Genetic Instabilities and Neurological Diseases

Genetic Instabilities and Neurological Diseases

Second Edition

Editors

ROBERT D. WELLS
Institute of Biosciences and Technology
Texas A&M University Health Sciences Center
The Texas Medical Center
Houston, Texas 77030-3303

TETSUO ASHIZAWA
Department of Neurology
University of Texas Medical Branch
Galveston, Texas 77555-0539

AMSTERDAM • BOSTON • HEIDELBERG • LONDON • NEW YORK • OXFORD
PARIS • SAN DIEGO • SAN FRANCISCO • SINGAPORE • SYDNEY • TOKYO
Academic Press is an imprint of Elsevier

Academic Press is an imprint of Elsevier
30 Corporate Drive, Suite 400, Burlington, MA 01803, USA
525 B Street, Suite 1900, San Diego, California 92101-4495, USA
84 Theobald's Road, London WC1X 8RR, UK

This book is printed on acid-free paper. ∞

Copyright © 2006, Elsevier Inc. All rights reserved.

No part of this publication may be reproduced or transmitted in any form or by any means, electronic or mechanical, including photocopy, recording, or any information storage and retrieval system, without permission in writing from the publisher.

Permissions may be sought directly from Elsevier's Science & Technology Rights Department in Oxford, UK: phone: (+44) 1865 843830, fax: (+44) 1865 853333, E-mail: permissions@elsevier.com. You may also complete your request on-line via the Elsevier homepage (http://elsevier.com), by selecting "Customer Support" and then "Obtaining Permissions."

Library of Congress Cataloging-in-Publication Data
Application submitted.

British Library Cataloguing in Publication Data
A catalogue record for this book is available from the British Library

ISBN 13: 978-0-12-369462-1
ISBN 10: 0-12-369462-0

For all information on all Elsevier Academic Press publications visit our Web site at www.books.elsevier.com

Printed in the United States of America
06 07 08 09 10   9 8 7 6 5 4 3 2 1

Working together to grow
libraries in developing countries

www.elsevier.com | www.bookaid.org | www.sabre.org

ELSEVIER   BOOK AID International   Sabre Foundation

Contents

Contributors ix
Preface xv

PART I Overview

Chapter 1 Overview of the Field 3
TETSUO ASHIZAWA AND ROBERT D. WELLS

PART II Myotonic Dystrophy

Chapter 2 Myotonic Dystrophies: An Overview 21
TETSUO ASHIZAWA AND PETER S. HARPER

Chapter 3 The RNA-Mediated Disease Process in Myotonic Dystrophy 37
CHARLES A. THORNTON, MAURICE S. SWANSON, AND THOMAS A. COOPER

Chapter 4 cis Effects of CTG Expansion in Myotonic Dystrophy Type 1 55
SITA REDDY AND SHARAN PAUL

Chapter 5 Normal and Pathophysiological Significance of Myotonic Dystrophy Protein Kinase 79
DERICK G. WANSINK, RENÉ E. M. A. VAN HERPEN, AND BÉ WIERINGA

Chapter 6 Biochemistry of Myotonic Dystrophy Protein Kinase 99
RAM SINGH AND HENRY F. EPSTEIN

Chapter 7 Clinical and Genetic Features of Myotonic Dystrophy Type 2 115
JAMIE M. MARGOLIS, LAURA P. W. RANUM, AND JOHN W. DAY

Chapter 8 Myotonic Dystrophy Type 2: Clinical and Genetic Aspects 131
RALF KRAHE, LINDA L. BACHINSKI, AND BJARNE UDD

Chapter 9 The Subtelomeric D4Z4 Repeat Instability in Facioscapulohumeral Muscular Dystrophy 151
SILVÈRE VAN DER MAAREL, RUNE R. FRANTS, AND GEORGE W. PADBERG

PART III Fragile X Syndrome

Chapter 10 Fragile X Syndrome and Fragile X-Associated Tremor/Ataxia Syndrome 165
RANDI J. HAGERMAN AND PAUL J. HAGERMAN

Chapter 11 Animal Models of Fragile X Syndrome: Mice and Flies 175
BEN A. OOSTRA AND DAVID L. NELSON

Chapter 12 Chromosomal Fragile Sites: Mechanisms of Cytogenetic Expression and Pathogenic Consequences 195
ROBERT I. RICHARDS

PART IV Kennedy's Disease

Chapter 13 Clinical Features and Molecular Biology of Kennedy's Disease 211
CHEUNJU CHEN AND KENNETH H. FISCHBECK

PART V Huntington's Disease

Chapter 14 Molecular Pathogenesis and Therapeutic Targets in Huntington's Disease 223
JOHN S. BETT, GILLIAN P. BATES, AND EMMA HOCKLY

Chapter 15 Molecular Pathogenesis of Huntington's Disease: The Role of Excitotoxicity 251
MAHMOUD A. POULADI, ILYA BEZPROZVANNY, LYNN A. RAYMOND, AND MICHAEL R. HAYDEN

v

Chapter 16 Huntington's Disease-like 2   261
RUSSELL L. MARGOLIS, SUSAN E. HOLMES,
DOBRILA D. RUDNICKI, ELIZABETH O'HEARN,
CHRISTOPHER A. ROSS, OLGA PLETNIKOVA, AND
JUAN C. TRONCOSO

PART VI   Friedreich's Ataxia

Chapter 17 Friedreich's Ataxia   277
MASSIMO PANDOLFO

Chapter 18 Experimental Therapeutics for Friedreich's Ataxia   297
ROBERT B. WILSON

Chapter 19 Evolution and Instability of the GAA Triplet-Repeat Sequence in Friedreich's Ataxia   305
IRENE DE BIASE, ASTRID RASMUSSEN, AND
SANJAY I. BIDICHANDANI

Chapter 20 Mouse Models for Friedreich's Ataxia   321
HÉLÈNE PUCCIO

Chapter 21 Triplexes, Sticky DNA, and the (GAA·TTC) Trinucleotide Repeat Associated with Friedreich's Ataxia   327
LESLIE S. SON AND ROBERT D. WELLS

PART VII   Spinocerebellar Ataxias

Chapter 22 Phosphorylation of Ataxin-1: A Link Between Basic Research and Clinical Application in Spinocerebellar Ataxia Type 1   339
KERRI M. CARLSON AND HARRY T. ORR

Chapter 23 Spinocerebellar Ataxia Type 2   351
STEFAN M. PULST

Chapter 24 Machado–Joseph Disease/Spinocerebellar Ataxia Type 3   363
HENRY PAULSON

Chapter 25 Spinocerebellar Ataxia Type 6   379
HIDEHIRO MIZUSAWA AND KINYA ISHIKAWA

Chapter 26 Pathogenesis of Spinocerebellar Ataxia Type 7: New Insights from Mouse Models and Ataxin-7 Function   387
DOMINIQUE HELMLINGER AND DIDIER DEVYS

Chapter 27 Spinocerebellar Ataxia Type 7: Clinical Features to Cellular Pathogenesis   399
GWENN A. GARDEN, RAY TRUANT, LISA M. ELLERBY, AND ALBERT R. LA SPADA

Chapter 28 Molecular Genetics of Spinocerebellar Ataxia Type 8   417
YOSHIO IKEDA, KATHERINE A. DICK, JOHN W. DAY, AND LAURA P. W. RANUM

Chapter 29 Spinocerebellar Ataxia Type 10: A Disease Caused by an Expanded $(ATTCT)_n$ Pentanucleotide Repeat   433
TETSUO ASHIZAWA

Chapter 30 DNA Structures and Genetic Instabilities Associated with Spinocerebellar Ataxia Type 10 $(ATTCT)_n \cdot (AGAAT)_n$ Repeats Suggest a DNA Amplification Model for Repeat Expansion   447
VLADIMIR N. POTAMAN, MALGORZATA J. PYTLOS, VERA I. HASHEM, JOHN J. BISSLER, MICHAEL LEFFAK, AND RICHARD R. SINDEN

Chapter 31 Spinocerebellar Ataxia Type 12   461
SUSAN E. HOLMES, ELIZABETH O'HEARN, NATIVIDAD CORTEZ-APREZA, H. S. HWANG, CHRISTOPHER A. ROSS, S. STRACK, AND RUSSELL L. MARGOLIS

Chapter 32 Spinocerebellar Ataxia 17 and Huntington's Disease-like 4   475
GIOVANNI STEVANIN AND ALEXIS BRICE

PART VIII   Other Polyamino Acid Repeats

Chapter 33 Polyalanine and Polyglutamine Diseases: Possible Common Mechanisms?   487
AIDA ABU-BAKER AND GUY A. ROULEAU

PART IX   Biophysics of PolyQ

Chapter 34 Chemical and Physical Properties of Polyglutamine Repeat Sequences   517
RONALD WETZEL

PART X   In Vivo Instability Studies

Chapter 35 Somatic Mosaicism of Expanded CAG·CTG Repeats in Humans and Mice: Dynamics, Mechanisms, and Consequences   537
PEGGY F. SHELBOURNE AND DARREN G. MONCKTON

Chapter 36 Transgenic Mouse Models of Unstable Trinucleotide Repeats: Toward an Understanding of Disease-Associated Repeat Size Mutation   563
MÁRIO GOMES-PEREIRA, LAURENT FOIRY, AND GENEVIÈVE GOURDON

PART XI   Insect Models

Chapter 37 *Drosophila* Models of Polyglutamine Disorders   587
GEORGE R. JACKSON, TZU-KANG SANG, AND J. PAUL TAYLOR

PART XII   Instability Mechanisms in Vivo and in Vitro

Chapter 38 Involvement of Genetic Recombination in Microsatellite Instability   597
RUHEE DERE, MICHEAL L. HEBERT, AND MAREK NAPIERALA

Chapter 39 Bending the Rules: Unusual Nucleic Acid
           Structures and Disease Pathology in the Repeat
           Expansion Diseases   617
           KAREN USDIN

Chapter 40 Replication of Expandable DNA Repeats   637
           SERGEI M. MIRKIN

Chapter 41 Error-Prone Repair of Slipped (CTG)·(CAG)
           Repeats and Disease-Associated
           Expansions   645
           GAGAN B. PANIGRAHI, RACHEL LAU, S. ERIN
           MONTGOMERY, MICHELLE R. LEONARD, JULIEN L.
           MARCADIER, MARIANA KEKIS, CAROLINE VOSCH,
           ANDREA TODD, AND CHRISTOPHER E. PEARSON

Chapter 42 DNA Repair Models for Understanding Triplet
           Repeat Instability   667
           YUAN LIU, RAJENDRA PRASAD, AND
           SAMUEL H. WILSON

Chapter 43 Models of Repair Underlying Trinucleotide
           DNA Expansion   679
           IRINA V. KOVTUN AND CYNTHIA T. MCMURRAY

Chapter 44 Transcription and Triplet Repeat
           Instability   691
           YUNFU LIN, VINCENT DION, AND JOHN H. WILSON

Chapter 45 Structural Characteristics of Trinucleotide
           Repeats in Transcripts   705
           WLODZIMIERZ J. KRZYZOSIAK, KRZYSZTOF
           SOBCZAK, AND MAREK NAPIERALA

PART XIII Mutations in Flanking Sequences

Chapter 46 Gross Rearrangements Caused by Long Triplet
           and Other Repeat Sequences   717
           ALBINO BACOLLA, MARZENA WOJCIECHOWSKA,
           BEATA KOSMIDER, JACQUELYNN E. LARSON,
           AND ROBERT D. WELLS

PART XIV Cancer and Genetic Instability

Chapter 47 Microsatellite Instability in Cancer   737
           MICHAEL J. SICILIANO

Index   749

Contributors

The number in parentheses indicates the chapter to which the author contributed.

Aida Abu-Baker (33)
Center for the Study of Brain Diseases, CHUM Research Center—Notre Dame Hospital, JA de Sève Pavillion, Montreal Quebec, Canada

Tetsuo Ashizawa (1, 2, 29)
Department of Neurology, University of Texas Medical Branch, Galveston, Texas

Linda L. Bachinski (8)
Department of Molecular Genetics, University of Texas M. D. Anderson Cancer Center, Houston, Texas

Albino Bacolla (46)
Institute of Biosciences and Technology, Center for Genome Research, Texas A&M University Health Science System, Texas Medical Center, Houston, Texas

Gillian P. Bates (14)
Department of Medical and Molecular Genetics, King's College London School of Medicine, Guy's Tower, Guy's Hospital, London, United Kingdom

John S. Bett (14)
Department of Medical and Molecular Genetics, King's College London School of Medicine, Guy's Tower, Guy's Hospital, London, United Kingdom

Ilya Bezprozvanny (15)
Department of Physiology, University of Texas Southwestern Medical Center, Dallas, Texas

Sanjay Bidichandani (19)
Departments of Biochemistry and Molecular Biology, and Pediatrics, University of Oklahoma Health Sciences Center, Oklahoma City, Oklahoma

John J. Bissler (30)
Division of Nephrology and Hypertension, Cincinnati Children's Hospital Medical Center, Cincinnati, Ohio

Alexis Brice (32)
Department of Genetics, Cytogenetics and Embryology APHP, Federation of Neurology APHP, INSERM U679 and Salpêtrière Hospital, University Paris-VI Medical School, Paris, France

Kerri M. Carlson (22)
Department of Laboratory Medicine and Pathology, and Institute of Human Genetics, University of Minnesota, Minneapolis, Minnesota

CheunJu Chen (13)
Neurogenetics Branch, National Institutes of Neurological Disorders and Strokes, National Institutes of Health, Bethesda, Maryland

Thomas A. Cooper (3)
Departments of Pathology and Molecular and Cellular Biology, Baylor College of Medicine, Houston, Texas

Natividad Cortez-Apreza (31)
Laboratory of Genetic Neurobiology, Division of Neurobiology, Department of Psychiatry, Johns Hopkins University School of Medicine, Baltimore, Maryland

John W. Day (7, 28)
Department of Neurology and the Institute of Human Genetics, University of Minnesota, Minneapolis, Minnesota

Irene De Biase (19)
Departments of Biochemistry and Molecular Biology, and Pediatrics, University of Oklahoma Health Sciences Center, Oklahoma City, Oklahoma

Ruhee Dere (38)
Institute of Biosciences and Technology, Center for Genome Research, Texas A&M University System Health Science Center, Texas Medical Center, Houston, Texas

Didier Devys (26)
IGBMC, CNRS/INSERM/ULP, Illkirch, France

Katherine A. Dick (28)
Departments of Genetics, Cell Biology, and Development, and Institute of Human Genetics, University of Minnesota, Minneapolis Minnesota

Vincent Dion (44)
Verna and Marrs McLean Department of Biochemistry and Molecular Biology, Baylor College of Medicine, Houston, Texas

Lisa M. Ellerby (27)
Buck Institute for Age Research, Novato, California

Henry F. Epstein (6)
Department of Neuroscience and Cell Biology, University of Texas Medical Branch at Galveston, Galveston, Texas

Kenneth H. Fischbeck (13)
Neurogenetics Branch, National Institutes of Neurological Disorders and Strokes, National Institutes of Health, Bethesda, Maryland

Laurent Foiry (36)
INSERM U781, Clinique Maurice Lamy, Hopital Necker Enfants Malades, Paris, France

Rune R. Frants (9)
Center for Human and Clinical Genetics, Leiden University Medical Center, Leiden, The Netherlands

Gwenn A. Garden (27)
Departments of Neurology and Center for Neurogenetics and Neurotherapeutics, University of Washington, Seattle, Washington

Mário Gomes-Pereira (36)
INSERM U781, Clinique Maurice Lamy, Hopital Necker Enfants Malades, Paris, France

Geneviève Gourdon (36)
INSERM U781, Clinique Maurice Lamy, Hopital Necker Enfants Malades, Paris, France

Paul J. Hagerman (10)
Department of Biochemistry and Molecular Medicine, University of California at Davis, Davis, California

Randi J Hagerman (10)
M. I. N. D. Institute and Department of University of California at Davis, Sacramento California

Peter S. Harper (2)
Department of Medical Genetics, University of Wales College of Medicine, Cardiff, Wales, United Kingdom

Vera I. Hashem (30)
Department of Human and Molecular Genetics, Baylor College of Medicine, Houston, Texas

Michael R. Hayden (15)
Center for Molecular Medicine and Therapeutics, Department of Medical Genetics, Children's and Women's Hospital, University of British Columbia, Vancouver, British Columbia, Canada

Micheal L. Hebert (38)
Institute of Biosciences and Technology, Center for Genome Research, Texas A&M University System Health Science Center, Texas Medical Center, Houston, Texas

Dominique Helmlinger (26)
IGBMC, CNRS/INSERM/ULP, Illkirch, France

Emma Hockly (14)
Department of Medical and Molecular Genetics, King's College London School of Medicine, Guy's Tower, Guy's Hospital, London, United Kingdom

Susan E. Holmes (16, 31)
Laboratory of Genetic Neurobiology, Division of Neurobiology, Department of Psychiatry, Johns Hopkins University School of Medicine, Baltimore, Maryland

H. S. Hwang (31)
Laboratory of Genetic Neurobiology, Division of Neurobiology, Department of Psychiatry, Johns Hopkins University School of Medicine, Baltimore, Maryland

Yoshio Ikeda (28)
Departments of Genetics, Cell Biology, and Development, and the Institute of Human Genetics, University of Minnesota, Minneapolis

Kinya Ishikawa (25)
Department of Neurology and Neurological Science, Graduate School, Tokyo Medical and Dental University, Bunkyo-ku, Tokyo, Japan

George R. Jackson (37)
Department of Neurology, Brain Research Institute, Center for Neurobehavioral Genetics, Semel Institute for Neuroscience and Human Behavior, David Geffen School of Medicine, University of California at Los Angeles, Los Angeles, California

Mariana Kekis (41)
Program of Genetics and Genomic Biology, The Hospital for Sick Children, and Department of Molecular and Medical Genetics, University of Toronto, Toronto, Ontario, Canada

Beata Kosmider (46)
Institute of Biosciences and Technology, Texas A&M University System Health Science Center, Center for Genome Research, Texas Medical Center, Houston, Texas

Irina V. Kovtun (43)
Departments of Molecular Pharmacology and Experimental therapeutics, and Biochemistry and Molecular Biology, Mayo Clinic Rochester, Rochester, Minnesota

Ralf Krahe (8)
Section of Cancer Genetics, Department of Molecular Genetics, University of Texas M. D. Anderson Cancer Center, Houston, Texas

Wlodzimierz J. Krzyzosiak (45)
Institute of Bioorganic Chemistry, Polish Academy of Sciences, Poznan, Poland

Albert R. La Spada (27)
Departments of Neurology, Laboratory Medicine, and Medicine, and Center for Neurogenetics and Neurotherapeutics, University of Washington, Seattle, Washington

Jacquelynn E. Larson (46)
Institute of Biosciences and Technology, Texas A&M University System Health Science Center, Center for Genome Research, Texas Medical Center, Houston, Texas

Rachel Lau (41)
Program of Genetics and Genomic Biology, The Hospital for Sick Children, Toronto, Ontario, Canada

Michael Leffak (30)
Department of Biochemistry and Molecular Biology, Wright State University, Dayton Ohio

Michelle R. Leonard (41)
Program of Genetics and Genomic Biology, The Hospital for Sick Children, and Department of Molecular and Medical Genetics, University of Toronto, Toronto, Ontario, Canada

Yunfu Lin (44)
Verna and Marrs McLean Department of Biochemistry and Molecular Biology, Baylor College of Medicine, Houston, Texas

Yuan Liu (42)
Laboratory of Structural Biology, National Institute of Environmental Health Sciences, National Institutes of Health, Research Triangle Park, North Carolina

Julien L. Marcadier (41)
Program of Genetics and Genomic Biology, The Hospital for Sick Children, and Department of Molecular and Medical Genetics, University of Toronto, Toronto, Ontario, Canada

Jamie M. Margolis (7)
Departments of Genetics, Cell Biology, and Development, and the Institute of Human Genetics, University of Minnesota, Minneapolis, Minnesota

Russell L. Margolis (16, 31)
Laboratory of Genetic Neurobiology, Division of Neurobiology, Department of Psychiatry; Department of Neurology; Program in Cellular and Molecular Medicine, Johns Hopkins University School of Medicine, Baltimore, Maryland

Cynthia T. McMurray (43)
Departments of Molecular Pharmacology and Experimental therapeutics, and Biochemistry and Molecular Biology, Mayo Clinic Rochester, Rochester, Minnesota

Sergei M. Mirkin (40)
Department of Biochemistry and Molecular Genetics, College of Medicine, University of Illinois at Chicago, Chicago, Illinois

Hidehiro Mizusawa (25)
Department of Neurology and Neurological Science, Graduate School, Tokyo Medical and Dental University, Bunkyo-ku, Tokyo, Japan

Darren G. Monckton (35)
Institute of Biomedical and Life Sciences, University of Glasgow, Anderson College, Glasgow, Scotland, United Kingdom

S. Erin Montgomery (41)
Program of Genetics and Genomic Biology, The Hospital for Sick Children; Department of Molecular and Medical Genetics, University of Toronto, Toronto, Ontario, Canada; and Albany Medical College, Albany, New York

Marek Napierala (38, 45)
Institute of Biosciences and Technology, Center for Genome Research, Texas A&M University System Health Science Center, Texas Medical Center, Houston, Texas

David L. Nelson (11)
Department of Molecular and Human Genetics, Baylor College of Medicine, Houston, Texas

Elizabeth O'Hearn (16, 31)
Departments of Neurology and Neuroscience, Johns Hopkins University School of Medicine, Baltimore, Maryland

Ben A. Oostra (11)
Department of Clinical Genetics, Erasmus M.C. University Rotterdam, Rotterdam, The Netherlands

Harry T. Orr (22)
Institute of Human Genetics, University of Minnesota, Minneapolis, Minnesota

George W. Padberg (9)
Department of Neurology, University Medical Center Nijmegen, Nijmegen, The Netherlands

Massimo Pandolfo (17)
Department of Neurology, Free University of Bruxelles, Erasme Hospital, Bruxelles, Belgium

Gagan B. Panigrahi (41)
Program of Genetics and Genomic Biology, The Hospital for Sick Children, Toronto, Ontario, Canada

Sharan Paul (4)
Institute for Genetic Medicine, Keck School of Medicine, University of Southern California, Los Angeles, California

Henry Paulson (24)
Department of Neurology, Carver College of Medicine, University of Iowa, Iowa City, Iowa

Christopher E. Pearson (41)
Program of Genetics and Genomic Biology, The Hospital for Sick Children, and Department of Molecular and Medical Genetics, University of Toronto, Toronto, Ontario, Canada

Olga Pletnikova (16)
Division of Neuropathology, Department of Pathology, Johns Hopkins University School of Medicine, Baltimore, Maryland

Vladimir N. Potaman (30)
Laboratory for DNA Structure and Mutagenesis, Center for Genome Research, Institute of Biosciences and Technology, Texas A&M University System Health Science Center, Houston, Texas

Mahmoud A. Pouladi (15)
Center for Molecular Medicine and Therapeutics, Department of Medical Genetics, Children's and Women's Hospital, University of British Columbia, Vancouver, British Columbia, Canada

Rajendra Prasad (42)
Laboratory of Structural Biology, National Institute of Environmental Health Sciences, National Institutes of Health, Research Triangle Park, North Carolina

Hélène Puccio (20)
Department of Molecular Pathology, Institute of Genetics and Molecular and Cellular Biology, CNRS/INSERM/ULP, Illkirch, France

Stefan M. Pulst (23)
Division of Neurology, Cedars-Sinai Medical Center, Los Angeles, California

Malgorzata J. Pytlos (30)
Laboratory for DNA Structure and Mutagenesis, Center for Genome Research, Institute of Biosciences and Technology, Texas A&M University System Health Science Center, Houston, Texas

Laura P. W. Ranum (7, 28)
Departments of Genetics, Cell Biology, and Development, and the Institute of Human Genetics, University of Minnesota, Minneapolis, Minnesota

Astrid Rasmussen (19)
Departments of Biochemistry and Molecular Biology, and Pediatrics University of Oklahoma Health Sciences Center, Oklahoma City, Oklahoma; Department of Neurogenetics and Molecular Biology, Instituto Nacional de Neurología y Neurocirugía Manuel Velasco Suárez, Mexico City, Mexico

Lynn A. Raymond (15)
Department of Psychiatry and Brain Research Center, University of British Columbia, Vancouver, British Columbia, Canada

Sita Reddy (4)
Institute for Genetic Medicine, Keck School of Medicine, University of Southern California, Los Angeles, California

Robert I. Richards (12)
ARC Special Research Center for the Molecular Genetics of Development, ARC/NHMRC Research Network in Genes and Environment in Development, School of Molecular and Biomedical Sciences, The University of Adelaide, Adelaide, South Australia

Christopher A. Ross (16, 31)
Laboratory of Genetic Neurobiology, Division of Neurobiology, Department of Psychiatry; Departments of Neurology and Neuroscience; Program in Cellular and Molecular Medicine, Johns Hopkins University School of Medicine, Baltimore, Maryland

Guy Rouleau (33)
Center for the Study of Brain Diseases, CHUM Research Center—Notre Dame Hospital, JA de Sève Pavillion, Montreal Quebec, Canada

Dobrila D. Rudnicki (16)
Laboratory of Genetic Neurobiology, Division of Neurobiology, Department of Psychiatry; Johns Hopkins University School of Medicine, Baltimore, Maryland

Tzu-Kang Sang (37)
Department of Neurology, David Geffen School of Medicine, University of California at Los Angeles, Los Angeles, California

Peggy F. Shelbourne (35)
Institute of Biomedical and Life Sciences, University of Glasgow, Anderson College, Glasgow, Scotland, United Kingdom

Michael J. Siciliano (47)
Department of Molecular Genetics, University of Texas M. D. Anderson Cancer Center, Houston, Texas

Richard R. Sinden (30)
Laboratory of DNA Structure and Mutagenesis, Center for Genome Research, Institute of Biosciences and Technology, Texas A&M University System Health Science Center, Houston, Texas

Ram Singh (6)
Department of Neuroscience and Cell Biology, University of Texas Medical Branch at Galveston, Galveston, Texas

Krzysztof Sobczak (45)
Institute of Bioorganic Chemistry, Polish Academy of Sciences, Poznan, Poland

Leslie S. Son (21)
Institute of Biosciences and Technology, Texas A&M University System Health Science Center, Center for Genome Research, Houston, Texas

Giovanni Stevanin (32)
Department of Genetics, Cytogenetics and Embryology APHP, Federation of Neurology APHP, INSERM U679 and Salpêtrière Hospital, University Paris-VI Medical School, Paris, France

S. Strack (31)
Department of Pharmacology, University of Iowa Carver College of Medicine, Iowa City, Iowa

Maurice S. Swanson (3)
Department of Molecular Genetics and Microbiology and Powell Gene Therapy Center, University of Florida College of Medicine, Gainesville, Florida

J. Paul Taylor (37)
Department of Neurology, University of Pennsylvania, Philadelphia, Pennsylvania

Charles A. Thornton (3)
Department of Neurology, University of Rochester Medical Center, Rochester, New York

Andrea Todd (41)
Program of Genetics and Genomic Biology, The Hospital for Sick Children, and Department of Molecular and Medical Genetics, University of Toronto, Toronto, Ontario, Canada

Juan Troncoso (16)
Department of Neurology; Division of Neuropathology, Department of Pathology, Johns Hopkins University School of Medicine, Baltimore, Maryland

Ray Truant (27)
Department of Biochemistry and Biomedical Science, McMaster University, Hamilton, Ontario, Canada

Bjarne Udd (8)
Department of Neurology, Vasa Central Hospital, Vasa; Tampere University Hospital, Finland; and Folkhälsan Institute of Genetics, University of Helsinki, Helsinki, Finland

Karen Usdin (39)
Laboratory of Molecular and cellular Biology, National Institute of Diabetes and Digestive and Kidney Diseases, National Institutes of Health, Bethesda, Maryland

Silvère M. van der Maarel (9)
Center for Human and Clinical Genetics, Leiden University Medical Center, Leiden, The Netherlands

René E. M. A. van Herpen (5)
Department of Cell Biology, Nijmegen Center for Molecular Life Sciences, Radboud University Nijmegen Medical Center, Nijmegen, The Netherlands

Caroline Vosch (41)
Program of Genetics and Genomic Biology, The Hospital for Sick Children, and Department of Molecular and Medical Genetics, University of Toronto, Toronto, Ontario, Canada

Derick G. Wansink (5)
Department of Cell Biology, Nijmegen Center for Molecular Life Sciences, Radboud University Nijmegen Medical Center, Nijmegen, The Netherlands

Robert D. Wells (1, 21, 46)
Institute of Biosciences and Technology, Texas A&M University System Health Science Center, Center for Genome Research, Texas Medical Center, Houston, Texas

Ronald Wetzel (34)
Graduate School of Medicine, University of Tennessee, Knoxville, Tennessee

Bé Wieringa (5)
Department of Cell Biology, Nijmegen Center for Molecular Life Sciences, Radboud University Nijmegen Medical Center, Nijmegen, The Netherlands

Robert B. Wilson (18)
Department of Pathology and Laboratory Medicine, University of Pennsylvania, Stellar-Chance Laboratories, Philadelphia, Pennsylvania

Samuel H. Wilson (42)
Laboratory of Structural Biology, National Institute of Environmental Health Sciences, National Institutes of Health, Research Triangle Park, North Carolina

John H. Wilson (44)
Verna and Marrs McLean Department of Biochemistry and Molecular Biology, Baylor College of Medicine, Houston, Texas

Marzena Wojciechowska (46)
Institute of Biosciences and Technology, Texas A&M University System Health Science Center, Center for Genome Research, Texas Medical Center, Houston, Texas

Preface

This book describes advances in our clinical and biomedical knowledge of hereditary neurological diseases caused by genomic instability. Volume I, published in 1998, reviewed the status of discoveries since the revelation in 1991 that the disease genes had expanded triplet repeat sequences as their responsible mutations. The current volume is focused on reviewing the status of each of the diseases and their mechanisms with an emphasis on discoveries since 1998. Several new diseases are added in this book. While some of them represent a new class of triplet repeat expansions, others include diseases caused by an expansion of tetra- and pentanucleotide repeats and those caused by a deletion of a minisatellite repeat.

The efforts of a large number of scientists in the fields of human genetics, cell biology, biochemistry, biophysics, chemistry and clinical sciences have focused their attention on the non-Mendelian expansion process as well as their protein products and their pathophysiology. This book is intended to serve as a comprehensive treatise of many aspects of this broad based attack on these neurological diseases. This book should be of interest to clinicians who wish to become better informed about these diseases and their etiology as well as scientists who desire to update and widen their knowledge base of the field. We hope that this book will facilitate communications between bench and bedside, which are the key to success in multidisciplinary translational research as emphasized in the recent NIH Roadmap.

This book serves as the second volume of the authoritative review of all neurological diseases that are related to repeat expansions. Due to the maturing nature of the field, more emphasis is placed on clinical and pathophysiological studies as well as mechanisms in complex organisms and cells compared to Volume I.

Despite the remarkable developments identified above, we are still a long way from developing effective therapeutic strategies for these diseases. This book is dedicated to those patients afflicted with these diseases and to their families with the conviction that future generations will not experience their pain and suffering.

A large number of dedicated and talented scientists have willingly contributed chapters to this book. Their prompt and dedicated participation has made publication of this treatise possible in a timely fashion.

We thank Ms. Peggy Weinshilboum and Ms. Lorrie Adams for their expert assistance and Ms. Hilary Rowe and Ms. Erin Labonte-McKay of Elsevier-Academic Press for their encouragement and participation.

ROBERT D. WELLS

TETSUO ASHIZAWA

PART I

Overview

CHAPTER 1

Overview of the Field

TETSUO ASHIZAWA AND ROBERT D. WELLS

Department of Neurology, The University of Texas Medical Branch, Galveston, Texas 77555
Institute of Biosciences and Technology, Texas A&M University System Health Science Center, Center for Genome Research, Houston, Texas 77030

I. Introduction
II. The Current Spectrum of Expanded Repeats in Human Diseases
III. Instability of Microsatellite Repeats
   A. Repeat Instability in Patient-Derived Tissues
   B. Molecular Mechanisms of Genetic Instability
   C. Microsatellite Repeat Instability and Population Genetics
IV. Pathogenic Mechanism of Neurological Diseases Caused by Expanded Microsatellites Repeats
   A. Genotype–Phenotype Correlation
   B. Pathogenic Mechanisms
V. Future Directions for Research on Repeats and Genomic Instabilities in Neurological Disorders
VI. Concluding Remarks
References

The objective of this introductory chapter is to provide an overview of this book by reviewing the current status of research in the molecular mechanisms of microsatellite repeat instability and in the pathogenic mechanisms of human diseases resulting from the expansion of microsatellite repeats. As the list of human diseases in this category grew longer in the past decade, we observed a wider variety of repeat motifs, which came with highly variable lengths of repeat tracts in different genomic locations. Studies of *in vitro, Escherichia coli*, and yeast models provided multiple molecular mechanisms of repeat instability. Data from human tissues and mouse models suggested that some of these mechanisms are particularly relevant to the expansion of specific repeats in human diseases. As for the pathogenic mechanism by which the repeat expansion leads to the disease, major models include a loss of function of the gene and a gain of function by the mutant RNA transcript or protein product. However, the pathogenic mechanism varies from one disease to another, depending on the motif, length, and intragenic location of the repeat. Understanding the molecular mechanisms of repeat instability and pathogenic process is not only of scientific interest but also essential for the development of rational treatment of these diseases.

I. INTRODUCTION

In 1991, expanded CGG·CCG and CAG·CTG triplet repeat sequence (TRS) were almost simultaneously identified as the mutations responsible for fragile X syndrome and Kennedy's disease, respectively. Since then, 17 additional neurological diseases have been found to be caused by expanded simple tandem repeats (microsatellite repeats), which have become an important class of mutations in human genetics. This book focuses on the instability of microsatellite repeats and neurological diseases caused by the expansion of these repeats. Substantial pages are devoted to the molecular mechanisms of repeat instability and the pathogenic mechanisms of these diseases. Moreover, clinical phenotype and genotype–phenotype correlation are reviewed in detail in appropriate chapters. We have accomplished these objectives by updating the topics covered in the prior edition of this book published in 1998 [10] and by adding new chapters on diseases, repeats, and molecular mechanisms that were not recognized or included in the previous edition. In this introductory chapter we intend to provide an overview of the field by reviewing the spectrum of expanded microsatellite repeats in human diseases, characteristics and molecular mechanisms of repeat instability involved in human diseases, and pathogenic mechanisms of neurological diseases caused by expanded repeats.

II. THE CURRENT SPECTRUM OF EXPANDED REPEATS IN HUMAN DISEASES

Since the discovery of expanded CAG and CGG repeats in SBMA and fragile X syndrome in 1991, the number of hereditary human neurological diseases caused by expansions of microsatellites has kept increasing (Table 1-1). In addition to the expansions of TRS, expanded tetra- and pentanucleotide repeats have been shown to cause myotonic dystrophy type 2 (DM2) [1] and spinocerebellar ataxia type 10 (SCA 10) [2], respectively. Unstable minisatellites have also been shown to cause human neurological diseases. Most mutations of myoclonic epilepsy of Unverricht and Lundborg (ULD, or EPM1) have been an expansion of a dodecamer repeat [3]. Furthermore, we should note that facioscapulohumeral muscular dystrophy (FSHD) was found with decreased length of the D4Z4 repeat, which consists of tandemly aligned 3.3 kb units, even before the discoveries of SBMA and fragile X syndrome [4].

Among the expanded TRS, the sequence of the repeat unit is variable, and some of the expanded sequences show interruptions of a pure repeat tract with cryptic units. The location of the repeat expansion within the respective gene is also variable. In the majority of the diseases of expanded TRS, the repeat resides in the coding region, whereas in other diseases expanded repeats are located in the 5′-untranslated region (UTR), the 3′-UTR or an intron. The length of the disease-causing expansion is also highly variable depending on the locus. It ranges from just 21 bp in oculopharyngeal muscular dystrophy (OPMD) [5] to 44,000 bp in DM2 [1]. In addition to these repeat expansions causing neurological disorders, a handful of human developmental diseases have been found to be caused by small expansions of polyalanine-coding trinucleotide repeats (see Part VIII) [6]. Thus, the repeat unit sequence, the presence of interrupting repeat units, the length of the repeat tract and the location of the repeat within the gene are readily recognizable variables in pathogenic repeats of human diseases.

III. INSTABILITY OF MICROSATELLITE REPEATS

Microsatellite repeats have been found to be polymorphic in the human population [7, 8]. The variable number of repeat units makes this class of polymorphism highly informative as genetic markers, and thus useful in establishing the human genetic linkage map and in positional cloning of genes responsible for human diseases. The repeat length polymorphism is thought to have resulted from instability of the repeat. Actual instability in the copy number of these repeat is frequently found in patients with hereditary nonpolyposis colon cancer [9], and is detectable in normal individuals with lower mutation rates (see Part XIV).

The repeat instability became a subject of greater interest in human genetics when expanded TRS were found as the mutation responsible for human hereditary neurological diseases. By the mid-1990s, the repeat instability in these disorders was characterized as follows: (a) the size of the expanded repeat frequently change when transmitted from parent to child, (b) the intergenerational changes are typically biased toward further expansion, (c) the gender of the transmitting parent often influences the extent and the direction of intergenerational changes, and (d) the repeat size often increases with age. These observations are attributable to instability of the expanded allele in somatic and germline tissues.

The exact mechanism of the instability of expanded repeats in human diseases is unknown. However, it has been recognized that the likelihood of repeat instability correlates with the copy number of the perfect repeat in these diseases, suggesting that the repeat itself plays a major role in the mechanism of instability. Studies *in vitro* and in simple organisms provided important insights in understanding the basic molecular mechanisms of the instability (see Part XII). In these studies, the expansions and deletions were shown to be mediated by DNA replication, repair, and recombination, probably acting in concert (reviewed in [10–13]). Almost all models derived from these experimental studies

TABLE 1-1 Summary of Neurological Diseases Caused by Microsatellite Repeat Expansions

Disorder	Inheritance	Gene/Locus	Chromosomal Localization	Protein Product	Expansion Size Normal	Expansion Size Mutant	Repeat Location	Mutation Type	Parental Gender Bias
Fragile X syndrome	X-linked dominant	FMR1 (FRAXA)	Xq27.3	FMRP	$(CGG)_{5-54}$ (45–54: "gray zone")	$(CGG)_{55-200}$ (premutation) $(CGG)_{230-1000}$ (full)	5'-UTR	LOF Fragile site	maternal
Fragile XE mental retardation	X-linked ?dominant	FMR2 (FRAXE)	Xq28	FMR2 protein	$(GCC)_{3-42}$	$(GCC)_{130-150}$ (pre) $(GCC)_{200-750}$ (full)	5'-UTR	LOF Fragile site	ND
Friedreich's ataxia	autosomal recessive	FRDA	9q13-21.1	frataxin	$(GAA)_{6-29}$	$(GAA)_{30-65}$ (pre) $(GAA)_{66-1700}$ (full)	intron 1	LOF (partial)	maternal
Myotonic dystrophy type 1	autosomal dominant	DMPK	19q13	myotonic dystrophy protein kinase	$(CTG)_{5-37}$	$(CTG)_{38-50}$ (premutation) $(CTG)_{50-3000}$ (full)	3'-UTR	RNA GOF LOF (partial)	paternal for small alleles and maternal large alleles
Myotonic dystrophy type 2	autosomal dominant	ZNF9	3q21	zinc finger protein 9	$(CCTG)_{<27}$	$(CCTG)_{75-11000}$	Intron 1	RNA GOF	ND
Oculopharyngeal muscular dystrophy	autosomal dominant@	OPMD	14q11.2-q13	poly(A)-binding protein-2	$(GCG)_6$	$(GCG)_{8-13}$	coding	GOF	ND
Spinobulbar muscular atrophy (Kennedy disease)	X-linked recessive	AR	Xq13-21	androgen receptor	$(CAG)_{6-35}$	$(CAG)_{36-66}$	coding	GOF LOF (partial)	ND
Huntington's disease	autosomal dominant	HD	4p16.3	huntingtin	$(CAG)_{6-35}$	$(CAG)_{36-39}$ (↓penetrance) $(CAG)_{40-121}$ (full)	coding	GOF	paternal
Huntington's disease like 2	autosomal dominant	HDL2 (JPH3)	16q24.3	junctophilin-3	$(CTG)_{6-28}$	$(CTG)_{40-58}$	coding/3'UTR	GOF	ND
Dentatorubral-pallidoluysian atrophy/Haw River syndrome	autosomal dominant	DRPLA (B37)	12p13.31	atrophin-1 (drplap)	$(CAG)_{3-36}$	$(CAG)_{49-88}$	coding	GOF	paternal
Spinocerebellar ataxia type 1	autosomal dominant	SCA1	6p23	ataxin-1	$(CAG)_{6-44}$*	$(CAG)_{39-82}$	coding	GOF	paternal
Spinocerebellar ataxia type 2	autosomal dominant	SCA2	12q24.1	ataxin-2	$(CAG)_{14-31}$	$(CAG)_{34-64}$	coding	GOF	paternal

(continued)

TABLE 1-1 (continued)

Disorder	Inheritance	Gene/Locus	Chromosomal Localization	Protein Product	Expansion Size Normal	Expansion Size Mutant	Repeat Location	Mutation Type	Parental Gender Bias
Spinocerebellar ataxia type 3 / Machado-Joseph disease	autosomal dominant	SCA3 (MJD1)	14q32.1	ataxin-3 (MJDp)	$(CAG)_{12-\sim 43}$	$(CAG)_{60->200}$	coding	GOF	paternal
Spinocerebellar ataxia type 6	autosomal dominant	CACNA1A	19p13	α_{1A}-voltage dependent calcium channel subunit	$(CAG)_{5-18}$	$(CAG)_{20-33}$	coding	ND	ND
Spinocerebellar ataxia type 7	autosomal dominant	SCA7	3p12-21.2	ataxin-7	$(CAG)_{7-34}$	$(CAG)_{37->250}$	coding	GOF	paternal
Spinocerebellar ataxia type 8	autosomal dominant	SCA8	19q13	ataxin-8	$(CTG)_{16-34}$	$(CTG)_{100-250}$**	3'UTR	ND	maternal (*en mass* contraction in sperm)
Spinocerebellar ataxia type 10	autosomal dominant	ATXN10 (SCA10)	22q13	ataxin-10	$(ATTCT)_{10-29}$	$(ATTCT)_{>800(full)}$ $(ATTCT)_{280-800(pre)}$	Intron 1	ND	paternal
Spinocerebellar ataxia type 12	autosomal dominant	PPP2R2B	5q32	Phosphatase PP2ABb	$(CAG)_{4-32}$	$(CAG)_{51-78}$	5'UTR	ND	ND
Spinocerebellar ataxia type 17	autosomal dominant	TBP	6q27	TATA binding protein	$(CAG)_{25-44}$	$(CAG)_{45-66}$	Coding	GOF	ND

Abbreviations: GOF, gain of function; LOF, loss of function; ND, not determined; *: Normal alleles 39–44 are interrupted by one or two CAT's while disease alleles 39–44 are not.; **: reduced penetrance with $(CTG)>107$; @: autosomal recessive with $(GCG)7$.

hypothesize the formation of non-B DNA secondary structures at specific motifs of DNA repeats [12–18] (see Fig. 46-1 in Part XIII).

While simple experimental strategies are powerful for studies on molecular mechanisms, data in these systems should be carefully evaluated when related to the mechanism of instability of expanded repeats in human diseases. The machinery involved in DNA replication, transcription, recombination, and repair differ somewhat in humans, and some of the disease-causing repeats are much larger than those studied in lower organisms. One of the most striking features of expanded microsatellite repeats in human diseases is the propensity to further expand in most situations, while repeats in lower organisms tend to undergo deletions, although use of yeast models have alleviated some of these problems [19]. Transgenic mice carrying expanded repeats and cell culture derived from patients with expanded repeats offer useful models by enabling experimental manipulations of various determinants in the mammalian/human background (see Part X). Together, these experimental systems should be able to effectively test various hypotheses of the repeat instability mechanism.

A. Repeat Instability in Patient-Derived Tissues

The correlation between perfect repeat copy number and instability is a common property of microsatellite repeats in general, and of those specifically involved in human diseases. The expanded $(CAG)_{21-30}$ repeat of spinocerebellar ataxia type 6 (SCA6) [20] is short and shows notable stability. Conversely, larger repeat expansions seen in fragile X syndrome [21], FRAXE mental retardation [22], DM1 [23, 24] and DM2 [1], Friedreich's ataxia [25], and SCA10 [2] show remarkable instabilities in somatic or germline tissues, or both. The relatively stable expansion of the SCA17 repeat is attributable to the interrupted repeat structure [26, 27]. At the *SCA1* and *SCA2* loci, some individuals have a CAG repeat allele in the "intermediate" range, which often contains interruptions within the repeats [28, 29]. Interrupted intermediate alleles have a shortened length of the pure repeat tract and increased repeat-size stability. These features are presented in chapters describing individual diseases in Parts II–VII. In accord with these observations, TRS in polyalanine expansion diseases, which are generally short (<35 repeats) and interrupted, are notably stable (see Part VIII) [6].

However, the perfect repeat copy number is not the only determinant of the repeat stability. The presence of *cis*- and *trans*-acting regulators has been predicted. The C/G content of the genomic region surrounding a disease-causing CTG·CAG repeat has been shown to correlate with the repeat instability [30]. Regulation of repeat-size stability by other *cis*-acting elements has been implicated in the CAG repeat of MJD [31] and the CGG repeat of fragile X syndrome [32]. As to the *trans*-acting factors, mutations in mismatch repair (MMR) genes have been shown to cause micro- and minisatellite repeat instability in patients with hereditary non-polyposis colon cancer (HNPCC) [9]. Transgenic mouse models of DM1 and HD have shown that various MMR deficiencies modify the stability of expanded CTG·CAG repeats (See Part X) [33–36].

Repeats located within the coding region of the gene generally show relatively small expansions compared with noncoding repeats. In SCA7, it has been postulated that very large expansions of coding repeats may induce embryonic lethality [37]. In HD and SCA1, mosaic alleles in the striatum have been shown to reach far beyond the apparent expansion limit found in blood DNA [38, 39], suggesting that the restricted expansion size of coding repeats is not an intrinsic nature of these repeats. Thus, the upper limit of repeat expansion size appears to be regulated by confounded factors in the cellular environment.

B. Molecular Mechanisms of Genetic Instability

Data from studies of double strand breaks, recombination, replication repair, mismatch repair, and nucleotide excision repair will be discussed in this section. Details of these topics are further discussed mainly in Part XII, and relevant issues are also discussed in Parts X and XIII.

1. Genetic Recombination and Double-Strand Break Repair

Double-strand breaks (DSB) are very potent initiators of genetic recombination [40]. Proteins involved in the repair of the DSB also participate in the recombination processes. It has been demonstrated, in different model systems, that TRS such as CTG·CAG and CGG·CCG efficiently induce DNA strand discontinuities that are repaired by the recombination machinery (reviewed in [41–44]). Therefore, the role of DSB repair and recombination pathways in generating repeat instabilities will be reviewed together.

Replication slippage was the dominant model in the 1990s to explain microsatellite instability, since recombination (as defined by the reciprocal crossing-over exchange) was not revealed by studies of human cases [10]. Three important facts argue against reciprocal exchange as a mechanism of TRS instability in patients; first, the lack of evidence supporting the exchange of the flanking sequences, second, no corresponding length changes of the second allele during the expansion/deletion event occurring at the other allele, and third, a simple reciprocal exchange cannot explain extremely large expansions observed in some of the diseases. On the other hand, data from the patients pointed toward gene conversion (a nonreciprocal event) in the instability of CGG·CCG repeats in the fragile X syndrome and CTG·CAG tracts in the myotonic dystrophy type 1 cases (reviewed in [43, 44]). Thus, gene conversion could be, in principle, the recombination pathway involved in CTG·CAG, CGG·CCG repeat instabilities (Fig. 1-1A). It should be pointed out that the small expansions in the polyalanine tracts (often encoded by different GCN triplets), recently shown to cause at least nine human diseases, are likely to arise from unequal crossing-over as the predominant mechanism [45–47]. Since polyalanine tracts are usually encoded by imperfect TRS (i.e., by variants of the alanine codons),

a simple replication slippage model cannot explain their instability [45–47].

a. Recombination Studies in *E. coli*

In model systems from bacteria to mammalian cells, recombination, including both gene conversion and crossing-over events, has been shown to be involved in TRS instability. Additionally, repetitive sequences also promoted homologous recombination in both prokaryotic and the eukaryotic systems presumably by virtue of forming unusual, non-B DNA structures. Different di-, tetra-, and pentanucleotide repetitive sequences have previously been shown to stimulate recombination [48–52]. The results of intermolecular and intramolecular studies in *E. coli* revealed that the frequency of crossing-over between long DM 1 CTG·CAG repeats was significantly elevated when compared to the nonrepeating controls [53, 54]. Stimulation of recombination was also observed for GAA·TTC repeats from the Friedreich's ataxia gene [14], however, the intramolecular process between long repeats was significantly hampered by formation of sticky DNA (see above). In the case of the CTG·CAG repeats, the recombination frequency was

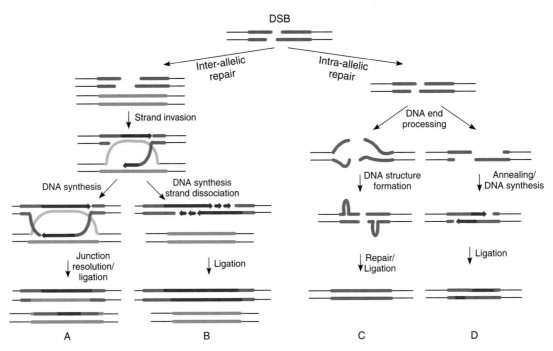

FIGURE 1-1 Inter-allelic and intra-allelic pathways in DSB repair leading to repeat instabilities. (A) Gene conversion without exchange of the flanking sequences. (B) Synthesis-dependent strand annealing (SDSA). (C) Nonhomologous end joining (NHEJ). (D) Single strand annealing (SSA). Only pathways, which do not involve or do not alter the sequence of the second allele, are presented. Note that a strong likelihood exists for the repeats to form stable secondary structures (e.g., hairpins) at any stage during the processes of DNA synthesis and annealing of the single-stranded DNA ends. Thus, non-B DNA structures formed by tandem repeats, besides being an important cause of the DSB formation, are also a direct source of the repeat instabilities. Thicker red, gray, and blue lines: repeat regions. The blue line designates the newly synthesized DNA tracts. See CD-ROM for color image.

dependent on the orientation of the repeat tract relative to the unidirectional origin of replication. When the CTG repeats were present on the lagging strand template, the frequency of recombination was substaintially higher in both inter- and intramolecular assays. The CTG·CAG tracts (as well as CGG·CCG and GAA·TTC) are known to arrest replication fork progression *in vitro* and *in vivo* [16, 55–57] due to their capabilities to adopt non-B DNA conformations. In the case of CTG·CAG repeats, this occurs predominantly when the CTG strand is located on the lagging strand template for replication [56]. In the model proposed to explain the orientation effect on recombination, stalling of the replication fork at the secondary structures led to the formation of nicks and/or DSB in the repeating tracts which stimulated their mutagenic repair *via* recombination [53, 54]. These studies also demonstrated a high level of TRS instability resulting from the recombination process in *E. coli*. A pronounced influence of DSB repair on TRS instability was also detected in the experiments, with transformation of break-containing plasmids into *E. coli* [58]. Repair of the DSB located in the CTG·CAG and CGG·CCG repeats resulted in dramatic increase of TRS deletions. Recently, Hashem et al. [59] showed using a genetic system in bacteria that mutations in *recA* and *recB*, which decrease the rate of recombination, had a stabilizing effect on CTG·CAG repeats lowering the high rates of deletion seen in recombination proficient cells. Thus, the recombination proficiency also correlated with the high rates of genetic instability in the triplet repeats.

b. TRS Instability During Mitotic and Meiotic Recombination in Yeast

Eukaryotic model systems, especially yeast, have been proven to be an excellent tool for the analysis of the involvement of recombination in the TRS instability, since mitotic and meiotic events can be analyzed separately in different genetic backgrounds [41, 42, 60]. Independent analyses of the mitotic and meiotic processes may be crucial in order to understand the timing of the events leading to the TRS expansion in humans.

Recently, several studies in yeast have been aimed towards understanding the role of DSB repair and recombination in the instability of TRS tracts, primarily CTG·CAG repeats (reviewed in [41–44]). Initial results obtained with relatively short TRS did not reveal a significant role of recombination in generating TRS instabilities [42]. It has been speculated that short CTG·CAG tracts may not be very efficient in generating DSB in yeast or that the breaks induced in the shorter repeats are repaired by pathways other than homologous recombination [42, 61]. In addition, experiments with *RAD52* mutants suggested that TRS instability is due to defects in replication rather than in recombination [42].

However, elegant experiments with long CTG·CAG tracts (up to 250 repeats) definitively implicated DSBs and recombination as important mechanisms of the repeat instability [62]. Freudenreich et al. [62] showed using both pulsed field electrophoresis of the yeast chromosomes and genetic assays, that long CTG·CAG tracts (130–250 repeats) induce DSB in a length-dependent manner and that these sequences have a high propensity for expansions during yeast transformation when the recombination event is initiated next to the repeat tract. The expansions of the CTG·CAG sequences in yeast were even more pronounced during meiosis when compared to the mitotic division [63–66]. Jankowski et al. [65] attributed these instabilities to DSB-induced recombination. Sequences as short as $(CTG·CAG)_{64}$ induced the *spo11* dependent DSB formation during meiosis [66]. Their repair resulted in deletions as well as in expansions of the CTG·CAG tract.

DSB were also artificially induced in yeast *in vivo* by use of the homing endonuclease (HO) [67]. In a study with short CTG·CAG tracts, almost 20% of DSB-induced gene conversion events led to TRS deletions (almost exclusively in the recipient locus) [68]. When a longer repeat $(CTG·CAG)_{98}$ was used, gene conversion resulted in frequent expansions (~30% of events) [69]. In the absence of the HO endonuclease, only contractions were observed. Interestingly, no expansions were detected when CTG·CAG repeats were replaced with the $(CAA·TTG)_{87}$ tract, substantiating the role of non-B DNA structures in the instability processes since neither CAA nor TTG repeat tracts have been shown to form stable hairpin structures [70].

Recently, Richard et al. [60] proposed a unifying model for CTG·CAG instabilities observed during both mitotic and meiotic gene conversion in yeast. This model is based on the synthesis-dependent strand annealing (SDSA) pathway [71], modified for the specificity of the repetitive sequences. Four crucial considerations are accommodated in this model: (i) the initial formation of the DSB in one of the TRS tracts; (ii) the importance of the unusual DNA structures in generation of the repeat instability; (iii) the absence of evidence for crossing-over exchange; and (iv) no change in the sequence of the donor/template DNA. The initial event of the SDSA pathway is an invasion of one or both DNA strands of the processed DSB ends into the DNA template followed by DNA synthesis and dissociation of the newly synthesized strands from the template (Fig. 1-1B). Out-of-register re-annealing of the unwound DNA strands together with hairpin structure formation on either of the strands results in the expansions or deletions of the TRS tract. Hence, the role of the unusual DNA structures in the recombinational instability of TRS tracts is not only limited to the initiation of the recombination event (via DSB induction) but it is also important at each of the subsequent synthesis and annealing steps, where slipped structures can be formed.

c. TRS Recombination and DSB Repair in Mammalian Models

The involvement of recombination and DSB repair in TRS instability has not been extensively studied in mammalian cells. Results of recent experiments in CHO cells demonstrated the influence of long CTG·CAG repeats (98 and 183 repeats) but not (CTG·CAG)$_{17}$ on the recombination between two copies of the *APRT* gene [72]. Meservy et al. [72] examined the changes in the CTG·CAG repeats initiated by homologous recombination between nearby *APRT* sequences. Long repeats underwent frequent large deletions (tenfold increase due to recombination). The frequency of the recombination-associated rearrangements extending outside of the CTG·CAG region was also increased over fiftyfold. The presence of the CTG·CAG repeats also had a reciprocal effect on the types of the recombination events observed. In the cell lines harboring (CTG·CAG)$_{183}$ repeats, the rate of the gene conversion events between the *APRT loci* was three- to fourfold lower and, in contrast, the rate of crossing-over was two- to threefold higher when compared to the control cell lines lacking the repeats [72].

Homologous recombination is a primary pathway of DSB repair in bacteria and lower eukaryotes including yeast [73]. In mammalian cells, nonhomologous end joining (NHEJ) is the primary means of DSB repair [40, 74]. The influence of the mammalian DSB repair on the stability of CTG·CAG tracts was studied in COS1 cells [75]. The DNA breaks were artificially introduced into the repeat region prior to transfection. The vast majority of the DSB repair events resulted in deletion of the TRS tracts, perhaps due to the structure formation at the repeat-containing DNA ends (Fig. 1-1C). It would be interesting to analyze the TRS instability after *in vivo* induction of the DSB in mammalian cells, since different pathways are known to participate in the repair of the breaks generated *in vitro* compared to those induced *in vivo* [76].

d. TRS Instability and DSB Repair in Transgenic Mouse Models

Mouse genetic experiments support the involvement of DSB repair in CTG·CAG repeats instability. Savouret et al. [77] tested the influence of several genetic products in both homologous recombination (HR) and NHEJ (*Rad52*, *Rad54*, and *DNA-PKcs*) on intergenerational and somatic instability of CTG·CAG repeats in transgenic mice. No change in the repeat stability was observed in *Rad54* and *DNA-PKcs* knockouts eliminating DSBR-HR as a likely mechanism of TRS expansions in their system.

However, lack of *Rad52* led to a significant decrease in the size of the expansions during intergenerational transmission. This implicated the contribution of the single-strand annealing (SSA, Fig. 1-1D) pathway in the CTG·CAG repeats instability in mice.

In summary, substantial evidence has accumulated to support the following general model of DSB/recombination-mediated TRS instability. Structures formed by TRS related to the human neurological diseases are capable of blocking DNA replication. They can also be recognized and subjected to repair by endonuclease excision. These processes (arrest of the replication fork progression as well as nucleolytic repair of the "structural lesion") may induce DNA strand discontinuities (nicks/breaks) which are very efficient substrates for recombinational repair. The repair of the DSB by the intra-allelic as well the inter-allelic (or ectopic) processes can lead to substantial TRS instability. It will be interesting to learn in the future how these processes are conducted in humans. A portion of this review has been presented previously [78]; also, the reader is directed to the chapter in this volume by Napierala et al. [79] which reviews other aspects of recombination-repair as a mechanism for genetic instabilities.

2. REPLICATION-REPAIR

DNA replication slippage was considered in the 1990s to be the major factor influencing the genetic instabilities of various TRS including CTG·CAG, GAA·TTC, CGG·CCG, and GAC·GTC, which are implicated in several hereditary diseases [10, 12, 42, 56, 62, 80–84]. Recently, replication was shown to influence the genetic instabilities of the CCTG·CAGG tetranucleotide repeats associated with DM2 [85]. The most interesting components related to replication include the pausing of the DNA polymerase at several non-B DNA structures, which causes the replication fork to collapse and then involves repair and recombination machinery to help restart replication (reviewed in [78]). In addition, the orientation of the repeat sequences relative to the direction of replication, the location of the repeat sequences, and the endonuclease involvement are also fascinating components of this process (reviewed in [78]).

3. MISMATCH REPAIR (MMR)

Plasmids harboring expanded CTG·CAG repeats in *E. coli* were found to be more stable when grown in methyldirected MMR-deficient strains than in the parental background [86]. It was subsequently shown that human MSH2 preferentially bind MMR proteins to looped-out secondary structures formed by CTG·CAG repeats [87]. The role of MMR was later refined by experiments that showed that active mismatch repair stabilized small instabilities (>8 repeats), but increased the occurrence of large deletions [88–91].

Studies in mouse model systems have suggested that MMR plays a significant role in regulation of the instability of CTG·CAG repeats. Somatic tissues of transgenic mice showed a substantial increase in instability in the MSH6-deficient background, whereas the somatic instability was completely blocked in the MSH3-deficient background [36]. This differential effect may be attributable to the competitive binding of MSH3 and MSH6 to MSH2 to form a functional complex. Somatic instability and germline expansions were found to be dependent upon MSH2 in Huntington's disease transgenic mice [33, 92, 93]. In experiments with DM1 transgenic mice with >300 repeats, a strong dependence was found on MSH2 for contractions during germline transmission and in spermatogonia [34, 94]. Most recently, a Mutl homologue, Pms2, was determined to increase the somatic mosaicism of CTG·CAG repeats [35]. These results support the hypothesis that mismatch repair is a key player in genetic instabilities, possibly through MSH2 or other downstream proteins, but its precise role that results in the expansions observed in the human diseases is still unclear.

4. Nucleotide Excision Repair (NER)

NER recognizes helical distortions in DNA generally created by bulky adducts such as pyrimidine dimers. NER also results in the recognition and removal of DNA loops often associated with the secondary structures formed by trinucleotide repeats. The UvrA protein preferentially bind to repeat loops of 1, 2 or 17 CAG repeats, and mutations in UvrA in *E. coli* affect the instability of the CTG·CAG repeats harbored on the plasmid [95, 96]. However, experiments in yeast have shown that deletion of Rad1, which is partially involved in NER, does not stimulate the instability of CTG·CAG repeats [19]. Hence, further work will be required to clarify the influence of NER on TRS instability.

C. Microsatellite Repeat Instability and Population Genetics

This edition touches the origin and population genetics of the mutant chromosome in some diseases such as DM1, DM2, Friedreich's ataxia, and some SCAs (see Parts II, VI, and VII). In most repeat-expansion diseases, families from different populations share the ancestral haplotype(s). An increased repeat size to the upper normal range or the "premutation" range and loss of interrupting sequences are considered to be steps toward the expansion to full mutation alleles. Data from these studies provide additional insights in the mechanism of repeat instability.

IV. PATHOGENIC MECHANISMS OF NEUROLOGICAL DISEASES CAUSED BY EXPANDED MICROSATELLITE REPEATS

A. Genotype–Phenotype Correlation

There appear to be several interesting genotype–phenotype correlations in human diseases caused by microsatellite expansions. The primary target organ of these diseases is either the central nervous system (CNS) (see Parts III–VII) or muscle (see Part II). Moreover, certain types of expansion mutations have shown typical phenotypic characteristics. For example, polyglutamine-coding (CAG)n expansions have consistently been shown to cause degenerative CNS diseases (see Parts IV, V, and VII), while large (CGG)n expansions in 5′ UTR have been seen in mental retardation syndrome with folate-sensitive fragile sites (see Part III). Large expansions of noncoding CTG-containing repeats appear to cause the phenotype of myotonic dystrophies (see Part II). Additionally, expansions of polyalanine-coding triplets cause developmental disorders with dysmorphic features, although the OPMD mutation leading to a polyalanine expansion in the *PABPN1* gene is an exception (see Part VIII) [6]. Since each mutation type has been shown to share a similar disease-causing molecular mechanism, the prediction of phenotype based on the type of a repeat expansion mutation may be justified, while prediction of the mutation from the phenotype remains to be difficult because different types of mutations have caused diseases with similar phenotypes.

Anticipation, a clinical phenomenon defined as progressively earlier onset of the disease with increasing severity in successive generations (see Part II), has become a hallmark of diseases with microsatellite expansions. However, it should be noted that anticipation may not be present in some diseases caused by repeat expansions, especially when the repeat size does not increase in successive generations. In recessive disorders, like Friedreich's ataxia, anticipation is automatically precluded because of the lack of vertical transmission of the disease phenotype. Conversely, when anticipation is present, the repeat instability may not always show increases of the repeat size in successive generations. This discrepancy may be attributable to a high degree of somatic instability in DM2 [97] and in some cases of DM1 [98].

However, these molecular explanations are not applicable to anticipation reported in SCA6 families where the repeat expansion is stable [99]. Whether anticipation is a true biological event or a result of ascertainment bias needs to be carefully investigated in each disease [100].

Conspicuous anticipation is often dependent on the gender of the transmitting parent. In many diseases, especially those with polyglutamine expansions, infantile- and juvenile-onset cases (caused by very large expansions) and *de novo* mutation cases (resulted from expansion of alleles in the upper normal range into alleles in the disease range) are found with paternal transmission of the repeat [101, 102]. In these diseases, expanded repeats in sperm show a greater degree of repeat-size mosaicism than those in blood cells, with the size distribution skewed toward expansion. In fragile X syndrome and DM1, the striking anticipation giving rise to children with mental retardation is typically seen with maternal transmission. The paucity of very large alleles in sperm of affected males may account for the almost exclusive maternal origin of these cases [103–105].

B. Pathogenic Mechanisms

When the identification of expanded microsatellite repeats defined the genetic mutations of these diseases, the molecular mechanisms by which repeat expansions cause the disease phenotypes became a major target of research. Histopathological examinations of affected tissues of patients, cell culture derived from patients' tissues, and genetic animal models provided substantial progresses in our understanding of the pathogenic mechanism of each disease. These studies led to three major models of the pathogenic mechanisms: (a) loss of function of the gene(s) by the repeat expansion, (b) gain of toxic function by the mutant protein product, and (c) gain of toxic function by the mutant RNA transcript. In diseases caused by repeat expansions, the pathogenic mechanism of each disease is different; however, diseases caused by expanded repeats of similar characteristics, such as repeat unit sequence, the location of the repeat in the gene, and the repeat expansion size, tend to share pathogenic mechanisms. It should also be noted that these mechanisms may not necessarily be mutually exclusive in some diseases.

1. Gain of Function by Expansion of Coding Repeats

All diseases caused by expanded CAG·CTG repeats coding for polyglutamines have been neurodegenerative diseases involving various parts of the central nervous system. These diseases include spinobulbar muscular atrophy (SBMA; Kennedy's disease; see Part IV), Huntington's disease (HD; see Part V), dentatorubral pallidoluysian atrophy (DRPLA), and several SCAs (see Part VII), which are discussed in detail in this book. The gene where the expanded CAG repeat is located was known for its function in SBMA (*AR*, androgen receptor), SCA6 (*CACNA1A*, P/Q type calcium channel), and SCA17 (*TBP*, TATA binding protein). The functions of the genes of the remaining disorders were unknown at the time of the identification of the mutations, and many of these are still being investigated.

The pathogenic mechanism of most diseases in this category has been attributed to a gain of toxic function by the respective mutant proteins, which contain expanded polyglutamines. The exact nature of gain of function may differ from one disease to another, depending on the biochemical function, expression pattern, subcellular localization, and catabolism of the respective mutant protein products, and may be altered by genetic and environmental modifiers. These processes have been studied particularly in detail in HD [106, 107] and SCA1 [108, 109]. Oligomers and insoluble aggregates of mutant proteins or protein fragments containing expanded polyglutamines have been detected in cells of patients with these diseases. The aggregates contain polymers of the mutant proteins in β-sheet structures, and contain ubiquitin and other proteins. However, the pathogenic role of aggregates is controversial with three very different views, in which aggregates are thought to be (a) pathogenic, (b) nonpathogenic (i.e., "innocent bystanders"), or (c) protective (i.e., eliminating toxic soluble mutant proteins into aggregates) [106, 107].

Although a gain of function is an attractive pathogenic mechanism for dominantly inherited diseases, it may not be the only problem in diseases with an expansion of a coding (CAG)n repeat. First, the polyglutamine expansion may alter the primary function of the protein, such as androgen receptor in SBMA [110] and P/Q type calcium channel in SCA6 [20]. Second, dominant negative effects of the mutant protein may lead to a loss of function of the protein [106, 107, 111]. Furthermore, a gain of function by the mutant mRNA, which contains an expanded CAG repeat, may need to be investigated as a potential contributor to the pathogenic mechanism of these diseases [112].

It should be noted that these polyglutamine-coding (CAG)n repeats are not the only coding repeats that cause human neurological diseases. The mutation of OPMD is an expansion of polyalanine-coding GCG repeat in the poly(A) binding protein nuclear 1 (*PABPN1*) gene (see Part VIII) [5]. The expansion has been thought to induce the formation of the filamentous nuclear inclusions. In Huntington disease-like 2 (HDL2; see Part V) [113] an expanded CTG repeat is located in an alternatively spliced exon (exon 2A) of the junctophilin-3 (*JPH3*) gene. Two of the three mRNA isoforms resulting from the alternative splicing place the repeat in the coding region in two different reading frames: one codes for polyalanine and the other for polyleucine. The third isoform put the repeat in the noncoding 3' UTR. Although no genes have been found on the opposite strand of *JPH3*, a brain from a patient with HDL2 showed neuronal intranuclear inclusions immunoreactive for expanded polyglutamine repeats [114]. Pathogenic roles of aggregated proteins in these diseases are intriguing, especially for the potential overlap of the pathogenic mechanism with the polyglutamine diseases.

2. SIMPLE LOSS OR GAIN OF FUNCTION BY EXPANSIONS OF UNTRANSLATED REPEATS

Expanded repeats can lead to a simple loss of function by interfering with the transcription of the gene in different ways. Large expansions of CGG·CCG repeats in the 5' UTR cause fragile X syndrome [21] and FRAXE Mental Retardation [22] by decreasing the transcription of the *FMR1* and *FMR2* genes through DNA methylation (see Part III). In Friedreich's ataxia, an expansion of an intronic GAA·TTC repeat hinders the transcription of the *FRDA1* gene by forming the "sticky" DNA structure (see Part VI) [17, 25]. Conversely, a simple gain of function has been postulated in SCA12, in which an expansion of the (CAG)n repeat in the 5' UTR increases the transcription of a protein phosphatase subunit gene, *PPP2RB* (see Part VII) [115, 116].

3. TRANS-DOMINANT RNA GAIN OF FUNCTION

In DM1 the mRNA of the *DMPK* gene contains an expanded (CTG)n repeat in the 3' UTR. Data from genetic mouse models, cell culture models, and patient-derived tissues strongly suggest that the mutant mRNA with the expanded (CUG)n repeat accumulates in the nuclear foci and sequesters proteins, especially those known as the muscleblind proteins, which regulate the processing of a wide variety of gene transcripts (see Part II) [117–119]. A huge expansion of a CCTG·CAGG repeat in intron 1 of the *ZNF9* gene has been postulated to cause DM2 by a mechanism similar to DM1 [118, 120]. In SCA8, the disease mechanism involving an expansion of a (CTG)n repeat in 3' UTR is still unclear. It has been postulated that the SCA8 mRNA encodes a sequence antisense to the *KLH1* gene on the opposite strand, and the expansion of the (CUG)n repeat perturbs the regulation of *KLH1* expression (see Part VII) [121, 122].

Small (CGG)n expansions (in the intermediate range) of *FMR1* (*FRAXA*) causes an adult-onset neurodegenerative disease known as "fragile X associated tremor and ataxia syndrome (FXTAS)" (see Part III) [123]. In this disease, the FMR1 mRNA level has been shown to be elevated, raising a possibility of a gain of function by the mutant FMR1 mRNA. Another disease whose pathogenic mechanism may involve a *trans*-dominant gain of function by RNA is SCA10, in which an expanded AUUCU repeat in the *ATXN10* transcript shows a very large expansion (see Part VII).

V. FUTURE DIRECTIONS FOR RESEARCH ON REPEATS AND GENOMIC INSTABILITIES IN NEUROLOGICAL DISORDERS

The spectrum of the expanded repeat motif that causes human neurological diseases may become even broader. Not only new TRS but also new tetra-, penta-, and other oligo-nucleotide repeats may be discovered as pathogenic repeats in the future. No human diseases have been reported with an expansion of a mono- or dinucleotide repeat. However, it is conceivable that their instability could also cause diseases or disease susceptibilities by altering gene functions. Polymorphic dinucleotide repeats in the promoter region have been shown to regulate the transcription by changing the DNA structure in some genes [124, 125].

Substantial advances have been made in understanding mechanisms of the repeat instability and expansion. Future studies on expanded repeats in experimental systems using *in vitro*, *E. coli*, yeast, cell culture, and transgenic animal models should shed light on the mechanism of the repeat instability in human diseases. Some of the transgenic animals closely recapitulate most of the characteristics of the instability of expanded repeats in human diseases (see Part X).

We anticipate that further elucidation of the pathogenic mechanisms of diseases caused by expanded repeats would lead to the development of rational treatments based on the disease mechanisms. However, the pathophysiology of these diseases may be complex and variable, involving multiple downstream pathways, which may vary in different cells or tissues. Therefore, the ideal treatment should be to normalize the mutation rather than correcting the downstream consequences of the mutation, and understanding mechanisms of repeat expansion is critical for development of treatment strategies to reduce the repeat expansion size. Adverse genome-wide effects of such treatments may have relevance to microsatellite instability observed in some cancers (see Part XIV).

VI. CONCLUDING REMARKS

Substantial advances in studies of the repeat instability and pathogenic mechanisms have been made since expanded microsatellite repeats were identified as disease-causing mutations. These studies are important for understanding the biological significance of the repeat instability and for future development of rational therapeutics for diseases caused by the repeat expansions.

Acknowledgments

This work was supported by National Institutes of Health grants ES11347 (RDW) and NS041547 (TA), the Robert A. Welch Foundation (RDW), the Friedreich's Ataxia Research Alliance (RDW), the Seek a Miracle Foundation (RDW), and the Muscular Dystrophy Association (TA). In addition, the authors thank our past and present coworkers for their numerous helpful suggestions.

References

1. Liquori, C. L., Ricker, K., Moseley, M. L., Jacobsen, J. F., Kress, W., Naylor, S. L., Day, J. W., and Ranum, L. P. (2001). Myotonic dystrophy type 2 caused by a CCTG expansion in intron 1 of ZNF9. *Science* **293**, 864–867.
2. Matsuura, T., Yamagata, T., Burgess, D. L., Rasmussen, A., Grewal, R. P., Watase, K., Khajavi, M., McCall, A. E., Davis, C. F., Zu, L., Achari, M., Pulst, S. M., Alonso, E., Noebels, J. L., Nelson, D. L., Zoghbi, H. Y., and Ashizawa, T. (2000). Large expansion of the ATTCT pentanucleotide repeat in spinocerebellar ataxia type 10. *Nat. Genet.* **26**, 191–194.
3. Virtaneva, K., D'Amato, E., Miao, J., Koskiniemi, M., Norio, R., Avanzini, G., Franceschetti, S., Michelucci, R., Tassinari, C. A., Omer, S., Pennacchio, L. A., Myers, R. M., Dieguez-Lucena, J. L., Krahe, R., de la Chapelle, A., and Lehesjoki, A. E. (1997). Unstable minisatellite expansion causing recessively inherited myoclonus epilepsy, EPM1. *Nat. Genet.* **15**, 393–396.
4. van Deutekom, J. C., Wijmenga, C., van Tienhoven, E. A., Gruter, A. M., Hewitt, J. E., Padberg, G. W., van Ommen, G. J., Hofker, M. H., and Frants, R. R. (1993). FSHD associated DNA rearrangements are due to deletions of integral copies of a 3.2 kb tandemly repeated unit. *Hum. Mol. Genet.* **2**, 2037–2042.
5. Brais, B., Bouchard, J. P., Xie, Y. G., Rochefort, D. L., Chretien, N., Tome, F. M., Lafreniere, R. G., Rommens, J. M., Uyama, E., Nohira, O., Blumen, S., Korczyn, A. D., Heutink, P., Mathieu, J., Duranceau, A., Codere, F., Fardeau, M., and Rouleau, G. A. (1998). Short GCG expansions in the PABP2 gene cause oculopharyngeal muscular dystrophy. *Nat. Genet.* **18**, 164–167.
6. Albrecht, A., and Mundlos, S. (2005). The other trinucleotide repeat, polyalanine expansion disorders. *Curr. Opin. Genet. Dev* **15**, 285–293.
7. Litt, M., and Luty, J. A. (1989). A hypervariable microsatellite revealed by in vitro amplification of a dinucleotide repeat within the cardiac muscle actin gene. *Am. J. Hum. Genet.* **44**, 397–401.
8. Weber, J. L., and May, P. E. (1989). Abundant class of human DNA polymorphisms which can be typed using the polymerase chain reaction. *Am. J. Hum. Genet.* **44**, 388–396.
9. Leach, F. S., Nicolaides, N. C., Papadopoulos, N., Liu, B., Jen, J., Parsons, R., Peltomaki, P., Sistonen, P., Aaltonen, L. A., Nystrom-Lahti, M., Guan, X.-Y., Zhang, J., Meltzer, P. S., Yu, J.-W., Kao, F.-T., Chen, D. J., Cerosaletti, K. M., Keith Fournier, R. E., Todd, S., Lewis, T., Leach, R. J., Naylor, S. L., Weissenbach, J., Mecklin, J.-P., Jarvinen, H., Petersen, G. M., Hamilton, S. R., Green, J., Jass, J., Watson, P., Lynch, H. T., Trent, J. M., de la Chapelle, A., Kinzler, K. W., and Vogelstein, B. (1993). Mutations of a mutS homolog in hereditary nonpolyposis colorectal cancer. *Cell* **75**, 1215–1225.
10. Wells, R. D., and Warren, S. T. (1998). Genetic Instabilities and Hereditary Neurological Diseases. San Diego, Academic Press.
11. Bowater, R. P., and Wells, R. D. (2001). The intrinsically unstable life of DNA triplet repeats associated with human hereditary disorders. *Prog. Nucleic Acid Res. Mol. Biol.* **66**, 159–202.
12. Sinden, R. R., Potaman, V. N., Oussatcheva, E. A., Pearson, C. E., Lyubchenko, Y. L., and Shlyakhtenko, L. S. (2002). Triplet repeat DNA structures and human genetic disease, dynamic mutations from dynamic DNA. *J. Biosci.* **27**, 53–65.
13. Rubinsztein, D. C., and Hayden, M. R. (1998). Analysis of Triplet Repeat Disorders. Oxford, Bios Scientific Publisher.
14. Napierala, M., Dere, R., Vetcher, A., and Wells, R. D. (2004). Structure-dependent recombination hot spot activity of GAA·TTC sequences from intron 1 of the Friedreich's ataxia gene. *J. Biol. Chem.* **279**, 6444–6454.
15. Potaman, V. N., Bissler, J. J., Hashem, V. I., Oussatcheva, E. A., Lu, L., Shlyakhtenko, L. S., Lyubchenko, Y. L., Matsuura, T., Ashizawa, T., Leffak, M., Benham, C. J., and Sinden, R. R. (2003). Unpaired structures in SCA10 (ATTCT)n·(AGAAT)n repeats. *J. Mol. Biol.* **326**, 1095–1111.
16. Siianova, E., and Mirkin, S. M. (2001). Expansion of trinucleotide repeats. *Mol. Biol. (Mosk)* **35**, 208–223.
17. Sakamoto, N., Chastain, P. D., Parniewski, P., Ohshima, K., Pandolfo, M., Griffith, J. D., and Wells, R. D. (1999). Sticky DNA, self-association properties of long GAA·TTC repeats in R.R.Y triplex structures from Friedreich's ataxia. *Mol. Cell* **3**, 465–475.
18. Darlow, J. M., and Leach, D. R. (1998). Secondary structures in d(CGG) and d(CCG) repeat tracts. *J. Mol. Biol.* **275**, 3–16.
19. Freudenreich, C. H., Kantrow, S. M., and Zakian, V. A. (1998). Expansion and length-dependent fragility of CTG repeats in yeast. *Science* **279**, 853–856.
20. Zhuchenko, O., Bailey, J., Bonnen, P., Ashizawa, T., Stockton, D. W., Amos, C., Dobyns, W. B., Subramony, S. H., Zoghbi, H. Y., and Lee, C. C. (1997). Autosomal dominant cerebellar ataxia (SCA6) associated with small polyglutamine expansions in the alpha 1A-voltage-dependent calcium channel. *Nat. Genet.* **15**, 62–69.
21. Fu, Y. H., Kuhl, D. P., Pizzuti, A., Pieretti, M., Sutcliffe, J. S., Richards, S., Verkerk, A. J., Holden, J. J., Fenwick, R. G., Jr., Warren, S. T., Oostra, B. A., Nelson, D. L., and Caskey, C. T. (1991). Variation of the CGG repeat at the fragile X site results in genetic instability, resolution of the Sherman paradox. *Cell* **67**, 1047–1058.
22. Knight, S. J., Flannery, A. V., Hirst, M. C., Campbell, L., Christodoulou, Z., Phelps, S. R., Pointon, J., Middleton-Price, H. R., Barnicoat, A., Pembrey, M. E., Holland, J., Oostra, B. A., Bobrow, M., and Davies, K. E. (1993). Trinucleotide repeat amplification and hypermethylation of a CpG island in FRAXE mental retardation. *Cell* **74**, 127–134.
23. Fu, Y-H., Pizzuti A., Fenwick, R., King, J., Rajnarayan, S., Dunne, P. W., Dubel, J., Nasser, G. A., Ashizawa, T., de Jong, P., Wieringa, B., Korneluk, R., Perryman, M. B., Epstein, H. F., and Caskey, C. T. (1992) An unstable triplet repeat in a gene related to myotonic muscular dystrophy. *Science* **225**, 1256–1258.

24. Brook, J. D., McCurrach, M. E., Harley, H. G., Buckler, A. J., Church, D., Aburatani, H., Hunter, K., Stanton, V. P., Thirion, J. P., Hudson, T., Sohn, R., Zemelman, B., Snell, R. G., Rundle, S. A., Crow, S., Davies, J., Shelbourne, P., Buxton, J., Jones, C., Juvonen, V., Johnson, K., Harper, P. S., Shaw, D. J., and Housman, D. E. (1992). Molecular basis of myotonic dystrophy, expansion of a trinucleotide (CTG) repeat at the 3′ end of a transcript encoding a protein kinase family member. *Cell* **68**, 799–808.
25. Campuzano, V., Montermini, L., Molto, M. D., Pianese, L., Cossee, M., Cavalcanti, F., Monros, E., Rodius, F., Duclos, F., Monticelli, A., Zara, F., Canizares, J., Koutnikova, H., Bidichandani, S. I., Gellera, C., Brice, A., Trouillas, P., De Michele, G., Filla, A., De Frutos, R., Palau, F., Patel, P. I., Di Donato, S., Mandel, J. L., Cocozza, S., Koenig, M., and Pandolfo, M. (1996). Friedreich's ataxia, autosomal recessive disease caused by an intronic GAA triplet repeat expansion. *Science* **271**, 1423–1427.
26. Nakamura, K., Jeong, S. Y., Uchihara, T., Anno, M., Nagashima, K., Nagashima, T., Ikeda, S., Tsuji, S., and Kanazawa, I. (2001). SCA17, a novel autosomal dominant cerebellar ataxia caused by an expanded polyglutamine in TATA-binding protein. *Hum. Mol. Genet.* **10**, 1441–1448.
27. Fujigasaki, H., Martin, J. J., De Deyn, P. P., Camuzat, A., Deffond, D., Stevanin, G., Dermaut, B., Van Broeckhoven, C., Durr, A., and Brice, A. (2001). CAG repeat expansion in the TATA box-binding protein gene causes autosomal dominant cerebellar ataxia. *Brain* **124**, 1939–1947.
28. Chung, M. Y., Ranum, L. P., Duvick, L. A., Servadio, A., Zoghbi, H. Y., and Orr, H. T. (1993). Evidence for a mechanism predisposing to intergenerational CAG repeat instability in spinocerebellar ataxia type I. *Nat. Genet.* **5**, 254–258.
29. Choudhry, S., Mukerji, M., Srivastava, A. K., Jain, S., and Brahmachari, S. K. (2001). CAG repeat instability at SCA2 locus, anchoring CAA interruptions and linked single nucleotide polymorphisms. *Hum. Mol. Genet.* **10**, 2437–2446.
30. Brock, G. J., Anderson, N. H., and Monckton, D. G. (1999). Cis-acting modifiers of expanded CAG/CTG triplet repeat expandability, associations with flanking GC content and proximity to CpG islands. *Hum. Mol. Genet.* **8**, 1061–1067.
31. Maciel, P., Gaspar, C., Guimaraes, L., Goto, J., Lopes-Cendes, I., Hayes, S., Arvidsson, K., Dias, A., Sequeiros, J., Sousa, A., and Rouleau, G. A. (1999). Study of three intragenic polymorphisms in the Machado-Joseph disease gene (MJD1) in relation to genetic instability of the (CAG)n tract. *Eur. J. Hum. Genet.* **7**, 147–156.
32. Crawford, D. C., Wilson, B., and Sherman, S. L. (2000). Factors involved in the initial mutation of the fragile X CGG repeat as determined by sperm small pool PCR. *Hum. Mol. Genet.* **9**, 2909–2918.
33. Manley, K., Shirley, T. L., Flaherty, L., and Messer, A. (1999). Msh2 deficiency prevents in vivo somatic instability of the CAG repeat in Huntington disease transgenic mice. *Nat. Genet.* **23**, 471–473.
34. Savouret, C., Brisson, E., Essers, J., Kanaar, R., Pastink, A., te Riele, H., Junien, C., and Gourdon, G. (2003). CTG repeat instability and size variation timing in DNA repair-deficient mice. *Embo. J.* **22**, 2264–2273.
35. Gomes-Pereira, M., Fortune, M. T., Ingram, L., McAbney, J. P., and Monckton, D. G. (2004). Pms2 is a genetic enhancer of trinucleotide CAG.CTG repeat somatic mosaicism, implications for the mechanism of triplet repeat expansion. *Hum. Mol. Genet.* **13**, 1815–1825.
36. van den Broek, W. J., Nelen, M. R., Wansink, D. G., Coerwinkel, M. M., te Riele, H., Groenen, P. J., and Wieringa, B. (2002). Somatic expansion behaviour of the (CTG)n repeat in myotonic dystrophy knock-in mice is differentially affected by Msh3 and Msh6 mismatch-repair proteins. *Hum. Mol. Genet.* **11**, 191–198.
37. Monckton, D. G., Cayuela, M. L., Gould, F. K., Brock, G. J., Silva, R., and Ashizawa, T. (1999). Very large (CAG)(n) DNA repeat expansions in the sperm of two spinocerebellar ataxia type 7 males. *Hum. Mol. Genet.* **8**, 2473–2478.
38. Kennedy, L., and Shelbourne, P. F. (2000). Dramatic mutation instability in HD mouse striatum, does polyglutamine load contribute to cell-specific vulnerability in Huntington's disease? *Hum. Mol. Genet.* **9**, 2539–2544.
39. Watase, K., Venken, K. J., Sun, Y., Orr, H. T., and Zoghbi, H. Y. (2003). Regional differences of somatic CAG repeat instability do not account for selective neuronal vulnerability in a knock-in mouse model of SCA1. *Hum. Mol. Genet.* **12**, 2789–2795.
40. Bishop, A. J., and Schiestl, R. H. (2000). Homologous recombination as a mechanisms for genome rearrangements: Environmental, and genetic effects. *Hum. Mol. Genet.* **9**, 2427–2434.
41. Lahue, R. S., and Slater, D. L. (2003). DNA repair and trinucleotide repeat instability. *Front. Biosci.* **8**, 653–665.
42. Lenzmeier, B. A., and Freudenreich, C. H. (2003). Trinucleotide repeat instability: A hairpin curve at the crossroads of replication, recombination, and repair. *Cytogenet. Genome Res.* **100**, 7–24.
43. Jakupciak, J. P., and Wells, R. D. (2000). Genetic instabilities of triplet repeat sequences by recombination. *IUBMB Life* **50**, 355–359.
44. Cleary, J. D., and Pearson, C. E. (2003). The contribution of cis-elements to disease-associated repeat instability: Clinical and experimental evidence. *Cytogenet. Genome Res.* **100**, 25–55.
45. Warren, S. T. (1997). Polyalanine expansion in synpolydactyly might result from unequal crossing-over of HOXD13. *Science* **275**, 408–409.
46. Amiel, J., Trochet, D., Clement-Ziza, M., Munnich, A., and Lyonnet, S. (2004). Polyalanine expansions in human. *Hum. Mol. Genet.* **13**, 235–243.
47. Brown, L. Y., and Brown, S. A. (2004). Alanine tracts: The expanding story of human illness and trinucleotide repeats. *Trends. Genet.* **20**, 51–58.
48. Sargent, R. G., Merrihew, R. V., Nairn, R., Adair, G., Meuth, M., and Wilson, J. H. (1996). The influence of a $(GT)_{29}$ microsatellite sequence on homologous recombination in the hamster APRT gene. *Nucleic Acids Res.* **24**, 746–753.
49. Majewski, J., and Ott, J. (2000). GT repeats are associated with recombination on human chromosome 22. *Genome Res.* **10**, 1108–1114.
50. Benet, A., Molla, G., and Azorin, F. (2000). d(GA·TC)(n) microsatellite DNA sequences enhance homologous DNA recombination in SV40 minichromosomes. *Nucleic Acids Res.* **28**, 4617–4622.
51. Shiroishi, T., Sagai, T., and Moriwaki, K. (1993). Hotspots of meiotic recombination in the mouse major histocompatibility complex. *Genetica* **88**, 187–196.
52. Kirkpatrick, D. T., Wang, Y. H., Dominska, M., Griffith, J. D., and Petes, T. D. (1999). Control of meiotic recombination and gene expression in yeast by a simple repetitive DNA sequence that excludes nucleosomes. *Mol. Cell. Biol.* **19**, 7661–7671.
53. Napierala, M., Parniewski, P. P., Pluciennik, A., and Wells, R. D. (2002). Long CTG·CAG repeat sequences markedly stimulate intramolecular recombination. *J. Biol. Chem.* **277**, 34087–34100.
54. Pluciennik, A., Iyer, R. R., Napierala, M., Larson, J. E., Filutowicz, M., and Wells, R. D. (2002). Long CTG·CAG repeats from myotonic dystrophy are preferred sites for intermolecular recombination. *J. Biol. Chem.* **277**, 34074–34086.
55. Krasilnikova, M. M., and Mirkin, S. M. (2004). Replication stalling at Friedreich's ataxia (GAA)n repeats in vivo. *Mol. Cell. Biol.* **24**, 2286–2295.
56. Samadashwily, G. M, Raca, G., and Mirkin, S. M. (1997). Trinucleotide repeats affect DNA replication *in vivo*. *Nat. Genet.* **17**, 298–304.

57. Ohshima, K., Kang, S., Larson, J. E., and Wells, R. D. (1996). Cloning, characterization, and properties of seven triplet repeat DNA sequences. *J. Biol. Chem.* **271**, 16773–16783.
58. Hebert, M. L., Spitz, L. A., and Wells, R. D. (2004). DNA double-strand breaks induce deletion of CTG·CAG repeats in an orientation-dependent manner in Escherichia coli. *J. Mol. Biol.* **336**, 655–672.
59. Hashem, V. I., Roshce, W. A., and Sinden, R. R. (2004). Genetic recombination destabilizes (CTG)n·(CAG)n repeats in E. coli. *Mutat. Res.* **554**, 95–109.
60. Richard, G. F., Cyncynatus, C., and Dujon, B. (2003). Contractions and expansions of CAG·CTG trinucleotide repeats occur during ectopic gene conversion in yeast, by a MUS81-independent mechanism. *J. Mol. Biol.* **326**, 769–782.
61. Freudenreich, C. H., Stavenhagen, J. B., and Zakian, V. A. (1997). Stability of a CTG·CAG trinucleotide repeat in yeast is dependent on its orientation in the genome. *Mol. Cell. Biol.* **17**, 2090–2098.
62. Freudenreich, CH., Kantrow, S. M., and Zakian, V. A. (1998). Expansion and length-dependent fragility of CTG repeats in yeast. *Science* **279**, 853–856.
63. Arai, N., Akiyama, R., Niimi, N., Nakatsubo, H., and Inoue, T. (1999). Meiotic contraction of CAG repeats in Saccharomyces cerevisiae. *Genes Genet. Syst.* **74**, 159–167.
64. Cohen, H., Sears, D. D., Zenvirth, D., Hieter, P., and Simchen, G. (1999). Increased instability of human CTG repeat tracts on yeast artificial chromosomes during gametogenesis. *Mol. Cell. Biol.* **19**, 4153–4158.
65. Jankowski, C., Nasar, F., and Nag, D. K. (2000). Meiotic instability of CAG repeat tracts occurs by double-strand break repair in yeast. *Proc. Natl. Acad. Sci. USA* **97**, 2134–2139.
66. Jankowski, C., and Nag, D. K. (2002). Most meiotic CAG repeat tract-length alterations in yeast are SPO11 dependent. *Mol. Genet. Genome* **267**, 64–70.
67. Richard, G. F., and Paques, F. (2000). Mini- and microsatellite expansions: the recombination connection. *EMBO Rep*, **1**, 122–126.
68. Richard, G. F., Dujon, B., and Haber, J. E. (1999). Double-strand break repair can lead to high frequencies of deletions within short CAG·CTG trinucleotide repeats. *Mol. Gen. Genet.* **261**, 871–882.
69. Richard, G. F., Goellner, G. M., McMurray, C. T., and Haber, J. E. (2000). Recombination-induced CAG trinucleotide repeat expansions in yeast involve the MRE11-RAD50-XRS2 complex. *EMBO J.* **19**, 2381–2390.
70. Gacy, A. M., Goellner, G. Juranic, N., Macura, S., and McMurray, C. T. (1995). Trinucleotide repeats that expand in human disease form hairpin structures in vitro. *Cell* **81**, 533–540.
71. Nassif, N., Penney, J., Pal, S., Engels, W. R., and Gloor, G. B. (1994). Efficient copying of nonhomologous sequences from ectopic sites via P-element-induced gap repair. *Mol. Cell. Biol.* **14**, 1613–1625.
72. Meservy, J. L., Sargent, R. G., Iyer, R. R., Chan, F., McKenzie, G. J., Wells, R. D., and Wilson, J. H. (2003). Long CTG tracts from the myotonic dystrophy gene induce deletions and rearrangements during recombination at the APRT locus in CHO cells. *Mol. Cell. Biol.* **23**, 3152–3162.
73. Lieber, M. R., Ma, Y., Pannicke, U., and Schwarz, K. (2003). Mechanism and regulation of human non-homologous DNA end-joining. *Nat. Rev. Mol. Cell Biol.* **4**, 712–720.
74. Kanaar, R., Hoeijmakers, J. H., and van Gent, D. C. (1998). Molecular mechanisms of DNA double strand break repair. *Trends Cell Biol*, **8**, 483–489.
75. Marcadier, J. L., and Pearson, C. E. (2003). Fidelity of primate cell repair of a double-strand break within a (CTG)·(CAG) tract. Effect of slipped DNA structures. *J. Biol. Chem.* **278**, 33848–33856.
76. van Heemst, D., Brugmans, L., Verkaik, N. S., and van Gent, D. C. (2004). End-joining of blunt DNA double-strand breaks in mammalian fibroblasts is precise and requires DNA-PK and XRCC4. *DNA Repair (Amst)* **3**, 43–50.
77. Savouret, C., Brisson, E., Essers, J., Kanaar, R. Pastink, A., te Riele, H., Junien, C., and Gourdon, G. (2003). CTG repeat instability and size variation timing in DNA repair-deficient mice. *EMBO J* **22**, 2264–2273.
78. Wells, R. D., Dere, R., Hebert, M., Napierala, M., and Son, L. S. (2005). Advances in mechanisms of genetic instability related to hereditary neurological diseases. Survey and Summary for *Nucl. Acids Res.* **33**, 3785–3798.
79. Napierala, M., Dere, R., Hebert, M., and Wells, R. D. (2006). Involvement of genetic recombination in microsatellite instability in *Genetic Instabilities and Hereditary Neurological Diseases*, Second Edition, (Wells, R. D., and Ashizawa, T., eds.), Elsevier-Academic Press Inc.,
80. Kang, S. Jaworski, A., Ohshima, K., and Wells, R. D. (1995). Expansion and deletion of CTG repeats from human disease genes are determined by the direction of replication in E. coli. *Nat. Genet.* **10**, 213–218.
81. Mochmann, L. H., and Wells, R. D. (2004). Transcription influences the types of deletion and expansion products in an orientation-dependent manner from GAC·GTC repeats. *Nucleic Acids Res.* **32**, 4469–4479.
82. Shimizu, M., Gellibolian, R., Oostra, B. A., and Wells, R. D. (1996). Cloning, characterization and properties of plasmids containing CGG triplet repeats from the FMR-1 gene. *J. Mol. Biol.* **258**, 614–626.
83. Maurer, D. J., O'Callaghan, B. L., and Livingston, D. M. (1996). Orientation dependence of trinucleotide CAG repeat instability in *Saccharomyces cerevisiae*. *Mol. Cell. Biol.* **16**, 6617–6622.
84. Cleary, J. D., Nichol, K., Wang, Y. H., and Pearson, C. E. (2002). Evidence of cis-acting factors in replication-mediated trinucleotide repeat instability in primate cells. *Nat. Genet.* **31**, 37–46.
85. Dere, R., Napierala, M., Ranum, L. P., and Wells, R. D. (2004). Hairpin structure-forming propensity of the (CCTG·CAGG) tetranucleotide repeats contributes to the genetic instability associated with myotonic dystrophy type 2. *J. Biol. Chem.* **279**, 41715–41726.
86. Jaworski, A., Rosche, W. A., Gellibolian, R., Kang, S., Shimizu, M., Bowater, R. P., Sinden, R. R., and Wells, R. D. (1995). Mismatch repair in Escherichia coli enhances instability of (CTG)n triplet repeats from human hereditary diseases. *Proc. Natl. Acad. Sci. U S A* **92**, 11019–11023.
87. Pearson, C. E., Ewel, A., Acharya, S., Fishel, R. A., and Sinden, R. R. (1997). Human MSH2 binds to trinucleotide repeat DNA structures associated with neurodegenerative diseases. *Hum. Mol. Genet.* **6**, 1117–1123.
88. Wells, R. D., Parniewski, P., Pluciennik, A., Bacolla, A., Gellibolian, R., and Jaworski, A. (1998). Small slipped register genetic instabilities in Escherichia coli in triplet repeat sequences associated with hereditary neurological diseases. *J. Biol. Chem.* **273**, 19532–19541.
89. Schumacher, S., Fuchs, R. P., and Bichara, M. (1998). Expansion of CTG repeats from human disease genes is dependent upon replication mechanisms in Escherichia coli, the effect of long patch mismatch repair revisited. *J. Mol. Biol.* **279**, 1101–1110.
90. Schmidt, K. H., Abbott, C. M., and Leach, D. R. (2000). Two opposing effects of mismatch repair on CTG repeat instability in Escherichia coli. *Mol. Microbiol.* **35**, 463–471.
91. Parniewski, P., Jaworski, A., Wells, R. D., and Bowater, R. P. (2000). Length of CTG·CAG repeats determines the influence of mismatch repair on genetic instability. *J. Mol. Biol.* **299**, 865–874.
92. Kovtun, I. V., and McMurray, C. T. (2001). Trinucleotide expansion in haploid germ cells by gap repair. *Nat. Genet.* **27**, 407–411.

93. Wheeler, V. C., Lebel, L. A., Vrbanac, V., Teed, A., te Riele, H., and MacDonald, M. E. (2003). Mismatch repair gene Msh2 modifies the timing of early disease in Hdh(Q111) striatum. *Hum. Mol. Genet.* **12**, 273–281.
94. Savouret, C., Garcia-Cordier, C., Megret, J., te Riele, H., Junien, C., and Gourdon, G. (2004). MSH2-dependent germinal CTG repeat expansions are produced continuously in spermatogonia from DM1 transgenic mice. *Mol. Cell Biol* **24**, 629–637.
95. Oussatcheva, E. A., Hashem, V. I., Zou, Y., Sinden, R. R., and Potaman, V. N. (2001). Involvement of the nucleotide excision repair protein UvrA in instability of CAG*CTG repeat sequences in Escherichia coli. *J. Biol. Chem.* **276**, 30878–30884.
96. Parniewski, P., Bacolla, A., Jaworski, A., and Wells, R. D. (1999). Nucleotide excision repair affects the stability of long transcribed (CTG*CAG) tracts in an orientation-dependent manner in Escherichia coli. *Nucleic Acids Res.* **27**, 616–623.
97. Day, J. W., Ricker, K., Jacobsen, J. F., Rasmussen, L. J., Dick, K. A., Kress, W., Schneider, C., Koch, M. C., Beilman, G. J., Harrison, A. R., Dalton, J. C., and Ranum, L. P. (2003). Myotonic dystrophy type 2, molecular, diagnostic and clinical spectrum. *Neurology* **60**, 657–664.
98. Monckton, D. G., Wong, L. J., Ashizawa, T., and Caskey, C. T. (1995). Somatic mosaicism, germline expansions, germline reversions and intergenerational reductions in myotonic dystrophy males, small pool PCR analyses. *Hum. Mol. Genet.* **4**, 1–8.
99. Matsuyama, Z., Kawakami, H., Maruyama, H., Izumi, Y., Komure, O., Udaka, F., Kameyama, M., Nishio, T., Kuroda, Y., Nishimura, M., and Nakamura, S. (1997). Molecular features of the CAG repeats of spinocerebellar ataxia 6 (SCA6). *Hum. Mol. Genet.* **6**, 1283–1287.
100. Ashizawa, T., and Conneally, P. M. (1999). Repeats may not be everything in anticipation. *Neurology* **53**, 1164–1165.
101. Kremer, B., Almqvist, E., Theilmann, J., Spence, N., Telenius, H., Goldberg, Y. P., and Hayden, M. R. (1995). Sex-dependent mechanisms for expansions and contractions of the CAG repeat on affected Huntington disease chromosomes. *Am. J. Hum. Genet.* **57**, 343–350.
102. Goldberg, Y. P., Kremer, B., Andrew, S. E., Theilmann, J., Graham, R. K., Squitieri, F., Telenius, H., Adam, S., Sajoo, A., Starr, E., Heiberg, A., Wolff, G., and Hayden, M. R. (1993). Molecular analysis of new mutations for Huntington's disease, intermediate alleles and sex of origin effects. *Nat. Genet.* **5**, 174–179.
103. Reyniers, E., Vits, L., De Boulle, K., Van Roy, B., Van Velzen, D., de Graaff, E., Verkerk, A. J., Jorens, H. Z., Darby, J. K., Oostra, B., and Willems, P. J. (1993). The full mutation in the FMR-1 gene of male fragile X patients is absent in their sperm. *Nat. Genet.* **4**, 143–146.
104. Lavedan, C., Hofmann-Radvanyi, H., Rabes, J. P., Roume, J., and Junien, C. (1993). Different sex-dependent constraints in CTG length variation as explanation for congenital myotonic dystrophy. *Lancet* **341**, 237.
105. Martorell, L., Gamez, J., Cayuela, M. L., Gould, F. K., McAbney, J. P., Ashizawa, T., Monckton, D. G., and Baiget, M. (2004). Germline mutational dynamics in myotonic dystrophy type 1 males, allele length and age effects. *Neurology* **62**, 269–274.
106. Landles, C., and Bates, G. P. (2004). Huntingtin and the molecular pathogenesis of Huntington's disease. Fourth in molecular medicine review series. *EMBO Rep.* **5**, 958–963.
107. Ross, C. A. (2004). Huntington's disease, new paths to pathogenesis. *Cell* **118**, 4–7.
108. Humbert, S., and Saudou, F. (2002). Toward cell specificity in SCA1. *Neuron* **34**, 669–670.
109. Orr, H. T., and Zoghbi, H. Y. (2001). SCA1 molecular genetics, a history of a 13 year collaboration against glutamines. *Hum. Mol. Genet.* **10**, 2307–2311.
110. Katsuno, M., Adachi, H., Tanaka, F., and Sobue, G. (2004). Spinal and bulbar muscular atrophy, ligand-dependent pathogenesis and therapeutic perspectives. *J. Mol. Med.* **82**, 298–307.
111. van Roon-Mom, W. M., Reid, S. J., Faull, R. L., and Snell, R. G. (2005). TATA-binding protein in neurodegenerative disease. *Neuroscience* **133**, 863–872.
112. Galvao, R., Mendes-Soares, L., Camara, J., Jaco, I., and Carmo-Fonseca, M. (2001). Triplet repeats, RNA secondary structure and toxic gain-of-function models for pathogenesis. *Brain. Res. Bull.* **56**, 191–201.
113. Holmes, S. E., O'Hearn, E., Rosenblatt, A., Callahan, C., Hwang, H. S., Ingersoll-Ashworth, R. G., Fleisher, A., Stevanin, G., Brice, A., Potter, N. T., Ross, C. A., and Margolis, R. L. (2001). A repeat expansion in the gene encoding junctophilin-3 is associated with Huntington disease-like 2. *Nat. Genet.* **29**, 377–378.
114. Walker, R. H., Rasmussen, A., Rudnicki, D., Holmes, S. E., Alonso, E., Matsuura, T., Ashizawa, T., Davidoff-Feldman, B., and Margolis, R. L. (2003). Huntington's disease–like 2 can present as chorea-acanthocytosis. *Neurology* **61**, 1002–1004.
115. Holmes, S. E., O'Hearn, E. E., McInnis, M. G., Gorelick-Feldman, D. A., Kleiderlein, J. J., Callahan, C., Kwak, N. G., Ingersoll-Ashworth, R. G., Sherr, M., Sumner, A. J., Sharp, A. H., Ananth, U., Seltzer, W. K., Boss, M. A., Vieria-Saecker, A. M., Epplen, J. T., Riess, O., Ross, C. A., and Margolis, R. L. (1999). Expansion of a novel CAG trinucleotide repeat in the 5′ region of PPP2R2B is associated with SCA12. *Nat. Genet.* **23**, 391–392.
116. Holmes, S. E., O'Hearn, E., and Margolis, R. L. (2003). Why is SCA12 different from other SCAs? *Cytogenet Genome Res.* **100**, 189–197.
117. Mankodi, A., Logigian, E., Callahan, L., McClain, C., White, R., Henderson, D., Krym, M., and Thornton, C. A. (2000). Myotonic dystrophy in transgenic mice expressing an expanded CUG repeat. *Science* **289**, 1769–1773.
118. Kanadia, R. N., Johnstone, K. A., Mankodi, A., Lungu, C., Thornton, C. A., Esson, D., Timmers, A. M., Hauswirth, W. W., and Swanson, M. S. (2003). A muscleblind knockout model for myotonic dystrophy. *Science* **302**, 1978–1980.
119. Jiang, H., Mankodi, A., Swanson, M. S., Moxley, R. T., and Thornton, C. A. (2004). Myotonic dystrophy type 1 is associated with nuclear foci of mutant RNA, sequestration of muscleblind proteins and deregulated alternative splicing in neurons. *Hum. Mol. Genet.* **13**, 3079–3088.
120. Mankodi, A., Teng-Umnuay, P., Krym, M., Henderson, D., Swanson, M., and Thornton, C. A. (2003). Ribonuclear inclusions in skeletal muscle in myotonic dystrophy types 1 and 2. *Ann. Neurol.* **54**, 760–768.
121. Koob, M. D., Moseley, M. L., Schut, L. J., Benzow, K. A., Bird, T. D., Day, J. W., and Ranum, L. P. (1999). An untranslated CTG expansion causes a novel form of spinocerebellar ataxia (SCA8). *Nat. Genet.* **21**, 379–384.
122. Nemes, J. P., Benzow, K. A., Moseley, M. L., Ranum, L. P., and Koob, M. D. (2000). The SCA8 transcript is an antisense RNA to a brain-specific transcript encoding a novel actin-binding protein (KLHL1). *Hum. Mol. Genet.* **9**, 1543–1551.
123. Hagerman, P. J., and Hagerman, R. J. (2004). Fragile X-associated tremor/ataxia syndrome (FXTAS). *Ment. Retard. Dev. Disabil. Res. Rev.* **10**, 25–30.
124. Exner, M., Minar, E., Wagner, O., and Schillinger, M. (2004). The role of heme oxygenase-1 promoter polymorphisms in human disease. *Free Radic. Biol. Med.* **37**, 1097–1104.
125. Rothenburg, S., Koch-Nolte, F., and Haag, F. (2001). DNA methylation and Z-DNA formation as mediators of quantitative differences in the expression of alleles. *Immunol. Rev.* **184**, 286–298.

PART II

Myotonic Dystrophy

Myotonic Dystrophies: An Overview

TETSUO ASHIZAWA AND PETER S. HARPER

Department of Neurology, The University of Texas Medical Branch, Galveston, Texas; and Institute of Medical Genetics, Cardiff University, Cardiff, UK

I. Introduction
II. Clinical Phenotype
   A. Multisystemic Phenotype
   B. Anticipation
   C. Congenital Myotonic Dystrophy
III. The Mutation Responsible for Myotonic Dystrophies
   A. Mapping and Identification of the DM1 and DM2 Mutations
   B. Molecular Explanations for Anticipation, Congenital Myotonic Dystrophy, and Parental Gender Effects
   C. Instability of $(CTG)_n$ and $(CCTG)_n$ Repeats in Myotonic Dystrophies
IV. Pathogenic Mechanisms of Myotonic Dystrophies
   A. Pathogenic Mechanism of DM1
   B. Molecular Pathogenesis of DM2
   C. Molecular Basis of Phenotypic Differences between DM1 and DM2
V. Impact of Advanced Knowledge in Myotonic Dystrophy on Diagnostics and Therapeutics
VI. Concluding Remarks
   Acknowledgment
   References

Myotonic dystrophies constitute a group of disorders characterized by myotonia, weakness, and atrophy of the muscle, and are associated with variable multisystemic phenotypes. To date, two genetic mutations have been identified, including the $(CTG)_n$ repeat expansion in myotonic dystrophy type 1 (DM1) and the $(CCTG)_n$ repeat expansion in DM2. These repeats are located in a noncoding region of the respective genes, *DMPK* and *ZNF9*. The currently available evidence suggests that a *trans*-dominant gain of function by the mutant RNA transcripts containing expanded repeats plays a central role in the pathogenic mechanism. Experimental models of these repeats have provided valuable insights into the molecular mechanism of the repeat instability observed in patients with these diseases. These research advances may open the field of novel experimental therapeutics.

I. INTRODUCTION

Myotonic dystrophies are autosomal dominant disorders characterized by myotonia, weakness, and atrophy of the skeletal muscle, and variable multisystemic phenotypes [1]. To date, only two genetic mutations have been identified for myotonic dystrophies: an expansion of the $(CTG)_n$ repeat in myotonic dystrophy type 1 (DM1) and an expansion of the (CCTG) repeat in type 2 (DM2) [2]. Myotonic dystrophies belong to a group of diseases caused by expansion of unstable microsatellite repeats. While DM1 and DM2 have many features similar to those of other repeat expansion diseases, they are also distinct from others, especially from the perspective of the clinical phenotype and the molecular pathogenic mechanism. Details of the disease phenotype of DM2 and mechanism of DM1 and DM2 are covered by other chapters in this book. Here, we provide an overview of myotonic dystrophies in comparison with other disorders of repeat instability.

DM1 was identified in 1909 by Steinert [3] and also by Batten and Gibb [4], and then became known as *myotonic dystrophy, Steinert's disease, dystrophia myotonica*, or *myotonica atrophica* [1]. Myotonic dystrophy was identified several decades after the discoveries of Friedreich's ataxia (FA) in 1863 [5] and Huntington's disease (HD) in 1872 [6]. However, DM1 has been known to clinicians much longer than most other diseases now recognized as repeat expansion disorders, such as spinal and bulbar muscular atrophy (SBMA or Kennedy's disease) identified in 1968 [7], fragile X syndrome in 1969 [8], Machado–Joseph disease (MJD, now also known as SCA3) in 1975 [9], and dentatorubral pallidoluysian atrophy (DRPLA) in 1982 [10]. Although autosomal dominant forms of ataxic disorders were reported earlier by Menzel, Holmes, Marie, and others before Steinert's report of myotonic dystrophy, these autosomal dominant crebellar ataxias (ADCAs) were not recognized as individual disease entities until respective genetic mutations started to be identified in 1993 [11].

In 1992, an expansion of an unstable CTG trinucleotide repeat $((CTG)_n)$ in the 3′ untranslated region (UTR) of the *dystrophia myotonica protein kinase* (*DMPK*) gene was identified as the genetic mutation of myotonic dystrophy [12–14]. In 2001, the second locus was identified with an expansion of the $(CCTG)_n$ repeat in the first intron of the *zinc finger 9* (*ZNF9*) gene in patients who carried the diagnosis of myotonic dystrophy type 2, proximal myotonic myopathy (PROMM), and proximal myotonic dystrophy (PDM) (see Section IIIA) [15]. Shortly thereafter, the International Myotonic Dystrophy Consortium (IDMC) published the following guidelines for the nomenclature [16]:

1. All multisystemic myotonic disorders including DM, PROMM, PDM, and DM2 are collectively called "myotonic dystrophies."
2. The loci for these diseases will be consecutively named *DM* followed by a number (*DMn*), such as *DM1, DM2, DM3, . . . , regardless of the clinical phenotype*.
3. To accommodate this nomenclature system, the chromosome 19q13.3 locus for myotonic dystrophy (OMIM# 160900) was changed from "*DM*" to "*DM1*."
4. If a new allelic disease is discovered, it will not be assigned to a new locus; instead, it will be assigned to the previously known locus. The allelic disease will be assigned to a new OMIM number with "#" in front, indicating that it is an allelic disease.
5. If diseases previously assigned to one locus turn out to be caused by mutations in two different genes located close to each other, the disease assigned to the locus more recently will be assigned to a new locus using the "*DMn*" system.

The guidelines further state, "The nomenclature does not preclude the use of traditional clinical terms such as 'PROMM' and 'PDM' for clinical diagnosis. Although the term 'myotonic dystrophy' may still be used as a clinical diagnosis of the disease caused by the CTG repeat expansion at the *DM1* locus, the preferred terminology is 'myotonic dystrophy type 1' or 'DM1,' which is easily distinguishable from DM2 and other myotonic dystrophies."

Subsequently, Le Ber et al. [17] reported a family with a non-DM1, non-DM2 multisystem myotonic disorder with cataracts and frontotemporal dementia and assigned the disease locus to chromosome 15q21-24. In Chapter 8, Krahe and his colleagues describe mapping of another locus to chromosome 16p in families with PROM/DM2-like phenotype. The official OMIM nomenclatures of *DM3* and *DM4* have not been given at present.

II. CLINICAL PHENOTYPE

A. Multisystemic Phenotype

Several repeat-expansion diseases show multisystemic phenotypes. In FA, hypertrophic cardiomyopathy and insulin resistance accompany the neurodegenerative phenotype of the central and peripheral nervous systems [18]. Fragile X syndrome is characterized by facial dysmorphism and cryptoorchidism in addition to the mental retardation [19]. Patients with SBMA have endocrine dysfunctions in addition to the motor neuron disease [20]. In HD, metabolic abnormalities have been found in peripheral tissues [21, 22]. In SCA7 and SCA10, some involvement of organs outside the nervous system has been postulated [23, 24]. However, the multisystemic phenotypes of myotonic dystrophies are much more robust and involve far greater numbers of organs compared with these disorders.

1. MULTISYSTEMIC PHENOTYPE OF DM1

Initial reports of myotonic dystrophy (presumably DM1) by Steinert and also Batten and Gibbs in 1909 included the description of myotonia; weakness and atrophy of distal limb, face, neck, and bulbar muscles; ptosis; and testicular atrophy [3, 4]. In 1911, Greenfield [25] first described cataract, and Curschmann (1912) [26] recognized the significance of cataract and testicular atrophy as systemic features of this disease. He also described endocrine and mental disturbances. Bradycardia was already noted by Steinert in his initial paper, but cardiac conduction defects were not systematically evaluated until the electrocardiogram came into regular use in the 1940s. Subsequent studies widened the spectrum of the multisystemic manifestations by documenting the involvement of smooth muscles, respiration, peripheral nerve, bone, and skin (reviewed in [1, 2]). The smooth muscle phenotype includes dysmotility of the gastrointestinal tract (manifested as dysphagia, emesis, diarrhea, constipation, megacolon, anal incontinence, and gallstone), incoordinate contraction of the uterus in labor, hypotension, and abnormal ciliary body and low intraocular pressure in eyes. Respiratory problems are obstructive and central sleep apnea, hypersomnolence, alveolar hypoventilation, and aspiration pneumonia secondary to dysphagia. Distal sensory loss and areflexia are also often seen in patients with DM1. The skeletal changes include cranial hyperostosis, enlargement of air sinuses, and a small pituitary fossa. Skin of DM1 patients often shows multiple pilomatricoma (hair follicle tumors) and frontal baldness, suggesting hair follicle abnormalities. Immunologically, the serum level of gamma globulin is decreased and granulocyte functions are defective. While eyes of DM1 patients typically show iridescent lens opacities in the posterior subcapsular region, which later mature into ordinary dense cataracts, retinopathy is also frequently observed. In addition to testicular atrophy, endocrine abnormalities involve other organs, resulting in insulin insensitivity, increased levels of follicle-stimulating hormone (FSH) and luteinizing hormone (LH), increased luteinizing hormone-releasing hormone (LRH) response, and adrenal and, potentially ovarian, dysfunctions.

Among these multisystemic abnormalities, skeletal muscle, cardiorespiratory, and brain disorders are of particular clinical importance because the disability and prognosis of DM1 patients heavily depend on them [2, 27, 28]. The skeletal muscle is affected primarily in distal limbs, face, and neck (especially sternocleidomastoid muscles) with weakness and atrophy. In the advanced stage, proximal muscle weakness often leads to an inability to rise from a chair and, eventually, to loss of ambulation. Bulbar muscle weakness may cause not only dysarthria but also severe dysphagia, which may lead to life-threatening aspiration pneumonia. Respiratory muscle weakness contributes to hypercapnia, respiratory insufficiency, and respiratory failure. Myotonia can be easily induced by making a tight fist or by direct percussion of the thener muscle in the hands, but it is also present in many other muscles including those in the tongue. Myotonia is readily detectable by electromyography as spontaneous prolonged discharges of compound muscle action potentials with "dive bomber" or "motorcycle revving" sounds. Myotonia in DM1 shows the warm-up phenomenon, in which repeated contraction of the muscle gradually alleviates the severity of myotonia. Cardiopulmonary problems are important because they may become life-threatening. Cardiac conduction block, especially at the His bundle, may cause complete atrioventricular block with a sudden death. Ventricular tachyarrhythmias may also threaten the life of the patient. Furthermore, congestive heart failure may result from dilated cardiomyopathy. General anesthesia is often complicated with postoperative respiratory failure. Although mental dysfunction of DM1 patients was already recognized by Curschmann [26] 3 years after Steinert's report, the disability stems from brain dysfunction and has been underestimated except in the cases of congenital DM1 (see Section IIC). Indeed, hypersomnia, abnormal frontal executive function, apathy, and avoidance behaviors often severely limit the social fitness of the patients, precluding them from adequate employment and social life. In congenital and childhood DM1, cognitive dysfunction often is the major functional limitation of the patient. Furthermore, dysfunction of the central respiratory center also plays important roles in sleep apnea and anesthesia-related respiratory complications. Thus, skeletal muscle, cardiorespiratory, and brain manifestations are the most important clinical problems in DM1.

2. MULTISYSTEMIC PHENOTYPE OF DM2

The clinical phenotype of DM2 is extensively reviewed in Chapters 7 and 8. Briefly, the phenotype of DM2 closely resembles that of DM1, although there are important differences [2, 29]. The most striking difference is the lack of congenital phenotype in DM2. In adult-onset cases, the multisystemic phenotype of DM2 is generally milder than that of DM1. Clinically, DM2 can be differentiated from DM1 by prominent proximal muscle weakness, generally milder and inconsistent myotonia, calf hypertrophy, muscle aching, and lack of ocular abnormalities other than cataracts. Smooth muscle abnormalities have not been well documented in DM2. Among these differences, predominantly proximal muscle weakness is the hallmark of DM2; hence DM2 is also widely known as PROMM by clinicians.

B. Anticipation

Anticipation is a genetic term for a clinical phenomenon defined as progressively earlier onset of the disease with increasing severity in successive generations. Fleischer (1918) [30] documented that cataract can be a sole sign of myotonic dystrophy in obligate heterozygotes in the early generation of affected families, providing the foundation for the concept of anticipation. Thus, the concept of anticipation, which is now recognized in many other diseases caused by expanded repeats, was first identified in myotonic dystrophy. However, anticipation remained controversial for the next several decades. With the state of knowledge in human genetics at that era, no one could provide a convincing biological explanation for anticipation. Thus, Penrose (1948) [31], who was a leading geneticist, argued that anticipation is attributable to ascertainment biases. Consequently, anticipation was not accepted as a biological phenomenon until Höweler (1989) [32] showed that ascertainment biases cannot adequately explain anticipation. Anticipation in DM1 was finally accepted as a true biological phenomenon when the genetic mutation was identified in 1992. However, anticipation had also been recognized in other diseases before 1992. For example, in Huntington's disease, subjects suffering from the juvenile-onset rigid form (Westphal variant) were identified in the latest generation of affected families [33], and in fragile X syndrome, anticipation was known as the Sherman Paradox [34]. Subsequently, anticipation has been recognized in many diseases caused by repeat expansions, including DM2. Molecular correlates of anticipation are discussed in Section IIIB.

C. Congenital Myotonic Dystrophy

Vanier [35] gave the first clear description of the severe congenital form of myotonic dystrophy in 1960. Now, the congenital form is recognized as a dramatic endpoint of anticipation in DM1. There are three important facts regarding this congenital form. First, the congenital form has been found only in DM1 families; despite the report of anticipation in some DM2 families, no cases of congenital DM2 have been identified. Second, in DM1 families, the congenital form is almost exclusively found in children who are born to affected mothers, although a handful of paternally transmitted cases have been reported [36]. Third, the congenital phenotype is characterized by clinical manifestations distinct from those of adult DM1. Neonates with the congenital form suffer from generalized hypotonia with respiratory and feeding difficulties, which are often associated talipes and polyhydroamnios, and if they survive the critical neonatal period, they develop a variable degree of mental retardation [1]. The infants with the congenital form do not exhibit myotonia and other clinical features characteristic of adult DM1 until they grow into early childhood. Cases with infantile onset have also been reported in SCA2 and SCA7, with very large CAG repeat expansions transmitted from the affected father [23–40]. Multiple cases of the congenital form of SCA7 have been reported with severe hypotonia, congestive heart failure, patent ductus arteriosus, cerebral and cerebellar atrophy, and visual loss [23, 39, 40]. Neonates with SCA2 present with neonatal hypotonia, developmental delay, and dysphagia. Ocular findings consist of retinitis pigmentosa [37–39]. It should be noted that neonatal hypotonia appears to be a common phenotype among these cases.

There was no dispute about the biological nature of congenital DM1; however, this effect of the gender of the transmitting parent was a puzzle to every clinician. The congenital DM1 phenotype has been attributed to a very large expansion size of the $(CTG)_n$ repeat, and this and the lack of congenital DM2 are discussed in Section IIIB.

III. THE MUTATION RESPONSIBLE FOR MYOTONIC DYSTROPHIES

A. Mapping and Identification of the DM1 and DM2 Mutations

1. DM1 MUTATION

The myotonic dystrophy locus (then designated *DM*) was one of the first human disease loci assigned to a chromosome by linkage. Linkage to the secretor and Lutheran blood group was found in 1971 [41, 42], and subsequently, the assignment of this locus to chromosome 19 was accomplished by linkage of *DM* to C3, which was known to be on chromosome 19 from somatic cell hybrid studies [43, 44]. This preceded assignments of the *SBMA* and fragile X syndrome loci to the X chromosome in 1978 [45] and 1969 [8], the *SCA1* locus to chromosome 6 by linkage to HLA in 1977 [46], and the *HD* locus to chromosome 4 in 1983 [47], which was the first human disease assigned to a chromosome by linkage to an anonymous restriction fragment length polymorphism (RFLP) marker. These assignments were considerably earlier than others such as FA [48], DRPLA, and other SCAs, which were assigned to the respective chromosomes after 1988 [49].

In 1991, the mutations of fragile X syndrome and SBMA were identified as expansions of trinucleotide repeats [50, 51]. Changes in the size of the $(CGG)_n$

repeat expansion in families with fragile X syndrome explained the Sherman Paradox, providing a hint that a similar mutation with an expansion of an unstable microsatellite might explain anticipation of myotonic dystrophy. Within a year, variably expanded mutant alleles were identified in DM patients [52–54], and a few months later, the $(CTG)_n$ expansion mutation was identified [12, 13]. Although normal individuals showed 37 or fewer CTGs, myotonic dystrophy patients had more than 50 CTGs, often reaching thousands of CTGs.

2. DM2 MUTATION

After identification of the DM1 mutation, up to 2% of patients with a clinical diagnosis of myotonic dystrophy were found to have no expansion allele of the $(CTG)_n$. In 1995, Ricker and his colleagues identified autosomal dominant German families with such patients [55]. Because these patients showed proximal muscle weakness as opposed to the distal muscle weakness of Steinert's myotonic dystrophy, the disease was named proximal myotonic myopathy (PROMM). Day and colleagues [56] and Udd and colleagues [57] reported families with slightly different myotonic dystrophy phenotypes in the absence of the $(CTG)_n$ expansion. They called these diseases myotonic dystrophy type 2 (DM2) and proximal myotonic dystrophy (PDM). Subsequently, Ranum and her colleagues [58] mapped the DM2 locus to chromosome 3q, and this was followed by a report that PROMM shares the same genetic locus [59]. In the end, however, PROMM, DM2, and PDM were found to be the same disease that is caused by an expansion of $(CCTG)_n$ in intron 1 of the *zinc finger 9* (*ZNF9*) gene [15].

B. Molecular Explanations for Anticipation, Congenital Myotonic Dystrophy, and Parental Gender Effects

In DM1, puzzling genetic phenomena, such as anticipation and the congenital form, which were difficult to explain by conventional Mendelian genetics, are largely attributable to the "dynamic" mutation. The size of the expanded repeat inversely correlates with the age at onset, and the repeat size increases in successive generations in DM1, providing the molecular basis for anticipation [60–62]. Paternal mutant repeats usually do not exceed 1000 CTGs in the offspring, whereas maternal transmission frequently gives rise to further expansions of the mutant repeats beyond 1000 CTGs in children with congenital myotonic dystrophy [63, 64]. Premutation alleles (38–50 CTGs) tend to expand into the full-mutation range more frequently with paternal transmission than maternal transmission [65, 66]. This accounts for the paternal origin of the *de novo* mutations of myotonic dystrophy. These observations provided important insights into the effects of the gender of the transmitting parent on the degree of anticipation in other diseases caused by trinucleotide repeat expansions, such as HD [67], DRPLA [68], SCA1 [69], SCA2 [70, 71], MJD/SCA3 [72, 73], and SCA7 [74, 75], and on the *de novo* mutations in HD [76] and SCA2 [70].

The conspicuous lack of the severe congenital form of DM2 and the mechanism of reported anticipation in DM2, however, remain puzzling. The genotype–phenotype correlation and characteristics of the $(CCTG)_n$ repeat instability in DM2 are vastly different from those in DM1 [29, 77]. First, the expansion size does not appear to correlate with the age at onset in DM2. Second, the size of the expanded $(CCTG)_n$ allele does not consistently increase as it is transmitted from generation to generation. Instead, expanded $(CCTG)_n$ repeats tend to show intergenerational contractions of the repeat size. It was also noted that expanded $(CCTG)_n$ alleles are highly unstable in somatic tissues. The high degree of somatic instability could explain the apparent lack of inverse correlation between the age at onset and the repeat expansion size. Anticipation appears to be present in DM2 despite the intergenerational contraction of $(CCTG)_n$ expansion size [77, 78]. While ascertainment biases need to be considered as the source of the observed anticipation in DM2, a continuous increase in $(CCTG)_n$ repeat size with age of the patient may be the key to solving this apparent dilemma. The lack of congenital DM2 is also an unsolved mystery. The $(CCTG)_n$ expansion should be potent enough to cause the severe phenotype in the "*trans*-dominant RNA-gain-of-function" theory of the pathogenic mechanism of myotonic dystrophies [79] (see Chapters 37–44 and Section IVC of this Chapter). The intergenerational contraction of the expanded $(CCTG)_n$ repeat may keep the repeat size small enough to preclude the occurrence of congenital DM2. However, other mechanisms may be necessary to explain the discordance in the congenital form between DM1 and DM2.

C. Instability of $(CTG)_n$ and $(CCTG)_n$ Repeats in Myotonic Dystrophies

The instability of the expanded repeat provided molecular explanations for anticipation and the congenital form. However, the mechanism of the repeat instability is still not fully understood. Investigations of the repeat instability have been done not only in patient-derived tissues and cells, but also in a variety of experimental systems, including *in vitro*, bacteria, yeast, and transgenic animal models.

1. THE MECHANISM OF (CTG)$_n$ REPEAT INSTABILITY

Cell cultures and tissues derived from patients with DM1 and transgenic mouse lines recapitulated the expansion-prone instability of the expanded (CTG)$_n$ repeat. The degree of repeat instability correlates with the repeat size. In tissues from myotonic dystrophy patients, the expanded (CTG)$_n$ repeat shows extensive and variable somatic and germline instability with a strong bias toward expansion. The size of the expanded repeat is variable in different somatic tissues (smaller in the cerebellum and larger in the muscle and most other somatic tissues than in the blood) [80–84]. Interestingly, it has been demonstrated that the expanded CAG repeats in the cerebellum exhibit the smallest degree of somatic mosaicism in HD, SCA1, MJD, and DRPLA [85]. The bias toward further expansion has been documented in the blood, muscle, and other tissues [86, 87]. The expanded repeat continues to increase in size as the patient ages in the blood and, perhaps, other tissues. As mature blood and muscle cells are "postmitotic" cells that are terminally and irreversibly differentiated and constantly replaced with cells newly differentiated from the stem cells, the age-dependent increase in repeat size is likely to be due to continuous expansions of the repeat size in the stem cell population. The age-dependent expansion may occur in other repeats in human diseases and their mouse models [85, 88–90]. Assuming the length-dependent potency of these repeats in the cummulative pathogenic process, the severity of the disease phenotype in a given tissue at a given age of the patient may be determined by both age and age-dependent repeat size of stem cells. Thus, the severity index of the tissue phenotype (P) may be roughly calculated with the formula

$$P = \int_{t_0}^{t_1} f(t)dt \times C,$$

where t_0 is the age at which cells start expressing the *DMPK* gene, t_1 is age, $f(t)$ is repeat size at age t, and C is a constant. This formula is by no means perfect. For example, as an additional variable, there may also be repeat instability in postmitotic cells [91]. The instability of the CTG repeat before establishing tissue-specific stem cells may be negligible because the repeat does not show striking instability during the early fetal development [92].

In the male germline, the expanded repeats show a greater degree of size variability [80, 86]. The size distribution of expanded alleles in sperm spans from an upper limit higher than that in blood to a lower limit in the range overlapping with normal allele distribution. The expanded CTG repeat in sperm does not manifest an age-dependent increase in repeat size [66]. In transgenic mouse models of DM1, the expanded (CTG)$_n$ repeat shows increases in size during the premeiotic stages of the spermatogenesis [93]. Further studies are needed to elucidate the determinants that account for the different instability among tissues.

In lymphoblastoid cells derived from DM1 patients, cells with larger repeats have a faster growth rate, enhancing the expansion bias of the repeat instability at the tissue level ("mitotic drive") [94, 95].

In transgenic mice, the expanded (CTG)$_n$ repeat in the transgene exhibits most of the characteristics of the expanded repeat in patients' tissues, including expansion bias, the parental sex effect, intergenerational instability, age-dependent increase, and intertissue variability of the repeat size [96–100]. Studies of transgenic mice with expanded CTG·CAG repeats in the background of mismatch repair gene deficiencies revealed that *Msh2*, *Msh3*, *Msh6*, and Pms2 play important roles in the repeat instability [93, 101, 102].

Studies *in vitro* provided evidence that CTG·CAG repeats form non-B DNA structures such as hairpins with slipped strands [103–107]. In *Escherichia coli*, (CTG)$_n$ repeats are unstable but deletion-prone, although expansions do occur depending on the repeat length and the direction of replication, and the mechanism of instability includes slippage of the strands at the replication folk, gene conversion-like events, and recombination [108]. Studies on the instability of (CTG)$_n$ repeats in yeast elucidated the roles of molecules involved in DNA replication and repair. The characteristics of CTG·CAG repeats differ from those of other repeats such as CGG·CCG and GAA·TTC [109, 110]. Differential processing of slipped repeats by mismatch repair mechanisms may explain the differences in mutation patterns between various disease loci or tissues [111, 112]. The data obtained in bacteria and yeasts should be carefully interpreted with considerations of interspecies differences when they are used as models of the repeat instability in humans. Nevertheless, these studies provided important insights into the mechanism of the instability.

2. MECHANISM OF (CCTG)$_n$ INSTABILITY

The instability of the (CCTG)$_n$ repeat in DM2 patients is reviewed in Chapter 7. The expanded (CCTG)$_n$ repeat exhibits intergenerational contraction (on average, about −17 kb per generation) [77]. Thus, the mechanism of the (CCTG)$_n$ repeat instability in DM2 may be very different from that of the (CTG)$_n$ repeat instability in DM1. Expanded DM2 alleles also show extensive somatic mosaicism in peripheral tissues obtained from DM2 patients with the putative age-dependent increase in expansion size [77]. However, studies of the mechanism of $(CCTG)_n$ instability have been limited to those done *in vitro* and in *E. coli*. A $(CAGG)_n$ repeat, but not the complementary $(CCTG)_n$ repeat, formed a defined base-paired hairpin structure *in vitro*. The orientation of the $(CCTG)_n$ repeat in reference to the replication origin of the vector influences the stability of the repeats [113]. These data may provide insights into the *in vivo* behaviors of the expanded $(CCTG)_n$ repeat in DM2 patients.

3. ORIGIN OF THE MUTATIONS OF DM1 AND DM2

Mytonic dystrophy is one of the most common inherited neuromuscular disorders, and has been described in global populations except for most sub-Saharan ethnic populations [114, 115]. The prevalence varies but generally ranges from 1/8000 to 1/50,000 in European and Japanese populations [116, 117]. High prevalences have been reported in different regions of the world, such as 1/475 in Charlevoix and Saguenay-Lac-Saint-Jean (SLSJ) in Northeastern Quebec, Canada [118]; 1/2114 in Yemenite Jews [119]; 1/5524 in Istria, Croatia [120], with a founder effect. Based on the paucity of myotonic dystrophy in sub-Saharan ethnic populations, it was postulated that the myotonic dystrophy mutation occurred after the human migration out of Africa [115]. When the $(CTG)_n$ repeat expansion was identified as the myotonic dystrophy mutation, the mutation was found to be associated with the Alu 1-kb insertion (Alu+) allele located 5 kb upstream of the $(CTG)_n$ repeat within the *DMPK* gene [54]. Since then, the $(CTG)_n$ repeat expansion has always been found on the Alu+ background in European and Asian populations [121, 122]. The only exception, in which the Alu− allele was associated with the $(CTG)_n$ repeat expansion, was found in a Nigerian Yoruba family, which is the sole reported DM1 family in the sub-Saharan ethnic population [123]. Subsequent analyses of $(CTG)_n$ and the Alu+/− polymorphism in worldwide populations appear to point to the consensus that $(CTG)_5$-Alu+ is the ancestral haplotype for all observed haplotypes and the $(CTG)_n$ expansion alleles have derived from this ancestral haplotype through larger $(CTG)_n$ alleles of the upper normal range [124]. Analyses of haplotypes of the *DMPK* region showed that most European and Asian DM1 $(CTG)_n$ expansion alleles are on one haplotype (haplotype A) background, despite the existence of diverse haplotypes among worldwide populations. The Nigerian myotonic dystrophy mutation was found to be on a different haplotype background [123].

DM2 appears to be most common in the German population [55, 56]. Most DM2 families in the United States are clearly Northern Europeans [56]. However, DM2 families have been identified in Afghanistan [125]. Haplotype analyses showed a common haplotype of at least 132 kb among families with DM2, suggesting a common founder, which was estimated to be established 200 to 540 generations ago [125, 126]. The $(CCTG)_n$ tract is interrupted on normal alleles, but these interruptions are lost on affected alleles, suggesting that the loss of interruptions may predispose alleles to further expansion [125].

4. MAINTAINING THE STEADY PREVALENCE OF DM1 IN THE PRESENCE OF ANTICIPATION

The terminal event of anticipation in DM1 is the congenital form, which accompanies severely compromised nuptial and reproductive capability. Consequently, anticipation is expected to gradually deplete the DM1 patient population. However, the prevalence of the disease has been relatively steady, and areas of high prevalence have been identified with a founder effect as described before. In part, this can be explained by the considerable pool of normal individuals who have premutation alleles (38–50 CTGs) or alleles just below the premutation range. These individuals can act as a reservoir for the future origin of new cases through genetic instability [65]. Because premutation alleles are unstable and quickly expand to the full mutation range, expansion into the upper range of normal repeat size, instead of expansion into the premutation range, must be considered the original mutational event of DM1 that took place after the migration out of Africa.

An additional factor to be considered to offset anticipation in DM1 is a phenomenon called *meiotic drive*. This is a segregation distortion with a preferential transmission at meiosis of a particular allele, in this case, a larger one. Two studies of normal individuals showed preferential transmission of the chromosome carrying the larger normal allele at the DM1 locus [127, 128]. However, the data are inconsistent with respect to the sex of the transmitting parent, and reanalysis has not shown any evidence of abnormal segregation. Similar suggestions of abnormal segregation of the expanded mutant allele in DM1 families [129–131] and in transgenic mouse lines [96] have been reported. However, a recent study using data from prenatal molecular studies, which are not subject to ascertainment bias, showed no evidence of meiotic drive [132]. Although segregation distortion has been reported in MJD [133–135], more recent studies have disputed this [136]. While non-Mendelian phenomena, such as anticipation and congenital DM1, have been documented in DM1, segregation distortion should be carefully examined with samples of larger size.

IV. PATHOGENIC MECHANISMS OF MYOTONIC DYSTROPHIES

The currently available data suggest that the pathogenic mechanisms of DM1 and DM2 are based primarily on a gain of toxic function by RNA transcripts containing the expanded repeats [137]. This novel pathogenic mechanism has not been previously recognized in human diseases, but is gaining increasing importance by being recognized as a potential mechanism for other diseases, such as fragile X-associated tremor/ataxia syndrome (FXTAS) caused by intermediate expansions of FMR1 $(CGG)_n$ repeats [138, 139] and SCA10 caused by expanded intronic pentanucleotide $(ATTCT)_n$ repeats [140]. However, there are some observations in DM1 and DM2 that cannot be explained by the current RNA gain-of-function model. It may be necessary to modify the RNA gain-of-function model or to consider additional pathogenic models to explain all features of DM1 and DM2.

A. Pathogenic Mechanism of DM1

The mechanism by which the expanded CTG repeat leads to the multisystemic clinical phenotype of DM1 is not fully understood. Because of the location of the $(CTG)_n$ repeat in the 3′ UTR, the coding information of this gene remains intact in the mutant *DMPK* gene. However, the $(CTG)_n$ repeat is transcribed into the messenger RNA (mRNA) as a $(CUG)_n$ repeat. Recent studies have led to two major models of the disease mechanism for DM1: (1) a gain of toxic function by the expanded CUG repeat in the mutant *DMPK* mRNA, and (2) a loss of function of the genes in the vicinity of the CTG repeat, including *DMWD*, *SIX5*, and *DMPK* itself [137]. Currently available data strongly support that the toxic gain-of-function model is the major pathogenic mechanism, whereas the loss of function of *DMPK* and adjacent genes could also contribute to the DM1 phenotype. Thus, the pathogenic mechanisms of DM1 and DM2 are distinct from those of other diseases caused by trinucleotide repeat expansions, such as fragile X syndrome (methylation and loss of function) [141]; FRDA1 (sticky DNA structure and loss of function) [18, 142]; SCA12 (increased transcription) [143, 144]; HD, DRPLA, SBMA, SCA1, SCA2, SCA3/MJD, SCA7, SCA17 (polyglutamine expansion and gain of function) [145, 146]; and OPMD (polyalanine expansion and gain of function) [147].

1. Gain of Function by the Mutant DMPK mRNA

Because the CTG repeat is located in the 3′ UTR, the gain-of-function model would not work for DM1 through the DMPK protein, but it could through the expanded CUG repeat in the mutant DMPK mRNA. In DM1 cells, $(CUG)_n$ repeats of the mutant DMPK mRNA have been shown to accumulate in nuclear foci [148]. Transgenic mice expressing noncoding $(CUG)_n$ repeats also develop $(CUG)_n$ nuclear foci and clinical and histological phenotypes closely resembling DM1 [149]. $(CUG)_n$ repeats have been shown to bind several proteins. Some of these proteins, such as muscleblind-like (MBNL) proteins, colocalize in the nuclear foci of $(CUG)_n$ repeats in DM1 cells [150–152]. Furthermore, mice deficient in *Mbnl1* develop a DM1-like phenotype [153]. Taken together, the sequestration of the MBNL1 protein into the nuclear $(CUG)_n$ repeat foci appears to be the major pathogenic process of human DM1. Another CUG-binding protein, CUG-BP1, is increased in the nucleus of DM1 cells, although it does not colocalize in the nuclear $(CUG)_n$ foci [79, 154–156]. Both MBNL1 and CUG-BP1 have been shown to regulate splicing of various gene transcripts, some of which have been shown to be altered in cells/tissues from patients with DM1 and transgenic mouse model of DM1 [149, 153, 157–160]. Mice overexpressing CUG-BP1 show splicing abnormalities [161]. Furthermore, although transfected cells expressing MBNL proteins and either $(CUG)_n$ or $(CAG)_n$ RNA repeats show colocalization of MBNL with both $(CUG)_n$ and $(CAG)_n$ repeat foci, only cells with $(CUG)_n$ foci exhibited splicing abnormalities, suggesting that colocalization of MBNL in the nuclear foci is separable from the splicing misregulation [162]. Recent studies demonstrated that CUG-BP1 and other CELF proteins regulate the equilibrium of splice site selection by antagonizing the facilitatory activity of MBNL proteins in the regulation of splicing [158, 163]. Thus, CUG-BP1 may also play a key role in the pathogenic mechanism of DM1.

2. Loss of DMPK Function

Because of decreased levels of *DMPK* mRNA in adult DM1 tissue, DMPK deficiency was proposed as the pathogenic mechanism of DM1 soon after identification of the DM1 mutation. As the transcription of the mutant DMPK gene appears unaltered, the loss of the DMPK mRNA has been attributed to the retention of mutant DMPK transcripts in nuclear foci in DM1 cells [164–166]. A dominant negative effect of the mutant DMPK mRNA on the wild-type DMPK mRNA has also been postulated [167].

Initial observations of DMPK-deficient mice showed no robust abnormalities [168, 169] except for some muscle weakness. However, more recent studies of homozygous and heterozygous *DMPK*-deficient mice suggested that haploinsufficiency of *DMPK* does cause skeletal and cardiac muscle abnormalities through alterations of sodium and calcium channels [170–172]. Furthermore, the DMPK isoforms may add a twist to the haploinsufficiency theory if the $(CTG)_n$ repeat expansion alters the isoforms [173, 174].

3. Effects on Adjacent Genes

The CTG repeat expansion affects the expression of DMPK and neighboring genes, *SIX5* and *DMWD*, in DM1 [175]. The expanded $(CTG)_n$ repeat alters the chromotin structure and hinders the transcription of the *SIX5* gene immediately downstream of *DMPK* by impeding an access of an enhancer element for the *SIX5* gene [176–178]. Although the actual reduction in the *SIX5* expression level in DM1 tissues may vary [179, 180], *SIX5* deficiency causes cataracts in both homozygous and heterozygous "knockout" mice [181, 182]. Cardiac conduction system abnormalities were also found in *SIX5*-deficient mice [183]. However, these mice do not exhibit histopathological, contractile, or electrophysiological abnormalities in the skeletal muscle [184, 185]. Further studies are needed to determine whether other organs show abnormalities compatible with the DM1 phenotype in these "knockout" mice. Although decreased *DMWD* mRNA levels were reported in the cytoplasm of DM1 cell lines and adult DM1 skeletal muscle samples [186], clinical consequences of the *DMWD* deficiency remain unknown.

B. Molecular Pathogenesis of DM2

The close similarity in clinical phenotype between DM1 and DM2 suggests that some part of pathogenic mechanism is shared by these two diseases. Like the $(CTG)_n$ repeat in DM1, the $(CCTG)_n$ repeat in the first intron of the *ZNF9* gene [15] is transcribed but not translated in DM2. Intranuclear RNA foci containing $(CCUG)_n$ repeats and MBNL proteins are detected in DM2 cells [150, 151]. Thus, an RNA gain-of-function model similar to that of DM1 may be applicable to DM2 and explains both the dominant inheritance and multisystemic nature of this disease. An alternative mechanism may be a loss of the *ZNF9* function. Interestingly, the DM-like phenotype of heterozygous *Znf9*-deficient mice is discussed in Chapters 7 and 8. However, *ZNF9* has no functional similarities to *DMPK* or adjacent genes in the *DM1* region. It is noteworthy that DM2 patients homozygous for the $(CCTG)_{>4000}$ expansion reported in an Afghan family exhibited the phenotype of ordinary heterozygous DM2 patients, including clinical course, muscle histology, and anti-MBNL1 staining and brain imaging features [151]. Thus, DM2 appears to be a true autosomal dominant disease.

C. Molecular Basis of Phenotypic Differences between DM1 and DM2

We should be reminded that there are clinical differences between DM1 and DM2, most notably the lack of the congenital form in DM2 [187]. We postulate the following possibilities to explain these differences: (1) the lack of loss of function of DMWD/DMPK/SIX5 in DM2, (2) differences in CTCF sites and the DNA methylation status between the DM1 and DM2 loci, (3) differential binding of RNA-binding proteins other than MBNL to $(CUG)_n$ and $(CCUG)_n$ repeats and their flanking regions, (4) the intergenerational contraction in DM2 preventing expression of the congenital phenotype, and (5) differences in tissue-dependent expression of *DMPK* and *ZNF9* during development. Stronger binding of $(CUG)_n$ repeats than of $(CCUG)_n$ by MBNL proteins repeats could explain the generally more severe phenotype observed in DM1 compared with DM2. However, studies of patient-derived cells demonstrated that RNA repeats form larger foci with more intense MBNL colocalization in DM2 cells than in DM1 cells [79]. $(CCTG)_n$ repeats in DM2 patients are generally longer than $(CTG)_n$ repeats in DM1 patients, and $(CCTG)_n$ repeats bind to MBNL proteins more strongly than do $(CTG)_n$ repeats *in vitro* [188]. Furthermore, homozygote DM2 patients do not exhibit a more severe phenotype than heterozygous DM2 patients [189]. These observations suggest that the extent of the sequestration of MBNL proteins in the nuclear RNA repeat foci may not be the most important determinant of phenotypic severity. Further studies are necessary to unravel the mechanism of the phenotypic differences between DM1 and DM2.

V. IMPACT OF ADVANCED KNOWLEDGE IN MYOTONIC DYSTROPHY ON DIAGNOSTICS AND THERAPEUTICS

DNA testing offers accurate diagnosis of DM1 and DM2. For DM1, polymerase chain reaction (PCR) across the repeat region can identify normal individuals by detecting two $(CTG)_n$ alleles in the normal repeat range. Detection of a single normal allele by PCR analysis indicates that the subject is either a normal individual homozygous for the normal allele or a patient heterozygous for the expanded allele, which cannot be amplified by PCR. To differentiate them, Southern blot analysis is used to determine whether the expanded allele is present. However, this approach may not work in some DM2 patients because the somatic mosaicism often obscures the expanded allele on Southern blot anlysis. In such cases, an additional DM2 repeat assay (RA) that consists of amplifying the CCTG repeat by repeat-primed PCR and probing the resultant product with an internal probe is recommended [77]. This approach ensures >99% specificity and sensitivity for known expansion [190]. Alternative approaches are based on the *in situ* hybridization of the repeat probe to the nuclear foci [191]. Antibodies against muscleblind proteins may also be useful. However, the major limitation of this approach is the requirement of muscle biopsy.

Potential therapeutic targets are expanded repeats in the DMPK gene (DNA) and the DMPK mRNA. Drugs that can manipulate the functions of MMR and other DNA repair genes might be able to direct the repeat instability toward contractions. Chemotherapeutic agents that cause double-strand breaks and other DNA damage, or inhibit DNA replication, may promote deletions of the repeats [192–194]. However, these approaches may introduce significant adverse effects on the genomic integrity. Targeting the expanded repeats may alleviate such problems, and this may be more readily achievable if the target is RNA. Antisense oligonucleosides and ribozymes targeting the expanded repeats in RNA have been tried in cell culture and mouse models with promising results [195, 196].

VI. CONCLUDING REMARKS

Substantial progress has been made in understanding the disease-causing mechanism of myotonic dystrophies. The *trans*-dominant gain-of-function models mediated by RNA transcripts containing expanded repeats appear to play a central role in the mechanisms of DM1 and DM2, and explain the complex multisystemic phenotypes. Thus, DM1 and DM2 belong to a class of genetic disorders not only phenotypically but also molecularly distinct from other disorders including those caused by an expansion of most other microsatellite repeats. While phenotypic similarities between DM1 and DM2 endorse the idea of a common mechanism based on the RNA gain of function, further studies are needed to explain phenotypic differences between these two diseases, especially the lack of the congenital form in DM2. Whether another multisystemic myotonic disorder, recently reported as DM3, shares a molecular mechanism with DM1 and DM2 is yet to be determined. These advances in the field of DM1 and DM2 may allow us to develop new experimental therapeutics based on the molecular mechanisms.

Acknowledgment

This work was supported by a grant (3801) from the Muscular Dystrophy Association (T.A.).

References

1. Harper, P. S. (2001). "Myotonic Dystrophy," 3rd ed. Saunders, London.
2. Ashizawa, T., and Harper, P. S. (2005) Myotonic dystrophies. *In* "Emery and Rimoin's Principles and Practice of Medical Genetics" (D. L. Rimoin, J. M. Connor, R. E. Pyeritz, and B. R. Korf, Eds.), 4th ed. Churchill Livingstone, New York.
3. Steinert, H. H. W. (1909). Myopathologische Beiträge: I. Über das klinische und anatomische Bilde des Muskelschwunds des Myotoniker. *Dtsch. Z. Nervenheilkd.* **37**, 58–104.
4. Batten, F. E., and Gibb, H. P. (1909). Myotonia atrophica. *Brain* **32**, 187–205.
5. Friedreich, N. (1863). Ueber degenerative atrophie der spinalen Hinterstränge. *Virchow's Arch. Pathol. Anat.* **26**, 391–419.
6. Huntington, G. (1872). On chorea. *Med. Surg. Rep.*, 317–321.
7. Kennedy, W. R., Alter, M., and Sung, J. H. (1968). Progressive proximal spinal and bulbar muscular atrophy of late onset: A sex-linked recessive trait. *Neurology* **18**, 671–680.
8. Lubs, H. A. (1969). A marker X chromosome. *Am. J. Hum. Genet.* **21**, 231–244.
9. Nakano, K. K., Dawson, D. M., and Spence, A. (1972). Machado disease: A hereditary ataxia in Portuguese emigrants to Massachusetts. *Neurology* **22**, 49–55.
10. Naito, H., and Oyanagi, S. (1982). Familial myoclonus epilepsy and choreoathetosis: Hereditary dentatorubral-pallidoluysian atrophy. *Neurology* **32**, 798–807.
11. Rosenberg, R. N. (1995). Autosomal dominant cerebellar phenotypes: The genotype has settled the issue. *Neurology* **45**, 1–5.
12. Fu, Y. H., Pizzuti, A., Fenwick, R. G., Jr., King, J., Rajnarayan, S., Dunne, P. W., Dubel, J., Nasser, G. A., Ashizawa, T., de Jong, P., et al. (1992). An unstable triplet repeat in a gene related to myotonic muscular dystrophy. *Science* **255**, 1256–1258.
13. Brook, J. D., McCurrach, M. E., Harley, H. G., Buckler, A. J., Church, D., Aburatani, H., Hunter, K., Stanton, V. P., Thirion, J. P., Hudson, T., et al. (1992). Molecular basis of myotonic dystrophy: Expansion of a trinucleotide (CTG) repeat at the 3' end of a transcript encoding a protein kinase family member. *Cell* **68**, 799–808.
14. Mahadevan, M., Tsilfidis, C., Sabourin, L., Shutler, G., Amemiya, C., Jansen, G., Neville, C., Narang, M., Barcelo, J., O'Hoy, K., et al. (1992). Myotonic dystrophy mutation: An unstable CTG repeat in the 3' untranslated region of the gene. *Science* **255**, 1253–1255.
15. Liquori, C. L., Ricker, K., Moseley, M. L., Jacobsen, J. F., Kress, W., Naylor, S. L., Day, J. W., and Ranum, L. P. (2001). Myotonic dystrophy type 2 caused by a CCTG expansion in intron 1 of ZNF9. *Science* **293**, 864–867.
16. International Myotonic Dystrophy Consortium (IDMC) (2000). New nomenclature and DNA testing guidelines for myotonic dystrophy type 1 (DM1). *Neurology* **54**, 1218–1221.

17. Le Ber, I., Martinez, M., Campion, D., Laquerriere, A., Betard, C., Bassez, G., Girard, C., Saugier-Veber, P., Raux, G., Sergeant, N., Magnier, P., Maisonobe, T., Eymard, B., Duyckaerts, C., Delacourte, A., Frebourg, T., and Hannequin, D. (2004). A non-DM1, non-DM2 multisystem myotonic disorder with frontotemporal dementia: Phenotype and suggestive mapping of the DM3 locus to chromosome 15q21-24. *Brain* **127**, 1979–1992.
18. Pandolfo, M. (2003). Friedreich ataxia. *Semin. Pediatr. Neurol.* **10**, 163–172.
19. Hagerman, R. J., and Hagerman, P. J. (2002). "Fragile X Syndrome: Diagnosis, Treatment and Research," 3rd ed. Johns Hopkins Univ. Press, Baltimore.
20. Sinnreich, M., and Klein, C. J. (2004). Bulbospinal muscular atrophy: Kennedy's disease. *Arch. Neurol.* **61**, 1324–1326.
21. Lodi, R., Schapira, A. H., Manners, D., Styles, P., Wood, N. W., Taylor, D. J., and Warner, T. T. (2000). Abnormal in vivo skeletal muscle energy metabolism in Huntington's disease and dentatorubropallidoluysian atrophy. *Ann. Neurol.* **48**, 72–76.
22. Andreassen, O. A., Dedeoglu, A., Stanojevic, V., Hughes, D. B., Browne, S. E., Leech, C. A., Ferrante, R. J., Habener, J. F., Beal, M. F., and Thomas, M. K. (2002). Huntington's disease of the endocrine pancreas: Insulin deficiency and diabetes mellitus due to impaired insulin gene expression. *Neurobiol. Dis.* **11**, 410–424.
23. Benton, C. S., de Silva, R., Rutledge, S. L., Bohlega, S., Ashizawa, T., and Zoghbi, H. Y. (1998). Molecular and clinical studies in SCA-7 define a broad clinical spectrum and the infantile phenotype. *Neurology* **51**, 1081–1086.
24. Rasmussen, A., Matsuura, T., Ruano, L., Yescas, P., Ochoa, A., Ashizawa, T., and Alonso, E. (2001). Clinical and genetic analysis of four Mexican families with spinocerebellar ataxia type 10. *Ann. Neurol.* **50**, 234–239.
25. Greenfield, J. G. (1911). Notes on family of "myotonia atrophica" and early cataract, with a report on an additional case of myotonia atrophica. *Rev. Neurol. Psychol.* **9**, 169–181.
26. Curschmann, H. (1912). Über familiare atrophische Myononie. *Dtsch. Z. Nervenheilkd.* **45**, 161–202.
27. Harper, P. S., van Engelen, B., Eymard, B., and Wilcox, D. E. (2004). Myotonic dystrophy: Present management and future therapy. Oxford Univ. Press, Oxford.
28. Harper, P. S., and Johnson, K. J. (2001). Myotonic dystrophy. *In* Scriver, C. R., Beaudet, A. L., Sly, W. S., Valle, D., eds., "The Metabolic and Molecular Bases of Inherited Diseases," 8th ed., pp. 5525–5550. McGraw–Hill Medical, New York.
29. Day, J. W., and Ranum, L. P. (2005). RNA pathogenesis of the myotonic dystrophies. *Neuromuscul. Disord.* **15**, 5–16.
30. Fleischer, B. (1918). Über myotonische Dystrophie mit Katarakt. *Albrecht von Graefes Arch. Klin. Ophthalmol.* **96**, 91–133.
31. Penrose, L. S. (1948). The problem af anticipation in pedigrees of dystrophia myotonica. *Ann. Eugen.* **14**, 125–232.
32. Höweler, C. J., Busch, H. F., Geraedts, J. P., Niermeijer, M. F., and Staal, A. (1989). Anticipation in myotonic dystrophy: Fact or fiction? *Brain* **112** (Pt. 3), 779–797.
33. Campbell, A. M., Corner, B., Norman, R. M., and Urich, H. (1961). The rigid form of Huntington's disease. *J. Neurol. Neurosurg. Psychiatry* **24**, 71–77.
34. Sherman, S. L., Jacobs, P. A., Morton, N. E., Froster-Iskenius, U., Howard-Peebles, P. N., Nielsen, K. B., Partington, M. W., Sutherland, G. R., Turner, G., and Watson, M. (1985). Further segregation analysis of the fragile X syndrome with special reference to transmitting males. *Hum. Genet.* **69**, 289–299.
35. Vanier, T. M. (1960). Dystrophia myotonica in childhood. *Br. Med. J.* **5208**, 1284–1288.
36. Zeesman, S., Carson, N., and Whelan, D. T. (2002). Paternal transmission of the congenital form of myotonic dystrophy type 1: A new case and review of the literature. *Am. J. Med. Genet.* **107**, 222–226.
37. Moretti, P., Blazo, M., Garcia, L., Armstrong, D., Lewis, R. A., Roa, B., and Scaglia, F. (2004). Spinocerebellar ataxia type 2 (SCA2) presenting with ophthalmoplegia and developmental delay in infancy. *Am. J. Med. Genet. A* **124**, 392–396.
38. Babovic-Vuksanovic, D., Snow, K., Patterson, M. C., and Michels, V. V. (1998). Spinocerebellar ataxia type 2 (SCA 2) in an infant with extreme CAG repeat expansion. *Am. J. Med. Genet.* **79**, 383–387.
39. Mao, R., Aylsworth, A. S., Potter, N., Wilson, W. G., Breningstall, G., Wick, M. J., Babovic-Vuksanovic, D., Nance, M., Patterson, M. C., Gomez, C. M., and Snow, K. (2002). Childhood-onset ataxia: Testing for large CAG-repeats in SCA2 and SCA7. *Am. J. Med. Genet.* **110**, 338–345.
40. Hsieh, M., Lin, S. J., Chen, J. F., Lin, H. M., Hsiao, K. M., Li, S. Y., Li, C., and Tsai, C. J. (2000). Identification of the spinocerebellar ataxia type 7 mutation in Taiwan: Application of PCR-based Southern blot. *J. Neurol.* **247**, 623–629.
41. Harper, P. S., Rivas, M. L., Bias, W. B., Hutchinson, J. R., Dyken, P. R., and McKusick, V. A. (1972). Genetic linkage confirmed between the locus for myotonic dystrophy and the ABH-secretion and Lutheran blood group loci. *Am. J. Hum. Genet.* **24**, 310–316.
42. Renwick, J. H., and Bolling, D. R. (1971). An analysis procedure illustrated on a triple linkage of use for prenatal diagnosis of myotonic dystrophy. *J. Med. Genet.* **8**, 399–406.
43. Whitehead, A. S., Solomon, E., Chambers, S., Bodmer, W. F., Povey, S., and Fey, G. (1982). Assignment of the structural gene for the third component of human complement to chromosome 19. *Proc. Natl. Acad. Sci. USA* **79**, 5021–5025.
44. McAlpine, P. J., Mohandas, T., Ray, M., Wang, H., and Hamerton, J. L. (1976). Assignment of the peptidase D gene locus (PEPD) to chromosome 19 in man. *Birth Defects* **12**, 204–205.
45. Pearn, J., and Hudgson, P. (1978). Anterior-horn cell degeneration and gross calf hypertrophy with adolescent onset: A new spinal muscular atrophy syndrome. *Lancet* **1**, 1059–1061.
46. Jackson, J. F., Currier, R. D., Terasaki, P. I., and Morton, N. E. (1977). Spinocerebellar ataxia and HLA linkage: Risk prediction by HLA typing. *N. Engl. J. Med.* **296**, 1138–1141.
47. Gusella, J. F., Wexler, N. S., Conneally, P. M., Naylor, S. L., Anderson, M. A., Tanzi, R. E., Watkins, P. C., Ottina, K., Wallace, M. R., Sakaguchi, A. Y., et al. (1983). A polymorphic DNA marker genetically linked to Huntington's disease. *Nature* **306**, 234–238.
48. Chamberlain, S., Shaw, J., Rowland, A., Wallis, J., South, S., Nakamura, Y., von Gabain, A., Farrall, M., and Williamson, R. (1988). Mapping of mutation causing Friedreich's ataxia to human chromosome 9. *Nature* **334**, 248–250.
49. Manto, M. U. (2005). The wide spectrum of spinocerebellar ataxias (SCAs). *Cerebellum* **4**, 2–6.
50. Fu, Y. H., Kuhl, D. P., Pizzuti, A., Pieretti, M., Sutcliffe, J. S., Richards, S., Verkerk, A. J., Holden, J. J., Fenwick, R. G., Jr., Warren, S. T., et al. (1991). Variation of the CGG repeat at the fragile X site results in genetic instability: Resolution of the Sherman paradox. *Cell* **67**, 1047–1058.
51. La Spada, A. R., Wilson, E. M., Lubahn, D. B., Harding, A. E., and Fischbeck, K. H. (1991). Androgen receptor gene mutations in X-linked spinal and bulbar muscular atrophy. *Nature* **352**, 77–79.
52. Aslanidis, C., Jansen, G., Amemiya, C., Shutler, G., Mahadevan, M., Tsilfidis, C., Chen, C., Alleman, J., Wormskamp, N. G., Vooijs, M., et al. (1992). Cloning of the essential myotonic dystrophy region and mapping of the putative defect. *Nature* **355**, 548–551.
53. Buxton, J., Shelbourne, P., Davies, J., Jones, C., Van Tongeren, T., Aslanidis, C., de Jong, P., Jansen, G., Anvret, M., Riley, B., et al. (1992). Detection of an unstable fragment of DNA specific to individuals with myotonic dystrophy. *Nature* **355**, 547–548.

54. Harley, H. G., Brook, J. D., Rundle, S. A., Crow, S., Reardon, W., Buckler, A. J., Harper, P. S., Housman, D. E., and Shaw, D. J. (1992). Expansion of an unstable DNA region and phenotypic variation in myotonic dystrophy. *Nature* **355**, 545–546.
55. Ricker, K., Koch, M. C., Lehmann-Horn, F., Pongratz, D., Speich, N., Reiners, K., Schneider, C., and Moxley, R. T., 3rd (1995). Proximal myotonic myopathy: Clinical features of a multisystem disorder similar to myotonic dystrophy. *Arch. Neurol.* **52**, 25–31.
56. Day, J. W., Roelofs, R., Leroy, B., Pech, I., Benzow, K., and Ranum, L. P. (1999). Clinical and genetic characteristics of a five-generation family with a novel form of myotonic dystrophy (DM2). *Neuromuscul. Disord.* **9**, 19–27.
57. Udd, B., Krahe, R., Wallgren-Pettersson, C., Falck, B., and Kalimo, H. (1997). Proximal myotonic dystrophy—a family with autosomal dominant muscular dystrophy, cataracts, hearing loss and hypogonadism: Heterogeneity of proximal myotonic syndromes? *Neuromuscul. Disord.* **7**, 217–228.
58. Ranum, L. P., Rasmussen, P. F., Benzow, K. A., Koob, M. D., and Day, J. W. (1998). Genetic mapping of a second myotonic dystrophy locus. *Nat. Genet.* **19**, 196–198.
59. Ricker, K., Grimm, T., Koch, M. C., Schneider, C., Kress, W., Reimers, C. D., Schulte-Mattler, W., Mueller-Myhsok, B., Toyka, K. V., and Mueller, C. R. (1999). Linkage of proximal myotonic myopathy to chromosome 3q. *Neurology* **52**, 170–171.
60. Harper, P. S., Harley, H. G., Reardon, W., and Shaw, D. J. (1992). Anticipation in myotonic dystrophy: New light on an old problem. *Am. J. Hum. Genet.* **51**, 10–16.
61. Harley, H. G., Rundle, S. A., Reardon, W., Myring, J., Crow, S., Brook, J. D., Harper, P. S., and Shaw, D. J. (1992). Unstable DNA sequence in myotonic dystrophy. *Lancet* **339**, 1125–1128.
62. Ashizawa, T., Dubel, J. R., Dunne, P. W., Dunne, C. J., Fu, Y. H., Pizzuti, A., Caskey, C. T., Boerwinkle, E., Perryman, M. B., Epstein, H. F., et al. (1992). Anticipation in myotonic dystrophy: II. Complex relationships between clinical findings and structure of the GCT repeat. *Neurology* **42**, 1877–1883.
63. Lavedan, C., Hofmann-Radvanyi, H., Shelbourne, P., Rabes, J. P., Duros, C., Savoy, D., Dehaupas, I., Luce, S., Johnson, K., and Junien, C. (1993). Myotonic dystrophy: Size- and sex-dependent dynamics of CTG meiotic instability, and somatic mosaicism. *Am. J. Hum. Genet.* **52**, 875–883.
64. Ashizawa, T., Dunne, P. W., Ward, P. A., Seltzer, W. K., and Richards, C. S. (1994). Effects of the sex of myotonic dystrophy patients on the unstable triplet repeat in their affected offspring. *Neurology* **44**, 120–122.
65. Martorell, L., Monckton, D. G., Sanchez, A., Lopez De Munain, A., and Baiget, M. (2001). Frequency and stability of the myotonic dystrophy type 1 premutation. *Neurology* **56**, 328–335.
66. Martorell, L., Gamez, J., Cayuela, M. L., Gould, F. K., McAbney, J. P., Ashizawa, T., Monckton, D. G., and Baiget, M. (2004). Germline mutational dynamics in myotonic dystrophy type 1 males: Allele length and age effects. *Neurology* **62**, 269–274.
67. Snell, R. G., MacMillan, J. C., Cheadle, J. P., Fenton, I., Lazarou, L. P., Davies, P., MacDonald, M. E., Gusella, J. F., Harper, P. S., and Shaw, D. J. (1993). Relationship between trinucleotide repeat expansion and phenotypic variation in Huntington's disease. *Nat. Genet.* **4**, 393–397.
68. Komure, O., Sano, A., Nishino, N., Yamauchi, N., Ueno, S., Kondoh, K., Sano, N., Takahashi, M., Murayama, N., Kondo, I., et al. (1995). DNA analysis in hereditary dentatorubral-pallidoluysian atrophy: Correlation between CAG repeat length and phenotypic variation and the molecular basis of anticipation. *Neurology* **45**, 143–149.
69. Chung, M. Y., Ranum, L. P., Duvick, L. A., Servadio, A., Zoghbi, H. Y., and Orr, H. T. (1993). Evidence for a mechanism predisposing to intergenerational CAG repeat instability in spinocerebellar ataxia type I. *Nat. Genet.* **5**, 254–258.
70. Schols, L., Gispert, S., Vorgerd, M., Menezes Vieira-Saecker, A. M., Blanke, P., Auburger, G., Amoiridis, G., Meves, S., Epplen, J. T., Przuntek, H., Pulst, S. M., and Riess, O. (1997). Spinocerebellar ataxia type 2: Genotype and phenotype in German kindreds. *Arch. Neurol.* **54**, 1073–1080.
71. Riess, O., Laccone, F. A., Gispert, S., Schols, L., Zuhlke, C., Vieira-Saecker, A. M., Herlt, S., Wessel, K., Epplen, J. T., Weber, B. H., Kreuz, F., Chahrokh-Zadeh, S., Meindl, A., Lunkes, A., Aguiar, J., Macek, M., Jr., Krebsova, A., Macek, M., Sr., Burk, K., Tinschert, S., Schreyer, I., Pulst, S. M., and Auburger, G. (1997). SCA2 trinucleotide expansion in German SCA patients. *Neurogenetics* **1**, 59–64.
72. Igarashi, S., Takiyama, Y., Cancel, G., Rogaeva, E. A., Sasaki, H., Wakisaka, A., Zhou, Y. X., Takano, H., Endo, K., Sanpei, K., Oyake, M., Tanaka, H., Stevanin, G., Abbas, N., Durr, A., Rogaev, E. I., Sherrington, R., Tsuda, T., Ikeda, M., Cassa, E., Nishizawa, M., Benomar, A., Julien, J., Weissenbach, J., Wang, G. X., Agid, Y., St George-Hyslop, P. H., Brice, A., and Tsuji, S. (1996). Intergenerational instability of the CAG repeat of the gene for Machado–Joseph disease (MJD1) is affected by the genotype of the normal chromosome: Implications for the molecular mechanisms of the instability of the CAG repeat. *Hum. Mol. Genet.* **5**, 923–932.
73. Maruyama, H., Nakamura, S., Matsuyama, Z., Sakai, T., Doyu, M., Sobue, G., Seto, M., Tsujihata, M., Oh-i, T., Nishio, T., et al. (1995). Molecular features of the CAG repeats and clinical manifestation of Machado–Joseph disease. *Hum. Mol. Genet.* **4**, 807–812.
74. David, G., Giunti, P., Abbas, N., Coullin, P., Stevanin, G., Horta, W., Gemmill, R., Weissenbach, J., Wood, N., Cunha, S., Drabkin, H., Harding, A. E., Agid, Y., and Brice, A. (1996). The gene for autosomal dominant cerebellar ataxia type II is located in a 5-cM region in 3p12-p13: Genetic and physical mapping of the SCA7 locus. *Am. J. Hum. Genet.* **59**, 1328–1336.
75. David, G., Abbas, N., Stevanin, G., Durr, A., Yvert, G., Cancel, G., Weber, C., Imbert, G., Saudou, F., Antoniou, E., Drabkin, H., Gemmill, R., Giunti, P., Benomar, A., Wood, N., Ruberg, M., Agid, Y., Mandel, J. L., and Brice, A. (1997). Cloning of the SCA7 gene reveals a highly unstable CAG repeat expansion. *Nat. Genet.* **17**, 65–70.
76. Goldberg, Y. P., Andrew, S. E., Theilmann, J., Kremer, B., Squitieri, F., Telenius, H., Brown, J. D., and Hayden, M. R. (1993). Familial predisposition to recurrent mutations causing Huntington's disease: Genetic risk to sibs of sporadic cases. *J. Med. Genet.* **30**, 987–990.
77. Day, J. W., Ricker, K., Jacobsen, J. F., Rasmussen, L. J., Dick, K. A., Kress, W., Schneider, C., Koch, M. C., Beilman, G. J., Harrison, A. R., Dalton, J. C., and Ranum, L. P. (2003). Myotonic dystrophy type 2: Molecular, diagnostic and clinical spectrum. *Neurology* **60**, 657–664.
78. Schneider, C., Ziegler, A., Ricker, K., Grimm, T., Kress, W., Reimers, C. D., Meinck, H., Reiners, K., and Toyka, K. V. (2000). Proximal myotonic myopathy: Evidence for anticipation in families with linkage to chromosome 3q. *Neurology* **55**, 383–388.
79. Mankodi, A., Teng-Umnuay, P., Krym, M., Henderson, D., Swanson, M., and Thornton, C. A. (2003). Ribonuclear inclusions in skeletal muscle in myotonic dystrophy types 1 and 2. *Ann. Neurol.* **54**, 760–768.
80. Jansen, G., Willems, P., Coerwinkel, M., Nillesen, W., Smeets, H., Vits, L., Howeler, C., Brunner, H., and Wieringa, B. (1994). Gonosomal mosaicism in myotonic dystrophy patients: Involvement of mitotic events in $(CTG)_n$ repeat variation and selection against extreme expansion in sperm. *Am. J. Hum. Genet.* **54**, 575–585.
81. Ashizawa, T., Dubel, J. R., and Harati, Y. (1993). Somatic instability of CTG repeat in myotonic dystrophy. *Neurology* **43**, 2674–2678.
82. Thornton, C. A., Johnson, K., and Moxley, R. T., 3rd (1994). Myotonic dystrophy patients have larger CTG expansions in skeletal muscle than in leukocytes. *Ann. Neurol.* **35**, 104–107.

83. Kinoshita, M., Takahashi, R., Hasegawa, T., Komori, T., Nagasawa, R., Hirose, K., and Tanabe, H. (1996). (CTG)$_n$ expansions in various tissues from a myotonic dystrophy patient. *Muscle Nerve* **19**, 240–242.
84. Ishii, S., Nishio, T., Sunohara, N., Yoshihara, T., Takemura, K., Hikiji, K., Tsujino, S., and Sakuragawa, N. (1996). Small increase in triplet repeat length of cerebellum from patients with myotonic dystrophy. *Hum. Genet.* **98**, 138–140.
85. Takano, H., Onodera, O., Takahashi, H., Igarashi, S., Yamada, M., Oyake, M., Ikeuchi, T., Koide, R., Tanaka, H., Iwabuchi, K., and Tsuji, S. (1996). Somatic mosaicism of expanded CAG repeats in brains of patients with dentatorubral-pallidoluysian atrophy: Cellular population-dependent dynamics of mitotic instability. *Am. J. Hum. Genet.* **58**, 1212–1222.
86. Monckton, D. G., Wong, L. J., Ashizawa, T., and Caskey, C. T. (1995). Somatic mosaicism, germline expansions, germline reversions and intergenerational reductions in myotonic dystrophy males: Small pool PCR analyses. *Hum. Mol. Genet.* **4**, 1–8.
87. Wong, L. J., Ashizawa, T., Monckton, D. G., Caskey, C. T., and Richards, C. S. (1995). Somatic heterogeneity of the CTG repeat in myotonic dystrophy is age and size dependent. *Am. J. Hum. Genet.* **56**, 114–122.
88. Sato, T., Oyake, M., Nakamura, K., Nakao, K., Fukusima, Y., Onodera, O., Igarashi, S., Takano, H., Kikugawa, K., Ishida, Y., Shimohata, T., Koide, R., Ikeuchi, T., Tanaka, H., Futamura, N., Matsumura, R., Takayanagi, T., Tanaka, F., Sobue, G., Komure, O., Takahashi, M., Sano, A., Ichikawa, Y., Goto, J., Kanazawa, I., et al. (1999). Transgenic mice harboring a full-length human mutant DRPLA gene exhibit age-dependent intergenerational and somatic instabilities of CAG repeats comparable with those in DRPLA patients. *Hum. Mol. Genet.* **8**, 99–106.
89. Ishiguro, H., Yamada, K., Sawada, H., Nishii, K., Ichino, N., Sawada, M., Kurosawa, Y., Matsushita, N., Kobayashi, K., Goto, J., Hashida, H., Masuda, N., Kanazawa, I., and Nagatsu, T. (2001). Age-dependent and tissue-specific CAG repeat instability occurs in mouse knock-in for a mutant Huntington's disease gene. *J. Neurosci. Res.* **65**, 289–297.
90. Kennedy, L., Evans, E., Chen, C. M., Craven, L., Detloff, P. J., Ennis, M., and Shelbourne, P. F. (2003). Dramatic tissue-specific mutation length increases are an early molecular event in Huntington disease pathogenesis. *Hum. Mol. Genet.* **12**, 3359–3367.
91. Kennedy, L., and Shelbourne, P. F. (2000). Dramatic mutation instability in HD mouse striatum: Does polyglutamine load contribute to cell-specific vulnerability in Huntington's disease? *Hum. Mol. Genet.* **9**, 2539–2544.
92. Martorell, L., Johnson, K., Boucher, C. A., and Baiget, M. (1997). Somatic instability of the myotonic dystrophy (CTG)$_n$ repeat during human fetal development. *Hum. Mol. Genet.* **6**, 877–880.
93. Savouret, C., Garcia-Cordier, C., Megret, J., te Riele, H., Junien, C., and Gourdon, G. (2004). MSH2-dependent germinal CTG repeat expansions are produced continuously in spermatogonia from DM1 transgenic mice. *Mol. Cell. Biol.* **24**, 629–637.
94. Ashizawa, T., Monckton, D. G., Vaishnav, S., Patel, B. J., Voskova, A., and Caskey, C. T. (1996). Instability of the expanded (CTG)$_n$ repeats in the myotonin protein kinase gene in cultured lymphoblastoid cell lines from patients with myotonic dystrophy. *Genomics* **36**, 47–53.
95. Khajavi, M., Tari, A. M., Patel, N. B., Tsuji, K., Siwak, D. R., Meistrich, M. L., Terry, N. H., and Ashizawa, T. (2001). "Mitotic drive" of expanded CTG repeats in myotonic dystrophy type 1 (DM1). *Hum. Mol. Genet.* **10**, 855–863.
96. Monckton, D. G., Coolbaugh, M. I., Ashizawa, K. T., Siciliano, M. J., and Caskey, C. T. (1997). Hypermutable myotonic dystrophy CTG repeats in transgenic mice. *Nat. Genet.* **15**, 193–196.
97. Gourdon, G., Radvanyi, F., Lia, A. S., Duros, C., Blanche, M., Abitbol, M., Junien, C., and Hofmann-Radvanyi, H. (1997). Moderate intergenerational and somatic instability of a 55-CTG repeat in transgenic mice. *Nat. Genet.* **15**, 190–192.
98. Fortune, M. T., Vassilopoulos, C., Coolbaugh, M. I., Siciliano, M. J., and Monckton, D. G. (2000). Dramatic, expansion-biased, age-dependent, tissue-specific somatic mosaicism in a transgenic mouse model of triplet repeat instability. *Hum. Mol. Genet.* **9**, 439–445.
99. Seznec, H., Lia-Baldini, A. S., Duros, C., Fouquet, C., Lacroix, C., Hofmann-Radvanyi, H., Junien, C., and Gourdon, G. (2000). Transgenic mice carrying large human genomic sequences with expanded CTG repeat mimic closely the DM CTG repeat intergenerational and somatic instability. *Hum. Mol. Genet.* **9**, 1185–1194.
100. Seznec, H., Agbulut, O., Sergeant, N., Savouret, C., Ghestem, A., Tabti, N., Willer, J. C., Ourth, L., Duros, C., Brisson, E., Fouquet, C., Butler-Browne, G., Delacourte, A., Junien, C., and Gourdon, G. (2001). Mice transgenic for the human myotonic dystrophy region with expanded CTG repeats display muscular and brain abnormalities. *Hum. Mol. Genet.* **10**, 2717–2726.
101. Van den Broek, W. J., Nelen, M. R., Wansink, D. G., Coerwinkel, M. M., te Riele, H., Groenen, P. J., and Wieringa, B. (2002). Somatic expansion behaviour of the (CTG)$_n$ repeat in myotonic dystrophy knock-in mice is differentially affected by Msh3 and Msh6 mismatch-repair proteins. *Hum. Mol. Genet.* **11**, 191–198.
102. Savouret, C., Brisson, E., Essers, J., Kanaar, R., Pastink, A., te Riele, H., Junien, C., and Gourdon, G. (2003). CTG repeat instability and size variation timing in DNA repair-deficient mice. *EMBO J.* **22**, 2264–2273.
103. Pearson, C. E., and Sinden, R. R. (1996). Alternative structures in duplex DNA formed within the trinucleotide repeats of the myotonic dystrophy and fragile X loci. *Biochemistry* **35**, 5041–5053.
104. Pearson, C. E., Wang, Y. H., Griffith, J. D., and Sinden, R. R. (1998). Structural analysis of slipped-strand DNA (S-DNA) formed in (CTG)$_n$: (CAG)$_n$ repeats from the myotonic dystrophy locus. *Nucleic Acids Res.* **26**, 816–823.
105. Ohshima, K., Kang, S., Larson, J. E., and Wells, R. D. (1996). Cloning, characterization, and properties of seven triplet repeat DNA sequences. *J. Biol. Chem.* **271**, 16773–16783.
106. Bacolla, A., Gellibolian, R., Shimizu, M., Amirhaeri, S., Kang, S., Ohshima, K., Larson, J. E., Harvey, S. C., Stollar, B. D., and Wells, R. D. (1997). Flexible DNA: Genetically unstable CTG·CAG and CGG·CCG from human hereditary neuromuscular disease genes. *J. Biol. Chem.* **272**, 16783–16792.
107. Wojciechowska, M., Bacolla, A., Larson, J. E., and Wells, R. D. (2005). The myotonic dystrophy type 1 triplet repeat sequence induces gross deletions and inversions. *J. Biol. Chem.* **280**, 941–952.
108. Hashem, V. I., Klysik, E. A., Rosche, W. A., and Sinden, R. R. (2002). Instability of repeated DNAs during transformation in *Escherichia coli*. *Mutat. Res.* **502**, 39–46.
109. Pelletier, R., Krasilnikova, M. M., Samadashwily, G. M., Lahue, R., and Mirkin, S. M. (2003). Replication and expansion of trinucleotide repeats in yeast. *Mol. Cell. Biol.* **23**, 1349–1357.
110. Shimizu, M., Fujita, R., Tomita, N., Shindo, H., and Wells, R. D. (2001). Chromatin structure of yeast minichromosomes containing triplet repeat sequences associated with human hereditary neurological diseases. *Nucleic Acids Res. Suppl.*, 71–72.
111. Panigrahi, G. B., Lau, R., Montgomery, S. E., Leonard, M. R., and Pearson, C. E. (2005). Slipped (CTG)*(CAG) repeats can be correctly repaired, escape repair or undergo error-prone repair. *Nat. Struct. Mol. Biol.* **12**, 654–662.
112. Owen, B. A., Yang, Z., Lai, M., Gajek, M., Badger, J. D., Hayes, J. J., Edelmann, W., Kucherlapati, R., Wilson, T. M., and

McMurray, C. T. (2005). (CAG)(*n*)-hairpin DNA binds to Msh2–Msh3 and changes properties of mismatch recognition. *Nat. Struct. Mol. Biol.* **12**, 663–670.

113. Dere, R., Napierala, M., Ranum, L. P., and Wells, R. D. (2004). Hairpin structure-forming propensity of the (CCTG·CAGG) tetranucleotide repeats contributes to the genetic instability associated with myotonic dystrophy type 2. *J. Biol. Chem.* **279**, 41715–41726.

114. Dada, T. O. (1973). Dystrophia myotonica in a Nigerian family. *East Afr. Med. J.* **50**, 213–228.

115. Ashizawa, T., and Epstein, H. F. (1991). Ethnic distribution of myotonic dystrophy gene. *Lancet* **338**, 642–643.

116. Harper, P. S. (1989). "Myotonic dystrophy." Saunders, Philadelphia.

117. Osame, M., and Furusho, T. (1983). [Genetic epidemiology of myotonic dystrophy in Kagoshima and Okinawa districts in Japan]. *Rinsho Shinkeigaku* **23**, 1067–1071.

118. Bouchard, G., Roy, R., Declos, M., Mathieu, J., and Kouladjian, K. (1989). Origin and diffusion of the myotonic dystrophy gene in the Saguenay region (Quebec). *Can. J. Neurol. Sci.* **16**, 119–122.

119. Segel, R., Silverstein, S., Lerer, I., Kahana, E., Meir, R., Sagi, M., Zilber, N., Korczyn, A. D., Shapira, Y., Argov, Z., and Abeliovich, D. (2003). Prevalence of myotonic dystrophy in Israeli Jewish communities: Inter-community variation and founder premutations. *Am. J. Med. Genet. A* **119**, 273–278.

120. Medica, I., Logar, N., Mileta, D. L., and Peterlin, B. (2004). Genealogical study of myotonic dystrophy in Istria (Croatia). *Ann. Genet.* **47**, 139–146.

121. Krndija, D., Savic, D., Mladenovic, J., Rakocevic-Stojanovic, V., Apostolski, S., Todorovic, S., and Romac, S. (2005). Haplotype analysis of the DM1 locus in the Serbian population. *Acta Neurol. Scand.* **111**, 274–277.

122. Pan, H., Lin, H. M., Ku, W. Y., Li, T. C., Li, S. Y., Lin, C. C., and Hsiao, K. M. (2001). Haplotype analysis of the myotonic dystrophy type 1 (DM1) locus in Taiwan: Implications for low prevalence and founder mutations of Taiwanese myotonic dystrophy type 1. *Eur. J. Hum. Genet.* **9**, 638–641.

123. Krahe, R., Eckhart, M., Ogunniyi, A. O., Osuntokun, B. O., Siciliano, M. J., and Ashizawa, T. (1995). De novo myotonic dystrophy mutation in a Nigerian kindred. *Am. J. Hum. Genet.* **56**, 1067–1074.

124. Tishkoff, S. A., Goldman, A., Calafell, F., Speed, W. C., Deinard, A. S., Bonne-Tamir, B., Kidd, J. R., Pakstis, A. J., Jenkins, T., and Kidd, K. K. (1998). A global haplotype analysis of the myotonic dystrophy locus: Implications for the evolution of modern humans and for the origin of myotonic dystrophy mutations. *Am. J. Hum. Genet.* **62**, 1389–1402.

125. Liquori, C. L., Ikeda, Y., Weatherspoon, M., Ricker, K., Schoser, B. G., Dalton, J. C., Day, J. W., and Ranum, L. P. (2003). Myotonic dystrophy type 2: Human founder haplotype and evolutionary conservation of the repeat tract. *Am. J. Hum. Genet.* **73**, 849–862.

126. Bachinski, L. L., Udd, B., Meola, G., Sansone, V., Bassez, G., Eymard, B., Thornton, C. A., Moxley, R. T., Harper, P. S., Rogers, M. T., Jurkat-Rott, K., Lehmann-Horn, F., Wieser, T., Gamez, J., Navarro, C., Bottani, A., Kohler, A., Shriver, M. D., Sallinen, R., Wessman, M., Zhang, S., Wright, F. A., and Krahe, R. (2003). Confirmation of the type 2 myotonic dystrophy (CCTG)$_n$ expansion mutation in patients with proximal myotonic myopathy/proximal myotonic dystrophy of different European origins: A single shared haplotype indicates an ancestral founder effect. *Am. J. Hum. Genet.* **73**, 835–848.

127. Carey, N., Johnson, K., Nokelainen, P., Peltonen, L., Savontaus, M. L., Juvonen, V., Anvret, M., Grandell, U., Chotai, K., Robertson, E., et al. (1994). Meiotic drive at the myotonic dystrophy locus? *Nat. Genet.* **6**, 117–118.

128. Chakraborty, R., Stivers, D. N., Deka, R., Yu, L. M., Shriver, M. D., and Ferrell, R. E. (1996). Segregation distortion of the CTG repeats at the myotonic dystrophy locus. *Am. J. Hum. Genet.* **59**, 109–118.

129. Magee, A. C., and Hughes, A. E. (1998). Segregation distortion in myotonic dystrophy. *J. Med. Genet.* **35**, 1045–1046.

130. Gennarelli, M., Dallapiccola, B., Baiget, M., Martorell, L., and Novelli, G. (1994). Meiotic drive at the myotonic dystrophy locus. *J. Med. Genet.* **31**, 980.

131. Zatz, M., Cerqueira, A., Vainzof, M., and Passos-Bueno, M. R. (1997). Segregation distortion of the CTG repeats at the myotonic dystrophy (DM) locus: New data from Brazilian DM families. *J. Med. Genet.* **34**, 790–791.

132. Zunz, E., Abeliovich, D., Halpern, G. J., Magal, N., and Shohat, M. (2004). Myotonic dystrophy—no evidence for preferential transmission of the mutated allele: A prenatal analysis. *Am. J. Med. Genet. A* **127**, 50–53.

133. Ikeuchi, T., Igarashi, S., Takiyama, Y., Onodera, O., Oyake, M., Takano, H., Koide, R., Tanaka, H., and Tsuji, S. (1996). Non-Mendelian transmission in dentatorubral–pallidoluysian atrophy and Machado–Joseph disease: The mutant allele is preferentially transmitted in male meiosis. *Am. J. Hum. Genet.* **58**, 730–733.

134. Rubinsztein, D. C., and Leggo, J. (1997). Non-Mendelian transmission at the Machado–Joseph disease locus in normal females: Preferential transmission of alleles with smaller CAG repeats. *J. Med. Genet.* **34**, 234–236.

135. Takiyama, Y., Sakoe, K., Soutome, M., Namekawa, M., Ogawa, T., Nakano, I., Igarashi, S., Oyake, M., Tanaka, H., Tsuji, S., and Nishizawa, M. (1997). Single sperm analysis of the CAG repeats in the gene for Machado–Joseph disease (MJD1): Evidence for non-Mendelian transmission of the MJD1 gene and for the effect of the intragenic CGG/GGG polymorphism on the intergenerational instability. *Hum. Mol. Genet.* **6**, 1063–1068.

136. Grewal, R. P., Cancel, G., Leeflang, E. P., Durr, A., McPeek, M. S., Draghinas, D., Yao, X., Stevanin, G., Alnot, M. O., Brice, A., and Arnheim, N. (1999). French Machado–Joseph disease patients do not exhibit gametic segregation distortion: A sperm typing analysis. *Hum. Mol. Genet.* **8**, 1779–1784.

137. Monckton, D. G., and Ashizawa, T. (2004). Molecular aspects of myotonic dystrophy: Our current understanding. *In* "Myotonic Dystrophy: Present Management and Future Therapy" (P. S. Harper, B. van Engelen, B. Eymard, and D. E. Wilcox, Eds.). Oxford Univ. Press, Oxford.

138. Oostra, B. A., and Willemsen, R. (2003). A fragile balance: FMR1 expression levels. *Hum. Mol. Genet.* 12 (Spec. No. 2), R249–R257.

139. Hagerman, P. J., and Hagerman, R. J. (2004). Fragile X-associated tremor/ataxia syndrome (FXTAS). *Ment. Retard. Dev. Disabil. Res. Rev.* **10**, 25–30.

140. Lin, X., and Ashizawa, T. (2005). Recent progress in spinocerebellar ataxia type-10 (SCA10). *Cerebellum* **4**, 37–42.

141. Jin, P., Alisch, R. S., and Warren, S. T. (2004). RNA and microRNAs in fragile X mental retardation. *Nat. Cell. Biol.* **6**, 1048–1053.

142. Voncken, M., Ioannou, P., and Delatycki, M. B. (2004). Friedreich ataxia: Update on pathogenesis and possible therapies. *Neurogenetics* **5**, 1–8.

143. Holmes, S. E., O'Hearn, E., Rosenblatt, A., Callahan, C., Hwang, H. S., Ingersoll-Ashworth, R. G., Fleisher, A., Stevanin, G., Brice, A., Potter, N. T., Ross, C. A., and Margolis, R. L. (2001). A repeat expansion in the gene encoding junctophilin-3 is associated with Huntington disease-like 2. *Nat. Genet.* **29**, 377–378.

144. Holmes, S. E., O'Hearn, E., and Margolis, R. L. (2003). Why is SCA12 different from other SCAs? *Cytogenet. Genome. Res.* **100**, 189–197.

145. Landles, C., and Bates, G. P. (2004). Huntingtin and the molecular pathogenesis of Huntington's disease: Fourth in Molecular Medicine Review Series. *EMBO Rep.* **5**, 958–963.
146. Everett, C. M., and Wood, N. W. (2004). Trinucleotide repeats and neurodegenerative disease. *Brain* **127**, 2385–2405.
147. Amiel, J., Trochet, D., Clement-Ziza, M., Munnich, A., and Lyonnet, S. (2004). Polyalanine expansions in human. *Hum. Mol. Genet.* **13** (Spec. No. 2), R235–R243.
148. Taneja, K. L., McCurrach, M., Schalling, M., Housman, D., and Singer, R. H. (1995). Foci of trinucleotide repeat transcripts in nuclei of myotonic dystrophy cells and tissues. *J. Cell Biol.* **128**, 995–1002.
149. Mankodi, A., Logigian, E., Callahan, L., McClain, C., White, R., Henderson, D., Krym, M., and Thornton, C. A. (2000). Myotonic dystrophy in transgenic mice expressing an expanded CUG repeat. *Science* **289**, 1769–1773.
150. Mankodi, A., Urbinati, C. R., Yuan, Q. P., Moxley, R. T., Sansone, V., Krym, M., Henderson, D., Schalling, M., Swanson, M. S., and Thornton, C. A. (2001). Muscleblind localizes to nuclear foci of aberrant RNA in myotonic dystrophy types 1 and 2. *Hum. Mol. Genet.* **10**, 2165–2170.
151. Fardaei, M., Rogers, M. T., Thorpe, H. M., Larkin, K., Hamshere, M. G., Harper, P. S., and Brook, J. D. (2002). Three proteins, MBNL, MBLL and MBXL, co-localize in vivo with nuclear foci of expanded-repeat transcripts in DM1 and DM2 cells. *Hum. Mol. Genet.* **11**, 805–814.
152. Jiang, H., Mankodi, A., Swanson, M. S., Moxley, R. T., and Thornton, C. A. (2004). Myotonic dystrophy type 1 is associated with nuclear foci of mutant RNA, sequestration of muscleblind proteins and deregulated alternative splicing in neurons. *Hum. Mol. Genet.* **13**, 3079–3088.
153. Kanadia, R. N., Johnstone, K. A., Mankodi, A., Lungu, C., Thornton, C. A., Esson, D., Timmers, A. M., Hauswirth, W. W., and Swanson, M. S. (2003). A muscleblind knockout model for myotonic dystrophy. *Science* **302**, 1978–1980.
154. Timchenko, L. T., Timchenko, N. A., Caskey, C. T., and Roberts, R. (1996). Novel proteins with binding specificity for DNA CTG repeats and RNA CUG repeats: Implications for myotonic dystrophy. *Hum. Mol. Genet.* **5**, 115–121.
155. Timchenko, L. T., Miller, J. W., Timchenko, N. A., DeVore, D. R., Datar, K. V., Lin, L., Roberts, R., Caskey, C. T., and Swanson, M. S. (1996). Identification of a $(CUG)_n$ triplet repeat RNA-binding protein and its expression in myotonic dystrophy. *Nucleic Acids Res.* **24**, 4407–4414.
156. Roberts, R., Timchenko, N. A., Miller, J. W., Reddy, S., Caskey, C. T., Swanson, M. S., and Timchenko, L. T. (1997). Altered phosphorylation and intracellular distribution of a $(CUG)_n$ triplet repeat RNA-binding protein in patients with myotonic dystrophy and in myotonin protein kinase knockout mice. *Proc. Natl. Acad. Sci. USA* **94**, 13221–13226.
157. Mankodi, A., Takahashi, M. P., Jiang, H., Beck, C. L., Bowers, W. J., Moxley, R. T., Cannon, S. C., and Thornton, C. A. (2002). Expanded CUG repeats trigger aberrant splicing of ClC-1 chloride channel pre-mRNA and hyperexcitability of skeletal muscle in myotonic dystrophy. *Mol. Cell* **10**, 35–44.
158. Ho, T. H., Charlet, B. N., Poulos, M. G., Singh, G., Swanson, M. S., and Cooper, T. A. (2004). Muscleblind proteins regulate alternative splicing. *EMBO J.* **23**, 3103–3112.
159. Savkur, R. S., Philips, A. V., Cooper, T. A., Dalton, J. C., Moseley, M. L., Ranum, L. P., and Day, J. W. (2004). Insulin receptor splicing alteration in myotonic dystrophy type 2. *Am. J. Hum. Genet.* **74**, 1309–1313.
160. Charlet, B. N., Savkur, R. S., Singh, G., Philips, A. V., Grice, E. A., and Cooper, T. A. (2002). Loss of the muscle-specific chloride channel in type 1 myotonic dystrophy due to misregulated alternative splicing. *Mol. Cell* **10**, 45–53.
161. Ho, T. H., Bundman, D., Armstrong, D. L., and Cooper, T. A. (2005). Transgenic mice expressing CUG-BP1 reproduce splicing mis-regulation observed in myotonic dystrophy. *Hum. Mol. Genet.* **14**, 1539–1547.
162. Ho, T. H., Savkur, R. S., Poulos, M. G., Mancini, M. A., Swanson, M. S., and Cooper, T. A. (2005). Colocalization of muscleblind with RNA foci is separable from misregulation of alternative splicing in myotonic dystrophy. *J. Cell Sci.* **118**, 2923–2933.
163. Dansithong, W., Paul, S., Comai, L., and Reddy, S. (2004). MBNL1 is the primary determinant of focus formation and aberrant IR splicing in DM1. *J. Biol. Chem.* **280**, 5773–5780.
164. Davis, B. M., McCurrach, M. E., Taneja, K. L., Singer, R. H., and Housman, D. E. (1997). Expansion of a CUG trinucleotide repeat in the 3' untranslated region of myotonic dystrophy protein kinase transcripts results in nuclear retention of transcripts. *Proc. Natl. Acad. Sci. USA* **94**, 7388–7393.
165. Furling, D., Lemieux, D., Taneja, K., and Puymirat, J. (2001). Decreased levels of myotonic dystrophy protein kinase (DMPK) and delayed differentiation in human myotonic dystrophy myoblasts. *Neuromuscul. Disord.* **11**, 728–735.
166. Krahe, R., Ashizawa, T., Abbruzzese, C., Roeder, E., Carango, P., Giacanelli, M., Funanage, V. L., and Siciliano, M. J. (1995). Effect of myotonic dystrophy trinucleotide repeat expansion on DMPK transcription and processing. *Genomics* **28**, 1–14.
167. Wang, Y. H., and Griffith, J. (1995). Expanded CTG triplet blocks from the myotonic dystrophy gene create the strongest known natural nucleosome positioning elements. *Genomics* **25**, 570–573.
168. Jansen, G., Groenen, P. J., Bachner, D., Jap, P. H., Coerwinkel, M., Oerlemans, F., van den Broek, W., Gohlsch, B., Pette, D., Plomp, J. J., Molenaar, P. C., Nederhoff, M. G., van Echteld, C. J., Dekker, M., Berns, A., Hameister, H., and Wieringa, B. (1996). Abnormal myotonic dystrophy protein kinase levels produce only mild myopathy in mice. *Nat. Genet.* **13**, 316–324.
169. Reddy, S., Smith, D. B., Rich, M. M., Leferovich, J. M., Reilly, P., Davis, B. M., Tran, K., Rayburn, H., Bronson, R., Cros, D., Balice-Gordon, R. J., and Housman, D. (1996). Mice lacking the myotonic dystrophy protein kinase develop a late onset progressive myopathy. *Nat. Genet.* **13**, 325–335.
170. Mounsey, J. P., Mistry, D. J., Ai, C. W., Reddy, S., and Moorman, J. R. (2000). Skeletal muscle sodium channel gating in mice deficient in myotonic dystrophy protein kinase. *Hum. Mol. Genet.* **9**, 2313–2320.
171. Reddy, S., Mistry, D. J., Wang, Q. C., Geddis, L. M., Kutchai, H. C., Moorman, J. R., and Mounsey, J. P. (2002). Effects of age and gene dose on skeletal muscle sodium channel gating in mice deficient in myotonic dystrophy protein kinase. *Muscle Nerve* **25**, 850–857.
172. Benders, A. A., Groenen, P. J., Oerlemans, F. T., Veerkamp, J. H., and Wieringa, B. (1997). Myotonic dystrophy protein kinase is involved in the modulation of the Ca^{2+} homeostasis in skeletal muscle cells. *J. Clin. Invest.* **100**, 1440–1447.
173. Wansink, D. G., van Herpen, R. E., Coerwinkel-Driessen, M. M., Groenen, P. J., Hemmings, B. A., and Wieringa, B. (2003). Alternative splicing controls myotonic dystrophy protein kinase structure, enzymatic activity, and subcellular localization. *Mol. Cell. Biol.* **23**, 5489–5501.
174. van Herpen, R. E., Oude Ophuis, R. J., Wijers, M., Bennink, M. B., van de Loo, F. A., Fransen, J., Wieringa, B., and Wansink, D. G. (2005). Divergent mitochondrial and endoplasmic reticulum association of DMPK splice isoforms depends on unique sequence arrangements in tail anchors. *Mol. Cell. Biol.* **25**, 1402–1414.

175. Saveliev, A., Everett, C., Sharpe, T., Webster, Z., and Festenstein, R. (2003). DNA triplet repeats mediate heterochromatin-protein-1-sensitive variegated gene silencing. *Nature* **422**, 909–913.
176. Frisch, R., Singleton, K. R., Moses, P. A., Gonzalez, I. L., Carango, P., Marks, H. G., and Funanage, V. L. (2001). Effect of triplet repeat expansion on chromatin structure and expression of DMPK and neighboring genes, SIX5 and DMWD, in myotonic dystrophy. *Mol. Genet. Metab.* **74**, 281–291.
177. Klesert, T. R., Otten, A. D., Bird, T. D., and Tapscott, S. J. (1997). Trinucleotide repeat expansion at the myotonic dystrophy locus reduces expression of DMAHP. *Nat. Genet.* **16**, 402–406.
178. Thornton, C. A., Wymer, J. P., Simmons, Z., McClain, C., and Moxley, R. T., 3rd (1997). Expansion of the myotonic dystrophy CTG repeat reduces expression of the flanking DMAHP gene. *Nat. Genet.* **16**, 407–409.
179. Eriksson, M., Ansved, T., Edstrom, L., Anvret, M., and Carey, N. (1999). Simultaneous analysis of expression of the three myotonic dystrophy locus genes in adult skeletal muscle samples: The CTG expansion correlates inversely with DMPK and 59 expression levels, but not DMAHP levels. *Hum. Mol. Genet.* **8**, 1053–1060.
180. Hamshere, M. G., Newman, E. E., Alwazzan, M., Athwal, B. S., and Brook, J. D. (1997). Transcriptional abnormality in myotonic dystrophy affects DMPK but not neighboring genes. *Proc. Natl. Acad. Sci. USA* **94**, 7394–7399.
181. Klesert, T. R., Cho, D. H., Clark, J. I., Maylie, J., Adelman, J., Snider, L., Yuen, E. C., Soriano, P., and Tapscott, S. J. (2000). Mice deficient in Six5 develop cataracts: Implications for myotonic dystrophy. *Nat. Genet.* **25**, 105–109.
182. Sarkar, P. S., Appukuttan, B., Han, J., Ito, Y., Ai, C., Tsai, W., Chai, Y., Stout, J. T., and Reddy, S. (2000). Heterozygous loss of Six5 in mice is sufficient to cause ocular cataracts. *Nat. Genet.* **25**, 110–114.
183. Wakimoto, H., Maguire, C. T., Sherwood, M. C., Vargas, M. M., Sarkar, P. S., Han, J., Reddy, S., and Berul, C. I. (2002). Characterization of cardiac conduction system abnormalities in mice with targeted disruption of Six5 gene. *J. Intervent. Card Electrophysiol.* **7**, 127–135.
184. Personius, K. E., Nautiyal, J., and Reddy, S. (2004). Myotonia and muscle contractile properties in mice with SIX5 deficiency. *Muscle Nerve.* **31**, 503–505.
185. Mistry, D. J., Moorman, J. R., Reddy, S., and Mounsey, J. P. (2001). Skeletal muscle Na currents in mice heterozygous for Six5 deficiency. *Physiol. Genom.* **6**, 153–158.
186. Alwazzan, M., Newman, E., Hamshere, M. G., and Brook, J. D. (1999). Myotonic dystrophy is associated with a reduced level of RNA from the DMWD allele adjacent to the expanded repeat. *Hum. Mol. Genet.* **8**, 1491–1497.
187. Moxley, R. T., 3rd, Meola, G., Udd, B., and Ricker, K. (2002). Report of the 84th ENMC Workshop: PROMM (proximal myotonic myopathy) and other myotonic dystrophy-like syndromes. 2nd workshop, 13–15th October 2000, Loosdrecht, The Netherlands. *Neuromuscul. Disord.* **12**, 306–317.
188. Kino, Y., Mori, D., Oma, Y., Takeshita, Y., Sasagawa, N., and Ishiura, S. (2004). Muscleblind protein, MBNL1/EXP, binds specifically to CHHG repeats. *Hum. Mol. Genet.* **13**, 495–507.
189. Schoser, B. G., Kress, W., Walter, M. C., Halliger-Keller, B., Lochmuller, H., and Ricker, K. (2004). Homozygosity for CCTG mutation in myotonic dystrophy type 2. *Brain* **127**, 1868–1877.
190. Udd, B., Meola, G., Krahe, R., Thornton, C., Ranum, L., Day, J., Bassez, G., and Ricker, K. (2003). Report of the 115th ENMC workshop: DM2/PROMM and other myotonic dystrophies. 3rd Workshop, 14–16 February 2003, Naarden, The Netherlands. *Neuromuscul. Disord.* **13**, 589–596.
191. Sallinen, R., Vihola, A., Bachinski, L. L., Huoponen, K., Haapasalo, H., Hackman, P., Zhang, S., Sirito, M., Kalimo, H., Meola, G., Horelli-Kuitunen, N., Wessman, M., Krahe, R., and Udd, B. (2004). New methods for molecular diagnosis and demonstration of the $(CCTG)_n$ mutation in myotonic dystrophy type 2 (DM2). *Neuromuscul. Disord.* **14**, 274–283.
192. Yang, Z., Lau, R., Marcadier, J. L., Chitayat, D., and Pearson, C. E. (2003). Replication inhibitors modulate instability of an expanded trinucleotide repeat at the myotonic dystrophy type 1 disease locus in human cells. *Am. J. Hum. Genet.* **73**, 1092–1105.
193. Gomes-Pereira, M., and Monckton, D. G. (2004). Chemically induced increases and decreases in the rate of expansion of a CAG∗CTG triplet repeat. *Nucleic Acids Res.* **32**, 2865–2872.
194. Hashem, V. I., Pytlos, M. J., Klysik, E. A., Tsuji, K., Khajavi, M., Ashizawa, T., and Sinden, R. R. (2004). Chemotherapeutic deletion of CTG repeats in lymphoblast cells from DM1 patients. *Nucleic Acids Res.* **32**, 6334–6346.
195. Furling, D., Doucet, G., Langlois, M. A., Timchenko, L., Belanger, E., Cossette, L., and Puymirat, J. (2003). Viral vector producing antisense RNA restores myotonic dystrophy myoblast functions. *Gene Ther.* **10**, 795–802.
196. Langlois, M. A., Lee, N. S., Rossi, J. J., and Puymirat, J. (2003). Hammerhead ribozyme-mediated destruction of nuclear foci in myotonic dystrophy myoblasts. *Mol. Ther.* **7**, 670–680.

The RNA-Mediated Disease Process in Myotonic Dystrophy

CHARLES A. THORNTON, MAURICE S. SWANSON, AND THOMAS A. COOPER

Department of Neurology, University of Rochester Medical Center, Rochester, New York 14642; Department of Molecular Genetics and Microbiology and Powell Gene Therapy Center, University of Florida College of Medicine, Gainesville, Florida 32610; and Department of Pathology and Department of Molecular and Cellular Biology, Baylor College of Medicine, Houston, Texas 77030

I. Introduction
II. Evidence for an RNA-Mediated Disease Mechanism in DM1
   A. Evidence against Conventional Mechanisms for Genetic Dominance
   B. Evidence for the RNA-Dominant Genetic Mechanism
   C. Effects of Expanded Poly(CUG) on Intracellular Transcript Localization
   D. Cell Culture Models of DM1 Involving Overexpression of CUG Expansion RNA
   E. Transgenic Mouse Models of DM1 Involving Overexpression of CUG Expansion RNA
   F. DM2 Results from Expression of Untranslated CCUG Repeats
III. Biochemical Basis for RNA-Mediated Disease
   A. Mutant DMPK and ZNF9 RNAs Are Retained in the Nucleus
   B. Alternative Splicing of Pre-mRNAs
   C. Alternative Splicing Is Disrupted in DM
   D. Mechanisms for Misregulated Alternative Splicing: Role for CELF and MBNL Proteins
   E. CELF Proteins Are RNA Splicing Regulators
   F. Muscleblind Sequestration Model for DM
IV. Unanswered Questions for the RNA-Mediated Disease Process in DM
   A. What Is the Molecular Basis for the Postnatal Splicing Switch Affected in DM?
   B. Are Other Biochemical Pathways Altered in DM?
   C. Why Do Existing Mouse Models Fail to Recapitulate Congenital DM Phenotypes?
References

Myotonic dystrophy (DM) is a dominantly inherited degenerative disease which is caused by the expansion of unstable $(CTG)_n$ and $(CCTG)_n$ microsatellites in the non-coding regions of two unrelated genes. In this review, we will discuss the molecular basis of DM disease pathogenesis with particular emphasis on DM type 1 (DM1). DM is the first example of an RNA-mediated disease. Mutant DMPK mRNAs are toxic because they accumulate in the nucleus and alter the activities of pre-mRNA alternative splicing factors during post-natal development. This targeted interference with the normal pathway of alternative splicing results in the retention of a distinct group of fetal exons in specific mRNAs during post-natal development. Ultimately, these mRNAs are translated into protein isoforms that are incompatible with the physiological demands of adult tissues. These studies have provided important new insights into the regulation of RNA splicing during development and suggest the possibility that additional RNA-mediated diseases exist.

I. INTRODUCTION

In myotonic dystrophy, the transcription products of a mutant gene interfere with cell function and trigger a heritable disease state, independently of the protein they encode. The weight of current evidence now supports this novel mechanism, and indicates several interesting features about the disease process. These include observations that genetic dominance can result from a toxic gain of function by mutant RNA, that repetitive sequences in noncoding RNA have biological activity, and that complex phenotypes may result from misregulated alternative splicing of pre-mRNA. Indeed, myotonic dystrophy has compelled us to revise our notions about the role of RNA in human genetics and the potential for *trans* effects among mRNAs. It is remarkable that no close precedent for this disease process has been recognized in other species thus far. Owing to the lack of biological precedent, and to the limited range of experimental systems that are currently available, our understanding of the biochemical mechanisms underlying this RNA-mediated disease process is still incomplete. Furthermore, the extent to which it may explain the signs and symptoms of myotonic dystrophy remains uncertain, though current evidence would indicate that it is the predominant pathophysiological mechanism. In this chapter, we summarize the evidence for an RNA-mediated disease process in myotonic dystrophy type 1 (DM1) and then review current information about the molecular mechanisms.

II. EVIDENCE FOR AN RNA-MEDIATED DISEASE MECHANISM IN DM1

Investigation of molecular pathogenesis in DM1 began with the 1992 discovery of an expanded CTG repeat segregating with disease in the affected kindreds [1–4]. Despite 90 years of clinical and physiological observations up to that point, few connections, if any, had been established with the cellular pathways that now appear most pertinent to the disease process. A very perplexing aspect of the CTG expansion mutation, when it was initially unveiled, was its unusual position within the *dystrophia myotonica protein kinase* (*DMPK*) gene. DM1 was one of the first instances in which the mutation causing a dominantly inherited human disorder was localized to the 3′ untranslated region of a gene [5]. Lacking information about the function of DMPK protein, or any obvious conclusion about how its activity could be affected by this mutation, the gene discovery did not provide any immediate insight into the disease mechanism. To address this question, the initial investigations were focused on the conventional mechanisms for genetic dominance, namely, the potential ways in which this mutation could affect DMPK or other proteins encoded at the DM1 locus, and the cellular consequences of altered protein activity. The investigation of a novel, RNA-mediated mechanism gathered momentum only after the initial results failed to establish a clear connection between DMPK activity and signs of DM1. The case for an RNA-mediated disease mechanism, therefore, currently rests on the cumulative weight of evidence against a conventional mechanism, together with observations supporting a deleterious gain of function by mutant RNA. In reviewing this topic, it is important to note that these two mechanistic alternatives, one conventional and involving the DMPK protein, the other unconventional and involving repeat expansion RNA, are not mutually exclusive. It remains entirely possible, if not likely, that both alternatives contribute to the extraordinarily complex phenotype of DM1. The focus of this chapter, however, is on RNA-mediated pathogenesis.

A. Evidence against Conventional Mechanisms for Genetic Dominance

Classic mechanisms for genetic dominance all involve the effects of mutations on proteins encoded by mutant alleles (reviewed in [6]). One potential effect of a point mutation or gene rearrangement is that the function of the protein encoded by a mutant allele is reduced or eliminated. In the case of autosomal genes, most mutations that eliminate function of a single allele are well tolerated, but for select genes this can result in dominantly inherited disease. This mechanism, which is particularly important for developmental disease, is referred to as haploinsufficiency (i.e., a single functional allele is not sufficient to protect against disease). By contrast, certain mutations, especially those that involve regulatory elements or gene duplication, may instead result in increased protein activity. However, the most common mechanism, in terms of number of individuals affected, is that point mutations or small rearrangements (deletions or insertions) causing dominantly inherited disease have some effect other than simple loss or increase of protein activity. For instance, a mutant protein may interfere with the function of its wild-type counterpart, creating a dominant-negative effect. Alternatively, a mutant protein may acquire a deleterious property that is unrelated to the natural function of its wild-type counterpart, such as, a change in conformational state that leads to illicit protein interactions or aggregation.

In the case of DM1, each of these conventional mechanisms for genetic dominance has been considered and found wanting. Precise measurement of DMPK protein levels in DM1 tissue has been problematic. In view of observations that DMPK mRNA from the mutant allele is retained in the nucleus [7], as discussed below, it would be expected that cellular levels of DMPK protein are reduced, and most data would support this conclusion [8, 9]. However, there are only a few indications that partial loss of DMPK activity can explain symptoms of DM1.

For example, disruption of the murine *Dmpk* gene has failed to reproduce major signs of DM1 in mice [10, 11]. An important exception is that homozygous *Dmpk* knockout mice show abnormalities of cardiac conduction [12], and, like the cardiac involvement in DM1, the defect is progressive [13]. However, the conduction defect in heterozygous *Dmpk* knockout mice, the condition that would most likely approximate the circumstance in human DM1 heart, was mild. Furthermore, DM1 is characterized by progressive cardiac fibrosis and fatty infiltration with preferential involvement of the specialized conduction tissue [14]. A similar degenerative process has not been described in *Dmpk* knockout mice. Perhaps the most persuasive argument against *DMPK* haploinsufficiency as a core mechanism for DM1, however, is the massive amount of genetic data that has accumulated from genetic analysis of families with the disease. This analysis has failed to uncover a single DM1 family having a point mutation or deletion within *DMPK* or any other gene at the DM1 locus. Given that DM1 analysis is one of the genetic tests that is most commonly performed, this cumulative experience would indicate that no mutation at the DM1 locus, other than an expansion of CTG repeats in *DMPK*, can produce signs of DM1.

Although most studies would suggest reduced DMPK protein expression in DM1 patients, the possibility of overexpression in particular tissues, or at certain stages of development, has not been eliminated. In this regard, it is noteworthy that transgenic mice that overexpress human DMPK develop a multisystem degenerative phenotype [15]. While these mice display histopathologic evidence of myopathy and prolonged insertional activity by electromyography, the histological changes do not closely resemble human DM1, and the duration and character of the electromyographic discharge do not qualify it as myotonia. Furthermore, the level of *DMPK* overexpression observed in this transgenic model far exceeds what has been reported in DM1 tissue. These findings indicate that marked overexpression of DMPK has deleterious effects on cardiac and skeletal muscle, but it is unlikely that these observations have a direct bearing on human DM1.

While the expanded repeat cannot directly impact DMPK protein because it lies beyond the termination codon in the final exon of *DMPK*, it could indirectly alter the DMPK translation product if it influenced splicing at upstream exons of the DMPK pre-mRNA. However, expansion of the CUG repeat had no noticeable impact on splicing of upstream DMPK exons [16]. Of note, the CTG repeat in *DMPK* is transcribed, yet the CUG repeat is excluded in a small fraction of DMPK transcripts, because of an infrequent splice event that uses an alternative 3' splice site located downstream of the CUG tract in the terminal exon [17]. The mRNA from the mutant allele, when spliced in this fashion, is not retained in the nucleus. Preferential export of these transcripts could theoretically skew the population of DMPK splice products available in the cytoplasm for translation. However, there is no evidence that this occurs, and the protein encoded by this low-abundance spliced product has not been detected. Thus, there is no evidence that expansion of the CTG repeat leads to production of mutant DMPK protein that could engender dominant-negative or gain-of-function effects. Taken together, these observations support the conclusion that effects of the CTG expansion on DMPK protein expression cannot provide a unitary explanation for the complex phenotype of DM1.

Finally, the effects of the CTG expansion on genes at the DM1 locus are not limited to *DMPK*. Expansion of the repeat induces changes in chromatin structure [18]. These changes, in turn, lead to partial silencing of the neighboring *SIX5* and *DMWD* genes [19–21]. However, this effect may not have a significant impact on steady-state levels of SIX5 or DMWD mRNA in DM1 tissue [22]. Furthermore, mice heterozygous for a *Six5* null allele fail to show signs of DM1 other than a modest increase in the frequency of cataracts [23, 24]. As cataracts are a common finding in inbred mouse strains, the specificity of this effect is uncertain. These results provide further evidence against haploinsufficiency as the major mechanism leading to signs of DM1.

B. Evidence for the RNA-Dominant Genetic Mechanism

Current models of RNA-mediated pathogenesis in DM1 envisage that transcripts from the mutant *DMPK* allele, which contain a tract of expanded CUG repeats, accumulate in foci and compromise one or more functions of the nucleus, such as regulated alternative splicing of pre-mRNA [25, 26]. Although there is no close precedent for this mechanism in other organisms, the current view of RNA-mediated pathogenesis took shape at a time when new biological roles of RNA were coming to light. For example, the discovery of micro-RNAs indicated that regulatory interactions among genes could occur at the level of their respective RNAs (reviewed in [27]). Similarly, modular properties of mRNA were revealed by bacterial transcripts in which genetic information for protein synthesis, conveyed in the coding region, was displayed alongside RNA structure-dependent regulatory elements ("riboswitches"), located in noncoding regions [28]. These parallel developments in RNA biology gave strength to the rationale for considering an unconventional solution to the dilemma posed by DM1: If the pathogenic effect does not involve DMPK protein, what about the mutant RNA? More specifically, would it be reasonable to consider the possibility that an expanded CUG repeat in the noncoding region has endowed the mutant DMPK transcript with a disease-causing property? In fact, an early study proposed a dominant-negative effect of the *DMPK* expansion mutation on polyadenylation [29]. Circumstantially, this idea would fit with observations that tissues most affected by DM1, such as skeletal muscle, heart, and brain, are tissues that express DMPK mRNA most highly [5]. However, this possibility did not rise above the level of interesting speculation until the key observation that RNA transcribed from the mutant *DMPK* allele was retained in the nucleus in discrete foci [7]. Subsequent support for an RNA-mediated disease process came from experiments in which various aspects of the DM1 phenotype were reproduced in model systems by expression of an untranslated expanded CUG repeat [30, 31]. Additional support came from the important discovery that a second form of myotonic dystrophy, now known as myotonic dystrophy type 2 (DM2), also resulted from expansion of an untranslated repeat [32]. Finally, as discussed later in Section III, definitive support came when specific models led to testable predictions, and these predictions were fulfilled [31, 33].

C. Effects of Expanded Poly(CUG) on Intracellular Transcript Localization

Elements controlling cytoplasmic localization, stability, and translation are often located in the 3' untranslated region (UTR) of mRNA [34]. Mutations in these elements may lead to mislocalization of mRNA in the cytoplasm. To determine if an expanded CUG repeat could have this effect, the location of DMPK mRNA was examined in fibroblasts and muscle cells using fluorescence *in situ* hybridization (FISH) [7]. In addition, the transcriptional output of the mutant *DMPK* allele was specifically examined by using a CAG repeat oligonucleotide probe that was complementary to the expanded tract of CUG repeats. This approach led to the surprising observation that mutant DMPK transcripts were not mislocalized in the cytoplasm; instead, they were retained in the nucleus in multiple foci. Furthermore, the site of retention was clearly separate from the site of transcription, suggesting a failure of nucleocytoplasmic transport. Biochemical fractionation confirmed that the transcripts from the mutant *DMPK* allele were quantitatively retained in nuclei of nonmitotic cells [35], and subsequent studies have shown that nuclear RNA foci also are present in DM1 brain [36]. Several experimental systems have now shown that placement of an expanded CUG repeat in the 3' UTR of mRNA consistently leads to nuclear retention of the repeat-bearing transcript [30, 31,

37]. These findings provided the first solid evidence for biological activity of the expanded CUG repeat and pointed to the nucleus as the likely location for an RNA-dominant pathogenic effect. The mechanism for nuclear retention is discussed further later.

D. Cell Culture Models of DM1 Involving Overexpression of CUG Expansion RNA

Congenital DM1 (CDM) is characterized by weakness and structural abnormalities of skeletal muscle at the time of birth [38]. This most severe form of DM1 is usually associated with CTG expansion lengths greater than 1000 repeats. Although the pathogenesis of CDM remains poorly understood, the structural abnormalities in skeletal muscle, including hypoplasia of muscle fibers and a superabundance of muscle precursor cells, suggested that muscle maturation during the second half of fetal development is delayed [39]. While there is no exact marker of this phenotype in cell culture systems, DM1 muscle precursor cells display a reduced ability to withdraw from the cell cycle and undergo fusion into multinucleate cells [40] (myotubes), an alteration that may relate to the developmental derangements in CDM. It is therefore noteworthy that the ability of myoblasts to undergo fusion and myogenic differentiation was suppressed by overexpression of $(CUG)_{200}$ in the 3' UTR of mRNAs encoding β-galactosidase or green fluorescent protein [30, 41]. Repeat tracts of this length triggered nuclear retention of the transgene mRNA, and correspondingly, they also repressed activity of the reporter protein. By comparison, similar transcripts containing $(CUG)_5$ did not accumulate in nuclear foci, nor did they inhibit myoblast fusion. Interestingly, these effects on differentiation were not induced by expression of expanded poly(CUG) unless it was also accompanied by upstream flanking sequences from the *DMPK* 3' UTR, suggesting that the repeat tract was necessary but not sufficient to disrupt skeletal myogenesis [41]. Taken together, these results supported an RNA-dominant mechanism for inhibition of myogenic differentiation in muscle cell cultures.

E. Transgenic Mouse Models of DM1 Involving Overexpression of CUG Expansion RNA

Another strategy to test the hypothesis of RNA-dominant disease was to place an expanded CTG repeat in the 3' UTR of a heterologous gene that was highly expressed in skeletal muscle of mice. To accomplish this, a $(CTG)_{250}$ tract was inserted in the 3' UTR of a genomic fragment containing the entire human skeletal actin (*HSA*) gene, including its flanking regulatory sequences [31]. This fragment was subsequently used to derive lines of transgenic mice. To control for the effects of overexpressing skeletal actin, parallel sets of transgenic lines were produced using *HSA* with (CTG)$_5$ inserted in the identical position. These lines of transgenic mice were designated *HSA* long repeat (*HSA*^{LR}) and *HSA* short repeat (*HSA*^{SR}). Of note, the amino acid sequences of murine and human skeletal actin are identical, and previous studies had shown no deleterious effects of overexpressing human skeletal actin in mouse muscle [42]. This experimental design did not replicate the circumstance of human DM1 in several important respects. For example, expression of the transgene was limited to skeletal muscle, and the *HSA* gene is turned on somewhat later than *Dmpk* during development. Thus, these mice provided a single-tissue test of the putative RNA-dominant disease mechanism. Also, the length of the expanded repeat in *HSA* transgenic mice, 0.75 kb, was smaller than the 5- to 12-kb CTG expansions that are typically found in muscle tissue of adults with symptomatic DM1 [43, 44]. In practice, the length of the expanded repeat in transgenic mice was limited by the tendency of large CTG tracts to undergo spontaneous internal deletions in *Escherichia coli* cloning vectors, coupled with the reduced germline and somatic instability of expanded CTG repeats in murine transgenes as compared with the human DM1 locus. However, *HSA* is the most abundant nuclear-encoded transcript in skeletal muscle, accounting for around 2% of mRNA at steady state [45]. Thus, if pathogenic effects depended on the mass of CUG expansion RNA in the nucleus, rather than length of individual repeat tracts, higher expression of a shorter repeat would be expected to achieve a DM1-like effect.

The phenotype in *HSA*^{LR} mice resembled DM1 in several important respects [31]. The hallmark of myotonic dystrophy is an impairment of muscle relaxation after voluntary activity. This phenomenon, known as myotonia, is caused by involuntary runs of repetitive action potentials that are generated within muscle fibers in the absence of ongoing neural activation. *HSA*^{LR} transgenic mice displayed noticeable hindlimb myotonia after rapid movement. Physiological recordings from the muscle fibers, using extracellular electrodes inserted during general anesthesia, confirmed the presence of prolonged myotonic discharges. Also, histological examination of skeletal muscle showed structural changes similar to those of human DM1. Mammalian muscle fibers are multinucleated cells in which the number of nuclei in relation to the volume of cytoplasm remains constant. The earliest structural alteration in human DM1 is an increase in the number of nuclei per muscle fiber and migration of nuclei from their normal position beneath the muscle membrane to the interior "central" region of the muscle fiber [46]. *HSA*^{LR} mice show a marked increase in the number of nuclei per fiber and the number of nuclei in a central position. The structure of the contractile apparatus is also disrupted, with some muscle fibers showing bands of myofibrils incorrectly oriented perpendicular to the axis of muscle contraction ("ring fibers") and regions of muscle cytoplasm devoid of myofibrils ("sarcoplasmic masses"). Both changes are characteristic of human DM1 muscle.

As expected, the *HSA*^{LR} transgene was highly expressed in skeletal muscle in several independent founder lines. While the expanded CUG did not interfere with either the polyadenylation or processing of transgene mRNA, it did have a dramatic effect on nucleocytoplasmic transport. The *HSA*^{LR} transcripts formed multiple foci in muscle nuclei (Fig. 3-1). In contrast to human DM1 muscle, where the number of foci is usually fewer than 5, there were dozens of foci per nucleus in *HSA*^{LR} mice. The reason for this difference is uncertain, but it may relate to shorter length of the expanded CUG or higher rates of *HSA* transcription. Notably, the CTG repeat was the only element from *DMPK* inserted into the *HSA*^{LR} transgene, indicating

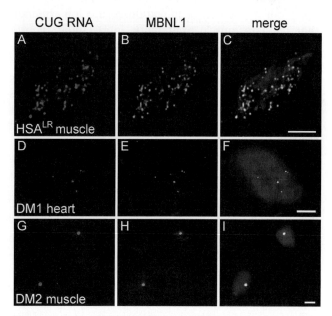

FIGURE 3-1 Muscleblind-like 1 (MBNL1) is sequestered in nuclear foci of CUG or CCUG expansion RNA. Shown are frozen sections of *HSA*^{LR} transgenic mouse skeletal muscle (A–C), DM1 cardiac muscle (D–F), and DM2 skeletal muscle (G–I). Nuclear foci are identified by RNA FISH using a Texas red-labeled CAG or CAGG probe (A,D,G). Colocalization of MBNL1 is shown by immunofluorescence (IF) using a polyclonal antibody (B,E,H). Merged images (C,F,I) also show the position of the nucleus. Bar = 5μm.

that this repetitive sequence was sufficient to induce a DM1-like pathogenic effect, independent of any other flanking sequence from the *DMPK* 3' UTR. The phenotypic consequences in *HSA*LR mice did not result from overexpression of actin protein, as indicated by the normal muscle structure and function in *HSA*SR mice that overexpress actin and normal levels of actin protein in *HSA*LR muscle. Likewise, the phenotypic consequence did not result from genomic integration of an expanded CTG repeat, because the lines of transgenic mice that harbored the transgene, but did not express the mRNA, remained normal. Indeed, a dose–response relationship between CUG RNA accumulation and muscle disease was apparent. For instance, a founder line that integrated a single copy of the *HSA*LR transgene showed mild histological changes in muscle that were never accompanied by myotonia, despite having conspicuous nuclear foci of CUG expansion RNA. However, when this line was bred to homozygosity, doubling the number of transgene copies, the result was a modest increase in *HSA*LR mRNA, prominent myotonia, and enhancement of muscle histological abnormalities.

Despite close scrutiny of muscle excitation in patients with DM1, the physiological basis for the myotonia was unknown. The *HSA*LR transgenic mice provided an opportunity to address this question by carrying out a detailed examination of the ion channels that regulate muscle excitability. Intracellular recordings from excised muscle tissue revealed an 80% reduction of chloride ion (Cl^-) conductance in the *HSA*LR muscle membrane [47]. Based on previous studies using inhibitors of Cl^- conduction [48], this alteration was clearly sufficient to account for the myotonia. This effect on ion conductance was demonstrated to result from a decrease in the level of ClC-1, the major Cl^- channel in skeletal muscle. When expression of this channel protein was assessed in human DM1, a similar defect was uncovered. Moreover, as discussed below, the loss of Cl^- conduction, in both *HSA*LR mice and human DM1, can be attributed to misregulated alternative splicing of the ClC-1 pre-mRNA [47, 49]. Taken together, these observations indicate that specific features of DM1 are induced in skeletal muscle by expression of RNA containing an expanded CUG repeat.

A second transgenic mouse model was derived using a large genomic fragment encompassing the entire *DMPK* gene [50]. By this approach, an expanded $(CUG)_{300}$ repeat was expressed, in the natural context of the DMPK transcript, and under the control of native elements regulating *DMPK* expression. Despite having fairly low levels of transgene expression, these mice displayed myotonia and histological abnormalities in muscle. Furthermore, they showed abnormal expression of microtubule-binding protein tau in brain, similar to observations in human DM1 [51]. By contrast, lines of transgenic mice expressing DMPK with a small $(CUG)_{55}$ expansion are normal [52]. These results raised the possibility that an RNA-dominant disease mechanism underlies the manifestations of DM1 in brain as well as in muscle.

F. DM2 Results from Expression of Untranslated CCUG Repeats

Once the DM1 mutation was identified, permitting genetic confirmation of the diagnosis, it became apparent that a subgroup of families affected by myotonic dystrophy did not, in fact, have DM1. The clinical features in these individuals were distinctive [53, 54]. For example, these individuals tended to be less severely affected, and the earliest weakness occurred in muscle groups that controlled movement of the shoulders and hips, as opposed to early involvement of the finger and ankle muscles in classic DM1. Despite these atypical features, these individuals often carried the diagnosis of DM1 because their progressive muscle weakness, histological changes in muscle, myotonia, cataracts, testicular atrophy, cardiac arrhythmia, and autosomal dominant inheritance overlapped with the broad phenotypic spectrum of DM1. The subsequent discovery that this phenocopy of DM1, now known as DM2, resulted from an expanded CCTG repeat in intron 1 of the *zinc finger 9* (*ZNF9*) gene, and that mutant RNA accumulates in nuclear foci in both disorders [32], had an obvious implication: the overlapping features of DM1 and DM2 can be attributed to the deleterious effects of poly(CUG) or poly(CCUG) RNA in the nucleus. The clinical and genetic features of DM2 are discussed further in this volume [55, 56].

III. BIOCHEMICAL BASIS FOR RNA-MEDIATED DISEASE

A. Mutant DMPK and ZNF9 RNAs Are Retained in the Nucleus

As noted in the previous section, early RNA fluorescence *in situ* hybridization (FISH) studies demonstrated that transcripts from mutant *DMPK* alleles were retained in the nucleus of DM1 fibroblasts and myoblasts [7, 35]. While the mechanism(s) involved in blocked nucleocytoplasmic export of mutant transcripts is unknown, there appears to be a repeat length threshold for nuclear retention. Subcellular fractionation and quantitative reverse transcription polymerase chain reaction (RT-PCR) analysis indicates that this threshold lies somewhere between 80 and 400 CUG repeats [57]. In agreement, ribonuclear foci are not detectable by RNA FISH in DM1 fibroblasts carrying *DMPK* alleles with 50–80 CTG repeats [30]. The number and morphology of ribonuclear foci in DM1 are also quite different in cell types and tissues. In proliferating DM1 cells in culture, these RNA-rich accumulations range from a few small foci in fibroblasts to dozens of larger foci in myoblasts, which express the *DMPK* gene at a higher level [7, 26, 35, 58]. Although only a few nuclear foci are observed in postmitotic cells such as myofibers and cortical neurons, these foci are comparatively large [36]. As assayed by quantitative RNA FISH, ribonuclear foci are also 8- to 13-fold more intense in DM2 skeletal muscle compared with DM1 skeletal muscle, perhaps due to expression differences between the *DMPK* and *ZNF9* genes, and they are structurally distinct with spheroidal foci predominating in DM1 and rodlike structures in DM2 [59]. Interestingly, the processing of both DMPK and ZNF9 pre-mRNAs into mature mRNAs does not appear to be affected by the corresponding expansion mutations. For DM1, ribonuclear foci contain fully processed and intact DMPK mRNAs [7, 35]. ZNF9 transcripts are also processed efficiently in DM2 cells but because the CCUG expansion is in an intron, the mRNA is efficiently exported, whereas ZNF9 intron 1 RNA, or possibly just the repeat region alone, is retained in nuclear foci [32] (L. Ranum, personal communication).

Why do ribonuclear foci form? One possibility is that DM-associated nuclear retention of specific RNAs results from the formation of a ribonucleoprotein complex that is designed to retain toxic mRNAs. This mechanism would be analogous to the retention of hyperedited dsRNAs by the p54nrb/PSF/Matrin 3 complex [60, 61]. Although A-I editing is not relevant to dsCUG or dsCCUG hairpins, it is intriguing that ribonuclear foci form in *Drosophila* expressing a GFP-DMPK-3′UTR(CTG)$_{162}$ transgene and these foci colocalize with non-A, the *Drosophila* p54nrb homolog [60, 62]. Interestingly, PSF and p54nrb have been implicated in several pathways, including DNA repair and transcription, but they are also components of the spliceosome, where they interact with Stem 1b of U5 snRNA [63, 64]. In DM, mutant dsCUG and dsCCUG RNAs bind preferentially to the muscleblind-like (MBNL) proteins, and the association of MBNL1 with these RNAs may be a critical step in ribonuclear foci formation [36, 59, 65, 66]. As described below, the splicing of specific exons is affected in DM, and the MBNL proteins are involved in this splicing regulation. Finally, current evidence indicates that ribonuclear foci do not colocalize with known nuclear structures, including Cajal bodies, nucleoli, PML bodies and splicing factor compartments (SFCs, also know as speckles) [36, 59]. Therefore, DM-associated foci may localize to a novel nuclear structure or they may be random accumulations of nuclear retained *DMPK* and *ZNF9* transcripts. Although ribonuclear foci are a characteristic feature of DM cells, it is noteworthy that the focal RNA population may not be the toxic component. Indeed, similar to the role of polyglutamine inclusions in Huntington's disease, nuclear RNA accumulations may be protective rather than pathogenic [67]. Instead, the intranuclear level of more diffusely distributed DMPK and ZNF9 mutant RNAs may correlate with toxicity.

B. Alternative Splicing of Pre-mRNAs

The vast majority of metazoan genes are split into exons and introns. Exons contain the sequence information that is ultimately incorporated into mRNA, and introns are the intervening sequences between the exons that are removed and discarded. Each gene is transcribed into a precursor mRNA (pre-mRNA) and exons are spliced together to form an mRNA, which is exported to the cytoplasm to be translated. Splicing occurs in the nucleus, except in the case of anucleate platelets [68], and is carried out by a large multicomponent complex called the spliceosome. Until recently, it was thought that most human genes each generate one mRNA by constitutive splicing of all exons. It is now known that up to 74% of human genes undergo alternative splicing in which exons or parts of exons can be skipped during pre-mRNA processing, resulting in the expression of multiple variant mRNAs [69]. There are several different patterns of alternative splicing (Fig. 3-2). Alternatively spliced mRNAs from individual genes are identical except for what are usually relatively small regions that vary between the different splice variants. Eighty percent of the time, this variability is within the coding regions of mRNAs, resulting in expression of divergent proteins [70]. Functional differences between these protein isoforms can range from undetectable to dramatic. As a result, the generation of alternatively spliced mRNAs can have a major impact on cell physiology. The frequency of alternative splicing explains how a large and diverse human proteome is generated from a surprisingly limited number of genes. In addition, up to 30% of alternative splicing events are predicted to introduce or remove regions that change the reading frame of an mRNA. These events most often result in the introduction of a premature termination codon, which ultimately results in the degradation of the mRNA by the nonsense mediated decay (NMD) pathway [71]. Therefore, in addition to generating different protein isoforms, alternative splicing is also responsible for controlling on/off gene expression decisions.

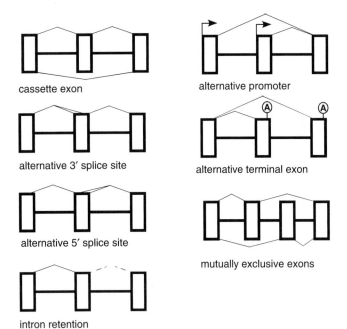

FIGURE 3-2 Alternative splicing pathways. Diagrams show exons as boxes and introns as thick lines between the boxes. Thin lines indicate the possible splicing patterns. Arrows indicate transcription initiation sites. Circled uppercase 'A' indicates a polyadenylation signal that lies 20–25 nucleotides upstream of what will become the 3′ end of the mRNA. For intron retention, the dotted line indicates that the intron can be either removed or retained in the mature mRNA.

Alternative splicing is often regulated according to cell type or developmental stage or in response to an acute stimulation, and plays a major role in the appropriate temporal and spatial expression of specific protein isoforms that can be critical for cell function. Regulation of alternative splicing involves interactions between RNA-binding proteins with specific binding sites within the pre-mRNA. These binding sites are usually within a few hundred nucleotides of the regulated exon. A large number of proteins that regulate splicing of a variety of pre-mRNAs have been identified [72]. Tissue-specific regulation of an alternative splicing event typically involves modulation of the activities of antagonistic regulatory factors rather than modulation of one tissue-specific factor [73]. Therefore, tissue-specific splicing regulation is often dependent on the ratio of the nuclear activities of different regulatory factors rather than the specific concentration of a single splicing regulator.

C. Alternative Splicing Is Disrupted in DM

An early premise of the RNA gain-of-function hypothesis as a mechanism of DM pathogenesis was that repeat-containing RNA expressed from the expanded allele developed a toxic gain of function by disrupting the normal function of RNA-binding proteins [74]. A prominent molecular feature of the disease is disruption of alternative splicing regulation [25]. Currently, 13 alternative splicing events are known to be disrupted in several different tissues from individuals with DM (Table 3-1). In all cases, there is a failure to express the splicing pattern that is characteristic of adult tissues. Instead, embryonic splicing patterns are retained in the adult. Therefore, a hallmark of DM is the disruption of developmentally regulated alternative splicing. The consequences and

TABLE 3-1 Misregulated Alternative Splicing in Myotonic Dystrophy

Gene	Exon/intron	Tissue	Pattern	Ref.
Cardiac troponin T (*TNNT2*)	Exon 5	Heart	Exon inclusion	[25]
Insulin receptor (*IR*)	Exon 11	Skeletal muscle	Exon exclusion	[79, 80]
Chloride channel (*CLCN-1*)	Intron 2	Skeletal muscle	Intron retention	[47, 49]
Chloride channel (*CLCN-1*)	Exon 7a	Skeletal muscle	Exon inclusion	[47, 49]
Tau (*MAPT*)	Exons 2 and 3	Brain	Exon exclusion	[36, 53]
Tau (*MAPT*)	Exon 10	Brain	Exon exclusion	[36]
Myotubularin-related protein 1 (*MTMR1*)	Exons 2.1 and 2.3	Skeletal muscle/heart	Exon exclusion	[88, 89]
Fast skeletal troponin T (*TNNT3*)	Fetal exon	Skeletal muscle	Exon inclusion	[33]
N-Methyl-D-aspartate receptor (*NMDAR1*)	Exon 5	Brain	Exon inclusion	[36]
Amyloid precursor protein (*APP*)	Exon 7	Brain	Exon exclusion	[36]
Ryanodine receptor (*RyR*)	Exon 70 (AS I)	Skeletal muscle	Exon exclusion	[86]
Sarcoplasmic/endoplasmic reticulum Ca^{2+} ATPase 1 (*SERCA1*)	Exon 22	Skeletal muscle	Exon exclusion	[86]
Sarcoplasmic/endoplasmic reticulum Ca^{2+} ATPase 2 (*SERCA2*)	Intron 19	Skeletal muscle	Intron retention	[86]

potential consequences of these misregulated splicing events are described below.

A large number of diseases are caused by mutations that disrupt sequence elements that are required for the normal cut-and-paste reactions of splicing [73, 75]. The result is aberrant splicing in which whole exons are skipped or cryptic splice sites are used, ultimately generating nonnatural mRNAs. In most cases, these RNAs contain premature termination codons and are degraded by the NMD pathway. This aberrant splicing should be distinguished from the aberrant regulation of splicing that is a hallmark of DM. The *cis*-acting mutations that cause aberrant splicing most often cause disease through loss of function of the mutated allele due to degradation of nonnatural mRNAs. In contrast, natural mRNA splice variants are expressed in DM; however, these mRNAs are expressed at inappropriate developmental stages. Embryonic or fetal splicing patterns reappear or persist in adult tissues, resulting in the expression of mRNA and protein isoforms that do not provide the necessary functional properties.

The first evidence for the disruption of alternative splicing in myotonic dystrophy was the identification of the fetal isoform of cardiac troponin T (cTNT) in DM1 cardiac tissue [25]. Troponin T is one of three dissimilar subunits of the troponin complex, which confers calcium sensitivity to the contractile apparatus and regulates the actin–myosin interactions that result in contraction. Different cTNT splice variants have distinct functional properties; specifically, the embryonic isoforms are less sensitive to Ca^{2+}, resulting in lower contractility in embryonic heart [76]. The altered splicing of cTNT has unknown consequences to heart function. Individuals with DM often develop arrhythmias [77], as well as a loss of myocardial function (78). However, it is unclear whether the expression of the fetal cTNT isoform plays a role in cardiac pathogenesis.

Two major clinical features of DM, insulin resistance and myotonia, directly correlate with misregulated splicing of the insulin receptor (IR) and muscle-specific chloride channel (ClC-1), respectively. Individuals with DM have an unusual form of insulin resistance that is due to a defect of skeletal muscle [79]. One particularly consistent molecular abnormality identified in individuals with DM is a failure to express the muscle-specific isoform of the IR in skeletal muscle tissue. The IR isoform expressed in DM muscle tissue and muscle cells grown in culture is a lower signaling isoform, which directly correlates with lower responsiveness of DM skeletal muscle to insulin stimulation [79]. IR splicing abnormalities were also identified in patients with DM2 consistent with a common mechanism of pathogenesis in DM1 and DM2 [80]. Skeletal muscle biopsies obtained 8 years apart from the same individual with DM2 demonstrated progression of the abnormal IR splicing pattern consistent with the observed progressive symptoms of the disease [80].

Another prominent feature of DM is myotonia, which correlates with the failure of the muscle-specific chloride channel (*ClC-1/CLCN1*) gene to switch from the embryonic to the adult splicing pattern. In the embryonic splicing pattern, one or both of two alternative exons is included, putting the mRNA out of frame and preventing expression of the full-length ClC-1 protein. Inclusion of these alternative exons introduces premature termination codons that are likely to cause loss of ClC-1 mRNAs by NMD [47, 49]. ClC-1 protein levels are reduced to 10% or less of normal skeletal muscle, which is sufficient to cause myotonia based on loss-of-function mutations of the *ClC-1* gene [47, 49]. It is interesting to note that ClC-1 protein expression is strongly induced soon after birth. This developmentally regulated expression is likely to be controlled at least in part by an alternative splicing switch that puts the mRNA in frame, and failure of this developmental splicing transition in individuals with myotonic dystrophy results in myotonia.

Tau, a microtubule-associated protein that functions in the polymerization and stability of microtubules, is encoded by the *MAPT* gene. Tau is the major component of neurofibrillary tangles (NFTs) that accumulate in neuronal cell soma and neuronal processes, and is implicated in the pathogenesis of several neurological disorders including Alzheimer's disease, frontotemporal dementia with parkinsonism linked to chromosome 17 (FTDP-17), Pick's disease, progressive supranuclear palsy (PSP), and corticobasal degeneration (CBD) [81]. The tau gene contains 16 exons; exons 2, 3, 6, and 10 undergo developmentally regulated alternatively splicing [36, 51, 82]. It is the regulation of exon 10, which encodes one of four microtubule binding domains, that is associated with a propensity of tau to aggregate. The normal ratio of mRNAs including or excluding exon 10, and therefore the ratio of proteins containing three and four microtubule binding domains, is one to one. Disruption of this ratio can promote the formation of insoluble tau aggregates, NFTs. NFTs can be detected in brains of individuals with DM1 [83, 84], and altered splicing of the tau alternative exons 2 and 10 has been demonstrated in DM CNS tissues [36, 51]. It remains to be determined whether these abnormalities are causative of the changes in brain function in adults with DM1; however, there is a suggestive correlation between aberrant splicing of tau, NFT formation, and age-related cognitive impairment [85].

The skeletal muscle ryanodine receptor (*RyR1*) and sarcoplasmic/endoplasmic reticulum Ca^{2+}-ATPase (*SERCA*) genes are primarily responsible for regulating intracellular Ca^{2+} homeostasis. During skeletal muscle sarcolemma depolarization, Ca^{2+} is released from the sarcoplasmic reticulum through RyR1 channels to initiate contraction. SERCA pumps Ca^{2+} back into the sarcoplasmic reticulum in an ATP-dependent manner to lower the cytoplasmic Ca^{2+} concentration for relaxation. RyR1 is the skeletal muscle isoform of three ryanodine receptor genes. Of the three *SERCA* genes, *SERCA1* and *SERCA2* are the major isoforms in fast and slow skeletal muscle, respectively. *RyR1* and *SERCA1* express immature isoforms in skeletal muscles from individuals with DM1, as well as the *HSA*^{LR} mouse model for DM1 [86]. In addition, expression of a previously unidentified isoform of SERCA2 was shown to be reduced in DM1 skeletal muscle.

The *RyR1* gene contains two cassette alternative exons, ASI and ASII, that undergo alternative splicing transitions later and earlier in development, respectively [87]. Interestingly, only ASI is affected in DM1 skeletal muscle [86], indicating that not all developmentally regulated splicing is altered. These results strongly suggest that different sets of *trans*-acting factors regulate the timing of ASI and ASII splicing transitions, and only those factors regulating splicing of ASI are affected in DM. The total abundance of RyR1 and SERCA1 and SERCA2 did not differ between DM1 and unaffected tissue samples, only the ratio of the isoforms expressed [86]. It is possible that misexpression of RyR1 and SERCA isoforms alters Ca^{2+} homeostasis in skeletal muscle and therefore plays a role in skeletal muscle wasting, one of the major debilitating features of the disease.

The myotubularin-related 1 (*MTMR1*) gene is one of a conserved family of phosphatidylinositol 3-phosphate phosphatases that are involved in regulating trafficking of intracellular vesicles. The alternative exons 2.1, 2.2, and 2.3 are normally included during heart and skeletal muscle postnatal development; however, there is a failure to complete this transition in DM1 heart and skeletal muscle [88, 89]. Although enzymatic assays indicate that the phosphatase activities of fetal and adult isoforms are similar, mutations within the *MTMR1* gene result in disease characterized by hypotonia and respiratory insufficiency which resemble congenital DM [90].

The *TNNT3* gene encodes the troponin T isoform that is expressed in fast skeletal muscle fibers. The premRNA expressed from this gene undergoes extensive alternative splicing; however, one exon that is expressed primarily in fetal skeletal muscle is retained in skeletal muscle from adults with DM1 [33]. The effect of this fetal isoform in adult skeletal muscle function remains to be determined.

Amyloid precursor protein (APP), a large transmembrane glycosylated protein, is a precursor of amyloid, which is the major component of senile plaques found in brains affected by Alzheimer's disease. At least eight isoforms of APP are generated via alternative splicing of exons 7, 8, and 15 [91]. In adult DM brain, the fetal APP isoform containing exon 7 is expressed inappropriately [36]. This exon encodes a serine protease inhibitor domain [92], but the consequences of inclusion of this domain in protein expressed in adult brain are unknown.

The *N*-methyl-D-aspartate receptors (NMDARs) are predominant mediators of excitatory synaptic transmission and play important roles in long-term potentiation that affect learning and memory. Functional diversity of the NMDA receptors is generated by expression of three different subunits, alternative splicing, and assembly as multimeric complexes, possibly tetramers [93]. One subunit, NR1, generates eight isoforms by alternative splicing differing in physiological properties as well as subcellular distribution [93]. Inclusion of NMDA NR1 exons 5 and 21 was shown to be affected in brain tissue from individuals with DM1 [36], raising a potential relationship between aberrant regulation of NMDA alternative splicing and CNS features in DM1.

D. Mechanisms for Misregulated Alternative Splicing: Role for CELF and MBNL Proteins

How does nuclear accumulation of CUG or CCUG repeat RNAs induce splicing changes? An early hypothesis for the RNA gain-of-function model proposed that the RNA expressed from the expanded allele disrupted the function of RNA-binding proteins having a *trans*-dominant effect on RNA processing of transcripts from multiple genes [29, 74]. The most straightforward model is one in which repeat-containing RNA binds and sequesters RNA-binding proteins, resulting in their loss of function. In pursuit of this hypothesis, investigators have identified five proteins that bind to CUG repeat-containing RNA: CUG-binding protein (CUG-BP, also called CUG-BP1), ETR-3 (also called CUG-BP2 and NAPOR), muscleblind-like protein (MBNL), the double-stranded RNA protein kinase, PKR, and hnRNP H [26, 74, 94–96]. Of these, the CELF and MBNL proteins have emerged as the factors that regulate the splicing events that are affected in DM1 and DM2.

The first factors implicated in DM were the CELF proteins, CUG-BP1 and ETR-3, which are encoded by two of six genes that make up the CELF family (CUG-BP1 and ETR-3-like factors). These proteins have been shown to have roles in mRNA translation, RNA editing, mRNA stability, and alternative splicing [97–101]. MBNL proteins have been demonstrated to regulate alternative splicing by binding to specific sequence motifs within pre-mRNAs [102]. hnRNP H also has roles in regulating alternative splicing and polyadenylation [103, 104, 105]. PKR is a dsRNA-binding protein with a protein kinase domain that is activated by binding to dsRNA.

Early studies demonstrated that PKR bound to and was activated by double-stranded CUG expanded repeats. However, subsequent genetic experiments in which a mouse model for DM1 expressing expanded CUG repeats was put into a mutant PKR background showed no effect on the disease phenotype [59].

Importantly, MBNL and CELF proteins have been demonstrated to regulate splicing of cTNT and IR alternative exons, two developmentally regulated alternative splicing events that are disrupted in DM. Furthermore, CELF and MBNL proteins have antagonistic activities on the cTNT and IR alternative exons. The splicing patterns observed in DM striated muscle are consistent with a loss of MBNL activity and a gain of CUG-BP1 activity. It is currently unclear whether the splicing defects are driven primarily by the loss of MBNL, gain of CELF, or a combination of both. There is strong evidence supporting a role for the loss of MBNL function as the determinative event in alternative splicing misregulation. All three MBNL proteins colocalize with foci containing CUG-repeat RNA [106]. Nucleoplasmic (nonfoci) MBNL1 is reduced 2.3-fold in DM1 cells compared with unaffected cells based on immunofluorescence measurements, consistent with either sequestration or increased degradation [36]. Furthermore, a mouse line ($Mbnl1^{\Delta E3/\Delta E3}$) in which the predominant isoforms of MBNL1 have been knocked out reproduces the misregulated splicing that is observed in DM heart and skeletal muscle tissues, and the phenotype of the mice is consistent with several DM symptoms [33]. In fact, the HSA^{LR} and $Mbnl1^{\Delta E3/\Delta E3}$ mouse models for DM are remarkably similar in skeletal muscle phenotype in that both exhibit myotonia, histological changes, and splicing abnormalities that are highly consistent with DM. It is interesting to note that neither model exhibits the striking skeletal muscle degeneration that is primarily responsible for mortality in DM. It is currently unclear whether this is due to the incomplete reproduction of the pathogenic mechanism or to species differences between human and mouse.

An alternative but not mutually exclusive view is that in addition to sequestering MBNL proteins, the accumulated repeat-containing RNAs interfere with nuclear signaling events, ultimately resulting in altered CELF and MBNL activities as downstream events. Evidence suggests that CELF and MBNL proteins play a role in developmentally regulated alternative splicing of specific pre-mRNAs. CELF protein abundance decreases postnatally, correlating with the timing of the splicing changes of known pre-mRNA targets of these proteins (cTNT and ClC-1 alternative exons) in heart and skeletal muscle [107]. Developmentally regulated splicing events (cTNT, ClC-1, and fast skeletal TNT) are disrupted in the $Mbnl1^{\Delta E3/\Delta E3}$ mouse model, indicative of a role for MBNL in these developmental transitions. The fact that splicing of both cTNT and IR is controlled by CELF and MBNL proteins and the activities of these proteins are antagonistic suggests linkage of these two families in developmental regulation of some alternative splicing events. The repeat-containing RNA could induce an aberrant signaling cascade characteristic of embryonic tissue or block a signaling event that is required for the transition of developmentally regulated splicing to later patterns, ultimately resulting in splicing patterns that are characteristic of earlier developmental stages in adult tissues.

E. CELF Proteins Are RNA Splicing Regulators

CUG-BP1, the first CELF protein to be identified, was purified from HeLa cytoplasmic extracts based on an assay to identify proteins that bound to $(CUG)_8$ RNA [74]. A total of six CELF paralogs have subsequently been identified [100]. All CELF proteins have similar protein structures (Fig. 3-3) containing three RNA recognition motifs (RRMs); two are at the N terminus and one near the C terminus separated by a 180- to 220-amino-acid domain that contains regions of the protein required for splicing activation and repression [108]. All CELF paralogs are expressed in brain, and three, CUG-BP1, ETR-3, and CELF6, are expressed in adult nonbrain tissues as well (Fig. 3-3). Soon after their discovery, CELF proteins were shown to regulate alternative splicing via binding to relatively short (4- to 6-nucleotide) intronic UG-rich motifs located adjacent to the alternative exons [100, 109]. CUG-BP1 and ETR-3 have been shown to regulate several of the pre-mRNAs whose splicing is misregulated in DM

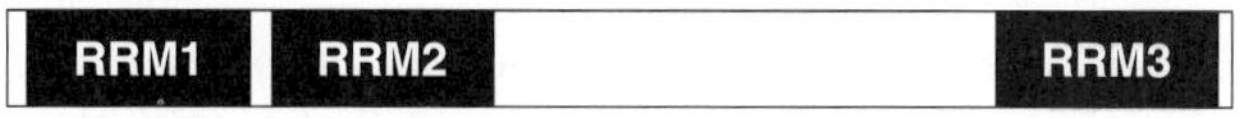

| RRM1 | RRM2 | | RRM3 |

NAMES			EXPRESSION
CUG-BP1	BRUNOL2		widely expressed
ETR-3	BRUNOL3	CUG-BP2 NAPOR	brain, striated muscle
CELF3	BRUNOL1		brain
CELF4	BRUNOL4		brain
CELF5	BRUNOL5		brain
CELF6	BRUNOL6		brain, kidney

FIGURE 3-3 CELF protein domain structure and expression. The diverse CELF nomenclature is indicated. Expression refers to expression in adult tissues for protein for ETR-3 and CUG-BP1 and for mRNA of CELF3-CELF6. RRM is the RNA recognition motif.

tissues (cTNT, IR, ClC-1, NMDAR1, and MTMR1) [25, 49, 79, 109, 110].

The evidence that CUG-BP1 and other CELF proteins play a role in DM pathogenesis comes from several results. First, CUG-BP1 binds to CUG repeat RNA [74], although it does not colocalize with repeat-containing RNA foci in DM cells [59]. Second, CUG-BP1 steady-state protein levels are increased in DM1 tissues and DM1 cells in culture [65, 74, 79]. Third, the splicing patterns in DM tissues of the three well-characterized targets of CUG-BP1 are consistent with increased CUG-BP1 activity [25, 49, 79]. Fourth, CUG-BP1 protein expression is developmentally downregulated, consistent with the switch to adult splicing patterns observed for cTNT, IR, and ClC1 in adult striated muscle [107]. Furthermore, the reversion to embryonic splicing patterns in adult DM tissue is consistent with the increased expression of CUG-BP1 observed in adult DM tissue. Fifth, to test directly whether increased expression of CUG-BP1 could contribute to a DM phenotype in mice, transgenic mice expressing CUG-BP1 in heart and skeletal muscle were generated. Both heart and skeletal muscle tissues expressed the aberrant splicing patterns for cTNT, IR, and MTMR1 that are observed in DM tissues. In addition, levels of expression greater than fourfold above endogenous levels result in lethality at birth, likely to be due to inability of pups to breathe due to muscle weakness [89].

The role of CUG-BP1 in DM pathogenesis is intertwined with that of MBNL proteins. As noted above, at least two splicing events that are misregulated in DM are also regulated by MBNL proteins, and regulation by MBNL is antagonistic to that of CELF proteins. Current investigations are directed toward determining the relative roles for a gain of CELF activity and a loss of MBNL activity in the splicing abnormalities observed in DM.

F. Muscleblind Sequestration Model for DM

In cells, RNAs exist as ribonucleoprotein (RNP) complexes. Thus, the demonstration that DMPK transcripts accumulated in ribonuclear foci led to the protein sequestration model for DM pathogenesis [26, 74]. This model proposes that DMPK and ZNF9 transcripts, which contain CUG and CCUG repeat expansions that fold into lengthy and stable RNA hairpins [93, 111, 112], sequester specific dsRNA-binding proteins. In other words, pathogenesis results from an RNA gain of function leading to a protein loss of function. Two predictions of the RNA hypothesis are that a factor so affected would colocalize with nuclear foci, and that its concentration elsewhere in the nucleus would be diminished. Although CUG-BP1 was first proposed as a candidate for this type of sequestered factor, subsequent observations argue against this possibility. First, CUG-BP1 is a ssRNA-binding protein and its binding to CUG repeats is not proportional to CUG repeat length [74, 111]. As DM disease severity is roughly proportional to repeat length, then the protein loss-of-function model requires that the activity of the sequestered factor be more affected by expression of $(CUG)_{750}$ versus $(CUG)_{75}$. Second, CUG-BP1 does not colocalize with ribonuclear foci in DM cells or tissues [66, 106, 113]. Third, in contrast to a key prediction of the sequestration model, CUG-BP1 splicing activity does not decrease, but actually increases, in DM cells and tissues [25, 74, 79]. Instead, the MBNL proteins have emerged as the sequestered factors that likely play a direct role in DM pathogenesis. As noted in the previous section, CUG-BP1 and the other CELF proteins play an important role in DM even though they are not the factors sequestered by mutant DMPK and ZNF9 RNAs.

The human MBNL proteins were first identified by their ability to bind preferentially to CUG expanded repeats [26]. Although they were initially named triplet repeat expansion, or EXP, proteins, subsequent characterization demonstrated that they were homologs of the *Drosophila* muscleblind (mbl) proteins. Intriguingly, loss of *mbl* gene expression in flies results in impairment of relatively late steps in the differentiation of both photoreceptor and muscle tissues [114, 115]. For muscle, loss of *mbl* gene expression results in abnormal Z-band differentiation and tendon matrix deposition at sites of indirect muscle attachment to the epidermis [114]. This was the first evidence that loss of muscleblind gene expression leads to developmental defects in particular muscle structures.

In contrast to *Drosophila*, there are three human *MBNL* genes that vary in their expression patterns (Fig. 3-4). The *MBNL1* gene, which is the best characterized *MBNL* family member, is expressed in all tissues surveyed, although the highest expression levels are in heart and skeletal muscle. The expression of *MBNL2* is also highest in skeletal muscle, whereas MBNL3 RNAs are most abundant in placenta, although the sizes of the major RNA species (~2 and ~11 kb) are quite different than that of either MBNL1 or MBNL2. Interestingly, these large and small MBNL3 RNAs are also detectable in liver and pancreas, whereas in muscle and placenta, intermediate-size RNAs (~6 and 7.5 kb) are present. These mRNAs code for a number of protein isoforms. For example, there are at least 13 mouse Mbnl1 proteins that arise from alternative splicing and that use

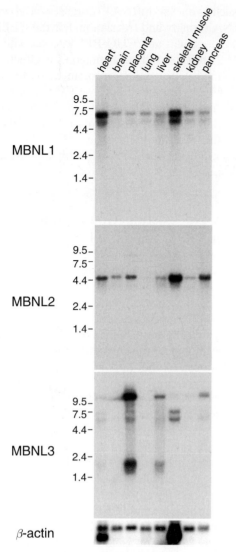

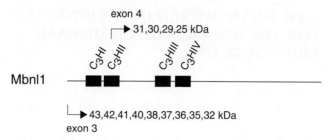

FIGURE 3-4 Expression of MBNL1, MBNL2, and MBNL3 in various human tissues. RNA blots were prepared from various tissues using 2 μg poly(A)+ RNA per lane. Blots were probed sequentially using MBNL1, MBNL2, and MBNL3 cDNAs under conditions that minimized cross-hybridization and finally probed with a β-actin cDNA as a loading control (performed by C. R. Urbinati, Loyola Marymount University). Size standards are indicated in kilobases.

FIGURE 3-5 Domain structure of mouse Mbnl1 proteins. A major feature of the Mbnl1 protein structure (line) is the four CCCH (C_3H) motifs (black boxes) and the various isoforms that initiate (arrows) in either exon 3 (below line) or exon 4 (above line). Note that the protein isoforms that initiate in exon 4 contain only two functional C_3H motifs.

different initiation codons located in either exon 3 or exon 4 (Fig. 3-5).

The protein sequestration hypothesis suggests that the MBNL proteins preferentially recognize CUG and CCUG expansion RNAs. Indeed, the RNA binding characteristics of the CUG-BP1, MBNL1, and PKR proteins have been compared using the yeast three-hybrid system, and only MBNL1 shows a strong interaction with dsCUG and dsCCUG in this system [116]. Interestingly, MBNL1 does not interact with dsRNA composed of completely paired CUG/CAG duplexes, suggesting that the U mismatch in dsCUG is important for binding specificity. It is hoped that future studies will reveal the molecular basis for the specificity of MBNL interactions with dsCUG and dsCCUG.

If DM is caused by MBNL protein sequestration by mutant DMPK and ZNF9 RNAs, then *Mbnl1* knockout mice should phenocopy characteristic features of DM disease. To test this possibility, $Mbnl1^{\Delta E3/\Delta E3}$ knockout mice have been generated so that only those proteins that use the initiation codon in exon 3 (E3) (see Fig. 3-5) are eliminated, because these are the isoforms that bind to CUG and CCUG repeat expansions [33]. In agreement with the sequestration hypothesis, these mice develop many of the characteristic multisystemic features of adult-onset DM, including skeletal muscle myotonia with associated split fibers and centralized myonuclei, subcapsular particulate ocular cataracts, and atrioventricular (AV) heart conduction block [33] (Kanadia et al., unpublished data). Of particular interest is the presence of the unusual particulate cataracts, which are common features of DM1 and DM2. These cataracts are not the more frequently observed age-related central opacity, and they develop even in the mildly affected DM1 population. Because the epidermal cell population, which lies just under the lens capsule, continues to undergo differentiation postnatally, it is possible that missplicing of specific pre-mRNAs is also responsible for DM-associated cataract formation, although this has not been demonstrated to date. In contrast, the missplicing of Clcn1/ClC-1, Tnnt2/cTNT, and Tnnt3 pre-mRNAs seen in adult DM tissues has been documented in adult $Mbnl1^{\Delta E3/\Delta E3}$ mice. As noted previously, there is a striking concordance between the skeletal muscle splicing patterns seen for HSA^{LR} mice expressing a $(CUG)_{250}$ expansion and $Mbnl1^{\Delta E3/\Delta E3}$ knockout mice [33] (Lin et al., unpublished data). Therefore, overexpression of CUG repeat expansion RNA is equivalent to loss of MBNL1 splicing activity.

IV. UNANSWERED QUESTIONS FOR THE RNA-MEDIATED DISEASE PROCESS IN DM

A. What Is the Molecular Basis for the Postnatal Splicing Switch Affected in DM?

In this chapter, we have reviewed the evidence supporting the RNA-mediated pathogenesis model for DM. Remarkably, several of the most characteristic features of this neuromuscular disease, including myotonia and insulin resistance, are caused by the failure to exclude specific fetal exons during postnatal development. This failure arises because the expression of the *DMPK* and *ZNF9* mutant alleles leads to MBNL sequestration and upregulation of the fetal splicing pattern promoted by the CELF proteins. What are the normal biochemical steps that result in fetal exon exclusion in adult tissues? A recent study reported that CUG-BP1 levels decline during postnatal development and are quite low in most adult tissues [105]. This would suggest that normal fetal exon skipping in the adult is promoted by unopposed MBNL splicing activity. This is a reasonable mechanistic explanation for most somatic tissues. However, the level of CUG-BP1 does not decrease postnatally in brain even though APP and NMDA NR1 RNA splicing patterns are altered in DM1 neurons. Thus, these splicing decisions may not be regulated by changes in the steady-state levels of the CELF and MBNL proteins. Alternatively, the activities of these splicing factors might be modulated by developmentally controlled interactions with other, yet unidentified, neuronal proteins and/or by specific posttranslational modifications (e.g., phosphorylation). Another intriguing aspect of potential CELF–MBNL interactions in the CNS is highlighted by the fact that most of the *CELF* genes are preferentially expressed in the brain (Fig. 3-3), whereas *MBNL* expression does not appear to show this bias (Fig. 3-4). Thus, it is possible that the antagonistic interactions between the MBNL and CELF proteins are not the primary events that regulate the splicing of the APP and NMDA NR1 exons affected in DM.

B. Are Other Biochemical Pathways Altered in DM?

In addition to RNA splicing, are other biochemical pathways affected by the DM1 and DM2 mutations and the associated changes in the activities of the CELF and MBNL proteins? In support of this possibility, splicing factors often play roles in other posttranscriptional events, such as nuclear mRNA export, as well as cytoplasmic mRNA stability and translation. For the CELF family, both calreticulin and CUG-BP1 interact with the 5′ UTR of p21 mRNA and have antagonistic effects on p21 translation [117]. The ETR-3 protein binds to AU-rich elements (AREs) in the 3′ UTR of COX-2 mRNA and stabilizes this mRNA while inhibiting its translation [101]. Therefore, it is likely that sequestration of the MBNL proteins and the associated alteration of CELF activity will have important implications for multiple steps in gene expression.

C. Why Do Existing Mouse Models Fail to Recapitulate Congenital DM Phenotypes?

Why is there no congenital form of DM2 and why do $Mbnl1^{\Delta E3/\Delta E3}$ knockout and HSA^{LR} transgenic mice fail to reproduce the severe developmental abnormalities seen in CDM? As mentioned in a previous section, this failure may reflect mouse–human differences in the function and development of skeletal muscle and the central nervous system. Another possibility is that the temporal and spatial expression pattern of the *ZNF9* gene may be incompatible with a congenital onset, although the mouse *Znf9* gene is turned on relatively early during embryogenesis [118] and nuclear foci of CCUG expansion RNA are clearly present in DM2 muscle precursor cells [59]. Because both *ZNF9* and *DMPK* are expressed *in utero*, but only in DM1 does this result in abnormal muscle and brain development, the biochemical consequences of nuclear retained CCUG-expanded ZNF9 and CUG-expanded DMPK transcripts may be quite different. If this conclusion is correct, it may also explain why degenerative phenotypes in DM2 are generally less severe than those in DM1, despite observations that the nuclear burden of repeat expansion RNA [59], and the extent of MBNL1 sequestration (Lin *et al.*, unpublished data), appears to be greater in the DM2 muscle nucleus. It remains to be determined whether these differences reflect distinct biochemical consequences of the CUG-versus-CCUG expansion sequences themselves or the added effects of DMPK versus ZNF9 RNA flanking sequences that are co-retained in ribonuclear foci [41].

References

1. Harley, H. G., Brook, J. D., Rundle, S. A., Crow, S., Reardon, W., Buckler, A. J., Harper, P. S., Housman, D. E., and Shaw, D. J. (1992). Expansion of an unstable DNA region and phenotypic variation in myotonic dystrophy. *Nature* 355, 545–546.
2. Buxton, J., Shelbourne, P., Davies, J., Jones, C., Van Tongeren, T., Aslanidis, C., de Jong, P., Jansen, G., Anvret, M., Riley, B., *et al.*

(1992). Detection of an unstable fragment of DNA specific to individuals with myotonic dystrophy. *Nature* **355**, 547–548.
3. Mahadevan, M., Tsilfidis, C., Sabourin, L., Shutler, G., Amemiya, C., Jansen, G., Neville, C., Narang, M., Barcelo, J., O'Hoy, K., et al. (1992). Myotonic dystrophy mutation: An unstable CTG repeat in the 3′ untranslated region of the gene. *Science* **255**, 1253–1255.
4. Fu, Y. H., Pizzuti, A., Fenwick, R. G., Jr., King, J., Rajnarayan, S., Dunne, P. W., Dubel, J., Nasser, G. A., Ashizawa, T., de Jong, P., et al. (1992). An unstable triplet repeat in a gene related to myotonic muscular dystrophy. *Science* **255**, 1256–1258.
5. Brook, J. D., McCurrach, M. E., Harley, H. G., Buckler, A. J., Church, D., Aburatani, H., Hunter, K., Stanton, V. P., Thirion, J. P., Hudson, T., et al. (1992). Molecular basis of myotonic dystrophy: Expansion of a trinucleotide (CTG) repeat at the 3′ end of a transcript encoding a protein kinase family member. *Cell* **69**, 385.
6. Wilkie, A. O. (1994). The molecular basis of genetic dominance. *J. Med. Genet.* **31**, 89–98.
7. Taneja, K. L., McCurrach, M., Schalling, M., Housman, D., and Singer, R. H. (1995). Foci of trinucleotide repeat transcripts in nuclei of myotonic dystrophy cells and tissues. *J. Cell Biol.* **128**, 995–1002.
8. Maeda, M., Taft, C. S., Bush, E. W., Holder, E., Bailey, W. M., Neville, H., Perryman, M. B., and Bies, R. D. (1995). Identification, tissue-specific expression, and subcellular localization of the 80- and 71-kDa forms of myotonic dystrophy kinase protein. *J. Biol. Chem.* **270**, 20246–20249.
9. Furling, D., Lemieux, D., Taneja, K., and Puymirat, J. (2001). Decreased levels of myotonic dystrophy protein kinase (DMPK) and delayed differentiation in human myotonic dystrophy myoblasts. *Neuromuscul. Disord.* **11**, 728–735.
10. Jansen, G., Groenen, P. J., Bachner, D., Jap, P. H., Coerwinkel, M., Oerlemans, F., van den Broek, W., Gohlsch, B., Pette, D., Plomp, J. J., et al. (1996). Abnormal myotonic dystrophy protein kinase levels produce only mild myopathy in mice. *Nat. Genet.* **13**, 316–324.
11. Reddy, S., Smith, D. B., Rich, M. M., Leferovich, J. M., Reilly, P., Davis, B. M., Tran, K., Rayburn, H., Bronson, R., Cros, D., et al. (1996). Mice lacking the myotonic dystrophy protein kinase develop a late onset progressive myopathy. *Nat. Genet.* **13**, 325–335.
12. Berul, C. I., Maguire, C. T., Aronovitz, M. J., Greenwood, J., Miller, C., Gehrmann, J., Housman, D., Mendelsohn, M. E., and Reddy, S. (1999). DMPK dosage alterations result in atrioventricular conduction abnormalities in a mouse myotonic dystrophy model. *J. Clin. Invest.* **103**, R1–7.
13. Berul, C. I., Maguire, C. T., Gehrmann, J., and Reddy, S. (2000). Progressive atrioventricular conduction block in a mouse myotonic dystrophy model. *J. Intervent. Card. Electrophysiol.* **4**, 351–358.
14. Nguyen, H. H., Wolfe, J. T., 3rd, Holmes, D. R., Jr., and Edwards, W. D. (1988). Pathology of the cardiac conduction system in myotonic dystrophy: A study of 12 cases. *J. Am. Coll. Cardiol.* **11**, 662–671.
15. O'Cochlain, D. F., Perez-Terzic, C., Reyes, S., Kane, G. C., Behfar, A., Hodgson, D. M., Strommen, J. A., Liu, X. K., van den Broek, W., Wansink, D. G., et al. (2004). Transgenic overexpression of human DMPK accumulates into hypertrophic cardiomyopathy, myotonic myopathy and hypotension traits of myotonic dystrophy. *Hum. Mol. Genet.* **13**, 2505–2518.
16. Krahe, R., Ashizawa, T., Abbruzzese, C., Roeder, E., Carango, P., Giacanelli, M., Funanage, V. L., and Siciliano, M. J. (1995). Effect of myotonic dystrophy trinucleotide repeat expansion on DMPK transcription and processing. *Genomics* **28**, 1–14.
17. Tiscornia, G. and Mahadevan, M. S. (2000). Myotonic dystrophy: The role of the CUG triplet repeats in splicing of a novel DMPK exon and altered cytoplasmic DMPK mRNA isoform ratios. *Mol. Cell.* **5**, 959–967.
18. Otten, A. D., and Tapscott, S. J. (1995). Triplet repeat expansion in myotonic dystrophy alters the adjacent chromatin structure. *Proc. Natl. Acad. Sci. USA* **92**, 5465–5469.
19. Klesert, T. R., Otten, A. D., Bird, T. D., and Tapscott, S. J. (1997). Trinucleotide repeat expansion at the myotonic dystrophy locus reduces expression of DMAHP. *Nat. Genet.* **16**, 402–406.
20. Thornton, C. A., Wymer, J. P., Simmons, Z., McClain, C., and Moxley, R. T., 3rd. (1997). Expansion of the myotonic dystrophy CTG repeat reduces expression of the flanking DMAHP gene. *Nat. Genet.* **16**, 407–409.
21. Alwazzan, M., Newman, E., Hamshere, M. G., and Brook, J. D. (1999). Myotonic dystrophy is associated with a reduced level of RNA from the DMWD allele adjacent to the expanded repeat. *Hum. Mol. Genet.* **8**, 1491–1497.
22. Eriksson, M., Ansved, T., Edstrom, L., Anvret, M., and Carey, N. (1999). Simultaneous analysis of expression of the three myotonic dystrophy locus genes in adult skeletal muscle samples: The CTG expansion correlates inversely with DMPK and 59 expression levels, but not DMAHP levels. *Hum. Mol. Genet.* **8**, 1053–1060.
23. Klesert, T. R., Cho, D. H., Clark, J. I., Maylie, J., Adelman, J., Snider, L., Yuen, E. C., Soriano, P., and Tapscott, S. J. (2000). Mice deficient in Six5 develop cataracts: Implications for myotonic dystrophy. *Nat. Genet.* **25**, 105–109.
24. Sarkar, P. S., Appukuttan, B., Han, J., Ito, Y., Ai, C., Tsai, W., Chai, Y., Stout, J. T., and Reddy, S. (2000). Heterozygous loss of Six5 in mice is sufficient to cause ocular cataracts. *Nat. Genet.* **25**, 110–114.
25. Philips, A. V., Timchenko, L. T., and Cooper, T. A. (1998). Disruption of splicing regulated by a CUG-binding protein in myotonic dystrophy. *Science* **280**, 737–741.
26. Miller, J. W., Urbinati, C. R., Teng-Umnuay, P., Stenberg, M. G., Byrne, B. J., Thornton, C. A., and Swanson, M. S. (2000). Recruitment of human muscleblind proteins to (CUG)(n) expansions associated with myotonic dystrophy. *EMBO J.* **19**, 4439–4448.
27. Zamore, P. D., and Haley, B. (2005). Ribo-gnome: The big world of small RNAs. *Science* **309**, 1519–1524.
28. Mandal, M., Boese, B., Barrick, J. E., Winkler, W. C., and Breaker, R. R. (2003). Riboswitches control fundamental biochemical pathways in *Bacillus subtilis* and other bacteria. *Cell* **113**, 577–586.
29. Wang, J., Pegoraro, E., Menegazzo, E., Gennarelli, M., Hoop, R. C., Angelini, C., and Hoffman, E. P. (1995). Myotonic dystrophy: Evidence for a possible dominant-negative RNA mutation. *Hum. Mol. Genet.* **4**, 599–606.
30. Amack, J. D., Paguio, A. P., and Mahadevan, M. S. (1999). Cis and trans effects of the myotonic dystrophy (DM) mutation in a cell culture model. *Hum. Mol. Genet.* **8**, 1975–1984.
31. Mankodi, A., Logigian, E., Callahan, L., McClain, C., White, R., Henderson, D., Krym, M., and Thornton, C. A. (2000). Myotonic dystrophy in transgenic mice expressing an expanded CUG repeat. *Science* **289**, 1769–1773.
32. Liquori, C. L., Ricker, K., Moseley, M. L., Jacobsen, J. F., Kress, W., Naylor, S. L., Day, J. W., and Ranum, L. P. (2001). Myotonic dystrophy type 2 caused by a CCTG expansion in intron 1 of ZNF9. *Science* **293**, 864–867.
33. Kanadia, R. N., Johnstone, K. A., Mankodi, A., Lungu, C., Thornton, C. A., Esson, D., Timmers, A. M., Hauswirth, W. W., and Swanson, M. S. (2003). A muscleblind knockout model for myotonic dystrophy. *Science* **302**, 1978–1980.
34. Grzybowska, E. A., Wilczynska, A., and Siedlecki, J. A. (2001). Regulatory functions of 3′UTRs. *Biochem. Biophys. Res. Commun.* **288**, 291–295.

35. Davis, B. M., McCurrach, M. E., Taneja, K. L., Singer, R. H., and Housman, D. E. (1997). Expansion of a CUG trinucleotide repeat in the 3' untranslated region of myotonic dystrophy protein kinase transcripts results in nuclear retention of transcripts. *Proc. Natl. Acad. Sci. USA* **94**, 7388–7393.
36. Jiang, H., Mankodi, A., Swanson, M. S., Moxley, R. T., and Thornton, C. A. (2004). Myotonic dystrophy type 1 is associated with nuclear foci of mutant RNA, sequestration of muscleblind proteins and deregulated alternative splicing in neurons. *Hum. Mol. Genet.* **13**, 3079–3088.
37. Ho, T. H., Savkur, R. S., Poulos, M. G., Mancini, M. A., Swanson, M. S., and Cooper, T. A. (2005). Colocalization of muscleblind with RNA foci is separable from mis-regulation of alternative splicing in myotonic dystrophy. *J. Cell Sci.* **118**, 2923–2933.
38. Harper, P. S. (1975). Congenital myotonic dystrophy in Britain: I. Clinical aspects. *Arch. Dis. Child.* **50**, 505–513.
39. Sarnat, H. B., and Silbert, S. W. (1976). Maturational arrest of fetal muscle in neonatal myotonic dystrophy: A pathologic study of four cases. *Arch. Neurol.* **33**, 466–474.
40. Furling, D., Coiffier, L., Mouly, V., Barbet, J. P., St Guily, J. L., Taneja, K., Gourdon, G., Junien, C., and Butler-Browne, G. S. (2001). Defective satellite cells in congenital myotonic dystrophy. *Hum. Mol. Genet.* **10**, 2079–2087.
41. Amack, J. D., and Mahadevan, M. S. (2001). The myotonic dystrophy expanded CUG repeat tract is necessary but not sufficient to disrupt C2C12 myoblast differentiation. *Hum. Mol. Genet.* **10**, 1879–1887.
42. Brennan, K. J., and Hardeman, E. C. (1993). Quantitative analysis of the human alpha-skeletal actin gene in transgenic mice. *J. Biol. Chem.* **268**, 719–725.
43. Wong, L. J., Ashizawa, T., Monckton, D. G., Caskey, C. T., and Richards, C. S. (1995). Somatic heterogeneity of the CTG repeat in myotonic dystrophy is age and size dependent. *Am. J. Hum. Genet.* **56**, 114–122.
44. Thornton, C. A., Johnson, K., and Moxley, R. T., 3rd (1994). Myotonic dystrophy patients have larger CTG expansions in skeletal muscle than in leukocytes. *Ann. Neurol.* **35**, 104–107.
45. Welle, S., Bhatt, K., and Thornton, C. A. (1999). Inventory of high-abundance mRNAs in skeletal muscle of normal men. *Genome Res.* **9**, 506–513.
46. Vassilopoulos, D., and Lumb, E. M. (1980). Muscle nuclear changes in myotonic dystrophy. *Eur. Neurol.* **19**, 237–240.
47. Mankodi, A., Takahashi, M. P., Jiang, H., Beck, C. L., Bowers, W. J., Moxley, R. T., Cannon, S. C., and Thornton, C. A. (2002). Expanded CUG repeats trigger aberrant splicing of ClC-1 chloride channel pre-mRNA and hyperexcitability of skeletal muscle in myotonic dystrophy. *Mol. Cell.* **10**, 35–44.
48. Furman, R. E., and Barchi, R. L. (1978). The pathophysiology of myotonia produced by aromatic carboxylic acids. *Ann. Neurol.* **4**, 357–365.
49. Charlet, B. N., Savkur, R. S., Singh, G., Philips, A. V., Grice, E. A., and Cooper, T. A. (2002). Loss of the muscle-specific chloride channel in type 1 myotonic dystrophy due to misregulated alternative splicing. *Mol. Cell.* **10**, 45–53.
50. Seznec, H., Agbulut, O., Sergeant, N., Savouret, C., Ghestem, A., Tabti, N., Willer, J. C., Ourth, L., Duros, C., Brisson, E., *et al.* (2001). Mice transgenic for the human myotonic dystrophy region with expanded CTG repeats display muscular and brain abnormalities. *Hum. Mol. Genet.* **10**, 2717–2726.
51. Sergeant, N., Sablonniere, B., Schraen-Maschke, S., Ghestem, A., Maurage, C. A., Wattez, A., Vermersch, P., and Delacourte, A. (2001). Dysregulation of human brain microtubule-associated tau mRNA maturation in myotonic dystrophy type 1. *Hum. Mol. Genet.* **10**, 2143–2155.
52. Gourdon, G., Radvanyi, F., Lia, A. S., Duros, C., Blanche, M., Abitbol, M., Junien, C., and Hofmann-Radvanyi, H. (1997). Moderate intergenerational and somatic instability of a 55-CTG repeat in transgenic mice. *Nat. Genet.* **15**, 190–192.
53. Thornton, C. A., Griggs, R. C., and Moxley, R. T., 3rd (1994). Myotonic dystrophy with no trinucleotide repeat expansion. *Ann. Neurol.* **35**, 269–272.
54. Ricker, K., Koch, M. C., Lehmann-Horn, F., Pongratz, D., Otto, M., Heine, R., and Moxley, R. T., 3rd (1994). Proximal myotonic myopathy: A new dominant disorder with myotonia, muscle weakness, and cataracts. *Neurology* **44**, 1448–1452.
55. Margolis, J. M., Ranum, L. P. W., and Day, J. W. (2006). Clinical and genetic features of myotonic dystrophy type 2. *In* "Genetic Instabilities and Neurological Diseases" (R. D. Wells and T. Ashizawa, Eds.), Ch. 6. Elsevier, Burlington, MA.
56. Krahe, R., and Bachinski, L. L. (2006). Myotonic dystrophy type 2: Clinical and genetic aspects. *In* "Genetic Instabilities and Neurological Diseases" (R. D. Wells and T. Ashizawa, Eds.), Ch. 7. Elsevier, Burlington, MA.
57. Hamshere, M. G., Newman, E. E., Alwazzan, M., Athwal, B. S., and Brook, J. D. (1997). Transcriptional abnormality in myotonic dystrophy affects DMPK but not neighboring genes. *Proc. Natl. Acad. Sci. USA* **94**, 7394–7399.
58. Taneja, K. L. (1998). Localization of trinucleotide repeat sequences in myotonic dystrophy cells using a single fluorochrome-labeled PNA probe. *Biotechniques* **24**, 472–476.
59. Mankodi, A., Teng-Umnuay, P., Krym, M., Henderson, D., Swanson, M., and Thornton, C. A. (2003). Ribonuclear inclusions in skeletal muscle in myotonic dystrophy types 1 and 2. *Ann. Neurol.* **54**, 760–768.
60. DeCerbo, J., and Carmichael, G. G. (2005). Retention and repression: Fates of hyperedited RNAs in the nucleus. *Curr. Opin. Cell Biol.* **17**, 302–308.
61. Kumar, M., and Carmichael, G. G. (1997). Nuclear antisense RNA induces extensive adenosine modifications and nuclear retention of target transcripts. *Proc. Natl. Acad. Sci. USA* **94**, 3542–3547.
62. Houseley, J. M., Wang, Z., Brock, G. J., Soloway, J., Artero, R., Perez-Alonso, M., O'Dell, K. M., and Monckton, D. G. (2005). Myotonic dystrophy associated expanded CUG repeat muscleblind positive ribonuclear foci are not toxic to *Drosophila*. *Hum. Mol. Genet.* **14**, 873–883.
63. Peng, R., Dye, B. T., Perez, I., Barnard, D. C., Thompson, A. B., and Patton, J. G. (2002). PSF and p54nrb bind a conserved stem in U5 snRNA. *RNA* **8**, 1334–1347.
64. Shav-Tal, Y., and Zipori, D. (2002). PSF and p54(nrb)/NonO: Multi-functional nuclear proteins. *FEBS Lett.* **531**, 109–114.
65. Dansithong, W., Paul, S., Comai, L., and Reddy, S. (2005). MBNL1 is the primary determinant of focus formation and aberrant insulin receptor splicing in DM1. *J. Biol. Chem.* **280**, 5773–5780.
66. Mankodi, A., Urbinati, C. R., Yuan, Q. P., Moxley, R. T., Sansone, V., Krym, M., Henderson, D., Schalling, M., Swanson, M. S., and Thornton, C. A. (2001). Muscleblind localizes to nuclear foci of aberrant RNA in myotonic dystrophy types 1 and 2. *Hum. Mol. Genet.* **10**, 2165–2170.
67. Arrasate, M., Mitra, S., Schweitzer, E. S., Segal, M. R., and Finkbeiner, S. (2004). Inclusion body formation reduces levels of mutant huntingtin and the risk of neuronal death. *Nature* **431**, 805–810.
68. Denis, M. M., Tolley, N. D., Bunting, M., Schwertz, H., Jiang, H., Lindemann, S., Yost, C. C., Rubner, F. J., Albertine, K. H., Swoboda, K. J., *et al.* (2005). Escaping the nuclear confines: Signal-dependent pre-mRNA splicing in anucleate platelets. *Cell* **122**, 379–391.

69. Johnson, J. M., Castle, J., Garrett-Engele, P., Kan, Z., Loerch, P. M., Armour, C. D., Santos, R., Schadt, E. E., Stoughton, R., and Shoemaker, D. D. (2003). Genome-wide survey of human alternative pre-mRNA splicing with exon junction microarrays. *Science* **302**, 2141–2144.
70. Modrek, B., Resch, A., Grasso, C., and Lee, C. (2001). Genome-wide detection of alternative splicing in expressed sequences of human genes. *Nucleic Acids Res.* **29**, 2850–2859.
71. Lewis, B. P., Green, R. E., and Brenner, S. E. (2003). Evidence for the widespread coupling of alternative splicing and nonsense-mediated mRNA decay in humans. *Proc. Natl. Acad. Sci. USA* **100**, 189–192.
72. Black, D. L. (2003). Mechanisms of alternative pre-messenger RNA splicing. *Annu. Rev. Biochem.* **72**, 291–336.
73. Faustino, N. A., and Cooper, T. A. (2003). Pre-mRNA splicing and human disease. *Genes Dev.* **17**, 419–437.
74. Timchenko, L. T., Miller, J. W., Timchenko, N. A., DeVore, D. R., Datar, K. V., Lin, L., Roberts, R., Caskey, C. T., and Swanson, M. S. (1996). Identification of a $(CUG)_n$ triplet repeat RNA-binding protein and its expression in myotonic dystrophy. *Nucleic Acids Res.* **24**, 4407–4414.
75. Cartegni, L., Chew, S. L., and Krainer, A. R. (2002). Listening to silence and understanding nonsense: Exonic mutations that affect splicing. *Nat. Rev. Genet.* **3**, 285–298.
76. McAuliffe, J. J., Gao, L. Z., and Solaro, R. J. (1990). Changes in myofibrillar activation and troponin C Ca^{2+} binding associated with troponin T isoform switching in developing rabbit heart. *Circ. Res.* **66**, 1204–1216.
77. Phillips, M. F., and Harper, P. S. (1997). Cardiac disease in myotonic dystrophy. *Cardiovasc. Res.* **33**, 13–22.
78. Vinereanu, D., Bajaj, B. P., Fenton-May, J., Rogers, M. T., Madler, C. F., and Fraser, A. G. (2004). Subclinical cardiac involvement in myotonic dystrophy manifesting as decreased myocardial Doppler velocities. *Neuromuscul. Disord.* **14**, 188–194.
79. Savkur, R. S., Philips, A. V., and Cooper, T. A. (2001). Aberrant regulation of insulin receptor alternative splicing is associated with insulin resistance in myotonic dystrophy. *Nat. Genet.* **29**, 40–47.
80. Savkur, R. S., Philips, A. V., Cooper, T. A., Dalton, J. C., Moseley, M. L., Ranum, L. P., and Day, J. W. (2004). Insulin receptor splicing alteration in myotonic dystrophy type 2. *Am. J. Hum. Genet.* **74**, 1309–1313.
81. D'Souza, I., and Schellenberg, G. D. (2005). Regulation of tau isoform expression and dementia. *Biochim. Biophys. Acta* **1739**, 104–115.
82. Gao, L., Tucker, K. L., and Andreadis, A. (2005). Transcriptional regulation of the mouse microtubule-associated protein tau. *Biochim. Biophys. Acta* **1681**, 175–181.
83. Vermersch, P., Sergeant, N., Ruchoux, M. M., Hofmann-Radvanyi, H., Wattez, A., Petit, H., Dwailly, P., and Delacourte, A. (1996). Specific tau variants in the brains of patients with myotonic dystrophy. *Neurology* **47**, 711–717.
84. Kiuchi, A., Otsuka, N., Namba, Y., Nakano, I., and Tomonaga, M. (1991). Presenile appearance of abundant Alzheimer's neurofibrillary tangles without senile plaques in the brain in myotonic dystrophy. *Acta Neuropathol. (Berlin)* **82**, 1–5.
85. Modoni, A., Silvestri, G., Pomponi, M. G., Mangiola, F., Tonali, P. A., and Marra, C. (2004). Characterization of the pattern of cognitive impairment in myotonic dystrophy type 1. *Arch. Neurol.* **61**, 1943–1947.
86. Kimura, T., Nakamori, M., Lueck, J. D., Pouliquin, P., Aoike, F., Fujimura, H., Dirksen, R. T., Takahashi, M. P., Dulhunty, A. F., and Sakoda, S. (2005). Altered mRNA splicing of the skeletal muscle ryanodine receptor and sarcoplasmic/endoplasmic reticulum Ca^{2+}-ATPase in myotonic dystrophy type 1. *Hum. Mol. Genet.* **14**, 2189–2200.
87. Futatsugi, A., Kuwajima, G., and Mikoshiba, K. (1995). Tissue-specific and developmentally regulated alternative splicing in mouse skeletal muscle ryanodine receptor mRNA. *Biochem. J.* **305 (Pt. 2)**, 373–378.
88. Buj-Bello, A., Furling, D., Tronchere, H., Laporte, J., Lerouge, T., Butler-Browne, G. S., and Mandel, J. L. (2002). Muscle-specific alternative splicing of myotubularin-related 1 gene is impaired in DM1 muscle cells. *Hum. Mol. Genet.* **11**, 2297–2307.
89. Ho, T. H., Bundman, D., Armstrong, D. L., and Cooper, T. A. (2005). Transgenic mice expressing CUG-BP1 reproduce splicing mis-regulation observed in myotonic dystrophy. *Hum. Mol. Genet.* **14**, 1539–1547.
90. Copley, L. M., Zhao, W. D., Kopacz, K., Herman, G. E., Kioschis, P., Poustka, A., Taudien, S., and Platzer, M. (2002). Exclusion of mutations in the MTMR1 gene as a frequent cause of X-linked myotubular myopathy. *Am. J. Med. Genet.* **107**, 256–258.
91. Sandbrink, R., Hartmann, T., Masters, C. L., and Beyreuther, K. (1996). Genes contributing to Alzheimer's disease. *Mol. Psychiatry*, **1**, 27–40.
92. Johnson, S. A., Pasinetti, G. M., May, P. C., Ponte, P. A., Cordell, B., and Finch, C. E. (1988). Selective reduction of mRNA for the beta-amyloid precursor protein that lacks a Kunitz-type protease inhibitor motif in cortex from Alzheimer brains. *Exp. Neurol.* **102**, 264–268.
93. Cull-Candy, S. G., and Leszkiewicz, D. N. (2004). Role of distinct NMDA receptor subtypes at central synapses. *Sci STKE* **2004**, re16.
94. Lu, X., Timchenko, N. A., and Timchenko, L. T. (1999). Cardiac elav-type RNA-binding protein (ETR-3) binds to RNA CUG repeats expanded in myotonic dystrophy. *Hum. Mol. Genet.* **8**, 53–60.
95. Tian, B., White, R. J., Xia, T., Welle, S., Turner, D. H., Mathews, M. B., and Thornton, C. A. (2000). Expanded CUG repeat RNAs form hairpins that activate the double-stranded RNA-dependent protein kinase PKR. *RNA* **6**, 79–87.
96. Kim, D. H., Langlois, M. A., Lee, K. B., Riggs, A. D., Puymirat, J., and Rossi, J. J. (2005). HnRNP H inhibits nuclear export of mRNA containing expanded CUG repeats and a distal branch point sequence. *Nucleic Acids Res.* **33**, 3866–3874.
97. Timchenko, N. A., Welm, A. L., Lu, X., and Timchenko, L. T. (1999). CUG repeat binding protein (CUGBP1) interacts with the 5′ region of C/EBPbeta mRNA and regulates translation of C/EBPbeta isoforms. *Nucleic Acids Res.* **27**, 4517–4525.
98. Anant, S., Henderson, J. O., Mukhopadhyay, D., Navaratnam, N., Kennedy, S., Min, J., and Davidson, N. O. (2001). Novel role for RNA-binding protein CUGBP2 in mammalian RNA editing. CUGBP2 modulates C to U editing of apolipoprotein B mRNA by interacting with apobec-1 and ACF, the apobec-1 complementation factor. *J. Biol. Chem.* **276**, 47338–47351.
99. Paillard, L., Omilli, F., Legagneux, V., Bassez, T., Maniey, D., and Osborne, H. B. (1998). EDEN and EDEN-BP, a cis element and an associated factor that mediate sequence-specific mRNA deadenylation in Xenopus embryos. *EMBO J.* **17**, 278–287.
100. Ladd, A. N., Charlet, N., and Cooper, T. A. (2001). The CELF family of RNA binding proteins is implicated in cell-specific and developmentally regulated alternative splicing. *Mol. Cell. Biol.* **21**, 1285–1296.
101. Mukhopadhyay, D., Houchen, C. W., Kennedy, S., Dieckgraefe, B. K., and Anant, S. (2003). Coupled mRNA stabilization and translational silencing of cyclooxygenase-2 by a novel RNA binding protein, CUGBP2. *Mol. Cell.* **11**, 113–126.
102. Ho, T. H., Charlet, B. N., Poulos, M. G., Singh, G., Swanson, M. S., and Cooper, T. A. (2004). Muscleblind proteins regulate alternative splicing. *EMBO J.* **23**, 3103–3112.

103. Arhin, G. K., Boots, M., Bagga, P. S., Milcarek, C., and Wilusz, J. (2002). Downstream sequence elements with different affinities for the hnRNP H/H' protein influence the processing efficiency of mammalian polyadenylation signals. *Nucleic Acids Res.* **30**, 1842–1850.
104. Chen, C. D., Kobayashi, R., and Helfman, D. M. (1999). Binding of hnRNP H to an exonic splicing silencer is involved in the regulation of alternative splicing of the rat beta-tropomyosin gene. *Genes Dev.* **13**, 593–606.
105. Chou, M. Y., Rooke, N., Turck, C. W., and Black, D. L. (1999). hnRNP H is a component of a splicing enhancer complex that activates a c-src alternative exon in neuronal cells. *Mol. Cell. Biol.* **19**, 69–77.
106. Fardaei, M., Rogers, M. T., Thorpe, H. M., Larkin, K., Hamshere, M. G., Harper, P. S., and Brook, J. D. (2002). Three proteins, MBNL, MBLL and MBXL, co-localize in vivo with nuclear foci of expanded-repeat transcripts in DM1 and DM2 cells. *Hum. Mol. Genet.* **11**, 805–814.
107. Ladd, A. N., Stenberg, M. G., Swanson, M. S., and Cooper, T. A. (2005). Dynamic balance between activation and repression regulates pre-mRNA alternative splicing during heart development. *Dev. Dyn.* **233**, 783–793.
108. Han, J., and Cooper, T. A. (2005). Identification of CELF splicing activation and repression domains in vivo. *Nucleic Acids Res.* **33**, 2769–2780.
109. Faustino, N. A., and Cooper, T. A. (2005). Identification of putative new splicing targets for ETR-3 using sequences identified by systematic evolution of ligands by exponential enrichment. *Mol. Cell. Biol.* **25**, 879–887.
110. Zhang, W., Liu, H., Han, K., and Grabowski, P. J. (2002). Region-specific alternative splicing in the nervous system: Implications for regulation by the RNA-binding protein NAPOR. *RNA* **8**, 671–685.
111. Michalowski, S., Miller, J. W., Urbinati, C. R., Paliouras, M., Swanson, M. S., and Griffith, J. (1999). Visualization of double-stranded RNAs from the myotonic dystrophy protein kinase gene and interactions with CUG-binding protein. *Nucleic Acids Res.* **27**, 3534–3542.
112. Sobczak, K., de Mezer, M., Michlewski, G., Krol, J., and Krzyzosiak, W. J. (2003). RNA structure of trinucleotide repeats associated with human neurological diseases. *Nucleic Acids Res.* **31**, 5469–5482.
113. Fardaei, M., Larkin, K., Brook, J. D., and Hamshere, M. G. (2001). In vivo co-localisation of MBNL protein with DMPK expanded-repeat transcripts. *Nucleic Acids Res.* **29**, 2766–2771.
114. Artero, R., Prokop, A., Paricio, N., Begemann, G., Pueyo, I., Mlodzik, M., Perez-Alonso, M., and Baylies, M. K. (1998). The muscleblind gene participates in the organization of Z-bands and epidermal attachments of *Drosophila* muscles and is regulated by Dmef2. *Dev. Biol.* **195**, 131–143.
115. Begemann, G., Paricio, N., Artero, R., Kiss, I., Perez-Alonso, M., and Mlodzik, M. (1997). *muscleblind*, a gene required for photoreceptor differentiation in Drosophila, encodes novel nuclear Cys3His-type zinc-finger-containing proteins. *Development* **124**, 4321–4331.
116. Kino, Y., Mori, D., Oma, Y., Takeshita, Y., Sasagawa, N., and Ishiura, S. (2004). Muscleblind protein, MBNL1/EXP, binds specifically to CHHG repeats. *Hum. Mol. Genet.* **13**, 495–507.
117. Iakova, P., Wang, G. L., Timchenko, L., Michalak, M., Pereira-Smith, O. M., Smith, J. R., and Timchenko, N. A. (2004). Competition of CUGBP1 and calreticulin for the regulation of p21 translation determines cell fate. *EMBO J.* **23**, 406–417.
118. Shimizu, K., Chen, W., Ashique, A. M., Moroi, R., and Li, Y. P. (2003). Molecular cloning, developmental expression, promoter analysis and functional characterization of the mouse CNBP gene. *Gene* **307**, 51–62.

cis Effects of CTG Expansion in Myotonic Dystrophy Type 1

SITA REDDY AND SHARAN PAUL

Institute for Genetic Medicine, Room 240, Keck School of Medicine, University of Southern California,
Los Angeles, California 90033

I. Introduction
II. Myotonic Dystrophy Type 1
   A. Genetics of DM1
   B. Mechanism of CTG Repeat Instability
   C. Age at Onset and Disease Course of DM1
   D. Clinical Features of DM1
III. Myotonic Dystrophy Type 2
   A. Genetics of DM2
   B. Age at Onset and Disease Course in DM2
   C. Clinical Features of DM2
IV. What Do the Genetics of DM1 and DM2 Tell Us about the Etiology of These Disorders?
V. Dominant RNA Effects Contribute to DM1 Skeletal Muscle Disease
VI. What Is the Mechanistic Basis of the Toxicity Associated with CUG Repeat Expression?
VII. Proteins That Interact with CUG Repeat Sequences
VIII. Is the Toxicity of CUG/CCUG Different?
IX. *cis* Effects of CTG Expansion at the DM1 Locus
X. Role of Decreased *DMPK* Levels in the Etiology of DM1
   A. *DMPK*
   B. Targeted Inactivation of *Dmpk* in Mice
   C. Loss of *Dmpk* Does Not Result in Gonadal Dysfunction, Cataracts, or Features of Congenital DM1
   D. Loss of *Dmpk* Results in Decreased Twitch and Tetanic Force Development in the Sternomastoid
   E. Depolarization-Mediated Calcium Efflux from the Sarcoplasmic Reticulum Is ~40% Smaller in $Dmpk^{-/-}$ Myotubes
   F. *Dmpk*-Deficient Mice Have Altered Sodium Channel Gating, with Reopenings Leading to Persistent Depolarizing Current in Skeletal Muscle
   G. *Dmpk* Deficiency May Contribute to Skeletal Muscle Weakness and Myotonia in DM1
   H. Inactivation of *Dmpk* Results in Cardiac Conduction Disorders
   I. *Dmpk*-Deficient Mice Have Altered Sodium Channel Gating in Cardiac Muscle
   J. *Dmpk*-Deficient Mice Show Decreased Phosphorylation of Phospholamban
   K. *Dmpk* Loss Alters Hippocampal Function
XI. Role of Decreased *SIX5* Levels in the Etiology of DM1
   A. *SIX5*
   B. Targeted Deletion of *Six5* Sequences in Mice
   C. Decreased *Six5* Levels Do Not Result in Skeletal Muscle Defects
   D. Decreased *Six5* Levels Result in Infrahisian Conduction Disease and Ventricular Hypertrophy
   E. *Six5* Deficiency Results in Nuclear Cataracts
   F. *Six5* Loss Results in Elevated FSH Levels, Testicular Atrophy, Leydig Cell Hyperproliferation, and Aberrant Spermiogenesis
XII. Possible Contribution of Other *cis* Effects at the DM1 Locus
XIII. Concluding Remarks
References

The myotonic dystrophies, DM1 and DM2, result from expanded CTG or CCTG repeat tracts located on chromosomes 19q and 3q, respectively. The two disorders demonstrate several similar features; however, DM1 is the more serious disorder, both exhibiting unique features that are not observed in DM2 and showing an increased incidence and severity of several symptoms that are shared between the two diseases. Although an RNA-dominant mechanism has been shown to underlie the development of several pathological features common to DM1 and DM2, locus-specific *cis* effects of CTG expansion have been hypothesized to explain both the increased severity and complexity of the symptoms exhibited in DM1. Consistent with this hypothesis CTG expansion has been shown to cause stochastic decreases in the steady-state levels of three genes, *DMPK*, *SIX5*, and *DMWD*, located in the vicinity of the CTG tract. If *cis* effects of CTG expansion influence the severity of DM1, it is predicted that inactivation of genes that demonstrate reduced steady-state levels in DM1 would result in partial DM1 phenotypes in model animals. To test this hypothesis, we and others have developed mice in which *Dmpk* and *Six5* have been functionally inactivated. Analyses of these mouse strains demonstrate that decreased levels of *Dmpk* and *Six5* result in a unique set of pathophysiological features that are observed in DM1 patients. Specifically, reduced *Dmpk* levels result in skeletal muscle weakness, calcium and sodium channel defects, cardiac conduction disease, characterized by expanded P–R intervals and atrioventricular conduction blocks, and hippocampal dysfunction. Inactivation of *Six5* in mice shows that *Six5* loss can contribute to congenital cataracts, infrahisian conduction blocks, ventricular hypertrophy, and pituitary and testicular dysfunction. Thus, a combination of both the dominant RNA effect and the locus-specific *cis* effects can explain both the severity and enormous variability that characterize DM1.

I. INTRODUCTION

Myotonic muscular dystrophy was established as a separate disease entity early in the 20th century. Independent observations made by Steinert (1909) and Batten and Gibb (1909) emphasized that a constellation of symptoms occurring in congruence with skeletal muscle myotonia, muscle weakness, and wasting constituted this fascinating disorder [1, 2]. Clinical evaluation now defines myotonic dystrophy 1 (DM1) as a multisystem disorder characterized by myotonia, muscle weakness, muscle atrophy, cardiac conduction disorders, ocular cataracts, endocrine defects, gonadal dysfunction, neuropathy, and psychiatric disease [3–5]. A striking feature of myotonic dystrophy is the enormous variation in expressivity and its pleiomorphic presentation. As early as 1917, Fleischer observed the phenomenon of anticipation in DM1 [6]. *Genetic anticipation* refers to the increase in the severity of symptoms and a decrease in the age at disease onset observed in successive generations of a disease pedigree. The existence of genetic anticipation in DM1 aroused considerable controversy over the years, as it was not clear whether this phenomenon resulted from observational and ascertainment biases or reflected a fundamental biological mechanism [7]. It was only with the discovery of inherited unstable repeat sequences that the biological basis for the phenomenon of genetic anticipation was unequivocally established.

II. MYOTONIC DYSTROPHY TYPE 1

A. Genetics of DM1

DM1 prevalence worldwide is approximately 1 in 8000. DM1 is an autosomal dominant disease, and results from the expansion of a CTG repeat tract on chromosome 19q13.3. The repeat expansion is located in the 3′ untranslated region of a protein kinase, *DMPK*, and is found immediately 5′ of a homeodomain encoding gene *SIX5* (Fig. 4-1) [8–11]. In the normal population the CTG tract varies in size from 5 to 37 repeats. At a threshold of 50 repeats, a striking increase in repeat instability manifests. Several elegant studies demonstrate that both the frequency and amplitude of repeat expansion increase as the CTG tract progressively expands in size [12, 13]. Thus, when a threshold of 50 repeats is reached, a powerful bias for the repeat tract to expand sets in, such that with each

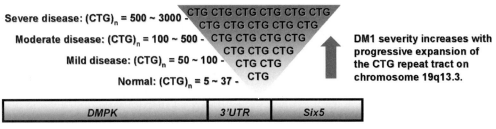

FIGURE 4-1 DM1 locus on chromosome 19q13.3. See CD-ROM for color image.

successive generation the CTG tract becomes progressively larger, reaching enormous sizes that can approach 4000 repeats [12, 13]. As the severity of symptoms roughly corresponds to CTG tract length [12, 13], the propensity of the unstable CTG tract to expand with each generation in a DM1 pedigree provides a molecular explanation for the phenomenon of genetic anticipation.

B. Mechanism of CTG Repeat Instability

The molecular mechanism underlying repeat instability has yet to be completely elucidated. However, examination of CTG tract length in DM1 patients demonstrates that there is both somatic and intergenerational instability. Interestingly, intergenerational instability at the DM1 locus appears to occur in two modes: At the low end of the repeat range associated with DM1 ($n = 50–80$), repeat expansion occurs as a function of length, such that relatively large ($n = \sim200$) expansions grow more frequent with increasing repeat number. When CTG tracts exceed a threshold of 80 repeats, the pattern of instability changes starkly, resulting in a high frequency of saltatory expansions ($n = 120–1250$). These large expansions are the basis for the marked genetic anticipation observed in DM1 pedigrees [12]. Studies in several model systems, including bacteria, yeast, and mice, have been used to throw light on this intriguing phenomenon and are discussed elsewhere in this volume. Our studies on CTG repeat instability in *Escherichia coli* demonstrate that CTG expansion requires the loss of SbcC, a protein that modulates cleavage of single-stranded DNA and degradation of duplex DNA from double-strand breaks. Our data are consistent with the hypothesis that noncanonical single strand-containing secondary structures in Okazaki fragments and/or double-strand breaks in repeat tracts are intermediates in CTG expansion [14].

C. Age at Onset and Disease Course of DM1

A rough correlation exists between the CTG repeat tract size and both the age at onset and disease course in DM1. DM1 occurs primarily at four different ages of onset [13, 15]:

1. Late Onset/Asymptomatic

Expansions of 50–100 repeats at best produce a late-onset and mild form of DM1, with cataracts being the most common finding and weakness and myotonia being rare.

2. Adult Onset

This form of the disease manifests with multisystem involvement, with the severity of the symptoms roughly increasing with the size of the repeat tract. However, the expressivity can be very variable, and presentation can include one or several features of DM1, including myotonia, muscle weakness, cardiac rhythm abnormalities, smooth muscle dysfunction, respiratory failure, endocrine and gonadal abnormalities, cataracts, and hypersomnolence. The disease progresses insidiously but can become debilitating in the fifth and sixth decades of life. Respiratory failure and cardiac disease are often responsible for death.

3. Childhood Onset

Facial weakness is evident as is myotonia. Low IQ, psychiatric disease, and early cardiac involvement characterize these patients.

4. Congenital Onset

The longest tracts, 500 to ~4000 repeats in length, often result in the severe congenital form of the disease. Congenital DM1 is usually inherited maternally [13]. The affected infants show widespread involvement of skeletal muscle. The most common symptoms are generalized hypotonia and weakness, pharyngeal weakness, and arthrogryposis involving predominantly the lower extremities. Less constant features include polyhydramnios, facial diplegia, diaphragmatic paralysis, respiratory failure, decreased motility of the gastrointestinal tract, congenital cataracts, and electrocardiographic abnormalities. Surviving infants show delayed motor development and are often mentally retarded. Interestingly, clinical myotonia is not observed in the first years of life. Congenitally affected infants who survive early childhood develop significant cardiorespiratory disease, which can lead to death in the third and fourth decades of life [3, 4, 5, 16].

D. Clinical Features of DM1

1. DM1 Skeletal Muscle Pathology and Skeletal Defects

Skeletal muscle disease in myotonic dystrophy manifests with myotonia or abnormal muscle relaxation, muscle pain, weakness, and atrophy. Most symptomatic adults demonstrate myotonia, which is usually more prominent earlier in the progression of the disease. Myotonia is most commonly observed as a difficulty in relaxing the grip. Facial muscles, the tongue, and other bulbar muscles can also exhibit myotonia. The pattern of muscles that demonstrate weakness is largely unwavering. Muscles of the face and neck show weakness and wasting early in the disease course. In the limbs, weakness is usually distal, with proximal weakness becoming demonstrable later in the course of the disorder. In addition to weakness, DM1 is associated with atrophy of the involved muscles. Variation in fiber size and fiber type content and presence of central nuclei, fibrosis, sarcoplasmic masses, and ringed fibers are characteristic histological findings. Electron microscopy demonstrates dissolution of myofibers, irregularities in the Z line, and degenerative changes in the mitochondria, sarcoplasmic reticulum, and transverse tubular system [3–5, 15].

Skeletal defects include cranial and facial abnormalities. Hyperostosis of the skull, small sela turcica, large sinuses, and micrognathia are reported in DM1. Defective bone development and thin ribs are sometimes observed in congenital DM1 [3–5].

2. DM1 Cardiac Pathology

Sudden cardiac failure is one of the main causes of death in DM1 patients and occurs with a high incidence (~30%). The predominant defects are conduction blocks that occur throughout the cardiac conduction system on electrophysiological testing. First-degree atrioventricular (AV) block and intraventricular conduction disorders are observed in 75% of DM1 patients [17]. Progressive deterioration of the conduction system, resulting in complete AV block or ventricular arrhythmias, is primarily responsible for sudden cardiac death [18]. Echocardiographic abnormalities include mitral valve prolapse, depressed left ventricular systolic function, reductions in ejection fraction, fractional shortening, and reduced stroke volume [19–22]. In some patients, impaired regional left ventricular relaxation has been attributed to "cardiac myotonia" [22]. Both hypertrophic cardiac atrophy and dilated cardiac myopathy are documented in DM1 [23–27]. Histological abnormalities include myocyte hypertrophy, myofibrillar loss, fibrosis, and fatty infiltration of the myocardium and conduction system [3–5]. Electron microscopy demonstrates dissolution of myofibers, irregularities in the Z line, and degenerative changes in the sarcoplasmic reticulum and transverse tubular system [3–5].

3. DM1 Smooth Muscle Pathology

Smooth muscle dysfunction is widespread. Clinical effects are seen principally in the gastrointestinal tract and result in disordered esophageal and gastric peristalsis. Irritable bowel-like symptoms are extremely common. Upper gastrointestinal tract involvement is usually invariable in later stages of DM1, and dysphagia and aspiration contribute to chest infections, which are a major cause of mortality [3–5, 15].

4. DM1 Eye and Lens Pathology

The incidence of lens opacities in DM1 is very high and manifests as posterior subcapsular, iridescent, multicolored cataracts in adults. Congenital cataracts are associated with congenital DM1. Other ocular defects include decreased vision and decreased intraocular pressure [3–5, 16].

5. DM1 Endocrine Pathology

Endocrine defects include abnormal glucose tolerance with elevated insulin levels. Pituitary and gonadal defects are prominent in DM1. Testicular atrophy, oligospermia, and hyperplasia of Leydig cells are frequently noted. Elevated follicle-stimulating hormone (FSH) and slightly increased luteinizing hormone (LH) levels are often observed in DM1 patients. Ovarian dysfunction is more variable. In severely affected women, pregnancy is rare. On average, female DM1 fertility is 75% that of normal controls [3–5]. Primary ovarian failure, ovarian atrophy, and abnormal ovarian structure are features reported in DM1 [3, 5]. Interestingly, the average age at which menopause is reached is earlier in DM1 patients (40 + 9.42) than in controls (48 + 4.19) [3].

6. DM1 CNS Pathology

CNS dysfunction manifests with mental retardation, hypersomnolence, depression, and anxiety disorders [3–5, 28–34]. It is of interest to note that visuospatial ability appears to be specifically compromised in DM1 patients [31]. In the severe congenital form of the disease, CNS abnormalities are developmental in their origin, whereas in the adult-onset form of DM1, CNS changes appear to be degenerative. In congenital DM1, ventricular enlargement and widening of the interhemispheric fissures suggest aberrant cerebral development or an abnormal buildup of spinal fluid in the brain [35]. In adult-onset DM1, reduction in brain weight, progressive white matter abnormalities, enlargement of the ventricles, and neuronal loss point to degenerative changes in the brain [36–39]. Significantly, presenile accumulation of Alzheimer's neurofibrillary tangles (NFTs) are reported in the cerebral neocortex, brain stem, olfactory nuclei, and limbic and paralimbic systems [40–42]. Gliosis and reactive astrocytosis are also features reported in DM1 [40–42]. Neuronal loss and NFT formation in the limbic region and cerebral cortex may contribute to abnormalities in learning, memory, depression, and anxiety in DM1. Similarly, neuronal loss and development of NFTs in the nuclei of the brainstem may play an important role in DM1-associated hypersomnolence.

III MYOTONIC DYSTROPHY TYPE 2

A. Genetics of DM2

Although the dynamic instability of CTG tracts provides a molecular explanation for the genetic anticipation observed in DM1 pedigrees, the mechanism whereby CTG expansion results in DM1 pathology has yet to be completely elucidated. An important clue was provided by the identification of DM2, a clinical disorder that shares many features with DM1, with the caveat that DM2 is, in general, a milder disease [43–46].

DM2, like DM1, is inherited as an autosomal dominant disorder [43]. Significantly, the genetic mutation in DM2 was discovered to be an enormous CCTG repeat expansion located in the intron of *ZNF9* (zinc finger protein 9) on chromosome 3q21 [47] (Fig. 4-2). An uninterrupted $(CCTG)_{20}$ is considered to be the premutation allele. DM2 patients encode CCTG repeat expansions of 75 to ~11,000 repeats, with a mean of ~5000 repeats. The smallest pathogenic repeat size is unclear, as DM2 demonstrates dramatic CCTG repeat instability in somatic tissues [47, 48].

Genetic anticipation is not as apparent in DM2 as in DM1. Singularly, affected children often have repeat lengths that are shorter than that observed in the parents, and longer repeat expansions are not always observed in patients with earlier disease onset. The size of the repeat correlates most closely with the age at which the blood was obtained. However, a positive correlation between repeat tract length and disease severity may be masked by the massive somatic instability of the CCTG repeat tract in DM2. It is also conceivable that all expansions greater than a given threshold exert similar pathological effects or, alternatively, that smaller repeats may be more toxic than larger CCTG repeat sequences [48].

B. Age at Onset and Disease Course in DM2

On average, the age of DM2 onset is greater than that observed in the adult-onset form of DM1. Day and colleagues report the median age of DM2 onset as 48 years (range, 13–67 years) [46]. Thus, in terms of disease onset, DM2 symptoms manifest much later in life when compared with DM1. Importantly, even though the repeat tracts can be four times larger in DM2 (~14,000 repeats) when compared with DM1 (~4000), both the severe congenital and early childhood onset forms of the disease are notably absent.

C. Clinical Features of DM2

DM2 patients develop several symptoms observed in DM1, including skeletal muscle disease, cardiac pathology, cataracts, testicular atrophy, and endocrine dysfunction [48]. However, DM2 is a less severe disease when compared with DM1, and there are enough differences in both the number and the severity of the symptoms that an experienced clinician would rarely confuse the two disorders. DM2 features that are both similar to and distinct from those of DM1 are now discussed.

1. DM2 Skeletal Muscle Pathology

Skeletal muscle disease in DM2 manifests with myotonia, pain, and muscle weakness. The pattern of muscle weakness involves the neck flexors, elbow extensors, finger flexors, and hip flexors and extensors. Variation in fiber size and presence of central nuclei are characteristic histological findings. Necrotic fibers, mild fibrosis, and adipose deposition are also noted.

Myotonia is more symptomatic in DM1 than in DM2. The incidence of myotonia in a large cohort of DM2 patients was 90 in 234 patients, or 38%; these numbers are almost certainly smaller than those observed in DM1. Significantly, weakness in DM2 often manifests much later in life. Ricker and colleagues observed the onset of weakness in the fourth or fifth

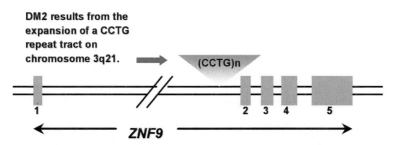

FIGURE 4-2 DM2 locus on chromosome 3q21. See CD-ROM for color image.

decade of life, whereas Day and colleagues noted that 30% of patients developed hip muscle weakness after 50. Muscle atrophy is usually mild and is present in only 9% of DM2 patients [43–46].

2. DM2 Cardiac Pathology

Cardiac arrhythmias are observed in DM2; however, in many DM2 families cardiac manifestations are more benign and the prognosis is less severe when compared with DM1. Although more longitudinal studies are required, the incidence of cardiac disease also appears to be lower in DM2, with only 20% of DM2 patients, compared with as many as 75% of DM1 patients, demonstrating either atrioventricular or intraventricular blocks [46, 49].

3. DM2 Eye and Lens Pathology

Lens opacities in DM2 are similar to those observed in DM1 and manifest as posterior subcapsular, iridescent, multicolored cataracts. The relative incidence of cataracts in DM1 and DM2 is currently unclear. Other ocular defects including decreased vision and decreased intraocular pressure have not been reported in DM2 [46].

4. DM2 Endocrine Pathology

Endocrine dysfunction in DM2, as in DM1, results in abnormal glucose tolerance. In fact, the prevalence of diabetes is higher in DM2 when compared with DM1. Testicular atrophy and increased FSH levels are also common in DM2. Hyperhydrosis appears to be more common in DM2 than in DM1 [43–46].

5. DM2 CNS Pathology

CNS dysfunction is much less evident in DM2 than in DM1. Mental retardation, depression, anxiety disorders, and compromised visuospatial ability are not reported in DM2. The developmental defects of the CNS observed in congenital DM1 are also absent in DM1. However, some DM2 patients have been shown to demonstrate white matter abnormalities on MRI scanning [50].

6. Symptoms That Are Not Observed or Are Not Prominent in DM2

Thus, a comparison of the clinical features of the two disorders demonstrates that:

1. Myotonia is more symptomatic in DM1 compared with DM2.
2. Muscle weakness manifests much later in life in DM2 patients compared with DM1 patients.
3. Muscle atrophy is mild in DM2 compared with DM1.
4. The incidence and severity of cardiac disease are lower in DM2 than in DM1.
5. Gastrointestinal involvement and respiratory insufficiency are rarely reported in DM2, whereas both features occur frequently and are severe in DM1.
6. Similarly, skeletal defects are also not prominent in DM2.
7. CNS dysfunction is more severe in DM1.
8. A severe congenital form of the disease is absent in DM2 [15] (Fig. 4-3).

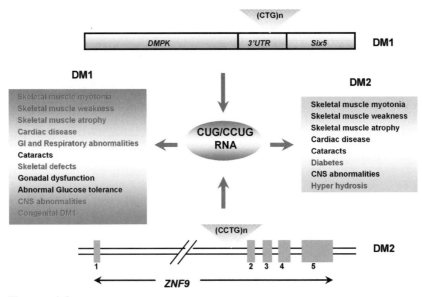

Figure 4-3 Clinical features of DM1 and DM2. Features shaded are unique to DM1 or show increased incidence and severity in DM1 when compared with DM2. Features that are more severe or are unique to DM2 are shaded. See CD-ROM for color image.

IV. WHAT DO THE GENETICS OF DM1 AND DM2 TELL US ABOUT THE ETIOLOGY OF THESE DISORDERS?

The congruence of the genetics and pathology in DM1 and DM2 supports the hypothesis that dominant effects associated with the expression of CUG or CCUG repeat-containing RNAs underlie the development of features that are common to both DM1 and DM2. However, evoking a single dominant RNA mechanism does not appear to be entirely satisfactory because:

1. There are pathological features not seen in DM2 that are prominent in DM1. Thus, what is the mechanism that underlies the development of features unique to DM1?
2. Features that are common to both DM1 and DM2 vary significantly in both severity and rate of incidence, with DM2 being the milder of the two diseases. Therefore, what is the molecular mechanism(s) in DM1 that is responsible for the increased severity and rate of incidence of the symptoms that are common to both disorders?

With respect to the latter point, it is of interest to recognize that even though DM2 repeat tracts can be as much as four times longer than the repeat tracts in DM1, the pathology observed in DM2 is less severe than that in DM1. Furthermore, DM2 patients homozygous for very large repeat tracts do not have a more severe disease, suggesting that the toxic effects of the CCTG repeats are saturatable [48]. Thus, even in the event that DM2 pathology results solely from a dominant RNA mechanism, the maximum toxicity resulting from a dominant RNA mechanism must be observed in DM2. However, as DM2 is a relatively mild disorder when compared with DM1, one must evoke a second mechanism that explains the development of both the unique features of DM1 and the increased severity and rate of incidence in DM1 of features that are common to both diseases.

What is responsible for the larger spectrum of defects and the increased severity of the shared symptoms in DM1? Three nonmutually exclusive possibilities may serve to explain the severity and multifactorial nature of DM1 pathology:

1. The CUG tract may be intrinsically more toxic that the CCUG repeat tract.
2. *DMPK* RNA may be expressed in a wider variety of tissues than *ZNF9* RNA.
3. Locus-specific effects may influence the severity and multifactorial nature of DM1.

V. DOMINANT RNA EFFECTS CONTRIBUTE TO DM1 SKELETAL MUSCLE DISEASE

To test the dominant RNA hypothesis, mice expressing mutant RNAs encoding expanded CUG repeats were developed by both the Thornton and Gourdon laboratories [51, 52]. Thornton and colleagues, who ectopically expressed CUG repeats in skeletal muscle, observed robust myotonia and several histopathological features observed in DM1 skeletal muscle, including central nuclei, fiber size variation, ring fibers, and an increase in the proportion of oxidative fibers. The Gourdon laboratory, which developed transgenic mice using human YACs that encode the mutant DM1 locus, observed lower body weight in postnatal adult animals and abnormal dentition. Myotonia was less prominent in these animals, and histological features observed included central nuclei, fiber size variation, fibrosis, foci of degeneration, and abnormalities in mitochondrial morphology. It is, however, unclear if either mouse model develops significant skeletal muscle weakness, as detailed structure–function analyses were not carried out in these studies. Visible muscle atrophy was also not observed in either mouse model.

These data therefore support the hypothesis that expression of mutant RNAs encoding expanded CUG repeats plays a causal role in the development of myotonia and several characteristic histopathological features of DM1 skeletal muscle disease. The relative contribution of the dominant RNA mechanism to the development of other features that are common to DM1 and DM2, in the context of the whole animal, is currently unknown and has yet to be established.

VI. WHAT IS THE MECHANISTIC BASIS OF THE TOXICITY ASSOCIATED WITH CUG REPEAT EXPRESSION?

To explain the toxicity of the CUG repeat encoding RNA, Caskey and colleagues proposed the protein sequestration model, in which expanded CUG repeat tracts are hypothesized to bind and sequester CUG repeat-specific RNA-binding proteins within the nucleus [53]. Sequestration of RNA-binding proteins by the CUG repeat expansions would therefore result in their progressive depletion from other RNA transcripts that may require such proteins for processing, turnover, transport, or translation. Thus expression of large CUG repeat expansions is hypothesized to result in loss of function of CUG-specific RNA-binding proteins, abnormal RNA processing of their target mRNAs, and the development of DM1 pathology. Importantly, consistent with this hypothesis, RNA localization studies demonstrate that mutant RNAs encoding expanded CUG or CCUG RNA form aberrant RNA foci within DM nuclei [47, 54].

VII. PROTEINS THAT INTERACT WITH CUG REPEAT SEQUENCES

Several laboratories have attempted to dissect the biochemical composition of the RNA foci both by purifying proteins that bind to CUG repeat sequences and by directly examining the localization of candidate proteins to DM foci *in vivo*. Two groups of proteins have been shown to bind to CUG repeats. In this context it is of interest to note that when CUG repeat tracts expand, the mutant transcript folds into stable hairpins [55]. RNA-binding proteins that bind specifically to CUG repeats can therefore belong to two classes: CUG RNA-binding proteins that bind to small single-stranded CUG repeats in the normal range, and a second class of proteins that bind specifically to expanded disease-associated CUG tracts that form double-stranded CUG hairpins. CUG-BP is a candidate for RNA-binding proteins that bind to single-stranded UG or CUG repeats [56]. CUG-BP does not bind to double-stranded CUG hairpins *in vitro* [57] nor does it appear associated with expanded CUG repeats in DM1 patient cells to any significant degree [58]. However, the steady-state levels of this protein are upregulated by the expression of the CUG repeat sequences [59].

The muscleblind family of proteins represents the second class. These proteins have been shown to bind specifically to disease-associated double-stranded CUG hairpins [60]. Significantly, binding is proportional to the length of the CUG hairpin [60]. Consistent with these *in vitro* data, in both DM1 and DM2 tissues mutant RNAs encoding CUG/CCUG repeats complex with the muscleblind family of RNA-binding proteins to form stable nuclear inclusions [58, 61]. Three proteins, MBNL, MBLL, and MBXL, make up the muscleblind family of proteins [58]. The sequestration of the muscleblind family of proteins in DM foci predict that this set of proteins is functionally inactivated in DM cells.

A second piece of evidence that supports the hypothesis that aberrant sequestration or regulation of physiologically important RNA-binding proteins by CUG repeats plays a mechanistic role in DM1 is the growing recognition that abnormal RNA function is an important biochemical event that may underlie one or more features of DM1 pathology [62–68]. Specifically, both the Thornton and Cooper laboratories have shown that abnormal chloride channel RNA splicing and turnover are observed in both DM1 and DM2 patient muscle cells and in transgenic mice expressing CUG repeats in skeletal muscle [62, 63]. Importantly, these studies link the aberrant processing of chloride channel RNA to the development of myotonia both in DM patients and in transgenic mice expressing CUG repeats in skeletal muscle. Studies by Cooper and colleagues have shown that the splicing of the insulin receptor and the cardiac troponin T RNA is also aberrant in DM1 and DM2 cells [64–66]. Abnormal splicing of myotubularin, NMDA NR1 receptor, amyloid-β precursor protein, and microtubule-associated protein tau RNAs has also been reported in DM1 [52, 67, 68].

Upregulation of CUG-BP levels both in cell culture and in mice has been shown to reproduce the aberrant RNA splice patterns observed in DM1 cells [62–67, 70]. Both the Timchenko and Cooper laboratories have shown that ectopic expression of high levels of CUG-BP in skeletal muscle and heart results in neonatal lethality, whereas moderate overexpression of CUG-BP results in central nuclei and fiber size and type variation [69, 70]. Cooper and colleagues and our laboratory have shown that short interfering RNA siRNA-mediated inactivation of MBNL in myoblasts also recapitulates the aberrant splice patterns observed in DM1 cells [71, 72]. Thus, both MBNL inactivation and CUG-BP overexpression appear to independently establish the aberrant splice patterns observed in DM1.

To test the relative importance of MBNL inactivation and CUG-BP overexpression in establishing the aberrant splice patterns observed in DM1, we carried out rescue experiments in which either MBNL was overexpressed or CUG-BP was silenced, in DM1 myoblasts. Significant rescue of aberrant IR splicing was achieved by MBNL overexpression but not by CUG-BP inactivation. Thus, these data demonstrate that although both functional inactivation of MBNL and upregulation of steady-state CUG-BP levels occur as a consequence of CUG repeat expression, it is the inactivation of MBNL that is the key event, whereas the overexpression of CUG-BP appears to play a secondary role, in establishing the aberrant IR splice pattern in DM1 myoblasts [72].

Consistent with these ideas, important data from the Swanson laboratory demonstrate that inactivation of *Mbnl* in mice is sufficient to result in robust myotonia, development of central nuclei, fiber splitting, and abnormal splicing of RNAs in a manner analogous to

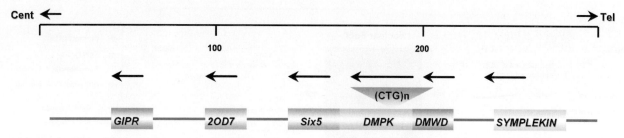

FIGURE 4-4 Schematic of the DM1 locus. Six independent transcripts map within a 200-kb region of the expanded CTG tract. Cent, centromeric region; Tel, telomeric region. Adapted, with permission, from Alwazzan et al. [78]. See CD-ROM for color image.

that observed in DM patients and transgenic mice overexpressing expanded CUG repeats [73]. Significantly, mice in which *Mbnl* is inactivated develop subcapsular dustlike opacities similar to those observed in DM patients [73]. The entire spectrum of pathological features resulting from inactivation of the muscleblind family of proteins in DM has, however, yet to be completely established.

In addition to the muscleblind proteins, Junghans and colleagues have demonstrated that several transcription factors sequester in DM1 foci [74]. The possible pathophysiological consequences of this observation and its relative importance in DM etiology have, however, yet to be established. Thus, taken together, these data demonstrate that toxic effects associated with CUG repeat expression can result in abnormal RNA splicing, myotonia, histopathological changes in skeletal muscle, and formation of subcapsular cataracts (Fig. 4-4).

VIII. IS THE TOXICITY OF CUG/CCUG DIFFERENT?

One possible reason that could explain the increased severity of DM1 compared with DM2 is the greater toxicity of CUG repeat sequences compared with CCUG repeats. Although the intrinsic toxicity of either repeat RNA is not completely understood, the two sequences appear to behave in a similar fashion by all currently established parameters.

First, both CUG and CCUG repeat-containing RNA form aberrant intranuclear RNA foci [47, 54]. In fact, it is likely that CCUG repeat tracts form more numerous RNA foci when compared with CUG repeat sequences [47]. Second, the muscleblind family of proteins sequester within foci formed by both CUG and CCUG RNA foci [58, 61]. Third, the pathological consequences of CUG and CCUG repeat expression appear to be similar insofar as the expression of either repeat sequence appears to result in aberrant RNA splicing [62–67].

A second possibility that could explain the difference in DM1 and DM2 pathology could emerge from the pattern of CUG and CCUG repeat expression *in vivo*. The patterns of *DMPK* and *ZNF9* in humans and mice demonstrate widespread expression of both genes [47, 75–77]. However, comparative analyses of the expression patterns of *DMPK* and *ZNF9* has not been carried out to date. Completion of such analyses will help establish if a restricted pattern of *ZNF9* expression, in key tissues/cell types, is responsible for the smaller spectrum of defects observed in DM2 patients.

IX. *CIS* EFFECTS OF CTG EXPANSION AT THE DM1 LOCUS

Alternatively, locus-specific effects may underlie the severity and multifactorial nature of DM1 when compared with DM2. The DM1 locus on chromosome 19q13.3 is a gene-dense region in which six independent transcripts, *GIPR*, *20D7*, *DMWD*, *DMPK*, *SIX5*, and *SYMPLEKIN*, map within a 200-kb region of the expanded CTG tract [78]. A schematic of this region and the direction of transcription of the six genes are shown in Fig. 4-5. To test the hypothesis that locus-specific effects may contribute to DM1 pathology, the expression levels of genes located in the vicinity of the CTG tract were studied in DM1 patient cells.

CTG expansion changes the transcript levels of three genes at the DM1 locus by different mechanisms:

1. Several reports demonstrate that *DMPK* mRNA and protein levels can decrease by as much as ~70 to 80% in DM1 patients when compared with controls [79, 80]. Part of the decrease in DMPK protein levels must result from the sequestration of the mutant *DMPK* RNA within the nucleus. However, other effects must also be at play, because more than a twofold decrease in *DMPK* transcript levels has been reported in DM1. Interestingly, both my laboratory and the Tapscott laboratory have demonstrated

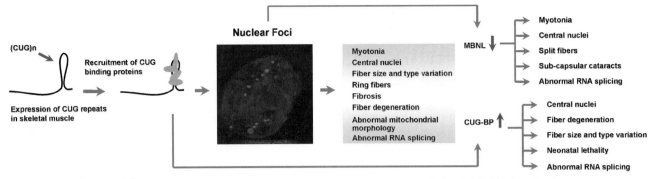

FIGURE 4-5 Dominant RNA model for DM1 pathogenesis in skeletal muscle. See CD-ROM for color image.

that inactivation of *Six5* in mice decreases *Dmpk* steady-state RNA levels *in vivo* (Section XIB). As SIX5 levels are lower in DM1 patients (see later), decreased SIX5 levels may also serve to reduce *DMPK* levels in DM1 patient cells.

2. Expanded CTG repeat sequences have been shown to have an increased affinity for histones *in vitro* [81]. Consistent with this observation, Tapscott and colleagues have observed that CTG tract expansion results in the loss of a DNase I hypersensitive site immediately 5' of *SIX5* [82]. These observations predict that CTG expansion results in local heterochromatin formation and the transcriptional downregulation of the *SIX5* allele, which is linked to the repeat expansion. In support of this hypothesis, both *SIX5* RNA and protein levels have been demonstrated to decrease two- to fourfold in DM1 patients [83, 84].

3. *DMWD*, a gene found immediately 5' of *DMPK*, demonstrates a ~20 to 50% decrease in steady-state cytoplasmic RNA levels of the *DMWD* allele linked to the CTG expansion [85]. However, as nuclear *DMWD* levels are not altered, these data demonstrate that the export of *DMWD* transcripts may be abnormal in DM1 patients. Interestingly, some *DMWD* transcripts have been found to extend beyond the weak polyadenylation signals that are found at the 3' end of the coding region of this gene [86]. Thus, if *DMWD* transcripts include the expanded repeat, such transcripts are also predicted to sequester within DM1 nuclei. The reason for the greater than two fold decrease in *SIX5* and *DMWD* transcript levels in DM1 patients is currently unknown.

It is important to note that CTG expansions cause probabilistic or stochastic changes in gene expression as a function of CTG tract size that do not manifest as all-or-none effects. Importantly, as the levels of *DMPK* and *SIX5* can drop below 50% of normal levels, the pathology resulting from the perturbations of these genes is expected to vary in severity from that of heterozygous loss to a phenotype that lies intermediate to that of a heterozygote and a complete null.

If *cis* effects of CTG expansion increase the severity and the complexity of the symptoms exhibited by DM1 patients, it is predicted that inactivation of genes that demonstrate reduced steady-state levels in DM1 would result in partial DM1 phenotypes in model animals. To test this hypothesis, we and others have developed mice in which *Dmpk* and *Six5* have been functionally inactivated. Consistent with this model, analyses of these mouse strains demonstrate that decreased levels of *Dmpk* and *Six5* result in a unique set of pathophysiological features that are observed in DM1 patients. Most of the current information on *cis* effects resulting from CTG expansion is derived from these animal models. Thus, the analyses pertaining to these mouse strains are presented here.

X. ROLE OF DECREASED *DMPK* LEVELS IN THE ETIOLOGY OF DM1

A. *DMPK*

DMPK is a serine–threonine kinase, which is the archetype of the myotonic dystrophy family of protein kinases. Members of this family include *Neurospora Cot1*, *Drosophila Wts*, *Drosophila Genghis Khan*, *Caenorhabditis elegans LET-502*, rat *ROCKα*, human *p160ROCK*, human *PK418*, murine *CRIK*, and rat *MRCK* [87, 88]. Several members of this family of serine–threonine kinases have been shown to interact with the Rho family of small GTPases. Epstein and colleagues have demonstrated that DMPK binds preferentially to Rac-1 of the Rho family, and that their coexpression results in the GTP-sensitive activation of DMPK [88, 89]. To test the pathophysiology of decreased steady-state levels of *Dmpk* in DM1, both our laboratory and the Wieringa laboratory developed mouse strains in which this gene was inactivated.

B. Targeted Inactivation of *Dmpk* in Mice

The mouse and human genes contain 15 exons, with the serine–threonine kinase domain spanning exons 2–8 [90, 91]. The consensus sequences for the ATP binding site and the serine–threonine kinase are located in exons 2 and 3 and exons 5 and 6, respectively [92, 93]. Thus, to functionally inactivate the *Dmpk* gene in mice, we replaced the 5′ UTR and the first 7 exons of *Dmpk* with a cassette encoding the neomycin phosphotransferase gene under the transcriptional control of the mouse phosphoglycerate kinase promoter. Inactivation of *Dmpk* was confirmed by Northern hybridization analyses, which showed the absence of *Dmpk* transcripts in tissue derived from $Dmpk^{-/-}$ mice [94]. In a parallel study, targeted inactivation of *Dmpk* was carried out using a similar strategy in the Wieringa laboratory [86].

C. Loss of *Dmpk* Does Not Result in Gonadal Dysfunction, Cataracts, or Features of Congenital DM1

$Dmpk^{+/-}$ and $Dmpk^{-/-}$ animals are fertile, and transmission of the mutation followed the normal segregation pattern for a mendelian gene; thus, negative selection against the mutant allele was not apparent. *Dmpk* mutant pups were healthy and showed no overt signs of hypotonia, respiratory distress, or gross anatomical abnormalities. Loss of *Dmpk* therefore did not result in either overt gonadal dysfunction or the development of features of congenital DM1 [86, 94]. Longitudinal studies on $Dmpk^{+/-}$ and $Dmpk^{-/-}$ animals from 3 months to ~2 years of age did not demonstrate an increased incidence of ocular cataract formation (our unpublished data).

D. Loss of *Dmpk* Results in Decreased Twitch and Tetanic Force Development in the Sternomastoid

To test the consequence of *Dmpk* loss on skeletal muscle structure and function, we carried out detailed structure–function analyses on the sternomastoid in *Dmpk* mutant animals. The sternomastoid was chosen as this muscle shows significant weakness and wasting even when involvement of other muscles is relatively mild in DM1 patients. As DM1 is a progressive disorder, longitudinal analyses were carried out and two age groups, 3–4 months and 7–11 months, were sampled [94].

Muscle twitch and tetanic force development were measured using direct stimulation of the muscle and a tension transducer to measure force. In these analyses muscles from 3- to 4-month-old *Dmpk* mutant animals did not show functional impairment. Importantly, a 30 to 50% decrease in both twitch and tetanic force was observed in 7- to 11-month-old $Dmpk^{-/-}$ mice. $Dmpk^{+/-}$ mice demonstrated an intermediate phenotype, showing a greater variability in force measurements, with 30% of the animals tested demonstrating substantial decreases in force, which were in the range observed for $Dmpk^{-/-}$ muscles. Thus, these data demonstrate that muscle force production is sensitive to changes in *Dmpk* dosage and that a progressive decrease in *Dmpk* levels results in an increase both in the incidence and in the severity of skeletal muscle weakness.

As muscle fiber degeneration and regeneration can contribute to progressive weakness, we tested the degree of muscle regeneration in wild-type and *Dmpk* mutant skeletal muscles by measuring the levels of *MyoD*, which is a marker for satellite cell activation. *MyoD* levels are known to closely correlate with the numbers of activated satellite cells both in crush injury models of muscle regeneration in wild-type mice and during the cycles of degeneration and regeneration observed in the *mdx* mouse model of Duchenne muscular dystrophy. Thus, steady-state *MyoD* RNA levels were measured by Northern blot analyses of RNA from limb and neck muscles from wild-type and *Dmpk* mutant animals. Consistent with the force measurements, we observe that *MyoD* levels are not upregulated in skeletal muscle from 3- to 4-month-old $Dmpk^{-/-}$ mice; however a three- to fourfold increase in *MyoD* was observed in 7- to 11-month-old $Dmpk^{-/-}$ muscles. Intermediate levels of *MyoD* were observed in $Dmpk^{+/-}$ muscles.

To further characterize the incidence of fiber degeneration and regeneration with time, wild-type and *Dmpk* mutant sternomastoid muscle sections were stained with antibodies against embryonic myosin heavy chain. These data demonstrate an increased incidence of eMHC-positive fibers in $Dmpk^{-/-}$ (1%) mouse muscles when compared with wild-type (0.4%) muscles at 7–11 months of age. $Dmpk^{+/-}$ muscles demonstrated a phenotype that was in-between those observed in wild-type and $Dmpk^{-/-}$ muscles. Although the percentages of eMHC-positive fibers are small in *Dmpk* mutant mice, these numbers are likely to be significant as eMHC is expressed transiently in regenerating muscle. Thus, these data demonstrate that there is a small but significant increase in muscle regeneration and degeneration with time as a function of decreasing *Dmpk* levels.

Morphometric analyses of hematoxylin and eosin-stained sections of the sternomastoid demonstrated that the frequency distribution of $Dmpk^{-/-}$ fiber sizes showed a wide range of fiber cross-sectional area when compared with wild-type controls. Both sternomastoid and forelimb muscles of $Dmpk^{-/-}$ muscles also demonstrated foci of degeneration, which included empty basal lamina ghosts associated with small-diameter fibers with some central nuclei and fibrosis. Ultrastructural analyses of 7- to 11-month-old $Dmpk^{-/-}$ muscles showed abnormal muscle structure in 20 to 60% of randomly selected fields. Abnormalities that were observed included one or more of the following features: bending and disintegration of the Z line, myofibrillar distortion, mitochondria that had lost their structural integrity, and dilated profiles of sarcoplasmic reticulum.

To test the mechanism underlying the decreased force production in $Dmpk^{-/-}$ skeletal muscle, we carried out sequential analyses that examined consecutive events that initiate skeletal muscle force generation. Specifically, we tested the integrity of nerve conduction, signal transduction at the neuromuscular junction, and excitation of the muscle membrane. The relatively normal results from these analyses demonstrate that muscle weakness in $Dmpk$ mutant mice results primarily from a myopathy rather than a neuropathy. Muscle fatigue, which is assessed as an indicator of the integrity of the oxidative and glycolytic pathways for energy production, were normal in $Dmpk^{-/-}$ muscles. Lastly, analyses of the relationship between direct muscle stimulation frequency and force generation (force–frequency relationships) demonstrated that maximum force generation occurred in the same range and rate of stimulation in wild-type and $Dmpk^{-/-}$ muscles. Thus, these data suggest that the kinetics of calcium release from the sarcoplasmic reticulum are similar in $Dmpk^{-/-}$ and wild-type muscle. Thus, structure–function analyses of $Dmpk$ mutant muscles make the following points:

1. Skeletal muscle force production is sensitive to changes in $Dmpk$ dosage. A stepwise decrease in $Dmpk$ levels in $Dmpk^{+/-}$ and $Dmpk^{-/-}$ mice results in a progressive increase in both the incidence and the severity of skeletal muscle weakness.

2. Although structural changes or cycles or regeneration and degeneration may contribute to skeletal muscle weakness in $Dmpk$-deficient mice, the relatively modest nature of these changes does not appear to completely explain the substantial decreases in skeletal muscle force production observed.

3. These data therefore support the hypothesis that functional abnormalities in excitation–contraction coupling may play a significant role in the skeletal muscle weakness in $Dmpk$-deficient muscles.

4. As maximum force generation occurred in the same range and rate of stimulation in wild-type and $Dmpk^{-/-}$ muscles, the kinetics of calcium release appear to be similar in $Dmpk^{-/-}$ and wild-type muscles. Therefore, defects that may underlie muscle weakness in $Dmpk$ mutant mice may include changes in the absolute levels of calcium released from the sarcoplasmic reticulum and abnormal actin–myosin cross-bridge formation.

E. Depolarization-Mediated Calcium Efflux from the Sarcoplasmic Reticulum Is ~40% Smaller in $Dmpk^{-/-}$ Myotubes

Features of excitation–contraction coupling can be studied *in vitro* using cultured muscle cells. Although structure–function analyses were not carried out in $Dmpk$-deficient mice generated in the Wieringa laboratory, their analyses of depolarization-mediated calcium efflux from the sarcoplasmic reticulum provide important insights into the molecular mechanisms that may underlie the skeletal muscle weakness that results as a consequence of $Dmpk$ loss [95].

Excitation–contraction coupling in skeletal muscle involves a set of sequential steps. First, a synaptic potential stimulates an action potential in the surface membrane. Subsequently, transmission of that signal into the transverse tubule system stimulates calcium release from the sarcoplasmic reticulum. Reaction of calcium released from the sarcoplasmic reticulum with troponin is the signal linking electrical excitation and contraction.

In skeletal muscle, the neuromuscular junction contains a ligand-gated receptor/channel that is opened by acetylcholine, which allows cations to diffuse through to create a local depolarization. This stimulates a rapidly propagating surface action potential based on the opening of membrane sodium channels. When a muscle cell is depolarized by an action potential, calcium ions enter the cell through the voltage-sensing dihydropyridine receptors (DHPRs), which are L-type calcium channels, located on transverse tubules. This calcium triggers a release of calcium, which is stored in the sarcoplasmic reticulum (SR), through the ryanodine receptors, which are calcium release channels located in the terminal cisternae of the sarcoplasmic reticulum.

In vitro differentiated $Dmpk^{-/-}$ myotubes exhibit higher resting Ca^{2+} levels (185 nM) compared with wild-type myotubes (122 nM) [95]. To test the etiology of this defect, $Dmpk^{-/-}$ myotubes were treated with tetrodotoxin, nifedipine, and ryanodine, which inhibit terodotoxin-sensitive voltage-operated sodium channels (TTXRs), DHPRs, and ryanodine receptors respectively. Ryanodine does not rescue this defect; however, tetrodotoxin and nifedipine, partially and completely, normalize the increased resting calcium levels in $Dmpk^{-/-}$ myotubes. These data therefore demonstrate that TTXRs and DHPRs have aberrant open probabilities that result in elevated resting Ca^{2+} levels in $Dmpk$-deficient myotubes.

Importantly, when depolarization was triggered with acetylcholine or KCl, calcium amplitudes were reduced ~40% in $Dmpk$-deficient myotubes. However, when resting calcium levels were normalized by prior treatment with tetrodotoxin, both the amplitude and the kinetics of calcium responses evoked by depolarization with KCl were similar in $Dmpk^{-/-}$ and wild-type myotubes. Interestingly, in this study, cultured DM1 muscle cells exhibited defects similar to those reported by Wieringa and colleagues [95–99]. Thus, these data are consistent with our findings in $Dmpk^{-/-}$ skeletal muscles and demonstrate that:

1. The absolute levels of calcium efflux from the sarcoplasmic reticulum are ~40% smaller in $Dmpk$-deficient myotubes.
2. The dampening effect on the release of calcium in $Dmpk^{-/-}$ myotubes results from elevated resting calcium levels, which is a consequence of the aberrant open configurations of TTXRs and DHPRs.
3. As depolarization-mediated calcium amplitudes and kinetics of calcium release are normal when the elevated levels of resting calcium are normalized by prior treatment with tetrodotoxin, these data demonstrate there are no intrinsic defects in the kinetics of calcium release from the sarcoplasmic reticulum.

F. *Dmpk*-Deficient Mice Have Altered Sodium Channel Gating, with Reopenings Leading to Persistent Depolarizing Current in Skeletal Muscle

Franke and colleagues have found repeated action potentials and abnormal gating of sodium channels in skeletal muscle biopsies of DM1 patients [99]. To follow up on the work of the Weiringa laboratory and to test if $Dmpk$ deficiency results in sodium channel defects similar to those observed by Franke and colleagues, we measured membrane potentials and sodium currents in skeletal muscle from wild-type, $Dmpk^{+/-}$ and $Dmpk^{-/-}$ skeletal muscle [100, 101]. These studies make the following points:

1. Intracellular membrane recordings demonstrate repetitive action potentials in both $Dmpk^{+/-}$ and $Dmpk^{-/-}$ muscles, induced by a single stimulus. These repetitive action potentials were blocked by lidocaine, a sodium channel blocking agent, and were not observed in wild-type mice.

2. Sodium channels in $Dmpk$-deficient muscles demonstrate more frequent and longer openings and longer bursts of openings with sustained depolarization. This recapitulates the sodium channel defect in DM1, thus providing a link between DM1-associated sodium channel defects and $Dmpk$ deficiency.
3. The sodium channel defect was identical in both $Dmpk^{+/-}$ and $Dmpk^{-/-}$ mice, demonstrating that partial deficiency of $Dmpk$ is sufficient to result in the sodium channel dysfunction.
4. The sodium channel defects were more prominent with increasing age in $Dmpk$ mutant muscles.

As skeletal muscle defects in $Dmpk^{-/-}$ mice are more severe than those in $Dmpk^{+/-}$ mice, whereas sodium channel lesions are similar in their manifestation in both $Dmpk^{+/-}$ and $Dmpk^{-/-}$ skeletal muscle, these data demonstrate that defects other than those stemming from sodium channel defects described by us must also contribute to skeletal muscle weakness in $Dmpk$-deficient muscles. Such defects may include: alterations in other ion channels including DHPRs as identified by Weiringa and colleagues, and functional defects in the contractile apparatus. With respect to the latter point, it is of interest to note that *Let-502*, a *Caenorhabditis elegans* gene that shows homology to *DMPK*, has been demonstrated to act in a pathway linking signals generated by the GTP-binding protein Rho to the myosin-based contractile apparatus [102].

The mechanism by which $Dmpk$ deficiency alters ion channel function is likely to involve phosphorylation. Significantly, Timchenko and colleagues have demonstrated that the β subunit of DHPR is phosphorylated by recombinant DMPK *in vitro* [103]. Similarly, although a direct demonstration of DMPK phosphorylation of sodium channels is not reported, previous studies by Moorman and colleagues demonstrated that the physiological effects of DMPK expression on sodium channel function in *Xenopus* oocytes are lost in a sodium channel mutant in which a phosphorylation site is lost due to mutation of a serine residue to an alanine [104].

G. *Dmpk* Deficiency May Contribute to Skeletal Muscle Weakness and Myotonia in DM1

Taken together these data demonstrate that $Dmpk$ deficiency can contribute to the increased skeletal muscle weakness and, potentially, to the higher incidence and severity of myotonia observed in DM1 patients compared with DM2 patients. Specifically:

1. Reduction in *Dmpk* levels is predicted to increase the incidence and severity of skeletal muscle weakness in DM1 patients by decreasing the depolarization-mediated efflux of calcium from the sarcoplasmic reticulum.

2. Sodium channel defects resulting from *Dmpk* deficiency may serve to exacerbate the incidence and severity of myotonia observed in DM1, as repetitive action potentials induced by a single stimulus in $Dmpk^{+/-}$ and $Dmpk^{-/-}$ mice are blocked by lidocaine, a sodium channel blocking agent.

H. Inactivation of Dmpk Results in Cardiac Conduction Disorders

In the heart, the electrical stimulus originates from the sinoatrial (SA) node, which is located in the upper part of the right atrium. As the stimulus proceeds away from the SA node in all directions and the atria are triggered to contract, the wave of depolarization sweeping through the atria is recorded as a P wave on the electrocardiogram (ECG). This wave of depolarization subsequently reaches the AV node, located between the atria and the ventricles. Depolarization slows within the AV node, and in this brief pause the blood flows from the atria into the ventricles. Depolarization continues down the His bundle, which extends down from the AV node, divides into the right and left bundle branches within the ventricular septum, and terminates in fine Purkinje fibers that contact the ventricular myocardial cells. This allows depolarization of the myocardial cells of the ventricles and begins ventricular contraction. This series of events produces the QRS complex on the ECG. There is a pause after the QRS complex, which is followed by a T wave, which represents repolarization of the ventricle. This set of events represents a single cardiac cycle. To test if *Dmpk* is required to maintain the integrity of cardiac conduction, we carried out the following electrophysiological studies [105–107].

1. Dmpk1/2 and Dmpk2/2 Mice Demonstrate First-Degree AV Block

In these experiments we sampled wild-type, $Dmpk^{+/-}$ and $Dmpk^{-/-}$ animals at four different age levels: 1–2 months, 4–6 months, 12–15 months, and 18–21 months. In these experiments, the mean sinus cycle length (beat-to-beat heart rate), P-wave duration (atrial conduction time), P–R interval (atrial and AV nodal conduction time), QRS interval (ventricular depolarization time), J–T interval (ventricular repolarization time), and Q–T interval (surrogate of action potential duration) were measured. Prolonged P–R intervals were observed in both $Dmpk^{+/-}$ (mean ± SD = 48 ± 8 ms) and $Dmpk^{-/-}$ (48 ± 7 ms) animals compared with wild-type controls (34 ± 5 ms) ($P < 0.001$ for all measurements). Elongation of the P–R interval or first-degree heart block was apparent in both $Dmpk^{+/-}$ and $Dmpk^{-/-}$ mice both under anesthesia and during ambulation. Other ECG intervals were unaltered in wild-type and *Dmpk* mutant animals [105]. As *Dmpk*-deficient mice demonstrated P–R prolongation with normal P-wave duration and QRS interval, these data indicate AV node dysfunction. Furthermore, a lack of conduction delay through the working atrial muscle and ventricular myocardium indicates specific dysfunction at the level of the specialized conduction tissue.

2. $Dmpk^{-/-}$ Mice Demonstrate Severe AV Conduction Disturbances Including Second- and Third-Degree AV Block

Sinus node, atrial, AV, and ventricular conduction parameters and refractoriness were studied using *in vivo* electrophysiology in wild-type, $Dmpk^{+/-}$, and $Dmpk^{-/-}$ mice. Rate-corrected sinus node recovery time (CSNRT) was evaluated for indirect estimation of sinus node function. Although CSNRT measurements were more variable in $Dmpk^{-/-}$ mice, the mean CSNRTs for mutant and control animals were similar, suggesting that sinus node function was not compromised in *Dmpk*-deficient mice. Programmed atrial stimulation was used to study AV nodal physiology, and in these experiments $Dmpk^{-/-}$ mice demonstrated more severe AV conduction disturbances during atrial pacing, which included both second- and third-degree AV blocks [105]. In a second independent study, His bundle recordings in $Dmpk^{-/-}$ mice demonstrated both increased A–H intervals (the atrial-His interval, which makes up most of the P–R interval on the surface ECG; 36.7 ± 4 ms versus 31.6 ± 4.8 ms; $P = 0.037$) and increased H–V intervals (the His–ventricular interval represents time from His bundle depolarization to the beginning of ventricular depolarization; 14.7 ± 2 ms versus 10.3 ± 0.8 ms; $P = 0.001$) compared with controls [106]. Thus, in $Dmpk^{-/-}$ mice, AV conduction abnormalities are located both in the suprahisian and infrahisian conduction tissue, with a higher incidence in the latter, a finding similar to that in DM1 patients.

3. Conduction Disease Severity Increases with Age in $Dmpk^{-/-}$ Mice

Conduction disease is not apparent in *Dmpk* mutant mice 1–2 months old; however, conduction disease is clearly established in mice 4 months and older. There is a trend toward elongation of the P–R interval with time, such that P–R intervals >50 ms were not observed in 4- to 6-month-old mice but became apparent at 16–17 months of age. Furthermore, the incidence of second- and third-degree AV block increased in 16- to 17-month-old mice, which primarily demonstrated P–R intervals that were ≥45 ms [105, 107]. The increase in the incidence of conduction disease with age and the correlation of severe conduction disorders with length of the P–R interval are also reminiscent of DM1 cardiac disease.

4. Sympatholytic and Cholinergic Effects Are Not Responsible for AV Block in *Dmpk* Mutant Animals

Programmed stimulation and pacing were performed both at baseline and in conjunction with isoproterenol administration. However, catecholamine stimulation did not alter the AV conduction parameters differentially in *Dmpk* mutant mice when and compared with controls. Following isoproterenol, atropine was administered to the mice. Atropine treatment shortened the sinus cycle length without changing P–R duration [105]. Thus, these data demonstrate that sympatholytic effects are not directly implicated in the AV block in *Dmpk* mutant mice. As the data show a lack of anticholinergic reversal of the AV block, they demonstrate that AV node dysfunction is not secondary to vagal action.

5. Gross Structural Differences Are Not Observed in *Dmpk* Mutant Hearts

Histological sections stained with hematoxylin and eosin and Gomori trichrome did not demonstrate significant atrophy or fibrosis in $Dmpk^{+/-}$ and $Dmpk^{-/-}$ hearts [105]. Thus, taken together, these studies demonstrate that:

1. *Dmpk* loss results in AV conduction delay. The data therefore demonstrate a critical role for *Dmpk* in AV node function.
2. The degree of P–R prolongation on the ECG is similar in $Dmpk^{+/-}$ and $Dmpk^{-/-}$ mice. $Dmpk^{-/-}$ mice, however, demonstrate more severe AV conduction disturbances. These studies therefore demonstrate that cardiac conduction is sensitive to *Dmpk* dosage. The data link *Dmpk* haploinsufficiency with AV conduction disturbances, which characterize DM1.
3. As gross structural changes do not manifest as a consequence of *Dmpk* loss, our data suggest that functional defects in impulse propagation may underlie the cardiac pathology observed in *Dmpk* mutant mice. The data do not, however, rule out subtle changes in cell–cell communication occurring as a consequence of structural changes in the conduction system.
4. As the AV node is a slow-conducting calcium channel-dependent tissue, it is possible that calcium current inhibition could underlie the prolonged AV conduction time in *Dmpk* mutant mice. Thus, if calcium amplitudes are dampened in a manner similar to that observed in skeletal muscle, such changes could contribute to the abnormally slow impulse propagation in the AV node.
5. It is also possible that alterations in sodium channel function in the fast-conducting tissues of the AV node and His–Purkinje system may contribute to the conduction disorders that result from *Dmpk* loss.

I. *Dmpk*-Deficient Mice Have Altered Sodium Channel Gating in Cardiac Muscle

To test the integrity of cardiac sodium channel function in *Dmpk* mutant mice, we used whole-cell and cell-attached patch-clamp recordings of ventricular cardiomyocytes enzymatically isolated from wild-type, $Dmpk^{+/-}$, and $Dmpk^{-/-}$ mice [108]. The primary findings of this study are:

1. Recordings from membrane patches containing one or a few sodium channels showed multiple sodium channel reopenings after the macroscopic current had subsided in both $Dmpk^{+/-}$ and $Dmpk^{-/-}$ mice when compared with controls (greater than threefold differences; $P < 0.05$). Thus *Dmpk* deficiency results in a sodium channel abnormality comprising frequent, long bursts of sodium channel reopenings during sustained depolarization, which results in a plateau of noninactivating late sodium current.
2. Macroscopic sodium current density was similar in *Dmpk* mutant and wild-type cardiomyocytes.
3. Action potential duration was significantly prolonged in both $Dmpk^{+/-}$ and $Dmpk^{-/-}$ mice.
4. $Dmpk^{+/-}$ and $Dmpk^{-/-}$ mice demonstrated similar cardiac sodium channel gating abnormalities.

These data demonstrate that similar sodium channel defects result from *Dmpk* deficiency both in skeletal muscle and in ventricular cardiomyocytes. As noted earlier, such defects could contribute to the conduction disease observed in $Dmpk^{+/-}$ and $Dmpk^{-/-}$ mice.

J. *Dmpk*-Deficient Mice Show Decreased Phosphorylation of Phospholamban

Recent studies by Ruiz-Lozano demonstrate that *Dmpk* colocalizes and coimmunoprecipitates with phospholamban, a muscle-specific sarcoplamic reticulum calcium ATPase (SERCA2a) inhibitor *in vivo* [109].

These authors have also shown that purified *Dmpk* phosphorylates phospholamban *in vitro*. Consistent with these data, phospholamban appears to be hypophosphorylated, and calcium uptake by the sarcoplamic reticulum in $Dmpk^{-/-}$ ventricular homogenates is impaired. These results therefore suggest that decreased phospholamban phosphorylation may alter calcium levels within $Dmpk^{-/-}$ cardiomyocytes.

Thus, electrophysiological studies clearly demonstrate an important role for *Dmpk* in maintaining the functional integrity of the cardiac conduction system. The mechanism whereby *Dmpk* loss precipitates cardiac rhythm disorders has yet to be completely elucidated. Our data are consistent with *Dmpk* being part of a tightly regulated signaling pathway in which small changes in *Dmpk* dosage result in hypophosphorylation of key targets. However, the identity and function of such targets have yet to be proven conclusively.

K. *Dmpk* Loss Alters Hippocampal Function

Epstein and colleagues hypothesized that loss of *Dmpk* could alter hippocampal function. This hypothesis is of special interest, as functional defects in the CNS are prominent in DM1 patients. Previous work by the Epstein laboratory had shown that DMPK can modify the actin cytoskeleton. Specifically, overexpression of DMPK in lens epithelial cells has been demonstrated to result in significant rearrangements of the actin cytoskeleton and plasma membrane [89]. DMPK has also been shown to phosphorylate and inactivate myosin phosphatase, which is predicted to alter the assembly and contractility of the actin cytoskeleton [110]. As changes in the actin cytoskeleton can influence both synaptic shape and the shape of dendritic spines, two events that are important for long-term potentiation, a form of synaptic plasticity that is believed to contribute to the cellular basis of memory storage, these authors examined hippocampal function in $Dmpk^{-/-}$ mice.

No changes in basal synaptic transmission in the CA1 area of the hippocampus were detected in $Dmpk^{-/-}$ mice. Furthermore, no changes in long-term potentiation were detected in $Dmpk^{-/-}$ mice 3 h after induction. However, *Dmpk* knockout mice demonstrated decreased decremental potentiation [111]. These studies therefore demonstrate that functional defects exist in the hippocampus of $Dmpk^{-/-}$ mice. However, the relevance of this finding to DM1 CNS dysfunction has yet to be completely understood.

Thus, when taken together, these studies demonstrate that DMPK loss can serve to increase both the severity and incidence of skeletal muscle disease and

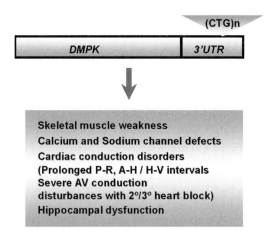

FIGURE 4-6 Contribution of *DMPK* loss to DM1 pathology. See CD-ROM for color image.

cardiac conduction disorders in DM1. The role of DMPK deficiency in the development of other DM1 features has yet to be clarified. As noted above, understanding the possible role of DMPK loss in the development of mental retardation and psychiatric disease in DM1 is an area of particular interest (Fig. 4-6).

XI. ROLE OF DECREASED *SIX5* LEVELS IN THE ETIOLOGY OF DM1

A. *SIX5*

The *Six* series of genes encode evolutionarily conserved transcription factors characterized by the *Six* domain and the *Six* homeodomain, both of which necessary for specific DNA binding. Ectopic expression of *Six* genes has been shown to alter cell fate. These data therefore suggest that the *Six* gene family members may play a critical role in organogenesis [112]. Abnormal function or decreased levels of several *SIX* genes are known to result in human genetic diseases. Specifically, mutations in *SIX3* result in holoprosencephaly [113], whereas haploinsufficiency of *SIX6* is responsible for bilateral anophthalmia [114]. *SIX5* levels are decreased two- to fourfold in cells of DM1 patient [83, 84]. To study the role of *SIX5* in DM1 etiology, both our laboratory and the Tapscott laboratory constructed and analyzed *Six5* mutant mice [115, 116].

B. Targeted Deletion of *Six5* Sequences in Mice

In the mouse strain developed in our laboratory, *Six5* sequences were replaced by a cassette encoding the neomycin phosphotransferase gene under the transcriptional control of the mouse phosphoglycerate kinase promoter. Specifically, the region deleted in this mouse strain includes 412 bp 5′ of the *Six5* ATG codon, *Six5* coding sequences, and ~180 bp 3′ of the termination codon [115]. A different mutation was developed by Tapscott and colleagues, who replaced *Six5* exon 1 with sequences coding for β-galactosidase [116]. Interestingly, steady-state *Dmpk* RNA levels decreased by ~15–25% in $Six5^{+/-}$ mice and ~50% in $Six5^{-/-}$ mice [115, 116]. These results therefore demonstrate that either Six5 function is required for the maintenance of normal steady-state *Dmpk* RNA levels or, alternatively, that the genomic sequences deleted in both strains of mice may contain regulatory elements that are required for normal *Dmpk* transcription.

C. Decreased *Six5* Levels Do Not Result in Skeletal Muscle Defects

To prevent the effect of reduced *Dmpk* levels from confounding the identification of possible skeletal muscle defects resulting from *Six5* loss, we studied young wild-type, $Six5^{+/-}$ and $Six5^{-/-}$ mice, which were 3–4 months of age, a point at which *Dmpk* deficiency does not result in a demonstrable skeletal muscle phenotype in $Dmpk^{+/-}$ and $Dmpk^{-/-}$ mice. In these analyses, no significant changes in skeletal muscle structure or function were observed, except that $Six5^{-/-}$ skeletal muscle showed more variability in force development when compared with wild-type or $Six5^{+/-}$ skeletal muscle [117]. The possible effects of *Six5* loss on skeletal muscle structure or function with increasing age are currently unknown. We studied the possible role of *Six5* haploinsufficiency on sodium channel function. However, unlike $Dmpk^{+/-}$ mice, $Six5^{+/-}$ mice did not demonstrate skeletal muscle sodium channel dysfunction [118]. Consistent with these results, histological analyses of the *Six5* mutant mice developed by the Tapscott laboratory at 3 months of age did not show significant abnormalities in skeletal muscle structure [116]. In these studies, electromyography of several muscles at 3 and 10 months of age did not demonstrate myotonia or abnormal calcium-activated potassium channel function in *Six5* mutant animals [116].

D. Decreased *Six5* Levels Result in Infrahisian Conduction Disease and Ventricular Hypertrophy

To test the possible role of Six5 haploinsufficiency in the etiology of DM1 cardiac disease we studied cardiac function in wild type and $Six5^{+/-}$ animals at both 2–3 months and 15–16 months of age [119].

1. $Six5^{+/-}$ Mice Demonstrate Longer QRS and HV Intervals

$Six5^{+/-}$ mice in both age groups did not show prolonged P–R intervals. However, ECG recordings on anesthetized 15- to 16-month-old $Six5^{+/-}$ mice demonstrated longer QRS intervals (19 ± 2 ms) when compared with wild-type controls (15 ± 2 ms) ($P = 0.0001$). Consistent with these results, intracardiac electrophysiological studies showed prolonged H–V intervals (18 ± 2 ms versus 15 ± 2 ms; $P = 0.002$) in 15- to 16-month-old $Six5^{+/-}$ mice. These differences were not, however, recorded on ambulation, demonstrating that the infrahisian conduction delay in $Six5^{+/-}$ mice is subtle and exacerbated by anesthesia.

2. $Six5^{+/-}$ Mice Demonstrate Increased Left Ventricular End-Diastolic Dimension and Ventricular Hypertrophy

Echocardiography showed that overall heart function is preserved in $Six5^{+/-}$ mice; however, the left ventricular end-diastolic dimension (LVEDD) was significantly larger in $Six5^{+/-}$ mice than in controls (3.0 ± 0.6 mm versus 2.4 ± 0.6 mm; $P = 0.039$). As wall thickness did not significantly differ between $Six5^{+/-}$ and control animals, total left ventricular mass must be increased. The increased chamber size most likely reflects early ventricular remodeling, secondary to diastolic dysfunction and cardiomyopathy. Thus, these data demonstrate that *Six5* haploinsufficiency can result in ventricular hypertrophy.

Exercise tolerance testing was carried out on a multilane graded treadmill machine designed for mice. In this exercise, mice are required to maintain running at a constant speed (100 m/min) at a 15° slope. This experiment demonstrated that only 25% of the $Six5^{+/-}$ mice could continue the test for longer than 9 min, whereas 86% of the control mice could last longer than 9 min. However, P–R intervals remained within normal limits and no arrhythmias were provoked by exercise. Thus, these data demonstrate that the reduced exercise ability demonstrated by $Six5^{+/-}$ mice may result from mild heart failure. We cannot, however, rule out that skeletal muscle defects could also contribute to the reduced exercise ability, as skeletal muscle function was not studied at 15–16 months of age in $Six5^{+/-}$ animals. These data support the following points:

1. In humans, a prolonged H–V interval has been demonstrated in 54% of adult myotonic dystrophy patients [120]. It is has also been demonstrated that

prolonged QRS and abnormal H–V interval are strongly correlated in DM1 patients [120]. It is therefore conceivable that *Six5* deficits contribute to this feature of DM1. It is less likely that the ~25% decrease in *Dmpk* levels, which is observed in *Six5*$^{+/-}$ mice, is responsible for development of the prolonged H–V interval, as infrahisian delays have only been recorded by us in *Dmpk*$^{-/-}$ mice.

2. Ventricular hypertrophy is observed in DM1 patients [23–27]; however, it occurs at lower frequency when compared with the incidence of conduction disease. It is likely that SIX5 loss contributes to this feature of DM1 cardiac disease. Our data do not rule out that ventricular hypertrophy may underlie the mild His–Purkinje delay observed in *Six5*$^{+/-}$ animals.

E. *Six5* Deficiency Results in Nuclear Cataracts

Slit-lamp examination of wild-type, *Six5*$^{+/-}$ and *Six5*$^{-/-}$ mice on a *129Sv* background, at 3 and 8 weeks of age, demonstrated the presence of lenticular cataracts. We graded lens opacities as small, intermediate, or advanced, when they obscured less than 10%, 10–50%, or greater than 50% of the visual axis, respectively. In our study, at 3 weeks of age, ~80 and ~20% of the *Six5*$^{-/-}$ mice demonstrated either intermediate or advanced opacities. At this time point, ~10, 20, and 30% of the *Six5*$^{+/-}$ showed small, intermediate, or advanced opacities, respectively. At 8 weeks, the percentage of advanced opacities increased in *Six5*$^{-/-}$ mice to 40%. Similarly, the severity and incidence of lens opacities increased in *Six5*$^{+/-}$ mice, such that 20, 30, and 50% of the mice showed small, intermediate, and advanced opacities, respectively. Wild-type animals did not show any lens defects at either time point. Thus, both the incidence and severity of cataracts increase as a function of both decreasing *Six5* dosage and increasing age ($P < 0.001$, two-sided Fisher exact test) [115].

When ocular sections were stained with hematoxylin and eosin to examine the histology of the lens, we observed that tissue destruction originated in the nucleus of the lens, spreading outward toward the cortex, as cataract formation progressed temporally. As the mitotic index of the lens epithelial layer was unaltered in *Six5*$^{+/-}$ and *Six5*$^{-/-}$ mice, progressive changes in the lens fibers may account for cataract formation in *Six5* mutant mice. *Six5*$^{+/-}$ and *Six5*$^{-/-}$ mice were not microphthalmic, and both the structural integrity and functional integrity of the retina were preserved in these animals [115].

Previous studies by Kawakami and colleagues have demonstrated the binding of SIX5 to the Na, K-ATPase α1 gene [121]. Thus, we tested whether steady-state levels of Na,K-ATPase α1 RNA were altered in *Six5*$^{+/-}$ and *Six5*$^{-/-}$ eyes. We observe that steady-state levels of Na,K-ATPase α1 RNA increase as a function of decreasing Six5 dosage. Therefore it is possible that abnormal ion homeostasis could contribute to the progressive breakdown of the lens tissue in *Six5*$^{+/-}$ and *Six5*$^{-/-}$ animals.

Six5 mutant mice developed in the Tapscott laboratory were studied on a *C57/BL/6/129Sv* background. These animals demonstrate a milder phenotype, where lens opacities were observed primarily in *Six5*$^{-/-}$ mice at 8 to 10 months of age. Alterations in Na,K-ATPase α1 RNA were also not detected in this mouse strain [116]. It is currently unclear why these differences in the severity of cataract formation are observed; one possible contributing factor could be the differences in the mouse backgrounds used in the two studies.

The cataracts observed in *Six5* mutant animals are nuclear in origin and do not resemble the subcapsular iridescent cataracts observed in adult-onset DM1. Recent studies by the Swanson laboratory demonstrate the formation of dustlike subcapsular cataracts in mice lacking *Mbnl1* [73]. Thus, sequestration of the muscleblind proteins may play the primary role in cataract formation in adult-onset DM1. However, as congenital DM1 can be associated with congenital cataracts [16], it is possible that *Six5* loss may contribute to this feature of DM1.

F. *Six5* Loss Results in Elevated FSH Levels, Testicular Atrophy, Leydig Cell Hyperproliferation, and Aberrant Spermiogenesis

Six5$^{-/-}$ mice are sterile and demonstrate progressive testicular atrophy. Specifically, testis size is normal at birth; by 12 weeks of age, however, the average size of the *Six5*$^{-/-}$ testis is ~30% that of wild-type controls [122]. To establish the mechanism that underlies this striking loss of testicular tissue and sterility, we studied the development and maintenance of the testis in wild-type, *Six5*$^{+/-}$, and *Six5*$^{-/-}$ animals. Consistent with the normal size of the testis at birth, no significant differences were observed in testis development during embryogenesis in *Six5*$^{+/-}$ and *Six5*$^{-/-}$ mice.

In wild-type animals, spermatogonia continue to undergo mitotic divisions for ~10 days after birth, at which time meiosis commences. Formation of haploid spermatids occurs between 2 and 3 weeks of age in the normal mouse testis. In wild-type mice, the first cycle of spermatogenesis is completed at ~6 weeks of age, when terminally differentiated spermatozoa are released from the Sertoli cells into the lumen of the seminiferous tubules. To study the integrity of this series of events we examined wild-type, $Six5^{+/-}$, and $Six5^{-/-}$ testis sections at 2 and 6 weeks of age. At 2 weeks, when meiosis has just commenced, histological observation of testis sections showed ~5 fold and ~25 fold increases in apoptotic cell death in $Six5^{+/-}$ and $Six5^{-/-}$ testis when compared with controls ($P = 0.004$ for a three-way comparison between wild type, $Six5^{+/-}$, and $Six5^{-/-}$ mice). Electron microscopic analyses at this time demonstrated that while although Sertoli cells were spared, cells of the spermatogenic series were selectively destroyed in $Six5^{+/-}$ and $Six5^{-/-}$ testis. At 6 weeks of age, when the first cycle of spermatogenesis is complete, spermatozoa were clearly visible in wild-type testes but were conspicuously absent in $Six5^{-/-}$ testes. However, FACS analyses of $Six5^{-/-}$ testicular cells at 2 and 12 weeks of age demonstrated the presence of haploid cells in $Six5^{-/-}$ testis. Taken together, these data demonstrate that $Six5$ is required for both spermatogenic cell viability and the successful completion of spermiogenesis, a process by which haploid spermatids develop into mature spermatozoa.

$Six5^{+/-}$ mice demonstrate an intermediate phenotype and show oligozoospermia. Specifically, sperm counts in $Six5^{+/-}$ testis were ~60% of that observed in wild-type controls ($P = 0.001$ for a three-way comparison between wild-type, $Six5^{+/-}$, and $Six5^{-/-}$ mice; $P = 0.03$ for a two-way comparison between wild-type and $Six5^{+/-}$ mice). Tubular atrophy was observed to increase as a function of decreasing $Six5$ dosage. Progressive Leydig cell hyperproliferation was observed in $Six5^{-/-}$ mice, and by 10 months of age, the intertubular spaces are filled with Leydig cells in $Six5^{-/-}$ testes.

Significantly, FSH levels were elevated ~1.5- and ~2-fold in $Six5^{+/-}$ and $Six5^{-/-}$ mice when compared with controls ($P = 0.03$ for a three-way comparison between wild type, $Six5^{+/-}$, and $Six5^{-/-}$ mice). However, both serum testosterone levels and inhibin α and βB RNA levels, within the testis, were unaltered in $Six5$ mutant mice. As testosterone and inhibin B, which a dimer of inhibin α and inhibin βB, feed back to negatively regulate FSH secretion by the pituitary, these data suggest that $Six5$ loss may result in pituitary dysfunction.

As Sertoli cell–germ cell signaling plays a key role in germ cell viability, we counted the numbers of terminally differentiated Sertoli cells in wild-type and $Six5$ mutant testes. Terminally differentiated Sertoli cell numbers were slightly lower in $Six5^{-/-}$ testis when compared with controls. However, this decrease in Sertoli cell number was not sufficient to explain the complete absence of spermatozoa in $Six5^{-/-}$ testis. Thus, it is likely that functional abnormalities in Sertoli cells or germ cells may play a significant role in the increased germ cell death and the absence of spermiogenesis observed in $Six5^{-/-}$ testis. To test if Sertoli cell-specific expression of paracrine factors that act on germ cells is altered, we studied the expression of desert hedgehog (Dhh) and SCF, which are required for germ cell viability and normal spermatogenesis [123, 124]. However, neither Dhh levels nor SCF levels were significantly altered in $Six5^{+/-}$ and $Six5^{-/-}$ testis. We therefore assessed germ cell integrity by studying the expression of c-Kit, which is the receptor for SCF [125]. As c-Kit levels were decreased in $Six5^{-/-}$ testis, these data suggest that abnormal signaling between germ cells and Sertoli cells may play a significant role in the increased germ cell death observed in $Six5$ mutant testis.

Six5 mutant mice developed by the Tapscott laboratory do not demonstrate gonadal dysfunction [116]. It is unclear if a subtle testicular defect manifests in these animals as detailed analyses of the gonads were not carried out. As noted earlier, mouse background strain differences could account for these phenotypic differences.

Drosophila $Six4$ (d-$Six4$) is homologous to $Six5$. Loss of d-$Six4$ has been shown to result in infertility, testicular atrophy, and gamete loss. Consistent with our results, abnormal gamete–soma interactions have been hypothesized to underlie the aberrant gametogenesis in d-$Six4$ mutant flies [126].

These studies, taken together, suggest that SIX5 loss may contribute to the elevated FSH levels and testicular atrophy observed in DM1 patients. Decreased SIX5 levels may also increase the incidence of ventricular hypertrophy, infrahisian conduction blocks, and development of congenital cataracts in DM1 (Fig. 4-7).

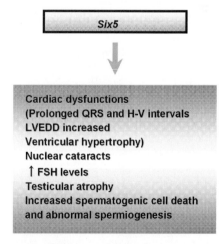

FIGURE 4-7 Contribution of $Six5$ loss to DM1 pathology. See CD-ROM for color image.

XII. POSSIBLE CONTRIBUTION OF OTHER *CIS* EFFECTS AT THE DM1 LOCUS

The boundaries to which heterochromatin spreads when the CTG repeat tract expands in size are unknown. It is currently also unclear if a functional relationship exists between the CTG repeat tract size and the distance to which heterochromatin spreads. It is also of interest to determine if CCTG expansions have similar effects on hetrochromatin formation on chromosome 3q21.

XIII. CONCLUDING REMARKS

Recognition that both *cis* effects resulting from CTG expansion and dominant RNA effects of CUG repeat expression contribute in varying degrees to the pathophysiology of DM1 provides important insights into the molecular mechanisms that govern both the multitude of symptoms and the variable expressivity of DM1. Importantly, expansion of the CTG repeat tract appears to cause stochastic changes in the levels of gene expression at the DM1 locus. Furthermore, the dysregulation of physiologically important RNA processing proteins by the expression of expanded CUG repeats may also demonstrate probabilistic behavior, rather than all-or-none effects. Thus, a summation of several stochastic events that occur as a consequence of CTG expansion can explain both the pleiomorphic expression and enormous variability that characterize DM1 (Fig. 4-8). Understanding the nature of the molecular defects that contribute to the unique set of symptoms that manifest in this fascinating disorder will provide a rational framework both for the clinical management of the symptoms and, ultimately, for the design of a therapy for DM1.

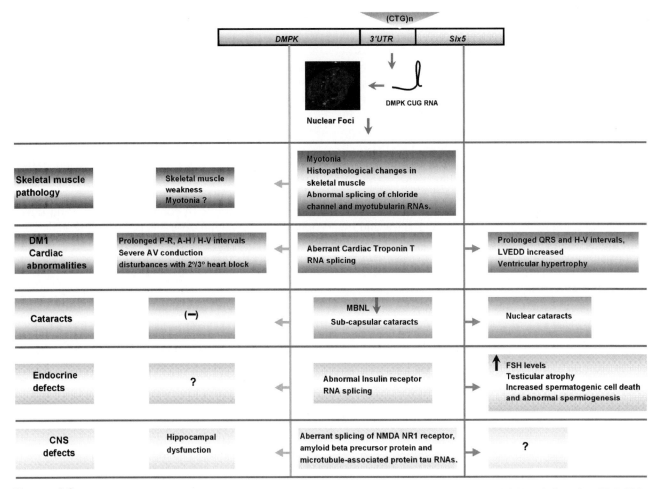

FIGURE 4-8 Relative contribution of *cis* effects of CTG expansion and the dominant RNA mechanism to DM1 pathology. See CD-ROM for color image.

References

1. Steinert, H. (1909). Myopathologische Beitrage1. Uber das klinischeund anatomische Bild des Muskelschwunds der Myotoniker. *Dtsch. Z. Nervenheilkd.* **37**, 58–104.
2. Batten, F. E., and Gibb, H. P. (1909). Myotonia atrophica. *Brain* **32**, 187–205.
3. Harper, P. S. (1989). "Myotonic Dystrophy." 2nd ed., Saunders, Philadelphia.
4. Harper, P. S. (2001). "Myotonic dystrophy," Vol. 37. Saunders, London.
5. Roses, A. D., Harper, P. S., and Bossen, E. H. (1979). Myotonic muscular dystrophy. *In* "The Handbook of Clinical Neurology" (P. J. Vinken and G. W. Bruyn, Eds.), Ch. 13, pp. 485–532. North-Holland, Amsterdam.
6. Fleischer, B. (1917). Uber myotonische Dystrophie. *Munch. Med. Wochenschr.* **64**, 1630.
7. Howeler, C. J., Busch, H. F., Geraedts, J. P., Niermeijer, M. F., and Staal, A. (1989). Anticipation in myotonic dystrophy: Fact or fiction? *Brain* **112**, 779–797.
8. Brook, J. D., McCurrach, M. E., Harley, H. G., Buckler, A. J., Church, D., Aburtani, H., Hunter, K., Stanton, V. P., Thirion, J.-P., Hudson, T., Sohn, R., Zemelman, B., Snell, R. G., Rundle, S. A., Crow, S., Davies, J., Selbourne, P. J. B., Jones, C., Juxonen, V., Johnson, K., Harper, P. S., Duncan, D. J., and Housman, D. E. (1992). Molecular basis of myotonic dystrophy: Expansion of a trinucleotide (CTG) repeat at the 3' end of a transcript encoding a protein kinase family member. *Cell* **68**, 799–808.
9. Fu, Y.-H., Pizutti, A., Fenwick, R. G., King, J., Rajnarayan, S., Dunne, P. W., Dubel, J., Nasser, G. A., Ashizawa, T., De Jong, P., Wieringa, B., Korneluk, R., Perryman, M. B., Epstein, H. F., and Caskey, C. T. (1992). An unstable triplet repeat in a gene related to myotonic muscular dystrophy. *Science* **255**, 1256–1258.
10. Mahadevan, M., Tsilfidis, C., Sabourin, L., Shutler, G., Amemiya, C., Jansen, G., Neville, C., Narang, M., Barcelo, J., O'Hoy, K., Leblond, S., Earle-Macdonald, J., de Jong, P. J., Wieringa, B., and Korneluk, R. G. (1992). Myotonic dystrophy mutation: An unstable CTG repeat in the 3' untranslated region of the gene. *Science* **255**, 1253–1255.
11. Boucher, C. A., King, S. K., Carey, N., Krahe, R., Winchester, C. L., Rahman, S., Creavin, T., Meghji, P., Bailey, M.E.S., and Chartier, F. L. (1995). A novel homeodomain-encoding gene is associated with a large CpG island interrupted by the myotonic dystrophy unstable (CTG)n repeat. *Hum. Mol. Genet.* **4**, 1919–1925.
12. Barcelo, J. M., Mahadevan, M. S., Tsilfidis, C., MacKenzie, A. E., and Korneluk, R. G. (1993). Intergenerational stability of the myotonic dystrophy protomutation. *Hum. Mol. Genet.* **2**, 705–709.
13. Harley, H., Rundle, S. A., MacMillan, J. C., Myring, J., Brook, J. D., Crow, S., Reardon, W., Fenton, I., Shaw, D. J., and Harper, P. S. (1993). Size of the unstable CTG repeat sequence in relation to phenotype and parental transmission in myotonic dystrophy. *Am. J. Hum. Genet.* **52**, 1164–1174.
14. Sarkar, P. S., Chang, H-C, Boudi, B. F., and Reddy, S. (1998). CTG repeats show bimodal amplification in *E.coli*. *Cell* **95**, 531–540.
15. Machuca-Tzili, L, Brook, D., and Hilton-Jones, D. (2005). Clinical and molecular aspects of the myotonic dystrophies: A review. *Muscle Nerve* **32**, 1–18.
16. Sarnat, H. B., O'Connor, T., and Byrne, P. A. (1976). Clinical effects of myotonic dystrophy on pregnancy and the neonate. *Arch. Neurol.* **7**, 459–465.
17. Fragola, P. V., Luzi, M., Calo, L., Antonini, G., Borzi, M., Frongillo, D., and Cannata, D. (1994). Cardiac involvement in myotonic dystrophy. *Am. J. Cardiol.* **74**, 1070–1072.
18. Hawley, R. J., Milner, M. R., Gottdiener, J. S., and Cohen, A. (1991). Myotonic heart disease: A clinical follow-up. *Neurology* **41**, 259–262.
19. Melacini, P., Villanova, C., Menegazzo, E., Novelli, G., Danieli, G., Rizzoli, G., Fasoli, G., Angelini, C., Buja, G., Miorelli, M., Dallapiccola, B., and Dalla Volta, S. (1995). Correlation between cardiac involvement and CTG trinucleotide repeat length in myotonic dystrophy. *J. Am. Coll. Cardiol.* **25**, 239–245.
20. Tokgozoglu, L. S., Ashizawa, T., Pacifico, A., Armstrong, R. M., Epstein, H. F., and Zoghbi, W. A. (1995). Cardiac involvement in a large kindred with myotonic dystrophy. *JAMA* **274**, 813–819.
21. Forsberg, H., Olofsson, B-O, Eriksson, A., and Andersson, S. (1990). Cardiac involvement in congenital myotonic dystrophy. *Br. Heart J.* **63**, 119–121.
22. Sonaglioni, G., Curatola, L., Bollettini, G., Agostini, L., Franchetta, G., Palestini, N., Pignotti, M., and Floris, B. (1984). Echocardiographic findings in dystrophia myotonica (Steinert's disease). *G. Ital. Cardiol.* **14**, 551–556.
23. Lin, A. E., Mitchell, F. M., Fitz, R. W., and Doyle, J. J. (1989). Dilated cardiomyopathy in myotonic dystrophy. *J. Am. Coll. Cardiol.* **1**, 262–263.
24. Premawardhana, L. D., and Thirunavakarasu, G. (1992). Myotonia dystrophica: First presentation as severe left ventricular failure complicating dilated cardiomypathy. *Postgrad. Med. J.* **795**, 67.
25. Fall, F. H., Young, W. W., Power, J. A., Faulkner, C. S., Hettleman, B. D., and Robb, J. F. (1990). Severe congestive heart failure and cardiomyopathy as a complication of myotonic dystrophy in pregnancy. *Obstet. Gynecol.* **76**, 481–485.
26. Pentimone, F., Del Corso, L., Vannini, A., Mori, L., and Moruzzo, D. (1990). Dilated cardiomyopathy and visceral anomalies in myotonic dystrophy. *Minerva Cardioangiol.* **38**, 231–234.
27. Igarashi, H., Momoi, M. Y., Yamagata, T., Shiraishi, H., and Eguchi, I. (1998). Hypertrophic cardiomyopathy in congenital myotonic dystrophy. *Pediatr. Neurol.* **18**, 366–369.
28. Calderon, R. (1966). Myotonic dystrophy: A neglected cause of mental retardation. *J Pediatr.* **68**, 423–431.
29. Palmer, B. W., Boone, K. B., Chang, L., Lee, A., and Black, S. (1994). Cognitive deficits and personality patterns in maternally versus paternally inherited myotonic dystrophy. *J. Clin Exp Neuropsychol.* **16**, 784–795.
30. Colombo, G., Perini, G. I., Miotti, M. V., Armani, M., and Angelini, C. (1992). Cognitive and psychiatric evaluation of 40 patients with myotonic dystrophy. *Ital. J. Neurol. Sci.* **13**, 53–58.
31. Censori, B., Danni, M., Del Pesce, M., and Provinciali, L. (1990). Neuropsychological profile in myotonic dystrophy. *J. Neurol.* **237**, 251–256.
32. Turnpenny, P., Clark, C., and Kelly, K. (1994). Intelligence quotient profile in myotonic dystrophy, intergenerational deficit, and correlation with CTG amplification. *J. Med. Genet.* **31**, 300–305.
33. Goossens, E., Steyaert, J., De Die-Smulders, C., Willekens, D., and Fryns, J. P. (2000). Emotional and behavioral profile and child psychiatric diagnosis in the childhood type of myotonic dystrophy. *Genet. Couns.* **11**, 317–327.
34. Giubilei, F., Antonini, G., Bastianello, S., Morino, S., Paolillo, A., Fiorelli, M., Ferretti, C., and Fieschi, C. (1999). Excessive daytime sleepiness in myotonic dystrophy. *J. Neurol Sci.* **164**, 60–63.
35. Garcia-Alix, A., Cabanas, F., Morales, C., Pellicer, A., Echevarria, J., Paisan, L., and Ouero, J. (1991). Cerebral abnormalities in congenital myotonic dystrophy. *Pediatr. Neurol.* **1**, 28–32.
36. Ono, S., Kanda, F., Takahashi, K., Fukuoka, Y., Jinnai, K., Kurisaki, H., Mitake, S., Inagaki, T., and Nagao, K. (1995). Neuronal cell loss in the dorsal raphe nucleus and the superior central nucleus in myotonic dystrophy: A clinicopathological correlation. *Acta Neuropathol. (Berlin)* **89**, 122–125.

37. Ono, S., Kanda, F., Takahashi, K., Fukuoka, Y., Jinnai, K., Kurisaki, H., Mitake, S., Inagaki, T., and Nagao, K. (1996). Neuronal loss in the medullary reticular formation in myotonic dystrophy: A clinicopathological study. *Neurology* **46**, 228–231.
38. Ono, S., Takahashi, K., Kanda, F., Jinnai, K., Fukuoka, Y., Mitake, S., Inagaki, T., Kurisaki, H., Nagao, K., and Shimizu, N. (2001). Decrease of neurons in the medullary arcuate nucleus in myotonic dystrophy. *Acta Neuropathol. (Berlin)* **102**, 89–93.
39. Di Costanzo, A., Di Salle, F., Santoro, L., Bonavita, V., and Tedeschi, G. (2002). Brain MRI features of congenital and adult-form myotonic dystrophy type 1: Case–control study. *Neuromuscul. Disord.* **12**, 476–483.
40. Kiuchi, A., Otsuka, N., Namba, Y., Nakano, I., and Tomonaga, M. (1991). Presenile appearance of abundant Alzheimer's neurofibrillary tangles without senile plaques in the brain in myotonic dystrophy. *Acta Neuropathol. (Berlin)* **1**, 1–5.
41. Yoshimura, N., Otake, M., Igarashi, K., Matsunaga, M., Takebe, K., and Kudo, H. (1990). Topography of Alzheimer's neurofibrillary change distribution in myotonic dystrophy. *Clin. Neuropathol.* **5**, 234–239.
42. Yoshimura, N. (1990). Alzheimer's neurofibrillary changes in the olfactory bulb in myotonic dystrophy. *Clin. Neuropathol.* **9**, 240–243.
43. Ricker, K., Koch, M. C., Lehmann-Horn, F., Pongratz, D., Otto, M., and Heine, R. (1994). Proximal myotonic myopathy: A new dominant disorder with myotonia, muscle weakness and cataracts. *Neurology* **44**, 1448–1452.
44. Ricker, K., Koch, M. C., Lehmann-Horn, F. Pongratz, D., Speich, N., and Reiners, K. (1995). Proximal myotonic myopathy: Clinical features of amultisystem disorder similar to myotonic dystrophy. *Arch. Neurol.* **52**, 25–31.
45. Moxley, R. T., III. (1996). Proximal myotonic myopathy: Mini-review of a recently delineated clinical disorder. *Neuromuscul. Disord.* **6**, 87–93.
46. Day, J. W., Ricker, K., Jacobsen, J. F., Rasmussen, L. J., Dick, K. A., Kress, W., Schneider, C., Koch, M. C., Beilman, G. J., Harrison, A. R., Dalton, J. C., and Ranum L. P. W. (2003). Myotonic dystrophy type 2. *Neurology* **60**, 657–664.
47. Liquori, C. L., Ricker, K., Moseley, M. L., Jacobsen, J. F., Kress, W., Naylor, S. L., Day, J. W., and Ranum, L. P. (2001). Myotonic dystrophy type 2 caused by a CCTG expansion in intron 1 of ZNF9. *Science* **293**, 864–867.
48. Day, J. W., and Ranum, L. P. W. (2005). RNA pathogenesis of the myotonic dystrophies. *Neuromuscul. Disord.* **15**, 5–16.
49. Meola, G., Sansone, V., Marinou, K., Cotelli, M., Moxley, R.T., III., Thornton, C. A., and De Ambroggi, L. (2002). Proximal myotonic myopathy: A syndrome with favourable prognosis? *J. Neurol Sci.* **193**, 89–96.
50. Hund, E., Jansen, O., Koch, M. C., Ricker, K., Fogel, W., Niedermaier, N., Otto, M., Kuhn, E., and Meinck, H. M. (1997). Proximal myotonic myopathy with MRI white matter abnormalities of the brain. *Neurology* **48**, 33–37.
51. Mankodi, A., Logigian, E., Callahan, L., McClain, C., White, R., Henderson, D., Krym, M., and Thornton, C. A. (2000). Myotonic dystrophy in transgenic mice expressing an expanded CUG repeat. *Science* **289**, 1769–1773.
52. Seznec, H., Agbulut, O., Sergeant, N., Savouret, C., Ghestem, A., Tabti, N., Willer, J. C., Ourth, L., Duros, C., Brisson, E., Fouquet, C., Butler-Browne, G., Delacourte, A., Junien, C., and Gourdon, G. (2001). Mice transgenic for the human myotonic dystrophy region with expanded CTG repeats display muscular and brain abnormalities. *Hum. Mol. Genet.* **10**, 2717–2726.
53. Caskey, C. T., Swanson, M. S., and Timchenko, L. T. (1996). Myotonic dystrophy: Discussion of molecular mechanism. *Cold Spring Harb. Symp. Quant. Biol.* **61**, 607–614.
54. Taneja, K. L., McCurrach, M. E., Shalling, M., Housman, D., and Singer, R. (1995). Foci of trinucleotide repeat transcripts in nuclei of myotonic dystrophy cells and tissues. *J. Cell Biol.* **128**, 995–1002.
55. Mariappan, S. V., Garcoa, A. E., and Gupta, G. (1996). Structure and dynamics of the DNA hairpins formed by tandemly repeated CTG triplets associated with myotonic dystrophy. *Nucleic Acids Res.* **15**, 775–783.
56. Timchenko, L. T., Miller, J. W., Timchenko, N. A., DeVore, D. R., Datar, K. V., Lin, L., Roberts, R., Caskey, C. T., and Swanson, M. S. (1996). Identification of a (CUG)n triplet repeat RNA-binding protein and its expression in myotonic dystrophy. *Nucleic Acids Res.* **24**, 4407–4414.
57. Michalowski, S., Miller, J. W., Urbinati, C. R., Paliouras, M., Swanson, M. S., and Griffith, J. (1999). Visualization of double-stranded RNAs from the myotonic dystrophy protein kinase gene and interactions with CUG-binding protein. *Nucleic Acids Res.* **27**, 3534–3542.
58. Fardaei, M., Rogers, M. T., Thorpe, H. M., Larkin, K., Hamshere, M. G., Harper, P. S., and Brook, J. D. (2002). Three proteins, MBNL, MBLL and MBXL, co-localize in vivo with nuclear foci of expanded-repeat transcripts in DM1 and DM2 cells. *Hum. Mol. Genet.* **11**, 805–814.
59. Timchenko N. A., Cai, Z. J., Welm, A. L., Reddy, S., Ashizawa, T., and Timchenko, L. T. (2001). RNA CUG repeats sequester CUG-BP1 and alter protein levels and activity of CUG-BP1. *J. Biol. Chem.* **276**, 7820–7826.
60. Miller J. W., Urbinati, C. R., Teng-Umnuay, P., Stenberg, M. G., Byrne, B. J., Thornton, C. A., and Swanson, M. S. (2000). Recruitment of human muscleblind proteins to (CUG)(n) expansions associated with myotonic dystrophy. *EMBO J.* **19**, 4439–4448.
61. Mankodi, A., Urbinati, C. R., Yuan, Q. P., Moxley R. T., Sansone, V., Krym, M., Henderson, D., Schalling, M., Swanson, M. S., and Thornton, C. A. (2001). Muscleblind localizes to nuclear foci of aberrant RNA in myotonic dystrophy types 1 and 2. *Hum. Mol. Genet.* **10**, 2165–2170.
62. Mankodi, A., Takahashi, M. P., Jiang, H., Beck, C. L., Bowers, W. J., Moxley, R. T., Cannon, S. C., and Thornton, C. A. (2002). Expanded CUG repeats trigger aberrant splicing of ClC-1 chloride channel pre-mRNA and hyperexcitability of skeletal muscle in myotonic dystrophy. *Mol. Cell.* **1**, 35–44.
63. Charlet, B. N., Savkur, R. S., Singh, G., Philips, A. V., Grice, E. A., and Cooper, T. A. (2002). Loss of the muscle-specific chloride channel in type I myotonic dystrophy due to misregulated alternative splicing. *Mol. Cell.* **10**, 45–53.
64. Philips, A. V., Timchenko, L. T., and Cooper, T. A. (1998). Disruption of splicing regulated by a CUG-binding protein in myotonic dystrophy. *Science* **280**, 737–741.
65. Savkur, R. S., Philips, A. V., and Cooper, T. A. (2001). Aberrant regulation of insulin receptor alternative splicing is associated with insulin resistance in myotonic dystrophy. *Nat. Genet.* **29**, 40–47.
66. Savkur, R. S, Philips, A. V, Cooper, T. A, Dalton, J. C, Moseley, M. L, Ranum L. P. W., and Day, J. W. (2004). Insulin receptor splicing alteration in Myotonic dystrophy type 2. *Am. J. Hum. Genet.* **74**, 1309–1313.
67. Buj-Bello, A., Furling, D., Tronchere, H., Laporte, J., Lerouge, T., Butler-Browne, G. S., and Mandel, J-L. (2002). Muscle-specific alternative splicing of myotubularin-related 1 gene is impaired in DM1 cells. *Hum. Mol. Genet.* **11**, 2297–2307.
68. Jiang, H., Mankodi, A., Swanson, M. S., Moxley, R. T., and Thornton, C. A. (2004). Myotonic dystrophy type 1 is associated with nuclear foci of mutant RNA, sequestration of muscleblind proteins and deregulated alternative splicing in neurons. *Hum. Mol. Genet.* **13**, 3079–3088.
69. Timchenko, N. A., Patel, R., Iakova, P., Cai, Z-J., Quan, L., and Timchenko L. T. (2004). Overexpression of CUG triplet repeat

binding protein, CUG-BP1, in mice inhibits myogenesis. *J. Biol. Chem.* **279**, 13129–13139.

70. Ho, T. H., Bundman, D., Armstrong, D. L., and Cooper, T. A. (2005). Transgenic mice expressing CUG-BP1 reproduce splicing mis-regulation observed in myotonic dystrophy. *Hum. Mol. Genet.* **14**, 1539–1547.

71. Ho, T. H., Charlet-B, N., Poulos, M. G., Singh, G., Swanson, M. S., and Cooper, T. A. (2004). Muscleblind proteins regulate alternative splicing. *EMBO. J.* **23**, 3103–3112.

72. Dansithong, W., Paul, S., Comai, L., and Reddy, S. (2005). MBNL1 is the primary determinant of focus formation and aberrant IR splicing in DM1. *J. Biol. Chem.* **280**, 5773–5780.

73. Kanadia, R. N., Johnstone, K. A., Mankodi, A., Lungu, C., Thornton, C. A., Esson, D., Timmers, A. M., Hauswirth, W. W., and Swanson, M. S. (2003). A muscleblind knockout model for myotonic dystrophy. *Science* **302**, 1978–1980.

74. Ebralidze, A., Wang, Y., Petkova, K., Ebralidse, R., and Junghans, R. P. (2004). RNA leaching of transcription factors disrupts transcription in myotonic dystrophy. *Science* **303**, 383–387.

75. Shimizu, K., Chen, W., Ashique, A. M., Moroi, R., and Li, Y-P. (2003). Molecular cloning, developmental expression, promoter analysis and functional characterization of the mouse CNBP gene. *Gene* **307**, 51–62.

76. Lam, L. T., Pham, Y. C. N., Man, N. T., and Morris, G. E. (2000). Characterization of the monoclonal antibody panel shows that the myotonic dystrophy protein kinase, DMPK, is expressed almost exlusively in muscle heart. *Hum. Mol. Genet.* **9**, 2167–2173.

77. Sarkar, P. S., Han, J., and Reddy, S. (2004). In situ hybridization analysis of Dmpk mRNA in adult mouse tissues. *Neuromuscul. Disord.* **14**, 497–506.

78. Alwazzan, M., Hamshere, M. G., Lennon, G. G., and Brook, J. D. (1998). Six transcripts map within 200 kilobases of the myotonic dystrophy expanded repeat. *Mamm. Genome* **9**, 485–487.

79. Fu, Y.-H., Friedman, D. L., Richards, S., Pearlman, J. A., Gibbs, R. A., Pizutti, A., Ashizawa, T., Perryman, M. B., Scarlato, G., Fenwick, R. G. J., and Caskey, C. T. (1993). Decreased expression of myotonin-protein kinase messenger RNA and protein in adult form of myotonic dystrophy. *Science* **260**, 235–237.

80. Wang, J., Pegoraro, E., Menegazzo, E., Gennarelli, M., Hoop, R. C., Angelini, C., and Hofmann, E. (1995). Myotonic dystrophy: Evidence for a possible dominant-negative RNA mutation. *Hum. Mol. Genet.* **4**, 599–606.

81. Wang, Y. H., Amirhaeri, S., Kang, S., Wells, R. D., Griffith, J. D. (1994). Preferential nucleosome assembly at DNA triplet repeats from myotonic dystrophy gene. *Science* **265**, 669–671.

82. Otten, A. D., and Tapscott, S. J. (1995). Triplet repeat expansion in myotonic dystrophy alters the adjacent chromatin structure. *Proc. Natl. Acad. Sci. USA* **92**, 5465–5469.

83. Klesert, T. R., Otten, A. D., Bird, T. D., and Tapscott, S. J. (1997). Trinucleotide repeat expansion at the myotonic dystrophy locus reduces expression of DMAHP. *Nat. Genet.* **16**, 402–407.

84. Thornton, C. A., Wymer, J. P., Simmons, Z., McClain, C., and Moxley, R. T., III. (1997). Expansion of the myotonic dystrophy CTG repeat reduces expression of the flanking DMAHP gene. *Nat. Genet.* **16**, 407–409.

85. Alwazzan, M., Newman, E., Hamshere, M. G., and Brook, J. D. (1999). Myotonic dystrophy is associated with a reduced level of RNA from the DMWD allele adjacent to the expanded repeat. *Hum. Mol. Genet.* **8**, 1491–1497.

86. Jansen, G., Groenen, P. J. T. A., Bachner, D., Jap, P. H. K., Coerwinkel, M., Oerlemans, F., Broek, W., Gohlsch, B., Pette, D., Plomp, J. J., Molenaar, P. C., Nederhoff, M. G. J., Echteld, C., Dekker, M., Berns, A., Hameister, H., and Wieringa, B. (1996). Abnormal myotonic dystrophy protein kinase levels produce only mild myopathy in mice. *Nat. Genet.* **13**, 423–442.

87. Zhao, Y., Loyer, P., Li, H., Valentine, V., Kidd, V., and Kraft, A. S. (1997). Cloning and chromosomal location of a novel member of the myotonic dystrophy family of protein kinases. *J. Biol. Chem.* **272**, 10013.

88. Shimizu, M., Wang, W., Walch, E. T., Dunne, P. W., and Epstein, H. F. (2000). Rac-1 and Raf-1 kinases, components of distinct signaling pathways, activate myotonic dystrophy protein kinase. *FEBS Lett.* **475**, 273–277.

89. Jin, S., Shimizu, M., Balasubramanyam, A., and Epstein H. F. (2000). Myotonic dystrophy protein kinase (DMPK) induces actin cytoskeletal reorganization and apoptotic-like blebbing in lens cells. *Cell Motil. Cytoskel.* **45**, 133–148.

90. Jansen, G., Mahadevan, M., Amemiya, C., Wormskamp, N., Segers, B., Hendriks, W., O'Hoy, K., Baird, S., Sabourin, L., Lennon, G., Jap, P. L., Iles, D., Coerwinkel, M., Hofker, M., Carrano, A. V., de Jong, P. J., Korneluk, R. G., and Wieringa, B. (1992). Characterization of the myotonic dystrophy region predicts multiple protein isoform-encoding mRNAs. *Nat. Genet.* **1**, 261–266.

91. Mahadevan, M. S., Amemiya, C., Jansen, G., Sabourin, L., Baird, S., Neville, C. E., Wormskamp, N., Segers, B., Batzer, M., Lamerdin, J., et al. (1993). Structure and genomic sequence of the myotonic dystrophy (DM kinase) gene. *Hum. Mol. Genet.* **2**, 299–304.

92. Kamps, M. P., Taylor, S. S., and Sefton, B. M. (1984). Direct evidence that oncogenic tyrosine kinases and cyclic AMP-dependent protein kinases have homologous ATP binding sites. *Nature* **310**, 589–592.

93. Hanks, S. K., Quinn, A. M., Hunter, T. (1988). The protein kinase family: Conserved features and deduced phylogeny of the catalytic domains. *Science* **241**, 42–52.

94. Reddy, S., Smith, D. B. J., Rich, M. M., Leferovich, J. M., Reilly, P., Davis, B. D., Tran, K., Rayrurn, H., Bronson, R., Cros, D., Balice-Gordon, R. J., and Housman, D. (1996). Mice lacking the myotonic dystrophy kinase develop a late onset myopathy. *Nat. Genet.* **13**, 423–442.

95. Benders, A. A., Groenen, P. J., Oerlemans, F. T., Veerkamp, J. H., and Wieringa B. (1997). Myotonic dystrophy protein kinase is involved in the modulation of the Ca^{2+} homeostasis in skeletal muscle cells. *J. Clin. Invest.* **100**, 1440–1447.

96. Jacobs, A. E. M., Benders, A. A., Oosterhof, A., Veerkamp, J. H., Van Mier, P, Wevers, R. A., and Joosten, E. M. G. (1990). The calcium homeostasis and membrane potential of cultured muscle cells from patients with myotonic dystrophy. *Biochim. Biophys. Acta* **1096**, 14–19.

97. Benders, A. A., Wevers, R. A., and Veerkamp, J. H. (1996). Ion transport I human skeletal muscle cells: Disturbances in myotonic dystrophy and Brody's disease. *Acta Physiol. Scand.* **156**, 355–367.

98. Rudel, R., Ruppersberg, J. P., and Spittelmeister, W. (1989). Abnormalities of the fast sodium current in myotonic dystrophy, recessive generalized myotonia and adynamia episodica. *Muscle Nerve* **12**, 281–287.

99. Franke, C., Hatt, H., Iaizzo, P. A., and Lehmann-Horn, F. (1990). Characterization of Na^+ channels and Cl^- conductance in resealed muscle fiber segments from patients with myotonic dystrophy. *J. Physiol. London* **425**, 391–405.

100. Mounsey, J. P., Mistry, D. J., Ai, C.W., Reddy, S., and Moorman, J. R. (2000). Skeletal muscle sodium channel gating in mice deficient in myotonic dystrophy protein kinase. *Hum. Mol. Genet.* **9**, 2313–2320.

101. Reddy, S., Mistry, D. J., Wang, Q. C., Geddis, L. M., Kutchai, H. C., Moorman, J. R., and Mounsey, J. P. (2002). Effects of age and gene dose on skeletal muscle sodium channel gating in mice deficient in myotonic dystrophy protein kinase. *Muscle Nerve*, **25**, 850–857.

102. Wissman, A., Ingles, J., McGee, J. D., and Mains, P. E. (1997). *Caenorhabditis elegans* LET-502 is related to Rho-binding kinases and human myotonic dystrophy kinase and interacts genetically with a homolog of the regulatory subunit of smooth muscle myosin phosphatase to affect cell shape. *Genes Dev.* **11**, 409–422.

103. Timchenko, L., Nastainczyk, W., Schneider, T., Patel, B., Hofman, F., and Caskey C. T. (1995). Full-length myotonin protein kinase (72 kDa) displays serine kinase activity. *Proc. Natl. Acad. Sci. USA* **92**, 5366–5370.

104. Mounsey, J. P., Xy, P., John, J. E., Horne, L. T., Gilbert, J., Roses, A. D., and Moorman, J. R. (1995). Modulation of skeletal muscle sodium channels by human myotonin protein kinase. *J. Clin. Invest.* **95**, 2379–2384.

105. Berul, C. I., Aronovitz, M. J., Saba, S., Housman, D., Mendelsohn, M., and Reddy, S. (1999). Atrioventricular conduction abnormalities are observed in mice lacking the myotonic dystrophy kinase. *J. Clin Invest.* **103**, R1–7.

106. Saba, S., VanderBrink, B. S., Luciano, B., Aronovitz, M. J., Berul, C. I., Reddy, S., Housman, D., Mendelsohn, M. E., Estes, N. A. M., and Wang, P. (1999). Localization of the site of conduction abnormality in a mouse model of myotonic dystrophy. *J. Cardiovasc. Electrophysiol.* **10**, 1214–1220.

107. Berul, C. I., Maguire, C. T., Gehrmann, J., Ai, C., and Reddy. S. (2000). Progressive atrioventricular conduction block in a mouse myotonic dystrophy model. *J. Intern. Cardiovasc. Electrophysiol.* **4**, 351–358.

108. Lee, H. C., Patel, M. K., Mistry, D. J., Wang, Q., Reddy, S., Moorman, J. R., and Mounsey, J. P. (2003). Abnormal Na channel gating in murine cardiac myocytes deficient in myotonic dystrophy protein kinase. *Physiol. Genom.* **12**, 147–157.

109. Kaliman, P., Catalucci, D., Lam, J. T., Kondo, R., Gutierrez, J. C. P., Reddy, S., Palacin, M., Zorzano, A., Chien, K. R., and Ruiz-Lozano, P. (2005). Myotonic dystrophy protein kinase phosphorylates phospholamban and regulates calcium uptake in cardiomyocte sarcoplasmic reticulum. *J. Biol. Chem.* **280**, 8016–8021.

110. Muranyi, A., Zhang, R., Liu, F., Hirano, K., Ito, M., Epstein, H. F., and Hartshorne, D. J. (2001). Myotonic dystrophy protein kinase phosphorylates the myosin phosphatase targeting subunit and inhibits myosin myosin phosphatase activity. *FEBS Lett.* **493**, 80–84.

111. Schulz, P. E., Mcintosh, A. D., Kasten, M. R., Wieringa, B., and Epstein, H. F. (2003). A role for myotonic dystrophy protein kinase in synaptic plasticity. *J. Neurophysiol.* **89**, 1177–1186.

112. Kawakami, K., Sato, S., Ozaki, H., and Ikeda, K. (2000). Six family genes-structure and function as transcription factors and their role in development. *Bioessays* **7**, 616–626.

113. Wallis, D. E., Roessler, E., Hehr, U., Nanni, L., Wiltshire, T., Richieri-Costa, A., Gillessen-Kaesbach, G., Zackai, E. H., Rommens, J., and Muenke, M. (1999). Mutations in the homeodomain of the human SIX3 gene cause holoprosencephaly. *Nat. Genet.* **22**, 196–198.

114. Gallardo, M. E., Lopez-Rios, J., Fernaud-Espinosa, I., Granadino, B., Sanz. R., Ramos, C., Ayuso, C., Seller, M. J., Brunner, H. G., Bovolenta, P., and Rodriguez, de Cordoba, S. (1999). Genomic cloning and characterization of the human homeobox gene SIX6 reveals a cluster of SIX genes in chromosome 14 and associates SIX6 hemizygosity with bilateral anophthalmia and pituitary anomalies. *Genomics* **61**, 82–91.

115. Sarkar, P. S., Appukuttan, B., Han, J., Ai, C., Tsai, W., Stout, J. T., and Reddy, S. (2000). Heterozygous loss of Six5 in mice is sufficient to cause ocular cataracts. *Nat. Genet.* **25**, 110–114.

116. Klesert, T. R., Cho, D. H., Clark, J. I., Maylie, J., Adelman, J., Snider, L., Yuen, E. C., Soriano, P., and Tapscott, S. J. (2000). Mice deficient in Six5 develop cataracts: Implications for myotonic dystrophy. *Nat. Genet.* **25**, 105–109.

117. Personius, K. E., Nautiyal, J., and Reddy, S. (2004). Loss of Six5 in mice is insufficient to cause myotonia or contractile abnormalities. *Muscle Nerve* **4**, 503–505.

118. Mistry, D. J., Moorman, J. R., Reddy, S., and Mounsey J. P. (2001). Skeletal muscle Na currents in mice heterozygous for Six5 deficiency. *Physiol. Genom.* **6**, 153–158.

119. Wakimoto, H., Maguire, C. T., Sherwood, M. C., Vargas, M. M., Sarkar, P. S., Han, J., Reddy, S., and Berul, C. I. (2002). Characterization of cardiac conduction system abnormalities in mice with targeted disruption of Six5 gene. *J. Interv Cardiovasc. Electrophysiol.* **7**, 127–135.

120. Babuty, D., Fauchier, L., Tena-Carbi, D., Poret, P., Leche, J., Raynaud, M., Fauchier, J. P., and Cosnay, P. (1999). Is it possible to identify infrahissian cardiac conduction abnormalities in myotonic dystrophy by non-invasive methods? *Heart* **82**, 634–637.

121. Ohto, H., Kamada, S., Tago, K., Tominaga, S. I., Ozaki, H., Sato, S., Kawakami, K. (1999). Cooperation of Six and Eya in activation of their target genes through nuclear translocation of Eya. *Mol. Cell. Biol.* **19**, 6815–6824.

122. Sarkar, P. S., Paul, S., Han, J., and Reddy, S. (2004). Six5 is required for spermatogenic cell survival and spermiogenesis. *Hum. Mol. Genet.* **13**, 1421–1431.

123. Bitgood, M. J., Shen, L., and McMahon, A. P. (1996). Sertoli cell signaling by desert hedgehog regulates the male germline. *Curr. Biol.* **6**, 298–304.

124. Loveland, K. L., and Schlatt, S. (1997). Stem cell factor and c-kit in the mammalian testis: Lessons originating from mother nature's gene knockouts. *J. Endocrinol.* **153**, 337–344.

125. Packer, A. L., Besmer, P., and Bachvarova, R. F. (1995). Kit ligand mediates survival of type A speramtogonia and dividing spermatocytes in postnatal mouse testis. *Mol. Reprod. Dev.* **42**, 303–310.

126. Kirby, R. J., Hamilton, G. M., Finnegan, D. J., Johnson, K. J., and Jarman, A. P. (2001). Drosophila homolog of the myotonic dystrophy-associated gene, SIX5, is required for muscle and gonad development. *Curr. Biol.* **11**, 1044–1049.

Normal and Pathophysiological Significance of Myotonic Dystrophy Protein Kinase

DERICK G. WANSINK, RENÉ E. M. A. VAN HERPEN, AND BÉ WIERINGA
Department of Cell Biology, Nijmegen Centre for Molecular Life Sciences, Radboud University Nijmegen Medical Centre, Nijmegen, The Netherlands

I. Introduction
II. Myotonic Dystrophy Protein Kinase
   A. Tissue Expression and *in Situ* Localization
   B. Alternative Splicing
   C. AGC Serine/Threonine Protein Kinase Group
III. The Role of Individual Protein Domains
   A. N-Terminal Leucine-Rich Domain
   B. Serine/Threonine Protein Kinase Domain
   C. VSGGG Motif
   D. Coiled-Coil Region
   E. C-Terminal Tails
IV. Substrates and Function
   A. DMPK and Ion Homeostasis
   B. DMPK and the Actomyosin Cytoskeleton
   C. SRF, CUG-BP, and MKBP
V. Transgenic Mice
   A. DMPK Knockout Mice
   B. Tg26-hDMPK
VI. Concluding Remarks
Acknowledgments
References

Currently, no single hypothesis can explain the wide variety and variability of symptoms involved in the complex clinical manifestation of myotonic dystrophy type 1 (DM1). Although there is strong support for a toxic-RNA gain-of-function effect of DMPK pre-mRNAs with repeat expansion, the possibility that an abnormal balance in protein products expressed from the mutant DMPK gene contributes to the DM1 phenotype should be investigated in more detail. We review here knowledge of DMPK biology, citing others' and our own work. DMPK is a member of the group of AGC kinases, which is mainly expressed in heart, skeletal and smooth muscle, and brain. Alternative splicing results in cell-type dependent expression of distinct isoforms, which partition in a species-dependent manner across cytosol, endoplasmic reticulum (ER), and the mitochondrial outer membrane (MOM). All DMPKs share a Leu-rich N-terminal leader segment, a typical kinase domain with Lys/Arg directed serine/threonine substrate specificity, and a coiled-coil domain involved in protein multimerization. Isoforms differ in absence/presence of an internal VSGGG motif and tails with motifs that serve as specific C-terminal anchors for ER and MOM. Cell biological and bioinformatic studies, including functional and structural homology comparison to related kinases, suggest that DMPKs may be regulators of mitochondrial dynamics, actin-cytoskeleton dynamics and ion homeostasis. Clearly, these are processes with specific relevance for muscle and brain, the main targets of disease in DM1 patients. Moreover, transgenic mouse and cell-model studies point out that strict control of DMPK expression is important for cellular viability and function.

I. INTRODUCTION

Myotonic dystrophy (DM) is known to occur as two distinct types of genetic disorders with remarkable similar symptomology but widely different prevalence, classified as DM type 1 (DM1, also known as Steinert's disease) and DM type 2 (DM2, also known as proximal myotonic myopathy, PROMM). The clinical manifestation of DM1 and DM2 is complex, and both disorders can be best characterized as multisystemic, neuromuscular diseases with problems involving skeletal muscle, heart, smooth muscle, central nervous system, eyes, and the endocrine system (summarized in Fig. 5-1) [1]. A very severe form, not found in DM2 families, is congenital DM1, characterized by high neonatal mortality, hypotonia, mental retardation, and respiratory distress [1].

Two different genetic mutations are associated with DM: an expanding $(CTG)_n$ repeat tract in the *DMPK*

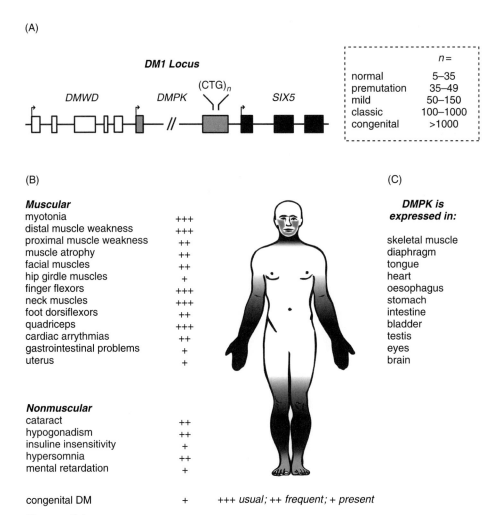

FIGURE 5-1 DM1 molecular pathogenesis: Organization of the gene locus, manifestation of symptoms, and tissue distribution of DMPK expression. (A) Schematic representation of the gene organization in the DM1 locus. The $(CTG)_n$ repeat tract is located in a gene-dense region on human chromosome 19q13, in exon 15 (i.e., the last exon) of the *DMPK* gene. This DNA segment also overlaps with the promoter region of the *SIX5* gene. *DMWD* is located upstream of the *DMPK* gene. Rectangles indicate exons: white, *DMWD* exons; gray *DMPK* exons (only exons 1 and 15 are indicated; see Fig. 4-2); black, *SIX5* exons. The straight line represents introns and intergenic sequences. Arrows denote start and direction of transcription for each gene. The disease classification with the corresponding length range for the somatic and intergenerational hypermutable $(CTG)_n$ repeat is indicated in the box. (B) Short summary of characteristic symptoms in DM1. The severity with which various organs and tissues (muscles) from various parts of the body were affected is indicated by + to +++. Typical skeletal muscle features are illustrated in the drawing; the darkest shading indicates the most severely affected muscle areas. (C) Global tissue type distribution of DMPK expression; data are from human and mouse studies (see text for details).

gene, mapped on chromosome 19q13, giving rise to DM1 (Fig. 5-1A) (MIM 160900) [2–4], and an unstable $(CCTG)_n$ repeat in the *ZNF9* gene on chromosome 3q21, which results in DM2 (MIM 602668) [5, 6]. In this chapter, we concentrate on DM1 as the most abundant form of DM and one of the most frequent genetic disorders in adults. After a brief summary of current knowledge about the pathophysiology of DM1, an overview is given on existing knowledge and new data regarding the protein products of the *DMPK* gene.

In 1991–1992, it was discovered simultaneously by several groups that the mutation in DM1, that is, the unstable $(CTG)_n$ repeat, is located in the 3′ untranslated region of the *DMPK* gene and, hence, does not encode any protein information, unlike, for example, $(CAG)_n$ repeats in polyglutamine diseases and $(GCN)_n$ repeats in polyalanine diseases [7, 8]. Soon thereafter, it was found that the $(CTG)_n$ repeat is also located close to or even in the promoter region of the downstream neighboring *SIX5* gene (Fig. 5-1A). Experimental data on mRNA and protein levels, although still somewhat equivocal, strongly suggested that expansion of the $(CTG)_n$ tract leads to haploinsufficiency of both *DMPK* and *SIX5* (summarized in [9]). Gene knockout mouse models for *DMPK* or *SIX5*, however, displayed only mild myopathies, cardiac defects, cataract, mild memory and cognition abnormalities, or defects in spermatogenesis [10–17], but certainly not the complex and highly variable multisystemic features seen in DM1 patients. Also, in transgenic DMPK overexpressor mice, the disease characteristics were not fully reproduced, albeit that similarities in heart problems were seen [12, 18]. This suggested that DM1 is not caused by a simple loss or gain of *DMPK* or *SIX5* function alone, and led to the understanding that the mere presence of the expanded $(CTG)_n$ tract itself might be crucial and could cause the pleiotropic toxic effects in patients. Although most people in the field were well aware that findings in genetically "homogeneous" animal models cannot always be reliably used for explanation of disease etiology in humans [e.g., 19, 20], the hypothesis about the direct involvement of repeats was given further support by an elegant transgenic study with mice bearing a transgene composed of a $(CTG)_n$ element driven from the actin promoter [21], and by several studies at the cellular level (summarized in [22]).

The currently favored explanation for the molecular pathophysiology of DM1 is a toxic gain of function at the RNA level, whereby long $(CUG)_n$ tracts in DMPK transcripts perturb RNA metabolism in the nucleus, including altered transcription and alternative modes of splicing (reviewed in [6]). After binding to members of the muscleblind (MBNL) family of splicing factors, DMPK transcripts carrying long $(CUG)_n$ repeats are retained in ribonuclear inclusions, seen as aggregates in nuclei of cultivated muscle cells from DM1 patients or directly in muscle or brain tissue material [23–26]. Moreover, an interaction of $(CUG)_n$ tracts with the CUG-binding protein (CUG-BP) has been proposed [27–29]. CUG-BP is a splicing factor and serves as modulator of alternative splice events with properties that act antagonistically to MBNL function [30].

What remains unexplained is how DMPK $(CUG)_n$ mRNA protein inclusions can cause both the change in cell function (i.e., myotonia, insulin resistance) and the tissue wasting and cell loss phenomena (muscle degeneration, brain white matter loss) of the disease. Also, the dosage effects of repeat length have not been clarified. Current thoughts are that $(CUG)_n$ RNA protein aggregates could be globally cytotoxic as in protein folding disorders, involving the ubiquitin proteasome system [31]. Conversely, RNA protein aggregates could also cause changes in the activity of transcription and splice factors acting in dynamic domains or hubs of the nucleus, thus interfering *in trans* with production, processing, or transport of mRNAs for specific proteins like the insulin receptor, chloride channel, and troponin T, explaining distinct DM1 features like insulin resistance and contractile defects in the myotonia. Both modes of RNA-driven toxicity may also coexist.

As no single model explains the complete plethora of clinical phenotypes observed in DM1, we consider it important to remain focused also on the pathobiological contribution of the DMPK protein, that is, the direct product of the mutated gene. The biological function of DMPK has been a topic of study ever since the discovery of the expansion mutation [9, 32, 33]. Considerable progress has been made by (1) the recognition of DMPK as a distinct and interesting member of the AGC group of serine/threonine protein kinases, (2) the identification of targets for its catalytic activity, and (3) a better description of the biological significance of distinct domains in the protein. Here, we summarize knowledge about the normal structure–function relationships in DMPK and speculate on how the protein could be involved in DM1 pathophysiology.

II. MYOTONIC DYSTROPHY PROTEIN KINASE

A. Tissue Expression and *in Situ* Localization

Detailed knowledge on the tissue and cell type distribution of DMPK expression is important because it will help us to generate a clearer picture of the putative role for DMPK protein in DM1 pathogenesis. Furthermore, it helps to provide us with a better understanding of RNA dominance effects in different cell types, given that only those cells that produce mRNA or protein products from the DMPK gene suffer from the deleterious effects of long $(CUG)_n$ tracts. For obvious reasons, most DMPK expression data, based on either mRNA (i.e., Northern blots, *in situ* hybridization, or reverse transcription polymerase chain reaction) or protein detection, originate from studies of mouse tissues. However, we have no reason to assume that the tissue and cell type patterns of gene expression in mouse differ very much from those in humans (Fig. 5-1C).

Already during the early years of study, DMPK turned out to be a very difficult target for the production of polyclonal and monoclonal antibodies. With the use of either peptides or bacterially expressed DMPK protein as immunogen, many antibodies were produced against various parts of the protein, but almost invariably strong cross-reactivity toward other proteins was seen [32, 34, 35]. DMPK antibodies identified protein products varying from ~42 to 84 kDa in size [32, 34–37] and also DMPK homologs MRCKα and -β [34]. It is now generally recognized that the smaller 42- to 55-kDa proteins are not genuine products of the *DMPK* gene [34–36]. Because many expression and localization experiments were based on use of these first-generation antibodies, there is still considerable controversy about the true distribution and subcellular localization of DMPK. Caution should therefore be taken in the interpretation of early publications.

Overall, DMPK protein and mRNA expression appear strongly correlated, suggesting that transcriptional regulation is predominant. Highest levels of gene products are found in tissues containing smooth muscle cells, like stomach, bladder, and intestine [18, 38, 39; van Herpen et al., unpublished]. DMPK is also prominently expressed in heart—in both ventricles and atria—and in skeletal muscle, diaphragm, and tongue [18, 34, 36, 38]. Protein expression was found in slow and fast skeletal muscle types, in both type I and type II fibers [40–42; van Herpen et al., unpublished]. DMPK expression increased during myoblast differentiation *in vitro* [43]. DMPK is also identified in testis and several regions of the brain, but global levels in the latter tissue are intrinsically low [18, 38, 39]. Whether brain expression is confined to either neurons or glial cells, or even specialized cell derivatives thereof, is still not clearly resolved [44, 45] (note that antibodies used in these studies recognize predominantly ~50-kDa proteins). This is, however, a particularly interesting issue, and more knowledge is urgently needed to explain the central nervous system problems observed in DM1 patients, including the mental problems, character changes, hypersomnia, and progressive loss of white matter [1, 46, 47]. DMPK is also present in the eye, but expression in the lens is controversial [38, 48, 49]. Sarkar et al. report strong expression in the liver [38], but we have not been able to reproduce this finding [18].

On a subcellular level, DMPK has been localized to the sarcoplasmic reticulum (SR) and close to SERCA IIa ATPase located in the T tubules of type I skeletal muscle fibers [32]. DMPK was also seen together with the acetylcholine receptor at neuromuscular junctions [50, 51], although this observation could not be confirmed by others [35]. In cardiac tissue, DMPK was found localized to intercalated discs [32, 35, 50–52] and associated with gap junctions and the SR [32, 53]. In the rat central nervous system, DMPK was seen associated with the endoplasmic reticulum (ER) and dendritic microtubules within adult spinal motor neurons [45]. All in all, these studies using many cell types and different mammalian species did not yield a uniform picture of where DMPK is located. The fact that DMPK has been described as a membrane-bound protein [40, 52, 54, 55], but also as a soluble protein [35], added to the confusion. We know now that the reason for this lack of clarity may be that DMPK actually consists of a mixture of isoforms, each with a distinct location (see below) [36, 56, 57]. Furthermore, we also know now that mouse–human differences in location of individual DMPK isoforms do exist [56].

B. Alternative Splicing

The *DMPK* gene consists of 15 exons (Fig. 5-2A) and the corresponding primary transcripts are subject to extensive alternative splicing in both human and mouse (Fig. 5-2B) [36, 39, 58]. In both species, two constitutive alternative splice modes and one regulated alternative splice mode prevail and specify a total of six major isoforms (Fig. 5-2B) [36]. All DMPK protein isoforms have in common a leucine-rich N terminus, a serine/threonine protein kinase domain specified by exons 2 to 8, and an α-helical coiled-coil region encoded by exons 10 to 12 (Fig. 5-2C) [3, 36, 39].

An alternative 5′ splice site in exon 8 competes with the "normal" 5′ splice site of intron 8 and determines inclusion of a 15-nucleotide stretch encoding the five-amino-acid sequence Val–Ser–Gly–Gly–Gly. This results in DMPK isoforms with (A, C, and E) or without (B, D, and F) this VSGGG motif (Figs. 5-2B and C). Exons 13 through 15 encode diverse C termini, the presence and nature of which are determined by usage of either the "normal" 3′ splice site of intron 13 or the skipping over of the first four nucleotides of exon 14 and use of an alternative 3′ splice in this exon. Finally,

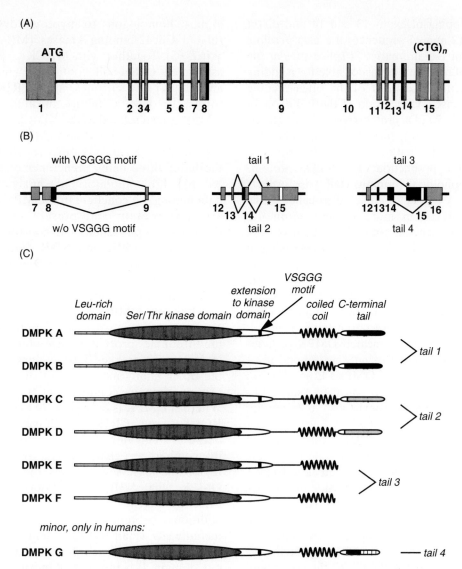

FIGURE 5-2 The *DMPK* gene encodes several alternatively spliced serine/threonine protein kinases. (A) The *DMPK* gene is strongly conserved between human and mouse and consists of 15 exons (shown is the human gene on scale; only intron sizes are different in the mouse gene). Rectangles indicate exons; the straight line represents introns. Exonic parts that are subject to alternative splicing, i.e., alternative use of 5' and 3' sites, in exons 8 and 14 are indicated in black. (B) Detailed illustration of alternative splice modes in the *DMPK* gene. Alternatively spliced exons or exonic fragments are indicated in black. (left) Inclusion of the last 15 nucleotides of exon 8 results in the presence of the VSGGG motif. (middle) Alternative skipping of the first four nucleotides of exon 14 produces two different open reading frames (ORFs), encoding proteins with C-terminal tail 1 or 2. Asterisks indicate the location of stop codons in the ORFs formed. (right top) Skipping of exons 13 and 14 by direct joining of exon 12–15 sequences is a smooth muscle-specific event, which results in the formation of an ORF with stop codon in the beginning of exon 15 encoding a DMPK isoform with a two-amino-acid C-terminal tail 3. (right bottom) A rare splice mode present only in human. A cryptic splice acceptor site in exon 15 defines exon 16. Exon 14-to-exon 16 fusion results in removal of the 5' part of exon 15, including the $(CUG)_n$ repeat. This splice mode encodes C-terminal tail 4. (C) Schematic representation of the protein domain organization of the six major DMPK isoforms A through F and minor human isoform DMPK G (all drawn to scale). All isoforms have a leucine-rich N terminus followed by a serine/threonine protein kinase domain, including an extension to the kinase domain with or without VSGGG motif, an α-helical coiled-coil domain, and the different C-terminal tail segments. Alternative splice modes shown under (B) define the presence or absence of the VSGGG motif and the nature of the C-terminal tail.

the complete skipping of exons 13 and 14 and direct joining of exon 12 and 15 sequences are also possible (Fig. 5-2B). Use of the alternative 3′ splice sites at the intron 13–exon 14 boundary is responsible for a coding shift in the 3′-terminal segment of the open reading frame of DMPK mRNA. Inclusion of the full exon 14 segment generates ~70-kDa proteins with a hydrophobic C terminus (Fig. 5-2C; tail 1, DMPK isoforms A and B), whereas skipping of the first four nucleotides from this exon results in the production of ~70-kDa proteins with a less hydrophobic C terminus (tail 2, isoforms C and D). DMPKs A through D are the so-called "long isoforms" and coexist in more or less equimolar amounts found primarily, but not exclusively, in skeletal muscle, heart, and brain [18, 36]. Complete skipping of exons 13 and 14 by direct joining of exon 12 and 15 sequences is the predominant event in smooth muscle cells, for example, stomach and bladder. This splice mode results in yet another frameshift and a new open reading frame with almost direct translation termination in exon 15, generating truncated DMPK proteins with a C terminus composed of only two amino acids (Fig. 5-2B; tail 3). The resultant DMPKs E and F (with and without VSGGG) are the so-called "short isoforms," ~60 kDa in size, and coexist also in equimolar amounts (Fig. 5-2C) [36].

Only in humans, just 3′ of the $(CUG)_n$ repeat in exon 15, a rarely used cryptic splice acceptor site is located, which defines a new terminal exon, called exon 16 [59]. Use of this splice site excises a large portion of exon 15, including the $(CUG)_n$ repeat (Fig. 5-2B). The resulting "E16+" transcripts encode a ~69-kDa DMPK isoform with a unique C-terminal tail. Based on analogy with our own nomenclature system for DMPK proteins, we termed this rare isoform *DMPK G* and the corresponding C terminus tail 4 (Fig. 5-2C) [57, 59]. It may be interesting to note here that Tiscornia and Mahadevan reported that, in contrast to mature DMPK transcripts containing a long $(CUG)_n$ tract, mature DMPK G transcripts can freely exit the nucleus, and DMPK G protein may therefore be present in relatively large amounts in the cytosol of cells from DM1 patients [59]. Follow-up studies are necessary, however, to fully understand the physiological and also potentially pathophysiological contribution of this rare splice event in humans.

C. AGC Serine/Threonine Protein Kinase Group

Based on sequence homology between individual serine/threonine protein kinase domains, DMPK is a member of the group of AGC kinases [60, 61]. DMPK is most homologous to myotonic dystrophy kinase-related Cdc42-binding kinases (MRCKs) α, β, and γ [62, 63]. Together with MRCKα, -β, and -γ, Rho-associated coiled-coil containing kinase (ROCK) I and II [64] and citron kinase (CRIK) [65], DMPK forms the DMPK family of kinases (Fig. 5-3A). DMPK is also related to the NDR family, consisting of NDR1 and -2 and Lats1 and -2 (Fig. 5-3A), and, more distantly, to PKA, PKB (also called Akt), and PKC, the latter three being archetypes of the AGC group [60, 61]. From bioinformatic analyses and genomic data mining, we conclude that DMPK is an evolutionary new protein and present only in mammals (Wansink, unpublished). In contrast, homologs for most other DMPK and NDR family members are present in *Xenopus laevis* and *Danio rerio* and also in nonvertebrates such as *Caenorhabditis elegans* and *Drosophila melanogaster*.

Although DMPK family members all have an N-terminal leucine-rich domain followed by a serine/threonine protein kinase domain in common, important differences between especially DMPK and the other family members are their size and overall protein domain composition (Fig. 5-3B) [64, 66]. Whereas DMPK is only ~60–70 kDa in size, MRCKs, ROCKs, and CRIK are huge proteins varying between ~160 and 240 kDa. Furthermore, these family members are characterized by a large coiled-coil region, at least seven times the length of that in DMPK (Fig. 5-3B; see below). Most importantly, MRCKs, ROCKs, and CRIK contain a number of additional protein domains with distinct functions, including a small GTPase binding domain, a PH domain, a cysteine-rich domain, and a citron homology domain. It is this lack of motifs with known function that renders DMPK unique among its family members and in the AGC group as a whole. Unfortunately, this feature is clearly not of much help in the elucidation of DMPK protein function.

III. THE ROLE OF INDIVIDUAL PROTEIN DOMAINS

A. N-Terminal Leucine-Rich Domain

As most protein kinases act in signaling networks together with other proteins, control of the phosphotransfer reaction by conformational changes within the kinase domain itself is often coupled to protein domains flanking the catalytic domain [60]. Such adjacent domains can link to signaling modules by providing a binding scaffold, help to localize the kinase to

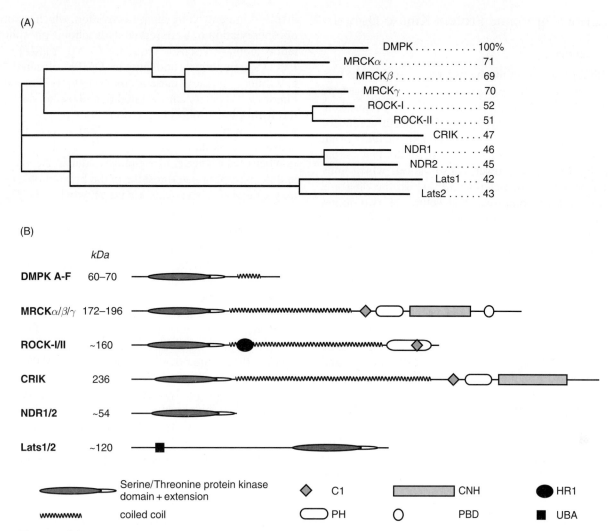

FIGURE 5-3. DMPK and NDR families of serine/threonine protein kinases. (A) Phylogenetic tree of the DMPK family (DMPK, MRCKα, -β, and -γ; ROCK-I and -II; and CRIK) and the NDR family (NDR1 and -2 and Lats1 and -2). The tree is based on homology comparison in the serine/threonine protein kinase domain using ClustalW [132]. Mouse sequences were used, and percentage sequence identity relative to the DMPK kinase domain is given. The tree is drawn to scale. (B) Domain organization of all mouse DMPK and NDR family members based on the SMART program [133] and coils program [134]. The N-terminal leucine-rich domain is not depicted, because it is not recognized as a specific protein domain, but exists in all four members of the DMPK family. The different domains are drawn to scale and symbols are explained at the bottom. C1, protein kinase C conserved region 1 domain or cysteine-rich domain; PH, pleckstrin homology domain; CNH, CNH domain or citron homology domain; PBD, P21-Rho-binding domain or Cdc42/Rac interactive binding (CRIB) region; HR1, HR1 or Rho effector domain; UBA, ubiquitin-associated domain. The following mouse sequences were used: NDR1, Q91VJ4; NDR2, Q7TSE6; Lats1, Q8BYR2; Lats2, Q7TSJ6; MRCKα, several ESTs; MRCKβ, NP898837; MRCKγ, XP140553; ROCK-I, P70335; ROCK-II, P70336; CRIK, P49025; DMPK, P54265.

specific subcellular compartments or control the oligomerization state of the protein [60]. Flanking the serine/threonine protein kinase domain in all DMPK splice isoforms is an N-terminal leucine-rich domain of ~70 amino acids, where almost every fourth amino acid is a leucine (Fig. 5-2C). Part of this segment forms a leucine zipper motif. A similar region is found in all DMPK family members, with 20–37% sequence identity compared with DMPK. The N-terminal region of MRCKα is known to mediate dimerization of the kinase domain [67]. Also in the case of ROCK-II, dimerization may be driven by the N-terminal region [68]. In contrast, the N terminus of DMPK seems not to be involved in multimerization [69, 70; van Herpen et al., unpublished]. Our own work has suggested that the leucine-rich domain may regulate DMPK kinase activity, but further study is necessary to clarify this [Wansink and van Herpen, unpublished].

B. Serine/Threonine Protein Kinase Domain

1. STRUCTURE

Although different mechanisms have evolved to regulate kinase activity, all kinases catalyze the same type of reaction, which is the transfer of the γ-phosphate from ATP to the hydroxyl group of serine, threonine, or tyrosine [71]. Typical of a kinase domain, including that in DMPK, are 11 major conserved subdomains separated by regions of lower conservation, which may contain insertions or deletions [3, 71]. In the *on* state, the kinase fold composes a structure of two lobes. The N-terminal lobe consists mainly of β sheets and a single prominent α helix, termed the αC helix, whereas the C-terminal, larger lobe is mainly α-helical [71–74]. Between the two lobes the kinase cleft is formed, in which ATP is bound by a highly conserved loop, called the phosphate-binding loop, or *P loop*. The P loop is a glycine-rich sequence containing a GXGXØG consensus, with Ø usually being a tyrosine or phenylalanine [71]. An invariant lysine in the ATP binding region, Lys100 in DMPK, makes contact with the α- and β-phosphates in ATP and, together with residues in the P loop, positions the phosphate groups of ATP for phosphotransfer [72, 74]. Positioning of the lysine in the interlobe cleft is stabilized and mediated via an ion pair formed with an invariant glutamate in the αC helix. This Lys–Glu pair is crucial to kinase activity, and mutation of the lysine renders a kinase inactive, as we also demonstrated for DMPK [57].

The conformation of the activation loop in the *off* state varies between kinases, although a similar mode of regulation is employed. In the nonphosphorylated state the activation loop folds into the active site, blocking the binding of ATP and peptide substrate [72, 75, 76]. Activation loop phosphorylation and autophosphorylation induce a conformational change that stabilizes the kinase domain in an open conformation permissive for substrate binding, leading to kinase activation. DMPK is subject to autophosphorylation, but the actual phosphorylation site in the protein has not been determined yet [40, 57, 77–79]. Besides positioning of ATP via the Lys–Glu ion pair, the αC helix also makes direct contact with the N-terminal region of the activation loop. Therefore, regulatory mechanisms that modulate kinase activity often involve structural alterations of the αC helix, as this domain is important to conformational changes that take place within the catalytic center [72–74].

Around 40 kinases, among which are all members of the DMPK family, contain a so-called protein kinase C-terminal domain (also called "extension to serine/threonine protein kinase") [60]. This accessory domain is involved in controlling kinase activation, which involves phosphorylation of a conserved hydrophobic phosphorylation motif—consensus sequence FXX[F/Y][S/T][F/Y], FVGYSY in mouse and human DMPK—located just C-terminal of the kinase domain [80–84]. For some kinases it has been demonstrated that phosphorylation of the hydrophobic phosphorylation motif recruits PDK1, which then phosphorylates the activation loop, thereby activating the kinase [72, 74, 81, 82]. In addition, interaction between a conserved arginine residue in the αC helix and the phosphorylated residue in the hydrophobic motif may synergistically activate kinase activity [73, 74, 82].

2. ACTIVITY

Reports on the enzymatic activity and substrate specificity of DMPK are relatively scarce and often based on the use of bacterially expressed protein, often improperly folded, or standard inefficient substrates like histone H1 or MBP [77–79]. Using a peptide library as substrate, our group was able to show that DMPK is actually a very active serine/threonine kinase that favors threonine over serine [57]. DMPK is a lysine/arginine-directed kinase and prefers at least three arginines or lysines among the five residues N-terminal to the phosphoacceptor site. Also, Bush and coworkers showed that DMPK prefers positively charged amino acids N-terminal to the phosphoacceptor, albeit that their consensus sequence was somewhat different from ours [70]. Not much is known about the consensus substrate sites of DMPK homologs, although it has been reported that also NDR1 and ROCK-II are lysine/arginine-directed kinases [85, 86].

A small number of natural DMPK substrates have been identified, and are discussed in Section IV. Screening of databases with the consensus phosphoacceptor sequence revealed numerous putative DMPK substrates [57]. However, similarity to the consensus per se may have only very limited predictive power. In the same way as has been documented for PKB [87], amino acids surrounding the arginine, lysine, and serine/threonine residues at the phosphorylation site affect the kinase's phosphorylation ability. This is illustrated by absence of a match to the consensus in DMPK itself, even though DMPK exhibits autophosphorylation [57, 77–79]. Additional criteria for putative substrates need to be used, such as evolutionary conservation of the phosphoacceptor site, cell and tissue distribution, and subcellular localization corresponding to that of DMPK itself. Knowing the subcellular localization of individual isoforms (see below) will thus ultimately help us to identify new substrates.

How is the kinase activity of DMPK regulated? Different mechanisms controlling kinase activity are known, among which are allosteric regulation, pseudo-substrate inhibition, oligomerization, and use of docking motifs and extended domains, like, for example, the aforementioned protein kinase C-terminal domain. Whether the last domain plays a role in the regulation of DMPK activity similar to its role for PKB [80], NDR1 [88], and MRCKα [67] has not yet been investigated. When peptides were used as substrate in an *in vitro* kinase assay, no fundamental differences in activity between DMPK isoforms were observed. Full-length myosin phosphatase targeting subunit 1 (MYPT1), however, was only significantly phosphorylated by mDMPKs E, F, and G [57; see also 89]. As E and F can be regarded "clipped" forms of mDMPKs A–D, this would be in accordance with an autoregulatory function for C-terminal tails 1 and 2, similar to what has been reported for MRCKα [67], ROCK-II [90], and human DMPK A/B [70]. Nevertheless, the putative pseudo-substrate autoinhibitory sequence in human DMPK A, as defined by Bush et al. [70], is not conserved in mouse. Also in tail 2, better conserved between mouse and human, no pseudo-substrate site could be identified. We therefore assume that tails 1 and 2 do not act as pseudo-substrates, but exert their effect via a different mechanism, possibly steric hindrance [57]. Next to the C terminus, the VSGGG motif also plays a role in the catalytic activity of DMPK, in both transphosphorylation and autophosphorylation (see below) [57].

C. VSGGG Motif

As detailed earlier, presence of the pentapeptide VSGGG motif depends on a constitutive mode of alternative splicing and is observed in 50% of the DMPK isoforms (Figs. 5-2B and C). Blast searches indicate that this sequence is unique in the entire protein sequence database. What then is the function of this tiny motif? A conspicuous in-gel mobility shift of DMPK isoforms containing a VSGGG motif suggested that the serine in the VSGGG motif (Ser379) might be a site for posttranslational modification with a strong effect on the conformational state of the protein [57]. Using a serine-to-alanine mutant, we have been able to demonstrate that Ser379 is not a site for glycosaminoglycan addition (as predicted by a posttranslational modification program) [57]. In contrast, preliminary data from our own laboratory now suggest that Ser379 may be a site for phosphorylation, most likely autophosphorylation (van Herpen and Wansink, unpublished).

Isoforms containing a VSGGG motif displayed higher phosphorylation and autophosphorylation activity than those without this motif [57]. Knowing that the VSGGG motif interrupts the protein kinase C-terminal domain, we postulate that Ser379 is a regulatory site for DMPK kinase activity. DMPK isoforms lacking the VSGGG motif may use a "flanking" phosphoacceptor (Thr375 or Thr379), or may be subject to a different type of conformational regulation.

D. Coiled-Coil Region

In all DMPK isoforms, the kinase domain is followed by an α-helical coiled-coil region (Fig. 5-2C) [3, 33, 39]. Coiled-coil regions occur in ~10% of all proteins in higher eukaryotes [60, 91], including signaling enzymes, transcription factors, and motor proteins [92, 93]. The coiled-coil region was first described as the main structural element of fibrous proteins including keratin, myosin, tropomyosin, and fibrinogen [92]. Typically, a coiled coil consists of two to five amphipathic α helices that wrap around one another into a left-handed helix to form a supercoil [92, 94]. At the interface of the two helices within the supercoil, distinctive repetition of amino acids allows packing of side chains into a "knobs-into-holes" manner [93].

In a polypeptide chain making up a coil sequence a periodicity of seven amino acids, called the heptad repeat, can be recognized. Two turns within one heptad sequence result in discrete positioning of similar amino acids along the same side of the α helix. The heptad repeat is designated as $(a-b-c-d-e-f-g)_n$ in one helix and $(a'-b'-c'-d'-e'-f'-g')_n$ in the opposite one and contains nonpolar core residues at positions *a* and *d*, which occupy the helical interface, whereas amino acids *e* and *g* are usually polar residues [92–94]. Amino acids at *a* and *d* positions—usually leucine, valine, and isoleucine—stabilize coiled-coil formation via hydrophobic interactions, whereas charged residues, for example, glutamate and lysine, at positions *e* and *g* form interhelical electrostatic interactions [92, 94]. *a–a'* and *d–d'* interactions between helices drive the "hydrophobic collapse" and formation of the supercoil, whereas *e–g'* and *g–e'* ionic interactions determine the specificity of the interhelical interaction [94]. Residues *b*, *c*, and *f* are hydrophilic, as they are positioned at the surface of the coiled coil and exposed to the solvent [92, 94].

The coiled-coil region in DMPK is relatively small and consists of ~9 heptad repeats, containing ~65 amino acids in total. For comparison, the coiled-coil region in ROCK-I is predicted to consist of more than 650 residues. The coiled-coil region in DMPK is involved in aggregation of the protein in large multimeric complexes of variable size, perhaps containing up to 10 DMPK molecules (van Herpen et al., unpublished) [69, 70]. These complexes can contain different isoforms, via homomeric and heteromeric interactions, and the multimerization behavior is dependent solely on the coiled-coil region (van Herpen et al., unpublished). As was already mentioned, the coiled-coil region in ROCK-II is also involved in multimerization, but there, unlike in DMPK, the N-terminal leucine-rich domain plays an important role [68].

The relevance of multimerization to the cell biological function of DMPK is not entirely clear. What has been shown, mainly using C-terminally truncated mutants, is that multimerization is important for kinase activity (van Herpen et al., unpublished) [69, 70]. Likewise, the kinase activities of ROCK-II and MRCKα are partly dependent on the presence of a coiled-coil region [67, 90, 95]. We have experimental evidence that the coiled-coil region in DMPK is also involved in substrate binding and subcellular localization [van Herpen et al., unpublished; 56].

E. C-Terminal Tails

The most important alternatively spliced domain in DMPK is the C terminus (Fig. 5-2C). As specified earlier, two long C termini containing 96 and 97 amino acids exist, named tail 1 and tail 2, respectively. Tail 1 and tail 2 are unique polypeptide structures as they show no appreciable homology to any other protein in the database. Tail 3 is only two amino acids long and may be considered a "clipped" form of tails 1 and 2. The variation in DMPK C-terminal tails leads to differences in kinase activity and substrate specificity (see Section IIIB2), but, above all, to diverse, isoform-specific subcellular locations [56, 57].

Expression of the six major mouse DMPK isoforms individually in various cell lines (Fig. 5-2C) consistently demonstrated that tail 1 directs DMPK to the ER, whereas tail 2 is responsible for association with the mitochondrial outer membrane (MOM) (Figs. 5-4A and B) [56, 57]. All information necessary for targeting to the ER is contained within tail 1, as YFP–tail 1 fusion proteins (containing the entire 96-amino acid-stretch) are located at the ER. In contrast, presence of the coiled-coil region is essential for tail 2 to drive proper localization of DMPKs C and D to the MOM. The two amino acids of tail 3 contain no targeting information, hence DMPK E and F behave as cytosolic proteins (Fig. 5-4C). Also, the minor human isoform DMPK G is a cytosolic protein [57]. The VSGGG motif does not influence localization for any of the DMPK variants.

Mouse DMPK truncation mutants in combination with site-directed mutagenesis indicated that the information

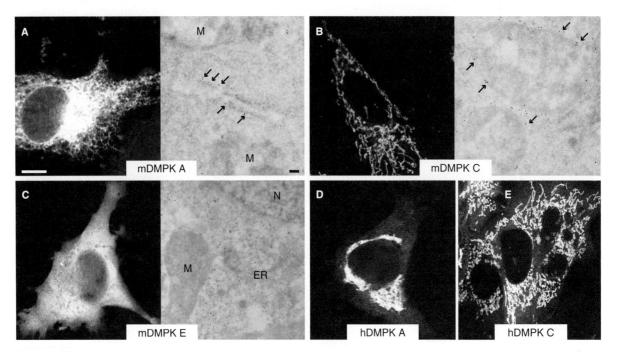

FIGURE 5-4. The subcellular localization of DMPK differs between isoforms and species. N-terminal YFP fusion proteins of mouse DMPK (mDMPK) and human DMPK (hDMPK) isoforms were expressed in Neuro2A cells and visualized by confocal scanning laser microscopy or by immunogold labeling using anti-YFP antibodies for immunoelectron microscopy (A) YFP–mDMPK A is associated with the endoplasmic reticulum. (B) YFP–mDMPK C is located at the cytosolic phase of the mitochondrial outer membrane. (C) YFP–mDMPK E is located in the cytosol. (D) YFP–hDMPK A and (E) YFP–hDMPK C are both associated with the mitochondrial outer membrane, but only YFP–hDMPK A induces strong mitochondrial aggregation around the nucleus. M, mitochondrion; N, nucleus; ER, endoplasmic reticulum. Bar for CSLM panels = 5 μm; bar for EM panels = 100 nm.

for ER or MOM association is confined in the ultimate C-terminal ~45 amino acids [56]. This, together with the observation that membrane association of mDMPKs A and C is resistant to alkaline sodium carbonate extraction, places DMPK in the family of C-terminally anchored proteins or tail-anchored (TA) proteins [96, 97]. In fact, DMPK is the first TA protein kinase characterized to date [56]. TA proteins are preferentially located at the ER or MOM, or both, and are characterized by a single C-terminal hydrophobic stretch, which anchors them in the membrane. In DMPK, computer analysis of the linear sequence information predicts a putative transmembrane domain (TMD) in tail 1 as well as in tail 2. Based on this, DMPK membrane association via insertion of its C terminus had already been suggested in the past [70, 77]. Still, further study is needed to resolve whether DMPK is at (directly or indirectly bound to other proteins) or in the membrane, as for most TA proteins the exact mechanism and mode of membrane association and topology are not yet clear.

Not much is known about the relationship between tail anchor structure and membrane specificity, that is, avidity for lipid composition. The positioning of charged amino acids flanking the putative TMD seem to define whether a protein anchors in the ER and/or in the MOM, although the exact mechanism and underlying rules are still not well understood [96]. The distribution of charged amino acids surrounding the TMD in DMPK tail 1 and tail 2 is different from that of other known TA proteins, rendering them unique tail anchors.

To our surprise, we observed that the human isoform carrying tail 1, hDMPK A, does not locate to the ER, as does its mouse ortholog, but to the MOM. We have demonstrated that this is the result of sequence differences between the human and mouse genomes [56]. In particular, a crucial arginine (Arg600) seems to disrupt the putative TMD in the human isoform. When this arginine is mutated to an alanine, the ER localization is "restored." Binding of DMPK tail 1 to membranes seems to have its own special features, as expression of the hDMPK A protein reproducibly resulted in clustering of mitochondria around the nucleus (Fig. 5-4D). This was only rarely seen for hDMPK C (Fig. 5-4E). On close inspection, we found that mDMPK A gave rise to stacked ER membrane structures, so-called organized smooth ER (OSER). It is presently not known whether similar molecular mechanisms play a role in the clustering of mitochondria or ER, or whether species specific effects of tail 1 may be involved.

What have we learned with respect to the function of membrane-associated DMPK isoforms? Variation in guidance motifs in the C termini may provide an adaptive mechanism for matching intracellular location to function during evolution. Note that the crucial R600 is found only in human and chimpanzee and not in mouse, rat, cow, and pig (Wansink, unpublished). The difference in location preference may render functions of mouse and human DMPK A orthologs completely divergent. On the other hand, it is conceivable that for proper function, DMPK simply needs to be positioned close to sites of ER–MOM contact. In reality, this could be achieved by association to either the ER or MOM side of the membranous interphase, and hence, mutational change could possibly be functionally neutral. No such speculation about an evolutionary shift in function for the DMPK C isoforms is needed, as both DMPK C isoforms show an identical distribution at the MOM in human and mouse cells.

Perinuclear clustering of mitochondria can have diverse roles, including the initiation of apoptosis [98] or local modulation of intracellular Ca^{2+} transients, as observed in parotid acinar cells [99]. Fission and fusion behavior and movement of mitochondria are dependent on integrity of the cytoskeletal network, and we know that DMPK may have a role in actomyosin dynamics by controlling myosin phosphatase activity via MYPT1 phosphorylation (see later text) [57, 89, 100]. Thus, a functional link between DMPK activity on the cytosolic face of the MOM and interaction with the actin cytoskeleton could well affect mitochondrial distribution. Whether similar associations could also play a role in OSER formation is not known.

IV. SUBSTRATES AND FUNCTION

What has determination of structural properties of protein domains and structure of isoforms taught us about DMPK's physiological role? To answer this question, we need to combine the new cell biological and biochemical data with already existing data from electrophysiological studies of cells with and without DMPK. Also, the analogy with known functions of closely related members from the AGC group of kinases is of great relevance to guidance of future research. When combined, all experimental data point to a role for DMPK in controlling ion homeostasis and actin cytoskeleton dynamics. This is summarized in Fig. 5-5, which shows all known DMPK activators and substrates in one scheme, together with relevant substrates and activators of DMPK family members MRCKs, ROCKs, and CRIK.

A. DMPK and Ion Homeostasis

Combined evidence from studies conducted over more than a decade have indicated that DMPK must be

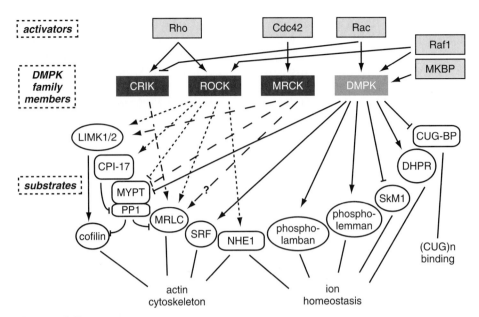

FIGURE 5-5 DMPK and family members, their activators, and substrates. Scheme summarizing the different signaling pathways upstream and downstream of the DMPK family members and their (putative) substrates for phosphorylation. Nearly all substrates can be functionally categorized into two groups and are proteins active in actomyosin dynamics of the cytoskeleton or in ion homeostasis. For the sake of clarity, arrows depicting a kinase substrate relationship are different for each kinase. Furthermore, only relevant ROCK substrates are shown (see additional list below). LIMK1/2, LIM kinase 1 and 2; CPI-17, PKC-potentiated inhibitor of 17 kDa; MYPT, myosin phosphatase targeting subunit; PP1, protein phosphatase 1; MRLC, myosin regulatory light chain; SRF, serum response factor; NHE1, Na^+–H^+ exchanger 1; SkM1, skeletal muscle voltage-gated sodium channel 1; DHPR, dihydropyridine receptor; CUG-BP, CUG-binding protein. The following ROCK substrates are not depicted: the ERM family members ezrin, radixin, moesin; adducin, calponin, MARCKS, EF-1α; intermediate filament proteins vimentin, GFAP, and NF-L; CD44, IP3 receptor, CRMP1/2, Tau, MAP2 [64]. MRCK may also phosphorylate ERM proteins. See CD-ROM for color image.

involved in the regulation of ion homeostasis, specifically via direct or indirect regulation of plasma membrane and ER-resident Ca^{2+} channels and voltage-gated Na^+ channels. When human DMPK was coexpressed in *Xenopus* oocytes with voltage-gated Na^+ channel proteins of human and rat, only the skeletal-specific, but not the cardiac-specific, channel current appeared altered [101, 102]. This was a surprising observation, because the putative DMPK phosphorylation sites in the skeletal muscle and cardiac isoforms are similar. Measurement of single-channel Na^+ currents in skeletal myocytes derived from $DMPK^{+/-}$ and $DMPK^{-/-}$ mice replicated Na^+ gating abnormalities [103, 104]. Late reopening of Na^+ channels similar to those seen in $DMPK^{-/-}$ skeletal muscle has also been described for $DMPK^{-/-}$ cardiac myocytes [105]. These reports, however, did not resolve the question whether the number of Na^+ channels in $DMPK^{-/-}$ cells could also be reduced because of adaptational effects [104, 106]. Modulation of Na^+ channel activity may provide a mechanism explaining the abnormal membrane excitability observed in patients with DM1, thus providing an alternative explanation to the hypothesis about involvement of the chloride channel ClC-1 [6]. Direct evidence showing a relation between DMPK kinase activity and modulation of Na^+ gating as an explanation for myotonia is still lacking and requires further investigation.

A role for DMPK in voltage-dependent Ca^{2+} release has also been proposed. In cultured DM1 muscle cells, $[Ca^{2+}]_i$ is high due to altered calcium influx through voltage-dependent Ca^{2+} channels, that is, the dihydropyridine receptor (DHPR) [107]. Furthermore, *in vitro* differentiated myotubes derived from $DMPK^{-/-}$ mice exhibit a higher resting $[Ca^{2+}]_i$ compared with wild-type myotubes, which is likely due to an altered open probability of voltage-dependent Ca^{2+} channels [108]. An interesting observation in this respect is that recombinant DMPK *in vitro* phosphorylates a peptide corresponding to a sequence in the β subunit of DHPR [78].

In cardiac myocytes, DMPK physically associates with and phosphorylates phospholamban, a muscle-specific SR Ca^{2+} ATPase (SERCA2a) inhibitor [109]. In ventricular homogenates from $DMPK^{-/-}$ mice, Ca^{2+} uptake in the SR was found to be impaired, suggesting a role for DMPK in cardiac contractility [109]. Taken together, these findings suggest that DMPK may be involved in initial events governing excitation–contraction coupling in skeletal or cardiac muscle [108, 109], but mechanistic details about dynamic positioning of channels or phosphorylation modifications of channel proteins or protein level effects are lacking.

Another recently published target for DMPK kinase activity is phospholemman [110]. Phospholemman is expressed in heart and skeletal muscle and regulates the Na^+/Ca^{2+} exchanger and Na^+/K^+ ATPase. When expressed in *Xenopus* oocytes, phospholemman induces chloride currents. Coexpression with DMPK resulted in a reduced appearance of phospholemman in oocyte membranes and lowered chloride conductivity. DMPK–phospholemman interaction may be yet another mechanism important for normal chloride homeostasis, and, if changed, there may be consequences with respect to altered chloride currents in patients with DM1 [110].

B. DMPK and the Actomyosin Cytoskeleton

DMPK relatives MRCK$\alpha/\beta/\gamma$, ROCK-I/II, and CRIK are involved in aspects of actomyosin remodeling that play a central role in muscle contraction, neurite outgrowth, and cell motility [64, 66]. Rho, Cdc42, and Rac, members of the Rho family of small GTPases, regulate the activity of these family members (Fig. 5-5) [111]. As mentioned earlier, according to pattern and profile search programs, DMPK does not contain a conventional small GTPase binding domain. Nevertheless, Rac-1 (and not Rho or Cdc42) has been shown to bind and activate DMPK, which also implicates involvement of DMPK in modulation of the actin cytoskeleton [112]. In the same report it was shown that also Raf1 kinase activates DMPK by direct phosphorylation. Unfortunately, the Rac-1 and Raf1 binding sites in DMPK were not clarified. Furthermore, the Epstein group reported that DMPK can induce actin cytoskeletal reorganization and apoptotic-like blebbing in lens cells [113]. We have observed similar phenomena, but further research is needed to recognize their true physiological significance and make sure they are not the result of DMPK overexpression (van Herpen et al. unpublished; see also [114]).

If DMPK function can be predicted by analogy examination, what then could be the downstream effectors in a DMPK-regulated signaling pathway? One of the best candidates is the myosin phosphatase targeting subunit (MYPT), a protein that acts as a subunit of myosin phosphatase (PP1) and is involved in the phosphorylation status of myosin regulatory light chain (MRLC) (Fig. 5-5). A large body of experimental evidence has convincingly demonstrated that MRCK$\alpha/\beta/\gamma$, ROCK-I/II, and CRIK are involved in the regulation of phosphorylation of MRLC [63, 64, 100, 115, 116]. This regulation is achieved either by direct interaction with MRLC or, indirectly, by inactivation of myosin phosphatase, through an inhibitory phosphorylation of MYPT (Fig. 5-5). Studies initiated in our group, based on functional analogy prediction and the identification of physical interaction of DMPK and MYPT in the yeast two-hybrid assay (mentioned in [33]), have now demonstrated that DMPK phosphorylates and inhibits MYPT1, but is probably not an efficient kinase for MRLC [57, 89]. Still, DMPK and the atrial form of MRLC do interact *in vitro* (Wansink et al., unpublished), corroborating a tripartite functional link between DMPK, MYPT, and MRLC. From the literature, we know that phosphorylation of MRLC is a critical step in the regulation of myosin II and affects, depending on the cell type, smooth muscle contraction, cell shape, cytokinesis, neuronal outgrowth, stress fiber formation, cell size regulation, apoptosis, and more [64, 117, 118].

The second set of downstream effectors for DMPK may be the LIM kinases (LIMK1/2) and the actin-binding protein cofilin (Fig. 5-5). Although virtually no experimental data are available to support this hypothetical proposal, we do know that LIMK1/2 are activated through phosphorylation by both ROCKs and MRCKs and that the corresponding phosphorylation site also matches the phosphoacceptor consensus for DMPK. Interestingly, Lats1 negatively regulates LIMK1 activity (not shown in Fig. 5-5) [119]. DMPK may modulate cofilin phosphorylation also through its interaction with myosin phosphatase [89, 120]. Further study is necessary to reveal if DMPK, ROCK-I/II, MRCK$\alpha/\beta/\gamma$, and CRIK function is redundant, or if they all have their own distinct roles in the phosphorylation and regulation of cofilin and MRLC [e.g., see 121].

C. SRF, CUG-BP, and MKBP

Three DMPK-interacting proteins do not fall into the preceding two categories. The first is the serum response factor (SRF). DMPK has been implicated in the regulation of skeletal and cardiac α-actin gene transcription [122]. DMPK phosphorylates SRF and enhances SRF-mediated promoter activity of α-actin genes in C_2C_{12} myoblasts and in nonmyogenic cells. We do not know whether there is any relationship with actomyosin dynamics as discussed earlier or if more transcription factors are substrates of DMPK.

DMPK phosphorylates CUG-BP, thereby decreasing its nuclear concentration [123]. In addition, transcripts from expanded DMPK alleles sequester CUG-BP, which affects regulation of splicing, transport, or translation of cellular mRNAs (see Section I). So a complex regulatory feedback mechanism has been proposed wherein decreased levels of DMPK, as in DM1 patients and DMPK knockout mice, lead to an elevation of the hypophosphorylated form of CUG-BP and an increase in nuclear concentration, thereby adversely affecting processing of pre-mRNAs [123–125]. More clarity is expected from reductionistic studies in CUG-BP knockout or knock-down model systems, to assess its pathobiological significance in isolation. Importantly, overexpression of CUG-BP in mouse models resulted in skeletal muscle features and heart features typical of DM1 [126, 127].

Apart from Rac-1 and Raf1, the only other protein reported to be involved in activation of DMPK is myotonin kinase-binding protein (MKBP) (Fig. 5-5). MKBP belongs to the family of small heat shock proteins and binds and activates DMPK *in vitro* and protects DMPK from heat-induced inactivation [128]. Although MKBP is upregulated in skeletal muscle of DM patients, it is at present unclear whether MKBP, together with DMPK, is involved in a stress response in muscle [128]. It is also not clear how activities of both proteins are hierarchically ordered.

V. TRANSGENIC MICE

A. DMPK Knockout Mice

To examine effects of a lowered DMPK dosage, two DMPK knockout mouse lines were independently generated [12, 13]. Both DMPK$^{-/-}$ models completely lack DMPK protein, but show only relatively mild phenotypes, including a mild myopathy with late onset. Increased fiber degeneration suggests a role for DMPK in maintenance of muscle fiber structural integrity [13]. Reddy's group also found cardiac conduction abnormalities in DMPK$^{-/-}$ and DMPK$^{+/-}$ mice similar to what is observed in DM1 patients [14]. In isolated DMPK$^{-/-}$ myoblasts and differentiated DMPK$^{-/-}$ myotubes, an altered Ca^{2+} response and an abnormal Na$^+$ channel opening were recorded, suggesting that DMPK is indeed involved in maintenance of ion homeostasis [104, 108]. Involvement of DMPK in ion homeostasis may also underlie its role in synaptic plasticity in the brain [17], which is relevant for the cognitive dysfunction associated with DM1, but this could also be due to a function in actomyosin dynamics, for example, in neuronal outgrowth.

It may be important to note that also knockout models for DMPK homologs ROCK-I [129] and ROCK-II [130] display milder phenotypes than anticipated, based on their crucial roles in Rho signaling and actocytoskeleton remodeling [64, 66]. Mice lacking CRIK display defective neurogenesis in the brain due to altered cytokinesis and massive apoptosis and die before adulthood [131]. Given the sequence and structural homology between DMPK family members, we expect that the lowered dosage of DMPK protein is partially compensated by the presence and activity of MRCKs, ROCKs, and maybe also CRIK.

B. Tg26-hDMPK

In the Tg26-hDMPK model, our group investigated the physiological consequences of *DMPK* overexpression [12]. Tg26-hDMPK mice carry ~25 copies of the complete human *DMPK* gene, including an intact promoter region, which results in a continuous surplus of hDMPK transcripts with a (CUG)$_{11}$ repeat and hDMPK protein isoforms (two- to eightfold higher protein expression) [12, 18]. In aged mice, cumulated through development, growth, and aging, *hDMPK* overexpression displayed a series of key systemic muscle features seen in DM1 patients [18]. Old Tg26-hDMPK animals revealed reduced workload tolerance, cardiomyopathy, myotonic myopathy, and arterial tone deficit, to name a few [18]. Although this does not necessarily support a role for DMPK in DM1 pathophysiology, these data underscore a vital role for proper *DMPK* expression in securing adequate muscle operation.

VI. CONCLUDING REMARKS

The protein level and stoichiometric composition of the DMPK isoform repertoire in distinct expressing cells are probably strictly regulated and must have biological significance. This notion has motivated us to investigate in detail the biochemical and cell biological characteristics of individual isoforms. In various studies, we have reported that structure, enzymatic activity, and, above all, subcellular localization of DMPK protein isoforms are strongly dictated by alternative splicing [36, 56, 57]. According to the current view, a splice imbalance of protein isoforms with relevant physiological functions (e.g., chloride channel, insulin receptor, tau protein, troponin T) is at the heart of the molecular pathogenesis of DM1 [6]. In this light, we urgently need to study if the dosage of expanded (CUG)$_n$ *DMPK* mRNA has effects on its own splicing and if the DM1 mutation could thus cause an imbalance in DMPK isoform level or composition. Strict control of isoform ratios of tail-anchored DMPKs and their levels and location may very well be important for the regulation of ion homeostasis, dynamics of the actomyosin cytoskeleton, or mitochondrial performance and, thus ultimately, the well-being and viability of muscle cells, neuronal cells, smooth muscle cells, and other cell types in which *DMPK* is expressed. For better judgment of the contribution of DMPK to DM1 pathology, we therefore still need to improve our understanding of how individual DMPK isoforms contribute to proper cell functioning.

Acknowledgments

Myotonic dystrophy research in the Wieringa laboratory was supported by grants from the Prinses Beatrix Fonds, the Stichting Spieren voor Spieren, the Dutch Cancer Society (KWF Kankerbestrijding), the Muscular Dystrophy Association (MDA), and l'Association Française contre les Myopathies (AFM).

References

1. Harper, P. S. (2001). "Myotonic Dystrophy." Saunders, London.
2. Fu, Y.-H., Pizzuti, A., Fenwick Jr, R. G., King, J., Rajnarayan, S., Dunne, P. W., Dubel, J., Nasser, G. A., Ashizawa, T., de Jong, P., Wieringa, B., Korneluk, R. G., Perryman, M. B., Epstein, H. F., and Caskey, C. T. (1992). An unstable triplet repeat in a gene related to myotonic muscular dystrophy. *Science* **255**, 1256–1258.
3. Brook, J. D., McCurrach, M. E., Harley, H. G., Buckler, A. J., Church, D., Aburatani, H., Hunter, K., Stanton, V. P., Thirion, J-P., Hudson, T., Sohn, R., Zemelman, B., Snell, R. G., Rundle, S. A., Crow, S., Davies, J., Shelbourne, P., Buxton, J., Jones, C., Juvonen, V., Johnson, K., Harper, P. S., Shaw, D. J., and Housman, D. E. (1992). Molecular basis of myotonic dystrophy: Expansion of a trinucleotide (CTG) repeat at the 3′ end of a transcript encoding a protein kinase family member. *Cell* **68**, 799–808.
4. Mahadevan, M., Tsilfidis, C., Sabourin, L., Shutler, G., Amemiya, C., Jansen, G., Neville, C., Narang, M., Barcelo, J., O'Hoy, K., Lebond, S., Earle-Macdonald, J., de Jong, P. J., Wieringa, B., and Korneluk, R. G. (1992). Myotonic dystrophy mutation: An unstable CTG repeat in the 3′ untranslated region of the gene. *Science* **255**, 1253–1255.
5. Liquori, C. L., Ricker, K., Moseley, M. L., Jacobsen, J. F., Kress, W., Naylor, S. L., Day, J. W., and Ranum, L. P. W. (2001). Myotonic dystrophy type 2 caused by a CCTG expansion in intron 1 of ZNF9. *Science* **293**, 864–867.
6. Ranum, L. P., and Day, J. W. (2004). Myotonic dystrophy: RNA pathogenesis comes into focus. *Am. J. Hum. Genet.* **74**, 793–804.
7. Amiel, J., Trochet, D., Clement-Ziza, M., Munnich, A., and Lyonnet, S. (2004). Polyalanine expansions in human. *Hum. Mol. Genet.* **13**, R235–R243.
8. Everett, C. M., and Wood, N. W. (2004). Trinucleotide repeats and neurodegenerative disease. *Brain* **127**, 2385–2405.
9. Wansink, D. G., and Wieringa, B. (2003). Transgenic mouse models for myotonic dystrophy type 1 (DM1). *Cytogenet. Genome Res.* **100**, 230–242.
10. Personius, K. E., Nautiyal, J., and Reddy, S. (2005). Myotonia and muscle contractile properties in mice with SIX5 deficiency. *Muscle Nerve* **31**, 503–505.
11. Sarkar, P. S., Paul, S., Han, J., and Reddy, S. (2004). Six5 is required for spermatogenic cell survival and spermiogenesis. *Hum. Mol. Genet.* **13**, 1421–1431.
12. Jansen, G., Groenen, P. J., Bachner, D., Jap, P. H., Coerwinkel, M., Oerlemans, F., van den Broek, W., Gohlsch, B., Pette, D., Plomp, J. J., Molenaar, P. C., Nederhoff, M. G., van Echteld, C. J., Dekker, M., Berns, A., Hameister, H., and Wieringa, B. (1996). Abnormal myotonic dystrophy protein kinase levels produce only mild myopathy in mice. *Nat. Genet.* **13**, 316–324.
13. Reddy, S., Smith, D. B. J., Rich, M. M., Leferovich, J. M., Reilly, P., Davis, B. M., Tran, K., Rayburn, H., Bronson, R., Cros, D., Balice-Gordon, R. J., and Housman, D. (1996). Mice lacking the myotonic dystrophy protein kinase develop a late onset progressive myopathy. *Nat. Genet.* **13**, 325–335.
14. Berul, C. I., Maguire, C. T., Aronovitz, M. J., Greenwood, J., Miller, C., Gehrmann, J., Housman, D., Mendelsohn, M. E., and Reddy, S. (1999). DMPK dosage alterations result in atrioventricular conduction abnormalities in a mouse myotonic dystrophy model. *J. Clin. Invest.* **103**, R1–R7.
15. Klesert, T. R., Cho, D. H., Clark, J. I., Maylie, J., Adelman, J., Snider, L., Yuen, E. C., Soriano, P., and Tapscott, S. J. (2000). Mice deficient in Six5 develop cataracts: Implications for myotonic dystrophy. *Nat. Genet.* **25**, 105–109.
16. Sarkar, P. S., Appukuttan, B., Han, J., Ito, Y., Ai, C., Tsai, W., Chai, Y., Stout, J. T., and Reddy, S. (2000). Heterozygous loss of Six5 in mice is sufficient to cause ocular cataracts. *Nat. Genet.* **25**, 110–114.
17. Schulz, P. E., McIntosh, A. D., Kasten, M. R., Wieringa, B., and Epstein, H. F. (2003). A role for myotonic dystrophy protein kinase in synaptic plasticity. *J. Neurophysiol.* **89**, 1177–1186.
18. O'Cochlain, D. F., Perez-Terzic, C., Reyes, S., Kane, G. C., Behfar, A., Hodgson, D. M., Strommen, J. A., Liu, X. K., van den Broek, W., Wansink, D. G., Wieringa, B., and Terzic, A. (2004). Transgenic overexpression of human DMPK accumulates into hypertrophic cardiomyopathy, myotonic myopathy and hypotension traits of myotonic dystrophy. *Hum. Mol. Genet.* **13**, 2505–2518.
19. Watchko, J. F., O'Day, T. L., and Hoffman, E. P. (2002). Functional characteristics of dystrophic skeletal muscle: Insights from animal models. *J. Appl. Physiol.* **93**, 407–417.
20. Durbeej, M., and Campbell, K. P. (2002). Muscular dystrophies involving the dystrophin–glycoprotein complex: An overview of current mouse models. *Curr. Opin. Genet. Dev.* **12**, 349–361.
21. Mankodi, A., Logigian, E., Callahan, L., McClain, C., White, R., Henderson, D., Krym, M., and Thornton, C. A. (2000). Myotonic dystrophy in transgenic mice expressing an expanded CUG repeat. *Science* **289**, 1769–1772.
22. Amack, J. D., and Mahadevan, M. S. (2004). Myogenic defects in myotonic dystrophy. *Dev. Biol.* **265**, 294–301.
23. Mankodi, A., Teng-Umnuay, P., Krym, M., Henderson, D., Swanson, M., and Thornton, C. A. (2003). Ribonuclear inclusions in skeletal muscle in myotonic dystrophy types 1 and 2. *Ann. Neurol.* **54**, 760–768.
24. Jiang, H., Mankodi, A., Swanson, M. S., Moxley, R. T., and Thornton, C. A. (2004). Myotonic dystrophy type 1 is associated with nuclear foci of mutant RNA, sequestration of muscleblind proteins and deregulated alternative splicing in neurons. *Hum. Mol. Genet.* **13**, 3079–3088.
25. Fardaei, M., Rogers, M. T., Thorpe, H. M., Larkin, K., Hamshere, M. G., Harper, P. S., and Brook, J. D. (2002). Three proteins, MBNL, MBLL and MBXL, co-localize in vivo with nuclear foci of expanded-repeat transcripts in DM1 and DM2 cells. *Hum. Mol. Genet.* **11**, 805–814.
26. Mankodi, A., Urbinati, C. R., Yuan, Q. P., Moxley, R. T., Sansone, V., Krym, M., Henderson, D., Schalling, M., Swanson, M. S., and Thornton, C. A. (2001). Muscleblind localizes to nuclear foci of

aberrant RNA in myotonic dystrophy types 1 and 2. *Hum. Mol. Genet.* **10**, 2165–2170.

27. Michalowski, S., Miller, J. W., Urbinati, C. R., Paliouras, M., Swanson, M. S., and Griffith, J. (1999). Visualization of double-stranded RNAs from the myotonic dystrophy protein kinase gene and interactions with CUG-binding protein. *Nucleic Acids Res.* **27**, 3534–3542.

28. Timchenko, N. A., Cai, Z. J., Welm, A. L., Reddy, S., Ashizawa, T., and Timchenko, L. T. (2001). RNA CUG repeats sequester CUGBP1 and alter protein levels and activity of CUGBP1. *J. Biol. Chem.* **276**, 7820–7826.

29. Philips, A. V., Timchenko, L. T., and Cooper, T. A. (1998). Disruption of splicing regulated by a CUG-binding protein in myotonic dystrophy. *Science* **280**, 737–740.

30. Ho, T. H., Charlet, B. N., Poulos, M. G., Singh, G., Swanson, M. S., and Cooper, T. A. (2004). Muscleblind proteins regulate alternative splicing. *EMBO J.* **23**, 3103–3112.

31. La Spada, A. R., Richards, R. I., and Wieringa, B. (2004). Dynamic mutations on the move in Banff. *Nat. Genet.* **36**, 667–670.

32. Ueda, H., Ohno, S., and Kobayashi, T. (2000). Myotonic dystrophy and myotonic dystrophy protein kinase. *Progr. Histochem. Cytochem.* **35**, 187–251.

33. Groenen, P., and Wieringa, B. (1998). Expanding complexity in myotonic dystrophy. *BioEssays* **20**, 901–912.

34. Lam, L. T., Pham, Y. C., Nguyen, T. M., and Morris, G. E. (2000). Characterization of a monoclonal antibody panel shows that the myotonic dystrophy protein kinase, DMPK, is expressed almost exclusively in muscle and heart. *Hum. Mol. Genet.* **9**, 2167–2173.

35. Pham, Y. C., Man, N., Lam, L. T., and Morris, G. E. (1998). Localization of myotonic dystrophy protein kinase in human and rabbit tissues using a new panel of monoclonal antibodies. *Hum. Mol. Genet.* **7**, 1957–1965.

36. Groenen, P. J. T. A., Wansink, D. G., Coerwinkel, M., van den Broek, W., Jansen, G., and Wieringa, B. (2000). Constitutive and regulated modes of splicing produce six major myotonic dystrophy protein kinase (DMPK) isoforms with distinct properties. *Hum. Mol. Genet.* **9**, 605–616.

37. van der Ven, P. F., Jansen, G., van Kuppevelt, T. H., Perryman, M. B., Lupa, M., Dunne, P. W., ter Laak, H. J., Jap, P. H., Veerkamp, J. H., Epstein, H. F., et al. (1993). Myotonic dystrophy kinase is a component of neuromuscular junctions. *Hum. Mol. Genet.* **2**, 1889–1894.

38. Sarkar, P. S., Han, J., and Reddy, S. (2004). In situ hybridization analysis of Dmpk mRNA in adult mouse tissues. *Neuromuscul. Disord.* **14**, 497–506.

39. Jansen, G., Mahadevan, M., Amemiya, C., Wormskamp, N., Segers, B., Hendriks, W., O'Hoy, K., Baird, S., Sabourin, L., Lennon, G. G., Jap, P. L., Iles, D., Coerwinkel, M., Hofker, M., Carrano, A. V., de Jong, P. J., Korneluk, R. G., and Wieringa, B. (1992). Characterization of the myotonic dystrophy region predicts multiple protein isoform-encoding mRNAs. *Nat. Genet.* **1**, 261–266.

40. Dunne, P. W., Ma, L., Casey, D. L., Harati, Y., and Epstein, H. F. (1996). Localization of myotonic dystrophy protein kinase in skeletal muscle and its alteration with disease. *Cell Motil. Cytoskeleton* **33**, 52–63.

41. Furling, D., Lam le, T., Agbulut, O., Butler-Browne, G. S., and Morris, G. E. (2003). Changes in myotonic dystrophy protein kinase levels and muscle development in congenital myotonic dystrophy. *Am. J. Pathol.* **162**, 1001–1009.

42. Eriksson, M., Hedberg, B., Carey, N., and Ansved, T. (2001). Decreased DMPK transcript levels in myotonic dystrophy 1 type IIA muscle fibers. *Biochem. Biophys. Res. Commun.* **286**, 1177–1182.

43. Furling, D., Lemieux, D., Taneja, K., and Puymirat, J. (2001). Decreased levels of myotonic dystrophy protein kinase (DMPK) and delayed differentiation in human myotonic dystrophy myoblasts. *Neuromuscul. Disord.* **11**, 728–735.

44. Endo, A., Motonaga, K., Arahata, K., Harada, K., Yamada, T., and Takashima, S. (2000). Developmental expression of myotonic dystrophy protein kinase in brain and its relevance to clinical phenotype. *Acta Neuropathol. (Berlin)* **100**, 513–520.

45. Balasubramanyam, A., Iyer, D., Stringer, J. L., Beaulieu, C., Potvin, A., Neumeyer, A. M., Avruch, J., and Epstein, H. F. (1998). Developmental changes in expression of myotonic dystrophy protein kinase in the rat central nervous system. *J. Comp. Neurol.* **394**, 309–325.

46. Kornblum, C., Reul, J., Kress, W., Grothe, C., Amanatidis, N., Klockgether, T., and Schroder, R. (2004). Cranial magnetic resonance imaging in genetically proven myotonic dystrophy type 1 and 2. *J. Neurol.* **251**, 710–714.

47. Modoni, A., Silvestri, G., Pomponi, M. G., Mangiola, F., Tonali, P. A., and Marra, C. (2004). Characterization of the pattern of cognitive impairment in myotonic dystrophy type 1. *Arch. Neurol.* **61**, 1943–1947.

48. Winchester, C. L., Ferrier, R. K., Sermoni, A., Clark, B. J., and Johnson, K. J. (1999). Characterization of the expression of DMPK and SIX5 in the human eye and implications for pathogenesis in myotonic dystrophy. *Hum. Mol. Genet.* **8**, 481–492.

49. Dunne, P. W., Ma, L., Casey, D. L., and Epstein, H. F. (1996). Myotonic protein kinase expression in human and bovine lenses. *Biochem. Biophys. Res. Commun.* **225**, 281–288.

50. Maeda, M., Taft, C. S., Bush, E. W., Holder, E., Bailey, W. M., Neville, H., Perryman, M. B., and Bies, R. D. (1995). Identification, tissue-specific expression, and subcellular localization of the 80-and 71-kDa forms of myotonic dystrophy kinase protein. *J. Biol. Chem.* **270**, 20246–20249.

51. Whiting, E. J., Waring, J. D., Tamai, K., Somerville, M. J., Hincke, M., Staines, W. A., Ikeda, J.-E., and Korneluk, R. G. (1995). Characterization of myotonic dystrophy kinase (DMK) protein in human and rodent muscle and central nervous system. *Hum. Mol. Genet.* **4**, 1063–1072.

52. Salvatori, S., Biral, D., Furlan, S., and Marin, O. (1997). Evidence for localization of the myotonic dystrophy protein kinase to the terminal cisternae of the sarcoplasmic reticulum. *J. Muscle Res. Cell Motil.* **18**, 429–440.

53. Mussini, I., Biral, D., Marin, O., Furlan, S., and Salvatori, S. (1999). Myotonic dystrophy protein kinase expressed in rat cardiac muscle is associated with sarcoplasmic reticulum and gap junctions. *J. Histochem. Cytochem.* **47**, 383–392.

54. Shimiokawa, M., Ishiura, S., Kameda, N., Yamamoto, M., Sasagawa, N., Saitoh, N., Sorimachi, H., Ueda, H., Ohno, S., Suzuki, K., and Kobayashi, T. (1997). Novel isoform of myotonin protein kinase-gene product of myotonic dystrophy is localized in the sarcoplasmic reticulum of skeletal muscle. *Am. J. Pathol.* **150**, 1285–1295.

55. Kameda, N., Ueda, H., Ohno, S., Shimokawa, M., Usuki, F., Ishiura, S., and Kobayashi, T. (1998). Developmental regulation of myotonic dystrophy protein kinase in human muscle cells in vitro. *Neuroscience* **85**, 311–322.

56. van Herpen, R. E. M. A., Oude Ophuis, R. J. A., Wijers, M., Bennink, M. B., van de Loo, F. A., Fransen, J., Wieringa, B., and Wansink, D. G. (2005). Divergent mitochondrial and endoplasmic reticulum association of DMPK splice isoforms depends on unique sequence arrangements in tail anchors. *Mol. Cell. Biol.* **25**, 1402–1414.

57. Wansink, D. G., van Herpen, R. E. M. A., Coerwinkel-Driessen, M. M., Groenen, P. J. T. A., Hemmings, B. A., and Wieringa, B. (2003). Alternative splicing controls myotonic dystrophy protein kinase structure, enzymatic activity and subcellular localization. *Mol. Cell. Biol.* **23**, 5489–5501.

58. Mahadevan, M. S., Amemiya, C., Jansen, G., Sabourin, L., Baird, S., Neville, C. E., Wormskamp, N., Segers, B., Batzer, M., Lamerdin, J., de Jong, P., Wieringa, B., and Korneluk, R. G. (1993). Structure and genomic sequence of the myotonic dystrophy (DM kinase) gene. *Hum. Mol. Genet.* **2**, 299–304.
59. Tiscornia, G., and Mahadevan, M. S. (2000). Myotonic dystrophy: The role of the CUG triplet repeats in splicing of a novel *DMPK* exon and altered cytoplasmic *DMPK* mRNA isoform ratios. *Mol. Cell* **5**, 959–967.
60. Manning, G., Whyte, D. B., Martinez, R., Hunter, T., and Sudarsanam, S. (2002). The protein kinase complement of the human genome. *Science* **298**, 1912–1934.
61. Caenepeel, S., Charydczak, G., Sudarsanam, S., Hunter, T., and Manning, G. (2004). The mouse kinome: Discovery and comparative genomics of all mouse protein kinases. *Proc. Natl. Acad. Sci. USA* **101**, 11707–11712.
62. Ng, Y., Tan, I., Lim, L., and Leung, T. (2004). Expression of the human myotonic dystrophy kinase-related Cdc42-binding kinase {gamma} is regulated by promoter DNA methylation and Sp1 binding. *J. Biol. Chem.* **279**, 34156–34164.
63. Leung, T., Chen, X.-Q., Tan, I., Manser, E., and Lim, L. (1998). Myotonic dystrophy kinase-related Cdc42-binding kinase acts as a Cdc42 effector in promoting cytoskeletal reorganization. *Mol. Cell. Biol.* **18**, 130–140.
64. Riento, K., and Ridley, A. J. (2003). Rocks: Multifunctional kinases in cell behaviour. *Nat. Rev. Mol. Cell Biol.* **4**, 446–456.
65. Madaule, P., Eda, M., Watanabe, N., Fujisawa, K., Matsuoka, T., Bito, H., Ishizaki, T., and Narumiya, S. (1998). Role of citron kinase as a target of the small GTPase Rho in cytokinesis. *Nature* **394**, 491–494.
66. Zhao, Z. S., and Manser, E. (2005). PAK and other Rho-associated kinases: Effectors with surprisingly diverse mechanisms of regulation. *Biochem J.* **2**, 201–214.
67. Tan, I., Seow, K. T., Lim, L., and Leung, T. (2001). Intermolecular and intramolecular interactions regulate catalytic activity of myotonic dystrophy kinase-related Cdc42-binding kinase α. *Mol. Cell. Biol.* **21**, 2767–2778.
68. Doran, J. D., Liu, X., Taslimi, P., Saadat, A., and Fox, T. (2004). New insights into the structure–function relationships of Rho-associated kinase: A thermodynamic and hydrodynamic study of the dimer-to-monomer transition and its kinetic implications. *Biochem. J.* **384**, 255–262.
69. Zhang, R., and Epstein, H. F. (2003). Homodimerization through coiled-coil regions enhances activity of the myotonic dystrophy protein kinase. *FEBS Lett.* **546**, 281–287.
70. Bush, E. W., Helmke, S. M., Birnbaum, R. A., and Perryman, M. B. (2000). Myotonic dystrophy protein kinase domains mediate localization, oligomerization, novel catalytic activity, and autoinhibition. *Biochemistry* **39**, 8480–8490.
71. Hanks, S. K., and Hunter, T. (1995). The eukaryotic protein kinase superfamily: Kinase (catalytic) domain structure and classification. *FASEB J.* **9**, 576–596.
72. Nolen, B., Taylor, S., and Ghosh, G. (2004). Regulation of protein kinases: Controlling activity through activation segment conformation. *Mol. Cell* **15**, 661–675.
73. Krupa, A., Preethi, G., and Srinivasan, N. (2004). Structural modes of stabilization of permissive phosphorylation sites in protein kinases: Distinct strategies in Ser/Thr and Tyr kinases. *J. Mol. Biol.* **339**, 1025–1039.
74. Huse, M., and Kuriyan, J. (2002). The conformational plasticity of protein kinases. *Cell* **109**, 275–282.
75. Johnson, L. N., Noble, M. E., and Owen, D. J. (1996). Active and inactive protein kinases: Structural basis for regulation. *Cell* **85**, 149–158.
76. Adams, J. A. (2003). Activation loop phosphorylation and catalysis in protein kinases: Is there functional evidence for the autoinhibitor model? *Biochemistry* **42**, 601–607.
77. Waring, J. D., Haq, R., Tamai, K., Sabourin, L. A., Ikeda, J.-E., and Korneluk, R. G. (1996). Investigation of myotonic dystrophy kinase isoform translocation and membrane association. *J. Biol. Chem.* **271**, 15187–15193.
78. Timchenko, L., Nastainczyk, W., Schneider, T., Patel, B., Hofmann, F., and Caskey, C. T. (1995). Full-length myotonin protein kinase (72 kDa) displays serine kinase activity. *Proc. Natl. Acad. Sci. USA* **92**, 5366–5370.
79. Dunne, P. W., Walch, E. T., and Epstein, H. F. (1994). Phosphorylation reactions of recombinant human myotonic dystrophy protein kinase and their inhibition. *Biochemistry* **33**, 10809–10814.
80. Yang, J., Cron, P., Thompson, V., Good, V. M., Hess, D., Hemmings, B. A., and Barford, D. (2002). Molecular mechanism for the regulation of protein kinase B/Akt by hydrophobic motif phosphorylation. *Mol. Cell* **9**, 1227–1240.
81. Biondi, R. M., and Nebreda, A. R. (2003). Signalling specificity of Ser/Thr protein kinases through docking-site-mediated interactions. *Biochem. J.* **372**, 1–13.
82. Frodin, M., Antal, T. L., Dummler, B. A., Jensen, C. J., Deak, M., Gammeltoft, S., and Biondi, R. M. (2002). A phosphoserine/threonine-binding pocket in AGC kinases and PDK1 mediates activation by hydrophobic motif phosphorylation. *EMBO J.* **21**, 5396–5407.
83. Sarbassov, D. D., Guertin, D. A., Ali, S. M., and Sabatini, D. M. (2005). Phosphorylation and regulation of Akt/PKB by the rictor–mTOR complex. *Science* **307**, 1098–1101.
84. Gao, T., Furnari, F., and Newton, A. C. (2005). PHLPP: A phosphatase that directly dephosphorylates Akt, promotes apoptosis, and suppresses tumor growth. *Mol. Cell* **18**, 13–24.
85. Millward, T. A., Heizmann, C. W., Schafer, B. W., and Hemmings, B. A. (1998). Calcium regulation of Ndr protein kinase mediated by S100 calcium-binding proteins. *EMBO J.* **17**, 5913–5922.
86. Turner, M. S., Fen Fen, L., Trauger, J. W., Stephens, J., and LoGrasso, P. (2002). Characterization and purification of truncated human Rho-kinase II expressed in Sf-21 cells. *Arch. Biochem. Biophys.* **405**, 13–20.
87. Obata, T., Yaffe, M. B., Leparc, G. G., Piro, E. T., Maegawa, H., Kashiwagi, A., Kikkawa, R., and Cantley, L. C. (2000). Peptide and protein library screening defines optimal substrate motifs for AKT/PKB. *J. Biol. Chem.* **275**, 36108–36115.
88. Millward, T. A., Hess, D., and Hemmings, B. A. (1999). Ndr protein kinase is regulated by phosphorylation on two conserved sequence motifs. *J. Biol. Chem.* **274**, 33847–33850.
89. Muranyi, A., Zhang, R., Liu, F., Hirano, K., Ito, M., Epstein, H. F., and Hartshorne, D. J. (2001). Myotonic dystrophy protein kinase phosphorylates the myosin phosphatase targeting subunit and inhibits myosin phosphatase activity. *FEBS Lett.* **493**, 80–84.
90. Amano, M., Chihara, K., Nakamura, N., Kaneko, T., Matsuura, Y., and Kaibuchi, K. (1999). The COOH terminus of Rho-kinase negatively regulates Rho-kinase activity. *J. Biol. Chem.* **274**, 32418–32424.
91. Rost, B. (2002). Did evolution leap to create the protein universe? *Curr. Opin. Struct. Biol.* **12**, 409–416.
92. Burkhard, P., Stetefeld, J., and Strelkov, S. V. (2001). Coiled coils: A highly versatile protein folding motif. *Trends Cell Biol.* **11**, 82–88.
93. Lupas, A. (1996). Coiled coils: New structures and new functions. *Trends Biochem. Sci.* **21**, 375–382.
94. Mason, J. M., and Arndt, K. M. (2004). Coiled coil domains: Stability, specificity, and biological implications. *ChemBioChem* **5**, 170–176.

95. Shimizu, T., Ihara, K., Maesaki, R., Amano, M., Kaibuchi, K., and Hakoshima, T. (2003). Parallel coiled-coil association of the RhoA-binding domain in Rho-kinase. *J. Biol. Chem.* **278**, 46046–46051.
96. Borgese, N., Colombo, S., and Pedrazzini, E. (2003). The tale of tail-anchored proteins: Coming from the cytosol and looking for a membrane. *J. Cell Biol.* **161**, 1013–1019.
97. High, S., and Abell, B. M. (2004). Tail-anchored protein biosynthesis at the endoplasmic reticulum: The same but different. *Biochem. Soc. Trans.* **32**, 659–662.
98. Takada, S., Shirakata, Y., Kaneniwa, N., and Koike, K. (1999). Association of hepatitis B virus X protein with mitochondria causes mitochondrial aggregation at the nuclear periphery, leading to cell death. *Oncogene* **18**, 6965–6973.
99. Bruce, J. I. E., Giovannucci, D. R., Blinder, G., Shuttleworth, T. J., and Yule, D. I. (2004). Modulation of $[Ca^{2+}]_i$ signaling dynamics and metabolism by perinuclear mitochondria in mouse parotid acinar cells. *J. Biol. Chem.* **279**, 12909–12917.
100. Amano, M., Fukata, Y., and Kaibuchi, K. (2000). Regulation and functions of Rho-associated kinase. *Exp. Cell Res.* **261**, 44–51.
101. Mounsey, J. P., Xu, P., John, J. E., 3rd, Horne, L. T., Gilbert, J., Roses, A. D., and Moorman, J. R. (1995). Modulation of skeletal muscle sodium channels by human myotonin protein kinase. *J. Clin. Invest.* **95**, 2379–2384.
102. Chahine, M., and George, A. L., Jr. (1997). Myotonic dystrophy kinase modulates skeletal muscle but not cardiac voltage-gated sodium channels. *FEBS Lett.* **412**, 621–624.
103. Mistry, D. J., Moorman, J. R., Reddy, S., and Mounsey, J. P. (2001). Skeletal muscle Na currents in mice heterozygous for Six5 deficiency. *Physiol. Genom.* **6**, 153–158.
104. Mounsey, J. P., Mistry, D. J., Ai, C. W., Reddy, S., and Moorman, J. R. (2000). Skeletal muscle sodium channel gating in mice deficient in myotonic dystrophy protein kinase. *Hum. Mol. Genet.* **9**, 2313–2320.
105. Lee, H. C., Patel, M. K., Mistry, D. J., Wang, Q., Reddy, S., Moorman, J. R., and Mounsey, J. P. (2003). Abnormal Na channel gating in murine cardiac myocytes deficient in myotonic dystrophy protein kinase. *Physiol. Genom.* **12**, 147–157.
106. Reddy, S., Mistry, D. J., Wang, Q. C., Geddis, L. M., Kutchai, H. C., Moorman, J. R., and Mounsey, J. P. (2002). Effects of age and gene dose on skeletal muscle sodium channel gating in mice deficient in myotonic dystrophy protein kinase. *Muscle Nerve* **25**, 850–857.
107. Jacobs, A. E., Benders, A. A., Oosterhof, A., Veerkamp, J. H., van Mier, P., Wevers, R. A., and Joosten, E. M. (1990). The calcium homeostasis and the membrane potential of cultured muscle cells from patients with myotonic dystrophy. *Biochim. Biophys. Acta* **1096**, 14–19.
108. Benders, A. G. M., Groenen, P. J. T. A., Oerlemans, F. T. J. J., Veerkamp, J. H., and Wieringa, B. (1997). Myotonic dystrophy protein kinase is involved in the modulation of the Ca^{2+} homeostasis in skeletal muscle cells. *J. Clin. Invest.* **100**, 1440–1447.
109. Kaliman, P., Catalucci, D., Lam, J. T., Kondo, R., Gutierrez, J. C. P., Reddy, S., Palacin, M., Zorzano, A., Chien, K. R., and Ruiz-Lozano, P. (2005). Myotonic dystrophy protein kinase (DMPK) phosphorylates phospholamban and regulates calcium uptake in cardiomyocyte sarcoplasmic reticulum. *J. Biol. Chem.*, **280**, 8016–8021.
110. Mounsey, J. P., John, J. E., 3rd, Helmke, S. M., Bush, E. W., Gilbert, J., Roses, A. D., Perryman, M. B., Jones, L. R., and Moorman, J. R. (2000). Phospholemman is a substrate for myotonic dystrophy protein kinase. *J. Biol. Chem.* **275**, 23362–23367.
111. Etienne-Manneville, S., and Hall, A. (2002). Rho GTPases in cell biology. *Nature* **420**, 629–635.
112. Shimizu, M., Wang, W., Walch, E. T., Dunne, P. W., and Epstein, H. F. (2000). Rac-1 and Raf-1 kinases, components of distinct signaling pathways, activate myotonic dystrophy protein kinase. *FEBS Lett.* **475**, 273–277.
113. Jin, S., Shimizu, M., Balasubramanyam, A., and Epstein, H. F. (2000). Myotonic dystrophy protein kinase (DMPK) induces actin cytoskeletal reorganization and apoptotic-like blebbing in lens cells. *Cell Motil. Cytoskeleton* **45**, 133–148.
114. Sasagawa, N., Kino, Y., Takeshita, Y., Oma, Y., and Ishiura, S. (2003). Overexpression of human myotonic dystrophy protein kinase in *Schizosaccharomyces pombe* induces an abnormal polarized and swollen cell morphology. *J. Biochem. (Tokyo)* **134**, 537–542.
115. Piekny, A. J., and Mains, P. E. (2002). Rho-binding kinase (LET-502) and myosin phosphatase (MEL-11) regulate cytokinesis in the early *Caenorhabditis elegans* embryo. *J. Cell Sci.* **115**, 2271–2282.
116. Yamashiro, S., Totsukawa, G., Yamakita, Y., Sasaki, Y., Madaule, P., Ishizaki, T., Narumiya, S., and Matsumura, F. (2003). Citron kinase, a rho-dependent kinase, induces di-phosphorylation of regulatory light chain of myosin II. *Mol. Biol. Cell* **14**, 1745–1756.
117. Hartshorne, D. J., Ito, M., and Erdodi, F. (2004). Role of protein phosphatase type 1 in contractile functions: Myosin phosphatase. *J. Biol. Chem.*
118. Somlyo, A. P., and Somlyo, A. V. (2003). Ca^{2+} sensitivity of smooth muscle and nonmuscle myosin II: Modulated by G proteins, kinases, and myosin phosphatase. *Physiol. Rev.* **83**, 1325–1358.
119. Yang, X., Yu, K., Hao, Y., Li, D. M., Stewart, R., Insogna, K. L., and Xu, T. (2004). LATS1 tumour suppressor affects cytokinesis by inhibiting LIMK1. *Nat. Cell Biol.* **6**, 609–617.
120. Samstag, Y., and Nebl, G. (2003). Interaction of cofilin with the serine phosphatases PP1 and PP2A in normal and neoplastic human T lymphocytes. *Adv. Enzyme Regul.* **43**, 197–211.
121. Wilkinson, S., Paterson, H. F., and Marshall, C. J. (2005). Cdc42-MRCK and Rho-ROCK signalling cooperate in myosin phosphorylation and cell invasion. *Nat. Cell Biol.* **7**, 255–261.
122. Iyer, D., Belaguli, N., Fluck, M., Rowan, B. G., Wei, L., Weigel, N. L., Booth, F. W., Epstein, H. F., Schwartz, R. J., and Balasubramanyam, A. (2003). Novel phosphorylation target in the serum response factor MADS box regulates alpha-actin transcription. *Biochemistry* **42**, 7477–7486.
123. Roberts, R., Timchenko, N. A., Miller, J. W., Reddy, S., Caskey, C. T., Swanson, M. S., and Timchenko, L. T. (1997). Altered phosphorylation and intracellular distribution of a $(CUG)_n$ triplet repeat RNA-binding protein in patients with myotonic dystrophy and in myotonin protein kinase knockout mice. *Proc. Natl. Acad. Sci. USA* **94**, 13221–13226.
124. Savkur, R. S., Philips, A. V., and Cooper, T. A. (2001). Aberrant regulation of insulin receptor alternative splicing is associated with insulin resistance in myotonic dystrophy. *Nat. Genet.* **29**, 40–47.
125. Ladd, A. N., Charlet-B., N., and Cooper, T. A. (2001). The CELF family of RNA binding proteins is implicated in cell-specific and developmentally regulated alternative splicing. *Mol. Cell. Biol.* **21**, 1285–1296.
126. Timchenko, N. A., Patel, R., Iakova, P., Cai, Z.-J., Quan, L., and Timchenko, L. T. (2004). Overexpression of CUG triplet repeat-binding protein, CUGBP1, in mice inhibits myogenesis. *J. Biol. Chem.* **279**, 13129–13139.
127. Ho, T. H., Bundman, D., Armstrong, D. L., and Cooper, T. A. (2005). Transgenic mice expressing CUG-BP1 reproduce splicing mis-regulation observed in myotonic dystrophy. *Hum. Mol. Genet.* **14**, 1539–1547.
128. Suzuki, A., Sugiyama, Y., Hayashi, Y., Nyu-i, N., Yoshida, M., Nonaka, I., Ishiura, S., Arahata, K., and Ohno, S. (1998). MKBP, a novel member of the small heat shock protein family, binds and

activates the myotonic dystrophy protein kinase. *J. Cell Biol.* **140**, 1113–1124.

129. Shimizu, Y., Thumkeo, D., Keel, J., Ishizaki, T., Oshima, H., Oshima, M., Noda, Y., Matsumura, F., Taketo, M. M., and Narumiya, S. (2005). ROCK-I regulates closure of the eyelids and ventral body wall by inducing assembly of actomyosin bundles. *J. Cell Biol.* **168**, 941–953.

130. Thumkeo, D., Keel, J., Ishizaki, T., Hirose, M., Nonomura, K., Oshima, H., Oshima, M., Taketo, M. M., and Narumiya, S. (2003). Targeted disruption of the mouse rho-associated kinase 2 gene results in intrauterine growth retardation and fetal death. *Mol. Cell. Biol.* **23**, 5043–5055.

131. Di Cunto, F., Imarisio, S., Hirsch, E., Broccoli, V., Bulfone, A., Migheli, A., Atzori, C., Turco, E., Triolo, R., Dotto, G. P., Silengo, L., and Altruda, F. (2000). Defective neurogenesis in citron kinase knockout mice by altered cytokinesis and massive apoptosis. *Neuron* **28**, 115–127.

132. Thompson, J. D., Higgins, D. G., and Gibson, T. J. (1994). CLUSTAL W: Improving the sensitivity of progressive multiple sequence alignment through sequence weighting, position-specific gap penalties and weight matrix choice. *Nucleic Acids Res.* **22**, 4673–4680.

133. Letunic, I., Copley, R. R., Schmidt, S., Ciccarelli, F. D., Doerks, T., Schultz, J., Ponting, C. P., and Bork, P. (2004). SMART 4.0: Towards genomic data integration. *Nucleic Acids Res.* **32**, D142–D144.

134. Lupas, A., Van Dyke, M., and Stock, J. (1991). Predicting coiled coils from protein sequences. *Science* **252**, 1162–1164.

CHAPTER 6

Biochemistry of Myotonic Dystrophy Protein Kinase

RAM SINGH AND HENRY F. EPSTEIN
Department of Neuroscience and Cell Biology, University of Texas Medical Branch, Galveston, Texas 77555-0620

I. Introduction
II. Structure of *Dm-1* Locus and Region
III. DMPK Structural Domains
   A. Leucine-Rich Repeat: Amino-Terminal Region
   B. Catalytic Domain
   C. Carboxy-Terminal Region
   D. Coiled-Coil Region
IV. Alternative Splicing and DMPK Isoforms
V. Functional Biochemical Properties of DMPK
   A. Homodimerization through the Coiled-Coil Region
   B. Interaction with Other Regulatory Proteins
   C. Substrate Specificity
VI. DMPK Family of Protein Kinases
VII. Tissue Expression of DMPK
   A. Heart
   B. Lens
   C. Skeletal Muscle
   D. Brain
VIII. Subcellular Localization of DMPK
   A. Carboxy-Terminal Membrane Anchoring
   B. Endoplasmic Reticulum, Mitochondrial, and Cytosolic Localization
IX. DMPK Function in Heart and Brain
   A. Excitability of Heart
   B. Synaptic Plasticity in Brain
X. Multiple Mechanisms in *Dm-1* Pathogenesis
   A. Effects of Local Chromatin Perturbation on the Expression of Neighboring Genes and DMPK
   B. Perturbation of Alternative RNA Splicing
   C. Haploinsufficiency of DMPK
XI. Conclusion
   Acknowledgments
   References

Myotonic dystrophy protein kinase (DMPK) and its isoforms are the protein products of the *Dm-1* locus on chromosome 19q13.3, and play role in skeletal weakness and cardiac myopathy. DMPK, a serine-threonine protein kinase, consists of four distinct regions; a leucine-rich repeats (LRR), a protein kinase (PK), a hydrophobic coiled-coil (H) and a putative transmembrane (T). The protein kinase phosphorylates and regulates myosin phosphatase (MYPT1), serum response factor (SRF), and phospholamban (PLN). Additionally, DMPK interacts with Rac-1, a member of the Rho family of small GTPases. DMPK knockout mice exhibit atrioventricular arrhythmias, impaired skeletal muscle development and delayed relaxation, cataracts, and diminished synaptic plasticity. The haploinsufficiency of DMPK, decreased transcription of DMWD and SIX5, and sequestration of muscleblind and related proteins involved in alternative splicing by CUG expansion in DMPK RNA appear to represent distinct pathogenic mechanisms in DM-1.

I. INTRODUCTION

Myotonic dystrophy (DM), a multisystem disorder was first identified in 1909 [1, 2], and is now documented as one of the most common neuromuscular diseases. It is an autosomal dominant neuromuscular disorder with a global occurrence of 1 per 8000. Two types of myotonic dystrophy are now recognized, DM1 and DM2, which are due to mutations in the *Dm-1* and *Dm-2* loci, respectively.

The *Dm-1* mutation was identified as an expansion of a $(CTG)_n$ in the 3′ untranslated region (3′ UTR) of a gene on chromosome 19q 13.3 locus, encoding a serine/threonine protein kinase DMPK [3–7] This myotonic dystrophy was the first neuromuscular disease recognized to be an inherited trait [8].

II. STRUCTURE OF *Dm-1* LOCUS AND REGION

Dm-1 comprises 15 exons, and encodes a protein of 629 amino acids that shares regions of homology with the DMPK family of protein kinases [3–7, 9, 10]. The human gene is about 13 kb in length and is transcribed in the telomere-to-centromere orientation (Fig. 6-1). Exons 1–8 constitute the amino terminus of the protein, with exons 2–8 showing homology to the serine/threonine protein kinase family. Exons 9–12 show homology to the coiled-coil domain of myosin. An Alu repeat is located in intron 8, and the CTG repeat is in the last exon (exon 15) downstream of the translation stop signal. The CTG repeat lies within the 3′ UTR of the sequences encoding the protein kinase (DMPK), approximately 500 bp upstream of the poly(A) signal [7, 11].

III. DMPK STRUCTURAL DOMAINS

DMPK is a serine/threonine protein kinase [12] with a catalytic domain of 43 kDa, followed by a helical region of 12 kDa and a nonpolar region homologous to known transmembrane domains (Fig. 6-2) [5, 7, 11, 13].

A. Leucine-Rich Repeat: Amino-Terminal Region

DMPK contains a single leucine-rich repeat (LRR) of about 69 amino acids amino-terminal to the catalytic domain. In human DMPK, amino acid residues 9–38 constitute a highly nonpolar region with similarities to previously studied LRRs [14]. The LRR in fact binds ubiquitin-1, which protects against polyubiquinylation and degradation [N. Nagamatsu, H. F. Epstein, and T. Ashizawa, unpublished results]. The DMPK amino-terminal region shows 93% identity between human and mouse, suggesting that this LRR region may be functionally significant [15, 16].

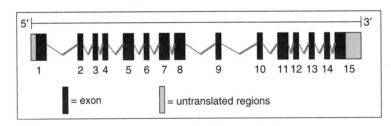

FIGURE 6-1 Human *Dm-1* gene structure including exon–intron information. See CD-ROM for color image.

```
  1 MGGHFWPPEP YTVFMWGSPW EADSPRVKLR GREKGRQTEG GAFPLVSSAL SGDPRFFSPT
 61 TPPAEPIVVR LKEVRLQRDD FEILKVIGRG AFSEVAVVKM KQTGQVYAMK IMNKWDMLKR
121 GEVSCFREER DVLVNGDRRW ITQLHFAFQD ENYLYLVMEY YVGGDLLTLL SKFGERIPAE
181 MARFYLAEIV MAIDSVHRLG YVHRDIKPDN ILLDRCGHIR LADFGSCLKL RADGTVRSLV
241 AVGTPDYLSP EILQAVGGGP GTGSYGPECD WWALGVFAYE MFYGQTPFYA DSTAETYGKI
301 VHYKEHLSLP LVDEGVPEEA RDFIQRLLCP PETRLGRGGA GDFRTHPFFF GLDWDGLRDS
361 VPPFTPDFEG ATDTCNFDLV EDGLTAMVSG GGETLSDIRE GAPLGVHLPF VGYSYSCMAL
421 RDSEVPGPTP MEVEAEQLLE PHVQAPSLEP SVSPQDETAE VAVPAAVPAA EAEAEVTLRE
481 LQEALEEEVL TRQSLSREME AIRTDNQNFA SQLREAEARN RDLEAHVRQL QERMELLQAE
541 GATAVTGVPS PRATDPPSHL DGPPAVAVGQ CPLVGPGPMH RRHLLLPARV PRPGLSEALS
601 LLLFAVVLSR AAALGCIGLV AHAGQLTAVW RRPGAARAP
```

FIGURE 6-2 Human DMPK amino acid sequence; color variation shows the different domain sequences: blue indicates amino-terminal, brown protein kinase domain, red coiled-coil, and dark blue transmembrane region sequences. See CD-ROM for color image.

B. Catalytic Domain

The kinase function is carried out by a catalytic domain the structure and catalytic residues of which are highly conserved among the serine/threonine protein kinase family [14]. The 43-kDa catalytic domain between residues 70 and 349 of human DMPK is related to many members of the serine/threonine protein kinase family [17–19]. DMPK shares catalytic domain sequences with other protein kinases, with consequent cross-reaction of antisera, a possible reason for the quite large size variations [20]. All 11 functional motifs characteristic of serine/threonine protein kinases are present [3].

C. Carboxy-Terminal Region

Residues 352–629 in human DMPK, which constitute the carboxy-terminal region, are 78.2% identical to mouse DMPK. Residues 461–538 form the putative α-helical region that is predicted by the stringent PAIRCOIL algorithm [21] to form a coiled coil. Furthermore, residues 503–530 are compatible with a leucine zipper [22]. Such a zipper represents a shorter, less stable version of a true coiled coil and might be the basis of readily reversible associations between protein molecules [23]. Like the similarities in the catalytic domain, the RhoA-binding domain within this region is a common characteristic of the DMPK group of protein kinases.

D. Coiled-Coil Region

A coiled-coil domain follows the catalytic domain [14]. The coiled-coil regions are 88.6% identical between human and mouse DMPKs. The α-helical coiled-coil region (residues 461–538) is necessary for the formation of DMPK homodimers, and the dimeric form enhances the catalytic efficiency of the kinase active centers [24]. An interesting structural feature of the broader myotonic dystrophy family of protein kinase (MDFPK) is the presence of varying lengths of α helices. These structural and functional findings in DMPK are likely to be significant in understanding the structure–function relationships of other important MDFPK members such as Rho kinase.

IV. ALTERNATIVE SPLICING AND DMPK ISOFORMS

Six major DMPK mRNA isoforms (Fig. 6-3) are conserved between humans and mice and are produced by a combination of three alternative splice events, one of

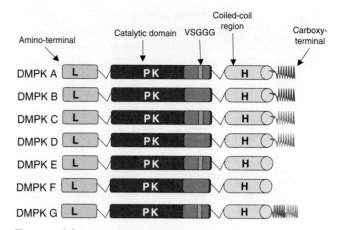

FIGURE 6-3 Six major (A-F) and one minor (G) spliceoforms of DMPK (11, 25). See CD-ROM for color image.

which is cell type specific [11]. All isoforms share an amino-terminal domain, a kinase domain, and a coiled-coil region, while alternative splicing determines the presence or absence of a five-amino-acid VSGGG motif and the nature of the carboxy terminus (three cell type-dependent variants). A new human DMPK isoform has been reported [25]. This minor isoform, designated DMPK G here, carries yet another carboxy terminus, but, more importantly, its mRNA lacks the $(CUG)_n$ repeat in its 3′ UTR. As a result, unlike DMPK transcripts bearing long $(CUG)_n$ repeats, DMPK G transcripts may more efficiently leave the nucleus, thus creating an altered DMPK isoform profile in the cytoplasm of cells of DM1 patients in whom the DMPK gene is expressed.

The differences in mobility behavior on sodium dodecyl sulfate (SDS)–polyacrylamide gels and on Western blots of the various DMPK isoforms agree with the predictions based on RNA splicing (Fig. 6-4). In lanes with long DMPK isoforms (A and C) containing the VSGGG motif (lanes A and C), there are four major bands of ~78, ~73, ~70, and ~67 kDa. In products without this motif, only two major bands in each lane are observed, namely, ~72 and 66 kDa for DMPK B (lane B), and ~74 and ~68 kDa for DMPK D (lane D). Strikingly, products corresponding to the smooth muscle isoforms (with or without the VSGGG motif) are less heterogeneous (lanes E and F) than those in other tissues [11, 26].

V. FUNCTIONAL BIOCHEMICAL PROPERTIES OF DMPK

DMPK is known to undergo autophosphorylation and to phosphorylate the general protein kinase substrate histone H1 *in vitro* [12]. But, in contrast to other

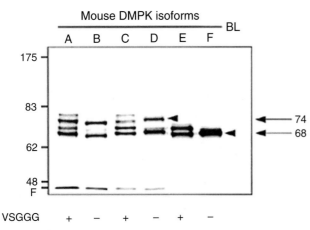

FIGURE 6-4 Major spliceoforms of DMPK with different carboxy terminals. A–F refer to the spliceoform constructs. They were individually expressed in COS-1 cells, and Western blots of cell lysates were reacted with the B-79 anti-DMPK polyclonal antiserum. Reprinted with permission from *Human Molecular Genetics*. See CD-ROM for color image.

serine/threonine protein kinases such as PKA and PKC, recombinant DMPK is resistant to several known inhibitors of serine/threonine kinases such as Y-27632 and HA-1077 at concentrations up to 10 μM [27]. These protein kinase inhibitors either stabilize autoinhibitory domains or block the ATP binding site, so that ATP binding is no longer possible.

A. Homodimerization through the Coiled-Coil Region

All of the serine/threonine kinases, which show strong conservation in their catalytic protein kinase domains to DMPK, also have varying lengths of α-helical coiled-coil forming sequence (Fig. 6-5). Sequences homologous to this region have been shown to be necessary for the binding of Rho, a Ras superfamily GTPase [28]. Addition of the coiled-coil domain cause DMPK to elute as a much higher than predicted molecular weight species, consistent with the formation of oligomers [29]. The *Arabidopsis thaliana* gene *TOUSLED* encodes a nuclear serine/ threonine kinase that also requires a coiled-coil domain for oligomerization and enzymatic activity [30], raising the possibility that kinase activity can be regulated by coiled coil-mediated oligomerization [29]. Indeed, enzyme kinetic analysis of the dimeric LPKH subfragment of DMPK reveals a threefold greater k_{cat}/K_m and a ten-fold greater V_{max} than for monomeric LPK [24].

B. Interaction with Other Regulatory Proteins

DMPK may play an important role in the pathophysiology of DM1 [31–35], probably via interaction with multiple signal molecules in different signaling pathways (Fig. 6-6). The actin cytoskeleton-linked GTPase Rac-1 physically binds to DMPK, and coexpression of Rac-1 with DMPK activates the transphosphorylation activity of DMPK in a GTP-sensitive manner. DMPK also may be phosphorylated by Raf-1 [36]. It has not been established whether phosphorylation by Raf-1 activates or inhibits DMPK activity. Raf-1 kinase is thought to be activated by the small GTPase Ras, which is a key element of the signaling pathway linked to the mitogen-activated protein (MAP) kinase cascade [37]. Rac1 and possibly Raf-1 may regulate the transphosphorylation of target proteins by DMPK and permit "cross-talk" between different signaling pathways [14, 36]. The interactions of Rac-1 and Raf-1 kinase with DMPK may be functionally significant because they both show binding, enzymatic activation, and sensitivity to known regulatory interactions. However, other GTPase family members including Ras, RhoA, and Cdc42 do not bind to DMPK [36].

C. Substrate Specificity

Recent findings revealed that DMPK may participate in a variety of cellular processes. Phospholamban [38], the β subunit of the dihydropyridine receptor [39],

FIGURE 6-5 Sequence of predicted coiled coil in H region of DMPK.

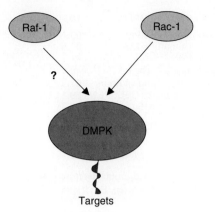

FIGURE 6-6 Model of activation of DMPK by Rac-1 and by Raf-1 kinase. See CD-ROM for color image.

mouse kallikrin-binding protein [40], CUG-binding protein/hNab50 [41], and the myosin phosphatase targeting subunit 1 (MYPT1) [27] have been identified as potential substrates for DMPK. The latter finding suggests a role for DMPK in cytoskeletal movement or intracellular transport dynamics, similar to the function of ROCK/Rho-kinase/ROK and the myotonic dystrophy kinase-related Cdc42-binding kinase (MRCK) as effectors of RhoA and Cdc42, respectively, in reorganization of the actin-based cytoskeleton.

1. Phosphorylation of Myosin Phosphatase Target Subunit 1

Myosin phosphatase (MP) is composed of three subunits [42]: a protein phosphatase 1 catalytic subunit, δ isoform (PP1cδ), and two noncatalytic subunits. One of the noncatalytic subunits of the δ isoform has a proposed targeting role and is termed myosin phosphatase targeting subunit (MYPT) [42, 43]. It is also known as the myosin-binding subunit (MBS). The two genes for MYPT located on chromosome 12 and chromosome 1 express MYPT1 and MYPT2, respectively [42, 44]. MYPT1 has a wide tissue distribution, with the highest levels found in smooth muscle and nonmuscle cells [42], whereas MYPT2 is restricted to brain and cardiac muscle [44].

Phosphorylation of MYPT1 by the DMPK homolog Rho kinase inhibits phosphatase activity [45] and increases the level of myosin phosphorylation and activity. Ca^{2+} sensitization in smooth muscle [42] and cytoskeletal rearrangements in nonmuscle cells are likely consequences of this interaction [45]. The similarity of the sequences of the catalytic domains of DMPK and Rho kinase [46] suggests that the both kinases may be related functionally as evidenced by their phosphorylation of MYPT1 and the consequent inhibition of MP activity.

2. Phosphorylation of Phospholamban

Phospholamban (PLN) is a muscle-specific sarcoplasmic reticulum (SR) Ca-ATPase inhibitor, highly expressed in cardiac muscle [38]. It has two adjacent residues, S16 and T17, identified as the phosphorylation sites for protein kinase A and Ca/calmodulin-dependent kinase II, respectively [47]. Co-immunoprecipitation studies show that DMPK and PLN can physically interact, and purified wild-type DMPK phosphorylates PLN *in vitro*, but not DMPK dead (mouse K110A) [38]. Moreover, PLN is underphosphorylated (twofold decreased) in SR vesicles from DMPK$^{-/-}$ mice compared with wild-type both *in vitro* and *in vivo*. Under physiological conditions, PLN phosphorylation at S16 by PKA leads to proportional increases in the rate of Ca uptake into SR and accelerates ventricular relaxation [48, 49]. DMPK appears to have a similar function.

3. Serum Response Factor

The serum response factor (SRF) belongs to the MCM1–agamous–ARG80–deficiens–SRF (MADS) box transcription factor family [50]. It is required for expression of MyoD, the skeletal myogenic factor, in both dividing and differentiating myoblasts [51]. SRF is regulated by changes in actin dynamics [52]. The highly conserved MADS box comprises the DNA binding domain and part of the dimerization domain in SRF-like transcription factors [53].

SRF may function as a target of DMPK [54]. Phosphorylation of SRF at T159 in the DNA binding domain of the MADS box by DMPK enhances transcription of the skeletal and cardiac α-actin genes in heterologous cells. Modulation of α-actin expression is consistent with the putative role of DMPK in skeletal muscle differentiation [33, 35, 55].

SRF can be detected in both nuclear and cytosolic fractions of skeletal myocytes [56], and its regulated translocation between cytosolic and nuclear compartments is important for gene expression in smooth muscle [57]. Thus, phosphorylation of SRF by DMPK could occur in a cytosolic location, with subsequent translocation of activated SRF to the nucleus. The putative regulation of SRF by DMPK and the significance of SRF in myogenesis may be related to the skeletal muscle phenotype in DM1.

VI. DMPK FAMILY OF PROTEIN KINASES

In humans, the catalytic domain of DMPK is closely related to those of PK428 and p160ROCK, 78 and 66.1% identity, respectively (Fig. 6-7). Both DMPK and

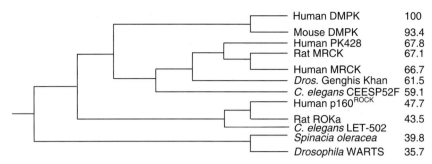

FIGURE 6-7 Myotonic dystrophy family of protein kinases. Reprinted from "Genetic Instabilities and Hereditary Neurological Diseases" (R. D. Wells and S. T. Warren, Eds.), with permission of Academic Press, (14).

PK428 have a leucine-rich amino-terminal domain and a hydrophobic region in their carboxy-terminal domains [14]. Additionally, both are expressed in skeletal muscle, heart, and nervous system. Possible functional redundancy between these enzymes has not been experimentally demonstrated, but this possibility may be necessary for the complete evaluation of DMPK function. Human p160^{ROCK}, another potentially redundant kinase, is activated by the 21-kDa GTPase RhoA. It shows significant homology to DMPK in its catalytic and Rho binding domains.

Several functional and structural characteristics of DMPK are related to the properties of other protein kinases in diverse organisms including *Neurospora* Cot-1, *Drosophila* Warts, rat ROK [58], *Caenorhabditis elegans* LET-502 [59], *Drosophila* Genghis Khan [60], citron Rho-interacting kinase [61], and rat DM kinase-related Cdc42-binding kinase [62]. Most importantly, these kinases, like DMPK and Rho kinase, have been implicated in the regulation of the actin cytoskeleton.

VII. TISSUE EXPRESSION OF DMPK

DMPK is expressed in a wide range of tissues, with the highest expression in cardiac, skeletal (most prominent in tongue, esophagus, and diaphragm), and smooth (stomach and colon) muscle. In total brain, a moderate level of expression is found, but the level appears to be higher in certain subregions of the central nervous system and lower in others. DMPK is not detected in the ovary, kidney, and pancreas [63, 64].

A. Heart

High levels of DMPK mRNA are detected in the epicardium, myocardium, and endocardium [63]. The 78- to 80-kDa isoform is expressed mainly in cardiac muscle. Specific monoclonal antibodies localized DMPK to intercalated disks in human heart. The 70- to 72-kDa isoform is expressed widely, including the heart [20].

B. Lens

Human and bovine lenses contain DMPK mRNA. DMPK has been localized by polyclonal antibody to the cuboidal lens epithelial cell layer [65–67].

C. Skeletal Muscle

DMPK is found in type I muscle fibers by immunocytochemistry. It is localized in neuromuscular junctions *in vivo* and *in vitro* [68] and in muscle spindles. Immunolocalization studies also show that anti-DMPK antibodies bind to the neuromuscular junction [15, 65, 69, 70] and SR in human skeletal muscle [71, 72]. DMPK levels increase during the period of human muscle fiber formation, both *in vitro* and *in vivo*. In normal adult muscle, DMPK is expressed in both fast and slow muscle fibers. In congenital myotonic dystrophy (CDM), DMPK levels are reduced throughout skeletal muscle development, consistent with decreased DMPK expression from the mutant allele [73].

Based on reaction with monoclonal anti-DMPK antibodies, DMPK expression is found to significantly increase in skeletal muscle between 9 and 16 weeks of human muscle development. The large increase in DMPK accumulation correlates with the formation of second-generation muscle fibers and the major period of muscle formation. After 20 weeks, no new fibers form and the muscle fibers undergo a process of maturation. DMPK levels remain high during this time.

This is consistent with evidence that DMPK transcription in muscle cells is under the control of muscle-specific regulatory elements located in the promoter and the first intron of the *Dm-1* gene [73, 74]. The increase in DMPK expression during *in vitro* myogenic differentiation is in good agreement with the appearance of DMPK during muscle formation and the persistence of high levels of DMPK during muscle growth *in vivo* [73].

D. Brain

The locations of DMPK in the central nervous system, and its developmental pattern of expression in rat brain and spinal cord have been studied using a monospecific rabbit antiserum (Fig. 6-8) [67]. This study demonstrates that DMPK expression begins after birth and increases gradually to peak at Postnatal Day 21 in many brain regions. After Postnatal Day 21 and proceeding to the adult, the pattern of expression becomes restricted to certain regions or cell groups in the central nervous system. Electron microscopy reveals expression within adult spinal motor neurons to the endoplasmic reticulum and dendritic microtubules [67].

Immunohistochemistry of the developmental expression of DMPK in the central nervous system using a specific antibody shows that the immunoreactive neurons appear in the early fetal frontal cortex and cerebellar granule cell layer, persist through 29 weeks of gestation, and then disappear [75]. Anti-DMPK antibodies bind to synaptic glomeruli, dendritic processes, and cytoplasm in rodent brain [20, 64].

DMPK mRNA is expressed in several regions of the brain including the hippocampus proper, dentate gyrus, and subiculum [63]. In the hippocampus, significant expression occurs in the pyramidal cell layers of the CA1, CA2, and CA3 fields. Expression is scattered in the stratum lacunosum and stratum oriens of the hippocampus. In the dentate gyrus, the granular layer and the polymorphic layers show strong expression. In the cerebellum, the Purkinje and granular cell layers express significant levels, while the molecular layer shows sparse and scattered expression. Thalamus, hypothalamus, and midbrain regions show low levels of expression. Within the ventricles, DMPK mRNA is detected in cells of the choroid plexus [63].

These studies of protein and mRNA suggest that DMPK may function in membrane trafficking and secretion within neurons associated with cognition, memory, and motor control.

VIII. SUBCELLULAR LOCALIZATION OF DMPK

The varying specificities of antibodies to either recombinant DMPK or DMPK peptides and the proposed localizations of DMPK have been a matter of debate. Detailed localization by immunofluorescence

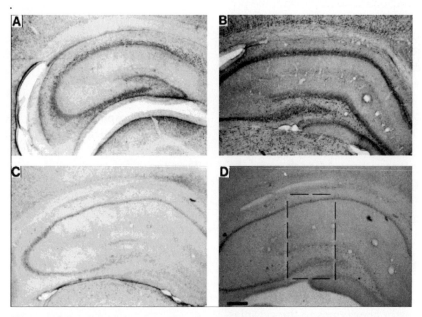

FIGURE 6-8 Developmental series of rat hippocampus (low power). Light microscopy of coronal sections from the dorsal hippocampus showing developmental changes in DMPK expression at postnatal stages P7 (A), P21 (B), P28 (C), and adult (D). Reprinted with permission of the *Journal of Comparative Neurology*, (64).

microscopy or immunohistochemistry may be complicated by the simultaneous expression of multiple isoforms in myoblasts, neuronal cells, or other cell types in which DMPK is expressed. As DMPK protein is in low abundance, it can be detected only by highly sensitive techniques in which nonspecificity may be a major problem. The criterion for specific anti-DMPK is: antibodies must react only with bonafide DMPK isoforms and do not react with DMPK knockout mouse tissues by localization and Western blot.

Subcellular fractionation and immunohistochemical studies indicate that DMPK localizes at the neuromuscular and myotendinous junctions [72] and terminal cisternae of the SR [76] in skeletal muscle. Furthermore, DMPK is found at intercalated disks and the tubular and junctional SR in heart muscle [76].

A. Carboxy-Terminal Membrane Anchoring

The anchoring proteins are multivalent and allow the assembly of several signaling proteins. The mechanisms by which anchoring proteins assemble at distinct subcellular sites are diverse. Structural membrane proteins, transmembrane receptors, and cytoskeletal proteins may provide such anchoring.

Full-length DMPK is a carboxy-terminal anchored serine/threonine protein kinase. DMPK isoforms A and C are strongly associated with membranes via their carboxy termini. Most mitochondrial outer membrane proteins have a transmembrane domain near the carboxy terminus and an amino-terminal cytosolic moiety. Carboxy-terminal anchor proteins constitute a group of proteins that specifically insert into intracellular membranes, using a single membrane-spanning region located close to the carboxy terminus which plays an important role [26, 29, 77–79].

B. Endoplasmic Reticulum, Mitochondrial, and Cytosolic Localization

Previous functional experiments suggest that the full-length DMPK binds to the endoplasmic reticulum (ER) [14, 26, 77, 78]. The amino acid sequence of the carboxy-terminal α helix in DMPK is very similar to those of the transmembrane domains of human HMG CoA reductase and rat microsomal aldehyde dehydrogenase that anchor these enzymes to the cytoplasmic face of the ER [80]. DMPK is predominantly cytoplasmic; only a very low signal is observed in the nucleus [26, 79].

Various DMPK isoforms may be associated with the ER mitochondria and cytosol. The A and B isoforms appear to be associated with the ER. The C and D isoforms may be located at the outer membrane of mitochondria. The E, F, and G isoforms are found in the cytosol. Membrane-associated DMPK isoforms are resistant to alkaline conditions. Mutagenic analysis shows that proper anchoring is differentially dependent on basic residues flanking putative transmembrane domains [79]. The combination of enhanced expression and knockout phenotypes in mutant mice suggests that DMPK may perform significant roles in heart and brain.

IX. DMPK FUNCTION IN HEART AND BRAIN

A. Excitability of Heart

Cardiac arrythmias are the major life-threatening problem in adult DM1. The significance of DMPK in this problem is supported by the finding that DMPK-deficient mice develop cardiac conduction defects including first-, second-, and third-degree atrioventricular block. These results suggest that the atrioventricular node and the His-Purkinje regions of the conduction systems are specifically compromised by DMPK [81]. Importantly DMPK$^{-/-}$ and DMPK$^{-/+}$ mice develop first-degree heart block in an age-dependent manner similarly to DM1 patients. These mice also show functionally abnormal cardiac Na channels consistent with DMPK being a possible regulator of muscle Na channels [82]. Phosphorylation of Na channels by protein kinases is recognized as an important mechanism for modulation of Na currents [83–86].

DMPK phosphorylation of PLN appears to be a new mechanism implicated in the regulation of cardiac contractility. DMPK knockouts result in a significant decrease in SR Ca^{2+} uptake activity, pointing toward PLN phosphrylation by DMPK as a physiologically relevant event [38]. The cardiac function can be regulated through DMPK phosphorylation of PLN, which provides a molecular mechanism for the cardiac dysfunction in DM1.

B. Synaptic Plasticity in Brain

DMPK is involved in the process of development of many neuronal types in several regions of the central nervous systems (CNS), and it correlates with the developmentally related CNS defects in neonatal myotonic dystrophy [20, 64, 75, 87].

Remodeling of the actin cytoskeleton and the postsynaptic dendritic spines is important for mechanisms underlying long-term potentiation (LTP), a use-dependent form of synaptic plasticity. LTP is considered a model for the cellular basis of memory storage and other cognitive functions. Changes in synaptic shape [88, 89] and the shape of dendritic spines [89] may be related to cytoskeletal remodeling. DMPK can alter the actin cytoskeleton [90]. It may play a role in the cytoskeletal modulation of dendritic spines, as DMPK mRNA and protein are present in the hippocampus [87], a structure that is associated with learning and memory.

Electrophysiological experiments on null mouse brain slices demonstrate the involvement of DMPK in synaptic plasticity. These mice show no changes in baseline synaptic transmission in hippocampal area CA1. DMPK$^{-/-}$ versus DMPK$^{+/+}$ or DMPK$^{+/-}$ mice show significant decreases (67%) in the decremental phase of LTP with a duration of 30–180 min (Fig. 6-9) [91]. These results suggest a role for DMPK in synaptic plasticity that could be relevant to the cognitive dysfunction associated with DM1.

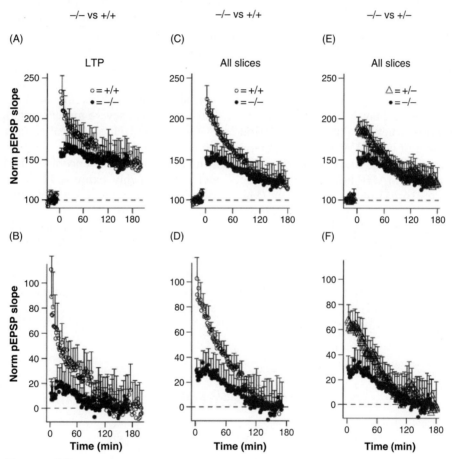

FIGURE 6-9 Normal sustained potentiation, but decreased decremental long-term synaptic potentiation (LTP) in DMPK null slices. (A) Sustained potentiation compared between wild (E) and knockout (F) slices. (B) The decremental potentiation from (A) on an expanded y axis in (B) after subtracting sustained potentiation at 150 min after high-frequency stimulation (HFS). There is a 64.9% reduction in decremental potentiation after 5 to 20 min HFS. (C) Averaging all DMPK$^{-/-}$ slices, including those that show sustained potentiation after HFS and those that return to baseline, demonstrates that sustained potentiation is identical in DMPK$^{-/-}$ and DMPK$^{+/+}$ slices. (D) Expanding the y axis from (C) and subtracting sustained potentiation demonstrates a 68.7% reduction in decremental potentiation. (E) Averaging all DMPK$^{-/-}$ and DMPK$^{+/-}$ slices, including those that showed long-term potentiation after HFS and those that showed only decremental potentiation, demonstrates that sustained potentiation is identical in the two groups. (F) There was a significant decrease in decremental potentiation between the two groups. Reprinted with permission of the *Journal of Neurophysiology*, (91).

X. MULTIPLE MECHANISMS IN *Dm-1* PATHOGENESIS

A decade after the discovery of the *Dm-1* mutation, understanding the disease mechanisms is still the object of ongoing research. The normal *Dm-1* gene contains 5–37 CTG repeats, whereas DM1 patients have the repeats expanded from 50 to thousands in the 3′ UTR. These triplet expansion mutations may produce the characteristic findings of DM1 by a combination of several mechanisms: chromatin perturbations leading to impaired transcription, sequestration of proteins regulating alternative RNA splicing, and decreased expression of proper DMPK isoforms (Fig. 6-10).

A. Effects of Local Chromatin Perturbation on the Expression of Neighboring Genes and DMPK

Expanded CTG repeats may be able to interfere with local chromatin structure. This perturbation could affect expression of both DMPK and neighboring genes. The enhancer and the first exon of the adjacent, more centromeric DMAHP homeobox gene are in close proximity to exon 15 of the *Dm-1* gene (Fig. 6-10), making them a vulnerable target for regional chromatin changes. A number of studies demonstrate that expansion of the CTG repeat alters the conformation of chromatin in the vicinity of the DMPK gene [66, 92, 93]. One consequence of this change is that the expression of neighboring genes *SIX5* and *DMWD* may be partly suppressed [93]. Gene *SIX5* is responsible for eye development and facial morphology and is expressed in corneal epithelium and endothelium, lens epithelium, the cellular layer of the retina, and sclera. *DMWD* is strongly expressed in brain and testes [92, 93].

B. Perturbation of Alternative RNA Splicing

Pathogenesis in skeletal muscle may be mediated predominantly by a gain of function in the mutant mRNA. It appears likely that transcripts from the mutant DMPK gene accumulate as foci in the nuclei of both cultured and biopsied tissues [18, 94] and that splicing and possibly other cellular functions are affected [95–99]. Strong experimental support for the involvement of transcript gain of function in skeletal muscle pathology comes from a transgenic mouse model [100] in which expanded CUG repeats were expressed at high levels in skeletal muscle. However, despite expanded transcripts in muscle nuclei, the mice did not develop severe muscle wasting.

Genetic and physiological studies have shown that defects in ion channels may cause myotonia [101, 102]. Mutations in the gene encoding ClC-1, a chloride channel that is highly expressed and specific to skeletal muscle, cause generalized myotonia (myotonia congenita). Alternative splicing of the ClC-1 transcript is disrupted so that the mature form of the chloride channel is not produced in DM1 biopsy cells or in mice transgenic for overexpression of expanded CUG. Myotonia may be also caused by mutations in *SCN4A*, the gene encoding the α subunit of the skeletal muscle sodium channel [103].

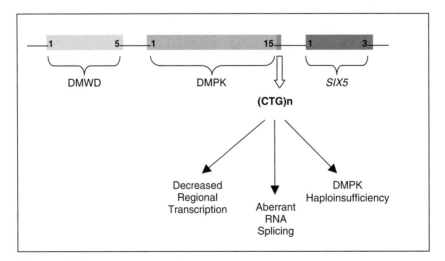

FIGURE 6-10 Multiple mechanisms in *Dm-1* pathogenesis. See CD-ROM for color image.

The reduced insulin response in skeletal muscle predisposes DM1 patients to diabetes [99]. Alternative splicing of the insulin receptor pre-mRNA is aberrantly regulated in DM1 skeletal muscle tissue and in expanded CUG overexpression, resulting in persistent production of the lower-affinity nonmuscle isoform (Fig. 6-10) [99].

The *muscleblind*-like protein family in mice contains three genes: *MBNL*, *MBLL*, and *MBXL*. The sequestration of MBNL proteins has been proposed as the major pathogenic mechanism resulting from mRNA gain of function [104, 105]. All three *muscleblind* gene proteins bind to dsCUG RNA, which suggests a link between expanded expression of mutant DM1 transcripts and nuclear sequestration of the *muscleblind* proteins. Immunofluorescence and histological analysis of *muscleblind* knockout mice [105, 106] reveals muscle myotonia, cataracts, and RNA splicing defects, which are characteristic of DM1 [105, 107]. These findings are consistent with the expansion of CUG in DMPK RNA transcripts leading to sequestration of *muscleblind* and interference with its role in the regulation of alternative RNA splicing as a major pathogenic mechanism in DM1.

C. Haploinsufficiency of DMPK

Early studies indicated that the expression of DMPK mRNA and protein is reduced in patient muscle biopsies and muscle cell culture [5, 108]. However, DMPK knockout mice do not have the totality of characteristic findings in DM1 skeletal muscle. Initial reports on these mice showed only a very mild, late-onset myopathy that is not typical of DM1 [33, 35]. Significantly, only these knockouts of the several mouse models show significant weakness in skeletal muscle. In the normal condition, the DMPK gene is transcribed, and the RNAs are processed in the nucleus, transported to the cytoplasm, and then translated into the DMPK. Conversely, in the DM1 patient, the DMPK gene is transcribed but the transcripts are largely retained within the nucleus, leading to decreased levels of DMPK mRNA and producing haploinsufficiency of DMPK [99, 109].

XI. CONCLUSION

DMPK and its isoforms are the established protein products of the *Dm-1* gene. DMPK acts as a serine/threonine kinase and can alter the functions of myosin phosphatase, phospholamban, and the serum response factor by their phosphorylation. These putative substrate proteins play important roles in the regulation of the actin cytoskeleton, cardiac excitation–contraction coupling, and muscle development, respectively. DMPK is closely related structurally to other protein kinases that are known to function in the regulation of the actin cytoskeleton in multiple cell types. DMPK knockout mice demonstrate cardiac arrhythmias, skeletal muscle weakness, and diminished synaptic plasticity consistent with clinical findings in DM1 patients. It is likely that haploinsufficiency of DMPK together with sequestration of proteins involved in alternative splicing by the expanded CUG RNA tracts and perturbation of chromatin structure leading to decreased transcription of regional genes may all play roles in the pathogenesis of *Dm-1*.

Acknowledgments

We thank our colleagues whose data we have presented as figures in this review. This research was supported, in part, by Grant 5RO1 NS35071-07 from the NIH, a grant from the Muscular Dystrophy Association, and the Cecil H. and Ida M. Green Endowment at the University of Texas.

References

1. Steinert, H. (1909). Myopathologische Beitrage: 1. Uber das klinischeund anatomische Bild des Muskelschwunds der Myotoniker. *Dtsch. Z. Nervenheilkd.* **37**, 58–104.
2. Batten, F., and Gibb, H. (1909). Myotonia atrophica. *Brain* **32**, 187–205.
3. Brook, J. D., McCurrah, M. E., Harley, H. G., Buckler, A. J., Church, D., Aburatani, H., Hunter, K., Stanton, V. P., Thirion, J. P., Hudson, T., Sohn, R., Zemelman, B., Snell, R. G., Rundle, S. A., Crow, S., Davies, J., Shelbourne, P., Buxton, J., Jones, C., Juvonen, V., Johnson, K., Harper, P. S., Shaw, D. J., and Housman, D. E. (1992). Molecular basis of myotonic dystrophy: Expansion of a trinucleotide (CTG) repeat at the 3' end of a transcript encoding a protein kinase family member. *Cell* **68**, 799–808.
4. Buxton, J., Shelbourne, P., Davies, J., Jones, C., Van-ongeren, T., Aslanidis, C., de-Jong, P., Jansen, G., Anvret, M., Riley, B., Williamson, R., and Johnson, K. (1992). Detection of an unstable fragment of DNA specific to individuals with myotonic dystrophy. *Nature* **355**, 547–548.
5. Fu, Y.-H., Pizzuti, A., Fenwick, R. G. J., King, J., Rajnarayan, S., Dunne, P. W., Dubel, J., Nasser, G. A., Ashizawa, T., De Jong, P., Wieringa, B., Korneluk, R., Perryman, M. B., Epstein, H. F., and Caskey, C. T. (1992). An unstable triplet repeat in a gene related to myotonic muscular dystrophy. *Science* **255**, 1256–1258.
6. Harley, H. G., Brook, J. D., Rundle, S. A., Crow, S., Reardon, W., Buckler, A. J., Harper, P. S., Houseman, D. E., and Shaw, D. (1992). Expansion of an unstable DNA region and phenotypic variation in myotonic dystrophy. *Nature* **355**, 545–546.
7. Mahadevan, M., Tsilfidis, C., Sabourin, L., Shutler, G., Amemiya, C., Jansen, G., Neville, C., Narang, M., Barcelo, J., O'Hoy, K., Leblond, S., Earle-MacDonald, J., De Jong, P. J., Wieringa, B., and Koneluk, R. G. (1992). Myotonic dystrophy mutation: An unstable CTG repeat in the 3' untranslated region of the gene. *Science* **255**, 1253–1255.
8. Curschmann, H. (1912). Uber familiare atrophische myotonie. *Dtsch. Z. Nervenheilkd.* **45**, 161–202.

9. Jansen, G., Mahadevan, M., Amemiya, C., Wormskamp, N., Segers, B., Hendriks, W., O'Hoy, K., Baird, S., Sabourin, L., Lennon, G., Jap, P. L., Iles, D., Coerwinkel, M., Hofker, M., Carrano, A. V., de Jong, P. J., Korneluk, R. G., and Wieringa, B. (1992). Characterization of the myotonic dystrophy region predicts multiple protein isoform-encoding mRNAs. *Nat. Genet.* **1**, 261–266.
10. Shaw, D. J., McCurrach, M., Rundle, S. A., Harley, H. G., Crow, S. R., Sohn, R., Thirion, J. P., Hamshere, M. G., Buckler, A. J., Harper, P. S., Housman, D. E., and Brook, J. D. (1993). Genomic organization and transcriptional units at the myotonic dystrophy locus. *Genomics* **18**, 673–679.
11. Groenen, P. J. T. A., Wansink, D. G., Coerwinkel, M., van den Broek, W., Jansen, G., and Wieringa, B. (2000). Constitutive and regulated modes of splicing produce six major myotonic dystrophy protein kinase (DMPK) isoforms with distinct properties. *Hum. Mol. Genet.* **9**, 605–616.
12. Dunne, P. W., Walch, E. T., and Epstein, H. F. (1994). Phosphorylation reactions of recombinant human myotonic dystrophy protein kinase and their inhibition. *Biochemistry* **33**, 10809–10814.
13. Mahadevan, M. S., Amemiya, C., Jansen, G., Sabourin, L., Baird, S., Neville, C. E., Wormskamp, N., Segers, B., Batzer, M., and Lamerdin, J. (1993). Structure and genomic sequence of the myotonic dystrophy (DM kinase). *Hum. Mol. Genet.* **2**, 299–304.
14. Epstein, H. F., and Jin, S. (1998). Biochemical studies of DM protein kinase (DMPK). *In* "Genetic Instabilities and Hereditary Neurological Diseases," pp. 147–167. Academic Press, San Diego.
15. van der Ven, P. F. M., Jensen, G., van Kuppevelt, T. H. M. S. M., Perryman, M. B., Lpa, M., Dunne, P. W., ter Laak, H. J., Jap, P. H. K., veerkamp, J. H., Epstein, H. F., and Wieringa, B. (1993). Myotonic dystrophy kinase is a component of neuromuscular junction. *Hum. Mol. Genet.* **2**, 1889–1894.
16. Waring, J. D., Haq, R., Tamai, K., Sabourin, L. A., Ikeda, J. E., and Korneluk, R. G. (1996). Investigation of myotonic dystrophy kinase isoform translocation and membrane association. *J. Biol. Chem.* **217**, 15187–15193.
17. Amack, J. D., Paguio, A. P., and Mahadevan, M. S. (1999). *cis* and *trans* effects of the myotonic dystrophy (DM) mutation in a cell culture model. *Hum. Mol. Genet.* **8**, 1975–1984.
18. Davis, B. M., McCurrach, M. E., Taneja, K. L., Singer, R. H., and Housman, D. E. (1997). Expansion of a CUG trinucleotide repeat in the 3′ untranslated region of myotonic dystrophy protein kinase transcripts results in nuclear retention of transcripts. *Proc. Natl. Acad. Sci. USA* **94**, 7388–7393.
19. Lam, L. T., Pham, Y. C. N., Nguyen, T. M., and Morris, G. E. (2000). Characterization of a monoclonal antibody panel shows that the myotonic dystrophy protein kinase, DMPK, is expressed almost exclusively in muscle and heart. *Hum. Mol. Genet.* **9**, 2167–2173.
20. Pham, Y. C. N., Man-Nguyen, T., Lam-Le, T., and Morris, G. E. (1998). Localization of myotonic dystrophy protein kinase in human and rabbit tissues using a new panel of monoclonal antibodies. *Hum. Mol. Genet.* **7**, 1957–1965.
21. Berger, B., Wilson, D., Wolf, E., Tonchev, T., Milla, M., and Kim, P. (1995). Predicting coiled coils by use of pairwise residue correlations. *Proc. Natl. Acad. Sci. USA* **92**, 8259–8263.
22. Woolfson, D. N., and Alber, T. (1995). Predicting oligomerization states of coiled coils. *Protein Sci.* **4**, 1596–1607.
23. Alber, T. (1992). Structure of the leucine zipper. *Curr. Opin. Genet. Dev.* **2**, 205–210.
24. Zhang, R., and Epstein, H. F. (2003). Homodimerization through coiled-coil regions enhances activity of the myotonic dystrophy protein kinase. *FEBS Lett.* **546**, 281–287.
25. Tiscornia, G., and Mahadevan, M. S. (2000). Myotonic dystrophy: The role of the CUG triplet repeats in splicing of a novel *DMPK* exon and altered cytoplasmic *DMPK* mRNA isoform ratios. *Mol. Cell* **5**, 959–967.
26. Wansink, D. G., van Herpen, R. E. M. A., Coerwinkel-Driessen, M. M., Groenen, P. J. T. A., Hemmings, B. A., and Wieringa, B. (2003). Alternative splicing controls myotonic dystrophy protein kinase structure, enzymatic activity, and subcellular localization. *Mol. Cell. Biol.* **23**, 5489–5501.
27. Muranyi, A., Zhang, R., Liu, F., Hirano, K., Ito, M., Epstein, H. F., and Hartshorne, D. J. (2001). Myotonic dystrophy protein kinase phosphorylates the myosin phosphatase targeting subunit and inhibits myosin phosphatase activity. *FEBS Lett.* **493**, 80–84.
28. Leung, T., Manser, E., Tan, L., and Lim, L. (1995). A novel serine/threonine kinase binding the Ras-related RhoA GTPase which translocates the kinase to peripheral membranes. *J. Biol. Chem.* **270**, 29051–29054.
29. Bush, E. W., Helmke, S. M., Birnbaum, R. A., and Perryman, M. B. (2000). Myotonic dystrophy protein kinase domains mediate localization, oligomerization, novel catalytic activity, and autoinhibition. *Biochemistry* **39**, 8480–8490.
30. Roe J. L., Durfee, T., Zupan, J. R., Repetti, P. P., McLean, B. G., and Zambryski, P. C. (1997). *TOUSLED* is a nuclear serine/threonine protein kinase that requires a coiled-coil region for oligomerization and catalytic activity. *J. Biol. Chem.* **272**, 5838–5845.
31. Berul, C. I., Maguire, C. T., Aronovitz, M. J., Greenwood, J., Miller, C., Gehrmann, J., Housman, D., Mendelsohn, M. E., and Reddy, S. (1999). DMPK dosage alterations result in atrioventricular conduction abnormalities in a mouse myotonic dystrophy model. *J. Clin. Invest.* **103**, 1–7.
32. Benders, A. A. G. M., Groenen, P. J. T. A., Oerlemans, F. T. J. J., Veerkamp, J. H., and Wieringa, B. (1997). Myotonic dystrophy protein kinase is involved in the modulation of the Ca^{2+} homeostasis in skeletal muscle cells. *J. Clin. Invest.* **100**, 1440–1447.
33. Jansen, G., Groenen, P. J. T. A., Bachner, D., Jap, P. H. K., Coerwinkel, M., Oerlemans, F., Van den Broek, W., Gohlsch, B., Pette, D., Plomp, J. J., Molenaar, P. C., Nederhoff, M. G. J., Van Echteld, C. J. A., Dekker, M., Berns, A., Hameister, H., and Wieringa, B. (1996). Abnormal myotonic dystrophy protein kinase levels produce only mild myopathy in mice. *Nat. Genet.* **13**, 316–324.
34. Krahe, R., Ashizawa, T., Abbruzzese, C., Roeder, E., Carango, P., Giacanelli, M., Funanage, V. L., and Siciliano, M. J. (1995). Effect of myotonic dystrophy trinucleotide repeat expansion on *DMPK* transcription and processing. *Genomics* **28**, 1–14.
35. Reddy, S., Smith, D. B. J., Rich, M. M., Leferovich, J. M., Reilly, P., Davis, B. M., Tran, K., Rayburn, H., Bronson, R., Cros, D., Balice-Gordon, R. J., and Housman, D. (1996). Mice lacking the myotonic dystrophy protein kinase develop a late onset progressive myopathy. *Nat. Genet.* **13**, 325–335.
36. Shimizu, M., Wang, W., Walch, E. T., Dunne, P. W., and Epstein, H. F. (2000). Rac-1 and Raf-1 kinases, components of distinct signaling pathways, activate myotonic dystrophy protein kinase. *FEBS Lett.* **23**, 273–277.
37. Marshall, C. J. (1995). Specificity of receptor tyrosine kinase signaling: Transient versus sustained extracellular signal-regulated kinase activation. *Cell* **80**, 179–185.
38. Kaliman, P., Catalucci, D., Lam, J. T., Kondo, R., Gutierrez, J. C., Reddy, S., Palacin, M., Zorzano, A., Chien, K. R., and Ruiz-Lozano, P. (2005). Myotonic dystrophy protein kinase phosphorylates phospholamban and regulates calcium uptake in cardiomyocyte sarcoplasmic reticulum. *J. Biol. Chem.* **280**, 8016–8021.
39. Timchenko, L., Nastainczyk, W., Schneider, T., Patel, B., Hofmann, F., and Caskey, C. T. (1995). Full-length myotonin protein kinase (72 kDa) displays serine kinase activity. *Proc. Natl. Acad. Sci. USA* **92**, 5366–5370.
40. Suzuki, A., Sugiyama, Y., Hayashi, Y., Nyu-i, N., Yoshida, M., Nonaka, I., Ishiura, S., Arahata, K., and Ohno, S. (1998). MKBP, a novel member of the small heat shock protein family, binds and

activates the myotonic dystrophy protein kinase. *J. Cell Biol.* **140**, 1113–1124.

41. Roberts, R., Timchenko, N. A., Miller, J. W., Reddy, S., Caskey, C. T., Swanson, M. S., and Timchenko, L. T. (1997). Altered phosphorylation and intracellular distribution of a (CUG)$_n$ triplet repeat RNA-binding protein in patients with myotonic dystrophy and in myotonin protein kinase knockout mice. *Proc. Natl. Acad. Sci. USA* **94**, 13221–13226.

42. Hartshorne, D. J., Ito M., and Erdodi, F. (1998). Myosin light chain phosphatase: Subunit composition, interactions and regulation. *J. Muscle Res. Cell Motil.* **19**, 325–341.

43. Alessi, D., MacDougall, L. K., Sola, M. M., Ikebe, M., and Cohen, P. (1992). The control of protein phosphatase-1 by targeting subunits: The major myosin phosphatase in avian smooth muscle is a novel form of protein phosphatase-1. *Eur. J. Biochem.* **210**, 1023–1035.

44. Fujioka, M., Takahashi, N., Odai, H., Araki, S., Ichikawa, K., Feng, J., Nakamura, M., Kaibuchi, K., Hartshorne, D. J., Nakano, T., and Ito, M. (1998). A new isoform of human myosin phosphatase targeting/regulatory subunit: cDNA cloning, tissue expression, and chromosomal mapping. *Genomics* **49**, 59–68.

45. Kaibuchi, K., Kuroda, S., and Amano, M. (1999). Regulation of the cytoskeleton and cell adhesion by the Rho family GTPases in mammalian cells. *Annu. Rev. Biochem.* **68**, 459–486.

46. Matsui, T., Amano, M., Yamamoto, T., Chihara, K., Nakafuku, M., Ito M., Nakano, T., Okawa, K., Iwamatsu, A., and Kaibuchi, K. (1996). Rho-associated kinase, a novel serine/threonine kinase, as a putative target for the small GTP binding protein Rho. *EMBO J.* **15**, 2208–2216.

47. Hagemann, D., Xiao, R. P. (2002). Dual site phospholamban phosphorylation and its physiological relevance in the heart. *Trends Cardiovasc. Med.* **12**, 51–56.

48. Luo, W., Chu, G., Sato, Y., Zhou, Z., Kadambi, V. J., and Kranias, E. G. (1998). Transgenic approaches to define the functional role of dual site phospholamban phosphorylation. *J. Biol. Chem.* **273**, 4734–4739.

49. Arai, M. (2000). Function and regulation of sarcoplasmic reticulum Ca^{2+}-ATPase: Advances during the past decade and prospects for the coming decade. *Jpn. Heart J.* **41**, 1–13.

50. Nurrish, S. J., and Treisman, R. (1995). DNA binding specificity determinants in MADS-box transcription factors. *Mol. Cell. Biol.* **15**, 4076–4085.

51. Gauthier-Rouviere, C., Vandromme, M., Tuil, D., Lautreou, N., Morris, M., Soulez, M., Kahn, A., Fernandez, A., and Lamb, N. (1996). Expression and activity of serum response factor is required for expression of the muscle-determining factor myoD in both dividing and differentiating mouse C2C12 myoblasts. *Mol. Boil. Cell.* **7**, 719–729.

52. Schratt, G., Philippar, U., Berger, J., Schwarz, H., Heidenreich, O., and Nordheim, A. (2002). Serum response factor is crucial for actin cytoskeletal organization and focal adhesion assembly in embryonic stem cells. *J. Cell Biol.* **156**, 737–750.

53. Johansen, F. E., and Prywes, R. (1995). Serum response factor: Transcriptional regulation of genes induced by growth factors and differentiation. *Biochim. Biophys. Acta* **1242**, 1–10.

54. Iyer, D., Belaguli, N., Flueck, M., Rowan, B. G., Wei, L., Weigel, N. L., Booth, F. W., Epstein, H. F., Schwartz, R. J., and Balasubramanyam, A., (2003). Novel phosphorylation target in the serum response factor MADS box regulates a-actin transcription. *Biochemistry* **42**, 7477–7486.

55. Bush, E. W., Taft, C. S., Meixell, G. E., and Perryman, M. B. (1996). Overexpression of myotonic dystrophy kinase in BC3H1 cells induces the skeletal muscle phenotype. *J. Biol. Chem.* **271**, 548–552.

56. Fluck, M., Carson, J. A., Schwartz, R. J., and Booth, F. W. (1999). SRF protein is upregulated during stretch-induced hypertrophy of rooster ALD muscle. *J. Appl. Physiol.* **86**, 1793–1799.

57. Camoretti-Mercado, B., Liu, H. W., Halayko, A. J., Forsythe, S. M., Kyle, J. W., Li, B., Fu, Y., McConville, J., Kogut, P., Vieira, J. E., Patel, N. M., Hershenson, M. B., Fuchs, E., Sinha, S., Miano, J. M., Parmacek, M. S., Burkhardt, J. K., and Solway, J. (2000). Physiological control of smooth muscle-specific gene expression through regulated nuclear translocation of serum response factor. *J. Biol. Chem.* **275**, 30387–30393.

58. Zhao, Y., Loyer, P., Li, H., Valentine, V., Kidd, V., and Kraft, A. S. (1997). Cloning and chromosomal location of a novel member of the myotonic dystrophy family of protein kinases. *J. Biol. Chem.* **272**, 10013–10020.

59. Wissmann, A., Ingles, J., McGhee, J. D., and Mains, P. E. (1997). *Caenorhabditis elegans* LET-502 is related to Rho-binding kinases and human myotonic dystrophy kinase and interacts genetically with a homology of the regulatory submit of smooth muscle myosin phosphatase to affect cell shape. *Genes Dev.* **11**, 409–422.

60. Luo, L., Lee, T., Tsai, L., Tang, G., Jan, L. Y., and Jan, Y. N. (1997). Genghis Khan (Gek) as a putative effector for Drosophila Cdc42 and regulator of actin polymerization. *Proc. Natl. Acad. Sci. USA* **94**, 12963–12968.

61. Di Cunto, F., Calautti, E., Hsiao, J., Ong, L., Topley, G., Turco, E., and Dotto, G. P. (1998). Citron Rho-interacting kinase, a novel tissue-specific ser/thr kinase encompassing the Rho-binding protein citron. *J. Biol. Chem.* **273**, 29706–29711.

62. Ridley, A. J., and Hall, A. (1992). The small GTP-binding protein rho regulates the assembly of focal adhesions and actin stress fibers in response to growth factors. *Cell* **70**, 389–399.

63. Sarkar, P. S., Han, J., Reddy, S. (2004). In-situ hybridization analysis of Dmpk mRNA in adult mouse tissues. *Neuromusc. Disord.* **14**, 497–506.

64. Balasubramanyam, A., Iyer, D., Stringer, J. L., Beaulieu, C., Potvin, A., Neumeyer, A. M., Avruch, J., and Epstein, H. F. (1998). Developmental changes in expression of myotonic dystrophy protein kinase in the rat central nervous system. *J. Comp. Neurol.* **394**, 309–325.

65. Dunne, P. W., Ma, L., Casey, D. L., and Epstein, H. F. (1996). Myotonic protein kinase expression in human and bovine lenses. *Biochem. Biophys. Res. Commun.* **225**, 281–288.

66. Winchester, C. L., Ferrier, R. K., Sermoni, A., Clark, B. J., and Johnson, K. J. (1999). Characterization of the expression of DMPK and SIX5 in the human eye and implications for pathogenesis in myotonic dystrophy. *Hum. Mol. Genet.* **8**, 481–492.

67. Jin, S., Shimizu, M., Balasubramanyam, A., and Epstein, H. F. (2000). Myotonic dystrophy protein kinase (DMPK) induces actin cytoskeletal reorganization and apoptotic-like blebbing in lens cells. *Cell Motil. Cytoskel.* **45**, 133–148.

68. Kobayashi, T., Shimokawa, M., Ishiura, S., Yamamoto, M., Kameda, N., Tanaka, H., Mizusawa, H., Ueda, H., and Ohno, S. (1997). Myotonic dystrophy protein kinase is a sarcoplasmic reticulum protein specifically localized in type I muscle fibers without colocalization of SERCA IIATPase. *Basic Appl. Myol.* **7**, 311–316.

69. Tachi, N., Kozuka, N., Ohya, K., Chiba, S., and Kikuchi, K. (1995). Expression of myotonic dystrophy protein kinase in biopsied muscles. *J. Neurol. Sci.* **132**, 61–64.

70. Salvatori, S., Biral, D., Furlan, S., and Marin, O. (1994). Identification and localization of the myotonic dystrophy gene product in skeletal and cardiac muscles. *Biochem. Biophys. Res. Commun.* **203**, 1365–1370.

71. Ueda, H., Shimokawa, M., Yamamoto, M., Kameda, N., Mizusawa, H., Baba, T., Terada, N., Fujii, Y., Ohno, S., Ishiura, S., and Kobayashib, T. (1999). Decreased expression of myotonic dystrophy protein kinase and disorganization of sarcoplasmic reticulum in skeletal muscle of myotonic dystrophy. *J. Neurol. Sci.* **162**, 38–50.

72. Shimokawa, M., Ishiura, S., Kameda, N., Yamamoto, M., Sasagawa, N., Saitoh, N., Sorimachi, H., Ueda, H., Ohno, S., Suzuki, K., and Kobayashi, T. (1997) Novel isoform of myotonin protein kinase. *Am. J. Pathol.* **150**, 1285–1295.
73. Furling, D., Lam, L. T., Agbulut, O., Butler-Browne, G. S., and Morris, G. E. (2003). Changes in myotonic dystrophy protein kinase levels and muscle development in congenital myotonic dystrophy. *Am. J. Pathol.* **162**, 1001–1009.
74. Storbeck, C. J., Sabourin, L. A., Waring, J. D., and Korneluk, R. G. (1998). Definition of regulatory sequence elements in the promoter region and the first intron of the myotonic dystrophy protein kinase gene. *J. Biol. Chem.* **273**, 9139–9147.
75. Endo, A., Motonaga, K., Arahata, K., Harada, K., Yamada, T., and Takashima, S. (2000). Developmental expression of myotonic dystrophy protein kinase in brain and its relevance to clinical phenotype. *Acta Neuropathol.* **100**, 513–520.
76. Ueda, H., Shimokawa, M., Yamamoto, M., Kameda, N., Mizusawa, H., Baba, T., Terada, N., Fujii, Y., Ohno, S., Ishiura, S., and Kobayashib, T. (1999). Decreased expression of myotonic dystrophy protein kinase and disorganization of sarcoplasmic reticulum in skeletal muscle of myotonic dystrophy. *J. Neurol. Sci.* **162**, 38–50.
77. Borgese, N., Gazzoni, I., Barberi, M., Colombo, S., and Pedrazzini, E. (2001). Targeting of a tail-anchored protein to endoplasmic reticulum and mitochondrial outer membrane by independent but competing pathways. *Mol. Biol. Cell* **12**, 2482–2496.
78. Borgese, N., Colombo, S., and Pedrazzini, E. (2003). The tale of tail-anchored proteins: Coming from the cytosol and looking for a membrane. *J. Cell Biol.* **161**, 1013–1019.
79. Herpen, R. E. M. A., Oude, R. J. A., Mietske, O., Miranda, W., Bennink, B., Loo, A. J. d., Fransen, J., Wieringa, B., and Wansink, D. G. (2005). Divergent mitochondrial and endoplasmic reticulum association of DMPK splice isoforms depends on unique sequence arrangements in tail anchors. *Mol. Cell. Biol.* **25**, 1402–1414.
80. Masaki, R., Yamamoto, A., and Tashiro, Y. (1994). Microsomal aldehyde dehydrogenase is localized to the endoplasmic reticulum via its carboxy-terminal 35 amino acids. *J. Cell. Biol.* **126**, 1407–1420.
81. Pall, G. S., Johnson, K. J., and Smith, G. L. (2003). Abnormal contractile activity and calcium cycling in cardiac myocytes isolated from dmpk knockout mice. *Physiol. Genom.* **13**, 139–146.
82. Lee, H. C., Patel, M. K., Mistry, D. J., Wang, Q., Reddy, S. Moorman, J. R., and Mounsey, J. P. (2003). Abnormal Na channel gating in murine cardiac myocytes deficient in myotonic dystrophy protein kinase. *Physiol. Genom.* **12**, 147–157.
83. Bendahhou, S., Cummins, T. R., Potts, J. F., Tong, J., and Agnew, W. S. (1995). Serine-1321-independent regulation of the adult skeletal muscle Na channel by protein kinase C. *Proc. Natl. Acad. Sci. USA* **92**, 12003–12007.
84. Frohnwieser, B., Weigl, L., and Schreibmayer, W. (1995). Modulation of cardiac sodium channel isoform by cyclic AMP dependent protein kinase does not depend on phosphorylation of serine 1504 in the cytosolic loop interconnecting transmembrane domains III and IV. *Pflueger's Arch.* **430**, 751–753.
85. Murray, K. T., Hu, N. N., Daw, J. R., Shin, H. G., Watson, M. T., Mashburn, A. B., and George, A. L. J. (1997) Functional effects of protein kinase C activation on the human cardiac Na channel. *Circ. Res.* **80**, 370–376.
86. Qu, Y., Rogers, J., Tanada, T., Scheuer, T., and Catterall, W. A. (1994). Modulation of cardiac Na channels expressed in a mammalian cell line and in ventricular myocytes by protein kinase C. *Proc. Natl. Acad. Sci. USA* **91**, 3289–3293.
87. Whiting, E. J., Waring, J. D., Tamai, K., Somerville, J., Hincke, M., Staines, W. A., Ikeda, J. E., and Korneluk, R. G. (1995). Characterization of myotonic dystrophy kinase (DMK) protein in human and rodent muscle and central nervous tissue. *Hum. Mol. Genet.* **4**, 1063–1072.
88. Kim, C. H., and Lisman, J. E. (1999). A role of actin filament in synaptic transmission and long-term potentiation. *J. Neurosci.* **19**, 4314–4324.
89. Engert, F., and Bonhoeffer, T. (1999). Dendritic spine changes associated with hippocampal long-term synaptic plasticity. *Nature* **399**, 66–70.
90. Fischer, M., Kaech, S., Wagner, U., Brinkhaus, H., and Matus, A. (2000). Glutamate receptors regulate actin-based plasticity in dendritic spines. *Nat. Neurosci.* **3**, 887–894.
91. Schulz, P. E., McIntosh, A. D., Kasten, M. R., Wieringa, B., and Epstein, H. F. (2003). A role for myotonic dystrophy protein kinase in synaptic plasticity. *J. Neurophysiol.* **89**, 1177–1186.
92. Eriksson, M., Ansved, T., Edstroem, L., Anvret, M., and Carey, N. (1999). Simultaneous analysis of expression of the three myotonic dystrophy locus genes in adult skeletal muscle samples: The CTG expansion correlates inversely with DMPK and 59 expression levels, but not DMAHP levels. *Hum. Mol. Genet.* **8**, 1053–1060.
93. Frisch, R., Singleton, K. R., Moses, P. A., Gonzalez, I. L., Carango, P., Marks, H. G., and Funanage, V. L. (2001). Effect of triplet repeat expansion on chromatin structure and expression of *DMPK* and neighboring genes, *SIX5* and *DMWD*, in myotonic dystrophy. *Mol. Genet. Metab.* **74**, 281–291.
94. Taneja, K. L., McCurrach, M., Schalling, M., Housman, D., and Singer, R. H. (1995) Foci of trinucleotide repeat transcripts in nuclei of myotonic dystrophy cells and tissues. *J. Cell Biol.* **128**, 995–1002.
95. Timchenko, L. T., Miller, J. W., Timchenko, N. A., DeVore, D. R., Datar, K. V., Lin, L., Roberts, R., Caskey, C. T., and Swanson, M. S. (1996). Identification of a $(CUG)_n$ triplet repeat RNA-binding protein and its expression in myotonic dystrophy. *Nucleic Acids Res.* **24**, 4407–4414.
96. Philips, A. V., Timchenko, L. T., and Cooper, T. A. (1998). Disruption of splicing regulated by a CUG-binding protein in myotonic dystrophy. *Science* **280**, 737–741.
97. Lu, X., Timchenko, N. A., and Timchenko, L. T. (1999). Cardiac elav-type RNA-binding protein (ETR-3) binds to RNA CUG repeats expanded in myotonic dystrophy. *Hum. Mol. Genet.* **8**, 53–60.
98. Miller, J. W., Urbinati, C. R., Teng-Umnuay, P., Stenberg, M. G., Byrne, B. J., Thornton, C. A., and Swanson, M. S. (2000). Recruitment of human muscleblind proteins to $(CUG)_{(n)}$ expansions associated with myotonic dystrophy. *EMBO J.* **19**, 4439–4448.
99. Savkur, R. S., Philips, A. V., and Cooper, T. A. (2001). Aberrant regulation of insulin receptor alternative splicing is associated with insulin resistance in myotonic dystrophy. *Nat. Genet.* **29**, 40–47.
100. Mankodi, A., Logigian, E., Callahan, L., McClain, C., White, R., Henderson, D., Krym, M., and Thornton, C. A. (2000). Myotonic dystrophy in transgenic mice expressing an expanded CUG repeat. *Science* **289**, 1769–1773.
101. Mankodi, A., and Thornton, C. A. (2002). Myotonic syndromes. *Curr. Opin. Neurol.* **15**, 545–552.
102. Koch, M. C., Steinmeyer, K., Lorenz, C., Ricker, K., Wolf, F., Otto, M., Zoll, B., Lehmann-Horn, F., Grzeschik, K.-H., and Jentsch, T. J. (1992). The skeletal muscle chloride channel in dominant and recessive human myotonia. *Science* **257**, 797–800.
103. Wu, F. F., Takahashi, M. P., Pegoraro, E., Angelini, C., Colleselli, P., Cannon, S. C., and Hoffman, E. P. (2001). A new mutation in a family with cold-aggravated myotonia disrupts Na channel inactivation. *Neurology* **56**, 878–884.
104. Fardaei, M., Rogers, M. T., Thorpe, H. M., Larkin, K., Hamshere, M. G., Harper, P. S., and Brook, J. D. (2002). Three proteins, *MBNL*, *MBLL* and *MBXL*, co-localize *in-vivo* with nuclear foci

of expanded-repeat transcripts in DM1 and DM2 cells. *Hum. Mol. Genet.* **11**, 805–814.
105. Kanadia, R. N., Johnstone, K. A., Mankodi, A., Lungu, C., Thornton, C. A., Esson, D., Timmers, A. M., Hauswirth, W. W., and Swanson, M. S. (2003). A muscleblind knockout model for myotonic dystrophy. *Science* **302**, 1978–1980.
106. Kanadia, R. N., Urbinatia, C. R., Crussellea, V. J., Luob, D., Leec, Y., Harrisonb, J. K., Ohc, S. P., and Swansona, M. S., (2003). Developmental expression of mouse muscleblind genes *Mbnl1*, *Mbnl2* and *Mbnl3*. *Gene Express. Patterns* **3**, 459–462.
107. Mankodi, A., Urbinati, C. R., Yuan, Q. P., Moxley, R. T., Sansone, V., Krym, M., Henderson, D., Schalling, M., Swanson, M. S., and Thornton, C. A. (2001). Muscleblind localizes to nuclear foci of aberrant RNA in myotonic dystrophy types 1 and 2. *Hum. Mol. Genet.* **10**, 2165–2170.
108. Hoffmann-Radvanyi, H., Lavedan, C., Rabes, J. P., Savoy, D., Duros, C., Johnson, K., and Junien, C. (1993). Myotonic dystrophy: Absence of CTG enlarged transcript in congenital forms, and low expression of the normal allele. *Hum. Mol. Genet.* **2**, 1263–1266.
109. Timchenko, N. A., Cai, Z. J., Welm, A. L., Reddy, S., Ashizawa, T., and Timchenko, L. T. (2001). RNA CUG repeats sequester CUGBP1 and alter protein levels and activity of CUGBP1. *J. Biol. Chem.* **276**, 7820–7826.

CHAPTER 7

Clinical and Genetic Features of Myotonic Dystrophy Type 2

JAMIE M. MARGOLIS, LAURA P. W. RANUM, AND JOHN W. DAY

Department of Genetics, Cell Biology, and Development, Department of Neurology, and the Institute of Human Genetics, University of Minnesota, Minneapolis, Minnesota 55455

I. Introduction
II. Genetic Features of DM2
   A. History
   B. DM2 Gene Identification
   C. Haplotype Analysis and Conservation
   D. Somatic Mosaicism
   E. Diagnostic Methods
   F. Intergenerational Changes
   G. Genotype–Phenotype Correlation
III. Clinical Features of Myotonic Dystrophy
   A. Muscle Pathology
   B. Multisystemic Features
   C. Central Nervous System Involvement
IV. Pathophysiological Models
   A. DM Pathogenic Models prior to DM2
   B. Identification of DM2 Indicates Breadth of RNA Effects in DM Pathogenesis
   C. Gain-of-Function RNA Model
V. CUG-BP and Muscleblind
   A. CUG-BP
   B. Muscleblind
   C. Downstream Targets of CUG-BP and Muscleblind
VI. Potential Causes of Clinical Distinctions between DM1 and DM2
VII. Conclusions
   References

Myotonic dystrophy (DM), the most common form of muscular dystrophy in adults, is a dominantly inherited disease characterized by a complex array of unusual multisystemic clinical features. After a CTG expansion in the 3′ untranslated region of the *DMPK* gene on chromosome 19 was found to cause the first identified form of DM (DM1), some affected families were found that did not carry that mutation, demonstrating the existence of a second genetic form of the disease (DM2). In 2001, we showed that DM2 is caused by a CCTG expansion in intron 1 of the *ZNF9* gene on chromosome 3. Detailed molecular and clinical comparisons of DM1 and DM2 have helped to simplify the pathogenic models of DM, because the striking genotypic and phenotypic parallels between the two diseases suggest a common disease mechanism. The fact that both mutations involve repeat tracts that are transcribed but not translated strongly supports a pathogenic model in which toxic effects of RNA containing either CUG or CCUG expansions lead to the peculiar multisystemic features of this disease.

I. INTRODUCTION

Myotonic dystrophy (DM), the most common form of adult-onset muscular dystrophy, is a dominantly inherited multisystemic disorder characterized by seemingly unrelated clinical features including myotonia, muscular dystrophy, cataracts, cardiac conduction defects, and specific endocrine abnormalities [1–3]. When the mutation that causes myotonic dystrophy type 1 (DM1) was identified as a CTG expansion in the 3′ untranslated region of the *myotonia dystrophia protein kinase (DMPK)* gene in 1992 [4, 5], it led to the awareness that a second genetic form existed, because some individuals clinically diagnosed with DM did not have the DM1 mutation [6–8]. In 2001, the mutation responsible for myotonic dystrophy type 2 (DM2) was found to be a tetranucleotide CCTG expansion in the first intron of the *zinc finger protein 9* gene *(ZNF9)* [9]. As both DM1 and DM2 are caused by markedly expanded microsatellite repeats that are transcribed but not translated, the comparative study of these two diseases strongly supported the hypothesis that they are both caused by toxic effects of RNA containing CUG or CCUG repeat expansions. Additional investigations of these two diseases will further clarify the pathophysiological features that constitute the complex multisystemic DM phenotype.

II. GENETIC FEATURES OF DM2

A. History

In 1909, Steinert, Batten, and Gibb identified myotonic dystrophy as a multisystemic disorder that is now recognized as the most common form of adult muscular dystrophy [10, 11]. However, it was not until 1992 that the genetic mutation responsible for the first form of DM (DM1) was identified as a CTG trinucleotide repeat expansion in the 3′ untranslated region of the serine–threonine kinase *myotonia dystrophica protein kinase* (*DMPK*) gene and in the promoter region of the neighboring homeodomain gene *SIX5* on chromosome 19q13 [4, 5, 12–14]. DM1 was the first dominantly inherited disease found to be caused by an untranslated trinucleotide repeat expansion; the repeat is transcribed into RNA but is not translated into a protein. Affected individuals have repeat sizes ranging between 50 and 4000 CTGs, whereas unaffected individuals' repeat sizes range between 5 and 37 CTGs.

B. DM2 Gene Identification

When the DM1 mutation was identified, some families diagnosed with DM were shown not to carry a CTG repeat expansion in *DMPK* [8], indicating that at least one additional mutation could cause this multisystemic phenotype. Several terms were coined to describe this syndrome, including proximal myotonic myopathy (PROMM), proximal myotonic dystrophy (PDM), and myotonic dystrophy type 2 (DM2) [6, 7, 15, 16]. In 1998, DM2 was mapped to chromosome 3q21 [16], and subsequently most families with PROMM and PDM were linked to the DM2 locus [17]. The location was further refined by analysis of recombinant chromosomes and with a linkage disequilibrium approach using ~100 newly developed short tandem repeat markers [9]. One of the markers in linkage disequilibrium segregated abnormally: affected individuals had only a single observable allele by polymerase chain reaction (PCR), and affected children often did not appear to inherit an allele from their affected parent. Southern analysis showed that the aberrant segregation pattern was due to the presence of a large expansion that could not be amplified by PCR. This large expansion was subsequently found by Southern analysis in all affected individuals, but not in 1360 control chromosomes [9].

The DM2 mutation is a tetranucleotide CCTG repeat expansion located in the first intron of the *zinc finger protein 9* (*ZNF9*) gene on chromosome 3q21 (Fig. 7-1) [9]. The DM2 expansion is the largest microsatellite expansion reported, with affected individuals carrying between 75 and 11,000 CCTG repeats (mean, 5000 CCTGs), with normal alleles having ≤26 CCTG repeats [9]. The DM2 CCTG repeat is part of the complex repeat motif $(TG)_n(TCTG)_n(CCTG)_n$, but on pathogenic alleles only the CCTG repeat is expanded. On normal alleles, the CCTG repeat tract is generally interrupted; interruptions are generally lost on expanded DM2 alleles (Figs. 7-2 and 7-4) [9].

C. Haplotype Analysis and Conservation

To assess the ancestral origins of the DM2 expansion, haplotype analysis of 71 DM2 families was performed (Fig. 7-3) [18]. The majority of the families used in one study were of German or European descent, along with a single family of Afghan descent. Near the DM2 repeat tract, there is a single 127-kb haplotype for all of these families, indicating a common founder. Over a broader region of ~700 kb encompassing the repeat, all of these families have one of three haplotypes, denoted as A, B, or C, in which the B and C haplotypes are variants of the consensus A haplotype. The B and C haplotypes appear to be derived from the A haplotype by a limited number

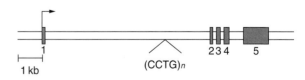

FIGURE 7-1 Genomic organization of the *ZNF9* gene. The position of the DM2 expansion in intron 1 is shown. The gene spans 11.3 kb of genomic sequence with an open reading frame of 1.5 kb. Figure reproduced with permission from *Science* 293, 864–867 (2001).

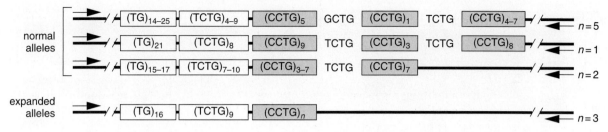

FIGURE 7-2 Schematic diagram of the DM2 expansion region, showing sequence configurations of normal and expanded repeat tracts. Interruptions present within the CCTG repeat of normal alleles are absent in expanded alleles. Figure reproduced with permission from *Science*, **293**, 864–867 (2001).

of ancestral recombination and microsatellite instability events. Similar to DM1, the haplotype conservation among the DM2 families suggests that DM2 arose from a single founder, at least in patients of European or Afghan descent [18]. Additionally, the conservation of haplotype A in the Afghan family allows the speculation that the DM2 mutation is an old mutation that was introduced into the Afghan gene pool sometime between 2000 and 1000 BC, when the ancient Aryan tribes of Indo-European extraction settled in Aryana (ancient Afghanistan). Similar haplotype results were found in a separate large panel of DM2 families [19].

Haplotype analysis of 228 normal chromosomes indicated that there are only rare perfect matches to the three major DM2 haplotypes in control samples [18]. A single control chromosome with a haplotype identical to a large portion of the B haplotype, the most common haplotype in DM2 families, was identified. This chromosome contains an uninterrupted CCTG tract of 20 repeats [18]. The largest normal DM2 repeat tract previously sequenced contains 26 CCTGs with two interruptions, giving an overall repeat tract of $(CCTG)_{12}$ *GCTG CCTG TCTG* $(CCTG)_{11}$ [9]. Because sequence interruptions are normally observed in controls and are thought to stabilize the repeat tract in other diseases [20–23], the lack of interruptions on this normal allele from the DM2 haplotype suggests that this allele represents a pool of premutation alleles for DM2 that may serve as a pool for further expansions (Fig. 7-4).

To gain insight into the function of the repeat, the evolutionary conservation of the repeat tract was analyzed in chimpanzee, gorilla, mouse, and rat. The repeat tracts of the animals that were tested were similar, but not identical, to the human DM2 repeat tract (Fig. 7-5) [18]. The TG portion of the repeat is found in human, chimpanzee, gorilla, mouse, and rat. The TCTG portion of the DM2 repeat is found in human, chimpanzee, gorilla, and mouse. The CCTG tract is interrupted in normal human and gorilla, whereas the chimpanzee has an uninterrupted CCTG tract [18]. Additionally, a 200-bp sequence 3' of the repeat tract is conserved between the human and all other mammals tested [18], suggesting a biological function for the repeat and 3' sequence.

D. Somatic Mosaicism

The DM2 mutation is the largest microsatellite expansion reported, with a mean of 5000 and a range of 75 to more than 11,000 CCTG repeats in affected individuals [9]. Distinct from DM1, there is marked repeat length heterogeneity in almost all individuals, with Southern analysis of DNA isolated from blood showing either broad smears of repeat lengths or several discrete bands (Fig. 7-6A). Further evidence of somatic instability comes from a pair of monozygotic twins who, at 31 years of age, had repeat expansions that differed in size by 11 kb (13 and 24 kb) (Fig. 7-6B) [9]. The size of the CCTG repeat tract can increase in a single individual over time, as was shown by one individual who had an increase of 2 kb over 3 years (Fig. 7-6C). This time-dependent somatic mosaicism leads to a positive correlation between the age of the individual at the time of the blood draw and the number of CCTG repeats observed on Southern analysis (Fig. 7-6D). The somatic instability and the tendency for expansions to increase in length over time have made the pathogenic threshold of the DM2 expansion difficult to define because affected individuals with the smallest expansions also have large repeats, which prevent an assessment of the effects of the smaller alleles.

E. Diagnostic Methods

The somatic instability and the unprecedented size of the DM2 expansion pose diagnostic problems not previously encountered with other expansion disorders, and complicate the molecular methods and interpretation of genetic testing. Consequently, to achieve the specificity and sensitivity required for clinically accurate detection of DM2 expansions, a reliable battery of assays includes PCR, Southern analysis, and a PCR-based repeat assay.

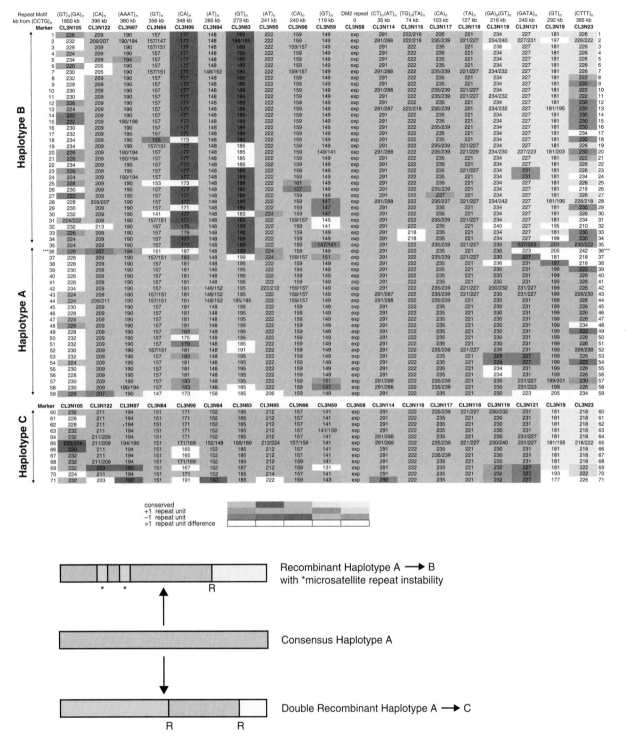

FIGURE 7-3 Haplotypes of 71 DM2 families. The three major affected haplotypes found in 71 DM2 families analyzed (A, B, and C) are shown. The haplotype between CL3N95 and CL3N118 is conserved in all affected alleles. The consensus haplotype A is indicated in gold. Minor deviations in repeat size are indicated by alternative colors with a color key located below the figure. The markers span 2.2 Mb, and the distance of each marker from the DM2 CCTG expansion is denoted at the top of the figure. The STR marker name and the repeat motif associated with each are designated. A schematic diagram of the proposed ancestral origin of DM2 haplotypes is shown at the bottom of the figure. The proposed ancestral relationship between the major haplotype variants is shown, with haplotypes B and C related to haplotype A by a small number of ancestral recombination and microsatellite instability events. See CD-ROM for color image. Permission requested to reproduce figure from University of Chicago Press and Liquori et al., *Am. J. Hum. Genet.* **73**, 849–862 (2003). © 2003 by The American Society of Human Genetics. All rights reserved.

Chapter 7 Clinical and Genetic Features of Myotonic Dystrophy Type 2

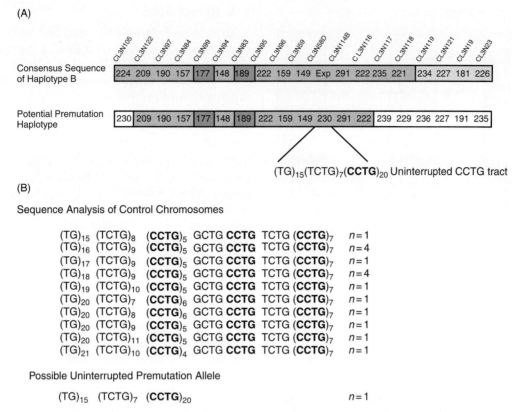

FIGURE 7-4 Identification of potential premutation allele in the general population. (A) The consensus between haplotype B and a possible premutation allele in an unaffected control is depicted. Identical coloring indicates the regions where the haplotypes are shared. The sequence of the DM2 expansion region in the possible premutation allele is indicated below the haplotype. (B) Schematic diagram showing the repeat configurations of the DM2 expansion region on 24 normal control alleles, 3 expanded affected alleles, and the putative premutation allele. See CD-ROM for color image. Permission requested to reproduce figure from University of Chicago Press and Liquori et al., *Am. J. Hum. Genet.* **73**, 849–862 (2003). © 2003 by the American Society of Human Genetics. All rights reserved.

1. PCR Assay

Most unaffected individuals can be genetically identified by the presence of two normal-sized DM2 alleles on routine PCR. Because CCTG expansions are too large to amplify by PCR, expansion-positive individuals have a single band on PCR representing their normal allele and, thus, are indistinguishable from the 15% of unaffected controls who are truly homozygous [24]. However, expansion carriers can be distinguished from homozygotes via family studies (Fig. 7-7A), as the affected offspring often do not appear to inherit an allele from their affected parent. This apparent non-mendelian inheritance pattern, referred to as the presence of a "blank allele," occurs because the expanded allele fails to amplify. Although it can occur in other circumstances, such as misidentified paternity, the presence of a blank allele provides strong support that a family carries the DM2 mutation [24].

FIGURE 7-5 Evolutionary comparisons of the DM2 repeat tract. Schematic diagram of DM2 repeat region, showing consensus sequence configurations found in human, chimpanzee, gorilla, mouse, and rat. Permission requested to reproduce figure from University of Chicago Press and Liquori et al., *Am. J. Hum. Genet.* **73**, 849–862 (2003). © 2003 by The American Society of Human Genetics. All rights reserved.

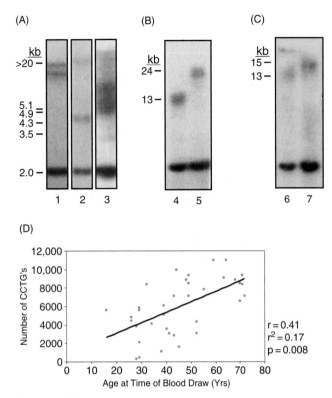

FIGURE 7-6 Instability of the DM2 expansion. (A) Somatic heterogeneity in blood. Southern blots of *BsoB*I-digested genomic DNA from blood revealed multiple expanded alleles in some affected individuals, some discrete in size (lanes 1 and 2) and others broad (lane 3). (B) Southern blots of *Eco*RI-digested genomic DNA from blood of monozygotic twins (lanes 4 and 5). (C) Expanded alleles increase in length over time. Southern blot of *Eco*RI-digested genomic DNA samples from blood taken from a single subject at 28 (lane 6) and 31 (lane 7) years of age, respectively. (D) Correlation between the size of the expanded allele in individuals with a single allele and age at the time the blood sample was taken. Figure reproduced with permission from *Science* **293**, 864–867 (2001).

2. Southern Analysis

The DM2 expansion is usually detectable by Southern analysis, but its unparalleled size and dramatic instability diminish the reliability of this diagnostic method. The size heterogeneity is so great that 20% of DM2 expansions are not detectable by Southern analysis [24]. The expanded alleles that are detectable by Southern analysis can appear as a single discrete band, multiple bands, or a broad smear containing many different repeat sizes (Fig. 7-7B). For other expansion disorders, such as DM1 and SCA8, somatic mosaicism is limited, and the expansion and control alleles are detected as equally intense bands by Southern analysis (Fig. 7-7B, lane 8). In contrast, for DM2, the detectable expansions are almost always less intense than the normal allele, indicating that even when a discrete band can be visualized, the rest of the expanded allele is varied in size and migrates as a diffuse smear containing many different repeat sizes [24].

3. Repeat Assay

To detect expansions not seen on Southern analysis, a repeat assay was created for DM2 using methods previously developed for DM1 and SCA10. This assay uses a PCR from a unique sequence flanking the DM2 expansion, with reverse primers at various sites within the CCTG tract, to generate a broad smear with molecular weights invariably higher than those from control DNA (Figs. 7-7C and D) [24]; specificity of the reaction is obtained by probing the product with a specific internal oligonucleotide probe. Although it provides no information about overall repeat expansion size, when performed properly, the repeat assay is sensitive and specific, and increases the rate of DM2 expansion detection to 99%, compared with the 80% detection rate with Southern analysis [24], with no false-positive results in 320 control chromosomes.

4. Other Diagnostic Methods

Two forms of *in situ* hybridization have been reported to detect the DM2 repeat [25]. Chromogenic *in situ* hybridization visualized both the genomic expansion and mutant transcripts in muscle biopsies, and fluorescence *in situ* hybridization visualized the DM2 expansion and allowed an estimation of repeat size in extended DNA fibers [25]. Clarification of the sensitivity and specificity of these methods may allow them to be used as diagnostic assays. Although the latter method provides a measure of expansion size of individual repeat tracts, the known somatic heterogeneity of repeat size for individual patients means that individual molecules from a given patient will dramatically vary in size.

F. Intergenerational Changes

The phenomenon of anticipation, the lowering of the age of onset or an increase in disease severity in successive generations, is a striking clinical hallmark of DM1. Earlier ages of onset for subsequent generations, based on clinical criteria, have also been reported in DM2 [24, 26]. However, in a large study, the expected trend of longer expansions in patients with an earlier age of disease onset was not observed. In fact, affected offspring had shorter expansions than affected parents, after either maternal or paternal transmission, raising the possibility that observed clinical differences between generations in DM2 families represented ascertainment biases. Furthermore, the somatic heterogeneity of the DM2 expansion and the fact that repeat size increases with age complicate this analysis and may mask meaningful biological effects of repeat length on disease onset and severity. However, it is also possible that there is a

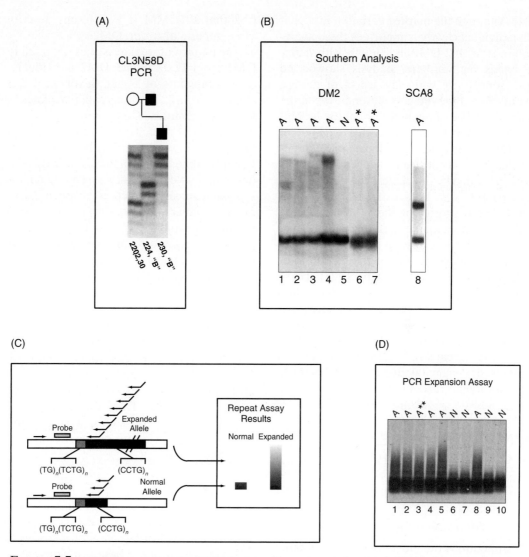

FIGURE 7-7 (A) PCR analysis of the CL3N58 marker. The genotype of each individual is shown in base pairs. Alleles too large to amplify by PCR, which are refered to as "blank alleles" and indicated by a "B," make the segregation of the markers appear nonmendelian. (B) Expansion detection by genomic Southern analysis. DM2 Southern analysis of genomic DNA from control (N) and affected individuals with detectable (A) and nondetectable (A*) expanded alleles is shown (lanes 1–7). In contrast to DM2, an SCA8 Southern blot (lane 8) shows equally intense signals for the normal and expanded alleles. (C) Schematic diagram of the PCR-based repeat assay. The straight arrow represents the flanking primer CL3N58-D R. Tailed arrows represent the JJP4CAGG primer. A third primer (JJP3, not shown), used to make the PCR more robust, has the same sequence as the hanging tail of JJP4CAGG. The primer used to probe Southern blots of the PCR products is CL3N58-E R. (D) Repeat assay results for affected individuals with expansions that were detected (A) or not detected (A**) by genomic Southern analysis are shown in lanes 1–5 and 8. Negative results from unaffected controls (N) are shown in lanes 6, 7, 9, and 10. Permission requested to reproduce figure from Lippincott, Williams and Wilkins and Day et al. (2003). Myotonic dystrophy type 2: Molecular, diagnostic and clinical spectrum. *Neurology* **60** (4), 657–664.

ceiling effect, wherein expansions of all sizes greater than a maximal pathogenic size exert similar effects irrespective of their length. Similarly, in adult-onset DM1, the most significant correlations between repeat length and disease onset are for repeats less than 400 CTGs [27], with occasional dramatic disparities in clinical presentations of patients with larger repeat expansions.

G. Genotype–Phenotype Correlation

In other microsatellite expansion disorders, the size of the repeat expansion is inversely correlated with the age of onset, with larger repeats associated with an earlier age of onset and greater disease severity. However, for DM2 this classic feature of repeat disorders does not hold true [24]. Although the marked variation in repeat size for each individual clearly complicates these analyses, information from 102 DM2 individuals with definable single bands on Southern analysis showed no correlation of repeat size with any aspect of disease onset or severity. The observed correlation between the DM2 repeat size and the individual's age at the time of the blood draw further demonstrates the tendency of repeats to increase over time. An independent demonstration that the DM2 repeat size is independent of disease severity is provided by the clinical characterization of siblings of parents who both carried DM2 expansions, in which homozygosity of the DM2 expansion [28] resulted in no discernable difference in age of onset, pattern of symptoms, disease progression, brain imaging, muscle pathology, or size of the repeat compared with heterozygous family members [28]. These data support the interpretation that there is a ceiling effect for DM2 expansions, and that further increases in size of the DM2 expansion, or presence of more copies of the DM2 expansion, do not result in greater pathogenicity. The role of repeat size in DM1 is complicated by the occurrence of congenital effects, which are more likely to occur for longer repeat expansions.

III. CLINICAL FEATURES OF MYOTONIC DYSTROPHY

A. Muscle Pathology

1. Pattern of Weakness

Muscle dysfunction is the most common complaint in DM1 and DM2. For both diseases, manual motor testing reveals early involvement of the neck flexors and lateral deep finger flexors [1, 8, 24, 29, 30]. Facial weakness, accompanied by ptosis, a common feature in DM1, is also seen in DM2. Although finger flexor weakness is an early sign of DM2 on neurological examination, this weakness is not a common concern or complaint of DM2 patients. Compared with DM1, which commonly causes early functional problems related to loss of dexterity and grip, initial symptoms of weakness in DM2 subjects more typically involve proximal lower extremity weakness that causes problems arising from chairs or climbing stairs. This symptomatic hip girdle weakness led some investigators to refer to this disorder as proximal myotonic myopathy (PROMM) [7], but neurological examination of asymptomatic family members shows that neck flexor and finger flexor weakness typically precedes difficulties with proximal lower extremities; collaborative studies of genetically confirmed subjects have clarified that PROMM is a different description of the same genetic disorder, DM2.

The pattern of weakness does not reliably distinguish DM1 from DM2. Some DM2 individuals do present with complaints of finger flexor weakness at a time when they have no discernable weakness of proximal lower extremities. Comparably, some subjects with adult-onset DM1 come to medical attention because of difficulty climbing stairs without complaints of finger flexor weakness, though they, like DM2 patients, always have finger flexor weakness on examination. A more reliable clinical distinction between DM1 and DM2 might be volar forearm atrophy, or hypotrophy, which occurs to a variable degree in DM1 but not in DM2, and is possibly related to the congenital features of DM1 that do not appear to be significantly recapitulated in DM2 [1, 7, 24, 31, 32].

2. Myotonia

Electrical myotonia, the persistent spontaneous reactivation of muscle fibers that is a hallmark of DM, is readily detected by electromyography in almost all DM1 and DM2 patients [1, 24]. Although seen in both DM1 and DM2, the myotonia found in DM1 tends to be more symptomatic. Even though nearly 75% of 234 DM2 patients tested also displayed grip and percussion myotonia [24], grip and percussion myotonia is more frequent in DM1.

3. Histology

The histological features of muscle biopsies obtained from both DM1 and DM2 patients are strikingly similar with a stereotypical presentation, allowing the diagnosis of DM to be suggested based on biopsy alone [1, 24, 33]. Atrophic fibers, extreme fiber size variation, as well as severely atrophic fibers with pyknotic myonuclei are found in both DM1 and DM2. Additionally, an extensive proliferation of centrally located nuclei exists in both diseases, with some fibers showing chains of central nuclei. Depending on the extent of muscle involvement, necrosis, fibrosis, and adipose deposition are present to variable degrees in both diseases. Type 1 fiber atrophy is a recognized feature of DM1, but not DM2, and the severely atrophic fibers in DM2 are predominantly of type 2 [33, 34].

B. Multisystemic Features

1. Cataracts

Posterior subcapsular iridescent multicolored opacities are seen in both DM1 and DM2 and may appear very early in the disease process, with an onset in the second decade or later, and can be detected on slit-lamp examination [1, 24]. When seen in profusion bilaterally, as they are in almost all individuals with DM1 and DM2 older than 20 years when studied by slit-lamp examination, they are essentially pathognomonic for DM. Surgical intervention may be required to restore visual acuity as early as the third decade of life [24].

2. Heart

Cardiac features of DM1 and DM2 include atrioventricular and intraventricular conduction defects, atrial fibrillation, and ventricular arrhythmias [1, 24, 35]. The development of potentially lethal arrhythmias cannot be easily predicted by routine electrocardiograms and does not correlate with development or severity of conduction defects [36]. Several subjects with DM2 also demonstrate congestive heart failure, suggesting that a primary cardiomyopathy may be a feature of this disease [24].

3. Endocrine and Other Systemic Features

Endocrine abnormalities reported in both DM1 and DM2 include hypotestosteronism, with elevated follicle-stimulating hormone (FSH) levels [1, 24], and insulin insensitivity (hyperinsulinemia, hyperglycemia) leading to a predisposition for type 2 diabetes in both DM1 and DM2 [37, 38]. Oligospermia with testicular atrophy and reduced male fertility also occurs in both diseases. Additional serological abnormalities include hypogammaglobulinemia, which results in reduction of both IgG and IgM, but not IgA, in both DM1 and DM2 [1, 24]. Hyperhidrosis is seen in both DM1 and DM2 [8, 39], and can sometimes be severe.

C. Central Nervous System Involvement

Central nervous system (CNS) involvement in DM includes both developmental and degenerative changes [24, 40–44]. The developmental CNS effects of DM1 are characterized by mental retardation and other cognitive and personality changes. Although there have been isolated reports of DM2 patients with mental retardation, mental retardation has not been causally associated with that disorder [8], but cognitive abnormalities, most prominently in visual–spatial and frontal lobe function, have been reported for DM2 [45]. Psychological dysfunction and abnormal white matter changes on MRI are characteristic of the degenerative changes seen in DM1 [41, 43], and similar white matter abnormalities have been seen in DM2 [46]. Longitudinal quantitative studies are required to better define the etiology and significance of degenerative CNS features of both diseases.

IV. PATHOPHYSIOLOGICAL MODELS

A. DM Pathogenic Models prior to DM2

After its identification in 1992, the pathogenic effects of the DM1 mutation were unclear because of its location in a noncoding region of the *DMPK* gene. Some of the proposed disease mechanisms included (1) haploinsufficiency of *DMPK*; (2) haploinsufficiency of *SIX5* and neighboring genes; and (3) RNA pathogenesis. *DMPK* knockout mice showed late-onset myopathy [47, 48] and cardiac conduction defects; a *SIX5* knockout mouse developed cataracts [49], though without the features specific to DM; and an RNA gain-of-function model with 250 CTG repeats in the 3′ untranslated region of the muscle-specific *human skeletal actin (HSA)* gene [50] developed myotonia and myopathic features characteristic of DM. Taken together, these data suggested an additive model of DM1 pathogenesis, which combined the pathogenic effects found in each individual model [51, 52].

B. Identification of DM2 Indicates Breadth of RNA Effects in DM Pathogenesis

Clinically, DM1 and DM2 share a strikingly similar set of unusual and multisystemic features. These clinical similarities, and the discovery that both diseases are caused by microsatellite repeat expansions in transcribed but untranslated genetic regions, strongly suggested a common pathogenic mechanism was responsible for both diseases. The normal function of *ZNF9* as a nucleic acid-binding protein bears no homology to *DMPK* or any of the other genes in the DM1 locus [53, 54]. Similarly, the genes within the DM2 locus (*KIAA1160, Rab 11B*, glycoprotein IX, *FLJ11631*, and *FLJ12057*) do not share any obvious relationship with those at the DM1 locus (*DMPK, SIX5, DMWD, FCGRT*), suggesting that effects of the DM1 and DM2 repeat expansions on expression of regional genes, as was proposed in the additive model of DM1, would not be likely to cause the phenotypic similarities of these diseases. Alternatively, the clinical and molecular parallels between DM1 and DM2 strongly suggested that the RNA gain-of-function pathogenic mechanism causes the wide variety of clinical features seen in both diseases.

C. Gain-of-Function RNA Model

The role of RNA was initially suggested by fluorescence *in situ* hybridization studies of DM1 muscle that demonstrated CUG repeats within ribonuclear

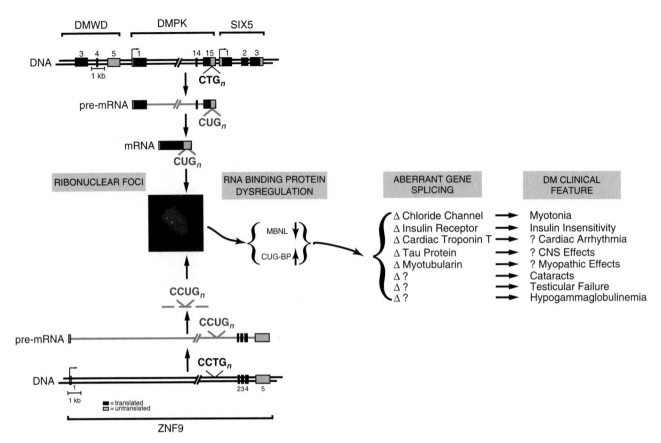

FIGURE 7-8 Pathogenic model of DM1 and DM2. The model of RNA pathogenesis in DM1 and DM2 is due to the untranslated expansions in each disease. Both expansions are transcribed; the DMPK mRNA containing the CUG expansion is incorporated into the ribonuclear inclusions; the CCUG expansion from the DM2 transcript is incorporated into ribonuclear inclusions, though it remains unclear whether any other elements of the ZNF9 transcript are also contained within the inclusions. Muscleblind protein (MBNL) binds to the ribonuclear inclusions; CUG-BP is increased by unclear mechanisms. Decreased MBNL and increased CUG-BP activity alter splicing of transcripts involved in DM pathogenesis, for example, transcripts encoding the chloride channel and insulin receptor. Although the genes responsible for some clinical features have not yet been identified (e.g., testicular failure and hypogammaglobulinemia), the occurrence of these abnormalities in both DM1 and DM2 indicates that they are likely to be caused by the toxic effects of repeat expansions in RNA, possibly involving the resultant decrease in MBNL and increase in CUG-BP. See CD-ROM for color image. Permission to reproduce figure from Elsevier, *Neuromuscular Disorders*. 15, 5–6 (2005).

inclusions [55], with parallel studies after the identification of the DM2 mutation showing CCUG repeat-containing transcripts in DM2 ribonuclear foci [9]. Additional evidence supporting an RNA gain-of-function model included studies in which a CUG repeat expansion in the 3′ UTR of *DMPK* mRNA was shown to inhibit myoblast differentiation [56]. Subsequently, compelling evidence was presented in the mouse model of Mankodi et al., who showed that CUG repeats expressed in the 3′ UTR of the *human skeletal actin* gene resulted in ribonuclear inclusions and myotonia [50]. The ability of DM1 and DM2 transcripts to bind or otherwise alter a specific set of RNA-binding proteins, including CUG-BP and three forms of muscleblind, further demonstrated the role of RNA in DM pathogenesis [57, 58]. These RNA-binding proteins are normally involved in the regulation of alternative splicing of various genes [59–62], and the current model of pathogenesis is that changes in activity of CUG-BP and MBNL result in aberrant splicing of their target genes, which, in turn, leads to the classic multisystemic features of DM1 and DM2 (Fig. 7-8).

V. CUG-BP AND MUSCLEBLIND

A. CUG-BP

The initial investigations of RNA binding proteins in DM pathogenesis focused on CUG-binding protein (CUG-BP), a member of the CELF family of proteins that regulate alternative splicing by binding to UG-rich sequences [63–65]. CUG-BP is normally downregulated in adult compared with embryonic cardiomyocytes [66], consistent with the observation that abnormally heightened levels of CUG-BP expression in DM adults result in aberrant splicing of one of its target genes, *cardiac troponin T* gene (*TNNT2*), with preservation of the fetal isoform containing exon 5 [60]. Transgenic mice overexpressing CUG-BP in skeletal and cardiac muscle [67] develop aberrant splice patterns that are also present in DM patients, including changes in cardiac troponin T, chloride channel, and myotubularin.

B. Muscleblind

Muscleblind (MBNL), a double-stranded RNA-binding protein, regulates alternative splicing in a pattern that is functionally antagonist to CUG-BP and other CELF proteins [65, 68]. MBNL proteins colocalize with the CUG-containing nuclear foci in DM1 and CCUG-containing foci in DM2 [57, 58, 62, 69]. As opposed to the fetal expression of CUG-BP, muscleblind is expressed continually throughout development and regulates splicing by binding intronic sites flanking an exon [66]. An MBNL knockout mouse model [70] exhibits many DM features, including myotonia, cataracts, and aberrant splicing of genes [70]. Although the ribonuclear inclusion in DM1 and DM2 do bind MBNL, this colocalization is not likely to be responsible for functionally significant downregulation, as the binding of MBNL to RNA foci and the disruption of splicing by MBNL are independent events [71].

C. Downstream Targets of CUG-BP and Muscleblind

Taken together, the data on RNA-binding protein alterations in DM1 and DM2 suggest that increased activity of CUG-BP or decreases in MBNL result in aberrant splicing of their target genes, which often results in the inappropriate expression of fetal isoforms of various proteins in adult tissues, leading to the multisystemic features of DM. Some of the genes abnormally spliced include cardiac troponin T, the insulin receptor, the skeletal muscle chloride channel, tau protein, and myotubularin.

1. Cardiac Troponin T

Cardiac troponin T (*TNNT2*), a thin contractile protein, was the first gene shown to be aberrantly spliced because of a *trans*-dominant effect of the CTG expansion in DM1 [60]. In heart tissue from DM1 adults, CUG-BP activity is increased above normal, resulting in aberrant splicing and retention of exon 5 in *TNNT2*, causing the presence in adult DM1 myocardium of the fetal isoform that is normally only found in developing heart [60, 66, 72]. Because *TNNT2* mutations cause 15% of all hypertrophic cardiomyopathies [73, 74], the aberrant splicing pattern of *TNNT2* in DM may contribute to the cardiac features of this multisystemic disease.

2. Insulin Receptor

The insulin receptor (*INSR*), a tetrameric structure composed of two α and two β subunits, has two predominant isoforms that either include or exclude exon 11 in the α subunit. The insulin-responsive IR-B isoform containing exon 11 is highly expressed in muscle, but the IR-A isoform excluding exon 11 is less responsive to insulin and is normally expressed at low levels in muscle [37, 75]. Both DM1 and DM2 skeletal muscle have aberrant *INSR* splicing, which results in the predominance of the insulin-insensitive IR-A isoform [37, 38], consistent with the insulin insensitivity seen in DM1 and DM2 patients [24, 76].

3. Chloride Channel

The skeletal muscle chloride channel (*CLC-1*) is a voltage-gated channel that modulates muscle excitability by increasing chloride conductance across the surface membrane [77]. Skeletal muscle from both DM1 and DM2 patients shows a variety of *CLC-1* splicing alterations, including exon skipping and use of alternative splice donor sites. These aberrant *CLC-1* splice forms cause truncation and loss of *CLC-1* protein [2, 65]. Because more than 60 *CLC-1* mutations cause inherited forms of myotonia congenita (MC) [78–80], the aberrant *CLC-1* splice pattern and loss of chloride channel protein in DM1 and DM2 are likely responsible for the clinical and electrical myotonia that are a hallmark of both diseases.

4. Tau and Myotubularin

Other proteins shown to be abnormally spliced in DM1, and likely to underlie features common to DM1 and DM2, are the tau protein and myotubularin. Tau proteins are microtubule-associated proteins [81] that have been associated with neurodegenerative disease. Like other targets of CELF RNA-binding proteins, abnormal expression of tau protein isoforms is evident in cDM1 CNS tissue [82, 83], which may contribute to the cognitive deficiencies that are a characteristic feature of DM. The protein myotubularin (*MTM1*) is involved in muscle differentiation, and directly affected in infantile myotubular myopathy. In skeletal muscle and cultured myocytes from congenitally affected DM1 patients, there is marked impairment of generation of the adult isoform [84], consistent with the known effects of DM on other downstream target proteins and consistent with the abnormal muscle fiber maturation in DM1.

VI. POTENTIAL CAUSES OF CLINICAL DISTINCTIONS BETWEEN DM1 AND DM2

In addition to the clinical features that have been associated with specific genetic splicing changes, DM1 and DM2 share additional multisystemic elements including testicular failure, hypogammaglobulinemia, and iridescent posterior subcapsular cataracts; the occurrence of these unusual abnormalities in both diseases strongly suggests that they are caused by the CUG and CCUG repeat expansions expressed in RNA, though the exact molecular mechanism has not yet been elucidated. Despite the remarkable clinical similarities of DM1 and DM2, they are not identical disorders. The primary clinical difference between the two is that DM2 does not have the congenital form that can occur in DM1 [24]. Although the similarities of the two diseases has helped to clarify the role of CUG and CCUG RNA repeat expansions in DM1 and DM2 pathogenesis, the clinical differences in the two diseases could reflect modulation of this RNA mechanism, or could be caused by distinct processes. One possible mechanism for modulating the toxic RNA effect could be that CUG-BP and MBNL affinity for CUG differs from their affinity for transcripts containing CCUG. Also, clinical differences in the two diseases could reflect variation in the RNA effects produced by dissimilarities in temporal and cell-specific expression patterns of *DMPK* and *ZNF9*. Alternatively, a separate mechanism could account for the phenotypic differences between DM1 and DM2, such as alterations in the expression of locus-specific genes including *DMPK* and *SIX5* for DM1 or *ZNF9* for DM2.

To explain the increased severity of congenital DM1, CTG repeat length has commonly been invoked, with longer repeat tracts postulated to result in earlier age of onset and greater disease severity. However, several lines of evidence now suggest that model is inadequate: some individuals with very long DM1 expansions have adult-onset disease; DM2 expansions are much longer than DM1 expansions but do not cause a severe congenital syndrome; DM2 subjects homozygous for large expansions have a disease similar to that of affected individuals who are heterozygous for an expansion. Disease severity in congenital DM1 may be attributed to hypermethylation at the DM1 locus [85]. CTCF sites that flank the repeat, along with the repeat itself, form an insulator element between *DMPK* and the *SIX5* enhancer. In congenital DM1, methylation of these CTCF sites may inactivate the insulator and increase expression of DMPK transcripts containing the expanded CTG repeat. Rather than the size of the repeat being directly responsible for disease severity, impaired regulation of DMPK expression caused by hypermethylation, with resultant overexpression of the repeat, may cause the increased disease severity typical of congenital DM1.

VII. CONCLUSIONS

Identification of a second example of adult-onset multisystemic myotonic dystrophy has helped clarify the molecular pathogenesis of both DM1 and DM2. There are several unusual genetic features of DM2: (1) it is the only identified tetranucleotide repeat disorder; (2) DM2 expansions are larger than any other identified repeat expansion, exceeding 11,000 repeats with a mean of 5000 repeats; (3) the degree of somatic mosaicism is unparalleled, with most subjects having dramatic heterogeneity of repeat length in blood; (4) expansion length can increase over time, in one instance growing by 2 kb over 3 years; (5) there is no discernable correlation between repeat length and disease severity. The size and somatic mosaicism have complicated genetic testing, but the currently available battery of a PCR-based repeat assay and Southern analysis is reliable, sensitive, and specific. The lack of correlation between repeat length and severity suggests that the pathogenic mechanism is saturable, with no further increase in pathogenicity as repeat lengths expand beyond a size with maximal effect; this ceiling effect is further reflected by the fact that individuals homozygous for DM2 expansions are clinically indistinguishable from those with only one expanded allele.

The clinical and molecular parallels between DM1 and DM2 have simplified the molecular model of DM pathogenesis, and strongly indicated that an RNA gain-of-function mechanism causes the multisystemic features present in both diseases. Although the molecular changes responsible for all of the multisystemic features of DM have not been fully defined, and the molecular causes of the clinical differences between DM1 and DM2 are yet to be determined, clarification of the primary pathophysiological process now provides a target for therapeutic intervention.

References

1. Harper, P. S. (2001). "Myotonic Dystrophy." Saunders, London.
2. Mankodi, A., Takahashi, M. P., Jiang, H., Beck, C. L., Bowers, W. J., Moxley, R. T., Cannon, S. C., and Thornton, C. A. (2002). Expanded CUG repeats trigger aberrant splicing of ClC-1 chloride channel pre-mRNA and hyperexcitability of skeletal muscle in myotonic dystrophy. *Mol. Cell* **10**, 35–44.
3. Ranum, L. P., and Day, J. W. (2002). Dominantly inherited, noncoding microsatellite expansion disorders. *Curr. Opin. Genet. Dev.* **12**, 266–271.

4. Brook, J. D., McCurrach, M. E., Harley, H. G., Buckler, A. J., Church, D., Aburatani, H., Hunter, K., Stanton, V. P., Thirion, J. P., Hudson, T., et al. (1992). Molecular basis of myotonic dystrophy: Expansion of a trinucleotide (CTG) repeat at the 3′ end of a transcript encoding a protein kinase family member. *Cell* **69**, 385.
5. Fu, Y. H., Pizzuti, A., Fenwick, R. G., Jr., King, J., Rajnarayan, S., Dunne, P. W., Dubel, J., Nasser, G. A., Ashizawa, T., de Jong, P., et al. (1992). An unstable triplet repeat in a gene related to myotonic muscular dystrophy. *Science* **255**, 1256–1258.
6. Thornton, C. A., Griggs, R. C., and Moxley, R. T. (1994). Myotonic dystrophy with no trinucleotide repeat expansion. *Ann. Neurol.* **35**, 269–272.
7. Ricker, K., Koch, M. C., Lehmann-Horn, F., Pongratz, D., Otto, M., Heine, R., and Moxley R. T., III (1994). Proximal myotonic myopathy: A new dominant disorder with myotonia, muscle weakness, and cataracts. *Neurology* **44**, 1448–1452.
8. Day, J. W., Roelofs, R., Leroy, B., Pech, I., Benzow, K., and Ranum, L. P. W. (1999). Clinical and genetic characteristics of a five-generation family with a novel form of myotonic dystrophy (DM2). *Neuromuscul. Disord.* **9**, 19–27.
9. Liquori, C., Ricker, K., Moseley, M. L., Jacobsen, J. F., Kress, W., Naylor, S., Day, J. W., and Ranum, L. P. W. (2001). Myotonic dystrophy type 2 caused by a CCTG expansion in intron 1 of ZNF9. *Science* **293**, 864–867.
10. Steinert, H. (1909). Myopathologische Beitrage 1. Uber das klinischeund anatomische Bild des Muskelschwunds der Myotoniker. *Dtsch. Z. Nervenheilkd.* **37**, 58–104.
11. Batten, F., and Gibb, H. (1909). Myotonia atrophica. *Brain* **32**, 187–205.
12. Buxton, J., Shelbourne, P., Davies, J., Jones, C., Van Tongeren, T., Aslanidis, C., de Jong, P., Jansen, G., Anvret, M., Riley, B., et al. (1992). Detection of an unstable fragment of DNA specific to individuals with myotonic dystrophy. *Nature* **355**, 547–548.
13. Harley, H. G., Brook, J. D., Rundle, S. A., Crow, S., Reardon, W., Buckler, A. J., Harper, P. S., Houseman, D. E., and Shaw, D. (1992). Expansion of an unstable DNA region and phenotypic variation in myotonic dystrophy. *Nature* **355**, 545–546.
14. Mahadevan, M., Tsilfidis, C., Sabourin, L., Shutler, G., Amemiya, C., Jansen, G., Neville, C., Narang, M., Barcelo, J., O'Hoy, K., Leblond, S., Earle-MacDonald, J., De Jong, P. J., Wieringa, B., Koneluk, R. G., et al. (1992). Myotonic dystrophy mutation: An unstable CTG repeat in the 3′ untranslated region of the gene. *Science* **255**, 1253–1255.
15. Udd, B., Krahe, R., Wallgren-Petterson, C., Falck, B., and Kalimo, H. (1997). Proximal myotonic dystrophy—a family with autosomal dominant muscular dystrophy, cataracts, hearing loss and hypogonadism: Heterogeneity of proximal myotonic syndromes? *Neuromuscul. Disord.* **7**, 217–228.
16. Ranum, L., Rasmussen, P., Benzow, K., Koob, M., and Day, J. (1998). Genetic mapping of a second myotonic dystrophy locus. *Nat. Genet.* **19**, 196–198.
17. Ricker, K., Grimm, T., Koch, M. C., Schneider, C., Kress, W., Reimers, C. D., Schulte-Mattler, W., Mueller-Myhsok, B., Toyka, K. V., and Mueller, C. R. (1999). Linkage of proximal myotonic myopathy to chromosome 3q. *Neurology* **52**, 170–171.
18. Liquori, C. L., Ikeda, Y., Weatherspoon, M., Ricker, K., Schoser, B. G., Dalton, J. C., Day, J. W., and Ranum, L. P. (2003). Myotonic dystrophy type 2: Human founder haplotype and evolutionary conservation of the repeat tract. *Am. J. Hum. Genet.* **73**, 849–862.
19. Udd, B., Meola, G., Krahe, R., Thornton, C., Ranum, L. P. W., Day, J., Bassez, G., and Ricker, K. (2003). Report of the 115th ENMC workshop: DM2/PROMM and other myotonic dystrophies. 3rd Workshop, 14–16 February 2003, Naarden, The Netherlands. *Neuromuscul. Disord.* **13**, 589–596.
20. Chung, M.-Y., Ranum, L. P. W., Duvick, L. A., Servadio, A., Zoghbi, H. Y., and Orr, H. T. (1993). Evidence for a mechanism predisposing to intergenerational CAG repeat instability in spinocerebellar ataxia type 1. *Nat. Genet.* **5**, 254–258.
21. Gunter, C., Paradee, W., Crawford, D. C., Meadows, K. A., Newman, J., Kunst, C. B., Nelson, D. L., Schwartz, C., Murray, A., Macpherson, J. N., Sherman, S. L., and Warren, S. T. (1998). Re-examination of factors associated with expansion of CGG repeats using a single nucleotide polymorphism in FMR1. *Hum. Mol. Genet.* **7**, 1935–1946.
22. Kunst, C. B., and Warren, S. T. (1994). Cryptic and polar variation of the fragile X repeat could result in predisposing normal alleles. *Cell* **77**, 853–861.
23. Pulst, S.-M., Nechiporuk, A., Nechiporuk, T., Gispert, S., Chen, X. N., Lopes-Cendes, I., Pearlman, S., Starkman, S., Orozco-Diaz, G., Lunkes, A., DeJong, P., Rouleau, G. A., Auburger, G., Korenberg, J. R., Figueroa, C., and Sahba, S. (1996). Moderate expansion of a normally biallelic trinucleotide repeat in spinocerebellar ataxia type 2. *Nat. Genet.* **14**, 269–276.
24. Day, J. W., Ricker, K., Jacobsen, J. F., Rasmussen, L. J., Dick, K. A., Kress, W., Schneider, C., Koch, M. C., Beilman, G. J., Harrison, A. R., Dalton, J. C., and Ranum, L. P. (2003). Myotonic dystrophy type 2: Molecular, diagnostic and clinical spectrum. *Neurology* **60**, 657–664.
25. Sallinen, R., Vihola, A., Bachinski, L. L., Huoponen, K., Haapasalo, H., Hackman, P., Zhang, S., Sirito, M., Kalimo, H., Meola, G., Horelli-Kuitunen, N., Wessman, M., Krahe, R., and Udd, B. (2004). New methods for molecular diagnosis and demonstration of the $(CCTG)_n$ mutation in myotonic dystrophy type 2 (DM2). *Neuromuscul. Disord.* **14**, 274–283.
26. Schneider, C., Ziegler, A., Ricker, K., Grimm, T., Kress, W., Reimers, C. D., Meinck, H., Reiners, K., and Toyka, K. V. (2000). Proximal myotonic myopathy: Evidence for anticipation in families with linkage to chromosome 3q. *Neurology* **55**, 383–388.
27. Hamshere, M. G., Harley, H., Harper, P., Brook, J. D., and Brookfield, J. F. (1999). Myotonic dystrophy: The correlation of (CTG) repeat length in leucocytes with age at onset is significant only for patients with small expansions. *J. Med. Genet.* **36**, 59–61.
28. Schoser, B. G., Kress, W., Walter, M. C., Halliger-Keller, B., Lochmuller, H., and Ricker, K. (2004). Homozygosity for CCTG mutation in myotonic dystrophy type 2. *Brain* **127**, 1868–1877.
29. Mathieu, J., Allard, P., Potvin, L., Prevost, C., and Begin, P. (1999). A 10-year study of mortality in a cohort of patients with myotonic dystrophy. *Neurology* **52**, 1658–1662.
30. Thornton, C. (1999). The myotonic dystrophies. *Semin. Neurol.* **19**, 25–33.
31. Ricker, K., Koch, M., Lehmann-Horn, F., Pongratz, D., Speich, N., Reiners, K., Schneider, C., and Moxley, R. T., 3rd (1995). Proximal myotonic myopathy: Clinical features of a multisystem disorder similar to myotonic dystrophy. *Arch. Neurol.* **52**, 25–31.
32. Moxley, R. (1996). Proximal myotonic myopathy: Mini-review of a recently delineated clinical disorder. *Neuromuscul. Disord.* **6**, 87–93.
33. Schoser, B. G., Schneider-Gold, C., Kress, W., Goebel, H. H., Reilich, P., Koch, M. C., Pongratz, D. E., Toyka, K. V., Lochmuller, H., and Ricker, K. (2004). Muscle pathology in 57 patients with myotonic dystrophy type 2. *Muscle Nerve* **29**, 275–281.
34. Vihola, A., Bassez, G., Meola, G., Zhang, S., Haapasalo, H., Paetau, A., Mancinelli, E., Rouche, A., Hogrel, J. Y., Laforet, P., Maisonobe, T., Pellissier, J. F., Krahe, R., Eymard, B., and Udd, B. (2003). Histopathological differences of myotonic dystrophy type 1 (DM1) and PROMM/DM2. *Neurology* **60**, 1854–1857.
35. Phillips, M. F., and Harper, P. S. (1997). Cardiac disease in myotonic dystrophy. *Cardiovasc. Res.* **33**, 13–22.

36. Colleran, J. A., Hawley, R. J., Pinnow, E. E., Kokkinos, P. F., and Fletcher, R. D. (1997). Value of the electrocardiogram in determining cardiac events and mortality in myotonic dystrophy. *Am. J. Cardiol.* **80**, 1494–1497.
37. Savkur, R. S., Philips, A. V., and Cooper, T. A. (2001). Aberrant regulation of insulin receptor alternative splicing is associated with insulin resistance in myotonic dystrophy. *Nat. Genet.* **29**, 40–47.
38. Savkur, R. S., Philips, A. V., Cooper, T. A., Dalton, J. C., Moseley, M. L., Ranum, L. P., and Day, J. W. (2004). Insulin receptor splicing alteration in myotonic dystrophy type 2. *Am. J. Hum. Genet.* **74**, 1309–1313.
39. Aminoff, M., Beckley, D., and McIlroy, M. (1985). Autonomic function in myotonic dystrophy. *Arch. Neurol.* **42**, 16.
40. Rubinsztein, J. S., Rubinsztein, D. C., McKenna, P. J., Goodburn, S., and Holland, A. J. (1997). Mild myotonic dystrophy is associated with memory impairment in the context of normal general intelligence. *J. Med. Genet.* **34**, 229–233.
41. Bungener, C., Jouvent, R., and Delaporte, C. (1998). Psychopathological and emotional deficits in myotonic dystrophy. *J. Neurol. Neurosurg. Psychiatry* **65**, 353–356.
42. Delaporte, C. (1998). Personality patterns in patients with myotonic dystrophy. *Arch. Neurol.* **55**, 635–640.
43. Ogata, A., Terae, S., Fujita, M., and Tashiro, K. (1998). Anterior temporal white matter lesions in myotonic dystrophy with intellectual impairment: An MRI and neuropathological study. *Neuroradiology* **40**, 411–415.
44. Wilson, B. A., Balleny, H., Patterson, K., and Hodges, J. R. (1999). Myotonic dystrophy and progressive cognitive decline: A common condition or two separate problems? *Cortex* **35**, 113–121.
45. Meola, G., Sansone, V., Perani, D., Colleluori, A., Cappa, S., Cotelli, M., Fazio, F., Thornton, C. A., and Moxley, R. T. (1999). Reduced cerebral blood flow and impaired visual–spatial function in proximal myotonic myopathy. *Neurology* **53**, 1042–1050.
46. Hund, E., Jansen, O., Koch, M., Ricker, K., Fogel, W., Niedermaier, N., Otto, M., Kuhn, E., and Meinck, H. (1997). Proximal myotonic myopathy with MRI white matter abnormalities of the brain. *Neurology* **48**, 33–37.
47. Jansen, G., Groenen, P. J. T. A., Bachner, D., Jap, P. H. K., Coerwinkel, M., Oerlemans, F., van den Broek, W., Gohlsch, B., Pette, D., Plomp, J. J., Molenaar, P. C., Nederhof, M. G. J., van Echted, C. J. A., Dekker, M., Berns, A., Hameister, H., and Wieringa, B. (1996). Abnormal myotonic dystrophy protein kinase levels produce only mild myopathy in mice. *Nat. Genet.* **13**, 316–324.
48. Reddy, S., Smith, D. B., Rich, M. M., Leferovich, J. M., Reilly, P., Davis, B. M., Tran, K., Rayburn, H., Bronson, R., Cros, D., Balice-Gordon, R. J., and Housman, D. (1996). Mice lacking the myotonic dystrophy protein kinase develop a late onset progressive myopathy. *Nat. Genet.* **13**, 325–335.
49. Klesert, T. R., Cho, D. H., Clark, J. I., Maylie, J., Adelman, J., Snider, L., Yuen, E. C., Soriano, P., and Tapscott, S. J. (2000). Mice deficient in Six5 develop cataracts: Implications for myotonic dystrophy. *Nat. Genet.* **25**, 105–109.
50. Mankodi, A., Logigian, E., Callahan, L., McClain, C., White, R., Henderson, D., Krym, M., and Thornton, C. A. (2000). Myotonic dystrophy in transgenic mice expressing an expanded CUG repeat. *Science* **289**, 1769–1773.
51. Tapscott, S. J. (2000). Deconstructing myotonic dystrophy. *Science* **289**, 1701–1702.
52. Groenen, P., and Wieringa, B. (1998). Expanding complexity in myotonic dystrophy. *Bioessays* **20**, 901–912.
53. Pellizzoni, L., Lotti, F., Maras, B., and Pierandrei-Amaldi, P. (1997). Cellular nucleic acid binding protein binds a conserved region of the 5′ UTR of *Xenopus laevis* ribosomal protein mRNAs. *J. Mol. Biol.* **267**, 264–275.
54. Pellizzoni, L., Lotti, F., Rutjes, S. A., and Pierandrei-Amaldi, P. (1998). Involvement of the *Xenopus laevis* Ro60 autoantigen in the alternative interaction of La and CNBP proteins with the 5′UTR of L4 ribosomal protein mRNA. *J. Mol. Biol.* **281**, 593–608.
55. Taneja, K. L., McCurrach, M., Schalling, M., Housman, D., and Singer, R. H. (1995). Foci of trinucleotide repeat transcripts in nuclei of myotonic dystrophy cells and tissues. *J. Cell Biol.* **128**, 995–1002.
56. Amack, J. D., Paguio, A. P., and Mahadevan, M. S. (1999). Cis and trans effects of the myotonic dystrophy (DM) mutation in a cell culture model. *Hum. Mol. Genet.* **8**, 1975–1984.
57. Mankodi, A., Urbinati, C. R., Yuan, Q. P., Moxley, R. T., Sansone, V., Krym, M., Henderson, D., Schalling, M., Swanson, M. S., and Thornton, C. A. (2001). Muscleblind localizes to nuclear foci of aberrant RNA in myotonic dystrophy types 1 and 2. *Hum. Mol. Genet.* **10**, 2165–2170.
58. Fardaei, M., Rogers, M. T., Thorpe, H. M., Larkin, K., Hamshere, M. G., Harper, P. S., and Brook, J. D. (2002). Three proteins, MBNL, MBLL and MBXL, co-localize in vivo with nuclear foci of expanded-repeat transcripts in DM1 and DM2 cells. *Hum. Mol. Genet.* **11**, 805–814.
59. Timchenko, L. T., Miller, J. W., Timchenko, N. A., DeVore, D. R., Datar, K. V., Lin, L., Roberts, R., Caskey, C. T., and Swanson, M. S. (1996). Identification of a $(CUG)_n$ triplet repeat RNA-binding protein and its expression in myotonic dystrophy. *Nucleic Acids Res.* **24**, 4407–4414.
60. Philips, A. V., Timchenko, L. T., and Cooper, T. A. (1998). Disruption of splicing regulated by a CUG-binding protein in myotonic dystrophy. *Science* **280**, 737–741.
61. Lu, X., Timchenko, N. A., and Timchenko, L. T. (1999). Cardiac elav-type RNA-binding protein (ETR-3) binds to RNA CUG repeats expanded in myotonic dystrophy. *Hum. Mol. Genet.* **8**, 53–60.
62. Miller, J. W., Urbinati, C. R., Teng-Umnuay, P., Stenberg, M. G., Byrne, B. J., Thornton, C. A., and Swanson, M. S. (2000). Recruitment of human muscleblind proteins to $(CUG)_n$ expansions associated with myotonic dystrophy. *EMBO J.* **19**, 4439–4448.
63. Ladd, A. N., Charlet, N., and Cooper, T. A. (2001). The CELF family of RNA binding proteins is implicated in cell-specific and developmentally regulated alternative splicing. *Mol. Cell. Biol.* **21**, 1285–1296.
64. Ladd, A. N., Nguyen, N. H., Malhotra, K., and Cooper, T. A. (2004). CELF6, a member of the CELF family of RNA-binding proteins, regulates muscle-specific splicing enhancer-dependent alternative splicing. *J. Biol. Chem.* **279**, 17756–17764.
65. Charlet, B. N., Savkur, R. S., Singh, G., Philips, A. V., Grice, E. A., and Cooper, T. A. (2002). Loss of the muscle-specific chloride channel in type 1 myotonic dystrophy due to misregulated alternative splicing. *Mol. Cell* **10**, 45–53.
66. Ladd, A. N., Stenberg, M. G., Swanson, M. S., and Cooper, T. A. (2005). Dynamic balance between activation and repression regulates pre-mRNA alternative splicing during heart development. *Dev. Dyn.* **233**, 783–793.
67. Ho, T. H., Bundman, D., Armstrong, D. L., and Cooper, T. A. (2005). Transgenic mice expressing CUG-BP1 reproduce splicing mis-regulation observed in myotonic dystrophy. *Hum. Mol. Genet.* **14**, 1539–1547.
68. Ho, T. H., Charlet, B. N., Poulos, M. G., Singh, G., Swanson, M. S., and Cooper, T. A. (2004). Muscleblind proteins regulate alternative splicing. *EMBO J.* **23**, 3103–3112.
69. Fardaei, M., Larkin, K., Brook, J. D., and Hamshere, M. G. (2001). In vivo co-localisation of MBNL protein with DMPK expanded-repeat transcripts. *Nucleic Acids Res.* **29**, 2766–2771.

70. Kanadia, R. N., Johnstone, K. A., Mankodi, A., Lungu, C., Thornton, C. A., Esson, D., Timmers, A. M., Hauswirth, W. W., and Swanson, M. S. (2003). A muscleblind knockout model for myotonic dystrophy. *Science* **302**, 1978–1980.
71. Ho, T. H., Savkur, R. S., Poulos, M. G., Mancini, M. A., Swanson, M. S., and Cooper, T. A. (2005). Colocalization of muscleblind with RNA foci is separable from mis-regulation of alternative splicing in myotonic dystrophy. *J. Cell. Sci.* **118**, 2923–2933.
72. Cooper, T. A., and Ordahl, C. P. (1985). A single cardiac troponin T gene generates embryonic and adult isoforms via developmentally regulated alternate splicing. *J. Biol. Chem.* **260**, 11140–11148.
73. Seidman, J. G., and Seidman, C. (2001). The genetic basis for cardiomyopathy: From mutation identification to mechanistic paradigms. *Cell* **104**, 557–567.
74. Thierfelder, L., Watkins, H., MacRae, C., Lamas, R., McKenna, W., Vosberg, H. P., Seidman, J. G., and Seidman, C. E. (1994). Alpha-tropomyosin and cardiac troponin T mutations cause familial hypertrophic cardiomyopathy: A disease of the sarcomere. *Cell* **77**, 701–712.
75. Moller, D. E., Yokota, A., Caro, J. F., and Flier, J. S. (1989). Tissue-specific expression of two alternatively spliced insulin receptor mRNAs in man. *Mol. Endocrinol.* **3**, 1263–1269.
76. Moxley, R. T., Griggs, R. C., and Goldblatt, D. (1980). Muscle insulin resistance in myotonic dystrophy: Effect of supraphysiologic insulinization. *Neurology* **30**, 1077–1083.
77. Dutzler, R., Campbell, E. B., Cadene, M., Chait, B. T., and MacKinnon, R. (2002). X-ray structure of a ClC chloride channel at 3.0 A reveals the molecular basis of anion selectivity. *Nature* **415**, 287–294.
78. Grunnet, M., Jespersen, T., Colding-Jorgensen, E., Schwartz, M., Klaerke, D. A., Vissing, J., Olesen, S. P., and Duno, M. (2003). Characterization of two new dominant ClC-1 channel mutations associated with myotonia. *Muscle Nerve* **28**, 722–732.
79. Sasaki, R., Ito, N., Shimamura, M., Murakami, T., Kuzuhara, S., Uchino, M., and Uyama, E. (2001). A novel CLCN1 mutation: P480T in a Japanese family with Thomsen's myotonia congenita. *Muscle Nerve* **24**, 357–363.
80. Steinmeyer, K., Lorenz, C., Pusch, M., Koch, M. C., and Jentsch, T. J. (1994). Multimeric structure of ClC-1 chloride channel revealed by mutations in dominant myotonia congenita (Thomsen). *Embo. J.* **13**, 737–743.
81. Buee, L., Bussiere, T., Buee-Scherrer, V., Delacourte, A., and Hof, P. R. (2000). Tau protein isoforms, phosphorylation and role in neurodegenerative disorders. *Brain Res Brain Res Rev* **33**, 95–130.
82. Andreadis, A., Brown, W. M., and Kosik, K. S. (1992). Structure and novel exons of the human tau gene. *Biochemistry* **31**, 10626–10633.
83. Sergeant, N., Sablonniere, B., Schraen-Maschke, S., Ghestem, A., Maurage, C. A., Wattez, A., Vermersch, P., and Delacourte, A. (2001). Dysregulation of human brain microtubule-associated tau mRNA maturation in myotonic dystrophy type 1. *Hum. Mol. Genet.* **10**, 2143–2155.
84. Buj-Bello, A., Laugel, V., Messaddeq, N., Zahreddine, H., Laporte, J., Pellissier, J. F., and Mandel, J. L. (2002). The lipid phosphatase myotubularin is essential for skeletal muscle maintenance but not for myogenesis in mice. *Proc. Natl. Acad. Sci. USA* **99**, 15060–15065.
85. Filippova, G. N., Thienes, C. P., Penn, B. H., Cho, D. H., Hu, Y. J., Moore, J. M., Klesert, T. R., Lobanenkov, V. V., and Tapscott, S. J. (2001). CTCF-binding sites flank CTG/CAG repeats and form a methylation-sensitive insulator at the DM1 locus. *Nat. Genet.* **28**, 335–343.

Myotonic Dystrophy Type 2: Clinical and Genetic Aspects

RALF KRAHE, LINDA L. BACHINSKI, AND BJARNE UDD

Section of Cancer Genetics, Department of Molecular Genetics, University of Texas M. D. Anderson Cancer Center, Houston, Texas 77030; Department of Neurology, Vasa Central Hospital, 65130 Vasa, Finland; Department of Neurology, Tampere University Hospital, Tampere, Finland; and Folkhälsan Institute of Genetics, University of Helsinki, 00014 Helsinki, Finland

I. Introduction
II. Clinical Phenotype
   A. Symptoms and Findings in DM2
   B. Age of Onset and Anticipation in DM2
   C. Homozygosity for DM2 Mutation
   D. What Is the Full Phenotypic Spectrum of DM2?
   E. Muscle Biopsy and Morphological Findings
III. Molecular Genetics
   A. Characteristics of the DM2 Repeat
   B. Evolutionary Conservation of the DM2 Repeat
   C. Population Studies and the Origin of the DM2 Repeat Expansion
   D. Molecular Diagnosis of DM2
   E. How Many Myotonic Dystrophies Are There?
IV. Molecular Pathophysiology
V. Concluding Remarks
   Acknowledgments
   Note Added in Proof
   References

Myotonic dystrophy (DM) is the most common muscular dystrophy in adults. Clinically and genetically, DM is a heterogeneous group of neuromuscular disorders, which is characterized by autosomal dominant inheritance, muscular dystrophy, myotonia, and multisystem involvement. To date, mutations for myotonic dystrophy type 1 (DM1) and type 2 (DM2) have been identified. DM1 and DM2 are caused by similar unstable microsatellite repeat expansions—in DM1 a $(CTG)_n$ expansion in *DMPK* in chromosome 19q13.3, in DM2 a $(CCTG)_n$ expansion in *ZNF9* in chromosome 3q21.3. The developing paradigm is that DM is an RNA disease, mediated by the mutant expansion of normally polymorphic microsatellite repeats with a $(CTG)_n$-like repeat motif. Transcription of the mutant repeats into $(CUG)_n/(CCUG)_n$ RNA is both necessary and sufficient to cause disease. Mutant RNA species accumulate in ribonuclear inclusions and interfere with proper RNA splicing, transcription and/or translation of a number of effecter genes, resulting in the pleiotropic phenotype characteristic of this disease. This interference may be due, in part, to sequestration of various proteins involved in these cellular processes, such as MBNL and CUGBP1. Additional "atypical" DM kindreds not segregating either the DM1 or DM2 mutation have been reported, and two additional loci have recently been mapped to chromosomes 15q21-q24 and 16p. Given the phenotypic similarities between patients with DM1 and DM2 and those with linkage to the DM3 or DM4 loci, it is tempting to speculate that the underlying mutation(s) may also be expansion of a repeat with a similar motif. The clinical, population, and molecular genetic aspects of DM2 are discussed in the context of DM1 and the newly mapped loci for DM3 and DM4.

I. INTRODUCTION

Myotonic dystrophy (DM) is the most common muscular dystrophy in adults, with an incidence of more than 1 in 8000. Clinically, DM is a heterogeneous group of neuromuscular disorders characterized by autosomal dominant inheritance, muscular dystrophy, myotonia, and multisystem involvement [1]. The discovery of an unstable $(CTG)_n$ trinucleotide repeat expansion in the 3' untranslated region of the *DM protein kinase* (*DMPK*) gene in chromosome 19q13.3 as the underlying genetic cause of myotonic dystrophy type 1 (DM1) in 1992 [2–4] made systematic diagnostic testing of patients possible. As a result, it was found that not all patients with a myotonic dystrophy phenotype had the DM1 mutation, indicating genetic heterogeneity [4–6]. However, as early as the late 1970s and early 1980s, physicians (Dr. Kenneth Ricker in Germany and Dr. Bjarne Udd in Finland) had already identified patients who, in contrast to DM1 patients, presented with a distinct pattern of proximal muscle weakness, along with myotonia, cataracts, and an autosomal dominant pattern of inheritance [7]. Thus, genetic heterogeneity was already suspected on clinical grounds, and once DNA testing for the DM1 mutation became available, these "atypical" DM patients proved to be negative.

Ricker, Moxley, and their colleagues presented abstracts describing these families at the American Academy of Neurology annual meeting in 1994 and, later the same year, at the International Congress on Neuromuscular Disorders in Kyoto. In three kindreds the pedigree structure permitted linkage analysis [5], and linkage to *DMPK* in 19q13.3 and the loci for the nondystrophic myotonias caused by mutations in the skeletal muscle chloride channel (*CLCN1* in 7q34) [8] or sodium channel (*SCN4A* in 17q23.3) [9] gene was excluded. Muscle biopsy studies in two patients showed that muscle fibers had intrinsic myotonic contractions and that the channel current properties were distinct from those seen in DM1 and the nondystrophic chloride or sodium channel myotonias [5]. The 15 patients in the three families presented shared a core of common features: (1) myotonia, especially in the grip and thighs; (2) proximal muscle weakness, apparent on arising from a squat; (3) posterior capsular, iridescent lens opacities, identical to those in DM1 [7]; and, by definition, a normally sized $(CTG)_n$ repeat in *DMPK* [5, 7]. Ricker, Moxley, and colleagues named this disorder proximal myotonic myopathy, or PROMM for short [5].

In 1994 Thornton et al. identified three patients with an atypical form of myotonic dystrophy associated with proximal weakness, calf hypertrophy, mild myotonia, cataracts, cardiac conduction disturbance, and gonadal insufficiency. The patients had normally sized $(CTG)_n$ repeats, and no other mutations were identified in *DMPK* [6]. Unlike those with DM1, these patients had normal distal muscle strength, normal tendon reflexes, only mild facial weakness, and no evidence of anticipation on family history. Soon after, there were new reports that helped to define the clinical spectrum of PROMM [10, 11]. Some of these patients had a peculiar type of muscle pain, some had a tremor [10], and the myotonia varied in severity within the same individual, sometimes even disappearing entirely [11]. Following the report by Ricker et al. in 1995 [11], others in Germany [12, 13], Italy [14], Spain [15], and the United States [16–20] identified patients with PROMM. A somewhat different clinical phenotype was described in an extended pedigree from Finland, in which patients presented with late-onset proximal weakness and severe wasting, cataracts, hearing loss, and male hypogonadism, but with myotonia detected on EMG only without clinical manifestation. The family was described as having proximal myotonic dystrophy (PDM) [21]. At this stage in 1997, the European Neuromuscular Center (ENMC) was instrumental in facilitating the first international workshop on PROMM (Proximal Myotonic Myopathies) and other Proximal Myotonic Syndromes [7]. The workshop report included extensive evaluation of all published families and diagnostic criteria [7].

Important advances in the molecular genetics of PROMM emerged in the late 1990s. Day and Ranum had been studying a large non-DM1 family from Minnesota (MN-1) described as having myotonic dystrophy type 2 (DM2) [22]. They had evaluated 62 family members, 22 of whom had an adult-onset progressive myopathy, which included highly penetrant cataracts typical of DM1 and myotonia on EMG. A large percentage of the patients had selective weakness of the long flexors for the distal phalanx of the fingers. In 1998, Ranum et al. published definite linkage of the disorder in their MN-1 family to chromosome 3q21 [23]. Subsequent linkage analysis in nine German families with PROMM confirmed linkage in eight of these to the same locus 3q21 [24]. The Finnish family previously described with a more severe proximal myotonic dystrophy (PDM) [21] also turned out to be linked to the DM2 locus [25]. Taken together, these observations strongly suggested that most patients and families with PROMM were linked to the novel DM2 locus [26]. The consensus at the 2nd ENMC Workshop on PROMM was that patients with PROMM, PDM, and DM2—initially considered three distinct diseases based on clinical grounds [7]—and those with linkage to 3q21 had the same disorder. Thus, the term *myotonic dystrophy type 2* (DM2) was adopted for all of the progressive myotonic multiorgan disorders linked to the DM2 locus [26, 27].

In 2001 the mutation underlying the disorder was identified [28]. By use of 23 Minnesota and 52 German families to identify shared markers and refine the location of the DM2 locus, it was found that the microsatellite marker CL3N58 in 3q21 showed the highest degree of linkage disequilibrium. Moreover, the marker showed an unusual nonmendelian segregation pattern. Affected offspring appeared to have inherited only one allele, which was always transmitted by the healthy parent. The CL3N58 marker was found to be located within the first intron of ZNF9, a gene encoding a zinc-finger protein, and to have a complex repeat structure: $(TG)_{14-25}(TCTG)_{4-10}(CCTG)_{11-26}$. When mutated, the $(CCTG)_n$ portion of the repeat tract is expanded to repeat lengths between ~75 and ~11,000 repeats (mean, ~5000, or ~300 to ~44 kb), and the mutant allele cannot be amplified by polymerase chain reaction (PCR). All of the 3q21-linked DM2, PROMM, and PDM kindreds proved to be segregating a mutant expansion of this repeat [28–31]. Interestingly, not only the linked families, but also several families previously published as nonlinked, proved to have this expansion, indicating considerable variation of phenotypic expression and corresponding difficulties in determining the correct affection status for linkage analyses [29, 32–34]. However, several families presenting with a clinical diagnosis of proximal myotonic syndrome were neither linked to nor harbored the mutated DM2 repeat, thus indicating additional genetic heterogeneity [31, 32] (see Section IIIE).

II. CLINICAL PHENOTYPE

Most of the clinical findings on DM2 have been reported in connection with publication of families identified with the disorder. Extensive evaluation of the phenotypic spectrum was conducted at the 3rd ENMC workshop in 2003 (Table 8-1) [31]. To date, only one larger study on the clinical phenotype in DM2 mutation-verified patients has been published by Day et al. [35].

A. Symptoms and Findings in DM2

Frequent major complaints and signs, or core features, of DM2 are: proximal muscle weakness, muscle pain and/or stiffness, cataracts, myotonia, tremor, cardiac disturbance, endocrinological abnormalities, and elevated γ-glutamyl transferase (γ-GT) [31, 36]. However, in any individual patient, any of these symptoms may be absent, and myotonia may be variable over time.

The most consistent muscle weakness is that of neck flexors, which may become severe. Mild ptosis is present in a minority of patients, combined with mild facial weakness. The only findings of distal muscle involvement are represented by weakness of deep finger flexors observed on specific testing in about half of the patients. Visible muscle atrophy is not a feature in the majority of the patients. When present, the grip myotonia has been characterized as jerky.

In addition to the aforementioned core features, a number of findings are occasionally associated with the disorder. Why these findings are less consistent is not yet clear. Cataracts are DM1-type posterior, subcapsular, iridescent cataracts or lens opacities. The typical cataracts may develop very late, not at all, or occasionally in presymptomatic patients before their thirties [35]. Brain functions in DM2 have been studied by Meola et al., who showed no major cognitive impairment but an avoidant personality change [37]. In MRI studies, minor to moderate white matter changes have been encountered, whereas assessment of atrophy was inconsistent [38, 39]. In the original MN-1 family, excessive sweating was reported in many patients but has not been a prominent finding in other cohorts [22]. Manifest diabetes mellitus is infrequent, whereas insulin resistance on testing is found in the majority of patients. Clinical male hypogonadism was marked in the Finnish PDM family, although subclinical follicle-stimulating hormone (FSH) elevations are a more regular finding. Cardiac complications have recently received more attention, including sudden cardiac deaths [40, 41], and may be the major cause of DM2-associated mortality. Whether sudden syncopal spells are related to cardiac arrhythmia or other malfunctions is not well understood. Occasionally, the severity of weakness showed extreme variation, mimicking even periodic paralyses, with patients presenting at emergency facilities. The generalized weakness always resolved spontaneously. A similar but less dramatic variation has repeatedly been reported regarding aggravation of stiffness and myotonia during pregnancy with spontaneous relief after delivery. In contrast to DM1, no specific obstetric problems have been reported with the exception of one family of Jewish descent [42], nor have complications during general anesthesia been observed. Other reports have extended the clinical spectrum of DM2-associated manifestations to include vestibular symptoms [43], parkinsonism [44], schizophrenia [45], or hyper-creatine kinasemia as the sole manifestation [46]. Overall, the clinical symptoms and findings found to be associated with the DM2 phenotype appear to be more varied and generally milder than those of DM1.

TABLE 8-1 Comparison of Clinical and Genetic Features in DM1 and DM2

	DM1	DM2
Core features*		
Myotonia	++[a]	(+) to +, on EMG
Muscle weakness	++	(+) to ++
Cataracts	++	− to ++
Localization of muscle weakness		
Facial weakness, jaw muscles	++	− to +
Distal limb muscle weakness	++	− to +
Proximal limb muscle weakness	(+)	+ to ++
Sternocleidomastoid muscle	++	+ to ++
Muscle symptoms		
Muscle/joint pain and stiffness	−	− to ++
Muscle strength variations	−	− to +
Muscle atrophy	++, distal	− to +
Muscle cramps	− to (+)	− to +
Calf hypertrophy	−	− to ++
Muscle biopsy		
Fiber atrophy	− to +, type 1 fibers	+ to ++, type 2 fibers
Cardiac arrhythmias	++	− to ++
Elevated serum creatine kinase levels	(+) to ++	(+) to ++
γ-Glutamyl transferase elevation	+	− to +
Hypoimmunoglobulinemia IgG	+	− to +
Hyperhidrosis	−	− to +
Brain		
Tremors	−	− to ++
Late change in mental state	++	− to (+)
Hypersomnia	+	− to (+)
Mental retardation	+, congenital form	−
Insulin resistance/glucose intolerance/diabetes	+	− to (+)
Male hypogonadism	+	− to +
Frontal baldness	++	− to (+)
Genetics		
Inheritance	AD	AD
Anticipation	++	− to (+)
Locus	*DMPK*	*ZNF9*
Chromosome	19q13.3	3q21.3
Expansion mutation	(CTG)$_n$	(CCTG)$_n$
Congenital form	+	−

[a]+, present; ++, pronounced; (+), variably present; −, absent. AD, autosomal dominant.

B. Age of Onset and Anticipation in DM2

As in DM1, specific DM2 symptoms and signs may have very different times of onset. Proximal muscle weakness is usually noted after age 30, and frequently it is markedly present only late in life, after age 60. As indicated above, cataracts have rarely been observed much earlier at presymptomatic stages. In rare instances, pain and stiffness have also been reported as early as the teens, but in a retrospective anamnestic survey, the latter features may be difficult to fully substantiate. However, in stark contrast to DM1, neither congenital nor childhood-onset DM2 cases with central nervous system symptoms have been identified or described to date [31, 35].

In the study by Day et al. [35], the most common first symptom identified by the patients was myotonia (39.7%), which is generally milder than that in DM1 adult-onset patients. However, the age range of first myotonia was very large, varying from 13 to 67. Equally common was muscle weakness observed as the first symptom (38.7%), with a similar age range of onset (18–66). Stiffness and/or pain were less often the first symptom (15.7%, 20–54 years); and in some cases muscle pain may be the most disabling problem for DM2 patients [36]. Cataracts were infrequently the first symptom of disease (8.1%).

In DM1, anticipation, that is, the increasing severity of phenotypic symptoms and/or the progressively earlier age of onset through successive generations, is a regular and typical feature, which at the molecular genetic level correlates with the increasing size of the mutant $(CTG)_n$ repeat expansion in *DMPK* from one generation to the next [47–56]. In contrast, in DM2, anticipation is not a regular feature, although it has been demonstrated statistically in a few German families [31, 57]. In DM1, the pronounced meiotic and mitotic instability and preponderance toward expansion seem to be the major effectors of and the molecular basis for the observed anticipation [48–53, 58, 59]. In contrast, in DM2, no direct correlation between disease severity and size of the mutant expansion could be shown [28, 29, 35]. DM2 repeat instability also appears to go in both directions, with approximately equal frequencies for expansions and contractions in available parent–offspring pairs [28, 29, 31], whereas over time, in individual patients, the expansion appears to increase [28]. It is, however, worth noting that all published studies to date, which are complicated by the enormous size of the DM2 repeat expansion and the associated technical difficulties in accurately sizing the $(CCTG)_n$ repeat, have relied on the analysis of DNA extracted from peripheral blood leukocytes or immortalized lymphoblastoid cell lines [28, 29, 31, 35]. So far, no study has investigated the size of mutant expansions in different affected tissues (including sperm) of a single or multiple patients or through successive generations. Thus, the observed lack of a genotype–phenotype correlation may be a function of the material analyzed.

C. Homozygosity for DM2 Mutation

In 2004, Schoser and colleagues reported the only known family with homozyogosity for the DM2 mutation, a large consanguineous family from Afghanistan with three homozygous sisters [60]. Compared with her two younger homozygous sisters, the oldest homozygous patient was clinically more severely affected and has since died (B. Schoser, personal communication). For the clinical course of symptoms, all three homozygotes were within the range expected for heterozygotes. Moreover, DM2 repeat length analysis, muscle histology, brain imaging studies, and muscleblind 1 (MBNL1) immunohistochemistry showed no appreciable differences between heterozygotes and homozygotes. Thus, homozygosity for the DM2 expansion does not seem to alter the disease phenotype as compared with the heterozygous state, and the DM2 repeat appears to be a truly autosomal dominant mutation. In addition to the observed homozygosity, the kindred was unique with respect to its ethnicity: to date, this Afghan kindred is the only reported non-European kindred from Asia [30].

D. What Is the Full Phenotypic Spectrum of DM2?

Phenotypic expression of DM2 is extremely variable, even within families, much more so than in DM1. Thus, the question arises as to whether mild and aberrant phenotypes can exist as sporadic manifestations and never be considered for diagnosis of DM2. This question is not yet answered, but in a pilot study in Finland on 82 consecutive patients with undetermined myopathy, no fewer than 17 (20%) proved to have DM2 expansions (B. Udd, personal communication). The inclusion criteria were very wide but had a preference for findings of myotonia or myotonia-like features on EMG, which makes the cohort not entirely random, but biased toward patients with some signs of muscle membrane disturbance. Nonetheless, only 3 of the 17 DM2-confirmed patients had a phenotype qualifying for diagnosis of PROMM/DM2 with mandatory core findings. These findings suggest that the DM2 mutation may underlie a wider spectrum of phenotypic manifestations than heretofore appreciated and that molecular genetic analysis should not be restricted to patients with clinically obvious DM2 disease and a family history.

E. Muscle Biopsy and Morphological Findings

Early muscle biopsy data in DM2 suggested the findings were similar to those in DM1 [5, 7]. The similarities consist of a highly increased number of internalized nuclei and variation of fiber size. However, later studies described clear differences, in that sarcoplasmic masses and ring fibers are not regular findings in DM2 as they are in DM1 [61, 62]. The use of myosin-specific immunohistochemistry allowed for a detailed examination of fiber types in DM2 patients. In a collaborative study Finnish, French, and Italian DM2 mutation-verified patients with biopsy specimens from vastus lateralis, deltoid, and biceps brachii muscles were analyzed [62]. Routine stains showed frequent nuclear clump fibers, without other features indicating neurogenic alteration, in early stages when the muscle was clinically unaffected. These and other extremely small atrophic fibers were identified as type 2 fibers, in contrast to established preferential type 1 fiber atrophy in DM1. The extremely small type 2 fibers in DM2 are not well detected on conventional ATPase staining. Other general histopathological findings consisted of frequent small group angulated fibers, rimmed vacuolated fibers, and moth-eaten fibers. Internal myonuclei were significantly more prevalent in type 2 fibers in DM2, whereas they occurred mainly in type 1 fibers in DM1. Morphometrical data from the study showed remarkable type 2 fiber atrophy in a subset of type 2 fibers, with other type 2 fibers showing hypertrophy. The reason for the rather specific type 2 fiber atrophy in DM2 is not known, but the clear difference from DM1 muscle morphology indicates that the findings on muscle biopsy and the differences between DM2 and DM1 regarding distribution of muscle weakness may have a common denominator. These findings were confirmed in a second study of 57 DM2-verified German patients with biopsies from deltoid, biceps, vastus lateralis, and tibialis anterior muscles [63]. In end-stage patients, fibrosis and fatty infiltration, starting in the interfascicular space and not within the fascicles, occurs. Results of electron microscopic studies showed no specific findings, especially no nuclear inclusions [31].

III. MOLECULAR GENETICS

A. Characteristics of the DM2 Repeat

The mutant DM2 repeat is located within intron 1 of the *ZNF9* gene in chromosome 3q21.3; it is composed of a complex motif with several polymorphic elements and was first described as $(TG)_n(TCTG)_n(CCTG)_n$ [28, 30], but variations such as $(TG)_{14–25}(CCTG)_5(GCTG)_1(CCTG)_1(TCTG)_1(CCTG)_{5–20}$ [28, 30] and $(TG)_{20–24}(TCTG)_{6–10}(CCTG)_{11–16}$ [29] have been reported. It has been proposed that in DM2 patients, it is the loss of the short intervening cryptic repeats (either TCTG and/or GCTG) in the tetranucleotide repeat array that causes the mutant expansion into large pure DM2 $(CCTG)_n$ repeat [28, 30], similar to other diseases associated with dynamic mutations [64–66]. Similar to the DM1 $(CTG)_n$ repeat, the DM2 $(CCTG)_n$ repeat is normally polymorphic. PCR across the repeat in normal individuals has shown a unimodal distribution of allele sizes, with most alleles differing by only 2 bp, indicating that the $(TG)_n$ motif is contributing heavily to the polymorphism at this locus. In the tails of the distribution, however, alleles are generally 4-bp apart in size, indicating additional variation in one or more of the 4-bp motifs of the repeat array. In contrast to Caucasian individuals, study of a group of 94 independent African-American samples

showed a bimodal allele distribution with a wider range of allele sizes, both smaller and larger, including some that cannot be accounted for by variation in the number of 2- or 4-bp units (L. Bachinski and R. Krahe, unpublished data) (see Figs. 8-1A and B). However, without further sequence analysis of the alleles, it is not possible to determine exactly which part of the repeat is responsible for this variation. Because the vast majority of normal individuals are heterozygous for the complex $(TG)_n(TCTG)_n(CCTG)_n$ DM2 repeat and because of

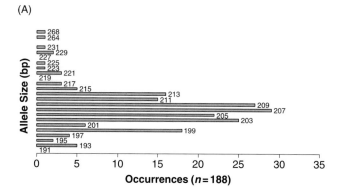

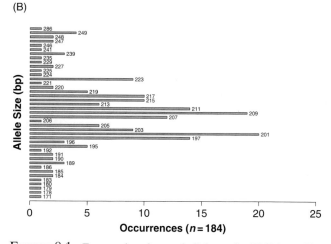

FIGURE 8-1 Frequencies of normal alleles at the DM2 locus. The DM2 repeat is a complex repeat array of at least three polymorphic repeats, $(TG)_n(TCTG)_n(CCTG)_n$, all of which can be polymorphic and contribute to the overall variability of alleles at the locus, such that alleles can differ by multiples of 2 or 4 bp. All sizes have been corrected for primer tail length and represent actual genomic length of amplified fragments. (A) Alleles observed in 188 European chromosomes. A total of 20 different alleles were observed, and the largest allele seen was 268 bp. The distribution of allele sizes appears to be unimodal, with 207 bp the most frequent (15.4%). Most alleles differ in size by 2 bp, reflecting variation in the $(TG)_n$ tract of the repeat array. (B) Alleles from 184 African-American chromosomes. A total of 41 alleles were observed, with the largest being 286 bp. Distribution appears to be bimodal, with 201 bp (10.9%) and 209 bp (10.3%) being most frequent. It is worth noting that a number of alleles having 1-bp size differences are seen, suggesting additional variation in the complex repeat array in sub-Saharan populations.

the high GC content of the locus, it is technically difficult to determine the exact sequence of normal alleles without further subcloning. The greater allelic diversity of normal DM2 alleles seen in African-American individuals is consistent with the generally observed greater genetic diversity for sub-Saharan populations.

B. Evolutionary Conservation of the DM2 Repeat

Evolutionary conservation of the repeat was studied by sequencing the repeat-containing region in three chimpanzees, two gorillas, one mouse, and one rat [30]. Both nonhuman primate species have complex repeat motifs similar, but not identical, to those of humans. Sequence in the regions approximately 500 bp in both directions from the repeat was more than 98% conserved between human and nonhuman primates. In contrast, only about 50% conservation was seen between rodent species and humans in this region. The rat sequence contained only the $(TG)_n$ element, whereas the mouse contained $(TG)_n$ and $(TCTG)_n$ elements, but not $(CCTG)_n$ elements, at the homologous intronic site. The apparent evolutionary conservation of the intronic sequences between humans and nonhuman as well as rodent species was interpreted as evidence for functional importance of intronic sequences, possibly in RNA processing of the pre-mRNA [30].

C. Population Studies and the Origin of the DM2 Repeat Expansion

Similar to DM1 [67], DM2 appears to be more prevalent in populations of European descent and, to date, has not been reported in other populations [26, 31]. In DM1, there is striking linkage disequilibrium (LD) around the $(CTG)_n$ expansion mutation [68–70]. With the exception of one sub-Sahara kindred [71], a single haplotype within and flanking *DMPK* has been shown to be in complete LD with the DM1 mutation. This has been interpreted as indicating that either predisposition for $(CTG)_n$ instability resulted from a founder effect, which occurred only once or a few times in human evolution, or *cis*-acting elements within the disease haplotype predispose the $(CTG)_n$ repeat to instability [69, 72].

LD studies, using both microsatellite markers (STRs) and single-nucleotide polymorphisms (SNPs), have been conducted to investigate the origin of the DM2 expanded repeat [29, 30]. Bachinski et al. [29] studied 17 DM2-confirmed kindreds of both Northern and Southern European origins, along with families that provided phase information on normal chromosomes. A total of 15 unique haplotypes on 160 independent normal chromosomes were observed using 7 polymorphic STR and 22 SNP markers, of which only 6 were polymorphic (Figs. 8-2A and B). Only one of these haplotypes—the most common in a Caucasian population sample, at 45.6%—was observed in the DM2 patients. The shared interval extended for a total distance of ~132–162 kb. These data suggest a single founding mutation in DM2 patients of diverse European populations, a situation reminiscent of that seen in DM1 [68, 69]. The estimated mutation age depended somewhat on the assumed historical population growth rate, with 90% credible intervals of 380–465 generations (for a growth rate of 5% per generation) and 515–540 generations (growth rate of 2% per generation). Using an average generation time of 20 years, the estimates correspond to a mutation age range of approximately 4000–11,000 years. Taken together, the data are consistent with a European origin of the DM2 mutation, but ancient enough to be present in several European subpopulations. The more extensive LD seen in DM2 relative to DM1 [29, 69] and apparent lack of DM2 patients in other non-Caucasian populations, such as the Japanese with a reported incidence for DM1 of 1 in 18,000 [1, 70, 73–80], are consistent with a more recent origin for the DM2 mutation relative to the DM1 mutation, after the divergence of the European and Asian lineages after the migration out of Africa some 80,000–100,000 years ago [67].

Liquori et al. [30] used 71 families with genetically confirmed DM2, all but one being of Northern European/German descent, and genotyped 12 STR markers. The common interval that was shared by all families with DM2 immediately flanks the repeat, extending up to 216 kb telomeric and 119 kb centromeric of the $(CCTG)_n$ expansion. The longest observed uninterrupted repeat in a normal individual was $(CCTG)_{20}$; they therefore suggested that, because this allele occurred on a conserved haplotype that extended ~470 kb, it may be representative of a pool of premutations for DM2. The more extensive LD seen by Liquori et al. in their DM2 families of Northern European/German descent relative to the geographically more diverse kindreds from Northern and Southern Europe studied by Bachinski et al. is simply a reflection of the larger LD commonly seen in geographically more refined populations.

Because of the observed pronounced LD, it has been suggested that similar to DM1 and other trinucleotide repeat diseases [81], an unknown *cis*-acting element contributes to the DM2 expansion [29]. In DM1, the complete allelic association with a single

(A)

Haplotype	rs762570	rs2342285	TSC175998	rs922834	TSC0280015	rs2128342	rs2342286	rs959487	rs1351596	TSC1065149	886191	886192	886189	886183	886187	DM2	TSC548022	TSC873597	309B5-AT1	814L21-GT2	rs2101155	TSC1443126	685P2-GT1	Obs/N
DM2	C	C	T	G	C	G	G	T	C	T	A	C	T	A	C	EXPANDED	A	C	293	123	A	A	221	17/17
A	C	C	T	G	C	G	G	T	C	T	A	C	T	A	C	193–229	A	C	293	123	A	A	221	73/160
B	C	C	T	G	C	G	G	T	C	T	A	C	T	A	C	193–231	A	**A**	293	123	A	A	221	64/160
C	C	C	T	G	C	G	G	T	C	T	**A**	C	T	A	C	203, 211, 213	A	**A**	293	123	A	A	221	4/160
D	C	C	T	G	C	G	G	T	C	T	A	C	T	A	C	203–205	A	**A**	293	123	A	A	**217**	3/160
E	C	C	T	G	C	G	G	T	C	T	A	C	T	A	C	197, 199, 209	A	C	293	**125**	A	A	221	3/160
F	C	C	T	G	C	G	G	T	C	T	A	C	T	A	C	221, 223	**G**	C	293	123	A	A	221	3/160
G	C	C	T	G	C	G	G	T	C	T	A	C	T	A	C	207, 211	A	**A**	293	**125**	A	A	221	2/160
H	C	C	T	G	C	G	G	T	C	T	A	C	T	A	C	209	A	C	293	123	A	A	**217**	1/160
I	C	C	T	G	C	G	G	T	C	T	A	C	T	A	C	213	A	C	293	123	A	A	**219**	1/160
J	C	C	T	G	C	G	G	T	C	T	**A**	C	T	A	C	205	A	**A**	293	123	A	A	**217**	1/160
K	C	C	T	G	C	G	G	T	C	T	**A**	C	T	A	C	199	A	C	293	123	A	A	221	1/160
L	C	C	T	G	C	G	G	T	C	T	**A**	C	T	A	C	199	A	C	293	**125**	A	A	221	1/160
M	C	C	T	G	C	G	G	T	C	T	A	C	T	A	C	213	A	**A**	293	**121**	A	A	221	1/160
N	C	C	**C**	G	C	G	G	T	C	T	A	C	T	A	C	211	A	C	293	123	A	A	221	1/160
O	C	C	**C**	G	C	G	G	T	C	T	A	C	T	A	C	205	A	C	295	123	A	A	221	1/160

(B)

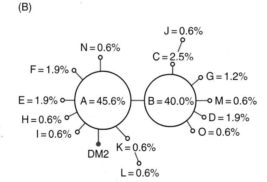

FIGURE 8-2 (A) Haplotypes observed across the DM2 region including 19 SNP and 3 microsatellite markers. Fifteen normal European haplotypes (A–O) are shown along with the DM2 haplotype, which is identical to the most common haplotype, A, except for the presence of the DM2 $(CCTG)_n$ expansion. (B) Haplotype evolutionary network based on parsimony. The areas of circles representing each haplotype are roughly proportional to the frequency of the haplotype being represented. Each haplotype is one mutational step from all of the haplotypes to which it is connected by a line.

haplotype in Eurasian populations has been interpreted as evidence that specific chromosomal context may be associated with the instability [68, 69] and that cis-acting elements may be required for expansion to occur. Additional evidence for the involvement of flanking sequences comes from the variable stability of similar $(CAG)_n$ repeats at different disease loci [81–83] and from transgenic mouse models of trinucleotide repeat instability [84, 85]. While the nature of such a cis-element(s) is currently unknown, recent in vitro studies suggest the distance of the repeat to the replication origin may be an important factor [83]. In this context, it is noteworthy that the DM2 region is relatively gene rich and lies within an early replicating region.

D. Molecular Diagnosis of DM2

The extreme variability of the phenotypes of both DM1 and DM2 make molecular diagnostic methods essential for correct diagnosis. For DM2, a three-step diagnostic procedure has been recommended [31]. First, standard PCR across the repeat serves to exclude individuals with two normal alleles [29, 35]. Depending on the population sampled, about 80–95% of individuals are heterozygous at this locus for the normally polymorphic repeat; therefore, individuals who amplify only a single allele have a high probability of having the expansion [31]. To confirm whether individuals showing a single allele have a second expanded allele that is resistant to PCR, a repeat-primed PCR (RP-PCR) assay is recommended, analogous to similar assays for DM1 [86, 87] and SCA10 [88]. Several variations on this assay have been published, with or without the addition of a blotting/hybridization step [29, 35]. Assays of this type are very sensitive and reliable, even without the blotting step, when compared with other methods, such as conventional Southern blotting. However, the size of the DM2 expansion cannot be determined using the RP-PCR mutation assay.

Because of the large size of the $(CCTG)_n$ expansions seen in this disease (300 bp–44 kb; mean, 20 kb), Southern blotting after conventional electrophoresis, the method of choice for the diagnosis of DM1, is unsatisfactory and inadequate to properly size the mutation. Pulsed-field electrophoresis can obviate this difficulty, and many repeats can be approximately sized by this method [29]. However, this method is laborious, and because of the size heterogeneity of the repeat even in cells of the same tissue (reflective of somatic instability), its sensitivity is estimated at only 70–80% [89], making it less than ideal for diagnostic purposes. Chromogenic *in situ* hybridization (CISH) has been used to detect both the genomic expansions and the mutant transcripts in muscle biopsy sections and can be used in routine DM2 diagnostics in laboratories where these procedures are in routine use [89]. Similar diagnostic approaches have been successfully implemented by other groups [90, 91]. Fluorescence *in situ* hybridization (FISH) on extended DNA fibers (fiber-FISH) can likewise be used to directly visualize the DM2 mutation and to estimate the repeat expansion sizes, but only on DNA extracted from cultured cells [89]. To date, no other methods have been developed for estimating the expansion size in DM2. Both pulsed-field electrophoresis and fiber-FISH are specialized techniques, best suited to a research environment rather than a diagnostic laboratory. Because of this, large-scale studies to estimate the repeat expansion sizes in DM2 patients have not been carried out, and the relationship, if any, between repeat size and disease severity or age of onset remains undetermined. The utility of the different diagnostic methods discussed here is highlighted in Fig. 8-3.

Recently, two additional methods were published, including a long-range PCR method on single genome equivalents, followed by agarose gel electrophoresis and repeat-specific oligohybridization [92], and another pulsed-field gel electrophoresis method followed by semiquantitative Southern blot analysis with a novel hybridization probe [93], both with reportedly high sensitivity.

E. How Many Myotonic Dystrophies Are There?

Le Ber and colleagues [94] recently described a large French kindred with a phenotype consistent with myotonic dystrophy, however with the addition of frontotemporal dementia. Patients were exhaustively characterized for essentially all known manifestations of myotonic myopathies, and many additional possible causes of the clinical observations were ruled out. Two-point LOD (2.23) and multipoint LOD (2.38) scores were obtained with several markers, including D15S153, under an "affecteds only" model. This was comparable to the maximum LOD score obtained by simulation. NPL analysis also produced significant results with a score of 12.82 ($P < 10^{-5}$). These results, although not conclusive, strongly suggested a third DM locus (DM3) in chromosome 15q21-q24 [94].

Additional genetic heterogeneity exists within the spectrum of the autosomal dominant myotonic myopathies, and identification of additional loci is to be expected [Bachinski et al., manuscript in preparation]. During the course of a 10-cM genome scan using approximately 400 microsatellite markers that was conducted to identify the DM2 locus, several families meeting the diagnostic criteria for PROMM or DM2 were identified who had no expansions in *DMPK* or *ZNF9* and whose disease did not segregate with these loci [29, 31]. Among these families was one that showed a LOD score of >1.3 in chromosome 16. To demonstrate whether this LOD score represented a novel locus, we genotyped a number of additional markers on chromosome 16 in this and other families segregating autosomal dominant progressive myotonic myopathy who had also tested negative for expansions at both the DM1 and DM2 loci. Using nine such families (four German, four Spanish, and one Brazilian), we obtained a LOD score of 3.83 in the p arm of chromosome 16. As most of these families are small, each contributes only a small positive amount to the total LOD score. However, in all cases, the LOD scores observed are near the simulated $ELOD_{max}$ for these families. Thus, it appears that chromosome 16 harbors a fourth DM locus (DM4) that is responsible for DM in most of these kindreds. Efforts are underway to identify the causative mutation. The similarities of the clinical manifestations observed in the DM3 and DM4

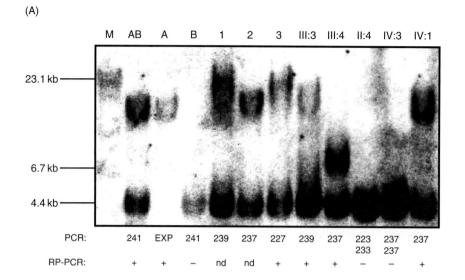
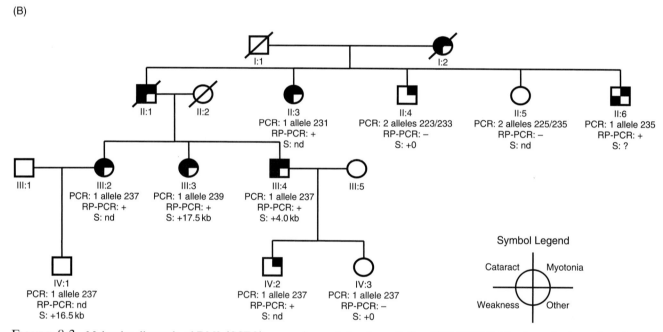

FIGURE 8-3 Molecular diagnosis of DM2 (CCTG)$_n$ expansion mutation in an Italian (I01) PROMM family and other selected cases. (A) FIGE Southern blot showing expanded (CCTG)$_n$ alleles at the *DM2* locus. Shown beneath each lane are the PCR allele sizes across the DM2 repeat and the results (+/−) of the repeat-primed PCR (RP-PCR) assay. Pedigree numbers in the lane headers refer to those shown in the pedigree (B). Lanes A and B are haploid hybrids for the normal and mutant chromosome 3, respectively; lane AB is the donor patient from which the hybrid cell lines were established. M is the size marker lane. Cases in lanes 1–3 are individual, unrelated patients. This representative autoradiograph demonstrates the size and spectrum (~4–19 kb) and the high level of somatic heterogeneity mutations within individuals and between individuals of the same kindred of the mutant expansions commonly encountered. (B) Pedigree of PROMM/DM2 kindred I01 [14], illustrating commonly encountered diagnostic problems and the importance of using multiple molecular diagnostic tests. Individual II:4 exemplifies a false-positive diagnosis. He had myotonia, but had two normal alleles and showed no expansion by either RP-PCR or Southern analysis. For individual II:6, who has myotonia and muscle weakness, Southern analysis gave an unclear molecular diagnosis, whereas RP-PCR clearly identified him as a mutation carrier. Individual IV:1 is a false-negative diagnosis. Although phenotypically normal, he has an expansion of +16.5 kb. Individual IV:3 is a true homozygote. Both RP-PCR and Southern analysis confirm the absence of the expansion, despite the presence of only a single allele.

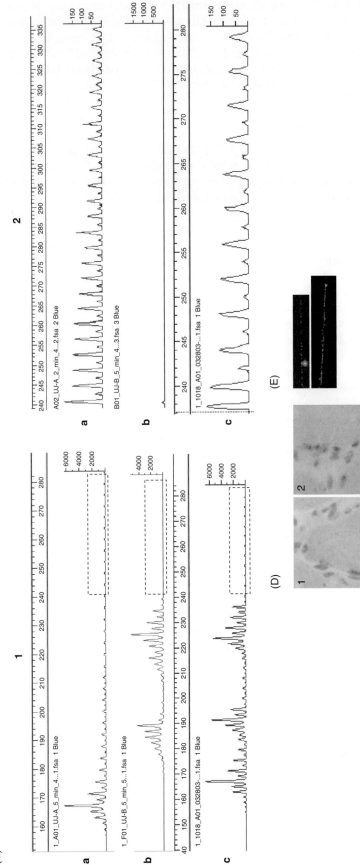

FIGURE 8-3. (C) RP-PCR assay for molecular diagnosis of DM2 (CCTG)$_n$ expansion mutation. Diploid–haploid conversion somatic cell hybrid cell lines established from a DM2 patient with the (CCTG)$_n$ expansion of approximately +18.5, with the two homologs of chromosome 3 separated, are used to illustrate the results of the RP-PCR assay. Panel 1 is the full view and panel 2 is the zoomed-in view of the areas in panel 1 corresponding to the dashed boxes. Sample a is a somatic cell hybrid cell line with the homolog chromosome 3 carrying the mutant DM2 alleles, sample b is the somatic cell hybrid cell line with the homolog chromosome 3 carrying the wild-type allele, and sample c is the original patient sample containing both chromosomes and alleles. The PCR products from the expanded (CCTG)$_n$ are visible in full view as small "bumps" on the baseline (panel 1, a and c), but can be better visualized in closeup as a repeating pattern of peaks (panel 2, a and c). (D) Visualization of the DM2 (CCTG)$_n$ expansion mutation by CISH on DM2 patient frozen muscle sections. The specific label (DAB) is seen as dark brown spots within the myonuclei. (1) A single spot signal representing the genomic DM2 mutation is obtained with the (CCTG)$_8$ sense oligonucleotide. (2) In the same patient, ribonuclear inclusions containing accumulated mutant RNAs are detected with the (CCAG)$_8$ antisense oligonucleotide as larger multifocal accumulations. (E) Direct visualization of the DM2 (CCTG)$_n$ expansion mutation by fiber-FISH on stretched chromatin fibers prepared from cultured peripheral leukocytes of a confirmed DM2 patient. A BAC clone from the region, 814L21 (red), covers the DM2 locus and identifies extended DNA fibers containing the DM2 region. Cohybridization of the (CCTG)$_8$ sense oligonucleotide (green) highlights the expanded (CCTG)$_n$ on the DNA fibers of affected individuals. The size of different DM2 expansions is determined by relating the relative length of hybridization measurements to the known size of the BAC as an internal standard. Representative fiber-FISH images: (top) A patient with a mean expansion of 7.4 kb (range = 3.1–15.9 kb, n = 15 images; ~18.0 kb by FIGE-S); (bottom) a patient with a mean expansion of 12.2 kb (range = 3.2–22.5 kb, n = 15 images; ~18.5 kb by FIGE-S). See CD-ROM for color image. (A) and (B) are reprinted with permission from Bachinski et al. (2003) *Am. J. Hum. Genet.* **73**, 835–848. (D) and (E) are reprinted with permission from Sallinen et al. (2004) *Neuromuscul. Disord.* **14**, 274–283.

patients to those of DM1 and DM2 suggest that the underlying mutations may also be $(CTG)_n$- or $(CCTG)_n$-like repeats, but it remains to be seen whether any or all of these additional DM loci are due to the expansion of unstable microsatellite repeats.

IV. MOLECULAR PATHOPHYSIOLOGY

There are many parallels between DM2 and DM1; the picture that has emerged in recent years is that both are RNA-mediated diseases. Currently, most of the data on the pathophysiological mechanisms in DM come from studies on DM1. In DM1 and DM2 the mutant repeats are located in untranslated regions of their respective genes—the 3′ UTR of *DMPK* in 19q13.3 and intron 1 of *ZNF9* in 3q21.3, respectively; they are transcribed but not translated [95, 96]. The DM1 and DM2 repeats are different from other noncoding repeat expansions, identified in other repeat diseases. For example, in fragile-X syndrome (FRAXA), a $(CCG)_n$ repeat expansion in the 5′ promoter region leads to the transcriptional silencing of the associated gene, *fragile-X mental retardation 1* (*FMR1*) [97]; in Friedreich's ataxia (FRDA), expansion of a $(GAA)_n$ repeat in intron 1 of the *frataxin 1* (*FXN*) gene interferes with transcription and leads to loss of expression [98]. For both types of repeats, the resulting loss of function is consistent with the paradigm that recessive diseases are most often due to loss of functional protein. In contrast, in $(CAG)_n$ repeat diseases, including Huntington's disease (HD) and various spinocerebellar ataxias (SCAs), the expansion in the coding regions of the associated genes is translated into polyglutamine tracts that confer a dominant gain of function [99].

Initially, three distinct hypotheses were proposed to explain the complex pattern of multisystemic manifestations caused by the DM1 $(CTG)_n$ repeat expansion [100, 101]: (1) the expanded repeat in the mutant transcript, present as $(CUG)_n$ repeats, affects processing of *DMPK* transcripts, resulting in DMPK protein haploinsufficiency; (2) the $(CTG)_n$ expansion alters the local chromatin structure, resulting in a field or position effect variegation and misexpression of genes flanking the expansion; and (3) expanded $(CUG)_n$ repeats in the mutant transcript interact with RNA-binding proteins by sequestering them and interfering with their normal activity, ultimately leading to disruption of cellular metabolism.

Initial evidence against DMPK haploinsufficiency in DM1 came from the demonstration that the DM1 expansion does not interfere with transcriptional initiation and transcription of either normal or mutant *DMPK* alleles, while affecting posttranscriptional processing of mutant allele transcripts [102]. As early as 1995, we hypothesized that the dominant inheritance of DM stems from a gain of function of a new, as-yet-unidentified function of the mutant DMPK RNA with the expanded $(CUG)_n$ repeat. We postulated that the 3′ UTR of the mutant mRNA would function in *trans* and inhibit such processes as muscle differentiation or interfere with regulation of multiple other genes, resulting in the pleiotropic DM phenotype [102]. In this model, the DM phenotype would be strictly dependent on the presence of the expanded $(CTG)_n$. Additional evidence came from two different *Dmpk* knockout mice [103, 104]. Heterozygous $Dmpk^{+/-}$ mice showed no evidence of muscle pathology, whereas older homozygous $Dmpk^{-/-}$ mice developed a mild myopathy. Neither mouse model showed the characteristic electrophysiological and pathological changes characteristic of DM. However, subsequent analysis of cardiac muscle conduction in heterozygous and homozygous knockout mice revealed cardiac conduction defects similar to those seen in human DM [105]. Mutant DMPK transcripts were shown to be retained in the nucleus where they accumulate in nuclear foci [106–109], interfering with the splicing and processing of its own transcripts [110] as well as other transcripts [107, 111] and preventing export to the cytoplasm where they would normally be translated into protein.

Consistent with the second hypothesis, altered expression of genes flanking the $(CTG)_n$ expansion has been observed for at least the immediate neighboring genes, including *SIX5*, which is associated with a large CpG island that is interrupted by the expanded DM1 repeat [112]. The altered local chromatin structure [113–115] suppresses the expression of *SIX5* [116–118]. Moreover, heterozygous and homozygous *Six5* knockout mice develop cataracts, suggesting that the cataracts characteristic of human DM might be caused by decreased expression of *SIX5* [119, 120]. A similar effect on the expression of *DMWD*, the closest telomeric gene, was also reported [121], which, because its expression in the testis and brain had been suggested as a candidate gene for the testicular atrophy and cognitive disturbances seen in DM1 [122]. In addition, in congenital DM (CDM) patients, the CpG island within *DMPK* and *SIX5* is hypermethylated [123], suggesting that epigenetic events might contribute to specific aspects of the DM phenotype.

The third hypothesis of an RNA-mediated pathogenesis for the mutant DMPK transcripts has been supported by *in vitro* studies [124–127] and a mouse model described by Mankodi and colleagues [128]. Mutant transcripts with expanded $(CUG)_n$ repeats inhibit the differentiation of myoblasts to myotubes in culture [124–126]. Transgenic mice expressing an untranslated $(CUG)_n$ expansion in an unrelated RNA, the *human skeletal actin*

Chapter 8 Myotonic Dystrophy Type 2: Clinical and Genetic Aspects

(*HSA*) gene, show nuclear foci of the mutant transgene and developed myotonia and myopathy [128]. Mice expressing a nonexpanded repeat did not, indicating that transcripts with expanded $(CUG)_n$ repeats are both necessary and sufficient to generate a DM phenotype [128].

The accumulation of mutant transcripts in nuclear foci suggested an effect on proteins that interact with the repeats. Several such proteins have now been identified. CUGBP1 was implicated in regulating the processing of mRNAs and translational initiation [129–134]. Miller and colleagues identified a novel family of triplet repeat expansion RNA-binding proteins, homologous to the *Drosophila* muscleblind (mbl) proteins, that specifically bind dsCUG RNAs, with binding proportional to the size of the expansion [135]. In *Drosophila*, mbl protein is required for terminal muscle and photorecptor cell differentiation. In DM1 cell lines, MBNL proteins accumulate in nuclear foci [135–139].

In 2001, the identification of an $(CCTG)_n$ expansion mutation in 3q21.3 for DM2 without obvious similarity of flanking genes to those surrounding the DM1 locus in 19q13.3, provided an additional boost for the toxic RNA gain-of-function model [28]. This was further underscored by the finding of similar accumulations of mutant RNAs in ribonuclear inclusions of DM2 muscle biopsies [140, 141]. Like $(CUG)_n$ ribonuclear inclusions in DM1, $(CCUG)_n$ inclusions in DM2 bind nuclear proteins involved in the posttranscriptional processing and splicing of other genes, including members of the muscleblind protein family, MBNL1 and MBNL2, and possibly other family members [135–139].

The almost complete recapitulation of the DM1 phenotype in a *Mbnl1* knockout mouse model, which also showed aberrant splicing of Clcn1, Tnnt1, and Tnnt3 transcripts, provided further evidence for the toxic RNA model and the aberrant splicing of "effector" genes [142]. To date, at least 12 genes have been identified as aberrantly spliced in DM1: cardiac troponin T (*TNNT2*) [138, 139, 143], insulin receptor (*INSR*) [137, 144], skeletal muscle chloride channel (*CLCN1*) [142, 145–147], microtubule-associated protein tau (*MAPT*) [148–150], myotubularin-related 1 phosphatase (*MTMR1*) [151], fast skeletal muscle troponin T (*TNNT3*) [142], NMDA receptor 1 (*NMDAR1*) [149], amyloid precursor protein (*APP*) [149], a number of sarcomeric proteins, including skeletal muscle ryanodine receptor 1 (*RYR1*) [152], sarcomeric/endoplasmic reticulum, fast twitch skeletal muscle Ca^{2+}-ATPase (*SERCA1* or *ATP2A1*) [152], the Z-disk-associated protein (*ZASP* or *LDB3*) (C. Thornton, personal communication), and titin (*TTN*) (C. Thornton, personal communication). For most of the affected gene transcripts, their splicing pattern is altered in such a way that the embryonic or fetal isoforms become the DM1-associated form. The observed splicing defects in the aforementioned genes are the molecular basis for such varied manifestations as the observed skeletal and cardiac muscle weakness, the cardiac conduction problems, insulin resistance, glucose intolerance and diabetes, myotonia, impaired muscle differentiation, CNS impairment, and elevated Ca^{2+} concentrations.

To date, only 3 of the 12 aberrantly spliced genes have also been confirmed in DM2: *CLCN1* [145], *INSR* [153], and *MAPT* (Maurage et al., in press). However, given the sequestration of MBLN1 and MBLN2 proteins in both $(CUG)_n$ and $(CCUG)_n$ ribonuclear inclusions and the overlap of phenotypes, it is likely that if not all, at least a subset of the genes affected in DM1 is also affected by aberrant splicing in DM2. Interestingly, splicing defects in DM patient muscle and in transgenic mice that express the mutant $(CUG)_n$ repeat are strikingly similar to those seen in *Mbnl1* knockout mice (C. Thornton, personal communication).

The inappropriate redistribution or leaching of various transcription factors, both general and differentiation factors, such as Sp1, by mutant RNA species in DM1 has recently been suggested as yet another pathogenic mechanism [154]. Sp1 leaching appeared to result in transcriptional downregulation of *CLCN1* transcription among other genes. It is possible that similar mechanisms are also operative in DM2.

We have used global gene expression profiling with microarrays to test the hypothesis that DM1 and DM2 share common pathological pathways. We globally compared expression in skeletal muscle biopsies of normal, individuals, those with DM1, and those with DM2 (manuscript in preparation). Comparison of gene expression profiles of sets of normal skeletal muscle biopsies with DM1 and DM2 biopsies showed considerable overlap in the genes down- and upregulated among DM1 and DM2 patients. Significant dysregulation of several functional gene categories, including muscle, myogenesis, calcium channel, DNA repair, ribosomal proteins, RNA binding, and proteasome, was observed. Numerous skeletal muscle-specific genes (e.g., several myosin heavy chains, tropomyosins, and troponins, as well as sarcomeric proteins) were specifically affected. Similar changes were seen *in vitro* with primary myoblast cultures established from skeletal muscle of congenital and adult-onset DM1 and DM2 patients for either total RNA or nuclear and cytoplasmic fractions separated. Expression profiling of nuclear and cytoplasmic fractions of DM1 myoblasts that displayed an inability to differentiate into multinucleated myotubes identified sets of genes downregulated in both fractions (e.g., β-tropomyosin) and downregulated in only the cytoplasmic fraction (e.g., troponin I, the slow-twitch/type I fiber skeletal muscle isoform; myomesin 2, a titin-associated protein). Taken together, these data suggest a global *trans* effect of the transcribed expansion on the DM1 or DM2 transcriptome.

CUGBP1 mRNA and protein overexpression, which has been demonstrated in DM1 patients, has been linked to the translational block of various gene products, including general and muscle-specific transcription factors as well as cell cycle regulatory genes [129, 131–134, 155, 156]. Skeletal muscle cells from DM patients fail to induce cytoplasmic levels of CUGBP1, whereas normal differentiated cells accumulate CUGBP1 in the cytoplasm. In normal cells, CUGBP1 upregulates p21 protein during differentiation through induction of p21 translation via binding to a GC-rich sequence located within the 5' region of p21 mRNA. Failure of DM cells to accumulate CUGBP1 in the cytoplasm leads to a significant reduction of p21 and to alterations of other proteins responsible for the cell cycle withdrawal. Thus, CUGBP1-dependent altered expression and activity of the proteins responsible for cell cycle withdrawal appear to lead to impaired muscle differentiation in DM muscle cells [132, 133, 156].

Interestingly, in the nucleus, MBNL1 and CUGBP1 protein activities appear to antagonize each other's functions for splicing. For exon 11 of INSR, CUGBP1 regulates the equilibrium of splice site selection by antagonizing the facilitatory activity of MBNL1 and MBNL2 splicing in a dose-dependent manner [137]. In DM1 cells, CUGBP1 protein levels are elevated by mechanisms that are independent of MBNL1 and MBNL2 loss. However, loss of MBNL1 function appears to be the key event, whereas CUGBP1 overexpression seems to play a secondary role in the aberrant INSR mRNA splicing [137].

Using a modified UV crosslinking assay to isolate proteins bound to mutant DMPK-derived RNA, Kim and colleagues recently identified hnRNP H as a nuclear RNA-binding protein capable of binding and possibly modulating nuclear retention of mutant DMPK mRNA [157]. Another interesting possibility is that mutant DM transcripts cause *trans*-RNA interference, through the direct interaction of mutant $(CUG)_n$ repeat containing transcripts with transcripts of other genes containing long $(CAG)_n$ tracts, such as *SCA1* and *TFIID*, and possibly others [127].

Thus, the emerging picture for the molecular pathophysiological mechanisms in DM1 and DM2 is that the mutant $(CTG)_n/(CUG)_n$ and $(CCTG)_n/(CCUG)_n$ transcripts form ribonuclear inclusions that sequester *trans*-acting factors that are involved in a variety of cellular processes, including transcriptional regulation, posttranscriptional processing and splicing, nuclear export, and translation. In conclusion, it may be the highly varied nature of the different pathogenic mechanisms that explains the pleiotropic nature of the DMs.

Unlike for DM1 where a collection of important mouse models has been generated over the last 12 years [158], only one mouse model currently exists for DM2 (Y.-P. Li, personal communication and manuscript submitted). Li and colleagues generated *Znf9* knockout mice [159]. Interestingly, $Znf9^{+/-}$ mice appear to recapitulate multiple features of the DM2 phenotype found in human patients, including myopathy, myotonia, cardiac hypertrophy, and cataracts, suggesting that Znf9/ZNF9 haploinsufficiency may at least in part account for certain features of DM2. *Znf9* is highly expressed in skeletal and heart muscle. In $Znf9^{+/-}$ mice, *Znf9* expression appears to be significantly decreased, and *Znf9* transgenic mice can rescue the myotonic dystrophy phenotype in $Znf9^{+/-}$ mice. However, in the light of recent findings in *Drosophila*, where reporter gene constructs with 3' UTR $(CTG)_n$ repeats in the human disease range did not produce a pathological phenotype, these findings are noteworthy. Thus, in *Drosophila*, neither expanded $(CUG)_n$ repeat RNAs nor their ribonuclear foci appear to be directly toxic, in contrast to expanded polyglutamine-containing proteins [160].

V. CONCLUDING REMARKS

The overlap of clinical manifestations and molecular pathogenic mechanisms between DM1 and DM2 has now substantiated an important role of an RNA-mediated toxic gain-of-function mechanism as a third possible mechanism for unstable repeat mutations, in addition to loss of gene expression or gain of function of novel pathogenic protein products. However, several important issues remain. For example, what is the molecular basis for the observed differences between DM1 and DM2, such as the distal versus proximal muscle weakness, the fiber type atrophy, the lack of a severe congenital form with mental retardation, and the apparent lack of genotype/phenotype correlations for DM2, with $(CCTG)_n$ expansions considerably larger than the $(CTG)_n$ expansions commonly seen in DM1 and stronger binding of MBLN proteins to $(CCUG)_n$ than $(CUG)_n$ ribonuclear inclusions? The latter would suggest that sequestration of MBLN splice factors cannot explain all of the clinical features for either DM1 or DM2. The role of CUGBP1, and possibly other proteins involved in RNA splicing and translation, provides another interesting avenue for further research. Is there a difference in the toxicity of $(CUG)_n$ and $(CCUG)_n$ mutant repeat expansions, or is the toxicity modulated by other flanking sequences in the mutant RNA as has been suggested [125]? Similarly, what features account for the observed differences between DM1 (and DM2) and SCA8, the only other known unstable $(CTG)_n$ repeat disease [95, 161].

Splicing is a very complex process, facilitated by several *trans*-acting multiprotein complexes that form the splice machinery and mediated through specific *cis* sequence consensus elements in downstream genes. Together with the extreme variability of the clinical manifestations in DM2, the complexity of the posttranscriptional processing of DM "effector" genes raises the interesting possibility of additional modifier polymorphisms in the splice sites, including splice enhancer and silencer elements, of affected genes. In this context, genomewide profiling approaches hold great promise in further clarifying the effects of the respective mutations on the expression and splicing of downstream genes. Given the pleotropic nature of the phenotype of the different DMs, the effects of the mutant expansions are likely not limited to just the 12 genes affected by aberrant splicing identified to date.

In light of the possible role for *Znf9* haploinsufficiency in $Znf9^{+/-}$ mice, the role of *ZNF9* in the pathogenesis of DM2 patients needs to be clarified. Clearly, mouse models such as $(CCUG)_n$ transgenic and knockin mice are needed to further dissect the functional role of the mutant DM2 repeat expansion.

Currently, at least four genetically distinct loci for DM have been mapped; for two, DM1 and DM2, the causative mutations have been identified. Given the phenotypic similarities between patients with DM1 and DM2 and those with linkage to the DM3 or DM4 locus, it is tempting to speculate that the underlying mutation(s) may also be $(CTG)_n$- or $(CCTG)_n$-like repeats. Whether or not that is in fact the case remains to be seen; however, if it is the case, it would provide additional support for the RNA-mediated pathogenesis for the DMs.

Acknowledgments

R. K. and L. L. B. thank the members of the Krahe laboratory for valuable discussions and Keith A. Baggerly and E. Lin for assistance with bioinformatics analyses of DM gene expression profiling data. We thank the European Neuro-Muscular Centre (ENMC) for their continued support of the International Working Group on DM2/PROMM and Other Myotonic Dystrophies and the numerous, participating collaborators of the group. R. K. was supported by grants from the NIH (AR48171); B. U. was supported by Medicinska understödsföreningen Liv och Hälsa r.f., the Tampere University Hospital Research Funds, and the Folkhälsan Institute of Genetics.

NOTE ADDED IN PROOF

Recent molecular genetic and clinical studies on the large French family presenting with a combined phenotype of PROMM and fronto-temporal dementia and putative linkage (LODmax=2.23) to chromosome 15q21-q24 (Le Ber et al. 2004) by the original authors have firmly identified the segregating disease as hereditary inclusion body myopathy with Paget disease of bone and fronto-temporal dementia (IBMPFD) due to a mutation in the *VCP* gene in 9p13.3-p12 (MIM 601023) (Didier Hannequin, University of Rouen, personal communication). With this clarification on the status of the putative 15q DM locus, we have designated our novel DM locus mapping to 16p as DM3 (manuscript submitted).

References

1. Harper, P. S. (2001). "Myotonic Dystrophy." Saunders, London.
2. Brook, J. D., McCurrach, M. E., Harley, H. G., Buckler, A. J., Church, D., Aburatani, H., Hunter, K., Stanton, V. P., Thirion, J. P., Hudson, T., Sohn, R., Zemelman, B., Snell, R. S., Rundle, S. A., Crow, S., Davies, J., Shelbourne, P., Buxton, J., Jones, C., Juvonen, V., Johnson, K., Harper, P. S., Shaw, D. J., and Housman, D. E. (1992). Molecular basis of myotonic dystrophy: Expansion of a trinucleotide (CTG) repeat at the 3′ end of a transcript encoding a protein kinase family member. *Cell* **68**, 799–808.
3. Fu, Y. H., Pizzuti, A., Fenwick, R. G., Jr., King, J., Rajnarayan, S., Dunne, P. W., Dubel, J., Nasser, G. A., Ashizawa, T., de Jong, P., Wieringa, B., Korneluk, R., Perryman, M. B., Epstein, H. F., and Caskey, C. T. (1992). An unstable triplet repeat in a gene related to myotonic muscular dystrophy. *Science* **255**, 1256–1258.
4. Mahadevan, M., Tsilfidis, C., Sabourin, L., Shutler, G., Amemiya, C., Jansen, G., Neville, C., Narang, M., Barcelo, J., O'Hoy, K., Leblond, S., Earle-Macdonald, J., de Jong, P., Wierenga, B., and Korneluk, R. G. (1992). Myotonic dystrophy mutation: An unstable CTG repeat in the 3′ untranslated region of the gene. *Science* **255**, 1253–1255.
5. Ricker, K., Koch, M. C., Lehmann-Horn, F., Pongratz, D., Otto, M., Heine, R. and Moxley, R. T., 3rd (1994). Proximal myotonic myopathy: A new dominant disorder with myotonia, muscle weakness, and cataracts. *Neurology* **44**, 1448–1452.
6. Thornton, C. A., Griggs, R. C., and Moxley, R. T., III (1994). Myotonic dystrophy with no trinucleotide repeat expansion. *Ann. Neurol.* **35**, 269–272.
7. Moxley, R. T., III, Udd, B., and Ricker, K. (1998). 54th ENMC International Workshop: PROMM (proximal myotonic myopathies) and other proximal myotonic syndromes. *Neuromuscul. Disord.* **8**, 508–518.
8. Abdalla, J. A., Casley, W. L., Cousin, H. K., Hudson, A. J., Murphy, E. G., Cornelis, F. C., Hashimoto, L., and Ebers, G. C. (1992). Linkage of Thomsen disease to the T-cell-receptor beta (TCRB) locus on chromosome 7q35. *Am. J. Hum. Genet.* **51**, 579–584.
9. McClatchey, A. I., Trofatter, J., McKenna-Yasek, D., Raskind, W., Bird, T., Pericak-Vance, M., Gilchrist, J., Arahata, K., Radosavljevic, D., Worthen, H. G., et al. (1992). Dinucleotide repeat polymorphisms at the SCN4A locus suggest allelic heterogeneity of hyperkalemic periodic paralysis and paramyotonia congenita [published erratum appears in Am J Hum Genet 1992 Oct;51(4):942] [see comments]. *Am. J. Hum. Genet.* **50**, 896–901.
10. Lacomis, D., Chad, D. A., and Smith, T. W. (1994). Proximal weakness as the primary manifestation of myotonic dystrophy in older adults [see comments]. *Muscle Nerve* **17**, 687–688.
11. Ricker, K., Koch, M. C., Lehmann-Horn, F., Pongratz, D., Speich, N., Reiners, K., Schneider, C., and Moxley, R. T., 3rd (1995). Proximal myotonic myopathy: Clinical features of a multisystem disorder similar to myotonic dystrophy. *Arch. Neurol.* **52**, 25–31.

12. Stoll, G., von Giesen, H. J., Koch, M. C., Arendt, G., and Benecke, R. (1995). Proximal myotonic myopathy syndrome in the absence of trinucleotide repeat expansions. *Muscle Nerve* **18**, 782–783.
13. Hund, E., Jansen, O., Koch, M. C., Ricker, K., Fogel, W., Niedermaier, N., Otto, M., Kuhn, E., and Meinck, H. M. (1997). Proximal myotonic myopathy with MRI white matter abnormalities of the brain. *Neurology* **48**, 33–37.
14. Meola, G., Sansone, V., Radice, S., Skradski, S., and Ptacek, L. (1996). A family with an unusual myotonic and myopathic phenotype and no CTG expansion (proximal myotonic myopathy syndrome): A challenge for future molecular studies. *Neuromuscul. Disord.* **6**, 143–150.
15. Gamez, J., Cervera, C., Fernandez, J., Martinez, J., Martorell, L., Teijeira, S., and Navarro, C. (1996). Proximal myotonic myopathy report of three families [abstract]. *Neuromuscul. Disord.* **6** (Suppl.), S46.
16. Sansone, V., Cappa, S., Cotelli, C., Rognone, F., Thornton, C. A. and Moxley, R. T. (1997). Magnetic resonance imaging and cognitive alterations in proximal myotonic myopathy (PROMM). *Neurology* **48** (Suppl.), A231.
17. Rowland, L. P. (1994). Thornton-Griggs-Moxley disease: Myotonic dystrophy type 2 [letter]. *Ann. Neurol.* **36**, 803–804.
18. Sander, H. W., Tavoulareas, G. P., and Chokroverty, S. (1996). Heat-sensitive myotonia in proximal myotonic myopathy. *Neurology* **47**, 956–962.
19. Griggs, R., Sansone, V., Lifton, A., and Moxley, R. T. (1997). Hypothyroidism unmasking proximal myotonic myopathy (PROMM). *Neurology* **48** (Suppl.), A267.
20. Moxley, R. T., Sansone, V., Lifton, A., and Thornton, C. A. (1997). Insulin resistance in proximal myotonic myopathy (PROMM). *Neurology* **48** (Suppl.), A229.
21. Udd, B., Krahe, R., Wallgren-Pettersson, C., Falck, B., and Kalimo, H. (1997). Proximal myotonic dystrophy—a family with autosomal dominant muscular dystrophy, cataracts, hearing loss and hypogonadism: Heterogeneity of proximal myotonic syndromes? *Neuromuscul. Disord.* **7**, 217–228.
22. Day, J. W., Roelofs, R., Leroy, B., Pech, I., Benzow, K., and Ranum, L. P. (1999). Clinical and genetic characteristics of a five-generation family with a novel form of myotonic dystrophy (DM2). *Neuromuscul. Disord.* **9**, 19–27.
23. Ranum, L. P., Rasmussen, P. F., Benzow, K. A., Koob, M. D., and Day, J. W. (1998). Genetic mapping of a second myotonic dystrophy locus. *Nat. Genet.* **19**, 196–198.
24. Ricker, K., Grimm, T., Koch, M. C., Schneider, C., Kress, W., Reimers, C. D., Schulte-Mattler, W., Mueller-Myhsok, B., Toyka, K. V., and Mueller, C. R. (1999). Linkage of proximal myotonic myopathy to chromosome 3q. *Neurology* **52**, 170–171.
25. Meola, G., Udd, B., Sansone, V., Ptacek, L., Lee, D., and Krahe, R. (1999). Dominant multi-system proximal myotonic myopathy syndromes: Clinical and genetic heterogeneity in three families. *Neurology* **52** (Suppl.), A95.
26. Moxley, R. T., Meola, G., Udd, B., and Ricker, K. (2002). Report of the 84th ENMC Workshop: PROMM (Proximal Myotonic Myopathy) and Other Myotonic Dystrophy-Like Syndromes: 2nd Workshop. *Neuromuscul. Disord.* **12**, 306–317.
27. International-Myotonic-Dystrophy-Consortium. (2000). New nomenclature and DNA testing guidelines for myotonic dystrophy type 1 (DM1). *Neurology* **54**, 1218–1221.
28. Liquori, C. L., Ricker, K., Moseley, M. L., Jacobsen, J. F., Kress, W., Naylor, S. L., Day, J. W., and Ranum, L. P. (2001). Myotonic dystrophy type 2 caused by a CCTG expansion in intron 1 of ZNF9. *Science* **293**, 864–867.
29. Bachinski, L. L., Udd, B., Meola, G., Sansone, V., Bassez, G., Eymard, B., Thornton, C. A., Moxley, R. T., Harper, P. S., Rogers, M. T., Jurkat-Rott, K., Lehmann-Horn, F., Wieser, T., Gamez, J., Navarro, C., Bottani, A., Kohler, A., Shriver, M. D., Sallinen, R., Wessman, M., Zhang, S., Wright, F. A., and Krahe, R. (2003). Confirmation of the type 2 myotonic dystrophy (CCTG)$_n$ expansion mutation in patients with proximal myotonic myopathy/proximal myotonic dystrophy of different European origins: A single shared haplotype indicates an ancestral founder effect. *Am. J. Hum. Genet.* **73**, 835–848.
30. Liquori, C. L., Ikeda, Y., Weatherspoon, M., Ricker, K., Schoser, B. G., Dalton, J. C., Day, J. W., and Ranum, L. P. (2003). Myotonic dystrophy type 2: Human founder haplotype and evolutionary conservation of the repeat tract. *Am. J. Hum. Genet.* **73**, 849–862.
31. Udd, B., Meola, G., Krahe, R., Thornton, C., Ranum, L., Day, J., Bassez, G., and Ricker, K. (2003). Report of the 115th ENMC workshop: DM2/PROMM and other myotonic dystrophies. 3rd Workshop, 14–16 February 2003, Naarden, The Netherlands. *Neuromuscul. Disord.* **13**, 589–596.
32. Kress, W., Mueller-Myhsok, B., Ricker, K., Schneider, C., Koch, M. C., Toyka, K. V., Mueller, C. R., and Grimm, T. (2000). Proof of genetic heterogeneity in the proximal myotonic myopathy syndrome (PROMM) and its relationship to myotonic dystrophy type 2 (DM2). *Neuromuscul. Disord.* **10**, 478–480.
33. Sun, C., Henriksen, O. A., and Tranebjaerg, L. (1999). Proximal myotonic myopathy: Clinical and molecular investigation of a Norwegian family with PROMM. *Clin. Genet.* **56**, 457–461.
34. Wieser, R., Volz, A., Vinatzer, U., Gardiner, K., Jager, U., Mitterbauer, M., Ziegler, A., and Fonatsch, C. (2000). Transcription factor GATA-2 gene is located near 3q21 breakpoints in myeloid leukemia. *Biochem. Biophys. Res. Commun.* **273**, 239–245.
35. Day, J. W., Ricker, K., Jacobsen, J. F., Rasmussen, L. J., Dick, K. A., Kress, W., Schneider, C., Koch, M. C., Beilman, G. J., Harrison, A. R., Dalton, J. C., and Ranum, L. P. (2003). Myotonic dystrophy type 2: Molecular, diagnostic and clinical spectrum. *Neurology* **60**, 657–664.
36. George, A., Schncider-Gold, C., Zier, S., Reiners, K., and Sommer, C. (2004). Musculoskeletal pain in patients with myotonic dystrophy type 2. *Arch. Neurol.* **61**, 1938–1942.
37. Meola, G., Sansone, V., Perani, D., Scarone, S., Cappa, S., Dragoni, C., Cattaneo, E., Cotelli, M., Gobbo, C., Fazio, F., Siciliano, G., Mancuso, M., Vitelli, E., Zhang, S., Krahe, R., and Moxley, R. T. (2003). Executive dysfunction and avoidant personality trait in myotonic dystrophy type 1 (DM-1) and in proximal myotonic myopathy (PROMM/DM-2). *Neuromuscul. Disord.* **13**, 813–821.
38. Kassubek, J., Juengling, F. D., Hoffmann, S., Rosenbohm, A., Kurt, A., Jurkat-Rott, K., Steinbach, P., Wolf, M., Ludolph, A. C., Lehmann-Horn, F., Lerche, H., and Weber, Y. G. (2003). Quantification of brain atrophy in patients with myotonic dystrophy and proximal myotonic myopathy: A controlled 3-dimensional magnetic resonance imaging study. *Neurosci. Lett.* **348**, 73–76.
39. Kornblum, C., Reul, J., Kress, W., Grothe, C., Amanatidis, N., Klockgether, T., and Schroder, R. (2004). Cranial magnetic resonance imaging in genetically proven myotonic dystrophy type 1 and 2. *J. Neurol.* **251**, 710–714.
40. Schneider-Gold, C., Beer, M., Kostler, H., Buchner, S., Sandstede, J., Hahn, D., and Toyka, K. V. (2004). Cardiac and skeletal muscle involvement in myotonic dystrophy type 2 (DM2): A quantitative ^{31}P-MRS and MRI study. *Muscle Nerve* **30**, 636–644.
41. Schoser, B. G., Ricker, K., Schneider-Gold, C., Hengstenberg, C., Durre, J., Bultmann, B., Kress, W., Day, J. W., and Ranum, L. P. (2004). Sudden cardiac death in myotonic dystrophy type 2. *Neurology* **63**, 2402–2404.
42. Newman, B., Meola, G., O'Donovan, D. G., Schapira, A. H., and Kingston, H. (1999). Proximal myotonic myopathy (PROMM) presenting as myotonia during pregnancy. *Neuromuscul. Disord.* **9**, 144–149.

43. Grewal, R. P., Zhang, S., Ma, W., Rosenberg, M., and Krahe, R. (2004). Clinical and genetic analysis of a family with PROMM. *J. Clin. Neurosci.* **11**, 603–605.
44. Chu, K., Cho, J. W., Song, E. C., and Jeon, B. S. (2002). A patient with proximal myotonic myopathy and parkinsonism. *Can. J. Neurol. Sci.* **29**, 188–190.
45. Schneider, C., Pedrosa Gil, F., Schneider, M., Anetseder, M., Kress, W., and Muller, C. R. (2002). Intolerance to neuroleptics and susceptibility for malignant hyperthermia in a patient with proximal myotonic myopathy (PROMM) and schizophrenia. *Neuromuscul. Disord.* **12**, 31–35.
46. Merlini, L., Sabatelli, P., Columbaro, M., Bonifazi, E., Pisani, V., Massa, R., and Novelli, G. (2005). Hyper-CK-emia as the sole manifestation of myotonic dystrophy type 2. *Muscle Nerve* **31**, 764–767.
47. Harley, H. G., Rundle, S. A., Reardon, W., Myring, J., Crow, S., Brook, J. D., Harper, P. S., and Shaw, D. J. (1992). Unstable DNA sequence in myotonic dystrophy. *Lancet* **339**, 1125–1128.
48. Lavedan, C., Hofmann-Radvanyi, H., Shelbourne, P., Rabes, J. P., Savoy, D., Dehaupas, I., Luce, S., Johnson, K., and Junien, C. (1993). Myotonic dystrophy: Size- and sex-dependent dynamics of CTG instability, and somatic mosaicism. *Am. J. Hum. Genet.* **52**, 875–883.
49. Ashizawa, T., Dubel, J. R., Dunne, P. W., Dunne, C. J., Fu, Y.-H., Caskey, C. T., Boerwinkle, E., Perryman, M. B., Epstein, H. F., and Hejtmancik, J. F. (1992). Anticipation in myotonic dystrophy: II. Complex relationships between clinical findings and structure of the GCT repeat. *Neurology* **42**, 1877–1883.
50. Ashizawa, T., Dunne, C. J., Dubel, J. M., Perryman, M. B., Boerwinkle, E., and Hejtmancik, J. F. (1992). Anticipation in myotonic dystrophy: I. Statistical verification based on clinical and haplotype findings. *Neurology* **42**, 1871–1877.
51. Barcelo, J. M., Mahadevan, M. S., Tsilfidis, C., MacKenzie, A. E., and Korneluk, R. G. (1993). Intergenerational stability of the myotonic dystrophy. *Hum. Mol. Genet.* **2**, 705–709.
52. Ashizawa, T., Dunne, P. W., Ward, P. A., Seltzer, W. K., and Richards, C. S. (1994). Effects of sex of myotonic dystrophy patients on the unstable triplet repeat in their affected offspring. *Neurology* **44**, 120–122.
53. Ashizawa, T., Anvret, M., Baiget, M., Barcelo, J. M., Brunner, H., Cobo, A. M., Dallapiccola, B., Fenwick, R. G., Jr., Grandell, U., Harley, H., Junien, C., Koch, M. C., Korneluk, R. G., Lavedan, C., Miki, T., Muley, J. C., Lopez de Munain, A., Novelli, G., Roses, A. D., Seltzer, W. K., Shaw, D. J., Smeets, H., Sutherland, G. R., Yamagata, H., and Harper, P. S. (1994). Characteristics of intergenerational contractions of the CTG repeat in myotonic dystrophy. *Am. J. Hum. Genet.* **54**, 414–423.
54. Hamshere, M. G., Harley, H., Harper, P., Brook, J. D., and Brookfield, J. F. (1999). Myotonic dystrophy: The correlation of (CTG) repeat length in leucocytes with age at onset is significant only for patients with small expansions. *J. Med. Genet.* **36**, 59–61.
55. Tsilfidis, C., MacKenzie, A. E., Mettler, G., Barcelo, J., and Korneluk, R. G. (1992). Correlation between CTG trinucleotide repeat length and frequency of severe congenital myotonic dystrophy. *Nat. Genet.* **1**, 192–195.
56. Logigian, E. L., Moxley, R. T. 3rd, Blood, C. L., Barbieri, C. A., Martens, W. B., Wiegner, A. W., and Thornton, C. A. (2004). Leukocyte CTG repeat length correlates with severity of myotonia in myotonic dystrophy type 1. *Neurology* **62**, 1081–1089.
57. Schneider, C., Ziegler, A., Ricker, K., Grimm, T., Kress, W., Reimers, C. D., Meinck, H., Reiners, K., and Toyka, K. V. (2000). Proximal myotonic myopathy: Evidence for anticipation in families with linkage to chromosome 3q. *Neurology* **55**, 383–388.
58. Tsilfidis, C., MacKenzie, A. E., Mettler, G., and Barcelo, J. (1992). Correlation between CTG trinucleotide repeat length and severe congenital myotonic dystrophy. *Nat. Genet.* **1**, 192–195.
59. Monckton, D. G., Wong, L. J., Ashizawa, T., and Caskey, C. T. (1995). Somatic mosaicism, germline expansions, germline reversions and intergenerational reductions in myotonic dystrophy males: Small pool PCR analyses. *Hum. Mol. Genet.* **4**, 1–8.
60. Schoser, B. G., Kress, W., Walter, M. C., Halliger-Keller, B., Lochmuller, H., and Ricker, K. (2004). Homozygosity for CCTG mutation in myotonic dystrophy type 2. *Brain* **127**, 1868–1877.
61. Bassez, G., Attarian, S., Laforet, P., Azulay, J. P., Rouche, A., Ferrer, X., Urtizberea, J. A., Pellissier, J. F., Duboc, D., Fardeau, M., Pouget, J., and Eymard, B. (2001). [Proximal myotonial myopathy (PROMM): Clinical and histology study]. *Rev. Neurol. (Paris)* **157**, 209–218.
62. Vihola, A., Bassez, G., Meola, G., Zhang, S., Haapasalo, H., Paetau, A., Mancinelli, E., Rouche, A., Hogrel, J. Y., Laforet, P., Maisonobe, T., Pellissier, J. F., Krahe, R., Eymard, B., and Udd, B. (2003). Histopathological differences of myotonic dystrophy type 1 (DM1) and PROMM/DM2. *Neurology* **60**, 1854–1857.
63. Schoser, B. G., Schneider-Gold, C., Kress, W., Goebel, H. H., Reilich, P., Koch, M. C., Pongratz, D. E., Toyka, K. V., Lochmuller, H., and Ricker, K. (2004). Muscle pathology in 57 patients with myotonic dystrophy type 2. *Muscle Nerve* **29**, 275–281.
64. Richards, R. I. (2001). Dynamic mutations: A decade of unstable expanded repeats in human genetic disease. *Hum. Mol. Genet.* **10**, 2187–2194.
65. Richards, R. I., and Sutherland, G. R. (1992). Dynamic mutations: A new class of mutations causing human. *Cell* **70**, 709–712.
66. Ashley, C. T., Jr., and Warren, S. T. (1995). Trinucleotide repeat expansion and human disease. *Annu. Rev. Genet.* **29**, 703–728.
67. Ashizawa, T., and Epstein, H. F. (1991). Ethnic distribution of myotonic dystrophy gene. *Lancet* **338**, 642–643.
68. Imbert, G., Kretz, C., Johnson, K., and Mandel, J. L. (1993). Origin of the expansion mutation in myotonic dystrophy. *Nat. Genet.* **4**, 72–76.
69. Neville, C. E., Mahadevan, M. S., Barcelo, J. M., and Korneluk, R. G. (1994). High resolution genetic analysis suggests one ancestral predisposing haplotype for the origin of the myotonic dystrophy mutation. *Hum. Mol. Genet.* **3**, 45–51.
70. Yamagata, H., Miki, T., Nakagawa, M., Johnson, K., Deka, R., and Ogihara, T. (1996). Association of CTG repeats and the 1-kb Alu insertion/deletion polymorphism at the myotonin protein kinase gene in the Japanese population suggests a common Eurasian origin of the myotonic dystrophy mutation. *Hum. Genet.* **97**, 145–147.
71. Krahe, R., Eckhart, M., Ogunniyi, A. O., Osuntokun, B. O., Siciliano, M. J., and Ashizawa, T. (1995). De novo myotonic dystrophy mutation in a Nigerian kindred. *Am. J. Hum. Genet.* **56**, 1067–1074.
72. Mahadevan, M. S., Foitzik, M. A., Surh, L. C., and Korneluk, R. G. (1993). Characterization and polymerase chain reaction (PCR) detection Alu deletion polymorphism in total linkage disequilibrium with myotonic dystrophy. *Genomics* **15**, 446–448.
73. Yamagata, H., Miki, T., Ogihara, T., Nakagawa, M., Higuchi, I., Shelbourne, P., Davies, J., and Johnson, K. (1992). Expansion of unstable DNA region in Japanese myotonic dystrophy patients. *Lancet* **339**, 692.
74. Yamagata, H., Miki, T., Yamanaka, N., Takemoto, Y., Kanda, F., Takahashi, K., Inui, T., Kinoshita, M., Nakagawa, M., Higuchi, I., Osame, M., and Ogihara, T. (1994). Characteristics of dynamic mutation in Japanese myotonic dystrophy. *Jpn. J. Hum. Genet.*
75. Yamagata, H., Miki, T., Sakoda, S.-I., Yamanaka, N., Davies, J., Shelbourne, P., Kubota, R., Takenaga, S., Nakagawa, M., Ogihara, T., and Johnson, K. (1994). Detection of a premutation in Japanese myotonic dystrophy. *Hum. Mol. Genet.* **3**, 819–820.
76. Davies, J., Yamagata, H., Shelbourne, P., Buxton, J., Ogihara, T., Nokelainen, P., Nakagawa, M., Williamson, R., Johnson, K., and Miki, T. (1992). Comparison of the myotonic dystrophy associated CTG repeat in European and Japanese populations. *J. Med. Genet.* **29**, 766–769.

77. Goldman, A., Ramsay, M., and Jenkins, T. (1994). Absence of myotonic dystrophy in southern African Negroids is associated with a significantly lower number of CTG trinucleotide repeats. *J. Med. Genet.* **31**, 37–40.
78. Goldman, A., Ramsay, M., and Jenkins, T. (1995). New founder haplotypes at the myotonic dystrophy locus in southern Africa. *Am. J. Hum. Genet.* **56**, 1373–1378.
79. Goldman, A., Ramsay, M., and Jenkins, T. (1996). Ethnicity and myotonic dystrophy: A possible explanation for its absence in sub-Saharan Africa. *Ann. Hum. Genet.* **60**, 57–65.
80. Yamagata, H., Nakagawa, M., Johnson, K., and Miki, T. (1998). Further evidence for a major ancient mutation underlying myotonic dystrophy from linkage disequilibrium studies in the Japanese population. *J. Hum. Genet.* **43**, 246–249.
81. Brock, G. J., Anderson, N. H., and Monckton, D. G. (1999). cis-acting modifiers of expanded CAG/CTG triplet repeat expandability: Associations with flanking GC content and proximity to CpG islands. *Hum. Mol. Genet.* **8**, 1061–1067.
82. Richards, R. I., Crawford, J., Narahara, K., Mangelsdorf, M., Friend, K., Staples, A., Denton, M., Easteal, S., Hori, T. A., Kondo, I., Jenkins, T., Goldman, A., Panich, V., Ferakova, E., and Sutherland, G. R. (1996). Dynamic mutation loci: Allele distributions in different populations. *Ann. Hum. Genet.* **60**, 391–400.
83. Cleary, J. D., Nichol, K., Wang, Y. H., and Pearson, C. E. (2002). Evidence of *cis*-acting factors in replication-mediated trinucleotide repeat instability in primate cells. *Nat. Genet.* **31**, 37–46.
84. Gourdon, G., Radvanyi, F., Lia, A. S., Duros, C., Blanche, M., Abitbol, M., Junien, C., and Hofmann-Radvanyi, H. (1997). Moderate intergenerational and somatic instability of a 55-CTG repeat in transgenic mice. *Nat. Genet.* **15**, 190–192.
85. Monckton, D. G., Coolbaugh, M. I., Ashizawa, K. T., Siciliano, M. J., and Caskey, C. T. (1997). Hypermutable myotonic dystrophy CTG repeats in transgenic mice. *Nat. Genet.* **15**, 193–196.
86. Warner, J. P., Barron, L. H., Goudie, D., Kelly, K., Dow, D., Fitzpatrick, D. R., and Brock, D. J. (1996). A general method for the detection of large CAG repeat expansions by fluorescent PCR. *J. Med. Genet.* **33**, 1022–1026.
87. Sermon, K., Seneca, S., De Rycke, M., Goossens, V., Van de Velde, H., De Vos, A., Platteau, P., Lissens, W., Van Steirteghem, A., and Liebaers, I. (2001). PGD in the lab for triplet repeat diseases: myotonic dystrophy, Huntington's disease and fragile-X syndrome. *Mol. Cell. Endocrinol.* **183** Suppl. 1, S77–S85.
88. Matsuura, T., and Ashizawa, T. (2002). Polymerase chain reaction amplification of expanded ATTCT repeat in spinocerebellar ataxia type 10. *Ann. Neurol.* **51**, 271–272.
89. Sallinen, R., Vihola, A., Bachinski, L. L., Huoponen, K., Haapasalo, H., Hackman, P., Zhang, S., Sirito, M., Kalimo, H., Meola, G., Horelli-Kuitunen, N., Wessman, M., Krahe, R., and Udd, B. (2004). New methods for molecular diagnosis and demonstration of the (CCTG)$_n$ mutation in myotonic dystrophy type 2 (DM2). *Neuromuscul. Disord.* **14**, 274–283.
90. Cardani, R., Mancinelli, E., Sansone, V., Rotondo, G., and Meola, G. (2004). Biomolecular identification of (CCTG)$_n$ mutation in myotonic dystrophy type 2 (DM2) by FISH on muscle biopsy. *Eur. J. Histochem.* **48**, 437–442.
91. Meola, G. (2005). Advanced microscopic and histochemical techniques: Diagnostic tools in the molecular era of myology. *Eur. J. Histochem.* **49**, 93–96.
92. Bonifazi, E., Vallo, L., Giardina, E., Botta, A., and Novelli, G. (2004). A long PCR-based molecular protocol for detecting normal and expanded ZNF9 alleles in myotonic dystrophy type 2. *Diagn. Mol. Pathol.* **13**, 164–166.
93. Jakubiczka, S., Vielhaber, S., Kress, W., Kupferling, P., Reuner, U., Kunath, B., and Wieacker, P. (2004). Improvement of the diagnostic procedure in proximal myotonic myopathy/myotonic dystrophy type 2. *Neurogenetics* **5**, 55–59.
94. Le Ber, I., Martinez, M., Campion, D., Laquerriere, A., Betard, C., Bassez, G., Girard, C., Saugier-Veber, P., Raux, G., Sergeant, N., Magnier, P., Maisonobe, T., Eymard, B., Duyckaerts, C., Delacourte, A., Frebourg, T., and Hannequin, D. (2004). A non-DM1, non-DM2 multisystem myotonic disorder with frontotemporal dementia: Phenotype and suggestive mapping of the DM3 locus to chromosome 15q21-24. *Brain* **127**, 1979–1992.
95. Ranum, L. P., and Day, J. W. (2004). Pathogenic RNA repeats: An expanding role in genetic disease. *Trends Genet* **20**, 506–512.
96. Day, J. W., and Ranum, L. P. (2005). RNA pathogenesis of the myotonic dystrophies. *Neuromuscul. Disord.* **15**, 5–16.
97. Jin, P., and Warren, S. T. (2000). Understanding the molecular basis of fragile X syndrome. *Hum. Mol. Genet.* **9**, 901–908.
98. Pandolfo, M. (1998). Molecular genetics and pathogenesis of Friedreich ataxia. *Neuromuscul. Disord.* **8**, 409–415.
99. Ross, C. A. (2002). Polyglutamine pathogenesis: Emergence of unifying mechanisms for Huntington's disease and related disorders. *Neuron* **35**, 819–822.
100. Timchenko, L. T. (1999). Myotonic dystrophy: The role of RNA CUG triplet repeats. *Am. J. Hum. Genet.* **64**, 360–364.
101. Tapscott, S. J. (2000). Deconstructing myotonic dystrophy. *Science* **289**, 1701–1702.
102. Krahe, R., Ashizawa, T., Abbruzzese, C., Roeder, E., Carango, P., Giacanelli, M., Funanage, V. L., and Siciliano, M. J. (1995). Effect of myotonic dystrophy trinucleotide repeat expansion on DMPK transcription and processing. *Genomics* **28**, 1–14.
103. Jansen, G., Groenen, P. J., Bachner, D., Jap, P. H., Coerwinkel, M., Oerlemans, F., van den Broek, W., Gohlsch, B., Pette, D., Plomp, J. J., Molenaar, P. C., Nederhoff, M. G., van Echteld, C. J., Dekker, M., Berns, A., Hameister, H., and Wieringa, B. (1996). Abnormal myotonic dystrophy protein kinase levels produce only mild myopathy in mice [see comments]. *Nat. Genet.* **13**, 316–324.
104. Reddy, S., Smith, D. B., Rich, M. M., Leferovich, J. M., Reilly, P., Davis, B. M., Tran, K., Rayburn, H., Bronson, R., Cros, D., Balice-Gordon, R. J., and Housman, D. (1996). Mice lacking the myotonic dystrophy protein kinase develop a late onset progressive myopathy. *Nat. Genet.* **13**, 325–335.
105. Berul, C. I., Maguire, C. T., Aronovitz, M. J., Greenwood, J., Miller, C., Gehrmann, J., Housman, D., Mendelsohn, M. E., and Reddy, S. (1999). DMPK dosage alterations result in atrioventricular conduction abnormalities in a mouse myotonic dystrophy model. *J. Clin. Invest.* **103**, R1–R7.
106. Taneja, K. L., McCurrach, M., Schalling, M., Housman, D., and Singer, R. H. (1995). Foci of trinucleotide repeat transcripts in nuclei of myotonic dystrophy cells and tissues. *J. Cell Biol.* **128**, 995–1002.
107. Wang, J., Pegoraro, E., Menegazzo, E., Gennarelli, M., Hoop, R. C., Angelini, C., and Hoffman, E. P. (1995). Myotonic dystrophy: Evidence for a possible dominant-negative RNA mutation. *Hum. Mol. Genet.* **4**, 599–606.
108. Davis, B. M., McCurrach, M. E., Taneja, K. L., Singer, R. H., and Housman, D. E. (1997). Expansion of a CUG trinucleotide repeat in the 3′ untranslated region of myotonic dystrophy protein kinase transcripts results in nuclear retention of transcripts. *Proc. Nat. Acad. Sci. USA* **94**, 7388–7393.
109. Koch, K. S., and Leffert, H. L. (1998). Giant hairpins formed by CUG repeats in myotonic dystrophy messenger RNAs might sterically block RNA export through nuclear pores. *J. Theor. Biol.* **192**, 505–514.

110. Tiscornia, G., and Mahadevan, M. S. (2000). Myotonic dystrophy: The role of the CUG triplet repeats in splicing of a novel DMPK exon and altered cytoplasmic DMPK mRNA isoform ratios. *Mol. Cell* **5**, 959–967.
111. Morrone, A., Pegoraro, E., Angelini, C., Zammarchi, E., Marconi, G., and Hoffman, E. P. (1997). RNA metabolism in myotonic dystrophy: Patient muscle shows decreased insulin receptor RNA and protein consistent with abnormal insulin resistance. *J. Clin. Invest.* **99**, 1691–1698.
112. Boucher, C. A., King, S. K., Carey, N., Krahe, R., Winchester, C. L., Rahman, S., Creavin, T., Meghji, P., Bailey, M. E., Chartier, F. L., et al. (1995). A novel homeodomain-encoding gene is associated with a large CpG island interrupted by the myotonic dystrophy unstable $(CTG)_n$ repeat. *Hum. Mol. Genet.* **4**, 1919–1925.
113. Wang, Y.-H., Amirhaeri, S., Kang, S., Wells, R. D., and Griffith, J. D. (1994). Preferential nucleosome assembly at DNA triplet repeats from the myotonic dystrophy gene. *Science* **265**, 669–671.
114. Wang, Y.-H., and Griffith, J. (1995). Expanded CTG triplet repeat blocks from the myotonic dystrophy gene create the strongest known natural nucleosome positioning element. *Genomics* **25**, 570–573.
115. Michalowski, S., Miller, J. W., Urbinati, C. R., Paliouras, M., Swanson, M. S., and Griffith, J. (1999). Visualization of double-stranded RNAs from the myotonic dystrophy protein kinase gene and interactions with CUG-binding protein. *Nucleic Acids Res.* **27**, 3534–3542.
116. Klesert, T. R., Otten, A. D., Bird, T. D., and Tapscott, S. J. (1997). Trinucleotide repeat expansion at the myotonic dystrophy locus reduces expression of DMAHP. *Nat. Genet.* **16**, 402–406.
117. Thornton, C. A., Wymer, J. P., Simmons, Z., McClain, C., and Moxley, R. T., 3rd (1997). Expansion of the myotonic dystrophy CTG repeat reduces expression of the flanking DMAHP gene. *Nat. Genet.* **16**, 407–409.
118. Korade-Mirnics, Z., Tarleton, J., Servidei, S., Casey, R. R., Gennarelli, M., Pegoraro, E., Angelini, C., and Hoffman, E. P. (1999). Myotonic dystrophy: Tissue-specific effect of somatic CTG expansions on allele-specific DMAHP/SIX5 expression. *Hum. Mol. Genet.* **8**, 1017–1023.
119. Klesert, T. R., Cho, D. H., Clark, J. I., Maylie, J., Adelman, J., Snider, L., Yuen, E. C., Soriano, P., and Tapscott, S. J. (2000). Mice deficient in Six5 develop cataracts: Implications for myotonic dystrophy. *Nat. Genet.* **25**, 105–109.
120. Sarkar, P. S., Appukuttan, B., Han, J., Ito, Y., Ai, C., Tsai, W., Chai, Y., Stout, J. T., and Reddy, S. (2000). Heterozygous loss of Six5 in mice is sufficient to cause ocular cataracts. *Nat. Genet.* **25**, 110–114.
121. Alwazzan, M., Newman, E., Hamshere, M. G., and Brook, J. D. (1999). Myotonic dystrophy is associated with a reduced level of RNA from the DMWD allele adjacent to the expanded repeat. *Hum. Mol. Genet.* **8**, 1491–1497.
122. Jansen, G., Bachner, D., Coerwinkel, M., Wormskamp, N., Hameister, H., and Wieringa, B. (1995). Structural organization and developmental expression pattern of the mouse WD-repeat gene DMR-N9 immediately upstream of the myotonic dystrophy locus. *Hum. Mol. Genet.* **4**, 843–852.
123. Steinbach, P., Glaser, D., Vogel, W., Wolf, M., and Schwemmle, S. (1998). The DMPK gene of severely affected myotonic dystrophy patients is hypermethylated proximal to the largely expanded CTG repeat. *Am. J. Hum. Genet.* **62**, 278–285.
124. Usuki, F., Ishiura, S., Saitoh, N., Sasagawa, N., Sorimachi, H., Kuzume, H., Maruyama, K., Terao, T., and Suzuki, K. (1997). Expanded CTG repeats in myotonin protein kinase suppresses myogenic differentiation. *NeuroReport* **8**, 3749–3753.
125. Amack, J. D., and Mahadevan, M. S. (2001). The myotonic dystrophy expanded CUG repeat tract is necessary but not sufficient to disrupt C2C12 myoblast differentiation. *Hum. Mol. Genet.* **10**, 1879–1887.
126. Bhagwati, S., Shafiq, S. A., and Xu, W. (1999). $(CTG)_n$ repeats markedly inhibit differentiation of the C2C12 myoblast cell line: Implications for congenital myotonic dystrophy. *Biochim. Biophys. Acta* **1453**, 221–229.
127. Sasagawa, N., Takahashi, N., Suzuki, K. and Ishiura, S. (1999). An expanded CTG trinucleotide repeat causes trans RNA interference: A new hypothesis for the pathogenesis of myotonic dystrophy. *Biochem. Biophys. Res. Commun.* **264**, 76–80.
128. Mankodi, A., Logigian, E., Callahan, L., McClain, C., White, R., Henderson, D., Krym, M., and Thornton, C. A. (2000). Myotonic dystrophy in transgenic mice expressing an expanded CUG repeat. *Science* **289**, 1769–1773.
129. Timchenko, L. T., Miller, J. W., Timchenko, N. A., DeVore, D. R., Datar, K. V., Lin, L., Roberts, R., Caskey, C. T., and Swanson, M. S. (1996). Identification of a $(CUG)_n$ triplet repeat RNA-binding protein and its expression in myotonic dystrophy. *Nucleic Acids Res.* **24**, 4407–4414.
130. Timchenko, L. T., Timchenko, N. A., Caskey, C. T., and Roberts, R. (1996). Novel proteins with binding specificity for DNA CTG repeats and RNA CUG repeats: Implications for myotonic dystrophy. *Hum. Mol. Genet.* **5**, 115–121.
131. Timchenko, N. A., Cai, Z. J., Welm, A. L., Reddy, S., Ashizawa, T., and Timchenko, L. T. (2001). RNA CUG repeats sequester CUGBP1 and alter protein levels and activity of CUGBP1. *J. Biol. Chem.* **267**, 7820–7826.
132. Timchenko, N. A., Iakova, P., Cai, Z. J., Smith, J. R., and Timchenko, L. T. (2001). Molecular basis for impaired muscle differentiation in myotonic dystrophy. *Mol. Cell. Biol.* **21**, 6927–6938.
133. Timchenko, N. A., Wang, G. L., and Timchenko, L. T. (2005). RNA CUG-binding protein 1 increases translation of 20-kDa isoform of CCAAT/enhancer-binding protein beta by interacting with the alpha and beta subunits of eukaryotic initiation translation factor 2. *J. Biol. Chem.* **280**, 20549–20557.
134. Timchenko, N. A., Welm, A. L., Lu, X., and Timchenko, L. T. (1999). CUG repeat binding protein (CUGBP1) interacts with the 5′ region of C/EBPbeta mRNA and regulates translation of C/EBPbeta isoforms. *Nucleic Acids Res.* **27**, 4517–4525.
135. Miller, J. W., Urbinati, C. R., Teng-Umnuay, P., Stenberg, M. G., Byrne, B. J., Thornton, C. A., and Swanson, M. S. (2000). Recruitment of human muscleblind proteins to $(CUG)_n$ expansions associated with myotonic dystrophy. *EMBO J.* **19**, 4439–4448.
136. Fardaei, M., Larkin, K., Brook, J. D., and Hamshere, M. G. (2001). In vivo co-localisation of MBNL protein with DMPK expanded-repeat transcripts. *Nucleic Acids Res.* **29**, 2766–2771.
137. Dansithong, W., Paul, S., Comai, L., and Reddy, S. (2005). MBNL1 is the primary determinant of focus formation and aberrant insulin receptor splicing in DM1. *J. Biol. Chem.* **280**, 5773–5780.
138. Ho, T. H., Charlet, B. N., Poulos, M. G., Singh, G., Swanson, M. S., and Cooper, T. A. (2004). Muscleblind proteins regulate alternative splicing. *EMBO J.* **23**, 3103–3112.
139. Ho, T. H., Savkur, R. S., Poulos, M. G., Mancini, M. A., Swanson, M. S., and Cooper, T. A. (2005). Colocalization of muscleblind with RNA foci is separable from mis-regulation of alternative splicing in myotonic dystrophy. *J. Cell. Sci.* **118**, 2923–2933.
140. Mankodi, A., Urbinati, C. R., Yuan, Q. P., Moxley, R. T., Sansone, V., Krym, M., Henderson, D., Schalling, M., Swanson, M. S., and Thornton, C. A. (2001). Muscleblind localizes to nuclear foci of aberrant RNA in myotonic dystrophy types 1 and 2. *Hum. Mol. Genet.* **10**, 2165–2170.
141. Mankodi, A., Teng-Umnuay, P., Krym, M., Henderson, D., Swanson, M., and Thornton, C. A. (2003). Ribonuclear inclusions in skeletal muscle in myotonic dystrophy types 1 and 2. *Ann. Neurol.* **54**, 760–768.

142. Kanadia, R. N., Johnstone, K. A., Mankodi, A., Lungu, C., Thornton, C. A., Esson, D., Timmers, A. M., Hauswirth, W. W., and Swanson, M. S. (2003). A muscleblind knockout model for myotonic dystrophy. *Science* **302**, 1978–1980.
143. Philips, A. V., Timchenko, L. T., and Cooper, T. A. (1998). Disruption of splicing regulated by a CUG-binding protein in myotonic dystrophy. *Science* **280**, 737–741.
144. Savkur, R. S., Philips, A. V., and Cooper, T. A. (2001). Aberrant regulation of insulin receptor alternative splicing is associated with insulin resistance in myotonic dystrophy. *Nat. Genet.* **29**, 40–47.
145. Mankodi, A., Takahashi, M. P., Jiang, H., Beck, C. L., Bowers, W. J., Moxley, R. T., Cannon, S. C., and Thornton, C. A. (2002). Expanded CUG repeats trigger aberrant splicing of ClC-1 chloride channel pre-mRNA and hyperexcitability of skeletal muscle in myotonic dystrophy. *Mol. Cell* **10**, 35–44.
146. Charlet, B. N., Savkur, R. S., Singh, G., Philips, A. V., Grice, E. A., and Cooper, T. A. (2002). Loss of the muscle-specific chloride channel in type 1 myotonic dystrophy due to misregulated alternative splicing. *Mol. Cell* **10**, 45–53.
147. Berg, J., Jiang, H., Thornton, C. A., and Cannon, S. C. (2004). Truncated ClC-1 mRNA in myotonic dystrophy exerts a dominant-negative effect on the Cl current. *Neurology* **63**, 2371–2375.
148. Sergeant, N., Sablonniere, B., Schraen-Maschke, S., Ghestem, A., Maurage, C. A., Wattez, A., Vermersch, P., and Delacourte, A. (2001). Dysregulation of human brain microtubule-associated tau mRNA maturation in myotonic dystrophy type 1. *Hum. Mol. Genet.* **10**, 2143–2155.
149. Jiang, H., Mankodi, A., Swanson, M. S., Moxley, R. T., and Thornton, C. A. (2004). Myotonic dystrophy type 1 is associated with nuclear foci of mutant RNA, sequestration of muscleblind proteins and deregulated alternative splicing in neurons. *Hum. Mol. Genet.* **13**, 3079–3088.
150. Wang, Y., Wang, J., Gao, L., Lafyatis, R., Stamm, S., and Andreadis, A. (2005). Tau exons 2 and 10, which are misregulated in neurodegenerative diseases, are partly regulated by silencers which bind a SRp30c.SRp55 complex that either recruits or antagonizes htra2beta1. *J. Biol. Chem.* **280**, 14230–14239.
151. Buj-Bello, A., Furling, D., Tronchere, H., Laporte, J., Lerouge, T., Butler-Browne, G. S., and Mandel, J. L. (2002). Muscle-specific alternative splicing of myotubularin-related 1 gene is impaired in DM1 muscle cells. *Hum. Mol. Genet.* **11**, 2297–2307.
152. Kimura, T., Nakamori, M., Lueck, J. D., Pouliquin, P., Aoike, F., Fujimura, H., Dirksen, R. T., Takahashi, M. P., Dulhunty, A. F., and Sakoda, S. (2005). Altered mRNA splicing of the skeletal muscle ryanodine receptor and sarcoplasmic/endoplasmic reticulum Ca^{2+}-ATPase in myotonic dystrophy type 1. *Hum. Mol. Genet.* **14**, 2189–2200.
153. Savkur, R. S., Philips, A. V., Cooper, T. A., Dalton, J. C., Moseley, M. L., Ranum, L. P., and Day, J. W. (2004). Insulin receptor splicing alteration in myotonic dystrophy type 2. *Am. J. Hum. Genet.* **74**, 1309–1313.
154. Ebralidze, A., Wang, Y., Petkova, V., Ebralidse, K., and Junghans, R. P. (2004). RNA leaching of transcription factors disrupts transcription in myotonic dystrophy. *Science* **303**, 383–387.
155. Timchenko, L. T., Iakova, P., Welm, A. L., Cai, Z. J., and Timchenko, N. A. (2002). Calreticulin interacts with C/EBPalpha and C/EBPbeta mRNAs and represses translation of C/EBP proteins. *Mol. Cell. Biol.* **22**, 7242–7257.
156. Timchenko, N. A., Patel, R., Iakova, P., Cai, Z. J., Quan, L., and Timchenko, L. T. (2004). Overexpression of CUG triplet repeat-binding protein, CUGBP1, in mice inhibits myogenesis. *J. Biol. Chem.* **279**, 13129–13139.
157. Kim, D. H., Langlois, M. A., Lee, K. B., Riggs, A. D., Puymirat, J., and Rossi, J. J. (2005). HnRNP H inhibits nuclear export of mRNA containing expanded CUG repeats and a distal branch point sequence. *Nucleic Acids Res.* **33**, 3866–3874.
158. Wansink, D. G., and Wieringa, B. (2003). Transgenic mouse models for myotonic dystrophy type 1 (DM1). *Cytogenet. Genome. Res.* **100**, 230–242.
159. Chen, W., Liang, Y., Deng, W., Shimizu, K., Ashique, A. M., Li, E., and Li, Y. P. (2003). The zinc-finger protein CNBP is required for forebrain formation in the mouse. *Development* **130**, 1367–1379.
160. Houseley, J. M., Wang, Z., Brock, G. J., Soloway, J., Artero, R., Perez-Alonso, M., O'Dell, K. M., and Monckton, D. G. (2005). Myotonic dystrophy associated expanded CUG repeat muscleblind positive ribonuclear foci are not toxic to Drosophila. *Hum. Mol. Genet.* **14**, 873–883.
161. Koob, M. D., Moseley, M. L., Schut, L. J., Benzow, K. A., Bird, T. D., Day, J. W., and Ranum, L. P. (1999). An untranslated CTG expansion causes a novel form of spinocerebellar ataxia (SCA8). *Nat. Genet.* **21**, 379–384.

CHAPTER 9

The Subtelomeric D4Z4 Repeat Instability in Facioscapulohumeral Muscular Dystrophy

SILVÈRE M. VAN DER MAAREL, RUNE R. FRANTS, AND GEORGE W. PADBERG

Center for Human and Clinical Genetics, Leiden University Medical Center, Leiden, The Netherlands; and Department of Neurology, University Medical Center Nijmegen, Nijmegen, The Netherlands

I. Introduction
II. Clinical Characteristics
III. Ancillary Investigations
IV. Linkage Analysis
V. Genetic/Linkage Heterogeneity
VI. Genetic Diagnosis of FSHD
VII. Timing and Origin of the D4Z4 Rearrangement
VIII. Candidate Genes
IX. Chromatin Remodeling
X. Myoblast Studies
XI. Concluding Remarks
Acknowledgments
References

Facioscapulohumeral muscular dystrophy (FSHD) is the third most common muscular dystrophy, with an incidence of 1 in 20,000. FSHD usually starts with facial weakness, followed by progression to shoulder girdle and upper arm muscles, often in an asymmetric fashion. Extramuscular symptoms such as retinovasculopathy and deafness can be part of the syndrome. FSHD has an autosomal pattern of inheritance, with a high *de novo* mutation frequency. Age of onset typically is the second decade of life, with nearly complete penetrance (95%) by age 20. FSHD is genetically associated with a contraction of a large polymorphic repeat array in the subtelomere of chromosome 4q. This array contains 11–100 copies of a 3.3-kb tandem repeat unit termed D4Z4 in healthy individuals, whereas FSHD patients carry one array of 1–10 units. It has become increasingly evident that the chromatin structure in this subtelomeric region is disrupted by the D4Z4 contraction, initiating a cascade of epigenetic events eventually leading to FSHD. At present, the challenge is to determine the precise epigenetic disease mechanism, the genes affected by this epigenetic mechanism and the consequences of their dysregulation.

I. INTRODUCTION

Facioscapulohumeral muscular dystrophy (FSHD) was first described by Landouzy and Dejerine in the late 19th century. Postmortem studies enabled them to prove the myopathic nature of the disease and the authors emphasized onset of the disease in the facial and shoulder muscles and the clearly heritable pattern to distinguish "their" disease from Duchenne dystrophy, which was the only recognized myopathic condition at the time [1, 2]. Subsequent generations of neurologists tried to apply mendelian knowledge to muscle disease. In all studies, early facial weakness served as an important item for clinical differential diagnosis. Still, landmark articles in the sixties postulated autosomal dominant, recessive, and X-linked modes of inheritance in FSHD [3]. The genetic localization of the causative deletion on 4q35 finally established the dominant mode of inheritance and explained the other frequently observed patterns as the result of the high mutation rate [4, 5].

II. CLINICAL CHARACTERISTICS

The majority of patients present with a rather characteristic clinical picture of asymmetrical facial and shoulder girdle muscle weakness. Mild weakness may go unnoticed, leading to reports of 20–30% asymptomatic (but clinically recognizable) gene carriers in completely investigated families. By the same mechanism, patients present occasionally with foot-extensor weakness, for which the term *scapuloperonal syndrome*, or pelvic girdle weakness, was coined, in which case limb–girdle syndrome is diagnosed. Facial weakness may be very mild and requires some clinical experience for recognition. Asymmetric weakness of the scapula fixators with relative sparing of the deltoid and forearm muscles and early involvement of the upper-arm muscles constitutes the classic picture. More than 70% of all gene carriers progress to abdominal muscle weakness and foot extensor weakness, whereas in 50%, pelvic girdle muscles become involved. Approximately 20% require the use of a wheelchair [6].

The age at onset usually is the second decade. FSHD is gradually progressive, but the rate of progression is extremely variable, and both rapid progression and a long standstill have been reported. Progression seems to slow after ages 50–60. In general, women appear to have a later age at onset and a somewhat slower rate of progression.

Fatigue and muscle pain are reported frequently, and a subclinical vascular retinopathy appears to be present in 60% of all gene carriers. Contractures, severe scoliosis, pectus excavatum, hearing loss, and visual loss are part of the clinical spectrum but occur infrequently. Respiratory failure requiring ventilatory support and dysphagia are rare complications, and cardiac conduction defects remain a topic of discussion [6].

The condition just described is usually observed in gene carriers with "average" residual FSHD repeat sizes that leave four to eight D4Z4 repeats present (see Section VI). Larger repeat arrays probably result in milder cases and even in nonpenetrant gene carriers, but this impression requires corroborating studies. Small repeat arrays with a one- to three-residues are highly associated with severe, early-onset FSHD, defined as facial weakness before the age of 5 and symptomatic shoulder weakness before the age of 10. These patients tend to have a higher rate of progression and a higher frequency of high-tone hearing loss and symptomatic retinal disease. Also, and particularly in Japan, epilepsy and mental retardation are reported as part of infantile FSHD. These cases tend to be sporadic and often reflect new mutations [7]. As the mutation rate is high, fitness is estimated to be reduced. Population differences in frequency and severity of FSHD have been suggested, but further studies are required.

The picture painted so far, and particularly the presence of other cases in the family suggestive of autosomal dominant inheritance, leaves little room for a diagnosis other than FSHD. Yet, there are cases and sometimes small families with exactly this clinical picture that lack a D4Z4 deletion. In some of these phenocopy cases, the molecular mechanism was shown also to act through D4Z4, but, at present, it is unclear if this holds true for all these patients [8] (see Section IX).

III. ANCILLARY INVESTIGATIONS

Apart from DNA diagnostics, there are no laboratory tests specific to FSHD. Creatine kinase is often mildly elevated, can be normal, and also, occasionally, is significantly elevated. EMG usually reveals a myopathic pattern, but in rare instances neuropathic features can be found. A biopsy from a clinically normal muscle might show few or no abnormalities. When affected, a muscle will show dystrophic features: small angular (regenerating) fibers, perivascular infiltrates, and lobulated fibers. None of these changes are specific, but the frequent observation of infiltrates has triggered discussions on possible vascular origins of FSHD and secondary immunological events [9]. The story of FSHD is rife with anecdotes of patients who report losing strength and bulk in a particular muscle in a matter of days. Unfortunately, muscle biopsies in such instances have not been described. These reports balance those of patients who report long periods of arrest of progression of the disease.

IV. LINKAGE ANALYSIS

Identification of the FSHD causative gene was not facilitated by generous gifts of Nature. No unequivocal biomarker had been identified to target a disease-specific metabolic or structural mechanism. Equally, no chromosomal rearrangement, like a translocation, had been reported to cosegregate with FSHD.

During the 1980s several groups embarked on the identification, characterization, and collection of biomaterial from large multigeneration families for linkage analysis. Although the intra- and interfamily variability in clinical phenotype was considerable, the inheritance pattern was accepted as autosomal dominant. Still, to minimize the complications of possible genetic heterogeneity, families large enough to independently yield significant linkage results were ascertained whenever feasible.

To extract optimal information from the clinical and genetic studies in these early days, an International Consortium was established in 1988. Through the use of restriction fragment length polymorphisms (RFLPs) and serological markers, 95% of the genome was excluded from harboring the FSHD locus by 1990 [10, 11]. In 1989, Weber and May reported a new type of highly informative genetic marker, the microsatellite markers, among which $(CA)_n$ is the prototype [12]. In collaboration with Dr. Weber, the first genomewide linkage scan with this type of marker was performed in a few highly informative Dutch FSHD pedigrees. Compared with the Southern blot technique used to investigate two-allele RFLPs, the multiallelic microsatellite markers were attractive because of their rich abundance in the genome to provide a dense genetic map, high polymorphism information content, and the fact that they could be analyzed by polymerase chain reaction (PCR)-based technology on minimal amounts of DNA.

One of the microsatellite markers, Mfd22, showed linkage with the FSHD locus at a distance of 13 cM with a highly significant LOD score above 6, without evidence of genetic (locus) heterogeneity [4]. The microsatellite marker Mfd22, corresponding to the locus D4S171, was assigned to chromosome 4 by using a somatic cell hybrid panel [13].

The saturation of the linkage region with additional markers and the subsequent generation of a physical map were considerable challenges in the early 1990s. Cosmid clone 13E was isolated from a chromosome 4 cosmid library (Los Alamos) by hybridization with a homeodomain probe [14]. Multicolor *in situ* hybridization to interphase nuclei demonstrated that this cosmid was the most telomeric probe available, because it mapped distal to D4S139 and D4F35S1. An almost single-copy probe, p13E-11, could be isolated from the cosmid. On Southern blots, this probe recognized a rather complicated polymorphic fragment pattern with different restriction enzymes. Interestingly, this probe consistently revealed a short *Eco*RI fragment, in the range of 10 to ca. 40 kb, in FSHD patients. As a single recombinant was not seen in the complete Dutch family material, p13E-11 mapped very close to the FSHD locus [15].

Could the short "FSHD fragment" be indicative of a chromosomal rearrangement? In a collection of eight sporadic patients and their parents from various parts of The Netherlands, it could clearly be shown that most of the sporadic patients had a novel fragment shorter than 30 kb, not present in either parent [15]. Soon afterward, a small family was identified with an apparently sporadic patient transmitting FSHD to one of the sons. Southern blot analysis confirmed the father to be a new mutation (mosaic), transmitting the rearranged fragment to his son—the top of a new FSHD pedigree [16].

This guilt by association was a strong indication that the rearrangement had hit the FSHD gene. Intriguingly, probe p13E-11 was subcloned from a cosmid clone, identified through a screen for homeodomain-containing sequences. Mapping the cosmid revealed that it contained multiple copies of this homeodomain-containing sequence. The restriction enzyme *Kpn*I, conveniently excised each repeat unit into a 3.3-kb fragment [17], designated D4Z4. Restriction mapping and sequencing of *Eco*RI fragments cloned from patients and controls showed that the proximal and distal parts of the *Eco*RI fragment are identical in patients and controls. Moreover, exact sizing of the fragments in FSHD families showed that the short fragments were of different size in the different families and differed by some 3.3 kb, supporting the hypothesis that the FSHD rearrangement was caused by a (recurrent) homologous recombination-based mechanism, thereby deleting an integral number of repeat units [17]. Analysis by pulsed-field gel electrophoresis (PFGE), allowing separation of the long *Eco*RI fragments, demonstrated that probe p13E-11 recognizes two highly polymorphic loci. Haplotype analysis unambiguously assigned one of the loci to chromosome 4q35. Bakker et al. were able to map the second p13E-11 locus to the tip of the long arm of chromosome 10 [18].

In the original article, an *Eco*RI fragment length of 28 kb or shorter, corresponding to six repeat units, was observed in FSHD patients. In subsequent larger patient series, a cutoff at 35 kb was proposed. In a population-based sample, PFGE revealed a multimodal length distribution from a few kilobase pairs up to more than 300 kb [19]. This means that FSHD patients have a chromosome 4 with 1 to 10 repeat units, whereas unaffected individuals may harbor repeat arrays of 11 or more units. At least one D4Z4 repeat seems to be required, as patients with a monosomy for the region do not manifest FSHD [20].

In many studies, a rough and inverse relationship between repeat unit number and clinical severity, for example, age at onset, has been established [21, 22]. However, this association is quite loose and has only limited prognostic value in patient management and counseling. Most large mendelian FSHD families transmit fragments between four and eight D4Z4 units. One to three repeat units are typically found in sporadic, severely affected patients. Above eight units, the clinical phenotype becomes less clear-cut, with frequent cases of nonpenetrance. The poor predictability of the natural course of FSHD, seen as variable expressivity and reduced penetrance, probably reflects the complex epigenetic mechanisms and the presence of phenotype-modifying genes.

V. GENETIC/LINKAGE HETEROGENEITY

Several FSHD families showing no linkage to chromosome 4 have been reported [23]. After identification of the causal repeat contraction, and careful reevaluation of diagnoses, a limited number of families, supporting the presence of a second FSHD locus, remain. Some of the apparent non-4q-linked families could be explained by a deletion of the p13E-11 probe region, compromising the visualization of the contracted repeat array present [24, 25]. The remaining non-4q families form a valuable resource for future studies. Identification of the putative FSHD2 locus/gene can reveal important clues with respect to the pathogenic mechanisms of FSHD; it can give direction to the metabolic pathways involved and identify intervention targets. Unfortunately, linkage studies in these families have been unsuccessful so far. The most obvious candidate locus for FSHD2, the highly homologous region on chromosome 10q, has been excluded [18]. Van Overveld et al. were able to show that the D4Z4 repeat array in such phenotypic FSHD patients shows hypomethylation, as also seen in standard chromosome 4q-linked FSHD patients [8]. This finding gives strong support for a common epigenetic pathway. Equally, it indicates in which direction candidate genes could be sought. Patients with ICF syndrome (immunodeficiency, centromere instability, facial abnormalities) with a genetic defect in the DNMT3B gene show hypomethylation of D4Z4, in addition to multiple other repeat loci (see Section IX).

VI. GENETIC DIAGNOSIS OF FSHD

The D4Z4 repeat is commonly detected by Southern blotting of genomic DNA digested with *Eco*RI and hybridized with probe p13E-11, which does not recognize the D4Z4 repeat proper, but a region just proximal to D4Z4 [15]. The unusually large size of the D4Z4 repeat requires a PFGE-based approach to completely separate these repeat arrays. Because of the positions of the *Eco*RI sites flanking the D4Z4 repeat, in FSHD patients,

this Southern blot analysis yields fragments of 10 kb (1 unit) to 38 kb (10 units), which is diagnostic of FSHD. However, a highly homologous and equally polymorphic repeat is also present in the subtelomere of chromosome 10q as a result of an ancient duplication of both chromosome ends [18, 26, 27]. Intriguingly, contraction of this repeat on chromosome 10q has never been reported to be associated with FSHD (Fig. 9-1). Thus, probe p13E-11 recognizes a total of four *Eco*RI fragments: two derived from chromosomes 4 and two from chromosomes 10. To discriminate between 4-derived and 10-derived fragments, additional restriction enzymes can be used: chromosome 4-derived units are sensitive to *Xap*I, whereas chromosome 10-derived units are sensitive to *Bln*I [28, 29]. In most cases, a combination of these enzymes is sufficient for the DNA diagnosis of FSHD.

There are, however, several complications. Subtelomeres are dynamic structures characterized by a high frequency of homologous and nonhomologous exchanges [30]. As a consequence of the subtelomeric localization of D4Z4, exchanges between 4-derived and 10-derived D4Z4 repeats are observed in some 20% of the population, with some variation between different populations [19, 25, 31, 32]. These exchanged repeats, which occur equally on both chromosome ends, can be either homogeneous or heterogeneous (i.e., consisting of a combination of 4-derived and 10-derived units). However, irrespective of the composition of the repeat, D4Z4 repeat contractions in FSHD are reported only on chromosome 4.

FIGURE 9-1 Schematic representation of the FSHD locus (not drawn to scale). The D4Z4 repeat (block arrows) is located in the subtelomeric domain of chromosome 4, at a distance of approximately 40 kb to the telomere. This polymorphic D4Z4 array may vary between 11 and 100 units, each 3.3 kb in size. In telomeric direction, a biallelic variation of chromosome 4, designated A and B, can be encountered with almost equal frequency. Patients with FSHD carry a repeat array of 1–10 units in association with a 4qA variation. In the proximal direction, several genes have been identified as potential candidate genes for FSHD. These include *FRG2* at 35 kb, *FRG1* at 120 kb, and *ANT1* at 3.5 Mb. Within each repeat unit, a potential homeobox gene (*DUX4*) is located (not drawn). The location of probe p13E-11, commonly used for diagnosis, is also indicated. As a result of an ancient duplication, the distal end of chromosome 10q is highly homologous to 4qA. This homology extends 40 kb in the proximal direction and ends within an inverted copy of D4Z4. At chromosome 10q, the D4Z4 array can vary between 1 and 100 units without pathological consequences. In one-fifth of the population, repeat exchanges between chromosomes 4 and 10 can be observed. However, FSHD is strictly linked to the 4qA subtelomere, as contracted repeats on chromosome 10, irrespective of the composition, have never been observed to cause FSHD.

CHAPTER 9 The Subtelomeric D4Z4 Repeat Instability in Facioscapulohumeral Muscular Dystrophy

These subtelomeric exchanges may, however, complicate the DNA diagnosis in some cases, for example, when a 10-derived repeat residing on chromosome 4 is contracted (false negative), or in case a 4-derived repeat on chromosome 10 is <38 kb (false positive). For these cases, additional PFGE-based diagnostic tests have been developed making use of chromosome 4-specific probes and allowing the correct chromosomal assignment of each of the four arrays [33].

A second complication arises from the observation of the biallelic variation of chromosome 4qter. Two allelic variants of chromosome 4q exist, designated 4qA and 4qB, which differ from each other in the region distal to D4Z4 [27]. Although both variants are almost equally common in the population, only D4Z4 contractions on 4qA chromosomes have been identified in FSHD [34]. In contrast, contracted D4Z4 repeats on 4qB chromosomes do not seem to be associated with FSHD, or have a strongly reduced penetrance [35]. As repeats on both chromosome ends seem to have an equal propensity to rearrange, it is hypothesized that additional *cis* factors are required to cause (4qA) or prevent (4qB) FSHD. For the diagnosis, it is thus imperative to include probes specific for 4qA and 4qB, respectively. These probes also allow the identification of patients with so-called proximally extended deletions [24, 25]. In these patients, estimated to be some 1–2% of the FSHD population, the partial deletion of D4Z4 is extended in the proximal direction and includes the probe region p13E-11. These patients have a normal spectrum of disease and would thus be misdiagnosed based on the use of probe p13E-11 only. As the probes 4qA and 4qB recognize a region distal to D4Z4, inclusion of these probes in the standard diagnosis of FSHD allows identification of these proximally extended deletion cases.

VII. TIMING AND ORIGIN OF THE D4Z4 REARRANGEMENT

In about 40% of new cases, a mitotic origin of the D4Z4 contraction is observed. Mitotic contractions of D4Z4 are observed in affected *de novo* patients, as well as clinically unaffected parents of nonmosaic *de novo* patients [36]. A marked gender difference is observed in mosaic carriers for the D4Z4 contractions. While mosaic females are more often the unaffected parent of a nonmosaic affected child, mosaic males are usually affected. Mosaic patients are often less severely affected, and a relationship has been reported between the severity and a combination of the residual repeat size and the proportion of cells carrying the contracted D4Z4 allele [36]. Mitotic alleles are usually detected as a "fifth" allele when using PFGE. Typically, in these cases, the ancestral and *de novo* alleles involved in the mitotic contraction show reduced signal intensity. However, in most diagnostic settings, conventional linear gel electrophoresis is employed for the diagnosis of FSHD. As this technique normally does not allow complete separation of all alleles, detection of mosaicism in FSHD patients is fully dependent on the reduced signal intensity of the disease fragment. A recent survey showed that in approximately 90% of mosaic FSHD patients, the linear gel-based diagnosis of FSHD had failed to reveal the mosaicism [37].

The presence of mosaicism for the FSHD allele in blood and the germline provides evidence for an early timing of the somatic rearrangement before the separation of the somatic and gonadal lineages. This also has important consequences for counseling, as nonaffected carriers of a mosaic FSHD allele have an increased recurrence risk, depending on the fraction of mosaic cells. Conversely, affected carriers of a mosaic FSHD allele have a reduced risk of transmitting the disease allele compared with nonmosaic carriers, but their affected (nonmosaic) children are likely to be more severely affected.

Detailed analysis of mosaic cases has provided evidence for the timing and mechanism of the mitotic D4Z4 contraction. Most contractions seem to evolve through a gene conversion mechanism without crossover, although a significant proportion of patients show evidence for the occurrence of a crossover. In these patients, three cell populations can be identified, as the donor and acceptor alleles are both changed by this mechanism. Moreover, it was demonstrated that the sister chromatid is the preferred partner for this rearrangement [38]. Gonadal mosaicism for FSHD has also been demonstrated by the observation of more than one affected sibling with the same FSHD allele in *de novo* families in which, in the parents, no evidence could be found for a mitotic (somatic) rearrangement of D4Z4. However, the frequency of gonadal mosaicism has not been studied in great detail.

VIII. CANDIDATE GENES

At the time the genetic defect for FSHD was first identified, it was postulated that D4Z4 might encode (part of) a gene that, when (partly) deleted, would cause FSHD. Initial efforts were therefore aimed at precise characterization of the D4Z4 repeat. Although each D4Z4 unit does contain a single open reading frame, termed *DUX4*, and encodes a putative double homeodomain protein, *in vivo* expression has never been demonstrated [39–41]. Recently, some evidence was presented for the presence of a DUX4 protein specifically in myoblast cultures of FSHD patients, but the identity and origin of this protein are not yet fully established [42]. Overexpression studies in C2C12 myoblasts of DUX4 fused to green fluorescent protein show that the protein is actively transported to the nucleus by virtue of nuclear localization signals within the homeodomains of the protein [43].

Next, efforts focused on the region centromeric to D4Z4 to search for the presence of genes that may play a role in the pathogenesis of FSHD, assuming that contraction of D4Z4 would lead to the transcriptional deregulation of one or more genes near D4Z4, similar to position effects as described in *Drosophila* [40, 44]. The region distal to D4Z4 was discarded for further analysis as it was believed to be relatively small (20 kb) and composed of repetitive sequences.

Interestingly, the region immediately adjacent, and up to several megabases centromeric, to D4Z4 is remarkably gene poor and characterized by a high density of repetitive elements and pseudogenes [45]. So far, only three genes have been identified in the immediate vicinity of D4Z4. These include FSHD candidate region gene 1 (*FRG1*) at 120-kb distance, a tubulin-related gene (*TUBB4q*) at 80-kb distance, and *FRG2* at 35-kb distance [46–48]. *FRG1* and *FRG2* both encode partially characterized proteins, whereas *TUBB4q* likely represents a pseudogene.

FRG2 encodes a putative nuclear protein of unknown function. Although *FRG2* is transcriptionally upregulated specifically in differentiating myoblast cultures of FSHD patients, a role in FSHD pathogenesis is controversial, as it is deleted at the disease allele in some patients with a proximally extended deletion [46]. Nevertheless, in reporter assays, it was demonstrated that the activity of its putative promoter is sensitive to increasing units of D4Z4, in agreement with the hypothesis that D4Z4 may have a silencing effect on neighboring genes. Moreover, in FSHD myoblast cultures, the *FRG2* copies from chromosomes 4 and 10 were transcriptionally upregulated which, in view of the enhanced pairing of 4qter and 10qter chromosome ends in FSHD cells [49], led to the hypothesis that a transvection mechanism (i.e., that pairing is essential for the upregulation of the copies on chromosomes 4 and 10) might operate at these loci [46].

FRG1 also encodes a protein of unknown function but is, in contrast to *FRG2*, highly conserved in vertebrates and nonvertebrates [50]. Stable and transient overexpression of the protein demonstrated it to be localized primarily in nucleoli, nuclear speckles, and Cajal bodies, all consistent with a role in RNA biogenesis [51]. Its transcriptional deregulation in FSHD is highly controversial, varying from upregulation, to no change, to downregulation [47, 52–54].

One other promising candidate gene, *ANT1*, is located approximately 3–5 Mb in the proximal direction [55]. The adenine nucleotide translocator ANT1 facilitates the export of ATP over the mitochondrial membrane. *ANT1* is expressed predominantly in terminally differentiated tissue and is highly expressed in heart and skeletal muscle [56, 57]. The transcriptional deregulation in FSHD is highly controversial for this gene too, but evidence has been presented supporting upregulation of ANT1 at the protein level in both affected and unaffected FSHD muscle [58]. Proteins involved in mitochondrial function and protection from oxidative stress also appeared to be modified in FSHD, and comprehensive follow-up studies are necessary to explore whether mitochondrial dysfunction may be an early event in FSHD pathogenesis.

IX. CHROMATIN REMODELING

Effort has also been focused on studying the mechanism by which contraction of D4Z4 can cause the transcriptional deregulation of essential genes. A protein complex consisting of YY1, nucleolin, and HMGB2 was identified that binds to D4Z4 and, when depleted in cell lines, causes the transcriptional upregulation of *FRG2* [52]. Therefore, it was proposed that this protein complex acts as a repressor complex for nearby genes and that with reduction of D4Z4, insufficient repression is achieved.

Evidence of a local change in chromatin structure came from the observation that D4Z4 repeats are hypomethylated in FSHD [8]. D4Z4 is very GC-rich, with no less than 290 CpG dinucleotides within a single unit. By a Southern blot-based methylation assay making use of two methylation-sensitive restriction enzymes, it was demonstrated that the D4Z4 repeat array is hypomethylated at the disease chromosome. This hypomethylation was also observed in nonpenetrant gene carriers, but not in patients suffering from a nonrelated muscular dystrophy, suggesting that hypomethylation is specific for FSHD but not sufficient to cause FSHD. Moreover, in patients clinically identical to those with FSHD but who do not carry the typical repeat contraction, hypomethylation was also observed, indicating that in these patients too, the disease mechanism acts through chromosome 4. In contrast to the 4q-linked patients, in which the hypomethylation is restricted to the contracted array only, in these non-4q-linked patients, the hypomethylation is observed on both chromosomes 4 and, possibly, also on the homologous repeats on chromosome 10.

In a survey, the question of whether there exists a relationship between D4Z4 methylation, residual repeat size, and clinical severity was asked [59]. In a small subset of patients in whom the methylation of D4Z4 at the disease chromosome could be measured specifically without interference of the normal chromosome, no direct relationship could be established between residual methylation, repeat size, and severity. However, analogous to the rough relationship between residual repeat size and severity, the most severely affected patients with repeat sizes in the range of 1–3 units all displayed pronounced hypomethylation of D4Z4. However, in the patient group carrying repeat sizes of 4–10 units, the hypomethylation was very variable, as is also observed for the clinical severity. This analysis was challenged by two issues: the relatively small patient group that could be analyzed and the fact that only 2 of the 290 CpG dinucleotides within D4Z4 were measured. Therefore, additional and more comprehensive methylation assays need to be developed for D4Z4 to address the question of whether differential methylation of D4Z4 can explain the overt clinical variation in disease presentation,

Interestingly, hypomethylation of D4Z4 was first reported in the unrelated ICF syndrome [60]. This very rare autosomal recessive disorder shows abnormalities of the juxtacentromeric heterochromatin of chromosomes 1 and 16 in mitogen-stimulated lymphocytes. ICF syndrome is caused by mutations in the DNA methyltransferase 3B (*DNMT3B*) gene, although there is evidence of genetic heterogeneity [61, 62]. As a consequence, the genome of these patients shows regions of hypomethylated DNA, often involving pericentromeric and subtelomeric repeat structures such as the Sat2, Sat3, and NBL2 repeats and D4Z4 [63]. Analogous to the non-4q-linked FSHD patients, in ICF patients, the hypomethylation is very pronounced and affects the D4Z4 repeats on both chromosomes 4 and 10 [8]. This suggests that a similar genetic defect underlies non-4q-linked FSHD, although mutations in *DNMT3B* have not been detected in non-4q-linked FSHD.

It is unknown why hypomethylation of D4Z4 is not associated with muscular dystrophy in ICF patients. Although these patients usually die young, this cannot satisfactorily explain the lack of muscular dystrophy. As described, a substantial proportion of FSHD cases have a very early onset, and these are typically carriers of a very small residual repeat array (1–3 units), which is strongly hypomethylated. Moreover, mildly affected ICF cases have been reported surviving beyond the age of 20 without a dystrophic phenotype. A role for the distal variation has also been excluded, as 4qA and 4qB chromosomes have been observed in ICF patients (S. M. van der Maarel, unpublished data).

The chromatin structure was further investigated by analysis of different chromatin markers [53, 64]. Histone H4 acetylation levels, colocalization studies with DAPI–bright foci or with regions enriched in heterochromatin protein -1a or histone H3 trimethylated at lysine 9 were consistent with 4qter being unexpressed euchromatin rather than heterochromatin. In agreement, the replication timing of 4qter was found to be very close to that of unexpressed euchromatin as well. For all of these parameters, no differences could be observed between fibroblasts, lymphoblastoid cell lines, and mononuclear blood cells of controls and FSHD patients. Moreover, studies of histone H4 acetylation levels along chromosome 4qter, in combination with semiquantitative expression analyses of genes on chromosome 4qter, provided no evidence in support of a spreading of heterochromatinization emanating from the D4Z4 repeat.

The focus has also shifted to the nuclear organization of 4qter. The mammalian nucleus is highly compartmentalized, with individual chromosomes occupying distinct territories most likely reflecting their gene density, transcriptional activity, replication timing, and chromosome size [65, 66]. In earlier studies, the relative localization of chromosome 4qter and 10qter with respect to the nucleolus and nuclear periphery was analyzed in lymphocytes of control individuals and FSHD patients [49, 67]. Although no evidence was found for a specific nuclear localization of 4qter, in one study a small but significantly enhanced frequency of chromosome pairing between chromosomes 4qter and 10qter was observed [49]. More recently, two independent studies reevaluated the organization of 4qter in the nucleus [68, 69]. It was demonstrated that 4qter largely occupies a peripheral territory in the nucleus, independent of cell type and chromosome territory effects. This predominant perinuclear localization was not observed for other chromosome ends, including the highly homologous 10qter, making it an important feature discriminating between these two chromosome ends. Detailed analysis of this peripheral localization by making use of probes spread over the distal end of chromosome 4q and by analysis of a X;4 translocation in which the derivative chromosome X contains 4 Mb of distal 4q showed that it is not D4Z4 itself, but sequences proximal to D4Z4, that are necessary and sufficient for this perinuclear localization. As 4q and 10q share only 40 kb of sequence homology proximal to D4Z4, it is likely that the upstream sequences may explain the differential localization of both chromosome ends.

Interestingly, this nuclear localization seems to be dependent on the integrity of the nuclear lamina, as this peripheral localization of 4qter is lost in cells deficient in lamin A/C [68]. Several neuromuscular disorders, designated as laminopathies, are caused by defects in nuclear lamina proteins, including lamin A/C and emerin (reviewed by Maraldi et al. [70]). Based on the aforementioned studies, different disease models have been proposed for FSHD (Fig. 9-2). All models imply a structural chromatin modulating function for D4Z4; it

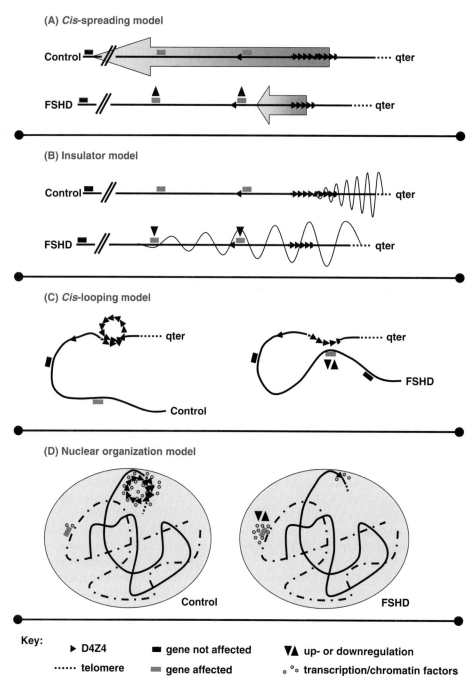

FIGURE 9-2 Different models have been proposed to explain FSHD. (A) In the *cis*-spreading model, it was hypothesized that D4Z4 may impose a closed chromatin structure on 4q and that, with contraction of D4Z4, the chromatin structure would become more open in a distance-dependent manner. Although initially there was some evidence supporting this model based on the distance-dependent upregulation of several candidate genes [52], recent studies do not support a simple *cis*-spreading model [53, 54]. (B) It was proposed that D4Z4 may act as an insulator separating euchromatic chromosome domains proximal to D4Z4 from heterochromatic chromatin distal to D4Z4. (C) It was proposed that D4Z4 normally forms intra-array loops when >11 units and, when contracted, interacts with other domains on chromosome 4 [53]. (D) Recent studies suggest that 4qter occupies a distinct perinuclear territory where it interacts with the nuclear lamina, in which chromatin and transcription factors are tethered [54, 69]. Perturbation of this interaction in FSHD may cause subtle misbalances of chromatin and transcription factors at 4qter and unrelated loci.

may either impose a chromatin structure to 4qter, act as an insulator, or participate in nuclear positioning along with other, more proximally located sequences. Irrespective of its exact function, in FSHD the chromatin structure seems to be impaired as evidenced by the reduced methylation levels.

X. MYOBLAST STUDIES

Relatively few studies have been performed on primary myoblast cell cultures to explore FSHD pathogenesis. One study reported normal calcium homeostasis as judged by calcium current properties, voltage dependency of contractile responses, and amplitude of evoked calcium transients [71]. In a second study, a vacuolar/necrotic phenotype was observed in myoblasts derived from affected FSHD muscle. Undifferentiated FSHD myoblast cultures manifested an increased susceptibility to oxidative stress, which was no longer seen on differentiation to myotubes [72]. Global gene expression profiling showed that in FSHD myoblasts, several cellular processes including oxidative stress were dysregulated. A recent study addressed the potential of primary FSHD myoblast cultures for autologous cell transplantation [73]. To this end, myoblasts were isolated from nonaffected muscle of FSHD patients, and in contrast to the previous study, these myoblasts did not reveal any morphological differences compared with control myoblasts. Also, other characteristics, including doubling time, differentiation parameters, and telomere size, were indistinguishable from those of control myoblasts. Moreover, these myoblasts participated in muscle structures when injected into immunodeficient mice, providing a potential opportunity for autologous cell transplantation. The apparent discrepancy between the studies may be related to differences in cell culture parameters, site of muscle biopsy, and degree of muscular dystrophy. It is of note that the major difference between the studies is the site of biopsy. The earlier study used affected muscle as a source of myoblast cultures, whereas the more recent study started from clinically unaffected muscle of FSHD patients.

XI. CONCLUDING REMARKS

The discovery of the D4Z4 repeat array contraction as the genetic defect in the vast majority of FSHD patients provided the first molecular basis for understanding FSHD pathogenesis. Most importantly for the FSHD community, it allowed reliable genetic diagnosis of FSHD, which is probably the most beneficial achievement so far for the FSHD community. Moreover, it provided an explanation for many of the observations made in the pre-repeat era. For example, the recurrent observation of a mildly affected parent with severely affected offspring [74] can now be explained by the presence of gonadosomal mosaicism for the disease allele in the parent. In addition, mosaicism can also in part explain some claims of anticipation. Partial deletions of D4Z4 and extension proximally into the probe region have now been recognized and can confirm some 1–2% as 4q-linked families that were previously suggestive for genetic heterogeneity.

On the other hand, many findings have obscured our view of FSHD pathogenesis. Unlike most monogenic disorders, it is likely that the rearrangement does not compromise the function of the disease gene, but rather causes the transcriptional deregulation of one or more disease genes *in cis* or *in trans* that (indirectly) cause FSHD. This makes the identification of the disease gene(s) more difficult and necessitates functional analysis of potential candidate genes. Functional analysis should not be restricted to transcription studies, as evidence is emerging that based on the method employed and the biomaterial sources (site of biopsy, duration of disease, pathology, etc.), different RNA or protein levels of candidate genes on 4qter or elsewhere in the genome have been reported.

Currently, there is no validated natural or transgenic animal model for FSHD. The development of a faithful animal model is seriously hampered by the primate specificity of the FSHD locus. D4Z4 is reported only in primates and the subtelomeric localization of FRG1 is also observed only in primates [67, 75]. It is conceivable that this unique combination of sequences and subtelomeric localization is essential for key steps in FSHD pathogenesis and that introducing only individual elements of the FSHD locus in the mouse genome (e.g., D4Z4 or one of the candidate genes) will, at best, reproduce only parts of the complex etiology of FSHD. Therefore, the feasibility and design of animal models and their preference to (human-derived) cellular model systems need to be carefully reconsidered. Despite these serious constraint, generation of faithful animal models is instrumental to the development of evidence-based therapeutic strategies.

Acknowledgments

We thank the Dutch FSHD patients for their continuous and active involvement in our studies. FSHD research is supported by grants from The Netherlands Organization for Scientific Research (NWO), the Prinses Beatrix Fonds, the Stichting Spieren voor Spieren, the Muscular Dystrophy Association USA, the FSH Society, the Stichting FSHD, the Shaw family, and the National Institute of Arthritis and Musculoskeletal and Skin Diseases (National Institutes of Health).

References

1. Landouzy, L., and Dejerine, J. (1885). De la myopathy atrophique progressive. *Rev. Med.* **5**, 81–117.
2. Landouzy, L., Dejerine, J. (1885). De la myopathie atrophique progressive. *Rev. Med.* **5**, 253–366.
3. Walton, J. N., Race, R. R., and Philip, U. (1955). On the inheritance of muscular dystrophy: With a note on the blood groups, and a note on colour vision and linkage studies. *Ann. Hum. Genet.* **20**, 1–38.
4. Wijmenga, C., Frants, R. R., Brouwer, O. F., Moerer, P., Weber, J. L., and Padberg, G. W. (1990). Location of facioscapulohumeral muscular dystrophy gene on chromosome 4. *Lancet* **336**, 651–653.
5. Wijmenga, C., van Deutekom, J. C., Hewitt, J. E., Padberg, G. W., van Ommen, G. J., Hofker, M. H., and Frants, R. R. (1994). Pulsed-field gel electrophoresis of the D4F104S1 locus reveals the size and the parental origin of the facioscapulohumeral muscular dystrophy (FSHD)-associated deletions. *Genomics* **19**, 21–26.
6. Padberg, G. W. (2004) *In* "Facioscapulohumeral Muscular Dystrophy: Clinical Medicine and Molecular Cell Biology" (M. Upadhyaya and D. N. Cooper, Eds.), pp. 41–54. Garland Science/BIOS Scientific, Oxon.
7. Funakoshi, M., Goto, K., and Arahata, K. (1998). Epilepsy and mental retardation in a subset of early onset 4q35- facioscapulohumeral muscular dystrophy. *Neurology* **50**, 1791–1794.
8. Van Overveld, P. G., Lemmers, R. J., Sandkuijl, L. A., Enthoven, L., Winokur, S. T., Bakels, F., Padberg, G. W., van Ommen, G. J., Frants, R. R., and van der Maarel, S. M. (2003). Hypomethylation of D4Z4 in 4q-linked and non-4q-linked facioscapulohumeral muscular dystrophy. *Nat. Genet.* **35**, 315–317.
9. Arahata, K., Ishihara, T., Fukunaga, H., Orimo, S., Lee, J. H., Goto, K., and Nonaka, I. (1995). Inflammatory response in facioscapulohumeral muscular dystrophy (FSHD): Immunocytochemical and genetic analyses. *Muscle Nerve* **2**, 56–66.
10. Lunt, P. W. (1989). A workshop on facioscapulohumeral (Landouzy–Dejerine) disease, Manchester, 16 to 17 November 1988. *J. Med. Genet.* **26**, 535–537.
11. Sarfarazi, M., Upadhyaya, M., Padberg, G., Pericak-Vance, M., Siddique, T., Lucotte, G., and Lunt, P. (1989). An exclusion map for facioscapulohumeral (Landouzy–Dejerine) disease. *J. Med. Genet.* **26**, 481–484.
12. Weber, J. L., and May, P. E. (1989). Abundant class of human DNA polymorphisms which can be typed using the polymerase chain reaction. *Am. J. Hum. Genet.* **44**, 388–396.
13. Weber, J. L., and May, P. E. (1990). Dinucleotide repeat polymorphism at the D4S171 locus. *Nucleic. Acids. Res.* **18**, 2202.
14. Burglin, T. R., Finney, M., Coulson, A., and Ruvkun, G. (1989). *Caenorhabditis elegans* has scores of homoeobox-containing genes. *Nature* **341**, 239–243.
15. Wijmenga, C., Hewitt, J. E., Sandkuijl, L. A., Clark, L. N., Wright, T. J., Dauwerse, H. G., Gruter, A. M., Hofker, M. H., Moerer, P., Williamson, R., van Ommen, G. J., Padberg, G. W., and Frants, R. R. (1992). Chromosome 4q DNA rearrangements associated with facioscapulohumeral muscular dystrophy. *Nat. Genet.* **2**, 26–30.
16. Wijmenga, C., Brouwer, O. F., Padberg, G. W., and Frants, R. R. (1992). Transmission of de-novo mutation associated with facioscapulohumeral muscular dystrophy. *Lancet* **340**, 985–986.
17. Van Deutekom, J. C., Wijmenga, C., van Tienhoven, E. A., Gruter, A. M., Hewitt, J. E., Padberg, G. W., van Ommen, G. J., Hofker, M. H., and Frants, R. R. (1993). FSHD associated DNA rearrangements are due to deletions of integral copies of a 3.2 kb tandemly repeated unit. *Hum. Mol. Genet.* **2**, 2037–2042.
18. Bakker, E., Wijmenga, C., Vossen, R. H., Padberg, G. W., Hewitt, J., van der Wielen, M., Rasmussen, K., and Frants, R. R. (1995). The FSHD-linked locus D4F104S1 (p13E-11) on 4q35 has a homologue on 10qter. *Muscle Nerve* **2**, 39–44.
19. Van Overveld, P. G., Lemmers, R. J., Deidda, G., Sandkuijl, L., Padberg, G. W., Frants, R. R., and van der Maarel, S. M. (2000). Interchromosomal repeat array interactions between chromosomes 4 and 10: A model for subtelomeric plasticity. *Hum. Mol. Genet.* **9**, 2879–2884.
20. Tupler, R., Berardinelli, A., Barbierato, L., Frants, Hewitt J. E., Lanzi, G., Maraschio, P., and Tiepolo, L. (1996). Monosomy of distal 4q does not cause facioscapulohumeral muscular dystrophy. *J. Med. Genet.* **33**, 366–370.
21. Lunt, P. W., Jardine, P. E., Koch, M. C., Maynard, J., Osborn, M., Williams, M., Harper, P. S., and Upadhyaya, M. (1995). Correlation between fragment size at D4F104S1 and age at onset or at wheelchair use, with a possible generational effect, accounts for much phenotypic variation in 4q35- facioscapulohumeral muscular dystrophy (FSHD). *Hum. Mol. Genet.* **4**, 951–958.
22. Tawil, R., Forrester, J., Griggs, R. C., Mendell, J., Kissel, J., McDermott, M., King, W., Weiffenbach, B., and Figlewicz, D., for the FSH-DY Group. (1996). Evidence for anticipation and association of deletion size with severity in facioscapulohumeral muscular dystrophy. *Ann. Neurol.* **39**, 744–748.
23. Gilbert, J. R., Stajich, J. M., Wall, S., Carter, S. C., Qiu, H., Vance, J. M., Stewart, C. S., Speer, M. C., Pufky, J., Yamaoka, L. H., Rozear, M., Samson, F., Fardeau, M., Roses, A. D., and Pricak-Vance, M. A. (1993). Evidence for heterogeneity in facioscapulohumeral muscular dystrophy (FSHD). *Am. J. Hum. Genet.* **53**, 401–408.
24. Lemmers, R. J., Osborn, M., Haaf, T., Rogers, M., Frants, R. R., Padberg, G. W., Cooper, D. N., van der Maarel, S. M., and Upadhyaya, M. (2003). D4F104S1 deletion in facioscapulohumeral muscular dystrophy: Phenotype, size, and detection. *Neurology* **61**, 178–183.
25. Lemmers, R. J. L. F., van der Maarel, S. M., van Deutekom, J. C. T., van der Wielen, M. J. R., Deidda, G., Dauwerse, H. G., Hewitt, J., Hofker, M., Bakker, E., Padberg, G. W., and Frants, R. R. (1998). Inter- and intrachromosomal subtelomeric rearrangements on 4q35: Implications for facioscapulohumeral muscular dystrophy (FSHD) aetiology and diagnosis. *Hum. Mol. Genet.* **7**, 1207–1214.
26. Deidda, G., Cacurri, S., Grisanti, P., Vigneti, E., Piazzo, N., and Felicetti, L. (1995). Physical mapping evidence for a duplicated region on chromosome 10qter showing high homology with the facioscapulohumeral muscular dystrophy locus on chromosome 4qter. *Eur. J. Hum. Genet.* **3**, 155–167.
27. Van Geel, M., Dickson, M. C., Beck, A. F., Bolland, D. J., Frants, R. R., van der Maarel, S. M., de Jong, P. J., and Hewitt, J. E. (2002). Genomic analysis of human chromosome 10q and 4q telomeres suggests a common origin. *Genomics* **79**, 210–217.
28. Deidda, G., Cacurri, S., Piazzo, N., and Felicetti, L. (1996). Direct detection of 4q35 rearrangements implicated in facioscapulohumeral muscular dystrophy (FSHD). *J. Med. Genet.* **33**, 361–365.
29. Lemmers, R. J. L., de Kievit, P., van Geel, M., van der Wielen, M. J., Bakker, E., Padberg, G. W., Frants, R. R., and van der Maarel, S. M. (2001). Complete allele information in the diagnosis of facioscapulohumeral muscular dystrophy by triple DNA analysis. *Ann. Neurol.* **50**, 816–819.
30. Mefford, H. C., Trask, B. J. (2002). The complex structure and dynamic evolution of human subtelomeres. *Nat. Rev. Genet.* **3**, 91–102.

31. Van Deutekom, J. C., Bakker, E., Lemmers, R. J., van der Wielen, M. J., Bik, E., Hofker, M. H., Padberg, G. W., and Frants, R. R. (1996). Evidence for subtelomeric exchange of 3.3 kb tandemly repeated units between chromosomes 4q35 and 10q26: Implications for genetic counselling and etiology of FSHD1. *Hum. Mol. Genet.* **5**, 1997–2003.
32. Wu, Z. Y., Wang, Z. Q., Murong, S. X., and Wang, N. (2004). FSHD in Chinese population: Characteristics of translocation and genotype–phenotype correlation. *Neurology* **63**, 581–583.
33. Lemmers, R. J. L. F., van der Wielen, M. J., Bakker, E., and van der Maarel, S. M. (2004). *In* "Facioscapulohumeral Muscular Dystrophy: Clinical Medicine and Molecular Cell Biology" (M. Upadhyaya and D. N. Cooper, Eds.), pp. 211–234. Garland Science/BIOS Scientific, Oxon.
34. Lemmers, R. J., de Kievit, P., Sandkuijl, L., Padberg, G. W., van Ommen, G. J., Frants, R. R., and van der Maarel, S. M. (2002). Facioscapulohumeral muscular dystrophy is uniquely associated with one of the two variants of the 4q subtelomere. *Nat. Genet.* **32**, 235–236.
35. Lemmers, R. J., Wohlgemuth, M., Frants, R. R., Padberg, G. W., Morava, E., and van der Maarel, S. M. (2004). Contractions of D4Z4 on 4qB subtelomeres do not cause facioscapulohumeral muscular dystrophy. *Am. J. Hum. Genet.* **75**, 1124–1130.
36. van der Maarel, S. M., Deidda, G., Lemmers, R. J., van Overveld, P. G., van der Wielen, M., Hewitt, J. E., Sandkuijl, L., Bakker, B., van Ommen, G. J., Padberg, G. W., and Frants, R. R. (2000). De novo facioscapulohumeral muscular dystrophy: Frequent somatic mosaicism, sex-dependent phenotype, and the role of mitotic transchromosomal repeat interaction between chromosomes 4 and 10. *Am. J. Hum. Genet.* **66**, 26–35.
37. Lemmers, R. J. L. F., van der Wielen, M. J. R., Bakker, E., Padberg, G. W., Frants, R. R., and van der Maarel, S. M. (2004). Somatic mosaicism in FSHD often goes undetected. *Ann. Neurol.* **55**, 845–850.
38. Lemmers, R. J., van Overveld, P. G., Sandkuijl, L. A., Vrieling, H., Padberg, G. W., Frants, R. R., and van der Maarel, S. M. (2004). Mechanism and timing of mitotic rearrangements in the subtelomeric D4Z4 repeat Involved in facioscapulohumeral muscular dystrophy. *Am. J. Hum. Genet.* **75**, 44–53.
39. Gabriels, J., Beckers, M. C., Ding, H., De Vriese, A., Plaisance, S., van der Maarel, S. M., Padberg, G. W., Frants, R. R., Hewitt, J. E., Collen, D., and Belayew, A. (1999). Nucleotide sequence of the partially deleted D4Z4 locus in a patient with FSHD identifies a putative gene within each 3.3 kb element. *Gene* **236**, 25–32.
40. Hewitt, J. E., Lyle, R., Clark, L. N., Valleley, E. M., Wright, T. J., Wijmenga, C., van Deutekom, J. C., Francis, F., Sharpe, P. T., Hofker, M., Frants, R. R., and Williamson, R. (1994). Analysis of the tandem repeat locus D4Z4 associated with facioscapulohumeral muscular dystrophy. *Hum. Mol. Genet.* **3**, 1287–1295.
41. Lyle, R., Wright, T. J., Clark, L. N., and Hewitt, J. E. (1995). The FSHD-associated repeat, D4Z4, is a member of a dispersed family of homeobox-containing repeats, subsets of which are clustered on the short arms of the acrocentric chromosomes. *Genomics* **28**, 389–397.
42. Coppee, F., Matteotti, C., Ansseau, E., Sauvage, S., Leclercq, I., Leroy, A., Marcowycz, A., Gerbaux, C., Figlewicz, D., Ding, H., and Belayew, A. (2004) *In* "Facioscapulohumeral Muscular Dystrophy: Clinical Medicine and Molecular Cell Biology" (M. Upadhyaya and D. N. Cooper, Eds.), pp. 117–134. Garland Science/BIOS Scientific, Oxon.
43. Ostlund, C., Garcia-Carrasquillo, R. M., Belayew, A., and Worman, H. J. (2005). Intracellular trafficking and dynamics of double homeodomain proteins. *Biochemistry* **44**, 2378–2384.
44. Winokur, S. T., Bengtsson, U., Feddersen, J., Mathews, K. D., Weiffenbach, B., Bailey, H., Markovich, R. P., Murray, J. C., Wasmuth, J. J., Altherr, M. R., and Schutte, B. C. (1994). The DNA rearrangement associated with facioscapulohumeral muscular dystrophy involves a heterochromatin-associated repetitive element: Implications for a role of chromatin structure in the pathogenesis of the disease. *Chromosome Res.* **2**, 225–234.
45. van Geel, M., Heather, L. J., Lyle, R., Hewitt, J. E., Frants, R. R., and de Jong, P. J. (1999). The FSHD region on human chromosome 4q35 contains potential coding regions among pseudogenes and a high density of repeat elements. *Genomics* **61**, 55–65.
46. Rijkers, T., Deidda, G., van Koningsbruggen, S., van Geel, M., Lemmers, R. J., van Deutekom, J. C., Figlewicz, D., Hewitt, J. E., Padberg, G. W., Frants, R. R., and van der Maarel, S. M. (2004). FRG2, an FSHD candidate gene, is transcriptionally upregulated in differentiating primary myoblast cultures of FSHD patients. *J. Med. Genet.* **41**, 826–836.
47. van Deutekom, J. C. T., Lemmers, R. J. L. F., Grewal, P. K., van Geel, M., Romberg, S., Dauwerse, H. G., Wright, T. J., Padberg, G. W., Hofker, M. H., Hewitt, J. E., and Frants, R. R. (1996). Identification of the first gene (*FRG1*) from the FSHD region on human chromosome 4q35. *Hum. Mol. Genet.* **5**, 581–590.
48. van Geel, M., van Deutekom, J. C., van Staalduinen, A., Lemmers, R. J., Dickson, M. C., Hofker, M. H., Padberg, G. W., Hewitt, J. E., de Jong, P. J., and Frants, R. R. (2000). Identification of a novel beta-tubulin subfamily with one member (TUBB4Q) located near the telomere of chromosome region 4q35. *Cytogenet. Cell Genet.* **88**, 316–321.
49. Stout, K., van der Maarel, S., Frants, R. R., Padberg, G. W., Ropers, H.-H., and Haaf, T. (1999). Somatic pairing between subtelomeric regions: Implications for human genetic disease? *Chrom. Res.* **7**, 323–329.
50. Grewal, P. K., Carim Todd, L., van der Maarel, S., Frants, R. R., and Hewitt, J. E. (1998). FRG1, a gene in the FSH muscular dystrophy region on human chromosome 4q35, is highly conserved in vertebrates and invertebrates. *Gene* **216**, 13–19.
51. van Koningsbruggen, S., Dirks, R. W., Mommaas, A. M., Onderwater, J. J., Deidda, G., Padberg, G. W., Frants, R. R., and van der Maarel, S. M. (2004). FRG1P is localised in the nucleolus, Cajal bodies, and speckles. *J. Med. Genet.* **41**.
52. Gabellini, D., Green, M., and Tupler, R. (2002). Inappropriate gene activation in FSHD: A repressor complex binds a chromosomal repeat deleted in dystrophic muscle. *Cell* **110**, 339–248.
53. Jiang, G., Yang, F., van Overveld, P. G., Vedanarayanan, V., van der Maarel, S. M., and Ehrlich, M. (2003). Testing the position-effect variegation hypothesis for facioscapulohumeral muscular dystrophy by analysis of histone modification and gene expression in subtelomeric 4q. *Hum. Mol. Genet.* **12**, 2909–2921.
54. Winokur, S. T., Chen, Y. W., Masny, P. S., Martin, J. H., Ehmsen, J. T., Tapscott, S. J., van der Maarel, S. M., Hayashi, Y., and Flanigan, K. M. (2003). Expression profiling of FSHD muscle supports a defect in specific stages of myogenic differentiation. *Hum. Mol. Genet.* **12**, 2895–2907.
55. Li, K., Warner, C. K., Hodge, J. A., Minoshima, S., Kudoh, J., Fukuyama, R., Maekawa, M., Shimizu, Y., Shimizu, N., and Wallace, D. C. (1989). A human muscle adenine nucleotide translocator gene has four exons, is located on chromosome 4, and is differentially expressed. *J. Biol. Chem.* **264**, 13998–14004.
56. Doerner, A., Pauschinger, M., Badorff, A., Noutsias, M., Giessen, S., Schulze, K., Bilger, J., Rauch, U., and Schultheiss, H. P. (1997). Tissue-specific transcription pattern of the adenine nucleotide translocase isoforms in humans. *FEBS Lett.* **414**, 258–262.
57. Stepien, G., Torroni, A., Chung, A. B., Hodge, J. A., and Wallace, D. C. (1992). Differential expression of adenine nucleotide translocator isoforms in mammalian tissues and during muscle cell differentiation. *J. Biol. Chem.* **267**, 14592–14597.

58. Laoudj-Chenivesse, D., Carnac, G., Bisbal, C., Hugon, G., Bouillot, S., Desnuelle, C., Vassetzky, Y., and Fernandez, A. (2005). Increased levels of adenine nucleotide translocator 1 protein and response to oxidative stress are early events in facioscapulohumeral muscular dystrophy muscle. *J. Mol. Med.* **83**, 216–224.
59. van Overveld, P. G., Enthoven, L., Ricci, E., Felicetti, L., Jeanpierre, M., Winokur, S. T., Frants, R. R., Padberg, G. W., and van der Maarel, S. M. (2005). *Ann. Neurol.*, in press.
60. Kondo, T., Bobek, M. P., Kuick, R., Lamb, B., Zhu, X., Narayan, A., Bourc'his, D., Viegas-Pequignot, E., Ehrlich, M., and Hanash, S. M. (2000). Whole-genome methylation scan in ICF syndrome: Hypomethylation of non-satellite DNA repeats D4Z4 and NBL2. *Hum. Mol. Genet.* **9**, 597–604.
61. Wijmenga, C., Hansen, R. S., Gimelli, G., Bjorck, E. J., Davies, E. G., Valentine, D., Belohradsky, B. H., van Dongen, J. J., Smeets, D. F., van den Heuvel, L. P., Luyten, J. A., Strengman, E., Weemaes, C., and Pearson, P. L. (2000). Genetic variation in ICF syndrome: Evidence for genetic heterogeneity. *Hum. Mutat.* **16**, 509–517.
62. Xu, G. L., Bestor, T. H., Bourc'his, D., Hsieh, C. L., Tommerup, N., Bugge, M., Hulten, M., Qu, X., Russo, J. J., and Viegas-Pequignot, E. (1999). Chromosome instability and immunodeficiency syndrome caused by mutations in a DNA methyltransferase gene. *Nature* **402**, 187–191.
63. Ehrlich, M. (2003). The ICF syndrome, a DNA methyltransferase 3B deficiency and immunodeficiency disease. *Clin. Immunol.* **109**, 17–28.
64. Yang, F., Shao, C. B., Vedanarayanan, V., and Ehrlich, M. (2004). Cytogenetic and immuno-FISH analysis of the 4q subtelomeric region, which is associated with facioscapulohumeral muscular dystrophy. *Chromosoma* **112**, 350–359.
65. Sun, H. B., Shen, J., and Yokota, H. (2000). Size-dependent positioning of human chromosomes in interphase nuclei. *Biophys. J.* **79**, 184–190.
66. Tanabe, H., Habermann, F. A., Solovei, I., Cremer, M., and Cremer, T. (2002). Nonrandom radial arrangements of interphase chromosome territories: Evolutionary considerations and functional implications. *Mutat. Res.* **504**, 37–45.
67. Winokur, S. T., Bengtsson, U., Vargas, J. C., Wasmuth, J. J., and Altherr, M. R. (1996). The evolutionary distribution and structural organization of the homeobox-containing repeat D4Z4 indicates a functional role for the ancestral copy in the FSHD region. *Hum. Mol. Genet.* **5**, 1567–1575.
68. Masny, P. S., Bengtsson, U., Chung, S. A., Martin, J. H., van Engelen, B., van der Maarel, S. M., and Winokur, S. T. (2004). Localization of 4q35.2 to the nuclear periphery: Is FSHD a nuclear envelope disease? *Hum. Mol. Genet.* **13**, 1857–1871.
69. Tam, R., Smith, K. P., and Lawrence, J. B. (2004). The 4q subtelomere harboring the FSHD locus is specifically anchored with peripheral heterochromatin unlike most human telomeres. *J. Cell Biol.* **167**, 269–279.
70. Maraldi
71. Vandebrouck, C., Imbert, N., Constantin, B., Duport, G., Raymond, G., and Cognard, C. (2002). Normal calcium homeostasis in dystrophin-expressing facioscapulohumeral muscular dystrophy myotubes. *Neuromuscul. Disord.* **12**, 266–272.
72. Winokur, S. T., Barrett, K., Martin, J. H., Forrester, J. R., Simon, M., Tawil, R., Chung, S. A., Masny P. S., and Figlewicz, D. A. (2003). Facioscapulohumeral muscular dystrophy (FSHD) myoblasts demonstrate increased susceptibility to oxidative stress. *Neuromuscul. Disord.* **13**, 322–333.
73. Vilquin, J. T., Marolleau, J. P., Sacconi, S., Garcin, I., Lacassagne, M. N., Robert, I., Ternaux, B., Bouazza, B., Larghero, J., and Desnuelle, C. (2005). Normal growth and regenerating ability of myoblasts from unaffected muscles of facioscapulohumeral muscular dystrophy patients. *Gene Ther.*
74. Brooke, M. H. (1977). "A Clinician's View of Neuromuscular Diseases." Williams & Wilkins, Baltimore.
75. Grewal, P. K., Bolland, D. J., Todd, L. C., and Hewitt, J. E. (1998). High-resolution mapping of mouse chromosome 8 identifies an evolutionary chromosomal breakpoint. *Mamm. Genome* **9**, 603–607.

PART III

Fragile X Syndrome

CHAPTER 10

Fragile X Syndrome and Fragile X-Associated Tremor/Ataxia Syndrome

RANDI J. HAGERMAN AND PAUL J. HAGERMAN

M.I.N.D. Institute and Department of Pediatrics, University of California, Davis, School of Medicine, Sacramento, California 95817; and Department of Biochemistry and Molecular Medicine, University of California, Davis, School of Medicine, Davis, California 95616

I. Introduction
II. Fragile X Syndrome
   A. Spectrum of Clinical Involvement
   B. Clinical Involvement in Premutation Carriers
III. Fragile X-Associated Tremor/Ataxia Syndrome
   A. Overview
   B. Clinical Features
   C. Epidemiology
   D. Neuropathology
   E. Molecular Pathogenesis
IV. Concluding Remarks
   Acknowledgments
   References

Fragile X syndrome, the leading heritable form of mental retardation, and fragile X-associated tremor/ataxia syndrome (FXTAS), a newly discovered neurodegenerative disorder, are both caused by CGG repeat expansions in the fragile X mental retardation 1 (*FMR1*) gene; however, the two disorders involve separate groups of individuals, and occur by entirely different molecular mechanisms. Although fragile X syndrome is caused by silencing of the *FMR1* gene for large (>200 CGG repeats; full mutation) expansions, FXTAS appears to be caused by increased transcriptional activity of smaller (55 to 200 CGG repeats; premutation) expansion. Furthermore, FXTAS is likely to be one of the few genetic disorders that are caused by a "toxic" gain-of-function of the *FMR1* mRNA. The clinical and molecular pathogenesis of both disorders will be addressed in this chapter.

I. INTRODUCTION

The fragile X mental retardation 1 (*FMR1*) gene (OMIM +309550) gives rise to several distinct clinical syndromes, which are determined in part by the size of the CGG repeat expansion in the 5' untranslated region of the *FMR1* gene. Based on the size of the CGG repeat, individuals are classified as having normal (5–44 CGG repeats), gray zone (45–54 CGG repeats), premutation (55–200 CGG repeats), or full mutation (>200 CGG repeats) alleles. Because the CGG repeat element is in a noncoding portion of the gene, none of the expanded forms of the gene affects the sequence of the *FMR1* protein (FMRP) product, and therefore, none of the disorders caused by the CGG repeat expansions is due to an altered protein product. Most full mutation alleles are hypermethylated in the promoter and adjacent CGG repeat regions, and this epigenetic transformation leads to transcriptional silencing and a deficiency or absence of FMRP [1]. Thus, it is the lack of FMRP that is responsible for the neurodevelopmental disorder fragile X syndrome, the leading heritable form of cognitive impairment, and the leading single gene associated with autism. Fragile X syndrome is the subject of the first half of this chapter.

Although full mutation expansions (>200 CGG repeats) usually result in fragile X syndrome, CGG repeat expansions in the premutation range lead to three distinct forms of clinical involvement, including behavioral/cognitive difficulties in some children [2], premature ovarian failure (POF) in about 20% of female carriers [3], and, among older adult carriers, the neurodegenerative disorder fragile X-associated tremor/ataxia syndrome (FXTAS) [4, 5]. The neurodevelopmental (behavioral/cognitive) features among some premutation carriers are likely to be due to the moderate reductions in FMRP in the upper end of the premutation range and, therefore, may be considered to be on the fragile X spectrum [6–8]. However, notwithstanding being caused by the same gene, fragile X syndrome and FXTAS are caused by entirely distinct molecular mechanisms, affect different groups of individuals, and involve carriers in entirely different age ranges. Whereas fragile X syndrome is fundamentally a protein deficiency disorder due to gene silencing, FXTAS is caused by abnormal expression of an expanded repeat mRNA; that is, FXTAS is not caused by lowered levels of FMRP. The clinical manifestations and pathogenesis of FXTAS are addressed in the second half of this chapter.

Because other chapters in this volume focus on the more molecular aspects of fragile X syndrome, including animal models, the current chapter focuses more on the clinical aspects of the disorder, although both clinical and molecular aspects are addressed.

II. FRAGILE X SYNDROME

A. Spectrum of Clinical Involvement

There is a broad range of clinical involvement in fragile X syndrome, from those with severe mental retardation and autism to those who have a normal IQ. Individuals with normal or near-normal IQs generally have learning disabilities and psychiatric and/or emotional problems [2]. Approximately 70% of females with fragile X syndrome do not have mental retardation (i.e., IQ > 70) [9, 10]; however, these individuals frequently present with learning disabilities. Even with a normal IQ, these girls commonly experience difficulties with math in school or suffer language deficits in addition to shyness and social anxiety [2, 11, 12]. The anxiety can be severe, resulting in selective mutism in certain environments, such as school [13].

Approximately 15% of boys with fragile X syndrome and full mutation alleles have an IQ within the normal to borderline range; however, these individuals typically are learning disabled, particularly with math and language deficits [10, 14]. Attention deficit hyperactivity disorder (ADHD) is seen in the majority of boys with fragile X syndrome, including those who do not have mental retardation [15, 16]. Males who are mosaic for *FMR1* allele size (i.e., with both full mutation and premutation alleles, the latter with or without hypermethylation) are more likely to present with learning disabilities without mental retardation [2, 14, 17]. Loesch et al. [18] found that the overall level of cognitive involvement correlates strongly with the level of FMRP, with individuals having higher IQs also tending to have higher FMRP levels as measured by the immunocytochemical method of Willemsen et al. [19].

The physical features of FXS often involve prominent ears with cupping of the superior aspect of the pinna, a long or narrow face, a high arched palate, hyperextensible finger joints, double jointed thumbs, and flat feet with significant pronation [2]. The level of involvement from a physical perspective also relates to FMRP levels; individuals with higher levels of FMRP are less likely to demonstrate the characteristic physical features associated with fragile X syndrome [18, 20]. Many of the physical features appear to be associated with connective tissue dysplasia, which is likely to be related to connective tissue genes that are dysregulated by the absence or deficiency of FMRP. The skin of individuals with FXS is typically soft and velvetlike, particularly on the dorsum of the hands.

Young children with fragile X syndrome usually do not have a long face, and approximately 30% do not have prominent ears; therefore, the behavioral features associated with fragile X syndrome are most helpful diagnostically in the younger age range [2, 17]. Such features include ADHD, impulsivity, dysinhibition, mood lability often leading to tantrum behavior, and anxiety. On the positive side, individuals with fragile X typically have a good sense of humor, an interest in social interactions, and a good memory for trivia that is of interest to them [2].

Approximately 30% of children with fragile X syndrome have autism [21–23]. The presence of autism is associated with a lower IQ and more severe language deficits than found in individuals who have fragile X syndrome without autism [21–24]. It is important to recognize the presence of autism, because these children require early, intensive intervention appropriate for autism for a better clinical outcome [2, 25]. The language features associated with fragile X syndrome include perseveration (repetition of common phrases such as "get out of here"), occasional echolalia, and sometimes a lack of language development [26]. Approximately 10% of children with fragile X syndrome are nonverbal; these individuals are typically the ones who have autism [2, 24]. Other children with fragile X syndrome may be very verbal, with speech that is both cluttered and repetitive. They may often ask the same question numerous times, even after an appropriate answer is given.

Finally, the diagnosis of fragile X syndrome is usually precipitated at 2–3 years of age by delays in normal language milestones. However, additional behavioral features, such as poor eye contact, hand flapping, hand biting, tactile sensitivity, and sensory integration problems, are common; when noticed, they should stimulate testing for fragile X syndrome by the clinician. All individuals with mental retardation or autism of unknown etiology should be tested for fragile X syndrome by *FMR1* DNA testing.

A variety of medications have been helpful for patients with fragile X syndrome, and are now a part of the standard treatment regimen. Such medications include stimulants for treating ADHD, selective serotonin reuptake inhibitors (SSRIs) for treating anxiety and obsessive–compulsive behavior, and atypical antipsychotics such as respiridone and aripiprazole for treating mood instability, aggression, and ADHD [27]. Other medications that have also been helpful are clonidine and guanfacine for anxiety and ADHD, melatonin for sleep disturbances, and anticonvulsants such as valproic acid and lamotrigine for seizures and mood instability [27].

B. Clinical Involvement in Premutation Carriers

Over the past two decades, there have been a number of reports regarding clinical involvement in premutation carriers, initially in females and more recently in males. The earlier reports commented on both emotional difficulties and subtle physical features, which occur in approximately 25% of carriers [12, 28–31]. More recently, studies of neuropsychological deficits in premutation carriers have documented executive function difficulties and memory problems in addition to features of dysinhibition [18, 32, 33]. Neuroimaging studies in premutation males, by Moore et al. [34], demonstrated a significant reduction in the volume of prefrontal white matter tracts, and reductions of gray and white matter volumes of the cerebellum, amygdala, hippocampal complex, and thalamus. Murphy et al. [35] found similar structural and metabolic changes in females in the same brain regions that were reported as abnormal in male premutation carriers. Most recently, Cornish et al. [36] have demonstrated deficits in social cognition in male premutation carriers from adolescence through adulthood. These findings are similar to previous reports by Loesch et al. [28] and Dorn et al. [37].

More severe clinical involvement in children with premutation alleles manifests as severe learning disabilities, often with mental retardation and/or autism [6, 7, 38, 39]. The more severe involvement in a minority of premutation carriers is often associated with both a deficit of FMRP and some features of fragile X syndrome, such as hand flapping, hand biting, hyperactivity, mood instability, and anxiety. These observations stimulated the studies of *FMR1* gene expression in the premutation range that led to the finding of elevated *FMR1* mRNA in premutation carriers [40].

In the late 1990s, we became increasingly aware of older males with the premutation who were presenting with tremor and ataxia. The early observations were initially presented at the National Fragile X Foundation Conference in 2000, and were subsequently published in 2001 [4]. As the number of older males with the premutation who were evaluated for neurological problems increased, we designated this "fragile X-associated tremor/ataxia syndrome" (FXTAS) [41]. FXTAS is the subject of the next section of this chapter.

III. FRAGILE X-ASSOCIATED TREMOR/ATAXIA SYNDROME

A. Overview

As noted earlier, carriers of premutation alleles can experience various forms of clinical involvement. Some children may display both physical (e.g., hyperextensible finger joints, prominent ears) and behavioral/emotional features that are within the spectrum of involvement of fragile X syndrome [12, 28, 30, 31, 42]. Because this form of clinical involvement generally affects individuals with larger premutation alleles [43], the underlying mechanism is likely to be the same as that of fragile X syndrome, namely, lowered levels of FMRP, which are observed for larger alleles within the premutation range [40, 44, 45]. Two other forms of clinical involvement associated with premutation alleles, premature ovarian failure [3, 46–49] and FXTAS [4, 5, 50–54], both appear to be unique to the premutation range and, therefore, are likely to be caused by a different mechanism than the neurodevelopmental disorder on the fragile X spectrum.

The recently discovered neurodegenerative disorder FXTAS appears to affect primarily older (>50) male carriers of premutation alleles [4, 5, 50–54], although some female carriers are clearly affected by FXTAS [51, 54]. The major clinical features of FXTAS include gait ataxia, progressive intention tremor, parkinsonism, and peripheral neuropathy [55–57]. Roughly 60% of premutation carriers who have clinical features of FXTAS also display characteristic (symmetric) increased intensity of the middle cerebellar peduncles and deep white matter on T2-weighted magnetic resonance images [50, 53]. In addition, analysis of postmortem brain tissue from several premutation carriers has revealed ubiquitin-positive inclusions in both neurons and astrocytes throughout the brain and brainstem [58a, 58b].

B. Clinical Features

The most prominent findings among adult premutation carriers with FXTAS are gait ataxia and intention tremor; both are present in 70–90% of cases [4, 5, 50–54] (Table 10-1). More variable symptoms include parkinsonism, numbness/pain in the lower extremities, and various forms of autonomic dysfunction (impotence, urinary and/or bowel incontinence). Gait ataxia most often begins with problems with balance, particularly with tandem gait. Walking difficulties generally progress to the point where successive use of a cane, walker, and/or wheelchair is required. The tremor is also progressive and is apparent with purposeful actions (e.g., writing, pouring water); however, a resting component is sometimes present as well. The tremor usually involves both upper extremities, and is often first noticed in the dominant hand. Cognitive problems are generally characterized by impairments of working memory, with relative sparing of language capabilities. Psychiatric symptoms are also common, and include anxiety, hostility, irritability, and apathy [59].

The progression of clinical involvement in FXTAS is highly variable, both in the age at onset of neurological symptoms and in the rate of progression of the neurological and cognitive deficits. Most individuals do not experience symptoms before the age of 50 (average age of onset, ~60), although a few have reported earlier onset. Onset may occur in the seventies or eighties, and some premutation carriers who are in their nineties exhibit no symptoms of FXTAS. Some individuals with FXTAS remain stable (i.e., without disease progression) for many years, whereas others experience a rapid downhill course (5–6 years). The reasons for the broad variability in penetrance, age at onset, rate of progression, severity of symptoms, and specific clinical manifestations are not understood at present. The influence of FXTAS on life expectancy has not yet been studied in a systematic fashion.

C. Epidemiology

Screening for tremor and ataxia among adult premutation carriers over 50 years of age, who were ascertained through known fragile X families (i.e., at least one child affected with fragile X syndrome), has revealed that at least one-third of such carriers have both ataxia and tremor [54]; this figure rises to at least one-half for male carriers older than 70. If a carrier frequency of ~1/800 is used for males in the general population [60], an estimated one in 3000 males in the general population has a lifetime risk of developing FXTAS. However, this estimate for prevalence may fall if the risk of developing FXTAS increases with increasing number of CGG repeats, as carriers ascertained through children or grandchildren would tend to have larger repeats, due to the increasing probability of transmission of larger full mutation alleles [61]. The number of female premutation carriers with features of FXTAS is not known, although it does occur in a small number of cases [56, 57, 62, 63]. The reduced numbers of female carriers with FXTAS may reflect the effects of random X inactivation; those who have a substantial fraction of their normal *FMR1* allele as the active allele do not suffer from the neurodegenerative disorder. Additional epidemiological studies are needed to resolve this issue. In this regard, although fragile X syndrome has been identified in nearly all of the major populations throughout the world, similar studies have not yet been performed for FXTAS.

To better assess the prevalence of FXTAS within the general population, screening studies for *FMR1* premutation alleles have been performed on patient groups, who are seen in adult neurology clinics, for features that overlap with FXTAS (e.g., essential tremor, ataxia, parkinsonism, multiple system atrophy). In more than a dozen studies reported to date (Table 10-2), several patterns are beginning to emerge. First, screens within ataxia populations over 50 years of age have identified approximately 2–5% as carriers of premutation alleles; the screen of 269 males in the German study represents an interesting exception [63]. Screens of Parkinson's disease

TABLE 10-1 Diagnostic Criteria for FXTAS[a,b]

Definite FXTAS	Clinical and Radiological or Neuropathological	Intention tremor or gait ataxia MCP sign Intranuclear inclusions
Probable FXTAS	Clinical or Radiological and Clinical (minor)	Intention tremor and gait ataxia MCP sign Parkinsonism
Possible FXTAS	Clinical and Radiological (minor)	Intention tremor or gait ataxia Cerebral white matter lesions; atrophy

[a] Mandatory criterion: number of CGG repeats between 55 and 200.
[b] Reference [55].

TABLE 10-2 Screening Studies of Various Neurological Populations for the Presence of Expanded (Premutation) *FMR1* Alleles

Screened population[a]	Sample size	Gender	Number (%) with premutation alleles[b]	Number (%) with gray zone alleles[c]	Cases all over 50 years	Study
Ataxia	59	Males	2 (3.4)	1 (1.7)	Yes	MacPherson et al., 2003 [74]
Ataxia	122	Males	5 (4.1)	0	Yes	Van Esch et al., 2005 [75]
Ataxia	131	Females	0	?	Yes	Van Esch et al., 2005 [76]
Ataxia	275	Males	6 (2.2; 4.2 for >50 years)	0	No	Brussino et al., 2005 [77]
Ataxia	27	Males, females	1 (3.7)	0	No	Biancalana et al., 2005 [78]
Ataxia	55	Males, females	0	0	No	Tan et al., 2004 [79]
Ataxia	269	Males	0	6 (2.2)	Yes	Zuhlke et al., 2004 [63]
Ataxia	241	Females	1 (0.41)	2 (0.83)	Yes	Zuhlke et al., 2004 [63]
Ataxia	51	Males	2 (4; 7.1 for >50 years)	0	No	Di Maria et al., 2003 [80]
Ataxia	167	Males	1 (0.6)	0	Yes	Milunsky and Maher, 2004 [81]
ET	81	Males	0	0	Yes	Garcia Arocena et al., 2004 [82]
ET	196	Males	0	3 (1.5)	No	Deng et al., 2004 [83]
ET	114	Males	0	0	No	Di Maria et al., 2003 [80]
ET	71	Males, females	0	0	No	Tan et al., 2004 [79]
Idiopathic PD	216	Males	0	4 (1.9)	No	Deng et al., 2004 [83]
Parkinsonism/PD	25/389	Males	0/0	<5	No	Toft and Farrer, 2005 [84]
Parkinsonism	265	Males	1 (0.4; 1 for >50 years)	5 (1.9; 4.7 for >50 years)	No	Hedrich et al., 2005 [85]
Parkinsonism	208	Females	1 (0.5; 1)	9 (4.3; 11)	No	Hedrich et al., 2005 [85]
Parkinsonism	26	Males, females	0	0	No	Tan et al., 2004 [79]
MSA	65	Males, females	0	3 (4.6)	Yes	Garland et al., 2004 [86]
MSA/OPCA	77/19	Males, females	1 (1.3)/0	4 (5)/0	No	Biancalana et al., 2005 [78]
MSA	15	Males, females	0	0	No	Tan et al., 2004 [79]
Probable MSA/MSA-C	223/76	Males, females	4 (1.8)/3 (4.0)	14 (6.3)/5 (6.6)	?	Kamm et al., 2005 [87]
Basal ganglia–cerebellar disease[d]	93	Males	1 (1.1; 1.7 for >50 yr)	0	No	Seixas et al., 2005 [88]
Basal ganglia–cerebellar disease[d]	140	Females	0	0	No	Seixas et al., 2005 [88]

[a] Principal form of clinical involvement.
[b] Premutation range, 55–200 CGG repeats.
[c] Gray zone range, 45–54 CGG repeats.
[d] Predominantly parkinsonism cases excluded.

or multiple system atrophy cases have not identified a significant number of premutation carriers. The absence of cases among those with Parkinson's disease or parkinsonism is interesting in view of the fact that about one-quarter of FXTAS cases were initially diagnosed with disorders with parkinsonism as a prominent component (e.g., idiopathic Parkinson's disease, multiple system atrophy) [64]. However, many of the screens of groups with parkinsonism and multiple system atrophy include substantial numbers of individuals who are younger than 50; thus, the low detection rates may be at least partially confounded by the younger age ranges.

Second, for those screening studies in which premutation alleles were detected (mostly among ataxia patients), most alleles possessed 80 to 100 CGG repeats, contrary to the general population, in which most premutation alleles have fewer than 80 repeats. The latter observation does argue for a size bias, and supports the aforementioned speculation that the true prevalence of FXTAS may be less than the 1/3000 figure.

D. Neuropathology

Postmortem examination of the brains of individuals who were clinically involved with FXTAS generally reveals variable degrees of global brain atrophy and both cerebellar and subcortical cerebral white matter disease [58a, 58b]. The spongiosis of the deep cerebellar white matter, which extends into the middle cerebellar peduncles (MCPs), is the neuropathological basis for the increased signal intensities observed in that region with T2-weighted MRI [5, 50]. This MRI finding ("MCP sign"), found in approximately 60% of carriers with tremor and/or ataxia, is currently used as a criterion for establishing a diagnosis of "definite" FXTAS (Table 10-1) [55]. Other findings include substantial Purkinje cell dropout with Bergmann gliosis [58a, 58b].

The cardinal feature of FXTAS is the presence of ubiquitin-positive, eosinophilic, intranuclear inclusions that are present in both neurons and astrocytes throughout the brain and brainstem [58a, 58b] in all FXTAS cases analyzed to date (~12 cases). The percentage of inclusion-bearing neural cells is highest in the hippocampus (~40–50% of neuronal nuclei), with lower counts (~5–10%) in the cerebral cortex. The inclusions themselves are solitary and spherical in shape (2–5 μm), and are PAS, tau, silver, polyglutamine, and synuclein negative [58; 58b]. These results demonstrate that the inclusions found in FXTAS brain tissue are distinct from the glial cytoplasmic inclusions associated with multiple system atrophy, from the intranuclear inclusions of the CAG repeat (polyglutamine) disorders, and from the cytoplasmic Lewy bodies found in the Lewy body dementias and Parkinson's disease. The presence of inclusions is currently used as a postmortem diagnostic criterion for FXTAS (Table 10-1).

The inclusions found in FXTAS cases have been recapitulated in a knockin mouse model in which the expanded (~100 CGG) repeat has been placed in the context of the mouse *Fmr1* gene [65]. However, unlike the situation in FXTAS, where numerous inclusions are found within astrocytic nuclei, no inclusions are detected in astrocytes in the mouse. This difference may explain the absence of either significant neurodegeneration or neurological dysfunction (at least thus far) in the mouse model, particularly if the pathogenesis in FXTAS involves astrocytic as well as neuronal dysfunction.

It should be noted that the neuropathology of FXTAS is quite different from that observed with fragile X syndrome: there is moderate to severe loss of brain volume with FXTAS, in contrast to the overall increase in brain volume observed in fragile X syndrome; no intranuclear inclusions have been detected in the brains of older adult males with fragile X syndrome [66].

E. Molecular Pathogenesis

Although caused by the same (*FMR1*) gene, fragile X syndrome and FXTAS are caused by different molecular mechanisms. The two disorders affect different groups of individuals (full mutation versus premutation alleles) in different age ranges (childhood versus adult onset). Fragile X syndrome is clearly a protein deficiency syndrome, with the absence of FMRP due to silencing of full mutation alleles of the gene. However, in the premutation range, the *FMR1* gene is not methylated, and is transcriptionally fully active. On the basis of this distinction, we proposed that FXTAS is likely to be caused by an RNA "toxic" gain of function [4, 50, 55, 58].

There are at least four arguments that support this RNA gain-of-function model for FXTAS. First, the neurodegenerative disorder is largely, perhaps exclusively, confined to carriers of premutation alleles, never having been reported in older adults with full mutation alleles. The absence of FXTAS among adults with full mutation alleles rules out mechanisms of disease pathogenesis that involve either the absence of FMRP or large CGG repeat expansions acting at the DNA level. Thus, FXTAS requires transcriptional activity of the *FMR1* gene. Second, it is now well established that expression of the *FMR1* gene is dysregulated in the premutation range, with elevated transcription [40, 44] and altered sites of transcriptional initiation [67]. Third, a *Drosophila melanogaster* model system for the neuropathology of FXTAS, which expresses an expanded (~90 CGG) repeat upstream of a reporter gene, results in both neurodegeneration (within the eye) and the presence of inclusions [68]. In the fly model, the neuropathology is present even when the expanded CGG repeat is transcribed upstream of a heterologous reporter. Therefore, the expanded repeat RNA is capable of inducing at least two features of the neuropathology of FXTAS. There are, however, important differences between the inclusions of FXTAS and those of the fly model. In particular, the distribution and physical properties of the inclusions are different between fly and human. In the fly, there are substantial numbers of cytoplasmic inclusions; however, no cytoplasmic inclusions have been detected in human cases of FXTAS. Finally, *FMR1* mRNA has now been detected within the inclusions [69]. This fourth observation provides an important parallel with the RNA gain-of-function model in myotonic dystrophy, where the intranuclear foci also contain the expanded (CUG or CCUG) repeat RNAs [70–72].

A protocol has recently been developed for the isolation of microgram quantities of purified inclusions from postmortem human brain tissue of FXTAS cases using automated sorting of immunofluorescence-tagged inclusions [73]. Mass spectrometric analysis of the proteins constituting the inclusions has identified more than 30 proteins [73]. A number of these proteins are of potential interest to the pathogenesis of FXTAS, including heat shock proteins (Hsp27, αB-crystallin); two RNA-binding proteins, heterogeneous nuclear ribonuclear protein A2 (hnRNP A2) and the protein homolog of *Drosophila* muscleblind (MBNL1); and several neurofilament proteins. Of particular interest is the finding of the intermediate filament protein lamin A/C (OMIM *150330) within the inclusions, as mutations in the *LMNA* gene (lamin A/C) are responsible for an autosomal recessive form of Charcot–Marie–Tooth disease of the axonal type (designated CMT2B1; OMIM 605588), and peripheral neuropathy is a common feature among FXTAS patients.

IV. CONCLUDING REMARKS

Over the past 5 years, our understanding of both the clinical manifestations and the molecular pathogenic consequences of expanded (CGG repeat) alleles of the *FMR1* gene have advanced substantially. A model has now been advanced to position FMRP in the pathway of glutamate-regulated synaptic growth, stability, and activity. Furthermore, there is a much greater understanding of the proteins that directly interact with FMRP and those that control *FMR1* gene expression at the transcriptional level. During this same period, a new, mechanistically distinct disorder (FXTAS) has been discovered that specifically affects older carriers of premutation expansions, and a "toxic" RNA model has been advanced to explain its pathogenesis. Based on the current pace of scientific discovery in the fragile X field, it is not unreasonable to expect that targeted therapies will be developed for both of these disorders within the next 5 years.

Acknowledgments

The authors acknowledge the input by, and many helpful discussions with, their collaborators. This work was supported in part by grants from the National Institutes of Health (HD36071, R.J.H.; HD40661 and NS43532, P.J.H.), by the UC Davis M.I.N.D. Institute (R.J.H., P.J.H.), and through support of the Boory Family Fund (P.J.H.).

References

1. Tassone, F., Hagerman, R. J., Iklé, D. N., Dyer, P. N., Lampe, M., Willemsen, R., Oostra, B. A., and Taylor, A. K. (1999). FMRP expression as a potential prognostic indicator in fragile X syndrome. *Am. J. Med. Genet.* **84**, 250–261.
2. Hagerman, R. J. (2002). Physical and behavioral phenotype. In "Fragile X Syndrome: Diagnosis, Treatment and Research" (R. J. Hagerman and P. J. Hagerman, Eds.), 3rd ed., pp. 3–109. Johns Hopkins Univ. Press, Baltimore.
3. Allingham-Hawkins, D. J., Babul-Hirji, R., Chitayat, D., Holden, J. J., Yang, K. T., Lee, C., Hudson, R., Gorwill, H., Nolin, S. L., Glicksman, A., Jenkins, E. C., Brown, W. T., Howard-Peebles, P. N., Becchi, C., Cummings, E., Fallon, L., Seitz, S., Black, S. H., Vianna-Morgante, A. M., Costa, S. S., Otto, P. A., Mingroni-Netto, R. C., Murray, A., Webb, J., Vieri, F., et al. (1999). Fragile X premutation is a significant risk factor for premature ovarian failure: The International Collaborative POF in Fragile X Study—preliminary data. *Am. J. Med. Genet.* **83**, 322–325.
4. Hagerman, R. J., Leehey, M., Heinrichs, W., Tassone, F., Wilson, R., Hills, J., Grigsby, J., Gage, B., and Hagerman, P. J. (2001). Intention tremor, parkinsonism, and generalized brain atrophy in male carriers of fragile X. *Neurology* **57**, 127–130.
5. Jacquemont, S., Hagerman, R. J., Leehey, M., Grigsby, J., Zhang, L., Brunberg, J. A., Greco, C., Des Portes, V., Jardini, T., Levine, R., Berry-Kravis, E., Brown, W. T., Schaeffer, S., Kissel, J., Tassone, F., and Hagerman, P. J. (2003). Fragile X premutation tremor/ataxia syndrome: Molecular, clinical, and neuroimaging correlates. *Am. J. Hum. Genet.* **72**, 869–878.
6. Tassone, F., Hagerman, R. J., Taylor, A. K., Mills, J. B., Harris, S. W., Gane, L. W., and Hagerman, P. J. (2000). Clinical involvement and protein expression in individuals with the *FMR1* premutation. *Am. J. Med. Genet.* **91**, 144–152.
7. Goodlin-Jones, B., Tassone, F., Gane, L. W., and Hagerman, R. J. (2004). Autistic spectrum disorder and the fragile X premutation. *J. Dev. Behav. Pediatr.* **25**, 392–398.
8. Hagerman, R. J. (2005). Lessons from fragile X regarding neurobiology, autism, and neurodegeneration. *J. Dev. Behav. Pediatr.*, in press.
9. de Vries, B. B., Wiegers, A. M., Smits, A. P., Mohkamsing, S., Duivenvoorden, H. J., Fryns, J. P., Curfs, L. M., Halley, D. J., Oostra, B. A., van den Ouweland, A. M., and Niermeijer, M. F. (1996). Mental status of females with an FMR1 gene full mutation. *Am. J. Med. Genet.* **58**, 1025–1032.
10. Bennetto, L., and Pennington, B. F. (2002). Neuropsychology. In "Fragile X Syndrome: Diagnosis, Treatment and Research" (R. J. Hagerman and P. J. Hagerman, Eds.), 3rd ed., pp. 206–298. Johns Hopkins Univ. Press, Baltimore.
11. Freund, L. S., Reiss, A. L., and Abrams, M. T. (1993). Psychiatric disorders associated with fragile X in the young female. *Pediatrics* **91**, 321–329.
12. Sobesky, W. E., Taylor, A. K., Pennington, B. F., Bennetto, L., Porter, D., Riddle, J., and Hagerman, R. J. (1996). Molecular–clinical correlations in females with fragile X. *Am. J. Med. Genet.* **64**, 340–345.
13. Hagerman, R. J., Hills, J., Scharfenaker, S., and Lewis, H. (1999). Fragile X syndrome and selective mutism. *Am. J. Med. Genet.* **83**, 313–317.
14. Hagerman, R. J., Hull, C. E., Safanda, J. F., Carpenter, I., Staley, L. W., RA, O. C., Seydel, C., Mazzocco, M. M., Snow, K., Thibodeau, S. N., et al. (1994). High functioning fragile X males: Demonstration of an unmethylated fully expanded FMR-1 mutation associated with protein expression. *Am. J. Med. Genet.* **51**, 298–308.
15. Munir, F., Cornish, K. M., and Wilding, J. (2000). A neuropsychological profile of attention deficits in young males with fragile X syndrome. *Neuropsychologia* **38**, 1261–1270.
16. Cornish, K. M., Turk, J., Wilding, J., Sudhalter, V., Munir, F., Kooy, F., and Hagerman, R. (2004). Annotation: Deconstructing the attention deficit in fragile X syndrome: A developmental neuropsychological approach. *J. Child. Psychol. Psychiatry* **45**, 1042–1053.
17. Merenstein, S. A., Sobesky, W. E., Taylor, A. K., Riddle, J. E., Tran, H. X., and Hagerman, R. J. (1996). Molecular–clinical correlations in males with an expanded FMR1 mutation. *Am. J. Med. Genet.* **64**, 388–394.

18. Loesch, D. Z., Huggins, R. M., and Hagerman, R. J. (2004). Phenotypic variation and FMRP levels in fragile X. *Ment. Retard. Dev. Disabil. Res. Rev.* **10**, 31–41.
19. Willemsen, R., Smits, A., Mohkamsing, S., van Beerendonk, H., de Haan, A., de Vries, B., van den Ouweland, A., Sistermans, E., Galjaard, H., and Oostra, B. A. (1997). Rapid antibody test for diagnosing fragile X syndrome: A validation of the technique. *Hum. Genet.* **99**, 308–311.
20. Loesch, D. Z., Huggins, R. M., Bui, Q. M., Taylor, A. K., and Hagerman, R. J. (2003). Relationship of deficits of FMR1 gene specific protein with physical phenotype of fragile X males and females in pedigrees: A new perspective. *Am. J. Med. Genet.* **118A**, 127–134.
21. Bailey, D. B., Jr., Hatton, D. D., Skinner, M., and Mesibov, G. B. (2001). Autistic behavior, FMR1 protein, and developmental trajectories in young males with fragile X syndrome. *J. Autism Dev. Disord.* **31**, 165–174.
22. Rogers, S. J., Wehner, E. A., and Hagerman, R. J. (2001). The behavioral phenotype in fragile X: Symptoms of autism in very young children with fragile X syndrome, idiopathic autism, and other developmental disorders. *J. Dev. Behav. Pediatr.* **22**, 409–417.
23. Kaufmann, W. E., Cortell, R., Kau, A. S., Bukelis, I., Tierney, E., Gray, R. M., Cox, C., Capone, G. T., and Stanard, P. (2004). Autism spectrum disorder in fragile X syndrome: Communication, social interaction, and specific behaviors. *Am. J. Med. Genet.* **129A**, 225–234.
24. Philofsky, A., Hepburn, S. L., Hayes, A., Hagerman, R. J., and Rogers, S. J. (2004). Linguistic and cognitive functioning and autism symptoms in young children with fragile X syndrome. *Am. J. Ment. Retard.* **109**, 208–218.
25. Braden, M. (2002). Academic interventions in fragile X. In "Fragile X Syndrome: Diagnosis, Treatment and Research" (R. J. Hagerman and P. J. Hagerman, Eds.), 3rd ed., pp. 428–464. Johns Hopkins Univ. Press, Baltimore.
26. Sudhalter, V., Scarborough, H. S., and Cohen, I. L. (1991). Syntactic delay and pragmatic deviance in the language of fragile X males. *Am. J. Med. Genet.* **38**, 493–497.
27. Hagerman, R. J. (2002). Medical follow-up and pharmacotherapy. In "Fragile X Syndrome: Diagnosis, Treatment and Research" (R. J. Hagerman and P. J. Hagerman, Eds.), 3rd ed., pp. 287–338. Johns Hopkins Univ. Press, Baltimore.
28. Loesch, D. Z., Hay, D. A., and Mulley, J. (1994). Transmitting males and carrier females in fragile X—revisited. *Am. J. Med. Genet.* **51**, 392–399.
29. Franke, P., Maier, W., Hautzinger, M., Weiffenbach, O., Gansicke, M., Iwers, B., Poustka, F., Schwab, S. G., and Froster, U. (1996). Fragile-X carrier females: Evidence for a distinct psychopathological phenotype? *Am. J. Med. Genet.* **64**, 334–339.
30. Franke, P., Leboyer, M., Gansicke, M., Weiffenbach, O., Biancalana, V., Cornillet-Lefebre, P., Croquette, M. F., Froster, U., Schwab, S. G., Poustka, F., Hautzinger, M., and Maier, W. (1998). Genotype–phenotype relationship in female carriers of the premutation and full mutation of FMR-1. *Psychiatry Res.* **80**, 113–127.
31. Riddle, J. E., Cheema, A., Sobesky, W. E., Gardner, S. C., Taylor, A. K., Pennington, B. F., and Hagerman, R. J. (1998). Phenotypic involvement in females with the FMR1 gene mutation. *Am. J. Ment. Retard.* **102**, 590–601.
32. Loesch, D. Z., Bui, M. Q., Grigsby, J., Butler, E., Epstein, J., Huggins, R. M., Taylor, A., and Hagerman, R. J. (2003). Effect of the fragile X status categories and the FMRP levels on executive functioning in fragile X males and females. *Neuropsychology* **17**, 646–657.
33. Moore, C. J., Daly, E. M., Schmitz, N., Tassone, F., Tysoe, C., Hagerman, R. J., Hagerman, P. J., Morris, R. G., Murphy, K. C., and Murphy, D. G. (2004). A neuropsychological investigation of male premutation carriers of fragile X syndrome. *Neuropsychologia* **42**, 1934–1947.
34. Moore, C. J., Daly, E. M., Tassone, F., Tysoe, C., Schmitz, N., Ng, V., Chitnis, X., McGuire, P., Suckling, J., Davies, K. E., Hagerman, R. J., Hagerman, P. J., Murphy, K. C., and Murphy, D. G. (2004). The effect of pre-mutation of X chromosome CGG trinucleotide repeats on brain anatomy. *Brain* **127**, 2672–2681.
35. Murphy, D. G. M., Mentis, M. J., Pietrini, P., Grady, C. L., Moore, C. J., Horwitz, B., Hinton, V., Dobkin, C. S., Schapiro, M. B., and Rapoport, S. I. (1999). Premutation female carriers of fragile X syndrome: A pilot study on brain anatomy and metabolism. *J. Am. Acad. Child Adolesc. Psychiatry* **38**, 1294–1301.
36. Cornish, K. M., Kogan, C., Turk, J., Manly, T., James, N., Mills, A., and Dalton, A. (2005). The emerging fragile X premutation phenotype: Evidence from the domain of social cognition. *Brain Cogn.* **57**, 53–60.
37. Dorn, M. B., Mazzocco, M. M., and Hagerman, R. J. (1994). Behavioral and psychiatric disorders in adult male carriers of fragile X. *J. Am. Acad. Child Adolesc. Psychiatry* **33**, 256–264.
38. Hagerman, R. J., Staley, L. W., O'Connor, R., Lugenbeel, K., Nelson, D., McLean, S. D., and Taylor, A. (1996). Learning-disabled males with a fragile X CGG expansion in the upper premutation size range. *Pediatrics* **97**, 122–126.
39. Aziz, M., Stathopulu, E., Callias, M., Taylor, C., Turk, J., Oostra, B., Willemsen, R., and Patton, M. (2003). Clinical features of boys with fragile X premutations and intermediate alleles. *Am. J. Med. Genet. B* **121**, 119–127.
40. Tassone, F., Hagerman, R. J., Taylor, A. K., Gane, L. W., Godfrey, T. E., and Hagerman, P. J. (2000). Elevated levels of *FMR1* mRNA in carrier males: A new mechanism of involvement in fragile X syndrome. *Am. J. Hum. Genet.* **66**, 6–15.
41. Jacquemont, S., Hagerman, R. J., Leehey, M., Grigsby, J., Zhang, L., Brunberg, J. A., Greco, C., des Portes, V., Jardini, T., Levine, R., Berry-Kravis, E., Brown, W. T., Schaeffer, S., Kissel, J., Tassone, F., and Hagerman, P. J. (2003). Fragile X premutation tremor/ataxia syndrome: Molecular, clinical, and neuroimaging correlates. *Am. J. Hum. Genetics.* **72**, 869–878.
42. Hagerman, R. J., and Hagerman, P. J. (2002). The fragile X premutation: Into the phenotypic fold. *Curr. Opin. Genet. Dev.* **12**, 278–283.
43. Johnston, C., Eliez, S., Dyer-Friedman, J., Hessl, D., Glaser, B., Blasey, C., Taylor, A., and Reiss, A. (2001). Neurobehavioral phenotype in carriers of the fragile X premutation. *Am. J. Med. Genet.* **103**, 314–319.
44. Tassone, F., Hagerman, R. J., Chamberlain, W. D., and Hagerman, P. J. (2000). Transcription of the *FMR1* gene in individuals with fragile X syndrome. *Am. J. Med. Genet. (Semin. Med. Genet.)* **97**, 195–203.
45. Kenneson, A., Zhang, F., Hagedorn, C. H., and Warren, S. T. (2001). Reduced FMRP and increased *FMR1* transcription is proportionally associated with CGG repeat number in intermediate-length and premutation carriers. *Hum. Mol. Genet.* **10**, 1449–1454.
46. Cronister, A., Schreiner, R., Wittenberger, M., Amiri, K., Harris, K., and Hagerman, R. J. (1991). Heterozygous fragile X female: Historical, physical, cognitive, and cytogenetic features. *Am. J. Med. Genet.* **38**, 269–274.
47. Schwartz, C. E., Dean, J., Howard Peebles, P. N., Bugge, M., Mikkelsen, M., Tommerup, N., Hull, C., Hagerman, R., Holden, J. J., and Stevenson, R. E. (1994). Obstetrical and gynecological complications in fragile X carriers: A multicenter study. *Am. J. Med. Genet.* **51**, 400–402.
48. Murray, A., Webb, J., Grimley, S., Conway, G., and Jacobs, P. (1998). Studies of FRAXA and FRAXE in women with premature ovarian failure. *J. Med. Genet.* **35**, 637–640.
49. Marozzi, A., Vegetti, W., Manfredini, E., Tibiletti, M. G., Testa, G., Crosignani, P. G., Ginelli, E., Meneveri, R., and Dalpra, L. (2000). Association between idiopathic premature ovarian failure and fragile X premutation. *Hum. Reprod.* **15**, 197–202.

50. Brunberg, J. A., Jacquemont, S., Hagerman, R. J., Berry-Kravis, E. M., Grigsby, J., Leehey, M. A., Tassone, F., Brown, W. T., Greco, C. M., and Hagerman, P. J. (2002). Fragile X premutation carriers: Characteristic MR imaging findings of adult male patients with progressive cerebellar and cognitive dysfunction. *AJNR Am. J. Neuroradiol.* **23**, 1757–1766.
51. Berry-Kravis, E., Lewin, F., Wuu, J., Leehey, M., Hagerman, R., Hagerman, P., and Goetz, C. G. (2003). Tremor and ataxia in fragile X premutation carriers: Blinded videotape study. *Ann. Neurol.* **53**, 616–623.
52. Leehey, M. A., Munhoz, R. P., Lang, A. E., Brunberg, J. A., Grigsby, J., Greco, C., Jacquemont, S., Tassone, F., Lozano, A. M., Hagerman, P. J., and Hagerman, R. J. (2003). The fragile X premutation presenting as essential tremor. *Arch. Neurol.* **60**, 117–121.
53. Jacquemont, S., Farzin, F., Hall, D., Leehey, M., Tassone, F., Gane, L., Zhang, L., Grigsby, J., Jardini, T., Lewin, F., Berry-Kravis, E., Hagerman, P. J., and Hagerman, R. J. (2004). Aging in individuals with the FMR1 mutation. *Am. J. Ment. Retard.* **109**, 154–164.
54. Jacquemont, S., Hagerman, R. J., Leehey, M. A., Hall, D. A., Levine, R. A., Brunberg, J. A., Zhang, L., Jardini, T., Gane, L. W., Harris, S. W., Herman, K., Grigsby, J., Greco, C. M., Berry-Kravis, E., Tassone, F., and Hagerman, P. J. (2004). Penetrance of the fragile X-associated tremor/ataxia syndrome in a premutation carrier population. *JAMA* **291**, 460–469.
55. Hagerman, P. J., and Hagerman, R. J. (2004). The fragile-X premutation: A maturing perspective. *Am. J. Hum. Genet.* **74**, 805–816. [Epub 2004 Mar 2029.]
56. Berry-Kravis, E., Potanos, K., Weinberg, D., Zhou, L., and Goetz, C. G. (2005). Fragile X-associated tremor/ataxia syndrome in sisters related to X-inactivation. *Ann. Neurol.* **57**, 144–147.
57. Jacquemont, S., Orrico, A., Galli, L., Sahota, P. K., Brunberg, J. A., Anichini, C., Leehey, M., Schaeffer, S., Hagerman, R. J., Hagerman, P. J., and Tassone, F. (2005). Spastic paraparesis, cerebellar ataxia, and intention tremor: A severe variant of FXTAS? *J. Med. Genet.* **42**, e14.
58a. Greco, C. M., Hagerman, R. J., Tassone, F., Chudley, A. E., Del Bigio, M. R., Jacquemont, S., Leehey, M., and Hagerman, P. J. (2002). Neuronal intranuclear inclusions in a new cerebellar tremor/ataxia syndrome among fragile X carriers. *Brain* **125**, 1760–1771.
58b. Greco, C. M., Berman, R. F., Martin, R. M., Tassone, F., Schwortz, P. H., Chang, A., Trapp, B. D., Iwahashi, C., Brunberg, J., Grigsby, J., Hessi, D., Becker, E. J., Papazian, J., Leehey, M. A., Hagerman, R. J. and Hagerman, P. J. (2006). Neuropathology of fragile X–associated tremor/ataxia syndrome (FXTAS). *Brain* **129**, 243–255. Epub 2005 Dec 2005.
59. Bacalman, S., Farzin, F., Bourgeois, J., Cogswell, J., Goodlin-Jones, B., Gane, L. W., Grigsby, J., Leehey, M., Tassone, F., and Hagerman, R. J. (2005). Psychiatric phenotype of the fragile X–associated tremor/ataxia syndrome (FXTAS) in males: Newly described fronto–subcortical dementia. *J. Clin. Psychiatry*, **67**, 87–94.
60. Dombrowski, C., Levesque, S., Morel, M. L., Rouillard, P., Morgan, K., and Rousseau, F. (2002). Premutation and intermediate-size *FMR1* alleles in 10572 males from the general population: Loss of an AGG interruption is a late event in the generation of fragile X syndrome alleles. *Hum. Mol. Genet.* **11**, 371–378.
61. Nolin, S. L., Brown, W. T., Glicksman, A., Houck, G. E., Jr., Gargano, A. D., Sullivan, A., Biancalana, V., Brondum-Nielsen, K., Hjalgrim, H., Holinski-Feder, E., Kooy, F., Longshore, J., Macpherson, J., Mandel, J. L., Matthijs, G., Rousseau, F., Steinbach, P., Vaisanen, M. L., von Koskull, H., and Sherman, S. L. (2003). Expansion of the fragile X CGG repeat in females with premutation or intermediate alleles. *Am. J. Hum. Genet.* **72**, 454–464.

62. Hagerman, R. J., Leavitt, B. R., Farzin, F., Jacquemont, S., Greco, C. M., Brunberg, J. A., Tassone, F., Hessl, D., Harris, S. W., Zhang, L., Jardini, T., Gane, L. W., Ferranti, J., Ruiz, L., Leehey, M. A., Grigsby, J., and Hagerman, P. J. (2004). Fragile-X-associated tremor/ataxia syndrome (FXTAS) in females with the *FMR1* premutation. *Am. J. Hum. Genet.* **74**, 1051–1056. [Epub 2004 Apr 1052.]
63. Zuhlke, C., Budnik, A., Gehlken, U., Dalski, A., Purmann, S., Naumann, M., Schmidt, M., Burk, K., and Schwinger, E. (2004). *FMR1* premutation as a rare cause of later onset ataxia: Evidence for FXTAS in female carriers. *J. Neurol.* **251**, 1418–1419.
64. Hall, D. A., Berry-Kravis, E., Jacquemont, S., Rice, C. D., Cogswell, J., Zhang, L., Hagerman, R. J., Hagerman, P. J., and Leehey, M. A. (2005). Initial diagnoses given to persons with the fragile X associated tremor/ataxia syndrome (FXTAS). *Neurology* **65**, 299–301.
65. Willemsen, R., Hoogeveen-Westerveld, M., Reis, S., Holstege, J., Severijnen, L. A., Nieuwenhuizen, I. M., Schrier, M., Van Unen, L., Tassone, F., Hoogeveen, A. T., Hagerman, P. J., Mientjes, E. J., and Oostra, B. A. (2003). The *FMR1* CGG repeat mouse displays ubiquitin-positive intranuclear neuronal inclusions: Implications for the cerebellar tremor/ataxia syndrome. *Hum. Mol. Genet.* **12**, 949–959.
66. Sabaratnam, M. (2000). Pathological and neuropathological findings in two males with fragile X syndrome. *J. Intell. Disabil. Res.* **44**, 81–85.
67. Beilina, A., Tassone, F., Schwartz, P. H., Sahota, P., and Hagerman, P. J. (2004). Redistribution of transcription start sites within the *FMR1* promoter region with expansion of the downstream CGG-repeat element. *Hum. Mol. Genet.* **13**, 543–549. [Epub 2004 Jan 2013.]
68. Jin, P., Zarnescu, D. C., Zhang, F., Pearson, C. E., Lucchesi, J. C., Moses, K., and Warren, S. T. (2003). RNA-mediated neurodegeneration caused by the fragile X premutation rCGG repeats in *Drosophila*. *Neuron* **39**, 739–747.
69. Tassone, F., Iwahashi, C., and Hagerman, P. J. (2004). *FMR1* RNA within the intranuclear inclusions of fragile X-associated tremor/ataxia syndrome (FXTAS). *RNA Biol.* **1**, 103–105.
70. Mankodi, A., and Thornton, C. A. (2002). Myotonic syndromes. *Curr. Opin. Neurol.* **15**, 545–552.
71. Ranum, L. P., and Day, J. W. (2004). Myotonic dystrophy: RNA pathogenesis comes into focus. *Am. J. Hum. Genet.* **74**, 793–804. [Epub 2004 Apr 2002.]
72. Day, J. W., and Ranum, L. P. (2005). RNA pathogenesis of the myotonic dystrophies. *Neuromuscul. Disord.* **15**, 5–16. [Epub 2004 Nov 2026.]
73. Iwahashi, C. K., Yasui, D. H., An, H.-J., Greco, C. M., Tassone, F., Nannen, K., Babineau, B., Lebrilla, C. B., Hagerman, R. J., and Hagerman, P. J. (2005). Protein composition of the intranuclear inclusions of FXTAS. *Brain*, **129**, 256–271. [Epub 2005 Oct 2004.]
74. Macpherson, J., Waghorn, A., Hammans, S., and Jacobs, P. (2003). Observation of an excess of fragile-X premutations in a population of males referred with spinocerebellar ataxia. *Hum. Genet.* **112**, 619–620. [Epub 2003 Feb 2027.]
75. Van Esch, H., Dom, R., Bex, D., Salden, I., Caeckebeke, J., Wibail, A., Borghgraef, M., Legius, E., Fryns, J. P., and Matthijs, G. (2005). Screening for FMR-1 premutations in 122 older Flemish males presenting with ataxia. *Eur. J. Hum. Genet.* **13**, 121–123.
76. Van Esch, H., Matthijs, G., and Fryns, J. P. (2005). Should we screen for *FMR1* premutations in female subjects presenting with ataxia? *Ann. Neurol.* **57**, 932–933.
77. Brussino, A., Gellera, C., Saluto, A., Mariotti, C., Arduino, C., Castellotti, B., Camerlingo, M., de Angelis, V., Orsi, L., Tosca, P., Migone, N., Taroni, F., and Brusco, A. (2005). FMR1 gene premutation is a frequent genetic cause of late-onset sporadic cerebellar ataxia. *Neurology* **64**, 145–147.

78. Biancalana, V., Toft, M., Le Ber, I., Tison, F., Scherrer, E., Thibodeau, S., Mandel, J. L., Brice, A., Farrer, M. J., and Durr, A. (2005). FMR1 premutations associated with fragile X-associated tremor/ataxia syndrome in multiple system atrophy. *Arch. Neurol.* **62**, 962–966.
79. Tan, E. K., Zhao, Y., Puong, K. Y., Law, H. Y., Chan, L. L., Yew, K., Tan, C., Shen, H., Chandran, V. R., Teoh, M. L., Yih, Y., Pavanni, R., Wong, M. C., and Ng, I. S. (2004). Fragile X premutation alleles in SCA, ET, and parkinsonism in an Asian cohort. *Neurology* **63**, 362–363.
80. Di Maria, E., Grasso, M., Pigullo, S., Faravelli, F., Abbruzzese, G., Barone, P., Martinelli, P., Ratto, S., Sciolla, R., Bellone, E., Dagna-Bricarelli, F., Ajmar, F., and Mandich, P. (2003). Further evidence that a tremor/ataxia syndrome may occur in fragile X premutation carriers. Paper presented at: American Society of Human Genetics, 53rd Annual Meeting; Los Angeles, CA.
81. Milunsky, J. M., and Maher, T. A. (2004). Fragile X carrier screening and spinocerebellar ataxia in older males. *Am. J. Med. Genet. A* **125**, 320.
82. Garcia Arocena, D., Louis, E. D., Tassone, F., Gilliam, T. C., Ottman, R., Jacquemont, S., and Hagerman, P. J. (2004). Screen for expanded FMR1 alleles in patients with essential tremor. *Mov. Disord.* **19**, 930–933.
83. Deng, H., Le, W., and Jankovic, J. (2004). Premutation alleles associated with Parkinson disease and essential tremor. *JAMA* **292**, 1685–1686.
84. Toft, M., and Farrer, M. (2005). Premutation alleles and fragile X-associated tremor/ataxia syndrome. *JAMA* **293**, 296; author reply 296–297.
85. Hedrich, K., Pramstaller, P. P., Stubke, K., Hiller, A., Kabakci, K., Purmann, S., Kasten, M., Scaglione, C., Schwinger, E., Volkmann, J., Kostic, V., Vieregge, P., Martinelli, P., Abbruzzese, G., Klein, C., and Zuhlke, C. (2005). Premutations in the FMR1 gene as a modifying factor in Parkin-associated Parkinson's disease? *Mov. Disord.* **20**, 1060–1062.
86. Garland, E. M., Vnencak-Jones, C. L., Biaggioni, I., Davis, T. L., Montine, T. J., and Robertson, D. (2004). Fragile X gene premutation in multiple system atrophy. *J. Neurol. Sci.* **227**, 115–118.
87. Kamm, C., Healy, D. G., Quinn, N. P., Wullner, U., Moller, J. C., Schols, L., Geser, F., Burk, K., Borglum, A. D., Pellecchia, M. T., Tolosa, E., del Sorbo, F., Nilsson, C., Bandmann, O., Sharma, M., Mayer, P., Gasteiger, M., Haworth, A., Ozawa, T., Lees, A. J., Short, J., Giunti, P., Holinski-Feder, E., Illig, T., Wichmann, H. E., Wenning, G. K., Wood, N. W., and Gasser, T. (2005). The fragile X tremor ataxia syndrome in the differential diagnosis of multiple system atrophy: Data from the EMSA Study Group. *Brain* **128**, 1855–1860. [Epub 2005 Jun 1859.]
88. Seixas, A. I., Maurer, M. H., Lin, M., Callahan, C., Ahuja, A., Matsuura, T., Ross, C. A., Hisama, F. M., Silveira, I., and Margolis, R. L. (2005). FXTAS, SCA10, and SCA17 in American patients with movement disorders. *Am. J. Med. Genet. A* **136**, 87–89.

CHAPTER 11

Animal Models of Fragile X Syndrome: Mice and Flies

BEN A. OOSTRA AND DAVID L. NELSON

Department of Clinical Genetics, Erasmus MC, Rotterdam, The Netherlands; and Department of Molecular and Human Genetics, Baylor College of Medicine, Houston, Texas

I. Introduction
II. Mouse Models
   A. *Fmr 1* Knockout Mice
   B. Macroorchidism
   C. Neuroanatomy and Physiology of the Knockout Brain
   D. Structural Abnormalities
   E. LTP/LTD
   F. Behavior
   G. Environmental Effects
   H. Instability of the CGG Repeat in Mice
   I. Mouse Model for FXTAS
   J. Knockout Mice for *Fmr 1* Paralogs *Fxr 1* and *Fxr 2*
III. Flies
   A. Behavioral Phenotypes
   B. Neuronal Phenotypes
   C. Modifying Phenotypes with Genes and Drugs
   D. Biochemistry
   E. FXTAS and CGG Models
IV. Conclusion
   Acknowledgments
   References

Fragile X syndrome, one of the most common forms of inherited mental retardation, is caused by an expansion of a polymorphic CGG repeat upstream of the coding region in the *FMR1* gene. These expansions block expression of the *FMR1* gene due to methylation of the promoter. Functional studies on the FMR1 protein have shown that the protein can bind RNA and might be involved in transport of RNAs from the nucleus to the cytoplasm. A role for FMR1 protein in translation of certain mRNAs has been suggested. Available tissues from fragile X patients, such as blood cells, cultured skin fibroblasts, and (rarely) postmortem material, do not easily allow studies on the molecular pathogenesis of the disease. An animal model could help in understanding the effects of the lack of FMR1 protein expression or the effects of the expression of a mutated FMR1 protein (FMRP). Numerous animal models for fragile X syndrome have been developed in both mice and flies, and these exhibit features that mimic the human phenotype. This review focuses on insights into the pathogenesis of fragile X syndrome that have resulted from the various animal models that perturb the functions of *FMR1* and related genes.

I. INTRODUCTION

Fragile X syndrome is one of the most common forms of inherited mental retardation. The disorder is characterized by some physical anomalies, including elongated facial structure, large protruding ears, hyperextensible joints, and, in males specifically, macroorchidism. In addition, the phenotype is characterized by cognitive impairment ranging from mild to severe and by other variable neurobehavioral indications, including hyperarousal, attention deficit, anxiety, social withdrawal, and seizure susceptibility (for review see [1]). In virtually all cases, the genetic cause of the syndrome is an expansion of a polymorphic CGG repeat upstream of the coding region in the *FMR1 gene* [2, 3]. Expansion of the repeat leads to methylation [4] and deacetylation [5, 6] of the repeat and surrounding sequences, including the promoter of the *FMR1* gene, resulting in a lack of transcription [7] and loss of the encoded protein, FMRP.

FMRP is a ubiquitously expressed RNA-binding protein, and several common alternatively spliced isoforms are found in approximately equal abundance. FMRP has five known functional domains, including two RNA binding KH domains, an RGG box (also involved in RNA binding), a nuclear localization signal (NLS), and a nuclear export signal (NES), as well as two coiled coils involved in protein–protein interaction (Fig. 11-1). The protein aggregates with multiple RNAs and proteins to form a messenger ribonucleic protein (mRNP). The aggregate is found predominantly associated with the actively translating ribosomes in the cell. It was shown that FMRP can bind to RNAs containing a G-quartet structure, with the aid of its RGG box [8, 9]. Darnell et al. identified the RNA target for the KH2 domain as a sequence-specific element within a complex tertiary structure termed the *FMRP kissing complex* [10]. They demonstrated that the association of FMRP with brain polyribosomes is abrogated by competition with the FMRP kissing complex RNA, but determination of the relationship of RNA binding to FMRP function and mental retardation requires more studies. Although FMRP is predominantly a cytoplasmatic protein, it does shuttle between nucleus and cytoplasm, perhaps transporting mRNAs in a selective manner. FMR1 protein has been suggested to play a role in control of translation of certain mRNAs. Two closely related genes, *FXR1* and *FXR2* [11, 12], have been described in all vertebrates analyzed to date [13]. These paralogous proteins have been shown to interact with FMRP, and are frequently found to be coexpressed, suggesting the potential for compensation in the absence of FMR1 function.

Individuals with a premutation do not show the classic phenotype of the fragile X syndrome and were initially thought to be asymptomatic, although a number of studies have reported mild learning disabilities and social phobias or anxiety disorders in a small subgroup of premutation carriers (see Chapter 10 [14]). In addition, approximately 20% of female premutation carriers manifest premature ovarian failure [15]. Recent studies have reported male individuals with alleles in the premutation range with increased *FMR1* mRNA levels that are up to eightfold higher than normal and with (mildly) reduced FMRP levels [16–18]. The elevated *FMR1* transcript levels were positively correlated with the number of CGG repeats [19]. The question whether these elevated *FMR1* mRNA levels and slightly reduced FMRP levels result in a mild fragile X phenotype was challenged by the recent description of older males carrying a premutation

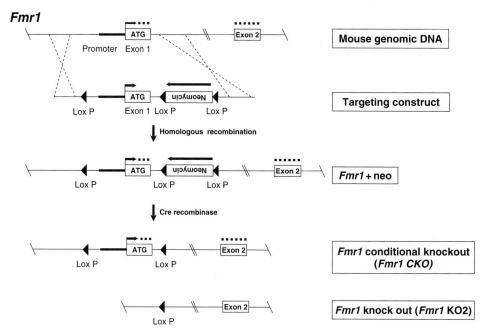

FIGURE 11-1 Schematic representation of the *Fmr1* alleles obtained after homologous recombination between the targeting construct and the mouse genomic DNA in ES cells.

(ranging between 71 and 135 CGGs) who exhibit a unique neurodegenerative syndrome characterized by progressive intention tremor and ataxia (fragile X-associated tremor/ataxia syndrome) [20–23].

Tissues available from fragile X patients, such as blood cells, cultured skin fibroblasts, and (rarely) postmortem material, do not easily allow studies on the molecular pathogenesis of the disease. Animal models have the potential to assist in understanding the effects of the lack of FMRP expression or the effects of the expression of a mutated FMR1 protein. Genes involved in development and differentiation are widely studied in the nematode worm *Caenorhabditis elegans* and the fruit fly *Drosophila melanogaster*. The development of new technologies, such as RNA interference, provides an opportunity to use additional nonmammalian vertebrates such as the frog, *Xenopus laevis*, and the zebra fish, *Danio rerio*, to study genes involved in early development of the vertebrate, despite the lack of methods allowing facile genome manipulation in these organisms.

Although lower vertebrates and also worms and flies can serve as good models, there are many reasons to use a mammalian system when specific details and aspects need to be studied. The laboratory mouse is, for practical and technical reasons, the mammal of choice for functional studies, and is arguably the model most likely to closely resemble human fragile X syndrome in the absence of *Fmr1*.

A murine model for fragile X syndrome has been generated, and these mice show differences in testis size and in some parts of the brain [24]. At the same time, these mice show some behavioral difficulties mimicking the human phenotype. The newest animal model of fragile X syndrome, the fruit fly *Drosophila*, has revealed several novel mechanistic insights into the disease. In this review we focus on fragile X syndrome with specific emphasis on the two different model organisms for this syndrome, the fruit fly *Drosophila melanogaster* and the mouse, and we review specifically what we have learned from the animal models. Much of our current understanding of the function(s) of FMR1 derives from analysis of mouse and fly models. These have been invaluable to the study of this common cause of mental retardation.

II. MOUSE MODELS

A. *Fmr1* Knockout Mice

Clues to the mechanisms that cause the abnormalities observed in fragile X syndrome are limited. To gain more insight into the pathological and physiological processes, a mouse model for the disease was generated [24]. The development of an animal model has major advantages. First, the unlimited supply of tissues provides an opportunity to study the effects of the lack of FMRP expression on the morphological and molecular levels. Second, the behavior of the knockout mice can be studied to understand the mechanisms involved in learning. Third, the knockout mice can be crossed with other transgenic mice carrying different mutations or with different expression patterns to study genetic interactions. The *FMR1* gene is highly conserved among species, and the murine homolog *Fmr1* shows 97% homology in amino acid sequence [25]. The expression patterns at the mRNA and protein levels are very similar in humans and mice [24, 26–29]. The function of the *FMR1* gene is therefore likely to be very similar in both species.

The original knockout mice (KO1) were generated by replacing the wild-type murine *Fmr1* gene with a nonfunctional *Fmr1* gene in which a neomycin resistance cassette was placed in exon 5, using homologous recombination in embryonic stem (ES) cells employing conventional transgenic ES technology. These ES cells were injected into blastocysts and transferred to pseudopregnant females. Highly chimeric males were crossed with wild-type C57Bl6 females to give birth to females heterozygous for the knockout mutation. Breeding those females with wild-type males resulted in knockout males. As a result of the integration of the neomycin cassette in the *Fmr1* gene, the mutant mice are no longer able to make normal *Fmr1* mRNA. Although the knockout mutation in the animal model is different from the mutation present in human fragile X patients, both mutations lead to an absence of the FMR1 protein in adults. Weight and light microscopic appearance of kidney, heart, spleen, liver, lung, or brain were not different between knockouts and normal littermates [24]. Other phenotypic characteristics, such as long face, prominent ears, high-arched palate, flat feet, hand calluses, and hyperextensible finger joints, have not been observed in fragile X mice [30]. No macroscopic or microscopic abnormalities could be detected in complete autopsies of knockout mice. The weights of several organs of mutant mice, except testis, did not significantly differ from those of control mice [31]. No significant statistical difference in body weight was found between age-matched groups of control and mutant mice. The absence of FMRP did not influence the reproduction or viability of the knockout mice. This is similar to the situation in humans, where human fragile X patients have a normal life span.

The original *Fmr1* knockout mouse (KO1) was limited by the requirement of studying the effect of the lack of Fmrp expression in the entire animal. To be able to dissect out the function of Fmrp further, regulation of its expression, both spatially and temporally, is needed. To create a new more versatile *Fmr1* in vivo knockout model, conditional *Fmr1* knockout mice (*Fmr1* CKO) were generated by flanking the promoter and first exon of *Fmr1* with bacteriophage P1-derived *lox* P sites [32]. A targeting construct containing the floxed mouse *Fmr1* promoter and exon 1, in addition to a floxed neomycin cassette in intron 1, was used to transfect ES cells to enable homologous recombination (Fig 11-2). An ES clone was picked and injected into blastocysts and an *Fmr1* + neo mouse line was obtained. The neomycin cassette still present in the ES cells was excised *in vitro* after transiently transfecting the ES clones with a cre-expressing plasmid. A mouse containing the floxed mouse *Fmr1* promoter and exon 1 (termed *Fmr1* CKO) was generated. The expression of Fmrp in *Fmr1* CKO mice is at the wild-type level [32]. The *Fmr1* KO2 line was obtained by crossing *Fmr1* CKO and cag–cre-expressing mice. The *Fmr1* KO2 line produces no detectable *Fmr1* mRNA or FMRP [32]. A limited number of experiments have been carried out with *Fmr1* KO2 mice, and thus far, no differences from *Fmr1* KO1 have been identified (unpublished results).

Male knockout mice do not express Fmrp in any of their cells; in contrast, in males with fragile X, the Fmr1 protein is absent in all cells except germ cells [33]. It was therefore originally suggested that FMRP might be essential for gametogenesis. The observation that both male and female knockout mice without any protein expression are fertile and have the same size progeny as controls indicates that Fmr1 is not necessary for spermatogenesis and oogenesis in mice.

B. Macroorchidism

One prominent phenotypic characteristic of fragile X patients is macroorchidism; sometimes this is manifested in childhood, but it is present in almost all fragile X patients after puberty [1]. Testicular weight was significantly higher in knockout mice than in normal littermates (Fig. 11-3). By 15 days of age, *FMR1* knockout mice have larger testes than their wild-type littermates [34]. In adult mice the differences become much more pronounced (Fig. 11-3). As in human males with fragile X, macroorchidism is present in >90% of adult knockout mice [29]. Microscopic examination of the testes of mutant mice revealed no structural differences when compared with controls, including a normal pattern of tubule size, a normal amount of interstitial mass, and normal spermatogenesis. The size of the testis is determined mainly by the number of Sertoli cells that support the proliferation and differentiation of the germ cells [35]. In the knockout mice, it appeared that the proliferative activity of the Sertoli cells was significantly higher, resulting in an increase in spermatogenic cell number and testicular weight. Follicle-stimulating hormone (FSH) plays an important, but not essential, role in Sertoli cell proliferation [36]. Sertoli cells are most sensitive to the mitogenic activity of FSH at the end of the fetal period and shortly after birth [37]. The circulating FSH level was measured in knockout mice, but not found to be elevated compared with that in wild-type littermates. This observation is similar to the situation in humans with fragile X, who show no evidence of an increased FSH level [38, 39]. Determination of the level of FSH receptor mRNA in the testis indicated a

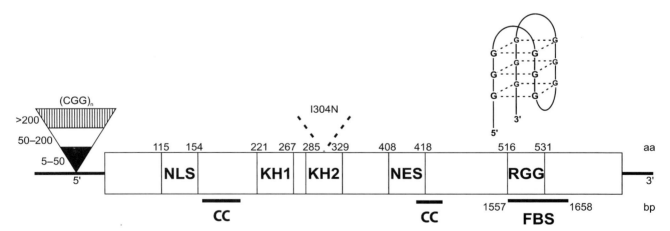

FIGURE 11-2 Schematic representation of FMR1 mRNA and protein. The known domains are indicated: NLS, nuclear localization signal; CC, coiled coils; KH, K-protein homology domain; NES, nuclear export signal; RGG box, Arg–Gly–Gly triplet; FBS, FMRP binding site. In addition, the Ile304Asn missense mutation found in a single, most severely affected patient is indicated. In the 5' UTR the different repeat classes are represented: 5–50, normal; 50–200, premutation; >200, full mutation. The G-quartet structure present in the mRNA is depicted above the FBS that overlaps with the RGG box.

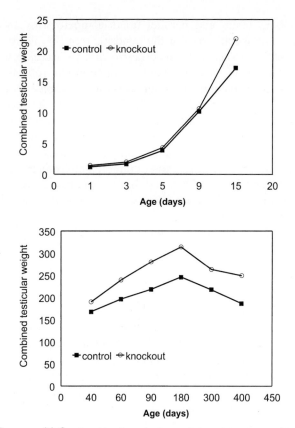

FIGURE 11-3 Combined testicular weight of knockout (filled symbols) and control (open symbols) mice at different ages.

slight increase, which, however, was not significant [34]. These findings suggest that elements of the FSH signal transduction pathway are not involved in development of macroorchidism in the fragile X syndrome.

Absence of FMRP in developing germ cells as a primary cause of development of macroorchidism is not very likely, because in affected human male fetuses, FMRP expression could be detected in the primordial germ cells [40]. The primary cause of the increased testis size in *Fmr1* knockout mice may be found in Sertoli cells. Sertoli cells of the wild-type neonatal mouse show a high expression of Fmrp on Postnatal Day 3. This expression disappears on Postnatal Day 14, suggesting a function for Fmrp in Sertoli cells during early postnatal life [28]. In knockout mice, and also in humans with fragile X, this high, early postnatal expression is not present. This might indicate that the absence of Fmrp during this postnatal period leads to dysregulation of Sertoli cell maturation and proliferation and, consequently, to development of macroorchidism in adult life.

A transgenic mouse has been generated with a yeast artificial chromosome (YAC) containing the human *FMR1* gene to determine whether the *Fmr1* knockout mouse phenotype could be rescued [41]. Macroorchidism was absent in knockout mice carrying the YAC transgene, indicating functional rescue by the human protein.

C. Neuroanatomy and Physiology of the Knockout Brain

Macroscopic and microscopic examination of different brain regions did not reveal differences between knockout and normal mice [42]. Staining with a number of antibodies did not show alterations in protein levels with the exception of Fmrp.

Metabolic rates were found to be altered in the knockout mouse. Qin et al. [43] measured the cerebral metabolic rates for glucose, CMR_{glc}, as an indicator of the level of functional activity in 38 brain regions of fragile X knockout mice and wild-type littermates. In 26 of the 38 regions, the CMR_{glc} level was significantly increased in the *Fmr1* knockout mice, particularly in the regions of the limbic system and primary sensory and posterior parietal cortical areas. Interestingly, these areas are involved in motor activity and exploration, processes that are affected in the knockout mice (see below). Thus, the increased level of energy metabolism in cortical structures correlates with abnormal dendritic spine morphology (below) and an increased excitatory state. The same group [44] determined *in vivo* regional rates of cerebral protein synthesis (rCPS) in adult wild-type and *Fmr1* null mice. A substantial decrease in rCPS was noted in all brain regions examined between the ages of 4 and 6 months in both wild-type and *Fmr1* null mice. Superimposed on the age-dependent decline in rCPS, a regionally selective elevation in rCPS was demonstrated in *Fmr1* null mice. The results suggest that the process of synaptic pruning during young adulthood may be reflected in decreased rCPS. These findings support the hypothesis that FMRP is a suppressor of translation in brain *in vivo*.

Significant region-specific differences in basal neurotransmitter and metabolite levels were reported between wild-type and *Fmr1* knockout animals [45, 46]. Particularly with respect to the emotional deficits that have been described in fragile X patients, it was postulated [45] that specifically limbic brain regions may suffer from imbalances in neurotransmission mediated by monoamines and amino acids. Comparing male *Fmr1* knockout mice with their wild-type counterparts revealed age- and region-specific differences in amino acids and monoamines and their metabolites. In juvenile knockout mice, aspartate and taurine were especially increased in cortical regions, striatum, hippocampus, cerebellum, and brainstem. In addition, juveniles showed an altered balance between excitatory and inhibitory amino acids in the caudal cortex, hippocampus, and brainstem. The precise causal relationship between the neurochemical and the morphological as well as behavioral abnormalities in *Fmr1* knockout mice requires further investigations.

D. Structural Abnormalities

A study of the hippocampus did reveal possible subtle neuroanatomical abnormalities. Sections through the hippocampus revealed excessive sprouting of the intra- and infrapyramidal mossy fibers in subfield CA3 in knockout mice [47]. This suggests increased axonal branching and synaptogenesis in the knockout mouse, perhaps related to the increased sensitivity to seizures. It has been shown that these hippocampal fibers are involved in spatial learning tasks [48–50]. However, a decrease in size of the mossy fibers in the same hippocampal area was reported in an independent study [51]. The reason for the discrepancy between the studies is not clear, but could be related to strain differences. FVB mice, which have an unusually small CA3 subfield, were used in the first study, whereas C57BL/6 mice with a relatively large CA3 subfield were used in the second study.

The pathological cellular mechanisms that may underlie the cortical behavioral and cognitive deficits described later are probably closely related to dysfunctions at the level of dendritic spines and their input. The dendritic spines of pyramidal cells of both fragile X patients and *Fmr1* knockout mice are unusually long and irregular [52–54]. Impaired spine morphology in the barrel region of somatosensory cortex has also been reported in young knockout mice [55]. As these spines appear morphologically immature, it has been suggested that FMRP is involved in spine maturation and pruning as well as synaptogenesis [53]. In a recent study, Koekkoek et al. showed that in the cerebellum Purkinje cells of knockout mice show elongated spines and enhanced long-term depression (LTD) induction at the parallel fiber synapses that innervate these spines [56]. In contrast to Nimchinsky et al., who reported that in knockout mice the phenotype/morphology seems to be transitory [55], Koekkoek et al. showed morphological changes also at an older age [56]. A similar morphological effect was noted in Purkinje cell-only knockout mice. The abnormalities of dendritic spines that Koekkoek et al. observed in cerebellar Purkinje cells of both global and cell-specific *Fmr1* null-mutants mimic only partly those that have been described for pyramidal cells in the cerebral cortex [53, 57]. They follow the same pattern in that the individual spines appear as immaturely shaped processes with elongated necks and heads, but they differ in that their density is normal. Apparently, the density of spines in Purkinje cells is more tightly regulated by compensatory mechanisms than that in pyramidal cells. The spine density in Purkinje cells is largely subject to a well-regulated process in which the climbing fibers and parallel fibers compete with each other for specific sites at the dendritic tree [58, 59]. It is therefore attractive to hypothesize that the accelerated elimination of multiple climbing fiber inputs reflects a mechanism to compensate for a slowdown in spine maturation. FMRP and *Fmr1* mRNA are present in spines and/or dendrites, and FMRP is translated in response to activation of metabotropic glutamate receptor (mGluR) type 1 in synaptoneurosomes [60]. The function of FMRP as an inhibitor of translation of targeted mRNAs *in vitro*, including its own mRNA and that of proteins involved in microtubule-dependent synapse growth and function, indicates that FMRP may act as a regulator of activity-dependent translation in synapses.

E. LTP/LTD

Considering that fragile X patients and *Fmr1* knockout mice display hippocampal-related memory deficits, the involvement of FMRP in transport of mRNAs, and differences in the dendritic mRNA pool in absence of FMRP, FMRP jumps out as a key candidate as a mediator of translation-dependent long-living forms of plasticity and memory. In fact, data has been gathered in recent years to strengthen a metabotropic glutamate receptor (mGluR)-dependent role of FMRP in synaptic plasticity. Weiler et al. showed that increased *Fmr1* mRNA associated with translational complexes in response to activation of mGluRs in synaptosomal fractions [60]. While no function of FMRP has been related to long-term potentiation (LTP) in electrophysiological studies in hippocampal sections of the knockout mouse [61] and also late-phase hippocampal LTP could not be proven to be affected in *Fmr1* knockout mice [62], Li et al. succeeded in demonstrating that LTP was reduced in the cortex of *Fmr1* knockout mice [63]. However, no studies have been performed to test for a possible LTD-related function of FMRP until recently. Huber and colleagues succeeded in demonstrating an enhancement of mGluR5-dependent LTD in the absence of the FMRP protein [64]. As a result of activation of postsynaptic group 1 mGluRs (predominantly mGluR5), AMPA and NMDA receptors are internalized and FMRP is synthesized. The negative regulatory function of FMRP on mRNA translation ensures limited expression of the proteins required for permanent receptor endocytosis. Thus, FMRP regulates the degree of LTD.

In the absence of FMRP, receptor intake is not negatively regulated, and consequently an exaggerated number of receptors are internalized. The decrease in receptor numbers on the postsynaptic membrane weakens the synapse and changes the morphology of the spines. The increase in the number of elongated and immature-appearing spines that has been observed in fragile X patients is suggested to be a result of incomplete synapse elimination; the mental retardation is thought to result from exaggerated LTD [64]. Thus, altered hippocampal LTD in fragile X patients may interfere with the normal formation and maintenance of synapses required for particular cognitive functions. The significance of FMRP-related LTD beyond the hippocampus has been emphasized in an elaborate effort by Koekkoek and colleagues [56]. The Purkinje cells of *Fmr1*-deficient mice have elongated spines and enhanced LTD induction at the parallel fiber synapses that innervate these spines. The observed cerebellar deficits are probably independent of further developmental aberrations downstream, as bilateral lesions of the cerebellar nuclei affected wild-type and *Fmr1* knockout mice alike.

F. Behavior

Many tests comparing behavior of Fmr1 KO mice with that of normal control littermates have yielded contrasting observations, and in some cases, these have been found to result from variation in genetic background, which might also explain part of the phenotypic variability in humans [65]. Abnormalities in synaptic processes in the cerebral cortex and hippocampus contribute to cognitive deficits in fragile X patients and Fmr1 knockout mice. For example, the Fmr1 knockout mouse that we have generated exhibits behavioral abnormalities and learning deficits and has audiogenic seizures that are consistent with the human syndrome [66, 67]. Fmr1 knockout mice also show increased exploratory and motor activity, deficits in spatial learning ability, and decreased anxiety-related responses [24, 41, 68].

1. MORRIS WATER MAZE TEST

The Morris water maze tests impairments in visual short-term memory and visual–spatial abilities in small rodents [69, 70]. In the water maze test, the mice are placed in a large circular pool filled with opaque water and are given the task to swim to a platform that can be either visible or hidden. No gross impairment in swimming ability was observed in either the knockout mice or the control group. Knockout mice and normal littermates were able to locate the hidden platform as a result of training. An example of a training trial of a normal littermate is illustrated in Fig. 11-4. Using video tracking the successive swims of an animal are recorded. During successive swims, both the time to find

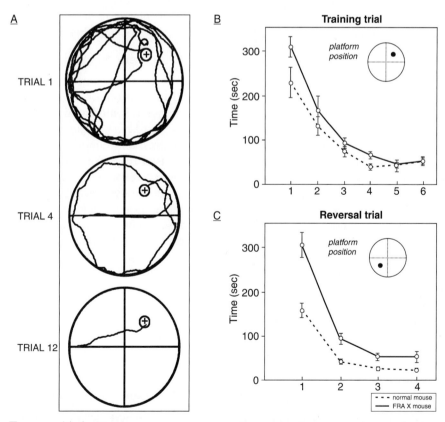

FIGURE 11-4 Morris water maze test performance of mutant and control mice. (A) Swimming trajectory of a control mouse during three training trials on a hidden-platform water maze. (B) Escape latency to find a hidden platform: training trials. (C) Escape latency during reversal trials. Filled circles: knockout mice; open circles: control littermates.

the platform and the distance traveled are dramatically reduced. Similar results were obtained with the knockout animals. Thus, the hidden platform test showed that mutant mice reached levels of performance equal to those of controls (Fig. 11-4), suggesting no impairment in learning the location of the platform. Following 12 training trials, the position of the hidden platform was changed and four reversal trials were carried out. Now the knockout mice experienced more difficulty than normal littermates in learning the new position, which was apparent both in increased escape latency and in increased path length. This appears to be due a relative inability of knockout mice to alter a learned spatial strategy. Spatial memory abilities like those in the hidden platform learning test have been shown to be highly dependent on hippocampal function.

The original Morris water maze tests were carried out in a mixed C57Bl6/129 background [24]. Near-normal performance was observed in a knockout line that had been backcrossed to C57BL/6 for more than 15 generations [62]. They examined F_1 siblings of C57BL/6 knockout and 129 crosses. Here, significant but subtle increased swim latencies in reversal trials were observed. These data suggest that strain differences between C57BL/6 and 129 influence the *Fmr1* knockout phenotype in behavioral experiments. Also, the knockout allele was bred into a background of FvB that was screened for normal visual abilities. These animals were tested in the Morris water maze test and similar results were obtained as described by Bakker et al. (unpublished results).

A mild learning deficit was also observed in the terrestrial radial arm maze test [51], which also tests spatial learning.

2. Conditioned and Contextual Fear Test

In this test, fear is measured by observation of "freezing" behavior (voluntary immobility of the mouse) when a tone is followed by a foot shock. After a brief training phase, the responses of the mice to the test environment (hippocampus dependent) and to the tone (amygdala dependent) are registered. Although fragile X mice reportedly froze in fear more frequently in response to both the test environment (context) and the tone in one experimental setup [62, 71, 72], no differences between knockout mice and controls were found by two other laboratories [41, 73]. Differences in performance between genotypes may be very mild and only measurable under some testing conditions.

3. Eyeblink Conditioning Test

Most studies have focused on hippocampal brain regions, and the potential roles of cerebellar deficits have not been investigated until recently. In a recent study [56] it was demonstrated that both global and Purkinje cell-specific knockouts of *Fmr1* show deficits in classic delay eyeblink conditioning in that the percentage of conditioned responses, as well their peak amplitude and peak velocity, is reduced. This eyeblink conditioning test is a well-described test for studying cerebellar function in associative motor learning, indicating that cerebellar deficits can contribute to motor learning deficits in fragile X patients. Moreover, fragile X patients display the same cerebellar deficits in eyeblink conditioning as the mutant mice [56]. These data indicate that a lack of FMRP leads to cerebellar deficits at both the cellular and behavioral levels, and they raise the possibility that cerebellar dysfunctions can contribute to motor learning deficits both in fragile X patients and in *Fmr1* knockout mice.

4. Motor Activity and Anxiety

Fragile X patients are reported to be hyperactive and anxious. Fragile X knockout mice are more active than their control littermates. However, the difference in activity between the groups is mild, and variance within groups is high. A seemingly contradictory observation is that *Fmr1* knockout mice displayed increased anxiety-related responses in the exploratory behavior test, whereas lesser anxiety is reported when measured in the open-field test [41]. However, no difference in anxiety was registered in the elevated plus maze, considered to be the most suitable test for measuring anxiety [72].

5. Acoustic Startle Reflex

The acoustic startle response is a behavioral tool used to assess brain mechanisms of sensorimotor integration and is mediated by neurons in the lower brainstem. The knockout mice exhibited increased auditory startle response amplitudes to low-intensity stimuli and decreased responses to high-intensity stimuli [41]. The increased response to low-intensity stimuli is compatible with the hyperarousal of fragile X patients, whereas the decreased reactivity of the fragile X knockout mouse to high-amplitude stimuli is compatible with the decreased functioning of the neuronal connections in fragile X patients.

A different setup was chosen by Koekkoek et al. [56], who tested enhanced startle responses to auditory stimuli. They analyzed the initial 60-ms periods of the eyeblink responses following the onset of the tone. The percentage of startle responses was significantly increased during all sessions in the *Fmr1* knockout mutants.

6. Audiogenic Seizures

Spontaneous seizures are observed in more than 20% of fragile X patients. These spontaneous seizures are never observed in the knockout mouse, but an increased sensitivity to epileptic seizures in response to auditory stimuli is detected in the knockout mouse [66]. This may indicate that absence of Fmrp results in increased cortical excitability and is consistent with the sensory hypersensitivity of fragile X patients.

It was proposed by Bear et al. that a lack of FMRP may lead to uncontrolled protein synthesis at the synapse and that reducing mGluR activation may (partly) reverse the effects due to the lack of FMRP [74]. Interestingly, recent preliminary data from Bauchwitz' group showed that partially blocking mGluR5 in the hippocampus with a specific antagonist (MPEP) indeed prevents the induction of audiogenic seizures in Fmr1 knockout mice [75]. These data demonstrate that at least some aspects of the phenotype in mice can be reversed by treatment. A large number of group I mGluR (ant)agonists are available that can be used to study the blocking and/or activation of mGluR receptors in mouse models for fragile X syndrome.

G. Environmental Effects

The ability to influence FMRP expression by experience has been demonstrated by Todd and Mack, who showed that FMRP levels increase in the somatosensory cortex of the rat in response to unilateral whisker stimulation [76]. In an effort to shed more light onto the experience-dependent production of FMRP, the same group showed that inhibiting translation in barrel cortex synaptic fractions suppressed the whisker-induced production of FMRP. Also, the levels of Fmr1 mRNA remained unchanged in this scenario. Furthermore, FMRP production depended on the activation of both NMDA receptors and mGluR1s [77]. Additional evidence that experience regulates the expression levels of FMRP in vivo at the level of translation was presented by Gabel et al. visual experience modulates the production of FMRP. Exposure of dark-reared rats to light rapidly (within 15 minutes of exposure) increases FMRP levels in the cell bodies and dendrites of the visual cortex. The upregulation occurs posttranscriptionally and can be inhibited by NMDA receptor antagonists [78].

A number of articles have described the effect of enriched environmental conditions (EECs) on FMRP expression. Increased FMRP immunoreactivity was demonstrated in visual cortex of rats exposed to complex environments for 20 days and in the motor cortex of rats trained on motor skill tasks for 7 days, as compared with appropriate controls [79]. Rats exposed to EECs for 20 days exhibited increased FMRP immunoreactivity in visual cortex compared with animals housed in standard laboratory caging. Animals exposed to EECs for 20 days had higher dentate gyrus FMRP levels than animals exposed 5 or 10 days [80]. These results provide further evidence of behaviorally induced alteration of FMRP expression and suggest regulation of FMRP abundance by synaptic activity.

Restivo et al. reared Fmr1 knockout mice in a C57BL/6 background under EECs and examined the possibility that experience-dependent stimulation alleviates their behavioral and neuronal abnormalities [81]. Fmr1 knockout mice kept in standard cages were hyperactive, displayed an altered pattern of open-field exploration, and did not show habituation. Quantitative morphological analyses revealed a reduction in basal dendrite length and branching together with more immature-appearing spines along apical dendrites of layer 5 pyramidal neurons in the visual cortex. Enrichment largely rescued these behavioral and neuronal abnormalities. Enrichment did not, as in rats, affect FMRP levels in the wild-type mice. These data suggest that EECs can in part rescue the phenotype seen in Fmr1 knockout mice, and suggest parallels with effects of early intervention in treating young children with fragile X syndrome.

H. Instability of the CGG Repeat in Mice

Still, very little is known about the mechanism and timing of the CGG repeat amplification in FMR1. Studies in yeast and Escherichia coli suggest that direction of replication, genetic background (including repair systems), transcription, and growth conditions can influence repeat instability. Several models for the repeat amplification have been proposed; the most likely models are discussed by Wells [82]. To study the mechanism and timing of repeat instability an animal model is required. Only in an animal model will it be possible to study gametogenesis and early embryogenesis at specific time points. Transgenic mouse models with a maximum track of 97 uninterrupted CGG repeats in the human FMR1 promoter region have been constructed. Despite the fact that premutations of this size are inherited unstably and expand to full mutations greater than 200 repeats in nearly 100% of human transmissions from mother to offspring, no repeat instability was observed in hundreds of murine meioses analyzed [83–85]. As factors other than repeat length, including the genomic environment, might have an influence on the rate of repeat expansion, in subsequent experiments the endogenous $(CGG)_8$ repeat of the mouse was replaced by a $(CGG)_{98}$ repeat [86]. In its natural genomic environment, the "knock-in" CGG triplet mouse shows moderate CGG repeat instability on both maternal and paternal transmission [87]. A different approach was taken by Peier and Nelson, who generated a transgenic mouse with a YAC containing the human *FMR1* gene with an elongated (CGG) repeat of different sizes [88]. Length-dependent instabilities in the form of small expansions and contractions were observed in both male and female transmissions over five generations. Alterations in tract length were found to occur exclusively in the 3' uninterrupted CGG tract. Large expansion events indicative of a transition from a premutation to a full mutation were not observed.

Although the thresholds of repeat expansion might differ between mouse and human, these murine models could enable the study of both the timing of repeat expansion and the mechanism of methylation-induced expression silencing in a mouse model.

I. Mouse Model for FXTAS

Males with alleles in the premutation range with increased *FMR1* mRNA levels that are up to eightfold higher than normal and with (mildly) reduced FMRP levels have been described [16]. The elevated *FMR1* transcript levels were positively correlated with the number of CGG repeats [19]. The question of whether these elevated *FMR1* mRNA levels and slightly reduced FMRP levels result in a mild fragile X phenotype was challenged by the recent description of older males carrying a premutation (PM), who exhibit a unique neurodegenerative syndrome characterized by progressive intention tremor and ataxia, termed FXTAS (fragile X-associated tremor/ataxia syndrome) [14, 20]. At autopsy, these patients have been shown to carry inclusion bodies in their neurons that include markers of protein degradation, such as ubiquitin [89].

Mice created to study repeat instability are potentially useful models for FXTAS as they carry CGG permutation length alleles expressed in the human *FMR1* or mouse *Fmr1* gene. The brains of expanded CGG repeat mice were analyzed neurohistologically and biochemically at different ages from neonate to the latest stage of life (1–72 weeks) [87]. Biochemically, elevated *Fmr1* mRNA levels (two- to fourfold) were already detectable in the first week of life. Neuropathological analysis of the brains showed the presence of neuronal ubiquitin-positive intranuclear inclusions throughout the brain (Fig. 11-5). The inclusions became visible at 30 weeks of age, and an increase was observed in both the number and the size of the inclusions during the course of life, which correlates with the progressive character of FXTAS. Next to ubiquitin, Hsp40 and the 20S catalytic core complex of the proteasome could be demonstrated as constituents of the inclusions. Strikingly, in contrast to brains from symptomatic FXTAS patients, inclusions were totally absent in astrocytes and no cell

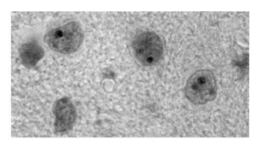

FIGURE 11-5 Distribution of neuronal ubiquitin-positive intranuclear inclusions in the colliculus inferior from the expanded repeat mouse at the age of 72 weeks by indirect immunoperoxidase staining. See CD-ROM for color image.

loss could be observed. Furthermore, a correlation was found between the occurrence of inclusions within specific brain regions from the mouse and the clinical features in symptomatic premutation carriers. Very recently, the expanded CGG repeat mouse was assessed for cognitive, behavioral, and neuromotor performance at different ages (20, 52, and 72 weeks). The results clearly indicate an age-dependent decline in visual–spatial learning capacities, a potential increase in anxiety levels, and mild neuromotor disturbances in the expanded CGG repeat mouse model [90]. This model should prove to be valuable for additional characterization of the mechanisms of development of FXTAS and, possibly, for testing potential treatments.

J. Knockout Mice for *Fmr1* Paralogs *Fxr1* and *Fxr2*

For the paralogs of *FMR1, FXR1, and FXR2*, no disease causing mutations has yet been described in humans. Like FMRP, both FXR1P and FXR2P are expressed in high levels in adult neurons and especially in Purkinje cells. In testes, FMRP, FXR1P, and FXR2P are all expressed but at various levels and in different cell types [91]. In murine skeletal muscles, Fxr1p is expressed in high levels and is localized within the muscle contractile bands [28, 92] with only very weak signals visible for Fmrp and Fxr2p [28]. *Fxr1* exhibits significant alternative splicing, and generates isoforms that vary from tissue to tissue [92, 93]. In contrast, no alternative splicing has been reported for *Fxr2*, and the several isoforms produced by *Fmr1* do not appear to be altered in relative abundance when comparing tissues [93, 94].

The absence of Fmrp in the *Fmr1* knockout mouse does not influence the expression of Fxr1p or Fxr2p detectably [28]. Knockout mice for *Fxr1* and *Fxr2* may reveal a phenotype and could shed light on the functions of Fxr1p and Fxr2p. A possible explanation for the relatively mild phenotype of the fragile X syndrome is that the protein products of two autosomal paralogs of *FMR1* (*FXR1* and *FXR2*) partially compensate for FMRP. This hypothesis predicts that double or triple knockouts of *FMR1*, *FXR1*, and *FXR2* will show a much more severe phenotype than knockouts of each individual gene.

An *Fxr2* knockout has been created, and the mice show no evidence of pathological abnormalities in brain or testes [95]. When tested for cognitive and behavioral characteristics, they show impaired Morris water maze learning and increased locomotor activity comparable to that of the fragile X knockout mouse. In addition, they have decreased rotarod performance, a delayed hindlimb response in the hotplate test, and less contextual fear. In contrast to the fragile X knockout mouse, the *Fxr2* knockout has a lower prepulse inhibition of the acoustic startle reflex and no significant difference in the acoustic startle response. *Fmr1/Fxr2* double mutants have been generated. *Fmr1/Fxr2* double knockout mice showed increased hyperactivity relative to *Fmr1* KO knockout and *Fxr2* knockout mice (unpublished results). On the other hand, *Fmr1/Fxr2* double knockout mice did not differ from *Fmr1* knockout mice or *Fxr2* knockout mice in anxiety-related responses. These findings suggest that both *Fmr1* and *Fxr2* genes contribute in an additive/cooperative manner to pathways controlling exploratory activity, but not to pathways involved in anxiety-like behavior.

After observation of circadian rhythm defects in *Drosophila* lacking *dFMR*, *Fmr1/Fxr2* double knockout mice were tested for circadian rhythm of activity. They were found to lack normal day/night cycling in wheel running activity altogether, whether they were exposed to a normal light/dark environment or complete darkness (unpublished results). Both single knockout models show normal rhythm in a light/dark environment, but a shorter circadian period in total darkness, and the *Fxr2* single knockout shows higher levels of sporadic activity during normal rest periods. These data provide further evidence that Fmrp and Fxr2p function in a common pathway, and that loss of function of either gene alone is insufficient to expose all the functions in which the gene participates. Additional study of double (and possibly triple) knockout animals should provide better insight into the role of Fmrp in the mouse.

Homozygous *Fxr1* knockout neonates die shortly after birth most likely due to cardiac or respiratory failure [96]. Histochemical analyses carried out on both skeletal and cardiac muscles show a disruption of cellular architecture and structure in E19 *Fxr1* neonates compared with wild-type littermates. In wild-type E19 skeletal and cardiac muscles, Fxr1p is localized to the costameric regions within the muscles. In E19 *Fxr1* knockout littermates, in addition to the absence of Fxr1p, costameric proteins vinculin, dystrophin, and α-actinin were found to be delocalized. A second mouse model (*Fxr1* 1 neo), which expresses strongly reduced levels of Fxr1p relative to wild-type littermates, does not display the neonatal lethal phenotype seen in the *Fxr1* knockouts, but does display a strongly reduced limb musculature and has a reduced life span of 18 weeks. No data are available on behavioral abnormalities in the *Fxr1* 1 neo mouse. For these studies an inducible *Fxr1* knockout mouse is needed in which the *Fxr1* gene is inactivated in the brain only.

III. FLIES

The fruit fly *Drosophila melanogaster* is an extremely valuable model organism for study of gene function. It has numerous advantages over mice, including rapid analysis due to short breeding cycles, ability to generate large numbers of animals at low cost, and exquisite genetics allowing precise expression of genes and facile definition of genetic modifiers of phenotypes. The primary disadvantage for studying a human genetic disease is the evolutionary distance from humans, which can result in less similarity of function for genes with sequence similarity.

In the case of *FMR1*, the fly genome contains a single gene with significant sequence similarity to FMR1, which is likely to represent the common ancestral sequence of the three paralogs found in vertebrates (Fig. 11-6 [97]). The fly gene has been termed either dFXR or dFMR; dFMR is used here. dFMR is 35–37% identical to the human FXRs at the DNA sequence level and between 56 and 65% identical at the protein level [98]. Functional domains (NLS, NES, KH1, KH2, and RGG) are conserved in location as well as in sequence, and the amino terminus is more highly conserved than the carboxy end of the protein, which is also the case among the three vertebrate proteins. The fly dFMR gene is clearly the best sequence match in the *Drosophila* genome [98–100]. The pattern of expression of dFMR is likewise similar, with higher levels of expression in neurons and reduced levels in glia [97–101]. It may be the case that the fly protein carries out functions that have been divided among the three vertebrate genes. In this view, it is likely that models in *Drosophila* may represent effects of loss or gain of function of all three genes in the mouse. This would predict more prominent phenotypes in fly models, particularly if the three vertebrate genes are capable of compensating for one another. In general, fly models have shown stronger phenotypes, further indicating that compensation by *FXR1* and *FXR2* may be relevant to fragile X syndrome.

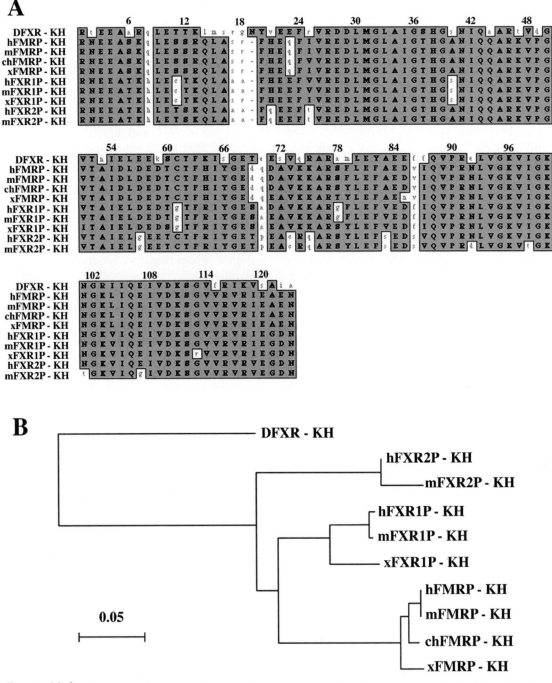

FIGURE 11-6 Comparison of the *Drosophila* dFMR protein sequence with other members of the fragile X-related gene family. (A) Amino acid sequence alignment of KH domains 1 and 2 of *Drosophila* dFXR; human and mouse FXR2P; human, mouse, and *Xenopus* FXR1P; and human, mouse, chicken, and *Xenopus* FMR1P. Identities and similarities are boxed and shaded. (B) Neighbor joining phylogenetic tree depicting the distance between pairs of sequences. The two most similar sequences were joined first. The other sequences were added one by one in order of decreasing similarity.

Several loss-of-function alleles for dFMR have been described, derived from imperfect excision of one of two P elements that were found to have inserted into the 5′ end of the gene. These all appear to be molecular nulls based on absence of dFMRP [98–100, 102]. In addition, reduced dFMRP is found in flies with the inserted P elements [99]. Dockendorff and colleagues generated transgenic lines that carry genomic fragments covering dFMR to demonstrate rescue of phenotypes [100]. Two lines were developed, one with a frameshift in the dFMR coding sequence that allows clear demonstration of the effect of dFMR in complementation experiments. Although mutant flies are viable and anatomically normal, loss of function of dFMR results in numerous phenotypes that can be detected by more detailed study. Effects are seen on behavior, neuronal and neuromuscular architecture, sperm structure, and abundance of some protein products. Some of these phenotypes have proven useful for screening for genetic and environmental modifiers.

A. Behavioral Phenotypes

Initial studies of loss of FMRP function found that mutant flies lacked normal circadian rhythms in both activity and hatching from the pupae (eclosion) [99, 100, 102]. Lack of rhythm is found when flies are placed in total darkness and this phenotype can be rescued by the intact dFMR transgene [100]. However, it is not found in a hypomorphic mutant fly [102]. While disrupted rhythm might suggest a role for dFMR in the function of the central oscillator, efforts to identify defects in clock function in these mutants have shown normal oscillation of clock component RNAs and proteins [99, 100, 102]. Indeed, CREB (cAMP response element-binding protein) oscillation is reduced in dFMR mutants in constant darkness, whereas PDF (pigment dispersing factor) is unaffected [100]. Unpublished data from *Fmr1/Fxr2* double knockout mice described earlier indicate that some central clock components cycling behavior or abundance are altered. While the conclusion from the fly models is that the central clock is intact, and that clock output to activity and eclosion is disrupted by loss of dFMR, mammalian studies may offer a different perspective. Because the fly data stimulated efforts to identify defects in circadian behavior in mice, it will be interesting to use data from the mouse clock studies to reexamine potential effects on homologous components in flies. Flies lacking dFMR also show impaired flight [98] and locomotor defects as larvae [103]. An RNA target for dFMR, Ppk1, which encodes a component of the epithelial sodium transport channel, is overexpressed in mutant flies and appears to play a role in generating this behavioral phenotype.

Courtship behavior in mutant flies has been studied by the Jongens group. Normal male flies attend female flies during courtship, and carry out stereotypical behaviors that include tapping, extending and vibrating one wing, and licking prior to copulation. dFMR mutant males were found to have a significantly reduced length of useful attention to females, and often failed to advance to later stages of courtship during a 10-min period of observation [100]. In a later study, this group used courtship behavior to assess learning and memory in the dFMR mutants [104]. Normal flies learn (after repeated courtship advances are rebuffed) that recently mated females are unreceptive, and their courtship attempts are reduced even when presented with virgin females for a period of 2–3 h. This finding has been used as a memory learning task. It was found that dFMR mutants, like wild-type flies, learn that previously mated females are unreceptive; however, when placed in a chamber with a virgin fly, the mutants reinitiated courtship attempts much more vigorously than did wild-type flies, suggesting a defect in immediate recall memory. The ability to perform behavioral assays and learning tasks in flies offers parallels to the human disease and the possibility of using fly models for studying modifiers, both genetic and pharmacological. Human fragile X patients clearly have learning disabilities, but are also found to have sleep abnormalities, which may result from altered circadian clock function.

B. Neuronal Phenotypes

Several studies have described changes in neuronal processes in different areas of the fly peripheral and central nervous systems in the absence of dFMR. Although overall maintenance of neuronal number and location is preserved, loss of dFMR results in overgrowth of the neuromuscular junction (NMJ), with increased neurite branching, enlarged boutons [98], and increased arborization and branching of central neurons such as the lateral and DC neurons, as well as mushroom bodies [99, 100, 105]. Together, these studies suggest a role for dFMR in negatively regulating neuronal process growth and complexity. In a recent study, Pan et al. demonstrated the cell specificity of the effects by mosaic analysis where individual neurons can be mutated and marked with a fluorescent marker [106]. This method allowed the conclusions that both axons and dendrites are affected in mushroom body neurons and that excess neurites can arise from the cell body in the absence of dFMR.

The overgrowth of neuronal processes in the fly loss-of-function models is highly reminiscent of the dendritic spine alterations found in mouse models and in some human autopsy specimens, further underscoring the similarity of the fly model. As alluded to earlier, this phenotype is less subtle in flies than it is in mice, which again could be the result in flies of the loss of the equivalent of *Fmr1* and its two paralogs in mice.

Another structural phenotype found in the dFMR null flies can be seen in sperm derived from male flies. In addition to having reduced courtship attention, male flies lacking dFMR are hypofertile. Zhang and coworkers investigated possible reasons for this and found that the regular 9 + 2 arrangement of microtubules in the sperm tail is disrupted, with the middle pair of microtubules absent in an increasing number of sperm during spermatogenesis [107]. The progressive loss suggests instability of the microtubules and a role for dFMR in stabilizing them. Although macroorchidism (enlarged testes) is a feature of human fragile X syndrome and the mouse knockout, both appear to have normal fertility. However, Zhang et al. found some abnormalities in knockout testes and sperm examined by electron microscopy [107]. Thus, the fly model is offering additional phenotypes that may prompt renewed examination of the mouse model and human patients.

Studies in *Drosophila* can also include rather precise overexpression of genes of interest using tissue-specific drivers of gene expression (Gal4 UAS drivers) introduced by breeding into a standardized transgenic fly or into one carrying a specialized P element insertion into the dFMR gene. Phenotypes can then be assessed in the specified tissue. A number of groups have used this approach to study the consequences of overexpression of dFMR. Widespread overexpression of dFMR is lethal in flies, and some targeted overexpression leads to cell death through apoptosis [97]. This is in accord with observations in cell culture of translational suppression by FMRP and lethality; there appears to be a limited quantity of FMRP that can be tolerated by a cell or organism. The Gal4 UAS system is temperature sensitive, allowing some modulation to be carried out by growth of the flies at different temperatures. This can allow fine tuning of phenotypes.

As with the mouse, the observation of opposing phenotypes in loss-of-function versus overexpression models provides strong evidence for a direct role of the fly FMR gene in the generation of the phenotype. Overexpression of dFMR lengthens circadian period length [100], reduces NMJ bouton sizes [98], and reduces neurite extension and branching [99, 100, 106], offering greater confidence in the specificity of these phenotypes. Overexpression also leads to phenotypes without counterparts in loss of function. A rough eye is found when dFMR is overexpressed using eye-specific drivers [108], and other structures can likewise show cell loss or disorganization [97]. Eye phenotypes are particularly useful because they can provide a readily assayable marker for carrying out screens for enhancers or suppressors of the phenotype. Overexpression of mutant versions of proteins is also more easily carried out in fly models. The widely studied KH2 domain point mutation I304N has been introduced into fly expression vectors (where it is I307N due to the slight difference in amino acid numbers). Overexpression of the fly I307N mutation leads to less severe phenotypes than overexpression of the wild-type dFMR, suggesting that I307N is likely to be a loss-of-function, rather than gain-of-function, mutation.

C. Modifying Phenotypes with Genes and Drugs

One of the most significant advantages of working with fly models is the ability to rapidly screen for genetic modifiers of a mutant phenotype. This can be carried out as a candidate screen, with genes that are hypothesized to function in concert with the gene of interest, or as a genomewide screen using the large collection of mutant fly strains available for coverage of the *Drosophila* genome. Zhang and coworkers, noting alterations at the NMJ in dFMR mutants that appeared similar to overexpression of the MAP1B gene Futsch, demonstrated that dFMR negatively regulates Futsch expression [98]. They could also show that Futsch mutations can reduce the NMJ phenotype in the dFMR mutant fly back to normal levels. Futsch mRNA is bound by dFMRP, just as mammalian MAP1B RNA is bound by FMRP, and Futsch protein levels appear to be under dFMR control as they increase some twofold in dFMR mutants and are decreased on overexpression of dFMR. Significantly, Futsch mutations cannot modify all dFMR fly phenotypes. The sperm alterations are not changed in a Futsch mutant background, for example [107]. MAP1B levels in mutant mice have been studied and found to be similarly dysregulated in the Fmr1 knockout [109]. Genes identified as potential partners for FMRP by biochemical interactions (such as co-immunoprecipitation or yeast two-hybrid assays) can be validated through genetic interactions and other biochemical and cell biological means in the fly. Schenk and colleagues identified CyFIP in this way and could demonstrate functional interaction in control of Rac1 in fly models, along with colocalization of the proteins [110]. Rac1 was found to be an RNA target of dFMR by Lee et al., who also showed genetic interaction [111]. Reeve et al. used a different approach, comparing protein levels for 32 candidate genes in dFMR mutant flies with normal pharate brain extracts [112]. They identified profilin as a consistently upregulated protein, and were able to demonstrate that dFMRP binds the profilin mRNA, and that reduction of profilin levels can reduce the dFMR mutant phenotype in neurite extension and branching. These studies increasingly point to a role for FMRP in maintenance of cytoskeletal components, and suggest mechanisms for disease pathology.

Identification of genetic modifiers can also be carried out in a nondirected fashion, using random mutagenesis of the fly genome to complement mutant phenotypes.

Zarenescu et al. used the rough eye phenotype caused by overexpression of dFMR to search for dominant enhancers and suppressors after chemical mutagenesis of flies [108]. They screened more than 50,000 flies, finding several complementation groups. One of these offered 19 mutant alleles that acted as dominant suppressors of the rough eye phenotype. The gene affected was found to be lethal (2) giant larvae (dlgl), a cytoskeletal protein involved in cell polarity and cytoplasmic transport. From this identification, they were able to demonstrate enhancement by dlgl of the dFMR loss-of-function phenotype at the NMJ and physical interaction between the dlgl protein and dFMR, in a complex that includes mRNA. Specific RNAs could also be isolated and identified. This study offers an example of the utility of fly models for carrying out hypothesis-independent screens on large mutant collections, an approach that is not currently feasible in mammalian models. Although such screens remain daunting even in the fly, they can provide highly valuable clues to functions not suggested by other lines of evidence.

The fly also offers the potential for using mutant phenotypes as targets for drug therapies. The observation of enhanced long-term depression in the *Fmr1* knockout mouse [64], coupled with current understanding of the requirement for local protein synthesis in group 1 metabotropic glutamate receptor (mGluR) signaling that mediates LTD, has led to the proposal that mGluR signaling is overresponsive in the absence of FMRP. This mGluR hypothesis for fragile X pathogenesis, proposed by Huber et al. [74], predicts that downregulation of mGluR signaling would suppress mutant phenotypes in FMR1 loss of function. Antagonists of mGluR were used by McBride and colleagues to test this possibility in the fly model [104]. They found that both the courtship behavior and learning abnormalities and the mushroom body neuron lobe defects could be corrected by antagonists such as MPEP and by LiCl, thought to also act to reduce mGluR signaling. The circadian activity defect was not corrected by these treatments. This study offers the possibility of using fly models for demonstrating potential efficacy of a variety of drug treatments, and allows consideration of development of screens for drug treatments if robust phenotypes can be developed that might be easily assayed.

D. Biochemistry

Because of their small size, flies are not typically considered for biochemical analyses. However, in the case of FMRP, the fly has provided some very interesting biochemical insights that had not been identified in larger organisms. Chief among these is participation of FMRP in the microRNA-mediated regulation of RNA abundance. dFMRP was found in complexes with other components of this system, principally Argonaute 2 and Dicer in S2 cells from *Drosophila* [113]. This led to further studies demonstrating a similar association in mammals [114]. Studies to identify proteins altered in abundance by the absence of dFMR with both a targeted [112] and a more global approach [107, 115] have been described. The latter studies use two-dimensional protein gels to demonstrate alterations of protein levels in specific fly tissues, comparing mutant and normal flies. Similar studies have been carried out in mice and mammalian cells in culture with limited success, perhaps owing to the compensatory effects of Fxr1 and/or Fxr2.

E. FXTAS and CGG Models

The fly has proven valuable in testing the concept that CGG sequences expressed in RNA are toxic. Jin and colleagues developed a fly model for neurodegeneration by expressing riboCGG in the context of a green fluorescent protein gene [116]. Directed expression in the eye demonstrated neurodegenerative effects, showing the ability of the CGG sequences to cause a phenotype outside of the FMR1 gene context. Moreover, the model showed similarities to human neurons in FXTAS, with nuclear inclusions that contain ubiquitin and HSP70. It was readily demonstrated in this model that overexpression of HSP70 could suppress the phenotype, showing a genetic interaction in addition to presence in the inclusion bodies. These flies are now a useful model for testing additional potential modifiers of this neurodegenerative disorder, allowing consideration of potential mechanisms of pathogenesis.

The fly has been a welcome and important addition to the armamentarium for determining the function of the *FMR1* gene and consequences of CGG repeat expansions. While caution is warranted in comparing the various consequences of loss of the single fly gene with loss of one of the three human or mouse paralogs, it is likely that this caveat is greatly outweighed by the improved phenotypes and rapid analyses of candidate modifiers, both genetic and environmental, offered by the fly. The fly is a particularly good model for study of the developmental and plasticity-related alterations in neurons, as this has been the focus of a large and very productive area of fly research. Because this represents the area most relevant to understanding and treating patients with fragile X syndrome, the fly offers great potential for helping to develop rational therapies for this common disorder.

IV. CONCLUSION

Regulation of translation of specific mRNAs might be an important role of FMRP as a response of neurons to stimuli. Disturbance of translation through the lack of FMRP might have great effects on the normal functioning of the brain. The disturbance might lead, in fragile X patients, to learning and memory deficits. Animal models might help to learn more about the function of the *FMR1* gene and the effect that the lack of the protein has on brain functioning. Furthermore, animal models might help in studying the timing and mechanism of the repeat amplification.

Acknowledgments

We thank our colleagues in Rotterdam, Antwerp, and Houston for their stimulating collaboration and discussions. This work was supported in part by grants from the FRAXA Foundation; The Netherlands Organization of Scientific Research; and the National Institutes of Health (R01 HD38038 to B.A.O. and D.L.N. and HD29256 to D.L.N.).

References

1. Hagerman, R. J. (2002). The physical and behavioural phenotype. *In* "Fragile X Syndrome: Diagnosis, Treatment and Research" (R. J. Hagerman, and P. Hagerman, Eds.), pp. 3–109. Johns Hopkins Univ. Press, Baltimore.
2. Verkerk, A. J., Pieretti, M., Sutcliffe, J. S., Fu, Y. H., Kuhl, D. P., Pizzuti, A., Reiner, O., Richards, S., Victoria, M. F., Zhang, F. P., et al. (1991). Identification of a gene (FMR-1) containing a CGG repeat coincident with a breakpoint cluster region exhibiting length variation in fragile X syndrome. *Cell* **65**, 905–914.
3. Oberlé, I., Rousseau, F., Heitz, D., Kretz, C., Devys, D., Hanauer, A., Boue, J., Bertheas, M. F., and Mandel, J. L. (1991). Instability of a 550-base pair DNA segment and abnormal methylation in fragile X syndrome. *Science* **252**, 1097–1102.
4. Sutcliffe, J. S., Nelson, D. L., Zhang, F., Pieretti, M., Caskey, C. T., Saxe, D., and Warren, S. T. (1992). DNA methylation represses FMR-1 transcription in fragile X syndrome. *Hum. Mol. Genet.* **1**, 397–400.
5. Chiurazzi, P., Pomponi, M. G., Pietrobono, R., Bakker, C. E., Neri, G., and Oostra, B. A. (1999). Synergistic effect of histone hyperacetylation and DNA demethylation in the reactivation of the FMR1 gene. *Hum. Mol. Genet.* **8**, 2317–2323.
6. Coffee, B., Zhang, F., Warren, S. T., and Reines, D. (1999). Acetylated histones are associated with FMR1 in normal but not fragile X-syndrome cells. *Nat. Genet.* **22**, 98–101.
7. Pieretti, M., Zhang, F. P., Fu, Y. H., Warren, S. T., Oostra, B. A., Caskey, C. T., and Nelson, D. L. (1991). Absence of expression of the FMR-1 gene in fragile X syndrome. *Cell* **66**, 817–822.
8. Darnell, J. C., Jensen, K. B., Jin, P., Brown, V., Warren, S. T., and Darnell, R. B. (2001). Fragile X mental retardation protein targets G quartet mRNAs important for neuronal function. *Cell* **107**, 489–499.
9. Schaeffer, C., Bardoni, B., Mandel, J. L., Ehresmann, B., Ehresmann, C., and Moine, H. (2001). The fragile X mental retardation protein binds specifically to its mRNA via a purine quartet motif. *EMBO J.* **20**, 4803–4813.
10. Darnell, J. C., Fraser, C. E., Mostovetsky, O., Stefani, G., Jones, T. A., Eddy, S. R., and Darnell, R. B. (2005). Kissing complex RNAs mediate interaction between the fragile-X mental retardation protein KH2 domain and brain polyribosomes. *Genes Dev.* **19**, 903–918.
11. Siomi, M. C., Siomi, H., Sauer, W. H., Srinivasan, S., Nussbaum, R. L., and Dreyfuss, G. (1995). FXR1, an autosomal homolog of the fragile X mental retardation gene. *EMBO J.* **14**, 2401–2408.
12. Zhang, Y., Oconnor, J. P., Siomi, M. C., Srinivasan, S., Dutra, A., Nussbaum, R. L., and Dreyfuss, G. (1995). The fragile X mental retardation syndrome protein interacts with novel homologs FXR1 and FXR2. *EMBO J.* **14**, 5358–5366.
13. Kirkpatrick, L. L., McIlwain, K. A., and Nelson, D. L. (2001). Comparative genomic sequence analysis of the FXR gene family: FMR1, FXR1, and FXR2. *Genomics* **78**, 169–177.
14. Hagerman, R. J., and Hagerman, P. J. (2006). Clinical and genetic features of myotonic dystrophy type 2. *In* "Genetic Instabilities and Neurological Diseases" (R. D. Wells and T. Ashizawa, Eds.), Ch. 10. Elsevier, Burlington, MA.
15. Sherman, S. L. (2000). Premature ovarian failure among fragile X premutation carriers: Parent-of-origin effect? *Am. J. Hum. Genet.* **67**, 11–13.
16. Tassone, F., and Hagerman, P. J. (2003). Expression of the FMR1 gene. *Cytogenet. Genome Res.* **100**, 124–128.
17. Kenneson, A., Zhang, F., Hagedorn, C. H., and Warren, S. T. (2001). Reduced FMRP and increased FMR1 transcription is proportionally associated with CGG repeat number in intermediate-length and premutation carriers. *Hum. Mol. Genet.* **10**, 1449–1454.
18. Primerano, B., Tassone, F., Hagerman, R. J., Hagerman, P., Amaldi, F., and Bagni, C. (2002). Reduced FMR1 mRNA translation efficiency in fragile X patients with premutations. *RNA* **8**, 1–7.
19. Allen, E. G., He, W., Yadav-Shah, M., and Sherman, S. L. (2004). A study of the distributional characteristics of FMR1 transcript levels in 238 individuals. *Hum. Genet.* **114**, 439–447.
20. Hagerman, P. J., and Hagerman, R. J. (2004). The fragile-X premutation: A maturing perspective. *Am. J. Hum. Genet.* **74**, 805–816.
21. Hagerman, P. J., and Hagerman, R. J. (2004). Fragile X-associated tremor/ataxia syndrome (FXTAS). *Ment. Retard. Dev. Disabil. Res. Rev.* **10**, 25–30.
22. Jacquemont, S., Hagerman, R. J., Leehey, M., Grigsby, J., Zhang, L., Brunberg, J. A., Greco, C., Des Portes, V., Jardini, T., Levine, R., et al. (2003) Fragile X premutation tremor/ataxia syndrome: Molecular, clinical, and neuroimaging correlates. *Am. J. Hum. Genet.* **72**, 869–878.
23. Hagerman, P. J., Greco, C. M., and Hagerman, R. J. (2003). A cerebellar tremor/ataxia syndrome among fragile X premutation carriers. *Cytogenet. Genome Res.* **100**, 206–212.
24. Bakker, C. E., Verheij, C., Willemsen, R., Vanderhelm, R., Oerlemans, F., Vermey, M., Bygrave, A., Hoogeveen, A. T., Oostra, B. A., Reyniers, E., et al. (1994). Fmr1 knockout mice: A model to study fragile X mental retardation. *Cell* **78**, 23–33.
25. Ashley, C. T., Sutcliffe, J. S., Kunst, C. B., Leiner, H. A., Eichler, E. E., Nelson, D. L., and Warren, S. T. (1993). Human and murine FMR-1: Alternative splicing and translational initiation downstream of the CGG-repeat. *Nat. Genet.* **4**, 244–251.
26. Hinds, H. L., Ashley, C. T., Sutcliffe, J. S., Nelson, D. L., Warren, S. T., Housman, D. E., and Schalling, M. (1993). Tissue specific expression of FMR-1 provides evidence for a functional role in fragile X syndrome. *Nat. Genet.* **3**, 36–43.

27. Khandjian, E. W., Fortin, A., Thibodeau, A., Tremblay, S., Cote, F., Devys, D., Mandel, J. L., and Rousseau, F. (1995). A heterogeneous set of FMR1 proteins is widely distributed in mouse tissues and is modulated in cell culture. *Human Mol Genet*, **4**, 783–790.
28. Bakker, C. E., de Diego Otero, Y., Bontekoe, C., Raghoe, P., Luteijn, T., Hoogeveen, A. T., Oostra, B. A., and Willemsen, R. (2000). Immunocytochemical and biochemical characterization of FMRP, FXR1P, and FXR2P in the mouse. *Exp. Cell Res.* **258**, 162–170.
29. De Diego Otero, Y., Bakker, C. E., Raghoe, P., Severijnen, L. W. F. M., Hoogeveen, A., Oostra, B. A., and Willemsen, R. (2000). Immunocytochemical characterization of FMRP, FXR1P and FXR2P during embryonic development in the mouse. *Gene Funct. Dis.* **1**, 28–37.
30. Kooy, R. F., Dhooge, R., Reyniers, E., Bakker, C. E., Nagels, G., Deboulle, K., Storm, K., Clincke, G., Dedeyn, P. P., Oostra, B. A., et al. (1996). Transgenic mouse model for the fragile X syndrome. *Am. J. Med. Genet.* **64**, 241–245.
31. Willems, P. J., Reyniers, E., and Oostra, B. A. (1995). An animal model for fragile X syndrome. *Mental Retard. Dev. Disabil. Res. Rev.* **1**, 298–302.
32. Mientjes, E. J., Nieuwenhuizen, I., Kirkpatrick, L., Zu, T., Hoogeveen-Westerveld, M., Severijnen, L., Willemsen, R., Nelson, D. L., and Oostra, B. A. The generation of a conditional Fmr1 knock out mouse model to study Fmrp function in vivo. *Neurobiol. Dis.*, Epub Oct 2005.
33. Reyniers, E., Vits, L., De Boulle, K., Van Roy, B., Van Velzen, D., de Graaff, E., Verkerk, A. J. M. H., Jorens, H. Z., Darby, J. K., Oostra, B. A., et al. (1993). The full mutation in the FMR-1 gene of male fragile X patients is absent in their sperm. *Nat. Genet.* **4**, 143–146.
34. Slegtenhorst-Eegdeman, K. E., van de Kant, H. J. G., Post, M., Ruiz, A., Uilenbroek, J. T. J., Bakker, C. E., Oostra, B. A., Grootegoed, J. A., de Rooij, D. G., and Themmen, A. P. N. (1998). Macro-orchidism in *FMR1* knockout mice is caused by increased Sertoli cell proliferation during testis development. *Endocrinology* **139**, 156–162.
35. Sharpe, R. M. (1993). "Experimental Evidence for Sertoli–Germ Cell and Sertoli–Leydig Cell Interactions." In Russell LD, Griswold MD (eds.), *The Sertoli Cell*. Clearwater, FL: Cache River Press; 1993: 391–418.
36. Kumar, T. R., Wang, Y., Lu, N., and Matzuk, M. M. (1997). Follicle stimulating hormone is required for ovarian follicle maturation but not male fertility. *Nat. Genet.* **15**, 201–204.
37. Orth, J. M. (1982). Proliferation of Sertoli cells in fetal and postnatal rats: A quantitative autoradiographic study. *Anat. Rec.* **203**, 485–492.
38. Moore, P. S., Chudley, A. E., and Winter, J. S. (1991). Pituitary–gonadal axis in prepubertal boys with the fragile X syndrome. *Am. J. Med. Genet.* **39**, 374–375.
39. Nielsen, K. B., Tommerup, N., Dyggve, H. V., and Schou, C. (1982). Macroorchidism and fragile X in mentally retarded males: Clinical, cytogenetic, and some hormonal investigations in mentally retarded males, including two with the fragile site at Xq28, fra(X)(q28). *Hum. Genet.* **61**, 113–117.
40. Malter, H. E., Iber, J. C., Willemsen, R., De Graaff, E., Tarleton, J. C., Leisti, J., Warren, S. T., and Oostra, B. A. (1997). Characterization of the full fragile X syndrome mutation in fetal gametes. *Nat. Genet.* **15**, 165–169.
41. Peier, A. M., McIlwain, K. L., Kenneson, A., Warren, S. T., Paylor, R., and Nelson, D. L. (2000). (Over)correction of FMR1 deficiency with YAC transgenics: Behavioral and physical features. *Hum. Mol. Genet.* **9**, 1145–1159.
42. Bakker, C. E., and Oostra, B. A. (2003). Understanding fragile X syndrome: Insights from animal models. *Cytogenet. Genome. Res.* **100**, 111–123.
43. Qin, M., Kang, J., and Smith, C. B. (2002). Increased rates of cerebral glucose metabolism in a mouse model of fragile X mental retardation. *Proc. Natl. Acad. Sci. USA* **99**, 15758–15763.
44. Qin, M., Kang, J., Burlin, T. V., Jiang, C., and Smith, C. B. (2005). Postadolescent changes in regional cerebral protein synthesis: An in vivo study in the FMR1 null mouse. *J. Neurosci.* **25**, 5087–5095.
45. Gruss, M., and Braun, K. (2001). Alterations of amino acids and monoamine metabolism in male Fmr1 knockout mice: A putative animal model of the human fragile X mental retardation syndrome. *Neural Plast.* **8**, 285–298.
46. Gruss, M., and Braun, K. (2004). Age- and region-specific imbalances of basal amino acids and monoamine metabolism in limbic regions of female Fmr1 knock-out mice. *Neurochem. Int.* **45**, 81–88.
47. Ivanco, T. L., and Greenough, W. T. (2002). Altered mossy fiber distributions in adult Fmr1 (FVB) knockout mice. *Hippocampus* **12**, 47–54.
48. Crusio, W. E., and Schwegler, H. (1987). Hippocampal mossy fiber distribution covaries with open-field habituation in the mouse. *Behav. Brain Res.* **26**, 153–158.
49. Schwegler, H., and Crusio, W. E. (1995). Correlations between radial-maze learning and structural variations of septum and hippocampus in rodents. *Behav. Brain Res.* **67**, 29–41.
50. Schwegler, H., Crusio, W. E., and Brust, I. (1990). Hippocampal mossy fibers and radial-maze learning in the mouse: A correlation with spatial working memory but not with non-spatial reference memory. *Neuroscience* **34**, 293–298.
51. Mineur, Y. S., Sluyter, F., de, W. S., Oostra, B. A., and Crusio, W. E. (2002). Behavioral and neuroanatomical characterization of the Fmr1 knockout mouse. *Hippocampus* **12**, 39–46.
52. Rudelli, R. D., Brown, W. T., Wisniewski, K., Jenkins, E. C., Laure-Kamionowska, M., Connell, F., and Wisniewski, H. M. (1985). Adult fragile X syndrome: Clinico-neuropathologic findings. *Acta Neuropathol.* **67**, 289–295.
53. Comery, T. A., Harris, J. B., Willems, P. J., Oostra, B. A., Irwin, S. A., Weiler, I. J., and Greenough, W. T. (1997). Abnormal dendritic spines in fragile X knockout mice: Maturation and pruning deficits. *Proc. Natl. Acad. Sci. USA* **94**, 5401–5404.
54. Irwin, S. A., Patel, B., Idupulapati, M., Harris, J. B., Crisostomo, R. A., Larsen, B. P., Kooy, F., Willems, P. J., Cras, P., Kozlowski, P. B., et al. (2001). Abnormal dendritic spine characteristics in the temporal and visual cortices of patients with fragile-X syndrome: A quantitative examination. *Am. J. Med. Genet.* **98**, 161–167.
55. Nimchinsky, E. A., Oberlander, A. M., and Svoboda, K. (2001). Abnormal development of dendritic spines in fmr1 knock-out mice. *J. Neurosci.* **21**, 5139–5146.
56. Koekkoek, S. K., Yamaguchi, K., Milojkovic, B. A., Dortland, B. R., Ruigrok, T. J., Maex, R., De Graaf, W., Smit, A. E., Vanderwerf, F., Bakker, C. E., et al. (2005). Deletion of FMR1 in Purkinje cells enhances parallel fiber LTD, enlarges spines, and attenuates cerebellar eyelid conditioning in fragile X syndrome. *Neuron* **47**, 339–352.
57. Irwin, S. A., Idupulapati, M., Gilbert, M. E., Harris, J. B., Chakravarti, A. B., Rogers, E. J., Crisostomo, R. A., Larsen, B. P., Mehta, A., Alcantara, C. J., et al. (2002). Dendritic spine and dendritic field characteristics of layer V pyramidal neurons in the visual cortex of fragile-X knockout mice. *Am. J. Med. Genet.* **111**, 140–146.
58. Kakizawa, S., Yamada, K., Iino, M., Watanabe, M., and Kano, M. (2003). Effects of insulin-like growth factor I on climbing fibre synapse elimination during cerebellar development. *Eur. J. Neurosci.* **17**, 545–554.
59. Cesa, R., Morando, L., and Strata, P. (2003). Glutamate receptor delta2 subunit in activity-dependent heterologous synaptic competition. *J. Neurosci.* **23**, 2363–2370.

60. Weiler, I. J., Irwin, S. A., Klintsova, A. Y., Spencer, C. M., Brazelton, A. D., Miyashiro, K., Comery, T. A., Patel, B., Eberwine, J., and Greenough, W. T. (1997). Fragile X mental retardation protein is translated near synapses in response to neurotransmitter activation. *Proc. Natl. Acad. Sci. USA* **94**, 5395–5400.
61. Godfraind, J. M., Reyniers, E., Deboulle, K., Dhooge, R., Dedeyn, P. P., Bakker, C. E., Oostra, B. A., Kooy, R. F., and Willems, P. J. (1996). Long-term potentiation in the hippocampus of fragile X knockout mice. *Am. J. Med. Genet.* **64**, 246–251.
62. Paradee, W., Melikian, H. E., Rasmussen, D. L., Kenneson, A., Conn, P. J., and Warren, S. T. (1999). Fragile X mouse: Strain effects of knockout phenotype and evidence suggesting deficient amygdala function. *Neuroscience* **94**, 185–192.
63. Li, J., Pelletier, M. R., Perez Velazquez, J. L., and Carlen, P. L. (2002). Reduced cortical synaptic plasticity and GluR1 expression associated with fragile X mental retardation protein deficiency. *Mol. Cell. Neurosci.* **19**, 138–151.
64. Huber, K. M., Gallagher, S. M., Warren, S. T., and Bear, M. F. (2002). Altered synaptic plasticity in a mouse model of fragile X mental retardation. *Proc. Natl. Acad. Sci. USA* **99**, 7746–7750.
65. Dobkin, C., Rabe, A., Dumas, R., El Idrissi, A., Haubenstock, H., and Ted Brown, W. (2000). Fmr1 knockout mouse has a distinctive strain-specific learning impairment. *Neuroscience* **100**, 423–429.
66. Musumeci, S. A., Bosco, P., Calabrese, G., Bakker, C., De Sarro, G. B., Elia, M., Ferri, R., and Oostra, B. A. (2000). Audiogenic seizures susceptibility in transgenic mice with fragile X syndrome. *Epilepsia* **41**, 19–23.
67. Musumeci, S. A., Hagerman, R. J., Ferri, R., Bosco, P., Dalla Bernardina, B., Tassinari, C. A., De Sarro, G. B., and Elia, M. (1999). Epilepsy and EEG findings in males with fragile X syndrome. *Epilepsia* **40**, 1092–1099.
68. D'Hooge, R., Nagels, G., Franck, F., Bakker, C. E., Reyniers, E., Storm, K., Kooy, R. F., Oostra, B. A., Willems, P. J., and Dedeyn, P. P. (1997). Mildly impaired water maze performance in male Fmr1 knockout mice. *Neuroscience* **76**, 367–376.
69. Morris, R. G. M. (1981). Spatial localization does not require the presence of local clues. *Learn. Motiv.* **12**, 239–260.
70. Morris, R. G., Garrud, P., Rawlins, J. N., and O'Keefe, J. (1982). Place navigation impaired in rats with hippocampal lesions. *Nature* **297**, 681–683.
71. Chen, L., and Toth, M. (2001). Fragile X mice develop sensory hyperreactivity to auditory stimuli. *Neuroscience* **103**, 1043–1050.
72. Nielsen, D. M., Derber, W. J., McClellan, D. A., and Crnic, L. S. (2002). Alterations in the auditory startle response in Fmr1 targeted mutant mouse models of fragile X syndrome. *Brain Res.* **927**, 8–17.
73. Van Dam, D., D'Hooge, R., Hauben, E., Reyniers, E., Gantois, I., Bakker, C. E., Oostra, B. A., Kooy, R. F., and De Deyn, P. P. (2000). Spatial learning, contextual fear conditioning and conditioned emotional response in Fmr1 knockout mice. *Behav. Brain Res.* **117**, 127–136.
74. Bear, M. F., Huber, K. M., and Warren, S. T. (2004). The mGluR theory of fragile X mental retardation. *Trends Neurosci.* **27**, 370–377.
75. Yan, Q. J., Rammal, M., Tranfaglia, M., and Bauchwitz, R. P. (2005). Suppression of two major fragile X syndrome mouse model phenotypes by the mGluR5 antagonist MPEP. *Neuropharmacology* **49**, 1053–1066.
76. Todd, P. K., and Mack, K. J. (2000). Sensory stimulation increases cortical expression of the fragile X mental retardation protein in vivo. *Brain Res. Mol. Brain Res.* **80**, 17–25.
77. Todd, P. K., Malter, J. S., and Mack, K. J. (2003). Whisker stimulation-dependent translation of FMRP in the barrel cortex requires activation of type I metabotropic glutamate receptors. *Brain Res. Mol. Brain Res.* **110**, 267–278.
78. Gabel, L. A., Won, S., Kawai, H., McKinney, M., Tartakoff, A. M., and Fallon, J. R. (2004). Visual experience regulates transient expression and dendritic localization of fragile X mental retardation protein. *J. Neurosci.* **24**, 10579–10583.
79. Irwin, S. A., Swain, R. A., Christmon, C. A., Chakravarti, A., Weiler, I. J., and Greenough, W. T. (2000). Evidence for altered fragile-X mental retardation protein expression in response to behavioral stimulation. *Neurobiol. Learn. Mem.* **73**, 87–93.
80. Irwin, S. A., Christmon, C. A., Grossman, A. W., Galvez, R., Kim, S. H., Degrush, B. J., Weiler, I. J., and Greenough, W. T. (2005). Fragile X mental retardation protein levels increase following complex environment exposure in rat brain regions undergoing active synaptogenesis. *Neurobiol. Learn. Mem.* **83**, 180–187.
81. Restivo, L., Ferrari, F., Passino, E., Sgobio, C., Bock, J., Oostra, B. A., Bagni, C., and Ammassari-Teule, M. (2005). Enriched environment promotes behavioral and morphological recovery in a mouse model for the fragile X syndrome. *Proc. Natl. Acad. Sci. USA* **102**, 11557–11562.
82. Son, L. S., and Wells, R. D. (2006). Triplexes, sticky DNA, and the (GAA TTC) trinucleotide repeat associated with Friedrich's ataxia. *In* "Genetic Instabilities and Neurological Diseases" (R. D. Wells and T. Ashizawa, Eds.), Ch. 20. Elsevier, Burlington, MA.
83. Bontekoe, C. J. M., de Graaff, E., Nieuwenhuizen, I. M., Willemsen, R., and Oostra, B. A. (1997). FMR1 premutation allele is stable in mice. *Eur. J. Hum. Genet.* **5**, 293–298.
84. Lavedan, C., Grabczyk, E., Usdin, K., and Nussbaum, R. L. (1998). Long uninterrupted CGG repeats within the first exon of the human FMR1 gene are not intrinsically unstable in transgenic mice. *Genomics* **50**, 229–240.
85. Lavedan, C. N., Garrett, L., and Nussbaum, R. L. (1997). Trinucleotide repeats (CGG)22TGG(CGG)43TGG(CGG)21 from the fragile X gene remain stable in transgenic mice. *Hum. Genet.* **100**, 407–414.
86. Bontekoe, C. J., Bakker, C. E., Nieuwenhuizen, I. M., van Der Linde, H., Lans, H., de Lange, D., Hirst, M. C., and Oostra, B. A. (2001). Instability of a (CGG)(98) repeat in the Fmr1 promoter. *Hum. Mol. Genet.* **10**, 1693–1699.
87. Willemsen, R., Hoogeveen-Westerveld, M., Reis, S., Holstege, J., Severijnen, L., Nieuwenhuizen, I., Schrier, M., VanUnen, L., Tassone, F., Hoogeveen, A., Hagerman, P., Mientjes, E., and Oostra, B. A. (2003). The FMR1 CGG repeat mouse displays ubiquitin-positive intranuclear neuronal inclusions: Implications for the cerebellar tremor/ataxia syndrome. *Hum. Mol. Genet.* **12**, 949–959.
88. Peier, A., and Nelson, D. (2002). Instability of a premutation-sized CGG repeat in FMR1 YAC transgenic mice. *Genomics* **80**, 423–432.
89. Greco, C. M., Hagerman, R. J., Tassone, F., Chudley, A. E., Del Bigio, M. R., Jacquemont, S., Leehey, M., and Hagerman, P. J. (2002). Neuronal intranuclear inclusions in a new cerebellar tremor/ataxia syndrome among fragile X carriers. *Brain* **125**, 1760–1771.
90. Van Dam, D., Errijgers, V., Kooy, R. F., Willemsen, R., Mientjes, E., Oostra, B. A., and De Deyn, P. P. (2005). Cognitive decline, neuromotor and behavioural disturbances in a mouse model for fragile-X-associated tremor/ataxia syndrome (FXTAS). *Behav. Brain Res.* **162**, 233–239.
91. Tamanini, F., Willemsen, R., van Unen, L., Bontekoe, C., Galjaard, H., Oostra, B. A., and Hoogeveen, A. T. (1997). Differential expression of FMR1, FXR1 and FXR2 proteins in human brain and testis. *Hum. Mol. Genet.* **6**, 1315–1322.
92. Dube, M., Huot, M. E., and Khandjian, E. W. (2000). Muscle specific fragile X related protein 1 isoforms are sequestered in the nucleus of undifferentiated myoblast. *BMC Genet.* **1**, 1–4.
93. Kirkpatrick, L. L., McIlwain, K. A., and Nelson, D. L. (1999). Alternative splicing in the murine and human FXR1 genes. *Genomics* **59**, 193–202.

94. Khandjian, E. W., Bardoni, B., Corbin, F., Sittler, A., Giroux, S., Heitz, D., Tremblay, S., Pinset, C., Montarras, D., Rousseau, F., et al. (1998). Novel isoforms of the fragile X related protein FXR1P are expressed during myogenesis. *Hum. Mol. Genet.* **7**, 2121–2128.
95. Bontekoe, C. J., McIlwain, K. L., Nieuwenhuizen, I. M., Yuva-Paylor, L. A., Nellis, A., Willemsen, R., Fang, Z., Kirkpatrick, L., Bakker, C. E., McAninch, R., et al. (2002). Knockout mouse model for Fxr2: A model for mental retardation. *Hum. Mol. Genet.* **11**, 487–498.
96. Mientjes, E. J., Willemsen, R., Kirkpatrick, L. L., Nieuwenhuizen, I. M., Hoogeveen-Westerveld, M., Verweij, M., Reis, S., Bardoni, B., Hoogeveen, A. T., Oostra, B. A., et al. (2004). Fxr1 knockout mice show a striated muscle phenotype: Implications for Fxr1p function in vivo. *Hum. Mol. Genet.* **13**, 1291–1302.
97. Wan, L., Dockendorff, T. C., Jongens, T. A., and Dreyfuss, G. (2000). Characterization of dFMR1, a *Drosophila melanogaster* homolog of the fragile X mental retardation protein. *Mol. Cell. Biol.* **20**, 8536–8547.
98. Zhang, Y. Q., Bailey, A. M., Matthies, H. J., Renden, R. B., Smith, M. A., Speese, S. D., Rubin, G. M., and Broadie, K. (2001). *Drosophila* fragile X-related gene regulates the MAP1B homolog Futsch to control synaptic structure and function. *Cell* **107**, 591–603.
99. Morales, J., Hiesinger, P. R., Schroeder, A. J., Kume, K., Verstreken, P., F. Rob Jackson, F. R., Nelson, D. L., and Hassan, B. A. (2002). *Drosophila* fragile X protein, DFXR, regulates neuronal morphology and function in the brain. *Neuron* **34**, 961–972.
100. Dockendorff, T. C., Su, H. S., McBride, S. M. J., Yang, Z., Choi, C. H., Siwicki, K. K., Sehgal, A., and Jongens, T. A. (2002). *Drosophila* lacking dfmr1 activity show defects in circadian output and fail to maintain courtship interest. *Neuron* **34**, 973–984.
101. Schenck, A., Van de Bor, V., Bardoni, B., and Giangrande, A. (2002). Novel features of dFMR1, the *Drosophila* orthologue of the fragile X mental retardation protein. *Neurobiol. Dis.* **11**, 53–63.
102. Inoue, S., Shimoda, M., Nishinokubi, I., Siomi, M., Okamura, M., Nakamura, A., Kobayashi, S., Ishida, N., and Siomi, H. (2002). A role for the *Drosophila* fragile x-related gene in circadian output. *Curr. Biol.* **12**, 1331.
103. Xu, K., Bogert, B. A., Li, W., Su, K., Lee, A., and Gao, F. B. (2004). The fragile X-related gene affects the crawling behavior of *Drosophila* larvae by regulating the mRNA level of the DEG/ENaC protein Pickpocket1. *Curr. Biol.* **14**, 1025–1034.
104. McBride, S. M., Choi, C. H., Wang, Y., Liebelt, D., Braunstein, E., Ferreiro, D., Sehgal, A., Siwicki, K. K., Dockendorff, T. C., Nguyen, H. T., et al. (2005). Pharmacological rescue of synaptic plasticity, courtship behavior, and mushroom body defects in a *Drosophila* model of fragile X syndrome. *Neuron* **45**, 753–764.
105. Michel, C. I., Kraft, R., and Restifo, L. L. (2004). Defective neuronal development in the mushroom bodies of *Drosophila* fragile X mental retardation 1 mutants. *J. Neurosci.* **24**, 5798–5809.
106. Pan, L., Zhang, Y. Q., Woodruff, E., and Broadie, K. (2004). The *Drosophila* fragile X gene negatively regulates neuronal elaboration and synaptic differentiation. *Curr. Biol.* **14**, 1863–1870.
107. Zhang, Y. Q., Matthies, H. J., Mancuso, J., Andrews, H. K., Woodruff, E., 3rd, Friedman, D., and Broadie, K. (2004). The *Drosophila* fragile X-related gene regulates axoneme differentiation during spermatogenesis. *Dev. Biol.* **270**, 290–307.
108. Zarnescu, D. C., Jin, P., Betschinger, J., Nakamoto, M., Wang, Y., Dockendorff, T. C., Feng, Y., Jongens, T. A., Sisson, J. C., Knoblich, J. A., et al. (2005). Fragile X protein functions with lgl and the par complex in flies and mice. *Dev. Cell* **8**, 43–52.
109. Lu, R., Wang, H., Liang, Z., Ku, L., O'Donnell W, T., Li, W., Warren, S. T., and Feng, Y. (2004). The fragile X protein controls microtubule-associated protein 1B translation and microtubule stability in brain neuron development. *Proc. Natl. Acad. Sci. USA* **101**, 15201–15206.
110. Schenck, A., Bardoni, B., Langmann, C., Harden, N., Mandel, J. L., and Giangrande, A. (2003). CYFIP/Sra-1 controls neuronal connectivity in *Drosophila* and links the Rac1 GTPase pathway to the fragile X protein. *Neuron* **38**, 887–898.
111. Lee, A., Li, W., Xu, K., Bogert, B. A., Su, K., and Gao, F. B. (2003). Control of dendritic development by the *Drosophila* fragile X-related gene involves the small GTPase Rac1. *Development* **130**, 5543–5552.
112. Reeve, S. P., Bassetto, L., Genova, G. K., Kleyner, Y., Leyssen, M., Jackson, F. R., and Hassan, B. A. (2005). The *Drosophila* fragile X mental retardation protein controls actin dynamics by directly regulating profilin in the brain. *Curr. Biol.* **15**, 1156–1163.
113. Caudy, A. A., Myers, M., Hannon, G. J., and Hammond, S. M. (2002). Fragile X-related protein and VIG associate with the RNA interference machinery. *Genes Dev.* **16**, 2491–2496.
114. Jin, P., Alisch, R. S., and Warren, S. T. (2004). RNA and microRNAs in fragile X mental retardation. *Nat. Cell Biol.* **6**, 1048–1053.
115. Zhang, Y. Q., Friedman, D. B., Wang, Z., Woodruff, E., 3rd, Pan, L., O'Donnell, J., and Broadie, K. (2005). Protein expression profiling of the drosophila fragile X mutant brain reveals upregulation of monoamine synthesis. *Mol. Cell. Proteomics* **4**, 278–290.
116. Jin, P., Zarnescu, D. C., Zhang, F., Pearson, C. E., Lucchesi, J. C., Moses, K., and Warren, S. T. (2003). RNA-mediated neurodegeneration caused by the fragile X premutation rCGG repeats in *Drosophila*. *Neuron* **39**, 739–747.

CHAPTER 12

Chromosomal Fragile Sites: Mechanisms of Cytogenetic Expression and Pathogenic Consequences

ROBERT I. RICHARDS

ARC Special Research Centre for the Molecular Genetics of Development, ARC/NHMRC Research Network in Genes and Environment in Development, School of Molecular and Biomedical Sciences, The University of Adelaide, Adelaide, SA 5005, Australia

I. Introduction
II. Historical Aspects of Chromosomal Fragile Sites
III. "Rare" Fragile Sites
   A. Folate-Sensitive Rare Fragile Sites
   B. Nonfolate-Sensitive Rare Fragile Sites
IV. "Common" or Constitutive Fragile Sites
   A. Mechanism of Cytogenetic Formation
   B. Contribution to Cancer
V. Conclusions
   Acknowledgments
   References

The comparative analysis of different chromosomal fragile sites has shed light on the molecular mechanisms responsible for their cytogenic manifestations and their various contributions to human pathology. A clear relationship has emerged, at least for the common fragile sites, between chromosomal breakage induced *in vitro* and DNA instability observed *in vivo*. This is particularly relevant given the finding that genes associated with some of these sites, have roles to play in cancer. The highly conserved relationships between the distinct classes of fragile sites and their associated genes suggest that normal functional roles are played by these relationships. For the rare, folate sensitive fragile sites the responsible CCG repeat is invariably located in the 5' untranslated region of the associated gene suggesting a normal role for these sequences in the RNA. For the common fragile sites, *FRA3B* and *FRA16D*, the protective function of their respective genes, *FHIT* and *WWOX/FOR* suggests that these sites and their associated genes are part of the cell's normal response to environmental conditions that cause replicative stress.

I. INTRODUCTION

Chromosomal fragile sites are reproducibly located nonstaining gaps or breaks in chromosomes that can be induced to appear under certain conditions of cell culture. More than 100 such sites have been identified in human chromosomes. The main criterion by which they have been classified is their frequency in the population. They are further classified according to the chemicals used in cell culture to induce their cytogenetic expression.

"Rare" fragile sites are seen only in the chromosomes of certain individuals (less than 5% of the population). The most thoroughly characterized of the rare fragile sites is *FRAXA*, and this locus is responsible for one of the most common forms of inherited mental retardation that consequently bears its name—fragile X syndrome. Detailed descriptions of the molecular basis of *FRAXA* and its clinical consequences appear in other chapters of this volume.

"Common" fragile sites are found in most human chromosomes and in all individuals in the population. Interest has been generated in the molecular basis of common fragile sites because of their reported association with regions of DNA instability in cancer [1, 2]. A growing body of evidence shows that certain common chromosomal fragile sites are regions of DNA instability in some forms of cancer.

A third, somewhat enigmatic, group of chromosomal fragile sites is associated with adenovirus 12 infection [3, 4]. This group of fragile sites appears distinct from the other classes in that their cytogenetic expression is associated with defects in transcription- and/or transcription-associated repair rather than DNA replication [4].

II. HISTORICAL ASPECTS OF CHROMOSOMAL FRAGILE SITES

Chromosomal fragile sites have attracted a great deal of speculation and controversy over both their molecular basis and their contribution to disease. The association of the *FRAXA* fragile site with its namesake, fragile X syndrome, together with a growing list of associations between various fragile sites and disease, has provided the impetus for intense investigation of both the chromosomal structures themselves and the pathways by which they contribute to human pathology.

Much speculation surrounded the molecular basis of the fragile X mutation prior to its characterization. In an attempt to explain the unusual segregation characteristics of the disease in affected families, a mechanism was proposed that involved a "local block to the reactivation" of the inactivated X chromosome [5]. The finding that methylation was associated with the *FRAXA* mutation [6] was seen by some to support this proposal; however, it was dispelled with the molecular characterization of the autosomal rare fragile site *FRA16A* [7]. This fragile site was found to have the same molecular basis as *FRAXA* (an expanded CGG repeat [8]), and revealed that the associated methylation was a consequence of the mutation, not a cause of it, as the CpG island at the *FRA16A* locus is not a normal site of imprinting of non-fragile site-expressing chromosomes [7].

The relationship between common chromosomal fragile sites and DNA instability in cancer has also been a subject of controversy. Although an association was proposed, based on statistical grounds [1, 2], it was soon challenged on the basis of both an individual example [9] and additional statistical analysis [10]. The detailed characterization of the *FRA3B* common fragile site and, more recently, the *FRA16D* common fragile site has demonstrated that not only are some common fragile sites the location of DNA instability in cancer but their *in vitro* fragility and *in vivo* breakage occur predominantly within very large genes that both span the region and indeed have the ability to contribute to cancer progression.

Therefore, the molecular characterization of numerous chromosomal fragile site loci has now confirmed a variety of types of contributions to human pathology, at least for certain of these fragile sites. However, although the comparative analysis of fragile sites from within one particular classification has been instructive in identifying some common properties, generalizing the findings from one fragile site locus to others can be misleading, as different fragile sites can also have distinct properties. It is also clear that distinguishing cause from consequence in regard to the role that chromosomal fragile sites play in biology has been difficult.

III. "RARE" FRAGILE SITES

The rare class of fragile sites is further classified according to their induction chemistry, with those sensitive to the level of folate (including *FRAXA*) being the most numerous ($n = 22$ out of a total of 29).

A. Folate-Sensitive Rare Fragile Sites

Six of the folate-sensitive chromosomal fragile sites (*FRAXA*, *FRAXE*, *FRAXF*, *FRA11B*, *FRA16A*, and *FRA10A*) have now been positionally cloned, and the DNA sequences responsible determined [7, 8, 11–14]. In each case the fragile site is due to the expansion of the longer (usually uninterrupted) alleles of a polymorphic CGG repeat beyond what appears to be a common threshold of ~230 copies. Interruptions in the repeat (such as a single AGG instead of one of the CGG copies) tend to stabilize the allele, probably by reducing slippage. In most cases, genes have been associated with the fragile site, and for each of these cases the repeat is located within the 5′ untranslated region of the gene transcript, suggesting a normal role for the CGG repeat in the RNA [15]. When the repeats are expanded beyond the threshold of ~230 copies, they are typically associated with silencing of transcription of the respective gene and methylation of its promoter region. However, expansions below this number have also been associated with disease, premutation alleles at the *FRAXA* locus giving rise to a premature ovarian failure in females [16] and late-onset tremor ataxia syndrome in males [17]. Because the pathogenic mechanism is due to an RNA dominant gain of function involving a widely expressed gene, there is a distinct possibility that gene transcripts from premutation alleles of other folate-sensitive fragile site loci will also give rise to late-onset tremor ataxia or premature ovarian failure. The detailed pathogenic pathways responsible for the consequences of *FRAXA* CGG repeat expansion that result in fragile X syndrome, fragile X-associated tremor/ataxia syndrome, and premature ovarian failure are described in other chapters of this volume. Another X-linked folate-sensitive fragile site, *FRAXE*, has also been associated with mental retardation [11], but no association of ataxia with premutation alleles of this locus has been reported.

Of the autosomal folate-sensitive fragile sites, only *FRA11B* has been associated with disease, in this case 11q− or Jacobsen syndrome [13]. This association was the first clear demonstration that the chromosome fragility seen *in vitro* following chemical induction had a counterpart in DNA instability *in vivo* (in this case, chromosome breakage). Although *FRA11B*-associated breakage is responsible for only a minority of 11q− cases, other CGG repeats in the vicinity have been implicated in non-*FRA11B* 11q− deletions [18].

This group of fragile sites is sensitive not only to folate levels, but also to thymidylate levels; however, the molecular mechanism by which the level of either of these agents controls cytogenetic appearance of chromosomal fragility at an expanded CGG repeat is yet to be determined [19]. One possibility is that the induction chemicals are affecting pyrimidine biosynthesis and, consequently, the level of cytosine available for replication of the expanded CCG repeat, which may be particularly sensitive because of its extreme cytosine content.

B. Nonfolate-Sensitive Rare Fragile Sites

Other chemicals (e.g., distamycin A and bromodeoxyuridine) have also been found to induce distinct rare fragile site loci, and two of these (*FRA10B* and *FRA16B*) have been characterized at the molecular level [20, 21]. In both cases, expansions of normally polymorphic AT-rich minisatellite repeats give rise to alleles that are able to cytogenetically express the respective chromosomal fragile site. Therefore, while the DNA sequence composition of the expanded repeats is clearly different for folate-sensitive and nonfolate-sensitive rare fragile sites, it is a distinct possibility that similar mechanisms exist for both cytogenetic expression of and mutation leading to rare fragile site alleles. The molecular process(es) by which either distamycin A or BrdU chemical induction results in cytogenetically observable chromosome fragility is yet to be determined.

IV. "COMMON" OR CONSTITUTIVE FRAGILE SITES

A. Mechanism of Cytogenetic Formation

Unlike "rare" fragile sites where expanded repeat sequences appear to coincide with the site of inducible chromosome breakage, attempts at identifying the necessary and sufficient conditions for the cytogenetic formation of "common" fragile sites have been far more problematic.

"Common" fragile sites are found on most human chromosomes and in all individuals in the population. The majority of these sites were initially observed under the conditions of thymidylate or folate stress used to induce rare folate-sensitive fragile sites. Subsequently it was found that aphidicolin, an inhibitor of DNA polymerase α, is a more effective inducing agent for this group of fragile sites [22]. A small additional group of common fragile sites are induced by distinct agents (5-azacytidine and bromodeoxyuridine). The exact number of common fragile sites present in the human genome depends (to some extent) on how the sites are defined. Although more than 70 aphidicolin-sensitive common fragile sites have been identified (see [23] for locations), these different fragile site loci exhibit quite markedly different frequency of cytogenetic expression with a typical hierarchy (i.e., *FRA3B* > *FRA16D* > *FRA6E* > *FRA7G* > *FRAXB* > others [24]). The greater the replicative stress placed on cells in culture, the greater the number of cells that exhibit fragile sites and the greater the number of distinct fragile sites that are seen in individual cells. The most readily observed common fragile sites (i.e., *FRA3B* and *FRA16D*) are therefore the most sensitive of multiple sites in the genome that manifest as nonstaining gaps or breaks in the chromosome in response to specific environmental conditions that typically impact on replication.

When induced in cultured cells, common fragile sites exhibit various forms of associated DNA instability, including sister chromatid exchanges, translocations, and deletions [25–27]. They have also been found to be sites of plasmid [28] and DNA virus [29] integration, in addition to acting as boundaries for gene amplification events both in cultured cells subject to selection and in cancer cells. The likelihood that each of these forms of DNA instability is associated with double-strand DNA breaks has led to the conclusion that double-strand breaks are likely to constitute part of the mechanism of their cytogenetic expression.

There is evidence that both genetic and environmental factors contribute to the level of common fragile site expression [26]. Studies in twins and sib pairs indicate a genetic component of control. Presumably this reflects variation between individuals in DNA sequences (cis-acting elements) in the vicinity of the fragile site loci or in the level of some cellular component (trans-acting factor) that is rate limiting for fragile site expression [30, 31]. Evidence of a role for environmental factors in common fragile site expression extends beyond the known cell culture-inducing agents to include the finding that cigarette smokers have consistently higher levels of fragile site expression [32, 33], whereas ethanol and caffeine [1] have been found to affect expression levels rather than act as inducing agents *per se*. In addition, a diverse array of mutagens and carcinogens have been reported to induce fragile sites [34].

1. CIS-ACTING ELEMENTS

Cytogenetic gaps or breaks at common fragile sites have been mapped across extensive distances (i.e., 4 Mb for *FRA3B* [35], >1 Mb for *FRA16D* [36]); however, the vast majority of breaks at the *FRA16D* locus localize to a significantly narrower interval of ~270 kb [37]. The DNA sequences that span the most frequent regions of *in vitro* induced and cytogenetically observed chromosome breakage at either the *FRA3B* or *FRA16D* fragile sites have been determined, and the locations of *in vivo* deletions in cancer cells mapped [38–42]. There is a good general correlation between *in vitro* fragility and *in vivo* DNA instability, suggesting a causal relationship. This causal relationship is strengthened by the observation that the relative frequency of *in vivo* deletions at common fragile site loci in cancer cells roughly parallels the relative frequency of *in vitro* cytogenetic expression at different fragile sites [37, 42, 43]. There are no specific sequences (e.g., expanded simple tandem repeats (STRs)) located at the boundaries of deletions. However within the commonly deleted regions are a greater number of "flexible" sequences identified by the FlexStab computer algorithm [44]. A higher frequency of these flexible sequences has also been found in the vicinity of the *FRA7H* common fragile site [44]. These FlexStab sequences can, however, be deleted from either the *FRA3B* or *FRA16D* fragile sites in cancer cells and the regions still cytogenetically express a fragile site [42, 45]. So, although these flexible sequences might render the region more likely to break at a particular point, they do not appear essential for fragile site expression. Indeed, the highest-scoring peak of flexibility (FlexStab1) at the *FRA16D* locus is missing from the otherwise very highly conserved orthologous sequence in mouse. Furthermore, one of the deletion endpoints of a *FRA16D* cancer cell deletion exhibits the *de novo* generation of an AT-rich 9-bp repeat, suggesting that repeats at fragile site loci might be a consequence rather than an obligate cause of fragility [42]. This finding is consistent with the reported coincident location of two rare fragile site loci (*FRA16B* and *FRA10B*) with the common fragile sites *FRA16C* and *FRA10E*, respectively [46].

The sequence conservation between human *FRA3B* and *FRA16D* and their orthologous mouse *Fra14A2* and *Fra8E1* loci is striking [47–49]. Orthologous genes (*FHIT/Fhit* or *WWOX/Wwox*) span the respective fragile sites in each species [47–49]. For *FRA16D*, this conservation is particularly remarkable given that most of the region is located within a single huge *WWOX* gene intron (778,855 bases in length in human and 640,482 bases in mouse). This is (as yet) the longest identified intron in humans, and assuming a transcription rate of 50 ms per base across the entire intron, it would take >10 h to transcribe.

The overall characteristics of the spread of breakpoints and the lack of specific repeat elements, unlike the rare fragile sites, suggest that the "common" and "rare" fragile sites are likely to have quite distinct DNA sequence requirements and cytogenetic expression mechanisms.

2. TRANS-ACTING FACTORS

The observation of simultaneous cytogenetic expression of multiple fragile sites in a single cell suggests that the induction process involves one or more cellular factors that, in certain circumstances, are rate-limiting for the chromosomal integrity of fragile site loci. Several such factors have now been identified. The first of these was ATR [24]. ATR deficiency in cells renders fragile site expression independent of inducing agent. Some cases of Seckel syndrome (*SCKL1*) have been found to carry mutations in the *ATR* gene. Cells from patients with these mutations show increased chromosomal breakage following replication stress [50]. Because ATR functions as a replication checkpoint kinase, it has been proposed that fragile sites are underreplicated chromosomal regions resulting from stalled replication forks that have escaped the ATR checkpoint. Furthermore, the fact that caffeine is an inhibitor of ATR kinase may explain the ability of caffeine to enhance common fragile site expression levels.

BRCA1 is a downstream target of ATR, and Arlt et al. [51] demonstrated that its G_2/M checkpoint function is required for common fragile site stability. The significance of this observation is clear for those cases of breast and other cancers involving *BRCA1* loss-of-function mutations, as it would be expected that the common fragile site loci in these cells would be more likely to exhibit DNA instability *in vivo*.

Exploring the known interaction between *BRCA1* and the Fanconi anemia (FA) pathway in the cellular DNA damage response, Howlett et al. [52] were able to demonstrate that the FA pathway is also required for regulation of common fragile site stability. Similar to *BRCA1* mutation carriers, FA patients are at increased risk of cancer, and therefore, their increased fragile site instability suggests a possible mechanistic pathway to explain this increased cancer susceptibility.

Finally, Musio et al. [53] explored the possible involvement of SMC1 in common fragile site expression because of the known role of this protein in the cellular response to DNA damage. SMCs are the structural maintenance of chromosome family of proteins that form part of the complexes that regulate the higher-order dynamics of chromosomes, including such processes as chromosome condensation and sister chromatid cohesion. Of this protein family, SMC1 and SMC3 were of particular interest to Musio et al. [53] as these proteins have roles in promoting repair of gaps and deletions, postreplicative double-strand break (DSB) repair, and SMC1 is part of the DNA damage response.

Absence of SMC1, in particular, was found to significantly increase breaks at the most frequently observed common fragile site loci (i.e., *FRA3B*, *FRA16D*, *FRAXB*), especially in the presence of aphidicolin [53]. SMC1 appears to act by preventing the collapse of stalled replication forks. Further connecting the relevant factors, it was found that aphidicolin induction results in an increase in SMC1 levels, an association between ATR and SMC1, and the ATR-mediated phosphorylation of SMC1. Cells deficient in ATR do not exhibit SMC1-dependent chromosome fragility, indicating the dependency of this function of SMC1 on interaction with and/or phosphorylation by ATR [53].

Just how each of these *trans*-acting factors contributes to the molecular events involved in elaborating a chromosomal fragile site awaits further, yet more detailed investigation (Fig. 12-1). In particular, none of the described *trans*-acting factors accounts for the reproducible location of common fragile sites. Although certain DNA sequences almost certainly contribute to this specificity, the recognition of these sequences by *trans*-acting factors is likely to play a role. One source of this specificity is likely to be replication timing.

Various studies have addressed the relationship between chromosomal fragile sites and replication timing. At the rare fragile site loci, these studies have

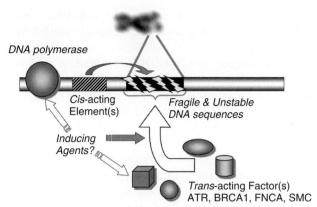

FIGURE 12-1 Components in common chromosomal fragile site expression. Various components for which there is experimental evidence of a contribution to chromosomal fragile site expression have been identified, although the mechanics of their action is as yet speculative. *cis*-Acting elements are DNA sequences in the vicinity of the fragile site but, at least for some common fragile sites (i.e., *FRA3B* and *FRA16D*), are not at the site of most frequent deletion in cancer cells. Fragile and unstable DNA sequences are expanded DNA repeats in the case of rare fragile sites, but for common fragile sites, the location of the break endpoints is quite variable and the specificity of these sequences unknown. Sequences of higher flexibility such as AT dinucleotide repeats may be more prone to breakage; however, variation in AT copy number in the population does not appear to correlate with variation in fragile site expression levels, at least for the *FRA16D* locus. *trans*-Acting factors are normal cellular components (probably all proteins) that by their presence or absence contribute to the genomewide expression of fragile sites, presumably by contributing to or interfering with normal chromosome condensation along with or following the process of DNA replication. Inducing agents are the dietary or environmental factors that might exert their rate-limiting effect on fragile site appearance by facilitating or inhibiting the interaction between *trans*-acting factors and the DNA sequence elements of the fragile site locus. ATR, ataxia telangiectasia-related; BRCA1, breast cancer 1; FA, Fanconi anemia pathway; SMC, structural maintenance of chromosome (1 and 3). See CD-ROM for color image.

sought to determine whether fragile site alleles were later in replicating than their nonfragile site-expressing allelic counterparts. Late replication and further delay of fragile site alleles have both been observed for the *FRAXA* and *FRAXE* loci [54–56]. However, the *FRA10B* locus is not located within a late-replicating region, nor are fragile site-expressing alleles of this locus delayed in their replication. Therefore, late and further delayed replication may not be a necessary condition for rare fragile site expression [57].

On the other hand, common fragile site loci consistently exhibit late replication, which is further delayed on induction, suggesting that this is a necessary component of the cytogenetic expression of this class of fragile site [58–61]. It may be that the timing of replication of the common fragile site loci is sufficient to expose these chromosomal regions to at least some of the specific conditions, including particular *trans*-acting factors, required for the manifestation of a cytogenetic break.

The third group of chromosomal fragile sites, associated with adenovirus 12 infection, are clearly distinct from the rare and common classes. Their cytogenetic expression is associated not with DNA replication, but with defects in transcription-associated repair [3], a process that has been referred to as *transcriptional healing* [62]. This process is dependent on the Cockayne syndrome B DNA repair–transcription coupling factor (CSB), and this protein has a role to play in the cytogenetic expression of this class of fragile sites [4]. Interestingly, the p53 protein also has a role to play in the expression of these fragile sites. Cytogenetic manifestation of fragile sites on chromosomes can therefore be due to quite different pathways involving distinct *trans*-acting factors.

B. Contribution to Cancer

1. DNA Instability in Cancer Cells

Common chromosomal fragile sites are associated with a variety of forms of DNA instability in cancer cells. These fragile site-associated DNA instabilities are also able to make genetic contributions to cancer in various ways, for example, through loss-of-function type mutations (typically associated with tumor suppressors) or through gain of function via amplification (typically associated with oncogenes).

a. Translocations

Prior to the identification of common fragile sites, a translocation breakpoint, t(3;8), in a familial case of renal carcinoma was mapped to 3p14 in what subsequently turned out to be the same location as the *FRA3B* fragile site [63]. *FRA3B* is also the site of t(3;16) and t(3;4) translocations in esophageal adenocarcinoma [64].

Translocations have also been mapped in the vicinity of the *FRA16D* fragile site both in cancer cells [65] and as germline mutations [66]. In both instances disturbances to the expression of the *MAF* proto-oncogene located ~1 Mb distal have been implicated in contributing either to the multiple myeloma phenotype, in the case of the former [65], or to ocular developmental abnormalities, for the latter [66]. Although the instability of this region would appear to be associated with the presence of the *FRA16D* fragile site, in neither case has disruption of the *WWOX* gene, within which most of the translocations are occurring, been implicated in the phenotype.

b. Deletions and Insertions

Both homozygous and heterozygous deletions have been found within or spanning common fragile site regions. The boundaries of the two types of deletions do not coincide with each other, suggesting that the mutation mechanisms are distinct [42]. Cells that exhibit homozygous deletion at one common fragile site locus can sometimes be found to have a homozygous deletion at another [37, 42, 43]. The frequency with which these deletions are observed correlates closely with the hierarchy of cytogenetic expression of the fragile site, providing strong evidence that there is a causal relationship between *in vitro* chromosome fragility and *in vivo* DNA instability. Therefore, understanding the molecular mechanism of cytogenetic fragility (progress described earlier) will provide good insight into the processes involved in this form of DNA instability in cancer [42, 43].

In addition to the clear correlation between *in vitro* chromosome fragility and *in vivo* DNA instability described earlier, various properties of DNA instability at common fragile site loci in cancer cells have been described [42]. Principal among these is the finding that the instability appears to be a fairly early event in cancer cell progression and that, once mutated, these regions can then remain quite stable. This is despite the finding that cells with substantial deletions in the *FRA3B* and/or *FRA16D* loci are still able to manifest fragile sites at these loci [42, 45]. There is also evidence that multiple deletion events have occurred at the one locus, suggesting that the inducing agent(s) and/or cellular environmental conditions responsible for the deletions persisted for some time or were present on multiple occasions.

Human keratinocytes immortalized with HPV16 have been found to frequently have the HPV16 site of integration coincide with the location of common chromosomal fragile sites [67]. Similarly, more than half of the sites of insertion of HPV16 DNA in cervical cancer coincide with common fragile sites [68, 69]. Although the large number and variety of nonfragile site insertions suggest that the relevant gene disruptions might not be essential for transformation, these observations together with the finding that transfected plasmid DNA has integrated at the *FRA3B* locus [70] suggest that common fragile site loci are susceptible to the insertion of foreign DNA.

c. Amplifications

Common fragile sites have been found to act as the boundaries of gene amplification via the breakage–fusion bridge (BFB) mechanism [71]. The fragile sites flanking the amplicon serve two roles: the more telomeric is involved in initiation of the BFB cycle, and the more centromeric determines the length of the amplicon. In addition, initiation of the BFB cycle is correlated with induction of fragile site expression [71]. Furthermore, hypoxia, which is a typical state for cells within an adenocarcinoma, is able to both induce fragile sites and initiate intrachromosomal amplification [72]. A functional contribution by the fragile site-mediated BFB gene amplification process to oncogenesis is evident from the *FRA7G*-associated intrachromosomal amplification of the *MET* oncogene in a human gastric carcinoma [73]. Similarly, BFB amplification in T47D breast carcinoma cells at the *FRA7I* fragile site is associated with increased expression of the *PIP* gene located within the amplicon [74].

2. Genes at Common Fragile Site Loci

The genes spanning or located near those common fragile sites that have been more thoroughly characterized do not exhibit any clear sequence or functional similarities. However, some of these genes do share the remarkable property of their extraordinary length, particularly as the majority of the sequences from primary transcripts from these genes are introns (Table 12-1). These long transcripts are conserved through vertebrate evolution, with the longest intron in the WWOX gene exceeding 63 kb in *fugu*, which is normally noteworthy for its short introns. A biological basis (if any) for these exceptionally long genes is yet to emerge.

The *FHIT* and *WWOX* genes span their respective fragile sites (*FRA3B* and *FRA16D*). Functional data suggest that aberrant expression of each of these genes makes a contribution to cancer, as follows.

a. *FHIT* Gene Spans *FRA3B*

A familial t(3;8) translocation associated with renal carcinoma was identified at 3p14 and presumed to locate a tumor suppressor gene, prior to this region being identified as the location of the *FRA3B* common chromosomal fragile site. A great deal of effort was therefore expended in both the characterization of the *FRA3B* fragile site and the identification of a gene or genes that might contribute to cancer. Since its identification in 1996 [77, 78] as spanning the *FRA3B* fragile site, there have been in excess of 500 publications concerning the *FHIT* gene and its encoded protein. Many of these publications report an association of some form with cancer. This is despite some initial concerns as to whether such an association functionally contributed to cancer or whether the proximity of the *FHIT* gene to the fragile site was merely a coincidence [79].

The *FHIT* gene was found to be frequently mutated (usually deletions) in cancer cells and to give rise to abnormal transcripts [77, 78], although some of these deletions leave the exons intact [80]. Sequence analysis of the *FRA3B/FHIT* region and cancer breakpoints gave some insight into the possible mechanism of deletion and identified the greater abundance of flexible sequences in the fragile site region [39, 40]. The murine *Fhit* gene was also found to map to a fragile site [81], whereas the sequence comparison of human *FHIT/FRA3B* and mouse *Fhit/Fra14A2* revealed that these orthologs are conserved but highly recombinogenic [49, 82]. Like its human counterpart, the murine *Fhit* locus gives rise to abnormal transcripts in cancer [83].

Genetic linkage analysis has provided evidence of a role for the *FHIT* gene in prostate cancer [84].

A substantial body of research has accumulated in support of a role for *FHIT* as a tumor suppressor. Replacement of *FHIT* in cancer cells deficient for the protein induces apoptosis and suppresses tumorigenicity [85, 86]. *FHIT*-rescued tumor cells exhibit increased apoptosis and G_0/G_1 arrest [87, 88]. Transgenic mice ablated for *Fhit* exhibit Muir–Torre-like syndrome and an increased sensitivity to chemical mutagens [89] that can be rescued with gene therapy [87].

Identification of the normal and cancer pathway(s) to which *FHIT* contributes has been problematic. Fhit is a diadenosine (Ap3A) hydrolase [90]. Fhit-induced apoptosis in cancer cells has been found to be limited by substrate binding rather than by its hydrolysis, suggesting that it is likely that some property of the substrate-bound form of the Fhit protein is crucial to apoptosis rather than Ap3A cleavage products [91]. A clue to the pathway in which *FHIT* might normally participate was revealed by the surprising finding that in *Drosophila* and *Caenorhabditis elegans*, *Fhit* is found in a "fused" gene with *Nitrilase*, as components of fused proteins might have related function [92]. FHIT has also been found to have a novel interaction with tubulin [93] and it is the physiological target of src [94].

b. *FOR/WWOX* gene spans *FRA16D*

The *FOR* (fragile site *FRA16D* oxidoreductase) or WWOX (WW-containing oxidoreductase) gene was identified by three different laboratories (1) on the basis of its spanning the *FRA16D* fragile site [41], (2) as a possible tumor suppressor due to its location within a loss of heterozygosity smallest region of overlap in breast cancer [95], or (3) as a gene that was induced by hyaluronidase [96]. Transcripts for this gene are alternatively spliced, mainly toward its 3' end, resulting

TABLE 12-1 Size of Common FRA-Associated Genes

Fragile site	Gene	Transcript (Mb)	mRNA length (kb)
FRA3B	FHIT	0.8	1.1
FRA16D	WWOX(FOR)	1.1	2.2
FRA4	GRID2	1.49	2.7
FRA6E	PARKIN	0.6	3.0

Source. References [75, 76].

in protein products with common N termini (containing two WW domains) and variable-length C termini where sequences necessary for oxidoreductase activity are located. A nuclear localization signal is also located between the WW domains. The most abundant spliced form encodes a protein that has been named FORII, WWOX, and WOX1 [41, 95, 96]. A high degree of conservation has been found (49% amino acid identity between human and *Drosophila*), with orthologs from various species forming a distinct branch of the oxidoreductase superfamily [97]. Since its discovery there has been mounting evidence for a role for WWOX in cancer cell biology.

WWOX has been described as a mitochondrial apoptogenic protein and an essential partner of p53 in cell death [96]. There is, however, disagreement over its cytoplasmic location, as Bednarek et al. [95] locate the protein in the Golgi. Ectopically expressed WWOX acts as an inhibitor of tumor growth in nude mice and, therefore, has been proposed to act as a tumor suppressor [98]. Analysis of a large number of breast cancer tissue samples and cell lines identified one of the alternative splice forms (known as WWOX variant 4 or FORIII) as an abundant transcript in ~50% breast cancer samples and cell lines [99]. There have been several reports describing interactions between WWOX and a variety of important yet functionally diverse proteins including JNK1 [100–104]. Biological significance for most of these interactions is yet to be verified.

Despite the reports of WWOX functioning as a tumor suppressor, Watanabe et al. [105] have challenged the notion that WWOX is a classic tumor suppressor based on the observations that a coding region mutation of WWOX is rarely found in cancer cells and that WWOX protein levels are elevated in gastric and breast carcinoma. Although the precise function of WWOX is yet to be determined, a growing list of reports suggest a likely association with cancer based on frequently observed aberrant expression [41, 99, 106–113].

To gain some insight into the normal role of WWOX and how perturbation of this role might contribute to cancer cell biology, a functional analysis of WWOX was recently undertaken using the *Drosophila* genetic model [97]. *Drosophila* either overexpressing or ablated for WWOX protein were viable and fertile, indicating that the protein is not essential for survival. However, the reported association of WWOX with p53 prompted an analysis of sensitivity to ionizing radiation, as p53-ablated *Drosophila* have increased sensitivity to gamma irradiation compared with their wild-type counterparts [114, 115]. While it was reported that *Drosophila* deficient for WWOX were significantly more sensitive to gamma irradiation [97], it now appears that this sensitivity is due to background genetic changes brought about by the homologous recombination mutagenesis approach utilised [O'Keefe et al., *unpublished observations*]. Further analysis will therefore be required in order to determine the role that this protein normally plays and therefore how perturbation of this role might contribute to cancer cell biology.

c. Other Common Fragile Sites and Associated Genes

i. *FRA6E/Parkin*

The *FRA6E* fragile site has been reported to span an extensive region of 3.6 Mb [116]. *Parkin* and eight other genes are located within this region, with *Parkin* being notable for its length (1.5 Mb). *Parkin* encodes an E3 ubiquitin ligase, and although deletions have previously been associated with autosomal recessive juvenile parkinsonism [117], recently mutations have also been identified in *Parkin* in a variety of cancers [116, 118–121]. Parkin has the characteristics of a tumor suppressor [118].

ii. *FRA4F/GRID2*

FRA4F is a common chromosomal fragile site located at 4q22, with a syntenic site located at 6C1 in mouse [122]. The ionotropic glutamate receptor delta2 gene, *GRID2*, is located within the *FRA4F* region and is again a large gene (~1.4 Mb in mouse, ~1.5 Mb in human) [122]. 4q34-q35 deletions that indicate the likely presence of a tumor suppressor have been identified in hepatocellular carcinoma, and these correspond to the *FRA4F* region [123]. Two independent translocations within the mouse *grid2* gene suggest that this region is unstable [124].

iii. Other Common Fragile Site Loci

Other common fragile site loci (*FRAXB* [43], *FRA7E* [46], *FRA7H* [44], *FRA7I* [74], *FRA7G* [125–127], *FRA9E* [128] and *FRA2G* [129]) have been characterized to varying extents at the molecular level. Genes have been identified at most of these loci, some of which show homozygous deletion and/or loss of expression in a variety of cancers. Evidence for the presence of a tumor suppressor has also been provided for the *FRA7G* locus [130].

As yet, no genes have been identified in association with the 161-kb region where fragility at *FRA7H* has been located [44]. Given the extreme length of genes found to span other common fragile site loci, it is possible that this entire 161-kb region is intronic to a gene that spans this site. In this regard, the computer prediction programs used to annotate the human genome had difficulty in identifying *WWOX* as a single gene because of its length and the presence of some largely noncoding exons. It is therefore possible that similar, large (as yet cryptic) genes may in fact be spanning some of these characterized fragile regions.

V. CONCLUSIONS

Chromosomal fragile sites are clearly intriguing structures with important roles in biology. The comparative analysis of different fragile site loci, both within and between different classes, has been particularly instructive of their mechanisms of cytogenetic expression and the pathogenic consequences of their presence on the chromosome.

A number of pressing questions remain in regard to fragile sites and their contribution to biology. Foremost among these is why it is that common fragile sites are located within genes that can contribute to cancer. These unstable DNA sequences are located within genes that play protective roles against cancer and, therefore, would appear to be something that would be selected against. Perhaps the answer to this dilemma lies in a normal role that this relationship serves. The sensitivity of the common fragile site regions to environmental damage suggests that they (and the genes that span them) might be part of the cell's mechanism for normal response to environmental factors that cause replicative stress. Initial challenge to the fragile site regions might be part of such a response through altered or increased expression that is transmitted to daughter cells as part of their increased resilience to the environmental stress. Persistent damage to the fragile sites would then see such a protective mechanism eventually turned into part of the problem of DNA damage contributing to cancer cell progression.

Another matter that requires clarification is the molecular mechanism(s) by which inducing agents actually cause chromosomes to exhibit cytogenetic fragile sites, and indeed, what is it that distinguishes fragile site loci from the rest of the genome?

Finally, for the rare folate-sensitive fragile sites it will be intriguing to see whether premutation alleles at the non-*FRAXA* loci are also responsible for cases of late-onset tremor/ataxia syndrome or similar symptoms.

Acknowledgments

This work was supported by Grant 207809 from the National Health and Medical Research Council (NHMRC) of Australia, the Australian Research Council (ARC) Special Research Centre for the Molecular Genetics of Development, and the ARC/NHMRC Research Network in Genes and the Environment in Development. Apologies to the literally hundreds of authors whose valuable and significant contributions to this field were not cited in this review because of space constraints.

Sincere thanks to Louise O'Keefe, Sonia Dayan, Donna Crack Yinghong Liu, Tanya Henshall, and Grant Booker for helpful comments on drafts of this work.

References

1. Yunis, J. J., and Soreng, A. L. (1984). Constitutive fragile sites and cancer. *Science* **226**, 1199–1204.
2. Hecht F., and Glover T. W. (1984). Cancer chromosome breakpoints and common fragile sites induced by aphidicolin. *Cancer Genet. Cytogenet.* **13**, 185–188.
3. Yu, A., Bailey, A. D., and Weiner, A. M. (1998). Metaphase fragility of the human *RNU1* and *RNU2* loci is induced by actinomycin D through a p53-dependant pathway. *Hum. Mol. Genet.* **7**, 609–617.
4. Yu, A., Fan, Y. H., Liao, D., Bailey, A. D., and Weiner, A. M. (2000). Activation of p53 or loss of the Cockayne syndrome group B repair protein causes metaphase fragility of human U1, U2, and 5S genes. *Mol. Cell* **5**, 801–810.
5. Laird, C. D. (1987). Proposed mechanism of inheritance and expression of the human fragile-X syndrome of mental retardation. *Genetics* **117**, 587–599.
6. Oberle, I., Rousseau, F., Heitz, D., Kretz, C., Devys, D., Hanauer, A., Boue, J., Bertheas, M. F., and Mandel, J. L. (1991). Instability of a 550-base pair DNA segment and abnormal methylation in fragile X syndrome. *Science* **252**, 1097–1102.
7. Nancarrow, J. K., Kremer, E., Holman, K., Eyre, H., Doggett, N., Le Paslier, D., Callen, D. F., Sutherland, G. R., and Richards, R. I. (1994). Implications of *FRA16A* structure for the mechanism of chromosomal fragile site genesis. *Science* **264**, 1938–1941.
8. Kremer, E., Pritchard, M., Lynch, M., Yu, S., Holman, K., Warren, S., Schlessinger, D., Sutherland, G. R., and Richards, R. I. (1991). DNA instability at the fragile X maps to a trinucleotide repeat sequence p(CCG)n. *Science* **252**, 1711–1714.
9. Simmers, R. N., Sutherland, G. R., West, A., and Richards, R. I. (1987). Fragile sites at 16q22 are not at the breakpoint of the chromosomal rearrangement in AMMoL. *Science* **236**, 92–94.
10. Sutherland, G. R., and Simmers, R. N. (1988). No statistical association between common fragile sites and nonrandom chromosome breakpoints in cancer cells. *Cancer Genet. Cytogenet.* **31**, 9–15.
11. Knight, S. J., Flannery, A. V., Hirst, M. C., Campbell, L., Christodoulou, Z., Phelps, S. R., Pointon, J., Middleton-Price, H. R., Barnicoat, A., Pembrey, M. E., et al. (1993). Trinucleotide repeat amplification and hypermethylation of a CpG island in FRAXE mental retardation. *Cell* **74**, 127–134.
12. Parrish, J. E., Oostra, B. A., Verkerk, A. J., Richards, C. S., Reynolds, J., Spikes, A. S., Shaffer, L. G., and Nelson, D. L. (1994). Isolation of a GCC repeat showing expansion in FRAXF, a fragile site distal to FRAXA and FRAXE. *Nat Genet.* **8**, 229–235.
13. Jones, C., Penny, L., Mattina, T., Yu, S., Baker, E., Voullaire, L., Langdon, W. Y., Sutherland, G. R., Richards, R. I., and Tunnacliffe, A. (1995). Association of a chromosome deletion syndrome with a fragile site within the proto-oncogene *CBL2*. *Nature* **376**, 145–149.
14. Sarafidou, T., Kahl, C., Martinez-Garay, I., Mangelsdorf, M., Gesk, S., Baker, E., Kokkinaki, M., Talley, P., Maltby, E. L., French, L., Harder, L., Hinzmann, B., Nobile, C., Richkind, K., Finnis, M., Deloukas, P., Sutherland, G. R., Kutsche, K., Moschonas, N. K., Siebert, R., and Gecz, J, for the European Collaborative Consortium for the Study of ADLTE (2004). Folate-sensitive

fragile site *FRA10A* is due to an expansion of a CGG repeat in a novel gene, *FRA10AC1*, encoding a nuclear protein. *Genomics* **84**, 69–81.

15. Chiang, P-W., Carpenter, L. E., and Hagerman, P. J. (2001). The 5′-untranslated region of the *FMR1* message facilitates translation by internal ribosome entry. *J. Biol. Chem.* **276**, 37916–37921.
16. Vianna-Morgante A. M,. Costa, S. S., Pares, A. S., and Verreschi, I. T. (1996). FRAXA premutation associated with premature ovarian failure. *Am. J. Med. Genet.* **64**, 373–375.
17. Hagerman, R. J., and Hagerman, P. J. (2002). The fragile X premutation: Into the phenotypic fold. *Curr. Opin. Genet. Dev.* **12**, 278–283.
18. Jones, C., Mullenbach, R., Grossfeld, P., Auer, R., Favier, R., Chien, K., James, M., Tunnacliffe, A., and Cotter, F. (2000). Co-localisation of CCG repeats and chromosome deletion breakpoints in Jacobsen syndrome: Evidence for a common mechanism of chromosome breakage. *Hum. Mol. Genet.* **9**, 1201–1208.
19. Glover, T. W. (1981). FUdR induction of the X chromosome fragile site: Evidence for the mechanism of folic acid and thymidine inhibition. *Am. J. Hum. Genet.* **33**, 234–242.
20. Yu, S., Mangelsdorf, M., Hewett, D., Hobson, L., Baker, E., Eyre, H., Lapsys, N., Le Paslier, D., Doggett, N., Sutherland, G. R., and Richards, R. I. (1997). Human chromosomal fragile site *FRA16B* is an amplified AT-rich minisatellite repeat. *Cell* **88**, 367–374.
21. Hewett, D. R., Handt, O., Mangelsdorf, M., Hobson, L., Eyre, H., Baker, E., Sutherland, G. R., Schuffenhauer, S., Mao, J., and Richards, R. I. (1998). Structure of *FRA10B* reveals common elements in repeat expansion and chromosomal fragile site genesis. *Mol. Cell* **1**, 773–781.
22. Glover, T. W., Berger, C., Coyle, J., and Echo, B. (1984). DNA polymerase α inhibition by aphidicolin induces gaps and breaks at common fragile sites in human chromosomes. *Hum. Genet.* **67**, 136–142.
23. Richards, R. I. (2001). Fragile and unstable chromosomes in cancer: Causes and consequences *Trends Genet.* **17**, 339–345.
24. Casper, A. M., Nghiem, P., Arlt, M. F., and Glover, T. W. (2002). ATR regulates fragile site stability. *Cell* **111**, 779–789.
25. Glover, T. W., and Stein, C. K. (1987). Induction of sister chromatid exchanges at common fragile sites. *Am. J. Hum. Genet.* **41**, 882–890.
26. Glover, T. W., and Stein, C. K. (1988). Chromosome breakage and recombination at fragile sites. *Am. J. Hum. Genet.* **43**, 265–273.
27. Wang, L., Paradee, Mullins, C., Shridhar, R., Rosati, R., Wilke, C. M., Glover, T. W., and Smith, D. I. (1997). Aphidicolin-induced FRA3B breakpoints cluster to two distinct regions. *Genomics* **41**, 485–488.
28. Rassool, F. V., McKeithen, T. W., Neilly, M. E., van Melle, E., Esponisa, R., III, and Le Beau, M. M. (1991). Preferential integration of marker DNA into the chromosomal fragile site at 3p14: An approach to cloning fragile sites. *Proc. Natl. Acad. Sci USA* **88**, 6657–6661.
29. Thorland, E. C., Myers, S. L., Persing, D. H., Sarkar, G., McGovern, R. M., Gostout, B. S., and Smith, D. I. (2000). Human papillomavirus type 16 integrations in cervical tumours frequently occur in common fragile sites. *Cancer Res.* **60**, 5916–5921.
30. Austin, M. J., Collins, J. M., Corey, L. A., Nance, W. E., Neale, M. C., Schieken, R. M., and Brown, J. A. (1992). Aphidicolin-inducible common fragile site expression: results from a population survey of twins *Am. J. Hum. Genet.* **50**, 76–83.
31. Tedeschi, B., Vernole, P., Sanna, M. L., and Nicoletti, B. (1992). Population genetics of aphidicolin-induced fragile sites *Hum. Genet.* **89**, 543–547.
32. Ban, S., Cologne, J. B., and Neriishi, K. (1995). Effect of radiation and cigarette smoking on expression of FudR-inducible common fragile sites in human peripheral lymphocytes. *Mutat. Res.* **334**, 197–203.
33. Stein, C. K., Glover, T. W., Palmer, J. L., and Glisson, B. S. (2002). Direct correlation between *FRA3B* expression and cigarette smoking. *Genes Chromosom. Cancer* **34**, 333–340.
34. Yunis, J. J., Soreng, A. L., and Bowe, A. E. (1987). Fragile sites are targets of diverse mutagens and carcinogens. *Oncogene* **1**, 59–69.
35. Becker, N. A, Thorland, E. C., Denison, S. R., Phillips, L. A., and Smith, D. I. (2002). Evidence that instability within the *FRA3B* region extends four megabases. *Oncogene* **21**, 8713–8722.
36. Krummel, K. A., Roberts, L. R., Kawakami, M., Glover, T. W., and Smith, D. I. (2000). The characterization of the common fragile site FRA16D and its involvement in multiple myeloma translocations. *Genomics* **69**, 37–46.
37. Mangelsdorf, M., Ried, K., Woollatt, E., Dayan, S., Eyre, H., Finnis, M., Hobson, L., Nancarrow, J., Venter, D., Baker, E., and Richards, R. I. (2000). Chromosomal fragile site *FRA16D* and DNA instability in cancer. *Cancer Res.* **60**, 1683–1689.
38. Boldog, F., Gemmill, R. M., West, J., Robinson, M., Robinson, L., Li, E., Roche, J., Todd, S., Waggoner, B., Lundstrom, R., Jacobson, J., Mullokandov, M. R., Klinger, H., and Drabkin, H. A. (1997). Chromosome 3p14 homozygous deletions and sequence analysis of *FRA3B*. *Hum. Mol. Genet.* **6**, 193–203.
39. Inoue, H., Ishii, H., Alder, H., Snyder, E., Druck, T., Huebner, K., and Croce, C. M. (1997). Sequence of the *FRA3B* common fragile region: Implications for the mechanism of FHIT deletion. *Proc. Natl. Acad. Sci. USA*. **94**, 14584–14589.
40. Mimori, K., Druck, T., Inoue, H., Alder, H., Berk, L., Mori, M., Huebner, K., and Croce, C. M. (1999). Cancer-specific chromosome alterations in the constitutive fragile region FRA3B. *Proc. Natl. Acad. Sci. USA* **96**, 7456–7461.
41. Ried, K., Finnis, M., Hobson, L., Mangelsdorf, M., Dayan, S., Nancarrow, J. K., Woollatt, E., Kremmidiotis, G., Gardner, A., Venter, D., Baker, E., and Richards, R. I. (2000). Common chromosomal fragile site *FRA16D* DNA sequence: Identification of the *FOR* gene spanning *FRA16D* and homozygous deletions and translocation breakpoints in cancer cells. *Hum. Mol. Genet.* **9**, 1651–1663.
42. Finnis, M., Dayan, S., Hobson, L., Chenevix-Trench, G., Friend, K., Ried, K., Venter, D., Woollatt, E., Baker, E., and Richards, R. I (2005). Common chromosomal fragile site *FRA16D* mutation in cancer cells. *Hum. Mol. Genet.* **14**, 1341–1349.
43. Arlt, M. F., Miller, D. E., Beer, D. G., and Glover, T. W. (2002). Molecular characterization of *FRAXB* and comparative common fragile site instability in cancer cells. *Genes Chromosom. Cancer* **33**, 82–92.
44. Mishmar, D., Rahat, A., Scherer, S. W., Nyakatura, G., Hinzmann, B., Kohwi, Y., Mandel-Gutfroind, Y., Lee, J. R., Drescher, B., Sas, D. E., Margalit, H., Platzer, M., Weiss, A., Tsui, L. C., Rosenthal, A., and Kerem, B. (1998). Molecular characterization of a common fragile site (FRA7H) on human chromosome 7 by the cloning of a simian virus 40 integration site. *Proc. Natl. Acad. Sci. USA* **95**, 8141–8146.
45. Corbin, S., Neilly, M. E., Espinosa, R. 3rd, Davis, E. M., McKeithan, T. W., and Le Beau, M. M. (2002). Identification of unstable sequences within the common fragile site at 3p14.2: Implications for the mechanism of deletions within fragile histidine triad gene/common fragile site at 3p14.2 in tumors. *Cancer Res.* **62**, 3477–3484.
46. Zlotorynski, E., Rahat, A., Skaug, J., Ben-Porat, N., Ozeri, E., Hershberg, R., Levi, A., Scherer, S. W., Margalit, H., and Kerem, B. (2003). Molecular basis for expression of common and rare fragile sites. *Mol. Cell. Biol.* **23**, 7143–7151.
47. Glover, T. W., Hoge, A. W., Miller, D. E., Ascara-Wilke, J. E., Adam, A. N., Dagenais, S. L., Wilke, C. M., Dierick, H. A., and Beer, D. G. (1998). The murine Fhit gene is highly similar to its human orthologue and maps to a common fragile site region. *Cancer Res.* **58**, 3409–3414.
48. Krummel, K. A., Denison, S. R., Calhoun, E., Phillips, L. A., and Smith, D. I. (2002). The common fragile site *FRA16D* and its

associated gene WWOX are highly conserved in the mouse at *Fra8E1*. *Genes Chromosom. Cancer* **34**, 154–167.

49. Matsuyama, A., Shiraishi, T., Trapasso, F., Kuroki, T., Alder, H., Mori, M., Huebner, K., and Croce, C. M. (2003). Fragile site orthologs FHIT/FRA3B and Fhit/Fra14A2: Evolutionarily conserved but highly recombinogenic. *Proc. Natl. Acad. Sci. USA* **100**, 14988–14993.

50. Casper, A. M., Durkin, S. G., Arlt, M. F., and Glover, T. W. (2004). Chromosomal instability at common fragile sites in Seckel syndrome. *Am. J. Hum. Genet.* **75**, 654–660.

51. Arlt, M. F., Xu, B., Durkin, S. G., Casper, A. M., Kastan, M. B., and Glover, T. W. (2004). BRCA1 is required for common-fragile-site stability via Its G/M checkpoint function. *Mol. Cell. Biol.* **24**, 6701–6709.

52. Howlett, N. G., Taniguchi, T., Durkin, S. G., D'Andrea A. D., and Glover, T. W. (2005). The Fanconi anemia pathway is required for the DNA replication stress response and for the regulation of common fragile site stability. *Hum. Mol. Genet.* **14**, 693–701.

53. Musio, A., Montagna, C., Mariani, T., Tilenni, M., Focarelli, M. L., Brait, L., Indino, E., Benedetti, P. A., Chessa, L., Albertini, A., Ried, T., and Vezzoni P. (2005). *SMC1* involvement in fragile site expression. *Hum. Mol. Genet.* **14**, 525–533.

54. Hansen, R. S., Canfield, T. K., Lamb, M. M., Gartler, S. M., and Laird, C. D. (1993). Association of fragile X syndrome with delayed replication of the FMR1 gene. *Cell* **73**, 1403–1409.

55. Hansen, R. S., Canfield, T. K., Fjeld, A. D., Mumm, S., Laird, C. D., and Gartler S. M. (1997). A variable domain of delayed replication in FRAXA fragile X chromosomes: X inactivation-like spread of late replication. *Proc. Natl. Acad. Sci. USA* **94**, 4587–4592.

56. Subramanian, P. S., Nelson, D. L., and Chinault, A. C. (1996). Large domains of apparent delayed replication timing associated with triplet expansion at *FRAXA* and *FRAXE*. *Am. J. Hum. Genet.* **59**, 407–416.

57. Handt, O., Baker E., Dayan, S., Gartler, S. M., Woollatt, E., Richards, R. I., and Hansen R. S. (2000). Analysis of replication timing at the *FRA10B* and *FRA16B* fragile site loci. *Chromosome Res.* **8**, 677–688.

58. Le Beau, M. M., Rassool, F. V., Neilly, M. E., Espinosa, R., 3rd, Glover, T. W., Smith, D. I., and McKeithan, T. W. (1998). Replication of a common fragile site, *FRA3B*, occurs late in S phase and is delayed further upon induction: Implications for the mechanism of fragile site induction. *Hum. Mol. Genet.* **7**, 755–761.

59. Wang, L., Darling, J., Zhang, J. S., Huang, H., Liu, W., and Smith, D. I. (1999). Allele-specific late replication and fragility of the most active common fragile site, *FRA3B*. *Hum. Mol. Genet.* **8**, 431–437.

60. Hellman, A., Rahat, A., Scherer, S. W., Darvasi, A., Tsui, L. C., and Kerem, B. (2000). Replication delay along *FRA7H*, a common fragile site on human chromosome 7, leads to chromosomal instability. *Mol. Cell. Biol.* **20**, 4420–4427.

61. Palakodeti, A., Han, Y., Jiang, Y., and Le Beau, M. M. (2004). The role of late/slow replication of the *FRA16D* in common fragile site induction. *Genes Chromosom. Cancer* **39**, 71–76.

62. Citterio, E., Vermeulen, W., and Hoeijmakers, J. H. (2000). Transcriptional healing. *Cell* **101**, 447–450.

63. Glover, T. W., Coyle-Morris, J. F., Li, F. P., Brown, R. S., Berger, C. S., Gemmill, R. M., and Hecht, F. (1988). Translocation t(3;8)(p14.2;q24.1) in renal cell carcinoma affects expression of the common fragile site at 3p14 (*FRA3B*) in lymphocytes. *Cancer Genet. Cytogenet.* **31**, 69–73.

64. Fang, J. M., Arlt, M. F., Burgess, A. C., Dagenais, S. L., Beer, D. G., and Glover, T. W. (2001). Translocation breakpoints in FHIT and FRA3B in both homologs of chromosome 3 in an esophageal adenocarcinoma. *Genes Chromosom. Cancer* **30**, 292–298.

65. Chesi, M., Bergsagel, P. L., Shonukan, O. O., Martelli, M. L., Brents, L. A., Chen, T., Schrok, E., Ried, T., and Kuehl, W. M. (1998). Frequent dysregulation of the c-*MAF* proto-oncogene at 16q23 by translocation to an Ig locus in multiple myeloma. *Blood* **91**, 4457–4463.

66. Jamieson, R. V., Perveen, R., Kerr, B., Carette, M., Yardley, J., Heon, E., Wirth, M. G., van Heyningen, V., Donnai, D., Munier, F., and Black, G. C. (2002). Domain disruption and mutation of the bZIP transcription factor, MAF, associated with cataract, ocular anterior segment dysgenesis and coloboma. *Hum. Mol. Genet.* **11**, 33–42.

67. Popescu, N. C., and DiPaolo, J. A. (1990). Integration of human papillomavirus 16 DNA and genomic rearrangements in immortalized human keratinocyte lines. *Cancer Res.* **50**, 1316–1323.

68. Thorland E. C., Myers, S. L., Persing, D. H., Sarkar, G., McGovern, R. M., Gostout, B. S., and Smith, D. I. (2000). Human papillomavirus type 16 integrations in cervical tumors frequently occur in common fragile sites. *Cancer Res.* **60**, 5916–5921.

69. Yu, T., Ferber, M. J., Cheung, T. H., Chung, T. K., Wong, Y. F., and Smith, D. I. (2005). The role of viral integration in the development of cervical cancer. *Cancer Genet. Cytogenet.* **158**, 27–34.

70. Rassool, F. V., McKeithan, T. W., Neilly, M. E., van Melle, E., Espinosa, R., 3rd, and Le Beau, M. M. (1991). Preferential integration of marker DNA into the chromosomal fragile site at 3p14: An approach to cloning fragile sites. *Proc. Natl. Acad. Sci. USA* **88**, 6657–6661.

71. Coquelle, A., Pipiras, E., Toledo, F., Buttin, G., and Debatisse, M. (1997). Expression of fragile sites triggers intrachromosomal mammalian gene amplification and sets boundaries to early amplicons. *Cell* **89**, 215–225.

72. Coquelle, A., Toledo, F., Stern, S., Bieth, A., and Debatisse, M. (1998). A new role for hypoxia in tumor progression: Induction of fragile site triggering genomic rearrangements and formation of complex DMs and HSRs. *Mol. Cell* **2**, 259–265.

73. Hellman, A., Zlotorynski, E., Scherer, S. W., Cheung, J., Vincent, J. B., Smith, D. I., Trakhtenbrot, L., and Kerem, B. (2002). A role for common fragile site induction in amplification of human oncogenes. *Cancer Cell.* **1**, 89–97.

74. Ciullo, M., Debily, M. A., Rozier, L., Autiero, M., Billault, A., Mayau, V., El Marhomy, S., Guardiola, J., Bernheim, A., Coullin, P., Piatier-Tonneau, D., and Debatisse, M. (2002). Initiation of the breakage-fusion-bridge mechanism through common fragile site activation in human breast cancer cells: The model of PIP gene duplication from a break at *FRA7I*. *Hum. Mol. Genet.* **11**, 2887–2894.

75. http://www.ensembl.org/index.html.

76. http://www.dsi.univ-paris5.fr/genatlas/.

77. Ohta, M., Inoue, H., Cotticelli, M. G., Kastury, K., Baffa, R., Palazzo, J., Siprashvilli, Z., Mori, M., McCue, P., Druck, T., Croce, C. M., and Huebner, K. (1996). The *FHIT* gene, spanning the chromosome 3p14.2 fragile site and renal carcinoma-associated t(3;8) breakpoint, is abnormal in digestive tract cancers. *Cell* **84**, 587–597.

78. Sozzi, G., Veronese, M. L., Negrini, M., Baffa, R., Cotticelli, M. G., Inoue, H., Tornielli, S., Pilotti, S., De Gregorio, L., Pastorino, U., Pierotti, M. A., Ohta, M., Huebner, K., and Croce, C. M. (1996). The FHIT gene 3p14.2 is abnormal in lung cancer. *Cell* **85**, 17–26.

79. Le Beau, M. M., Drabkin, H., Glover, T. W., Gemmill, R., Rassool, F. V., McKeithan, T. W., and Smith, D. I. (1998). An FHIT tumor suppressor gene? *Genes Chromosom. Cancer* **21**, 281–289.

80. Wang, L., Darling, J., Zhang, J. S., Qian, C. P., Hartmann, L., Conover, C., Jenkins, R., and Smith, D. I. (1998). Frequent homozygous deletions in the FRA3B region in tumor cell lines still leave the FHIT exons intact. *Oncogene* **16**, 635–642.

81. Glover, T. W., Hoge, A. W., Miller, D. E., Ascara-Wilke, J. E., Adam, A. N., Dagenais, S. L., Wilke, C. M., Dierick, H. A., and Beer, D. G. (1998). The murine Fhit gene is highly similar to its human orthologue and maps to a common fragile site region. *Cancer Res.* **58**, 3409–3414.

82. Shiraishi, T., Druck, T., Mimori, K., Flomenberg, J., Berk, L., Alder, H., Miller, W., Huebner, K., and Croce, C. M. (2001). Sequence conservation at human and mouse orthologous common fragile regions, FRA3B/FHIT and Fra14A2/Fhit. *Proc. Natl. Acad. Sci. USA* **98**, 5722–5727.
83. Pekarsky, Y., Druck, T., Cotticelli, M. G., Ohta, M., Shou, J., Mendrola, J., Montgomery, J. C., Buchberg, A. M., Siracusa, L. D., Manenti, G., Fong, L. Y., Dragani, T. A., Croce, C. M., and Huebner, K. (1998). The murine Fhit locus: Isolation, characterization, and expression in normal and tumor cells. *Cancer Res.* **58**, 3401–3408.
84. Larson, G. P., Ding, Y., Cheng, L. S., Lundberg, C., Gagalang, V., Rivas, G., Geller, L., Weitzel, J., MacDonald, D., Archambeau, J., Slater, J., Neuberg, D., Daly, M. B., Angel, I., Benson, A. B. 3rd, Smith, K., Kirkwood, J. M., O'Dwyer, P. J., Raskay, B., Sutphen, R., Drew, R., Stewart, J. A., Werndli, J., Johnson, D., Ruckdeschel, J. C., Elston, R. C., and Krontiris, T. G. (2005). Genetic linkage of prostate cancer risk to the chromosome 3 region bearing *FHIT*. *Cancer Res.* **65**, 805–814.
85. Siprashvili, Z., Sozzi, G., Barnes, L. D., McCue, P., Robinson, A. K., Eryomin, V., Sard, L., Tagliabue, E., Greco, A., Fusetti, L., Schwartz, G., Pierotti, M. A., Croce, C. M., and Huebner, K. (1997). Replacement of Fhit in cancer cells suppresses tumorigenicity. *Proc. Natl. Acad. Sci. USA* **94**, 13771–13776.
86. Sevignani, C., Calin, G. A., Cesari, R., Sarti, M., Ishii, H., Yendamuri, S., Vecchione, A., Trapasso, F., and Croce, C. M. (2003). Restoration of fragile histidine triad (FHIT) expression induces apoptosis and suppresses tumorigenicity in breast cancer cell lines. *Cancer Res.* **63**, 1183–1187.
87. Dumon, K. R., Ishii, H., Fong, L. Y., Zanesi, N., Fidanza, V., Mancini, R., Vecchione, A., Baffa, R., Trapasso, F., During, M. J., Huebner, K., and Croce, C. M. (2001). FHIT gene therapy prevents tumor development in Fhit-deficient mice. *Proc. Natl. Acad. Sci. USA* **98**, 3346–3351.
88. Sard, L., Accornero, P., Tornielli, S., Delia, D., Bunone, G., Campiglio, M., Colombo, M. P., Gramegna, M., Croce, C. M., Pierotti, M. A., and Sozzi, G. (1999). The tumor-suppressor gene FHIT is involved in the regulation of apoptosis and in cell cycle control. *Proc. Natl. Acad. Sci. USA* **96**, 8489–8492.
89. Fong, L. Y., Fidanza, V., Zanesi, N., Lock, L. F., Siracusa, L. D., Mancini, R., Siprashvili, Z., Ottey, M., Martin, S. E., Druck, T., McCue, P. A., Croce, C. M., and Huebner, K. (2000). Muir–Torre-like syndrome in Fhit-deficient mice. *Proc. Natl. Acad. Sci. USA* **97**, 4742–4747.
90. Barnes, L. D., Garrison, P. N., Siprashvili, Z., Guranowski, A., Robinson, A. K., Ingram, S. W., Croce, C. M., Ohta, M., and Huebner, K. (1996). Fhit, a putative tumor suppressor in humans, is a dinucleoside 5′,5‴-P1,P3-triphosphate hydrolase. *Biochemistry* **35**, 11529–11535.
91. Trapasso, F., Krakowiak, A., Cesari, R., Arkles, J., Yendamuri, S., Ishii, H., Vecchione, A., Kuroki, T., Bieganowski, P., Pace, H. C., Huebner, K., Croce, C. M., and Brenner, C. (2003). Designed FHIT alleles establish that Fhit-induced apoptosis in cancer cells is limited by substrate binding. *Proc. Natl. Acad. Sci. USA* **100**, 1592–1597.
92. Pekarsky, Y., Campiglio, M., Siprashvili, Z., Druck, T., Sedkov, Y., Tillib, S., Draganescu, A., Wermuth, P., Rothman, J. H., Huebner, K., Buchberg, A. M., Mazo, A., Brenner, C., and Croce, C. M. (1998). Nitrilase and Fhit homologs are encoded as fusion proteins in *Drosophila melanogaster* and *Caenorhabditis elegans*. *Proc. Natl. Acad. Sci. USA* **95**, 8744–8749.
93. Chaudhuri, A. R., Khan, I. A., Prasad, V., Robinson, A. K., Luduena, R. F., and Barnes, L. D. (1999). The tumor suppressor protein Fhit: A novel interaction with tubulin. *J. Biol. Chem.* **274**, 24378–24382.
94. Pekarsky, Y., Garrison, P. N., Palamarchuk, A., Zanesi, N., Aqeilan, R. I., Huebner, K., Barnes, L. D., and Croce, C. M. (2004). Fhit is a physiological target of the protein kinase Src. *Proc. Natl. Acad. Sci. USA* **101**, 3775–3779.
95. Bednarek, A. K., Laflin, K. J., Daniel, R. L., Liao, Q., Hawkins, K. A., and Aldaz, C. M. (2000). WWOX, a novel WW domain-containing protein mapping to human chromosome 16q23.3–24.1, a region frequently affected in breast cancer. *Cancer Res.* **60**, 2140–2145.
96. Chang, N. S., Pratt, N., Heath, J., Schultz, L., Sleve, D., Carey, G. B., and Zevotek, N. (2001). Hyaluronidase induction of a WW domain-containing oxidoreductase that enhances tumor necrosis factor cytotoxicity. *J. Biol. Chem.* **276**, 3361–3370.
97. O'Keefe, L., Liu, Y-H., Perkins, A., Dayan, S., Saint, R. B., and Richards, R. I. (2005). *FRA16D* common chromosomal fragile site oxido-reductase (FOR/WWOX) protects against the effects of ionising radiation in *Drosophila*. *Oncogene*. **24**, 6590–6596.
98. Bednarek, A. K., Keck-Waggoner, C. L., Daniel, R. L., Laflin, K. J., Bergsagel, P. L., Kiguchi, K., Brenner, A. J., and Aldaz, C. M. (2001). WWOX, the FRA16D gene, behaves as a suppressor of tumor growth. *Cancer Res.* **61**, 8068–8073.
99. Driouch, K., Prydz, H., Monese, R., Johansen, H., Lidereau, R., and Frengen, E. (2002). Alternative transcripts of the candidate tumor suppressor gene, WWOX, are expressed at high levels in human breast tumors. *Oncogene* **21**, 1832–1840.
100. Chang, N. S., Doherty, J., Ensign, A., Lewis, J., Heath, J., Schultz, L., Chen, S. T., and Oppermann, U. (2003). Molecular mechanisms underlying WOX1 activation during apoptotic and stress responses. *Biochem. Pharmacol.* **66**, 1347–1354.
101. Sze, C. I., Su, M., Pugazhenthi, S., Jambal, P., Hsu, L. J., Heath, J., Schultz, L., and Chang, N. S. (2004). Down-regulation of WW domain-containing oxidoreductase induces Tau phosphorylation in vitro: A potential role in Alzheimer's disease. *J. Biol. Chem.* **279**, 30498–30506.
102. Ludes-Meyers, J. H., Kil, H., Bednarek, A. K., Drake, J., Bedford, M. T., and Aldaz, C. M. (2004). WWOX binds the specific proline-rich ligand PPXY: Identification of candidate interacting proteins. *Oncogene* **23**, 5049–5055.
103. Aqeilan, R. I., Pekarsky, Y., Herrero, J. J., Palamarchuk, A., Letofsky, J., Druck, T., Trapasso, F., Han, S. Y., Melino, G., Huebner, K., and Croce, C. M. (2004). Functional association between WWOX tumor suppressor protein and p73, a p53 homolog. *Proc. Natl. Acad. Sci. USA* **101**, 4401–4406.
104. Aqeilan, R. I., Palamarchuk, A., Weigel, R. J., Herrero, J. J., Pekarsky, Y., and Croce, C. M. (2004). Physical and functional interactions between the WWOX tumor suppressor protein and the AP-2gamma transcription factor. *Cancer Res.* **64**, 8256–8261.
105. Watanabe, A., Hippo, Y., Taniguchi, H., Iwanari, H., Yashiro, M., Hirakawa, K., Kodama, T., and Aburatani, H. (2003). An opposing view on WWOX protein function as a tumor suppressor. *Cancer Res.* **63**, 8629–8633.
106. Paige, A. J., Taylor, K. J., Taylor, C., Hillier, S. G., Farrington, S., Scott, D., Porteous, D. J., Smyth, J. F., Gabra, H., and Watson, J. E. (2001). WWOX: A candidate tumor suppressor gene involved in multiple tumor types. *Proc. Natl. Acad. Sci. USA* **98**, 11417–11422.
107. Yakicier, M. C., Legoix, P., Vaury, C., Gressin, L., Tubacher, E., Capron, F., Bayer, J., Degott, C., Balabaud, C., and Zucman-Rossi, J. (2001). Identification of homozygous deletions at chromosome 16q23 in aflatoxin B1 exposed hepatocellular carcinoma. *Oncogene* **20**, 5232–5238.
108. Kuroki, T., Trapasso, F., Shiraishi, T., Alder, H., Mimori, K., Mori, M., and Croce, C. M. (2002). Genetic alterations of the tumor suppressor gene WWOX in esophageal squamous cell carcinoma. *Cancer Res.* **62**, 2258–2260.

109. Ishii, H., Vecchione, A., Furukawa, Y., Sutheesophon, K., Han, S. Y., Druck, T., Kuroki, T., Trapasso, F., Nishimura, M., Saito, Y., Ozawa, K., Croce, C. M., Huebner, K., and Furukawa, Y. (2003). Expression of FRA16D/WWOX and FRA3B/FHIT genes in hematopoietic malignancies. *Mol. Cancer Res.* **1**, 940–947.
110. Yendamuri, S., Kuroki, T., Trapasso, F., Henry, A. C., Dumon, K. R., Huebner, K., Williams, N. N., Kaiser, L. R., and Croce, C. M. (2003). WW domain containing oxidoreductase gene expression is altered in non-small cell lung cancer. *Cancer Res.* **63**, 878–881.
111. Aqeilan, R. I., Kuroki, T., Pekarsky, Y., Albagha, O., Trapasso, F., Baffa, R., Huebner, K., Edmonds, P., and Croce, C. M. (2004). Loss of WWOX expression in gastric carcinoma. *Clin. Cancer Res.* **10**, 3053–3058.
112. Guler, G., Uner, A., Guler, N., Han, S. Y., Iliopoulos, D., Hauck, W. W., McCue, P., and Huebner K. (2004). The fragile genes FHIT and WWOX are inactivated coordinately in invasive breast carcinoma. *Cancer* **100**, 1605–1614.
113. Kuroki, T., Yendamuri, S., Trapasso, F., Matsuyama, A., Aqeilan, R. I., Alder, H., Rattan, S., Cesari, R., Nolli, M. L., Williams, N. N., Mori, M., Kanematsu, T., and Croce, C. M. (2004). The tumor suppressor gene WWOX at FRA16D is involved in pancreatic carcinogenesis. *Clin. Cancer Res.* **10**, 2459–2465.
114. Sogame, N., Kim, M., and Abrams, J. M. (2003). Drosophila p53 preserves genomic stability by regulating cell death. *Proc. Natl. Acad. Sci. USA* **100**, 4696–4701.
115. Lee, J. H., Lee, E., Park, J., Kim, E., Kim, J., and Chung J. (2003). In vivo p53 function is indispensable for DNA damage-induced apoptotic signaling in *Drosophila*. *FEBS Lett.* **550**, 5–10.
116. Denison, S. R., Callahan, G., Becker, N. A, Phillips, L. A, and Smith, D. I. (2003). Characterization of FRA6E and its potential role in autosomal recessive juvenile parkinsonism and ovarian cancer. *Genes Chromosom. Cancer* **38**, 40–52.
117. Kitada, T., Asakawa, S., Hattori, N., Matsumine, H., Yamamura, Y., Minoshima, S., Yokochi, M., Mizuno, Y., and Shimizu, N. (1998). Mutations in the parkin gene cause autosomal recessive juvenile parkinsonism. *Nature* **392**, 605–608.
118. Cesari, R., Martin, E. S., Calin, G. A., Pentimalli, F., Bichi, R., McAdams, H., Trapasso, F., Drusco, A., Shimizu, M., Masciullo, V., D'Andrilli, G., Scambia, G., Picchio, M. C., Alder, H., Godwin, A. K., and Croce, C. M. (2003). Parkin, a gene implicated in autosomal recessive juvenile parkinsonism, is a candidate tumor suppressor gene on chromosome 6q25-q27. *Proc. Natl. Acad. Sci. USA* **100**, 5956–5961.
119. Denison, S. R., Wang, F., Becker, N. A., Schule, B., Kock, N., Phillips, L. A., Klein, C., and Smith, D. I. (2003). Alterations in the common fragile site gene Parkin in ovarian and other cancers. *Oncogene* **22**, 8370–8378.
120. Wang, F., Denison, S., Lai, J. P., Philips, L. A., Montoya, D., Kock, N., Schule, B., Klein, C., Shridhar, V., Roberts, L. R., and Smith, D. I. (2004). Parkin gene alterations in hepatocellular carcinoma. *Genes Chromosom. Cancer* **40**, 85–96.
121. Picchio, M. C., Martin, E. S., Cesari, R., Calin, G. A., Yendamuri, S., Kuroki, T., Pentimalli, F., Sarti, M., Yoder, K., Kaiser, L. R., Fishel, R., and Croce, C. M. (2004). Alterations of the tumor suppressor gene Parkin in non-small cell lung cancer. *Clin Cancer Res.* **10**, 2720–2724.
122. Rozier, L., El-Achkar, E., Apiou, F., and Debatisse, M. (2004). Characterization of a conserved aphidicolin-sensitive common fragile site at human 4q22 and mouse 6C1: Possible association with an inherited disease and cancer. *Oncogene* **23**, 6872–6880.
123. Bluteau, O., Beaudoin, J. C., Pasturaud, P., Belghiti, J., Franco, D., Bioulac-Sage, P., Laurent-Puig, P., and Zucman-Rossi J. (2002). Specific association between alcohol intake, high grade of differentiation and 4q34-q35 deletions in hepatocellular carcinomas identified by high resolution allelotyping. *Oncogene* **21**, 1225–1232.
124. Robinson, K. O., Petersen, A. M., Morrison, S. N., Elso, C. M., and Stubbs, L. (2005). Two reciprocal translocations provide new clues to the high mutability of the Grid2 locus. *Mamm. Genome* **16**, 32–40.
125. Huang, H., Qian, J., Proffit, J., Wilber, K., Jenkins, R., and Smith, D. I. (1998). FRA7G extends over a broad region: Coincidence of human endogenous retroviral sequences (HERV-H) and small polydispersed circular DNAs (spcDNA) and fragile sites. *Oncogene* **16**, 2311–2319.
126. Engelman, J. A., Zhang, X. L., and Lisanti, M. P. (1998). Genes encoding human caveolin-1 and -2 are co-localized to the D7S522 locus (7q31.1), a known fragile site (FRA7G) that is frequently deleted in human cancers. *FEBS Lett.* **436**, 403–410.
127. Huang, H., Reed, C. P., Mordi, A., Lomberk, G., Wang, L., Shridhar, V., Hartmann, L., Jenkins, R., and Smith, D. I. (1999). Frequent deletions within FRA7G at 7q31.2 in invasive epithelial ovarian cancer. *Genes Chromosom. Cancer* **24**, 48–55.
128. Callahan, G., Denison, S. R., Phillips, L. A., Shridhar, V., and Smith, D. I. (2003). Characterization of the common fragile site FRA9E and its potential role in ovarian cancer. *Oncogene* **22**, 590–601.
129. Limongi, M. Z., Pelliccia, F., and Rocchi, A. (2003). Characterization of the human common fragile site FRA2G. *Genomics* **81**, 93–97.
130. Zenklusen, J. C., Hodges, L. C., LaCava, M., Green, E. D., and Conti, C. J. (2000). Definitive functional evidence for a tumor suppressor gene on human chromosome 7q31.1 neighboring the Fra7G site. *Oncogene* **19**, 1729–1733.

Part IV

Kennedy's Disease

CHAPTER 13

Clinical Features and Molecular Biology of Kennedy's Disease

CHEUNJU CHEN AND KENNETH H. FISCHBECK

Neurogenetics Branch, National Institutes of Neurological Disorders and Stroke, National Institutes of Health, Bethesda, Maryland 20892

I. Introduction
II. Clinical Features of SBMA
III. Laboratory Studies
IV. Differential Diagnosis
V. Management
VI. Genetics of SBMA
VII. The Androgen Receptor
    A. Structure of the Androgen Receptor Gene and Protein
    B. Androgen Receptor Activation
    C. The Androgen Receptor Ligands
VIII. Androgen Receptor Function in the Nervous System
IX. Pathological Mechanisms in SBMA
    A. Toxic Gain of Function
    B. Inclusions and Aggregates
    C. Ligand-Dependent Effects in SBMA Models
X. Therapeutic Approaches in SBMA
    A. Histone Deacetylase Inhibitors
    B. Anti-androgens
References

Spinal and bulbar muscular atrophy (SBMA), or Kennedy's disease, is an X-linked motor neuron disease caused by CAG repeat expansion in the androgen receptor gene, resulting in polyglutamine tract expansion in the receptor protein. Affected males develop a chronic, progressive neuromuscular deficit and also may show signs of androgen insensitivity. As in other polyglutamine diseases, a pathological feature is inclusion formation in cells expressing the mutant protein. Unlike most other polyglutamine disorders, the normal function of the mutant protein in SBMA is well known. The mutation leads to both a toxic gain of function in affected cells and a loss of normal receptor function. Lower motor neurons in the spinal cord and brainstem express high levels of the androgen receptor, and these are the cells most susceptible to degeneration in SBMA. In cell culture and animal models, expression of mutant receptor leads to motor neuron dysfunction and cell death. Binding of androgen to the mutant receptor protein has been shown to be important in the pathogenesis of SBMA. There is currently no specific treatment for this disease. However, anti-androgen treatment has been found to be effective in animal models and is currently being tested in patients with SBMA.

I. INTRODUCTION

Spinal and bulbar muscular atrophy (SBMA), or Kennedy's disease, is a chronic and slowly progressive, X-linked, adult-onset motor neuron disease affecting males in mid- to late adulthood [1, 2]. It has been reported primarily in individuals of Asian and European descent. SBMA appears to be more common in Japanese and Finnish populations than other ethnic groups, which is likely due to a founder effect [3–6]. The disease was described by Dr. Hiroshi Kawahara in 1897 in two brothers and a maternal uncle. In 1968, Drs. William Kennedy, Milton Alter, and Joo Ho Sung published a detailed description of the disease, including the pathological and electrodiagnostic findings [1]. Since then, the disease entity has widely been known as Kennedy– Alter–Sung syndrome or Kennedy's disease.

II. CLINICAL FEATURES OF SBMA

Affected males often have early muscle cramps and fasciculations, but they first seek medical attention for progressive weakness of bulbar and limb muscles. Weakness usually starts in the proximal muscles of the lower extremities or the shoulder girdle. Distal muscles can also be affected, causing hand and ankle weakness. Patients often develop atrophy in the affected muscles. The weakness results in a progressive gait impairment, and patients may become wheelchair dependent two to three decades after the onset of symptoms.

Bulbar involvement may lead to weakness and atrophy of the face, tongue, and throat muscles. Patients often have facial fasciculations, particularly noticeable around the mouth and chin, and in the tongue. Weakness of the jaw and tongue muscles may lead to difficulty in jaw closure, and wasting and fissuring of the tongue [1, 7, 8]. Bulbar weakness often results in dysarthria. Choking is a common problem in affected individuals who have pharyngeal involvement and may lead to aspiration in the late stages of the disease. Dysphagia may also occur, but the severity seldom prevents adequate nutritional intake. Extraocular muscles are spared.

Other commonly associated neurological features include postural hand tremor, depressed or absent deep tendon reflexes, and sensory loss. Patients with SBMA rarely complain of sensory symptoms. However, on careful testing, mild distal sensory loss to vibration and pinprick may be observed [2, 7, 9–11]. There is no upper motor neuron involvement.

Affected males often display signs of androgen insensitivity, which may have onset before the neurological signs and symptoms. Gynecomastia is present in about half of patients. Other endocrinological abnormalities include reduced fertility, erectile dysfunction, and testicular atrophy [11, 12].

Female carriers do not develop the full clinical manifestations of the disease. They may experience muscle cramps or a mild hand tremor. On laboratory examination, they may have elevated serum creatine kinase levels and electromyographic abnormalities [1, 13, 14], but they do not develop the progressive weakness in bulbar and limb muscles that is seen in affected males.

III. LABORATORY STUDIES

SBMA patients have elevated serum creatine kinase levels, usually two to five times the upper limit of normal [1, 10, 15, 16]. Other associated conditions that have been reported include impaired glucose tolerance and hyperlipidemia [10, 16, 17]. Testosterone levels in patients with SBMA vary from slightly decreased or normal to elevated [2, 15, 18, 19].

Electromyography and nerve conduction studies show evidence of motor and sensory neuron loss. Nerve conduction studies often show reduced or unelicitable compound motor action potentials and prolonged distal motor latencies. Sensory nerve action potentials are also often reduced or absent. Electromyography of both proximal and distal muscles shows signs of chronic denervation and partial renervation. Motor units have reduced recruitment and action potentials with large amplitudes and prolonged duration. Signs of acute denervation, such as fasciculations and fibrillation potentials, may also be seen [1, 16, 20].

Pathologically, there is a reduced number of motor neurons in the brainstem and spinal cord, as well as a decreased number of sensory neurons in the dorsal root ganglia [10, 12]. Nuclear inclusions may be seen in the remaining motor neurons with appropriate immunohistochemical staining [21]. Sural nerve biopsy shows a loss of large-diameter axons [12, 22]. Skeletal muscle biopsy shows evidence of chronic denervation, with muscle fiber atrophy, small angulated fibers, and grouped atrophy. Myopathic changes are uncommon [1, 23].

IV. DIFFERENTIAL DIAGNOSIS

SBMA patients are often misdiagnosed as having other neuromuscular disorders. Approximately 1 in 25 individuals diagnosed with amyotrophic lateral sclerosis (ALS) have SBMA on genetic testing [24]. However, in contrast to ALS, SBMA is gradually progressive and patients usually have a normal life span. In addition, patients with ALS have upper motor neuron signs such as hyperreflexia and spasticity, in addition to lower motor neuron signs. Another distinguishing feature of SBMA may be the presence of gynecomastia and other signs of androgen insensitivity. Other misdiagnoses of

SBMA include other disorders that present with proximal muscle weakness such as myasthenia gravis, chronic inflammatory neuropathy, and inflammatory or metabolic myopathy. The diagnosis of SBMA is made through genetic testing, that is, by determining the length of the CAG (cytosine–adenine–guanine) repeat in the androgen receptor (AR) gene by use of polymerase chain reaction.

V. MANAGEMENT

There is currently no specific treatment for SBMA; therefore, management is supportive. To maintain ambulation, physical therapy and assistive devices may be helpful. Individuals with loss of arm or hand strength or function may benefit from occupational therapy. Some patients with SBMA have breast reduction surgery for the gynecomastia [25]. Unlike ALS, patients usually do not have such severe involvement of the bulbar muscles to require feeding tubes or communication devices. Respiratory muscles are usually spared.

Genetic counseling is helpful. The daughters of affected males are obligate carriers. Affected males pass the mutant allele to each of their daughters. Heterozygous female carriers have a 50% chance of passing on the mutation to each child; thus, they have a 50% chance of having an affected son and a 50% chance of having a daughter who is a carrier. Carrier testing for at-risk female relatives and prenatal testing for female carriers are available.

VI. GENETICS OF SBMA

The causative defect in SBMA is expansion of a CAG trinucleotide repeat in the first exon of the AR gene on the X chromosome [26]. The CAG repeat encodes a polyglutamine tract in the amino-terminal domain of the protein. In normal individuals, the AR gene has 5 to 36 CAGs [26], and patients with SBMA have 40 to 66 CAGs [26]. SBMA was the first polyglutamine expansion disorder discovered. Currently, this group of disorders consists of nine diseases, including Huntington's disease, dentatorubral pallidoluysian atrophy (DRPLA), and six of the dominantly inherited spinocerebellar ataxias (SCAs 1, 2, 3, 6, 7, and 17). All of these disorders except SBMA follow an autosomal dominant pattern of inheritance. In each disorder, the length of the expanded repeat correlates inversely with the age at onset [15, 27, 28]. However, patients with identical repeat lengths may have different clinical manifestations and disease severity [19]. Thus, it is believed that other environmental and genetic factors may influence the disease process.

VII. THE ANDROGEN RECEPTOR

A. Structure of the Androgen Receptor Gene and Protein

The AR gene is located on the long arm of the X chromosome at Xq11-q12. Thus, males have a single copy of the AR gene, and females have two copies of the gene. In females, one allele undergoes random X inactivation. The open reading frame is encoded by eight exons, and the protein is composed of 919 amino acids, which are organized in well-defined regions: (1) the amino-terminal transactivation domain; (2) the DNA binding domain; (3) the hinge region; and (4) the ligand binding domain. The amino-terminal transactivation domain is encoded by exon 1 and contains polyglutamine, polyproline, and polyglycine repeats. The DNA binding domain is encoded by exons 2 and 3, and each exon codes for a zinc finger. The first zinc finger determines the specificity of AR binding to the hormone responsive element in the genomic DNA of target genes. The second zinc finger stabilizes the DNA–protein binding by AR dimerization. The hinge region has a nuclear localization signal and is encoded by the 5′ end of exon 4. The ligand binding domain is encoded by the 3′ region of exon 4 and exons 5–8 [29, 30]. In SBMA, there is abnormal expansion of the polyglutamine tract in the transactivation domain encoded by exon 1 (Fig. 13-1).

B. Androgen Receptor Activation

The AR is a member of the nuclear receptor family, and regulates gene expression in response to the binding of androgens (testosterone and dihydrotestosterone). In the absence of ligand, the AR is located in the cytoplasm as a multimeric complex with heat shock proteins. Once bound to ligand, the androgen receptor dissociates from the accessory proteins, dimerizes, and translocates into the nucleus [29, 31]. In the nucleus, the AR binds as a dimer to a specific recognition sequence in the promoter regions of androgen responsive genes and functions as a transcription factor.

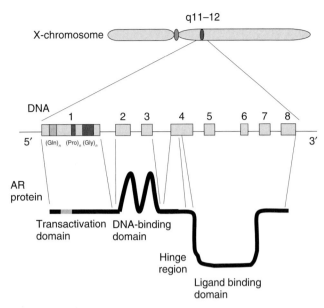

FIGURE 13-1 Structure of the androgen receptor gene and protein. The androgen receptor gene is located on the X chromosome at site q11–12. The gene consists of eight exons. Exon 1 encodes the transactivation domain. Exons 2 and 3 encode the DNA binding domain. The 5' end of exon 4 encodes the hinge region. The 3' end of exon 4 and exons 5 through 8 code the ligand binding domain. See CD-ROM for color image.

C. The Androgen Receptor Ligands

The AR is activated by two hormones that exert both androgenic and anabolic actions: testosterone and dihydrotestosterone (DHT). Ninety-five percent of testosterone is synthesized by Leydig cells in the testes; 5% is synthesized by cortical cells of the adrenal gland. Testosterone is secreted and constitutes the majority of circulating androgen. In tissues that contain the enzyme 5α-reductase, testosterone is converted to DHT. DHT has a higher affinity for the AR than testosterone, and once DHT is bound, the complex is more stable and exerts a more potent stimulus on gene expression than does testosterone [32–34]. The conversion of testosterone to DHT also prevents the conversion of testosterone to the less potent androstenedione. The conversion to DHT thus amplifies the effects of testosterone.

Androgens play an essential role in male sexual development and behavior. In certain tissues, such as the prostate, epididymis, seminal vesicle, and skin, the AR binds primarily DHT, while in other tissues such as the skeletal muscles or testes, where the activity of 5α-reductase is low or absent, testosterone serves as the primary ligand. Testosterone functions in the stimulation of bone mass, muscle mass, sexual dimorphism, and spermatogenesis. Virilization of external genitalia during development depends on DHT. In adult men, DHT plays an important role in the growth of facial and body hair, acne, male pattern baldness, and prostatic enlargement [34, 35]. In addition, testosterone and DHT alter serum lipid profiles and fat distribution [35].

VIII. ANDROGEN RECEPTOR FUNCTION IN THE NERVOUS SYSTEM

The role of androgens in male sexual development and the function of the AR in skeletal muscle and bone growth have been well studied. However, AR function in the nervous system is less well understood. The AR is expressed in many areas of the brain and spinal cord, including both sexually dimorphic and nondimorphic neurons. In the cerebrum, ARs are found within the periventricular nucleus of the hypothalamus and the amygdala [36], where they are believed to be involved in the control of reproductive function and sexual behavior. High concentrations of AR are found within the nucleus ambiguus, hypoglossal nucleus, facial nucleus, and trigeminal motor nuclei [37]. In the spinal cord, the AR is expressed mainly in motor neurons in the anterior horns [38]. The AR is also expressed in sensory neurons in the dorsal root ganglia [21]. In pathological specimens from patients with SBMA, there is degeneration of these AR-containing brainstem and spinal cord motor neurons as well as the sensory neurons in the dorsal root ganglia.

In addition to controlling reproductive function and behavior, androgens exert a trophic response on neurons. One major area of androgen action in the human spinal cord is Onuf's nucleus, which corresponds to the spinal nucleus of the bulbocavernosus (SNB) in rodents. Motor neurons in this nucleus innervate perineal muscles and are involved in copulatory behavior. This is a sexually dimorphic nucleus, in that males have a greater number of motor neurons in the nucleus compared with females. When females are treated with androgens, they develop a similar number of motor neurons as males [39]. Androgen deprivation in males decreases motor neuron cell size, dendritic length, and number and size of gap junctions; whereas androgen replacement therapy reverses this process [40]. Similar effects are seen in motor neuronal cell lines that express the AR. In the presence of androgens, the cells develop larger cell bodies, broader and longer neuritic processes, and enhanced neurite branching [41–43]. Androgens also play a trophic role in peripheral nerves. After resection of the facial, hypoglossal, and sciatic nerves, androgens increase the rate of axonal regeneration [43–45]. At the molecular level, androgens are linked to an increase in mRNA expression of structural proteins such as β-actin, β-tubulin, neuritin, and connexins [40, 44, 46].

IX. PATHOLOGICAL MECHANISMS IN SBMA

A. Toxic Gain of Function

The expanded polyglutamine tract of the mutant AR protein does not interfere with normal cellular localization or ligand binding [41], but the expanded polyglutamine repeat does lead to a decrease in receptor expression and altered transcriptional activation [41, 47–49]. This loss of normal AR function may account for the signs of androgen insensitivity in SBMA patients. However, it does not appear to be the principal cause of motor neuron toxicity. The AR knockout (Tfm) mouse has signs of androgen insensitivity but normal motor function [50]. Patients with loss of AR function (androgen insensitivity syndrome) also have signs of feminization without loss of motor neuron function. Women who are heterozygous or homozygous for the expanded repeat mutation have subclinical or mild expression of the disease phenotype [13, 51]. These observations have led to the belief that the mutant AR in SBMA causes motor neuron dysfunction primarily by a toxic gain of function rather than by a loss of function.

In cell culture models, the expression of mutant AR leads to abnormalities in cellular functions such as the development of dystrophic neurites and cell bodies and, ultimately, leads to cell death [43, 47, 52]. In transgenic *Drosophila* and mouse models in which the aberrant AR protein has been inserted into the genome, motor neuron dysfunction and cell death occur [53–55]. These findings support the hypothesis that the mutant AR protein is toxic to motor neurons.

B. Inclusions and Aggregates

A common pathological feature of the polyglutamine diseases is the presence of intracellular inclusions of mutant protein in the neurons that are susceptible to degeneration. These inclusions may be found in the nucleus, cytoplasm, or neuronal processes. Inclusions are seen in lower motor neurons in SBMA patients, as well as in mouse and cell culture models [10, 21, 52, 56]. In affected patients, inclusions are also found in nonneuronal tissues that express AR, such as the testis, scrotal skin, and skeletal muscle [57].

The inclusions are composed of fibrils that may be stabilized by hydrogen bonding and glutamyl-lysine crosslinks [58]. The inclusions stain positively for ubiquitin and amino-terminal epitopes of the AR [21, 59]. In addition to the mutant protein itself, the aggregates are composed of proteins normally involved in the ubiquitin–proteosome pathway, molecular chaperones, and transcriptional factors. Sequestration and depletion of these cellular proteins may compromise their normal function and lead to cellular toxicity.

Both the ubiquitin–proteosome pathway and the molecular chaperones play an important function in protein folding, trafficking, and degradation. In cell culture systems of SBMA, overexpression of such molecular chaperones as heat shock proteins (Hsp70 and Hsp40) decreases the amount of aggregation and toxicity due to the mutant AR [60, 61]. When transgenic mouse models of SBMA were cross-bred with those that overexpressed Hsp70, there was a reduction in nuclear inclusion formation and an amelioration in the behavioral phenotype [62]. These studies support the hypothesis that the sequestration of these factors hinders the cell's ability to properly repair and degrade protein, and that their overexpression reverses the toxicity.

Transcriptional cofactors such as cAMP response element-binding protein (CBP) have also been found to be present within the nuclear inclusions [63, 64]. CBP functions as a histone acetyltransferase and is essential in activating transcription. At the level of transcription, the acetylation and deacetylation of histones play an important role in the modulation of chromatin structure and in the regulation of gene expression. Transcriptionally active genes are associated with highly acetylated histones, whereas genes that are less active are associated with low levels of acetylation. Hypoacetylated histones bind tightly to DNA, inhibit the access of transcription factors and RNA polymerases, and thus maintain a transcriptionally silent state. When histones are acetylated, the chromatin takes on a more open conformation, thus enhancing the access of transcription factors. In SBMA, sequestration of CBP may lead to changes in gene expression that result in cellular toxicity. Overexpression of CBP has been shown to rescue the functional and morphological phenotype in a *Drosophila* model of SBMA [65].

Inclusions are also found in neurites in models of SBMA and other polyglutamine diseases. These inclusions may mechanically disrupt axonal transport and lead to mitochondrial redistribution and dysfunction [66, 67].

These findings have led to the hypothesis that the aggregation of the mutant protein is important to the mechanism of polyglutamine-mediated neurodegeneration. However, in some models of SBMA and other

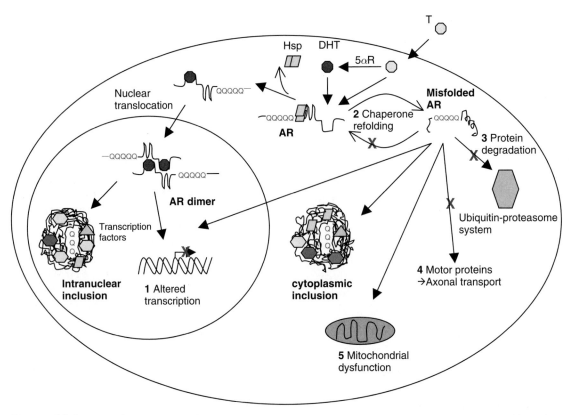

FIGURE 13-2 Model for cellular pathogenesis in SBMA. Testosterone (T) is converted to dihydrotestosterone (DHT) by the enzyme 5α-reductase (5αR). On T or DHT binding to AR, heat shock proteins (Hsp) dissociate from the AR, and the receptor translocates into the nucleus and binds to specific sequences in regulatory DNA. The mutant AR has an altered confirmation that results in the formation of intranuclear or cytoplasmic inclusions. Among the pathological mechanisms that may be involved in SBMA are (1) transcriptional dysregulation, (2) altered chaperone function, (3) altered ubiquitin–proteasome activity, (4) altered axonal transport, and (5) mitochondrial dysfunction. Intranuclear and cytoplasmic inclusions may be toxic, but recent studies indicate a cytoprotective role. See CD-ROM for color image.

polyglutamine diseases, the presence of intracellular inclusions has not been associated with toxicity [56, 64, 68, 69]. There is evidence that although the polyglutamine expansion in the AR in SBMA leads to inclusion formation, aggregation, and cytotoxicity, the inclusions may be protective by sequestering toxic proteins, facilitating their degradation, and increasing cell survival [68, 69] (Fig. 13-2).

C. Ligand-Dependent Effects in SBMA Models

Cell culture and transgenic animal models have shown that ligand binding is important in the development of the pathological phenotype of SBMA. The formation of intracellular inclusions is dependent on the degree of AR activation by ligands. In neuronal cell culture systems of SBMA, nuclear inclusions form only in the presence of androgen [43, 52, 56].

In transgenic mouse models, motor impairment is gender-related, with males more severely affected than females [55, 64, 70, 71]. The SBMA phenotype may be produced with the administration of testosterone to transgenic females [54]. Depletion of androgens in male mice by castration or by the administration of leuprorelin, a lutenizing hormone-releasing hormone agonist, prevents motor neuron toxicity [54, 71, 72]. Similar results have been obtained in the *Drosophila* SBMA model [53]. This ligand dependency is also supported clinically. Female carriers of the SBMA mutation may have abnormal electromyograms and slightly elevated creatine kinase. However, except for occasional muscle cramps and minimal tremors in the hands, they are clinically asymptomatic [14, 73]. It had been hypothesized that females remain relatively unaffected due to the silencing of the mutated allele by random inactivation of the X chromosome. However, women who are homozygous for the mutant SBMA AR gene have been reported to be minimally asymptomatic [74]. These findings support the importance of androgens in the development of the disease in humans, as well as in mice and flies. The variability in the age at onset and clinical progression in males with SBMA may be due not only to the length of the polyglutamine expansion, but also to the individual variation in levels of circulating androgens.

X. THERAPEUTIC APPROACHES IN SBMA

There is no treatment currently available for SBMA. However, based on animal studies and the pathogenic mechanisms believed to be involved in SBMA, there are at least two potential therapeutic approaches that may be promising.

A. Histone Deacetylase Inhibitors

At the level of transcription, the acetylation and deacetylation of histones play an important role in the level of transcription, such that active genes are associated with acetylated histones. The extent of histone acetylation is controlled by histone acetyltransferases and histone deacetlyases (HDACs). A proposed mechanism of neuronal toxicity in SBMA and other polyglutamine diseases is reduction in histone acetylase activity. In animal studies, HDAC inhibitors have been shown to improve the behavioral and histopathological phenotype in SBMA and other polyglutamine diseases [75–77]. Phenylbutyrate and suberoyl anilide hydroxaminic acid (SAHA) are HDAC inhibitors that have been used in the treatment of cancer and other disorders. Valproic acid is another drug with HDAC inhibitory activity that has been in widespread clinical use for the treatment of epilepsy and bipolar disorder. Phenylbutyrate and valproic acid are well-tolerated drugs with known safety profiles. The use of HDAC inhibitors in spinal muscular atrophy, another genetically inherited motor neuron disease, is currently being investigated in clinical trials. These pharmacological agents may be good candidates for investigation as therapy in SBMA and other polyglutamine disorders, as well.

B. Anti-androgens

Unlike other polyglutamine disorders, the function of the mutant protein in SBMA is known. It has been shown that the presence of ligand is important for the formation of nuclear inclusions in several model systems. And in transgenic animal models, the removal of ligand ameliorates the disease phenotype. Clinical trials are being conducted with the androgen antagonist leuprorelin [78]. On the basis of animal studies, GnRH agonists such as leuprorelin are promising candidates for treatment in SBMA. Studies are also currently being done to examine the differential effects of testosterone and DHT on mutant AR toxicity (D. Merry, personal communication). High levels of 5α-reductase have been found in spinal cord motor neurons [79]. Similarly, DHT was shown to bind to motor neurons in the lower pons, medulla oblongata, and spinal cord, the cells most susceptible to degeneration in SBMA [38]. Previous studies have also shown that there are low levels of this enzyme in skeletal muscle [80]. This indicates that although DHT is likely the primary ligand for AR in motor neurons, testosterone may be the primary ligand in skeletal muscle. This leads to the possibility of using pharmacologically available 5α-reductase inhibitors as potential therapy for SBMA. Currently, two pharmacological agents (finasteride and dutasteride) are approved by the Food and Drug Administration for the treatment of benign prostatic hypertrophy and male pattern baldness. Both these agents have low toxicity profiles. It is possible that selective suppression of DHT with a 5α-reductase inhibitor would preserve motor neuron function without having the adverse effect of reducing the anabolic effects of testosterone in muscle. This may lead to stabilization or amelioration of the neurological symptoms experienced by patients with SBMA.

References

1. Kennedy, W. R., Alter, M., and Sung, J. H. (1968). Progressive proximal spinal and bulbar muscular atrophy of late onset: A sex-linked recessive trait. *Neurology* **18**, 671–680.
2. Harding, A. E., Thomas, P. K., Baraitser, M., Bradbury, P. G., Morgan-Hughes, J. A., and Ponsford, J. R. (1982). X-linked recessive bulbospinal neuronopathy: A report of ten cases. *J. Neurol. Neurosurg. Psychiatry* **45**, 1012–1019.
3. Tanaka, F., Doyu, M., Ito, Y., Matsumoto, M., Mitsuma, T., Abe, K., Aoki, M., Itoyama, Y., Fischbeck, K. H., and Sobue, G. (1996). Founder effect in spinal and bulbar muscular atrophy (SBMA). *Hum. Mol. Genet.* **5**, 1253–1257.
4. Udd, B., Juvonen, V., Hakamies, L., Nieminen, A., Wallgren-Pettersson, C., Cederquist, K., and Savontaus, M. L. (1998). High prevalence of Kennedy's disease in Western Finland: Is the syndrome underdiagnosed? *Acta Neurol. Scand.* **98**, 128–133.
5. Lund, A., Udd, B., Juvonen, V., Andersen, P. M., Cederquist, K., Davis, M., Gellera, C., Kolmel, C., Ronnevi, L. O., Sperfeld, A. D., Sorensen, S. A., Tranebjaerg, L., Van Maldergem, L., Watanabe, M., Weber, M., Yeung, L., and Savontaus, M. L. (2001). Multiple founder effects in spinal and bulbar muscular atrophy (SBMA, Kennedy disease) around the world. *Eur. J. Hum. Genet.* **9**, 431–436.
6. Lund, A., Udd, B., Juvonen, V., Andersen, P. M., Cederquist, K., Ronnevi, L. O., Sistonen, P., Sorensen, S. A., Tranebjaerg, L., Wallgren-Pettersson, C., and Savontaus, M. L. (2000). Founder effect in spinal and bulbar muscular atrophy (SBMA) in Scandinavia. *Eur. J. Hum. Genet.* **8**, 631–636.

7. Olney, R. K., Aminoff, M. J., and So, Y. T. (1991). Clinical and electrodiagnostic features of X-linked recessive bulbospinal neuronopathy. *Neurology* **41**, 823–828.
8. Suzuki, T., Endo, K., Igarashi, S., Fukuda, M., and Tanaka, M. (1997). Isolated bilateral masseter atrophy in X-linked recessive bulbospinal neuronopathy. *Neurology* **48**, 539–540.
9. Barkhaus, P. E., Kennedy, W. R., Stern, L. Z., and Harrington, R. B. (1982). Hereditary proximal spinal and bulbar motor neuron disease of late onset: A report of six cases. *Arch. Neurol.* **39**, 112–116.
10. Sobue, G., Hashizume, Y., Mukai, E., Hirayama, M., Mitsuma, T., and Takahashi, A. (1989). X-linked recessive bulbospinal neuronopathy: A clinicopathological study. *Brain* **112** (Pt. 1), 209–232.
11. Arbizu, T., Santamaria, J., Gomez, J. M., Quilez, A., and Serra, J. P. (1983). A family with adult spinal and bulbar muscular atrophy, X-linked inheritance and associated testicular failure. *J. Neurol. Sci.* **59**, 371–382.
12. Nagashima, T., Seko, K., Hirose, K., Mannen, T., Yoshimura, S., Arima, R., Nagashima, K., and Morimatsu, Y. (1988). Familial bulbo-spinal muscular atrophy associated with testicular atrophy and sensory neuropathy (Kennedy–Alter–Sung syndrome): Autopsy case report of two brothers. *J. Neurol. Sci.* **87**, 141–152.
13. Sobue, G., Doyu, M., Kachi, T., Yasuda, T., Mukai, E., Kumagai, T., and Mitsuma, T. (1993). Subclinical phenotypic expressions in heterozygous females of X-linked recessive bulbospinal neuronopathy. *J. Neurol. Sci.* **117**, 74–78.
14. Mariotti, C., Castellotti, B., Pareyson, D., Testa, D., Eoli, M., Antozzi, C., Silani, V., Marconi, R., Tezzon, F., Siciliano, G., Marchini, C., Gellera, C., and Donato, S. D. (2000). Phenotypic manifestations associated with CAG-repeat expansion in the androgen receptor gene in male patients and heterozygous females: A clinical and molecular study of 30 families. *Neuromuscul. Disord.* **10**, 391–397.
15. Igarashi, S., Tanno, Y., Onodera, O., Yamazaki, M., Sato, S., Ishikawa, A., Miyatani, N., Nagashima, M., Ishikawa, Y., and Sahashi, K. (1992). Strong correlation between the number of CAG repeats in androgen receptor genes and the clinical onset of features of spinal and bulbar muscular atrophy. *Neurology* **42**, 2300–2302.
16. Hui, A. C., Cheung, P. T., Tang, A. S., Fu, M., Wong, L., and Kay, R. (2004). Clinical and electrophysiological features in Chinese patients with Kennedy's disease. *Clin. Neurol. Neurosurg.* **106**, 309–312.
17. Adachi, H., Katsuno, M., Minamiyama, M., Waza, M., Sang, C., Nakagomi, Y., Kobayashi, Y., Tanaka, F., Doyu, M., Inukai, A., Yoshida, M., Hashizume, Y., and Sobue, G. (2005). Widespread nuclear and cytoplasmic accumulation of mutant androgen receptor in SBMA patients. *Brain* **128**, 659–670.
18. MacLean, H. E., Choi, W. T., Rekaris, G., Warne, G. L., and Zajac, J. D. (1995). Abnormal androgen receptor binding affinity in subjects with Kennedy's disease (spinal and bulbar muscular atrophy). *J. Clin. Endocrinol. Metab.* **80**, 508–516.
19. Dejager, S., Bry-Gauillard, H., Bruckert, E., Eymard, B., Salachas, F., LeGuern, E., Tardieu, S., Chadarevian, R., Giral, P., and Turpin, G. (2002). A comprehensive endocrine description of Kennedy's disease revealing androgen insensitivity linked to CAG repeat length. *J. Clin. Endocrinol. Metab.* **87**, 3893–3901.
20. Ferrante, M. A., and Wilbourn, A. J. (1997). The characteristic electrodiagnostic features of Kennedy's disease. *Muscle Nerve* **20**, 323–329.
21. Li, M., Miwa, S., Kobayashi, Y., Merry, D. E., Yamamoto, M., Tanaka, F., Doyu, M., Hashizume, Y., Fischbeck, K. H., and Sobue, G. (1998). Nuclear inclusions of the androgen receptor protein in spinal and bulbar muscular atrophy. *Ann. Neurol.* **44**, 249–254.
22. Guidetti, D., Vescovini, E., Motti, L., Ghidoni, E., Gemignani, F., Marbini, A., Patrosso, M. C., Ferlini, A., and Solime, F. (1996). X-linked bulbar and spinal muscular atrophy, or Kennedy disease: Clinical, neurophysiological, neuropathological, neuropsychological and molecular study of a large family. *J. Neurol. Sci.* **135**, 140–148.
23. Ringel, S. P., Lava, N. S., Treihaft, M. M., Lubs, M. L., and Lubs, H. A. (1978). Late-onset X-linked recessive spinal and bulbar muscular atrophy. *Muscle Nerve* **1**, 297–307.
24. Parboosingh, J. S., Figlewicz, D. A., Krizus, A., Meininger, V., Azad, N. A., Newman, D. S., and Rouleau, G. A. (1997). Spinobulbar muscular atrophy can mimic ALS: The importance of genetic testing in male patients with atypical ALS. *Neurology* **49**, 568–572.
25. Sperfeld, A. D., Karitzky, J., Brummer, D., Schreiber, H., Haussler, J., Ludolph, A. C., and Hanemann, C. O. (2002). X-linked bulbospinal neuronopathy: Kennedy disease. *Arch. Neurol.* **59**, 1921–1926.
26. La Spada, A. R., Wilson, E. M., Lubahn, D. B., Harding, A. E., and Fischbeck, K. H. (1991). Androgen receptor gene mutations in X-linked spinal and bulbar muscular atrophy. *Nature* **352**, 77–79.
27. La Spada, A. R., Roling, D. B., Harding, A. E., Warner, C. L., Spiegel, R., Hausmanowa-Petrusewicz, I., Yee, W. C., and Fischbeck, K. H. (1992). Meiotic stability and genotype–phenotype correlation of the trinucleotide repeat in X-linked spinal and bulbar muscular atrophy. *Nat. Genet.* **2**, 301–304.
28. Doyu, M., Sobue, G., Mukai, E., Takahashi, A., and Mitsuma, T. (1992). DNA diagnosis of X-linked recessive bulbospinal muscular atrophy by androgen receptor gene mutations. *Rinsho Shinkeigaku* **32**, 336–339.
29. Poletti, A. (2004). The polyglutamine tract of androgen receptor: From functions to dysfunctions in motor neurons. *Front. Neuroendocrinol.* **25**, 1–26.
30. MacLean, H. E., Warne, G. L., and Zajac, J. D. (1997). Localization of functional domains in the androgen receptor. *J. Steroid Biochem. Mol. Biol.* **62**, 233–242.
31. Jenster, G., Trapman, J., and Brinkmann, A. O. (1993). Nuclear import of the human androgen receptor. *Biochem. J.* **293** (Pt. 3), 761–768.
32. Wilson, E. M., and French, F. S. (1976). Binding properties of androgen receptors: Evidence for identical receptors in rat testis, epididymis, and prostate. *J. Biol. Chem.* **251**, 5620–5629.
33. George, F. W., and Noble, J. F. (1984). Androgen receptors are similar in fetal and adult rabbits. *Endocrinology* **115**, 1451–1458.
34. Zhou, Z. X., Lane, M. V., Kemppainen, J. A., French, F. S., and Wilson, E. M. (1995). Specificity of ligand-dependent androgen receptor stabilization: Receptor domain interactions influence ligand dissociation and receptor stability. *Mol. Endocrinol.* **9**, 208–218.
35. Mooradian, A. D., Morley, J. E., and Korenman, S. G. (1987). Biological actions of androgens. *Endocr. Rev.* **8**, 1–28.
36. Sheridan, P. J. (1978). Localization of androgen- and estrogen-concentrating neurons in the diencephalon and telencephalon of the mouse. *Endocrinology* **103**, 1328–1334.
37. Yu, W. H., and McGinnis, M. Y. (2001). Androgen receptors in cranial nerve motor nuclei of male and female rats. *J. Neurobiol.* **46**, 1–10.
38. Sar, M., and Stumpf, W. E. (1977). Androgen concentration in motor neurons of cranial nerves and spinal cord. *Science* **197**, 77–79.
39. Breedlove, S. M. (1986). Cellular analyses of hormone influence on motoneuronal development and function. *J. Neurobiol.* **17**, 157–176.
40. Matsumoto, A. (1997). Hormonally induced neuronal plasticity in the adult motoneurons. *Brain Res. Bull.* **44**, 539–547.
41. Brooks, B. P., Paulson, H. L., Merry, D. E., Salazar-Grueso, E. F., Brinkmann, A. O., Wilson, E. M., and Fischbeck, K. H. (1997). Characterization of an expanded glutamine repeat androgen receptor in a neuronal cell culture system. *Neurobiol. Dis.* **3**, 313–323.
42. Brooks, B. P., Merry, D. E., Paulson, H. L., Lieberman, A. P., Kolson, D. L., and Fischbeck, K. H. (1998). A cell culture model for androgen effects in motor neurons. *J. Neurochem.* **70**, 1054–1060.

43. Simeoni, S., Mancini, M. A., Stenoien, D. L., Marcelli, M., Weigel, N. L., Zanisi, M., Martini, L., and Poletti, A. (2000). Motoneuronal cell death is not correlated with aggregate formation of androgen receptors containing an elongated polyglutamine tract. *Hum. Mol. Genet.* **9**, 133–144.
44. Jones, K. J. (1994). Androgenic enhancement of motor neuron regeneration. *Ann. N. Y. Acad. Sci.* **743**, 141–161; discussion 161–144.
45. Jones, K. J., Oblinger, M. M. (1994). Androgenic regulation of tubulin gene expression in axotomized hamster facial motoneurons. *J. Neurosci.* **14**, 3620–3627.
46. Marron, T. U., Guerini, V., Rusmini, P., Sau, D., Brevini, T. A., Martini, L., and Poletti, A. (2005). Androgen-induced neurite outgrowth is mediated by neuritin in motor neurones. *J. Neurochem.* **92**, 10–20.
47. Lieberman, A. P., Harmison, G., Strand, A. D., Olson, J. M., and Fischbeck, K. H. (2002). Altered transcriptional regulation in cells expressing the expanded polyglutamine androgen receptor. *Hum. Mol. Genet.* **11**, 1967–1976.
48. Mhatre, A. N., Trifiro, M. A., Kaufman, M., Kazemi-Esfarjani, P., Figlewicz, D., Rouleau, G., and Pinsky, L. (1993). Reduced transcriptional regulatory competence of the androgen receptor in X-linked spinal and bulbar muscular atrophy. *Nat. Genet.* **5**, 184–188.
49. Nakajima, H., Kimura, F., Nakagawa, T., Furutama, D., Shinoda, K., Shimizu, A., and Ohsawa, N. (1996). Transcriptional activation by the androgen receptor in X-linked spinal and bulbar muscular atrophy. *J. Neurol. Sci.* **142**, 12–16.
50. Sato, T., Matsumoto, T., Yamada, T., Watanabe, T., Kawano, H., and Kato, S. (2003). Late onset of obesity in male androgen receptor-deficient (AR KO) mice. *Biochem. Biophys. Res. Commun.* **300**, 167–171.
51. Schmidt, B. J., Greenberg, C. R., Allingham-Hawkins, D. J., Spriggs, E. L. (2002). Expression of X-linked bulbospinal muscular atrophy (Kennedy disease) in two homozygous women. *Neurology* **59**, 770–772.
52. Piccioni, F., Simeoni, S., Andriola, I., Armatura, E., Bassanini, S., Pozzi, P., and Poletti, A. (2001). Polyglutamine tract expansion of the androgen receptor in a motoneuronal model of spinal and bulbar muscular atrophy. *Brain Res. Bull.* **56**, 215–220.
53. Takeyama, K., Ito, S., Yamamoto, A., Tanimoto, H., Furutani, T., Kanuka, H., Miura, M, Tabata, T., and Kato, S. (2002). Androgen-dependent neurodegeneration by polyglutamine-expanded human androgen receptor in *Drosophila*. *Neuron* **35**, 855–864.
54. Katsuno, M., Adachi, H., Kume, A., Li, M., Nakagomi, Y., Niwa, H., Sang, C., Kobayashi, Y., Doyu, M., and Sobue, G. (2002). Testosterone reduction prevents phenotypic expression in a transgenic mouse model of spinal and bulbar muscular atrophy. *Neuron* **35**, 843–854.
55. McManamny, P., Chy, H. S., Finkelstein, D. I., Craythorn, R. G., Crack, P. J., Kola, I., Cheema, S. S., Horne, M. K., Wreford, N. G., O'Bryan, M. K., De Kretser, D. M., and Morrison, J. R. (2002). A mouse model of spinal and bulbar muscular atrophy. *Hum. Mol. Genet.* **11**, 2103–2111.
56. Walcott, J. L., and Merry, D. E. (2002). Ligand promotes intranuclear inclusions in a novel cell model of spinal and bulbar muscular atrophy. *J. Biol. Chem* **277**, 50855–50859.
57. Li, M., Nakagomi, Y., Kobayashi, Y., Merry, D. E., Tanaka, F., Doyu, M., Mitsuma, T., Hashizume, Y., Fischbeck, K. H., and Sobue, G. (1998). Nonneural nuclear inclusions of androgen receptor protein in spinal and bulbar muscular atrophy. *Am. J. Pathol.* **153**, 695–701.
58. Kahlem, P., Terre, C., Green, H., and Djian, P. (1996). Peptides containing glutamine repeats as substrates for transglutaminase-catalyzed cross-linking: Relevance to diseases of the nervous system. *Proc. Natl. Acad. Sci. USA* **93**, 14580–14585.
59. Ellerby, L. M., Hackam, A. S., Propp, S. S., Ellerby, H. M., Rabizadeh, S., Cashman, N. R., Trifiro, M. A., Pinsky, L., Wellington, C. L., Salvesen, G. S., Hayden, M. R., and Bredesen, D. E. (1999). Kennedy's disease: Caspase cleavage of the androgen receptor is a crucial event in cytotoxicity. *J. Neurochem.* **72**, 185–195.
60. Bailey, C. K., Andriola, I. F. M., Kampinga, H. H., and Merry, D. E. (2002). Molecular chaperones enhance the degradation of expanded polyglutamine repeat androgen receptor in a cellular model of spinal and bulbar muscular atrophy. *Hum. Mol. Genet.* **11**, 515–523.
61. Kobayashi, Y., Kume, A., Li, M., Doyu, M., Hata, M., Ohtsuka, K., and Sobue, G. (2000). Chaperones Hsp70 and Hsp40 suppress aggregate formation and apoptosis in cultured neuronal cells expressing truncated androgen receptor protein with expanded polyglutamine tract. *J. Biol. Chem.* **275**, 8772–8778.
62. Adachi, H., Katsuno, M., Minamiyama, M., Sang, C., Pagoulatos, G., Angelidis, C., Kusakabe, M., Yoshiki, A., Kobayashi, Y., Doyu, M., and Sobue, G. (2003). Heat shock protein 70 chaperone overexpression ameliorates phenotypes of the spinal and bulbar muscular atrophy transgenic mouse model by reducing nuclear-localized mutant androgen receptor protein. *J. Neurosci.* **23**, 2203–2211.
63. McCampbell, A., Taylor, J. P., Taye, A. A., Robitschek, J., Li, M., Walcott, J., Merry, D., Chai, Y., Paulson, H., Sobue, G., and Fischbeck, K. H. (2000). CREB-binding protein sequestration by expanded polyglutamine. *Hum. Mol. Genet.* **9**, 2197–2202.
64. Abel, A., Walcott, J., Woods, J., Duda, J., and Merry, D. E. (2001). Expression of expanded repeat androgen receptor produces neurologic disease in transgenic mice. *Hum. Mol. Genet.* **10**, 107–116.
65. Taylor, J. P., Taye, A. A., Campbell, C., Kazemi-Esfarjani, P., Fischbeck, K. H., and Min, K. T. (2003). Aberrant histone acetylation, altered transcription, and retinal degeneration in a *Drosophila* model of polyglutamine disease are rescued by CREB-binding protein. *Genes Dev.* **17**, 1463–1468.
66. Szebenyi, G., Morfini, G. A., Babcock, A., Gould, M., Selkoe, K., Stenoien, D. L., Young, M., Faber, P. W., MacDonald, M. E., McPhaul, M. J., and Brady, S. T. (2003). Neuropathogenic forms of huntingtin and androgen receptor inhibit fast axonal transport. *Neuron* **40**, 41–52.
67. Piccioni, F., Pinton, P., Simeoni, S., Pozzi, P., Fascio, U., Vismara, G., Martini, L., Rizzuto, R., and Poletti, A. (2002). Androgen receptor with elongated polyglutamine tract forms aggregates that alter axonal trafficking and mitochondrial distribution in motor neuronal processes. *FASEB J.* **16**, 1418–1420.
68. Taylor, J. P., Tanaka, F., Robitschek, J., Sandoval, C. M., Taye, A., Markovic-Plese, S., and Fischbeck, K. H. (2003). Aggresomes protect cells by enhancing the degradation of toxic polyglutamine-containing protein. *Hum. Mol. Genet.* **12**, 749–757.
69. Arrasate, M., Mitra, S., Schweitzer, E. S., Segal, M. R., and Finkbeiner, S. (2004). Inclusion body formation reduces levels of mutant huntingtin and the risk of neuronal death. *Nature* **431**, 805–810.
70. La Spada, A. R., Peterson, K. R., Meadows, S. A., McClain, M. E., Jeng, G., Chmelar, R. S., Haugen, H. A., Chen, K., Singer, M. J., Moore, D., Trask, B. J., Fischbeck, K. H., Clegg, C. H., and McKnight, G. S. (1998). Androgen receptor YAC transgenic mice carrying CAG 45 alleles show trinucleotide repeat instability. *Hum. Mol. Genet.* **7**, 959–967.
71. Katsuno, M., Adachi, H., Inukai, A., and Sobue, G. (2003). Transgenic mouse models of spinal and bulbar muscular atrophy (SBMA). *Cytogenet. Genome. Res.* **100**, 243–251.
72. Chevalier-Larsen, E. S., O'Brien, C. J., Wang, H., Jenkins, S. C., Holder, L., Lieberman, A. P., and Merry, D. E. (2004). Castration restores function and neurofilament alterations of aged symptomatic

males in a transgenic mouse model of spinal and bulbar muscular atrophy. *J. Neurosci.* **24**, 4778–4786.

73. Ishihara, H., Kanda, F., Nishio, H., Sumino, K., Chihara, K. (2001). Clinical features and skewed X-chromosome inactivation in female carriers of X-linked recessive spinal and bulbar muscular atrophy. *J. Neurol.* **248**, 856–860.

74. Schmidt, B. J., Greenberg, C. R., Allingham-Hawkins, D. J., and Spriggs, E. L. (2002). Expression of X-linked bulbospinal muscular atrophy (Kennedy disease) in two homozygous women. *Neurology* **59**, 770–772.

75. Minamiyama, M., Katsuno, M., Adachi, H., Waza, M., Sang, C., Kobayashi, Y., Tanaka, F., Doyu, M., Inukai, A., and Sobue, G. (2004). Sodium butyrate ameliorates phenotypic expression in a transgenic mouse model of spinal and bulbar muscular atrophy. *Hum. Mol. Genet.* **13**, 1183–1192.

76. Ferrante, R. J., Kubilus, J. K., Lee, J., Ryu, H., Beesen, A., Zucker, B., Smith, K., Kowall, N. W., Ratan, R. R., Luthi-Carter, R., and Hersch, S. M. (2003). Histone deacetylase inhibition by sodium butyrate chemotherapy ameliorates the neurodegenerative phenotype in Huntington's disease mice. *J. Neurosci.* **23**, 9418–9427.

77. Hockly, E., Richon, V. M., Woodman, B., Smith, D. L., Zhou, X., Rosa, E., Sathasivam, K., Ghazi-Noori, S., Mahal, A., Lowden, P. A., Steffan, J. S., Marsh, J. L., Thompson, L. M., Lewis, C. M., Marks, P. A., and Bates, G. P. (2003). Suberoylanilide hydroxamic acid, a histone deacetylase inhibitor, ameliorates motor deficits in a mouse model of Huntington's disease. *Proc. Natl. Acad. Sci. USA* **100**, 2041–2046.

78. Banno, H., Adachi, H., Katsuno, M., Suzuki, K., Atsuta, N., Watanabe, H., Tanaka, F., Doyu, M., and Sobue, G. (2005, Dec 15). Mutant androgen receptor accumulation in spinal and bulbar muscular atrophy scrotal skin: A pathogenic marker. *Ann. Neurol.* [Epub ahead of print].

79. Pozzi, P., Bendotti, C., Simeoni, S., Piccioni, F., Guerini, V., Marron, T. U., Martini, L., and Poletti, A. (2003). Androgen 5-alpha-reductase type 2 is highly expressed and active in rat spinal cord motor neurones. *J. Neuroendocrinol.* **15**, 882–887.

80. Thigpen, A. E., Silver, R. I., Guileyardo, J. M., Casey, M. L., McConnell, J. D., and Russell, D. W. (1993). Tissue distribution and ontogeny of steroid 5 alpha-reductase isozyme expression. *J. Clin. Invest.* **92**, 903–910.

PART V

Huntington's Disease

Molecular Pathogenesis and Therapeutic Targets in Huntington's Disease

JOHN S. BETT, GILLIAN P. BATES, AND EMMA HOCKLY

King's College, London, Department of Medical and Molecular Genetics, GKT School of Medicine, 8th Floor Guy's Tower, Guy's Hospital, London SE1 9RT, UK

I. Introduction
II. Mechanisms of HD Pathogenesis
   A. Mutant Huntingtin Aggregation and Inclusion Body Formation
   B. Toxicity of Huntingtin N-Terminal Fragments
   C. Cellular Protein Quality Control Mechanisms
   D. The Autophagy–Lysosome Pathway in HD
   E. Impairment of Axonal Transport
   F. Transcriptional Dysregulation
III. Experimental Therapeutics in Models of HD
   A. Models Used for Identification of Potential Therapeutic Agents
   B. Preclinical Testing in Mouse Models of HD
IV. Conclusion
Acknowledgments
References

Huntington's disease (HD) is an autosomal dominant, late onset neurodegenerative disease that is caused by a CAG/polyglutamine repeat expansion. Since the cloning of the HD gene in 1993, great progress has been made in understanding its molecular pathogenesis and uncovering potential therapeutic targets. In addition, a wide range of excellent genetic models have been generated that include, yeast, *C. elegans*, *D. melanogaster*, mammalian cell culture models, mouse and rat. These are being used to further unravel and validate the mechanisms by which cellular function becomes disrupted in HD and to develop a drug discovery pipeline through which promising drugs can be tested in a variety of genetic systems. This will enable the translation of basic research into the clinic and hopefully the eventual development of an effective treatment or cure for HD.

I. INTRODUCTION

Huntington's disease (HD) is an autosomal dominant progressive neurodegenerative disorder for which there is currently no effective therapy. Individuals with HD usually remain unaffected until midlife, when they manifest the typical motor and emotional symptoms, including chorea, rigidity, bradykinesia, irritability, and chronic depression [1].

The gene causing HD was cloned in 1993, and the mutation was found to be an expanded CAG repeat in the first exon, which is translated into an abnormally long polyglutamine tract in the N-terminus of a large 348-kDa protein named *huntingtin* [2]. Individuals with 35 or fewer repeats will remain unaffected, whereas individuals with repeats of 40 or more develop HD within a normal life span. The disease shows incomplete penetrance in the range of 36 to 39 contiguous glutamines, and the likelihood of developing the disorder with repeats in this range most likely depends on other genetic and environmental factors. There is an inverse correlation between the length of the polyglutamine tract and age of onset, whereby the most extreme cases present as a juvenile form of the disease. These are normally caused by repeats of more than 50 glutamines, although the exact number required for juvenile onset is difficult to define. Juvenile cases of HD normally result from paternal transmission of the mutant allele, which is associated with large expansions of the CAG trinucleotide.

HD is one of a group of at least nine neurodegenerative diseases known as the polyglutamine diseases, all of which are caused by the expansion of a polyglutamine tract in unrelated proteins. These include several spinocerebellar ataxias (SCA types 1, 2, 3, 6, 7 and 17), spinal and bulbar muscular atrophy (SBMA), and dentatorubral pallydoluysian atrophy (DRPLA). All present with autosomal dominant inheritance with the exception of SBMA, which is sex-linked.

The normal function of huntingtin remains to be firmly established, but it has been implicated in a variety of cellular processes including vesicular transport, cytoskeletal anchoring, and clathrin-mediated endocytosis [3]. Although there is evidence that a loss of function in mutant huntingtin contributes to HD [4, 5], and knocking out huntingtin in the forebrain of mice leads to neurodegeneration [6], this is unlikely to be the major route of toxicity for several reasons. First, knocking out normal huntingtin causes mice to die early in embryogenesis [7], and patients homozygous for the mutation are not easily distinguished from heterozygotes [8, 9]. In addition, ectopic expression of a mutant N-terminal huntingtin fragment in the presence of two functional copies of the *HD* gene is sufficient to cause neurological disease in mice [10]. Taken together, this suggests that although loss of function in mutant huntingtin can contribute to HD pathogenesis, the primary mechanism of toxicity in HD stems from a gain of function in mutant huntingtin.

II. MECHANISMS OF HD PATHOGENESIS

Over the past 10 years or so, major progress has been made in identifying pathogenic mechanisms and potential therapeutic targets in HD. These include aggregation of mutant huntingtin [11], processing of mutant huntingtin to generate a toxic N-terminal fragment [12–14], impairment of the ubiquitin–proteasome system [15], axonal transport defects [16], and transcriptional dysregulation [17]. It is hoped that the availability of many excellent cellular and animal models of HD will allow major progress to be made in elucidating and validating these underlying pathogenic mechanisms. The development of a high-throughput drug discovery pipeline where promising drugs are tested in a variety of genetic systems from baker's yeast to the mouse will greatly assist in transforming basic research from the laboratory into the clinic [18]. Ultimately, it is hoped this will aid in finding an effective treatment or cure for HD.

A. Mutant Huntingtin Aggregation and Inclusion Body Formation

The first transgenic mouse models of HD were generated to express an N-terminal fragment of mutant human huntingtin with expanded CAG repeats [10]. The most extensively studied of these is the R6/2 line, originally expressing around 150 repeats, which displays a neurological phenotype consisting of progressive motor dysfunction, abnormal gait, mild resting tremor, and hindlimb clasping. Examination of brains from these mice and other HD mouse models revealed the presence of inclusion bodies in neuronal nuclei, which stain positive for N-terminal huntingtin and ubiquitin [19]. In addition, analysis of postmortem HD brains demonstrated that inclusion bodies are present in neuronal nuclei and neuronal processes of both juvenile and adult-onset patients [20]. The presence of inclusion bodies in HD brains immediately suggested that the elongated polyglutamine tract may confer an altered biophysical property of mutant huntingtin that causes misfolding and aggregation of the huntingtin protein. This hypothesis was tested *in vitro*, where the aggregation of purified GST-tagged N-terminal huntingtin containing either pathogenic or nonpathogenic polyglutamine tracts was assessed [21]. Strikingly, pathogenic proteins with 53 contiguous glutamines formed fibrillar amyloid-like structures on cleavage of the GST moiety, whereas nonpathogenic proteins with only 30 glutamines did not. This was in keeping with work from Max Perutz, who previously hypothesized and showed that long polyglutamine tracts can form β-sheet structures that are stabilized by H bonding between main chain and side chain amides [22]. Therefore, because of the close correlation between the threshold of huntingtin aggregation and the disease-causing length of the CAG repeat, the aggregation of mutant huntingtin has become a very strong candidate as a major toxic mechanism in HD. Toxicity from aggregation could be exerted by aberrant protein interactions of mutant huntingtin and subsequent coaggregation of important cellular proteins, especially those naturally harboring a polyglutamine tract [23]. Further, it has been shown that the formation of large insoluble aggregates *in vitro* is dependent on both protein concentration and time [24], which may explain the delay in both inclusion formation and clinical manifestation of the disease. This finding is also consistent with the aggregation of huntingtin being a nucleation-dependent event, whereby an aggregated nucleus of a certain size or shape is necessary for the process to begin. Indeed, addition of a preformed "seed" markedly enhances the rate of mutant huntingtin aggregation. Although mutant huntingtin can form spherical, annular, and fibrillar aggregates *in vitro* [25], it is currently unclear which of these aggregation pathways is likely to be most toxic or which most closely represents aggregates in the human HD brain.

The propensity of mutant polyglutamine proteins to aggregate appears to be a universal property, as inclusion bodies can be artificially induced in eukaryotes such as the baker's yeast *Saccharomyces cerevisiae* [26], the nematode *Caenorhabditis elegans* [27–29], the fruitfly *Drosophila melanogaster* [30–32], and the common house mouse *Mus musculus* [19]. However, the exact relationship of inclusion bodies to HD pathology is still unclear. It has been shown that directing preformed polyglutamine inclusions to the nucleus is highly toxic to cells [33]. In addition, it appears extremely likely that spherical aggregates with a diameter of 1 μm or greater would interfere with the tightly controlled workings of the nucleus and the transport of important cellular cargoes along the axon. However, it is still unclear if the presence of an inclusion body is a direct prerequisite for cellular pathology or death. For example, it has been shown in HD cell models that apoptotic cell death does not correlate with the presence of a visible inclusion body, and that inclusion body formation may even predict survival for cultured neurons [34, 35]. To this end, many researchers believe inclusion bodies exert a neuroprotective effect by sequestering potentially more toxic soluble or oligomeric aggregated forms of the mutant protein [36]. There is yet another possibility that the presence of inclusion bodies represents the culmination of a toxic aggregation process, whereby it is the process of aggregation that drives the recruitment and dysfunction of essential cellular proteins. In support of this, it has been shown that the recruitment of the polyglutamine tract-bearing TATA box-binding protein (TBP) to inclusions requires ongoing synthesis of mutant huntingtin [37]. When transcription of mutant huntingtin has been stopped, inclusion bodies are recruitment-incompetent. In this case, an inclusion body may not be as pathogenic as soluble huntingtin or oligomeric huntingtin aggregates.

Nevertheless, inclusion bodies serve as a marker of disease whereby their formation in the nucleus and neuropil correlates with disease pathology in transgenic mice [38]. It has also been shown in a tetracycline-inducible conditional mouse model of HD that shutting off expression of mutant huntingtin reverses neuropathology and causes the disappearance of inclusions, further providing a link between the presence of inclusion bodies and HD pathogenesis [39]. In addition, inhibiting aggregation of mutant huntingtin using a synthetic bivalent binding peptide comprising nonpathogenic polyglutamine tracts was sufficient to rescue photoreceptor degeneration in a *Drosophila* model of HD, as well as reduce inclusion formation [40].

B. Toxicity of Huntingtin N-Terminal Fragments

It is currently unknown why, despite widespread expression throughout the body, mutant huntingtin appears to exert its toxic effect chiefly in neurons. In addition, it is not understood why some populations of neurons may be more vulnerable to toxicity than others, and several explanations have been proposed to account for this. These include neuronal population-specific expansions of the CAG repeat in affected brain regions [41], differences in huntingtin-interacting proteins between different neurons [3], and the fact that mutant huntingtin has distinct aggregation properties in different neurons [42]. However, another attractive explanation is that differential cleavage of full-length huntingtin in different neuron types generates a more toxic N-terminal fragment of the mutant protein in specific populations of neurons. Indeed, full-length huntingtin has been reported to contain several cleavage sites within the first 500 or so amino acids (Fig. 14-1) [3]. These include cas-

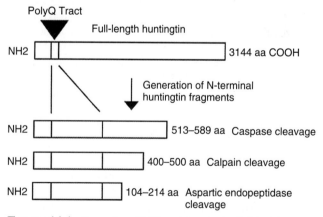

FIGURE 14-1 Processing of full-length huntingtin. Full-length mutant huntingtin appears to undergo proteolytic processing to generate various N-terminal fragments. Proposed cleavage sites include caspase, calpain, and aspartic endopeptidase sites. N-terminal fragments are thought to have increased propensity to aggregate and thus are potentially more toxic than full-length huntingtin.

pase-3, calpain, and aspartic endopeptidase sites, and mutational disruption of these sites or the use of specific protease inhibitors has been shown to decrease both cleavage and cellular toxicity [12, 14, 43]. In addition, N-terminal fragments of mutant huntingtin are more toxic to mammalian cells than full-length huntingtin [44, 45], which may be related to their increased ability to aggregate and accumulate in the nucleus [46–48].

Consistent with this idea, analyses of HD patient brains and brains from mouse models reveal that inclusion bodies from affected regions are composed mainly of an N-terminal fragment of mutant huntingtin [13, 20, 49]. Thus, the generation of an N-terminal fragment with increased toxicity and aggregation properties would represent a neuron type-specific event that could explain the differential neuronal dysfunction and death observed in HD. In support of this, transgenic HD mice expressing only an N-terminal fragment of the *HD* gene with around 150 repeats have a much wider pathology and severe neurological phenotype [10] than full-length knock-in mouse models [50, 51]. Therefore, a more detailed understanding of the precise events that generate an N-terminal fragment in HD brains is an important undertaking. The ability to manipulate or prevent these events will then represent an exciting and promising therapeutic window.

C. Cellular Protein Quality Control Mechanisms

The presence of inclusion bodies throughout HD patient brains, mice, and other genetic systems suggests that pathogenic polyglutamines escape the normal rigid protein quality-control mechanisms that exist within the cell. After a protein is translated, it must reach its native structural conformation to become a fully functional protein, a process that is greatly assisted by several classes of molecular chaperones. For example, two major classes of chaperones are the Hsp40 and Hsp70 families, which act in concert to recognize and promote the refolding of exposed hydrophobic polypeptide stretches in an ATP-dependent manner. If the protein cannot be refolded, it is tagged for destruction by the ubiquitin–proteasome system (UPS), a major route of cellular protein clearance. In this system, four or more ubiquitin monomers are attached to the misfolded protein through a series of enzymatic steps that signal the protein's destruction by the 26S proteasome, a barrel-shaped proteolytic machine.

The formation of inclusion bodies in HD and other polyglutamine diseases suggests that these elaborate cellular quality control mechanisms have difficulty in dealing with mutant polyglutamine proteins such as huntingtin. Elucidating the exact nature of the involvement of both molecular chaperones and the UPS in HD is therefore of extreme interest, as this could lead to promising therapeutic targets.

1. MOLECULAR CHAPERONES AS A THERAPEUTIC TARGET IN HD

The finding that various classes of molecular chaperones colocalize with inclusion bodies in HD transgenic mice [52, 53] and in polyglutamine disease patient brains [54–56] suggests that chaperones can recognize misfolded and aggregated polyglutamine proteins. Therefore, major effort has been expended to investigate the effect of modulating chaperone levels in models of HD and other polyglutamine diseases. Although many different chaperones have a noticeable effect on mutant huntingtin-induced phenotypes [26, 28, 57], members of the Hsp40 and Hsp70 families have sparked the most interest and shown the most promise as therapeutic targets for HD and the polyglutamine diseases. For example, overexpression of members of the Hsp70 and Hsp40 families of chaperones can interfere with polyglutamine aggregation and decrease mutant polyglutamine-associated toxicity in cell models [52, 54, 58–60]. In addition, overexpression of Hsp70 in a *Drosophila* model of polyglutamine disease can dramatically suppress neurodegeneration [61], and this effect is more pronounced when Hsp40 and Hsp70 family members are overexpressed simultaneously [62]. The expression of Hsp40 and Hsp70 appears to alter the biochemical properties of mutant polyglutamine proteins, which renders them more soluble [62]. Indeed, it has been shown in yeast and in a cell-free system that Hsp40 and Hsp70 can act synergistically to increase the detergent solubility of polyglutamine aggregates, and retard the formation of ordered fibrils in favor of amorphous aggregates (Fig. 14-2) [63]. Interestingly, addition of these chaperones is effective at modifying the *in vitro* aggregation of polyglutamines only if added during the lag phase of aggregation [25, 63], suggesting they interfere with early aggregate precursors but do not affect preformed inclusion bodies. The combination of Hsp40 and Hsp70 has also proved effective at inhibiting the negative interaction of mutant huntingtin with the polyglutamine tract-containing TBP. [37]. It was shown *in vitro* that a GST–N-terminal mutant huntingtin fusion protein undergoes spontaneous self-association on cleavage of GST, and subsequently interacts with other monomers of mutant huntingtin and TBP [37]. Incubation of mutant huntingtin in the presence of Hsp40 and Hsp70, however, markedly decreases the self-association step and subsequently decreases the interaction of mutant huntingtin monomers with TBP. Because transcriptional dysregulation and the recruitment of TBP to inclusions

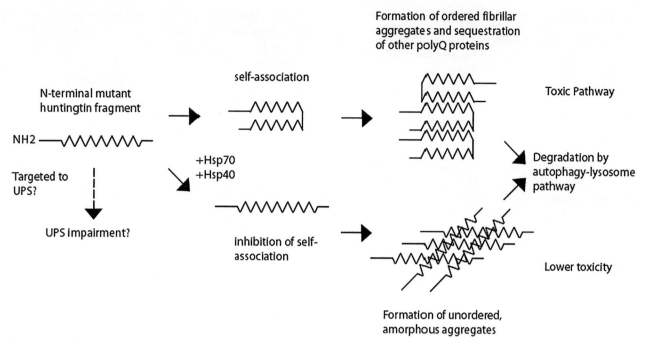

FIGURE 14-2 Protective effect of molecular chaperones on the aggregation of mutant huntingtin and the role of the ubiquitin–proteasome system (UPS) and autophagy in clearance of huntingtin. N-terminal fragments of mutant huntingtin can self-associate via interactions of the mutant polyglutamine tract. This step may be critical in the formation of highly ordered fibrillar aggregates of mutant huntingtin, and is thought to drive the recruitment and aggregation of other cellular polyglutamine proteins. Overexpression of the molecular chaperones Hsp70 and Hsp40 appears to inhibit the self-association step of mutant huntingtin, and is associated with the formation of unordered, amorphous aggregates and decreased cellular toxicity. Mutant huntingtin is targeted to the UPS for degradation, but it is unknown if this targeting occurs before or after it has aggregated. This could potentially cause dysfunction of the UPS, or alternatively, an age-dependent decline in UPS activity may enhance the aggregation of mutant huntingtin. The autophagy–lysosome pathway may be responsible for degrading aggregated polyglutamine proteins, as it is capable of degrading organelles and larger protein complexes.

constitute a common feature in HD and the polyglutamine diseases (see later), preventing the co-aggregation of TBP by overexpression of Hsp40 and Hsp70 could in part explain the beneficial effects of these chaperones.

To validate the potential of overexpressing chaperones as a therapeutic avenue in HD and other polyglutamine diseases, it is important to show that this approach has a beneficial effect in polyglutamine mouse models. Indeed, the overexpression of Hsp70 has improved the neurological phenotype in some mouse models of polyglutamine disease [64, 65], but has shown a more modest effect in others [66]. It has also been reported that overexpression of Hsp70 can delay early aggregate formation in the R6/2 transgenic HD mouse brain without significantly affecting the neurological phenotype [53]. Thus, because overexpression of Hsp70 in polyglutamine mouse models has less effect than in other systems, inducing the expression of several chaperones together may prove more effective in ameliorating mutant polyglutamine-induced neurological phenotypes. To this end, there have been major efforts to identify compounds that can activate the mammalian heat shock response. One such compound that has been identified is geldanamycin, a benzoquinone ansamycin that activates Hsp40, Hsp70, and Hsp90 and suppresses aggregation of N-terminal mutant huntingtin in mammalian cells [60]. Geldanamycin and radicicol, a fungicidal antibiotic, have also been shown to induce and maintain the heat shock response in HD transgenic brain slices over 3 weeks in culture, which results in a 1-week delay in aggregate formation [53]. Another interesting compound that has been identified by several laboratories as a suppressor of mutant huntingtin-induced neurotoxic phenotypes is celastrol [67]. This compound has been found to elicit a robust heat shock response in mammalian cells, involving the upregulation of Hsp70, Hsp40, and Hsp27, which may explain its observed suppression of neurotoxic phenotypes.

Results with inducers of the heat shock response so far look promising as therapeutic agents in HD, and current efforts are aimed at identifying compounds that can both cross the blood–brain barrier and activate the heat shock response in neurons. It is hoped that testing these compounds in mouse models for their ability to suppress neurological phenotypes will bring us one step closer to finding a promising therapy for HD patients.

2. The Ubiquitin–Proteasome System in HD

The UPS is responsible for degrading damaged or misfolded cellular proteins, and is also essential for normal physiological protein turnover. The association of ubiquitylated inclusion bodies in polyglutamine disease with components of the proteasomal machinery [54, 56] suggests the UPS has indeed targeted the aggregated proteins for destruction. However, once long polyglutamine proteins have aggregated, they appear to be resistant to degradation by the UPS [68], suggesting the effort exerted by the UPS to degrade aggregated proteins in HD could potentially compromise its normal cellular housekeeping function, leading to neuronal dysfunction. Direct evidence of a UPS impairment in HD came from a cell model stably expressing green fluorescent protein fused to a strong UPS degradation signal [15]. In this UPS reporter cell line, the accumulation of green fluorescence in the cell acts as an indicator of UPS inhibition as judged by treatment with proteasome inhibitors. Remarkably, it was found that mutant huntingtin, but not normal huntingtin, caused a marked accumulation of fluorescence, indicating that the expression of a pathogenic polyglutamine-containing protein directly impairs the UPS. Furthermore, there was a correlation between the level of proteasome impairment and the presence and size of inclusion bodies. It has been proposed that this impairment may be the result of pathogenic polyglutamines becoming stably trapped within the proteasome core, thus limiting normal proteasome activity [69, 70]. The close association of proteasomes with cellular inclusion bodies and in the brains of transgenic HD mice suggests that these proteasomes are indeed likely to be "clogged" with aggregated polyglutamine proteins [71, 72]. However, data show that the cellular impairment in UPS function is not caused primarily by a direct interaction between the proteasome and mutant huntingtin [73], and so other factors must account for the observed impairment.

Although a robust impairment of the UPS is observed in cell models of polyglutamine disease [15, 74], no evidence of proteasome impairment has been detected in polyglutamine disease mouse models to date [49, 75, 76]. In contrast, adult-onset HD postmortem brains show a decrease of in vitro proteasome activity, whereas juvenile HD brains actually show an upregulation in activity [77]. These results suggest that in vivo, UPS impairment may be a secondary contributor to polyglutamine pathology and not a primary mechanism of toxicity. Nevertheless, the ability to alter UPS function pharmacologically represents an attractive therapeutic target for several reasons. First, modulating the UPS in various polyglutamine disease models is known to alter inclusion body formation and toxicity [78–80]. Also, although polyglutamine tracts are not easily degraded by eukaryotic proteasomes [70, 72], directing mutant huntingtin to the UPS is still sufficient to decrease inclusion formation [81, 82]. This suggests that increasing the proteasomal processing of mutant huntingtin in HD brains may delay huntingtin aggregation and disease progression, and further investigation of the therapeutic potential of this strategy is definitely worth pursuing.

D. The Autophagy–Lysosome Pathway in HD

It has been shown that terminating transcription of mutant huntingtin in a conditional mouse model of HD can reverse the neurological phenotype and cause the disappearance of inclusions [39]. Therefore, cellular mechanisms must exist that are capable of degrading aggregated polyglutamine proteins. In addition to the UPS, a major process involved in the clearance of cellular proteins is the autophagy–lysosome pathway. In this process, double-membraned vacuoles engulf proteins and organelles, fuse with primary lysosomes, and release their contents to be degraded by proteases in the acidic environment of the lysosome. The observation that the UPS cannot degrade aggregated huntingtin [68] (presumably because aggregated regions are too tightly bound to be unfolded and fed through the 26S proteasome catalytic core) suggests that the autophagy-lysosome pathway may be involved in the degradation of aggregated polyglutamine proteins (Fig. 14-2).

The autophagy–lysosome pathway has therefore been investigated as a potential route of aggregate clearance. Both normal huntingtin and mutant huntingtin have been observed in autophagic vacuoles from cultured mouse neurons [83], and it has been shown that inhibiting autophagy causes the accumulation of aggregates and inhibits the clearance of mutant huntingtin [84, 85]. Therefore, enhancing autophagy represents an exciting therapeutic approach to HD. Indeed, data show that inducing autophagy with rapamycin improves neurodegeneration in a *Drosophila* HD model, and treating transgenic HD mice with the rapamycin analog CCI-779 reduces aggregation and improves the neurological phenotype [86]. Rapamycin acts by binding to the mammalian target of rapamycin (mTOR) protein kinase, a negative regulator of the autophagic pathway. This interaction inhibits its kinase activity and stimulates autophagy. These results suggest that finding compounds to cross the blood–brain barrier and induce autophagy in the HD brain may well prove to have a positive therapeutic outcome.

E. Impairment of Axonal Transport

An interesting pathogenic mechanism recently implicated in HD is disruption of axonal transport. Axonal transport is a microtubule-based process necessary for the distribution of essential protein complexes from the cell body to the nerve endings, and disruption of this in certain neuronal processes has been proposed to play a role in HD pathogenesis [16]. Indeed, the formation of inclusion bodies in neuronal processes has been described in human HD patient brains [20], validating the possibility that defects in the transport of axonal cargoes may be involved in the molecular pathology of HD. In support of this idea, it has been shown that *Drosophila* expressing mutant N-terminal huntingtin form inclusion bodies in axonal processes and show signs of organelle accumulations in the axon, characteristic of an axonal transport blockage [87]. This dysfunction could occur by a direct blockage of axons by large aggregates and also by the sequestration of polyglutamine-containing proteins with vital axonal transport functions [88]. Interestingly, knockdown of normal huntingtin causes a defective axonal transport phenotype in *Drosophila*, suggesting that the loss of normal huntingtin function could potentially contribute to axonal transport phenotypes [87]. This is supported by work that shows that huntingtin is involved in the vesicular transport of brain-derived neurotrophic factor (BDNF) along the axon [5]. Thus, it is possible that mutant huntingtin could sequester normal huntingtin through aberrant interaction with its polyglutamine tract, reducing normal levels of soluble huntingtin. The loss of function of huntingtin and other important proteins, combined with a physical blockage in the axonal processes, could therefore contribute to an axonal transport defect in HD.

Interestingly, it has been shown that mutations in dynein, an essential molecular motor in axonal transport, can enhance the disease phenotype of HD transgenic *Drosophila* and mice [89]. The proposed mechanism for this exacerbation of phenotype is an impairment of the lysosome–autophagy pathway, which is dependent on microtubule-based transport. As a key clearance mechanism of mutant huntingtin, reducing autophagic activity speeds up the process of aggregate formation in the cell, which is likely to account for the observed increase in toxicity.

It has yet to be firmly established that axonal transport defects exist and contribute to disease in HD mouse models or patients. Nonetheless, current strategies aimed at inhibiting the aggregation of mutant huntingtin would, in theory, prove beneficial in preventing axonal transport blockage by large aggregates. Other strategies aimed at pharmacologically enhancing axonal transport in HD may also be effective therapies.

F. Transcriptional Dysregulation

Over the past several years, a body of evidence has implicated transcriptional dysregulation as a likely contributor to HD pathogenesis. The correct temporal and spatial regulation of gene expression is one of the most important processes to an organism. Therefore, disturbing transcriptional balance in a biological system is likely to have dramatic consequences. It has become clear that numerous transcriptional pathways are altered in the HD mouse model and in patients with HD [90, 91]. Thus, identifying which genes are modulated by disease progression may provide new insights into disease pathogenesis, and methods to counteract transcriptional dysregulation in HD patients represent an exciting therapeutic strategy.

A major transcriptional program implicated in HD pathogenesis is the expression of cAMP response element (CRE)-regulated genes. CRE promoter elements are bound by cAMP response element-binding protein (CREB), which, on phosphorylation by protein kinase A, recruits the polyglutamine tract-bearing coactivator CREB-binding protein (CBP) to execute transcription of the relevant genes. The finding that CBP coaggregates into inclusion bodies in a polyglutamine-dependent manner in HD cell and mouse models, as well as HD patient brains, suggests that the expression of CRE-regulated genes may be disturbed in HD [23, 92, 93]. Indeed, microarray analysis of HD cell models confirms that transcription of CRE-regulated genes is downregulated [94], and expanded polyglutamine tracts are shown to interfere with CREB-dependent transcription [95]. Alterations in CRE-dependent transcription have also been observed *in vivo*, where CRE-regulated genes are actually upregulated [96]. The expression of CRE-regulated genes is responsible for many important neuronal processes, highlighted by the fact that genetic ablation of CREB in the mouse forebrain causes striatal and hippocampal degeneration [97]. Thus, it is easy to imagine how dysfunction of CBP via aberrant polyglutamine-dependent interactions could contribute to HD pathogenesis. In support of this, it has been shown that mutant N-terminal huntingtin interacts with and inhibits the acetyltransferase activity of CBP and p300/CBP-associated factor (P/CAF) [98]. The acetylation of histones by acetyltransferases causes a relaxation of DNA packaging, allowing easier access for transcription factors to DNA sequence elements. Therefore, the reduction of histone acetylation observed in a *Drosophila* HD model may be a result of decreased CBP function [98]. Furthermore, it was found that treating *Drosophila* HD models with the histone deacetylase (HDAC) inhibitor SAHA can increase the acetylation of histones and reduce lethality [98]. In addition, SAHA has been found to slow down the neurological symptoms in R6/2 HD transgenic mice [99], further validating HDAC inhibitors as promising therapeutic candidates for the treatment of HD.

It has also been shown that the polyglutamine tract-bearing TBP can interact with mutant N-terminal huntingtin in a polyglutamine-dependent manner *in vitro*, and is recruited to cellular polyglutamine inclusions [37]. It is also reported to be present in inclusions of HD patient brains [100]. TBP is part of the general transcription machinery and because of its vital cellular importance, aberrant interactions with mutant huntingtin may contribute to the molecular pathology of HD. In fact, mutant huntingtin inhibits the binding of TBP to its native DNA sequence *in vitro*, and toxicity in a yeast HD model has been shown to be elicited through polyglutamine-dependent interactions of huntingtin with TBP [37]. Strikingly, this effect was found to be attenuated in the presence of the heat shock proteins Hsp40 and Hsp70, indicating that induction of the heat shock response could potentially improve transcriptional dysregulation, as well as exert other protective effects (discussed in earlier section).

Many other interesting transcriptional pathways have been implicated in HD pathogenesis [101–103]. Particularly interesting is the finding that normal huntingtin interacts with the REST-NRSF (repressor element 1 transcription factor–neuron-restrictive silencer factor) transcriptional repressor in the cytoplasm, limiting its availability to bind to neuron-restrictive silencer elements (NRSEs) in the nucleus [104]. This promotes the transcription of genes with NRSEs, which encode for various neuroprotective proteins including BDNF. Mutant huntingtin, however, fails to bind as well as normal huntingtin, resulting in the accumulation of REST-NRSF in the nucleus and, ultimately, the repression of genes regulated by NRSEs. This also suggests that loss of huntingtin function contributes to HD pathology.

The success of HDAC inhibitors in fly and mouse models of HD validates the targeting of transcriptional impairment as an important therapeutic pursuit in HD. Further preclinical mouse trials, if successful, will open the door for this class of compounds to be tested in clinical trials for their ability to attenuate symptoms of HD patients.

III. EXPERIMENTAL THERAPEUTICS IN MODELS OF HD

Numerous genetic models of HD have been developed including cell models [46, 105, 106], yeast [26, 107], *Caenorhabditis elegans* [27–29], *Drosophila melanogaster* [30–32], transgenic [10, 108–111] and knock-in [50, 51, 112, 113] mouse models, and a transgenic rat model [114]. Also, viral vector models [115] have been developed that can be applied to numerous species including primates.

Clearly, a vital aim of research in these models is to identify targets for therapeutic intervention, and the use of these models to validate these targets and to screen potential agents for their ability to modify disease progression is an ongoing and expanding area in HD research. The concept of a drug discovery pipeline has gained currency, wherein simpler models are used to screen libraries of compounds, while more complex models are used to verify findings and sift positive "hits" for further testing (Fig. 14-3) [116] Compounds that show efficacy in a number of different models, which mimic various aspects of the disease, would be considered for clinical testing in HD patients. The utility of this pipeline is crucially dependent on the availability of good models that accurately recapitulate aspects of HD, and comparing results obtained in different models can help to validate their predictive power.

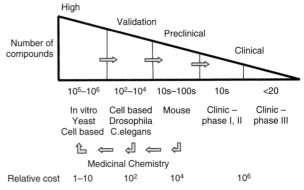

FIGURE 14-3 The drug discovery pipeline for potential therapeutics for the treatment of polyglutamine disease. A process of drug discovery from high-throughput screening of random chemical libraries to clinical trials of the most promising candidates flows from left to right. At each segment, the number of compounds screened decreases, while the cost of screening each candidate increases. Adapted, with permission, from Hughes and Olson [116].

A. Models Used for Identification of Potential Therapeutic Agents

1. In Vitro Filter Retardation Assay

The simplest model used to screen compounds as therapeutic targets depends on the tendency of huntingtin containing expanded polyglutamine to aggregate *in vitro*. This model uses an automated filter retardation assay to screen compounds for their ability to inhibit protein aggregation in a wholly *in vitro* system. The first study published using this model screened 184,880 compounds and identified the benzothiazoles as a potentially interesting class of compounds [117]. Although this model is very simple and cannot mimic the complexity of the situation *in vivo*, it is a very high-throughput assay, which can identify targets based on very well-defined biochemical criteria. However, as it is a fully *in vitro* model, it is capable of detecting only molecules that interact directly with the huntingtin protein, and does not pick up molecules that could influence aggregation by modulating other aspects of the cell machinery.

The utility of such high-throughput screens may ultimately depend on the quality of the library they are used to screen. The assay in itself does not distinguish between molecules on the basis of their druglike properties, merely on their ability to inhibit aggregation. Therefore, molecules will be identified that are capable of disrupting aggregation, but are not viable targets for further investigation *in vivo*, due to poor pharmacological properties. Dependent on the molecules identified by this primary screen, medicinal chemistry can be used to synthesize anologs with similar properties, but better-predicted pharmacological profiles. These new compounds can themselves be screened for inhibitory activity. Any compound identified can then be assessed in more sophisticated and biologically relevant assays. Lipinski's "rule of five" [118] (Table 14-1) provides a guideline as to which compounds are likely to be viable drugs, based on how likely they are to be absorbed. Compounds that violate more than two of the rules, or that are wildly beyond the range for any one rule, are unlikely to be useful therapeutically. Other considerations also influence whether a compound has therapeutic potential. For example, certain structural motifs may be chemically or metabolically unstable, leading to very rapid degradation and, therefore, a short duration of tissue exposure to the agents. Others may have inherent toxicity.

2. Mammalian Cell Culture and Yeast-Based Models

Yeast [107] and cell-based [94] models represent the next level of complexity. These models are still relatively high-throughput, and can potentially be used to screen thousands of compounds [119]. A number of mammalian cell models are available, based on neuronal PC12 cells [23, 94], COS-7 or HEK293 cells [120], stably transfected with polyglutamine-encoding genes. These models have more biological relevance than wholly *in vitro* systems, as the cellular machinery is intact. Therefore, they can be used to screen molecules with a much wider range of target mechanisms.

A variety of readouts are possible, with aggregation inhibition and polyglutamine-induced cell death being obvious first-line candidates. The use of cell death as an endpoint has the advantage of making no *a priori* assumptions about viable molecular targets, but does make the assumption that cell death, rather than neuronal dysfunction, is the dominant biologically important cellular outcome in HD, an assumption that may not be valid [121] as severe symptoms are noted in R6/2 mouse in the absence of cell death.

Several cell-based screens for potential therapeutic agents have recently been published [119, 122, 123]. A library of 1040 FDA-approved compounds, compiled by NINDS, was tested by 26 independent laboratories in a number of models: In one study, compounds were assesed in PC12 cells at concentrations ranging from 0.16 to 100 μM for their ability to delay or prevent polyQ-induced cell death [119]. Eighteen compounds prevented cell death and 51 delayed it. Among the most active compounds were caspase inhibitors and cannabinoids. In an independent study, compounds were assesed in PC12 cells at concentrations ranging from 0.16 to 100 μM for their ability to reduce polyQ-induced lactate dehydrogenase (LDH) release, a marker of cell death. [123]. Five compounds attenuated LDH release. Of these, three (acivicin, nipecotic acid, and mycophenolic acid) also reduced aggregate formation in the cell model.

In an alternative strategy [122], these 1040 compounds were assesed in a filter retardation assay for their ability to inhibit aggregation of an N-terminal

TABLE 14-1 Lipinski's Rule of Five for Prediction of Good Compound Permeability/Absorption Properties [118]

Parameter	Limits
H-bond donors	<5
H-bond acceptors	<10
Molecular weight	<500
Log P^a	<5

^{a} Log P is the log ratio of the compound's solubility in octanol to its solubility in water.

fragment of huntingtin containing 58 glutamines. Ten compounds were found to inhibit aggregation with an IC$_{50}$ below 15 μM. The six most effective compounds were then tested at concentrations ranging from 0.01 to 500 μM for their ability to delay or prevent abnormal huntingtin localization in a striatal cell line derived from the *HdhQ111* knock-in mouse. Two compounds, celastrol and juglone, showed efficacy in this model. Celastrol showed efficacy in six independent laboratories as an aggregation inhibitor and suppressor of neurotoxicity [67, 122]. It is a quinone methide triterpene and an active component of Chinese herbal medicines [67]. It is a potent inducer of the heat shock response, a promising therapeutic target in HD, as discussed earlier.

An alternative approach [124] aims to identify suppressors of polyglutamine aggregation in a yeast model [125]. If effective, these compounds are then tested in a filter retardation assay [126], and for their ability to reduce inclusion formation and ameliorate toxicity in HD models such as PC12 cells [127] and *Drosophila* [32, 127], before proceeding to preclinical trials in the mouse. This approach has identified a small molecule, C2-8, that has the ability to inhibit aggregation in yeast and mammalian cells [124]. It was also found to suppress neurodegeneration in *Drosophila* and reduce the size of inclusions in an organotypic R6/2 hippocampal slice culture model [128]. Full preclinical mouse trials with C2-8 are currently underway. A high-throughput cell-based fluorescence resonance energy transfer (FRET) assay for polyQ aggregation [120] has been used to screen 2800 compounds in HEK293 cells. A number of active compounds were identified, and one, Y-27632, was further characterized. Y-27632, an inhibitor of the Rho-associated kinase p160ROCK, inhibited aggregation in HEK293 cells with an EC$_{50}$ of ≈5 μM, and reduced photoreceptor degeneration in a *Drosophila* [98] model of HD.

HDAC inhibitors were first shown to have therapeutic potential in yeast [107] and neuronal cell culture [129], and these findings have been confirmed in *Drosophila* [98] and murine models of HD [99, 130, 131].

3. INVERTEBRATE MODELS

Although yeast and cell models can yield vital data on the action of compounds in living systems with intact cellular machinery, in no way can they mimic the complexity of a functional nervous system with its myriad intercellular connections. Therefore, researchers have sought to use more complex models to elucidate the activity of compounds in whole organisms.

One of the most comprehensively studied and simplest multicellular organisms is the nematode worm, *Caenorhabditis elegans*. Models in which fluorescent polyglutamine proteins are expressed in the muscle of this worm have been developed [28]. Proteins with a polyglutamine tract of about 35 or longer show a length-dependent propensity to aggregate in this model, and this aggregation is paralleled by a loss of motility (Fig. 14-4). These characteristics can be used as the basis for medium-throughput assays of therapeutic compounds. An alternative model in which a polyQ protein is neuronally expressed has also been described, and may have greater relevance to the disease processes in HD [29]. This model has been used to show that Sir2 activation by resveratrol can protect against polyQ toxicity [132], both in flies and in neuronal cells derived from the *HdhQ111* knock-in mouse [106].

Drosophila melanogaster is a well-studied model organism, with considerable complexity and homologs to many mammalian genes. Numerous *Drosophila* models of HD and polyglutamine disease are available (for reviews see Marsh et al. [133, 134]). These models strikingly display many of the features of human polyglutamine disease, such as neurodegeneration, apparent in the loss of photoreceptor neurons, which can be used as the basis of an assay of drug effectiveness (Fig. 14-5). They also parallel the movement disorder with a loss of climbing ability, and have reduced survival or failure to eclose (emerge from the pupa). These characteristics can also be quantified to form the basis of screens for the effectiveness of therapeutic interventions.

Numerous therapeutic studies have been published on *Drosophila* models of HD [40, 98, 124, 135], with compounds including HDAC inhibitors and aggregation inhibitors rescuing lethality and neurodegeneration in dose-dependent manner. The relatively high throughput has made it possible to test combinations of compounds [135] at a range of doses, and the best rescue has been achieved by using several drugs (e.g., SAHA, cystamine, and Congo red) in combination at much lower doses than would be required for maximal effect as a single treatment. It may be that such a cocktail approach may pay dividends in HD, as it already has in AIDS and cancer. The drugs may interact to enhance the beneficial effects of each, while reducing toxic side effects seen at higher doses. This is particularly important in diseases such as HD, where it can be predicted that a patient will need to be on therapy for many years.

Studies in invertebrates are obviously far more time-consuming and expensive than studies *in vitro* and in cell models. However, by use of these systems, it is possible to monitor processes with striking homologies to human symptomology, such as neurodegeneration, loss of motility, and reduced survival. Many fundamental cellular processes are conserved throughout the

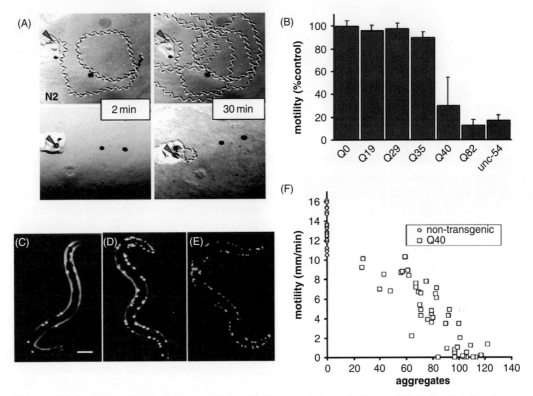

FIGURE 14-4 Expression of polyQ expansions in *C. elegans* muscle results in a motility defect that directly corresponds to aggregate formation. (A) Time-lapse micrographs illustrating tracks left by 5-day-old wild-type (N2) and Q82 animals 2 and 30 min after being placed at the position marked by the red arrow. (B) Quantitation of motility index for 4- to 5-day-old Q0, Q19, Q29, Q35, Q40, Q82, and *unc-54*(r293) animals. Data are means ± SD for at least 50 animals of each type as a percentage of N2 motility. (C–E) Epifluoresence micrographs of late larval/young adult Q40 animals illustrating various numbers of aggregates in different animals. Bar = 0.1 mm. (F) Comparison of motility and aggregate number in adult Q40 animals (squares) and nontransgenic siblings (circles, no aggregates). Aggregate number is representative of number of muscle cells affected, as the body wall muscle cells had, on average, 1.3 ± 0.6 aggregates per cell (n = 212). Reprinted, with permission, from Morley et al. [198]. See CD-ROM for color image.

animal kingdom, and there is hope that much of the dysfunction that can be successfully targeted in these systems can be translated to mammalian and, ultimately, human disease.

4. RODENT MODELS

Although *Drosophila* and *C. elegans* models have a degree of complexity far in excess of that seen in single-celled organisms, they are still a long way from humans in terms of biology. From the time the HD gene was cloned, it was a goal of researchers to develop mammalian models of HD with clear biological relevance to human disease. All mammals, including mice and humans, share a basic body plan and nearly all anatomical features, including similar digestive, circulatory, and nervous systems and the presence of a blood–brain barrier. Metabolism of drugs can differ markedly between species, due largely to allotropic factors [136, 137], but also species-specific effects. However, these models have the most clear-cut similarity to human disease, and it is probable that any hits determined by the higher-throughput screens described earlier will have to be verified in mammalian models before proceeding to the clinic.

a. Neurotoxic Lesion Models

Before the mutation was discovered, the only murine models available to study HD were lesion models, in which striatal cells were killed by neurotoxins such as 3-nitropropionic acid (3-NP) [138] and quinolinic acid [139]. Although these recapitulated some of the symptoms of HD, there is little reason to believe that they could accurately reflect the molecular events that lead to HD, and it is therefore highly questionable that they could have predictive power for assessment of therapeutic strategies. Furthermore, they clearly fail to reproduce the progressive nature of HD.

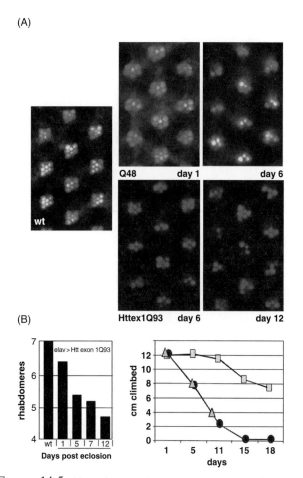

FIGURE 14-5 Neurodegeneration in *Drosophila* expressing polyQ proteins is progressive. (A) The rhabdomeres (photorecptor cells) at different ages are shown for flies expressing a pure polyQ peptide (Q48) and expressing a mutant exon1 fragment of a human Htt gene with 93Qs (Httex1Q93). The rhabdomere constellations get progressively worse. The severity of the effect is greater with the pure polyQ than with the pathogenic human Htt protein fragment. (B) The progressive loss of rhabdomeres in flies expressing Httex1Q93 over 12 days is shown in comparison with wild-type eyes, which exhibit seven rhabdomeres throughout life. Motor function is also impaired and is progressively lost as shown by the climbing assay. Flies exhibit negative geotropism. The distance climbed in 20 days was measured for flies expressing Q48 (circles) and Httex1Q93 (triangles) under the control of elav-Gal4 and compared with that of the nonexpressing sibs Q48/CyO (squares). Note that the climbing ability progressively declines for both genotypes. See CD-ROM for color image. Reprinted, with permission, from Marsh et al. [133, 134].

Lesion models are relatively convenient to work with, as lesions can be created to order, and despite the advent of more sophisticated models, some researchers have chosen to continue to work with lesion models to perform preclinical trials [140–146]. Until recently, lesion models were the only option for creating mammalian models in species other than mouse [147, 148], and these models continue to be popular for first-line testing of interventions with a surgical component [148, 149] due in part to the technical difficulties of performing surgery on mice. It is to be hoped, however, that the availability of rat transgenics [114] and lentiviral models [115] that can be applied to other mammals will soon supersede the lesion models as the models of choice for assessment of surgical approaches. Typically, such approaches may target the regeneration of atrophied brain regions (specifically the striatum) [149], and researchers in the field hope that many questions of neural plasticity and neurogenesis can be answered in animals lacking a functional striatum following toxic lesion. However, because the presence of mutant huntingtin might be predicted to exert an influence on graft survival, researchers in this field have risen to the challenge of performing striatal grafts in transgenic mouse models of HD [150].

b. Genetic Models

Since the development of the first genetic mammalian models, the transgenic R6 mouse lines [10], in 1996, a large variety of murine models of HD have been developed to model various aspects of the disease and to answer mechanistic questions [39, 50, 51, 108–110, 112, 113]. However, the original R6/2 line, with its rapidly progressing and well-defined phenotype, has remained the model of choice for most preclinical trials of therapeutic strategems. The reasons for this are manifold, but among the most compelling is its very rapid disease course, such that it is possible to complete a behavioral trial within 3 months. The R6/2 line expresses exon 1 of human huntingtin with >150 CAG repeats, under the control of the huntingtin promoter.

In some cases, the very rapidity of disease progression in R6/2 has been perceived to be a disadvantage, in that researchers have worried over whether the degeneration in R6/2 might be so aggressive as to preclude treatments with agents that might be predicted to have optimal effects in the early stages of disease. In these cases, researchers have typically opted for the similar, but slower-progressing R6/1 model [10], or the N171-82Q model [108], which is, similarly, a transgenic fragment model, but expresses a slightly longer fragment of the gene and, again, progresses to mortality somewhat more slowly than R6/2. A range of other transgenic models are available [10, 108, 110], but no therapeutic trials in these models have thus far been published.

Researchers are extremely keen to use full-length knock-in models in therapeutic tests. These models have a pathological-length CAG repeat inserted into the endogenous mouse huntingtin [50, 51, 112, 113], and it could be predicted that they might more accurately mimic some aspects of human disease. However, these models are slow to develop symptoms and do not die prematurely, and thus far, no therapeutic trial in these models has been published. Identification of novel early biological markers of disease that show clear-cut and reproducible effects in full-length models may aid in the development of tests that can use these mice to confirm findings from transgenic models. As no model can provide an exact replica of human Huntington's disease, it is vital to test therapeutic strategies in a variety of models before proceeding to human trials. An alternative strategy might be to use the nonmurine mammalian models that are becoming available, to rule out species-specific effects.

B. Preclinical Testing in Mouse Models of HD

The complexity of the mouse models has enabled a large range of outcome measures to become established, covering almost the entire spectrum of disease. Preference is given to measures that clearly mirror symptoms in humans, as the ultimate aim is to cure these symptoms. A useful outcome measure should be quantifiable and progressive and should show a clear difference between wild-type and diseased animals. We have previously discussed the requirements of such measures in more detail [18]. Most therapeutic studies have used survival time as an intuitively understood, quantitative measure that is amenable to statistics. Weight loss recapitulates the wasting seen in HD, whereas rotarod performance (Fig. 14-6A) can be related to the motor dysfunction, and grip strength decline (Fig. 14-6B) reflects the muscle wastage symptomatic of the disease. Hindlimb clasping is an oft-described but unquantifiable feature of the phenotype in these animals that has sometimes been included as an outcome measure. Other behavioral tests such as open-field and bar maneuver tests are less frequently included components of the behavioral test battery used in therapeutic trials. Cognitive and affective impairments have been described, but these are more operator-dependent, time-consuming, and difficult to quantify, and have thus been used less frequently as outcome measures in therapeutic trials. Diabetes is common in R6/2 mice, and reduction of blood glucose has been used as an additional outcome measure.

Besides symptomatic outcome measures, many studies also investigate biological markers of relevance to the disease and/or the process targeted. For example, many studies include a postmortem analysis of brain weight, neuronal morphology, and aggregate load in the brain.

A large number of therapeutic trials in mouse models have been published (Table 14-2). The targets approached fall into a number of more-or-less well defined groups. In some cases, the target is barely understood, but follows from observations in HD and other diseases.

(A)

(B)
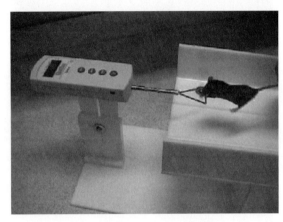

FIGURE 14-6 (A) Rotarod apparatus for analysis of motor deterioration in mice. The rotating drum is set to accelerate over a period of 600 s. The ability of R6/2 mice to walk on the rotaing drum progressively declines from 5 weeks of age. (B) Grip strength apparatus for analysis of forelimb muscular strength and/or motor impersistence. The measured strength of R6/2 mice to walk on the rotaing drum progressively declines from 10 weeks of age.

1. MODULATION OF THE ENVIRONMENT

Observations that monozygotic twins may vary in their disease presentation [151] point to a role for the environment in disease progression. Several independent groups [152–155] have noted that differences in environment, including nutrition, social interaction, cage size, and availability of running wheels, toys, and bedding, can alter the phenotype in R6/2, R6/1, and N171-82Q mice.

Environmental enrichment can enhance neurogenesis and neuroplasticity, even in adult mice [156–158], and recent studies have begun to elucidate the mechanisms by which environmental stimulation could influence behavior in the R6/1 mouse [159–161], enhancing BDNF levels and delaying the loss of cannabinoid receptors. These mechanistic studies may help to define new targets for pharmacological intervention. As unenriched

TABLE 14-2 Published Studies of Therapeutic Approaches in Mammalian Models of Huntington's Disease[a]

Treatment	Putative target	Model	Start (weeks)	Dosage	Outcome	Ref.
Striatal transplantation	Striatal repair	R6/1	10	IS	↑ Open field	[150]
RNA interference	Huntingtin expression	N171-82Q	4	IS, IC	↑ Rotarod, ↑ stride length, ↓ inclusions	[185]
DNA vaccine	Toxic protein	R6/2	5, 7	100 μg plasmid 2x	↓ Diabetes	[186]
Cystamine	Transglutaminase	R6/2	7	1 μM/d IP	↑ Survival, ↑ weight, ↓ abnormal movements	[178]
		R6/2	3	112 mg/kg IP	↑ Survival, ↑ weight, ↑ rotarod, ↑ brain morphology, ↓ aggregation	[187]
		R6/2	3	225 mg/kg IP	↑ Survival, ↑ weight, ↓ aggregation	[180]
		R6/2	Prenatal	112 mg/kg in water	↑ Survival, ↓ motor dysfunction	
		R6/2/TG−/−	3	112 mg/kg in water	↑ Survival, ↓ motor dysfunction	
Rapamycin ester	Autophagy	N171-82Q	4	20 mg/kg IP	↑ Rotarod, ↓ weight, ↑ grip strength, ↑ wire maneuver, ↓ brain weight, ↓ aggregation	[86]
Congo red	Aggregation	R6/2	9	0.5 mg IP/48 h	↑ Rotarod, ↑ weight, ↓ aggregation	[166]
				168 mg/28 d IC	↑ Rotarod, ↑ weight, ↓ aggregation	
Trehalose	Aggregation	R6/2	3	0.2% in water	No effect	[162]
				2% in water	↑ Survival, ↑ rotarod, ↑ weight, ↓ brain atrophy, ↓ inclusions	
				5% in water	No effect	
Minocycline	Aggregation/caspase inhibition	R6/2	6	5 mg/kg IP	↑ Rotarod, ↑ survival	[164]
		R6/2	4	1 g/L in water	No effect	[163]
			4	5 g/L in water	↓ Weight	
		3NP	20	45 mg/kg IP	↑ Survival, ↓ motor tests, ↓ striatal morphology	[144]
		3NP-rat			No effect	[145]
		QA rat			↓ Lesion size	[165]
						[184]
Doxycycline	Aggregation	R6/2	4	2 g/L in water	No effect	[163]
			4	6 g/L in water	No effect	
Riluzole	Aggregation/glutamate antagonism	R6/2	3	10 mg/kg PO	↑ Survival, ↑ weight, ↓ hyperactivity, ↓ aggregation	[188]
		R6/2	5	0.3 g/L in water	No effect	Submitted
				0.6 g/L in water	↓ Weight	
Lithium	Excitotoxicity	R6/2	5	15 mg/kg	↓ Weight	[189]
			10	15 mg/kg	↓ Weight, ↑ rotarod	

[a] All doses daily unless specified. Arrows indicate direction of effect. For example, "↑ rotarod" implies a longer latency on the rotarod test, which is a positive outcome, and "↓ Diabetes" implies a reduction in diabetes, which is also a positive outcome. PO, oral dosing by gavage; IP, intraperitoneal injection; SC, subcutaneous injection; ICV, intra-cerebroventricular injection. Doses in food or water are given as a percentage or concentration, not as the calculated dose in mg/kg, as this depends on many variables, and different researchers may use different calculations.

Compound	Mechanism	Model	n	Dose	Effect	Ref
GDNF	Excitotoxicity	R6/2	4–5	IS	No effect	[190]
Gabapentin	Excitotoxicity	R6/2	6	100 mg/kg/d SC pump	No effect	[191]
Gabapentin-lactam	Excitotoxicity	R6/2	6	100 mg/kg/d SC pump	↑ Beam walking, ↓ aggregation	[191]
MPEP	Glutamate-mediated excitotoxicity	R6/2	3.5	100 mg/kg	↑ Survival, ↑ rotarod, ↑ open field	[192]
LY379268	Glutamate-mediated excitotoxicity	R6/2	3.5	1.2 mg/kg	↑ Survival, ↑ open field	[192]
Coenzyme Q10	Excitotoxicity/antioxidant	R6/2	4	0.2% in diet	↑ Survival, ↑ rotarod, ↑ weight, ↑ brain morphology	[170]
		N171-82Q	4	0.2% in diet	↑ Survival, ↑ weight	[169]
		N171-82Q	8	0.2% in diet	↑ Rotarod, ↑ weight	[155]
		N171-82Q	8	0.2% in diet	↑ Survival, ↑ rotarod	[170]
Remacemide	Excitotoxicity/antioxidant	R6/2	4	0.007% in diet	↑ Survival, ↑ rotarod, ↑ weight, ↑ brain morphology	[169]
		N171-82Q	4	0.007% in diet	↑ Survival, ↑ weight	[155]
		N171-82Q	8	0.007% in diet	↑ Rotarod, ↑ weight	[170]
		N171-82Q	8	0.007% in diet	↑ Survival	
Coenzyme Q10/remacemide (combination)	Excitotoxicity/antioxidant	R6/2	4	0.2% CoQ 0.007% Rem	↑ Survival, ↑ rotarod, ↑ weight, ↑ brain morphology	[169]
		N171-82Q	4	"	↑ Survival, ↑ weight	
		N171-82Q	8	"	↑ Rotarod, ↑ weight	
Lipoic acid	Antioxidant	R6/2	4	0.05% in diet	↑ Survival, ↑ weight	[171]
		N171-82Q	4	0.05% in diet	↑ Survival, ↑ weight	[171]
2-Sulfo-tert.-phenylbutyinitrone	Antioxidant	R6/2	4	0.06% in diet	No effect	[172]
		N171-82Q	4	0.06% in diet	No effect	
BN82451	Oxidative damage	R6/2	4	0.15% in diet	↑ Survival, ↑ rotarod, ↑ brain morphology, ↓ aggregation	[181]
Creatine	Energy homeostasis	R6/2	4	1% in diet	↑ Survival, ↑ weight, ↑ rotarod	[179]
		R6/2	4	2% in diet	↑ Survival, ↑ weight, ↑ rotarod (all maximal), ↑ brain morphology, ↓ aggregation, ↓ diabetes	
		R6/2	4	3% in diet	↑ Survival	
		R6/2	6	2% in diet	↑ Survival, ↑ weight, ↑ rotarod, ↑ brain morphology	[182]
		R6/2	8	2% in diet	↑ Survival, ↑ weight, ↑ rotarod	
		R6/2	10	2% in diet	No effect	
		N171-82Q	4	2% in diet	↑ Survival, ↑ weight, ↑ rotarod, ↑ brain morphology, ↓ aggregation, ↓ diabetes	
Tacrine, moclobemide, creatine:	Neurotransmitter levels (combination therapy)	R6/2	5	TMC	↑ Survival, ↑ weight, ↑ T maze, ↑ water mazes, ↓ diabetes; normalizes gene expression	[174]

(continued)

TABLE 14-2 (continued)

Treatment	Putative target	Model	Start (weeks)	Dosage	Outcome	Ref.
3 mg/kg Tacrine IP (T)			5	TM	—	
10 mg/kg moclobemide (M)			5	TC	—	
1% creatine in diet (C)			5	MC	—	
			5	T	↑ Survival	
			5	M	—	
			5	C	—	
Celecoxib		N171-82Q	8	0.15% in diet	↓ Survival	[155]
Chlorpromazine		N171-82Q	13	1 mg/kg SC Timed release	No effect	[155]
Dichloroacetate	Mitochondrial dysfunction	R6/2	4	100 mg/kg in water	↑ Survival, ↑ weight, ↑ rotarod, ↑ brain morphology, ↓ diabetes	[175]
		N171-82Q	4		↑ Survival, ↑ weight, ↑ rotarod, ↑ brain morphology	
SAHA	Transcriptional dysregulation	R6/2	5	0.67 g/L in water	↑ Rotarod, ↓ weight, ↑ neuronal morphology	[99]
Sodium butyrate	Transcriptional dysregulation	R6/2	3	100 mg/kg	No effect	[131]
				200 mg/kg	↑ Survival, ↑ rotarod	
				400 mg/kg	↑ Survival, ↑ rotarod	
				600 mg/kg	↑ Survival, ↑ rotarod	
				1200 mg/kg	↑ Survival (maximal), ↑ rotarod, ↑ brain morphology	
				5000 mg/kg	↓ Survival	
				10,000 mg/kg	↓ Survival	
Sodium phenylbutyrate	Transcriptional dysregulation	N171-82Q	10–11	100 mg/kg IP	↑ Survival, ↑ brain morphology	[130]
Mithramycin	Transcriptional dysregulation	R6/2	3	150 µg/kg IP	↑ Survival, ↑ rotarod, ↑ brain morphology	[167]
Taurosodeoxycholic acid	Apoptosis	R6/2	6	0.5 mg/kg SC	↑ Rotarod, ↓ TUNEL+ cells, ↓ striatal atrophy	[176]
zVad-fmk	Caspase inhibition	R6/2	7	5 g/kg/28 d ICV pump	↑ Survival, ↑ rotarod	[177]
					↑ Survival	[164]
YVad-cmk	Caspase inhibition	R6/2	7	2.5 g/kg/28 d ICV pump	No effect	[164]

Treatment	Mechanism	Model	Age (wk)	Dose/intervention	Effect	Ref
DEVD-fmk	Caspase inhibition	R6/2	7	2.5 g/kg/28 d ICV pump	No effect	[164]
YVad-cmk and DEVD-fmk	Caspase inhibition	R6/2	7	2.5 g/kg/28 d ICV pump	↑ Survival, ↑ rotarod	[164]
CGS21680	Adenosine receptor	R6/2	7	0.5 mg/kg IP	No effect	[173]
				2.5 mg/kg IP	↑ Rotarod, ↑ open field, ↓ inclusions	
				5.0 mg/kg IP	↑ Rotarod, ↑ open field, ↓ inclusions	
Paroxetine	Serotonin reuptake inhibition	N171-82Q	8	5 mg/kg SC	↑ Survival, ↑ rotarod, ↑ weight	[193]
			13		↑ Survival	
EFAs	Membrane protein regulation	R6/1	Prenatal	254 mg/48 h	↑ Survival, ↑ behavioral battery	[194]
Environmental enrichment	Unknown: neuronal plasticity, neurogenesis, etc.	R6/2	4	Enhanced diet	↑ Survival, ↑ weight, ↑ behavioral battery	[152]
			4	+WT in cage	↑ Survival, ↑ weight	
			2.5	+early weaning and behavioral testing	↑ Survival, ↑ weight	
			4	+breeding	↑ Survival, ↑ weight	
		R6/2	4	Tube, nesting, food on floor	↑ Rotarod,	[154]
			4	+larger cage, more social interaction, toys, wheels	↑ Rotarod, ↑ peristriatal volume	
		R6/1	4	Toys, etc.	↑ Bar maneuver, ↓ abnormal movements, ↑ peristriatal volume; normalize growth factors	[159] [160] [195]
		N171-82Q	8	Toys, wheels, social interaction	↑ Rotarod, ↑ weight	[155]
Dietary restriction	Unknown: growth factors, chaperones, energy homeostasis	N171-82Q	8	Fast every second day	↑ Survival, ↑ rotarod, ↑ weight, ↓ tremor, ↓ aggregation	[196]
Human umbilical cord blood	Unknown	R6/2		$\sim 7 \times 10^7$	↑ Survival, ↑ weight	[197]
				$\sim 10^8$	↑ Survival, ↑ weight	

laboratory conditions represent a very unnaturalsparseness of experience, and enriched conditions are probably simply going some way to removing the deficit, many researchers, ourselves included, take the view that a pharmacological agent should be expected to improve the phenotype beyond the level that could be achieved by good husbandry alone, and therefore perform all studies on enriched mice. However, care must be taken to ensure that environment is standardized in all cages.

2. Modulating Polyglutamine Aggregation

Aggregation remains a rational target for intervention. A number of approaches may be useful. First, one could use a small molecule to directly inhibit aggregation by interacting with the mutant protein. Second, one could use molecules that stimulate the cellular machinery to deal with misfolded protein. This could involve using molecules that are capable of upregulating chaperones involved in protein folding, or enhancing the clearance of aggregated or misfolded protein, for example, by stimulating autophagy or the ubiquitin–proteasome system.

A number of compounds have been tested due to their ability to inhibit aggregation *in vitro*, in cell culture, in an organotypic slice culture or in flies. These include Congo red [126, 127, 128], trehalose [162], minocycline [163], doxycycline [163], and riluzole (submitted). Many of these compounds have other biological effects besides inhibiting polyQ aggregation. Minocycline, for example, is a caspase inhibitor [164], and it is in that context it was first tested. Chen et al. [164] reported phenotypic improvements, but these have not been replicated [145, 163], for reasons that remain unclear. It has even been reported to have deleterious effects in some models [144], and its use should be approached with caution [165]. It is surprising that Congo red is reported to be as effective whether administered by intraperitoneal injection (IP) or by intracerebroventricular pumps (ICV), given that it would not be expected to cross the blood–brain barrier [166]. Meanwhile, an improvement in neurodegeneration in *Drosophila* and in the N171-82Q behavioral phenotype was seen with CCI-779, an analog of rapamycin used to stimulate the autophagy pathway [86]. Rapamycin acts by binding to the mTOR protein kinase, a negative regulator of the autophagic pathway. This interaction inhibits its kinase activity and stimulates autophagy.

It is as yet unclear whether drugs that interact directly with polyglutamine-containing proteins have clinical potential. It may well be that it is not feasible to achieve high enough concentrations *in vivo* to have efficacy via such a simple mechanism. A more fruitful approach may be to investigate compounds that interfere with the aggregation process through the folding and clearance pathways. Cell and yeast-based assays for aggregation inhibitors have the capacity to identify compounds that act both directly and indirectly, whereas wholly *in vitro* assays can identify only compounds that interact directly with the polyQ-containing protein.

3. Redressing Transcriptional Dysregulation

Severe transcriptional dysregulation is a hallmark of polyQ neurodegeneration in all the models used. Transcriptional repression is the dominant result, and many attempts have been made to alleviate repression using HDAC inhibitors and the transcriptional activator mithramycin. These have been successful in models ranging from yeast to mouse [98, 99, 107, 129–131], and have been among the most successful agents so far tested in mice. R6/2 mice treated with SAHA [99] showed the greatest improvement in rotarod performance, amounting to a 4-week delay in symptoms (Fig. 14-7A), whereas mithramycin elicited

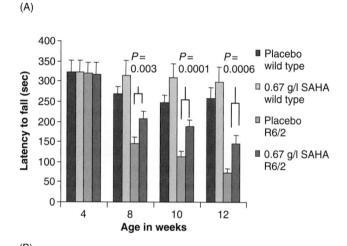

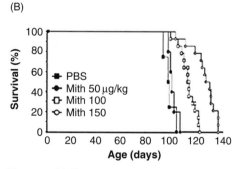

Figure 14-7 (A) Improvement in rotarod performance due to SAHA treatment. Twelve-week-old R6/2 mice treated with SAHA perform approximately at the level of untreated 8-week-old R6/2 mice. (B) Improvement in survival due to mithramycin treatment. Mice treated with 150 μM mithramycin survive approximately 29% longer than untreated mice. This is the greatest reported increase in survival in any published study. See CD-ROM for color image.

a 29% increase in survival [167], the highest yet reported in R6/2 (Fig. 14-7B). Meanwhile, sodium butyrate has also shown efficacy in a mouse model of SBMA [168].

Other successful interventions have included targeting excitotoxicity and/or oxidative damage [169–172], upregulating neuroreceptors and neurotransmitters [173, 174], redressing inbalances in energy homeostasis [175], and inhibiting apoptosis [164, 176, 177].

4. FOLLOWING UP POSITIVE RESULTS WITH MECHANISTIC STUDIES

Although the use of models to elucidate mechanisms has undoubtedly led to the definition of targets for therapeutic intervention, the reverse process is equally important. The choice to test a particular compound in models of HD has typically been made on logical grounds due to a known property of that molecule. For example, it may have been demonstrated to inhibit aggregation *in vitro* or to interfere with the transcriptional machinery. However, even if the compound is convincingly shown to have beneficial effects *in vivo*, it is often by no means clear that it is acting on the proposed target. Many of the compounds tested to date are relatively simple compounds with multiple biological targets. For example, minocycline has antibiotic properties, is a caspase inhibitor, and has also been shown to inhibit aggregation. If a compound has multiple known targets, it is often unclear which, if any, is responsible for a given effect. Follow-up studies can be used to demonstrate that the compound does indeed have the expected biological effects, or to bring to light unexpected effects that could indicate that a molecule is working "off target." They can also be used to further dissect mechanisms, for example, by investigating the subsets of genes differentially regulated by treatment with HDAC inhibitors.

These studies can be invaluable for target validation, and can form the basis for compound refinement, such that analogs may be developed that interact more strongly with the "true" target, but have fewer off-target interactions, which may be responsible for toxicity, undesirable side effects, and poor tolerability. These studies can also help to elucidate causes of toxicity. If toxicity is found to be "on target," that is to say, a compound that has beneficial effects at one dose has deleterious effects at a higher dose through the same mechanism, this may limit the therapeutic window of all compounds working via this mechanism. If, on the other hand, the toxicity is "off target," that is, working via a mechanism other than that which causes the beneficial effects, it may be possible to design molecules that minimize the side effect while maximizing the therapeutic benefits.

5. INTERPRETATION AND REPLICATION OF PRECLINICAL TRIALS

In July 2002, a workshop was organized by the Hereditary Disease Foundation in Cardiff, involving researchers from many of the laboratories conducting preclinical trials in mouse models of HD. The aim of this meeting was to reach a consensus on what would be required from such trials to have sufficient evidence to take a given therapy into human clinical trials. Chief among the recommendations were that results found in one laboratory should be replicated by fully independent researchers using the same mouse model, and that positive effects found in a rapidly progressing fragment model such as R6/2 should be repeated in a full-length model such as the *Hdh150* line [50].

Few of the studies published have been replicated by fully independent groups. The transglutaminase inhibitor cystamine [178–180] and creatine [179, 181, 182], used to redress imbalances in energy homeostasis, have both been reported to have efficacy in R6/2 and N171-82Q; in the latter case, the articles are published by groups who collaborate closely with one another. Another study found no effect of creatine in R6/2 [174]. Results in a genetic cross of R6/2 and a transglutaminase knockout mouse [180] suggest that cystamine does not exert its beneficial effects primarily through transglutaminase inhibition, whereas PET imaging and histological studies [183] support its neuroprotective potential in R6/2. Coenzyme Q and remacemide have also been reported to have positive effects in R6/2 and N171-82Q [169, 170], but have more recently been reported to reduce survival in N171-82Q [155]. Conflicting reports exist for minocycline in various models [144, 145, 163–165, 184].

Negative publication bias, however has prevented the publication of numerous studies that have failed to replicate published findings. False positives may occur for any number of reasons, including small sample size, poor baseline matching, operator bias, and simple statistic variation. It is worth noting that at a *P* value of 0.05, one in 20 tests will come out statistically significant, and most studies include multiple outcome measures, enhancing the chances of finding false positives. False negatives can also occur due to small sample size, poor statistical power of tests, and random variation. It is therefore vital that findings be replicated independently before proceeding to costly and time-consuming clinical trials in HD patients. Furthermore, we should remember that *all* the models of HD discussed remain invalidated until such time as an agent that has shown efficacy crosses successfully into clinical use. Until that time, we must be careful to make use of all available data to determine which candidates are most promising for crossing into clinical trials.

Differences in disease phenotype occur between and even within laboratories. Most striking is the range of life expectancies published for R6/2 by different laboratories. Many laboratories publish a mean survival of <100 days, while we regularly keep cohorts of mice beyond this age with no mortality. Furthermore, the untreated mean survival times seem to vary widely even within a laboratory, such that, within the same laboratory, mice treated with an agent reported to improve survival actually survive a shorter time than mice forming the placebo group in a second study within the same publication [164]. Clearly, these discrepancies are cause for some concern, and studies claiming improvements in survival in which the placebo group has an unusually low mean survival should be carefully scrutinized.

Despite strenuous efforts to develop the *Hdh150* line [50] as a supplementary model for preclinical testing, it has, as yet, not proven possible to conduct trials in this model. This is chiefly because the phenotype of this model is so subtle that enormous numbers of mice would be required to be able to detect any improvement due to treatment. Furthermore the lack of a reduction in life span makes survival studies impossible, and years, rather than months are required to monitor the slow development of the muscle atrophy and the behavioral phenotype.

IV. CONCLUSION

Clearly, the discovery of the causal mutation for Huntington's disease and the subsequent development of many excellent disease models have led to an enormous improvement of our understanding of the disease and great hopes for the development of effective therapies that can delay the onset, cure the disease, or ameliorate symptoms. The concerted efforts of researchers to gather data from across the spectrum of disease models, and to use information gathered from each study to inform, continually refine and improve tests at all levels of complexity will surely eventually lead to the ultimate goal of successful treatment of Huntington's disease in the clinical setting.

Acknowledgments

Work in the author's laboratory is funded by the Wellcome Trust, the Huntington's Disease Society of America, and the Hereditary Disease Foundation. J.S.B. is a Wellcome Trust Prize Student.

References

1. Bates, G. P., Harper, P. S., and Jones, A. L. (2002). Huntington's Disease, Oxford University Press, Oxford.
2. Huntington's Disease Collaborative Research Group. (1993). A novel gene containing a trinucleotide repeat that is expanded and unstable on Huntington's disease chromosomes. *Cell* **72**, 971–983.
3. Li, S. H., and Li, X. J. (2004). Huntingtin–protein interactions and the pathogenesis of Huntington's disease. *Trends Genet.* **20**, 146–154.
4. Zuccato, C., Ciammola, A., Rigamonti, D., Leavitt, B. R., Goffredo, D., Conti, L., MacDonald, M. E., Friedlander, R. M., Silani, V., Hayden, M. R., Timmusk, T., Sipione, S., and Cattaneo, E. (2001). Loss of huntingtin-mediated BDNF gene transcription in Huntington's disease. *Science* **293**, 493–498.
5. Gauthier, L. R., Charrin, B. C., Borrell-Pages, M., Dompierre, J. P., Rangone, H., Cordelieres, F. P., De Mey, J., MacDonald, M. E., Lessmann, V., Humbert, S., and Saudou, F. (2004). Huntingtin controls neurotrophic support and survival of neurons by enhancing BDNF vesicular transport along microtubules. *Cell* **118**, 127–138.
6. Dragatsis, I., Levine, M. S., and Zeitlin, S. (2000). Inactivation of Hdh in the brain and testis results in progressive neurodegeneration and sterility in mice. *Nat. Genet.* **26**, 300–306.
7. Duyao, M. P., Auerbach, A. B., Ryan, A., Persichetti, F., Barnes, G. T., McNeil, S. M., Ge, P., Vonsattel, J. P., Gusella, J. F., Joyner, A. L., et al. (1995). Inactivation of the mouse Huntington's disease gene homolog Hdh. *Science* **269**, 407–410.
8. Wexler, N. S., Young, A. B., Tanzi, R. E., Travers, H., Starosta-Rubinstein, S., Penney, J. B., Snodgrass, S. R., Shoulson, I., Gomez, F., Ramos Arroyo, M. A., et al. (1987). Homozygotes for Huntington's disease. *Nature* **326**, 194–197.
9. Myers, R. H., Leavitt, J., Farrer, L. A., Jagadeesh, J., McFarlane, H., Mastromauro, C. A., Mark, R. J., and Gusella, J. F. (1989). Homozygote for Huntington disease. *Am. J. Hum. Genet.* **45**, 615–618.
10. Mangiarini, L., Sathasivam, K., Seller, M., Cozens, B., Harper, A., Hetherington, C., Lawton, M., Trottier, Y., Lehrach, H., Davies, S. W., and Bates, G. P. (1996). Exon 1 of the HD gene with an expanded CAG repeat is sufficient to cause a progressive neurological phenotype in transgenic mice. *Cell* **87**, 493–506.
11. Bates, G. (2003). Huntingtin aggregation and toxicity in Huntington's disease. *Lancet* **361**, 1642–1644.
12. Wellington, C. L., Singaraja, R., Ellerby, L., Savill, J., Roy, S., Leavitt, B., Cattaneo, E., Hackam, A., Sharp, A., Thornberry, N., Nicholson, D. W., Bredesen, D. E., and Hayden, M. R. (2000). Inhibiting caspase cleavage of huntingtin reduces toxicity and aggregate formation in neuronal and nonneuronal cells. *J. Biol. Chem.* **275**, 19831–19838.
13. Kim, Y. J., Yi, Y., Sapp, E., Wang, Y., Cuiffo, B., Kegel, K. B., Qin, Z. H., Aronin, N., and DiFiglia, M. (2001). Caspase 3-cleaved N-terminal fragments of wild-type and mutant huntingtin are present in normal and Huntington's disease brains, associate with membranes, and undergo calpain-dependent proteolysis. *Proc. Natl. Acad. Sci. USA* **98**, 12784–12789.
14. Lunkes, A., Lindenberg, K. S., Ben-Haiem, L., Weber, C., Devys, D., Landwehrmeyer, G. B., Mandel, J. L., and Trottier, Y. (2002). Proteases acting on mutant huntingtin generate cleaved products that differentially build up cytoplasmic and nuclear inclusions. *Mol. Cell* **10**, 259–269.
15. Bence, N. F., Sampat, R. M., and Kopito, R. R. (2001). Impairment of the ubiquitin–proteasome system by protein aggregation. *Science* **292**, 1552–1555.

16. Gunawardena, S., and Goldstein, L. S. (2005). Polyglutamine diseases and transport problems: Deadly traffic jams on neuronal highways. *Arch. Neurol.* **62**, 46–51.
17. Luthi-Carter, R., and Cha, J.-H. J. (2003). Mechanisms of transcriptional dysregulation in Huntington's disease. *Clin. Neurosci. Res.* **3**, 165–177.
18. Hockly, E., Woodman, B., Mahal, A., Lewis, C. M., and Bates, G. (2003). Standardization and statistical approaches to therapeutic trials in the R6/2 mouse. *Brain Res. Bull.* **61**, 469–479.
19. Davies, S. W., Turmaine, M., Cozens, B. A., DiFiglia, M., Sharp, A. H., Ross, C. A., Scherzinger, E., Wanker, E. E., Mangiarini, L., and Bates, G. P. (1997). Formation of neuronal intranuclear inclusions underlies the neurological dysfunction in mice transgenic for the HD mutation. *Cell* **90**, 537–548.
20. DiFiglia, M., Sapp, E., Chase, K. O., Davies, S. W., Bates, G. P., Vonsattel, J. P., and Aronin, N. (1997). Aggregation of huntingtin in neuronal intranuclear inclusions and dystrophic neurites in brain. *Science* **277**, 1990–1993.
21. Scherzinger, E., Lurz, R., Turmaine, M., Mangiarini, L., Hollenbach, B., Hasenbank, R., Bates, G. P., Davies, S. W., Lehrach, H., and Wanker, E. E. (1997). Huntingtin-encoded polyglutamine expansions form amyloid-like protein aggregates in vitro and in vivo. *Cell* **90**, 549–558.
22. Perutz, M. F., Johnson, T., Suzuki, M., and Finch, J. T. (1994). Glutamine repeats as polar zippers: Their possible role in inherited neurodegenerative diseases. *Proc. Natl. Acad. Sci. USA* **91**, 5355–5358.
23. Kazantsev, A., Preisinger, E., Dranovsky, A., Goldgaber, D., and Housman, D. (1999). Insoluble detergent-resistant aggregates form between pathological and nonpathological lengths of polyglutamine in mammalian cells. *Proc. Natl. Acad. Sci. USA* **96**, 11404–11409.
24. Scherzinger, E., Sittler, A., Schweiger, K., Heiser, V., Lurz, R., Hasenbank, R., Bates, G. P., Lehrach, H., and Wanker, E. E. (1999). Self-assembly of polyglutamine-containing huntingtin fragments into amyloid-like fibrils: Implications for Huntington's disease pathology. *Proc. Natl. Acad. Sci. USA* **96**, 4604–4609.
25. Wacker, J. L., Zareie, M. H., Fong, H., Sarikaya, M., and Muchowski, P. J. (2004). Hsp70 and Hsp40 attenuate formation of spherical and annular polyglutamine oligomers by partitioning monomer. *Nat. Struct. Mol. Biol.* **11**, 1215–1222.
26. Krobitsch, S., and Lindquist, S. (2000). Aggregation of huntingtin in yeast varies with the length of the polyglutamine expansion and the expression of chaperone proteins. *Proc. Natl. Acad. Sci. USA* **97**, 1589–1594.
27. Faber, P. W., Alter, J. R., MacDonald, M. E., and Hart, A. C. (1999). Polyglutamine-mediated dysfunction and apoptotic death of a *Caenorhabditis elegans* sensory neuron. *Proc. Natl. Acad. Sci. USA* **96**, 179–184.
28. Satyal, S. H., Schmidt, E., Kitagawa, K., Sondheimer, N., Lindquist, S., Kramer, J. M., and Morimoto, R. I. (2000). Polyglutamine aggregates alter protein folding homeostasis in *Caenorhabditis elegans*. *Proc. Natl. Acad. Sci. USA* **97**, 5750–5755.
29. Parker, J. A., Connolly, J. B., Wellington, C., Hayden, M., Dausset, J., and Neri, C. (2001). Expanded polyglutamines in *Caenorhabditis elegans* cause axonal abnormalities and severe dysfunction of PLM mechanosensory neurons without cell death. *Proc. Natl. Acad. Sci. USA* **98**, 13318–13323.
30. Jackson, G. R., Salecker, I., Dong, X., Yao, X., Arnheim, N., Faber, P. W., MacDonald, M. E., and Zipursky, S. L. (1998). Polyglutamine-expanded human huntingtin transgenes induce degeneration of *Drosophila* photoreceptor neurons. *Neuron* **21**, 633–642.
31. Kazemi-Esfarjani, P., and Benzer, S. (2000). Genetic suppression of polyglutamine toxicity in *Drosophila*. *Science* **287**, 1837–1840.
32. Marsh, J. L., Walker, H., Theisen, H., Zhu, Y. Z., Fielder, T., Purcell, J., and Thompson, L. M. (2000). Expanded polyglutamine peptides alone are intrinsically cytotoxic and cause neurodegeneration in *Drosophila*. *Hum. Mol. Genet.* **9**, 13–25.
33. Yang, W., Dunlap, J. R., Andrews, R. B., and Wetzel, R. (2002). Aggregated polyglutamine peptides delivered to nuclei are toxic to mammalian cells. *Hum. Mol. Genet.* **11**, 2905–2917.
34. Saudou, F., Finkbeiner, S., Devys, D., and Greenberg, M. E. (1998). Huntingtin acts in the nucleus to induce apoptosis but death does not correlate with the formation of intranuclear inclusions. *Cell* **95**, 55–66.
35. Arrasate, M., Mitra, S., Schweitzer, E. S., Segal, M. R., and Finkbeiner, S. (2004). Inclusion body formation reduces levels of mutant huntingtin and the risk of neuronal death. *Nature* **431**, 805–810.
36. Kopito, R. R. (2000). Aggresomes, inclusion bodies and protein aggregation. *Trends Cell Biol.* **10**, 524–530.
37. Schaffar, G., Breuer, P., Boteva, R., Behrends, C., Tzvetkov, N., Strippel, N., Sakahira, H., Siegers, K., Hayer-Hartl, M., and Hartl, F. U. (2004). Cellular toxicity of polyglutamine expansion proteins: Mechanism of transcription factor deactivation. *Mol. Cell* **15**, 95–105.
38. Li, H., Li, S. H., Cheng, A. L., Mangiarini, L., Bates, G. P., and Li, X. J. (1999). Ultrastructural localization and progressive formation of neuropil aggregates in Huntington's disease transgenic mice. *Hum. Mol. Genet.* **8**, 1227–1236.
39. Yamamoto, A., Lucas, J. J., and Hen, R. (2000). Reversal of neuropathology and motor dysfunction in a conditional model of Huntington's disease. *Cell* **101**, 57–66.
40. Kazantsev, A., Walker, H. A., Slepko, N., Bear, J. E., Preisinger, E., Steffan, J. S., Zhu, Y. Z., Gertler, F. B., Housman, D. E., Marsh, J. L., and Thompson, L. M. (2002). A bivalent Huntingtin binding peptide suppresses polyglutamine aggregation and pathogenesis in *Drosophila*. *Nat. Genet.* **30**, 367–376.
41. Kennedy, L., Evans, E., Chen, C. M., Craven, L., Detloff, P. J., Ennis, M., and Shelbourne, P. F. (2003). Dramatic tissue-specific mutation length increases are an early molecular event in Huntington disease pathogenesis. *Hum. Mol. Genet.* **12**, 3359–3367.
42. Tagawa, K., Hoshino, M., Okuda, T., Ueda, H., Hayashi, H., Engemann, S., Okado, H., Ichikawa, M., Wanker, E. E., and Okazawa, H. (2004). Distinct aggregation and cell death patterns among different types of primary neurons induced by mutant huntingtin protein. *J. Neurochem.* **89**, 974–987.
43. Gafni, J., Hermel, E., Young, J. E., Wellington, C. L., Hayden, M. R., and Ellerby, L. M. (2004). Inhibition of calpain cleavage of huntingtin reduces toxicity: Accumulation of calpain/caspase fragments in the nucleus. *J. Biol. Chem.* **279**, 20211–20220.
44. Hackam, A. S., Singaraja, R., Wellington, C. L., Metzler, M., McCutcheon, K., Zhang, T., Kalchman, M., and Hayden, M. R. (1998). The influence of huntingtin protein size on nuclear localization and cellular toxicity. *J. Cell Biol.* **141**, 1097–1105.
45. Martindale, D., Hackam, A., Wieczorek, A., Ellerby, L., Wellington, C., McCutcheon, K., Singaraja, R., Kazemi-Esfarjani, P., Devon, R., Kim, S. U., Bredesen, D. E., Tufaro, F., and Hayden, M. R. (1998). Length of huntingtin and its polyglutamine tract influences localization and frequency of intracellular aggregates. *Nat. Genet.* **18**, 150–154.
46. Lunkes, A., and Mandel, J. L. (1998). A cellular model that recapitulates major pathogenic steps of Huntington's disease. *Hum. Mol. Genet.* **7**, 1355–1361.
47. Hackam, A. S., Singaraja, R., Zhang, T., Gan, L., and Hayden, M. R. (1999). In vitro evidence for both the nucleus and cytoplasm as subcellular sites of pathogenesis in Huntington's disease. *Hum. Mol. Genet.* **8**, 25–33.

48. Li, H., Li, S. H., Johnston, H., Shelbourne, P. F., and Li, X. J. (2000). Amino-terminal fragments of mutant huntingtin show selective accumulation in striatal neurons and synaptic toxicity. *Nat. Genet.* **25**, 385–389.
49. Zhou, H., Cao, F., Wang, Z., Yu, Z. X., Nguyen, H. P., Evans, J., Li, S. H., and Li, X. J. (2003). Huntingtin forms toxic NH$_2$-terminal fragment complexes that are promoted by the age-dependent decrease in proteasome activity. *J. Cell Biol.* **163**, 109–118.
50. Lin, C. H., Tallaksen-Greene, S., Chien, W. M., Cearley, J. A., Jackson, W. S., Crouse, A. B., Ren, S., Li, X. J., Albin, R. L., and Detloff, P. J. (2001). Neurological abnormalities in a knock-in mouse model of Huntington's disease. *Hum. Mol. Genet.* **10**, 137–144.
51. Shelbourne, P. F., Killeen, N., Hevner, R. F., Johnston, H. M., Tecott, L., Lewandoski, M., Ennis, M., Ramirez, L., Li, Z., Iannicola, C., Littman, D. R., and Myers, R. M. (1999). A Huntington's disease CAG expansion at the murine Hdh locus is unstable and associated with behavioural abnormalities in mice. *Hum. Mol. Genet.* **8**, 763–774.
52. Jana, N. R., Tanaka, M., Wang, G., and Nukina, N. (2000). Polyglutamine length-dependent interaction of Hsp40 and Hsp70 family chaperones with truncated N-terminal huntingtin: Their role in suppression of aggregation and cellular toxicity. *Hum. Mol. Genet.* **9**, 2009–2018.
53. Hay, D. G., Sathasivam, K., Tobaben, S., Stahl, B., Marber, M., Mestril, R., Mahal, A., Smith, D. L., Woodman, B., and Bates, G. P. (2004). Progressive decrease in chaperone protein levels in a mouse model of Huntington's disease and induction of stress proteins as a therapeutic approach. *Hum. Mol. Genet.* **13**, 1389–1405.
54. Cummings, C. J., Mancini, M. A., Antalffy, B., DeFranco, D. B., Orr, H. T., and Zoghbi, H. Y. (1998). Chaperone suppression of aggregation and altered subcellular proteasome localization imply protein misfolding in SCA1. *Nat. Genet.* **19**, 148–154.
55. Chai, Y., Koppenhafer, S. L., Bonini, N. M., and Paulson, H. L. (1999). Analysis of the role of heat shock protein (Hsp) molecular chaperones in polyglutamine disease. *J. Neurosci.* **19**, 10338–10347.
56. Schmidt, T., Lindenberg, K. S., Krebs, A., Schols, L., Laccone, F., Herms, J., Rechsteiner, M., Riess, O., and Landwehrmeyer, G. B. (2002). Protein surveillance machinery in brains with spinocerebellar ataxia type 3: Redistribution and differential recruitment of 26S proteasome subunits and chaperones to neuronal intranuclear inclusions. *Ann. Neurol.* **51**, 302–310.
57. Wyttenbach, A., Sauvageot, O., Carmichael, J., Diaz-Latoud, C., Arrigo, A. P., and Rubinsztein, D. C. (2002). Heat shock protein 27 prevents cellular polyglutamine toxicity and suppresses the increase of reactive oxygen species caused by huntingtin. *Hum. Mol. Genet.* **11**, 1137–1151.
58. Zhou, H., Li, S. H., and Li, X. J. (2001). Chaperone suppression of cellular toxicity of huntingtin is independent of polyglutamine aggregation. *J. Biol. Chem.* **276**, 48417–48424.
59. Kobayashi, Y., Kume, A., Li, M., Doyu, M., Hata, M., Ohtsuka, K., and Sobue, G. (2000). Chaperones Hsp70 and Hsp40 suppress aggregate formation and apoptosis in cultured neuronal cells expressing truncated androgen receptor protein with expanded polyglutamine tract. *J. Biol. Chem.* **275**, 8772–8778.
60. Sittler, A., Lurz, R., Lueder, G., Priller, J., Lehrach, H., Hayer-Hartl, M. K., Hartl, F. U., and Wanker, E. E. (2001). Geldanamycin activates a heat shock response and inhibits huntingtin aggregation in a cell culture model of Huntington's disease. *Hum. Mol. Genet.* **10**, 1307–1315.
61. Warrick, J. M., Chan, H. Y., Gray-Board, G. L., Chai, Y., Paulson, H. L., and Bonini, N. M. (1999). Suppression of polyglutamine-mediated neurodegeneration in *Drosophila* by the molecular chaperone HSP70. *Nat. Genet.* **23**, 425–428.
62. Chan, H. Y., Warrick, J. M., Gray-Board, G. L., Paulson, H. L., and Bonini, N. M. (2000). Mechanisms of chaperone suppression of polyglutamine disease: Selectivity, synergy and modulation of protein solubility in *Drosophila*. *Hum. Mol. Genet.* **9**, 2811–2820.
63. Muchowski, P. J., Schaffar, G., Sittler, A., Wanker, E. E., Hayer-Hartl, M. K., and Hartl, F. U. (2000). Hsp70 and hsp40 chaperones can inhibit self-assembly of polyglutamine proteins into amyloid-like fibrils. *Proc. Natl. Acad. Sci. USA* **97**, 7841–7846.
64. Cummings, C. J., Sun, Y., Opal, P., Antalffy, B., Mestril, R., Orr, H. T., Dillmann, W. H., and Zoghbi, H. Y. (2001). Overexpression of inducible HSP70 chaperone suppresses neuropathology and improves motor function in SCA1 mice. *Hum. Mol. Genet.* **10**, 1511–1518.
65. Adachi, H., Katsuno, M., Minamiyama, M., Sang, C., Pagoulatos, G., Angelidis, C., Kusakabe, M., Yoshiki, A., Kobayashi, Y., Doyu, M., and Sobue, G. (2003). Heat shock protein 70 chaperone overexpression ameliorates phenotypes of the spinal and bulbar muscular atrophy transgenic mouse model by reducing nuclear-localized mutant androgen receptor protein. *J. Neurosci.* **23**, 2203–2211.
66. Hansson, O., Nylandsted, J., Castilho, R. F., Leist, M., Jaattela, M., and Brundin, P. (2003). Overexpression of heat shock protein 70 in R6/2 Huntington's disease mice has only modest effects on disease progression. *Brain Res.* **970**, 47–57.
67. Westerheide, S. D., Bosman, J. D., Mbadugha, B. N., Kawahara, T. L., Matsumoto, G., Kim, S., Gu, W., Devlin, J. P., Silverman, R. B., and Morimoto, R. I. (2004). Celastrols as inducers of the heat shock response and cytoprotection. *J. Biol. Chem.* **279**, 56053–56060.
68. Verhoef, L. G., Lindsten, K., Masucci, M. G., and Dantuma, N. P. (2002). Aggregate formation inhibits proteasomal degradation of polyglutamine proteins. *Hum. Mol. Genet.* **11**, 2689–2700.
69. Goellner, G. M., and Rechsteiner, M. (2003). Are Huntington's and polyglutamine-based ataxias proteasome storage diseases? *Int. J. Biochem. Cell Biol.* **35**, 562–571.
70. Venkatraman, P., Wetzel, R., Tanaka, M., Nukina, N., and Goldberg, A. L. (2004). Eukaryotic proteasomes cannot digest polyglutamine sequences and release them during degradation of polyglutamine-containing proteins. *Mol. Cell* **14**, 95–104.
71. Jana, N. R., Zemskov, E. A., Wang, G., and Nukina, N. (2001). Altered proteasomal function due to the expression of polyglutamine-expanded truncated N-terminal huntingtin induces apoptosis by caspase activation through mitochondrial cytochrome c release. *Hum. Mol. Genet.* **10**, 1049–1059.
72. Holmberg, C. I., Staniszewski, K. E., Mensah, K. N., Matouschek, A., and Morimoto, R. I. (2004). Inefficient degradation of truncated polyglutamine proteins by the proteasome. *EMBO J.* **23**, 4307–4318.
73. Bennett, E. J., Bence, N. F., Jayakumar, R., and Kopito, R. R. (2005). Global impairment of the ubiquitin–proteasome system by nuclear or cytoplasmic protein aggregates precedes inclusion body formation. *Mol. Cell* **17**, 351–365.
74. Park, Y., Hong, S., Kim, S. J., and Kang, S. (2005). Proteasome function is inhibited by polyglutamine-expanded ataxin-1, the SCA1 gene product. *Mol. Cell* **19**, 23–30.
75. Diaz-Hernandez, M., Hernandez, F., Martin-Aparicio, E., Gomez-Ramos, P., Moran, M. A., Castano, J. G., Ferrer, I., Avila, J., and Lucas, J. J. (2003). Neuronal induction of the immunoproteasome in Huntington's disease. *J. Neurosci.* **23**, 11653–11661.
76. Bowman, A. B., Yoo, S. Y., Dantuma, N. P., and Zoghbi, H. Y. (2005). Neuronal dysfunction in a polyglutamine disease model occurs in the absence of ubiquitin–proteasome system impairment and inversely correlates with the degree of nuclear inclusion formation. *Hum. Mol. Genet.* **14**, 679–691.
77. Seo, H., Sonntag, K. C., and Isacson, O. (2004). Generalized brain and skin proteasome inhibition in Huntington's disease. *Ann. Neurol.* **56**, 319–328.

78. Cummings, C. J., Reinstein, E., Sun, Y., Antalffy, B., Jiang, Y., Ciechanover, A., Orr, H. T., Beaudet, A. L., and Zoghbi, H. Y. (1999). Mutation of the E6-AP ubiquitin ligase reduces nuclear inclusion frequency while accelerating polyglutamine-induced pathology in SCA1 mice. *Neuron* **24**, 879–892.
79. Nollen, E. A., Garcia, S. M., van Haaften, G., Kim, S., Chavez, A., Morimoto, R. I., and Plasterk, R. H. (2004). Genome-wide RNA interference screen identifies previously undescribed regulators of polyglutamine aggregation. *Proc. Natl. Acad. Sci. USA* **101**, 6403–6408.
80. Warrick, J. M., Morabito, L. M., Bilen, J., Gordesky-Gold, B., Faust, L. Z., Paulson, H. L., and Bonini, N. M. (2005). Ataxin-3 suppresses polyglutamine neurodegeneration in *Drosophila* by a ubiquitin-associated mechanism. *Mol. Cell* **18**, 37–48.
81. Kaytor, M. D., Wilkinson, K. D., and Warren, S. T. (2004). Modulating huntingtin half-life alters polyglutamine-dependent aggregate formation and cell toxicity. *J. Neurochem.* **89**, 962–973.
82. Michalik, A., and Van Broeckhoven, C. (2004). Proteasome degrades soluble expanded polyglutamine completely and efficiently. *Neurobiol. Dis.* **16**, 202–211.
83. Kim, M., Lee, H. S., LaForet, G., McIntyre, C., Martin, E. J., Chang, P., Kim, T. W., Williams, M., Reddy, P. H., Tagle, D., Boyce, F. M., Won, L., Heller, A., Aronin, N., and DiFiglia, M. (1999). Mutant huntingtin expression in clonal striatal cells: Dissociation of inclusion formation and neuronal survival by caspase inhibition. *J. Neurosci.* **19**, 964–973.
84. Ravikumar, B., Duden, R., and Rubinsztein, D. C. (2002). Aggregate-prone proteins with polyglutamine and polyalanine expansions are degraded by autophagy. *Hum. Mol. Genet.* **11**, 1107–1117.
85. Qin, Z. H., Wang, Y., Kegel, K. B., Kazantsev, A., Apostol, B. L., Thompson, L. M., Yoder, J., Aronin, N., and DiFiglia, M. (2003). Autophagy regulates the processing of amino terminal huntingtin fragments. *Hum. Mol. Genet.* **12**, 3231–3244.
86. Ravikumar, B., Vacher, C., Berger, Z., Davies, J. E., Luo, S., Oroz, L. G., Scaravilli, F., Easton, D. F., Duden, R., O'Kane, C. J., and Rubinsztein, D. C. (2004). Inhibition of mTOR induces autophagy and reduces toxicity of polyglutamine expansions in fly and mouse models of Huntington disease. *Nat. Genet.* **36**, 585–595.
87. Gunawardena, S., Her, L. S., Brusch, R. G., Laymon, R. A., Niesman, I. R., Gordesky-Gold, B., Sintasath, L., Bonini, N. M., and Goldstein, L. S. (2003). Disruption of axonal transport by loss of huntingtin or expression of pathogenic polyQ proteins in Drosophila. *Neuron* **40**, 25–40.
88. Lee, W. C., Yoshihara, M., and Littleton, J. T. (2004). Cytoplasmic aggregates trap polyglutamine-containing proteins and block axonal transport in a *Drosophila* model of Huntington's disease. *Proc. Natl. Acad. Sci. USA* **101**, 3224–3229.
89. Ravikumar, B., Acevedo-Arozena, A., Imarisio, S., Berger, Z., Vacher, C., O'Kane C, J., Brown, S. D., and Rubinsztein, D. C. (2005). Dynein mutations impair autophagic clearance of aggregate-prone proteins. *Nat. Genet.* **37**, 771–776.
90. Luthi-Carter, R., Strand, A., Peters, N. L., Solano, S. M., Hollingsworth, Z. R., Menon, A. S., Frey, A. S., Spektor, B. S., Penney, E. B., Schilling, G., Ross, C. A., Borchelt, D. R., Tapscott, S. J., Young, A. B., Cha, J. H., and Olson, J. M. (2000). Decreased expression of striatal signaling genes in a mouse model of Huntington's disease. *Hum. Mol. Genet.* **9**, 1259–1271.
91. Luthi-Carter, R., Hanson, S. A., Strand, A. D., Bergstrom, D. A., Chun, W., Peters, N. L., Woods, A. M., Chan, E. Y., Kooperberg, C., Krainc, D., Young, A. B., Tapscott, S. J., and Olson, J. M. (2002). Dysregulation of gene expression in the R6/2 model of polyglutamine disease: Parallel changes in muscle and brain. *Hum. Mol. Genet.* **11**, 1911–1926.
92. Steffan, J. S., Kazantsev, A., Spasic-Boskovic, O., Greenwald, M., Zhu, Y. Z., Gohler, H., Wanker, E. E., Bates, G. P., Housman, D. E., and Thompson, L. M. (2000). The Huntington's disease protein interacts with p53 and CREB-binding protein and represses transcription. *Proc. Natl. Acad. Sci. USA* **97**, 6763–6768.
93. Nucifora, F. C., Jr., Sasaki, M., Peters, M. F., Huang, H., Cooper, J. K., Yamada, M., Takahashi, H., Tsuji, S., Troncoso, J., Dawson, V. L., Dawson, T. M., and Ross, C. A. (2001). Interference by huntingtin and atrophin-1 with cbp-mediated transcription leading to cellular toxicity. *Science* **291**, 2423–2428.
94. Wyttenbach, A., Swartz, J., Kita, H., Thykjaer, T., Carmichael, J., Bradley, J., Brown, R., Maxwell, M., Schapira, A., Orntoft, T. F., Kato, K., and Rubinsztein, D. C. (2001). Polyglutamine expansions cause decreased CRE-mediated transcription and early gene expression changes prior to cell death in an inducible cell model of Huntington's disease. *Hum. Mol. Genet.* **10**, 1829–1845.
95. Shimohata, T., Nakajima, T., Yamada, M., Uchida, C., Onodera, O., Naruse, S., Kimura, T., Koide, R., Nozaki, K., Sano, Y., Ishiguro, H., Sakoe, K., Ooshima, T., Sato, A., Ikeuchi, T., Oyake, M., Sato, T., Aoyagi, Y., Hozumi, I., Nagatsu, T., Takiyama, Y., Nishizawa, M., Goto, J., Kanazawa, I., Davidson, I., Tanese, N., Takahashi, H., and Tsuji, S. (2000). Expanded polyglutamine stretches interact with TAFII130, interfering with CREB-dependent transcription. *Nat. Genet.* **26**, 29–36.
96. Obrietan, K., and Hoyt, K. R. (2004). CRE-mediated transcription is increased in Huntington's disease transgenic mice. *J. Neurosci.* **24**, 791–796.
97. Mantamadiotis, T., Lemberger, T., Bleckmann, S. C., Kern, H., Kretz, O., Martin Villalba, A., Tronche, F., Kellendonk, C., Gau, D., Kapfhammer, J., Otto, C., Schmid, W., and Schutz, G. (2002). Disruption of CREB function in brain leads to neurodegeneration. *Nat. Genet.* **31**, 47–54.
98. Steffan, J. S., Bodai, L., Pallos, J., Poelman, M., McCampbell, A., Apostol, B. L., Kazantsev, A., Schmidt, E., Zhu, Y. Z., Greenwald, M., Kurokawa, R., Housman, D. E., Jackson, G. R., Marsh, J. L., and Thompson, L. M. (2001). Histone deacetylase inhibitors arrest polyglutamine-dependent neurodegeneration in *Drosophila*. *Nature* **413**, 739–743.
99. Hockly, E., Richon, V. M., Woodman, B., Smith, D. L., Zhou, X., Rosa, E., Sathasivam, K., Ghazi-Noori, S., Mahal, A., Lowden, P. A., Steffan, J. S., Marsh, J. L., Thompson, L. M., Lewis, C. M., Marks, P. A., and Bates, G. P. (2003). Suberoylanilide hydroxamic acid, a histone deacetylase inhibitor, ameliorates motor deficits in a mouse model of Huntington's disease. *Proc. Natl. Acad. Sci. USA* **100**, 2041–2046.
100. Huang, C. C., Faber, P. W., Persichetti, F., Mittal, V., Vonsattel, J. P., MacDonald, M. E., and Gusella, J. F. (1998). Amyloid formation by mutant huntingtin: Threshold, progressivity and recruitment of normal polyglutamine proteins. *Somat. Cell. Mol. Genet.* **24**, 217–233.
101. Boutell, J. M., Thomas, P., Neal, J. W., Weston, V. J., Duce, J., Harper, P. S., and Jones, A. L. (1999). Aberrant interactions of transcriptional repressor proteins with the Huntington's disease gene product, huntingtin. *Hum. Mol. Genet.* **8**, 1647–1655.
102. Dunah, A. W., Jeong, H., Griffin, A., Kim, Y. M., Standaert, D. G., Hersch, S. M., Mouradian, M. M., Young, A. B., Tanese, N., and Krainc, D. (2002). Sp1 and TAFII130 transcriptional activity disrupted in early Huntington's disease. *Science* **296**, 2238–2243.
103. Holbert, S., Denghien, I., Kiechle, T., Rosenblatt, A., Wellington, C., Hayden, M. R., Margolis, R. L., Ross, C. A., Dausset, J., Ferrante, R. J., and Neri, C. (2001). The Gln–Ala repeat transcriptional activator CA150 interacts with huntingtin: Neuropathologic and genetic evidence for a role in Huntington's disease pathogenesis. *Proc. Natl. Acad. Sci. USA* **98**, 1811–1816.

104. Zuccato, C., Tartari, M., Crotti, A., Goffredo, D., Valenza, M., Conti, L., Cataudella, T., Leavitt, B. R., Hayden, M. R., Timmusk, T., Rigamonti, D., and Cattaneo, E. (2003). Huntingtin interacts with REST/NRSF to modulate the transcription of NRSE-controlled neuronal genes. *Nat. Genet.* **35**, 76–83.

105. Igarashi, S., Morita, H., Bennett, K. M., Tanaka, Y., Engelender, S., Peters, M. F., Cooper, J. K., Wood, J. D., Sawa, A., and Ross, C. A. (2003). Inducible PC12 cell model of Huntington's disease shows toxicity and decreased histone acetylation. *NeuroReport* **14**, 565–568.

106. Trettel, F., Rigamonti, D., Hilditch-Maguire, P., Wheeler, V. C., Sharp, A. H., Persichetti, F., Cattaneo, E., and MacDonald, M. E. (2000). Dominant phenotypes produced by the HD mutation in STHdh(Q111) striatal cells. *Hum. Mol. Genet.* **9**, 2799–2809.

107. Hughes, R. E., Lo, R. S., Davis, C., Strand, A. D., Neal, C. L., Olson, J. M., and Fields, S. (2001). Altered transcription in yeast expressing expanded polyglutamine. *Proc. Natl. Acad. Sci. USA* **98**, 13201–13206.

108. Schilling, G., Becher, M. W., Sharp, A. H., Jinnah, H. A., Duan, K., Kotzuk, J. A., Slunt, H. H., Ratovitski, T., Cooper, J. K., Jenkins, N. A., Copeland, N. G., Price, D. L., Ross, C. A., and Borchelt, D. R. (1999). Intranuclear inclusions and neuritic aggregates in transgenic mice expressing a mutant N-terminal fragment of huntingtin. *Hum. Mol. Genet.* **8**, 397–407.

109. Laforet, G. A., Sapp, E., Chase, K., McIntyre, C., Boyce, F. M., Campbell, M., Cadigan, B. A., Warzecki, L., Tagle, D. A., Reddy, P. H., Cepeda, C., Calvert, C. R., Jokel, E. S., Klapstein, G. J., Ariano, M. A., Levine, M. S., DiFiglia, M., and Aronin, N. (2001). Changes in cortical and striatal neurons predict behavioral and electrophysiological abnormalities in a transgenic murine model of Huntington's disease. *J. Neurosci.* **21**, 9112–9123.

110. Slow, E. J., van Raamsdonk, J., Rogers, D., Coleman, S. H., Graham, R. K., Deng, Y., Oh, R., Bissada, N., Hossain, S. M., Yang, Y. Z., Li, X. J., Simpson, E. M., Gutekunst, C. A., Leavitt, B. R., and Hayden, M. R. (2003). Selective striatal neuronal loss in a YAC128 mouse model of Huntington disease. *Hum. Mol. Genet.* **12**, 1555–1567.

111. Reddy, P. H., Williams, M., Charles, V., Garrett, L., Pike-Buchanan, L., Whetsell, W. O., Jr., Miller, G., and Tagle, D. A. (1998). Behavioural abnormalities and selective neuronal loss in HD transgenic mice expressing mutated full-length HD cDNA. *Nat. Genet.* **20**, 198–202.

112. Wheeler, V. C., White, J. K., Gutekunst, C. A., Vrbanac, V., Weaver, M., Li, X. J., Li, S. H., Yi, H., Vonsattel, J. P., Gusella, J. F., Hersch, S., Auerbach, W., Joyner, A. L., and MacDonald, M. E. (2000). Long glutamine tracts cause nuclear localization of a novel form of huntingtin in medium spiny striatal neurons in HdhQ92 and HdhQ111 knock-in mice. *Hum. Mol. Genet.* **9**, 503–513.

113. Levine, M. S., Klapstein, G. J., Koppel, A., Gruen, E., Cepeda, C., Vargas, M. E., Jokel, E. S., Carpenter, E. M., Zanjani, H., Hurst, R. S., Efstratiadis, A., Zeitlin, S., and Chesselet, M. F. (1999). Enhanced sensitivity to N-methyl-D-aspartate receptor activation in transgenic and knockin mouse models of Huntington's disease. *J. Neurosci. Res.* **58**, 515–532.

114. von Horsten, S., Schmitt, I., Nguyen, H. P., Holzmann, C., Schmidt, T., Walther, T., Bader, M., Pabst, R., Kobbe, P., Krotova, J., Stiller, D., Kask, A., Vaarmann, A., Rathke-Hartlieb, S., Schulz, J. B., Grasshoff, U., Bauer, I., Vieira-Saecker, A. M., Paul, M., Jones, L., Lindenberg, K. S., Landwehrmeyer, B., Bauer, A., Li, X. J., and Riess, O. (2003). Transgenic rat model of Huntington's disease. *Hum. Mol. Genet.* **12**, 617–624.

115. de Almeida, L. P., Ross, C. A., Zala, D., Aebischer, P., and Deglon, N. (2002). Lentiviral-mediated delivery of mutant huntingtin in the striatum of rats induces a selective neuropathology modulated by polyglutamine repeat size, huntingtin expression levels, and protein length. *J. Neurosci.* **22**, 3473–3483.

116. Hughes, R. E., and Olson, J. M. (2001). Therapeutic opportunities in polyglutamine disease. *Nat. Med.* **7**, 419–423.

117. Heiser, V., Engemann, S., Brocker, W., Dunkel, I., Boeddrich, A., Waelter, S., Nordhoff, E., Lurz, R., Schugardt, N., Rautenberg, S., Herhaus, C., Barnickel, G., Bottcher, H., Lehrach, H., and Wanker, E. E. (2002). Identification of benzothiazoles as potential polyglutamine aggregation inhibitors of Huntington's disease by using an automated filter retardation assay. *Proc. Natl. Acad. Sci. USA* **99** (Suppl. 4), 16400–16406.

118. Lipinski, C. A., Lombardo, F., Dominy, B. W., and Feeney, P. J. (2001). Experimental and computational approaches to estimate solubility and permeability in drug discovery and development settings. *Adv. Drug. Deliv. Rev.* **46**, 3–26.

119. Aiken, C. T., Tobin, A. J., and Schweitzer, E. S. (2004). A cell-based screen for drugs to treat Huntington's disease. *Neurobiol. Dis.* **16**, 546–555.

120. Pollitt, S. K., Pallos, J., Shao, J., Desai, U. A., Ma, A. A., Thompson, L. M., Marsh, J. L., and Diamond, M. I. (2003). A rapid cellular FRET assay of polyglutamine aggregation identifies a novel inhibitor. *Neuron* **40**, 685–694.

121. Tobin, A. J., and Signer, E. R. (2000). Huntington's disease: The challenge for cell biologists. *Trends Cell Biol.* **10**, 531–536.

122. Wang, J., Gines, S., MacDonald, M. E., and Gusella, J. F. (2005). Reversal of a full-length mutant huntingtin neuronal cell phenotype by chemical inhibitors of polyglutamine-mediated aggregation. *BMC Neurosci.* **6**, 1.

123. Wang, W., Duan, W., Igarashi, S., Morita, H., Nakamura, M., and Ross, C. A. (2005). Compounds blocking mutant huntingtin toxicity identified using a Huntington's disease neuronal cell model. *Neurobiol Dis.* **20**, 500–508.

124. Zhang, X., Smith, D. L., Meriin, A. B., Engemann, S., Russel, D. E., Roark, M., Washington, S. L., Maxwell, M. M., Marsh, J. L., Thompson, L. M., Wanker, E. E., Young, A. B., Housman, D. E., Bates, G. P., Sherman, M. Y., and Kazantsev, A. G. (2005). A potent small molecule inhibits polyglutamine aggregation in Huntington's disease neurons and suppresses neurodegeneration in vivo. *Proc. Natl. Acad. Sci. USA* **102**, 892–897.

125. Meriin, A. B., Zhang, X., He, X., Newnam, G. P., Chernoff, Y. O., and Sherman, M. Y. (2002). Huntington toxicity in yeast model depends on polyglutamine aggregation mediated by a prion-like protein Rnq1. *J. Cell Biol.* **157**, 997–1004.

126. Heiser, V., Scherzinger, E., Boeddrich, A., Nordhoff, E., Lurz, R., Schugardt, N., Lehrach, H., and Wanker, E. E. (2000). Inhibition of huntingtin fibrillogenesis by specific antibodies and small molecules: Implications for Huntington's disease therapy. *Proc. Natl. Acad. Sci. USA* **97**, 6739–6744.

127. Apostol, B. L., Kazantsev, A., Raffioni, S., Illes, K., Pallos, J., Bodai, L., Slepko, N., Bear, J. E., Gertler, F. B., Hersch, S., Housman, D. E., Marsh, J. L., and Thompson, L. M. (2003). A cell-based assay for aggregation inhibitors as therapeutics of polyglutamine-repeat disease and validation in Drosophila. *Proc. Natl. Acad. Sci. USA* **100**, 5950–5955.

128. Smith, D. L., Portier, R., Woodman, B., Hockly, E., Mahal, A., Klunk, W. E., Li, X. J., Wanker, E., Murray, K. D., and Bates, G. P. (2001). Inhibition of polyglutamine aggregation in R6/2 HD brain slices-complex dose–response profiles. *Neurobiol. Dis.* **8**, 1017–1026.

129. McCampbell, A., Taye, A. A., Whitty, L., Penney, E., Steffan, J. S., and Fischbeck, K. H. (2001). Histone deacetylase inhibitors reduce polyglutamine toxicity. *Proc. Natl. Acad. Sci. USA* **98**, 15179–15184.

130. Gardian, G., Browne, S. E., Choi, D. K., Klivenyi, P., Gregorio, J., Kubilus, J. K., Ryu, H., Langley, B., Ratan, R. R., Ferrante, R. J.,

and Beal, M. F. (2005). Neuroprotective effects of phenylbutyrate in the N171–82Q transgenic mouse model of Huntington's disease. *J. Biol. Chem.* **280**, 556–563.
131. Ferrante, R. J., Kubilus, J. K., Lee, J., Ryu, H., Beesen, A., Zucker, B., Smith, K., Kowall, N. W., Ratan, R. R., Luthi-Carter, R., and Hersch, S. M. (2003). Histone deacetylase inhibition by sodium butyrate chemotherapy ameliorates the neurodegenerative phenotype in Huntington's disease mice. *J. Neurosci.* **23**, 9418–9427.
132. Parker, J. A., Arango, M., Abderrahmane, S., Lambert, E., Tourette, C., Catoire, H., and Neri, C. (2005). Resveratrol rescues mutant polyglutamine cytotoxicity in nematode and mammalian neurons. *Nat. Genet.* **37**, 349–350.
133. Marsh, J. L., Pallos, J., and Thompson, L. M. (2003). Fly models of Huntington's disease. *Hum. Mol. Genet.* **12** (Spec. No. 2), R187–R193.
134. Marsh, J. L., and Thompson, L. M. (2004). Can flies help humans treat neurodegenerative diseases? *Bioessays* **26**, 485–496.
135. Agrawal, N., Pallos, J., Slepko, N., Apostol, B. L., Bodai, L., Chang, L. W., Chiang, A. S., Thompson, L. M., and Marsh, J. L. (2005). Identification of combinatorial drug regimens for treatment of Huntington's disease using *Drosophila*. *Proc. Natl. Acad. Sci. USA* **102**, 3777–3781.
136. Mahmood, I. (1999). Allometric issues in drug development. *J. Pharm. Sci.* **88**, 1101–1106.
137. Mahmood, I. (1999). Prediction of clearance, volume of distribution and half-life by allometric scaling and by use of plasma concentrations predicted from pharmacokinetic constants: A comparative study. *J. Pharm. Pharmacol.* **51**, 905–910.
138. Beal, M. F., Brouillet, E., Jenkins, B. G., Ferrante, R. J., Kowall, N. W., Miller, J. M., Storey, E., Srivastava, R., Rosen, B. R., and Hyman, B. T. (1993). Neurochemical and histologic characterization of striatal excitotoxic lesions produced by the mitochondrial toxin 3-nitropropionic acid. *J. Neurosci.* **13**, 4181–4192.
139. Beal, M. F., Kowall, N. W., Ellison, D. W., Mazurek, M. F., Swartz, K. J., and Martin, J. B. (1986). Replication of the neurochemical characteristics of Huntington's disease by quinolinic acid. *Nature* **321**, 168–171.
140. de Lago, E., Urbani, P., Ramos, J. A., Di Marzo, V., and Fernandez-Ruiz, J. (2005). Arvanil, a hybrid endocannabinoid and vanilloid compound, behaves as an antihyperkinetic agent in a rat model of Huntington's disease. *Brain Res.* **1050**, 210–216.
141. Kim, J. H., Kim, S., Yoon, I. S., Lee, J. H., Jang, B. J., Jeong, S. M., Lee, B. H., Han, J. S., Oh, S., Kim, H. C., Park, T. K., Rhim, H., and Nah, S. Y. (2005). Protective effects of ginseng saponins on 3-nitropropionic acid-induced striatal degeneration in rats. *Neuropharmacology* **48**, 743–756.
142. Rodriguez, A. I., Willing, A. E., Saporta, S., Cameron, D. F., and Sanberg, P. R. (2003). Effects of Sertoli cell transplants in a 3-nitropropionic acid model of early Huntington's disease: A preliminary study. *Neurotoxicol. Res.* **5**, 443–450.
143. McBride, J. L., During, M. J., Wuu, J., Chen, E. Y., Leurgans, S. E., and Kordower, J. H. (2003). Structural and functional neuroprotection in a rat model of Huntington's disease by viral gene transfer of GDNF. *Exp. Neurol.* **181**, 213–223.
144. Diguet, E., Fernagut, P. O., Wei, X., Du, Y., Rouland, R., Gross, C., Bezard, E., and Tison, F. (2004). Deleterious effects of minocycline in animal models of Parkinson's disease and Huntington's disease. *Eur. J. Neurosci.* **19**, 3266–3276.
145. Diguet, E., Rouland, R., and Tison, F. (2003). Minocycline is not beneficial in a phenotypic mouse model of Huntington's disease. *Ann. Neurol.* **54**, 841–842.
146. Keene, C. D., Rodrigues, C. M., Eich, T., Linehan-Stieers, C., Abt, A., Kren, B. T., Steer, C. J., and Low, W. C. (2001). A bile acid protects against motor and cognitive deficits and reduces striatal degeneration in the 3-nitropropionic acid model of Huntington's disease. *Exp. Neurol.* **171**, 351–360.
147. Roitberg, B. Z., Emborg, M. E., Sramek, J. G., Palfi, S., and Kordower, J. H. (2002). Behavioral and morphological comparison of two nonhuman primate models of Huntington's disease. *Neurosurgery* **50**, 137–145; discussion 145–136.
148. Kendall, A. L., Rayment, F. D., Torres, E. M., Baker, H. F., Ridley, R. M., and Dunnett, S. B. (1998). Functional integration of striatal allografts in a primate model of Huntington's disease. *Nat. Med.* **4**, 727–729.
149. Dunnett, S. B. (1999). Striatal reconstruction by striatal grafts. *J. Neural. Transm. Suppl.* **55**, 115–129.
150. Dunnett, S. B., Carter, R. J., Watts, C., Torres, E. M., Mahal, A., Mangiarini, L., Bates, G., and Morton, A. J. (1998). Striatal transplantation in a transgenic mouse model of Huntington's disease. *Exp. Neurol.* **154**, 31–40.
151. Georgiou, N., Bradshaw, J. L., Chiu, E., Tudor, A., O'Gorman, L., and Phillips, J. G. (1999). Differential clinical and motor control function in a pair of monozygotic twins with Huntington's disease. *Mov. Disord.* **14**, 320–325.
152. Carter, R. J., Hunt, M. J., and Morton, A. J. (2000). Environmental stimulation increases survival in mice transgenic for exon 1 of the Huntington's disease gene. *Mov. Disord.* **15**, 925–937.
153. Van Dellen, A., and Hannan, A. J. (2004). Genetic and environmental factors in the pathogenesis of Huntington's disease. *Neurogenetics* **5**, 9–17.
154. Hockly, E., Cordery, P. M., Woodman, B., Mahal, A., van Dellen, A., Blakemore, C., Lewis, C. M., Hannan, A. J., and Bates, G. P. (2002). Environmental enrichment slows disease progression in R6/2 Huntington's disease mice. *Ann. Neurol.* **51**, 235–242.
155. Schilling, G., Savonenko, A. V., Coonfield, M. L., Morton, J. L., Vorovich, E., Gale, A., Neslon, C., Chan, N., Eaton, M., Fromholt, D., Ross, C. A., and Borchelt, D. R. (2004). Environmental, pharmacological, and genetic modulation of the HD phenotype in transgenic mice. *Exp. Neurol.* **187**, 137–149.
156. Van Praag, H., Kempermann, G., and Gage, F. H. (1999). Running increases cell proliferation and neurogenesis in the adult mouse dentate gyrus. *Nat. Neurosci.* **2**, 266–270.
157. Van Praag, H., Christie, B. R., Sejnowski, T. J., and Gage, F. H. (1999). Running enhances neurogenesis, learning, and long-term potentiation in mice. *Proc. Natl. Acad. Sci. USA* **96**, 13427–13431.
158. Brown, J., Cooper-Kuhn, C. M., Kempermann, G., Van Praag, H., Winkler, J., Gage, F. H., and Kuhn, H. G. (2003). Enriched environment and physical activity stimulate hippocampal but not olfactory bulb neurogenesis. *Eur. J. Neurosci.* **17**, 2042–2046.
159. Spires, T. L., Grote, H. E., Garry, S., Cordery, P. M., Van Dellen, A., Blakemore, C., and Hannan, A. J. (2004). Dendritic spine pathology and deficits in experience-dependent dendritic plasticity in R6/1 Huntington's disease transgenic mice. *Eur. J. Neurosci.* **19**, 2799–2807.
160. Spires, T. L., Grote, H. E., Varshney, N. K., Cordery, P. M., van Dellen, A., Blakemore, C., and Hannan, A. J. (2004). Environmental enrichment rescues protein deficits in a mouse model of Huntington's disease, indicating a possible disease mechanism. *J. Neurosci.* **24**, 2270–2276.
161. Glass, M., van Dellen, A., Blakemore, C., Hannan, A. J., and Faull, R. L. (2004). Delayed onset of Huntington's disease in mice in an enriched environment correlates with delayed loss of cannabinoid CB1 receptors. *Neuroscience* **123**, 207–212.
162. Tanaka, M., Machida, Y., Niu, S., Ikeda, T., Jana, N. R., Doi, H., Kurosawa, M., Nekooki, M., and Nukina, N. (2004). Trehalose alleviates polyglutamine-mediated pathology in a mouse model of Huntington disease. *Nat. Med.* **10**, 148–154.
163. Smith, D. L., Woodman, B., Mahal, A., Sathasivam, K., Ghazi-Noori, S., Lowden, P. A., Bates, G. P., and Hockly, E. (2003). Minocycline

and doxycycline are not beneficial in a model of Huntington's disease. *Ann. Neurol.* **54**, 186–196.

164. Chen, M., Ona, V. O., Li, M., Ferrante, R. J., Fink, K. B., Zhu, S., Bian, J., Guo, L., Farrell, L. A., Hersch, S. M., Hobbs, W., Vonsattel, J. P., Cha, J. H., and Friedlander, R. M. (2000). Minocycline inhibits caspase-1 and caspase-3 expression and delays mortality in a transgenic mouse model of Huntington disease. *Nat. Med.* **6**, 797–801.

165. Diguet, E., Gross, C. E., Tison, F. and Bezard, E. (2004). Rise and fall of minocycline in neuroprotection: Need to promote publication of negative results. *Exp. Neurol.* **189**, 1–4.

166. Sanchez, I., Mahlke, C., and Yuan, J. (2003). Pivotal role of oligomerization in expanded polyglutamine neurodegenerative disorders. *Nature* **421**, 373–379.

167. Ferrante, R. J., Ryu, H., Kubilus, J. K., D'Mello, S., Sugars, K. L., Lee, J., Lu, P., Smith, K., Browne, S., Beal, M. F., Kristal, B. S., Stavrovskaya, I. G., Hewett, S., Rubinsztein, D. C., Langley, B., and Ratan, R. R. (2004). Chemotherapy for the brain: The antitumor antibiotic mithramycin prolongs survival in a mouse model of Huntington's disease. *J. Neurosci.* **24**, 10335–10342.

168. Minamiyama, M., Katsuno, M., Adachi, H., Waza, M., Sang, C., Kobayashi, Y., Tanaka, F., Doyu, M., Inukai, A., and Sobue, G. (2004). Sodium butyrate ameliorates phenotypic expression in a transgenic mouse model of spinal and bulbar muscular atrophy. *Hum. Mol. Genet.* **13**, 1183–1192.

169. Schilling, G., Coonfield, M. L., Ross, C. A., and Borchelt, D. R. (2001). Coenzyme Q10 and remacemide hydrochloride ameliorate motor deficits in a Huntington's disease transgenic mouse model. *Neurosci. Lett.* **315**, 149–153.

170. Ferrante, R. J., Andreassen, O. A., Dedeoglu, A., Ferrante, K. L., Jenkins, B. G., Hersch, S. M., and Beal, M. F. (2002). Therapeutic effects of coenzyme Q10 and remacemide in transgenic mouse models of Huntington's disease. *J. Neurosci.* **22**, 1592–1599.

171. Andreassen, O. A., Ferrante, R. J., Dedeoglu, A., and Beal, M. F. (2001). Lipoic acid improves survival in transgenic mouse models of Huntington's disease. *NeuroReport* **12**, 3371–3373.

172. Klivenyi, P., Ferrante, R. J., Gardian, G., Browne, S., Chabrier, P. E., and Beal, M. F. (2003). Increased survival and neuroprotective effects of BN82451 in a transgenic mouse model of Huntington's disease. *J. Neurochem.* **86**, 267–272.

173. Chou, S. Y., Lee, Y. C., Chen, H. M., Chiang, M. C., Lai, H. L., Chang, H. H., Wu, Y. C., Sun, C. N., Chien, C. L., Lin, Y. S., Wang, S. C., Tung, Y. Y., Chang, C., and Chern, Y. (2005). CGS21680 attenuates symptoms of Huntington's disease in a transgenic mouse model. *J. Neurochem.* **93**, 310–320.

174. Morton, A. J., Hunt, M. J., Hodges, A. K., Lewis, P. D., Redfern, A. J., Dunnett, S. B., and Jones, L. (2005). A combination drug therapy improves cognition and reverses gene expression changes in a mouse model of Huntington's disease. *Eur. J. Neurosci.* **21**, 855–870.

175. Andreassen, O. A., Ferrante, R. J., Huang, H. M., Dedeoglu, A., Park, L., Ferrante, K. L., Kwon, J., Borchelt, D. R., Ross, C. A., Gibson, G. E., and Beal, M. F. (2001). Dichloroacetate exerts therapeutic effects in transgenic mouse models of Huntington's disease. *Ann. Neurol.* **50**, 112–117.

176. Keene, C. D., Rodrigues, C. M., Eich, T., Chhabra, M. S., Steer, C. J., and Low, W. C. (2002). Tauroursodeoxycholic acid, a bile acid, is neuroprotective in a transgenic animal model of Huntington's disease. *Proc. Natl. Acad. Sci. USA* **99**, 10671–10676.

177. Ona, V. O., Li, M., Vonsattel, J. P., Andrews, L. J., Khan, S. Q., Chung, W. M., Frey, A. S., Menon, A. S., Li, X. J., Stieg, P. E., Yuan, J., Penney, J. B., Young, A. B., Cha, J. H., and Friedlander, R. M. (1999). Inhibition of caspase-1 slows disease progression in a mouse model of Huntington's disease. *Nature* **399**, 263–267.

178. Karpuj, M. V., Becher, M. W., Springer, J. E., Chabas, D., Youssef, S., Pedotti, R., Mitchell, D., and Steinman, L. (2002). Prolonged survival and decreased abnormal movements in transgenic model of Huntington disease, with administration of the transglutaminase inhibitor cystamine. *Nat. Med.* **8**, 143–149.

179. Dedeoglu, A., Kubilus, J. K., Yang, L., Ferrante, K. L., Hersch, S. M., Beal, M. F., and Ferrante, R. J. (2003). Creatine therapy provides neuroprotection after onset of clinical symptoms in Huntington's disease transgenic mice. *J. Neurochem.* **85**, 1359–1367.

180. Bailey, C. D., and Johnson, G. V. (2005). The protective effects of cystamine in the R6/2 Huntington's disease mouse involve mechanisms other than the inhibition of tissue transglutaminase. *Neurobiol. Aging* in press.

181. Ferrante, R. J., Andreassen, O. A., Jenkins, B. G., Dedeoglu, A., Kuemmerle, S., Kubilus, J. K., Kaddurah-Daouk, R., Hersch, S. M., and Beal, M. F. (2000). Neuroprotective effects of creatine in a transgenic mouse model of Huntington's disease. *J. Neurosci.* **20**, 4389–4397.

182. Andreassen, O. A., Dedeoglu, A., Ferrante, R. J., Jenkins, B. G., Ferrante, K. L., Thomas, M., Friedlich, A., Browne, S. E., Schilling, G., Borchelt, D. R., Hersch, S. M., Ross, C. A., and Beal, M. F. (2001). Creatine increase survival and delays motor symptoms in a transgenic animal model of Huntington's disease. *Neurobiol. Dis.* **8**, 479–491.

183. Wang, X., Sarkar, A., Cicchetti, F., Yu, M., Zhu, A., Jokivarsi, K., Saint-Pierre, M., and Brownell, A. L. (2005). Cerebral PET imaging and histological evidence of transglutaminase inhibitor cystamine induced neuroprotection in transgenic R6/2 mouse model of Huntington's disease. *J. Neurol. Sci.* **231**, 57–66.

184. Bantubungi, K., Jacquard, C., Greco, A., Pintor, A., Chtarto, A., Tai, K., Galas, M. C., Tenenbaum, L., Deglon, N., Popoli, P., Minghetti, L., Brouillet, E., Brotchi, J., Levivier, M., Schiffmann, S. N., and Blum, D. (2005). Minocycline in phenotypic models of Huntington's disease. *Neurobiol. Dis.* **18**, 206–217.

185. Harper, S. Q., Staber, P. D., He, X., Eliason, S. L., Martins, I. H., Mao, Q., Yang, L., Kotin, R. M., Paulson, H. L., and Davidson, B. L. (2005). RNA interference improves motor and neuropathological abnormalities in a Huntington's disease mouse model. *Proc. Natl. Acad. Sci. USA* **102**, 5820–5825.

186. Miller, T. W., Shirley, T. L., Wolfgang, W. J., Kang, X., and Messer, A. (2003). DNA vaccination against mutant huntingtin ameliorates the HDR6/2 diabetic phenotype. *Mol. Ther.* **7**, 572–579.

187. Dedeoglu, A., Kubilus, J. K., Jeitner, T. M., Matson, S. A., Bogdanov, M., Kowall, N. W., Matson, W. R., Cooper, A. J., Ratan, R. R., Beal, M. F., Hersch, S. M., and Ferrante, R. J. (2002). Therapeutic effects of cystamine in a murine model of Huntington's disease. *J. Neurosci.* **22**, 8942–8950.

188. Schiefer, J., Landwehrmeyer, G. B., Luesse, H. G., Sprunken, A., Puls, C., Milkereit, A., Milkereit, E., and Kosinski, C. M. (2002). Riluzole prolongs survival time and alters nuclear inclusion formation in a transgenic mouse model of Huntington's disease. *Mov. Disord.* **17**, 748–757.

189. Wood, N. I., and Morton, A. J. (2003). Chronic lithium chloride treatment has variable effects on motor behaviour and survival of mice transgenic for the Huntington's disease mutation. *Brain Res. Bull.* **61**, 375–383.

190. Popovic, N., Maingay, M., Kirik, D., and Brundin, P. (2005). Lentiviral gene delivery of GDNF into the striatum of R6/2 Huntington mice fails to attenuate behavioral and neuropathological changes. *Exp. Neurol.* **193**, 65–74.

191. Zucker, B., Ludin, D. E., Gerds, T. A., Lucking, C. H., Landwehrmeyer, G. B., and Feuerstein, T. J. (2004). Gabapentin-lactam, but not gabapentin, reduces protein aggregates and

improves motor performance in a transgenic mouse model of Huntington's disease. *Naunyn Schmiedeberg's Arch. Pharmacol.* **370**, 131–139.

192. Schiefer, J., Sprunken, A., Puls, C., Luesse, H. G., Milkereit, A., Milkereit, E., Johann, V., and Kosinski, C. M. (2004). The metabotropic glutamate receptor 5 antagonist MPEP and the mGluR2 agonist LY379268 modify disease progression in a transgenic mouse model of Huntington's disease. *Brain Res.* **1019**, 246–254.

193. Duan, W., Guo, Z., Jiang, H., Ladenheim, B., Xu, X., Cadet, J. L., and Mattson, M. P. (2004). Paroxetine retards disease onset and progression in huntington mutant mice. *Ann. Neurol.* **55**, 590–594.

194. Clifford, J. J., Drago, J., Natoli, A. L., Wong, J. Y., Kinsella, A., Waddington, J. L., and Vaddadi, K. S. (2002). Essential fatty acids given from conception prevent topographies of motor deficit in a transgenic model of Huntington's disease. *Neuroscience* **109**, 81–88.

195. Van Dellen, A., Blakemore, C., Deacon, R., York, D., and Hannan, A. J. (2000). Delaying the onset of Huntington's in mice. *Nature* **404**, 721–722.

196. Duan, W., Guo, Z., Jiang, H., Ware, M., Li, X. J., and Mattson, M. P. (2003). Dietary restriction normalizes glucose metabolism and BDNF levels, slows disease progression, and increases survival in huntingtin mutant mice. *Proc. Natl. Acad. Sci. USA* **100**, 2911–2916.

197. Ende, N., and Chen, R. (2001). Human umbilical cord blood cells ameliorate Huntington's disease in transgenic mice. *J. Med.* **32**, 231–240.

198. Morley, J. F., Brignull, H. R., Weyers, J. J., and Morimoto, R. I. (2002). The threshold for polyglutamine-expansion protein aggregation and cellular toxicity is dynamic and influenced by aging in *Caenorhabditis elegans*. *Proc. Natl. Acad. Sci. USA* **99**, 10417–10422.

CHAPTER 15

Molecular Pathogenesis of Huntington's Disease: The Role of Excitotoxicity

MAHMOUD A. POULADI, ILYA BEZPROZVANNY,
LYNN A. RAYMOND, MICHAEL R. HAYDEN

Center for Molecular Medicine and Therapeutics, Department of Medical Genetics, Children's and Woman's Hospital, University of British Columbia, Vancouver, BC, Canada V6T 1Z4; Department of Physiology, University of Texas Southwestern Medical Center, Dallas, Texas 75390; and Department of Psychiatry and Brain Research Centre, University of British Columbia, Vancouver, BC, Canada V6T 1Z3

I. Introduction
II. Glutamate and Neurotransmission
III. Glutamate and Excitotoxicity
   A. NMDA Receptors Play a Key Role in Calcium-Induced Excitotoxicity
   B. Disruption of Calcium Signaling Activates Neurotoxic Processes
   C. Role of the Mitochondria
IV. Excitotoxicity and HD
   A. NMDA Receptors
   B. mGluR5 and InsP3R1 Receptors
   C. Mitochondria
V. Implications for Therapy
VI. Concluding Remarks
Acknowledgments
References

I. INTRODUCTION

Huntington's disease (HD) is a progressive neurological disorder characterized by involuntary movements, emotional disturbances, and dementia. The underlying genetic lesion is an expansion of a CAG trinucleotide repeat in the *HD* gene that results in an expanded polyglutamine (polyQ) stretch at the N terminus of the huntingtin protein (htt). The cardinal neuropathological feature of HD is a selective loss in the striatum of medium-sized spiny neurons. A number of pathogenic mechanisms contributing to the observed neuronal loss have been identified. Recent evidence strongly implicates aberrant glutamate signaling, disrupted neuronal calcium handling, and the accompanying excitotoxicity in the pathogenesis of HD.

II. GLUTAMATE AND NEUROTRANSMISSION

Glutamate is the principal excitatory neurotransmitter in the vertebrate central nervous system (CNS), and its distribution is widespread throughout the CNS. During normal physiological synaptic transmission, glutamate is released from presynaptic termini of glutamatergic neurons into the synaptic cleft, resulting in a brief and localized rise in glutamate concentration and binding to postsynaptic glutamate receptors. Activation of these glutamate receptors leads to depolarization of the postsynaptic neuron, increasing the probability of firing an action potential. The excitatory action of glutamate is then terminated by the efficient and rapid removal of glutamate from the synaptic cleft by high-affinity glutamate uptake systems in neuronal and glial nerve termini [1].

The postsynaptic actions of glutamate are mediated by biochemically distinct glutamate receptors that are broadly classified into GTP-binding protein-coupled metabotropic receptors (mGluRs) and ion channel-forming ionotropic receptors (iGluRs) (Table 15-1). mGluRs are are divided into three groups based on sequence homology, pharmacological properties, and signaling pathways. Group I mGluRs (mGluR1 and mGluR5) mediate their action via phospholipase C (PLC)-linked hydrolysis of phosphoinositides. Group II (mGluR2 and mGluR3) and group III (mGluR4, mGluR6–mGluR8) mGluRs, on the other hand, are either negatively linked to adenyl cyclases or linked to ion channels.

iGluRs are classified into three subfamilies according to their specific affinity to the agonists N-methyl D-aspartate (NMDA), α-amino-3-hydroxy-5-methyl-4-isoxazole propionate (AMPA), and kainate. AMPA receptors are heteromeric structures formed by association of a combination of the four subunits GluR1–GluR4. AMPA receptors are permeable to potassium and sodium ions but, due to modifications to the transcripts encoding the GluR2 subunit, are normally impermeable to calcium ions. AMPA receptors are involved in rapid excitatory transmission at glutamatergic synapses.

Kainate receptors are formed by combination of kainate receptor subunits in a fashion similar to AMPA and NMDA receptors, and are divided into two groups according to their affinity for binding to kainate. The KA1 and KA2 subunits bind with high affinity to kainate, whereas the GluR5–GluR7 subunits exhibit lower-affinity binding to kainate. Similar to AMPA receptors, some kainate subunits (GluR5 and GluR6) are modified on the RNA level to make them impermeable to calcium. Kainate receptor subunits are present at both the pre- and postsynaptic neuronal termini. In comparison to NMDA and AMPA receptors, little is known about the role of kainate receptors in normal physiology and disease, although studies have shown them to play an important role in neurotransmitter release from presynaptic termini at both excitatory and inhibitory synapses.

NMDA receptor channels comprise distinct subunits with differential tissue distribution and temporal expression. Different combinations of subunits form tetrameric structures with unique pharmacological and signaling properties. The subunits are classified into three subfamilies based on sequence homology: the NR1 subunit, NR2 (with four subtypes, A–D), and NR3 (with two subtypes, A and B). Functional NMDA receptors are formed predominantly by combination of the NR1 subunit with NR2 and, in some cases, NR3 subunits. NMDA receptors are unique in that they allow entry of monovalent ions such as sodium and potassium in addition to the divalent calcium ions. NMDA receptors are also unique in that efficient activation and opening of the channel require the simultaneous occurrence of three events: (1) binding of the agonist glutamate to the NR2 subunit; (2) binding of glycine, which acts as a co-agonist, to the NR1 subunit; and (3) release in response to membrane depolarization of the magnesium ion block that prevents ion entry in the resting state.

It is now widely accepted that the excitotoxicity associated with aberrant glutamate signaling is attributable largely to iGluRs, particularly NMDA receptors.

TABLE 15-1 Glutamate Receptor Families and Subtypes

Ionotropic	NMDA	NR1
		NR2A, NR2B, NR2C, NR2D
		NR3A, NR3B
	AMPA	GluR1–GluR4
	Kainate	GluR5, GluR6
		KA1, KA2
Metabotropic	Group I	mGluR1, mGluR5
	Group II	mGluR2, mGluR3
	Group III	mGluR4, mGluR6, mGluR7, mGluR8

III. GLUTAMATE AND EXCITOTOXICITY

A. NMDA Receptors Play a Key Role in Calcium-Induced Excitotoxicity

NMDA receptors have the highest affinity for glutamate among the glutamate receptors, and numerous studies have shown them to be the primary agents of glutamate-mediated neurotoxicity. Indeed, *in vitro* studies have demonstrated that blockade of NMDA receptor activation prevents neuronal death in excitotoxicity assays [2]. Furthermore, transfection of NMDA receptor subunits into nonneuronal cells in the presence of glutamate-containing culture media leads to cell death that is prevented by NMDA receptor antagonists [3]. This property of NMDA receptors is attributed largely to their high permeability to calcium ions and their lower agonist-induced desensitization characteristics, giving NMDA receptors the capacity to cause sustained increases in intracellular calcium levels on prolonged or enhanced activation, and leading to disturbed neuronal calcium signaling and cell death.

Despite widespread expression of NMDA receptors throughout the CNS, distinct neuronal populations are lost in the different neuropathological conditions in which a role for excitotoxicity has been implicated. This selective susceptibility of the various neuronal populations is thought to reflect the differential tissue distribution of NMDA receptor subtypes, an idea supported by *in vitro* transfection studies. For example, NMDA receptor-mediated cell death was found to be significantly higher in cells transfected with the NR1/NR2A and NR1/NR2B subtypes, compared with cells transfected with the NR1/NR2C and NR1/NR2D subtypes [4, 5], a pattern that closely parallels open channel probabilities and calcium permeabilities of the different subtypes [6–8]. That the ability of a given subtype to induce cell death directly relates to the extent to which it is permeable to calcium underscores the importance of calcium signaling in excitotoxicity.

B. Disruption of Calcium Signaling Activates Neurotoxic Processes

Rises in intracellular calcium ions as a result of excessive or enhanced activation of specific calcium channels act as the intracellular mediators of excitotoxicity. This is accomplished through the disruption of numerous calcium-dependent enzymes and processes, ultimately leading to cell death. For example, calcium-induced activation of nucleases causes degradation of nuclear DNA and disruption of genomic organization. Protease activation leads to digestion of cytoskeletal and other essential cellular organelles [9, 10], as well as release of toxic fragments as a result of cleavage of proteins such as huntingtin [11], causing further cellular damage. Activated lipases break down cellular membranes, often releasing components that could cause further damage, such as arachidonic acid, the metabolism of which leads to release of oxygen free radicals [12].

Are intracellular calcium levels the sole determinant of glutamate neurotoxicity? Some studies seemed to indicate that mere accumulation of intracellular calcium beyond a certain threshold was sufficient to induce neurotoxicity and that levels of intracellular calcium, therefore, could be used as predictors of cell death [13, 14]. Other studies, however, reported contradictory results, finding no consistent correlation between increases in intracellular calcium levels and induced neurodegeneration [15, 16]. Further studies employing *in vitro* models of anoxia provided even more compelling findings demonstrating that although calcium channel blockers prevent increases in intracellular calcium levels, they fail to prevent neuronal degeneration [17]. These observations suggest that intracellular calcium levels, *per se*, are not the primary determinants of excitotoxicity, although the topic is still a matter of debate [18].

An alternative hypothesis put forth to explain the apparent dissociation observed under certain conditions between intracellular calcium levels and neurotoxicity holds that glutamate-induced, calcium-mediated neurotoxicity is not merely a function of the magnitude of intracellular calcium increases, but more importantly the source of calcium influx (reviewed in [19]). This "source specificity" hypothesis is supported by studies showing that although increases in intracellular calcium levels mediated by NMDA receptors are toxic, similar increases mediated by voltage-sensitive calcium channels fail to induce cell death [20, 21]. A corollary of this hypothesis in the context of NMDA receptor subtypes is that the capacity of a given subtype to induce cell death is dependent not only on its calcium permeability properties, but also its associated signal transduction pathways. The significance of this corollary becomes apparent when considered in the context of a recent study by Hardingham and coworkers, who demonstrated differential effects on cell survival of synaptic and extrasynaptic NMDA receptors: whereas calcium entry through synaptic NMDA receptors induced prosurvival events, calcium entry through extrasynaptic NMDA receptors led to cell death, presumably due to differences in the associated signaling complexes between synaptic and extrasynaptic NMDA receptors [22].

C. Role of the Mitochondria

In addition to their role in energy production and metabolism, mitochondria play an important role in calcium handling and homeostasis, and have been implicated in a number of neurodegenerative diseases [23, 24]. In the context of NMDA receptor activation, increases in intracellular calcium concentrations beyond a certain threshold level induce mitochondrial uptake and sequestration of calcium ions to maintain a set level of intracellular calcium and sustain calcium homeostasis [25–27]. However, prolonged calcium overload in conjunction with rising oxidative stress ultimately leads to mitochondrial damage and activation of the mitochondrial permeability transition, causing release of calcium from the mitochondria back into the cytosol [28, 29]. Activation of the permeability transition and release of mitochondrial calcium content are found to coincide with the cellular events mediated by disrupted calcium regulation that lead to cell death.

IV. EXCITOTOXICITY AND HD

The involvement of excitotoxicity in the pathogenesis of HD was first suggested by rodent studies in which intrastriatal injections of kainic acid (KA) or quinolinic acid (QA, an NMDA receptor agonist) produced lesions that mimicked many of the neurochemical and histopathological features of HD [30–35] and were associated with HD-like behavioral deficits [36–38]. A number of human and animal studies have since identified defects in NMDA and mGluR5 signaling, as well as mitochondrial calcium handling in HD patients and animal models of HD. Collectively, these studies give rise to a coherent, multifactorial model of mutant huntingtin-mediated alteration of glutamate receptor activity and calcium signaling as a primary contributor to neuronal degeneration in HD [39].

A. NMDA Receptors

Involvement of NMDA receptors in the pathology of HD was initially inferred from two lines of studies. The first set of studies demonstrated that injection of glutamate agonists into the striatum of rodents results in HD-like neuronal lesions. For example, injection of kainate into the striatum of rats was found to induce HD-like neuropathological changes. This effect of kainate was dependent on glutamate release, as it was prevented by prior decortication and removal of the corticostriatal afferents [40]. Further studies showed that injection of the NMDA receptor agonist QA into the striatum of rodents resulted in the most accurate reproduction of the histological and neuropathological changes seen in HD [32, 35], and was associated with HD-like behavioral changes in lesioned animals as well [36–38]. The second set of studies showed that neurons expressing NMDA receptors seem to be preferentially lost in HD, suggesting a role for NMDA receptors in enhancing the susceptibility to cell death.

For example, analysis of postmortem brain tissues from patients with HD showed that NMDA receptor binding was reduced by 93% in the putamen from HD brains compared with binding in normal brains [41]. Furthermore, *in situ* hybridization histochemistry studies of rat striatum showed that striatal projection neurons, the population selectively lost in HD patients, displayed enhanced expression of NR1/NR2B-type NMDA receptors compared with the spared interneurons [42]. The difference in NMDA receptor subtype expression was suggested to contribute to the relative vulnerability and resistance of striatal projections and interneurons, respectively, to NMDA receptor-mediated excitotoxicity.

Although these studies demonstrated the capacity of NMDA receptor overactivation to cause HD-like neuropathological and behavioral changes and provided correlative evidence from brains of HD patients, no evidence of a direct modulation of NMDA receptor function by mutant huntingtin was provided. With the use of huntingtin and NMDA receptor cotransfected HEK293 cells, the first such evidence demonstrated that mutant (Htt-138Q), but not wild-type (Htt-15Q), huntingtin enhances NMDA receptor currents [43], an effect that is specific to the NR1/NR2B NMDA receptor subtype and not the NR1/NR2A subtype. By use of the same cotransfection system, it was subsequently shown that compared with wild-type huntingtin (Htt-15Q), mutant huntingtin (Htt-138Q) leads to increased susceptibility to NMDA receptor-mediated cell death [44]. Significantly, this increase in NMDA receptor-mediated excitotoxic death is markedly diminished when an N-terminal fragment of mutant huntingtin is used in place of the full-length mutant protein. Furthermore, the enhancement of NMDA receptor-mediated cell death by mutant huntingtin is greater in cells transfected with the NR1/NR2B NMDA receptor subtype than in those transfected with the NR1/NR2A subtype. That the potentiation of NMDA-induced currents and enhancement of sensitivity to NMDA receptor-mediated cell death by mutant huntingtin are NR2B-specific is intellectually satisfying, as NR1/NR2B is the principal NMDA receptor subtype expressed in medium spiny neurons of the striatum [42].

These observations of enhanced NMDA receptor activity in the presence of mutant huntingtin were further validated in the YAC72 transgenic mouse model of HD. YAC72 transgenic mice express the entire human *HD* gene, with 72 CAG repeats under the control of the endogenous huntingtin promoter and regulatory elements, and recapitulate many of the behavioral and neuropathological features of the human condition [45]. By use of whole-cell patch clamp recordings, it was demonstrated that NMDA receptor peak current amplitudes and current density are significantly larger in medium spiny neurons from YAC72 mice than in those from wild-type mice [46]. It was further demonstrated that medium spiny neurons from YAC72 mice show enhanced susceptibility to QA- and NMDA-induced cell death compared with those from wild-type mice. This enhancement is specific to medium spiny neurons and is not observed in cerebellar granule neurons from YAC72 mice [46], an observation consistent with the pathology of HD in which no apparent cerebellar degeneration is observed. Corroborating the subtype specificity reported in HEK293 cotransfection studies, treatment of medium spiny neurons from YAC72 with ifenprodil, an NR2B-specific NMDA antagonist, prevents excitotoxic cell death, further implicating the NR1/NR2B NMDA receptor subtype in HD.

The enhanced excitotoxic cell death mediated by NMDA receptors was shown to occur via the intrinsic apoptotic pathway using primary medium spiny neuronal cultures from YAC46 and YAC72 mice [47], an observation that was also validated using primary medium spiny neurons isolated from YAC128 animals [48]. Furthermore, although defects in mitochondrial function were shown to contribute to the enhancement in NMDA receptor-mediated cell death, the difference in the extent to which mitochondrial stressors alone enhance cell death compared with NMDA receptor-mediated cell death indicates that NMDA receptor function and/or NMDA receptor-specific downstream signaling partners are also altered by mutant huntingtin [49].

B. mGluR5 and InsP3R1 Receptors

Several lines of evidence suggest a role for the group I mGluR5 receptors in pathogenesis of HD. First, mGluR5 is preferentially expressed in the striatal medium spiny neurons that are lost in HD compared with the large aspiny interneurons, which are largely spared. Through the use of *in situ* hybridization, the presence of mGluR5 mRNA in the striatum has been demonstrated, with preferential expression in medium spiny neurons compared with interneurons [50, 51]. It has been suggested that the higher levels of mGluR5 expression in medium spiny neurons compared with interneurons is a potential factor contributing to their selective loss in HD.

Second, mGluR5 has been shown to enhance NMDA receptor activity. Given the established role of NMDA receptor overactivation in mediating excitotoxic cell death, factors potentiating NMDA receptor signaling are expected to result in enhanced excitotoxicity. Combined electrophysiological and microfluorometric recordings helped to demonstrate that application of the group I agonist 3,5-dihydroxyphenylglycine (3,5-DHPG) strongly enhances NMDA-induced membrane depolarization and intracellular calcium accumulation in striatal neurons [52]. This effect is specific to spiny neurons and is not observed in large aspiny interneurons. Similarly, Pisani and coworkers demonstrated that NMDA-induced membrane depolarization and inward currents in mouse striatal slices are potentiated in the presence of the mGluR5 agonist 2-chloro-5-hydroxyphenylglycine (CHPG), an effect that is absent in neurons treated with the mGluR5 antagonist 2-methyl-6-(phenylethynyl)-pyridine (MPEP), and in neurons from mGluR5-deficient mice [53]. The impact of the mGluR5-mediated enhancement of NMDA receptor signaling on excitotoxicity has been examined by several groups. Using Selective noncompetitive antagonists of mGluR5 (MPEP, SIB-1757, and SIB-1893) were used to demonstrate that inhibition of mGluR5 signaling is neuroprotective against NMDA-mediated excitotoxicity in cortical cultures [54]. Furthermore, treatment of mice with MPEP or SIB-1893 protects against neurodegeneration induced by intrastriatal injections of NMDA or QA. Similar observations were made in rats, where intrastriatal injection of the group I antagonist 1-aminoindan-1,5-dicarboxylic acid (AIDA) or (S)-4-chloro-3-hydroxyphenylglycine (4C3HPG) protected against NMDA- and QA-induced striatal lesions [55]. Finally, treating rats with MPEP, a mGluR5 antagonist, was shown to protect against QA-induced striatal lesions by a mechanism that involves both pre- and postsynaptic components [56]. Furthermore, MPEP treatment was shown to reduce the body weight loss, electroencephalographic alterations, and spatial memory impairment that accompany QA-induced lesions.

Third, and perhaps most importantly, activation of G-protein-coupled mGluR5 receptors may contribute to the pathology of HD due to alterations in its downstream signaling partners. mGluR5 functions by stimulating phospholipase C (PLC)-mediated hydrolysis of phosphatidylinositol 4,5-bisphosphate to generate the second messengers diacylglycerol (DAG) and inositol 1,4,5-trisphosphate (InsP3). The latter binds and activates endoplasmic reticulum (ER)-bound type I InsP3R1 receptors, leading to calcium release from the ER and increasing neuronal calcium load. Using a yeast two-hybrid system with InsP3R1 carboxy-terminal fragment as bait, Tang et al. identified an interaction between InsP3R1 and huntingtin-associated protein 1A (HAP1A), and subsequently demonstrated the formation of a tertiary complex between InsP3R1, HAP1A, and huntingtin [57]. Huntingtin was found to interact with InsP3R1 in the absence HAP1A, although with lower affinity. Furthermore, mutant huntingtin was shown to sensitize InsP3R1 to InsP3, an effect that is not observed with wild-type huntingtin. The authors further demonstrated in primary cultures of rat medium spiny neurons that mutant huntingtin sensitizes InsP3R1 to activation by InsP3 and facilitates ER calcium release in response to threshold concentrations of the group I mGluR agonist DHPG, effects that are not observed with wild-type huntingtin. The functional consequences of these effects were demonstrated in medium spiny neuron cultures obtained from transgenic HD YAC128 mice expressing full-length mutant huntingtin with 128 polyQ repeats, where mutant huntingtin-mediated increase in InsP3R1 activity translates into increased glutamate-induced excitotoxicity compared with wild-type, an effect that is prevented by treatment with the InsP3R1 blockers 2-APB and enoxaparin [48].

C. Mitochondria

The involvement of mitochondria in neurodegenerative diseases in general, and HD in particular, has been long recognized. Clues to the involvement of the mitochondria in HD were provided by animal studies in which treatment with mitochondrial toxins led to neuronal degeneration that mimicked neurodegenerative changes observed in human HD [58, 59]. Several subsequent studies identified striatal-specific mitochondrial defects in postmortem brains of HD patients [60–62]. Mitochondria from lymphoblasts of HD patients were less resistant to induction of the mitochondrial permeability transition on calcium challenge, compared with normal individuals [63]. Similarly, mitochondria from brains of YAC72 transgenic mice are less resistant to calcium challenge than mitochondria from brains of wild-type or YAC18 control animals. This has the effect of activating the mitochondrial permeability transition and causing the release of calcium and apoptotic factors, effectively facilitating calcium dysregulation and the induction of cell death. In addition, huntingtin is found to associate directly with mitochondria, providing a potential mechanism for the disruption of calcium handling by the mitochondria [63–65]. For example, it was suggested that mutant huntingtin may influence mitochondrial calcium handling directly by forming ion channels in the mitochondrial membrane [39].

More direct evidence for a mitochondrial role in potentiating NMDA receptor-mediated excitotoxicity was demonstrated in transgenic YAC models of HD. Using primary medium spiny neurons from YAC46 mice, it was demonstrated that inhibition of the mitochondrial permeability transition with cyclosporine A or bongkrekic acid or boosting of mitochondrial function with coenzyme Q10 substantially diminishes NMDA receptor-mediated cell death [47] and abolishes the observed difference in NMDA receptor-mediated cell death between YAC46 and wild-type. Similar results were obtained from medium spiny neurons from YAC128 mice [48].

V. IMPLICATIONS FOR THERAPY

Collectively, the studies described give rise to a coherent, multifactorial model of mutant huntingtin-mediated alteration of glutamate receptor activity and calcium signaling as a primary contributor to neuronal degeneration in HD (Fig. 15-1), and suggest that modulation of excitotoxicity and factors contributing to it may be a viable therapeutic strategy. In this regard, Slow and colleagues have recently shown that medium spiny neurons from shortstop, a transgenic YAC mouse model expressing exon 1 and exon 2 of mutant huntingtin with 128 polyQ repeats, do not show enhanced susceptibility to NMDA receptor-mediated excitotoxicity *in vitro* or QA-induced striatal lesions *in vivo* compared with wild-type [66]. Remarkably, this attenuation of the enhancement in excitotoxicity was associated with an absence of neurodegeneration and behavioral abnormalities in these animals, suggesting that treatments that attenuate the enhanced excitotoxicity implicated in HD may be therapeutically useful.

In considering assessment of candidate therapeutics for HD, a number of factors have to be taken into account. Substantial resources are required to develop novel compounds from initial screening and identification assays to eventual efficacy testing and approval for use in humans. Thus, selecting compounds with known mechanisms of action that are currently being used in the treatment of other disorders is of great advantage in terms of more rapid applications to humans with HD. Indeed, choosing compounds with proven track records known to target mechanisms implicated in HD will likely increase the chances of success. In addition, many of the disrupted processes implicated in the pathogenesis of HD are readily assessed using *in vitro* assays, and the ability of candidate compounds to offer neuroprotection in such assays should be part of the criteria used in selecting candidate compounds. The compounds chosen for assessment must also be shown to cross the blood–brain barrier to a significant extent, and offer neuroprotection in rodent models of HD. Furthermore, given the considerable amount of time for which the treatment would need to be administered to patients, ease of administration and favorable tolerability profiles are additional criteria that need to be considered in selecting candidate compounds for assessment for the treatment of HD.

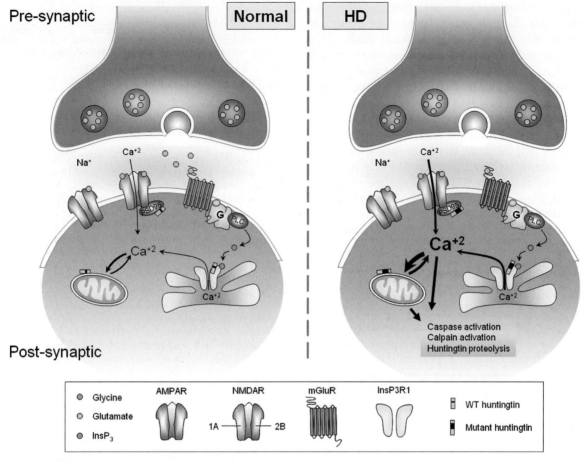

FIGURE 15-1 Glutamate receptor signaling and calcium handling are altered in HD. Under normal conditions, glutamate released from corticostriatal neuron projections binds to NMDA receptors on medium spiny neurons, leading to the opening of the ionotropic receptors and influx of calcium ions from the extracellular milieu. Binding of glutamate to the G-protein-coupled mGluR5 leads to phospholipase C (PLC)-mediated generation of inositol triphosphate (IP3) and the activation of inositol triphosphate receptors (InsP3Rs) and release of calcium ions from the ER. Influx of calcium from the extracellular milieu or release from intracellular stores such as the ER leads to the activation of a variety of pathways that are essential for the normal functioning of neurons, and as such, neuronal calcium levels are tightly regulated by a number of mechanisms, including uptake and buffering by the mitochondria. In HD, NMDA receptor activity is altered by mutant huntingtin, allowing excess calcium influx into the cell. Moreover, mutant huntingtin sensitizes InsP3R1, causing enhanced activity and leading to excess calcium release from the ER into the cytosol in response to mGluR5-mediated activation. Furthermore, mutant huntingtin is associated with defective mitochondrial functioning and increased susceptibility to the activation of mitochondrial permeability transition, leading to defective calcium handling. Collectively, mutant huntingtin causes the disruption of glutamate signaling and calcium handling pathways, leading to the activation of the apoptotic machinery and eventual cell death. See CD-ROM for color image.

Viable pharmacological modulators that meet many of these criteria exist for a number of therapeutically tractable targets nominated by the described studies. For example, glutamate release may be modulated to normalize glutamate receptor activity and abrogate the excitotoxic cell death observed. A number of agents that inhibit glutamate release exist including lamotrigine, a drug shown to be effective in the treatment of a number of conditions including epilepsy and bipolar disorder [67, 68]. Conversely, glutamate signaling may be normalized by upregulation of glutamate uptake processes, a strategy recently shown experimentally to be amenable to therapy in other neurodegenerative conditions [69]. Compounds capable of modulating glutamate uptake by affecting glutamate transporter expression or localization include beta-lactams such as ceftriaxone, a drug used in the treatment of meningitis, and citicoline, a compound originally developed for the treatment of stroke and which is currently being tested for Alzheimer's and Parkinson's diseases [70].

In addition to glutamate release and uptake, modulation of NMDA receptor activity may be therapeutically useful. In this regard, memantine, an activity-dependent NMDA receptor antagonist recently approved for the treatment of Alzheimer's disease, may be of value. Memantine has been shown to be neuroprotective in numerous *in vitro* and *in vivo* excitotoxicity studies [71], and its use in HD may prevent the observed neuronal degeneration by normalizing NMDA receptor activity. For the same reasons, use of mGluR5 receptor and InsP3R1 receptor antagonists may be of therapeutic value in HD, and a number of experimental compounds modulating these targets exist. The experimental compound MPEP, for example, has been shown to inhibit mGluR5 receptor activity, and its use led to a decline in disease progression in the R6/2 transgenic mouse model of HD [72]. Finally, use of compounds that target the mitochondria either by inhibiting the mitochondrial permeability transition or by boosting mitochondrial functions may be therapeutic. Such compounds include rasagiline, a drug recently approved for Parkinson's disease [73], and coenzyme Q10 [74].

Evidence in support of the validity of these targets for therapy is provided by a recent study that demonstrated in primary medium spiny neurons from YAC128 mice that inhibitors of NMDA receptors, mGluR5, InsP3R1, and mitochondrial permeability transition effectively attenuate the enhanced glutamate-mediated excitotoxicity observed *in vitro* [48], providing for the possibility of a pharmaceutically friendly avenue for the modulation of excitotoxicity.

VI. CONCLUDING REMARKS

Despite considerable advances in understanding the pathogenesis of HD, no effective treatments are available and the disease remains fatal to individuals afflicted with it. The described studies strongly implicate deranged neuronal calcium signaling and the accompanying excitotoxicity in the pathogenesis of HD, and nominate a number of targets the modulation of which may cure the disease.

Acknowledgments

M.R.H. and L.A.R. are supported by grants from the Canadian Institutes for Health Research and Michael Smith Foundation for Health Research. M.R.H. and I.B. are supported by grants from the Huntington Disease Society of America and the Hereditary Disease Foundation. M.R.H., L.A.R., and I.B. are supported by grants from the HighQ Foundation. M.R.H. is supported by the Jack and Doris Brown Foundation. I.B. is supported by the National Institutes of Health. M.A.P. is supported by funding from Canadian Institutes for Health Research. M.R.H. holds a Canada Research Chair in Human Genetics.

References

1. Kandel, E. R., Schwartz, J. H., and Jessell, T. M. (2000). "Principles of Neural Science". McGraw-Hill Medical, New York.
2. Choi, D. W., Koh, J. Y., and Peters, S. (1988). Pharmacology of glutamate neurotoxicity in cortical cell culture: attenuation by NMDA antagonists. *J. Neurosci.* **8**, 185–196.
3. Cik, M., Chazot, P. L., and Stephenson, F. A. (1993). Optimal expression of cloned NMDAR1/NMDAR2A heteromeric glutamate receptors: a biochemical characterization. *Biochem. J.* **296 (Pt 3)**, 877–83.
4. Lynch, D. R., and Guttmann, R. P. (2001). NMDA receptor pharmacology: perspectives from molecular biology. *Curr. Drug Targets.* **2**, 215–231.
5. Lynch, D. R., and Guttmann, R. P. (2002). Excitotoxicity: perspectives based on N-methyl-D-aspartate receptor subtypes. *J. Pharmacol. Exp. Ther.* **300**, 717–723.
6. Chen, N., Luo, T., and Raymond, L. A. (1999). Subtype-dependence of NMDA receptor channel open probability. *J. Neurosci.* **19**, 6844–6854.
7. Burnashev, N., Zhou, Z., Neher, E., and Sakmann, B. (1995). Fractional calcium currents through recombinant GluR channels of the NMDA, AMPA and kainate receptor subtypes. *J. Physiol.* **485 (Pt 2)**, 403–418.
8. Grant, E. R., Bacskai, B. J., Pleasure, D. E., Pritchett, D. B., Gallagher, M. J., Kendrick, S. J., Kricka, L. J., and Lynch, D. R. (1997). N-methyl-D-aspartate receptors expressed in a nonneuronal cell line mediate subunit-specific increases in free intracellular calcium. *J. Biol. Chem.* **272**, 647–656.
9. Siman, R., and Noszek, J. C. (1988). Excitatory amino acids activate calpain I and induce structural protein breakdown in vivo. *Neuron* **1**, 279–287.
10. Mills, L. R., and Kater, S. B. (1990). Neuron-specific and state-specific differences in calcium homeostasis regulate the generation and degeneration of neuronal architecture. *Neuron* **4**, 149–63.
11. Goldberg, Y. P., Nicholson, D. W., Rasper, D. M., Kalchman, M. A., Koide, H. B., Graham, R. K., Bromm, M., Kazemi-Esfarjani, P., Thornberry, N. A., Vaillancourt, J. P., and Hayden, M. R. (1996). Cleavage of huntingtin by apopain, a proapoptotic cysteine protease, is modulated by the polyglutamine tract. *Nat. Genet.* **13**, 442–449.
12. Lazarewicz, J. W., Salinska, E., and Wroblewski, J. T. (1992). NMDA receptor-mediated arachidonic acid release in neurons: role in signal transduction and pathological aspects. *Adv. Exp. Med. Biol.* **318**, 73–89.
13. Hartley, D. M., Kurth, M. C., Bjerkness, L., Weiss, J. H., and Choi, D. W. (1993). Glutamate receptor-induced 45Ca2+ accumulation in cortical cell culture correlates with subsequent neuronal degeneration. *J. Neurosci.* **13**, 1993–2000.
14. Eimerl, S., and Schramm, M. (1994). The quantity of calcium that appears to induce neuronal death. *J. Neurochem.* **62**, 1223–6.
15. Michaels, R. L., and Rothman, S. M. (1990). Glutamate neurotoxicity in vitro: antagonist pharmacology and intracellular calcium concentrations. *J. Neurosci.* **10**, 283–292.
16. Dubinsky, J. M., and Rothman, S. M. (1991). Intracellular calcium concentrations during "chemical hypoxia" and excitotoxic neuronal injury. *J. Neurosci.* **11**, 2545–2551.
17. Marcoux, F. W., Weber, M. L., Probert, A. W., Jr., and Dominick, M. A. (1992). Hypoxic neurodegeneration in culture: calcium influx, electron microscopy, and neuroprotection with excitatory amino acid antagonists. *Ann. N. Y. Acad. Sci.* **648**, 303–305.
18. Stout, A. K., and Reynolds, I. J. (1999). High-affinity calcium indicators underestimate increases in intracellular calcium concentrations associated with excitotoxic glutamate stimulations. *Neuroscience* **89**, 91–100.

19. Sattler, R., and Tymianski, M. (2000). Molecular mechanisms of calcium-dependent excitotoxicity. *J. Mol. Med.* **78**, 3–13.
20. Tymianski, M., Charlton, M. P., Carlen, P. L., and Tator, C. H. (1993). Source specificity of early calcium neurotoxicity in cultured embryonic spinal neurons. *J. Neurosci.* **13**, 2085–2104.
21. Sattler, R., Charlton, M. P., Hafner, M., and Tymianski, M. (1998). Distinct influx pathways, not calcium load, determine neuronal vulnerability to calcium neurotoxicity. *J. Neurochem.* **71**, 2349–2364.
22. Hardingham, G. E., Fukunaga, Y., and Bading, H. (2002). Extrasynaptic NMDARs oppose synaptic NMDARs by triggering CREB shut-off and cell death pathways. *Nat. Neurosci.* **5**, 405–414.
23. Nicholls, D. G., Budd, S. L., Ward, M. W., and Castilho, R. F. (1999). Excitotoxicity and mitochondria. *Biochem. Soc. Symp.* **66**, 55–67.
24. Manfredi, G., and Beal, M. F. (2000). The role of mitochondria in the pathogenesis of neurodegenerative diseases. *Brain Pathol.* **10**, 462–472.
25. Kiedrowski, L., and Costa, E. (1995). Glutamate-induced destabilization of intracellular calcium concentration homeostasis in cultured cerebellar granule cells: role of mitochondria in calcium buffering. *Mol. Pharmacol.* **47**, 140–147.
26. Wang, G. J., and Thayer, S. A. (1996). Sequestration of glutamate-induced Ca2+ loads by mitochondria in cultured rat hippocampal neurons. *J. Neurophysiol.* **76**, 1611–1621.
27. Peng, T. I., and Greenamyre, J. T. (1998). Privileged access to mitochondria of calcium influx through N-methyl-D-aspartate receptors. *Mol. Pharmacol.* **53**, 974–980.
28. Bernardi, P., and Petronilli, V. (1996). The permeability transition pore as a mitochondrial calcium release channel: a critical appraisal. *J. Bioenerg. Biomembr.* **28**, 131–138.
29. Ichas, F., and Mazat, J. P. (1998). From calcium signaling to cell death: two conformations for the mitochondrial permeability transition pore. Switching from low- to high-conductance state. *Biochim. Biophys. Acta.* **1366**, 33–50.
30. Coyle, J. T., and Schwarcz, R. (1976). Lesion of striatal neurones with kainic acid provides a model for Huntington's chorea. *Nature* **263**, 244–246.
31. McGeer, E. G., and McGeer, P. L. (1976). Duplication of biochemical changes of Huntington's chorea by intrastriatal injections of glutamic and kainic acids. *Nature* **263**, 517–519.
32. Beal, M. F., Kowall, N. W., Ellison, D. W., Mazurek, M. F., Swartz, K. J., and Martin, J. B. (1986). Replication of the neurochemical characteristics of Huntington's disease by quinolinic acid. *Nature* **321**, 168–171.
33. Beal, M. F., Kowall, N. W., Swartz, K. J., Ferrante, R. J., and Martin, J. B. (1988). Systemic approaches to modifying quinolinic acid striatal lesions in rats. *J. Neurosci.* **8**, 3901–3908.
34. DiFiglia, M. (1990). Excitotoxic injury of the neostriatum: a model for Huntington's disease. *Trends Neurosci.* **13**, 286–289.
35. Beal, M. F., Ferrante, R. J., Swartz, K. J., and Kowall, N. W. (1991). Chronic quinolinic acid lesions in rats closely resemble Huntington's disease. *J. Neurosci.* **11**, 1649–1659.
36. Popoli, P., Pezzola, A., Domenici, M. R., Sagratella, S., Diana, G., Caporali, M. G., Bronzetti, E., Vega, J., and Scotti de Carolis, A. (1994). Behavioral and electrophysiological correlates of the quinolinic acid rat model of Huntington's disease in rats. *Brain Res. Bull.* **35**, 329–335.
37. Furtado, J. C., and Mazurek, M. F. (1996). Behavioral characterization of quinolinate-induced lesions of the medial striatum: relevance for Huntington's disease. *Exp. Neurol.* **138**, 158–168.
38. Shear, D. A., Dong, J., Gundy, C. D., Haik-Creguer, K. L., and Dunbar, G. L. (1998). Comparison of intrastriatal injections of quinolinic acid and 3-nitropropionic acid for use in animal models of Huntington's disease. *Prog. Neuropsychopharmacol. Biol. Psychiatry* **22**, 1217–1240.
39. Bezprozvanny, I., and Hayden, M. R. (2004). Deranged neuronal calcium signaling and Huntington disease. *Biochem. Biophys. Res. Commun.* **322**, 1310–1317.
40. McGeer, E. G., McGeer, P. L., and Singh, K. (1978). Kainate-induced degeneration of neostriatal neurons: dependency upon corticostriatal tract. *Brain Res.* **139**, 381–383.
41. Young, A. B., Greenamyre, J. T., Hollingsworth, Z., Albin, R., D'Amato, C., Shoulson, I., and Penney, J. B. (1988). NMDA receptor losses in putamen from patients with Huntington's disease. *Science* **241**, 981–983.
42. Landwehrmeyer, G. B., Standaert, D. G., Testa, C. M., Penney, J. B., Jr., and Young, A. B. (1995). NMDA receptor subunit mRNA expression by projection neurons and interneurons in rat striatum. *J. Neurosci.* **15**, 5297–5307.
43. Chen, N., Luo, T., Wellington, C., Metzler, M., McCutcheon, K., Hayden, M. R., and Raymond, L. A. (1999). Subtype-specific enhancement of NMDA receptor currents by mutant huntingtin. *J. Neurochem.* **72**, 1890–1898.
44. Zeron, M. M., Chen, N., Moshaver, A., Lee, A. T., Wellington, C. L., Hayden, M. R., and Raymond, L. A. (2001). Mutant huntingtin enhances excitotoxic cell death. *Mol. Cell Neurosci.* **17**, 41–53.
45. Hodgson, J. G., Agopyan, N., Gutekunst, C. A., Leavitt, B. R., LePiane, F., Singaraja, R., Smith, D. J., Bissada, N., McCutcheon, K., Nasir, J., Jamot, L., Li, X. J., Stevens, M. E., Rosemond, E., Roder, J. C., Phillips, A. G., Rubin, E. M., Hersch, S. M., and Hayden, M. R. (1999). A YAC mouse model for Huntington's disease with full-length mutant huntingtin, cytoplasmic toxicity, and selective striatal neurodegeneration. *Neuron* **23**, 181–192.
46. Zeron, M. M., Hansson, O., Chen, N., Wellington, C. L., Leavitt, B. R., Brundin, P., Hayden, M. R., and Raymond, L. A. (2002). Increased sensitivity to N-methyl-D-aspartate receptor-mediated excitotoxicity in a mouse model of Huntington's disease. *Neuron* **33**, 849–860.
47. Zeron, M. M., Fernandes, H. B., Krebs, C., Shehadeh, J., Wellington, C. L., Leavitt, B. R., Baimbridge, K. G., Hayden, M. R., and Raymond, L. A. (2004). Potentiation of NMDA receptor-mediated excitotoxicity linked with intrinsic apoptotic pathway in YAC transgenic mouse model of Huntington's disease. *Mol. Cell. Neurosci.* **25**, 469–479.
48. Tang, T. S., Slow, E., Lupu, V., Stavrovskaya, I. G., Sugimori, M., Llinas, R., Kristal, B. S., Hayden, M. R., and Bezprozvanny, I. (2005). Disturbed Ca2+ signaling and apoptosis of medium spiny neurons in Huntington's disease. *Proc. Natl. Acad. Sci. U. S. A.* **102**, 2602–2607.
49. Shehadeh, J., Fernandes, H. B., Zeron Mullins, M. M., Graham, R. K., Leavitt, B. R., Hayden, M. R., and Raymond, L. A. (2006). Striatal neuronal apoptosis is preferentially enhanced by NMDA receptor activation in YAC transgenic mouse model of Huntington disease. *Neurobiol. Dis.* **21**, 392–403.
50. Kerner, J. A., Standaert, D. G., Penney, J. B., Jr., Young, A. B., and Landwehrmeyer, G. B. (1997). Expression of group one metabotropic glutamate receptor subunit mRNAs in neurochemically identified neurons in the rat neostriatum, neocortex, and hippocampus. *Brain Res. Mol. Brain Res.* **48**, 259–269.
51. Tallaksen-Greene, S. J., Kaatz, K. W., Romano, C., and Albin, R. L. (1998). Localization of mGluR1a-like immunoreactivity and mGluR5-like immunoreactivity in identified populations of striatal neurons. *Brain Res.* **780**, 210–217.
52. Calabresi, P., Centonze, D., Pisani, A., and Bernardi, G. (1999). Metabotropic glutamate receptors and cell-type-specific vulnerability in the striatum: implication for ischemia and Huntington's disease. *Exp. Neurol.* **158**, 97–108.
53. Pisani, A., Gubellini, P., Bonsi, P., Conquet, F., Picconi, B., Centonze, D., Bernardi, G., and Calabresi, P. (2001). Metabotropic glutamate receptor 5 mediates the potentiation of N-methyl-D-aspartate responses in medium spiny striatal neurons. *Neuroscience* **106**, 579–587.

54. Bruno, V., Ksiazek, I., Battaglia, G., Lukic, S., Leonhardt, T., Sauer, D., Gasparini, F., Kuhn, R., Nicoletti, F., and Flor, P. J. (2000). Selective blockade of metabotropic glutamate receptor subtype 5 is neuroprotective. *Neuropharmacology* **39**, 2223–2230.
55. Orlando, L. R., Alsdorf, S. A., Penney, J. B., Jr., and Young, A. B. (2001). The role of group I and group II metabotropic glutamate receptors in modulation of striatal NMDA and quinolinic acid toxicity. *Exp. Neurol.* **167**, 196–204.
56. Popoli, P., Pintor, A., Tebano, M. T., Frank, C., Pepponi, R., Nazzicone, V., Grieco, R., Pezzola, A., Reggio, R., Minghetti, L., De Berardinis, M. A., Martire, A., Potenza, R. L., Domenici, M. R., and Massotti, M. (2004). Neuroprotective effects of the mGlu5R antagonist MPEP towards quinolinic acid-induced striatal toxicity: involvement of pre- and post-synaptic mechanisms and lack of direct NMDA blocking activity. *J. Neurochem.* **89**, 1479–1489.
57. Tang, T. S., Tu, H., Chan, E. Y., Maximov, A., Wang, Z., Wellington, C. L., Hayden, M. R., and Bezprozvanny, I. (2003). Huntingtin and huntingtin-associated protein 1 influence neuronal calcium signaling mediated by inositol-(1,4,5) triphosphate receptor type 1. *Neuron* **39**, 227–239.
58. Beal, M. F., Brouillet, E., Jenkins, B., Henshaw, R., Rosen, B., and Hyman, B. T. (1993). Age-dependent striatal excitotoxic lesions produced by the endogenous mitochondrial inhibitor malonate. *J. Neurochem.* **61**, 1147–1150.
59. Brouillet, E., Hantraye, P., Ferrante, R. J., Dolan, R., Leroy-Willig, A., Kowall, N. W., and Beal, M. F. (1995). Chronic mitochondrial energy impairment produces selective striatal degeneration and abnormal choreiform movements in primates. *Proc. Natl. Acad. Sci. U. S. A.* **92**, 7105–7109.
60. Gu, M., Gash, M. T., Mann, V. M., Javoy-Agid, F., Cooper, J. M., and Schapira, A. H. (1996). Mitochondrial defect in Huntington's disease caudate nucleus. *Ann. Neurol.* **39**, 385–389.
61. Browne, S. E., Bowling, A. C., MacGarvey, U., Baik, M. J., Berger, S. C., Muqit, M. M., Bird, E. D., and Beal, M. F. (1997). Oxidative damage and metabolic dysfunction in Huntington's disease: selective vulnerability of the basal ganglia. *Ann. Neurol.* **41**, 646–653.
62. Tabrizi, S. J., Cleeter, M. W., Xuereb, J., Taanman, J. W., Cooper, J. M., and Schapira, A. H. (1999). Biochemical abnormalities and excitotoxicity in Huntington's disease brain. *Ann. Neurol.* **45**, 25–32.
63. Panov, A. V., Gutekunst, C. A., Leavitt, B. R., Hayden, M. R., Burke, J. R., Strittmatter, W. J., and Greenamyre, J. T. (2002). Early mitochondrial calcium defects in Huntington's disease are a direct effect of polyglutamines. *Nat. Neurosci.* **5**, 731–736.
64. Gutekunst, C. A., Li, S. H., Yi, H., Ferrante, R. J., Li, X. J., and Hersch, S. M. (1998). The cellular and subcellular localization of huntingtin-associated protein 1 (HAP1): comparison with huntingtin in rat and human. *J. Neurosci.* **18**, 7674–7686.
65. Choo, Y. S., Johnson, G. V., MacDonald, M., Detloff, P. J., and Lesort, M. (2004). Mutant huntingtin directly increases susceptibility of mitochondria to the calcium-induced permeability transition and cytochrome c release. *Hum. Mol. Genet.* **13**, 1407–1420.
66. Slow, E. J., Graham, R. K., Osmand, A. P., Devon, R. S., Lu, G., Deng, Y., Pearson, J., Vaid, K., Bissada, N., Wetzel, R., Leavitt, B. R., and Hayden, M. R. (2005). Absence of behavioral abnormalities and neurodegeneration in vivo despite widespread neuronal huntingtin inclusions. *Proc. Natl. Acad. Sci. U. S. A* **102**, 11402–11407.
67. Richens, A., and Yuen, A. W. (1991). Overview of the clinical efficacy of lamotrigine. *Epilepsia* **32 Suppl 2**, S13–S16.
68. Hurley, S. C. (2002). Lamotrigine update and its use in mood disorders. *Ann. Pharmacother.* **36**, 860–873.
69. Rothstein, J. D., Patel, S., Regan, M. R., Haenggeli, C., Huang, Y. H., Bergles, D. E., Jin, L., Dykes, H. M., Vidensky, S., Chung, D. S., Toan, S. V., Bruijn, L. I., Su, Z. Z., Gupta, P., and Fisher, P. B. (2005). Beta-lactam antibiotics offer neuroprotection by increasing glutamate transporter expression. *Nature* **433**, 73–77.
70. Adibhatla, R. M., and Hatcher, J. F. (2005). Cytidine 5′-diphosphocholine (CDP-choline) in stroke and other CNS disorders. *Neurochem. Res.* **30**, 15–23.
71. Parsons, C. G., Danysz, W., and Quack, G. (1999). Memantine is a clinically well tolerated N-methyl-D-aspartate (NMDA) receptor antagonist–a review of preclinical data. *Neuropharmacology* **38**, 735–767.
72. Schiefer, J., Sprunken, A., Puls, C., Luesse, H. G., Milkereit, A., Milkereit, E., Johann, V., and Kosinski, C. M. (2004). The metabotropic glutamate receptor 5 antagonist MPEP and the mGluR2 agonist LY379268 modify disease progression in a transgenic mouse model of Huntington's disease. *Brain Res.* **1019**, 246–254.
73. Mandel, S., Weinreb, O., Amit, T., and Youdim, M. B. (2005). Mechanism of neuroprotective action of the anti-Parkinson drug rasagiline and its derivatives. *Brain Res. Brain Res. Rev.* **48**, 379–387.
74. Beal, M. F. (2004). Therapeutic effects of coenzyme Q10 in neurodegenerative diseases. *Methods Enzymol.* **382**, 473–487.

Huntington's Disease-like 2

RUSSELL L. MARGOLIS, SUSAN E. HOLMES, DOBRILA D. RUDNICKI,
ELIZABETH O'HEARN, CHRISTOPHER A. ROSS, OLGA PLETNIKOVA,
JUAN C. TRONCOSO

Laboratory of Genetic Neurobiology, Division of Neurobiology, Department of Psychiatry; Department of Neurology; Division of Neuropathology, Department of Neuroscience; Department of Pathology; and Program in Cellular and Molecular Medicine, Johns Hopkins University School of Medicine, Baltimore, Maryland 21287

I. Introduction
II. Detection of the HDL2 Expansion Mutation
III. HDL2 at the Bedside
IV. HDL2 and Neuroacanthocytes
V. Neuropathology of HDL2
VI. Epidemiology of HDL2
VII. *JPH3* and HDL2: Phenotype–Genotype Relationship
VIII. HDL2 Is Not a Polyglutamine Disease
IX. The HDL2 Locus and Junctophilin-3: Structure and Function
X. HDL2 and Toxic Transcripts
XI. HDL2 and Intranuclear Protein Aggregates
XII. Conclusion
Acknowledgments
References

Huntington's disease-like 2 (HDL2), first described in 2001, is a rare autosomal dominant adult-onset progressive neurodegenerative disorder. Clinically and pathologically, HDL2 is nearly indistinguishable from Huntington's disease (HD), and is characterized, like HD, by selective striatal degeneration and neuronal intranuclear inclusions. Like HD, the cause of HDL2 is an expansion of a polymorphic CAG/CTG repeat expansion with the threshold for disease at about 40 triplets. The HDL2 repeat is located on chromosome 16q24.3, in a variably spliced exon of the gene *junctophilin-3* (*JPH3*). Unlike HD, the HDL2 repeat does not encode polyglutamine, but instead is variably in-frame to encode polyalanine or polyleucine, or falls within the 3' untranslated region (UTR). The mechanism by which the HDL2 expansion mutation leads to nearly the same phenotype as the HD polyglutamine expansion is unknown, but may have similarities to that of myotonic dystrophy types 1 and 2.

I. INTRODUCTION

The HD gene was discovered 12 years ago, and since then, intensive investigations in dozens of laboratories have generated multiple hypotheses of HD pathogenesis. The key issue now is how to choose among these hypotheses to select potential targets for therapeutic agents. One approach is to find other diseases that mimic the HD phenotype, and then determine pathways that are abnormal in both diseases. This approach has proven of great value in Alzheimer's disease (AD). Mutations in genes encoding presenilin-1, presenilin-2, and amyloid-precursor protein (APP) cause rare autosomal dominant forms of AD, and the pathway defined by these genes is now a prime target for the development of therapeutic agents [1].

What would be the characteristics of a disorder similar enough to HD to apply this type of comparative approach? The most obvious answer is a disorder that is also caused by expression of a protein with an expanded polyglutamine repeat. Indeed, one important clue to HD pathogenesis is the fact that all eight other known polyglutamine diseases (spinocerebellar ataxia [SCA] types 1, 2, 3, 6, 7, and 17; dentatorubropallidoluyisan atrophy; spinal and bulbar atrophy) are primarily, if not exclusively, diseases of neurodegeneration. The presence of nuclear polyglutamine-containing protein aggregates in brain in most of these disorders stimulated additional investigations into common mechanisms [2, 3]. The general hypothesis that has emerged is that expanded polyglutamine tracts are selectively toxic to neurons. Nonetheless, the pattern of neurodegeneration in each of the polyglutamine diseases is distinct. In the case of HD, the outstanding pathological feature is selective neurodegeneration of striatal medium spiny neurons [4–6], yet why these neurons are so prominently lost in HD remains unknown. The most frequently invoked mechanism is excitotoxicity, a form of cell death induced by excess activation of glutamate receptors [7].

A potential role of excitotoxicity in HD pathogenesis has been postulated, but not proven, for more than 25 years. Medium spiny neurons receive substantial glutamatergic input from corticostriatal projection neurons (which themselves receive glutamatergic input). High levels of huntingtin have been detected in at least some of these projecting neurons (particularly in cortical layer V), suggesting that the primary damage could be to these neurons, leading to excess glutamate release at their striatal terminals and subsequent secondary striatal neuronal loss [8]. Many investigations have shown that injection of glutamate agonists into striatum damages striatal neurons with some selectivity [8–10], forming the basis of animal models of HD prior to the discovery of the HD gene. *N*-Methyl-D-aspartate (NMDA) receptors appear selectively decreased in presymptomatic HD patients [11], providing more direct evidence of HD-related changes to striatal glutamatergic signal transduction. HD transgenic mice also have altered glutamate receptor expression [12]. Cell studies have demonstrated that medium spiny neurons have increased sensitivity to NMDA receptor-mediated neurotoxicity [13] and that mutant huntingtin may affect the functional properties of glutamate receptors and glutamate uptake mechanisms [14, 15]. As predicted from the excitotoxicity and bioenergetic hypotheses, both the mitochondrial cofactor coenzyme Q (CoQ) and the NMDA antagonist remacemide slowed disease progression in HD transgenic mice [16], though little effect of these agents was detected in a large HD clinical trial [17]. The glutamatergic antagonistic riluzole [18] similarly slowed disease progression in HD transgenic mice.

Another possibility is that medium spiny neurons may be particularly vulnerable to inhibition of mitochondrial respiratory processes [19], potentially reflecting the high metabolic demand required to maintain their large membrane potential [5]. Systemic administration of mitochondrial inhibitors to animals can result, depending on the species and the choice of inhibitor, in selective striatal neurodegeneration [20–22]. Experiments using corticostriatal slices have demonstrated that medium spiny neurons respond to pharmacologically induced loss of mitochondrial membrane potential with massive increases in cytosolic calcium [23]. There is also evidence of mitochondrial abnormalities in the striatum of HD patients [24]. Other factors that have been implicated in selective striatal neuronal vulnerability include dopamine, through modulation of intrastriatal glutamate levels and conversion to toxic metabolites [25, 26]; stress and corticosteroids [5]; and variations in the level of expression of calcium buffering proteins [27, 28]. The difference in neuronal loss between HD and the other polyglutamine diseases [29–31] also strongly suggests that an aspect of striatal vulnerability in HD derives from properties of the huntingtin protein distinct from the polyglutamine expansion. These properties might include higher or lower levels of huntingtin expression in vulnerable neurons, an issue still unresolved [8, 32]; a mutation-induced loss of a striatal-specific function of the normal protein, such as induction of a trophic factor [33]; or interaction with other proteins that are themselves differentially expressed in striatal medium spiny neurons, similar to the interaction of ataxin-1 with leucine-rich acidic nuclear protein (LANP), a protein expressed predominately in cerebellar Purkinje cells [34].

Do true HD-like disorders exist? Relatively few diseases are characterized by autosomal dominant adult-onset progressive degeneration of medium spiny neurons. Overall, in carefully examined individuals with an HD-like syndrome or pathology in Europe and North America, as few as 1% do not have the *HD* mutation [35–38]. Nonetheless, we were able to identify four individuals from our HD clinic with a familial progressive neurodegenerative disorder clinically indistinguishable from HD who had neither the HD mutation nor any other mutation known to confer an HD-like syndrome [39].

II. DETECTION OF THE HDL2 EXPANSION MUTATION

To determine the causative mutation in HDL2, we screened genomic DNA of these four individuals with a modified version of the repeat expansion detection (RED) assay [40]. This assay detects the presence of CAG/CTG repeats longer than about 40 triplets in genomic DNA. Interpretation is complicated by the existence of expanded but not pathological CAG/CTG repeats in 20–25% of the population, and the RED assay itself does not provide information on the locus of the expanded repeat. However, strategies have been developed in which the RED assay can be used to determine the precise repeat within the genome that is expanded without resorting to positional cloning [41]. It has been used to find the expansions that cause spinocerebellar ataxia type 8 [42] and type 12 [43]. In the first of the four probands investigated, a CAG/CTG expansion was detected by the RED assay that could not be explained by any known repeat expansion. All affected members of the proband's family also carried the expansion [44]. We then used the RED assay and genomic library screening to find the locus of the repeat, enabling us to develop a simple polymerase chain reaction (PCR) assay to determine the presence or absence of the repeat expansion in any given individual [45]. The PCR results confirmed that the repeat expansion completely segregated with disease in the family. We used the PCR assay to determine that one of the other four individuals from our clinic with an HD-like familial disorder, who had also shown an expansion by the RED assay, was from another branch of the proband's family.

III. HDL2 AT THE BEDSIDE

In any given patient, HDL2 is clinically indistinguishable from HD, and the range of clinical presentations of HDL2 and HD is the same. The index HDL2 family [46] and some members of a second American HDL2 family [47, 48] present in the fourth decade with prominent weight loss, poor coordination, rigidity, dysarthria, hyperreflexia, bradykinesia, dystonia, and tremor. Cerebellar signs are absent, and chorea and eye movement abnormalities are minimal. Psychiatric disturbances, including depression, irritability, and apathy, are common, and dementia is universal. Over 10–15 years, affected individuals become profoundly demented, rigid, and virtually immobile, with death from nonspecific complications following thereafter (Table 16-1).

With detection of additional HDL2 families, it has become clear that most individuals outside of the index

TABLE 16-1 Signs and Symptoms in the HDL2 Index Family

Motor signs	Frequency	Nonmotor signs	Frequency
Dysarthria	100%	Weight loss	100%
Rigidity	100%	Dementia	100%
Hyperreflexia	100%	Psychiatric syndromes	100%
Action tremor	88%		
Bradykinesia or reduced movement	88%		
Gait abnormality	88%		
Chorea	62%		
Dystonia	62%		

Source: Derived from Margolis et al. [46].

pedigree have a clinical syndrome that corresponds to typical HD. Onset tends to be at a somewhat older age than in the index family, with prominent chorea and dysmetric saccades and less prominent dystonia, bradykinesia, tremor, hyperreflexia, and dysarthria. Dementia and psychiatric disturbances are clearly still a part of the syndrome, but at least the dementia, and perhaps the entire clinical presentation, appears to evolve more slowly. Video clips of HDL2 patients have been deposited with the Movement Disorder Society [49].

IV. HDL2 AND NEUROACANTHOCYTES

Chorea-acanthocytosis is an autosomal recessive disorder caused by mutations in *VPS13A*, and characterized by chorea, dystonia, seizures, cognitive and psychiatric abnormalities, myopathy, and acanthocytosis (abnormally spiculated red blood cells). Neuroimaging and neuropathological findings demonstrate degeneration of the caudate and putamen. The McLeod syndrome is a multisystem X-linked disorder caused by mutations in the gene *Xk*, which is also characterized in part by basal ganglia degeneration and acanthocytosis [50]. Interestingly, acanthocytes were detected in all three individuals of the HDL2 family described by Walker and colleagues, in one of three affected family members from a Mexican HDL2 family, and in neither of two individuals from the index family nor an individual from a fourth pedigree [48]. The detection of acanthocytes in some HDL2 cases may be coincidental or secondary to nonspecific factors. However, the *JPH3* protein product (see Section VII), like the products of

VPS13 and *Xk*, is associated with plasma membranes, and can be detected in red blood cells (D. Rudnicki, unpublished data). It is therefore conceivable that the expansion mutation in *JPH3* might disrupt membranes and lead to acanthocytosis.

V. NEUROPATHOLOGY OF HDL2

HDL2 is indistinguishable from HD on MRI scan (Fig. 16-1). As in HD, images of moderately advanced cases show extensive caudate atrophy, moderate cortical atrophy, and little evidence of atrophy elsewhere in the brain.

Four HDL2 brains have now been examined pathologically, with consistent findings. The most thoroughly examined HDL2 brain was from a member of the index family. Disease onset was in midlife, followed by a typical disease course ending in death in a nursing home about 20 years after disease onset. Repeat length was 53 triplets (see below). On gross examination (Fig. 16-2A), brain weight was 1064 g, with mild atrophy of frontal, temporal, mesial parietal, and mesial occipital gyri. There were no abnormalities of basal blood vessels or of the cranial nerves. Viewed in coronal sections, mild atrophy of cortical gray matter and dilation of the ventricular system were evident, with no evidence of white matter demyelination. Of particular importance, and as anticipated from MRI scans, the head of the caudate and the putamen were severely atrophic, with only mild atrophy of the globus pallidus. The pons, medulla, cerebellum, thalamus, hypothalamus, hippocampus, amygdala, entorhinal cortex, substantia nigra, and subthalamic nucleus were normal.

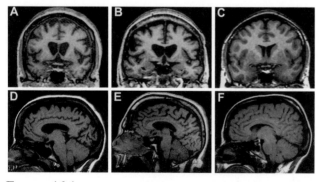

FIGURE 16-1 HDL2 is identical to HD on MRI scans. (A, D) HDL2. MRI was performed at age 36, at a disease duration of 10 years. (B, E) HD. MRI performed at age 48, at a disease duration of 12 years. (C, F) Normal control. MRI performed at age 43. Note the atrophy of the striatum and cerebral cortex in the HDL2 and HD cases, with relative sparing of the cerebellum and brainstem. Reprinted, with permission, from [46], *Annals of Neurology*, copyright Wiley–Liss Inc, 2001.

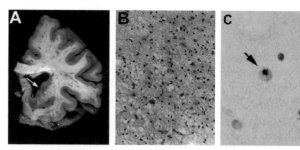

FIGURE 16-2 HDL2 pathology. (A) Gross pathology. Prominent striatal atrophy (arrow) and moderate cortical atrophy. (B) Microscopic pathology of the caudate. Neuronal degeneration, astrocytic gliosis, and vacuolization. (C) Intranuclear inclusions. Staining by 1C2 antibody (arrow). Reprinted, with permission, from [46], *Annals of Neurology*, copyright Wiley–Liss Inc, 2001.

Microscopic examination of the caudate revealed severe neuronal degeneration affecting small neurons more than large neurons, reactive astrocytosis, and neuropil vacuolation, with a dorsal-to-ventral gradient of decreasing severity similar to that observed in most cases of HD (Fig. 16-2B). The putamen was affected, though less severely, with the same gradient noted in the caudate, and moderate neuronal loss and astrocytosis were noted in the globus pallidum. In the substantia nigra, which was grossly intact and not depigmented, microscopic examination showed moderate neuronal degeneration accompanied by pigment incontinence but no Lewy bodies. Neither β-amyloid deposits nor neurofibrillary tangles were detected in neocortex, entorhinal cortex, or hippocampus using Hirano silver stains.

Perhaps the most striking finding on pathological examination was the presence of intranuclear inclusions, more common in cortex than in striatum, stained by anti-ubiquitin antibodies and 1C2 antibodies (Fig. 16-2C), but not by anti-huntingtin antibodies. Staining with 1C2 antibodies was of particular interest, as this reagent has been used as a selective stain for proteins or polypeptide fragments containing expanded tracts of polyglutamine, both in blots and in tissue. TorsinA, a molecular chaperone in which mutations cause dystonia [47], and TATA-binding protein (TBP) (Rudnicki, unpublished data) have also been detected in these inclusions. Recently, RNA inclusions have also been detected in HDL2 brain sections (see below).

VI. EPIDEMIOLOGY OF HDL2

In North America, 28 genetically documented cases of HDL2 have been detected in 12 pedigrees, with the index pedigree accounting for nearly half of all cases.

In 538 individuals from North America with an HD-like syndrome who did not have the HD mutation, 6 cases of HDL2 were detected [51], suggesting that as few as 1% of all HD-like cases have HDL2. This estimate may be somewhat low, as 300 of the HD-like cases were anonymous samples sent to a commercial laboratory for HD testing. It is likely that many, if not most, of these cases had tardive dyskinesia or other syndromes only remotely resembling HD. In Japan, no cases of HDL2 have been detected [51, 52], though the frequency of HDL2 in Asia outside of Japan is unknown. No cases have been detected in screens of Europeans, with the exception of one individual from Morocco [53, 54]. Nonetheless, it is clear that HDL2 is rare in North America and rare or nonexistent in Europe and Japan.

Two striking aspects of HDL2 epidemiology have emerged. First, the mutation has been detected only in individuals of African ancestry. All American HDL2 families of known ethnicity either completely or partially identify themselves as African American. The Moroccan patient with HDL2 was originally from the southern part of Morocco, an area populated primarily by individuals of African descent. The single Mexican pedigree with HDL2 is from a region originally colonized by Africans. Second, HDL2 is nearly as common as HD among South Africans of African descent. Seventy-eight of ninety-three South Africans of European ancestry (84%) referred for HD genetic testing to the University of Witwatersand had the HD mutation, and none had the HDL2 mutation. On the other hand, only 21 of 56 patients with African or mixed African-European ancestry referred for HD testing had the HD mutation (38%), whereas 15 of 56 (27%) had the HDL2 mutation. Preliminary evidence also suggests that North American and South African HDL2 cases share a common haplotype [51; A. Krause, personal communication]. This finding is consistent with an African, and potentially a West African, origin for the HDL2 mutation.

VII. *JPH3* AND HDL2: PHENOTYPE–GENOTYPE RELATIONSHIP

The range of the *JPH3* repeat length in the general population and in individuals with HDL2 is strikingly similar to the range of the CAG repeat length in *huntingtin (htt)*. In unaffected individuals, the normal length of the *JPH3* repeat varies from 6 to 28 triplets with a mode of 14 triplets, compared with the normal range in *htt* of 4–35. *JPH3* repeat lengths associated with HDL2 range from 40 to 58, whereas htt repeat lengths associated with HD range from 36 to >200 triplets, with the vast majority about 40–60 triplets in length. In HD, repeats in the range 36–39 are incompletely penetrant. Similarly, *JPH3* repeat expansions in the range 40–45 may also be incompletely penetrant, based on the evidence of one individual from an HDL2 pedigree with a repeat expansion of 44 and no clear signs of HDL2 by age 65 [51].

It is not possible to exclude phenotypic manifestations of shorter *JPH3* repeats [51]. A 48-year-old woman with *JPH3* alleles containing 12 and 33 triplets acutely developed a nonprogressive cerebellar disorder after a hospitalization for uncontrolled diabetes mellitus type II. Her son, with alleles of 35 and 14 triplets, developed Cogan's syndrome, an autoimmune disease characterized by optic and audiovestibular findings, at age 25. A detailed neurological examination at age 30 revealed jerky horizontal and vertical gaze and saccades, possible dysdiadochokinesis, and moderate unsteadiness with tandem walk. It is unclear whether the HDL2 repeat expansion has any relationship to the neurological syndromes developed by these two patients. One intriguing possibility is that *JPH3* repeats of intermediate length do not themselves cause neurological damage, but may increase the vulnerability to damage from other insults. However, the shift in repeat length from mother to son does provide evidence for unstable vertical transmission of repeat lengths as short as 33 triplets [51]. Similarly, *htt* repeat lengths of 27 to 35 triplets have never been associated with disease, but are unstable on vertical transmission, typically from father to child.

Repeat instability during vertical transmission and the correlation between longer repeat length and younger age of onset are characteristic features of repeat expansion diseases. The five-triplet variability of expanded repeat length in two sibships of the index HDL2 family demonstrates that expanded repeats are not stable on vertical transmission. The few examples in which the length of an HDL2 expanded repeat is known in both parent and child show a tendency toward a modest increase in repeat length during vertical transmission, consistent with the suggestion of anticipation in the index pedigree, though insufficient data are available to meaningfully quantify the change or determine if, as in HD, the likelihood of expansion during transmission varies according to the sex of the transmitting parent [44, 51]. It is, however, clearly apparent that longer *JPH3* repeat lengths strongly correlate with a younger age of onset, with a slope similar to that observed in HD (Fig. 16-3) [51]. We predict that evidence for anticipation will emerge with the accumulation of data on age of onset and repeat length in parent–offspring pairs.

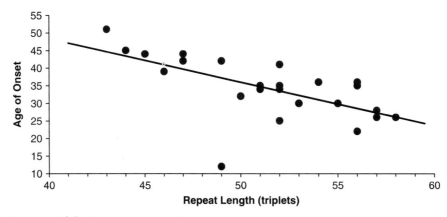

FIGURE 16-3 Younger age of HDL2 onset is associated with longer repeat lengths. $N = 24$, $R = 0.62$, $r^2 = 0.39$, $P = 0.0011$. The relationship is quantitatively similar to that observed in HD. Reprinted, with permission, from [51], *Annals of Neurology*, copyright Wiley–Liss Inc, 2004.

VIII. HDL2 IS NOT A POLYGLUTAMINE DISEASE

Eleven diseases in addition to HDL2 are caused by CAG/CTG expansions. In eight, the repeat encodes polyglutamine, and it is the expanded glutamine that is thought to be pathogenic. All of these diseases are characterized by adult onset and selective neurodegeneration. Of the remaining three CAG/CTG expansion diseases, myotonic dystrophy type 1 (DM1) is a multisystem disease with a phenotypic presentation fundamentally different from those of the other disorders (though developmental and degenerative changes in the central nervous system may be present). The repeat is located in the 3′ untranslated region of the gene *DMPK*, and most repeat expansions are much longer than in the polyglutamine diseases. The expanded range of the SCA8 repeat is also much longer than the polyglutamine disease repeats, though the phenotype of the disease is similar. SCA12 repeat expansions are longer than those in HD, but similar in length to typical SCA3 expansions, though there is no convincing evidence that the repeat encodes polyglutamine. Our initial prediction, based on the precedent of other CAG/CTG genes and the detection of intranuclear inclusions staining with 1C2 antibodies, was therefore that HDL2 would be a polyglutamine disease.

Consistent with this prediction, an open reading frame (ORF) does exist at the HDL2 locus on 16q24.3 in which the repeat is in-frame to encode polyglutamine. The ORF begins 345 bp 5′ to the repeat, with a stop codon 81 bp 3′ to the repeat, and is predicted to encode 142 residues in addition to the repeat. However, there is no gene, experimentally identified or predicted through bioinformatic algorithms, that includes this ORF. No expressed sequence tag (EST) corresponds to the ORF, the predicted protein sequence is not homologous to any known proteins, and the region 5′ to the repeat is poorly conserved in the mouse. Our experimental efforts to identify a transcript containing this open reading frame have similarly failed. Reverse transcription (RT) PCR, using cDNA derived from human brain and one primer in the CAG repeat ORF and a second primer in other nearby potential exons, did not yield a product. Multitissue Northern blots or cDNA library screens using oligonucleotide probes antisense to the regions of the ORF flanking either side of the CAG repeat similarly failed to detect a CAG repeat-containing transcript. Finally, we generated a polyclonal antibody against an antigenic epitope encoded by the putative ORF. Although this antibody was able to detect a protein product of the right size on Western blots of protein extracted from cells transiently transfected with the ORF, no bands were detected on Western blots of human brain (Rudnicki et al., unpublished data). These antibodies do not stain the intranuclear inclusions present in HDL2 brain. We tentatively conclude, based on these negative experiments, that there is no expression of a transcript from the HDL2 locus in which the reading frame is in the CAG orientation.

IX. THE HDL2 LOCUS AND JUNCTOPHILIN-3: STRUCTURE AND FUNCTION

Although there is no evidence that the HDL2 repeat is expressed in the CAG orientation, there are convincing data that the repeat is expressed in the CTG orientation. On the CTG strand, the repeat is located 760 nucleotides downstream of the 3′ end of exon 1 and more than 36 kb upstream of exon 2B of *JPH3* [45] (Fig. 16-4). A polyadenylation signal is located 281 nucleotides 3′ to

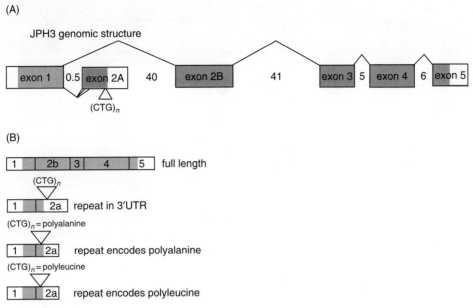

FIGURE 16-4 *JPH3* structure. (A) Genomic structure. Gray regions indicate open reading frame. Small numbers between exons indicate length of introns in kilobases. Note that one or more additional alternate exons may exist between exon 2B and exon 3. Additional transcripts missing one or more middle exons also appear to be expressed, but rarely. Not to scale. See text for details. (B) *JPH3* transcripts. Note that the primary transcript, encoding the full-length *JPH3* protein, does not include the CTG repeat.

the end of the repeat (in human sequence, not mouse), GENSCAN predicts a transcript in which exon 1 of *JPH3* is spliced to an exon containing the CTG repeat, and multiple ESTs exist in which exon 1 is spliced to an alternate terminal exon containing the repeat, which we have termed *2A*. We have experimentally shown the existence of exon 1–exon 2A transcripts in human cerebral cortex by RT-PCR. These experiments, and different ESTs, indicate that multiple different splice acceptor sites are used in the exon 1–exon 2A junction. The different splice sites change the reading frame, so that the exon 2A repeat can encode polyalanine or polyleucine or reside in the 3′ untranslated region. Preliminary experimental and EST analysis suggests that the polyalanine encoding variant is predominant, but even that variant is expressed at a much lower level than the full-length transcript, which includes exons 1 and 2B-5, but not 2A.

The junctophilins were initially described by Takeshima and colleagues in 2000 [86, 89] as part of an effort to understand the junctional complexes formed between sarcoplasmic reticulum (SR) and plasma membrane (PM) in myocytes and cardiocytes and between endoplasmic retiuclum (ER) and PM in neurons. Four members of the family have been identified (Table 16-2), each characterized by three motifs. Each type of junctophilin contains eight membrane occupation and recognition nexus (MORN) motifs, which anchor the N-terminal third of the protein to

TABLE 16-2 Differential Expression of the Junctophilins[a]

Junctophilin	Expression (mouse)				Phenotype of homozygote knockout
	Brain	Skel[b]	Cardiac	Testes	
Type 1	+	+++	++	−	Early postnatal death, skeletal muscle abnormalities
Type 2	−	++	+++	−	Embryonic lethality (cardiac)
Type 3	+++	−	−	+	Mild motor, shortened life span
Type 4	+++	−	−	−	Not available

[a] Expression analysis determined by Northern blot. See text for references.
[b] Skeletal muscle.

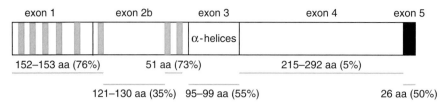

FIGURE 16-5 Junctophilin protein structure. The gray area indicates MORN motifs; the black region is the ER/SR transmembrane insertion domain. The exons encoding each part of the protein are indicated above the depiction of the protein. Predicted functional domains, along with percentage identity among human junctophilin types 1, 2, and 3, are indicated by lines below the protein depiction. Derived from Nishi et al. [89].

the plasma membrane, probably via interaction with phospholipids. Each junctophilin also includes a central region that is predicted to form α helices and a C-terminal region that encodes an SR/ER transmembrane domain (Fig. 16-5).

The functional significance of the junctophilins derives primarily from amphibian cell models and from knockout mice missing both copies of junctophilin-1 or junctophilin-2 (Fig. 16-6). Amphibian embryos injected with cRNA of junctophilin-1 developed junctional complexes between ER and PM [86]. Mice with no junctophilin-2 expression died by about embryonic day 9.5. Cardiocytes from these embryos showed a 90% loss of the primary functional SR–PM junction. The cardiocytes displayed random, asynchronous, and prolonged calcium transients during spontaneous heart oscillations. Abnormal transients were abolished by depleting internal calcium stores, but not by eliminating external calcium, suggesting that loss of junctophilin-2 leads to abnormal release of SR calcium through ryanodine receptors [86]. Consistent with the findings in junctophilin-2 knockout mice, ultrastructural examination of myocytes in junctophilin-1 knockout postnatal mice revealed reduced numbers and abnormal structure of the triad junction that normally forms between PM (formed into transverse tubules) and SR. Functionally, the muscle did not contract normally when electrically stimulated and responded excessively to increases in external calcium concentration [55].

The putative role of junctophilins in modulating internal calcium flux leads to the attractive hypothesis that loss of expression of the predominant, full-length form of *JPH3* could be central to HDL2 pathogenesis. How a repeat in a variably spliced exon would influence expression of the full-length transcript is not clear. However, it is conceivable that the repeat expansion could alter *JPH3* transcript splicing to favor the shortened exon 1–exon 2A transcript over the full-length transcript. Although there is so far no evidence for such a mechanism in HDL2, variable lengths of an intronic dinucleotide repeat alter splicing of *eNOS* [56], and insertion of a dodecamer repeat into exon 9 of bovine *c-myb* leads to exclusion of this exon in the full-length transcript [57]. More importantly, direct demonstration of a decrease in expression of full-length *JPH3* transcript in HDL2 brain has proven elusive. *JP3* (the mouse ortholog of *JPH3*) knockout mice show motor incoordination at 6–12 weeks, but no gross brain abnormalities and no electrophysiological abnormalities of cerebellar function [87]. We have examined these mice at ages of more than a year. $JP3^{+/-}$ mice have motor abnormalities that are somewhat progressive and have a shorter life span than

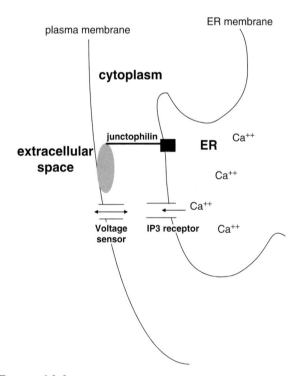

FIGURE 16-6 Junctophilin function. Putative role of junctophilin in formation of the junctional complex between plasma membrane and endoplasmic reticulum, bringing plasma membrane voltage sensors into the proximity of IP3 receptors that modulate calcium release from the endoplasmic reticulum. Gray oval = MORN motif region. Black box = ER/SR insertion domain. Adapted, in part, from Takeshima et al. [86].

littermate controls. The phenotype in *JP3*$^{-/-}$ mice is slightly more pronounced (Rudnicki et al., unpublished data). Overall, however, the phenotype, especially of the *JP3*$^{+/-}$ mice, appears relatively mild compared with the severity of HDL2. The nonlethal effects of JP3 loss, compared with the severe phenotype resulting from knockout of *JP1* and *JP2*, may reflect redundancy in spatial distribution and function between *JP3* and *JP4* [88]. It seems reasonable to propose that loss of *JPH3* expression may contribute to the HDL2 phenotype, but that it is not sufficient to cause the disease.

X. HDL2 AND TOXIC TRANSCRIPTS

If loss of expression alone is unlikely to account for the pathogenesis of HDL2, then the alternative hypothesis is that the repeat is expressed and in some way toxic. As noted earlier, it is unlikely that this toxicity is mediated by a polyglutamine tract, but the short version of the *JPH3* transcript is expressed in brain, predominantly in a form to encode polyalanine. Overexpression of expanded polyalanine and polyleucine tracts in other genes or gene fragments is toxic to mammalian cells, and aggregates form in these model systems that stain with the 1C2 antibody [58, 59]. Oculopharyngeal muscular dystrophy (OPMD), an adult-onset neuromuscular degenerative disorder, is caused by an alanine expansion [60], and inclusions formed by this disease have also been successfully stained with the 1C2 antibody [61]. Expression of expanded polyalanine and/or polyleucine tracts could therefore explain both HDL2 pathogenesis and the intranuclear aggregates observed in HDL2 brain. However, we have thus far been unable to stain intranuclear inclusions in HDL2 brain with antibodies directed against epitopes flanking the JPH3 repeat expressed in either the polyalanine or the polyleucine reading frame, nor have we detected expression of JPH3 with an expanded polyalanine or polyleucine tract on Western blots (Rudnicki et al., unpublished data). Although not definitive, these data suggest that neither polyalanine nor polyleucine is a major factor in HDL2 pathogenesis.

Several other possibilities remain. The repeat could cause other abnormalities of splicing more complicated than a simple shift to the exon 1–exon 2A version of *JPH3*. Increased expression of a splice variant could exert a dominant negative effect on normal *JPH3* function. Another possibility is toxicity at the level of the transcript, as has been detected in myotonic dystrophy types 1 and 2 (DM1, DM2) [62]. DM1 is caused by a CTG repeat expansion in the 3′ untranslated region of the gene *myotonic dystrophy protein kinase 1* (*DMPK1*) [63–65], whereas DM2, with a phenotype very similar to that of DM1, is caused by an intronic CCTG repeat expansion in *ZNF9* [66]. Overexpression of DMPK1 with an expanded CTG repeat results in the formation of small nuclear RNA aggregates containing the abnormal transcript. Similar-appearing aggregates are also detected in brain and muscle from DM1 patients [67–71] and muscle from DM2 patients [66]. In DM1 and DM2, CUG-binding protein 1 level is increased, while the protein muscleblind is less available, perhaps from sequestration into RNA inclusions [72, 73]. Both proteins alter splicing of other RNA species, with the result that splicing is abnormal in multiple genes in DM. The altered function of some of these abnormally spliced genes directly corresponds to DM manifestations. Preliminary data suggest that the HDL2 expansion can lead to RNA inclusions, with some properties similar to those seen in DM1 and DM2 (Rudnicki et al., unpublished data). This finding provides tantalizing evidence that toxic transcripts may play a role in HDL2 pathogenesis.

XI. HDL2 AND INTRANUCLEAR PROTEIN AGGREGATES

How is it possible to account for the intranuclear protein aggregates in HDL2? It appears unlikely that the aggregates result from proteins with expanded tracts of polyglutamine, polyalanine, or polyleucine. One clue is that 1C2 staining is not specific for expanded polyglutamine tracts. In addition to polyalanine and polyleucine aggregates, 1C2 stains inclusions that characterize hyaline intranuclear inclusion disease [74], aggregates in Purkinje cells in cerebellum from SCA6 patients even though the polyglutamine tracts in SCA6 are no more than about 30 residues in length [75], and aggregations of the androgen receptor with a normal-length polyglutamine tract but mutations of selected lysines [76]. It is therefore likely that the 1C2 antibody detects abnormal protein conformations that are not specific to long polyglutamine tracts, and that such conformations are present in many different types of aggregates, potentially including HDL2 aggregates. In addition, the 1C2 antibody was originally designed to detect TBP [77], and bands corresponding to TBP are frequently discernable on Western blots stained with 1C2. We have detected TBP in HDL2 aggregates (Rudnicki et al., unpublished data); it is possible that this accounts for 1C2 staining. Whether HDL2 protein aggregates and RNA foci are in some way related remains an open question; preliminary examination did not reveal costaining of the two types of aggregates, and the RNA foci are much smaller than the protein aggregates. It is possible that RNA foci are the nidus around which protein aggregates form, and that the RNA core is not detectable once surrounded by proteins.

The protein aggregates in HDL2 brain may therefore represent a nonspecific response, particularly because inclusions are common in the aging and degenerating brain. For instance, Marinesco bodies are ubiquitinated intranuclear inclusions found in the pigmented region of the substantia nigra [78]. They generally increase in number with aging and do not necessarily correlate with neurodegeneration. They have been shown to stain for vacuole-creating protein (VCP), a protein putatively associated with neurodegeneration [79], as well as the polyglutamine-containing protein ataxin-3 and at least part of the proteasome complex [80–82]. Lewy bodies, associated with Parkinson's disease and Lewy body dementia, are inclusions that contain, among other proteins, α-synuclein and ubiquitin. Experiments in which rats were given the mitochondrial complex I inhibitor rotenone suggest that metabolic impairment can give rise to these structures [83]. A neurodegenerative disorder associated with fragile X permutation carrier status has been reported [84]; these individuals also appear to have neuronal intranuclear inclusions that stain for ubiquitin [85].

XII. CONCLUSION

HDL2 is a disease at the crossroads of HD and DM. On the one hand, the clinical and pathological phenotypes of HDL2 are very similar to those of HD. On the other hand, the mutation is a CTG expansion with a number of molecular features similar to the CTG expansion found in *DMPK1*. The precise mechanism by which the CTG repeat expansion causes disease is uncertain, though it is now possible to speculate that the pathogenic process may involve toxic RNA species, as in DM, and perhaps a loss of normal *JPH3* expression (Fig. 16-7). We anticipate that one or more of the pathogenic processes thought to lead to selective striatal vulnerability in HD, such as excitotoxicity, mitochondrial dysfunction, and altered regulation of transcription, may also be part of the pathogenic process of HDL2. We further predict that the upstream processes by which these pathways are activated in HDL2 may resemble the pathogenic processes, such as abnormal regulation of splicing, previously detected in DM. The availability of HDL2 brain tissue and the development of cell and animal models of HDL2 should facilitate elucidation of the pathogenesis of HDL2 and perhaps, in the process, lead to insights into HD and DM.

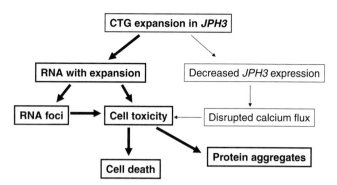

FIGURE 16-7 Hypothetical pathogenic pathway. A simplified conception of HDL2 pathogenesis that ties together the available preliminary data. Loss of expression of *JPH3* is depicted as a less important aspect of pathogenesis than toxic transcripts.

Acknowledgments

The authors thank Dr. Nancy Sachs, Dr. Ruth Walker, Dr. Amanda Krause, Dr. Adam Rosenblatt, and Dr. Mark Moliver for valuable insight and guidance; Amira Pavlova, Abdul Bachani, John Hwang, Abhijit Agarwal, and Alka Ahuja for research assistance; Ms. Marie Sonderman for technical assistance; and the individuals with HDL2 and their families for their patience and cooperation. This work was supported by the Hereditary Disease Foundation and NIH Grants NS016375 and NS38054.

References

1. Hardy, J., and Selkoe, D. J. (2002). The amyloid hypothesis of Alzheimer's disease: Progress and problems on the road to therapeutics. *Science* **297**, 353–356.
2. Ross, C. A. (1997). Intranuclear neuronal inclusions: A common pathogenic mechanism for glutamine-repeat neurodegenerative diseases? *Neuron* **19**, 1147–1150.
3. Margolis, R. L. (2002). The spinocerebellar ataxias: Order emerges from chaos. *Curr. Neurol. Neurosci. Rep.* **2**, 447–456.
4. Sieradzan, K. A., and Mann, D. M. (2001). The selective vulnerability of nerve cells in Huntington's disease. *Neuropathol. Appl. Neurobiol.* **27**, 1–21.
5. Mitchell, I. J., Cooper, A. J., and Griffiths, M. R. (1999). The selective vulnerability of striatopallidal neurons. *Prog. Neurobiol.* **59**, 691–719.
6. Ross, C. (2002). Polyglutamine pathogenesis: Emergence of unifying mechanisms for Huntington's disease and related disorders. *Neuron* **35**, 819.
7. Mattson, M. P. (2003). Excitotoxic and excitoprotective mechanisms: Abundant targets for the prevention and treatment of neurodegenerative disorders. *Neuromol. Med.* **3**, 65–94.
8. Fusco, F. R., Chen, Q., Lamoreaux, W. J., Figueredo-Cardenas, G., Jiao, Y., Coffman, J. A., Surmeier, D. J., Honig, M. G., Carlock, L. R., and Reiner, A. (1999). Cellular localization of huntingtin in striatal and cortical neurons in rats: Lack of correlation with neuronal vulnerability in Huntington's disease. *J. Neurosci.* **19**, 1189–1202.
9. Coyle, J. T., and Schwarcz, R. (1976). Lesion of striatal neurons with kainic acid provides a model for Huntington's chorea. *Nature* **263**, 244–246.

10. Beal, M. F. (1992). Does impairment of energy metabolism result in excitotoxic neuronal death in neurodegenerative illnesses? *Ann. Neurol.* **31**, 119–130.
11. Albin, R. L., Young, A. B., Penney, J. B., Handelin, B., Balfour, R., Anderson, K. D., Markel, D. S., and Tourtellotte, W. W. (1990). Abnormalities of striatal projection neurons and N-methyl-D-aspartate receptors in presymptomatic Huntington's disease. *N. Engl. J. Med.* **322**, 1293–1298.
12. Cha, J. H., Kosinski, C. M., Kerner, J. A., Alsdorf, S. A., Mangiarini, L., Davies, S. W., Penney, J. B., Bates, G. P., and Young, A. B. (1998). Altered brain neurotransmitter receptors in transgenic mice expressing a portion of an abnormal human huntington disease gene. *Proc. Natl. Acad. Sci. USA* **95**, 6480–6485.
13. Zeron, M. M., Hansson, O., Chen, N., Wellington, C. L., Leavitt, B. R., Brundin, P., Hayden, M. R., and Raymond, L. A. (2002). Increased sensitivity to N-methyl-D-aspartate receptor-mediated excitotoxicity in a mouse model of Huntington's disease. *Neuron* **33**, 849–860.
14. Li, H., Li, S. H., Johnston, H., Shelbourne, P. F., and Li, X. J. (2000). Amino-terminal fragments of mutant huntingtin show selective accumulation in striatal neurons and synaptic toxicity. *Nat. Genet.* **25**, 385–389.
15. Chen, N., Luo, T., Wellington, C., Metzler, M., McCutcheon, K., Hayden, M. R., and Raymond, L. A. (1999). Subtype-specific enhancement of NMDA receptor currents by mutant huntingtin. *J. Neurochem.* **72**, 1890–1898.
16. Ferrante, R. J., Andreassen, O. A., Dedeoglu, A., Ferrante, K. L., Jenkins, B. G., Hersch, S. M., and Beal, M. F. (2002). Therapeutic effects of coenzyme Q10 and remacemide in transgenic mouse models of Huntington's disease. *J. Neurosci.* **22**, 1592–1599.
17. Huntington's Disease Study Group (2001). A randomized, placebo-controlled trial of coenzyme Q10 and remacemide in Huntington's disease. *Neurology* **57**, 397–404.
18. Schiefer, J., Landwehrmeyer, G. B., Luesse, H. G., Sprunken, A., Puls, C., Milkereit, A., Milkereit, E., and Kosinski, C. M. (2002). Riluzole prolongs survival time and alters nuclear inclusion formation in a transgenic mouse model of Huntington's disease. *Mov. Disord.* **17**, 748–757.
19. Beal, M. F. (2000). Energetics in the pathogenesis of neurodegenerative diseases. *Trends Neurosci.* **23**, 298–304.
20. Guyot, M. C., Hantraye, P., Dolan, R., Palfi, S., Maziere, M., and Brouillet, E. (1997). Quantifiable bradykinesia, gait abnormalities and Huntington's disease-like striatal lesions in rats chronically treated with 3-nitropropionic acid. *Neuroscience* **79**, 45–56.
21. Alexi, T., Hughes, P. E., Knusel, B., and Tobin, A. J. (1998). Metabolic compromise with systemic 3-nitropropionic acid produces striatal apoptosis in Sprague–Dawley rats but not in BALB/c ByJ mice. *Exp. Neurol.* **153**, 74–93.
22. Brouillet, E., Hantraye, P., Ferrante, R. J., Dolan, R., Leroy-Willig, A., Kowall, N. W., and Beal, M. F. (1995). Chronic mitochondrial energy impairment produces selective striatal degeneration and abnormal choreiform movements in primates. *Proc. Natl. Acad. Sci. USA* **92**, 7105–7109.
23. Pisani, A., Bonsi, P., Bernardi, G., and Calabresi, P. (2002). Impairment of mitochondrial metabolism differentially affects striatal neuronal subtypes. *NeuroReport* **13**, 641–644.
24. Browne, S. E., Bowling, A. C., MacGarvey, U., Baik, M. J., Berger, S. C., Muqit, M. M. K., Bird, E. D., and Beal, M. F. (1997). Oxidative damage and metabolic dysfunction in Huntington's disease: Selective vulnerability of the basal ganglia. *Ann. Neurol.* **41**, 646–653.
25. Chapman, A. G., Durmuller, N., Lees, G. J., and Meldrum, B. S. (1989). Excitotoxicity of NMDA and kainic acid is modulated by nigrostriatal dopaminergic fibres. *Neurosci. Lett.* **107**, 256–260.
26. Cheng, N., Maeda, T., Kume, T., Kaneko, S., Kochiyama, H., Akaike, A., Goshima, Y., and Misu, Y. (1996). Differential neurotoxicity induced by L-DOPA and dopamine in cultured striatal neurons. *Brain Res.* **743**, 278–283.
27. Cicchetti, F., and Parent, A. (1996). Striatal interneurons in Huntington's disease: Selective increase in the density of calretinin-immunoreactive medium-sized neurons. *Mov. Disord.* **11**, 619–626.
28. Burke, R. E., and Baimbridge, K. G. (1993). Relative loss of the striatal striosome compartment, defined by calbindin-D28k immunostaining, following developmental hypoxic–ischemic injury. *Neuroscience* **56**, 305–315.
29. Ross, C. A. (1995). When more is less: Pathogenesis of glutamine repeat neurodegenerative diseases. *Neuron* **15**, 493–496.
30. Zoghbi, H. Y., and Orr, H. T. (2000). Glutamine repeats and neurodegeneration. *Annu. Rev. Neurosci.* **23**, 217–247.
31. Zoghbi, H. Y., and Orr, H. T. (2000). Glutamine repeats and neurodegeneration [in process citation]. *Annu. Rev. Neurosci.* **23**, 217–247.
32. Ferrante, R. J., Gutekunst, C.-A., Persichetti, F., McNeil, S. M., Kowall, N. W., Gusella, J. F., MacDonald, M. E., Beal, M. F., and Hersh, S. M. (1997). Heterogeneous topographic and cellular distribution of huntingtin expression in the normal human neostriatum. *J. Neurosci.* **17**, 3052–3063.
33. Zuccato, C., Ciammola, A., Rigamonti, D., Leavitt, B. R., Goffredo, D., Conti, L., MacDonald, M. E., Friedlander, R. M., Silani, V., Hayden, M. R., Timmusk, T., Sipione, S., and Cattaneo, E. (2001). Loss of huntingtin-mediated BDNF gene transcription in Huntington's disease. *Science* **293**, 493–498.
34. Matilla, A., Koshy, B., Cummings, C. J., Isobe, T., Orr, H. T., and Zoghbi, H. Y. (1997). The cerebellar leucine rich acidic nuclear protein (LANP) interacts with ataxin-1. *Nature* **389**, 974–978.
35. Andrew, S. E., Goldberg, Y. P., Kremer, B., Squitieri, F., Theilmann, J., Zeisler, J., Telenius, H., Adam, S., Almquist, E., Anvret, M., Lucotte, G., Stoessl, A. J., Campanella, G., and Hayden, M. R. (1994). Huntington disease without CAG expansion: Phenocopies or errors in assignment? *Am. J. Hum. Genet.* **54**, 852–863.
36. Persichetti, F., Srinidhi, J., Kanaley, L., Ge, P., Myers, R. H., D'Arrigo, K., Barnes, G. T., MacDonald, M. E., Vonsattel, J. P., and Gusella, J. F. (1994). Huntington's disease CAG trinucleotide repeats in pathologically confirmed post-mortem brains. *Neurobiol. Dis.* **1**, 159–166.
37. Xuereb, J. H., MacMillan, J. C., Snell, R., Davies, P., and Harper, P. S. (1996). Neuropathological diagnosis and CAG repeat expansion in Huntington's disease. *J. Neurol. Neurosurg. Psychiatry* **60**, 78–81.
38. Stevanin, G., Fujigasaki, H., Lebre, A. S., Camuzat, A., Jeannequin, C., Dode, C., Takahashi, J., San, C., Bellance, R., Brice, A., and Durr, A. (2003). Huntington's disease-like phenotype due to trinucleotide repeat expansions in the TBP and JPH3 genes. *Brain* **126**, 1599–1603.
39. Rosenblatt, A., Ranen, N. G., Rubinsztein, D. C., Stine, O. C., Margolis, R. L., Wagster, M. V., Becher, M. W., Rosser, A. E., Leggo, J., Hodges, J. R., ffrench-Constant, C. K., Sherr, M., Franz, M. L., Abbott, M. H., and Ross, C. A. (1998). Patients with features similar to Huntington's disease, without CAG expansion in huntingtin. *Neurology* **51**, 215–220.
40. Schalling, M., Hudson, T. J., Buetow, K. W., and Housman, D. E. (1993). Direct detection of novel expanded trinucleotide repeats in the human genome. *Nat. Genet.* **4**, 135–139.
41. Koob, M. D., Benzow, K. A., Bird, T. D., Day, J. W., Moseley, M. L., and Ranum, L. P. (1998). Rapid cloning of expanded trinucleotide repeat sequences from genomic DNA. *Nat. Genet.* **18**, 72–75.
42. Koob, M. D., Moseley, M. L., Schut, L. J., Benzow, K. A., Bird, T. D., Day, J. W., and Ranum, L. P. (1999). An untranslated CTG expansion causes a novel form of spinocerebellar ataxia (SCA8). *Nat. Genet.* **21**, 379–384.

43. Holmes, S. E., O'Hearn, E. E., McInnis, M. G., Gorelick-Feldman, D. A., Kleiderlein, J. J., Callahan, C., Kwak, N. G., Ingersoll-Ashworth, R. G., Sherr, M., Sumner, A. J., Sharp, A. H., Ananth, U., Seltzer, W. K., Boss, M. A., Vieria-Saecker, A. M., Epplen, J. T., Riess, O., Ross, C. A., and Margolis, R. L. (1999). Expansion of a novel CAG trinucleotide repeat in the 5' region of PPP2R2B is associated with SCA12. *Nat. Genet.* **23**, 391–392.

44. Margolis, R. L., and Ross, C. A. (2001). Expansion explosion: New clues to the pathogenesis of repeat expansion neurodegenerative diseases. *Trends Mol. Med.* **7**, 479–482.

45. Holmes, S. E., O'Hearn, E., Rosenblatt, A., Callahan, C., Hwang, H. S., Ingersoll-Ashworth, R. G., Fleisher, A., Stevanin, G., Brice, A., Potter, N. T., Ross, C. A., and Margolis, R. L. (2001). A repeat expansion in the gene encoding junctophilin-3 is associated with Huntington disease-like 2. *Nat. Genet.* **29**, 377–378.

46. Margolis, R. L., O'Hearn, E., Rosenblatt, A., Willour, V., Holmes, S. E., Franz, M. L., Callahan, C., Hwang, H. S., Troncoso, J. C., and Ross, C. A. (2001). A disorder similar to Huntington's disease is associated with a novel CAG repeat expansion. *Ann. Neurol.* **50**, 373–380.

47. Walker, R. H., Morgello, S., Davidoff-Feldman, B., Melnick, A., Walsh, M. J., Shashidharan, P., and Brin, M. F. (2002). Autosomal dominant chorea-acanthocytosis with polyglutamine-containing neuronal inclusions. *Neurology* **58**, 1031–1037.

48. Walker, R. H., Jankovic, J., O'Hearn, E., and Margolis, R. L. (2003). Phenotypic features of Huntington's disease-like 2. *Mov. Disord.* **18**, 1527–1530.

49. Walker, R. H., Rasmussen, A., Rudnicki, D., Holmes, S. E., Alonso, E., Matsuura, T., Ashizawa, T., Davidoff-Feldman, B., and Margolis, R. L. (2003). Huntington's disease-like 2 can present as chorea-acanthocytosis. *Neurology* **61**, 1002–1004.

50. Danek, A., Jung, H. H., Melone, M. A., Rampoldi, L., Broccoli, V., and Walker, R. H. (2005). Neuroacanthocytosis: New developments in a neglected group of dementing disorders. *J. Neurol. Sci.* **229/230**, 171–186.

51. Margolis, R. L., Holmes, S. E., Rosenblatt, A., Gourley, L., O'Hearn, E., Ross, C. A., Seltzer, W. K., Walker, R. H., Ashizawa, T., Rasmussen, A., Hayden, M., Almqvist, E. W., Harris, J., Fahn, S., MacDonald, M. E., Mysore, J., Shimohata, T., Tsuji, S., Potter, N., Nakaso, K., Adachi, Y., Nakashima, K., Bird, T., Krause, A., and Greenstein, P. (2004). Huntington's disease-like 2 (HDL2) in North America and Japan. *Ann. Neurol.* **56**, 670–674.

52. Shimohata, T., Onodera, O., Honma, Y., Hirota, K., Nunomura, Y., Kimura, T., Kawachi, I., Sanpei, K., Nishizawa, M., and Tsuji, S. (2004). Gene diagnosis of patients with chorea [Japanese]. *Rinsho Shinkeigaku* **44**, 149–153.

53. Stevanin, G., Camuzat, A., Holmes, S. E., Julien, C., Sahloul, R., Dode, C., Hahn-Barma, V., Ross, C. A., Margolis, R. L., Durr, A., and Brice, A. (2002). CAG/CTG repeat expansions at the Huntington's disease-like 2 locus are rare in Huntington's disease patients. *Neurology* **58**, 965–967.

54. Bauer, I., Gencik, M., Laccone, F., Peters, H., Weber, B. H., Feder, E. H., Weirich, H., Morris-Rosendahl, D. J., Rolfs, A., Gencikova, A., Bauer, P., Wenning, G. K., Epplen, J. T., Holmes, S. E., Margolis, R. L., Ross, C. A., and Riess, O. (2002). Trinucleotide repeat expansions in the junctophilin-3 gene are not found in Caucasian patients with a Huntington's disease-like phenotype. *Ann. Neurol.* **51**, 662.

55. Ito, K., Komazaki, S., Sasamoto, K., Yoshida, M., Nishi, M., Kitamura, K., and Takeshima, H. (2001). Deficiency of triad junction and contraction in mutant skeletal muscle lacking junctophilin type 1. *J. Cell Biol.* **154**, 1059–1067.

56. Bilbao, D., and Valcarcel, J. (2003). Getting to the heart of a splicing enhancer. *Nat. Struct. Biol* **10**, 6–7.

57. Shinagawa, T., Ishiguro, N., Horiuchi, M., Matsui, T., Okada, K., and Shinagawa, M. (1997). Deletion of c-myb exon 9 induced by insertion of repeats. *Oncogene* **14**, 2775–2783.

58. Rankin, J., Wyttenbach, A., and Rubinsztein, D. C. (2000). Intracellular green fluorescent protein–polyalanine aggregates are associated with cell death. *Biochem. J* **348** (Pt. 1), 15–19.

59. Dorsman, J. C., Pepers, B., Langenberg, D., Kerkdijk, H., Ijszenga, M., Den Dunnen, J. T., Roos, R. A., and van Ommen, G. J. (2002). Strong aggregation and increased toxicity of polyleucine over polyglutamine stretches in mammalian cells. *Hum. Mol. Genet* **11**, 1487–1496.

60. Brais, B., Bourchard, J. P., Xie, Y. G., Rochefort, D. L., Chretien, N., Tome, F. M., LaFreniere, R. G., Rommens, J. M., Uyama, E., Nohira, O., Blumen, S., Korcyn, A. D., Heutink, P., Mathieu, J., Duranceau, A., Codere, F., Fardeau, M., and Rouleau, G. A. (1998). Short GCG expansions in the PABP2 gene cause oculopharyngeal muscular dystrophy. *Nat. Genet.* **18**, 164–167.

61. Sugaya, K., Matsubara, S., Miyamoto, K., Kawata, A., and Hayashi, H. (2003). An aggregate-prone conformational epitope in trinucleotide repeat diseases. *NeuroReport* **14**, 2331–2335.

62. Ranum, L. P., and Day, J. W. (2004). Pathogenic RNA repeats: An expanding role in genetic disease. *Trends Genet.* **20**, 506–512.

63. Brook, J. D., McCurrach, M. E., Harley, H. G., Buckler, A. J., Church, D., Aburatani, H., Hunter, K., Davies, J., Shelbourne, P., Buxton, J., Jones, C., Juvonen, V., Johnson, K., Harper, P. S., Shaw, D. J., and Housman, D. E. (1992). Molecular basis of myotonic dystrophy: Expansion of a trinucleotide (CTG) repeat at the 3' end of a transcript encoding a protein kinase family member. *Cell* **68**, 799–808.

64. Mahadevan, M., Tsilfidis, C., Sabourin, L., Shutler, G., Amemiya, C., Jansen, G., Nelville, C., Narang, M., Barcelo, J., O'Hoy, K., Leblond, S., Earle-MacDonald, J., DeJong, P. J., Wieringa, B., and Korneluk, R. G. (1992). Myotonic dystrophy mutation: An unstable CTG repeat in the 3' untranslated region of the gene. *Science* **255**, 1253–1255.

65. Fu, Y.-H., Pizzuti, A., Fenwick, R. G., King, J., Rajnarayan, S., Dunne, P. W., Dubel, J., Nasser, G. A., Ashizawa, T., De Jong, P., Wieringa, B., Korneluk, R., Perryman, M. B., Epstein, H. F., and Caskey, C. T. (1992). An unstable triplet repeat in a gene related to myotonic muscular dystrophy. *Science* **255**, 1256–1258.

66. Liquori, C. L., Ricker, K., Moseley, M. L., Jacobsen, J. F., Kress, W., Naylor, S. L., Day, J. W., and Ranum, L. P. (2001). Myotonic dystrophy type 2 caused by a CCTG expansion in intron 1 of ZNF9. *Science* **293**, 864–867.

67. Taneja, K. L., McCurrach, M., Schalling, M., Housman, D., and Singer, R. H. (1995). Foci of trinucleotide repeat transcripts in nuclei of myotonic dystrophy cells and tissues. *J. Cell Biol.* **128**, 995–1002.

68. Davis, B. M., McCurrach, M. E., Taneja, K. L., Singer, R. H., and Housman, D. E. (1997). Expansion of a CUG trinucleotide repeat in the 3' untranslated region of myotonic dystrophy protein kinase transcripts results in nuclear retention of transcripts. *Proc. Natl. Acad. Sci. USA* **94**, 7388–7393.

69. Jiang, H., Mankodi, A., Swanson, M. S., Moxley, R. T., and Thornton, C. A. (2004). Myotonic dystrophy type 1 is associated with nuclear foci of mutant RNA, sequestration of muscleblind proteins and deregulated alternative splicing in neurons. *Hum. Mol. Genet* **13**, 3079–3088.

70. Mankodi, A., Teng-Umnuay, P., Krym, M., Henderson, D., Swanson, M., and Thornton, C. A. (2003). Ribonuclear inclusions in skeletal muscle in myotonic dystrophy types 1 and 2. *Ann. Neurol.* **54**, 760–768.

71. Mankodi, A., Logigian, E., Callahan, L., McClain, C., White, R., Henderson, D., Krym, M., and Thornton, C. A. (2000). Myotonic dystrophy in transgenic mice expressing an expanded CUG repeat. *Science* **289**, 1769–1773.

72. Timchenko, L. T., Timchenko, N. A., Caskey, C. T., and Roberts, R. (1996). Novel proteins with binding specificity for DNA CTG repeats and RNA CUG repeats: Implications for myotonic dystrophy. *Hum. Mol. Genet* **5**, 115–121.
73. Miller, J. W., Urbinati, C. R., Teng-Umnuay, P., Stenberg, M. G., Byrne, B. J., Thornton, C. A., and Swanson, M. S. (2000). Recruitment of human muscleblind proteins to (CUG)(n) expansions associated with myotonic dystrophy. *EMBO J.* **19**, 4439–4448.
74. Takahashi, J., Fukuda, T., Tanaka, J., Minamitani, M., Fujigasaki, H., and Uchihara, T. (2000). Neuronal intranuclear hyaline inclusion disease with polyglutamine-immunoreactive inclusions. *Acta Neuropathol. (Berl.)* **99**, 589–594.
75. Ishikawa, K., Owada, K., Ishida, K., Fujigasaki, H., Shun, L. M., Tsunemi, T., Ohkoshi, N., Toru, S., Mizutani, T., Hayashi, M., Arai, N., Hasegawa, K., Kawanami, T., Kato, T., Makifuchi, T., Shoji, S., Tanabe, T., and Mizusawa, H. (2001). Cytoplasmic and nuclear polyglutamine aggregates in SCA6 Purkinje cells. *Neurology* **56**, 1753–1756.
76. Thomas, M., Dadgar, N., Aphale, A., Harrell, J. M., Kunkel, R., Pratt, W. B., and Lieberman, A. P. (2004). Androgen receptor acetylation site mutations cause trafficking defects, misfolding, and aggregation similar to expanded glutamine tracts. *J. Biol. Chem.* **279**, 8389–8395.
77. Trottier, Y., Lutz, Y., Stevanin, G., Imbert, G., Devys, D., Cancel, G., Saudou, F., Weber, C., David, G., Tora, L., Agid, Y., Brice, A., and Mandel, J.-L. (1995). Polyglutamine expansion as a pathological epitope in Huntington's disease and four dominant cerebellar ataxias. *Nature* **378**, 403–406.
78. Dickson, D. W., Wertkin, A., Kress, Y., Ksiezak-Reding, H., and Yen, S. H. (1990). Ubiquitin immunoreactive structures in normal human brains: Distribution and developmental aspects. *Lab. Invest.* **63**, 87–99.
79. Mizuno, Y., Hori, S., Kakizuka, A., and Okamoto, K. (2003). Vacuole-creating protein in neurodegenerative diseases in humans. *Neurosci. Lett.* **343**, 77–80.
80. Fujigasaki, H., Verma, I. C., Camuzat, A., Margolis, R. L., Zander, C., Lebre, A. S., Jamot, L., Saxena, R., Anand, I., Holmes, S. E., Ross, C. A., Durr, A., and Brice, A. (2001). SCA12 is a rare locus for autosomal dominant cerebellar ataxia: A study of an Indian family. *Ann. Neurol.* **49**, 117–121.
81. Kumada, S., Uchihara, T., Hayashi, M., Nakamura, A., Kikuchi, E., Mizutani, T., and Oda, M. (2002). Promyelocytic leukemia protein is redistributed during the formation of intranuclear inclusions independent of polyglutamine expansion: An immunohistochemical study on Marinesco bodies. *J. Neuropathol. Exp. Neurol.* **61**, 984–991.
82. Kettner, M., Willwohl, D., Hubbard, G. B., Rub, U., Dick, E. J., Jr., Cox, A. B., Trottier, Y., Auburger, G., Braak, H., and Schultz, C. (2002). Intranuclear aggregation of nonexpanded ataxin-3 in Marinesco bodies of the nonhuman primate substantia nigra. *Exp. Neurol.* **176**, 117–121.
83. Betarbet, R., Sherer, T. B., MacKenzie, G., Garcia-Osuna, M., Panov, A. V., and Greenamyre, J. T. (2000). Chronic systemic pesticide exposure reproduces features of Parkinson's disease. *Nat. Neurosci.* **3**, 1301–1306.
84. Hagerman, R. J., Leehey, M., Heinrichs, W., Tassone, F., Wilson, R., Hills, J., Grigsby, J., Gage, B., and Hagerman, P. J. (2001). Intention tremor, parkinsonism, and generalized brain atrophy in male carriers of fragile X. *Neurology* **57**, 127–130.
85. Greco, C. M., Hagerman, R. J., Tassone, F., Chudley, A. E., Del Bigio, M. R., Jacquemont, S., Leehey, M., and Hagerman, P. J. (2002). Neuronal intranuclear inclusions in a new cerebellar tremor/ataxia syndrome among fragile X carriers. *Brain* **125**, 1760–1771.
86. Takeshima, H., Komazaki, S., Nishi, M., Iino, M., and Kangawa, K. (2000). Junctophilins: A novel family of junctional membrane complex proteins. *Mol. Cell* **6**, 11–22.
87. Nishi, M., Hashimoto, K., Kuriyama, K., Komazaki, S., Kano, M., Shibata, S., and Takeshima, H. (2002) Motor discoordination in mutant mice lacking junctophilin type 3. *Biochem. Biophys. Res. Commun.* **292**, 318–324.
88. Nishi, M., Sakagami, H., Komazaki, S., Kondo, H., and Takeshima, H. (2003) Coexpression of junctophilin type 3 and type 4 in brain. *Brain Res. Mol. Brain Res.* **118**, 102–110.
89. Nishi, M., Mizushima, A., Nakagawara, K., and Takeshima, H. (2000) Characterization of human junctophilin subtype genes. *Biochem. Biophys. Res. Commun.* **273**, 920–927.

Part VI

Friedreich's Ataxia

CHAPTER 17

Friedreich's Ataxia

MASSIMO PANDOLFO
Department of Neurology, Free University of Brussels, Erasone Hospital, Brussels, Belgium

I. Introduction
II. Clinical and Pathological Aspects of Friedreich's Ataxia
   A. Epidemiology
   B. Pathology
   C. Clinical Aspects
   D. Prognosis
III. Isolation and Analysis of the Friedreich Ataxia Gene
   A. Mapping and Cloning of the FRDA Gene
   B. Structure of the FRDA Gene
   C. Expression of the FRDA Gene
IV. Gene Mutations in FRDA
   A. Point Mutations
   B. GAA Trinucleotide Repeat Expansion
   C. Detection and Diagnostic Value of Expanded GAA Repeats
   D. Instability of the Expanded GAA Triplet Repeat
V. Origin of the Expanded GAA Repeat
VI. Pathogenic Mechanism of the GAA Expansion
   A. Effect of the Expanded GAA Repeat on Frataxin Gene Expression
   B. Properties of the Expanded GAA Repeat
VII. Phenotype–Genotype Correlation
VIII. Conclusion and Perspectives
References

Friedreich's ataxia, the most frequent cause of inherited ataxia in Caucasians, is due in most cases to a large expansion of an intronic GAA repeat, resulting in decreased expression of the target frataxin gene. The autosomal recessive inheritance of the disease gives this triplet-repeat mutation some unique features of natural history and evolution. The frataxin gene encodes a mitochondrial protein that has homologs in all eukaryotes and in gram-negative bacteria. Studies in yeast and mouse models and biochemical investigations indicate a role for frataxin in the assembly of iron–sulfur clusters in the mitochondrion. Frataxin deficiency leads to abnormal mitochondrial iron metabolism, decreased actvities of iron–sulfur cluster-containing enzymes, reduced oxidative phosphorylation, and possibly increased oxidative stress. The basis of the specific cell vulnerability observed in Friedreich's ataxia is still unclear. Increasing knowledge of pathogenetic mechanisms allows the proposal of novel treatment approaches.

I. INTRODUCTION

In 1853, Nicholaus Friedreich, Professor of Medicine in Heidelberg, described a "degenerative atrophy of the posterior columns of the spinal cord" leading to progressive ataxia, sensory loss, and muscle weakness, often associated with scoliosis, foot deformity, and cardiopathy [1, 2]. The disease might afflict several individuals in a sibship, but parents were never affected. Some critics, particularly Charcot, suspected that Friedreich's patients had tabes, a form of neurosyphilis. However, after Friedreich published additional cases in 1866 and 1867 [3, 4], it became generally accepted that he had described a new disease entity. Friedreich was able to pinpoint all the essential clinical and pathological features of the disease. He just missed the loss of deep tendon reflexes, which Erb later described in 1885. The new disease was given the name *Friedreich's ataxia* (currently abbreviated as FRDA) in 1882 by Brousse [5]. By 1890, Ladame [6] had already reported more than 100 cases. These pioneering studies were done in the context of the late 19th- and early 20th-century flourishing of neuropathology and clinical neurology that gave shape to the modern nosological structure of these disciplines. Inevitably, progress occurred through debates and disputes. It is not surprising that the subsequent identification of clinically similar diseases and the presence of cases that could not be easily classified somewhat blurred the definition of FRDA for many years [7]. Only in the late 1970s did renewed interest in the disease prompt a reevaluation of the literature and analysis of large series of patients to establish clear diagnostic criteria. The landmark studies were done first by the Québec Collaborative Group [8], then by Harding [9]. Recessive inheritance was firmly established as an essential feature of FRDA [8-11]. The Québec Collaborative Group, after evaluating 50 patients of French Canadian ancestry, subdivided them into four groups: typical FRDA (33 patients), incomplete FRDA (3 patients, all from the same sibship), atypical FRDA (6 patients), and non-FRDA (8 patients). The clinical features of the typical FRDA group were identified

TABLE 17-1 Diagnostic Criteria for FRDA According to Harding [9]

Autosomal recessive inheritance
Onset before age 25
Within 5 years from onset:
Limb and truncal ataxia
Absent tendon reflexes in the legs
Extensor plantar responses
Motor NCV >40 m/s in upper limbs with small or absent SAPs
After 5 years since onset:
As above plus dysarthria
Additional criteria, not essential for diagnosis, present in >2/3 of cases:
Scoliosis
Pyramidal weakness of the legs
Absent reflexes in upper limbs
Distal loss of joint position and vibration sense in lower limbs
Abnormal EKG
Other features, present in <50% of cases:
Nystagmus
Optic atrophy
Deafness
Distal weakness and wasting
Pes cavus
Diabetes

and proposed as diagnostic criteria. Harding [9], however, felt that such criteria, although appropriate for advanced cases, were too strict to allow diagnosis of early cases. She analyzed 115 patients from 90 families, some at an early stage of the disease, and proposed the diagnostic criteria listed in Table 17-1, which include certain signs and symptoms that may not be present at the onset, but have to manifest as the diseases evolves. These studies, as well as more recent ones [12, 13], identified a degree of variability in the clinical features of FRDA, including age at onset, rate of progression, and severity and extent of disease involvement. As remarked by Harding [9], such variability, sometimes occurring even within the same sibship [14, 15], is greater than that found in most other recessive neurological diseases. Patients may be confined to a wheelchair in their early teens or still be ambulatory in their late thirties. Cardiac complications may be minimal, or absent, or so severe as to cause premature death. Only some patients develop skeletal abnormalities, optic atrophy, diabetes mellitus, and sensorineural deafness. Atypical cases, with an overall FRDA-like phenotype but missing one or more essential features of typical FRDA, are sometimes observed in sibships along with typical cases [15], indicating that they represent extreme examples of the clinical spectrum of FRDA. But other atypical cases cluster in families, clouding classification. Examples include Acadian FRDA, observed in a specific population of French origin living in North America, which has a milder course than classic FRDA and is rarely accompanied by a cardiomyopathy [16, 17]; late-onset Friedreich's ataxia (LOFA), a disease with all the features of FRDA but with an onset after 25 years of age [18, 19]; and Friedreich's ataxia with retained reflexes (FARR), a variant in which tendon reflexes in the lower limbs are preserved [20]. After identification of the FRDA gene and of its most common mutation, the unstable hyperexpansion of a GAA triplet-repeat polymorphism [21], genotype–phenotype correlations became possible, clarifying these issues and having important consequences on the diagnostic criteria for FRDA.

II. CLINICAL AND PATHOLOGICAL ASPECTS OF FRIEDREICH'S ATAXIA

A. Epidemiology

The GAA triplet-repeat expansion causing FRDA is found only in individuals from Europe, North Africa, the Middle East, and India [22]. As detailed later, this specific distribution is likely to be the consequence of unique mutational events. In these populations, FRDA is the most common of the hereditary ataxias. Conversely, the disease is not found in individuals whose ancestors belong entirely to other ethnic groups. In the populations involved, except in areas where other ataxic disorders are exceptionally prevalent because of a founder effect, FRDA generally accounts for half of the overall heredodegenerative ataxia cases and for three-quarters of those with onset before age 25 [23]. The disease seems to have a fairly similar prevalence of about 2×10^{-5} in almost all Caucasian populations studied [10, 23–26], with local clusters due to a founder effect, as those observed in Rimouski, Québec [27], and Kathikas-Arodhes, Cyprus [28].

B. Pathology

1. Central Nervous System

FRDA causes a characteristic pattern of central nervous system (CNS) pathology [29, 30]. Friedreich [1–4, 31] identified the degeneration of the posterior columns of the spinal cord as a hallmark of the disease. The posterior columns contain the central branches of the axons of large dorsal root ganglia (DRG) sensory neurons. These axons extend without interruption to the brainstem, forming the gracile (Goll) and cuneate (Burdach) tracts, the former originating at the lumbosacral level, the latter from cervicothoracic segments. The posterior columns appear shrunken, grayish, and translucent. Demyelination, loss of fibers, and fibrillary gliosis are more severe in the Goll than in the Burdach tract; that is, the fibers originating more caudally are more severely affected. Atrophy is also observed in the spinocerebellar tracts, the dorsal being more affected than the ventral. Clarke's column, where the spinocerebellar tracts originate, shows severe loss of neurons. Therefore, the sensory systems providing information to the brain and cerebellum about the position and speed of body segments, particularly the lower limbs, are severely compromised in FRDA. Motor neurons in the ventral horns are fairly well preserved, with minimal losses. Conversely, the long crossed and uncrossed corticospinal motor tracts are atrophied. The pattern of atrophy of the long tracts of fibers suggests a "dying back" process [32], particularly clear in the case of corticospinal fibers. These are severely atrophic at the lumbar level, much less so in the cervical cord and brainstem, and of normal appearance in the cerebral peduncula. Such a process implies that, at least in some types of neurons, the primary biochemical defect in FRDA acts first or more severely on axons than on cell bodies, leading to more severe involvement of the long sensory and motor pathways.

In the brainstem, what appears to be transsynaptic degeneration, with intense gliosis, can be observed in the gracile and cuneate nuclei, where the dorsal column tracts terminate. The medial lemnisci, which continue the central sensory pathway after these nuclei, often show shrinkage and loss of myelin, particularly in their ventral portion, deriving from the gracile nuclei. In most cases, the sensory pathways originating from the cranial nerves also exhibit myelin pallor and loss of fibers, including the entering roots of nerves V, IX, and X, the descending trigeminal tracts, and the solitary tracts. The auditory and vestibular systems are sometimes affected, with demyelination and loss of entering fibers from nerve VIII and atrophy of the lateral vestibular nuclei, cochlear nuclei, superior olives, and inferior colliculi. The accessory cuneate nuclei, corresponding to Clarke's column in the spinal cord, are markedly atrophic. Cranial motor nuclei and the medial longitudinal bundles appear instead to be spared. Therefore, also in the brainstem the sensory system is selectively affected, with primary sensory neuron loss and transsynaptic degeneration in relay nuclei.

Cerebellar atrophy is not a major characteristic of FRDA. The cerebellar cortex shows only mild loss of Purkinje cells and occasional axonal torpedoes, particularly in the superior lamellae, usually late in the disease course. Quantitative analysis of synaptic terminals, however, suggests a loss of contacts over Purkinje cell bodies and proximal dendrites, probably reflecting the loss of cerebellar afferents [33]. Conversely, the deep cerebellar nuclei, where cerebellar efferents originate, are severely affected, with marked neuronal loss and gliosis in the dentate nucleus. The presence of additional iron accumuation in this normally iron-rich structure is somewhat controversial. Magnetic resonance imaging (MRI) data support an increased iron content compared with that of age-matched controls [34], but pathological findings are less clear [33]. As a consequence of dentate atrophy, the superior cerebellar pedunculi appear markedly atrophic, and the red nuclei, where many fibers from the dentate terminate, may show some gliosis. The basal ganglia may, in some cases, exhibit moderate cell loss in the external pallidus and subthalamic nuclei. The thalamus, striatum, and substantia nigra do not appear to be directly involved by the disease. The cerebral cortex appears normal, with the exception of a loss of large pyramidal cells in the primary motor areas when pyramidal tract atrophy is advanced. Optic nerves and tracts show a variable degree of involvement. When present, atrophy of the visual system may extend to the lateral geniculate bodies [35].

In summary, the neuropathology of FRDA is characterized by atrophy of the central sensory pathways carrying to the brain and cerebellum crucial information for the correct execution of movement and for equilibrium, atrophy of the cerebellar efferent pathway, and atrophy of the distal portion of the corticospinal motor tracts. As described later, the consequent malfunction of each of these systems contributes to the characteristic clinical picture of FRDA.

Finally, as many patients with FRDA die as a consequence of heart disease, it is not uncommon to observe widespread hypoxic changes and focal infarcts in the CNS.

2. Peripheral Nervous System

Sensory axonal polyneuropathy is a hallmark of FRDA [8, 9, 32, 36–39]. The motor component of peripheral nerves is better preserved. In the DRG, loss of large primary sensory neurons is an invariable and early finding. Proliferation of capsule cells, forming clumps called *Residualknötchen of Nageotte*, and irregular appearance of intragangliar fibers are also observed. Atrophy of the central branches of the axons of the affected primary sensory cells causes thinning of the dorsal roots, particularly at the lumbosacral level, and atrophy of the peripheral branches causes a marked loss of large myelinated fibers from peripheral nerves. The fine, unmyelinated fibers are involved to a limited extent [40], and interstitial connective tissue is increased.

3. Heart

The heart is clinically or subclinically affected in the vast majority of FRDA patients [8, 9, 12, 13, 29, 41]. Enlargement of the heart is the typical finding, with thickening of the ventricular walls and interventricular septum [41, 42]. After a long disease course, however, the gross appearance becomes more that of a dilatative cardiomyopathy [43]. Pericardial adhesions are often observed. Microscopically, hypertrophic cardiomyocytes are intermingled with fibers undergoing atrophy or granular degeneration and normal-appearing fibers. Connective tissue is increased, with diffuse and focal inflammatory cell infiltration. Intracellular iron deposits in cardiomyocytes [44, 45] are a specific finding in FRDA cardiomyopathy [46].

4. Other Organs

About 10% of FRDA patients have diabetes mellitus and show a loss of islet cells without the signs of autoimmune aggression found in type I diabetes [47].

Skeletal abnormalities are very common in FRDA. Scoliosis is observed in more than half of the patients, mostly as a double thoracolumbar curve [48, 49]. As a rule, a kyphotic curve is associated with FRDA. The kyphoscoliosis of FRDA resembles the idiopathic form of scoliosis [50]. Although it is often suggested that it derives from muscular imbalance due to the neurological disease, this has never been convincingly demonstrated. Pes cavus, pes equinovarus, and clawing of the toes are also very common and thought to be secondary to nerve degeneration [8, 9].

Amyotrophy of small hand muscles and of distal leg and foot muscles is a common feature of the disease, which may be observed relatively early, before the appearance of disuse atrophy [9], reflecting some involvement of spinal motor neurons.

C. Clinical Aspects

1. Onset

Typical FRDA is said to have its onset around puberty, but wide variations are observed [8, 9, 12, 13, 51, 52]. Some patients show symptoms well before puberty, even at ages 2–3 [53, 54], and a few others have a very late onset in adult life. In the "typical FRDA" patients described by the Québec Collaborative Group [8], onset was before age 20. In her diagnostic criteria, Harding [9] extended this limit to age 25. The status of later-onset cases (LOFA) remained uncertain until the molecular defect was identified, when it became clear that LOFA patients carry mutations in the same gene as do typical FRDA cases. Age of onset is more variable among families than within families [9, 14], but large variations within a sibship are occasionally observed. The dynamic nature of the causative mutation, a GAA triplet-repeat expansion in the frataxin gene [21], explains part of the variability, probably around 50% [55]. The effect of modifier genes or environmental factors should account for the remaining variability in age of onset and severity. Rare patients with frataxin point mutations may also have unusual features, including a very early or late age of onset (see below).

Gait instability (65%) and generalized clumsiness (25%) are the usual initial symptoms. Occasionally, nonneurological manifestations, such as scoliosis (5%) and cardiomyopathy (5%) precede the onset of ataxia [8, 9, 12, 13].

2. Neurological Signs and Symptoms

The cardinal neurological feature of FRDA is a progressive, unremitting trunk and limb ataxia (Fig. 17-1). As already noted by Friedreich, it is a mixed cerebellar–sensory ataxia, the pathological basis of which is the degeneration of both sensory pathways and cerebellar afferent and efferent pathways. Most commonly, it begins with clumsiness in gait and frequent falls (truncal ataxia). Limb incoordination, dysmetria, and intention tremor then follow. Progression is steady, until patients lose the ability to perform fine-motor activities, to walk, to stand, and eventually to sit without support. Speech is affected within 5-years of onset [8, 9, 12, 13], and becomes more and more indistinct as disease progresses. Dysarthria in FRDA patients consists of slow, jerky speech with sudden utterances [56, 57].

Muscle tone is most often decreased, sometimes normal, rarely increased. Muscular weakness is common and

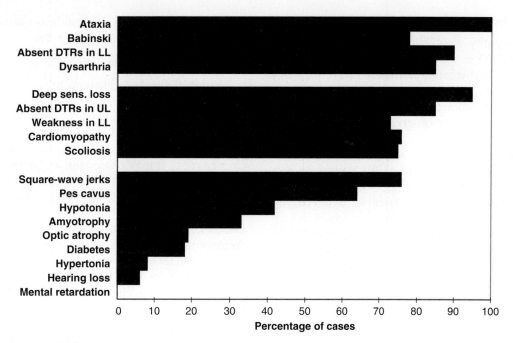

FIGURE 17-1 Clinical findings in an unselected series of 27 consecutive patients with a positive FRDA molecular test. Abbreviations: DTR: deep tendon reflexes; LL: lower limbs.

progressive, particularly at the lower limbs, usually affecting the proximal muscles first [58]. It is thought to derive, at least in part, from the degeneration of the corticospinal tracts. However, when patients become wheelchair-bound, they still have significant strength in their lower limbs, on average 70% of normal, clearly indicating that ataxia and not weakness is the primary cause for loss of ambulation in FRDA [58]. Early atrophy of distal lower limb muscles and of small hand muscles may cause some additional weakness [9]. In the later stages, when patients are wheelchair-bound, disuse atrophy occurs.

Sensory loss is another cardinal manifestation of FRDA. It is the consequence of the degeneration of large sensory fibers in peripheral nerves and in the central sensory pathways. Loss of position and vibration sense commonly occurs at onset and, in most cases, after 2 years. Vibration sense loss may sometimes precede the appearance of ataxia. Perception of light touch, pain, and temperature is initially normal, but tends to decrease in most patients with advanced disease, as even small sensory neurons and fibers become affected.

Another consequence of the sensory axonal neuropathy, the loss of deep tendon reflexes (DTRs), at least in the lower limbs, was considered a diagnostic criterion [8, 9]. DTRs in all four limbs are indeed lost by most patients within a few years of onset. However, a minority of patients with a proven molecular diagnosis of FRDA have elicitable tendon reflexes for several years after diagnosis (FARR); a few even have exaggerated reflexes with spasticity. Some rare cases have a very prominent spastic phenotype with relatively little ataxia that is clinically indistinguishable from a hereditary spastic paraplegia [59]. Despite loss of DTRs, most patients have extensor plantar responses as a consequence of pyramidal tract involvement, a characteristic combination of neurological signs in FRDA. However, a minority of patients have flexor plantar responses. Clearly, the relative weight of sensory neuropathy, causing areflexia and hypotonia, and of pyramidal tract degeneration, causing extensor plantar responses and spasticity, varies from patient to patient. In most cases, the typically mixed picture results, but sometimes one component may obscure the other.

The most common abnormality of ocular movements in FRDA is fixation instability with square-wave jerks [60]. In addition, patients may show various combinations of cerebellar, vestibular, and brain-stem oculomotor signs. Ophthalmoparesis is not observed and nystagmus is uncommon.

About 30% of patients develop optic atrophy, with or without visual impairment, particularly in the later stages of disease [8, 9, 13, 13, 61–63]. Sensorineural hearing loss, also more common with advanced disease, affects about 20% of patients [64, 65]. Optic atrophy and sensorineural hearing loss tend to be associated in the same patients, and with diabetes, more often than expected by chance alone [9, 66], probably as a sign of more severe, widespread disease. These symptoms can progress to the point of rendering the patient functionally blind and deaf.

3. Heart Disease

Cardiomyopathy may significantly contribute to disability in FRDA and even cause premature death as a consequence of arrhythmias and heart failure [67]. In a significant portion of FRDA patients, however, heart disease remains asymptomatic [8, 9, 12, 13, 68–73]. Cardiac involvement is more often observed in patients with earlier age of onset [74]. When symptoms occur, the most common complaints are shortness of breath (40% of patients) and palpitations (11%) [9]. Typical electrocardiographic abnormalities include widespread T-wave inversions and signs of ventricular hypertrophy. Conduction disturbances and supraventricular ectopic beats (in 10% of patients) are occasional findings [12, 13, 72, 73]. Atrial fibrillation is a negative prognostic sign [38], as it precedes fatal cardiac complications by 6 months or less in almost a quarter of the cases in which these occur. Electrocardiographic changes vary in time, with occasional normal recordings: this has led to underestimation of their frequency. If repeated recordings are obtained, however, electrocardiography seems to be the most sensitive test for FRDA cardiomyopathy. Echocardiography and Doppler–echocardiography can demonstrate concentric hypertrophy of the ventricles (62%) or asymmetric septal hypertrophy (29%), along with diastolic function abnormalities [9, 75]. As the disease progresses, the transition from a hypertrophic to a dilatative type of cardiomyopathy has been documented [43].

4. Diabetes Mellitus

About 10% of FRDA patients have diabetes mellitus, and an additional 20% have carbohydrate intolerance. A detailed study of glucose and insulin metabolism in FRDA patients revealed a deficiency in arginine-stimulated insulin secretion in all cases, including normotolerant individuals [76], suggesting that beta cells are invariably affected by the primary genetic defect of FRDA. In addition, peripheral insulin resistance may also contribute to the development of diabetes in FRDA [77]. Diabetes usually appears after the neurological symptoms. It may be revealed by the appearance of characteristic symptoms, like polyuria and polydipsia, or by glucose testing. Although some cases may be initially controlled with oral hypoglycemic drugs, insulin dependence is eventually the rule. Diabetes may aggravate the neurological picture by adding diabetic neuropathic complications, such as distal sensory loss. It is therefore important to regularly check FRDA patients for the development of diabetes, as it may increase the burden of disease and even promote potentially fatal complications. Family studies have suggested that having a sib with diabetes increases the risk for an individual with FRDA of developing diabetes up to 45% [14]. These subjects should be tested for glucose tolerance every 6 months.

5. Other Clinical Manifestations

Most patients have skeletal abnormalities, even in the early phase of the disease. Kyphoscoliosis affects 85–100% of patients and may be severe in 10%, particularly when onset is before puberty. It is usually slowly progressive, bearing little relation to the degree of muscle weakness. It may cause pain and cardio-respiratory problems. Pes cavus and pes equinovarus, although found in more than 50% of patients, are not typical of FRDA, being even more common in other neuropathic diseases, such as Charcot–Marie–Tooth (CMT) disease. Autonomic disturbances, most commonly cold and cyanotic legs and feet, occur with increasing frequency as the disease advances [78]. Parasympathetic abnormalities, including decreased heart rate variability parameters, have been reported [79]. Urgency of micturition is rare.

6. Laboratory Investigations

DNA testing is currently used to establish the diagnosis of FRDA. No other laboratory test provides diagnostic results. Lipoproteins, vitamin E, lactate, pyruvate, urinary organic acids, serum very long chain fatty acids (VLCFAs), serum phytanic acid, and leukocyte and/or fibroblast lysosomal enzymes are all normal in FRDA. These tests are occasionally done to exclude some differential diagnoses: α- and β-lipoproteinemia (Bassen–Kornzweig and Tangier diseases), isolated vitamin E deficiency, mitochondrial disorders, adrenoleukodystrophy, adrenomyeloneuropathy, Refusum's disease, and a number of lysosomal storage diseases. Markers of oxidative stress have been found to be increased in FRDA patients, including plasma malondialdeheyde [80], a marker of lipid peroxidation; urinary 8-hydroxydeoxyguanosine [81], a marker of oxidative damage to nucleic acids; and reduced plasma free glutathione [82]. These tests, however, are performed only in a research setting, and the observed abnormalities are not specific to this disease.

7. Neuroimaging

The current structural and functional CNS imaging technologies not only can reveal neuropathological details with extraordinary definition, but can also detect metabolic dysfunctions associated with the disease process.

CNS structural imaging in FRDA patients, by MRI or CT scanning, closely reflects the findings of postmortem neuropathological studies. The spinal cord is the most severely affected structure, appearing much more compromised than any other part of the CNS, including the cerebellum. This pattern of involvement contrasts with patterns of other inherited degenerative ataxias [83], in particular the group of early-onset ataxias with retained reflexes (EOCAs), which have a similar age of onset and are also recessively inherited. In EOCAs, contrary to FRDA, the cerebellum is much more atrophic then the cervical spinal cord. Thinning of the cervical spinal cord can be detected on sagittal and axial images in almost all FRDA patients [83–87]. Magnetic resonance signal abnormalities in the posterior and lateral columns can also be observed, consistent with the degenerative process affecting these structures [87]. As stated, intracranial structures, including brainstem, cerebellum, and cerebrum, are less evidently affected, but they are not completely spared. Accurate assessment of regional atrophy on CT scans, however, indicates that cerebral hemispheres, brainstem, and cerebellum are all smaller in FRDA patients compared with healthy controls, and the extent of this diffuse atrophy correlates with clinical severity [86]. In addition, mild vermian and lobar cerebellar atrophy clearly occurs in more severe and more advanced cases [85, 88]. A morphometric study of the CNS in FRDA patients with current MRI-based approaches has, however, not yet been performed. Such a study would be important to better define the involvement of CNS structures in the disease, and, if performed longitudinally and prospectively, it could provide a measure of its natural history. In addition, recent advances in structural imaging, such as diffusion tensor imaging, allow the direct assessment of the integrity of fiber tracts and may be a way of following the progressive degeneration of long axons. They are now being evaluated in pilot studies.

With respect to functional imaging studies in the CNS, blood flow in the cerebellum, as assessed by technetium-99m-labeled hexamethylpropyleneamine oxime single-photon-emission computed tomography (^{99m}Tc-HMPAO SPECT), appears to be markedly decreased, more than expected for the degree of atrophy [88]. Interestingly, positron emission tomography (PET) scans reveal increased glucose metabolism in the brain of FRDA patients who are still ambulatory. This finding appears to be specific to FRDA [89]. As disease progresses and patients lose their ability to walk, glucose consumption decreases and eventually becomes subnormal [86, 89]. Although not yet fully explained, this abnormality may relate to mitochondrial dysfunction. Mitochondrial dysfunction could be more directly revealed by magnetic resonance spectroscopy (MRS). This technique allows a direct estimate of some chemicals in a given volume of tissue. Phosphorus-31 MRS analysis (^{31}P MRS), in particular, provides levels of phosphorus-containing compounds that are involved in energy metabolism, such as adenosine triphosphate (ATP) and phosphocreatine. ^{31}P-MRS studies have detected deficits in *in vivo* oxidative phosphorylation in the skeletal muscle and the myocardium of FRDA patients [90]. Furthermore, a relationship has been demonstrated between the GAA triplet-repeat expansion and the extent of *in vivo* energy metabolism deficit. In particular, in skeletal muscle there is an inverse linear correlation between the size of the expansion and the post-exercise rate of ATP synthesis [91]. Similar findings are being obtained from the central nervous system (R. Lodi, personal communication).

8. NEUROPHYSIOLOGICAL INVESTIGATIONS

Electromyographic and electroneurographic studies in FRDA reveal the underlying dying-back axonal sensory neuropathy. Sensory action potentials (SAPs) in peripheral nerves are severely reduced or absent, even early in the course of the disease [92–94]. Motor and sensory nerve conduction velocities (NCVs) are within or just below the normal range, a feature that helps in distinguishing an early case of FRDA from a case of demyelinating hereditary sensorimotor neuropathy, such as CMT disease. Dispersion and delay of somatosensory evoked potentials (SEPs), observed in all patients, indicate degeneration of both peripheral and central sensory fibers [13]. Abnormalities in SEPs are not significantly related to disease duration, but rather to the size of the GAA expansion, suggesting that DRG degeneration is an early and scarcely progressive event in FRDA [96]. Brainstem auditory evoked potentials (BAEPs) instead progressively deteriorate in all patients, beginning from the most rostral component, wave V [13, 95]. Visual evoked potentials (VEPs) are commonly (50–90%) reduced in amplitude, but P-100 latency is not increased [13, 61–63, 95]. Analysis of motor evoked potentials by magnetic stimulation reveals slowing of central motor conduction in all cases [97], which is progressive and related to disease duration [98], as expected if pyramidal tract degeneration is a later and progressive feature of the disease.

D. Prognosis

FRDA is a progressive disease that inevitably leads to increasing disability. Patients lose their ability to walk, on average, 15 years after onset, but variability is very large [9, 99]. Early onset and left ventricular hypertrophy appear to be predictors of a faster rate of progression of the disease [66, 99]. The burden of neurological impairment, cardiomyopathy, and occasionally diabetes result in a shortened life expectancy [99]. Older studies found that most patients died in their thirties, but survival may be significantly prolonged by treatment of cardiac symptoms, particularly arrhythmias, by antidiabetic treatment, and by prevention and control of complications resulting from prolonged disability. Carefully assisted patients may live several more decades. No cure is yet available. Thanks to our increasingly better understanding of the pathogenesis of

FRDA, treatments that may affect the degenerative process are now being developed and tested in preclinical and clinical studies. Any success would dramatically change the outlook for these patients, for whom even symptomatic pharmacological treatment has so far been disappointing. In this regard, attempts were made to use drugs that affect neurotransmitters acting in the cerebellar circuitry, the levels of which are thought to be modified, usually reduced, by the degenerative process. Drugs tested include cholinergic agonists, such as physostigmine [100] and choline; neuropeptides, such as TRH [101, 102]; serotoninergic agonists, such as 5-hydroxytryptophan [103, 104] and buspirone; and the dopaminergic drug amantadine [105–107]. No convincingly positive result has been obtained.

Physical therapy can help patients to deal with their neurological deficit. Rehabilitation programs should include exercises aimed at maximizing the residual capacity of motor control. Orthopedic interventions are sometimes necessary, including surgical correction of severe scoliosis and, in patients who can still walk, surgical correction of foot deformity.

III. ISOLATION AND ANALYSIS OF THE FRIEDREICH ATAXIA GENE

A. Mapping and Cloning of the FRDA Gene

During the 1970s and early 1980s, many attempts were made to identify the primary biochemical defect in FRDA [108–110]. Several abnormalities were proposed as the primary defect, including lipoamide dehydrogenase deficiency [111–114], pyruvate carboxylase deficiency [115], mitochondrial malic enzyme deficiency [116–120], increased urinary taurine excretion [121, 122], and abnormal lipids [123]. None of these observations has been confirmed. A positional cloning approach was then undertaken to identify the defective gene in FRDA [124–127]. The primary mapping of the locus was accomplished in 1988 by Chamberlain and her group at St. Mary's Hospital in London [128]. They examined 20 families with at least three affected sibs according to Harding's criteria, and demonstrated linkage without recombination to the anonymous marker D9S15, on chromosome 9. A critical interval of 240 kb was subsequently defined by analysis of recombination events and homozygosity analysis [129–132].

B. Structure of the FRDA Gene

Initially called X25, the FRDA gene was found by starting from a putative exon independently identified by exon amplification and by computer analysis of random cosmid sequences [21]. The gene is composed of seven exons spread over 85 kb of genomic DNA. The first five exons, numbered 1 to 5a, are localized within a 40-kb interval; then a large intron of 30 kb precedes the sixth exon, exon 5b; the seventh exon, exon 6, is an additional 15 kb downstream. Transcription goes in the centromere to telomere direction. The 5' end of the gene, including its first exon, is associated with an unmethylated CpG island. Six of the seven exons of the FRDA gene contain protein coding sequence. Exons 1 to 4 are invariably joined together during the processing of the FRDA transcript. Variability is then introduced in the 3' end of the mature mRNA by the alternative usage of exon 5a or 5b as fifth exon, leading to the synthesis of different protein isoforms differing at their carboxy terminus. On the basis of Northern blot analysis, exon 5a-containing transcripts are much more common in all examined tissues; they encode for a novel protein called *frataxin* [21]. Further variability is introduced by the possible addition of exon 6 to exon 5b-containing transcripts, which occurs when a donor splice site consensus sequence in exon 5b is used instead of a polyadenylation site a few base pairs downstream. However, no changes in the protein sequence are introduced by the usage of exon 6, which is entirely noncoding.

C. Expression of the FRDA Gene

A 1.3-kb major transcript can be identified in all tissues, in agreement with the predicted size of an exon 5a-containing mRNA. Fainter bands of 1.05, 2.0, 2.8, and 7.3 kb are also detected in heart. The 1.05- and 2.0-kb bands correspond to exon 5b- and exon 5b + 6-containing isoforms, respectively. The larger bands derive from a transcribed pseudogene (M. Pandolfo, unpublished results). The FRDA gene shows tissue-specific levels of expression [21]. Among human adult tissues, the highest RNA levels are found in heart and CNS, particularly in the spinal cord; intermediate levels are found in liver, skeletal muscle, and pancreas; and minimal levels in other tissues. Overall, expression of frataxin appears to be high in the primary sites of degeneration in FRDA, both within and outside the CNS.

The developmental expression of the gene has been investigated in mouse embryo and adult by RNA *in situ* hybridization [133, 134]. Expression is negligible until Embryonic Day 10 (E10); then it progressively increases, becoming very clear at E14. Frataxin RNA is found within the CNS, both in proliferating cells of the ependymal layer (periventricular zone) and in more mature cells in the developing forebrain. In the adult, brain expression is restricted to the ependymal layer, choroid plexus, and granular layer of the cerebellum. The spinal cord exhibits a characteristic pattern of expression, highest in the thoracolumbar region, starting around E12.5. Surprisingly with respect to the pathology, expression is higher in the anterior horns than in the posterior horns. The frataxin gene is prominently transcribed in large neuronal cells in the DRG by E12.5 up to adult life. In the mouse developing embryo, the frataxin gene is also prominently expressed in extraneural tissues, such as heart and in tissues that are apparently not affected in FRDA patients, such as liver, muscle, thymus, skin, developing teeth, and brown fat [21, 133, 134] (the last tissue, present in newborns, is particularly rich in mitochondria). Frataxin is highly expressed in mouse adult and fetal kidney, but little is found in human adult kidney, which is not a site of pathology in FRDA [21, 133]. Apart from differences between mouse and human, the discrepancy between frataxin patterns of expression, which is broader than the FRDA sites of pathology, might be accounted for, in part, by the nondividing nature of the affected cell types (neurons, cardiomyocytes, beta cells of the pancreas), meaning that they cannot be replaced when they die, unlike liver, skeletal muscle, and cells expressing frataxin during fetal development.

IV. GENE MUTATIONS IN FRDA

A. Point Mutations

Point mutations in the frataxin gene are a rare cause of FRDA [21]. Only about 2% of FRDA chromosomes carry sequence changes resulting in the premature truncation of frataxin or in an amino acid change of likely functional significance. However, the identification of these point mutations has been essential in identifying the FRDA gene. XX point mutations have been fully characterized so far (Table 17-2). In all cases, affected individuals are heterozygous for their point mutation, with a normal frataxin coding sequence on the other homolog of chromosome 9. As a rule, patients with truncating mutations have a typical FRDA phenotype, while some missense mutations give rise to atypical, usually milder phenotypes. For example, the G130V mutation, found in patients from several different countries, is usually accompanied by a mild phenotype of truncal ataxia with little or no limb ataxia and no dysarthria. Other missense mutations, such as W155R, behave as null mutations and result in typical FRDA. Functional studies on mutant frataxins suggest that milder phenotypes correlate with some residual activity, as evaluated by their ability to complement frataxin deletion in yeast [135].

B. GAA Trinucleotide Repeat Expansion

In the vast majority of FRDA chromosomes (98%), an abnormal GAA repeat expansion occurs within the first intron of the gene encoding frataxin [21]. This intron is 12 kb in size, and the triplet repeat is localized about 1.4 kb after exon 1, in the middle of a repetitive sequence of the Alu-Sx family (Fig. 17-2). The GAA repeat apparently is derived from a polyA expansion of the canonical A5TACA6 sequence linking the two

TABLE 17-2 Frequencies (in %) of Haplotypes Associated with Small Normal (SN), Large Normal (LN), and Expanded (E) GAA Repeat Alleles (from Reference [159])

Haplotype[a]	GAA alleles		
	SN	LN	E
AT2CC	0.7	45.6	50.9
AT3CC	0.0	8.8	20.8
AT2CT	2.9	21.1	14.2
CT1CC	10.8	0.0	2.8
xTxCx	31.7	93.0	96.2
xCxCx	17.3	3.6	3.8
xCxTx	50.0	3.6	0.0
xTxTx	0.7	0.0	0.0
n[b]	139	57	106

[a] Haplotypes are constructed from 5 differents markers with alleles A or C, C or T, 1 to 6 and C or T, respectively. Position of the markers along the genomic map is shown in Fig. 3. Full haplotypes are only shown for some of them. Partial haplotypes, where x means any alleles, are shown in the remaining cases.
[b] n indicates the number of independent analyzed chromosomes.

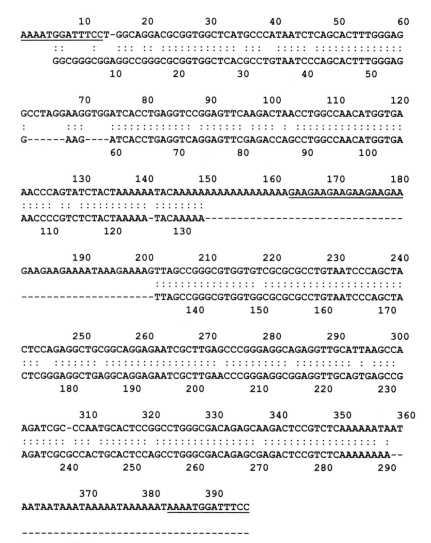

FIGURE 17-2 Genomic sequence flanking the GAA triplet repeat. The GAA-Alu sequence is aligned to the Alu Sx consensus sequence. The most frequent allele (GAA)$_9$ is shown. The GAA repeat is preceeded by an (A)$_{18}$ polyA tract and followed by the AATAAAGAAAAG sequence, which are not part of the Alu consensus sequence. The duplicated sequences flanking the Alu repeat are underlined.

halves of the Alu sequence. Normal chromosomes bear fewer than 40 triplets, whereas FRDA chromosomes have from 66 to >1000. The triplet expansion in FRDA results in a prolonged tract of purines on one strand and pyrimidines on its complement (R·Y sequence).

C. Detection and Diagnostic Value of Expanded GAA Repeats

Two techniques may be used to detect the expanded GAA repeats associated with FRDA: Southern blot (SB) analysis and polymerase chain reaction (PCR). SB is optimally sensitive in demonstrating heterozygotes as well as homozygotes for expanded GAA repeats. By digestion of genomic DNA with the restriction enzyme *Bsi*HKA I, a normal fragment of about 2400 bp containing the GAA repeat can be detected by a frataxin exon 1 probe [136]. Expanded repeats generate larger bands. The main disadvantages of the SB technique are, compared with PCR, the larger amount of DNA required for the analysis and the longer procedure. Several primer pairs have been developed to amplify the GAA repeat [21, 137]. In all cases, the amplification protocol requires special precautions to enhance the efficiency and specificity of the reaction, particularly for heterozygote detection. However, despite some technical difficulties, PCR remains the method of choice to test for expanded GAA repeats. PCR is effective in accurately determining the size of expansions in homo- and heterozygotes, requires little DNA, and is rapidly carried out.

Detection of the expansion mutation thereby provides a most useful diagnostic test. The length of expansion has, however, limited value for individual prognosis (see later text). Finding of a heterozygous mutation in a patient should lead to the search for a point mutation in the other allele, but it should be kept in mind that the frequency of expansion heterozygotes in the normal Caucasian population is about 1 in 90 [138, 139]. Based on the frequency of point mutations in FRDA chromosomes in Caucasians, in the absence of stratification, only 4 in 10,000 FRDA patients would be expected to carry a point mutation on both alleles. However, functional studies and animal models indicate that homozygosity for a mutation totally inactivating the frataxin gene is lethal. Absence of the expansion therefore excludes the diagnosis of FRDA.

D. Instability of the Expanded GAA Triplet Repeat

During examination of a family for the FRDA expanded repeat, it is common to see it changing in size when transmitted from parent to child [21, 66, 136, 137]. Instability during parent–offspring transmission can also be indirectly demonstrated by the detection of two distinct alleles in affected children of consanguineous parents, who are expected to be homozygous by descent at the FRDA locus [21]. Both expansions and contractions of expanded GAA repeats can be observed. In triplet-repeat diseases with dominant inheritance, the dynamic nature of the mutation is reflected in anticipation, where increase in expansion sizes in successive generations correlates with earlier onset and, usually, more rapid progression of disease. Recessive inheritance precludes the occurrence of anticipation, but variation in the size of FRDA expanded alleles does underlie phenotypic variation, regardless of anticipation, as detailed later. Differences in stability between paternal and maternal transmission of the expanded alleles cannot be studied in patients, because establishing the parental origin of each expanded allele would require complex manipulations, such as the isolation of each chromosome 9 homolog in a separate somatic cell hybrid. Such differences are best studied in carrier children, for whom parental origin of expanded allele can easily be determined by linkage analysis. In FRDA, fully expanded alleles most often contract during paternal transmission, but are equally likely to further expand or contract during maternal transmission [140, 141], a result also supported by sperm analysis [140]. In this regard, FRDA resembles the other diseases associated with very large expansions in noncoding regions, such as fragile X syndrome and myotonic dystrophy, whereas smaller expansions of CAG repeats in coding regions, found in dominant ataxias or Huntington's disease, tend to undergo size increases during paternal transmission.

Mitotic instability, leading to somatic mosaicism for expansion sizes, adds to meiotic instability in some diseases associated with large triplet-repeat expansions, such as myotonic dystrophy [142]. The phenomenon can be observed in FRDA as well [66]. Analysis of GAA expansions reveals ample variations in different cell types or tissues from the same patient. Furthermore, heterogeneity among cells occurs at a variable degree in different tissues. For instance, cultured fibroblasts and cerebellar cortex show very little heterogeneity in expansion sizes among cells, lymphocytes are more heterogeneous, and most brain regions show a quite complex pattern of allele sizes, indicating extensive cellular heterogeneity [143]. Although some of these differences could be accounted for by a major period of instability during the first weeks of embryonic development, one is also prompted to conclude that the GAA expanded repeats are inherently more stable in some cell types [143]. In general, it is clear that determining the size of a patient's expansions in peripheral blood lymphocytes, from which DNA is usually obtained, provides only a single sample of the overall repeat size distribution occurring within that patient and, therefore, only an approximate estimate of expansion sizes in affected tissues.

V. ORIGIN OF THE EXPANDED GAA REPEAT

The estimated frequency of GAA expansion carriers is about 1 in 90 in the Caucasian population [10, 24, 25, 27, 28, 138, 139], making it the most common triplet-repeat expansion identified to date. FRDA heterozygotes do not appear to have any specific health problem that could cause reduced fitness [9, 144]. This makes the natural history of the mutation at the population level strikingly different from that of dominant or X-linked diseases due to trinucleotide expansions. Large expansions in these diseases are newly formed from unstable alleles of intermediate sizes, resulting in the phenomenon of anticipation. In FRDA, large expanded alleles are transmitted by asymptomatic carriers, and new expansion events in heterozygotes would have no consequence at the phenotypic level. Only the much rarer homozygotes, who have FRDA, are less likely to reproduce. GAA expansions are therefore maintained in the population with minimal negative selective pressure.

Alleles at the GAA repeat site can be subdivided into three classes depending on their length: short normal (SN) alleles (~82% in Europeans), long normal (LN) alleles (~17% in Europeans), pathological expanded (E) alleles (~1% in Europeans) [145, 146]. The length polymorphism of the GAA repeat in normal alleles suggests that it was generated by two types of events. Small changes, plus or minus one trinucleotide, may have caused limited size heterogeneity. Such small changes were likely to be the consequence of occasional events of polymerase "stuttering" during DNA replication, that is, slippage followed by mis-realignment of the newly synthesized strand by one or, rarely, a few repeat units [147]. This basic polymorphism-generating mechanism has been postulated for all simple sequence repeats [148]. By comparison, the jump from the SN to the LN group was probably a singular event. Linkage disequilibrium (LD) studies were carried out in European, but also Yemenite and North African families, with single-nucleotide polymorphisms (SNPs) spanning the FRDA gene and with polymorphisms of the polyA sequence adjacent to the GAA repeat. These studies indicate that E and LN alleles appear genetically homogeneous and likely related, whereas SN alleles represent a more heterogeneous class of alleles [145, 149].

Possibly, the event that created LN alleles was the sudden duplication of an SN allele containing 8 or 9 GAA triplets, creating an LN allele with 16 or 18 GAA triplets. This occurred presumably in Africa, leading to a population of chromosomes with LN alleles sharing the same background haplotype. Single repeat insertion/deletions, resulting from DNA polymerase "stuttering," gave rise to the spectrum of stable GAA repeats ranging from 12 to about 25 triplets. One or a few of these chromosomes subsequently migrated to Europe and/or to the Middle East, but not to East Asia, where no LN (or E) alleles are found. It is hard to speculate about the mechanism leading to such a sudden doubling of the repeat; however, similar events have been shown to occur in triplet repeats cloned into bacterial plasmids [150]. Recombination-based mechanisms such as unequal sister-chromatid exchange and gene conversion have been proposed as generators of variability in tandem repeats [148] and in microsatellites [151], but alternative hypotheses, such as the occurrence of an exceptionally large slippage event, cannot be excluded. The passage from LN to E alleles probably involved a second genetic event of the same kind that generated "very long" LN alleles containing 32–36 GAA triplets, still on the same haplotype background as the "shorter" LN alleles from which they derived. By reaching the instability threshold, estimated as 34 GAA triplets [146], they form a reservoir for expansions. The occurrence of a second duplication event is suggested by the lack of both E and LN alleles, with more than 21 GAA triplets alleles in sub-Saharan Africans. The ethnic–geographic distribution of FRDA could be explained if the second event occurred prior to the divergence of Indo-Europeans and Afro-Asiatic speakers. According to this scenario, the extent of LD between LN alleles and linked marker loci on chromosomes of African descent is expected to be lower than that between LN and E alleles and the same marker in Europeans [152, 153], as in fact has been observed [22]. Accordingly, LN chromosomes in Africa appear to be 3.2 times older than the LN chromosomes in Europe, and these appear to be 1.27 times older than E chromosomes. Assuming the age of LN African chromosomes to be in the range of 100,000 years, one would date the origin of European LN chromosomes at about 30,000 years ago and that of E chromosomes at about 25,000 years ago, that is, following the Upper Paleolithic population expansion [154].

It was possible to directly observe the hyperexpansion of premutant "very long" LN alleles containing more than 34 GAA triplets. This length is close to the instability threshold for other triplet-repeat-associated disorders, such as those involving CGG and CAG repeats [155].

Strand displacement during DNA replication is thought to be the mechanism that leads to reiterative synthesis and expansion. For this phenomenon to occur, the displaced strand has to form some kind of secondary structure [156]. A single DNA strand containing a GAA repeat is able to form different types of secondary structures [157], which may be involved in instability. A single CTT strand seems structureless [157], and this difference may play a role in determining whether deletions or expansions are favored according to the direction of the replicating fork. Strand displacement is promoted by stalling of DNA polymerase caused by an alternate DNA structure, or by tightly bound proteins, or both [148]. The triplex-forming ability of long FRDA GAA repeats, discussed later, may be involved in repeat instability by causing DNA polymerase stalling as well as by forming a target for protein binding.

VI. PATHOGENIC MECHANISM OF THE GAA EXPANSION

A. Effect of the Expanded GAA Repeat on Frataxin Gene Expression

The expanded GAA repeat has been shown to exert its disease-causing effect by suppressing FRDA gene expression [21, 145]. This loss-of-function pathogenetic mechanism is in accordance with the recessive nature of the disease. FRDA can therefore be defined as a deficiency of frataxin, the protein for which it encodes.

Evidence of frataxin deficiency has been obtained at both the RNA and protein levels. Individuals with FRDA show a severe reduction in the level of mature frataxin mRNA, when tested by ribonuclease (RNase) protection [145], by real-time reverse transcription PCR [158], and by use of microarrays [159]. Heterozygous carriers show levels between those of affected individuals and healthy controls. All portions of frataxin mRNA are reduced in abundance to the same extent, and no evidence of partially processed transcripts is found. These data suggest inhibition of transcription as the most likely mechanism leading to frataxin deficiency, but other possibilities, such as abnormal excision of the expansion-containing intron, cannot be excluded at this time. Reduction in abundance of the transcripts is not linked to abnormal methylation of the CpG island containing exon 1, as indicated by analysis with methylation-sensitive restriction enzymes, contrary to the case of fragile X syndrome [21]. Western blot analysis of tissue samples from FRDA patients confirms a severe but not complete frataxin deficiency [160]. The residual amount of frataxin mRNA and protein is inversely proportional to expansion sizes. A significant amount (20–30% of normal) of mRNA and protein is produced by expanded alleles containing fewer than 300 triplets [160], whereas larger expansion results in smaller amounts of frataxin, down to less than 5% of normal. The existence of a graded effect of expansions on the residual frataxin level provides a biological basis for the correlation between expansion sizes and phenotypic features.

B. Properties of the Expanded GAA Repeat

R·Y sequences such as the FRDA GAA repeat are known to adopt non-B-DNA structures, particularly intramolecular triple helices (Fig. 17-3) [161]. This was shown to occur for GAA repeats containing 38 or more triplets in supercoiled plasmids, both at pH 4.5 and at pH 8.1 [162]. Under physiological conditions, the most likely conformation is an R–R·Y intramolecular triplex structure, where the purine-rich strand of the Watson–Crick R·Y DNA duplex dissociates and winds back down the major groove of the DNA helix, pairing in an antiparallel orientation with the central purine-rich strand, via reverse Hoogsteen hydrogen bonds. Triplex structures have been shown to effectively inhibit transcription [163–165], making this mechanism a particularly attractive possibility. The spontaneous formation of an intramolecular triplex by an R·Y sequence, however, requires DNA supercoiling [161]. A wave of local negative supercoiling is known to be generated during transcriptional elongation *in vivo*, behind the polymerase [166, 167]. This could trigger the formation of an intramolecular triplex [168], potentially blocking further transcription. An effect on transcription was first suggested by transfection experiments in which the expression of a two-exon reporter gene was inhibited by the insertion of a GAA repeat of pathological length in the intron. Those experiments did not reveal splicing abnormalities and provided evidence in favor of a transcription block between the two exons [169]. The GAA repeat is a tract of oligopurines (R) and oligopyrimidines (Y). It has been proposed that the pathological structure adopted by disease-causing lengths of this repeat is a triplex [137, 162, 169, 170–172]. Triplexes are three-stranded nucleic acid structures that can form at such R·Y sequences [148]. The third strand occupies the major groove of the DNA double helix, forming Hoogsteen pairs between R or Y bases of the Watson–Crick base pairs. In intramolecular triplexes, as can be observed *in vitro* in supercoiled plasmid DNA, the R·Y DNA folds back onto itself to form the triple-helical structure. Four different isomers may form, two based on R·R·Y and two on Y·R·Y structures. Intermolecular triplexes are formed between oligo- or polynucleotides (DNA or RNA) and target R·Y sequences on duplex DNA. Thorough investigations of triplexes conducted in the 1980s and 1990s provided substantial information on the type of sequence required, the effects of pH and methylation of C residues, the effect of interposing non-R·Y sequences, the influence of environmental factors on the stabilization of the four triplex isomers, the effect of stabilization by intercalating agents, and related factors [161, 162, 173–185]. R·R·Y triplexes are more versatile than Y·R·Y triplexes because they will tolerate more diverse pairing schemes and because their stability does not depend on lower pH, but depends on the presence of divalent metal ions. This is the type of triplex formed by the FRDA GAA repeat at neutral pH [186].

A new type of DNA structure that implies intramolecular triplex formation was shown to be adopted by lengths of GAA as found in FRDA. This structure was called "sticky DNA" and was first demonstrated in

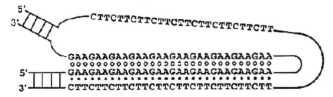

FIGURE 17-3 Model of an intramolecular R-R·Y triple helix formed by a GAA trinucleotide repeat (courtesy of Dr. K. Ohshima).

plasmids containing long tracts of GAA [187]. Sticky DNA was discovered as an anomalously retarded band in agarose gels in which linearized plasmids containing GAA repeats were separated [188]. Such a slow-migrating band was shown to have a number of physicochemical properties that are typical of intramolcular R·R·Y triplexes. In particular, the retarded band appeared only if the plasmid was negatively supercoiled prior to linearization, and it was sensitive to divalent ion concentration and temperature as is typical for R·R·Y triplexes. The possible intermolecular nature of the structure was suggested by the correlation between its abundance and plasmid DNA concentration. This was proven by electron microscopy analysis, which revealed bimolecular complexes formed by joining two plasmids through the region containing the GAA repeat. An excellent correlation was found between the lengths of GAA and the formation of this novel conformation: FRDA patients have 66 or more repeats; sticky DNA was found only for repeats longer than 59 units. *In vitro* transcription studies of $(GAA)_n$ repeats (where n = 9–150) using T7 or SP6 RNA polymerase showed that, when a gel-isolated sticky DNA template was transcribed, the amount of full-length RNA synthesized was significantly reduced compared with the transcription of the linear template. Surprisingly, transcriptional inhibition was observed not only for the sticky DNA template, but also for another DNA molecule used as an internal control in an orientation-independent manner. The molecular mechanism of transcriptional inhibition by sticky DNA was sequestration of the RNA polymerases by direct binding to the complex DNA structure [189]. A $(GAAGGA)_{65}$ sequence, also found in intron 1 of the frataxin gene, does not form sticky DNA, nor does it inhibit transcription *in vivo* and *in vitro* or associate with the FRDA disease state [190]. This finding suggests that interruptions in the GAA sequence may destabilize its structure and facilitate transcription. A systematic analysis of the effects of introducing interruptions into a $(GAA·TTC)_{150}$ repeat by substituting an increasing number of A's with G's has confirmed that the sticky DNA/triplex structure is progressively destabilized and it fails to form when the sequence becomes $(GAAGGA)_{75}$. As the tendency to form a sticky DNA/triplex structure decreases, less and less inhibition of transcription is observed *in vivo* and *in vitro* [191].

VII. PHENOTYPE–GENOTYPE CORRELATION

In FRDA, as in other repeat disorders, regardless of their pathogenetic mechanisms, the size of the expanded repeat is inversely related to age of onset and disease severity [65, 136, 137, 141, 192]. As FRDA patients have two expanded alleles, correlation is more complex than that in dominant or X-linked triplet-repeat diseases, in which patients have only one expanded allele. Earlier age of onset, earlier age when confined to a wheelchair, more rapid rate of disease progression, and presence and severity of cardiomyopathy [136, 193], scoliosis [136], and diabetes [137] best correlate with the size of the smaller repeat (GAA-1). Figure 17-4 is a scattergram showing the relationship between age at onset and GAA-1 in a sample of 140 FRDA patients [136]. The contribution of the size of the larger allele (GAA-2) to the phenotype is more difficult to evaluate, as GAA-2 is not independent of GAA-1 in many respects (size limit imposed by GAA-1, consanguinity, founder events, etc.).

In all studies, correlation coefficients between disease severity parameters and expansion sizes in peripheral blood lymphocytes have not been very high ($r = -0.69$ to -0.75), indicating the existence of additional sources of phenotypic variation. Somatic mosaicism for expansion sizes may be one of these factors. As analysis of lymphocytes provides only one sample of the repeat size distribution occurring within a patient; correlations with the phenotype can only be approximate. However, some studies show striking examples of clinical variability in FRDA that cannot be accounted for by GAA repeat length variations [66, 194]. For example, we observed a FRDA family with three affected siblings with markedly different phenotypic presentations, including one with spastic paraplegia. Molecular analysis showed midsize GAA repeat expansion sizes in all three individuals. The effect of genetic and environmental modifiers remains the most likely hypothesis to explain non-GAA-related variability. Discovering these modifiers would provide important information to better understand pathogenesis and

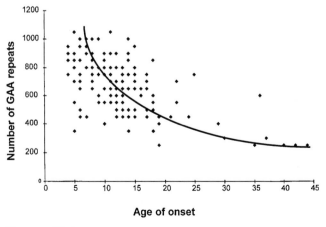

FIGURE 17-4 Scattergram of age of onset versus GAA-1 size in 140 FRDA patients (from Reference [66]).

develop treatment, but unfortunately none has so far been identified.

VIII. CONCLUSION AND PERSPECTIVES

The peculiar nature of the mutation causing FRDA, the expanded GAA triplet repeat, does not affect the coding sequence of frataxin. Patients with this disease therefore make an insufficient amount of an otherwise normal protein. This situation offers a target to develop a therapy, If it were possible to increase frataxin production of FRDA patients, even to levels that are similar to those of healthy carriers, the course of the disease could possibly be stopped and maybe even improved. Increased frataxin production could be achieved in a number of ways.

1. Production could be increased through gene replacement therapy, that is, by introducing a frataxin gene without the GAA expansion into the patient cells.
2. The level of frataxin could be increased directly by administration of frataxin. The protein should, however, be modified in such a way that it is able to reach the nerve cells affected by the disease and the mitochondria inside these cells.
3. Production could be increased by intervening in the GAA expansion with molecules that can destabilize the triple-helical structure and shift the equilibrium toward the physiological double helix that allows frataxin expression.

Though still in their infancy, all these approaches are under study. Recently, encouraging results were obtained with gene replacement therapy, with partial correction of the oxidative stress hypersensitivity of FRDA fibroblasts by frataxin-encoding adeno-associated virus and lentivirus vectors (Fleming et al., 2005).

Other ways to treat the disease may come from studies on the function of frataxin. On the basis of these findings, therapeutic approaches aimed at free radical control and respiratory chain activation may be proposed. With respect to anti-oxidant molecules and respiratory chain stimulants, some coenzyme Q derivatives (idebenone, coenzyme Q-10) have already yielded promising results, not only in experimental models [193], but also in clinical trials, at least on FRDA cardiomyopathy [194, 195]. Automated high-throughput tests to evaluate a large number of molecules for their ability to correct the functional consequences of frataxin deficiency are underway. An intriguing possibility would be the identification of small molecules capable of effectively replacing frataxin by binding mitochondrial iron and increasing its bio-availability. One other possibility in treating FRDA may come from cellular therapies, in particular, the use of stem cells. However, the widespread nature of the neurodegeneration in FRDA is a major obstacle to this approach, because it would require the diffuse delivery of cells to the CNS of the patients.

The most positive note is the remarkable progress that has been made in understanding the pathogenesis of FRDA since the gene responsible for it was discovered in 1996. In addition, investigating the pathogenesis of FRDA has stimulated research on numerous basic areas of biology, from DNA structure and biochemistry to iron metabolism. But most exciting is the now realistic perspective of developing a treatment for this so far incurable neurodegenerative disease.

References

1. Friedrich, N. (1863). Über degenerative Atrophie der spinalen Hinterstränge. *Virchow's Arch. Pathol. Anat.* **27**, 1–26.
2. Friedrich, N. (1863). Über degenerative Atrophie der spinalen Hinterstränge. *Virchow's Arch. Pathol. Anat.* **26**, 433–459.
3. Friedrich, N. (1876). Über ataxie mit besonderer berücksichtigung der hereditären formen. *Virchow's Arch. Pathol. Anat.* **68**, 145–245.
4. Friedrich, N. (1877). Über ataxie mit besonderer berücksichtigung der hereditären formen. *Virchow's Arch. Pathol. Anat.* **70**, 140–142.
5. Brousse, M. (1872). De l'ataxie héréditaire. Thèse de Montpellier.
6. Ladame, P. (1880). Friedreich's disease. *Brain* **13**, 467–537.
7. Bell, J., and Carmichael, E. A. (1929). On hereditary ataxia and spastic paraplegia. *Treas. Hum. Inherit.* **4**, 141–281.
8. Geoffroy, G., Barbeau, A., Breton, G., Lemieux, B., Aube, M., Leger, C., and Bouchard, J. P. (1976). Clinical description and roentgenologic evaluation of patients with Friedreich ataxia. *Can. J. Neurol. Sci.* **3**, 279–286.
9. Harding, A. E. (1981). Friedreich's ataxia: A clinical and genetic study of 90 families with an analysis of early diagnosis criteria and intrafamilial clustering of clinical features. *Brain* **104**, 589–620.
10. Skre, H. (1975). Friedreich's ataxia in western Norway. *Clin. Genet.* **7**, 287–298.
11. Harding, A. E., and Zilkha, K. J. (1981). 'Pseudo-dominant' inheritance in Friedreich's ataxia. *J. Med. Genet.* **18**, 285–287.
12. Filla, A., De Michele, G., Caruso, G., Marconi, R., and Campanella, G. (1990). Genetic data and natural history of Friedreich's disease: A study of 80 Italian patients. *J. Neurol.* **237**, 345–351.
13. Muller-Felber, W., Rossmanith, T., Spes, C., Chamberlain, S., Pongratz, D., and Deufel, T. (1993). The clinical spectrum of Friedreich's ataxia in German families showing linkage to the FRDA locus on chromosome 9. *Clin. Invest.* **71**, 109–114.
14. Winter, R. M., Harding, A. E., Baraitser, M., and Bravery, M. B. (1981). Intrafamilial correlation in Friedreich's ataxia. *Clin. Genet* **20**, 419–427.
15. Filla, A., De Michele, G., Cavalcanti, F., Santorelli, F., Santoro, L., and Campanella, G. (1991). Intrafamilial phenotype variation in

Friedreich's disease: Possible exceptions to diagnostic criteria. *J. Neurol.* **238**, 147–150.

16. Barbeau, A., Roy, M., Sadibelouiz, M., and Wilensky, M. A. (1984). Recessive ataxia in Acadians and "Cajuns." *Can. J. Neurol. Sci* **11**, 526–533.

17. Richter, A., Poirier, J., Mercier, J., Julien, D., Morgan, K., Roy, M., Gosselin, F., Bouchard, J. P., and Melancon, S. B. (1996). Friedreich ataxia in Acadian families from eastern Canada: Clinical diversity with conserved haplotypes. *Am. J. Med. Genet.* **64**, 594–601.

18. Klockgether, T., Chamberlain, S., Wullner, U., Fetter, M., Dittmann, H., Petersen, D., and Dichgans, J. (1993). Late-onset Friedreich's ataxia: Molecular genetics, clinical neurophysiology, and magnetic resonance imaging. *Arch. Neurol.* **50**, 803–806.

19. De Michele, G., Filla, A., Cavalcanti, F., Di Maio, L., Pianese, L., Castaldo, I., Calabrese, O., Monticelli, A., Varrone, S., Campanella, G., and Coccoza, S. (1994). Late onset Friedreich's disease: Clinical features and mapping of mutation to the FRDA locus. *J. Neurol. Neurosurg. Psychiatry* **57**, 977–979.

20. Palau, F., De Michele, G., Vilchez, J. J., Pandolfo, M., Monros, E., Cocozza, S., Smeyers, P., Lopez-Arlandis, J., Campanella, G., Di Donato, S., and Filla, A. (1997). Early-onset ataxia with cardiomyopathy and retained tendon reflexes maps to Friedreich's ataxia locus on chromosome 9q. *Ann. Neurol.* **37**, 359–362.

21. Campuzano, V., Montermini, L., Moltó, M. D., Pianese, L., Cossée, M., Cavalcanti, F., Monros, E., Rodius, F., Duclos, F., Monticelli, A., Zara, F., Cañizares, J., Koutnikova, H., Bidichandani, S., Gellera, C., Brice, A., Trouillas, P., De Michele, G., Filla, A., de Frutos, R., Palau, F., Patel, P. I., Di Donato, S., Mandel, J.-L., Cocozza, S., Koenig, M., and Pandolfo, M. (1996). Friedreich ataxia: Autosomal recessive disease caused by an intronic GAA triplet repeat expansion. *Science* **271**, 1423–1427.

22. Labuda, M., Labuda, D., Miranda, C., Poirier, J., Soong, B., Barucha, N. E., and Pandolfo, M. (2000) Unique origin and specific ethnic distribution of the Friedreich ataxia GAA expansion. *Neurology* **54**, 2322–2324.

23. Harding, A. E. (1983). Classification of the hereditary ataxias and paraplegias. *Lancet* **1**, 1151–1155.

24. Romeo, G., Menozzi, P., Ferlini, A., Fadda, S., Di Donato, S., Uziel, G., and Lucci, B. (1983). Incidence of Friedreich ataxia in Italy estimated from consanguinous marriages. *Am. J. Hum. Genet.* **35**, 523–529.

25. Leone, M., Brignolio, F., Rosso, M. G., Curtoni, E. S., Moroni, A., Tribolo, A., and Schiffer, D. (1990). Friedreich's ataxia: A descriptive epidemiological study in an Italian population. *Clin. Genet.* **38**, 161–169.

26. Lopez-Arlandis, J. M., Vilchez, J. J., Palau, F., and Sevilla, T. (1995). Friedreich's ataxia: An epidemiological study in Valencia, Spain, based on consanguinity analysis. *Neuroepidemiology* **14**, 14–19.

27. Bouchard, J. P., Barbeau, A., Bouchard, R., Paquet, M., and Bouchard, R. W. (1979). A cluster of Friedreich's ataxia in Rimouski, Quebec. *Can. J. Neurol. Sci* **6**, 205–208.

28. Dean, G., Chamberlain, S., and Middleton, L. (1988). Friedreich's ataxia in Kathikas-Arodhes, Cyprus, *Lancet* **1**, 587.

29. Hewer, R. L. (1968). Study of fatal cases of Friedreich's ataxia. *Br. Med. J.* **3**, 649–652.

30. Lamarche, J. B., Lemieux, B., and Lieu, H. B. (1984). The neuropathology of "typical" Friedreich's ataxia in Quebec. *Can. J. Neurol. Sci* **11**, 592–600.

31. Friedreich, N. (1863). Uber degenerative Atrophie der spinalen Hinterstränge. *Virchow's Arch. Pathol. Anat.* **26**, 391–419.

32. Said, G., Marion, M. H., Selva, J., and Jamet, C. (1986). Hypotrophic and dying-back nerve fibers in Friedreich's ataxia. *Neurology* **36**, 1292–1299.

33. Koeppen A. (2001). The neuropathology of inherited ataxias. *In* "The Cerebellum and Its Disorders" (M. Manto and M. Pandolfo, Eds.). Cambridge Univ. Press, London/New York.

34. Waldvogel, D., van Gelderen, P., and Hallett, M. (1999). Increased iron in the dentate nucleus of patients with Friedreich ataxia. *Ann. Neurol.* **46**: 123–125.

35. Carroll, W. M., Kriss, A., Baraitser, M., Barrett, G., and Halliday, A. M. (1980). The incidence and nature of visual pathway involvement in Friedreich's ataxia: A clinical and visual evoked potential study of 22 patients. *Brain* **103**, 413–434.

36. Hughes, J. T., Brownell, B., and Hewer, R. L. (1968). The peripheral sensory pathway in Friedreich's ataxia: An examination by light and electron microscopy of the posterior nerve roots, posterior root ganglia, and peripheral sensory nerves in cases of Friedreich's ataxia. *Brain* **91**, 803–818.

37. Ouvrier, R. A., McLeod, J. G., and Conchin, T. E. (1982). Friedreich's ataxia: Early detection and progression of peripheral nerve abnormalities. *J. Neurol. Sci* **55**, 137–145.

38. Harding, A. E. (1984). "The Hereditary Ataxias and Related Disorders." Churchill Livingstone, Edinburgh/New York.

39. Jitpimolmard, S., Small, J., King, R. H., Geddes, J., Misra, P., McLaughlin, J., Muddle, J. R., Cole, M., Harding, A. E., and Thomas, P. K. (1993). The sensory neuropathy of Friedreich's ataxia: An autopsy study of a case with prolonged survival. *Acta Neuropathol. (Berl.).* **86**, 29–35.

40. Nolano, M., Provitera, V., Crisci, C., Saltalamacchia, A. M., Wendelschafer-Crabb, G., Kennedy, W. R., Filla, A., Santoro, L., and Caruso, G. (2001) Small fibers involvement in Friedreich's ataxia. *Ann. Neurol.* **50**, 17–25.

41. Gottdiener, J. S., Hawley, R. J., Maron, B. J., Bertorini, T. F., and Engle, W. K. (1982). Characteristics of the cardiac hypertrophy in Friedreich's ataxia. *Am. Heart J.* **103**, 525–531.

42. Pasternac, A., Krol, R., Petitclerc, R., Harvey, C., Andermann, E., and Barbeau, A. (1980). Hypertrophic cardiomyopathy in Friedreich's ataxia: Symmetric or asymmetric? *Can. J. Neurol. Sci* **7**, 379–382.

43. Casazza, F., and Morpurgo, M. (1996). The varying evolution of Friedreich's ataxia cardiomyopathy. *Am. J. Cardiol.* **77**, 895–898.

44. Lamarche, J. B., Cote, M., and Lemieux, B. (1980). The cardiomyopathy of Friedreich's ataxia: Morphological observations in 3 cases. *Can. J. Neurol. Sci* **7**, 389–396.

45. Lamarche, J. B., Shapcott, D., Côté, M., and Lemieux, B. (1993). Cardiac iron deposits in Friedreich's ataxia. *In* "Handbook of Cerebellar Diseases" (R. Lechtenberg, Ed.), pp. 453–458. Marcel Dekker, New York.

46. Babcock, M., de Silva, D., Oaks, R., Davis-Kaplan, S., Jiralerspong, S., Montermini, L., Pandolfo, M., and Kaplan, J. (1997). Regulation of mitochondrial iron accumulation by Yfh1, a putative homolog of frataxin. *Science* **276**, 1709–1712.

47. Schoenle, E. J., Boltshauser, E. J., Baekkeskov, S., Landin Olsson, M., Torresani, T., and von Felten, A. (1989). Preclinical and manifest diabetes mellitus in young patients with Friedreich's ataxia: No evidence of immune process behind the islet cell destruction. *Diabetologia* **32**, 378–381.

48. Allard, P., Dansereau, J., Thiry, P. S., Geoffroy, G., Raso, J. V., and Duhaime, M. (1982). Scoliosis in Friedreich's ataxia. *Can. J. Neurol. Sci* **9**, 105–111.

49. Labelle, H., Tohme, S., Duhaime, M., and Allard, P. (1986). Natural history of scoliosis in Friedreich's ataxia. *J. Bone Joint Surg. [Am]* **68**, 564–572.

50. Aronsson, D. D., Stokes, I. A., Ronchetti, P. J., and Labelle, H. B. (1994). Comparison of curve shape between children with cerebral palsy, Friedreich's ataxia, and adolescent idiopathic scoliosis. *Dev. Med. Child. Neurol.* **36**, 412–418.

51. Campanella, G., Filla, A., De Falco, F., Mansi, D., Durivage, A., and Barbeau, A. (1980). Friedreich's ataxia in the south of Italy: A clinical and biochemical survey of 23 patients. *Can. J. Neurol. Sci.* **7**, 351–357.
52. D'Angelo, A., Di Donato, S., Negri, G., Beulche, F., Uziel, G., and Boeri, R. (1980). Friedreich's ataxia in northern Italy: I. Clinical, neurophysiological and in vivo biochemical studies. *Can. J. Neurol. Sci* **7**, 359–365.
53. Ulku, A., Arac, N., and Ozeren, A. (1988). Friedreich's ataxia: A clinical review of 20 childhood cases. *Acta Neurol. Scand.* **77**, 493–497.
54. De Michele, G., Di Maio, L., Filla, A., Majello, M., Cocozza, S., Cavalcanti, F., Mirante, E., and Campanella, G. (1996). Childhood onset of Friedreich ataxia: A clinical and genetic study of 36 cases. *Neuropediatrics* **27**, 3–7.
55. Pandolfo, M. (2003) Friedreich ataxia. *Semin. Pediatr. Neurol.* **10**, 163–172.
56. Gentil, M. (1990). Dysarthria in Friedreich disease. *Brain. Lang.* **38**, 438–448.
57. Cisneros, E., and Braun, C. M. (1995). Vocal and respiratory diadochokinesia in Friedreich's ataxia: Neuropathological correlations. *Rev. Neurol. (Paris)* **151**, 113–123.
58. Beauchamp, M., Labelle, H., Duhaime, M., and Joncas, J. (1995). Natural history of muscle weakness in Friedreich's ataxia and its relation to loss of ambulation. *Clin. Orthop.* 270–275.
59. Badhwar A, Jansen A, Andermann F, Pandolfo M, and Andermann E. (2004) Striking intrafamilial phenotypic variability and spastic paraplegia in the presence of similar homozygous expansions of the FRDA1 gene. *Mov. Disord.* **19**, 1424–1431.
60. Spieker, S., Schulz, J. B., Petersen, D., Fetter, M., Klockgether, T., and Dichgans, J. (1995). Fixation instability and oculomotor abnormalities in Friedreich's ataxia. *J. Neurol.* **242**, 517–521.
61. Kirkham, T. H., and Coupland, S. G. (1981). An electroretinal and visual evoked potential study in Friedreich's ataxia. *Can. J. Neurol. Sci* **8**, 289–294.
62. Livingstone, I. R., Mastaglia, F. L., Edis, R., and Howe, J. W. (1981). Visual involvement in Friedreich's ataxia and hereditary spastic ataxia: A clinical and visual evoked response study. *Arch. Neurol.* **38**, 75–79.
63. Rabiah, P. K., Bateman, J. B., Demer, J. L., and Perlman, S. (1997). Ophthalmologic findings in patients with ataxia. *Am. J. Ophthalmol.* **123**, 108–117.
64. Ell, J., Prasher, D., and Rudge, P. (1984). Neuro-otological abnormalities in Friedreich's ataxia. *J. Neurol. Neurosurg. Psychiatry* **47**, 26–32.
65. Cassandro, E., Mosca, F., Sequino, L., De Falco, F. A., and Campanella, G. (1986). Otoneurological findings in Friedreich's ataxia and other inherited neuropathies. *Audiology* **25**, 84–91.
66. Montermini, L., Richter, A., Morgan, K., Justice, C. M., Julien, D., Castelloti, B., Mercier, J., Poirier, J., Capazzoli, F., Bouchard, J. P., Lemieux, B., Mathieu, J., Vanasse, M., Seni, M. H., Graham, G., Andermann, F., Andermann, E., Melançon, S., Keats, B. J. B., Di Donato, S., and Pandolfo, M. (1997). Phenotypic variability in Friedreich ataxia: Role of the associated GAA triplet repeat expansion. *Ann. Neurol.* **41**, 675–682.
67. Leone, M., Rocca, W. A., Rosso, M. G., Mantel, N., Schoenberg, B. S., and Schiffer, D. (1988). Friedreich's disease: Survival analysis in an Italian population. *Neurology.* **38**, 1433–1438.
68. Hartman, J. M., and Booth, R. W. (1960). Friedreich's ataxia: A neurocardiac disease. *Am. Heart J.* **60**, 716–720.
69. Boyer, S. H., Chisholm, A. W., and McKusick, V. A. (1962). Cardiac aspects of Friedreich's ataxia. *Circulation* **25**, 493–505.
70. Harding, A. E., and Hewer, R. L. (1983). The heart disease of Friedreich's ataxia: A clinical and electrocardiographic changes in 30 cases. *Q. J. Med.* **52**, 489–502.
71. Pentland, B., and Fox, K. A. (1983). The heart in Friedreich's ataxia. *J. Neurol. Neurosurg. Psychiatry* **46**, 1138–1142.
72. Child, J. S., Perloff, J. K., Bach, P. M., Wolfe, A. D., Perlman, S., and Kark, R. A. (1986). Cardiac involvement in Friedreich's ataxia: A clinical study of 75 patients. *J. Am. Coll. Cardiol.* **7**, 1370–1378.
73. Alboliras, E. T., Shub, C., Gomez, M. R., Edwards, W. D., Hagler, D. J., Reeder, G. S., Seward, J. B., and Tajik, A. J. (1986). Spectrum of cardiac involvement in Friedreich's ataxia: Clinical, electrocardiographic and echocardiographic observations. *Am. J. Cardiol.* **58**, 518–524.
74. Maione, S., Giunta, A., Filla, A., De Michele, G., Spinelli, L., Liucci, G. A., Campanella, G., and Condorelli, M. (1997). May age of onset be relevant in the occurrence of left ventricular hypertrophy in Friedreich's ataxia? *Clin. Cardiol.* **20**, 141–145.
75. Morvan, D., Komajda, M., Doan, L. D., Brice, A., Isnard, R., Seck, A., Lechat, P., Agid, Y., and Grosgogeat, Y. (1992). Cardiomyopathy in Friedreich's ataxia: A Doppler-echocardiographic study. *Eur. Heart J.* **13**, 1393–1398.
76. Finocchiaro, G., Baio, G., Micossi, P., Pozza, G., and Di Donato, S. (1988). Glucose metabolism alterations in Friedreich's ataxia. *Neurology* **38**, 1292–1296.
77. Fantus, I. G., Seni, M. H., and Andermann, E. (1993). Evidence for abnormal regulation of insulin receptors in Friedreich's ataxia. *J. Clin. Endocrinol. Metab.* **76**, 60–63.
78. Margalith, D., Dunn, H. G., Carter, J. E., and Wright, J. M. (1984). Friedreich's ataxia with dysautonomia and labile hypertension. *Can. J. Neurol. Sci* **11**, 73–77.
79. Pousset, F., Kalotka, H., Durr, A., Isnard, R., Lechat, P., Le Heuzey, J. Y., Thomas, D., and Komajda, M. (1996). Parasympathetic activity in Friedreich's ataxia. *Am. J. Cardiol.* **78**, 847–850.
80. Emond, M., Lepage, G., Vanasse, M., and Pandolfo, M. (2000) Increased levels of plasma malondialdehyde in Friedreich ataxia. *Neurology* **55**, 1752–1753.
81. Schulz, J. B., Dehmer, T., Schöls, L., Mende, H., Hardt, C., Vorgerd, M., Bürk, K., Matson, W., Dichgans, J., Beal, M. F., and Bogdanov, M. B. (2000). Oxidative stress in patients with Friedreich ataxia. *Neurology* **55**, 1719–1721.
82. Piemonte, F., Pastore, A., Tozzi, G., Tagliacozzi, D., Santorelli, F. M., Carrozzo, R., Casali, C., Damiano, M., Federici, G., Bertini, E. Glutathione in blood of patients with Friedreich's ataxia. (2001). *Eur. J. Clin. Invest.* **31**, 1007–1011.
83. Riva, A., and Bradac, G. B. (1995). Primary cerebellar and spinocerebellar ataxia an MRI study on 63 cases. *J. Nuroradiol.* **22**, 71–76.
84. Wessel, K., Schroth, G., Diener, H. C., Muller-Forell, W., and Dichgans, J. (1989). Significance of MRI-confirmed atrophy of the cranial spinal cord in Friedreich's ataxia. *Eur. Arch. Psychiatry Neurol. Sci.* **238**, 225–230.
85. Wullner, U., Klockgether, T., Petersen, D., Naegele, T., and Dichgans, J. (1993). Magnetic resonance imaging in hereditary and idiopathic ataxia [see comments]. *Neurology* **43**, 318–325.
86. Junck, L., Gilman, S., Gebarski, S. S., Koeppe, R. A., Kluin, K. J., and Markel, D. S. (1994). Structural and functional brain imaging in Friedreich's ataxia. *Arch. Neurol.* **51**, 349–355.
87. Mascalchi, M., Salvi, F., Piacentini, S., and Bartolozzi, C. (1994). Friedreich's ataxia: MR findings involving the cervical portion of the spinal cord. *AJR Am. J. Roentgenol.* **163**, 187–190.
88. Giroud, M., Septien, L., Pelletier, J. L., Dueret, N., and Dumas, R. (1994). Decrease in cerebellar blood flow in patients with Friedreich's ataxia: A TC-HMPAO SPECT study of three cases. *Neurol. Res.* **16**, 342–344.
89. Gilman, S., Junck, L., Markel, D. S., Koeppe, R. A., and Kluin, K. J. (1990). Cerebral glucose hypermetabolism in Friedreich's ataxia detected with position emission tomography. *Ann. Neurol.* **28**, 750–757.

90. Schapira, A., and Lodi, R. (2004). Assessment of in vitro and in vivo mitochondrial function in Friedreich's ataxia and Huntington's disease. *Methods Mol. Biol.* **277**, 293–307.
91. Lodi, R., Cooper, J. M., Bradley, J. L., Manners, D., Styles, P., Taylor, D. J., and Schapira, A. H. (1999). Deficit of in vivo mitochondrial ATP production in patients with Friedreich ataxia. *Proc. Natl. Acad. Sci. USA* **96**, 11492–11495.
92. McLeod, J. G. (1971). An electrophysiological and pathological study of peripheral nerves in Friedreich's ataxia. *J. Neurol. Sci* **12**, 333–349.
93. Peyronnard, J. M., Bouchard, J. P., and Lapointe, M. (1976). Nerve conduction studies and electromyography in Friedreich's ataxia. *Can. J. Neurol. Sci.* **3**, 313–317.
94. Ackroyd, R. S., Finnegan, J. A., and Green, S. H. (1984). Friedreich's ataxia. A clinical review with neurophysiological and echocardiographic findings. *Arch. Dis. Child.* **59**, 217–221.
95. Vanasse, M., Garcia-Larrea, L., Neuschwander, P., Trouillas, P., and Mauguiere, F. (1988). Evoked potential studies in Friedreich's ataxia and progressive early onset cerebellar ataxia. *Can. J. Neurol. Sci* **15**, 292–298.
96. Santoro, L., De Michele, G., Perretti, A., Crisci, C., Cocozza, S., Cavalcanti, F., Ragno, M., Monticelli, A., Filla, A., and Caruso, G. (1999). Relation between trinucleotide GAA repeat length and sensory neuropathy in Friedreich's ataxia. *J. Neurol. Neurosurg. Psychiatry.* **66**, 99–96.
97. Mondelli, M., Rossi, A., Scarpini, C., and Guazzi, G. C. (1995). Motor evoked potentials by magnetic stimulation in hereditary and sporadic ataxia. *Electromyogr. Clin. Neurophysiol.* **35**, 415–424.
98. Santoro, L., Perretti, A., Lanzillo, B., Coppola, G., De Joanna, G., Manganelli, F., Cocozza, S., De Michele, G., Filla, A., and Caruso, G. (2000). Influence of GAA expansion size and disease duration on central nervous system impairment in Friedreich's ataxia: Contribution to the understanding of the pathophysiology of the disease. *Clin. Neurophysiol.* **111**, 1023–1030.
99. De Michele, G., Perrone, F., Filla, A., Mirante, E., Giordano, M., De Placido S., and Campanella, G. (1996). Age of onset, sex, and cardiomyopathy as predictors of disability and survival in Friedreich's disease: A retrospective study on 119 patients. *Neurology* **47**, 1260–1264.
100. Kark, R. A., Budelli, M. M., and Wachsner, R. (1981). Double-blind, triple-crossover trial of low doses of oral physostigmine in inherited ataxias. *Neurology* **31**, 288–292.
101. Le Witt, P. A., and Ehrenkranz, J. R. (1982). TRH and spinocerebellar degeneration [letter]. *Lancet* **2**, 981.
102. Filla, A., De Michele, G., Di Martino, L., Mengano, A., Iorio, L., Maggio, M. A., and Campanella, G. (1989). Chronic experimentation with TRH administered intramuscularly in spinocerebellar degeneration: Double-blind cross-over study in 30 subjects. *Riv. Neurol.* **59**, 83–88.
103. Wessel, K., Hermsdorfer, J., Deger, K., Herzog, T., Huss, G. P., Kompf, D., Mai, N., Schimrigk, K., Wittkamper, A., and Ziegler, W. (1995). Double-blind crossover study with levorotatory form of hydroxytryptophan in patients with degenerative cerebellar diseases. *Arch. Neurol.* **52**, 451–455.
104. Trouillas, P., Serratrice, G., Laplane, D., Rascol, A., Augustin, P., Barroche, G., Clanet, M., Degos, C. F., Desnuelle, C., Dumas, R., et al. (1995). Levorotatory form of 5-hydroxytryptophan in Friedreich's ataxia: Results of a double-blind drug-placebo cooperative study. *Arch. Neurol.* **52**, 456–460.
105. Peterson, P. L., Saad, J., and Nigro, M. A. (1988). The treatment of Friedreich's ataxia with amantadine hydrochloride [see comments]. *Neurology* **38**, 1478–1480.
106. Filla, A., De Michele, G., Orefice, G., Santorelli, F., Trombetta, L., Banfi, S., Squitieri, F., Napolitano, G., Puma, D., and Campanella, G. (1993). A double-blind cross-over trial of amantadine hydrochloride in Friedreich's ataxia. *Can. J. Neurol. Sci.* **20**, 52–55.
107. Botez, M. I., Botez-Marquard, T., Elie, R., Pedraza, O. L., Goyette, K., and Lalonde, R. (1996). Amantadine hydrochloride treatment in heredodegenerative ataxias: A double blind study. *J. Neurol. Neurosurg. Psychiatry* **61**, 259–264.
108. Barbeau, A. (1976). Friedreich's ataxia 1976: A overview. *Can. J. Neurol. Sci.* **3**, 389–397.
109. Barbeau, A. (1978). Friedreich's ataxia 1978: An overview. *Can. J. Neurol. Sci.* **5**, 161–165.
110. Barbeau, A. (1982). Friedreich's disease 1982: Etiologic hypotheses—a personal analysis. *Can. J. Neurol. Sci.* **9**, 243–263.
111. Blass, J. P., Kark, R. A. P., and Menon, N. K. (1976). Low activities of the pyruvate and oxoglutarate dehydrogenase complexes in five patients with Friedreich's ataxia. *N. Engl. J. Med.* **295**, 62–67.
112. Kark, R. A., and Rodriguez-Budelli, M. (1979). Pyruvate dehydrogenase deficiency in spinocerebellar degenerations. *Neurology* **29**, 126–131.
113. Kark, R. A., Budelli, M. M., Becker, D. M., Weiner, L. P., and Forsythe, A. B. (1981). Lipoamide dehydrogenase: Rapid heat inactivation in platelets of patients with recessively inherited ataxia. *Neurology* **31**, 199–202.
114. Robinson, B. H., Sherwood, W. G., Kahler, S., O'Flynn, M. E., and Nadler, H. (1981). Lipoamide dehydrogenase deficiency. *N. Engl. J. Med.* **304**, 53–54.
115. Dijkstra, U. J., Willems, J. L., Joosten, E. M., and Gabreels, F. J. (1983). Friedreich ataxia and low pyruvate carboxylase activity in liver and fibroblasts. *Ann. Neurol.* **13**, 325–327.
116. Stumpf, D. A., Parks, J. K., Eguren, L. A., and Haas, R. (1982). Friedreich ataxia: III. Mitochondrial malic enzyme deficiency. *Neurology* **32**, 221–227.
117. Stumpf, D. A., Parks, J. K., and Parker, W. D. (1983). Friedreich's disease: IV. Reduced mitochondrial malic enzyme activity in heterozygotes. *Neurology* **33**, 780–783.
118. Chamberlain, S., and Lewis, P. D. (1983). Normal mitochondrial malic enzyme levels in Friedreich's ataxia fibroblasts. *J. Neurol. Neurosurg. Psychiatry* **46**, 1050–1051.
119. Gray, R. G., and Kumar, D. (1985). Mitochondrial malic enzyme in Friedreich's ataxia: Failure to demonstrate reduced activity in cultured fibroblasts. *J. Neurol. Neurosurg. Psychiatry* **48**, 70–74.
120. Fernandez, R. J., Civantos, F., Tress, E., Maltese, W. A., and De Vivo, D. C. (1986). Normal fibroblast mitochondrial malic enzyme activity in Friedreich's ataxia. *Neurology* **36**, 869–872.
121. Lemieux, B., Barbeau, A., Beroniade, V., Shapcott, D., Breton, G., Geoffroy, G., and Melancon, S. (1976). Aminoacid metabolism in Friedreich's ataxia. *Can. J. Neurol. Sci.* **3**, 373–378.
122. Lemieux, B., Giguère, R., Barbeau, A., Melancon, S., and Shapcott, D. (1978). Taurine in cerebrospinal fluid in Friedreich's ataxia. *Can. J. Neurol. Sci.* **5**, 125–129.
123. Walker, J. L., Chamberlain, S., and Robinson, N. (1980). Lipids and lipoproteins in Friedreich's ataxia. *J. Neurol. Neurosurg. Psychiatry* **43**, 111–117.
124. Chamberlain, S., Walker, J. L., Sachs, J. A., Wolf, E., and Festenstein, H. (1979). Non-association of Friedreich's ataxia and HLA based on five families. *Can. J. Neurol. Sci.* **6**, 451–452.
125. Koeppen, A. H., Goedde, H. W., Hirth, L., Benkmann, H. G., and Hiller, C. (1980). Genetic linkage in hereditary ataxia. *Lancet* **1**, 92–93.
126. Chamberlain, S., Worrall, C. S., South, S., Shaw, J., Farrall, M., and Williamson, R. (1987). Exclusion of the Friedreich ataxia gene from chromosome 19. *Hum. Genet.* **76**, 186–189.
127. Keats, B. J., Ward, L. J., Lu, M., Krieger, S., Wilensky, M. A., Forster-Gibson, C. J., Roy, M., Monte, M., Barbeau, A., Simpson, N. E., Eiberg, H., Tippett, P., Williamson, R., and Chamberlain, S. (1987). Linkage studies of Friedreich ataxia by

means of blood-group and protein markers. *Am. J. Hum. Genet.* **41**, 627–634.

128. Chamberlain, S., Shaw, J., Rowland, A., Wallis, J., South, S., Nakamura, Y., von Gabain, A., Farrall, M., and Williamson, R. (1988). Mapping of mutation causing Friedreich' ataxia to human chromosome 9. *Nature* **334**, 248–250.

129. Rodius, F., Duclos, F., Wrogemann, K., Le Paslier, D., Ougen, P., Billault, A., Belal, S., Musenger, C., Brice, A., Dürr, A., Mignard, C., Sirugo, G., Weissenbach, J., Cohen, D., Hentati, F., Ben Hamida, M., Mandel, J. L., and Koenig, M. (1994). Recombinations in individuals homozygous by descent localize the Friedreich ataxia locus in a cloned 450-kb interval. *Am. J. Hum. Genet.* **54**, 1050–1059.

130. Monros, E., Smeyers, P., Rodius, F., Cañizares, J., Moltó, M. D., Vilchez, J., Pandolfo, M., Lopez-Arlandis, J., de Frutos, R., Prieto, F., Koenig, M., and Palau, F. (1994). Refined mapping of Friedreich ataxia locus by identification of recombinant events in patients homozygous by descent. *Eur. J. Hum. Genet.* **2**, 291–299.

131. Duclos, F., Rodius, F., Wrogemann, K., Mandel, J.-L., and Koenig, M. (1994). The Friedreich ataxia region: Characterization of two novel genes and reduction of the critical region to 300 kb. *Hum. Mol. Genet.* **3**, 909–914.

132. Montermini, L., Rodius, F., Pianese, L., Moltó, M. D., Cossée, M., Campuzano, V., Cavalcanti, F., Monticelli, A., Palau, F., Gyapay, G., Wenhert, M., Zara, F., Patel, P. I., Cocozza, S., Koenig, M., and Pandolfo, M. (1995). The Friedreich ataxia critical region spans a 150-kb interval on chromosome 9q13. *Am. J. Hum. Genet.* **57**, 1061–1067.

133. Koutnikova, H., Campuzano, V., Foury, F., Dollé, P., Cazzalini, O., and Koenig, M. (1997). Studies of human, mouse and yeast homologues indicate a mitochondrial function for frataxin. *Nat. Genet.* **16**, 345–351.

134. Jiralerspong, S., Liu, Y., Montermini, L., Stifani, S., and Pandolfo, M. (1997). Frataxin shows developmentally regulated tissue-specific expression in the mouse embryo. *Neurobiol. Dis.* **4**, 103–113.

135. Cavadini, P., Gellera, C., Patel, P. I., and Isaya, G. (2000). Human frataxin maintains mitochondrial iron homeostasis in Saccharomyces cerevisiae. *Hum. Mol. Genet.* **12**, 2523–2530.

136. Dürr, A., Cossée, M., Agid, Y., Campuzano, V., Mignard, C., Penet, C., Mandel, J.-L., Brice, A., and Koenig, M. (1996). Clinical and genetic abnormalities in patients with Friedreich's ataxia. *N. Engl. J. Med.* **335**, 1169–1175.

137. Filla, A., De Michele, G., Cavalcanti, F., Pianese, L., Monticelli, A., Campanella, G., and Cocozza, S. (1996). The relationship between trinucleotide (GAA) repeat length and clinical features in Friedreich ataxia. *Am. J. Hum. Genet.* **59**, 554–560.

138. Cossée, M., Schmitt, M., Campuzano, V., Reutenauer, L., Moutou, C., Mandel, J.-L., and Koenig, M. (1997). Evolution of the Friedreich's ataxia trinucleotide repeat expansion: Founder effect and premutations. *Proc. Natl. Acad. Sci. USA* **94**, 7452–7457.

139. Epplen, C., Epplen, J. T., Frank, G., Miterski, B., Santos, E. J. M., and Schöls, L. (1997). Differential stability of the $(GAA)_n$ tract in the Friedreich ataxia gene. *Hum. Genet.* **99**, 834–836.

140. Pianese, L., Cavalcanti, F., De Michele, G., Filla, A., Campanella, G., Calabrese, O., Castaldo, I., Monticelli, A., and Cocozza, S. (1997). The effect of parental gender on the GAA dynamic mutation in the FRDA gene. *Am. J. Hum. Genet.* **60**, 463–466.

141. Monros, E., Moltó, M. D., Martinez, F., Cañizares, J., Blanca, J., Vilchez, J. J., Prieto, F., de Frutos, R., and Palau, F. (1997). Phenotype correlation and intergenerational dynamics of the Friedreich ataxia GAA trinucleotide repeat. *Am. J. Hum. Genet.* **61**, 101–110.

142. Anvret, M., Ahlberg, G., Grandell, U., Hedberg, B., Johnson, K., and Edstrom, L. (1993). Larger expansions of the CTG repeat in muscle compared to lymphocytes from patients with myotonic dystrophy. *Hum. Mol. Genet.* **2**, 1397–1400.

143. Montermini, L., Kish, S. J., Jiralerspong, S., Lamarche, J. B., and Pandolfo, M. (1997). Somatic mosaicism for the Friedreich's ataxia GAA triplet repeat expansions in the central nervous system. *Neurology* **49**, 606–610.

144. Harding, A. E. (1994). "The Hereditary Ataxias and Related Disorders," Churchill Livingstone, Edinburgh/New York.

145. Cossée, M. Campuzano, V., Koutnikova, H., Fischbeck, K. H., Mandel, J.-L., Koenig, M., Bidichandani, S. Patel, P. I., Moltó, M. D., Cañizares, J., de Frutos, R., Pianese, L., Cavalcanti, F., Monticelli, A., Cocozza, S., Montermini, L., and Pandolfo, M. (1997). Frataxin fracas. *Nat. Genet.* **15**, 337–338.

146. Montermini, L., Andermann, E., Richter, A., Pandolfo, M., Cavalcanti, F., Pianese, L., Iodice, L., Farina, G., Monticelli, A., Turano, M., Filla, A., De Michele, G., and Cocozza, S. (1997). The Friedreich ataxia GAA triplet repeat: Premutation and normal alleles. *Hum. Mol. Genet.* **6**, 1261–1266.

147. Richards, R. I., and Sutherland, G. R. (1994). Simple repeat DNA is not replicated simply. *Nat. Genet.* **6**, 114–116.

148. Wells, R. D. (1996) Molecular basis of genetic instability of triplet repeats. *J. Biol. Chem.* **271**, 2875–2878.

149. Monticelli, A., Giacchetti, M., De Biase, I., Pianese, L., Turano, M., Pandolfo, M., and Cocozza, S. (2004). New clues on the origin of the Friedreich ataxia expanded alleles from the analysis of new polymorphisms closely linked to the mutation. *Hum. Genet.* **114**, 458–463.

150. Pluciennik, A., Iyer, R. R., Parniewski, P., and Wells, R. D. (2000). Tandem duplication: A novel type of triplet repeat instability. *J. Biol. Chem.* **275**, 28386–28397.

151. Jakupciak, J. P., and Wells, R. D. (2000). Gene conversion (recombination) mediates expansions of CTG·CAG repeats. *J. Biol. Chem.* **275**, 40003–40013.

152. Labuda, D., Labuda, M., and Zietkiewicz, E. (1997). The genetic clock and the age of the founder effect in growing populations: A lesson from French Canadians and Ashkenazim. *Am. J. Hum. Genet.* **61**, 768–771.

153. Harpending, H. C., Batzer, M. A., Gurven, M., Jorde, L. B., Rogers, A. R., and Sherry, S. T. (1998). Genetic traces of ancient demography. *Proc. Natl. Acad. Sci. USA* **95**, 1961–1967.

154. Geschwind, D. H., Perlman, S., Grody, W., Telatar, M., Montermini, L., Pandolfo, M., and Gatti, R. A. (1997). The Friedreich's ataxia GAA repeat expansion in patients with recessive or sporadic ataxia. *Neurology* **49**, 1004–1009.

155. Eichler, E. E., Holden, J. J. A., Popovich, B. W., Reiss, A. L., Snow, K., Thibodeau, S. N., Richards, C. S., Ward, P. A., and Nelson, D. L. (1994). Length of uninterrupted CGG repeats determines instability in the FMR1 gene. *Nat. Genet.* **8**, 88–94.

156. Parniewski, P., and Staczek, P. (2002). Molecular mechanisms of TRS instability. *Adv. Exp. Med. Biol.* **516**, 1–25.

157. LeProust, E. M., Pearson, C. E., Sinden, R. R., and Gao, X. (2000). Unexpected formation of parallel duplex in GAA and TTC trinucleotide repeats of Friedreich's ataxia. *J. Mol. Biol.* **302**, 1063–1080.

158. Pianese, L., Turano, M., Lo Casale, M. S., De Biase, I., Giacchetti, M., Monticelli, A., Criscuolo, C., Filla, A., and Cocozza, S. (2004). Real time PCR quantification of frataxin mRNA in the peripheral blood leucocytes of Friedreich ataxia patients and carriers. *J. Neurol. Neurosurg. Psychiatry* **75**, 1061–1063.

159. Coppola, G., Choi, S-H., Santos, M., Tentler, D., Wexler, E. M., Koeppen, A. H., Pandolfo, M., and Geschwind, D. H. Gene expression profiling in frataxin-deficient mice: Microarray evidence for significant expression changes without detectable neurodegeneration. *Neurobiol. Dis.* 2006, Jan. 29. [Epub ahead of print].

160. Campuzano, V., Montermini, L., Lutz, Y., Cova, L., Hindelang, C., Jiralerspong, S., Trottier, Y., Kish, S. J., Faucheux, B., Trouillas, P., Authier, F. J., Dürr, A., Mandel, J.-L., Vescovi, A. L., Pandolfo, M.,

and Koenig, M. (1997). Frataxin is reduced in Friedreich ataxia patients and is associated with mitochondrial membranes. *Hum. Mol. Genet.* **6**, 1771–1780.

161. Wells, R. D., Collier, D. A., Hanvey, J. C., Shimizu, M., and Wohlrab, F. (1988). The chemistry and biology of unusual DNA structures adopted by oligopurine·oligopyrimidine sequences. *FASEB. J.* **2**, 2939–2949.

162. Ohshima, K., Kang, S., Larson, J. E., and Wells, R. D. (1996). Cloning, characterization, and properties of seven triplet repeat DNA sequences. *J. Biol. Chem.* **271**, 16773–16783.

163. Grabczyk, E., and Fishman, M. C. (1995). A long purine–pyrimidine homopolymer acts as a transcriptional diode. *J. Biol. Chem.* **270**, 1791–1797.

164. Cooney, M., Czernuszewicz, G., Postel, E. H., Flint, S. J., and Hogan, M. E. (1988). Site-specific oligonucleotide binding represses transcription of the human c-myc gene in vitro. *Science* **241**, 456–459.

165. Duval-Valentin, G., Thuong, N. T., and Helene, C. (1992). Specific inhibition of transcription by triple helix-forming oligonucleotides. *Proc. Natl. Acad. Sci. USA* **89**, 504–508.

166. Liu, L. F., and Wang, J. C. (1987). Supercoiling of the DNA template during transcription. *Proc. Natl. Acad. Sci. USA* **84**, 7024–7027.

167. Wu, H. Y., Shyy, S. H., Wang, J. C., and Liu, L. F. (1988). Transcription generates positively and negatively supercoiled domains in the template. *Cell* **53**, 433–440.

168. Htun, H., and Dahlberg, J. E. (1989). Topology and formation of triple-stranded H-DNA. *Science* **243**, 1571–1576.

169. Ohshima, K., Montermini, L., Wells, R. D., and Pandolfo, M. (1998). Inhibitory effects of expanded GAA·TTC triplet repeats from intron 1 of Friedreich's ataxia gene on transcription and replication in vivo. *J. Biol. Chem.* **273**, 14588–14595.

170. Bidichandani, S. I., Ashizawa, T., and Patel, P. I. (1998). The GAA triplet-repeat expansion in Friedreich's ataxia interferes with transcription and may be associated with an unusual DNA structure. *Am. J. Hum. Genet.* **62**, 111–121.

171. Grabczyk, E., and Usdin, K. (2000). The GAA*TTC triplet repeat expanded in Friedreich's ataxia impedes transcription elongation by T7 RNA polymerase in a length and supercoil dependent manner. *Nucleic Acids Res.* **28**, 2815–2822.

172. Gacy, A. M., Goellner, G. M., Spiro, C., Dyer, R., Mikesell, M., Yao, J. Z., Johnson, A. J., Juranic, N., Macura, S., Richter, A., Melançon, S. B., and McMurray C. T. (1997). "DNA Structures Associated with Class I Expansion of GAA in Friedreich's Ataxia." Poster CS3-103, presented at Santa Fe meeting on "Unstable Triplets, Microsatellites, and Human Disease;" April 1–6, 1997 (J. Griffith, R. D. Wells, and D. L. Nelson, organizers).

173. Frank-Kamenetskii, M. D., and Mirkin, S. M. (1995). Triplex DNA structures. *Annu. Rev. Biochem.* **64**, 65–95.

174. Soyfer, V. N., and Potaman, V. N. (1996). *In* "Triple-Helical Nucleic Acids. (R. C. Garber, Ed.), Springer-Verlag, New York.

175. Sinden, R. R. (1994). "DNA Structure and Function." Academic Press, San Diego.

176. Guieysse, A-L., Praseuth, D., Grigoriev, M., Harel-Bellan, A., and Helene, C. (1996). Detection of covalent triplex with human cells. *Nucleic Acids Res.* **24**, 4210–4216.

177. Bacolla, A., Ulrich, M. J., Larson, J. E., Ley, T. J., and Wells, R. D. (1995). An Intramolecular triplex in the human gamma-globin 5′-flanking region is altered by point mutations associated with hereditary persistence of fetal hemoglobin. *J. Biol. Chem.* **270**, 24556–24563.

178. Xu, G., and Goodridge, A. G. (1996). Characterization of a polypyrimidine/polypurine tract in the promoter of the gene for chicken malic enzyme. *J. Biol. Chem.* **271**, 16008–16019.

179. Hanvey, J. C., Shimizu, M., and Wells, R. D. (1989). Site-specific inhibition of *Eco*RI restriction/modification enzymes via DNA triple helix. *Nucleic Acids Res.* **18**, 157–161.

180. Shimizu, M., Hanvey, J. C., and Wells, R. D. (1989). Intramolecular DNA triplexes in supercoiled plasmids: I. Effect of loop size on formation and stability. *J. Biol. Chem.* **264**, 5944–5949.

181. Hanvey, J. C., Shimizu, M., and Wells, R. D. (1989). Intramolecular DNA triplexes in supercoiled plasmids: II. Effect of base composition and non-central interruptions on formation and stability. *J. Biol. Chem.* **264**, 5950–5956.

182. Hanvey, J. C., Shimizu, M., and Wells R. D. (1989). Site-specific inhibition of *Eco*RI restriction/modification enzymes via DNA triple helix. *Nucleic Acids Res.* **18**, 157–161.

183. Shimizu, M., Hanvey, J. C., and Wells, R. D. (1990). Multiple non-B-DNA conformations of polypurine·polypyrimidine sequences in plasmids. *Biochemistry* **29**, 4704–4713.

184. Kang, S., Wohlrab, F., and Wells, R. D. (1992). GC rich flanking tracts decrease the kinetics of intramolecular DNA triplex formation. *J. Biol. Chem.* **267**, 19435–19442.

185. Kang, S., Wohlrab, F., and Wells, R. D. (1992). Metal ions cause the isomerization of certain intramolecular triplexes. *J. Biol. Chem.* **267**, 1259–1264.

186. Potaman, V. N., Oussatcheva, E. A., Lyubchenko, Y. L., Shlyakhtenko, L. S., Bidichandani, S. I., Ashizawa, T., and Sinden, R.R. (2004). Length-dependent structure formation in Friedreich ataxia $(GAA)_n*(TTC)_n$ repeats at neutral pH. *Nucleic Acids Res.* **32**, 1224–1231.

187. Sakamoto, N., Chastain, P. D., Parniewski, P., Ohshima, K., Pandolfo, M., Griffith, J. D., and Wells, R. D. (1999). Sticky DNA: self-association properties of long GAA·TTC repeats in R·R·Y triplex structures from Friedreich ataxia. *Mol. Cell* **3**: 465–475.

187. Sakamoto, N., Ohshima, K., Montermini, L., Pandolfo, M., and Wells, R. D. (2001). Sticky DNA, a self-associated complex formed at long GAA·TTC repeats in intron 1 of the frataxin gene, inhibits transcription. *J. Biol. Chem.* **276**, 27171–27177.

188. Ohshima, K., Sakamoto, N., Labuda, M., Poirier, J., Moseley, M. L., Montermini, L., Ranum, L. P., Wells, R. D., and Pandolfo, M. (1999). A nonpathogenic GAAGGA repeat in the Friedreich gene: Implications for pathogenesis. *Neurology* **53**, 1854–1857.

189. Sakamoto, N., Larson, J. E., Iyer, R. R., Montermini, L., Pandolfo, M., and Wells, R. D. (2001). GGA*TCC-interrupted triplets in long GAA·TTC repeats inhibit the formation of triplex and sticky DNA structures, alleviate transcription inhibition, and reduce genetic instabilities. *J. Biol. Chem.* **276**, 27178–27187.

190. Lamont, P. J., Davis, M. B., and Wood, N. W. (1997). Identification and sizing of the GAA trinucleotide repeat expansion of Friedreich's ataxia in 56 patients: Clinical and genetic correlates. *Brain* **120**, 673–680.

191. Isnard, R., Kalotka, H., Dürr, A., Cossée, M., Schmitt, M., Pousset, F., Thomas, D., Brice, A., Koenig, M., and Komajda, M. (1997). Correlation between left ventricular hypertrophy and GAA trinucleotide repeat length in Friedreich's ataxia. *Circulation* **95**, 2247–2249.

192. Fleming, J., Spinoulas, A., Zheng, M., et al. Partial correction of sensitivity to oxidant stress in friedreich ataxia patient fibroblasts by frataxin-encoding adeno-associated virus and lentivirus vectors. *Hum Gene Ther*. 2005; **16**: 947–956

193. Seznec, H., Simon, D., Monassier, L., et al. Idebenone delays the onset of cardiac functional alteration without correction of Fe-S enzymes deficit in a mouse model for Friedreich ataxia. *Hum Mol Genet* 2004; **13**: 1017–1024.

194. Lodi, R., Rajagopalan, B., Blamire, A. M., et al. Cardiac energetics are abnormal in Friedreich ataxia patients in the absence of cardiac dysfunction and hypertrophy: an in vivo 31P magnetic resonance spectroscopy study. *Cardiovasc Res*. 2001; **52**: 111–119.

195. Buyse, G., Mertens, L., Di Salvo, G., et al. Idebenone treatment in Friedreich's ataxia: Neurological, cardiac, and biochemical monitoring. *Neurology* 2003; **60**: 1679–1681.

CHAPTER 18

Experimental Therapeutics for Friedreich's Ataxia

ROBERT B. WILSON

Department of Pathology and Laboratory Medicine, University of Pennsylvania,
Stellar-Chance Laboratories, Philadelphia, Pennsylvania

I. Introduction
II. Parabenzoquinones
   A. Rationale
   B. Idebenone
   C. Coenzyme Q10
   D. MitoQ
   E. EPI-A0001
III. Selenium and Glutathione Peroxidase Mimetics
IV. Creatine and Carnitine
V. *FRDA* Upregulation
VI. Summary
   References

Friedreich's ataxia (FRDA) research has now entered the treatment era. Preliminary data on the use of the parabenzoquinones idebenone and coenzyme Q10 (CoQ10) indicate that they are safe and well tolerated, and suggest possible benefit for the cardiomyopathy of FRDA. Phase II trials of these compounds are in progress, phase III trials are in the planning stages, and additional parabenzoquinones and parabenzoquinone analogs are in development. Development of improved ataxia scales for phase III trials is in progress. Positive trends were noted in mitochondrial ATP production after 4 months of treatment with L-carnitine. Selenium and glutathione peroxidase mimetics enhanced cell viability in tissue culture models of FRDA, and high-throughput drug screens using cell-based models of FRDA are in progress. The presence of intact exonic sequences in most disease alleles opens up the possibility of upregulating expression of the disease gene for FRDA; several investigators are pursuing this possibility. One or more of these approaches, perhaps in combination, could potentially halt, and perhaps even reverse, aspects of the disorder.

I. INTRODUCTION

Since the breakthrough identification of the disease gene for Friedreich's ataxia (FRDA) in 1996, progress in understanding the nature of the genetic defect and in elucidating the function of the encoded protein, frataxin, has been so rapid that, less than 10 years later, a large number of treatment possibilities are being pursued. The clinical presentation, genetics, and pathology of FRDA are discussed in detail in Chapter 17 [1]. This chapter describes some of the current experimental therapeutics for FRDA, from approaches designed to mitigate the effects of decreased frataxin function to approaches designed to increase frataxin expression. The furthest advanced of these approaches, with clinical trials in progress, is the use of parabenzoquinones, which are described first.

II. PARABENZOQUINONES

A. Rationale

The rationale for the use of parabenzoquinones such as coenzyme Q10 (CoQ10) in the treatment of FRDA derives from the function of the disease protein, frataxin, which chaperones iron in the mitochondrial matrix and promotes iron utilization, particularly incorporation into iron–sulfur clusters [2–7]. Iron–sulfur clusters are prosthetic groups important for the functions of many mitochondrial proteins, including mitochondrial respiratory complexes I, II, and III. Impaired iron–sulfur cluster assembly would be expected to decrease mitochondrial function by impeding the normal flux of electrons through the electron transport chain, thereby increasing the production of free radicals. FRDA cells also accumulate mitochondrial iron [8–10], likely as a secondary consequence of the iron–sulfur cluster assembly defect. Ferrous iron generates toxic reactive oxygen species (ROS) by reducing oxygen to the superoxide radical and reducing hydrogen peroxide to the hydroxyl radical (the Fenton reactions) [11]. Yeast depleted of Yfh1p, the yeast frataxin homolog, exhibit impaired iron–sulfur cluster assembly, decreased activities of mitochondrial iron–sulfur cluster enzymes, impaired mitochondrial respiration, sensitivity to oxidative stress, and mitochondrial iron accumulation [6–8, 12–15]. The myocardium of patients with FRDA exhibits specific decreases in the activities of mitochondrial iron–sulfur cluster enzymes, including respiratory complexes I, II, and III, as well as impaired mitochondrial bioenergetics [16–18]. Urinary 8-hydroxy-2′-deoxyguanosine (8OH2′dG) and plasma malondialdehyde, both markers of oxidative stress, are increased in patients with FRDA [19, 20].

Contradictory evidence regarding the role of oxidative stress in FRDA derives from striated-muscle frataxin-deficient mice and neuron/cardiac frataxin-deficient mice, which were generated using a conditional knockout approach [10]. Together, these mice recapitulate the major pathophysiological and biochemical characteristics of FRDA, including progressive large sensory neuronal dysfunction, hypertrophic cardiomyopathy, decreased activities of iron–sulfur cluster enzymes, and a secondary accumulation of mitochondrial iron [10]. Yet a variety of well-designed experiments indicated that oxidative stress in the striated-muscle frataxin-deficient mice was not increased relative to that of normal controls, and superoxide dismutase mimetics did not enhance survival [21]. A possible explanation for these results is that individual cells in the conditional knockout models lack frataxin completely, whereas individual cells from patients with FRDA still express residual frataxin. By 7 weeks of age, respiratory chain enzyme activities in the cardiac tissue of the striated-muscle frataxin-deficient mice were decreased to only 20% of normal; mitochondria in individual cardiomyocytes may have shut down completely. Partially dysfunctional mitochondria, which are likely present for a longer period in humans with FRDA than in the conditional knockout models, would be expected to increase intracellular oxidative stress. Nevertheless, how significantly oxidative stress contributes to the signs and symptoms of FRDA in humans remains an open question.

B. Idebenone

Idebenone is a short-chain parabenzoquinone derivative with a structure very similar to that of CoQ10. Like CoQ10, idebenone transfers electrons from complexes I and II to complex III in the mitochondrial electron transport chain [22]. When incubated with intact mitochondria *in vitro*, idebenone is rapidly reduced in the presence of NADH or succinate; this reduction is more rapid than for exogenous CoQ10 [22]. Reduced idebenone stimulates state 3 respiration, an effect that is blocked by the complex III inhibitor antimycin A [21]. Idebenone also decreases lipid peroxidation in the presence of $FeCl_3$, apparently by interacting with complex I and accepting electrons otherwise destined for oxygen to form superoxide radicals [21]. Idebenone was marketed in Japan from 1986 to 1998 for the treatment of cognitive difficulties following stroke, with an estimated eight million patients treated in this period. In clinical trials for Alzheimer's disease, Huntington's disease, and multi-infarct dementia, in which a variety of doses and dosing regimens were employed, idebenone was found to be safe and well tolerated.

In heart homogenates from patients with valvular stenosis, Fe^{2+}, but not Fe^{3+}, decreased the activity of respiratory complex II and increased lipoperoxidation [23]. Addition of the water-soluble antioxidant ascorbate increased lipoperoxidation by reducing Fe^{3+} to Fe^{2+}. Although the addition of the iron chelator desferrioxamine protected complex II from iron injury, the activity of the Krebs cycle enzyme aconitase was significantly decreased. Idebenone protected complex II, lipids, and aconitase from iron injury in the heart homogenates. Based on these results, idebenone was given to three individuals with FRDA at 5 mg/kg/day. Left ventricular mass index decreased in all three patients after 4–9 months of therapy [23].

Hausse et al. gave 5 mg/kg/day idebenone to 38 patients with FRDA in a 6-month, open-label trial [24].

Ataxia scale measurements, using the International Cooperative Ataxia Rating Scale (ICARS), did not improve significantly. However, ataxia scales such as the ICARS are relatively insensitive to change, and anecdotal evidence of improvement in ataxia was noted. For example, in several patients, parents or teachers noted an improvement in fine-motor coordination (such as in handwriting), more fluent speech, and a decrease in swallowing difficulties. Left ventricular mass index (LVMI), as measured by echocardiography, decreased 14% ($P < 0.001$ by paired t test) after 6 months of treatment. One patient whose left ventricular mass index remained unchanged after 6 months of 5 mg/kg/day idebenone responded after 6 months of 10 mg/kg/day, suggesting a dose–response effect.

Fournier et al. gave 5 mg/kg/day idebenone to 11 patients with FRDA in a 1-year, open-label trial and obtained similar results, with the LVMI decreasing significantly ($P = 0.001$), but ICARS measurements, and measurements of cardiac function, remaining unchanged [25]. Leber et al. gave 5 mg/kg/day idebenone to 31 adults and 17 children with FRDA in a 3-year, open-label trial [26]. LVMI decreased significantly in both the adults and the children after 12 months, and this decrease was sustained after 3 years. ICARS scores were stable in the adults but worsened significantly in the children, and ocular square wave jerks worsened significantly in both groups.

In a double-blind, placebo-controlled trial of idebenone for 28 patients with FRDA, Mariotti et al. found favorable differences in LVMI ($P = 0.048$) and interventricular septal (IVS) thickness ($P = 0.05$) after 6 months of 5 mg/kg/day treatment [27]. After 12 months of treatment the differences in LVMI and IVS thickness increased ($P = 0.01$ and $P = 0.004$, respectively). Only patients with IVS or posterior wall (PW) thicknesses >12 mm were included. No changes in ejection fraction (EF) or ataxia measures (ICARS) were seen.

In an open trial of idebenone for eight patients with FRDA, Buyse et al. noted improvements in LVMI ($P = 0.03$), longitudinal cardiac strain ($P < 0.001$) and strain rate ($P < 0.001$), and radial cardiac strain rate ($P < 0.05$) after 1 year of 5 mg/kg/day treatment [28]. Scores on the Cooperative Ataxia Group Rating Scale (CAGRS) worsened slightly ($P = 0.016$) over the course of the year. Because cardiac strain and strain rate specifically worsen with fibrosis, the improvements in cardiac strain and strain rate reported by Buyse et al. argue strongly against the hypothesis that the decrease in cardiac hypertrophy with idebenone treatment represents shrinkage due to fibrosis.

In an open trial of idebenone for nine patients with FRDA, Artuch et al. reported improvements in ICARS scores after 3 months ($P = 0.017$), 6 months ($P = 0.012$), and 12 months ($P = 0.007$) relative to baseline, as well as improvements in ICARS scores between 3 and 6 months ($P = 0.011$), with 5 mg/kg/day treatment [29]. LVMI did not change significantly; however, most of the patients in the cohort studied lacked cardiac hypertrophy at baseline.

The inconsistent effects of idebenone on the ataxia scale scores of patients with FRDA may derive in part from the insensitivity to change of rating scales, such as the ICARS and CAGRS, based largely on the neurological examination. In addition, long sensory neurons undergo apoptosis during disease progression in FRDA. Given the irreversibility of this neuronal loss, one might expect that a beneficial effect of idebenone would be reflected more by neuroprotection, and might thus be subtle, than by a dramatic reversal of neurological signs and symptoms. It is also possible that higher doses of idebenone are required to achieve effective concentrations in neuronal cells.

The data of Artuch et al. [29] provide another possible explanation for the inconsistent effects of idebenone on ataxia scale scores in the clinical trials described earlier. All of the patients studied by Artuch et al. improved in ICARS subscores measuring fine manipulation, nystagmus, and eye movements, whereas only patients with the lowest number of GAA repeats in their FRDA alleles improved in ICARS subscores measuring kinetic function, posture, and gait. The apparently more favorable response of patients with the lowest number of GAA repeats could be explained by the significant negative correlations observed between serum idebenone concentrations (which ranged from 0.04 to 0.37 μmol/L) and GAA repeats ($r = -0.736$, $P = 0.024$) and the ICARS scores after 12 months of treatment ($r = -0.817$, $P = 0.007$). These data suggest that patients with larger GAA repeat expansions may absorb idebenone less well and, therefore, may respond neurologically less well. Buyse et al. [28] did not report GAA repeat sizes. However, the study by Hausse et al. [24] included only patients with more than 300 GAA repeats in each *FRDA* allele, and the study by Mariotti et al. [27] included only patients with 500 or more GAA repeats in each *FRDA* allele. Taken together, these studies, none of which reported any adverse effects of idebenone at 5 mg/kg/day, support the need for testing higher doses in patients with FRDA.

Dr. Paul Taylor and colleagues conducted an open-label, dose escalation trial to determine the maximum tolerated one-day dose of idebenone in children, adolescents, and adults with FRDA (personal communication). Three patients in each of the three age cohorts received oral idebenone starting at 0.83 mg/kg three times daily (2.5 mg/kg/day, children and adolescents) or 1.7 mg/kg three times daily (5.0 mg/kg/day, adults). Patients were monitored by physical examination, electrocardiograms, blood tests, and frequent checks of vital signs. All cohorts (adult, adolescent, and children) completed dose escalation to the maximum dose level allowed under the protocol (75 mg/kg). Adverse events were graded in accordance with the Common Toxicity Criteria, Version 2.0, as developed by the Cancer Therapy Evaluation Program of the National Cancer Institute (http://ctep.cancer.gov/forms/CTCv20-4-30-992.pdf). Grade 2 was considered dose-limiting toxicity (DLT), and no DLT was observed. Adverse events were infrequent and mild (most commonly nausea) and occurred over a range of dose levels; hence it is uncertain whether any of the adverse events were drug related.

Dr. Nicholas Di Prospero and colleagues conducted an open-label trial to examine the toxicity and tolerability of "high-dose" idebenone administered as a multiple-dose regimen (60 mg/kg/day) for a short inpatient course, and then over 1 month, to patients with FRDA (personal communication). Of the 14 patients who completed the trial, compliance with the three-times-daily regimen was excellent, with greater than 95% compliance with all scheduled doses. Open-ended questioning of patient's health during biweekly phone inquiries revealed that more than 60% of patients reported subjective improvement in their overall condition. These observations included decreased fatigue, improved balance and stability, improved peripheral sensation, and improved fine-motor tasks such as handwriting. These results suggest that higher doses of idebenone are relatively well tolerated and should be explored for efficacy in patients with FRDA.

The only double-blind, placebo-controlled trial of idebenone for FRDA of significant length was that of Mariotti et al. [27], described earlier, who found a statistically significant decrease in cardiac hypertrophy. Taken together with the concordant results from open trials, there can be little question that idebenone decreases the cardiac hypertrophy associated with hypertrophic cardiomyopathy in FRDA. In addition to the lack of any evidence that the decrease in cardiac hypertrophy after treatment with idebenone for 6 months or longer is secondary to a deleterious effect of the drug on the heart, the study of Buyse et al. [28] provides reassurance that the decrease in hypertrophy is in fact secondary to a beneficial effect on cardiac function. In particular, the study of Buyse et al. essentially rules out that the decrease in cardiac hypertrophy is secondary to fibrosis.

There are, however, insufficient data on the natural history of FRDA cardiomyopathy to say for certain that a decrease in hypertrophy, or an improvement in any other cardiac index, represents a true surrogate for clinical efficacy. Cardiac scales such as the New York Heart Association functional classification are of little use in a population whose activities are limited primarily by ataxia. Clinical endpoints related to cardiac function could include a decreased incidence or severity of congestive heart failure, a decreased incidence or severity of cardiac arrhythmias, or decreased mortality, which is usually associated with end-stage cardiac disease. Although such endpoints would likely be acceptable in phase III studies, a very long and/or large trial might still be required. Particular subpopulations of patients with FRDA (late stage, symptomatic, large repeat expansions, etc.) would likely facilitate such a trial. Alternatively, unpublished data suggest that ataxia scales based on performance measures are sufficiently sensitive to disease progression that they could be used as primary endpoints in a phase III trial of idebenone for the ataxia of FRDA (Dr. David Lynch, personal communication), though hard evidence for a significant effect of idebenone on the ataxia of FRDA is currently lacking.

C. Coenzyme Q10

Lodi et al. gave 400 mg/day CoQ10 and 2100 IU/day vitamin E to 10 patients with FRDA in a 6-month, open-label trial [18]. After 3 months, cardiac phosphocreatine-to-ATP ratios and the maximum rate of skeletal muscle mitochondrial ATP production improved ($P = 0.03$ and $P = 0.01$, respectively) by ^{31}P magnetic resonance spectroscopy (^{31}P-MRS). These improvements were sustained after 6 months of treatment. Echocardiographic measures, including LVMI, were unchanged. ICARS scores worsened in six patients, but failed to reach statistical significance ($P = 0.06$) overall. Posture and gait scores deteriorated in eight of the patients, which did reach statistical significance ($P = 0.02$) [18].

In a follow-up study, Hart et al. continued giving 400 mg/day CoQ10 and 2100 IU/day vitamin E to the same 10 patients for a total of 4 years [30]. The improvements in cardiac phosphocreatine-to-ATP ratios and the maximum rate of skeletal muscle mitochondrial ATP production noted at 3 months were sustained after 4 years. In addition, ICARS scores and left ventricular hypertrophy did not worsen, and cardiac fractional shortening improved ($P = 0.02$). Hart et al. used cross-sectional data from 77 untreated patients with FRDA to assess whether clinical scores declined less rapidly in the 10 patients receiving 400 mg/day CoQ10 and 2100 IU/day vitamin E. Taking into account GAA repeat expansion sizes, the rate of change in total ICARS scores, and in kinetic scores, for

6 of the 10 patients was better than predicted by the cross-sectional data. The decline in posture and gait scores was generally similar to, and the decline in dominant-hand dexterity was equal to or greater than, the decline predicted by the cross-sectional data.

D. MitoQ

MitoQ is an idebenone derivative targeted to mitochondria by covalent attachment to a lipophilic triphenylphosphonium cation [31]. Because of the large membrane potential of mitochondria, MitoQ accumulates several hundredfold within mitochondria in cultured cells [32]. Significant doses were fed safely to mice over long periods, with steady-state distributions within heart, brain, liver, and muscle, suggesting that oral administration will be straightforward [32]. Primary FRDA fibroblasts in cell culture lose viability when depleted of glutathione using L-buthionine-(S,R)-sulfoximine (BSO), which blocks the rate-limiting enzyme in glutathione synthesis; MitoQ protected against this loss of viability at concentrations several-hundredfold lower than those of idebenone [33]. Preliminary safety studies of MitoQ in humans indicate that it is safe and well-tolerated (Martin Delatycki, personal communication). Clinical trials of MitoQ for FRDA are being planned, but are not yet underway as of this writing.

E. EPI-A0001

EPI-A0001 is a bio-isostere of CoQ10 being developed by Edison Pharmaceuticals for the treatment of FRDA and other diseases affecting the mitochondrial respiratory chain. The structure of EPI-A0001 is similar to that of CoQ10, except the redox properties of the parabenzoquinone ring system have been altered to improve the efficiency of electron transfer between complexes I and III, particularly in disease states associated with defects in complex I or III. EPI-A0001 was evaluated in primary FRDA fibroblasts depleted of glutathione using BSO, and analogs of EPI-A0001 were evaluated in yeast depleted of Yfh1p. EPI-A0001 protected the primary FRDA fibroblasts at concentrations more than a hundredfold lower than those of idebenone, and the analogs of EPI-A0001 improved indirect measures of mitochondrial function in the yeast model system more than eightfold. Because EPI-A0001 is plant-derived, it is likely to be well-tolerated and nontoxic. Clinical trials of EPI-A0001 for FRDA are being planned, but are not yet underway as of this writing.

III. SELENIUM AND GLUTATHIONE PEROXIDASE MIMETICS

The viability of FRDA fibroblasts depleted of glutathione using BSO was enhanced by preincubation with selenium in a dose-dependent manner [34]. Selenium supplementation simultaneous with BSO addition was ineffective. Selenium supplementation had no effect on glutathione levels, but did increase glutathione peroxidase (GPX) activity. Based on these results, Jauslin et al. tested 17 GPX mimetics and found that all of them rescued FRDA cell viability from the BSO effect, with diselenides, in general, more potent than monoselenides [34]. The GPX mimetic ebselen was active with an EC_{50} of ~10 µM and was also protective in a rat model of cerebral ischemia [35], suggesting that GPX mimetics could be considered as a treatment strategy for FRDA.

IV. CREATINE AND CARNITINE

The rationale for treating individuals with FRDA with creatine and carnitine derives from the mitochondrial dysfunction underlying the disorder. Supplementation with creatine might increase phosphocreatine (PCr), which is generated by phosphorylation of creatine by creatine kinase and serves as a cellular energy reservoir for the generation of ATP. Mitochondrial dysfunction is often associated with impaired beta-oxidation and a secondary carnitine deficiency [36]. Schols et al. conducted a randomized, placebo-controlled, three-period crossover trial of 6.75 g/day creatine and 3 g/day L-carnitine for 15 ambulatory patients with FRDA [37]. The treatment periods were 4 months each, with 1-month washouts between. Outcome measures included ^{31}P-MRS, echocardiography, the ICARS, and a pegboard test. Relative to placebo and creatine treatments, positive trends were noted in the ^{31}P-MRS measurements of patients after 4 months of L-carnitine, though these trends did not reach statistical significance ($P = 0.06$, F test). The other outcome measures remained essentially unchanged with each treatment. The data suggest that larger trials of L-carnitine for patients with FRDA may be warranted.

V. *FRDA* UPREGULATION

FRDA is unusual among genetic disorders in that the exonic sequences of the disease gene are intact. FRDA is caused by intronic GAA repeat expansions in the disease gene (discussed in detail in Chapter 21 [38]). Campuzano et al. identified the FRDA disease gene, *FRDA* (originally designated *X25*), on chromosome 9q13 [39]. Most patients with FRDA (~97%) have expansions of a GAA repeat in the first intron of both *FRDA* alleles [39]. Normal alleles have 40 or fewer GAA repeats, whereas disease alleles have from approximately 100 to more than 1700 repeats [40]. The GAA repeat expansions interfere with mRNA transcription by forming triplex DNA structures termed *sticky DNA*, and the size of the expansion correlates inversely with expression of the encoded protein, frataxin [41, 42].

Grabczyk and Usdin used oligodeoxyribonucleotides to block triplex formation and achieved concentration-dependent increases in full-length *FRDA* transcripts [4]. Napierala et al. used Novobiocin to repress the formation of sticky DNA by lowering negative superhelical density [43]. Sarsero et al. constructed stable cell lines expressing enhanced green fluorescent protein (eGFP) fusions of *FRDA* in BAC clones; *FRDA* expression was increased by treatment of the cells with hemin, an iron-containing compound synthesized in mitochondria, and by treatment with butyric acid, a histone deacetylase inhibitor [44]. Gottesfeld, Dervan, and their colleagues are studying pyrrole-imidazole polyamides, which are cell-permeable small molecules that bind to specific DNA sequences [45, 46]; they are currently testing GAA-targeted polyamides, which have the potential to relieve *FRDA* transcriptional repression. An advantage of these compounds over others for *FRDA* upregulation is that in experiments in mouse models of human disease, they have been shown to be bioavailable and nontoxic [47].

VI. SUMMARY

FRDA research has now entered the treatment era. Parabenzoquinones and other antioxidants are generally safe and well-tolerated and may slow the progression of the disease. High-throughput drug screens using cell-based models of FRDA are in progress. The presence of intact exonic sequences in most disease alleles opens up the possibility of upregulating *FRDA* expression. One or more of these approaches, perhaps in combination, could potentially halt, and perhaps even reverse, aspects of the disorder.

References

1. Pandolfo, M. (2006). Friedreich's ataxia. In "Genetic Instabilities and Neurological Diseases" (R. D. Wells and T. Ashizawa, Eds.), 2nd ed., Ch. 17. Elsevier, Burlington, MA.
2. Patel, P. I., and Isaya, G. (2001). Friedreich ataxia: From GAA triplet-repeat expansion to frataxin deficiency. *Am. J. Hum. Genet.* **69**, 15–24.
3. Adamec, J., Rusnak, F., Owen, W. G., Naylor, S., Benson, L. M., Gacy, A. M., and Isaya, G. (2000). Iron-dependent self-assembly of recombinant yeast frataxin: Implications for Friedreich ataxia. *Am. J. Hum. Genet.* **67**, 549–562.
4. Cavadini, P., O'Neill, H. A., Benada, O., and Isaya, G. (2002). Assembly and iron-binding properties of human frataxin, the protein deficient in Friedreich ataxia. *Hum. Mol. Genet.* **11**, 217–227.
5. Gakh, O., Adamec, J., Gacy, A. M., Twesten, R. D., Owen, W. G., and Isaya, G. (2002). Physical evidence that yeast frataxin is an iron storage protein. *Biochemistry* **41**, 6798–6804.
6. Muhlenhoff, U., Richhardt, N., Ristow, M., Kispal, G., and Lill, R. (2002). The yeast frataxin homolog Yfh1p plays a specific role in the maturation of cellular Fe/S proteins. *Hum. Mol. Genet.* **11**, 2025–2036.
7. Muhlenhoff, U., Gerber, J., Richhardt, N., and Lill, R. (2003). Components involved in assembly and dislocation of iron–sulfur clusters on the scaffold protein Isu1p. *EMBO J.* **22**, 4815–4825.
8. Babcock, M., de Silva, D., Oaks, R., Davis-Kaplan, S., Jiralerspong, S., Montermini, L., Pandolfo, M., and Kaplan, J. (1997). Regulation of mitochondrial iron accumulation by Yfh1p, a putative homolog of frataxin. *Science* **276**, 1709–1712.
9. Delatycki, M. B., Camakaris, J., Brooks, H., Evans-Whipp, T., Thorburn, D. R., Williamson, R., and Forrest, S. M. (1999). Direct evidence that mitochondrial iron accumulation occurs in Friedreich ataxia. *Ann. Neurol.* **45**, 673–675.
10. Puccio, H., Simon, D., Cossee, M., Criqui-Filipe, P., Tiziano, F., Melki, J., Hindelang, C., Matyas, R., Rustin, P., and Koenig, M. (2001). Mouse models for Friedreich ataxia exhibit cardiomyopathy, sensory nerve defect and Fe–S enzyme deficiency followed by intramitochondrial iron deposits. *Nat. Genet.* **27**, 181–186.
11. Horton, A. A., and Fairhurst, S. (1987). Lipid peroxidation and mechanisms of toxicity. *CRC Crit. Rev. Toxicol.* **18**, 27–79.
12. Wilson, R. B., and Roof, D. M. (1997). Respiratory deficiency due to loss of mitochondrial DNA in yeast lacking the frataxin homologue. *Nat. Genet.* **16**, 352–357.
13. Foury, F., and Cazzalini, O. (1997). Deletion of the yeast homologue of the human gene associated with Friedreich's ataxia elicits iron accumulation in mitochondria. *FEBS Lett.* **411**, 373–377.
14. Radisky, D. C., Babcock, M. C., and Kaplan, J. K. (1999). The yeast frataxin homologue mediates mitochondrial iron efflux: Evidence for a mitochondrial iron cycle. *J. Biol. Chem.* **274**, 4497–4499.
15. Foury, F. (1999). Low iron concentration and aconitase deficiency in a yeast frataxin homologue deficient strain. *FEBS Lett.* **456**, 281–284.
16. Rotig, A., de Lonlay, P., Chretien, D., Foury, F., Koenig, M., Sidi, D., Munnich, A., and Rustin, P. (1997). Aconitase and mitochondrial iron–sulphur protein deficiency in Friedreich's ataxia. *Nat. Genet.* **17**, 215–217.
17. Bradley, J. L., Blake, J. C., Chamberlain, S., Thomas, P. K., Cooper, J. M., and Schapira, A. H. (2000). Clinical, biochemical and molecular genetic correlations in Friedreich's ataxia. *Hum. Mol. Genet.* **9**, 275–282.
18. Lodi, R., Hart, P. E., Rajagopalan, B., Taylor, D. J., Crilley, J. G., Bradley, J. L., Blamire, A. M., Manners, D., Styles, P., Schapira, A. H. V., and Cooper, J. M. (2001). Antioxidant treatment improves in vivo cardiac and skeletal muscle bioenergetics in patients with Friedreich's ataxia. *Ann. Neurol.* **49**, 590–596.
19. Schulz, J. B., Dehmer, T., Schols, L., Mende, H., Hardt, C., Vorgerd, M., Burk, K., Matson, W., Dichgans, J., Beal, M. F., and Bogdanov, M. B. (2000). Oxidative stress in patients with Friedreich ataxia [see comments]. *Neurology* **55**, 1719–1721.
20. Emond, M., Lepage, G., Vanasse, M., and Pandolfo, M. (2000). Increased levels of plasma malondialdehyde in Friedreich ataxia [see comments]. *Neurology* **55**, 1752–1753.

21. Seznec, H., Simon, D., Bouton, C., Reutenauer, L., Hertzog, A., Golik, P., Procaccio, V., Patel, M., Drapier, J. C., Koenig, M., and Puccio, H. (2005). Friedreich ataxia: The oxidative stress paradox. *Hum. Mol. Genet.* **14**, 463–474.
22. Sugiyama, Y., Fujita, T., Matsumoto, M., Okamoto, K., and Imada, I. (1985). Effects of idebenone (CV-2619) and its metabolites on respiratory activity and lipid peroxidation in brain mitochondria from rats and dogs. *J. Pharmacobiodyn.* **8**, 1006–1017.
23. Rustin, P., von Kleist-Retzow, J. C., Chantrel-Groussard, K., Sidi, D., Munnich, A., and Rotig, A. (1999). Effect of idebenone on cardiomyopathy in Friedreich's ataxia: A preliminary study. *Lancet* **354**, 477–479.
24. Hausse, A. O., Aggoun, Y., Bonnet, D., Sidi, D., Munnich, A., Rotig, A., and Rustin, P. (2002). Idebenone and reduced cardiac hypertrophy in Friedreich's ataxia [see comments.]. *Heart* **87**, 346–349.
25. Fournier A., Therien J., Vanasse M., Émond M., and Pandolfo M. (2001). Effect of idebenone on cardiomyopathy in Friedreich's ataxia. *Can. J. Cardiol.* **17** Suppl C:89C–288C.
26. Leber, I., Tanguy, M. L., Pousset, F., Rivaud, S., Jedynak, P., Welter, M. L., Mallet, A., Brice, A., Durr, A. (2003). Long term follow-up of patients with Friedreich's ataxia treated with Idebenone: A French open trial for over 3 years. *Friedreich's Ataxia Research Alliance.* http://www.faresearchalliance.org/docs/abstracts2003.pdf (accessed November 16, 2005).
27. Mariotti, C., Solari, A., Torta, D., Marano, L., Fiorentini, C., and Di Donato, S. (2003). Idebenone treatment in Friedreich patients: One-year-long randomized placebo-controlled trial. *Neurology* **60**, 1676–1679.
28. Buyse, G., Mertens, L., Di Salvo, G., Matthijs, I., Weidemann, F., Eyskens, B., Goossens, W., Goemans, N., Sutherland, G. R., and Van Hove, J. L. (2003). Idebenone treatment in Friedreich's ataxia: Neurological, cardiac, and biochemical monitoring. *Neurology* **60**, 1679–1681.
29. Artuch, R., Aracil, A., Mas, A., Colome, C., Rissech, M., Monros, E., and Pineda, M. (2002). Friedreich's ataxia: Idebenone treatment in early stage patients. *Neuropediatrics* **33**, 190–193.
30. Hart, P. E., Lodi, R., Rajagopalan, B., Bradley, J. L., Crilley, J. G., Turner, C., Blamire, A. M., Manners, D., Styles, P., Schapira, A. H., and Cooper, J. M. (2005). Antioxidant treatment of patients with Friedreich ataxia: Four-year follow-up. *Arch. Neurol.* **62**, 621–626.
31. Kelso, G. F., Porteous, C. M., Coulter, C. V., Hughes, G., Porteous, W. K., Ledgerwood, E. C., Smith, R. A., and Murphy, M. P. (2001). Selective targeting of a redox-active ubiquinone to mitochondria within cells: Antioxidant and antiapoptotic properties. *J. Biol. Chem.* **276**, 4588–4596.
32. Smith, R. A., Porteous, C. M., Gane, A. M., and Murphy, M. P. (2003). Delivery of bioactive molecules to mitochondria in vivo. *Proc. Natl. Acad. Sci. USA* **100**, 5407–5412.
33. Jauslin, M. L., Meier, T., Smith, R. A., and Murphy, M. P. (2003). Mitochondria-targeted antioxidants protect Friedreich ataxia fibroblasts from endogenous oxidative stress more effectively than untargeted antioxidants. *FASEB J.* **17**, 1972–1974.
34. Jauslin, M. L., Wirth, T., Meier, T., and Schoumacher, F. (2002). A cellular model for Friedreich ataxia reveals small-molecule glutathione peroxidase mimetics as novel treatment strategy. *Hum. Mol. Genet.* **11**, 3055–3063.
35. Dawson, D. A., Masayasu, H., Graham, D. I., and Macrae, I. M. (1995). The neuroprotective efficacy of ebselen (a glutathione peroxidase mimic) on brain damage induced by transient focal cerebral ischaemia in the rat. *Neurosci. Lett.* **185**, 65–69.
36. Infante, J. P., and Huszagh, V. A. (2000). Secondary carnitine deficiency and impaired docosahexaenoic (22:6n-3) acid synthesis: A common denominator in the pathophysiology of diseases of oxidative phosphorylation and beta-oxidation. *FEBS Lett.* **468**, 1–5.
37. Schols, L., Zange, J., Abele, M., Schillings, M., Skipka, G., Kuntz-Hehner, S., van Beekvelt, M. C., Colier, W. N., Muller, K., Klockgether, T., Przuntek, H., and Vorgerd, M. (2005). L-Carnitine and creatine in Friedreich's ataxia: A randomized, placebo-controlled crossover trial. *J. Neural Transm.* **112**, 789–796.
38. Son, L. S., and Wells, R. D. (2006). Triplexes, sticky DNA, and the (GAA·TTC) trinucleotide repeat associated with Friedreich's ataxia. *In* "Genetic Instabilities and Neurological Diseases" (R. D. Wells and T. Ashizawa, Eds.), 2nd ed., Ch. 21. Elsevier, Burlington, MA.
39. Campuzano, V., Montermini, L., Molto, M. D., Pianese, L., Cossee, M., Cavalcanti, F., Monros, E., Rodius, F., Duclos, F., and Monticelli, A. (1996). Friedreich's ataxia: Autosomal recessive disease caused by an intronic GAA triplet repeat expansion [see comments]. *Science* **271**, 1423–1427.
40. Pandolfo, M. (1999). Molecular pathogenesis of Friedreich's ataxia. *Neurol. Rev.* **56**, 1201–1208.
41. Ohshima, K., Montermini, L., Wells, R. D., and Pandolfo, M. (1998). Inhibitory effects of expanded GAA–TTC triplet repeats from intron I of the Friedreich ataxia gene on transcription and replication in vivo. *J. Biol. Chem.* **273**, 14588–14595.
42. Sakamoto, N. (1999). Sticky DNA: Self-association properties of long GAA·TTC repeats in R·R·Y triplex structures from Friedreich's ataxia. *Mol. Cell* **3**, 465–475.
43. Grabczyk, E., and Usdin, K. (2000). Alleviating transcript insufficiency caused by Friedreich's ataxia triplet repeats. *Nucleic Acids Res.* **28**, 4930–4937.
44. Sarsero, J. P., Li, L., Wardan, H., Sitte, K., Williamson, R., and Ioannou, P. A. (2003). Upregulation of expression from the FRDA genomic locus for the therapy of Friedreich ataxia. *J. Gene Med.* **5**, 72–81.
45. Gottesfeld, J. M., Turner, J. M., and Dervan, P. B. (2000). Chemical approaches to control gene expression. *Gene Expr.* **9**, 77–91.
46. Dervan, P. B., and Edelson, B. S. (2003). Recognition of the DNA minor groove by pyrrole-imidazole polyamides. *Curr. Opin. Struct. Biol.* **13**, 284–299.
47. Dickinson, L. A., Burnett, R., Melander, C., Edelson, B. S., Arora, P. S., Dervan, P. B., and Gottesfeld, J. M. (2004). Arresting cancer proliferation by small-molecule gene regulation. *Chem. Biol.* **11**, 1583–1594.

CHAPTER 19

Evolution and Instability of the GAA Triplet-Repeat Sequence in Friedreich's Ataxia

IRENE DE BIASE, ASTRID RASMUSSEN, AND SANJAY I. BIDICHANDANI

Departments of Biochemistry and Molecular Biology, and Pediatrics, University of Oklahoma Health Sciences Center, Oklahoma City, Oklahoma, and Department of Neurogenetics and Molecular Biology, Instituto Nacional de Neurología y Neurocirugía Manuel Velasco Suárez, Mexico City, Mexico

I. Introduction
II. Origin and Evolution of Polymorphic GAA-TR Alleles at the *FXN* Locus
III. Geographic Distribution and Population Genetics of the GAA-TR Mutation
IV. Intergenerational Instability of GAA-TR Alleles at the *FXN* Locus
V. Somatic Instability of GAA-TR Alleles at the *FXN* Locus
VI. Is the GAA-TR Sequence at the *FXN* Locus Unique?
VII. Concluding Remarks
References

Patients with Friedreich's ataxia are homozygous for expanded GAA triplet-repeat alleles (E alleles) in intron 1 of the *FXN* gene. Although it is the most common recessive ataxia among Europeans, Friedreich's ataxia is far less common or even nonexistent in other populations. This is largely reflective of the variable prevalence of E alleles (and/or premutation intermediates) in different populations. Patients have a wide range of E allele lengths, the size of which correlates with phenotypic severity. In this chapter, we describe the evolution of the GAA triplet-repeat sequence at the *FXN* locus, specifically addressing how E alleles arose in a stepwise fashion and how that correlates with the variable geographic prevalence of Friedreich's ataxia. We describe the instability of E alleles during intergenerational transmission, specifically the role of the sex of the transmitting parent and the genetic status of the offspring in determining the size distribution of E alleles encountered in nature. Recent discoveries are highlighted that show that somatic instability of E alleles in mammalian systems is tissue-specific and age-dependent, observations that challenge the prevailing models for the molecular mechanism underlying GAA triplet-repeat instability. We discuss the potential phenotypic impact associated with somatic instability of alleles at the cusp of normal and E alleles ("borderline" alleles). We also present data indicating that the *FXN* locus may not be unique in its ability to foster large expansions of the GAA triplet-repeat sequence. We conclude with suggestions for future avenues of research to further delineate the properties and determinants of GAA triplet-repeat instability.

I. INTRODUCTION

Of the three billion bases in the human genome, about 10% are composed of a million copies of a quasi-randomly inserted, ~300-bp primate-specific sequence called the *Alu* element. The human *FXN* gene, located on the long arm of chromosome 9, has an inordinate share of 39 *Alu* elements inserted in several locations. This chapter is concerned with the consequence of one of these *Alu* insertions, located in intron 1, within which evolved a GAA triplet-repeat (GAA-TR) sequence. Its stepwise expansion eventually led to the creation of a hypermutable (unstable) GAA-TR sequence, abnormally large expansions of which constitute the most common mutation in the etiology of Friedreich's ataxia. Indeed, with the carrier frequency of this particular mutation approaching 0.5–1% in some populations, it is presently the most prevalent of all disease-causing, unstable triplet repeats.

Friedreich's ataxia is an autosomal recessive disease and the vast majority of patients are homozygous for expanded GAA-TR alleles [1]. Friedreich's ataxia is characterized by slowly progressive ataxia with onset usually before the age of 25 years, depressed tendon reflexes, dysarthria, Babinski responses, and loss of position and vibration senses [2, 3]. The neurological manifestations result from degeneration of the dorsal root ganglia, posterior columns, corticospinal tracts, and dorsal spinocerebellar tracts. Cerebellar pathology may be seen in late stages. About two-thirds of individuals develop cardiomyopathy, and diabetes mellitus occurs in 10% of individuals. The length of the expanded GAA-TR alleles correlates with several objective parameters of disease severity, including age of onset, presence of leg muscle weakness/wasting, duration until wheelchair use, and prevalence and severity of cardiomyopathy and secondary skeletal complications [4–8]. The expanded GAA-TR sequence interferes with transcription in a length-dependent manner and is associated with a severe deficiency of frataxin protein in individuals homozygous for the GAA expansion [9–11]. Given the severe deficiency associated with very long repeats, the total residual mRNA and protein in patients are determined as a function of the shorter of the two repeats. It is for this reason there is a strong phenotypic correlation associated with the size of the shorter of the two expanded GAA-TR alleles. As the length of the expanded GAA-TR allele determines the severity of disease, this makes the understanding of the mechanisms of GAA-TR instability a worthwhile endeavor. Deciphering the molecular mechanisms involved in intergenerational and somatic instability is likely to have important implications for understanding disease pathogenesis and for accurate genotype–phenotype correlation. Moreover, at least in principle, reversal of the expanded GAA-TR allele in somatic cells of patients could serve as a potential therapeutic strategy.

As indicated in the title, we focus on the evolution and genetic instability of this GAA-TR sequence. We describe the evolution of the GAA-TR sequence into polymorphic alleles, their geographic and racial distribution, and the genetic instability displayed by the different GAA-TR alleles during intergenerational transmission and within the soma. We also highlight recent discoveries regarding the behavior of expanded GAA-TR alleles in human/mammalian development and aging. Finally, given that Friedreich's ataxia is currently the only known disease caused by expansion of a GAA-TR sequence, we explore whether the *FXN* locus is unique in its ability to support such a hypermutable sequence.

II. ORIGIN AND EVOLUTION OF POLYMORPHIC GAA-TR ALLELES AT THE *FXN* LOCUS

The expanded GAA-TR sequence underlying Friedreich's ataxia is located in intron 1 of the *FXN* gene on the long arm of chromosome 9 (Fig. 19-1). It originated within an *Alu*Sx element following modification of the consensus $(A)_n TAC(A)_n$ motif at the center [1, 12]. Consequently, this particular GAA-TR sequence is primate-specific, and consistent with the age of *Alu*Sx elements, it is observed in the *FXN* genes of the great apes, Old World and New World primates, indicating that the GAA-TR sequence is more than 30 million years old. Throughout most of primate genomic evolution, the length of the GAA-TR sequence has remained ≤5 triplets [13, 14]. Even though the GAA-TR sequence is polymorphic and somewhat longer in the great apes, and shows almost the same sequence context as in humans, expansion of the repeat tract to >6 triplets occurred specifically in humans (Fig. 19-2).

In humans the allelic distribution reveals a significant degree of polymorphism [1, 12, 15, 16]. Most human chromosomes have <12 triplets (short normal [SN] alleles). Among Europeans, SN alleles account for 80–85% of all chromosomes. Approximately 15% of chromosomes contain 12–30 triplets (long normal [LN] alleles). Disease-causing, expanded (E) alleles contain 66–1700 triplets, and account for 0.65% of European chromosomes. Alleles containing 34–100 triplets are

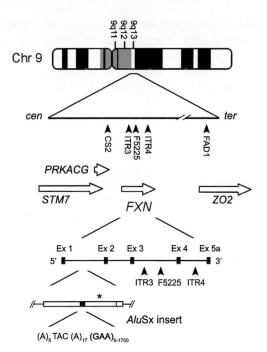

FIGURE 19-1 GAA-TR sequence at the *FXN* locus. The five coding exons (1 through 5a) of the *FXN* gene are shown in the context of other transcriptional units that map within chromosome 9q13. All transcriptional units are depicted as arrows to indicate their relative direction of transcription. The location of linked (CS2 and FAD1) and intragenic (ITR3, F5225, and ITR4) polymorphic markers used for linkage disequilibrium studies at the *FXN* locus are indicated by vertical filled arrowheads. The physical distance between CS2 and FAD1 is approximately 200 kb. Also shown are the *Alu*Sx insert in intron 1 of the *FXN* gene and the location of the GAA-TR sequence at its center. The asterisk denotes the closest SNP, located 212 bp downstream of the GAA-TR sequence.

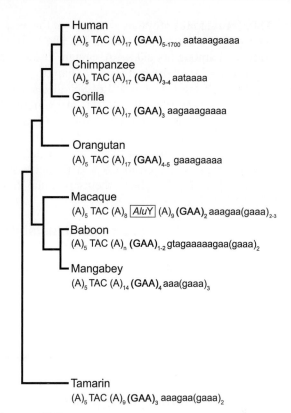

FIGURE 19-2 Phylogenetic analysis of the GAA-TR sequence at the center of the *Alu*Sx element in various primate *FXN* genes. Data are shown for humans (*Homo sapiens*), the great apes [chimpanzee (*Pan troglodytes*), gorilla (*Gorilla gorilla*), and orangutan (*Pongo pygmaeus*)], Old World monkeys [macaque (*Macaca mulatta*), baboon (*Mandrillus leucophaeus*), and mangabey (*Cerocebus torquatus*)], and a New World monkey [tamarin (*Sanguinus oedipus*)]. Only the sequence at the center of the *Alu*Sx element is shown, which includes the central $(A)_n TAC(A)_n$ consensus sequence and G/A-rich sequences located immediately downstream. The GAA-TR sequence is shown in boldface, and the number of repeating units is indicated accordingly. Note that the GAA-TR sequence is located 3' of the central $(A)_n TAC(A)_n$ consensus sequence in all primate genomes (the macaque genome has an additional *Alu*Y element located just 5' of the GAA-TR sequence). The G/A-rich sequence located 3' of the repeat is somewhat conserved. The human–orangutan genetic distance, that is, the maximum distance within the great apes as a group, is estimated to be 13 million years (range, 12–15). The estimated genetic distances between human and Old World monkey and between human and New World monkey are 23 (range, 21–25) and 33 (range, 32–36) million years, respectively [56]. The data are compiled from Refs. [13, 14] and our unpublished data.

called premutation (PM) alleles, given their propensity to undergo hyperexpansion and produce E alleles upon intergenerational transmission [12, 15–17]. The exact prevalence of PM alleles is unknown, but it is less than that of E alleles.

Even before the discovery of the *FXN* gene, significant linkage disequilibrium was detected with markers linked to the *FXN* locus in German [18] and Italian [19] patients. This indicated that there was a strong "founder effect," at least among the European patients; that is, the same mutation was present in all/most patients. Following the discovery of the GAA-TR mutation, several independent studies using intragenic and tightly linked markers confirmed that the same mutation was found in all patients [15, 20–22].

These studies of linkage disequilibrium also revealed important clues to the generation of the various polymorphic allele classes. In a study of Spanish patients, Monrós et al. [20] found strong linkage disequilibrium using a biallelic single-nucleotide polymorphism (SNP), FAD1 (see Fig. 19-1) located ~120 kb telomeric to the *FXN* gene; the A allele was seen in 83% of mutant chromosomes, but in only 23% of normal chromosomes. In a more comprehensive study of predominantly French patients Cossée et al. [15] found, by analysis of five intragenic and tightly linked markers, that most E alleles shared the same major haplotype (A-T-2-C-[C/T] at tel-FAD1-ITR4-F5225-ITR3-(GAA)$_n$-CS2-cen) (see

Fig. 19-1). More significantly, they showed that this major haplotype was also commonly seen in LN alleles, but it was rarely seen among SN alleles. Although 75% of LN alleles and 86% of E alleles had a similar haplotype, this haplotype only accounted for <4% of SN alleles. These data indicate not only that all E alleles are genetically related and share the same origin, but also that E alleles most likely originated from LN alleles. In contrast, SN alleles revealed a broad array of haplotypes, including in rare cases the same one seen in LN and E alleles, supporting the notion that LN alleles themselves may have arisen from an ancient SN allele. Given that the most prevalent SN allele contains 8 or 9 triplets and alleles with ~18 triplets constitute the most common LN allele, a chance duplication of an SN allele could explain the creation of LN alleles. Monticelli et al. [22] studied the polymorphic poly(A) tract immediately preceding the GAA-TR sequence and another SNP located 212 bp downstream (Fig. 19-1) and found that SN alleles with 9 triplets shared the same haplotype with LN and E alleles, supporting the hypothesis that LN alleles arose via duplication of an SN allele with 9 triplets (Fig. 19-3).

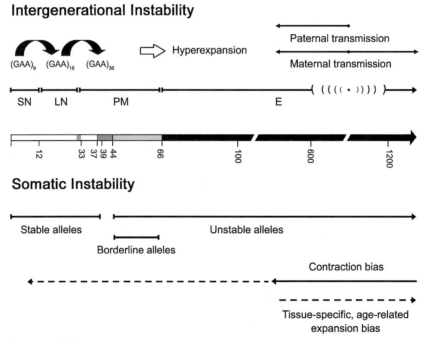

FIGURE 19-3 Genetic instability of the GAA-TR sequence at the *FXN* locus. The thick horizontal arrow at the center represents the length of GAA-TR alleles (in triplets). All other arrows represent observed length variations (instability), with their directions representing contractions (to the left) or expansions (to the right). Arrows above and below the central thick arrow represent intergenerational and somatic instability, respectively. Intergenerational instability: The various alleles and their lengths are depicted by horizontal lines. SN = short normal (<12 triplets); LN = long normal (12–30 triplets); PM = premutation (34–65 triplets); E = expanded/disease-causing (66 to >1200 triplets). See text for discussion of overlap between both PM and LN alleles and PM and E alleles. The parentheses located between 600 and 1200 triplets represent a "sink" seen in the distribution of E alleles in Friedreich's ataxia pedigrees (see text for details). The multistep hypothesis for the generation of E alleles is depicted at the top; the two curved arrows indicate historic duplications resulting in the transition from SN to LN and from LN to PM alleles, respectively. The unfilled horizontal arrow depicts the "hyperexpansion" of PM alleles seen in some intergenerational transmissions, resulting in the formation of E alleles in offspring. Horizontal arrows indicate the tendency of paternal transmissions of E alleles to result in contractions, and for maternal transmissions to produce contractions or further expansions with equal frequency. Somatic instability: Horizontal lines depict the different alleles with respect to their behavior in somatic cells *in vivo*. Alleles with ≤39 triplets are stable and those with ≥44 triplets are unstable. See text for discussion of "borderline alleles" (shown here to span 44–65 triplets). The arrows at the bottom depict the behavior of E alleles seen in mammalian cells *in vivo*. Most tissues show a strong contraction bias (left-pointing arrow), with the interrupted line depicting the occasional complete reversion of E alleles back to the normal size range. The right-pointing interrupted arrow represents the tendency of E alleles to undergo further expansion in a tissue-specific and age-related manner (see text for details).

These data support the following "multistep hypothesis" for the generation of the various allele classes at the human *FXN* locus. The initial insertion of an *Alu*Sx element into intron 1 of an ancient primate *FXN* gene was followed by formation of a GAA-TR sequence at its center. This repeat remained under 6 triplets in length throughout primate evolution, but increased into the SN allele range in humans. A single SN allele duplication event constituted the critical step: formation of LN alleles. LN alleles serve as a reservoir for E alleles, most likely via premutation intermediates, which are known to undergo hyperexpansion during intergenerational transmission (see Section IV). The multistep hypothesis has been further refined through the study of the allelic distribution and susceptibility to Friedreich's ataxia in various human populations (see Section III).

III. GEOGRAPHIC DISTRIBUTION AND POPULATION GENETICS OF THE GAA-TR MUTATION

Prior to the identification of the *FXN* gene and the GAA-TR mutation, Friedreich's ataxia was often quoted as the most prevalent recessive ataxia, without any further qualification. However, soon after the cloning of the *FXN* gene, it became apparent that although Friedreich's ataxia was indeed the most frequent recessive ataxia in Europeans (accounting for ~75% of all recessive/sporadic ataxia patients), North Africans, and the people of the Middle East and the Indian subcontinent, it was almost nonexistent in China, Japan, and Southeast Asia [4, 14, 20, 21, 23, 24]. Labuda et al. [21] noted that despite testing patients of multiple races from all over the United States and Canada, they never observed E alleles in African Americans, Southeast Asians, or Native Americans. They also found that LN and E alleles from North Africa, the Middle East, and the Indian subcontinent shared the same European founder haplotype, indicating that the generation of LN and E alleles likely occurred only once in human history, and that Friedreich's ataxia is limited to speakers of Indo-European and Afro-Asiatic language groups.

The population distribution of LN alleles provides additional clues to the origin of E alleles (Figs. 19-4 and 19-5). All populations that have E alleles (Indo-European and Afro-Asiatic speakers) also have LN alleles [12, 15, 21, 25]. Native Americans and East and Southeast Asians who do not have E alleles also do not have LN alleles [14, 21, 25]. However, LN alleles are seen in sub-Saharan Africans, with almost the same prevalence as in Europeans, despite the absence of E alleles [21]. Indeed, most of the LN alleles in sub-Saharan Africans also showed the same haplotype core as the European chromosomes (T-C at ITR4 and ITR3). Together, these data further support and refine the multistep hypothesis. They indicate that the SN-to-LN transition most likely preceded the migration out of Africa. The branch of humans that migrated into East Asia and subsequently into the Americas had only the SN allele [21, 25] (Fig. 19-5). However, in the people who populated the globe from Europe in the north, through the Middle East, to the Indian subcontinent in the south, a second critical transition from LN to PM alleles must have occurred to explain the additional existence of E alleles (Fig. 19-5). Based on linkage disequilibrium data from several European studies, Colombo and Carobene [26] estimated that the origin of PM alleles (or at least their sudden rise in prevalence

FIGURE 19-4 Distribution of GAA-TR alleles in various human populations. The entire allelic size range, from SN to E alleles, is seen in Europeans, North Africans, and the people of the Middle East and the Indian subcontinent. Only SN alleles are found among East and Southeast Asians and Native Americans. Sub-Saharan Africans are characterized by the presence of SN and LN alleles and the absence of PM and E alleles.

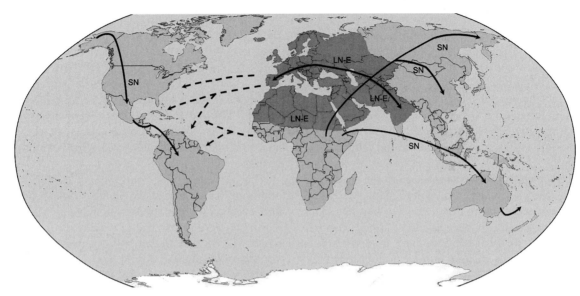

FIGURE 19-5 Geographic distribution and migration of GAA-TR alleles in human history. All solid lines depict ancient migrations of humans. Interrupted lines depict recent human migrations to the Americas. (Note: Other recent human migration events, for example, to Australia, are not shown.) Geographic areas colored in orange show the extent of the world populated by speakers of Indo-European and Afro-Asiatic language groups, that is, those who carry LN and E alleles and are predisposed to the development of Friedreich's ataxia. East and Southeast Asians, aboriginal Australians, and Native Americans have only SN alleles and do not develop Friedreich's ataxia. The peopling of the Americas (depicted by interrupted lines) is used as an example of more recent spread of alleles predisposing to the development of Friedreich's ataxia. The various lines depict the migration of Europeans to North America (largely supplanting the local native population), the migration of Europeans to Mexico and Central and South America (with variable, but significant, levels of genetic admixture with native populations), and the human migration from sub-Saharan Africa to the Americas. See CD-ROM for color image.

and/or their spread) occurred about 700 generations ago, consistent with the Upper Paleolithic population expansion. Labuda et al. [21] also arrived at a similar conclusion using a different method. Together these data indicate that the transition from SN to LN allele occurred once in human history and most likely occurred in Africa, but the critical transition to PM allele and the ensuing susceptibility to Friedreich's ataxia via E alleles developed (and spread) exclusively among Indo-European and Afro-Asiatic speakers. Local population structure and dynamics nevertheless result in variations in the prevalence of Friedreich's ataxia among the susceptible populations. For example, there is a higher frequency of disease in Cyprus [27] and northern Finland [28], and a lower frequency in the more populated southern Finland [28] and in some parts of India [29, 30].

It follows that populations with significant genetic admixture with susceptible populations would also have some risk of developing Friedreich's ataxia. The Americas with their waves of recent migration provide a useful test case (Fig. 19-5). Native Americans have only SN alleles and are therefore not at risk of developing Friedreich's ataxia [21, 25]. In the United States and Canada, however, the native populations have largely been supplanted by Europeans and other nonnative populations, and the national incidences of Friedreich's ataxia are comparable to those in Europe, with Labuda et al. [21] noting positive cases only among those of European and Middle Eastern ancestry despite a multiracial testing pool. No data are available on Friedreich's ataxia in African Americans, but Labuda et al. [21] reported a slightly higher prevalence of LN alleles compared with sub-Saharan Africans (17.3% versus 11.9%), possibly reflecting some genetic admixture with European *FXN* genes.

The history of the peopling of present-day Mexico is very different from the U.S. and Canadian examples. Gómez et al. [25] reported an eightfold lower prevalence of Friedreich's ataxia among Mexican mestizos compared with published European data. They found that all E alleles in Mexico shared the same haplotype as European chromosomes. They found that the level of genetic admixture of European and native *FXN* genes in the mestizo gene pool was commensurate with the observed lower prevalence of Friedreich's ataxia, thus demonstrating that the disease was introduced to Mexico via European genetic admixture with the native population. These data further support the notion that LN and E alleles arose only once in human history.

IV. INTERGENERATIONAL INSTABILITY OF GAA-TR ALLELES AT THE *FXN* LOCUS

Two forms of intergenerational instability characterize the behavior of GAA-TR alleles, and they depend on the size of the allele being transmitted: hyperexpansion of PM alleles and the intergenerational instability of E alleles. SN and LN alleles are stable during intergenerational transmission [8, 12, 15, 31]. Although intergenerational instability has obvious clinical implications, a careful analysis of this phenomenon also helps to explain the reason for the particular size distribution of E alleles frequently seen in Friedreich's ataxia pedigrees.

Premutation alleles may undergo hyperexpansion, expanding severalfold in size, upon parental transmission [12, 15–17]. What we know about this form of instability is limited because it is rarely observed. It should be noted that this is in marked contrast to other diseases caused by triplet-repeat expansions, all of which are dominantly inherited, and frequently undergo transition from premutation to mutation length. In the case of Friedreich's ataxia, the vast majority of patients inherit E alleles from asymptomatic heterozygous parents. However, from the few reports describing hyperexpansions of PM alleles at the *FRDA* locus (including two unpublished cases from our laboratory), we catalogued 11 hyperexpansions. Despite such a small number, the following observations can be made:

1. Premutation alleles have ranged in size from 34 to 100 triplets (mode = 38, mean = 47.6, 95% confidence interval [CI] = 34.5–60.7). (Note: Not all reported PM alleles were sequenced and consequently some sizes merely reflect estimates by gel electrophoresis.)
2. Hyperexpansion is possible via both maternal and paternal transmission with no obvious sex bias among donor parents or recipient offspring.
3. Hyperexpansion can result in an increase of the original PM allele length by more than tenfold (mean = 9.2-fold; range = 1.6- to 19-fold; 95% CI = 5.3- to 13.1-fold).
4. It is not clear if every PM allele undergoes hyperexpansion, or if an allele that hyperexpands once will (or is more likely to) do so in subsequent intergenerational transmissions.

The phenomenon of genetic anticipation, wherein the onset of disease occurs at progressively earlier ages and with increased severity in subsequent generations, is a hallmark of dominantly inherited diseases caused by unstable triplet-repeat expansions. The molecular basis for this phenomenon is the stepwise, intergenerational increase in size of the expanded allele. Anticipation is not seen in Friedreich's ataxia because it is an autosomal recessive disease in which the vast majority of patients inherit from both parents E alleles that usually span a narrow size range. Even though a wide range of E allele sizes have been observed in those with Friedreich's ataxia, ranging from 66 to 1700 triplets, the median length is between 800 and 1000 triplets, and it is very uncommon to encounter alleles with <600 or >1200 triplets [4–8]. However, the E alleles seen in offspring of PM allele carriers via hyperexpansion have ranged in size from 62 to 683 triplets (mean = 410 triplets, 95% CI = 263–558 triplets). This indicates that subsequent expansions may be occurring in these pedigrees to attain the length of commonly observed E alleles; that is, the rare short E alleles merely represent a transient state before their expansion to the commonly observed E allele sizes. Although this has not been directly tested in Friedreich's ataxia pedigrees, it is plausible that analysis of preceding generations of carriers of short E alleles may reveal the initial transition from PM to E alleles.

Given that all E alleles share a common origin and are therefore ultimately related, the observed range of alleles among Friedreich's ataxia pedigrees must represent the changes that have occurred via multiple intergenerational transmissions. A careful analysis of the behavior of E alleles during intergenerational transmission may therefore help to explain the size distribution of alleles observed in nature. Intergenerational transmission of E alleles almost always results in length variation, and this tends to depend on various factors [8, 31, 32], including the sex and age of the transmitting parent, the length of the E allele being transmitted, and, interestingly, the genetic status of the offspring with respect to being a heterozygote (carrier) or homozygote (patient). Maternal transmissions result in nearly equal frequency of further expansion or contraction, whereas paternal transmissions almost always result in contractions. These changes are commonly on the order of 100–150 triplets, representing an alteration of approximately 10–20% in the original E allele length. It should be noted that a complete reversion of an E allele to the normal size has never been observed via intergenerational transmission in Friedreich's ataxia pedigrees. In maternal transmission, longer E alleles tend to contract and shorter alleles more frequently expand [31], whereas in paternal transmission, the degree of contraction increases with longer E alleles [8, 31]. With increasing parental age, there is a corresponding increase in the magnitudes of both paternal contractions and maternal expansions [31]. Paternally derived contractions were shown to result from the correspondingly shorter length of E alleles in sperm [31, 32]. Furthermore, as suggested by the impact of age on paternal contractions, the sperm–blood difference was found to be more pronounced in older males, suggesting an age-dependent decrease in E allele length in sperm [31]. Lastly, heterozygous offspring inherit larger E alleles than their homozygous counterparts [8]. More than half the transmissions to homozygotes were contractions, while their heterozygous carrier sibs tended to inherit more expansions. Furthermore, maternal expansions were larger and paternal contractions were smaller in magnitude when E alleles were inherited by heterozygotes [8].

The observed size range of E alleles in Friedreich's ataxia pedigrees indicates that there are two phenomena that regulate allelic size. First, there is an upper limit to the length of E alleles. This is especially puzzling in the case of heterozygotes, where cellular function is not expected to be affected due to the half-normal levels of frataxin derived from the normal allele, and so there is no apparent reason why very long E alleles could not exist in nature. Second, there seems to be a preferred range of E allele size, such that there seems to be a "sink" in the triplet-repeat range 600–1200 (Fig. 19-3). Both these observations, the "sink" phenomenon and the apparent upper size limit, may not be entirely independent, as the mechanisms regulating the "sink" might automatically enforce an upper limit. Perhaps these are controlled prior to fertilization via dynamics involved in the germline, or via mechanisms acting at the level of fertilization or in early embryonic development. The observation that E alleles in heterozygous carriers are longer than those in homozygotes (patients) implies that there may be a postfertilization mechanism for selection of slightly smaller E alleles in the absence of a normal allele [8]. This is conceivably related to the minimum amount of frataxin required in early embryonic development, as it is known that *frda* null mice suffer embryonic lethality [33]. Thus, homozygotes may play a role in preventing E alleles from expanding indefinitely. However, this does not explain why an upper limit seems to apply in heterozygotes, and why the E alleles would not simply keep expanding in length from one generation to the other. However, Monrós et al. [8] directly observed that there was no overall size difference in the E alleles across generations by comparing E alleles in heterozygous parents versus heterozygous offspring. One possibility in the prevention of continuous expansion may stem from the normalizing effect of passing through the male germline; contractions occur at a high rate via paternal transmission and larger contractions occur when fathers transmit large E alleles [8, 31].

Furthermore, what makes a PM allele behave differently from an E allele during intergenerational transmission? It may be more approachable to view this question as follows: Is there a clear distinction between LN and PM alleles, on the one hand, and between PM and E alleles, on the other? Given our lack of knowledge involving the dynamics of intergenerational transmission of PM alleles, unfortunately, the diagnosis of a PM allele is made only after the fact, that is, after actually observing a hyperexpansion. We do not yet know the distribution of allele lengths in the postmeiotic, germ cell compartment of carriers of PM alleles. For example, is there a gradual increase in the proportion of germ cells with E alleles as the PM allele length increases, or is there a clear threshold length below which they are stable in intergenerational transmission and above which they are clearly PM alleles that undergo frequent hyperexpansion? On the other end of the size spectrum, what causes PM alleles to undergo multiplicative increases in length with no apparent sex bias, whereas E alleles change by only a fraction of their length with significant parental sex bias determining the type of length variation? It should be noted that when strictly defined, the term *E allele* refers to an allele that causes disease, whereas a *PM allele* actually refers to its property of being unstable during intergenerational transmission. Notwithstanding the lack of clarity implicit in the nomenclature, it is tempting to speculate that the peculiar behavior of PM alleles, that is, hyperexpansion via both maternal and paternal transmission, stems from a strict size limitation. Delatycki et al. [17] demonstrated a stepwise expansion by analyzing a carrier of a PM allele with 100 triplets in his blood. The subject's sperm showed 320 triplets, which further expanded to 540 triplets in his offspring. This contrasts with the observation that E alleles are almost always shorter in sperm compared with the blood sample from the same subject [31, 32]. It is plausible that PM alleles have to fit within a defined size window, above and below which hyperexpansion may not be mechanistically possible. Another factor that may be crucial is the purity of the GAA-TR sequence. Alleles with >27 triplets are rarely seen at the *FXN* locus, and Montermini et al. [12] reported that alleles of this length are frequently interrupted with one or more (GAGGAA) hexanucleotide motifs, which tend to stabilize them during intergenerational transmission.

V. SOMATIC INSTABILITY OF GAA-TR ALLELES AT THE *FXN* LOCUS

Somatic instability refers to the variability in repeat length observed among different tissues/cell types of the same individual or among different cells within the same tissue/organ. By definition, these are postzygotic changes, which may occur during growth and development, differentiation, and aging of the individual. Early studies following the discovery of the *FXN* gene found that E alleles were variable in length in DNA isolated from various tissue sources from the same patient [6, 34–37]. E alleles were found to vary among different subregions of the brain, with developmentally related regions showing similar allele sizes [34]. Lymphoblastoid cell lines, generated by Epstein–Barr virus-mediated transformation of B lymphocytes from patients, almost always show shorter E alleles compared with peripheral blood DNA from the same individual [37]. However, in all of these studies, E alleles were detected by polymerase chain reaction (PCR) amplification or by genomic Southern blot analysis of the *FXN* gene, in which the expanded "band" is usually generated from analysis of 0.06–12 μg genomic DNA, or the equivalent of 10^4 to 2×10^6 nucleated cells. These commonly used assays, therefore, estimate the repeat size of only the "constitutional" or "most common" allele and, thereby, reflect only gross somatic variability.

Techniques such as small-pool PCR (SP–PCR) and single-genome PCR (SG-PCR) have been employed to detect GAA-TR variability among individual cells or genomes, thus revealing changes in small proportions of the cells being analyzed (Fig. 19-6) [38–40]. A major advantage of these techniques is that they reflect the true distribution of allele sizes *in vivo*, starting from genomic DNA, that is, without the need for sophisticated methods of collection and storage of tissues/cells [41].

There are many reasons for investigating somatic instability. Besides providing mechanistic clues that could be surmised from the observed pattern of somatic instability, systematic studies of somatic instability may explain why some tissues are pathologically affected, and may help to refine genotype–phenotype correlations. Current ideas for the mechanism of GAA-TR instability, derived from simple model systems, focus on DNA replication and how length variation may be accomplished via altered DNA conformations [10, 39, 42]. DNA recombination is another mechanism that has been proposed [43]. Here, we focus on studies of somatic instability in mammalian systems. We describe data that challenge the existing models and indicate that, although erroneous replication and recombination are indeed critical for instability in simpler life forms, somatic instability of the GAA-TR sequence in human (mammalian) cells is likely to stem primarily from nonreplication-dependent, postmitotic mechanisms that are tissue-specific and age-dependent. We also comment on the initiation of somatic instability, in terms of both the threshold length of repeat tract required for this phenomenon to manifest and the timing of its development.

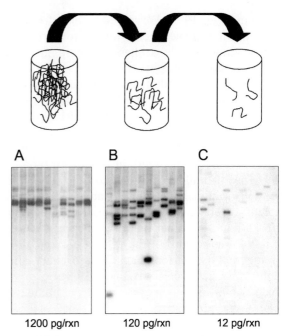

FIGURE 19-6 Small-pool PCR and single-genome PCR for the analysis of GAA-TR instability. PCR analysis of serial dilutions of genomic DNA reveal variable bands that, within a defined range of DNA concentrations, represent repeat lengths contained in individual DNA molecules. Gels showing the results of PCR using 1200, 120, and 12 pg per reaction (i.e., per lane) indicate the importance of selecting the correct DNA concentration for analysis. (A) The use of 1200 pg/reaction tends to overrepresent the "constitutional" or "most common" allele in the sample (equivalent to the repeat length detected by conventional PCR). (B) The use of 120 pg/reaction is more representative of the true distribution of repeat lengths (see Refs. [38, 41] for a discussion of experimental and theoretical details) and is referred to as small-pool PCR because it allows the analysis of a collection of approximately 5–50 molecules per reaction. (C) The use of 12 pg/reaction shows only some lanes with products of amplification (because only some aliquots of the diluted DNA sample are expected to contain amplifiable molecules). At this concentration, it is possible to analyze repeat lengths in individual molecules, hence the name single-genome PCR. Given the absence of competition between alleles of widely varying sizes, single-genome PCR is often used to accurately measure the frequency of large expansions.

Al-Mahdawi et al. [44] created a transgenic mouse model with a yeast artificial chromosome (YAC) carrying the entire human *FXN* locus containing a $(GAA)_{190}$ E allele in intron 1. They found an age-dependent increase in "smearing" involving the adult mouse cerebellum that suggested a tissue-specific, age-dependent increase in somatic instability. Miranda et al. [45] created another transgenic "knock-in" mouse model carrying a $(GAA)_{230}$ insert at the mouse *frda* locus, which also seemed to show a similar result in their cerebellar sample. We performed SP-PCR analysis of the transgenic mice described by Al-Mahdawi et al. [44] and confirmed an age-dependent, tissue-specific increase in somatic instability, specifically in adult cerebellum (Fig. 19-7). We noted a bias for further expansion. Other tissues, most with greater proliferative potential, showed only a slight increase in somatic instability or no change over time.

SP-PCR analysis of adult human tissues derived from autopsies of patients with Friedreich's ataxia revealed that the high mutation load of E alleles showed no obvious correlation with the proliferative capacity of the tissues analyzed (Fig. 19-8). Large contractions were seen in every tissue analyzed. In peripheral blood, 5% of the genes had alleles that had contracted by >50% of the constitutional allele length, and 0.3% contained complete reversions to the normal or premutation length [38]. On the other hand, large expansions were seen in various regions of the central nervous system, including the dorsal root ganglia, a primary site of pathology in Friedreich's ataxia (Fig. 19-8).

Lymphoblastoid cell lines derived from homozygous and heterozygous carriers of a wide range of E allele sizes, serially passaged in culture to investigate the effect of DNA replication in the context of the human chromosome, demonstrated that E alleles were remarkably stable (Fig. 19-8). Even after 30–50 population doublings in culture, E alleles in lymphoblastoid cells displayed very low levels of instability, estimated to be a hundredfold lower compared with blood samples from the same individuals [38]. Sharma et al. [38] estimated that peripheral leukocytes, in contrast, had undergone fewer than 15 binary divisions from their bone marrow stem cells. This disparity in level of somatic instability, therefore, cannot simply be explained by a replication-based mechanism.

When does somatic instability initiate in human tissues? The defined E allele lengths in a fertilized oocyte or in the early postzygotic stage could deteriorate rapidly during the burst of cell division characterizing early embryonic development or, indeed, during cellular differentiation accompanying organogenesis in the first trimester of intrauterine life. However, our

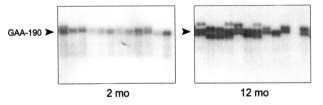

FIGURE 19-7 Age-dependent increase in somatic instability of an expanded GAA-TR allele in the cerebellum of a transgenic mouse model. Small-pool PCR shows low-level instability in the cerebellum of a 2-month-old mouse and significantly higher somatic instability in the cerebellum of a 12-month-old mouse littermate. This mouse model contains a transgenic YAC with the entire human *FXN* locus containing a GAA-TR allele with 190 triplets [44].

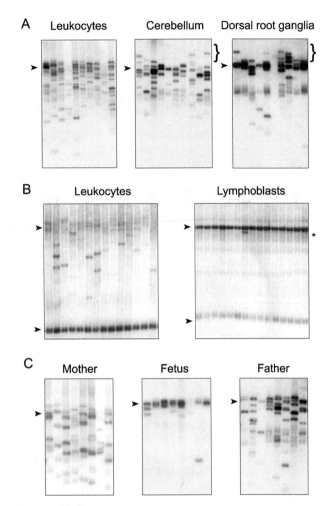

FIGURE 19-8 Tissue-specific and age-dependent somatic instability of E alleles at the human *FXN* locus. All results are from small-pool PCR experiments, and arrowheads indicate the location of the constitutional allele. (A) Small-pool PCR of human tissues (derived from the same individual) shows large contractions in all tissues. Large expansions are specifically seen in the cerebellum and dorsal root ganglia (indicated by braces). (B) A significantly higher level of instability is seen in peripheral leukocytes compared with a lymphoblastoid cell line derived from the same individual (the asterisk indicates an occasional somatic mutation). Similar results were obtained for cell lines derived from multiple subjects carrying a wide range of E allele sizes, despite serial passaging of lymphoblastoid cell lines over 30–50 population doublings [38]. (C) Small-pool PCR analysis of peripheral blood samples of an 18-week fetus (from a pregnancy terminated after testing positive for Friedreich's ataxia) and from both parents. The E allele shows a high level of somatic instability in adult tissues, whereas the fetal sample shows significantly less instability. The high level of instability in the parental blood samples (which is comparable to that in blood samples from other adults) rules out the cause of the low level of instability observed in fetal blood as being due to any *cis*-acting factors affecting either of the parental E alleles.

SP-PCR results on tissues of an 18-week fetus (derived from a pregnancy terminated for Friedreich's ataxia) reveal a significantly lower level of somatic instability (mutation load) compared with adult tissues (Fig. 19-8).

These data indicate that somatic instability is not caused predominantly by the rapid cell division and organogenesis associated with early human development. Indeed, it is temporally regulated, and is probably related to the biochemical processes involved in, but not limited to, cellular aging.

Whereas all the above data suggest a role for *trans*-acting factors in the etiology of somatic instability, an obvious question to ask is whether there are any intraallelic, *cis*-acting factors that are essential for DNA instability to manifest *in vivo*. Given the wide array of GAA-TR alleles present in nature, Pollard et al. [39] sought to determine if there is a minimum threshold length for the initiation of somatic instability *in vivo* and if the purity of the repeat tract affected instability. SP-PCR analyses showed that alleles with ≤39 triplets were stable and alleles with ≥44 triplets were unstable, thus establishing a threshold length for somatic instability between 40 and 44 triplets in human cells *in vivo* (Fig. 19-9). The implications of this finding are manifold. It indicates that the minimum allele length required for somatic instability is greater than that required for intergenerational instability of PM alleles (Fig. 19-9), suggesting that different cellular mechanisms are responsible for these two forms of instability. It is likely that a structural transition underlies the initiation of somatic instability *in vivo*. The minimum length for the formation of sticky DNA (59 triplets) [46] is greater than the 40- to 44-repeat threshold for the initiation of somatic instability. However, Potaman et al. [47] showed that the GAA-TR sequence undergoes a transition from a stable intramolecular triplex at 9–23 triplets to an intramolecular bi-triplex structure at 42 triplets, coinciding in length with the threshold for the initiation of somatic instability in human cells. Pollard et al. [39] also demonstrated that a $(GAGGAA)_n$ hexanucleotide interruption located at the 3′ end of the repeat tract completely stabilized a GAA-TR allele *in vivo* despite the presence of an adjacent pure GAA-TR sequence of 76 triplets. They speculated that the interruption could interfere with a structural transition that is required for somatic instability within human cells.

Does somatic instability have any phenotypic consequences? Sharma et al. [40] proposed the concept of "borderline alleles," that is, alleles not conventionally in the E allele size range, but nevertheless long enough to display somatic instability; the instability *per se* confers pathogenic allele status (Figs. 19-3 and 19-9). They used SP-PCR and SG-PCR to analyze a variety of somatic tissues derived from individuals who were carriers of one large E allele and one borderline allele. They found that even though the "constitutional" allele (determined by conventional PCR and sequencing) was not in the E allele range, instability of the borderline allele

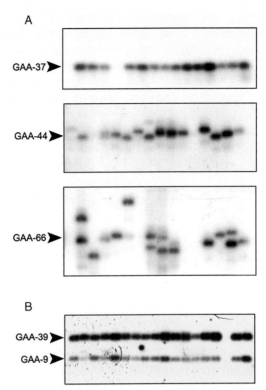

FIGURE 19-9 The threshold length for the initiation of somatic instability at the *FXN* locus in human cells *in vivo* is between 40 and 44 triplets. Small-pool PCR analysis of peripheral leukocytes shows that alleles with ≤39 triplets are stable and those with ≥44 triplets are unstable. (A) The somatically unstable GAA-44 and GAA-66 alleles are derived from symptomatic individuals, and the stable GAA-37 allele is from an asymptomatic sib of the carrier of the GAA-44 allele. (Note: The unstable E alleles are not shown in each of the three individuals.) (B) The GAA-39 allele is so far the longest allele shown to be stable in somatic cells *in vivo*. Despite its stability in somatic cells, this GAA-39 allele underwent hyperexpansion in two intergenerational transmissions, producing E alleles in two of three offspring.

resulted in a large number of somatic cells that were homozygous for alleles with >66 triplets. An asymptomatic older sib of one of the patients was shown to have a large E allele and a somatically stable $(GAA)_{37}$ allele, further supporting the hypothesis that somatic instability *per se* was essential for the generation of the disease phenotype. They speculated that expansions of the borderline alleles would accumulate over time in particular somatic tissues, thus resulting in the late-onset, slowly progressive, Friedreich's ataxia phenotype seen in such individuals. Given the stochastic nature of this process, it is plausible that borderline alleles may result in nonpenetrance; that is, not all carriers of borderline alleles (coupled with a conventional E allele) would necessarily manifest a disease phenotype.

Altogether, the observations of somatic instability in Friedreich's ataxia seem to indicate that it plays a "modifying" role in disease pathogenesis. For example, somatic instability could explain the phenotypic disparity seen among individuals with similar constitutional E alleles or in cases where the disease severity is in marked contrast to what could be reasonably expected from the E allele lengths [36]. The observation of large expansions in dorsal root ganglia could explain their early involvement in the disease process. Similarly, an age-dependent increase in cerebellar instability, as seen in mice, could be the molecular basis for the cerebellar degeneration seen in later stages of Friedreich's ataxia.

VI. IS THE GAA-TR SEQUENCE AT THE *FXN* LOCUS UNIQUE?

Of the three triplet-repeat motifs that are involved in unstable expansions and human disease, $(CGG \cdot GCC)_n$ and $(CAG \cdot CTG)_n$ expansions have been detected at multiple loci, but the *FXN* gene is the only known location of a large $(GAA \cdot TCC)_n$ expansion (the latter motif has been referred to as GAA-TR throughout this chapter). Is the *FXN* locus unique in its ability to foster such an unstable sequence or are there other loci with similar GAA-TR expansions? In a comprehensive search of the entire human genome, Clark et al. [48, 49] made the following observations:

1. Of the 10 nonredundant, triplet-repeat motifs, the GAA-TR sequence shows the most instances of large expanded tracts. There are nearly a 1000 loci that have $(GAA)_{8+}$ sequences, 29 of which are $(GAA)_{30+}$ sequences, analogous to PM alleles at the *FXN* locus.
2. GAA-TR sequences preferentially map within the 3′ poly(A) tails of *Alu* elements. In fact, the central *Alu* location of the GAA-TR sequence at the *FXN* locus is highly unusual, with only two other sequences showing such a location. One of these two loci also showed the pattern of allele distribution like the *FXN* locus, suggesting a similarity in the evolution of these alleles.
3. The vast majority of GAA-TR sequences mapped within large tracts of G/A-rich sequences (termed G/A islands).
4. Evolutionarily older *Alu* elements had significantly longer GAA-TR sequences, indicating continued expansion throughout primate genomic evolution.
5. The inordinate expansion of GAA-TR sequences is seen in, and is specific to, all mammalian genomes.
6. Large expansions, including $(GAA)_{100+}$ alleles (analogous to E alleles at the *FXN* locus), were observed at 9 of the 29 loci initially detected as having alleles analogous to PM alleles. Two of these sequences are located within introns of genes of unknown

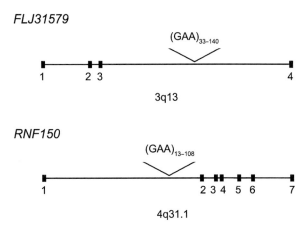

FIGURE 19-10 Human genomic loci with $(GAA)_{100+}$ alleles located within introns of transcriptional units. A diagrammatic representation of the transcriptional units [*FLJ31579* located at 3q13 and the gene for ring finger protein *RNF150* located at 4q31.1] is shown along with the relative positions of the GAA-TR sequence.

function (Fig. 19-10), and they could conceivably serve as disease-causing or disease-susceptibility loci.

All of these data indicate that the *FXN* locus is not unique in its ability to host an unstable GAA-TR sequence, and it is perhaps only a matter of time before other traits are linked to unstable GAA-TR sequences. Furthermore, given the ability of long GAA trinucleotide repeats to adopt non-B-DNA structures [9, 10, 46, 50–53], to interfere with DNA replication [10, 39, 42, 53] and gene transcription [9, 10, 52], to mediate position effect variegation [54], and to cause flanking mutagenesis [37], these sequences are likely to have important implications for mammalian genomic structure and function.

These genomic studies also yielded some clues to the genesis of GAA-TR sequences. On observing the preferential association of GAA-TR sequences with 3′ poly(A) tails of *Alu* elements, Chauhan et al. [55] also noted that the length of the GAA-TR sequence was inversely proportional to the length of the pure poly(A) tail immediately upstream (5′) of the triplet repeat. Extending this analysis to all GAA-TR sequences in the human genome, whether in *Alu* elements or not, we found that it was very common to have poly(A) tails immediately upstream of GAA-TR sequences, whereas they were very rarely seen on the 3′ side of the triplet repeat. These observations indicate that not only do GAA-TR sequences preferentially form within poly(A) tails (either in *Alu* elements or other transposon relics), but also, interestingly, they display a strong polarity in their formation. It is presently unknown if this property is related to the underlying mechanism(s) of GAA-TR instability.

VII. CONCLUDING REMARKS

Expansion of the GAA-TR sequence in Indo-Europeans and North Africans led to the genetic susceptibility to Friedreich's ataxia. Starting from the critical steps of formation of LN and PM alleles, it remains unknown how the E allele attained the prevalence it has. Indeed, Friedreich's ataxia is the most common inherited ataxia among Europeans. In this chapter we have summarized the salient features of GAA-TR instability at the human *FXN* locus. Clearly, a lot remains unknown. Much more data need to be gathered to understand the properties and determinants of genetic instability of PM and E alleles at the *FXN* locus. For example, a direct way of measuring the frequency of hyperexpansion of PM alleles would be to perform SP-PCR or single-sperm PCR of semen samples obtained from male PM carriers. Whether there is a defined size window for PM allele hyperexpansion to be mechanistically possible via intergenerational transmission could be addressed in carefully designed transgenic mouse experiments. Similarly, additional investigations are required to determine the mechanism(s) underlying the tissue-specific, age-dependent increase in somatic instability of E alleles. Perhaps, the mismatch repair system plays a role in mediating GAA-TR instability in the postmitotic cells of the cerebellum. Moreover, careful experiments need to be performed to identify the exact cell type(s) involved in the enhanced instability. Another intriguing line of investigation involves the study of other unstable GAA-TR sequences in the human genome. For example, SP-PCR experiments could identify if the threshold for the initiation of somatic instability varies in different genomic locations. Furthermore, the occurrence of intergenerational hyperexpansions could be investigated at the loci that show PM-length alleles.

Lastly, strategies need to be designed to accelerate the natural contraction bias that exists in somatic cells *in vivo*. The demonstration that DNA replication is not necessarily required for somatic instability suggests that a strategy that could work in postmitotic tissues may be possible. This could be developed as a potential therapeutic strategy to slow the disease process in Friedreich's ataxia.

References

1. Campuzano, V., Montermini, L., Moltó, M. D., Pianese, L., Cossée, M., Cavalcanti, F., Monrós, E., Rodius, F., Duclos, F., Monticelli, A., Zara, F., Cañizares, J., Koutnikova, H., Bidichandani, S. I., Gellera, C., Brice, A., Trouillas, P., De Michele, G., Filla, A., de Frutos, R., Palau, F., Patel, P. I., Di Donato, S., Mandel, J-L., Cocozza, S., Koenig, M., and Pandolfo, M. (1996). Friedreich's ataxia: Autosomal recessive disease caused by an intronic GAA triplet repeat expansion. *Science* **271**, 1423–1427.
2. Harding, A. E. (1981). Friedreich's ataxia: A clinical and genetic study of 90 families with an analysis of early diagnosis criteria and intrafamilial clustering of clinical features. *Brain* **104**, 589–620.
3. Bidichandani, S. I., and Ashizawa, T. Friedreich ataxia. (Updated 2004) In: "GeneReviews: Genetic Disease Online Reviews at GeneTests-GeneClinics." Copyright, University of Washington, Seattle. Available at *http://www.geneclinics.org/*.
4. Dürr, A., Cossée, M., Agid, Y., Campuzano, V., Mignard, C., Penet, C., Mandel, J. L., Brice, A., and Koenig, M. (1996). Clinical and genetic abnormalities in patients with Friedreich's ataxia. *N. Engl. J. Med.* **335**, 1169–1175.
5. Filla, A., De Michele, G., Cavalcanti, F., Pianese, L., Monticelli, A., Campanella, G., and Cocozza, S. (1996). The relationship between trinucleotide (GAA) repeat length and clinical features in Friedreich ataxia. *Am. J. Hum. Genet.* **59**, 554–560.
6. Montermini, L., Richter, A., Morgan, K., Justice, C. M., Julien, D., Castellotti, B., Mercier, J., Poirier, J., Capozzoli, F., Bouchard, J. P., Lemieux, B., Mathieu, J., Vanasse, M., Seni, M. H., Graham, G., Andermann, F., Andermann, E., Melancon, S. B., Keats, B. J., Di Donato, S., and Pandolfo, M. (1997). Phenotypic variability in Friedreich ataxia: Role of the associated GAA triplet repeat expansion. *Ann. Neurol.* **41**, 675–682.
7. Delatycki, M. B., Paris, D. B., Gardner, R. J., Nicholson, G. A., Nassif, N., Storey, E., MacMillan, J. C., Collins, V., Williamson, R., and Forrest, S. M. (1999). Clinical and genetic study of Friedreich ataxia in an Australian population. *Am. J. Med. Genet.* **87**, 168–174.
8. Monrós, E., Moltó, M. D., Martinez, F., Cañizares, J., Blanca, J., Vílchez, J. J., Prieto, F., de Frutos, R., and Palau, F. (1997). Phenotype correlation and intergenerational dynamics of the Friedreich ataxia GAA trinucleotide repeat. *Am. J. Hum. Genet.* **61**, 101–110.
9. Bidichandani, S. I., Ashizawa, T., and Patel, P. I. (1998). The GAA triplet-repeat expansion in Friedreich ataxia interferes with transcription and may be associated with an unusual DNA structure. *Am. J. Hum. Genet.* **62**, 111–121.
10. Ohshima, K., Montermini, L., Wells, R. D., and Pandolfo, M. (1998). Inhibitory effects of expanded GAA·TTC triplet repeats from intron I of the Friedreich ataxia gene on transcription and replication in vivo. *J. Biol. Chem.* **273**, 14588–14595.
11. Campuzano, V., Montermini, L., Lutz, Y., Cova, L., Hindelang, C., Jiralerspong, S., Trottier, Y., Kish, S. J., Faucheux, B., Trouillas, P., Authier, F. J., Durr, A., Mandel, J. L., Vescovi, A., Pandolfo, M., and Koenig, M. (1997). Frataxin is reduced in Friedreich ataxia patients and is associated with mitochondrial membranes. *Hum. Mol. Genet.* **6**, 1771–1780.
12. Montermini, L., Andermann, E., Labuda, M., Richter, A., Pandolfo, M., Cavalcanti, F., Pianese, L., Iodice, L., Farina, G., Monticelli, A., Turano, M., Filla, A., De Michele, G., and Cocozza, S. (1997). The Friedreich ataxia GAA triplet repeat: Premutation and normal alleles. *Hum. Mol. Genet.* **6**, 1261–1266.
13. González-Cabo, P., Sánchez, M. I., Cañizares, J., Blanca, J. M., Martinez-Arias, R., De Castro, M., Bertranpetit, J., Palau, F., Moltó, M. D., and de Frutos, R. (1999). Incipient GAA repeats in the primate Friedreich ataxia homologous genes. *Mol. Biol. Evol.* **16**, 880–883.
14. Justice, C. M., Den, Z., Nguyen, S. V., Stoneking, M., Deininger, P. L., Batzer, M. A., and Keats, B. J. (2001). Phylogenetic analysis of the Friedreich ataxia GAA trinucleotide repeat. *J. Mol. Evol.* **52**, 23223–23228.
15. Cossée, M., Schmitt, M., Campuzano, V., Reutenauer, L., Moutou, C., Mandel, J. L., and Koenig, M. (1997). Evolution of the

Friedreich's ataxia trinucleotide repeat expansion: Founder effect and premutations. *Proc. Natl. Acad. Sci. USA* **94**, 7452–7457.

16. Epplen, C., Epplen, J. T., Frank, G., Miterski, B., Santos, E. J., and Schöls, L. (1997). Differential stability of the (GAA)$_n$ tract in the Friedreich ataxia (STM7) gene. *Hum. Genet.* **99**, 834–836.
17. Delatycki, M. B., Paris, D., Gardner, R. J., Forshaw, K., Nicholson, G. A., Nassif, N., Williamson, R., and Forrest, S. M. (1998). Sperm DNA analysis in a Friedreich ataxia premutation carrier suggests both meiotic and mitotic expansion in the FRDA gene. *J. Med. Genet.* **35**, 713–716.
18. Zühlke, C., Gehlken, U., Purmann, S., Kunisch, M., Mülller-Myhsok, B., Kreuz, F., and Laccone, F. (1999). Linkage disequilibrium and haplotype analysis in German Friedreich ataxia families. *Hum. Hered.* **49**, 90–96.
19. Pianese, L., Cocozza, S., Campanella, G., Castaldo, I., Cavalcanti, F., De Michele, G., Filla, A., Monticelli, A., Munaro, M., Redolfi, E., Varrone, S., and Pandolfo, M. (1994). Linkage disequilibrium between FD1-D9S202 haplotypes and the Friedreich's ataxia locus in a central-southern Italian population. *J. Med. Genet.* **31**, 133–135.
20. Monrós, E., Cañizares, J., Moltó, M. D., Rodius, F., Montermini, L., Cossée, M., Martínez, F., Prieto, F., de Frutos, R., Koenig, M., Pandolfo, M., Bertranpetit, J., and Palau, F. (1996). Evidence for a common origin of most Friedreich ataxia chromosomes in the Spanish population. *Eur. J. Hum. Genet.* **4**, 191–198.
21. Labuda, M., Labuda, D., Miranda, C., Poirier, J., Soong, B. W., Barucha, N. E., and Pandolfo, M. (2000). Unique origin and specific ethnic distribution of the Friedreich ataxia GAA expansion. *Neurology* **54**, 2322–2324.
22. Monticelli, A., Giacchetti, M., De Biase, I., Pianese, L., Turano, M., Pandolfo, M., and Cocozza, S. (2004). New clues on the origin of the Friedreich ataxia expanded alleles from the analysis of new polymorphisms closely linked to the mutation. *Hum. Genet.* **114**, 458–463.
23. Sasaki, H., Yabe, I., Yamashita, I., and Tashiro, K. (2000). Prevalence of triplet repeat expansion in ataxia patients from Hokkaido, the northernmost island of Japan. *J. Neurol. Sci.* **175**, 45–51.
24. Mori, M., Adachi, Y., Kusumi, M., and Nakashima, K. (2001). A genetic epidemiological study of spinocerebellar ataxias in Tottori prefecture, Japan. *Neuroepidemiology* **20**, 144–149.
25. Gómez, M., Clark, R. M., Nath, S. K., Bhatti, S., Sharma, R., Alonso, E., Rasmussen, A., and Bidichandani, S. I. (2004). Genetic admixture of European *FRDA* genes is the cause of Friedreich ataxia in the Mexican population. *Genomics* **84**, 779–784.
26. Colombo, R., and Carobene, A. (2000). Age of the intronic GAA triplet repeat expansion mutation in Friedreich ataxia. *Hum. Genet.* **106**, 455–458.
27. Dean, G., Chamberlain, S., and Middleton, L. (1988). Friedreich's ataxia in Kathikas-Arodhes, Cyprus. *Lancet* **1**, 8585–8587.
28. Juvonen, V., Kulmala, S. M., Ignatius, J., Penttinen, M., and Savontaus, M. L. (2002). Dissecting the epidemiology of a trinucleotide repeat disease: Example of FRDA in Finland. *Hum. Genet.* **110**, 36–40.
29. Chattopadhyay, B., Gupta, S., Gangopadhyay, P. K., Das, S. K., Roy, T., Mukherjee, S. C., Sinha, K. K., Singhal, B. S., and Bhattacharya, N. P. (2004). Molecular analysis of GAA repeats and four linked bi-allelic markers in and around the frataxin gene in patients and normal populations from India. *Ann. Hum. Genet.* **68**, 189–195.
30. Mukerji, M., Choudhry, S., Saleem, Q., Padma, M. V., Maheshwari, M. C., and Jain, S. (2000). Molecular analysis of Friedreich's ataxia locus in the Indian population. *Acta Neurol. Scand.* **102**, 227–229.
31. De Michele, G., Cavalcanti, F., Criscuolo, C., Pianese, L., Monticelli, A., Filla, A., and Cocozza, S. (1998). Parental gender, age at birth and expansion length influence GAA repeat intergenerational instability in the X25 gene: Pedigree studies and analysis of sperm from patients with Friedreich's ataxia. *Hum. Mol. Genet.* **7**, 1901–1906.
32. Pianese, L., Cavalcanti, F., De Michele, G., Filla, A., Campanella, G., Calabrese, O., Castaldo, I., Monticelli, A., and Cocozza, S. (1997). The effect of parental gender on the GAA dynamic mutation in the FRDA gene. *Am. J. Hum. Genet.* **60**, 460–463.
33. Cossée, M., Puccio, H., Gansmuller, A., Koutnikova, H., Dierich, A., LeMeur, M., Fischbeck, K., Dolle, P., and Koenig, M. (2000). Inactivation of the Friedreich ataxia mouse gene leads to early embryonic lethality without iron accumulation. *Hum. Mol. Genet.* **9**, 1219–1226.
34. Montermini, L., Kish, S. J., Jiralerspong, S., Lamarche, J. B., and Pandolfo, M. (1997). Somatic mosaicism for Friedreich's ataxia GAA triplet repeat expansions in the central nervous system. *Neurology* **49**, 606–610.
35. Machkhas, H., Bidichandani, S. I., Patel, P. I., and Harati, Y. (1998). A mild case of Friedreich ataxia: Lymphocyte and sural nerve analysis for GAA repeat length reveals somatic mosaicism. *Muscle Nerve* **21**, 390–393.
36. Bidichandani, S. I., Garcia, C. A., Patel, P. I., and Dimachkie, M. M. (2000). Very late-onset Friedreich ataxia despite large GAA triplet repeat expansions. *Arch. Neurol.* **57**, 246–251.
37. Bidichandani, S. I., Purandare, S. M., Taylor, E. E., Gumin, G., Machkhas, H., Harati, Y., Gibbs, R. A., Ashizawa, T., and Patel, P. I. (1999). Somatic sequence variation at the Friedreich ataxia locus includes complete contraction of the expanded GAA triplet repeat, significant length variation in serially passaged lymphoblasts and enhanced mutagenesis in the flanking sequence. *Hum. Mol. Genet.* **8**, 2425–2436.
38. Sharma, R., Bhatti, S., Gómez, M., Clark, R. M., Murray, C., Ashizawa, T., and Bidichandani, S. I. (2002). The GAA triplet-repeat sequence in Friedreich ataxia shows a high level of somatic instability in vivo, with a significant predilection for large contractions. *Hum. Mol. Genet.* **11**, 2175–2187.
39. Pollard, L. M., Sharma, R., Gómez, M., Shah, S., Delatycki, M. B., Pianese, L., Monticelli, A., Keats, B. J., and Bidichandani, S. I. (2004). Replication-mediated instability of the GAA triplet repeat mutation in Friedreich ataxia. *Nucleic Acids Res.* **32**, 5962–5971.
40. Sharma, R., De Biase, I., Gómez, M., Delatycki, M. B., Ashizawa, T., and Bidichandani, S. I. (2004). Friedreich ataxia in carriers of unstable borderline GAA triplet-repeat alleles. *Ann. Neurol.* **56**, 898–901.
41. Gomes-Pereira, M., Bidichandani, S. I., and Monckton, D. G. (2004). Analysis of unstable triplet repeats using small-pool polymerase chain reaction. *Methods Mol. Biol.* **277**, 61–76.
42. Krasilnikova, M. M., and Mirkin, S. M. (2004). Replication stalling at Friedreich's ataxia (GAA)n repeats in vivo. *Mol. Cell. Biol.* **24**, 2286–2295.
43. Napierala, M., Dere, R., Vetcher, A., and Wells, R. D. (2004). Structure-dependent recombination hot spot activity of GAA·TTC sequences from intron 1 of the Friedreich's ataxia gene. *J. Biol. Chem.* **279**, 6444–6454.
44. Al-Mahdawi, S., Pinto, R. M., Ruddle, P., Carroll, C., Webster, Z., and Pook, M. (2004). GAA repeat instability in Friedreich ataxia YAC transgenic mice. *Genomics* **84**, 301–310.
45. Miranda, C. J., Santos, M. M., Ohshima, K., Smith, J., Li, L., Bunting, M., Cossée, M., Koenig, M., Sequeiros, J., Kaplan, J., and Pandolfo, M. (2002). Frataxin knockin mouse. *FEBS Lett.* **512**, 291–297.
46. Sakamoto, N., Chastain, P. D., Parniewski, P., Ohshima, K., Pandolfo, M., Griffith, J. D., and Wells, R. D. (1999). Sticky DNA: Self-association properties of long GAA·TTC repeats in R.R.Y triplex structures from Friedreich's ataxia. *Mol. Cell* **3**, 465–475.

47. Potaman, V. N., Oussatcheva, E. A., Lyubchenko, Y. L., Shlyakhtenko, L. S., Bidichandani, S. I., Ashizawa, T., and Sinden, R. R. 92004). Length-dependent structure formation in Friedreich ataxia $(GAA)_n*(TTC)_n$ repeats at neutral pH. *Nucleic Acids Res.* **32**, 1224–1231.
48. Clark, R. M., Dalgliesh, G. L., Endres, D., Gómez, M., Taylor, J., and Bidichandani, S. I. (2004). Expansion of GAA triplet repeats in the human genome: Unique origin of the FRDA mutation at the center of an Alu. *Genomics* **83**, 373–383.
49. Clark, R. M., Bhaskar, S. S., Miyahara, M., Dalgliesh, G. L., and Bidichandani, S. I. (2006). Expansion of GAA trinucleotide repeats in mammals. *Genomics*, **87**, 57–67.
50. Wells, R. D., Collier, D. A., Hanvey, J. C., Shimizu, M., and Wohlrab, F. (1988). The chemistry and biology of unusual DNA structures adopted by oligopurine·oligopyrimidine sequences. *FASEB J.* **2**, 2939–2949.
51. Sakamoto, N., Larson, J. E., Iyer, R. R., Montermini, L., Pandolfo, M., and Wells, R. D. (2001). GGA*TCC-interrupted triplets in long GAA*TTC repeats inhibit the formation of triplex and sticky DNA structures, alleviate transcription inhibition, and reduce genetic instabilities. *J. Biol. Chem.* **276**, 27178–27187.
52. Grabczyk, E., and Usdin, K. (2000). The GAA*TTC triplet repeat expanded in Friedreich's ataxia impedes transcription elongation by T7 RNA polymerase in a length and supercoil dependent manner. *Nucleic Acids Res.* **28**, 2815–2822.
53. Gacy, A. M., Goellner, G. M., Spiro, C., Chen, X., Gupta, G., Bradbury, E. M., Dyer, R. B., Mikesell, M. J., Yao, J. Z., Johnson, A. J., Richter, A., Melancon, S. B., and McMurray, C. T. (1998). GAA instability in Friedreich's ataxia shares a common, DNA-directed and intraallelic mechanism with other trinucleotide diseases. *Mol. Cell* **1**, 583–593.
54. Saveliev, A., Everett, C., Sharpe, T., Webster, Z., and Festenstein, R. (2003). DNA triplet repeats mediate heterochromatin-protein-1-sensitive variegated gene silencing. *Nature* **422**, 909–913.
55. Chauhan, C., Dash, D., Grover, D., Rajamani, J., and Mukerji, M. J. (2002). Origin and instability of GAA repeats: Insights from Alu elements. *Biomol. Struct. Dyn.* **20**, 253–263.
56. Glazko, G. V., and Nei, M. (2003). Estimation of divergence times for major lineages of primate species. *Mol. Biol. Evol.* **20**, 424–434.

Mouse Models for Friedreich's Ataxia

HÉLÈNE PUCCIO

Institut de Génétique et de Biologie Moléculaire et Cellulaire (IGBMC), CNRS/INSERM/Université Louis Pasteur, 67404 Illkirch cedex, CU de Strasbourg, France

I. Introduction
II. Mouse Models for Friedreich's Ataxia
   A. Conditional Knockout Models
   B. Inducible Conditional Knockout Models
   C. Models that Reproduce the Molecular Defect
References

Friedreich ataxia (FRDA), the most common recessive ataxia, is characterized by degeneration of the large sensory neurons and spinocerebellar tracts and cardiomyopathy. It is caused by severely reduced levels of frataxin, a mitochondrial protein involved in iron–sulfur cluster (ISC) biosynthesis. Mouse models have been important tools in dissecting the steps of pathogenesis in FRDA. Furthermore, animal models that reproduce some of the key events in a pathology are essential for the development of effective therapies, both pharmacological and gene therapy approaches. This chapter presents an overview of the current mouse models that have been developed for FRDA.

I. INTRODUCTION

Friedreich's ataxia (FRDA) is the most common hereditary ataxia in Caucasians, with an estimated incidence of 1 in 30,000 [1]. This neurodegenerative disease is characterized by degeneration of the large sensory neurons and spinocerebellar tracts, cardiomyopathy, and an increased incidence of diabetes [2, 3]. FRDA is caused by the partial loss of frataxin, a nuclear-encoded mitochondrial protein thought to be involved in Fe–S protein synthesis. The majority of patients are homozygous for a $(GAA)_n$ triplet expansion within the first intron of the gene, leading to inhibition of transcriptional elongation [4]. A few patients (4%) are compound heterozygotes for the triplet expansion and for a point mutation in the frataxin gene. Frataxin, although highly conserved throughout evolution, has no homology with other proteins, thus preventing a prediction of its functional domain based on its sequence. However, it is its highly conserved nature that has helped to define its role in Fe–S biogenesis [5].

Our understanding of disease pathogenesis and frataxin function comes from evidence from differents models (yeast, cell, and mouse) and from human samples accumulated over the years. Yeast in which the frataxin homolog gene *YFH1* has been deleted harbor a petite phenotype, suggesting a mitochondrial defect [6]. Moreover, *YFH1* was identified as a high-copy-number suppressor of a yeast mutant deficient in intracellular iron usage, thereby linking the yeast frataxin homolog to iron metabolism [7]. Analysis of patient cardiac autopsies and biopsies revealed iron deposits, as well as a selective deficit in the activity of a specific set of mitochondrial proteins bearing Fe–S clusters (complexes I, II, and III of the respiratory chain and aconitases) [8]. These Fe–S proteins are extremely sensitive to free radicals, and their deficit was initially thought to be a consequence of increased oxidative stress generated through the Fenton reaction by mitochondrial iron accumulation. However, the characterization of mice deficient in frataxin has shown that the primary deficit in the disease is Fe–S protein deficiency followed by secondary mitochondrial iron accumulation [9, 10], and that this mitochondrial iron accumulation does not seem to generate oxidative stress [11]. The most recent data from yeast models demonstrate a role for frataxin as an iron chaperone closely involved in ISC assembly/protection and heme biosynthesis. Indeed, reconstitutional studies, as well as *in vivo* studies using yeast strain from which frataxin has been deleted, demonstrate that frataxin is required, although not essential, for Fe–S cluster biosynthesis [12]. A direct interaction of frataxin with ISU1 (yeast)/ISCU (mammals), the scaffolding protein involved in Fe–S cluster synthesis, was demonstrated, suggesting that frataxin might serve as an iron donor protein in the biosynthesis of Fe–S clusters [13, 14]. Mammalian frataxin has also been shown to interact *in vitro* with ferrochelatase, an Fe–S cluster-containing enzyme involved in the last step of heme synthesis [15]. Lastly, both YFH1 and mammalian frataxin have been shown *in vivo* to interact with the mitochondrial aconitase, an Fe–S cluster-containing enzyme of the Krebs cycle [16]. Taken together, these data strongly suggest that frataxin is involved in delivering iron either for Fe–S cluster biogenesis or for heme synthesis, as well as in protecting the Fe–S cluster of aconitase.

II. MOUSE MODELS FOR FRIEDREICH'S ATAXIA

To study further the mechanism of the disease and to test pharmacological therapy, several mouse models have been generated. Our group generated a classic mouse model by constitutive inactivation of frataxin by homologous recombination [17]. Homozygous deletion of frataxin causes embryonic lethality a few days after implantation, demonstrating an important role for frataxin during early development. These results suggest that the milder phenotype in humans is due to residual frataxin expression associated with the expansion mutations. No iron accumulation was observed during embryonic resorption, suggesting that cell death might be due to a mechanism independent of iron accumulation.

A. Conditional Knockout Models

To circumvent embryonic lethality, our group generated, in parallel, two different conditional knockout models, based on the Cre-lox system, in which frataxin was deleted either specifically in skeletal and cardiac muscle (using a transgenic mice expressing the recombinase Cre under the muscle creatine kinase promoter) or in a more generalized frataxin-deficient line including neuronal tissues (neuron-specific enolase (NSE) promoter) [9]. Both models are viable and reproduce some morphological and biochemical features observed in FRDA patients, including cardiac hypertrophy without skeletal muscle involvement in the heart and striated muscle frataxin-deficient line, large sensory neuron dysfunction without alteration of the small sensory and motor neurons in the more generalized frataxin-deficient line, and deficient activities of complexes I–III of the respiratory chain and of the aconitases in both lines. These animals provide an important resource for pathophysiological studies and testing of new treatments.

Murine FRDA cardiomyopathy is characterized by the early onset of dilation with development of left ventricular hypertrophy followed by reduced systolic function. Detailed time course experiments in the cardiac model revealed that the Fe–S enzyme deficiencies begin in the initial phase of the pathology, at the onset of the cardiac dysfunction, whereas intramitochondrial iron accumulation occurs at the end stage of the disease. Moreover, NSE mutant animals do not manifest any iron deposit, but do have a deficit of Fe–S enzymes. Both models therefore indicate that the Fe–S deficiency and cardiomyopathy are independent of mitochondrial iron accumulation. These results support the necessary role of frataxin for efficient Fe–S cluster synthesis, although it is nonessential, because, despite the absence of detectable frataxin at birth, there is still 50% Fe–S enzyme activity at 4 weeks of age in the cardiac model [10]. Therefore, in agreement with recently published results in the yeast model, in the absence of frataxin, Fe–S cluster biosynthesis would occur at a very reduced rate [12]. Several reports suggest that continuous oxidative damage resulting from hampered superoxide dismutase (SOD) signaling participates in the mitochondrial deficiency and, ultimately, neuronal and cardiac cell death. Indeed, the SOD activity is abnormally low in the diseased mouse [18], however, SOD mimetics, contrary to idebenone (see later), have no effect on the survival of animals, and no detectable oxidative stress can be measured [11]. These results suggest that free radical production in the murine FRDA model is a minor component of the pathophysiology. Although these results at first appear to contradict the general agreement, evidence of oxidative damage in FRDA patients is contradictory. These mutant mice therefore represent the first mammalian models in which treatment strategies for the human disease can be evaluated.

As previously mentioned, these animal models have been used for therapeutic trials. Idebenone is a short-chain synthetic analog of coenzyme Q10 that can function as an electron carrier in the mitochondrial respiratory chain and acts as a potent free radical scavenger [19, 20]. During the last 3–5 years, many therapeutic trials have assessed the clinical value of idebenone, but the results remain controverial, mainly because of the clinical heterogeneity of the disease and the lack of randomized placebo-controlled studies [21]. In view of the main methodological difficulties in evaluating the effects of Idebenone in patients, mouse models have become a powerful tool. We found that idebenone effectively delays progressive cardiac hypertrophy and dilation and preserves ventricular contractility by 1 week, thus increasing the life span of the animal by 10% [10]. However, in contrast with the recent observation made on one FRDA patient heart biopsy, idebenone does not restore Fe–S enzyme activity. The results from the placebo-controlled double-blind trial therefore strengthen the results obtained in patient trials, as most of them were open trials without placebo controls, and support the use of idebenone for the human disease.

The neuron-specific mouse model we generated developed a movement disorder characterized by gait abnormalities and loss of proprioception [9]. Furthermore, electrophysiological studies revealed a specific large sensory nerve conduction defect with normal motor nerve conduction. Although these features mimic the neurological symptoms in FRDA patients, our mouse model is extremely severe, with a life expectancy of 24 days. The severity of this model makes any therapeutic approaches very difficult. Furthermore, this model exhibits lesions not seen in the human disease (liver and spongiform cortical lesions), preventing cell-specific degeneration mechanism studies.

B. Inducible Conditional Knockout Models

To obtain specific and progressive neurological models for FRDA, we generated inducible knockout mouse models using two transgenic lines (28.4 and 28.6, having distinct neuronal specificities) expressing the tamoxifen-dependent recombinase (Cre-ER^T) under the mouse prion protein (Prp) promoter, thus enabling us to spatiotemporally control somatic mutagenesis of conditional alleles of the targeted genes. Both Prp-CRE-ER^T lines express the Cre-ER^T recombinase in the nervous system, but whereas the 28.4 line has a wide expression pattern, the Cre-ER^T expression of the 28.6 line is mostly restricted to the hippocampus, the cerebellum, and the dorsal root ganglia (DRG). Both lines developed the most prominent features of the human disease: a slowly progressive mixed cerebellar and sensory ataxia associated with a progressive loss of proprioception and absence of motor involvement [22]. These mouse models also parallel the human disease at the histopathological level. The models have degeneration of the posterior columns of the spinal cord that appear translucent, because of demyelination and loss of fibers, and severe lesions of neurons in Clarke's columns, hallmarks of FRDA. In addition, one of the mutant lines had specific damage to the large sensory neuron cell bodies in the DRG, another distinctive feature of FRDA. The time of occurrence of these lesions suggests that as in patients, the anomalies observed in the neuronal cell bodies of the DRG are a primary event, whereas the neuronal loss in Clarke's column and the degeneration in the posterior column might be secondary events.

The progressive neurodegeneration of the DRG represents an excellent model for unraveling the pathological cascade leading to neuronal death in FRDA. Several cell death pathways can be activated during neurodegeneration, including apoptosis and the more recently accepted mechanism involving autophagy [23]. Although oxidative insult to cultured cells from FRDA patients results mostly in apoptotis for FRDA [24, 25], no evidence for such a cell death mechanism *in vivo* has been provided so far. In particular, apoptosis could not be detected in either the complete frataxin knockout [17] or the conditional mouse [9] model. Surprisingly, we have observed that the degenerative mechanism involved in the DRG neurons is an autophagic process, leading to removal and degradation of damaged cytosolic proteins and organelles. Autophagy is characterized by the presence of autophagic vacuoles and autophagosomes that are formed by rearrangement of subcellular membranes (rough endoplasmic reticulum or trans-Golgi system) to sequester cytosolic constituents and organelles and traffic them to lysosomes for degradation [26]. Different steps of the autophagic process were clearly observed in the large myelineated DRG neurons of the Cb mutants. In addition to autophagosomes, lipofuscin accumulation was also observed in these neurons. Lipofuscin is composed of proteins, lipids, carbohydrates, and metals, particularly iron [27] derived from mitochondria and metalloproteins that are incompletely autophagocytosed and degraded by lysosomes, thereby leading to their accumulation. As lipofuscin accumulation has been reported in both the DRG and cardiomyocytes of FRDA patients [28–30], this cellular response is certainly a close consequence of frataxin deficiency. Moreover, both processes appear to be cell autonomous and progressive, as the ganglion cell neurons that show autophagosomes and/or lipofuscin accumulation are deleted for frataxin. Furthermore, the concurrent activation of both processes could explain the lack of detectable iron deposits within the DRG.

Finally, both the inducible neurological mouse models [22] and the previous FRDA mouse models that we generated [9] clearly show accumulation of damaged mitochondria as a direct consequence of frataxin deficiency. Interestingly, the cardiomyocytes in the cardiac model appear to compensate for lack of frataxin by mitochondrial proliferation and cellular hypertrophy without entering an autophagic process. In contrast, DRG neurons appear to survive without frataxin for a longer period by intracellular removal of damaged organelles and proteins through the autophagic and lysosomal pathways. Therefore, the slowly progressive phenotype of the inducible neurological models would reflect the progressive nature of the autophagic process, proposed as a protective mechanism for the elimination of defective mitochondria with dysfunctional inner membranes [31].

In conclusion, these models represent excellent tools with which to unravel the pathological cascade of FRDA and to test compounds that interfere with the degenerative process, such as antioxidants, which are good pharmacological candidates. Furthermore, these models should prove useful for the investigation of neurodegenerative mechanisms characterized by delayed onset and slow progression over years or decades. The novel spatiotemporally controlled conditional gene-targeting approach that we have used is particularly adapted to study the mechanisms of late-onset and slowly progressive neurodegeneration and is amenable to large experimental flexibility through modulation at will of the timing of induction.

C. Models that Reproduce the Molecular Defect

Although the mouse models we have developed are excellent tools for gaining an understanding of the pathophysiology of the disease, and for assessing some therapeutic protocols, they still do not mimic the situation in the human disease, because conditional gene targeting leads to complete loss of frataxin in some cells at a specific time in development, whereas FRDA is characterized by partial frataxin deficiency in all cells throughout life. Therefore, the need to develop new animal models of the disease remains.

Pandolfo's group has attempted to generate a mouse model by introducing a $(GAA)_{230}$ repeat within the mouse frataxin gene to mirror the chronically reduced levels of frataxin expression found in the human disease [32]. Bred with the *Frda* knockout, the authors obtained animals expressing 25–36% of wild-type frataxin levels, an expression level associated with mildly affected FRDA patients. Unfortunately, these mice did not develop abnormalities of motor coordination, cardiomyopathy, iron metabolism, or response to iron loading up to the age of 1 year. Thus, frataxin levels 25–30% those of the wild type seem to be compatible with normal neurological function and iron metabolism in mice. Furthermore, the GAA repeat is meiotically and mitotically stable in the mouse strain they investigated. Chamberlain's group was able to overcome the embryonic lethality of the *Frda* knockout by generating a transgenic mouse that contains the entire frataxin gene within a human YAC clone on the null mouse background [33]. The human frataxin was expressed in the appropriate tissues at levels comparable to those of endogenous mouse frataxin, and was correctly processed and localized to mitochondria. Biochemical analysis of heart tissue demonstrated preservation of mitochondrial respiratory chain function. Therefore, it is possible to envision the generation of a mouse model expressing low levels of frataxin by using a human YAC derived from a patient, thereby having a repeat within the first intron inhibiting the expression of frataxin. Toward this goal, Pook's group has generated two lines of human FRDA YAC transgenic mice that contain GAA repeat expansions within the appropriate genomic context [34]. Both lines show intergenerational instability. The authors have not yet determined if crossing these mice with the frataxin knockout mice will produce an effective mouse model for FRDA pathology. The levels of human frataxin mRNA and protein they have measured appear too high. However, because they have found an instability of the GAA expansions, these transgenics might be useful in generating an appropriate model for FRDA. It is clearly important to generate, in addition to the already existing models, a mouse model that molecularly mirrors the human disease. This would be particularly useful for the study of potential GAA repeat-based therapies.

References

1. Cossee, M., Schmitt, M., Campuzano, V., Reutenauer, L., Moutou, C., Mandel, J. L., and Koenig, M. (1997). Evolution of the Friedreich's ataxia trinucleotide repeat expansion: Founder effect and premutations. *Proc. Natl. Acad. Sci. USA.* **94**, 7452–7457.
2. Harding, A. E. (1981). Friedreich's ataxia: A clinical and genetic study of 90 families with an analysis of early diagnostic criteria and intrafamilial clustering of clinical features. *Brain* **104**, 589–620.
3. Pandolfo, M. (1998). Molecular genetics and pathogenesis of Friedreich ataxia. *Neuromuscul. Disord.* **8**, 409–415.
4. Campuzano, V., Montermini, L., Moltó, M. D., Pianese, L., Cossée, M., Cavalcanti, F., Monrós, E., Rodius, F., Duclos, F., Monticelli, A., Zara, F., Cañizares, J., Koutnikova, H., Bidichandani, S. I., Gellera, C., Brice, A., Trouillas, P., De Michele, G., Filla, A., de Frutos, R., Palau, F., Patel, P. I., Di Donato, S., Mandel, J-L., Cocozza, S., Koenig, M., and Pandolfo, M. (1996). Friedreich's ataxia: Autosomal recessive disease caused by an intronic GAA triplet repeat expansion. *Science* **271**, 1423–1427.
5. Huynen, M. A., Snel, B., Bork, P., and Gibson, T. J. (2001). The phylogenetic distribution of frataxin indicates a role in iron-sulfur cluster protein assembly. *Hum. Mol. Genet.* **10**, 2463–2468.
6. Wilson, R. B., and Roof, D. M. (1997). Respiratory deficiency due to loss of mitochondrial DNA in yeast lacking the frataxin homologue. *Nat. Genet.* **16**, p. 352–357.
7. Babcock, M., de Silva, D., Oaks, R., Davis-Kaplan, S., Jiralerspong, S., Montermini, L., Pandolfo, M., and Kaplan, J. (1997). Regulation of mitochondrial iron accumulation by Yfh1p, a putative homolog of frataxin. *Science* **276**, 1709–1712.
8. Rotig, A., de Lonlay, P., Chretien, D., Foury, F., Koenig, M., Sidi, D., Munnich, A., and Rustin, P. (1997). Aconitase and mitochondrial iron–sulphur protein deficiency in Friedreich's ataxia. *Nat. Genet.* **17**, 215–217.
9. Puccio, H., Simon, D., Cossee, M., Criqui-Filipe, P., Tiziano, F., Melki, J., Hindelang, C., Matyas, R., Rustin, P., and Koenig, M. (2001). Mouse models for Friedreich ataxia exhibit cardiomyopathy, sensory nerve defect and Fe–S enzyme deficiency followed by intramitochondrial iron deposits. *Nat. Genet.* **27**, 181–186.
10. Seznec, H., Simon, D., Monassier, L., Criqui-Filipe, P., Gansmuller, A., Rustin, P., Koenig, M., Puccio, H., Carelle, N., Weber, P., and Metzger, D. (2004). Idebenone delays the onset of cardiac functional alteration without correction of Fe–S enzymes deficit in a mouse model for Friedreich ataxia mouse models with progressive cerebellar and sensory ataxia reveal autophagic neurodegeneration in dorsal root ganglia. *Hum. Mo. Genet.* **13**, 1017–1024. [Epub 2004 Mar 17.]
11. Seznec, H., Simon, D., Bouton, C., Reutenauer, L., Hertzog, A., Golik, P., Procaccio, V., Patel, M., Drapier, J. C., Koenig, M., and Puccio, H. (2005). Friedreich ataxia: The oxidative stress paradox. *Hum. Mol. Genet.* **14**, 463–474.
12. Muhlenhoff, U., Richhardt, N., Ristow, M., Kispal, G., and Lill, R. (2002). The yeast frataxin homolog Yfh1p plays a specific role in the maturation of cellular Fe/S proteins. *Hum. Mol. Genet.* **11**, 2025–2036.
13. Gerber, J., Muhlenhoff, U., and Lill, R. (2003). An interaction between frataxin and Isu1/Nfs1 that is crucial for Fe/S cluster synthesis on Isu1. *EMBO Rep.* **4**, 906–911. [Epub 2003 Aug 15.]
14. Yoon, T., and Cowan, J. A. (2003). Iron–sulfur cluster biosynthesis: Characterization of frataxin as an iron donor for assembly of [2Fe–2S] clusters in ISU-type proteins. *J. Am. Chem. Soc.* **125**, 6078–6084.
15. Yoon, T., and Cowan, J. A. (2004). Frataxin-mediated iron delivery to ferrochelatase in the final step of heme biosynthesis. *J. Biol. Chem.* **27**, 27.
16. Bulteau, A. L., O'Neill, H. A., Kennedy, M. C., Ikeda-Saito, M., Isaya, G., and Szweda, L. I. (2004). Frataxin acts as an iron chaperone protein to modulate mitochondrial aconitase activity. *Science.* **305**, 242–245.
17. Cossée, M., Puccio, H., Gansmuller, A., Koutnikova, H., Dierich, A., LeMeur, M., Fischbeck, K., Dolle, P., and Koenig, M. (2000). Inactivation of the Friedreich ataxia mouse gene leads to early embryonic lethality without iron accumulation. *Hum. Mol. Genet.* **9**, 1219–1226.
18. Chantrel-Groussard, K., Geromel, V., Puccio, H., Koenig, M., Munnich, A., Rotig, A., and Rustin, P. (2001). Disabled early recruitment of antioxidant defenses in Friedreich's ataxia. *Hum. Mol. Genet.* **10**, 2061–2067.
19. Mordente, A., Martorana, G. E., Minotti, G., and Giardina, B. (1998). Antioxidant properties of 2,3-dimethoxy-5-methyl-6-(10-hydroxydecyl)-1,4-benzoquinone (idebenone). *Chem. Res. Toxicol.* **11**, 54–63.
20. Gillis, J. C., Benefield, P., and McTavish, D. (1994). Idebenone: A review of its pharmacodynamic and pharmacokinetic properties, and therapeutic use in age-related cognitive disorders. *Drugs Aging* **5**, 133–152.
21. Rustin, P. (2003). The use of antioxidants in Friedreich's ataxia treatment. *Expert Opin. Invest. Drugs* **12**, 569–575.
22. Simon, D., Seznec, H., Gansmuller, A., Carelle, N., Weber, P., Metzger, D., Rustin, P., Koenig, M., and Puccio, H. (2004). Friedreich ataxia mouse models with progressive cerebellar and sensory ataxia reveal autophagic neurodegeneration in dorsal root ganglia. *J. Neurosci.* **24**, 1987–1995.
23. Xue, L., Fletcher, G. C., and Tolkovsky, A. M. (1999). Autophagy is activated by apoptotic signalling in sympathetic neurons: An alternative mechanism of death execution. *Mol. Cell. Neurosci.* **14**, 180–198.
24. Santos, M. M., Ohshima, K., and Pandolfo, M. (2001). Frataxin deficiency enhances apoptosis in cells differentiating into neuroectoderm. *Hum. Mol. Genet.* **10**, 1935–1944.
25. Wong, A., Yang, J., Cavadini, P., Gellera, C., Lonnerdal, B., Taroni, F., and Cortopassi, G. (1999). The Friedreich's ataxia mutation confers cellular sensitivity to oxidant stress which is rescued by chelators of iron and calcium and inhibitors of apoptosis. *Hum. Mol. Genet.* **8**, 425–340.
26. Reggiori, F., and Klionsky, D. J. (2002). Autophagy in the eukaryotic cell. *Eukaryot. Cell* **1**, 11–21.
27. Brunk, U. T., and Terman, A. (2002). Lipofuscin: Mechanisms of age-related accumulation and influence on cell function. *Free Radic. Biol. Med.* **33**, 611–619.
28. Lamarche, J., Luneau, C., and Lemieux, B. Ultrastructural observations on spinal ganglion biopsy in Friedreich's ataxia: A preliminary report. *Can. J. Neurol. Sci.* 1982. **9**(2): 137–139.
29. Lamarche, J. B., Cote, M., and Lemieux, B. (1980). The cardiomyopathy of Friedreich's ataxia morphological observations in 3 cases. *Can. J. Neurol. Sci.* **7**, 389–396.
30. Larnaout, A., Belal, S., Zouari, M., Fki, M., Ben Hamida, C., Goebel, H. H., Ben Hamida, M., and Hentati, F. (1997). Friedreich's ataxia with isolated vitamin E deficiency: A neuropathological study of a Tunisian patient. *Acta. Neuropathol. (Berl.).* **93**, 633–637.
31. Lemasters, J. J., Nieminen, A. L., Qian, T., Trost, L. C., Elmore, S. P., Nishimura, Y., Crowe, R. A., Cascio, W. E., Bradham, C. A., Brenner, D. A., and Herman, B. (1998). The mitochondrial permeability transition in cell death: A common mechanism in necrosis, apoptosis and autophagy. *Biochim. Biophys. Acta.* **1366**, 177–196.

32. Miranda, C. J., Santos, M. M., Ohshima, K., Smith, J., Li, L., Bunting, M., Cossée, M., Koenig, M., Sequieros, J., Kaplan, J., and Pandolfo, M. (2002). Frataxin knockin mouse. *FEBS Lett.* **512**, 291–297.
33. Pook, M. A., Al-Mahdawi, S., Carroll, C. J., Cossee, M., Puccio, H., Lawrence, L., Clark, P., Lowrie, M. B., Bradley, J. L., Cooper, J. M., Koenig, M., and Chamberlain, S. (2001). Rescue of the Friedreich's ataxia knockout mouse by human YAC transgenesis. *Neurogenetics*. **3**, 185–193.
34. Al-Mahdawi, S., Pinto, R. M., Ruddle, P., Carroll, C., Webster, Z., and Pook, M. (2004). GAA repeat instability in Friedreich ataxia YAC transgenic mice. *Genomics* **84**, 301–310.

CHAPTER 21

Triplexes, Sticky DNA, and the (GAA·TTC) Trinucleotide Repeat Associated with Friedreich's Ataxia

LESLIE S. SON AND ROBERT D. WELLS

Institute of Biosciences and Technology, Center for Genome Research, Texas A&M University System Health Science Center, Texas Medical Center, Houston, Texas 77030

I. Introduction
II. DNA Structures Associated with (R·Y) Sequences
   A. Triplexes
   B. Sticky DNA
III. Sticky DNA Properties and Detection
   A. Requirements for Sticky DNA Formation
   B. Assays for Detection of Sticky DNA
IV. Effect of Sticky DNA on Cellular Mechanisms
   A. Transcription
   B. Replication
   C. Recombination
V. Concluding Remarks
   Acknowledgments
   References

DNA polypurine·polypyrimidine (R·Y) sequences have the ability to form triplexes and sticky DNA. The properties of triplexes have been studied since the late 1950s. Triple helix formation may function in the regulation of cellular processes from transcription to genome integrity. Sticky DNA is a novel non-B-DNA structure (a type of triplex) that is formed by long tracts of the trinucleotide repeat sequence (GAA·TTC) that is associated with the hereditary neurological disease Friedreich's ataxia. The formation of this conformation is dependent on the lengths and extents of polymorphisms of the repeating tracts, the orientation of the repeats, negative supercoiling, and the presence of divalent metal ions. These requirements for formation and stability have been measured by several *in vitro* methods including gel mobility, chemical probing, and electron microscopy. Also, long (GAA·TTC) repeats that are prone to form sticky DNA cause replication pausing and inhibit transcription. Both sticky DNA and triplexes exist and function *in vivo*. The structural properties of triplexes and sticky DNA and their role in cellular mechanisms are discussed.

I. INTRODUCTION

This chapter focuses on the existence and characteristics of the novel non-B-DNA structure, sticky DNA that forms with the trinucleotide repeating sequence (TRS) (GAA·TTC). This TRS is found as an expansion product in patients who have Friedreich's ataxia (FRDA), a severe hereditary neurodegenerative disease. The expansion is found within the first intron of the *X25* gene that codes for the protein frataxin and can reach lengths of >1000 repeats [1–4]. The possible molecular biological mechanisms leading to the expansion of the repeat, as well as the pathophysiological and clinical aspects of the disease, are discussed in other chapters of this book written by Pandolfo [5], Wilson [6], De Biase [7], Mirkin [8], and their colleagues.

Long (GAA·TTC) repeats stall replication and inhibit transcription *in vivo* in *Escherichia coli* and *in vitro* [9–11]. Also, the long stretches of (GAA·TTC) have the propensity to form non-B-DNA structures (triplexes and sticky DNA) [12–14]. Due to these findings concerning the purine·pyrimidine (R·Y)-rich sequence, it is hypothesized that non-B-DNA structure formation of the long (GAA·TTC) tracts may be responsible for the aberrant cellular mechanisms.

The following sections further elaborate on the existence of the non-B-DNA structures that are associated with the (GAA·TTC) TRS. The requirements for formation of these structures and assays used for their detection are reviewed. Also, the effect of sticky DNA on cellular mechanisms and the possible role in the loss-of-function manifestations of Friedreich's ataxia are discussed.

II. DNA STRUCTURES ASSOCIATED WITH (R·Y) SEQUENCES

Non-B-DNA structures can form as a consequence of cellular mechanistic activities such as replication, transcription, and DNA recombination and repair [1, 15–17]. DNA sequence composition, supercoiling, single-strandedness of the helix, and even intracellular environmental factors such as metal ions and the presence of proteins all contribute to the propensity for non-B structure formation [1, 16–18]. The (GAA·TTC) repeating sequence, associated with FRDA, is an example of an (R·Y)-rich sequence that has mirror repeat symmetry, allowing for the formation of both inter- and intramolecular triplexes with either two purine strands and one pyrimidine strand or two pyrimidine strands and one purine-rich tract. (GAA·TTC) repeats also have direct repeat symmetry, which is conducive to sticky DNA formation in the presence of supercoiling and metal ions. Since the discovery of triplexes in 1957 [19], triplex structures, also referred to as triple helices or H-DNA, have been observed both *in vitro* and *in vivo* using a variety of techniques and model systems [1, 14, 16–18, 20–24]. Sticky DNA, a more recent DNA structural discovery, has also been observed *in vitro* and *in vivo* in *E. coli* [13, 25–27].

A. Triplexes

Triplexes are formed by polypurine·polypyrimidine (R·Y) DNA sequences that contain a mirror repeat symmetry [18, 23, 24]. These triple-stranded conformations form at (R·Y) regions where the purine strand of the Watson–Crick duplex associates with a third strand through Hoogsteen hydrogen bonds in the major groove of the original B conformation duplex.

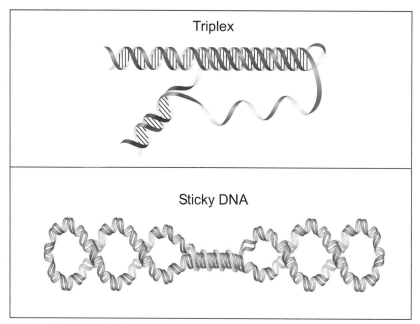

FIGURE 21-1 Models of an intramolecular triplex and sticky DNA structures. See CD-ROM for color image.

This third strand may be purine- or pyrimidine-rich depending on the pH and ionic conditions. For this reason, we denote two kinds of triplex structures: one is a purine·purine·pyrimidine (R·R·Y) conformation and the other is a pyrimidine · purine · pyrimidine (Y·R·Y) conformation [23]. The third strand of the (R·R·Y) triplex is in an antiparallel orientation with respect to the duplex, and the third strand of the (Y·R·Y) triplex is in a parallel orientation. Furthermore, the third strand of a triplex may originate from a single DNA (R·Y) tract (intramolecular) or between two separate DNA sequences (intermolecular). The propensity for one conformation to form over the other is based on the pH and metal ion conditions. pH values below 5.5 cause cytosine residues to be protonated, therefore promoting the pyrimidine strand to Hoogsteen bond with the purine strand of the duplex. More neutral pH conditions promote the formation of the (R·R·Y) triple helices [23]. Metal ions are known to have triplex-stabilizing effects as well. Divalent metal ions such as magnesium (Mg^{2+}), manganese (Mn^{2+}), and zinc (Zn^{2+}) have been shown to stabilize the (R·R·Y) conformation by binding to the negatively charged phosphate backbone of the DNA [28–30]. Both the (Y·R·Y) triple helix and the (R·R·Y) are stabilized by polyamines like spermine and spermidine [31–34].

Triplex structures have been identified both *in vitro* and in *in vivo* model systems using a variety of analyses [18, 23]. Chemical and enzymatic probes such as psoralen, P1 and S1 nucleases, and methyl-directed restriction site modification have proven to be quite effective in detection of H-DNA conformations [30, 35–37]. Another assay that has been useful in triplex identification is the application of anti-DNA antibodies [38, 39]. Base-specific DNA antibody binding has demonstrated not only the ability to distinguish triple helices from double helices, but also to distinguish specific triplex conformations in systems from *E. coli* to human cells [38]. Other detection methods include chemical probing, several different methods of spectroscopy, two-dimensional gel analysis, electron microscopy, triplex affinity binding assays, and X-ray diffraction [23].

The conclusion that triple helices exist in various conformations, that is, (R·R·Y) or (Y·R·Y) helices and intra- or intermolecular formation, may be an indication that these different conformations hold different biological roles in living systems. Although the biological significance of triplex structures remains to be firmly elucidated, we know that naturally occurring mirror repeat sequences that have the ability to form triplexes are abundant in the mammalian genome [40] and are most predominant in promoters and exons of genes [41, 42]. Numerous studies on proteins that preferentially bind to H-DNA structures [43–46] indicate that triplex formation within these regions might play a role in the regulation of cellular mechanisms such as replication, transcription, chromosomal folding, and nucleosome exclusion [18, 23]. Other studies show that triplexes can inhibit DNA replication [47, 48] and induce DNA recombination and repair [49–51], further supporting the idea that creation of triplex structures within these significant regions of a gene plays a fundamental role in the function of gene expression.

Most recently, studies on a poly (R·Y) sequence from the human PKD1 gene identified large genomic rearrangements within the sequence caused by double-strand breaks [52]. Interestingly, the breakpoints were localized to regions prone to form triple helices and other non-B-DNA structures. This study concluded that non-B-DNA conformations (triplexes in this case) trigger genomic rearrangements through recominbination– repair activities. Furthermore, naturally occurring DNA sequences that formed triplexes (H-DNA) induced double-strand breaks at the locus of the H-DNA, which caused subsequent mutagenesis in mammalian cells [53].

B. Sticky DNA

Sticky DNA was discovered by Sakamoto et al. in 1999 [13]. This non-B-DNA structure is thought to be a specific type of triplex that forms under certain environmental conditions. Although many of the requirements are similar to those of previously characterized triplex structures, sticky DNA has unique properties in the field of non-B-DNA conformations. The following section discusses the characteristics and biological implications of sticky DNA in Friedreich's ataxia.

III. STICKY DNA PROPERTIES AND DETECTION

A. Requirements for Sticky DNA Formation

As mentioned previously, many of the requirements for sticky DNA formation are similar to those of previously characterized triplex structures. However, sticky DNA is distinctive in that it forms only intramolecularly and only in the presence of two long duplex tracts of (GAA·TTC) or (GAA·TTC) in one duplex and

(GAAGGA·TCCTTC) in the other duplex [21-1]. Also, pairs of long tracts of (GA·TC) and (GGA·TCC) form sticky DNA [27], as expected. The minimum repeat length for each of these sequences for structure formation varies depending on the sequences involved, but in most cases must be >60 repeats [13, 25, 27]. The conformation of sticky DNA was originally thought to be a bi-triplex [13] formed from the association of two separate plasmids, each containing a single TRS. Each (R·Y)-rich sequence formed an individual triplex by donating a portion of single-stranded DNA to the other plasmid's duplex DNA, creating a very stable bi-triplex conformation. However, further experiments on the requirements for sticky DNA formation have modified this proposed model. Vetcher et al. tested monomer and dimer DNA containing one and two (GAA·TTC) tracts, respectively, for the ability to form sticky DNA [26]. Sticky DNA was found only in the dimer species but not in the monomer, proving that the structure must have both (GAA·TTC) tracts in the same closed circular plasmid, forming a single long triplex rather than a bi-triplex structure. Additionally, this result was confirmed by the observation that plasmids harboring a single long tract of (GAA·TTC)$_{150}$ had a greater propensity to form dimers over monomers *in vivo* compared with the control vector [25]. The results were dependent on the length of the TRS, the direct orientation of the repeat inserts, *in vivo* negative supercoil density, and the extent of nucleotide interruptions within the repeating tracts.

Sticky DNA can form at both neutral and slightly acidic pH values. However, the amount of structure formation is 3.5 times higher at pH 8.0 than at pH 5.5 [13]. Prior work has shown that below pH 6, (Y·R·Y) triplexes are stabilized [23, 24, 54], whereas (R·R·Y) triple helices are stabilized at a neutral pH [55] in the presence of divalent metal ions. The greater propensity for stabilization of sticky DNA at neutral pH with magnesium (Mg^{2+}) may indicate that the conformation of sticky DNA is an (R·R·Y) triplex.

Supercoiling is also required for two repeating (GAA·TTC) tracts to interact with each other to form sticky DNA. Sakamoto et al. found that when plasmid DNA was relaxed with topoisomerase I followed by the introduction of supercoiling with ethidium bromide (EtBr), the higher the amount of supercoiling introduced, the faster the migration with negative supercoil densities ($-\sigma$) ranging from 0 to 0.042. However, a slower migration was observed for topoisomers with $-\sigma = 0.051$–0.074, suggesting that relaxation occurred at these densities as a result of sticky DNA formation [13]. At even higher densities, the topoisomers again migrated faster. Other studies revealed a similar correlation between the amount of negative supercoiling in a closed circular plasmid with the length of the DNA/DNA interacting regions between the two (GAA·TTC) repeating tracts. Vetcher et al. determined this length by observing plasmids harboring the long (GAA·TTC) repeats with different amounts of negative supercoiling under electron microscopy (EM) [26]. Supercoil density values between -0.05 and -0.075 revealed the greatest amount of DNA/DNA interaction. Currently, studies are being done to test the requirement for negative supercoiling to form sticky DNA *in vitro* (L. S. Son and R. D. Wells, unpublished data).

Many studies have addressed the effect of metal ions on triplexes [28, 29, 56–58]. It is hypothesized that if sticky DNA is, in fact, a type of (R·R·Y) triplex, it would be stabilized by similar divalent metal ions, that is, Mg^{2+}, Mn^{2+}, Zn^{2+}, or Co^{2+} [23]. According to prior studies, this stabilization is also dependent on time and temperature of incubation of the DNA with the metal ion. Previously, experiments were conducted on the ability of mono- and divalent metal ions to reform sticky DNA [13]. Two linear plasmids containing a single tract of (GAA·TTC)$_{150}$ were used for this experiment in which no sticky DNA formation was observed. More extensive work is currently being done on the propensity for sticky DNA formation by supercoiled plasmids containing two tracts of (GAA·TTC) in the presence of various divalent metal ions (L. S. Son and R. D. Wells, unpublished data).

B. Assays for Detection of Sticky DNA

The methods used to detect sticky DNA are similar to those used for other non-B-DNA structures. The first observations that long (GAA·TTC) repeats formed a novel non-B conformation occurred by gel electrophoresis mobility shifts [13]. On digestion of plasmid DNA harboring two (GAA·TTC)$_{150}$ repeating tracts (a biologically formed dimer in these studies), a linear fragment that ran at the appropriate length (7.1 kb) and a DNA band migrating seven times slower (42 kb) (retarded band) than the linear DNA were observed. To interpret the nature of the retarded band species, the DNA band was isolated from the gel and treated with EDTA at 80°C for 10 min, chelating any Mg^{2+} that might stabilize a non-B structure formed by the repeating tracts. The result was the loss of the RB species and complete restoration of linear monomer DNA product. Interestingly, the RB species appeared to be stable when incubated at 80°C for 1 h in the absence of EDTA. This revealed the extreme thermal stability and confirmed the significance of divalent metal ions for the maintenance of the structure. The extent of gel retardation of sticky DNA compared with the linear monomer appears to be a result of the length of the digested duplex arms outside of the structure, as well as the extent of association of the two (GAA·TTC) tracts. This is discussed later in this chapter.

Chemical and enzymatic probing has also been performed on plasmids that can form the sticky DNA structure. Studies using P1 nuclease, which cleaves DNA at partially unpaired or single-stranded regions, as well as OsO_4 modification of single-stranded thymine residues, reveal an interaction of the two (GAA·TTC) tracts with single-stranded characteristics both at the duplex–duplex junctions just outside of the repeats and also a single-stranded pyrimidine tract within the TRS [13]. Another chemical that has been used to detect sticky DNA is nitrogen mustard [25]. Nitrogen mustard (HN_2) has the ability to crosslink guanine residues when the guanines are in close proximity to each other. The strategy was that if sticky DNA formed *in vivo*, the HN_2 would crosslink the guanine nucleotides that were associating within the structure. Vetcher et al. treated plasmids containing different lengths of (GAA·TTC) repeats with this chemical in an *in situ* experiment to attempt to lock the sticky DNA in its conformation while still inside *E. coli* cells. Results showed that treatment of the DNA with HN_2 followed by restriction digestion displayed the same RB product as those plasmids that were not treated with HN_2. Interestingly, when the HN_2-treated plasmids were incubated at 80°C for 10 min in the presence of EDTA, the retarded band still remained intact. Hence, the HN_2 did trap the sticky DNA conformation by cross-linking the guanine resides [25]. However, it could be argued that either the preexistence of sticky DNA *in vivo* allowed for the HN_2 crosslinking or the crosslinking itself played a role in the association of the two (GAA·TTC) tracts. Obviously, the latter criticism can be raised with all studies involving an exogenous chemical probe. Also, although nitrogen mustard can penetrate the *E. coli* cell wall, it is highly toxic to the organism and eventually results in cell death. This leads to another consideration as to whether sticky DNA was formed *in vivo* and locked into the conformation inside the cell or sticky DNA formation is simply a structural artifact of restriction digestion with (GAA·TTC)-containing plasmids.

Direct visual observations of sticky DNA have been made through the use of electron microscopy (EM) (Fig. 21-2). Sakamoto et al. digested the dimer DNA, which harbored two long (GAA·TTC) tracts, with several restriction enzymes [13]. The results of EM on the sticky DNA retarded band that resulted from these individual enzymatic digestions showed that the distance of the restriction site cleavage from the two interacting repeat tracts determined the length of the duplex arms and hence the size of the X-shaped structure. The gel mobility of these different-sized X-shaped structures was also determined by cleavage with the different restriction enzymes followed by agarose gel electrophoresis [13]. The longer the duplex arms of the plasmid (based on the location of the restriction site), the slower the mobility on

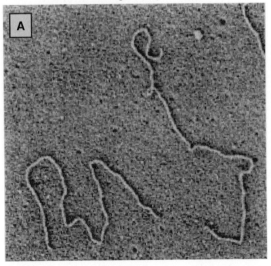
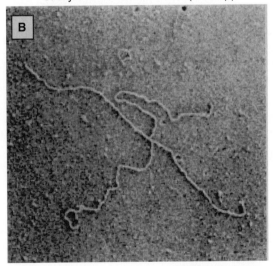

FIGURE 21-2 Electron microscopy of plasmid DNA harboring two tracts of (GAA·TTC)$_{150}$. The dimer of pRW3822 was digested with *Xmn*I to yield both linear monomer and a retarded band. Both species were isolated from an agarose gel. (*A*) Linear monomeric DNA of length 7.1 kb. (*B*) Retarded band DNA that migrates up to six times more slowly (42 kb) than the linear DNA on an agarose gel. The sticky DNA is observed as an X-shaped structure where the (GAA·TTC) tracts are associated at the intersection. Reprinted, with permission, from Sakamoto et al. [9].

the gel. Furthermore, EM has been applied to sticky DNA to confirm the hypothesis that the conformation is a single long triplex versus the original idea that it is an intermolecular bi-triplex model [26]. A single, long triplex model of sticky DNA would be very rigid when compared with the duplex DNA. Conversely, an intermolecular bi-triplex model would be more flexible due to the extent of unpaired bases between the two triplexes. Through electron microscopy, the degree of bending of the DNA/DNA interacting TRS region was determined, confirming the rigidity of the sticky DNA region, hence a single long triplex conformation.

The question of sticky DNA's existence *in vivo* has been an arduous task, as there are no direct assays to analyze the presence of the structure inside the cell. However, because we know that the formation of some non-B-DNA structures, specifically triplexes and sticky DNA, is dependent on negative supercoiling [13, 18, 23, 26], it is possible to measure variations of superhelical density occurring inside the cell as a result of structure formation. Due to the inherent ability of a cell to maintain a certain level of supercoiling using topoisomerases, the formation of a non-B-DNA structure alters the natural supercoiling capacity of the molecule, thus promoting topoisomerases either to relax or to incorporate negative supercoil turns into the DNA. This relaxation or increase in the negative superhelicity, resulting in a family of topoisomer species, can be measured on agarose gels on isolation of the DNA. Previously, this method has been applied to the determination of left-handed Z-DNA and cruciforms in *E. coli* [59–61]. Studies are currently being conducted on the *in vivo* existence of sticky DNA with the use of this assay. From our current knowledge, it appears that sticky DNA is unique compared with other non-B-DNA structures, both in its characteristics for formation and stability and in its detailed conformational features (L. S. Son and R. D. Wells, unpublished work).

IV. EFFECT OF STICKY DNA ON CELLULAR MECHANISMS

A. Transcription

Friedreich's ataxia is associated with the expansion of the (GAA·TTC) TRS in the frataxin gene. As mentioned previously, the expansion results in reduced levels of X25 mRNA transcript and mature protein in patients. Investigations were conducted on the mechanism that produces this effect. Plasmids containing (GAA·TTC) repeat lengths of 9–270 triplets were transfected into eukaryotic COS-7 cells to observe the length and orientation effect of the TRS on transcription, gene expression, and replication [9]. RNase protection assays and Northern blot analyses demonstrated very low levels of mature mRNA with increasing lengths of TRS. Also, an orientation dependence was shown to affect transcription, i.e., the mRNA was much lower when transcription used $(TTC)_n$ as the template than when the $(GAA)_n$ was transcribed. Furthermore, analysis of β-galactosidase activity resulting from *lacZ* expression in these cells significantly decreased with respect to the TRS length and orientation.

Also, Sakamoto et al. studied the effect of sticky DNA on bacterial and eukaryotic *in vitro* transcription [11]. For these studies, plasmids harboring (GAA·TTC) repeat lengths of 9–150 were transformed into the *E. coli* SURE strain, followed by DNA isolation and digestion with a restriction endonuclease. The DNA was run on an agarose gel, and both linear and sticky DNA retarded bands were isolated. *In vitro* transcription of the linear and retarded bands was carried out by either the T7 or SP6 RNA polymerase. The results showed that the amount of full-length transcript was greatly reduced for the retarded band containing sticky DNA compared with the transcript of the linear template.

Furthermore, results demonstrated that the transcriptional inhibition was due to sequestration of the RNA polymerases by direct binding to the DNA structure. Other sequences such as the (GAAGGA·TCCTTC) repeat were also used for this study in which no inhibition of transcription was observed. Previous work identified the hexanucleotide repeat (GAAGGA·TCCTTC) in FRDA patients [62]. This repeating sequence was found to be nonpathogenic in FRDA. A complementary body of work researched the effect of interruptions within pure (GAA·TTC) repeats on the amount of transcriptional activity [12]. It was found that the incorporation of >11% of the (GGA·TCC) triplet into a pure $(GAA·TTC)_{150}$ repeating sequence resulted in the inhibition of sticky DNA retarded band and also alleviated the inhibition of transcription. In summary, these data suggest that the influence of (GAA·TTC) and sticky DNA in FRDA may be at the transcriptional level.

B. Replication

Analysis of the amount of replicated plasmid in COS-7 cells was analyzed by Southern blot hybridization. Results showed that the hybridization signal for the longest (GAA·TTC) repeats was greatly reduced compared with that for the shorter tracts. Compared with control plasmids of similar general length and plasmids with smaller TRS, the plasmids harboring a longer (GAA·TTC) TRS were recovered in much lower concentrations. This indicated a specific effect of the TRS on replication, rather than a lower efficiency of replication due to plasmid length. Other studies on the effect of the (GAA·TTC) repeats on replication pausing *in vivo* have been done [47] (also, see Chapter 40 by Mirkin [8]). These studies concluded that long tracts of (GAA·TTC) repeats have profound effects on cellular mechanisms. Furthermore, it was hypothesized that DNA secondary structures such as triplexes and/or sticky DNA may be involved in the etiology of FRDA.

C. Recombination

Sticky DNA has also been implicated in repair and recombination mechanisms. Napierala et al. demonstrated that the presence of two long (GAA·TTC) tracts within a single plasmid reduces the frequency of recombination, which is opposite to the results found for the (CTG·CAG) TRS associated with myotonic dystrophy type 1 [63]. The (GAA·TTC) repeat sequence and recombination are discussed further in detail in Chapter 38 by Dere et al. [64].

Also, the possible involvement of other proteins that participate in DNA metabolism with sticky DNA were studied [25]. In this work, the relationship of plasmids harboring long (GAA·TTC) tracts to form dimer molecules over monomers (D/M ratio) with the ability to adopt the sticky DNA structure was characterized. The length and orientation of the single repeating tract influenced the dimerization of the plasmids. As mentioned earlier, only plasmids with two long (GAA·TTC) tracts in the direct repeat orientation can form sticky DNA. The enhancement of the D/M with increasing tract length reflects the interaction of the TRS tracts. Because RecA and Uvr proteins are involved in DNA metabolism, the effect of mutations in these genes on the D/M ratios were tested. A parental strain (RR1), $recA^-$ strain (HB101), and $uvrA^-$, $uvrB^-$, $uvrC^-$, and $uvrD^-$ strains were evaluated. The results showed that mutations in $uvrA^-$, $uvrB^-$, $uvrC^-$, and $uvrD^-$ had no effect on the D/M for the vector control (pSPL3) compared with the parental strain. However, with the plasmid containing (GAA·TTC)$_{150}$, a very significant difference in the D/M ratio was observed in the $uvrA^-$ and $uvrB^-$ strains, but the ratio appeared similar to that of the parental strain in the $uvrC^-$ and $uvrD^-$ strains. This indicates that the UvrA$_2$/UvrB complex that recognizes DNA helical distortions is important in the enhancement of the D/M ratio that is correlated with sticky DNA formation [25]. The $recA^-$ strain was not prone to form dimers; therefore, a D/M ratio could not be determined.

V. CONCLUDING REMARKS

The role of non-B-DNA structures in the onset of human disease has become a critical topic of research [14, 52, 65]. The function of sticky DNA in the etiology of FRDA has attracted great interest in the past 7 years, as it appears to be a unique structure specific for the (R·Y) repeat expansions found in FRDA patients. Although we do not fully understand the mechanism by which sticky DNA leads to the eventual loss of frataxin protein, this conformation clearly stalls replication and inhibits transcriptional activity *in vivo* in *E. coli* and HeLa extracts. Thus, it is imperative to move these experimental methodologies into human cells to bridge the gap between the initial expansion of the repeat tract and the manifestation of the disease.

Acknowledgments

This work was supported by grants from the National Institutes of Health (NS37554 and ES11347), the Robert A. Welch Foundation, the Friedreich's Ataxia Research Alliance, and the Seek-a-Miracle Foundation (Muscular Dystrophy Foundation).

References

1. Bowater, R. P., and Wells, R. D. (2001). The intrinsically unstable life of DNA triplet repeats associated with human hereditary disorders. *Prog. Nucleic Acid Res. Mol. Biol.* **66**, 159–202.
2. Pandolfo, M. (2002). The molecular basis of Friedreich ataxia. *Adv. Exp. Med. Biol.* **516**, 99–118.
3. Pandolfo, M. (1999). Molecular pathogenesis of Friedreich ataxia. *Arch. Neurol.* **56**, 1201–1208.
4. Pandolfo, M., and Koenig, M. (1998). Freidreich's ataxia. *In* "Genetic Instabilities and Hereditary Neurological Diseases" (R. D. Wells, and S. T. Warren, Eds.), pp. 373–398. Academic Press, San Diego.
5. Pandolfo, M. (2006). Friedreich's ataxia. *In* "Genetic Instabilities and Neurological Diseases" (R. D. Wells, and T. Ashizawa, Eds.), 2nd ed., Ch. 17. Elsevier, Burlington, MA.
6. Wilson, R. B. (2006). Experimental therapeutics for Friedreich's ataxia. *In* "Genetic Instabilities and Neurological Diseases" (R. D. Wells, and T. Ashizawa, Eds.), 2nd ed., Ch. 18. Elsevier, Burlington, MA.
7. De Biase, I., Rasmussen, A., and Bidichandani, S. (2006). Evolution and instability of GAA triplet-repeat sequence in Friedreich's ataxia. *In* "Genetic Instabilities and Neurological Diseases" (R. D. Wells, and T. Ashizawa, Eds.), 2nd ed., Ch. 19. Elsevier, Burlington, MA.
8. Mirkin, S. (2006). Replication of expandable DNA repeats. *In* "Genetic Instabilities and Neurological Diseases" (R. D. Wells, and T. Ashizawa, Eds.), 2nd ed., Ch. 41. Elsevier, Burlington, MA.
9. Ohshima, K., Montermini, L., Wells, R. D., and Pandolfo, M. (1998). Inhibitory effects of expanded GAA·TTC triplet repeats from intron I of the Friedreich ataxia gene on transcription and replication *in vivo*. *J. Biol. Chem.* **275**, 14588–14595.

10. Bidichandani, S. I., Ashizawa, T., and Patel, P. I. (1998). The GAA triplet-repeat expansion in Friedreich ataxia interferes with transcription and may be associated with an unusual DNA structure. *Am. J. Hum. Genet.* **62**, 111–121.
11. Sakamoto, N., Ohshima, K., Montermini, L., Pandolfo, M., and Wells, R. D. (2001). Sticky DNA, a self-associated complex formed at long GAA∗TTC repeats in intron 1 of the frataxin gene, inhibits transcription. *J. Biol. Chem.* **276**, 27171–27177.
12. Sakamoto, N., Larson, J. E., Iyer, R. R., Montermini, L., Pandolfo, M., and Wells, R. D. (2001). GGA∗TCC-interrupted triplets in long GAA∗TTC repeats inhibit the formation of triplex and sticky DNA structures, alleviate transcription inhibition, and reduce genetic instabilities. *J. Biol. Chem.*, **276**, 27178–27187.
13. Sakamoto, N., Chastain, P. D., Parniewski, P., Ohshima, K., Pandolfo M., Griffith, J. D., and Wells, R. D. (1999). Sticky DNA: Self-association properties of long GAA·TTC repeats in R·R·Y triplex structures from Friedreich's ataxia. *Mol. Cell* **3**, 465–475.
14. Potaman, V. N., Oussatcheva, E. A., Lyubchenko, Y. L., Shlyakhtenko, L. S., Bidichandani, S. I., Ashizawa, T., and Sinden, R. R. (2004). Length-dependent structure formation in Friedreich ataxia $(GAA)_n·(TTC)_n$ repeats at neutral pH. *Nucleic Acids Res.* **32**, 1224–1231.
15. Wells, R. D., Dere, R., Hebert, M. L., Napierala, M., and Son, L. S. (2005). Advances in mechanisms of genetic instability related to hereditary neurological diseases. *Nucleic Acids Res.* **33**, 3785–3798.
16. Bacolla, A., and Wells, R. D. (2004). Non-B DNA conformations, genomic rearrangements, and human disease. *J. Biol. Chem.* **279**, 47411–47414.
17. Wells, R. D. (1996). Molecular basis of genetic instability of triplet repeats. *J. Biol. Chem.* **271**, 2875–2878.
18. Sinden, R. R. (1994). "DNA Structure and Function." Academic Press, San Diego.
19. Felsenfeld, G., Davies, D. R., and Rich, A. (1957). Formation of a three-stranded polynucleotide molecule. *J. Am. Chem. Soc.* **79**, 2023–2024.
20. Tiner, W. J., Sr., Potaman, V. N., Sinden, R. R., and Lyubchenko, Y. L. (2001). The structure of intramolecular triplex DNA: Atomic force microscopy study. *J. Mol. Biol.* **314**, 353–357.
21. LeProust, E. M., Pearson, C. E., Sinden, R. R., and Gao, X. (2000). Unexpected formation of parallel duplex in GAA and TTC trinucleotide repeats of Friedreich's ataxia. *J. Mol. Biol.* **302**, 1063–1080.
22. Mariappan, S. V., Catasti, P., Silks, L. A., 3rd, Bradbury, E. M., and Gupta, G. (1999). The high-resolution structure of the triplex formed by the GAA·TTC triplet repeat associated with Friedreich's ataxia. *J. Mol. Biol.* **285**, 2035–2052.
23. Soyfer, V. N., and Potaman, V. N. (1996). "Triple-Helical Nucleic Acids." Springer-Verlag, New York.
24. Frank-Kamenetskii, M. D., and Mirkin, S. M. (1995). Triplex DNA structures. *Annu. Rev. Biochem.* **64**, 65–95.
25. Vetcher, A. A., and Wells, R. D. (2004). Sticky DNA formation in vivo alters the plasmid dimer/monomer ratio. *J. Biol. Chem.* **279**, 6434–6443.
26. Vetcher, A. A., Napierala, M., Iyer, R. R., Chastain, P. D., Griffith, J. D., and Wells, R. D. (2002). Sticky DNA, a long GAA·GAA·TTC triplex that is formed intramolecularly, in the sequence of intron 1 of the frataxin gene. *J. Biol. Chem.* **277**, 39217–39227.
27. Vetcher, A. A., Napierala, M., and Wells, R. D. (2002). Sticky DNA: Effect of the polypurine·polypyrimidine sequence. *J. Biol. Chem.* **277**, 39228–39234.
28. Malkov, V. A., Soyfer, V. N., and Frank-Kamenetskii, M. D. (1992). Effect of intermolecular triplex formation on the yield of cyclobutane photodimers in DNA. *Nucleic Acids Res.* **20**, 4889–4895.
29. Kang, S. M., Wohlrab, F., and Wells, R. D. (1992). Metal ions cause the isomerization of certain intramolecular triplexes. *J. Biol. Chem.* **267**, 1259–1264.
30. Ussery, D. W., and Sinden, R. R. (1993). Environmental influences on the in vivo level of intramolecular triplex DNA in *Escherichia coli*. *Biochemistry* **32**, 6206–6213.
31. Beal, P. A., and Dervan, P. B. (1991). Second structural motif for recognition of DNA by oligonucleotide-directed triple-helix formation. *Science* **251**, 1360–1363.
32. Thomas, T., and Thomas, T. J. (1993). Selectivity of polyamines in triplex DNA stabilization. *Biochemistry* **32**, 14068–14074.
33. Hampel, K. J., Crosson, P., and Lee, J. S. (1991). Polyamines favor DNA triplex formation at neutral pH. *Biochemistry* **30**, 4455–4459.
34. Hampel, K. J., Burkholder, G. D., and Lee, J. S. (1993). Plasmid dimerization mediated by triplex formation between polypyrimidine–polypurine repeats. *Biochemistry* **32**, 1072–1077.
35. Guieysse, A. L., Praseuth, D., Grigoriev, M., Harel-Bellan, A., and Helene, C. (1996). Detection of covalent triplex within human cells. *Nucleic Acids Res.* **24**, 4210–4216.
36. Michel, D., Chatelain, G., Herault, Y., and Brun, G. (1992). The long repetitive polypurine/polypyrimidine sequence $(TTCCC)_{48}$ forms DNA triplex with PU–PU–PY base triplets in vivo. *Nucleic Acids Res.* **20**, 439–443.
37. Parniewski, P., Kwinkowski, M., Wilk, A., and Klysik, J. (1990). Dam methyltransferase sites located within the loop region of the oligopurine–oligopyrimidine sequences capable of forming H-DNA are undermethylated in vivo. *Nucleic Acids Res.* **18**, 605–611.
38. Stollar, B. D. (1992). Immunochemical analyses of nucleic acids. *Prog. Nucleic Acid Res. Mol. Biol.* **42**, 39–77.
39. Lee, J. S., Latimer, L. J., Haug, B. L., Pulleyblank, D. E., Skinner, D. M., and Burkholder, G. D. (1989). Triplex DNA in plasmids and chromosomes. *Gene* **82**, 191–199.
40. Schroth, G. P., and Ho, P. S. (1995). Occurrence of potential cruciform and H-DNA forming sequences in genomic DNA. *Nucleic Acids Res.* **23**, 1977–1983.
41. Kinniburgh, A. J. (1989). A cis-acting transcription element of the c-myc gene can assume an H-DNA conformation. *Nucleic Acids Res.* **17**, 7771–7778.
42. Pestov, D. G., Dayn, A., Siyanova, E., George, D. L., and Mirkin, S. M. (1991). H-DNA and Z-DNA in the mouse c-Ki-ras promoter. *Nucleic Acids Res.* **19**, 6527–6532.
43. Thoma, B. S., Wakasugi, M., Christensen, J., Reddy, M. C., and Vasquez, K. M. (2005). Human XPC-hHR23B interacts with XPA-RPA in the recognition of triplex-directed psoralen DNA interstrand crosslinks. *Nucleic Acids Res.* **33**, 2993–3001.
44. Shigemori, Y., and Oishi, M. (2004). Specific cleavage of DNA molecules at RecA-mediated triple-strand structure. *Nucleic Acids Res.* **32**, e4.
45. Vasquez, K. M., Christensen, J., Li, L., Finch, R. A., and Glazer, P. M. (2002). Human XPA and RPA DNA repair proteins participate in specific recognition of triplex-induced helical distortions. *Proc. Natl. Acad Sci. USA* **99**, 5848–5853.
46. Guieysse, A. L., Praseuth, D., and Helene, C. (1997). Identification of a triplex DNA-binding protein from human cells. *J. Mol. Biol.* **267**, 289–298.
47. Krasilnikova, M. M., and Mirkin, S. M. (2004). Replication stalling at Friedreich's ataxia $(GAA)_n$ repeats in vivo. *Mol. Cell. Biol.* **24**, 2286–2295.
48. Dayn, A., Samadashwily, G. M. and Mirkin, S. M. (1992). Intramolecular DNA triplexes: Unusual sequence requirements and influence on DNA polymerization. *Proc. Natl. Acad. Sci. USA* **89**, 11406–11410.
49. Wang, G., Seidman, M. M., and Glazer, P. M. (1996). Mutagenesis in mammalian cells induced by triple helix formation and transcription-coupled repair. *Science* **271**, 802–805.

50. Seidman, M. M., and Glazer, P. M. (2003). The potential for gene repair via triple helix formation. *J. Clin. Invest.* **112**, 487–494.
51. Faruqi, A. F., Datta, H. J., Carroll, D., Seidman, M. M., and Glazer, P. M. (2000). Triple-helix formation induces recombination in mammalian cells via a nucleotide excision repair-dependent pathway. *Mol. Cell. Biol.* **20**, 990–1000.
52. Bacolla, A., Jaworski, A., Larson, J. E., Jakupciak, J. P., Chuzhanova, N., Abeysinghe, S. S., O'Connell, C. D., Cooper, D. N., and Wells, R. D. (2004). Breakpoints of gross deletions coincide with non-B DNA conformations. *Proc. Natl. Acad. Sci. USA* **101**, 14162–14167.
53. Wang, G., and Vasquez, K. M. (2004). Naturally occurring H-DNA-forming sequences are mutagenic in mammalian cells. *Proc. Natl. Acad. Sci. USA* **101**, 13448–13453.
54. Ohshima, K., Kang, S., Larson, J. E., and Wells, R. D. (1996). TTA·TAA triplet repeats in plasmids form a non-H bonded structure. *J. Biol. Chem.* **271**, 16784–16791.
55. Faucon, B., Mergny, J. L., and Helene, C. (1996). Effect of third strand composition on the triple helix formation: Purine versus pyrimidine oligodeoxynucleotides. *Nucleic Acids Res.* **24**, 3181–3188.
56. Panyutin, I. G., and Wells, R. D. (1992). Nodule DNA in the $(GA)_{37} \cdot (CT)_{37}$ insert in superhelical plasmids. *J. Biol. Chem.* **267**, 5495–5501.
57. Lyamichev, V. I., Voloshin, O. N., Frank-Kamenetskii, M. D., and Soyfer, V. N. (1991). Photofootprinting of DNA triplexes. *Nucleic Acids Res.* **19**, 1633–1638.
58. Kohwi, Y., and Kohwi-Shigematsu, T. (1988). Magnesium ion-dependent triple-helix structure formed by homopurine–homopyrimidine sequences in supercoiled plasmid DNA. *Proc. Natl. Acad. Sci. USA* **85**, 3781–3785.
59. Jaworski, A., Hsieh, W. T., Blaho, J. A., Larson, J. E., and Wells, R. D. (1987). Left-handed DNA in vivo. *Science* **238**, 773–777.
60. Jaworski, A., Blaho, J. A., Larson, J. E., Shimizu, M., and Wells, R. D. (1989). Tetracycline promoter mutations decrease non-B DNA structural transitions, negative linking differences and deletions in recombinant plasmids in *Escherichia coli*. *J. Mol. Biol.* **207**, 513–526.
61. Zacharias, W., Jaworski, A., Larson, J. E., and Wells, R. D. (1988). The B- to Z-DNA equilibrium in vivo is perturbed by biological processes. *Proc. Natl. Acad. Sci. USA* **85**, 7069–7073.
62. Ohshima, K., Sakamoto, N., Labuda, M., Poirier, J., Moseley, M. L., Montermini, L., Ranum, L. P., Wells, R. D., and Pandolfo, M. (1999). A nonpathogenic GAAGGA repeat in the Friedreich gene: Implications for pathogenesis. *Neurology* **53**, 1854–1857.
63. Napierala, M., Dere, R., Vetcher, A., and Wells, R. D. (2004). Structure-dependent recombination hot spot activity of GAA·TTC sequences from intron 1 of the Friedreich's ataxia gene. *J. Biol. Chem.* **279**, 6444–6454.
64. Dere, R., and Hebert, M. L., Napierala, M. (2006). Involvement of genetic recombination in microsatellite instability. In "Genetic Instabilities and Neurological Diseases" (R. D. Wells and T. Ashizawa, Eds.), 2nd ed., Ch. 38. Elsevier, Burlington, MA.
65. Wells, R. D., and Warren, S. T. (Eds.) (1998). "Genetic Instabilities and Hereditary Neurological Diseases." Academic Press, San Diego.

PART VII

Spinocerebellar Ataxias

CHAPTER 22

Phosphorylation of Ataxin-1: A Link Between Basic Research and Clinical Application in Spinocerebellar Ataxia Type 1

KERRI M. CARLSON AND HARRY T. ORR

Department of Laboratory Medicine and Pathology, and Institute of Human Genetics University of Minnesota, Mayo Mail Code 206, Minneapolis, Minnesota 55455

I. Introduction
II. Insights into Normal Ataxin-1 Function
III. Factors Mediating SCA1 Pathogenesis
IV. Phosphorylation of Ataxin-1: A Mediator of SCA1 Pathogenesis
   A. Phosphorylation Sites in Ataxin-1
   B. Localization of Phosphorylated S776-Ataxin-1
   C. S776 and Ataxin-1 Aggregation
   D. S776 and SCA1 Pathogenesis *in Vivo*
   E. 14-3-3: An Ataxin-1 Interactor
   F. 14-3-3 Stabilizes Ataxin-1
   G. 14-3-3 and Neurodegeneration *in Vivo*
V. AKT Signaling: A Role in SCA1 Pathogenesis?
   A. AKT and Ataxin-1 Phosphorylation
   B. AKT Mediates the 14-3-3/Ataxin-1 Interaction
   C. AKT-Dependent Ataxin-1 Phosphorylation and SCA1 Pathogenesis
   D. PI3K/AKT Pathway and SCA1 Pathogenesis
VI. The Search for Modifiers of Ataxin-1 S776 Phosphorylation
   A. Development of a Cell-Based Assay for Ataxin-1 Phosphorylation
   B. Inhibitors of Ataxin-1 Phosphorylation
   C. Ataxin-1 Signaling Pathways
VII. Concluding Remarks
References

Spinocerebellar ataxia type 1 (SCA1) is an autosomal dominant neurodegenerative disease caused by an expanded polyglutamine tract in ataxin-1. Although the mechanism by which mutant ataxin-1 causes selective neuronal degeneration remains unknown, studies have demonstrated that if mutant ataxin-1 expression is decreased, Purkinje cells can recover normal function. These studies suggest that therapies aimed at decreasing mutant ataxin-1 protein levels may be lead candidates for SCA1 treatment. Phosphorylation of ataxin-1 at S776 is an important mediator of ataxin-1 turnover and SCA1 pathogenesis. Mice expressing a *SCA1 [82Q]*-A776 transgene are behaviorally indistinguishable from wild-type mice. Pathologically, these mice develop a very mild SCA1 phenotype late in life. On a molecular level, S776 has been shown to mediate the interaction between ataxin-1 and 14-3-3. 14-3-3 binding to ataxin-1 stabilizes the ataxin-1 protein. A cell-based assay has been used to identify two signaling pathways important for regulating S776 phosphorylation. One pathway appears to work in a calcium-dependent manner to phosphorylate both mutant and wild-type ataxin-1. The second pathway is mutant specific and involves PI3K/AKT signaling. Lead candidates for SCA1 treatment may include therapies aimed at decreasing ataxin-1 phosphorylation, resulting in an increase in ataxin-1 turnover in the cell. The discovery of S776 phosphorylation as an important regulator of ataxin-1 turnover, as well as the identification of two signaling pathways involved in this regulation, opens up a new area to explore for therapeutic development in SCA1.

I. INTRODUCTION

Spinocerebellar ataxia type I (SCA1) is an autosomal-dominant neurodegenerative disorder caused by an expansion of a CAG repeat in the *SCA1* gene. The CAG repeat encodes for a polyglutamine tract in the *SCA1* gene product, ataxin-1 [1]. Mutant *SCA1* alleles range in size from 39 to 82 uninterrupted CAG repeats [2]. SCA1 is a member of a group of polyglutamine disorders that includes spinocerebellar ataxia types 2 [3–5], 3 [6], 6 [7], 7 [8], and 17 [9], Huntington's disease (HD) [10], spinobulbar muscular atrophy [11], and dentatorubropallidoluysian atrophy (DRPLA) [12–14].

Clinically, SCA1 is characterized by ataxia, dysphagia, dysarthria, and progressive motor dysfunction. These symptoms usually present within the third to fourth decade of life and progress throughout the next 10 to 30 years, ultimately ending in death. The age of onset and severity of the disease are inversely correlated with the polyglutamine repeat length. Pathologically, SCA1 is characterized by a loss of Purkinje cells in the cerebellar cortex, inferior olive neurons, and deep cerebellar neurons [15].

Currently, effective therapeutics are not available to treat patients with SCA1 or related polyglutamine diseases. This is due largely to the lack of understanding about the cellular function of these proteins, as well as the pathogenic mechanisms associated with the diseases. Recent research on SCA1 has begun to reveal the role of ataxin-1 in the cell and to identify factors that are important in disease pathogenesis. In this chapter, we briefly summarize current knowledge concerning ataxin-1 function, as well as new insights into SCA1 pathogenesis. We focus specifically on the role of ataxin-1 phosphorylation for the disease process and how this information sets the stage for identifying potential therapeutics for SCA1.

II. INSIGHTS INTO NORMAL ATAXIN-1 FUNCTION

When ataxin-1 was first identified as the protein linked to SCA1, its function was not apparent from its sequence. The *SCA1* transcript is ubiquitously expressed, although the site of SCA1 pathogenesis is specific to the Purkinje cells and certain neuronal populations in the brainstem [16, 17]. Structurally, the ataxin-1 protein is characterized by a polyglutamine tract, a functional nuclear localization signal, and an AXH domain that has been implicated in RNA binding and protein/protein interaction [18].

Recent studies aimed at elucidating the pathogenic mechanism of SCA1 are pointing to a role for ataxin-1 in RNA metabolism. Both wild-type and mutant ataxin-1 localize to the nucleus of Purkinje cells, a primary site of SCA1 pathogenesis [16]. Within the nucleus, ataxin-1 has been shown to associate with PML (a nuclear matrix-associated protein) and to fractionate with the nuclear matrix *in vivo* [19]. PML and the nuclear matrix are associated with RNA metabolism, including mRNA stability and transport [20]. In addition to its colocalization with PML and the nuclear matrix, ataxin-1 has RNA binding capabilities [21]. This interaction is most likely mediated through the AXH domain of ataxin-1 [18]. *In vitro* RNA binding ability of ataxin-1 is inversely related to the length of its polyglutamine tract [21].

Finally, recent data indicate that, in tissue culture cells, ataxin-1 colocalizes in nuclear bodies with the nuclear export factor TAP/NXF1 in a RNA-dependent manner. Moreover, with the use of fluorescence recovery after photobleaching (FRAP), wild-type ataxin-1, but not mutant ataxin-1, has been shown to shuttle out of the cell nucleus [22]. Although ataxin-1 does not seem to have a specific export signal, it is possible that its ability to shuttle out of the nucleus is dependent on its interaction with another protein, such as TAP/NXF1, in an export complex or on its association with RNA. Collectively, these studies point to a role for ataxin-1 in RNA processing and, perhaps, RNA export from the nucleus.

In addition to a role in nuclear export, studies also link ataxin-1 to RNA transcription. In a genetic screen for modifiers of ataxin-1-induced neurodegeneration using a drosophila model of SCA1, multiple transcriptional cofactors such as Sin3A were identified as modulators of ataxin-1-induced pathogenesis [23]. Subsequently, it was demonstrated that ataxin-1 interacts with the silencing mediator of retinoid and thyroid hormone receptors (SMRT), a transcriptional corepressor, and with histone deacetylase 3 (HDAC3). In this study, it was also shown that ataxin-1 is able to bind chromosomes and, when linked to DNA, is able to mediate transcriptional repression of a reporter gene [24].

Support for the role of ataxin-1 in both RNA export and transcriptional regulation comes from a recent study that showed that ataxin-1 is covalently modified by the protein small ubiquitin-like modifier (SUMO) [25]. The modification of proteins by addition of small polypeptides such as SUMO is an important mechanism for controlling protein events within the cell. Other proteins modified by SUMO, including PML and Sp100, are also involved in both transcriptional regulation and nuclear transport [26, 27].

III. FACTORS MEDIATING SCA1 PATHOGENESIS

All evidence indicates that the polyglutamine expansion in ataxin-1 imparts a toxic, gain-of-function mutation. Transgenic mice expressing a *SCA1* allele with 82 glutamines (*SCA1[82Q]*) present with Purkinje cell degeneration and progressive ataxia, similar to human patients [28]. This pathology is not observed in *SCA1* knockout mice, supporting the idea that SCA1 is not due to loss of ataxin-1 function [29].

In a genetic screen for modifiers of SCA1 pathogenesis, genes involved in pathways including oxidative stress, RNA processing, protein folding/heat shock response, the ubiquitin–proteolytic pathway, and transcriptional regulation were all identified as modifiers of SCA1 pathogenesis [23]. These data suggest that SCA1 pathogenesis is a complex process affecting multiple pathways in target cells. Studies implicate both inherent properties of the ataxin-1 protein, as well as misregulation of cellular pathways in SCA1 pathogenesis.

Intragenic properties of the mutant ataxin-1 protein are important for SCA1 pathogenesis (Fig. 22-1). Examination of both human patients and transgenic models demonstrates that the length of the polyglutamine tract is inversely correlated with both the age of onset and the severity of SCA1 [16, 28]. Although polyglutamine expansion is necessary for SCA1, it is not sufficient for pathogenesis. Data have shown that ataxin-1 localization also plays an important role in SCA1 pathogenesis. The ataxin-1 protein is localized to the nucleus of neurons. Transgenic mice expressing a *SCA1 [82Q]* allele with a mutated nuclear localization signal do not present with an ataxia phenotype [30]. These data suggest that nuclear localization of ataxin-1 is another important factor in ataxin-1-induced pathogenesis. Finally, recent data identified ataxin-1 phosphorylation as another intragenic modifier of pathogenesis by altering ataxin-1 turnover in the cell [31, 32].

In addition to intragenic factors necessary for SCA1 pathogenesis, multiple lines of evidence support a role for protein misfolding and degradation pathways in the disease process. Mutations in six genes involved in protein folding/heat shock response or ubiquitin–proteolytic pathways were identified as enhancers of ataxin-1-induced pathogenesis in a genetic screen using a drosophilia model of SCA1 [23]. Likewise, a hallmark of polyglutamine disease, such as SCA1, is the presence of large nuclear inclusions of the mutant protein [19, 33, 34]. The role of inclusions in pathogenesis is still debated, but in SCA1 these aggregates stain positive for ubiquitin, the chaperone HDJ-2/HSDJ, and the 20S proteasome [35]. This suggests that protein misfolding is a characteristic of mutant ataxin-1 and that cellular protein folding and clearance pathways modify disease severity.

Support for the importance of protein folding and clearance in the SCA1 disease process comes from transgenic mouse studies. These studies have shown that *SCA1 [82Q]* transgenic mice that overexpress the HSP70 chaperone have a somewhat improved SCA1 phenotype compared with *SCA1 [82Q]* mice [36]. Alternatively, transgenic mice expressing mutant ataxin-1 but lacking *Ube3a*, a ubiquitin–protein ligase, show increased Purkinje cell pathology despite fewer nuclear inclusions [37].

Finally, both the amount of mutant ataxin-1 and the length of exposure to the mutant protein are important *in vivo* mediators of SCA1 pathogenesis. The age of onset and severity of the ataxia phenotype vary in transgenic mice expressing different levels of an *SCA1 [82Q]* transgene. Increasing levels of ataxin-1 expression result in a more severe neuropathological phenotype at an earlier age [28]. Likewise, when ataxin-1 levels are decreased in the cerebellum of *SCA1 [82Q]* transgenic mice using siRNA, the motor function of these mice significantly improves. The recovery of motor function in these mice is accompanied by both improvement of the SCA1 cerebellar pathology and the disappearance of nuclear inclusions in the Purkinje cells [38].

Although the siRNA experiments demonstrate that decreased levels of ataxin-1 protein expression can result in an improvement in the SCA1 phenotype *in vivo*, they do not address the role of age in this phenotypic recovery. This question has been addressed through the use of a conditional mouse model of SCA1. The length of exposure to the mutant protein in these mice is inversely correlated to the ability of these mice to recover once the

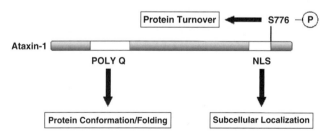

FIGURE 22-1 Intragenic modulators of SCA1 pathogenesis. Three characteristics of the ataxin-1 protein itself have been linked to SCA1 pathogenesis. First, the polyglutamine tract length is an important determinant of protein conformation and misfolding. Second, the nuclear localization sequence (NLS) directs the protein to the necessary subcellular localization. Finally, phosphorylation of ataxin-1 at S776 is an important determinant of ataxin-1 protein turnover in the cell.

expression of the ataxin-1 protein is halted. At an early disease stage, both the Purkinje cell pathology and the motor dysfunction associated with SCA1 are completely reversible after cessation of transgene expression. Only partial recovery is observed when transgene expression is halted at a later disease stage [39]. Taken together, these experiments suggest that strategies aimed at decreasing ataxin-1 protein levels in SCA1 patients may be effective at both halting the disease progression and leading to at least a partial recovery from the disease phenotype.

IV. PHOSPHORYLATION OF ATAXIN-1: A MEDIATOR OF SCA1 PATHOGENESIS

A. Phosphorylation Sites in Ataxin-1

Phosphorylation is an important mechanism for regulating protein function, interaction, and clearance [40]. For this reason, phosphorylation was hypothesized to be an important regulator of ataxin-1 function and SCA1 pathogenesis. An analysis of the ataxin-1 primary sequence indicated the presence of many potential phosphorylation sites in the ataxin-1 protein [17]. Subsequently, by the use of mass spectrometry, two novel phosphorylation sites have been confirmed: S776 [31] and S239 [41]. Although the role of S239 remains unknown, S776 has been demonstrated to be an important factor in both SCA1 pathogenesis and regulation of ataxin-1/protein interactions [31, 32].

B. Localization of Phosphorylated S776-Ataxin-1

To understand the role of pS776 in ataxin-1 biology, it was important to determine the subcellular localization of pS776-ataxin-1. Ataxin-1 localizes strongly to the nucleus of Purkinje cells, the prominent cellular site of SCA1 pathogenesis. Ataxin-1 expression is also detected in the cell body and dendrites of the Purkinje cells [16]. By use of an antibody specific to pS776-ataxin-1 (PN1168), the subcellular location of pS776-ataxin-1 was examined.

Immunohistochemical analysis of cerebellar slices from transgenic mice expressing either an *SCA1 [82Q]* or an *SCA1 [30Q]* transgene was conducted using both the phospho-specific antibody PN1168 and an antibody that recognizes both phosphorylated and nonphosphorylated ataxin-1 (11750). In both transgenic mouse lines, staining with the 11750 antibody revealed strong ataxin-1 signal in the nucleus, as well as staining in the cell body and dendrites, consistent with previous studies. On the other hand, staining using PN1168 revealed pS776-ataxin-1 predominantly in the nucleus, with light staining in the cell body and no detectable staining in the dendrites. Likewise, no PN1168 staining was detected in Purkinje cells from transgenic mice with a mutated nuclear localization signal (*SCA1 [82Q]*-T772) [30]. These data suggest that phosphorylation of ataxin-1 is linked in a yet to be determined fashion to the transport of the protein into the nucleus of the cell [31].

C. S776 and Ataxin-1 Aggregation

One hallmark of SCA1 and other polyglutamine diseases is the formation of nuclear inclusions [19]. To investigate the role of S776 in the formation of ataxin-1 inclusions, two stably transfected CHO cell lines were generated. The first line expressed mutant ataxin-1 with S776 (ataxin-1 [82Q]-S776) and the second expressed mutant ataxin-1 with the serine mutated to an alanine (ataxin-1 [82Q]-A776), eliminating phosphorylation at this residue. More than 60% of ataxin-1 [82Q]-S776 CHO cells contained large nuclear inclusions visualized by 11750 antibody staining. In contrast, less than 0.1% of the ataxin-1 [82Q]-A776 CHO cells contained nuclear inclusions. This change in nuclear inclusion formation was accompanied by a tenfold decrease in the amount of ataxin-1 [82Q]-A776 protein detected in the insoluble fraction of cell lysates, compared with cells expressing ataxin-1 [82Q]-S776. On the basis of these data, it appears that S776 is an important factor affecting the solubility of ataxin-1 and, therefore, its ability to form nuclear inclusions [31].

D. S776 and SCA1 Pathogenesis in Vivo

The cell culture experiments suggested that S776 was an important mediator of ataxin-1 solubility; however, they did not address the role of S776 in SCA1 pathogenesis. To address this question, transgenic mice overexpressing *SCA1 [82Q]*-A776 in a Purkinje cell-specific manner were generated (Fig. 22-2A). Consistent with the cell culture data, *SCA1 [82Q]*-A776 localized to the nucleus of the Purkinje cells in these animals; however, the rate of nuclear inclusion formation was decreased considerably compared with that of *SCA1 [82Q]*-S776-expressing mice. At 5 weeks of age, no nuclear inclusions were observed in the Purkinje cells of *SCA1 [82Q]*-A776 mice. By 56 weeks of age, only 32% of the *SCA1 [82Q]*-A776-expressing Purkinje cells contained nuclear inclusions. In *SCA1 [82Q]*-S776 mice, 50% of the Purkinje cells contained nuclear inclusions by 18 weeks of age. By 37 weeks

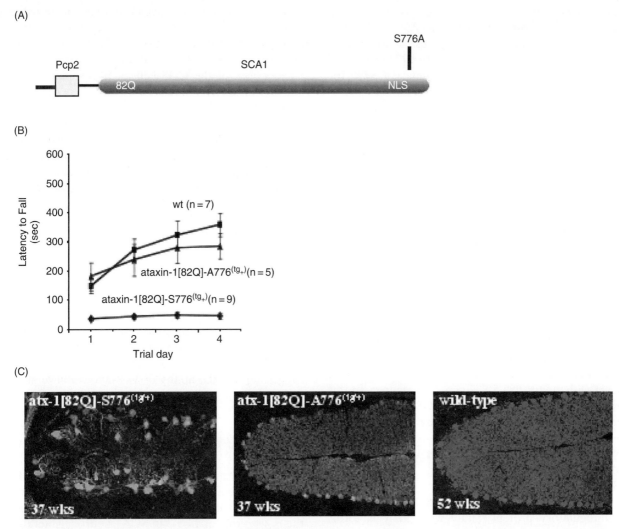

FIGURE 22-2 Transgenic mice expressing ataxin-1-A776 do not show signs of SCA1 pathology. (A). A diagram of the ataxin-1 [82Q]-A776 transgene. The Purkinje cell specific regulatory element *Pcp2* was used to drive transgene expression. (B). Rotarod analysis was used to assess neurological behavior in the ataxin-1 [82Q]-A776 mice. Behaviorally, 46 week-old ataxin-1 [82Q]-A776 transgenic mice were indistinguishable from 1 year-old wild type mice. This was in contrast to ataxin-1-S776 transgenic mice that did not perform well on the test. (C). Calbindin immunofluroescence was used to assess Purkinje cell morphology in the cerebellum of a 37 week old ataxin-1 [82Q]-A776 mouse, a 37 week old ataxin-1 [82Q]-S776 mouse and a 52 week old wild type mouse. 37 week-old ataxin-1 [82Q]-A776 transgenic mice showed very little SCA1 pathology compared to age-matched ataxin-1 [82Q]-A776 mice.

of age, 100% of the Purkinje cells in these animals contained nuclear inclusions. Therefore, by mutation of S776 to A776 in the ataxin-1 protein, a significant reduction in nuclear inclusion formation was detected *in vivo*, consistent with the cell culture data [31].

Overall, *SCA1 [82Q]*-A776 mice displayed a very mild SCA1 phenotype late in life. By homecage behavior, *SCA1 [82Q]*-A776 animals were indistinguishable from their wild type littermates at all ages examined, even when the transgene was bred to homozygosity. Similarly, by the accelerating rotarod, 19-week-old *SCA1 [82Q]*-A776 mice performed as well as age-matched FVB controls. This is in contrast to *SCA1 [82Q]*-S776 mice, which demonstrate rotarod deficits as early as 5 weeks of age (Fig. 22-2B).

Although, behaviorally, *SCA1 [82Q]*-A776 mice were indistinguishable from wild-type animals, pathologically they displayed a mild SCA1 phenotype. At 18 weeks of age, the Purkinje cells of *SCA [82Q]*-A776 showed no signs of the dendritic thinning and Purkinje cell heterotopia observed in 18-week-old *SCA1 [82Q]*-S776 mice. By 46 weeks of age, however, slight thinning of the molecular layer was observed, although the Purkinje cell body layer remained intact. This differed from the Purkinje cell pathology observed in *SCA1 [82Q]*-S776 mice, where by 37 weeks of age, complete atrophy of the dendritic tree and extensive disorganization of the Purkinje cell layer were observed (Fig. 22-2C) [31]. The mild Purkinje cell pathology and lack of behavioral abnormalities noted in the *SCA1 [82Q]*-A776 transgenic mice demonstrate the *in vivo* importance of S776 for mutant ataxin-1 pathogenesis.

Both polyglutamine length and nuclear localization have been shown to be important intragenic modulators of SCA1 pathogenesis. On the basis of the aforementioned studies, S776 appears to be another important intragenic modulator of disease, perhaps by mediating ataxin-1 solubility. Although the presence of S776 is not sufficient to cause SCA1, it is necessary for the pathogenesis to progress. This leads to the question: By what mechanism does phorphorylation of S776 impact disease progression?

E. 14-3-3: An Ataxin-1 Interactor

As a first step toward elucidating the potential mechanism by which S776 mediates ataxin-1-induced pathogenesis, a screen was conducted to identify proteins that interact with ataxin-1 in a S776-specific manner. In this screen, immunoprecipitation from COS cell lysates followed by mass spectrometry was used to identify proteins that interact with ataxin-1 [82Q]-S776 but not with ataxin-1 [82Q]-A776. A yeast two-hybrid screen was used to confirm the interaction.

By use of this approach, 14-3-3 was identified as an ataxin-1-S776-specific interactor. 14-3-3 is a multifunctional regulatory protein family with many isoforms [42]. Several of these isoforms, including β, ε, and ζ, interact with ataxin-1 in a Neuro2A cell line. 14-3-3 has been shown to interact with both mutant and wild-type ataxin-1-S776; however, the strength of this interaction appears to be mediated by the length of the polyglutamine tract. In COS cells, the relative level of 14-3-3 bound to ataxin-1 was found to increase with increasing lengths of the polyQ tract, independent of the phosphorylation state of ataxin-1 [32].

F. 14-3-3 Stabilizes Ataxin-1

As described earlier, S776 is an important mediator of both ataxin-1 solubility and the formation of nuclear inclusions. Because the interaction of 14-3-3 with ataxin-1 is dependent on the presence of S776, it is possible that 14-3-3 may play an active role in determining ataxin-1 solubility and nuclear inclusion formation. To address this question, the localization of 14-3-3 in COS cells was examined.

The subcellular localization of 14-3-3 in transfected COS cells was altered by the presence of ataxin-1 [82Q]-S776, but not ataxin-1 [82Q]-A776. When expressed alone, 14-3-3 distributed to the cytoplasm and nucleoplasm of the COS cells. Cotransfection of both ataxin-1 [82Q]-S776 and 14-3-3 resulted in the redistribution of the 14-3-3 protein to the nuclear inclusions formed by ataxin-1 [82Q]-S776. This redistribution of 14-3-3 was not seen when 14-3-3 was cotransfected with ataxin-1 [82Q]-A776 (a protein that rarely forms nuclear inclusions when transfected alone).

One explanation for the observed aggravated nuclear inclusion formation in the presence of 14-3-3 is that 14-3-3 is acting to stabilize the ataxin-1 protein. To test this hypothesis, the steady-state level of ataxin-1 in HeLa cells was determined in the presence or absence of 14-3-3. When ataxin-1 and 14-3-3 were cotransfected into HeLa cells, the steady-state level of ataxin-1 increased in the presence of 14-3-3. This increase in protein levels was dependent on polyglutamine tract length, supporting the data suggesting that interaction of 14-3-3 with ataxin-1 is dependent on polyglutamine tract length. 14-3-3 did not increase the levels of ataxin-1 [82Q]-A776, suggesting that the ability of 14-3-3 to stabilize ataxin-1 is the result of its direct interaction with S776 [32].

G. 14-3-3 and Neurodegeneration in Vivo

The experiments just described suggest that 14-3-3, through its ability to bind to and stabilize ataxin-1, might be an important player in SCA1 pathogenesis. To look at the *in vivo* role of 14-3-3 in SCA1 neurodegeneration, a drosophilia model of SCA1 was used. Double transgenic flies were generated that expressed both 14-3-3 and ataxin-1 [82Q]-S776 in a retinal-specific manner. These flies presented with a thin, disorganized retinal layer, highly abnormal rhabdomeres, and disordered ommatidia. This phenotype was much more severe than that observed in transgenic flies expressing ataxin-1 [82Q]-S776 alone. 14-3-3 transgenic flies did not have a visible phenotype. These data support a role for 14-3-3 in SCA1 pathogenesis, perhaps by binding to and stabilizing the mutant ataxin-1 protein [32].

V. AKT SIGNALING: A ROLE IN SCA1 PATHOGENESIS?

A. AKT and Ataxin-1 Phosphorylation

By the use of both cell culture and a transgenic mouse model, the importance of the S776 residue in ataxin-1 to SCA1 pathogenesis has been demonstrated.

Online database searches identified AKT as one possible kinase for phosphorylating S776. AKT is a serine/threonine protein kinase also known as protein kinase B (PKB) [43]. An *in vitro* kinase assay showed that ataxin-1 [30Q] is a substrate for AKT1. This relationship was confirmed in HeLa cells cotransfected with ataxin-1 [2Q] and either a constitutively active (CA) or dominant negative (DN) AKT. Coexpression of ataxin-1 [2Q] and CA AKT, but not DN AKT, resulted in an increased level of phosphorylated ataxin-1 [2Q]-S776 [32]. These results support the hypothesis that AKT acts as a kinase for ataxin-1-S776 phosphorylation.

B. AKT Mediates the 14-3-3/Ataxin-1 Interaction

Because S776 of ataxin-1 is both a substrate for AKT phosphorylation and important for 14-3-3 binding to ataxin-1, it is likely that AKT is able to mediate 14-3-3 binding to ataxin-1. This question was addressed using an *in vitro* binding assay. It was demonstrated that 14-3-3 binding of ataxin-1 was dependent on S776 phosphorylation by AKT. Likewise, in cell culture, interaction of ataxin-1 with 14-3-3 was increased in cells cotransfected with ataxin-1 and AKT-CA. This interaction was decreased in cells cotransfected with ataxin-1 and AKT-DN [32]. Overall, it appears that AKT phosphorylation of S776 is able to mediate 14-3-3 binding to ataxin-1.

C. AKT-Dependent Ataxin-1 Phosphorylation and SCA1 Pathogenesis

Double transgenic flies expressing both 14-3-3 and ataxin-1 [82Q] have an aggravated SCA1 phenotype. Because the interaction between 14-3-3 and ataxin-1 is mediated by AKT activity *in vitro*, the question becomes whether increased AKT activity can also enhance the SCA1 phenotype. Similar to the data obtained for the 14-3-3/ataxin-1 transgenic flies, double transgenic flies expressing ataxin-1 [82Q]-S776 and AKT1 in a retinal-specific manner showed a much more severe phenotype than flies expressing either AKT1 or ataxin-1 [82Q]-S776 alone [32].

To confirm the relationship between AKT and ataxin-1 neurodegeneration *in vivo*, transgenic flies expressing ataxin-1 [82Q] were crossed to flies with either a chromosomal duplication of AKT1 or a loss-of-function allele of AKT1. The SCA1 phenotype observed in ataxin-1 [82Q] transgenic flies was suppressed by the presence of a loss-of-function AKT1 allele and exacerbated by the presence of multiple AKT1 alleles, consistent with a role for AKT1 in SCA1 pathogenesis [32].

D. PI3K/AKT Pathway and SCA1 Pathogenesis

The PI3K/AKT signaling pathway is involved in a variety of cellular processes including survival pathways. In this signaling pathway, PI3K activation results in the activation of AKT and downstream effectors of AKT [43]. Because of its role in AKT activation, the role of PI3K in SCA1-induced neurodegeneration was also explored.

Transgenic flies expressing ataxin-1 [82Q]-S776 were crossed to flies expressing PI3K. A dramatic increase in severity of the ataxin-1-induced neurodegeneration was observed in the offspring. This increase in phenotypic severity was accompanied by an increase in ataxin-1 [82Q] protein levels as observed by immunofluoresence [32].

Overall, data demonstrate that the protein 14-3-3 interacts with ataxin-1 in a phospho-S776-specific manner. This interaction is mediated by AKT, a kinase that is able to phosphorylate S776 in ataxin-1. Interaction of 14-3-3 with ataxin-1 stabilizes the ataxin-1 protein and enhances the formation of nuclear inclusions [32]; however, S776 phosphorylation is not a mutant-specific event. Phosphorylation of S776 occurs on both mutant and wild-type protein, suggesting that phosphorylation not only is important for SCA1 pathogenesis but also plays a role in the normal cellular function of ataxin-1. One way to gain an understanding of the role of phosphorylation in the cellular function of ataxin-1 is to identify the signaling pathways that are responsible for the phosphorylation event. Besides helping to gain an understanding of the biology of ataxin-1, knowledge of these signaling pathways may also prove useful for the identification of potential therapeutic targets for SCA1.

VI. THE SEARCH FOR MODIFIERS OF ATAXIN-1 S776 PHOSPHORYLATION

A. Development of a Cell-Based Assay for Ataxin-1 Phosphorylation

To aid in the discovery of the signaling pathways involved in ataxin-1 phosphorylation, a cell-based assay was developed to screen for chemical modifiers of ataxin-1 phosphorylation (Fig. 22-3A). By identification of compounds that either inhibit or increase S776 phosphorylation, a signaling pathway responsible for S776 phosphorylation may be pieced together. In this phosphorylation assay, stably transfected CHO cells expressing green fluorescent protein (GFP)-tagged ataxin-1

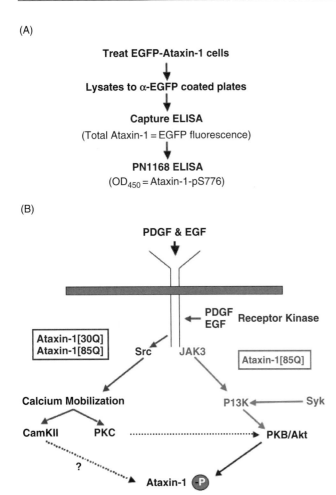

FIGURE 22-3 The identification of modifiers of ataxin-1 phosphorylation. (A). A flowchart detailing the cell-based assay developed in order to identify compounds that modulate ataxin-1 phosphorylation. (B). Screening of a library of kinase and phosphatase inhibitors using the cell-based phosphorylation assay identified two ataxin-1 signaling pathways. One pathway (blue), mediated through calcium signaling, is involved in phosphorylation of both mutant and wild type ataxin-1. The second pathway (red) preferentially phosphorylates mutant ataxin-1 and involves the PI3K/AKT signaling pathway. See CD-ROM for color image.

protein are exposed to a compound of interest. Following compound exposure, lysates are prepared from the cells, and GFP-tagged ataxin-1 is captured using 96-well plates coated with GFP antibody. The level of GFP fluorescence is quantified to represent the total amount of ataxin-1 protein in the lysates. Next, a sandwich ELISA is performed using the PN1168 antibody (specific for ataxin-1 pS776 protein). The relative level of phosphorylation is determined by dividing the PN1168 ELISA results by the GFP fluorescence for each compound [44].

The phosphorylation assay described above was used to screen both mutant and wild-type protein against a compound library containing known kinase and phosphatase inhibitors. Two main types of compounds were identified in the screen. The first type of compound modulated the phosphorylation of both mutant and wild-type ataxin-1, whereas the second type of compound acted in a mutant-specific manner [44].

B. Inhibitors of Ataxin-1 Phosphorylation

Inhibitors that affected both mutant and wild-type ataxin-1 phosphorylation were associated with kinases involved with intracellular calcium signaling. These kinases included PKC, CaMKII, ERK2, TRKs, and Src family tyrosine kinases. The activity of these compounds against both mutant and wild-type protein suggests that they most likely affect the normal physiological signaling pathway of ataxin-1. This dual activity does not make them strong candidates for therapeutic agents; however, they may prove to be useful tools for further dissecting the cellular role of ataxin-1 [44].

A second group of compounds may prove to be more useful toward the development of an effective SCA1 therapy. These compounds mediate phosphorylation of ataxin-1 in a mutant-specific manner, thereby targeting a unique signaling pathway that may be activated in response to mutant ataxin-1. Kinases targeted by these mutant-specific compounds include JAK tyrosine kinases, cyclin-dependent kinases, PKA, p56, Syk, and the PI3K/AKT kinases. Each kinase identified can be linked to the PI3K/AKT signaling pathway already implicated in SCA1 pathogenesis (see Section V) [44].

C. Ataxin-1 Signaling Pathways

The data collected in the kinase/phosphatase inhibitor screen suggest that the normal signaling pathway linked with ataxin-1 phosphorylation is associated with calcium homeostasis in the cell. The presence of mutant ataxin-1 results in a misregulation of this normal signaling pathway, leading to activation of the PI3K/AKT signaling pathway. The activation of this pathway is supported by data that demonstrated increased activated AKT levels in the brains of transgenic mice expressing ataxin-1 [82Q], compared with the brains of mice expressing wild-type ataxin-1 [30Q] [44] (Fig. 22-3B).

Because phosphorylation of ataxin-1 is necessary for SCA1 pathogenesis, a potential therapy might include a compound that could interfere with this phosphorylation step. The identification of a mutant-specific signaling pathway regulating ataxin-1 phosphorylation may prove to be an exciting discovery in terms of therapeutic development. Therapies that target this mutant-specific pathway may be able to reduce SCA1 pathogenesis without interfering with the wild-type function of the protein.

VII. CONCLUDING REMARKS

The causative mutation in SCA1 is a polyglutamine repeat tract expansion in ataxin-1. The consequences of this expansion are complex, involving multiple pathways including protein folding and degradation pathways, RNA processing, transcriptional regulation, and oxidative stress [23, 36]. Although the polyglutamine tract expansion is necessary for SCA1 pathogenesis, it is not sufficient to cause disease. Other properties of the protein have also been implicated in pathogenesis, including the ability of the protein to enter the nucleus [30] and, more recently, the phosphorylation of S776 (the focus of this chapter) [31, 32, 44].

Protein phosphorylation is a well-known regulatory mechanism in the cell [40]. S776 is phorphorylated on both wild-type and mutant ataxin-1, pointing to a role for phosphorylation in the normal cellular function of ataxin-1 as well as pathogenesis. On a molecular level, the phosphorylation of S776 is an important mediator of the interaction of ataxin-1 with 14-3-3, leading to a stabilization of the ataxin-1 protein. This interaction increases with increasing polyglutamine length, favoring the stabilization of mutant over wild-type ataxin-1 and, perhaps, resulting in a toxic accumulation of mutant protein in the cell. Two cell signaling pathways are linked to ataxin-1 phosphorylation. One pathway appears to work in a calcium-dependent manner to phosphorylate both mutant and wild-type ataxin-1. This pathway is most likely involved in the regulation of wild-type ataxin-1 in the cell. The second pathway involves PI3K/AKT signaling and appears to be specific to mutant ataxin-1.

One model of ataxin-1-induced pathogenesis involves the accumulation of mutant ataxin-1 due to 14-3-3 stabilization. This increase in mutant protein may either directly or indirectly lead to activation of the PI3K/AKT pathway. Activation of this pathway, besides leading to downstream changes in the cell, would also lead to increasing levels of phosphorylated mutant ataxin-1 and an even greater accumulation of mutant protein. In addition, although levels of mutant ataxin-1 are increasing, changing calcium homeostasis in the cell may concurrently be leading to a downregulation of wild-type ataxin-1. In support of this hypothesis, calcium handling in SCA1-affected cells is altered [45, 46]. In transgenic mice that express mutant ataxin-1, downregulation of many genes involved in calcium handling, including IP3R1 (an intracellular calcium channel) and SERCA2 (a calcium pump), has been observed [45]. Downregulation of these genes during the SCA1 disease process may result in perturbation of the normal ataxin-1 signaling pathway and lead to a shift in ataxin-1 phosphorylation favoring mutant ataxin-1.

Until recently, a lack of knowledge surrounding the cellular function of ataxin-1, as well as the pathways involved in pathogenesis, has made the development of effective therapeutics for SCA1 challenging. Both siRNA studies [38] and studies using a conditional SCA1 mouse model [39] have demonstrated that strategies aimed at decreasing ataxin-1 protein levels may be effective for treating SCA1. Likewise, the observation that SCA1 knockout mice do not develop a disease phenotype suggests that therapies targeting ataxin-1 protein levels may not be detrimental to patients. The discovery of phosphorylation as an important regulator of ataxin-1 protein turnover opens up a new avenue to explore for therapeutic development in SCA1. Compounds that decrease ataxin-1 phosphorylation, resulting in an increase in ataxin-1 turnover in the cell, may be lead candidates for further studies aimed at identifying effective therapeutics for SCA1.

References

1. Orr, H. T., Chung, M. Y., Banfi, S., Kwiatkowski, T. J., Jr., Servadio, A., Beaudet, A. L., McCall, A. E., Duvick, L. A., Ranum, L. P., and Zoghbi, H. Y. (1993). Expansion of an unstable trinucleotide CAG repeat in spinocerebellar ataxia type 1. *Nat. Genet.* **4**, 221–226.
2. Chung, M. Y., Ranum, L. P., Duvick, L. A., Servadio, A., Zoghbi, H. Y., and Orr, H. T. (1993). Evidence for a mechanism predisposing to intergenerational CAG repeat instability in spinocerebellar ataxia type I. *Nat. Genet.* **5**, 254–258.
3. Sanpei, K., Takano, H., Igarashi, S., Sato, T., Oyake, M., Sasaki, H., Wakisaka, A., Tashiro, K., Ishida, Y., Ikeuchi, T., Koide, R., Saito, M., Sato, A., Tanaka, T., Hanyu, S., Takiyama, Y., Nishizawa, M., Shimizu, N., Nomura, Y., Segawa, M., Iwabuchi, K., Eguchi, I., Tanaka, H., Takahashi, H., and Tsuji, S. (1996). Identification of the spinocerebellar ataxia type 2 gene using a direct identification of repeat expansion and cloning technique, DIRECT. *Nat. Genet.* **14**, 277–284.
4. Pulst, S. M., Nechiporuk, A., Nechiporuk, T., Gispert, S., Chen, X. N., Lopes-Cendes, I., Pearlman, S., Starkman, S., Orozco-Diaz, G., Lunkes, A., DeJong, P., Rouleau, G. A., Auburger, G., Korenberg, J. R., Figueroa, C., and Sahba, S. (1996). Moderate expansion of a normally biallelic trinucleotide repeat in spinocerebellar ataxia type 2. *Nat. Genet.* **14**, 269–276.
5. Imbert, G., Saudou, F., Yvert, G., Devys, D., Trottier, Y., Garnier, J. M., Weber, C., Mandel, J. L., Cancel, G., Abbas, N., Durr, A., Didierjean, O., Stevanin, G., Agid, Y., and Brice, A. (1996). Cloning of the gene for spinocerebellar ataxia 2 reveals a locus with high sensitivity to expanded CAG/glutamine repeats. *Nat. Genet.* **14**, 285–291.

6. Kawaguchi, Y., Okamoto, T., Taniwaki, M., Aizawa, M., Inoue, M., Katayama, S., Kawakami, H., Nakamura, S., Nishimura, M., Akiguchi, I., et al. (1994). CAG expansions in a novel gene for Machado–Joseph disease at chromosome 14q32.1. *Nat. Genet.* **8**, 221–228.
7. Zhuchenko, O., Bailey, J., Bonnen, P., Ashizawa, T., Stockton, D. W., Amos, C., Dobyns, W. B., Subramony, S. H., Zoghbi, H. Y., and Lee, C. C. (1997). Autosomal dominant cerebellar ataxia (SCA6) associated with small polyglutamine expansions in the alpha 1A-voltage-dependent calcium channel. *Nat. Genet.* **15**, 62–69.
8. David, G., Abbas, N., Stevanin, G., Durr, A., Yvert, G., Cancel, G., Weber, C., Imbert, G., Saudou, F., Antoniou, E., Drabkin, H., Gemmill, R., Giunti, P., Benomar, A., Wood, N., Ruberg, M., Agid, Y., Mandel, J. L., and Brice, A. (1997). Cloning of the SCA7 gene reveals a highly unstable CAG repeat expansion. *Nat. Genet.* **17**, 65–70.
9. Nakamura, K., Jeong, S. Y., Uchihara, T., Anno, M., Nagashima, K., Nagashima, T., Ikeda, S., Tsuji, S., and Kanazawa, I. (2001). SCA17, a novel autosomal dominant cerebellar ataxia caused by an expanded polyglutamine in TATA-binding protein. *Hum. Mol. Genet.* **10**, 1441–1448.
10. The Huntington's Disease Collaborative Research Group (1993). A novel gene containing a trinucleotide repeat that is expanded and unstable on Huntington's disease chromosomes. *Cell* **72**, 971–983.
11. La Spada, A. R., Wilson, E. M., Lubahn, D. B., Harding, A. E., and Fischbeck, K. H. (1991). Androgen receptor gene mutations in X-linked spinal and bulbar muscular atrophy. *Nature* **352**, 77–79.
12. Koide, R., Ikeuchi, T., Onodera, O., Tanaka, H., Igarashi, S., Endo, K., Takahashi, H., Kondo, R., Ishikawa, A., Hayashi, T., et al. (1994). Unstable expansion of CAG repeat in hereditary dentatorubral-pallidoluysian atrophy (DRPLA). *Nat. Genet.* **6**, 9–13.
13. Nagafuchi, S., Yanagisawa, H., Sato, K., Shirayama, T., Ohsaki, E., Bundo, M., Takeda, T., Tadokoro, K., Kondo, I., Murayama, N., et al. (1994). Dentatorubral and pallidoluysian atrophy expansion of an unstable CAG trinucleotide on chromosome 12p. *Nat. Genet.* **6**, 14–18.
14. Burke, J. R., Wingfield, M. S., Lewis, K. E., Roses, A. D., Lee, J. E., Hulette, C., Pericak-Vance, M. A., and Vance, J. M. (1994). The Haw River syndrome: Dentatorubropallidoluysian atrophy (DRPLA) in an African-American family. *Nat. Genet.* **7**, 521–524.
15. Subramony, S. H., and Vig, P. J., (1998). *In* "Genetic Instabilities and Hereditary Neurological Diseases" (R. D. Wells and S. T. Warren, Eds.), pp. 231–239. Academic Press, San Diego.
16. Servadio, A., Koshy, B., Armstrong, D., Antalffy, B., Orr, H. T., and Zoghbi, H. Y. (1995). Expression analysis of the ataxin-1 protein in tissues from normal and spinocerebellar ataxia type 1 individuals. *Nat. Genet.* **10**, 94–98.
17. Banfi, S., Servadio, A., Chung, M. Y., Kwiatkowski, T. J., Jr., McCall, A. E., Duvick, L. A., Shen, Y., Roth, E. J., Orr, H. T., and Zoghbi, H. Y. (1994). Identification and characterization of the gene causing type 1 spinocerebellar ataxia. *Nat. Genet.* **7**, 513–520.
18. de Chiara, C., Giannini, C., Adinolfi, S., de Boer, J., Guida, S., Ramos, A., Jodice, C., Kioussis, D., and Pastore, A. (2003). The AXH module: An independently folded domain common to ataxin-1 and HBP1. *FEBS Lett.* **551**, 107–112.
19. Skinner, P. J., Koshy, B. T., Cummings, C. J., Klement, I. A., Helin, K., Servadio, A., Zoghbi, H. Y., and Orr, H. T. (1997). Ataxin-1 with an expanded glutamine tract alters nuclear matrix-associated structures. *Nature* **389**, 971–974.
20. Borden, K. L. (2002). Pondering the promyelocytic leukemia protein (PML) puzzle: Possible functions for PML nuclear bodies. *Mol. Cell. Biol.* **22**, 5259–5269.
21. Yue, S., Serra, H. G., Zoghbi, H. Y., and Orr, H. T. (2001). The spinocerebellar ataxia type 1 protein, ataxin-1, has RNA-binding activity that is inversely affected by the length of its polyglutamine tract. *Hum. Mol. Genet.* **10**, 25–30.
22. Irwin, S., Vandelft, M., Pinchev, D., Howell, J. L., Graczyk, J., Orr, H. T., and Truant, R. (2005). RNA association and nucleocytoplasmic shuttling by ataxin-1. *J. Cell Sci.* **118**, 233–242.
23. Fernandez-Funez, P., Nino-Rosales, M. L., de Gouyon, B., She, W. C., Luchak, J. M., Martinez, P., Turiegano, E., Benito, J., Capovilla, M., Skinner, P. J., McCall, A., Canal, I., Orr, H. T., Zoghbi, H. Y., and Botas, J. (2000). Identification of genes that modify ataxin-1-induced neurodegeneration. *Nature* **408**, 101–106.
24. Tsai, C. C., Kao, H. Y., Mitzutani, A., Banayo, E., Rajan, H., McKeown, M., and Evans, R. M. (2004). Ataxin 1, a SCA1 neurodegenerative disorder protein, is functionally linked to the silencing mediator of retinoid and thyroid hormone receptors. *Proc. Natl. Acad. Sci. USA* **101**, 4047–4052.
25. Riley, B. E., Zoghbi, H. Y., and Orr, H. T. (2005). SUMOylation of the polyglutamine repeat protein, ataxin-1, is dependent on a functional nuclear localization signal. *J. Biol. Chem.* **280**, 21942–21948.
26. Melchior, F., Schergaut, M., and Pichler, A. (2003). SUMO: Ligases, isopeptidases and nuclear pores. *Trends Biochem. Sci.* **28**, 612–618.
27. Muller, S., Ledl, A., and Schmidt, D. (2004). SUMO: A regulator of gene expression and genome integrity. *Oncogene* **23**, 1998–2008.
28. Burright, E. N., Clark, H. B., Servadio, A., Matilla, T., Feddersen, R. M., Yunis, W. S., Duvick, L. A., Zoghbi, H. Y., and Orr, H. T. (1995). SCA1 transgenic mice: A model for neurodegeneration caused by an expanded CAG trinucleotide repeat. *Cell* **82**, 937–948.
29. Matilla, A., Roberson, E. D., Banfi, S., Morales, J., Armstrong, D. L., Burright, E. N., Orr, H. T., Sweatt, J. D., Zoghbi, H. Y., and Matzuk, M. M. (1998). Mice lacking ataxin-1 display learning deficits and decreased hippocampal paired-pulse facilitation. *J. Neurosci.* **18**, 5508–5516.
30. Klement, I. A., Skinner, P. J., Kaytor, M. D., Yi, H., Hersch, S. M., Clark, H. B., Zoghbi, H. Y., and Orr, H. T. (1998). Ataxin-1 nuclear localization and aggregation: Role in polyglutamine-induced disease in SCA1 transgenic mice. *Cell* **95**, 41–53.
31. Emamian, E. S., Kaytor, M. D., Duvick, L. A., Zu, T., Tousey, S. K., Zoghbi, H. Y., Clark, H. B., and Orr, H. T. (2003). Serine 776 of ataxin-1 is critical for polyglutamine-induced disease in SCA1 transgenic mice. *Neuron* **38**, 375–387.
32. Chen, H. K., Fernandez-Funez, P., Acevedo, S. F., Lam, Y. C., Kaytor, M. D., Fernandez, M. H., Aitken, A., Skoulakis, E. M., Orr, H. T., Botas, J., and Zoghbi, H. Y. (2003). Interaction of Akt-phosphorylated ataxin-1 with 14–3–3 mediates neurodegeneration in spinocerebellar ataxia type 1. *Cell* **113**, 457–468.
33. DiFiglia, M., Sapp, E., Chase, K. O., Davies, S. W., Bates, G. P., Vonsattel, J. P., and Aronin, N. (1997). Aggregation of huntingtin in neuronal intranuclear inclusions and dystrophic neurites in brain. *Science* **277**, 1990–1993.
34. Holmberg, M., Duyckaerts, C., Durr, A., Cancel, G., Gourfinkel-An, I., Damier, P., Faucheux, B., Trottier, Y., Hirsch, E. C., Agid, Y., and Brice, A. (1998). Spinocerebellar ataxia type 7 (SCA7): A neurodegenerative disorder with neuronal intranuclear inclusions. *Hum. Mol. Genet.* **7**, 913–918.
35. Cummings, C. J., Mancini, M. A., Antalffy, B., DeFranco, D. B., Orr, H. T., and Zoghbi, H. Y. (1998). Chaperone suppression of aggregation and altered subcellular proteasome localization imply protein misfolding in SCA1. *Nat. Genet.* **19**, 148–154.
36. Cummings, C. J., Sun, Y., Opal, P., Antalffy, B., Mestril, R., Orr, H. T., Dillmann, W. H., and Zoghbi, H. Y. (2001). Over-expression of

inducible HSP70 chaperone suppresses neuropathology and improves motor function in SCA1 mice. *Hum. Mol. Genet.* **10**, 1511–1518.
37. Cummings, C. J., Reinstein, E., Sun, Y., Antalffy, B., Jiang, Y., Ciechanover, A., Orr, H. T., Beaudet, A. L., and Zoghbi, H. Y. (1999). Mutation of the E6-AP ubiquitin ligase reduces nuclear inclusion frequency while accelerating polyglutamine-induced pathology in SCA1 mice. *Neuron* **24**, 879–892.
38. Xia, H., Mao, Q., Eliason, S. L., Harper, S. Q., Martins, I. H., Orr, H. T., Paulson, H. L., Yang, L., Kotin, R. M., and Davidson, B. L. (2004). RNAi suppresses polyglutamine-induced neurodegeneration in a model of spinocerebellar ataxia. *Nat. Med.* **10**, 816–820.
39. Zu, T., Duvick, L. A., Kaytor, M. D., Berlinger, M. S., Zoghbi, H. Y., Clark, H. B., and Orr, H. T. (2004). Recovery from polyglutamine-induced neurodegeneration in conditional SCA1 transgenic mice. *J. Neurosci.* **24**, 8853–8861.
40. Hunter, T. (2000). Signaling—2000 and beyond. *Cell* **100**, 113–127.
41. Vierra-Green, C. A., Orr, H. T., Zoghbi, H. Y., and Ferrington, D. A. (2005). Identification of a novel phosphorylation site in ataxin-1. *Biochim. Biophys. Acta* **1744**, 11–18.
42. Bridges, D., and Moorhead, G. B. (2004). 14–3–3 proteins: A number of functions for a numbered protein. *Sci STKE* **2004**, re10.
43. Brazil, D. P., Yang, Z. Z., and Hemmings, B. A. (2004). Advances in protein kinase B signalling: AKTion on multiple fronts. *Trends Biochem. Sci.* **29**, 233–242.
44. Kaytor, M. D., Byam, C. E., Tousey, S. K., Stevens, S. D., Zoghbi, H. Y., and Orr, H. T. (2005). A cell-based screen for modulators of ataxin-1 phosphorylation. *Hum. Mol. Genet.* **14**, 1095–1105.
45. Lin, X., Antalffy, B., Kang, D., Orr, H. T., and Zoghbi, H. Y. (2000). Polyglutamine expansion down-regulates specific neuronal genes before pathologic changes in SCA1. *Nat. Neurosci.* **3**, 157–163.
46. Vig, P. J., Subramony, S. H., and McDaniel, D. O. (2001). Calcium homeostasis and spinocerebellar ataxia-1 (SCA-1). *Brain Res. Bull.* **56**, 221–225.

CHAPTER 23

Spinocerebellar Ataxia Type 2

STEFAN-M. PULST
Division of Neurology, Cedars–Sinai Medical Center, Los Angeles, California

I. Introduction
II. Identification of the SCA2 Gene
III. Repeat Range
IV. Anticipation and Meiotic Instability of the SCA2 Repeat
V. Genetic Modifiers of Age of Disease Onset
VI. Frequency and Phenotype
   A. Ataxia
   B. Eye Movements and Retinal Changes
   C. Movement Disorders
   D. Neuropathy
   E. Dementia
VII. Neuropathology
VIII. Function
   A. Sequence Homologies and Protein Domains
   B. Expression Patterns
   C. Function
IX. Mouse Models
X. Outlook
References

I. INTRODUCTION

What is now known as spinocerebellar ataxia type 2 (SCA2) was first described clinically as a distinct genetic entity in a large, homogeneous population of patients with dominantly inherited cerebellar ataxia from the Holguin province of Cuba [1]. Genotypic differentiation from the then only known ataxia locus, the SCA1 locus, on chromosome 6p did not occur until 1990 [2]. In 1971, Wadia and Swami had already pointed to the importance of slowed saccadic eye movements in a subset of patients with inherited ataxias in India [3]. Subsequent genotyping after identification of the SCA2 gene indicated that six of these families with slow eye movements carried mutations in the SCA2 gene [4]. In East Indian pedigrees, however, slow eye movements do not appear to be a distinguishing feature [5].

The prevalence of SCA2 in Holguin was 41 per 100,000, much higher than in the western part of Cuba or in other parts of the world [6]. It was initially speculated that the high incidence might be related to a founder effect or the interaction of a disease gene with an environmental toxin. Eighty-one patients from 20 different pedigrees were clinically examined. The majority were of Caucasian Spanish ancestry. In addition to ataxic gait and other cerebellar findings, many patients had slow saccadic eye movements, which in some had progressed to ophthalmoparesis. Tendon reflexes were brisk during the first years of life, but absent several years later. Seven autopsied cases showed a marked reduction of Purkinje cells.

II. IDENTIFICATION OF THE SCA2 GENE

In 1993, the SCA2 locus was mapped to Chromosome 12 in two ethnic populations. Using a genomewide screen, Gispert et al. [7] identified a 20-cM interval on Chromosome 12q24.1 that contained the SCA2 locus. Pulst et al. [8] confirmed this location in a second pedigree of Southern Italian descent and demonstrated that SCA2 showed marked anticipation of disease onset. Of 15 parent–child pairs, 14 showed earlier disease onset by at least 1 year. By the use of closely linked genetic markers, it could be demonstrated that anticipation was not due to biased ascertainment based on later onset in asymptomatic gene carriers [8]. This observation strongly suggested that SCA2 was caused by an unstable DNA repeat. Additional pedigrees from Italy, Tunisia, Austria, French Canada, Martinique, and France [9, 10] were also shown to have linkage to the new SCA2 locus. Several crucial recombination events in the Cuban pedigree finally limited SCA2 location to a 1-cM interval between markers D12S1328 and D12S1329 [11].

The SCA2 gene was independently identified by groups using three different approaches. Pulst et al. [12] employed a positional cloning strategy and constructed a physical map of the critical region using P1 artificial chromosomes (PACs) and bacterial artificial chromosomes (BACs). The contigs were then searched for CAG repeat-containing sequences. Using a genomically based assay to detect expanded polyglutamines, designated DIRECT, Sanpei et al. [13] analyzed genomic DNAs from several patients with SCA2. The DNAs were digested with several restriction enzymes, and Southern blots were hybridized under high-stringency conditions with a (CAG)$_{55}$ oligonucleotide. Using the 1C2 monoclonal antibody that recognizes long stretches of glutamines, Imbert et al. [14] identified clones in expression libraries generated from lymphoblastoid cDNA generated from SCA2 and SCA7 patients. Of several positive clones, one clone contained 22 glutamines. Interestingly, this clone, which was later shown to encode the SCA2 cDNA, was identified in the library made from patients with SCA7 and contained the normal 22 glutamines.

III. REPEAT RANGE

The normal alleles are not highly polymorphic. The two normal alleles, which account for >90% of alleles in most studies, have 22 and 23 CAG repeats [12, 13, 15]. Normal alleles typically show one or two CAA interruptions. In contrast to the CAT interruptions coding for histidine that are found in SCA1, the CAA interruptions do not interrupt the glutamine tract at the protein level (Fig. 23-1). Rare normal alleles ranging from 15 to 32 repeats have also been identified [13–15].

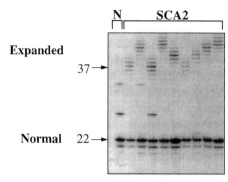

FIGURE 23-1 Expansion of the SCA2 repeat in chromosome 12-linked ataxia patients (SCA2). The DNA from a normal control (N) shows homozygosity for 22 repeats.

Intermediate alleles have yet to be convincingly associated with an abnormal phenotype, but can manifest meiotic instability by resulting in a pathological expansion in a subsequent generation. Alleles with reduced penetrance range from 32 to 35 repeats. These may or may not be associated with phenotypic changes depending on the age of the proband. An allele of 34 repeats was seen in the asymptomatic mother of a woman with SCA2. An allele with 32 repeats has also been seen in an asymptomatic 19-year-old whose symptomatic father carried an allele of 40 repeats [16]. The contracted allele had no CAA interruptions (Fig. 23-2).

Fernandez-Funez et al. [17] described a family segregating 33-CAG-repeat alleles in the SCA2 gene. These alleles were associated with disease onset as late as 86 years in one patient. In the Holguin SCA2 population, a woman with a typical SCA2 phenotype and an age of onset of 48 years carried an SCA2 allele with 32 repeats [18, 19].

Several asymptomatic individuals in SCA2 pedigrees with 34 and 35 repeats on chromosomes carrying the disease haplotype have been identified [13, 14, 20]. However, all of these individuals were still younger than the mean age of onset plus one standard deviation expected for the respective repeat length.

IV. ANTICIPATION AND MEIOTIC INSTABILITY OF THE SCA2 REPEAT

As in other diseases caused by unstable CAG DNA repeats, there is a clear inverse correlation between age of onset and repeat length. The best correlation is obtained with a negative exponential fit [12, 19]. The widest range of age of onset is observed for fewer than

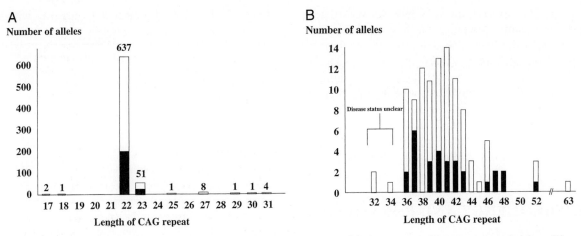

FIGURE 23-2 Distribution of SCA2 alleles on normal (A) and disease (B) chromosomes. Data were compiled from Riess et al. [15] (open boxes) and Pulst et al. [12] (solid boxes). For the discussion of individuals with 32 and 34 repeats, see text.

40 repeats. For example, the presence of 37 repeats was associated with ages of onset ranging from 15 to 65 [12, 19]. For larger repeat sizes, the variability is less, and repeat sizes of >45 are almost always associated with disease onset under 20 years of age [12–16, 20]. Homozygosity for an expanded SCA2 allele does not appear to influence age of onset [13].

Initial observations in the Cuban pedigrees [21] and in the FS pedigree from Southern Italy [8] did not point to consistent differences in the degree of anticipation depending on paternal or maternal inheritance. However, analysis of changes in CAG repeat sizes has indicated that large expansions are almost exclusively observed, when the repeat is passed through the paternal germline [15, 16, 20]. In contrast to these reports, Cancel et. al. [16] did not find a paternal bias.

Choudry et al. [22] examined repeat structure of the normal and expanded SCA2 alleles in Indian families. Similar to previous studies [12], they found that most normal alleles carried two interruptions with a 8 + 4 + 8 structure. All pathological repeats were uninterrupted. They identified two single-nucleotide polymorphisms (SNPs) 177 and 106 bp upstream of the CAG repeat. The first SNP also results in an amino acid change from valine to leucine. The CC haplotype showed complete association with the expansion mutation in the Indian study population. Absence of the 5′ CAA interruption in chromosomes with the CC haplotype appeared to be one of the factors predisposing to repeat expansion and suggested a polar variation in the SCA2 repeat.

V. GENETIC MODIFIERS OF AGE OF DISEASE ONSET

Common to all polyQ diseases is a significant inverse correlation between repeat length and age of onset, albeit with tremendous variability within each repeat length. This variation points to the likely existence of a number of *cis*- and *trans*-acting genetic factors, nonallelic genetic modifiers, and stochastic and environmental factors influencing age of onset (AO), in addition to the pathogenic allele itself. Although genetic modifiers have been identified in cell culture and model systems, relatively little work has been done in humans. Understanding these factors is not trivial, but may have important implications for understanding pathogenesis and improving counseling of presymptomatic individuals.

Using the Cuban SCA2 founder population, Pulst et al. recently estimated the amount of age of onset (AO) variance attributable to shared genetic and environmental factors (familiality) by examining the coefficient of intraclass correlation in siblings. They found that 55% of the residual variance in AO (after correction of the effect of the SCA2 CAG repeat) was familial [19].

Several groups have examined the effects of candidate modifier genes on residual AO variance. Hayes et al. [23] identified the polyQ tract in the RAI1 gene as modifying AO in a set of SCA2 families from different ethnic backgrounds. The effect was specific to SCA2 and not seen in SCA3 patients. RAI1 as a candidate modifier was also confirmed in Indian SCA2 patients [24].

Pulst et al. [19] examined the effect of the polyQ tract in all known polyQ disease genes on SCA2 onset.

Only the polyQ repeat in the CACNA1A (SCA6) gene showed significant association both with the longest allele and with the genotype determined by summing the repeats on both alleles. They used an approach of allelic association in two groups highly discordant for AO after correction for the SCA2 repeat. No effect was observed for the RAI1 gene or the ApoE gene, but allelic diversity for these two genes was reduced as a result of the founder population, and power to detect an effect was thus limited (Pulst, unpublished).

VI. FREQUENCY AND PHENOTYPE

SCA2 has a worldwide distribution [1, 15, 16, 20, 25, 26]. Its highest prevalence lies in the Cuban province of Holguin [1, 18, 19, 21]. Most patients in Holguin are of Caucasian Spanish ancestry and migrated from the Canary Islands. In this founder population, ataxic gait and other cerebellar findings are universal, and many patients have slow saccadic eye movements. Tendon reflexes are usually brisk during the first years of life, but absent several years later. SCA2 is also particularly common in India [3–5]. Analysis of a large number of SCA2 pedigrees has indicated a wide range of phenotypic manifestations [15, 16, 20, 25–27] that make SCA2 indistinguishable from other SCAs in the individual patient, although some findings such as slow saccades, peripheral neuropathy, and dementia are particularly common in SCA2 [25].

Most SCA2 mutations have arisen on different founder chromosomes [20, 28]. In Gunma prefecture, at least two founder haplotypes were identified with different CCG or CCGCCG interruptions of the CAG repeat [29]. In contrast, some German, Serbian, and French families shared the same haplotype, suggesting a common founder or a recurrent mutation on an at-risk chromosome [28]. An identical core haplotype established by alleles in the loci D12S1672 and D12S1333 in pedigrees of diverse ethnic origin from India, Japan, and England probably represents a haplotype common in these populations, rather than indicating a common founder (Table 23-1). [30].

A. Ataxia

Ataxia is universally present and is usually a presenting sign, although some Cuban patients may present with muscle cramps. A subclinical neuropathy may be identified before any other clinical signs [31]. Ataxia involves gait and stance, but is also prominent in appendicular functions. In the Cuban population a prominent

TABLE 23-1 SCA2 Phenotype Compared with SCA1 and SCA3[a]

	SCA1	SCA2[1]	SCA2[2]	SCA3
Cerebellar dysfunction	100	100	100	100
Reduced saccadic velocity	50	71	92	10
Myoclonus	0	40	0	4
Dystonia or chorea	20	0	38	8
Pyramidal involvement	70	29	31	70
Peripheral neuropathy	100	94	44	80
Intellectual impairment	20	31	37	5

[a]Percentages of patients with a specific sign are indicated. Percentages for SCA1, SCA2[1], and SCA3 were modified from those of Riess et al. [15]; those for SCA2[2], from Geschwind et al. [20].

truncal oscillation was observed when patients were standing with their eyes open. Quantitative assessments have recently been used in the evaluation of these patients and may provide an important addition to ataxia rating scales [32].

B. Eye Movements and Retinal Changes

Abnormal eye movements have been identified in all clinical studies of SCA2 [reviewed 33, 34–37]. Burk et al. [34] found that SCA2 patients had significantly slower saccadic speed (138°/s) than patients with SCA1 (244°/s) or SCA3 (347°/s). All eight SCA2 patients had saccadic velocities 2 SD below the mean of a control group. In a follow-up study, SCA2 patients were characterized by reduced saccadic velocity and the absence of square-wave jerks and gaze-evoked nystagmus [35].

Buttner et al. [37] compared patients with SCA1, SCA2, SCA3, and SCA6 identified by direct mutation analysis. Patients with SCA2 had the slowest peak saccadic velocity, ranging from 80 to 295°/s (normal, >400°/s). Saccades were also slowed in SCA1 patients, but patients with SCA3 or SCA6 had normal saccades. A recent analysis of Cuban SCA2 patients indicated that saccade velocity was inversely related to CAG repeat length [38].

Retinal degeneration is common in SCA7 and was thought to be exclusive to that ataxia. In SCA2 it has been described in the setting of infantile SCA2 associated with >200 repeats [39]. Retinal pigmentary degeneration was also noted in a 48-year-old woman with 41 CAG repeats in the SCA2 gene who had developed night blindness at age 28, four years before the onset of ataxia [40].

C. Movement Disorders

Movement disorders are common in SCA2. Parkinsonism and even L-DOPA-responsive parkinsonism without significant ataxia have recently been recognized as prominent aspects of the SCA2 phenotype. Sasaki et al. [41] described parkinsonism in a man homozygous for the SCA2 mutation. Recently, Gwinn-Hardy et al. [42] described a Taiwanese family with several members who displayed prominent parkinsonian signs that were responsive to L-DOPA. Shan et al. [43] confirmed this observation in another two Taiwanese patients, including reduction of ^{18}F-DOPA distribution in both the putamen and caudate nuclei. The SCA2 PD phenotype has recently been reviewed [44].

Geschwind et al. [20] found dystonia or chorea in 38% of their patients, and Cancel et al. [16] described dystonia in 9%. Patients with dystonia had longer repeats than those without. Sasaki et al. [41] point to the presence of choreiform movements in their patients in Japan. Myoclonus is prominent in Cuban SCA2 patients, especially in those with early onset (Pulst, personal observation). Cancel et al. [16] found that patients with myoclonus had longer repeats than those without. Schols et al. [45] reported postural tremor as the most common extrapyramidal sign in SCA2.

D. Neuropathy

In most studies, hyperreflexia due to upper motor neuron dysfunction is followed by hyporeflexia, indicating the presence of peripheral nerve dysfunction. In Cuban SCA2 patients, amplitude of sensory nerve potentials was decreased, even in the absence of other clinical signs [31]. Subclinical involvement by a sensory neuropathy was also confirmed in two Japanese SCA2 pedigrees [46]. Eighty percent of French SCA2 patients had a neuropathy [47]. Cancel et al. [16] observed fasciculations in 25% of SCA2 patients. Both CAG length and duration influenced the frequency of decreased reflexes and vibration sense in the lower extremities, amyotrophy, and fasciculations [16].

E. Dementia

Even in nondemented subjects, verbal and executive dysfunction can be detected with a frequency of 25 to 37% [27, 48, 49]. In a Northern Italian SCA2 pedigree, five of six individuals displayed frontal executive dysfunction despite a Mini-Mental Status score in the nondemented range [50]. Gambardella et al. [51] observed early and selective impairment of conceptual reasoning ability, as shown by abnormalities in the Wisconsin Card Sorting Test.

VII. NEUROPATHOLOGY

Several postmortem examinations have been reported in the Holguin population of Cuba [1]. There was a marked reduction in the number of cerebellar Purkinje cells. In silver preparations, Purkinje cell dendrites had poor arborization and torpedo-like formation of their axons as they passed through the granular layer. Parallel fibers were scanty. Granule cells were decreased in number, whereas Golgi and basket cells were well preserved, as were neurons in the dentate and other cerebellar nuclei. In the brainstem, there was marked neuronal loss in the inferior olive and pontocerebellar nuclei. Six of seven brains also had marked loss in the substantia nigra. In five spinal cords that were available for analysis, marked demyelination was present in the posterior columns and, to a lesser degree, in the spinocerebellar tracts. Motor neurons and neurons in Clarke's column were reduced in size and number. Especially in lumbar and sacral segments, anterior and posterior roots were partially demyelinated

Dürr et al. [49] reported autopsy findings in two patients from Martinican families. In addition to the findings reported by Orozco et al. [1], they also noted severe gyral atrophy most prominent in the frontotemporal lobes. The cerebral cortex was thinned, but without neuronal rarefaction. The cerebral white matter was atrophic and gliotic. Degeneration in the nigroluyso-pallidal system again mainly involved the substantia nigra. One brain showed patchy loss in parts of the third-nerve nuclei. Adams and Pulst [52] reported similar findings in one member of the FS pedigree (see Fig. 23-3). Nerve biopsy has shown moderate loss of large myelinated fibers [53].

Brains of Cubans and Germans with SCA2 have recently been examined using unconventionally thick serial sections through the brainstem and thalamus [54–56]. Sections were stained for lipofuscin pigment and Nissl material. These analyses showed that all of the pre-cerebellar nuclei (red, pontine, arcuate, prepositus hypoglossal, superior vestibular, lateral vestibular, medial vestibular, interstitial vestibular, spinal vestibular, vermiform, lateral reticular, external cuneate, subventricular, paramedian reticular, intercalate, interfascicular hypoglossal, and conterminal nuclei, pontobulbar body, reticulotegmental nucleus of the pons, inferior olive, and nucleus of Roller) were involved in the degenerative processes. The thalamus was also consistently affected.

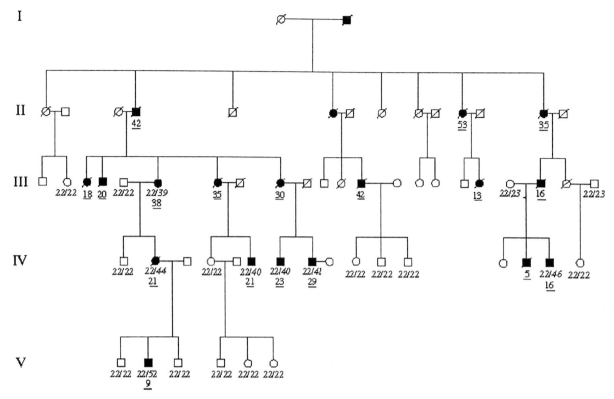

FIGURE 23-3 Anticipation of age of onset and meiotic SCA2 CAG repeat length instability in the FS pedigree. Age of onset is underlined. Reprinted, with permission, from Adams and Pulst [52].

There was consistent involvement of the lateral geniculate body; the lateral posterior, ventral anterior, ventral lateral, ventral posterior lateral, and ventral posterior medial thalamic nuclei; and the extraterritorial reticular nucleus. Significant involvement of cranial nerve nuclei and tracts was also reported.

VIII. FUNCTION

A. Sequence Homologies and Protein Domains

Using genomic clones as probes, both Pulst et al. [12] and Sanpei et al. [13] identified several SCA2 cDNA clones from a fetal brain and a frontal cortex cDNA library, respectively. Both groups could not unequivocally identify the 5' end of the SCA2 cDNA and used 5' RACE and cloned reverse transcription polymerase chain reaction products to identify additional cDNA sequence. Both groups predict the identical amino acid sequence with 1312 amino acids (22 glutamines) [12] and 1313 amino acids (23 glutamines) [13] with the CAG repeat coding for polyglutamine.

Sequence analysis of the 4-kb cDNA clone isolated by Imbert et al. [14] predicted a much shorter open reading frame (ORF). However, a second ORF in a different frame partially overlapped the first and predicted a larger protein if ribosomal frameshifting were to occur. Otherwise, the cDNA sequence agreed with the sequence obtained by the other two groups. It is likely that the 4.0-kb cDNA clone represents a rare cDNA, as none of the other cDNA clones isolated by Pulst et al. [12] or Sanpei et al. [13] contained the second ORF. Furthermore, the protein recognized by the 1CE antibody and by SCA2-specific antibodies is more consistent with a protein of larger molecular weight [57, 58].

The 5' sequence of the SCA2 cDNA is extremely GC-rich and two potential ATG initiation codons can be identified. The most 5' ATG is located 78 bp downstream of an in-frame stop codon. Usage of this translation initiation site predicts a protein of 140.1 kDa. The second ATG, which has a better Kozak consensus sequence, is located just 5' to the CAG repeat and would result in a protein with relative molecular weight of 125 kDa. Proteins observed by Western blot analysis and conservation of the 5' ATG in the mouse [59] suggest that the 5' ATG is the predominant site of translation initiation.

In analogy to the other SCA gene products, the SCA2 gene product has been designated ataxin-2. Homology searches using cDNA and amino acid sequences have identified several homologies with proteins of known function. Significant sequence homology was detected with a protein designated ataxin-2-related protein (A2RP) and the mouse SCA2 protein [12, 60] Despite the significant homologies, the polyglutamine tract in human ataxin-2 is not present in A2RP or in mouse ataxin-2, suggesting that it may not be important for ataxin-2 function. However, all acidic amino acids constituting a highly acidic domain adjacent to the human polyglutamine domain are conserved in A2RP and mouse ataxin-2.

B. Expression Patterns

The SCA2 gene is widely expressed. On Northern blots, a 4.5-kb transcript is recognized in RNAs isolated from brain, heart, placenta, liver, skeletal muscle, and pancreas [12–14]. Little or no expression was seen in lung or kidney. The transcript is expressed throughout the brain. In RNAs isolated from SCA2 lymphoblastoid cell lines, expression of both the normal and expanded alleles is seen using reverse transcription polymerase chain reaction [12].

The SCA2 transcript in the mouse is of identical size [59]. Expression during mouse embryonic development with strong expression on Days E11 and E12 suggests that ataxin-2 may have a role in normal embryogenesis.

Antibodies to ataxin-2 recognize a 145-kDa protein in mouse and human brains [59]. In addition, a larger protein of approximately 200 kDa and several smaller proteins are recognized by these antibodies, suggesting that ataxin-2 may undergo posttranslational processing and aggregation. In normal and SCA2 brains, ataxin-2 has a cytoplasmic localization.

By use of antibodies to ataxin-2, the expression pattern of ataxin-2 in normal and SCA2 brains has been studied [57, 58]. In normal brains, ataxin-2 has a cytoplasmic localization. In SCA2 brain, antibodies to ataxin-2 or to expanded polyglutamine repeats show intense cytoplasmic staining that appears significantly stronger than in simultaneously stained control brains. Instead of a finely granular staining pattern, the entire cytoplasm is strongly immunoreactive in SCA2 brains. Although Huynh et al. [57] did not identify intranuclear inclusion bodies in three cerebella from SCA2 patients, Koyano et al. [61] reported ubiquitinated intranuclear inclusions in about 1 to 2% of neurons in affected areas. Purkinje neurons, however, did not show inclusions in Japanese patients, suggesting that the formation of intranuclear inclusions is a late event and not necessary for pathogenesis.

Pang et al. [62] detected intranuclear inclusions in a larger number of neurons in the brainstem and the cortex in two brains from SCA2 patients including neurons that are normally not involved in the neurodegenerative process. Unfortunately, the authors did not use an antibody to ataxin-2 for the detection of the intranuclear inclusions, relying instead on an antibody to polyglutamine repeats. These authors also reported on the presence of intranuclear inclusions in glial cells, confirming the observations by Huynh et al. [57] that ataxin-2 staining in glial cells was increased in SCA2 brains.

C. Function

A potential function for ataxin-2 was suggested by the identification of protein interactors. Using the yeast two-hybrid system, Shibata et al. [63] identified a 403-amino acid protein that interacted with the C-terminal half of ataxin-2. This protein, designated A2BP1 for ataxin-2-binding protein 1, contains two RNP RNA binding motifs. The fact that ataxin-2 contains SM motifs, which are found in proteins involved in RNA splicing, suggested that ataxin-2 and A2BP1 function in RNA transport or processing.

Recent support for a role for ataxin-2 in RNA metabolism has come from two different lines of experiments. Ralser et al. [64] discovered that in yeast, human ataxin-2 can functionally substitute for deletion mutants for poly(A)-binding protein 1 (Pabp1). In mammalian cells subjected to heat shock, ataxin-2 is found in stress granules, which represent complexes of untranslated mRNAs and ribonuclear proteins.

In the fly, changes in ataxin-2 expression result in severe phenotypic changes [65]. In *Caenorhabditis elegans* knock-downs of the ataxin-2 and A2BP1 orthologs result in reduced egg masses and embryonic lethality [66]. Ciosk et al. [67] followed up on these initial observations and determined that in the absence of ataxin-2, the germline was abnormally masculinized. In the worm, ataxin-2 physically interacted with PAB-1, one of the two *C. elegans* poly(A)-binding proteins. The developmental defects appeared to result from inappropriate translational regulation, which is normally mediated by the conserved KH-domain protein GLD-1 and a second KH-domain protein, MEX-3.

Given the potential involvement of ataxin-2 in mRNA regulation, it is quite surprising that by biochemical subcellular fractionation and immunocytochemical localization, ataxin-2 localized to the Golgi [68]. As ataxin-2 lacks a leader sequence, it is likely that it localizes to the cytoplasmic side of the Golgi, although definite proof for this notion is lacking. In addition to the perinuclear Golgi localization, endogenous and GFP-tagged exogenous ataxin-2 localize to puncta in the cytoplasm. Exogenously expressed mutant ataxin-2[Q58] and ataxin-2[Q104] have lost the Golgi-predominant location and result in increased cell death.

IX. MOUSE MODELS

Huynh et al. [58] expressed full-length human ataxin-2 under the control of the Purkinje cell-specific Pcp2 regulatory element in C57BL/6JxDBA/2J mice. Three lines were generated that expressed mutant ataxin-2 with 58 glutamine repeats (ataxin-2[Q58]) and two that expressed wild-type ataxin-2[Q22].

Clasping, footprinting, and rotarod analysis were performed to determine the effect of transgene expression on Purkinje cell function. Clasping was observed at 4 to 12 months of age depending on the transgenic line. Stride length was significantly altered in mice expressing mutant ataxin-2 compared with wild-type mice or mice expressing ataxin-2[Q22]. Alterations began at 8 to 16 weeks depending on the transgenic line. Rotarod testing confirmed these functional deficits. At 6 weeks, motor performance of transgenic animals was not different from that of wild-type mice. At 16 weeks, double transgenic Q58–11 mice already performed poorly on rotarod testing, whereas heterozygous Q58–11 animals performed as well as wild-type animals. Although mice from the Q58–5B line exhibited stride length deficits later than those from the Q58–11 line, their rotarod performance matched that of the Q58–11 mice. Functional deficits were progressive. At 26 weeks, both heterozygous and homozygous Q58–11 animals showed severely impaired motor performance. The rotarod performance of animals expressing ataxin-2[Q22] was not significantly different from that of wild-type animals.

Functional changes were accompanied by morphological alterations. Labeling with antibodies to either ataxin-2 or 1C2, which recognizes expanded polyQ repeats, did not reveal intranuclear inclusions, although Purkinje cell cytoplasm labeled intensely with these antibodies in animals expressing ataxin-2[Q58]. Ubiquitin labeling was undetectable in Purkinje cells of wild-type and transgenic animals. To characterize the dendritic morphology of Purkinje cells, an antibody to calbindin 28K was used. All three transgenic lines expressing ataxin-2[Q58] showed a progressive reduction in calbindin staining. At 24–27 weeks, Purkinje cell number was reduced by 50 to 53% in lines Q58–5B, Q58–11, and Q58–19.

Gain of function or gain of toxic function, as relevant in SCA2 pathogenesis, has been supported by results with an ataxin-2-deficient mouse line. Similar to SCA1$^{-/-}$ mice, these mice exhibit only a mild phenotype. Ataxin-2-deficient mice do not develop apparent neurodegeneration and have no gross tissue or developmental changes. These observations support the hypothesis that polyQ expansion does not lead to a loss of normal ataxin-2 function in patients with SCA2 [69].

X. OUTLOOK

Identification of the SCA2 gene has led to a definition of the phenotype that includes not only the typical ataxia phenotype, but also rarer parkinsonian phenotypes that may at times be indistinguishable from idiopathic Parkinson's disease. The normal function of ataxin-2 remains poorly understood, although several studies support a function in mRNA translational regulation or transport. Animal models of the disease are consistent with a gain of function of mutant ataxin-2. It remains to be seen if polyQ expansion leads to a simple gain of function or if mutant ataxin-2 acquires toxic or novel functions.

References

1. Orozco, G., Estrada, R., Perry, T. L., Arana, J., Fernandez, R., Gonzalez-Quevedo, A., Galarraga, J., and Hansen, S. (1989). Dominantly inherited olivopontocerebellar atrophy from eastern Cuba: Clinical, neuropathological, and biochemical findings. *J. Neurol. Sci.* **93**, 37–50.
2. Auburger, G., Orozco, G., Capote, R. F., Sanchez, S. G., Perez, R., Chamberlain, S., and Baute, L. H. (1990). Autosomal dominant ataxia: Genetic evidence for locus heterogeneity from a Cuban founder-effect population. *Am. J. Hum. Genet.* **46**, 1163–1177.
3. Wadia, N. H., and Swami, R. K. (1971). A new form of heredofamilial spinocerebellar degeneration with slow eye movements (nine families). *Brain.* **94**, 359–374.
4. Wadia, N., Pang, J., Desai, J., Mankodi, A., Desai, M., and Chamberlain, S. (1998). A clinicogenetic analysis of six Indian spinocerebellar ataxia (SCA2) pedigrees: The significance of slow saccades in diagnosis [review]. *Brain* **121**, 2341–2355.
5. Chakravarty, A., and Mukherjee, S. C., (2002). Autosomal dominant cerebellar ataxias in ethnic Bengalees in West Bengal—an Eastern Indian state. *Acta Neurol. Scand.* **105**, 202–208.
6. Gudmundsson, K. (1969). The prevalence and occurrence of some rare neurological diseases in Iceland. *Acta Neurol. Scand.* **45**, 114–118.
7. Gispert, S., Twells, R., Orozco, G., Brice, A., Weber, J., Herdero, L. Scheufler, K., Riley, B., Allotey, R. I., Nothers, C., Hillerman, R., Lunkes, A., Khati, C., Stevanin, G., Hernandez, A., Magariuno, C., Klockgether, T., Durr, A., Chneiweiss, H., Enczmann, J., Farrall, M., Beckmann, J., Mullan, M., Wernet, P., Agid, Y., Freund, H. J., Williamson, R., Auburger, G., and Chamberlain, S. (1994). Chromosomal assignment of the second (Cuban) locus for autosomal dominant cerebellar ataxia (SCA2) to chromosome 12q23–24.1 *Nat. Genet.* **4**, 295–299.
8. Pulst, S.-M., Nechiporuk, A., and Starkman, S. (1993). Anticipation in spinocerebellar ataxia type 2. *Nat. Genet.* **5**, 8–10.

9. Belal, S., Cancel, G., Stevanin, G., Hentati, F., Khati, C., Ben Hamida, C., Auburger, G., Agid, Y., Ben Hamida, M., and Brice, A. (1994). Clinical and genetic analysis of a Tunisian family with autosomal dominant cerebellar ataxia type 1 linked to the SCA2 locus. *Neurology* **44**, 1423–1426.
10. Nechiporuk, A., Lopes-Cendes, I., Nechiporuk, T., Starkman, S., Andermann, E., Rouleau, G. A., Weissenbach, J. S., Kort, E., and Pulst, S. M. (1996). Genetic mapping of spinocerebellar ataxia type 2 gene on human chromosome 12. *Neurology* **46**, 1731–1735.
11. Gispert, S., Lunkes, A., Santos, N., Orozco, G., Ha-Hao, D., Ratzlaff, T., Agiar, J., Torrens, I., Heredero, L., Brice, A., Cancel, G., Stevanin, G., Vernant, J.-C., Durr, A., Lepage-Lezin, A., Belal, S., Ben-Hamida, M., Pulst, S.-M., Rouleau, G., Weissenbach, H., LePaslier, D., Kucherlapati, R., Montgomery, K., Fukui, K., and Auburger, G. (1995). Localization of the candidate gene D-amino acid oxidase outside the refined 1-cM region of spinocerebellar ataxia 2. *Am. J. Hum. Genet.* **57**, 972–975.
12. Pulst, S. M., Nechiporuk, A., Nechiporuk, T., Gispert, S., Chen, X. N., Lopes-Cendes, I., Pearlman, S., Starkman, S., Orozco-Diaz, G., Lunkes, A., DeJong, P., Rouleau, G. A., Auburger, G., Korenberg, J. R., Figueroa, C., and Sahba, S. (1996). Moderate expansion of a normally biallelic trinucleotide repeat in spinocerebellar ataxia type 2. *Nat. Genet.* **14**, 269–276.
13. Sanpei, K., Takano, H., Igarashi S., Sato, T., Oyake, M., Sasaki, H., Wakisaka, A., Tashiro, K., Ishida. Y., Ikeuchi, T., Koide, R., Saito, M., Sato, A., Tanaka. T., Hanyu, S., Takiyama, Y., Nishizawa, M., Shimizu, N., Nomura, Y., Segawa, M., Iwabuchi, K., Eguchi, I., Tanaka, H., Takahashi, H., and Tsuji, S. (1996). Identification of the spinocerebellar ataxia type 2 gene using a direct identification of repeat expansion and cloning technique, DIRECT. *Nat. Genet.* **14**, 277–284.
14. Imbert, G., Saudou, F., Yvert, G., Devys, D., Trottier, Y., Garnier, J. M., Weber, C., Mandel, J., Cancel, G., Abbas, N., Durr, A., Didierjean, O., Stevanin, G., Agid, Y., and Brice, A. (1996). Cloning of the gene for spinocerebellar ataxia 2 reveals a locus with high sensitivity to expanded CAG glutamine repeats. *Nat. Genet.* **14**, 285–291.
15. Riess, O., Laccone, F., Gispert, S., Schols, L., Zuhlke, C., Vieira-Saecker, A. M., Herlt, S., Wessel, K., Epplen, J. T., Weber, B. H., Kreuz, F., Chahrokh-Zadeh, S., Meindl, A., Lunkes, A., Aguiar, J., Macek, M., Jr., Krebsova, A., Macek, M., Sr., Burk, K., Tinschert, S., Schreyer, I., Pulst, S. M., and Auburger, G. (1997). SCA2 trinucleotide expansion in German SCA patients. *Neurogenetics* **1**, 59–64.
16. Cancel, G., Dürr, A., Didierjean, O., Imbert, G., Burk, K., Lezin, A., Belal, S., Benomar, A., Abada-Bendib, M., Vial, C., Guimaraes, J., Chneiweiss, H., Stevanin, G., Yvert, G., Abbas, N., Saudou, F., Lebre, A. S., Yahyaoui, M., Hentati, F., Vernant, J. C., Klockgether, T., Mandel, J. L, Agid, Y., and Brice, A. (1997). Molecular and clinical correlations in spinocerebellar ataxia 2: A study of 32 families. *Hum. Mol. Genet.* **6**, 709–715.
17. Fernandez-Funez, P., Nino-Rosales, M. L., de Gouyon, B., She, W. C., Luchak, J. M., Martinez, P., Turiegano, E., Benito, J., Capovilla, M., Skinner, P. J., McCall, A., Canal, I., Orr, H. T., Zoghbi, H. Y., and Botas, J. (2000). Identification of genes that modify ataxin-1-induced neurodegeneration. *Nature* **408**, 101–106.
18. Santos, N., Aguiar, J., Fernandez, J., et al. (1999). Molecular diagnosis of a sample of the Cuban population with spinocerebellar ataxia type 2. *Biotecnol. Apli.* **16**, 219–221.
19. Pulst, S. M., Santos, N., Wang, D., Yang, H., Huynh, D., Velazquez, L., and Figueroa, K. P. (2005) Spinocerebellar ataxia type 2: PolyQ repeat variation in the CACNA1A calcium channel modifies age of onset (2005). *Brain* **128**, 2297–2303. Epub 2005 Jul 6.
20. Geschwind, D. H., Perlman, S. B., Figueroa, C. P., Treiman, L. J., and Pulst, S. M. (1997). The prevalence and wide clinical spectrum of the spinocerebellar ataxia type 2 trinucleotide repeat in patients with autosomal dominant cerebellar ataxia. *Am. J Hum. Genet* **60**, 842–850.
21. Orozco, G., Fleites, A., Cordoves Sagaz, R., and Auburger, G. (1990). Autosomal dominant cerebellar ataxia: Clinical analysis of 263 patients from a homogeneous population in Holguin, Cuba. *Neurology* **40**, 1369–1375.
22. Choudhry, S., Mukerji, M., Srivastava, A. K., Jain, S., and Brahmachari, S. K. (2001). CAG repeat instability at SCA2 locus: Anchoring CAA interruptions and linked single nucleotide polymorphisms. *Hum. Mol. Genet.* **10**, 2437–2446.
23. Hayes, S., Turecki, G., Brisebois, K., Lopes-Cendes, I., Gaspar, C., Riess, O., Ranum, L. P., Pulst S. M., and Rouleau, G. A. (2000). CAG repeat length in RAI1 is associated with age at onset variability in spinocerebellar ataxia type 2 (SCA2). *Hum. Mol. Genet.* **9**, 1753–1758.
24. Chattopadhyay, B., Ghosh, S., Gangopadhyay, P. K., Das, S. K., Roy, Sinha, K. K., Jha, D. K., Mukherjee, S. C., Chakraborty, A., Singhal, B. S., Bhattacharya, A. K., and Bhattacharyya N. P. (2003). Modulation of age at onset in Huntington's disease and spinocerebellar ataxia type with patients originated from eastern India. *Neurosci. Lett.* **345**, 93–96
25. Maschke, M., Oehlert, G., Xie, T. D., Perlman, S., Subramony S. H., Kumar, N., Ptacek, L. J., and Gomez, C. M. (2005). Clinical feature profile of spinocerebellar ataxia type 1–8 predicts genetically defined subtypes. *Mov. Disord.* **21**. [Epub ahead of print.]
26. Rosa, A. L., Molina, I., Kowaljow, V., and Conde, C. B. (2005). Brisk deep-tendon reflexes as a distinctive phenotype in an Argentinean spinocerebellar ataxia type 2 pedigree. *Mov. Disord.* **17**. [Epub ahead of print.]
27. Armstrong, J., Bonaventura, I., Rojo, A., Gonzalez, G., Corral, J., Nadal, N., Volpini, V., and Ferrer, I. (2005). Spinocerebellar ataxia type 2 (SCA2) with white matter involvement. *Neurosci. Lett.* **381**, 247–251.
28. Didierjean, O., Cancel, G., Stevanin, G., Durr, A. K., Benomar, A., Lezin A., Belal, S., Abada-Bendid, M., Klockgether, T., and Brice, A. (1999). Linkage disequilibrium at the SCA2 locus. *J Med. Genet.* **36**, 415–417.
29. Mizushima, K., Watanabe, M., Kondo, I., Okamoto, K., Shizuka, M., Abe, K., Aoki, M., and Shoji, M. (1999). Analysis of spinocerebellar ataxia type 2 gene and haplotype analysis: (CCG)1-2 polymorphism and contribution to founder effect. *Med. Genet.* **36**, 112-114.
30. Pang, J., Allotey, R., Wadia, N., Sasaki, H., Bindoff, L., and Chamberlain, S. A. (1999). Common disease haplotype segregating in spinocerebellar ataxia 2 (SCA2) pedigrees of diverse ethnic origin. *Eur. J. Hum. Genet*. **7**, 841–845.
31. Velazquez, L., and Medina, E. E. (1998). Neurophysiological evaluation of patients with spinocerebellar ataxia type 2. *Rev. Neurol.* **160**, 921–926. [In Spanish.]
32. Velazquez-Perez, L., de la Hoz-Oliveras, J., Perez-Gonzalez, R., Hechavarria, P. R., and Herrera-Dominguez, H. (2001). Quantitative evaluation of disorders of coordination in patients with Cuban type 2 spinocerebellar ataxia. *Rev Neurol.* **32**, 601–606. [In Spanish.]
33. Pulst, S. M., and Perlman, S. (2000). Hereditary Ataxias. *In* "Neurogenetics" (S. M. Pulst, Ed.). Oxford Univ. Press, London/ New York pp. 231–264.
34. Burk K., Abele M., Fetter M., Dichgans J., Skalej M., Laccone F., Didierjean O., Brice A., and Klockgether, T. (1996). Autosomal dominant cerebellar ataxia type I clinical features and MRI in families with SCA1, SCA2 and SCA3. *Brain* **119**, 1497–1505.
35. Burk K., Fetter M., Abele M., Laccone F., Brice A., Dichgans J., and Klockgether, T. (1999). Autosomal dominant cerebellar ataxia type I: Oculomotor abnormalities in families with SCA1, SCA2, and SCA3. J Neurol. **246**, 789–797.

36. Rivaud-Pechoux, S., Durr, A., Gaymard, B., Cancel, G., Ploner, C. J., Agid, Y., Brice. A., and Pierrot-Deseilligny, C. (1998). Eye movement abnormalities correlate with genotype autosomal dominant cerebellar ataxia type I. *Ann. Neurol.* **43**, 297–302.
37. Buttner N., Geschwind, D., Jen, J. C., Perlman, S., Pulst, S. M., and Baloh, R. W. (1998). Oculomotor phenotypes in autosomal dominant ataxias. *Arch. Neurol.* **55**, 1353–1357.
38. Velazquez-Perez, L., Seifried, C., Santos-Falcon, N., Abele, M., Ziemann, U., Almaguer, L. E., Martinez-Gongora, E., Sanchez-Cruz, G., Canales, N., Perez-Gonzalez, R., Velazquez-Manresa, M., Viebahn, B., von Stuckrad-Barre, S., Fetter, M., Klockgether, T., and Auburger, G. (2004). Saccade velocity is controlled by polyglutamine size in spinocerebellar ataxia 2. *Ann. Neurol.* **56**, 444–447.
39. Babovic-Vuksanovic, D., Snow, K., Patterson, M., and Michels, V. V. (1998). Spinocerebellar ataxia type 2 (SCA 2) in an infant with extreme CAG repeat expansion. *Am. J. Med. Genet.* **79**, 383–387.
40. Rufa, A., Dotti, M. T., Galli, L., Orrico, A., Sicurelli, F., and Federico, A. (2002) Spinocerebellar ataxia type 2 (SCA2) associated with retinal pigmentary degeneration. *Eur. Neurol.* **47**, 128–129.
41. Sasaki, H., Wakisaka, A., Sanpei, K., Takano, H., Igarashi, S., Ikeuchi, T., Iwabuchi, K., Fukazawa, T., Hamada, T., Yuasa, T., Tsuji, S., and Tashiro K. (1998). Phenotype variation correlates with CAG repeat length in SCA2-a study of 28 Japanese patients. *J. Neurol. Sci* **159**, 202–208.
42. Gwinn-Hardy, K., Chen, J. Y., Liu, H. C., Liu, T. Y., Boss, M., Seltzer, W., Adam, A., Singleton, A., Koroshetz, W., Waters, C., Hardy, J., and Farrer, M. (2000). Spinocerebellar ataxia type 2 with parkinsonism in ethnic Chinese. *Neurology* **55**, 800–805.
43. Shan, D. E., Soon, B. W., Sun, C. M., Lee, S. J., Liao, K. K., and Liu, R. S. (2001). Spinocerebellar ataxia type 2 presenting as familial levodopa-responsive parkinsonism. *Ann. Neurol.* **50**, 812–815.
44. Furtado, S., Payami, H., Lockhart, P. J., Hanson, M., Nutt, J. G., Singleton, A. A., Singleton, A., Bower, J., Utti, R. J., Bird, T. D., de la Fuente-Fernandez, R., Tsuboi Y, Klimek M. L., Suchowersky, O., Hardy, J., Calne, D. B., Wszolek, Z. K., Farrer, M., Gwinn-Hardy, K., and Stoessl, A. J. (2004). Profile of families with parkinsonism-predominant spinocerebellar ataxia type 2 (SCA2) [review]. *Mov. Disord.* **19**, 622–629.
45. Schols, L., Peters, S., Szymanski, S., Kruger, R., Lange, S., Hardt, C., Riess, O., and Przuntek, H. (2000). Extrapyramidal motor signs in degenerative ataxias. *Arch. Neurol.* **57**, 1495–1500.
46. Ueyama, H., Kumamoto, T., Nagao, S., et al. (1998). Clinical and genetic studies of spinocerebellar ataxia type 2 in Japanese kindreds. *Acta Neurol. Scand.* **98**, 427–432.
47. Kubis, N., Durr, A., Gugenheim, M., Chneiweiss, H., Mazzetti, P., Brice, A., and Bouche, P. (1999). Polyneuropathy in autosomal dominant cerebellar ataxias: Phenotype–genotype correlation. *Muscle Nerve* **22**, 712–717.
48. Burk, K., Globas, C., Bosch, S., Graber, S., Abele, M., Brice, A., Dichgans, J., Daum, I., and Klockgether T. (1999). Cognitive deficits in spinocerebellar ataxia 2. *Brain* **19**, 769–777.
49. Durr, A., Smadja, D., Cancel, G., Lezin, A., Stevanin, G., Mikol, J., Bellance, R., Buisson, G. G., Chneiweiss, H., Dellanave, J., Agid, Y., Brice, A., and Vernant, J. D. (1995). Autosomal dominant cerebellar ataxia type 1 *in* Martinique (French West Indies): Clinical and neuropathological analysis of 53 patients from three unrelated SCA2 families. *Brain* **118**, 1573–1581.
50. Storey, E., Forrest, S. M., Shaw, J. H., Mitchell, P., and Gardner, R. J. (1999). Spinocerebellar ataxia type 2: Clinical features of a pedigree displaying prominent frontal-executive dysfunction. *Arch. Neurol.* **56**, 43–50.
51. Gambardella, A., Annesi, G., Bono, F., Spadafora, P., Valentino, P., Pasqua, A. A., Mazzei, R., Montesanti, R., Conforti, F. L., Oliveri, R. L., Zappia, M., Aguglia, U., and Quattrone A. (1998). CAG repeat length and clinical features in three Italian families with spinocerebellar ataxia type 2 (SCA2): Early impairment of Wisconsin Card Sorting Test and saccade velocity. *J. Neurol.* **245**, 647–652.
52. Adams, C., and Pulst, S. M. (1997). Clinical and molecular analysis of a pedigree of southern Italian ancestry with spinocerebellar ataxia type 2. *Neurology* **49**, 1163–1166.
53. Filla, A., DeMichele, G., Banfi, S., Santoro, L., Perretti, A., Cavalcanti, F., Pianese, L., Castaldo, I., Barrieri, F., Campanella, G., and Cocozza, S. (1995) Has spinocerebellar ataxia type 2 a distinct phenotype? Genetic and clinical study of an Italian family. *Neurology* **45**, 793–796.
54. Rub, U., Del Turco, D., Burk, K., Diaz, G. O., Auburger, G., Mittelbronn, M., Gierga, K., Ghebremedhin, E., Schultz, C., Schols, L., Bohl, J., Braak, H., and Deller T. (2005) Extended pathoanatomical studies point to a consistent affection of the thalamus in spinocerebellar ataxia type 2. *Neuropathol. Appl. Neurobiol.* **31**, 127–140.
55. Rub, U., Gierga, K., Brunt, E. R., de Vos, R. A., Bauer, M., Schols, L., Burk, K., Auburger, G., Bohl, J., Schultz, C., Vuksic, M., Burbach, G. J., Braak, H., and Deller, T. (2005). Spinocerebellar ataxias types 2 and 3: Degeneration of the precerebellar nuclei isolates the three phylogenetically defined regions of the cerebellum. *J. Neural Transm.* **112**, 1523–1545. Epub 2005 Mar 23.
56. Gierga, K., Burk, K., Bauer, M., Orozco Diaz, G., Auburger, G., Schultz, C., Vuksic, M., Schols, L., de Vos, R. A., Braak, H., Deller, T., and Rub, U. (2005). Involvement of the cranial nerves and their nuclei in spinocerebellar ataxia type 2 (SCA2). *Acta Neuropathol.* (Berl.) **109**, 617–631.
57. Huynh, D. P., Del Bigio, M. R., Ho, D. H., and Pulst S. M. (1999). Expression of ataxin-2 in brains from normal individuals and patients with Alzheimer's disease and spinocerebellar ataxia 2. *Ann. Neurol* **45**, 232–241.
58. Huynh, D. P., Figueroa, K. P., Hoang, N., and Pulst, S. M. (2000). Nuclear localization or inclusion body formation are not necessary for SCA2 pathogenesis in man or mouse. *Nat. Genet.* **2**, 44–45.
59. Nechiporuk, T., Huynh, D. P., Figueroa, K., Sahba, S., Nechiporuk, A., and Pulst, S. M. (1998). The mouse SCA2 gene: cDNA sequence, alternative splicing and protein expression. *Hum. Mol. Genet.* **8**, 1301–1309.
60. Figueroa, K. P., and Pulst, S. M. (2003). Identification and expression of the gene for human ataxin-2-related protein on chromosome 16. *Exp. Neurol.* **184**, 669–678.
61. Koyano, S., Uchihara, T., Fujigasaki, H., Nakamura, A., Yagishita, S., and Iwabuchi, K. (1999). Neuronal intranuclear inclusions in spinocerebellar ataxia type 2: Triple-labeling immunofluorescent study. *Neurosci. Lett.* **273**, 117–120.
62. Pang, J. T., Giunti, P., Chamberlain, S., An, S. F., Vitaliani, R., Scaravilli, T., Martinian, L., Wood, N. W., Scaravilli, F., and Ansorge, O. (2002). Neuronal intranuclear inclusions in SCA2: A genetic, morphological and immunohistochemical study of two cases. *Brain* **125**, 656–663.
63. Shibata, H., Huynh, D. P., and Pulst, S. M. (2000) A novel protein with RNA-binding motifs interacts with ataxin-2. *Hum. Mol. Genet.* **9**, 1303–1313.
64. Ralser, M., Albrecht, M., Nonhoff, U., Lengauer, T., Lehrach, H., and Krobitsch, S. (2005). An integrative approach to gain insights into the cellular function of human ataxin-2 *J. Mol. Biol.* **364**, 203–214.
65. Satterfield, T. F., Jackson, S. M., and Pallanck, L. J. (2002). A drosophila homolog of the polyglutamine disease gene SCA2 is a

dosage-sensitive regulator of actin filament formation. *Genetics* **162**, 1687–1702.
66. Kiehl, T. R., Shibata, H., and Pulst, S. M. (2000). The ortholog of human ataxin-2 is essential for early embryonic patterning in *C. elegans*. *J. Mol. Neurosci.* **15**, 231–241.
67. Ciosk, R., DePalma, M., and Priess, J. R. (2004). ATX-2, the *C. elegans* ortholog of ataxin 2, functions in translational regulation in the germline. *Development* **131**, 4831–4841.
68. Huynh, D. P., Yang, H. T., Vakharia, H., Nguyen, D., and Pulst, S. M. (2003). Expansion of the polyQ repeat in ataxin-2 alters its Golgi localization, disrupts the Golgi complex and causes cell death. *Hum. Mol. Genet.* **12**, 1485–1496.
69. Kiehl, T.-R., Nechiporuk, A., Figueroa, K. P., Keatings, M. T., Huynh, D. P., and Pulst, S. M. (2006). Generation and characterization of Sca 2 (ataxin-2) Knockout mice. *Biochem. Biophys. Rec. Commun.* **339**, 17–24. Epub 2005 Nov 8.

Machado–Joseph Disease/Spinocerebellar Ataxia Type 3

HENRY PAULSON

Associate Professor, Department of Neurology, University of Iowa Carver College of Medicine, Iowa City, Iowa 52242

I. Clinical Features
II. Neuropathological Features
III. Molecular Genetic Features
IV. The MJD1 Gene Product, Ataxin-3
V. Protein Misfolding as a Central Feature of Pathogenesis
VI. Further Insights from Animal Models
VII. Toward Therapy
References

This dominantly inherited ataxia was first described among immigrants from the Portuguese Azorean islands. Clinical studies in the 1970s revealed a remarkably heterogeneous, familial neurodegenerative disorder among Azoreans and Azorean immigrants [1, 2]. Affected persons typically displayed progressive ataxia, often with parkinsonian features and motor neuron involvement. The clinical features of this disease, which came to be known as Machado–Joseph disease (MJD), sometimes varied greatly even within the same family. In the early 1990s the genetic locus for MJD was narrowed to chromosome 14q, and the pathogenic mutation was soon discovered to be an unstable expansion of a CAG repeat in the MJD1 gene [3]. The pathogenic repeat encodes an expanded stretch of the amino acid glutamine in the disease protein, known as ataxin-3 or MJD1p. Thus, MJD joined the growing group of polyglutamine neurodegenerative diseases, now numbering nine [4]. At the same time the MJD1 gene defect was being identified, European scientists separately mapped what had been thought to be an unrelated ataxia, spinocerebellar ataxia type 3 (SCA3), to the same chromosomal region [5]. But once the pathogenic expansion underlying MJD was discovered it soon became clear that SCA3 was caused by the same mutation. Thus, MJD and SCA3 are the same genetic disorder. The official Human Genome Organization (HUGO) designation for this disorder is MJD, but it is also referred to as SCA3 or MJD/SCA3 in the literature.

I. CLINICAL FEATURES

In many countries MJD is the most common dominant ataxia, representing about 20% in most U.S. studies and approaching 50% or more in German, Japanese, Portuguese, and Chinese series (Table 24-1) [7–12, 114]. The relatively high prevalence of MJD in Japan reflects a higher frequency of large normal alleles in the Japanese population, providing a "reservoir" for new expansions [13]. In most countries, however, the MJD mutation is rarely detected in patients with the diagnosis of "sporadic" ataxia.

The core clinical feature in MJD is progressive ataxia due to cerebellar and brainstem dysfunction. Ataxia, however, never occurs in isolation. Numerous other clinical problems reflect progressive dysfunction in the brainstem, oculomotor system, pyramidal and extrapyramidal pathways, lower motor neurons, and peripheral nerves [12, 14–29, 114]. Most commonly, affected persons present in the young-adult to midadult years with progressive gait imbalance, often accompanied by vestibular disturbance and speech difficulties. Over time a wide range of visual and oculomotor problems may surface, including nystagmus, jerky ocular pursuits, slowing of saccades, dysconjugate eye movements, and ophthalmoplegia with lid retraction and apparent bulging eyes. In advanced stages of disease, patients are wheelchair-bound and have severe dysarthria and dysphagia. Patients may also develop facial and temporal atrophy, dystonia, spasticity, and amyotrophy. Dementia is not a typical feature, even in advanced disease, though there is evidence for subcortical dysfunction and mild cognitive impairment in MJD [30–32]). Survival after disease onset ranges from ~20 to 25 years [33].

The age of onset varies widely in MJD, reflecting differences in the size of the pathogenic repeat; disease symptom onset has been reported in persons as young as 5 and those older than 70. Likewise, the phenotype of MJD varies markedly, sometimes even within the same family [34]. Researchers have classified MJD into several types based on this variability [18, 26]. "Type I" disease begins earlier (mean age of onset, about 25 years) and is characterized by prominent spasticity and rigidity, bradykinesia, and minimal ataxia. "Type II" disease, the most common form of disease, begins in young- to midadult years (mean age of onset, ~38 years) and is characterized by progressive ataxia and upper motor neuron signs. "Type III" disease, the late-onset form of disease (mean age of onset, ~50 years), is characterized by ataxia and significant peripheral nerve involvement resulting in amyotrophy and generalized areflexia. Some researchers have even described a "type IV" form of disease characterized by parkinsonism [14, 35, 36].

More than most other ataxias, MJD can manifest with features suggesting extrapyramidal disease. As indicated above, in some cases bradykinesia, rigidity, dystonia, and tremor can closely resemble parkinsonism and may even respond to dopaminergic therapy [37, 38]. Parkinsonism appears to be more common among African Americans than Caucasians [36].

Schols and colleagues [8, 39] noted that severe spasticity and pronounced peripheral neuropathy are more frequently associated with MJD than other dominant ataxias. Particularly in late-adult-onset MJD, peripheral axonal neuropathy with areflexia can be a prominent feature, sometimes accompanied by amyotrophy and limb muscle fasciculations [6, 40]. Intriguingly, the degree of neuropathy correlates more closely with the patient's age than with the CAG repeat length. This suggests that normal age-related attenuation of peripheral nerve function is accelerated in MJD.

Sleep problems are common in MJD. Schols and colleagues [1] found that MJD patients had greater trouble falling asleep and more nocturnal awakening, with sleep impairment being more common in older patients and in those with prominent brainstem involvement. REM sleep behavior disorder is common [15, 16, 41], and central sleep apnea has been documented in some patients. Restless leg syndrome occurs in more than half of MJD patients, and can be the only manifestation of disease in the rare patients with intermediate-sized repeat expansions [42].

TABLE 24-1 Prevalence of MJD among Dominant Ataxias in Different Populations

Study	Ethnicity	% MJD
Schols (1995) [8]	German	49
Durr et al. (1996) [6]	European, North African	28
Inoue et al. (1996)	Japanese	56
Watanabe et al. (1998) [12]	Japanese	34
Moseley et al. (1998) [7]	United States (African American, German)	21
Takano et al. (1998) [13]	Caucasian	30
	Japanese	43
Pujana et al. (1999)	Spanish	15
Nagaoka et al. (1999)	Japanese	20
Saleem et al. (2000) [134]	Indian	5
Soong et al. (2001) [10]	Chinese	47
Zhou et al. (2001) [60]	Chinese	35
Silveira et al. (2002) [9]	Portuguese/Brazilian	63

Not surprisingly, the phenotypic heterogeneity of MJD is mirrored by various abnormalities noted on brain imaging [43, 44]. The most common feature on MRI of the brain is pontocerebellar atrophy with a dilated fourth ventricle, reflecting loss of tissue in the superior and middle cerebellar peduncles, the pons, and the cerebellar vermis. Other MRI abnormalities can include atrophy of the globus pallidus and midbrain, with sparing of the cerebral cortex. Consistent with the frequent extrapyramidal features of MJD, MRI abnormalities of the basal ganglia are detected more often in MJD than in most other SCAs [45]. The severity of imaging abnormalities correlates with CAG repeat length: patients with longer repeats and thus earlier-onset disease are more likely to show progressive MRI abnormalities at an earlier age. But the severity of MRI findings also correlates with the patient's age independent of the age of onset and repeat length. For instance, elderly patients with late-onset MJD can show surprisingly pronounced atrophy on MRI [44]. Single-photon-emission computed tomography (SPECT) imaging studies also have documented diffuse central nervous system (CNS) abnormalities, including reduced dopamine transporter density in the striatum [46].

II. NEUROPATHOLOGICAL FEATURES

Befitting the clinical heterogeneity of MJD, signs of neurodegeneration are widespread and variable [17, 19–23, 26, 47, 48]. Typically, neurodegeneration involves deeper structures of the basal ganglia, numerous brainstem nuclei, and the cerebellum. Neuropathological findings are not limited to these regions, however. Atrophy and neuronal loss in the following regions have been described: globus pallidus, various thalamic nuclei, subthalamic nucleus, substantia nigra, red nucleus, medial longitudinal fasciculus, various pontine nuclei and cranial motor nerve nuclei, superior and middle cerebellar peduncles, cerebellar dentate nucleus, Clarke's column and spinocerebellar tracts, vestibular nucleus, anterior horn cells, posterior columns, and posterior root ganglia. The cerebral cortex, olivary nuclei, and corticospinal tracts are often relatively spared in MJD, and, in a subset of patients, the cerebellar cortex can be relatively spared.

As in other polyglutamine disease proteins, the disease protein in MJD accumulates within proteinacous inclusions in various brain regions. These inclusions are often found inside the cell nuclei of specific populations of neurons [47, 49–52]. Neuronal intranuclear inclusions are spherical, ubiquitinated structures that also contain many other proteins. In MJD, nuclear inclusions (NIs) are abundant in pontine neurons but have also been observed in other brainstem neuronal populations, thalamus, substantia nigra, and, rarely, striatum. Extranuclear accumulations also occur in MJD brain [47, 53]. Indeed, ubiquitin-positive deposits in the cytoplasm of motor neurons were described in MJD before the discovery of NIs. This feature of disease, which has been less well characterized than NIs, warrants further investigation given the frequency of motor neuron involvement in MJD.

III. MOLECULAR GENETIC FEATURES

Kawaguchi et al. [3] discovered the mutation in MJD to be an unstable CAG repeat expansion in the coding region of the MJD1 gene. This 11-exon gene encodes the protein ataxin-3 or MJDp [39, 54]. The CAG repeat resides in the 10th exon, in-frame to encode a polyglutamine tract near the carboxyl terminus of this ~42-kDa protein.

Since its discovery, the MJD mutation has been identified worldwide as a cause of hereditary ataxia in diverse ethnic groups [3, 8–11, 55–60]. In many regions of the world, MJD is the most common autosomal dominant ataxia. In the United States, MJD, SCA2, and SCA6 are the three most common dominant ataxias.

MJD differs from other CAG/polyglutamine diseases in that the others have contiguous or narrowly separated ranges for normal and expanded alleles. In contrast, MJD shows a comparatively wide gap between normal repeat lengths and fully penetrant expanded repeat lengths, though recently discovered intermediate alleles have begun to narrow this gap (~51–59 repeats). Normal alleles range from 12 to ~43 repeats, whereas expanded alleles are nearly always 60 repeats or longer, ranging up to ~87 in length [14, 18, 24, 55–57, 114]. More than 90% of normal alleles have fewer than 31 repeats, with the most common repeat lengths being 14, 21–24, and 27 [61–63]. More recently, rare intermediate expansions with repeats of ~51–59 [42, 63, 64] and possibly as low as 45 [65], have been reported to manifest as a milder form of disease or, in some cases, simply as restless leg syndrome with or without peripheral neuropathy.

Sporadic MJD due to *de novo* expansions is quite rare, much less common than in several other polyglutamine diseases including SCA2, SCA6, and Huntington's disease (HD). The rather large jump in repeat size that a normal allele would need to undergo to expand into the disease range reduces the likelihood of such *de novo* expansions. The low frequency of intermediate-sized alleles contributes to the rarity of sporadic MJD.

As in all CAG/polyglutamine diseases, the size of the expansion correlates inversely with age of symptom onset: a correlation coefficient of between roughly 0.7 and 0.9 has been observed in many patient series. On average, larger repeats cause earlier disease onset and may be associated with faster disease progression [66]. The size of the expanded repeat length also determines, in part, the clinical phenotypes mentioned above (Table 24-2). Patients with type I disease, the early-onset dystonic form of disease, typically have repeats larger than those of patients with type II and III phenotypes. In one study, Sasaki et al. [24] found that patients with type I, II, and III phenotypes had mean repeat sizes of 80, 76, and 73, respectively. The largest CAG repeats are often associated with pyramidal signs and dystonia [67], whereas the smallest repeat size (<73) is associated with peripheral neuropathy [25, 114]. Many clinical signs, including amyotrophy, ophthalmoplegia, and dysphagia, correlate better with disease duration than with CAG repeat size [114]. Indeed, the peripheral nervous system findings of MJD seem to have a strong, age-dependent component.

Anticipation, the tendency for disease to worsen from generation to generation, occurs in MJD as in other polyQ diseases. Durr and colleagues [6a] and Takiyama and colleagues [63] each noted a mean anticipation of approximately 10 years for age of onset from one generation to the next. The molecular basis for anticipation in MJD is clear: the tendency for repeat sizes to expand from one generation to the next, coupled with the fact that longer repeats cause earlier-onset disease, leads to earlier manifestation of disease symptoms in successive generations. More than half of parent–child transmissions of the MJD repeat are unstable, with ~75% further expanding in the next generation [55, 63]. Anticipation seems to be slightly more prominent with paternal than with maternal inheritance, though not nearly as profound as it is with Huntington's disease. Takiyama and colleagues [68] also reported anticipation with maternal transmission despite little change in repeat size, suggesting that other unknown genetic or environmental factors may also contribute to anticipation.

Although MJD is a dominant disorder, there may be a gene dosage effect. In a Yemenese family, several individuals homozygous for MJD expansions were found to have earlier onset and more rapid progression than heterozygous individuals [69]. The expansions in this family were relatively small (66–72), and some heterozygotes remained asymptomatic well into their seventh decade. This limited anecdotal evidence suggests a dose dependency to the toxicity of the expansion. Consistent with a gene dosage effect, homozygous mice are more severely impaired than heterozygous mice in two recently developed mouse models [70, 71], and the severity of the phenotype in a *Drosophila* model of MJD also correlated with the level of transgene expression [72].

A single-nucleotide polymorphism (SNP) exists immediately after the CAG repeat. This SNP is usually guanine in normal alleles, whereas cytosine is present in all expanded alleles. Whether this nucleotide difference contributes in any way to pathogenesis is unknown, but it has been suggested to play a role in repeat instability [73, 74]. This SNP was recently exploited in RNA interference (RNAi) studies that selectively suppressed expression of the mutant allele while sparing expression of the normal allele [75].

The normal repeat is stably transmitted and shows no somatic mosaicism in repeat size. In contrast, expanded MJD1 repeats display a modest degree of somatic mosaicism, most notably in the cerebellar cortex [76]. In the cerebellum this leads to smaller, not larger, expanded repeats. Thus somatic mosaicism cannot explain the preferential involvement of cerebellum in disease.

IV. THE MJD1 GENE PRODUCT, ATAXIN-3

The MJD1 gene encodes a 42-kDa protein, ataxin-3 or MJDp (Fig. 24-1). Ataxin-3 is widely expressed in the CNS and elsewhere in the body [39, 51, 77–81]. Indeed the author has found ataxin-3 to be expressed in every cell line or tissue examined in his laboratory [79; unpublished observations]. The disconnect between widespread expression of ataxin-3 and selective neuronal degeneration in MJD is reminiscent of other polyglutamine neurodegenerative diseases, in which selective tissue vulnerability cannot be explained by selective protein expression.

TABLE 24-2 MJD Subtypes[a]

Type	Age of onset	Repeat length	Major features
I	<25	>75	Dystonia, rigidity, ataxia
II	~20–50	>73	Ataxia, pyramidal and bulbar signs
III	>40	<72	Ataxia, peripheral signs

[a] CAG repeat lengths represent the approximate transition points at which the clinical subtype is increasingly likely to occur.

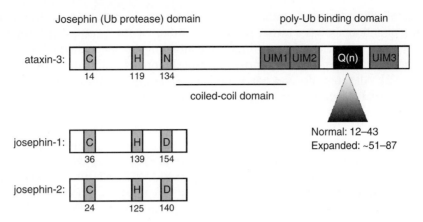

FIGURE 24-1 Ataxin-3 and related Josephin domain-containing proteins. Ataxin-3 is a relatively small protein with a glutamine repeat near the carboxyl terminus. The approximate ranges for normal and disease repeats are shown. The amino-terminal conserved "josephin" domain, a predicted coiled-coil domain, and several ubiquitin interacting motifs (UIMs) are also shown. Josephin-1 and josephin-2 represent two smaller proteins containing similar josephin domains without flanking motifs.

Ataxin-3 exists in at least two major splice forms that differ only in their carboxyl termini. Both contain the polyglutamine domain encoded by exon 10. The originally published isoform [3] is generated by a translational read-through from exon 10 into the subsequent intron, whereas the second isoform represents the product of a transcript correctly spliced from exons 10 to 11. Both splice variants are believed to be expressed in disease brain [51], but whether one is more toxic to neurons than the other is not known.

Ataxin-3 is a small, soluble protein that can shuttle in and out of the nucleus. The amount of ataxin-3 in the cytoplasm versus nucleus can vary from cell type to cell type and may also depend on the particular splice variant expressed [82]. In many cells a fraction of the cellular pool of ataxin-3 is intranuclear, bound to the nuclear matrix [80, 83]. In unaffected brain and in normal neurons, ataxin-3 appears to be largely cytoplasmic, but in MJD brain the protein localizes primarily to the nucleus of neurons, both in humans and in mouse models [50, 51, 70, 71, 78, 79]. Interestingly, in cell models nuclear ataxin-3 is more easily recognized by the polyglutamine-specific antibody 1C2 than is cytoplasmic ataxin-3, suggesting that the tertiary or quaternary conformation of the protein differs depending on its subcellular location [83]. Why ataxin-3 preferentially localizes to neuronal cell nuclei in disease is uncertain. This same phenomenon, however, has been described in other polyQ diseases including HD.

The rather large and polymorphic polyQ tract present in normal human ataxin-3 is unique to human ataxin-3. Other vertebrate ataxin-3 species have a much smaller polyglutamine domain. For example, a stretch of only four glutamine residues interrupted by a histidine residue is found in rodents. Clearly, a substantial glutamine repeat is not essential for the core functions of ataxin-3, although the longer normal repeat in humans may modulate details of ataxin-3 function.

Close orthologs of ataxin-3 exist in other mammals and fish, and more distant orthologs are present in plants, flies, and worms. A protein motif common to all ataxin-3 orthologs is the evolutionarily conserved, N-terminal "josephin" domain [84]. Two other human genes encode proteins with josephin domains, josephin-1 and josephin-2 (Swiss-Prot Accession Nos. Q15040 and Q8TAC2, respectively). These two proteins are ~50% identical to one another and ~25% identical to the josephin domain of ataxin-3. They lack the other motifs of ataxin-3: in addition to its josephin and polyQ domains, ataxin-3 has a predicted a coiled-coil domain, two closely spaced ubiquitin-interacting motifs (UIMs) upstream of the polyglutamine domain, and, in one of two major splice variants, a third UIM downstream of the polyglutamine tract.

What is the function of ataxin-3? Protein informatics revealed that the josephin domain contains the predicted catalytic amino acid triad found in ubiquitin-specific cysteine proteases [84]. Recent studies confirmed that ataxin-3 can cleave poly-ubiquitin chains from test substrates and from free poly-ubiquitin chains *in vitro* [85–88]. When the predicted active site cysteine residue in ataxin-3 is mutated, this de-ubiquitinating activity is lost. And when overexpressed in cells, catalytically inactive ataxin-3 causes a buildup of poly-ubiquitinated proteins [89]. The recent solution structure for the josephin domain of ataxin-3 reveals a papain-like fold to the protease domain [87, 88]. Structural studies further suggest that the josephin domain has a rather rigid globular structure, whereas the carboxy-terminal, polyglutamine-containing half of the protein comprises a more flexible tail [90–92]. Although no one has yet identified specific physiological substrates for ataxin-3, all available evidence indicates that ataxin-3 is indeed a de-ubiquitinating enzyme (DUB).

The precise physiological roles of this rather unusual DUB remain unclear. A clue to its potential roles comes from the fact that ataxin-3 has multiple UIMs, either two or three depending on the splice variant. UIMs are protein motifs that mediate ubiquitin binding to various proteins implicated in traditional (i.e., proteasomal) and nontraditional roles of ubiquitin pathways [93]. The first two UIMs mediate ataxin-3 binding to poly-ubiquitin chains [85, 94–96]. If these UIMs are mutated, ubiquitin binding by ataxin-3 is lost. In the test tube, ataxin-3 recognizes ubiquitin chains that are linked through either lysine 48 or lysine 63 (K48- and K63-linked chains, respectively). Ataxin-3 preferentially recognizes chains of three or more ubiquitin molecules, with a binding affinity for K48-linked tetra-ubiquitin in the submicromolar range [94]. When the proteasome is pharmacologically inhibited, poly-ubiquitinated proteins that accumulate in cells bind to, and coprecipitate with, ataxin-3 [89, 94, 95]. Which poly-ubiquitin chain type, K48-linked or K63-linked, is the preferred ubiquitin substrate *in vivo* remains unknown. Of the dozens of DUBs expressed in humans, ataxin-3 is unique in having several UIMs flanking a polyglutamine tract. Comparative analysis of the two related human josephin proteins lacking these additional motifs, josephin-1 and josephin-2, may provide additional clues to ataxin-3 function.

An attractive model for ataxin-3 action is that its UIMs regulate, or restrict, the ubiquitinated species on which it can act. Most DUBs do not have UIMs. Because ataxin-3 does have UIMs that bind poly-ubiquitin chains, it seems likely that heavily ubiquitinated substrates are favored for recognition by ataxin-3. Distal ubiquitin residues in these chains might then be presented to the protease site for cleavage. In this manner, ataxin-3 could serve to trim heavily ubiquitinated species before their presentation to the proteolytic core of the proteasome. Substrate delivery to the proteasome typically requires that the substrate be conjugated to a K48-linked poly-ubiquitin chain at least four ubiquitin molecules long. Excessively long or branched chains, however, might even impede this delivery. It is appealing to speculate that ataxin-3 acts to trim branched or excessively long ubiquitin chains on substrates destined for proteasomal degradation. Through this action, ataxin-3 would increase the efficiency of substrate delivery to the proteolytic chamber of the proteasome. Interestingly, as a UIM-containing protein, ataxin-3 itself can undergo post-translational modification by the addition of one or more ubiquitin molecules [89]. Ubiquitination of ataxin-3 could serve to regulate its own DUB activity.

Mounting evidence suggests that ataxin-3 does, in fact, function in the ubiquitin–proteasome degradation pathway. Ataxin-3 cosediments with proteasomes in yeast and interacts with two other ubiquitin-binding proteins, VCP/p97 and HHR23 [96], both of which have already been implicated in substrate delivery to the proteasome. Presentation of ubiquitinated substrates to the 26S proteasome is a complex process mediated by a range of cellular factors, one of which may prove to be ataxin-3. In addition, ataxin-3 regulates aggresome formation through its interactions with dynein [86]. Because aggresomes are a major pathway through which cells process and eliminate abnormal proteins, ataxin-3 is strategically placed to influence cellular protein surveillance.

Regardless of whether ataxin-3 directly acts in proteasomal function, its ubiquitin-associated activities link ataxin-3 to protein quality control pathways already implicated in polyglutamine disease pathogenesis. Its involvement in ubiquitin-dependent processes leads this author to speculate that the primary pathway through which mutant ataxin-3 causes neuronal dysfunction and degeneration is perturbation of protein surveillance pathways. With the wealth of information that has recently surfaced on ataxin-3 structure and function, it should not take researchers long to determine how polyglutamine expansion alters this disease protein's activities in cellular protein homeostasis.

V. PROTEIN MISFOLDING AS A CENTRAL FEATURE OF PATHOGENESIS

A common link among polyQ diseases is the presence of neuronal inclusions containing the mutant protein. These inclusions are observed in select brain regions, differing among the various polyQ diseases [97–99]. Their discovery in MJD and other polyQ diseases suggested that protein misfolding is central to pathogenesis. Consistent with this view, recombinant mutant polyQ proteins adopt amyloid-like conformations and form aggregates *in vitro* in the absence of other proteins [100–102]. Importantly, the repeat threshold required for aggregation *in vitro* largely mirrors the threshold length for human disease. Furthermore, recombinant mutant ataxin-3 adopts increased beta-sheet secondary structure and forms aggregates *in vitro*, supporting an altered conformation for the disease protein [103]. These and other lines of evidence suggest that conformational abnormalities of mutant ataxin-3 underlie its pathogenic nature.

Though inclusions are found in various polyQ disease brains, their distribution correlates imperfectly with regions of degeneration, both in affected humans and in transgenic mouse models. This is the case in MJD as well, where degeneration in many deep basal ganglionic, thalamic, brainstem, and spinal cord regions is not always accompanied by significant inclusions.

To directly address aggregate toxicity, Wetzel and colleagues delivered preformed, fibrillar aggregates to cultured cells [104]. Aggregates generated from expanded or normal repeats were toxic to cells, provided that the aggregates were transported into the nucleus. In contrast, monomeric forms of normal and expanded polyQ had no effect. Although this result suggests that aggregates are directly toxic, pure polyQ complexes of this sort may differ fundamentally from the macroscopic inclusions found in neurons. The latter are heterogeneous collections of diverse proteins, the accumulation of which is almost certainly regulated by cellular processes. Recently, Finkbeiner and colleagues assessed the toxicity of these cell-derived inclusions by following transfected neurons over time to determine the temporal and causal relationship of inclusion formation to cell death [105]. Neurons containing visible inclusions were *less* likely to die than neurons in which the mutant protein remained diffuse. Based on this result, neuronal inclusions would seem to play a protective role.

The seemingly contradictory findings of these two studies are compatible when one realizes that the biochemical process of aggregation and the formation of mature inclusions are not one and the same. Results in a *Drosophila* model of MJD shed light on the issue. In transgenic flies, overexpression of specific Hsp chaperones markedly suppresses polyglutamine toxicity without noticeably changing inclusions [106]. This suppression, however, is accompanied by a marked increase in the solubility of mutant ataxin-3 [107]. Importantly, this result dissociates inclusions from toxicity, yet also implies that misfolding and aggregation of disease protein are central to pathogenesis. Genetic screens in *Caenorhabditis elegans*, yeast, and flies have confirmed a role for chaperones in buffering the toxicity of polyQ proteins [108, 109]. It is fair to say that Hsp chaperones are the class of proteins most consistently identified as modulators of polyglutamine toxicity and aggregation in various model systems (reviewed in [97, 98, 108]).

Together, Hsp chaperones and the ubiquitin–proteasome pathway (UPP) constitute a key cellular defense against mutant polyglutamine disease proteins. Cells must ensure that proteins damaged by physiological stress or mutations (expanded polyglutamine, for example) are dealt with efficiently. If a protein's native conformation cannot be achieved, refolding or disaggregating efforts by Hsp chaperones ensue or the protein is targeted for degradation. For many misfolded proteins, the principal route for protein destruction is the UPP. If the concentration of misfolded proteins exceeds cellular folding and degradative capacity, these proteins can form insoluble aggregates, which may, in turn, become sequestered in inclusions.

Because inclusions in MJD and other polyQ disorders are ubiquitinated and sequester proteasome components [50, 110, 111], a failure of the UPP has been suggested to underlie disease. Proteasomes cannot fully digest peptides containing expanded polyQ repeats *in vitro* [112], overexpression of pathogenic polyQ proteins impairs the UPP in cell-based reporter assays [113], and ataxin-3 is more prone to aggregate when the proteasome is pharmacologically inhibited in transfected cells [110]. These results suggest that attempts by the cell to digest pathogenic polyQ tracts might cause the release of aggregation-prone fragments, or that polyQ proteins themselves might stall the proteasome, rendering it unavailable for normal quality control functions. Despite these *in vitro* and cell-based data, however, there is no compelling *in vivo* evidence that mutant polyQ proteins inhibit the proteasome in affected tissues [e.g., see 114]. Moreover, in one study employing transfected cells, mutant ataxin-3 proved to be as easily degraded as expanded ataxin-3 [89]. Still, some studies have documented decreased proteasome function in the aging brain, suggesting that the age-dependent nature of MJD may reflect age-related impairment of neuronal protein quality control (QC).

In summary, although the hallmark inclusions of MJD are probably not directly toxic, they are clearly not a normal feature of neurons. At a minimum, inclusions represent biological markers of a failure in aberrant protein clearance, and may represent a successful cellular response to the problem of polyQ protein aggregation. Current evidence further suggests that the intellectually appealing hypothesis that mutant polyQ proteins choke off, or overwhelm, neuronal QC will not explain pathogenesis in MJD or any polyQ disease. Instead, more subtle impairment of QC may be one of several insults the neuron faces when chronically exposed to expanded polyQ. Regardless, the intriguing connection of ataxin-3 to ubiquitin pathways through its DUB and ubiquitin binding activities means that the theory that proteasomal function is perturbed in MJD will need to be looked at with a fresh perspective, taking advantage of newly developed transgenic mouse models of disease [70, 71].

VI. FURTHER INSIGHTS FROM ANIMAL MODELS

The first animal model of MJD was produced by Ikeda et al. [115], who expressed ataxin-3 in mice using a Purkinje cell-specific promoter. Whereas mice expressing full-length mutant ataxin-3 were normal, mice expressing a truncated fragment developed progressive ataxia with massive Purkinje cell degeneration. This constituted the first compelling evidence that protein context critically influences the toxicity of expanded polyglutamine. This important point has since been corroborated by many studies in other polyQ diseases [116]. The results obtained in this model (which is no longer available) also suggested that physiological or stress-induced cleavage of ataxin-3 in the brain might unleash a particularly toxic species that initiates or accelerates disease.

Recent results in a *Drosophila* model expressing full-length ataxin-3 [117] shed light on how the protein context of ataxin-3 may influence pathogenesis. Several years ago, Bonini and colleagues used a truncated fragment of mutant ataxin-3 to model polyglutamine neurodegeneration in the fruit fly [72]. This model recapitulated many features of polyglutamine diseases, including inclusion formation and neurodegeneration. The ataxin-3 fragment proved to be highly toxic. In contrast, full-length expanded ataxin-3 with an even longer repeat causes a much milder, selectively neurotoxic phenotype. Intriguingly, *normal* ataxin-3, which causes no problems when overexpressed in flies, actually suppresses the toxicity of other polyglutamine disease proteins including expanded ataxin-3 fragment. This suppression requires the ubiquitin-linked functions of ataxin-3; if the UIMs or the active site cysteine is mutated, ataxin-3 loses much or all of this activity. Thus, mutant ataxin-3 seems to possess counterposing activities: both a toxic property of expanded polyglutamine and an intrinsic suppressor activity to mitigate this toxicity. Consistent with this, full-length expanded ataxin-3 becomes severely toxic when its active site cysteine residue is mutated. This finding is interesting in light of the fact that the threshold repeat length to cause MJD in humans is larger than that for other polyQ diseases. Perhaps, in humans, mutant ataxin-3 counteracts its own toxicity through its ubiquitin-related functions as a DUB. Indeed, expanded ataxin-3 binds poly-ubiquitin chains as well as normal ataxin-3 *in vitro* [94] and, in flies, seems to retain some residual suppressor function.

The study of MJD has been aided by the recent development of two transgenic mouse models [70, 71]. Both models express full-length ataxin-3. Cemal and colleagues [71] generated mice expressing normal (Q15) or expanded (Q64 to Q84) alleles of the human *MJD1* gene contained within a yeast artificial chromosome (YAC). These mice express ataxin-3 widely in the brain and body, presumably in the normal distribution of the *MJD1* gene. Mice expressing expanded ataxin-3 (Q84) develop a mild cerebellar phenotype with neuronal cell loss, gliosis, and abundant inclusion formation. Neurodegenerative changes are most notable in the pons, deep cerebellar nuclei, and cerebellar cortex, though older mice also show peripheral nerve demyelination and axonal loss. The phenotype shows a clear dose dependency, as homozygous YAC mice are more severely affected than hemizygous mice. No proteolytic fragment of ataxin-3 was noted by the authors, but several higher-molecular-weight ataxin-3 bands detected on gels could represent insoluble oligomers or post-translationally modified ataxin-3. Because this model expresses the full *MJD1* gene in its normal genomic context, it is well suited for many important studies: (1) the molecular basis of regional neuronal selectivity; (2) the mechanism underlying repeat instability; (3) the role of splice variants in normal and disease brain; and (4) the potential contribution to pathogenesis of a CAG repeat-induced, translational frameshift in the MJD1 protein, recently described by Rouleau and colleagues [118]. It also represents a valid animal model in which to explore the potential therapeutic utility of RNAi suppression for MJD. To date, this model has seen limited use by other researchers, though one group reported increased neuronal levels of the Hsp chaperone Hsp27 in affected brain regions of Q84 YAC transgenic mice [119].

Colomer and colleagues recently created a MJD transgenic mouse model that expresses full-length ataxin-3 cDNA behind the prion promoter [70]. The ataxin-3 cDNA corresponds to the originally published splice variant, MJD1a [3]. Mice expressing normal ataxin-3 (Q20) are entirely normal. In contrast, mice expressing mutant ataxin-3 (Q71) above a critical concentration develop a progressive motor phenotype with ataxia, weight loss, and early death. Neuronal inclusions are observed and neurons of the substantia nigra, which are a known target in MJD, are reduced in number. As with the YAC mouse, the dose dependency in this mouse is striking: whereas hemizygous Q71 mice have very little phenotype, homozygous Q71 mice have a severe, early-onset phenotype. The severity of the phenotype in homozygous mice may make this model suitable for clinical trials of potential preventive compounds. Because the same prion promoter has been used to drive expression of at least two other polyglutamine disease genes, yielding somewhat different phenotypes in these other models, the MJD Q71 mice may also prove useful in exploring the protein context of polyglutamine toxicity. Finally, it will be a valuable model in which to explore the role of neuronal dysfunction, as the severe behavioral phenotype in this model appears to be out of proportion to the apparently minimal degree of brain atrophy.

Research into disease mechanisms in MJD would benefit from the creation of still more animal models. To date, for example, no one has reported either a MJD1 knockout or knock-in mouse. Both would help to answer fundamental questions about ataxin-3 and MJD. Nevertheless, the existence now of two complementary transgenic mouse models of MJD should enhance investigations into potential pathogenic mechanisms. Four postulated mechanisms in MJD and other polyglutamine diseases are the production of a toxic proteolytic fragment, transcriptional dysregulation, perturbation of axonal transport, and mitochondrial impairment. Here I briefly discuss each of these hypotheses in light of the existing animal models.

An intriguing feature of the MJD Q71 mice is the accumulation in brain of a putative, C-terminal ataxin-3 fragment. A similar-sized, ataxin-3 species was detected by Goti and colleagues [70] in human MJD brains and proved to be toxic when transiently expressed in neuroblastoma cells. The protease responsible for this apparent cleavage event has not been identified, but Berke and colleagues [120] independently reported that a similar ataxin-3 fragment was generated by caspases in cells undergoing apoptosis. In contrast to Goti et al. [70], however, this second group [120] did not detect a similar fragment in MJD brain. As shown in MJD models in cells, mice, and flies [50, 72, 115, 121], truncation of ataxin-3 can accelerate aggregation and cytotoxicity. Still, full-length ataxin-3 can form inclusions in the absence of any detectable cleavage, though less efficiently than truncated ataxin-3 [e.g., 122]. Proving that such a proteolytic event is central to pathogenesis will be difficult, but clearly worth pursuing if compounds can be found to block proteolysis. If generation of a toxic fragment is a key event in pathogenesis, one would expect to find a similar fragment in the YAC Q84 mouse model. To the author's knowledge, this has not been investigated.

Several cell-based and disease tissue-based studies are consistent with the hypothesis that expanded polyQ proteins trigger disease by perturbing gene transcription through aberrant protein–protein interactions. In MJD brain, the basal transcription factor TATA-binding protein (which is also the SCA17 disease protein) localizes to some nuclear inclusions [123].

Ataxin-3 also can repress CREB-binding protein (CBP)-dependent transcription in transfected cells [124], and mutant ataxin-3 can recruit essentially the entire cellular pool of CBP into inclusions [125]. Evert and colleagues have shown that mutant ataxin-3 upregulates cytokine genes in a neural cell line and that normal and mutant ataxin-3 have divergent effects on transcriptional profiles in cells [126]. It is important to caution that most of the aforementioned work was performed in cell models that may not accurately reflect what happens in a chronically diseased neuron in the brain. It is hoped that gene expression analyses and directed studies of specific transcription-associated factors similar to those performed in other polyglutamine disease models [e.g., 127] will now be performed in the recently published mouse models of MJD. Because nontraditional roles for ubiquitin include transcription factor activation [97], the function of ataxin-3 in ubiquitin pathways could link directly to transcriptional dysregulation.

The frequent occurrence of motor neuron loss and peripheral neuropathy in MJD suggests that axonal dysfunction plays a significant role in disease. Histopathological analyses of some polyQ disease brains show widespread neuritic inclusions, raising the possibility that perturbation of axonal transport contributes to pathogenesis [128]. Aggregated disease protein may physically block transport and titrate motor proteins away from their normal functions. Expression of a mutant ataxin-3 fragment in *Drosophila*, for example, was found to cause axonal blockages [129]. In addition, ataxin-3 interacts with the cytosolic histone deacetylase 6 and dynein during transport of misfolded proteins in cultured cells, thus linking ataxin-3 to cytoskeletal transport processes [86]. Despite these promising leads, the role of axonal pathology has not been studied rigorously in MJD brain. Clearly this aspect of disease warrants further study, both in human MJD tissue and in transgenic mouse models.

Finally, impairment of mitochondria could trigger neuronal dysfunction and death in MJD and other polyQ diseases. In neurons, mitochondria play a central, integrating role in energy production, neuronal health, and apoptotic cell death pathways. Although most evidence implicating mitochondrial involvement in polyQ toxicity has been derived from other diseases, a few studies suggest mutant ataxin-3 has effects on mitochondria. For example, Tsai et al. [130] reported decreased Bcl-2 and increased cytochrome C levels in cells expressing mutant ataxin-3; moreover, polyglutamine-induced death in these cells was attenuated by exogenously expressed Bcl-2. One way expanded polyQ proteins could disrupt normal mitochondrial function is through the formation of aberrant ion channels [131]. For this to happen, however, the polyQ domain would likely need to become separated from most of its surrounding polypeptide through a proteolytic event. Although ataxin-3 can be cleaved in apoptotic paradigms [120] and may undergo proteolysis in Q71 transgenic mice [70], there is no evidence as yet that ataxin-3 fragments can form such channels. The availability of transgenic mouse models with significant neuronal dysfunction and selective degeneration will now allow scientists to determine whether mutant ataxin-3 directly binds mitochondria and/or alters their activity *in vivo*.

VII. TOWARD THERAPY

MJD is a relentlessly progressive disease for which there is currently no cure. It is important, however, to remember that numerous symptoms occurring in MJD can respond to symptomatic therapy. For example, some patients have parkinsonian signs that may improve with dopaminergic agents [37, 38]. The restless leg syndrome experienced by many MJD patients can also respond to dopaminergic agents. Aberrant sleep patterns and daytime sleepiness, a common complaint in MJD, can benefit from modafanil or other sympathomimetic agents or, if it reflects REM behavior disorder, clonazepam. Some patients may find their level of alertness benefited by amantadine, though the pharmacological basis of this response is unclear. Unfortunately, none of these medications has been studied in a double-bind, placebo-controlled clinical trial. This is due, in part, to the difficulty of performing multicenter trials for a disease as rare and as clinically heterogeneous as MJD,

A major goal of current research in MJD is to understand disease mechanisms in order that the disease might be slowed or prevented altogether. Currently there is no proven preventive medicine for MJD or any other polyglutamine disease. It is hoped that some elements of pathogenesis will be shared across this class of diseases. If so, ongoing studies in HD may identify useful compounds to slow disease. Clinicians caring for patients with MJD will need to keep a close eye on current HD trials testing compounds that enhance bioenergetics or inhibit histone deacetyltransferases.

One potentially powerful route to therapy is to turn off the disease gene. Because the mutant *MJD1* allele acts through a dominant toxic mechanism, suppressing its expression should be beneficial regardless of the precise cellular route(s) by which ataxin-3 is toxic to neurons. RNA interference (RNAi) has recently been used successfully to silence ataxin-3 in cell models [75] and other polyglutamine disease genes in mouse models [132, 133]. The exquisite sequence specificity of RNAi has even permitted silencing of the mutant alleles while sparing the normal allele [75]. With the recent development of transgenic mouse models of MJD that recapitulate important disease features, RNAi therapy for MJD now can be tested *in vivo*.

References

1. Schols, L., Haan, J., Riess, O., Amoiridis, G., and Przuntek, H. (1998). Sleep disturbance in spinocerebellar ataxias: Is the SCA3 mutation a cause of restless legs syndrome? *Neurology* **51**, 1603–1607.
2. Sudarsky, L., and Coutinho, P. (1995). Machado–Joseph disease. *Clin. Neurosci.* **3**, 17–22.
3. Kawaguchi Y., Okamoto T., Taniwaki M., Aizawa, M., Inoue, M., katayama, S., Kawakami, H., Nakamura, S., Nishimura, M., and Akiguchi, I. (1994). CAG expansions in a novel gene for Machado–Joseph disease at chromosome 14q32.1. *Nat. Genet.* **8**, 221–228.
4. Zoghbi, H. Y., and Orr, H. T. (2000). Glutamine repeats and neurodegeneration. *Annu. Rev. Neurosci.* **23**, 217–247.
5. Stevanin, G., Cancel, G., Durr, A., Chneiweiss, H., Dubourg, O., Weissenbach, J., Cann, H. M., Agid, Y., and Brice, A. (1995). The gene for spinal cerebellar ataxia 3 (SCA3) is located in a region of 3 cM on chromosome 14q24.3-q32.2. *Am. J. Hum. Genet.* **56**, 193–201.
6. Durr A., Stevanin G., Cancel G., Duyckaerts C., Abbas N., Didierjean O., Chneiweiss H., Benomar A., Lyon-Caen O., Julien J., Serdaru M., Penet C., Agid Y., and Brice A. (1996). Spinocerebellar ataxia 3 and Machado–Joseph disease: Clinical, molecular and neuropathological features. *Ann. Neurol.* **39**, 490–499.
7. Moseley, M. L., Benzow, K. A., Schut, L. J., Bird, T. D., Gomez, C. M., Barkhaus, P. E., Blindauer, K. A., Labuda, M., Pandolfo, M., Koob, M. D., and Ranum, L. P. (1998). Incidence of dominant spinocerebellar and Friedreich triplet repeats among 361 ataxia families. *Neurology* **51**, 1666–1671.
8. Schols, L. (1995). Machado–Joseph disease mutation as the genetic basis of most spinocerebellar ataxias in Germany. *J. Neurol. Neurosurg. Psychiatry* **59**, 49–50.
9. Silveira, I., Miranda, C., Guimaraes, L., Moreira, M. C., Alonso, I., Mendonca, P., Ferro, A., Pinto-Basto, J., Coelho, J., Ferreirinha, F., Poirier, J., Parreira, E., Vale, J., Januario, C., Barbot, C., Tuna, A., Barros, J., Koide, R., Tsuji, S., Holmes, S. E., Margolis, R. L., Jardim, L., Pandolfo, M., Coutinho, P., and Sequeiros, J. (2002). Trinucleotide repeats in 202 families with ataxia: A small expanded (CAG)$_n$ allele at the SCA17 locus. *Arch. Neurol.* **59**, 623–629.
10. Soong, B., Lu, Y., Choo, K., and Lee H. (2001). Frequency analysis of autosomal dominant cerebellar ataxias in Taiwanese patients and clinical and molecular characterization of spinocerebellar ataxia type 6. *Arch. Neurol.* **58**, 1105–1109.
11. Tsai, H. F., Liu, C. S., Leu, T. M., Wen, F. C., Lin, S. J., Liu, C. C., Yang, D. K., Li, C., and Hsieh, M. (2004). Analysis of trinucleotide repeats in different SCA loci in spinocerebellar ataxia patients and in normal population of Taiwan. *Acta. Neurol. Scand.* **109**, 355–360.
12. Watanabe, H., Tanaka, F., Matsumoto, M., Doyu, M., Ando, T., Mitsuma, T., and Sobue, G. (1998). Frequency analysis of autosomal dominant cerebellar ataxias in Japanese patients and clinical characterization of spinocerebellar ataxia 6. *Clin. Genet.* **53**, 13–19.
13. Takano, H., Cancel, G., Ikeuchi, T., Lorenzetti, D., Mawad, R., Stevanin, G., Didierjean, O., Durr, A., Oyake, M., Shimohata, T., Sasaki, R., Koide, R., Igarashi, S., Hayashi, S., Takiyama, Y.,

Nishizawa, M., Tanaka, H., Zoghbi, H. Y., Brice, A., and Tsuji, S. (1998). Close association between prevalence of dominantly inherited spinocerebellar ataxias with CAG-repeat expansions and frequencies of large normal CAG alleles in Japanaese and Caucasian populations. *Am. J. Hum. Genet.* **63**, 1060–1066.

14. Cancel G., Abbas N., Stevanin G., Durr A., Chneiweiss H., Neri C., Duyckaerts C., Penet C., Cann H. M., Agid Y., and Brice A., (1995). Marked phenotypic heterogeneity associated with expansion of a CAG repeat sequence at the spinocerebellar ataxia 3/Machado–Joseph disease locus. *Am. J. Hum. Genet.* **57**, 809–816.
15. Friedman J. H. (2002). Presumed rapid eye movement behavior disorder in Machado-Joseph disease (spinocerebellar ataxia type 3). *Mov. Disord.* **17**, 1350–1353.
16. Friedman J. H., Fernandez H. H., and Sudarsky L. R. (2003). REM behavior disorder and excessive daytime somnolence in Machado-Joseph disease (SCA-3). *Mov. Disord.* **18**, 1520–1522.
17. Lin K. P., and Soong B. W. (2002). Peripheral neuropathy of Machado–Joseph disease in Taiwan: A morphometric and genetic study. *Eur. Neurol.* **48**, 210–217.
18. Matsumura, R., Takayanagi, T., Fujimoto, Y., Murata, K., Mano, Y., Horikawa, H., and Chuma, T. (1996). The relationship between trinucleotide repeat length and phenotypic variation in Machado–Joseph disease. *J. Neurol. Sci.* **139**, 52–57.
19. Rub, U., de Vos, R. A., Schultz, C., Brunt, E. R., Paulson, H., and Braak, H. (2002). Spinocerebellar ataxia type 3 (Machado–Joseph disease): Severe destruction of the lateral reticular nucleus. *Brain.* **125** (Pt. 9), 2115–2124.
20. Rub, U., de Vos, R. A., Brunt, E. R., Schultz, C., Paulson, H., Del Tredici, K., and Braak, H. (2002). Degeneration of the external cuneate nucleus in spinocerebellar ataxia type 3 (Machado–Joseph disease). *Brain Res.* **953**, 126–134.
21. Rub, U., Brunt, E. R., Gierga, K., Schultz, C., Paulson, H., de Vos, R. A., and Braak, H. (2003). The nucleus raphe interpositus in spinocerebellar ataxia type 3 (Machado–Joseph disease). *J. Chem. Neuroanat.* **25**, 115–127.
22. Rub, U., Brunt, E. R., de Vos, R. A., Del Turco, D., Del Tredici, K., Gierga, K., Schultz, C., Ghebremedhin, E., Burk, K., Auburger, G., and Braak, H. (2004). Degeneration of the central vestibular system in spinocerebellar ataxia type 3(SCA3) patients and its possible clinical significance. *Neuropathol. Appl. Neurobiol.* **4**, 402–414.
23. Rub, U., Burk, K., Schols, L., Brunt, E. R., de Vos, R. A., Diaz, G. O., Gierga, K., Ghebremedhin, E., Schultz, C., Del Turco, D., Mittelbronn, M., Auburger, G., Deller, T., and Braak, H. (2004). Damage to the reticulotegmental nucleus of the pons in spinocerebellar ataxiatype 1, 2, and 3. *Neurology* **63**, 1258–1263.
24. Sasaki, H., Wakisaka, A., Fukazawa, T., Iwabuchi, K., Hamada, T., Takada, A., Mukai, E., Matsuura, T., Yoshiki, T., and Tashiro, K. (1995). CAG repeat expansion of Machado–Joseph disease in the Japanese: Analysis of the repeat instability for parental transmission, and correlation with disease phenotype. *J. Neurol. Sci.* **133**, 128–133.
25. Schols, L., Amoiridis, G., Epplen, J. T., Langkafel, M., Przuntek, H., and Riess, O. (1996). Relations between genotype and pheotype in German patients with the Machado–Joseph disease mutation. *J. Neurol. Neurosurg. Psychiatry* **61**, 466–470.
26. Sequeiros, J., and Coutinho, P. (1993). Epidemiology and clinical aspects of Machado–Joseph disease [review; no abstract available]. *Adv. Neurol.* **61**, 139–153.
27. Soong, B. W., Cheng, C. H., Liu, R. S., and Shan, D. E. (1997). Machado–Joseph disease: Clinical, molecular, and metabolic characterization in Chinese kindreds. *Ann. Neurol.* **41**, 446–452.
28. Takiyama, Y., Okynagi, S., Kawashima, S., Sakamoto, H., Saito, K., Yoshida, M., Tsuji, S., Mizuno, Y., and Nishizawa, M. (1994). A clinical and pathologic study of a large Japanese family with Machado–Joseph disease tightly linked to the DNA markers on chromosome 14q. *Neurology* **44**, 1302–1308.
29. Zhou, L. S., Fan, M. Z., Yang, B. X., Weissenbach, J., Wang, G. X., and Tsuji, S. (1997). Machado-Joseph disease in four Chinese pedigrees: Molecular analysis of 15 patients including two juvenile cases and clinical correlations. *Neurology* **48**, 482–485.
30. Kawai Y., Takeda A., Abe Y., Washimi Y., Tanaka F., and Sobue G. (2004). Cognitive impairments in Machado–Joseph disease. *Arch. Neurol.* **61**, 1757–1760.
31. Maruff, P., Tyler, P., Burt, T., Currie, B., Burns, C., and Currie, J. (1996). Cognitive deficits in Machado–Joseph disease. *Ann. Neurol.* **40**, 421–427.
32. Zawacki, T. M., Grace, J., Friedman, J. H., and Sudarsky, L. (2002). Executive and emotional dysfunction in Machado-Joseph disease. *Mov. Disord.* **17**, 1004–1010.
33. Klockgether, T., Ludtke R., Kramer B., Abela, M., Burk, K., Schols, L., Riess, O., Laccone, F., Boesch, S., Lopes-Cendes, I., Brice, A., Inzelberg, R., Zilber, N., and Dichgans, J. (1998). The natural history of degenerative ataxia: A retrospective study in 466 patients. *Brain* **121**, 589–600.
34. Subramony, S. H., and Currier, R. D. (1996). Intrafamilial variability in Machado–Joseph disease. *Mov. Disord.* **11**, 741–743.
35. Gwinn-Hardy, K., Singleton, A., O'Suilleabhain, P., Boss, M., Nicholl, D., Adam, A., Hussey, J., Critchley, P., Hardy, J., and Farrer, M. (2001). Spinocerebellar ataxia type 3 phenotypically resembling Parkinson disease in a black family. *Arch. Neurol.* **58**, 296–299.
36. Subramony, S. H., Hernandez, D., Adam, A., Smith-Jefferson, S., Hussey, J., Gwinn-Hardy, K., Lynch, T., McDaniel, O., Hardy, J., Farrer, M., and Singleton, A. (2002). Ethnic differences in the expression of neurodegenerative disease:Machado-Joseph disease in Africans and Caucasians. *Mov. Disord.* **17**, 1068–1071.
37. Buhmann C., Bussopulos A., and Oechsner M. (2003). Dopaminergic response in parkinsonian phenotype of Machado–Joseph disease. *Mov. Disord.* **18**, 219–221.
38. Tuite, P. J., Rogaeva, E. A., St. George-Hyslop, P. H., and Lang, A. E. (1995). Dopa-responsive parkinsonism phenotype of Machado–Joseph disease: Confirmation of 14q CAG expansion. *Ann. Neurol.* **38**, 684–687.
39. Schmitt, I., Evert, B. O., Khazneh, H., Klockgether, T., and Wuellner, U. (2003). The human MJD gene: Genomic structure and functional characterization of the promoter region. *Gene* **314**, 81–88.
40. Soong, B. W., and Lin, K. P. (1998). An electrophysiologic and pathologic study of peripheral nerves in individuals with Machado–Joseph disease. *Chin. Med.* **61**, 181–187.
41. Iranzo A., Munoz E., Santamaria J., Vilaseca I., Mila M., and Tolosa E. (2003). REM sleep behavior disorder and vocal cord paralysis in Machado-Joseph disease. *Mov. Disord.* **18**, 1179–1183.
42. Van Alfen, N., Sinke, R., Zwarts, M., Gabreels-Festen, A., Praamstra, P., Kremer, B. P., and Horstink, M. W. (2001). Intermediate CAG repeat lengths (53, 54) for MJD/SCA3 are associated with an abnormal phenotype. *Ann. Neurol.* **49**, 805–808.
43. Murata, Y., Yamaguchi, S., Kawakami, H., Imon, Y., Maruyama, H., Sakai, T., Kazuta, T., Ohtake, T., Nishimura, M., Saida, T., Chiba, S., Oh-I, T., and Nakamura, S. (1998). Characteristic magnetic resonance imaging findings in Machado–Joseph disease. *Arch. Neurol.* **55**, 33–37.
44. Onodera, O., Idezuka, J., Igarashi, S., Takiyama, Y., Endo, K., Takano, H., Oyake, M., Tanaka, H., Inuzuka, T., Hayashi, T., Yuasa, T., Ito, J., Miyatake, T., and Tsuji, S. (1998). Progressive atrophy of cerebellum and brainstem as a function of age and size of the expanded CAG repeats in the MJD1 gene in Machado–Joseph disease. *Ann. Neurol.* **43**, 288–296.
45. Klockgether T., Skalej M., Wedekind D., Luft A. R., Welte D., Schulz J. B., Abele M., Burk K., Laccone F., Brice A., and Dichgans J.

45. (1998). Autosomal dominant cerebellar ataxia type I: MRI based volumetry of posterior fossa structures and basal ganglia in spinocerebellar ataxia type 1, 2 and 3. *Brain* **121**, 1687–1693.
46. Etchebehere E., Cendes, F., Lopes-Cendes I., Pereira, J. A., Lima, M. C., Sansana, C. R., Silva, C. A., Camargo, M. F., Santos, A. O., Ramos, C. D., Camargo, E. E. (2001). Brain single-photon emission computed tomography and magnetic resonance imaging in Machado–Joseph disease. *Arch. Neurol.* **58**, 1257–1263.
47. Hayashi M., Kobayashi K., and Furuta H. (2003). Immunohistochemical study of neuronal intranuclear and cytoplasmic inclusions in Machado–Joseph disease. *Psychiatry Clin. Neurosci.* **57**, 205–213.
48. Koeppen, A. H., Dickson, A. C., Lamarche, J. B., and Robitaille, Y. (1999). Synapses on the hereditary ataxias. *J. Neuropathol. Exp. Neurol.* **58**, 748–764.
49. Munoz, E., Rey, M. J., Mila, M., Cardozo, A., Ribalta, T., Tolosa, E., and Ferrer, I. (2002). Intranuclear inclusions, neuronal loss and CAG mosaicism in two patients with Machado–Joseph disease. *J. Neurol. Sci.* **200**, 19–25.
50. Paulson, H. L., Perez, M. K., Trottier, Y., Trojanowski, J. Q., Subramony, S. H., Das, S. S., Vig, P., Mandel, J. L., Fischbeck, K. H. and Pittman, R. N. (1997). Intranuclear inclusions of expanded polyglutamine protein in spinocerebellar ataxia type 3. *Neuron* **19**, 333–344.
51. Schmidt, T., Landwehrmeyer, B., Schmitt, I., Trottier, Y., Auburger, G., Laccone, F., Klockgether, T., Volpel, M., Epplen, J.T., Schols, L., and Riess, O. (1998). An isoform of ataxin-3 accumulates in the nucleus of neuronal cells in affected brain regions of SCA3 patients. *Brain Pathol.* **8**, 669–679.
52. Uchihara, T., Iwabuchi, K., Funata, N., and Yagishita, S. (2002). Attenuated nuclear shrinkage in neurons with nuclear aggregates: A morphometric study on pontine neurons of Machado–Joseph disease brains. *Exp. Neurol.* **178**, 124–128.
53. Suenaga, T., Matsushima, H., Nakamura, S., Akiguchi, I., and Kimura, J. (1993). Ubiquitin-immunoreactive inclusions in anterior horn cells and hypoglossal neurons in a case with Joseph's disease. *Acta Neuropathol.* **85**, 341–344.
54. Ichikawa Y., Goto J., Hattori M., Toyoda A., Ishii K., Jeong S. Y., Hashida H., Masuda N., Ogata K., Kasai F., Hirai M., Maciel P., Rouleau G. A., Sakaki Y., and Kanazawa I. (2001). The genomic structure and expression of MJD, the Machado–Joseph disease gene. *J. Hum. Genet.* **46**, 413–422.
55. Maciel, P., Gaspar, C., DeStefano, A. L., Silveira, I., Coutinho, P., Radvany, J., Dawson, D. M., Sudarsky, L., Guimaraes, J., Loureiro, J. E. L., Nezarati, M. M., Corwin, L. I., Lopes-Cendes, I., Rooke, K., Rosenberg, R., MacLeod, P., Farrer, L. A., Sequeiros, J., and Rouleau, G. A. (1995). Correlation between CAG repeat length and clinical features in Machado–Joseph disease. *Am. J. Hum. Genet.* **57**, 54–61.
56. Matilla, T., McCall, A., Subramony, S. H., and Zoghbi, H. Y. (1995). Molecular clinical correlations in spinocerebellar ataxia type 3 and Machado–Joseph disease. *Ann. Neurol.* **38**, 68–72.
57. Ranum, L. P. W., Lundgren, J. K., Schut, L. J., Ahrens, M. J., Perlman, S., Aita, J., Bird, T. D., Gomez, C., and Orr, H. T. (1995). Spinocerebellar ataxia type 1 and Machado–Joseph disease: Incidence of CAG expansions among adult-onset ataxia patients from 311 families with dominant, recessive or sporadic ataxia. *Am. J. Hum. Genet.* **57**, 603–608.
58. Silveira, I., Lopes-Cendes, I., Kish, S., Maciel, P., Gaspar, C., Coutinho, P., Botez, M. I., Teive, H., Arruda, W., Steiner, C. E., Pinto-Junior, W., Maciel, J. A., Jerin, S., Sack, G., Andermann, E., Sudarsky, L., Rosenberg, R., MacLeod, P., Chitayat, D., Babul, R., Sequeiros, J., Rouleau, G. A. (1996). Frequency of spinocerebellar ataxia type I, dentatorubral-pallidoluysian atrophy and Machado–Joseph disease mutations in a large group of spinocerebellar ataxia patients. *Neurology* **46**, 214–218.
59. Storey, E., du Sart, D., Shaw, J., Lorentzos, P., Kelly, L., McKinley Gardner, R. J., Forrest. S. M., Biros, I., and Nicholson, G. A. (2000). Frequency of spinocerebellar ataxia types 1, 2, 3, 6, and 7 in Australian patients with spinocerebellar ataxia. *Am. J. Med. Genet.* **95**, 351–357.
60. Zhou, X. Y., Takiyama, Y., Igarashi, S., Li, Y. F., Zhou, B. Y., Gui, D. C., Endo, K., Tanaka, H., Chen, Z. H., Zhou, Y., Qiao, W., and Gu, W. (2001). Spinocerebellar ataxia type 1 in China. *Arch. Neurol.* **58**, 789–794.
61. Gaspar, C., Lopes-Cendes, I., Hayes, S., Goto, J., Arvidsson, K., Dias, A., Silveira, I., Maciel, P., Coutinho, P., Lima, M., Zhou, Y. X., Soong, B. W., Watanabe, M., Giunti, P., Stevanin, G., Riess, O., Sasaki, H., Hsieh, M., Nicholson, G. A., Brunt, E., Higgins, J. J., Lauritzen, M., Tranebjaerg, L., Volpini, V., Wood, N., Ranum, L., Tsuji, S., Brice, A., Sequeiros, J., and Rouleau, G. A. (2001). Ancestral origins of the Machado–Joseph disease mutation: A worldwide haplotype study. *Am. J. Hum. Genet.* **68**, 523–528.
62. Rubinsztein, C., Leggo, J., Coetzee, G. A., Irvine, R. A., Buckley, M., and Ferguson-Smith, A. (1995). Sequence variation and size ranges of CAG repeats in the Machado–Joseph disease, spinocerebellar ataxia type 1 and androgen receptor genes. *Hum. Mol. Genet.* **4**, 1585–1590.
63. Takiyama, Y., Igarashi, S., Rogaeva, E. A., Endo, K., Rogaev, E. I., Tanaka, H., Sherrington, R., Sanpei, K., Liang, Y., Saito, M., Tsuda, T., Takano, H., Ikeda, M., Lin, C., Chi, H., Kennedy, J. L., Lang, A. E., Wherrett, J. R., Segawa, M., Nomura, Y., Yuasa, T., Weissenbach, J., Yoshida, M., Nishizawa, M., and Kidd, K. K. (1995). Evidence for intergenerational instability in the CAG repeat in the MJD1 gene and for conserved haplotypes at flanking markers amongst Japanese and Caucasian subjects with Machado–Joseph disease. *Hum. Mol. Genet.* **4**, 1137–1146.
64. Gu W., Ma H., Wang K., Jin M., Zhou Y., Liu X., Wang G., and Shen Y. (2004). The shortest expanded allele of the MJD1 gene in a Chinese MJD kindred with autonomic dysfunction. *Eur. Neurol.* **52**, 107–111. [Epub 2004 Aug 13.]
65. Padiath, Q. S., Srivastava, A. K., Roy, S., Jain, S., and Brahmachari, S. K. (2005). Identification of a novel 45 repeat unstable allele associated with a disease phenotype at the MJD1/SCA3 locus. *Am. J. Med. Genet. B. Neuropsychiatr. Genet.* **133**, 124–126.
66. Maruyama, H., Izumi, Y., Morino, H., Oda, M., Toji, H., Nakamura, S., and Kawakami, H. (2002). Difference in disease-free survival curve and regional distribution according to subtype of spinocerebellar ataxia: A study of 1,286 Japanese patients. *Am. J. Med. Genet.* **114**, 578–583.
67. Jardim, L., Pereira, M., Silveira, I., Ferro, A., Sequeiros, J., and Giugliani, R. (2001). Neurologic findings in Machado–Joseph disease. *Arch. Neurol.* **58**, 899–904.
68. Takiyama, Y., Shimazaki, H., Morita, M., Soutome, M., Sakoe, K., Esumi, E., Muramatsu, S. I., Yoshida, M. (1998). Maternal anticipation in Machado–Joseph disease (MJD): Some maternal factors independent of the number of CAG repeat units may play a role in genetic anticipation in a Japanese MJD family. *J. Neurol. Sci.* **155**, 141–145.
69. Lerer I., Merims D., Abeliovich D., Zlotogora J., and Gadoth N. (1996). Machado–Joseph disease: Correlation between clinical features, the CAG repeat length and homozygosity for the mutation. *Eur. J. Hum. Genet.* **4**, 3–7.
70. Goti, D., Katzen, S. M., Mez, J., Kurtis, N., Kiluk, J., Ben-Haiem, L., Jenkins N. A., Copeland, N. G., Kakizuka, A., Sharp, A. H., Ross, C. A., Mouton, P. R., and Colomer, V. (2004). A mutant ataxin-3 putative-cleavage fragment in brains of Machado–Joseph disease patients and transgenic mice is cytotoxic above a critical concentration. *J. Neurosci.* **24**, 10266–10279.

71. Cemal, C. K., Carroll, C. J., Lawrence, L., Lowrie, M. B., Ruddle, P., Al-Mahdawi, S., King, R. H., Pook, M. A., Huxley, C., and Chamberlain, S. (2002). YAC transgenic mice carrying pathological alleles of the MJD1 locus exhibit a mild and slowly progressive cerebellar deficit. *Hum. Mol. Genet.* **11**, 1075–1094.
72. Warrick, J. M., Paulson, H. L., Gray-Board, G. L., Bui, Q. T., Fischbeck, K. H., Pittman, R. N., and Bonini, N. M. (1998). Expanded polyglutamine protein forms nuclear inclusions and causes neural degeneration in drosophila. *Cell* **93**, 1–20.
73. Igarashi S., Takiyama Y., Cancel G., Rogaeva E. A., Sasaki H., Wakisaka A., Zhou X-Y., Takano H., Endo, K., Sanpei, K., Oyake, M., Tanaka, H. Stevanin, G., Abbas, N., Durr, A., Rogaev, E, I., Sherrington, R., Tsuda, T., Ideda, M., Cassa, E., Nishizawa, M., Benomar, A., Julien, J., Weissenbach, J., Want, G. X., Agid, Y., St. George-Hyslop, Ph. H., Brice, A., and Tsuji, S. (1996). Intergenerational instability of the CAG repeat of the gene for Machado–Joseph disease (MJD1) is affected by the genotype of the normal chromosome: Implications for the molecular mechanisms of the instability of the CAG repeat. *Hum. Mol. Genet.* **5**, 923–932.
74. Matsumura, R., Takayanagi, T., Murata, K., Futamura, N., Hirano, M., and Ueno, S. (1996). Relationship between (CAG)$_n$ C configuration to repeat instability of the Machado–Joseph disease gene. *Hum. Genet.* **98**, 643–645.
75. Miller, V. M., Xia, H., Marrs, G. L., Gouvion, C. M., Lee, G., Davidson, B. L., and Paulson, H. L. (2003). Allele-specific silencing of dominant disease genes. *Proc. Natl. Acad. Sci. USA* **100**, 7195–7200.
76. Maciel, P., Lopes-Cendes, I., Kish, S., Sequeiros, J., and Rouleau, G. A. (1997) Mosaicism of the CAG repeat in CNS tissue in relation to age at death in spinocerebellar ataxia type 1 and Machado–Joseph disease patients. *Am. J. Hum. Genet.* **60**, 993–996.
77. do Carmo Costa M., Gomes-da-Silva J., Miranda C. J., Sequeiros J., Santos M. M., and Maciel P. (2004). Genomic structure, promoter activity, and developmental expression of the mouse homologue of the Machado–Joseph disease (MJD) gene. *Genomics* **84**, 361–373.
78. Nihiyama, K., Murayama, S., Goto, J., Watanabe, M., Hashida, H., Katayama, S., Numura, Y., Nakamura, S., and Kanazawa, I. (1996). Regional and cellular expression of the Machado–Joseph disease gene in brains of normal and affected individuals. *Ann. Neurol.* **40**, 776–781
79. Paulson, H. L., Das, S. S., Crino, P. B., Perez, M. K., Patel, S. C., Gotsdiner, D., Fischbeck, K. H., and Pittman, R. N. (1997). Machado–Joseph disease gene product is a cytoplasmic protein widely expressed in brain. *Ann. Neurol.* **41**, 453–462.
80. Tait, D., Riccio, M., Sittler, A., Scherzinger, E., Santi, S., Ognibene, A., Maraldi, N. M., Lehrach, H., and Wanker, E. E. (1998). Ataxin-3 is transported into the nucleus and associates with the nuclear matrix. *Hum. Mol. Genet.* **7**, 991–997.
81. Wang, G., Keiko, I., Nukina, N., Goto J, Ichikawa, Y., Uchida, K., Sakamoto, T., and Kanazawa, I. (1997). Machado–Joseph disease gene product identified in lymphocytes and brain. *Biochem. Biophys. Res. Commun.* **233**, 476–479.
82. Trottier, Y., Cancel, G., An-Gourfinkel, I., Lutz, Y., Weber, C., Brice, A., Hirsch, E., and Mandel, J. L. (1998). Heterogeneous intracellular localization and expression of ataxin-3. *Neurobiol. Dis.* **5**, 335–347.
83. Perez, K. P., Paulson, H. L., and Pittman, R. N. (1999). Ataxin-3 with an altered conformation that exposes the polyglutamine domain is associated with the nuclear matrix. *Hum. Mol. Genet.* **8**, 2377–2385.
84. Scheel, H., Tomiuk, S., and Hofmann, K. (2003). Elucidation of ataxin-3 and ataxin-7 function by integrative bioinformatics. *Hum. Mol. Genet.* **12**, 2845–2852.

85. Burnett B., Li F., and Pittman R. N. (2003). The polyglutamine neurodegenerative protein ataxin-3 binds polyubiquitylated proteins and has ubiquitin protease activity. *Hum. Mol. Genet.* **12**, 3195–3205. [Epub 2003 Oct 14.]
86. Burnett B. G., and Pittman R. N. (2005). The polyglutamine neurodegenerative protein ataxin 3 regulates aggresome formation. *Proc. Natl. Acad. Sci. USA* **102**, 4330–4335. [Epub 2005 Mar 14.]
87. Mao, Y., Senic-Matuglia, F., Di Fiore, P. P., Polo, S., Hodsdon, M. E., and De Camilli, P. (2005). Deubiquitinating function of ataxin-3: Insights from the solution structure of the Josephin domain. *Proc. Natl. Acad. Sci. USA* **102**, 12700–12705. [Epub 2005 Aug 23.]
88. Nicastro, G., Menon, R. P., Masino, L., Knowles, P. P., McDonald, N. Q., and Pastore, A. (2005). The solution structure of the Josephin domain of ataxin-3: Structural determinants for molecular recognition. *Proc. Natl. Acad. Sci. USA.* **102**, 10493–10498. [Epub 2005 Jul 14.]
89. Berke, S. J., Chai, Y., Marrs, G. L., Wen, H., and Paulson, H. L. (2005). Defining the role of ubiquitin-interacting motifs in the polyglutamine disease protein, ataxin-3. *J. Biol. Chem.* **280**, 32026–32034. [Epub 2005 Jul 21.]
90. Chow M. K., Mackay J. P., Whisstock J. C., Scanlon M. J., and Bottomley S. P. (2004). Structural and functional analysis of the Josephin domain of the polyglutamine protein ataxin-3. *Biochem. Biophys. Res. Commun.* **322**, 387–394.
91. Masino, L., Musi, V., Menon, R. P., Fusi, P., Kelly, G., Frenkiel, T. A, Trottier, Y., and Pastore, A. (2003). Domain architecture of the polyglutamine protein ataxin-3: A globular domain followed by a flexible tail. *FEBS Lett.* **549**, 21–25.
92. Masino, L., Nicastro, G., Menon, R. P., Dal Piaz, F., Calder, L., and Pastore, A. (2004). Characterization of the structure and the amyloidogenic properties of the Josephin domain of the polyglutamine-containing protein ataxin-3. *J. Mol. Biol.* **344**, 1021–1035.
93. Hofmann K., and Falquet L. (2001). A ubiquitin-interacting motif conserved in components of the proteasomal and lysosomal protein degradation systems. *Trends Biochem. Sci.* **6**, 347–350.
94. Chai Y., Berke S. S., Cohen R. E., and Paulson H. L. (2004). Polyubiquitin binding by the polyglutamine disease protein ataxin-3 links its normal function to protein surveillance pathways. *J. Biol. Chem.* **279**, 3605–311. [Epub 2003 Nov 5.]
95. Donaldson K. M., Li W., Ching K. A., Batalov S., Tsai C. C., and Joazeiro C. A. (2003). Ubiquitin-mediated sequestration of normal cellular proteins into polyglutamine aggregates. *Proc. Natl. Acad. Sci. USA* **100**, 8892–8897. [Epub 2003 Jul 11.]
96. Doss-Pepe E. W., Stenroos E. S., Johnson W. G., and Madura K. (2003). Ataxin-3 interactions with rad23 and valosin-containing protein and its associations with ubiquitin chains and the proteasome are consistent with a role in ubiquitin-mediated proteolysis. *Mol. Cell. Biol.* **23**, 6469–6483.
97. Berke, S. J., and Paulson, H. L. (2003). Protein aggregation and the ubiquitin proteasome pathway: Gaining the upper hand on neurodegeneration. *Curr. Opin. Genet. Dev.* **13**, 253–261.
98. Ross, C. A., and Poirier, M. A. (2004). Protein aggregation and neurodegenerative disease. *Nat. Med.* **10** (Suppl.), S10–S17.
99. Taylor, J. P., Hardy, J., and Fischbeck, K. H. (2002). Toxic proteins in neurodegenerative disease. *Science* **296**, 1991–1995.
100. Chen, S., Ferrone, F. A., and Wetzel, R. (2002). Huntington's disease age-of-onset linked to polyglutamine aggregation nucleation. *Proc, Natl, Acad, Sci, USA* **99**, 11884–11889.
101. Kayed, R., Head, E., Thompson, J. L., McIntire, T. M., Milton, S. C., Cotman, C. W., and Glabe, C. G. (2003). Common structure of soluble amyloid oligomers implies common mechanism of pathogenesis. *Science* **300**, 486–489.
102. Scherzinger, E., Lurz, R., Turmaine, M., Mangiarini, L., Hollenbach, B., Hasenbank, R., Bates, G. P., Davies, S. W.,

Lehrach, H., Wanker, E. E. (1997). Huntingtin-encoded polyglutamine expansions form amyloid-like protein aggregates in vitro and in vivo. *Cell* **90**, 549–558.

103. Bevivino A. E., and Loll P. J. (2001). An expanded glutamine repeat destabilizes native ataxin-3 structure and mediates formation of parallel beta -fibrils. *Proc. Natl. Acad. Sci. USA* **98**, 11955–11960.

104. Yang, W., Dunlap, J. R., Andrews, R. B., and Wetzel, R. (2002). Aggregated polyglutamine peptides delivered to nuclei are toxic to mammalian cells. *Hum. Mol. Genet.* **11**, 2905–2917.

105. Arrasate, M., Mitra, S., Schweitzer, E. S., Segal, M. R., and Finkbeiner, S. (2004). Inclusion body formation reduces levels of mutant huntington and the risk of neuronal death. *Nature* **431**, 805–810.

106. Warrick, J. M., Chan, H. Y., Gray-Board, G. L., Chai, Y., Paulson, H. L., and Bonini, N. M. (1999). Suppression of polyglutamine-mediated neurodegeneration in *Drosophila* by the molecular chaperone HSP70. *Nat. Genet.* **23**, 425–428.

107. Chan, H. Y., Warrick, J. M., Gray-Board, G. L., Paulson, H. L., and Bonini, N. M. (2000). Mechanisms of chaperone suppression of polyglutamine disease: Selectivity, synergy and modulation of protein solubility in *Drosophila. Hum. Mol. Genet.* **9**, 2811–2820.

108. Muchowski, P. J., and Wacker, J. L. (2005). Modulation of neurodegeneration by molecular chaperones. *Nat. Rev. Neurosci.* **6**, 11–22.

109. Willingham, S., Outeiro, T. F., DeVit, M. J., Lindquist, S. L., and Muchowski, P. J. (2003). Yeast genes that enhance the toxicity of a mutant huntington fragment or alpha-synuclein. *Science* **302**, 1769–1772.

110. Chai Y., Koppenhafer S. L., Shoesmith S. J., Perez M. K., and Paulson H. L. (1999). Evidence of proteasome involvement in polyglutamine disease: Localization to nuclear inclusions in SCA3/MJD and suppression of polyglutamine aggregation in vitro. *Hum. Mol. Genet.* **8**, 673–682.

111. Schmidt, T., Lindenberg, K. S., Krebs, A., Schols, L., Laccone, F., Herms, J., Rechsteiner, M., Riess, O., and Landwehrmeyer, G. B. (2002). Protein surveillance machinery in brains with spinocerebellar ataxia type 3: Redistribution and differential recruitment of 26S proteasome subunits and chaperones to neuronal intranuclear inclusions. *Ann. Neurol.* **51**, 302–310.

112. Venkatraman, P., Wetzel, R., Tanaka, M., Nukina, N., and Goldberg, A. L. (2004). Eukaryotic proteasomes cannot digest polyglutamine sequences and release them during degradation of polyglutamine-containing proteins. *Mol. Cell* **14**, 95–104.

113. Bence N. F., Sampat R. M., and Kopito R. R. (2001). Impairment of the ubiquitin–proteasome system by protein aggregation. *Science* **292**, 1552–1555.

114. Bowman, A. B., Yoo, S. Y., Dantuma, N. P., and Zoghbi, H. Y. (2005). Neuronal dysfunction in a polyglutamine disease model occurs in the absence of ubiquitin–proteasome system impairment and inversely correlates with the degree of nuclear inclusion formation. *Hum. Mol. Genet.* **14**, 679–691.

115. Ikeda H., Yamaguchi M., Sugai, S., Aze Y., Narumiya S., and Kakizuka A. (1996). Expanded polyglutamine in the Machado–Joseph disease protein induces cell death in vitro and in vivo. *Nat. Genet.* **13**, 196–202.

116. La Spada A. R., and Taylor J. P. (2003). Polyglutamines placed into context. *Neuron.* **38**, 681–684.

117. Warrick, J. M., Morabito, L. M., Bilen, J., Gordesky-Gold, B., Faust, L. Z., Paulson, H. L., and Bonini, N. M. (2005). Ataxin-3 suppresses polyglutamine neurodegeneration in *Drosophila* by a ubiquitin-associated mechanism. *Mol. Cell.* **18**, 37–48.

118. Toulouse, A., Au-Yeung, F., Gaspar, C., Roussel, J., Dion, P., and Rouleau, G. A. (2005). Ribosomal frameshifting on MJD-1 transcripts with long CAG tracts. *Hum. Mol. Genet.* **14**, 2649–2660. [Epub 2005 Aug 8.]

119. Chang W. H., Cemal C. K., Hsu Y. H., Kuo C. L., Nukina N., Chang M. H., Hu H. T., Li C., and Hsieh M. (2005). Dynamic expression of Hsp27 in the presence of mutant ataxin-3. *Biochem. Biophys. Res. Commun.* **336**, 258–267.

120. Berke S. J., Schmied F. A., Brunt E. R., Ellerby L. M., and Paulson H. L. (2004). Caspase-mediated proteolysis of the polyglutamine disease protein ataxin-3. *J. Neurochem.* **89**, 908–918.

121. Jana N. R., and Nukina N. (2004). Misfolding promotes the ubiquitination of polyglutamine-expanded ataxin-3, the defective gene product in SCA3/MJD. *Neurotoxicol. Res.* **6**, 523–533.

122. Evert, B. O., and Wullne, U., Schulz, J. B., Weller, M., Groscurth, P., Trottier, Y., Brice, A., and Klockgether, T. (1999). High level expression of expanded full-length ataxin-3 in vitro causes cell death and formation of intranuclear inclusions in neuronal cells. *Hum. Mol. Genet.* **8**, 1169–1176.

123. Perez, M. K., Paulson, H. L., Pendse, S. J., Saionz, S. J., Bonini, N. M., and Pittman, R. N. (1998). Recruitment and the role of nuclear localization in polyglutamine-mediated aggregation. *J. Cell. Biol.* **143**, 1457–1470.

124. Chai, Y., Wu, L., Griffin, J. D., and Paulson, H. L. (2001). The role of protein composition in specifying nuclear inclusion formation in polyglutamine disease. *J. Biol. Chem.* **276**, 44889–44897.

125. Chai, Y., Shao, J., Miller, V. M., Williams, A., and Paulson, H. L. (2002). Live-cell imaging reveals divergent intracellular dynamics of polyglutamine disease proteins and supports a sequestration model of pathogenesis. *Proc. Natl. Acad. Sci.* USA **99**, 9310–9315.

126. Evert B. O., Vogt I. R., Vieira-Saecker A. M., Ozimek L., de Vos R. A., Brunt E. R., Klockgether T., and Wullner U. (2003). Gene expression profiling in ataxin-3 expressing cell lines reveals distinct effects of normal and mutant ataxin-3. *J. Neuropathol. Exp. Neurol.* **62**, 1006–1018.

127. Sugars, K. L., and Rubinsztein, D. C. (2003). Transcriptional abnormalities in Huntington disease. *Trends Genet.* **19**, 233–238.

128. Gunawardena, S., and Goldstein, L. S. (2005). Polyglutamine diseases and transport problems: Deadly traffic jams on neuronal highways. *Arch. Neurol.* **62**, 46–51.

129. Gunawardena, S., Her, L. S., Brusch, R. G., Laymon, R. A., Niesman, I. R., Gordesky-Gold, B., Sintasath, L., Bonini, N. M., and Goldstein, L. S. (2003). Disruption of axonal transport by loss of huntingtin or expression of pathogenic polyQ proteins in Drosophila. *Neuron* **40**, 25–40.

130. Tsai, H. F., Tsai, H. J., and Hsieh, M. (2004). Full-length expanded ataxin-3 enhances mitochondrial-mediated cell death and decreases Bcl-2 expression in human neuroblastoma cells. *Biochem. Biophys. Res. Commun.* **324**, 1274–1282.

131. Monoi, H., Futaki, S., Kugimiya, S., Minakata, H., and Yoshihara, K. (2000). Poly-L-glutamine forms cation channels: Relevance to the pathogenesis of the polyglutamine diseases. *Biophys. J.* **78**, 2892–2899.

132. Harper S. Q., Staber P. D., He X., Eliason S. L., Martins I. H., Mao Q., Yang L, Kotin R. M., Paulson H. L., and Davidson B. L. (2005). RNA interference improves motor and neuropathological abnormalities in a Huntington's disease mouse model. *Proc. Natl. Acad. Sci. USA.* **102**, 5820–5825. [Epub 2005 Apr 5.]

133. Xia, H., Mao, Q., Eliason, S. L., Harper, S. Q., Martins, I. H., Orr, H. T., Paulson, H. L., Yang, L., Kotin, R. M., and

Davidson, B. L. (2004). RNAi suppresses polyglutamine-induced neurodegeneration in a model of spinocerebellar ataxia. *Nat. Med.* **10**, 816–820. [Epub 2004 Jul 4.]
134. Saleem, Q., Choudhry, S., Mukerji, M., Bashyam, L., Padma, M. V., Chakravarthy, A., Maheshwari, M. C., Jain, S., and Brahmachari, S. K. (2000). Molecular analysis of autosomal dominant hereditary ataxias in the Indian population: High frequency of SCA2 and evidence for a common founder mutation. *Hum. Genet.* **106**, 179–187.
135. Inoue, K., Hanihara, T., Yamada, Y., Kosaka, K., Katsuragi, T., and Iwabuchi, K. (1996). Clinical and genetic evaluation of Japanese autosomal dominant cerebellar ataxias; is Machado-Joseph disease common in the Japanese? *J. Neurol. Neurosurg. Psychiatry.* **60**: 697–698.
136. Pujana, M. A., Corral, J., Gratacos, M., Combarros, O., Berciano, J., Genis, D., Banchs, I., Estivill, X., Volpini, V. (1999). Spinocerebellar ataxias in Spanish patients: genetic analysis of familial and sporadic cases. The Ataxia Study Group. *Hum. Genet.* **104**: 516–522.
137. Nagaoka, U., Suzuki, Y., Kawanami, T., Kurita, K., Shikama, Y., Honda, K., Abe, K., Nakajima, T., and Kato, T. (1999). Regional differences in genetic subgroup frequency in hereditary cerebellar ataxia, and a mophometrical study of brain MR images in SCA1, MJD and SCA6. *J. Neurol. Sci.* **164**: 187–194.

Spinocerebellar Ataxia Type 6

HIDEHIRO MIZUSAWA AND KINYA ISHIKAWA

Department of Neurology and Neurological Science, Graduate School, Tokyo Medical and Dental University, 1-5-45 Yushima, Bunkyo-ku, Tokyo 113-8519, Japan

I. Introduction
II. Etiology, Pathogenesis, and Neuropathology
III. Clinical Manifestation
IV. Examination and Diagnosis
V. Treatment, Prognosis, and Perspective
References

Spinocerebellar ataxia type 6 (SCA6), one of the autosomal dominant neurodegenerative diseases, is caused by small expansions of CAG repeat that encodes polyglutamine tract for the α_{1A} (P/Q type; alpha 12.1) subunit of the voltage-gated calcium channel (*CACNA1A*, $Ca_v2.1$). Among the CAG repeats and their expansions known to cause human diseases, the length and expansion in SCA6 patients are the smallest and even the expanded length is within a range of normal sizes in other repeats. Clinically, patients with SCA6 show progressive, and rather "pure", cerebellar dysfunctions including gait ataxia, nystagmus, dysarthria and incoordination of the limbs at an average age-of-onset around 45 years. Vertigo with or without down beat nystagmus, typically experienced with rapid change in the head position, is also a characteristic clinical feature that would be important to clinically differentiate SCA6 from other autosomal dominant cerebellar ataxias. Despite that the expansion is small in SCA6, the mutation causes cerebellar dysfunction by the mechanism not clearly known yet. The expanded polyglutamine causes alteration in calcium channel function in cultured cells, suggesting that the mutation causes channel dysfunction other than a simple loss-of-function. In addition, formation of aggregation containing calcium channel protein has been found in Purkinje cells of SCA6 brains. However, protein aggregation is not within the nucleus as in most other polyglutamine diseases. Further studies are needed to elucidate molecular pathomechanism underlying SCA6 and to develop an effective treatment.

I. INTRODUCTION

Spinocerebellar ataxia type 6 (SCA6) is one of the autosomal dominant cerebellar ataxias (ADCAs), and its main clinical feature is slowly progressive ataxia consistent with ADCA type III, namely, pure cerebellar ataxia [1]. SCA6 is caused by the expansion of CAG repeats in the gene for the α_{1A} (P/Q type, $\alpha12.1$) subunit of the voltage-gated calcium channel (*CACNA1A*, $Ca_v2.1$), which is located on the short arm of chromosome 19 [2]. Before the report of SCA6, a nationwide linkage analysis for patients with pure cerebellar ataxia was performed in Japan, and the locus 19p13 was identified in half of the families collected [3]. This group of patients turned out to have SCA6. A very characteristic feature of SCA6 is the small size of both normal and expanded CAG repeats. Even the expanded repeats are usually within normal range in other polyglutamine diseases. In Japan, no case of Friedreich's ataxia has been idendified by gene analysis, and SCA6 seems to be either the most common or the second most common disease of the hereditary spinocerebellar ataxias (SCAs) [4–7]. The prevalence of SCA6 is lower in Europe and North America [8–11], and other SCAs with pure cerebellar ataxias such as SCA5, SCA10, SCA11, SCA14, SCA15, SCA16, and SCA22 appear uncommon in Japan [12–18]. There is also another SCA with pure cerebellar ataxia linked to chromosome 16q, the associated gene of which has been identified very recently [19–23]. The remainder of the initial families just mentioned have this type of SCA [3].

II. ETIOLOGY, PATHOGENESIS, AND NEUROPATHOLOGY

SCA6 is caused by small expansions of CAG repeats in exon 47 of *CACNA1A* (Ca$_v$2.1) [2, 3]. Although the normal range is 5 to 18 repeats, SCA6 patients have 20 to 33 repeats. The 19-CAG repeat has been reported as the lower limit of expansion causing SCA6 [24, 25], although this repeat has been observed in normal alleles including our report [3]. Our three asymptomatic cases with 19 repeats harbored 13 repeats in the other normal chromosome. In contrast, among four individuals with 19/19, 19/13, 19/11, and 19/11 repeats, ataxia was seen in only one case with homozygous 19/19 repeats [25]. Therefore, the total number may have some influence. However, because a SCA6 patient with 19/7 repeats was also reported [24], the significance of a single 19 repeat should be elucidated in the future. The repeat number 20 is definitely pathological because a neuropathological study of a case with 20 repeats revealed specific inclusions in Purkinje cells (unpublished data).

Although there is an inverse correlation between age at onset and repeat length, the age-at-onset range is 36 years, for example, in patients with 22 CAG repeats in the expanded allele (Fig. 25-1) [26]. Factors other than the CAG repeat length of *CACNA1A* may be influencing the age at onset. In contrast to apparent genetic anticipation in other polyglutamine diseases, the repeat length is usually the same within a given SCA6 family, probably due to the small size of the expansion. The repeat length is also the same among various parts of a given SCA6 brain [27]. These facts suggest that the CAG repeat in *CACNA1A* is very stable in both meiotic and mitotic divisions.

Mutations other than the repeat expansion of the gene *CACNA1A* cause familial hemiplegic migraine (FHM: point mutations, deletions, splice abnormalities) and episodic ataxia type 2 (EA2: point mutations) [28]. The mechanism of phenotypic variations between the repeat expansion and other mutations has not yet been clarified. Furthermore, some mutant ataxic mice have resulted from point mutations or a splice abnormality of *CACNA1A*, namely, tottering (*tg*), leaner (*tg^{la}*), rolling mouse Nagoya (*tg^{rol}*), and rocker (*rkr*). Although, in humans, SCA6, FHM, and EA2 are autosomal dominant diseases, these mouse ataxias are transmitted as autosomal recessive traits.

Voltage-gated Ca channels are composed of several subunits including α_1, β, and α_2/δ. Because the α_{1A} subunit forms the pore for Ca ions and appears very important for channel functions, mutations of the gene might cause impairment of channel functions. Indeed, a few studies using cultured cells reported alterations of Ca channel functions, but the degree of change was not that large, and the quality of the alteration was different in different reports [29–32]. Therefore, although there may be a functional abnormality of the Ca channel, the contribution to the pathogenesis of SCA6 would not be large enough. On the other hand, neuropathologically, SCA6 brain shows loss of Purkinje cells, milder degeneration of granule cells, and thinning of the molecular layer with astrogliosis [27] (Fig. 25-2). Because α_{1A} Ca

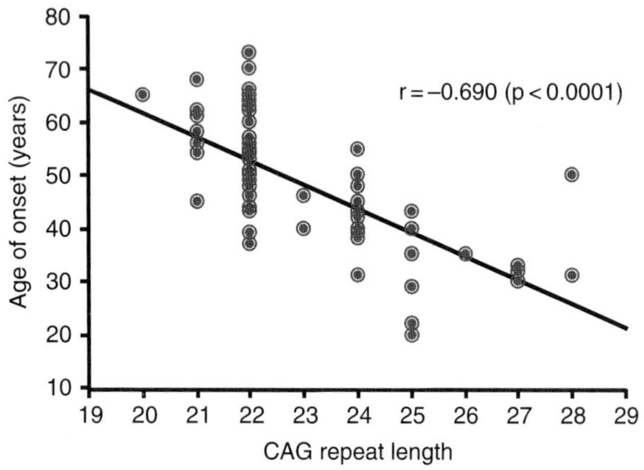

FIGURE 25-1 Age at onset correlates inversely with the number of the CAG repeats of the *CACNA1A* gene, suggesting the expansion is the cause of the illness. Note the wide range of age at onset with the same length of repeats. For example, the age-at-onset range is more than 30 years for a CAG repeat length of 22 [26]. See CD-ROM for color image.

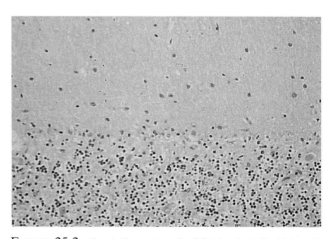

FIGURE 25-2 Cerebellar cortex of a SCA6 patient. Purkinje cells are decreased in number, and the remaining cells are atrophic, with milder loss of granule cells. There is astrogliosis of Bergman's glia in the Purkinje cell layer and astogliosis in the molecular layer. Hematoxylin and eosin stain. ×80. See CD-ROM for color image.

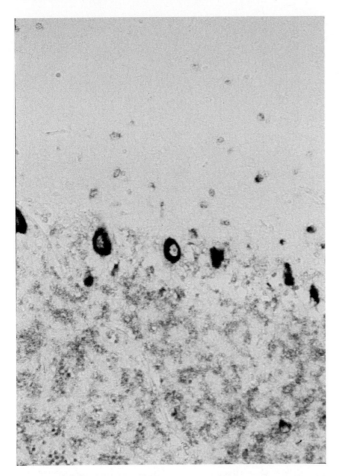

FIGURE 25-3 *In situ* hybridization of mRNA of the *CACNA1A* gene. The cerebellum, particularly Purkinje cells, shows a very strong reaction followed by a less marked reaction of granule cells. ×100. See CD-ROM for color image.

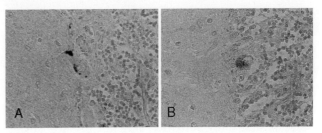

FIGURE 25-4 Two types of cytoplasmic inclusions specific to both SCA6 and Purkinje cells. Remaining Purkinje cells contain α_{1A} Ca channel protein-positive inclusions in the cytoplasm (A). Finer cytoplasmic inclusions positive with 1C2 antibody suggesting the presence of elongated polyglutamine tracts are also seen in Purkinje cells (B). ×150. See CD-ROM for color image.

channel mRNA and protein are widely expressed in the central nervous system, particularly in cerebellar Purkinje cells [33] (Fig. 25-3), it seems quite reasonable that the Punkinje cell, which seems to be the strongest expressor of the causative gene *CACNA1A*, would be affected most profoundly. In fact, there are two types of cytoplasmic inclusion bodies specific to both Purkinje cells and SCA6 (Fig. 25-4) [33, 34]. Larger inclusions are immunostained by antibodies against α_{1A} Ca channel protein and appear to be composed of the channel protein. Finer inclusions are positively immunostained with antibody 1C2, suggesting that the latter finer inclusions contain expanded polyglutamine tracts with conformational changes. Interestingly, intranuclear inclusions, which are a hallmark common to other polyglutamine diseases, are not apparent in SCA6, at least by light microscopy. However, the presence of the inclusions implies that the mutant protein may have toxic effects on Purkinje cells, as in other polyglutamine diseases. Our study on cultured HEK cells further suggested that the α_{1A} Ca channel with expanded polyglutamine may be prone to fragmentation and cell death [35]. This is a striking feature because the result may suggest that the proteolytic cleavage seen in other polyglutamine diseases could operate with small polyglutamine expansions. Matsuyama et al. also showed that a Ca channel with expanded polyglutamine repeats may disturb the antiapoptotic role of normal Ca channels [36]. The neuropathological changes were usually pronounced in the superior vermis and milder in the hemisphere in most patients [27, 37–40]. When cerebellar pathology is marked, the inferior olivary nucleus exhibits neuronal loss to some extent, with some cases exhibiting the pathology of cerebello-olivary atrophy. The fact that the two types of inclusions are found exclusively in Purkinje cells indicates the Purkinje cell may be the primary target in SCA6.

III. CLINICAL MANIFESTATION

Clinically, SCA6 is a slowly progressive cerebellar impairmant including nystagmus, cerebellar dysarthria, incoordination of the extremities, ataxic gait, and muscle hypotonia. According to our study of 140 patients, the mean age at onset is 47.2 ± 11.5, the latest age at onset of the various spinocerebellar ataxias [26]. It is, however, important to note that some patients become ill in their twenties. Unsteadiness of gait or gait ataxia was the most frequent initial symptom in our and other reports [8, 9, 37, 41]. The second most frequent initial symptom was "vertigo" or oscillopsia. Yabe et al. (2003) reported that vertigo or oscillopsia was experienced by 68% of their patients [42], whereas only 12% of the patients in our cohort exhibited these symptoms [26]. The difference could be due to the confusion of vertigo with unsteadiness: patients might complain of "vertigo" when they suffer from unsteady gait and vice versa. Vertigo, however is one of the clinical features characteristic of SCA6 among the various SCAs [37]. In particular, vertigo and oscillopsia are induced by changes in head position (positioning vertigo) [8, 37, 42–45]. Symptomatic vertigo was present in only 8.8% of the subjects, but neuro-otological testing revealed positioning vertigo in 38.9% of the subjects examined. This difference suggests that daily activities may not provide strong enough stimulation to induce vertigo, or patients may unconsciously avoid stimulation inducing vertigo. Vertigo/oscillopsia and positioning vertigo/oscillopsia tend to be accompanied by downbeat positioning nystagmus (DPN) with or without downbeat gaze nystagmus. Various episodic features, including marked vertigo and episodic ataxia, were reported in previous studies of SCA6 [8, 37, 42], but only a small number of our subjects had true and significant episodic symptoms as in others [4]. Episodic symptoms may be frequent in foreign reports, and this variability may be due to differences in modifying gene's effect among various ethnic groups.

Although extracerebellar signs such as pyramidal tract signs, abnormal involuntary movements, parkinsonism, hyporeflexia, sensory disturbances, intellectual impairment, and urinary incontinence have also been reported, the frequencies of these signs are very low in our cohort and not constant among various reports [4, 5, 8–10, 37, 41, 46]. In addition, there were no anatomical substrates responsible for these extracerebellar signs. Extracerebellar signs do not appear to be significant at present.

IV. EXAMINATION AND DIAGNOSIS

Routine laboratory examinations of blood, urine, and feces, as well as cerebrospinal fluid, are all normal. Brain MRI of patients with SCA6 is the most important examination and reveals atrophy of the cerebellum, particularly the superior vermis, without atrophy of the brainstem and cerebrum (Fig. 25-5) [47, 48]. Rarely, brainstem atrophy may be observed [49]. These features on MRI, however, do not differentiate SCA6 from other SCAs presenting with pure cerebellar ataxia, such as sporadic late cortical cerebellar atrophy and autosomal dominat cerebellar ataxia linked to chromosome 16 [19].

Neuro-otological examination reveals DPN, particularly in those with positioning vertigo [26, 42, 50, 51]. In contrast, DPN is uncommon or variable in multiple-system atrophy [42, 51, 52] and other autosomal dominant SCAs, particularly Machado–Joseph disease or SCA3 [42, 51]. Positioning vertigo and DPN tend to

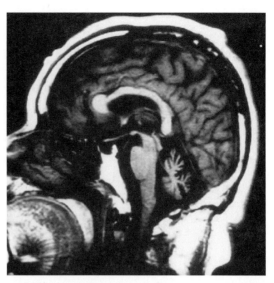

FIGURE 25-5 Brain MRI scan of a SCA6 patient. The cerebellum, particularly the vermis, is atrophied, with widening of sulci and the fourth ventricle. The brainstem and the cerebrum appear normal. T1WI, midsagittal view.

be present at a later stage in SCA6. DPN was reported to be related to impaired cancellation of the vestibulo-ocular reflex (VOR); namely, the inhibitory effect of Purkinje cells in the flocculus and nodulus on the vestibular nuclei is impaired, on the basis of a hyperactive VOR [53–55]. Extensive Purkinje cell loss including the flocculus has been reported in SCA6 [37]. These findings may be accounted for by the pathology in the vermis, flocculus, paraflocculus, or nodulus, with sparing of the paramedian pontine reticular formation, in consonance with previous reports in the literature [56–63].

Although the neuro-otological findings are characteristic of SCA6, they are not specific to the disease. Many autosomal dominant SCAs present with pure cerebellar ataxia, as do sporadic cases with SCA6 gene mutation. Furthermore, some SCA patients with involvement of multiple systems, as in Machado–Joseph disease/SCA3, may present with pure cerebellar phenotype [64]. The diagnosis of SCA6 should be confirmed by gene analysis.

V. TREATMENT, PROGNOSIS, AND PERSPECTIVE

In Japan, the TRH derivatives taltirelin hydrate and protirelin tartate have been approved by the Ministry of Health, Welfare and Labor as effective for ataxic patients with various spinocerebellar ataxias including olivopontocerebellar atrophy or MSA-C and are widely used. In addition, taltirelin hydrate may have other effects on the nervous system, such as a neurotrophic effect [65]. However, general improvement with use of taltirelin was observed in only 22.6% of the patients, twice that in control. In some, episodic symptoms of SCA6 as well as EA2 may be improved by acetazolamide, but the effect is usually limited. The present status of treatment of SCA6 is quite insufficient, although SCA6 patients tend to be ambulatory until relatively late ages, and life span does not seem to be shortened [66].

SCA6 is one of a few neurological diseases in which the functions of causative proteins have been explored, for example, SOD1 causing ALS1, androgen receptor causing spinal and bulbar muscular atrophy, and protein kinase Cγ (PKCγ) causing SCA14 [15]. These diseases may be of some advantage in elucidation of the pathomechanisms. However, research related to Purkinje cells, the main target of SCA6, appears to suffer from the fact that cultured cells into which α_{1A} Ca channels are introduced show only Q-type properties, whereas Purkinje cells exhibit exclusively P-type channel properties. To overcome this difficulty, a suitable animal model is essential. A great deal of effort should be expended in elucidation of the SCA6 pathomechanism, and the development of new therapies based on the mechanism should be continuously pursued.

References

1. Harding, A. E. (1982). The clinical features and classification of the late onset autosomal dominant cerebellar ataxias: A study of 11 families, including descendants of 'Drew family of Walworth.' *Brain* **105**, 1–28.
2. Zhuchenko, O., Bailey, J., Bonnen, P., Ashizawa, T., Stockton, D. W., Amos, C., Dobyns, W. B., Subramony, S. H., Zoghbi, H. Y., and Lee, C. C. (1997). Autosomal dominant cerebellar ataxia (SCA6) associated with small polyglutamine expansions in the α1A voltage-dependent calcium channel. *Nat. Genet.* **15**, 62–69.
3. Ishikawa, K., Tanaka, H., Saito, M., Ohkoshi, N., Fujita, T., Yoshizawa, K., Ikeuchi, T., Watanabe, M., Hayashi, A., Takiyama, Y., Nishizawa, M., Nakano, I., Matsubayashi, K., Miwa, M., and Shoji, S. (1997). Japanese families with autosomal dominant pure cerebellar ataxia map to chromosome 19p13.1-p13.2 and are strongly associated with mild CAG expansions in the spinocerebellar ataxia type 6 gene in chromosome 19p13.1. *Am. J. Hum. Genet.* **61**, 336–346.
4. Matsumura, R., Futamura, N., Fujimoto, Y., Yanagimoto, S., Horikawa, H., Suzumura, A., and Takayanagi, T. (1997). Spinocerebellar ataxia type 6. Molecular and clinical features of 35 Japanese patients including one homozygous for the CAG repeat expansion. *Neurology* **49**, 1238–1243.
5. Watanabe, H., Tanaka, F., Matsumoto, M., Doyu, M., Ando, T., Mitsuma, T., and Sobue, G. (1998). Frequency analysis of autosomal dominant cerebellar ataxias in Japanese patients and clinical characterization of spinocerebellar ataxia type 6. *Clin. Genet.* **53**, 13–19.
6. Sasaki, H., Yabe, I., Yamashita, I., and Tashiro, K. (2000). Prevalence of triplet repeat expansion in ataxia patients from Hokkaido, the northernmost island of Japan. *J. Neurol. Sci.* **175**, 45–51.
7. Maruyama, H., Izumi, Y., Morino, H., Oda, M., Toji, H., Nakamura, S., and Kawakami, H. (2002). Difference in disease-free survival curve and regional distribution according to subtype of spinocerebellar ataxia: A study of 1,286 Japanese patients. *Am. J. Med. Genet.* **114**, 578–583.
8. Geschwind, D. H., Perlman, S., Figueroa, K. P., Karrim, J., Baloh, R. W., and Pulst, S. M. (1997). Spinocerebellar ataxia type 6: Frequency of the mutation and genotype–phenotype correlations. *Neurology* **49**, 1247–1251.
9. Stevanin, G., Durr, A., David, G., Didierjean, O., Cancel, G., Rivaud, S., Tourbah, A., Warter, J. M., Agid, Y., and Brice, A. (1997). Clinical and molecular features of spinocerebellar ataxia type 6. *Neurology* **49**, 1243–1246.
10. Schöls, L., Amoiridis, G., Buttner, T., Przuntek, H., Epplen, J. T., and Riess, O. (1997). Autosomal dominant cerebellar ataxia: Phenotypic differences in genetically defined subtypes? *Ann. Neurol.* **42**, 924–932.
11. Schöls, L., Kruger, R., Amoiridis, G., Przuntek, H., Epplen, J. T., and Riess, O. (1998). Spinocerebellar ataxia type 6: Genotype and phenotype in German kindreds. *J. Neurol. Neurosurg. Psychiatry* **64**, 67–73.
12. Holmberg, M., Johansson, J., Forsgren, L., Heijbel, J., Sandgren, O., and Holmgren, G. (1995). Localization of autosomal dominant cerebellar ataxia associated with retinal degeneration and anticipation to chromosome 3p12-p21.1. *Hum. Mol. Genet.* **4**, 1441–1445.
13. Zu, L., Figueroa, K., Grewal, R., and Pulst, S. (1999). Mapping of a new autosomal dominant spinocerebellar ataxia to chromosome 22. *Am. J. Hum. Genet.* **64**, 594–599.
14. Worth, P. F., Giunti, P., Gardner-Thorpe, C., Dixon, P. H., Davis, M. B., and Wood, N. W. (1999). Autosomal dominant cerebellar ataxia type III: Linkage in a large British family to a 7.6-cM region on chromosome 15q14-21.3. *Am. J. Hum. Genet.* **65**, 420–426.
15. Chen, D. H., Brkanac, Z., Verlinde, C. L., Tan, X. J., Bylenok, L., Nochlin, D., Matsushita, M., Lipe, H., Wolff, J., Fernandez, M., Cimino, P. J., Bird, T. D., and Raskind, W. H. (2003). Missense mutations in the regulatory domain of PKC gamma: A new mechanism for dominant nonepisodic cerebellar ataxia. *Am. J. Hum. Genet.* **72**, 839–849.
16. Knight, M. A., Kennerson, M. L., Anney, R. J., Matsuura, T., Nicholson, G. A., Salimi-Tari, P., Gardner, R. J., Storey, E., and Forrest, S. M. (2003). Spinocerebellar ataxia type 15 (SCA15) maps to 3p24.2–3pter: Exclusion of the ITPR1 gene, the human orthologue of an ataxic mouse mutant. *Neurobiol. Dis.* **13**, 147–157.
17. Miyoshi, Y., Yamada, T., Tanimura, M., Taniwaki, T., Arakawa, K., Ohyagi, Y., Furuya, H., Yamamoto, K., Sakai, K., Sasazuki, T., and Kira, J. (2001). A novel autosomal dominant spinocerebellar ataxia (SCA16) linked to chromosome 8q22.1-24.1. *Neurology* **57**, 96–100.
18. Chung, M. Y., Lu, Y. C., Cheng, N. C., and Soong, B. W. (2003). A novel autosomal dominant spinocerebellar ataxia (SCA22) linked to chromosome 1p21-q23. *Brain* **126**, 1293–1299.
19. Nagaoka, U., Takashima, M., Ishikawa, K., Yoshizawa, K., Yoshizawa, T., Ishikawa, M., Yamawaki, T., Shoji, S., and Mizusawa, H. (2000). A gene on SCA4 locus causes dominantly inherited pure cerebellar ataxia. *Neurology* **54**, 1971–1975.
20. Takashima, M., Ishikawa, K., Nagaoka, U., Shoji, S., and Mizusawa, H. (2001). A linkage disequilibrium at the candidate gene locus for 16q-linked autosomal dominant cerebellar ataxia type III in Japan. *J. Hum. Genet.* **46**, 167–171.
21. Li, M., Ishikawa, K., Toru, S., Tomimitsu, H., Takashima, M., Goto, J., Takiyama, Y., Sasaki, H., Imoto, I., Inazawa, J., Toda, T., Kanazawa, I., and Mizusawa, H. (2003). Physical map and haplo-

type analysis of 16q-linked autosomal dominant cerebellar ataxia (ADCA) type III in Japan. *J. Hum. Genet.* **48**, 111–118.

22. Owada, K., Ishikawa, K., Toru, S., Ishida, G., Gomyoda, M., Tao, O., Noguchi, Y., Kitamura, K., Kondo, I., Noguchi, E., Arinami, T., and Mizusawa, H. (2005). A clinical, genetic and neuropathologic study in a family with 16q-linked ADCA type III. *Neurology*, in press.

23. Ishikawa, K., Toru, S., Tsunemi, T., Li, M., Kobayashi, K., Yokota, T., Amino, T., Owada, K., Fujigasaki, H., Sakamoto, M., Tomimitsu, H., Takashima, M., Kumagai, J., Noguchi, Y., Kawashima, Y., Ohkoshi, N., Ishida, G., Gomyoda, M., Yoshida, M., Hashizume, Y., Saito, Y., Murayama, S., Yamanouchi, H., Mizutani, T., Kondo, I., Toda, T., and Mizusawa, H. (2005). Autosomal dominant cerebellar ataxia linked to chromosome 16q22.1 is associated with a single nucleotide substitution in the 5′-untranslated region of the gene encoding a protein with spectrin repeat and Rho guanine-nucleotide exchange factor domains. *Am. J. Hum. Genet.* **77**, 280–296.

24. Katayama, T., Ogura, Y., Aizawa, H., Kuroda, H., Suzuki, Y., Kuroda, K., and Kikuchi, K. (2000). Nineteen CAG repeats of the SCA6 gene in a Japanese patient presenting with ataxia. *J. Neurol.* **247**, 711–712.

25. Mariotti, C., Gellera, C., Grisoli, M., Mineri, R., Castucci, A., and Di Donato, S. (2001). Pathogenic effect of an intermediate-size SCA-6 allele (CAG)(19) in a homozygous patient. *Neurology* **57**, 1502–1504.

26. Takahashi, H., Ishikawa, K., Tsutsumi, T., Fujigasaki, H., Kawata, A., Okiyama, R., Fujita, T., Yoshizawa, K., Yamaguchi, S., Tomiyasu, H., Yoshii, F., Mitani, K., Shimizu, N., Yamazaki, M., Miyamoto, T., Orimo, T., Shoji, S., Kitamura, K., and Mizusawa, H. (2004). A clinical and genetic study in a large cohort of patients with spinocerebellar ataxia type 6. *J. Hum. Genet.* **49**, 256–264.

27. Ishikawa, K., Watanabe, M., Yoshizawa, K., Fujita, T., Iwamoto, H., Yoshizawa, T., Harada, K., Nakamagoe, K., Komatsuzaki, Y., Satoh, A., Doi, M., Ogata, T., Kanazawa, I., Shoji, S., and Mizusawa, H. (1999). Clinical, neuropathological, and molecular study in two families with spinocerebellar ataxia type 6 (SCA6). *J. Neurol. Neurosurg. Psychiatry* **67**, 86–89.

28. Ophoff, R. A., Terwindt, G. M., Vergouwe, M. N., van Eijk, R., Oefner, P. J., Hoffman, S. M., Lamerdin, J. E., Mohrenweiser, H. W., Bulman, D. E., Ferrari, M., Haan, J., Lindhout, D., van Ommen, G. J., Hofker, M. H., Ferrari, M. D., and Frants, R. R. (1996). Familial hemiplegic migraine and episodic ataxia type-2 are caused by mutations in the Ca^{2+} channel gene CACNL1A4. *Cell* **87**, 543–552.

29. Matsuyama, Z., Minoru, W., Mori, Y., Kawakami, H., Nakamura, S., and Imoto, K. (1999). Direct alteration of the P/Q-type Ca^{2+} channel property by polyglutamine expansion in spinocerebellar ataxia 6. *J. Neurosci.* **19** (RC-14), 1–5.

30. Toru, S., Murakoshi, T., Ishikawa, K., Saegusa, H., Fujigasaki, H., Uchihara, T., Nagayama, S., Osanai, M., Mizusawa, H., and Tanabe, T. (2000). Spinocerebellar ataxia type 6 mutation alters P-type calcium channel function. *J. Biol. Chem.* **275**, 10893–10898.

31. Restituito, S., Thompson, R. M., Eliet, J., Raike, R. S., Riedl, M., Charnet, P., and Gomez, C. M. (2000). The polyglutamine expansion in spinocerebellar ataxia type 6 causes a beta subunit-specific enhanced activation of P/Q-type calcium channels in *Xenopus* oocytes. *J. Neurosci.* **20**, 6394–6403.

32. Piedras-Renteria, E. S., Watase, K., Harata, N., Zhuchenko, O., Zoghbi, H. Y., Lee, C. C., and Tsien, R. W. (2001). Increased expression of alpha 1A Ca^{2+} channel currents arising from expanded trinucleotide repeats in spinocerebellar ataxia type 6. *J. Neurosci.* **21**, 9185–9193.

33. Ishikawa, K., Fujigasaki, H., Saegusa, H., Ohwada, K., Fujita, T., Iwamoto, H., Komatsuzaki, Y., Toru, S., Toriyama, H., Watanabe, M., Ohkoshi, N., Shoji, S., Kanazawa, I., Tanabe, T., and Mizusawa, H. (1999). Abundant expression and cytoplasmic aggregations of α1A voltage-dependent calcium channel protein associated with neurodegeneration in spinocerebellar ataxia type 6. *Hum. Mol. Gene* **8**, 1185–1193.

34. Ishikawa, K., Owada, K., Ishida, K., Fujigasaki, H., Shun, L. M., Tsunemi, T., Ohkoshi, N., Toru, S., Mizutani, T., Hayashi, M., Arai, N., Hasegawa, K., Kawanami, T., Kato, T., Makifuchi, T., Shoji, S., Tanabe, T., and Mizusawa, H. (2001). Cytoplasmic and nuclear polyglutamine aggregates in SCA6 Purkinje cells. *Neurology* **56**, 1753–1756.

35. Kubodera, T., Yokota, T., Ohwada, K., Ishikawa, K., Miura, H., Matsuoka, T., and Mizusawa, H. (2003). Proteolytic cleavage and cellular toxicity of the human alpha1A calcium channel in spinocerebellar ataxia type 6. *Neurosci. Lett.* **341**, 74–78.

36. Matsuyama, Z., Yanagisawa, N. K., Aoki, Y., Black, J. L., 3rd, Lennon, V. A., Mori, Y., Imoto, K., and Inuzuka, T. (2004). Polyglutamine repeats of spinocerebellar ataxia type 6 impair the cell-death-preventing effect of CaV2.1 Ca^{2+} channel: Loss-of-function cellular model of SCA6. *Neurobiol. Dis.* **17**, 198–204.

37. Gomez, C. M., Thompson, R. M., Gammack, J. T., Perlman, S. L., Dobyns, W. B., Truwit, C. L., Zee, D. S., Clark, H. B., and Anderson, J. H. (1997). Spinocerebellar ataxia type 6: Gaze-evoked and vertical nystagmus, Purkinje cell degeneration, and variable age of onset. *Ann. Neurol.* **42**, 933–950.

38. Sasaki, H., Kojima, H., Yabe, I., Tashiro, K., Hamada, T., Sawa, H., Hiraga, H., and Nagashima, K. (1998). Neuropathological and molecular studies of spinocerebellar ataxia type 6 (SCA6). *Acta Neuropathol.* **95**, 199–204.

39. Takahashi, H., Ikeuchi, T., Honma, Y., Hayashi, S., and Tsuji, S. (1998). Autosomal dominant cerebellar ataxia (SCA6): Clinical, genetic and neuropathological study in a family. *Acta Neuropathol.* **95**, 333–337.

40. Tsuchiya, K., Ishikawa, K., Watabiki, S., Tone, O., Taki, K., Haga, C., Takashima, M., Ito, U., Okeda, R., Mizusawa, H., and Ikeda, K. (1998). A clinical, genetic, neuropathological study in a Japanese family with SCA 6 and a review of Japanese autopsy cases of autosomal dominant cortical cerebellar atrophy. *J. Neurol. Sci.* **160**, 54–59.

41. Ikeuchi, T., Takano, H., Koide, R., Horikawa, Y., Honma, Y., Onishi, Y., Igarashi, S., Tanaka, H., Nakao, N., Sahashi, K., Tsukagoshi, H., Inoue, K., Takahashi, H., and Tsuji, S. (1997). Spinocerebellar ataxia type 6: CAG repeat expansion in a1A voltage-dependent calcium channel gene and clinical variations in Japanese population. *Ann. Neurol.* **42**, 879–884.

42. Yabe, I., Sasaki, H., Takeichi, N., Takei, A., Hamada, T., Fukushima, K., and Tashiro, K. (2003). Positional vertigo and macroscopic downbeat positioning nystagmus in spinocerebellar ataxia type 6 (SCA6). *J. Neurol.* **250**, 440–443.

43. Harada, H., Tamaoka, A., Watanabe, M., Ishikawa, K., and Shoji, S. (1998). Downbeat nystagmus in two siblings with spinocerebellar ataxia type 6 (SCA 6). *J. Neurol. Sci.* **160**, 161–163.

44. Sinke, R. J., Ippel, E. F., Diepstraten, C. M., Beemer, F. A., Wokke, J. H., van Hilten, B. J., Knoers, N. V., van Amstel, H. K., and Kremer, H. P. (2001). Clinical and molecular correlations in spinocerebellar ataxia type 6: A study of 24 Dutch families. *Arch. Neurol.* **58**, 1839–1844.

45. Durig, J. S., Jen, J. C., and Demer, J. L. (2002). Ocular motility in genetically defined autosomal dominant cerebellar ataxia. *Am. J. Ophthalmol.* **133**, 718–721.

46. Yabe, I., Sasaki, H., Matsuura, T., Takada, A., Wakisaka, A., Suzuki, Y., Fukazawa, T., Hamada, T., Oda, T., Ohnishi, A., and Tashiro, K. (1998). SCA6 mutation analysis in a large cohort of the Japanese patients with late-onset pure cerebellar ataxia. *J. Neurol. Sci.* **156**, 89–95.

47. Murata, Y., Kawakami, H., Yamaguchi, S., Nishimura, M., Kohriyama, T., Ishizaki, F., Matsuyama, Z., Mimori, Y., and

Nakamura, S. (1998). Characteristic magnetic resonance imaging findings in spinocerebellar ataxia 6. *Arch. Neurol.* **55**, 1348–1352.
48. Satoh, J. I., Tokumoto, H., Yukitake, M., Matsui, M., Matsuyama, Z., Kawakami, H., Nakamura, S., and Kuroda, Y. (1998). Spinocerebellar ataxia type 6: MRI of three Japanese patients. *Neuroradiology* **40**, 222–227.
49. Sugawara, M., Toyoshima, I., Wada, C., Kato, K., Ishikawa, K., Hirota, K., Ishiguro, H., Kagaya, H., Hirata, Y., Imota, T., Ogasawara, M., and Masamune, O. (2000). Pontine atrophy in spinocerebellar ataxia type 6. *Eur. Neurol.* **43**, 17–22.
50. Jen, J. C., Yue, Q., Karrim, J., Nelson, S. F., and Baloh, R. W. (1998). Spinocerebellar ataxia type 6 with positional vertigo and acetazolamide responsive episodic ataxia. *J. Neurol. Neurosurg. Psychiatry* **65**, 565–568.
51. Tsutsumi, T., Kitamura, K., Tsunoda, A., Noguchi, Y., and Mitsuhashi, M. (2001). Electronystagmographic findings in patients with cerebral degenerative disease. *Acta Otolaryngol. (Suppl.)* **545**, 136–139.
52. Bertholon, P., Bronstein, A. M., Davies, R. A., Rudge, P., and Thilo, K. V. (2002). Positional down beating nystagmus in 50 patients: Cerebellar disorders and possible anterior semicircular canalithiasis. *J. Neurol. Neurosurg. Psychiatry* **72**, 366–372.
53. Takemori, S. (1975). Visual suppression of vestibular nystagmus after cerebellar lesions. *Ann. Otol. Rhinol. Laryngol.* **84**, 318–326.
54. Halmagyi, G. M., Rudge, P., Gresty, M. A., and Sanders, M. D. (1983). Downbeating nystagmus: A review of 62 cases. *Arch. Neurol.* **40**, 777–784.
55. Thurston, S. E., Leigh, R. J., Abel, L. A., and Dell'Osso, L. F. (1987). Hyperactive vestibulo-ocular reflex in cerebellar degeneration: Pathogenesis and treatment. *Neurology* **37**, 53–57.
56. Zee, D. S., Yamazaki, A., Butler, P. H., and Gucer, G. (1981). Effects of ablation of flocculus and paraflocculus of eye movements in primate. *J. Neurophysiol.* **46**, 878–899.
57. Henn, V., Lang, W., Hepp, K., and Reisine, H. (1984). Experimental gaze palsies in monkeys and their relation to human pathology. *Brain* **107**, 619–636.
58. Fetter, M., Klockgether, T., Schulz, J. B., Faiss, J., Koenig, E., and Dichgans, J. (1994). Oculomotor abnormalities and MRI findings in idiopathic cerebellar ataxia. *J. Neurol.* **241**, 234–241.
59. Moschner, C., Perlman, S., and Baloh, R. W. (1994). Comparison of oculomotor findings in the progressive ataxia syndromes. *Brain* **117**, 15–25.
60. Buttner, U., and Grundei, T. (1995). Gaze-evoked nystagmus and smooth pursuit deficits: Their relationship studied in 52 patients. *J. Neurol.* **242**, 384–389.
61. Buttner, N., Geschwind, D., Jen, J. C., Perlman, S., Pulst, S. M., and Baloh, R. W. (1998). Oculomotor phenotypes in autosomal dominant ataxias. *Arch. Neurol.* **55**, 1353–1357.
62. Burk, K., Fetter, M., Abele, M., Laccone, F., Brice, A., Dichgans, J., and Klockgether, T. (1999). Autosomal dominant cerebellar ataxia type I: Oculomotor abnormalities in families with SCA1, SCA2, and SCA3. *J. Neurol.* **246**, 789–797.
63. Lin, C. Y., and Young, Y. H. (1999). Clinical significance of rebound nystagmus. *Laryngoscope* **109**, 1803–1805.
64. Ishikawa, K., Mizusawa, H., Igarashi, S., Takiyama, Y., Tanaka, H., Ohkoshi, N., Shoji, S., and Tsuji, S. (1996). Pure cerebellar ataxia phenotype in Machado–Joseph disease. *Neurology* **46**, 1776–1777.
65. Iwasaki, Y., Ikeda, K., Shiojima, T., and Kinoshita, M. (1992). TRH analogue, TA-0910 (3-methyl-(s)-5,6-dihydroorotyl-L-histidyl-L-prolinamide) enhances neurite outgrowth in rat embryo ventral spinal cord in vitro. *J. Neurol. Sci.* **112**, 147–151.
66. Ishikawa, K., Mizusawa, H., Saito, M., Tanaka, H., Nakajima, N., Kondo, N., Kanazawa, I., Shoji, S., and Tsuji, S. (1996). Autosomal dominant pure cerebellar ataxia: A clinical and genetic analysis of eight Japanese families. *Brain* **119**, 1173–1182.

Chapter 26

Pathogenesis of Spinocerebellar Ataxia Type 7: New Insights from Mouse Models and Ataxin-7 Function

DOMINIQUE HELMLINGER AND DIDIER DEVYS

IGBMC, CNRS/INSERM/ULP, BP 10142, 67404 Illkirch Cedex, France

I. Introduction
II. Ataxin-7, the Protein Mutated in SCA7
   A. Expression Levels and Subcellular Localization
   B. Ataxin-7 Function
   C. Ataxin-7 Paralogs
   D. Ataxin-7 Incorporation into TFTC-Type Complexes Is Not Affected by PolyQ Expansion
   E. Future Directions: Effect of PolyQ-Expanded ATXN7 on TFTC/STAGA Function
III. SCA7 Pathogenesis
   A. Phenotypic and Neuropathological Abnormalities of SCA7 Mouse Models
   B. Retinal Pathology of SCA7 Mice
   C. Aggregation of Mutant Ataxin-7 into Nuclear Inclusions
   D. Proteolytic Processing of Mutant Ataxin-7
   E. Polyglutamine Expansion Stabilizes Ataxin-7
   F. Chaperones and the Ubiquitin–Proteasome System
   G. Transcriptional Dysregulation in SCA7 Retinal Dysfunction
IV. Concluding Remarks
Acknowledgments
References

Spinocerebellar ataxia type 7 (SCA7) is an adult-onset neurodegenerative disorder caused by a polyglutamine (polyQ) expansion in ataxin-7 (ATXN7). SCA7 is unique among all polyQ disorders as it is the only one affecting the retina, which is particularly suitable for phenotypic and molecular analysis. Furthermore, SCA7 is the only polyQ disorder in which infantile cases with extremely large expansions are found. This appeared favorable for generating a detectable neurological phenotype over the mouse life span, using either very long CAG/polyQ tracts or cell type-specific overexpression of cDNA constructs with modest repeat lengths. Many SCA7 mouse models have been generated and have provided significant new insights into SCA7 pathogenesis that have advanced our understanding of the mechanisms potentially implicated in polyQ disorders. Moreover, the recent identification of ATXN7 normal function as a subunit of transcriptional coactivator complexes offers new opportunities to understand the mechanism of transcriptional alterations underlying neuronal dysfunction in SCA7 mouse models. Emerging studies propose a link between SCA7 pathogenesis in the retina and ATXN7 normal function.

I. INTRODUCTION

Spinocerebellar ataxia type 7 (SCA7) is an adult-onset neurodegenerative disorder belonging to the clinically and genetically heterogeneous group of autosomal dominant cerebellar ataxias (ADCAs) [1, 2]. In ADCA type II (SCA7), cerebellar ataxia is associated with progressive pigmentary macular degeneration and loss of visual acuity [3–6]. Although SCA7 patients typically present with gait and limb ataxia associated with dysarthria, those with an earlier disease onset tend to develop visual impairment before ataxia [7]. SCA7 is characterized by a marked clinical anticipation, and infantile cases, with an onset below 2 years of age, show a more rapid progression, manifest a spectrum of phenotypes broader than that of adult-onset patients, and die within a few months [7–9].

Pathologically, SCA7 presents as olivocerebellar atrophy with massive neuronal loss and gliosis in the cerebellum, affecting mainly Purkinje cells and the dentate nucleus, and in inferior olive nuclei and mild loss of neurons in the basis pontis and substantia nigra. Degenerative changes in the retina initially affect cone photoreceptors, but progress toward a cone–rod dystrophy [4–6]. Infantile SCA7 cases have been reported with limited cell loss throughout the central nervous system, which is probably due to the short duration of the disease [10–13].

SCA7 is caused by the expansion of a CAG trinucleotide repeat within the coding region of the *SCA7* gene [14–16]. Mutant SCA7 alleles contain between 38 and 70 repeats, but extreme expansions have been documented in infantile cases, up to 460 repeats [7, 8, 17]. The SCA7 mutation is the most unstable among polyQ disorders, as CAG repeats expand at almost every maternal or paternal transmission. The largest CAG repeats found in juvenile and infantile cases are always transmitted by affected fathers carrying moderate expanded repeats [8, 17], in agreement with the extreme instability of the SCA7 CAG repeat in the male germline [18]. The importance of genomic context for SCA7 CAG repeat instability has been suggested by studies comparing transgenic mice carrying a SCA7 mutation in either a genomic fragment or a cDNA construct. The CAG repeat was highly unstable specifically in its proper genomic context [19].

SCA7 is unique among all polyQ disorders as it is the only one that affects the retina, and is particularly suitable for phenotypic and molecular analysis. Furthermore, SCA7 is the only polyQ disorder in which infantile cases with extremely large expansions are found. This appeared favorable for generating a detectable neurological phenotype over the mouse life span, using either very long CAG/polyQ tracts or cell type-specific overexpression of cDNA constructs with modest repeat lengths (Table 26-1). Many SCA7 mouse models have been generated and have provided significant new insights into SCA7 pathogenesis that advanced our understanding of the mechanisms potentially implicated in polyQ disorders. Moreover, the recent identification of ataxin-7 (ATXN7) normal function as a subunit of TFTC/STAGA transcriptional coactivator complexes provides new opportunities to understand the mechanism of transcriptional alterations underlying neuronal dysfunction in SCA7 mouse models. Emerging studies propose a link between SCA7 pathogenesis in the retina and ATXN7 normal function.

II. ATAXIN-7, THE PROTEIN MUTATED IN SCA7

A. Expression Levels and Subcellular Localization

Northern blot analysis has shown that *SCA7* expression is ubiquitous, including different brain regions and nonneuronal tissues [14, 20]. In particular, Lindenberg

TABLE 26-1 SCA7 Mouse Models

Model design			Expression		Phenotype			
Promoter	Protein length	PolyQ length	Level	Cell type	Onset (months)	Severity[a]	Death	Line Ref.
Transgenic SCA7 mice								
Prp	Full-length	92Q	Moderate	CNS	3	+++	5 months	6076[b]
Rhodopsin	Full-length	90Q	High	Rods	1	+++	No	R7E[b]
Pcp-2	Full-length	90Q	Moderate	Purkinje	11	+	No	P7E[b]
PDGF-B	Full-length	128Q	Moderate	CNS	5	+/++	8 months/No	B7E2[b]
Knock-in SCA7 mice								
Endogenous	Full-length	266Q	Endogenous		1	+++	5 months	H10[b]

[a] The number of plus signs denotes the severity of the phenotype, from + (mild) to +++ (severe).
[b] Data summarized in this table were obtained from the most characterized lines in each study.

et al. showed by *in situ* hybridization that *SCA7* mRNA is found in both affected and unaffected neurons [21]. Regional distribution of ATXN7 immunoreactivity does not differ between patients and controls. Several studies using different antibodies suggest that endogenous ATXN7 subcellular localization varies between neuronal types, being either nuclear, cytoplasmic, or even both [20–24]. However, ATXN7 is not preferentially nuclear in neurons affected in SCA7 pathology.

ATXN7 sequence analysis predicted a bipartite nuclear localization signal (NLS, position 378–393), which was shown to be functional in transfected cells [25, 26]. Another study suggested that ATXN7 nuclear localization relied on two putative NLSs located in the C-terminal part of the protein [27]. Recombinant normal ATXN7 was found to be nuclear in all neuronal types from different transgenic mice [28, 29], and Yoo et al. showed that endogenous mouse ATXN7 is predominantly nuclear in the cerebellum using fractionation experiments [30].

An ATXN7 isoform (ATXN7b), generated by incorporation of a previously uncharacterized *SCA7* exon, is expressed predominantly in the central nervous system (CNS) and exhibits cytosolic localization [31]. It still has not been determined whether incorporation of an alternative 58-amino acid C-terminal extension in a poorly conserved region of ATXN7 modifies its function or plays a specific role in SCA7 pathogenesis.

B. Ataxin-7 Function

Ataxin-7 is a 892-amino acid protein, for which the normal function was recently discovered. Bioinformatic analyses identified a significant block of homology between human ATXN7 and a yeast open reading frame, *YGL066w*, that was identified as encoding a novel subunit of the SAGA complex (Spt/Ada/Gcn5 acetylase) and therefore named SAGA-associated factor 73 (Sgf73) [32, 33].

The yeast SAGA is a multisubunit coactivator complex required for transcription of a subset of RNA polymerase II-dependent genes [34, 35]. We and others have shown that human ATXN7 is an integral component of mammalian SAGA-like complexes, the TATA-binding protein-free TAF-containing complex (TFTC), and the SPT3/TAF9/GCN5 acetyltransferase complex (STAGA) [36, 37] TFTC/STAGA complexes contain the GCN5 histone acetyltransferase (HAT), TRRAP, SPT, and ADA proteins, and a subset of TATA-binding protein (TBP)-associated factors (TAFs). These complexes were shown to preferentially acetylate histone H3 in both free and nucleosomal contexts and to activate transcription on chromatin templates [38–40]. ATXN7 is the fourth member of the polyQ proteins, in addition to the androgen receptor, TBP, and atrophin-1, the normal function of which is involved in transcriptional regulation.

C. Ataxin-7 Paralogs

We also defined a *SCA7* gene family [36]. Four ATXN7 paralogs were identified in vertebrate genomes. Sequence comparison of ATXN7 family members notably revealed three conserved blocks: block I (residues 126–176 in human ATXN7), block II (residues 341–400), and block III (residues 508–565) (Fig. 26-1). The expandable polyQ motif, located N-terminal to block I, is present in human ATXN7 and in its vertebrate orthologs, but is absent in ATXN7 paralogs. In invertebrate and yeast genomes, a single ATXN7 sequence was found and characterized by a weakly conserved block I (16% identity between *Homo sapiens* and *Saccharomyces cerevisiae*) located N-terminal to a highly conserved block II (42% identity between *H. sapiens* and *S. cerevisiae*), and by absence of block III [36, 41, 42]. Two glutamine-rich regions (11 and 8 Q's) are found in the C-terminal part of ATXN7 yeast ortholog (Sgf73) (Fig. 26-1).

Many TAF paralogs can participate in TFTC/STAGA assembly and have redundant and complementary roles in transcriptional regulation [43]. It would thus be interesting to analyze ATXN7 paralog expression profiles, to determine whether these different proteins can be present in TFTC/STAGA complexes and what their respective roles are *in vivo*. In particular, ATXN7 family members harboring block III might have a specific function as this domain is not present in Sgf73 and thus not required for SAGA complex assembly.

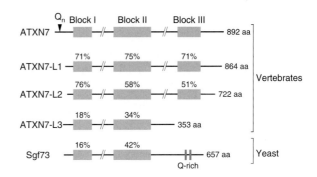

FIGURE 26-1 ATXN7 gene family. Schematic representation of ATXN7 paralogs found in vertebrates (human, mouse, and fish). A single ATXN7 ortholog is found in yeast and in invertebrates. For each conserved block (gray boxes), sequence identities of human paralogs and Sgf73 are indicated relative to human ATXN7. Adapted from Animal Models of Movement Disorders, Helmlinger, D., and Devys, D., SCA7 mouse models, (2005), with permission from Elsevier.

D. Ataxin-7 Incorporation into TFTC-Type Complexes Is Not Affected by PolyQ Expansion

Recent work from our laboratory showed that both normal and mutant ATXN7 can incorporate into TFTC/STAGA complexes immunopurified from SCA7 patient cell lines or transfected HEK293 cells [36]. A recent study in yeast demonstrated that Sgf73 is required for SAGA complex integrity [44]. The study further showed that expression of human ATXN7 in yeast carrying a Sgf73 deletion restored a functional SAGA complex. However, under the same conditions, polyQ-expanded ATXN7 assembled a SAGA complex with reduced level of different subunits (Ada2, Ada3, and TAF12). Similar observations were reported in an independent study using stable cell lines expressing normal or mutant ATXN7 [37]. It remains to be determined, however, whether such effects can be observed in mammals, in particular, by using existing SCA7 mouse models (see Sections IIIE and IIIG).

E. Future Directions: Effect of PolyQ-Expanded ATXN7 on TFTC/STAGA Function

Both normal and mutant ATXN7 can be incorporated into TFTC/STAGA complexes. It will thus be of major interest to assess HAT activity in TFTC/STAGA complexes containing either normal or mutant ATXN7. Two recent studies using yeast and stable cell lines indicate that polyQ-expanded ATXN7 inhibits the nucleosomal histone acetylation function of SAGA or STAGA *in vitro* [37, 44]. Furthermore, polyQ-expanded ATXN7 inhibits HAT activity of STAGA complexes containing normal ATXN7, indicating a dominant negative mechanism for mutant ATXN7. To confirm this hypothesis, it is crucial to determine whether inactivation of STAGA-dependent nucleosomal acetylation occurs *in vivo* and is responsible for mutant ATXN7-induced neuronal toxicity (Section IIIG).

Alternatively, a polyQ expansion in TFTC/STAGA complexes could lead to deregulated transcriptional activity of the TFTC/STAGA complexes. The presence of the polyQ expansion within a coactivator complex would serve as an aberrant interaction domain with the chromatin, sequestering TFTC/STAGA at aberrant sites. In this model, deregulation of TFTC/STAGA function would lead to aberrant histone modifications and a diversion of activators away from the genes they normally regulate, leading to a loss of a tissue-specific gene expression profile and, eventually, to SCA7 pathogenesis.

III. SCA7 PATHOGENESIS

A. Phenotypic and Neuropathological Abnormalities of SCA7 Mouse Models

All SCA7 mouse models that have been generated express full-length ATXN7, but differ in the size of the polyQ expansion and in the neuronal types targeted (Table 26-1). Four SCA7 transgenic models have been generated using heterologous promoters. Platelet-derived growth factor chain B (PDGF-B) and prion (Prp) promoters led to widespread overexpression of mutant ATXN7 with 90 to 128 Q's, whereas rhodopsin (Rho) and Pcp2 promoters led to restricted overexpression of the transgene product in Purkinje cells and rod photoreceptors, respectively [28, 45–47]. Yoo et al. generated SCA7 knock-in mice by introducing a 266-CAG repeat into the *Sca7* locus by homologous recombination, leading to expression of the mutant protein at endogenous levels and in the proper spatiotemporal pattern [30].

In knock-in and transgenic models, widespread expression of mutant ATXN7 in the CNS causes an adult-onset, progressive neurological phenotype leading to premature death. These mice typically present with gait ataxia and motor incoordination. As disease progresses, symptoms become more complex and variable between the different models, although tremor, weight loss, and hypoactivity are consistently observed. Phenotypic differences are likely brought out by variable expression patterns and by the distinct genetic backgrounds used to generate these mice. Also, differences in mutant ATXN7 expression levels and polyQ length likely account for the specific severity reported in the different models. Knock-in mice present symptoms strikingly similar to those reported in the infantile patient from whom the 266-CAG expansion was cloned, with comparable disease onset (1 month) and progression (death around 5 months). Finally, the enhanced toxicity reported in homozygous Sca7 266Q/266Q mice and in homozygous R7E mice indicates a gene dosage effect [29, 30].

As demonstrated for other polyQ disorders, prolonged exposure of neurons to mutant proteins is necessary for neuronal dysfunction. Thus, either high overexpression of mutant ATXN7 or very long polyQ tracts were shown to be required to produce a phenotype during the short mouse life span. Indeed, no ataxic phenotype was detected in one Prp line (line 6529) that expresses mutant ATXN7 at 0.75-fold the endogenous level. A severe neurological phenotype and premature death were produced by overexpression of ATXN7 with 92 Q's (2.5-fold the endogenous level) in another Prp line (line 6561) and by an endogenous level of ATXN7 with an extremely large polyQ expansion (266 Q's).

Reduced-size Purkinje cell bodies have been described in two transgenic models (P7E and Prp lines) and in the knock-in model [28, 30, 47]. Purkinje cell pathology correlated with the onset of neurological signs and occurred without significant cell loss, as suggested by normal cell counts and by the absence of apoptotic cells. Interestingly, SCA1 knock-in mice exhibit Purkinje cell pathology, with marked reduction of dendritic arborization and significant cell loss [48]. Although Purkinje cells are affected in both SCA1 and in SCA7 knock-in mice, the dissimilar neuronal pathology highlights the importance of protein context in polyQ-induced toxicity.

B. Retinal Pathology of SCA7 Mice

SCA7 is unique among all polyQ disorders, as it is the only one commonly affecting the retina in addition to the brain. By the use of different SCA7 mouse models, numerous studies have attempted to decipher the specific mechanisms that underlie SCA7-restricted retinal involvement. Furthermore, mouse retina is a valuable tissue in which to study neurodegeneration, as neuronal dysfunction can be reproducibly quantified in living animals using electroretinography (ERG). An early-onset decrease in ERG recordings was observed in all SCA7 models and eventually led to the complete absence of photoreceptor responses [28, 30, 47]. In R7E mice, in which transgene expression is restricted to rod photoreceptors, rod response progressively decreased up to flat recordings from 1 year of age. In mouse models expressing mutant ATXN7 in all retinas (knock-in and Prp lines) cone-mediated responses were shown to be affected before rod-mediated responses, raising the possibility that mouse cone photoreceptors may be more sensitive to mutant ATXN7, similarly to the human pathology. However, mouse cone photoreceptors may simply be particularly sensitive to any polyQ expansion, as suggested by occurrence of a cone–rod dystrophy in a Huntington's disease (HD) mouse model [49].

The most prominent pathological feature of the retina from all SCA7 mouse models is progressive thinning of photoreceptor outer and inner segments, in which phototransduction takes place. Significant photoreceptor cell loss was consistently evidenced by cell counting and TUNEL staining in the retinas from SCA7 mice. However, photoreceptor degeneration was limited, as knock-in and R7E mice retained 70–80% of photoreceptors until terminal stages of the disease. Furthermore, ERG abnormalities appeared prior to photoreceptor cell loss, which was detectable at 15 weeks of age in knock-in mice and from 1 year of age in R7E mice.

In R7E mice, repression of rhodopsin promoter activity led to early and severe downregulation of transgene expression. Although residual expression of mutant ATXN7 was found to be negligible from 9 weeks of age, SCA7 transgenic mice showed a progressive decline in photoreceptor activity, leading to complete loss of ERG responses from 1 year of age [29]. These results suggest that mutant ATXN7-induced toxicity might not be reversible beyond a given pathological threshold. In R7E mice, disease progression despite long-term extinction of mutant ATXN7 expression could be due to a combination of irreversible transcriptional abnormalities and partial persistence of aggregates (see Section IIIE).

C. Aggregation of Mutant Ataxin-7 into Nuclear Inclusions

Like other polyglutamine disease-causing proteins, mutant ATXN7 aggregates and forms nuclear inclusions (NIs), which were first detected by immunohistology analyses of patient brains. Several studies using different anti-ATXN7 antibodies showed that NIs were not particularly enriched in SCA7-vulnerable neurons [22–24, 50, 51]. However, the regional distribution of NIs varied between these different studies.

All SCA7 mouse models that have been generated display numerous NIs, both in regions that are affected in patients (e.g., cerebellum, inferior olive nucleus, retina) and in nonaffected areas (e.g., cerebral cortex, hippocampus) [28, 30, 46]. Interestingly, in SCA7 knock-in mice, aggregate formation occurred after the onset of neuronal dysfunction [30]. However, NIs developed earlier in cone than in rod photoreceptors from these mice, agreeing with an earlier onset of cone dysfunction as assessed by specific ERG recordings. In the early stages of SCA7 knock-in mouse pathogenesis, mutant ATXN7 could only be detected as insoluble material by Western blot analysis of cerebellar extracts, before the appearance of visible NIs in Purkinje cells. Together with the observation that expanded polyQ impedes ATXN7 turnover (Section IIIE), these results strongly suggest that mutant ATXN7 is accumulating more rapidly in affected neurons. Insoluble microaggregates that cannot be detected by conventional microscopy were recently identified in other polyQ models and could themselves be pathogenic (reviewed in [52]).

In 2-years-old R7E mice, a significant number of rods retained aggregates despite early loss of mutant ATXN7 expression and normal proteasomal activity (D. Helmlinger, unpublished results). This observation suggests that, in this model, aggregate formation is rapid whereas aggregate clearance is partial and much slower than reported previously in an inducible HD model [53]. In addition, fluorescence imaging of living cells expressing polyQ expansions revealed the coexistence of stable and dynamic aggregates [54], which could explain the observation that aggregate formation is not fully reversible in R7E mice.

Many studies have reported abnormal accumulations of a variety of essential proteins, including chaperones and proteasome subunits, into polyQ aggregates from patients and animal models. It has been proposed that recruitment and sequestration of transcription factors, particularly those containing a glutamine-rich motif (e.g., CREB-binding protein [CBP]), would interfere with their normal function, thereby contributing to pathogenesis. In all SCA7 transgenic models, CBP was found associated with aggregates but without any evidence of depletion from its normal localization. Moreover, Yoo et al. reported that CBP was found associated with NIs late in pathogenesis (18 weeks of age, i.e., just before death), whereas these knock-in mice displayed severe transcriptional defects as early as 4 weeks of age [30]. So far, no transcription factor for which localization has been analyzed in SCA7 mouse models was found to be depleted in the nucleoplasm, even when it was associated with NIs [46]. Furthermore, there was no correlation between the presence of a glutamine-rich motif in the respective transcription factors and their likelihood of being recruited into NIs.

D. Proteolytic Processing of Mutant Ataxin-7

Specific proteolytic processing of expanded polyQ proteins has been reported for HD, DRPLA, and SCA3, but not for SCA1 (reviewed in [55]). It has been hypothesized that this cleavage would lead to nuclear translocation and aggregation of polyQ protein fragments and thus initiate toxic events within the nucleus. Immunohistochemical analysis of SCA7 transgenic mice revealed that ATXN7-formed NIs are not immunoreactive for antibodies recognizing C-terminal epitopes of the protein [47, 56]. This cleavage event affects specifically mutant ATXN7 as the normal form remains full-length and was consistently observed in all neuronal types expressing mutant ATXN7, in particular, both SCA7-vulnerable and spared neurons [28, 46]. However, it is still not known whether this corresponds to specific cleavage of soluble mutant ATXN7 preceding the formation of aggregates or to nonspecific trimming of aggregated proteins.

E. Polyglutamine Expansion Stabilizes Ataxin-7

Several pieces of experimental evidence indicate that there is differential handling of mutant versus normal ATXN7 by neurons *in vivo*. Normal recombinant ATXN7 is barely detectable in P7E and B7E2 transgenic mice, whereas mutant ATXN7 progressively accumulates, becoming easily detectable as homogenous nuclear staining, and then gradually aggregates to form NIs [46]. This observation was then confirmed in SCA7 knock-in mice, indicating that selective stabilization of mutant ATXN7 is a crucial pathogenic mechanism [30]. Mutant ATXN7 selective accumulation probably results from a reduced turnover of expanded polyQ ATXN7, as mutant SCA7 mRNA levels were constant with age and identical to wild-type SCA7 mRNA levels. In knock-in mice, more rapid accumulation of mutant ATXN7 was observed in brain regions exhibiting marked dysfunction [30]. These regions include the cerebellum and retina, which are known to degenerate in SCA7 patients, but intriguingly also the hippocampus, which is not classically affected in SCA7. Thus, the mechanisms by which polyQ expansion stabilizes ATXN7 and whether this observation accounts for the regional specific degeneration in SCA7 need to be further investigated.

F. Chaperones and the Ubiquitin–Proteasome System

Nuclear inclusions formed by polyQ-expanded proteins accumulate molecular chaperones and components of the ubiquitin–proteasome system. Manipulation of chaperone levels in various cellular and invertebrate polyQ models and *in vitro* proteasome assays strongly suggested that impairment of the protein folding and degradation machineries may be a common pathogenic mechanism in these diseases [57]. However, overexpression of HSP70 chaperone led to a modest improvement in neurological symptoms in mouse models [58–60]. In mouse models of SCA7 and HD, the levels of several chaperones, including HSP70 and the HSP40 chaperones, were found to be reduced [49, 60]. HSP40 cochaperones recognize abnormally folded polypeptides and present them to HSP70, while at the same time stimulating HSP70 ATPase activity.

Two recent studies in SCA7 transgenic and knock-in mice directly challenged the involvement of chaperone and proteasome impairment in polyQ pathogenesis [56, 61]. Using newly generated transgenic mice that overexpress both HSP70 and HDJ2 specifically in rod photoreceptors, we demonstrated that high-level expression of these molecular chaperones, alone or in combination, did not modulate SCA7 retinal phenotype. Rod dysfunction was not overcome in R7E mice overexpressing both HSP70 and HSP40, as assessed by the similar reduction of ERG responses and the comparable histological and transcriptional abnormalities. Surprisingly, we found that although coexpression of HSP70 with its cofactor efficiently suppressed mutant

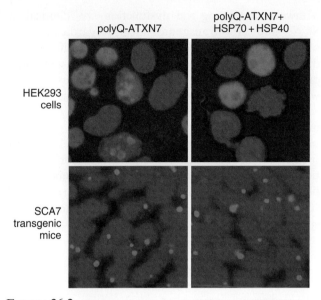

FIGURE 26-2 Overexpression of both HSP70 and HSP40 chaperones suppresses aggregate formation in transfected cells (top) but not in SCA7 transgenic mice (bottom). Mutant ataxin-7 was detected using an anti-N-terminal antibody in HEK293T cells or in rod photoreceptors from R7E mice. See CD-ROM for color image. Reprinted from Animal Models of Movement Disorders, Helmlinge, D. and Devys, D., SCA7 mouse models, 637–648 (2005), with permission from Elsevier.

protein aggregation in transfected cells (Fig. 26-2), the chaperones did not prevent neuronal toxicity or aggregate formation in SCA7 mice [56]. These contradictory findings substantiate the value of using *in vivo* models when studying polyQ pathogenesis.

Bowman et al. assessed the contribution of the ubiquitin–proteasome system in SCA7 knock-in mouse retinopathy, using transgenic mice that express a green fluorescent protein-based reporter substrate (UbG76V-GFP [62]). The levels of the reporter remained low during the initial phase of disease and were even increased, through an expression-dependent mechanism, suggesting that neuronal dysfunction occurs in the presence of a functional ubiquitin–proteasome system [61]. An *in vitro* assay also showed normal proteasome proteolytic activity in the retina. Altogether, these data exclude significant impairment of the protein folding and degradation machineries as a necessary step for SCA7 toxicity in the retina.

G. Transcriptional Dysregulation in SCA7 Retinal Dysfunction

Several studies have suggested that transcriptional dysregulation may be a primary pathogenic process in polyQ diseases (reviewed in [63, 64]). In agreement with this hypothesis, increasing evidence points to the nucleus as the prime site of polyQ toxicity. In particular, several studies have shown that nuclear localization of expanded polyQ proteins is necessary and sufficient to induce toxicity, at least in SCA1 and SBMA models [65–67].

Expanded polyQ-containing proteins have been proposed to interfere with the normal function of several transcription factors, such as CBP and Sp1 [68–72]. Early and severe transcriptional alterations were detected in various polyQ eukaryotic models [72–76]. Finally, histone deacetylase inhibitors, which stimulate gene transcription, are protective in HD mouse and *Drosophila* models [71, 77, 78]. However, the exact molecular mechanims underlying polyQ-induced transcriptional dysregulation and the extent to which altered expression of these genes contributes to neuronal dysfunction are yet unclear.

In retinas from both SCA7 transgenic (R7E) and knock-in mice, mutant ATXN7 induces an early and dramatic downregulation of several photoreceptor-specific genes, particularly those encoding components of the phototransduction pathway, such as rhodopsin, cone opsins, and transducins (Fig. 26-3C). Time-course analysis revealed that both models develop comparable abnormalities as revealed by progressive reduction of rhodopsin mRNA levels concomitant with decreased rod ERG responses and thinning of their segments (Fig. 26-3) [29, 30]. These results support the hypothesis that the polyQ-induced retinal dysfunction observed in mice and perhaps also in SCA7 patients may be caused by reduced expression of photoreceptor-specific genes.

Crx has a major role in the transcriptional regulation of photoreceptor-specific genes such as rhodopsin [79, 80]. Using *in vitro* experiments, La Spada and coworkers reported an interaction between the Gcn5 HAT and the Crx activator through ATXN7, suggesting that Crx recruits STAGA and Gcn5 to retinal-specific genes [27, 37, 45]. These authors also showed that expanded ATXN7 impairs both Gcn5 histone H3 acetylation activity and Crx DNA binding *in vitro*. Presymptomatic SCA7 transgenic mice presented very mild transcriptional abnormalities at an age when Crx recruitment to its promoters was significantly reduced.

Finally, Gcn5 occupancy at these promoters was unchanged in SCA7 transgenic mice, showing that STAGA can be recruited to retina-specific genes in a Crx-independent mechanism [37]. By contrast, Yoo et al. reported early and robust transcriptional alterations without modification of Crx levels and activity in their SCA7 knock-in model and found that some retina-specific genes that are not regulated by Crx were also severely downregulated [30]. Furthermore, in all SCA7 retinal models, Crx nuclear distribution was found unaltered by aggregation of mutant ATXN7 into NIs, precluding abnormal sequestration of Crx by mutant

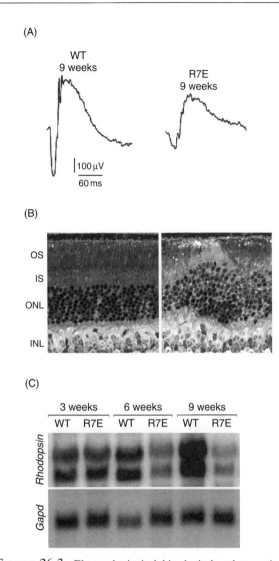

FIGURE 26-3 Electrophysiogical, histological, and transcriptional abnormalities in retinas from SCA7 transgenic mice (R7E mice). (A) Scotopic electroretinographic (ERG) responses from wild-type and R7E mice at 9 weeks of age. The decrease in ERG response was concomitant with progressive thinning of the segment layer. (B) The photoreceptor nuclear layer (ONL) is completely disrupted by numerous whorls that correspond to focal thinning of the outer segment (OS) layer. OS, outer segment; IS, inner segment; ONL, outer nuclear layer; INL, inner nuclear layer. (C) Northern blot of total RNA extracted from wild-type and R7E mice revealed that rhodopsin mRNA levels are progressively and severely reduced in SCA7 mice. See CD-ROM for color image. Reprinted from Animal Models of Movement Disorders, Helmlinge, D. and Devys, D., SCA7 mouse models, 637–648 (2005), with permission from Elsevier.

ATXN7. Finally, R6/2 HD transgenic mice develop a retinopathy comparable to that of SCA7 transgenic mice and a dramatic downregulation of rhodopsin expression, demonstrating that rhodopsin promoter activity can be repressed *in vivo* by polyQ expansion in two unrelated proteins, ATXN7 and huntingtin [29, 49]. Further investigations are needed to assess whether the molecular pathways underlying rod dysfunction are identical in SCA7 and HD transgenic mice. Regardless, these observations together indicate that the specific retinal dysfunction in SCA7 patients cannot be explained exclusively by specific interference of mutant ATXN7 with Crx-dependent transcriptional regulation.

In summary, study of the retinopathy of SCA7 mouse models has provided new insights into the molecular mechanisms underlying neuronal dysfunction in this tissue. Mutant ATXN7-induced photoreceptor dysfunction, as assessed by a progressive decline in their electrophysiological activity, may be due primarily to early and severe transcriptional dysregulation of photoreceptor-specific genes.

IV. CONCLUDING REMARKS

All SCA7 mice models that have been generated express full-length ATXN7 with a polyQ expansion ranging from 90 to 266 repeats and together faithfully reproduce several key neurological features reported in SCA7 patients. These models have unraveled some of the early molecular events involved in SCA7 pathogenesis. Expansion of the polyQ tract in ATXN7 markedly reduces the turnover of the mutant protein, thus increasing mutant ATXN7 levels in neurons and presumably inducing its aggregation in the nucleus. Nuclear accumulation of mutant ATXN7 parallels the onset of functional deficits in affected neurons, indicating that mutant ATXN7 compromises essential nuclear functions. In retinas from these mice, expression of genes essential for normal photoreceptor function is repressed early and dramatically, thereby triggering a progressive decline in photoreceptor activity leading to a complete loss of electrophysiological responses and, eventually, limited cell loss (Fig. 26-2). The retinal phenotype in SCA7 mice thus primarily results from severe and long-lasting neuronal dysfunction.

These SCA7 retinal models are particularly attractive to explore further the therapeutic potential of drugs that could improve the transcriptional changes observed in polyQ disorders. Indeed, mouse retina is a suitable tissue for drug delivery, and SCA7 models present functional abnormalities that can be easily assessed in living animals and that are directly linked to dramatic transcriptional alterations.

Comparison of different polyQ mouse models reveals that either very long CAG repeats or significant overexpression of transgenes with modest repeat lengths is required to generate a phenotype during the short life span of the mouse. SCA7 knock-in (266 Q's) and transgenic (90 Q's) mice developed a similar phenotype and showed comparable molecular alterations, such as selective stabilization of mutant ATXN7 and transcriptional dysregulation. Both models are thus valuable tools with which to further dissect the molecular mechanisms underlying SCA7 pathogenesis.

Acknowledgments

The authors thank Jean-Louis Mandel and Làszlò Tora for continuous support and critical reading of the manuscript. The work on SCA7 in the authors' laboratory is funded by the Institut National de la Santé et de la Recherche Médicale, the Centre National de la Recherche Scientifique, the Hôpital Universitaire de Strasbourg (HUS), the Collège de France, and the European Community (EUROSCA, PL 503304), and the National Organization of Rare Disorders.

References

1. Harding, A. E. (1993). Clinical features and classification of inherited ataxias. *Adv. Neurol.* **61**, 1–14.
2. Konigsmark, B. W., and Weiner, L. P. (1970). The olivopontocerebellar atrophies: A review. *Medicine (Baltimore)* **49**, 227–241.
3. Benomar, A., Le Guern, E., Durr, A., Ouhabi, H., Stevanin, G., Yahyaoui, M., Chkili, T., Agid, Y., and Brice, A. (1994). Autosomal-dominant cerebellar ataxia with retinal degeneration (ADCA type II) is genetically different from ADCA type I. *Ann. Neurol.* **35**, 439–444.
4. Enevoldson, T. P., Sanders, M. D., and Harding, A. E. (1994). Autosomal dominant cerebellar ataxia with pigmentary macular dystrophy: A clinical and genetic study of eight families. *Brain* **117**, 445–460.
5. Gouw, L. G., Digre, K. B., Harris, C. P., Haines, J. H., and Ptacek L. J. (1994). Autosomal dominant cerebellar ataxia with retinal degeneration: Clinical, neuropathologic, and genetic analysis of a large kindred. *Neurology* **44**, 1441–1447.
6. Martin, J. J., Van Regemorter, N., Krols, L., Brucher, J. M., de Barsy, J., Szliwowski, H., Evrard, P., Ceuterick, C., Tassignon, M. J., and Smet-Dieleman, H. (1994). On an autosomal dominant form of retinal-cerebellar degeneration: An autopsy study of five patients in one family. *Acta Neuropathol.* **88**, 277–286.
7. Johansson, J., Forsgren, L., Sandgren, O., Brice, A., Holmgren, G. and Holmberg, M. (1998). Expanded CAG repeats in Swedish spinocerebellar ataxia type 7 (SCA7) patients: Effect of CAG repeat length on the clinical manifestation. *Hum. Mol. Genet.* **7**, 171–176.
8. Benton, C. S., de Silva, R., Rutledge, S. L., Bohlega, S., Ashizawa, T., and Zoghbi, H. Y. (1998). Molecular and clinical studies in SCA-7 define a broad clinical spectrum and the infantile phenotype. *Neurology* **51**, 1081–1086.
9. David, G., Durr, A., Stevanin, G., Cancel, G., Abbas, N., Benomar, A., Belal, S., Lebre, A. S., Abada-Bendib, M., Grid, D., Holmberg, M., Yahyaoui, M., Hentati, F., Chkili, T., Agid, Y., and Brice, A. (1998). Molecular and clinical correlations in autosomal dominant cerebellar ataxia with progressive macular dystrophy (SCA7). *Hum. Mol. Genet.* **7**, 165–170.
10. Carpenter, S., and Schumacher, G. A. (1966). Familial infantile cerebellar atrophy associated with retinal degeneration. *Arch. Neurol.* **14**, 82–94.
11. de Jong, P. T., de Jong, J. G., de Jong-Ten Doeschate, J. M., and Delleman, J. W. (1980). Olivopontocerebellar atrophy with visual disturbances: An ophthalmologic investigation into four generations. *Ophthalmology* **87**, 793–804.
12. Ryan, S. J., Jr., Knox, D. L. Green, W. R. and Konigsmark, B. W. (1975). Olivopontocerebellar degeneration: Clinicopathologic correlation of the associated retinopathy. *Arch. Ophthalmol.* **93**, 169–172.
13. Traboulsi, E. I., Maumenee, I. H., Green, W. R., Freimer, M. L., and Moser, H. (1988). Olivopontocerebellar atrophy with retinal degeneration: A clinical and ocular histopathologic study. *Arch. Ophthalmol.* **106**, 801–806.
14. David, G., Abbas, N., Stevanin, G., Durr, A., Yvert, G., Cancel, G., Weber, C., Imbert, G., Saudou, F., Antoniou, E., Drabkin, H., Gemmill, R., Giunti, P., Benomar, A., Wood, N., Ruberg, M., Agid, Y., Mandel, J. L., and Brice, A. (1997). Cloning of the SCA7 gene reveals a highly unstable CAG repeat expansion. *Nat. Genet.* **17**, 65–70.
15. Del-Favero, J., Krols, L., Michalik, A., Theuns, J., Lofgren, A., Goossens, D., Wehnert, A., Van den Bossche, D., Van Zand, K., Backhovens, H., van Regenmorter, N., Martin, J. J., and Van Broeckhoven, C. (1998). Molecular genetic analysis of autosomal dominant cerebellar ataxia with retinal degeneration (ADCA type II) caused by CAG triplet repeat expansion. *Hum. Mol. Genet.* **7**, 177–186.
16. Koob, M. D., Benzow, K. A., Bird, T. D., Day, J. W., Moseley, M. L., and Ranum, L. P. (1998). Rapid cloning of expanded trinucleotide repeat sequences from genomic DNA. *Nat. Genet.* **18**, 72–75.
17. van de Warrenburg, B. P., Frenken, C. W., Ausems, M. G., Kleefstra, T., Sinke, R. J., Knoers, N. V., and Kremer, H. P. (2001). Striking anticipation in spinocerebellar ataxia type 7: The infantile phenotype. *J. Neurol.* **248**, 911–914.
18. Monckton, D. G., Cayuela, M. L., Gould, F. K., Brock, G. J., Silva, R., and Ashizawa, T. (1999). Very large (CAG)(n) DNA repeat expansions in the sperm of two spinocerebellar ataxia type 7 males. *Hum. Mol. Genet.* **8**, 2473–2478.
19. Libby, R. T., Monckton, D. G., Fu, Y. H., Martinez, R. A., McAbney, J. P. Lau, R., Einum, D. D., Nichol, K., Ware, C. B., Ptacek, L. J., Pearson, C. E., and La Spada, A. R. (2003). Genomic context drives SCA7 CAG repeat instability, while expressed SCA7 cDNAs are intergenerationally and somatically stable in transgenic mice. *Hum. Mol. Genet.* **12**, 41–50.
20. Strom, A. L., Jonasson, J., Hart, P., Brannstrom, T., Forsgren, L., and Holmberg, M. (2002). Cloning and expression analysis of the murine homolog of the spinocerebellar ataxia type 7 (SCA7) gene. *Gene* **285**, 91–99.
21. Lindenberg, K. S., Yvert, G., Muller, K., and Landwehrmeyer, G. B. (2000). Expression analysis of ataxin-7 mRNA and protein in human brain: Evidence for a widespread distribution and focal protein accumulation. *Brain Pathol.* **10**, 385–394.
22. Cancel, G., Duyckaerts, C., Holmberg, M., Zander, C., Yvert, G., Lebre, A. S., Ruberg, M., Faucheux, B., Agid, Y., Hirsch, E., and Brice, A. (2000). Distribution of ataxin-7 in normal human brain and retina. *Brain* **123**, 2519–2530.
23. Einum, D. D., Townsend, J. J., Ptacek, L. J., and Fu, Y. H. (2001). Ataxin-7 expression analysis in controls and spinocerebellar ataxia type 7 patients. *Neurogenetics* **3**, 83–90.
24. Jonasson, J., Strom, A. L., Hart, P., Brannstrom, T., Forsgren, L., and Holmberg, M. (2002). Expression of ataxin-7 in CNS and non-CNS tissue of normal and SCA7 individuals. *Acta Neuropathol. (Berl).* **104**, 29–37.
25. Kaytor, M. D., Duvick, L. A., Skinner, P. J., Koob, M. D., Ranum, L. P., and Orr. H. T. (1999). Nuclear localization of the spinocerebellar ataxia type 7 protein, ataxin-7. *Hum. Mol. Genet.* **8**, 1657–1664.
26. Zander, C., Takahashi, J., El Hachimi, K. H., Fujigasaki, H., Albanese, V., Lebre, A. S., Stevanin, G., Duyckaerts, C., and Brice, A. (2001). Similarities between spinocerebellar ataxia type 7 (SCA7) cell models and human brain: Proteins recruited in inclusions and activation of caspase-3. *Hum. Mol. Genet.* **10**, 2569–2579.

27. Chen, S., Peng, G. H., Wang, X., Smith, A. C., Grote, S. K., Sopher, B. L., and La Spada, A. R. (2004). Interference of Crx-dependent transcription by ataxin-7 involves interaction between the glutamine regions and requires the ataxin-7 carboxy-terminal region for nuclear localization. *Hum. Mol. Genet.* **13**, 53–67.
28. Garden, G. A., Libby, R. T., Fu, Y. H., Kinoshita, Y., Huang, J., Possin, D. E., Smith, A. C., Martinez, R. A., Fine, G. C., Grote, S. K., Ware, C. B., Einum, D. D., Morrison, R. S., Ptacek, L. J., Sopher, B. L., and La Spada, A. R. (2002). Polyglutamine-expanded ataxin-7 promotes non-cell-autonomous Purkinje cell degeneration and displays proteolytic cleavage in ataxic transgenic mice. *J. Neurosci.* **22**, 4897–4905.
29. Helmlinger, D., Abou-Sleymane, G., Yvert, G., Rousseau, S., Weber, C., Trottier, Y., Mandel, J. L., and Devys, D. (2004). Disease progression despite early loss of polyglutamine protein expression in SCA7 mouse model. *J. Neurosci.* **24**, 1881–1887.
30. Yoo, S. Y., Pennesi, M. E., Weeber, E. J., Xu, B., Atkinson, R., Chen, S., Armstrong, D. L., Wu, S. M., Sweatt, J. D., and Zoghbi, H. Y. (2003). SCA7 knockin mice model human SCA7 and reveal gradual accumulation of mutant ataxin-7 in neurons and abnormalities in short-term plasticity. *Neuron* **37**, 383–401.
31. Einum, D. D., Clark, A. M., Townsend, J. J., Ptacek, L. J., and Fu, Y. H. (2003). A novel central nervous system-enriched spinocerebellar ataxia type 7 gene product. *Arch. Neurol.* **60**, 97–103.
32. Gavin, A. C., Bosche, M., Krause, R., Grandi, P., Marzioch, M., Bauer, A., Schultz, J., Rick, J. M., Michon, A. M., Cruciat, C. M., Remor, M., Hofert, C., Schelder, M., Brajenovic, M., Ruffner, H., Merino, A., Klein, K., Hudak, M., Dickson, D., Rudi, T., Gnau, V., Bauch, A., Bastuck, S., Huhse, B., Leutwein, C., Heurtier, M. A., Copley, R. R., Edelmann, A., Querfurth, E., Rybin, V., Drewes, G., Raida, M., Bouwmeester, T., Bork, P., Seraphin, B., Kuster, B., Neubauer, G., and Superti-Furga, G. (2002). Functional organization of the yeast proteome by systematic analysis of protein complexes. *Nature* **415**, 141–147.
33. Sanders, S. L., Jennings, J., Canutescu, A., Link, A. J., and Weil, P. A. (2002). Proteomics of the eukaryotic transcription machinery: Identification of proteins associated with components of yeast TFIID by multidimensional mass spectrometry. *Mol. Cell. Biol.* **22**, 4723–4738.
34. Grant, P. A., Duggan, L., Cote, J., Roberts, S. M., Brownell, J. E., Candau, R., Ohba, R., Owen-Hughes, T., Allis, C. D., Winston, F., Berger, S. L., and Workman, J. L. (1997). Yeast Gcn5 functions in two multisubunit complexes to acetylate nucleosomal histones: Characterization of an Ada complex and the SAGA (Spt/Ada) complex. *Genes Dev.* **11**, 1640–1650.
35. Grant, P. A., Schieltz, D., Pray-Grant, M. G., Steger, D. J., Reese, J. C., Yates, J. R. 3rd, and Workman, J. L. (1998). A subset of TAF(II)s are integral components of the SAGA complex required for nucleosome acetylation and transcriptional stimulation. *Cell* **94**, 45–53.
36. Helmlinger, D., Hardy, S., Sasorith, S., Klein, F., Robert, F., Weber, C., Miguet, L., Potier, N., Van-Dorsselaer, A., Wurtz, J. M., Mandel, J. L., Tora, L., and Devys, D. (2004). Ataxin-7 is a subunit of GCN5 histone acetyltransferase-containing complexes. *Hum. Mol. Genet.* **13**, 1257–1265.
37. Palhan, V. B., Chen, S., Peng, G. H., Tjernberg, A., Gamper, A. M., Fan, Y., Chait, B. T., La Spada, A. R., and Roeder, R. G. (2005). Polyglutamine-expanded ataxin-7 inhibits STAGA histone acetyltransferase activity to produce retinal degeneration. *Proc. Natl. Acad. Sci. USA* **102**, 8472–8477.
38. Brand, M., Yamamoto, K., Staub, A., and Tora, L. (1999). Identification of TATA-binding protein-free TAFII-containing complex subunits suggests a role in nucleosome acetylation and signal transduction. *J. Biol. Chem.* **274**, 18285–18289.
39. Martinez, E., Palhan, V. B., Tjernberg, A., Lymar, E. S., Gamper, A. M., Kundu, T. K., Chait, B. T., and Roeder, R. G. (2001). Human STAGA complex is a chromatin-acetylating transcription coactivator that interacts with pre-mRNA splicing and DNA damage-binding factors in vivo. *Mol. Cell. Biol.* **21**, 6782–6795.
40. Wieczorek, E., Brand, M., Jacq, X., and Tora, L. (1998). Function of TAF(II)-containing complex without TBP in transcription by RNA polymerase II. *Nature* **393**, 187–191.
41. Mushegian, A. R., Vishnivetskiy, S. A., and Gurevich, V. V. (2000). Conserved phosphoprotein interaction motif is functionally interchangeable between ataxin-7 and arrestins. *Biochemistry* **39**, 6809–6813.
42. Scheel, H., Tomiuk, S., and Hofmann, K. (2003). Elucidation of ataxin-3 and ataxin-7 function by integrative bioinformatics. *Hum. Mol. Genet.* **12**, 2845–2852.
43. Tora, L. (2002). A unified nomenclature for TATA box binding protein (TBP)-associated factors (TAFs) involved in RNA polymerase II transcription. *Genes Dev.* **16**, 673–675.
44. McMahon, S. J., Pray-Grant, M. G., Schieltz, D., Yates, J. R., 3rd, and Grant, P. A. (2005). Polyglutamine-expanded spinocerebellar ataxia-7 protein disrupts normal SAGA and SLIK histone acetyltransferase activity. *Proc. Natl. Acad. Sci. USA* **102**, 8478–8482.
45. La Spada, A. R., Fu, Y. H., Sopher, B. L., Libby, R. T., Wang, X., Li, L. Y., Einum, D. D., Huang, J., Possin, D. E., Smith, A. C., Martinez, R. A., Koszdin, K. L., Treuting, P. M., Ware, C. B., Hurley, J. B., Ptacek, L. J., and Chen, S. (2001). Polyglutamine-expanded ataxin-7 antagonizes CRX function and induces cone-rod dystrophy in a mouse model of SCA7. *Neuron* **31**, 913–927.
46. Yvert, G., Lindenberg, K. S., Devys, D., Helmlinger, D, Landwehrmeyer, G. B., and Mandel, J. L. (2001). SCA7 mouse models show selective stabilization of mutant ataxin-7 and similar cellular responses in different neuronal cell types. *Hum. Mol. Genet.* **10**, 1679–1692.
47. Yvert, G., Lindenberg, K. S., Picaud, S., Landwehrmeyer, G. B., Sahel, J. A., and Mandel, J. L. (2000). Expanded polyglutamines induce neurodegeneration and transneuronal alterations in cerebellum and retina of SCA7 transgenic mice. *Hum. Mol. Genet.* **9**, 2491–2506.
48. Watase, K., Weeber, E. J., Xu, B., Antalffy, B., Yuva-Paylor, L., Hashimoto, K., Kano, M., Atkinson, R., Sun, Y., Armstrong, D. L., Sweatt, J. D., Orr, H. T., Paylor, R., and Zoghbi, H. Y. (2002). A long CAG repeat in the mouse Sca1 locus replicates SCA1 features and reveals the impact of protein solubility on selective neurodegeneration. *Neuron* **34**, 905–919.
49. Helmlinger, D., Yvert, G., Picaud, S., Merienne, K., Sahel, J., Mandel, J. L., and Devys, D. (2002). Progressive retinal degeneration and dysfunction in R6 Huntington's disease mice. *Hum. Mol. Genet.* **11**, 3351–3359.
50. Ansorge, O., Giunti, P., Michalik, A., Van Broeckhoven, C., Harding, B., Wood, N., and Scaravilli, F. (2004). Ataxin-7 aggregation and ubiquitination in infantile SCA7 with 180 CAG repeats. *Ann. Neurol.* **56**, 448–452.
51. Mauger, C., Del-Favero, J., Ceuterick, C., Lubke, U., Van Broeckhoven, C., and Martin, J. (1999). Identification and localization of ataxin-7 in brain and retina of a patient with cerebellar ataxia type II using anti-peptide antibody. *Brain Res. Mol. Brain Res.* **74**, 35–43.
52. Michalik, A., and Van Broeckhoven, C. (2003). Pathogenesis of polyglutamine disorders: Aggregation revisited. *Hum. Mol. Genet.* **12**, 173–186.
53. Martin-Aparicio, E., Yamamoto, A., Hernandez, F., Hen, R., Avila, J., and Lucas, J. J. (2001). Proteasomal-dependent aggregate reversal and absence of cell death in a conditional mouse model of Huntington's disease. *J. Neurosci.* **21**, 8772–8781.

54. Stenoien, D. L., Mielke, M., and Mancini, M. A. (2002). Intranuclear ataxin1 inclusions contain both fast- and slow-exchanging components. *Nat. Cell. Biol.* **4**, 806–810.
55. Tarlac, V., and Storey, E. (2003). Role of proteolysis in polyglutamine disorders. *J. Neurosci. Res.* **74**, 406–416.
56. Helmlinger, D., Bonnet, J., Mandel, J. L., Trottier, Y., and Devys, D. (2004). Hsp70 and Hsp40 chaperones do not modulate retinal phenotype in SCA7 mice. *J. Biol. Chem.* **279**, 55969–55977.
57. Opal, P., and Zoghbi, H. Y. (2002). The role of chaperones in polyglutamine disease. *Trends Mol. Med.* **8**, 232–236.
58. Cummings, C. J., Sun, Y., Opal, P., Antalffy, B., Mestril, R., Orr, H. T., Dillmann, W. H., and Zoghbi, H. Y. (2001). Over-expression of inducible HSP70 chaperone suppresses neuropathology and improves motor function in SCA1 mice. *Hum. Mol. Genet.* **10**, 1511–1518.
59. Hansson, O., Nylandsted, J., Castilho, R. F., Leist, M., Jaattela, M., and Brundin, P. (2003). Overexpression of heat shock protein 70 in R6/2 Huntington's disease mice has only modest effects on disease progression. *Brain Res.* **970**, 47–57.
60. Hay, D. G., Sathasivam, K., Tobaben, S., Stahl, B., Marber, M., Mestril, R., Mahal, A., Smith, D. L., Woodman, B., and Bates, G. P. (2004). Progressive decrease in chaperone protein levels in a mouse model of Huntington's disease and induction of stress proteins as a therapeutic approach. *Hum. Mol. Genet.* **13**, 1389–1405.
61. Bowman, A. B., Yoo, S. Y., Dantuma, N. P., and Zoghbi, H. Y. (2005). Neuronal dysfunction in a polyglutamine disease model occurs in the absence of ubiquitin–proteasome system impairment and inversely correlates with the degree of nuclear inclusion formation. *Hum. Mol. Genet.* **14**, 679–691.
62. Lindsten, K., Menendez-Benito, V., Masucci, M. G., and Dantuma, N. P. (2003). A transgenic mouse model of the ubiquitin/proteasome system. *Nat. Biotechnol.* **21**, 897–902.
63. Cha, J. H. (2000). Transcriptional dysregulation in Huntington's disease. *Trends Neurosci.* **23**, 387–392.
64. Sugars, K. L., and Rubinsztein, D. C. (2003). Transcriptional abnormalities in Huntington disease. *Trends Genet.* **19**, 233–238.
65. Katsuno, M., Adachi, H., Kume, A., Li, M., Nakagomi, Y., Niwa, H., Sang, C., Kobayashi, Y., Doyu, M., and Sobue, G. (2002). Testosterone reduction prevents phenotypic expression in a transgenic mouse model of spinal and bulbar muscular atrophy. *Neuron* **35**, 843–854.
66. Klement, I. A., Skinner, P. J., Kaytor, M. D., Yi, H., Hersch, S. M., Clark, H. B., Zoghbi, H. Y., and Orr, H. T. (1998). Ataxin-1 nuclear localization and aggregation: Role in polyglutamine-induced disease in SCA1 transgenic mice. *Cell* **95**, 41–53.
67. Takeyama, K., Ito, S., Yamamoto, A., Tanimoto, H., Furutani, T., Kanuka, H., Miura, M., Tabata, T., and Kato, S. (2002). Androgen-dependent neurodegeneration by polyglutamine-expanded human androgen receptor in *Drosophila*. *Neuron* **35**, 855–864.
68. Dunah, A. W., Jeong, H., Griffin, A., Kim, Y. M., Standaert, D. G., Hersch, S. M., Mouradian, M. M., Young, A. B., Tanese, N., and Krainc, D. (2002). Sp1 and TAFII130 transcriptional activity disrupted in early Huntington's disease. *Science* **296**, 2238–2243.
69. Li, S. H., Cheng, A. L., Zhou, H., Lam, S., Rao, M., Li, H., and Li, X. J. (2002). Interaction of Huntington disease protein with transcriptional activator Sp1. *Mol. Cell. Biol.* **22**, 1277–1287.
70. Nucifora, F. C., Jr., Sasaki, M., Peters, M. F., Huang, H., Cooper, J. K., Yamada, M., Takahashi, H., Tsuji, S., Troncoso, J., Dawson, V. L., Dawson, T. M., and Ross, C. A. (2001). Interference by huntingtin and atrophin-1 with cbp-mediated transcription leading to cellular toxicity. *Science* **291**, 2423–2428.
71. Steffan, J. S., Bodai, L., Pallos, J., Poelman, M., McCampbell, A., Apostol, B. L., Kazantsev, A., Schmidt, E., Zhu, Y. Z., Greenwald, M., Kurokawa, R., Housman, D. E., Jackson, G. R., Marsh, J. L., and Thompson, L. M. (2001). Histone deacetylase inhibitors arrest polyglutamine-dependent neurodegeneration in *Drosophila*. *Nature* **413**, 739–743.
72. Taylor, J. P., Taye, A. A., Campbell, C., Kazemi-Esfarjani, P., Fischbeck, K. H., and Min, K. T. (2003). Aberrant histone acetylation, altered transcription, and retinal degeneration in a *Drosophila* model of polyglutamine disease are rescued by CREB-binding protein. *Genes Dev.* **17**, 1463–1468.
73. Hughes, R. E., Lo, R. S., Davis, C., Strand, A. D., Neal, C. L., Olson, J. M., and Fields, S. (2001). Altered transcription in yeast expressing expanded polyglutamine. *Proc. Natl. Acad. Sci. USA* **98**, 13201–13206.
74. Lin, X., Antalffy, B., Kang, D., Orr, H. T., and Zoghbi, H. Y. (2000). Polyglutamine expansion down-regulates specific neuronal genes before pathologic changes in SCA1. *Nat. Neurosci.* **3**, 157–163.
75. Luthi-Carter, R., Strand, A., Peters, N. L., Solano, S. M., Hollingsworth, Z. R., Menon, A. S., Frey, A. S., Spektor, B. S., Penney, E. B., Schilling, G., Ross, C. A., Borchelt, D. R., Tapscott, S. J., Young, A. B., Cha, J. H., and Olson, J. M. (2000). Decreased expression of striatal signaling genes in a mouse model of Huntington's disease. *Hum. Mol. Genet.* **9**, 1259–1271.
76. Luthi-Carter, R., Strand, A. D., Hanson, S. A., Kooperberg, C., Schilling, G., La Spada, A. R., Merry, D. E., Young, A. B., Ross, C. A., Borchelt, D. R., and Olson, J. M. (2002). Polyglutamine and transcription: Gene expression changes shared by DRPLA and Huntington's disease mouse models reveal context-independent effects. *Hum. Mol. Genet.* **11**, 1927–1937.
77. Hockly, E., Richon, V. M., Woodman, B., Smith, D. L., Zhou, X., Rosa, E., Sathasivam, K., Ghazi-Noori, S., Mahal, A., Lowden, P. A., Steffan, J. S., Marsh, J. L., Thompson, L. M., Lewis, C. M., Marks, P. A., and Bates, G. P. (2003). Suberoylanilide hydroxamic acid, a histone deacetylase inhibitor, ameliorates motor deficits in a mouse model of Huntington's disease. *Proc. Natl. Acad. Sci. USA* **100**, 2041–2046.
78. Ferrante, R. J., Kubilus, J. K., Lee, J., Ryu, H., Beesen, A., Zucker, B., Smith, K., Kowall, N. W., Ratan, R. R., Luthi-Carter, R., and Hersch, S. M. (2003). Histone deacetylase inhibition by sodium butyrate chemotherapy ameliorates the neurodegenerative phenotype in Huntington's disease mice. *J. Neurosci.* **23**, 9418–9427.
79. Chen, S., Wang, Q. L., Nie, Z., Sun, H., Lennon, G., Copeland, N. G., Gilbert, D. J., Jenkins, N. A., and Zack, D. J. (1997). Crx, a novel Otx-like paired-homeodomain protein, binds to and transactivates photoreceptor cell-specific genes. *Neuron* **19**, 1017–1030.
80. Furukawa, T., Morrow, E. M., and Cepko, C. L. (1997). Crx, a novel otx-like homeobox gene, shows photoreceptor-specific expression and regulates photoreceptor differentiation. *Cell* **91**, 531–541.

CHAPTER 27

Spinocerebellar Ataxia Type 7: Clinical Features to Cellular Pathogenesis

GWENN A. GARDEN, RAY TRUANT, LISA M. ELLERBY, AND ALBERT R. LA SPADA

Department of Neurology, Department of Laboratory Medicine, Department of Medicine, and Center for Neurogenetics and Neurotherapeutics, University of Washington, Seattle, Washington; Department of Biochemistry and Biomedical Science, McMaster University, Hamilton, Ontario, Canada; and Buck Institute for Age Research, Novato, California

I. Introduction
II. Neuropathology
III. Disease Gene Identification, Size Ranges, and Spectrum of Severity
IV. Repeat Instability in SCA7
V. SCA7 Neurodegeneration: Models and Mechanisms
VI. SCA7 Retinal Degeneration: Identification of Molecular Players and Pathways
VII. Ataxin-7 Normal Function Provides Clues to SCA7 Disease Pathogenesis
VIII. Ataxin-7 Proteolytic Cleavage and Turnover
IX. Unanswered Questions and Future Directions
Acknowledgments
References

Of the various CAG/polyglutamine repeat diseases, spinocerebellar ataxia type 7 (SCA7) is unique for a number of reasons. The clinical severity in SCA7 is rather broad, with cases ranging from infantile onset with early death due to nonneurological involvement to elderly presentations of isolated ataxia that progress extremely slowly. The breadth of clinical presentation and natural history stems from the marked CAG repeat instability at the SCA7 locus, making this repeat disorder the most unstable mutation of the CAG/polyglutamine repeat disease category. The pronounced anticipation (i.e., worsening disease severity as manifested by earlier age of disease onset and more rapid disease progression with familial transmission of a disease mutation) in afflicted pedigrees made SCA7 a highly likely candidate for a causal triplet repeat mutation. Indeed, evidence for a polyglutamine expansion as the cause of SCA7 was found 2 years before the SCA7 disease gene was cloned [1]. Advances in our understanding of SCA7's molecular basis and genetic instability have led to important insights into cell type-specific neurodegeneration and *cis*-element control of repeat instability. There are many reasons to expect that studies of the SCA7 gene's normal function, investigations into the molecular basis of SCA7 neurodegeneration, and the development of therapeutic interventions for SCA7 will greatly influence related endeavors directed at the other CAG/polyglutamine repeat diseases.

I. INTRODUCTION

SCA7 is an autosomal dominant inherited neurodegenerative syndrome of progressive cerebellar ataxia and retinal degeneration that affects people on every continent [2–7]. The clinical association of hereditary ataxia with macular dystrophy has long been reported [8]. Patients with dominantly inherited ataxia in combination with retinal dystrophy were initially differentiated from the larger category of autosomal dominant cerebellar ataxia (ADCA) as ADCA type II, based on the presence of macular degeneration [9]. Currently, the only identified gene defect associated with ADCA type II is an unstable CAG repeat expansion in the coding region of a gene that was designated as ataxin-7 in 1997 [10]. Almost all patients with ADCA type II have more than 36 CAG repeats on one ataxin-7 allele and, thus, carry the diagnosis of "SCA7." In a tiny subset of patients who have been followed with a provisional diagnosis of "ADCA type II," ataxin-7 CAG repeat expansion and linkage to the SCA7 locus are not observed [11]. The precise genetic basis of such admittedly rare SCA7 phenocopies is yet to be established.

Ataxia (i.e., impaired coordination manifested as difficulty with speech, walking, manual dexterity) is the most common clinical feature of SCA7 and is often the first reported symptom [11]. Patients with SCA7 can develop early loss of color discrimination, and visual impairment may deteriorate to blindness [12]. A subset of patients (usually juvenile-onset or early adult onset) present initially with vision loss; however, both features may develop concurrently at the time of symptom onset [11]. SCA7 patients may display more extensive neurological deficits including dysarthria, dysphagia, hypoacusis, and eye movement abnormalities (slow saccades; staring) that can progress to frank ophthalmoplegia. Involvement of the corticospinal tracts resulting in exaggerated deep tendon reflexes, spasticity, and extensor plantar reflexes can also be present [11, 13]. Abnormalities of the extrapyramidal motor system are less commonly observed, but have been reported in patients with very long repeat lengths [11]. Dysfunction of the peripheral nervous system, including sensory abnormalities (primarily loss of vibrioception) and evidence of motor neuronopathy, has been reported in severely affected SCA7 patients [13]. As in other diseases secondary to DNA repeat expansions, families with SCA7 demonstrate the pattern of genetic anticipation, where subsequent generations develop more severe forms of the disease after inheriting repeat expansions longer than those of their parents. When very long repeats are inherited, SCA7 may present with an early-onset, rapidly progressive juvenile form or a very severe infantile form [14]. Infantile-onset SCA7 is remarkable for its widespread disease pathology that includes organ systems outside the central nervous system (CNS). Indeed, severe cardiac anomalies in SCA7 infantile-onset cases indicate that the largest ataxin-7 CAG repeat expansions can produce nonneurological disease and exceedingly rapid progression, as one such infant succumbed to disease at 6 weeks of age [14].

Clinical evaluation of patients with SCA7 is often augmented by neuroimaging studies. Neuroradiology examination typically reveals marked atrophy of the cerebellum and pons, whereas a subset of patients may also demonstrate high T2 signal intensity in the transverse pontine fibers [15]. MRI-based volumetric analysis revealed that SCA7 patients demonstrate significantly more pontine atrophy than patients with other SCAs (Fig. 27-1). Interestingly, it appears that volume loss (as measured by these methods) may develop in the pons at an earlier stage of disease than atrophy of the cerebellum, suggesting that the primary site of disease onset may be in brainstem structures rather than in the cerebellar folia [16].

Ophthalmological evaluation is critical for the diagnosis of SCA7, as retinal degeneration is the single clinical feature that permits differentiation of SCA7 from the more than 25 other SCAs. On funduscopic examination, degeneration of the macula is noted, suggesting that SCA7 is purely a macular dystrophy. However, full-field electroretinograms (ERGs) of SCA7 patients reveal dysfunction of cone photoreceptor cells throughout the retina prior to any rod photoreceptor abnormality [17]. Such widespread loss of cone photoreceptor function ahead of rod photoreceptor dysfunction indicates that SCA7 is actually a cone–rod dystrophy type of retinal degeneration. As cone photoreceptors are responsible for color vision, the earliest sign of SCA7 retinal degeneration is impaired blue–yellow discrimination (dyschromatopsia) [18]. Functional retinal abnormalities, including progressive loss of visual acuity and abnormalities on ERG, thus temporally precede the development of detectable anatomic retinal degeneration, and support the hypothesis that SCA7 first impacts the function of central cones [19, 20]. SCA7 patients initially complain of problems with central vision instead of with peripheral vision, and they typically develop

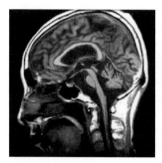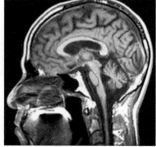

FIGURE 27-1 Neuroimaging of SCA7 patients reveals marked pontine atrophy. Here we see MRI scans of the brains of two SCA7 patients, one with advanced disease (left) and one with mild cerebellar atrophy (right). A pronounced but similar degree of atrophy (~35%) of the pons (circled) is apparent in both scans, emphasizing the early onset of pontine atrophy in this form of SCA. Reprinted from Bang et al., 2004, *J. Neurol. Neurosurg. Psychiatry* **75**, 1452. Used with permission.

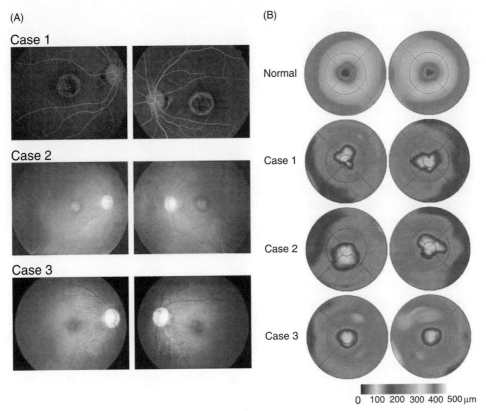

FIGURE 27-2 Funduscopic and optical coherence tomography appearance of SCA7 retinal degeneration. Ophthalmological examination of three siblings with adult-onset SCA7 and pronounced visual impairment was performed. (A) Fundus photographs reveal marked symmetrical foveal atrophy in cases 1 and 2, and less severe foveal atrophy in case 3. (B) Optical coherence tomography analysis demonstrates centrifugal thinning of the retina in a concentric or "bull's eye" pattern. Note again the less severe retinal degeneration in case 3. See CD-ROM for color image. Reprinted from Ahn et al., 2005, *Am. J. Ophthalmol.* **139**, 923. Used with permission.

central scotomas [21]. As retinal disease progresses in SCA7, the rod photoreceptor cells become involved and affected patients ultimately become completely blind. As the centrally located macula is particularly cone-rich, SCA7 patients develop an atrophic maculopathy detectable by fundoscopy. The appearance of macular dystropy has been described as bull's-eye maculopathy [19]. Concentric rings of retinal thinning can also be appreciated by optical coherence tomography (Fig. 27-2).

II. NEUROPATHOLOGY

Autopsy studies of patients with SCA7 have demonstrated atrophy of the cerebellar cortex, dentate nucleus, inferior olive, subthalamic nucleus, and olivocerebellar, spinocerebellar, and pyramidal tracts [22, 23]. Histological examinations of tissue from SCA7 patients have revealed extensive loss of cerebellar Purkinje cells, as well as prominent neuronal loss and gliosis in the inferior olive, with only mild changes in the cerebellar granule cell layer [23]. Neuronal loss has been reported in brainstem cranial nerve motor nuclei (III, IV, and XII), spinal motor neurons, and regions of the basal ganglia, in particular, substantia nigra [24]. Demyelination of the pyramidal tracts and the posterior columns of the spinal cord has also been observed [24].

A common pathological feature of neurodegenerative diseases associated with polyglutamine-expanded proteins is the presence of nuclear inclusions (NIs) that contain the mutant protein together with a host of other proteins, including ubiquitin–proteasome components and transcription factors. Numerous studies of SCA7 neuropathology in human patients and in SCA7 mouse models have delineated the pattern of NI formation and the representation of various factors and proteins within SCA7 NIs [25–29]. Although extensive NI formation is present in specific neuronal populations that are vulnerable to neurodegeneration in SCA7, the presence of NIs extends well beyond the vulnerable regions to areas of the neuraxis not subject to prominent degeneration. For example, numerous NIs are observed in the inferior olivary complex and basis pontis regions in postmortem SCA7 brains; however, nonvulnerable regions of the CNS, such as the cerebral cortex, also display frequent NIs. The reverse is also true, as certainly vulnerable neuronal populations in SCA7 do not always contain NIs [30]. Indeed, in juvenile-onset SCA7 associated with a CAG repeat length of 180, prominent NIs were detected in multiple organ systems that remained disease-free and in various neuronal populations (e.g., hippocampus) that did not demonstrate neuronal loss [31]. By documenting a disconnection between NI formation and disease pathology, such early studies on SCA7 patient material contributed to the emerging view that inclusion body formation is not required for molecular pathology in polyglutamine diseases. Later studies of NI formation in mouse models of SCA7 then demonstrated that inclusion of certain protein factors in NIs may shed light on disease pathogenesis, but, in many cases, recruitment and sequestration of protein factors could not be predicted based on the function of the protein or the presence of a polyglutamine tract [28, 32].

III. DISEASE GENE IDENTIFICATION, SIZE RANGES, AND SPECTRUM OF SEVERITY

Localization of the gene responsible for the ADCA type II phenotype revealed that diverse pedigrees from disparate geographic regions all demonstrated linkage to chromosome 3p12–21.1 [33, 34], allowing the SCA7 designation to be applied to such patients. Further mapping of this region then permitted investigators to test the highly likely hypothesis that a CAG/polyglutamine repeat expansion might be involved, especially after the expanded polyglutamine epitope-directed 1C2 antibody was developed and found to detect an abnormal protein product in the lymphocytes of SCA7 patients [1]. Screening of CAG repeat sequences from a YAC contig spanning a 5-cM critical region thus yielded a novel gene that was named *ataxin-7* [10]. Initial studies of the ataxin-7 CAG repeat indicated that expansions beyond 37 CAGs produced the SCA7 disease phenotype, and normal individuals possess ataxin-7 repeats ranging in size from 7 to 34 CAGs [10]. The ataxin-7 gene encoded a novel protein of (then) unknown function and typically contains a 10-CAG repeat starting at codon 30 that is translated into a polyglutamine tract in the ataxin-7 protein. This CAG repeat expands to 37 to >250 triplets in SCA7 patients, and there is a

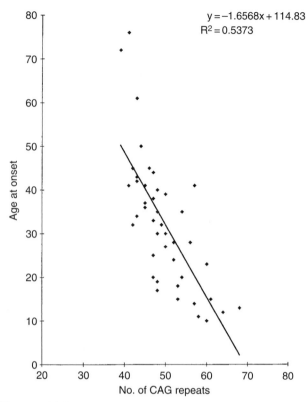

FIGURE 27-3 Genotype – phenotype correlation in SCA7. Here, age at onset is plotted as a function of CAG repeat length for 45 SCA7 patients. A strong inverse correlation between allele size and age at onset was observed ($r = 0.74$, $P < 0.0001$), accounting for the prominent anticipation in this CAG/polyglutamine repeat disorder. Reprinted from Giunti et al., 1999, *Am. J. Hum. Genet.* **64**, 1594. Used with permission.

strong inverse correlation between the age of disease onset and the length of the repeat mutation (Fig. 27-3), such that longer repeats tend to produce earlier ages of onset [6, 13, 14]. There is also a correlation between repeat length and type of clinical presentation in SCA7, with repeat mutations of <59 CAGs often yielding initial cerebellar findings and those ≥59 CAGs typically producing visual impairment as the harbinger of disease status [6].

IV. REPEAT INSTABILITY IN SCA7

A striking similarity between SCA7 and other genetic diseases caused by CAG repeat expansion is the strong negative correlation between the length of the expanded repeat and the age of symptom onset. A very profound form of genetic "anticipation" (i.e., worsening clinical severity as a disease is transmitted from parent to offspring), however, accentuates the unique clinical picture of SCA7. In SCA7, the age of disease onset in children who inherit the gene defect is, on average, 20 years earlier than the age of disease onset in their affected parent [35]. The SCA7 mutation is more unstable in male gametogenesis than female gametogenesis, with the most severe infantile form due to very long repeat expansions that are transmitted paternally [14]. In one large study of SCA7 repeat instability, intergenerational CAG repeat length changes ranging from −13 to +62 repeats were observed in 44 instances of parent-to-child transmission [35]. A significant difference between paternal expansion tendency and maternal expansion tendency was noted, as paternally transmitted SCA7 alleles displayed an average repeat length increase of +17 CAGs, whereas maternally transmitted SCA7 alleles showed an average repeat length increase of +4 CAGs. Furthermore, intermediate alleles (28–35 CAGs) that do not result in a disease phenotype can expand via paternal transmission into mutant alleles that result in *de novo* cases of SCA7 [36]. Despite the inherent difference in repeat instability between male and female gametogenesis, SCA7 is actually more frequently inherited by maternal transmission [35]. This phenomenon appears secondary to the fact that repeat expansions in sperm can be so long that they impact the survival of gametes as well as early embryos, accounting for an increased miscarriage rate among female partners of affected male SCA7 patients [37].

The marked CAG repeat instability at the SCA7 locus stands in sharp contrast to the rather modest CAG repeat instability observed at the androgen receptor (AR) locus, where CAG repeat expansion underlies X-linked spinal and bulbar muscular atrophy (SBMA), another polyglutamine disorder. For example, a CAG40 tract at the AR locus in SBMA changes length only about one-quarter of the time that it is transmitted, and when it does so, the length changes are small, ranging from −4 to +7 repeats [38]. In SCA7, however, a CAG40 tract expands by 7 repeats on average with each transmission, and most alterations involve jumps of >10, sometimes >20, and occasionally >100 repeats [35]. The fact that CAG tracts of identical length behave quite differently depending on the locus at which they reside strongly supports the existence of *cis*-acting DNA elements that promote repeat instability at certain CAG/CTG repeat loci. Studies of the role of *cis*-acting elements in modulating trinucleotide repeat instability suggest that two types of *cis*-elements exist: those that are haplotype-specific and those that are locus-specific. In the case of haplotype-specific *cis*-elements, certain chromosomal haplotypes are tightly associated with the transition from a normal-length repeat to a disease-causing allele [39, 40]. For example, all myotonic dystrophy CTG expansion chromosomes in Europeans share an identical nine-marker haplotype [41]. Specific chromosomal backgrounds, on which repeat expansion has occurred, have also been documented for SCA2, SCA3, and *de novo* Huntington's disease (HD) cases arising from so-called "intermediate alleles" at the upper end of the normal range [42–44]. Although these are compelling associations, dissection of cause from effect is difficult if not impossible in these circumstances. Furthermore, the role of the original chromosomal background in perpetuating polar repeat expansion once in the disease length range is unclear.

To identify *cis*-acting factors responsible for CAG expansion, the molecular basis of repeat instability at the SCA7 locus has been studied. To model instability, expanded CAG tracts from the human SCA7 locus were introduced into mice either on 13.5-kb genomic fragments or out of genomic context on ataxin-7 cDNAs [45]. Comparison of the transmission of the SCA7 CAG repeats revealed that genomic context drives repeat instability with an obvious bias toward expansion, whereas SCA7 CAG repeats introduced on ataxin-7 cDNAs showed few length alterations (Fig. 27-4). Deletion of the genomic region 3′ to the SCA7 repeat stabilized the transmission of the CAG tract relative to that observed with longer genomic fragments in additional lines of mice, suggesting that *cis*-information conferring an instability potential to the SCA7 triplet resides within the 3′ region. What sequences within the 3′ region are responsible for promoting CAG/CTG repeat instability at the SCA7 locus? Analysis of DNA sequence adjacent to CAG/CTG repeats with marked instability suggests that such regions possess a high percentage GC content [46]. In the case of SCA7 and a number of other highly unstable CAG/CTG repeat loci, binding sites for a protein known as CTCF (i.e., the "CTCCC-binding factor") are found within such GC-rich regions. CTCF functions as a negative regulator of transcription [47], and can mediate a variety of other regulatory functions, including promoter activation, silencing, and enhancer blocking/insulator boundary element creation [48]. As CTCF binding yields a change in the DNA structure of the region to which it binds, CTCF emerged as a logical candidate for involvement in trinucleotide repeat instability. One of us thus tested the hypothesis that an intact 3′ CTCF binding site is necessary for SCA7 repeat instability by re-deriving SCA7 CAG repeat expansion genomic fragment transgenic mice with a mutant 3′ CTCF binding site. Independent lines of such SCA7 transgenic mice displayed ablation of CAG/CTG repeat instability on intergenerational transmission, suggesting that an intact CTCF binding site is required for repeat instability at the SCA7 locus (A. R. La Spada et al., unpublished results).

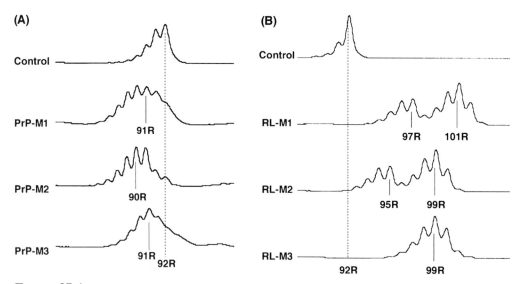

FIGURE 27-4 Genomic context drives CAG repeat instability at the SCA7 locus. Comparison of intergenerational CAG repeat instability between SCA7 "PrP" cDNA transgenic mice (A) and SCA7 "RL" genomic fragment mice (B). As shown here, the SCA7 "PrP" cDNA transgenic mice (A) display only minor repeat contractions on ABI capillary gel electrophoresis analysis of the PCR-amplified SCA7 CAG repeat tract. However, the SCA7 "RL" genomic fragment transgenic mice show numerous repeat increases with a significant tendency to repeat expansion. Reprinted from Libby et al., 2003, *Hum. Mol. Genet.* **12**, 41. Used with permission.

V. SCA7 NEURODEGENERATION: MODELS AND MECHANISMS

Our current understanding of polyglutamine neurodegeneration is based on a variety of biochemical, molecular, genetic, and organismal studies. Although the CAG/polyglutamine repeat disorders share many features, the different disease-causing proteins are unrelated except for the fact that they all contain an expanded polyglutamine tract. This observation suggested that there is something inherently toxic about long polyglutamine stretches. The pathological thresholds for almost all the CAG/polyglutamine repeat diseases are nearly identical (between 35 and 40); and in all cases, the age of onset and severity of the disease are inversely related to the repeat length. What is the molecular basis of this threshold effect? In 1994, Perutz et al. proposed the idea that once the number of glutamines crosses a threshold of 35, then the polyglutamine tract acquires an altered conformation composed of beta-pleated sheets folding back on one another—the so-called "polar zipper" hypothesis [49]. Biochemical validation of altered conformation came with production of an antibody that could specifically detect expanded polyglutamine tracts [1], indicating that a gain of structure at >36 repeats may correspond to the dominant gain of function observed in these diseases. Indeed, structural biologists have defined polyglutamine tracts as being either water-filled nanotubes or beta-sheets with cross-beta crystallites by crystallography [50, 51]. Nuclear magnetic resonance (NMR) studies have found polyglutamine expansions to be highly mobile and thus incapable of polar side-group interactions in proteins [52]. Other NMR studies have revealed that short polyglutamine expansions are not structured at all and form flexible loops [53], but, on expansion to disease-length repeats, become beta-pleated sheets [54]. These data are consistent with the concept of gain of function by gain of structure. The production of abnormally folded polyglutamine-containing proteins as a key event in polyglutamine disease pathogenesis became clear when analysis of human brain material from patients with SCA3 and studies of a mouse model of HD detected NIs [55, 56]. On the basis of this work, the theory has emerged that polyglutamine expansion tracts misfold, become resistant to degradation, and then accumulate as protein aggregates in nuclei and cytoplasm. The ability of Congo red to prevent polyglutamine neurotoxicity by inhibiting oligomerization and the toxicity of "diffuse" mutant polyglutamine-expanded protein (and not "inclusions") to produce cell death in cultured neurons support a model in which adoption of an altered conformation is the crucial event in polyglutamine disease pathogenesis [57, 58]. Visible aggregates (or "inclusions") are not toxic *per se*, but instead may be correlated with the coexistence of misfolded oligomeric intermediates. Thus, the toxicity of the polyglutamine expansion tract relates to its propensity to misfold and to the formation of these toxic intermediates, which are not visible at the light microscope level, but may be visualized by transmission electron microscopy, atomic force microscopy, and Fourier transform infrared spectroscopy [59].

Ataxin-7 is widely expressed throughout the central nervous system (CNS) and in many other nonneuronal organ systems, yet disease pathology in adult-onset SCA7 is specifically restricted to only certain populations of neurons. Thus, although an appreciation of the general mechanism of polyglutamine neurotoxicity is important to our understanding of SCA7 disease pathogenesis, how do we approach this issue of cell type specificity in SCA7? To accomplish this, investigators have generated cell culture and animal models of SCA7 with the polyglutamine expansion expressed in the context of full-length ataxin-7 protein. To model SCA7 in mice, a number of groups have used heterologous promoter-expression systems (including the rhodopsin promoter, Purkinje cell protein-2 promoter, and murine prion protein promoter) [26, 28, 29, 32], as the regulatory elements controlling ataxin-7 gene expression are not well characterized. These models have provided useful insights into the nature of SCA7 neurodegeneration, emphasizing the importance of trans neuronal or noncell-autonomous effects in polyglutamine-expanded ataxin-7-mediated neuropathology. The murine prion promoter (MoPrP) SCA7 mouse model, in particular, has highlighted the importance of cell–cell interactions in polyglutamine disease neurodegeneration. Comparison of MoPrP SCA7 transgenic mice expressing similar levels of ataxin-7–92Q and ataxin-7–24Q protein revealed onset of gait ataxia at 12 weeks of age in the so-called PrP-SCA7-c92Q mice, whereas PrP-SCA7-c24Q mice are visibly normal beyond 2 years of age [26]. Immunostaining of cerebellar sections from such PrP-SCA7-c92Q mice around this age revealed abundant NIs in neurons of the molecular layer and granule cell layer of the cerebellum. Due to a deletion of MoPrP flanking sequence in the creation of the final MoPrP vector, expression of the transgene is typically not observed in Purkinje cells of the cerebellum [60]. Thus, in the MoPrP SCA7 mouse model, it was possible to determine if lack of expression of polyglutamine-expanded ataxin-7 in Purkinje cells would protect them. Immunohistochemical analysis of cerebellar sections from the PrP-SCA7-c92Q line 6076 mice indicated that Purkinje cells are quite severely affected, with loss of calbindin staining becoming dramatically apparent by 8 weeks of age, long before the onset of a visible phenotype [26]. Indeed, by 13 weeks of age, when most of these mice are just becoming visibly ataxic, almost all of the calbindin staining is lost and marked degeneration of Purkinje cell dendritic arbors is apparent (Fig. 27-5). The dramatic Purkinje cell

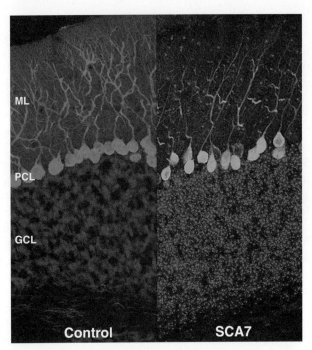

FIGURE 27-5 Noncell-autonomous Purkinje cell degeneration in a mouse model of SCA7. Confocal microscopy analysis of cerebellar sections from a SCA7 transgenic mouse created by inserting an ataxin-7 CAG-92 into the murine prion protein expression vector (SCA7) and from an age- and sex-matched nontransgenic littermate (Control). Staining with an anti-ataxin-7 antibody (magenta), a calbindin antibody (green), and DAPI (blue) reveals a healthy, normal-appearing cerebellum characterized by properly oriented Purkinje cells with extensive dendritic arborization in the control mice. However, SCA7 transgenic mice display pronounced Purkinje cell degeneration as evidenced by decreased dendritic arborization and displacement of Purkinje cell bodies. Interestingly, although numerous neurons in the granule cell layer (GCL) and the molecular layer (ML) display aggregates of ataxin-7, there is no accumulation of mutant ataxin-7 in the degenerating Purkinje cells due to lack of appreciable expression there. As the Purkinje cells degenerate without expressing the mutant protein, the degeneration is described as noncell-autonomous. See CD-ROM for color image. Adapted from Garden et al., 2002, *J. Neurosci.* **22**, 4897. Used with permission.

degeneration is accompanied by extensive ataxin-7 antibody labeling of NIs in all neurons of the cerebellum, except for the Purkinje cells. Furthermore, ultrastructural analysis of the Purkinje cells reveals marked histopathology of the Purkinje cell soma with accompanying degenerative changes. These findings led to the conclusion that the Purkinje cell degeneration in the MoPrP SCA7 mice is noncell-autonomous, and thus that noncell-autonomous processes could be operating in this and other polyglutamine repeat diseases. In another mouse model of SCA7 where expression of ataxin-7 90Q was restricted to photoreceptors, postsynaptic degenerative changes were observed in bipolar neurons, similarly supporting a role for noncell-autonomous "transneuronal" pathology in SCA7 [29].

What is the molecular basis of the noncell-autonomous Purkinje cell degeneration observed in the MoPrP SCA7 mouse model? The Purkinje cells are the only neurons in the cerebellum not expressing the expanded ataxin-7, and yet they are one of the most severely affected populations. In SCA7 patients, the Purkinje cells are also severely affected, and yet NIs are not prominent in these cells [27]. Indeed, examination of cerebellar sections from patients with SCA1, SCA2, SCA3, and DRPLA indicated that NIs are uncommon in Purkinje cells [61]. Purkinje cell degeneration in the MoPrP SCA7 mouse model may involve withdrawal of trophic factors. In the case of SCA7, brain-derived neurotrophic factor (BDNF) or glia-derived neurotrophic factor (GDNF) could be involved, as Purkinje cells possess an enormous number of synaptic contacts and glial support, from cells both within the cerebellum and outside the cerebellum (Fig. 27-6). BDNF, which is important for granule cell survival via the PI-3 kinase/Akt pathway [62], may be normally supplied at high levels by these numerous contacts, and could diminish in the face of mutant ataxin-7 induced neurotoxicity. GDNF, produced by the neighboring Bergmann glia, can also support the development and survival of cerebellar Purkinje cells [63]. Interestingly, in the MoPrP SCA7 mouse model, the surrounding Bergmann glia cells display extensive ultrastructural pathology. As Bergmann glia are the cells that maintain glutamate and amino acid homeostasis in the molecular layer of the cerebellum, we hypothesized that the noncell-autonomous Purkinje cell injury observed in the PrP-SCA7-c92Q mice might be secondary to Bergmann glia pathology. To test this hypothesis, transgenic mice with polyglutamine-expanded ataxin-7 under the control of a promoter (Gfa2) that drives expression in only Bergmann glia in the cerebellum were generated [64]. Such Gfa2-SCA7-92Q mice do develop ataxia and Purkinje cell degeneration, suggesting that Bergmann glial dysfunction may contribute to the extra cellular alterations that produce Purkinje cell degeneration in SCA7. Analysis of the Gfa2-SCA7–92Q mice indicated that mutant ataxin-7 is interfering with the ability of Bergmann glia to take up glutamate by impairing expression of the Bergmann glia-specific glutamate transporter GLAST. In this way, accumulation of glutamate in Purkinje cell synaptic clefts could produce glutamate excitotoxicity, resulting in Purkinje cell dysfunction and degeneration. However, as Gfa2-92Q-ataxin-7 mice demonstrate only a moderate phenotype with late onset and do not exhibit early mortality, Bergmann glial dysfunction is

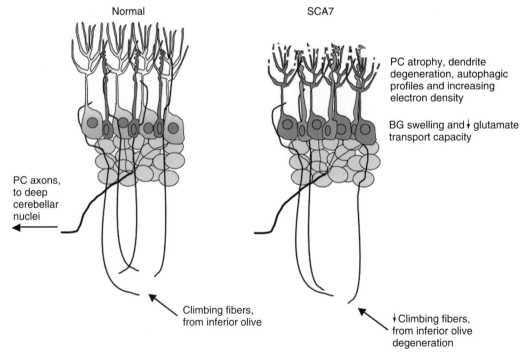

FIGURE 27-6 Model for Purkinje cell degeneration in SCA7. Purkinje cells with extensive dendritic arborization receive trophic support from synaptic contacts, including climbing fibers from the inferior olive, and from Bergmann glia whose cell bodies are adjacent to Purkinje cells and whose radial processes ensheath Purkinje cell arbors in a normal cerebellum. In SCA7 cerebellum, however, Purkinje cells display dendritic atrophy and reduced dendritic arborization. This is accompanied by Bergmann glia pathology and may involve impaired glutamate transport function of the adjacent Bergmann glia. See CD-ROM for color image.

VI. SCA7 RETINAL DEGENERATION: IDENTIFICATION OF MOLECULAR PLAYERS AND PATHWAYS

likely not the only cause of Purkinje cell degeneration or the broader neurodegenerative phenotype in SCA7.

SCA7 is classified as an ADCA type II because affected patients display retinal involvement in addition to cerebellar neurodegeneration. SCA7 is unique among the CAG/polyglutamine diseases, as it is the only such disorder with a prominent retinal degeneration phenotype. One key question in the polyglutamine disease field is why certain neurons degenerate in one disorder, whereas other neuron populations degenerate in other diseases, despite the fact that all the polyglutamine proteins appear to be expressed throughout the neuraxis. A likely explanation for selective vulnerability is that the protein context of each disease polypeptide accounts for these differences, due to disease-specific protein–protein interactions. The MoPrP SCA7 mouse model provided the first glimpse of a possible mechanistic underpinning for cell type-specific retinal degeneration in SCA7. In this model, transgenic expression of ataxin-7 in all three nuclear layers of the retina was achieved, permitting accurate recapitulation of the SCA7 cone–rod dystrophy phenotype in PrP-SCA7-c92Q mice [28]. When the retinas of such transgenic mice were examined for histological changes, periodic thinning of the photoreceptor cell layer was observed beginning at around 11 weeks of age. This periodic thinning produced areas where the number of nuclei eventually decreased from a normal width of 12 nuclei to as few as 6 nuclei. To determine what cells were primarily being lost, retinal sections were immunostained with antibodies specific for cone pigments, and cone photoreceptors were dramatically decreased in number (Fig. 27-7). ERGs were then obtained, and cone cell responses were noted to degrade prior to rod photoreceptor responses. Importantly, SCA7 c92Q mice became blind long before an appreciable number of photoreceptors were lost, indicating that neuronal dysfunction accounted for the retinal phenotype. Thus, histological and electrophysiological studies indicated that PrP-SCA7-c92Q transgenic mice faithfully recapitulate the cone–rod dystrophy type of retinal degeneration seen in SCA7 patients [28].

To explain the selective vulnerability of the retina in SCA7 and to account for the rather unusual cone–rod dystrophy type of retinal degeneration seen in this disorder, a role for the cone–rod homeobox protein (CRX) in SCA7 retinal degeneration was examined.

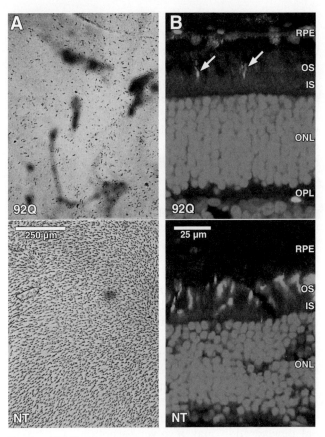

FIGURE 27-7 Recapitulation of the SCA7 cone–rod dystrophy retinal degeneration phenotype in transgenic mice. (A) Green cone-specific antibody immunostaining of retinal whole mounts from PrP-SCA7-c92Q transgenic mice (92Q) and age-matched nontransgenic littermate control (NT) reveals a dramatic loss of cone photoreceptors. (B) Retinal sections from PrP-SCA7-c92Q mice (92Q) and age-matched nontransgenic littermate control (NT) were immunostained with a cone-specific antibody and counterstained with propidium iodide (red). Although numerous cone photoreceptors are apparent in the outer segments of the NT retina, only two cone photoreceptors (marked by arrows) can be seen in the 92Q retina. RPE, retinal pigmented epithelium; OS, outer segments; IS, inner segments; ONL, outer nuclear layer; OPL, outer plexiform layer. See CD-ROM for color image. Adapted from La Spada et al., 2001, Neuron 31, 913. Used with permission.

CRX is a 299-amino acid transcription factor with a homeobox domain and glutamine-rich region, and its expression is highest in the retina and the pineal gland [65]. As glutamine tracts often interact with one another when in transcription factors and coactivators, various investigators have focused on the likely interaction of polyglutamine-expanded disease proteins with transcription factors containing a glutamine tract, such as CREB-binding protein (CBP) [66, 67]. Mutations in the CRX gene cause an autosomal dominant form of cone–rod dystrophy (CORD2) [68, 69], suggesting that impairment in CRX function could produce cone–rod dystrophy in SCA7. After demonstrating that ataxin-7 and CRX engage in a physical interaction by yeast two-hybrid assay and *in vitro* co-immunoprecipitation, the functional significance of the interaction was tested by determining if polyglutamine-expanded ataxin-7 could prevent CRX from activating transcription from its target consensus-binding element [28]. These studies found that polyglutamine-expanded ataxin-7 could interfere with CRX's transactivation function *in vitro*, and subsequent quantitative real-time reverse transcription polymerase chain reaction (RT-PCR) studies on the retinas of presymptomatic PrP-SCA7-c92Q transgenic mice revealed significant reductions in the levels of cone-expressed genes and rod-expressed genes that are targets of CRX regulation. These findings suggested that polyglutamine-expanded ataxin-7 interference of CRX function accounts for the selective cell type-specific retinal degeneration seen in SCA7 patients. The importance of transcription interference for SCA7 retinal degeneration has been further underscored by studies of a SCA7 knock-in mouse model that expresses ataxin-7 with 266 glutamines [70]. These mice also undergo a retinal degeneration process and display significant reductions in a wide range of cone-expressed and rod-expressed genes, some of which are not targets of CRX regulation. Recent advances in our understanding of the normal function of ataxin-7 indicate that SCA7 retinal degeneration may involve disruption of a broad range of transcription-related processes in addition to CRX transcription interference.

VII. ATAXIN-7 NORMAL FUNCTION PROVIDES CLUES TO SCA7 DISEASE PATHOGENESIS

One of the first goals of genetic research subsequent to the identification of a disease gene is to understand the normal biological function of the encoded protein. Understanding the normal biological function of a mutated protein can provide insights into the molecular basis of disease pathogenesis and can guide researchers to pathways of interest for drug development. For six of nine of the polyglutamine diseases, the normal biological function of the disease protein was unknown at the time of its discovery. Ataxin-7, the SCA7 disease protein, was among such proteins, the gene products of which had been unrecognized, until their respective discoveries as the cause of a human disease. Ataxin-7 is an 892-amino acid protein that displays both nuclear and cytosolic localization [30, 71]. At least three nuclear localization signals (NLSs) have been proposed for ataxin-7, one of which resides at amino acids 378–393 and two of which are located in the carboxy terminus at amino acid positions 705–708 and 835–838 [72, 73]. Very recent data have shown that ataxin-7 can shuttle into and out of the nucleus via a classic leucine-type nuclear export signal (NES) [74]. The coexistence of a functional NLS and NES in ataxin-7 suggests that it may be shuttling another protein or regulating subcellular localization of a key factor. Related to these domains, therefore, is a 50-amino acid motif that resembles the phosphate binding site of arrestin proteins [31]. Arrestins are transmembrane receptor proteins that constitute the largest family of membrane receptors, and are involved in the attenuation of activated receptors. In the retina, a highly abundant protein, arrestin-1 (visual arrestin), potentiates the receptor-mediated inhibition of rhodopsin. The phosphate binding domains of visual arrestin and β-arrestin show considerable sequence similarity to ataxin-7. Indeed, swapping these domains can result in an ataxin-7–arrestin chimera that can functionally attenuate rhodopsin or the β_2-adrenergic receptor [75]. Although the nuclear localization of ataxin-7 has been emphasized, ataxin-7 is also prominent in the cytoplasm on immunofluorescence or immunohistochemistry analysis [30, 71], suggesting that a membrane-associated activity is not inconceivable. As arrestins not only mediate the attenuation of G-protein coupled receptor signaling but can also act as signal transducers themselves by localizing to the nucleus [76], the shuttling activity of ataxin-7 implies that ataxin-7 may function analogously and could be involved in integrating cell signaling from a plasma membrane receptor to events in the nucleus.

Although the prospect of ataxin-7 regulating cell signaling—akin to an arrestin—remains theoretical, recent studies have definitively characterized ataxin-7 as a core component of STAGA, a transcription coactivator complex. Transcription coactivator complexes are large multi-subunit protein complexes that mediate protein–protein and enzymatic interactions between upstream DNA-bound activator proteins and the RNA polymerase II transcription complex [77, 78]. Unlike bacteria that possess a simple system of a single DNA sequence-specific binding protein interacting with the critical sigma subunit of bacterial RNA polymerase [79], eukaryotic transcription involves multiple analogs of these proteins in higher-order 100+ protein factor complexes that permit cell type-specific signaling and orchestrate chromatin remodeling [80]. One well-characterized complex that can function as a mediator between DNA-bound transcription activators and the RNA polymerase II machinery is the SAGA complex, originally defined in the yeast *Saccharomyces cerevisiae* [81]. SAGA contains histone acetyltransferase (HAT) activity, mediated by the Gcn5 enzyme. The 1.8-MDa SAGA complex at its core comprises Ada1, -2, -3, and -5 proteins, defined as genetic interactors with the herpes simplex virus VP16 transactivator, in combination with Gcn5. SAGA also contains a set of TATA-binding proteins (TBPs) or TATA-associated factors (TAFs), specifically Taf5, -6, -9, -10, and -12, and a set of TBP-like proteins, namely, Spt3, -7, -8, and -20. A recently described component of SAGA is Sgf73, a protein of unknown, but essential function in yeast that turns out to be a homolog of human ataxin-7 [82]. Although the identity between human ataxin-7 and yeast Sgf73 is only 25%, human ataxin-7 can actually complement Sgf73 null yeast strains, indicating that the two proteins are functionally interchangeable [83].

Although the characterization of Sgf73 as a homolog of ataxin-7 prompted one group to determine if ataxin-7 is a subunit of the mammalian equivalent of the SAGA complex known as STAGA [84], another group's unbiased mass spectrometry sequencing analysis of STAGA-associated factors identified a 110-kDa protein as ataxin-7 [85]. Independent biochemical studies of the STAGA complex by each of these two teams of investigators subsequently confirmed that ataxin-7 is indeed a core component of STAGA [84, 85]. The identification of ataxin-7 as a core component of the STAGA complex raises questions about the effect of mutant ataxin-7 on STAGA complex function. After demonstrating that both normal ataxin-7 and polyglutamine-expanded ataxin-7 are incorporated into the STAGA complex in HEK293T cells stably expressing ataxin-7, HAT assays with ataxin-7 immunoprecipitates were performed and revealed that incorporation of polyglutamine-expanded ataxin-7 dramatically reduces HAT activity [85] (Fig. 27-8). This observation led to the hypothesis that CRX is a STAGA-dependent transcription factor and that polyglutamine-expanded ataxin-7 might be interfering with CRX transactivation by reducing STAGA complex function. Chromatin immunoprecipitation assays performed on retinal samples from presymptomatic and early symptomatic MoPrP SCA7 transgenic mice that develop cone–rod dystrophy retinal degeneration found evidence for progressively impaired chromatin remodeling at CRX-regulated genes in the retinas of PrP-SCA7-c92Q mice [85], supporting a model of altered STAGA complex function in SCA7 retinal degeneration (Fig. 27-9). These results suggested for the first time that the normal function of a polyglutamine disease protein intersects with its pathogenic mechanism, an observation with important implications for the molecular basis of all polyglutamine disorders. Parallel studies in yeast support this model, as polyglutamine-expanded ataxin-7 is incapable of rescuing *Sgf73* null yeast, but normal ataxin-7 does rescue the *Sgf73* deletion strain [83]. This observation paves the way for the powerful

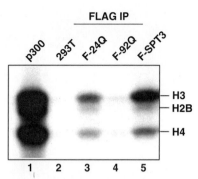

FIGURE 27-8 Polyglutamine-expanded ataxin-7 inactivates the HAT activity of the STAGA complex on incorporation. HEK293T cells that were either untransfected (293T), stably transfected with FLAG-ataxin-7-24Q (F-24Q), stably transfected with FLAG-ataxin-7-92Q (F-92Q), or stably transfected with Spt3 (F-SPT3) were subjected to immunoprecipitation (IP) with anti-FLAG antibody, and the IP material was tested for HAT activity on an oligonucleosome substrate. Although ataxin-7-24Q IP material and Spt3 IP material displayed HAT activity for histone 3 and histone 4, ataxin-7-92Q IP material showed no detectable HAT activity. Recombinant p300-expressing HEK293T cells serve a positive control, whereas IP material from untransfected HEK293T cells is the negative control. Adapted from Palhan et al., 2005, *Proc. Natl. Acad. Sci. USA* **102**, 8472. Used with permission.

genetic system of yeast to further elucidate the function of ataxin-7. Interestingly, of the nine known polyglutamine disease proteins, the only other polyglutamine disease protein with a closely related yeast homolog is TBP, the causal gene product for SCA17 and a target of the STAGA complex [86]. Although current data are consistent with polyglutamine expansion of ataxin-7 having a trans-dominant effect on STAGA complex function in SCA7, the *in vivo* necessity of ataxin-7 for STAGA complex function and stability remains unclear and must await generation of ataxin-7 knockout mice.

Although the exact function of ataxin-7 in STAGA is unknown, polyglutamine expansion in ataxin-7 can disrupt HAT activity of SAGA in yeast [83] and HAT activity of STAGA in retinal cells [85]. This would indicate that in the context of SAGA/STAGA, the polyglutamine tract expansion in ataxin-7 culminates in a loss of Gcn5 function, but such a finding appears contrary to the genetics of SCA7. However, in the global context of transcription activation, STAGA can essentially be considered a switching or balancing mechanism, and thus, its disruption is consistent with a gain-of-function phenotype. An analogy would be a broken light switch keeping the lights on in an otherwise dark room: the broken switch has a loss of function, but the light constantly on in the room is perceived as a gain of function. How is STAGA function affected by polyglutamine expansion in ataxin-7? The SAGA complex, much like the RNA polymerase II initiation complex, is defined as

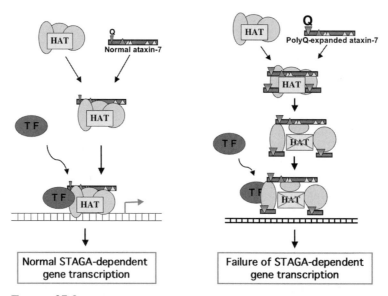

FIGURE 27-9 SCA7 retinal degeneration may involve impaired STAGA coactivator complex function. Here we see a model for how polyglutamine-expanded ataxin-7 produces retinal degeneration in SCA7 patients. Ataxin-7 is incorporated into the STAGA complex regardless of polyglutamine length. The STAGA co-activator complex possesses intrinsic HAT activity because of the presence of Gcn5 typically, and associates with certain transcription factors (TF) to permit chromatin remodeling at target genes. In this way, STAGA-dependent TFs transactivate their target genes. In SCA7 retina, however, incorporation of polyglutamine-expanded ataxin-7 into the STAGA complex leads to inhibition of Gcn5-mediated HAT activity, resulting in failure of STAGA-dependent gene transcription for TFs that rely on STAGA for target gene transactivation. See CD-ROM for color image.

a scaffold of protein–protein interactions. This scaffold has recently been visualized by electron microscopy down to 30-nm resolution to reveal a structure of two saddle shapes that may define large surface areas for essential protein–protein interactions to regulate RNA polymerase II transcription by HAT activity on chromatin [87]. Therefore, polyglutamine expansion of ataxin-7 may disrupt the scaffold and thereby disallow the Gcn5 HAT to be properly directed in space. Of course, one conundrum remains: Why would disruption of a member of a general transcription complex that is expressed in most cell types produce a specific pattern of pathology? The answer may require that we identify other involved STAGA-dependent transcription factors in the retina and cerebellum whose functional impairment underlies the production of the SCA7 phenotype.

VIII. ATAXIN-7 PROTEOYLTIC CLEAVAGE AND TURNOVER

Another important feature of polyglutamine disease pathogenesis that may account for cell type specificity in the various diseases is proteolytic cleavage. In HD, SBMA, DRPLA, SCA3, and SCA7, evidence exists for the preferential nuclear accumulation of truncated polyglutamine-containing fragments of the respective disease proteins [26, 56, 88–90]. One hypothesis of polyglutamine disease pathogenesis (the so-called "toxic fragment hypothesis") posits the production of truncated polyglutamine disease protein as a key step in the disease process. The importance of proteolytic cleavage for disease pathogenesis is most convincing for HD, where protease cleavage processes have been carefully studied. Both *in vitro* and *in vivo* studies of HD are consistent with caspase- and calpain-mediated cleavage of huntingtin protein [91–95]. Ultimately, truncation of the huntingtin protein to the final toxic peptide product may involve cleavage by an as yet unidentified aspartyl protease [96]. Whether huntingtin protein is transported to the nucleus via interaction with an NLS-containing shuttle protein and cleaved in the nucleus or diffuses into the nucleus as a truncated fragment after cleavage in the cytosol is unclear.

Studies of ataxin-7 in mouse models and cell culture support the existence of an amino-terminal truncated fragment in SCA7. In the MoPrP SCA7 mouse model, an ~55-kDa ataxin-7 fragment is detected with both an amino-terminal directed ataxin-7 antibody and the

1C2 antibody [26]. Immunostaining studies of SCA7 mice generated with the rhodopsin promoter similarly support the existence of an amino-terminal truncation fragment [29]. Similar to huntingtin, ataxin-3, AR, and atrophin-1, ataxin-7 can be cleaved by caspase-7 [97], and caspase cleavage can modulate ataxin-7 cell toxicity *in vitro* [98]. Cleavage of ataxin-7 by caspase-7 occurs at two conserved caspase consensus sites in the amino-terminal region of the protein at positions 266 and 344. Cleavage at either site would generate a short fragment containing the amino terminus with the polyglutamine tract but without the NES signal, potentially resulting in accumulation of mutant ataxin-7 fragment in the nucleus. Expression of a mutant ataxin-7 construct, with an amino-terminal fragment similar in size to the truncation product found *in vivo*, and lacking the NES, showed enhanced nuclear localization and cell toxicity when compared with full-length mutant ataxin-7 protein containing the NES [74]. The proteases responsible for *in vivo* generation of an amino-terminal nuclear 50- to 60-kDa ataxin-7 fragment are unknown; therefore, it will be important to determine if caspase-7 cleavage of ataxin-7 occurs *in vivo* and is relevant to disease pathogenesis.

It is not currently understood how the polyglutamine-expanded form of ataxin-7 causes degeneration in specific neuronal populations. Several cell biology studies have attempted to determine the molecular mechanism by which polyglutamine-expanded ataxin-7 expression leads to cellular toxicity. Studies of polyglutamine-expanded ataxin-7 in cultured neuronal cells lines demonstrate that the activated form of the proapoptotic protease caspase-3 is sequestered into these aggregates, and similar findings were observed in cortical neurons from SCA7 patients [99]. Overexpression of polyglutamine-expanded ataxin-7 in cultured cerebellar cells led to activation of the intrinsic apoptotic pathway with increased expression of the proapoptotic protein Bax, decreased expression of antiapoptotic bcl-2, release of cytochrome c and Smac/DIABLO from mitochondria, activation of caspase-9 and caspase-3, and ultimately DNA degradation consistent with apoptotic cell death [100]. However, a component of the neurological dysfunction in SCA7 mice is likely due to reversible physiological alterations that do not involve neuronal loss. Electrophysiological examination of hippocampal slice preparations from SCA7 knock-in mice demonstrated normal synaptic connectivity within the hippocampus but mild impairment of the immediate phase of long-term potentiation (LTP) due in part to an impairment in posttetanic potentiation (PTP) [70]. Thus, whether and how activation of apoptotic or autophagic cell death pathways contributes to SCA7 disease pathogenesis remain unknown. Interestingly, although polyglutamine-expanded ataxin-7 accumulation might be envisioned to inhibit proteasome function and thereby produce a toxic cascade, *in vivo* studies of the SCA7 knock-in mouse model have indicated that neuronal dysfunction actually can occur in the absence of ubiquitin–proteasome system impairment [101]. The interrelationship between subcellular localization, cell death pathway activation, proteolytic cleavage, ataxin-7 turnover, and SCA7 disease pathogenesis thus needs to be clarified.

IX. UNANSWERED QUESTIONS AND FUTURE DIRECTIONS

The realization that ataxin-7 is a transcription coactivator subunit further adds to the increasing body of evidence that many polyglutamine disease proteins are transcription factors or co-regulators. Indeed, of the nine polyglutamine disease proteins, seven are associated with prominent nuclear inclusions, and for all seven of nine, evidence for a role in transcription regulation has been uncovered. AR and TBP are well-known transcription factors, the functions of which had been well established prior to the discovery of CAG repeat expansions in the respective diseases that they cause: X-linked spinal and bulbar muscular atrophy and SCA17. The huntingtin protein has been shown to regulate transcription through an interaction with repressor element-1 transcription factor (REST) in neurons, whereas a short amino-terminal fragment of the huntingtin protein interferes with transcription regulators such as CBP, Sp1, and TAFII-130 [102–104]. The association of huntingtin with such factors supports a role for it in the regulation of transcription. A close homolog of atrophin-1 (the gene responsible for dentatorubral pallidoluysian atrophy) in the fruit fly can function as a transcription co-repressor [105]. Human atrophin-1 protein, like the huntingtin protein, can interfere with CBP-mediated transcription [103]. Ataxin-1 may modulate RNA polymerase II-mediated transcription through an interaction with PQBP-1 [106]. Independent data have also suggested that ataxin-1 may function as a transcription co-repressor through its interaction with SMRT, a repressor of transcription [107]. Ataxin-3, by binding to histone proteins, has been reported to interfere with the chromatin remodeling activities of transcription coactivators with HAT activity, in particular CBP and p300, and may therefore function as a transcription co-repressor [108]. Given these findings, the tendency of polyglutamine disease proteins to interact with polyglutamine-containing transcription factors, and the documentation of gene transcription alterations as an early event in disease pathogenesis in both mouse models and human patient material [109, 110], it has been suggested that many polyglutamine diseases may be considered "transcriptionopathies" [111]. That polyglutamine expansions expressed in yeast yield transcriptional changes that resemble those observed in SAGA subunit mutants would seem to support this view [112]. Similarly compelling has been the demonstration that histone deacteylase (HDAC) inhibitors are potentially useful therapeutic compounds for the treatment of polyglutamine diseases such as HD and SBMA [113–115]. As polyglutamine expansion of ataxin-7 can impair STAGA-dependent GCN5 HAT activity and HDAC inhibitors can rescue this defect, it will be important to determine if HDAC inhibitors hold therapeutic promise for the treatment of SCA7 retinal degeneration.

Recent studies of SCA1 and HD mouse models have shown that RNA interference (RNAi) may be a viable approach for therapeutic intervention [116, 117]. In the case of SCA1, intracerebellar delivery of a short hairpin RNA (shRNA) directed against human ataxin-1 completely eliminated expression of the mutant transgene and thus prevented Purkinje cell degeneration in the classic SCA1 B05 mouse model. SCA7 is a particularly excellent candidate for such approaches and their translation for use in humans, as the requisite mouse models exist and the SCA7 retinal degeneration phenotype offers an optimal proving ground for the evaluation of RNAi-mediated interventions. The rationale for selection of SCA7 as an ideal test case for RNAi-mediated therapy for polyglutamine neurodegeneration is based on the accessibility of the eye and the isolation of its vasculature from the systemic circulation, limiting potential side effects from nonspecific effects of shRNAs. In a number of retinal degeneration animal models ranging from rodents to dogs, methods for delivery and successful long-term expression of therapeutic gene products have been developed and refined [118, 119]. Adeno-associated virus serotype 2 (AAV2) vectors have been shown to efficiently transduce retinal photoreceptors when delivered by subretinal injection [120–122]. If knock-down of a polyglutamine-expanded protein is to be considered as a potential therapy for such diseases, then the available SCA7 mouse models offer an opportunity to develop successful human ataxin-7-directed shRNAs and to derive AAV2-ataxin-7 shRNA vectors that could one day be tested in humans via subretinal injection. Future studies of the molecular basis of SCA7 should provide other exciting opportunities for therapy development for this disorder. Indeed, an exciting prospect is that therapeutic work on SCA7 may prove to be applicable to the rest of the polyglutamine diseases.

Acknowledgments

Our research on SCA7 is supported by grants from the National Institutes of Health (EY14061 and GM59356 to A.R.L., NS40251A to L.M.E.); the Canadian Institutes of Health Research (MOP 36518 and a New Scientist Award to R.T.); and the National Organization for Rare Disorders (G.A.G.).

References

1. Trottier, Y., Lutz, Y., Stevanin, G., Imbert, G., Devys, D., Cancel, G., Saudou, F., Weber, C., David, G., Tora, L., et al. (1995). Polyglutamine expansion as a pathological epitope in Huntington's disease and four dominant cerebellar ataxias. *Nature* **378**, 403–406.
2. Abe, T., Tsuda, T., Yoshida, M., Wada, Y., Kano, T., Itoyama, Y., and Tamai, M. (2000). Macular degeneration associated with aberrant expansion of trinucleotide repeat of the SCA7 gene in 2 Japanese families. *Arch. Ophthalmol.* **118**, 1415–1421.
3. Bryer, A., Krause, A., Bill, P., Davids, V., Bryant, D., Butler, J., Heckmann, J., Ramesar, R., and Greenberg, J. (2003). The hereditary adult-onset ataxias in South Africa. *J. Neurol. Sci.* **216**, 47–54.
4. Gu, W., Wang, Y., Liu, X., Zhou, B., Zhou, Y., and Wang, G. (2000). Molecular and clinical study of spinocerebellar ataxia type 7 in Chinese kindreds. *Arch. Neurol.* **57**, 1513–1518.
5. Jardim, L. B., Silveira, I., Pereira, M. L., Ferro, A., Alonso, I., do Ceu Moreira, M., Mendonca, P., Ferreirinha, F., Sequeiros, J., and Giugliani, R. (2001). A survey of spinocerebellar ataxia in South Brazil: 66 new cases with Machado–Joseph disease, SCA7, SCA8, or unidentified disease-causing mutations. *J. Neurol.* **248**, 870–876.
6. Johansson, J., Forsgren, L., Sandgren, O., Brice, A., Holmgren, G., and Holmberg, M. (1998). Expanded CAG repeats in Swedish spinocerebellar ataxia type 7 (SCA7) patients: Effect of CAG repeat length on the clinical manifestation. *Hum. Mol. Genet.* **7**, 171–176.
7. Storey, E., du Sart, D., Shaw, J. H., Lorentzos, P., Kelly, L., McKinley Gardner, R. J., Forrest, S. M., Biros, I., and Nicholson, G. A. (2000). Frequency of spinocerebellar ataxia types 1, 2, 3, 6, and 7 in Australian patients with spinocerebellar ataxia. *Am. J. Med. Genet.* **95**, 351–357.
8. Duinkerke-Eerola, K. U., Cruysberg, J. R., and Deutman, A. F. (1980). Atrophic maculopathy associated with hereditary ataxia. *Am. J. Ophthalmol.* **90**, 597–603.
9. Harding, A. E. (1983). Classification of the hereditary ataxias and paraplegias. *Lancet* **1**, 1151–1155.
10. David, G., Abbas, N., Stevanin, G., Durr, A., Yvert, G., Cancel, G., Weber, C., Imbert, G., Saudou, F., Antoniou, E., Drabkin, H., Gemmill, R., Giunti, P., Benomar, A., Wood, N., Ruberg, M., Agid, Y., Mandel, J. L., and Brice, A. (1997). Cloning of the SCA7 gene reveals a highly unstable CAG repeat expansion. *Nat. Genet.* **17**, 65–70.
11. Giunti, P., Stevanin, G., Worth, P. F., David, G., Brice, A., and Wood, N. W. (1999). Molecular and clinical study of 18 families with ADCA type II: Evidence for genetic heterogeneity and de novo mutation. *Am. J. Hum. Genet.* **64**, 1594–1603.
12. Hamilton, S. R., Chatrian, G. E., Mills, R. P., Kalina, R. E., and Bird, T. D. (1990). Cone dysfunction in a subgroup of patients with autosomal dominant cerebellar ataxia. *Arch. Ophthalmol.* **108**, 551–556.
13. David, G., Durr, A., Stevanin, G., Cancel, G., Abbas, N., Benomar, A., Belal, S., Lebre, A. S., Abada-Bendib, M., Grid, D., Holmberg, M., Yahyaoui, M., Hentati, F., Chkili, T., Agid, Y., and Brice, A. (1998).

Molecular and clinical correlations in autosomal dominant cerebellar ataxia with progressive macular dystrophy (SCA7). *Hum. Mol. Genet.* **7**, 165–170.

14. Benton, C. S., de Silva, R., Rutledge, S. L., Bohlega, S., Ashizawa, T., and Zoghbi, H. Y. (1998). Molecular and clinical studies in SCA-7 define a broad clinical spectrum and the infantile phenotype. *Neurology* **51**, 1081–1086.

15. Bang, O. Y., Huh, K., Lee, P. H., and Kim, H. J. (2003). Clinical and neuroradiological features of patients with spinocerebellar ataxias from Korean kindreds. *Arch. Neurol.* **60**, 1566–1574.

16. Bang, O. Y., Lee, P. H., Kim, S. Y., Kim, H. J., and Huh, K. (2004). Pontine atrophy precedes cerebellar degeneration in spinocerebellar ataxia 7: MRI-based volumetric analysis. *J. Neurol. Neurosurg. Psychiatry* **75**, 1452–1456.

17. To, K. W., Adamian, M., Jakobiec, F. A., and Berson, E. L. (1993). Olivopontocerebellar atrophy with retinal degeneration. An electroretinographic and histopathologic investigation. *Ophthalmology* **100**, 15–23.

18. Gouw, L. G., Digre, K. B., Harris, C. P., Haines, J. H., and Ptacek, L. J. (1994). Autosomal dominant cerebellar ataxia with retinal degeneration: Clinical, neuropathologic, and genetic analysis of a large kindred. *Neurology* **44**, 1441–1447.

19. Ahn, J. K., Seo, J. M., Chung, H., and Yu, H. G. (2005). Anatomical and functional characteristics in atrophic maculopathy associated with spinocerebellar ataxia type 7. *Am. J. Ophthalmol.* **139**, 923–925.

20. Aleman, T. S., Cideciyan, A. V., Volpe, N. J., Stevanin, G., Brice, A., and Jacobson, S. G. (2002). Spinocerebellar ataxia type 7 (SCA7) shows a cone–rod dystrophy phenotype. *Exp. Eye Res.* **74**, 737–745.

21. Enevoldson, T. P., Sanders, M. D., and Harding, A. E. (1994). Autosomal dominant cerebellar ataxia with pigmentary macular dystrophy: A clinical and genetic study of eight families. *Brain* **117**, 445–460.

22. Martin, J. J., Van Regemorter, N., Krols, L., Brucher, J. M., de Barsy, T., Szliwowski, H., Evrard, P., Ceuterick, C., Tassignon, M. J., Smet-Dieleman, H., et al. (1994). On an autosomal dominant form of retinal-cerebellar degeneration: An autopsy study of five patients in one family. *Acta Neuropathol. (Berl.)* **88**, 277–286.

23. Michalik, A., Martin, J. J., and Van Broeckhoven, C. (2004). Spinocerebellar ataxia type 7 associated with pigmentary retinal dystrophy. *Eur. J. Hum. Genet.* **12**, 2–15.

24. Martin, J., Van Regemorter, N., Del-Favero, J., Lofgren, A., and Van Broeckhoven, C. (1999). Spinocerebellar ataxia type 7 (SCA7): Correlations between phenotype and genotype in one large Belgian family. *J. Neurol. Sci.* **168**, 37–46.

25. Einum, D. D., Townsend, J. J., Ptacek, L. J., and Fu, Y. H. (2001). Ataxin-7 expression analysis in controls and spinocerebellar ataxia type 7 patients. *Neurogenetics* **3**, 83–90.

26. Garden, G. A., Libby, R. T., Fu, Y. H., Kinoshita, Y., Huang, J., Possin, D. E., Smith, A. C., Martinez, R. A., Fine, G. C., Grote, S. K., Ware, C. B., Einum, D. D., Morrison, R. S., Ptacek, L. J., Sopher, B. L., and La Spada, A. R. (2002). Polyglutamine-expanded ataxin-7 promotes non-cell-autonomous Purkinje cell degeneration and displays proteolytic cleavage in ataxic transgenic mice. *J. Neurosci.* **22**, 4897–4905.

27. Holmberg, M., Duyckaerts, C., Durr, A., Cancel, G., Gourfinkel-An, I., Damier, P., Faucheux, B., Trottier, Y., Hirsch, E. C., Agid, Y., and Brice, A. (1998). Spinocerebellar ataxia type 7 (SCA7): A neurodegenerative disorder with neuronal intranuclear inclusions. *Hum. Mol. Genet.* **7**, 913–918.

28. La Spada, A. R., Fu, Y. H., Sopher, B. L., Libby, R. T., Wang, X., Li, L. Y., Einum, D. D., Huang, J., Possin, D. E., Smith, A. C., Martinez, R. A., Koszdin, K. L., Treuting, P. M., Ware, C. B., Hurley, J. B., Ptacek, L. J., and Chen, S. (2001). Polyglutamine-expanded ataxin-7 antagonizes CRX function and induces cone–rod dystrophy in a mouse model of SCA7. *Neuron* **31**, 913–927.

29. Yvert, G., Lindenberg, K. S., Picaud, S., Landwehrmeyer, G. B., Sahel, J. A., and Mandel, J. L. (2000). Expanded polyglutamines induce neurodegeneration and trans-neuronal alterations in cerebellum and retina of SCA7 transgenic mice. *Hum. Mol. Genet.* **9**, 2491–2506.

30. Lindenberg, K. S., Yvert, G., Muller, K., and Landwehrmeyer, G. B. (2000). Expression analysis of ataxin-7 mRNA and protein in human brain: Evidence for a widespread distribution and focal protein accumulation. *Brain Pathol.* **10**, 385–394.

31. Ansorge, O., Giunti, P., Michalik, A., Van Broeckhoven, C., Harding, B., Wood, N., and Scaravilli, F. (2004). Ataxin-7 aggregation and ubiquitination in infantile SCA7 with 180 CAG repeats. *Ann. Neurol.* **56**, 448–452.

32. Yvert, G., Lindenberg, K. S., Devys, D., Helmlinger, D., Landwehrmeyer, G. B., and Mandel, J. L. (2001). SCA7 mouse models show selective stabilization of mutant ataxin-7 and similar cellular responses in different neuronal cell types. *Hum. Mol. Genet.* **10**, 1679–1692.

33. Benomar, A., Krols, L., Stevanin, G., Cancel, G., LeGuern, E., David, G., Ouhabi, H., Martin, J. J., Durr, A., Zaim, A., et al. (1995). The gene for autosomal dominant cerebellar ataxia with pigmentary macular dystrophy maps to chromosome 3p12-p21.1. *Nat. Genet.* **10**, 84–88.

34. Holmberg, M., Johansson, J., Forsgren, L., Heijbel, J., Sandgren, O., and Holmgren, G. (1995). Localization of autosomal dominant cerebellar ataxia associated with retinal degeneration and anticipation to chromosome 3p12-p21.1. *Hum. Mol. Genet.* **4**, 1441–1445.

35. Gouw, L. G., Castaneda, M. A., McKenna, C. K., Digre, K. B., Pulst, S. M., Perlman, S., Lee, M. S., Gomez, C., Fischbeck, K., Gagnon, D., Storey, E., Bird, T., Jeri, F. R., and Ptacek, L. J. (1998). Analysis of the dynamic mutation in the SCA7 gene shows marked parental effects on CAG repeat transmission. *Hum. Mol. Genet.* **7**, 525–532.

36. Stevanin, G., Giunti, P., Belal, G. D., Durr, A., Ruberg, M., Wood, N., and Brice, A. (1998). De novo expansion of intermediate alleles in spinocerebellar ataxia 7. *Hum. Mol. Genet.* **7**, 1809–1813.

37. Monckton, D. G., Cayuela, M. L., Gould, F. K., Brock, G. J., Silva, R., and Ashizawa, T. (1999). Very large (CAG)(n) DNA repeat expansions in the sperm of two spinocerebellar ataxia type 7 males. *Hum. Mol. Genet.* **8**, 2473–2478.

38. La Spada, A. R., Roling, D. B., Harding, A. E., Warner, C. L., Spiegel, R., Hausmanowa Petrusewicz, I., Yee, W. C., and Fischbeck, K. H. (1992). Meiotic stability and genotype–phenotype correlation of the trinucleotide repeat in X-linked spinal and bulbar muscular atrophy. *Nat. Genet.* **2**, 301–304.

39. Cleary, J. D., and Pearson, C. E. (2003). The contribution of cis-elements to disease-associated repeat instability: Clinical and experimental evidence. *Cytogenet. Genome Res.* **100**, 25–55.

40. La Spada, A. R. (1997). Trinucleotide repeat instability: Genetic features and molecular mechanisms. *Brain Pathol.* **7**, 943–963.

41. Imbert, G., Kretz, C., Johnson, K., and Mandel, J. L. (1993). Origin of the expansion mutation in myotonic dystrophy. *Nat. Genet.* **4**, 72–76.

42. Choudhry, S., Mukerji, M., Srivastava, A. K., Jain, S., and Brahmachari, S. K. (2001). CAG repeat instability at SCA2 locus: Anchoring CAA interruptions and linked single nucleotide polymorphisms. *Hum. Mol. Genet.* **10**, 2437–2446.

43. Goldberg, Y. P., McMurray, C. T., Zeisler, J., Almqvist, E., Sillence, D., Richards, F., Gacy, A. M., Buchanan, J., Telenius, H., and Hayden, M. R. (1995). Increased instability of intermediate alleles in families with sporadic Huntington disease compared to similar sized intermediate alleles in the general population. *Hum. Mol. Genet.* **4**, 1911–1918.

44. Limprasert, P., Nouri, N., Heyman, R. A., Nopparatana, C., Kamonsilp, M., Deininger, P. L., and Keats, B. J. (1996). Analysis of CAG repeat of the Machado–Joseph gene in human, chimpanzee and monkey populations: A variant nucleotide is associated with the number of CAG repeats. *Hum. Mol. Genet.* **5**, 207–213.

45. Libby, R. T., Monckton, D. G., Fu, Y. H., Martinez, R. A., McAbney, J. P., Lau, R., Einum, D. D., Nichol, K., Ware, C. B., Ptacek, L. J., Pearson, C. E., and La Spada, A. R. (2003). Genomic context drives SCA7 CAG repeat instability, while expressed SCA7 cDNAs are intergenerationally and somatically stable in transgenic mice. *Hum. Mol. Genet.* **12**, 41–50.

46. Brock, G. J., Anderson, N. H., and Monckton, D. G. (1999). *cis*-acting modifiers of expanded CAG/CTG triplet repeat expandability: Associations with flanking GC content and proximity to CpG islands. *Hum. Mol. Genet.* **8**, 1061–1067.

47. Lobanenkov, V. V., Nicolas, R. H., Adler, V. V., Paterson, H., Klenova, E. M., Polotskaja, A. V., and Goodwin, G. H. (1990). A novel sequence-specific DNA binding protein which interacts with three regularly spaced direct repeats of the CCCTC-motif in the 5′-flanking sequence of the chicken c-myc gene. *Oncogene* **5**, 1743–1753.

48. Ohlsson, R., Renkawitz, R., and Lobanenkov, V. (2001). CTCF is a uniquely versatile transcription regulator linked to epigenetics and disease. *Trends Genet.* **17**, 520–527.

49. Perutz, M. F., Johnson, T., Suzuki, M., and Finch, J. T. (1994). Glutamine repeats as polar zippers: Their possible role in inherited neurodegenerative diseases. *Proc. Natl. Acad. Sci. USA* **91**, 5355–5358.

50. Perutz, M. F., Finch, J. T., Berriman, J., and Lesk, A. (2002). Amyloid fibers are water-filled nanotubes. *Proc. Natl. Acad. Sci. USA* **99**, 5591–5595.

51. Sikorski, P., and Atkins, E. (2005). New model for crystalline polyglutamine assemblies and their connection with amyloid fibrils. *Biomacromolecules* **6**, 425–432.

52. Gordon-Smith, D. J., Carbajo, R. J., Stott, K., and Neuhaus, D. (2001). Solution studies of chymotrypsin inhibitor-2 glutamine insertion mutants show no interglutamine interactions. *Biochem. Biophys. Res. Commun.* **280**, 855–860.

53. Masino, L., Musi, V., Menon, R. P., Fusi, P., Kelly, G., Frenkiel, T. A., Trottier, Y., and Pastore, A. (2003). Domain architecture of the polyglutamine protein ataxin-3: A globular domain followed by a flexible tail. *FEBS Lett.* **549**, 21–25.

54. Tanaka, M., Morishima, I., Akagi, T., Hashikawa, T., and Nukina, N. (2001). Intra- and intermolecular beta-pleated sheet formation in glutamine-repeat inserted myoglobin as a model for polyglutamine diseases. *J. Biol. Chem.* **276**, 45470–45475.

55. Davies, S. W., Turmaine, M., Cozens, B. A., DiFiglia, M., Sharp, A. H., Ross, C. A., Scherzinger, E., Wanker, E. E., Mangiarini, L., and Bates, G. P. (1997). Formation of neuronal intranuclear inclusions underlies the neurological dysfunction in mice transgenic for the HD mutation. *Cell* **90**, 537–548.

56. Paulson, H. L., Perez, M. K., Trottier, Y., Trojanowski, J. Q., Subramony, S. H., Das, S. S., Vig, P., Mandel, J. L., Fischbeck, K. H., and Pittman, R. N. (1997). Intranuclear inclusions of expanded polyglutamine protein in spinocerebellar ataxia type 3. *Neuron* **19**, 333–344.

57. Arrasate, M., Mitra, S., Schweitzer, E. S., Segal, M. R., and Finkbeiner, S. (2004). Inclusion body formation reduces levels of mutant huntingtin and the risk of neuronal death. *Nature* **431**, 805–810.

58. Sanchez, I., Mahlke, C., and Yuan, J. (2003). Pivotal role of oligomerization in expanded polyglutamine neurodegenerative disorders. *Nature* **421**, 373–379.

59. Poirier, M. A., Li, H., Macosko, J., Cai, S., Amzel, M., and Ross, C. A. (2002). Huntingtin spheroids and protofibrils as precursors in polyglutamine fibrilization. *J. Biol. Chem.* **277**, 41032–41037.

60. Fischer, M., Rulicke, T., Raeber, A., Sailer, A., Moser, M., Oesch, B., Brandner, S., Aguzzi, A., and Weissmann, C. (1996). Prion protein (PrP) with amino-proximal deletions restoring susceptibility of PrP knockout mice to scrapie. *EMBO J.* **15**, 1255–1264.

61. Koyano, S., Iwabuchi, K., Yagishita, S., Kuroiwa, Y., and Uchihara, T. (2002). Paradoxical absence of nuclear inclusion in cerebellar Purkinje cells of hereditary ataxias linked to CAG expansion. *J. Neurol. Neurosurg. Psychiatry.* **73**, 450–452.

62. Dudek, H., Datta, S. R., Franke, T. F., Birnbaum, M. J., Yao, R., Cooper, G. M., Segal, R. A., Kaplan, D. R., and Greenberg, M. E. (1997). Regulation of neuronal survival by the serine-threonine protein kinase Akt. *Science* **275**, 661–665.

63. Mount, H. T., Dean, D. O., Alberch, J., Dreyfus, C. F., and Black, I. B. (1995). Glial cell line-derived neurotrophic factor promotes the survival and morphologic differentiation of Purkinje cells. *Proc. Natl. Acad. Sci. USA* **92**, 9092–9096.

64. Custer, S. K., Garden, G. A., Gill, N., Libby, R. T., Guyenet, S. J., Westrum, L. E., Sopher, B. L., and La Spada, A. R. (2006). Bergmann glia expression of polyglutamine-expanded ataxin-7 produces Purkinje cell degeneration and implicates glial-induced impairment of glutamate transport in SCA7. Submitted for publication.

65. Chen, S., Wang, Q. L., Nie, Z., Sun, H., Lennon, G., Copeland, N. G., Gilbert, D. J., Jenkins, N. A., and Zack, D. J. (1997). Crx, a novel Otx-like paired-homeodomain protein, binds to and transactivates photoreceptor cell-specific genes. *Neuron* **19**, 1017–1030.

66. McCampbell, A., Taylor, J. P., Taye, A. A., Robitschek, J., Li, M., Walcott, J., Merry, D., Chai, Y., Paulson, H., Sobue, G., and Fischbeck, K. H. (2000). CREB-binding protein sequestration by expanded polyglutamine. *Hum. Mol. Genet.* **9**, 2197–2202.

67. Steffan, J. S., Kazantsev, A., Spasic-Boskovic, O., Greenwald, M., Zhu, Y. Z., Gohler, H., Wanker, E. E., Bates, G. P., Housman, D. E., and Thompson, L. M. (2000). The Huntington's disease protein interacts with p53 and CREB-binding protein and represses transcription. *Proc. Natl. Acad. Sci. USA* **97**, 6763–6768.

68. Freund, C. L., Gregory-Evans, C. Y., Furukawa, T., Papaioannou, M., Looser, J., Ploder, L., Bellingham, J., Ng, D., Herbrick, J. A., Duncan, A., Scherer, S. W., Tsui, L. C., Loutradis-Anagnostou, A., Jacobson, S. G., Cepko, C. L., Bhattacharya, S. S., and McInnes, R. R. (1997). Cone–rod dystrophy due to mutations in a novel photoreceptor-specific homeobox gene (CRX) essential for maintenance of the photoreceptor. *Cell* **91**, 543–553.

69. Swain, P. K., Chen, S., Wang, Q. L., Affatigato, L. M., Coats, C. L., Brady, K. D., Fishman, G. A., Jacobson, S. G., Swaroop, A., Stone, E., Sieving, P. A., and Zack, D. J. (1997). Mutations in the cone–rod homeobox gene are associated with the cone–rod dystrophy photoreceptor degeneration. *Neuron* **19**, 1329–1336.

70. Yoo, S. Y., Pennesi, M. E., Weeber, E. J., Xu, B., Atkinson, R., Chen, S., Armstrong, D. L., Wu, S. M., Sweatt, J. D., and Zoghbi, H. Y. (2003). SCA7 knockin mice model human SCA7 and reveal gradual accumulation of mutant ataxin-7 in neurons and abnormalities in short-term plasticity. *Neuron* **37**, 383–401.

71. Cancel, G., Duyckaerts, C., Holmberg, M., Zander, C., Yvert, G., Lebre, A. S., Ruberg, M., Faucheux, B., Agid, Y., Hirsch, E., and Brice, A. (2000). Distribution of ataxin-7 in normal human brain and retina. *Brain* **123** (Pt. 12), 2519–2530.

72. Chen, S., Peng, G. H., Wang, X., Smith, A. C., Grote, S. K., Sopher, B. L., and La Spada, A. R. (2004). Interference of Crx-dependent transcription by ataxin-7 involves interaction between the glutamine regions and requires the ataxin-7 carboxy-terminal region for nuclear localization. *Hum. Mol. Genet.* **13**, 53–67.

73. Kaytor, M. D., Duvick, L. A., Skinner, P. J., Koob, M. D., Ranum, L. P., and Orr, H. T. (1999). Nuclear localization of the spinocerebellar ataxia type 7 protein, ataxin-7. *Hum. Mol. Genet.* **8**, 1657–1664.
74. Taylor, J., Grote, S. K., Xia, J., Vandelft, M., Gracyzk, J., Ellerby, L. M., and Truant, R. (2006). Ataxin-7 can export from the nucleus via a conserved exportin-dependent signal. In press.
75. Mushegian, A. R., Vishnivetskiy, S. A., and Gurevich, V. V. (2000). Conserved phosphoprotein interaction motif is functionally interchangeable between ataxin-7 and arrestins. *Biochemistry* **39**, 6809–6813.
76. Scott, M. G., Le Rouzic, E., Perianin, A., Pierotti, V., Enslen, H., Benichou, S., Marullo, S., and Benmerah, A. (2002). Differential nucleocytoplasmic shuttling of beta-arrestins: Characterization of a leucine-rich nuclear export signal in beta-arrestin2. *J. Biol. Chem.* **277**, 37693–37701.
77. Blazek, E., Mittler, G., and Meisterernst, M. (2005). The mediator of RNA polymerase II. *Chromosoma* **113**, 399–408.
78. Conaway, J. W., Florens, L., Sato, S., Tomomori-Sato, C., Parmely, T. J., Yao, T., Swanson, S. K., Banks, C. A., Washburn, M. P., and Conaway, R. C. (2005). The mammalian mediator complex. *FEBS Lett.* **579**, 904–908.
79. Borukhov, S., and Nudler, E. (2003). RNA polymerase holoenzyme: Structure, function and biological implications. *Curr. Opin. Microbiol.* **6**, 93–100.
80. Yang, X. J. (2005). Multisite protein modification and intramolecular signaling. *Oncogene* **24**, 1653–1662.
81. Timmers, H. T., and Tora, L. (2005). SAGA unveiled. *Trends Biochem. Sci.* **30**, 7–10.
82. Sanders, S. L., Jennings, J., Canutescu, A., Link, A. J., and Weil, P. A. (2002). Proteomics of the eukaryotic transcription machinery: Identification of proteins associated with components of yeast TFIID by multidimensional mass spectrometry. *Mol. Cell. Biol.* **22**, 4723–4738.
83. McMahon, S. J., Pray-Grant, M. G., Schieltz, D., Yates, J. R., 3rd, and Grant, P. A. (2005). Polyglutamine-expanded spinocerebellar ataxia-7 protein disrupts normal SAGA and SLIK histone acetyltransferase activity. *Proc. Natl. Acad. Sci. USA* **102**, 8478–8482.
84. Helmlinger, D., Hardy, S., Sasorith, S., Klein, F., Robert, F., Weber, C., Miguet, L., Potier, N., Van-Dorsselaer, A., Wurtz, J. M., Mandel, J. L., Tora, L., and Devys, D. (2004). Ataxin-7 is a subunit of GCN5 histone acetyltransferase-containing complexes. *Hum. Mol. Genet.* **13**, 1257–1265.
85. Palhan, V. B., Chen, S., Peng, G. H., Tjernberg, A., Gamper, A. M., Fan, Y., Chait, B. T., La Spada, A. R., and Roeder, R. G. (2005). Polyglutamine-expanded ataxin-7 inhibits STAGA histone acetyltransferase activity to produce retinal degeneration. *Proc. Natl. Acad. Sci. USA* **102**, 8472–8477.
86. Nakamura, K., Jeong, S. Y., Uchihara, T., Anno, M., Nagashima, K., Nagashima, T., Ikeda, S., Tsuji, S., and Kanazawa, I. (2001). SCA17, a novel autosomal dominant cerebellar ataxia caused by an expanded polyglutamine in TATA-binding protein. *Hum. Mol. Genet.* **10**, 1441–1448.
87. Wu, P. Y., Ruhlmann, C., Winston, F., and Schultz, P. (2004). Molecular architecture of the *S. cerevisiae* SAGA complex. *Mol. Cell* **15**, 199–208.
88. Li, M., Miwa, S., Kobayashi, Y., Merry, D. E., Yamamoto, M., Tanaka, F., Doyu, M., Hashizume, Y., Fischbeck, K. H., and Sobue, G. (1998). Nuclear inclusions of the androgen receptor protein in spinal and bulbar muscular atrophy. *Ann. Neurol.* **44**, 249–254.
89. Schilling, G., Becher, M. W., Sharp, A. H., Jinnah, H. A., Duan, K., Kotzuk, J. A., Slunt, H. H., Ratovitski, T., Cooper, J. K., Jenkins, N. A., Copeland, N. G., Price, D. L., Ross, C. A., and Borchelt, D. R. (1999). Intranuclear inclusions and neuritic aggregates in transgenic mice expressing a mutant N-terminal fragment of huntingtin. *Hum. Mol. Genet.* **8**, 397–407.
90. Schilling, G., Wood, J. D., Duan, K., Slunt, H. H., Gonzales, V., Yamada, M., Cooper, J. K., Margolis, R. L., Jenkins, N. A., Copeland, N. G., Takahashi, H., Tsuji, S., Price, D. L., Borchelt, D. R., and Ross, C. A. (1999). Nuclear accumulation of truncated atrophin-1 fragments in a transgenic mouse model of DRPLA. *Neuron* **24**, 275–286.
91. Gafni, J., and Ellerby, L. M. (2002). Calpain activation in Huntington's disease. *J. Neurosci.* **22**, 4842–4849.
92. Goldberg, Y. P., Nicholson, D. W., Rasper, D. M., Kalchman, M. A., Koide, H. B., Graham, R. K., Bromm, M., Kazemi Esfarjani, P., Thornberry, N. A., Vaillancourt, J. P., and Hayden, M. R. (1996). Cleavage of huntingtin by apopain, a proapoptotic cysteine protease, is modulated by the polyglutamine tract. *Nat. Genet.* **13**, 442–449.
93. Kim, Y. J., Yi, Y., Sapp, E., Wang, Y., Cuiffo, B., Kegel, K. B., Qin, Z. H., Aronin, N., and DiFiglia, M. (2001). Caspase 3-cleaved N-terminal fragments of wild-type and mutant huntingtin are present in normal and Huntington's disease brains, associate with membranes, and undergo calpain-dependent proteolysis. *Proc. Natl. Acad. Sci. USA* **98**, 12784–12789.
94. Wellington, C. L., Ellerby, L. M., Hackam, A. S., Margolis, R. L., Trifiro, M. A., Singaraja, R., McCutcheon, K., Salvesen, G. S., Propp, S. S., Bromm, M., Rowland, K. J., Zhang, T., Rasper, D., Roy, S., Thornberry, N., Pinsky, L., Kakizuka, A., Ross, C. A., Nicholson, D. W., Bredesen, D. E., and Hayden, M. R. (1998). Caspase cleavage of gene products associated with triplet expansion disorders generates truncated fragments containing the polyglutamine tract. *J. Biol. Chem.* **273**, 9158–9167.
95. Wellington, C. L., Singaraja, R., Ellerby, L., Savill, J., Roy, S., Leavitt, B., Cattaneo, E., Hackam, A., Sharp, A., Thornberry, N., Nicholson, D. W., Bredesen, D. E., and Hayden, M. R. (2000). Inhibiting caspase cleavage of huntingtin reduces toxicity and aggregate formation in neuronal and nonneuronal cells. *J. Biol. Chem.* **275**, 19831–19838.
96. Lunkes, A., Lindenberg, K. S., Ben-Haiem, L., Weber, C., Devys, D., Landwehrmeyer, G. B., Mandel, J. L., and Trottier, Y. (2002). Proteases acting on mutant huntingtin generate cleaved products that differentially build up cytoplasmic and nuclear inclusions. *Mol. Cell* **10**, 259–269.
97. Ellerby, L. M., Hackam, A. S., Propp, S. S., Ellerby, H. M., Rabizadeh, S., Cashman, N. R., Trifiro, M. A., Pinsky, L., Wellington, C. L., Salvesen, G. S., Hayden, M. R., and Bredesen, D. E. (1999). Kennedy's disease: Caspase cleavage of the androgen receptor is a crucial event in cytotoxicity. *J. Neurochem.* **72**, 185–195.
98. Young, J. E., Gouw, L. G., Propp, S. S., Lin, A., Hermel, E., Logvinova, A., Chen, S. F., Bredesen, D. E., Sopher, B. L., Chen, S., Ptacek, L. J., Truant, R., Fu, Y. H., La Spada, A. R., and Ellerby, L. M. (2005). Proteolytic cleavage of ataxin-7 by caspase-7 modulates cellular toxicity and transcriptional dysregulation. Submitted for publication.
99. Zander, C., Takahashi, J., El Hachimi, K. H., Fujigasaki, H., Albanese, V., Lebre, A. S., Stevanin, G., Duyckaerts, C., and Brice, A. (2001). Similarities between spinocerebellar ataxia type 7 (SCA7) cell models and human brain: Proteins recruited in inclusions and activation of caspase-3. *Hum. Mol. Genet.* **10**, 2569–2579.
100. Wang, H. L., Yeh, T. H., Chou, A. H., Kuo, Y. L., Luo, L. J., He, C. Y., Huang, P. C. and Li, A. H. (2005). Polyglutamine-expanded ataxin-7 activates mitochondrial apoptotic pathway of cerebellar neurons by upregulating Bax and downregulating Bcl-x(L). *Cell Signal.* **18**, 541–552.

101. Bowman, A. B., Yoo, S. Y., Dantuma, N. P., and Zoghbi, H. Y. (2005). Neuronal dysfunction in a polyglutamine disease model occurs in the absence of ubiquitin–proteasome system impairment and inversely correlates with the degree of nuclear inclusion formation. *Hum. Mol. Genet.* **14**, 679–691.
102. Dunah, A. W., Jeong, H., Griffin, A., Kim, Y. M., Standaert, D. G., Hersch, S. M., Mouradian, M. M., Young, A. B., Tanese, N., and Krainc, D. (2002). Sp1 and TAFII130 transcriptional activity disrupted in early Huntington's disease. *Science* **296**, 2238–2243.
103. Nucifora, F. C., Jr., Sasaki, M., Peters, M. F., Huang, H., Cooper, J. K., Yamada, M., Takahashi, H., Tsuji, S., Troncoso, J., Dawson, V. L., Dawson, T. M., and Ross, C. A. (2001). Interference by huntingtin and atrophin-1 with cbp-mediated transcription leading to cellular toxicity. *Science* **291**, 2423–2428.
104. Zuccato, C., Tartari, M., Crotti, A., Goffredo, D., Valenza, M., Conti, L., Cataudella, T., Leavitt, B. R., Hayden, M. R., Timmusk, T., Rigamonti, D., and Cattaneo, E. (2003). Huntingtin interacts with REST/NRSF to modulate the transcription of NRSE-controlled neuronal genes. *Nat. Genet.* **35**, 76–83.
105. Zhang, S., Xu, L., Lee, J., and Xu, T. (2002). Drosophila atrophin homolog functions as a transcriptional corepressor in multiple developmental processes. *Cell* **108**, 45–56.
106. Okazawa, H., Rich, T., Chang, A., Lin, X., Waragai, M., Kajikawa, M., Enokido, Y., Komuro, A., Kato, S., Shibata, M., Hatanaka, H., Mouradian, M. M., Sudol, M., and Kanazawa, I. (2002). Interaction between mutant ataxin-1 and PQBP-1 affects transcription and cell death. *Neuron* **34**, 701–713.
107. Tsai, C. C., Kao, H. Y., Mitzutani, A., Banayo, E., Rajan, H., McKeown, M., and Evans, R. M. (2004). Ataxin 1, a SCA1 neurodegenerative disorder protein, is functionally linked to the silencing mediator of retinoid and thyroid hormone receptors. *Proc. Natl. Acad. Sci. USA* **101**, 4047–4052.
108. Li, F., Macfarlan, T., Pittman, R. N., and Chakravarti, D. (2002). Ataxin-3 is a histone-binding protein with two independent transcriptional corepressor activities. *J. Biol. Chem.* **277**, 45004–45012.
109. Lin, X., Antalffy, B., Kang, D., Orr, H. T., and Zoghbi, H. Y. (2000). Polyglutamine expansion down-regulates specific neuronal genes before pathologic changes in SCA1. *Nat. Neurosci.* **3**, 157–163.
110. Luthi-Carter, R., Strand, A., Peters, N. L., Solano, S. M., Hollingsworth, Z. R., Menon, A. S., Frey, A. S., Spektor, B. S., Penney, E. B., Schilling, G., Ross, C. A., Borchelt, D. R., Tapscott, S. J., Young, A. B., Cha, J. H., and Olson, J. M. (2000). Decreased expression of striatal signaling genes in a mouse model of Huntington's disease. *Hum. Mol. Genet.* **9**, 1259–1271.
111. La Spada, A. R., and Taylor, J. P. (2003). Polyglutamines placed into context. *Neuron* **38**, 681–684.
112. Hughes, R. E., Lo, R. S., Davis, C., Strand, A. D., Neal, C. L., Olson, J. M., and Fields, S. (2001). Altered transcription in yeast expressing expanded polyglutamine. *Proc. Natl. Acad. Sci. USA* **98**, 13201–13206.
113. Hockly, E., Richon, V. M., Woodman, B., Smith, D. L., Zhou, X., Rosa, E., Sathasivam, K., Ghazi-Noori, S., Mahal, A., Lowden, P. A., Steffan, J. S., Marsh, J. L., Thompson, L. M., Lewis, C. M., Marks, P. A., and Bates, G. P. (2003). Suberoylanilide hydroxamic acid, a histone deacetylase inhibitor, ameliorates motor deficits in a mouse model of Huntington's disease. *Proc. Natl. Acad. Sci. USA* **100**, 2041–2046.
114. Minamiyama, M., Katsuno, M., Adachi, H., Waza, M., Sang, C., Kobayashi, Y., Tanaka, F., Doyu, M., Inukai, A., and Sobue, G. (2004). Sodium butyrate ameliorates phenotypic expression in a transgenic mouse model of spinal and bulbar muscular atrophy. *Hum. Mol. Genet.* **13**, 1183–1192.
115. Steffan, J. S., Bodai, L., Pallos, J., Poelman, M., McCampbell, A., Apostol, B. L., Kazantsev, A., Schmidt, E., Zhu, Y. Z., Greenwald, M., Kurokawa, R., Housman, D. E., Jackson, G. R., Marsh, J. L., and Thompson, L. M. (2001). Histone deacetylase inhibitors arrest polyglutamine-dependent neurodegeneration in *Drosophila*. *Nature* **413**, 739–743.
116. Harper, S. Q., Staber, P. D., He, X., Eliason, S. L., Martins, I. H., Mao, Q., Yang, L., Kotin, R. M., Paulson, H. L., and Davidson, B. L. (2005). RNA interference improves motor and neuropathological abnormalities in a Huntington's disease mouse model. *Proc. Natl. Acad. Sci. USA* **102**, 5820–5825.
117. Xia, H., Mao, Q., Eliason, S. L., Harper, S. Q., Martins, I. H., Orr, H. T., Paulson, H. L., Yang, L., Kotin, R. M., and Davidson, B. L. (2004). RNAi suppresses polyglutamine-induced neurodegeneration in a model of spinocerebellar ataxia. *Nat. Med.* **10**, 816–820.
118. Acland, G. M., Aguirre, G. D., Ray, J., Zhang, Q., Aleman, T. S., Cideciyan, A. V., Pearce-Kelling, S. E., Anand, V., Zeng, Y., Maguire, A. M., Jacobson, S. G., Hauswirth, W. W., and Bennett, J. (2001). Gene therapy restores vision in a canine model of childhood blindness. *Nat. Genet.* **28**, 92–95.
119. Ali, R. R., Sarra, G. M., Stephens, C., Alwis, M. D., Bainbridge, J. W., Munro, P. M., Fauser, S., Reichel, M. B., Kinnon, C., Hunt, D. M., Bhattacharya, S. S., and Thrasher, A. J. (2000). Restoration of photoreceptor ultrastructure and function in retinal degeneration slow mice by gene therapy. *Nat. Genet.* **25**, 306–310.
120. Ali, R. R., Reichel, M. B., De Alwis, M., Kanuga, N., Kinnon, C., Levinsky, R. J., Hunt, D. M., Bhattacharya, S. S., and Thrasher, A. J. (1998). Adeno-associated virus gene transfer to mouse retina. *Hum. Gene Ther.* **9**, 81–86.
121. Ali, R. R., Reichel, M. B., Thrasher, A. J., Levinsky, R. J., Kinnon, C., Kanuga, N., Hunt, D. M., and Bhattacharya, S. S. (1996). Gene transfer into the mouse retina mediated by an adeno-associated viral vector. *Hum. Mol. Genet.* **5**, 591–594.
122. Flannery, J. G., Zolotukhin, S., Vaquero, M. I., LaVail, M. M., Muzyczka, N., and Hauswirth, W. W. (1997). Efficient photoreceptor-targeted gene expression in vivo by recombinant adeno-associated virus. *Proc. Natl. Acad. Sci. USA* **94**, 6916–6921.

Molecular Genetics of Spinocerebellar Ataxia Type 8

YOSHIO IKEDA,[1,3] KATHERINE A. DICK,[1,3] JOHN W. DAY,[2,3] AND LAURA P. W. RANUM[1,3]

Departments of [1]Genetics, Cell Biology, and Development and [2]Neurology and the [3]Institute of Human Genetics, University of Minnesota, Minneapolis, Minnesota 55455

I. Introduction
II. Identification of the SCA8 CTG Repeat Expansion
   A. RAPID Cloning
   B. Organization of the SCA8 Disease Gene
III. The CTG Repeat Expansion Cosegregates with a Novel Form of Ataxia
   A. SCA8 Expansion Screening in Families with Unknown Forms of Ataxia
   B. Clinical Features of SCA8
   C. Neuropathology of SCA8
IV. Reduced Penetrance Commonly Found in SCA8
   A. Disease Penetrance in the MN-A Family Is Affected by CTG Length
   B. Reduced Penetrance in Other Families
   C. SCA8 Expansions on Control Chromosomes
V. Genetic Analysis of SCA8 Expansion Chromosomes
   A. Haplotype Analysis of SCA8 Expansion Chromosomes
   B. SCA8 Expansions Cosegregate with Ataxia in Small Families
VI. SCA8 Repeat Instability: Possible Influences on Disease Penetrance
   A. Length of the Polymorphic CTA Repeat Tract Preceding the CTG Expansion
   B. Duplicating Interruptions within the CTG Expansion
   C. Repeat Instability during Transmission
   D. En Masse CTG Repeat Contractions in Sperm
   E. Germline Instability of the CTG Repeat Expansion by Small Pool PCR
VII. Molecular Parallels with Myotonic Dystrophy
VIII. Modeling SCA8 Pathogenesis Using the Mouse and Fly
IX. Conclusions
   References

Spinocerebellar ataxia type 8 (SCA8) is a dominantly inherited, slowly progressive neurodegenerative disorder caused by a CTG repeat expansion. Affected individuals show relatively pure cerebellar symptoms affecting gait and limb coordination, speech, and eye movements. The SCA8 CTG expansion, which shows dramatic genetic instability, was the first apparently non-coding triplet repeat reported to cause ataxia. A puzzling and controversial feature of the disease is the high degree of reduced penetrance that is found in all SCA8 families, including a large dominant family (MN-A) with a logarithm of the odds (LOD) score of 6.8 at $\theta = 0.00$. Haplotype analysis of 37 different SCA8 ataxia families indicates that the SCA8 expansion associated with ataxia has arisen independently at least three times. Molecular parallels between mutations in SCA8 and myotonic dystrophy types 1 and 2, along with the known toxic properties of transcripts containing expanded CUG repeats, suggest the possibility that an RNA gain of function mechanism may play a role in disease pathogenesis.

I. INTRODUCTION

Spinocerebellar ataxia type 8 (SCA8) is a slowly progressive, mainly cerebellar ataxia characterized by dramatic repeat instability and a high degree of reduced penetrance. Among the SCAs, SCA8 was the first example of a dominant spinocerebellar ataxia not caused by the expansion of a CAG polyglutamine encoding repeat tract [1] and similar to myotonic dystrophy type 1 (DM1), the only other disease reported to be caused by a noncoding CTG expansion mutation. SCA8 can now be classified with a growing group of diseases in which microsatellite repeat expansions are transcribed but do not appear to be translated, including SCA10, SCA12, DM1, DM2, and fragile X tremor ataxia syndrome (FXTAS) [1–6]. The pathogenic mechanism underlying several of these disorders points to an RNA gain of function mechanism. This chapter will discuss the clinical and molecular features of SCA8 and proposed models of disease pathogenesis.

II. IDENTIFICATION OF THE SCA8 CTG REPEAT EXPANSION

A. RAPID Cloning

In 1999, the authors used the repeat analysis, pooled isolation and detection (RAPID) cloning method was used to identify the mutation for a previously unknown form of spinocerebellar ataxia, SCA8, which is caused by a noncoding CTG repeat expansion within a gene of unknown function [1, 7]. RAPID cloning allowed the isolation of the candidate CAG/CTG expansion mutation directly from the DNA of a single individual, eliminating the need for linkage analysis or extensive clinical data [1, 8]. By removing biases, such as the need for large families, identifying a mutation that is characterized by a high degree of reduced disease penetrance.

To identify candidate expansion mutations, the repeat expansion detection (RED) assay was performed on DNA samples from patients with dominant but unknown forms of ataxia [1, 7, 9]. This screen identified a previously uncharacterized expansion of 80 repeats in an affected mother and daughter. To clone the expansion and unique sequence flanking the repeat, the RAPID cloning procedure was performed and then the genomic insert was sequenced, revealing an expansion of 80 uninterrupted CTG/CAG repeats preceded by 11 CTA/TAG repeats [1, 7]. Although identifying a polyglutamine encoding ataxia gene was expected, the only polyglutamine open reading frame (ORF) contained a single methionine followed by a polyglutamine stretch. Furthermore, sequence analysis did not reveal splice donor or acceptor signals that would allow a polyglutamine ORF to extend through the expansion as part of a spliced transcript, and no transcripts spanning the repeat in the polyglutamine direction have been detected. In summary, these observations indicate that it is unlikely that the SCA8 expansion can be translated into a polyglutamine tract [1]. Since the identification of the SCA8 gene, RAPID cloning has been used to directly clone disease-causing expansion mutations for Huntington disease like 2 (HDL2) and spinocerebellar ataxia type 12 (SCA12) [10, 11].

B. Organization of the SCA8 Disease Gene

The CTG repeat tract, which is conserved in chimpanzees, gorillas, and orangutans [12, 13], is located at the 3' end of a highly alternatively spliced transcript that is expressed at low levels in the CTG direction (Fig. 28-1). The low-steady-state transcript levels in

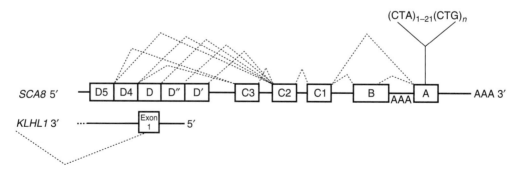

FIGURE 28-1 SCA8 gene organization. Exons are shown as boxes, and alternate splice forms are indicated by dashed lines. The SCA8 repeat tract is located in exon A and is transcribed in the CTG orientation. The region of the SCA8 gene that overlaps the 5' end of KLHL1 is shown. Various alternative splice forms of the SCA8 transcript are indicated. Figure reproduced with permission from S. Karger AG, Basel, and *Cytogenet. Genome Res.* **100**, 175–183 (2003).

the central nervous system (CNS), detectable by reverse transcription-polymerase chain reaction (RT-PCR) but not by Northern or *in situ* hybridization analysis, have made SCA8 expression and alternative splicing analysis difficult [1, 14]. The SCA8 gene overlaps the 5' end of Kelch like 1 (KLHL1), which encodes an actin binding protein that is transcribed in the opposite direction [1, 15]. Although no functional relationship between the two transcripts has been demonstrated, a possible function of the SCA8 transcript suggested by the genomic organization may be to regulate KLHL1 transcripts through an antisense mechanism [15]. A much shorter version of the SCA8 transcript is found in mice, but this truncated version of the gene does not contain the CTG repeat tract [16]. Although RT-PCR shows that SCA8 is transcribed in the CTG orientation, and sequence analysis shows a short ORF with 41 amino acids plus the CTG expansion, this ORF appeared unlikely to be translated because of the relatively large number of upstream start and stop codons, leading to the hypothesis that SCA8 is mediated by an RNA mechanism similar to that of DM1 [1, 8].

III. THE CTG REPEAT EXPANSION COSEGREGATES WITH A NOVEL FORM OF ATAXIA

A. SCA8 Expansion Screening in Families with Unknown Forms of Ataxia

PCR analysis of the expansion was performed on DNA samples from the kindred from which the expansion had originally been cloned. Both of the affected individuals and two at-risk family members carried the expansion [1], and subsequent screening of the ataxia family collection [17] identified probands from 11 additional ataxia kindreds who also carried expansions. To investigate the pathogenicity of the repeat expansion, we collected and examined additional members from these kindreds; the largest of these is a seven-generation kindred (MN-A), from which 92 members were evaluated (Fig. 28-2). All of the affected individuals in the family have an expanded allele, and linkage analysis between ataxia and the expansion in this kindred gave a maximum LOD score of 6.8 at $\theta = 0.00$.

B. Clinical Features of SCA8

On the basis of clinical evaluations of over 200 patients from 25 separate families, it is apparent that SCA8 presents as a slowly progressive ataxia that largely spares brainstem and cerebral functions [1, 8, 18–23]. In the MN-A family disease onset ranges from 13 to 60 years, with gait incoordination the most frequently reported initial symptom. Consistent with slow disease progression, the need for mobility aids usually occurs 20 years after the presentation of initial symptoms. Speech is dysarthric with ataxic and spastic components for all examined individuals [18]. Oculomotor involvement is commonly found in moderate to severely affected patients [18, 20, 24]. Occasionally, intermittent low-amplitude myoclonic jerks in the fingers and arms and mild athetotic movements of extended fingers are detected. Hyperreflexia is a common finding, with an elicitable Babinski sign sometimes observed in severely affected individuals [18, 20]. Occasionally, mild sensory loss is observed, indicated by decreased vibratory perception [18].

Magnetic resonance imaging (MRI) analysis shows atrophy of the cerebellar hemispheres and vermis in affected SCA8 individuals [18, 19, 23], with no apparent involvement of the brainstem, cerebral hemispheres, and basal ganglia. An affected MN-A family member tracked over a 9-year period with MRI revealed little change, consistent with the slowly progressive disease course (Fig. 28-3) [18]. In contrast, Zeman et al. reported a patient having had two MRI scans: the initial scan was determined to be normal, whereas a second scan 4 years later showed clear cerebellar atrophy [25]. Of note, MRI analysis of a 71-year-old individual with an SCA8 expansion, who is clinically unaffected, showed mild cerebellar atrophy, indicating that asymptomatic individuals may still have cerebellar atrophy [26].

C. Neuropathology of SCA8

Autopsy tissue from two SCA8 patients showed consistent neuropathological features. The first patient, from the MN-A family, had an expansion of 110 combined repeats with disease duration of 9 years. Moderate to severe loss of Purkinje cells, mild granule cell loss, and neuronal loss with gliosis in the olivary nuclei were found, with no evidence of neurodegenerative abnormalities in other parts of the brain. Similar neuropathological changes were found in autopsy tissue from a second, unrelated SCA8 expansion positive individual with 140 repeats. This second case was characterized by more severe loss of Purkinje cells, granule cells, and inferior olivary neurons. These findings from two unrelated, genetically confirmed SCA8 patients can be summarized as cerebellar cortical degeneration with accompanying olivary degeneration [27].

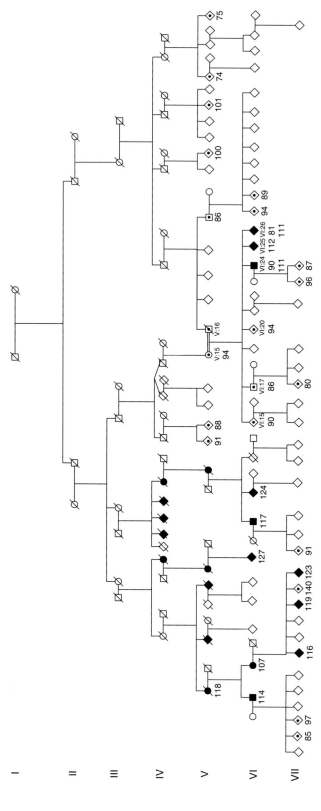

FIGURE 28-2 The large SCA8 kindred (MN-A family). Filled symbols indicate individuals with ataxia; symbols with a dot indicate individuals who inherited the CTG expansion but are not clinically affected by ataxia. The CTG repeat lengths of expanded alleles are indicated below the symbols. Haplotype analyses using five short tandem repeat markers confirm that both branches of the family inherited the expanded repeat from a common founder. Family members homozygous for the SCA8 expansion and their affected heterozygous siblings (individuals VI, 24–26) had similar clinical features, with comparable ages of onset and rates of disease progression. Figure reproduced with permission from *Nature Genetics* **21**, 379–384 (1999) (http://www.nature.com).

CHAPTER 28 Molecular Genetics of Spinocerebellar Ataxia Type 8

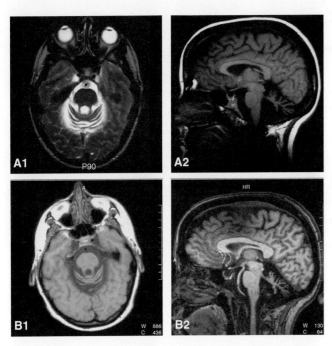

FIGURE 28-3 Serial MRI scans of an affected individual. Horizontal (A1, B1) and sagittal (A2, B2) MRI scans from an affected individual at ages 26 (A) and 35 (B) years. The earlier image is 9 years after onset (17 years). There is marked cerebellar atrophy, minimal brainstem atrophy, and no evidence of cerebral involvement. Very little change occurs over the 9-year period between scans, which is consistent with the slow progression of the disease. Figure reproduced with permission from Day et al., Spinocerebellar ataxia type 8: clinical features in a large family, *Neurology* **55**(5), 649–657 (2000).

IV. REDUCED PENETRANCE COMMONLY FOUND IN SCA8

A. Disease Penetrance in the MN-A Family Is Affected by CTG Length

Within the large MN-A family there are 17 individuals with an expanded repeat who were not clinically affected at the time of evaluation. The ages at exam of these asymptomatic carriers ranged from 14 to 74 years, with a mean (43 ± 17 years) that was comparable to the mean age at exam of the affected family members. The repeat lengths among these asymptomatic carriers are significantly ($p < 10^{-8}$) shorter than those in affected individuals (means of 90 and 116 repeats, respectively), indicating that disease penetrance in the MN-A family is influenced by CTG repeat length. All but one of the individuals with a combined CTA/CTG repeat tract of over 110 repeats are clinically affected. Although this 42-year-old individual, with 143 combined CTA/CTG repeats, has exhibited no signs of ataxia during repeated neurological exams, SCA8 is an adult onset disorder with a documented age of onset as old as 65 years; therefore, an asymptomatic status for this individual is not unanticipated. These data demonstrate that disease penetrance is affected by CTG repeat length in the MN-A family [1].

B. Reduced Penetrance in Other Families

In the MN-A family, SCA8 is transmitted in an autosomal dominant pattern with alleles less than 110 combined repeats showing reduced penetrance. However, other SCA8 families show a complex inheritance pattern in which only a subset of expansion carriers from a given family are affected [1, 18, 23, 26, 28]. Representative SCA8 pedigrees that have presented with apparently distinct inheritance patterns are shown in Fig. 28-4 [29]. Families A and B appear to have transmitted ataxia in a dominant pattern with affected individuals in multiple generations. In contrast, families C and D appear recessive with multiple affected individuals in a single generation, and the affected individuals in families E and F appear as sporadic cases with no other affected family members.

In contrast to the relatively large number of affected patients in the MN-A family ($n = 13$), the 36 smaller SCA8 ataxia families have significantly fewer affected individuals with only 2 families having 3 affected individuals, 9 families having 2 affected individuals, and 25 families with only a single affected individual. Although only a subset of the expansion carriers in the MN-A family develop ataxia (13/35), disease penetrance is significantly higher in the MN-A pedigree than in the 36 smaller ataxia families we have studied as well as families reported by other groups worldwide [1, 18, 20, 23, 26, 28, 29].

Within the additional SCA8 ataxia families reported, repeat sizes among affected and unaffected expansion carriers overlap and often exceed the pathogenic threshold found in the MN-A family [29]. The tight correlation between repeat size and pathogenesis found in the MN-A family is not present in these other ataxia families: SCA8 expansions vary dramatically in size, and the presence of an SCA8 expansion cannot be used to predict whether an asymptomatic individual will develop ataxia [29–32].

C. SCA8 Expansions on Control Chromosomes

Unexpectedly, SCA8 expansions have also been found in control samples that our lab and others have screened [29, 32, 33]. Ten SCA8 alleles (0.4%) larger than 74 combined CTA/CTG repeats were identified, which is the smallest expansion found in an ataxia patient, in a screen of 2626 unrelated control chromosomes that were analyzed in Minnesota and Canada [29]. One of these control expansions was from a Centre d'Etude du Polymorphisme Humain (CEPH) grandmother (family 1416) (Fig. 28-4b). Medical histories indicate that neither this

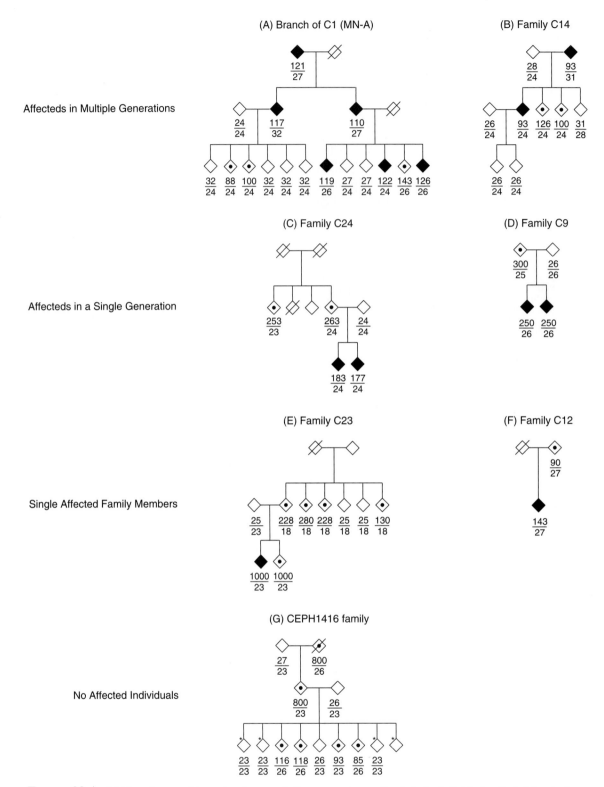

FIGURE 28-4 SCA8 pedigrees with varying degrees of disease penetrance. Symbols for individuals affected by ataxia are blackened, and unaffected expansion carriers are indicated by symbols with a dot inside them. A diagonal line through a symbol denotes an individual who is deceased. The size of the expanded and unexpanded SCA8 alleles is shown below the individuals. The numbers designating each family correspond to those that appear in Fig. 28-5A. Individuals indicated with an asterisk are negative for CTG expansion by Southern analysis. Figure reproduced with permission from University of Chicago Press and Ikeda et al., Am. J. Hum. Genet. **75**, 3–16 (2004). © 2004 by The American Society of Human Genetics. All rights reserved.

woman nor her son (54 years, 800 repeats) was affected by ataxia. All 6 of the SCA8 expansion carriers in this family were asymptomatic at the time of clinical evaluation; however, given that the individuals in generation III were children when they were last clinically evaluated, it is not known whether they have or will develop ataxia [29].

Expansions containing more than 74 combined repeats occurred on 12/292 (4%) independent ataxia chromosomes in our original collection of probands from genetically undefined ataxia families [29]. Although the frequency of expansions greater than 74 combined repeats among unrelated ataxia probands is significantly higher than in the general population (10/2626 chromosomes, $p = 4 \times 10^{-25}$), the relative frequency of alleles with more than 74 combined repeats in the general population (~0.4%) is higher than that of all forms of ataxia (~1/10,000). Screens of other ataxia collections have generated similar findings [19–22, 25, 28, 34–37].

It has been reported that populations with a high frequency of large normal CAG repeat alleles often have a higher prevalence of pathogenic expanded repeats that cause disease [38]. Although the distribution of SCA8 alleles with less than 50 repeats in the Finnish population is similar to that of other ethnic groups, Finnish ataxia and control populations show surprisingly high frequencies of SCA8 expansions [12, 39]. Juvonen *et al.* found SCA8 expansions in 2.9% of controls tested and at significantly higher frequencies in ataxia patients: 15.1%, 18.4%, and 5.5% for families classified with hereditary, dominant, and sporadic ataxia, respectively [20, 40]. These findings suggest that for SCA8 high frequencies of the expansion in controls (more than 80 repeats) may predict the disease frequency better than high frequencies of alleles at the upper end of the normal range (but less than 50 CTGs). In addition, this study and our own investigations strongly suggest that homozygous SCA8 expansion carriers are at increased risk of developing ataxia [1, 29, 34, 36, 37, 39, 41–43].

Taken together, the human genetics studies performed by us and worldwide indicate that the SCA8 CTG repeat can cause ataxia, but that additional environmental or genetic modifiers, including repeat length, affect disease penetrance [1, 29, 32, 34, 39].

V. GENETIC ANALYSIS OF SCA8 EXPANSION CHROMOSOMES

A. Haplotype Analysis of SCA8 Expansion Chromosomes

To better understand the origin of the SCA8 expansion and the reduced penetrance of the disease, haplotype analysis was performed on a panel of 37 SCA8 ataxia families from the United States, Canada, Japan, and Mexico, with 13 SCA8 expansion positive samples sent to Athena Diagnostics for ataxia testing, 7 control samples with expansions, and 14 expansion carriers with psychiatric diseases [29]. Seventeen polymorphic short tandem repeat (STR) markers spanning ~1 Mb of DNA flanking the SCA8 CTG repeat were analyzed. Two ancestrally related haplotypes (A and A′) were observed in the Caucasian population, which included 31 SCA8 families as well as groups of psychiatric patients and controls, indicating a common origin for the pathogenic and nonpathogenic expansions within the Caucasian population (Fig. 28-5A) [29]. Additionally, the Japanese and Mexican ataxia families showed two other distinct haplotypes (B and C, respectively) (Figs. 28-5A and 28-5B). The identification of independently arising SCA8 expansions among ataxia families with various ethnic backgrounds further supports the direct role of the CTG expansion in disease pathogenesis.

B. SCA8 Expansions Cosegregate with Ataxia in Small Families

Genetic studies demonstrate that the cosegregation of the SCA8 expansion and ataxia is highly significant in the MN-A family (LOD = 6.8, $\theta = 0.00$) [1]. To determine whether the SCA8 expansions found in the other 36 smaller ataxia families are simply found by chance or whether the expansions predispose carriers to ataxia, the incidence that the expansion cosegregated with ataxia in family members other than the probands was examined [29]. If the SCA8 expansions do not cause ataxia, then it would be expected that the frequency of SCA8 expansions in additional affected first degree relatives would be ~50%. In contrast, 12 of the 13 affected first degree relatives also inherited the SCA8 expansion, indicating that the expansion also cosegregates with ataxia in these families ($p = 0.0038$). The only exception was found in a family in which two sisters were affected with a clinically distinct form of ataxia characterized by rapid disease progression, choreiform movements, a severe sensory neuronopathy, and neuromyotonic discharges seen by electromyography. Because the striking clinical differences in this family suggest the presence of a different disease in the family, this family was excluded from further analysis. Linkage analysis was performed on the remaining 10 small families with multiple affected individuals; the LOD scores were consistently positive and when combined exceeded 2.0 (LOD = 2.02), the threshold considered significant for testing linkage to a single specific locus [29, 44]. The cosegregation of the SCA8 expansion among additional affected relatives in the group of

(A)

Group I: Haplotype analysis of 37 SCA8 families

Group II: Haplotype analysis of 13 diagnostic ataxia samples

Group III: Haplotype analysis of 5 normal controls and 2 CEPH families

Group IV: Haplotype analysis of 14 major psychosis patients

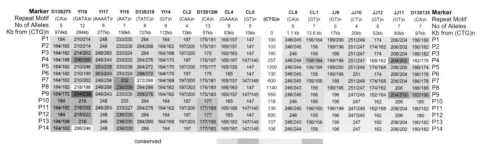

FIGURE 28-5 (continued)

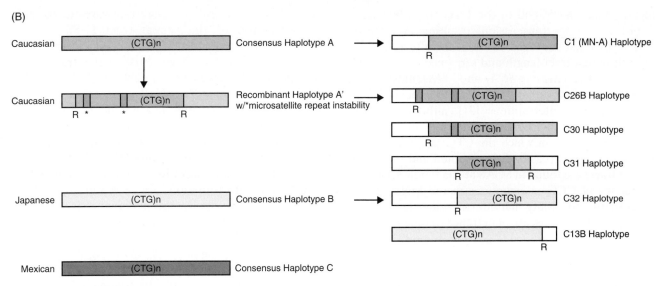

FIGURE 28-5 (A) Haplotype analysis of several SCA8 expansion positive families with and without ataxia. The markers in this figure are ordered according to their physical distance from the CTG repeat. Group I: Haplotypes of 37 SCA8 positive families in which at least one member has been diagnosed with ataxia. Two predominant and probably related SCA8 expansion haplotypes (A and A') were found in 18 and 13 families, respectively. Six families, including 4 from Japan, have a clearly distinct second haplotype (B), and a Mexican family shows evidence for a third independent haplotype (C). The consensus haplotype A is indicated in yellow. Minor deviations in repeat size of ±1 repeat unit flanked by markers with conserved allele sizes are indicated by alternative colors, with the color key located below the figure. Recombinant regions that are not conserved among families are uncolored. Two families (C13 and C26) presented with homozygous expansion positive patients, and the separate SCA8 haplotypes for these families are indicated. The microsatellite marker name, its repeat motif, and its distance from the SCA8 CTG repeat expansion are shown at the top of the figure. The size range of the combined CTA/CTG repeat expansions in the family (or in some case a single expansion carrier) are shown. Group II: SCA8 expansion haplotypes of 13 samples sent to Athena Diagnostics for testing have either haplotype A or haplotype A', except for a single subject (AD13) with haplotype B. Group III: Haplotypes of 7 normal control families, including 2 CEPH families with SCA8 expansions. Four of these families had haplotype A, and 3 had haplotype A'. Group IV: Haplotypes of 14 major psychosis patients with CTG expansions had either haplotype A ($n = 8$) or haplotype A' ($n = 6$). (B) Proposed summary of the ancestral origins based on the analysis of 37 SCA8 ataxia families. The current haplotypes are likely to have arisen from a small number of ancestral recombination and microsatellite instability events. "R" indicates a recombination event, and the asterisk symbolizes an area with microsatellite repeat instability. See CD-ROM for color image. Figure reproduced with permission from University of Chicago Press and Ikeda et al., *Am. J. Hum. Genet.* **75**, 3–16, 2004. © 2004 by The American Society of Human Genetics. All rights reserved.

small ataxia families further indicates that the SCA8 expansion directly predisposes individuals to developing ataxia [29].

VI. SCA8 REPEAT INSTABILITY: POSSIBLE INFLUENCES ON DISEASE PENETRANCE

A. Length of the Polymorphic CTA Repeat Tract Preceding the CTG Expansion

As in other disorders, the SCA8 repeat tract is a compound containing 2 repeat motifs. The first repeat motif is a CTA, containing 1–21 repeats, that is stably transmitted within families, but polymorphic at the population level. The CTA repeat tract is directly followed by the unstable CTG expansion, forming a compound repeat tract with the overall configuration $(CTA)_n(CTG)_{exp}$ [1, 8, 31, 34]. Generally, the overall length of the SCA8 repeat tract has been reported without indicating the respective lengths of the CTA and CTG repeat tracts. A notable molecular difference between the MN-A family and other families with lower disease penetrance is that the MN-A family has only 3 CTA repeats, whereas the CTA repeat tracts in other families are much longer, varying between 8 and 15 CTAs [29]. These data suggest that variations in CTA repeat length may contribute to differences in disease penetrance.

B. Duplicating Interruptions within the CTG Expansion

Despite the common ancestral origin of Caucasian SCA8 expansions, expanded alleles often have different triplet interruptions, with one or more CCG, CTA, CTC, CCA, or CTT motifs found within the CTG expansion [8, 31]. Surprisingly, these interruptions, which are often located near the 5′ end of the CTG tract, frequently duplicate during transmission, resulting in offspring with alleles that vary from those of the affected parent in both repeat tract length and sequence configuration (Fig. 28-6). In contrast to *SCA1* and *FMR1*, most normal SCA8 repeat tracks with less than 50 repeats do not have sequence interruptions [31], although Sobrido et al. [35] have reported a single normal allele with 23 combined repeats in which the CTG tract has a CAG interruption. Whereas both interrupted and pure CTG repeat tracts are found in SCA8 ataxia families, the high frequency of CCG interruptions in the MN-A family suggests the possibility that interruptions play a role in the relatively high disease penetrance found in this family [8, 31].

C. Repeat Instability during Transmission

In addition to the high mutability of the SCA8 sequence interruptions discussed previously, expansion alleles also show dramatic intergenerational instability [1, 8]. The changes in SCA8 expansion size are large compared to other dominantly inherited SCAs, but are typically not as large as for DM1 [1, 45–52]. In general, paternal transmissions result in a contraction of the repeat tract (−86 to +7), whereas maternal transmissions result in expansions (−11 to +900). The largest maternally transmitted alleles include increases of +250, +375, +600, and +900 CTG repeats [1, 41]. The maternal bias for SCA8 repeat expansions has not been observed in other SCAs, but it is similar to the maternal expansion biases for two other noncoding expansion disorders: fragile X syndrome and DM1 [1, 8, 53, 54]. Examples of intergenerational changes for maternal and paternal transmissions are shown in Fig. 28-7.

In the MN-A family, the maternal expansion and paternal deletion biases clearly affect disease penetrance, with 90% of maternal transmissions resulting in ataxia and the remaining 10% involving the transmission of expanded alleles from both parents (Fig. 28-2) [1, 8]. In contrast, 16 of the 19 asymptomatic individuals who carried repeat expansions received the SCA8 expansion from the father. The maternal penetrance bias observed in the MN-A family is likely caused by the transmission of expansions above the pathogenic threshold of approximately 110 combined repeats, whereas paternal transmissions tend to result in contractions below the pathogenic threshold for the MN-A

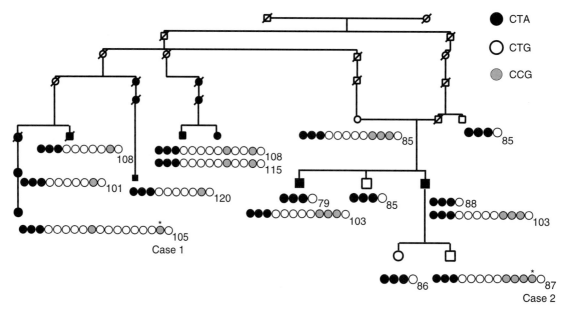

FIGURE 28-6 An abbreviated pedigree for the large SCA8 kindred (MN-A) showing the different allele configurations found within this family. The two cases in which allele configurations changed over a single generation are noted as case 1 and case 2. The CTA, CTG, and CCG triplets are indicated by black, white, and gray circles, respectively. The number of CTG repeats at the 3′ end of each allele is indicated by a subscript number. Filled pedigree symbols indicate individuals with ataxia. Haplotype analysis confirms that both branches of the family inherited the expanded repeat from a common, closely related founder. Figure reproduced with permission from Oxford University Press and Moseley et al., SCA8 CTG repeat: en masse contractions in sperm and intergenerational sequence changes may play a role in reduced penetrance. *Hum. Mol. Genet.* **9**(14), 2125–2130 (2000).

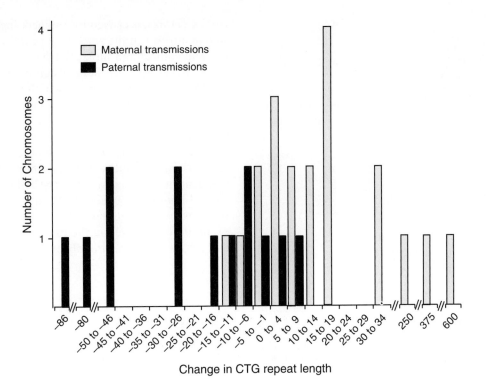

FIGURE 28-7 Intergenerational variation in repeat number for maternal and paternal transmissions. Repeat variation is shown as a decrease or an increase in CTG repeat units. Maternal and paternal transmissions are represented by white and black bars, respectively. Figure reproduced with permission from *Nature Genetics* **21**, 379–384 (1999) (http://www.nature.com).

family [1, 18]. Although the maternal penetrance bias is striking for the MN-A family, this bias is not found in most of the other SCA8 families that have been reported [8, 20, 29].

D. En Masse CTG Repeat Contractions in Sperm

We examined sperm to further investigate SCA8 repeat instability and the paternal repeat contraction bias [8, 31]. Southern analysis on sperm DNA from two unrelated individuals showed that each expanded allele underwent a massive contraction into a size range less often associated with ataxia (from 500 to ~80 and from 800 to ~100) (Fig. 28-8A). Similar trends were also found in individuals with smaller somatic expansions, with these individuals also showing contractions in sperm below approximately 100 repeats (Fig. 28-8B). The equal intensities of the bands representing the normal and expanded alleles suggest that all or nearly all of the expanded allele in the sperm had contracted. The tendency for the SCA8 expanded allele to contract in sperm is likely to contribute to the reduced penetrance in some SCA8 families [21, 31].

E. Germline Instability of the CTG Repeat Expansion by Small Pool PCR

The variation in repeat sizes found in sperm samples from SCA8 patients was analyzed in more detail by small pool PCR. Extracted sperm DNA was diluted to as few as 50, 15, and 1 genome equivalent for PCR. SCA8 expansion samples extracted from blood served as size controls. Small pool PCR of the expanded and unexpanded alleles from blood showed no instability. In contrast, small pool PCR of sperm samples from two unrelated SCA8 patients, either heterozygous or homozygous for the expansion (samples A and B: 119/26 and 345/110 repeats in blood, respectively), showed deletions from the original allele size in blood into a narrow window of contracted allele sizes (Fig. 28-9). The results shown for patient A are consistent with previous Southern analysis of sperm DNA showing repeat contractions into a relatively narrow size window, with the vast majority of contracted alleles varying by less than 10 repeat units. For the homozygous patient, two clusters of contracted alleles are found in sperm, suggesting that the upper and lower size clusters correspond to allele specific contractions. It is surprising that, although mean allele sizes in sperm vary by only approximately 20

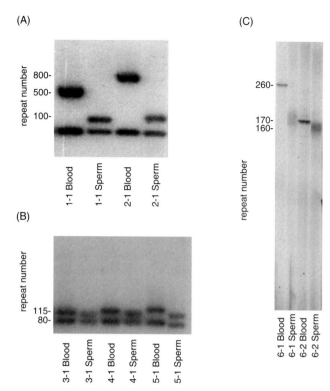

FIGURE 28-8 En masse contraction of SCA8 alleles in sperm. (A) Dramatic repeat length changes in patients 1 and 2 detected by Southern blotting. The repeat lengths of patients 1 and 2 contract from 500 and 800 repeats in blood to ~80 and ~100 repeats in sperm, respectively. The probe used did not contain the CTG repeat. (B) Southern blots of blood and sperm DNA from patients with smaller expansions in their blood reveal the same trend in contractions of the expanded allele in sperm to much smaller allele sizes, usually below approximately 100 repeats. Again, the equal intensities of the bands representing the normal and expanded alleles indicate that repeat contractions occurred in all or nearly all of the sperm with expanded alleles. (C) PCR analysis of SCA8 contractions in two patients from a family with paternal disease transmission. Although contraction of repeats in sperm is again observed, the resulting alleles remain within a more penetrant size range (more than 100 CTGs). Approximate repeat numbers are shown to the left of each figure. Figure reproduced with permission from Oxford University Press and Moseley et al., SCA8 CTG repeat: en masse contractions in sperm and intergenerational sequence changes may play a role in reduced penetrance, *Hum. Mol. Genet.* **9**(14), 2125–2130 (2000).

repeats, the original allele expansions in blood were dramatically different in size (110 versus 345 repeats).

VII. MOLECULAR PARALLELS WITH MYOTONIC DYSTROPHY

SCA8 was the second disorder shown to be caused by a noncoding CTG expansion; DM1 is caused by a CTG expansion in the 3' UTR of the *DMPK* gene. In 2001, we showed that a second form of myotonic dystrophy (DM2), is caused by a noncoding CCTG expansion in intron 1 of the zinc finger protein 9 (*ZNF9*) gene [5]. The clinical and molecular parallels between DM1 and DM2, as well as other cell culture and murine models [55], now strongly support a pathogenic RNA gain of function model in which the CUG and CCUG transcripts themselves cause trans dominant changes in the regulation of alternative splicing of other genes. Molecular parallels between SCA8, DM1, and DM2 suggest the possibility that an RNA gain of function mechanism may play a role in SCA8 [8]. The differing clinical features between SCA8 and the myotonic dystrophies are consistent with the fact that SCA8 is almost exclusively expressed in the brain, whereas the *DMPK* and *ZNF9* genes are broadly expressed [5].

VIII. MODELING SCA8 PATHOGENESIS USING THE MOUSE AND FLY

In an effort to further elucidate the molecular mechanisms underlying SCA8 pathogenesis, a transgenic mouse model was generated using human bacterial artificial chromosome (BAC) clones containing the expanded or normal SCA8 gene [56]. The SCA8–BAC expansion mice have a $(CTA)_3(CTG)_{118}$ repeat tract cloned from the affected MN-A family proband, whereas the control construct contains a $(CTA)_8(CTG)_{11}$ repeat track. The SCA8–BAC expansion, but not the BAC control lines, develops a progressive neurological phenotype that is fatal in high-copy-number lines. This model demonstrates that expression of the SCA8 CTG expansion, but not control transcripts, is pathogenic [56] and is currently being used to define the molecular mechanisms of the disease and the interesting reduced penetrance.

In addition, Mutsuddi *et al.* have developed a *Drosophila* model of SCA8 [57] and showed that retinal expression of normal or expanded SCA8 transcripts induces a late onset, progressive neurodegeneration. Modifier screens have identified mutations in four genes (*staufen, muscleblind, split ends*, and *CG3249*, which encodes a putative protein kinase A anchor protein) as potential genetic interactors. All four of these genes, which are conserved in *Drosophila* and humans, encode neuronally expressed RNA binding proteins [57]. Although both normal and expanded repeat tracks cause a phenotype in this model, the genetic interaction between *muscleblind* and SCA8 varies with repeat size. These data indicate that expansions in the SCA8 transcript alter interactions with RNA binding proteins in *Drosophila*, which could play a role in SCA8 pathogenesis [57].

CHAPTER 28 Molecular Genetics of Spinocerebellar Ataxia Type 8

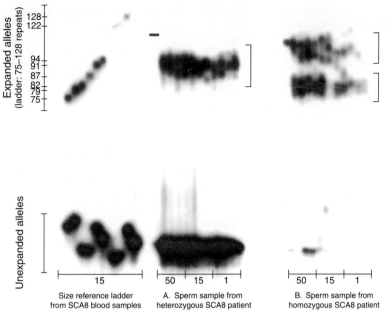

FIGURE 28-9 Small pool PCR analysis of repeat length variation in two unrelated SCA8 sperm samples. Sperm DNA is diluted to as few as 50, 15, and 1 genome equivalents as templates for a single PCR reaction. The results of heterozygous and homozygous SCA8 sperm samples (A and B: 119/26 and 345/110 repeats in blood, respectively) are shown (middle and right panels). The results of small pool PCR in blood DNA samples from unrelated SCA8 patients are also shown as a size reference ladder for expanded alleles (left panel). For comparison purposes, the bar in the middle panel shows approximately where the expansion of 119 repeats in blood would be expected to migrate.

IX. CONCLUSIONS

The data summarized here describe various pieces of a complicated genetic disorder, with an emphasis on the molecular genetic features of the SCA8 expansion and variations in disease penetrance. The reduced penetrance of the SCA8 CTG expansion differs from many of the other dominant SCAs identified to date. Positional cloning approaches that depend on the collection of large families were used to identify the SCA1, SCA2, SCA3, and SCA6 mutations. In contrast, the SCA8 mutation was isolated from a single ataxia patient using RAPID cloning [1, 7]. It is therefore not surprising that the genetic characteristics and disease penetrance of SCA8 do not follow the pattern of previously defined SCAs. Most genetic studies have reported overall repeat size but have not examined CTA versus CTG repeat lengths or sequence interruptions. Although this additional information would help to clarify the genetic complexities of SCA8, further analysis in cell culture and animal models will also be needed to understand the mechanisms of repeat instability, the potential role that sequence interruptions play in affecting disease penetrance, and the possible involvement of an RNA gain of function mechanism.

References

1. Koob, M. D., Moseley, M. L., Schut, L. J., Benzow, K. A., Bird, T. D., Day, J. W., and Ranum, L. P. W. (1999). An untranslated CTG expansion causes a novel form of spinocerebellar ataxia (SCA8). *Nat. Genet.* **21**, 379–384.
2. Holmes, S. E., Hearn, E. O., Ross, C. A., and Margolis, R. L. (2001). SCA12: An unusual mutation leads to an unusual spinocerebellar ataxia. *Brain Res. Bull.* **56**, 397–403.
3. Matsuura, T., Yamagata, T., Burgess, D. L., Rasmussen, A., Grewal, R. P., Watase, K., Khajavi, M., McCall, A. E., Davis, C. F., Zu, L., Achari, M., Pulst, S. M., Alonso, E., Noebels, J. L., Nelson, D. L., Zoghbi, H. Y., and Ashizawa, T. (2000). Large expansion of the ATTCT pentanucleotide repeat in spinocerebellar ataxia type 10. *Nat. Genet.* **26**, 191–194.
4. Hagerman, R. J., Leehey, M., Heinrichs, W., Tassone, F., Wilson, R., Hills, J., Grigsby, J., Gage, B., and Hagerman, P. J. (2001). Intention tremor, parkinsonism, and generalized brain atrophy in male carriers of fragile X. *Neurology* **57**, 127–130.
5. Liquori, C., Ricker, K., Moseley, M. L., Jacobsen, J. F., Kress, W., Naylor, S., Day, J. W., and Ranum, L. P. W. (2001). Myotonic dystrophy type 2 caused by a CCTG expansion in intron 1 of ZNF9. *Science* **293**, 864–867.

6. Jacquemont, S., Hagerman, R. J., Leehey, M., Grigsby, J., Zhang, L., Brunberg, J. A., Greco, C., Des Portes, V., Jardini, T., Levine, R., Berry-Kravis, E., Brown, W. T., Schaeffer, S., Kissel, J., Tassone, F., and Hagerman, P. J. (2003). Fragile X premutation tremor/ataxia syndrome: Molecular, clinical, and neuroimaging correlates. *Am. J. Hum. Genet.* **72**, 869–878.

7. Koob, M. D., Benzow, K. A., Bird, T. D., Day, J. W., Moseley, M. L., and Ranum, L. P. W. (1998). Rapid cloning of expanded trinucleotide repeat sequences from genomic DNA. *Nat. Genet.* **18**, 72–75.

8. Mosemiller, A. K., Dalton, J. C., Day, J. W., and Ranum, L. P. W. (2003). Molecular genetics of spinocerebellar ataxia type 8 (SCA8). *Cytogenet. Genome Res.* **100**, 175–183.

9. Schalling, M., Hudson, T., Buetow, K., and Housman, D. (1993). Direct detection of novel expanded trinucleotide repeats in the human genome. *Nat. Genet.* **4**, 135–139.

10. Holmes, S. E., O'Hearn, E. E., McInnis, M. G., Gorelick-Feldman, D. A., Kleiderlein, J. J., Callahan, C., Kwak, N. G., Ingersoll-Ashworth, R. G., Sherr, M., Sumner, A. J., Sharp, A. H., Ananth, U., Seltzer, W. K., Boss, M. A., Vieria-Saecker, A. M., Epplen, J. T., Riess, O., Ross, C. A., and Margolis, R. L. (1999). Expansion of a novel CAG trinucleotide repeat in the 5' region of PPP2R2B is associated with SCA12. *Nat. Genet.* **23**, 391–392.

11. Holmes, S. E., O'Hearn, E., Rosenblatt, A., Callahan, C., Hwang, H. S., Ingersoll-Ashworth, R. G., Fleisher, A., Stevanin, G., Brice, A., Potter, N. T., Ross, C. A., and Margolis, R. L. (2001). A repeat expansion in the gene encoding junctophilin-3 is associated with Huntington disease-like 2. *Nat. Genet.* **29**, 377–378.

12. Andres, A. M., Soldevila, M., Saitou, N., Volpini, V., Calafell, F., and Bertranpetit, J. (2003). Understanding the dynamics of spinocerebellar ataxia 8 (SCA8) locus through a comparative genetic approach in humans and apes. *Neurosci. Lett.* **336**, 143–146.

13. Andres, A. M., Soldevila, M., Lao, O., Volpini, V., Saitou, N., Jacobs, H. T., Hayasaka, I., Calafell, F., and Bertranpetit, J. (2004). Comparative genetics of functional trinucleotide tandem repeats in humans and apes. *J. Mol. Evol.* **59**, 329–339.

14. Janzen, M. A., Moseley, M. L., Benzow, K. A., Day, J. W., Koob, M. D., and Ranum, L. P. W. (1999). Limited expression of SCA8 is consistent with cerebellar pathogenesis and toxic gain of function RNA model. *Am. J. Hum. Genet.* **65**, A267.

15. Nemes, J. P., Benzow, K. A., Moseley, M. L., Ranum, L. P. W., and Koob, M. D. (2000). The SCA8 transcript is an antisense RNA to a brain-specific transcript encoding a novel actin-binding protein (KLHL1). *Hum. Mol. Genet.* **9**, 1543–1551 [Correction/Addition *Hum. Mol. Genet.* **1549**, 2777].

16. Benzow, K. A., and Koob, M. D. (2002). The KLHL1-antisense transcript (KLHL1AS) is evolutionarily conserved. *Mamm. Genome* **13**, 134–141.

17. Moseley, M. L., Benzow, K. A., Schut, L. J., Bird, T. D., Gomez, C. M., Barkhaus, P. E., Blindauer, K. A., Labuda, M., Pandolfo, M., Koob, M. D., and Ranum, L. P. W. (1998). Incidence of dominant spinocerebellar and Friedreich triplet repeats among 361 ataxia families. *Neurology* **51**, 1666–1671.

18. Day, J. W., Schut, L. J., Moseley, M. L., Durand, A. C., and Ranum, L. P. W. (2000). Spinocerebellar ataxia type 8: Clinical features in a large family. *Neurology* **55**, 649–657.

19. Ikeda, Y., Shizuka, M., Watanabe, M., Okamoto, K., and Shoji, M. (2000). Molecular and clinical analyses of spinocerebellar ataxia type 8 in Japan. *Neurology* **54**, 950–955.

20. Juvonen, V., Hietala, M., Paivarinta, M., Rantamaki, M., Hakamies, L., Kaakkola, S., Vierimaa, O., Penttinen, M., and Savontaus, M. L. (2000). Clinical and genetic findings in Finnish ataxia patients with the spinocerebellar ataxia 8 repeat expansion. *Ann. Neurol.* **48**, 354–361.

21. Silveira, I., Alonso, I., Guimaraes, L., Mendonca, P., Santos, C., Maciel, P., Fidalgo De Matos, J. M., Costa, M., Barbot, C., Tuna, A., Barros, J., Jardim, L., Coutinho, P., and Sequeiros, J. (2000). High germinal instability of the (CTG)n at the SCA8 locus of both expanded and normal alleles. *Am. J. Hum. Genet.* **66**, 830–840.

22. Brusco, A., Cagnoli, C., Franco, A., Dragone, E., Nardacchione, A., Grosso, E., Mortara, P., Mutani, R., Migone, N., and Orsi, L. (2002). Analysis of SCA8 and SCA12 loci in 134 Italian ataxic patients negative for SCA1–3, 6 and 7 CAG expansions. *J. Neurol.* **249**, 923–929.

23. Topisirovic, I., Dragasevic, N., Savic, D., Ristic, A., Keckarevic, M., Keckarevic, D., Culjkovic, B., Petrovic, I., Romac, S., and Kostic, V. S. (2002). Genetic and clinical analysis of spinocerebellar ataxia type 8 repeat expansion in Yugoslavia. *Clin. Genet.* **62**, 321–324.

24. Anderson, J. H., Yavuz, M. C., Kazar, B. M., Christova, P., and Gomez, C. M. (2002). The vestibulo-ocular reflex and velocity storage in spinocerebellar ataxia 8. *Arch. Ital. Biol.* **140**, 323–329.

25. Zeman, A., Stone, J., Porteous, M., Burns, E., Barron, L., and Warner, J. (2004). Spinocerebellar ataxia type 8 in Scotland: Genetic and clinical features in seven unrelated cases and a review of published reports. *J. Neurol. Neurosurg. Psychiatry* **75**, 459–465.

26. Ikeda, Y., Shizuka-Ikeda, M., Watanabe, M., Schmitt, M., Okamoto, K., and Shoji, M. (2000). Asymptomatic CTG expansion at the SCA8 locus is associated with cerebellar atrophy on MRI. *J. Neurol. Sci.* **182**, 76–79.

27. Ikeda, Y., Moseley, M. L., Dalton, J. C., Su, M. T., Hsieh-Li, H. M., Lee-Chen, G. J., Clark, H. B., Day, J. W., and Ranum, L. P. W. (2004). Purkinje cell degeneration and 1C2 positive neuronal intranuclear inclusions in SCA8 patients. *Am. J. Hum. Genet.* **Supplement**, 441.

28. Cellini, E., Nacmias, B., Forleo, P., Piacentini, S., Guarnieri, B. M., Serio, A., Calabro, A., Renzi, D., and Sorbi, S. (2001). Genetic and clinical analysis of spinocerebellar ataxia type 8 repeat expansion in Italy. *Arch. Neurol.* **58**, 1856–1859.

29. Ikeda, Y., Dalton, J. C., Moseley, M. L., Gardner, K. L., Bird, T. D., Ashizawa, T., Seltzer, W. K., Pandolfo, M., Milunsky, A., Potter, N. T., Shoji, M., Vincent, J. B., Day, J. W., and Ranum, L. P. W. (2004). Spinocerebellar ataxia type 8: Molecular genetic comparisons and haplotype analysis of 37 families with ataxia. *Am. J. Hum. Genet.* **75**, 3–16.

30. Ranum, L. P. W., Moseley, M. L., Leppert, M., Guan den Eng, M. F., La Spada, A. R., Koob, M. D., and Day, J. W. (1999). Massive CTG expansions and deletions reduce penetrance of spinocerebellar ataxia type 8. *Am. J. Hum. Genet.* **65**, A466.

31. Moseley, M. L., Schut, L. J., Bird, T. D., Koob, M. D., Day, J. W., and Ranum, L. P. W. (2000). SCA8 CTG repeat: En masse contractions in sperm and intergenerational sequence changes may play a role in reduced penetrance. *Hum. Mol. Genet.* **9**, 2125–2130.

32. Worth, P. F., Houlden, H., Giunti, P., Davis, M. B., and Wood, N. W. (2000). Large, expanded repeats in SCA8 are not confined to patients with cerebellar ataxia. *Nat. Genet.* **24**, 214–215.

33. Vincent, J. B., Neves-Pereira, M. L., Paterson, A. D., Yamamoto, E., Parikh, S. V., Macciardi, F., Gurling, H. M., Potkin, S. G., Pato, C. N., Macedo, A., Kovacs, M., Davies, M., Lieberman, J. A., Meltzer, H. Y., Petronis, A., and Kennedy, J. L. (2000). An unstable trinucleotide-repeat region on chromosome 13 implicated in spinocerebellar ataxia: A common expansion locus. *Am. J. Hum. Genet.* **66**, 819–829.

34. Stevanin, G., Herman, A., Durr, A., Jodice, C., Frontali, M., Agid, Y., and Brice, A. (2000). Are (CTG)n expansions at the SCA8 locus rare polymorphisms? *Nat. Genet.* **24**, 213.

35. Sobrido, M. J., Cholfin, J. A., Perlman, S., Pulst, S. M., and Geschwind, D. H. (2001). SCA8 repeat expansions in ataxia: A controversial association. *Neurology* **57**, 1310–1312.

36. Tazon, B., Badenas, C., Jimenez, L., Munoz, E., and Mila, M. (2002). SCA8 in the Spanish population including one homozygous patient. *Clin. Genet.* **62**, 404–409.

37. Izumi, Y., Maruyama, H., Oda, M., Morino, H., Okada, T., Ito, H., Sasaki, I., Tanaka, H., Komure, O., Udaka, F., Nakamura, S., and Kawakami, H. (2003). SCA8 repeat expansion: Large CTA/CTG repeat alleles are more common in ataxic patients, including those with SCA6. *Am. J. Hum. Genet.* **72**, 704–709.
38. Takano, H., Cancel, G., Ikeuchi, T., Lorenzetti, D., Mawad, R., Stevanin, G., Didierjean, O., Durr, A., Oyake, M., Shimohata, T., Sasaki, R., Koide, R., Igarashi, S., Hayashi, S., Takiyama, Y., Nishizawa, M., Tanaka, H., Zoghbi, H., Brice, A., and Tsuji, S. (1998). Close associations between prevalences of dominantly inherited spinocerebellar ataxias with CAG-repeat expansions and frequencies of large normal CAG alleles in Japanese and Caucasian populations. *Am. J. Hum. Genet.* **63**, 1060–1066.
39. Juvonen, V., Hietala, M., Kairisto, V., and Savontaus, M. L. (2005). The occurrence of dominant spinocerebellar ataxias among 251 Finnish ataxia patients and the role of predisposing large normal alleles in a genetically isolated population. *Acta Neurol. Scand.* **111**, 154–162.
40. Juvonen, V., Kairisto, V., Hietala, M., and Savontaus, M. L. (2002). Calculating predictive values for the large repeat alleles at the SCA8 locus in patients with ataxia. *J. Med. Genet.* **39**, 935–936.
41. Corral, J., Genis, D., Banchs, I., San Nicolas, H., Armstrong, J., and Volpini, V. (2005). Giant SCA8 alleles in nine children whose mother has two moderately large ones. *Ann. Neurol.* **57**, 549–553.
42. Brusco, A., Gellera, C., Cagnoli, C., Saluto, A., Castucci, A., Michielotto, C., Fetoni, V., Mariotti, C., Migone, N., Di Donato, S., and Taroni, F. (2004). Molecular genetics of hereditary spinocerebellar ataxia: Mutation analysis of spinocerebellar ataxia genes and CAG/CTG repeat expansion detection in 225 Italian families. *Arch. Neurol.* **61**, 727–733.
43. Schols, L., Bauer, I., Zuhlke, C., Schulte, T., Kolmel, C., Burk, K., Topka, H., Bauer, P., Przuntek, H., and Riess, O. (2003). Do CTG expansions at the SCA8 locus cause ataxia? *Ann. Neurol.* **54**, 110–115.
44. Ott, J. (1991). "Analysis of Human Genetic Linkage." The John's Hopkins University Press, Baltimore.
45. Tsilfidis, C., MacKenzie, A. E., Mettler, G., Barcelo, J., and Korneluk, R. G. (1992). Correlation between CTG trinucleotide repeat length and frequency of severe congenital myotonic dystrophy. *Nat. Genet.* **1**, 192–195.
46. Chung, M.-Y., Ranum, L. P. W., Duvick, L. A., Servadio, A., Zoghbi, H. Y., and Orr, H. T. (1993). Evidence for a mechanism predisposing to intergenerational CAG repeat instability in spinocerebellar ataxia type 1. *Nat. Genet.* **5**, 254–258.
47. Maciel, P., Gaspar, C., DeStefano, A. L., Silveira, I., Coutinho, P., Radvany, J., Dawson, D. M., Sudarsky, L., Guimaraes, J., Loureiro, J. E., et al. (1995). Correlation between CAG repeat length and clinical features in Machado–Joseph disease. *Am. J. Hum. Genet.* **57**, 54–61.
48. Maruyama, H., Nakamura, S., Matsuyama, Z., Sakai, T., Doyu, M., Sobue, G., Seto, M., Tsujihata, M., Oh-i, T., Nishio, T., Sunohara, N., Takahashi, R., Hayashi, M., Nishino, I., Ohtake, T., Oda, T., Nishimura, M., Saida, T., Matsumoto, H., Baba, M., Kawaguchi, Y., Kakizuka, A., and Kawakimi, H. (1995). Molecular features of the CAG repeats and clinical manifestation of Machado–Joseph disease. *Hum. Mol. Genet.* **4**, 807–812.
49. Cancel, G., Durr, A., Didierjean, O., Imbert, G., Burk, K., Lezin, A., Belal, S., Benomar, A., Abadabendib, M., Vial, C., Guimaraes, J., Chneiweiss, H., Stevanin, G., Yvert, G., Abbas, N., Saudou, F., Lebre, A., Yahyaoui, M., Hentati, F., Vernant, J., Klockgether, T., Mandel, J., Agrid, Y., and Brice, A. (1997). Molecular and clinical correlations in spinocerebellar ataxia 2—a study of 32 families. *Hum. Mol. Genet.* **6**, 709–715.
50. David, G., Abbas, N., Stevanin, G., Durr, A., Yvert, G., Cancel, G., Weber, C., Imbert, G., Saudou, F., Antoniou, E., Drabkin, H., Gemmill, R., Giunti, P., Benomar, A., Wood, N., Ruberg, M., Agid, Y., Mandel, J. L., and Brice, A. (1997). Cloning of the SCA7 gene reveals a highly unstable CAG repeat expansion. *Nat. Genet.* **17**, 65–70.
51. Jodice, C., Mantuano, E., Veneziano, L., Trettel, F., Sabbadini, G., Calandriello, L., Francia, A., Spadaro, M., Pierelli, F., Salvi, F., Ophoff, R., Frants, R., and Frontali, M. (1997). Episodic ataxia type 2 (EA2) and spinocerebellar atxia type 6 (SCA6) due to CAG repeat expansion in the CACNA1A gene on chromosome 19p. *Hum. Mol. Genet.* **6**, 1973–1978.
52. Zhuchenko, O., Bailey, J., Bonnen, P., Ashizawa, T., Stockton, D. W., Amos, C., Dobyns, W. B., Subramony, S. H., Zoghbi, H. Y., and Lee, C. C. (1997). Autosomal dominant cerebellar ataxia (SCA6) associated with small polyglutamine expansions in the alpha-1A-voltage-dependent calcium channel. *Nat. Genet.* **15**, 62–69.
53. Groenen, P., and Wieringa, B. (1998). Expanding complexity in myotonic dystrophy. *Bioessays* **20**, 901–912.
54. Jin, P., and Warren, S. T. (2000). Understanding the molecular basis of fragile X syndrome. *Hum. Mol. Genet.* **9**, 901–908.
55. Ranum, L. P. W., and Day, J. W. (2004). Myotonic dystrophy: RNA pathogenesis comes into focus. *Am. J. Hum. Genet.* **74**, 793–804.
56. Moseley, M. L., Ikeda, Y., Gao, W., Weatherspoon, M. R., Su, M. T., Hsieh-Li, H. M., Lee-Chen, G. J., Chen, G., Reinert, K. C., Ebner, T. J., Day, J. W., and Ranum, L. P. W. (2004). SCA8 BAC transgenic mice: Cerebellar dysfunction and 1C2/ubiquitin positive neuronal inclusions. *Am. J. Hum. Genet.* **Supplement**, 387.
57. Mutsuddi, M., Marshall, C. M., Benzow, K. A., Koob, M. D., and Rebay, I. (2004). The spinocerebellar ataxia 8 noncoding RNA causes neurodegeneration and associates with staufen in Drosophila. *Curr. Biol.* **14**, 302–308.

CHAPTER 29

Spinocerebellar Ataxia Type 10: A Disease Caused by an Expanded (ATTCT)$_n$ Pentanucleotide Repeat

TETSUO ASHIZAWA

Department of Neurology, The University of Texas Medical Branch (UTMB), Galveston, Texas

I. Introduction
II. Identification and Characterization of the SCA10 Mutation
   A. Identification of the (ATTCT)$_n$ Expansion as the SCA10 Mutation
   B. Structure of the (ATTCT)$_n$ Repeat
   C. Instability of the Expanded (ATTCT)$_n$ Repeat
III. Clinical Phenotype and Ethnicity
   A. Spectrum of the SCA10 Phenotype
   B. Ethnic Distribution of SCA10 and Phenotypic Variations
IV. Genotype–Phenotype Correlation
V. Diagnostic Utility of the Mutation
VI. Pathogenic Mechanisms
   A. *ATXN10* Gene and Function of the Ataxin-10 Protein
   B. Pathogenic Models for SCA10
   C. Other Pathogenic Considerations
VII. Conclusion
   References

Spinocerebellar ataxia type 10 (SCA10; OMIM *603516) is an autosomal dominant disease caused by a large (4–22.5 kb) expansion of an (ATTCT)$_n$ pentanucleotide repeat in intron 9 of the *ataxin-10* (*ATXN10*) gene on chromosome 22q13.3. SCA10 is the only human disease caused by an expansion of a pentanucleotide repeat and has been found only in limited populations on the American continents. The principal clinical phenotype of SCA10 is cerebellar ataxia with or without epilepsy. Epilepsy frequently afflicts Mexican families, but peculiarly spares Brazilian families. Expanded (ATTCT)$_n$ repeats in some SCA10 patients examined were interrupted by various penta-, hexa-, and heptanucleotide repeat units. Expanded repeats show repeat-size mosaicism in somatic cells and sperm; however, the pattern of mosaicism is stable over time in the blood cells of adult patients *in vivo*. Paternal transmission of an expanded allele frequently results in an increase or a decrease in the repeat size in the offspring. The instability observed with paternal transmission may be attributable to the repeat-size mosaicism in sperm. In contrast, the repeat size is remarkably stable with maternal transmission, suggesting relative stability of the expanded repeat in oocytes. *In vitro*, uninterrupted (ATTCT)$_n$ repeats form unpaired DNA structure. The pathogenic mechanism of SCA10 is unknown. However, because the (ATTCT)$_n$ repeat is located in an intron, the mutation is not translated into the protein sequence. Loss of *ATXN10* function and RNA-mediated gain of function are currently under investigation as potential pathogenic models of SCA10.

I. INTRODUCTION

Spinocerebellar ataxia type 10 (SCA10) is the only human disease caused by an expansion of a pentanucleotide repeat. Although an increasing number of human diseases have been found to be caused by expansions of microsatellite repeats, most involve trinucleotide repeats. Only the $(CCTG)_n$ tetranucleotide repeat in myotonic dystrophy type 2 (DM2) [1] and the $(CCCCGCCCCGCG)_n$ dodecamer repeat in the myoclonic epilepsy of Unverricht and Lundborg (EPM1) [2] are nontrinucleotide repeats whose expansion causes human neurological diseases.

SCA10 is caused by an $(ATTCT)_n$ repeat expanded in intron 9 of the *ATXN10* gene on chromosome 22q13.3. The repeat size at this locus is polymorphic and ranges from 10 to 29 ATTCTs in normal individuals; however, in patients with SCA10, the expanded repeat easily exceeds 4 kb (800 ATTCTs) and can reach 22.5 kb (4500 ATTCTs). The intronic location and large expansion size resemble the expanded repeats of $(GAA)_n$ in Friedreich's ataxia 1 (FRDA) and $(CCTG)_n$ in DM2. However, in addition to the differences in repeat-unit sequences, the SCA10 $(ATTCT)_n$ repeat differs from these repeats in terms of the pattern of repeat interruptions, the secondary DNA structure, the characteristics of repeat-size instability, and the resultant disease phenotype. The pathogenic mechanism triggered by the $(ATTCT)_n$ repeat expansion is unknown. This chapter reviews SCA10 by describing the identification and characterization of the mutation, clinical phenotype and ethnicity, genotype–phenotype correlation, repeat structures, repeat-length stability, diagnostic utility of the mutation, and potential pathogenic mechanisms.

II. IDENTIFICATION AND CHARACTERIZATION OF THE SCA10 MUTATION

A. Identification of the $(ATTCT)_n$ Expansion as the SCA10 Mutation

In 1998, a large pedigree of a Mexican-American family with dominantly inherited cerebellar ataxia was reported with an unknown genetic locus distinct from the known loci of spinocerebellar ataxias (SCAs) [3]. In 1999, the locus of the disease in this family [4] and another large Mexican-American family [5] was independently mapped to chromosome 22q13-qter by linkage, and this locus was designated *SCA10*. Two recombination events in these families indicated that the *SCA10* gene resides within a 3.8-cM interval between *D22S1140* and *D22S1160*. By using additional polymorphic markers, the SCA10 region was narrowed to a 2.7-cM region between *D22S1140* and *D22S1153* [6]. Because the two large families of SCA10 showed anticipation, it was postulated that the mutation responsible for SCA10 is an expansion of an unstable repeat, and the search began for the mutation using positional cloning strategies.

Although the DNA sequence of the entire euchromatic part of human chromosome 22 had just become available, 11 gaps still remained to be sequenced [7]. Because *D22S1160* and *D22S1153* were located in one of these gaps, the exact physical size of the SCA10 candidate region could not be determined. Nevertheless, two contigs composed of bacterial artificial chromosomes (BACs), phage-P1-derived artificial chromosomes (PACs), and cosmids were covering most of the neighboring regions. Fourteen trinucleotide repeats (≥ 4 repeats in length) were examined that were listed in this region of the chromosome 22 genome database at the Sanger Centre. None of these repeats showed expansions in the affected members of the SCA10 families. To explore the possibility that an expansion of a polyglutamine-coding repeat is the SCA10 mutation, genomic DNA samples from affected family members were examined for a CAG or CAA expansion using repeat expansion detection (RED) analysis [8]. There was no evidence for an expansion of such repeats outside of the *ERDA1* [9] and *SEF2-1* [10] loci, at which CAG repeats are frequently expanded in normal subjects. Western blot analysis of proteins obtained from patients' lymphoblastoid cells, using the 1C2 antibody (an antibody against the TATA box binding protein) that recognizes expanded polyglutamine tracts [11], also failed to detect proteins containing expanded polyglutamine tracts. These results reduced the likelihood that SCA10 is caused by a polyglutamine expansion and prompted the extension of the investigation to microsatellite repeats other than trinucleotide repeats.

The author and coworkers systematically screened the candidate genomic region for polymorphic microsatellite repeats by polymerase chain reaction (PCR) analysis and encountered a polymorphic $(ATTCT)_n$ pentanucleotide tandem repeat in intron 9 of the *E46L* gene, which is now known as *ATXN10* (Genbank accession no. AL050282) (Fig. 29-1). Normal alleles contained 10–22 repeats with 82.1% heterozygosity [12]. The distribution of normal alleles was unimodal and similar among Caucasian, Mexican, and Japanese populations (Fig. 29-2). The allele distributions in each of the three ethnically defined populations (including 127 persons from the Mexican population) were consistent with the Hardy–Weinberg equilibrium ($p \gg 0.05$). Subsequently, a normal individual was identified with an allele with 29 ATTCTs, which represents the current upper limit of normal allele size [13].

PCR analysis of this repeat demonstrated a uniform lack of heterozygosity in all affected individuals and carriers of the disease haplotype in SCA10 families

Chapter 29 Spinocerebellar Ataxia Type 10

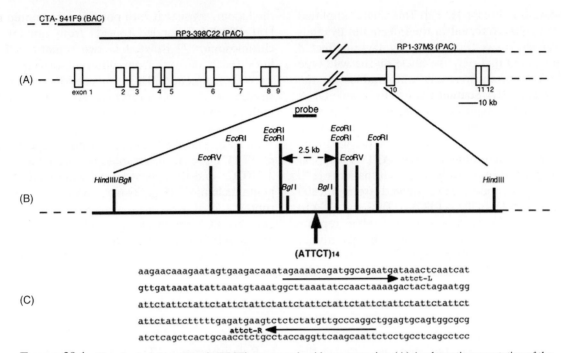

FIGURE 29-1 The physical map of the (ATTCT)$_n$ pentanucleotide repeat region. (A) A schematic presentation of the structure of *ATXN10*. *ATXN10* consists of 12 exons. The (ATTCT)$_n$ repeat is located in intron 9. The gap at the left of PAC 37M3 does not represent a missing sequence, but was introduced to preserve the scale. (B) A restriction map of the (ATTCT)$_n$ repeat region defined by flanking *Hin*dIII restriction sites (nt 17,023 and 34,567). The numbers are nucleotide positions in PAC37M3. "Probe" indicates the position of the probe used (nt 25,222–26,021) to detect the 2.5-kb *Eco*RI fragment shown in Fig. 29-3C in the Southern blot analysis. The (ATTCT)$_n$ repeat is located downstream of the probe within the 2.5-kb *Eco*RI fragment. (C) Nucleotide sequence of an (ATTCT)$_{14}$ repeat. Arrows underline PCR primer sequences (attct-L and attct-R) that were used for amplification of the ATTCT repeat region shown in Fig. 29-3C.

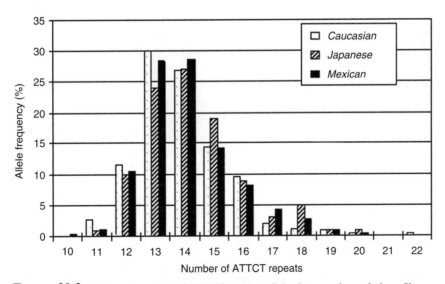

FIGURE 29-2 Distribution of the (ATTCT)$_n$ repeat alleles in normal populations. Shown is a histogram of the normal (ATTCT)$_n$ repeat alleles in European ($n = 250$), Japanese ($n = 100$), and Mexican ($n = 254$) chromosomes. The distribution of the alleles was unimodal with similar patterns among these populations. The (ATTCT) 29 allele (see text) is not shown.

(Figs. 29-3A and 29-3B) [5, 12]. The single amplified allele of the ATTCT repeat in the affected individuals (or the carriers) was transmitted by the unaffected parent, suggesting that only the allele on the wild-type (non-SCA10) chromosome is amplifiable in the offspring (Fig. 29-3B). Southern blots of genomic DNA showed a variably expanded allele, in addition to a normal allele, in all affected individuals, whereas all unaffected family members showed only the wild-type allele (Fig. 29-3C), indicating that the $(ATTCT)_n$ repeat region is expanded exclusively in SCA10 patients [12]. The size of expanded alleles ranged from 4 (800 ATTCTs) to 22.5 kb (4500 ATTCTs). Thus, the SCA10 expansion is one of the largest microsatellite repeats known to exist in the human genome. Subsequently, an ataxia patient and her asymptomatic mother, who both had a 280-repeat allele, were identified, raising the possibility that this allele is an intermediate allele with reduced penetrance [13, 14]. It was confirmed that the expanded $(ATTCT)_n$ repeat is absent from over 1000 normal chromosomes. It was also demonstrated that an inverse correlation exists between the size of the expanded ATTCT repeat and the age of onset ($n = 26$, $r^2 = 0.34, p = 0.018$) (Fig. 29-4) [12]. These observations support that this repeat expansion is the disease-causing mutation in SCA10.

B. Structure of the $(ATTCT)_n$ Repeat

As mentioned in Section IIA, the repeat number of normal alleles ranges from 10 to 29 ATTCTs [12–14]. Sequence analysis of the alleles obtained from 20 normal individuals with relatively small alleles showed tandem repeats of ATTCT without interruption [12, 13]. The large size of SCA10 patients' expanded ATTCT repeats had prevented their cloning and sequencing. The difficulty was primarily attributable to the preferential PCR amplification and cloning of the coexisting wild-type allele, which has a much shorter length. Repeat-primed PCR was developed to amplify the region containing the $(ATTCT)_n$ repeat [15]. This assay uses the forward primer that anneals to the flanking sequence upstream of the repeat and the reverse primer corresponding to the repeat sequence, which is attached to a hanging unique sequence that has no homology to any part of the genomic sequence (Fig. 29-5). Analysis of genomic DNA samples from normal subjects by this method showed an uninterrupted ladder of repeat-containing PCR products up to the size of the larger normal allele, whereas DNA from SCA10 patients showed the ladder expanded above the normal range (Fig. 29-5) [15]. However, the expanded allele often showed interrupted patterns at the top of the ladder, consistent with possible interrupted repeats [13]. Cloned expanded repeats from mutant human chromosomes 22 isolated in two somatic cell hybrid lines (each derived from different SCA10 patients) were also sequenced. The sequences obtained showed a stretch of 40–41 uninterrupted ATTCTs followed by multiple interruptions by two kinds of AT-rich heptanucleotide repeats, ATTTTCT and ATATTCT, in the 5′ end [16]. The 3′ end of the expanded repeat consisted of ATCCT repeats interrupted by a variable number of ATCCCs, and the repeat ended with a single ATTCT pentanucleotide (Fig. 29-6). Almost identical interrupted repeat structures were identified in different clones of PCR products of expanded repeats from these two SCA10 patients, suggesting that these interruptions are not artifacts introduced during PCR amplification, cloning, or sequencing. The author and coworkers speculate that the interrupting sequences may give clues to deciphering the expansion mechanism and may be important in the basic pathogenic mechanism of SCA10.

Relatively small ATTCT repeats (up to 46 ATTCTs) have been examined for the secondary DNA structure (see Chapter 30 in this volume). Under superhelical tension, two-dimensional (2D) agarose gel electrophoresis of topoisomers of plasmids containing $(ATTCT)_{11-46}$ showed evidence of DNA uncoiling. This was confirmed by atomic force microscopy, which showed unpairing of the two strands. Binding of chloroacetylaldehyde to the repeat region of the plasmid indicated the accessibility of this reagent to the uncoiled region [17]. The unpaired DNA structure may induce chromosome fragility and DNA methylation. However, in collaboration with Dr. Lisa G. Shaffer of Washington State University, chromosome 22 fragility was excluded by cytogenetic studies of leukocytes obtained from SCA10 patients (data not shown). Southern blot analysis of genomic SCA10 DNA digested with methylation-dependent enzymes showed no evidence of aberrant methylation in the SCA10 region (data not shown).

In Chapter 30 of this volume, Potaman et al. describe changes in the $(ATTCT)_n$ repeat sequences in plasmid. They found that $(ATTCT)_n$ repeats increased the length, although some deletions were also observed. The expansion in plasmid was not due simply to an increase in the number of ATTCT repeat unit, but involved complex events including inversion and transition. Potaman et al. summarized the characteristics of the expansions as follows: (1) the repeat tract increased in length from 23 or 24 repeats to as many as 46 repeats; (2) >96% of expansion mutations contained an inversion of the repeat, resulting in the sequence 5′-$(TATTC)_{5-11}(GAATA)_{9-35}$-3′; (3) all plasmids were

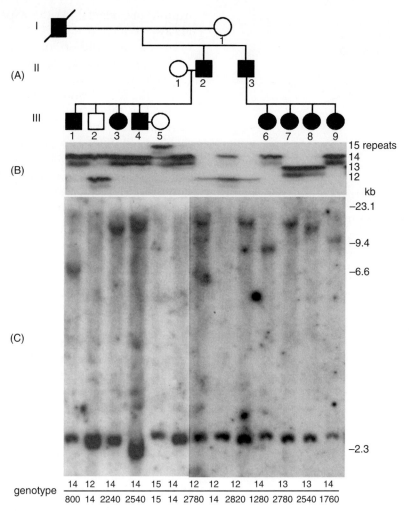

FIGURE 29-3 Expansion mutations in a representative SCA10 family. (A) Pedigrees of the representative Mexican-American family studied for the *SCA10* mutation. (B) PCR analysis of the ATTCT pentanucleotide repeat. The (ATTCT)$_n$ pentanucleotide repeat region was PCR-amplified from the genomic DNA samples of the family members using primers attct-L and attct-R. All affected individuals showed a single allele of variable size (note that each band is accompanied by a faint band underneath due to PCR artifact). Two unaffected individuals (I-1 and III-2) are heterozygous, and two spouses (II-1 and III-5) are homozygous for the (ATTCT)$_n$ repeat. In this family, affected individuals in the second generation (II-2 and II-3) failed to transmit their (ATTCT)$_{12}$ allele to the affected offspring (III-1, III-3, III-4, III-6, III-7, III-8, and III-9), whereas an unaffected offspring (III-2) received this allele from the affected father (II-2). The alleles of unaffected parents (I-1 and II-1) were passed on to the offspring in a pattern consistent with Mendelian inheritance. These data suggest that the affected individuals are apparently hemizygous for the (ATTCT) repeat. (C) Southern blot analysis of expansion mutations of the (ATTCT)$_n$ repeat region. Southern blots of genomic DNA samples revealed variably expanded alleles in affected members of the families shown. All individuals examined have a normal allele (2.5 kb). The apparent variability in normal allele size is attributable to gel-loading artifacts, as additional analyses using the same (*Eco*RI) and different (*Eco*RV, *Hin*dIII, and *Bgl*I) restriction endonucleases did not show consistent variability in normal allele size. The genotype of each individual is shown at the bottom of the figure, with an estimated number of pentanucleotide repeats for disease-associated chromosomes based on the fragment size.

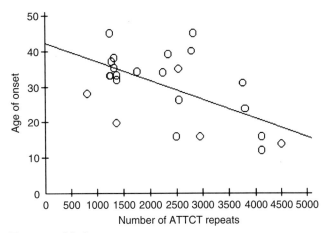

FIGURE 29-4 Correlation between the size of expanded $(ATTCT)_n$ repeat and the age of onset. A scatter plot shows an inverse correlation between the size of expansion and the age of onset in SCA10 patients ($r^2 = 0.34$, $p = 0.018$). Each symbol represents an SCA10 patient, and the linear regression line is shown.

head-to-tail dimers; (4) the expanded repeat product always contained 5'-$(TATTC)_n$ $(GAATA)_n$-3' [this occurred for either orientation of the repeat, $(ATTCT)_n \cdot (AGAAT)_n$ or $(AGAAT)_n \cdot (ATTCT)_n$]; and (5) the DNA sequence contained an A·T to G·C transition at the 3' end within the $(AGAAT)_n$ repeat, which resulted in the formation of a longer flanking direct repeat. They speculated that the inversion event involves an inter- or intramolecular strand switch. Whether these data are directly relevant to the expanded $(ATTCT)_n$ repeat in humans is not clear. However, inter- and intramolecular strand switches may offer attractive explanations for the complex interruptions of the expanded $(ATTCT)_n$ repeat observed in patients with SCA10.

C. Instability of the Expanded (ATTCT)n Repeat

The stability of microsatellite repeats generally decreases as the repeat size increases. Large expansion alleles found in some of these repeats, such as the $(CGG)_n$ of fragile X syndrome [18], the $(GAA)_n$ repeats of FRDA [19], the $(CTG)_n$ repeat of DM1, and the $(CCTG)_n$ repeat of DM2 [20], show conspicuous intergenerational instability and inter- and intratissue mosaicism. However, characteristics of instability vary among these repeats, and some of them mimic and others differ from those of SCA10 repeats.

Expanded $(ATTCT)_n$ repeats show intergenerational repeat-size changes, which are dependent on the gender of the transmitting parent. The repeat-size mosaicism of the expanded $(ATTCT)_n$ repeat is clearly detectable in the somatic tissues and sperm of SCA10 patients. However, data suggest that the size of expanded repeats appears to be relatively stable in somatic cells [21].

1. INTERGENERATIONAL CHANGES IN THE SIZE OF ATTCT REPEATS

It was found that normal alleles are stably transmitted from generation to generation. However, the expanded $(ATTCT)_n$ often showed changes in repeat size during parent-to-child transmission of the disease. The average size of the expanded $(ATTCT)_n$ repeat in affected individuals and mutation carriers from five Mexican families was 2120 repeats, with a range of 800–4500 repeats. The expansion size (mean ± SD) in the disease allele transmitted from the affected father was larger than that from the affected mother (2540 ± 960 repeats versus 1600 ± 960 repeats). Paternal transmissions of the expanded alleles lead to a high degree of instability, which results in either conspicuous

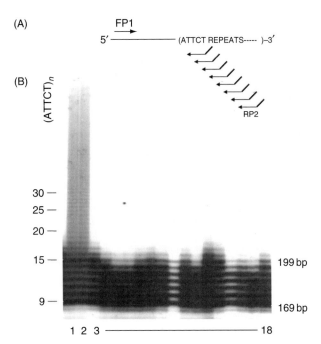

FIGURE 29-5 Repeat-primed PCR analysis of the $(ATTCT)_n$ repeat. (A) The strategy of repeat-primed PCR. The following PCR primers were used: forward primer (FP1), 5'-GAAGACAAATAGAAAACAGATGGCAGA-3'; reverse primer (RP2), 5'-TACGCATCCCAGTTTGAGACGG(**AATAG**)$_8$-3'. The forward primer anneals to the region upstream of the $(ATTCT)_n$ repeat, whereas the reverse primer anneals to the $(ATTCT)_n$ repeat. The hanging unique sequence assures the annealing of the reverse primer to the 3' end of the PCR products. (B) Amplified products show a ladder that goes up to the size corresponding to the larger allele with normal samples, whereas a much longer ladder beyond the upper limit of the normal range is detected with SCA10 samples. Lanes 1 and 2, SCA10 samples; lanes 3–18, normal samples.

5'-----------(attct)40-45 attttct attct (atattct attct)2 attttct (attct)11 atattct (attct)4-5 atattct (attct)10-13 atattct (attct)2 atattct (attct)11-12 (atattct attct)2 (attct)10-11 (atattct attct)4 (attct)2 atattct (attct)8-10 (atattct attct)5 (attct)7-8 atattct (attct)n -----------//---------- (atcct)4-103 (atccc)0-48 (atcct)2-3 attct-----------3'

FIGURE 29-6 The interrupted sequence of expanded ATTCT repeats in SCA10 patients. Genomic DNA was obtained from somatic cell hybrids and leukocytes derived from SCA10 patients. The expanded SCA10 repeat was amplified by PCR or repeat-primed PCR and cloned into the TOPO-TA vector for sequencing. Shown here is the consensus sequence obtained from the sequences of seven clones from three cell lines derived from two SCA10 patients (i.e., three clones from a somatic cell hybrid line containing 2300 ATTCT repeats, three clones from one with 800 ATTCT repeats, and one clone from a lymphoblast line with the same 800 ATTCT repeats). The insert size in these clones suggested that the repeat had contracted to ~2.5 kb or less during the cloning process. –//– indicates the middle region that could not be sequenced. Because of the repeat-rich region 3' to the ATTCT repeat, the downstream PCR primer was placed >500 bp away from the end of the ATTCT repeat, limiting the sequencing of the 3' end of the repeat. These data suggest that the expanded ATTCT repeats are interrupted by close variants of the AT-rich sequences, unlike uninterrupted short wild-type ATTCT repeats.

expansions or conspicuous contractions, whereas maternal transmissions show no or little change [21, 22]. The absolute repeat-size difference between parent and child was 1143 ± 1042 in 15 paternal transmissions and 20 ± 19 in 8 maternal transmissions ($p < 0.01$). The mean ± SD of the repeat-size change (i.e., the repeat size of the child minus the repeat size of the parent) was 247 ± 1555 in the paternal transmissions and 0 ± 28 in the maternal transmissions, suggesting that there is a mild expansion bias with the paternal transmission ($p < 0.01$) (Fig. 29-7). It is important to point out that the intergenerational changes in repeat size varied from family to family. For example, in one family all paternal transmissions of the expanded alleles resulted in contractions, whereas most paternal transmissions in other families led to further expansions. Thus, there may be family-specific factors that determine the degree and the direction of intergenerational repeat-size changes.

2. SOMATIC AND GERMLINE INSTABILITY OF THE ATTCT EXPANSION

The intergenerational changes in expanded allele size are attributable to instability in the parental germline or somatic cells, somatic instability in the child, or any combination thereof. Somatic instability of the expanded repeat can be detected as intra- and intertissue variability of repeat size. DNA was obtained from buccal cells and blood cells from four patients. All normal alleles of these samples and normal control subjects showed no evidence of instability. Conversely, expanded alleles showed intra- and intertissue variability in expansion size in some patients (Fig. 29-8). The intratissue mosaicism was presented as a smear or multiple discrete bands on Southern blot analysis. The intertissue mosaicism was detected as variability in the main allele size or differences in the pattern of intratissue mosaicism between tissues. These observations indicate that the expanded $(ATTCT)_n$ repeat in SCA10 is unstable in somatic tissues.

Sperm samples are difficult to obtain from patients with SCA10 who mostly belong to Catholic societies. A limited number of sperm samples, however, clearly showed repeat-size mosaicism, suggesting instability of the expanded $(ATTCT)_n$ repeat in male germline cells. The degree of repeat-size heterogeneity of the expanded allele in sperm appears to be greater than that in blood cells (Fig. 29-8). Although the sample size is small, these results suggest that the expanded alleles in male germline cells are highly unstable and may account for the instability observed with paternal transmissions.

3. STABILITY OVER TIME OF THE $(ATTCT)_n$ EXPANSION IN BLOOD CELLS FROM PATIENTS WITH SCA10

The existence of repeat-size mosaicism in SCA10 patients suggests that the expanded $(ATTCT)_n$ repeat was unstable some time in the past. However, it does not

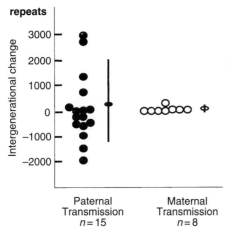

FIGURE 29-7 Effects of the gender of the transmitting parent on the repeat size of the offspring. Paternal transmission of the expanded allele (filled circles, $n = 15$) results in a greater degree of intergenerational change than maternal transmission (open circles, $n = 8$) (paternal change, +247 ± 1555 repeats; maternal, 0 ± 28 repeats, $p < 0.01$). Paternal transmissions gave rise to both elongations and contractions of the expanded alleles. Ovals and error bars indicate means and standard deviations, respectively.

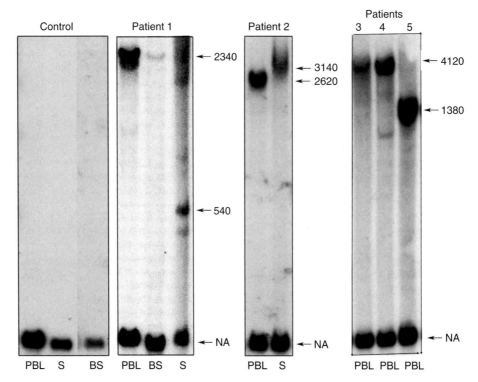

FIGURE 29-8 Repeat instability in somatic tissues and sperm. A normal control subject shows the 2.5-kb EcoRI fragment (normal allele). In patient 1, the sperm shows a smearlike pattern of multiple bands of expansion, whereas the other tissues show the major allele of a similar size (2340 repeats). PBLs of this patient show another band just underneath of the 2340-repeat allele. The expanded band of the buccal sample is substantially fainter; this is probably due to degraded DNA. In patient 2, the sperm shows a smearlike band larger than the expanded allele of PBLs. In patient 4, two distinct bands are seen in PBLs. In patient 5, the expanded band in PBLs shows a greater width than the normal band, suggesting repeat-size heterogeneity. Abbreviations: NA, normal allele; S, sperm; BS, buccal sample. Numbers on the right side of each panel indicate the number of ATTCT repeat units.

indicate whether the repeat instability is an ongoing process in patients. Samples of DNA were obtained from the blood of patients with SCA10 on two occasions separated by a 5-year interval. The size of the expanded allele did not show detectable changes in each sample set (Fig. 29-9). The majority of these patients showed repeat-size mosaicism with additional bands, which were variably smaller than the main band. Interestingly, the pattern of this mosaicism did not change in each patient after the 5-year interval. These data suggest that the expanded alleles in peripheral blood leukocytes (PBLs) are relatively stable, although rare instability events do occur perhaps in the bone marrow stem cell populations, as evidenced by stable somatic mosaicism.

4. Changes in ATTCT Expansion after Serial Passaging of SCA10 Lymphoblastoid Cell Lines in Culture

Cell culture provides a useful model to study the instability of repeats. Lymphoblastoid cell lines

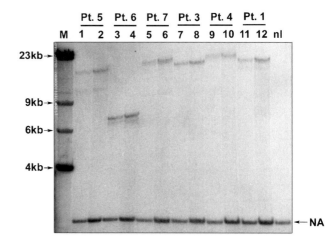

FIGURE 29-9 ATTCT expansion is stable in PBLs during a 5-year interval. The odd-numbered lanes represent initial PBL DNA, and the even-numbered lanes represent PBL DNA obtained 5 years later. The size of the expansion, including that of the mosaic allele, did not change in each patient. Abbreviations: M, λHindIII marker; NA, normal allele; nl, normal control individual.

(LBCLs) are of particular importance because they can be readily established from peripheral blood cells. The establishment of cultures of other cell types requires fresh surgical specimens. The establishment of LBCLs requires the transformation of lymphocytes by Epstein–Barr virus (EBV). Following EBV transformation, LBCLs showed 100–780 fewer repeats than the original alleles in PBLs in about one-third of the cases. This change may be due to the fact that EBV preferentially transforms B-lymphocytes, which represent only a subpopulation of PBLs; therefore, this observation may be explained by the selection of a subpopulation of PBLs that were already mosaic for the expansion size. Alternatively, this shift could be caused by mutational events in LBCLs during the EBV transformation or the subsequent culture period.

Expanded ATTCT repeats in some LBCLs showed a variable degree of instability during serial passes in culture (Fig. 29-10). In a LBCL with contracted repeats after transformation, the expanded alleles showed further contractions during passages (Fig. 29-10A). In some LBCLs, an apparent new contracted allele arose during the culture and gradually replaced the progenitor allele. Some LBCLs (Fig. 29-10B) showed a smear of expansion immediately after the EBV transformation, followed by a gradual loss of size heterogeneity resulting in a sharp, narrow band in subsequent passages. These changes are probably attributable to a shift in cell population due to clonal expansion, although repeat-size mutations during culture cannot be excluded. As predicted, the normal alleles of all LBCLs showed no instability. In most LBCLs, however, the expansion size remained stable after EBV transformation and serial passages (Fig. 29-10B).

5. INSTABILITY OF THE EXPANDED ATTCT REPEAT IN SERIALLY PASSED CLONAL LBCLS

To explore whether repeat-size mutation occurs in LBCLs, clonal LBCLs were established. Because clonal LBCLs are derived from a single cell [23], changes in repeat size in clonal lines are attributable to mutation events occurring after cloning (Fig. 29-11). One clonal line showed a contraction of the expanded allele during the first five passages, suggesting that a repeat-size mutation had occurred. However, most clonal lines showed no detectable changes in the size of the expanded allele, suggesting that the mutational event in culture is a relatively rare phenomenon.

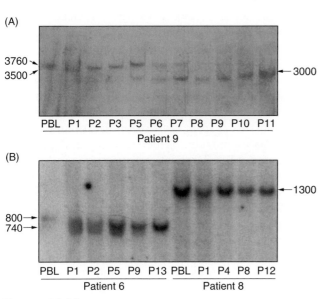

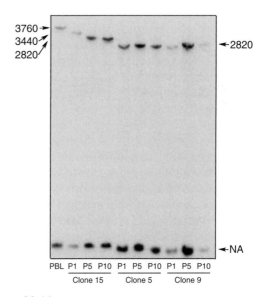

FIGURE 29-10 Repeat instability in LBCLs. Only expanded alleles are shown to highlight the instability. (A) The 3760-repeat expansion in PBLs contracted to 3500 through the EBV transformation at the first passage (P1). The expanded allele gradually increased in size to 3760 again during the subsequent passages (P1–P5). Moreover, a 3000-repeat allele appeared in P5 and gradually replaced the 3760 repeats. (B) The LBCLs on the left showed repeat-size heterogeneity through the transformation. Following the passages, the broad band converted to a tight one of 740 repeats. In the LBCLs on the right, the allele with 1300 repeats was stable throughout the passages and the transformation.

FIGURE 29-11 Repeat instability in clonal LBCLs. Clonal LBCLs (clones 5, 9, and 15) were established from an SCA10 patient who had a 3760-repeat allele in PBLs. Contraction was noted between passages 1 and 5 of clone 15. Abbreviations: P1, passage 1; P5, passage 5; P10, passage 10.

III. CLINICAL PHENOTYPE AND ETHNICITY

A. Spectrum of the SCA10 Phenotype

Since the identification of the $(ATTCT)_n$ expansion as the mutation for SCA10, more than 100 affected members have been identified and 67 patients from 18 families with SCA10 have been examined. In a series of Mexican SCA10 families, the phenotypic spectrum of SCA10 has been characterized, ranging from pure cerebellar ataxia to a more complex ataxia syndrome with epilepsy, neuropsychological disturbances, sensory polyneuropathy, and corticospinal tract signs [24]. Neuropsychological disturbances were most readily detected as subnormal or lower normal intelligence quotients. Depression and abnormal psychosocial behaviors have been observed in some patients, although it is unclear whether these are part of the SCA10 phenotype or are secondary to the chronic illness. Sensory polyneuropathy was clearly documented by nerve conduction studies, although sensory symptoms are rare complaints of patients. In one of these families, nonneurological phenotypes, including hepatic insufficiency or failure, normocytic normochromic anemia, and thrombocytopenia, were found with the SCA10 mutation [24]. Hepatic failure was the cause of death in some affected members. However, whether these nonneurological phenotypes were attributable to the *SCA10* mutation or whether they coincidentally segregated with SCA10 in the family is unclear.

Occurrences of epilepsy in SCA10 patients were noted in the initial report of this disease, but it was attributed to the coexisting cerebral cysticercosis [3]. However, subsequent studies clearly documented that epilepsy is by far the most conspicuous nonataxic phenotype in SCA10. Although ataxia was the initial symptom in most SCA10 patients, seizures often followed within a few years [22]. The prevalence of epilepsy differed from family to family, ranging from 25 to 100%; overall, 60% of Mexican SCA10 patients had epilepsy [22, 24]. Epilepsy consists of generalized motor seizures, complex partial seizures, or both, and the complex partial seizures often lead to secondary generalization. Anticonvulsive drug therapies are effective initially, but status epilepticus has been observed in two patients, one with a fatal outcome [22]. It was also noted that some patients became totally disabled from epilepsy in spite of relatively mild ataxia. These patients had complex partial seizures several times per day and generalized tonic–clonic seizures 2–3 times per week, despite thorough trials of anticovulsive treatments including implantation of a vagal nerve stimulator. Most SCA10 patients with epilepsy had abnormal EEGs, which frequently showed epileptiform discharges. Thus, SCA10 typically presents with cerebellar ataxia followed by epilepsy in the Mexican population, with variable other neurological disturbances, although some patients show only cerebellar ataxia.

B. Ethnic Distribution of SCA10 and Phenotypic Variations

A search for SCA10 among ataxia families from United States (non-Hispanic), Europe (including Spain), and Japan failed to identify the SCA10 mutation in these populations [25, 26]. Combination of these data with reports from other labs [27] led to the initial conclusion that SCA10 is relatively rare in non-Mexican populations. Unassigned SCA families from the United States and other countries were screened for the SCA10 mutation, using the repeat-primed PCR that can readily detect expansion alleles of the SCA10 repeat [15, 28]. Additional families of Mexican origin were identified. In 2003, this effort led to the discovery of the SCA10 mutation in 5 southeastern Brazilian families [29]. Like the Mexican SCA10 patients, all Brazilian patients had clinical signs of cerebellar ataxia with cerebellar atrophy on magnetic resonance imaging (MRI). Interestingly, epilepsy, which was frequently found in the Mexican families, was absent in all 43 patients from these Brazilian SCA10 families [29]. Other extracerebellar features were also absent from most of the Brazilian patients. There was no correlation between the epilepsy phenotype and ATTCT repeat expansion size in Mexican SCA10 families [22], and repeat expansion size could not explain the lack of epilepsy in Brazilian patients (Fig. 29-12) [29]. The basis of the phenotypic difference between the two ethnic origins remains unknown. Single nucleotide polymorphisms representing three regions of linkage disequilibrium were studied, as were additional short tandem repeat polymorphisms within the *ATXN10* gene. Preliminary data (not shown) suggest that mutant chromosomes of the Mexican and Brazilian families share the same haplotype within the *ATXN10* gene. Currently the haplotype is being expanded to regions flanking *ATXN10*, and haplotype frequencies in the general populations of Mexico and southeast Brazil are being determined. These studies may provide insights into the origins of SCA10 mutations and potential roles of *cis*-acting elements in the ethnicity-dependent

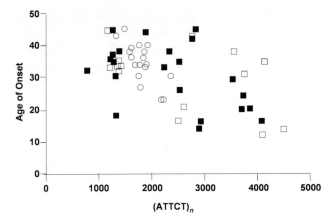

FIGURE 29-12 A lack of correlation of the epilepsy phenotype with the age of onset and (ATTCT)$_n$ repeat expansion size. The expansion size is inversely correlated with the age of onset in Brazilian patients with SCA10 (open circles, $n = 18$, $r^2 = 0.532$, $p < 0.01$) and in Mexican patients with (solid squares, $n = 22$) and without (open squares, $n = 14$) epilepsy (for all 54 patients, $r^2 = 0.264$, $p < 0.01$). Note that expansion size overlaps between the Brazilian patients and the Mexican patients, and the epilepsy phenotype was associated with a wide range of repeat size and age of onset among the Mexican patients. Thus, neither expansion size nor age of onset can explain the absence of epilepsy phenotype from Brazilian patients.

phenotypic variation. Trans-acting genetic and environmental factors may also contribute to the phenotypic disparity between these two SCA10 populations.

IV. GENOTYPE–PHENOTYPE CORRELATION

As previously mentioned, the age of onset shows an inverse correlation with the size of expanded (ATTCT)$_n$ repeat in SCA10 patients. However, the correlation is relatively weak and repeat size accounts for only ~35% of the variance in the age of onset. Somatic instability of the expanded repeat [21] may contribute to the low correlation coefficient (r^2). For example, the age of onset may show a tighter correlation with expansion size in the brain. Because no opportunity to examine the brain tissues from SCA10 patients has occurred, the difference in the expansion size between blood and brain tissues cannot be assessed.

Factors other than somatic instability may also explain the weak correlation. Different patterns of the interruptions in the expanded (ATTCT)$_n$ repeat may have different pathogenic potentials, which may lead to different ages of onset, disease severity, and phenotypic characteristics [13]. Other genetic and environmental factors may also modify the age of onset and other phenotypic expressions of SCA10.

V. DIAGNOSTIC UTILITY OF THE MUTATION

The molecular diagnosis of SCA10 is usually straightforward. First, PCR amplification of the (ATTCT)$_n$ repeat region is performed [12]. If two heterozygous alleles in the normal range (29 ATTCTs or smaller) are present, the diagnosis of SCA10 is excluded. Samples that show a single allele on PCR analysis undergo repeat-primed PCR, and those showing no extended ladder beyond the upper limit of the normal range are diagnosed as normal samples [15]. The samples that show expanded alleles are then subjected to Southern blot analysis to determine the expansion size. Expanded alleles of 4 kb (800 ATTCTs) or larger are considered SCA10 mutations.

The size of the expanded allele has no practical utility for predicting the age of onset in individual patients due to the weak correlation. The diagnostic significance of alleles larger than the upper normal limit but smaller than 4 kb is currently uncertain. As mentioned earlier, an allele of ~280 ATTCTs has been reported in an asymptomatic woman and her affected daughter [13]. This allele may represent the smallest pathogenic allele, which has reduced penetrance. However, the possibility that this is a normal allele coincidentally found in the daughter, who happened to have ataxia, cannot be excluded.

VI. PATHOGENIC MECHANISMS

The mechanism by which the expanded ATTCT repeat causes the SCA10 phenotype is unknown. The physiological function of the *ATAXN10* protein product remains to be investigated. At present all pathogenic mechanisms that have been postulated in diseases with microsatellite expansions and other novel mechanisms for SCA10 are being considered. However, the likelihood of each pathogenic model may be assessed on the basis of the intronic location of this pentanucleotide repeat and other clues found in preliminary studies.

A. *ATXN10* Gene and Function of the Ataxin-10 Protein

ATXN10 is a 172.8-kb gene consisting of 12 exons with an open reading frame (ORF) of 1428 bp, encoding 475 amino acids. The protein product of the *ATXN10* gene is a novel 53-kDa protein, E46L (like mouse brain protein E46, also known as ataxin-10; GenBank accession number NM_013236). There are orthologs of E46L, including mouse brain protein E46 (GenBank accession number P28658) and Drosophila CG4975-PA (GenBank accession number NP_611253), suggesting conservation among vertebrate and invertebrate species. Analysis of the amino acid sequence of the human E46L protein suggests that it does not contain any transmembrane domain, nuclear localization signal, or other known functional motifs [12]. However, the C-terminal end of E46L contains two armadillo repeat domains often found in membrane-associated proteins such as β-catenin. E46L mRNA was shown to be expressed throughout the brain [12]. März et al. [30] found that rat ataxin-10 is a cytoplasmic protein of ~55 kDa, which forms homotrimeric complexes via "tip-to-tip" contact with the concave sides of the molecules facing each other. Immunostaining of mouse and human brain sections with their antibody against rat ataxin-10 protein revealed a predominantly cytoplasmic and perinuclear localization, with a clear restriction to olivocerebellar regions. "Knock down" of *SCA10* in primary cerebellar and cortical neuronal cells in culture by small interfering RNAs (siRNAs) increased apoptosis in cerebellar and cortical neurons. These findings may argue for a loss-of-function phenotype in SCA10 patients.

B. Pathogenic Models for SCA10

1. Loss of Function

Among diseases caused by trinucleotide repeat expansions, a loss of function is the key pathogenic mechanism in FRDA [19], fragile X syndrome, and fragile XE (FRAXE) mental retardation syndrome [31, 32]. The expanded $(GAA)_n$ repeat hinders the transcription of the *FRDA* gene by forming the "sticky" DNA structure [33]. However, $(ATTCT)_n$ repeats do not form the "sticky" DNA structure; they form an unpaired structure, which is not likely to interfere with transcription [17]. In fragile X syndrome, the $(CGG)_n$ expansion silences the *FMR1* gene by methylating the nearby CpG islands in the promoter and the repeat itself [34]. Preliminary studies showed that the DNA methylation status in the *ATXN10* region is not altered in SCA10 (Ashizawa, unpublished data). Furthermore, it should be noted that FRDA is an autosomal recessive disorder and fragile X syndrome and FRAXE mental retardation are X-linked disorders, whereas SCA10 is an autosomal dominant disease. Loss of function generally is not an attractive pathogenic model for an autosomal dominant disease because haploinsufficiency usually is not sufficient to cause diseases. Nevertheless, there are some autosomal dominant disorders caused by haploinsufficiency, especially when the protein product is tightly regulated for its physiological function. In DM1, which is dominantly inherited, loss of function of *DMPK* by an expanded noncoding $(CTG)_n$ repeat has been postulated to contribute to the pathogenic mechanism [20, 35]. Thus, it is plausible that the expansion of $(ATTCT)_n$ induces a loss of *ATXN10* function that is sufficient to cause the disease. Preliminary data suggest that peripheral SCA10 cells are not deficient in *ATXN10* mRNA or protein [12] (Ashizawa, unpublished data). However, it is still possible that the expression of *ATXN10* may be compromised by expansion of the $(ATTCT)_n$ repeat in the brain, where the level of expression is substantially higher and the demand for the ataxin-10 protein might be more intense.

2. Gain of Function

The gain-of-function mechanism mediated by a mutant protein containing an expanded tract of tandemly repeated amino acid residues is highly unlikely in SCA10. First, the $(ATTCT)_n$ repeat is located in an intron of the *ATXN10* gene. Second, no other expressed sequences including this repeat are present on the same or opposite strand in the genome database. Third, this pentanucleotide repeat is polymorphic and would introduce a shift in the reading frame in normal alleles if it were located in the coding region. Furthermore, it may be noted that known translated trinucleotide repeats encoding polyglutamine or polyalanine repeats are short, and a large repeat expansion has not been found in the coding region of genes.

However, it is plausible that the RNA transcript containing expanded $(AUUCU)_n$ repeats gains a toxic function in SCA10. The RNA-mediated gain of function is an attractive model to explain the pathogenic mechanism of some autosomal dominant diseases caused by noncoding repeat expansions, and it is the most convincing model for the pathogenic mechanism of DM1 and DM2. DM1 is caused by a large $(CTG)_n$ repeat expansion in the 3' UTR of the *DMPK* gene [36], whereas a large expansion of a CCTG tetranucleotide repeat in the first intron of the zinc finger 9 (*ZNF9*) gene results in DM2 [1]. In DM1 and DM2, the mutant RNA transcripts containing expanded $(CUG)_n$ and $(CCUG)_n$ repeats colocalize with muscleblind proteins, respectively, and form RNA–protein complexes in intranuclear foci [37]. The functional inactivation of one of the muscleblind proteins (*Mbnl1*) was sufficient to cause several DM1 and DM2 phenotyes in mice, suggesting that sequestration of the muscleblind protein plays an important role in DM pathogenesis [38]. However, colocalization of the $(CUG)_n$-containing RNA and muscleblind proteins may not be relevant to the functional loss of muscleblind proteins [39]. The expanded $(CUG)_n$ repeat in the SCA8 mRNA could initiate a similar mechanism in the brain [40]. Furthermore, premutation alleles of the fragile X $(CGG)_n$ repeat causes fragile-X-associated tremor/ataxia syndrome (FXTAS) without decreasing the transcription of the *FMR1* gene. Instead, the *FMR1* mRNA level may be increased. Thus, it has been postulated that $(GCC)_n$-containing *FMR1* mRNA from the premutation allele gains a toxic function through interactions with ribo-$(GCC)_n$ binding proteins [35, 41]. Thus, RNA gain-of-function models have increasingly been recognized as the pathogenic mechanism of diseases with noncoding repeat expansions. Among these diseases, DM2 exhibits notable similarities with SCA10 regarding the following characteristics: (1) the mutation is an expansion of a nontriplet repeat; (2) the repeat is located in an intron of the gene; (3) the size of expansion is very large, reaching >20 kb; and (4) autosomal dominant inheritance. However, the sequence of the repeat unit (CCTG in DM2 and ATTCT in SCA10) is remarkably different. Work is in progress on the hypothesis that expanded $(AUUCU)_n$ repeats bind specific proteins and alter the functional availability of these proteins in SCA10 cells.

3. OTHER PATHOGENIC CONSIDERATIONS

Although a loss of function and the trans-dominant RNA gain of function may be plausible pathogenic mechanisms of SCA10, other possibilities should not be dismissed. The $(AUUCU)_n$ expansion in the transcript of *ATXN10* could alter the splicing pattern, leading to aberrant isoforms with toxic or dominant negative effects. The expanded $(ATTCT)_n$ repeat might have ripple effects on the transcription of the neighboring genes. Intron 9 of the *ATXN10* gene is large (>60 kb) and may contain noncoding RNA (ncRNA) genes, which are difficult to detect with current sequence analysis algorithms. It is also possible that the expanded $(ATTCT)_n$ repeat sequestrates important DNA-binding proteins. Any of these mechanisms could play a pathogenic role in SCA10 as a component of complex pathogenic mechanisms.

VII. CONCLUSION

SCA10 has several features distinct from those of other diseases of microsatellite instabilities. It is the only disease in which the expanded repeat is a pentanucleotide $(ATTCT)_n$ repeat. The expanded repeat is large and shows complex patterns of interruptions. Repeat size appears to be unstable in male germline cells but relatively stable in female germline and somatic cells. The repeat forms an unpaired DNA structure, which may function as an aberrant replication origin. SCA10 has been found only on the American continents. Phenotypically, SCA10 patients show relatively pure cerebellar ataxia often accompanied by epilepsy, although other mild extracerebellar phenotypes have also been documented. Further studies will be needed to elucidate the mechanism of the repeat expansion, the ethnic origin of the mutation, and the pathogenic mechanism of the disease.

Acknowledgment

This work is supported by a grant from the National Institutes of Health (NS41547).

References

1. Liquori, C. L., Ricker, K., Moseley, M. L., Jacobsen, J. F., Kress, W., Naylor, S. L., Day, J. W., and Ranum, L. P. (2001). Myotonic dystrophy type 2 caused by a CCTG expansion in intron 1 of ZNF9. *Science* **293**, 864–867.
2. Lalioti, M. D., Scott, H. S., and Antonarakis, S. E. (1997). What is expanded in progressive myoclonus epilepsy? *Nat. Genet.* **17**, 17.
3. Grewal, R. P., Tayag, E., Figueroa, K. P., Zu, L., Durazo, A., Nunez, C., and Pulst, S. M. (1998). Clinical and genetic analysis of a distinct autosomal dominant spinocerebellar ataxia. *Neurology* **51**, 1423–1426.
4. Zu, L., Figueroa, K. P., Grewal, R., and Pulst, S. M. (1999). Mapping of a new autosomal dominant spinocerebellar ataxia to chromosome 22. *Am. J. Hum. Genet.* **64**, 594–599.
5. Matsuura, T., Achari, M., Khajavi, M., Bachinski, L. L., Zoghbi, H. Y., and Ashizawa, T. (1999). Mapping of the gene for a novel spinocerebellar ataxia with pure cerebellar signs and epilepsy. *Ann. Neurol.* **45**, 407–411.
6. Matsuura, T., Watase, K., Nagamitsu, S., Zoghbi, H. Y., and Ashizawa, T. (1999). Fine mapping of the spinocerebellar ataxia type 10 region and search for a polyglutamine expansion. *Ann. Neurol.* **46**, 480.
7. Dunham, I., Shimizu, N., Roe, B. A., Chissoe, S., Hunt, A. R., Collins, J. E., Bruskiewich, R., Beare, D. M., Clamp, M., Smink, L. J., Ainscough, R., Almeida, J. P., Babbage, A., Bagguley, C., Bailey, J., Barlow, K., Bates, K. N., Beasley, O., Bird, C. P., Blakey, S., Bridgeman, A. M., Buck, D., Burgess, J., Burrill, W. D., O'Brien, K. P., et al. (1999). The DNA sequence of human chromosome 22. *Nature* **402**, 489–495.
8. Schalling, M., Hudson, T. J., Buetow, K. H., and Housman, D. E. (1993). Direct detection of novel expanded trinucleotide repeats in the human genome. *Nat. Genet.* **4**, 135–139.
9. Ikeuchi, T., Sanpei, K., Takano, H., Sasaki, H., Tashiro, K., Cancel, G., Brice, A., Bird, T. D., Schellenberg, G. D., Pericak-Vance, M. A., Welsh-Bohmer, K. A., Clark, L. N., Wilhelmsen, K., and Tsuji, S. (1998). A novel long and unstable CAG/CTG trinucleotide repeat on chromosome 17q. *Genomics* **49**, 321–326.
10. Breschel, T. S., McInnis, M. G., Margolis, R. L., Sirugo, G., Corneliussen, B., Simpson, S. G., McMahon, F. J., MacKinnon, D. F., Xu, J. F., Pleasant, N., Huo, Y., Ashworth, R. G., Grundstrom, C., Grundstrom, T., Kidd, K. K., DePaulo, J. R., and Ross, C. A. (1997). A novel, heritable, expanding CTG repeat in an intron of the SEF2-1 gene on chromosome 18q21.1. *Hum. Mol. Genet.* **6**, 1855–1863.
11. Trottier, Y., Lutz, Y., Stevanin, G., Imbert, G., Devys, D., Cancel, G., Saudou, F., Weber, C., David, G., Tora, L., et al. (1995). Polyglutamine expansion as a pathological epitope in Huntington's disease and four dominant cerebellar ataxias. *Nature* **378**, 403–406.
12. Matsuura, T., Yamagata, T., Burgess, D. L., Rasmussen, A., Grewal, R. P., Watase, K., Khajavi, M., McCall, A. E., Davis, C. F., Zu, L., Achari, M., Pulst, S. M., Alonso, E., Noebels, J. L.,

Nelson, D. L., Zoghbi, H. Y., and Ashizawa, T. (2000). Large expansion of the ATTCT pentanucleotide repeat in spinocerebellar ataxia type 10. *Nat. Genet.* **26**, 191–194.
13. Matsuura, T., Fang, P., Pearson, C. E., Jayakar, P., Ashizawa, T., Roa, B. B., and Nelson, D. L. (2006). Interruptions in the expanded ATTCT repeat of spinocerebellar ataxia type 10: repeat purity as a disease modifier? *Am. J. Hum. Genet.* **78**, 125–129.
14. Fang, P., Matsuura, T., Teive, H. A. G., Raskin, S., Jayakar, P., Schmitt, E., Ashizawa, T., and Roa, B. B. (2002). Spinocerebellar ataxia type 10 ATTCT repeat expansions in Brazilian patients and in a patient with early onset of ataxia. *Am. J. Hum. Genet.* **71**, 507.
15. Matsuura, T., and Ashizawa, T. (2002). Polymerase chain reaction amplification of expanded ATTCT repeat in spinocerebellar ataxia type 10. *Ann. Neurol.* **51**, 271–272.
16. Gao, R., Lin, X., Xu, W., Matsuura, T., and Ashizawa, T. (2004). The interrupted structure of expanded ATTCT pentanucleotide repeats in SCA10 patients. *Movement Disorders* **19**, 1123.
17. Potaman, V. N., Bissler, J. J., Hashem, V. I., Oussatcheva, E. A., Lu, L., Shlyakhtenko, L. S., Lyubchenko, Y. L., Matsuura, T., Ashizawa, T., Leffak, M., Benham, C. J., and Sinden, R. R. (2003). Unpaired structures in SCA10 (ATTCT)$_n$·(AGAAT)$_n$ repeats. *J. Mol. Biol.* **326**, 1095–1111.
18. Oostra, B. A., and Willemsen, R. (2002). The X chromosome and fragile X mental retardation. *Cytogenet. Genome Res.* **99**, 257–264.
19. Pandolfo, M. (2003). Friedreich ataxia. *Semin. Pediatr. Neurol.* **10**, 163–172.
20. Ashizawa, T., and Monckton, D. (2004). Molecular aspects of myotonic dystrophy: Our current understanding. *In* "Myotonic Dystrophy—Present Management, Future Therapy" (P. Harper, B. van Engelen, B. Eymard, and D. Wilcox, Eds.), pp. 14–36. Oxford University Press, Oxford.
21. Matsuura, T., Fang, P., Lin, X., Khajavi, M., Tsuji, K., Rasmussen, A., Grewal, R. P., Achari, M., Alonso, M. E., Pulst, S. M., Zoghbi, H. Y., Nelson, D. L., Roa, B. B., and Ashizawa, T. (2004). Somatic and germline instability of the ATTCT repeat in spinocerebellar ataxia type 10. *Am. J. Hum. Genet.* **74**, 1216–1224.
22. Grewal, R. P., Achari, M., Matsuura, T., Durazo, A., Tayag, E., Zu, L., Pulst, S. M., and Ashizawa, T. (2002). Clinical features and ATTCT repeat expansion in spinocerebellar ataxia type 10. *Arch. Neurol.* **59**, 1285–1290.
23. Khajavi, M., Tari, A. M., Patel, N. B., Tsuji, K., Siwak, D. R., Meistrich, M. L., Terry, N. H., and Ashizawa, T. (2001). "Mitotic drive" of expanded CTG repeats in myotonic dystrophy type 1 (DM1). *Hum. Mol. Genet.* **10**, 855–863.
24. Rasmussen, A., Matsuura, T., Ruano, L., Yescas, P., Ochoa, A., Ashizawa, T., and Alonso, E. (2001). Clinical and genetic analysis of four Mexican families with spinocerebellar ataxia type 10. *Ann. Neurol.* **50**, 234–239.
25. Matsuura, T., Ranum, L. P., Volpini, V., Pandolfo, M., Sasaki, H., Tashiro, K., Watase, K., Zoghbi, H. Y., and Ashizawa, T. (2002). Spinocerebellar ataxia type 10 is rare in populations other than Mexicans. *Neurology* **58**, 983–984.
26. Fujigasaki, H., Tardieu, S., Camuzat, A., Stevanin, G., LeGuern, E., Matsuura, T., Ashizawa, T., Durr, A., and Brice, A. (2002). Spinocerebellar ataxia type 10 in the French population. *Ann. Neurol.* **51**, 408–409.
27. Sasaki, H., Yabe, I., and Tashiro, K. (2003). The hereditary spinocerebellar ataxias in Japan. *Cytogenet. Genome Res.* **100**, 198–205.
28. Cagnoli, C., Michielotto, C., Matsuura, T., Ashizawa, T., Margolis, R. L., Holmes, S. E., Gellera, C., Migone, N., and Brusco, A. (2004). Detection of large pathogenic expansions in FRDA1, SCA10, and SCA12 genes using a simple fluorescent repeat-primed PCR assay. *J. Mol. Diagn.* **6**, 96–100.
29. Teive, H. A., Roa, B. B., Raskin, S., Fang, P., Arruda, W. O., Neto, Y. C., Gao, R., Werneck, L. C., and Ashizawa, T. (2004). Clinical phenotype of Brazilian families with spinocerebellar ataxia 10. *Neurology* **63**, 1509–1512.
30. Marz, P., Probst, A., Lang, S., Schwager, M., Rose-John, S., Otten, U., and Ozbek, S. (2004). Ataxin-10, the spinocerebellar ataxia type 10 neurodegenerative disorder protein, is essential for survival of cerebellar neurons. *J. Biol. Chem.* **279**, 35542–35550.
31. Pietrobono, R., Tabolacci, E., Zalfa, F., Zito, I., Terracciano, A., Moscato, U., Bagni, C., Oostra, B., Chiurazzi, P., and Neri, G. (2005). Molecular dissection of the events leading to inactivation of the FMR1 gene. *Hum. Mol. Genet.* **14**, 267–277.
32. Greene, E., Handa, V., Kumari, D., and Usdin, K. (2003). Transcription defects induced by repeat expansion: Fragile X syndrome, FRAXE mental retardation, progressive myoclonus epilepsy type 1, and Friedreich ataxia. *Cytogenet. Genome Res.* **100**, 65–76.
33. Sakamoto, N., Larson, J. E., Iyer, R. R., Montermini, L., Pandolfo, M., and Wells, R. D. (2001). GGA*TCC-interrupted triplets in long GAA*TTC repeats inhibit the formation of triplex and sticky DNA structures, alleviate transcription inhibition, and reduce genetic instabilities. *J. Biol. Chem.* **276**, 27178–27187.
34. Lim, J. H., Booker, A. B., and Fallon, J. R. (2005). Regulating fragile X gene transcription in the brain and beyond. *J. Cell Physiol.* **205**, 170–175
35. Frisch, R., Singleton, K. R., Moses, P. A., Gonzalez, I. L., Carango, P., Marks, H. G., and Funanage, V. L. (2001). Effect of triplet repeat expansion on chromatin structure and expression of DMPK and neighboring genes, SIX5 and DMWD, in myotonic dystrophy. *Mol. Genet. Metab.* **74**, 281–291.
36. Fu, Y. H., Pizzuti, A., Fenwick, R. G., Jr., King, J., Rajnarayan, S., Dunne, P. W., Dubel, J., Nasser, G. A., Ashizawa, T., de Jong, P., et al. (1992). An unstable triplet repeat in a gene related to myotonic muscular dystrophy. *Science* **255**, 1256–1258.
37. Mankodi, A., Teng-Umnuay, P., Krym, M., Henderson, D., Swanson, M., and Thornton, C. A. (2003). Ribonuclear inclusions in skeletal muscle in myotonic dystrophy types 1 and 2. *Ann. Neurol.* **54**, 760–768.
38. Kanadia, R. N., Johnstone, K. A., Mankodi, A., Lungu, C., Thornton, C. A., Esson, D., Timmers, A. M., Hauswirth, W. W., and Swanson, M. S. (2003). A muscleblind knockout model for myotonic dystrophy. *Science* **302**, 1978–1980.
39. Ho, T. H., Savkur, R. S., Poulos, M. G., Mancini, M. A., Swanson, M. S., and Cooper, T. A. (2005). Colocalization of muscleblind with RNA foci is separable from mis-regulation of alternative splicing in myotonic dystrophy. *J. Cell Sci.* **118**, 2923–2933.
40. Mosemiller, A. K., Dalton, J. C., Day, J. W., and Ranum, L. P. (2003). Molecular genetics of spinocerebellar ataxia type 8 (SCA8). *Cytogenet. Genome Res.* **100**, 175–183.
41. Hagerman, R. J., Leavitt, B. R., Farzin, F., Jacquemont, S., Greco, C. M., Brunberg, J. A., Tassone, F., Hessl, D., Harris, S. W., Zhang, L., Jardini, T., Gane, L. W., Ferranti, J., Ruiz, L., Leehey, M. A., Grigsby, J., and Hagerman, P. J. (2004). Fragile-X-associated tremor/ataxia syndrome (FXTAS) in females with the FMR1 premutation. *Am. J. Hum. Genet.* **74**, 1051–1056.

DNA Structures and Genetic Instabilities Associated with Spinocerebellar Ataxia Type 10 (ATTCT)$_n$·(AGAAT)$_n$ Repeats Suggest a DNA Amplification Model for Repeat Expansion

VLADIMIR N. POTAMAN, MALGORZATA J. PYTLOS, VERA I. HASHEM[1]
JOHN J. BISSLER, MICHAEL LEFFAK, AND RICHARD R. SINDEN

Laboratory of DNA Structure and Mutagenesis, Center for Genome Research, Institute of Biosciences and Technology, Texas A&M University System Health Science Center, Houston, Texas 77030-3303; Division of Nephrology and Hypertension, Cincinnati Children's Hospital Medical Center, Cincinnati, Ohio 45229-3039; and Department of Biochemistry and Molecular Biology, Wright State University, Dayton, Ohio 45435-0001

I. Introduction
  A. DNA Unwinding, Replication, and Transcription
  B. Eukaryotic DNA Replication and the Promiscuous Replication Hypothesis
II. The SCA10 Repeat Forms Unwound DNA
III. Complex Replication-Based Instability of SCA10 (ATTCT)$_n$·(AGAAT)$_n$ Repeats in *Escherichia coli*
  A. SCA10 Repeats Do Not Undergo Deletion at a High Rate in *Escherichia coli*
  B. Complex Expansion Mutations Occur in (ATTCT)$_n$·(AGAAT)$_n$ Repeats in *Escherichia coli*
IV. Models for Complex Expansion Mutations in *Escherichia coli* and Expansion and Instability in Humans
  A. Mechanisms Leading to the Formation of the Complex Expansion Mutation in *Escherichia coli*
  B. Amplification Leading to Repeat Expansion in Human Cells
V. Concluding Remarks
References

Many human hereditary diseases have been associated with the instability of DNA repeats in the genome. Spinocerebellar ataxia type 10 is associated with expansion from a normal range of 10 to 22 (ATTCT)$_n$·(AGAAT)$_n$ pentanucleotide repeats to as many as 4,500 repeats. While other unstable DNA repeats can form slipped strand structures, intramolecular triplex DNA, or quadruplex structures that might block DNA replication, the (ATTCT)$_n$·(AGAAT)$_n$ repeat is expected to form

[1]Present address: Department of Human and Molecular Genetics, Baylor College of Medicine, Houston, Texas 77030.

unwound or unpaired regions in negatively supercoiled DNA. Using two-dimensional gel electrophoresis, atomic force microscopy and chemical probe analysis we show that at moderate superhelical densities, the $(ATTCT)_n \cdot (AGAAT)_n$ repeat forms an unpaired region, which then extends into adjacent A+T-rich flanking sequences. For plasmids containing 29 repeats, above the normal human size range, a locally condensed structure formed at high superhelical densities. Although the structure is "condensed," the bases remained unpaired. Data are also presented showing that this pentanucleotide repeat exhibits unusual characteristics when grown in *E. coli*. The $(ATTCT)_n \cdot (AGAAT)_n$ repeats do not undergo deletion at a high rate and, unlike trinucleotide repeats that are rapidly deleted in *E. coli*, $(ATTCT)_n \cdot (AGAAT)_n$ repeats undergo expansion associated with a complex mutational event upon prolonged growth in *E. coli*. The complex mutations involved both plasmid dimerization and inversion of part of the repeat tract, forming an inverted repeat. The results are consistent with a high frequency of primer-template misalignment during DNA replication. Models for repeat expansion are discussed which involve aberrant DNA replication as a critical step in repeat expansion.

I. INTRODUCTION

A. DNA Unwinding, Replication, and Transcription

A DNA unwinding element (DUE) is an A+T-rich sequence ranging from 30 to >100 bp in length in which the duplex DNA is prone to unpairing (also called melting or unwinding). Kowalski et al. first showed that A+T-rich tracts possessed single-stranded character in supercoiled DNA, as they were susceptible to cleavage by the single-strand-specific mung bean nuclease [1, 2]. *In vitro*, a DUE can be maintained as a stable unwound structure in the presence of negative DNA supercoiling [3–5], and the spinocerebellar ataxia type 10 (SCA10) $(ATTCT)_n \cdot (AGAAT)_n$ repeat appears as a stable bubble in supercoiled plasmid DNA [6]. DNA unpairing *in vivo* may be promoted by torsional stress resulting from the equilibrium level of supercoiling or by transient supercoils introduced by the movement of an RNA transcription complex [7, 8].

DUEs are essential components of bacterial and yeast replication origins [9, 10] and some mammalian origins [11]. Most mammalian origins of DNA replication contain a DUE and/or other DNA sequences that could form cruciforms, intramolecular triplexes, or other alternative DNA structures that may present DNA in single-strand form [12, 13]. An unpaired region in DNA can provide a site where the replication complex assembles [12–14]. For example, initiation of replication at the *Escherichia coli* chromosome origin, *oriC*, involves binding of the initiator protein DnaA to sites (DnaA boxes) located within *oriC*, local unwinding of a DUE, and loading of DNA helicase, primase, and polymerases to the unpaired region [15, 16]. Bacterial *oriC* regions vary in size from about 200 up to 1000 bp, and they generally contain several DnaA boxes and some form of an A+T-rich region [16–18].

The chromosomes of higher eukaryotes have defined replication origins, but few replication origins have been characterized in human cells [19–24]. The initiation of DNA synthesis begins at multiple sites within a broad region called an initiation zone [14, 25–30]. Initiation zones in human cells have been characterized at origins associated with several genes, including the c-myc gene [23, 31–36] and the β-globin gene [32, 37]. A chromosomal origin of bidirectional replication was first identified 3.5 kb 5′ of the human c-myc gene [35, 36, 38–40]. A 2.4-kb fragment of this region, referred to as the c-myc core, was shown to control replication initiation (replicator activity) when transposed to an ectopic chromosomal site [23, 34] and contains sites for replication initiation [38–40]. The 2.4-kb c-myc core origin endows plasmids with autonomously replicating sequence activity in HeLa cells [38–40] and allows replication in human cell extracts of supercoiled plasmids free of RNA or DNA primers [11]. Several transcription factor consensus binding sites are present in the core origin [23, 41], as are a DNA unwinding element that is unwound *in vivo* and a potential triplex-forming region [11, 42–44]. Deletion analyses have shown that some of these regions of the c-myc origin, including the DUE, are essential for chromosomal replication activity [23, 34, 45]. The *c-myc* DUE-binding protein (DUE-B) plays a role in regulating replication initiation in HeLa cells [31].

The *c-myc* gene DUE is also known as the FUSE (far upstream element), which is a critical regulatory component of transcription control for the *c-myc* gene [44, 46, 47]. The human *c-myc* DUE presumably couples DNA topology with transcription, as a transcription factor (FUSE-binding protein, FBP) and the FBP antagonist, the FBP-interaction repressor (FIR), compete for binding at the DUE when single stranded [44, 47–50]. The DUE melts from superhelical energy generated by transcription, and thus the DUE provides a real-time feedback mechanism for sensing the level of transcription [46]. The observation that the eukaryotic replication initiator origin recognition complex (ORC) binds preferentially to supercoiled DNA [51] suggests that the topological equivalence between superhelical and unwound states could allow DUEs to act as replication switches.

Significantly, a link between transcription and replication initiation is emerging [31, 45, 46, 52–55].

Provocatively, it has been noted that disease-associated expanding repeats are found within genes that are transcribed. Transcription, alternative DNA conformational transitions, and instability by aberrant replication activities may thus be tightly linked.

B. Eukaryotic DNA Replication and the Promiscuous Replication Hypothesis

DNA replication poses at least two problems for the eukaryotic cell. The entire genome must be replicated in each cell cycle, but no part of the genome is ordinarily replicated more than once per cycle. Duplication of the genome only once per cell division is achieved with two mutually exclusive periods of the mitotic cycle, when either prereplicative complexes (pre-RCs) can assemble but not initiate or replication can initiate but pre-RCs cannot assemble. These periods of the cell cycle are determined by oscillation in the activities of the S-phase kinases Cdc7 and Cdk1 [56].

In yeast, origins are bound by the hexameric ORC [57, 58], which enlists essential members of the pre-RC, *Saccharomyces cerevisiae* Cdc6p (scCdc6p), scCdt1, and the minichromosome maintenance proteins MCM2–7 [56]. Although ORC is bound to origins through most or all of the cell cycle [59], it is during a window of opportunity in the G1 phase when S-phase kinase (SPK) activity is low that origins are licensed for replication by the deposition of Cdc6, Cdt1, and the MCMs to form the pre-RC. The functions of the replication proteins identified in yeast are highly conserved in frogs [60], humans [61], and other higher eukaryotes. Studies in *Xenopus* egg extracts show that subsequent binding of MCM10 and the Cdc45 loading factor Mus101 is required for the addition of Cdc45, unwinding of the template, association of the single-strand DNA binding replication protein A (RPA), and, ultimately, recruitment of DNA polymerase α [62, 63]. Thus, SPKs trigger the conversion of pre-RCs into active replication complexes to initiate DNA synthesis [64–68]. High levels of kinase activity from the beginning of S phase through metaphase lead to the phosphorylation of multiple pre-RC proteins and prevent the assembly of new pre-RCs [69].

1. THE PROMISCUOUS REPLICATION HYPOTHESIS

Amplification of specific regions of DNA by re-replication is well-documented in several systems, including the selective amplification of chorion genes during normal *Drosophila* development and puff II/9A in *Sciara* [70–73]. For those origins involved in amplification, the normal cellular controls limiting initiation to once per cell cycle are abrogated [70, 71]. In these cases, and in the establishment of drug resistance in tumor cells [74], amplification has been proposed to occur by an onion skin mechanism in which re-replication leads to multiple replication forks [70, 75–77], followed by recombination to generate linear tandem arrays of amplified DNA [78].

As an alternative to gene amplification via re-replication, DNA recombination has been proposed as a mechanism for producing autonomously replicating episomes, autonomously replicating double-minute chromosomes, and homogeneously staining regions. Karyotypic data in dividing clonal cell populations suggest that recombination-based amplification can result from the generation of DNA ends in the chromosome [79]. Amplification is frequently associated with replication origins, A+T-rich regions, inverted repeats, or polypurine·polypyrimidine tracts [70, 80, 81]. Fragile sites have been implicated as a causative factor in oncogene amplification [82]. Thus, although commonly thought to arise by strand breakage, reinitiation at an aberrant origin could also generate abnormal DNA ends, leading to recombinational amplification [78].

The authors propose that (ATTCT)·(AGAAT) repeats of sufficient length can form unwound DNA structures that lead to repeat amplification. One plausible mechanism is that a supercoiling-induced (e.g., unwound) structure bypasses the steps of pre-RC assembly that normally require the low CDK activity environment of G1 phase and allows pol α/primase to initiate replication in the high CDK environment of S phase without the association of origin-bound checkpoint proteins. This aberrant DNA replication origin may then initiate promiscuous rounds of replication. Alternatively, the supercoiling-induced structure may be recognized as a distortion by proteins involved in DNA repair (e.g., RPA, XPA, XPC, MSH2) [83–86] and generate a 3′-OH primer by strand breakage or enzymatic nicking. Repetitive rounds of slipped mispairing during replication could then lead to repeat amplification and recombination [87]. These events may occur even more frequently during early embryogenesis or germ cell development, where chromatin structure and replication differ from that in somatic cells and dynamic epigenetic modifications are occurring [88, 89].

Other situations exist where aberrant replication and even transcription occur at unpaired DNA bubbles. pol α/primase can initiate replication in the absence of accessory proteins if a single-strand template is available [90]. Aberrant transcription can also occur at unpaired bubbles. Transcription by RNA polymerase II can occur from the yeast *CUP1* promoter if DNA is sufficiently negatively supercoiled to open an unpaired region at the promoter [91]. Supercoiling presumably destabilizes the duplex, allowing the RNA pol II to bind to a single strand and initiate transcription. Though not the normal mechanism for transcriptional control, it may represent a minimal, possibly primitive, transcription pathway. Regulatory mechanisms have evolved to repress unscheduled, and potentially deleterious, transcription and replication. However, given the necessity of an unwound region for transcription and replication, and proteins designed to bind to unpaired regions, it may not be surprising that regulatory controls may be unable to prevent fractious events from occurring at the SCA10 DNA repeats that may be maintained as large unpaired regions.

This chapter will discuss the unwound alternative DNA structures formed by the SCA10 $(ATTCT)_n \cdot (AGAAT)_n$ repeat, complex expansion mutations associated with aberrant replication at this sequence in *E. coli*, models that explain the unusual mutations in *E. coli*, and replication-based models that might explain repeat expansion in human cells.

II. THE SCA10 REPEAT FORMS UNWOUND DNA

Unlike most of the trinucleotide and tetranucleotide repeats, which are G+C-rich, the only expanding pentanucleotide repeat, $(ATTCT)_n \cdot (AGAAT)_n$, is 80% A+T-rich. It lacks necessary symmetry elements to form hairpins, intramolecular triplexes, or quadruplexes. Because of the high A+T content, the $(ATTCT)_n \cdot (AGAAT)_n$ duplex is expected to be thermodynamically less stable than a mixed DNA sequence [3], and, therefore, various factors such as torsional stress, temperature, and protein binding may promote helix unpairing. Following strand separation, DNA within the repeats may form a slipped–mispaired conformation or persist in an unwound state. Slipped-strand DNA formation was not observed in the repeats after denaturation–renaturation [6], which may be explained by the preferential nucleation of the double-stranded structure in the flanking sequences with higher G+C content. Rather than forming slipped-strand structures, several lines of evidence presented here show that, under physiological conditions, $(ATTCT)_n \cdot (AGAAT)_n$ repeats form stable unpaired DNA structures in DNA with natural, physiological levels of supercoiling.

A two-dimensional (2D) agarose gel analysis of DNA unwinding is shown in Fig. 30-1. For this, plasmid DNA was treated with a topoisomerase I-containing nuclear extract in the presence of various concentrations of ethidium bromide. This provides a set of 8 DNA topoisomer populations with superhelical density in a range of $\sigma = 0$ to -0.08. These were then mixed together to provide the distribution of topoisomers from $L-L_0 = 0$ to -19 shown in Fig. 30-1. At 37°C a topological transition occurs in supercoiled plasmids containing $(ATTCT)_{11} \cdot (AGAAT)_{11}$ cloned with the flanking sequences from the human SCA10 locus. This transition is manifested as the retardation of DNA mobility in the first dimension beginning from a topoisomer number $L-L_0 = -14$, which corresponds to a superhelical density for the transition (σ_t) of $\sigma_t = -0.05$. The further decrease in mobility of topoisomers with higher superhelical density, which reaches a plateau in the 2D gel pattern at $\sigma_t = -15$, is

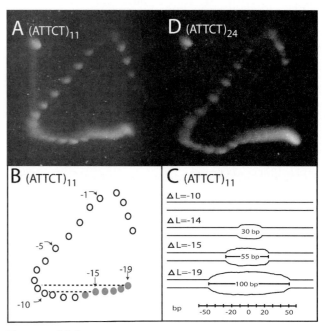

FIGURE 30-1 DNA unpairing detected by 2D agarose gel analysis. (A) 2D agarose gel with plasmid pVP5-11, containing $(ATTCT)_{11} \cdot (AGAAT)_{11}$, with 146-bp 5′ and 36-bp 3′ flanking human sequences from chromosome 22. Electrophoresis was from top to bottom in the first dimension and from left to right in the second. (B) Positions of individual topoisomers. The unpairing transition begins at topoisomer -14. Topoisomer -15 migrates as a molecule with 9.5 negative supercoils, as denoted by the dashed line, indicating unpairing of about 52 bp (10.4 bp/turn). Topoisomer -19 has unpaired about 105 bp. (C) Schematics of the strand conformations in the repeated sequence of different topoisomers of plasmid pVP5-11 (compare with 2D gel in part B). (D) 2D gel showing DNA unpairing in plasmid pVP5-29, containing $(ATTCT)_{29} \cdot (AGAAT)_{29}$.

consistent with the progressive DNA unpairing in the A+T-rich pentanucleotide region. Such a 2D gel pattern is typical of helix unpairing within DNA unwinding elements and is very different from the sharp drop in mobility that is characteristic of the all-or-none transitions seen for cruciform, Z-DNA, or intramolecular triplex DNA formation [3, 5]. Topoisomer -15 migrates as if it possesses 9.5 negative superhelical turns and, therefore, 5.5 supercoils are lost, consistent with the unpairing of 5.5 helical turns of duplex DNA, which may span the entire $(ATTCT)_{11} \cdot (AGAAT)_{11}$ sequence. At greater superhelical density, unpairing continues and reaches 105 total base pairs in topoisomer-19, which would include A+T-rich regions flanking the (ATTCT)·(AGAAT) repeats (Figs. 30-1B and 30-1C). Similar patterns were observed for plasmids containing longer lengths of repeats, including $(ATTCT)_{29} \cdot (AGAAT)_{29}$, as shown in Fig. 30-1C. The magnitude of relaxation for the plasmid with 29 repeats is greater than that observed for the plasmid with 11 repeats, indicating a longer unpaired DNA region (Fig. 30-1D).

DNA unpairing within the supercoiled plasmid was confirmed by atomic force microscopy (AFM). Supercoiled DNA samples were deposited onto aminopropyl mica at room temperature. Regions of DNA unpairing were clearly visible in plasmid DNA containing 23 and 29 repeats (Figs. 30-2A–D). For unpaired structures in the plasmid containing $(ATTCT)_{23} \cdot (AGAAT)_{23}$ (Fig. 30-2A), the length of the unwound region was determined to be 33 ± 5 nm ($n = 23$), which corresponds to 98 ± 15 bp of duplex DNA. This agrees well with the predicted length of 115 bp for complete unpairing, given that single-stranded and duplex DNA do not exhibit the same contour length in AFM [92]. At a superhelical density of $\sigma = -0.069$, visible unpaired structures were observed in 48% of the molecules counted. The DNA unpairing is temperature-dependent [5], and an even higher percentage of plasmids with unpaired structures is expected at 37°C. No such structures have ever been observed in AFM images of pUC8 or many other plasmids without these cloned ATTCT repeats [93–96]. Unpaired structures were not observed in AFM images of more than 100 molecules containing the $(ATTCT)_n \cdot (AGAAT)_n$ repeats ($n = 23, 29$) at superhelical densities less negative than $\sigma = -0.04$ (above that required for unpairing).

Additional and more critical evidence of the gradual strand separation in the repeats was provided by chemical probe analysis using chloroacetaldehyde (CAA), which reacts with unpaired cytosines and adenines. As the level of negative superhelical energy increased in consecutive topoisomer fractions, DNA

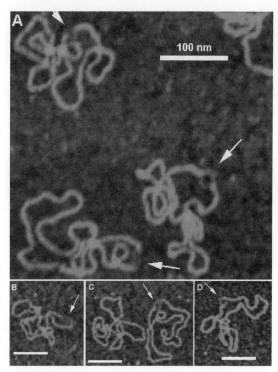

FIGURE 30-2 DNA unpairing detected by atomic force microscopy. (A, B) Tapping mode AFM images of supercoiled pEO5-23 DNA, containing $(ATTCT)_{23} \cdot (AGAAT)_{23}$. Arrows point to regions where the two complementary DNA strands are unpaired. A total of 48% of the molecules analyzed from topoisomer fractions 6 (panel A, $\sigma = -0.068$) and 7 (panel B, $\sigma = -0.075$) showed unpaired regions in which the individual strands were visible, as in panel A, or regions that appeared as discontinuities in the supercoiled duplex, as in panel B. (C, D) Unpaired structures in plasmid pVP5-29 containing $(ATTCT)_{29} \cdot (AGAAT)_{29}$ plus human flanking A+T-rich regions. (C) Topoisomer fraction number 4, $\sigma = -0.060$. (D) Topoisomer fraction number 3, $\sigma = -0.049$. See CD-ROM for color image.

unpairing was first detected in topoisomer population number 4 ($\sigma = -0.04$), where the bases at the center of $(ATTCT)_{11} \cdot (AGAAT)_{11}$ repeats were reactive with CAA (Fig. 30-3, lane 4). This is very close to the superhelical density value at which the transition on the 2D gel was observed (Figs. 30-1A and 30-1B). Upon further increase in the negative superhelical density up to $\sigma = -0.06$ in topoisomer fraction number 6, the entire repeat tract unwound and became strongly chemically reactive, mirroring the gradual unpairing observed in the 2D agarose gel. In topoisomer fractions 7 and 8 (the highest superhelical density of $\sigma = -0.08$), the chemically reactive region spanned about 106–135 bp, including flanking A+T-rich sequences. Thus, three independent experimental analyses show that $(ATTCT)_n \cdot (AGAAT)_n$ tracts form unpaired DNA structures in DNA containing physiologically relevant levels of supercoiling.

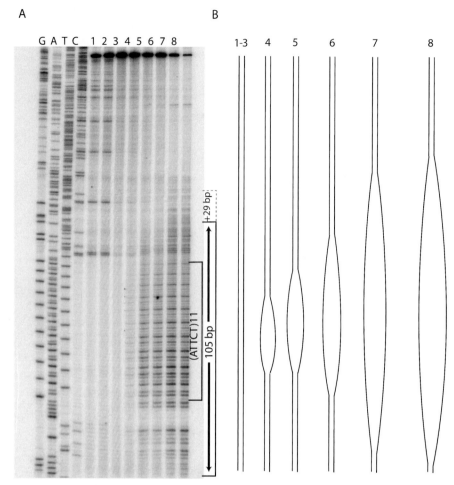

FIGURE 30-3 DNA unpairing in $(ATTCT)_{11} \cdot (AGAAT)_{11}$ tracts mapped by chemical probe analysis. (A) DNA sequence lanes are indicated by G, A, T, and C. The repeat sequence reads $(AGAAT)_{11}$ from bottom to top, from $(ATTCT)_{11}$ in the template strand. Lanes 1–8 contain individual DNA topoisomer fraction numbers 1–8, with 1 being relaxed DNA and 8 being the most negatively supercoiled. DNA topoisomers were incubated with CAA, and the sites of modification were detected by primer extension. Bands of radioactivity denote sites of chemical modification that block further DNA polymerization. Polymerization can stop one or two bases before the site of modification [121]. The region corresponding to the 11 ATTCT repeats (55 bp) is indicated, as is an extended A+T-rich region that unwinds at higher superhelical densities (105 bp). Superhelical densities of topoisomer fractions: 1, $\sigma = 0$; 2, $\sigma = -0.022$; 3, $\sigma = -0.033$; 4, $\sigma = -0.043$; 5, $\sigma = -0.054$; 6, $\sigma = -0.062$; 7, $\sigma = -0.069$; 8, $\sigma = -0.080$. (B) Schematic of unwinding for DNA topoisomer fractions 1–8.

To gain additional insight into the structural properties of unwound DNA, the repeat length dependence for DNA unpairing and alternative strand conformations were analyzed in unwound regions. In a range of repeat lengths $n = 5$ to 46, the threshold superhelical density for unpairing transition ($\sigma_t = -0.05$) was independent of the repeat length. A competing unwinding region (97 bp, 70% A+T) elsewhere in the plasmid opens up when the length of repeats is too short to support stable unpairing in the repeats ($n < 8$). Thus, DNA unpairing depends only on the thermodynamic stability of A+T-rich regions but not on their particular sequences. However, at the normal ($n = 14$) and longer repeat lengths, due to their extremely high A+T content the $(ATTCT)_n \cdot (AGAAT)_n$ regions are the first to unpair. At long repeat lengths, the fully unpaired single strands may not be a favorable DNA structure. Indeed, a "collapsed" structure forms in a region where more than 150 bp may be unpaired (29 repeats and the A+T-rich flanking sequence), as observed by AFM (Figs. 30-4A and 30-4B). The "collapsed" structure appears to be relatively loose, allowing penetration of small CAA molecules and their reaction with

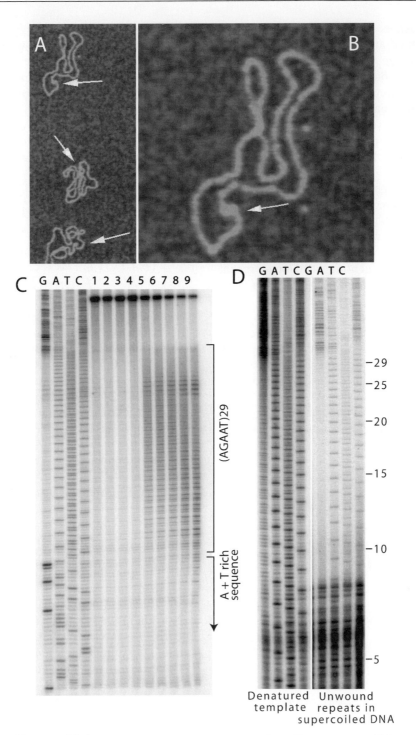

FIGURE 30-4 Long unpaired repeats form "collapsed" structures at high superhelical density. (A) AFM images of plasmid pVP5-29 containing 29 repeats under DNA unpairing conditions. The DNA sample had a superhelical density of $\sigma = -0.091$. Arrows point to the collapsed structures. (B) Zoomed image of one molecule. (C) Adenines in the (AGAAT)$_{29}$ strand are reactive with CAA. Superhelical densities of topoisomer fractions: 1, $\sigma = 0$; 2, $\sigma = -0.025$; 3, $\sigma = -0.035$; 4, $\sigma = -0.049$; 5, $\sigma = -0.060$; 6, $\sigma = -0.063$; 7, $\sigma = -0.070$; 8, $\sigma = -0.080$; 9, $\sigma = -0.091$. (D) Oligonucleotide hybridization to an unpaired repeated sequence in supercoiled plasmid allows enzymatic sequencing with Sequenase (right four sequence lanes). A DNA sequencing reaction obtained by oligonucleotide hybridization with a denatured template is shown in the left four lanes. See CD-ROM for color image.

adenines and cytosines (Fig. 30-4C). Moreover, oligonucleotides, which are much larger than CAA molecules, can hybridize with their complementary sequences in unpaired but "collapsed" regions and serve as primers for subsequent DNA polymerization (Fig. 30-4D). The nucleobase accessibility in the "collapsed" structure indicates the absence of stable alternative hydrogen-bonded structures in unpaired strands, which are probably loosely wrapped around each other.

In summary, several experimental techniques provide evidence of stable supercoil-induced DNA unpairing in the (ATTCT)·(AGAAT) repeats. Although DNA unpairing may occur in any A+T-rich sequence, due to the extremely high A+T content $(ATTCT)_n \cdot (AGAAT)_n$ regions preferentially unpair at normal and longer repeat lengths ($n > 14$). It appears that, at short lengths or moderate superhelical densities in longer repeat tracts, unpaired regions have well-separated strands that are visible in AFM as denatured bubbles. At high superhelical densities, long unpaired repeat tracts "collapse" and probably form structures with loosely intertwined strands. Unpaired single strands are accessible for normal interactions with small CAA and larger oligonucleotide molecules. It is feasible that they are also accessible for protein binding by helicase, primase, and DNA polymerase.

III. COMPLEX REPLICATION-BASED INSTABILITY OF SCA10 $(ATTCT)_n \cdot (AGAAT)_n$ REPEATS IN *ESCHERICHIA COLI*

A. SCA10 Repeats Do Not Undergo Deletion at a High Rate in *Escherichia coli*

To measure rates of SCA10 $(ATTCT)_n \cdot (AGAAT)_n$ repeat deletion in *E. coli*, a genetic assay that relies on the restoration of CAT gene activity to a Cm^r phenotype was utilized [97]. This assay can measure the deletion of long (CTG)·(CAG) repeats to tract lengths 24 or shorter, as well as detect complete deletions of repeats cloned into the CAT gene [97]. In the case of (ATTCT)·(AGAAT) repeats, the genetic system measures deletion to a single repeat plus an additional base pair [98].

Under chloramphenicol selection, the Cm^r reversion rates for plasmids pVHpenta23R and pVHpenta24, containing 23 and 24 repeats in opposite orientations [i.e., $(AGAAT)_{23} \cdot (ATTCT)_{23}$ and $(ATTCT)_{24} \cdot (AGAAT)_{24}$], were very low in the wild-type *E. coli* strain MG1655 in log phase growth. The rates were 3.9×10^{-9} and 7.1×10^{-9} revertants per cell per generation, respectively. Similar rates were also observed in two other *recA* strains, JTT1 and RS2 [98]. Revertants were analyzed by PCR and DNA sequencing, and 80–90% were deletions of all but one repeat plus an additional base pair. This contrasts with rates of deletion of (CAG)·(CTG) repeats, which typically range from 10^{-2} to 10^{-4} [97, 99–101].

B. Complex Expansion Mutations Occur in $(ATTCT)_n \cdot (AGAAT)_n$ Repeats in *Escherichia coli*

Using a biochemical assay for repeat instability [102–104], pVHpenta24 and pVHpenta23R were grown and recultivated for up to 15 days in *recA* derivatives of *E. coli* strains JTT1 and RS2 in the absence of selective pressure from chloramphenicol. The plasmid DNA was then purified and the repeat tract characterized. In 50–80% of all independent recultivation assays, expansion of the repeat tract was detected as a larger *Pvu*II–*Nco*I fragment [98]. The increase in length was consistent with an increase from 23 or 24 (ATTCT)·(AGAAT) repeats to about 46 (ATTCT)·(AGAAT) repeats. In some cell populations, deletions were also observed, but they occurred in conjunction with the expansion. This behavior is in contrast to that for (CAG)·(CTG) repeats, in which deletions rapidly occur and take over the population within several recultivations [103, 105, 106].

To understand the molecular basis for the complex expansion mutation, the plasmids were characterized and the repeat tracts were sequenced. Figure 30-5 summarizes the characteristics of the expansions. (1) The repeat tract increased in length from 23 or 24 repeats to as many as 46 repeats. (2) More than 96% of expansion mutations contained an inversion of the repeat, resulting in the sequence $5'$-$(TATTC)_{5-11}(GAATA)_{9-35}$-$3'$. (3) All plasmids were head-to-tail dimers. (4) The expanded repeat product always contained $5'$-$(TATTC)_n(GAATA)_n$-$3'$. This occurred for either orientation of the repeat: $(ATTCT)_n \cdot (AGAAT)_n$ or $(AGAAT)_n \cdot (ATTCT)_n$. (5) The DNA sequence contained an A·T to G·C transition at the $3'$ end within the $(AGAAT)_n$ repeat, which resulted in the formation of a longer flanking direct repeat.

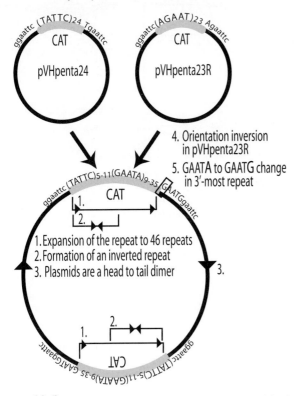

FIGURE 30-5 Complex mutational events associated with the expansion of $(ATTCT)_n \cdot (AGAAT)_n$ repeats. The initial pentanucleotide repeat sequences $(AGAAT)_{23} \cdot (ATTCT)_{23}$Agaattcc or $(ATTCT)_{24} \cdot (AGAAT)_{24}$Tgaattcc underwent spontaneous complex expansion mutations to $(TATTC)_{5-11}$ $(GAATA)_{9-35}$Ggaattcc. This figure summarizes the events associated with these complex mutations.

IV. MODELS FOR COMPLEX EXPANSION MUTATIONS IN *ESCHERICHIA COLI* AND EXPANSION AND INSTABILITY IN HUMANS

A. Mechanisms Leading to the Formation of the Complex Expansion Mutation in *Escherichia coli*

The formation of an inverted repeat region $5'-(TATTC)_{5-11}(GAATA)_{9-35}-3'$ must occur by an intramolecular strand switch when $(AGAAT)_{24}$ comprises the lagging template strand or by an intermolecular strand switch when $(AGAAT)_{23}$ comprises the leading template strand. This is required as the formation of a transient intramolecular hairpin in the pyrimidine-rich strand or a leading nascent to lagging template transient misalignment of the pyrimidine-rich strand must occur to produce the observed inverted repeat [98]. DNA synthesis following misalignment results in the sequence inversion. Hairpin, or intermolecular strand switch, templated synthesis of the purine-rich strand could not produce this sequence. This is basically a quasi-palindrome correction-type mutation, which can occur by a strand switch in either the leading or lagging strands in different mutational systems [107–111]. In the case of the SCA10 repeat, this type of mutation happens in both strands for opposite orientations of the insert, a situation not previously observed.

The model shown in Fig. 30-6 explains the main features of the most complex mutation associated with $(AGAAT)_{23} \cdot (ATTCT)_{23}$, and it is similar to models proposed by Lovett and colleagues for RecA-independent deletion and expansion events coupled with plasmid dimerization [112–114]. In this orientation, the repeat tract is inverted at the 5' end of the repeat (from 5'-AGAAT-3' to 5'-ATTCT-3'), concomitant with plasmid dimerization, repeat tract expansion, and formation of the inverted repeat. To explain this, leading strand synthesis of the pyrimidine-rich nascent strand must occur for 9–35 repeats before the intermolecular strand switch happens. Primer–template misalignment during replication is necessary to produce repeat tracts longer than 23 copies. Then, the primer–template must

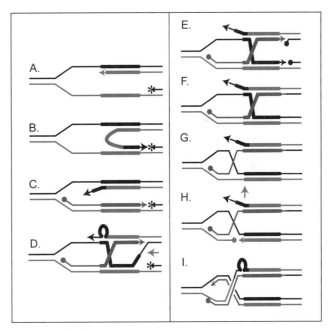

FIGURE 30-6 Model of the complex expansion mutation with inversion of the pentanucleotide repeat and plasmid dimerization. See text for details. Arrowheads at the ends of lines represent the 3' end of a nascent DNA strand. The asterisks at the ends of lines represent a 5' end of a DNA strand. The separate arrows at structures D and G denote sites of nicking. See CD-ROM for color image.

unpair, the strand switch must occur, and replication must continue on the lagging template strand to result in the inversion of the repeat (Fig. 30-6B). The nascent strand must then unpair from the lagging template strand (Fig. 30-6C). If synthesis proceeds through the repeat tract and into the flanking *Eco*RI direct repeat (used for cloning the repeats), this may facilitate realignment of the leading nascent strand to the leading template (Fig. 30-6D). At some point then, a strand exchange must occur (Fig. 30-6D) and a nick is then introduced (indicated by the arrow), which leads to a Holliday junction (Figs. 30-6E, 30-6F, and 30-6G). Following branch migration and the introduction of a second nick (Fig. 30-6H), the lagging template is ligated to the leading nascent strand (Fig. 30-6I). Continued synthesis from the 3' end restores a crossover replication fork that leads to plasmid dimerization (Fig. 30-6I).

The genetic instabilities of the SCA10 A+T-rich pentanucleotide repeat in *E. coli* are unlike those for $(CTG)_n \cdot (CAG)_n$ or $(CGG)_n \cdot (CCG)_n$ repeats associated with myotonic dystrophy (and many other neurodegenerative diseases) and fragile X syndrome, respectively. $(CTG)_n \cdot (CAG)_n$, $(CGG)_n \cdot (CCG)_n$, and $(GAA)_n \cdot (TTC)_n$ repeats can form DNA secondary structures that present strong blocks to DNA replication. In addition, they are deleted at high rates in *E. coli* in a length-dependent fashion. $(ATTCT)_n \cdot (AGAAT)_n$ repeats show very different properties. They do not form a DNA secondary structure that can block replication [6], and they undergo deletion at a very low rate, as presented previously. Repeat lengths greater than $(ATTCT)_{18} \cdot (AGAAT)_{18}$ undergo the complex expansion mutations in *E. coli* discussed earlier. The instabilities are likely due to the formation of stable unpaired DNA structures *in vivo*. Unpairing may facilitate or promote primer–template misalignment during replication, including the inter- and intramolecular strand switch events that can cause the sequence inversion forming the inverted repeat.

B. Amplification Leading to Repeat Expansion in Human Cells

It was shown that plasmid-containing $(ATTCT)_n \cdot (AGAAT)_n$ repeats supported initiation and replication in HeLa cell extracts without the addition of a specific initiation protein [6]. In cells, $(ATTCT)_n \cdot (AGAAT)_n$ repeats may unwind and support repetitive initiation of DNA replication, resulting in amplification of the repeat tract. Aberrant replication initiation could also be responsible for the instability observed in somatic cells. Unrestrained superhelical tension measured at

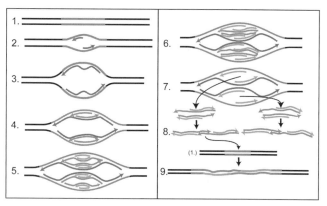

FIGURE 30-7 Model for repeat expansion based on aberrant replication origin activity. Step 1: Duplex DNA with SCA10 repeat shown in green. Step 2: The A+T-rich $(ATTCT)_n \cdot (AGAAT)_n$ repeat unwinds and replication initiates within the repeat. Nascent strands are shown in red. Step 3: DNA unpairing within the $(ATTCT)_n \cdot (AGAAT)_n$ repeats then reoccurs. Step 4: Replication reinitiates at the unpaired regions. The second nascent strand is shown in blue. Step 5: Onion skin replication continues by repeated initiation within the $(ATTCT)_n \cdot (AGAAT)_n$ repeat, leading to an 8-fold amplification. Step 6: Fork movement ceases or slows at the first and second forks while replication continues from the third fork (orange nascent strand). Step 7: Continued replication will lead to a displacement of four DNA molecules from the original DNA molecule. Step 8: Pairs of these molecules, from opposite template strands, possess complementary single-strand ends that can hybridize, ultimately leading to the formation of blunt-ended molecules. These can then be joined into very long strands. Step 9: The homologous recombination or non-homologous end-joining activities of mammalian cells can introduce the pure expanded repeat tracts back into the original SCA10 repeat tract, leading to massive expansion. See CD-ROM for color image.

active genes in living cells is sufficient to support DNA unwinding [8, 115, 116]. If $(ATTCT)_n \cdot (AGAAT)_n$ acts as a replication origin, fractious DNA replication may lead to repeat amplification and expansion, as shown in Fig. 30-7 and as discussed in Section IB.

V. CONCLUDING REMARKS

The complex expansion mutations seen in *E. coli* may not explain the specific etiology of the massive expansion associated with SCA10 or the high rate of expansion and contraction in SCA10 cells [117–120]. Partial sequences of expanded $(ATTCT)_n \cdot (AGAAT)_n$ repeats in patients have not detected the specific 5'-$(TATTC)_n(GAATA)_n$-3' inverted repeat formed in *E. coli* (T. Ashizawa, personal communication). However, the expansion in *E. coli* provides insight into mechanisms that may be responsible for repeat expansion in human cells. $(CAG) \cdot (CTG)$ and $(CGG) \cdot (CCG)$ repeats can block DNA replication, and these repeats undergo rapid deletion in *E. coli*. $(ATTCT)_n \cdot (AGAAT)_n$ repeats do not form a structure that can block replication, and they expand in a complex mutation in *E. coli*. Thus, a block to replication cannot be the only causative factor associated with repeat expansion or deletion in human cells, and models invoking replication blockage cannot be valid for all repeats. Potaman *et al.* suggested models, one of which is shown in Fig. 30-7, in which aberrant replication at DNA repeats is associated with repeat deletion and expansion. The model in Fig. 30-7 is supported by the complex expansion mutations observed in *E. coli*. In *E. coli*, the instability is associated with strand switching that can happen in both the leading and lagging strands during replication. Human genetic instabilities associated with other disease-associated triplet or tetranucleotide repeats may also result from aberrant replication. Various DNA repeats with assorted symmetry elements, which can form a variety of DNA secondary structures, are often associated with human replication origins [12, 13]. Thus, aberrant replication initiation, rather than replication blockage, may be a critical factor in repeat expansion.

Acknowledgments

This work was supported by NIH Grants ES05508 (RRS), GM53819 (ML), and DK061458 (JJB).

References

1. Sheflin, L. G., and Kowalski, D. (1984). Mung bean nuclease cleavage of dA+dT-rich sequence or an inverted repeat sequence in supercoiled PM2 DNA depends on ionic environment. *Nucleic Acids Res.* **12**, 7087–7104.
2. Sheflin, L. G., and Kowalski, D. (1985). Altered DNA conformations detected by mung bean nuclease occur in promoter and terminator regions of supercoiled pBR322 DNA. *Nucleic Acids Res.* **13**, 6137–6155.
3. Sinden, R. R. (1994). "DNA Structure and Function." Academic Press, San Diego.
4. Umek, R. M., and Kowalski, D. (1987). Yeast regulatory sequences preferentially adopt a non-B conformation in supercoiled DNA. *Nucleic Acids Res.* **15**, 4467–4480.
5. Kowalski, D., Natale, D. A., and Eddy, M. J. (1988). Stable DNA unwinding, not "breathing," accounts for single-strand-specific nuclease hypersensitivity of specific A+T-rich sequences. *Proc. Natl. Acad. Sci. USA* **85**, 9464–9468.
6. Potaman, V. N., Bissler, J. J., Hashem, V. I., Oussatcheva, E. A., Lu, L., Shlyakhtenko, L. S., Lyubchenko, Y. L., Matsuura, T., Ashizawa, T., Leffak, M., Benham, C. J., and Sinden, R. R. (2003). Unpaired structures in SCA10 (ATTCT)$_n$·(AGAAT)$_n$ repeats. *J. Mol. Biol.* **326**, 1095–1111.
7. Liu, L. F., and Wang, J. C. (1987). Supercoiling of the DNA template during transcription. *Proc. Natl. Acad. Sci. USA* **84**, 7024–7027.
8. Kramer, P. R., Fragoso, G., Pennie, W., Htun, H., Hager, G. L., and Sinden, R. R. (1999). Transcriptional state of the mouse mammary tumor virus promoter can affect topological domain size *in vivo*. *J. Biol. Chem.* **274**, 28590–28597.
9. Lin, S., and Kowalski, D. (1997). Functional equivalency and diversity of cis-acting elements among yeast replication origins. *Mol. Cell. Biol.* **17**, 5473–5484.
10. Miller, C. A., Umek, R. M., and Kowalski, D. (1999). The inefficient replication origin from yeast ribosomal DNA is naturally impaired in the ARS consensus sequence and in DNA unwinding. *Nucleic Acids Res.* **27**, 3921–3930.
11. Berberich, S., Trivedi, A., Daniel, D. C., Johnson, E. M., and Leffak, M. (1995). In vitro replication of plasmids containing human c-myc DNA. *J. Mol. Biol.* **245**, 92–109.
12. Pearson, C. E., Zorbas, H., Price, G. B., and Zannis-Hadjopoulos, M. (1996). Inverted repeats, stem-loops, and cruciforms: Significance for initiation of DNA replication. *J. Cell Biochem.* **63**, 1–22.
13. Boulikas, T. (1996). Common structural features of replication origins in all life forms. *J. Cell. Biochem.* **60**, 297–316.
14. Bogan, J. A., Natale, D. A., and DePamphilis, M. L. (2000). Initiation of eukaryotic DNA replication: Conservative or liberal? *J. Cell. Physiol.* **184**, 139–150.
15. Kornberg, A., and Baker, T. A. (1992). "DNA Replication." W. H. Freeman & Co., New York.
16. Leonard, A. C., and Grimwede, J. E. (2005). Building a bacterial orisome: Emergence of new regulatory features for replication unwinding. *Mol. Microbiol.* **55**, 978–985.
17. Mackiewicz, P., Zakrzewska-Czerwinska, J., Zawilak, A., Dudek, M. R., and Cebrat, S. (2004). Where does bacterial replication start? Rules for predicting the oriC region. *Nucleic Acids Res.* **32**, 3781–3791.
18. Zawilak-Pawlik, A., Kois, A., Majka, J., Jakimowicz, D., Smulczyk-Krawczyszyn, A., Messer, W., and Zakrzewska-Czerwinska, J. (2005). Architecture of bacterial replication initiation complexes: Orisomes from four unrelated bacteria. *Biochem. J.* **389**, 471–481.
19. Aladjem, M. I., Groudine, M., Brody, L. L., Dieken, E. S., Fournier, R. E., Wahl, G. M., and Epner, E. M. (1995). Participation of the human beta-globin locus control region in initiation of DNA replication. *Science* **270**, 815–819.
20. Little, R. D., Platt, T. H., and Schildkraut, C. L. (1993). Initiation and termination of DNA replication in human rRNA genes. *Mol. Cell. Biol.* **13**, 6600–6613.
21. Spradling, A. C., de Cicco, D. V., Wakimoto, B. T., Levine, J. F., Kalfayan, L. J., and Cooley, L. (1987). Amplification of the X-linked Drosophila chorion gene cluster requires a region upstream from the s38 chorion gene. *EMBO J.* **6**, 1045–1053.
22. Dijkwel, P. A., Vaughn, J. P., and Hamlin, J. L. (1991). Mapping of replication initiation sites in mammalian genomes by two-dimensional gel analysis: Stabilization and enrichment of replication intermediates by isolation on the nuclear matrix. *Mol. Cell. Biol.* **11**, 3850–3859.
23. Liu, G., Malott, M., and Leffak, M. (2003). Multiple functional elements comprise a mammalian chromosomal replicator. *Mol. Cell. Biol.* **23**, 1832–1842.
24. Paixao, S., Colaluca, I. N., Cubells, M., Peverali, F. A., Destro, A., Giadrossi, S., Giacca, M., Falaschi, A., Riva, S., and Biamonti, G. (2004). Modular structure of the human lamin B2 replicator. *Mol. Cell. Biol.* **24**, 2958–2967.
25. Gilbert, D. M. (2004). In search of the holy replicator. *Nat. Rev. Mol. Cell Biol.* **5**, 848–855.
26. Gilbert, D. M. (2001). Making sense of eukaryotic DNA replication origins. *Science* **294**, 96–100.
27. Dobbs, D. L., Shaiu, W. L., and Benbow, R. M. (1994). Modular sequence elements associated with origin regions in eukaryotic chromosomal DNA. *Nucleic Acids Res.* **22**, 2479–2489.
28. Benbow, R. M., Zhao, J., and Larson, D. D. (1992). On the nature of origins of DNA replication in eukaryotes. *Bioessays* **14**, 661–670.
29. DePamphilis, M. L. (2003). Eukaryotic DNA replication origins: Reconciling disparate data. *Cell* **114**, 274–275.

30. DePamphilis, M. L. (1998). Initiation of DNA replication in eukaryotic chromosomes. *J. Cell Biochem. Suppl.* **30–31**, 8–17.
31. Casper, J. M., Kemp, M. G., Ghosh, M., Randall, G. M., Vaillant, A., and Leffak, M. (2005). The c-myc DNA unwinding element binding protein modulates the assembly of DNA replication complexes in vitro. *J. Biol. Chem.* **280**, 13071–13083.
32. Kamath, S., and Leffak, M. (2001). Multiple sites of replication initiation in the human beta-globin gene locus. *Nucleic Acids Res.* **29**, 809–817.
33. Tao, L., Dong, Z., Leffak, M., Zannis-Hadjopoulos, M., and Price, G. (2000). Major DNA replication initiation sites in the c-myc locus in human cells. *J. Cell. Biochem.* **78**, 442–457.
34. Malott, M., and Leffak, M. (1999). Activity of the c-myc replicator at an ectopic chromosomal location. *Mol. Cell. Biol.* **19**, 5685–5695.
35. Trivedi, A., Waltz, S. E., Kamath, S., and Leffak, M. (1998). Multiple initiations in the c-myc replication origin independent of chromosomal location. *DNA Cell Biol.* **17**, 885–896.
36. Waltz, S. E., Trivedi, A. A., and Leffak, M. (1996). DNA replication initiates non-randomly at multiple sites near the c-myc gene in HeLa cells. *Nucleic Acids Res.* **24**, 1887–1894.
37. Wang, L., Lin, C. M., Brooks, S., Cimbora, D., Groudine, M., and Aladjem, M. I. (2004). The human beta-globin replication initiation region consists of two modular independent replicators. *Mol. Cell. Biol.* **24**, 3373–3386.
38. McWhinney, C., and Leffak, M. (1988). Episomal persistence of a plasmid containing human c-myc DNA. *Cancer Cells*, 467–471.
39. McWhinney, C., and Leffak, M. (1990). Autonomous replication of a DNA fragment containing the chromosomal replication origin of the human c-myc gene. *Nucleic Acids Res.* **18**, 1233–1242.
40. McWhinney, C., Waltz, S. E., and Leffak, M. (1995). Cis-acting effects of sequences within 2.4-kb upstream of the human c-myc gene on autonomous plasmid replication in HeLa cells. *DNA Cell Biol.* **14**, 565–579.
41. Marcu, K. B., Bossone, S. A., and Patel, A. J. (1992). myc function and regulation. *Annu. Rev. Biochem.* **61**, 809–860.
42. Kumar, S., and Leffak, M. (1991). Conserved chromatin structure in c-myc 5' flanking DNA after viral transduction. *J. Mol. Biol.* **222**, 45–57.
43. Kumar, S., and Leffak, M. (1989). DNA topology of the ordered chromatin domain 5' to the human c-myc gene. *Nucleic Acids Res.* **17**, 2819–2833.
44. Michelotti, G. A., Michelotti, E. F., Pullner, A., Duncan, R. C., Eick, D., and Levens, D. (1996). Multiple single-stranded cis elements are associated with activated chromatin of the human c-myc gene in vivo. *Mol. Cell. Biol.* **16**, 2656–2669.
45. Ghosh, M., Liu, G., Randall, G., Bevington, J., and Leffak, M. (2004). Transcription factor binding and induced transcription alter chromosomal c-myc replicator activity. *Mol. Cell. Biol.* **24**, 10193–10207.
46. Kouzine, F., Liu, J., Sanford, S., Chung, H. J., and Levens, D. (2004). The dynamic response of upstream DNA to transcription-generated torsional stress. *Nat. Struct. Mol. Biol.* **11**, 1092–1100.
47. He, L., Liu, J., Collins, I., Sanford, S., O'Connell, B., Benham, C. J., and Levens, D. (2000). Loss of FBP function arrests cellular proliferation and extinguishes c-myc expression. *EMBO J.* **19**, 1034–1044.
48. Duncan, R., Bazar, L., Michelotti, G., Tomonaga, T., Krutzsch, H., Avigan, M., and Levens, D. (1994). A sequence-specific, single-strand binding protein activates the far upstream element of c-myc and defines a new DNA-binding motif. *Genes Dev.* **8**, 465–480.
49. Liu, J., Akoulitchev, S., Weber, A., Ge, H., Chuikov, S., Libutti, D., Wang, X. W., Conaway, J. W., Harris, C. C., Conaway, R. C., Reinberg, D., and Levens, D. (2001). Defective interplay of activators and repressors with TFIH in xeroderma pigmentosum. *Cell* **104**, 353–363.
50. Liu, J., He, L., Collins, I., Ge, H., Libutti, D., Li, J., Egly, J. M., and Levens, D. (2000). The FBP interacting repressor targets TFIIH to inhibit activated transcription. *Mol. Cell* **5**, 331–341.
51. Remus, D., Beall, E. L., and Botchan, M. R. (2004). DNA topology, not DNA sequence, is a critical determinant for Drosophila ORC-DNA binding. *EMBO J.* **23**, 897–907.
52. Jenke, A. C., Stehle, I. M., Herrmann, F., Eisenberger, T., Baiker, A., Bode, J., Fackelmayer, F. O., and Lipps, H. J. (2004). Nuclear scaffold/matrix attached region modules linked to a transcription unit are sufficient for replication and maintenance of a mammalian episome. *Proc. Natl. Acad. Sci. USA* **101**, 11322–11327.
53. MacAlpine, D. M., Rodriguez, H. K., and Bell, S. P. (2004). Coordination of replication and transcription along a Drosophila chromosome. *Genes Dev.* **18**, 3094–3105.
54. Nieduszynski, C. A., Blow, J. J., and Donaldson, A. D. (2005). The requirement of yeast replication origins for pre-replication complex proteins is modulated by transcription. *Nucleic Acids Res.* **33**, 2410–2420.
55. Danis, E., Brodolin, K., Menut, S., Maiorano, D., Girard-Reydet, C., and Mechali, M. (2004). Specification of a DNA replication origin by a transcription complex. *Nat. Cell Biol* **6**, 721–730.
56. Bell, S. P., and Dutta, A. (2002). DNA replication in eukaryotic cells. *Annu. Rev. Biochem.* **71**, 333–374.
57. Aparicio, O. M., Weinstein, D. M., and Bell, S. P. (1997). Components and dynamics of DNA replication complexes in S. cerevisiae: Redistribution of MCM proteins and Cdc45p during S phase. *Cell* **91**, 59–69.
58. Bell, S. P., and Stillman, B. (1992). ATP-dependent recognition of eukaryotic origins of DNA replication by a multiprotein complex. *Nature* **357**, 128–134.
59. Diffley, J. F. (1994). Eukaryotic DNA replication. *Curr. Opin. Cell Biol.* **6**, 368–372.
60. Walter, J., and Newport, J. (2000). Initiation of eukaryotic DNA replication: Origin unwinding and sequential chromatin association of Cdc45, RPA, and DNA polymerase alpha. *Mol. Cell* **5**, 617–627.
61. Eward, K. L., Obermann, E. C., Shreeram, S., Loddo, M., Fanshawe, T., Williams, C., Jung, H. I., Prevost, A. T., Blow, J. J., Stoeber, K., and Williams, G. H. (2004). DNA replication licensing in somatic and germ cells. *J. Cell Sci.* **117**, 5875–5886.
62. Van Hatten, R. A., Tutter, A. V., Holway, A. H., Khederian, A. M., Walter, J. C., and Michael, W. M. (2002). The Xenopus Xmus101 protein is required for the recruitment of Cdc45 to origins of DNA replication. *J. Cell Biol.* **159**, 541–547.
63. Hashimoto, Y., and Takisawa, H. (2003). Xenopus Cut5 is essential for a CDK-dependent process in the initiation of DNA replication. *EMBO J.* **22**, 2526–2535.
64. Lei, M., and Tye, B. K. (2001). Initiating DNA synthesis: From recruiting to activating the MCM complex. *J. Cell Sci.* **114**, 1447–1454.
65. Nishitani, H., and Lygerou, Z. (2002). Control of DNA replication licensing in a cell cycle. *Genes Cells* **7**, 523–534.
66. Weinreich, M., Liang, C., Chen, H. H., and Stillman, B. (2001). Binding of cyclin-dependent kinases to ORC and Cdc6p regulates the chromosome replication cycle. *Proc. Natl. Acad. Sci. USA* **98**, 11211–11217.
67. Sclafani, R. A. (2000). Cdc7p-Dbf4p becomes famous in the cell cycle. *J. Cell Sci.* **113**, 2111–2117.
68. Lee, J. H., Xu, B., Lee, C. H., Ahn, J. Y., Song, M. S., Lee, H., Canman, C. E., Lee, J. S., Kastan, M. B., and Lim, D. S. (2003). Distinct functions of Nijmegen breakage syndrome in ataxia telangiectasia mutated-dependent responses to DNA damage. *Mol. Cancer Res.* **1**, 674–681.
69. Nguyen, V. Q., Co, C., and Li, J. J. (2001). Cyclin-dependent kinases prevent DNA re-replication through multiple mechanisms. *Nature* **411**, 1068–1073.

70. Spradling, A. C. (1999). ORC binding, gene amplification, and the nature of metazoan replication origins. *Genes Dev.* **13**, 2619–2623.
71. Calvi, B. R., and Spradling, A. C. (1999). Chorion gene amplification in Drosophila: A model for metazoan origins of DNA replication and S-phase control. *Methods* **18**, 407–417.
72. Liang, C., Spitzer, J. D., Smith, H. S., and Gerbi, S. A. (1993). Replication initiates at a confined region during DNA amplification in Sciara DNA puff II/9A. *Genes Dev.* **7**, 1072–1084.
73. Tower, J. (2004). Developmental gene amplification and origin regulation. *Annu. Rev. Genet.* **38**, 273–304.
74. Schimke, R. T. (1988). Gene amplification in cultured cells. *J. Biol. Chem.* **263**, 5989–5992.
75. Baran, N., Neer, A., and Manor, H. (1983). "Onion skin" replication of integrated polyoma virus DNA and flanking sequences in polyoma-transformed rat cells: Termination within a specific cellular DNA segment. *Proc. Natl. Acad. Sci. USA* **80**, 105–109.
76. Stark, G. R., Debatisse, M., Giulotto, E., and Wahl, G. M. (1989). Recent progress in understanding mechanisms of mammalian DNA amplification. *Cell* **57**, 901–908.
77. Schimke, R. T. (1992). Gene amplification: What are we learning? *Mutat. Res.* **276**, 145–149.
78. Syu, L. J., and Fluck, M. M. (1997). Site-specific in situ amplification of the integrated polyomavirus genome: A case for a context-specific over-replication model of gene amplification. *J. Mol. Biol.* **271**, 76–99.
79. Singer, M. J., Mesner, L. D., Friedman, C. L., Trask, B. J., and Hamlin, J. L. (2000). Amplification of the human dihydrofolate reductase gene via double minutes is initiated by chromosome breaks. *Proc. Natl. Acad. Sci. USA* **97**, 7921–7926.
80. Baran, N., Lapidot, A., and Manor, H. (1987). Unusual sequence element found at the end of an amplicon. *Mol. Cell. Biol.* **7**, 2636–2640.
81. Kirschner, L. S. (1996). De novo generation of simple sequence during gene amplification. *Nucleic Acids Res.* **24**, 2829–2834.
82. Hellman, A., Zlotorynski, E., Scherer, S., Cheung, J., Vincent, J., Smith, D. I., Trakhtenbrot, L., and Kerem, B. (2002). A role for common fragile site induction in amplification of human oncogenes. *Cancer Cell* **1**, 89–97.
83. Patrick, S. M., and Turchi, J. J. (1999). Replication protein A (RPA) binding to duplex cisplatin-damaged DNA is mediated through the generation of single-stranded DNA. *J. Biol. Chem.* **274**, 14972–14978.
84. Wakasugi, M., and Sancar, A. (1999). Order of assembly of human DNA repair excision nuclease. *J. Biol. Chem.* **274**, 18759–18768.
85. Volker, M., Mone, M. J., Karmakar, P., Van Hoffen, A., Schul, W., Vermeulen, W., Hoeijmakers, J. H., van Driel, R., Van Zeeland, A. A., and Mullenders, L. H. (2001). Sequential assembly of the nucleotide excision repair factors in vivo. *Mol. Cell* **8**, 213–224.
86. Pearson, C. E., Ewel, A., Acharya, S., Fishel, R. A., and Sinden, R. R. (1997). Human MSH2 binds to trinucleotide repeat DNA structures associated with neurodegenerative diseases. *Hum. Mol. Genet.* **6**, 1117–1123.
87. Cromie, G. A., Connelly, J. C., and Leach, D. R. (2001). Recombination at double-strand breaks and DNA ends: Conserved mechanisms from phage to humans. *Mol. Cell* **8**, 1163–1174.
88. Fuentes-Mascorro, G., Serrano, H., and Rosado, A. (2000). Sperm chromatin. *Arch. Androl.* **45**, 215–225.
89. Santos, F., Peters, A. H., Otte, A. P., Reik, W., and Dean, W. (2005). Dynamic chromatin modifications characterise the first cell cycle in mouse embryos. *Dev. Biol.* **280**, 225–236.
90. Collins, K. L., and Kelly, T. J. (1991). Effects of T antigen and replication protein A on the initiation of DNA synthesis by DNA polymerase alpha-primase. *Mol. Cell. Biol.* **11**, 2108–2115.
91. Leblanc, B. P., Benham, C. J., and Clark, D. J. (2000). An initiation element in the yeast CUP1 promoter is recognized by RNA polymerase II in the absence of TATA box-binding protein if the DNA is negatively supercoiled. *Proc. Natl. Acad. Sci. USA* **97**, 10745–10750.
92. Rivetti, C., Walker, C., and Bustamante, C. (1998). Polymer chain statistics and conformational analysis of DNA molecules with bends or sections of different flexibility. *J. Mol. Biol.* **280**, 41–59.
93. Lushnikov, A. Y., Brown, B. A., Oussatcheva, E. A., Potaman, V. N., Sinden, R. R., and Lyubchenko, Y. L. (2004). Interaction of the Z alpha domain of human ADAR1 with a negatively supercoiled plasmid visualized by atomic force microscopy. *Nucleic Acids Res.* **32**, 4704–4712.
94. Potaman, V. N., Oussatcheva, E. A., Lyubchenko, Y. L., Shlyakhtenko, L. S., Bidichandani, S. I., Ashizawa, T., and Sinden, R. R. (2004). Length-dependent structure formation in Friedreich ataxia (GAA)$_n$·(TTC)$_n$ repeats at neutral pH. *Nucleic Acids Res.* **32**, 1224–1231.
95. Tiner, Sr. W. J., Potaman, V. N., Sinden, R. R., and Lyubchenko, Y. L. (2001). The structure of intramolecular DNA triplex: Atomic force miroscopy study. *J. Mol. Biol.* **314**, 353–357.
96. Shlyakhtenko, L. S., Potaman, V. N., Sinden, R. R., and Lyubchenko, Y. L. (1998). Structure and dynamics of supercoil-stabilized DNA cruciforms. *J. Mol. Biol.* **280**, 61–72.
97. Hashem, V. I., Rosche, W. A., and Sinden, R. R. (2002). Genetic assays for measuring rates of (CAG)·(CTG) repeat instability in *Escherichia coli*. *Mutat. Res.* **502**, 25–37.
98. Hashem, V. I., Klysik, E. A., Edwards, S. F., and Sinden, R. R. (2005). Complex mutations associated with (ATTCT)$_n$·(AGAAT)$_n$ repeats in *Escherichia coli*. Submitted for publication.
99. Hashem, V. I., Klysik, E. A., Rosche, W. A., and Sinden, R. R. (2002). Instability of repeated DNAs during transformation in *Escherichia coli*. *Mutat. Res.* **502**, 39–46.
100. Hashem, V. I., and Sinden, R. R. (2002). Chemotherapeutically induced deletion of expanded triplet repeats. *Mutat. Res.* **508**, 107–119.
101. Hashem, V. I., Rosche, W. A., and Sinden, R. R. (2004). Genetic recombination destabilizes (CTG)$_n$(CAG)$_n$ repeats in E. coli. *Mutat. Res.* **554**, 95–109.
102. Kang, S., Jaworski, A., Ohshima, K., and Wells, R. D. (1995). Expansion and deletion of CTG triplet repeats from human disease genes are determined by the direction of replication in *E. coli*. *Nat. Genet.* **10**, 213–218.
103. Jaworski, A., Rosche, W. A., Gellibolian, R., Kang, S., Shimizu, M., Bowater, R. P., Sinden, R. R., and Wells, R. D. (1995). Mismatch repair in *Escherichia coli* enhances instability of (CTG)$_n$ repeats from human hereditary diseases. *Proc. Natl. Acad. Sci. USA* **92**, 11019–11023.
104. Shimizu, M., Gellibolian, R., Oostra, B. A., and Wells, R. D. (1996). Cloning, characterization and properties of plasmids containing CGG triplet repeats from the FMR-1 gene. *J. Mol. Biol.* **258**, 614–626.
105. Rosche, W. A., Jaworski, A., Kang, S., Kramer, S. F., Larson, J. E., Geidroc, D. P., Wells, R. D., and Sinden, R. R. (1996). Single-stranded DNA-binding protein enhances the stability of CTG triplet repeats in *Escherichia coli*. *J. Bacteriol.* **178**, 5042–5044.
106. Bowater, R. P., Rosche, W. A., Jaworski, A., Sinden, R. R., and Wells, R. D. (1996). Relationship between *Escherichia coli* growth and deletions of CTG·CAG triplet repeats in plasmids. *J. Mol. Biol.* **264**, 82–96.
107. Rosche, W. A., Trinh, T. Q., and Sinden, R. R. (1997). Leading strand specific spontaneous mutation corrects a quasipalindrome by an intermolecular strand switch mechanism. *J. Mol. Biol.* **269**, 176–187.

108. Rosche, W. A., Ripley, L. S., and Sinden, R. R. (1998). Primer-template misalignments during leading strand DNA synthesis account for the most frequent spontaneous mutations in a quasipalindromic region in Escherichia coli. *J. Mol. Biol.* **284**, 633–646.
109. Viswanathan, M., Lacirignola, J. J., Hurley, R. L., and Lovett, S. T. (2000). A novel mutational hotspot in a natural quasipalindrome in *Escherichia coli. J. Mol. Biol.* **302**, 553–564.
110. Yoshiyama, K., Higuchi, K., Matsumura, H., and Maki, H. (2001). Directionality of DNA replication fork movement strongly affects the generation of spontaneous mutations in Escherichia coli. *J. Mol. Biol.* **307**, 1195–1206.
111. Yoshiyama, K., and Maki, H. (2003). Spontaneous hotspot mutations resistant to mismatch correction in Escherichia coli: Transcription-dependent mutagenesis involving template-switching mechanisms. *J. Mol. Biol.* **327**, 7–18.
112. Morag, A. S., Saveson, C. J., and Lovett, S. T. (1999). Expansion of DNA repeats in *Escherichia coli*: Effects of recombination and replication functions. *J. Mol. Biol.* **289**, 21–27.
113. Bzymek, M., Saveson, C. J., Feschenko, V. V., and Lovett, S. T. (1999). Slipped misalignment mechanisms of deletion formation: In vivo susceptibility to nucleases. *J. Bacteriol.* **181**, 477–482.
114. Lovett, S. T., Drapkin, P. T., Sutera, V. A., Jr., and Gluckman-Peskind, T. J. (1993). A sister-strand exchange mechanism for *recA*-independent deletion of repeated DNA sequences in *Escherichia coli. Genetics* **135**, 631–642.
115. Kramer, P. R., and Sinden, R. R. (1997). Measurement of unrestrained negative supercoiling and topological domain size in living human cells. *Biochemistry* **36**, 3151–3158.
116. Ljungman, M., and Hanawalt, P. C. (1992). Localized torsional tension in the DNA of human cells. *Proc. Natl. Acad. Sci. USA* **89**, 6055–6059.
117. Matsuura, T., Yamagata, T., Burgess, D. L., Rasmussen, A., Grewal, R. P., Watase, K., Khajavi, M., McCall, A. E., Davis, C. F., Zu, L., Achari, M., Pulst, S. M., Alonso, E., Noebels, J. L., Nelson, D. L., Zoghbi, H. Y., and Ashizawa, T. (2000). Large expansion of the ATTCT pentanucleotide repeat in spinocerebellar ataxia type 10. *Nat. Genet.* **26**, 191–194.
118. Lin, X., and Ashizawa, T. (2003). SCA10 and ATTCT repeat expansion: Clinical features and molecular aspects. *Cytogenet. Genome Res.* **100**, 184–188.
119. Matsuura, T., Fang, P., Lin, X., Khajavi, M., Tsuji, K., Rasmussen, A., Grewal, R. P., Achari, M., Alonso, M. E., Pulst, S. M., Zoghbi, H. Y., Nelson, D. L., Roa, B. B., and Ashizawa, T. (2004). Somatic and germline instability of the ATTCT repeat in spinocerebellar ataxia type 10. *Am. J. Hum. Genet.* **74**, 1216–1224.
120. Grewal, R. P., Achari, M., Matsuura, T., Durazo, A., Tayag, E., Zu, L., Pulst, S. M., and Ashizawa, T. (2002). Clinical features and ATTCT repeat expansion in spinocerebellar ataxia type 10. *Arch. Neurol.* **59**, 1285–1290.
121. Tudek, B., Kowalczyk, P., and Ciesla, J. M. (1999). Localization of chloroacetaldehyde-induced DNA damage in human p53 gene by DNA polymerase fingerprint analysis. *IARC Sci Publ.* **150**, 279–293.

CHAPTER 31

Spinocerebellar Ataxia Type 12

Susan E. Holmes,[1] Elizabeth O'Hearn,[2,3] Natividad Cortez-Apreza,[1] H. S. Hwang,[1] Christopher A. Ross,[1–4] S. Strack,[5] and Russell L. Margolis[1,2,4]

[1]Laboratory of Genetic Neurobiology, Division of Neurobiology, Department of Psychiatry, [2]Departments of Neurology and [3]Neuroscience, and [4]Program in Cellular and Molecular Medicine, Johns Hopkins University School of Medicine, Baltimore, Maryland, and [5]Department of Pharmacology, University of Iowa Carver College of Medicine, Iowa City, Iowa

I. Introduction
II. SCA12 Epidemiology and Genetics
III. SCA12 Clinical Aspects
   A. Clinical Features
   B. Neuroimaging and Neuropathology
   C. Diagnostic Considerations
IV. Molecular Basis of SCA12
   A. SCA12 Repeat Lies in the 5′ Region of the Most Common *PPP2R2B* Transcript
   B. Alternately Spliced *PPP2R2B* Transcripts Encode Alternate Isoforms of Bβ
   C. SCA12 Is Not a Polyglutamine Disease
V. Function of *PPP2R2B* Gene Products
   A. Protein Phosphatase 2A
   B. Expression of B-Family Regulatory Subunits
   C. Structure of B-Family Regulatory Subunits
   D. Bβ Isoforms Determine Subcellular Localization of PP2A
VI. Possible Mechanisms of Disease Pathogenesis
References

Spinocerebellar ataxia type 12 (SCA12) is caused by a CAG repeat expansion in *PPP2R2B*, which encodes one of the brain-specific regulatory subunits of the trimeric phosphatase PP2A. SCA12 is the second most common SCA in India, accounting for approximately 8% of dominant ataxia cases; however, it is a rare disease in all other populations studied to date, having been found in only the single North American index pedigree. Clinically, SCA12 is the only inherited SCA in which action tremor is the presenting and most common sign. The SCA12 CAG repeat is found in the probable promoter region of the *PPP2R2B* variant encoding the predominant isoform (Bβ1) of the Bβ regulatory subunit. Repeat expansion appears to drive increased transcription from this promoter, suggesting that disease pathogenesis may involve overexpression of Bβ1 and lead to altered activity of PP2A, a ubiquitous enzyme implicated in multiple cellular processes including apoptosis. The SCA12 repeat is also within the intronic sequence of multiple alternately spliced transcripts with alternate promoters that encode additional Bβ isoforms, including Bβ2, which targets PP2A to the mitochondria and promotes apoptosis when overexpressed. Pathogenesis may also involve an expansion-induced shift in splicing or choice of promoters, leading to an increase in expression of Bβ2.

I. INTRODUCTION

Thirteen neurological disorders associated with expansions of CAG/CTG repeats have been described, beginning with spinal and muscular atrophy in 1991 [1–4]. One of the more recently identified CAG expansion disorders, spinocerebellar ataxia type 12 (SCA12) [5], belongs to the group of autosomal dominant, adult onset, progressive neurodegenerative ataxias, which includes more than 25 distinct forms with overlapping and variable phenotypes. Although clinical and even pathological findings cannot always distinguish among the subtypes of SCA, identification of causative mutations or linkage to unique chromosomal loci has allowed for genetic definition [2] and has also provided a window into the pathogenesis of these disorders. Three subgroups of SCA have been identified on the basis of current understanding of pathogenesis. In SCAs 1, 2, 3, 7, and 17, the CAG expanded repeat is within the coding sequence of the respective genes, leading to protein products containing abnormally long and toxic polyglutamine tracts [6]. In the second group, which includes SCA6 and episodic ataxia types 1 and 2 (EA1 and EA2, respectively), the expanded repeat appears to disrupt function of the encoded calcium or potassium channels [2]. SCA12 belongs to the third group, also including SCA10 [7] and potentially SCA8 [8], in which repeat expansions are localized within noncoding regions of the associated genes and may be affecting gene transcription or splicing.

This chapter will review SCA12, which has distinctive features at the clinical, pathological, and molecular levels. The pathogenic pathway has not been established, but aspects of the phenotype, genotype, and molecular genetics provide insight into possible mechanisms of pathogenesis.

II. SCA12 EPIDEMIOLOGY AND GENETICS

SCA12 is a rare disease found in only two ethnic groups to date. The pedigree initially described was American and said to be of German descent [5, 9]. Subsequently, SCA12 was identified in approximately 25 Indian pedigrees by two separate groups, accounting for roughly 8% of dominant ataxia cases in India, making it the second most common SCA in India [10–12]. Ascertainment has been predominantly in northern India; all cases come from a single endogamous ethnic group from Haryana State and share a common haplotype, which differs from that of the American index family [12].

No additional SCA12 pedigrees have been detected in North America [5], and no SCA12 cases have been identified in ataxia patients of European [5, 13–16], Japanese [17–19], or Chinese [20, 21] origins.

Normal repeat length at the SCA12 locus ranges from 4 to 32 triplets, with 10 triplets being the most common allele length in all populations examined (Fig. 31-1) [5, 10, 13–15, 20–23]. The proportion of alleles containing more than 12 triplets is greater in India than in other populations [10, 13]; the higher frequency of expansions in this population may reflect the frequency of larger normal alleles, as has been found for other SCAs [13]. Expanded alleles in clearly affected individuals range in length from 51 to 78 triplets [5, 12, 13].

The minimum length of repeat expansion sufficient to cause SCA12 has yet to be established. Alleles of 40 and 41 triplets were reported in two sporadic late onset ataxia patients of northern German origin [16], raising the possibility that alleles in this size range may be pathogenic, but the lack of additional affected family members precludes a definitive interpretation. An allele of 45 triplets was reported in a 28-year-old unaffected individual from India with no known family history of neurological disorder [13], a finding that is also of uncertain significance due to the young age of the subject. Alleles of intermediate length have also been associated with different phenotypes. A German patient with typical Creutzfeld–Jacob disease (CJD) had an SCA12 allele of 49 triplets; again the relationship between the repeat expansion and the disease is unknown. Although the initial symptoms resembled those of SCA12, the lack of additional family members precludes distinguishing a large, rare normal allele with no effect on phenotype from an unmasking of SCA12 at a subclinical stage by CJD [16]. Finally, an expansion of 53 triplets was identified in an Iranian

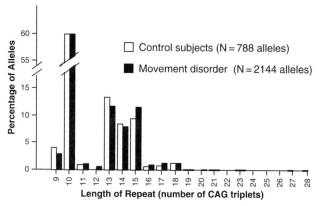

FIGURE 31-1 Distribution of normal allele lengths in control subjects and movement disorder patients (including subjects with sporadic ataxia, familial ataxia, and Parkinson's disease) of European descent. Reprinted with permission from *Brain Research Bulletin*, © 2001 [28].

woman with unipolar depression and her two monozygotic twin sons with schizophrenia [24]. Whereas this demonstrates that expanded alleles exist in the Iranian population, a clear relationship between the expanded allele and the psychiatric disorders seen in this family cannot be established. No SCA12 expansions have been detected in other sets of patients with bipolar disorder, schizophrenia, or Alzheimer's disease (Holmes, unpublished data).

Compared to the polyglutamine diseases (excepting SCA3), SCA12 is notable for relatively long repeat expansions and a relatively mild phenotype, consistent with molecular evidence (Section IVC) suggesting that SCA12 is unlikely to be caused by a polyglutamine expansion. The first known individual homozygous for an SCA12 expansion was identified in India [12]. This individual remains asymptomatic despite repeat lengths of 52 and 59 triplets, suggesting that homozygosity does not increase disease severity or dramatically decrease age of onset. There is no detectable anticipation in SCA12 [12]. Repeat length is only modestly unstable during vertical transmission, with slight contractions and slight expansions observed in both the index pedigree and various Indian pedigrees [10, 13, 24].

III. SCA12 CLINICAL ASPECTS

A. Clinical Features

Patients with SCA12 typically present in their third or fourth decade with action tremor (including both postural and kinetic tremors) of the arm and/or head that interferes with behaviors, such as writing and drinking [5, 9, 10, 13]. Age at onset of disease ranges from 8 to 55 years, with a mean of 34 years in the American kindred [5] and 37.2 years among Indian SCA12 patients [10]. On initial presentation, these patients are often thought to have essential tremor. β-blockers or primidone treatment reduces tremor in some SCA12 patients, and at least one patient reported that alcohol "calms" arm tremors (personal communication). Over the 20–30 years following presentation with tremor, most patients with SCA12 develop mild symptoms of slowly progressive gait unsteadiness, incoordination in hand and arm more than leg, and mildly slurred speech. The appearance of these symptoms suggests a spinocerebellar ataxia-like disorder rather than essential tremor. In addition to postural and kinetic tremors of the head and arms, examination usually reveals mild gait ataxia, upper greater than lower limb dysmetria, interrupted ocular pursuit, dysmetric saccades, subtle dysarthria, and hyperreflexia (see Table 31-1). A Babinski sign has been observed in some SCA12 patients, and both axial and limb dystonias have been reported. Mild and typically asymptomatic parkinsonism, including a paucity of spontaneous movement and subtle bradykinesia, has been seen in the American SCA12 kindred [9]. Anxiety and depression occur more commonly in SCA12 than in the general population. In late stages of SCA12 in the American kindred (Table 31-1), several patients developed cognitive abnormalities that included disorientation and memory loss, and primitive reflexes and *gegenhalten* have been detected on exam [9]. In addition to progressive cognitive decline, tremor prominence, arm dysmetria, and gait ataxia gradually worsen in SCA12 over time, and patients in late stages may lose the ability to walk or care for themselves independently (unpublished data). Also observed in late SCA12 are dysphagia and signs of autonomic dysfunction, including urinary incontinence and orthostatic hypotension.

Patients from India with SCA12 are clinically similar to North American SCA12 patients, but differences have been noted. Both groups usually present with arm and head tremors, followed by progressive gait ataxia and other signs of cerebellar dysfunction in addition to hyperreflexia [5, 9, 10, 13]. In two Indian patients, the onset of gait ataxia or dysarthria preceded tremor [10]. Unlike American SCA12 patients, cognitive dysfunction

TABLE 31-1 Main Clinical Features of SCA12

Early Features

Action tremor of arms and/or head

Mid-Course

Tremor: increased amplitude, distribution
Gait ataxia
Dysmetria, arm more than leg
Eye movement abnormalities
   Interrupted pursuit
   Ocular dysmetria
Hyperreflexia
Parkinsonism, mild
   Paucity of movement
   Bradykinesia
Focal dystonia
Anxiety or depression

Late Stage

Tremor: increased amplitude, distribution
Ataxia, dysmetria, eye signs (increased)
Primitive reflexes
Babinski sign (in some cases)
Gegenhalten
Dementia
Dysphagia
Autonomic dysfunction
   Urinary incontinence
   Orthostatic hypotension

has not been frequently reported in Indian patients with SCA12, although memory impairment was observed in two patients in their seventh or eighth decade [10, 13]. Parkinsonian features, such as paucity of movement and bradykinesia, that are common in American cases [9] have not been reported in Indian SCA12 patients. Subclinical sensory and motor axonal neuropathies have been detected in some Indian but no American patients with SCA12 [10]. These differences in phenotype suggest a role for additional genetic factors that may differentially modify the course of SCA12 in American and Indian families.

B. Neuroimaging and Neuropathology

Brain imaging [computed tomography (CT) or magnetic resonance imaging (MRI)] of American SCA12 patients reveals atrophy of the cerebral cortex and cerebellum with generalized ventriculomegaly (Fig. 31-2) [9]. Similar neuroimaging results are found in Indian SCA12 patients [10, 13]. The American proband with SCA12, who became progressively disoriented and forgetful in her sixth and seventh decades, died at 64 of urosepsis. Her brain CT scan (Fig. 31-2A), taken at age 60, revealed diffuse cerebral, cortical, and cerebellar atrophy with ventriculomegaly. The proband represents the only known case in which SCA12 neuropathology has been studied [25] (unpublished data). Examination of the brain revealed mild to moderate atrophy of the cerebral cortex and cerebellum, moderate Purkinje cell loss in the cerebellum, and occasional intraneuronal inclusions that were most frequent in substantia nigra neurons. Similar intraneuronal inclusions were rarely found in cerebellar Purkinje cells and cerebral cortical neurons. There was no evidence of Parkinson's or Alzheimer's disease. The neuronal intranuclear inclusions did not immunostain for polyglutamine (1C2 antibody staining was negative) [25] (unpublished data), supporting the hypothesis that the CAG expansion mutation in SCA12 does not lead to polyglutamine production (Section IVC).

C. Diagnostic Considerations

Although SCA12 is a rare disorder, the diagnosis should be considered in patients (especially of Indian descent) who exhibit upper extremity and/or head tremors in the presence of additional neurological signs, including gait ataxia, limb dysmetria, dysarthria, abnormal eye movements, hyperreflexia, parkinsonian features, autonomic dysfunction, dysphagia, and mood or cognitive abnormalities. The diagnosis of multiple system atrophy (MSA) may be suggested by the relatively late onset of

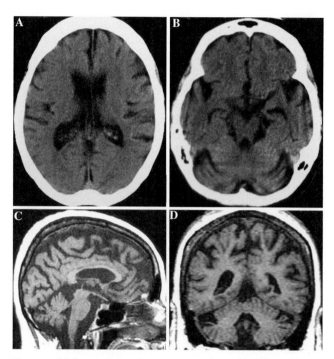

FIGURE 31-2 Neuroimages from two patients with SCA12 in the original American SCA12 kindred. (A, B) Head CT scans of SCA12 proband. This 60-year-old woman had action tremor of the head and arms beginning at age 35 and later developed progressive gait ataxia, limb dysmetria, eye movement abnormalities, hyperreflexia, mild parkinsonism, and dementia. The CT scans reveal moderate cerebellar and cerebrocortical atrophy with enlarged lateral ventricles. (C, D) T1-weighted MRI of a 59-year-old woman with SCA12. This patient had head and arm action tremors for 10 years and later showed mild gait ataxia, upper extremity dysmetria, abnormal eye movements, hyperreflexia, and mild parkinsonism. Sagittal (C) and coronal (D) MRIs reveal moderate cerebellar and cerebrocortical atrophy, with preservation of brainstem volume (C) and enlarged lateral ventricles (D). Reprinted with permission from *Brain Research Bulletin*, © 2001 [28].

progressive cerebellar abnormalities, parkinsonism, hyperreflexia, and autonomic dysfunction. However, a family history of relatives with similar neurologic and psychiatric disorders precludes the diagnosis of MSA, because the latter is typically sporadic. Neurologic dysfunction in SCA12 appears to progress more slowly than that in MSA, and many SCA12 patients remain employed and independent in their sixth and seventh decades.

SCA12 is the only SCA in which action tremor is the most common and earliest sign, although action tremor may be seen in patients with other inherited ataxias, including SCA2, SCA3, SCA6, SCA8, SCA14, SCA16, SCA19, and SCA21, ataxia due to a mutation in FGF14, dentatorubral–pallidolloysian atrophy (DRPLA), and fragile-X-associated tremor/ataxia syndrome (FXTAS) [26]. The diagnosis of SCA12 should be considered in any patient with action tremor in whom there is a family history that combines tremor plus another neurological or psychiatric sign or symptom. Familial essential tremor, in isolation, has not been associated with the SCA12 mutation [27]. β-blockers and primidone have reduced tremor prominence in some patients with SCA12. Benzodiazepines and antidepressants may be useful for syndromic anxiety or depression.

IV. MOLECULAR BASIS OF SCA12

A. SCA12 Repeat Lies in the 5′ Region of the Most Common *PPP2R2B* Transcript

The SCA12 repeat, in the CAG orientation, lies within the probable promoter region of the predominant variant of the gene *PPP2R2B* [5,28]. *PPP2R2B* encodes Bβ (also known as PR55β), one of multiple regulatory subunits of the ubiquitous phosphatase PP2A [29] (Section VA).

PPP2R2B is spaced over a region of approximately 493 kb on chromosome 5q32. The originally published cDNA (M64930) consisted of 9 exons (7A and 9–16). Subsequently, several lines of evidence demonstrated that the structure of the gene is complex, with multiple alternatively spliced transcripts identified by Reverse Transcriptase Polymerase Chain Reaction (RT-PCR) and bioinformatics approaches [30] (Fig. 31-3). The exons currently known, 1–16, are labeled. The predominant protein isoform, Bβ1, is derived primarily from the originally described transcript beginning with exon 7.

Exon 7 is itself proving to be quite complex, with two sources of variation: the use of multiple adjacent transcription start sites and an alternately spliced internal intron (IVS 7b/7c) (Fig. 31-3). A multiple-start-site element downstream-1 (MED-1) sequence (GCTCCC) [31] lies 65 nt 3′ to the CAG repeat [5] and is conserved in mouse and rat. This motif is typically found downstream of transcription initiation sites in TATA-less RNA polII promoter genes in which multiple initiation sites are used (MSS promoters). Available experimental and bioinformatics evidence suggests that the exon 7 promoter is an MSS promoter, using start sites 5′ to the repeat, within the repeat, and predominantly 3′ to the repeat (Fig. 31-3). 5′-Rapid Amplification of cDNA Ends (RACE) products and verified full-length cDNAs begin within 5–10 nt downstream from the CAG repeat [5], whereas a small subset of human and mouse brain Expressed Sequence Tags (ESTs) begin within or 5′ to the repeat. The Berkeley Drosophila Genome Project (BDGP) neural net promoter program [32] predicts a promoter spanning the repeat and a transcript start site (TSS) within the repeat. Thus, the CAG repeat, relative to the predominant Bβ1-encoding transcript, may be partially or entirely contained within the 5′ untranslated region (UTR), but is more often external to the transcript.

Removal of alternate intron 7b is seen in a handful of ESTs and by RT-PCR of a transfected reporter construct (H. Hwang, unpublished data). This transcript variant also encodes Bβ1, with a shortened 5′ UTR. The rare removal of intron 7c yields a transcript in which the open reading frame (ORF) of Bβ1 extends through the repeat (Section IVC).

Similar to other repeat-expansion disorders in which the expansion lies within noncoding regions, the SCA12 repeat may affect gene expression at the level of transcription, splicing, and/or translation. The position of the repeat within the probable promoter region of the predominant transcript variant suggests that the expansion may alter transcription levels, and reporter gene experiments demonstrate that the repeat-containing region does function as a promoter, with repeat expansion driving higher levels of reporter activity [33]. Although the positioning of the repeat within the 5′ UTR could lead to an effect of repeat expansion on translation, preliminary evidence suggests that increased protein levels are correlated with increased levels of transcript, indicating that the repeat may affect transcriptional efficiency. Sequences within UTRs may also affect mRNA stability, transport, and subcellular localization; therefore, it remains possible that expanded repeats alter the stability or location of *PPP2R2B* transcripts. It is conceivable, though less likely, that the expansion of a repeat in the MSS promoter region could affect the choice of transcription start site, increasing the proportion of repeat-containing transcripts and magnifying any effect of an expanded repeat within the 5′ UTR.

B. Alternately Spliced *PPP2R2B* Transcripts Encode Alternate Isoforms of Bβ

In addition to the predominant transcripts beginning with exon 7, alternately spliced variants consisting of multiple combinations of exons 1–6, 7D, and 8 spliced to exons 9–16 (Fig. 31-3) have been identified through experimental and bioinformatics approaches. The variable exons encode different N termini, yielding protein isoforms Bβ2–Bβ6, which have differing properties (Section VD). The CAG repeat is intronic in transcripts encoding these alternate protein isoforms. Some isoforms are encoded by several transcripts with alternate combinations of 5′-untranslated exons (e.g., Bβ3 is encoded by transcripts that include exons 1A, 2, and 5 or 3–5); the use of alternate promoters and alternate 5′ UTRs presumably provides the mechanism for the regulation of expression of the protein products (Section VD). Although it has not been demonstrated experimentally, it is conceivable that the SCA12 repeat expansion could disrupt regulation of splicing (as exemplified

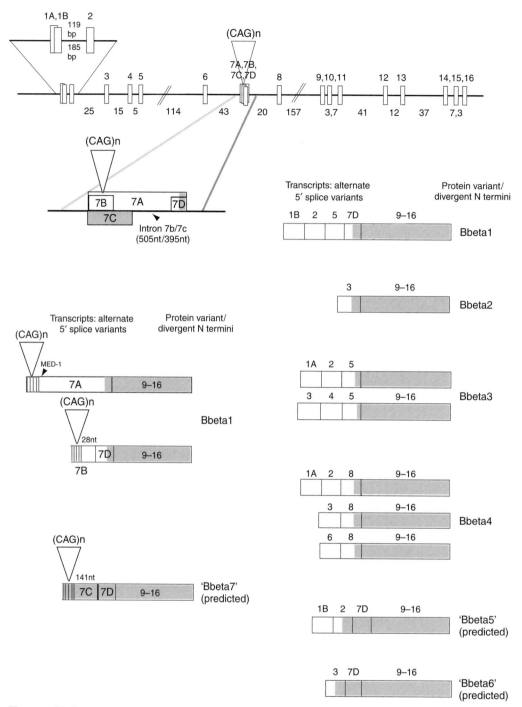

FIGURE 31-3 Genomic structure of *PPP2R2B* and alternately spliced transcripts. The gene is spaced over approximately 493 kb on chromosome 5q32. Exons are not shown to scale. Approximate intron sizes are indicated below the line. Filled regions represent ORFs beginning with Met. Alternate combinations of upstream exons lead to protein variants with divergent N termini and varied 5′ UTRs. Bβ2, Bβ3, and Bβ4 have been functionally tested (Section V), though Bβ3 and Bβ4 were not detected endogenously. All depicted transcript variants are represented by brain-derived clones in the human EST database and/or were identified by RT-PCR of human cortex. Experimental and bioinformatics evidence suggests the use of multiple transcription start sites for exon 7 upstream, within, and downstream from the SCA12 repeat in human and mouse (vertical gray lines in repeat-containing transcripts), with the majority of transcription from this promoter beginning downstream of the repeat. Exon 7 contains an alternately spliced intron (7b/7c), with two alternate donor sites (28 and 141 nt downstream from the repeat) spliced to a common acceptor site (127 nt upstream from the Bβ1 coding region). Use of the second donor site yields a transcript predicted to encode an alternate N terminus in which the repeat encodes polyserine (Bβ7 has not been functionally tested); only one of the known ESTs (human fetal kidney) corresponding to this splice variant begins 5′ to the repeat and includes an ATG codon upstream of the repeat. Use of only the first donor site (yielding a transcript with CAG repeat within 5′ UTR) was seen by RT-PCR of a reporter construct.

by the effect on exon skipping of a dodecamer repeat insertion in exon 9 of bovine *C-myb* [34]), potentially leading to a shift in the ratio of the Bβ isoforms.

C. SCA12 Is Not a Polyglutamine Disease

SCA12 belongs to the subset of CAG/CTG repeat expansion disorders that do not appear to be associated with expansions of polyglutamine tracts. SCA8 and DM1 are caused by expansions of CTG repeats within 3′ UTRs [8, 35–37], and Huntington's Disease-Like 2 (HDL2) is caused by expansion of a CTG repeat within different reading frames [38].

As described in Section IVA, the majority of transcripts derived from the exon 7 promoter do not include the repeat, or they include it within the 5′ UTR; however, several ESTs predict a repeat-containing ORF contiguous with the predominant protein variant (Bβ1). Three of these begin within the repeat and do not contain a start codon other than that of Bβ1, whereas one EST (from fetal kidney) begins 137 nt 5′ to the repeat (Fig. 31-3) and includes two possible start codons upstream of the repeat. The reading frame in the repeat region cannot be determined definitively, as the ESTs all contain insertions or deletions of one or two nucleotides relative to genomic sequence; however, the predicted N terminus (Bβ7) according to the genomic sequence encodes a polyserine stretch rather than polyglutamine. Similar transcripts were not detected by RT-PCR of frontal cortex, and no transcripts were detected using probes immediately 5′ and 3′ to the CAG repeat on Northern blots of mRNA from brain and the LAN-1 neuroblastoma cell line (which expresses *PPP2R2B* at a high level), suggesting that repeat-containing transcripts are not expressed at significant levels, at least in these tissues.

Experimental evidence at the protein level supports the conclusion that the SCA12 CAG repeat does not encode polyglutamine. Polyglutamine expansions were not detected on Western blots of protein from patient lymphoblastoid cell lines using the 1C2 antibody [24], and, as noted in Section IIIB, no 1C2-positive inclusions have been detected in SCA12 brains.

V. FUNCTION OF *PPP2R2B* GENE PRODUCTS

A. Protein Phosphatase 2A

Greater than 99% of protein phosphorylation in eukaryotic organisms occurs on serine and threonine residues, with the remainder occurring on tyrosines. The more than 300 Ser/Thr kinases in the human genome are opposed by fewer than 30 genes that encode recognizable Ser/Thr phosphatase catalytic domains. Most Ser/Thr dephosphorylation is carried out by two enzyme families: protein phosphatase 1 (PP1) and protein phosphatase 2A (PP2A). PP2A has been implicated in numerous cellular processes, including oncogenesis, DNA replication, growth, and differentiation, regulation of kinase cascades, morphogenesis, and cytokinesis, neurotransmitter release, ion channel function, microtubule assembly, and apoptosis [39–42].

PP2A exists in several dimeric and trimeric forms. The most common, heterotrimeric PP2A holoenzyme, consists of a core dimer of a catalytic (C) and scaffolding (A) subunit and a third variable (B) regulatory subunit (Fig. 31-4). Because the catalytic subunit is usually found in complexes with diverse regulatory subunits, cells typically express dozens of PP2A holoenzymes with distinct substrate specificities and subcellular distributions [43]. Methyl esterification of the C subunit promotes association with regulatory subunits. It has been suggested that inhibition of methyltransferase activity by increased levels of homocysteine found in Alzheimers disease brains is responsible for decreased PP2A activity and tau hyperphosphorylation [44–46].

Three gene families encode regulatory subunits of the heterotrimeric PP2A enzyme: *PPP2R2* (also known

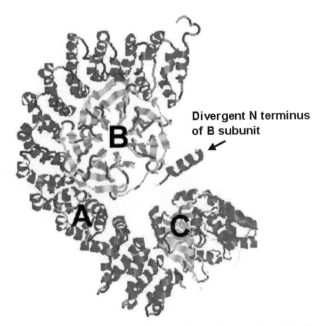

FIGURE 31-4 Model of the PP2A heterotrimer containing B-family regulatory subunits. A ribbon representation of the Aα-subunit structure [55] is arranged with model structures of catalytic (C) and B-family subunits. The edge of the B subunit's β-propeller interacts with the concave surface of the hook-shaped Aα molecule, leaving the divergent B subunit's divergent N terminus free to interact with substrates or anchoring proteins [59]. See CD-ROM for color image.

as B or PR55; this family includes the gene *PPP2R2B*, under discussion here), *PPP2R5* (B', PR61, B56), and *PPP2R3* (B''). In vertebrates, each family consists of 3–5 homologous genes, many of which are further diversified by alternative splicing. Except for a bipartite consensus motif that mediates interaction with the scaffolding subunit [47], B, B', and B'' subunits share little sequence and predicted structural similarity and may therefore employ different mechanisms to regulate phosphatase function. The B family of regulatory subunits regulates the dephosphorylation of substrates including histone-1 [48], vimentin [49], and tau [50].

B. Expression of B-Family Regulatory Subunits

Vertebrates have four genes encoding B-family regulatory subunits Bα–δ, with greater than 80% sequence identity. The Bα and Bδ isoforms are broadly expressed in a variety of tissues and cell types, whereas expression of the Bγ isoform and the various Bβ isoforms encoded by *PPP2R2B* is restricted to the nervous system (Fig. 31-5A) [51, 52]. *PPP2R2B* is unique among B-family subunit-encoding genes in giving rise through alternate splicing to several protein isoforms with variant N termini (Bβ1, Bβ2, etc. as described in Section IVB) [53, 54]. Even in the brain, however, Bα appears to be the most abundant B-family regulatory subunit, suggesting that Bβ and Bγ mediate specialized neuronal functions of PP2A, whereas Bα and Bδ may target PP2A to dephosphorylate more abundant and widely expressed substrates. The neuron-specific PP2A regulatory subunits also display interesting temporal expression profiles. Bβ1 mRNA levels drop during postnatal rat brain development, whereas both Bβ2 and Bγ expression increases dramatically over the same time course (Fig. 31-5B).

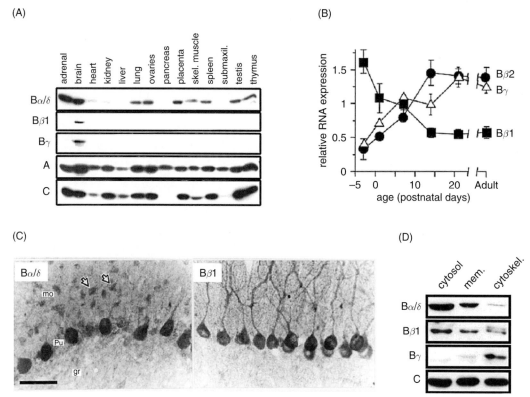

FIGURE 31-5 Differential expression of B-family subunits of PP2A. (A) Immunoblot of total extracts from the listed rat tissues with PP2A-subunit antibodies shows widespread expression of Bα/δ and brain specificity of Bβ1 and Bγ. (B) Ribonuclease protection analysis of RNA from a developmental series of rat forebrains demonstrates a drop in Bβ1 and a rise in Bβ2 and Bγ mRNA abundance (normalized to a cyclophilin internal control and to the average within each set). (C) Immunoperoxidase localization of Bα/δ and Bβ1 in rat cerebellum. Bα and/or Bδ are localized to the cytoplasm and nucleus of Purkinje cell (Pu) and to unidentified cell bodies in the molecular layer (mo). Granule cells are lightly stained. In contrast, Bβ1 is selectively localized to the cytoplasm and dendritic arbors, but not to the nuclei of Purkinje cells. (D) Cytosolic, membrane, and cytoskeletal fractions from rat brain were immunoblotted for B family and the catalytic subunit. Reprinted with permission from *Journal of Comparative Neurology*, © 1998 [51]; portions of panel B reprinted with permission from *Journal of Biological Chemistry*, © 2003 [54].

PP2A subunits are expressed in distinct, but overlapping neuron populations [51]. In the striatum, Bα/δ is present in small-to-medium size, presumably gamma amino butyric acid-ergic (GABA-ergic) neurons, as well as large cholinergic interneurons, whereas Bβ1 is detectable only in the latter class of neurons. Differences in the subcellular localization of B subunits are apparent throughout the brain: Bα/δ is detected in the cell body and soma of neurons, whereas Bβ1 is excluded from the nucleus and is more prominent in neuronal processes (Fig. 31-5C).

Fractionation experiments (Fig. 32-5D) show that Bα/δ is mostly cytosolic. Bβ1 is more evenly distributed, and Bγ is associated with cytoskeletal elements. On the basis of these data, the putative function of B-family regulatory subunits is to restrict the access of PP2A to as yet undetermined subsets of substrates; altered levels of Bβ1 due to the SCA12 repeat expansion (Section IVA) would therefore be predicted to shift PP2A target specificity.

C. Structure of B-Family Regulatory Subunits

The Aα scaffolding subunit is the only PP2A subunit for which the structure has been solved [55]. Aα is shaped like a bracket that brings regulatory and catalytic subunits in close proximity (Fig. 31-4). The PP2A catalytic subunit may adopt a structure similar to that of the related catalytic subunit of PP1, a globular protein with a C-terminal tail that is also involved in regulatory subunit binding [56–58]. B-family regulatory subunits contain up to seven recognizable tryptophan-aspartate (WD) repeats and therefore, like other similar proteins, probably adopt a seven-bladed β-propeller structure. Disruption of this structure by mutagenesis, or deletion of key residues in the domains that interact with the A-scaffolding subunit, disrupts association of Bγ with the rest of the PP2A holoenzyme [59]. Charge-reversal mutagenesis determined that residues along the edge of the Bγ β-propeller interact with the concave surface of the A subunit, positioning the divergent N terminus facing away from the holoenzyme where it can interact with specific substrates or anchoring proteins (Fig. 31-4) [59]. This arrangement is likely to be similar in all B-family subunits, including the Bβ isoforms encoded by *PPP2R2B*.

D. Bβ Isoforms Determine Subcellular Localization of PP2A

The four verified alternative isoforms of Bβ (Bβ1–Bβ4) share the C-terminal β-propeller domain, but they have different N termini of between 4 and 27 residues in length (Figs. 31-6A and 31-6B) encoded by alternately spliced transcripts of *PPP2R2B* as described in Section IVB. The putative isoforms Bβ5–Bβ7 are predicted to have longer N termini ranging from 82 to 166 amino acids and include the 24 amino acids encoded by the N terminus of Bβ1. *PPP2R2B* transcripts encoding Bβ1 and Bβ2 are found throughout the vertebrate subphylum [51]. Splice forms encoding the remaining isoforms have so far only been identified in RNA from primate brain. Bβ1–Bβ4 all form heterotrimers with A and C subunits of PP2A, and all tested heterotrimers retain enzymatic activity, even when lacking the N terminus (Fig. 31-6).

Striking differences in the subcellular localization of Bβ isoforms were observed after expression in PC6-3 cells, a subline of PC12 pheochromocytoma cells. Whereas Bβ1, Bβ3, Bβ4, and BβΔN-GFP were localized diffusely in the cytoplasm, Bβ2 localized to mitochondria (Fig. 31-6E) [54]. The unique N terminus of Bβ2 targets the PP2A holoenzyme to mitochondrial import receptors, but the C-terminal β-propeller acts as a "stop-transfer" domain, which resists the partial unfolding step required for passage through the narrow import channel [60]. In sum, the N-terminal domain of the Bβ variants provides site specificity, whereas the remainder of the protein interacts with other elements of PP2A and may also modulate localization.

Inducible and transient overexpression of Bβ2 has no effect on growth rate and viability under basal conditions, but it accelerates apoptosis when PC6-3 cells are stressed by serum removal. In contrast, overexpression of the Bβ1 splice form or the neuron-specific Bγ subunit to similar levels does not affect viability. Bβ2 appears to recruit PP2A activity to the outer mitochondrial membrane, sensitizing neuronal cells to apoptotic stimuli [60], apparently by modulating the mitochondrial apoptosis pathway upstream of Bcl-2 (Dagda et al., unpublished data).

VI. POSSIBLE MECHANISMS OF DISEASE PATHOGENESIS

Whereas the pathogenic pathway leading from repeat expansion to the SCA12 phenotype has not been established, the available evidence leads to some hypotheses. The relatively benign course and variability of the SCA12 phenotype distinguish it from most CAG/CTG expansion diseases, suggesting a possible gene dosage effect with potential susceptibility to modifying factors. The CAG repeat in *PPP2R2B* is unlikely to encode polyglutamine. Whereas a toxic gain of function of an alternate mutant protein product is possible, such as one

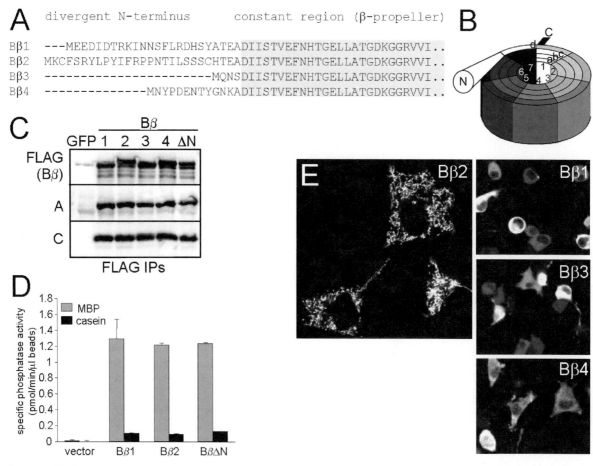

FIGURE 31-6 Bβ splicing affects subcellular targeting. (A) Alignment of N termini of Bβ splice isoforms. (B) Schematic structure model of B-family subunits showing WD repeats (1–7) and component β strands (a–d). (C) Immunoprecipitates of transiently expressed Bβ splice forms C-terminally tagged with a FLAG epitope and Green Flourescent Protein (GFP) contain similar amounts of scaffolding (A) and catalytic (C) subunits. The BβΔN mutant lacks the divergent N terminus. (D) *In vitro* phosphatase assays were carried out with Bβ-isoform immunoprecipitates (see C) and the model substrates myelin basic protein (MBP) and casein phosphorylated by PKA. (E) Transiently transfected PC6-3 cells show punctuate mitochondrial localization of Bβ2 and diffuse cytosolic distribution of Bβ1, Bβ3, and Bβ4-GFP fusion proteins. Panel D reproduced with permission from *Journal of Biological Chemistry*, © 2003 [54].

containing an expanded polyserine tract, transcripts encoding such proteins appear to be very uncommon and thus are unlikely to have a significant role in pathogenesis. The toxicity of RNA transcripts containing expanded CTG or CCTG repeats has been demonstrated in DM1, DM2 [61, 62], and preliminarily in HDL2 (Chapter 16 of this volume). Although there is no precedent for expanded CAG repeats leading to RNA toxicity, this possibility has not been excluded in SCA12.

As discussed in Section V, *PPP2R2B* encodes numerous isoforms of Bβ, a brain-specific regulatory subunit of PP2A. The regulatory subunits of PP2A determine substrate specificity and subcellular distribution of the enzyme. Preliminary evidence suggests that CAG repeat expansion increases expression of the predominant isoform of Bβ (Bβ1) in a reporter system using the exon 7 promoter. Increased levels of Bβ1 *in vivo* could result in shifts in PP2A target specificity. (Interestingly, *PPP2R2B* was markedly up-regulated in a cell model overexpressing the prion protein (PrPC) [63].) Microarray analysis followed by Northern blot confirmed the specific up-regulation of *PPP2R2B*. This finding lends support to the hypothesis that subclinical SCA12 may have been unmasked by CJD in a patient with an SCA12 allele of 49 triplets (Section II).

Alternately, CAG repeat expansion could alter upstream promoter use and/or splicing of alternate upstream exons, potentially leading to a shift in ratio between Bβ isoforms. A functional increase in Bβ2, which as noted earlier modulates the Bcl2 apoptotic pathway, would be predicted to have a considerable effect on cell function and survival.

SCA12 is the first disease shown to be associated with a mutation in a gene encoding a phosphatase subunit.

Functional studies demonstrate an important role for the *PPP2R2B*-encoded Bβ isoforms in the regulation of phosphatase activity in the brain, and preliminary evidence suggests that the SCA12 mutation may alter the expression levels of one or more isoforms. Cell and animal models will help determine the effect of the SCA12 expansion mutation on expression of *PPP2R2B*. Should the hypothesis of *PPP2R2B*-mediated changes in PP2A activity prove correct, elucidation of the pathways linking PP2A to neuronal death should yield novel insights into the role of phosphatases in neuronal survival and death.

Acknowledgments

The authors thank Drs. Achal K. Srivastava, Satish Jain, K. K. Sinha, and Mitali Mukerji for valuable insight and guidance; Abhijit Agarwal and Alka Ahuja for research assistance; Ms. Marie Sonderman for technical assistance; and family members who have participated in our investigations of SCA12 for their patience and cooperation. This work was supported by the National Ataxia Foundation and by NIH Grants NS42930, NS38054, and NS016375.

References

1. La Spada, A. R., Wilson, E. M., Lubahn, D. B., Harding, A. E., and Fischbeck, K. H. (1991). Androgen receptor gene mutations in X-linked spinal and bulbar muscular atrophy. *Nature* **352**, 77–79.
2. Margolis, R. L. (2002). The spinocerebellar ataxias: Order emerges from chaos. *Curr. Neurol. Neurosci. Rep.* **2**, 447–456.
3. Schols, L., Bauer, P., Schmidt, T., Schulte, T., and Riess, O. (2004). Autosomal dominant cerebellar ataxias: Clinical features, genetics, and pathogenesis. *Lancet Neurol.* **3**, 291–304.
4. van de Warrenburg, B. P., Sinke, R. J., and Kremer, B. (2005). Recent advances in hereditary spinocerebellar ataxias. *J. Neuropathol. Exp. Neurol.* **64**, 171–180.
5. Holmes, S. E., O'Hearn, E. E., McInnis, M. G., Gorelick-Feldman, D. A., Kleiderlein, J. J., Callahan, C., Kwak, N. G., Ingersoll-Ashworth, R. G., Sherr, M., Sumner, A. J., Sharp, A. H., Ananth, U., Seltzer, W. K., Boss, M. A., Vieria-Saecker, A. M., Epplen, J. T., Riess, O., Ross, C. A., and Margolis, R. L. (1999). Expansion of a novel CAG trinucleotide repeat in the 5′ region of PPP2R2B is associated with SCA12. *Nat. Genet.* **23**, 391–392.
6. Zoghbi, H. Y., and Orr, H. T. (2000). Glutamine repeats and neurodegeneration [in process citation]. *Annu. Rev. Neurosci.* **23**, 217–247.
7. Matsuura, T., Yamagata, T., Burgess, D. L., Rasmussen, A., Grewal, R. P., Watase, K., Khajavi, M., McCall, A. E., Davis, C. F., Zu, L., Achari, M., Pulst, S. M., Alonso, E., Noebels, J. L., Nelson, D. L., Zoghbi, H. Y., and Ashizawa, T. (2000). Large expansion of the ATTCT pentanucleotide repeat in spinocerebellar ataxia type 10. *Nat. Genet.* **26**, 191–194.
8. Koob, M. D., Moseley, M. L., Schut, L. J., Benzow, K. A., Bird, T. D., Day, J. W., and Ranum, L. P. (1999). An untranslated CTG expansion causes a novel form of spinocerebellar ataxia (SCA8). *Nat. Genet.* **21**, 379–384.
9. O'Hearn, E., Holmes, S. E., Calvert, P. C., Ross, C. A., and Margolis, R. L. (2001). SCA-12: Tremor with cerebellar and cortical atrophy is associated with a CAG repeat expansion. *Neurology* **56**, 299–303.
10. Srivastava, A. K., Choudhry, S., Gopinath, M. S., Roy, S., Tripathi, M., Brahmachari, S. K., and Jain, S. (2001). Molecular and clinical correlation in five Indian families with spinocerebellar ataxia 12. *Ann. Neurol.* **50**, 796–800.
11. Sinha, K. K., Worth, P. F., Jha, D. K., Sinha, S., Stinton, V. J., Davis, M. B., Wood, N. W., Sweeney, M. G., and Bhatia, K. P. (2004). Autosomal dominant cerebellar ataxia: SCA2 is the most frequent mutation in eastern India. *J. Neurol. Neurosurg. Psychiatry* **75**, 448–452.
12. Bahl, S., Virdi, K., Mittal, U., Sachdeva, M., Kalla, A., Holmes, S. E., O'Hearn, E., Margolis, R. L., Jain, S., Srivastava, A. K., and Mukerji, M. (2005). Evidence of a common founder for SCA12 in the Indian population. *Ann. Hum. Genet.* **69**, 1–7.
13. Fujigasaki, H., Verma, I. C., Camuzat, A., Margolis, R. L., Zander, C., Lebre, A. S., Jamot, L., Saxena, R., Anand, I., Holmes, S. E., Ross, C. A., Durr, A., and Brice, A. (2001). SCA12 is a rare locus for autosomal dominant cerebellar ataxia: A study of an Indian family. *Ann. Neurol.* **49**, 117–121.
14. Worth, P. F., and Wood, N. W. (2001). Spinocerebellar ataxia type 12 is rare in the United Kingdom. *Neurology* **56**, 419–420.
15. Brusco, A., Gellera, C., Cagnoli, C., Saluto, A., Castucci, A., Michielotto, C., Fetoni, V., Mariotti, C., Migone, N., Di Donato, S., and Taroni, F. (2004). Molecular genetics of hereditary spinocerebellar ataxia: Mutation analysis of spinocerebellar ataxia genes and CAG/CTG repeat expansion detection in 225 Italian families. *Arch. Neurol.* **61**, 727–733.
16. Hellenbroich, Y., Schulz-Schaeffer, W., Nitschke, M. F., Kohnke, J., Handler, G., Burk, K., Schwinger, E., and Zuhlke, C. (2004). Coincidence of a large SCA12 repeat allele with a case of Creutzfeld–Jacob disease. *J. Neurol. Neurosurg. Psychiatry* **75**, 937–938.
17. Maruyama, H., Izumi, Y., Morino, H., Oda, M., Toji, H., Nakamura, S., and Kawakami, H. (2002). Difference in disease-free survival curve and regional distribution according to subtype of spinocerebellar ataxia: A study of 1,286 Japanese patients. *Am. J. Med. Genet.* **114**, 578–583.
18. Matsumura, R., Futamura, N., Ando, N., and Ueno, S. (2003). Frequency of spinocerebellar ataxia mutations in the Kinki district of Japan. *Acta Neurol. Scand.* **107**, 38–41.
19. Sasaki, H., Yabe, I., and Tashiro, K. (2003). The hereditary spinocerebellar ataxias in Japan. *Cytogenet. Genome Res.* **100**, 198–205.
20. Tsai, H. F., Liu, C. S., Leu, T. M., Wen, F. C., Lin, S. J., Liu, C. C., Yang, D. K., Li, C., and Hsieh, M. (2004). Analysis of trinucleotide repeats in different SCA loci in spinocerebellar ataxia patients and in normal population of Taiwan. *Acta Neurol. Scand.* **109**, 355–360.
21. Zhao, Y., Tan, E. K., Law, H. Y., Yoon, C. S., Wong, M. C., and Ng, I. (2002). Prevalence and ethnic differences of autosomal-dominant cerebellar ataxia in Singapore. *Clin. Genet.* **62**, 478–481.
22. Cholfin, J. A., Sobrido, M. J., Perlman, S., Pulst, S. M., and Geschwind, D. H. (2001). The SCA12 mutation as a rare cause of spinocerebellar ataxia. *Arch. Neurol.* **58**, 1833–1835.
23. Sulek, A., Hoffman-Zacharska, D., Bednarska-Makaruk, M., Szirkowiec, W., and Zaremba, J. (2004). Polymorphism of trinucleotide repeats in non-translated regions of SCA8 and SCA12 genes: Allele distribution in a Polish control group. *J. Appl. Genet.* **45**, 101–105.
24. Holmes, S. E., O'Hearn, E., Brahmachari, S. K., Choudhry, S., Srivastava, A. K., Jain, S., Ross, C. A., and Margolis, R. L. (2003). *In* "Genetics of Movement Disorder" (S.-M. Pulst, Ed.), pp. 121–130. Academic Press, San Diego.
25. O'Hearn, E., Pletnikova, O., Holmes, S. E., Trojanowski, J. Q., and Margolis, R. L. (2004). SCA12 neuropathology: Cerebral cortical

25. and cerebellar atrophy, Purkinje cell loss, and neuronal intranuclear inclusions. *Mov. Disord.* **19**, 1124.
26. O'Hearn, E. (2005). *In* "Current Therapy in Neurologic Disease" (R. T. Johnson, J. W. Griffin, and J. C. McArthur, Eds.). Mosby, Inc., St. Louis, MO.
27. Nicoletti, G., Annesi, G., Carrideo, S., Tomaino, C., Di Costanzo, A., Zappia, M., and Quattrone, A. (2002). Familial essential tremor is not associated with SCA-12 mutation in southern Italy. *Mov. Disord.* **17**, 837–838.
28. Holmes, S. E., O' Hearn, E., Ross, C. A., and Margolis, R. L. (2001). SCA12: An unusual mutation leads to an unusual spinocerebellar ataxia. *Brain Res. Bull.* **56**, 397–403.
29. Mayer, R. E., Hendrix, P., Cron, P., Mattheis, R., Stone, S. R., Goris, J., Merlevede, W., Hofsteenge, J., and Hemmings, B. A. (1991). Structure of the 55-kDa regulatory subunit of protein phosphatase 2A: Evidence for a neuronal-specific isoform. *Biochem. J.* **30**, 3589–3597.
30. Holmes, S. E., O'Hearn, E., and Margolis, R. L. (2003). Why is SCA12 different from other SCAs? *Cytogenet. Genome Res.* **100**, 189–197.
31. Ince, T. A., and Scotto, K. W. (1995). A conserved downstream element defines a new class of RNA polymerase II promoters. *J. Biol. Chem.* **270**, 30249–30252.
32. Reese, M. G., Harris, N. L., and Eeckman, F. H. (1996). *In* "Biocomputing: Proceedings of the 1996 Pacific Symposium" (L. Hunter and T. Klein, Eds.), pp. 737–738. World Scientific Publishing Co., Singapore.
33. Holmes, S. E., Fujigasaki, H., O'Hearn, E., Antonarakis, S., Cooper, J. K., Callahan, C., Gorelick-Feldman, D. A., Verma, I. C., Saxena, R., Durr, A., Brice, A., Ross, C. A., and Margolis, R. L. (2000). Spinocerebellar ataxia type 12 (SCA12): Additional evidence for a causative role of the CAG repeat expansion in PPP2R2B. *Abstr./Am. Soc. Hum. Genet.* **67**, 200.
34. Shinagawa, T., Ishiguro, N., Horiuchi, M., Matsui, T., Okada, K., and Shinagawa, M. (1997). Deletion of c-myb exon 9 induced by insertion of repeats. *Oncogene* **14**, 2775–2783.
35. Brook, J. D., McCurrach, M. E., Harley, H. G., Buckler, A. J., Church, D., Aburatani, H., Hunter, K., Davies, J., Shelbourne, P., Buxton, J., Jones, C., Juvonen, V., Johnson, K., Harper, P. S., Shaw, D. J., and Housman, D. E. (1992). Molecular basis of myotonic dystrophy: Expansion of a trinucleotide (CTG) repeat at the 3' end of a transcript encoding a protein kinase family member. *Cell* **68**, 799–808.
36. Mahadevan, M., Tsilfidis, C., Sabourin, L., Shutler, G., Amemiya, C., Jansen, G., Nelville, C., Narang, M., Barcelo, J., O'Hoy, K., Leblond, S., Earle-MacDonald, J., DeJong, P. J., Wieringa, B., and Korneluk, R.G. (1992). Myotonic dystrophy mutation: An unstable CTG repeat in the 3' untranslated region of the gene. *Science* **255**, 1253–1255.
37. Fu, Y.-H., Pizzuti, A., Fenwick, R. G., King, J., Rajnarayan, S., Dunne, P. W., Dubel, J., Nasser, G. A., Ashizawa, T., De Jong, P., Wieringa, B., Korneluk, R., Perryman, M. B., Epstein, H. F., and Caskey, C. T. (1992). An unstable triplet repeat in a gene related to myotonic muscular dystrophy. *Science* **255**, 1256–1258.
38. Holmes, S. E., O'Hearn, E., Rosenblatt, A., Callahan, C., Hwang, H. S., Ingersoll-Ashworth, R. G., Fleisher, A., Stevanin, G., Brice, A., Potter, N. T., Ross, C. A., and Margolis, R. L. (2001). A repeat expansion in the gene encoding junctophilin-3 is associated with Huntington disease-like 2. *Nat. Genet.* **29**, 377–378.
39. Janssens, V., Goris, J., and Van Hoof, C. (2005). PP2A: The expected tumor suppressor. *Curr. Opin. Genet. Dev.* **15**, 34–41.
40. Price, N. E., and Mumby, M. C. (1999). Brain protein serine/threonine phosphatases. *Curr. Opin. Neurobiol.* **9**, 336–342.
41. Santoro, R., and Grummt, I. (2001). Molecular mechanisms mediating methylation-dependent silencing of ribosomal gene transcription. *Mol. Cell* **8**, 719–725.
42. Virshup, D. M. (2000). Protein phosphatase 2A: A panoply of enzymes. *Curr. Opin. Cell Biol.* **12**, 180–185.
43. Janssens, V., and Goris, J. (2001). Protein phosphatase 2A: A highly regulated family of serine/threonine phosphatases implicated in cell growth and signalling. *Biochem. J.* **353**, 417–439.
44. Vafai, S. B., and Stock, J. B. (2002). Protein phosphatase 2A methylation: A link between elevated plasma homocysteine and Alzheimer's disease. *FEBS Lett.* **518**, 1–4.
45. Sontag, E., Hladik, C., Montgomery, L., Luangpirom, A., Mudrak, I., Ogris, E., and White, C. L., III (2004). Downregulation of protein phosphatase 2A carboxyl methylation and methyltransferase may contribute to Alzheimer disease pathogenesis. *J. Neuropathol. Exp. Neurol.* **63**, 1080–1091.
46. Sontag, E., Luangpirom, A., Hladik, C., Mudrak, I., Ogris, E., Speciale, S., and White, C. L., III (2004). Altered expression levels of the protein phosphatase 2A ABalphaC enzyme are associated with Alzheimer disease pathology. *J. Neuropathol. Exp. Neurol.* **63**, 287–301.
47. Li, X., and Virshup, D. M. (2002). Two conserved domains in regulatory B subunits mediate binding to the A subunit of protein phosphatase 2A. *Eur. J Biochem.* **269**, 546–552.
48. Ferrigno, P., Langan, T. A., and Cohen, P. (1993). Protein phosphatase 2A1 is the major enzyme in vertebrate cell extracts that dephosphorylates several physiological substrates for cyclin-dependent protein kinases. *Mol. Biol. Cell* **4**, 669–677.
49. Turowski, P., Myles, T., Hemmings, B. A., Fernandez, A., and Lamb, N. J. (1999). Vimentin dephosphorylation by protein phosphatase 2A is modulated by the targeting subunit B55. *Mol. Biol. Cell* **10**, 1997–2015.
50. Sontag, E., Nunbhakdi-Craig, V., Lee, G., Bloom, G. S., and Mumby, M. C. (1996). Regulation of the phosphorylation state and microtubule-binding activity of tau by protein phosphatase 2A. *Neuron* **17**, 1201–1207.
51. Strack, S., Zaucha, J. A., Ebner, F. F., Colbran, R. J., and Wadzinski, B. E. (1998). Brain protein phosphatase 2A: Developmental regulation and distinct cellular and subcellular localization by B subunits. *J. Comp. Neurol* **392**, 515–527.
52. Strack, S., Chang, D., Zaucha, J. A., Colbran, R. J., and Wadzinski, B. E. (1999). Cloning and characterization of B delta, a novel regulatory subunit of protein phosphatase 2A. *FEBS Lett.* **460**, 462–466.
53. Schmidt, K., Kins, S., Schild, A., Nitsch, R. M., Hemmings, B. A., and Gotz, J. (2002). Diversity, developmental regulation and distribution of murine PR55/B subunits of protein phosphatase 2A. *Eur. J. Neurosci.* **16**, 2039–2048.
54. Dagda, R. K., Zaucha, J. A., Wadzinski, B. E., and Strack, S. (2003). A developmentally regulated, neuron-specific splice variant of the variable subunit Bbeta targets protein phosphatase 2A to mitochondria and modulates apoptosis. *J. Biol. Chem.* **278**, 24976–24985.
55. Groves, M. R., Hanlon, N., Turowski, P., Hemmings, B. A., and Barford, D. (1999). The structure of the protein phosphatase 2A PR65/A subunit reveals the conformation of its 15 tandemly repeated HEAT motifs. *Cell* **96**, 99–110.
56. Egloff, M. P., Cohen, P. T., Reinemer, P., and Barford, D. (1995). Crystal structure of the catalytic subunit of human protein phosphatase 1 and its complex with tungstate. *J. Mol. Biol.* **254**, 942–959.
57. Goldberg, J., Huang, H. B., Kwon, Y. G., Greengard, P., Nairn, A. C., and Kuriyan, J. (1995). Three-dimensional structure of the catalytic subunit of protein serine/threonine phosphatase-1. *Nature* **376**, 745–753.

58. Terrak, M., Kerff, F., Langsetmo, K., Tao, T., and Dominguez, R. (2004). Structural basis of protein phosphatase 1 regulation. *Nature* **429**, 780–784.
59. Strack, S., Ruediger, R., Walter, G., Dagda, R. K., Barwacz, C. A., and Cribbs, J. T. (2002). Protein phosphatase 2A holoenzyme assembly: Identification of contacts between B-family regulatory and scaffolding A subunits. *J. Biol. Chem.* **277**, 20750–20755.
60. Dagda, R. K., Barwacz, C. A., Cribbs, J. T., and Strack, S. (2005). Unfolding-resistant translocase targeting: A novel mechanism for outer mitochondrial membrane localization exemplified by the Bbeta 2 regulatory subunit of protein phosphatase 2A. *J. Biol. Chem.* **280**, 27375–27382.
61. Liquori, C. L., Ricker, K., Moseley, M. L., Jacobsen, J. F., Kress, W., Naylor, S. L., Day, J. W., and Ranum, L. P. (2001). Myotonic dystrophy type 2 caused by a CCTG expansion in intron 1 of ZNF9. *Science* **293**, 864–867.
62. Fardaei, M., Rogers, M. T., Thorpe, H. M., Larkin, K., Hamshere, M. G., Harper, P. S., and Brook, J. D. (2002). Three proteins, MBNL, MBLL and MBXL, colocalize in vivo with nuclear foci of expanded-repeat transcripts in DM1 and DM2 cells. *Hum. Mol. Genet.* **11**, 805–814.
63. Satoh, J., and Yamamura, T. (2004). Gene expression profile following stable expression of the cellular prion protein. *Cell Mol. Neurobiol.* **24**, 793–814.

CHAPTER 32

Spinocerebellar Ataxia 17 and Huntington's Disease-like 4

GIOVANNI STEVANIN AND ALEXIS BRICE

INSERM U679 (formerly U289) and Department of Genetics, Cytogenetics and Embryology APHP, Salpêtrière Hospital, 75013 Paris, France

I. Introduction
II. Genetic Aspects of SCA17–HDL4
   A. Size and Structure of the Normal Repeat
   B. Behavior of the Pathological Expansions: Range and Incomplete Penetrance
   C. Instability and Origin of the Expansions
   D. Epidemiology and Relative Frequency of SCA17
III. Phenotype of SCA17–HDL4
   A. Clinical Heterogeneity
   B. Phenotype–Genotype Correlations
   C. Neuropathology of SCA17–HDL4
IV. Physiopathological Consequences of the Expansion
V. Conclusion
   References

Spinocerebellar ataxia 17 (SCA17), also known as Huntington's disease-like 4 (HDL4), is the lastest member of the group of "polyglutaminopathies" to have been identified. SCA17 is caused by the expansion above 44 units of a CAG/CAA repeat in the coding region of the TATA box binding protein (TBP) gene, leading to an abnormal expansion of a polyglutamine stretch in the corresponding protein. Alleles with 43 and 44 repeats have been identified in sporadic cases and their pathogenicity remains uncertain. Furthermore, incomplete penetrance of pathological alleles up to 49 repeats has been suggested. The imperfect nature of the repeat makes intergenerational instability extremely rare, and *de novo* mutations are most probably the result of partial duplications. This is one of the rarer forms of autosomal dominant cerebellar ataxias, but the associated phenotype is often severe, involving various systems (cerebral cortex, striatum, and cerebellum) with an extremely variable age at onset (range: 3–75 years) and clinical presentation. A Huntington's disease-like phenotype has also been reported with SCA17 expansions, and this gene is thought to account for a small proportion of patients with a Huntington's disease-like phenotype and cerebellar signs. Parkinson's disease-like and Alzheimer's disease-like phenotypes have also been described with small SCA17 expansions.

In addition, a patient with Creutzfeldt–Jakob disease-like presentation has been reported. At the cellular level, the abnormal protein is expressed at the same level as its normal counterpart and forms neuronal intranuclear inclusions (NIIs) containing other proteins involved in protein folding or degradation. The increase in the size of the glutamine stretch enhances transcription *in vitro*, probably leading to transcription deregulation. More interestingly, the TBP mutated in SCA17 is recruited in the NIIs in other polyglutaminopathies, suggesting its direct involvement in the transcription down-regulation observed in these diseases.

I. INTRODUCTION

Spinocerebellar ataxia 17 (SCA17) or Huntington's disease-like 4 (HDL4) is a rare neurodegenerative disorder (MIM#607136) caused by a polyglutamine expansion [1]. It belongs to a group commonly referred to as "polyglutaminopathies," which also includes Huntington's disease (HD), spinal and bulbar muscular atrophy, dentatorubral pallidoluysian atrophy (DRPLA), and 6 forms of autosomal dominant cerebellar ataxias (ADCAs) [2]. In the absence of treatment, they can all lead to dramatic neurological dysfunction and ultimately to death. The number of glutamines observed in the pathological proteins varies from 21 to more than 400 in the 9 diseases currently identified, but in most cases the phenotype manifests above a repeat number varying between 35 and 40. This class of disorders also shares, with few exceptions, other common clinical, genetic, and physiopathological features, suggesting common pathological mechanisms: i.e., a negative correlation between the size of the repeat expansion and the age at onset, anticipation with parental sex bias, instability of the repeat on expanded alleles, and the formation of intranuclear inclusions in neurons.

SCA17 was first reported in a sporadic case of a complex neurological disorder with cerebellar ataxia, pyramidal signs, and severe intellectual impairment. This patient carried 63 trinucleotide repeats in the gene encoding the TATA binding protein (TBP) on chromosome 6q27 [3]. This new genetic entity was subsequently identified in familial ADCA cases and was shown to differ in several aspects from other polyglutaminopathies [4–6]. Pathological expansions were also found in patients with an HD-like phenotype (HDL4) or with clinical features compatible with Alzheimer's, Parkinson's, or Creutzfeldt–Jakob disease, highlighting the clinical heterogeneity of this genetic entity [7–9].

II. GENETIC ASPECTS OF SCA17–HDL4

TBP is an important and general transcription factor ubiquitously expressed from a single gene on chromosome 6q27, and it constitutes an integral component of the transcription initiation complexes of the three RNA polymerases [10–12]. TBP is the DNA-binding subunit of the RNA polymerase II transcription factor D (TFIID), a protein complex involved in mRNA transcription, and anchors the complex to the TATA box upstream of the first codon. TBP also plays a role in TATA-less promoter genes by correctly positioning the polymerase on the DNA.

A. Size and Structure of the Normal Repeat

The N terminus of the protein that modulates the DNA binding activity of the C terminus [12] contains a long stretch of glutamines, as in other transcription factors or homeobox proteins [13]. This repeat is impure and is encoded by 3 CAG stretches, interrupted by 1–3 CAA codons (Table 32-1). The glutamine stretch is polymorphic in the normal population, and large population studies (>5000 control chromosomes) have determined that the normal range is between 25 and 42 residues, with most alleles containing 32–39 repeats [3–6, 8, 11, 14–16]. The allelic distribution varies slightly, however, according to ethnic–geographical origins [14, 16]. In a more recent Japanese study, alleles of up to 45 repeats were detected in the control population [9].

B. Behavior of the Pathological Expansions: Range and Incomplete Penetrance

Given its high polymorphism and the mean size of the repeat, which is above the pathological threshold of most polyglutamine diseases, the TBP gene was considered a good candidate gene for neurodegenerative and psychiatric diseases in which anticipation was suspected [11, 16]. An abnormally expanded CAG repeat (63 repeats) was initially described in a 14-year-old child with signs of severe ataxia and cognitive impairment without a family history of neurological disorders [3]. A series of studies worldwide subsequently reported familial cases with other repeat expansions in this gene and allowed the pathological range to be defined. The threshold for pathological expansions varies according to the study from 43 to 45 repeats [7, 9, 17–19]. Sporadic patients carrying from 43 to 63 repeats have been identified, whereas affected members of families with dominant transmission of the disease carry between 45 and 66 CAG/CAA repeats [5–9, 18–28].

Oda et al. reported a family in which one patient carried 43 repeats and another carried normal alleles, suggesting that this repeat size may not be pathogenic in this family [9]. In addition, alleles with a similar number of repeats were reported in patients carrying known ADCA mutations and have been detected in controls, reinforcing the hypothesis that they are not pathogenic [9, 19]. In the absence of evidence of cosegregation of the allele with 44 repeats [7, 9], and because of the overlap between controls' and patients' ranges, caution is needed in diagnosis for such small expansions, which so far have only accounted for sporadic cases.

Determination of the pathological threshold is also complicated by the existence of an incomplete penetrance, which has been suggested for patients carrying 45–49 repeats because healthy carriers with 46, 48, and 49 CAG/CAA repeats aged 59, 69, and 76 years, respectively, have been reported [7, 23, 25].

TABLE 32-1 Structure of the CAG/CAA Repeat in Wild-Type and Expanded Alleles[a,b]

	Repeat number	Structure of the repeat							
Wild type	25–42	CAG_3	CAA_3	CAG_{8-11}	CAACAGCAA	CAG_{15-18}	CAACAG		
Koide et al., 1999 spo, *de novo*	63	CAG_3	CAA_3	CAG_9	CAACAGCAA	CAG_9	CAACAGCAA	CAG_{19}	CAACAG
Shatunov et al., 2004 spo, *de novo*	55	CAG_3	CAA_3	CAG_9	CAACAGCAA	CAG_{15}	CAACAGCAA	CAG_{17}	CAACAG
Nakamura et al., 2001	55	CAG_3	CAA_3	CAG_9	CAACAGCAA	CAG_{16}	CAACAGCAA	CAG_{16}	CAACAG
Fujigasaki et al., 2001	46	CAG_3	CAA_3	CAG_9	CAACAGCAA	CAG_{26}	CAACAG		
Nakamura et al., 2001	48	CAG_3	CAA_3	CAG_6	CAACAGCAA	CAG_{31}	CAACAG		
Nakamura et al., 2001	47	CAG_3	CAA_3	CAG_8	CAACAGCAA	CAG_{28}	CAACAG		
Nakamura et al., 2001 spo	47	CAG_3	CAA_3	CAG_6	CAACAGCAA	CAG_{30}	CAACAG		
Zuhlke et al., 2001	51	CAG_3	CAA_3	CAG_9	CAACAGCAA	CAG_{31}	CAACAG		
Silveira et al., 2002	43	CAG_3	CAA_3	CAG_9	CAACAGCAA	CAG_{23}	CAACAG		
Stevanin et al., 2003 spo, reduced penetrance	46	CAG_3	CAA_3	CAG_{11}	CAACAGCAA	CAG_{24}	CAACAG		
Stevanin et al., 2003 spo	44	CAG_3	CAA_3	CAG_9	CAACAGCAA	CAG_{24}	CAACAG		
Zuhlke et al., 2003 spo, homoZ	47	CAG_3	CAA_3	CAG_9	CAACAGCAA	CAG_{27}	CAACAG		
Zuhlke et al., 2003 Reduced penetrance	48	CAG_3	CAA_3	CAG_9	CAACAGCAA	CAG_{28}	CAACAG		
Oda et al., 2004 four spo, one homoZ	44–47	CAG_3	CAA_3	CAG_x	CAACAGCAA	CAG_y	CAACAG		
Brusco et al., 2004 spo, reduced penetrance?	45	CAG_3	CAA_3	CAG_9	CAACAGCAA	CAG_{25}	CAACAG		
Brusco et al., 2004	45	CAG_3	CAA_3	CAG_8	CAACAGCAA	CAG_{26}	CAACAG		
Toyoshima et al., 2004 homoZ	48	CAG_3	CAA_3	CAG_6	CAACAGCAA	CAG_{31}	CAACAG		
Bruni et al., 2004	52	CAG_3	CAA_3	CAG_9	CAACAGCAA	CAG_{32}	CAACAG		
Wu et al., 2004 spo	46	CAG_3	CAA_3	CAG_6	CAACAGCAA	CAG_{29}	CAACAG		
Maltecca et al., 2003 Paternally unstable	53–66	CAG_3	CAA_4	CAG_{44-57}	CAACAG				
Zuhlke et al., 2001 Maternally unstable	53–55	CAG_3	CAA_4	CAG_{45-47}	CAACAG				
Zuhlke et al., 2005 Paternally unstable, reduced penetrance	49–53	CAG_3	CAA_4	CAG_{41-45}	CAACAG				

[a] References [3–7, 9, 18, 21–23, 25–30].
[b] Abbreviations: spo, sporadic cases; homoZ, homozygous cases.

C. Instability and Origin of the Expansions

Elongation of repeated CAG elements, alone or as the result of the loss of CAA interruptions, is the major mechanism leading to abnormally elongated SCA17 repeats [3–7, 9, 18, 21–23, 25–30]. In three cases, including two *de novo* expansions, the expanded repeat is the result of partial duplication or insertion of repeats into the CAG/CAA stretch [3, 28]. Koide et al. excluded meiotic unequal crossover and suggested either a displacement of the 5' end of the Okasaki fragment, generating a flap endonuclease FEN1-resistant hairpin, or an unequal sister chromatid recombination, potentially leading to partial intramolecular duplication [3]. In SCA2, SCA6, SCA7, and HD, neomutations occur mostly on large normal paternal alleles that undergo the expansion of pure repeats to the pathological range [31–38]. In the two SCA17 *de novo* cases, expansions also occurred on paternal chromosomes carrying 37–39 repeats, although with a different mechanism.

The structure of the expanded repeat therefore varies according to (1) the number of repeats in internal CAG stretches, particularly the third one in which most expansions occur, (2) the loss of CAA interruptions, which could influence the stability of the region, as well as (3) the presence of internal duplications or insertions (Table 32-1).

Compared to other polyglutamine diseases, the SCA17 mutation is unusually stable during parent–child transmissions. The discontinuous distribution of repeat size in different populations reflects this stability of the normal repeat [15]. This probably results partially from the presence of CAA interruptions, as already observed in SCA1 or SCA2 normal alleles. This would make it unlikely for the multistep gradual expansion to be observed in SCA17. Indeed, the rare cases of intergenerational instability, with increases of +1 to +13, have only been reported in alleles lacking CAA interruptions, resulting in pure CAG stretches over 40 units (Table 32-1). In these instances, instability is observed in both paternal and maternal transmission [4, 23, 27].

D. Epidemiology and Relative Frequency of SCA17

To date, approximately 51 SCA17 families have been reported worldwide, representing approximately 120 patients. Most families–cases are from Japan ($n = 17$ families) [3, 5, 9, 21, 39] and Germany ($n = 19$ families) [4, 19, 20, 23, 25, 26, 40], a distribution reminiscent of that of SCA6, which is also found frequently in both of these countries. The remaining families are as follows: 7 from Italy [17, 24, 27, 29, 30], 3 from Taiwan [8, 22], 2 from France [7], and 1 each from Belgium [6], the United States [28], and Portugal [18]. The frequency of SCA17 ranges from 0.3 to 3% among studies of ADCA families [3, 6, 9, 18, 19, 29, 39], but represents less than 1% of HD-like patients [7, 20]. SCA17 was not found in other series of ADCA families [41, 42].

More than 10 SCA17 cases had no family history of neurological diseases, 2 of which were proven to result from *de novo* expansions [3, 28], whereas incomplete penetrance was observed in 4 others [7, 9, 23, 25], which emphasizes the importance of analyzing this gene in isolated cases with a compatible phenotype.

III. PHENOTYPE OF SCA17–HDL4

A series of studies attempted to establish the clinical spectrum of this particular and very heterogeneous form of ADCA [3–9, 18–28, 40, 43]. The clinical and neuropathological spectrum associated with this mutation appears to be broader than previously suspected.

A. Clinical Heterogeneity

The clinical signs in SCA17 patients reported in the literature are summarized in Table 32-2. The symptoms at onset, which occurs at a mean age of 34 ± 13 years (range: 3–75 years), are predominantly gait instability [18, 27, 29] or other movement disorders, such as focal dystonia [30, 40] or chorea [7, 20]. Psychiatric disturbances such as behavioral changes, psychosis, or depression, as well as dementia, can also be the presenting symptoms [3, 6, 7, 20].

At the time patients are examined, the most prevalent abnormality after cerebellar ataxia is dementia (77%). This is particularly important in clinical practice because overt and early dementia is rare in ADCAs, with the exception of DRPLA. Psychiatric alterations [6, 19, 20, 27, 30] and abnormal movements are frequent, especially chorea, choreoathetosis, and dystonia [5, 19, 27, 30, 40, 43]. Parkinsonism occurs in half of the

TABLE 32-2 Frequency of the Most Commonly Associated Neurological Signs in SCA17 Patients[a]

	Frequency of associated signs
No. of families	$n = 42$
No. of patients	$n = 90$
Age at onset	35% (range 6–75 years)
Cerebellar ataxia	97% (70/72)
Dementia	77% (53/69)
Psychiatric symptoms	66% (35/53)
Pyramidal signs	56% (22/39)
Abnormal movements	54% (39/72)
Dystonia	62% (26/42)
Chorea, choreoathetosis	36% (16/45)
Parkinsonism	50% (24/48)
Epilepsy	37% (15/40)

[a] References [3–9, 18–28, 40, 43]. Numbers in paranatheses corresponds to the number of cases obseved total.

patients, and epilepsy is commonly observed [5, 30]. Spasticity with brisk reflexes may also be found [3, 4, 9]. The cardinal features of SCA17 therefore are the association of cerebellar ataxia with dementia and other movement disorders. These signs are typically observed in a small proportion of families with an HD-like phenotype, and indeed SCA17 accounts for less than 1% of HD-like cases with cerebellar ataxia (range: 43–52 repeats) [7, 17, 20, 21]. Similarly, given the overlap of this clinical profile, patients with Parkinson's disease with dementia-like, Alzheimer's disease-like, or Creutzfeldt–Jakob disease-like phenotypes have been reported to carry expansions in the TBP gene (range: 46–55 repeats) [8, 22, 28]. SCA17 was not, however, found to be a common cause of Parkinson's disease, primary dystonia, epilepsy, bipolar disorder, or schizophrenia in large series [3, 16, 19, 44, 45].

After short disease durations (2–3 years), magnetic resonance imaging (MRI) findings vary greatly, from normal to moderate global atrophy or a focal atrophy of the cerebellum [4, 40]. After longer durations, the atrophy is always pronounced in the cerebellum and mild in the brain, with relative sparing of the brainstem.

Diagnosis is established by genetic testing, and other laboratory investigations are not necessary. Unfortunately, once the disease has manifested, it is unremittingly progressive, leading to loss of autonomy, with death occurring at a mean age of 39 ± 20 years after a mean disease duration of 19 ± 9 years. Treatment is purely symptomatic.

B. Phenotype–Genotype Correlations

The clinical picture and age at onset are variable even among patients of the same family carrying the same number of repeats, suggesting that the size of the repeat has a limited influence on the course of the pathology [6]. A correlation, although not as strong as in other ADCAs [46], exists between the age at onset and the size of the repeat (Fig. 32-1). Linear and exponential regressions show that only 64% and 88%, respectively, of the variance in the age at onset is explained by the repeat size. The influence of modifier genes could be crucial to explaining the remaining variance and could also account for the wide clinical heterogeneity. Because most of the SCA17 expansions are stable, it is unlikely that somatic mosaicism of the size of the expansion accounts for the interindividual variability, in contrast to other polyglutaminopathies. Alternatively, the very low correlation could be due to the difficulty in determining the exact age at onset because of the variability in the presenting sign(s), which is not always cerebellar ataxia. Interestingly, age at death is correlated with the size of the triplet repeat in a small cohort

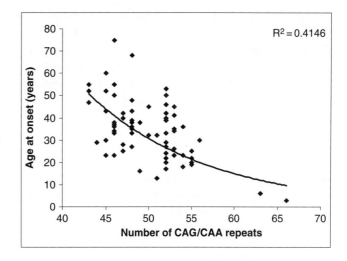

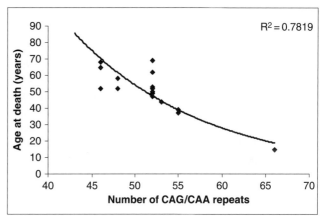

FIGURE 32-1 Correlation (exponential) between the age at onset ($n = 79$) or the age at death ($n = 16$) and the size of the repeat in SCA17-affected patients [3–7, 9, 18–23, 25, 27–30, 40, 43, and unpublished data].

of 16 patients of diverse origins (Fig. 32-1) [5, 6, 19, 27, 30, 43].

Anticipation does not appear to be a feature of this disease, which is in accordance with the very rare instability of the CAG/CAA repeat. Maltecca et al. reported, however, a familial case with a marked anticipation associated with a marked instability of the repeat [27].

Homozygosity has been shown to lower the age at onset in DRPLA, SCA3, and SCA6 but not in HD, which, however, shows increased severity [47–50]. Three SCA17 patients homozygous for repeat expansions in the TBP gene are not more severely affected than patients carrying one abnormal copy of the gene, supporting a gain of function hypothesis [9, 21, 25]. This might be due to incomplete penetrance of one of the expanded alleles. Neuropathological alterations did, however, extend to the hippocampus and brainstem in one homozygous patient [21].

C. Neuropathology of SCA17–HDL4

Although the clinical profile of SCA17 is reminiscent of that of DRPLA, its neuropathological basis differs. Neuropathological lesions are mild in the brainstem compared to other ADCA entities [6], but they are similarly marked in the cerebellum and the cerebral cortex.

Six patients' brains have been investigated neuropathologically and show similar features [5, 6, 19, 21, 30]. There is a mild global atrophy of the brain, predominantly in the cerebellum because of severe Purkinje cell loss and Bergman's gliosis. Neuronal loss is mild in the dentate nucleus [except in Rolfs et al. [19]] and the granular layer. Atrophy is moderate in the cerebral cortex, predominating in the motor cortex and visual areas, with abnormal arborization of neuronal dendrites and spongiosis [6, 21, 30]. In the brainstem, the pontine nuclei are spared but the locus coeruleus and the substantia nigra are mildly affected, with few deposits of free melanin pigments in the latter [6, 30].

However, differences occur according to the size of the repeat and/or disease duration. Atrophy of the substantia nigra is only observed in patients with the longest disease duration [19]. The inferior olivary nuclei are also atrophied in all but one patient, who carries the smallest repeat size ($n = 46$) and has a disease duration of only 10 years [6, 19, 30]. Similarly, the basal ganglia are spared in this patient [6], whereas atrophy is severe in the other patients, particularly in the caudate nucleus [21, 30]. Finally, the spinal cord is normal except in a patient with a 54-repeat expansion and a 24-year disease duration, who shows a loss of anterior horn cells [19]. There are no significant differences at the pathological level between heterozygous and homozygous patients, except additional atrophy of the hippocampus in one of the latter [21].

IV. PHYSIOPATHOLOGICAL CONSEQUENCES OF THE EXPANSION

The SCA17 gene product, TBP, has a well-known function and is widely expressed in the central nervous system and other tissues, which contrasts with the selective pattern of degeneration observed in patients.

As in other ADCAs, the pathological hallmarks of this disease are the presence of neuronal intranuclear inclusions (NIIs) containing the pathological proteins, as well as heat shock proteins and ubiquitin [5, 6], however with a lower frequency. Staining using specific antibodies or the 1C2 antibody [51, 52] is essentially nuclear, often diffuse, and focal in 0–3% of the neurons according to the structure [21, 30]. NIIs were not detected in visceral organs [21], but they were observed in various structures of the brain. They are found predominantly in the cerebral cortex, basal ganglia, and the midbrain reticular formation, but are also found in structures that are spared, such as the pontine nuclei, dentate nucleus, anterior horn, and inferior olives [6, 21, 30]. However, they are not detected in Purkinje cells, which severely degenerate [19]. In general, except for the cerebral cortex, there is an inverse correlation between the presence of NIIs and severe lesions in a given structure [6, 30], i.e., no NIIs are detected in the Purkinje and granule cell layers or in the locus coeruleus, all of which degenerate, but they are detected in the putamen, dentate nucleus, and pontine nuclei, which are unaffected or only mildly affected [6]. Moderate atrophy of the inferior olivary was, however, reported in one case, which had the highest density of NIIs [30]. More interestingly, Bruni et al. observed a link between the density of diffuse nuclear staining and cell loss, suggesting a more toxic effect of the nonaggregated protein [30].

The fact that the lack of TBP in TBP knockout mice is embryonic lethal, whereas homozygosity for SCA17 trinucleotide expansions in humans is not, indicates that the expansion is not responsible for a major loss of TBP function [53]. The absence of major differences between heterozygous and homozygous carriers suggests complete dominance, which is compatible with a gain of function mechanism or a dominant negative effect of the mutation.

In vitro, SCA17 models demonstrated that the increase in the size of the glutamine stretch inside full-length TBP enhances its insolubility and aggregation as well as the transcription of a Creb-responsive element-mediated luciferase reporter gene [15]. Interestingly, TBP is sequestered in other polyQ diseases, suggesting its direct involvement in the transcription deregulation, although this is an early down-regulation of transcription [54–56]. The exact mechanisms remain unknown but, given the role of TBP in anchoring the transcription machinery to the DNA, it can be postulated that alteration of the tertiary structure of this protein would lead to an alteration of its binding to the DNA and/or activation–binding–function of other TFIID components.

V. CONCLUSION

SCA17 is a rare neurodegenerative disorder in Asians and Caucasians. The cardinal features are the association of cerebellar ataxia, dementia, psychiatric features, and parkinsonism, with frequent occurrence of abnormal movements, such as chorea or dystonia. This clinical profile overlaps with those of other neurodegenerative diseases and patients with Alzheimer's disease-like,

Creutzfeldt–Jakob disease-like, or Parkinson's disease-like phenotypes or, more importantly, with a Huntington's disease-like phenotype (HDL4) that have been reported to carry SCA17 mutations.

SCA17 is caused by CAG expansions that can be associated with a loss of CAA interruptions or partial CAG/CAA repeat duplications in the TBP gene, leading to polyglutamine expansions above 43–44 repeats. Incomplete penetrance concerns repeats between 45 and 49 units, and the repeat is stable except in expanded alleles having lost CAA interruptions. The pathogenicity of repeats with 43 and 44 repeats remains unclear. These features have consequences for genetic counseling, particularly the existence of reduced penetrance for presymptomatic testing.

TBP and ubiquitin-positive NIIs are detected in the cerebral cortex, basal ganglia, pons, and dentate nucleus, but are absent from Purkinje cells, a pattern that does not match perfectly with neuronal loss. Atrophy mainly concerns the cerebellum, with global atrophy of the cortex, dentate nucleus, substantia nigra, and locus coeruleus.

TBP, which is also involved in other polyglutamine disorders, probably represents a key element of the pathology in these disorders, all of which are associated with transcription deregulation.

Acknowledgments

The authors' work is financially supported by grants from the European Community (EUROSCA), the Verum Foundation, and the Programme Hospitalier de Recherche Clinique.

References

1. Zoghbi, H. Y., and Orr, H. T. (2000). Glutamine repeats and neurodegeneration. *Annu. Rev. Neurosci.* **23**, 217–247.
2. Stevanin, G., Durr, A., and Brice, A. (2002). Spinocerebellar ataxias caused by polyglutamine expansions. *Adv. Exp. Med. Biol.* **516**, 47–77.
3. Koide, R., Kobayashi, S., Shimohata, T., Ikeuchi, T., Maruyama, M., Saito, M., Yamada, M., Takahashi, H., and Tsuji, S. (1999). A neurological disease caused by an expanded CAG trinucleotide repeat in the TATA-binding protein gene: A new polyglutamine disease? *Hum. Mol. Genet.* **8**, 2047–2053.
4. Zuhlke, C., Hellenbroich, Y., Dalski, A., Kononowa, N., Hagenah, J., Vieregge, P., Riess, O., Klein, C., and Schwinger, E. (2001). Different types of repeat expansion in the TATA-binding protein gene are associated with a new form of inherited ataxia. *Eur. J. Hum. Genet.* **9**, 160–164.
5. Nakamura, K., Jeong, S. Y., Uchihara, T., Anno, M., Nagashima, K., Nagashima, T., Ikeda, S., Tsuji, S., and Kanazawa, I. (2001). SCA17, a novel autosomal dominant cerebellar ataxia caused by an expanded polyglutamine in TATA-binding protein. *Hum. Mol. Genet.* **10**, 1441–1448.
6. Fujigasaki, H., Martin, J. J., De Deyn, P. P., Camuzat, A., Deffond, D., Stevanin, G., Dermaut, B., Van Broeckhoven, C., Durr, A., and Brice, A. (2001). CAG repeat expansion in the TATA box-binding protein gene causes autosomal dominant cerebellar ataxia. *Brain* **124**, 1939–1947.
7. Stevanin, G., Fujigasaki, H., Lebre, A. S., Camuzat, A., Jeannequin, C., Dode, C., Takahashi, J., San, C., Bellance, R., Brice, A., and Durr, A. (2003). Huntington's disease-like phenotype due to trinucleotide repeat expansions in the TBP and JPH3 genes. *Brain* **126**, 1599–1603.
8. Wu, Y. R., Fung, H. C., Lee-Chen, G. J., Gwinn-Hardy, K., Ro, L. S., Chen, S. T., Hsieh-Li, H. M., Lin, H. Y., Lin, C. Y., Li, S. N., and Chen, C. M. (2005). Analysis of polyglutamine-coding repeats in the TATA-binding protein in different neurodegenerative diseases. *J. Neural Transm.* **112**, 539–546.
9. Oda, M., Maruyama, H., Komure, O., Morino, H., Terasawa, H., Izumi, Y., Imamura, T., Yasuda, M., Ichikawa, K., Ogawa, M., Matsumoto, M., and Kawakami, H. (2004). Possible reduced penetrance of expansion of 44 to 47 CAG/CAA repeats in the TATA-binding protein gene in spinocerebellar ataxia type 17. *Arch. Neurol.* **61**, 209–212.
10. Rigby, P. W. (1993). Three in one and one in three: It all depends on TBP. *Cell* **72**, 7–10.
11. Imbert, G., Trottier, Y., Beckmann, J., and Mandel, J. L. (1994). The gene for the TATA binding protein (TBP) that contains a highly polymorphic protein coding CAG repeat maps to 6q27. *Genomics* **21**, 667–668.
12. Lescure, A., Lutz, Y., Eberhard, D., Jacq, X., Krol, A., Grummt, I., Davidson, I., Chambon, P., and Tora, L. (1994). The N-terminal domain of the human TATA-binding protein plays a role in transcription from TATA-containing RNA polymerase II and III promoters. *EMBO J.* **13**, 1166–1175.
13. Gerber, H. P., Seipel, K., Georgiev, O., Hofferer, M., Hug, M., Rusconi, S., and Schaffner, W. (1994). Transcriptional activation modulated by homopolymeric glutamine and proline stretches. *Science* **263**, 808–811.
14. Gostout, B., Liu, Q., and Sommer, S. S. (1993). "Cryptic" repeating triplets of purines and pyrimidines (cRRY(i)) are frequent and polymorphic: Analysis of coding cRRY(i) in the proopiomelanocortin (POMC) and TATA-binding protein (TBP) genes. *Am. J. Hum. Genet.* **52**, 1182–1190.
15. Reid, S. J., Rees, M. I., Roon-Mom, W. M., Jones, A. L., MacDonald, M. E., Sutherland, G., During, M. J., Faull, R. L., Owen, M. J., Dragunow, M., and Snell, R. G. (2003). Molecular investigation of TBP allele length: A SCA17 cellular model and population study. *Neurobiol. Dis.* **13**, 37–45.
16. Rubinsztein, D. C., Leggo, J., Crow, T. J., DeLisi, L. E., Walsh, C., Jain, S., and Paykel, E. S. (1996). Analysis of polyglutamine-coding repeats in the TATA-binding protein in different human populations and in patients with schizophrenia and bipolar affective disorder. *Am. J. Med. Genet.* **67**, 495–498.
17. Cellini, E., Forleo, P., Nacmias, B., Tedde, A., Bagnoli, S., Piacentini, S., and Sorbi, S. (2004). Spinocerebellar ataxia type 17 repeat in patients with Huntington's disease-like and ataxia. *Ann. Neurol.* **56**, 163–164.
18. Silveira, I., Miranda, C., Guimaraes, L., Moreira, M. C., Alonso, I., Mendonca, P., Ferro, A., Pinto-Basto, J., Coelho, J., Ferreirinha, F., Poirier, J., Parreira, E., Vale, J., Januario, C., Barbot, C., Tuna, A., Barros, J., Koide, R., Tsuji, S., Holmes, S. E., Margolis, R. L., Jardim, L., Pandolfo, M., Coutinho, P., and Sequeiros, J. (2002). Trinucleotide repeats in 202 families with ataxia: A small expanded (CAG)n allele at the SCA17 locus. *Arch. Neurol.* **59**, 623–629.
19. Rolfs, A., Koeppen, A. H., Bauer, I., Bauer, P., Buhlmann, S., Topka, H., Schols, L., and Riess, O. (2003). Clinical features and neuropathology of autosomal dominant spinocerebellar ataxia (SCA17). *Ann. Neurol.* **54**, 367–375.

20. Bauer, P., Laccone, F., Rolfs, A., Wullner, U., Bosch, S., Peters, H., Liebscher, S., Scheible, M., Epplen, J. T., Weber, B. H., Holinski-Feder, E., Weirich-Schwaiger, H., Morris-Rosendahl, D. J., Andrich, J., and Riess, O. (2004). Trinucleotide repeat expansion in SCA17/TBP in white patients with Huntington's disease-like phenotype. *J. Med. Genet.* **41**, 230–232.
21. Toyoshima, Y., Yamada, M., Onodera, O., Shimohata, M., Inenaga, C., Fujita, N., Morita, M., Tsuji, S., and Takahashi, H. (2004). SCA17 homozygote showing Huntington's disease-like phenotype. *Ann. Neurol.* **55**, 281–286.
22. Wu, Y. R., Lin, H. Y., Chen, C. M., Gwinn-Hardy, K., Ro, L. S., Wang, Y. C., Li, S. H., Hwang, J. C., Fang, X., Hsieh-Li, H. M., Li, M. L., Tung, L. C., Su, M. T., Lu, K. T., and Lee-Chen, G. J. (2004). Genetic testing in spinocerebellar ataxia in Taiwan: Expansions of trinucleotide repeats in SCA8 and SCA17 are associated with typical Parkinson's disease. *Clin. Genet.* **65**, 209–214.
23. Zuhlke, C., Dalski, A., Schwinger, E., and Finckh, U. (2005). Spinocerebellar ataxia type 17: Report of a family with reduced penetrance of an unstable Gln49 TBP allele, haplotype analysis supporting a founder effect for unstable alleles and comparative analysis of SCA17 genotypes. *B.M.C. Med. Genet.* **6**, 27.
24. De Michele, G., Maltecca, F., Carella, M., Volpe, G., Orio, M., De Falco, A., Gombia, S., Servadio, A., Casari, G., Filla, A., and Bruni, A. (2003). Dementia, ataxia, extrapyramidal features, and epilepsy: Phenotype spectrum in two Italian families with spinocerebellar ataxia type 17. *Neurol. Sci.* **24**, 166–167.
25. Zuhlke, C., Gehlken, U., Hellenbroich, Y., Schwinger, E., and Burk, K. (2003). Phenotypical variability of expanded alleles in the TATA-binding protein gene. Reduced penetrance in SCA17? *J. Neurol.* **250**, 161–163.
26. Zuhlke, C. H., Spranger, M., Spranger, S., Voigt, R., Lanz, M., Gehlken, U., Hinrichs, F., and Schwinger, E. (2003). SCA17 caused by homozygous repeat expansion in TBP due to partial isodisomy 6. *Eur. J. Hum. Genet.* **11**, 629–632.
27. Maltecca, F., Filla, A., Castaldo, I., Coppola, G., Fragassi, N. A., Carella, M., Bruni, A., Cocozza, S., Casari, G., Servadio, A., and De Michele, G. (2003). Intergenerational instability and marked anticipation in SCA-17. *Neurology* **61**, 1441–1443.
28. Shatunov, A., Fridman, E. A., Pagan, F. I., Leib, J., Singleton, A., Hallett, M., and Goldfarb, L. G. (2004). Small de novo duplication in the repeat region of the TATA-box-binding protein gene manifest with a phenotype similar to variant Creutzfeldt–Jakob disease. *Clin. Genet.* **66**, 496–501.
29. Brusco, A., Gellera, C., Cagnoli, C., Saluto, A., Castucci, A., Michielotto, C., Fetoni, V., Mariotti, C., Migone, N., Di Donato, S., and Taroni, F. (2004). Molecular genetics of hereditary spinocerebellar ataxia: Mutation analysis of spinocerebellar ataxia genes and CAG/CTG repeat expansion detection in 225 Italian families. *Arch. Neurol.* **61**, 727–733.
30. Bruni, A. C., Takahashi-Fujigasaki, J., Maltecca, F., Foncin, J. F., Servadio, A., Casari, G., D'Adamo, P., Maletta, R., Curcio, S. A., De Michele, G., Filla, A., El Hachimi, K. H., and Duyckaerts, C. (2004). Behavioral disorder, dementia, ataxia, and rigidity in a large family with TATA box-binding protein mutation. *Arch. Neurol.* **61**, 1314–1320.
31. Myers, R. H., MacDonald, M. E., Koroshetz, W. J., Duyao, M. P., Ambrose, C. M., Taylor, S. A., Barnes, G., Srinidhi, J., Lin, C. S., Whaley, W. L., Lazzarini, A. M., Schwarz, M., Wolff, G., Bird, E. D., Vonsattel, J. P., and Gusella, J. F. (1993). De novo expansion of a (CAG)n repeat in sporadic Huntington's disease. *Nature Genet.* **5**, 168–173.
32. Stevanin, G., Giunti, P., Belal, G. D. S., Durr, A., Ruberg, M., Wood, N., and Brice, A. (1998). De novo expansion of intermediate alleles in spinocerebellar ataxia 7. *Hum. Mol. Genet.* **7**, 1809–1813.
33. Mittal, U., Roy, S., Jain, S., Srivastava, A. K., and Mukerji, M. (2005). Post-zygotic de novo trinucleotide repeat expansion at spinocerebellar ataxia type 7 locus: Evidence from an Indian family. *J. Hum. Genet.* **50**, 155–157.
34. Giunti, P., Stevanin, G., Worth, P., David, G., Brice, A., and Wood, N. W. (1999). Molecular and clinical study of 18 families with ADCA type II: Evidence for genetic heterogeneity and de novo mutation. *Am. J. Hum. Genet.* **64**, 1594–1603.
35. Shizuka, M., Watanabe, M., Ikeda, Y., Mizushima, K., Okamoto, K., and Shoji, M. (1998). Molecular analysis of a de novo mutation for spinocerebellar ataxia type 6 and (CAG)n repeat units in normal elder controls. *J. Neurol. Sci.* **161**, 85–87.
36. Watanabe, M., Satoh, A., Kanemoto, M., Ohkoshi, N., and Shoji, S. (2000). De novo expansion of a CAG repeat in a Japanese patient with sporadic Huntington's disease. *J. Neurol. Sci.* **178**, 159–162.
37. Schols, L., Gispert, S., Vorgerd, M., Menezes Vieira-Saecker, A. M., Blanke, P., Auburger, G., Amoiridis, G., Meves, S., Epplen, J. T., Przuntek, H., Pulst, S. M., and Riess, O. (1997). Spinocerebellar ataxia type 2: Genotype and phenotype in German kindreds. *Arch. Neurol.* **54**, 1073–1080.
38. Bauer, P., Kraus, J., Matoska, V., Brouckova, M., Zumrova, A., and Goetz, P. (2004). Large de novo expansion of CAG repeats in patient with sporadic spinocerebellar ataxia type 7. *J. Neurol.* **251**, 1023–1024.
39. Maruyama, H., Izumi, Y., Morino, H., Oda, M., Toji, H., Nakamura, S., and Kawakami, H. (2002). Difference in disease-free survival curve and regional distribution according to subtype of spinocerebellar ataxia: A study of 1,286 Japanese patients. *Am. J. Med. Genet.* **114**, 578–583.
40. Hagenah, J. M., Zuhlke, C., Hellenbroich, Y., Heide, W., and Klein, C. (2004). Focal dystonia as a presenting sign of spinocerebellar ataxia 17. *Mov. Disord.* **19**, 217–220.
41. Alendar, A., Euljkovic, B., Savic, D., Djarmati, A., Keckarevic, M., Ristic, A., Dragasevic, N., Kosic, V., and Romac, S. (2004). Spinocerebellar ataxia type 17 in the Yugoslav population. *Acta Neurol. Scand.* **109**, 185–187.
42. Seixas, A. I., Maurer, M. H., Lin, M., Callahan, C., Ahuja, A., Matsuura, T., Ross, C. A., Hisama, F. M., Silveira, I., and Margolis, R. L. (2005). FXTAS, SCA10, and SCA17 in American patients with movement disorders. *Am. J. Med. Genet. A* **136A**, 87–89.
43. Filla, A., De Michele, G., Cocozza, S., Patrignani, A., Volpe, G., Castaldo, I., Ruggiero, G., Bonavita, V., Masters, C., Casari, G., and Bruni, A. (2002). Early onset autosomal dominant dementia with ataxia, extrapyramidal features, and epilepsy. *Neurology* **58**, 922–928.
44. Hernandez, D., Hanson, M., Singleton, A., Gwinn-Hardy, K., Freeman, J., Ravina, B., Doheny, D., Gallardo, M., Weiser, R., Hardy, J., and Singleton, A. (2003). Mutation at the SCA17 locus is not a common cause of parkinsonism. *Parkinsonism Relat. Disord.* **9**, 317–320.
45. Grundmann, K., Laubis-Herrmann, U., Dressler, D., Vollmer-Haase, J., Bauer, P., Stuhrmann, M., Schulte, T., Schols, L., Topka, H., and Riess, O. (2004). Mutation at the SCA17 locus is not a common cause of primary dystonia. *J. Neurol.* **251**, 1232–1234.
46. Stevanin, G., Durr, A., and Brice, A. (2000). Clinical and molecular advances in autosomal dominant cerebellar ataxias: From genotype to phenotype and physiopathology. *Eur. J. Hum. Genet.* **8**, 4–18.
47. Sato, K., Kashihara, K., Okada, S., Ikeuchi, T., Tsuji, S., Shomori, T., Morimoto, K., and Hayabara, T. (1995). Does homozygosity advance the onset of dentatorubral–pallidoluysian atrophy? *Neurology* **45**, 1934–1936.
48. Sobue, G., Doyu, M., Nakao, N., Shimada, N., Mitsuma, T., Maruyama, H., Kawakami, S., and Nakamura, S. (1996). Homozygosity for Machado–Joseph disease gene enhances phenotypic severity [letter]. *J. Neurol. Neurosurg. Psychiatry* **60**, 354–356.

49. Ikeuchi, T., Takano, H., Koide, R., Horikawa, Y., Honma, Y., Onishi, Y., Igarashi, S., Tanaka, H., Nakao, N., Sahashi, K., Tsukagoshi, H., Inoue, K., Takahashi, H., and Tsuji, S. (1997). Spinocerebellar ataxia type 6: CAG repeat expansion in alpha1A voltage-dependent calcium channel gene and clinical variations in Japanese population. *Ann. Neurol.* **42**, 879–884.
50. Squitieri, F., Gellera, C., Cannella, M., Mariotti, C., Cislaghi, G., Rubinsztein, D. C., Almqvist, E. W., Turner, D., Bachoud-Levi, A. C., Simpson, S. A., Delatycki, M., Maglione, V., Hayden, M. R., and Donato, S. D. (2003). Homozygosity for CAG mutation in Huntington disease is associated with a more severe clinical course. *Brain* **126**, 946–955.
51. Trottier, Y., Lutz, Y., Stevanin, G., Imbert, G., Devys, D., Cancel, G., Saudou, F., Weber, C., David, G., Laszlo, T., Agid, Y., Brice, A., and Mandel, J.-L. (1995). Polyglutamine expansion as a pathological epitope in Huntington's disease and four dominant cerebellar ataxias. *Nature* **378**, 403–406.
52. Stevanin, G., Trottier, Y., Cancel, G., Dürr, A., David, G., Didierjean, O., Bürk, K., Imbert, G., Saudou, F., Abada-Bendib, M., Gourfinkel-An, I., Benomar, A., Abbas, N., Klockgether, T., Grid, D., Agid, Y., Mandel, J.-L., and Brice, A. (1996). Screening for proteins with polyglutamine expansions in autosomal dominant cerebellar ataxias. *Hum. Mol. Genet.* **5**, 1887–1892.
53. Martianov, I., Viville, S., and Davidson, I. (2002). RNA polymerase II transcription in murine cells lacking the TATA binding protein. *Science* **298**, 1036–1039.
54. Uchihara, T., Fujigasaki, H., Koyano, S., Nakamura, A., Yagishita, S., and Iwabuchi, K. (2001). Non-expanded polyglutamine proteins in intranuclear inclusions of hereditary ataxias—triple-labeling immunofluorescence study. *Acta Neuropathol. (Berlin)* **102**, 149–152.
55. Roon-Mom, W. M., Reid, S. J., Jones, A. L., MacDonald, M. E., Faull, R. L., and Snell, R. G. (2002). Insoluble TATA-binding protein accumulation in Huntington's disease cortex. *Brain Res. Mol. Brain Res.* **109**, 1–10.
56. Perez, M. K., Paulson, H. L., Pendse, S. J., Saionz, S. J., Bonini, N. M., and Pittman, R. N. (1998). Recruitment and the role of nuclear localization in polyglutamine-mediated aggregation. *J. Cell Biol.* **143**, 1457–1470.

PART VIII

Other Polyamino Acid Repeats

CHAPTER 33

Polyalanine and Polyglutamine Diseases: Possible Common Mechanisms?

AIDA ABU-BAKER AND GUY A. ROULEAU

Center for the Study of Brain Diseases, CHUM Research Center—Notre Dame Hospital, J. A. de Sève Pavillion, Montreal, Quebec, H2L 4M1, Canada

I. Introduction
   A. Noncoding Diseases
   B. Coding Diseases
II. Polyglutamine Diseases
   A. Common Features
   B. Recent Molecular Mechanisms in PolyQ Disease Research: Lessons for PolyAla Disorders
   C. Recent Treatment Advances in the PolyQ Disease Field
III. Polyalanine Disorders
   A. Description
   B. PolyAla Disorders: Examples and Recent Molecular Mechanisms
IV. Comparison of PolyAla and PolyQ Diseases: Similarities and Differences
   A. Protein Aggregates
   B. β-Sheet Structure of Expanded PolyQ and PolyAla Tracts
   C. Gene Structure–Function
   D. Clinical Symptoms
   E. Expansion Length
   F. Mutational Mechanism
   G. Age of Onset
   H. Protein Intracellular Sublocalization
   I. Loss versus Gain of Function
   J. Cell Specificity
   K. Ala Tract versus Q Tract
V. Oculopharyngeal Muscular Dystrophy and PolyQ Diseases
   A. OPMD Background
   B. PolyA Binding Protein Nuclear 1: OPMD Gene Product
   C. Protein Aggregates in OPMD and PolyQ Diseases
   D. OPMD Recent Molecular Mechanisms
   E. Similar Molecular Mechanisms between OPMD and PolyQ Diseases
   F. Treatment in OPMD and PolyQ Diseases
   G. Intriguing Questions
VI. Conclusion
   References

Expansions of polyglutamine repeats are known to cause a variety of human neurodegenerative diseases, and more recently, expansions of alanine tracts have been shown to cause at least nine human conditions. Diseases of polyglutamine expansion are more common and have been the most thoroughly investigated. The observation that a number of polyalanine diseases form protein aggregates suggests a common mechanism with polyglutamine diseases. This chapter will review the rapidly growing body of knowledge concerning polyalanine disorders, with a comparison to polyglutamine disorders. First, the recent advances of molecular mechanisms underlying polyglutamine diseases will be outlined briefly. Second, the polyalanine disorders are reviewed. Third, polyglutamine and polyalanine disorders will be compared. Finally, oculopharyngeal muscular dystrophy, a polyalanine disorder, will be reviewed in more detail as it shares the most mechanisms with polyglutamine diseases.

1. INTRODUCTION

Trinucleotide repeat diseases are caused by expansion of trinucleotide repeats in a gene. Spinobulbar muscular dystrophy (SBMA) was the first triplet repeat disorder described in the early 1990s. Trinucleotide repeat diseases can be classified on the basis of the location of the trinucleotide repeats: diseases involving noncoding repeats (untranslated sequences) and diseases involving coding sequences (exonic). They can be also classified according to the amino acid for which they code: polyglutamine (polyQ) repeat diseases and polyalanine (polyAla) repeat diseases.

A. Noncoding Diseases

At least six diseases fall in the category of noncoding triplet repeat diseases. Because each disease involves a unique repeated sequence, these diseases show relatively little resemblance to one another. The noncoding-repeat diseases are typically multisystem disorders involving the dysfunction–degeneration of many different tissues. The size and variation of the repeat expansions are typically much greater in the noncoding-repeat diseases than in the coding-repeat diseases. Examples of noncoding-repeat diseases include fragile X syndrome (FRAXA), fragile XE mental retardation (FRAXE), Friedreich's ataxia (FRDA), mytotonic dystrophy (DM), spinocerebellar ataxia type 8 (SCA8), and spinocerebellar ataxia type 12 (SCA12) [1–6].

B. Coding Diseases

The second group of triplet repeat disorders are all caused by exonic repeat expansions. They include both polyglutamine (polyQ) and polyalanine (polyAla) diseases.

1. POLYGLUTAMINE DISEASES

PolyQ diseases that are caused by $(CAG)_n$ repeat expansions represent the largest group of trinucleotide repeat diseases (Table 33-1). PolyQ diseases include Huntington's disease (HD), spinobulbar muscular dystrophy (SBMA), spinocerebellar ataxia (SCA) types 1–3, 6, and 7, and dentatorubral pallidoluysian atrophy (DRPLA) [reviewed by Zoghbi and Orr [7]].

2. POLYALANINE DISEASES

PolyAla tract expansions have now been described in several human diseases. At least 9 disorders are associated with polyAla expansions (Table 33-2). Examples of PolyAla disorders include synpolydactyly [8, 9], cleidocranial dysplasia [10], oculopharyngeal muscular dystrophy (OPMD) [11], familial holoprosencephaly [12], hand–foot–genital syndrome [13], blepharophimosis–ptosis–epicanthus inversus syndrome type II [14], X-linked mental retardation and epilepsy [15], X-linked mental retardation with growth hormone deficiency [16], and congenital central hypoventilation syndrome [17].

The goal of this chapter is to compare and contrast the molecular pathogenesis of polyAla and polyQ diseases. First, recent advances of molecular pathogenesis and treatment in polyQ diseases are outlined. Second, similarities and differences between polyAla and polyQ diseases are discussed.

II. POLYGLUTAMINE DISEASES

A. Common Features

The polyQ disorders consist of a group of nine neurodegenerative diseases with overlapping phenotypes that affect distinct neuronal subsets, causing neuronal dysfunction and death. PolyQ diseases have much in common (Table 33-1). (1) With the exception of SBMA, these neurodegenerative disorders are dominantly inherited. (2) They all typically begin in adulthood, with degeneration causing progressive neuronal dysfunction and eventually neuronal loss 10–20 years after onset of symptoms. (3) The polyQ regions are the most important feature shared by the related proteins [18]. (4) Above a certain threshold, the greater the number of polyQ repeats in a protein, the earlier the onset of disease and the more severe the symptoms, suggesting that abnormally long polyQ tracts render their host protein toxic to nerve cells. Disease typically develops when the number of uninterrupted repeats exceeds approximately 35 glutamines, except for SCA6, which is associated with expansions above 21 repeats [19]. (5) The repeats show both somatic and germline instability. (6) Although there is widespread expression of the relevant protein throughout the brain and other tissues, only a certain subset of neurons are vulnerable to dysfunction in each of these diseases. (7) PolyQ repeats confer a gain of function when they are pathologically expanded.

TABLE 33-1 Polyglutamine Diseases

Disease	Gene	Protein function	Normal CAG tract size	Expanded polyQ tract size	Affected cells
Dentatorubral pallidoluysian atrophy (DRPLA)	DRPLA (atrophin-1) on chromosome 12	Insulin pathway	6–35	49–88	Loss of neurons from globus pallidus, basal ganglia, cerebellum, cerebral cortex
Huntington's disease (HD)	Huntington on chromosome 4	Axonal transport	11–34	36–121	Loss of neurons from striatum and cortex
Spinocerebellar ataxia type 1 (SCA1)	SCA1 (ataxin-1) on chromosome 6	Nerve cell connection learning	6–44	39–81	Atrophy of cerebellum and brainstem
Spinocerebellar ataxia type 2 (SCA2)	SCA2 (ataxin-2) on chromosome 12	Protein–protein interaction	14–31	36–64	Cerebellum, midbrain granule cells
Spinocerebellar ataxia type 3 (SCA3) or Machado–Joseph disease	SCA3 ataxin-3 on chromosome 14	Polyubiquitin binding protein	12–43	56–86	Basal ganglia, cerebellum, brainstem, spinal cord
Spinobulbar muscular atrophy (SBMA)	Androgen receptor (AR) on X chromosome	Transcription factor	9–36	38–62	Spinal cord, anterior horn, dorsal root ganglion, bulbar region of the brain, loss of lower motor neurons
Spinocerebellar ataxia type 6 (SCA6)	SCA6 on chromosome 19	Calcium channels	4–18	21–33	Cerebellum, Purkinje cells, granule cells
Spinocerebellar ataxia type 7 (SCA7)	SCA7 (ataxin-7) on chromosome 3	Subunit of GCN5 histone acetyltransferase	4–19	37–306	Visual cortex, cerebral cortex, optic tract

It is important to note that whereas the polyQ diseases are similar to each other, they are not identical. One of the most striking features of polyQ diseases is their multiple effects, i.e., their various effects on behavior, anatomy, and cellular and molecular physiology. Despite this fact, many scientists working on the polyQ diseases believe that finding a treatment for one CAG repeat disorder may help lead to a cure for the other eight as well.

A model of pathogenesis suggests that the expanded polyQ repeat confers toxic gain of function that results in neuronal dysfunction and death. PolyQ diseases are characterized by common features such as protein aggregates, toxic polyQ structures, apoptosis [20], and cell vulnerability.

B. Recent Molecular Mechanisms in PolyQ Disease Research: Lessons for PolyAla Disorders

Given the similarities described, the polyQ diseases are hypothesized to progress via common cellular mechanisms. The advances and developments in polyQ disease research over the last 5 years reflect the multifaceted complexity of these disorders. A striking array of cellular and molecular mechanisms has been proposed to contribute to polyQ pathogenesis. Multiple pathologic mechanisms have been discovered, each serving as a potential therapeutic target. Proposed mechanisms include sequestration of cellular factors by the abnormal protein [21–23], defects in the potential clearance pathways [i.e., chaperones, ubiquitin–proeasome pathway (UPP) [24], and lysosomal–autophagy pathway] [25], inappropriate induction of caspases and apoptosis [26], inhibition of neuron-specific functions such as axonal transport and maintenance of synaptic integrity [27], and transcriptional alterations. Other putative mechanisms include intracellular events such as disruption of cellular transport, mitochondrial dysfunction, altered calcium signaling, excitotoxicity [28], and oxidative stress. Important factors that have been shown to contribute to polyQ disease pathogenesis include the effects of posttranslational modifications on expanded polyQ protein and the importance of subcellular and tissue context.

TABLE 33-2 Polyalanine Diseases

Disease	Symptoms	Gene	Protein function	Normal polyAla tract size	Expanded polyAla tract size	Protein aggregates	Affected cells	Ref.
Synpolydactyly (SPD)	Hand–foot malformation with syndactyly and polydactyly, brachydactyly	HOXD13	Transcription factor patterning of dorsal axis limbs, genitals	15	22–29	Cytoplasmic	Skeletal	[8,9]
Hand–foot–genital syndrome (HFGS)	Hand–foot malformation with short thumbs–great toes, abnormal genitalia	HOXA13	Transcription factor patterning of dorsal axis limbs, genitals	18	24–26	Cytoplasmic	Skeletal	[13]
Cleidocranial dysplasia (CCD)	Skeletal dysplasia, open fontanelles, tooth abnormalities, hypoplastic clavicles, short stature	RUNX2 (CBFA1)	Transcription factor central role in morphogenesis of skeleton, osteoblast differentiation	17	27	Cytoplasmic	Skeletal	[10]
Congenital central hypoventilation syndrome (CCSH)	Loss of ventilary response and Hirschsprung disease	PHOX2B	Transcription factor, development of brain	20	25–29		Nervous system	[17]
Holoprosencephaly (HPE)	Malformation of midline structures of the forebrain and facial cranium	ZIC2	Transcription factor, development of brain and limbs	15	25		Nervous system	[12]
Blepharophimosis–ptosis–epicanthus inversus syndrome (BPES)	Blepharophimosis, ptosis, epicanthus inversus, ovarian failure	FOXL2	Helix–forkhead transcription factor expressed in developing eyes and ovaries	14	22–24	Nuclear and cytoplasmic	Nervous system	[14]
Infantile spasm syndrome X-linked (MR)	Mental retardation, epilepsy, dystonia	ARX	Transcription factor role in development of cerebral cortex and axonal guidance	16 12	18–23 20	Nuclear	Nervous system	[15]
X-linked mental retardation with growth hormone deficiency (MR and GH)	Mental retardation, short stature caused by growth hormone deficiency	SOX3	Transcription factor, neuronal differentiation in brain and spinal cord	15	26	Cytoplasmic	Nervous system	[16]
Oculopharyngeal muscular dystrophy (OPMD)	Progressive, late onset, muscular weakness of eyelid, pharynx	PABPN1, polyA binding protein	Regulates length of polyA mRNA tails	10 10	12–17 11	Nuclear	Muscles	[11]

1. Protein Misfolding

Cells have adapted sophisticated quality-control measures to protect against the accumulation of misfolded and aggregated proteins. For example, molecular chaperones promote proper protein folding and prevent aggregation of nonnative proteins [29]. Proteins that remain misfolded are tagged by ubiquitin and then degraded primarily by the ubiquitin–proteasome system, but also by the phagosome–lysosome system (autophagy).

Protein misfolding and altered solubility of proteins are fundamental defects conferred by polyQ expansion. Failure to detect and eliminate misfolded proteins may contribute to the pathogenesis of neurodegenerative disease. Conversely, it has been suggested that the ubiquitin–proeasome pathway itself may be a target for toxic proteins [24].

The fact that the cells are unable to efficiently degrade the misfolded and expanded polyQ proteins is also supported by the observation that aggresomes containing polyQ-expanded proteins are seen in many models. Aggresomes are the perinuclear inclusion bodies formed around the microtubule organizing center (MTOC) by active minus-end-directed transport of misfolded protein on microtubules [30].

Chaperones colocalize with polyQ aggregates in both cell-based polyQ expression systems and patient tissues. Human DNAJ (heat shock protein 40, HSP40) was found to associate with ataxin-1 aggregates in cell culture and in SCA1 patient tissue [31]. Heat shock proteins 40 and 70 (HSP40 and HSP70) were found to colocalize with aggregates of expanded ataxin-3 and androgen receptor in cell culture models [31–33]. Chaperone overexpression suppressed polyQ aggregation in cell culture, yeast, *Caenorhabditis elegans*, and *Drosophila* [34–40]. Most strikingly, chaperone overexpression has been demonstrated to suppress polyQ-mediated neuronal degeneration in three *Drosophila* models [39, 40].

2. Impairment of Ubiquitin–Proteasome Pathway and Autophagy

Ubiquitin and proteasome components are associated with polyQ inclusions in both disease models and patient tissues [41, 42]. The persistence of ubiquitinated forms of polyQ protein associated with proteasome components indicates that cells are attempting to destroy polyQ-containing proteins and that these substrates may be resistant to proteasomal degradation. Autophagy is a regulated process that is characterized by sequestration of bulk cytoplasm and organelles in autophagic vesicles and their delivery to the lysosome, where the materials are degraded and recycled to general nutrient stores to maintain essential cellular functions under starvation conditions [43].

A report demonstrated that the mammalian target of the drug rapamycin, the mTOR protein, was sequestered in polyQ aggregates in cell models, transgenic mice, and human brains [25]. Sequestration of mTOR impairs its kinase activity and induces autophagy, a key clearance pathway for mutant Huntington, htt [25]. Reduced mTOR activity protected against polyQ toxicity, as the specific mTOR inhibitor rapamycin attenuates aggregate formation and cell death in cell models of HD; inhibition of autophagy has the opposite effects [25].

3. Aberrant Protein Interactions

One possible molecular pathogenetic mechanism underlying polyQ diseases is that mutant polyQ protein binds to protein(s) differently than does normal protein, thereby altering cell function [44, 45].

4. Transcription Dysregulation and Histone Acetylation Alteration

Transcriptional dysregulation, an event downstream of polyQ aggregation, has been implicated in expanded polyQ disease pathogenesis [46]. Interaction of expanded polyQ repeat disease proteins with transcription factors such as TATA binding protein (TBP), CREB binding protein (CBP), p300, p53, mSin3A, nuclear receptor corepressor (N-CoR), Sp1, CA150, C-terminal binding protein (CtBP), cone–rod homeobox protein (CRX), TATA box binding protein associated factor 130 (TAFII130), and RNA polymerase II, both directly and through sequestration of these factors in polyQ aggregates, suggests possible mechanisms for the transcriptional repression caused by polyQ proteins [22, 23, 47–53].

On the basis of the physical interaction between polyQ-containing proteins and transcription factors, it is not surprising that levels of some mRNAs are altered in polyQ diseases [54]. Many of the transcription factors with altered activity in polyQ diseases directly or indirectly regulate histone acetylation, a process that helps regulate transcription through covalent modification of chromatin. This raises the possibility that perturbation of these pathways might be a primary cellular defect in polyQ diseases [55].

Sequestration of the transcriptional coactivator CREB binding protein (CBP), a histone acetyltransferase, has been implicated in the pathogenesis of polyQ expansion neurodegenerative diseases [56]. Direct interaction of htt *in vitro* with the acetyltransferase domain of CBP and p300/CBP-associated factor (PCAF), as well as inhibition of histone acetyltransferase activity, suggests that a loss of acetyltransferase activity may contribute to pathogenesis [55]. An increase in acetylation of nuclear histone proteins, facilitated by histone deacetylase inhibitors, reduces neurodegeneration in cell, fly, and mouse models of HD [21, 55, 57, 58]. Parker et al. reported the identification of sir-2.1, a histone deacetylase, as a genetic modifier of polyQ toxicity. Sir-2 activation through increased *sir-2.1* dosage or treatment with the sirtuin activator resveratrol specifically rescued mutant polyglutamine cytotoxicity in nematode and mammalian neurons [59].

5. Transport Defects

Interference with axonal transport has been implicated in HD and other polyQ diseases [60]. Gunawardena et al. showed that normal levels of htt are required for proper axonal transport in Drosophila [60]. On the other hand, expression of expanded polyQ proteins disrupts axonal transport in larval neurons. Szebenyi and colleagues found that polyQ proteins directly inhibit fast axonal transport and suggested that axonal transport defects may be a common feature of polyQ disease pathogenesis [27]. Cornett et al. provided evidence that polyQ expansion of htt impairs its nuclear export [61]. A more recent study showed that the key role of htt is to promote brain-derived neurotrophic factor (BDNF) transport and suggested that loss of this function might contribute to pathogenesis [62].

6. Mitochondrial Dysfunction

Mitochondrial dysfunction is another prominent feature of polyQ diseases [63, 64]. These diseases are characterized by impaired respiration, increased stress-induced mitochondrial depolarization, increased free radical production associated with oxidation damage, and abnormal energy metabolite concentrations and utilization. Mitochondria play an important role in the mediation of apoptotic pathways. Perturbation of the mitochondrial membrane results in the release of cytochrome *c* which may initiate a caspase-mediated apoptotic cascade [28, 65–67]. Ruan et al. have provided evidence that striatal cells from mutant htt knock-in mice are selectively vulnerable to mitochondrial complex II inhibitor-induced cell death through a nonapoptotic pathway [68].

7. Alteration in Ca^{2+} Signaling

Disruption of calcium homeostasis has been shown to contribute to the molecular basis of neurotoxicity observed in polyQ diseases [64, 69]. An intriguing model of htt toxicity that integrates various cellular components involved in calcium signaling, including N-methyl-d-aspartate (NMDA) receptors and mitochondria, was proposed by Tang et al. [70]. Glutamate receptor antagonists and drugs that enhance mitochondrial function were found to protect neurons in mouse, rat, and nonhuman primate models of neurodegenerative diseases [70].

8. Posttranslational Modifications (Phosphorylation, Ubiquitation, and SUMOylation) and the Importance of Context

The contribution of protein context to the modulation of polyQ toxicity in SCA1 has been reported [71, 72]. These studies suggest that phosphorylation of ataxin-1 at a site far removed from the polyQ tract has a critical role in pathogenesis, perhaps by promoting interaction with the multifunctional regulator protein, 14-3-3 [71].

Immunohistochemical studies show that both the ubiquitinated cytoplasmic inclusions and the nuclear membrane are aberrantly phosphorylated in DRPLA-affected neurons [73]. The small ubiquitin-related modifier (SUMO) posttranslationally modifies many proteins with roles in diverse processes, including regulation of transcription, chromatin structure, and DNA repair. Some proteins can be modified by SUMO and ubiquitin, but with distinct functional consequences. It is possible that the effects of ubiquitination and SUMOylation are largely due to the binding of proteins bearing specific interaction domains. Genetic manipulation of the SUMO conjugation pathway in *Drosophila* indicates that SUMOylation contributes to neurodegeneration in at least two polyQ repeat diseases, SBMA and HD [74–76].

The intracellular and extracellular contexts play key roles in determining the toxicity of mutant polyQ proteins. For example, cell–cell interactions seem to be critical in detrmining HD pathology in the cortex, as reported by Gu et al. [77]. Other cell types (e.g., glia) in addition to neurons may also contribute to HD neuronal pathology.

In addition to extracellular factors, the intracellular environment seems to play a key role in determining mutant polyQ protein's effects. Rousseau et al. showed that aggregation of a protein containing a polyQ stretch of pathological length is abolished when its expression is targeted to the endoplasmic reticulum [78]. When expressed in the mitochondria, a protein containing 73 Q is entirely soluble, whereas the nucleocytosolic equivalent has an extremely high tendency to aggregate. These results imply that polyglutamine aggregation is a property restricted to the nucleocytosolic compartment and suggest the existence of compartment-specific cofactors promoting or preventing the aggregation of pathological proteins [78].

C. Recent Treatment Advances in the PolyQ Disease Field

Compounds targeting a broad range of molecular pathways have been found to have beneficial effects in some polyQ models. Some drugs block very early events leading to polyQ diseases pathology, whereas others improve polyQ-associated symptoms. PolyQ disease models have been used to assess the efficacy of compounds that target specific molecular events, such as protein aggregates, protein misfolding, apoptosis, free radical damage, transcription, mitochondria dysfunction, excitotoxicity, and membrane integrity. Table 33-3 shows different potential therapeutic strategies tested on animal and/or cell models of polyQ diseases.

III. POLYALANINE DISORDERS

A. Description

An Ala expansion was first identified as the disease-causing mutation in synpolydactyly syndrome (SPD). Since then, similar mutations have been described in eight additional disorders. In contrast to polyQ diseases, which cause late onset neurodegenerative diseases, all polyAla disorders, except for OPMD, result in early developmental defects, such as malformations of the brain, digits, and other structures. All of the affected genes in polyAla diseases, except polyA binding protein 1 (PABPN1), which causes OPMD, code for transcription factors that play important roles in early development. Findings in OPMD, as well as other polyAla diseases, indicate that protein misfolding and aggregation may be common features shared with other polyQ diseases.

B. PolyAla Disorders: Examples and Recent Molecular Mechanisms

1. SYNPOLYDACTYLY: *HOXD13*

Synpolydactyly Syndrome (SPD) was the first human disorder discovered to be caused by an *HOX* gene mutation. Its hallmark is a dominantly inherited syndactyly involving the third and fourth fingers and the fourth and fifth toes [79]. The genetic cause is an expansion of an imperfect polyAla coding repeat in exon 1 of *HOXD13* [8]. The 15 normally occurring Ala residues in the N-terminal domain of HOXD13 are expanded by 7–14 alanines in affected cases [8, 9]. Imperfect repeat means that the polyAla repeat is encoded by different codons (four different codons code for alanine). SPD, also known as syndactyly type II, is characterized by incomplete penetrance and variable expressivity. The evolutionary highly conserved homeobox *HOX* genes encode for transcription factors that play key roles during limb and genital development [80]. Studies on affected families with different sized expansions showed that both penetrance and phenotype severity increase progressively with increasing expansion size [9].

The polyAla expansions have two effects. First they inactivate the mutant protein and second the mutant protein itself interferes with the function of wild-type HOXD13 and other 5'-HOXD proteins in the developing limb. These expansions are thus an example of a dominant negative mutation. Loss of function mutations in *HOXD13*, by contrast, produce subtly different distal limb abnormalities in humans [81]. In mice, inactivated *HOXD13* alleles have a very mild phenotype without polydactyly [82, 83], further indicating that the polyAla expansion SPD mutations are not simple loss of function mutations.

The first report to show that polyAla expansion in HOXD13 is necessary and sufficient for aggregation of the protein was published by Albrecht et al. [84]. The study presents evidence for mutant HOXD13 misfolding in cell culture. An increase of the Ala repeat above a certain length (22 Ala) was associated with a shift in the localization of HOXD13 from the nucleus to the cytoplasm, where it formed large amorphous aggregates. Cytoplasmic aggregation of mutant HOXD13 protein was influenced by the length of the repeat, the level of expression, and the efficacy of degradation by the proteasome. Heat shock proteins HSP70 and HSP40 were found to colocalize with the aggregates. Activation of the chaperone system by geldanamycin was found to reduce aggregate formation [84].

2. HAND–FOOT–GENITAL SYNDROME: *HOXA13*

Hand–foot–genital syndrome (HFGS) is a rare, dominantly inherited disorder characterized by shortening of the thumbs and big toes and distal genitourinary tract malformation [85, 86]. An expansion of a polyAla stretch from 18 to 24–26 alanines in the N terminus of *HOXA13* (homeobox gene 13) results in HFGS [87]. *HOXA13* has three alanine tracts that are 14, 12, and 18 alanines long. Expansions have been found in each of the three repeats in HFGS [81]. The phenotype caused by the Ala expansions is similar to that caused by presumed null alleles [13, 88]. A mouse mutant with a 10-Ala expansion in the third repeat of HOXA13 showed a phenotype similar to HFGS in humans [88]. Some evidence suggests that Ala tract expansions in *HOXA13* might confer a dominant negative activity. Expanded HOXA13 has

TABLE 33-3 Treatments in OPMD and Polyglutamine Diseases

Compound	Mode of Action	Effects	Reference no.
	Oculopharyngeal Muscular Dystrophy		
Small molecules			
Chaperones	Protein folding	Increased the solubility of mutant PABPN1-Ala17 and transfected cell survival rate in OPMD cell model	113
Human HSP70, HSP40	Antiamyloid inhibitor	Reduced aggregation and cell death in cell models of OPMD	115
Congo red	Antiprotein aggregation	Onset and severity of OPMD abnormalities were substantially delayed and attenuated by doxycycline treatment in transgenic mice	152
Doxycycline			
	PolyQ Diseases		
A. Small molecules			
1. Antiprotein aggregation:			
Congo red	Antiprotein aggregation (antiamyloid)	Reduced aggregation and cell death in HD cell models	202
Thioflavine S	Chemical disruption of aggregation	Decreased polyglutamine aggregates in cerebrum and liver	159
Chrysamine G	Antiprotein aggregation	Improved motor dysfunction and extended lifespan in a transgenic mouse model of HD	
Direct fast yellow			
Trehalose (disaccharides)		Reduced aggregate formation and suppressed apoptosis in several polyQ disease models	203
2. Transglutaminase inhibitors:			
Cystamine	Transglutamine inhibitor and antioxidant	Inhibited the activity of transglutaminase (TGase), an enzyme involved in the formation of protein aggregates	204
		Inhibited caspase-1 activity	124
		Increased levels of HSP40	
		Improved motor function in HD transgenic mice and decreased nerve cell death in HD of protein aggregates	204
3. Protein folding drugs			
Geldamycin	Protein folding	Retarded expanded poly ataxin-1 in cell culture	205
	Activates HSP90		
4. Histone deacetylase inhibitors:			
Mithramycin		Increased the histone acetylation level and gene transcription rescue effects in HD transgenic fly and mouse models	55
Sodium butyrate			
SAHA			
5. Antiapoptotic drugs:			
zVAD-fmk	Caspase inhibitors	Improved survival of R6/2 mice	26
YVAD-fmk	Inhibited caspase-1 and -3		
	Inhibited inflammation		
Minocyclin antibiotics (tetracycline)	Protein aggregation inhibitor permeability transition and mitochondria-induced cell death pathways	Delayed disease progression and death in HD transgenic mouse	206, 207
	Also blocked mitochondrial		

6. Mitochondria agents:

Creatine	Stimulated pyruvate dehydrogenase activity (improved mitochondrial function)	Inhibited aggregation, improved survival, and delayed onset of neuropathology in transgenic mice	208
Dichloroacetate		Effective in animal models of HD	

7. Excitotoxicity drugs:

Coenzyme Q_{10}	Cofactor of the electron transport chain, an important antioxidant	Neuroprotective effects	209
Riluzole, creatine	Reduced glutamate, inhibited polyQ aggregation Glutamate-release inhibitors	Useful in transgenic mouse of polyQ diseases	210
Remacemide ketamine	Antiapoptotic, antiexcitotoxic effects	Improved performance on the rotarod test, but had effect on survival of HD mice	211
Lithium chloride	Glutamate receptor antagonist	Enhanced mitochondrial function	212
Baclofen, tamotrigine	Glutamate-release inhibitors		
Tauroursodeoxycholic acid (TUDCA), a hydrophilic bile acid	Antioxidant Inhibited mitochondrial permeability transition Inhibited cytochrome c, Bax translocation, and caspase activation Neuroprotection against glutamate excitotoxicity	Slowed neurodegeneration and decreased the size and number of aggregates in HD mouse model	212
Lithium (anti-degressive drug)		Inactivation of NMDA receptors, decreased expression of proapoptotic proteins, p53 and Bax, enhanced expression of the cytoprotective protein, Bcl-2, and activation of the cell survival kinase, akt In a rat HD model, lithium significantly reduced brain lesions	213

8. Sirtuin-activating compound

Resveratrol	Sir-2 activation	Rescued mutant polyQ cytotoxicity in *C. elegans* and mammalian neurons	59
B. Antibodies or intrabodies (scFv)	Disrupted interactions of mutant protein with itself and other cellular targets	Inhibited aggregation of mutant htt in cell culture Altered cytotoxicity	214 215
C. siRNA	Inhibition of polyQ proteins by targeting the htt mRNA	Rescued polyQ-mediated cytotoxicity	216
	Selective targeting of mutant htt		217
D. Peptides	Short protein sequences that can bind mutant polyQ protein and inhibit aggregation	Suppressed aggregation and neurodegeneration	218

been shown to induce cytoplasmic aggregates in cell culture [84].

3. CLEIDOCRANIAL DYSPLASIA: *RUNX2*

PolyAla expansions in the transcription factor RUNX2 causes cleidocranial dysplasia (CCD) [10]. CCD is an autosomal dominant condition characterized by defective endochondral and intramembranous bone formation [89]. Typical features include hypoplasia–aplasia of clavicles, delayed closure of fontanelles, supernumerary teeth, short stature, and other changes in skeletal patterning and growth [10]. *RUNX2* (also known as *CBFA1*) is a member of the Runt family of transcription factors that is essential for osteoblast and chondrocyte differentiation [90]. Targeted inactivation of one allele of the *RUNX2* (*CBFAL*) gene in the mouse is sufficient to cause a similar constellation of skeletal malformations [90]. The human *RUNX2* contains 17 alanines in the N terminus, and expansion to 27 alanines causes CCD [10]. Expanded RUNX2 has been shown to induce cytoplasmic aggregates in cell culture [84].

4. CONGENITAL CENTRAL HYPOVENTILATION SYNDROME: *PHOX2B*

Congenital central hypoventilation syndrome (CCSH) is a life-threatening disorder involving an impaired ventilatory response to hypercarbia and hypoxemia. CCSH is an autosomal dominant disorder with incomplete penetrance [17]. *PHOX2B* is the main disease-causing gene for CCSH, encoding for a highly conserved paired box homeodomain transcription factor, which has two polyAla repeats in the C-terminal region of 9 and 20 residues. *PHOX2B* is expressed in the central and peripheral autonomic nervous systems during development and is essential for the formation of the medullary reflex circuits. Mutation screening of *PHOX2B* in CCHS patients has revealed in-frame duplications within the 20-Ala stretch, leading to expansions from +5 to +13 Ala residues, and frameshift mutations, leading to aberrant C-terminal regions [17, 91–93]. A correlation between the length of the expansion and the severity of the autonomic dysfunction symptoms was observed [93].

5. HOLOPROSENCEPHALY: *ZIC2*

Holoprosencephaly (HPE) is a term that was first used in 1963 to describe a spectrum of malformations that have failure of the proper formation of midline structures of the forebrain [94]. In HPE patients, the cerebral hemispheres fail to separate into distinct left and right halves. The majority of HPE cases are sporadic, although families with both autosomal dominant and autosomal recessive HPE have been described.

Genetic studies showed that heterozygous mutations in the *ZIC2* (*ZIC* gene family member 2) gene are one cause of HPE [12, 95]. A subsequent report confirms that partial loss of ZIC2 activity leads to a similar malformation in mice [12, 96]. The ZIC proteins are transcription factors that are expressed during brain development, and they play an important role in the development of the nervous system [97]. The Ala stretch expansion from 15 to 25 in the C terminus of ZIC2 also has been found to cause HPE [12, 98]. The phenotypes of HPE patients with Ala tract expansions are similar to those in patients with loss of function mutations, suggesting that Ala tract expansion leads to a loss of protein function. Cotransfection assays revealed that transcription factors carrying polyAla-expanded tracts impair transcriptional regulation of target genes. Ala-expanded ZIC2 was shown to repress the transcriptional activity of the *ApoE* regulatory region as compared to the wild-type protein [98].

6. BLEPHAROPHIMOSIS–PTOSIS–EPICANTHUS INVERSUS SYNDROME TYPE II: *FOXL2*

Expansion from 14 to 24 residues of a polyAla tract in the C terminus of *FOXL2* gene causes blepharophimosis–ptosis–epicanthus inversus syndrome (BPES) [14]. The gene has been identified as the putative forkhead transcription factor *FOXL2* (forkhead box-like 2) [14]. Forkhead proteins are found in all eukaryotes and serve important functions in the establishment of the body axis and the development of tissues from all three germ layers in animals [100]. FOXL2 is expressed in developing eyelids as well as in fetal and adult ovarian follicular cells. Unlike type I BPES, which is associated with premature ovarian failure (POF), type II affected individuals are fertile. Both types of BPES are autosomal dominant conditions characterized by eyelid dysplasia (blepharophimosis), drooping eyelids (ptosis), and a tiny skin fold running inward and upward from the lower lid (epicanthus inversus) [101]. It is likely that Ala tract expansion in this gene results in at least a partial loss of function [100, 102].

A study of BPES in a cell culture model reported that a polyAla expansion in the transcription factor FOXL2 leads to a shift in protein localization from the nucleus to the cytoplasm. The polyAla expansion was also found to induce the formation of intranuclear aggregates and cytoplasmic aggregation [103]. The role of aggregation in cell death requires additional exploration [103].

7. X-Linked Infantile Spasm Syndrome (West Syndrome) and Partington Syndrome (Mental Retardation and Epilepsy): ARX

PolyA tract expansions in the Aristaless-related homeobox (ARX) protein have been identified in a subset of patients with infantile spasms and mental retardation [104]. Expansions in two of the four Ala tracts of ARX were found in families with nonsyndromic mental retardation, as well as in X-linked infantile spasm syndrome (West syndrome, WS) and Partington syndrome [15, 105].

WS is composed of the triad of infantile spasms, an abnormal EEG pattern termed hypsarrhythmia, and mental retardation [104]. Patients with Partington syndrome show mental retardation and dystonic movements of the hands. The term "infantile spasms" refers to a type of seizure, a severe form of early onset epilepsy often associated with a poor prognosis. Infantile spasms are believed to reflect abnormal interactions between the cortex and brainstem structures. Ninety percent of infantile spasms begin in children younger than 12 months.

ARX belongs to the paired-type homeobox genes that are expressed in the forebrain and in the floorplate of the developing central nervous system [105]. The phenotypes associated with different ARX mutations demonstrate remarkable pleiotropy, but they also comprise a nearly continuous series of developmental disorders that begins with hydranencephaly, lissencephaly, and agenesis of the corpus callosum and ends with a series of overlapping syndromes with apparently normal brain structure [89–106]. The striking difference in phenotypes suggests either a partial loss of function or another specific mutational mechanism that results from the Ala repeat expansions.

In a recent model of infantile spasms and mental retardation, expression of expanded ARX in both cell culture and mouse brain slices results in filamentous nuclear inclusions and an increase in cell death [107]. These inclusions are ubiquitinated and contain HSP70. Coexpressing HSP70 decreased the percentage of cells with nuclear inclusions. These results suggest that expansions in one of the ARX polyAla tracts result in nuclear protein aggregation associated with cell death, leading to infantile spasms and mental retardation [107].

8. X-Linked Mental Retardation with Growth Hormone Deficiency: SOX3

An Ala expansion from 15 to 26 residues in the transcription factor SOX3 results in X-linked mental retardation with growth hormone deficiency (XLMR + GH) [16]. XLMR + GH results in mental retardation associated with facial or growth abnormalities and/or metabolic or neuromuscular disorders. SOX3 is highly expressed in the fetal brain and spinal cord, playing an important role in neuronal development. Specifically, SOX3 is expressed during neural and pituitary development. The polyAla-expansion-induced dysfunction of the SOX3 protein may disturb transcription pathways and the regulation of genes involved in cellular processes and functions required for cognitive and pituitary development [16]. Expanded SOX3 has been shown to induce cytoplasmic aggregates in cell culture [84].

9. Oculopharyngeal Muscular Dystrophy: PABPN1

Oculopharyngeal muscular dystrophy (OPMD) is an autosomal dominant, midlife onset (~45 years) disease affecting skeletal muscle. It is characterized by progressive swallowing difficulties, eyelid drooping, and proximal muscle weakness. OPMD is caused by expansions of the short (GCG) trinucleotide repeat in the coding sequence of the PABPN1 gene [11]. The normal PABPN1 gene has a $(GCG)_6$ repeat encoding a polyAla stretch at the 5' end, whereas in OPMD patients this repeat is expanded to $(GCG)_{8-13}$. Due to the presence of a GCA GCA GCA GCG coding sequence adjacent to the $(GCG)_6$ repeat, the wild-type PABPN1 protein has a 10-Ala stretch, whereas the mutated PABPN1 in dominant OPMD has 12–17 alanines in the N-terminal domain [11]. The highest prevalence of OPMD is in the French-Canadian population, where, 1/1000 people are at risk, though the disease is found worldwide. An OPMD project began in the authors lab in 1990 with the collection of large families segregating the disorder. In 1995, the mapping of the gene was reported, and in 1998 polyA binding protein nuclear 1 (PABPN1) was identified as the gene responsible for OPMD [11, 108]. Interestingly, homozygosity for the $(GCG)_7$ allele leads to an autosomal recessive form of OPMD. [99].

Filamentous intranuclear inclusions (IIs) are considered the pathological hallmark of OPMD [109]. Until relatively recently, among the family of polyAla expanded proteins, PABPN1 was the only one believed to form IIs. For at least five other polyAla diseases it has now been shown that mutant protein forms protein aggregates [84, 103, 107]. The mechanism by which polyAla expansions in transcription factors and the nuclear protein PABPN1 result in cellular dysfunction remains to be defined. The following section compares polyAla diseases to polyQ diseases.

IV. COMPARISON OF POLYALA AND POLYQ DISEASES: SIMILARITIES AND DIFFERENCES

A. Protein Aggregates

An important feature shared by polyQ diseases, and by many polyAla diseases, is the presence of protein aggregates (or inclusions). Mammalian models of polyQ and polyAla diseases indicate that mutant proteins are more prone to aggregate than their wild-type counterparts and cause significantly more cell death [110, 111]. In all of these diseases, the pathology and eventual death of specific cell populations are associated with the accumulation of distinct abnormal unfolded polypeptides. However, the role of protein aggregates in polyQ and/or polyAla disease pathogenesis is still not clear.

In a wide variety of systemic diseases, unfolded polypeptides accumulate in cells, largely as insoluble aggregates, and appear to play a critical role in disease pathogenesis. Intracellular inclusions of denatured proteins are characteristic features of many neurological diseases, including amyotrophic lateral sclerosis, Alzheimer's disease, Parkinson's disease, and polyQ diseases [reviewed in Sherman and Goldberg [112]].

OPMD IIs (as well as other protein aggregates that have been reported in other polyAla disorders [84, 103, 107]) are distinct, but they also share some properties with aggregates observed in inherited neurodegenerative diseases associated with expanded polyQ stretches. First, as with polyQ expanded proteins, the protein aggregates induced by the expression of mutant PABPN1 and mutant HOXD13 were found to sequester components of UPP and molecular chaperones [84, 113–115]. Second, both polyAla and polyQ protein aggregates are associated with cytotoxicity [107, 113, 115, 116]. Third, in polyQ diseases, the occurrence of protein aggregates in the nucleus and cytoplasm of the patient's brain is dependent on the length of the polyQ tract. Similarly, it was shown that the size of the Ala expansions correlates with the severity of the cytoplasmic aggregation of mutant HOXD13 [84]. The smallest Ala repeat expansion in HOXD13 was predominantly located in the nucleus, whereas the long expansions in HOXD13 were cytotoxic and mainly located in the cytoplasm [84]. Fourth, in both polyQ and polyAla diseases, the formation of protein aggregates is a time-dependent process [117].

It is important to note that the formation of polyAla aggregates was only detected in cell culture models of SPD, CCD, HFGS, and MRX + GH [84, 103]. Patient tissues need be examined for the presence of protein aggregates, though it is possible that cells that expressed the mutant protein died long before pathological material became available. Nonetheless, the potential effects of aggregation on cell viability should be also investigated in these polyAla disorders.

B. β-Sheet Structure of Expanded PolyQ and PolyAla Tracts

The expanded polyQ and polyAla protein repeats are thought to destabilize the native configuration of the mutant protein, resulting in the formation of protein aggregates within the cell. The propensity of expanded polyQ and polyAla to aggregate must derive from their ability to self-interact. The best known structure to explain the molecular mechanism of polyQ aggregates is the "polar zipper" described by Max Perutz [118]. In this modeling structure, polyQ tracts form extended antiparallel β strands held together by an extensive network of hydrogen bonds between both the main-chain and side-chain amides [118]. In polyQ diseases, inclusions appear to be involved in protein–protein interactions that are especially stable due to the extensive formation of β-sheet structures in polyQ repeats [119, 120]. In addition, some aggregate polypeptides may be covalently cross-linked by transglutaminase [121–123]. A large body of studies supports the role of transglutaminase in the aggregation of expanded polyQ. Transglutaminase is an enzyme that catalyzes the formation of cross-links between glutamine and lysine residues [124].

Hydrophobic polyAla peptides can also form β-pleated-sheet fibrillar macromolecules *in vitro* [125] and are resistant to chemical denaturation and enzymatic degradation [126]. It appears likely that an increase in Ala repeat length above 12–22 Ala results in misfolding and/or aggregation of the protein due to biophysical limitations [127]. Expression of a 19–37 Ala stretch fused to green fluorescent protein (GFP) in COS cells induces the formation of intracellular aggregates that are associated with cell death [116, 128]. The nuclear inclusions in Machado–Joseph disease (MJD) were reported to contain polyAla [129]. A transcriptional or translational frameshift occurring within the polyQ-encoding CAG repeats results in the production of polyAla-containing mutant proteins that are more harmful to the cell than polyQ [129].

C. Gene Structure–Function

In contrast to some polyQ proteins, the functions of all polyAla proteins are well-defined. In all polyAla disorders, except OPMD, the affected genes are transcription factors that function in the regulation of important early developmental processes (Table 33-2). In SBMA, a polyQ disease, the affected protein "androgen receptor" is known as a transcription factor. None of the other affected genes in polyQ disorders are transcription factors, but several reports have shown the involvement of transcription factors in these diseases. PolyQ proteins have different functions (Table 33-1).

It is important to mention that in some polyQ diseases, such as MJD, HD, and SCA7, the exact function of the affected gene was not known until more recently. Progress in the past 5 years has revealed important normal functions of these genes. For example, wild-type htt is suggested to be involved in the transport of vesicles [62] and to up-regulate the transcription of brain-derived neurotrophic factor (BDNF) [130]. The beneficial antiapoptotic effect property of wild-type htt has been also reported [131]. Wild-type htt encourages neuronal survival by blocking the activation of apoptotic cell cascades [132]. Ataxin-3, the protein mutated in Machado–Joseph disease (SCA3), belongs to a novel group of cysteine proteases and is predicted to be active against ubiquitin chains or related substrates [133]. Wild-type ataxin-3 has been shown to be a polyubiquitin binding protein that associates with the proteasome [133]. Furthermore, wild-type ataxin-3 was reported to be a histone binding protein with two independent transcriptional corepressor activities [52] and was found to be a component of the SAGA histone acetyltransferase complex [134]. Helmlinger et al. have demonstrated that wild-type ataxin-7 is a subunit of GCN5 histone acetyltransferase-containing complexes [135]. Increasing evidence indicates that loss of function may also be important in polyQ disease pathology [62].

D. Clinical Symptoms

Although all polyQ diseases are neurodegenerative, the symptoms and brain pathology differ in each polyQ disease. Expansion in the Ala tract in polyAla diseases also results in different abnormalities in the organs and tissues (Tables 33-1 and 33-2). PolyAla disorders are responsible for skeletal malformations (SPD, HFGS, CCD), abnormalities of the nervous system including mental retardation (HPE, WS, XLMR + GH, CCHS), and muscle disease (OPMD). Therefore, both polyQ and polyAla diseases can affect the nervous system (Tables 33-1 and 33-2). Analysis of repeat size and symptomatology in polyQ disease patients has clearly demonstrated an inverse relationship between the size of the repeat and the age of onset. The larger the polyQ expansion in size, the more severe the symptoms in polyQ diseases. A positive correlation between the size of the expansion and the severity of the phenotype has been demonstrated for HOXD13 in SPD [9]. Such correlations have been suggested but remain unconfirmed for PABPN1 expansions in OPMD [136–138] and PHOX2B in CCHS.

E. Expansion Length

It is striking that the length of the normal Ala tract is similar in all transcription factors (14–20 Ala residues) and so is the expanded tract that causes disease (18–29 Ala residues), suggesting a common underlying mechanism in these congenital malformation syndromes. An important difference between polyAla and polyQ diseases is that pathologically expanded polyAla tracts (18–29) are shorter than pathologically expanded polyQ tracts (35–85) (Tables 33-1 and 33-2). It is interesting that when the polyQ expansions are large, there is significant overlap in the phenotypes of these polyQ disorders.

F. Mutational Mechanism

Another contrast between polyAla and polyQ is the mutational mechanism. Polymerase slippage has long been assumed to be the mechanism of expansion in polyQ diseases. In contrast, for polyAla expansions unequal allelic homologus recombination during meiosis and/or mitosis is thought to be the mechanism. Expanded polyAla tracts are meiotically and mitotically stable, whereas expanded polyQ tracts tend to be meiotically and mitotically unstable. PolyQ repeats show both somatic and germline instability. PolyAla tracts show only a low degree of polymorphism [139].

G. Age of Onset

PolyQ diseases as well as OPMD are generally late onset diseases, whereas most of the polyAla diseases are congenital syndromes.

H. Protein Intracellular Sublocalization

All polyAla proteins are localized in the nucleus, mainly because of their functions as transcription factors or in mRNA polyadenylation. The strong association of long (>10) Ala repeats with nuclear proteins suggests a function related to the nucleus and regulation of transcription. On the other hand, different subcelullar (nuclear and/or cytoplasmic) localization for polyQ proteins has been reported (Tables 33-1 and 33-2).

I. Loss versus Gain of Function

Except for OPMD, the gene mutations in polyAla disorders function mainly by disrupting the activity of the protein and lead to loss of function or act as dominant negative alleles HOXD13. All polyQ pathogenesis models suggest that the expansion of CAG repeats confers a toxic gain of function, i.e., disease develops because the mutant form of the protein gains a new function, not because the protein loses its normal function. The presence of a normal copy of the gene will not counteract the deleterious effects of the abnormal protein. However, for some proteins, the expanded polyQ tract may cause a partial loss of function, as is the case with androgen receptor, whereas SBMA patients have some androgen insensitivity. Intriguing results have shown that the loss of function of wild-type proteins in other polyQ diseases may contribute to the pathogenesis.

J. Cell Specificity

Although all proteins known to have expanded polyQ repeats are expressed in many tissues in the human body, each is toxic to only a specific group of nerve cells (Table 33-1). The specific neurons that are affected differ from one disease to the next. Selective vulnerability is not absolute because different neuronal populations may be susceptible to cell dysfunction and/or death. On the other hand, skeletal malformations and abnormalities of the nervous system are presented in polyAla disorders (Table 33-2).

K. Ala Tract versus Q Tract

PolyAla tract residues have been found in 494 human proteins [139]. Interestingly, Ala stretches do not exceed 20 Ala in humans and are relatively short homopolymeric repeats when compared with polyQ tracts. Ala tracts have been shown to be necessary for transcription factor activity [140]. Albrecht et al. performed a genome-wide search for polyAla tracts longer than 8 Ala [84]. They identified 137 proteins, 7 of which could not be assigned to any functional category. A total of 102 (77%) were described to be located in the nucleus, and 32 (23%) were cytoplasmic, membrane-bound, or extracellular proteins. Of the nuclear proteins, 66 (50% of total) were transcription factors. The relative percentage of transcription factors increased with the length of the repeat. The results indicate that Ala repeats are likely to have a function related to the nucleus, presumably in the regulation of transcription [84].

The functional significance of polyQ tracts is not known. One possibility is that this tract plays a role in facilitating protein–protein interactions. A significant majority of polyQ tracts are found in proteins involved in transcription and signal transduction [141].

V. OCULOPHARYNGEAL MUSCULAR DYSTROPHY AND POLYQ DISEASES

Because more is known about OPMD than other polyAla disorders, and because OPMD has been shown to share several common mechanisms with polyQ diseases, the focus will be on the common mechanisms between OPMD and polyQ diseases that have been investigated so far.

A. OPMD Background

OPMD is a very unique disease sharing common features with other polyAla disorders, as well as with muscular dystrophic disorders and polyQ diseases. OPMD is an interesting disease for several reasons. The affected gene *PABPN1* is not a transcription factor. Wild-type PABPN1 is involved in pre-mRNA polyadenylation [142, 143]. The disease course and underlying mechanisms in OPMD in some ways appear to be similar to those seen in polyQ expansion disorders, as protein misfolding and subsequent aggregation appear to be central to the pathogenesis.

Both OPMD and polyQ diseases are coding triplet trinucleotide diseases and both are late onset diseases. Both polyQ and polyAla are hydrophobic amino acids that are able to form β-sheet structures and induce intracellular protein aggregation, as discussed previously. The most striking difference between OPMD and polyQ diseases is repeat length. Short expansions (>2) of the polyAla stretch in wild-type PABPN1 cause OPMD, whereas much longer expansions of polyQ (>35) are required to cause neurodegenerative diseases. It is proposed that polyAla is more harmful to the cell than polyQ [129]. Another difference is that in OPMD the pathogenic effects are mostly observed in skeletal muscle cells, as opposed to neuronal tissues in the case of polyQ disorders.

B. PolyA Binding Protein Nuclear 1 (PABPN1): OPMD Gene Product

Wild-type PABPN1 is an abundant nuclear protein, with widespread staining in the nucleoplasm [144]. It is mostly concentrated in discrete nuclear domains called "speckles." Wild-type PABPN1 is a well-studied protein with a predicted molecular weight of 32.8 kDa [145]. Wild-type PABPN1 is a multidomain protein of 306 amino acids [146] that comprise the polyAla stretch of 10 consecutive Ala and a proline-rich region in the acidic N terminus. The residues located between positions 125 and 161 are predicted to form an α-helix structure, which is required for the stimulation of the polyA polymerase [219]. The central region of wild-type PABPN1 also includes a putative ribonucleoprotein (RNP) type RNA binding domain (RBD) [146]. The basic C-terminal domain of wild-type PABPN1, which is rich in dimethylated arginine residues, contains a nuclear localization signal (NLS) that interacts with transportin (a nuclear transport receptor). The association of wild-type PABPN1 and transportin occurs in a Ras-related nuclear protein (Ran) GTP-sensitive manner, suggesting an active transport pathway [147].

All eukaryotic mRNA are posttranscriptionally modified at their 3' ends by the addition of a polyA tail, to which are bound two different proteins: PABPN1 (nuclear) and PABPC (cytoplasmic). Whereas wild-type PABPN1 protein stimulates the rapid and processive polymerization of the polyA tail in the nucleus [142, 148], PABPC protein in the cytoplasm functions in the initiation of translation and in the regulation of mRNA decay [149]. At steady state, wild-type PABPN1 is a nuclear protein [114, 144, 147], whereas PABPC is cytoplasmic [149]. Several studies suggested that PABPC is a shuttling protein, entering the nucleus and being exported to the cytoplasm in association with mRNA [150]. Wild-type PABPN1 is also able to perform nucleocytoplasmic shuttling, as demonstrated by heterokaryon assays [147].

C. Protein Aggregates in OPMD and PolyQ Diseases

Mutated PABPN1 is prone to form insoluble intranuclear inclusions (IIs), particularly in the nucleus where wild-type PABPN1 normally resides. OPMD was the first polyAla disorder identified that leads to IIs in the muscle nuclei of patients [109, 151]. Calado et al. showed that OPMD IIs were more resistant to salt extraction than the wild-type protein, which is dispersed in the nucleoplasm [114]. Viewed with the electron microscope, the OPMD IIs consist of collections of tubular filaments approximately 8.5 nm in external diameter, disposed in tangles or palisades decorated by anti-PABPN1 antibodies [114]. OPMD IIs were found to be associated with cell death [113, 115, 116, 152], which is similar to polyQ diseases. It is unknown how the presence of IIs disrupts neuronal or muscular function. Nevertheless, the fact that IIs are preferentially found in susceptible neurons and muscle fibers suggests that they play a critical role in pathogenesis.

D. OPMD Recent Molecular Mechanisms

The dominant pattern of inheritance in OPMD is more consistent with a gain of function, rather than a loss. The mechanism by which an expanded Ala tract leads to OPMD is unknown. Several lines of evidence support the idea that OPMD shares some commn molecular mechanisms with polyQ diseases. The role of the protein's oligomerization to OPMD pathogenicity was documented [116]. Among the models put forward to better understand OPMD pathogenesis, some hypothesize that a general disruption of either the ubiquitin–proteasome pathway (UPP) [113] or cellular RNA metabolism [114] appears to contribute to the disease. OPMD intranuclear aggregates were shown to sequester wild-type PABPN1 [114], polyA RNA [114], hnRNP proteins [153], and components of the UPP and heat shock protein (HSP) pathway [113–115]. The role of the nucleus in OPMD pathogenesis has been investigated [154]. The apoptosis pathway has been implicated in OPMD pathogenesis [152]. Similar mechanisms have also been well-documented in polyQ diseases.

E. Similar Molecular Mechanisms between OPMD and PolyQ Diseases

1. Oligomerization in OPMD and PolyQ Diseases

Wild-type PABPN1, when bound to polyA RNA, forms both linear filaments and discrete sized, compact oligomeric particles *in vitro* [145]. Oligomerization of wild-type PABPN1 is mediated via two potential oligomerization domains (ODs), one in the middle of the protein and the other at the C terminus [116].

Evidence has been provided, in cell culture, that oligomerization of mutant PABPN1-Ala17 is important in the formation of IIs and for cell death in OPMD. Inactivating oligomerization of mutant PABPN1-Ala17 by deletions from either of the ODs prevented nuclear protein aggregation and significantly reduced cell death [116]. ODs do not overlap with the Ala stretch of PABPN1. These findings suggest that oligomerization of mutant PABPN1 plays a crucial role in the formation of OPMD nuclear protein aggregation, whereas the expanded polyAla stretch is necessary but not sufficient to induce OPMD protein aggregation.

It seems surprising that extension of the Ala repeat by two residues in mutant PABPN1 should have such a drastic effect as causing OPMD. To find out whether Ala repeats tend to aggregate, Perutz et al. [127] made the peptide $D_2A_{10}K_2$ and measured its CD spectrum in solution as a function of pH. The spectra showed that the peptide forms α helices at all pHs, with no sign of aggregation. α helices can adhere to each other by forming coiled coils, but it is most unlikely that expansion from six to eight Ala would have this effect. Perutz and coworkers suggested the following interpretation: Ala repeats are hydrophobic and would, therefore, occupy internal positions in the protein. An additional Ala would be a misfit that lowers the free energy barrier to unfolding of the protein [127]. In a protein structural study, wild-type PABPN1 was shown to form filaments slowly, whereas expanded PABPN1 formed filaments more quickly. Structural analysis of the fibrils indicated antiparallel β sheets [155].

The facilitation of oligomerization to aggregate formation and toxicity in polyQ diseases has also been documented. One of the earliest oligomerization studies of polyQ domain proteins in mammalian neurons was carried out by Onodera et al. [156]. Varying lengths of polyQ–GFP fusion proteins were expressed in cultured COS-7 cells. The 19Q–GFP and 35Q–GFP proteins (nonpathologic length) distributed diffusely throughout the cytoplasm. In contrast, 56- and 80-Q fusion proteins (pathologic length) formed fibrillar arrays resembling those previously observed in neurons in HD and in a transgenic mouse model. These aggregates were intranuclear and intracytoplasmic [156]. Iuchi et al. [157] described oligomeric and polymeric aggregates formed in cells by expanded polyQ. These aggregates are not dissociated by concentrated formic acid, an extremely effective solvent for otherwise insoluble proteins [157].

Full-length human ataxin-3 was used to investigate the changes in secondary structure, aggregation behavior, and fibril formation associated with an increase from the normal length of 27 Q to a pathogenic length of 78 Q. 78-Q ataxin-3 was found to aggregate more strongly than 27-Q ataxin. A marked decrease in α-helical secondary structure accompanies expansion of the polyQ tract, suggesting destabilization of the native protein. Infrared spectroscopy of fibrils reveals a high β-sheet content and suggests a parallel, rather than an antiparallel, sheet conformation [158].

More evidence of polyQ oligomerization came from the use of antiamyloid inhibitors. The azo-dye Congo red binds preferentially to β sheets containing amyloid fibrils and can specifically inhibit oligomerization and disrupt preformed oligomers [159]. The study showed that inhibition of polyQ oligomerization by Congo red prevents ATP depletion and caspase activation, preserves normal cellular protein synthesis and degradation functions, and promotes the clearance of expanded polyQ repeats *in vivo* and *in vitro*. Infusion of Congo red into a transgenic mouse model of HD, well after the onset of symptoms, promotes the clearance of expanded repeats *in vivo* and exerts marked protective effects on survival, weight loss, and motor function. Therefore, oligomerization may be a crucial determinant of the biochemical properties of expanded polyQ tracts that are central to their chronic cytotoxicity [159].

Both Congo red and doxycycline (antiamyloid compounds) have been found to reduce aggregation and cell death in cell models of OPMD [160]. The parallel protection that Congo red affords against the cytotocity of expanded polyQ and polyAla is consistent with the idea that abnormal protein aggregation and accumulation may be deleterious in all intracellular amyloidoses, irrespective of the primary mutation. Work on a transgenic model of OPMD provided evidence that doxycycline (antiamyloid compound) can be used as a treatment in OPMD, suggesting that this drug is able to inhibit oligomerization induced by mutant PABPN1.

2. Ubiquitin–Proteasome Pathway and Molecular Chaperones in OPMD and PolyQ Diseases

The situation in OPMD appears to have many parallels with polyQ diseases, raising the possibility that misfolded, aggregate-prone proteins may perturb similar pathways, irrespective of the nature of the mutation or protein context. A unifying pathological hallmark of OPMD and polyQ diseases is the presence of ubiquitin-positive inclusions. The appearance of ubiquitinated aggregates implies an underlying incapability of the cellular chaperones and proteasome machinery that normally function to prevent the accumulation of misfolded proteins.

The II formation in mutant PABPN1-Ala17 may result from an imbalance between protein refolding and aggregation. Different observations converge to suggest that a gain of function of mutant PABPN1 may cause the accumulation of nuclear filaments observed in OPMD [109]. The involvement of the ubiquitin–proteasome pathway and molecular chaperones in OPMD is well-documented in several studies. First, Calado et al. showed that OPMD-specific nuclear inclusions in deltoid muscle from OPMD patients recruit, in addition to wild-type PABPN1, ubiquitin and the subunits of the proteasome [114]. Second, human, yeast, and chemical chaperones were shown to reduce both aggregation and cell death in an OPMD cell model without affecting the levels of mutant PABPN1 [115]. Third, Abu-Baker et al. [113] presented evidence that the UPP and molecular chaperones are part of the cellular response to mutant polyAla-containing PABPN1. Both HSP70 and ubiquitin were shown to be recruited into IIs of OPMD in cell-based models and human tissue. The proteasome inhibitor lactacystin resulted in an enhancement of both nuclear and perinuclear protein aggregation that is associated with cellular toxicity. Moreover, overexpression of molecular chaperones (HSP40 and HSP70) suppressed protein aggregation and toxicity by increasing the solubility of mutant PABPN1-Ala17 [113]. Ravikumar et al. also demonstrated that epoxomicin (another proteasome inhibitor) did increase the proportion of COS-7 cells expressing GFP-Ala19 aggregates [161]. These and other results are consistent with those obtained for polyQ diseases, where treatment of transfected cells with proteasome inhibitors resulted in a greater accumulation of polyQ polypeptides, an increase in the number of inclusion bodies, and enhanced apoptosis [32, 37, 162].

3. THE ROLE OF RNA IN OPMD AND POLYQ DISEASES

To find out whether a loss of function underlies OPMD, Calado et al. [114] measured the steady-state polyA tail length in both OPMD and normal myoblasts. Surprisingly, there was no significant differences between the steady-state polyA tail length in OPMD and in normal myoblasts. However, the nuclear inclusions were shown to sequester polyA RNA. This raises the possibility that in OPMD the polyAla expansions in the mutant PABPN1 protein may interfere with the cellular traffic of polyA RNA [114].

The authors have identified two wild-type PABPN1 interacting proteins; the heterogeneous nuclear ribonucleoproteins (hnRNPs) A1 and A/B [153]. These hnRNPs are mRNA binding proteins that are involved in both mRNA processing and mRNA export from the nucleus to cytoplasm [163]. Both hnRNPs A1 and A/B can interact (via the C terminus) with wild-type PABPN1, and they also colocalize with mutant PABPN1 in OPMD IIs [153]. Mutant PABPN1 nuclear inclusions were found to sequester other proteins involved in mRNA biogenesis, such as polyA polymerase (PAP) [164]. It has been proposed that wild-type PABPN1 plays a role in mRNA export [147]. The interaction between wild-type PABPN1 and hnRN proteins might be required for the packaging of mRNA for the export process. Together, these observations and a report of OPMD IIs sequestering mRNA [114] suggest that OPMD IIs may be toxic by acting as "mRNA traps" interfering with the mRNA nucleocytoplasmic export.

Work on OPMD revealed that mutant PABPN1 overexpression leads to up-regulation genes encoding proteins that are sequestered in OPMD nuclear inclusions [165]. Interestingly, many of the up-regulated genes encode proteins involved in RNA processing, mRNA splicing, packaging, and transport [165].

It has been shown that the formation of mutant PABPN1 inclusions requires binding to polyA, and interference with any of the protein domains required for stimulation of polyA polymerase prevents the formation of inclusions [164]. Interfering with the RNA binding domain (RBD) led to mutant protein not forming IIs. More work, however, is required to correlate RBD and the cytotoxicity associated in OPMD.

Alterations in RNA metabolism have been proposed to play a role in the pathogenesis of polyQ diseases [166]. At least two polyQ disease proteins (ataxin-2 and atrophin) contain motifs found in RNA splicing proteins [42]. Ataxin-2 contains two RNA splicing motifs, which are shared with snRNPs [167]. A novel protein (ataxin-2 binding protein 1, A2BP1) with RNA binding motifs was also found to interact with ataxin-2. RNA binding proteins with RNP motifs are involved in various RNA metabolic functions, such as RNA processing and intracellular RNA transport [168]. This evolutionary conservation suggests that proteins in this family may be involved in critical steps of RNA processing. One possible role of the ataxin-2–A2BP1 complex may be RNA transport in neurons. Neuronal cytotoxicity of ataxin-2 with expanded polyQ tracts might be due to altered RNA transport function. This hypothesis is consistent with the absence of intranuclear inclusions or aggregates from SCA2 [169]. A2BP1 binds to the ataxin-2 C terminus. The C-terminal ataxin-2 fragment does not include the polyQ tract.

Yue et al. demonstrated that ataxin-1 has RNA binding activity that is inversely affected by the length of its polyQ tract [170]. This suggests that ataxin-1 plays a role in RNA metabolism and that the expansion of the polyQ tract may alter this function.

Other work on SCA1 suggests that the normal role of ataxin-1 may be in RNA processing, perhaps nuclear RNA export [171]. Ataxin-1 nuclear inclusions were found to recruit the mRNA export factor, TAP/NXF1.

Wild-type, but not mutant, ataxin-1 was shown to be capable of nuclear export. These results suggest that nuclear retention of mutant ataxin-1 may be an important toxic gain of function in the disease [171].

4. THE ROLE OF THE NUCLEUS IN OPMD AND PolyQ DISEASES

The composition of intranuclear aggregates might reflect the process of intranuclear aggregate formation and contribute to the pathogenesis of OPMD. The authors showed that most mutant PABPN1-positive (Ala expanded form) intranuclear aggregates are structures distinct from intranuclear speckles (nuclear domains that are rich in splicing factors). In contrast, the promyelocytic leukemia (PML) protein, a major component of nuclear bodies, strongly colocalized to intranuclear aggregates of mutant PABPN1 [154]. Similar to these observations, mutant ataxin-1 was shown to cause a specific redistribution of the nuclear matrix-associated domain containing promyelocytic leukemia protein in transfected COS cells [172]. In human brains, intranuclear aggregates observed in SCA3, SCA7, and DRPLA colocalize with PML in a characteristic pattern [173–175]. This may suggest the involvement of PML in a process of intranuclear aggregate formation. Evidence that the nuclear environment is necessary for mutant PABPN1 inclusion formation and cellular toxicity was also presented [154]. This was achieved by inactivating the mutant PABPN1 nuclear localization signal (NLS) and generating full-length mutant PABPN1 fused to a strong nuclear export signal. Targeting of mutant PABPN1 to the cytoplasm resulted in significant suppression of both intranuclear aggregate formation and cellular toxicity, two hallmarks of OPMD [154].

It is becoming increasingly clear that, although the expanded polyAla stretch in mutant PABPN1 is crucial to initiate OPMD pathogenesis, residues outside the polyAla stretch also play important roles in the disease. Similar observations have also been reported in polyQ diseases [176].

Nucleocytoplasmic transport has become an important aspect of models explaining the pathogenesis of polyQ diseases [177–180]. In the case of SCA1, nuclear localization of the polyQ expanded ataxin-1 protein is critical to initiate pathogenesis. Transgenic mice carrying polyglutamine expanded ataxin-1 with an inactivated NLS do not develop neuropathology [179]. Yang et al. have demonstrated that aggregates of polyQ peptides containing an NLS are localized to nuclei and can lead to dramatic cell death [180]. Another finding suggested that mutant htt acts within the nucleus to induce neurodegeneration [177]. Blockage of the nuclear localization of mutant htt suppressed its ability to form intranuclear inclusions and induce neurodegeneration [177].

5. TRANSCRIPTION IN OPMD AND PolyQ DISEASES

Reports support wild-type PABPN1's involvement in transcription. First, wild-type PABPN1 directly interacts with the SKI-interacting protein (SKIP) [181], a muscle transcription regulatory protein that, along with MyoD, regulates muscle cell differentiation [181]. Second, normal PABPN1 associates with RNA polymerase II during transcription and accompanies the released transcript to the nuclear pore [182]. Modifications of these processes by expanded PABPN1 might play an important role in OPMD. A growing number of data suggest that transcription dysregulation may be an early event in polyQ pathogenesis (Section IIB).

6. APOPTOSIS IN OPMD AND PolyQ DISEASES

A study on OPMD transgenic mice demonstrated that doxycycline reduces the susceptibility of cells to proapoptotic insults. Doxycycline treatment influenced Bax levels, as this protein permeabilizes mitochondria, leading to cytochrome c release and subsequent activation of caspases 9 and 3 after various proapoptotic insults [152]. Bax levels were elevated in the muscle of Ala17 mice compared to wild-type littermates, and Bax levels in OPMD mice were reduced by doxycycline. It is becoming increasingly apparent that apoptosis, a highly regulated cell death pathway, is crucial to neurodegeneration in polyQ repeat diseases. The level and activity of caspases, which are crucial for the initiation and execution of apoptosis, are elevated in polyQ diseases [183].

7. NEUROLOGICAL INVOLVEMENT IN OPMD AND PolyQ DISEASES

OPMD is considered a primary myopathic disorder, and little morphologic evidence exists for a neurogenic etiology, though the peripheral nervous system (PNS) seems to be affected in some cases [184]. More evidence has emerged suggesting neurological involvement in OPMD. Boukriche et al. reported the case of an OPMD patient, who developed a severe chronic axonal neuropathy [185]. Nakashima et al. reported an OPMD patient with neurogenic features in the electrophysiological and pathological findings [186]. Analysis of mitotic and postmitotic tissues in OPMD transgenic animals revealed ubiquitinated PABPN1-positive IIs in neuronal cells. Similar IIs in postmortem brain sections from an OPMD patient were also observed. These results indicate that mutant PABPN1, presumably via the toxic effects of its polyAla tract, can lead to inclusion formation and neurodegeneration in both mice and the humans [187]. These findings raise the possibility that neurogenic changes, as in polyQ diseases, may contribute to the pathophysiology of OPMD [187].

F. Treatment in OPMD and PolyQ Diseases

1. OPMD

a. Surgical Options in OPMD

No medical treatment is available for OPMD. Surgical treatments are used to correct the ptosis [188] and improve swallowing [189, 190] in moderately to severely affected individuals (Table 33-3). However, ptosis will probably return within 5–15 years after the surgery.

b. Doxycycline (Antiprotein Aggregation)

The first potential therapeutic strategy for OPMD tested in animal models has been published [152]. The onset and severity of OPMD abnormalities were substantially delayed and attenuated by doxycycline treatment, which may exert its therapeutic effect by reducing aggregates and by distinct antiapoptotic properties. Doxycycline may represent a safe and feasible treatment for this disease [152]. Doxycycline and minocycline have been tested as potential treatments in mouse models of HD with conflicting results. It is possible that OPMD muscle may be a more accessible target for doxycycline, which would need to cross the blood–brain barrier to help treat afflictions such as HD.

2. PolyQ Diseases

Table 33-3 summarizes therapeutic strategies used in this field. Although beneficial effects were shown, it is unknown whether the compounds are acting at the target for which they were selected.

G. Intriguing Questions

1. Soluble versus Aggregates?

Which is the toxic form of the mutant protein: soluble, insoluble, or both? Does aggregation have a causative role in polyQ and polyAla cytotoxicty? Does it have a role in disease progression, or is this process in some way protective? Although the causative role of aggregation in polyQ and polyAla diseases remains controversial, blockage of the aggregation process in polyQ and OPMD diseases often results in a reduction in cytotoxicity *in vitro* and *in vivo*.

On the other hand, several studies support the hypothesis that aggregates are neither necessary nor sufficient for neuronal dysfunction in polyQ diseases, but rather that inclusion formation may be an active cellular defense mechanism. A study on the mutated htt protein showed that neuronal toxicity results primarily from the soluble protein and that II formation is protective for a certain period of time [191–193]. This suggests that protein aggregation represents a transient beneficial coping response to mutated proteins by reducing the amount of soluble protein in the nuclei of neurons. Some dissociation exists between the cellular distribution of polyQ protein aggregates and the selective pattern of cell loss in affected tissue [177, 179, 194, 195]. Moreover, interruption of inclusion formation results in enhanced toxicity, suggesting that inclusion formation may be a mechanism that assists in the clearance of misfolded proteins [177, 196]. However, a protective role for inclusions does not preclude the possibility that the tendency of the protein to aggregate, either with itself or with other proteins, is nonetheless important to its toxicity.

Inclusion formation may be required for disease progression as opposed to disease onset. The most likely explanation is that inclusions and other visible protein aggregates represent an end stage of a molecular cascade of several steps and that earlier steps in the cascade may be more directly tied to pathogenesis than the inclusions themselves. It would be interesting to establish the contribution of the soluble form of mutant PABPN1 to OPMD-associated toxicity.

2. Cell Specificity?

Why are neurons and muscle the tissues most affected in polyQ and OPMD diseases, respectively? Wirtschafter et al. proposed a hypothesis for the selective involvement of extraocular muscles affected in OPMD and the onset of symptoms in adulthood [197]. It is based on the phenomenon of continuous remodeling of normal adult extraocular muscles (EOMs). Unlike the EOMs, the myonuclei of other skeletal muscles are postmitotic in the adult unless the muscles are injured. In OPMD, the ongoing production of mutant PABPN1 in muscles undergoing continuous remodeling could result in a failure of accurate production of mRNA required for the maintenance of the myocytes [197]. Mouly et al. proposed that muscle satellite cells are implicated in OPMD pathogenesis [198]. The regenerative capacity of skeletal muscle depends on the number of available satellite cells and their proliferative capacity. In muscular dystrophies, the proliferative capacity of satellite cells is dramatically decreased [198].

One of the compelling mysteries in the field of polyQ diseases is the mechanism underlying selective neuronal degeneration in the setting of ubiquitously expressed expanded polyQ proteins. Factors contributing to selective vulnerability can partly be explained by different molecular mechanisms discussed in the following. For example, interacting proteins are likely to contribute to selective vulnerability. Certain interacting proteins may bind to disease proteins in a way that promotes misfolding and aggregation. Understanding this in molecular terms may be instrumental in defining cell-specific therapeutic targets [199, 200].

3. Loss or Gain of Function?

Until a few years ago, it was assumed that the expanded polyQ repeat confers a toxic gain of function that results in neuronal dysfunction and death. New data suggest that loss of the normal function of polyQ proteins may contribute to polyQ disease pathology [61, 62]. For example, the regional vulnerability that characterizes HD might be explained by a loss of htt's normal functions [201]. Mutant polyQ protein or mutant PABPN1 may mediate some of its toxic effects through the disruption of wild-type function. Although the primary function of wild-type PABPN1 was shown to be normal in the mutant protein (Section VE), it is not known what other functions this protein may have and if these are affected in the mutant protein.

4. Primary or Parallel Mechanisms?

One problem that investigators face is distinguishing primary from secondary events in polyQ disease pathogenesis. CAG expansion in polyQ diseases and GCG expansions in OPMD certainly trigger a cascade of complex and evolving events. However, on the other hand, the multiplicity of mechanisms suggests the existence of multiple therapeutic targets.

VI. CONCLUSION

Although 10 years ago no one could have predicted the biological complexities underlying polyQ and polyAla diseases, no one would have anticipated the remarkable progress in understanding these diseases or that they may share similar pathogenic mechanisms.

References

1. Abrams, M. T., Reiss, A. L., Freund, L. S., Baumgardner, T. L., Chase, G. A., and Denckla, M. B. (1994). Molecular–neurobehavioral associations in females with the fragile X full mutation. *Am. J. Med. Genet.* **51**, 317–327.
2. Mulley, J. C., Yu, S., Loesch, D. Z., Hay, D. A., Donnelly, A., et al. (1995). FRAXE and mental retardation. *J. Med. Genet.* **32**, 162–169.
3. Harding, A. E. (1981). Friedreich's ataxia: A clinical and genetic study of 90 families with an analysis of early diagnostic criteria and intrafamilial clustering of clinical features. *Brain* **104**, 589–620.
4. Harper, P. S. (1989). "Myotonic Dystrophy." Saunders, London.
5. Greenfield, J. G. (1954). "The Spino-cerebellar Degenerations," p. 112. C. C. Thomas, Springfield, IL.
6. Holmes, S. E., O'Hearn, E. E., McInnis, M. G., Gorelick-Feldman, D. A., Kleiderlein, J. J., et al. (1999). Expansion of a novel CAG trinucleotide repeat in the 5′ region of PPP2R2B is associated with SCA12. *Nat. Genet.* **23**, 391–392.
7. Zoghbi, H. Y., and Orr, H. T. (2000). Glutamine repeats and neurodegeneration. *Annu. Rev. Neurosci.* **23**, 217–247.
8. Muragaki, Y., Mundlos, S., Upton, J., and Olsen, B. R. (1996). Altered growth and branching patterns in synpolydactyly caused by mutations in HOXD13. *Science* **272**, 548–551.
9. Goodman, F. R., Mundlos, S., Muragaki, Y., Donnai, D., Giovannucci-Uzielli, M. L., Lapi, E., Majewski, F., McGaughran, J., McKeown, C., Reardon, W., et al. (1997). Synpolydactyly phenotypes correlate with size of expansions in HOXD13 polyalanine tract. *Proc. Natl. Acad. Sci. USA* **94**, 7458–7463.
10. Mundlos, S., Otto, F., Mundlos, C., Mulliken, J. B., Aylsworth, A. S., Albright, S., Lindhout, D., Cole, W. G., Henn, W., Knoll, J. H., Owen, M. J., Mertelsmann, R., Zabel, B. U., and Olsen, B. R. (1997). Mutations involving the transcription factor CBFA1 cause cleidocranial dysplasia. *Cell* **89**, 773–779.
11. Brais, B., Bouchard, J. P., Xie, Y. G., Rochefort, D. L., Chretien, N., Tome, F. M., Lafreniere, R. G., Rommens, J. M., Uyama, E., Nohira, O., Blumen, S., Korczyn, A. D., Heutink, P., Mathieu, J., Duranceau, A., Codere, F., Fardeau, M., and Rouleau, G. A. (1998). Short GCG expansions in the PABP2 gene cause oculopharyngeal muscular dystrophy. *Nat. Genet.* **18**, 164–167.
12. Brown, S. A., Warburton, D., Brown, L. Y., Yu, C.-Y., Roeder, E. R., Stengel-Rutkowski, S., Hennekam, R. C. M., and Muenke, M. (1998). Holoprosencephaly due to mutation in ZIC2 a homologue of Drosophila odd-paired. *Nat. Genet.* **20**, 180–183.
13. Mortlock, D. P., Post, L. C., and Innis, J. W. (1996). The molecular basis of hypodactyly (Hd): A deletion in Hoxa 13 leads to arrest of digital arch formation. *Nat. Genet.* **13**, 284–289.
14. Crisponi, L., Deiana, M., Loi, A., Chiappe, F., Uda, M., Amati, P., Bisceglia, L., Zelante, L., Nagaraja, R., Porcu, S., et al. (2001). The putative forkhead transcription factor FOXL2 is mutated in blepharophimosis/ptosis/epicanthus inversus syndrome. *Nat. Genet.* **27**, 159–166.
15. Stromme, P., Mangelsdorf, M. E., Shaw, M. A., Lower, K. M., Lewis, S. M., Bruyere, H., Lutcherath, V., Gedeon, A. K., Wallace, R. H., Scheffer, I. E., Turner, G., Partington, M., Frints, S. G., Fryns, J. P., Sutherland, G. R., Mulley, J. C., and Gecz, J. (2002). Mutations in the human ortholog of Aristaless cause X-linked mental retardation and epilepsy. *Nat. Genet.* **30**, 441–445.
16. Laumonnier, F., Ronce, N., Hamel, B. C., Thomas, P., Lespinasse, J., Raynaud, M., Paringaux, C., Van Bokhoven, H., Kalscheuer, V., Fryns, J. P., Chelly, J., Moraine, C., and Briault, S. (2002). Transcription factor SOX3 is involved in X-linked mental retardation with growth hormone deficiency. *Am. J. Hum. Genet.* **71**, 1450–1455.
17. Amiel, J., Laudier, B., Attie-Bitach, T., Trang, H., De Pontual, L., Gener, B., Trochet, D., Etchevers, H., Ray, P., Simonneau, M., et al. (2003). Polyalanine expansion and frameshift mutations of the paired-like homeobox gene PHOX2B in congenital central hypoventilation syndrome. *Nat. Genet.* **33**, 459–461.
18. Gusella, J. F., and MacDonald, M. E. (2000). Molecular genetics: Unmasking polyglutamine triggers in neurodegenerative disease. *Nat. Rev. Neurosci.* **1**, 109–115.
19. Cummings, C. J., and Zoghbi, H. Y. (2000). Fourteen and counting: Unraveling trinucleotide repeat diseases. *Hum. Mol. Genet.* **9**, 909–916.

20. Wellington, C. L., Ellerby, L. M., Hackam, A. S., Margolis, R. L., Trifiro, M. A., Singaraja, R., McCutcheon, K., Salvesen, G. S., Propp, S. S., Bromm, M., Rowland, K. J., Zhang, T., Rasper, D., Roy, S., Thornberry, N., Pinsky, L., Kakizuka, A., Ross, C. A., Nicholson, D. W., Bredesen, D. E., and Hayden, M. R. (1998). Caspase cleavage of gene products associated with triplet expansion disorders generates truncated fragments containing the polyglutamine tract. *J. Biol. Chem.* **273**, 9158–9167.
21. McCampbell, A., Taye, A. A., Whitty, L., Penney, E., Steffan, J. S., and Fischbeck, K. H. (2001). Histone deacetylase inhibitors reduce polyglutamine toxicity. *Proc. Natl. Acad. Sci. USA* **98**, 15179–15184.
22. Steffan, J. S., Kazantsev, A., Spasic-Boskovic, O., Greenwald, M., Zhu, Y. Z., Gohler, H., Wanker, E. E., Bates, G. P., Housman, D. E., and Thompson, L. M. (2000). The Huntington's disease protein interacts with p53 and CREB-binding protein and represses transcription. *Proc. Natl. Acad. Sci. USA* **97**, 6763–6768.
23. Nucifora, F. C., Jr., Sasaki, M., Peters, M. F., Huang, H., Cooper, J. K., Yamada, M., Takahashi, H., Tsuji, S., Troncoso, J., Dawson, V. L., Dawson, T. M., and Ross, C. A. (2001). Interference by huntingtin and atrophin-1 with cbp-mediated transcription leading to cellular toxicity. *Science* **291**, 2423–2428.
24. Bence, N. F., Sampat, R. M., and Kopito, R. R. (2001). Impairment of the ubiquitin–proteasome system by protein aggregation. *Science* **292**, 1552–1555.
25. Ravikumar, B., Vacher, C., Berger, Z., Davies, J. E., Luo, S., Oroz, L. G., Scaravilli, F., Easton, D. F., Duden, R., O'Kane, C. J., and Rubinsztein, D. C. (2004). Inhibition of mTOR induces autophagy and reduces toxicity of polyglutamine expansions in fly and mouse models of Huntington disease. *Nat. Genet.* **36**, 585–595.
26. Sanchez, I., Xu, C. J., Juo, P., Kakizaka, A., Blenis, J., and Yuan, J. (1999). Caspase-8 is required for cell death induced by expanded polyglutamine repeats. *Neuron* **22**, 623–633.
27. Szebenyi, G., Morfini, G. A., Babcock, A., Gould, M., Selkoe, K., Stenoien, D. L., Young, M., Faber, P. W., MacDonald, M. E., McPhaul, M. J., and Brady, S. T. (2003). Neuropathogenic forms of huntingtin and androgen receptor inhibit fast axonal transport. *Neuron* **40**, 41–52.
28. Tabrizi, S. J., Cleeter, M. W, Xuereb, J., Taanman, J. W., Cooper, J. M., and Schapira, A. H. (1999). Biochemical abnormalities and excitotoxicity in Huntington's disease brain. *Ann. Neurol.* **45**, 25–32.
29. Hartl, F. U., and Hayer-Hartl, M. (2002). Molecular chaperones in the cytosol: From nascent chain to folded protein. *Science* **295**, 1852–1858.
30. Johnston, J. A., Ward, C. L., and Kopito, R. R. (1998). Aggresomes: A cellular response to misfolded proteins. *J. Cell Biol.* **143**, 1883–1898.
31. Chai, Y., Koppenhafer, S. L., Bonini, N. M., and Paulson, H. L. (1999). Analysis of the role of heat shock protein (Hsp) molecular chaperones in polyglutamine disease. *J. Neurosci.* **19**, 10338–10347.
32. Chai, Y., Koppenhafer, S. L., Shoesmith, S. J., Perez, M. K., and Paulson, H. L. (1999). Evidence for proteasome involvement in polyglutamine disease: Localization to nuclear inclusions in SCA3/MJD and suppression of polyglutamine aggregation *in vitro*. *Hum. Mol. Genet.* **8**, 673–682.
33. Stenoien, D. L., Cummings, C. J., Adams, H. P., Mancini, M. G., Patel, K., DeMartino, G. N., Marcelli, M., Weigel, N. L., and Mancini, M. A. (1999). Polyglutamine-expanded androgen receptors form aggregates that sequester heat shock proteins, proteasome components and SRC-1, and are suppressed by the HDJ-2 chaperone. *Hum. Mol. Genet.* **8**, 731–741.
34. Kobayashi, Y., Kume, A., Li, M., Doyu, M., Hata, M., Ohtsuka, K., and Sobue, G. (2000). Chaperones Hsp70 and Hsp40 suppress aggregate formation and apoptosis in cultured neuronal cells expressing truncated androgen receptor protein with expanded polyglutamine tract. *J. Biol. Chem.* **275**, 8772–8778.
35. Muchowski, P. J., Schaffar, G., Sittler, A., Wanker, E. E., Hayer-Hartl, M. K., and Hartl, F. U. (2000). Hsp70 and Hsp40 chaperones can inhibit self-assembly of polyglutamine proteins into amyloid-like fibrils. *Proc. Natl. Acad. Sci. USA* **97**, 7841–7846.
36. Satyal, S. H., Schmidt, E., Kitagawa, K., Sondheimer, N., Lindquist, S., Kramer, J. M., and Morimoto, R. I. (2000). Polyglutamine aggregates alter protein folding homeostasis in *Caenorhabditis elegans*. *Proc. Natl. Acad. Sci. USA* **97**, 5750–5755.
37. Cummings, C. J., Mancini, M. A., Antalffy, B., DeFranco, D. B., Orr, H. T., and Zoghbi, H. Y. (1998). Chaperone suppression of aggregation and altered subcellular proteasome localization imply protein misfolding in SCA1. *Nat. Genet.* **19**, 148–154.
38. Fernandez-Funez, P., Nino-Rosales, M. L., de Gouyon, B., She, W. C., Luchak, J. M., Martinez, P., Turiegano, E., Benito, J., Capovilla, M., Skinner, P. J., McCall, A., Canal, I., Orr, H. T., Zoghbi, H. Y., and Botas, J. (2000). Identification of genes that modify ataxin-1-induced neurodegeneration. *Nature* **408**, 101–106.
39. Warrick, J. M., Chan, H. Y., Gray-Board, G. L., Chai, Y., Paulson, H. L., and Bonini, N. M. (1999). Suppression of polyglutamine-mediated neurodegeneration in Drosophila by the molecular chaperone HSP70. *Nat. Genet.* **23**, 425–428.
40. Kazemi-Esfarjani, P., and Benzer, S. (2000). Genetic suppression of polyglutamine toxicity in Drosophila. *Science* **287**, 1837–1840.
41. Alves-Rodrigues, A., Gregori, L., and Figueiredo-Pereira, M. E. (1998). Ubiquitin, cellular inclusions and their role in neurodegeneration. *Trends Neurosci.* **21**, 516–520.
42. Paulson, H. L. (1999). Protein fate in neurodegenerative proteinopathies: Polyglutamine diseases join the (mis)fold. *Am. J. Hum. Genet.* **64**, 339–345.
43. Klionsky, D. J., and Ohsumi, Y. (1999). Cytoplasm to vacuole protein transport. *Annu. Rev. Cell Dev. Biol.* **15**, 1–32.
44. Li, X. J., et al. (1995). A huntingtin-associated protein enriched in brain with implications for pathology. *Nature* **378**, 398–402.
45. Perez, M. K., Paulson, H. L., Pendse, S. J., Saionz, S. J., Bonini, N. M., and Pittman, R. N. (1998). Recruitment and the role of nuclear localization in polyglutamine-mediated aggregation. *J. Cell Biol.* **143**, 1457–1470.
46. Wyttenbach, A., Swartz, J., Kita, H., Thykjaer, T., Carmichael, J., Bradley, J., Brown, R., Maxwell, M., Schapira, A., Orntoft, T. F., Kato, K., and Rubinsztein, D. C. (2001). Polyglutamine expansions cause decreased CRE-mediated transcription and early gene expression changes prior to cell death in an inducible cell model of Huntington's disease. *Hum. Mol. Genet.* **10**, 1829–1845.
47. Huang, C. C., Faber, P. W., Persichetti, F., Mittal, V., Vonsattel, J. P., MacDonald, M. E., and Gusella, J. F. (1998). Amyloid formation by mutant huntingtin: Threshold, progressivity and recruitment of - normal polyglutamine proteins. *Somat. Cell Mol. Genet.* **24**, 217–233.
48. Boutell, J. M., Thomas, P., Neal, J. W., Weston, V. J., Duce, J., Harper, P. S., and Jones, A. L. (1999). Aberrant interactions of transcriptional repressor proteins with the Huntington's disease gene product, huntingtin. *Hum. Mol. Genet.* **9**, 1647–1655.
49. Shimohata, T., Nakajima, T., Yamada, M., Uchida, C., Onodera, O., Naruse, S., Kimura, T., Koide, R., Nozaki, K., Sano, Y., Ishiguro, H., Sakoe, K., Ooshima, T., Sato, A., Ikeuchi, T., Oyake, M., Sato, T., Aoyagi, Y., Hozumi, I., Nagatsu, T., Takiyama, Y., Nishizawa, M., Goto, J., Kanazawa, I., Davidson, I., Tanese, N., Takahashi, H., and Tsuji, S. (2000). Expanded polyglutamine stretches interact with TAFII130, interfering with CREB-dependent transcription. *Nat. Genet.* **26**, 29–36.

50. Suhr, S. T., Senut, M. C., Whitelegge, J. P., Faull, K. F., Cuizon, D. B., and Gage, F. H. (2001). Identities of sequestered proteins in aggregates from cells with induced polyglutamine expression. *J. Cell Biol.* **153**, 283–294.
51. Dunah, A. W., Jeong, H., Griffin, A., Kim, Y. M., Standaert, D. G., Hersch, S. M., Mouradian, M. M., Young, A. B., Tanese, N., and Krainc, D. (2002). Sp1 and TAFII130 transcriptional activity disrupted in early Huntington's disease. *Science* **296**, 2238–2243.
52. Mantamadiotis, T., Lemberger, T., Bleckmann, S. C., Kern, H., Kertz, O., Villalba, A. M., et al. (2002). Disruption of CREB function in brain leads to neurodegeneration. *Nat Genet.* **31**, 47–54.
53. Kegel, K. B., Meloni, A. R., Yi, Y., Kim, Y. J., Doyle, E., Cuiffo, B. G., Sapp, E., Wang, Y., Qin, Z. H., Chen, J. D., Nevins, J. R., Aronin, N., and DiFiglia, M. (2002). Huntingtin is present in the nucleus, interacts with the transcriptional corepressor C-terminal binding protein, and represses transcription. *J. Biol. Chem.* **277**, 7466–7476.
54. Cha, J. J. (2000). Transcriptional dysregulation in Huntington's disease. *Trends Neurosci.* **23**, 387–392.
55. Steffan, J. S., Bodai, L., Pallos, J., Poelman, M., McCampbell, A., Apostol, B. L., Kazantsev, A., Schmidt, E., Zhu, Y. Z., Greenwald, M., Kurokawa, R., Housman, D. E., Jackson, G. R., Marsh, J. L., and Thompson, L. M. (2001). Histone deacetylase inhibitors arrest polyglutamine-dependent neurodegeneration in Drosophila. *Nature* **413**, 739–743.
56. Taylor, J. P., Taye, A. A., Campbell, C., Kazemi-Esfarjani, P., Fischbeck, K. H., and Min, K. T. (2003). Aberrant histone acetylation, altered transcription, and retinal degeneration in a *Drosophila* model of polyglutamine disease are rescued by CREB-binding protein. *Genes Dev.* **17**, 1463–1468.
57. Hughes, R. E., Lo, R. S., Davis, C., Strand, A. D., Neal, C. L., Olson, J. M., and Fields, S. (2001). Altered transcription in yeast expressing expanded polyglutamine. *Proc. Natl. Acad. Sci. USA* **98**, 13201–13206.
58. Steffan, J., and Thompson, L. (2003). Targeting aggregation in the development of therapeutics for the treatment of Huntington's disease and other polyglutamine repeat diseases. *Expert Opin. Ther. Targets* **7**, 201–213.
59. Parker, J. A., Arango, M., Abderrahmane, S., Lambert, E., Tourette, C., Catoire, H., and Néri, C. (2005). Reserveratrol rescues mutant polyglutamine cytotoxicity in nematode and mammalian neurons. *Nat. Genet.* **37**, 349–350.
60. Gunawardena, S., Her, L. S., Brusch, R. G., Laymon, R. A., Niesman, I. R., Gordesky-Gold, B., Sintasath, L., Bonini, N. M., and Goldstein, L. S. (2003). Disruption of axonal transport by loss of huntingtin or expression of pathogenic polyQ proteins in Drosophila. *Neuron* **40**, 25–40.
61. Cornett, J., Cao, F., Wang, C. E., Ross, C. A., Bates, G. P., Li, S. H., and Li, X. J. (2005). Polyglutamine expansion of huntingtin impairs its nuclear export. *Nat. Genet.* **37**, 198–204.
62. Gauthier, L. R., Charrin, B. C., Borrell-Pages, M., Dompierre, J. P., Rangone, H., Cordelieres, F. P., De Mey, J., MacDonald, M. E., Lessmann, V., Humbert, S., and Saudou, F. (2004). Huntingtin controls neurotrophic support and survival of neurons by enhancing BDNF vesicular transport along microtubules. *Cell* **118**, 127–138.
63. Menke, T., Gille, G., Reber, F., Janetzky, B., Andler, W., Funk, R. H., and Reichmann, H. (2003). Coenzyme Q_{10} reduced the toxicity of retenone in neuronal cultures by preserving the mitochondrial membrane potential. *Biofactors* **18**, 65–72.
64. Panov, A. V., Gutekunst, C. A., Leavitt, B. R., Hayden, M. R., Burke, J. R., Strittmatter, W. J., and Greenamyre, J. T. (2002). Early mitochondrial calcium defects in Huntington's disease are a direct effect of polyglutamines. *Nat. Neurosci.* **5**, 731–736.
65. Sawa, A., Wiegand, G. W., Cooper, J., Margolis, R. L., Sharp, A. H., Lawler, J. F., Jr., Greenamyre, J. T., Snyder, S. H., and Ross, C. A. (1999). Increased apoptosis of Huntington disease lymphoblasts associated with repeat length-dependent mitochondrial depolarization. *Nat. Med.* **5**, 1194–1198.
65a. Sawa, A. (2001). Mechanisms for neuronal cell death and dysfunction in Huntington's disease: Pathological cross-talk between the nucleus and the mitochondria? *J. Mol. Med.* **79**, 375–381.
66. Gu, M., Gash, M. T., Mann, V. M., Javoy-Agid, F., Cooper, J. M., and Schapira, A. H. (1996). Mitochondrial defect in Huntington's disease caudate nucleus. *Ann. Neurol.* **39**, 385–389.
67. Browne, S. E., Bowling, A. C., MacGarvey, U., Baik, M. J., Berger, S. C., Muqit, M. M., Bird, E. D., and Beal, M. F. (1997). Oxidative damage and metabolic dysfunction in Huntington's disease: Selective vulnerability of the basal ganglia. *Ann. Neurol.* **41**, 646–653.
68. Ruan, Q., Lesort, M., MacDonald, M. E., and Johnson, G. V. W. (2004). Striatal cells from mutant huntingtin knock-in mice are selectively vulnerable to mitochondrial complex II inhibitor-induced cell death through a non-apoptotic pathway. *Hum. Mol. Genet.* **13**, 669–681.
69. Hansson, O., Guatteo, E., Mercuri, N. B., Bernardi, G., Li, X.-J., Castilho, R. F., and Brundin, P. (2001). Resistance to NMDA toxicity correlates with appearance of nuclear inclusions, behavioural deficits and changes in calcium homeostasis in mice transgenic for exon 1 of the huntington gene. *Eur. J. Neurosci.* **14**, 1492.
70. Tang, T. S., Slow, E., Lupu, V., Stavrovskaya, I. G., Sugimori, M., Llinas, R., Kristal, B. S., Hayden, M. R., and Bezprozvanny, I. (2005). Disturbed Ca^{2+} signaling and apoptosis of medium spiny neurons in Huntington's disease. *Proc. Natl. Acad. Sci. USA* **102**, 2602–2607.
71. Chen, H. K., Fernandez-Funez, P., Acevedo, S. F., Lam, Y. C., Kaytor, M. D., Fernandez, M. H., Aitken, A., Skoulakis, E. M., Orr, H. T., Botas, J., and Zoghbi, H. Y. (2003). Interaction of akt-phosphorylated ataxin-1 with 14-3-3 mediates neurodegeneration in spinocerebellar ataxia type 1. *Cell* **113**, 457–468.
72. Emamian, E. S., Kaytor, M. D., Duvick, L. A., Zu, T., Tousey, S. K., Zoghbi, H. Y., Clark, H. B., and Orr, H. T. (2003). Serine 776 of ataxin-1 is critical for polyglutamine-induced disease in SCA1 transgenic mice. *Neuron* **38**, 375–387.
73. Yazawa, I. (2000). Aberrant phosphorylation of dentatorubral-pallidoluysian atrophy (DRPLA) protein complex in brain tissue. *Biochem. J.* **351**, 587–593.
74. Chan, H. Y., Warrick, J. M., Andriola, I., Merry, D., and Bonini, N. M. (2002). Genetic modulation of polyglutamine toxicity by protein conjugation pathways in *Drosophila*. *Hum. Mol. Genet.* **11**, 2895–2904.
75. Li, Y., Wang, H., Wang, S., Quon, D., Liu, Y. W., and Cordell, B. (2003). Positive and negative regulation of APP amyloidogenesis by sumoylation. *Proc. Natl. Acad. Sci. USA* **100**, 259–264.
76. Steffan, J. S., Agrawal, N., Pallos, J., Rockabrand, E., Trotman, L. C., Slepko, N., Illes, K., Lukacsovich, T., Zhu, Y. Z., Cattaneo, E., et al. (2004). SUMO modification of Huntingtin and Huntington's disease pathology. *Science* **304**, 100–104.
77. Gu, X., Li, C., Wei, W., Lo, V., Gong, S., Li, S. H., Iwasato, T., Itohara, S., Li, X. J., Mody, I., Heintz, N., and Yang, X. W. (2005). Pathological cell–cell interactions elicited by a neuropathogenic form of mutant Huntingtin contribute to cortical pathogenesis in HD mice. *Neuron* **46**, 433–444.
78. Rousseau, E., Dehay, B., Ben-Haiem, L., Trottier, Y., Morange, M., and Bertolotti, A. (2004). Targeting expression of expanded polyglutamine proteins to the endoplasmic reticulum or mitochondria prevents their aggregation. *Proc. Natl. Acad. Sci. USA* **101**, 9648–9653.

79. Temtamy, S. A., and McKusick, V. A. (1978). The Genetics of Hand Malformations. New York: Alan R. Liss (pub). p. 465.
80. Izpisua-Belmonte and Duboule (1992). Homeobox genes and pattern formation in the vertebrate limb. *Dev. Biol.* **152**, 26–36.
81. Brown, L. Y., and Brown, S. A. (2004). Alanine tracts: The expanding story of human illness and trinucleotide repeats. *Trends Genet.* **20**, 51–58.
82. Dolle, P., Izpisua-Belmonte, J. C., Brown, J., Tickle, C., and Duboule, D. (1993). Hox genes and the morphogenesis of the vertebrate limb. *Prog. Clin. Biol. Res.* **383A**, 11–20.
83. Davies, A. P., and Capecchi, M. R. (1996). A mutational analysis of the 5′ HoxD genes: Dissection of genetic interactions during limb development in the mouse. *Development* **122**, 1175–1185.
84. Albrecht, A. N., Kornak, U., Boddrich, A., Suring, K., Robinson, P. N., Stiege, A. C., Lurz, R., Stricker, S., Wanker, E. E., and Mundlos, S. (2004). A molecular pathogenesis for transcription factor associated polyalanine tract expansions. *Hum. Mol. Genet.* **13**, 2351–2359.
85. Halal, F. (1988). The hand–foot–genital (hand–foot–uterus) syndrome: Family report and update. *Am. J. Med. Genet.* **30**, 793–803.
86. Utsch, B., Becker, K., Brock, D., Lentze, M. J., Bidlingmaier, F., and Ludwig, M. (2002). A novel stable polyalanine [poly(A)] expansion in the *HOXA13* gene associated with hand–foot–genital syndrome: Proper function of poly(A)-harbouring transcription factors depends on a critical repeat length? *Hum. Genet.* **110**, 488–494.
87. Goodman, F. R., Bacchelli, C., Brady, A. F., Brueton, L. A., Fryns, J. P., Mortlock, D. P., Innis, J. W., Holmes, L. B., Donnenfeld, A. E., Feingold, M., et al. (2000). Novel HOXA13 mutations and the phenotypic spectrum of hand–foot–genital syndrome. *Am. J. Hum. Genet.* **67**, 197–202.
88. Innis, J. W., Mortlock, D., Chen, Z., Ludwig, M., Williams, M. E., Williams, T. M., Doyle, C. D., Shao, Z., Glynn, M., Mikulic, D., Lehmann, K., Mundlos, S., and Utsch, B. (2004). Polyalanine expansion in HOXA13: Three new affected families and the molecular consequences in a mouse model. *Hum. Mol. Genet.* **13**, 2841–2851.
89. Jones (1997). "Smith's Recognizable Patterns of Human Malformation," 5th ed. W. B. Saunders Co., Philadelphia.
90. Otto, F., Thornell, A. P., Crompton, T., Denzel, A., Gilmour, K. C., Rosewell, I. R., Stamp, G. W., Beddington, R. S., Mundlos, S., Olsen, B. R., Selby, P. B., and Owen, M. J. (1997). Cbfa1, a candidate gene for cleidocranial dysplasia syndrome, is essential for osteoblast differentiation and bone development. *Cell* **89**, 765–771.
91. Sasaki, A., Kanai, M., Kijima, K., Akaba, K., Hashimoto, M., Hasegawa, H., Otaki, S., Koizumi, T., Kusuda, S., Ogawa, Y., Tuchiya, K., Yamamoto, W., Nakamura, T., and Hayasaka, K. (2003). Molecular analysis of congenital central hypoventilation syndrome. *Hum. Genet.* **114**, 22–26.
92. Weese-Mayer, D. E., Berry-Kravis, E. M., Zhou, L., Maher, B. S., Silvestri, J. M., Curran, M. E., and Marazita, M. L. (2003). Idiopathic congenital central hypoventilation syndrome: Analysis of genes pertinent to early autonomic nervous system embryologic development and identification of mutations in Phox2b. *Am. J. Med. Genet.* **123**, 267–278.
93. Matera, I., Bachetti, T., Puppo, F., Di Duca, M., Morandi, F., Casiraghi, G. M., Cilio, M. R., Hennekam, R., Hofstra, R., Schober, J. G., Ravazzolo, R., Ottonello, G., and Ceccherini, I. (2004). PHOX2B mutations and polyalanine expansions correlate with the severity of the respiratory phenotype and associated symptoms in both congenital and late onset central hypoventilation syndrome. *J. Med. Genet.* **41**, 373–380.
94. DeMyer, W. E., Zeman, W., and Palmer, C. G. (1963). Familial alobar holoprosencephaly (arhinencephaly) with median cleft lip and palate: Report of patient with 46 chromosomes. *Neurology* **13**, 913–918.
95. Brown, L. Y., Odent, S., David, V., Blayau, M., Dubourg, C., Apacik, C., Delgado, M. A., Hall, B. D., Reynolds, J. F., Sommer, A., Wieczorek, D., Brown, S. A., and Muenke, M. (2001). Holoprosencephaly due to mutations in ZIC2: Alanine tract expansion mutations may be caused by parental somatic recombination. *Hum. Mol. Genet.* **10**, 791–796.
96. Nagai, T., Aruga, J., Minowa, O., Sugimoto, T., Ohno, Y., Noda, T., and Mikoshiba, K. (2000). Zic2 regulates the kinetics of neurulation. *Proc. Natl. Acad. Sci. USA* **97**, 1618–1623.
97. Nagai, T., Aruga, J., Takada, S., Gunther, T., Sporle, R., Schughart, K., and Mikoshiba, K. (1997). The expression of the mouse Zic1, Zic2, and Zic3 gene suggests an essential role for Zic genes in body pattern formation. *Dev. Biol.* **182**, 299–313.
98. Brown, L., Paraso, M., Arkell, R., and Brown, S. (2004). In vitro analysis of partial loss-of-function ZIC-2 mutations in holoprosencephaly: Alanine tract expansion modulates DNA binding and transactivation. *Hum. Mol. Genet.* **14**, 411–420.
99. Blumen, S. C., Brais, B., Korczyn, A. D., Medinsky, S., Chapman, J., Asherov, A., Nisipeanu, P., Codere, F., Bouchard, J. P., Fardeau, M., Tome, F. M., and Rouleau, G. A. (1999). Homozygotes for oculopharyngeal muscular dystrophy have a severe form of the disease. *Ann. Neurol.* **46**, 115–118.
100. Prueitt and Zinn (2001). A fork in the road to fertility. *Nat. Genet.* **27**, 132–134.
101. Zlotogora, J., Sagi, M., and Cohen, T. (1983). The blepharophimosis, ptosis, and epicanthus inversus syndrome: Delineation of two types. *Am. J. Hum. Genet.* **35**, 1020–1027.
102. De Baere, E., Beysen, D., Oley, C., Lorenz, B., Cocquet, J., De Sutter, P., Devriendt, K., Dixon, M., Fellous, M., Fryns, J. P., Garza, A., Jonsrud, C., Koivisto, P. A., Krause, A., Leroy, B. P., Meire, F., Plomp, A., Van Maldergem, L., De Paepe, A., Veitia, R., and Messiaen, L. (2003). FOXL2 and BPES: mutational hotspots, phenotypic variability, and revision of the genotype–phenotype correlation. *Am. J. Hum. Genet.* **72**, 478–487.
103. Caburet, S., Demarez, A., Moumne, L., Fellows, M., De Baere, E., and Veitia, R. A. (2004). A recurrent polyalanine expansion in the transcription factor FOXL2 induces extensive nuclear and cytoplasmic protein aggregation. *J. Med. Genet.* **41**, 931–946.
104. Feinberg, A. P., and Leahy, W. R. (1977). Infantile spasms: Case report of sex-linked inheritance. *Dev. Med. Child Neurol.* **19**, 524–526.
105. Bienvenu, T., Poirier, K., Friocourt, G., Bahi, N., Beaumont, D., Fauchereau, F., Ben Jeema, L., Zemni, R., Vinet, M. C., Francis, F., Couvert, P., Gomot, M., Moraine, C., van Bokhoven, H., Kalscheuer, V., Frints, S., Gecz, J., Ohzaki, K., Chaabouni, H., Fryns, J. P., Desportes, V., Beldjord, C., and Chelly, J. (2002). ARX, a novel Prd-class-homeobox gene highly expressed in the telencephalon, is mutated in X-linked mental retardation. *Hum. Mol. Genet.* **11**, 981–991.
106. Kato, M., Das, S., Petras, K., Kitamura, K., Morohashi, K., Abuelo, D. N., Barr, M., Bonneau, D., Brady, A. F., Carpenter, N. J., Cipero, K. L., Frisone, F., Fukuda, T., Guerrini, R., Iida, E., Itoh, M., Lewanda, A. F., Nanba, Y., Oka, A., Proud, V. K., Saugier-Veber, P., Schelley, S. L., Selicorni, A., Shaner, R., Silengo, M., Stewart, F., Sugiyama, N., Toyama, J., Toutain, A., Vargas, A. L., Yanazawa, M., Zackai, E. H., and Dobyns, W. B. (2004). Mutations of ARX are associated with striking pleiotropy and consistent genotype–phenotype correlation. *Hum. Mutat.* **23**, 147–159.
107. Nasrallah, I. M., Minarcik, J. C., and Golden, J. A. (2004). A polyalanine tract expansion in Arx forms intranuclear inclusions and results in increased cell death. *J. Cell Biol.* **167**, 411–416.

108. Brais, B., Xie, Y. G., Sanson, M., Morgan, K., Weissenbach, J., Korczyn, A. D., Blumen, S. C., Fardeau, M., Tome, F. M., Bouchard, J. P., and Rouleau, G. A. (1995). The oculopharyngeal muscular dystrophy locus maps to the region of the cardiac alpha and beta myosin heavy chain genes on chromosome 14q11.2-q13. *Hum. Mol. Genet.* **4**, 429–434.
109. Tome, F. M., and Fardeau, M. (1980). Nuclear inclusions in oculopharyngeal dystrophy. *Acta Neuropathol. (Berlin)* **49**, 85–87.
110. Rubinsztein, D. C., Wyttenbach, A., and Rankin, J. (1999). Intracellular inclusions, pathological markers in diseases caused by expanded polyglutamine tracts? *J. Med. Genet.* **36**, 265–270.
111. Shanmugam, V., Dion, P., Rochefort, D., Laganiere, J., Brais, B., and Rouleau, G. A. (2000). PABP2 polyalanine tract expansion causes intranuclear inclusions in oculopharyngeal muscular dystrophy. *Ann. Neurol.* **48**, 798–802.
112. Sherman, M. Y., and Goldberg, A. L. (2001). Cellular defenses against unfolded proteins: A cell biologist thinks about neurodegenerative diseases. *Neuron* **29**, 15–32.
113. Abu-Baker, A., Messaed, C., Laganiere, J., Gaspar, C., Brais, B., and Rouleau, G. A. (2003). Involvement of the ubiquitin–proteasome pathway and molecular chaperones in oculopharyngeal muscular dystrophy. *Hum. Mol. Genet.* **12**, 2609–2623.
114. Calado, A., Tome, F. M. S., Brais, B., Rouleau, G. A., Kuhn, U., Wahle, E., and Carmo-Fonseca, M. (2000). Nuclear inclusions in oculopharyngeal muscular dystrophy consist of poly(A) binding protein 2 aggregates which sequester poly(A) RNA. *Hum. Mol. Genet.* **9**, 2321–2328.
115. Bao, Y. P., Cook, L. J., O'Donovan, D., Uyama, E., and Rubinsztein, D. C. (2002). Mammalian, yeast, bacterial, and chemical chaperones reduce aggregate formation and death in a cell model of oculopharyngeal muscular dystrophy. *J. Biol. Chem.* **277**, 12263–12269.
116. Fan, X., Dion, P., Laganiere, J., Brais, B., and Rouleau, G. A. (2001). Oligomerization of polyalanine expanded PABPN1 facilitates nuclear protein aggregation that is associated with cell death. *Hum. Mol. Genet.* **10**, 2341–2351.
117. Albrecht, A. N., and Mundlos, S. (2005). The other trinucleotide repeat: Polyalanine expansion disorders. *Curr. Opin. Genet. Dev.* **15**, 285–293.
118. Perutz, M. F., Johnson, T., Suzuki, M., and Finch, J. T. (1994). Glutamine repeats as polar zippers: Their possible role in inherited neurodegenerative diseases. *Proc. Natl. Acad. Sci. USA* **91**, 5355–5358.
119. Bates, G. P., Mangiarini, L., and Davies, S. W. (1998). Transgenic mice in the study of polyglutamine repeat expansion diseases. *Brain Pathol.* **6**, 699–714.
120. Lathrop, R. H., Casale, M., Tobias, D. J., Marsh, J. L., and Thompson, L. M. (1998). Modeling protein homopolymeric repeats: Possible polyglutamine structural motifs for Huntington's disease. *Proc. Int. Conf. Intell. Syst. Mol. Biol.* **6**, 105–114.
121. Cooper, A. J., Sheu, K. F., Burke, J. R., Onodera, O., Strittmatter, W. J., Roses, A. D., and Blass, J. P. (1997). Polyglutamine domains are substrates of tissue transglutaminase: Does transglutaminase play a role in expanded CAG/poly-Q neurodegenerative diseases? *J. Neurochem.* **69**, 431–434.
122. Igarashi, S., Koide, R., Shimohata, T., Yamada, M., Hayashi, Y., Takano, H., Date, H., Oyake, M., Sato, T., Sato, A., Egawa, S., Ikeuchi, T., Tanaka, H., Nakano, R., Tanaka, K., Hozumi, I., Inuzuka, T., Takahashi, H., and Tsuji, S. (1998). Suppression of aggregate formation and apoptosis by transglutaminase inhibitors in cells expressing truncated DRPLA protein with an expanded polyglutamine stretch. *Nat. Genet.* **18**, 1111–1117.
123. Kahlem, P., Green, H., and Dijan, P. (1998). Transglutaminase action imitates Huntington's disease: Selective polymerization of Huntingtin containing expanded polyglutamine. *Mol. Cell* **1**, 595–601.
124. Lesort, M., Chun, W., Tucholski, J., and Johnson, G. V. (2002). Does tissue transglutaminase play a role in Huntington's disease? *Neurochem. Int.* **40**, 37–52.
125. Blondelle, S. E., Forood, B., Houghten, R. A., and Perez-Paya, E. (1997). Polyalanine-based peptides as models for self-associated beta-pleated-sheet complexes. *Biochemistry* **36**, 8393–8400.
126. Forood, B., Perez-Paya, E., Houghten, R. A., and Blondelle, S. E. (1995). Formation of an extremely stable polyalanine beta-sheet macromolecule. *Biochem. Biophys. Res. Commun.* **211**, 7–13.
127. Perutz, M. F., Pope, B. J., Owen, D., Wanker, E. E., and Scherzinger, E. (2002). Aggregation of proteins with expanded glutamine and alanine repeats of the glutamine-rich and asparagine-rich domains of Sup35 and of the amyloid beta-peptide of amyloid plaques. *Proc. Natl. Acad. Sci. USA* **99**, 5596–5600.
128. Rankin, J., Wyttenbach, A., and Rubinsztein, D. C. (2000). Intracellular green fluorescent protein–polyalanine aggregates are associated with cell death. *Biochem. J.* **348**, 15–19.
129. Gaspar, C., Jannatipour, M., Dion, P., Laganiere, J., Sequeiros, J., Brais, B., and Rouleau, G. A. (2000). CAG tract of MJD-1 may be prone to frameshifts causing polyalanine accumulation. *Hum. Mol. Genet.* **9**, 1957–1966.
130. Zuccato, C., Ciammola, A., Rigamonti, D., Leavitt, B. R., Goffredo, D., Conti, L., MacDonald, M. E., Friedlander, R. M., Silani, V., Hayden, M. R., Timmusk, T., Sipione, S., and Cattaneo, E. (2001). Loss of Huntingtin-mediated BDNF gene transcription in Huntington's disease. *Science* **293**, 493–498.
131. Rangone, H., Humbert, S., and Saudou, F. (2004). Huntington's disease: How does huntingtin, an anti-apoptotic protein, become toxic? *Pathol. Biol. (Paris)* **52**, 338–342.
132. Rigamonti, D., Bauer, J. H., De Fraja, C., Conti, L., Sipione, S., Sciorati, C., Clementi, E., Hackam, A., Hayden, M. R., Li, Y., Cooper, J. K., Ross, C. A., Govoni, S., Vincenz, C., and Cattaneo, E. (2000). Wild-type huntingtin protects from apoptosis upstream of caspase-3. *J. Neurosci.* **20**, 3705–3713.
133. Burnett, B., Li, F., and Pittman, R. N. (2003). The polyglutamine neurodegenerative protein ataxin-3 binds polyubiquitinylated proteins and has ubiquitin protease activity. *Hum. Mol. Genet.* **12**, 3195–3205.
134. Scheel, H., Tomiuk, S., and Hofmann, K. (2003). Elucidation of ataxin-3 and ataxin-7 function by integrative bioinformatics. *Hum. Mol. Genet.* **12**, 2845–2852.
135. Helmlinger, D., Hardy, S., Sasorith, S., Klein, F., Robert, F., Weber, C., Miguet, L., Potier, N., Van Dorsselaer, A., Wurtz, J.-M., Mandel, J.-L., Tora, L, and Devys, D. (2004). Ataxin-7 is a subunit of GCN5 histone acetyltransferase-containing complexes. *Hum. Mol. Genet.* **13**, 1257–1265.
136. Hill, M. E., Creed, G. A., McMullan, T. F., Tyers, A. G., Hilton-Jones, D., Robinson, D. O., and Hammans, S. R. (2001). Oculopharyngeal muscular dystrophy: Phenotypic and genotypic studies in a UK population. *Brain* **124**, 522–526.
137. Schober, R., Kress, W., Grahmann, F., Kellermann, S., Baum, P., Gunzel, S., and Wagner, A. (2001). Unusual triplet expansion associated with neurogenic changes in a family with oculopharyngeal muscular dystrophy. *Neuropathology* **21**, 45–52.
138. Pou Serradell, A., Lloreta, J., Corominas, J., Hammouda, E., Urtizberea, J., Richard, P., and Brais, B. (2004). Oculopharyngeal muscular dystrophy: Study of patients from seven Spanish families with different GCG expansions in *PABP2* gene. *Neurologia* **19**, 239–247.
139. Lavoie, H., Debeane, F., Trinh, Q. D., Turcotte, J. F., Corbeil-Girard, L. P., Dicaire, M. J., Saint-Denis, A., Page, M., Rouleau, G. A., and Brais, B. (2003). Polymorphism, shared functions and convergent

evolution of genes with sequences coding for polyalanine domains. *Hum. Mol. Genet.* **12**, 2967–2979.
140. Han, K., and Manley, J. L. (1993). Functional domains of the *Drosophila* engrailed protein. *EMBO J.* **12**, 2723–2733.
141. Young, E. T., Sloan, J. S., and Van Riper, K. (2000). Trinucleotide repeats are clustered in regulatory genes in *Saccharomyces cerevisiae*. *Genetics* **154**, 1053–1068.
142. Wahle, E. (1991). A novel poly(A)-binding protein acts as specificity factor in the second phase of messenger RNA polyadenylation. *Cell* **66**, 759–768.
143. Wahle, E. (1991). Purification and characterization of a mammalian polyadenylate polymerase involved in the 3′ end processing of messenger RNA precursors. *J. Biol. Chem.* **266**, 3131–3139.
144. Krause, S., Fakan, S., Weis, K., and Wahle, E. (1994). Immunodetection of poly(A) binding protein II in the cell nucleus. *Exp. Cell Res.* **214**, 75–82.
145. Wahle, E., Lustig, A., Jeno, P., and Maurer, P. (1993). Mammalian poly(A)-binding protein II. Physical properties and binding to polynucleotides. *J. Biol. Chem.* **268**, 2937–2945.
146. Nemeth, A., Krause, S., Blank, D., Jenny, A., Jeno, P., Lusting, A., and Wahle, E. (1995). Isolation of genomic and cDNA clones encoding bovine poly(A) binding protein II. *Nucleic Acids Res.* **23**, 4034–4041.
147. Calado, U., Kutay, U., Kühn, E., Wahle, E., and Carmo-Fonseca, M. (2000). Deciphering the cellular pathway for transport of poly(A)-binding protein II. *RNA* **6**, 245–256.
148. Wahle, E. (1995). Poly(A) tail length control is caused by termination of processive synthesis. *J. Biol. Chem.* **270**, 2800–2808.
149. Gorlach, M., Burd, C. G., and Dreyfuss, G. (1994). The mRNA poly(A)-binding protein: Localization, abundance, and RNA-binding specificity. *Exp. Cell Res.* **211**, 400–407.
150. Afonina, E., Stauber, R., and Pavlakis, G. N. (1998). The human poly(A)-binding protein 1 shuttles between the nucleus and the cytoplasm. *J. Biol. Chem.* **273**, 13015–13021.
151. Tome, F. M., Chateau, D., Helbling-Leclerc, A., and Fardeau, M. (1997). Morphological changes in muscle fibers in oculopharyngeal muscular dystrophy. *Neuromuscul. Disord.* **7**, S63–S69.
152. Davies, J. E., Wang, L., Garcia-Oroz, L., Cook, L. J., Vacher, C., O'Donovan, D. G., and Rubinsztein, D. C. (2005). Doxycycline attenuates and delays toxicity of the oculopharyngeal muscular dystrophy mutation in transgenic mice. *Nat. Med.* **6**, 672–677.
153. Fan, X., Messaed, C., Dion, P., Laganiere, J., Brais, B., Karpati, G., and Rouleau, G. A. (2003). HnRNP A1 and A/B interaction with PABPN1 in oculopharyngeal muscular dystrophy. *Can. J. Neurol. Sci.* **30**, 244–251.
154. Abu-Baker, A., Laganiere, S., Fan, X., Laganiere, J., Brais, B., and Rouleau, G. A. (2005). Cytoplasmic targeting of mutant poly(A)-binding protein nuclear 1 suppresses protein aggregation and toxicity in oculopharyngeal muscular dystrophy. *Traffic* **6**, 1–14.
155. Scheuermann, T., Schulz, B., Blume, A., Wahle, E., Rudolph, R., and Schwarz, E. (2003). Trinucleotide expansions leading to an extended poly-L-alanine segment in the poly(A) binding protein PABPN1 cause fibril formation. *Protein Sci.* **12**, 2685–2692.
156. Onodera, O., Burke, J. R., Miller, S. E., Hester, S., Tsuji, S., Roses, A. D., and Strittmatter, W. J. (1997). Oligomerization of expanded-polyglutamine domain fluorescent fusion proteins in cultured mammalian cells. *Biochem. Biophys. Res. Commun.* **238**, 599–605.
157. Iuchi, G., Hoffner, P., Verbeke, P., Dijan, and Green, H. (2003). Oligomeric and polymeric aggregates formed by proteins containing expanded polyglutamine. *Proc. Natl. Acad. Sci. USA* **100**, 2409–2414.
158. Benvivino, A. E., and Loll, P. J. (2001). An expanded glutamine repeat destabilizes native ataxin-3 structure and mediates formation of parallel fibrils. *Proc. Natl. Acad. Sci. USA* **98**, 11955–11960.

159. Sanchez, I., Mahlke, C., and Yuan, J. (2003). Pivotal role of oligomerization in expanded polyglutamine neurodegenerative disorders. *Nature* **421**, 373–379.
160. Bao, Y. P., Sarkar, S., Uyama, E., and Rubinsztein, D. C. (2004). Congo red, doxycycline, and HSP70 overexpression reduce aggregate formation and cell death in cell models of oculopharyngeal muscular dystrophy. *J. Med. Genet.* **41**, 47–51.
161. Ravikumar, B., Duden, R., and Rubinsztein, D. C. (2002). Aggregate-prone proteins with polyglutamine and polyalanine expansions are degraded by autophagy. *Hum. Mol. Genet.* **11**, 1107–1117.
162. Wyttenbach, A., Carmichael, J., Swartz, J., Furlong, R. A., Narain, Y., Rankin, J., and Rubinsztein, D. C. (2000). Effects of heat shock, heat shock protein 40 (HDJ-2), and proteasome inhibition on protein aggregation in cellular models of Huntington's disease. *Proc. Natl. Acad. Sci. USA* **97**, 2898–2903.
163. Nakielny, S., and Dreyfuss, G. (1997). Nuclear export of proteins and RNAs. *Curr. Opin. Cell Biol.* **9**, 420–429.
164. Tavanez, J. P., Calado, P., Braga, J., Lafarga, M., and Carmo-Fonseca, M. (2005). In vivo aggregation properties of the nuclear poly(A)-binding protein PABPN1. *RNA* **5**, 752–762.
165. Corbeil-Girard, L. P., Klein, A. F., Sasseville, A. M., Lavoie, H., Dicaire, M. J., Saint-Denis, A., Page, M., Duranceau, A., Codere, F., Bouchard, J. P., Karpati, G., Rouleau, G. A., Massie, B., Langelier, Y., and Brais, B. (2005). PABPN1 overexpression leads to upregulation of genes encoding nuclear proteins that are sequestered in oculopharyngeal muscular dystrophy nuclear inclusions. *Neurobiol. Dis.* **18**, 551–567.
166. Deleault, N. R., Lucassen, R. W., and Supattapone, S. (2003). RNA molecules stimulate prion protein conversion. *Nature* **16**, 717–720.
167. Neuwald, A. F., and Koonin, E. V. (1998). Ataxin-2, global regulators of bacterial gene expression, and spliceosomal snRNP proteins share a conserved domain. *J. Mol. Med.* **76**, 3–5.
168. Burd, C. G., and Dreyfuss, G. (1994). Conserved structures and diversity of functions of RNA-binding proteins. *Science* **265**, 615–621.
169. Huynh, D. P., Del Bigio, M. R., Ho, D. H., and Pulst, S.-M. (1999). Expression of ataxin 2 in brains from normal individuals and patients with Alzheimer's disease and spinocerebellar ataxia 2 (SCA2). *Ann. Neurol.* **45**, 232–241.
170. Yue, S., Serra, H. G., Zoghbi, H. Y., and Orr, H. T. (2001). The spinocerebellar ataxia type 1 protein, ataxin-1, has RNA-binding activity that is inversely affected by the length of its polyglutamine tract. *Hum. Mol. Genet.* **10**, 25–30.
171. Irwin, S., Vandelft, M., Pinchev, D., Howell, J. L., Graczyk, J., Orr, H. T., and Truant, R. (2005). RNA association and nucleocytoplasmic shuttling by ataxin-1. *J. Cell Sci.* **118**, 233–242.
172. Skinner, P. J., Koshy, B. T., Cummings, C. J., Klement, I. A., Helin, K., Servadio, A., Zoghbi, H. Y., and Orr, H. T. (1997). Ataxin-1 with an expanded glutamine tract alters nuclear matrix-associated structures. *Nature* **389**, 971–974.
173. Dovey, C. L., Varadaraj, A., Wyllie, A. H., and Rich, T. (2004). Stress responses of PML nuclear domains are ablated by ataxin-1 and other nucleoprotein inclusions. *J. Pathol.* **203**, 877–883.
174. Yamada, M., Sato, T., Shimohata, T., Hayashi, S., Igarashi, S., Tsuji, S., and Takahashi, H. (2001). Interaction between neuronal intranuclear inclusions and promyelocytic leukemia protein nuclear and coiled bodies in CAG repeat diseases. *Am. J. Pathol.* **159**, 1785–1795.
175. Takahashi, J., Fujigasaki, H., Zander, C., El Hachimi, K. H., Stevanin, G., Durr, A., Lebre, A. S., Yvert, G., Trottier, Y., The, H., Hauw, J. J., Duyckaerts, C., and Brice, A. (2002). Two populations of neuronal intranuclear inclusions in SCA7 differ in size and promyelocytic leukaemia protein content. *Brain* **125**, 1534–1543.

176. Chai, Y., Wu, L., Griffin, J. D., and Paulson, H. L. (2001). The role of protein composition in specifying nuclear inclusion formation in polyglutamine disease. *J. Biol. Chem.* **276**, 44889–44897.
177. Saudou, F., Finkbeiner, S., Devys, D., and Greenberg, M. E. (1998). Huntingtin acts in the nucleus to induce apoptosis but death does not correlate with the formation of intranuclear inclusions. *Cell* **95**, 55–66.
178. Peters, M. F., Nucifora, F. C., Jr., Kushi, J., Seaman, H. C., Cooper, J. K., Herring, W. J., Dawson, V. L., Dawson, T. M., and Ross, C. A. (1999). Nuclear targeting of mutant Huntingtin increase toxicity. *Mol. Cell Neurosci.* **2**, 121–128.
179. Klement, I. A., Skinner, P. J., Kaytor, M. D., Yi, H., Hersch, S. M., Clark, H. B., Zoghbi, H. Y., and Orr, H. T. (1998). Ataxin-1 nuclear localization and aggregation: Role in polyglutamine-induced disease in SCA1 transgenic mice. *Cell* **95**, 41–53.
180. Yang, W., Dunlap, J. R., Andrews, R. B., and Wetzel, R. (2002). Aggregated polyglutamine peptides delivered to nuclei are toxic to mammalian cells. *Hum. Mol. Genet.* **11**, 2905–2917.
181. Kim, Y.-J., Noguchi, Y. K., Hayashi, T., Tsukahara, T., and Shimizu, A. K. (2001). The product of an oculopharyngeal muscular dystrophy gene, poly(A)-binding protein 2, interacts with SKIP and stimulates muscle-specific gene expression. *Hum. Mol. Genet.* **10**, 1129–1139.
182. Bear, D. G., Fomproix, N., Soop, T., Björkroth, B., Masich, S., and Daneholt, B. (2003). Nuclear poly(A)-binding protein PABPN1 is associated with RNA polymerase II during transcription and accompanies the released transcript to the nuclear pore. *Exp. Cell Res.* **286**, 332–344.
183. Wellington, C. L., Singaraja, R., Ellerby, L., Savill, J., Roy, S., Leavitt, B., et al. (2000). Inhibiting caspase cleavage of huntingtin reduces toxicity and aggregate formation in neuronal and nonneuronal cells. *J. Biol. Chem.* **275**, 19831–19838.
184. Probst, A., Tackmann, W., Stoeckli, H. R., Jerusalem, F., and Ulrich, J. (1982). Evidence for a chronic axonal atrophy in oculopharyngeal "muscular dystrophy." *Acta Neuropathol. (Berlin)* **57**, 209–216.
185. Boukriche, Y., Maisonobe, T., and Masson, C. (2002). Neurogenic involvement in a case of oculopharyngeal muscular dystrophy. *Muscle Nerve* **25**, 98–101.
186. Nakashima, D., Nakajima, H., Ishida, S., Sugino, M., Kimura, F., and Hanafusa, T. (2003). Preferential distal muscle involvement in case of oculopharyngeal muscular dystrophy with (GCG) 13 expansion. *Rinsho Shinkeigaku* **43**, 560–563.
187. Dion, P., Shanmugam, V., Gaspar, C., Messaed, C., Meijer, I., Toulouse, A., Laganiere, J., Roussel, J., Rochefort, D., Laganiere, S., Allen, C., Karpati, G., Bouchard, J. P., Brais, B., and Rouleau, G. A. (2005). Transgenic expression of an expanded (GCG) 13 repeat PABPN1 leads to weakness and coordination defects in mice. *Neurobiol. Dis.* **3**, 528–536.
188. Codere, F. (1993). Oculopharyngeal muscular dystrophy. *Can. J. Ophthalmol.* **28**, 1–2.
189. Duranceau, A. (1997). Cricopharyngeal myotomy in the management of neurogenic and muscular dysphagia. *Neuromuscul. Disord.* **7**, S85–S89.
190. Mathieu, J., Lapointe, G., Brassard, A., Tremblay, C., Brais, B., Rouleau, G. A., and Bouchard, J. P. (1997). A pilot study on upper esophageal sphincter dilatation for the treatment of dysphagia in patients with oculopharyngeal muscular dystrophy. *Neuromuscul. Disord.* **7**, S100–S104.
191. Sisodia, S. S. (1998). Nuclear inclusions in glutamine repeat disorders: Are they pernicious, coincidental, or beneficial? *Cell* **95**, 1.
192. Arrasate, M., Mitra, S., Schweitzer, E. S., Segal, M. R., and Finkbeiner, S. (2004). Inclusion body formation reduces levels of mutant huntingtin and the risk of neuronal death. *Nature* **431**, 805–810.
193. Orr, H. T. (2004). Neurodegenerative disease: Neuron protection agency. *Nature* **431**, 747–748.
194. Gutekunst, C. A., Li, S. H., Yi, H., Mulroy, J. S., Kuemmerle, S., Rye, D., Ferrante, R. J., Hersch, S. M., and Li, X. J. (1999). Nuclear and neuropil aggregates in Huntington's disease: Relationship to neuropathology. *J. Neurosci.* **19**, 2522–2534.
195. Kuemmerle, S., Gutekunst, C. A., Klein, A. M., Li, X. J., Li, S. H., Beal, M. F., Hersch, S. M., and Ferrante, R. J. (1999). Huntington aggregates may not predict neuronal death in Huntington's disease. *Ann. Neurol.* **46**, 842–849.
196. Muchowski, P. J. (2002). Protein misfolding, amyloid formation, and neurodegeneration: A critical role for molecular chaperones? *Neuron* **35**, 9–12.
197. Wirtschafter, J. D., Ferrington, D. A., and McLoon, L. K. (2004). Continuous remodeling of adult extraocular muscles as an explanation for selective craniofacial vulnerability in oculopharyngeal muscular dystrophy. *J. Neuroophthalmol.* **24**, 62–67.
198. Mouly, V., Aamiri, A., Bigot, A., Cooper, R. N., Di Donna, S., Furling, D., Gidaro, R., Jacquemin, V., Mamchaoui, K., Negroni, E., Perie, S., Renault, V., Silva-Barbosa, S. D., and Butler-Browne, G. S. (2005). *Acta Physiol. Scand.* **184**, 3–15.
199. Sittler, A., Wälter, S., Wedemeyer, N., Hasenbank, R., Scherzinger, E., Eickhoff, H., Bates, G., et al. (1998). SH3GL3 associates with the huntingtin exon 1 protein and promotes the formation of polyGln-containing protein aggregates. *Mol. Cell* **2**, 427–436.
200. Matilla, A., Koshy, B. T., Cummings, C. J., Isobe, T., Orr, H. T., and Zoghbi, H. Y. (1997). The cerebellar leucine-rich acidic nuclear protein interacts with ataxin-1. *Nature* **389**, 974–978.
201. Cattaneo, E., Rigamonti, D., Goffredo, D., Zuccato, C., Squitieri, F., and Sipione, S. (2001). Loss of normal huntingtin function: New developments in Huntington's disease research. *Trends Neurosci.* **24**, 1–8.
202. Poirier, M. A., Li, H., Macosko, J., Cai, S., Amzel, M., and Ross, C. A. (2002). Huntingtin spheroids and protofibrils as precursors in polyglutamine fibrilization. *J. Biol. Chem.* **277**, 41032–41037.
203. Tanaka, M., Machida, Y., Niu, S., Ikeda, T., Jana, N. R., Doi, H., Kurosawa, M., Nekooki, M., and Nukina, N. (2004). Trehalose alleviates polyglutamine-mediated pathology in a mouse model of Huntington disease. *Nat. Med.* **10**, 148–154.
204. Karpuj, M. V., Becher, M. W., Springer, J. E., Chabas, D., Youssef, S., Pedotti, R., Mitchell, D., and Steinman, L. (2002). Prolonged survival and decreased abnormal movements in transgenic model of Huntington disease, with administration of the transglutaminase inhibitor cystamine. *Nat. Med.* **8**, 143–149.
205. Sittler, A., Lurz, R., Lueder, G., Priller, J., Lehrach, H., Hayer-Hartl, M. K., Hartl, F. U., and Wanker, E. E. (2001). Geldanamycin activates a heat shock response and inhibits huntingtin aggregation in a cell culture model of Huntington's disease. *Hum. Mol. Genet.* **10**, 1307–1315.
206. Wang, X., Zhu, S., Drozda, M., Zhang, W., Stavrovskaya, I. G., Cattaneo, E., Ferrante, R. J., Kristal, B. S., and Friedlander, R. M. (2003). Minocycline inhibits caspase-independent and -dependent mitochondrial cell death pathways in models of Huntington's disease. *Proc. Natl. Acad. Sci. USA* **100**, 10483–10487.
207. Chen, M., Ona, V. O., Li, M., Ferrante, R. J., Fink, K. B., Zhu, S., Bian, J., Guo, L., Farrell, L. A., Hersch, S. M., Hobbs, W., Vonsattel, J. P., Cha, J. H., and Friedlander, R. M. (2000). Minocycline inhibits caspase-1 and caspase-3 expression and delays mortality in a transgenic mouse model of Huntington disease. *Nat. Med.* **6**, 797–801.

208. Ferrante, R. J., Andreassen, O. A., Jenkins, B. G., Dedeoglu, A., Kuemmerle, S., Kubilus, J. K., Kaddurah-Daouk, R., Hersch, S. M., and Beal, M. F. (2000). Neuroprotective effects of creatine in a transgenic mouse model of Huntington's disease. *J. Neurosci.* **20**, 4389–4397.
209. Beal, M. F., Henshaw, D. R., Jenkins, B. G., Rosen, B. R., and Schulz, J. B. (1994). Coenzyme Q_{10} and nicotinamide block striatal lesions produced by the mitochondrial toxin malonate. *Ann. Neurol.* **36**, 882–888.
210. Heiser, V., et al. (2002). Identification of benzothiazoles as potential polyglutamine aggregation inhibitors of Huntington's disease by using an automated filter retardation assay. *Proc. Natl. Acad. Sci. USA* **99**, 16400–16406.
211. Wood, N. I., and Morton, A. J. (2003). Chronic lithium chloride treatment has variable effects on motor behavior and survival of mice transgenic for the Huntington's disease mutation. *Brain Res. Bull.* **61**, 375–383.
212. Keene, C. D., Rodrigues, C. M. P., Eich, T., Chhabra, M. S., Steer, C. J., and Low, W. C. (2002). Taurousodeoxycholic acid, a bile acid, is neuroprotective in a transgenic animal model of Huntington's disease. *Proc. Natl. Acad. Sci. USA* **99**, 10671–10676.
213. Chuang, D. M., Chen, R. W., Chalecka-Franaszek, E., Ren, M., Hashimoto, R., Senatorov, V., Kanai, H., Hough, C., Hiroi, T., and Leeds, P. (2002). Neuroprotective effects of lithium in cultured cells and animal models of diseases. *Bipolar Disord.* **4**, 129–136.
214. Lecerf, J.-M., Shirley, T. L., Zhu, Q., Kazantsev, A., Amersdorfer, P., Housman, D. E., Messer, A., and Huston, J. S. (2001). Human single-chain Fv intrabodies counteract *in situ* huntingtin aggregation in cellular models of Huntington's disease. *Proc. Natl. Acad. Sci. USA* **98**, 4764–4769.
215. Khoshnan, A., Ko, J., and Patterson, P. H. (2002). Effects of intracellular expression of anti-huntingtin antibodies of various specificities on mutant huntingtin aggregation and toxicity. *Proc. Natl. Acad. Sci. USA* **99**, 1002–1007.
216. Xia, H., Mao, Q., Paulson, H. L., and Davidson, B. L. (2002). siRNA-mediated gene silencing in vitro and in vivo. *Nat. Biotechnol.*, 1008–1010.
217. Nagai, Y., Tucker, T., Ren, H., Kenan, D. J., Henderson, B. S., Keene, J. D., Strittmatter, W. J., and Burke, J. R. (2000). Inhibition of polyglutamine protein aggregation and cell death by novel peptides identified by phage display screening. *J. Biol. Chem.* **275**, 10437–10442.
218. Kazantsev, A., Walker, H. A., Slepko, N., Bear, J. E., Preisinger, E., Steffan, J. S., Zhu, Y. Z., Gertler, F. B., Housman, D. E., Marsh, J. L., and Thompson, L. M. (2002). A bivalent Huntingtin binding peptide suppresses polyglutamine aggregation and pathogenesis in Drosophila. *Nat. Genet.* **30**, 367–376.
219. Kerwitz, Y., Kühn, U., Lilie, H., Knoth, A., Scheuermann, T., Friedrich, H., Schwarz, E., and Wahle, E. (2003). Stimulation of poly(A) polymerase through a direct interaction with the nuclear poly(A) binding protein allosterically regulated by RNA. *EMBO J.* **22**, 3705–3714.

PART IX

Biophysics of PolyQ

CHAPTER 34

Chemical and Physical Properties of Polyglutamine Repeat Sequences

RONALD WETZEL

Graduate School of Medicine, University of Tennessee, Knoxville, Tennessee 37920

I. Introduction
II. Chemical and Physical Properties of the Amino Acid Glutamine
III. Solubility and Conformations of the Polyglutamine Sequence
    A. Conformations of Normal Repeat Length Polyglutamine
    B. Conformation of Expanded Polyglutamine
IV. Normal Roles of the Polyglutamine Sequence in Proteins
    A. Polyglutamine as a Flexible Linker between Secondary Structural Elements and Domains
    B. Polyglutamine as a Solubilizing Domain
    C. Polyglutamine as an Interaction or Aggregation Domain
V. Aggregation of Polyglutamine Sequences
    A. Varieties of Polyglutamine Aggregates
    B. Amyloid-like Aggregates with a Polyglutamine Core
    C. Polyglutamine Protein Aggregation Not Related to a Polyglutamine Amyloid Core
    D. Cellular Factors Influencing Aggregation
VI. Conclusions
References

In expanded CAG repeat diseases, the lengthening of the encoded polyglutamine sequence beyond a critical repeat length threshold introduces disease risk over a narrow repeat length range, and further expansions decrease age of onset. Though the mechanisms of these diseases remain obscure, pathology is presumably triggered by either a change in the properties of the polyglutamine itself or a change in some property of its host protein engendered by the expansion. This chapter reviews current understanding of the chemical and physical properties of polyglutamine sequences, focusing especially on conformation and aggregation. Also reviewed is the current limited knowledge of the normal role of polyglutamine repeats in proteins.

I. INTRODUCTION

An important subset of the trinucleotide repeat diseases is the group involving expansion of an in-frame CAG repeat in an open reading frame, leading to the expression of an expanded polyglutamine (polyGln) sequence. The sizes, predicted structures, cellular locations, and, where known, functions of the polyGln-containing proteins associated with these diseases vary. The diseases themselves exhibit overlapping neurological features, all are associated with neuronal polyGln aggregates, and all but one exhibit repeat length thresholds in the range of 40 Gln residues. It is generally accepted that the bulk of the pathology caused by polyGln expansion occurs via a toxic gain of function, although it appears possible that loss of function of the affected gene product might contribute in some cases [1]. There are two broad categories of mechanism that have been proposed to account for gains of function. In one class of mechanism, the root cause of disease is a common ability of the expanded polyGln sequence to subvert the normal function of each of these diverse proteins to generate new or modified activities with toxic consequences. Alternatively, there may be some common repeat length dependent property of the expanded polyGln sequence itself that plays an initiating role in the mechanisms of all these diseases. According to either model, it will be critical to understand the biophysical properties of the polyGln sequences embedded in these disease proteins and how these properties are affected by repeat length expansion, sequence context, and cellular context.

It is possible to imagine mechanisms by which polyGln expansion might alter the normal function of a protein in which it is embedded. The expanded polyGln might alter the structure and function of its host protein, either through a simple steric effect or due to a propagation of altered structure that is initiated by the polyGln segment. Alternatively, expanded polyGln might affect the trafficking and subcellular localization of the protein in which it occurs. If the polyGln sequence plays an important role in the proteins containing it, and especially if the nature of that role is similar for all the polyGln disease proteins, then one can imagine a common disease mechanism in which repeat expansion compromises that role. The challenge in elaborating such ideas has been to imagine mechanisms that, on the one hand, operate on a widely diverse set of protein structures, protein cellular locations, and protein functions, yet on the other hand produce overlapping disease features for all nine disease proteins.

There are also several ways in which repeat expansion might alter the properties of the polyGln segment itself. Repeat expansion might alter the polyGln folding thermodynamics such that the expanded polyGln segments take on a different three-dimensional structure that has toxic properties. Alternatively, toxic properties may come about via the ability of repeat expansion to magnify the intrinsic tendency of the polyGln sequence to form aggregates.

It has been surprisingly difficult to study the polyGln sequence both in isolation and in the context of a larger protein. As discussed here, glutamine exhibits what is generally considered to be a hydrophilic side chain, leading to the expectation that glutamine polymers likewise would be relatively hydrophilic and soluble. However, polyglutamine sequences exhibit a robust tendency to aggregate, greatly complicating efforts to study their properties in the monomeric state. Two broad classes of approach have emerged to deal with the aggregation problem: (a) studies with recombinant fusion proteins that feature solubilizing fusion partners [2] and (b) studies with simple, chemically synthesized polyGln peptides that can be rendered kinetically soluble by transient exposure to disaggregating solvents [3]. The story of the elucidation of some of the properties of these interesting homopolymers is intertwined with the story of increased understanding of how to manage the aggregation process. Despite some significant successes, great challenges remain, especially in studying native polyGln proteins and especially when they contain very long repeats. This chapter therefore must be viewed as a preliminary survey of a field in its infancy.

II. CHEMICAL AND PHYSICAL PROPERTIES OF THE AMINO ACID GLUTAMINE

Glutamine is one of two amino acids in the standard repertoire of 20 for ribosomal protein synthesis that contains a side chain amide group (the other being asparagine). This imparts a number of interesting properties onto this amino acid. First, the two amide hydrogens on the single, unsubstituted side chain amide nitrogen present the possibility of multiple H bonds. Second, despite the lack of charge in the side chain, a single Gln side chain in isolation is relatively hydrophilic due to its polarity and the ability of the side chain amide to interact with solvent water through hydrogen bonding. In scales of amino acid hydrophobicity, Gln is found with other amino acids (Glu, Asp, Asn, Ser, Thr, Lys, and Arg) whose side chains are compatible with solvent water [4]. In a study of surface accessibility in globular proteins, the vast majority of Gln residues are found to be at least partially surface exposed in the folded tertiary structure [5]. Third, consistent with Chou–Fasman and other parameters for predicting secondary structure from amino acid sequence [6], Gln is about equally likely to be located in α helix, β extended chain, or reverse turns in globular proteins [7]. An interesting and important caveat to the preceding discussion of solubility is that the amino acid Gln is not a particularly soluble amino acid [8], perhaps due to a strong self-association in the solid state. This self-interaction may in large part account for the tendency of polyGln to aggregate.

Although the Gln side chain is relatively stable, Gln residues are capable of undergoing a chemical deamidation reaction to generate glutamic acid [9]. Gln deamidation is much less favorable than Asn deamidation [9], however, and is unlikely to play a role in proteins that have a limited lifetime in the cell. Enzymatic hydrolysis of the side chain amide of Gln is also possible [10]. During the process of chemical hydrolysis of the side chain amide, isomerization to an isopeptide bond (in which the side chain carboxylate is substituted into the main chain amide bond) and racemization at the α carbon can both occur [11]; however, because these reactions occur at about the same relatively slow rate as hydrolysis itself, they are unlikely to have an impact on polyGln sequences in the cell.

The Gln side chain is a target for several enzymatic activities [10]. The Gln residue is the target of the transglutaminases that establish covalent, amide linkages between Gln side chains and primary amino groups both in the side chains of Lys residues in proteins and in small-molecule polyamines [12]. Several publications call attention to the possible role of transglutamine-dependent cross-linking of polyGln as a kind of nucleating event for polyGln aggregation, but results exploring this possibility have been mixed [12–16].

III. SOLUBILITY AND CONFORMATIONS OF THE POLYGLUTAMINE SEQUENCE

One group of hypotheses on expanded polyGln pathology posits that a toxic conformation of monomeric polyGln species exists that is only favored in sequences expanded beyond the pathological threshold. To address the feasibility of this hypothesis, one needs to know the preferred conformation of normal length polyGln peptides in monomeric proteins in solution, and then how the conformation might change in response to repeat length expansion.

A. Conformations of Normal Repeat Length Polyglutamine

Initial attempts to obtain structural information on the solution structure of polyGln by working with simple peptides gave varying results [17–19], which may have been due to difficulties in controlling and accounting for aggregation in the sample. Application of a disaggregation protocol involving transient exposure to organic solvents followed by high-speed centrifugation [3, 20], coupled with the use of flanking charged amino acids to promote kinetic solubility [17, 18], allowed the determination that aggregate-free, monomeric polyGln peptides in solution exhibit the signature circular dichroism (CD) spectrum of the random coil [21], in agreement with the earlier report of Altschuler et al. [18].

Perutz and coworkers initiated an approach to the study of polyGln segments in proteins, constructing model proteins by replacing the normal long, extended loop of the protease inhibitor CI2 (not to be confused with the antibody 1C2) with oligoGln segments [22]. Nuclear magnetic resonance (NMR) experiments subsequently showed that the oligoGln segments in these mutants are highly mobile and noninteracting [23]. Although these mutant proteins form aggregates, this proves to be mediated by domain–domain interactions within the rest of the protein and not by interactions of the oligoGln segments [24]. Interestingly, the oligoGln regions of these aggregates remain disordered and do not appear to interact with each other [23, 24]. This suggests that polyGln proteins can undergo a range of aggregation reactions, some mediated by polyGln and some by other sequence elements.

CD and NMR studies of recombinant protein fusions with either pure polyGln sequences [25] or huntingtin (htt) exon 1 fragment [26] also show that the normal length polyGln sequence in a protein host tends to favor disordered structure. At the same time, other studies suggest that polyGln can sometimes engage different, more ordered elements of secondary structure, depending on the protein context and repeat length. Tanaka et al. inserted polyGln sequences into a turn between two elements of α helix in myoglobin and reported that the inserts adopted a structure enriched in β turns [27]. Popiel et al. reported increased α helix in fusion constructs between thioredoxin and polyGln sequences [28]. Further analyses of these and other protein fusions will be required before some general rules for the impact of adjacent protein structure on polyGln structure can be elucidated. In all such experiments, it will be critical to insure that the proteins being analyzed are monomeric and not soluble aggregates, and to demonstrate that altered secondary structure is attributable to the polyGln and not to misfolding of the host protein–fusion partner.

An additional complication in the determination of the solution structure of normal length polyGln is the question of the meaning of the term "random coil." An emerging view of the denatured or unfolded state, whether globular proteins in denaturants or "intrinsically disordered" proteins in native buffer, is that such proteins are not really completely randomized statistical coils, but rather are fluctuating collections of small elements of locally organized secondary structure [29]. Such a state would be indistinguishable from a true random coil by many experimental tests, such as CD spectroscopy, but may be favored on theoretical grounds [29]. This opens the possibility that two proteins exhibiting similar CD spectra, for example, while harboring similar amounts of various secondary structural elements, might have different preferred arrangements of these fluctuating elements along the linear sequence. In addition, a secondary structural element like polyproline type II helix [30], which exhibits a CD spectrum very similar to that of classical random coil [31], could exist undetected in various amounts in the mix of flickering structures. In this regard, it is of interest that Creamer and coworkers have demonstrated that the Gln residue is one of the better amino acids in being able to adopt the polyproline type II helix conformation [30, 32].

An analysis of the conformational role of the oligoproline segment on the C-terminal side of the polyGln repeat of huntingtin illustrates some of the subtleties masked under the term "random coil" [33]. Although the CD spectrum of a Q_{40} peptide at 37°C exhibits a random coil signature, cooling of the peptide solution to 5°C uncovers a significant amount of α helix, consistent with the significant propensity for Gln residues to be found in α helix in globular proteins (Section II) and the observations of Popiel et al. [28] discussed earlier. However, a $Q_{40}P_{10}$ peptide solution at both 37 and 5°C exhibits CD spectra essentially identical to that of Q_{40} at 37°C (that is, indicating no appreciable α helix). The results suggest a tendency of polyGln to harbor nascent elements of not only α-helix, but also of some other structure that is (a) induced by oligoproline, (b) indistinguishable from random coil in the CD, and (c) capable of discouraging α-helix formation. It seems possible that this structure is polyproline type II helix [33].

B. Conformation of Expanded Polyglutamine

The ability to work with longer synthetic polyGln peptides conferred by an organic solvent disaggregation protocol [3, 20] allowed Chen et al. [21] to provide CD data showing that expansions of polyGln above the Huntington's disease repeat length, in analogy to normal repeat length polyGln, are disordered in solution. This was supported by subsequent studies using both CD and NMR on recombinant fusion proteins featuring either a simple expanded polyGln sequence [25] or an expanded polyGln in huntingtin exon 1 [26].

Other data, however, have been interpreted to indicate that repeat length expansion leads to an altered polyGln conformation. In particular, 1C2, a monoclonal antibody generated against the polyglutamine disease protein TBP (TATA box binding protein), has been shown to bind to proteins containing expanded polyGln, but not normal length polyGln [34]. However, 1C2 has also been shown to bind normal repeat length sequences as low as Q_{16} [26, 35]. In addition, analysis of the polyGln binding of monoclonal antibody MW1, which exhibits Western blot staining of polyGln fusion proteins very similar to that of 1C2, suggests that preferential binding of such antibodies to expanded polyGln might be explained not by an altered conformation, but by a "linear lattice" effect: the improved binding seen for a high local concentration of a repeat sequence of individual ligand elements, in this case a repeating pattern of short oligoGln segments, which may be the real antibody epitope [26].

There is much to learn about polyGln conformation and its response to repeat length expansion. The view that expanded polyGln sequences exist in an altered conformation within the monomer seems based largely on the use of the 1C2 antibody, whose epitope remains poorly understood. At the same time, one of the most challenging current problems in protein chemistry is the characterization of the disordered state, and tools like CD and even NMR may not be capable of detecting, let alone describing, subtle structural variations within the disordered state that might nonetheless be functionally significant.

IV. NORMAL ROLES OF THE POLYGLUTAMINE SEQUENCE IN PROTEINS

Even though the expanded CAG repeat diseases are primarily gain of function disorders, it seems likely that investigators would be better situated to speculate on how polyGln expansion might influence neurodegenerative disease if the roles of polyGln in the normal structures and functions of the proteins in which it is found were understood. Besides the 9 currently known expanded CAG repeat disease proteins [1], many other polyGln-encoding sequences exist in the human genome, both below and above the normal disease risk repeat threshold of about 35–40. Yeast and Drosophila possess a far greater number of proteins with significant polyGln sequences than do humans. In no case is there any clear indication of the function of the polyGln component in any of these proteins. At the same time, there are many observations on how varying the repeat length of the polyGln sequence affects protein stability, activity, and/or function. Continued work in this area may lead to an understanding of why polyGln sequences exist in proteins. Learning about how polyGln sequences influence protein folding, stability, trafficking, function, etc. in a general way may help to construct hypotheses for disease mechanisms.

A. Polyglutamine as a Flexible Linker between Secondary Structural Elements and Domains

1. EVOLUTIONARY TRENDS IN POLYGLUTAMINE REPEAT LENGTH

Considerations of protein structure, folding, and molecular evolution suggest that, at least for some of the disease proteins, the polyGln repeat must be in an interdomain segment where its variable length can be accommodated with minimal disruption of function. Across species, large differences in repeat length are commonly observed. Thus, the polyGln sequences beginning at residue 58 of the deposited sequences for the TATA box binding proteins (TBP) of human, hamster, mouse, chicken, and the habu viper have repeat lengths of 38, 14, 13, 6, and 4, respectively, whereas the rest of the protein sequences are identical or nearly identical. The glutamine repeat in the Clock gene product has an invariant length of 6 in the human, but is polymorphic, ranging from 25 to 33, among Drosophila species [36]. Murine interleukin 2 contains an N-terminal polyGln repeat ranging in length, in different mouse strains, up to 21 Glns, whereas the repeat is entirely absent from the human protein [37].

Given the chemical nature of Gln (see previous discussion), it would not be surprising to find polyGln repeats in solvated interdomain segments or in turn elements within folding units. Structural analysis of the p160 coactivator binding domain of CREB-binding protein (CBP), either alone [38] or in complex with its binding partner [39], shows that a pentaGln segment within this domain is located in a short, solvent-exposed loop between two α helices (Fig. 34-1). Although this segment is not known to undergo repeat expansion or to exhibit repeat length polymorphism, it is easy to see how such a polyGln segment located in a turn in a folded protein could expand without disrupting the fold of the rest of the molecule (although an increase in residues could diminish folding stability [40]). Human and mouse CBPs share this Q_5 element, whereas in *Caenorhaladitis elegans* and *Drosophila melanogaster* it is replaced by an entirely different sequence of 4–5 residues [39].

2. Repeat Length Effects on Protein Function

Variations in repeat lengths of polyGln repeats at interior positions in protein sequences generally modulate, rather than abrogate or completely alter, function.

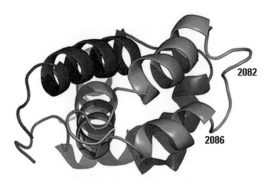

FIGURE 34-1 Cartoon ribbon diagram of the heterodimer of the complementary interaction domains of murine CBP (orange) and human ACTR (blue), with the Q_5 segment of CBP (residues 2082–2086) shown in green. The image was constructed in PyMOL (DeLano Scientific, LLC) from the deposited PDB file 1KBH from the NMR structure determination [39]. See CD-ROM for color image.

Perhaps the best documented example is the repeat sequence in the androgen receptor (AR). This protein contains an N-terminal domain (NTD) that includes the glutamine repeat, a central DNA binding domain, a hinge region, and a C-terminal ligand binding domain (LBD). Ligand binding to the LBD facilitates an interaction with the NTD that is required for binding to a transcriptional coactivator. Several studies consistently show that the repeat length of the NTD polyGln influences AR function. Thus, contraction from the normal wild-type range repeat length of 20 to a repeat length of 9 generated an AR with both enhanced ligand affinity and enhanced NTD–LBD interactions, leading to increased androgen-dependent growth of prostate cancer cells in culture [41], consistent with a clinical correlation of higher prostate cancer risk with decreased polyGln repeat length. Complete removal of the polyGln repeat has a similar effect of enhancing AR transcriptional activity, but only on androgen response elements that require an NTD–LBD interaction [42]. Likewise, expansion to a repeat length of 26 leads to a decrease in androgen-dependent transcriptional activity [43]. Although it is not clear how repeat length variation produces these effects, the monotonic decrease in activity with increasing repeat length, where activity depends on the productive interaction of two independent binding units at opposite ends of the polypeptide chain, suggests a simple chain entropy effect [40] whereby longer polyGlns increase the statistical barrier to the productive interactions of the LBD and NTD. In this system, one can think of the polyGln as a molecular rheostat, producing systematic gradations in the normal protein activity when it expands or contracts. This is, of course, in contrast to the effect of repeat length variation in expanded CAG repeat diseases, which appears to involve the generation of a new and toxic functionality.

PolyGln exhibits a similar molecular rheostat role in some other proteins that have been studied. Expansion of the polyGln repeat in ataxin-3 from 22 to 64 leads to a less than three-fold decrease in its binding constant to tetraubiquitin [44]. Loss of the N-terminal polyGln repeat from the Foxp1 transcription factor enhances its ability to repress transcription [45], although the structural basis of this effect has not been worked out.

Not all studies show quantitative changes in protein activity with repeat length. For example, expansion of polyGln from Q_{10} to Q_{60} in ataxin-7, discovered to have a role in transcriptional regulation, does not ablate its ability to form a transcriptional complex [46]. These studies were not designed to detect quantitative differences, however. It is also possible that repeats on the extreme termini of multidomain polypeptides will not play the same rheostat role as repeats located between domains. In a preliminary exploration of the possible role of repeat sequences in transcription factors, Gerber et al. inserted various repeat length polyGln or polyPro sequences into the DNA binding domain of GAL4 and observed complex repeat-length-dependent behavior in the activation of transcriptional activity *in vitro* and *in vivo* [47]. Although it is difficult to assess the meaning of experiments on unnatural repeat proteins, such experiments may be useful in exploring the kinds of roles homopolymeric repeats can play in the structure and function of transcription factors and other proteins.

The trend of domain–domain interactions becoming less favorable as polyGln repeat length increases makes sense from considerations of protein folding and from knowledge of the nature of the polyGln sequence in solution. It does not explain, however, the wide variations in repeat length among homologous proteins of different species and how the consequent variations in function might confer advantages onto the organism.

3. Effect of Repeat Length on Folding and Stability of the Host Protein

Many of the polyGln disease proteins are quite large, complex, and very difficult to either isolate intact from homologous tissue or produce intact by recombinant means. The relatively small AT3 protein is an exception. The human protein (with a Q_{22} repeat) has 376 residues, of which the N-terminal half constitutes the Josephin domain, whose structure was solved and shown to be in the cysteine protease family [48], consistent with the protein's activity as a ubiquitin hydrolase [49]. Expression of a construct lacking the C-terminal 170 residues, including the polyGln repeat, produces a Josephin domain that, compared to full length AT3, is prone to aggregate [48]. Like many globular proteins, intact AT3, as well as its isolated Josephin domain, undergoes unfolding-dependent aggregation to make amyloid-like fibrils [50, 51]. This might suggest a mechanism whereby repeat expansion directly destabilizes the folded structure of AT3 and through this destabilization facilitates aggregation of the Josephin domain (and not the polyGln domain). However, it appears that repeat expansion does not destabilize the folded structure of AT3 [52]. The key question would appear to be whether AT3 aggregates in the cell feature an H-bonded core region rich in Josephin sequences or, alternatively, rich in polyGln sequences. In fact, there is evidence for both. On the one hand, polyGln-mediated aggregation is supported by cell experiments with AT3 and other polyGln proteins demonstrating the importance of the polyGln repeat in the recruitment of different proteins into inclusions [53, 54]. On the other hand, some neuronal aggregates not rich in polyGln appear to be capable of recruiting polyGln proteins, including AT3 [55–57], and aggregates of differently truncated versions of AT3 have been shown to selectively recruit other cellular proteins [58]. The possibility that expansion of a polyGln repeat might destabilize and/or structurally perturb adjacent folded domains, which is supported by some studies [27, 121] but refuted by others [52], is an important concept. First, it could help account for the repeat-length dependent effects on protein activity reviewed in Section IV. A. 2. Second, it implies that expansion might, via different mechanisms, lead to aggregation where either the polyGln or the adjacent domain plays a central structural role and hence defines the seeding ability of the aggregate (Fig. 34-2). Further support for a role of polyGln repeat expansion in including structural perturbation and subsequent aggregation of an attached globular domain has been obtained in model studies on the small, beta-sheet rich protein CRABP I (cellular retinoic acid binding protein I) (Z. Ignatova and L. Gierasch, personal communication).

B. Polyglutamine as a Solubilizing Domain

The decreased solubility of AT3 with the removal of the polyGln-containing C terminus [48] suggests the possibility that polyGln might serve a solubilizing role for proteins containing this sequence. Whereas other amino acid sequence motifs might be capable of serving similar roles, polyGln may provide certain advantages, such as possible differences in solution structure (yet to be deduced) and the ability to confer solubility without radically changing charge characteristics or introducing other functionalities.

C. Polyglutamine as an Interaction or Aggregation Domain

The only consistently observed "function" of the polyGln sequence is its ability to form aggregates. As described in the next section, even relatively short polyGln peptides are "drawn in" to an ongoing aggregation process initiated by the spontaneous aggregation of a long repeat length polyGln. In fact, recruitment of cellular polyGln proteins in an ongoing aggregation reaction has been demonstrated in cell culture and animal models and has been proposed as a mechanism of toxicity (see Section VB). The ability of polyGln sequences to self-assemble was, in fact, the basis for early speculation about polyGln function, even before their aggregation potential and disease involvement were established. Noting that many polyGln sequences occur in transcription factors, Perutz and coworkers proposed that the polyGln sequence might be a hydrophilic, or

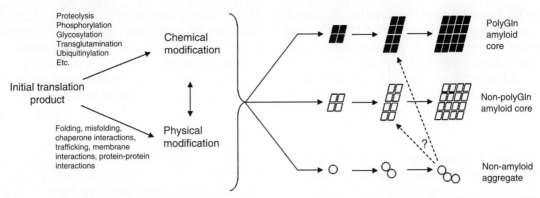

FIGURE 34-2 A general scheme for *in vivo* aggregation of polyGln proteins, showing how chemical and physical modifications of the protein can interact to modulate the tendency of the protein to aggregate. It also shows different possible aggregation modes and how the ability of most of these aggregates to grow inexorably leads to the possibility of a distribution of aggregate sizes in and around the cell. Not shown as a possible modifying factor is trafficking of the protein within the cell. The dotted lines show the possibility that protofibrils, a particular type of nonamyloid aggregate, under some conditions may be capable of transmuting directly into amyloid fibrils, which would effectively constitute an unusual mechanism of nucleation.

polar, version of the leucine zipper motifs well-known to mediate heterodimer formation within a class of transcription factors [59, 60]. Studies on simple polyGln peptides *in vitro* do not support this speculation, however, because, as a nucleated growth polymerization, organized (that is, amyloid-like) polyGln aggregation does not stop at the dimer stage but rapidly, and essentially irreversibly, goes on to form higher aggregates (see Section V). This rampant aggregation activity appears to be somewhat modulated when the polyGln sequence is embedded within some globular proteins, but there are no reports of reversible dimer formation. No evidence exists that polyGln segments can act as reversible dimerization domains for mixing and matching of protein modules in analogy to leucine zippers.

The preceding discussion notwithstanding, one can imagine other ways in which cells might exploit a more aggressive aggregation activity like that of polyGln. Some life forms have learned to exploit the intrinsic ability of polypeptides to form aggregates. For example, a number of yeast prions have been described that are basically amyloid-like self-propagating systems [61]. Amyloid-like structures have evolved to serve functional roles in bacteria [62, 63]. In the case of polyGln, it is known that various polyGln proteins and repeat lengths can participate in the elongation of an amyloid-like polyGln aggregate, once aggregation has been initiated or an aggregate seed provided (see Section V). Thus, it seems theoretically possible that an aggregation reaction confined specifically to polyGln sequences might have evolved in the cell, possibly as a regulatory mechanism or even as part of a cell death program. For example, aggregation followed by elimination (for example, by the ubiquitin–proteasome system) could serve as a means of rapid down-regulation of the family of polyGln proteins in a cell compartment. If polyGln aggregation serves such normal functions in the cell, its mediation of the neurotoxic response to repeat expansion would simply constitute an example of a disease mechanism that is an exaggeration of a normal cellular pathway. No evidence exists that cells exploit polyGln aggregation for this purpose, however, and it is difficult to imagine how such a system could be safely managed by the cell. As with the model for polyGln as a flexible separator of functional protein domains, this hypothesis also would not explain why homologous proteins from different organisms might have a long, short, or nonexistent polyGln repeat.

V. AGGREGATION OF POLYGLUTAMINE SEQUENCES

With the discovery of polyGln-containing cellular inclusions in patient material and in cell and animal models of expanded polyGln diseases [2, 64–66], work began in a number of laboratories to investigate details of the polyGln aggregation process. *In vitro* studies using recombinantly expressed huntingtin exon 1 fusion proteins showed that polyGln aggregation exhibits a concentration-dependent and repeat-length-dependent aggregation [2]. Work in cell models confirmed the enhanced ability of expanded polyGln sequences to make cellular inclusions and also provided the first glimpse of the promiscuity of polyGln aggregation, in showing that aggregates of one polyGln repeat length are capable of recruiting other polyGln proteins into the growing aggregate [66]. The strict repeat length dependence of the aggressiveness of polyGln aggregation was later confirmed using synthetic polyGln peptides [21] as well as other model systems [67]. Together, these studies show a remarkable correspondence between the repeat length threshold for disease risk seen in most polyGln diseases [1] with the repeat length dependence of the aggressiveness of aggregation, a relationship often invoked as supporting a primary role of aggregation in disease onset and disease mechanism.

A. Varieties of Polyglutamine Aggregates

In analogy to other neurodegenerative diseases associated with protein aggregation [68], a variety of aggregates of polyGln peptides and proteins have been reported. Physically, aggregates can differ in size, morphology, the degree and type of order (such as amyloid-like) defining the core structure of the aggregate, the peptide sequence responsible for making up that core, reactivity to antibodies, functionality such as the ability to seed elongation, ability to bind certain dyes, stability to SDS, and fundamental substructure as investigated by X-ray diffraction or Fourier transform infrared spectroscopy (FTIR). Aggregates *in vivo* can differ in their subcellular localization and their structural composition in terms of other cellular proteins. In addition, aggregates that have accumulated over years and decades in the brain appear to be chemically different from aggregates grown *in vitro* or only accumulated over several days in cell culture. Brain aggregates, for example, are stable to concentrated formic acid, suggesting chemical cross-links [69]. At the same time, rather amazingly, brain aggregates retain their ability to seed polyGln aggregation even after storage of the tissue for decades [70].

It appears likely, even probable, that many expanded CAG repeat disease proteins are capable of forming a variety of aggregates. The reader will recall several occasions in the earlier sections of this chapter (Section IIIA, CI2; Section IVA, AT3) where evidence was presented for polyGln proteins being capable of a number of different aggregation paths, some in which the polyGln is central to the assembly of the aggregate core and some in which it is peripheral. Figure 34-2 is an attempt to systematize at least some of the possibilities schematically and will be referred to throughout this section, in which the formation and properties of a number of aggregate types will be discussed.

1. Aggregate Nomenclature

Historically, the word aggregation has been used in protein chemistry to refer to any strong association between 2 or more polypeptide chains, even if the association involves native interactions holding together a well-folded multimeric protein [6]. More recently, the word has been applied to nonnative, undesirable interactions. Whether composed of 2 or 2000 polypeptide chains, such assemblies are considered to be nonnative aggregates.

As discussed earlier, these aggregates can differ significantly from each other. A few aggregates are sufficiently distinct to have acquired their own terminology, although these names are not always applied consistently. *Microaggregate* is a nonspecific term generically applied to very small aggregates, especially if they are only observable using particularly sensitive methods. Microaggregates in principle can be amyloid-like or more disordered. Very small aggregates containing relatively few polypeptide chains are often referred to as *oligomers* (from the Greek *oligos*, meaning few or little). One type of oligomeric structure is called *spherical oligomers*, spheroids, or globules, on the basis of their appearance in the electron microscope (EM) [71] or atomic force microscope (AFM) [72]. The individual spherical particles tend to be about 5 nm in diameter and are sometimes observed to be self-associated in linear or irregular arrays. The next largest aggregate often observed in studies of aggregation diseases is the *protofibril* (not to be confused with the *protofilament*, the hierarchical substructure of the amyloid fibril [73]). In some EM and AFM studies, protofibrils appear to be composed of linear or annular assemblies of spherical oligomers [74, 75]. Protofibrils tend to have diameters in the 5-nm range and lengths of no more than 50–100 nm. *Amyloid* fibrils are somewhat thicker than protofibrils, normally exhibiting diameters in the 8- to 12-nm range and lengths ranging from very short to over 1 μm. They often exhibit a twisted structure due to their composition of multiple protofilaments bundled together [73]. The phrase *amyloid-like* is sometimes used to refer to aggregates that are composed of long protofilaments that are, however, assembled somewhat differently than the classic amyloid fibril. Aggregates with no apparent regular structure are often called *amorphous aggregates*, although it is generally not known whether they are disordered at the molecular level. The word *inclusion* has been restricted to particularly large aggregates, visible in light microscopy, that in some cases can be seen to be composed of smaller, fibrillar structures [64].

B. Amyloid-like Aggregates with a Polyglutamine Core

Micrographs of nuclear inclusions in Huntington's disease (HD) brains reveal an underlying fibrous substructure resembling amyloid fibrils [64], and amyloid fibrils are formed from recombinantly expressed polyGln-containing fusion proteins [2]. Using the hypothesis that the properties of the polyGln sequence must be centrally important in the expanded CAG repeat diseases, a significant amount of work has been done to investigate the assembly and nature of the aggregates formed by the polyGln sequence itself [3, 18–21, 76–81]. Though of varied morphologies, these aggregates appear to possess a fundamental amyloid-like structure. In electron microscopy, the aggregates are shown to be either isolated protofilaments or protofilaments organized together to form ribbons or fibrils [80]. These aggregates exhibit the classical amyloid response to thioflavin T, strong β-sheet CD spectra, and binding to antibodies recognizing a generic conformational amyloid epitope [80].

1. Polyglutamine Aggregate Structure

The molecular structure of amyloid-like polyGln aggregates has drawn significant attention. Like other amyloid fibrils, high-resolution structural determination is beyond the reach of current methodology. Fibril diffraction data led Perutz et al. to propose a structural model related to conventional β sheets in which the H-bond network typical for β sheets was supplemented with H bonds contributed to by the amide groups of the Gln side chains [17]. Later, on the basis of the same diffraction data, Perutz and coworkers proposed an alternate model more resembling the parallel β-helix folding motif [82, 83], but with an open, solvated central channel [84]; subsequently, an independent analysis of this same diffraction data led to a more conventional, β-sheet-related model [85]. New diffraction data generated from aggregates of different repeat length polyGln peptides also support this more conventional model for an aggregate structure related to typical β sheets in globular proteins [81]. Whatever the details of the model, one interesting question about polyGln aggregates is how a fundamental structural motif accommodates different repeat lengths of polyGln. That is, if, for example, a Q_{40} peptide fits perfectly into a four-strand model for a building block of the β sheet network, what happens with a Q_{45} peptide? Does the exact replica of the four-strand building block form, with the five extra Gln residues extruded from the core structure? Or does the width of the β sheet expand to neatly take up the extra Gln residues and maximize the number of H bonds? The fibril diffraction data on aggregates of simple polyGln peptides support the latter model, in showing that the slabs of β sheet that make up the aggregate expand in width (in the strand direction) in aggregates of longer polyGln peptides [81].

This result was already anticipated by mutagenesis experiments in which proline–glycine (PG) pairs were inserted at regular intervals through a polyGln sequence [79]. In a series of peptides with differing length oligoGln segments between PG pairs, aggregation rate was optimal with Q_9 and Q_{10} segments, but it was also possible with shorter segments of Q_7 or Q_8. If all of these peptides, as required by the structural limitations of proline, place the PG pairs at turn positions in the aggregate, it would seem that the widths of β sheet in the shorter oligoGln segments must decrease in order to construct a viable aggregate. These mutagenesis data, however, appear to be consistent with both a standard antiparallel β sheet model and an irregular parallel β-helix model [79]. The aggregate structures and aggregation kinetics of the PG peptides are very similar to those of unbroken polyGln of similar repeat length [79]. Furthermore, huntingtin exon 1 fragments containing similar PG-interspersed polyGln sequences behave similarly to huntingtin exon 1 with unbroken polyGln, both in forming intracellular aggregates and in compromising cell viability [86].

No direct data are available to address the possible role of the side chain amide groups in stabilizing the structure of amyloid-like polyGln aggregates, as proposed in the polar zipper structural model of Perutz. However, a crystal structure of a Gln/Asn-rich segment of a yeast prion, thought to be a structure also represented in the amyloid fibrils made by this same peptide, reveals a packed structure in which side chain H bonding between adjacent strands within a β-sheet, plays a major role [87]. Structural data on other amyloids may well be relevant to the structure of amyloid-like polyGln aggregates [88].

2. Nucleation of Assembly of Amyloid-like Polyglutamine Aggregates

Detailed studies on simple synthetic polyGln peptides show that the assembly of amyloid-like aggregates occurs by a mechanism known as nucleated growth polymerization [77]. In this mechanism, initiation of the reaction is dependent on a highly energetically unfavorable nucleation event, whereas the subsequent fibril growth is driven by rounds of monomer additions [89, 90] (Fig. 34-3). Consistent with this mechanism, polyGln aggregation exhibits a lag phase with little or no observable aggregation but that can be aborted by providing seeds of preformed aggregate [77, 80]. Rigorous examination of the early stages of polyGln aggregation, according to a "thermodynamic" model for the nucleus, led to an unprecedented model in which the nucleus is a high-energy state of the monomer [77].

Because, as discussed earlier, polyGln monomers in solution have no strongly favored folded structure, the generation of the nucleus for aggregation is considered to be a kind of highly unfavorable protein folding reaction [77]. As a very rare species, the nucleus cannot be

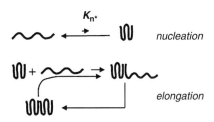

FIGURE 34-3 Nucleation mechanism proposed for polyGln amyloid formation based on aggregation kinetics and conformational analysis of monomeric polyGln [80]. In the mechanism, extended polyGln is in equilibrium (controlled by the nucleation equilibrium constant K_{n*}) with a very low concentration of the condensed nucleus. Elongation of the nucleus through an interaction with another molecule of extended polyGln, controlled by the second-order elongation rate constant, helps to commit the nucleus to a productive elongation reaction. Further turns of the elongation cycle follow to produce aggregates. Adapted from Ref. [80] with permission. Copyright (2002) National Academy of Sciences, U.S.A.

observed or studied by physical means; it can only be inferred by analysis of the reaction kinetics. One estimate places the equilibrium constant for aggregation nucleation of a Q_{47} peptide at about 10^{-9} [91]. This corresponds to a steady-state concentration of the nucleus of about 10 fM for a monomer concentration of 10 µM; to characterize such a concentration of a molecule, a physical technique would not only have to have the sensitivity to detect the nuclei, it would also have to have sufficient specificity that a 10^9 higher concentration of ground-state monomers would be largely opaque to the same technique—overall, a daunting set of conditions. In contrast to the poorly organized, fluctuating structure of the monomer (Section III), the nucleus is visualized as a folded, β-sheet-rich monomeric species that is capable of initiating aggregation when it encounters another (disordered) monomer.

Nucleation kinetics studies help to account for the repeat length dependence of the polyGln aggregation reaction. Parallel studies on peptides of repeat lengths Q_{28}, Q_{36}, and Q_{47} show that the critical nucleus, that is, the number of molecules involved in nucleus formation, is equal to one for all three repeat lengths. What makes Q_{47} aggregation relatively aggressive is that the nucleation equilibrium constant, small as it is for Q_{47}, becomes progressively smaller for Q_{36} and Q_{28} repeats [77]. To explore the magnitude of the effect of these differences on aggregation kinetics for physiologically relevant levels of polyGln, the kinetic parameters extracted from studies on micromolar concentrations of polyGln peptides were used to calculate approximate lag times for a polyGln concentration of 100 pM. For Q_{28}, the lag time is about 1300 years, for Q_{36} about 140 years, and for Q_{47} about 30 years [77].

The experimental challenges in characterizing the nucleus increase the importance of computational methods in understanding the unfavorable folding of the polyGln monomer. A simulation found that a relatively favorable state of the polyGln monomer in solution is an irregular parallel β helix (Fig. 34-4) in which main chain H bonds connect adjacent strands in the coil, and every other side chain is directed into the core of the helix to participate in H bonding [92]. This is consistent with a nucleation kinetics analysis of a polyGln sequence with regularly interspersed PG pairs, showing that such peptides aggregate through the same mechanism (nucleus of one) as unbroken polyGln; such a peptide can be accommodated into the irregular parallel β helix, where the PG residues are placed in turn positions [79]. This model can account for the increasing favorability of nucleation with increasing repeat length, because a Q_{40} peptide, for example, would neatly occupy about two full turns of helix [92]. Although

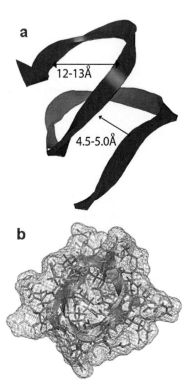

FIGURE 34-4 Computational modeling of the aggregation nucleus [92]. (A) Ribbon diagram of the proposed parallel β-helix nucleus for polyGln aggregation. The 4.5- to 5.0-Å distance shows the strand–strand separation between two turns of the helix, in agreement with the typical strand–strand distance for a parallel β sheet. The 12- to 13-Å distance is for the strand–strand separation across the diameter of the helix, corresponding to the space into which Gln side chains are thrust. (B) Packing of Gln side chains in a cross section of the model. The polypeptide main chain is shown as a ribbon, whereas the gray gridwork shows the molecular surface. Underlying the gridwork are the Gln side chains, which are directed both into the core of the helix (where they H-bond with each other and with the main chain) and out of the helix, where they can interact with solvent water. See CD-ROM for color image.

similar to the water-filled nanotube model of Perutz *et al.* [84], this model for the nucleus does not feature an interior water channel, which does not appear to be present in polyGln aggregates [81, 85] and which appears to be energetically unfavorable [83]. However, because the simulation was not conducted in explicit water, it does not directly address the possible role of water in the interior of the β sheet.

A computer-generated model for the repeat unit in the polyGln aggregate [93], which is also compatible with the PG mutagenesis data [79], features an antiparallel sheet arrangement such as those typically found in globular proteins. In fact, this may be more compatible with the X-ray fiber diffraction data [81, 85] than a parallel β-helix-based model.

Although it is important to study the homogeneous nucleation of polyGln peptides in order to arrive at an understanding of the fundamental folding and aggregation properties of polyGln under the simplest conditions of sequence context and environmental context, it should also be recognized that quite different conditions are obtained in the cell, where polyGln exists in the context of unique flanking sequences, and the environment is rich in many small and large molecules at high concentrations. It may be that the same mechanism of nucleation holds *in vivo*, but is operated on by a variety of accelerating and decelerating factors. It is also possible that the entire nucleation mechanism might be different *in vivo*, tending more toward some form of heterogeneous nucleation.

3. POLYGLUTAMINE ELONGATION

The second phase of the nucleated growth polymerization mechanism is elongation, the systematic addition of monomeric units to the growth face of the aggregate (Fig. 34-3). Fibril elongation is a second-order reaction dependent on a characteristic rate constant and on the molar concentrations of both the aggregate and the monomer. In general, amyloid fibril elongation is very specific, so that fibrils of one peptide are often only poor seeds for elongation of another amyloidogenic peptide [94]. This trend does not appear to apply to aggregates of polyGln proteins in which the polyGln segment is the core component of the fibril structure. In this case, any polyGln sequence, so long as it is exposed in the protein structure, can be efficiently recruited into a growing polyGln aggregate, with only a modest dependence of rate on repeat lengths. Thus, the pseudo-first-order rate constant for elongation of a Q_{47} aggregate by various monomeric polyGln peptides increases by about threefold over the repeat length range 10–40 (A. M. Bhattacharyya and R. Wetzel, unpublished results). The promiscuity of polyGln fibril elongation has a number of implications that will be explored in this section.

a. Recruitment of Polyglutamine Proteins into a Growing Aggregate

As implied by the *in vitro* data on simple peptides, aggregates of polyGln proteins are often very good at recruiting other polyGln proteins in cell and animal models [53, 66, 95, 96]. Two of the proteins often observed to be colocalized to aggregates are the polyGln-containing transcription factors CBP and TBP [53, 95–97]. Such observations led to a model for pathogenesis in which aggregation initiated by the expanded repeat form of the disease protein causes depletion, through recruitment, of other (even normal length) polyGln proteins important to cell survival. Loss of activity because of coaggregation might be due to simple sequestration [53, 95, 96], inactivation due to unfolding in the aggregation process, or the loss of protein molecules altogether by initiation of aggregate clearance mechanisms like proteolysis [98]. In a cell model, aggregation of expanded polyGln huntingtin leads to colocalization of CBP with the htt aggregates, depletion of CBP from the cytosol, and cytotoxicity; overexpression of CBP protects the cells against the toxic effect of htt aggregation [99].

Although protein recruitment mediated by the apparent high specificity of the formation of amyloid-like polyGln aggregates is a conceptually pleasing mechanism that has accumulated substantial supporting evidence, protein recruitment by misfolded protein aggregates may not be so universally simple. For example, using the fluorescence of GFP (green fluorescent protein) fused to CBP to follow recruitment in a mammalian cell culture system, it was found that inclusions of full length AT3 recruit CBP strongly, whereas htt exon 1 inclusions give variable degrees of recruitment and AT1 inclusions recruit CBP weakly or not at all [100]. It appears that not all aggregates exhibit a functional, recruitment-positive polyGln aggregation core, either because the polyGln core is present but masked or because of an alternative mode of aggregation (Fig. 34-2) that is not mediated by polyGln interactions at all (see Section VC).

b. Polyglutamine Amyloid Elongation as an Accelerator of Nucleation

One surprising consequence of the promiscuity of polyGln aggregation is the finding that normal length polyGln sequences in the environment can greatly enhance the already relatively favorable nucleation kinetics of an expanded repeat polyGln sequence. To understand this, it is necessary to return to the mechanism of nucleated growth polymerization in Fig. 34-3. In the mechanism, the nucleus is in a rapid preequilibrium with bulk phase disordered monomer, so that, effectively, the forward rate of nucleation depends on the thermodynamic stability of the nucleus, not on the microscopic rate constant by which it forms from disordered monomer. Because of the ability of normal length polyGln peptides to polymerize on a nucleus of an expanded polyGln, normal length polyGlns in the environment would be predicted to enhance the overall nucleation process by increasing the portion of transiently formed nuclei that proceed toward aggregate formation, rather than regress to disordered monomer. This theoretical possibility has been confirmed by *in vitro* experiments in which relatively low concentrations of short polyGln are added to a low concentration of expanded polyGln (A. M. Bhattacharyya, A. K. Thakur, and R. Wetzel, unpublished data). Although mechanistically related to the phenomenon described in Section VB, the ability of normal length polyGln to accelerate expanded polyGln nucleation is fundamentally different because it influences the rate-limiting initiation of polyGln aggregation, whereas polyGln-mediated protein recruitment has been understood to depend on the preexistence of an ongoing aggregation reaction.

These results have important implications for biology, because of the wide range of polyGln sequences of varying repeat lengths present in different cellular compartments. In fact, experiments in a Drosophila model confirm that overexpression of a Q_{20} peptide accelerates the aggregation and toxicity of an expanded polyGln form of htt (N. Slepko, J. L. Marsh, and L. M. Thompson, unpublished results). This effect may also be responsible for the reported linkage between repeat length (within the normal range) of the SCA6 protein and age of onset of SCA2 [101].

c. Effects of Flanking Sequences on Amyloid-like Polyglutamine Aggregation

There are many levels of complexity in trying to understand the cellular basis of polyGln protein behavior. One that is relatively unexplored is the role of flanking sequences in the ability of polyGln to take part in amyloid fibril growth and elongation [102]. As discussed earlier, the solution structure and aggregation behavior of polyGln itself have become relatively accessible through the use of chemically synthesized peptides. These studies reveal the baseline properties of polyGln sequences. How these properties may be modulated or masked when the polyGln element is part of a larger translation product appears to vary considerably depending on sequence. Furthermore, since flanking sequences may contain signals that transport the entire protein, with its polyGln, to different environments in the cell, and because some flanking sequences can be removed or modified differently in different cellular environments, the interaction between polyGln properties and flanking sequence can be quite dynamic and difficult to understand.

The influence of amino acids immediately adjacent to polyGln can be very dramatic. The pure polyGln sequence is extremely aggregation prone *in vitro*, so that even after the application of otherwise successful disaggregation protocols, and even in acid conditions that normally favor solubility, it is impossible to maintain the peptide in the monomeric state [80]. Addition of a few charged amino acids on either end of the sequence dramatically improves this initial solubility, while at the same time not compromising the ability of the peptide to ultimately aggregate. Lysine residues are particularly effective in providing this solubility [18, 80], perhaps because they retain a full positive charge at neutral pH. Acidic residues Glu and Asp are less effective, and Arg, in spite of its very basic side chain, is not nearly as effective as Lys; mixed acidic and basic residues [17, 80] are also less effective than an equal number of Lys residues (S. Chen and R. Wetzel, unpublished results), perhaps due to the lower net charge of these peptides.

All polyGln expansion disease proteins have a polyGln sequence exhibiting, at least in the full length protein, both N- and C-flanking sequences. Little is known about the possible effects of these more complex flanking sequences on polyGln aggregation. The one case that has been studied in detail suggests that the possible influence of these flanking sequences should not be ignored. A P_{10} oligoPro sequence following a Q_{40} sequence has been shown to slow down aggregation *in vitro* and, perhaps more importantly, decrease aggregate stability. The effect depends on the P_{10} being connected to the polyGln via a normal peptide bond and being located C-terminal, and not N-terminal, to the polyGln [33], suggesting that the effect is not mediated by a change in global hydrophobicity.

Formally, there are at least two ways in which a short element of sequence might influence polyGln aggregation. In one mode, as appears to be the case with oligoPro, it can modulate a fundamental property of the polyGln sequence. In another mode, it can introduce an alternative aggregation domain and pathway, which might effectively compete with polyGln for defining the structural basis of the aggregation reaction. The prototypical example of the second mode is from the work of the Weissman laboratory on yeast prions [103]. In these experiments, two amyloidogenic aggregation domains from different prions were fused in a single sequence. When this soluble protein was seeded with aggregates of either aggregation domain, it engaged in an aggregation reaction whose course was dictated by the identity of the seed [103]. These experiments provide a model for the kind of multiple and potentially mutually exclusive aggregation phenomena that might occur in more complex polyGln polypeptides, in analogy to the artificial polyGln-inserted CI2 discussed in Section IIIA. The high level of specificity of aggregation suggested by the Weissman experiments is likely to be limited to aggregation reactions mediated by highly ordered amyloid-like structures; it is not always clear whether particular cellular aggregation reactions are amyloid-like or are mediated by less specific interactions (see Section VC).

An additional mode by which the flanking sequence can modulate polyGln aggregation is when the flanking sequences are longer and thus have the potential to form independent domains that can sterically mask the polyGln element and inhibit its aggregation. It is a common observation that proteolytic fragmentation of polyGln proteins tends to make them more prone to aggregate. Artificial fusion proteins can be relatively stable to aggregation until the fusion partner is removed by proteolysis; in fact, this was exploited by the Wanker group in the design of a polyGln aggregation assay starting from recombinant protein [2, 104]. Characterization of inclusions isolated from cells often reveals a high degree of fragmentation in the deposited protein [64], and specific inhibitors of known proteolysis sites in a polyGln disease protein can suppress aggregate formation and toxicity [105]. Animal and cellular models based on protein fragments, such as htt exon 1, are often more aggressive, both in aggregation and in pathology, than models based on full length proteins [106]. The mechanisms by which flanking domains shield polyGln sequences from aggregation, or alter the course of aggregation, are undoubtedly complex and may prove to vary significantly, even for different fragments of the same protein. Alternative proteolytic cleavage events can influence not only the aggressiveness of aggregation, but also the cellular localization of the aggregates produced [107].

d. Exploiting the Specificity of Polyglutamine Amyloid Elongation

The specificity of amyloid growth *in vitro* [94] and *in vivo* [103] suggests that it might be possible to develop tissue stains capable of identifying particular aggregates by their ability to seed further elongation. Such a staining procedure has been developed in which viable polyGln aggregates in cells or tissue act as templates for the addition of biotin-labeled polyGln peptides, which can subsequently be identified by standard biotin-targeting reagents [70]. The existence of multiple types of aggregates of polyGln disease proteins in brain tissue is very clearly revealed when this method is used in parallel with other staining methods, such as antibodies to the polyGln or non-polyGln segments of the disease protein [70]. The method is also very sensitive, revealing the existence of polyGln–amyloid core microaggregates in material that does not exhibit aggregates by other methods and, by contrast, demonstrating the absence of polyGln recruitment activity in even large aggregates that are revealed by other methods. If the recruitment–sequestration model for polyGln toxicity serves as the molecular mechanism for at least some expanded polyGln pathology (Section VB), then it might be expected that cytotoxicity will correlate not with the total aggregate burden, as revealed, for example, by a polyclonal antibody against the disease protein, but rather with the seeding activity of particular subpopulations of aggregates. Recruitment-based stains will help in the full cataloging of the aggregation process in disease, and through this increased awareness of aggregate diversity should ultimately reveal which aggregates are toxic, which are protective scavengers of unaffiliated monomers, and which are neutral red herrings.

4. Polyglutamine Aggregate Stability

In principle, the stability of a polyGln aggregate may be just as important as its rate of formation. Although amyloid fibrils of many proteins are extremely stable, some amyloids exhibit more modest stability. A convenient measure of stability is the critical concentration, or C_r, which experimentally is simply the concentration of monomeric protein remaining in solution when an aggregation reaction comes to equilibrium. On the basis of this definition, the C_r is also the concentration of monomer below which aggregation cannot occur. The stability of simple polyGln aggregates tends to increase as repeat length increases. In phosphate-buffered saline, the C_r of the peptide $K_2Q_{15}K_2$ is on the order of 15 μM; longer polyGln repeats exhibit substantially lower, submicromolar C_r values [21]. If the same C_r were applicable to the cellular environment, the steady-state cellular concentration of Q_{15} peptides would have to exceed 15 μM before their aggregation would become an issue. However, molecular crowding in the cell [108] increases the rate of formation and decreases the C_r of polyGln aggregation [33], which presumably accounts for the ability of relatively short polyGln peptides to be recruited into aggregates *in vivo*. Sequences flanking the polyGln repeat can also modulate polyGln aggregate stability [33].

C. Polyglutamine Protein Aggregation Not Related to a Polyglutamine Amyloid Core

A number of observations cited in the preceding sections of this chapter can be interpreted to indicate the existence of multiple possible aggregation pathways for polyGln proteins *in vivo*. This idea is illustrated in Fig. 34-2, which shows schematically how the biophysical possibilities intrinsic to the polypeptide sequence of the disease protein make possible multiple aggregation pathways, and how various intracellular factors and processes, in addition to the intrinsic aggregation kinetics, can serve to influence the actual aggregation reactions that are obtained under particular conditions. Clearly, many *in vivo* aggregates of polyGln proteins are organized according to the ability of the polyGln to make an amyloid fibril core; these aggregates are perhaps most clearly recognized by their ability to respond to the recruitment stain described earlier. An example of the ability of another portion of an expanded polyGln disease protein to serve as the organizing, amyloid core of an aggregate is the Josephin domain of AT3, which can form amyloid in the absence of its polyGln sequence (Section IVA). Because amyloid formation by the Josephin domain appears to require significant destabilizing conditions, it is not yet clear whether its aggregation contributes to disease pathology. In the future, it may be possible to address such questions by characterizing cellular aggregates in disease tissue for the nature of their amyloid core, either by characterization of a proteolysis-resistant core or by the ability of the aggregates to seed elongation of peptide fragments representing other portions of the disease protein (Section VB).

Other, apparently nonamyloid, types of aggregates of polyGln proteins have also been described that might best be placed in the lower aggregation pathway in Fig. 34-2. *In vitro* incubation of purified, recombinantly expressed fragments of huntingtin leads to the generation of globular and protofibrillar aggregates [109] similar to those observed in the early stages of amyloid formation by other disease proteins and implicated in disease pathology [110]. The formation and fates of these oligomeric structures appear to be influenced by some molecular chaperones [111], as suggested in Fig. 34-2. Kopito and coworkers describe the isolation of globules and clusters of globules, in the 4- to 50-nm diameter range, from cells producing an expanded polyGln huntingtin exon 1 fragment [72]. These aggregates do not appear to contain fibrillar material, but they were not tested for their ability to act as seeds for polyGln elongation, which is potentially a more sensitive test for amyloid-like structures than microscopic methods. Although spherical oligomers and protofibrils of other disease proteins are often discussed as on-pathway intermediates in amyloid assembly [110], and although there is reason to believe that a structural relationship exists between protofibrils and mature amyloid fibrils of other disease proteins like Aβ [112, 113], much remains to be done before this precursor–product relationship is convincingly established for any amyloid. In fact, there is strong evidence from studies of some amyloidogenic proteins that protofibrillar structures are not on the assembly pathway but, at best, serve as off-pathway reservoirs for monomers during amyloid assembly [114]. The ambiguous relationship between oligomeric aggregates and mature amyloid fibrils is indicated by the dotted arrow with a question mark in Fig. 34-2. Similar uncertainty clouds the question of whether particular aggregates contribute to the disease process, a subject not addressed in this chapter.

D. Cellular Factors Influencing Aggregation

The focus of this chapter is on the biophysics and biochemistry of the polyglutamine sequence. Some excursions from the topic of pure polyGln behavior have been required in an effort to organize and clarify the voluminous literature touching on polyGln folding and aggregation. Another important aspect of polyGln behavior in the cell is how it might be impacted by the cellular environment. One example of this is found in the discussion of how other normal length polyGlns in the environment are expected to accelerate the nucleation of amyloid-like aggregate formation by an expanded polyGln segment (Section VB). The discussion of how other domains in a polyGln protein limit the ability of the polyGln to form amyloid-like aggregates (Section VB) brings attention to the very important role of proteolysis in the aggregation pathway. In this regard, limitations on the ability of the cell to break down polyGln sequences [115] are just as important as other factors, such as caspases and other specific endoproteinases, that appear to generate specific, aggregation-prone fragments [116]. Molecular chaperones have been shown to modulate both aggregation and toxicity of polyGln in cell and animal models [117, 118]. In yeast models of polyGln disease, a specific yeast prion is required to make the cell susceptible to polyGln aggregation and cell growth delay [119]. Complete coverage of this important and very active area is beyond the scope of this review, and the preceding publications are only cited to give the reader a flavor of the degree to which cellular physiology can filter the imperatives of polyglutamine biophysics. The many factors that can impact polyGln flux and aggregation in the cell may help to explain why some cells are functionally affected by expanded polyGln protein expression while others are not.

VI. CONCLUSIONS

The human genome contains many homopolymeric repeats, most commonly of hydrophilic amino acids like glutamine, which are often found in transcription factors [120]. The expanded polyGln repeat diseases have highlighted the existence of these repeat sequences and at the same time underscored scientists' ignorance of their roles in biology, as well as the incomplete understanding of their physical chemistry. PolyGln is a particularly complex sequence that has attracted an exceptional amount of attention because of its association with a fascinating and insidious group of diseases. It is reasonable to expect that, when the denouement in the search for the underlying mechanisms of expanded CAG repeat diseases is reached, a vastly improved knowledge of the physical chemistry and roles in normal biology of polyGln sequences will be achieved, and perhaps to some extent of other homopolymeric sequences as well.

Acknowledgment

The author gratefully acknowledges Rohit Pappu and Henry Paulson for ideas and helpful comments on the manuscript, Zoya Ignatova Lila Gierasch for sharing results prior to publication, and the National Institute on Aging (R01AG19322) for financial support.

References

1. Bates, G. P., and Benn, C. (2002). The polyglutamine diseases. *In* "Huntington's Disease," (G. P. Bates, P. S. Harper, and L. Jones, Eds.), pp. 429–472. Oxford University Press, Oxford, UK.
2. Scherzinger, E., Lurz, R., Turmaine, M., Mangiarini, L., Hollenbach, B., Hasenbank, R., Bates, G. P., Davies, S. W., Lehrach, H., and Wanker, E. E. (1997). Huntingtin-encoded polyglutamine expansions form amyloid-like protein aggregates in vitro and in vivo. *Cell* **90**, 549–558.
3. Wetzel, R. (2005). Protein folding and aggregation in the expanded polyglutamine repeat diseases. *In* "The Protein Folding Handbook, Part II," (J. Buchner and T. Kiefhaber, Eds.), pp. 1170–1214. Wiley-VCH, Weinheim.
4. Wolfenden, R., Andersson, L., Cullis, P. M., and Southgate, C. C. (1981). Affinities of amino acid side chains for solvent water. *Biochemistry* **20**, 849–855.
5. Chothia, C. (1976). The nature of the accessible and buried surfaces in proteins. *J. Mol. Biol.* **105**, 1–12.
6. Schulz, G. E., and Schirmer, R. H. (1979). "Principles of Protein Structure." Springer-Verlag, Berlin.
7. Levitt, M. (1978). Conformational preferences of amino acids in globular proteins. *Biochemistry* **17**, 4277–4285.
8. Amino acid information (1997). URL: http://prowl.rockefeller.edu/aainfo/solub.html.
9. Robinson, A. B., and Rudd, C. J. (1974). Deamidation of glutaminyl and asparaginyl residues in peptides and proteins. *Curr. Top. Cell. Regul.* **8**, 247–295.
10. Wold, F. (1985). Reactions of the amide side-chains of glutamine and asparagine in vivo. *Trends Biochem. Sci.* **10**, 4–6.
11. Geiger, T., and Clarke, S. (1987). Deamidation, isomerization, and racemization at asparaginyl and aspartyl residues in peptides. *J. Biol. Chem.* **262**, 785–794.
12. Lesort, M., Chun, W., Tucholski, J., and Johnson, G. V. (2002). Does tissue transglutaminase play a role in Huntington's disease? *Neurochem. Int.* **40**, 37–52.
13. Kahlem, P., Terre, C., Green, H., and Djian, P. (1996). Peptides containing glutamine repeats as substrates for transglutaminase-catalyzed cross-linking: Relevance to diseases of the nervous system [see comments]. *Proc. Natl. Acad. Sci. USA* **93**, 14580–14585.
14. Cooper, A. J. L., Sheu, K. R., Burke, J. R., Onodera, O., Strittmatter, W. J., Roses, A. D., and Blass, J. P. (1997). Transglutaminase-catalyzed inactivation of glyceraldehyde 3-phosphate dehydrogenase and alpha-ketoglutarate dehydrogenase complex by polyglutamine domains of pathological length. *Proc. Natl. Acad. Sci. USA* **94**, 12604–12609.
15. Igarashi, S., Koide, R., Shimohata, T., Yamada, M., Hayashi, Y., Takano, H., Date, H., Oyake, M., Sato, T., Sato, A., Egawa, S., Ikeuchi, T., Tanaka, H., Nakano, R., Tanaka, K., Hozumi, I., Inuzuka, T., Takahashi, H., and Tsuji, S. (1998). Suppression of aggregate formation and apoptosis by transglutaminase inhibitors in cells expressing truncated DRPLA protein with an expanded polyglutamine stretch. *Nat. Genet.* **18**, 111–117.
16. Chun, W., Lesort, M., Tucholski, J., Ross, C. A., and Johnson, G. V. (2001). Tissue transglutaminase does not contribute to the formation of mutant huntingtin aggregates. *J. Cell Biol.* **153**, 25–34.
17. Perutz, M. F., Johnson, T., Suzuki, M., and Finch, J. T. (1994). Glutamine repeats as polar zippers: Their possible role in inherited neurodegenerative diseases. *Proc. Natl. Acad. Sci. USA* **91**, 5355–5358.
18. Altschuler, E. L., Hud, N. V., Mazrimas, J. A., and Rupp, B. (1997). Random coil conformation for extended polyglutamine stretches in aqueous soluble monomeric peptides. *J. Pept. Res.* **50**, 73–75.
19. Sharma, D., Sharma, S., Pasha, S., and Brahmachari, S. K. (1999). Peptide models for inherited neurodegenerative disorders: Conformation and aggregation properties of long polyglutamine peptides with and without interruptions. *FEBS Lett.* **456**, 181–185.
20. Chen, S., and Wetzel, R. (2001). Solubilization and disaggregation of polyglutamine peptides. *Protein Sci.* **10**, 887–891.
21. Chen, S., Berthelier, V., Yang, W., and Wetzel, R. (2001). Polyglutamine aggregation behavior *in vitro* supports a recruitment mechanism of cytotoxicity. *J. Mol. Biol.* **311**, 173–182.
22. Stott, K., Blackburn, J. M., Butler, P. J., and Perutz, M. (1995). Incorporation of glutamine repeats makes protein oligomerize: Implications for neurodegenerative diseases. *Proc. Natl. Acad. Sci. USA* **92**, 6509–6513.
23. Gordon-Smith, D. J., Carbajo, R. J., Stott, K., and Neuhaus, D. (2001). Solution studies of chymotrypsin inhibitor-2 glutamine insertion mutants show no interglutamine interactions. *Biochem. Biophys. Res. Commun.* **280**, 855–860.
24. Chen, Y. W., Stott, K., and Perutz, M. F. (1999). Crystal structure of a dimeric chymotrypsin inhibitor 2 mutant containing an inserted glutamine repeat. *Proc. Natl. Acad. Sci. USA* **96**, 1257–1261.
25. Masino, L., Kelly, G., Leonard, K., Trottier, Y., and Pastore, A. (2002). Solution structure of polyglutamine tracts in GST-polyglutamine fusion proteins. *FEBS Lett.* **513**, 267–272.
26. Bennett, M. J., Huey-Tubman, K. E., Herr, A. B., West, A. P., Ross, S. A., and Bjorkman, P. J. (2002). A linear lattice model for polyglutamine in CAG expansion diseases. *Proc. Natl. Acad. Sci. USA* **99**, 11634–11639.
27. Tanaka, M., Morishima, I., Akagi, T., Hashikawa, T., and Nukina, N. (2001). Intra- and intermolecular beta-pleated sheet formation in glutamine-repeat inserted myoglobin as a model for polyglutamine diseases. *J. Biol. Chem.* **276**, 45470–45475.
28. Popiel, H. A., Nagai, Y., Onodera, O., Inui, T., Fujikake, N., Urade, Y., Strittmatter, W. J., Burke, J. R., Ichikawa, A., and Toda, T. (2004). Disruption of the toxic conformation of the expanded polyglutamine stretch leads to suppression of aggregate formation and cytotoxicity. *Biochem. Biophys. Res. Commun.* **317**, 1200–1206.

29. Fleming, P. J., and Rose, G. D. (2005). Conformational properties of unfolded proteins. In "Protein Folding Handbook, Part I," (J. Buchner and T. Kiefhaber, Eds.), pp. 710–736. Wiley-VCH, Weinheim.
30. Stapley, B. J., and Creamer, T. P. (1999). A survey of left-handed polyproline II helices. *Protein Sci.* **8**, 587–595.
31. Tiffany, M. L., and Krimm, S. (1972). Effect of temperature on the circular dichroism spectra of polypeptides in the extended state. *Biopolymers* **11**, 2309–2316.
32. Rucker, A. L., Pager, C. T., Campbell, M. N., Qualls, J. E., and Creamer, T. P. (2003). Host-guest scale of left-handed polyproline II helix formation. *Proteins* **53**, 68–75.
33. Bhattacharyya, A. M., Thakur, A. K., Hermann, V. M., Thiagarajan, G., Williams, A. D., Chellgren, B. W., Creamer, T. P., and Wetzel, R. (2005). Oligoproline effects on polyglutamine conformation and aggregation. *J. Mol. Biol.* **355**, 524–535.
34. Trottier, Y., Lutz, Y., Stevanin, G., Imbert, G., Devys, D., Cancel, G., Saudou, F., Weber, C., David, G., Tora, L., *et al.* (1995). Polyglutamine expansion as a pathological epitope in Huntington's disease and four dominant cerebellar ataxias. *Nature* **378**, 403–406.
35. Perez, M. K., Paulson, H. L., and Pittman, R. N. (1999). Ataxin-3 with an altered conformation that exposes the polyglutamine domain is associated with the nuclear matrix. *Hum. Mol. Genet.* **8**, 2377–2385.
36. Saleem, Q., Anand, A., Jain, S., and Brahmachari, S. K. (2001). The polyglutamine motif is highly conserved at the clock locus in various organisms and is not polymorphic in humans. *Hum. Genet.* **109**, 136–142.
37. Matesanz, F., and Alcina, A. (1996). Glutamine and tetrapeptide repeat variations affect the biological activity of different mouse interleukin-2 alleles. *Eur. J. Immunol.* **26**, 1675–1682.
38. Lin, C. H., Hare, B. J., Wagner, G., Harrison, S. C., Maniatis, T., and Fraenkel, E. (2001). A small domain of CBP/p300 binds diverse proteins: Solution structure and functional studies. *Mol. Cell* **8**, 581–590.
39. Demarest, S. J., Martinez-Yamout, M., Chung, J., Chen, H., Xu, W., Dyson, H. J., Evans, R. M., and Wright, P. E. (2002). Mutual synergistic folding in recruitment of CBP/p300 by p160 nuclear receptor coactivators. *Nature* **415**, 549–553.
40. Nagi, A. D., and Regan, L. (1997). An inverse correlation between loop length and stability in a four-helix-bundle protein. *Fold Des.* **2**, 67–75.
41. Wang, Q., Udayakumar, T. S., Vasaitis, T. S., Brodie, A. M., and Fondell, J. D. (2004). Mechanistic relationship between androgen receptor polyglutamine tract truncation and androgen-dependent transcriptional hyperactivity in prostate cancer cells. *J. Biol. Chem.* **279**, 17319–17328.
42. Callewaert, L., Christiaens, V., Haelens, A., Verrijdt, G., Verhoeven, G., and Claessens, F. (2003). Implications of a polyglutamine tract in the function of the human androgen receptor. *Biochem. Biophys. Res. Commun.* **306**, 46–52.
43. Lamb, D. J., Puxeddu, E., Malik, N., Stenoien, D. L., Nigam, R., Saleh, G. Y., Mancini, M., Weigel, N. L., and Marcelli, M. (2003). Molecular analysis of the androgen receptor in ten prostate cancer specimens obtained before and after androgen ablation. *J. Androl.* **24**, 215–225.
44. Chai, Y., Berke, S. S., Cohen, R. E., and Paulson, H. L. (2004). Polyubiquitin binding by the polyglutamine disease protein ataxin-3 links its normal function to protein surveillance pathways. *J. Biol. Chem.* **279**, 3605–3611.
45. Wang, B., Lin, D., Li, C., and Tucker, P. (2003). Multiple domains define the expression and regulatory properties of Foxp1 forkhead transcriptional repressors. *J. Biol. Chem.* **278**, 24259–24268.
46. Helmlinger, D., Hardy, S., Sasorith, S., Klein, F., Robert, F., Weber, C., Miguet, L., Potier, N., Van-Dorsselaer, A., Wurtz, J. M., Mandel, J. L., Tora, L., and Devys, D. (2004). Ataxin-7 is a subunit of GCN5 histone acetyltransferase-containing complexes. *Hum. Mol. Genet.* **13**, 1257–1265.
47. Gerber, H. P., Seipel, K., Georgiev, O., Hofferer, M., Hug, M., Rusconi, S., and Schaffner, W. (1994). Transcriptional activation modulated by homopolymeric glutamine and proline stretches. *Science* **263**, 808–811.
48. Nicastro, G., Menon, R. P., Masino, L., Knowles, P. P., McDonald, N. Q., and Pastore, A. (2005). The solution structure of the Josephin domain of ataxin-3: Structural determinants for molecular recognition. *Proc. Natl. Acad. Sci. USA* **102**, 10493–10498.
49. Burnett, B., Li, F., and Pittman, R. N. (2003). The polyglutamine neurodegenerative protein ataxin-3 binds polyubiquitylated proteins and has ubiquitin protease activity. *Hum. Mol. Genet.* **12**, 3195–3205.
50. Masino, L., Nicastro, G., Menon, R. P., Dal Piaz, F., Calder, L., and Pastore, A. (2004). Characterization of the structure and the amyloidogenic properties of the Josephin domain of the polyglutamine-containing protein ataxin-3. *J. Mol. Biol.* **344**, 1021–1035.
51. Chow, M. K., Paulson, H. L., and Bottomley, S. P. (2004). Destabilization of a nonpathological variant of ataxin-3 results in fibrillogenesis via a partially folded intermediate: A model for misfolding in polyglutamine disease. *J. Mol. Biol.* **335**, 333–341.
52. Chow, M. K., Ellisdon, A. M., Cabrita, L. D., and Bottomley, S. P. (2004). Polyglutamine expansion in ataxin-3 does not affect protein stability: Implications for misfolding and disease. *J. Biol. Chem.* **279**, 47643–47651.
53. Perez, M. K., Paulson, H. L., Pendse, S. J., Saionz, S. J., Bonini, N. M., and Pittman, R. N. (1998). Recruitment and the role of nuclear localization in polyglutamine-mediated aggregation. *J. Cell Biol.* **143**, 1457–1470.
54. Uchihara, T., Fujigasaki, H., Koyano, S., Nakamura, A., Yagishita, S., and Iwabuchi, K. (2001). Non-expanded polyglutamine proteins in intranuclear inclusions of hereditary ataxias—triple-labeling immunofluorescence study. *Acta Neuropathol (Berlin)* **102**, 149–152.
55. Lieberman, A. P., Trojanowski, J. Q., Leonard, D. G., Chen, K. L., Barnett, J. L., Leverenz, J. B., Bird, T. D., Robitaille, Y., Malandrini, A., and Fischbeck, K. H. (1999). Ataxin 1 and ataxin 3 in neuronal intranuclear inclusion disease. *Ann. Neurol.* **46**, 271–273.
56. Takahashi, J., Tanaka, J., Arai, K., Funata, N., Hattori, T., Fukuda, T., Fujigasaki, H., and Uchihara, T. (2001). Recruitment of nonexpanded polyglutamine proteins to intranuclear aggregates in neuronal intranuclear hyaline inclusion disease. *J. Neuropathol. Exp. Neurol.* **60**, 369–376.
57. Fujigasaki, H., Uchihara, T., Takahashi, J., Matsushita, H., Nakamura, A., Koyano, S., Iwabuchi, K., Hirai, S., and Mizusawa, H. (2001). Preferential recruitment of ataxin-3 independent of expanded polyglutamine: An immunohistochemical study on Marinesco bodies. *J. Neurol. Neurosurg. Psychiatry* **71**, 518–520.
58. Chai, Y., Wu, L., Griffin, J. D., and Paulson, H. L. (2001). The role of protein composition in specifying nuclear inclusion formation in polyglutamine disease. *J. Biol. Chem.* **276**, 44889–44897.
59. Perutz, M. F., Staden, R., Moens, L., and Baere, I. D. (1993). Polar zippers. *Curr. Biol.* **3**, 249–253.
60. Perutz, M. F. (1996). Glutamine repeats and inherited neurodegenerative diseases: Molecular aspects. *Curr. Opin. Struct. Biol.* **6**, 848–858.
61. Lindquist, S. (1997). Mad cows meet psi-chotic yeast: The expansion of the prion hypothesis. *Cell* **89**, 495–498.
62. Chapman, M. R., Robinson, L. S., Pinkner, J. S., Roth, R., Heuser, J., Hammar, M., Normark, S., and Hultgren, S. J. (2002). Role of *Escherichia coli* curli operons in directing amyloid fiber formation. *Science* **295**, 851–855.

63. Fukui, T., Shiraki, K., Hamada, D., Hara, K., Miyata, T., Fujiwara, S., Mayanagi, K., Yanagihara, K., Iida, T., Fukusaki, E., Imanaka, T., Honda, T., and Yanagihara, I. (2005). Thermostable direct hemolysin of vibrio parahaemolyticus is a bacterial reversible amyloid toxin. *Biochemistry* **44**, 9825–9832.
64. DiFiglia, M., Sapp, E., Chase, K. O., Davies, S. W., Bates, G. P., Vonsattel, J. P., and Aronin, N. (1997). Aggregation of huntingtin in neuronal intranuclear inclusions and dystrophic neurites in brain. *Science* **277**, 1990–1993.
65. Davies, S. W., Turmaine, M., Cozens, B. A., DiFiglia, M., Sharp, A. H., Ross, C. A., Scherzinger, E., Wanker, E. E., Mangiarini, L., and Bates, G. P. (1997). Formation of neuronal intranuclear inclusions underlies the neurological dysfunction in mice transgenic for the HD mutation. *Cell* **90**, 537–548.
66. Paulson, H. L., Perez, M. K., Trottier, Y., Trojanowski, J. Q., Subramony, S. H., Das, S. S., Vig, P., Mandel, J. L., Fischbeck, K. H., and Pittman, R. N. (1997). Intranuclear inclusions of expanded polyglutamine protein in spinocerebellar ataxia type 3. *Neuron* **19**, 333–344.
67. Morley, J. F., Brignull, H. R., Weyers, J. J., and Morimoto, R. I. (2002). The threshold for polyglutamine-expansion protein aggregation and cellular toxicity is dynamic and influenced by aging in *Caenorhabditis elegans*. *Proc. Natl. Acad. Sci. USA* **99**, 10417–10422.
68. Ross, C. A., and Poirier, M. A. (2005). What is the role of protein aggregation in neurodegeneration? *Nat. Rev. Mol. Cell Biol.*, **6**, 891–898.
69. Iuchi, S., Hoffner, G., Verbeke, P., Djian, P., and Green, H. (2003). Oligomeric and polymeric aggregates formed by proteins containing expanded polyglutamine. *Proc. Natl. Acad. Sci. USA* **100**, 2409–2414.
70. Osmand, A. P., Berthelier, V., and Wetzel, R. (2006). Imaging polyglutamine deposits in brain tissue. *Methods Enzymol.*, in press.
71. Nybo, M., Svehag, S.-E., and Nielsen, E. H. (1999). An ultrastructural study of amyloid intermediates in A beta (1–42) fibrillogenesis. *Scand. J. Immunol.* **49**, 219–223.
72. Mukai, H., Isagawa, T., Goyama, E., Tanaka, S., Bence, N. F., Tamura, A., Ono, Y., and Kopito, R. R. (2005). Formation of morphologically similar globular aggregates from diverse aggregation-prone proteins in mammalian cells. *Proc. Natl. Acad. Sci. USA* **102**, 10887–10892.
73. Sunde, M., and Blake, C. (1997). The structure of amyloid fibrils by electron microscopy and X-ray diffraction. *Adv. Protein Chem.* **50**, 123–159.
74. Kowalewski, T., and Holtzman, D. M. (1999). In situ atomic force microscopy study of Alzheimer's beta-amyloid peptide on different substrates: New insights into mechanism of beta-sheet formation. *Proc. Natl. Acad. Sci. USA* **96**, 3688–3693.
75. Harper, J. D., Wong, S. S., Lieber, C. M., and Lansbury, P. T., Jr. (1999). Assembly of A beta amyloid protofibrils: An in vitro model for a possible early event in Alzheimer's disease. *Biochemistry* **38**, 8972–8980.
76. Berthelier, V., Hamilton, J. B., Chen, S., and Wetzel, R. (2001). A microtiter plate assay for polyglutamine aggregate extension. *Anal. Biochem.* **295**, 227–236.
77. Chen, S., Ferrone, F., and Wetzel, R. (2002). Huntington's disease age-of-onset linked to polyglutamine aggregation nucleation. *Proc. Natl. Acad. Sci. USA* **99**, 11884–11889.
78. Yang, W., Dunlap, J. R., Andrews, R. B., and Wetzel, R. (2002). Aggregated polyglutamine peptides delivered to nuclei are toxic to mammalian cells. *Hum. Mol. Genet.* **11**, 2905–2917.
79. Thakur, A., and Wetzel, R. (2002). Mutational analysis of the structural organization of polyglutamine aggregates. *Proc. Natl. Acad. Sci. USA* **99**, 17014–17019.
80. Chen, S., Berthelier, V., Hamilton, J. B., O'Nuallain, B., and Wetzel, R. (2002). Amyloid-like features of polyglutamine aggregates and their assembly kinetics. *Biochemistry* **41**, 7391–7399.
81. Sharma, D., Shinchuk, L., Inouye, H., Wetzel, R., and Kirschner, D. A. (2005). Polyglutamine homopolymers having 8–45 repeats form slab-like β-crystallite assemblies. *Proteins: Struct., Funct., Bioinformat.*, **61**, 398–411.
82. Jenkins, J., and Pickersgill, R. (2001). The architecture of parallel beta-helices and related folds. *Prog. Biophys. Mol. Biol.* **77**, 111–175.
83. Wetzel, R. (2002). Ideas of order for amyloid fibril structure. *Structure* **10**, 1031–1036.
84. Perutz, M. F., Finch, J. T., Berriman, J., and Lesk, A. (2002). Amyloid fibers are water-filled nanotubes. *Proc. Natl. Acad. Sci. USA* **99**, 5591–5595.
85. Sikorski, P., and Atkins, E. (2005). New model for crystalline polyglutamine assemblies and their connection with amyloid fibrils. *Biomacromolecules* **6**, 425–432.
86. Poirier, M. A., Jiang, H., and Ross, C. A. (2005). A structure-based analysis of huntingtin mutant polyglutamine aggregation and toxicity: Evidence for a compact beta-sheet structure. *Hum. Mol. Genet.* **14**, 765–774.
87. Nelson, R., Sawaya, M. R., Balbirnie, M., Madsen, A. O., Riekel, C., Grothe, R., and Eisenberg, D. (2005). Structure of the cross-beta spine of amyloid-like fibrils. *Nature* **435**, 773–778.
88. Liebman, S. W. (2005). Structural clues to prion mysteries. *Nat. Struct. Mol. Biol.* **12**, 567–568.
89. Jarrett, J. T., and Lansbury, P. J. (1993). Seeding "one-dimensional crystallization" of amyloid: A pathogenic mechanism in Alzheimer's disease and scrapie? [review]. *Cell* **73**, 1055–1058.
90. Ferrone, F. (1999). Analysis of protein aggregation kinetics. *Methods Enzymol.* **309**, 256–274.
91. Bhattacharyya, A. M., Thakur, A., and Wetzel, R. (2005). Polyglutamine aggregation nucleation: Thermodynamics of a highly unfavorable protein folding reaction. *Proc. Natl. Acad. Sci. USA* **102**, 15400–15405.
92. Khare, S. D., Ding, F., Gwanmesia, K. N., and Dokholyan, N. V. (2005). Molecular origin of polyglutamine aggregation in neurodegenerative diseases. *Public Library of Science-Computational Biology* **1**, 230–235.
93. Starikov, E. B., Lehrach, H., and Wanker, E. E. (1999). Folding of oligoglutamines: a theoretical approach based upon thermodynamics and molecular mechanics. *J. Biomol. Struct. Dyn.* **17**, 409–427.
94. O'Nuallain, B., Williams, A. D., Westermark, P., and Wetzel, R. (2004). Seeding specificity in amyloid growth induced by heterologous fibrils. *J. Biol. Chem.* **279**, 17490–17499.
95. Huang, C. C., Faber, P. W., Persichetti, F., Mittal, V., Vonsattel, J. P., MacDonald, M. E., and Gusella, J. F. (1998). Amyloid formation by mutant huntingtin: threshold, progressivity and recruitment of normal polyglutamine proteins. *Somat. Cell Mol. Genet.* **24**, 217–233.
96. Kazantsev, A., Preisinger, E., Dranovsky, A., Goldgaber, D., and Housman, D. (1999). Insoluble detergent-resistant aggregates form between pathological and nonpathological lengths of polyglutamine in mammalian cells. *Proc. Natl. Acad. Sci. USA* **96**, 11404–11409.
97. Steffan, J. S., Kazantsev, A., Spasic-Boskovic, O., Greenwald, M., Zhu, Y. Z., Gohler, H., Wanker, E. E., Bates, G. P., Housman, D. E., and Thompson, L. M. (2000). The Huntington's disease protein interacts with p53 and CREB-binding protein and represses transcription. *Proc. Natl. Acad. Sci. USA* **97**, 6763–6768.
98. Jiang, H., Nucifora, F. C., Jr., Ross, C. A., and DeFranco, D. B. (2003). Cell death triggered by polyglutamine-expanded huntingtin in a neuronal cell line is associated with degradation of CREB-binding protein. *Hum. Mol. Genet.* **12**, 1–12.

99. Nucifora, F. C., Jr., Sasaki, M., Peters, M. F., Huang, H., Cooper, J. K., Yamada, M., Takahashi, H., Tsuji, S., Troncoso, J., Dawson, V. L., Dawson, T. M., and Ross, C. A. (2001). Interference by huntingtin and atrophin-1 with cbp-mediated transcription leading to cellular toxicity. *Science* **291**, 2423–2428.
100. Chai, Y., Shao, J., Miller, V. M., Williams, A., and Paulson, H. L. (2002). Live-cell imaging reveals divergent intracellular dynamics of polyglutamine disease proteins and supports a sequestration model of pathogenesis. *Proc. Natl. Acad. Sci. USA* **99**, 9310–9315.
101. Pulst, S.-M., Santos, N., Wang, D., Yang, H., Huynh, D., Velasquez, L., and Figueroa, K. P. (2005). Spinocerebellar ataxia type 2: PolyQ repeat variation in the CACNA1A calcium channel modifies age of onset. *Brain*, **128**, 2297–2303.
102. Nozaki, K., Onodera, O., Takano, H., and Tsuji, S. (2001). Amino acid sequences flanking polyglutamine stretches influence their potential for aggregate formation. *Neuroreport* **12**, 3357–3364.
103. Chien, P., and Weissman, J. S. (2001). Conformational diversity in a yeast prion dictates its seeding specificity. *Nature* **410**, 223–227.
104. Scherzinger, E., Sittler, A., Schweiger, K., Heiser, V., Lurz, R., Hasenbank, R., Bates, G. P., Lehrach, H., and Wanker, E. E. (1999). Self-assembly of polyglutamine-containing huntingtin fragments into amyloid-like fibrils: Implications for Huntington's disease pathology. *Proc. Natl. Acad. Sci. USA* **96**, 4604–4609.
105. Wellington, C. L., Singaraja, R., Ellerby, L., Savill, J., Roy, S., Leavitt, B., Cattaneo, E., Hackam, A., Sharp, A., Thornberry, N., Nicholson, D. W., Bredesen, D. E., and Hayden, M. R. (2000). Inhibiting caspase cleavage of huntingtin reduces toxicity and aggregate formation in neuronal and nonneuronal cells. *J. Biol. Chem.* **275**, 19831–19838.
106. Rubinsztein, D. C. (2002). Lessons from animal models of Huntington's disease. *Trends Genet.* **18**, 202–209.
107. Lunkes, A., Lindenberg, K. S., Ben-Haiem, L., Weber, C., Devys, D., Landwehrmeyer, G. B., Mandel, J. L., and Trottier, Y. (2002). Proteases acting on mutant huntingtin generate cleaved products that differentially build up cytoplasmic and nuclear inclusions. *Mol. Cell* **10**, 259–269.
108. Minton, A. P. (2000). Implications of macromolecular crowding for protein assembly. *Curr. Opin. Struct. Biol.* **10**, 34–39.
109. Poirier, M. A., Li, H., Macosko, J., Cai, S., Amzel, M., and Ross, C. A. (2002). Huntingtin spheroids and protofibrils as precursors in polyglutamine fibrilization. *J. Biol. Chem.* **277**, 41032–41037.
110. Caughey, B., and Lansbury, P. T. (2003). Protofibrils, pores, fibrils, and neurodegeneration: separating the responsible protein aggregates from the innocent bystanders. *Annu. Rev. Neurosci.* **26**, 267–298.
111. Wacker, J. L., Zareie, M. H., Fong, H., Sarikaya, M., and Muchowski, P. J. (2004). Hsp70 and Hsp40 attenuate formation of spherical and annular polyglutamine oligomers by partitioning monomer. *Nat. Struct. Mol. Biol.* **11**, 1215–1222.
112. Kheterpal, I., Lashuel, H. A., Hartley, D. M., Walz, T., Lansbury, P. T., Jr., and Wetzel, R. (2003). A beta protofibrils possess a stable core structure resistant to hydrogen exchange. *Biochemistry* **42**, 14092–14098.
113. Williams, A. D., Sega, M., Chen, M., Kheterpal, I., Geva, M., Berthelier, V., Kaleta, D. T., Cook, K. D., and Wetzel, R. (2005). Structural properties of Aβ protofibrils stabilized by a small molecule. *Proc. Natl. Acad. Sci. USA* **102**, 7115–7120.
114. Gosal, W. S., Morten, I. J., Hewitt, E. W., Smith, D. A., Thomson, N. H., and Radford, S. E. (2005). Competing pathways determine fibril morphology in the self-assembly of beta(2)-microglobulin into amyloid. *J. Mol. Biol.* **351**, 850–864.
115. Venkatraman, P., Wetzel, R., Tanaka, M., Nukina, N., and Goldberg, A. L. (2004). Eukaryotic proteasomes cannot digest polyglutamine sequences and release them intact during degradation of polyglutamine-containing proteins. *Mol. Cell* **14**, 95–104.
116. Kim, Y. J., Yi, Y., Sapp, E., Wang, Y., Cuiffo, B., Kegel, K. B., Qin, Z. H., Aronin, N., and DiFiglia, M. (2001). Caspase 3-cleaved N-terminal fragments of wild-type and mutant huntingtin are present in normal and Huntington's disease brains, associate with membranes, and undergo calpain-dependent proteolysis. *Proc. Natl. Acad. Sci. USA* **98**, 12784–12789.
117. Kazemi-Esfarjani, P., and Benzer, S. (2000). Genetic suppression of polyglutamine toxicity in Drosophila. *Science* **287**, 1837–1840.
118. Wyttenbach, A. (2004). Role of heat shock proteins during polyglutamine neurodegeneration: mechanisms and hypothesis. *J. Mol. Neurosci.* **23**, 69–96.
119. Osherovich, L. Z., and Weissman, J. S. (2001). Multiple Gln/Asn-rich prion domains confer susceptibility to induction of the yeast [PSI(+)] prion. *Cell* **106**, 183–194.
120. Oma, Y., Kino, Y., Sasagawa, N., and Ishiura, S. (2004). Intracellular localization of homopolymeric amino acid-containing proteins expressed in mammalian cells. *J. Biol. Chem.* **279**, 21217–21222.
121. Bevivino, A. E., and Loll, P. J. (2001). An expanded glutamine repeat destabilizes native ataxin-3 structure and mediates formation of parallel beta–fibrils. *Proc. Natl. Acad. Sci. USA* **98**, 11955–11960.

PART X

In Vivo Instability Studies

CHAPTER 35

Somatic Mosaicism of Expanded CAG·CTG Repeats in Humans and Mice: Dynamics, Mechanisms, and Consequences

PEGGY F. SHELBOURNE AND DARREN G. MONCKTON

Institute of Biomedical and Life Sciences, University of Glasgow, Anderson College Building, Glasgow G11 6NU, United Kingdom

I. Introduction
   A. The CAG·CTG Repeat Expansion Loci
   B. Why Are There So Many CAG·CTG Repeat Disorders?
II. Dynamics of Somatic Mosaicism in Humans
   A. Genotyping and Measurement of Somatic Mosaicism
   B. Somatic Mosaicism in Myotonic Dystrophy Type 1
   C. Somatic Mosaicism in Huntington Disease
   D. Somatic Mosaicism in Other CAG·CTG Repeat Disorders
   E. General Features of Somatic Mosaicism in Humans
III. Dynamics of Somatic Mosaicism in Human and Mouse Cell Models
   A. Dynamics of Somatic Mosaicism in Human Tissue Culture
   B. Dynamics and Tissue Specificity of Somatic Mosaicism in Transgenic Mice
   C. Dynamics of Somatic Mosaicism in Mouse Tissue Culture
IV. Consequences of Somatic Mosaicism
   A. Age-Dependent Allele Length Measurements
   B. Does Expansion-Biased Somatic Mosaicism Contribute to Disease Pathology?
V. *Cis*-Acting Modifiers of Somatic Mosaicism
VI. *Trans*-Acting Genetic Modifiers of Somatic Mosaicism
   A. DNA Mismatch Repair Genes Are Required To Generate Somatic Mosaicism
   B. Other DNA Repair Pathways
   C. Strain Effects
VII. What Is the Mechanism of Somatic Expansion?
   A. Replication Slippage Is Not the Major Mechanism of Expansion
   B. Are DNA Hairpins and/or Other Unusual Structures Mutation Intermediates?
   C. Somatic Expansion Could Occur by Inappropriate DNA Mismatch Repair
VIII. Somatic Mosaicism as a Therapeutic Target
   A. Can Repeat Expansion Be Modified by Drugs?
   B. Is DNA Mismatch Repair a Therapeutic Target?
   C. Somatic Mosaicism and Off-Target Drug Effects
IX. Concluding Remarks
   References

Since the discovery that expanded CAG·CTG repeat tracts are the primary defect in disorders such as myotonic dystrophy and Huntington disease, it has been apparent that the intergenerational instability of the expanded alleles underlies the unusual genetics of these disorders, notably anticipation and sex-of-parent biases. It has been demonstrated that the expanded repeats are also somatically unstable in a process that is age-dependent, tissue-specific, and expansion-biased, properties that very likely contribute to the tissue specificity and progressive nature of the symptoms. Unlike intergenerational instability, somatic expansion appears to be faithfully reproduced in transgenic mouse models and can also be replicated in tissue culture. In combination with detailed patient studies, these model systems have yielded considerable insight into the mechanism of expansion, which appears to be cell-division-independent, inappropriate DNA mismatch repair, rather than the more commonly assumed replication slippage. These new insights into somatic mosaicism reaffirm the CAG·CTG repeat as a therapeutic target. Indeed, evidence reveals that the rate of expansion can be suppressed by a number of agents. Excitingly, emerging knowledge of the mechanism of expansion raises new prospects for intervention based on rational drug design, with suppression of expansion as an intermediate aspiration and reversion as the ultimate goal.

I. INTRODUCTION

A. The CAG·CTG Repeat Expansion Loci

Of the growing list of disorders associated with the expansion of simple sequence repeats, most are associated with the expansion of a CAG·CTG repeat. This affords researchers a unique opportunity to compare and contrast a variety of data sets and to more fully understand the important factors mediating genetic instability at these loci. The diseases associated with CAG·CTG repeat expansions include the neurodegenerative disorders in which a CAG repeat encodes a polyglutamine tract in the resultant protein: spinal and bulbar muscular atrophy (SBMA) [100], Huntington disease (HD) [181], dentatorubral pallidoluysian atrophy (DRPLA) [93], Machado–Joseph disease [MJD, also known as spinocerebellar ataxia (SCA) type 3] [86], and the other spinocerebellar ataxias, types 1 [145], 2 [73], 7 [31], and 17 [94, 141]. In certain circumstances, SCA6 [206] is considered to be a member of this group on the basis of downstream pathology, because SCA6 shares the polyglutamine aggregates typical of these disorders [76]. However, the "CAG expansions" at this locus are short (the pathogenic range is ~20–30 repeats), well within what would generally be considered the normal range at any of the other CAG·CTG loci, and show very low levels of intergenerational instability and the absence of somatic instability [77, 82, 128, 172, 173, 206]. In the remaining disorders, the repeat tract is noncoding. In SCA12 a CAG repeat is located in the 5′ untranslated region of the *PPP2R2B* gene [68], whereas the CAG repeat is on the nontranscribed strand in myotonic dystrophy type 1 (DM1) [8, 19, 23, 46, 63, 115] and SCA8 [95]. Although the downstream pathology in these three disorders is very likely to differ from that observed in the polyglutamine repeat disorders [67, 161], current understanding of the mechanism of expansion suggests that there is no reason to believe that these loci will behave differently. Indeed, it would appear that the dynamics of expanded alleles at these loci overlap with those observed at the polyglutamine disease repeat loci. Two further loci that contain expanded CAG·CTG repeats are *CTG18.1* [15] and *ERDA1* [72, 140], for which expanded alleles have been observed at a detectable frequency in the general population, but for which no deleterious phenotype has been ascribed. Nonetheless, expanded alleles at these loci appear to be genetically unstable to a greater or lesser extent, in both the germline and soma [15, 35, 72, 140, 188], and their dynamics are presumably regulated by mechanisms similar to those of the other expanded CAG·CTG repeat loci.

Although it is probable that the factors mediating CAG·CTG repeat instability in the germline and soma will be similar, evidence suggests that the pathways are, at least partially, distinct [133, 135, 137]. Moreover, although somatic mosaicism appears to be fully reproducible in the mouse [45, 116, 171], very large germline expansions observed at some of the human loci have not yet been replicated in the mouse, thus depriving researchers of a definitive model system in which to further investigate this phenomenon.

B. Why Are There So Many CAG?CTG Repeat Disorders?

Most of the repeat expansion disorders are associated with trinucleotides, and most of these are CAG·CTG repeat expansions. The basis for this excess currently is not known, but may include the following: (1) an ascertainment bias of discovery, based on assumptions regarding the involvement of triplet repeats in disease pathology and, in particular, the involvement of polyglutamine expansions in neurodegenerative diseases; (2) an underlying biological propensity for these sequences to expand; (3) the relative toxicity of expanded polyglutamine tracts that the expanded CAG frequently encodes; or (4) the relative laxity with which small polyglutamine tracts, the precursors of pathogenic expansions, are tolerated within proteins. Regardless, the fact that so many diseases with CAG·CTG repeat expansions have been studied (and modeled), affords a unique opportunity for understanding the biology of the loci.

II. DYNAMICS OF SOMATIC MOSAICISM IN HUMANS

A. Genotyping and Measurement of Somatic Mosaicism

There are two principal approaches to genotyping and measuring the degree of somatic mosaicism in a given sample. First, traditional Southern blot hybridization of restriction-enzyme-digested genomic DNA and, second, methodologies based upon polymerase chain reaction (PCR) analysis. Detection of mutant alleles by Southern blot hybridization of genomic DNA requires ~5–10 μg of DNA, equivalent to that contained in >800,000 cells. Normally, a stable locus is visualized as a single, relatively sharp band after autoradiography. By contrast, extensive somatic mosaicism can result in the production of a diffuse, smeary band [196]. By using this approach it is relatively straightforward to determine the region of peak intensity and, thus, measure the modal allele length in a given sample. Definition of the extent of somatic mosaicism is more problematic, as the tails of the smear quickly run into the background signal of the autoradiograph. Measures such as the midpeak width provide a quantitative answer to this problem [196], but they inevitably fail to record the full extent of mosaicism. At best, it is likely that only those alleles that comprise >5% of the total population of cells will be detected. Moreover, a CAG·CTG repeat sequence cannot be used to probe a Southern blot of total genomic DNA due to the many cross-hybridizing fragments found throughout the genome, necessitating the use of restriction enzymes that release fragments with sufficient flanking DNA to permit the use of a locus-specific probe (typically on the order of 500 bp to 1 kb in length). These relatively large fragment sizes, coupled with the relatively poor resolving power of agarose gels, mean that it is typically not possible to detect small differences in allele size. Accordingly, this approach is relatively insensitive in terms of the amount of DNA required for the analysis, the fraction of variant alleles that can be defined, and the ability to precisely define allele length. Nonetheless, Southern blot hybridization of restriction-enzyme-digested genomic DNA does have the advantage that the use of low percentage agarose gels allows the resolution and detection of very large alleles (>5,000 repeats). Theoretically, pulsed field gel electrophoresis should permit the characterization of expansions with tens of thousands of repeats.

PCR-based approaches have the potential for far greater sensitivity in terms of the amount of input DNA required, the ability to accurately define allele length, and the ability to quantitatively define both the most common variants and those present in only a small fraction of cells. However, PCR has the disadvantages that very large alleles are refractory to efficient amplification and that, unless special care is taken, preferential amplification of smaller alleles can mask the presence of larger alleles. Moreover, polymerase stutter during PCR can complicate allele assignment. Consequently, many of the potential advantages of PCR are not realized using standard methods. Most of these approaches use 10–100 ng of input DNA as template. Although this is a relatively small amount compared to that employed in Southern blot hybridization of restriction-enzyme-digested genomic DNA, it still represents many thousands of cells worth of DNA. The DNA is amplified with oligonucleotide primers that incorporate either a radioactive label followed by resolution on denaturing polyacrylamide gels and detection by autoradiography or a fluorescent label and electrophoresis using an automated DNA sequencing apparatus. Data emanating from automated sequencers are commonly analyzed by using GeneScan software. The resolving power of such gels is very high, and theoretically alleles can be sized to within a single repeat unit. Although this high level of precision in defining allele length can be applied to alleles within the normal range and small expansions, the degree of PCR stutter generated in amplifying large expansions is such that, even when amplifying from single input molecules, a characteristic "hedgehog" pattern of peaks, differing by a single repeat unit, is observed [59, 205]. Of particular note, the PCR stutter is strongly biased toward generating contractions, and the hedgehog distribution shifts toward smaller products with increasing amplification cycle numbers [101, 205]. Thus, it is not always appropriate to assume that the most prominent peak represents the actual allele size. Moreover, the inclusion of only one labeling molecule in each amplified product (e.g., if oligonucleotide primers are labeled with a single radioactive or fluorescent molecule) can lead to preferential detection of smaller alleles. Thus, unless extreme care is taken, larger alleles may not be detected, particularly when they are present in a small fraction of cells. Indeed, it has been demonstrated that standard GeneScan-based approaches grossly underestimate the true degree of somatic mosaicism in highly variable samples [45, 89].

An alternative method that does capture much of the potential sensitivity of PCR is small pool PCR (SP-PCR) (see Figs. 35-1A–C) [52, 80, 135]. In this method, multiple small aliquots of DNA are amplified and resolved by using agarose gel electrophoresis. Subsequent detection by Southern blot hybridization is performed under conditions where it is possible to detect amplification products derived from single input molecules and, hence, from single cells. The effects of amplification bias against longer alleles are reduced by using small amounts of input DNA, low numbers of PCR cycles, and probes

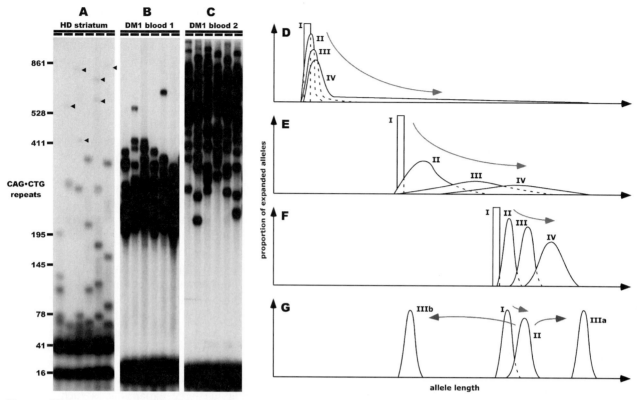

FIGURE 35-1 Dynamics of somatic mosaicism. Small pool PCR analysis of somatic mosaicism in striatal cell DNA from an HD patient (A) and blood DNA from two DM1 patients (B, C). Each lane contains the amplification products from ~20–40 cellular equivalents of DNA. The scale on the left shows the molecular weight marker converted to number of CAG·CTG repeats. In HD striatal DNA, most cells have acquired no, or only very small, somatic expansions. However, a subset of cells have acquired very large expansions, many in excess of several hundred repeats (some faint alleles have been highlighted with arrowheads). The level of variation in the blood DNA of DM1 patients inheriting larger alleles is greater with a more obvious increase in modal allele length. The distribution in DM1 patient 1's DNA is still highly skewed toward expansions. In DM1 patient 2, the level of variation is even greater and is more normally distributed. The histograms of allele length versus proportion of cells show schematic representations of the dynamics of somatic mosaicism over time (red arrows) *in vivo* with different progenitor allele lengths (D–F) and *in vitro* in tissue culture (G). In individuals inheriting relatively small expansions (D-I), the rate of expansion is low and the modal allele size increases slowly with time (D-II, D-III). A small subset of cells acquire larger expansions and generate a long tail of large expansions in some tissues in older individuals (D-IV) (e.g., in striatal cells in HD, see panel A). In individuals inheriting larger alleles (E-I), the rate of somatic expansion is much higher and highly skewed distributions may be generated (E-II) (e.g., in DM1 blood cells, see panel B), which eventually become more normally distributed (E-III, E-IV). In individuals inheriting very large expansions (F-I), the rate of somatic expansion is very high and the repeat expands rapidly in an apparently synchronous manner in all cells (F-II, F-III, F-IV) (e.g., in DM1 blood cells, see panel C). In tissue culture the progenitor allele (G-I) increases synchronously in size in all cells (G-II), but rare large expansions (G-IIIa) or rare large deletions (G-IIIb) followed by clonal expansion can lead to dramatic shifts in the repeat length distribution (blue arrows) [54, 90]. See CD-ROM for color image.

containing the appropriate repeats during subsequent Southern blot hybridization (despite being amplified less efficiently, larger alleles hybridize more efficiently). Thus, by using SP-PCR, it is possible to generate detailed, fully quantitative distributions of allele length variation in samples that contain repeat tracts of up to at least 1,000 units. Moreover, rare variants that are only present in a small fraction of cells can be sized and quantified with high accuracy. SP-PCR has proved to be highly effective in analyzing both germline instability and somatic mosaicism at the CAG·CTG repeat expansion loci and has revealed important insights missed by using standard approaches [88, 89, 124, 126, 133, 135].

Similar levels of sensitivity and accuracy can be achieved by using PCR amplification from physically selected single cells. This has proved particularly effective for analyzing germline variation in sperm [103, 104, 203, 204] and, to a lesser extent, oocytes [87], where mutation frequencies are often very high. However, such an approach has not been widely used to investigate somatic mosaicism because it is technically demanding, relatively labor-intensive, and not well-suited to the identification of rare variants. Another development that has great potential is laser capture microdissection. By using this technique, it is possible to select specific cell types from complex tissues and more precisely define patterns of cell-type-specific allele length variation within a given tissue [66, 190, 200].

B. Somatic Mosaicism in Myotonic Dystrophy Type 1

In myotonic dystrophy type 1 (DM1), it was apparent from the first Southern blot analyses of restriction-enzyme- digested genomic DNA that somatic mosaicism was prominent [8, 23, 46, 115]. Rather than the sharp band that was expected, the expanded allele in many patients presented as a heterogeneous smear, reflecting very high levels of CTG repeat length variability. It was soon established that the degree of heterogeneity observed in blood DNA was dependent on both the allele length and the age at sampling [196]. Patients sampled at an older age and/or with very large expansions presented more size heterogeneity than those with smaller alleles or those sampled at an early age. Similarly, it was quickly noted that the allele length observed in skeletal muscle was much larger than that observed in blood DNA [3, 6, 127, 182, 202]. Average allele lengths measured in adult skeletal muscle DNA are often very large, frequently in excess of 2,000 or more repeats, even in individuals with relatively small alleles in their blood DNA. Subsequently, it was shown that the heterogeneous smear observed by standard analyses could be resolved into its component alleles by procedures such as small pool PCR [135]. These analyses confirmed that somatic mosaicism was allele-length- and age-dependent [135, 196]. Intriguingly, patients with very small expansions did not show significant expansion in skeletal muscle, did not develop severe muscle weakness, and did not display the ribonuclear foci characteristic of DM1 pathology [118]. More commonly, in younger individuals or those with relatively small expansions, the distributions observed were highly skewed, with a tail of large expanded alleles and a relatively sharp lower boundary, below which mutant alleles were rarely observed (see Fig. 35-1B). This lower boundary was conserved between tissues of the same individual and was assumed to represent the size of the progenitor allele (i.e., the single allele transmitted in the egg or sperm from the affected parent), revealing that somatic mosaicism is highly biased toward expansion [135]. Mutational spectra in older individuals, particularly those with larger alleles, are highly diffuse and more normally distributed (see Fig. 35-1C). The expansion-biased nature of somatic mosaicism has been confirmed in serial analyses of blood samples from the same patient taken at different ages [123, 125]. In all cases investigated, the average repeat length was greater in blood DNA samples taken at a later age. Although similar studies of patients did not reveal an age-dependent increase in average allele length in skeletal muscle DNA [3, 182], they were likely compromised by the technical approach used, as the samples were migrating at the upper resolution limit of the gel. Examination of fetal and newborn material has confirmed that, even in individuals with very large alleles associated with congenital DM1 [6, 79, 84, 122, 196–198], the degree of somatic mosaicism in tissues during early development, including skeletal muscle, is very low compared to that observed in adults. Moreover, it has been observed that the difference in average allele length of blood and skeletal muscle DNA from the same patient is highly age-dependent [202]. These data confirm that somatic mosaicism in skeletal muscle cells of DM1 patients is age-dependent and expansion-biased and occurs much more rapidly than in blood cells. These findings are very provocative with regard to both tissue specificity of the symptoms and mechanism of expansion. Is DM1 primarily a skeletal-muscle-based disorder because of the tissue specificity of the expansion process? How does the repeat expand more rapidly in postmitotic skeletal muscle cells compared to rapidly dividing blood cell lineages?

Although it is clear that average allele lengths are much greater in DM1 skeletal muscle cells than in blood cells, the relative stability of the repeat in other tissues is less clear, with only a few studies having been performed and all using relatively insensitive bulk DNA analyses. Despite this, the data reveal that the repeat is also particularly unstable in cardiac muscle and other peripheral tissues, such as the liver, kidney, and pancreas (at levels comparable to those in skeletal muscle DNA), but relatively stable in lung and other tissues with a high proportion of white cells, for example, spleen and thymus [75, 79, 84, 91, 122, 197, 198]. Within the brain, the DM1 repeat appears to be highly unstable in all regions, with the striking exception of the cerebellar cortex, where the repeat is even more stable than that observed in blood cells [75, 91].

C. Somatic Mosaicism in Huntington Disease

Once HD was identified as a polyglutamine disease in 1993 [181], studies were initiated to investigate genetic instability in germline and somatic tissues. These demonstrated that the expanded CAG repeat tract is unstable in about 80% of intergenerational transmissions with additional increases more common, particularly in the paternal germline [138]. In contrast, early studies reported little evidence of somatic mutation length mosaicism [2, 113, 209], although later studies confirmed that the size of the expansion can vary both within and between tissues of individuals affected by HD [4, 32, 50, 74, 85,

180]. The greatest size variability was observed in the brain, particularly in regions such as the striatum and cortex that show the most neuropathological involvement [4, 74, 85, 178, 180]. Somatic mosaicism was most prominent in juvenile onset cases of HD [4, 180], where cells of the cortex/striatum had acquired expansions that were ~13 CAG repeats larger than cells of relatively spared regions of the brain, such as the cerebellum [180]. Sensitive SP-PCR methodologies have been employed to reveal brain-region-specific, expansion-biased HD mutation profiles, where the magnitude and variability of the mutation size changes observed were much greater [88] than previously documented (see Fig. 35-1A). Prior to widespread pathological cell death, a significant proportion (~10%) of the striatal cells have acquired expansions of >200 CAG repeats, whereas others have expansions that exceed 1,000 CAG repeats in length, some 25 times the size of the expansion inherited from the affected parent. Although expansion length changes appear to occur earlier in the striatum than in other regions of the brain, significant expansion-biased CAG repeat length gains are also seen in the cortex, but not in the hypothalamus or cerebellum. In end-stage brain tissue, very large somatic length gains (up to 500 CAG repeats) are still seen in the cortex, but not in the striatum, suggesting that striatal cells with the largest HD mutation expansions are lost during the disease process. Importantly, this study demonstrated that the HD mutation is capable of expansion-biased length changes in somatic tissues that are as great as, if not greater than, the size changes observed in the germline. Very large mutation length gains in the striatum are present before significant levels of neuronal loss and manifestation of overt clinical symptoms, and therefore they are an early molecular event in the disease process. Furthermore, the fact that mutation profiles in different regions of the brain correlate with the topography and temporal order of neuropathological involvement is consistent with the notion that mutation length variability in somatic tissues may contribute to both the progressive nature and cell-selective aspects of human HD pathogenesis.

D. Somatic Mosaicism in Other CAG?CTG Repeat Disorders

Somatic mosaicism has been demonstrated in tissues derived from individuals with SCA1 [27, 65, 92, 112, 114, 208], SCA2 [24, 129], MJD [25, 78, 112, 114, 178, 179], SCA7 [31, 60, 133], SBMA [2, 13, 78, 178, 179], and DRPLA [65, 66, 78, 144, 176, 178, 179, 184, 190]. In common with HD, many studies report greater levels of size mosaicism in the brain than in other peripheral tissues, although smaller and less varied CAG repeat sizes occur in the cerebellum [27, 65, 78, 112, 129, 144, 176, 179, 184]. The exception is SBMA where, although overall levels of mutation length variability in somatic tissues are lower than in other polyglutamine diseases [78, 179], mutation lengths appear to be more variable in tissues such as cardiac and skeletal muscle, skin, and prostate than in the central nervous system [178]. At the present time, the observed patterns of somatic mosaicism do not provide direct evidence that mutation length gains contribute to the distinct, tissue-specific cell death and dysfunction observed in the non-HD polyglutamine diseases. However, more sophisticated investigations of somatic length variability may help to resolve this issue. First, sensitive techniques, such as SP-PCR, unlike amplification of bulk genomic DNA, serve to reveal the full range of mutation lengths present in a tissue. Second, approaches like laser capture microdissection can be used to investigate the mutation profiles of defined cell types within the brain. For example, studies of laser-dissected cerebellar cells in DRPLA patients have demonstrated that the expansion lengths in granule cells are significantly less variable than in Purkinje cells and glia [66, 190]. Finally, caution must be applied when interpreting studies of tissue from patients who died after a long disease course, which is most commonly the case in the published literature cited here. In end-stage-disease tissue, the residual cell profile is much altered due to the loss of vulnerable neurons and the proliferation of cells such as astrocytes. Therefore, in order to more clearly define the relationship between somatic mutation length and pathological vulnerability in polyglutamine diseases, key evidence will most likely be provided by studies investigating defined cell types (isolated by techniques such as laser capture microdissection) from candidate brain regions in the rare, early disease cases where cell loss is minimal.

E. General Features of Somatic Mosaicism in Humans

It is clear that somatic mosaicism of many expanded CAG·CTG repeat loci is present to a greater or lesser extent and follows a number of general features: it is highly tissue-specific, expansion-biased, and age-dependent. Moreover, there appears to be a general conservation of tissue specificity with high levels of variation in regions of the brain, with the dramatic exception of the cerebellum. In all cases, the dynamics are consistent with a process of expansion that is mediated by multiple small length changes, for which the frequency is highly allele-length-dependent [135]. Although the expansion process appears to be highly stochastic at the level of the individual cell, it is likely that the mutation rates are high enough so that within a population of cells the process is, in fact, highly deterministic. Thus, relatively small alleles, for which the mutation rate is relatively low, expand only very slowly, and this leads to a very gradual increase in the modal allele length. However, as any one allele expands, its mutation frequency will increase and the chances of it expanding further are enhanced. This can lead to the highly skewed distributions with very long tails of expanded alleles observed in some tissues of HD patients, who inherit relatively short expansions (see Figs. 35-1A and 35-1D) [88]. The rate of expansion in individuals who inherit larger alleles is higher, such that a more processive increase in modal allele length is observed, leading to the skewed distributions commonly observed in DM1 patient blood DNA (see Figs. 35-1B and 35-1E) [135]. Very large alleles have a very high mutation frequency, so that the whole population of cells expands in an apparently synchronous manner in individuals who inherit them (see Figs. 35-1C and 35-1F). Molecular mechanisms that might give rise to such dynamics are discussed in Section VII.

III. DYNAMICS OF SOMATIC MOSAICISM IN HUMAN AND MOUSE CELL MODELS

The analysis of patient samples has yielded a great deal of information regarding the dynamics of somatic mosaicism in humans. However, such studies are ultimately limited by the availability of appropriate samples and confounding effects of patient age, genetic heterogeneity, and unknown environmental effects, coupled with the obvious inability to perform prospective experiments in humans. Thus, a great deal of effort has been invested in establishing experimental models that replicate the genetic instability observed in humans. This includes a vast amount of work in single cell model organisms such as bacteria and yeast, which will be covered in depth in other chapters in this volume. Here data from the analysis of somatic mosaicism in transgenic mice and mammalian cell tissue culture are reviewed.

A. Dynamics of Somatic Mosaicism in Human Tissue Culture

The most in-depth studies reported using human cell cultures have described the analysis of repeat dynamics in Epstein–Barr virus (EBV) immortalized lymphoblastoid cell lines (LBCL) derived from DM1 patients [7, 90]. These lines partially replicate the expansion-biased dynamics observed *in vivo*, with a gradual increase in average repeat length observed with increasing passage number (see Fig. 35-1G). However, they are also subject to two additional, probably not unrelated, phenomena that appear to be artifacts of the cell culture/immortalization process. First, large discontinuous shifts in the average allele length are periodically observed (see Fig. 35-1G). Second, a relatively high frequency of large contractions is detected. The large discontinuous shifts appear to be mediated by *in vitro* selection and clonal expansion of spontaneously arising LBCL cells that carry expanded alleles [90]. Indeed, it has been demonstrated directly that LBCL subclones with larger repeats divide more rapidly than those with smaller expansions. This growth advantage appears to be associated with decreased levels of $p21^{WAF1}$, a negative regulator of the Erk1,2 cell cycle progression protein [90]. Similar effects are not observed in transgenic mouse cell models of expanded DM1 CAG·CTG repeats in which the CTG repeat is not maintained within the context of the *DMPK* transcript (see Section IIIC) and is presumably mediated by the expanded CUG repeat toxic RNA effect [160] specific to myotonic dystrophy. It is also unclear whether such a repeat-length-mediated cell selection contributes somatic mosaicism in blood cells of DM1 patients or is an artifact of the severe selection effects created during *in vitro* culture. The former would appear unlikely, given the relative stability of the repeat in blood cell lineages relative to other postmitotic tissues (see Section IIB). Surprisingly, large contractions observed at high frequency *in vitro* probably occur in cells that do not persist in the culture perhaps due to associated negative selection effects, as the contractions do not accumulate with increasing passage number [7].

Analyses have also been performed on primary human cell lines predominantly derived from fetal tissues [198, 199]. Repeat instability in these lines appears to replicate that observed *in vivo* inasmuch as the alleles expand over time. Despite their potential utility, surprisingly little work has been performed with such lines. This may be due to the finite life span of primary human cell lines and the limited availability of donor samples from which to generate new lines.

B. Dynamics and Tissue Specificity of Somatic Mosaicism in Transgenic Mice

Despite the apparent failure to reproduce genetic instability in some of the early transgenic mouse models carrying expanded CAG·CTG arrays [14, 22, 51, 71], it is now clear that somatic mosaicism can be faithfully reproduced in mice [45, 57, 88, 106, 116, 167, 170, 186, 191, 194]. A number of transgenic lines incorporating flanking sequences from a variety of human disease loci, as well as from several knock-in mouse models, have been generated in which somatic mosaicism has been observed. As in humans, the basic dynamics of somatic mosaicism in all of these models are age-dependent, expansion-biased, and highly tissue-specific. Low levels of mosaicism are detected in newborns, confirming that the repeat is generally very stable during the highly proliferative phase of embryonic development. Expansions are typically detectable in 2- to 3-month-old mice and accumulate with age thereafter. In some tissues of older mice expansions of several hundred repeats are observed, and this is on a scale comparable to that observed in human adults [45, 89, 170, 191]. Interestingly, small deletions are prominent in younger mice [45]. This is consistent with the idea that the bias toward expansion may be relatively subtle, with both small deletions and small expansions common. Even a slight bias toward expansion will eventually lead to net length gains in most cells over time. Also, as in humans, no obvious relationship exists between the tissue specificity of somatic expansion and the proliferative capacity of the tissue [45, 88, 105, 116, 170]. In many different mouse models the patterns of somatic mosaicism are similar, suggesting a major role for tissue-specific *trans*-acting modifiers. Large expanded repeats are often most unstable in peripheral tissues, such as kidney and liver, and relatively stable in tissues such as heart, skeletal muscle, blood, and lung. They also appear to be unstable in many regions of the brain, particularly the striatum, but relatively stable in the cerebellum [45, 89, 105, 167, 170, 191, 194]. However, there are exceptions to these generalizations. Most strikingly, in a transgenic model of DM1 derived from human cosmid clones incorporating arrays of ~300 repeats, the tissue specificity of somatic mosaicism is highly dependent on the integration site [170]. The repeat is somatically unstable in all lines generated following the broad pattern indicated earlier, with one exception where the repeat appears very stable in both liver and kidney, but unstable in skeletal muscle. Thus, it is clear that the tissue specificity of somatic mosaicism is also influenced by *cis*-acting flanking sequence modifiers. The importance of *cis* factors is further highlighted in mouse models of HD. As in humans, a relatively small expansion (70–80 repeats) in a knock-in HD model is largely stable in peripheral tissues, such as liver and kidney, but grossly unstable in specific brain regions, particularly in the striatum and cortex, the two major affected brain regions in HD [89]. These data suggest that major *cis*-acting flanking sequence modifiers of tissue-specific instability at the *HD* locus are conserved between human and mouse. Unexpectedly, however, it has subsequently transpired that allele length is also a major modifier of the tissue specificity of somatic instability. Surprisingly, HD knock-in mice carrying alleles of ~150 repeats demonstrate very high levels of instability in peripheral tissues, such as liver. The magnitude of allele size increases is comparable to, if not greater than, the degree of instability observed in the brain [88]. Moreover, the region-specific patterns of instability within the brain are much less pronounced than those in HD mice carrying smaller alleles. These data suggest that, although tissue-specific somatic mosaicism of small expansions is highly influenced by *cis*-acting flanking sequence modifiers, larger alleles are less influenced by *cis* effects and more responsive to *trans*-acting tissue-specific modifiers [88]. These findings have major implications for interpreting data relating tissue specificity of somatic mosaicism to pathogenesis in humans and animal models, given that model systems typically have much larger inherited allele lengths than human patients. For example, SP-PCR analysis of a knock-in mouse model of SCA1 with an allele length of 154 CAG repeats has been used to reveal large somatic expansions in the striatums of older mice, compared to low levels of variation in the cerebellums of the same animals [191]. It is possible that the large somatic expansions observed in the striatums of these animals may be a product of the very large inherited repeat tract and may not reflect the mutational dynamics of this locus in SCA1 patients, who typically inherit alleles of 40–60 repeats.

A potential *cis*-acting modifier through which tissue-specific instability could be mediated is gene transcription. However, little supporting evidence exists in humans because the associated disease genes are ubiquitously expressed. Similarly, no correlation between expression levels of *DMPK* and repeat instability was observed in a DM1 transgenic mouse model [105]. Nonetheless, when considering relationships between the degree of somatic mosaicism and transcription levels or the proliferative capacity of the tissue (or indeed, any other factor), it must be kept in mind that most tissues are not a homogeneous collection of cells, but complex organs composed of multiple cell types with different transcription levels, proliferative capacities, and metabolic properties and, very likely, different repeat dynamics. In support of this supposition, a trimodal distribution of repeat lengths is observed within the kidney of one transgenic model and is assumed to reflect three distinct populations of cells that differ in their ability to metabolize expanded repeats [45]. Likewise, a bimodal distribution of repeat lengths in liver cell DNA is also observed in several other mouse models [88, 105, 106, 117, 170].

C. Dynamics of Somatic Mosaicism in Mouse Tissue Culture

Although it appears that the tissue-specific patterns of somatic mosaicism are not dictated by the proliferative capacity of the cells, it remains possible that such relationships have been masked by the complex nature of intact tissues. To address this possibility, cell lines have been established using transgenic mouse tissue, for example, from the DM1 *Dmt*-D mice [54]. Interestingly, tissue-specific repeat dynamics observed *in vivo* are replicated *in vitro*. Specifically, high levels of expansion are observed in cultured kidney cells, kidney being the tissue in which the highest levels of mosaicism are observed *in vivo* [45]. In contrast, despite having a proliferation rate that is similar to those of the kidney cell lines, very low levels of expansion were observed in cultured lung cells, again mirroring lung cell repeat profiles *in vivo*. Indeed, an allele of ~170 repeats was stably maintained in cultured lung cells for over 100 population doublings [54]. These data definitively establish that cell division is not sufficient to drive instability.

Although these cell lines have proved useful in investigating the dynamics of expanded repeats, the potential for clonal expansion events must be carefully considered when interpreting cell culture experiments [54–56]. Although the cell line repeats generally appear to expand in a manner replicating patterns observed *in vivo*, occasional large jumps in repeat length are observed [54–56], as in the human LBCL cell lines [7, 90]. These events appear to be mediated by the clonal expansion of cells that happen to contain an allele that differs in length from the remaining population of cells. Clonal expansion is most likely mediated by acquired mutations elsewhere in the genome that provide a selective advantage in the competitive environment of *in vitro* cell culture. This interpretation is supported by the fact that both large expansions and large deletions are observed [54–56]. Unfortunately, clonal expansion of this type can grossly compromise the assessment of repeat dynamics, particularly when selective pressure is increased by use of genotoxic agents (see Section VIIIA).

IV. CONSEQUENCES OF SOMATIC MOSAICISM

It has become apparent over the past few years that the degree of somatic mosaicism observed in the CAG·CTG repeat disorders is far greater than previously realized. Is this merely an interesting biological phenomenon, or does it have implications for how these disorders are diagnosed, experiments are interpreted, or the ramifications for downstream pathology are assessed?

A. Age-Dependent Allele Length Measurements

The molecular diagnosis of CAG·CTG repeat disorders and much downstream research are dependent on measuring the repeat length in a given patient, or sample experimental system. Clearly, the phenomenon of somatic mosaicism has the potential to complicate allele length measurements. Indeed, the phenomenon of mosaicism raises the question of which allele length measurements are appropriate. Should we measure the most abundant allele? The allele size inherited from the affected parent? The allele size present in blood? The allele size present in the affected tissue? In terms of molecular diagnosis and subsequent use of the data in determining genotype–phenotype relationships, it would seem most sensible to try to identify an individual's progenitor allele size, i.e., the single allele length transmitted through egg or sperm from the affected parent. However, to date, it has been common practice to record the most abundant allele, i.e., the highest peak/strongest band in a PCR assay or the midlength of the smear in a Southern blot analysis of restriction-enzyme-digested genomic blood DNA. However, if somatic mosaicism is age-dependent and expansion-biased, then such allele length measurements will be dependent on the age of sampling, i.e., progenitor allele lengths will be overestimated in individuals sampled at an older age. Moreover, because of anticipation, the age at sampling within a given family is often highly biased: more severely affected individuals who inherit larger alleles are typically sampled at a younger age than those who inherit smaller expansions and develop symptoms later in life.

So, do these concerns have practical implications for the prognostic diagnosis of these disorders? Evidence from DM1 cases suggests that for this disorder, where somatic mosaicism in blood DNA is prominent, the answer is yes. In particular, estimates of intergenerational length changes can be compromised. Experiments in which the measured average repeat lengths in the blood DNA of parent and offspring were compared by using standard methods, approximately 6.5% of transmissions were defined as contractions [5]. If such transmissions truly represented intergenerational contractions, then a reduction in disease severity in the child would be expected. However, in almost 50% of these cases, clinical anticipation was still observed and in none was a later age at onset in the child demonstrated, a phenomenon referred to as the Ashizawa paradox. Almost certainly, this paradox can be explained as an artifact of the expansion-biased nature of somatic mosaicism and age at sampling, i.e., the inherited allele length in the parent sampled at an older age was overestimated relative to that of the child sampled at an earlier age. Indeed, several detailed studies of parent–child pairs, in which somatic mosaicism has been used to estimate the progenitor allele length, are entirely consistent with this interpretation [121, 124, 135].

Although it is clear that there is a highly significant inverse correlation between measured allele length and age at onset in DM1, variation in measured allele length typically accounts for less than 50% of the variation in age at onset [36, 120, 156]. These data leave open the possibilities that allele length is not the major predictor of disease severity in DM1 and that some other factor is even more important. Indeed, data have been presented that support a threshold level of ~250–400 repeats, beyond which allele length has no effect on disease severity [62, 70, 168]. Despite this, it seems very likely that inherited repeat length is the major modifier of disease severity in DM1 and that there is no threshold effect because current genotype–phenotype relationships in DM1 are significantly compromised by a failure to account for age-at-sampling biases inherent in sampling patients. Consistent with this interpretation, the average allele length measured in skeletal muscle DNA shows a poorer correlation with age at onset than that provided by blood DNA data [202]. This is presumably because the degree of mosaicism in skeletal muscle DNA is even greater than that in blood DNA, and so the measured allele length in skeletal muscle is more prone to age-at-sampling effects.

Are similar age-at-sampling effects likely to have compromised genotype–phenotype relationships in the other CAG·CTG repeat disorders? In the other CAG·CTG repeat disorders, inherited allele lengths are typically much smaller (<100 repeats) than those observed in many patients with DM1. Although such alleles are somatically unstable, the pattern of mosaicism observed in blood DNA is such that the average allele length probably does not change greatly with time. Thus, it may be concluded that it is unlikely that age-at-sampling effects have greatly compromised genotype–phenotype relationships in the non-DM1 CAG·CTG repeat disorders.

B. Does Expansion-Biased Somatic Mosaicism Contribute to Disease Pathology?

It is clear that, with the exception of SCA8 [95], longer inherited repeat tracts result in increased disease severity and earlier age at onset in all the other CAG·CTG repeat expansion disorders. Similarly, allele length effects observed in cell culture and animal models indicate that larger alleles are inherently more toxic. Thus, the default expectation must be that cells acquiring somatic expansions are predicted to experience an increased pathological load compared to those cells that have not acquired somatic expansions. Therefore, simple logic dictates that the default expectation must be that age-dependent, expansion–biased somatic mosaicism plays a contributing role in the progressive nature of the symptoms. Likewise, the same arguments support a link between the tissue-specific patterns of somatic mosaicism and observed symptomatology. Thus, in disorders such as HD and DM1, where very large somatic expansions are observed in the primary sites of tissue pathology, the default expectation must be that expansion-biased somatic mosaicism is a factor in both the progressive nature and tissue specificity of the symptoms. Of course, although these arguments are logically irreproachable, it remains possible that the relative magnitude of the length changes mediated by somatic mosaicism is not sufficient to significantly influence downstream pathological processes. However, given the magnitude of the length changes observed, this seems profoundly unlikely. The average repeat length in the skeletal muscle cells of an adult onset DM1 patient is usually several thousand repeats [3, 6, 127, 182, 202]. This is at least as large as the allele typically inherited by congenitally affected children [183] and maybe larger, given the general failure to take into account age-dependent somatic mosaicism in genotyping. In HD, the arguments are even more compelling. Intergenerational increases of one repeat unit result in an average decrease in the age at onset of approximately 2 years [61], revealing that the phenotype is spectacularly sensitive to inherited repeat length. It has been demonstrated that the repeat can exceed 1,000 repeats in some striatal cells of patients inheriting 40–50 repeats [88]. It seems improbable that such massive somatic length gains do not have catastrophic consequences for the individual cells that contain them. Consistent with this view is the absence of such large expansions in residual striatal material from end-stage patients [88].

Thus, it seems very likely that somatic mosaicism influences the tissue specificity and progressive nature of DM1 and HD symptoms. If this is not the case, then explanations need to be invoked to account for the fact that somatic expansion has no pathological consequences. Such explanations may include the possibility that large expansions result in reduced levels of mutant transcripts in DM1 and HD and/or suppressed translation of the mutant protein in HD. However, given that the major pathogenic pathways in DM1 and HD appear to be mediated by dominant gain-of-function effects, it might be argued that inhibition of transcription/translation of the mutant transcript would be protective to cells. This does not appear to be the case because large DM1 expansions in skeletal muscle are not protective against muscle pathology, and striatal cells with large HD expansions do not appear to be spared [88]. Alternatively, DM1 transcripts/HD proteins containing very large expansions may be less toxic than those with smaller expansions. However, this suggestion both is counterintuitive and lacks experimental support. Thus, the balance of evidence favors the interpretation that somatic expansion is a contributory factor in the tissue specificity and progressive nature of both HD and DM1.

This conclusion raises the following question: is somatic expansion actually required for disease pathology? The degree of somatic mosaicism in newborns is generally very low, yet individuals inheriting very large expansions may present with symptoms at birth, e.g., congenital cases of DM1 [30, 183] and the infantile form of SCA7 [1, 12, 83]. Clearly, in these cases additional somatic expansion is not required to precipitate symptoms. Is the same likely to be true for adult onset forms of these disorders? At this point it simply is not known because the repeats tend to be somatically unstable in all patients. It is clear that genetic instability does not play a role in many inherited late onset progressive disorders. Nonetheless, in adult onset cases of HD and DM1, where somatic expansion may increase allele length by hundreds or even thousands of repeats in affected tissues [3, 6, 88, 182], it remains possible that somatic mosaicism is actually required to precipitate symptoms.

As discussed previously, a convincing case for somatic expansion playing a role in mediating the tissue specificity and progressive nature of symptoms can be made for DM1 and HD. Is this likely to be true for the other CAG·CTG repeat expansion disorders? This is difficult to assess because a detailed analysis of the somatic dynamics of most of the expanded loci has not been performed using technologies capable of revealing the true extent of mosaicism in humans. Nonetheless, one common finding in many studies is the relative stability of the repeat in the cerebellum compared to in other regions of the brain. This finding questions the role of somatic expansion in mediating the tissue-specific pathology of the other polyglutamine expansion disorders that show significant cerebellar involvement. One explanation is that the cerebellum is the brain region most susceptible to polyglutamine-induced neurotoxicity, and cerebellar ataxia therefore may be the default outcome of an expanded polyglutamine tract in a ubiquitously expressed protein. In this case, HD may not be associated with cerebellar ataxia because the extreme somatic expansions precipitate pathological events in striatal cells before cerebellar Purkinje cells. On this point it is worth noting that, in late-stage disease, most regions of the brain are affected in all disorders. This includes cerebellar degeneration in HD and striatal pathology in the SCAs. Likewise, individuals inheriting very large expansions at any of the polyglutamine repeat loci typically develop a severe juvenile or infantile clinical picture that is very similar across all of the disorders [207]. These cases show widespread patterns of neuropathology and little evidence of the cell-restrictive nature of pathology commonly observed in adult onset forms of the disorders. Indeed, very severe infantile cases (who typically inherit more than 100 repeats) often have pathological symptoms that extend beyond the central nervous system to peripheral tissues [1, 9, 12, 83, 136]. An analogous situation is probably present in many of the mouse models that express very large polyglutamine arrays. In these animals, the tissue-specific neuronal pathology observed in adult onset human patients is poorly conserved because large expansions may override cell-specific patterns of pathology and result in more generalized cytotoxicity.

V. *CIS*-ACTING MODIFIERS OF SOMATIC MOSAICISM

It is well-established that CAG·CTG allele length and repeat tract purity are major modifiers of intralocus genetic instability. Longer alleles are more unstable than shorter ones, and interrupted arrays are much more stable than pure repeat tracts. It is also clear that there are major locus-specific, flanking sequence modifiers of both germline and somatic instability. The most comprehensive supporting data relate to germline instability where clear locus-specific differences, corrected for allele length effects, have been carefully quantified [17]. These data have revealed that expanded repeats at the *HD*, *SCA7*, *SCA2*, and *DM1* loci are much more unstable than repeats at the *MJD*, *SBMA*, *CTG18.1*, and *DRPLA* loci. Intriguingly, these differences in genetic instability are strongly correlated with the GC content of the flanking sequence. The most unstable loci are located within CpG islands and have a high GC content in the immediate flanking DNA. The more stable loci are located in regions depleted of CpG sites and with low GC content. The nature of the link between these variables remains unresolved, but a cause-and-effect relationship between GC/CpG content and instability could be mediated by differences in methylation status and/or chromatin structure.

Unfortunately, equivalent data regarding the detailed quantitative dynamics of somatic mosaicism at the human CAG·CTG loci are not yet available. Nonetheless, preliminary findings indicate that germline and somatic instability of a locus is responsive to a number of similar factors. For instance, although longer alleles are inherited by MJD patients, the degree of somatic mosaicism observed in their brains is lower than that observed in SCA1 patients, who typically inherit shorter alleles [25, 114].

Similar evidence for *cis*-acting flanking sequence modifiers of genetic instability has come from the analysis of mouse models. Transgenes incorporating large expanded repeat arrays with small amounts of human genomic flanking sequence are highly susceptible to the position effects that modulate both germline and somatic instability [45, 116, 134, 205]. Conversely, transgenes that incorporate longer genomic flanking sequences are more resistant to position effects [106, 170]. However, the precise nature of the flanking sequence elements remains elusive. It is possible that, as in the case of intralocus tissue-specific patterns of instability, the interlocus variation in repeat stability could be mediated by differences in transcriptional activity. Support for this hypothesis is provided by the R6 HD transgenic mouse lines. Three lines in which the transgene is expressed show high levels of instability, whereas the one line in which the transgene is not expressed is somatically stable [116]. In contrast, repeats in high-expresser SCA7 cDNA transgenic mouse lines are much more stable than those in SCA7 genomic fragment mouse lines, in which the repeat tract does not appear to be expressed [106]. Thus, it would appear that repeat tract expression per se is neither necessary nor sufficient to drive somatic expansion.

VI. *TRANS*-ACTING GENETIC MODIFIERS OF SOMATIC MOSAICISM

The availability of transgenic mouse lines that model somatic mosaicism, combined with the array of genetically modified mice carrying defined mutations, has provided researchers the opportunity to directly test the role of candidate genes in the expansion pathway. Perhaps not surprisingly, genes involved in DNA repair pathways present as the most obvious candidates for a role in this process.

A. DNA Mismatch Repair Genes Are Required to Generate Somatic Mosaicism

DNA mismatch repair is the major pathway for correcting misincorporation errors that arise during DNA replication [21, 81, 132]. In addition to correcting base–base mismatches, the mismatch repair machinery can also correct small insertion/deletion loops that arise as a product of replication slippage. The critical importance of this pathway in protecting against replication slippage errors is demonstrated in the tumors of patients with hereditary nonpolyposis colon cancer (HNPCC), in which components of the mismatch repair pathway have been inactivated and genome-wide microsatellite destabilization ensues [18, 44, 102, 142, 148]. This microsatellite instability is also present in the tissues of genetically modified mice carrying loss-of-function mutations in a number of mammalian mismatch repair genes [10, 11, 33, 42, 108, 158, 163]. In both situations, microsatellite arrays, comprising ~5–30 repeats of 1–4 bp monomer units, are grossly destabilized and a high frequency of mutant alleles is observed. However, the progenitor and mutant alleles rarely differ in size by more than a few repeats, and both expansions and deletions are generated.

There are two major classes of mismatch repair genes present in mammalian somatic cells: the MutS homologues (*MSH2*, *MSH6*, and *MSH3*) and the MutL homologues (*MLH1*, *PMS1*, *PMS2*, and *MLH3*) [21, 81, 132]. The protein products of both classes function as heterodimers. DNA mismatches are initially recognized by either the MutS-α (MSH2–MSH6) [38] or MutS-β (MSH2–MSH3) heterodimer [49]. Although the absolute roles of MutS-α and MutS-β are not well-defined, they do appear to differ in their binding capacities [49]. MutS-α appears to be predominantly involved in the repair of base–base mismatches and MutS-β in the repair of insertion/deletion loops [81]. MutS complexes bound to mismatches then recruit MutL complexes. The major MutL complex, MutL-α, comprises MLH1 and PMS2. The other MutL complex identified in mammalian cells is MutL-β, comprising MLH1 and PMS1 [162]. MLH1 is also thought to interact with MLH3 to form a third MutL heterodimer in mammalian cells [109]. The precise roles of MutL-β and the putative MLH1/MLH3 heterodimers in somatic cell mismatch repair remain unclear. *Pms1* null mice are defective in some aspects of mismatch repair, presenting with instability of mononucleotide repeats, but not dinucleotide repeats [158]. Conflicting data have been presented on the role of Mlh3 in somatic mismatch repair. A dominant negative allele appears to increase microsatellite instability in tissue culture [109], but no such increase in microsatellite instability was detected in *Mlh3* null mice [108]. Although Mlh3 clearly has a role in meiosis [108], it is nonetheless expressed at high levels in many somatic tissues, consistent with a role in somatic mismatch repair [109].

A number of studies have demonstrated that *Msh2*, *Msh3*, and *Pms2* are all required to generate normal levels of CAG·CTG repeat somatic mosaicism in transgenic mice [53, 119, 169, 186, 195]. *Msh2* deficiency resulted in a complete suppression of expansion in R6/1 mice carrying an HD exon 1 transgene with ~90 repeats and HD knock-in mice with 111 repeats [119, 195]. Interestingly, lack of Msh2 in a transgenic model carrying a very large (>300 repeats) DM1 transgene resulted in complete loss of expansion but, in addition, induced a large increase in the frequency of somatic contractions [169]. The reason for this discrepancy remains unclear, but it could represent either an allele length effect, a background strain effect, or a role for cis-acting modifiers. These data clearly demonstrate that mismatch repair proteins are required to generate somatic expansions. Notably, the inclusion of the MutL homologue *Pms2* in this list suggests that their involvement is not limited to the MutS-mediated stabilization of slipped-strand structures as has previously been proposed [97, 119], but is more likely to implicate a full-scale mismatch repair reaction (see Section VIIC) [53].

B. Other DNA Repair Pathways

Other candidate genes have been tested using transgenic mouse models carrying expanded CAG·CTG repeats and null alleles in DNA repair genes. These include two genes involved in homologous recombination, *Rad52* and *Rad54*, and one gene involved in nonhomologous end-joining, *DNA-PKcs*. None of these genes had detectable effects on the somatic mutational dynamics of a large expanded CAG·CTG repeat [169]. The flap endonuclease FEN1 has also been implicated in mediating expansions in the germline, but no effect on somatic mosaicism of an HD transgene was observed in mice heterozygous for a *Fen1* null allele [174]. It is not possible to definitively determine whether Fen1 is required to generate somatic expansions because homozygous null *Fen1* mice are not viable [98].

C. Strain Effects

Few analyses have been performed to investigate the potential of background strain effects to modulate repeat instability observed in transgenic mouse models. Although modifier effects are less dramatic than those observed with null alleles in the mismatch repair pathway, clear evidence for the existence of strain-specific modifier loci has been reported [186]. These data reveal that natural variation can influence repeat dynamics in mice and may provide another route to identifying genes that play a critical role in the expansion process. Moreover, these data highlight the possibility that the human population may be segregating modifier gene variants that influence repeat expansion. If the hypothesis that proposes somatic mosaicism contributes to both the tissue specificity and progression of symptoms in the CAG·CTG repeat expansion disorders is correct, such modifier gene variants may also modify age at onset and disease severity.

VII. WHAT IS THE MECHANISM OF SOMATIC EXPANSION?

A. Replication Slippage Is Not the Major Mechanism of Expansion

It has long been assumed that the mechanism of repeat expansion is DNA replication slippage [99] (see Fig. 35-2). This is an attractive model, particularly when incorporating the idea that slippage might preferentially occur in Okazaki fragments during lagging strand synthesis [164]. This refinement introduces a mechanism whereby expansion might predominate over deletion and provides an explanation for the repeat instability threshold. Despite the attractiveness of this model and its widespread acceptance [20, 26, 28, 165, 193], no direct evidence exists that replication slippage plays a major role in the expansion of CAG·CTG repeats in humans or mice. Indeed, evidence is now mounting to suggest that DNA replication slippage is not the major pathway mediating genetic instability at the expanded CAG·CTG repeat loci. The replication slippage model makes two major predictions. First, as DNA replication is linked to the cell cycle, expansion should be a cell-cycle-dependent event. Second, slippage products must escape mismatch repair to become fixed as mutations (see Fig. 35-2). Tissue culture experiments have clearly demonstrated that cell division is not sufficient to drive expansion [54], and the accumulation of major expansions in postmitotic tissues, such as neurons and muscle cells in adult humans [3, 6, 88, 182] and mice [45, 89, 116], strongly argues against a requirement for DNA replication/cell division in the expansion pathway. Replication slippage errors should be repaired by the DNA mismatch repair pathway (see Fig. 35-2). Individuals with expanded repeats are fully proficient in DNA mismatch repair; thus, slippage events should be repaired. It is possible that so many DNA replication slippage errors are made at the expanded DNA repeat tracts that the DNA mismatch repair pathway is overwhelmed and/or a fraction of events is missed and that this leads to expansion. If this were the case, loss of function of the mismatch repair pathway should lead to an increase in the expansion rate analogous to the increase in mutation rate observed at microsatellite

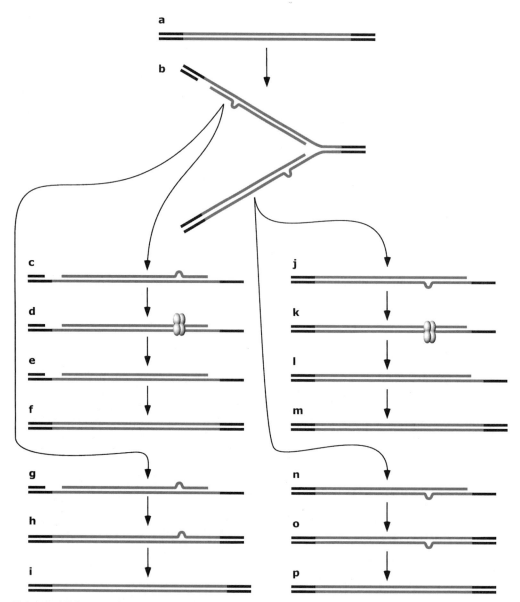

FIGURE 35-2 The DNA replication slippage model. The double-stranded duplex (a) comprises the repeat sequence (red) and unique sequence flanking DNA (black). During replication (b) polymerase slippage may occur, leading to a loop out on the either the daughter strand (c) or the template strand (j). Both slipped-strand structures should be recognized by the DNA mismatch repair pathway (d, k) and the error corrected (e, l), resulting in no net length change (f, m). If, however, the slipped-strand structures (g, n) escape mismatch repair and replication is completed (h, o), then during either a later round of nonreplicative mismatch repair or the next round of DNA replication, the error may be fixed to yield an expansion for daughter strand loop outs (i) or a deletion for template strand loop outs (p). The magnitude of the length changes will be equal to the size of the loop-out structures. Daughter strand loop outs may preferentially occur on the lagging strand when the repeat comprises the entire Okazaki fragment and slippage may occur at both ends [164]. See CD-ROM for color image.

repeats in tumors of HNPCC patients. However, loss of function of major components of the mismatch repair machinery results in suppression of expansion, indicating that DNA mismatch repair proteins are actually required to generate expansions [53, 119, 169, 186, 195]. Thus, it appears clear that a simple replication slippage model cannot account for all aspects of somatic expansion of CAG·CTG repeat loci. This is not to say that replication slippage does not occur, simply that the accumulated evidence does not support a major role for this pathway in mediating the high levels of somatic instability observed *in vivo*.

B. Are DNA Hairpins and/or Other Unusual Structures Mutation Intermediates?

The repetitive nature of expanded CAG·CTG repeats increases the propensity of such sequences to adopt nonstandard B-DNA duplex conformations [96, 149, 192]. Most notably, both single-stranded CAG and CTG repeat tracts tend to fold back on themselves and form hairpin structures [47, 131, 155, 201]. Such structures can present potent blocks to DNA polymerase *in vitro* [143] and have been widely assumed to be mutation intermediates. However, no direct evidence suggests that such structures exist *in vivo* in humans or mice. These mutation intermediates might exist only transiently and, if so, direct demonstration of their existence would be extremely challenging. Importantly though, the *in vivo* data describing mutational dynamics argue against a major role for hairpin structures. CAG and CTG hairpins contain destabilizing A·A and T·T mismatches, respectively, and stable hairpins are only likely to form at physiological temperatures when there are ~4–5 repeats in each arm of the stem. The addition of 1–2 repeats in the loop (one repeat is the minimum length required to make a 180° turn [146]) leads to a minimum total hairpin length of ~10 repeats. By assuming that the size of the hairpin relates to the size of the induced mutation (as it would in the replication slippage model), then CAG·CTG length changes of ~10 repeats should be observed as a result of the expansion process. In contrast, most intergenerational length changes at the CAG·CTG loci are much smaller than 10 repeats [17]. Likewise, the corresponding somatic length change mutations observed in young humans and mice are also less than 10 repeats [45, 89]. Although larger length change mutations accumulate with time, they are more likely to be the result of multiple, frequent small length changes [135] rather than less frequent, large length changes. Moreover, large hairpins are likely to be refractory to simple DNA mismatch repair, which is inefficient at repairing large loops/hairpins [110, 139]. On the other hand, it is possible that hairpin structures are mutation intermediates that are processed by a mechanism that eventually leads to the fixation of only small repeat length changes [147].

DNA hairpins are not the only unusual structures that CAG·CTG repeats can adopt, and it is very probable that nonstandard DNA conformations are adopted at some stage in the expansion pathway. One of the other unusual structures that can be formed is slipped-strand DNA (S-DNA) [151–153, 177] (see Fig. 35-3). S-DNA is a duplex structure in which both strands contain the same number of repeats but the repeat sequences are misaligned, thus generating complementary loops on opposing strands. Although no direct evidence shows

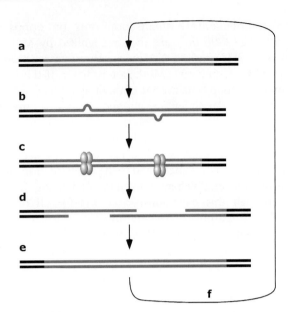

FIGURE 35-3 The inappropriate DNA mismatch repair model. This model posits that the repeat tract is replicated faithfully, yielding a double-stranded duplex (a) comprising the repeat sequence (red) and unique sequence flanking DNA (black). At some point the repeat sequences adopt a slipped-strand structure that comprises complementary loop outs of 1–3 repeats on opposite strands (b). Such loop outs should be recognized by DNA mismatch repair proteins (c). If the sequences opposite the loop outs are excised (d) and the resulting gap filled, a net expansion will result (e) equal in magnitude to the size of the loop out on one of the strands. Multiple small expansions (f) could lead to the accumulation of large expansions over time, even in nondividing cells [53]. See CD-ROM for color image.

that S-DNA exists *in vivo*, it can be readily formed *in vitro* by simply melting and reannealing a duplex repeat sequence [151]. The most common size of loop outs formed is 1–3 repeat units [153]. This falls squarely within the range that is recognized and corrected by the DNA mismatch repair pathway. Furthermore, it has been demonstrated that MSH2 can bind S-DNA [150].

C. Somatic Expansion Could Occur by Inappropriate DNA Mismatch Repair

Accumulating data from humans and mice reveal that somatic mosaicism accrues as the result of multiple small changes in an expansion-biased, length-dependent, DNA mismatch repair-dependent, and cell-division-independent process. The most parsimonious model that can account for these dynamics is inappropriate DNA mismatch repair [53] (see Fig. 35-3). In this model, it is proposed that expanded repeats are faithfully replicated by the cell during S phase, but at some other stage in the cell cycle the repeats adopt an S-DNA conformation with complementary small loop outs on opposing strands. The loop outs are then recognized by MutS-β (MSH2/MSH3) and a DNA mismatch repair reaction is initiated by recruiting a MutL heterodimer. If the complementary loop outs are far enough apart they will be repaired as independent events, and if repair is directed to incorporate the loop out on the opposite strand, a net gain of repeats, equivalent to the size of the original misalignment, will result. Multiple rounds of misalignment and mismatch repair could lead to the accumulation, over time, of large length changes, even in nondividing cells.

Although this mechanism accounts for all of the established features of somatic mosaicism *in vivo*, it is based upon a number of substantial, but testable, assumptions. First, it assumes that S-DNA structures are formed *in vivo*. This might occur during gene transcription, spontaneous DNA breathing, or chromatin remodeling. Data have revealed that chromatin is far from static and that DNA sequences are constantly being unwound and rewound on the histone core [130]. Such chromatin remodeling induces changes in upstream and downstream levels of superhelical tension. Negative superhelical tension can dramatically lower the melting temperature of duplex DNA [107] and might give rise to single-stranded domains within which S-DNA structures could form. Second, the model predicts that a full mismatch repair reaction takes place. It is already established that Msh2, Msh3, and Pms2 are required to generate normal levels of expansion [107]. This model predicts that the other mismatch repair proteins, Mlh1, Mlh3, and/or Pms1, will also be required, in addition to downstream factors such as an exonuclease (probably EXOI [48] or DNA polymerase δ or ϵ [189]), PCNA [185], human replication protein A [159], replication factor C [41], a DNA polymerase (probably DNA polymerase δ) [111], and a DNA ligase. Third, in order to generate a net expansion bias, the mismatch repair reaction must itself be biased toward using the strand containing the loop out as the "correct" template strand. If the strand complementary to the loop out was used as the template for repair, then deletions would ensue. During "normal" postreplication DNA mismatch repair, strand recognition is assumed to be mediated by direct physical linkage to the replication fork [185] and exonuclease activity directed to the newly synthesized DNA by either the free 3′ end on the leading strand or either of the free 3′ and 5′ ends present on Okazaki fragments on the lagging strand. Consistent with this, strand recognition during reconstituted *in vitro* mismatch repair reactions is entirely determined by the placement of a single-stranded nick within the substrate [41]. Although strand recognition during nonreplicative DNA mismatch repair remains an enigma, it could occur through the utilization of spontaneous single-stranded nicks or through the recruitment of an unknown endonuclease. However, it is possible that, in the absence of any strand recognition signals, the system may be inherently biased to incorporate loop outs, rather than deleting potentially important DNA sequences. This bias need only be very subtle to result in net expansions over time.

Whether the mechanism of expansion is really as simple as inappropriate DNA mismatch repair remains to be elucidated. However, progress achieved in understanding the important factors mediating somatic instability will be extended through the use of an impressive array of model systems now established to test competing models.

VIII. SOMATIC MOSAICISM AS A THERAPEUTIC TARGET

Regardless of downstream pathology, by definition, the primary cause of the repeat expansion disorders is the repeat expansion. Therefore, management of repeat length presents as a potential therapeutic target. In particular, the likely role of somatic mosaicism in exacerbating the tissue specificity and progressive nature of the symptoms suggests that suppression of the rate of somatic expansion is also likely to be therapeutically beneficial. Moreover, in all disorders, even those in which somatic mosaicism may not be critical, reversion of the repeat tract to a length that falls below the disease threshold would be predicted to be curative.

A. Can Repeat Expansion Be Modified by Drugs?

It is well-established that many genotoxic compounds affect DNA metabolism, with consequent effects on DNA mutation rates. However, normal mammalian cells are very sensitive to genotoxic assault, and DNA damage detection pathways can trigger an apoptotic response if high levels of DNA damage are induced [29]. Consequently, the levels of induced DNA damage that can be tolerated and repaired are relatively low. Thus, even with an extremely genotoxic agent, the frequency with which DNA damage is induced at a given locus is very low. This is exemplified by analysis of the *HPRT/hprt* locus, which is well-characterized in terms of its susceptibility to induced mutation [69]. This is because negative selection can applied for a functional gene and, consequently, very low mutation rates can be detected using drug selection. Indeed, mutation frequencies as low as 1 in 10^7 can be detected in normal cells under standard growth conditions [69].

However, although mutation frequencies can be enhanced by up to 1,000-fold by genotoxic assault, the highest levels of gene-specific mutation achieved are on the order of 1 in 10^4 cells [37, 58]. Such a low level of induced mutation at an expanded repeat locus would be of limited clinical utility, as it would be desirable to modulate repeat length in the vast majority of cells. Thus, evidence from unique sequence nuclear loci suggests that modulation of repeat dynamics by using genotoxic drugs might not be a productive approach. On the other hand, it is clear that hypervariable tandem repeat minisatellite loci (repeat units on the order of 10–100 bp and array sizes of 1–20 kb), used in DNA fingerprinting/DNA profiling, are particularly susceptible to radiation- and chemical-induced mutations in the mammalian germline [39, 40, 166, 187]. The minisatellite loci have very high spontaneous mutation rates and appear to mutate via a meiotic recombination-based process [80]. Induced mutations are assumed to arise not by direct induction of DNA damage at the minisatellite, but by a trans effect mediated by drug-induced differences in the complement of DNA metabolism/repair enzymes within exposed cells [16]. Thus, it remains possible that expanded CAG·CTG repeats, which have high basal mutation rates in both the germline and soma in normal mammalian cells, might also be particularly susceptible to induced mutations via indirect trans effects. Indeed, data using mammalian cell culture models have revealed that changes in the rate of expansion of expanded CAG·CTG repeats can be induced by using a variety of drugs [55, 64, 157, 199]. Importantly, this group includes chemicals that reduce the rate of expansion as well as those that accelerate the rate of expansion [55]. In addition, preliminary data suggest that some agents may be able to specifically induce deletions in the repeat tract [64]. However, most studies have been conducted with rapidly dividing cell cultures and chronic exposure to drugs over a period of many weeks. These conditions, and the presence of severe cell stress, greatly increase the likelihood that cell selection events may mask effects on basal DNA metabolism and/or generate false positive results [55]. Consequently, the true utility of the agents identified awaits the replication of their effectiveness in an *in vivo* mammalian model.

B. Is DNA Mismatch Repair a Therapeutic Target?

The studies highlighted in Section VIIIA have used known genotoxic agents in an empirical approach that has not generally been based on a detailed understanding of the expansion mechanism. However, as an in-depth understanding of this mechanism emerges, particularly the identification of specific DNA repair proteins that play a crucial role, the rational design of therapeutic agents may be possible. It has been established that the mismatch repair proteins, Msh2 and Msh3, are required for expansion. These proteins therefore present as excellent therapeutic targets, because drugs that disrupt their function may be clinically beneficial. Clearly, given the known role that defects in the mismatch repair pathway have in mediating a cancer-predisposition phenotype, e.g., HNPCC, gross disruption of the mismatch repair pathway would likely yield the highly undesirable side effect of increased cancer predisposition. However, nearly all HNPCC families have defects in either *MSH2* or *MLH1*, whereas a few families have mutations in *MSH6* and *PMS2* [154]. Interestingly, so far no HNPCC families have been identified with defects in *MSH3*. Likewise, in contrast to the gross cancer-predisposition phenotype observed in *Msh2-*, *Mlh1-*, and *Pms2*-deficient mice [10, 11, 33, 42, 108, 158, 163], *Msh3* nulls are not tumor-prone [34]. Therefore, specific inhibition of MSH3 activity, either directly or by preventing its interaction with mutation intermediates, MSH2, or other downstream effector proteins, might be therapeutically beneficial and associated with a low risk of significant side effects.

C. Somatic Mosaicism and Off-Target Drug Effects

The contention that somatic mosaicism may itself be a therapeutic target raises the possibility that a drug directed to another aspect of the pathogenic pathway may have unexpectedly negative consequences on somatic mosaicism. Although this may seem unlikely, understanding of the factors that mediate somatic mosaicism and the true mode of action of many drugs remains limited. Numerous ongoing screens are aimed at identifying drugs that target the pathological consequences of polyglutamine repeat expansion, with most using aggregate formation as a biological marker. The *in vivo* efficacy of candidate drugs is evaluated, not unreasonably, in animal models that carry very large expansions and have a rapid disease course. One such class of compounds identified as having some utility in ameliorating the effects of expanded polyglutamine repeat mediated toxicity is the histone deactylase (HDAC) inhibitors such as SAHA and sodium butyrate [43, 175]. These drugs have proved effective in both *Drosophila* and mouse models of HD. The presumptive mode of action of HDAC inhibitors in HD is to relieve transcriptional repression mediated by expanded polyglutamine inhibition of histone acetyltransferases such as CREB and CBP. Transcriptional repression in HDAC-inhibitor-treated cells is mediated by an opening of the chromatin structure. Given the putative role of open chromatin in facilitating repeat expansion, it is possible to foresee that this class of drugs may have potential side effects on repeat expansion. The potential negative effects on somatic mosaicism might be outweighed by the positive effects on downstream polyglutamine toxicity. Although mouse data support this view, the animal models in these studies carry expansions that are much larger than those typically inherited by HD patients and have a disease course that is orders of magnitude more rapid. It is possible that the relative dynamics of the positive and negative effects of drugs, like HDAC inhibitors, on polyglutamine toxicity and somatic mosaicism could be very different in the slowly progressive disease course in humans. Therefore, it will be important to determine the potential effects that HDAC inhibitors, and any other candidate drugs, have on somatic mosaicism *in vivo* before clinical trials in humans are initiated.

IX. CONCLUDING REMARKS

It is clear from the data summarized here that age-dependent, tissue-specific, expansion-biased somatic mosaicism is a major feature of the CAG·CTG repeat disorders, with significant implications for downstream pathology and the interpretation of animal model data. Importantly, insights into the dynamics of somatic mosaicism have revealed that the process of somatic expansion is, itself, a therapeutic target. Studies in the many excellent models now available confirm the feasibility of this approach and have identified *bona fide* new drug targets. Nonetheless, much remains to be done to determine the full cellular dynamics and role of somatic mosaicism in all of the CAG·CTG repeat expansion disorders, elucidate the precise mechanism of somatic expansion, and utilize this information to develop effective therapies.

Acknowledgments

The authors thank the Dynamic Mutation Group at the University of Glasgow for helpful discussions and the Association Française Contre Les Myopathies, Hereditary Disease Foundation, High Q Foundation, Huntington Disease Association of Great Britain, Lister Institute, Medical Research Council, Muscular Dystrophy Association, Myotonic Dystrophy Support Group, Muscular Dystrophy Campaign, and Wellcome Trust for support.

References

1. Ansorge, O., Giunti, P., Michalik, A., Van Broeckhoven, C., Harding, B., Wood, N., and Scaravilli, F. (2004). Ataxin-7 aggregation and ubiquitination in infantile SCA7 with 180 CAG repeats. *Ann. Neurol.* **56**, 448–452.
2. Ansved, T., Lundin, A., and Anvret, M. (1998). Larger CAG expansions in skeletal muscle compared with lymphocytes in Kennedy disease but not in Huntington disease. *Neurology* **51**, 1442–1444.
3. Anvret, M., Ahlberg, G., Grandell, U., Hedberg, B., Johnson, K., and Edstrom, L. (1993). Larger expansions of the CTG repeat in muscle compared to lymphocytes from patients with myotonic dystrophy. *Hum. Mol. Genet.* **2**, 1397–1400.
4. Aronin, N., Chase, K., Young, C., Sapp, E., Schwarz, C., Matta, N., Kornreich, R., Landwehrmeyer, B., Bird, E., Beal, M. F., et al. (1995). CAG expansion affects the expression of mutant Huntingtin in the Huntington's disease brain. *Neuron* **15**, 1193–1201.
5. Ashizawa, T., Anvret, M., Baiget, M., Barcelo, J. M., Brunner, H., Cobo, A. M., Dallapiccola, B., Fenwick, R. G., Jr., Grandell, U., Harley, H., et al. (1994). Characteristics of intergenerational contractions of the CTG repeat in myotonic dystrophy. *Am. J. Hum. Genet.* **54**, 414–423.
6. Ashizawa, T., Dubel, J. R., and Harati, Y. (1993). Somatic instability of CTG repeat in myotonic dystrophy. *Neurology* **43**, 2674–2678.
7. Ashizawa, T., Monckton, D. G., Vaishnav, S., Patel, B. J., Voskova, A., and Caskey, C. T. (1996). Instability of the expanded (CTG)n repeats in the myotonin protein kinase gene in cultured lymphoblastoid cell lines from patients with myotonic dystrophy. *Genomics* **36**, 47–53.
8. Aslanidis, C., Jansen, G., Amemiya, C., Shutler, G., Mahadevan, M., Tsilfidis, C., Chen, C., Alleman, J., Wormskamp, N. G. M., Vooijs, M., et al. (1992). Cloning of the essential myotonic dystrophy region and mapping of the putative defect. *Nature* **355**, 548–550.
9. Babovic-Vuksanovic, D., Snow, K., Patterson, M. C., and Michels, V. V. (1998). Spinocerebellar ataxia type 2 (SCA2) in an infant with extreme CAG repeat expansion. *Am. J. Med. Genet.* **79**, 383–387.
10. Baker, S., Bronner, C. E., Zhang, L., Plug, A. W., Robatzek, M., Warren, G., Elliott, E. A., Yu, J., Ashley, T., Arnheim, N., et al. (1995). Male mice defective in the DNA mismatch repair gene *PMS2* exhibit abnormal chromosome synapsis in meiosis. *Cell* **82**, 309–319.
11. Baker, S. M., Plug, A. W., Prolla, T. A., Bronner, C. E., Harris, A. C., Yao, X., Christie, D. M., Monell, C., Arnheim, N., Bradley, A., et al. (1996). Involvement of mouse Mlh1 in DNA mismatch repair and meiotic crossing over. *Nat. Genet.* **13**, 336–342.
12. Benton, C. S., de Silva, R., Rutledge, S. L., Bohlega, S., Ashizawa, T., and Zoghbi, H. Y. (1998). Molecular and clinical studies in SCA-7 define a broad clinical spectrum and the infantile phenotype. *Neurology* **51**, 1081–1086.
13. Biancalana, V., Serville, F., Pommier, J., Julien, J., Hanauer, A., and Mandel, J. L. (1992). Moderate instability of the trinucleotide repeat in spino bulbar muscular atrophy. *Hum. Mol. Genet.* **1**, 255–258.
14. Bingham, P. M., Scott, M. O., Wang, S., McPhaul, M. J., Wilson, E. M., Garben, J. Y., Merry, D. E., and Fischbeck, K. H. (1995). Stability of an expanded trinucleotide repeat in the androgen receptor gene in transgenic mice. *Nat. Genet.* **9**, 191–196.
15. Breschel, T. S., McInnis, M. G., Margolis, R. L., Sirugo, G., Corneliussen, B., Simpson, S. G., McMahon, F., MacKinnon, D. F., Xu, J. F., Pleasant, N., et al. (1997). A novel, heritable, expanding

CTG repeat in an intron of the SEF2 1 gene on chromosome 18q21.1. *Hum. Mol. Genet.* **6**, 1855–1863.
16. Bridges, B. A. (2001). Radiation and germline mutation at repeat sequences: Are we in the middle of a paradigm shift? *Radiat. Res.* **156**, 631–641.
17. Brock, G. J. R., Anderson, N. H., and Monckton, D. G. (1999). Cis-acting modifiers of expanded CAG/CTG triplet repeat expandability: Associations with flanking GC content and proximity to CpG islands. *Hum. Mol. Genet.* **8**, 1061–1067.
18. Bronner, C. E., Baker, S. M., Morrison, P. T., Warren, G., Smith, L. G., Lescoe, M. K., Kane, M., Earabino, C., Lipford, J., Lindblom, A., et al. (1994). Mutation in the DNA mismatch repair gene homologue hMLH1 is associated with hereditary non-polyposis colon cancer. *Nature* **368**, 258–261.
19. Brook, J. D., McCurrach, M. E., Harley, H. G., Buckler, A. J., Church, D., Aburatani, H., Hunter, K., Stanton, V. P., Thirion, J.-P., Hudson, T., et al. (1992). Molecular basis of myotonic dystrophy: Expansion of a trinucleotide (CTG) repeat at the 3′ end of a transcript encoding a protein kinase family member. *Cell* **68**, 799–808.
20. Brown, L. Y., and Brown, S. A. (2004). Alanine tracts: The expanding story of human illness and trinucleotide repeats. *Trends Genet.* **20**, 51–58.
21. Buermeyer, A. B., Deschenes, S. M., Baker, S. M., and Liskay, R. M. (1999). Mammalian DNA mismatch repair. *Annu. Rev. Genet.* **33**, 533–564.
22. Burright, E. N., Clark, H. B., Servadio, A., Matilla, T., Feddersen, R. M., Yunis, W. S., Duvick, L. A., Zoghbi, H. Y., and Orr, H. T. (1995). *SCA1* transgenic mice: A model for neurodegeneration caused by an expanded CAG trinucleotide repeat. *Cell* **82**, 937–948.
23. Buxton, J., Shelbourne, P., Davies, J., Jones, C., Tongeren, T. V., Aslanidis, C., de Jong, P., Jansen, G., Anvret, M., Riley, B., et al. (1992). Detection of an unstable fragment of DNA specific to individuals with myotonic dystrophy. *Nature* **355**, 547–548.
24. Cancel, G., Durr, A., Didierjean, O., Imbert, G., Burk, K., Lezin, A., Belal, S., Benomar, A., Abada-Bendib, M., Vial, C., et al. (1997). Molecular and clinical correlations in spinocerebellar ataxia 2: A study of 32 families. *Hum. Mol. Genet.* **6**, 709–715.
25. Cancel, G., Gourfinkel-An, I., Stevanin, G., Didierjean, O., Abbas, N., Hirsch, E., Agid, Y., and Brice, A. (1998). Somatic mosaicism of the CAG repeat expansion in spinocerebellar ataxia type 3/Machado–Joseph disease. *Hum. Mutat.* **11**, 23–27.
26. Chi, L. M., and Lam, S. L. (2005). Structural roles of CTG repeats in slippage expansion during DNA replication. *Nucleic Acids Res.* **33**, 1604–1617.
27. Chong, S. S., McCall, A. E., Cota, J., Subramony, S. H., Orr, H. T., Hughes, M. R., and Zoghbi, H. Y. (1995). Gametic and somatic tissue specific heterogeneity of the expanded *SCA1* CAG repeat in spinocerebellar ataxia type 1. *Nat. Genet.* **10**, 344–350.
28. Cleary, J. D., and Pearson, C. E. (2005). Replication fork dynamics and dynamic mutations: The fork-shift model of repeat instability. *Trends Genet.* **21**, 272–280.
29. Coates, P. J., Lorimore, S. A., and Wright, E. G. (2005). Cell and tissue responses to genotoxic stress. *J. Pathol.* **205**, 221–235.
30. Cobo, A. M., Poza, J. J., Martorell, L., Lopez de Munain, A., Emparanza, J. I., and Baiget, M. (1995). Contribution of molecular analyses to the estimation of the risk of congenital myotonic dystrophy. *J. Med. Genet.* **32**, 105–108.
31. David, G., Abbas, N., Stevanin, G., Dürr, A., Yvert, G., Cancel, G., Weber, C., Imbert, G., Saudou, F., Antoniou, E., et al. (1997). Cloning of the SCA7 gene reveals a highly unstable CAG repeat expansion. *Nat. Genet.* **17**, 65–70.
32. De Rooij, K. E., De Koning Gans, P. A., Roos, R. A., Van Ommen, G. J., and Den Dunnen, J. T. (1995). Somatic expansion of the (CAG)n repeat in Huntington disease brains. *Hum. Genet.* **95**, 270–274.
33. de Wind, N., Dekker, M., Berns, A., Radman, M., and Riele, H. (1995). Inactivation of the mouse *Msh2* gene results in mismatch repair deficiency, methylation tolerance, hyperrecombination and predisposition to cancer. *Cell* **82**, 321–330.
34. de Wind, N., Dekker, M., Claij, N., Jansen, L., van Klink, Y., Radman, M., Riggins, G., van der Valk, M., van't Wout, K., and te Riele, H. (1999). HNPCC-like cancer predisposition in mice through simultaneous loss of Msh3 and Msh6 mismatch-repair protein functions. *Nat. Genet.* **23**, 359–362.
35. Deka, R., Guangyun, S., Wiest, J., Smelser, D., Chunhua, S., Zhong, Y., and Chakraborty, R. (1999). Patterns of instability of expanded CAG repeats at the ERDA1 locus in general populations. *Am. J. Hum. Genet.* **65**, 192–198.
36. Di Leo, R., Rodolico, C., De Gregorio, C., Recupero, A., Coglitore, S., Annesi, G., Toscano, A., Messina, C., and Vita, G. (2004). Cardiovascular autonomic control in myotonic dystrophy type 1: A correlative study with clinical and genetic data. *Neuromuscul. Disord.* **14**, 136–141.
37. Diaz-Llera, S., Podlutsky, A., Osterholm, A. M., Hou, S. M., and Lambert, B. (2000). Hydrogen peroxide induced mutations at the HPRT locus in primary human T-lymphocytes. *Mutat. Res.* **469**, 51–61.
38. Drummond, J. T., Li, G. M., Longley, M. J., and Modrich, P. (1995). Isolation of an hMSH2-p160 heterodimer that restores DNA mismatch repair to tumor cells. *Science* **268**, 1909–1912.
39. Dubrova, Y. E., Jeffreys, A. J., and Malashenko, A. M. (1993). Mouse minisatellite mutations induced by ionizing radiation. *Nat. Genet.* **5**, 92–94.
40. Dubrova, Y. E., Nesterov, V. N., Krouchinsky, N. G., Ostapenko, V. A., Neumann, R., Neil, D. L., and Jeffreys, A. J. (1996). Human minisatellite mutation rate after the Chernobyl accident. *Nature* **380**, 683–686.
41. Dzantiev, L., Constantin, N., Genschel, J., Iyer, R. R., Burgers, P. M., and Modrich, P. (2004). A defined human system that supports bidirectional mismatch-provoked excision. *Mol. Cell.* **15**, 31–41.
42. Edelmann, W., Cohen, P. E., Kane, M., Lau, K., Morrow, B., Bennett, S., Umar, A., Kunkel, T., Cattoretti, G., Chaganti, R., et al. (1996). Meiotic pachytene arrest in MLH1-deficient mice. *Cell* **85**, 1125–1134.
43. Ferrante, R. J., Kubilus, J. K., Lee, J., Ryu, H., Beesen, A., Zucker, B., Smith, K., Kowall, N. W., Ratan, R. R., Luthi-Carter, R., et al. (2003). Histone deacetylase inhibition by sodium butyrate chemotherapy ameliorates the neurodegenerative phenotype in Huntington's disease mice. *J. Neurosci.* **23**, 9418–9427.
44. Fishel, R., Lescoe, M. K., Rao, M. R., Copeland, N. G., Jenkins, N. A., Garber, J., Kane, M., and Kolodner, R. (1993). The human mutator gene homolog MSH2 and its association with hereditary nonpolyposis colon cancer. *Cell* **75**, 1027–1038.
45. Fortune, M. T., Vassilopoulos, C., Coolbaugh, M. I., Siciliano, M. J., and Monckton, D. G. (2000). Dramatic, expansion-biased, age-dependent, tissue-specific somatic mosaicism in a transgenic mouse model of triplet repeat instability. *Hum. Mol. Genet.* **9**, 439–445.
46. Fu, Y. H., Pizzuti, A., Fenwick, R. G., King, J., Rajnarayan, S., Dunne, P. W., Dubel, J., Nasser, G. A., Ashizawa, T., de Jong, P., et al. (1992). An unstable triplet repeat in a gene related to myotonic muscular dystrophy. *Science* **255**, 1256–1258.
47. Gacy, A. M., Goellner, G., Juranic, N., Macura, S., and McMurray, C. T. (1995). Trinucleotide repeats that expand in human disease form hairpin structures *in vitro*. *Cell* **81**, 533–540.
48. Genschel, J., Bazemore, L. R., and Modrich, P. (2002). Human exonuclease I is required for 5′ and 3′ mismatch repair. *J. Biol. Chem.* **277**, 13302–13311.

49. Genschel, J., Littman, S. J., Drummond, J. T., and Modrich, P. (1998). Isolation of MutSbeta from human cells and comparison of the mismatch repair specificities of MutSbeta and MutSalpha. *J. Biol. Chem.* **273**, 19895–19901.
50. Giovannone, B., Sabbadini, G., Di Maio, L., Calabrese, O., Castaldo, I., Frontali, M., Novelleto, A., and Squitieri, F. (1997). Analysis of (CAG)n size heterogeneity in somatic and sperm cell DNA from intermediate and expanded Huntington disease gene carriers. *Hum. Mutat.* **10**, 458–464.
51. Goldberg, Y. P., Kalchman, M. A., Metzler, M., Nasir, J., Zeisler, J., Graham, R., Koide, H. B., O'Kusky, J., Sharp, A. H., Ross, C. A., et al. (1996). Absence of disease phenotype and intergenerational stability of the CAG repeat in transgenic mice expressing the human Huntington disease transcript. *Hum. Mol. Genet.* **5**, 177–185.
52. Gomes-Pereira, M., Bidichandani, S. I., and Monckton, D. G. (2004). Analysis of unstable triplet repeats using small-pool polymerase chain reaction. *Methods Mol. Biol.* **277**, 61–76.
53. Gomes-Pereira, M., Fortune, M. T., Ingram, L., McAbney, J. P., and Monckton, D. G. (2004). Pms2 is a genetic enhancer of trinucleotide CAG·CTG repeat somatic mosaicism: implications for the mechanism of triplet repeat expansion. *Hum. Mol. Genet.* **13**, 1815–1825.
54. Gomes-Pereira, M., Fortune, M. T., and Monckton, D. G. (2001). Mouse tissue culture models of unstable triplet repeats: *in vitro* selection for larger alleles, mutational expansion bias and tissue specificity, but no association with cell division rates. *Hum. Mol. Genet.* **10**, 845–854.
55. Gomes-Pereira, M., and Monckton, D. G. (2004). Chemically induced increases and decreases in the rate of expansion of a CAG*CTG triplet repeat. *Nucleic Acids Res.* **32**, 2865–2872.
56. Gomes-Pereira, M., and Monckton, D. G. (2004). Mouse tissue culture models of unstable triplet repeats. *Methods Mol. Biol.* **277**, 215–228.
57. Gourdon, G., Radvanyi, F., Lia, A. S., Duros, C., Blanche, M., Abitbol, M., Junien, C., and Hofmann Radvanyi, H. (1997). Moderate intergenerational and somatic instability of a 55 CTG repeat in transgenic mice. *Nat. Genet.* **15**, 190–192.
58. Greber, B., Lehrach, H., and Himmelbauer, H. (2003). Characterization of trimethylpsoralen as a mutagen for mouse embryonic stem cells. *Mutat. Res.* **525**, 67–76.
59. Grewal, R. P., Leeflang, E. P., Zhang, L., and Arnheim, N. (1998). The mutation properties of spinal and bulbar muscular atrophy disease alleles. *Neurogenetics* **1**, 249–252.
60. Gu, W., Wang, Y., Liu, X., Zhou, B., Zhou, Y., and Wang, G. (2000). Molecular and clinical study of spinocerebellar ataxia type 7 in Chinese kindreds. *Arch. Neurol.* **57**, 1513–1518.
61. Gusella, J. F., and MacDonald, M. E. (2000). Molecular genetics: Unmasking polyglutamine triggers in neurodegenerative disease. *Nat. Rev. Neurosci.* **1**, 109–115.
62. Hamshere, M. G., Harley, H., Harper, P., Brook, J. D., and Brookfield, J. F. (1999). Myotonic dystrophy: The correlation of (CTG) repeat length in leucocytes with age at onset is significant only for patients with small expansions. *J. Med. Genet.* **36**, 59–61.
63. Harley, H. G., Brook, J. D., Rundle, S. A., Crow, S., Reardon, W., Buckler, A. J., Harper, P. S., Housman, D., and Shaw, D. J. (1992). Expansion of an unstable DNA region and phenotypic variation in myotonic dystrophy. *Nature* **355**, 545–546.
64. Hashem, V. I., Pytlos, M. J., Klysik, E. A., Tsuji, K., Khajavi, M., Ashizawa, T., and Sinden, R. R. (2004). Chemotherapeutic deletion of CTG repeats in lymphoblast cells from DM1 patients. *Nucleic Acids Res.* **32**, 6334–6346.
65. Hashida, H., Goto, J., Kurisaki, H., Mizusawa, H., and Kanazawa, I. (1997). Brain regional differences in the expansion of a CAG repeat in the spinocerebellar ataxias: Dentatorubral pallidoluysian atrophy, Machado–Joseph disease, and spinocerebellar ataxia type 1. *Ann. Neurol.* **41**, 505–511.
66. Hashida, H., Goto, J., Suzuki, T., Jeong, S., Masuda, N., Ooie, T., Tachiiri, Y., Tsuchiya, H., and Kanazawa, I. (2001). Single cell analysis of CAG repeat in brains of dentatorubral–pallidoluysian atrophy (DRPLA). *J. Neurol. Sci.* **190**, 87–93.
67. Holmes, S. E., O'Hearn, E., and Margolis, R. L. (2003). Why is SCA12 different from other SCAs? *Cytogenet. Genome Res.* **100**, 189–197.
68. Holmes, S. E., O'Hearn, E. E., McInnis, M. G., Gorelick-Feldman, D. A., Kleiderlein, J. J., Callahan, C., Kwak, N. G., Ingersoll-Ashworth, R. G., Sherr, M., Sumner, A. J., et al. (1999). Expansion of a novel CAG trinucleotide repeat in the 5′ region of PPP2R2B is associated with SCA12. *Nat. Genet.* **23**, 391–392.
69. Hou, S. M., Van Dam, F. J., de Zwart, F., Warnock, C., Mognato, M., Turner, J., Podlutskaja, N., Podlutsky, A., Becker, R., Barnett, Y., et al. (1999). Validation of the human T-lymphocyte cloning assay—ring test report from the EU concerted action on HPRT mutation (EUCAHM). *Mutat. Res.* **431**, 211–221.
70. Hsiao, K. M., Chen, S. S., Li, S. Y., Chiang, S. Y., Lin, H. M., Pan, H., Huang, C. C., Kuo, H. C., Jou, S. B., Su, C. C., et al. (2003). Epidemiological and genetic studies of myotonic dystrophy type 1 in Taiwan. *Neuroepidemiology* **22**, 283–289.
71. Ikeda, H., Yamaguchi, M., Sugai, S., Aze, Y., Narumiya, S., and Kakizuka, A. (1996). Expanded polyglutamine in the Machado–Joseph disease protein induces cell death *in vitro* and *in vivo*. *Nat. Genet.* **13**, 196–202.
72. Ikeuchi, T., Sanpei, K., Takano, H., Sasaki, H., Tashiro, K., Cancel, G., Brice, A., Bird, T. D., Schellenberg, G. D., PericakVance, M. A., et al. (1998). A novel long and unstable CAG/CTG trinucleotide repeat on chromosome 17q. *Genomics* **49**, 321–326.
73. Imbert, G., Saudou, F., Yvert, G., Devys, D., Trottier, Y., Garnier, J. M., Weber, C., Mandel, J. L., Cancel, G., Abbas, N., et al. (1996). Cloning of the gene for spinocerebellar ataxia 2 reveals a locus with high sensitivity to expanded CAG/glutamine repeats. *Nat. Genet.* **14**, 285–291.
74. Ishiguro, H., Yamada, K., Sawada, H., Nishii, K., Ichino, N., Sawada, M., Kurosawa, Y., Matsushita, N., Kobayashi, K., Goto, J., et al. (2001). Age-dependent and tissue-specific CAG repeat instability occurs in mouse knock-in for a mutant Huntington's disease gene. *J. Neurosci. Res.* **65**, 289–297.
75. Ishii, S., Nishio, T., Sunohara, N., Yoshihara, T., Takemura, K., Hikiji, K., Tsujino, S., and Sakuragawa, N. (1996). Small increase in triplet repeat length of cerebellum from patients with myotonic dystrophy. *Hum. Genet.* **98**, 138–140.
76. Ishikawa, K., Owada, K., Ishida, K., Fujigasaki, H., Shun Li, M., Tsunemi, T., Ohkoshi, N., Toru, S., Mizutani, T., Hayashi, M., et al. (2001). Cytoplasmic and nuclear polyglutamine aggregates in SCA6 Purkinje cells. *Neurology* **56**, 1753–1756.
77. Ishikawa, K., Watanabe, M., Yoshizawa, K., Fujita, T., Iwamoto, H., Yoshizawa, T., Harada, K., Nakamagoe, K., Komatsuzaki, Y., Satoh, A., et al. (1999). Clinical, neuropathological, and molecular study in two families with spinocerebellar ataxia type 6 (SCA6). *J. Neurol. Neurosurg. Psychiatry* **67**, 86–89.
78. Ito, Y., Tanaka, F., Yamamoto, M., Doyu, M., Nagamatsu, M., Riku, S., Mitsuma, T., and Sobue, G. (1998). Somatic mosaicism of the expanded CAG trinucleotide repeat in mRNAs for the responsible gene of Machado Joseph disease (MJD), dentatorubral pallidoluysian atrophy (DRPLA), and spinal and bulbar muscular atrophy (SBMA). *Neurochem. Res.* **23**, 25–32.
79. Jansen, G., Willems, P., Coerwinkel, M., Nillesen, W., Smeets, H., Vits, L., Howeler, C., Brunner, H., and Wieringa, B. (1994). Gonosomal

mosaicism in myotonic dystrophy patients: Involvement of mitotic events in (CTG)n variation and selection against extreme expansion in sperm. *Am. J. Hum. Genet.* **54**, 575–585.
80. Jeffreys, A. J., Tamaki, K., MacLeod, A., Monckton, D. G., Neil, D. L., and Armour, J. A. L. (1994). Complex gene conversion events in germline mutation at human minisatellites. *Nat. Genet.* **6**, 136–145.
81. Jiricny, J. (2000). Mediating mismatch repair. *Nat. Genet.* **24**, 6–8.
82. Jodice, C., Mantuano, E., Veneziano, L., Trettel, F., Sabbadini, G., Calandriello, L., Francia, A., Spadaro, M., Pierelli, F., Salvi, F., et al. (1997). Episodic ataxia type 2 (EA2) and spinocerebellar ataxia type 6 (SCA6) due to CAG repeat expansion in the CACNA1A gene on chromosome 19p. *Hum. Mol. Genet.* **6**, 1973–1978.
83. Johansson, J., Forsgren, L., Sandgren, O., Brice, A., Holmgren, G., and Holmberg, M. (1998). Expanded CAG repeats in Swedish spinocerebellar ataxia type 7 (SCA7) patients: Effect of CAG repeat length on the clinical manifestation. *Hum. Mol. Genet.* **7**, 171–176.
84. Joseph, J. T., Richards, C. S., Anthony, D. C., Upton, M., Perez-Atayde, A. R., and Greenstein, P. (1997). Congenital myotonic dystrophy pathology and somatic mosaicism. *Neurology* **49**, 1457–1460.
85. Kahlem, P., and Djian, P. (2000). The expanded CAG repeat associated with juvenile Huntington disease shows a common origin of most or all neurons and glia in human cerebrum. *Neurosci. Lett.* **286**, 203–207.
86. Kawaguchi, Y., Okamoto, T., Taniwaki, M., Aizawa, M., Inoue, M., Katayama, S., Kawakami, H., Nakamura, S., Nishimura, M., Akiguchi, I., et al. (1994). CAG expansions in a novel gene for Machado-Joseph disease at chromosome 14q32.1. *Nat. Genet.* **8**, 221–227.
87. Kaytor, M. D., Burright, E. N., Duvick, L. A., Zoghbi, H. Y., and Orr, H. T. (1997). Increased trinucleotide repeat instability with advanced maternal age. *Hum. Mol. Genet.* **6**, 2135–2139.
88. Kennedy, L., Evans, E., Chen, C. M., Craven, L., Detloff, P. J., Ennis, M., and Shelbourne, P. F. (2003). Dramatic tissue-specific mutation length increases are an early molecular event in Huntington disease pathogenesis. *Hum. Mol. Genet.* **12**, 3359–3367.
89. Kennedy, L., and Shelbourne, P. F. (2000). Dramatic mutation instability in HD mouse striatum: Does polyglutamine load contribute to cell-specific vulnerability in Huntington's disease? *Hum. Mol. Genet.* **9**, 2539–2544.
90. Khajavi, M., Tari, A. M., Patel, N. B., Tsuji, K., Siwak, D. R., Meistrich, M. L., Terry, N. H., and Ashizawa, T. (2001). "Mitotic drive" of expanded CTG repeats in myotonic dystrophy type 1 (DM1). *Hum. Mol. Genet.* **10**, 855–863.
91. Kinoshita, M., Takahashi, R., Hasegawa, T., Komori, T., Nagasawa, R., Hirose, K., and Tanabe, H. (1996). (CTG)n expansions in various tissues from a myotonic dystrophy patient. *Muscle Nerve* **19**, 240–242.
92. Koefoed, P., Hasholt, L., Fenger, K., Nielsen, J. E., Eiberg, H., Buschard, K., and Sorensen, S. A. (1998). Mitotic and meiotic instability of the CAG trinucleotide repeat in spinocerebellar ataxia type 1. *Hum. Genet.* **103**, 564–569.
93. Koide, R., Ikeuchi, T., Onodera, O., Tanaka, H., Igarashi, S., Endo, K., Takahashi, H., Kondo, R., Ishikawa, A., Hayashi, T., et al. (1994). Unstable expansion of CAG repeat in hereditary dentatorubral-pallidoluysian atrophy (DRPLA). *Nat. Genet.* **6**, 9–13.
94. Koide, R., Kobayashi, S., Shimohata, T., Ikeuchi, T., Maruyama, M., Saito, M., Yamada, M., Takahashi, H., and Tsuji, S. (1999). A neurological disease caused by an expanded CAG trinucleotide repeat in the TATA-binding protein gene: A new polyglutamine disease? *Hum. Mol. Genet.* **8**, 2047–2053.
95. Koob, M. D., Moseley, M. L., Schut, L. J., Benzow, K. A., Bird, T. D., Day, J. W., and Ranum, L. P. (1999). An untranslated CTG expansion causes a novel form of spinocerebellar ataxia (SCA8). *Nat. Genet.* **21**, 379–384.
96. Kovtun, I. V., Goellner, G., and McMurray, C. T. (2001). Structural features of trinucleotide repeats associated with DNA expansion. *Biochem. Cell Biol.* **79**, 325–336.
97. Kovtun, I. V., and McMurray, C. T. (2001). Trinucleotide expansion in haploid germ cells by gap repair. *Nat. Genet.* **27**, 407–411.
98. Kucherlapati, M., Yang, K., Kuraguchi, M., Zhao, J., Lia, M., Heyer, J., Kane, M. F., Fan, K., Russell, R., Brown, A. M., et al. (2002). Haploinsufficiency of Flap endonuclease (Fen1) leads to rapid tumor progression. *Proc. Natl. Acad. Sci. USA* **99**, 9924–9929.
99. Kunkel, T. A. (1993). Nucleotide repeats. Slippery DNA and diseases. *Nature* **365**, 207–208.
100. La Spada, A. R., Wilson, E. M., Lubahn, D. B., Harding, A. E., and Fischbeck, K. H. (1991). Androgen receptor gene mutations in X-linked spinal and bulbar muscular atrophy. *Nature* **352**, 77–79.
101. Lai, Y., Shinde, D., Arnheim, N., and Sun, F. (2003). The mutation process of microsatellites during the polymerase chain reaction. *J. Comput. Biol.* **10**, 143–155.
102. Leach, F. S., Nicolaides, N. C., Papadopoulos, N., Liu, B., Jen, J., Parsons, R., Peltomaki, P., Sistonen, P., Aaltonen, L. A., Nystrom-Lahti, M., et al. (1993). Mutations of a mutS homolog in hereditary nonpolyposis colorectal cancer. *Cell* **75**, 1215–1225.
103. Leeflang, E. P., Tavaré, S., Marjoram, P., Neal, C. O. S., Srinidhi, J., MacDonald, M. E., de Young, M., Wexler, N. S., Gusella, J. F., and Arnheim, N. (1999). Analysis of germline mutation spectra at the Huntington's disease locus supports a mitotic mutation mechanism. *Hum. Mol. Genet.* **8**, 173–183.
104. Leeflang, E. P., Zhang, L., Tavaré, S., Hubert, R., Srinidhi, J., MacDonald, M. E., Myers, R. H., de Young, M., Wexler, N. S., Gusella, J. F., et al. (1995). Single sperm analysis of the trinucleotide repeats in the Huntington's disease gene: Quantification of the mutation frequency and spectrum. *Hum. Mol. Genet.* **4**, 1519–1526.
105. Lia, A. S., Seznec, H., Hofmann-Radvanyi, H., Radvanyi, F., Duros, C., Saquet, C., Blanche, M., Junien, C., and Gourdon, G. (1998). Somatic instability of the CTG repeat in mice transgenic for the myotonic dystrophy region is age dependent but not correlated to the relative intertissue transcription levels and proliferative capacities. *Hum. Mol. Genet.* **7**, 1285–1291.
106. Libby, R. T., Monckton, D. G., Fu, Y. H., Martinez, R. A., McAbney, J. P., Lau, R., Einum, D. D., Nichol, K., Ware, C. B., Ptacek, L. J., et al. (2003). Genomic context drives SCA7 CAG repeat instability, while expressed SCA7 cDNAs are intergenerationally and somatically stable in transgenic mice. *Hum. Mol. Genet.* **12**, 41–50.
107. Lilley, D. M. (1988). DNA opens up—supercoiling and heavy breathing. *Trends Genet.* **4**, 111–114.
108. Lipkin, S. M., Moens, P. B., Wang, V., Lenzi, M., Shanmugarajah, D., Gilgeous, A., Thomas, J., Cheng, J., Touchman, J. W., Green, E. D., et al. (2002). Meiotic arrest and aneuploidy in MLH3-deficient mice. *Nat. Genet.* **31**, 385–390.
109. Lipkin, S. M., Wang, V., Jacoby, R., Banerjee-Basu, S., Baxevanis, A. D., Lynch, H. T., Elliott, R. M., and Collins, F. S. (2000). MLH3: A DNA mismatch repair gene associated with mammalian microsatellite instability. *Nat. Genet.* **24**, 27–35.
110. Littman, S. J., Fang, W. H., and Modrich, P. (1999). Repair of large insertion/deletion heterologies in human nuclear extracts is directed by a 5′ single-strand break and is independent of the mismatch repair system. *J. Biol. Chem.* **274**, 7474–7481.
111. Longley, M. J., Pierce, A. J., and Modrich, P. (1997). DNA polymerase delta is required for human mismatch repair *in vitro*. *J. Biol. Chem.* **272**, 10917–10921.

112. Lopes-Cendes, I., Maciel, P., Kish, S., Gaspar, C., Robitaille, Y., Clark, H. B., Koeppen, A. H., Nance, M., Schut, L., Silveira, I., et al. (1996). Somatic mosaicism in the central nervous system in spinocerebellar ataxia type 1 and Machado–Joseph disease. *Ann. Neurol.* **40**, 199–206.

113. MacDonald, M. E., Barnes, G., Srinidhi, J., Duyao, M. P., Ambrose, C. M., Myers, R. H., Gray, J., Conneally, P. M., Young, A., Penney, J., et al. (1993). Gametic but not somatic instability of CAG repeat length in Huntington's disease. *J. Med. Genet.* **30**, 982–986.

114. Maciel, P., Lopes-Cendes, I., Kish, S., Sequeiros, J., and Rouleau, G. A. (1997). Mosaicism of the CAG repeat in CNS tissue in relation to age at death in spinocerebellar ataxia type 1 and Machado–Joseph disease patients. *Am. J. Hum. Genet.* **60**, 993–996.

115. Mahadevan, M., Tsilfidis, C., Sabourin, L., Shutler, G., Amemiya, C., Jansen, G., Neville, C., Narang, M., Barcelo, J., O'Hoy, K., et al. (1992). Myotonic dystrophy mutation: An unstable CTG repeat in the 3′ untranslated region of the gene. *Science* **255**, 1253–1255.

116. Mangiarini, L., Sathasivam, K., Mahal, A., Mott, R., Seller, M., and Bates, G. P. (1997). Instability of highly expanded CAG repeats in mice transgenic for the Huntington's disease mutation. *Nat. Genet.* **15**, 197–200.

117. Mangiarini, L., Sathasivam, K., Seller, M., Cozens, B., Harper, A., Hetherington, C., Lawton, M., Trottier, Y., Lehrach, H., Davies, S. W., et al. (1996). Exon 1 of the HD gene with an expanded CAG repeat is sufficient to cause a progressive neurological phenotype in transgenic mice. *Cell* **87**, 493–506.

118. Mankodi, A., Urbinati, C. R., Yuan, Q. P., Moxley, R. T., Sansone, V., Krym, M., Henderson, D., Schalling, M., Swanson, M. S., and Thornton, C. A. (2001). Muscleblind localizes to nuclear foci of aberrant RNA in myotonic dystrophy types 1 and 2. *Hum. Mol. Genet.* **10**, 2165–2170.

119. Manley, K., Shirley, T. L., Flaherty, L., and Messer, A. (1999). *Msh2* deficiency prevents *in vivo* somatic instability of the CAG repeat in Huntington disease transgenic mice. *Nat. Genet.* **23**, 471–473.

120. Marchini, C., Lonigro, R., Verriello, L., Pellizzari, L., Bergonzi, P., and Damante, G. (2000). Correlations between individual clinical manifestations and CTG repeat amplification in myotonic dystrophy. *Clin. Genet.* **57**, 74–82.

121. Martorell, L., Gamez, J., Cayuela, M. L., Gould, F. K., McAbney, J. P., Ashizawa, T., Monckton, D. G., and Baiget, M. (2004). Germline mutational dynamics in myotonic dystrophy type 1 males: Allele length and age effects. *Neurology* **62**, 269–274.

122. Martorell, L., Johnson, K. J., Boucher, C. A., and Baiget, M. (1997). Somatic instability of the myotonic dystrophy (CTG)$_n$ repeat during human fetal development. *Hum. Mol. Genet.* **6**, 877–880.

123. Martorell, L., Martinez, J. M., Carey, N., Johnson, K., and Baiget, M. (1995). Comparison of CTG repeat length expansion and clinical progression of myotonic dystrophy over a five year period. *J. Med. Genet.* **32**, 593–596.

124. Martorell, L., Monckton, D. G., Gamez, J., and Baiget, M. (2000). Complex patterns of male germline instability and somatic mosaicism in myotonic dystrophy type 1. *Eur. J. Hum. Genet.* **8**, 423–430.

125. Martorell, L., Monckton, D. G., Gamez, J., Johnson, K. J., Gich, I., Lopez de Munain, A., and Baiget, M. (1998). Progression of somatic CTG repeat length heterogeneity in the blood cells of myotonic dystrophy patients. *Hum. Mol. Genet.* **7**, 307–312.

126. Martorell, L., Monckton, D. G., Sanchez, A., Lopez De Munain, A., and Baiget, M. (2001). Frequency and stability of the myotonic dystrophy type 1 premutation. *Neurology* **56**, 328–335.

127. Massari, A., Gennarelli, M., Menegazzo, E., Pizzuti, A., Silani, V., Mastrogiacomo, I., Pagani, E., Angelini, C., Scarlato, G., Novelli, G., et al. (1995). Postzygotic instability of the myotonic dystrophy p[AGC] in repeat supported by larger expansions in muscle and reduced amplifications in sperm. *J. Neurol.* **242**, 379–383.

128. Matsumura, R., Futamura, N., Fujimoto, Y., Yanagimoto, S., Horikawa, H., Suzumura, A., and Takayanagi, T. (1997). Spinocerebellar ataxia type 6. Molecular and clinical features of 35 Japanese patients including one homozygous for the CAG repeat expansion. *Neurology* **49**, 1238–1243.

129. Matsuura, T., Sasaki, H., Yabe, I., Hamada, K., Hamada, T., Shitara, M., and Tashiro, K. (1999). Mosaicism of unstable CAG repeats in the brain of spinocerebellar ataxia type 2. *J. Neurol.* **246**, 835–839.

130. Mellor, J. (2005). The dynamics of chromatin remodeling at promoters. *Mol. Cell* **19**, 147–157.

131. Mitas, M., Yu, A., Dill, J., Kamp, T. J., Chambers, E. J., and Haworth, I. S. (1995). Hairpin properties of single-stranded DNA containing a GC-rich triplet repeat: (CTG)15. *Nucleic Acids Res.* **23**, 1050–1059.

132. Modrich, P. (1997). Strand-specific mismatch repair in mammalian cells. *J. Biol. Chem.* **272**, 24727–24730.

133. Monckton, D. G., Cayuela, M. L., Gould, F. K., Brock, G. J. R., de Silva, R., and Ashizawa, T. (1999). Very large (CAG)$_n$ DNA repeat expansions in the sperm of two spinocerebellar ataxia type 7 males. *Hum. Mol. Genet.* **8**, 2473–2478.

134. Monckton, D. G., Coolbaugh, M. I., Ashizawa, K., Siciliano, M. J., and Caskey, C. T. (1997). Hypermutable myotonic dystrophy CTG repeats in transgenic mice. *Nat. Genet.* **15**, 193–196.

135. Monckton, D. G., Wong, L.-J. C., Ashizawa, T., and Caskey, C. T. (1995). Somatic mosaicism, germline expansions, germline reversions and intergenerational reductions in myotonic dystrophy males: Small pool PCR analyses. *Hum. Mol. Genet.* **4**, 1–8.

136. Moretti, P., Blazo, M., Garcia, L., Armstrong, D., Lewis, R. A., Roa, B., and Scaglia, F. (2004). Spinocerebellar ataxia type 2 (SCA2) presenting with ophthalmoplegia and developmental delay in infancy. *Am. J. Med. Genet.* **124**, 392–396.

137. Moseley, M. L., Schut, L. J., Bird, T. D., Koob, M. D., Day, J. W., and Ranum, L. P. (2000). SCA8 CTG repeat: En masse contractions in sperm and intergenerational sequence changes may play a role in reduced penetrance. *Hum. Mol. Genet.* **9**, 2125–2130.

138. Myers, R. H., Marans, K. S., and MacDonald, M. E. (1998). Huntington's Disease. *In* "Genetic Instabilities and Hereditary Neurological Diseases" (R. D. Wells and S. T. Warren, Eds.), pp. 301–323. Academic Press, San Diego.

139. Nag, D. K., White, M. A., and Petes, T. D. (1989). Palindromic sequences in heteroduplex DNA inhibit mismatch repair in yeast. *Nature* **340**, 318–320.

140. Nakamoto, M., Takebayashi, H., Kawaguchi, Y., Narumiya, S., Taniwaki, M., Nakamura, Y., Ishikawa, Y., Akiguchi, I., Kimura, J., and Kakizuka, A. (1997). A CAG/CTG expansion in the normal population. *Nat. Genet.* **17**, 385–386.

141. Nakamura, K., Jeong, S. Y., Uchihara, T., Anno, M., Nagashima, K., Nagashima, T., Ikeda, S., Tsuji, S., and Kanazawa, I. (2001). SCA17, a novel autosomal dominant cerebellar ataxia caused by an expanded polyglutamine in TATA-binding protein. *Hum. Mol. Genet.* **10**, 1441–1448.

142. Nicolaides, N. C., Papadopoulos, N., Liu, B., Wei, Y. F., Carter, K. C., Ruben, S. M., Rosen, C. A., Haseltine, W. A., Fleischmann, R. D., Fraser, C. M., et al. (1994). Mutations of two PMS homologues in hereditary nonpolyposis colon cancer. *Nature* **371**, 75–80.

143. Ohshima, K., and Wells, R. D. (1997). Hairpin formation during DNA synthesis primer realignment *in vitro* in triplet repeat

sequences from human hereditary disease genes. *J. Biol. Chem.* **272**, 16798–16806.
144. Onodera, O., Oyake, M., Takano, H., Ikeuchi, T., Igarashi, S., and Tsuji, S. (1995). Molecular cloning of a full-length cDNA for dentatorubral-pallidoluysian atrophy and regional expressions of the expanded alleles in the CNS. *Am. J. Hum. Genet.* **57**, 1050–1060.
145. Orr, H. T., Chung, M.-Y., Banfi, S., Kwiatkowski, T. J., Jr., Servadio, A., Beaudet, A. L., McCall, A. E., Duvick, L. A., Ranum, L. P. W., and Zoghbi, H. Y. (1993). Expansion of an unstable CAG repeat in spinocerebellar ataxia type 1. *Nat. Genet.* **4**, 221–226.
146. Paiva, A. M., and Sheardy, R. D. (2004). Influence of sequence context and length on the structure and stability of triplet repeat DNA oligomers. *Biochemistry* **43**, 14218–14227.
147. Panigrahi, G. B., Lau, R., Montgomery, S. E., Leonard, M. R., and Pearson, C. E. (2005). Slipped (CTG)*(CAG) repeats can be correctly repaired, escape repair or undergo error-prone repair. *Nat. Struct. Mol. Biol.* **12**, 654–662.
148. Papadopoulos, N., Nicolaides, N. C., Wei, Y. F., Ruben, S. M., Carter, K. C., Rosen, C. A., Haseltine, W. A., Fleischmann, R. D., Fraser, C. M., Adams, M. D., et al. (1994). Mutation of a mutL homolog in hereditary colon cancer. *Science* **263**, 1625–1629.
149. Parniewski, P., and Staczek, P. (2002). Molecular mechanisms of TRS instability. *Adv. Exp. Med. Biol.* **516**, 1–25.
150. Pearson, C. E., Ewel, A., Acharya, S., Fishel, R. A., and Sinden, R. R. (1997). Human MSH2 binds to trinucleotide repeat DNA structures associated with neurodegenerative diseases. *Hum. Mol. Genet.* **6**, 1117–1123.
151. Pearson, C. E., and Sinden, R. R. (1996). Alternative structures in duplex DNA formed within the trinucleotide repeats of the myotonic dystrophy and fragile X loci. *Biochemistry* **35**, 5041–5053.
152. Pearson, C. E., Tam, M., Wang, Y. H., Montgomery, S. E., Dar, A. C., Cleary, J. D., and Nichol, K. (2002). Slipped-strand DNAs formed by long (CAG)·(CTG) repeats: Slipped-out repeats and slip-out junctions. *Nucleic Acids Res.* **30**, 4534–4547.
153. Pearson, C. E., Wang, Y. H., Griffith, J. D., and Sinden, R. R. (1998). Structural analysis of slipped-strand DNA (S-DNA) formed in (CTG)n·(CAG)n repeats from the myotonic dystrophy locus. *Nucleic Acids Res.* **26**, 816–823.
154. Peltomaki, P., and Vasen, H. (2004). Mutations associated with HNPCC predisposition—update of ICG-HNPCC/INSiGHT mutation database. *Disease Markers* **20**, 269–276.
155. Petruska, J., Arnheim, N., and Goodman, M. F. (1996). Stability of intrastrand hairpin structures formed by the CAG/CTG class of DNA triplet repeats associated with neurological diseases. *Nucleic Acids Res.* **24**, 1992–1998.
156. Phillips, M. F., Steer, H. M., Soldan, J. R., Wiles, C. M., and Harper, P. S. (1999). Daytime somnolence in myotonic dystrophy. *J. Neurol.* **246**, 275–282.
157. Pineiro, E., Fernandez-Lopez, L., Gamez, J., Marcos, R., Surralles, J., and Velazquez, A. (2003). Mutagenic stress modulates the dynamics of CTG repeat instability associated with myotonic dystrophy type 1. *Nucleic Acids Res.* **31**, 6733–6740.
158. Prolla, T. A., Baker, S. M., Harris, A. C., Tsao, J. L., Yao, X., Bronner, C. E., Zheng, B., Gordon, M., Reneker, J., Arnheim, N., et al. (1998). Tumour susceptibility and spontaneous mutation in mice deficient in Mlh1, Pms1 and Pms2 DNA mismatch repair. *Nat. Genet.* **18**, 276–279.
159. Ramilo, C., Gu, L., Guo, S., Zhang, X., Patrick, S. M., Turchi, J. J., and Li, G. M. (2002). Partial reconstitution of human DNA mismatch repair in vitro: Characterization of the role of human replication protein A. *Mol. Cell. Biol.* **22**, 2037–2046.
160. Ranum, L. P., and Day, J. W. (2004). Myotonic dystrophy: RNA pathogenesis comes into focus. *Am. J. Hum. Genet.* **74**, 793–804.
161. Ranum, L. P., and Day, J. W. (2004). Pathogenic RNA repeats: An expanding role in genetic disease. *Trends Genet.* **20**, 506–512.
162. Raschle, M., Marra, G., Nystrom-Lahti, M., Schar, P., and Jiricny, J. (1999). Identification of hMutLbeta, a heterodimer of hMLH1 and hPMS1. *J. Biol. Chem.* **274**, 32368–32375.
163. Reitmair, A. H., Schmits, R., Ewel, A., Bapat, B., Redston, M., Mitri, A., Waterhouse, P., Mittrucker, H. W., Wakeham, A., Liu, B., et al. (1995). MSH2 deficient mice are viable and susceptible to lymphoid tumours. *Nat. Genet.* **11**, 64–70.
164. Richards, R. I., and Sutherland, G. R. (1994). Simple DNA is not replicated simply. *Nat. Genet.* **6**, 114–116.
165. Ruggiero, B. L., and Topal, M. D. (2004). Triplet repeat expansion generated by DNA slippage is suppressed by human flap endonuclease 1. *J. Biol. Chem.* **279**, 23088–23097.
166. Sadamoto, S., Suzuki, S., Kamiya, K., Kominami, R., Dohi, K., and Niwa, O. (1994). Radiation induction of germline mutation at a hypervariable mouse minisatellite locus. *Int. J. Radiat. Biol.* **65**, 549–557.
167. Sato, T., Oyake, M., Nakamura, K., Nakao, K., Fukusima, Y., Onodera, O., Igarashi, S., Takano, H., Kikugawa, K., Ishida, Y., et al. (1999). Transgenic mice harboring a full-length human mutant DRPLA gene exhibit age-dependent intergenerational and somatic instabilities of CAG repeats comparable with those in DRPLA patients. *Hum. Mol. Genet.* **8**, 99–106.
168. Savic, D., Rakocvic-Stojanovic, V., Keckarevic, D., Culjkovic, B., Stojkovic, O., Mladenovic, J., Todorovic, S., Apostolski, S., and Romac, S. (2002). 250 CTG repeats in DMPK is a threshold for correlation of expansion size and age at onset of juvenile–adult DM1. *Hum. Mutat.* **19**, 131–139.
169. Savouret, C., Brisson, E., Essers, J., Kanaar, R., Pastink, A., te Riele, H., Junien, C., and Gourdon, G. (2003). CTG repeat instability and size variation timing in DNA repair-deficient mice. *EMBO J.* **22**, 2264–2273.
170. Seznec, H., Agbulut, O., Sergeant, N., Savouret, C., Ghestem, A., Tabti, N., Willer, J. C., Ourth, L., Duros, C., Brisson, E., et al. (2001). Mice transgenic for the human myotonic dystrophy region with expanded CTG repeats display muscular and brain abnormalities. *Hum. Mol. Genet.* **10**, 2717–2726.
171. Seznec, H., Lia-Baldini, A. S., Duros, C., Fouquet, C., Lacroix, C., Hofmann-Radvanyi, H., Junien, C., and Gourdon, G. (2000). Transgenic mice carrying large human genomic sequences with expanded CTG repeat mimic closely the DM CTG repeat intergenerational and somatic instability. *Hum. Mol. Genet.* **9**, 1185–1194.
172. Shizuka, M., Watanabe, M., Ikeda, Y., Mizushima, K., Kanai, M., Tsuda, T., Abe, K., Okamoto, K., and Shoji, M. (1998). Spinocerebellar ataxia type 6: CAG trinucleotide expansion, clinical characteristics and sperm analysis. *Eur. J. Neurol.* **5**, 381–387.
173. Sinke, R. J., Ippel, E. F., Diepstraten, C. M., Beemer, F. A., Wokke, J. H., van Hilten, B. J., Knoers, N. V., van Amstel, H. K., and Kremer, H. P. (2001). Clinical and molecular correlations in spinocerebellar ataxia type 6: A study of 24 Dutch families. *Arch. Neurol.* **58**, 1839–1844.
174. Spiro, C., and McMurray, C. T. (2003). Nuclease-deficient FEN-1 blocks Rad51/BRCA1-mediated repair and causes trinucleotide repeat instability. *Mol. Cell. Biol.* **23**, 6063–6074.
175. Steffan, J. S., Bodai, L., Pallos, J., Poelman, M., McCampbell, A., Apostol, B. L., Kazantsev, A., Schmidt, E., Zhu, Y. Z., Greenwald, M., et al. (2001). Histone deacetylase inhibitors arrest

polyglutamine-dependent neurodegeneration in *Drosophila*. *Nature* **413**, 739–743.

176. Takano, H., Onodera, O., Takahashi, H., Igarashi, S., Yamada, M., Oyake, M., Ikeuchi, T., Koide, R., Tanaka, H., Iwabuchi, K., et al. (1996). Somatic mosaicism of expanded CAG repeats in brains of patients with dentatorubral–pallidoluysian atrophy: Cellular population-dependent dynamics of mitotic instability. *Am. J. Hum. Genet.* **58**, 1212–1222.

177. Tam, M., Erin Montgomery, S., Kekis, M., Stollar, B. D., Price, G. B., and Pearson, C. E. (2003). Slipped (CTG)·(CAG) repeats of the myotonic dystrophy locus: Surface probing with anti-DNA antibodies. *J. Mol. Biol.* **332**, 585–600.

178. Tanaka, F., Reeves, M. F., Ito, Y., Matsumoto, M., Li, M., Miwa, S., Inukai, A., Yamamoto, M., Doyu, M., Yoshida, M., et al. (1999). Tissue specific somatic mosaicism in spinal and bulbar muscular atrophy is dependent on CAG repeat length and androgen receptor gene expression level. *Am. J. Hum. Genet.* **65**, 966–973.

179. Tanaka, F., Sobue, G., Doyu, M., Ito, Y., Yamamoto, M., Shimada, N., Yamamoto, K., Riku, S., Hshizume, Y., and Mitsuma, T. (1996). Differential pattern in tissue specific somatic mosaicism of expanded CAG trinucleotide repeat in dentatorubral pallidoluysian atrophy, Machado–Joseph disease, and X linked recessive spinal and bulbar muscular atrophy. *J. Neurol. Sci.* **135**, 43–50.

180. Telenius, H., Kremer, B., Goldberg, Y. P., Theilmann, J., Andrew, S. E., Zeisler, J., Adam, S., Greenberg, C., Ives, E. J., Clarke, L. A., et al. (1994). Somatic and gonadal mosaicism of the Huntington disease gene CAG repeat in brain and sperm. *Nat. Genet.* **6**, 409–414.

181. The Huntington's Disease Collaborative Research Group (1993). A novel gene containing a trinucleotide repeat that is expanded and unstable on Huntington's disease chromosomes. *Cell* **72**, 971–983.

182. Thornton, C. A., Johnson, K. J., and Moxley, R. T. (1994). Myotonic dystrophy patients have larger CTG expansions in skeletal muscle than in leukocytes. *Ann. Neurol.* **35**, 104–107.

183. Tsilfidis, C., MacKenzie, A. E., Mettler, G., Barcelo, J., and Korneluk, R. G. (1992). Correlation between CTG trinucleotide repeat length and frequency of severe congenital myotonic dystrophy. *Nat. Genet.* **1**, 192–195.

184. Ueno, S., Kondoh, K., Kotani, Y., Komure, O., Kuno, S., Kawai, J., Hazama, F., and Sano, A. (1995). Somatic mosaicism of CAG repeat in dentatorubral–pallidoluysian atrophy (DRPLA). *Hum. Mol. Genet.* **4**, 663–666.

185. Umar, A., Buermeyer, A. B., Simon, J. A., Thomas, D. C., Clark, A. B., Liskay, R. M., and Kunkel, T. A. (1996). Requirement for PCNA in DNA mismatch repair at a step preceding DNA resynthesis. *Cell* **87**, 65–73.

186. van Den Broek, W. J., Nelen, M. R., Wansink, D. G., Coerwinkel, M. M., te Riele, H., Groenen, P. J., and Wieringa, B. (2002). Somatic expansion behaviour of the $(CTG)_{(n)}$ repeat in myotonic dystrophy knock-in mice is differentially affected by Msh3 and Msh6 mismatch-repair proteins. *Hum. Mol. Genet.* **11**, 191–198.

187. Vilarino-Guell, C., Smith, A. G., and Dubrova, Y. E. (2003). Germline mutation induction at mouse repeat DNA loci by chemical mutagens. *Mutat. Res.* **526**, 63–73.

188. Vincent, J. B., Kovacs, M., Krol, R., Barr, C. L., and Kennedy, J. L. (1999). Intergenerational CAG repeat expansion at ERDA1 in a family with childhood-onset depression, schizoaffective disorder, and recurrent major depression. *Am. J. Med. Genet.* **88**, 79–82.

189. Wang, H., and Hays, J. B. (2002). Mismatch repair in human nuclear extracts. Time courses and ATP requirements for kinetically distinguishable steps leading to tightly controlled 5′ to 3′ and aphidicolin-sensitive 3′ to 5′ mispair-provoked excision. *J. Biol. Chem.* **277**, 26143–26148.

190. Watanabe, H., Tanaka, F., Doyu, M., Riku, S., Yoshida, M., Hashizume, Y., and Sobue, G. (2000). Differential somatic CAG repeat instability in variable brain cell lineage in dentatorubral pallidoluysian atrophy (DRPLA): A laser-captured microdissection (LCM)-based analysis. *Hum. Genet.* **107**, 452–457.

191. Watase, K., Venken, K. J., Sun, Y., Orr, H. T., and Zoghbi, H. Y. (2003). Regional differences of somatic CAG repeat instability do not account for selective neuronal vulnerability in a knock-in mouse model of SCA1. *Hum. Mol. Genet.* **12**, 2789–2795.

192. Wells, R. D. (1998). DNA structure, triplet repeats, and hereditary neurological diseases. *J. Biochem. Mol. Biol.* **31**, 2–19.

193. Wells, R. D., Dere, R., Hebert, M. L., Napierala, M., and Son, L. S. (2005). Advances in mechanisms of genetic instability related to hereditary neurological diseases. *Nucleic Acids Res.* **33**, 3785–3798.

194. Wheeler, V. C., Auerbach, W., White, J. K., Srinidhi, J., Auerbach, A., Ryan, A., Duyao, M. P., Vrbanac, V., Weaver, M., Gusella, J. F., et al. (1999). Length-dependent gametic CAG repeat instability in the Huntington's disease knock-in mouse. *Hum. Mol. Genet.* **8**, 115–122.

195. Wheeler, V. C., Lebel, L. A., Vrbanac, V., Teed, A., te Riele, H., and MacDonald, M. E. (2003). Mismatch repair gene *Msh2* modifies the timing of early disease in Hdh(Q111) striatum. *Hum. Mol. Genet.* **12**, 273–281.

196. Wong, L.-J. C., Ashizawa, T., Monckton, D. G., Caskey, C. T., and Richards, C. S. (1995). Somatic heterogeneity of the CTG repeat in myotonic dystrophy is age and size dependent. *Am. J. Hum. Genet.* **56**, 114–122.

197. Wong, L. J., and Ashizawa, T. (1997). Instability of the $(CTG)_n$ repeat in congenital myotonic dystrophy. *Am. J. Hum. Genet.* **61**, 1445–1448.

198. Wöhrle, D., Kennerknecht, I., Wolf, M., Enders, H., Schwemmle, S., and Steinbach, P. (1995). Heterogeneity of DM kinase repeat expansion in different fetal tissues and further expansion during cell proliferation *in vitro*: Evidence for a causal involvement of methyl-directed DNA mismatch repair in triplet repeat stability. *Hum. Mol. Genet.* **4**, 1147–1153.

199. Yang, Z., Lau, R., Marcadier, J. L., Chitayat, D., and Pearson, C. E. (2003). Replication inhibitors modulate instability of an expanded trinucleotide repeat at the myotonic dystrophy type 1 disease locus in human cells. *Am. J. Hum. Genet.* **73**, 1092–1105.

200. Yoon, S. R., Dubeau, L., de Young, M., Wexler, N. S., and Arnheim, N. (2003). Huntington disease expansion mutations in humans can occur before meiosis is completed. *Proc. Natl. Acad. Sci. USA* **100**, 8834–8838.

201. Yu, A., Dill, J., and Mitas, M. (1995). The purine-rich trinucleotide repeat sequences d(CAG)15 and d(GAC)15 form hairpins. *Nucleic Acids Res.* **23**, 4055–4057.

202. Zatz, M., Passos-Bueno, M. R., Cerqueira, A., Marie, S. K., Vainzof, M., and Pavanello, R. C. M. (1995). Analysis of the CTG repeat in skeletal muscle of young and adult myotonic dystrophy patients: When does the expansion occur? *Hum. Mol. Genet.* **4**, 401–406.

203. Zhang, L., Fischbeck, K. H., and Arnheim, N. (1995). CAG repeat length variation in sperm from a patient with Kennedy's disease. *Hum. Mol. Genet.* **4**, 303–305.

204. Zhang, L., Leeflang, E. P., Yu, J., and Arnheim, N. (1994). Studying human mutations by sperm typing: Instability of CAG trinucleotide repeats in the human androgen receptor gene. *Nat. Genet.* **7**, 531–535.

205. Zhang, Y., Monckton, D. G., Siciliano, M. J., Connor, T. H., and Meistrich, M. L. (2002). Age and insertion site dependence of

repeat number instability of a human DM1 transgene in individual mouse sperm. *Hum. Mol. Genet.* **11**, 791–798.

206. Zhuchenko, O., Bailey, J., Bonnen, P., Ashizawa, T., Stockton, D. W., Amos, C., Dobyns, W. B., Subramony, S. H., Zoghbi, H. Y., and Lee, C. C. (1997). Autosomal dominant cerebellar ataxia (SCA6) associated with small polyglutamine expansions in the alpha 1A-voltage-dependent calcium channel. *Nat. Genet.* **15**, 62–69.

207. Zoghbi, H. Y., and Orr, H. T. (2000). Glutamine repeats and neurodegeneration. *Annu. Rev. Neurosci.* **23**, 217–247.

208. Zuhlke, C., Hellenbroich, Y., Schaaff, F., Gehlken, U., Wessel, K., Schubert, T., CervosNavarro, J., Pickartz, H., and Schwinger, E. (1997). CAG repeat analyses in frozen and formalin fixed tissues following primer extension preamplification for evaluation of mitotic instability of expanded SCA1 alleles. *Hum. Genet.* **100**, 339–344.

209. Zuhlke, C., Riess, O., Bockel, B., Lange, H., and Thies, U. (1993). Mitotic stability and meiotic variability of the (CAG)n repeat in the Huntington disease gene. *Hum. Mol. Genet.* **2**, 2063–2067.

CHAPTER 36

Transgenic Mouse Models of Unstable Trinucleotide Repeats: Toward an Understanding of Disease-Associated Repeat Size Mutation

MÁRIO GOMES-PEREIRA, LAURENT FOIRY AND GENEVIÈVE GOURDON
Inserm U781, Clinique Maurice Lamy, Hôpital Necker Enfants Malades, 75015 Paris, France

I. Trinucleotide Repeats and Human Disease
II. Repeat Dynamics
   A. Intergenerational Instability
   B. Somatic Instability
III. Mouse Models of Repeat Instability
   A. Initial Attempts
   B. *Cis* Modifiers of Trinucleotide Repeat Instability
   C. *Trans*-Acting Modifiers of Trinucleotide Repeat Instability
   D. Timing of Repeat Instability
IV. Concluding Remarks
   A. Recreation of Intergenerational Instability in Transgenic Mice
   B. Recreation of Somatic Mosaicism in Transgenic Mice
References

Since the identification of unstable trinucleotide repeat expansions as the genetic cause of an increasing number of human diseases, a great effort has focused on the understanding of the molecular mechanisms of triplet repeat size mutation. While intergenerational trinucleotide repeat expansion accounts for the phenomenon of anticipation, tissue-specific, age-dependent, expansion-biased somatic mosaicism may contribute to the phenotypic variability and progressive nature of these disorders. Therefore, unravelling the molecular dynamics of unstable triplet repeats will provide important insight towards the understanding of disease genetics and pathology. Significant advances have been possible due to the development of mouse models of unstable trinucleotide tracts, either by transgene random insertion or homologous recombination. Transgenic mice have provided an excellent tool to investigate the metabolism of expanded trinucleotide sequences, such as the identification of *cis*- and *trans*-acting modifiers of repeat size mutation. Nonetheless, differences in terms of the nature of transgenic sequences inserted into the mouse genome, repeat tract length and insertion site made it difficult to compare the levels of repeat size variability between different mouse models. Indeed, each new animal model might have raised as many questions as it answered. In an attempt to rationalize some of the observations collected by many independent research groups, we provide an overview of the mouse models of expanded trinucleotide repeats generated to date. Overall, observations reported in this chapter strongly support a role for intratrinucleotide features, genomic context, sex, tissue- and cell type-specific modifiers in the control of intergenerational and somatic trinucleotide repeat instability.

I. TRINUCLEOTIDE REPEATS AND HUMAN DISEASE

Since the first identification of a trinucleotide repeat expansion as the genetic basis of a human disease in 1991 [1, 2], an increasing number of disorders have been associated with triplet repeat mutations. Most involve a CAG·CTG repeat expansion, as identified in spinocerebellar ataxia types 1 (SCA1) [3], 2 (SCA2) [4], 3 (SCA3, also known as Machado–Joseph Disease, MJD) [5], and 7 (SCA7) [6], Huntington's disease (HD) [7], dentatorubral–pallidoluysian atrophy (DRPLA) [8], spinal and bulbar muscular atrophy (SBMA) [2], and myotonic dystrophy type 1 (DM1) [9–11]. However, other triplet repeats have also been implicated in human disease: expanded CGG·CCG sequences cause fragile X syndrome (FRAXA) [1], whereas long GAA·TTC trinucleotides are responsible for Friedreich's ataxia (FRDA) [12].

Trinucleotide repeat expansion disorders may be mild to severely debilitating, or even fatal. This diverse clinical severity has led to intense investigation of the route from repeat expansion to disease development. Various molecular pathways are thought to be involved. These pathways include gain of function of polyglutamine tracts [13] and expanded CUG RNA transcripts [14], loss of function of associated genes due to RNA mislocalization [15], transcriptional interference of "sticky" DNA [16] or chromatin remodeling [17], promoter inactivation by DNA methylation [18], and other as yet uncharacterized pathogenic mechanisms [19].

Despite the diverse pathogenic pathways triggered by trinucleotide repeats, these mutations share important genetic properties, most of which reflect common dynamics and molecular mechanisms of repeat length mutation. Improvements in the understanding of any given disorder are therefore likely to have major implications for the understanding of other diseases of the same family. Transgenic mouse models with expanded trinucleotide sequences have been generated over the past decade. This chapter presents the significant advances obtained from intensive studies of these animal models. It also presents a brief parallel with human data to highlight the general principles underlying the intriguing dynamics of expanded trinucleotide repeats.

II. REPEAT DYNAMICS

Trinucleotide repeats are usually polymorphic in normal individuals, with up to 35 repeats. However, once the number of repeats increases into the expanded range associated with disease, the repeated tracts become highly unstable, displaying a marked tendency for further expansion [20, 21].

The instability of expanded trinucleotide repeats was initially revealed by the diffuse hybridization signals observed on Southern blot analysis of genomic DNA restriction fragments. The heterogeneous smears obtained were shown to consist of unresolved alleles of different sizes [22, 23]. The expansion mutation is strongly dependent upon repeat tract length: the probability of expansion increases with the number of tandem repeats, hence the term "dynamic mutations" [24]. Expansion-biased repeat length mutations are observed in the germline and in somatic cells. However, the molecular mechanisms of triplet repeat expansion are not fully understood, and the extent to which the dynamics of different expanded repeats shares common mechanisms of repeat size mutation in the soma and germline is unknown.

A. Intergenerational Instability

Repeats within the normal, nonpathogenic range are transmitted stably between generations, expanding only rarely in steps of one or a few repeats. However, as repeat size amplifies, the probability of further expansion increases dramatically [25]. Therefore, disease-associated alleles typically change in size during transmission from parent to offspring [21].

Expansion-biased intergenerational instability explains clinical anticipation, a feature of most trinucleotide repeat expansion disorders. Anticipation refers to the increased severity of the phenotype and the earlier age at onset observed in successive generations of an affected family. Genetic anticipation results from the tendency of the repeat to increase in length during transmission and from the association between longer repeats, more severe symptoms, and an earlier age at onset [26].

Intergenerational instability is also affected by the parental origin of the disease allele. Most of these disorders, especially those involving smaller repeat expansions (~35–80 repeats), display a greater risk of repeat expansion upon paternal transmission. Disorders with this pattern include SBMA [27], DRPLA [8], SCA7 [28,

29], and HD [30, 31]. In contrast, FRAXA and FRDA, which involve very large expansions (~200–2000 repeats), display mostly maternal transmission [1, 32, 33]. DM1 presents a more complex sex-related pattern of inheritance. Small expansions (from 37 to ~200 repeats) are more unstable when transmitted by the male parent, leading to an excess of transmitting grandfathers in DM1 pedigrees [34], whereas the largest expansions (up to several thousand repeats), associated with congenital DM1, are usually transmitted by the female parent [35].

Thus, trinucleotide repeats display expansion-biased intergenerational instability, dependent on repeat tract length and other unidentified factors associated with the sex of the transmitting parent.

B. Somatic Instability

Repeat size differences between somatic cells have been documented in most trinucleotide repeat disorders. Disease length alleles often show somatic mosaicism in an age-dependent, expansion-biased, and highly tissue-specific manner.

DM1 provides an outstanding example of somatic instability. The high prevalence of the disorder and the availability of large affected families make DM1 a particularly interesting disease for studies of the molecular mechanisms underlying somatic mosaicism. Extensive somatic instability of DM1 CTG repeats has been reported in a wide range of human tissues, including peripheral blood lymphocytes, liver, brain, and heart [35–38]. CAG tissue-specific somatic instability has also been reported in HD [39, 40], SCA1 [41], SCA3 [42], SCA7 [29, 43], DRPLA [44], and SBMA [45, 46] patients. Somatic repeat variability has also been described for expanded CGG·CCG repeats in FRAXA [47, 48] and disease length GAA·TTC alleles in FRDA [49–51]. Interestingly, *FRDA* alleles with large expansions display a strong bias toward contractions in peripheral blood cells [52] that is not observed in other trinucleotide expansion disorders.

In DM1 patients, expansion of the CTG repeat is positively correlated with disease severity [10, 53]. The repeat expansions observed in the most affected tissue—the skeletal muscle—are consistently much longer than those in peripheral blood leukocytes from the same individual [22, 54–56]. In addition, longer mean DM1 repeat lengths and broader ranges of variability have been found in older patients, who usually present with more severe clinical pictures [22, 57, 58]. Similarly, remarkably large repeat expansions accumulate specifically in the striatum of HD patients, the region of the brain most commonly affected in these patients [40]. Therefore, it is likely that somatic mosaicism contributes, at least in part, to the tissue specificity and progressive nature of many of the disorders associated with unstable DNA triplet repeats. Somatic expansion may be required to trigger late onset symptoms in individuals inheriting small alleles.

Neither the molecular mechanisms underlying repeat instability over successive generations nor the bases of tissue-specific repeat dynamics at any of the expanded repeat loci are understood. Further investigation in patients is limited by the availability of appropriate samples throughout the lifetime of an individual and is compromised by interindividual genetic and environmental variations. The development of trinucleotide repeat instability model systems provides a powerful tool for exploring the complex dynamics of simple repetitive DNA sequences.

III. MOUSE MODELS OF REPEAT INSTABILITY

The dynamics of expanded triplet repeats has been investigated by introducing trinucleotide sequences into simple model organisms such as *Escherichia coli* and *Saccharomyces cerevisiae*. Mammalian, bacterial, and yeast cells differ considerably in terms of DNA replication enzymology and chromatin structure. However, the use of genetically and biochemically defined systems in *E. coli* and *S. cerevisiae* has provided valuable insight into the factors affecting the instability of repetitive DNA sequences. Orientation with respect to the origin of replication [59–61], transcription [62], and DNA repair gene mutations [63–66] has been implicated in repeat dynamics. However, studies of trinucleotide repeat mutation in simple model organisms are compromised by the inherent tendency of such repeats to contract in bacteria and yeast, in contrast to the predominantly expansion-biased behavior observed at most loci in humans. In addition, simple model organisms cannot be used to study repeat instability in the germline or during somatic development. More complex models are therefore required for studies of the dynamics of disease-associated trinucleotide repeats. Transgenic mice have been generated to provide mammalian model systems for the assessment of repeat biology *in vivo*. The mouse is an ideal model organism as it has a short generation time, is genetically similar to humans, and can be manipulated genetically. The generation of transgenic animals has provided a unique opportunity to unravel important aspects of the molecular genetics of trinucleotide repeats.

A. Initial Attempts

Investigators first attempted to create murine models of trinucleotide repeat instability by introducing disease gene cDNAs containing expanded CAG·CTG repeat tracts into the mouse genome. Initial characterization of the resulting transgenic mice showed no trinucleotide repeat instability [67–70] and revealed that the expanded tracts were stable in mice, despite repeat lengths in the range known to display significant levels of intergenerational instability in humans. Intergenerational and somatic instability was reproduced in mice only if longer CAG·CTG tracts [71, 72], or moderately sized expansions within their native genomic DNA context [73], were introduced (Table 36-1).

The intergenerational changes in repeat length initially observed in these studies were smaller than those detected in humans. However, some important features of trinucleotide repeat instability were effectively reproduced in some of these mice: the sex of the transmitting parent and age at the time of

TABLE 36-1 Single-Copy Mouse Models of Trinucleotide Repeat Instability Studied Most Extensively

Mouse model	Year	Repeat sequence	Transgene origin and size	Type of transgenesis	Genetic background	Reference numbers
DM20 DM50 DM300	1997	$(CTG)_{20}$ $(CTG)_{50}$ $(CTG)_{300}$	Human *DM1* locus (45 kb)	Random integration	B6D2/F1 × C57BL/6	73, 79
Dmt-D *Dmt*-E	1997	$(CTG)_{162}$	Human *DMPK* 3′ UTR (1 kb)	Random integration	FVB	71
R6	1997	$(CAG)_{110-150}$	*HD* exon 1 (2 kb)	Random integration	CBA × C57BL/6	72
D02	1997	$(CAG)_{82}$	Human *SCA1* cDNA (3.2 kb)	Random integration	FVB/N	83
AR YAC TG1 *AR* YAC TG33	1998	$(CAG)_{45}$	Human *AR* gene (350 kb) Human *AR* gene (150 kb)	Random integration	C57BL/6 × C3H	84
Drm12 Drm21 Drm22	1999	$(CAG)_{78}$ $(CAG)_{76}$ $(CAG)_{77}$	Human *DRPLA* gene (36 kb)	Random integration	C57BL/6	96
Hdh6/Q72 Hdh4/Q80	1999	$(CAG)_{72}$ $(CAG)_{80}$	Human *HD* locus	Knock-in	129/Sv/ter × C57BL/6	86
Hdh^{Q20} Hdh^{Q50} Hdh^{Q92} Hdh^{Q111}	1999	$(CAG)_{20}$ $(CAG)_{50}$ $(CAG)_{92}$ $(CAG)_{111}$	Human *HD* locus	Knock-in	129SvEv/CD1	82
Sca1$^{78Q/2Q}$ Sca1$^{154/2Q}$	2000 2003	$(CAG)_{78}$ $(CAG)_{154}$	Human *SCA1* gene	Knock-in	129/SvEv C57BL/6 × 129/SvEv	88, 89
Fmr1$(CGG)_{98}$	2001	$(CGG)_{98}$	Human *FMR1* gene	Knock-in	FVB C57/B16J	77
Dmpk$(CTG)_{84}$	2002	$(CTG)_{84}$	Human *DMPK* gene	Knock-in	(C57BL/6 × 129/Ola) × C3H × FVB × Balb/C	81
TG296	2002	$(CGG)_9AGG$ $(CGG)_9AGG$ $(CGG)_{72}$	Human *FRM1* gene (5 kb)	Random integration	FVB/N	76
PrP-SCA7 92R RL-SCA7 92R α-SCA7 92R	2003	$(CAG)_{92}$	Human *SCA7* cDNA (4.5 kb) Human *SCA7* gene (13.5 kb) Human *SCA7* gene (8.3 kb)	Random integration	C57BL/6J × C3H/HeJ	85
FRDA YG22	2004	$(GAA)_{190}$	Human *FRDA* gene (370 kb)	Random integration	C57BL/6J	78

breeding were found to influence repeat instability [71, 72]. These encouraging results demonstrated that trinucleotide repeat dynamics could be recreated in mice, opening new avenues toward the understanding of the molecular mechanisms of repeat length mutations in mammalian cells. Numerous transgenic lines were subsequently generated and studied in detail (Table 36-1), providing important insights into complex triplet repeat biology and the role of modifiers affecting repeat stability.

Expanded trinucleotide sequences other than CAG·CTG tracts have been introduced in transgenic mice (Table 36-1). These sequences include expanded CGG·CCG repeats [74–77] and GAA·TTC tracts [78].

Although the instability of trinucleotide repeats is grounded on the intrinsic repetitive nature of the expanded DNA sequence, it is greatly affected by a variety of elements and factors. Triplet repeat dynamics is likely to depend on the tendency of a repetitive sequence to form secondary structures, the location of the repeat within the affected gene, cell- and tissue type-specific factors, parent of origin effects, sequences within and around the repeat, stage of embryonic development, state of epigenetic modification, and possibly other unidentified factors. Such modifiers can be classified as those directly associated with the expanding repeat (*cis*-acting elements) and those that interact with the repeat (*trans*-acting factors). The development of transgenic mouse models has most certainly contributed to the dissection of triplet repeat metabolism and the identification of modifiers.

B. *Cis* Modifiers of Trinucleotide Repeat Instability

Trinucleotide repeat instability is governed by important *cis* elements. Chromosomal position may account for interlocus variation in repeat mutability, whereas differences in mutation rates at a single locus (intralocus variation) may be associated with the number of repeats and the composition of the repeat tract.

1. Intra-trinucleotide Factors: Repeat Length and Sequence

Expanded CAG·CTG repeats were found to be stably transmitted across generations in mice transgenic for the disease-associated gene cDNA for the expanded *androgen receptor (AR)* gene containing 45 CAG repeats [67], the *HD* gene with 44 CAG triplet repeats [70], the *SCA3* gene harboring 79 CAG repeats [69], and the *SCA1* gene with 82 CAG repeats [68]. Taken together, these early results suggested that large repeat numbers, which are usually unstable in humans, are not sufficient to drive trinucleotide repeat dynamics in mice. It has even been proposed that fundamental aspects of triplet repeat biology may differ in mice and humans, with mice requiring longer repeat tracts than humans to display repeat size variation. Special efforts have been made to introduce larger CAG·CTG trinucleotide repeats in mice in order to recreate repeat size instability. Consistent with a greater length threshold for repeat size mutation in mice, significant levels of intergenerational instability were reported in animals carrying randomly integrated fragments of human genomic DNA containing longer trinucleotide sequences: a $(CAG)_{115-150}$ tract derived from the *HD* gene exon 1 in R6 mice [72] and a $(CTG)_{162}$ repeat derived from the 3' UTR of the *DMPK* gene in *Dmt* mice [71] (Table 36-1). In further support of the key role of trinucleotide repeat length in DNA dynamics, the introduction of a large fragment of human genomic DNA derived from the *DM1* locus and containing various numbers of CTG repeats (20, 55, or >300) into transgenic mice has revealed increasing levels of intergenerational and somatic trinucleotide instability with increasing repeat sizes [73, 79, 80]. In particular, no instability was detected in the DM20 line and a low mutation rate upon transmission of 2.8–6.5% has been reported in single-copy DM55 lines, whereas >90% of transmissions resulted in repeat length changes in DM300 mice. The existence of a repeat length threshold for repeat size mutation was also confirmed in a knock-in mouse model of DM1, in which an 11- or 84-CTG repeat sequence was introduced in the murine *Dmpk* gene [81]. The shorter repeat showed no instability, but the longer allele underwent expansion mutation in the soma. Knock-in models of HD with various numbers of CAG repeats (20, 50, 92, and 111) corroborated these results [82]. No intergenerational instability was observed in *Hdh^{Q20}* mice, whereas a low frequency (4%) of repeat length changes was detected in the *Hdh^{Q50}* lines. The highest levels of intergenerational repeat length mutation were observed for the longest transgenic CAG repeats in *Hdh^{Q92}* (55%) and *Hdh^{Q111}* (73%) pedigrees. Similarly, whereas an interrupted *SCA1* cDNA allele containing 30 CAG repeat units was stably transmitted across generations of transgenic mice, an expanded allele with 82 CAGs showed intergenerational repeat instability, with 67% of female transmissions being unstable [83]. Taken together, these results strengthen the hypothesis that trinucleotide repeat length itself is a major determinant of intergenerational instability.

2. Extra-trinucleotide Repeat Factors: The Genomic Context

a. cDNA versus Genomic Constructs

Early reports supported the idea that long transgenic repeats (>110 repeat units) showed higher levels of intergenerational repeat instability [71, 72, 79, 82] than shorter repeats. However, factors other than intra-trinucleotide repeat features have been strongly implicated in the control of DNA dynamics. CAG·CTG repeats of 45 or 55 units within human genomic DNA fragments carrying the entire *AR* [84] or *DMPK* [73] gene, respectively, appear to undergo repeat length changes in about 2–10% of transmissions. In contrast, a (CAG)$_{45}$ repeat tract within the *AR* cDNA environment is stably transmitted across generations [67]. Among all of the CAG·CTG repeat mice, the *AR* YAC (CAG)$_{45}$ [84] and DM55 mice [73] show germline instability with the smallest trinucleotide repeat expansions (Table 36-2), indicating that the length threshold for repeat instability in the mouse may be lowered by the inclusion of the appropriate flanking sequence elements from the human locus. The cDNA context may not be enough to drive repeat instability for an expanded trinucleotide sequence. Indeed, a cDNA construct containing an expanded *SCA1* allele carrying as many as 82 CAG units has been found to be highly stable [68]. The lack of instability in most of these cDNA-transgenic mice implies that the chromosomal context around the repeat plays a major role in controlling instability. The site of integration in the mouse genome may, however, compensate for the lack of important *cis* elements within the cDNA transgene, enhancing repeat length mutation and generating moderate levels of intergenerational instability as described in another *SCA1* cDNA-transgenic mouse line [83].

Of all the CAG polyglutamine repeat diseases, SCA7 shows the highest level of anticipation [43] and therefore provides a good model for studies of the biology of trinucleotide repeats. The importance of the chromatin environment in the control of CAG expansion was assessed

Table 36-2 Summary of Intergenerational Instability in Mouse Models of Expanded Trinucleotide Repeats[a]

Mouse model	Overall mutation rate	Female transmissions			Male transmissions			Reference number
		Mutation rate	Contraction rate and magnitude of length changes	Expansion rate and magnitude of length changes	Mutation rate	Contraction rate and magnitude of length changes	Expansion rate and magnitude of length changes	
AR YAC TG1	10%	16% (*) (**)	15% (1–3)	1% (1)	5% (*)	2% (18–20)	3% (1)	84
AR YAC TG33								
D02	49%	67% (*) (**)	67% (**) (1–9)	0%	3%	N/D	N/D	83
DM300	>90%	92%	15% (1–40)	77% (1–60)	97% (**)	3% (1–25)	94% (1–60)	79
Dmt-D	65%	58% (**)	50% (1–4)	8% (1)	68% (**)	4% (−11)	64% (1–7)	71
Dmt-E	31%	57% (**)	53% (1–7)	4% (2)	9%	0%	9% (1–2)	71
Drm12	17%	27% (**)	27% (1–3)	0%	7% (**)	1% (1–2)	6% (1)	96
Drm21								
Drm22								
Fmr1(CGG)$_{98}$	10%	6%	0%	6% (1–2)	19%	5% (11)	14% (2–6)	77
FRDA YG22	23%	24%	21% (1–5)	3% (1–5)	22% (**)	10% (1–5)	12% (1–5)	78
Hdh6/Q72	25%	27%	25% (1–4)	2% (2)	23%	0%	23% (1–8)	86
Hdh4/Q80								
HdhQ111	73%	87%	77% (1–8)	10% (1–2)	66%	13% (1–2)	53% (1–3)	82
PrP-SCA7 92R	0–17% (*)	10–16% (*)	N/D (1–4)	N/D (1–5)	9–24% (*)	0%	N/D (1–3)	85
RL-SCA7 92R	0–83% (*)	65–82% (*)	N/D (1–8)	N/D (1–10)	66–84% (*)	N/D (1–7)	N/D (1–11)	85
α-SCA7 92R	22–56% (*)	44–65% (*)	N/D (1)	N/D (1–2)	14–47% (*)	N/D (1–3)	N/D (1–2)	85
R6	~65%	N/D	N/D	N/D	N/D (**)	N/D	N/D	72
Sca78Q/2Q	37%	69% (**)	69% (1–6)	<1% (1)	9% (**)	5% (1–4)	4% (1)	88
Sca154Q/2Q	76%	86% (**)	86% (1–19)	0%	71%	44% (1–5)	27% (1–3)	89
TG296	17% (*)	14%	3% (1–2)	11% (1–2)	18%	4% (1–2)	14% (1–2)	76

[a] Abbreviations: N/D, not determined; (*), line-dependent; (**), age-dependent.

by monitoring repeat instability in the germline and soma of mice carrying SCA7 expansions of identical size within different flanking sequences [85]. A (CAG)$_{92}$ transgenic tract derived from the *SCA7* locus, carried by either a cDNA construct or a genomic fragment, was randomly integrated into the mouse genome to investigate the *cis* factors governing repeat dynamics. The cDNA construct included ~80 bp of upstream and ~210 bp of downstream genomic flanking DNA. In contrast, the genomic fragment was 13.5 kb long and encompassed ~5 kb of DNA upstream from the CAG repeat and ~8 kb downstream from the CAG repeat, including an entire CpG island. The intergenerational mutation frequencies of the expanded CAG repeats were 5–7 times higher when the repeats were flanked by their genomic context (RL-SCA7 92R line) than when in their cDNA environment (PrP-SCA7 92R line) (Table 36-2) [85]. In addition, the expanded SCA7 CAG repeat was found to be significantly more likely to expand further than to contract when in the genomic context [85], showing that positional factors influence not only the rate but also the direction of mutation events. Similarly, the genomic construct showed high levels of tissue-specific somatic mosaicism, whereas the cDNA construct remained stable in all tissues analyzed [85]. In conclusion, DNA flanking regions not only affect intergenerational instability but are also necessary for the accumulation of marked somatic mosaicism.

Some of the important *cis* elements required to drive repeat size mutation at human disease loci may have been conserved in the homologous mouse genes. Knock-in mouse models of HD [82, 86, 87], DM1 [81], and SCA1 [88, 89], in which homologous recombination was used to replace the endogenous murine repeat with an expanded CAG·CTG sequence, display trinucleotide repeat instability in the germline and somatic cells. Although one may argue that the mutation rates [81, 86–88] and/or magnitude of intergenerational repeat length changes [81, 82, 86–89] are lower in knock-in mouse models than expected for a similar sized allele at endogenous human loci, important *cis* factors necessary to drive CAG·CTG repeat instability are at least partially conserved in the mouse genome. Sex of parent effects on repeat length transmission [82, 86–89] and dramatic expansion-biased, age-dependent, tissue-specific somatic instability were reproduced, to some extent, in some of these mice [81, 82, 87, 89, 90]. Despite the overall homology between the mouse and human genomes, the *cis* factors influencing repeat instability may not necessarily be conserved between mice and humans for all trinucleotide repeat loci. An *Frda*^{230GAA} knock-in mouse failed to reproduce intergenerational or somatic repeat length variability [91], indicating that the mouse *Frda* locus lacks the key *cis* elements driving the instability of an expanded GAA trinucleotide sequence. Furthermore, very limited intergenerational instability has been reported in *Fmr1* (CGG)$_{98}$ knock-in mice [77]. Thus, not all of the endogenous mouse repeat loci seem to contain the *cis* elements involved in repeat dynamics.

Therefore, major *cis*-acting factors present in the human genome may be required to increase the rate and magnitude of repeat length mutations in mice, as previously suggested [73, 84, 92, 93]. Nonetheless, some trinucleotide repeats appropriately inserted into the homologous mouse genes may undergo repeat length mutation, but only at higher repeat numbers. cDNA constructs seem to lack sequences important for repeat instability, accounting for the absence of repeat length changes observed with most of these constructs.

b. Site of Integration of the Transgene

The role of flanking sequences in the dynamics of expanded trinucleotide repeats is demonstrated by the different levels of instability detected in mouse lines carrying the same transgene, integrated into different genomic locations.

Three single-copy *Dmt* transgenic lines were generated by random integration of an expanded (CTG)$_{162}$ repeat sequence derived from the 3' UTR of the *DMPK* gene, flanked by ~750 bp of human genomic DNA [71]. Two lines in particular, *Dmt*-D and *Dmt*-E (Table 36-1), differing in transgene integration site, display remarkably different levels of intergenerational and somatic instability. The repeat is highly unstable in the somatic tissues of *Dmt*-D mice, but very stable in the *Dmt*-E line [94]. Similarly, the mean mutation frequency in the sperm of *Dmt*-D mice (14.2%) is significantly higher than in *Dmt*-E mice (5.5%) at 8 months of age [95]. These results demonstrate the dependence of *Dmt* transgene mutation frequency on insertion site.

The random insertion of a much larger, 45-kb genomic fragment derived from the *DM1* locus containing either 55 or >300 CTG repeats revealed limited integration-site-dependent germline repeat instability [73, 79]. The remarkably high intergenerational mutation rates (>90%) of longer repeat sequences (>300 CTG) in DM300 mice appeared to be less affected by positional factors [79]. It therefore seems possible that the transmission of larger numbers of repeats, which are intrinsically more unstable, is less sensitive to genomic context. Alternatively, the intergenerational instability of triplet repeats lying within large genomic transgenes (carrying longer flanking human DNA sequences) is more likely to be minimally affected by the integration site. Evidence supporting this view is provided by three transgenic *Drm* lines generated by random integration of a 36-kb cosmid containing an expanded *DRPLA* (CAG)$_{76-78}$ allele; these lines display similar levels of germline instability, regardless of the site of transgene integration [96].

Different patterns of tissue-specific somatic mosaicism were also detected in different DM300 lines [79], suggesting that the dynamics of expanded repeats in the soma of these mice is affected by the genomic environment. When randomly inserted into R6 transgenic mice, a genomic fragment derived from *HD* exon 1, containing 115–150 CAG repeat units, displays different levels of instability, corroborating the hypothesis that positional effects are involved in repeat dynamics [72]. Interestingly, neither somatic nor intergenerational instability was observed in the only R6 line not expressing the transgene, suggesting that repeat length mutation frequencies and transcription may be correlated [72].

The relationship between somatic repeat instability and transgene expression was investigated by detailed analysis in DM55 mice [73]. No correlation was found between the degree of trinucleotide repeat length variability and the level of transgenic transcripts in somatic tissues [80]. Therefore, the association between repeat instability and transgene expression in some models may reflect a requirement for an open chromatin structure, rather than for transcription *per se*. An open chromatin environment may facilitate unusual DNA conformations acting as mutation intermediates, or it may provide access for the machinery involved in expansion of the repeat tract.

A more extensive analysis of the early cDNA-transgenic mouse models might shed light on a possible link between somatic mosaicism and transgene expression. Most triplet repeat-containing cDNA transgenes have been reported to display no significant somatic mosaicism, but detailed studies have not been performed in tissues in which the transgene is expressed under the control of tissue-specific promoters: the Purkinje-cell-specific *Pcp2-L7* gene promoter in *SCA1* [68] and *SCA3* [69] cDNA lines and the CNS-specific *NSE* gene promoter in *AR* cDNA mice [67].

c. Mapping *Cis*-Acting Modifiers of Repeat Instability

Differences in the mutation profiles of mouse lines carrying the same transgene integrated at different sites provide strong evidence for the involvement of positional elements in the control of repeat dynamics [72, 79, 94, 95]. However, it is extremely difficult to carry out a comprehensive study of *cis* modifiers of repeat instability based on a comparison of all of the mouse models generated due to differences in repeat number, nature of the trinucleotide sequence, and genetic background.

Two DM300 lines have been shown to differ in the amount of human genomic DNA at the 5′ end of the integrated fragment. A difference of ~15 kb at the 5′ end of the integrated fragment results in remarkably different profiles of tissue-specific somatic mosaicism, particularly in the liver and kidney [79]. This difference suggests that important sequences regulating trinucleotide repeat stability may be located within the 5′ sequences of the DM1 transgene. Surprisingly, whereas somatic mosaicism in the liver and kidney is modified extensively by these putative *cis* elements, no effect is observed in other tissues.

To identify elements increasing the instability of the genomic SCA7 fragment, most of the flanking DNA downstream from the CAG repeat was deleted from RL-SCA7 92R transgenic mice [85]. The resulting, shorter genomic fragment was inserted into the mouse genome to generate the deleted α-SCA7 92R transgenic line (Table 36-1) and was monitored during parent-to-offspring transmissions. The repeat within the modified transgene was significantly more stable than that in RL-SCA7 92R lines, not only in terms of mutation rate per transmission but also in terms of magnitude of repeat length changes. Although still prevalent, expansions were less frequent within the deleted α-SCA7 92R fragment than in the longer RL-SCA7 92R genomic fragment [85]. Nonetheless, the levels of repeat instability of the SCA7 3′-deleted α-SCA7 92R lines were still greater than those in the PrP-SCA7 92R cDNA lines. This elegant work clearly demonstrated the requirement for *cis* modifiers specific to the human SCA7 genomic context to drive trinucleotide repeat instability in the mouse genome and positioned putative *cis* elements, such as origins of replication, CpG islands, and CTCF binding sites (for a review, see ref. [97]), downstream from the CAG repeat.

d. G1C Context of the Flanking Sequences

Data from human and mouse models have revealed that genomic environment or the sequences flanking the repeat play a critical role in determining whether an expanded repeat is unstable on transmission from parent to offspring or in somatic tissues (for a review, see ref. [97]).

In humans, the best evidence for *cis* control is provided by the observation that, in CAG·CTG repeat expansion disorders, tracts of identical length behave differently, depending upon the locus at which they reside. For example, the length of a $(CAG)_{40}$ tract at the *AR* locus changes in only about 25% of human transmissions, mostly by −4 to +7 repeats [98]. In contrast, *SCA7* alleles containing a $(CAG)_{40}$ tract expand in most transmissions, and the changes may involve the addition of more than 100 repeats [6, 29]. The sequences immediately flanking the repeats are thought to have a major effect on intergenerational repeat instability at different chromosomal loci, but the precise nature of the *cis* modifiers within the flanking DNA sequences is still unknown. *Cis* instability elements may be binding sites for proteins regulating DNA conformation or may themselves participate in the formation of specific DNA structures during DNA replication or repair.

A comparative study of intergenerational instability in CAG·CTG expansion disorders has shown that the mutation rate and the size and direction of length changes differ between human loci, with some loci apparently more mutable and able to expand to greater lengths than others [93]. In particular, higher levels of instability of an expanded CAG·CTG repeat during transmission were found to be directly correlated with higher G+C content of flanking sequences [93]. Moreover, most unstable CAG·CTG loci were shown to be located within CpG islands, [92, 93]. Interestingly, it has been proposed that CpG islands could be initiation sites not only for transcription but also for replication [99, 100].

The basis of the relationship between G+C content and trinucleotide instability remains unclear. However, the GC-rich environment surrounding an unstable repeat may be more susceptible to changes in helical DNA dynamics, leading to an increase in mutation rate [93]. At the highly unstable *SCA7* locus, for instance, this effect may be mediated by the high density of replication-associated motifs, particularly in the region 3′ to the CAG repeat, which is necessary for marked repeat instability in the germline and somatic cells of transgenic mice [85]. In summary, nonrepetitive sequences immediately flanking trinucleotide repeats may influence the structural dynamics of DNA, accessibility for the machinery involved in size mutation, and consequently the level of repeat instability.

C. *Trans*-Acting Modifiers of Trinucleotide Repeat Instability

Experimental evidence has strongly implicated *trans*-acting factors in the dynamics of expanded trinucleotide repeat sequences both in the germline and in somatic tissues. However, the precise identity and mechanism of action of such modifiers remain elusive. Data collected from bacteria and yeast model systems have suggested that candidate modifiers may include proteins involved in DNA replication and repair [65, 66, 101, 102]. The development of mouse models of trinucleotide repeat instability has opened up new research avenues for the investigation of *trans*-acting modifiers of repeat dynamics.

In the next section, an overview of data from mice will be provided, supporting a role for genetic and environmental *trans*-acting modifiers of repeat instability.

1. Sex of Transmitting Parent Effect on Intergenerational Instability

The frequency, magnitude, and direction of trinucleotide repeat size mutation during transmission across generations depend on the sex of the transmitting parent. However, the identity of the factors involved in sex bias during transmission from parent to offspring remains unclear. Despite differences in the frequency of sex-specific mutations and in the magnitude and direction of repeat length changes between humans and mice, parent of origin effects have been reported in some mouse models of unstable trinucleotide repeats (Table 36-2).

In terms of the frequency of intergenerational mutations, higher mutation rates for expanded CAG·CTG tracts have mostly been reported for female germline transmission in numerous transgenic lines [71, 82–84, 88, 89, 96]. In contrast, other studies have found no clear effect of the sex of the transmitting parent on the frequency of intergenerational instability [79, 85, 86].

The sex of the transmitting parent affects both the frequency of intergenerational mutations and the direction of the repeat length changes (Table 36-2). Most of the CAG·CTG transgenic mouse models generated have shown an expansion bias (up to 10 repeats) when expanded CAG·CTG repeats are transmitted through the male germline [71, 77, 79, 82, 85–87], whereas a strong bias toward deletions is usually observed in maternal transmissions (up to 10 repeats) [71, 83, 84, 86–89, 96]. These observations suggest that fundamental aspects of mouse biology may lead to a greater tendency of the expanded CAG·CTG repeat to contract in the female germline of some mouse models.

However, not all mouse models carrying an expanded CAG·CTG tract show the same pattern of sex-specific intergenerational instability. Although female $Sca1^{154Q}$ knock-in mice display a prominent tendency toward the contraction of the repeat, a slight bias toward contraction is also reported in male transmission, in contrast to the models previously described [89]. Transgenic DM300 animals carrying the longest CAG·CTG repeat tracts introduced in mice exhibit an expansion bias in both male and female transmissions and the largest intergenerational length changes [expansions of up to 60 repeat units for a $(CTG)_{>300}$ tract] [79]. Mean intergenerational amplification is larger in male than in female transmission. However, this difference between male and female transmission is reduced for larger repeats due to the tendency toward smaller amplifications with increasing size of the transgenic trinucleotide in transmitting males [79]. Interestingly, in DM1 families, the expansion of repeats containing more than ~500 CTG units is greater after maternal transmission [35, 103], suggesting that DM300 mice may reproduce aspects of the length-dependent sex bias characteristic of DM1.

Models of expanded repeats other than CAG·CTG have provided different results in terms of sex bias (Table 36-2). A YAC transgenic model carrying an interrupted $(CGG)_{120}$ sequence showed no sex of transmitting parent effect, with expansions prevalent in the male and female germlines [76]. The *Fmr1* $(CGG)_{98}$ knock-in mouse model displays higher mutation rates in the male germline and an expansion bias in both male and female transmissions [77]. Transgenic lines carrying expanded GAA trinucleotides have been shown to display different sex biases, depending on the insertion site of the transgene. One line was found to show an expansion bias in both male and female transmissions, whereas another displayed a strong bias toward contractions in female transmissions and no overall bias in male transmissions [78]. Taken together, these results suggest that the site of integration and the associated genomic environment most certainly contribute to the difference in sex of transmitting parent effects between independent transgenic lines.

In conclusion, the existence of sex-specific differences in intergenerational mutational rates suggests that sex-specific *trans*-acting factors may be involved in the production of germinal mosaicism in males and females. Basic differences in male and female gametogenesis (e.g., number of cell divisions involved) may account for the sex-specific bias observed in humans and mice. Alternatively, fundamental differences in DNA metabolism between spermatogenesis and oogenesis may account for these effects.

2. Tissue- and Cell-Specific Somatic Instability

Transgenic mouse lines carrying trinucleotide repeat sequences either long enough or within the appropriate genomic context have recreated expansion-biased, age-dependent, tissue-specific somatic mosaicism of expanded CAG·CTG [72, 79–82, 87, 89, 90, 94] and GAA·TTC trinucleotide repeats [78] (Table 36-3).

Interestingly, in a comparison of independent mouse lines, a few tissues have frequently been found to display high levels of trinucleotide repeat variability, whereas others usually carry stable expanded repeats (Table 36-3). The kidney [79–81, 87, 94, 96], liver [72, 79, 82, 85, 87, 96], striatum [72, 82, 87, 89, 90], and pancreas [79, 87] show high levels of somatic mosaicism. Expanded repeats are generally more stable in the cerebellum [72, 79, 80, 82, 87, 89, 94, 96], heart [72, 78, 82, 90, 94, 96], and lung [79, 82, 90, 94] (Table 36-3). In contrast with most mouse models of unstable triplet repeats, *FRDA* $(GAA)_{190}$ transgenic mice present high levels of somatic instability in the cerebellum [78], a tissue in which the repeats are usually remarkably stable. This led to the suggestion that unique mechanisms of repeat size mutation for GAA repeats may operate in the cerebellum.

Three main conclusions can be drawn from these observations. First, the identical mutation profiles of animals of the same transgenic mouse line maintained in a common environment imply that somatic mosaicism is a highly deterministic phenomenon, controlled by a series of factors requiring further investigation. Second, the apparent limited conservation of intertissue patterns of somatic mosaicism between different transgenic lines (Table 36-3) suggests that the tissue specificity of repeat instability is affected by tissue-specific *trans*-acting modifiers. Third, levels of tissue-specific trinucleotide repeat instability do not appear to correlate with the reported rates of cell turnover in each somatic tissue [80, 81, 90, 94]. In one study, >80% of striatal cells displayed an increase in allele length of at least five CAG repeat units, whereas 1 in 2500 cells tripled the size of their repeats [90]. Given that >65% of cells in the striatum are postmitotic neurons and that no glial cell proliferation is observed in HD knock-in mice striatums [86], it is difficult to reconcile the high levels of somatic mosaicism in this tissue with successive replication-based expansions, as suggested by the replication slippage model [104]. Furthermore, the subventricular zone of the striatum does not preferentially contain the largest mutant alleles, and instability of the CAG repeat mutation was not detected during the period of greatest proliferation of neuronal precursors in the developing mouse brain [90]. These data suggest that age-dependent instability is not driven by rapid cell proliferation and that the accumulation of trinucleotide repeat size variation *in vivo* is not entirely dependent on cell division.

In addition to intertissue variability, triplet repeat instability also varies within individual tissues. Therefore, it appears that the repeat is not equally susceptible to mutation in all cells, leading to intratissue repeat size variation. For instance, in the kidney of *Dmt-D* mice, three major subpopulations of cells seem to exist with different mutational dynamics, resulting in a characteristic trimodal distribution [94]. Similarly, bimodal mutation profiles have been detected in the liver of different mouse models carrying unstable CAG·CTG trinucleotide repeats [72, 79–81, 85]. The multimodal nature of repeat size distributions suggests that important cell type-specific factors may mediate repeat size dynamics and, therefore, intratissue differences in mutation profiles. Both genetic and environmental modifiers may

TABLE 36-3 Summary of Somatic Mosaicism in Mouse Models of Expanded Trinucleotide Repeat Dynamics[a]

Mouse model	Somatic tissues showing various levels of somatic mosaicism			Reference numbers
	Low	Medium	High	
DM300	Blood	Eye	Kidney (*)	79
	Cerebellum	Heart	Liver (*)	
	Lung	Skeletal muscle	Pancreas	
		Tongue	Testis	
Dmpk(CTG)$_{84}$	Heart	Brain	Kidney	81
	Skeletal muscle	Eyes	Stomach mucosa	
	Skin	Small intestine		
Dmt-D	Cerebellum	Liver	Kidney	94
	Heart	Striatum		
	Lung			
Drm12	Cerebellum	Gonads	Cerebrum	96
Drm21	Heart	Tail	Kidney	
Drm22	Spleen		Liver	
FRDA YG22	Peripheral tissues	Brainstem	Cerebellum	78
		Cerebrum		
		Spinal cord		
Hdh6/72Q	Colon	Brain cortex	Striatum	90
Hdh4/Q40	Liver	Cerebellum		
	Lung	Eye		
	Heart	Hindbrain		
	Skeletal muscle	Hippocampus		
	Spleen	Kidney		
		Olfactory bulb		
		Spinal cord		
Hdh^{Q92}	Cerebellum	Brain cortex	Liver	82
Hdh^{Q111}	Heart	Kidney	Striatum	
	Lung			
R6	Cerebellum	Brain cortex	Liver	72
	Heart	Kidney	Striatum	
	Spleen			
RL-SCA7 92R	Heart	Kidney	Brain	85
	Skeletal muscle	Ovaries	Liver	
		Retina		
Sca1$^{154Q/2Q}$	Cerebellum	Brain cortex	Striatum	89
		Hippocampus		
		Kidney		
		Liver		
		Olfactory bulb		
		Pons		

[a] Abbreviation: (*), Line-dependent.

be responsible for a multimodal repeat distribution within a tissue. Given that the dynamics of expanded trinucleotide sequences is affected by environmental genotoxic factors [105], certain tissues or cell types present in areas exposed to the outside of the body may have distinct repeat profiles [81].

In summary, the identification of tissue- and cell-type-specific profiles of trinucleotide repeat instability should open new avenues toward the understanding of *trans*-acting modifiers of repeat dynamics and contribute to dissection of the molecular mechanisms of triplet repeat expansion.

3. Genetic Modifiers of Repeat Instability

a. DNA Mismatch Repair Genes and Proteins

DNA repair pathways, particularly the DNA mismatch repair (MMR) system, were rapidly identified as a plausible modifier of trinucleotide repeat instability, as they are required to maintain genomic integrity in both prokaryotes and eukaryotes and to stabilize the cellular genome, by correcting single mismatches, and short unpaired regions, such as small insertions and deletions [106]. The association between impaired MMR activity and high genetic instability of simple DNA repeat sequences is particularly strong in hereditary nonpolyposis colorectal cancer (HNPCC) [107], suggesting that MMR mutations may render repetitive DNA sequences unstable. However, unlike HNPCC, which is characterized by genome-wide instability, trinucleotide repeat diseases are associated with the expansion of a repetitive sequence within a single locus. The current mammalian model of MMR has shown that heterodimers formed between MSH2 and either MSH3 or MSH6 (MutSβ and MutSα, respectively) are critical for the initial recognition of single-base mismatches or small insertions and deletions. The recruitment of a second heterodimer consisting of MLH1 and either PMS2 (MutLα) or MLH3 (MutLβ) is thought to be essential for subsequent excision and resynthesis, resulting in DNA repair [106].

The breeding of transgenic mouse lines in MMR-deficient genetic backgrounds provided insight into the dynamics of expanded triplet repeats and the molecular mechanism of repeat length expansion. *Msh2* deficiency was initially reported to stabilize an expanded CAG·CTG tract in the somatic tissues of R6 mice transgenic for exon 1 of the *HD* gene [108]. Although one functional repeat allele was still sufficient to generate high levels of repeat size variability, large repeat expansions were not detected in the striatum of these mice if both *Msh2* alleles were mutated [108]. *Msh2* deficiency also abolishes trinucleotide repeat expansions in the male germline of these mice [109]. A lack of somatic repeat expansions or contractions in the absence of the *Msh2* gene product has also been reported in the striatum of Hdh^{Q111} knock-in mice, further implicating this gene in the control of repeat instability [110]. Moreover, stabilization of the transgenic CAG·CTG tract significantly delayed the manifestation of early disease symptoms, providing further evidence for a role of somatic mosaicism in triggering disease symptoms [110]. The effects of *Msh2* mutation on the intergenerational dynamics of expanded CAG·CTG repeats in Hdh^{Q111} knock-in mice proved to be more complex. *Msh2* deficiency did not alter the frequency of maternally or paternally inherited changes. Instead, the change in paternally inherited repeat sizes shifted from predominantly increase in an $Msh2^{+/+}$ background to exclusively contraction in $Msh2^{-/-}$ mice. In contrast, a preponderance of contractions and a few expansions were detected in female germline Hdh^{Q111} transmissions in both genetic backgrounds [110]. Therefore, although *Msh2* appears to be required for paternally inherited and somatic expansions, CAG repeat instability in Hdh^{Q111} knock-in mice must also involve *Msh2*-independent processes, generating maternally inherited expansion and contraction events and repeated loss through the male germline.

Control of the direction of repeat length mutation (either expansion or contraction) by *Msh2* has been reported in other mouse models. Breeding of transgenic DM300 mice, carrying >300 CAG·CTG repeats, in an *Msh2*-deficient background did not stabilize the repeat [111]. Both overall mutability and the magnitude of repeat length changes remained unchanged in the absence of the *Msh2* gene product. However, *Msh2* mutation resulted in a spectacular shift toward repeat contractions in the soma and in both male and female transmissions, indicating that functional *Msh2* alleles are required to drive repeat length changes toward expansion in these mice [111]. Furthermore, these observations confirm that, in the absence of Msh2, the repeat could be processed through an alternative pathway leading to contractions.

A second *MutS* homologue has been shown to act as an enhancer of expansion-biased trinucleotide repeat somatic instability *in vivo*. *Dmpk* (CTG)$_{84}$ knock-in mice bred in an *Msh3*-deficient background displayed a complete block of expansion-biased somatic repeat instability, without the accumulation of repeat size mutations over time [81]. However, the finding that *Msh6* mutations increase the frequency of somatic expansions in the same transgenic model [81] suggests that the dynamics of expanded trinucleotide repeats depends on competition between Msh3 and Msh6 for binding to Msh2. How MutS homologues Msh2 and

Msh3 mediate CAG·CTG repeat expansions has not yet been determined. It was initially suggested that this process might involve the recognition, stabilization, and protection of alternative trinucleotide repeat DNA structures by MMR proteins [81, 108, 112]. Consistent with this view, MSH2 was shown to bind slipped-stranded DNA structures *in vitro* [113]. However, *Pms2* deletion from *Dmt*-D transgenic mice carrying unstable CAG·CTG expanded repeats reduced the absolute rate of somatic expansion by about 50% [114]. The involvement of a MutL homologue in the expansion pathway strengthens the hypothesis that a full MMR reaction might also be required to generate expansions.

In conclusion, an increasing number of MMR proteins appear to be involved in the control of trinucleotide repeat dynamics upon transmissions and in the soma, strongly implicating the proteins of this DNA repair pathway in the generation of repeat size mutations.

b. Other Genetic Modifiers

The expansion of trinucleotide repeats by recombination has been reported in recombination-proficient *E. coli* strains [115, 116] and in *S. cerevisiae* [117, 118]. It has therefore been speculated that recombination events between sister chromatids or homologous chromosomes could play a role in generating somatic and germline repeat size mutations in humans, in a manner similar to the nonreciprocal transfer of repeats between alleles in the germline previously described for human minisatellites [119]. Recombination-mediated trinucleotide repeat instability requires DNA strand breaks. Interestingly, CAG·CTG repeats show a length-dependent susceptibility to strand breaks in yeast [120].

Homologous recombination is mediated by the Rad52 group of proteins, including Rad52 and Rad54 [121]. The involvement of homologous recombination-mediated repair in CAG·CTG repeat instability has been investigated in DM300 mice bred in a *Rad52*- or *Rad54*-deficient background. The loss of *Rad52* and *Rad54* did not affect the mutation frequencies of the CAG·CTG repeat tract or the frequency of expansions and contractions. As in control animals, a strong bias toward repeat expansion was detected in homologous recombination-deficient mice [111]. These observations, and the similar germline mutation profiles of old heterozygous and homozygous *HD* knock-in mice [87], rule out homologous recombination as the driving force behind mouse intergenerational repeat instability.

Interestingly, expansions were significantly smaller in the absence of Rad52, suggesting that this protein may be involved in triplet repeat metabolism [111]. Given the lack of effect of *Rad54* mutation, it was suggested that Rad52 may play a role in triplet repeat dynamics through its activity in single-strand annealing (SSA) repair mechanisms [111]. SSA is an error-prone double-strand-break repair pathway that does not depend on other Rad52 group proteins [122] or MutL homologues [123]. However, a MutL homologue, Pms2, increases somatic mosaicism in transgenic mice [114]. Therefore, one may hypothesize that multiple DNA metabolism pathways might participate in trinucleotide repeat size mutation in a nonexclusive manner.

Double strand-break repair may also occur via nonhomologous end-joining, which requires the DNA-dependent protein kinase catalytic subunit or DNA-PKcs [124]. However, *DNA-PKcs* knock-out mice, like control animals, have high levels of expansion-biased repeat instability, suggesting that nonhomologous end-joining probably is not involved in triplet repeat expansion [111].

Other proteins involved in several aspects of DNA metabolism may also modify the dynamics of expanded trinucleotide repeats. These modifiers may affect the mutation rates, direction of bias, and/or magnitude of repeat length changes and could be investigated by a systematic study of the effects of different mouse genetic backgrounds on the dynamics of expanded trinucleotide repeats. Circumstantial evidence points to the existence of strain-specific modifiers of repeat instability. Levels of somatic mosaicism are highest in *Dmpk* $(CGT)_{84}$ knock-in mice in which the contribution of the C3H mouse genetic background exceeds 50% [81], consistent with repeat dynamics being modified by strain-specific *trans*-acting factors. The identification of such modifiers should increase the understanding of interindividual differences in trinucleotide repeat instability and provide great insight into the mechanism of size mutation.

Finally, it has also been reported that the direction of expanded CAG·CTG mutations observed in intergenerational transmissions in R6 transgenic animals depends on the sex of the offspring [109]. The similar CAG repeat distributions in X- and Y-bearing parental sperm [125] suggest that unidentified sex-specific *trans*-acting factors may affect trinucleotide repeat dynamics during postzygotic development [125]. Exciting as they may appear, these results have not been replicated in other mouse models [76–78, 95, 111], suggesting that these observations may be specific to the R6 transgenic line.

D. Timing of Repeat Instability

The timing of repeat instability in gametogenesis and somatic development is unclear and requires further investigation. Identification of the time window during which repeat size variability accumulates will provide insight into the pathways of trinucleotide repeat dynamics.

The development of mouse models of unstable trinucleotide repeats has provided an important research tool for studies of the timing of repeat instability.

1. Timing of Intergenerational Instability

a. Male Transmission

The very low heterogeneity of repeat tract lengths in the somatic tissues of young adult mice suggests that intergenerational instability results primarily from parental germinal events. Triplet repeat dynamics has therefore been monitored throughout male and female intergenerational transmissions to determine the timing of expansion and contraction events.

In some transgenic mouse models, pedigree analyses have shown a parental age-dependent increase in the frequency [78, 87, 88, 95, 96] or magnitude of paternally inherited mutations [72, 79, 95, 96]. One report has shown that the magnitude of CAG repeat expansion in humans increases by a factor of up to five with the age of the transmitting father [96]. One explanation for the paucity of such reports is that human pedigree analyses involve small families and are confounded by somatic instability. The large age differences between parents and their offspring at the time of blood sample collection clearly hinder the analysis [22]. Only by studying instability in the sperm from a transmitting father can the distribution of allele lengths likely to be inherited by his offspring be predicted [126]. Indeed, the mutation profiles determined by polymerase chain reaction (PCR) for transgenic mouse sperm overlap the sizes inherited by the offspring [95], suggesting that instability during transmission results mainly from germinal instability in the father [43, 127].

In mice, mean spontaneous mutation frequencies in the sperm of *Dmt*-D and DM300 lines carrying expanded CAG·CTG repeats were found to increase considerably over a period of 20 months, with a marked increase in the size of expansions [95, 128], consistent with the positive correlation between the age of the transmitting father and the length of the repeat inherited by the offspring [79, 95].

The age dependence of expansion magnitude in the male germline implies that mutations accumulate over time in spermatogenic stem cells. In an attempt to determine the timing of repeat expansion in the male germline, repeat size variability has been assessed at various stages of spermatogenesis in germinal cell types sorted by fluorescence-activated cell sorting (FACS) analysis. In R6 HD transgenic mice carrying about 120 CAG repeats, male germinal expansions appear to be produced only in the latest stages of spermatogenesis (specifically during the transition from round spermatids to elongating spermatids) through an Msh2-dependent gap repair mechanism initiated by DNA strand breaks [112]. Although very attractive, these results are surprising as the gene encoding Msh2 is not expressed during the final stages of mouse spermatogenesis and no Msh2 protein is detected in round spermatids in men [129] or mice [130]. In contrast, a similar study on the DM300 transgenic mouse line has demonstrated that the median repeat size in spermatozoa is not significantly different from that in spermatogonia [128]. Thus, germinal expansions are already present at the first stages of spermatogenesis and are therefore probably generated through a meiosis-independent mechanism. In the same study, contractions were also detected in the spermatogonia of *Msh2* knock-out mice, indicating that they also occur before rather than after meiosis [128]. Single-molecule DNA analysis of testicular germ cells from two HD patients has also suggested that expansions may occur before the end of the first meiotic division [131]. These data suggest that male germline instability probably results from small additive expansion events in spermatogenic stem cells undergoing mitotic division and/or repeated DNA repair, except in R6 HD transgenic mice.

b. Female Germline

Five independent studies based on mouse pedigree analysis have shown an increase in the frequency of contractions during female transmissions with age [83, 84, 88, 89, 95, 96], whereas others have shown an increase in the magnitude of repeat length changes with age [83, 88, 89, 95, 96].

The stage of female gametogenesis at which the CAG·CTG repeat becomes unstable was investigated in unfertilized mature oocytes isolated from superovulated *SCA1* cDNA transgenic females, with single-cell PCR analysis used to monitor CAG repeat size variation. The analysis revealed a time window in oocyte development during which repeat instability occurs: oocytes collected at 7 weeks of age showed limited repeat size variability, whereas oocytes harvested at 20 weeks of age displayed overt contraction-biased repeat instability [83]. The lack of repeat instability in oocytes from young females suggests that germline mutations occur while the oocytes are arrested in meiosis I via a mechanism independent of DNA replication [83].

In summary, the monitoring of trinucleotide dynamics during mouse gametogenesis has provided evidence that, as in humans, the molecular mechanisms governing germline instability in male and female gametogenesis probably differ.

c. Postfertilization Event

In studies of the effect of *Msh2* deficiency on intergenerational instability, it has been reported that the distribution of expansions and contractions in *Msh2*[+/−] sperm overlaps with that in the corresponding *Msh2*[+/−]

offspring, but not with that in $Msh2^{-/-}$ offspring. A strong bias toward contractions was observed in the $Msh2^{-/-}$ progeny [111]. Although clearly demonstrating that the *Msh2* genotype of the offspring affects intergenerational instability, these results have wider implications. Contrary to the commonly accepted idea that intergenerational instability results simply from events taking place in the progenitor germline, a second event occurring soon after fertilization must contribute to the repeat length changes observed across generations of DM300 mice [111]. The timing and characterization of this event require further investigation.

2. Timing of Somatic Triplet Repeat Instability

Of all the trinucleotide expansion disorders, DM1 presents the widest range of expansion sizes and is the disease for which somatic mosaicism has been most extensively characterized.

Analyses of DM1 fetuses have shown a first wave of somatic instability in early embryogenesis, probably between 13 and 16 weeks of gestation, leading to major differences in repeat size variability between tissues [23, 132]. The accumulation of repeat length variability in the soma continues into adulthood, probably through the accumulation of small expansion-biased mutational changes [55, 57]. A detailed study of DM1 somatic instability over time in peripheral blood cells from patients has demonstrated the progression of expansion-biased repeat size heterogeneity and established a correlation between average repeat number, inherited repeat length, and age [57].

Many mouse models have recreated expansion-biased somatic mosaicism that, although weak or absent at a young age (<5 weeks), accumulates over time, resulting in high levels of repeat size variability in old animals [72, 78–82, 85, 89, 90, 94, 96]. Given the shorter life span of mice, the detected levels of somatic mosaicism in most of these models are at least comparable to, if not greater than, those observed in humans. A time-dependent increase in the somatic variability of repeat size suggested a continuous accumulation of stepwise repeat gains over time. It has been suggested that trinucleotide repeat expansions may occur at different rates in different tissues, thereby generating tissue-specific patterns of somatic mosaicism in mice and humans. Thus, repeat size mutation in the soma appears to occur throughout the adult life of an individual, although at different rates in different tissues.

Instability during mouse development has not been studied in detail. Mosaic animals, resulting from very rare mutation events in the first cell divisions following fertilization, have been reported in *Dmt* transgenic lines [71]. A detailed analysis showed that the size variability of an expanded trinucleotide repeat probably occurs very shortly after fertilization in transgenic R6 mice, with variability decreasing at later stages of development [133]. Further investigations of trinucleotide repeat dynamics during the early stages of embryonic development are required, as they may provide important clues for the understanding of the molecular mechanisms of somatic repeat instability.

IV. CONCLUDING REMARKS

The dynamics of the various disease-associated trinucleotide repeats (particularly CAG·CTG repeats) described has shown some degree of similarity. Beyond a repeat number threshold, expanded alleles show expansion-biased intergenerational and somatic instability, suggesting that the same mutation mechanisms may be shared by different repeat tracts. Advances in mouse genetics have made it possible to develop animal models, making a major contribution to dissection of the molecular pathways of repeat size mutation.

A. Recreation of Intergenerational Instability in Transgenic Mice

Trinucleotide repeat instability has been recreated in mice only when long repeat sequences (usually longer than 100 repeats) are introduced into the mouse genome [43, 72, 79]. Alternatively, shorter trinucleotide repeat tracts are unstable only if located in the appropriate genomic environment, within a large transgenic fragment of human genomic DNA [73, 76, 84, 85, 96], or at the endogenous locus in knock-in mice [77, 81, 82, 86–89]. This finding emphasizes the importance of repeat size and genomic flanking sequences in the control of mutation frequencies. Indeed, the transgenic mice with the longest trinucleotide repeat tracts within a large genomic fragment derived from the remarkably unstable human *DM1* locus [79] have recreated dramatic levels of intergenerational repeat instability. In other transgenic models the level of intergenerational instability seems to depend on the sex of the transmitting parent.

The sex of the transmitting parent has been shown to affect not only the frequency of mutation but also the direction (expansions versus contractions) and the magnitude of intergenerational changes in trinucleotide repeat sequences. This effect is probably mediated by the DNA context flanking the repeat and is well-illustrated by a comparison of DM300 [79] and *Dmt*-D [71] mice, carrying ~45 and ~1 kb of human genomic DNA flanking the repeat, respectively. DM300 mice display considerable germline instability in both male and female transmissions [79], whereas *Dmt*-D mice display contraction-biased transmissions through the female germline [71]. In PrP-SCA7 92R cDNA-transgenic lines, female transmissions display a bias toward contractions, whereas the frequency of expansions is higher if the repeat is located in its human genomic context in the RL-SCA7 92R line [85]. The direction of repeat size variation may also be controlled by unidentified sex-specific *trans*-acting factors. For example, the metabolism of trinucleotide repeats may be intrinsically different during female and male gametogenesis, as suggested by observations in humans and mice.

In most transgenic mouse models of unstable trinucleotide repeats, the magnitude of amplifications remains much smaller than that detected in patients. The molecular bases of these differences remain unclear, but they may include genomic environment, differences in mouse DNA metabolism, mouse age, and life span. A linear regression analysis has revealed that the mean increase in the size of CAG·CTG repeats during male transmission in DRPLA transgenic mice may be estimated at +0.31 per year and +0.0073 per spermatogenesis cycle in mice and +0.27 and +0.012, respectively, in DRPLA patients [96]. These results strongly suggest that the smaller intergenerational changes observed in mice are due to differences in reproductive life spans and that a common molecular mechanism underlies the age-dependent expansion of inherited CAG·CTG repeats in both species.

In summary, major differences in intergenerational instability have been observed between murine models and human repeat disease loci for most of the transgenic lines created: (1) rates of intergenerational repeat instability are lower in most of the mouse models; (2) large intergenerational repeat expansions have not yet been reported in mice; and (3) frequencies of parent-to-offspring contractions are often greater in mice than in humans. Despite the differences, trinucleotide repeat instability in transgenic models retains key properties previously described in humans: (1) the requirement of a length threshold for the detection of repeat instability; (2) an effect of the sex of the transmitting parent on intergenerational repeat instability; and (3) the modulation of repeat instability by flanking genomic sequences. Thus, mice provide an ideal model for studies of the biology of disease-associated expanded trinucleotide repeats.

B. Recreation of Somatic Mosaicism in Transgenic Mice

Transgenic models have also shed light on the molecular mechanisms driving repeat size mutation in the soma. Replication, transcription, and DNA repair pathways may account, at least partly, for the accumulation of repeat length variability in somatic cells. Differences between the levels of somatic mosaicism and intergenerational instability detected in mouse models indicate that the clear differences observed between the somatic and germinal pathways in humans [22] may be replicated in mice. Furthermore, the genomic context, which makes it possible to achieve high levels of somatic mosaicism in some transgenic mice, is not necessarily sufficient for intergenerational instability, which remains modest in these animals [71, 86, 90, 94].

In addition to posing a fascinating scientific question, somatic mosaicism may also account for the tissue specificity and progressive nature of the symptoms. The development of transgenic mice has led to the identification of MMR proteins as important modifiers of somatic mosaicism [81, 108, 111, 112, 114, 128] and to an understanding of the wider implications of the accumulation of repeat size variability in somatic tissues. *Hdh* knock-in transgenic lines carrying expanded CAG repeat sequences display the predominant accumulation of expansion-biased somatic mosaicism in the striatum [82, 87, 90]. The selective vulnerability of striatal neurons in HD may be caused by cell-specific determinants that directly or indirectly induce expansion-biased mutability of the CAG repeat over time. The cells in which the repeat is most unstable may succumb earlier because a critical polyglutamine concentration threshold is exceeded in these cells first [90]. Consistent with this hypothesis, the elimination of CAG expansion by Msh2 deficiency delays the timing of disease onset, but it does not affect the specific vulnerability of striatal neurons [110]. Thus, Msh2 and other genetic enhancers of somatic mosaicism may trigger disease symptoms without interfering with disease selectivity [110]. No causal relationship between the degree of somatic mosaicism and selective neurodegeneration was found in the *Sca1$^{154Q/2Q}$* mice carrying a longer repeat tract [89]. This observation suggests that selective neuronal dysfunction may be correlated to somatic instability specifically in HD or that longer repeat tracts [(CAG)$_{154}$ in *Sca1$^{154Q/2Q}$* versus (CAG)$_{70-111}$ in *Hdh* knock-in mice] may be associated with levels of somatic instability so high that they no longer correlate with tissue-specific neuronal dysfunction. The level of transcription of the disease-associated gene must also be taken into account as a factor determining the pattern of tissue-specific neurodegeneration.

In conclusion, transgenic mouse models have proved to be an excellent tool for unraveling the complex dynamics of expanded trinucleotide repeats, helping to dissect the molecular mechanisms of disease-associated repeat expansion in the germline and throughout the soma. Paradoxically, the diversity of mouse models may be a curse as well as a blessing, facilitating studies of the biology of triplet repeats but raising major questions requiring further investigation. In this review, a brief summary has been presented of the general principles controlling the dynamics of expansions of trinucleotide repeats, based on experimental data collected from various mouse models.

Acknowledgments

The authors thank the Association Française contre les Myopathies (AFM), INSERM, and the Université René-Descartes Paris V for financial support. MGP is supported by a fellowship from the Fundação para a Ciência e Tecnologia (FCT, Portugal). LF is supported by a grant from the Ministère Français de la Recherche et de la Technologie.

References

1. Fu, Y. H., Kuhl, D. P., Pizzuti, A., Pieretti, M., Sutcliffe, J. S., Richards, S., Verkerk, A. J., Holden, J. J., Fenwick, R. G., Jr., Warren, S. T., et al. (1991). Variation of the CGG repeat at the fragile X site results in genetic instability: Resolution of the Sherman paradox. *Cell* **67**, 1047–1058.
2. La Spada, A. R., Wilson, E. M., Lubahn, D. B., Harding, A. E., and Fischbeck, K. H. (1991). Androgen receptor gene mutations in X-linked spinal and bulbar muscular atrophy. *Nature* **352**, 77–79.
3. Orr, H. T., Chung, M. Y., Banfi, S., Kwiatkowski, T. J., Jr., Servadio, A., Beaudet, A. L., McCall, A. E., Duvick, L. A., Ranum, L. P., and Zoghbi, H. Y. (1993). Expansion of an unstable trinucleotide CAG repeat in spinocerebellar ataxia type 1. *Nat. Genet.* **4**, 221–226.
4. Imbert, G., Saudou, F., Yvert, G., Devys, D., Trottier, Y., Garnier, J. M., Weber, C., Mandel, J. L., Cancel, G., Abbas, N., Durr, A., Didierjean, O., Stevanin, G., Agid, Y., and Brice, A. (1996). Cloning of the gene for spinocerebellar ataxia 2 reveals a locus with high sensitivity to expanded CAG/glutamine repeats. *Nat. Genet.* **14**, 285–291.
5. Kawaguchi, Y., Okamoto, T., Taniwaki, M., Aizawa, M., Inoue, M., Katayama, S., Kawakami, H., Nakamura, S., Nishimura, M., Akiguchi, I., et al. (1994). CAG expansions in a novel gene for Machado-Joseph disease at chromosome 14q32.1. *Nat. Genet.* **8**, 221–228.
6. David, G., Abbas, N., Stevanin, G., Durr, A., Yvert, G., Cancel, G., Weber, C., Imbert, G., Saudou, F., Antoniou, E., Drabkin, H., Gemmill, R., Giunti, P., Benomar, A., Wood, N., Ruberg, M., Agid, Y., Mandel, J. L., and Brice, A. (1997). Cloning of the SCA7 gene reveals a highly unstable CAG repeat expansion. *Nat. Genet.* **17**, 65–70.
7. Huntington's Disease Research Collaborative Group (1993). A novel gene containing a trinucleotide repeat that is expanded and unstable on Huntington's disease chromosomes. *Cell* **72**, 971–983.
8. Koide, R., Ikeuchi, T., Onodera, O., Tanaka, H., Igarashi, S., Endo, K., Takahashi, H., Kondo, R., Ishikawa, A., Hayashi, T., et al. (1994). Unstable expansion of CAG repeat in hereditary dentatorubral–pallidoluysian atrophy (DRPLA). *Nat. Genet.* **6**, 9–13.
9. Fu, Y. H., Pizzuti, A., Fenwick, R. G., Jr., King, J., Rajnarayan, S., Dunne, P. W., Dubel, J., Nasser, G. A., Ashizawa, T., de Jong, P., et al. (1992). An unstable triplet repeat in a gene related to myotonic muscular dystrophy. *Science* **255**, 1256–1258.
10. Brook, J. D., McCurrach, M. E., Harley, H. G., Buckler, A. J., Church, D., Aburatani, H., Hunter, K., Stanton, V. P., Thirion, J. P., Hudson, T., et al. (1992). Molecular basis of myotonic dystrophy: Expansion of a trinucleotide (CTG) repeat at the 3' end of a transcript encoding a protein kinase family member. *Cell* **69**, 385.
11. Mahadevan, M., Tsilfidis, C., Sabourin, L., Shutler, G., Amemiya, C., Jansen, G., Neville, C., Narang, M., Barcelo, J., O'Hoy, K., et al. (1992). Myotonic dystrophy mutation: An unstable CTG repeat in the 3' untranslated region of the gene. *Science* **255**, 1253–1255.
12. Campuzano, V., Montermini, L., Molto, M. D., Pianese, L., Cossee, M., Cavalcanti, F., Monros, E., Rodius, F., Duclos, F., Monticelli, A., et al. (1996). Friedreich's ataxia: Autosomal recessive disease caused by an intronic GAA triplet repeat expansion. *Science* **271**, 1423–1427.
13. Gusella, J. F., and MacDonald, M. E. (2000). Molecular genetics: Unmasking polyglutamine triggers in neurodegenerative disease. *Nat. Rev. Neurosci.* **1**, 109–115.
14. Mankodi, A., Logigian, E., Callahan, L., McClain, C., White, R., Henderson, D., Krym, M., and Thornton, C. A. (2000). Myotonic dystrophy in transgenic mice expressing an expanded CUG repeat. *Science* **289**, 1769–1773.
15. Davis, B. M., McCurrach, M. E., Taneja, K. L., Singer, R. H., and Housman, D. E. (1997). Expansion of a CUG trinucleotide repeat in the 3' untranslated region of myotonic dystrophy protein kinase transcripts results in nuclear retention of transcripts. *Proc. Natl. Acad. Sci. USA* **94**, 7388–7393.
16. Sakamoto, N., Chastain, P. D., Parniewski, P., Ohshima, K., Pandolfo, M., Griffith, J. D., and Wells, R. D. (1999). Sticky DNA: Self-association properties of long GAA.TTC repeats in R.R.Y triplex structures from Friedreich's ataxia. *Mol. Cell* **3**, 465–475.
17. Klesert, T. R., Otten, A. D., Bird, T. D., and Tapscott, S. J. (1997). Trinucleotide repeat expansion at the myotonic dystrophy locus reduces expression of DMAHP. *Nat. Genet.* **16**, 402–406.
18. Jin, P., and Warren, S. T. (2000). Understanding the molecular basis of fragile X syndrome. *Hum. Mol. Genet.* **9**, 901–908.
19. Koob, M. D., Moseley, M. L., Schut, L. J., Benzow, K. A., Bird, T. D., Day, J. W., and Ranum, L. P. (1999). An untranslated CTG expansion causes a novel form of spinocerebellar ataxia (SCA8). *Nat. Genet.* **21**, 379–384.
20. Richards, R. I. (2001). Dynamic mutations: A decade of unstable expanded repeats in human genetic disease. *Hum. Mol. Genet.* **10**, 2187–2194.
21. Cummings, C. J., and Zoghbi, H. Y. (2000). Fourteen and counting: Unraveling trinucleotide repeat diseases. *Hum. Mol. Genet.* **9**, 909–916.
22. Monckton, D. G., Wong, L. J., Ashizawa, T., and Caskey, C. T. (1995). Somatic mosaicism, germline expansions, germline reversions and intergenerational reductions in myotonic dystrophy males: Small pool PCR analyses. *Hum. Mol. Genet.* **4**, 1–8.
23. Wörhle, D., Kennerknecht, I., Wolf, M., Enders, H., Schwemmle, S., and Steinbach, P. (1995). Heterogeneity of DM kinase repeat expansion in different fetal tissues and further expansion during cell proliferation in vitro: Evidence for a casual involvement of methyl-directed DNA mismatch repair in triplet repeat stability. *Hum. Mol. Genet.* **4**, 1147–1153.
24. Richards, R. I., and Sutherland, G. R. (1992). Dynamic mutations: A new class of mutations causing human disease. *Cell* **70**, 709–712.
25. Imbert, G., Kretz, C., Johnson, K., and Mandel, J. L. (1993). Origin of the expansion mutation in myotonic dystrophy. *Nat. Genet.* **4**, 72–76.
26. Harper, P. S., Harley, H. G., Reardon, W., and Shaw, D. J. (1992). Anticipation in myotonic dystrophy: New light on an old problem. *Am. J. Hum. Genet.* **51**, 10–16.
27. Biancalana, V., Serville, F., Pommier, J., Julien, J., Hanauer, A., and Mandel, J. L. (1992). Moderate instability of the trinucleotide repeat in spinobulbar muscular atrophy. *Hum. Mol. Genet.* **1**, 255–258.

28. David, G., Durr, A., Stevanin, G., Cancel, G., Abbas, N., Benomar, A., Belal, S., Lebre, A. S., Abada-Bendib, M., Grid, D., Holmberg, M., Yahyaoui, M., Hentati, F., Chkili, T., Agid, Y., and Brice, A. (1998). Molecular and clinical correlations in autosomal dominant cerebellar ataxia with progressive macular dystrophy (SCA7). *Hum. Mol. Genet.* **7**, 165–170.
29. Gouw, L. G., Castaneda, M. A., McKenna, C. K., Digre, K. B., Pulst, S. M., Perlman, S., Lee, M. S., Gomez, C., Fischbeck, K., Gagnon, D., Storey, E., Bird, T., Jeri, F. R., and Ptacek, L. J. (1998). Analysis of the dynamic mutation in the SCA7 gene shows marked parental effects on CAG repeat transmission. *Hum. Mol. Genet.* **7**, 525–532.
30. Duyao, M., Ambrose, C., Myers, R., Novelletto, A., Persichetti, F., Frontali, M., Folstein, S., Ross, C., Franz, M., Abbott, M., et al. (1993). Trinucleotide repeat length instability and age of onset in Huntington's disease. *Nat. Genet.* **4**, 387–392.
31. Zuhlke, C., Riess, O., Bockel, B., Lange, H., and Thies, U. (1993). Mitotic stability and meiotic variability of the (CAG)n repeat in the Huntington disease gene. *Hum. Mol. Genet.* **2**, 2063–2067.
32. De Michele, G., Cavalcanti, F., Criscuolo, C., Pianese, L., Monticelli, A., Filla, A., and Cocozza, S. (1998). Parental gender, age at birth and expansion length influence GAA repeat intergenerational instability in the X25 gene: Pedigree studies and analysis of sperm from patients with Friedreich's ataxia. *Hum. Mol. Genet.* **7**, 1901–1906.
33. Pianese, L., Cavalcanti, F., De Michele, G., Filla, A., Campanella, G., Calabrese, O., Castaldo, I., Monticelli, A., and Cocozza, S. (1997). The effect of parental gender on the GAA dynamic mutation in the FRDA gene. *Am. J. Hum. Genet.* **60**, 460–463.
34. Brunner, H. G., Bruggenwirth, H. T., Nillesen, W., Jansen, G., Hamel, B. C., Hoppe, R. L., de Die, C. E., Howeler, C. J., van Oost, B. A., Wieringa, B., et al. (1993). Influence of sex of the transmitting parent as well as of parental allele size on the CTG expansion in myotonic dystrophy (DM). *Am. J. Hum. Genet.* **53**, 1016–1023.
35. Lavedan, C., Hofmann-Radvanyi, H., Shelbourne, P., Rabes, J. P., Duros, C., Savoy, D., Dehaupas, I., Luce, S., Johnson, K., and Junien, C. (1993). Myotonic dystrophy: Size- and sex-dependent dynamics of CTG meiotic instability, and somatic mosaicism. *Am. J. Hum. Genet.* **52**, 875–883.
36. Jansen, G., Willems, P., Coerwinkel, M., Nillesen, W., Smeets, H., Vits, L., Howeler, C., Brunner, H., and Wieringa, B. (1994). Gonosomal mosaicism in myotonic dystrophy patients: Involvement of mitotic events in (CTG)n repeat variation and selection against extreme expansion in sperm. *Am. J. Hum. Genet.* **54**, 575–585.
37. Shelbourne, P., Winqvist, R., Kunert, E., Davies, J., Leisti, J., Thiele, H., Bachmann, H., Buxton, J., Williamson, B., and Johnson, K. (1992). Unstable DNA may be responsible for the incomplete penetrance of the myotonic dystrophy phenotype. *Hum. Mol. Genet.* **1**, 467–473.
38. Kinoshita, M., Takahashi, R., Hasegawa, T., Komori, T., Nagasawa, R., Hirose, K., and Tanabe, H. (1996). (CTG)n expansions in various tissues from a myotonic dystrophy patient. *Muscle Nerve* **19**, 240–242.
39. Telenius, H., Kremer, B., Goldberg, Y. P., Theilmann, J., Andrew, S. E., Zeisler, J., Adam, S., Greenberg, C., Ives, E. J., Clarke, L. A., et al. (1994). Somatic and gonadal mosaicism of the Huntington disease gene CAG repeat in brain and sperm. *Nat. Genet.* **6**, 409–414.
40. Kennedy, L., Evans, E., Chen, C. M., Craven, L., Detloff, P. J., Ennis, M., and Shelbourne, P. F. (2003). Dramatic tissue-specific mutation length increases are an early molecular event in Huntington disease pathogenesis. *Hum. Mol. Genet.* **12**, 3359–3367.
41. Chong, S. S., McCall, A. E., Cota, J., Subramony, S. H., Orr, H. T., Hughes, M. R., and Zoghbi, H. Y. (1995). Gametic and somatic tissue-specific heterogeneity of the expanded SCA1 CAG repeat in spinocerebellar ataxia type 1. *Nat. Genet.* **10**, 344–350.
42. Tanaka, F., Ito, Y., and Sobue, G. (1999). Somatic mosaicism of expanded CAG trinucleotide repeat in the neural and nonneural tissues of Machado–Joseph disease (MJD). *Nippon Rinsho* **57**, 838–842.
43. Monckton, D. G., Cayuela, M. L., Gould, F. K., Brock, G. J., Silva, R., and Ashizawa, T. (1999). Very large (CAG)(n) DNA repeat expansions in the sperm of two spinocerebellar ataxia type 7 males. *Hum. Mol. Genet.* **8**, 2473–2478.
44. Takano, H., Onodera, O., Takahashi, H., Igarashi, S., Yamada, M., Oyake, M., Ikeuchi, T., Koide, R., Tanaka, H., Iwabuchi, K., and Tsuji, S. (1996). Somatic mosaicism of expanded CAG repeats in brains of patients with dentatorubral–pallidoluysian atrophy: Cellular population-dependent dynamics of mitotic instability. *Am. J. Hum. Genet.* **58**, 1212–1222.
45. Tanaka, F., Reeves, M. F., Ito, Y., Matsumoto, M., Li, M., Miwa, S., Inukai, A., Yamamoto, M., Doyu, M., Yoshida, M., Hashizume, Y., Terao, S., Mitsuma, T., and Sobue, G. (1999). Tissue-specific somatic mosaicism in spinal and bulbar muscular atrophy is dependent on CAG-repeat length and androgen receptor–gene expression level. *Am. J. Hum. Genet.* **65**, 966–973.
46. Tanaka, F., Sobue, G., Doyu, M., Ito, Y., Yamamoto, M., Shimada, N., Yamamoto, K., Riku, S., Hshizume, Y., and Mitsuma, T. (1996). Differential pattern in tissue-specific somatic mosaicism of expanded CAG trinucleotide repeats in dentatorubral–pallidoluysian atrophy, Machado–Joseph disease, and X-linked recessive spinal and bulbar muscular atrophy. *J. Neurol. Sci.* **135**, 43–50.
47. Oberle, I., Rousseau, F., Heitz, D., Kretz, C., Devys, D., Hanauer, A., Boue, J., Bertheas, M. F., and Mandel, J. L. (1991). Instability of a 550-base pair DNA segment and abnormal methylation in fragile X syndrome. *Science* **252**, 1097–1102.
48. Rousseau, F., Heitz, D., Biancalana, V., Blumenfeld, S., Kretz, C., Boue, J., Tommerup, N., Van Der Hagen, C., DeLozier-Blanchet, C., Croquette, M. F., et al. (1991). Direct diagnosis by DNA analysis of the fragile X syndrome of mental retardation. *New Engl. J. Med.* **325**, 1673–1681.
49. Bidichandani, S. I., Purandare, S. M., Taylor, E. E., Gumin, G., Machkhas, H., Harati, Y., Gibbs, R. A., Ashizawa, T., and Patel, P. I. (1999). Somatic sequence variation at the Friedreich ataxia locus includes complete contraction of the expanded GAA triplet repeat, significant length variation in serially passaged lymphoblasts and enhanced mutagenesis in the flanking sequence. *Hum. Mol. Genet.* **8**, 2425–2436.
50. Hellenbroich, Y., Schwinger, E., and Zuhlke, C. (2001). Limited somatic mosaicism for Friedreich's ataxia GAA triplet repeat expansions identified by small pool PCR in blood leukocytes. *Acta Neurol. Scand.* **103**, 188–192.
51. Montermini, L., Kish, S. J., Jiralerspong, S., Lamarche, J. B., and Pandolfo, M. (1997). Somatic mosaicism for Friedreich's ataxia GAA triplet repeat expansions in the central nervous system. *Neurology* **49**, 606–610.
52. Sharma, R., Bhatti, S., Gomez, M., Clark, R. M., Murray, C., Ashizawa, T., and Bidichandani, S. I. (2002). The GAA triplet-repeat sequence in Friedreich ataxia shows a high level of somatic instability in vivo, with a significant predilection for large contractions. *Hum. Mol. Genet.* **11**, 2175–2187.
53. Hunter, A., Tsilfidis, C., Mettler, G., Jacob, P., Mahadevan, M., Surh, L., and Korneluk, R. (1992). The correlation of age of onset with CTG trinucleotide repeat amplification in myotonic dystrophy. *J. Med. Genet.* **29**, 774–779.
54. Anvret, M., Ahlberg, G., Grandell, U., Hedberg, B., Johnson, K., and Edstrom, L. (1993). Larger expansions of the CTG repeat in muscle compared to lymphocytes from patients with myotonic dystrophy. *Hum. Mol. Genet.* **2**, 1397–1400.
55. Ashizawa, T., Dubel, J. R., and Harati, Y. (1993). Somatic instability of CTG repeat in myotonic dystrophy. *Neurology* **43**, 2674–2678.
56. Thornton, C. A., Johnson, K., and Moxley, R. T., III (1994). Myotonic dystrophy patients have larger CTG expansions in skeletal muscle than in leukocytes. *Ann. Neurol.* **35**, 104–107.

57. Martorell, L., Monckton, D. G., Gamez, J., Johnson, K. J., Gich, I., de Munain, A. L., and Baiget, M. (1998). Progression of somatic CTG repeat length heterogeneity in the blood cells of myotonic dystrophy patients. *Hum. Mol. Genet.* **7**, 307–312.
58. Wong, L. J., Ashizawa, T., Monckton, D. G., Caskey, C. T., and Richards, C. S. (1995). Somatic heterogeneity of the CTG repeat in myotonic dystrophy is age and size dependent. *Am. J. Hum. Genet.* **56**, 114–122.
59. Kang, S., Jaworski, A., Ohshima, K., and Wells, R. D. (1995). Expansion and deletion of CTG repeats from human disease genes are determined by the direction of replication in *E. coli*. *Nat. Genet.* **10**, 213–218.
60. Freudenreich, C. H., Stavenhagen, J. B., and Zakian, V. A. (1997). Stability of a CTG/CAG trinucleotide repeat in yeast is dependent on its orientation in the genome. *Mol. Cell. Biol.* **17**, 2090–2098.
61. Miret, J. J., Pessoa-Brandao, L., and Lahue, R. S. (1998). Orientation-dependent and sequence-specific expansions of CTG/CAG trinucleotide repeats in *Saccharomyces cerevisiae*. *Proc. Natl. Acad. Sci. USA* **95**, 12438–12443.
62. Bowater, R. P., Jaworski, A., Larson, J. E., Parniewski, P., and Wells, R. D. (1997). Transcription increases the deletion frequency of long CTG. CAG triplet repeats from plasmids in *Escherichia coli*. *Nucleic Acids Res.* **25**, 2861–2868.
63. Schmidt, K. H., Abbott, C. M., and Leach, D. R. (2000). Two opposing effects of mismatch repair on CTG repeat instability in *Escherichia coli*. *Mol. Microbiol.* **35**, 463–471.
64. Parniewski, P., Bacolla, A., Jaworski, A., and Wells, R. D. (1999). Nucleotide excision repair affects the stability of long transcribed (CTG*CAG) tracts in an orientation-dependent manner in *Escherichia coli*. *Nucleic Acids Res.* **27**, 616–623.
65. Jaworski, A., Rosche, W. A., Gellibolian, R., Kang, S., Shimizu, M., Bowater, R. P., Sinden, R. R., and Wells, R. D. (1995). Mismatch repair in *Escherichia coli* enhances instability of (CTG)n triplet repeats from human hereditary diseases. *Proc. Natl. Acad. Sci. USA* **92**, 11019–11023.
66. Schweitzer, J. K., and Livingston, D. M. (1997). Destabilization of CAG trinucleotide repeat tracts by mismatch repair mutations in yeast. *Hum. Mol. Genet.* **6**, 349–355.
67. Bingham, P. M., Scott, M. O., Wang, S., McPhaul, M. J., Wilson, E. M., Garbern, J. Y., Merry, D. E., and Fischbeck, K. H. (1995). Stability of an expanded trinucleotide repeat in the androgen receptor gene in transgenic mice. *Nat. Genet.* **9**, 191–196.
68. Burright, E. N., Clark, H. B., Servadio, A., Matilla, T., Feddersen, R. M., Yunis, W. S., Duvick, L. A., Zoghbi, H. Y., and Orr, H. T. (1995). SCA1 transgenic mice: a model for neurodegeneration caused by an expanded CAG trinucleotide repeat. *Cell* **82**, 937–948.
69. Ikeda, H., Yamaguchi, M., Sugai, S., Aze, Y., Narumiya, S., and Kakizuka, A. (1996). Expanded polyglutamine in the Machado–Joseph disease protein induces cell death in vitro and in vivo. *Nat. Genet.* **13**, 196–202.
70. Goldberg, Y. P., Kalchman, M. A., Metzler, M., Nasir, J., Zeisler, J., Graham, R., Koide, H. B., O'Kusky, J., Sharp, A. H., Ross, C. A., Jirik, F., and Hayden, M. R. (1996). Absence of disease phenotype and intergenerational stability of the CAG repeat in transgenic mice expressing the human Huntington disease transcript. *Hum. Mol. Genet.* **5**, 177–185.
71. Monckton, D. G., Coolbaugh, M. I., Ashizawa, K. T., Siciliano, M. J., and Caskey, C. T. (1997). Hypermutable myotonic dystrophy CTG repeats in transgenic mice. *Nat. Genet.* **15**, 193–196.
72. Mangiarini, L., Sathasivam, K., Mahal, A., Mott, R., Seller, M., and Bates, G. P. (1997). Instability of highly expanded CAG repeats in mice transgenic for the Huntington's disease mutation. *Nat. Genet.* **15**, 197–200.
73. Gourdon, G., Radvanyi, F., Lia, A. S., Duros, C., Blanche, M., Abitbol, M., Junien, C., and Hofmann-Radvanyi, H. (1997). Moderate intergenerational and somatic instability of a 55-CTG repeat in transgenic mice. *Nat. Genet.* **15**, 190–192.
74. Lavedan, C., Grabczyk, E., Usdin, K., and Nussbaum, R. L. (1998). Long uninterrupted CGG repeats within the first exon of the human FMR1 gene are not intrinsically unstable in transgenic mice. *Genomics* **50**, 229–240.
75. Baskaran, S., Datta, S., Mandal, A., Gulati, N., Totey, S., Anand, R. R., and Brahmachari, V. (2002). Instability of CGG repeats in transgenic mice. *Genomics* **80**, 151–157.
76. Peier, A. M., and Nelson, D. L. (2002). Instability of a premutation-sized CGG repeat in FMR1 YAC transgenic mice. *Genomics* **80**, 423–432.
77. Bontekoe, C. J., Bakker, C. E., Nieuwenhuizen, I. M., van der Linde, H., Lans, H., de Lange, D., Hirst, M. C., and Oostra, B. A. (2001). Instability of a (CGG)98 repeat in the Fmr1 promoter. *Hum. Mol. Genet.* **10**, 1693–1699.
78. Al-Mahdawi, S., Pinto, R. M., Ruddle, P., Carroll, C., Webster, Z., and Pook, M. (2004). GAA repeat instability in Friedreich ataxia YAC transgenic mice. *Genomics* **84**, 301–310.
79. Seznec, H., Lia-Baldini, A. S., Duros, C., Fouquet, C., Lacroix, C., Hofmann-Radvanyi, H., Junien, C., and Gourdon, G. (2000). Transgenic mice carrying large human genomic sequences with expanded CTG repeat mimic closely the DM CTG repeat intergenerational and somatic instability. *Hum. Mol. Genet.* **9**, 1185–1194.
80. Lia, A. S., Seznec, H., Hoffmann-Radvany, H., Radvany, F., Duros, C., Saquet, C., Blanche, M., Junien, C., and Gourdon, G. (1998). Somatic instability of the CTG repeat in mice transgenic for the myotonic dystrophy region is age dependent but not correlated to the relative intertissue transcription levels and proliferative capacities. *Hum. Mol. Genet.* **7**, 1285–1291.
81. van Den Broek, W. J., Nelen, M. R., Wansink, D. G., Coerwinkel, M. M., te Riele, H., Groenen, P. J., and Wieringa, B. (2002). Somatic expansion behaviour of the (CTG)(n) repeat in myotonic dystrophy knock-in mice is differentially affected by Msh3 and Msh6 mismatch-repair proteins. *Hum. Mol. Genet.* **11**, 191–198.
82. Wheeler, V. C., Auerbach, W., White, J. K., Srinidhi, J., Auerbach, A., Ryan, A., Duyao, M. P., Vrbanac, V., Weaver, M., Gusella, J. F., Joyner, A. L., and MacDonald, M. E. (1999). Length-dependent gametic CAG repeat instability in the Huntington's disease knock-in mouse. *Hum. Mol. Genet.* **8**, 115–122.
83. Kaytor, M. D., Burright, E. N., Duvick, L. A., Zoghbi, H. Y., and Orr, H. T. (1997). Increased trinucleotide repeat instability with advanced maternal age. *Hum. Mol. Genet.* **6**, 2135–2139.
84. La Spada, A. R., Peterson, K. R., Meadows, S. A., McClain, M. E., Jeng, G., Chmelar, R. S., Haugen, H. A., Chen, K., Singer, M. J., Moore, D., Trask, B. J., Fischbeck, K. H., Clegg, C. H., and McKnight, G. S. (1998). Androgen receptor YAC transgenic mice carrying CAG 45 alleles show trinucleotide repeat instability. *Hum. Mol. Genet.* **7**, 959–967.
85. Libby, R. T., Monckton, D. G., Fu, Y. H., Martinez, R. A., McAbney, J. P., Lau, R., Einum, D. D., Nichol, K., Ware, C. B., Ptacek, L. J., Pearson, C. E., and La Spada, A. R. (2003). Genomic context drives SCA7 CAG repeat instability, while expressed SCA7 cDNAs are intergenerationally and somatically stable in transgenic mice. *Hum. Mol. Genet.* **12**, 41–50.
86. Shelbourne, P. F., Killeen, N., Hevner, R. F., Johnston, H. M., Tecott, L., Lewandoski, M., Ennis, M., Ramirez, L., Li, Z., Iannicola, C., Littman, D. R., and Myers, R. M. (1999). A Huntington's disease CAG expansion at the murine Hdh locus is unstable and associated with behavioural abnormalities in mice. *Hum. Mol. Genet.* **8**, 763–774.
87. Ishiguro, H., Yamada, K., Sawada, H., Nishii, K., Ichino, N., Sawada, M., Kurosawa, Y., Matsushita, N., Kobayashi, K., Goto, J., Hashida, H., Masuda, N., Kanazawa, I., and Nagatsu, T. (2001). Age-dependent and tissue-specific CAG repeat instability occurs

in mouse knock-in for a mutant Huntington's disease gene. *J. Neurosci. Res.* **65**, 289–297.

88. Lorenzetti, D., Watase, K., Xu, B., Matzuk, M. M., Orr, H. T., and Zoghbi, H. Y. (2000). Repeat instability and motor incoordination in mice with a targeted expanded CAG repeat in the Sca1 locus. *Hum. Mol. Genet.* **9**, 779–785.

89. Watase, K., Venken, K. J., Sun, Y., Orr, H. T., and Zoghbi, H. Y. (2003). Regional differences of somatic CAG repeat instability do not account for selective neuronal vulnerability in a knock-in mouse model of SCA1. *Hum. Mol. Genet.* **12**, 2789–2795.

90. Kennedy, L., and Shelbourne, P. F. (2000). Dramatic mutation instability in HD mouse striatum: Does polyglutamine load contribute to cell-specific vulnerability in Huntington's disease? *Hum. Mol. Genet.* **9**, 2539–2544.

91. Miranda, C. J., Santos, M. M., Ohshima, K., Smith, J., Li, L., Bunting, M., Cossee, M., Koenig, M., Sequeiros, J., Kaplan, J., and Pandolfo, M. (2002). Frataxin knockin mouse. *FEBS Lett.* **512**, 291–297.

92. Gourdon, G., Dessen, P., Lia, A. S., Junien, C., and Hofmann-Radvanyi, H. (1997). Intriguing association between disease associated unstable trinucleotide repeat and CpG island. *Ann. Genet.* **40**, 73–77.

93. Brock, G. J., Anderson, N. H., and Monckton, D. G. (1999). Cis-acting modifiers of expanded CAG/CTG triplet repeat expandability: Associations with flanking GC content and proximity to CpG islands. *Hum. Mol. Genet.* **8**, 1061–1067.

94. Fortune, M. T., Vassilopoulos, C., Coolbaugh, M. I., Siciliano, M. J., and Monckton, D. G. (2000). Dramatic, expansion-biased, age-dependent, tissue-specific somatic mosaicism in a transgenic mouse model of triplet repeat instability. *Hum. Mol. Genet.* **9**, 439–445.

95. Zhang, Y., Monckton, D. G., Siciliano, M. J., Connor, T. H., and Meistrich, M. L. (2002). Age and insertion site dependence of repeat number instability of a human DM1 transgene in individual mouse sperm. *Hum. Mol. Genet.* **11**, 791–798.

96. Sato, T., Oyake, M., Nakamura, K., Nakao, K., Fukusima, Y., Onodera, O., Igarashi, S., Takano, H., Kikugawa, K., Ishida, Y., Shimohata, T., Koide, R., Ikeuchi, T., Tanaka, H., Futamura, N., Matsumura, R., Takayanagi, T., Tanaka, F., Sobue, G., Komure, O., Takahashi, M., Sano, A., Ichikawa, Y., Goto, J., Kanazawa, I., Katsuki, M., and Tsuji, S. (1999). Transgenic mice harboring a full-length human mutant DRPLA gene exhibit age-dependent intergenerational and somatic instabilities of CAG repeats comparable with those in DRPLA patients. *Hum. Mol. Genet.* **8**, 99–106.

97. Cleary, J. D., Nichol, K., Wang, Y. H., and Pearson, C. E. (2002). Evidence of cis-acting factors in replication-mediated trinucleotide repeat instability in primate cells. *Nat. Genet.* **31**, 37–46.

98. La Spada, A. R., Roling, D. B., Harding, A. E., Warner, C. L., Spiegel, R., Hausmanowa-Petrusewicz, I., Yee, W. C., and Fischbeck, K. H. (1992). Meiotic stability and genotype–phenotype correlation of the trinucleotide repeat in X-linked spinal and bulbar muscular atrophy. *Nat. Genet.* **2**, 301–304.

99. Antequera, F. (2003). Structure, function and evolution of CpG island promoters. *Cell. Mol. Life Sci.* **60**, 1647–1658.

100. Delgado, S., Gomez, M., Bird, A., and Antequera, F. (1998). Initiation of DNA replication at CpG islands in mammalian chromosomes. *EMBO J.* **17**, 2426–2435.

101. Parniewski, P., Jaworski, A., Wells, R. D., and Bowater, R. P. (2000). Length of CTG.CAG repeats determines the influence of mismatch repair on genetic instability. *J. Mol. Biol.* **299**, 865–874.

102. Schweitzer, J. K., and Livingston, D. M. (1999). The effect of DNA replication mutations on CAG tract stability in yeast. *Genetics* **152**, 953–963.

103. Ashizawa, T., Dunne, P. W., Ward, P. A., Seltzer, W. K., and Richards, C. S. (1994). Effects of the sex of myotonic dystrophy patients on the unstable triplet repeat in their affected offspring. *Neurology* **44**, 120–122.

104. Richards, R. I., and Sutherland, G. R. (1994). Simple repeat DNA is not replicated simply. *Nat. Genet.* **6**, 114–116.

105. Gomes-Pereira, M., and Monckton, D. G. (2004). Chemically induced increases and decreases in the rate of expansion of a CAG·CTG triplet repeat. *Nucleic Acids Res.* **32**, 2865–2872.

106. Marti, T. M., Kunz, C., and Fleck, O. (2002). DNA mismatch repair and mutation avoidance pathways. *J. Cell. Physiol.* **191**, 28–41.

107. Peltomaki, P. (2001). Deficient DNA mismatch repair: A common etiologic factor for colon cancer. *Hum. Mol. Genet.* **10**, 735–740.

108. Manley, K., Shirley, T. L., Flaherty, L., and Messer, A. (1999). Msh2 deficiency prevents in vivo somatic instability of the CAG repeat in Huntington disease transgenic mice. *Nat. Genet.* **23**, 471–473.

109. Kovtun, I. V., Therneau, T. M., and McMurray, C. T. (2000). Gender of the embryo contributes to CAG instability in transgenic mice containing a Huntington's disease gene. *Hum. Mol. Genet.* **9**, 2767–2775.

110. Wheeler, V. C., Lebel, L. A., Vrbanac, V., Teed, A., te Riele, H., and MacDonald, M. E. (2003). Mismatch repair gene Msh2 modifies the timing of early disease in Hdh(Q111) striatum. *Hum. Mol. Genet.* **12**, 273–281.

111. Savouret, C., Brisson, E., Essers, J., Kanaar, R., Pastink, A., te Riele, H., Junien, C., and Gourdon, G. (2003). CTG repeat instability and size variation timing in DNA repair-deficient mice. *EMBO J.* **22**, 2264–2273.

112. Kovtun, I. V., and McMurray, C. T. (2001). Trinucleotide expansion in haploid germ cells by gap repair. *Nat. Genet.* **27**, 407–411.

113. Pearson, C. E., Ewel, A., Acharya, S., Fishel, R. A., and Sinden, R. R. (1997). Human MSH2 binds to trinucleotide repeat DNA structures associated with neurodegenerative diseases. *Hum. Mol. Genet.* **6**, 1117–1123.

114. Gomes-Pereira, M., Fortune, M. T., Ingram, L., McAbney, J. P., and Monckton, D. G. (2004). Pms2 is a genetic enhancer of trinucleotide CAG/CTG repeat somatic mosaicism: Implications for the mechanism of triplet repeat expansion. *Hum. Mol. Genet.* **13**, 1815–1825.

115. Jakupciak, J. P., and Wells, R. D. (1999). Genetic instabilities in (CTG.CAG) repeats occur by recombination. *J. Biol. Chem.* **274**, 23468–23479.

116. Jakupciak, J. P., and Wells, R. D. (2000). Gene conversion (recombination) mediates expansions of CTG·CAG repeats. *J. Biol. Chem.* **275**, 40003–40013.

117. Jankowski, C., Nasar, F., and Nag, D. K. (2000). Meiotic instability of CAG repeat tracts occurs by double-strand break repair in yeast. *Proc. Natl. Acad. Sci. USA* **97**, 2134–2139.

118. Richard, G. F., Goellner, G. M., McMurray, C. T., and Haber, J. E. (2000). Recombination-induced CAG trinucleotide repeat expansions in yeast involve the MRE11–RAD50–XRS2 complex. *EMBO J.* **19**, 2381–2390.

119. Jeffreys, A. J., Tamaki, K., MacLeod, A., Monckton, D. G., Neil, D. L., and Armour, J. A. (1994). Complex gene conversion events in germline mutation at human minisatellites. *Nat. Genet.* **6**, 136–145.

120. Freudenreich, C. H., Kantrow, S. M., and Zakian, V. A. (1998). Expansion and length-dependent fragility of CTG repeats in yeast. *Science* **279**, 853–856.

121. van Gent, D. C., Hoeijmakers, J. H., and Kanaar, R. (2001). Chromosomal stability and the DNA double-stranded break connection. *Nat. Rev. Genet.* **2**, 196–206.

122. Paques, F., and Haber, J. E. (1999). Multiple pathways of recombination induced by double-strand breaks in *Saccharomyces cerevisiae*. *Microbiol. Mol. Biol. Rev.* **63**, 349–404.

123. Sugawara, N., Paques, F., Colaiacovo, M., and Haber, J. E. (1997). Role of *Saccharomyces cerevisiae* Msh2 and Msh3 repair proteins in double-strand break-induced recombination. *Proc. Natl. Acad. Sci. USA* **94**, 9214–9219.

124. Haber, J. E. (2000). Partners and pathways repairing a double-strand break. *Trends Genet.* **16**, 259–264.

125. Kovtun, I. V., Welch, G., Guthrie, H. D., Hafner, K. L., and McMurray, C. T. (2004). CAG repeat lengths in X- and Y-bearing sperm indicate that gender bias during transmission of Huntington's disease gene is determined in the embryo. *J. Biol. Chem.* **279**, 9389–9391.

126. Martorell, L., Monckton, D. G., Gamez, J., and Baiget, M. (2000). Complex patterns of male germline instability and somatic mosaicism in myotonic dystrophy type 1. *Eur. J. Hum. Genet.* **8**, 423–430.

127. Leeflang, E. P., Tavare, S., Marjoram, P., Neal, C. O., Srinidhi, J., MacFarlane, H., MacDonald, M. E., Gusella, J. F., de Young, M., Wexler, N. S., and Arnheim, N. (1999). Analysis of germline mutation spectra at the Huntington's disease locus supports a mitotic mutation mechanism. *Hum. Mol. Genet.* **8**, 173–183.

128. Savouret, C., Garcia-Cordier, C., Megret, J., te Riele, H., Junien, C., and Gourdon, G. (2004). MSH2-dependent germinal CTG repeat expansions are produced continuously in spermatogonia from DM1 transgenic mice. *Mol. Cell. Biol.* **24**, 629–637.

129. Bocker, T., Barusevicius, A., Snowden, T., Rasio, D., Guerrette, S., Robbins, D., Schmidt, C., Burczak, J., Croce, C. M., Copeland, T., Kovatich, A. J., and Fishel, R. (1999). hMSH5: A human MutS homologue that forms a novel heterodimer with hMSH4 and is expressed during spermatogenesis. *Cancer Res.* **59**, 816–822.

130. Richardson, L. L., Pedigo, C., and Ann Handel, M. (2000). Expression of deoxyribonucleic acid repair enzymes during spermatogenesis in mice. *Biol. Reprod.* **62**, 789–796.

131. Yoon, S. R., Dubeau, L., de Young, M., Wexler, N. S., and Arnheim, N. (2003). Huntington disease expansion mutations in humans can occur before meiosis is completed. *Proc. Natl. Acad. Sci. USA* **100**, 8834–8838.

132. Martorell, L., Johnson, K., Boucher, C. A., and Baiget, M. (1997). Somatic instability of the myotonic dystrophy (CTG)n repeat during human fetal development. *Hum. Mol. Genet.* **6**, 877–880.

133. Kovtun, I. V., Thornhill, A. R., and McMurray, C. T. (2004). Somatic deletion events occur during early embryonic development and modify the extent of CAG expansion in subsequent generations. *Hum. Mol. Genet.* **13**, 3057–3068.

Part XI

Insect Models

Drosophila Models of Polyglutamine Disorders

GEORGE R. JACKSON, TZU-KANG SANG, and J. PAUL TAYLOR

Neurogenetics Program, Department of Neurology, Brain Research Institute, Center for Neurobehavioral Genetics, Semel Institute for Neuroscience and Human Behavior, David Geffen School of Medicine at University of California at Los Angeles, Los Angeles, California 90095; Department of Neurology, University of Pennsylvania School of Medicine, Philadelphia, Pennsylvania 19104

I. Introduction
II. Study of Neurodegeneration Using Genetic Approaches
III. Fly Models of Polyglutamine Diseases
   A. Spinocerebellar Ataxia Type 3 (Machado–Joseph Disease)
   B. Huntington's Disease
   C. Pure Polyglutamine Models
   D. Spinocerebellar Ataxia Type 1
   E. Spinobulbar Muscular Atrophy
IV. Modifiers of Fly Polyglutamine Models
References

Transgenic models in the fly using targeted misexpression of human polyglutamine disease-associated proteins have been created. These include models for spinocerebellar ataxias 1 and 3, Huntington's disease, and spinobulbar muscular atrophy. In many instances, robust neurodegeneration is observed in fly models. Drosophila models are sensitized genetic systems that permit the power of fly genetics to be harnessed in an effort to identify modifier genes or compounds. Here, we review the most widely used fly models and highlight key modifier genes and drugs.

I. INTRODUCTION

One powerful means of analyzing disease pathogenesis is the use of transgenic animal models, most notably in the mouse. However, the development and analysis of mouse models can be costly, time-consuming, and frustrating. Drosophila models of neurodegenerative disease show similarities to human disease. Since 1998, a number of models of polyglutamine disorders have been created by using the genetically tractable model organism Drosophila melanogaster. Furthermore, genetic screens using fly models have begun to provide insight into the pathogenesis of polyglutamine disorders. This chapter provides an overview of fly models pertinent to polyglutamine disorders and highlights key genetic and chemical modifiers that have been identified by using such models. This work is not intended to provide a comprehensive review; rather, the reader should be able to appreciate the contributions that Drosophila models have made toward an understanding of polyglutamine disorders. The chapter is also intended to serve as a starting point for those readers who would like to obtain a more detailed understanding of fly models of neurodegenerative disorders.

II. STUDY OF NEURODEGENERATION USING GENETIC APPROACHES

Many of the genes implicated in specific human diseases have at least one homologue in *Drosophila*. A database of human disease genes that possess homologues in the fly is available on line (http://super-fly.ucsd.edu/homophila/). Homologues of human neurodegenerative disease genes can be identified in the genome of the fly, and the function of these genes can be investigated by generating mutations in the *Drosophila* homologue and then analyzing the resulting phenotypes. This approach has been used to study the fly homologue of ataxin-2, the gene mutated in spinocerebellar ataxia type 2 [1]. Investigation of the *Drosophila* homologue of atrophin, a gene in which a CAG expansion gives rise to dentatorubral pallidoluysian atrophy (DRPLA), has identified its role as a transcription factor [2]. An alternative method employs RNA interference-mediated knock-down of gene expression. This approach has been used to identify a role for the *Drosophila* homologue of huntingtin in axonal transport and cell survival [3].

A pathogenic process of interest may be analyzed in *Drosophila* so long as it can be recapitulated in a manner that shares characteristics with pathology in humans. The toolbox of *Drosophila* genetics then may be used in order to study the pathogenic process. Mutations can be identified that affect this process without making biased assumptions about the relevant pathways. This potential for genetic approaches to shed light on disease pathogenesis makes flies a powerful model system in neurodegeneration research.

The assumption that key features underlying the biology of *Drosophila* and humans are conserved underlies invertebrate models of polyglutamine disorders. In general, fundamental aspects of cell biology in flies are surprisingly similar to those of humans in domains including membrane trafficking, the cytoskeleton, neuronal connectivity, synaptogenesis, and cell signaling and survival. Of course, important differences between flies and humans exist that must be taken into consideration in the interpretation of genetic models. The circulatory systems and cognitive processes of flies are much simpler than those of humans. However, in some cases the relative simplicity of *Drosophila* as compared to human gene organization proves beneficial in genetic analysis. Redundancy may exist in humans, such that duplicated versions of genes exist that are present in only one copy in flies; the study of biological processes in the fly may be simplified by this lack of redundancy. The most important aspect of invertebrate approaches is that a number of genetic manipulations can be carried out that are impossible or impractical to carry out in mammals. Large numbers of flies can be mutagenized and analyzed in a short period of time. Given the considerable success that invertebrate approaches have had in delineating processes such as cell cycle control, signal transduction, and pattern formation, similar approaches to the study of polyglutamine disorders are likely to continue to yield powerful insights into disease mechanisms.

In the early 20th century, a *Drosophila melanogaster* mutant was identified in the Morgan lab at Columbia when a white-eyed fly was noted among wild-type red-eyed flies. The eye continues to be the focus of research not only because adult eye phenotypes are easily detected but also because, unlike most organs in the fly, the eye is tolerant of genetic disruption of essential biological processes. In addition, the eye is dispensable for survival of the fly under laboratory conditions. A number of technologies can be utilized to generate, identify, and characterize mutations in the eye, giving it unrivaled potential for deciphering gene function.

III. FLY MODELS OF POLYGLUTAMINE DISEASES

Huntington's disease (HD) is the prototypic disease caused by an unstable CAG repeat expansion. HD is an autosomal dominant neurodegenerative disorder characterized by chorea, dementia, and affective changes [4]. It is caused by the expansion of an unstable CAG tract within exon 1 of the huntingtin gene, resulting in the expression of an expanded polyglutamine tract near the amino terminus of huntingtin [5, 6]. At least nine other neurodegenerative disorders are associated with expansion of polyglutamine, including autosomal dominant spinocerebellar ataxias (SCA1, SCA2, Machado–Joseph disease or SCA3, SCA6, SCA7, SCA12, and SCA17), spinobulbar muscular atrophy (SBMA), and dentatorubral pallidoluysian atrophy (DRPLA) [7]. Post mortem HD brain shows neuronal loss from the cortex and striatum, as well as cytoplasmic and nuclear protein aggregates containing amino-terminal fragments of mutant huntingtin [5]. Table 37-1 summarizes a number of glutamine repeat disorders that have been modeled in *Drosophila*. These models are described in this section, with more attention paid to those models that have become most widely used or with which the authors have experience.

A. Spinocerebellar Ataxia Type 3 (Machado–Joseph Disease)

In 1998, the first two *Drosophila* models of polyglutaminopathy appeared [8, 9]. The first of these [8] expressed a truncated form of ataxin-3/MJD, the mutant protein in

TABLE 37-1 Summary of Fly Polyglutamine Models

Protein context	Repeat length	Reference numbers	Promoters
Ataxin-3	Truncated Q78	8	glass, 24B, dpp, elav
		3	Appl, Da
		19	Loco, M1B
		18	30Y
	Full length Q78	8	glass
	Full length Q84	8	glass
Huntingtin	Truncated Q75 (1–171)	9	glass (direct fusion)
	Truncated Q120 (1–171)	9	glass (direct fusion)
	Truncated Q93 (exon 1)	21	elav
		26	OK107
		23	Rh1
		62	da
	Truncated Q97 (exon 1)	22	elav, glass
		26	OK107
	Truncated Q103deltaP	22	elav
	Truncated Q128 (1–548)	27	glass, elav, hsp70
"Pure" polyglutamine	Q48	63	elav
	Q108	28	glass, elav, dpp, sev
		3	Appl
	Q63	64	glass, Appl
	Q127	31	glass, Appl
	Q79	65	glass
	Q92	65	glass
Ataxin-1	Full length Q82	33	elav, ddc, glass
	Full length Q82	34	glass, ap
		66	Rh1, C5, C109–68
		18	chat
Androgen receptor	Full length Q52	39	glass
	Truncated Q52 (androgen binding domain)	39	glass
	Truncated Q112 (amino terminus)	43	glass

spinocerebellar ataxia type 3 (Machado–Joseph disease, MJD) [10]. Clinical features of SCA3 include ataxia, dystonia, tremor, pyramidal signs, and fasciculations. Bonini and coworkers used the binary GAL4/UAS expression system [11]. In the MJD protein, the unstable polyglutamine repeat lies toward the carboxyl terminus. A pathogenic repeat length of Q78 was engineered into the UAS vector. Expression of these constructs was driven by crossing with stocks expressing the yeast transcriptional activator GAL4 under control of various tissue-specific enhancers.

These enhancers included the panneuronal *elav (embryonic lethal abnormal vision)* GAL4, which is expressed in all neurons, from embryos to adults [12, 13]. Nonneural drivers included 24B-GAL4, which expresses in presumptive mesoderm and muscle [11], and *dpp (decapentaplegic)* GAL4 [14], which is expressed in all imaginal disks at the anterior–posterior boundary. Bonini and coworkers also used the GMR-GAL4 driver (*glass* multimer reporter) [15, 16]. *glass* is a zinc finger transcription factor expressed in the eye disk beginning in third instar larvae [17]. During this stage, the homogeneous eye epithelium differentiates into specialized cell types comprising the adult eye, including photoreceptor and mechanosensory neurons, pigment cells, and cone cells. Differentiation occurs as the morphogenetic furrow progresses, with successive differentiation of cells into ommatidial preclusters within and posterior to the furrow.

Weak and medium strength Q78 lines expressed in presumptive mesoderm and muscle cells were larval lethal with this 24B driver, whereas the strong Q78 insertion was embryonic–early larval lethal. Expression using the *dpp*-GAL4 driver yielded no phenotypes. Directed expression to all cells of the peripheral and central nervous systems (PNS and CNS) using the *elav*-GAL driver was lethal for the strongest insertion of Q78, whereas other Q78 insertions were viable but resulted in early adult death. Using the medium strength Q78 insertion line, abnormal retinal morphology resulted with the *elav*-GAL4 driver even at eclosion; this progressed to widespread degeneration by 4 days posteclosion.

Dramatic results were obtained by using GMR-GAL4. The weak Q78 line showed a mildly rough external appearance at eclosion with progressive depigmentation thereafter, presumably due to pigment cell degeneration. The medium and strong insertions showed a rough, depigmented eye at eclosion; histology showed cell loss that was progressively more severe with the stronger Q78 insertion lines examined.

Localization of MJD transgene products was examined in third instar larval eye disks by immunohistochemistry. The Q78 transgene product was cytoplasmic just posterior to the morphogenetic furrow but showed progressive nuclear localization in older cells farther from the furrow.

Other workers have analyzed phenotypes of Q78 in the embryo under control of *daughterless* [3], in the CNS by using different neuronal drivers such as *Appl* [3] and *choline acetyltransferase* [18], and in glia by using *loco* and *M1B* [19].

B. Huntington's Disease

The second fly model of polyglutamine disease to appear used amino-terminal fragments of huntingtin cDNA encoding 75 or 120 glutamine residues [9]. This fragment includes exons 2, 3, and a portion of 4. By using the numbering system of the published IT15 sequence encoding 23 repeats, these constructs encode the first 171 amino acids of human huntingtin. These cDNAs were subcloned directly into GMR [15], which uses a minimal HSP70 promoter and the *glass* enhancer to drive expression in all cells of the eye from third instar larval through adult life without the need for crosses with a GAL4 line. Q75 lines showed some late onset degeneration of photoreceptor neurons; however, demonstration of such effects required four copies of the transgene. Q120 lines bearing single copies of the transgene showed normal external morphology of the eye and normal retinal morphology at eclosion; however, a subset of Q120 lines showed photoreceptor neuron degeneration beginning at 4 days and progressing thereafter. This progressive degeneration can be assessed in living flies by using the optical neutralization or pseudo-pupil technique [20]. Histology at 10 days showed disruption of retinal morphology with intense staining in degenerating photoreceptor cell bodies and loss of rhabdomeres. Ultrastructural analysis of degenerating photoreceptors revealed some apoptotic features, including nuclear and cytoplasmic condensation and chromatin clumping, with preservation of mitochondria. Immunohistochemical analysis using an antibody recognizing the amino terminus of human huntingtin showed cytoplasmic localization in third instar larval eye disks regardless of repeat length. Huntingtin immunoreactivity in Q75 lines was cytoplasmic at eclosion but showed some nuclear staining, in addition, at 10 days. The Q120 product showed both cytoplasmic and diffuse nuclear staining at eclosion, but by 10 days showed some punctate immunoreactivity indicating aggregate formation.

Thompson and coworkers developed UAS-huntingtin lines by using exon 1 constructs including repeat lengths of Q93 [21] and Q97 [22]. The former does not produce clear external eye phenotypes under control of GMR-GAL4, although the latter has been reported to do so at elevated temperatures [22] (GAL4-driven phenotypes are highly temperature-sensitive). However, under control of GMR-GAL4, UAS-Q93 lines do produce marked thinning of the retina when viewed by using longitudinal sections [23]. The majority of reported work using Q93 lines has estimated photoreceptor loss by using optical neutralization under control of *elav*-GAL4 [21, 22, 24, 25]. Late onset loss of outer photoreceptors for UAS-Q93 under control of Rh1-GAL4, as well as formation of aggregates immunoreactive for the conformation-dependent epitope EM48, has been demonstrated [23]. Degeneration of CNS when UAS-Q93 is driven under control of the mushroom body-specific driver *OK107* also has been reported [26]. Phenotypes of expanded polyglutamine within larger fragments of huntingtin have been reported [27]. As of autumn 2005, no full length model of HD has appeared in a peer-reviewed publication.

C. Pure Polyglutamine Models

The protein context in which expanded polyglutamine tracts appear is critical to phenotypic effects observed in *Drosophila* models. Other investigators have addressed the importance of protein context by expressing quasi-pure polyglutamine peptides in transgenic animals. Thompson and coworkers expressed CAG tracts encoding 108 repeats flanked by six additional amino- and four carboxy-terminal amino acids by using the GAL4–UAS system [28]. Phenotypic effects were analyzed by using drivers including *elav*-GAL4, GMR-GAL4, *sev*-GAL4, and *dpp*-GAL4. Phenotypic effects of the Q108 transgene were generally more severe than those obtained with the truncated MJD construct; in many cases lethality was complete. By using the panneural *elav*-GAL4 driver to express UAS-Q108, incomplete lethality was observed; survivors showed early adult death. By using the *sev*-GAL4 driver [29], which drives expression beginning in third instar larval eye disks in a subset of photoreceptor precursors (R3, R4, and R7) as well as cone cell precursors [30], the UAS-Q108 transgene was lethal. Sensory bristles were absent from surviving eyes. By using GMR-GAL4, significant lethality was obtained, likely due to "leaky" expression outside the eye. Surviving eyes showed a severe external phenotype, including the absence of pigment and necrotic patches. Ultrastructural analysis of GMR-GAL4-driven UAS-Q108 has revealed features of autophagy and apoptosis [23]. In contrast to the benign effects of expanded repeats expressed within the MJD fragment, *dpp*-GAL4-driven UAS-Q108 resulted in pupal lethality. Dissected pupae showed abnormal morphology of the head and mesothorax, as well as missing third legs. Degenerative phenotypes of Q108 driven in outer photoreceptors only by using the Rh1-GAL4 driver have been analyzed by using confocal and transmission electron microscopy [23].

A number of other models using quasi-pure polyglutamine tracts have been reported. Among these is the Q127 reported by Kazemi-Esfarjani and Benzer [31]. These authors used a P element screen to identify a fly HSP40 protein as a suppressor of their GMR-GAL4-driven Q127 phenotype (see Section IV for further discussion of HSP modifiers).

D. Spinocerebellar Ataxia Type 1

SCA1 was the first of the autosomal dominant ataxias to be identified as a polyglutamine disorder [42]. Clinical features, apart from ataxia, may include pyramidal signs and neuropathy. Feany and Bender were the first to publish a fly model of SCA1, but they did not characterize it extensively [32]. Botas and coworkers developed a fly model of SCA1 and reported the results of P element screen for modifiers [34]. Full length ataxin-1 Q82 produced rough eye phenotypes under control of GMR-GAL4. Expression in interneurons using *apterous*-GAL4 also produced neurodegeneration.

E. Spinobulbar Muscular Atrophy

Spinobulbar muscular atrophy (SBMA), also known as Kennedy's disease, is an X-linked disorder characterized by slowly progressive weakness resulting from degeneration of motor neurons of the brainstem and spinal cord [35]. Patients with SBMA often exhibit mild signs of feminization such as gynecomastia and, less frequently, infertility despite normal serum androgen levels. The androgen insensitivity observed in this disorder provided clues to its underlying cause, and in 1991 the molecular basis of SBMA was found to be a trinucleotide (CAG) repeat expansion in the first exon of the androgen receptor (AR) gene [36]. Normal alleles exhibit polymorphism with repeat length ranging from 9 to 36, whereas patients with SBMA always have repeat sizes greater than 40 (the longest reported to date is 62 repeats). As with many other repeat expansion diseases, a correlation exists between repeat length and clinical manifestations in SBMA (37).

The AR is a ligand-dependent transcription factor belonging to the nuclear hormone receptor superfamily [46]. Similar to other nuclear hormone receptors, the AR becomes activated after binding by a ligand. The primary ligands for AR are testosterone and its more potent derivative dihydrotestosterone (DHT). Ligand binding to the C-terminal ligand binding domain is an important regulatory event in AR function, resulting in a conformational change that disrupts an intramolecular interaction between the amino terminus and the carboxy terminus, initiates posttranslational modifications, dissociates the AR from heat shock proteins (HSPs), and results in translocation of the active receptor–ligand complex to the nucleus, where it alters the transcription of target genes. The polyglutamine stretch is within the amino-terminal transactivation domain of the AR.

The clinical manifestations of SBMA are restricted to males, whereas female carriers are unaffected. This gender specificity has been difficult to reconcile with the toxic gain of function mechanism presumably shared by all polyglutamine diseases. Until recently, two hypotheses competed to explain gender specificity. The first proposed that females were protected from clinical manifestations by random X inactivation, resulting in only ~50% of the motor neurons expressing the mutant form of AR. This hypothesis presumed that resprouting from the remaining healthy motor neurons was sufficient to protect females from clinical signs and symptoms. The alternative hypothesis proposed that the toxicity of polyglutamine-expanded AR is ligand-dependent, restricting the phenotype to males because of their substantially higher androgen levels. Other studies, in part performed in *Drosophila*, have resolved this issue with surprising clarity.

Takeyama and colleagues successfully recapitulated SBMA in a fly model by expressing full length AR protein harboring a 52-glutamine stretch in the *Drosophila* eye [27]. Despite high-level expression, the eyes appeared normal. However, exposure of these flies to DHT, a potent ligand of the androgen receptor, resulted in substantial eye degeneration in a polyglutamine-length-dependent manner. These data indicate that a ligand-dependent event is required to convert polyglutamine-expanded AR to its toxic form. Consistent with this idea, flies expressing a truncated version of the AR consisting of the N-terminal transactivation domain, which includes the polyglutamine stretch, led to ligand-independent degeneration; and this phenotype was suppressed by coexpressing the unliganded C-terminal AR fragment. The nature of the conversion of AR to a toxic form remains elusive, but it conceivably may involve any of the ligand-dependent alterations to AR that are known to occur, including dissociation of the amino terminus and carboxy terminus, posttranslational modifications, dissociation from HSPs, or nuclear translocation. Degeneration likely is not caused by AR-dependent changes in transcription, however, because equivalent or greater ligand-dependent degeneration occurs with several AR antagonists [47]. The ligand-dependent nature of polyglutamine-expanded AR toxicity has been confirmed in two transgenic mouse models of SBMA [48, 49]. Bonini and coworkers also have reported a model using a truncated fragment of AR [28].

IV. MODIFIERS OF FLY POLYGLUTAMINE MODELS

Polyglutamine-containing aggregates in mouse and cell models contain a number of additional proteins, including molecular chaperones [50–53]. Fly models of polyglutamine disease also accumulate some of these components in inclusions [25, 54, 55]. Both candidate-based and unbiased genetic screens in the fly have begun to identify polyglutamine modifiers, which include chaperones. These include HSP70; overexpression of a human HSP70, HSPAL, suppresses the eye phenotype of MJD-Q78 and GMR-huntingtin-Q120 flies [54]. An unbiased transposon-based screen subsequently identified *Drosophila* HDJ1, an HSP40; misexpression of HDJ1 suppressed the quasi-pure Q127 phenotype [22]. Moderate rescue of the rough eye phenotype is observed when HSPA1L is expressed with Q108 [55]. In contrast, the eye phenotype of Q108 is dramatically suppressed by HDJ [55]. The eye phenotype of ataxin-1 is also suppressed by HSPA1L and HDJ1 [10].

The role of apoptosis in glutamine repeat disorders is controversial; however, in the case of HD, enough evidence has been derived from animal models implicating caspases-1 and -3 to support trials of minocycline, an antibiotic with caspase inhibitor activity [56–59]. In flies, expression of the baculoviral antiapoptotic protein P35 or the functionally similar fly protein DIAP1 suppresses phenotypes caused by cell death genes [34, 60]. Results of experiments testing suppression of polyglutamine phenotypes in the fly eye have been inconsistent. P35 suppresses ataxin-1 [10] and -3 [8, 10] phenotypes, but it enhances Q127 [10] and has no clear effect on Q108 [55] or htt-Q120 [12]. Similar inconsistency is observed for DIAP1: suppression for ataxins-1 and -3 [10], no effect on Q127 [10], Q108 [55], or htt-Q120 (Salecker and Jackson, unpublished data). These data indicate that the protein context in which polyglutamine is expressed determines its responsiveness to modifiers.

A candidate-based approach revealed striking suppression of Q108 and htt exon1 phenotypes by inactivation of *Drosophila* Apaf-1, Dark [15]. In a genetic background in which both copies of Dark had been mutated, strong suppression of Q108 and huntingtin exon 1 phenotypes occurred, as well as suppression of caspase activation, cell death, and aggregate formation. An example of suppression of the UAS-Q108 phenotype is illustrated in Fig. 37-1. The observation that virtually complete suppression of polyglutamine toxicity *in vivo* is accompanied by diminshed aggregate formation

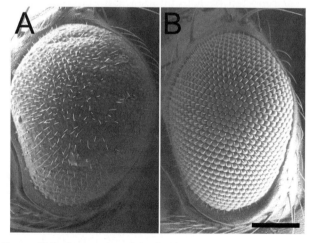

FIGURE 37-1 Suppression of the Q108 eye phenotype (A) by homozygous mutation of *dark* (B). Genotypes: (A) GMR-GAL4/+, +, UAS-Q108/+; (B) GMR-GAL4/+, *dark*^{CD4}, UAS-Q108/+. Scale bar = 100 μm.

could be considered evidence that aggregates are toxic, although this is highly controversial [51]. Misexpression of the *Drosophila* Bax-like protein Drob-1 suppresses retinal phenotypes of MJD-Q78, whereas overexpression of the Drob-1 antagonist Buffy enhances Q78 effects [56].

Drosophila models have also shown their worth in the validation of small compounds predicted to inhibit polyglutamine pathogenesis on the basis of cell-based studies. Studies of histone deacetylase inhibitors as suppressors of mutant huntingtin toxicity in the fly have led to validation in mouse models and clinical trials of phenylbutyrate in HD [1, 57, 58]. Rapamycin has been shown to suppress htt-Q120 toxicity, supporting a role for autophagy in polyglutamine disease [59]. Other studies have identified an inhibitor of the Rho-associated kinase p160ROCK as a suppressor of mutant htt exon 1 phenotypes in the fly [25]. Lithium was identified as a suppressor of GMR-htt-Q120 on the basis of the optical neutralization technique [60].

Acknowledgments

This work was supported by grants from the National Institutes of Health (NS002116, NS002116, and AG016570 to GRJ and NS044125 to JPT) and by a John J. Wasmuth Postdoctoral Fellowship of the Hereditary Disease Foundation to TKS.

References

1. Satterfield, T. F., Jackson, S. M., and Pallanck, L. J. (2002). A Drosophila homolog of the polyglutamine disease gene SCA2 is a dosage-sensitive regulator of actin filament formation. *Genetics* **162**, 1687–1702.
2. Zhang, S., Xu, L., Lee, J., and Xu, T. (2002). Drosophila atrophin homolog functions as a transcriptional corepressor in multiple developmental processes. *Cell* **108**, 45–56.
3. Gunawardena, S. et al. (2003). Disruption of axonal transport by loss of huntingtin or expression of pathogenic polyQ proteins in Drosophila. *Neuron* **40**, 25–40.
4. Folstein, S. E. (1989). "Huntington's Disease: A Disorders of Families," Johns Hopkins University Press, Baltimore.
5. Bates, G. (2003). Huntingtin aggregation and toxicity in Huntington's disease. *Lancet* **361**, 1642–1644.
6. Young, A. B. (2003). Huntingtin in health and disease. *J Clin Invest* **111**, 299–302.
7. Zoghbi, H. Y., and Orr, H. T. (2000). Glutamine repeats and neurodegeneration. *Annu. Rev. Neurosci.* **23**, 217–247.
8. Warrick, J. M., et al. (1998). Expanded polyglutamine protein forms nuclear inclusions and causes neural degeneration in Drosophila. *Cell* **93**, 939–949.
9. Jackson, G. R., et al. (1998). Polyglutamine-expanded human huntingtin transgenes induce degeneration of Drosophila photoreceptor neurons. *Neuron* **21**, 633–642.
10. Kawaguchi, Y., et al. (1994). CAG expansions in a novel gene for Machado-Joseph disease at chromosome 14q32.1. *Nat. Genet.* **8**, 221–228.
11. Brand, A. H., and Perrimon, N. (1993). Targeted gene expression as a means of altering cell fates and generating dominant phenotypes. *Development* **118**, 401–415.
12. Robinow, S., and White, K. (1988). The locus elav of Drosophila melanogaster is expressed in neurons at all developmental stages. *Developmental Biology* **126**, 294–303.
13. Lin, D. M., and Goodman, C. S. (1994). Ectopic and increased expression of Fasciclin II alters motoneuron growth cone guidance. *Neuron* **13**, 507–523.
14. Wilder, E. L., and Perrimon, N. (1995). Dual functions of wingless in the Drosophila leg imaginal disc. *Development* **121**, 477–488.
15. Hay, B. A., Wolff, T., and Rubin, G. M. (1994). Expression of baculovirus P35 prevents cell death in Drosophila. *Development* **120**, 2121–2129.
16. Freeman, M. (1996). Reiterative use of the EGF receptor triggers differentiation of all cell types in the Drosophila eye. *Cell* **87**, 651–660.
17. Ellis, M. C., O'Neill, E. M., and Rubin, G. M. (1993). Expression of Drosophila glass protein and evidence for negative regulation of its activity in non-neuronal cells by another DNA-binding protein. *Development* **119**, 855–865.
18. Ghosh, S., and Feany, M. B. (2004). Comparison of pathways controlling toxicity in the eye and brain in Drosophila models of human neurodegenerative diseases. *Hum. Mol. Genet.* **13**, 2011–2018.
19. Kretzschmar, D., et al. (2005). Glial and neuronal expression of polyglutamine proteins induce behavioral changes and aggregate formation in Drosophila. *Glia* **49**, 59–72.
20. Franceschini, N. (1972). Pupil and pseudopupil in the compound eye of *Drosophila*.
21. Steffan, J. S., et al. (2001). Histone deacetylase inhibitors arrest polyglutamine-dependent neurodegeneration in Drosophila. *Nature* **413**, 739–743.
22. Steffan, J. S., et al. (2004). SUMO modification of Huntingtin and Huntington's disease pathology. *Science* **304**, 100–104.
23. Sang, T. K., et al. (2005). Inactivation of Drosophila Apaf-1 related killer suppresses formation of polyglutamine aggregates and blocks polyglutamine pathogenesis. *Hum. Mol. Genet.* **14**, 357–372.
24. Steffan, J. S., and Thompson, L. M. (2003). Targeting aggregation in the development of therapeutics for the treatment of Huntington's disease and other polyglutamine repeat diseases. *Expert Opin. Ther. Targets* **7**, 201–213.
25. Pollitt, S. K., et al. (2003). A rapid cellular FRET assay of polyglutamine aggregation identifies a novel inhibitor. *Neuron* **40**, 685–694.
26. Agrawal, N., et al. (2005). Identification of combinatorial drug regimens for treatment of Huntington's disease using Drosophila. *Proc. Natl. Acad. Sci. USA* **102**, 3777–3781.
27. Lee, W. C., Yoshihara, M., and Littleton, J. T. (2004). Cytoplasmic aggregates trap polyglutamine-containing proteins and block axonal transport in a Drosophila model of Huntington's disease. *Proc. Natl. Acad. Sci. USA* **101**, 3224–3229.
28. Marsh, J. L., et al. (2000). Expanded polyglutamine peptides alone are intrinsically cytotoxic and cause neurodegeneration in Drosophila. *Hum. Mol. Genet.* **9**, 13–25.
29. Sun, X., and Artavanis-Tsakonas, S. (1997). Secreted forms of DELTA and SERRATE define antagonists of Notch signaling in Drosophila. *Development* **124**, 3439–3448.
30. Tomlinson, A., Bowtell, D. D., Hafen, E., and Rubin, G. M. (1987). Localization of the sevenless protein, a putative receptor for positional information, in the eye imaginal disc of Drosophila. *Cell* **51**, 143–50.
31. Kazemi-Esfarjani, P., and Benzer, S. (2000). Genetic suppression of polyglutamine toxicity in Drosophila. *Science* **287**, 1837–1840.
32. Orr, H. T., et al. (1993). Expansion of an unstable trinucleotide CAG repeat in spinocerebellar ataxia type 1. *Nat. Genet.* **4**, 221–226.

33. Feany, M. B., and Bender, W. W. (2000). A Drosophila model of Parkinson's disease. *Nature* **404**, 394–398.
34. Fernandez-Funez, P., et al. (2000). Identification of genes that modify ataxin-1-induced neurodegeneration. *Nature* **408**, 101–106.
35. Kennedy, W. R., Alter, M., and Sung, J. H. (1968). Progressive proximal spinal and bulbar muscular atrophy of late onset. A sex-linked recessive trait. *Neurology* **18**, 671–680.
36. La Spada, A. R., Wilson, E. M., Lubahn, D. B., Harding, A. E., and Fischbeck, K. H. (1991). Androgen receptor gene mutations in X-linked spinal and bulbar muscular atrophy. *Nature* **352**, 77–79.
37. Doyu, M., et al. (1992). Severity of X-linked recessive bulbospinal neuronopathy correlates with size of the tandem CAG repeat in androgen receptor gene. *Ann. Neurol.* **32**, 707–710.
38. Truss, M., and Beato, M. (1993). Steroid hormone receptors: interaction with deoxyribonucleic acid and transcription factors. *Endocr. Rev.* **14**, 459–479.
39. Takeyama, K., et al. (2002). Androgen-dependent neurodegeneration by polyglutamine-expanded human androgen receptor in Drosophila. *Neuron* **35**, 855–864.
40. Furutani, T., et al. (2005). Human expanded polyQ androgen receptor mutants in neurodegeneration as a novel ligand target. *J. Pharmacol. Exp. Ther.*
41. Katsuno, M., et al. (2002). Testosterone reduction prevents phenotypic expression in a transgenic mouse model of spinal and bulbar muscular atrophy. *Neuron* **35**, 843–854.
42. Chevalier-Larsen, E. S., et al. (2004). Castration restores function and neurofilament alterations of aged symptomatic males in a transgenic mouse model of spinal and bulbar muscular atrophy. *J. Neurosci.* **24**, 4778–4786.
43. Chan, H. Y., Warrick, J. M., Andriola, I., Merry, D., and Bonini, N. M. (2002). Genetic modulation of polyglutamine toxicity by protein conjugation pathways in Drosophila. *Hum Mol Genet* **11**, 2895–904.
44. Cummings, C. J., et al. (1998). Chaperone suppression of aggregation and altered subcellular proteasome localization imply protein misfolding in SCA1. *Nature Genetics* **19**, 148–154.
45. Chai, Y., Koppenhafer, S. L., Shoesmith, S. J., Perez, M. K., and Paulson, H. L. (1999). Evidence for proteasome involvement in polyglutamine disease: Localization to nuclear inclusions in SCA3/MJD and suppression of polyglutamine aggregation in vitro. *Human Molecular Genetics* **8**, 673–682.
46. Chai, Y., Koppenhafer, S. L., Bonini, N. M., and Paulson, H. L. (1999). Analysis of the role of heat shock protein (Hsp) molecular chaperones in polyglutamine disease [In Process Citation]. *J. Neurosci.* **19**, 10338–10347.
47. Suhr, S. T., et al. (2001). Identities of sequestered proteins in aggregates from cells with induced polyglutamine expression. *J. Cell Biol.* **153**, 283–294.
48. Warrick, J. M., et al. (1999). Suppression of polyglutamine-mediated neurodegeneration in Drosophila by the molecular chaperone HSP70. *Nat. Genet.* **23**, 425–428.
49. Sang, T.-K., and Jackson, G. R. (2005). Drosophila models of neurodegenerative disease. *NeuroRx* **2**, 438–446.
50. Ona, V. O., et al. (1999) Inhibition of caspase-1 slows disease progression in a mouse model of Huntington's disease. *Nature* **399**, 263–267.
51. Chen, M., et al. (2000). Minocycline inhibits caspase-1 and caspase-3 expression and delays mortality in a transgenic mouse model of Huntington disease. *Nat. Med.* **6**, 797–801.
52. Hersch, S., Fink, K., Vonsattel, J. P., and Friedlander, R. M. (2003). Minocycline is protective in a mouse model of Huntington's disease. *Ann. Neurol.* **54**, 841; author reply 842–843.
53. Thomas, M., Ashizawa, T., and Jankovic, J. (2004). Minocycline in Huntington's disease: A pilot study. *Mov. Disord.* **19**, 692–695.
54. Hay, B. A., Wassarman, D. A., and Rubin, G. M. (1995). Drosophila homologs of baculovirus inhibitor of apoptosis proteins function to block cell death. *Cell* **83**, 1253–1262.
55. Arrasate, M., Mitra, S., Schweitzer, E. S., Segal, M. R., and Finkbeiner, S. (2004). Inclusion body formation reduces levels of mutant huntingtin and the risk of neuronal death. *Nature* **431**, 805–810.
56. Senoo-Matsuda, N., Igaki, T., and Miura, M. (2005). Bax-like protein Drob-1 protects neurons from expanded polyglutamine-induced toxicity in Drosophila. *Embo. J.* **24**, 2700–2713.
57. Ferrante, R. J., et al. (2003). Histone deacetylase inhibition by sodium butyrate chemotherapy ameliorates the neurodegenerative phenotype in Huntington's disease mice. *J. Neurosci.* **23**, 9418–9427.
58. Hockly, E., et al. (2003). Suberoylanilide hydroxamic acid, a histone deacetylase inhibitor, ameliorates motor deficits in a mouse model of Huntington's disease. *Proc. Natl. Acad. Sci. USA* **100**, 2041–2046.
59. Ravikumar, B., et al. (2004). Inhibition of mTOR induces autophagy and reduces toxicity of polyglutamine expansions in fly and mouse models of Huntington disease. *Nat. Genet.* **36**, 585–595.
60. Berger, Z., et al. (2005). Lithium rescues toxicity of aggregate-prone proteins in Drosophila by perturbing Wnt pathway. *Hum. Mol. Genet.*
61. Warrick, J. M., et al. (2005). Ataxin-3 suppresses polyglutamine neurodegeneration in Drosophila by a ubiquitin-associated mechanism. *Mol. Cell* **18**, 37–48.
62. Lievens, J. C., Rival, T., Iche, M., Chneiweiss, H., and Birman, S. (2005). Expanded polyglutamine peptides disrupt EGF receptor signaling and glutamate transporter expression in Drosophila. *Hum. Mol. Genet.* **14**, 713–724.
63. Kazantsev, A., Walker, H. A., Slepko, N., Bear, J. E., Preisinger, E., Steffan, J. S., Zhu, Y.-Z., Gertler, F. B., Housman, D. E., Marsh, J. L., and Thompson, L. M. (2002). A bivalent Huntingtin binding peptide suppresses polyglutamine aggregation and pathogenesis in Drosophila. *Nature Genetics*.
64. Kazemi-Esfarjani, P., and Benzer, S. (2002). Suppression of polyglutamine toxicity by a Drosophila homolog of myeloid leukemia factor 1. *Hum. Mol. Genet.* **11**, 2657–2672.
65. Higashiyama, H., et al. (2002). Identification of ter94, Drosophila VCP, as a modulator of polyglutamine-induced neurodegeneration. *Cell Death Differ.* **9**, 264–273.
66. Tsuda, H., et al. (2005). The AXH domain of Ataxin-1 mediates neurodegeneration through its interaction with Gfi-1/Senseless proteins. *Cell* **122**, 633–644.

PART XII

Instability Mechanisms in Vivo and in Vitro

CHAPTER 38

Involvement of Genetic Recombination in Microsatellite Instability

RUHEE DERE, MICHEAL L. HEBERT, AND MAREK NAPIERALA

Institute of Biosciences and Technology, Center for Genome Research, Texas A&M University System Health Science Center, Texas Medical Center, Houston, Texas 77030-3303

I. Introduction
II. Trinucleotide Repeat Instability Associated with Recombination in Clinical Cases
III. Large CTG·CAG Expansions in *Escherichia coli* Are Caused by Gene Conversion
IV. Microsatellite Sequences Influence the Frequency of Recombination
   A. Role of Non-B-DNA Structures Formed by Microsatellite Sequences in Recombination
   B. CTG·CAG Repeats Expanded in Myotonic Dystrophy Type 1 Are Preferred Sites of Intramolecular and Intermolecular Recombination
   C. Recombination Properties of the CCTG·CAGG Tetranucleotide Repeats Associated with Myotonic Dystrophy Type 2
   D. Structure-Dependent Recombination Hot Spot Activity of GAA·TTC Sequences
V. Involvement of DNA Incisions (Breaks, Nicks, and Gaps) in Stimulation of Trinucleotide Repeat Sequence Instabilities
   A. Role of Double-Strand Breaks in Genetic Instability
   B. Effect of Single-Strand Nicks and Gaps on the Instability of CTG·CAG Repeats
VI. Concluding Remarks
References

This chapter reviews the involvement of genetic recombination in the instability of trinucleotide and tetranucleotide repeats. The role of nonreciprocal gene conversion and reciprocal crossing over in generating expansions and deletions of the repeat tracts in clinical cases, as well as in various model systems, is presented. The current status of studies related to microsatellite sequences as hot spots or cold spots of genetic recombination is also reviewed. Knowledge about the mechanisms of the expansions and the factors that influence these processes may help in the future to design therapeutic approaches aimed specifically to induce deletions of the already expanded pathogenic microsatellites or ultimately prevent the expansion from occurring.

I. INTRODUCTION

Microsatellites or short tandem repeats are DNA sequences composed of repetitive units of one to six nucleotides [1, 2]. These repetitive elements are highly abundant in many prokaryotic and especially in eukaryotic genomes [3, 4]. Microsatellites occupy approximately 3% of the human genome; they are highly polymorphic and can be found in all chromosomes in a variety of lengths [2–4].

Genetic instabilities of microsatellite sequences have been widely observed throughout the genomes of all organisms studied [5]. In the majority of cases, this phenomenon occurs without phenotypical consequences, but in some circumstances instabilities, mostly expansions of the repeat tracts, result in the development of disease [5, 6]. Expansions of tri-, tetra-, and pentanucleotide microsatellites have been identified as a cause of more than 20 neurological diseases, including myotonic dystrophy types 1 and 2 (DM1, DM2), fragile X syndrome, Friedreich's ataxia, and the spinocerebellar ataxias (SCA) [5, 6]. These expansions, which can be enormous (frequently reaching several thousands of repeats), are always limited to a single disease *locus* in the genome [6]. The length of the repeats in the other *loci*, even those containing repeated sequences identical to the expanded tract, remains unchanged. This *locus*-specific instability is a characteristic feature of the expansions causing neurological diseases and contrasts with the genomewide microsatellite instability observed in mismatch repair deficient cancer cells [7–9].

What is the principal mechanism underlying these expansions? This question has been a subject of numerous extensive studies since the discovery of the CTG·CAG repeat expansions in the myotonic dystrophy type 1 gene [10, 11] and the CGG·CCG repeat expansions in the fragile X syndrome gene [12].

In the 1990s, replication slippage was the dominant model to explain instability of the trinucleotide repeat sequence (TRS), because recombination (as defined by the reciprocal crossing over exchange, Fig. 38-1A) was not revealed by studies of human cases [13–15]. Three important facts argued at that time against a reciprocal crossing over event as a mechanism of TRS instability in patients: first, the lack of evidence supporting the exchange of the flanking sequences; second, no corresponding length changes of the second allele during the expansion–deletion event occurring at the other allele; and third, a simple reciprocal exchange could not explain the extremely large expansions observed in some of the diseases. On the other hand, data from the patients pointed toward gene conversion (a nonreciprocal event, Figs. 38-1B and 38-1C) in the instability of the CGG·CCG repeats in the fragile X syndrome and CTG·CAG tracts in myotonic dystrophy type 1 (reviewed in refs. [6] and [16]) and Machado–Joseph disease (MJD or SCA3, spinocerebellar ataxia type 3) cases [17, 18].

The aim of this chapter is to review the data related to the involvement of the recombination processes in the instability of trinucleotide (CTG·CAG, GAA·TTC) and tetranucleotide repeats (CCTG·CAGG) in *Escherichia coli* and other model systems. The role of nonreciprocal gene conversion and reciprocal crossing over processes

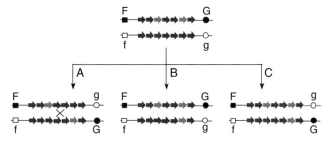

FIGURE 38-1 Genetic recombination between tandem repeats: (A) reciprocal crossing over exchange without any corresponding gene conversion; (B) gene conversion event not associated with crossing over; and (C) gene conversion accompanied by crossing over. Repeating sequences are shown as arrows. To differentiate between reciprocal and nonreciprocal events, two sequence interruptions (red arrows) were introduced in one of the homologous sequences. Markers flanking the repeats (F, f and G, g) are shown to indicate crossing over events. For simplicity, recombining tandem repeats are identical in size, and the processes do not lead to any changes in the sizes of the repeat tracts. See CD-ROM for color image.

in generating expansions and deletions of the repeat tracts in *E. coli* will be evaluated. Current knowledge about the role of microsatellite sequences as hot spots or cold spots in genetic recombination will also be reviewed. Finally, the effect of the repair of different types of DNA strand discontinuities including single-strand nicks and gaps or double-strand breaks (DSBs) and the principal role of the non-B-DNA structures formed by repetitive sequences in triggering genetic instability will be discussed.

II. TRINUCLEOTIDE REPEAT INSTABILITY ASSOCIATED WITH RECOMBINATION IN CLINICAL CASES

Genetic recombination was first demonstrated to be an important source of instability of minisatellites, a family of moderately sized (usually 6–100 bp), hypervariable repetitive elements [19–23]. In the case of microsatellites, recombination, mostly gene conversion, was repeatedly suggested to be a mechanism of the trinucleotide repeat instability in affected individuals.

The expansion of microsatellite sequences at a particular *locus* in the human genome is a complex process, often initiated by establishing the group of "at-risk" alleles harboring the longest, uninterrupted repeat tracts (founder effect). These alleles have a high potential for large expansions in subsequent generations, leading to the phenotype changes [24]. Somatic instability is the last, often disease-specific, step in the expansion process and occurs to various extents in different tissues. Genetic recombination may be involved in each of the previously mentioned events of microsatellite instability.

Detailed studies on the instability of the CTG·CAG repeats in the Machado–Joseph disease gene (MJD1) suggested that intermediate alleles, which later give rise to the large expanded repeats, arose through an interallelic gene conversion [18]. Similarly, in the case of the fragile X syndrome, Montagnon et al. proposed that the initial step in the transition from normal to the pathological repeat range can be attributed to unequal crossing over [25].

On the basis of haplotype analyses, gene conversion was postulated to be responsible for the reversion mutation, from the premutation size to the normal range, in the CGG·CCG repeat region of the fragile X mental retardation gene (FMR1) [26]. Similarly, a prenatal diagnosis performed in a female from a fragile X family carrying a large premutation showed a loss of expansion mutation due to the complex recombination pattern resembling either gene conversion or a double crossover [27]. Brown et al. have also indicated the role of gene conversion in the fragile X syndrome reversion mutations [28]. In addition, genetic recombination, in the form of either reciprocal crossing over or gene conversion without the crossing over exchange, was reported to influence the instability of CTG·CAG repeats in the myotonic dystrophy type 1 *locus* [11, 29]. Similar to the FMR1 gene, reversion mutations have also been reported for the CTG·CAG region of DM1 [30]. The analysis of the flanking markers excluded a single crossing over event and suggested a nonreciprocal gene conversion as a possible mechanism of these mutations [30].

Intergenerational instability of the SCA3 CTG·CAG repeats was also attributed to recombination. Interestingly, Igarashi et al. found that the genotype of the normal (not the expanded) chromosome substantially affects the degree of intergenerational instability of the TRS, suggesting interallelic gene conversion as a potential mechanism responsible for repeat expansion [17].

Thus, genetic recombination is involved in the TRS instability related to the diseases caused by relatively small repeat expansions in the coding regions of the mutated genes (so-called type I diseases [31]), as well as in the disorders associated with massive microsatellite expansions in noncoding regions of the mutated genes (termed type II diseases). Additionally, small expansions in the polyalanine tracts (often encoded by different GCN triplets), linked to at least nine human diseases, are likely to arise predominantly via an unequal crossing over mechanism [32, 33]. Because polyalanine tracts are usually encoded by imperfect trinucleotide repeats (i.e., by the variants of the alanine codons), a simple replication slippage model cannot explain their instability [32, 33].

Study of the mechanisms of the repeat instability directly in the human genome is very difficult. Often several different processes such as replication, recombination, or repair in concert with each other influence this instability. Moreover, it is likely that various mechanisms (e.g., slippage and gene conversion without crossing over) might lead to the same genetic outcome at the level of the DNA sequence, thus making the determination of the predominant pathway involved in the repeat instability impossible. Therefore, model systems, predominantly *E. coli* and *S. cerevisiae*, and also mammalian cell culture and transgenic animals are essential tools to study the involvement of different pathways in microsatellite instability.

III. LARGE CTG·CAG EXPANSIONS IN *ESCHERICHIA COLI* ARE CAUSED BY GENE CONVERSION

The potential role of recombination in the expansion of triplet repeat sequences using a two-plasmid recombination system in *E. coli* has been demonstrated [34, 35]. Biochemical and genetic analyses showed large, multiplefold expansions of CTG·CAG repeats *in vivo*, which occurred by homologous recombination. Two different plasmids were used in these experiments: the derivatives of pUC19, which contained the unidirectional ColE1 origin of replication together with the ampicillin resistance gene, and the derivatives of pACYC184, which harbored the tetracycline resistance gene. No significant homology exists between these vectors, thus providing an ideal system to observe the potential effects of genetic recombination on the different TRS cloned into these plasmids.

To analyze the recombinogenic behavior of the TRS, the experiments were conducted in three isogenic *E. coli* strains that differed in their recombination capacity [34, 35]. Cells were cotransformed with suitable plasmids to evaluate the effect of the types of TRS (CTG·CAG, GAA·TTC), their length, orientation relative to the origin of replication, and effect of the sequence heterology (interruptions). Extensive restriction mapping and DNA sequencing techniques were utilized to determine the size of the expansions as well as the mechanism that generated the instability. Multiplefold expansions were frequently found, which were dependent on the presence of repeat lengths greater than (CTG·CAG)$_{30}$ on both plasmids and RecA as well as RecBC functions. The largest expansions detected changed the length of the recipient tract from (CTG·CAG)$_{36}$ to the (CTG·CAG)$_{212}$ repeats. This increase in the size of the CTG·CAG tract corresponds in humans to the transformation from the normal to the full mutation DM1 allele [10, 36, 37].

Interestingly, cases of transfer of the G→A point mutation between homologous CTG·CAG tracts were also detected in the experiments with the pACYC derivative containing a tract of (CTG·CAG)$_{175}$ with two G→A interruptions (at the repeat positions 28 and 69) and the pUC19 derivative harboring an uninterrupted tract of (CTG·CAG)$_{80}$. The transfer of this polymorphism was not accompanied by the exchange of any DNA fragment between the two recombining sequences, which demonstrates the involvement of nonreciprocal gene conversion in this process (Fig. 38-1B).

To determine whether crossing over was associated with frequent expansion events, the flanking sequences of several recombination products were analyzed by DNA sequencing and restriction mapping. Results of these analyses demonstrated that gene conversion without the exchange of the flanking markers was the mechanism mediating TRS expansions. The effect of sequence heterology on the CTG·CAG expansions was also analyzed. The presence of the long CAG·CTG tracts harboring G→A point mutations resulted in rare expansion events, whereas the uninterrupted tracts of similar length readily recombined and expanded. Thus, as little as ~2% of sequence heterology was sufficient to disrupt the process of recombination-mediated expansions. These data are in agreement with the analyses performed in other model systems, as well as with human genetic studies demonstrating the stabilizing role of sequence interruptions on the TRS tracts (reviewed in ref. [6]).

These results showed, for the first time, that recombination pathways can be responsible for large expansions in an *E. coli* model system. However, no quantitative data regarding the frequency of the recombination events could be obtained from these experiments. The determination of the stimulatory or inhibitory effects of the TRS on recombination could help to establish the potential significance of the recombination-related mechanisms in repeat instability.

IV. MICROSATELLITE SEQUENCES INFLUENCE THE FREQUENCY OF RECOMBINATION

It was found that gene conversion could lead to the saltatory, massive increase in the size of the repeated sequences. If the TRS tracts, in contrast to the nonrepeating sequences, are capable of stimulating genetic recombination, repeat instabilities observed in patients and in the model organisms could result from multiple recombination events frequently initiated at the microsatellite *loci*. In this case, each individual recombination event within the TRS could lead to a small increase or decrease in the length of the repeated tract in a stepwise fashion.

A. Role of Non-B-DNA Structures Formed by Microsatellite Sequences in Recombination

Direct comparison between genetic and physical maps performed for genomes of different organisms showed that recombination events are more likely to occur in some regions of the chromosomes than in others. In many cases, specific sequences were implicated to stimulate (hot spots) or to suppress (cold spots) genetic recombination [38]. Often repeats including interspersed elements such as Alu repeats or LINE sequences were associated with recombination hot spots [39–41]. Also, tandem repeats, both minisatellites and microsatellites, have been shown to promote homologous recombination in prokaryotic as well as eukaryotic cells, presumably by adopting unusual DNA structures. The involvement of microsatellite sequences in genetic recombination is summarized in Table 38-1.

The best documented is the role of dinucleotide repeats, especially GT·AC, in stimulating recombination in prokaryotes, lower eukaryotes, and humans (Table 38-1). These sequences are among the most abundant microsatellites in the human genome and have been widely used as genetic markers in the construction of genetic maps of various genomes. The GT·AC repeat tracts are also capable of adopting a left-handed Z-DNA conformation (reviewed in ref. [42]). Majewski and Ott correlated the distribution of different microsatellites and mitotic recombination patterns along human chromosome 22 [43]. Significant association between increased recombination and the presence of GT·AC repeats was found. This effect was especially pronounced for the longer dinucleotide repeats (>45 bp), indicating that the hot spot activity is associated with longer tracts, whereas shorter repeats do not have an influence on recombination [43].

TABLE 38-1 Influence of Microsatellite Sequences on Genetic Recombination

Sequence[a, c]	Length	Structure	Effect	Organism	Reference numbers
(G·C)	<45	Triplex	Stimulation	E. coli	134
(CA·TG)	30–75	Z-DNA	Stimulation	E. coli, yeast, mammalian cells, human chromosome 22	42, 43, 135–140
(CG·CG)	32	Z-DNA	Stimulation	Mammalian cells	138
(GA·TC)	22	Triplex	Stimulation	Mammalian cells	44
(CTG·CAG)	<250	Hairpins–slipped structures	Stimulation	E. coli, yeast, mammalian cells	48, 49, 77, 107, 110, 141
(CGG·CCG)		Hairpins–slipped structures, tetraplex	Stimulation	Yeast	78
(GAA·TTC)	<176	Triplex–sticky DNA	Stimulation	E. coli	80
(CCTG·CAGG)	<200	Hairpins–slipped structures	Stimulation	E. coli, mouse	79, 142
(CCGAT·ATCGG)	12	ND, hairpins–slipped structures[b]	Stimulation	Yeast	45
(CCGNN·NNCGG)	12	ND	Stimulation		45
(CCGNN·NNCGG)	48	ND	Inhibition		45

[a] For nomenclature of the microsatellite sequences in this chapter, the orientation is always 5′ to 3′ for both designations of the antiparallel strands, separated by a center dot. For example, CTG·CAG designates a duplex sequence of repeating CTG, which may also be written TGC or GCT; CAG, the complementary strand, may also be written as AGC or GCA.

[b] Structure predicted using Mfold v.3.1 [143].

[c] Note: All direct repeat sequences can also form, at least in principle, slipped structures, and all inverted repeats can also form cruciforms [103, 144, 145].

The GA·TC repeats, another highly abundant dinucleotide microsatellite sequence in the human genome, were demonstrated to stimulate recombination in mammalian cells by as much as two orders of magnitude, when located on the SV40 episome [44]. Benet et al. proposed that the recombination potential of the GA·TC repeats could be linked to the propensity of these tracts to adopt triplex structures [44]. An interesting recombinational behavior of the pentanucleotide CCGNN·NNCGG sequence was described in yeast [45]. Depending on the number of repeating units, two opposite effects on recombination were observed: the (CCGNN·NNCGG)$_{12}$ demonstrated a meiotic hot spot activity elevating the recombination, whereas a tract four times longer exhibited a cold spot activity by repressing recombination. This seemingly contradictory phenomenon was most likely related to the competitive interactions between the four adjacent hot spots [(CCGNN·NNCGG)$_{48}$ = 4 × (CCGNN·NNCGG)$_{12}$].

In summary, the recombinational behavior of different microsatellites relies on their propensities to adopt non-B-DNA structures, because almost all of the microsatellites implicated in the stimulation of recombination are capable of the formation of triplex structures, Z-DNA conformation, hairpins, or slipped structures (Table 38-1). The length effect observed in several recombination studies conducted with tandem repeats might also be related to the increased stability of the unusual DNA structures formed by longer tracts. Our results obtained with tri- and tetranucleotide repeats fully corroborated this hypothesis. The studies on the recombinational capacity of the CTG·CAG and the GAA·TTC trinucleotide repeats as well as the CCTG·CAGG tetranucleotide tracts associated with the microsatellite expansion disorders will be presented in the following sections of this chapter.

B. CTG·CAG Repeats Expanded in Myotonic Dystrophy Type 1 Are Preferred Sites of Intramolecular and Intermolecular Recombination

Several studies on the frequency and mechanisms of recombination processes were conducted in different model organisms by using both episomal and chromosomal assays [46, 47]. A majority of these analyses utilized a pair of homologous sequences encoding a gene whose activity determined a specific phenotype (e.g., antibiotic resistance), and the recombinants were selected for or against the acquired phenotype. To determine the frequency of recombination between microsatellites, two independent intramolecular, as well as intermolecular, episomal recombination assays were designed. In these systems, crossing over between the repeating tracts frequently led to length changes of the microsatellite sequences and was also associated with a new phenotype of the *E. coli* harboring the recombinant plasmids [48, 49].

1. Long CTG·CAG Repeats Are Hot Spots of Intramolecular Recombination

In order to analyze the recombinational behavior of the CTG·CAG tracts in *E. coli*, an intramolecular recombination system was developed in which a crossover event between pairs of CTG·CAG tracts of different lengths and nonrepeating control sequences leads to the deletion of the intervening sequence, which contains the *GFP* gene [48]. After the transformation of the plasmids into the appropriate *E. coli* strain, the recombination-dependent deletion of the intervening sequence results in white colony formation. In contrast, when the plasmid containing the *GFP* gene (i.e., recombination substrate) is established in the host cell, the expression of this gene leads to the formation of a fluorescent green colony. Thus, the frequency of the recombination events was measured as the ratio of the number of white colonies to the total number of viable cells (Fig. 38-2A).

The experiments performed in various genetic backgrounds revealed that long CTG·CAG microsatellites are preferred sites of intramolecular recombination in *E. coli*. The frequency of recombination between two directly repeated CTG·CAG tracts was up to ~10 times higher than between two nonrepeating sequences of similar length in recombination proficient cells. The effect of length of the homologous CTG·CAG tracts on the recombination frequency was dramatic. Short TRS tracts (17 repeats) recombined with a frequency similar to that of controls containing nonrepeating homologous tracts. It was found that an increase in the length of the homologous sequences from $(CTG·CAG)_{17}$ to $(CTG·CAG)_{165}$ resulted in a 60-fold increase in the recombination frequency between direct repeats. In addition, a pronounced effect of the orientation, relative to the origin of replication, on the frequency of recombination was found. When the CTG repeats were located on the lagging strand template (termed orientation II [50]), recombination events occurred with a much higher frequency than in orientation I (CTG sequence is on the leading strand template).

Intramolecular recombination between CTG·CAG tracts led to high genetic instability (deletions and expansions) of the repeating sequences (Fig. 38-2C). The level of instability was strongly associated with the TRS length [48].

Thus, short CTG·CAG tracts neither induced recombination nor significantly stimulated repeat instability in the intramolecular system. Similarly, several analyses performed on the human genome revealed a lack of correlation between recombination rate and the presence of the short microsatellites containing less than ~20 repeats. In addition, recombination events did not have an influence on the instability of various short microsatellite repeats [43, 51, 52]. A very surprising and profound effect of the CTG·CAG repeat orientation, relative to the direction of the replication fork progression, on the frequency of recombination was also found in the intermolecular studies and will be discussed in the next section.

2. Expanded CTG·CAG Repeats Stimulate Intermolecular Recombination

Intermolecular recombination between two CTG·CAG tracts located on two independent episomes can be considered a model system for interallelic or ectopic recombination. In order to study the frequency of intermolecular recombination, derivatives of pBR322 (ColE1 plasmid encoding ampicillin and tetracycline resistance) and pFW25 (replicating using the R6K π-protein-dependent origin of replication and encoding chloramphenicol resistance) containing inserts with different lengths of the CTG·CAG repeats were used [49]. In the parental *E. coli* strain, lacking the gene encoding the π protein, cotransformed with both plasmids and cultured in the presence of all three antibiotics, only cointegrants resulting from the CTG·CAG-mediated recombination between the pBR322 and pFW25 derivatives could survive because parental cells cannot support replication of the R6K origin containing plasmids. In the isogenic *E. coli* harboring the *pir* gene (encoding the π protein), both plasmids can independently replicate; hence, the number of colonies obtained reflects the background level of cotransformants (Fig. 38-2B) [49].

Consistent with previous results from the intramolecular system, it was found that the frequency of homologous intermolecular recombination between two CTG·CAG tracts is up to 60-fold higher than between two nonrepeating sequences. The frequency of recombination is proportional to the length of the repeating tract: the longer the tracts, the higher the recombination frequency. The recombination events are more frequent when the TRSs are in orientation II than in orientation I relative to the origin of replication. Thus, investigations conducted in two different systems revealed a dramatic stimulation of recombination by long CTG·CAG repeats [49].

Several features of the CTG·CAG repeats may contribute to their recombinogenic behavior. During the recombination process, two TRS tracts can hybridize

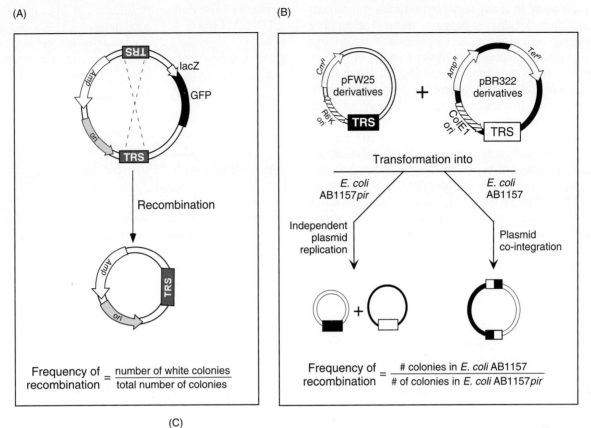

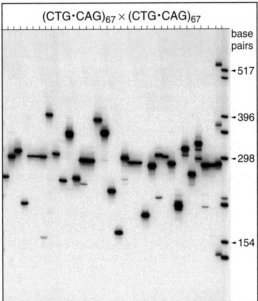

FIGURE 38-2 Systems to study recombinational potential of different TRSs in *E. coli*. (A) Intramolecular and (B) intermolecular assays to analyze the frequency of recombination. The approximate positions of the origins of replication, antibiotic resistance genes, TRS tracts, and the *GFP* gene are indicated. (C) Analysis of the genetic instability (deletions and expansions) of the TRS in the products of intramolecular recombination between a pair of $(CTG \cdot CAG)_{67}$ tracts. The autoradiogram shows the analysis of the length of the CTG·CAG inserts in the recombination products obtained from ~30 individual colonies. See CD-ROM for color image.

with each other in many registers, whereas nonrepeating sequences can align in only one frame. The number of possible alignments between two homologous CTG·CAG tracts increases with the number of repeats present in the recombining sequences, which can influence the kinetics of the synapsis step of homologous recombination. Interestingly, the frequency of both intramolecular and intermolecular recombination depends on the orientation of the CTG·CAG repeats relative to the unidirectional origin of replication. When the CTG repeats were present on the lagging strand template, the frequency of recombination was substantially higher. This strongly implicates a role for DNA replication in recombination between the TRS tracts. The CTG·CAG repeats are known to arrest replication fork progression *in vitro* and *in vivo*, due to their capabilities to adopt non-B-DNA conformations [53–55]. In the case of the CTG·CAG repeats, this occurs predominantly when the CTG strand is located on the lagging strand template for replication (i.e., orientation II). Stalling of the replication fork at the secondary structures leads to the formation of nicks and/or double-strand breaks (DSBs) in the repeating tracts, which stimulates their mutagenic repair via recombination [56–58]. In addition, structures formed by CTG·CAG repeats can be recognized and subjected to repair by endonuclease excision. The processes of nucleolytic repair of the "structural lesion" may induce DNA strand discontinuities (nicks, breaks), which are very efficient substrates for recombinational repair [59]. Thus, compared to the nonrepeating controls or short TRSs, recombination between two long CTG·CAG tracts is highly stimulated because the DNA discontinuities can be created independently in each of the tracts.

The recombination frequency between the CTG·CAG repeats containing a pair of G→A interruptions was lower than the frequency with pure uninterrupted tracts. Two distinct mechanisms can potentially be responsible for the strong effect of the interruptions on the frequency of recombination. First, the presence of interruptions results in the formation of imperfectly aligned heteroduplex recombination intermediates that contain mismatches at the sites of the interruptions. These mismatches may attract the mismatch repair proteins MutS and MutL, which can inhibit RecA-mediated strand transfer [60]. Thus, the recombination intermediates may be destabilized by the mismatch repair system, thereby diminishing the recombination frequency. Second, the interruptions can reduce the propensity of the CTG·CAG sequences to form stable secondary structures that may impede the progression of the replication fork and consequently the formation of DNA "structural lesions" [6]. It is possible that a combination of these two effects could account for the lower recombination frequency of the interrupted CTG·CAG tracts. Sequence heterology also had a profound effect on the extent of expansions observed in the gene conversion studies described in the previous section [34, 35].

C. Recombination Properties of the CCTG·CAGG Tetranucleotide Repeats Associated with Myotonic Dystrophy Type 2

The instability of nontriplet repeat sequences has been associated with three hereditary neurological disorders. These include a dodecamer sequence CCCCGCCC-CGCG·CGCGGGGCGGGG causative of progressive myoclonus epilepsy of the Unverricht–Lundborg type (EPM1) [61], a pentanucleotide ATTCT·AGAAT expansion that is implicated in spinocerebellar ataxia type 10 (SCA10) [62], and a tetranucleotide CCTG·CAGG sequence associated with myotonic dystrophy type 2 (DM2) [63].

The DM2 CCTG·CAGG expansion was mapped to intron 1 of the zinc finger protein 9 gene (ZNF9) on chromosome 3q21 [63]. The phenotypic characteristics of this disorder are very similar to those of myotonic dystrophy type 1, including myotonia, proximal weakness, frontal balding, cardiac arrhythmias, insulin-resistance-associated diabetes mellitus, polychromatic cataracts, and infertility [64–67]. The CCTG·CAGG repeats are unique in the length of the repeat expansion. Although certain triplet repeats are known to expand by a few hundred to a few thousand copies [5, 6, 68], the tetranucleotide repeats have been shown to expand to the enormous size of ~11,000 repeats (~44 kb) [63]. This is very interesting because unaffected individuals carry up to 26 frequently interrupted CCTG·CAGG repeats. Additionally, a time-dependent somatic variation in the number of repeats was described [63].

DNA structure has been implicated as one of the major factors influencing the genetic instabilities of triplet repeat disorders [5, 68–71]. In order to investigate the structural properties of the CCTG·CAGG tetranucleotide repeats, chemical and enzymatic probing studies were performed on $d(CAGG)_{26}$ and $d(CCTG)_{26}$ synthetic oligonucleotides [72]. These single-stranded oligonucleotides are useful as models because they resemble unpaired regions of the DNA. Studies with five different chemical and enzymatic probes revealed that $d(CAGG)_{26}$ formed a stable folded-back hairpin structure. This hairpin comprised a terminal loop containing six unpaired residues and a stem that was stabilized by Watson–Crick G·C and C·G pairs as well as non-Watson–Crick A·G and G·A interactions (Fig. 38-3). In contrast, unlike the d(CAGG)$_{26}$ oligomer, the complementary d(CCTG)$_{26}$ did not form any secondary structure that could be detected under the conditions used for these studies.

The molecular mechanisms underlying the massive expansion of the CCTG·CAGG tetranucleotide repeats are under investigation. A mammalian cell culture assay was used to demonstrate that replication-based slippage contributes to the genetic instability of these repeats [72]. Previous work with triplet repeats has shown replication slippage to influence their genetic instabilities [50, 68, 73, 74]. Additionally, the triplet repeats showed an orientation bias, where orientation II was more unstable when compared to the corresponding orientation I [5, 6, 50, 54, 68, 73–78]. Thus, CTG·CAG repeats are more unstable in the orientation prone to delete (orientation II). This is attributed to the ability of the CTG strand to form a more stable hairpin structure than the complementary CAG strand. In case of the CCTG·CAGG tetranucleotide sequence, orientation II (CCTG is on the lagging strand template) was more unstable than orientation I (CCTG is on the template for leading strand synthesis) [72], similar to the behavior observed with the DM1 sequence. However, the CCTG·CAGG repeats are unstable in the orientation prone to expand rather than the orientation prone to delete, as is the case for all TRSs (Fig. 38-3). This is because the CAGG strand is more prone to form a hairpin structure on the newly synthesized lagging strand as opposed to the CCTG strand. Thus, the tetranucleotide repeats tend to expand in orientation II and to delete in orientation I. The preferential ability of one of the two strands to form stable hairpin structures explains the instability phenomenon as for the TRS [72].

Because relatively small CCTG·CAGG expansions (<2 times the initial length of the repeat tract) were found by using a replication-based mammalian instability assay, the recombination properties of this sequence were analyzed. In order to study the recombination behavior of the CCTG·CAGG repeats, the intramolecular assay described earlier was used [79]. The tetranucleotide repeats were shown to be length-dependent

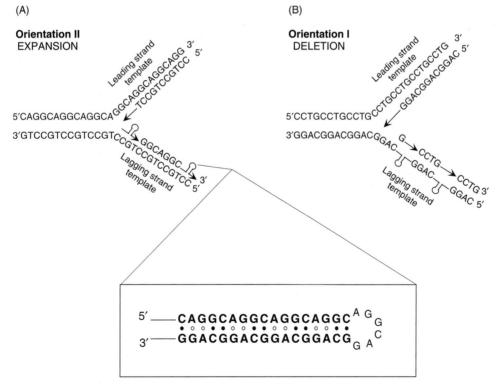

FIGURE 38-3 Structure-dependent replication slippage model for CCTG·CAGG tetranucleotide repeats. (A) Orientation II gives rise to expansions due to the propensity of the CAGG repeats on the newly synthesized Okazaki fragments to form folded-back hairpin structures. (B) Orientation I, on the other hand, can give rise to deletions as these non-B-DNA structures are formed on the lagging strand template. The proposed hairpin structure [72] has been magnified to show the most plausible, quasi-stable stem loop, containing both Watson–Crick C·G and G·C base pairs (filled circles) and nonstandard A·G and G·A interactions (open circles).

recombination hot spots. The recombination frequency was 27-fold higher for the longest tetranucleotide repeat than for the control containing two 354-bp non-repeating DNA fragments in recombination proficient E. coli cells. Moreover, a distinct orientation effect was observed, where orientation II gave higher frequencies than orientation I. This is similar to the results found in the intra- and intermolecular studies with the CTG·CAG sequence [48, 80]. This orientation effect could be attributed to the propensity for the CAGG strand to form hairpin structures on the newly synthesized Okazaki fragment [72]. These structures could indirectly stimulate DNA strand breaks, which would lead to the recombination between the two homologous tracts and the deletion of the intervening sequence.

DNA damage may occur due to the structure-forming propensity of the repeats, which could lead to induction of a number of repair events. The SOS response is elicited on extensive damage to the DNA, and in order to check whether this pathway was involved in the recombination between CCTG·CAGG repeats, the intramolecular recombination studies were carried out in E. coli SOS-repair mutants. In an SOS$^+$ strain, the frequencies of recombination were comparable to those of the parental strains. However, in a strain lacking the SOS response due to a mutation in the lexA gene, the frequencies were much lower than in the SOS$^+$ strain. This mutation in the lexA gene causes the LexA repressor to remain bound to the promoters of the SOS-controlled genes and renders the RecA protein unable to cleave the repressor, leading to inactivation of the SOS-repair pathway. Thus, an increase in the frequency of recombination in the SOS$^+$ strain when compared to the deficient strain could potentially indicate the participation of some gene product(s) under the control of the SOS-repair pathway in the recombination between the CCTG·CAGG repeats. The exact gene or set of genes involved in the repair of the breaks that are presumably occurring within the repeating tracts would be difficult to elucidate, because 40 or more genes have been identified to be under the SOS regulon [81], many of which are involved in the repair of double-strand breaks. Furthermore, the twentyfold increase in the amount of RecA protein itself observed upon the induction of the SOS response [82, 83] might be responsible for the increase in the frequency of recombination.

Analyses of the recombination products from the parental as well as the SOS$^+$ and SOS$^-$ strains showed that the percentages of expanded and deleted clones were similar in all of the strains. This indicates that the mechanism of the repair event occurring in all strains was similar and could simply involve a RecA-mediated homologous recombination pathway. This pathway might be further accentuated in the presence of the SOS response due to the higher expression of certain genes (e.g., ruvAB, umuDC, or recN) involved in homologous recombination [81].

The analyses of the tetranucleotide repeat size in the products of recombination between the two CCTG·CAGG tracts revealed a high percentage of expanded clones. This was in contrast to the studies with the CTG·CAG and GAA·TTC repeats that demonstrated the predominance of deletions in the recombination products [48, 80]. This propensity of the CCTG·CAGG repeats to expand could be attributed to the greater overall stability of these repeats as compared to the TRS tract. It is likely that the expanded products were formed even during recombination between long TRSs, but these were just not maintained during subsequent rounds of replication, which are required to analyze the repeat lengths in the recombinants.

D. Structure-Dependent Recombination Hot Spot Activity of GAA·TTC Sequences

The GAA·TTC repeats are one of the most common short tandem repeats among all of the 10 possible TRSs in the human genome [3, 4]. They are frequently associated with the 3' end of a polyA stretch present in the Alu elements [84]. Large expansions of the GAA·TTC sequence in intron 1 of the Friedreich's ataxia (FRDA) gene were found to be associated with this autosomal recessive disorder [85]. Friedreich's ataxia is almost exclusively caused by the GAA·TTC repeat expansions above the pathogenic threshold of ~70 trinucleotide repeats [86]. Only ~6% of patients were found to carry one expanded allele and a point mutation in the other copy of the FRDA gene, whereas in ~94% of cases, expansion of the GAA·TTC tracts was found on both chromosomes [87, 88].

GAA·TTC tracts are structurally polymorphic and may exist in the following distinctly different conformations: (1) an orthodox right-handed B-DNA duplex; (2) a folded-back intramolecular triplex or bitriplex composed of one consecutive stretch of R·Y repeats [89–91]; or (3) the sticky DNA conformation, a complex triplex structure formed by two directly repeated, relatively long GAA·TTC tracts (>60 repeats). Sticky DNA is a very stable conformation that can be present in neutral pH and requires negative supercoiling as well as Mg^{2+} ions [92–97]. The structural properties of the GAA·TTC repeats are related to their pathogenic potential.

Intramolecular and intermolecular recombination studies showed that the frequency of recombination between the GAA·TTC tracts was as much as 15 times higher than that of the nonrepeating control sequences. Analyses of the recombination products also demonstrated the expansions and deletions of the GAA·TTC repeats as a result of recombination [80].

Compared to the CTG·CAG or CCTG·CAGG repeats, the recombinational properties of the GAA·TTC sequence were unique, because they strongly depended on the secondary structure polymorphism of this sequence. Unexpectedly, it was found that an increase in the length of the GAA·TTC repeats decreased the intramolecular recombination frequency between these tracts [80]. Short GAA·TTC tracts that can exist in the cell as a B-DNA duplex or a triplex, but cannot form the sticky DNA structure due to their length [92, 95], were excellent substrates for intramolecular recombination and therefore recombine with the highest frequency. It is very likely that folded-back intramolecular triplex structures formed by short GAA·TTC tracts are responsible for the recombination hot spot activity of these repeats, because lowering of the *in vivo* negative superhelical density, using the DNA gyrase inhibitor novobiocin [98, 99], decreases the potential of these tracts to recombine. On the other hand, the frequency of intramolecular recombination between the tracts of $(GAA·TTC)_{176}$ was significantly hampered by the formation of sticky DNA [80]. Relaxation of the DNA *in vivo* using novobiocin repressed the formation of sticky DNA and, hence, increased the frequency of recombination between long tracts. Thus, the GAA·TTC sequence is the first example of a DNA-structure-dependent recombination hot spot [80].

The results obtained in intramolecular recombination studies together with the data from *in situ* cross-linking of the sticky DNA conformation show that this structure can exist in living cells [97]. Moreover, a substantial change can be made in the amount of the sticky DNA formed using novobiocin, influencing its biological functions. These findings may have profound consequences for new therapeutic strategies for Friedreich's ataxia. This autosomal recessive disease is caused by the reduced expression of frataxin as a result of the GAA·TTC expansion in the first intron of the FRDA gene [85, 100]. The formation of non-B-DNA structures by long GAA·TTC tracts inhibits the transcription of the FRDA gene and leads to a reduction in the amount of the *X25* mRNA in FRDA patients [100, 101]. Thus, potential therapies may be aimed at alleviating the inhibition of transcription. The use of specific pharmacological agents capable of interfering with the formation of unusual DNA structures adopted by GAA·TTC repeats may up-regulate the expression of the FRDA gene, thereby providing an effective therapeutic approach.

V. INVOLVEMENT OF DNA INCISIONS (BREAKS, NICKS, AND GAPS) IN STIMULATION OF TRINUCLEOTIDE REPEAT SEQUENCE INSTABILITIES

Different types of DNA incisions (double-strand breaks, single-strand nicks, and single-strand gaps) have been postulated to be potential instigators of the genetic instability of trinucleotide repeat sequences [5, 68, 102, 103]. These DNA strand discontinuities can stimulate instability by causing DNA polymerase pausing, resulting in the replication fork collapse [104] and initiation of recombinational repair [34, 35, 48, 49]. An intraallelic system was designed to study the effects of various DNA incisions on genetic instability. The lesions were introduced into a single TRS tract present in a plasmid, and the products arising from the repair of these lesions were analyzed. Investigations into the influence of different types of incisions have revealed that instability is stimulated by double-strand breaks and single-strand gaps.

A. Role of Double-Strand Breaks in Genetic Instability

Several studies have suggested that double-strand breaks (DSBs) are responsible for the expansions and deletions of the TRS [77, 78, 105–112]. First discovered in 1998 by Freudenreich et al. [77], the repeat tract of $(CTG·CAG)_{250}$ located on the yeast chromosome showed breakage within the TRS region. This breakage was hypothesized to cause the observed increase in recombination. DSBs between two short $(CTG·CAG)_5$ repeats interrupted by 18 bp of the I-*Sce*I recognition site led to high frequencies of the TRS deletions. Moreover, a sixtyfold higher frequency of rearrangements was observed when a triplet repeat tract was used as the template in the repair of the chromosomal DSBs [105]. The expansions of the CTG·CAG sequences in yeast were even more pronounced during meiosis when compared to the mitotic division. Jankowski et al. attributed these instabilities to DSB-induced recombination. Sequences as short as $(CTG·CAG)_{64}$ induced *spo11*-dependent DSB formation during meiosis. Their repair resulted in deletions as well as in expansions of the CTG·CAG tract. Expansions of the CAG·CTG trinucleotide repeats through recombination-induced DSB repair were dependent upon the integrity of the Mre11p-Rad50p-Xrs2p complex [109]. In addition to the CTG·CAG repeats, the CGG·CCG tracts of 81 and 160 repeats also induced recombination in an orientation-independent manner [78]. The authors concluded that the observed recombination was a measure of double-strand break repair.

Experiments in mammalian model systems have been conducted to evaluate the influence of double-strand break repair on genetic instability. Studies in CHO cells with the expanded CTG·CAG repeats located in the *APRT* gene displayed an increased rate of deletions and rearrangements during homologous recombination [110]. The influence of double-strand breaks introduced directly to one side of a CTG·CAG tract was also studied in COS-1 cells. The majority of DSB repair products resulted in deletions of the TRS, which was attributed to the formation of stable secondary structures [111]. In a DM1 mouse model with the expanded CTG·CAG tract of >300 repeats, mutations in Rad52 led to a slight decrease in the size of the expansions [112]. Interestingly, mutations in NHEJ and homologous recombination genes (Rad54 and DNA-PKcs) appeared to have no effect on instability.

A direct assay was developed to determine whether DSBs, within the TRS tracts of the premutation length, induced genetic instabilities [113]. Plasmids were created containing (CTG·CAG)$_{43}$ or (CTG·CAG)$_{70}$ repeats and (CGG·CCG)$_{43}$ repeats with point mutations such that unique restriction sites were generated within the TRS. By using these sites, DSBs were generated *in vitro* and the linearized plasmids were transformed into *E. coli* cells. The repair–replication processes resulted in expansions or deletions of the repeating tract (Fig. 38-4). The predominant products were deletions, and expansions remained small and infrequent. The insert orientation, length, sequence (CTG·CAG or CGG·CCG), absence of RecA, and deletion of the RecBCD exonuclease influenced the amount of instability generated during the repair process.

In the studies described previously, the plasmids used contained one tract of repetitive DNA. Because homologous sequences were not present in the cells at the time of the DSB repair, no interallelic repair events would be expected. In experiments conducted in the presence of a second plasmid, containing the (CTG·CAG)$_{30}$ tract as a repair template, the intermolecular repair events between these two plasmids were monitored [114]. In spite of the availability of the homologous CTG·CAG tract, a majority of DSB repair events occurred via an intramolecular mechanism. Thus, an intraalleic pathway seems to be the dominant mechanism to repair DSBs associated with TRS tracts.

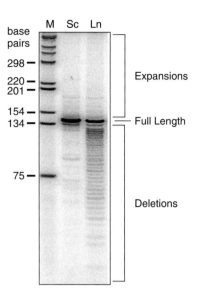

FIGURE 38-4 Effect of double-strand breaks on CTG·CAG instability. Supercoiled (Sc) and linear (Ln) forms of plasmids containing (CTG·CAG)$_{43}$ were transformed into the parental *E. coli* strain. Restriction analyses were performed to analyze the instability of the CTG·CAG repeat inserts [113]. TRS containing fragments were separated in nondenaturing polyacrylamide gel. Full length, deletion, and expansion bands are indicated on the right. Size markers from the 1-kb DNA ladder (M) are indicated on the left.

B. Effect of Single-Strand Nicks and Gaps on the Instability of CTG·CAG Repeats

Although the effects of DSBs on microsatellite instability have been studied extensively, the role of other types of DNA incisions (e.g., nicks or gaps) in TRS tracts has not been explored. Single-strand nicks and single-strand gaps have been hypothesized to be involved in expansions and deletions of the TRS [102, 115, 116]. These lesions can be generated *in vivo* through several pathways, such as DNA repair (mismatch repair, nucleotide excision repair, or base excision repair), discontinuous replication on the lagging strand, or recombination [103]. If unrepaired prior to replication, DNA strand discontinuities might lead to the DNA polymerase stalling and replication fork collapse that have been shown to influence genetic instabilities [104, 117, 118]. Nicks and one-base gaps induced CAG·CTG repeat expansions *in vitro*, due to the weak strand displacement activity of human DNA polymerase β [115]. Additionally, studies suggested that expansions in mouse haploid germ cells arise by gap repair, which is dependent on the Msh2 repair complex [116]. However, no detailed analyses have been reported that directly compare the effects of DSBs, nicks, and gaps on TRS instability.

Similar to the studies on DSBs, a direct assay was developed to determine the influence of nicks and single-stranded gaps within the CTG·CAG repeat sequence in *E. coli*. Plasmids containing (CTG·CAG)$_{68}$ with a single- or double-recognition site for the *Bbv*CI endonuclease within the TRS were constructed. Nicks and gaps in the TRS tracts were generated by *in vitro* cleavage with the nicking variants, *Bbv*CIA and *Bbv*CIB, of the *Bbv*CI enzyme [114]. Plasmids containing single-strand lesions were introduced into *E. coli* and the extent of instability was analyzed. Repair of the plasmid harboring a 30-nt gap generated a dominant 30-bp deletion product, whereas nicks and gaps of 15 nt did not induce any expansions or deletions. Formation of the precise 30-bp deletion product was dependent upon the presence of the repeating sequence and was stimulated by the incubation of the gap-containing plasmid with *E. coli* DNA ligase. The amount of the 30-nt deletion product was independent of the RecFOR gap repair pathway.

Small loop structures or mismatches arising from the slippage and misalignment of short tandem repeats can be excised by mismatch repair or nucleotide excision repair proteins, leading to the formation DNA gaps [77, 119–128]. If these gaps are sufficiently long, secondary structures, such as hairpins, may be formed in the single-stranded repeat regions. Most models for deletions propose that during DNA synthesis polymerase will bypass the hairpin structure located on the template strand and fill in the gap. However, a simpler model is suggested to account for these instabilities [114]. Upon formation of the 30-nt gap at the CTG·CAG tract, the free ends of the duplex (i.e., 5' and 3' ends) will be brought into close proximity by the folded-back structure formed within the repeat sequence, allowing for ligation across the gap (Fig. 38-5). This product of ligation will generate a deletion of the TRS during the next round of replication. The minimum length of the gap required to induce instability may vary depending on the thermodynamic stability of the structures formed by different repeat sequences at this gap.

Prior studies proposed that DNA gaps can promote expansions of the CTG·CAG repeats [115, 116]. On the contrary, data indicate that gaps in the DNA are not ideal

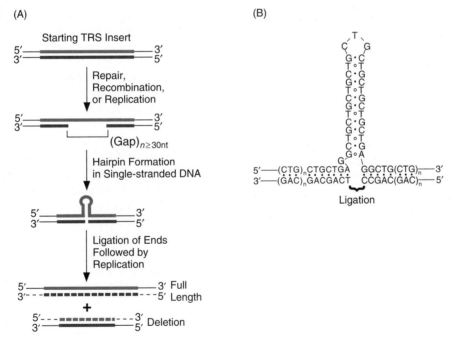

FIGURE 38-5 Involvement of gap repair in TRS instability. (A) Generation of a gap within the CTG·CAG tract can result from repair, recombination, or replication. The red line represents the CTG strand, whereas the blue line represents the CAG strand. Once the gap is formed, the single-stranded region will likely adopt a folded-back hairpin structure. The stability of this hairpin is determined by the length and the sequence composition of the gap region. Hairpin formation brings the free ends of the duplex DNA in close enough proximity for DNA ligase to join the 5' and 3' ends of the gap. The next round of DNA replication will result in two products: a plasmid containing a full length insert (dashed blue line with solid red line) and a plasmid with a deletion of the DNA fragment, corresponding precisely to the size of the gap (solid blue line with dashed red line). (B) Secondary structure model of the CTG·CAG region containing the 30-nt gap formed in the CAG strand. The filled circles represent Watson–Crick base pairs, whereas the open circles indicate T·T mispairs.

substrates to initiate expansion events in an *E. coli* model system. Due to the large number of DNA metabolic pathways that are capable of generating gaps, the gap repair mechanism may contribute to the bias toward deletions observed in *E. coli*. It is possible that the overwhelming number of DNA gaps repaired by ligation masks the rare expansion events generated by other pathways. This hypothesis is consistent with experiments that demonstrated that mutations in yeast DNA ligase (*cdc9-1* and *cdc9-2*) increased expansion events of (CTG·CAG)$_{78}$ repeats in an orientation-dependent manner [129]. It will be interesting to learn in the future how DNA nicks and gaps formed in the TRS tracts are repaired in humans.

VI. CONCLUDING REMARKS

This chapter focused on recombination and related pathways as powerful mechanisms inducing instability of microsatellite sequences. The current status of the studies connected to the recombinational capacity of these sequences was also reviewed.

In the past 15 years remarkable progress has been achieved in the understanding of the complex mechanisms of repeat instabilities. Studies have revealed that recombination, together with replication and repair, is one of the major processes responsible for instability of microsatellites (Fig. 38-6). Several *cis* elements as well as *trans*-acting factors, influencing genetic instability were discovered (reviewed in refs. [5], [6], [68], and [130]). The sequences of the repeats, their length, the presence of the interruptions in the repeating tract, the orientation of the repeats relative to the origin of replication, the distance from the origin, transcription through the repeats, and DNA methylation status are among the most important determinants of the extent of the instability [5, 6, 68, 130] (Fig. 38-6).

Several mechanisms can simultaneously act on repetitive sequences leading to their genetic instability. Recombination repair could play an especially important role in the circumstances where DNA replication cannot contribute to the instability, e.g., in the terminally differentiated tissues that no longer undergo cell division. Stimulation of DSBs by TRSs is well-documented in eukaryotic cells [130]. Interestingly, DSB distribution often reflects the distribution of crossover and gene conversion hot spots [131].

Considering the statistical overrepresentation of trinucleotide microsatellites in eukaryotes, the frequent recombination events between TRS tracts may be a source of mutations (deletions and inversions) leading to genetic diseases [132]. In addition, recombination between TRSs may have an important evolutionary role by promoting rearrangements of genetic information within different *loci*, leading to the formation of novel genes [133].

It is postulated that the formation of non-B-DNA structures by microsatellite sequences is the central and most important factor determining the instability of a particular repeat tract (Fig. 38-6). Practically all proposed models of microsatellite instability emphasize the ability of the repeat sequences to adopt unusual, non-B-DNA structures such as hairpins, slipped structures, triplexes, sticky DNA structure, tetraplexes, and unwound DNA conformations that can interfere with various processes of DNA metabolism (reviewed in ref. [103]). Moreover, microsatellites lacking strong structural potential are frequently used as negative controls in the studies of repeat instability [109]. Stable, non-B-DNA conformations may induce the formation of DSBs, nicks, or gaps, which are the initial events in the generation of repeat instabilities. In addition, during the repair of these lesions structure-promoted DNA slippage can lead to the genetic instability.

In summary, knowledge about the mechanisms of the expansions and the factors that influence these processes may help in the future to design therapeutic approaches aimed specifically to induce deletions of the already expanded pathogenic microsatellites or ultimately to prevent the expansion from occurring.

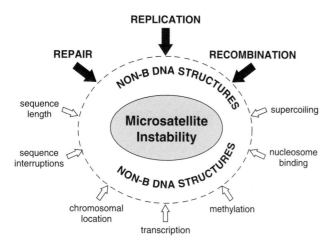

FIGURE 38-6 Major contributors to microsatellite instability *in vivo*. Replication, recombination, and repair in conjunction with several other processes and factors influence repeat instability predominantly via induction of the formation of various non-B-DNA structures.

Acknowledgments

We thank all members of this laboratory for their comments and helpful discussions. We are especially grateful to Drs. Albino Bacolla, Anna Pluciennik, John Jakupciak, Ravi Iyer, Pawel Parniewski, and

Alexandre Vetcher and Mrs. Jacquelynn Larson for their contribution to this work. This work was supported by grants from the National Institutes of Health (NS37554 and ES11347), the Robert A. Welch Foundation, the Muscular Dystrophy Foundation, and the Seek-a-Miracle Foundation to Robert D. Wells.

References

1. Weber, J. L., and Wong, C. (1993). Mutation of human short tandem repeats. *Hum. Mol. Genet.* **2**, 1123–1128.
2. Hearne, C. M., Ghosh, S., and Todd, J. A. (1992). Microsatellites for linkage analysis of genetic traits. *Trends Genet.* **8**, 288–294.
3. Subramanian, S., Madgula, V. M., George, R., Mishra, R. K., Pandit, M. W., Kumar, C. S., and Singh, L. (2003). Triplet repeats in human genome: Distribution and their association with genes and other genomic regions. *Bioinformatics* **19**, 549–552.
4. Subramanian, S., Mishra, R. K., and Singh, L. (2003). Genome-wide analysis of microsatellite repeats in humans: Their abundance and density in specific genomic regions. *Genome Biol.* **4**, R13.
5. Wells, R. D., and Warren, S. T., Eds. (1998). "Genetic Instabilities and Hereditary Neurological Diseases." Academic Press, San Diego.
6. Cleary, J. D., and Pearson, C. E. (2003). The contribution of cis-elements to disease-associated repeat instability: Clinical and experimental evidence. *Cytogenet. Genome Res.* **100**, 25–55.
7. Thibodeau, S. N., Bren, G., and Schaid, D. (1993). Microsatellite instability in cancer of the proximal colon. *Science* **260**, 816–819.
8. Eshleman, J. R., and Markowitz, S. D. (1996). Mismatch repair defects in human carcinogenesis. *Hum. Mol. Genet.* **5**, 1489–1494.
9. Janin, N. (2000). A simple model for carcinogenesis of colorectal cancers with microsatellite instability. *Adv. Cancer Res.* **77**, 189–221.
10. Brook, J. D., McCurrach, M. E., Harley, H. G., Buckler, A. J., Church, D., Aburatani, H., Hunter, K., Stanton, V. P., Thirion, J. P., Hudson, T., Sohn, R., Zemelman, B., Snell, R. G., Rundle, S. A., Crow, S., Davies, J., Shelbourne, P., Buxton, J., Jones, C., Juvonen, V., Johnson, K., Harper, P. S., Shaw, D. J., and Housman, D. E. (1992). Molecular basis of myotonic dystrophy: expansion of a trinucleotide (CTG) repeat at the 3' end of a transcript encoding a protein kinase family member. *Cell* **68**, 799–808.
11. Tsilfidis, C., MacKenzie, A. E., Mettler, G., Barcelo, J., and Korneluk, R. G. (1992). Correlation between CTG trinucleotide repeat length and frequency of severe congenital myotonic dystrophy. *Nat. Genet.* **1**, 192–195.
12. Kremer, E. J., Pritchard, M., Lynch, M., Yu, S., Holman, K., Baker, E., Warren, S. T., Schlessinger, D., Sutherland, G. R., and Richards, R. I. (1991). Mapping of DNA instability at the fragile X to a trinucleotide repeat sequence p(CCG)n. *Science* **252**, 1711–1714.
13. Richards, R. I., and Sutherland, G. R. (1992). Dynamic mutations: A new class of mutations causing human disease. *Cell* **70**, 709–712.
14. Kremer, B., Almqvist, E., Theilmann, J., Spence, N., Telenius, H., Goldberg, Y. P., and Hayden, M. R. (1995). Sex-dependent mechanisms for expansions and contractions of the CAG repeat on affected Huntington disease chromosomes. *Am. J. Hum. Genet.* **57**, 343–350.
15. Leeflang, E. P., Zhang, L., Tavare, S., Hubert, R., Srinidhi, J., MacDonald, M. E., Myers, R. H., de Young, M., Wexler, N. S., Gusella, J. F., and Arnheim, N. (1995). Single sperm analysis of the trinucleotide repeats in the Huntington's disease gene: Quantification of the mutation frequency spectrum. *Hum. Mol. Genet.* **4**, 1519–1526.
16. Jakupciak, J. P., and Wells, R. D. (2000). Genetic instabilities of triplet repeat sequences by recombination. *IUBMB Life* **50**, 355–359.
17. Igarashi, S., Takiyama, Y., Cancel, G., Rogaeva, E. A., Sasaki, H., Wakisaka, A., Zhou, Y. X., Takano, H., Endo, K., Sanpei, K., Oyake, M., Tanaka, H., Stevanin, G., Abbas, N., Durr, A., Rogaev, E. I., Sherrington, R., Tsuda, T., Ikeda, M., Cassa, E., Nishizawa, M., Benomar, A., Julien, J., Weissenbach, J., Wang, G. X., Agid, Y., StGeorgeHyslop, P. H., Brice, A., and Tsuji, S. (1996). Intergenerational instability of the CAG repeat of the gene for Machado–Joseph disease (MJD1) is affected by the genotype of the normal chromosome: Implications for the molecular mechanisms of the instability of the CAG repeat. *Hum. Mol. Genet.* **5**, 923–932.
18. Mittal, U., Srivastava, A. K., Jain, S., and Mukerji, M. (2005). Founder haplotype for Machado-Joseph disease in the Indian population: Novel insights from history and polymorphism studies. *Arch. Neurol.* **62**, 637–640.
19. Vergnaud, G., and Denoeud, F. (2000). Minisatellites: Mutability and genome architecture. *Genome Res.* **10**, 899–907.
20. Richard, G. F., and Paques, F. (2000). Mini- and microsatellite expansions: The recombination connection. *EMBO Rep.* **1**, 122–126.
21. Jeffreys, A. J., Barber, R., Bois, P., Buard, J., Dubrova, Y. E., Grant, G., Hollies, C. R., May, C. A., Neumann, R., Panayi, M., Ritchie, A. E., Shone, A. C., Signer, E., Stead, J. D., and Tamaki, K. (1999). Human minisatellites, repeat DNA instability and meiotic recombination. *Electrophoresis* **20**, 1665–1675.
22. Wahls, W. P., Wallace, L. J., and Moore, P. D. (1990). Hypervariable minisatellite DNA is a hotspot for homologous recombination in human cells. *Cell* **60**, 95–103.
23. Wahls, W. P., and Moore, P. D. (1998). Recombination hotspot activity of hypervariable minisatellite DNA requires minisatellite DNA binding proteins. *Somat. Cell Mol. Genet.* **24**, 41–51.
24. Richards, R. I. (2001). Dynamic mutations: A decade of unstable expanded repeats in human genetic disease. *Hum. Mol. Genet.* **10**, 2187–2194.
25. Montagnon, M., Bogyo, A., Deluchat, C., Jokic, M., Chateau, C., Taillandier, A., Thomas, F., Simon-Bouy, B., Boue, J., Serre, J. L., Serre, J. L., Boué, A., and Mornet, E. (1994). Transition from normal to premutated alleles in fragile X syndrome results from a multistep process. *Eur. J. Hum. Genet.* **2**, 125–131.
26. van den Ouweland, A. M., Deelen, W. H., Kunst, C. B., Uzielli, M. L., Nelson, D. L., Warren, S. T., Oostra, B. A., and Halley, D. J. (1994). Loss of mutation at the FMR1 locus through multiple exchanges between maternal X chromosomes. *Hum. Mol. Genet.* **3**, 1823–1827.
27. Losekoot, M., Hoogendoorn, E., Olmer, R., Jansen, C. C., Oosterwijk, J. C., van den Ouweland, A. M., Halley, D. J., Warren, S. T., Willemsen, R., Oostra, B. A., and Bakker, E. (1997). Prenatal diagnosis of the fragile X syndrome: Loss of mutation owing to a double recombinant or gene conversion event at the FMR1 locus. *J. Med. Genet.* **34**, 924–926.
28. Brown, W. T., Houck, G. E., Jr., Ding, X., Zhong, N., Nolin, S., Glicksman, A., Dobkin, C., and Jenkins, E. C. (1996). Reverse mutations in the fragile X syndrome. *Am. J. Med. Genet.* **64**, 287–292.
29. O'Hoy, K. L., Tsilfidis, C., Mahadevan, M. S., Neville, C. E., Barcelo, J., Hunter, A. G., and Korneluk, R. G. (1993). Reduction in size of the myotonic dystrophy trinucleotide repeat mutation during transmission. *Science* **259**, 809–812.
30. Brunner, H. G., Jansen, G., Nillesen, W., Nelen, M. R., de Die, C. E., Howeler, C. J., van Oost, B. A., Wieringa, B., Ropers, H. H., and Smeets, H. J. (1993). Brief report: Reverse mutation in myotonic dystrophy. *New Engl. J. Med.* **328**, 476–480.
31. Reddy, P. S., and Housman, D. E. (1997). The complex pathology of trinucleotide repeats. *Curr. Opin. Cell Biol.* **9**, 364–372.
32. Brown, L. Y., and Brown, S. A. (2004). Alanine tracts: The expanding story of human illness and trinucleotide repeats. *Trends. Genet.* **20**, 51–58.

33. Warren, S. T. (1997). Polyalanine expansion in synpolydactyly might result from unequal crossing-over of HOXD13. *Science* **275**, 408–409.
34. Jakupciak, J. P., and Wells, R. D. (1999). Genetic instabilities in (CTG·CAG) repeats occur by recombination. *J. Biol. Chem.* **274**, 23468–23479.
35. Jakupciak, J. P., and Wells, R. D. (2000). Gene conversion (recombination) mediates expansions of CTG·CAG repeats. *J. Biol. Chem.* **275**, 40003–40013.
36. Mahadevan, M., Tsilfidis, C., Sabourin, L., Shutler, G., Amemiya, C., Jansen, G., Neville, C., Narang, M., Barcelo, J., O'Hoy, K., Leblond, S., Earle-MacDonald, J., de Jong, P. J., Wieringa, B., and Korneluk, R. G. (1992). Myotonic dystrophy mutation: An unstable CTG repeat in the 3′ untranslated region of the gene. *Science* **255**, 1253–1255.
37. Fu, Y.-H., Pizzuti, A., Fenwick, R. G., Jr., King, J., Rajnarayan, S., Dunne, P. W., Dubel, J., Nasser, G. A., Ashizawa, T., de Jong, P. J., Wieringa, B., Korneluk, R. G., Perryman, M. B., Epstein, H. F., and Caskey, C. T. (1992). An unstable triplet repeat in a gene related to myotonic muscular dystrophy. *Science* **255**, 1256–1258.
38. Petes, T. D. (2001). Meiotic recombination hot spots and cold spots. *Nat. Rev. Genet.* **2**, 360–369.
39. Batzer, M. A., and Deininger, P. L. (2002). Alu repeats and human genomic diversity. *Nat. Rev. Genet.* **3**, 370–379.
40. Rowold, D. J., and Herrera, R. J. (2000). Alu elements and the human genome. *Genetica* **108**, 57–72.
41. Deininger, P. L., and Batzer, M. A. (1999). Alu repeats and human disease. *Mol. Genet. Metab.* **67**, 183–193.
42. Blaho, J. A., and Wells, R. D. (1989). Left-handed Z-DNA and genetic recombination. *Prog. Nucleic Acid Res. Mol. Biol.* **37**, 107–126.
43. Majewski, J., and Ott, J. (2000). GT repeats are associated with recombination on human chromosome 22. *Genome Res.* **10**, 1108–1114.
44. Benet, A., Molla, G., and Azorin, F. (2000). d(GA × TC)(n) microsatellite DNA sequences enhance homologous DNA recombination in SV40 minichromosomes. *Nucleic Acids Res.* **28**, 4617–4622.
45. Kirkpatrick, D. T., Wang, Y. H., Dominska, M., Griffith, J. D., and Petes, T. D. (1999). Control of meiotic recombination and gene expression in yeast by a simple repetitive DNA sequence that excludes nucleosomes. *Mol. Cell. Biol.* **19**, 7661–7671.
46. Paques, F., and Haber, J. E. (1999). Multiple pathways of recombination induced by double-strand breaks in *Saccharomyces cerevisiae*. *Microbiol. Mol. Biol. Rev.* **63**, 349–404.
47. Lambert, S., Saintigny, Y., Delacote, F., Amiot, F., Chaput, B., Lecomte, M., Huck, S., Bertrand, P., and Lopez, B. S. (1999). Analysis of intrachromosomal homologous recombination in mammalian cell, using tandem repeat sequences. *Mutat. Res.* **433**, 159–168.
48. Napierala, M., Parniewski, P. P., Pluciennik, A., and Wells, R. D. (2002). Long CTG·CAG repeat sequences markedly stimulate intramolecular recombination. *J. Biol. Chem.* **277**, 34087–34100.
49. Pluciennik, A., Iyer, R. R., Napierala, M., Larson, J. E., Filutowicz, M., and Wells, R. D. (2002). Long CTG·CAG repeats from myotonic dystrophy are preferred sites for intermolecular recombination. *J. Biol. Chem.* **277**, 34074–34086.
50. Kang, S., Jaworski, A., Ohshima, K., and Wells, R. D. (1995). Expansion and deletion of CTG repeats from human disease genes are determined by the direction of replication in *E. coli*. *Nat. Genet.* **10**, 213–218.
51. Kayser, M., Roewer, L., Hedman, M., Henke, L., Henke, J., Brauer, S., Kruger, C., Krawczak, M., Nagy, M., Dobosz, T., Szibor, R., de Knijff, P., Stoneking, M., and Sajantila, A. (2000). Characteristics and frequency of germline mutations at microsatellite loci from the human Y chromosome, as revealed by direct observation in father/son pairs. *Am. J. Hum. Genet.* **66**, 1580–1588.
52. Ellegren, H. (2004). Microsatellites: Simple sequences with complex evolution. *Nat. Rev. Genet.* **5**, 435–445.
53. Kang, S., Ohshima, K., Shimizu, M., Amirhaeri, S., and Wells, R. D. (1995). Pausing of DNA synthesis in vitro at specific loci in CTG and CGG triplet repeats from human hereditary disease genes. *J. Biol. Chem.* **270**, 27014–27021.
54. Ohshima, K., Kang, S., Larson, J. E., and Wells, R. D. (1996). Cloning, characterization, and properties of seven triplet repeat DNA sequences. *J. Biol. Chem.* **271**, 16773–16783.
55. Samadashwily, G. M., Raca, G., and Mirkin, S. M. (1997). Trinucleotide repeats affect DNA replication in vivo. *Nat. Genet.* **17**, 298–304.
56. Cox, M. M. (1998). A broadening view of recombinatorial DNA repair in bacteria. *Genes Cells* **3**, 65–78.
57. Haber, J. E. (1999). DNA recombination: The replication connection. *Trends. Biochem. Sci.* **24**, 271–275.
58. Michel, B., Flores, M. J., Viguera, E., Grompone, G., Seigneur, M., and Bidnenko, V. (2001). Rescue of arrested replication forks by homologous recombination. *Proc. Natl. Acad. Sci. USA* **98**, 8181–8188.
59. Bishop, A. J., and Schiestl, R. H. (2000). Homologous recombination as a mechanism for genome rearrangements: Environmental and genetic effects. *Hum. Mol. Genet.* **9**, 2427–2334.
60. Worth, L., Jr., Clark, S., Radman, M., and Modrich, P. (1994). Mismatch repair proteins MutS and MutL inhibit RecA-catalyzed strand transfer between diverged DNAs. *Proc. Natl. Acad. Sci. USA* **91**, 3238–3241.
61. Lalioti, M. D., Scott, H. S., Buresi, C., Rossier, C., Bottani, A., Morris, M. A., Malafosse, A., and Antonarakis, S. E. (1997). Dodecamer repeat expansion in cystatin B gene in progressive myoclonus epilepsy. *Nature* **386**, 847–851.
62. Matsuura, T., Yamagata, T., Burgess, D. L., Rasmussen, A., Grewal, R. P., Watase, K., Khajavi, M., McCall, A. E., Davis, C. F., Zu, L., Achari, M., Pulst, S. M., Alonso, E., Noebels, J. L., Nelson, D. L., Zoghbi, H. Y., and Ashizawa, T. (2000). Large expansion of the ATTCT pentanucleotide repeat in spinocerebellar ataxia type 10. *Nat. Genet.* **26**, 191–194.
63. Liquori, C. L., Ricker, K., Moseley, M. L., Jacobsen, J. F., Kress, W., Naylor, S. L., Day, J. W., and Ranum, L. P. (2001). Myotonic dystrophy type 2 caused by a CCTG expansion in intron 1 of ZNF9. *Science* **293**, 864–867.
64. Moxley, R. T., III, and Ricker, K. (1995). Proximal myotonic myopathy. *Muscle Nerve* **18**, 557–558.
65. Finsterer, J. (2002). Myotonic dystrophy type 2. *Eur. J. Neurol.* **9**, 441–447.
66. Mankodi, A., and Thornton, C. A. (2002). Myotonic syndromes. *Curr. Opin. Neurol.* **15**, 545–552.
67. Day, J. W., Ricker, K., Jacobsen, J. F., Rasmussen, L. J., Dick, K. A., Kress, W., Schneider, C., Koch, M. C., Beilman, G. J., Harrison, A. R., Dalton, J. C., and Ranum, L. P. W. (2003). Myotonic dystrophy type 2: Molecular, diagnostic and clinical spectrum. *Neurology* **60**, 657–664.
68. Bowater, R. P., and Wells, R. D. (2001). The intrinsically unstable life of DNA triplet repeats associated with human hereditary disorders. *Prog. Nucleic Acid Res. Mol. Biol.* **66**, 159–202.
69. Mitas, M. (1997). Trinucleotide repeats associated with human disease. *Nucleic Acids Res.* **25**, 2245–2254.
70. Darlow, J. M., and Leach, D. R. F. (1998). Secondary structures in d(CGG) and d(CCG): Repeat tracts. *J. Mol. Biol.* **275**, 3–16.
71. Sinden, R. R., Potaman, V. N., Oussatcheva, E. A., Pearson, C. E., Lyubchenko, Y. L., and Shlyakhtenko, L. S. (2002). Triplet repeat

DNA structures and human genetic disease: Dynamic mutations from dynamic DNA. *J. Biosci.* **27**, 53–65.
72. Dere, R., Napierala, M., Ranum, L. P., and Wells, R. D. (2004). Hairpin structure-forming propensity of the (CCTG·CAGG) tetranucleotide repeats contributes to the genetic instability associated with myotonic dystrophy type 2. *J. Biol. Chem.* **279**, 41715–41726.
73. Shimizu, M., Gellibolian, R., Oostra, B. A., and Wells, R. D. (1996). Cloning, characterization and properties of plasmids containing CGG triplet repeats from the FMR-1 gene. *J. Mol. Biol.* **258**, 614–626.
74. Maurer, D. J., O'Callaghan, B. L., and Livingston, D. M. (1996). Orientation dependence of trinucleotide CAG repeat instability in *Saccharomyces cerevisiae*. *Mol. Cell. Biol.* **16**, 6617–6622.
75. Freudenreich, C. H., Stavenhagen, J. B., and Zakian, V. A. (1997). Stability of a CTG·CAG trinucleotide repeat in yeast is dependent on its orientation in the genome. *Mol. Cell. Biol.* **17**, 2090–2098.
76. Miret, J. J., Pessoa-Brandao, L., and Lahue, R. S. (1998). Orientation-dependent and sequence-specific expansions of CTG/CAG trinucleotide repeats in *Saccharomyces cerevisiae*. *Proc. Natl. Acad. Sci. USA* **95**, 12438–12443.
77. Freudenreich, C. H., Kantrow, S. M., and Zakian, V. A. (1998). Expansion and length-dependent fragility of CTG repeats in yeast. *Science* **279**, 853–856.
78. Balakumaran, B. S., Freudenreich, C. H., and Zakian, V. A. (2000). CGG·CCG repeats exhibit orientation-dependent instability and orientation-independent fragility in *Saccharomyces cerevisiae*. *Hum. Mol. Genet.* **9**, 93–100.
79. Dere, R., and Wells, R. D. (2005). (CCTG·CAGG) repeats associated with DM2 are SOS-repair stimulated recombination hotspots which are prone to expansions. Manuscript in preparation.
80. Napierala, M., Dere, R., Vetcher, A., and Wells, R. D. (2004). Structure-dependent recombination hot spot activity of GAA·TTC sequences from intron 1 of the Friedreich's ataxia gene. *J. Biol. Chem.* **279**, 6444–6454.
81. Courcelle, J., Khodursky, A., Peter, B., Brown, P. O., and Hanawalt, P. C. (2001). Comparative gene expression profiles following UV exposure in wild-type and SOS-deficient *Escherichia coli*. *Genetics* **158**, 41–64.
82. Kowalczykowski, S. C., Dixon, D. A., Eggleston, A. K., Lauder, S. D., and Rehrauer, W. M. (1994). Biochemistry of homologous recombination in *Escherichia coli*. *Microbiol. Rev.* **58**, 401–465.
83. Cox, M. M. (1999). Recombinational DNA repair in bacteria and the RecA protein. *Prog. Nucleic Acid Res. Mol. Biol.* **63**, 311–366.
84. Chauhan, C., Dash, D., Grover, D., Rajamani, J., and Mukerji, M. (2002). Origin and instability of GAA repeats: Insights from Alu elements. *J. Biomol. Struct. Dyn.* **20**, 253–263.
85. Campuzano, V., Montermini, L., Molto, M. D., Pianese, L., Cossee, M., Cavalcanti, F., Monros, E., Rodius, F., Duclos, F., Monticelli, A., Zara, F., Canizares, J., Koutnikova, H., Bidichandani, S. I., Gellera, C., Brice, A., Trouillas, P., DeMichele, G., Filla, A., DeFrutos, R., Palau, F., Patel, P. I., DiDonato, S., Mandel, J. L., Cocozza, S., Koenig, M., and Pandolfo, M. (1996). Friedreich's ataxia: Autosomal recessive disease caused by an intronic GAA triplet repeat expansion. *Science* **271**, 1423–1427.
86. Bidichandani, S. I., Ashizawa, T., and Patel, P. I. (1998). The GAA triplet-repeat expansion in Friedreich ataxia interferes with transcription and may be associated with an unusual DNA structure. *Am. J. Hum. Genet.* **62**, 111–121.
87. Delatycki, M. B., Paris, D. B., Gardner, R. J., Nicholson, G. A., Nassif, N., Storey, E., MacMillan, J. C., Collins, V., Williamson, R., and Forrest, S. M. (1999). Clinical and genetic study of Friedreich ataxia in an Australian population. *Am. J. Med. Genet.* **87**, 168–174.
88. Delatycki, M. B., Knight, M., Koenig, M., Cossee, M., Williamson, R., and Forrest, S. M. (1999). G130V, a common FRDA point mutation, appears to have arisen from a common founder. *Hum. Genet.* **105**, 343–346.
89. Ohshima, K., Montermini, L., Wells, R. D., and Pandolfo, M. (1998). Inhibitory effects of expanded GAA·TTC triplet repeats from intron I of the Friedreich ataxia gene on transcription and replication *in vivo*. *J. Biol. Chem.* **275**, 14588–14595.
90. Grabczyk, E., and Usdin, K. (2000). The GAA·TTC triplet repeat expanded in Friedreich's ataxia impedes transcription elongation by T7 RNA polymerase in a length and supercoil dependent manner. *Nucleic Acids Res.* **28**, 2815–2822.
91. Potaman, V. N., Oussatcheva, E. A., Lyubchenko, Y. L., Shlyakhtenko, L. S., Bidichandani, S. I., Ashizawa, T., and Sinden, R. R. (2004). Length-dependent structure formation in Friedreich ataxia (GAA)n·(TTC)n repeats at neutral pH. *Nucleic Acids Res.* **32**, 1224–1231.
92. Sakamoto, N., Chastain, P. D., Parniewski, P., Ohshima, K., Pandolfo, M., Griffith, J. D., and Wells, R. D. (1999). Sticky DNA: self-association properties of long GAA·TTC repeats in R·R·Y triplex structures from Friedreich's ataxia. *Mol. Cell* **3**, 465–475.
93. Sakamoto, N., Larson, J. E., Iyer, R. R., Montermini, L., Pandolfo, M., and Wells, R. D. (2001). GGA*TCC-interrupted triplets in long GAA*TTC repeats inhibit the formation of triplex and sticky DNA structures, alleviate transcription inhibition, and reduce genetic instabilities. *J. Biol. Chem.* **276**, 27178–27187.
94. Sakamoto, N., Ohshima, K., Montermini, L., Pandolfo, M., and Wells, R. D. (2001). Sticky DNA, a self-associated complex formed at long GAA*TTC repeats in intron 1 of the frataxin gene, inhibits transcription. *J. Biol. Chem.* **276**, 27171–27177.
95. Vetcher, A. A., Napierala, M., Iyer, R. R., Chastain, P. D., Griffith, J. D., and Wells, R. D. (2002). Sticky DNA, a long GAA·GAA·TTC triplex that is formed intramolecularly, in the sequence of intron 1 of the frataxin gene. *J. Biol. Chem.* **277**, 39217–39227.
96. Vetcher, A. A., Napierala, M., and Wells, R. D. (2002). Sticky DNA: Effect of the polypurine·polypyrimidine sequence. *J. Biol. Chem.* **277**, 39228–39234.
97. Vetcher, A. A., and Wells, R. D. (2004). Sticky DNA formation in vivo alters the plasmid dimer/monomer ratio. *J. Biol. Chem.* **279**, 6434–6443.
98. Lewis, R. J., Singh, O. M., Smith, C. V., Skarzynski, T., Maxwell, A., Wonacott, A. J., and Wigley, D. B. (1996). The nature of inhibition of DNA gyrase by the coumarins and the cyclothialidines revealed by X-ray crystallography. *EMBO J.* **15**, 1412–1420.
99. Tsai, F. T., Singh, O. M., Skarzynski, T., Wonacott, A. J., Weston, S., Tucker, A., Pauptit, R. A., Breeze, A. L., Poyser, J. P., O'Brien, R., Ladbury, J. E., and Wigley, D. B. (1997). The high-resolution crystal structure of a 24-kDa gyrase B fragment from *E. coli* complexed with one of the most potent coumarin inhibitors, clorobiocin. *Proteins* **28**, 41–52.
100. Campuzano, V., Montermini, L., Lutz, Y., Cova, L., Hindelang, C., Jiralerspong, S., Trottier, Y., Kish, S. J., Faucheux, B., Trouillas, P., Authier, F. J., Durr, A., Mandel, J. L., Vescovi, A., Pandolfo, M., and Koenig, M. (1997). Frataxin is reduced in Friedreich ataxia patients and is associated with mitochondrial membranes. *Hum. Mol. Genet.* **6**, 1771–1780.
101. Cossee, M., Campuzano, V., Koutnikova, H., Fischbeck, K., Mandel, J. L., Koenig, M., Bidichandani, S. I., Patel, P. I., Molte, M. D., Canizares, J., De Frutos, R., Pianese, L., Cavalcanti, F., Monticelli, A., Cocozza, S., Montermini, L., and Pandolfo, M. (1997). Frataxin fracas. *Nat. Genet.* **15**, 337–338.
102. Siyanova, E. I., and Mirkin, S. M. (2001). Expansion of trinucleotide repeats. *Mol. Biol.* **35**, 208–223.

103. Wells, R. D., Dere, R., Hebert, M. L., Napierala, M., and Son, L. S. (2005). Advances in mechanisms of genetic instability related to hereditary neurological diseases. *Nucleic Acids Res.* in press.
104. Pelletier, R., Krasilnikova, M. M., Samadashwily, G. M., Lahue, R., and Mirkin, S. M. (2003). Replication and expansion of trinucleotide repeats in yeast. *Mol. Cell. Biol.* **23**, 1349–1357.
105. Richard, G. F., Dujon, B., and Haber, J. E. (1999). Double-strand break repair can lead to high frequencies of deletions within short CAG·CTG trinucleotide repeats. *Mol. Gen. Genet.* **261**, 871–882.
106. Jankowski, C., Nasar, F., and Nag, D. K. (2000). Meiotic instability of CAG repeat tracts occurs by double-strand break repair in yeast. *Proc. Natl. Acad. Sci. USA* **97**, 2134–2139.
107. Jankowski, C., and Nag, D. K. (2002). Most meiotic CAG repeat tract-length alterations in yeast are SPO11 dependent. *Mol. Genet. Genom.* **267**, 64–70.
108. Arai, N., Akiyama, R., Niimi, N., Nakatsubo, H., and Inoue, T. (1999). Meiotic contraction of CAG repeats in *Saccharomyces cerevisiae*. *Genes Genet. Syst.* **74**, 159–167.
109. Richard, G. F., Goellner, G. M., McMurray, C. T., and Haber, J. E. (2000). Recombination-induced CAG trinucleotide repeat expansions in yeast involve the MRE11-RAD50-XRS2 complex. *EMBO J.* **19**, 2381–2390.
110. Meservy, J. L., Sargent, R. G., Iyer, R. R., Chan, F., McKenzie, G. J., Wells, R. D., and Wilson, J. H. (2003). Long CTG tracts from the myotonic dystrophy gene induce deletions and rearrangements during recombination at the APRT locus in CHO cells. *Mol. Cell. Biol.* **23**, 3152–3162.
111. Marcadier, J. L., and Pearson, C. E. (2003). Fidelity of primate cell repair of a double-strand break within a (CTG)·(CAG) tract. Effect of slipped DNA structures. *J. Biol. Chem.* **278**, 33848–33856.
112. Savouret, C., Brisson, E., Essers, J., Kanaar, R., Pastink, A., te Riele, H., Junien, C., and Gourdon, G. (2003). CTG repeat instability and size variation timing in DNA repair-deficient mice. *EMBO J.* **22**, 2264–2273.
113. Hebert, M. L., Spitz, L. A., and Wells, R. D. (2004). DNA double-strand breaks induce deletion of CTG·CAG repeats in an orientation-dependent manner in *Escherichia coli*. *J. Mol. Biol.* **336**, 655–672.
114. Hebert, M. L., and Wells, R. D. (2005). Roles of double-strand breaks, nicks, and gaps in stimulating deletions of CTG·CAG repeats by intramolecular DNA repair. *J. Mol. Biol.* **353**, 961–979.
115. Hartenstine, M. J., Goodman, M. F., and Petruska, J. (2002). Weak strand displacement activity enables human DNA polymerase beta to expand CAG·CTG triplet repeats at strand breaks. *J. Biol. Chem.* **277**, 41379–41389.
116. Kovtun, I. V., and McMurray, C. T. (2001). Trinucleotide expansion in haploid germ cells by gap repair. *Nat. Genet.* **27**, 407–411.
117. Hashem, V. I., Pytlos, M. J., Klysik, E. A., Tsuji, K., Khajavi, M., Ashizawa, T., and Sinden, R. R. (2004). Chemotherapeutic deletion of CTG repeats in lymphoblast cells from DM1 patients. *Nucleic Acids Res.* **32**, 6334–6346.
118. Krasilnikova, M. M., and Mirkin, S. M. (2004). Analysis of triplet repeat replication by two-dimensional gel electrophoresis. *Methods Mol. Biol.* **277**, 19–28.
119. Parniewski, P., Jaworski, A., Wells, R. D., and Bowater, R. P. (2000). Length of CTG·CAG repeats determines the influence of mismatch repair on genetic instability. *J. Mol. Biol.* **299**, 865–874.
120. Rolfsmeier, M. L., and Lahue, R. S. (2000). Stabilizing effects of interruptions on trinucleotide repeat expansions in *Saccharomyces cerevisiae*. *Mol. Cell. Biol.* **20**, 173–180.
121. Rolfsmeier, M. L., Dixon, M. J., and Lahue, R. S. (2000). Mismatch repair blocks expansions of interrupted trinucleotide repeats in yeast. *Mol. Cell* **6**, 1501–1507.
122. Oussatcheva, E. A., Hashem, V. I., Zou, Y., Sinden, R. R., and Potaman, V. N. (2001). Involvement of the nucleotide excision repair protein UvrA in instability of CAG·CTG repeat sequences in *Escherichia coli*. *J. Biol. Chem.* **276**, 30878–30884.
123. Parniewski, P., Bacolla, A., Jaworski, A., and Wells, R. D. (1999). Nucleotide excision repair affects the stability of long transcribed (CTG·CAG) tracts in an orientation-dependent manner in *Escherichia coli*. *Nucleic Acids Res.* **27**, 616–623.
124. Schumacher, S., Fuchs, R. P., and Bichara, M. (1997). Two distinct models account for short and long deletions within sequence repeats in *Escherichia coli*. *J. Bacteriol.* **179**, 6512–6517.
125. Schumacher, S., Fuchs, R. P., and Bichara, M. (1998). Expansion of CTG repeats from human disease genes is dependent upon replication mechanisms in *Escherichia coli*: The effect of long patch mismatch repair revisited. *J. Mol. Biol.* **279**, 1101–1110.
126. Jaworski, A., Rosche, W. A., Gellibolian, R., Kang, S., Shimizu, M., Bowater, R. P., Sinden, R. R., and Wells, R. D. (1995). Mismatch repair in *Escherichia coli* enhances instability of (CTG)n triplet repeats from human hereditary diseases. *Proc. Natl. Acad. Sci. USA* **92**, 11019–11023.
127. Wells, R. D., Parniewski, P., Pluciennik, A., Bacolla, A., Gellibolian, R., and Jaworski, A. (1998). Small slipped register genetic instabilities in *Escherichia coli* in triplet repeat sequences associated with hereditary neurological diseases. *J. Biol. Chem.* **273**, 19532–19541.
128. Schmidt, K. H., Abbott, C. M., and Leach, D. R. (2000). Two opposing effects of mismatch repair on CTG repeat instability in *Escherichia coli*. *Mol. Microbiol.* **35**, 463–471.
129. Ireland, M. J., Reinke, S. S., and Livingston, D. M. (2000). The impact of lagging strand replication mutations on the stability of CAG repeat tracts in yeast. *Genetics* **155**, 1657–1665.
130. Lenzmeier, B. A., and Freudenreich, C. H. (2003). Trinucleotide repeat instability: A hairpin curve at the crossroads of replication, recombination, and repair. *Cytogenet. Genome Res.* **100**, 7–24.
131. Kauppi, L., Jeffreys, A. J., and Keeney, S. (2004). Where the crossovers are: Recombination distributions in mammals. *Nat. Rev. Genet.* **5**, 413–424.
132. Cox, R., and Mirkin, S. M. (1997). Characteristic enrichment of DNA repeats in different genomes. *Proc. Natl. Acad. Sci. USA* **94**, 5237–5242.
133. Young, E. T., Sloan, J. S., and Van Riper, K. (2000). Trinucleotide repeats are clustered in regulatory genes in *Saccharomyces cerevisiae*. *Genetics* **154**, 1053–1068.
134. Kohwi, Y., and Panchenko, Y. (1993). Transcription-dependent recombination induced by triple-helix formation. *Genes Dev.* **7**, 1766–1778.
135. Wahls, W. P., Wallace, L. J., and Moore, P. D. (1990). The Z-DNA motif d(TG)30 promotes reception of information during gene conversion events while stimulating homologous recombination in human cells in culture. *Mol. Cell. Biol.* **10**, 785–793.
136. Murphy, K. E., and Stringer, J. R. (1986). RecA independent recombination of poly[d(GT)-d(CA)] in pBR322. *Nucleic Acids Res.* **14**, 7325–7340.
137. Stringer, J. R. (1985). Recombination between poly[d(GT)·d(CA)] sequences in simian virus 40-infected cultured cells. *Mol. Cell. Biol.* **5**, 1247–1259.
138. Bullock, P., Miller, J., and Botchan, M. (1986). Effects of poly[d(pGpT)·d(pApC)] and poly[d(pCpG)·d(pCpG)] repeats on homologous recombination in somatic cells. *Mol. Cell. Biol.* **6**, 3948–3953.
139. Treco, D., and Arnheim, N. (1986). The evolutionarily conserved repetitive sequence d(TG·AC)n promotes reciprocal exchange and generates unusual recombinant tetrads during yeast meiosis. *Mol. Cell. Biol.* **6**, 3934–3947.
140. Gendrel, C. G., Boulet, A., and Dutreix, M. (2000). (CA/GT)(n) microsatellites affect homologous recombination during yeast meiosis. *Genes Dev.* **14**, 1261–1268.

141. Nag, D. K., Suri, M., and Stenson, E. K. (2004). Both CAG repeats and inverted DNA repeats stimulate spontaneous unequal sister-chromatid exchange in *Saccharomyces cerevisiae*. *Nucleic Acids Res.* **32**, 5677–5684.
142. Shiroishi, T., Sagai, T., and Moriwaki, K. (1993). Hotspots of meiotic recombination in the mouse major histocompatibility complex. *Genetica* **88**, 187–196.
143. Zuker, M. (2003). Mfold web server for nucleic acid folding and hybridization prediction. *Nucleic Acids Res.* **31**, 3406–3415.
144. Bacolla, A., and Wells, R. D. (2004). Non-B DNA conformations, genomic rearrangements, and human disease. *J. Biol. Chem.* **279**, 47411–47414.
145. Sinden, R. R. (1994). "DNA Structure and Function." Academic Press, San Diego.

Bending the Rules: Unusual Nucleic Acid Structures and Disease Pathology in the Repeat Expansion Diseases

KAREN USDIN

Laboratory of Molecular and Cellular Biology, National Institute of Diabetes and Digestive and Kidney Diseases, National Institutes of Health, Bethesda, Maryland, 20892

I. Introduction
II. Unusual Structures Formed by the DNA and RNA versions of the Disease-Associated Repeats
   A. DNA Structures
   B. RNA Structures
   C. Effects of Interruptions
   D. Evidence for Secondary Structures *in Vivo*
III. Potential Biological Consequences of These Unusual Structures
   A. Expansion
   B. Chromosome Fragility
   C. Effects on Gene Expression
IV. Concluding Remarks
References

The repeat expansion diseases are a group of genetic disorders with an unusual mutational mechanism: they are caused by expansion of a specific tandem repeat array. Many disorders in this category are also unusual in that the repeat tract is located outside of the open reading frame of the affected gene. Thus, in these cases pathology does not simply involve the generation of a malfunctioning or toxic protein. Rather, these disorders provide new paradigms for disease pathology that may be related to the repeats at the DNA or RNA level. The repeats responsible for these disorders all have the ability to form secondary structures, many of which involve a bend or fold. This chapter describes the structures formed by these repeats, how the properties of these structures inform current models of expansion, and different ways in which these structures could affect chromatin organization, transcription, and translation.

I. INTRODUCTION

Most known disease-causing mutations that do not result in early death or infertility can be stably transmitted to subsequent generations. The repeat expansion diseases are different in that they involve an increase in length or expansion of an unstable tandem repeat tract, with the resultant expanded allele being predisposed to further instability. This instability, which is sometimes referred to as a dynamic mutation, also differs from the general microsatellite instability characteristic of diseases, such as hereditary nonpolyposis colon cancer (HNPCC), which result from mutations in mismatch repair (MMR) genes [1]. In fact, as will be seen later in this chapter, mutations in MMR genes actually decrease repeat instability in this group of disorders [2–5]. Both increases (expansions) and decreases (contractions) in repeat number do occur, but expansions predominate, with increases above a certain threshold causing disease pathology. Expansion is thought to occur during either germline development or very early embryogenesis. In addition to instability occurring within this relatively narrow developmental window, some repeats also show significant somatic instability, the extent of which can vary between tissues. The risk of expansion is related to repeat number, the purity of the repeat tract, the sex of the transmitting parent, and the chromosomal context. In some diseases, expansion involves only a small increase in repeat number, whereas in others the expanded repeat tract can be many times larger than the allele from which it was derived. Although many factors that affect expansion risk are known, the actual mechanism of repeat expansion is not, nor is it known whether small expansions occur by the same mechanism as large expansions or how somatic and germline–early embryonic instability are related.

In general a direct relationship exists between the number of uninterrupted repeats in the expanded allele and disease severity, and in those disorders that are not congenital an inverse relationship exists between repeat number and age of onset. Current data suggest different ways in which the repeats can be pathological, depending on factors that include the sequence of the repeat, its location within the affected gene, the function of the gene itself, its chromosomal context, and the cells in which the gene is expressed.

In a subset of the repeat expansion diseases, expansion produces an increase in the length of a polyglutamine tract in the affected protein. It is the toxicity of this expanded amino acid stretch that is responsible for the neurodegeneration seen in these disorders. However, in the remaining expansion diseases the repeats are outside the coding region of the gene. Thus, in these instances, disease pathology is not simply a question of the production of an aberrant protein. Ten different human diseases belong to this category (Table 39-1). They are progressive myoclonus epilepsy type 1 (EPM1), myotonic dystrophy types 1 and 2 (DM1 and DM2), two dominantly inherited spinocerebellar ataxias (SCAs, SCA8 and SCA10), a recessively inherited ataxia, Friedreich's ataxia (FRDA), FRAXE mental retardation, and the three disorders that result from expansion at the FMR1 locus, namely, fragile X mental retardation syndrome (FXS), fragile X associated tremor and ataxia syndrome (FXTAS), and fragile X associated premature ovarian failure. Two other repeat expansion diseases potentially fall into this category: SCA12 and Huntington disease-like 2 (HDL2). Because these disorders are all described in greater detail in other chapters in this volume, just a few of their salient features will be described here.

In EPM1, an autosomal recessive disorder, the repeat C_4GC_4GCG is located in the promoter of the cystatin B gene [6, 7]. Expansion is thought to disrupt the optimal arrangement of transcription factor binding sites, thereby decreasing promoter activity. For the remaining diseases in this category, the repeats appear to do more than simply physically interrupt a regulatory unit. For example, in DM1, a dominantly inherited

TABLE 39-1 Human Repeat Expansion Diseases Where the Repeat Is or May Be Outside of the Coding Sequence[a]

Disease	Repeat unit[b]	Gene	Location	Inheritance	Reference numbers
Progressive myoclonus epilepsy type 1 (EPM1)	C_4GC_4GCG	Cystatin B	Promoter	AR	7
Fragile X syndrome, FXTAS, FXPOF	CGG	FMR1	5′ UTR	XLD	182, 183
FRAXE mental retardation	GCC	FMR2	5′ UTR	XLD	20, 21
Friedreich's ataxia	GAA	Frataxin	Intron	AR	18
Myotonic dystrophy type 1	CTG	DMPK	3′ UTR	AD	184
Myotonic dystrophy type 2	CCTG	Znf9	Intron	AD	12
SCA8	CTG	SCA8	Noncoding RNA	AD	15
SCA10	ATTCT	Ataxin-10	Intron	AD	17
SCA12	CAG	PPP2R2B	Promoter or 5′ UTR?	AD	43
Huntington disease-like 2	CTG	Junctophilin-3	3′ UTR?	AD	44

[a] Abbreviations: XLD, X-linked dominant; AD, autosomal dominant; AR, autosomal recessive; SCA, spinocerebellar ataxia.
[b] Where known, the sense strand sequence of the repeat is shown.

disorder and the most common form of muscular dystrophy in adults, the disease-causing repeat CTG·CAG is located in the 3′ UTR of the dystrophia myotonica protein kinase (DMPK) gene [8]. Although mice overexpressing DMPK or containing targeted disruptions of the DMPK gene do not show classic features of DM like myotonia [9, 10], mice expressing long, transcribed but untranslated CTG repeats do [11]. In addition a second form of DM, DM2, is caused by expansion of a related repeat, CCTG·CAGG, in intron 1 of the unrelated zinc finger protein 9 (ZNF9) gene [12]. These data support a role for the DM repeats at the RNA level in the pathology of both diseases. A congenital form of DM1 is also seen. In this form of DM1, methylation of the region flanking the repeat occurs [13]. This region contains binding sites for CTCF, a zinc finger protein that, together with the repeat, normally insulates the DMPK gene from the effects of a downstream enhancer [14]. Methylation abolishes CTCF binding and, therefore, its insulator function. This results in an increase in DMPK expression that might contribute to the more severe phenotype seen in this form of DM1 [14].

SCA8 and SCA10 have symptoms like those seen in SCAs resulting from expansion of a polyglutamine tract. However, the SCA8 repeat is located in a gene without an open reading frame (ORF) [15]. Transgenic mice containing human BAC clones with the expanded SCA8 gene develop a progressive neurological phenotype, and expression of SCA8 transcripts in the *Drosophila* retina induces a late onset, progressive neurodegeneration [16] that is also suggestive of an RNA-mediated pathology. DM and SCA8 may thus share an RNA gain of function mechanism, with the clinical differences perhaps arising from differences in when and where the individual transcripts are made. SCA10 results from expansion of a pentanucleotide repeat unit, ATTCT·AGAAT, in intron 9 of the ataxin-10 gene. This gene encodes a member of the armadillo repeat family of proteins [17]. Preliminary evidence suggests that SCA10 mRNA levels are similar in lymphoblastoid cells of unaffected and affected individuals and that the repeat is not immediately degraded [17]. If this is also true of cells in the brain, it might be that a gain of RNA function is responsible for this SCA as well. Friedreich's ataxia (FRDA), the most common recessively inherited ataxia, is caused by a GAA·TTC repeat tract in the first intron of the frataxin gene, which encodes a protein important for normal mitochondrial function [18]. Expansion leads to a deficit of frataxin mRNA. Thus, the relationship between expansion and disease pathology is likely to differ from that of the dominantly inherited ataxias. Evidence suggests that the frataxin deficit results not from aberrant pre-mRNA splicing but from a transcription defect [19].

FRAXE mental retardation is caused by expansion of a CCG·CGG repeat in the 5′ UTR of the FMR2 gene [20, 21], a putative transcription factor gene on the X chromosome [21, 22]. Like the large CTG·CAG expansions that give rise to congenital DM1, expansion in the FMR2 gene results in DNA hypermethylation. However, unlike what is seen in congenital DM1, hypermethylation decreases FMR2 transcription. A folate-sensitive fragile site is also seen that is coincident with the repeat. Expansion of the same repeat in the 5′ UTR of the fragile X mental retardation (FMR1) gene about 600 kb centromeric to FMR2 causes three clinically distinct disorders, depending on the repeat number in the expanded allele. Individuals with 59–200 repeats, known as premutation carriers, are at risk of fragile X associated tremor ataxia syndrome (FXTAS) [23–30], a late onset disorder characterized by progressive intention tremor, gait abnormalities, and gradual cognitive decline. Female carriers are also at risk of fragile X associated premature ovarian failure (FXPOF) [31–33]. In both men and women, a direct relationship exists between repeat number and the amount of FMR1 mRNA that is made [34–36]. In contrast to FMR2, the FMR1 repeat is oriented so that the CGG strand is present in the transcript. When transcribed but untranslated CGG repeats are expressed in the *Drosophila* eye, retinal degeneration is seen that is related to repeat number and mRNA level [37]. Expression of physiologically reasonable amounts of CGG-RNA in human cells is associated with a slightly elevated level of cell death, sensitivity to external apoptotic stimuli, decreased substrate attachment, and changes in the expression of a wide variety of different genes with potential effects on neuronal survival and fertility [38]. Evidence therefore points to an RNA-based mechanism of pathology in FXTAS and, in all likelihood, in FXPOF as well. Women with 59–200 repeats are also at risk of having a child with >200 repeats. These children are said to be carriers of full mutations, and most have a form of mental retardation known as FXS in which autistic-like symptoms, behavior problems, digestive difficulties and macroorchidism are frequent comorbid features. Expansion is also associated with the appearance of a folate-sensitive fragile site coincident with the expanded repeat, as well as heterochromatinization and methylation of the promoter [39]. In addition to the resultant decrease in the amount of transcript from these alleles, any residual mRNA made is also poorly translated. The net result is a deficit in the protein product of this gene, FMRP, a regulator of translation [40, 41] important in the response to synaptic activation [42].

In two other disorders, Huntington disease-like 2 (HDL2) and SCA12, both autosomal dominant disorders resulting from CTG·CAG expansions, it is not yet clear where in the gene the repeat is located. However, in both instances protein-mediated toxicity is an unlikely source of disease pathology [43]. The CTG strand of the HDL2 repeat is found in alternate transcripts of the junctophilin-3 gene [44]. In SCA12 the repeat is located in the PPP2R2B gene, which encodes a subunit of the protein phosphatase PPA. In this case it is the CAG-containing strand that is transcribed [43].

One nonhuman repeat expansion disease has been described to date, a form of canine progressive myoclonus epilepsy type 2. Like EPM1 in humans, this disorder is caused by expansion of a dodecamer repeat, this time GC_2GC_6GC, that is located in the only exon of the Epm2b gene [45]. Epm2 pathology is believed to arise not due to an effect of the expansion on the Epm2b protein but due to the 900-fold decrease in Epm2b mRNA that is seen in affected animals [45]. Because the Emp2b promoter is apparently unmethylated, the repeat may somehow interfere with transcription elongation [45].

II. UNUSUAL STRUCTURES FORMED BY THE DNA AND RNA VERSIONS OF THE DISEASE-ASSOCIATED REPEATS

A. DNA Structures

Since the genetic basis of the first of these disorders was uncovered in 1991, much effort has been expended to understand why the repeats are prone to expansion and how expansion can lead to disease symptoms. Much of this effort has been directed at understanding the biochemical properties of these repeats at both the DNA and RNA level.

All of the disease-associated repeats are somewhat unusual. Some repeats have unusual properties even when in duplexes with their complementary strand. For example, CTG·CAG, CGG·CCG, GAA·TTC, and C_4GC_4GCG· $CGCG_4CG_4$ are more flexible than normal DNA in some assays [46–50], with CTG·CAG repeats being the most flexible. However, although CGG·CCG repeats are more flexible than random duplex DNA, they also have a stable radius of curvature [47]. Although most of these duplexes are relatively stable, the A+T-rich SCA10 repeat shows a tendency to unpair when negatively supercoiled [51].

The two strands of the FRDA repeat can form parallel duplexes in addition to the typical antiparallel duplex [52]. The individual strands of some of these repeats can also form homoduplexes [53, 54], some of which have quite unusual conformations. For example, short tracts of the repeat CCG form an e-motif structure that comprises staggered antiparallel duplexes. These duplexes contain Watson–Crick (WC) base-paired G·C pairs separated by nonpaired extrahelical C's that are oriented diagonally across the minor groove, as illustrated in Fig. 39-1A [53].

Longer repeat tracts show a propensity to fold. For example, all of the repeats with a CNG motif form fairly stable hairpins [53, 55–63]. The stems of these hairpins consist of WC C·G base pairs interspersed with bases that are not WC base-paired in a 2:1 ratio. In some instances these non-WC base-paired bases are hydrogen-bonded and/or intrahelical, e.g., in the case of CGG and CTG hairpins [55, 60]. In other cases these bases can be extrahelical. For example, in the case of the CCG strand of the FXS repeat, an extended e-motif like that adopted by homoduplexes is formed. In this hairpin, the C's not involved in G·C base pairs are extrahelical. They lie in the minor groove of a highly distorted helix, where stacking interactions lead to their protonation [58]. Of the CNG hairpins associated with disease, those formed by repeats of the sequence CGG are by far the most stable. This is because they contain $G_{syn}·G_{anti}$ base pairs that are relatively stable and cause very little helix distortion [64]. The remaining hairpins show a hierarchy of stability CTG > CAG > CCG. The individual strands of the repeat responsible for FRDA also form hairpins [65]. However, although the CAGG strand in the repeat responsible for DM2 and the ATTCT strand of the repeat responsible for SCA10 also form hairpins, their complementary strands do not [66, 67].

In addition to hairpins, GC-rich repeats can also form a variety of other folded structures, as illustrated in Fig. 39-1. For example, under some circumstances the G-rich strand of repeats like the FXS repeat [68–71] and the EPM1 repeat [72] can also form folded hairpins or guanine tetraplexes (Fig. 39-1C). These tetraplexes are stabilized primarily by the formation of guanine tetrads, cyclic square-planar arrangements of four G's. In these tetrads, each G simultaneously donates two hydrogen bonds to a neighboring G and accepts two hydrogen bonds from its other neighbor. In addition to G_4 tetrads, two other tetrads have been proposed to exist in FXS tetraplexes, GCGC tetrads [70] and C_4 tetrads [71, 73]. Both GCGC- and C_4-tetrad-containing tetraplexes can be formed intra- or intermolecularly, in which case two or four different strands may be involved. In the latter case the strands can be either parallel or antiparallel. These structures can be extremely stable [69]. However, which structure predominates may depend on the circumstances: C_4-tetrads are

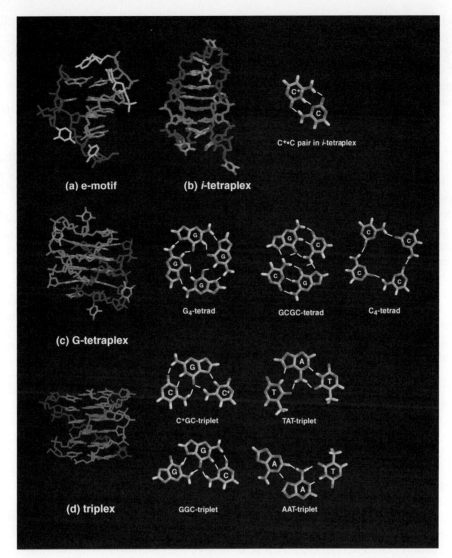

FIGURE 39-1 Graphic representations of some of the unusual types of DNA structures that can be formed by the disease-causing repeats: the e-motif structure formed by CCG repeats [53], a generic intramolecular i-tetraplex [179], intermolecular G-tetraplexes formed by model sequences with CGG repeats that form either GCGC-tetrads [70] or C_4-tetrads [71], and generic RRY [180] and YRY triplexes [181]. The information for each of these structures was derived from the X-ray or NMR coordinates found in the Molecular Modeling database (MMDB) at NCBI (http://www.ncbi.nlm.nih.gov:80/entrez/query.fcgi?CMD=search&DB=structure). Rasmol was used to highlight some of the key features of each structure. The individual strands of the e-motif structure are colored blue and green, and the extrahelical cytosines are shown in yellow. The backbone of the intramolecular i-tetraplex is shown in blue with two pairs of C·C base pairs shown in red and green, illustrating the intercalated nature of the hydrogen bonding. The two different backbones for the hairpin dimer form of a G-tetraplex are shown in blue and green. The Watson–Crick base-paired strands of the triplex are shown in red and blue, and the third strand is shown in green. See CD-ROM for color image.

stabilized by protonation and are thus stabilized by lower pHs, whereas tetraplexes with GCGC-tetrads may be formed more readily from the folding of very long hairpins comprising predominantly G·C base pairs. The C-rich strands of these repeats can form neutral tetraplexes consisting of parallel oriented hairpins [74]. In these structures, the C's that are not WC-base-paired can remain intrahelical or generate C·C+ pairs connecting the hairpins. At lower pHs, i-motif structures or i-tetraplexes can form. These structures consist of four

C-rich regions held together by intercalated C·C+ base pairs, as shown in Fig. 39-1B [74, 75].

The tendency of the repeats to fold can also produce slipped-DNA (S-DNA) structures when misalignment of the complementary strands occurs [76]. Misalignment produces looped out regions in different parts of the duplex that can adopt one or more of the folded conformations described earlier.

In addition to relatively unstable hairpins, the GAA·TTC tract can also form three-stranded structures called triplexes (Fig. 39-1D). These sorts of structures require mirror symmetry of the DNA and a strong strand bias with respect to the distribution of purines and pyrimidines. These triplexes can be either intra- or intermolecular, and the third strand can contain either TTC repeats forming a pyrimidine:purine:pyrimidine (YRY) triplex or GAA repeats forming a purine:purine:pyrimidine (RRY) triplex [77–81]. The purine-rich strand in both instances participates in normal WC hydrogen bonding with its complementary strand, while simultaneously forming Hoogsteen hydrogen bonds with the third strand lying in the major groove of the WC duplex. The YRY triplex is made up of C+GC and TAT triplets, whereas the RRY triplex contains a mixture of GGC and AAT triplets, as shown in Fig. 39-1. Because the C+GC triplets require the third strand C to be protonated, YRY triplexes are stabilized by low pH. Two RRY triplexes present on the same molecule can interact to form "sticky DNA" [82].

The structures formed by the repeat responsible for canine Epm2 have not been examined. However, the sequence of the repeat (GC_2GC_6GC) is very similar to that of the EPM1 repeat (C_4GC_4GCG or GC_4GC_4GC), which forms G-tetraplexes and *i*-tetraplexes. It is therefore very likely that the Epm2 repeat forms a similar set of structures.

B. RNA Structures

Perhaps not surprisingly, the CUG repeat [83] responsible for DM1, SCA8, and perhaps HDL2, the FXS repeat, CGG [84], the FRAXE MR repeat, CCG, and the SCA12 repeat, CAG [85], all form very stable RNA hairpins with conformations similar to those of their DNA counterparts. In spite of the fact that the CCTG repeat responsible for DM2 does not form DNA hairpins even at very low temperature, CCUG repeats form hairpins that are stable at 37°C [85]. Similarly, although the ATTCT repeat only forms a hairpin at low temperature, the AUUCU repeat present in the SCA10 transcript forms an RNA hairpin that is present at physiologically reasonable conditions. The most stable of all these hairpins, perhaps not surprisingly, are those formed by the GC-rich repeats [85].

C. Effects of Interruptions

Interruptions to the purity of the repeat tract affect expansion risk and possibly also the severity of the resultant disease symptoms. The effect that interruptions have on the type and stability of structures that are formed by the repeat mirrors these phenomena. For example, many FMR1 alleles have a CGG·CCG tract that contains AGG·CCT interruptions. These interruptions destabilize both the DNA hairpins [62] and tetraplexes [86] formed by the CGG strand. They also reduce the propensity of these repeats to form S-DNA [87]. AGG interruptions in CGG-RNA favor the formation of multiple smaller hairpins, each of lower stability, rather than the single large hairpin that is seen in the absence of interruptions [84]. Similar results have been reported for the RNA structures produced by other interrupted repeat tracts [88].

D. Evidence for Secondary Structures *in Vivo*

Direct proof that some of these structures form *in vivo* is lacking. However, indirect evidence for these structures comes from the vast body of literature for which structure formation is the most reasonable explanation. For example, CAG-RNA binds to a protein specific for certain double-stranded RNAs in yeast [89], supporting the idea that this repeat is able to form a duplex or hairpin *in vivo*. Because other disease-associated CNG repeats have a stability *in vitro* that is much higher than that of the CAG hairpin, it is reasonable to assume that they too can form hairpins inside cells. Direct evidence for triplex formation *in vivo* has been found for other sequences that form such structures *in vitro* [90, 91]. In addition, cross-linking studies with mustard gas demonstrate the formation of sticky DNA in bacteria [92]. Because sticky DNA involves the interaction of two GAA·TTC triplexes, the formation of GAA·TTC triplexes can be inferred. Finally, a large number of human proteins exist that are thought to play an important role in dealing with such structures [93–105]. This includes the product of the gene affected in FXS, FMRP, an RNA binding protein with a specific affinity for intramolecular tetraplexes [106] and the loop regions of a bihairpin "kissing complex" [107]. It is thus not unreasonable to think that these sorts of structures can form *in vivo* and have biological consequences.

III. POTENTIAL BIOLOGICAL CONSEQUENCES OF THESE UNUSUAL STRUCTURES

A. Expansion

These structures suggest ways in which the repeats may be predisposed to instability. However, data from bacteria, yeast, and mice have implicated a confusing array of genes involved DNA replication, repair, and recombination in expansion in these organisms. It is unclear which of these is actually responsible for expansion in humans. The situation is complicated by the fact that, unlike what is seen in humans, deletions predominate in bacteria and yeast. This is significant because expansions and contractions may occur by different mechanisms [108, 109]. In spite of this, a coherent picture is beginning to emerge in which secondary structure plays an important role.

Many of the genes implicated in instability are involved in DNA replication. For example, mutations in RAD27, the yeast homologue of human FEN-1, lead to dramatic increases in expansion frequency [110–112]. A small increase in expansion size is also seen in mice with relatively small CAG·CTG tracts in a FEN-1 heterozygous background [113]. FEN-1 is a structure-sensitive flap endonuclease whose activity is blocked by many of the disease-causing repeats [114, 115]. One of FEN-1's major tasks is the processing of the 5′ flaps of Okazaki fragments that are generated by strand displacement by an upstream DNA polymerase during lagging strand DNA synthesis. Secondary structures within the flap would prevent the removal of the redundant flap bases. These additional bases could then be incorporated into the nascent DNA strand, leading to expansions as illustrated in Fig. 39-2A. This is consistent with yeast

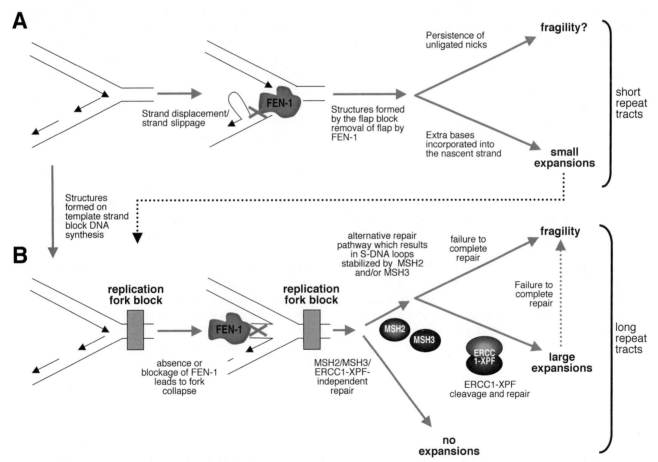

FIGURE 39-2 Models for events occurring during the replication of short and long repeat tracts. (A) Strand displacement or strand slippage during lagging strand DNA synthesis of short repeat tracts gives rise to the small expansions responsible for polyglutamine expansion diseases or to new "premutation" alleles for disorders involving longer repeat tracts and larger expansions. (B) Unresolved secondary structures in the template of longer repeat tracts lead to replication fork collapse. In the presence of MSH2, MSH3, and ERCC1-XPF, fork rescue is accomplished via a repair pathway that leads to expansion because of the way the resultant S-DNA is processed. Failure to complete this repair results in chromosome fragility. In the absence of any one of these proteins, an alternative pathway is favored that does not lead to expansion. See CD-ROM for color image.

data showing that expansions are more frequent when the strand forming the most stable secondary structure is the lagging strand daughter strand because this strand would be more resistant to FEN-1. However, expansions would be limited in size to the amount of DNA displaced. On the other hand, repeated strand slippage, which could easily occur during lagging strand synthesis, could give rise to large expansions relatively readily [116]. Slippage of a fragment entirely comprising the repeat could generate a 5′ flap that would also be resistant to FEN-1 cleavage [117]. This is particularly intriguing because the threshold for repeat expansion in humans is ~40–60 repeats, roughly the size of a mammalian Okazaki fragment. A similar situation might arise, as in the case of SCA10, where the repeat itself acts as an origin of replication [51]. *In vitro* experiments confirm that expansion can occur in the absence of strand displacement and that this effect is most marked when the primer is located entirely within the repeat [117]. The tendency to slip might be exacerbated by the formation of secondary structures in the template that block DNA synthesis and thus favor slippage over elongation.

However, in addition to lagging strand DNA synthesis, FEN-1 is important for other biological processes, including stalled replication fork rescue and DNA gap repair [118]. FEN-1 colocalizes with stalled replication forks in human cells where it acts to allow replication fork restart or lesion bypass [119]. The folded structures formed by disease-associated repeats block DNA polymerase *in vitro* [69, 77, 120, 121] and cause replication fork stalling in bacteria and yeast [122–124]. Moreover, inhibitors of replication elongation increase expansion frequencies in primary fibroblasts from DM1 patients, and these expansions are only seen in actively dividing cells [125]. Thus, some expansions could result from a problem with replication fork progression. If FEN-1 is missing or unable to successfully restart the fork or bypass the replication block, an alternative pathway may be used to rescue the fork that gives rise to expansions, as illustrated in Fig. 39-2B.

Data from yeast suggest that both strand slippage–strand displacement and stalled replication fork models may be valid, depending on the size of the repeat tract [126]. Changes in both long and short CAG·CTG repeat tracts are unaffected by mutations in Chk1, which in yeast is required for G2 arrest but not for the S-phase checkpoint [127, 128]. They are, however, both sensitive to mutations in Mrc1 [127, 128]. Mrc1, the yeast Claspin homologue, is specific to the S-phase checkpoint. Mrc1 travels with the replication fork, and its absence results in the uncoupling of replication fork progression and replication. Events confined to S phase, probably DNA replication, are thus likely to be responsible for the expansion of both long and short repeat tracts in yeast. However, these tracts show a differential sensitivity to mutations in other proteins active in S phase, consistent with different mechanisms being responsible for small and large expansions. For example, shorter repeat tracts are more sensitive to mutations in Rad17. This suggests a problem at the level of Okazaki fragment processing because Rad17, the yeast homologue of the human Rad1 gene, is part of a complex that recognizes unligated nicks [128]. Longer repeats, on the other hand, are more sensitive to Rad53 mutations. Because Rad53 is required for coordinating repair at stalled replication forks [129], a problem with the replication fork is more likely to be responsible for the expansion of larger repeat tracts. This is consistent with the greater dependence on Mrc1 for preventing the expansion of long repeat tracts in yeast [127].

However, in some mouse models, somatic expansions are not related to the rate of cell proliferation, and in some cases the products of expansion can be seen in postmitotic cells like brain and muscle [130]. Therefore, a replication-based process may not be responsible for this sort of expansion. The ability of S-DNA to bind the mismatch repair (MMR) protein Msh2p [131] suggests an alternative mechanism. Although Msh2p usually acts in MMR as a dimer with either Msh3p or Msh6p, recognition of S-DNA by a component of the MMR systems suggests that this pathway or proteins in it may be involved in expansion. In support of this contention, MSH2 and MSH3 [2–4], but not MSH6 [5], are required for the expansion of CTG·CAG repeats in mice. Mutation in Pms2p, another MMR protein, does not abolish expansion but reduces its incidence by about 50% [130]. One interpretation of these findings is that expansions arise from the mismatch process itself.

However, other explanations are again possible. For example, mutations in Rad1, the yeast homologue of XPF, also decrease expansion of a human minisatellite [132]. Because in yeast msh2 and msh3 mutations are epistatic with Rad1 mutations with respect to repeat expansion, this suggests that all three genes act in the same pathway to generate expansions [133]. Rad1p is involved in nucleotide excision repair (NER), however, because Msh2 and Msh3 mutations do not affect UV sensitivity, the effect of these proteins on expansion is probably not mediated via this pathway [133]. The Rad10–Rad1 complex, which corresponds to the ERCC1–XPF complex in humans, has single-stranded DNA endonuclease activity and together with Msh2p and Msh3p is thought play a role in the repair of heteroduplex loops generated during strand invasion [134]. Rad1p cleaves the strand opposite the extruded loop, rather than the loop itself [134]. Expansion would result if the loop were used as a template for repair synthesis.

In murine cells, strand breaks induced by H_2O_2 lead to an increased expansion of a CAG·CTG tract [108]. This would be consistent with a role for break repair in expansion. However, mutations in the homologous recombination (HR) genes RAD52 and RAD54 do not affect the expansion of a long CTG·CAG repeat in mice. This has led to the suggestions that HR is not involved and that expansion occurs via an intraallelic gap repair process [108]. However, the effect of rad52 and rad54 mutations does not necessarily rule out a HR-based mechanism. First, the Rad10–Rad1-dependent HR pathway in yeast is independent of Rad52 [135]. Additionally, Rad52 and Rad54 null mice show no overt phenotype [136–138], and inactivation of these genes affects only a subset of HR reactions [136, 139]. The fact that mutations in SPO11, which introduces double-strand breaks (DSBs) during meiosis, also reduce expansion frequency in yeast [140] supports a role for DSB repair in expansion, at least in some instances.

All of these disparate observations can be reconciled as follows: SPO11 causes DSBs during meiosis, and stalled replication forks are a frequent source of single- and double-strand breaks in cells that are actively dividing. It may be that in such cells these processes are responsible for initiating most of the events that lead to expansion. In postmitotic or quiescent cells exogenous sources of strand breaks may play a larger role. Break repair that involves strand invasion would provide an opportunity for strand misalignment or slippage within the repeat tract that would favor the formation of S-DNA. Because both strands of most of the disease-causing repeats can form folded structures, this might occur at high frequency. The combined action of MSH2, MSH3, and ERCC1-XPF on the S-DNA loops would then be responsible for the final steps in the expansion process. The absence of one of these proteins might block the use of this pathway, resulting in an alternative one being used that does not lead to expansion.

The effect of the SCA10 repeat on FEN-1 and DNA polymerases has not yet been tested. Because the repeat is prone to unpairing [51] and the ATTCT strand forms a hairpin [67], albeit one that is not very thermostable *in vitro*, it may be that expansion proceeds in the same way as suggested for the other diseases. However, the propensity of these repeats to unpair when negatively supercoiled suggests another way in which expansion could occur. The SCA10 repeat acts as an origin of replication in mammalian cells [51], presumably because it is easily unwound. Onion skin replication may result from repeated initiation within the repeat tract, with the resultant replication products somehow becoming incorporated into the nascent strand, thereby leading to the massive expansions that are a hallmark of this disease [51].

B. Chromosome Fragility

Chromosome fragility is an incompletely understood phenomenon seen in mammalian cells grown in the presence of specific inducers. Fragile sites appear as constrictions or nonstaining gaps in chromatin and undergo frequent chromosome breakage both *in vitro* and *in vivo*. Such sites are found coincident with the expanded CGG·CCG repeat tract in individuals with FXS and FRAXE MR. These sites are both induced by folate stress. Folate stress reduces the dCTP pools available for DNA synthesis: too little folate decreases the levels of the precursors for dCTP synthesis, whereas too much folate inhibits ribonucleotide reductase and the production of dCTP from TTP. Other classes of fragile sites are relatively insensitive to folate levels but can be induced by agents such as bromodeoxyuridine, distamycin-A, and aphidicolin. The underlying sequence responsible for these fragile sites differs from the sequence responsible for the folate-inducible ones. Distamycin- and bromodeoxyuridine-inducible fragile sites comprise A+T-rich minisatellite repeats that, like the CGG·CCG repeats, can form hairpins [141, 142]. Although the sequences responsible for aphidicolin-sensitive sites are unknown, it is known that these sites extend over megabases of DNA and have a complex sequence composition [143].

CGG·CCG repeats exclude nucleosomes *in vitro* [144] and in yeast [145], perhaps because they have a relatively stable radius of curvature [47]. The methylation that frequently occurs in expanded repeats exacerbates exclusion [146]. This suggests one possible cause of the microscopic appearance of the fragile site: repeat-induced nucleosome exclusion would inhibit the assembly of normal chromatin on the repeat, resulting in a chromatin gap. The fact that the repeat responsible for the distamycin-inducible fragile site at FRA16B also excludes nucleosomes in the presence of distamycin [147] supports this idea. However, it is unclear how folate stress would affect nucleosome exclusion in the case of the folate-sensitive sites. In addition, although no agent capable of inducing chromosome fragility at CTG·CAG repeats in mammalian cells has been identified, these repeats, like CGG·CCG repeats, are fragile in yeast [111]. Yet these repeats do not exclude nucleosomes; rather, they form some of the strongest known nucleosome positioning signals [148, 149]. Aphidicolin-sensitive sites are rich in flexible sequences [150], but just how this translates into chromosome fragility is not clear.

The one factor common to all known fragile sites is that the inducer in each case interferes with DNA replication by reducing nucleotide pools in the case of the folate stress, by inhibiting DNA polymerases like Pol α in the case of aphidicolin, and by binding to the A+T-rich regions thereby potentially affecting the efficiency of DNA replication in the cases of distamycin-A and bromodeoxyuridine. Moreover, the ability of agents like fluorodeoxyuridine, a thymidylate synthase inhibitor, to induce fragility at folate-sensitive sites [151] and for this effect to be reversed by thymidine but not folate suggests that nucleotide pool levels are critical for the expression of this group of fragile sites.

Yeast mutations that affect repeat instability also affect chromosome fragility [127, 128]. This adds weight to the idea that a common mechanism underlies both processes. In human cells, siRNA knockdown of ATR, the ataxia–telangiectasia and RAD3-related protein involved in the response to stalled replication forks, results in the expression of fragile sites in the absence of inducers [152]. In addition, humans with mutations in ATR (Seckel syndrome 1, OMIM no. 210600) have increased levels of chromosome breakage at fragile sites [153]. These observations illustrate that the connection between instability and fragility is not confined to yeast. Because in humans chromosome fragility is only seen when repeat numbers are large, it may be that fragility results from those replication forks stalled on long repeat tracts that cannot be either restarted or repaired (Fig. 39-2B).

C. Effects on Gene Expression

1. POTENTIAL DUPLEX DNA-MEDIATED EFFECTS

The unusual properties of DNA duplexes containing the repeats may contribute to the altered expression of certain expanded alleles.

Carriers of FMR1 premutation alleles produce higher than normal levels of FMR1 mRNA. This effect is not seen in an individual with a point mutation in FMR1 that causes a loss of FMRP function [156]. This suggests that the transcription increase is not due to feedback regulation of the gene in response to dropping FMRP levels. Premutation alleles also show a shift in the preferred transcription start site, with alleles with longer repeat tracts showing more transcription from upstream start sites than normal alleles [157]. Although it may be that the increase in transcription reflects an increase in the number of binding sites of a repeat-binding protein that promotes transcription initiation, it is tempting to think that the ability of long CGG·CCG repeat tracts to exclude nucleosomes [144] plays a role by making the entire region more accessible to factors that activate transcription.

2. POTENTIAL HAIRPIN DNA-MEDIATED EFFECTS

DNA hairpins formed by the repeats may also affect gene expression. Carriers of FXS full mutation alleles generally make less FMR1 mRNA than either premutation carriers or unaffected individuals. In those with the full mutation, the promoter tends to be heavily methylated and the histones bound to the repeats contain modifications characteristic of heterochromatin [39]. Carriers of FRAXE MR full mutation alleles also show methylation of the region upstream of the expanded repeat. The hairpins and S-DNA structures formed by the CGG·CCG repeats are very good substrates for DNA methyltransferases [158]. The conformation of the mispaired or extrahelical C's at CpG steps in the structures formed by the CCG strand of the repeat resembles the methylation intermediate in which the target C is flipped out of the double helix into the active site of the methylase [158]. The propensity of the repeats to be methylated *in vitro* is attributed to this conformational similarity and to the increased affinity of such enzymes for their transition state intermediates. Because DNA methylation can lead to the generation of heterochromatin and methylation is able to spread from one region to another, this seems a reasonable explanation for the hypermethylation and heterochromatinization of the FMR1 promoter.

3. POTENTIAL HAIRPIN RNA-MEDIATED EFFECTS

a. RNA Interference

In congenital DM1, increased DNA methylation 5' of the repeat occurs despite the absence of CpG residues in the repeat [13]. No information is available for the heterochromatinization or methylation status of the flanking regions of other disease-causing repeats. However, the GAA·TTC tracts responsible for FRDA, which also lack methylatable C's, and the CTG·CAG repeats responsible for DM1 are both associated with heterochromatinization of transgenes in mice [159]. Therefore, DNA methylation may not be the primary event in the heterochromatinization process. It may still be that hairpins are responsible, but they may be RNA rather than DNA hairpins. Evidence from plants and fungi suggests that RNA hairpins can lead to heterochromatinization and gene silencing via an RNA interference (RNAi) mechanism [160, 161]. Evidence from the silencing of transgenes in mammals suggests that a similar pathway exists in humans. It has previously been shown that CGG-RNA hairpins are substrates for human Dicer [84]. Dicer is a component of the RNAi machinery that converts certain double-stranded RNAs into the smaller RNAs that participate in the various RNAi pathways that affect gene expression. Participation in the pathway that leads to heterochromatinization may represent a general mechanism by which large transcribed repeat tracts can silence the genes from which they originate (illustrated in Fig. 39-3). The efficacy of this process may depend on the number of repeats, the amount of the transcript available to enter the RNAi pathway, how good a substrate the hairpins are for Dicer, the location of the repeat in the gene, and the sensitivity of the flanking regions to the spread of heterochromatin.

In addition to causing changes in chromatin modification, RNAi can lead to the targeted degradation of homologous mRNAs and the translational suppression of mRNAs with which it shares homology. Although the FMR1 mRNA and FMRP deficits in carriers of fragile X full mutation alleles are generally thought to reflect other problems, it may be that these sorts of processes also contribute to the mRNA and protein deficit in those individuals with full mutations who still make some mRNA. The ability of some repeats to participate in RNAi raises the possibility that expansion could also affect the expression of unrelated genes whose transcripts contain the same repeats.

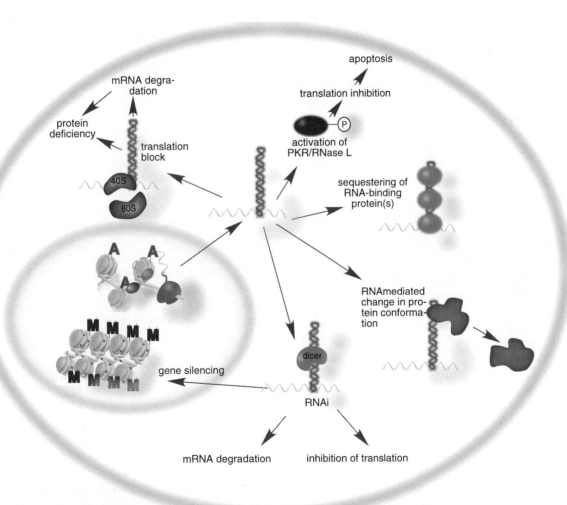

FIGURE 39-3 Diagrammatic representation of some of the potential biological effects of RNA hairpins formed by the disease-causing repeats. See CD-ROM for color image.

b. Activation of Other Responses to dsRNA

In addition to the RNAi mechanism, mammalian cells have other responses to the presence of dsRNA that may be relevant for disease-causing repeats that form RNA hairpins. For example, some dsRNAs are also able to bind and activate enzymes like RNase L or the interferon-inducible protein kinase, PKR. Activation of these enzymes can lead to decreased translation of a wide variety of different genes and ultimately to apoptosis, as illustrated in Fig. 39-3. CAG-RNA does bind PKR in a yeast 3-hybrid assay. This may have implications for diseases like SCA12, which may result from expansion of a transcribed CAG tract. However, although the CUG repeat binds and activates PKR *in vitro* [162], these molecules do not interact *in vivo* [89] and the fragile X RNA hairpin does not activate PKR at all [84]. Differences in the ability of different repeat RNA hairpins to activate PKR may be related to differences in the precise structural details of the individual hairpins.

c. Sequestering of RNA Binding Proteins

RNA hairpins have also been implicated in pathology resulting from sequestering proteins to which they bind. In a yeast 3-hybrid assay the DM1 and DM2 repeats bind to proteins from the muscleblind family in a length-dependent way [163–165]. This family, which consists of MBNL1/EXP, MBNL2/MBLL, and MBNL3/MBXL/MBLX/CHC, is required for the terminal differentiation of muscle and photoreceptor cells [89]. The length dependence of protein binding may indicate that secondary structure is important for binding because a certain minimum number of repeats is required for stable hairpin formation [89]. Because a fully Watson–Crick base-paired duplex containing the sequence CUG·CAG does not bind MBNL1 efficiently, MBNL1 may prefer bulge-containing dsRNAs [89]. Because mice with targeted disruptions of the MBNL1 gene show many of the symptoms of myotonic dystrophy [166], repeat-mediated sequestration of this group of proteins provides a compelling explanation of some aspects of disease pathology. However, Mbnl1 null mice are not obviously affected at birth. This suggests that some other factor(s) is important, at least for congenital DM1. It may be that repeat-induced alterations in chromatin structure that affect the expression of adjacent genes or an RNAi-mediated effect on the expression of other CUG repeat containing genes contribute to the severity of symptoms seen in this form of DM1.

MBNL1 binds to RNAs with a CHG or CHHG motif, where H = A, C, or U [89]. A similar mechanism may therefore account for disease symptoms not only in disorders resulting from expansions of CTG and CCTG but also in those like SCA12 that potentially result from the presence of an expanded CAG tract in the transcript.

d. RNA-Induced Protein Misfolding?

RNA hairpins may have other pathological consequences. The fact that MBNL1 does not bind to CGG repeats suggests that the mechanism of RNA toxicity in FXTAS and perhaps FXPOF differs from that responsible for DM1 and DM2, or at least that the protein(s) involved does. In the *Drosophila* model in which the FXS repeat is overexpressed in the eye, retinal degeneration can be rescued by overexpression of HSP70 [37]. This suggests that the repeat pathology may involve RNA-structure-mediated protein misfolding. This is interesting given the superficial similarities of the FXTAS inclusions to the inclusions seen in the polyglutamine repeat expansion disorders. Interesting precedents for such a mechanism exist, including the observations that a highly structured RNA molecule plays a role in prion conversion [167] and that RNA architecture is an important determinant of whether an arginine-rich peptide from HIV-1 Rev adopts an extended or α-helical conformation [168].

A variety of CGG repeat binding proteins have been described, including KU70/80 [169], CGGBP1/p20 [170], and members of the Pur family of proteins [171]. The Pur family, which consists of the proteins Purα, Purβ, and Purγ is particularly interesting because Purα null mice show neurological problems, including gait abnormalities and severe tremor [172]. Pur family members are DNA and RNA binding proteins involved in a diverse array of biological functions, including the regulation of transcription, translation, mRNA transport, replication, and growth. They bind to sequences with a NGG motif and show a binding preference to certain stable stem loop structures in which the Pur binding site is in the loop [173].

e. Ribosome Stalling

RNA hairpins can also affect the efficiency with which the mRNA is translated. The 40S ribosomal subunit stalls on the FMR1 mRNA during its scanning of the 5' UTR in search of the translation initiation site [174]. The very stable hairpins formed by the FXS repeats could well be responsible [84]. In bacteria, ribosome stalling leads to mRNA cleavage and degradation [175]. It may be that a similar phenomenon contributes to the FMR1 mRNA deficit in some individuals with long CGG repeat tracts, as well as to the mRNA deficit in other disorders like Epm2 where a ribosome may encounter a stable RNA hairpin.

f. Other Potential Effects of RNA Hairpins

When the repeat is in an intron, as it is in FRDA and SCA10, the possibility of ribosome stalling is moot unless the RNA hairpin also affects transcript splicing. Other potential effects of RNA hairpins include their ability to cause RNA polymerase to pause or to terminate transcription [176]. However, because preliminary evidence suggests that the mRNA is not affected in individuals with SCA10, this is apparently not the mechanism responsible for this disorder. It remains a possibility for FRDA, although, as will be discussed next, *in vitro* studies implicate another structure formed by these repeats in the mRNA deficit.

4. Potential Triplex-Mediated Effects on Transcription

Transcription inhibition by GAA·TTC tracts occurs in transfected COS-7 cells [121] and *in vitro* with both HeLa nuclear extracts and T7 RNA polymerase (RNAP) [78]. This suggests that the FRDA repeats have an intrinsic ability to affect transcription. This effect is not mediated at the level of transcription initiation, but upon elongation. It has been suggested by many that triplexes are responsible. A variety of different types of triplexes have been invoked, including DNA triplexes [78, 177], sticky DNA [82], and triplexes in which the nascent transcript forms the third strand [77]. Transcripts terminating at the 3' end of the repeat accumulate during transcription elongation *in vitro* [78]. This effect is exacerbated by template supercoiling. Similar truncated transcripts are seen when the repeat is transcribed from the opposite direction. However, in this case termination is also exacerbated by low pH [78]. The structure responsible for transcription termination in both directions may be an intramolecular triplex formed during transcription. An RRY triplex would explain the termination characteristics in the first instance and an YRY triplex in the second because this triplex involves protonated C's. Addition of RNA homologous to the transcript has no effect on the yield of transcript. This suggests that a triplex in which the RNA acts as the third strand is not responsible for the transcription defect. Premature termination is reduced when oligonucleotides that block triplex formation are added to the reaction [178], suggesting that the triplex forms during transcription elongation. According to this view, the triplex blocks the RNA polymerase at the triplex–duplex junction at the downstream side of the repeat. This leads to transcription termination with retention of the polymerase on the template.

The strand bias with respect to the distribution of purines and pyrimidines found in the Epm2b repeat suggests that triplex formation might also be possible. This could lead to a transcription elongation defect as proposed for FRDA.

IV. CONCLUDING REMARKS

The repeats responsible for the known repeat expansion diseases form a variety of different DNA and RNA secondary structures. Evidence suggests that these structures play a role both in the underlying mutational event responsible for these disorders and in the resultant pathophysiology in many instances. The favored mechanisms of disease pathology range from effects on chromatin structure to the sequestration of factors important for normal cell function.

Work on the known repeat expansion diseases has taught us much about what sort of unusual DNA and RNA structures can be formed by the disease-associated repeats. It has also broadened appreciation for what sort of other sequences may form similar structures and thus may be responsible for as yet unidentified repeat expansion diseases. Because ORFs constitute a tiny fraction of most genes, expansions occurring outside of these regions may be a much more significant problem than generally appreciated. Therefore, the disease mechanisms discussed in this chapter may not be isolated genetic curiosities but relatively common causes of disease pathology.

Acknowledgment

I thank my sister, Dr. Martine Usdin, for her thoughtful reading of this chapter and her helpful comments.

References

1. Fishel, R., Lescoe, M. K., Rao, M. R., Copeland, N. G., Jenkins, N. A., Garber, J., Kane, M., and Kolodner, R. (1993). The human mutator gene homolog MSH2 and its association with hereditary nonpolyposis colon cancer. *Cell* **75**, 1027–1038.
2. Kovtun, I. V., and McMurray, C. T. (2001). Trinucleotide expansion in haploid germ cells by gap repair. *Nat. Genet.* **27**, 407–411.
3. Manley, K., Shirley, T. L., Flaherty, L., and Messer, A. (1999). Msh2 deficiency prevents in vivo somatic instability of the CAG repeat in Huntington disease transgenic mice. *Nat. Genet.* **23**, 471–473.
4. Savouret, C., Garcia-Cordier, C., Megret, J., te Riele, H., Junien, C., and Gourdon, G. (2004). MSH2-dependent germinal CTG repeat expansions are produced continuously in spermatogonia from DM1 transgenic mice. *Mol. Cell Biol.* **24**, 629–637.
5. van den Broek, W. J., Nelen, M. R., Wansink, D. G., Coerwinkel, M. M., te Riele, H., Groenen, P. J., and Wieringa, B. (2002). Somatic expansion behaviour of the (CTG)n repeat in myotonic dystrophy knock-in mice is differentially affected by Msh3 and Msh6 mismatch-repair proteins. *Hum. Mol. Genet.* **11**, 191–198.

6. Lafreniere, R. G., Rochefort, D. L., Chretien, N., Rommens, J. M., Cochius, J. I., Kalviainen, R., Nousiainen, U., Patry, G., Farrell, K., Soderfeldt, B., Federico, A., Hale, B. R., Cossio, O. H., Sorensen, T., Pouliot, M. A., Kmiec, T., Uldall, P., Janszky, J., Pranzatelli, M. R., Andermann, F., Andermann, E., and Rouleau, G. A. (1997). Unstable insertion in the 5' flanking region of the cystatin B gene is the most common mutation in progressive myoclonus epilepsy type 1, EPM1. *Nat. Genet.* **15**, 298–302.
7. Lalioti, M. D., Mirotsou, M., Buresi, C., Peitsch, M. C., Rossier, C., Ouazzani, R., Baldy-Moulinier, M., Bottani, A., Malafosse, A., and Antonarakis, S. E. (1997). Identification of mutations in cystatin B, the gene responsible for the Unverricht–Lundborg type of progressive myoclonus epilepsy (EPM1). *Am. J. Hum. Genet.* **60**, 342–351.
8. Ranum, L. P., and Day, J. W. (2004). Myotonic dystrophy: RNA pathogenesis comes into focus. *Am. J. Hum. Genet.* **74**, 793–804.
9. Jansen, G., Groenen, P. J., Bachner, D., Jap, P. H., Coerwinkel, M., Oerlemans, F., van den Broek, W., Gohlsch, B., Pette, D., Plomp, J. J., Molenaar, P. C., Nederhoff, M. G., van Echteld, C. J., Dekker, M., Berns, A., Hameister, H., and Wieringa, B. (1996). Abnormal myotonic dystrophy protein kinase levels produce only mild myopathy in mice. *Nat. Genet.* **13**, 316–324.
10. Reddy, S., Smith, D. B., Rich, M. M., Leferovich, J. M., Reilly, P., Davis, B. M., Tran, K., Rayburn, H., Bronson, R., Cros, D., Balice-Gordon, R. J., and Housman, D. (1996). Mice lacking the myotonic dystrophy protein kinase develop a late onset progressive myopathy. *Nat. Genet.* **13**, 325–335.
11. Mankodi, A., Logigian, E., Callahan, L., McClain, C., White, R., Henderson, D., Krym, M., and Thornton, C. A. (2000). Myotonic dystrophy in transgenic mice expressing an expanded CUG repeat. *Science* **289**, 1769–1773.
12. Liquori, C. L., Ricker, K., Moseley, M. L., Jacobsen, J. F., Kress, W., Naylor, S. L., Day, J. W., and Ranum, L. P. (2001). Myotonic dystrophy type 2 caused by a CCTG expansion in intron 1 of ZNF9. *Science* **293**, 864–867.
13. Steinbach, P., Glaser, D., Vogel, W., Wolf, M., and Schwemmle, S. (1998). The DMPK gene of severely affected myotonic dystrophy patients is hypermethylated proximal to the largely expanded CTG repeat. *Am. J. Hum. Genet.* **62**, 278–285.
14. Filippova, G. N., Thienes, C. P., Penn, B. H., Cho, D. H., Hu, Y. J., Moore, J. M., Klesert, T. R., Lobanenkov, V. V., and Tapscott, S. J. (2001). CTCF-binding sites flank CTG/CAG repeats and form a methylation-sensitive insulator at the DM1 locus. *Nat. Genet.* **28**, 335–343.
15. Koob, M. D., Moseley, M. L., Schut, L. J., Benzow, K. A., Bird, T. D., Day, J. W., and Ranum, L. P. (1999). An untranslated CTG expansion causes a novel form of spinocerebellar ataxia (SCA8). *Nat. Genet.* **21**, 379–384.
16. Mutsuddi, M., Marshall, C. M., Benzow, K. A., Koob, M. D., and Rebay, I. (2004). The spinocerebellar ataxia 8 noncoding RNA causes neurodegeneration and associates with staufen in Drosophila. *Curr. Biol.* **14**, 302–308.
17. Matsuura, T., Yamagata, T., Burgess, D. L., Rasmussen, A., Grewal, R. P., Watase, K., Khajavi, M., McCall, A. E., Davis, C. F., Zu, L., Achari, M., Pulst, S. M., Alonso, E., Noebels, J. L., Nelson, D. L., Zoghbi, H. Y., and Ashizawa, T. (2000). Large expansion of the ATTCT pentanucleotide repeat in spinocerebellar ataxia type 10. *Nat. Genet.* **26**, 191–194.
18. Campuzano, V., Montermini, L., Molto, M. D., Pianese, L., Cossee, M., Cavalcanti, F., Monros, E., Rodius, F., Duclos, F., Monticelli, A., Zara, F., Canizares, J., Koutnikova, H., Bidichandani, S. I., Gellera, C., Brice, A., Trouillas, P., De Michele, G., Filla, A., De Frutos, R., Palau, F., Patel, P. I., Di Donato, S., Mandel, J. L., Cocozza, S., Koenig, M., and Pandolfo, M. (1996). Friedreich's ataxia: autosomal recessive disease caused by an intronic GAA triplet repeat expansion. *Science* **271**, 1423–1427.
19. Bidichandani, S. I., Ashizawa, T., and Patel, P. I. (1998). The GAA triplet-repeat expansion in Friedreich ataxia interferes with transcription and may be associated with an unusual DNA structure. *Am. J. Hum. Genet.* **62**, 111–121.
20. Gecz, J., Gedeon, A. K., Sutherland, G. R., and Mulley, J. C. (1996). Identification of the gene FMR2, associated with FRAXE mental retardation. *Nat. Genet.* **13**, 105–108.
21. Gu, Y., Shen, Y., Gibbs, R. A., and Nelson, D. L. (1996). Identification of FMR2, a novel gene associated with the FRAXE CCG repeat and CpG island. *Nat. Genet.* **13**, 109–113.
22. Gecz, J., Bielby, S., Sutherland, G. R., and Mulley, J. C. (1997). Gene structure and subcellular localization of FMR2, a member of a new family of putative transcription activators. *Genomics* **44**, 201–213.
23. Tassone, F., Hagerman, R. J., Taylor, A. K., Mills, J. B., Harris, S. W., Gane, L. W., and Hagerman, P. J. (2000). Clinical involvement and protein expression in individuals with the FMR1 premutation. *Am. J. Med. Genet.* **91**, 144–152.
24. Hagerman, R. J., Leehey, M., Heinrichs, W., Tassone, F., Wilson, R., Hills, J., Grigsby, J., Gage, B., and Hagerman, P. J. (2001). Intention tremor, parkinsonism, and generalized brain atrophy in male carriers of fragile X. *Neurology* **57**, 127–130.
25. Greco, C. M., Hagerman, R. J., Tassone, F., Chudley, A. E., Del Bigio, M. R., Jacquemont, S., Leehey, M., and Hagerman, P. J. (2002). Neuronal intranuclear inclusions in a new cerebellar tremor/ataxia syndrome among fragile X carriers. *Brain* **125**, 1760–1771.
26. Berry-Kravis, E., Lewin, F., Wuu, J., Leehey, M., Hagerman, R., Hagerman, P., and Goetz, C. G. (2003). Tremor and ataxia in fragile X premutation carriers: Blinded videotape study. *Ann. Neurol.* **53**, 616–623.
27. Hagerman, P. J., Greco, C. M., and Hagerman, R. J. (2003). A cerebellar tremor/ataxia syndrome among fragile X premutation carriers. *Cytogenet. Genome Res.* **100**, 206–212.
28. Jacquemont, S., Hagerman, R. J., Leehey, M. A., Hall, D. A., Levine, R. A., Brunberg, J. A., Zhang, L., Jardini, T., Gane, L. W., Harris, S. W., Herman, K., Grigsby, J., Greco, C. M., Berry-Kravis, E., Tassone, F., and Hagerman, P. J. (2004). Penetrance of the fragile X-associated tremor/ataxia syndrome in a premutation carrier population. *J. Am. Med. Assoc.* **291**, 460–469.
29. Hagerman, P. J., and Hagerman, R. J. (2004). Fragile X-associated tremor/ataxia syndrome (FXTAS). *Ment. Retard. Dev. Disabil. Res. Rev.* **10**, 25–30.
30. Hagerman, R. J., Leavitt, B. R., Farzin, F., Jacquemont, S., Greco, C. M., Brunberg, J. A., Tassone, F., Hessl, D., Harris, S. W., Zhang, L., Jardini, T., Gane, L. W., Ferranti, J., Ruiz, L., Leehey, M. A., Grigsby, J., and Hagerman, P. J. (2004). Fragile-X-associated tremor/ataxia syndrome (FXTAS) in females with the FMR1 premutation. *Am. J. Hum. Genet.* **74**, 1051–1056.
31. Machado-Ferreira, M. C., Costa-Lima, M. A., Boy, R. T., Esteves, G. S., and Pimentel, M. M. (2002). Premature ovarian failure and FMR1 premutation co-segregation in a large Brazilian family. *Int. J. Mol. Med.* **10**, 231–233.
32. Sherman, S. L. (2000). Premature ovarian failure in the fragile X syndrome. *Am. J. Med. Genet.* **97**, 189–194.
33. Murray, A. (2000). Premature ovarian failure and the FMR1 gene. *Semin. Reprod. Med.* **18**, 59–66.
34. Kenneson, A., Zhang, F., Hagedorn, C. H., and Warren, S. T. (2001). Reduced FMRP and increased FMR1 transcription is proportionally associated with CGG repeat number in intermediate-length and premutation carriers. *Hum. Mol. Genet.* **10**, 1449–1454.
35. Tassone, F., Hagerman, R. J., Loesch, D. Z., Lachiewicz, A., Taylor, A. K., and Hagerman, P. J. (2000). Fragile X males with unmethylated, full mutation trinucleotide repeat expansions have elevated levels of FMR1 messenger RNA. *Am. J. Med. Genet.* **94**, 232–236.

36. Tassone, F., Hagerman, R. J., Taylor, A. K., Gane, L. W., Godfrey, T. E., and Hagerman, P. J. (2000). Elevated levels of FMR1 mRNA in carrier males: A new mechanism of involvement in the fragile-X syndrome. *Am. J. Hum. Genet.* **66**, 6–15.
37. Jin, P., Zarnescu, D. C., Zhang, F., Pearson, C. E., Lucchesi, J. C., Moses, K., and Warren, S. T. (2003). RNA-mediated neurodegeneration caused by the fragile X premutation rCGG repeats in Drosophila. *Neuron* **39**, 739–747.
38. Handa, V., Goldwater, D., Stiles, D., Cam, M., Poy, G., Kumari, D., and Usdin, K. (2005). Long CGG-repeat tracts are toxic to human cells: implications for carriers of fragile X premutation alleles. *FEBS Lett.* **579**, 2702–2708.
39. Coffee, B., Zhang, F., Ceman, S., Warren, S. T., and Reines, D. (2002). Histone modifications depict an aberrantly heterochromatinized FMR1 gene in fragile X syndrome. *Am. J. Hum. Genet.* **71**, 923–932.
40. Laggerbauer, B., Ostareck, D., Keidel, E. M., Ostareck-Lederer, A., and Fischer, U. (2001). Evidence that fragile X mental retardation protein is a negative regulator of translation. *Hum. Mol. Genet.* **10**, 329–338.
41. Siomi, H., Ishizuka, A., and Siomi, M. C. (2004). RNA interference: A new mechanism by which FMRP acts in the normal brain? What can Drosophila teach us? *Ment. Retard. Dev. Disabil. Res. Rev.* **10**, 68–74.
42. Weiler, I. J., Irwin, S. A., Klintsova, A. Y., Spencer, C. M., Brazelton, A. D., Miyashiro, K., Comery, T. A., Patel, B., Eberwine, J., and Greenough, W. T. (1997). Fragile X mental retardation protein is translated near synapses in response to neurotransmitter activation. *Proc. Natl. Acad. Sci. USA* **94**, 5395–5400.
43. Holmes, S. E., O'Hearn, E. E., McInnis, M. G., Gorelick-Feldman, D. A., Kleiderlein, J. J., Callahan, C., Kwak, N. G., Ingersoll-Ashworth, R. G., Sherr, M., Sumner, A. J., Sharp, A. H., Ananth, U., Seltzer, W. K., Boss, M. A., Vieria-Saecker, A. M., Epplen, J. T., Riess, O., Ross, C. A., and Margolis, R. L. (1999). Expansion of a novel CAG trinucleotide repeat in the 5′ region of PPP2R2B is associated with SCA12. *Nat. Genet.* **23**, 391–392.
44. Holmes, S. E., O'Hearn, E., Rosenblatt, A., Callahan, C., Hwang, H. S., Ingersoll-Ashworth, R. G., Fleisher, A., Stevanin, G., Brice, A., Potter, N. T., Ross, C. A., and Margolis, R. L. (2001). A repeat expansion in the gene encoding junctophilin-3 is associated with Huntington disease-like 2. *Nat. Genet.* **29**, 377–378.
45. Lohi, H., Young, E. J., Fitzmaurice, S. N., Rusbridge, C., Chan, E. M., Vervoort, M., Turnbull, J., Zhao, X. C., Ianzano, L., Paterson, A. D., Sutter, N. B., Ostrander, E. A., Andre, C., Shelton, G. D., Ackerley, C. A., Scherer, S. W., and Minassian, B. A. (2005). Expanded repeat in canine epilepsy. *Science* **307**, 81.
46. Bacolla, A., Gellibolian, R., Shimizu, M., Amirhaeri, S., Kang, S., Ohshima, K., Larson, J. E., Harvey, S. C., Stollar, B. D., and Wells, R. D. (1997). Flexible DNA: Genetically unstable CTG·CAG and CGG·CCG from human hereditary neuromuscular disease genes. *J. Biol. Chem.* **272**, 16783–16792.
47. Chastain, P. D., and Sinden, R. R. (1998). CTG repeats associated with human genetic disease are inherently flexible. *J. Mol. Biol.* **275**, 405–411.
48. Baldi, P., Brunak, S., Chauvin, Y., and Pedersen, A. G. (1999). Structural basis for triplet repeat disorders: a computational analysis. *Bioinformatics* **15**, 918–929.
49. Jithesh, P. V., Singh, P., and Joshi, R. (2001). Molecular dynamics studies of trinucleotide repeat DNA involved in neurodegenerative disorders. *J. Biomol. Struct. Dyn.* **19**, 479–495.
50. Bhattacharyya, D., Kundu, S., Thakur, A. R., and Majumdar, R. (1999). Sequence directed flexibility of DNA and the role of cross-strand hydrogen bonds. *J. Biomol. Struct. Dyn.* **17**, 289–300.
51. Potaman, V. N., Bissler, J. J., Hashem, V. I., Oussatcheva, E. A., Lu, L., Shlyakhtenko, L. S., Lyubchenko, Y. L., Matsuura, T., Ashizawa, T., Leffak, M., Benham, C. J., and Sinden, R. R. (2003). Unpaired structures in SCA10 (ATTCT)n·(AGAAT)n repeats. *J. Mol. Biol.* **326**, 1095–1111.
52. LeProust, E. M., Pearson, C. E., Sinden, R. R., and Gao, X. (2000). Unexpected formation of parallel duplex in GAA and TTC trinucleotide repeats of Friedreich's ataxia. *J. Mol. Biol.* **302**, 1063–1080.
53. Zheng, M., Huang, X., Smith, G. K., Yang, X., and Gao, X. (1996). Genetically unstable CXG repeats are structurally dynamic and have a high propensity for folding. An NMR and UV spectroscopic study. *J. Mol. Biol.* **264**, 323–336.
54. Smith, G. K., Jie, J., Fox, G. E., and Gao, X. (1995). DNA CTG triplet repeats involved in dynamic mutations of neurologically related gene sequences form stable duplexes. *Nucleic Acids Res.* **23**, 4303–4311.
55. Mitas, M., Yu, A., Dill, J., and Haworth, I. S. (1995). The trinucleotide repeat sequence d(CGG)15 forms a heat-stable hairpin containing Gsyn·Ganti base pairs. *Biochemistry* **34**, 12803–12811.
56. Chen, X., Mariappan, S. V., Catasti, P., Ratliff, R., Moyzis, R. K., Laayoun, A., Smith, S. S., Bradbury, E. M., and Gupta, G. (1995). Hairpins are formed by the single DNA strands of the fragile X triplet repeats: Structure and biological implications. *Proc. Natl. Acad. Sci. USA* **92**, 5199–5203.
57. Mariappan, S. V., Garcoa, A. E., and Gupta, G. (1996). Structure and dynamics of the DNA hairpins formed by tandemly repeated CTG triplets associated with myotonic dystrophy. *Nucleic Acids Res.* **24**, 775–783.
58. Yu, A., Barron, M. D., Romero, R. M., Christy, M., Gold, B., Dai, J., Gray, D. M., Haworth, I. S., and Mitas, M. (1997). At physiological pH, d(CCG)15 forms a hairpin containing protonated cytosines and a distorted helix. *Biochemistry* **36**, 3687–3699.
59. Suen, I. S., Rhodes, J. N., Christy, M., McEwen, B., Gray, D. M., and Mitas, M. (1999). Structural properties of Friedreich's ataxia d(GAA) repeats. *Biochim. Biophys. Acta* **1444**, 14–24.
60. Mitas, M., Yu, A., Dill, J., Kamp, T. J., Chambers, E. J., and Haworth, I. S. (1995). Hairpin properties of single-stranded DNA containing a GC-rich triplet repeat: (CTG)15. *Nucleic Acids Res.* **23**, 1050–1059.
61. Yu, A., Dill, J., and Mitas, M. (1995). The purine-rich trinucleotide repeat sequences d(CAG)15 and d(GAC)15 form hairpins. *Nucleic Acids Res.* **23**, 4055–4057.
62. Gacy, A. M., Goellner, G., Juranic, N., Macura, S., and McMurray, C. T. (1995). Trinucleotide repeats that expand in human disease form hairpin structures in vitro. *Cell* **81**, 533–540.
63. Nadel, Y., Weisman-Shomer, P., and Fry, M. (1995). The fragile X syndrome single strand d(CGG)n nucleotide repeats readily fold back to form unimolecular hairpin structures. *J. Biol. Chem.* **270**, 28970–28977.
64. Mitas, M. (1997). Trinucleotide repeats associated with human disease. *Nucleic Acids Res.* **25**, 2245–2254.
65. Heidenfelder, B. L., Makhov, A. M., and Topal, M. D. (2003). Hairpin formation in Friedreich's ataxia triplet repeat expansion. *J. Biol. Chem.* **278**, 2425–2431.
66. Dere, R., Napierala, M., Ranum, L. P., and Wells, R. D. (2004). Hairpin structure-forming propensity of the (CCTG·CAGG) tetranucleotide repeats contributes to the genetic instability associated with myotonic dystrophy type 2. *J. Biol. Chem.* **279**, 41715–41726.
67. Handa, V., Yeh, H. J., McPhie, P., and Usdin, K. (2005). The AUUCU-repeats responsible for spinocerebellar ataxia type 10 form unusual RNA hairpins. *J. Biol. Chem.* **280**, 29340–29345.
68. Fry, M., and Loeb, L. A. (1994). The fragile X syndrome d(CGG)n nucleotide repeats form a stable tetrahelical structure. *Proc. Natl. Acad. Sci. USA* **91**, 4950–4954.

69. Usdin, K., and Woodford, K. J. (1995). CGG repeats associated with DNA instability and chromosome fragility form structures that block DNA synthesis in vitro. *Nucleic Acids Res.* **23**, 4202–4209.
70. Kettani, A., Kumar, R. A., and Patel, D. J. (1995). Solution structure of a DNA quadruplex containing the fragile X syndrome triplet repeat. *J. Mol. Biol.* **254**, 638–656.
71. Patel, P. K., Bhavesh, N. S., and Hosur, R. V. (2000). Cation-dependent conformational switches in d-TGGCGGC containing two triplet repeats of fragile X syndrome: NMR observations. *Biochem. Biophys. Res. Commun.* **278**, 833–838.
72. Saha, T., and Usdin, K. (2001). Tetraplex formation by the progressive myoclonus epilepsy type-1 repeat: Implications for instability in the repeat expansion diseases. *FEBS Lett.* **491**, 184–187.
73. Bhavesh, N. S., Patel, P. K., Karthikeyan, S., and Hosur, R. V. (2004). Distinctive features in the structure and dynamics of the DNA repeat sequence GGCGGG. *Biochem. Biophys. Res. Commun.* **317**, 625–633.
74. Fojtik, P., and Vorlickova, M. (2001). The fragile X chromosome (GCC) repeat folds into a DNA tetraplex at neutral pH. *Nucleic Acids Res.* **29**, 4684–4690.
75. Pataskar, S. S., Dash, D., and Brahmachari, S. K. (2001). Intramolecular i-motif structure at acidic pH for progressive myoclonus epilepsy (EPM1) repeat d(CCCCGCCCCGCG)n. *J. Biomol. Struct. Dyn.* **19**, 307–313.
76. Pearson, C. E., and Sinden, R. R. (1996). Alternative structures in duplex DNA formed within the trinucleotide repeats of the myotonic dystrophy and fragile X loci. *Biochemistry* **35**, 5041–5053.
77. Ohshima, K., Kang, S., Larson, J. E., and Wells, R. D. (1996). Cloning, characterization, and properties of seven triplet repeat DNA sequences. *J. Biol. Chem.* **271**, 16773–16783.
78. Grabczyk, E., and Usdin, K. (2000). The GAA*TTC triplet repeat expanded in Friedreich's ataxia impedes transcription elongation by T7 RNA polymerase in a length and supercoil dependent manner. *Nucleic Acids Res.* **28**, 2815–2822.
79. Jain, A., Rajeswari, M. R., and Ahmed, F. (2002). Formation and thermodynamic stability of intermolecular (R*R*Y) DNA triplex in GAA/TTC repeats associated with Freidreich's ataxia. *J. Biomol. Struct. Dyn.* **19**, 691–699.
80. Mariappan, S. V., Catasti, P., Silks, L. A., III, Bradbury, E. M., and Gupta, G. (1999). The high-resolution structure of the triplex formed by the GAA/TTC triplet repeat associated with Friedreich's ataxia. *J. Mol. Biol.* **285**, 2035–2052.
81. Potaman, V. N., Oussatcheva, E. A., Lyubchenko, Y. L., Shlyakhtenko, L. S., Bidichandani, S. I., Ashizawa, T., and Sinden, R. R. (2004). Length-dependent structure formation in Friedreich ataxia (GAA)n*(TTC)n repeats at neutral pH. *Nucleic Acids Res.* **32**, 1224–1231.
82. Vetcher, A. A., Napierala, M., Iyer, R. R., Chastain, P. D., Griffith, J. D., and Wells, R. D. (2002). Sticky DNA, a long GAA·GAA·TTC triplex that is formed intramolecularly, in the sequence of intron 1 of the frataxin gene. *J. Biol. Chem.* **277**, 39217–39227.
83. Napierala, M., and Krzyosiak, W. J. (1997). CUG repeats present in myotonin kinase RNA form metastable "slippery" hairpins. *J. Biol. Chem.* **272**, 31079–31085.
84. Handa, V., Saha, T., and Usdin, K. (2003). The fragile X syndrome repeats form RNA hairpins that do not activate the interferon-inducible protein kinase, PKR, but are cut by Dicer. *Nucleic Acids Res.* **31**, 6243–6248.
85. Sobczak, K., de Mezer, M., Michlewski, G., Krol, J., and Krzyzosiak, W. J. (2003). RNA structure of trinucleotide repeats associated with human neurological diseases. *Nucleic Acids Res.* **31**, 5469–5482.
86. Weisman-Shomer, P., Cohen, E., and Fry, M. (2000). Interruption of the fragile X syndrome expanded sequence d(CGG)(n) by interspersed d(AGG) trinucleotides diminishes the formation and stability of d(CGG)(n) tetrahelical structures. *Nucleic Acids Res.* **28**, 1535–1541.
87. Pearson, C. E., Eichler, E. E., Lorenzetti, D., Kramer, S. F., Zoghbi, H. Y., Nelson, D. L., and Sinden, R. R. (1998). Interruptions in the triplet repeats of SCA1 and FRAXA reduce the propensity and complexity of slipped strand DNA (S-DNA) formation. *Biochemistry* **37**, 2701–2708.
88. Sobczak, K., and Krzyzosiak, W. J. (2005). CAG repeats containing CAA interruptions form branched hairpin structures in spinocerebellar ataxia type 2 transcripts. *J. Biol. Chem.* **280**, 3898–3910.
89. Kino, Y., Mori, D., Oma, Y., Takeshita, Y., Sasagawa, N., and Ishiura, S. (2004). Muscleblind protein, MBNL1/EXP, binds specifically to CHHG repeats. *Hum. Mol. Genet.* **13**, 495–507.
90. Kohwi, Y., Malkhosyan, S. R., and Kohwi-Shigematsu, T. (1992). Intramolecular dG·dG·dC triplex detected in *Escherichia coli* cells. *J. Mol. Biol.* **223**, 817–822.
91. Karlovsky, P., Pecinka, P., Vojtiskova, M., Makaturova, E., and Palecek, E. (1990). Protonated triplex DNA in *E. coli* cells as detected by chemical probing. *FEBS Lett.* **274**, 39–42.
92. Vetcher, A. A., and Wells, R. D. (2004). Sticky DNA formation in vivo alters the plasmid dimer/monomer ratio. *J. Biol. Chem.* **279**, 6434–6443.
93. Weisman-Shomer, P., and Fry, M. (1993). QUAD, a protein from hepatocyte chromatin that binds selectively to guanine-rich quadruplex DNA. *J. Biol. Chem.* **268**, 3306–3312.
94. Weisman-Shomer, P., and Fry, M. (1994). Stabilization of tetrahelical DNA by the quadruplex DNA binding protein QUAD. *Biochem. Biophys. Res. Commun.* **205**, 305–311.
95. Frantz, J. D., and Gilbert, W. (1995). A novel yeast gene product, G4p1, with a specific affinity for quadruplex nucleic acids. *J. Biol. Chem.* **270**, 20692–20697.
96. Frantz, J. D., and Gilbert, W. (1995). A yeast gene product, G4p2, with a specific affinity for quadruplex nucleic acids. *J. Biol. Chem.* **270**, 9413–9419.
97. Sarig, G., Weisman-Shomer, P., Erlitzki, R., and Fry, M. (1997). Purification and characterization of qTBP42, a new single-stranded and quadruplex telomeric DNA-binding protein from rat hepatocytes. *J. Biol. Chem.* **272**, 4474–4482.
98. Musso, M., Nelson, L. D., and Van Dyke, M. W. (1998). Characterization of purine-motif triplex DNA-binding proteins in HeLa extracts. *Biochemistry* **37**, 3086–3095.
99. Weisman-Shomer, P., Naot, Y., and Fry, M. (2000). Tetrahelical forms of the fragile X syndrome expanded sequence d(CGG)(n) are destabilized by two heterogeneous nuclear ribonucleoprotein-related telomeric DNA-binding proteins. *J. Biol. Chem.* **275**, 2231–2238.
100. Lin, Y. C., Shih, J. W., Hsu, C. L., and Lin, J. J. (2001). Binding and partial denaturing of G-quartet DNA by Cdc13p of *Saccharomyces cerevisiae*. *J. Biol. Chem.* **276**, 47671–47674.
101. Weisman-Shomer, P., Cohen, E., and Fry, M. (2002). Distinct domains in the CArG-box binding factor A destabilize tetraplex forms of the fragile X expanded sequence d(CGG)n. *Nucleic Acids Res.* **30**, 3672–3681.
102. Hayashi, N., and Murakami, S. (2002). STM1, a gene which encodes a guanine quadruplex binding protein, interacts with CDC13 in *Saccharomyces cerevisiae*. *Mol. Genet. Genomics* **267**, 806–813.
103. Kang, S. G., and Henderson, E. (2002). Identification of non-telomeric G4-DNA binding proteins in human, *E. coli*, yeast, and Arabidopsis. *Mol. Cells* **14**, 404–410.
104. Todd, A., Cossons, N., Aitken, A., Price, G. B., and Zannis-Hadjopoulos, M. (1998). Human cruciform binding protein belongs to the 14–3–3 family. *Biochemistry* **37**, 14317–14325.

105. Rass, U., and Kemper, B. (2002). Crp1p, a new cruciform DNA-binding protein in the yeast *Saccharomyces cerevisiae*. *J. Mol. Biol.* **323**, 685–700.
106. Darnell, J. C., Jensen, K. B., Jin, P., Brown, V., Warren, S. T., and Darnell, R. B. (2001). Fragile X mental retardation protein targets G quartet mRNAs important for neuronal function. *Cell* **107**, 489–499.
107. Darnell, J. C., Fraser, C. E., Mostovetsky, O., Stefani, G., Jones, T. A., Eddy, S. R., and Darnell, R. B. (2005). Kissing complex RNAs mediate interaction between the fragile-X mental retardation protein KH2 domain and brain polyribosomes. *Genes Dev.* **19**, 903–918.
108. Kovtun, I. V., Thornhill, A. R., and McMurray, C. T. (2004). Somatic deletion events occur during early embryonic development and modify the extent of CAG expansion in subsequent generations. *Hum. Mol. Genet.* **13**, 3057–3068.
109. Fleming, K., Riser, D. K., Kumari, D., and Usdin, K. (2003). Instability of the fragile X syndrome repeat in mice: The effect of age, diet and mutations in genes that affect DNA replication, recombination and repair proficiency. *Cytogenet. Genome Res.* **100**, 140–146.
110. Schweitzer, J. K., and Livingston, D. M. (1998). Expansions of CAG repeat tracts are frequent in a yeast mutant defective in Okazaki fragment maturation. *Hum. Mol. Genet.* **7**, 69–74.
111. Freudenreich, C. H., Kantrow, S. M., and Zakian, V. A. (1998). Expansion and length-dependent fragility of CTG repeats in yeast. *Science* **279**, 853–856.
112. White, P. J., Borts, R. H., and Hirst, M. C. (1999). Stability of the human fragile X (CGG)(n) triplet repeat array in *Saccharomyces cerevisiae* deficient in aspects of DNA metabolism. *Mol. Cell Biol.* **19**, 5675–5684.
113. Spiro, C., and McMurray, C. T. (2003). Nuclease-deficient FEN-1 blocks Rad51/BRCA1-mediated repair and causes trinucleotide repeat instability. *Mol. Cell Biol.* **23**, 6063–6074.
114. Spiro, C., Pelletier, R., Rolfsmeier, M. L., Dixon, M. J., Lahue, R. S., Gupta, G., Park, M. S., Chen, X., Mariappan, S. V., and McMurray, C. T. (1999). Inhibition of FEN-1 processing by DNA secondary structure at trinucleotide repeats. *Mol. Cell* **4**, 1079–1085.
115. Henricksen, L. A., Tom, S., Liu, Y., and Bambara, R. A. (2000). Inhibition of flap endonuclease 1 by flap secondary structure and relevance to repeat sequence expansion. *J. Biol. Chem.* **275**, 16420–16427.
116. Richards, R. I., and Sutherland, G. R. (1994). Simple repeat DNA is not replicated simply. *Nat. Genet.* **6**, 114–116.
117. Ruggiero, B. L., and Topal, M. D. (2004). Triplet repeat expansion generated by DNA slippage is suppressed by human flap endonuclease 1. *J. Biol. Chem.* **279**, 23088–23097.
118. Qiu, J., Li, X., Frank, G., and Shen, B. (2001). Cell cycle-dependent and DNA damage-inducible nuclear localization of FEN-1 nuclease is consistent with its dual functions in DNA replication and repair. *J. Biol. Chem.* **276**, 4901–4908.
119. Sharma, S., Otterlei, M., Sommers, J. A., Driscoll, H. C., Dianov, G. L., Kao, H. I., Bambara, R. A., and Brosh, R. M., Jr. (2004). WRN helicase and FEN-1 form a complex upon replication arrest and together process branch migrating DNA structures associated with the replication fork. *Mol. Biol. Cell* **15**, 734–750.
120. Usdin, K. (1998). NGG-triplet repeats form similar intrastrand structures: Implications for the triplet expansion diseases. *Nucleic Acids Res.* **26**, 4078–4085.
121. Ohshima, K., Montermini, L., Wells, R. D., and Pandolfo, M. (1998). Inhibitory effects of expanded GAA·TTC triplet repeats from intron I of the Friedreich ataxia gene on transcription and replication in vivo. *J. Biol. Chem.* **273**, 14588–14595.
122. Krasilnikova, M. M., and Mirkin, S. M. (2004). Replication stalling at Friedreich's ataxia (GAA)n repeats in vivo. *Mol. Cell Biol.* **24**, 2286–2295.
123. Pelletier, R., Krasilnikova, M. M., Samadashwily, G. M., Lahue, R., and Mirkin, S. M. (2003). Replication and expansion of trinucleotide repeats in yeast. *Mol. Cell Biol.* **23**, 1349–1357.
124. Samadashwily, G. M., Raca, G., and Mirkin, S. M. (1997). Trinucleotide repeats affect DNA replication in vivo. *Nat. Genet.* **17**, 298–304.
125. Yang, Z., Lau, R., Marcadier, J. L., Chitayat, D., and Pearson, C. E. (2003). Replication inhibitors modulate instability of an expanded trinucleotide repeat at the myotonic dystrophy type 1 disease locus in human cells. *Am. J. Hum. Genet.* **73**, 1092–1105.
126. Freudenreich, C. H., and Lahiri, M. (2004). Structure-forming CAG/CTG repeat sequences are sensitive to breakage in the absence of Mrc1 checkpoint function and S-phase checkpoint signaling: Implications for trinucleotide repeat expansion diseases. *Cell Cycle* **3**, 1370–1374.
127. Lahiri, M., Gustafson, T. L., Majors, E. R., and Freudenreich, C. H. (2004). Expanded CAG repeats activate the DNA damage checkpoint pathway. *Mol. Cell* **15**, 287–293.
128. Callahan, J. L., Andrews, K. J., Zakian, V. A., and Freudenreich, C. H. (2003). Mutations in yeast replication proteins that increase CAG/CTG expansions also increase repeat fragility. *Mol. Cell Biol.* **23**, 7849–7860.
129. Cha, R. S., and Kleckner, N. (2002). ATR homolog Mec1 promotes fork progression, thus averting breaks in replication slow zones. *Science* **297**, 602–606.
130. Gomes-Pereira, M., Fortune, M. T., Ingram, L., McAbney, J. P., and Monckton, D. G. (2004). Pms2 is a genetic enhancer of trinucleotide CAG·CTG repeat somatic mosaicism: Implications for the mechanism of triplet repeat expansion. *Hum. Mol. Genet.* **13**, 1815–1825.
131. Pearson, C. E., Ewel, A., Acharya, S., Fishel, R. A., and Sinden, R. R. (1997). Human MSH2 binds to trinucleotide repeat DNA structures associated with neurodegenerative diseases. *Hum. Mol. Genet.* **6**, 1117–1123.
132. Jauert, P. A., Edmiston, S. N., Conway, K., and Kirkpatrick, D. T. (2002). RAD1 controls the meiotic expansion of the human HRAS1 minisatellite in *Saccharomyces cerevisiae*. *Mol. Cell Biol.* **22**, 953–964.
133. Saparbaev, M., Prakash, L., and Prakash, S. (1996). Requirement of mismatch repair genes MSH2 and MSH3 in the RAD1-RAD10 pathway of mitotic recombination in *Saccharomyces cerevisiae*. *Genetics* **142**, 727–736.
134. Kearney, H. M., Kirkpatrick, D. T., Gerton, J. L., and Petes, T. D. (2001). Meiotic recombination involving heterozygous large insertions in *Saccharomyces cerevisiae*: Formation and repair of large, unpaired DNA loops. *Genetics* **158**, 1457–1476.
135. Schiestl, R. H., and Prakash, S. (1988). RAD1, an excision repair gene of *Saccharomyces cerevisiae*, is also involved in recombination. *Mol. Cell Biol.* **8**, 3619–3626.
136. Rijkers, T., Van Den Ouweland, J., Morolli, B., Rolink, A. G., Baarends, W. M., Van Sloun, P. P., Lohman, P. H., and Pastink, A. (1998). Targeted inactivation of mouse RAD52 reduces homologous recombination but not resistance to ionizing radiation. *Mol. Cell Biol.* **18**, 6423–6429.
137. Essers, J., Hendriks, R. W., Swagemakers, S. M., Troelstra, C., de Wit, J., Bootsma, D., Hoeijmakers, J. H., and Kanaar, R. (1997). Disruption of mouse RAD54 reduces ionizing radiation resistance and homologous recombination. *Cell* **89**, 195–204.
138. Essers, J., Hendriks, R. W., Wesoly, J., Beerens, C. E., Smit, B., Hoeijmakers, J. H., Wyman, C., Dronkert, M. L., and Kanaar, R. (2002). Analysis of mouse Rad54 expression and its implications for homologous recombination. *DNA Repair (Amsterdam)* **1**, 779–793.

139. Dronkert, M. L., Beverloo, H. B., Johnson, R. D., Hoeijmakers, J. H., Jasin, M., and Kanaar, R. (2000). Mouse RAD54 affects DNA double-strand break repair and sister chromatid exchange. *Mol. Cell Biol.* **20**, 3147–3156.
140. Jankowski, C., and Nag, D. K. (2002). Most meiotic CAG repeat tract-length alterations in yeast are SPO11 dependent. *Mol. Genet. Genomics* **267**, 64–70.
141. Yu, S., Mangelsdorf, M., Hewett, D., Hobson, L., Baker, E., Eyre, H. J., Lapsys, N., Le Paslier, D., Doggett, N. A., Sutherland, G. R., and Richards, R. I. (1997). Human chromosomal fragile site FRA16B is an amplified AT-rich minisatellite repeat. *Cell* **88**, 367–374.
142. Hewett, D. R., Handt, O., Hobson, L., Mangelsdorf, M., Eyre, H. J., Baker, E., Sutherland, G. R., Schuffenhauer, S., Mao, J. I., and Richards, R. I. (1998). FRA10B structure reveals common elements in repeat expansion and chromosomal fragile site genesis. *Mol. Cell* **1**, 773–781.
143. Becker, N. A., Thorland, E. C., Denison, S. R., Phillips, L. A., and Smith, D. I. (2002). Evidence that instability within the FRA3B region extends four megabases. *Oncogene* **21**, 8713–8722.
144. Wang, Y. H., Gellibolian, R., Shimizu, M., Wells, R. D., and Griffith, J. (1996). Long CCG triplet repeat blocks exclude nucleosomes: A possible mechanism for the nature of fragile sites in chromosomes. *J. Mol. Biol.* **263**, 511–516.
145. Tomita, N., Fujita, R., Kurihara, D., Shindo, H., Wells, R. D., and Shimizu, M. (2002). Effects of triplet repeat sequences on nucleosome positioning and gene expression in yeast minichromosomes. *Nucleic Acids Res. Suppl.*, 231–232.
146. Wang, Y. H., and Griffith, J. (1996). Methylation of expanded CCG triplet repeat DNA from fragile X syndrome patients enhances nucleosome exclusion. *J. Biol. Chem.* **271**, 22937–22940.
147. Hsu, Y. Y., and Wang, Y. H. (2002). Human fragile site FRA16B DNA excludes nucleosomes in the presence of distamycin. *J. Biol. Chem.* **277**, 17315–17319.
148. Wang, Y. H., Amirhaeri, S., Kang, S., Wells, R. D., and Griffith, J. D. (1994). Preferential nucleosome assembly at DNA triplet repeats from the myotonic dystrophy gene. *Science* **265**, 669–671.
149. Wang, Y. H., and Griffith, J. (1995). Expanded CTG triplet blocks from the myotonic dystrophy gene create the strongest known natural nucleosome positioning elements. *Genomics* **25**, 570–573.
150. Zlotorynski, E., Rahat, A., Skaug, J., Ben-Porat, N., Ozeri, E., Hershberg, R., Levi, A., Scherer, S. W., Margalit, H., and Kerem, B. (2003). Molecular basis for expression of common and rare fragile sites. *Mol. Cell Biol.* **23**, 7143–7151.
151. Jacobs, P. A., Hunt, P. A., Mayer, M., Wang, J. C., Boss, G. R., and Erbe, R. W. (1982). Expression of the marker (X) (q28) in lymphoblastoid cell lines. *Am. J. Hum. Genet.* **34**, 552–557.
152. Casper, A. M., Nghiem, P., Arlt, M. F., and Glover, T. W. (2002). ATR regulates fragile site stability. *Cell* **111**, 779–789.
153. Casper, A. M., Durkin, S. G., Arlt, M. F., and Glover, T. W. (2004). Chromosomal instability at common fragile sites in Seckel syndrome. *Am. J. Hum. Genet.* **75**, 654–660.
154. Otten, A. D., and Tapscott, S. J. (1995). Triplet repeat expansion in myotonic dystrophy alters the adjacent chromatin structure. *Proc. Natl. Acad. Sci. USA* **92**, 5465–5469.
155. Klesert, T. R., Otten, A. D., Bird, T. D., and Tapscott, S. J. (1997). Trinucleotide repeat expansion at the myotonic dystrophy locus reduces expression of DMAHP. *Nat. Genet.* **16**, 402–406.
156. Tassone, F., Hagerman, R. J., Chamberlain, W. D., and Hagerman, P. J. (2000). Transcription of the FMR1 gene in individuals with fragile X syndrome. *Am. J. Med. Genet.* **97**, 195–203.
157. Beilina, A., Tassone, F., Schwartz, P. H., Sahota, P., and Hagerman, P. J. (2004). Redistribution of transcription start sites within the FMR1 promoter region with expansion of the downstream CGG-repeat element. *Hum. Mol. Genet.* **13**, 543–549.
158. Laayoun, A., and Smith, S. S. (1995). Methylation of slipped duplexes, snapbacks and cruciforms by human DNA(cytosine-5)methyltransferase. *Nucleic Acids Res.* **23**, 1584–1589.
159. Saveliev, A., Everett, C., Sharpe, T., Webster, Z., and Festenstein, R. (2003). DNA triplet repeats mediate heterochromatin-protein-1-sensitive variegated gene silencing. *Nature* **422**, 909–913.
160. Mette, M. F., Aufsatz, W., van der Winden, J., Matzke, M. A., and Matzke, A. J. (2000). Transcriptional silencing and promoter methylation triggered by double-stranded RNA. *EMBO J.* **19**, 5194–5201.
161. Volpe, T. A., Kidner, C., Hall, I. M., Teng, G., Grewal, S. I., and Martienssen, R. A. (2002). Regulation of heterochromatic silencing and histone H3 lysine-9 methylation by RNAi. *Science* **297**, 1833–1837.
162. Tian, B., White, R. J., Xia, T., Welle, S., Turner, D. H., Mathews, M. B., and Thornton, C. A. (2000). Expanded CUG repeat RNAs form hairpins that activate the double-stranded RNA-dependent protein kinase PKR. *RNA* **6**, 79–87.
163. Miller, J. W., Urbinati, C. R., Teng-Umnuay, P., Stenberg, M. G., Byrne, B. J., Thornton, C. A., and Swanson, M. S. (2000). Recruitment of human muscleblind proteins to (CUG)(n) expansions associated with myotonic dystrophy. *EMBO J.* **19**, 4439–4448.
164. Mankodi, A., Urbinati, C. R., Yuan, Q. P., Moxley, R. T., Sansone, V., Krym, M., Henderson, D., Schalling, M., Swanson, M. S., and Thornton, C. A. (2001). Muscleblind localizes to nuclear foci of aberrant RNA in myotonic dystrophy types 1 and 2. *Hum. Mol. Genet.* **10**, 2165–2170.
165. Fardaei, M., Rogers, M. T., Thorpe, H. M., Larkin, K., Hamshere, M. G., Harper, P. S., and Brook, J. D. (2002). Three proteins, MBNL, MBLL and MBXL, co-localize in vivo with nuclear foci of expanded-repeat transcripts in DM1 and DM2 cells. *Hum. Mol. Genet.* **11**, 805–814.
166. Kanadia, R. N., Johnstone, K. A., Mankodi, A., Lungu, C., Thornton, C. A., Esson, D., Timmers, A. M., Hauswirth, W. W., and Swanson, M. S. (2003). A muscleblind knockout model for myotonic dystrophy. *Science* **302**, 1978–1980.
167. Adler, V., Zeiler, B., Kryukov, V., Kascsak, R., Rubenstein, R., and Grossman, A. (2003). Small, highly structured RNAs participate in the conversion of human recombinant PrP(Sen) to PrP(Res) in vitro. *J. Mol. Biol.* **332**, 47–57.
168. Ye, X., Gorin, A., Frederick, R., Hu, W., Majumdar, A., Xu, W., McLendon, G., Ellington, A., and Patel, D. J. (1999). RNA architecture dictates the conformations of a bound peptide. *Chem. Biol.* **6**, 657–669.
169. Uliel, L., Weisman-Shomer, P., Oren-Jazan, H., Newcomb, T., Loeb, L. A., and Fry, M. (2000). Human Ku antigen tightly binds and stabilizes a tetrahelical form of the fragile X syndrome d(CGG)n expanded sequence. *J. Biol. Chem.* **275**, 33134–33141.
170. Deissler, H., Behn-Krappa, A., and Doerfler, W. (1996). Purification of nuclear proteins from human HeLa cells that bind specifically to the unstable tandem repeat (CGG)n in the human FMR1 gene. *J. Biol. Chem.* **271**, 4327–4334.
171. Johnson, E. M. (2003). The Pur protein family: Clues to function from recent studies on cancer and AIDS. *Anticancer Res.* **23**, 2093–2100.
172. Khalili, K., Del Valle, L., Muralidharan, V., Gault, W. J., Darbinian, N., Otte, J., Meier, E., Johnson, E. M., Daniel, D. C., Kinoshita, Y., Amini, S., and Gordon, J. (2003). Puralpha is essential for postnatal brain development and developmentally coupled cellular proliferation as revealed by genetic inactivation in the mouse. *Mol. Cell Biol.* **23**, 6857–6875.
173. Chepenik, L. G., Tretiakova, A. P., Krachmarov, C. P., Johnson, E. M., and Khalili, K. (1998). The single-stranded DNA binding protein, Pur-alpha, binds HIV-1 TAR RNA and activates HIV-1 transcription. *Gene* **210**, 37–44.

174. Feng, Y., Zhang, F., Lokey, L. K., Chastain, J. L., Lakkis, L., Eberhart, D., and Warren, S. T. (1995). Translational suppression by trinucleotide repeat expansion at FMR1. *Science* **268**, 731–734.
175. Sunohara, T., Jojima, K., Tagami, H., Inada, T., and Aiba, H. (2004). Ribosome stalling during translation elongation induces cleavage of mRNA being translated in *Escherichia coli. J. Biol. Chem.* **279**, 15368–15375.
176. Farnham, P. J., and Platt, T. (1981). Rho-independent termination: Dyad symmetry in DNA causes RNA polymerase to pause during transcription in vitro. *Nucleic Acids Res.* **9**, 563–577.
177. Bidichandani, S. I., Purandare, S. M., Taylor, E. E., Gumin, G., Machkhas, H., Harati, Y., Gibbs, R. A., Ashizawa, T., and Patel, P. I. (1999). Somatic sequence variation at the Friedreich ataxia locus includes complete contraction of the expanded GAA triplet repeat, significant length variation in serially passaged lymphoblasts and enhanced mutagenesis in the flanking sequence. *Hum. Mol. Genet.* **8**, 2425–2436.
178. Grabczyk, E., and Usdin, K. (2000). Alleviating transcript insufficiency caused by Friedreich's ataxia triplet repeats. *Nucleic Acids Res.* **28**, 4930–4937.
179. Han, X., Leroy, J. L., and Gueron, M. (1998). An intramolecular i-motif: The solution structure and base-pair opening kinetics of d(5mCCT3CCT3ACCT3CC). *J. Mol. Biol.* **278**, 949–965.
180. Bartley, J. P., Brown, T., and Lane, A. N. (1997). Solution conformation of an intramolecular DNA triplex containing a nonnucleotide linker: Comparison with the DNA duplex. *Biochemistry* **36**, 14502–14511.
181. Radhakrishnan, I., de los Santos, C., and Patel, D. J. (1993). Nuclear magnetic resonance structural studies of A·AT base triple alignments in intramolecular purine·purine·pyrimidine DNA triplexes in solution. *J. Mol. Biol.* **234**, 188–197.
182. Verkerk, A. J., Pieretti, M., Sutcliffe, J. S., Fu, Y. H., Kuhl, D. P., Pizzuti, A., Reiner, O., Richards, S., Victoria, M. F., Zhang, F. P., et al. (1991). Identification of a gene (FMR-1) containing a CGG repeat coincident with a breakpoint cluster region exhibiting length variation in fragile X syndrome. *Cell* **65**, 905–914.
183. Fu, Y. H., Kuhl, D. P., Pizzuti, A., Pieretti, M., Sutcliffe, J. S., Richards, S., Verkerk, A. J., Holden, J. J., Fenwick, R. G., Jr., Warren, S. T., et al. (1991). Variation of the CGG repeat at the fragile X site results in genetic instability: Resolution of the Sherman paradox. *Cell* **67**, 1047–1058.
184. Brook, J. D., McCurrach, M. E., Harley, H. G., Buckler, A. J., Church, D., Aburatani, H., Hunter, K., Stanton, V. P., Thirion, J. P., Hudson, T., et al. (1992). Molecular basis of myotonic dystrophy: Expansion of a trinucleotide (CTG) repeat at the 3′ end of a transcript encoding a protein kinase family member. *Cell* **68**, 799–808.

Replication of Expandable DNA Repeats

Sergei M. Mirkin

Department of Biochemistry and Molecular Genetics, College of Medicine,
University of Illinois at Chicago, Chicago, Illinois 60607

I. Introduction
II. Experimental Approach
III. Main Experimental Results
IV. Replication Model for Repeat Expansions
V. Concluding Remarks
    References

This chapter describes our structure-functional analysis of expandable DNA repeats during the last two decades. Our studies have, in fact, originated even before the discovery of repeat expansions in human disease, while we characterized an unusual three stranded DNA structure, H-DNA, formed by homopurine-homopyrimidine mirror repeats. Little did we know at a time that one of those repeats, $(GAA)_n$ $(TTC)_n$, will be implicated in the development of the hereditary human disorder, Friedreich's ataxia. We have subsequently found that formation of unusual DNA structures by H motifs during the DNA synthesis *in vitro* could block even the most processive DNA polymerases. Interestingly, the polymerase itself triggered the formation of unusual DNA structures that subsequently inhibited it. Simple DNA repeats including, but not limited to H motifs were, thus, called "suicidal sequences" for the DNA polymerization. To study whether this concept holds *in vivo*, we have analyzed the replication fork progression through different expandable repeats in bacterial, yeast and cultured mammalian cells, via 2-dimesional electrophoretic analysis of the replication intermediates. In all three systems, these repeats attenuated DNA replication, and there was a good agreement between the repeat instability and replication inhibition thresholds. Based on these observations we postulated the replication model for repeat expansions, assuming that expansions occur during multiple replication fork stalls and restarts within structure-prone repetitive runs. The validity of this model for the development and progression of human disease is discussed.

I. INTRODUCTION

The author's interest in simple DNA repeat replication began before the phenomenon of repeat expansions was even described. It was triggered by the discovery, in collaboration with Victor Lyamichev and Maxim Frank-Kamenetskii, of a novel DNA structure called H-DNA [1], which is represented as an intramolecular DNA triplex formed by homopurine–homopyrimidine mirror repeats (Fig. 40-1). This structure contains a substantial single-stranded DNA segment together with triplex-to-duplex and duplex-to-single strand junctions, making its nucleation energy rather high [2]. Remarkably, the two complementary DNA strands in this structure are not linked, i.e., formation of H-DNA is topologically equivalent to the unwinding of the entire homopurine–homopyrimidine run. It is, therefore, favored in negatively supercoiled DNA. Depending on the chemical nature of the third strand, homopyrimidine or homopurine, there are two subclasses of H-DNA, designated H-y or H-r,

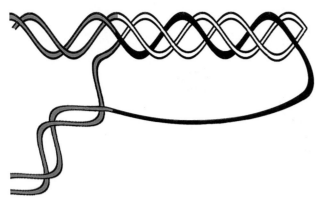

FIGURE 40-1 H-DNA. Black and white ribbons complementary strands of an H motif; gray ribbons, adjacent DNA. In the H-y conformation, the white ribbon corresponds to the homopyrimidine strands, whereas in the H-r conformation the white ribbon is the homopurine strand.

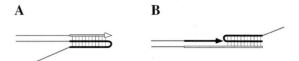

FIGURE 40-2 Models for triplex-caused polymerization arrest. (A) Polymerization on the single-stranded template. (B) Polymerization on the double-stranded template. White strips are homopyrimidine strands, black strips are homopyrimidine strands, gray lines are flanking DNA, and arrows are 3′ ends of nascent DNA chains.

respectively. The H-y structure is built from TA*T and CG*C+ triads. The requirement for cytosine protonation makes this structure favorable under acidic pH. The H-r structure is built of CG*G, TA*A, and occasionally TA*T triads. These triads are stable at physiological pH in the presence of divalent cations.

Upon moving to the United States in 1989, the focus of the author's research shifted from structural studies of H-DNA to establishment of its biological role(s). Two important considerations determined the direction of the research. First, it was assumed that an H-r conformation is more likely to form and function *in vivo*, given its pH independence. Second, there was some concern about significant DNA supercoiling, required for H-DNA formation. Although steady-state negative supercoiling is typical for bacterial DNA, it is hardly true for eukaryotic chromosomes [3]. Yet, H motifs are abundant in eukaryotes, but not in prokaryotes [4]. Although transcriptional DNA supercoiling [5] occurs in all kingdoms of life, it is probably too transient to outweigh the slow kinetics of triplex formation [6].

The author and coworkers began to wonder, therefore, whether H-DNA could be formed in the course of fundamental genetic transactions without direct impact from DNA supercoiling. A powerful boost for this consideration came from the pioneering studies of Manor and coworkers, who found that DNA polymerization was stopped in the middle of H motifs in single-stranded DNA templates [7, 8]. They suggested that when the newly synthesized DNA chain reaches the center of a repeat, its remaining single-stranded segment folds back, forming a triplex, which in turn traps DNA polymerase (Fig. 40-2A).

It has subsequently found that similar mechanisms are true for more biologically relevant double-stranded DNA templates [9]. DNA polymerase progression through H motifs was analyzed within nicked circular DNAs, where DNA supercoiling was irrelevant. DNA polymerization initiated at a nick and proceeded due to an intrinsic ability of DNA polymerases to displace the nontemplate DNA strand. It appeared that polymerization progressed smoothly when the purine strand served as a template and the pyrimidine strand was displaced. When the pyrimidine strand served as a template and the purine strand was displaced, in contrast, DNA polymerization was completely blocked at the middle of an H motif. The proposed mechanism is presented in Fig. 40-2B. During DNA synthesis, DNA polymerase runs into an H motif, displacing its nontemplate DNA strand. The displaced homopurine strand can fold back, forming a stable triplex downstream from the polymerase that, in turn, blocks further polymerization. If the displaced strand is homopyrimidine, stable triplex is not formed, because cytosines are not protonated, and polymerization proceeds normally.

These examples illustrate a crucial feature of the polymerase–triplex interplay. Triplexes did not exist in original DNA templates, but were formed in the course of DNA polymerase progression, blocking further DNA polymerization. Thus, H motifs were called suicidal sequences for the DNA polymerization [9]. The mechanisms of triplex-caused polymerization arrest were subsequently unraveled in studies of the DNA polymerase passage through triplex and duplex structures at different temperatures. It appeared that DNA polymerase can transverse triplexes at temperatures where they are about to melt, whereas duplexes are passed far below their T_m. Thus, DNA polymerase can actively unwind double-helical DNA, but lacks such ability for triplexes.

Are these findings applicable to repeat replication *in vivo*? On the one hand, an obvious analogue for the displaced nontemplate strand in the experiments is a transiently single-stranded portion of the lagging strand template, called the Okazaki initiation zone (OIZ) [10]. On the other hand, the real replication fork contains a plethora of accessory proteins besides DNA polymerases, including leading and lagging strand DNA helicases, a single-stranded DNA-binding protein, primases, topoisomerases, etc. Single-stranded DNA-binding proteins preclude folding of OIZ into stable secondary structures, including triplexes [11]. Furthermore, several DNA helicases were shown to untangle unusual DNA conformations in DNA templates [12, 13]. It was hypothesized, therefore, that an H motif could inhibit DNA replication *in vivo*, when its homopurine run is positioned in the lagging strand template, but this inhibition should be much more modest than that observed *in vitro*.

II. EXPERIMENTAL APPROACH

Evaluation of replication slowing at a repetitive run is, in fact, a challenging project. The main complication is that the replication fork progresses with an enormous speed, ranging from 100 bp/sec in eukaryotes up to 1000 bp/sec in bacteria [14]. Thus, if a 100-bp-long repeat within a 6-kb plasmid slows the replication fork progression down tenfold, the overall time of plasmid replication would increase by a mere 1 sec in bacteria and only 10 sec in eukaryotes. Therefore, most conventional methods of DNA replication analysis are not applicable to this problem.

To address this problem, an approach was chosen called two-dimensional neutral–neutral electrophoresis of replication intermediates. This technique was originally developed for mapping of the replication origins [15], but eventually it was expanded toward detecting replication blocks [16]. Various repeats have been cloned into bacterial, yeast, or mammalian vectors and their replication analyzed in the corresponding cell types.

The principles of this analysis for ColE1-derived bacterial plasmids [17] are presented in Fig. 40-3. These plasmids replicate unidirectionally, forming Θ-like intermediates. Those intermediates convert into bubbles upon restriction digest upstream of the replication origin. The size of such bubbles increases from 1 to 2 times during plasmid replication. The differences in sizes and shapes allow the separation of replication intermediates in two dimensions in the agarose gel, where the progression of the replication fork is manifested by the so-called bubble arc. If the replication fork progresses smoothly throughout the whole plasmid, the bubble arc is fairly smooth (Fig. 40-3A). Replication stalling at a repeat results in the preferential accumulation of an intermediate of defined size and shape, i.e., appearance of a distinct bulge on the otherwise smooth arc (Fig. 40-3B).

A modification of this approach allows the confirmation that replication stops coincide with the repeat. After the first dimension of electrophoresis, replication intermediates are in-gel digested with the restriction enzyme, cutting the plasmid either immediately upstream or downstream from the repeat. As a result, a fraction of bubble-shaped molecules converts into identical y-shaped products (Fig. 40-4). In the second dimension of the electrophoresis, these intermediates migrate similarly and are detected as a horizontal line. The right panel of Fig. 40-4 demonstrates that restriction digest upstream of the repeat (relative to the *ori*) leaves the bulge on the bubble arc, whereas the left panel shows that cleavage downstream of the repeat shifts the bulge onto the horizontal line.

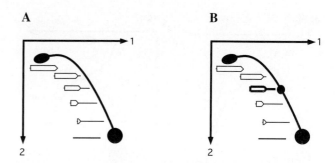

FIGURE 40-3 Two-dimensional electrophoretic analysis of bacterial plasmid replication. (A) Each dot on the bubble arc represents a replication intermediate, starting from the small bubble at the origin of replication (at the bottom) and ending at the largest, fully replicated bubble (at the top). The smooth bubble arc reflects the uniform speed of replication throughout the plasmid. (B) If replication progression is slowed down by a repeat within the plasmid (bold intermediate), a distinct bulge appears on the bubble arc.

The principles of the replication analysis for eukaryotic (yeast or mammalian) plasmids are illustrated in Fig. 40-5. These plasmids replicate bidirectionally; thus, each plasmid is divided into two domains replicated by different forks. Depending on the repeat's position and orientation, complementary repetitive runs could serve as templates for either leading or lagging DNA strand synthesis. Replication intermediates cleaved with appropriate restriction enzymes convert into Y-shaped structures. Those structures differ from nonreplicated

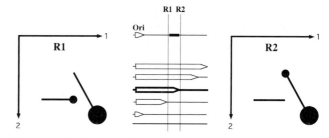

FIGURE 40-4 Mapping replication stops in bacterial plasmids using in-gel digest after the first dimension of two-dimensional gel electrophoresis. The central panel depicts plasmid replication. Vertical lines mark the positions of two restriction sites immediately upstream, R2, and downstream, R1, of the repeat (black rectangle). Replication intermediates are shown as bubbles, and the bold bubble reflects replication stalling. This stalled intermediate becomes y-shaped upon R1 digestion (left panel), but it stays on the bubble arc upon R2 digestion (right panel).

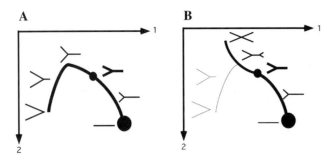

FIGURE 40-5 Two-dimensional electrophoretic analysis of plasmid replication in eukaryotes. (A) Each dot on the Y arc represents a replication intermediate, starting from the small Y at the origin of replication and ending at the fully replicated Y. If replication progression is slowed down, but not blocked altogether, by a repeat within the plasmid (bold intermediate), a distinct bulge appears on the Y arc. (B) If a repeat blocks the replication fork completely, a second replication fork approaches it from the opposite end, resulting in the appearance of double-Y intermediates and a spike on the Y arc.

DNA in their size and shape, which allows their separation by two-dimensional gel electrophoresis as a characteristic Y arc. If replication is attenuated by a repeat, but not blocked altogether, a bulge appears on the otherwise smooth Y arc due to the accumulation of intermediates of a specific size and shape (Fig. 40-5A). In the case of complete blockage, a second replication fork would approach the repeat from the opposite side, resulting in the appearance of double-Y intermediates, forming a spike on the Y arc (Fig. 40-5B).

III. MAIN EXPERIMENTAL RESULTS

By using the preceding approaches, the author and coworkers began looking for the effect of various H motifs on the replication fork progression *in vivo*. The first repeats studied, $d(GA)_n \cdot d(TC)_n$ and $d(G)_n \cdot d(C)_n$, gave evident replication stops in bacteria. The mechanisms of these replication blockages, however, appeared to be unrelated to unusual structures of template DNA strands. For the $d(GA)_n \cdot d(TC)_n$ repeat, replication stalling was caused by protein binding [18], whereas replication attenuation by the $d(G)_n \cdot d(C)_n$ run was mediated by transcription through this repeat [19]. It took almost 10 years to identify an H motif that seems to cause replication attenuation via triplex formation [20]. This happened to be the $(GAA)_n \cdot (TTC)_n$ repeat, which belongs to the class of expandable trinucleotide repeats responsible for severe hereditary neurological disorders in humans. Long before this last finding, the author and coworkers became involved in studying replication of various expandable repeats.

Several reasons determined this choice of studies. First, it was generally believed that a propensity for expansions correlates with the ability of repeats to form unusual secondary structures, such as imperfect hairpins, slipped-strand structures, unwound structures, G quartets, and triplexes [21]. Formation of these structures by expandable repeats was shown to stall DNA polymerization *in vitro* [22–24]. This stalling occasionally resulted in misalignment between nascent and template DNA strands, leading to repeat expansions or contractions *in vitro* [25]. These data, together with the fact that there is a massive accumulation of repeated DNA in the process of expansions, made it tempting to speculate that anomalous DNA replication of the repeats could be responsible for their expansions.

Most of the published data were, in fact, consistent with this hypothesis. First, stabilities of trinucleotide repeats in bacterial, yeast, and cultured mammalian cells appeared to depend on their orientation relative to replication origins [26–32]. Second, mutations in the replication apparatus of bacteria and yeast increased the instability of trinucleotide repeats. These included deletions in the yeast FLAP endonuclease (*rad27*), missense mutations in the yeast DNA polymerase δ and PCNA, proofreading mutants of *Escherichia coli* DNA polymerase III [33–39], missense mutations in the *RFC1* gene (*rfc1-1*) encoding the large subunit of the clamp loading complex [40], and mutations in the SRS2 DNA helicase [41].

Direct data on the character of the replication fork progression through expandable repeats were, however, lacking. Thus, replication fork progression has been studied through expandable trinucleotide triplet repeats $d(CGG)_n \cdot d(CCG)_n$, $d(CTG)_n \cdot d(CAG)_n$, and $d(GAA)_n \cdot d(TTC)_n$ in bacterial, yeast, and mammalian cells using electrophoretic analysis of replication intermediates, described in the previous section. In time it has been found that all of these repeats attenuate the replication fork progression in bacteria, yeast [40, 42], or mammals (unpublished results). Characteristic replication stalling, mediated by the $(CGG)_n$ repeats in various experimental systems, is shown in Fig. 40-6. Bulges

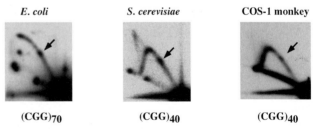

FIGURE 40-6 Experimental demonstration of replication stalling by $(CGG)_n$ repeats in *E. coli*, *S. cerevisiae*, and COS-1 monkey cells. In all instances, $(CGG)_n$ runs were situated in the lagging strand templates of the corresponding prokaryotic or eukaryotic plasmids. Arrows point to the replication stall sites.

are evident on either bubble or Y arcs of replication intermediates isolated from *E. coli* or eukaryotic cells, respectively. Replication stall sites were mapped to the repeats by using the in-gel digest technique described earlier [42]. Mapping experiments have additionally revealed an underreplication of the lagging DNA strand, pointing to the problem with lagging strand synthesis as the cause of replication attenuation.

In both prokaryotes and eukaryotes, repeat-caused replication inhibition was length-dependent. The strength of inhibition in all systems depended on the repeat's base composition in the following order: $(CGG)_n \cdot (CCG)_n > (GAA)_n \cdot (TTC)_n > (CAG)_n \cdot (CTG)_n$. Even in the strongest cases, however, replication was only modestly (two-to threefold) slowed down. Thus, only repeat-caused replication stalling was detected due to extraordinary resolution of the two-dimensional gel electrophoresis. Although all systems gave qualitatively similar results, substantial quantitative differences were observed between different cell types. For example, a relatively short $(CGG)_{10} \cdot (CCG)_{10}$ repeat inhibited DNA replication in yeast, whereas the corresponding inhibition in bacterial and mammalian cells was observed for the 4-times-longer repeat. This could be explained by the fact that the *Saccharomyces cerevisiae* genome is quite AT-rich, making 100% GC-rich runs, such as $(CGG)_n$ repeats, extremely foreign. Furthermore, replication stalling by the $(CGG)_n \cdot (CCG)_n$ repeat in yeast did not depend on its orientation in the replicon, unlike the orientation-dependent replication blockage in bacteria [42] and mammals (I. Voineagu, unpublished results).

Several arguments support the notion that replication attenuation at trinucleotide repeats is linked to their propensity to expand. First, normal alleles of repeat-containing genes in humans contain multiple interruptions that disappear in expandable alleles, resulting in homogeneous repetitive runs [43]. In normal *FMR1* alleles, for example, $(CGG)_n$ runs are interrupted by AGG triplets. It has been found that these AGG interruptions abolished the replication blockage in bacteria [42]. Second, it has also been found that replication attenuation caused by $(GAA)_n \cdot (TTC)_n$ repeats in yeast was strikingly orientation-dependent, as was evident when the repeat's homopurine strand served as the lagging strand template. Remarkably, length polymorphism of the $(GAA)_n \cdot (TTC)_n$ repeat in terms of both expansions and contractions drastically increased in the repeat's orientation, responsible for the replication stalling [20]. Finally, the data suggest that an impasse during lagging strand synthesis is the prime event resulting in replication fork stalling [40, 42]. Supporting this are studies of repeat stability in various replication mutants that implicate the lagging strand synthesis machinery [44].

At the same time, there are also unanswered questions about the replication–expansion connection. One uncertainty is the association between the expansion threshold in humans and replication inhibition thresholds in our systems. It is well-established that trinucleotide repeats tend to expand upon reaching the length of 30–40 uninterrupted units [45]. Replication stalling has been observed at similar lengths for $(CGG)_n \cdot (CCG)_n$ repeats in bacterial [42] and mammalian cells (I. Voineagu, unpublished results), as well as for $(GAA)_n \cdot (TTC)_n$ repeats in yeast [20]. Yet in the $(CAG)_n \cdot (CTG)_n$ case, only longer ($n = 80$) repeats stall replication in every system studied [40, 42]. Furthermore, replication attenuation for $(CAG)_n \cdot (CTG)_n$ repeats is quite weak except in special circumstances, such as treatment of bacterial cells with chloramphenicol. Another challenge is to explain the prevalence of repeat expansions over contractions in human pedigrees and transgenic mice [46] in the frame of the replication stall hypothesis. In studies of $(GAA)_n \cdot (TTC)_n$ repeats in yeast, it appeared that replication blockage leads to both expansions and contractions [20]. This is true for other model organisms as well. Which mechanisms determine expansion bias in humans? These and other questions should be addressed before the link between repeat replication and expansions is firmly established.

IV. REPLICATION MODEL FOR REPEAT EXPANSIONS

On the basis of the preceding data and discussions, it was suggested that expansions–contractions could occur during replication fork stalling and restart within trinucleotide repeats [20]. One possible model is presented in Fig. 40-7. When the leading DNA polymerase runs into an expandable repeat, an OIZ on the lagging strand template DNA becomes repetitive (Fig. 40-7A), folding into a stable conformation (Fig. 40-7B). This structure would prevent an orderly progression of the lagging DNA polymerase followed by the leading DNA polymerase, resulting in replication fork stalling and dissociation (Fig. 40-7C). For replication to restart, some accessory proteins should unwind the lagging strand template. In the meantime, the newly synthesized leading strand and its template could misalign within the repeat (Fig. 40-7D). Such misalignment can obviously generate repetitive slip outs on either the template or nascent DNA strand. Resumption of DNA synthesis would then lead to repeat contractions or expansions, respectively (Fig. 40-7D).

A problem with this model, however, is that slip outs between the leading DNA strand and its template should be energetically unfavorable and unstable. A provocative solution for this problem is suggested in Fig. 40-8. Collapse of the stalled replication fork

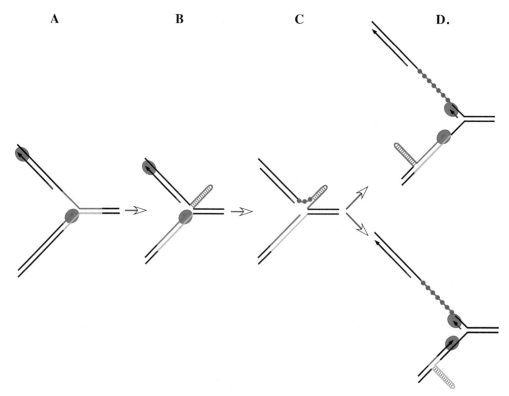

FIGURE 40-7 Replication model for repeat expansions–contractions. (A) Entrance of the leading strand DNA polymerase into a repeat exposes a repetitive run on the lagging strand template. (B) Repetitive lagging strand template folds into a stable secondary structure, blocking the lagging strand synthesis. (C) Replication fork collapses; unwinding of the lagging strand template begins. (D) Replication fork restart via polymerase reloading; repeat misalignment between the leading strand and its template leads to expansions (upper panel) or contractions (lower panel). Black lines are flanking DNA, red lines are structure-prone strand of a repeat, green lines are complementary strand of a repeat, purple lines are RNA primers, arrows are 3′ ends of nascent DNA chains, golden ovals are leading and lagging DNA polymerases, and gray circles are lagging strand guardians such as RPA, helicases, etc. See CD-ROM for color image.

commonly leads to its reversal [47, 48]. Reversal of a fork stalled within an expandable repeat would lead to the formation of a four-way junction with the single-stranded repetitive extension at the 3′ end of the leading

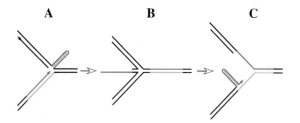

FIGURE 40-8 Repeat expansions upon the replication fork reversal. (A) Replication stalling within the repetitive tract. (B) Reversal of the collapsed replication fork exposes repetitive extension of the leading DNA strand, making it free to fold into stable secondary structures. (C) Flipping back of the reversed fork could result in repeat misalignments between the leading DNA strand and its template. Black lines are flanking DNA, red lines are structure-prone strand of a repeat, green lines are complementary strand of a repeat, and arrows are 3′ ends of nascent DNA chains. See CD-ROM for color image.

strand (Fig. 40-8B). Such a single-stranded repeat would tend to acquire a secondary structure. To restart replication, the reversed fork should be flipped back by the eukaryotic RuvAB homologues (Fig. 40-8C). Upon the restart, extra repeats will be added to the leading strand.

These models suggest that although initial blockage occurs during the lagging strand synthesis, repeat expansions and contractions occur on the leading strand. This distinguishes the model from other hypotheses on repeat instabilities, implicating lagging strand synthesis [49–51]. As previously mentioned, some experimental data show that repeat expansions can occur on the leading DNA strand [52, 53].

V. CONCLUDING REMARKS

As one can see, certain similarity exists between the early model for triplex-caused polymerization blockage (Fig. 40-2) and the more recent model for repeat expansions during replication fork stall–restart (Fig. 40-7). In both instances, single-stranded repetitive DNA segments are generated during DNA synthesis. Their propensity to form non-B-DNA structures, whether triplexes or others, would then block further replication. It is believed that repeat length polymorphism accumulates during the replication fork attempts to escape from this self-made trap. Thus, the term suicidal DNA sequences, which was coined in 1993 [9], appears to be more than adequate. Clearly, many more questions remain to be answered. One of the most important is what triggers initial repeat expansions in particular human families? A favorite suggestion, yet to be proved, is that this trigger could be a change in the position and/or orientation of a repeat within the replichore [54]. Experiments are under way to address this and other questions on the mechanisms of repeat expansions.

Acknowledgments

I thank current and former members of my lab, Randal Cox, Andrew Dayn, Andrei Krasilnikov, Maria Krasilnikova, Ekaterina Mirkin, Gordana Raca, George Samadashwily, and Irina Voineagu, for their invaluable experimental and intellectual contributions to this research. These studies were supported by grant no. GM60987 from the NIH and grant no. MCB-9723924 from the NSF.

References

1. Mirkin, S. M., Lyamichev, V. I., Drushlyak, K. N., Dobrynin, V. N., Filippov, S. A., and Frank-Kamenetskii, M. D. (1987). DNA H form requires a homopurine-homopyrimidine mirror repeat. *Nature* **330**, 495–497.
2. Lyamichev, V. I., Mirkin, S. M., Kumarev, V. P., Baranova, L. V., Vologodskii, A. V., and Frank-Kamenetskii, M. D. (1989). Energetics of the B-H transition in supercoiled DNA carrying d(CT)x·d(AG)x and d(C)n·d(G)n inserts. *Nucleic Acids Res*. **17**, 9417–9423.
3. Sinden, R. R., Carlson, J. O., and Pettijohn, D. E. (1980). Torsional tension in the DNA double helix measured with trimethylpsoralen in living *E. coli* cells: Analogous measurements in insect and human cells. *Cell* **21**, 773–783.
4. Cox, R., and Mirkin, S. M. (1997). Characteristic enrichment of DNA repeats in different genomes. *Proc. Natl. Acad. Sci. USA* **94**, 5237–5242.
5. Liu, L. F., and Wang, J. C. (1987). Supercoiling of the DNA template during transcription. *Proc. Natl. Acad. Sci. USA* **84**, 7024–7027.
6. Paes, H. M., and Fox, K. R. (1997). Kinetic studies on the formation of intermolecular triple helices. *Nucleic Acids Res*. **25**, 3269–3274.
7. Lapidot, A., Baran, N., and Manor, H. (1989). (dT-dC)n and (dG-dA)n tracts arrest single stranded DNA replication in vitro. *Nucleic Acids Res*. **17**, 883–900.
8. Baran, N., Lapidot, A., and Manor, H. (1991). Formation of DNA triplexes accounts for arrests of DNA synthesis at d(TC)$_n$ and d(GA)$_n$ tracts. *Proc. Natl. Acad. Sci. USA* **88**, 507–511.
9. Samadashwily, G. M., Dayn, A., and Mirkin, S. M. (1993). Suicidal nucleotide sequences for DNA polymerization. *EMBO J*. **12**, 4975–4983.
10. DePamphilis, M. L. (2002). Eukaryotic DNA replication fork. *Chemtracts—Biochem. Mol. Biol*. **15**, 313–325.
11. Samadashwily, G. M., and Mirkin, S. M. (1994). Trapping DNA polymerases using triplex-forming oligodeoxyribonucleotides. *Gene* **149**, 127–136.
12. Maine, I. P., and Kodadek, T. (1994). Efficient unwinding of triplex DNA by a DNA helicase. *Biochem. Biophys. Res. Comm*. **204**, 1119–1124.
13. Kopel, V., Pozner, A., Baran, N., and Manor, H. (1996). Unwinding of the third strand of a DNA triple helix, a novel activity of the SV40 large T-antigen helicase. *Nucleic Acids Res*. **24**, 330–335.
14. Kornberg, A., and Baker, T. (1992). "DNA Replication," 2nd ed. W. H. Freeman and Co., New York.
15. Brewer, B. J., and Fangman, W. L. (1987). The localization of replication origins on ARS plasmids in *S. cerevisiae*. *Cell* **51**, 463–471.
16. Brewer, B. J., and Fangman, W. L. (1988). A replication fork barrier at the 3′ end of yeast ribosomal RNA genes. *Cell* **55(4)**, 637–643.
17. Martin-Parras, L., Hernandez, P., Martinez-Robles, M., and Schvartzman, J. B. (1991). Unidirectional replication as visualised by two-dimensional agarose gel electrophoresis. *J. Mol. Biol*. **220**, 843–855.
18. Krasilnikova, M. M., Smirnova, E. V., Krasilnikov, A. S., and Mirkin, S. M. (2001). A new trick for an old dog: TraY binding to a homopurine-homopyrimidine run attenuates DNA replication. *J. Mol. Biol*. **313**, 271–282.
19. Krasilnikova, M. M., Samadashwily, G. M., Krasilnikov, A. S., and Mirkin, S. M. (1998). Transcription through a simple DNA repeat blocks replication elongation. *EMBO J*. **17**, 5095–5102.
20. Krasilnikova, M. M., and Mirkin, S. M. (2004). Replication stalling at Friedreich's ataxia (GAA)n repeats in vivo. *Mol. Cell. Biol*. **24**, 2286–2295.
21. McMurray, C. T. (1999). DNA secondary structure: A common and causative factor for expansion in human disease. *Proc. Natl. Acad. Sci. USA* **96**, 1823–1825.
22. Kang, S., Ohshima, K., Shimizu, M., Amirhaeri, S., and Wells, R. D. (1995). Pausing of DNA synthesis in vitro at specific loci in CTG and CGG triplet repeats from human hereditary disease genes. *J. Biol. Chem*. **270**, 27014–27021.
23. Usdin, K., and Woodford, K. J. (1995). CGG repeats associated with DNA instability and chromosome fragility from structures that block DNA synthesis in vitro. *Nucleic Acids Res*. **23**, 4202–4209.
24. Gacy, A. M., Goellner, G. M., Spiro, C., Chen, X., Gupta, G., Bradbury, E. M., Dyer, R. B., Mikesell, M. J., Yao, J. Z., Johnson, A. J., Richter, A., Melancon, S. B., and McMurray, C. T. (1998). GAA instability in Friedreich's ataxia shares a common, DNA-directed and intraallelic mechanism with other trinucleotide diseases. *Mol. Cell* **1**, 583–593.
25. Ohshima, K., and Wells, R. D. (1997). Hairpin formation during DNA synthesis primer realignment in vitro in triplet repeat sequences from human hereditary disease genes. *J. Biol. Chem*. **272**, 16798–16806.
26. Kang, S., Jaworski, A., Ohshima, K., and Wells, R. D. (1995). Expansion and deletion of CTG repeats from human disease genes are determined by the direction of replication in *E. coli*. *Nat. Genet*. **10**, 213–218.
27. Shimizu, M., Gellibolian, R., Oostra, B. A., and Wells, R. D. (1996). Cloning, characterization and properties of plasmids containing CGG triplet repeats from the FMR-1 gene. *J. Mol. Biol*. **258**, 614–626.

28. Freudenreich, C. H., Stavenhagen, J. B., and Zakian, V. A. (1997). Stability of a CTG/CAG trinucleotide repeat in yeast is dependent on its orientation in the genome. *Mol. Cell. Biol.* **17**, 2090–2098.
29. Miret, J. J., Pessoa-Brandao, L., and Lahue, R. S. (1998). Orientation-dependent and sequence-specific expansions of CTG/CAG trinucleotide repeats in *Saccharomyces cerevisiae*. *Proc. Natl. Acad. Sci. USA* **95**, 12438–12443.
30. Balakumaran, B. S., Freudenreich, C. H., and Zakian, V. A. (2000). CGG/CCG repeats exhibit orientation-dependent instability and orientation-independent fragility in *Saccharomyces cerevisiae*. *Hum. Mol. Genet.* **9**, 93–100.
31. Cleary, J. D., Nichol, K., Wang, Y. H., and Pearson, C. E. (2002). Evidence of cis-acting factors in replication-mediated trinucleotide repeat instability in primate cells. *Nat. Genet.* **31**, 37–46.
32. Panigrahi, G. B., Cleary, J. D., and Pearson, C. E. (2002). In vitro (CTG)*(CAG) expansions and deletions by human cell extracts. *J. Biol. Chem.* **277**, 13926–13934.
33. Freudenreich, C. H., Kantrow, S. M., and Zakian, V. A. (1998). Expansion and length-dependent fragility of CTG repeats in yeast. *Science* **279**, 853–856.
34. Schweitzer, J. K., and Livingston, D. M. (1998). Expansions of CAG repeat tracts are frequent in a yeast mutant defective in Okazaki fragment maturation. *Hum. Mol. Genet.* **7**, 69–74.
35. Schweitzer, J. K., and Livingston, D. M. (1999). The effect of DNA replication mutations on CAG tract stability in yeast. *Genetics* **152**, 953–963.
36. Spiro, C., Pelletier, R., Rolfsmeier, M. L., Dixon, M. J., Lahue, R. S., Gupta, G., Park, M. S., Chen, X., Mariappan, S. V. S., and McMurray, C. T. (1999). Inhibition of FEN-1 processing by DNA secondary structure at trinucleotide repeats. *Mol. Cell* **4**, 1079–1085.
37. White, P. J., Borts, R. H. and Hirst, M. C. (1999). Stability of the human fragile X (CGG)n triplet repeat array in *Saccharomyces cerevisiae* deficient in aspects of DNA metabolism. *Mol. Cell. Biol.* **19**, 5675–5684.
38. Ireland, M. J., Reinke, S. S., and Livingston, D. M. (2000). The impact of lagging strand replication mutations on the stability of CAG repeat tracts in yeast. *Genetics* **155**, 1657–1665.
39. Iyer, R. R., Pluciennik, A., Rosche, W. A., Sinden, R. R., and Wells, R. D. (2000). DNA polymerase III proofreading mutants enhance the expansion and deletion of triplet repeat sequences in *Escherichia coli*. *J. Biol. Chem.* **275**, 2174–2184.
40. Pelletier, R., Krasilnikova, M. M., Samadashwily, G. M., Lahue, R. S., and Mirkin, S. M. (2003). Replication and expansion of trinucleotide repeats in yeast. *Mol. Cell. Biol.* **23**, 1349–1357.
41. Bhattacharyya, S., and Lahue, R. S. (2004). *Saccharomyces cerevisiae* Srs2 DNA helicase selectively blocks expansions of trinucleotide repeats. *Mol. Cell. Biol.* **24**, 7324–7330.
42. Samadashwily, G. M., Raca, G., and Mirkin, S. M. (1997). Trinucleotide repeats affect DNA replication in vivo. *Nat. Genet.* **17**, 298–304.
43. Kunst, C. B., and Warren, S. T. (1994). Cryptic and polar variation of the fragile X repeat could result in predisposing normal alleles. *Cell* **77**, 853–861.
44. Mirkin, S. M. (2004). Molecular models for repeat expansions. *Chemtracts—Biochem. Mol. Biol.* **17**, 639–662.
45. Wells, R. D. (1996). Molecular basis of genetic instability of triplet repeats. *J. Biol. Chem.* **271**, 2875–2878.
46. Cleary, J. D., and Pearson, C. E. (2003). The contribution of cis-elements to disease-associated repeat instability: Clinical and experimental evidence. *Cytogenet. Genome Res.* **100**, 25–55.
47. Higgins, N. P., Kato, K., and Strauss, B. (1976). A model for replication repair in mammalian cells. *J. Mol. Biol.* **101**, 417–425.
48. Postow, L., Ullsperger, C., Keller, R. W., Bustamante, C., Vologodskii, A. V., and Cozzarelli, N. R. (2001). Positive torsional strain causes the formation of a four-way junction at replication forks. *J. Biol. Chem.* **276**, 2790–2796.
49. McMurray, C. T. (1995). Mechanisms of DNA expansion. *Chromosoma* **104**, 2–13.
50. Usdin, K., and Grabczyk, E. (2000). DNA repeat expansions and human disease. *Cell. Mol. Life Sci.* **57**, 914–931.
51. Bowater, R. P., and Wells, R. D. (2001). The intrinsically unstable life of DNA triplet repeats associated with human hereditary disorders. *Prog. Nucleic Acid Res. Mol. Biol.* **66**, 159–202.
52. Iyer, R. R., and Wells, R. D. (1999). Expansion and deletion of triplet repeat sequences in *Escherichia coli* occur on the leading strand of DNA replication. *J. Biol. Chem.* **274**, 3865–3877.
53. Hashem, V. I., and Sinden, R. R. (2005). Duplications between direct repeats stabilized by DNA secondary structure occur preferentially in the leading strand during DNA replication. *Mutat. Res.* **570**, 215–226.
54. Mirkin, S. M., and Smirnova, E. V. (2002). Positioned to expand. *Nat. Genet.* **31**, 5–6.

CHAPTER 41

Error-Prone Repair of Slipped (CTG)·(CAG) Repeats and Disease-Associated Expansions

GAGAN B. PANIGRAHI[1], RACHEL LAU[1], S. ERIN MONTGOMERY[1-3],
MICHELLE R. LEONARD[1,2], JULIEN L. MARCADIER[1,2], MARIANA KEKIS[1,2],
CAROLINE VOSCH[1,2], ANDREA TODD[1,2,4] AND CHRISTOPHER E. PEARSON[1,2]

[1]Program of Genetics & Genomic Biology, The Hospital for Sick Children, Toronto, Ontario, Canada M5G 1X8,
[2]Department of Molecular & Medical Genetics, University of Toronto, Toronto, Ontario, Canada, [3]Albany Medical College, Albany, New York, and [4]University of Toronto, Toronto, Ontario, Canada

I. Introduction
II. Slipped DNAs as Mutagenic Intermediates
   A. Structure of Slipped DNAs
   B. Repair of Heteroduplexes with Components of Slipped DNAs
III. Processing of Unpaired DNAs
   A. Nucleotide Excision Repair and Repeat Instability?
   B. Random-Sequence Heteroduplex Repair
   C. Mismatch Repair and (CTG)·(CAG) Instability
IV. Binding of MMR Proteins to Trinucleotide Repeats
V. Processing of Slipped (CTG)·(CAG) Repeats by Human Cell Extracts
   A. Three Repair Outcomes: Correct, Escaped, and Error-Prone Repair
   B. Slipped-DNA Repair Is Independent of MMR and NER Proteins
   C. Neuronlike Cell Extracts Process Slipped DNAs
   D. Mechanism of Correct Repair
   E. Mechanism of Escaped Repair
   F. Mechanism of Error-Prone Repair
   G. Comparison of Slipped (CTG)·(CAG) Processing with Base–Base Mismatches and Random-Sequence Heteroduplexes
   H. Role of Repair Proteins in (CTG)·(CAG) Instability?
   I. Slipped-DNA Processing and Repeat Instability in Human Disease
V. Summary and Future Studies
References

Slipped-DNAs, formed during DNA replication, repair or recombination may be mutagenic intermediates of disease-causing (CAG)·(CTG) instability. Curiously, (CAG)·(CTG) expansions in mice require mismatch repair proteins MSH2/3, which contrasts with their protective role against mutations. We determined the fidelity of slipped-DNA repair using human cell extracts and slipped (CAG)·(CTG) substrates. We observed correct repair, escaped repair and error-prone repair. Choice of repair path depended upon slip-out sequence (CAG or CTG) and whether substrates are expansion or deletion intermediates. Contrary to expectations that slipped-DNAs will escape repair, CAG slip-outs in deletion intermediates were correctly repaired with efficiencies comparable to a G-T mismatch, the most efficiently repaired mismatch. In contrast, CTG slip-outs escaped correct repair. Surprisingly, expansion intermediates with slip-outs of CAG or CTG yielded products of error-prone repair—where attempts at repair failed to remove all excess repeats. This lower fidelity of repair presents a novel path to generate repeat expansions. Human neuron-like cell extracts yielded each repair outcome, supporting a role for these processes in repeat instability in patient brains. Deficiencies of MSH2 or MSH3 did not affect repair outcome. Thus, the contribution of the MSH2/3 proteins to instability may precede or be coincident with structure formation/strand-slippage. Expansion biases in different diseases and patient tissues may be explained by slipped-DNA repair: Expansion intermediates were poorly repaired relative to deletion intermediates, and only expansion intermediates were subject to error-prone repair—yielding products that retained excess repeats. Thus, features of slipped-(CAG)·(CTG) processing may be involved in disease-associated repeat instability.

I. INTRODUCTION

Gene-specific instability of certain trinucleotide repeat tracts, such as (CTG)·(CAG) repeats, is the causative factor for at least 40 neurological, neurodegenerative, and neuromuscular diseases [1, 1a, 2]. Instability is prevalent in somatic tissues [1a, 2] as well as in the germline (parent-to-offspring transmissions) [1a, 2, 3]. The mutation patterns and levels vary between the different genetic loci or tissues [1–11]. For example, in diseases like myotonic dystrophy (DM1), Huntington's disease (HD), dentatorubral pallidoluysian atrophy (DRPLA), spinobulbar muscular atrophy (SBMA), and several types of spinocerebellar ataxia (SCA types 1–3, 6–8, 10, 12, and 17), repeat instability is found in both mitotic and postmitotic tissues such as the blood, muscle, and various brain regions. Generally, in nonaffected individuals the size of the repeat tract ranges from 5 to 25 repeat units. Repeat tracts that are approximately 34 repeats in length can become genetically unstable and can expand up to several hundreds or thousands of repeats, depending on the disease or tissue. An understanding of the mechanism of this important form of mutation has provided surprising insights into DNA metabolism, mutagenesis, and genetics. Although many researchers have demonstrated active contributions to repeat instability by processes of DNA replication and recombination [1a, 6, 11a, 12–23], this chapter will focus upon the contribution of DNA repair to repeat instability, specifically with respect to the proposed mutagenic intermediate slipped-strand DNAs.

II. SLIPPED DNAS AS MUTAGENIC INTERMEDIATES

The mechanism by which repeat expansion occurs is complex. Rather than one mechanism or metabolic process, a variety of processes seem to be involved. Invariably, slipped DNAs are thought to be mutagenic intermediates in the mutation process and could occur during DNA replication, repair, recombination, and/or transcription (Fig. 41-1). At replication forks, formation of slip outs of the repeats in the template (continuous) strands may allow these sequences to be bypassed during replication synthesis, resulting in repeat deletion intermediates [11a]. This deletion process may be facilitated on the lagging strand template, which is transiently single-stranded prior to the retrograde formation of each Okazaki fragment. Alternatively, if the slip outs were to form in the newly replicated nascent (nicked) strand of either the lagging or leading strand, the outcome would lead to expansion intermediates [11a]. In the absence of repair, these intermediates would, upon a subsequent round of replication, lead to deletion or expansion products. Nonreplicating DNAs may also incur the formation of slipped-DNA intermediates, for example, in brain tissues. Slippage may occur in regions of single-strandedness, which are known to occur in neural tissues [24–26]. Sites of DNA damage, such as nicks, or chemical or oxidative damage may facilitate slippage at the repeats. Slippage of the nicked or continuous strand undergoing the repair process could lead to expansions or deletion intermediates, respectively. Aberrant repair of these intermediates could result in repeat instability. Slipped DNA may also be formed during unequal crossover recombination between sister chromatids or homologous recombination or during recombination-mediated repair of double-strand breaks [1a]. Finally, formation of slipped DNAs in repeat sequences during transcription and could lead to length changes in the repeat tract. All disease-associated repeat sequences are actively transcribed. Unwinding of the repeat containing double-stranded DNA by moving transcription and complex introduces single-stranded regions in DNA, which may facilitate the formation of hairpins and unusual DNA structures. Such processes may be facilitated by the head-on collision of transcription complexes with progressing replication forks. Thus, most DNA metabolic processes, dependent and independent of replication, may contribute to the formation of mutagenic intermediates. If not properly repaired, slipped DNAs may pose a risk for repeat instability.

A. Structure of Slipped DNAs

To elucidate the mechanism of repeat instability and to understand the interaction of repair proteins and how the slipped-strand mutagenic intermediates

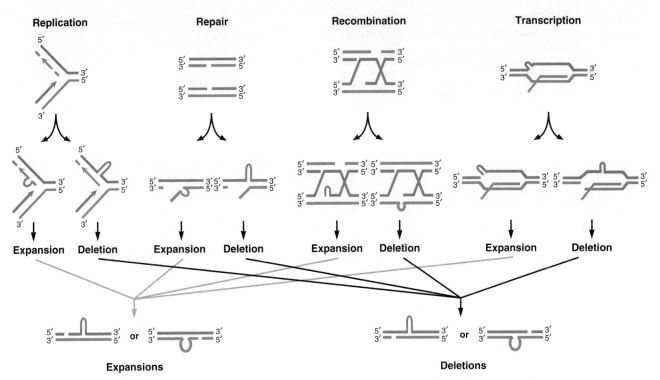

FIGURE 41-1 Slipped-DNA mutagenic intermediates may form during processes of DNA metabolism, including DNA replication, repair (nicked, chemically damaged, or unwound DNAs), recombination, and/or transcription. Slipped DNAs may form at replication forks on either the nascent (nicked) or template (continuous) strand, giving rise to expansion intermediates or deletion intermediates, respectively. Similarly, slipped DNAs formed in nonreplicating DNA during repair, recombination, or transcription stages. In all cases, slip outs in the nicked strand are expansion intermediates, whereas slip outs in the continuous strand are deletion intermediates. Nicks may arise either 5′ or 3′ of the slip out.

are processed, it is critical to understand the structural details of these intermediates. Just as an appreciation of the biophysical features of four-way DNA Holliday junctions has been informative to understanding DNA recombination [27], the characterization of slipped structures should provide insight into the mechanisms of repeat instability. Disease-associated repeats have been shown to adopt various unusual DNA conformations, such as quadruplexes [28] and parallel [29], triplex [21], and sticky DNA [30] structures. This chapter focuses upon slipped DNAs, long thought to be mutagenic intermediates of repeat instability (reviewed in refs. [1a,31] and [32]). Slipped DNAs, discovered in 1996, are formed by out-of-register pairings between complementary repeat strands [31]. Slipped intermediate DNAs (SI-DNAs) are heteroduplexes of $(CTG)_x \cdot (CAG)_y$, where $x \neq y$. When $x > y$ or $x < y$, it leads to slip outs of CTG or slip outs of CAG repeats, each having lengths equivalent to the difference between x and y. S-DNAs are homoduplex slipped DNAs where $x = y$. These were discovered and have been structurally characterized in detail by the Pearson lab [31, 33–37].

True slipped DNAs contain several critical components including slip outs, which can form intrastrand hairpins with hairpin tips or unpaired single-stranded loops, and slipped junctions, from whose points the slip outs extrude from the Watson–Crick duplex (Fig. 41-2A). Hairpins are major components of the slipped-strand DNA structures formed by the (CTG)·(CAG) repeats [35, 38–42]. Oligonucleotides containing $(CTG)_{25}$ or $(CAG)_{25}$ formed intrastrand hairpins with G·C base pairs and T–T or A–A mismatches [38]. These hairpins, despite their mismatches, are biophysically stable. In the $(CTG)_n$ hairpins, T–T mismatches are stacked in the helix and are further stabilized by forming two hydrogen bonds [39]. In contrast, the A–A mismatches in $(CAG)_n$ hairpins are poorly stacked in the helix and do not form base pairs [43]. Hairpin tips are an integral feature of intrastrand hairpins formed by CAG or CTG repeats. Two variant hairpin tip conformations are possible depending upon whether the number of repeats contained in the slip out is even or odd (Fig. 41-2A, see inset) [42, 44]. These distinct tip conformations may be recognized and/or cleaved differentially by proteins. In addition to intrastrand hairpins, slipped-out repeats, particularly long tracts of CAG, can

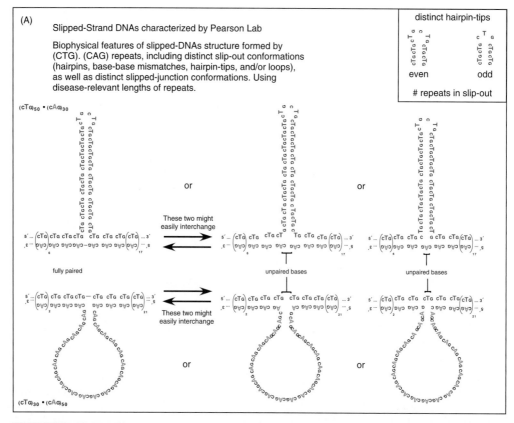

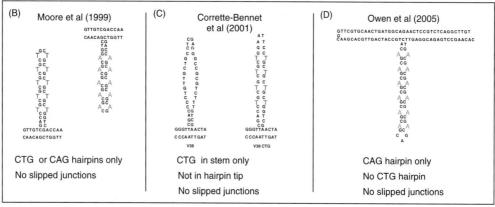

FIGURE 41-2 Comparison of the biophysical features of repeat-containing heteroduplexes. (A) In the Pearson laboratory, slipped intermediate heteroduplexes (SI-DNAs) were composed of complementary repeat strands containing an excess of 20 CTG repeats [$(CTG)_{50} \cdot (CAG)_{30}$] or an excess of 20 CAG repeats [$(CTG)_{30} \cdot (CAG)_{50}$]. These SI-DNAs harbored each of the components expected in a true slipped DNA: slipped junctions composed of three arms, two made of complementary repeat strands and the third being the CTG or CAG slip out. Slip outs may assume intrastrand hairpins with hairpin tips or single-stranded random coil conformations. CAG slip outs preferentially assume single-stranded loops, whereas CTG slip outs assume only hairpins. Hairpin tip conformation can vary depending upon the total number of repeats in the slip out (inset). These characterizations have been extensive [31, 33–37]. Slipped DNAs used in repair assays were structurally identical [71]. All of the 20 excess CTG repeats extruded together as an intrastrand hairpin from a single unique point off-center from the $(CAG)_{30}$, whereas all of the 20 excess CAG repeats extruded together as an unpaired random coil from a single unique point off-center from the $(CTG)_{30}$. Importantly, the repeats are present at the tip, stem, junction, and opposing strand of the heteroduplexes. (B) Moore et al. [52] have used either $(CTG)_{10}$ or $(CAG)_{10}$ intrastrand hairpins formed in the absence of their complementary repeat-containing strand; hence, these structures did not contain slipped junctions. (C) Corrette-Bennett et al. [55] used random-sequence heteroduplex loops or intrastrand pseudo-hairpins containing some CTG repeat units. However, CTG repeats were present only in the intrastrand stem, but not the hairpin tip; these also lacked slipped junctions. CAG repeats were not studied. (D) Owen et al. [57] used $(CAG)_{13}$ intrastrand hairpins in heteroduplexes, which lacked slipped junctions. CTG repeats were not studied.

adopt very stable single-stranded structures [31, 40, 45, 46]. The CTG slip outs predominantly formed intrastrand hairpins, whereas the CAG slip outs predominantly formed unpaired single-strand loops [36, 37] (Fig. 41-2A). Three-way slipped junctions are another critical feature of slipped DNAs. The sequences at four-way Holliday junctions can dramatically alter their structural conformation and ability to be recombined [27]; in a similar manner, the slipped-junction conformation may determine their repair outcome. Slip outs extrude from distinct points along the Watson–Crick repeat duplexes. Slipped-junction conformations are distinct for CTG and CAG slip outs (Fig. 41-2A). Slipped junctions may be fully base-paired or they may contain one or two mismatched bases on the strand opposite the slipped-out repeats (Fig. 41-2A).

Slipped (CTG)·(CAG) structures are biophysically stable and cannot branch migrate, consistent with intrastrand interactions at least at the slip-out junctions [31, 33]. A slip out that is unhindered by intrastrand base pairings may "creep" [47] or "inch" [48, 49] along the Watson–Crick duplex; such slip-out migration would lead to a net translational slippage of the two strands. In this instance, like branch migration of a Holliday junction, migration of a slipped-out loop would require only the energy necessary for the breaking and reforming of the interstrand hydrogen bonds ahead of and behind the slip out, respectively. However, if intrastrand interactions exist within the slipped-out strand (such as intrastrand hairpins or limited pairings at the junction), then migration of the slip out would be a thermodynamically costly process [50], requiring that both the intrastrand base pairs as well as the base pairs at the slip-out junction be broken and reformed.

Several features of slipped DNAs make them suitable models for mutagenic intermediates of repeat instability, namely, the propensity of slipped-structure formation correlated with both repeat tract length and repeat tract purity, both known determinants of genetic instability. The amount of S-DNA formed by a genetically stable length of 17 repeats increased from 2% to 70% for a diseased length of 50 or more repeat units [31, 34]. Interruptions in the repeat tracts of SCA1 and SCA2 are genetically stable relative to their pure tracts of the same length. Interruptions in the repeat tract dramatically reduce the propensity to form slipped DNAs, which suggests that interruptions may provide genetic stability by ablating the formation of mutagenic intermediates [35]. Moreover, slipped DNAs formed by trinucleotide repeats were remarkably stable, biophysically, and unlike other unusual DNA structures (cruciforms, Z-DNA, or triplexes) their extrusion and sustainment do not require superhelical tension, suggesting that they may form *in vivo* [31]. Thus, the structural features of true slipped DNAs, including conformations of the slip outs, and the slipped junctions may permit aberrant processing of these mutagenic intermediates in a manner that leads to repeat length mutations.

B. Repair of Heteroduplexes with Components of Slipped DNAs

Genome integrity is maintained by various DNA repair pathways. Repair machinery recognizes and processes varieties of DNA damage, including mismatched, unpaired, and chemically adducted DNA. Multiple repair pathways, including mismatch repair (MMR), nucleotide excision repair (NER), and base excision repair (BER), with distinct and overlapping mechanisms are responsible for repairing these damages. Determination of the contribution of DNA repair to trinucleotide repeat instability is crucial to understanding this disease-causing mutation. Most models of repeat instability presume that slipped-DNA intermediates escape repair, leading to expansions and deletions [5, 51–54]. On the other hand, the possibility may exist that slipped repeats carrying unique structural features are repaired in an error-prone fashion, resulting in products with lengths differing from either strand. It has been reported that heteroduplex DNAs composed of $(CTG)_{10}$ or $(CAG)_{10}$ escaped repair in a yeast model system [52]. However, the heteroduplexes used in this study did not contain all components of slipped DNAs, in that they contained only CAG or CTG repeats in the loop strand and lacked complementary repeats in the non-slipped strand (Fig. 41-2B). Another yeast study used a pseudo-inverted repeat with a stem composed of four sets of paired CTG units, but hairpin tips and junctions lacking either CTG or CAG units [55] (Fig. 41-2C). The study found that this structure, like a base-paired inverted repeat, escaped processing in either mitotic yeast or yeast extracts, relative to an unpaired loop composed of a random sequence. Similar DNA substrates were used to reveal binding by repair proteins [56, 57] (Fig. 41-2D). The absence of critical biophysical features of true slipped DNAs from the previous may have missed important recognition and or processing outcomes, for example, a true slipped junction rather than a three-way junction with only one arm composed of CNG repeats. A true slipped structure has three repeat-containing arms: two that are Watson–Crick and one that is only Crick (or only Watson). Incomplete base pairings might exist at junctions of slipped (CTG)·(CAG) repeats [58]. These mismatched base pairs may elicit different capacities for binding affinity and may alter downstream effects, such as levels of ADP→ATP exchange. The processing of true slipped DNAs that have been extensively characterized at the biophysical level has been reported and is described next.

III. PROCESSING OF UNPAIRED DNAs

A. Nucleotide Excision Repair and Repeat Instability?

The nucleotide excision repair (NER) pathway involves the formation of unwound regions of DNA encompassing long DNA stretches (~30 nt) and processing of these by structure-specific enodonucleases critical for NER [59]. The NER pathway is essential for the removal of bulky DNA lesions by ultraviolet (UV) light and certain DNA-distorting lesions induced by chemicals. It consists of damage recognition and unwinding of the distorted strand followed by incision of the damaged strand. The gap generated by the excision is restored through repair synthesis [60–62]. In addition to repair of chemical damage, NER proteins have been implicated in other DNA processes. In yeast, but not humans [63], processing of random-sequence heteroduplex loops required the NER protein Rad1 [64]. In human cells the processing of triplex DNAs requires functional NER proteins [65]. Few studies have implicated the NER pathway in repeat instability. Bacterial models suggest roles for NER in either protecting or enhancing large (CTG)·(CAG) deletions, phenomena that may depend upon transcription, replication direction, and repeat tract length [56, 66]. Bacterial models revealed that transcription-increased large deletions [56, 66–68] showed structure-specific binding of purified bacterial UvrA2 to CTG or CAG hairpin heteroduplexes with 1, 2, or 17 repeats. Binding was unaffected by hairpin length, suggesting that the bound determinant was common to each substrate, possibly the three-way junction. In contrast to bacteria, in yeast NER, deficiencies in Rad1 had no effect on the trinucleotide repeat (TNR) instability of CTG repeats [69]. Moreover, in humans, xeroderma pigmentosum (XP), a disease showing UV-induced skin neoplasms caused by defects in NER proteins, TNR instability at the DM1, SBMA, MJD, or HD loci was not observed [70]. Other results did not detect a role for the structure-specific endonucleases XPF or XPG in the processing of slipped (CTG)·(CAG) repeats [71], but this finding cannot exclude the contribution of NER proteins upstream of slip-out processing.

B. Random-Sequence Heteroduplex Repair

DNA loops or heteroduplexes formed at nonrepetitive DNA sequences and are thought to arise *in vivo* during deletion events or recombination events. If not repaired, loops act as mutagenic precursors to deletion or insertion mutations. Thus, repair of these structures is crucial for genome integrity. DNA loop repair has been studied by using *in vivo* assays [55, 72–75] and *in vitro* assays using cell-free extracts [55, 63, 76–80]. In these studies, loop-dependent and nick-dependent repair modes have been observed. In loop-directed repair, the loops are excised from the covalently closed circular repair substrates [81], whereas in nick-directed repair, the nick is used as a guideline for excision and subsequent synthesis used the nonnicked strand as a template for repair [55, 63, 81]. Loop repair can proceed from either the 3' or 5' to the loop [78, 82], and loop length plays a major role in the repair process [78]. Although the 12-nt loops are MMR-dependent, the larger, 216-nt loops are MMR- and NER-independent [55, 63, 81]. Because the repeat-containing loops or hairpins are structurally different from the nonrepeat heteroduplexes, it is of interest to know whether human repair machinery processes the slip out of CTG and CAG in the same way that it processes the nonrepeat structures.

C. Mismatch Repair and (CTG)·(CAG) Instability

Various bacterial, yeast, fly, and mouse model systems have been used to investigate the role of MMR proteins in (CTG)·(CAG) repeats. Conflicting conclusions have been drawn regarding the role of MMR in TNR instability. Bacterial systems have shown decreased [83] and increased [84] levels of deletion events. The difference between the two is likely the use of longer and shorter repeat tracts, respectively. Studies with yeast have also yielded mixed results, with some reports claiming no effect [69, 85, 86] and others claiming destabilization [87]. Strong evidence suggests that mammalian MMR is required for (CTG)·(CAG) expansions, and for this purpose the focus will be on mammalian systems.

Genomewide microsatellite instability in tumors from hereditary nonpolyposis colorectal carcinoma (HNPCC) patients is predominantly caused by mutations in the

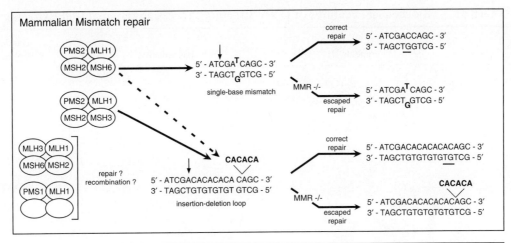

FIGURE 41-3 Mammalian MMR proteins and DNA lesion recognizing proteins. The upper panel shows the protein complex involved in repairing base–base mismatches and insertion–deletion loops. The lower panel summarizes the repair status in tumors or mice deficient in various mismatch repair proteins and their effect on genome integrity and repeat instability. For details see text.

hMSH2 and hMLH1 genes—mutations that result in a nonfunctional MMR system (Fig. 41-3). The mammalian repair pathway consists of a number of proteins, which work in concert to recognize and repair mismatched DNAs [88]. MMR proteins form two related protein complexes: MSH2-MSH3:MLH1-PMS2 and MSH2-MSH6:MLH1-PMS2 (Fig. 41-3). (Unless noted, nomenclature refers to the murine–human MMR genes.) The DNA damage-recognizing protein complexes MSH2-MSH3 and MSH2-MSH6 have overlapping but not identical binding specificity. Although both bind to single base insertion–deletion intermediates, multiple nucleotide insertion–deletion loops (up to 13 nt) were bound only by the hMSH2-hMSH3 complex [89, 90] and the G–T base–base mismatch was bound only by the hMSH2-hMSH3 complex [89]. The binding preferences of these distinct complexes reflect their roles in the repair of base–base mismatches and short insertion–deletion loops at most simple repeat tracts [88]. MSH2 and MLH1 are absolutely required for base–base and short loop repair (Fig. 41-3 and Table 41-1). It is widely believed that DNA lesion recognition complexes MSH2-MSH3 and MSH2-MSH6 function as a tetrameric complex with MLH1-PMS2, MLH1-PMS1, or MLH1-MLH3 [91]. Good evidence supports a role for PMS2 in the MMR pathway; the repair role of PMS1 and MLH3, both present in somatic and germline tissues, is less clear [92, 93]. In addition to mismatch repair, several of the eukaryotic MMR proteins participate in non-MMR DNA metabolic processes, such as damage surveillance, transcription-coupled repair, and recombinogenic and meiotic processes, as well as site-specific somatic hypermutation and class-switch recombination at immunoglobin genes [94].

TABLE 41-1 Biochemical Features of DNA Structure Processing (Human Cell Extracts)

	Slipped (CTG)·(CAG) repeats	Base–base mismatch	Random-sequence heteroduplex (2–429 nt)
Structure specific	Yes	Yes	Yes
Slip out sensitive	Yes (CAG > CTG)	G–T > G–G = A–A = C–T > A–C = A–G > T–T >> C–C	?
Length sensitive	?	Yes (1–12)	Yes (12–216)
Expansion versus contraction (insertion versus deletion)	Yes	No	?
Nick directed	Yes	Yes	Yes[a]
5′ > 3′	Yes	No	No
Bidirectional	Yes	Yes	Yes
Localized	Yes	Yes	Yes
Loop directed (loop excision)	No	N/A	Yes
Error-prone	Yes (only for slip out on nicked strand)	No	?
ATP dependent	Yes	Yes	Yes
ddTTP insensitive (polymerase β is not required)	Yes	Yes	Yes
Aphidicolin sensitive (polymerase α, δ, or ε is required)	Yes	Yes Polymerase δ	Yes
MMR–NER dependent	No, no	Yes, no	Yes, no[b]
SH-SY5Y	Yes, all outcomes	Yes	?
SH-SY5Y + retinoic acid neuronlike	Yes, all outcomes	Yes	?

[a] Also loop-directed.
[b] Up to 12 nucleotides MMR-dependent.

A requirement for MSH2 and MSH3 in (CTG)·(CAG) expansions (not deletions) has been provided by transgenic mice. The effect of MMR deficiencies on (CTG)·(CAG) instability in mice is summarized in Fig. 41-3 (lower panel). Spontaneous expansions of a tract of 112 (CTG)·(CAG) repeats were abolished (stabilized) when mice were crossed to an MSH2-deficient background, as shown by the Messer group [95]. Similarly, tracts of 84 (CTG)·(CAG) repeats were somatically stabilized and destabilized when crossed with MSH3- and MSH6-deficient mice, respectively [96]. These results suggest a role for MSH2 and MSH3, but not for MSH6, in the maintenance of long (CTG)·(CAG) repeat tracts. The enhanced CTG destabilization (expansions) in the MSH6-deficient mice might suggest that a complex of MSH2-MSH3 could be required for the expansions [96]. (This assumption was based upon the potential interdependence of MSH2-MSH3 levels on MSH6 levels.) In contrast to Wieringa and colleagues [96], a more recent study reported CAG stabilization, not destabilization, in an MSH6-deficient background [57], but confirmed the stabilization in the absence of MSH3.

Gourdon and colleagues [97] investigated the involvement of various DNA repair proteins in (CTG)·(CAG) repeat instability by using a DM1 transgenic mouse model, which carries a large repeat (>300 units). In contrast to the HD and DM1 knock-in mice [95, 96], the Gourdon group [97, 98] found that the absence of MSH2 did not result in the stabilization of CTG repeats. Rather, it changed from expansions to deletions of similar frequencies. This applied to both germline and somatic instability. Although Msh2−/− HD mice showed stable transmission of the TNR tracts, somatic instability, evident as deletions, was observed [99]. Similarly, the HD mice show a requirement of MSH2 for expansions but not for deletions [100]. So far, in four independent transgenic (CTG)·(CAG) mice it seems that MSH2, and in two cases MSH3, is required for expansions. In two of the four mice, repeat deletions are observed in the absence of MSH2 [97, 98, 101]. Thus, it seems that expansions are MSH2-dependent and the arising deletions may reflect the requirement of MSH2 for either correct repair of deletion events or protection from their occurrence.

Toward identifying a possible role of MutL homologues as modifiers of trinucleotide repeat dynamics, Monckton and colleagues [102] crossed Pms2-deficient mice with mice carrying a (CTG)·(CAG) tract with >160 units. In contrast to MSH2 and MSH3 deficiencies, mice deficient in PMS2 still displayed CTG expansions, but at reduced levels compared to PMS2-proficient mice. The PMS2-deficient mice also showed an increased frequency of very rare, but large deletions (>20 repeats) in a few cells (detectable only by small-pool polymerase chain reaction). These findings reveal that in addition to MMR enzymes that directly bind mismatched DNA, proteins that are subsequently recruited to the complex also play a central role in the accumulation of repeat length changes. Interestingly, complexes of hMLH1:hPMS2 but not hMLH1:hPMS1 can bind both double- and single-stranded DNA [93]. The reduced effect of a PMS2 deficiency upon TNR instability relative to a deficiency of either MSH2 or MSH3 may be due to a degree of redundancy, possibly through the participation of PMS1 or MLH3 in lieu of PMS2. Such functional redundancy may also account for the reduced effect of Pms2 deficiency on mononucleotide repeat instability relative to an Mlh1 deficiency [103]. Although some MMR proteins are required for (CTG)·(CAG) expansions, their role in this process is not evident from the *in vivo* data alone.

IV. BINDING OF MMR PROTEINS TO TRINUCLEOTIDE REPEATS

Repair of unpaired or damaged DNA typically begins with recognition of the unpaired site by a DNA structure-specific binding protein. For MMR, this role is filled by MSH2 and its associated partners. In humans, a strong correlation exists between exchange efficiencies (ADP→ATP) and ATPase activity of hMSH2 complexes [104] and repair efficiencies [105–107] of base–base mismatches and short insertion–deletion loops of random sequences. For example, the G–T mismatch is the best repaired and, when bound by MMR proteins, leads to efficient ADP→ATP exchange and ATPase activity of MSH2. In contrast the C–C mismatch is very poorly repaired; even though the C–C mismatch is bound by MMR, this binding does not yield any ADP→ATP exchange. Thus, although binding to a DNA substrate does not necessarily imply repair, exchange would, however, definitely commit to repair, *where repair outcome is known*. However, for mutation processes like (CTG)·(CAG) expansions, ADP→ATP exchange or lack thereof has undefined consequences. Notably, MMR proteins bind to fully duplexed DNA without any ADP→ATP exchange: could or should anything be inferred here? DNA binding and ATPase hydrolysis may be interpreted very differently for different DNA substrates, particularly for those substrates that might be processed to mutation products.

Several reports observed binding of MMR proteins to trinucleotide repeats. In 1997, MSH2 binding was shown to slipped DNAs but not to duplex repeats [33]. Furthermore, MSH2 bound preferentially to CAG slip outs (K_D = 50–200 nM), whereas CTG slip outs bound poorly. This implicates a strand asymmetry in MSH2 recognition. Furthermore, binding affinity increased with the length of the repeat sequence, which paralleled with the repeat instability in humans. This direct association with repeat number in the slip outs suggested that hMSH2 was binding to the A–A mismatches of the CAG slip outs [33]. A more recent study reported binding of the human hMSH2-hMSH3 proteins to CAG hairpin heteroduplexes, which lacked true slipped junctions [57] (Fig. 41-2D). Binding affinity for the CAG hairpin was high (K_D = 7–18 nM); a CTG hairpin was not tested. This study confirmed the previous observations made with hMSH2 alone [33], in that binding affinity increased with increasing CAG repeat length. hMSH2-hMSH3 binding protected the CAG hairpin; CTG hairpins were not studied. Could the presence of a true slipped junction, with the potential presence of unpaired repeats at the junction (Fig. 41-2A), affect the footprint or downstream effects? What might be the biological significance of MSH2 or MSH2-MSH3 binding to trinucleotide repeat DNAs?

Interestingly, binding of hMSH2-hMSH3 to a CAG hairpin that lacked a slipped junction showed reduced ATPase activity relative to a preferred short (CA)$_4$ loop [57]. MSH2-MSH3 directs the repair of insertion–deletion loops of up to 13 nucleotides, a process that is tightly linked with DNA-binding-dependent ATPase activity of MSH2-MSH3, which is regulated by a mismatch-stimulated ADP→ATP exchange [90]. One interpretation of this is that the ATPase activity of MSH2 is not required for its role in (CTG)·(CAG) expansions. A test of this hypothesis would be to cross (CTG)·(CAG) transgenic mice with mice harboring ATPase-defective MSH2 protein that retains full mismatch binding activity [108]. If binding but not ATPase activity were sufficient, these mismatch-repair-deficient mice should display spontaneous (CTG)·(CAG) expansions, unlike MSH2 nullizygous mice [95].

The significance of hMSH2 binding alone or complexed with hMSH3 to either repair or instability remains unclear, particularly because the correct or escaped processing of slipped repeats occurs independent of either protein [71] (see the following section). Although many possible interpretations exist, the lack of ATPase may reflect the possibility that binding does not lead to repair. However, one should be cautious of such interpretations with DNA substrates that lack critical components of slipped DNAs. Importantly, mechanistic interpretations of any kind would be best supported by repair data. The exact role that any of the MMR proteins is playing in causing (CTG)·(CAG) repeat instability is not known. Despite the evidence that MMR can modulate the (CTG)·(CAG) instability in bacteria, yeast, or transgenic mice [51], no direct study links the MMR repair proteins or pathways to the processing of slipped repeats.

V. PROCESSING OF SLIPPED (CTG)·(CAG) REPEATS BY HUMAN CELL EXTRACTS

An understanding of how slipped DNAs, which are likely mutagenic intermediates of repeat instability, are processed by human proteins may prove important to understanding the mechanism of this important mutation. The fidelity and efficiency of processing of slipped DNAs with slipped junctions and slip outs of $(CAG)_{20}$ or $(CTG)_{20}$ by human cell extracts were determined. Three repair outcomes were observed: correct repair, escaped repair, and error-prone repair. Each repair path depended upon nick location and slip-out composition (CAG or CTG). Interestingly, error-prone repair was observed only in the expansion substrates. In these substrates the excess repeats were incompletely excised, presenting a novel path to generate expansions but not deletions. Neuronlike cell extracts yielded same results, supporting the possibility that these processes are involved in (CTG)·(CAG) instability in patient postmitotic brain cells. Known mismatch and nucleotide excision repair proteins such as hMSH2, hMSH3, hMLH1, XPF, or XPG were not involved. These results may explain the different mutation patterns occurring between various disease loci or tissues.

Slipped DNAs modeled expansion and deletion intermediates (Fig. 41-4) that can arise at any one or more of the DNA metabolism stages, such as replication forks, nicked–damaged DNA, and/or recombination or transcription sites (Fig. 41-1). Briefly, slipped-out CAG or CTG structures formed between complementary strands with 30 and 50 repeats were used as substrates in an *in vitro* repair assay. Each substrate carries a nick uniquely placed proximal to the slip out and located

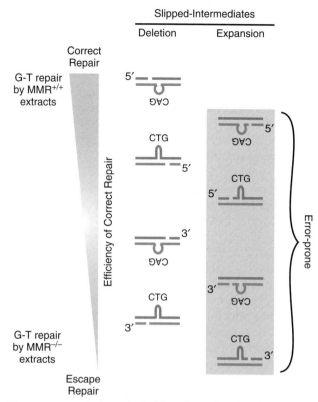

All processes are independent of the mismatch and nucleotide excision repair proteins (MSH2, MSH3, MLH1, XPF and XPG).

FIGURE 41-4 Summarized efficiency of correct repair of slipped DNAs containing an excess of 20 CAG repeats [$(CTG)_{30}$·$(CAG)_{50}$] or an excess of 20 CTG repeats [$(CTG)_{50}$·$(CAG)_{30}$] [71]. CAG and CTG slip outs assumed loop and hairpin conformations. In all cases, slip outs in the continuous strand are deletion intermediates, whereas slip outs in the nicked strand are expansion intermediates. Nicks were located either 5′ or 3′ of the slip out. The hierarchy of repair efficiency for the eight slipped substrates is shown. As a gauge, the high levels of G–T mismatch repair (28%) by MMR-proficient extracts and the poor repair of a G–T mismatch by MMR-defective cell extracts were used. Generally, the efficiency of correct repair was better for deletion intermediates (slip outs in the continuous strand) than for expansion intermediates (slip outs in the nicked strand). CAG slip outs were repaired with higher efficiencies than CTG slip outs. Slip outs with 5′ nicks were repaired better than slip outs with 3′-nicks. All expansion intermediates, but not deletion intermediates, gave rise to error-prone repair, as shown by the bracket. Neither MMR or NER proteins or their respective pathways played any role in the repair process. In the absence of hMSH2, hMSH3, hMLH1, XPF, or XPG or greatly reduced levels of hPMS2, hPMS1, or ERCC1, repair fidelities and efficiencies were indistinguishable from those obtained with the repair-proficient HeLa extracts. Similar outcomes were produced for human neuroblastoma cell extracts either before or after differentiation to postmitotic neuronlike cells.

either upstream (5′) or downstream (3′) of the slip out. Substrates carrying the slip outs on the continuous strand opposite the nick are deletion intermediates (Fig. 41-4), whereas substrates carrying slip outs in the same strand as the nick are expansion intermediates

(Fig. 41-4). Slipped structures were biophysically identical to those characterized previously (Fig. 41-2A) [37, 58, 109]. The difference of 20 repeats between the 30 and 50 complementary repeat strands mimicked the mutagenic length changes associated with Huntington's disease and myotonic dystrophy families and tissues [1a, 2, 3, 6, 8, 10, 11, 110].

Human cell extracts used as sources for slipped-DNA processing activities were competent in many forms of replication, repair, and recombination. Cell extracts are competent in base–base mismatch repair [106], large loop repair [82], base excision repair, psoralens cross-link repair, pyrimidine dimer and other UV damage repair, bulky base damage [111] double-strand-break repair [112, 113], homologous recombination [112], and triplex-mediated recombination.

At replication forks, slip outs of the newly replicated (nicked) or template (continuous) strands (Fig. 41-1) are intermediates of expansion or deletion events, respectively. Similarly, in nonreplicating DNAs (sites of damage or recombination), slippage of the nicked or continuous strand (Fig. 41-1) would lead to expansion or deletion intermediates, respectively. Correct repair of these intermediates would use the continuous (progenitor) strand as a template for repair. Most models of repeat instability presume that these intermediates escape repair, yielding expansions and deletions.

A. Three Repair Outcomes: Correct, Escaped, and Error-Prone Repair

Three repair outcomes were observed: correct repair, escaped repair, and error-prone repair. As outlined in Fig. 41-5 (top panel), correct repair involved slip-out recognition and DNA excision directed to the nicked strand that proceeded across the slip out (deletion intermediates) or excised the slip out (expansion intermediates), and the resulting gap was filled in and ligated, yielding a correct repair product with the same number of repeats as the continuous strand. In contrast, escaped repair involved only minimal processing of the slipped DNA, limited to ligation of the nick (Fig. 41-5, middle two panels). Error-prone repair occurred only for expansion intermediates (slip outs on the same strand as the nick) (Fig. 41-5, lower panel). Correct repair of expansion intermediates requires excision of the excess repeats contained in the slip out, followed by gap filling and ligation. However, many attempts to excise the slipped-out repeats were aborted, but the resulting gap was filled and ligated to yield products that retained some excess repeats. These error-prone repair products were a series of slipped intermediates much like the starting slipped DNA, but they contained a range of excess repeats. The number of excess repeats varied from 1 through to the total number of repeats in the starting slip out. For example, a starting slip out of $(CTG)_{50} \cdot (CAG)_{30}$ having an excess of 20 CTG repeats yielded a series of error-prone repair products containing an excess of 19, 18, 17, 16, 15, 14, ..., 31 CTG repeats. Correct repair products would contain 30 repeats in both strands, whereas escaped repair would retain the full slip out of 20 CTG repeats. These bands were error-prone repair products, where the excision step of the slipped-out repeats is faulty while the detection, strand selection, synthesis, and ligation steps are intact. Although the excision step is incomplete, the subsequent steps through to ligation are completed. Thus, this was termed *error-prone* repair, rather than *incomplete* or *aborted* repair. In this manner, a novel path to expansions is revealed: error-prone repair only arose from expansion intermediates (slip outs in the nicked strand) and products retained an excess of repeats, which may lead to expansions.

A hierarchy of repair efficiencies emerged based upon the sequence composition of the slip out, the location of the nick relative to the strand harboring the slip out, and the nick polarity to the slip out (Fig. 41-4). CAG slip outs were correctly repaired better than CTG slip outs. Deletion intermediates (slip outs in the continuous strand) were correctly repaired better than expansion intermediates (slip outs in the nicked strand). Slip outs with 5′ nicks were correctly repaired more efficiently than slip outs with 3′ nicks. The most efficiently repaired slip outs were processed at efficiencies as great as G–T mismatches by MMR-competent cell extracts, and the most poorly repaired substrates were processed at levels as low as G–T mismatches by MMR-deficient cell extracts. The G–T mismatch is the most efficiently repaired base–base mismatch (Fig. 41-4) [105, 106]. The differential repair of CAG slip outs relative to CTG slip outs may be due to their different biophysical features (unpaired random coils versus intrastrand hairpins) and/or to the distinct slip-out junction conformation each can assume (Fig. 41-2A) [58]. It has been shown previously that single-strand binding protein preferentially binds to CAG slip outs [58], a preference that may reflect the repair outcomes. The poor repair of perfectly paired hairpin substrates [114] may be related to the escaped repair observed here. The increased repair efficiency of deletion substrates over expansion substrates may explain the expansion bias of these sequences in diseased families. The error-prone repair of expansion substrates, but not of deletion substrates, also serves as an explanation for the expansion bias. Thus, several features of slipped $(CTG) \cdot (CAG)$ processing support the involvement of these processes in disease-associated repeat instability.

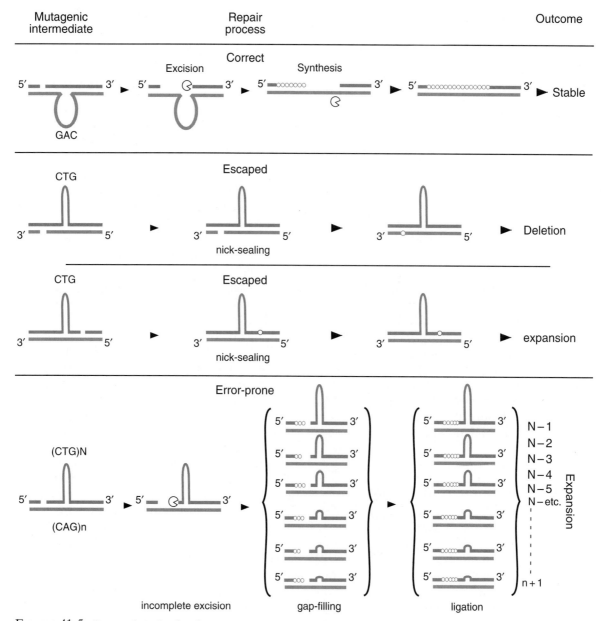

FIGURE 41-5 Proposed mechanism for correct, escaped, or error-prone repair of slipped-DNA intermediates, with newly incorporated regions of specific and nonspecific synthesis indicated by dots. Correct repair is directed to the continuous strands, using the continuous strand as a template for repair, and would stably maintain the repeat. Escaped repair is limited to the sealing of the nick and could lead to deletion or expansion products. Error-prone repair, where the fault is incomplete removal of excess repeats, with strand discrimination, gap-filling, and ligation steps left intact, could lead to expansions. In error-prone repair, the number of excess repeats retained ranges from $n + 1$ to the total amount of 20 repeats (N) present in the starting slip out. Error-prone repair has been detected only for expansion substrates (slip out in nicked strand). All processes are independent of hMSH2, hMSH3, hMLH1, XPF, and XPG. Similar outcomes were produced for human neuroblastoma cell extracts either before or after differentiation to postmitotic neuronlike cells.

B. Slipped-DNA Repair Is Independent of MMR and NER Proteins

Repair of slipped DNAs was independent of MMR proteins because extracts of human cells deficient in specific mismatch repair proteins processed slipped repeats with equal efficiency and fidelity (Fig. 41-4, Table 41-1). Similarly, NER played no role in the repair process. In the absence of hMSH2, hMSH3, hMLH1, XPF, or XPG or greatly reduced levels of hPMS2, hPMS1, or ERCC1, repair outcomes were indistinguishable from those obtained with the repair-proficient

HeLa extracts (Fig. 41-4, Table 41-1). Thus, neither MMR nor NER proteins are involved in the processing of slipped (CTG)·(CAG) repeats.

C. Neuronlike Cell Extracts Process Slipped DNAs

Repeat instability in patient tissues occurs in both proliferating and nonproliferating tissues, such as neurons. Somatic CAG expansions actively occur in the brains (striatum) of patients with Huntington's disease, dentatorubral pallidoluysian atrophy, and spinocerebellar ataxia types 1, 2, and 3, as well as in several MMR-proficient, but not MMR-deficient, transgenic mice with expanded CAG tracts [(reviewed in 1a), 96, 97, 101, 102, 110, 115]. This ongoing mutation may contribute to the neuronal specificity and progressive nature of pathology for some diseases. The cerebellum in many CAG expansion diseases shows repeat deletions. By using extracts of the human neuroblastoma (SH-SY5Y) cells before and after terminal differentiation by neuronlike cells, it was shown that correct, escaped, and error-prone repair outcomes can occur at the same efficiencies [71]. Thus, both proliferating and postmitotic cells harbor the same DNA activities to process slipped DNAs. This reveals that, even with the repair capacity changes associated with neural differentiation [116, 117], repair fidelity and efficiency seemed unaffected. Thus, the aberrant processing of slipped (CTG)·(CAG) repeats can contribute to their instability in neuronal tissues and may in this manner augment pathogenesis.

D. Mechanism of Correct Repair

The correct repair of slipped DNAs overturns the widely held presumption that they will escape repair [57]. The levels of correct repair were comparable to the high efficiency of G–T mismatch repair performed using the same extracts (~27% versus ~28%) (Fig. 41-4). It is noteworthy that the efficiency of slipped-DNA repair was high, similar to G–T mismatch repair, revealing the importance of this process. Thus, mutagenic events may constantly be occurring but are rarely detected as they are frequently repaired with high fidelity. Correct repair of slipped DNAs is independent of MMR and NER proteins and, hence, is independent of these repair pathways (Fig. 41-6). Which proteins or repair pathways are involved in slipped-DNA repair is not known, but may have some overlap with the unknown factors involved in random-sequence heteroduplex repair—a process that is independent of several MMR proteins.

E. Mechanism of Escaped Repair

Slipped repeats have long been expected to escape repair; however, only some slipped DNAs escaped repair. CTG slip outs are biophysically different from CAG slip outs. One might expect that they would be processed in a different way. CTG slip outs in deletion intermediates are poorly repaired (less than 5%) at six- to eightfold less compared to the CAG slip-out substrate (Fig. 41-4, lower left). This low level of repair was comparable to that of a G–T mismatch by MMR-defective cell extracts. Particularly poor repair occurred for CTG slip outs with 3' nicks (Fig. 41-4, lower right and left). This poor repair does not require binding of mismatch repair proteins hMSH2 or hMSH3, indicating that if escaped repair is the result of protection by a bound protein those proteins are not MMR proteins. Slipped (CTG)·(CAG) repeats may be masked from repair by bound proteins, but these are not MSH2/3. Thus, the *in vitro* binding of slipped DNAs or CAG hairpins by MSH2 or MSH2-MSH3 is unlikely to reflect a masking function [33, 57] (Fig. 41-6). If human MMR proteins are involved in (CTG)·(CAG) expansions their role is distinct from the processing of slipped intermediates, but may be involved in their formation (Fig. 41-6). Furthermore, the structure-specific NER endonucleases, XPF or XPG, also are not required for escaped repair of slipped DNAs.

F. Mechanism of Error-Prone Repair

Why did error-prone repair arise only from expansion intermediates and not from deletion intermediates? Error-prone repair arising only from processing of expansion intermediates (slip outs on the nicked strand) was distinct from processing of deletion intermediates (slip outs on the continuous strand). Excision of DNA across a slip out may be dramatically different from excision of a slip out. These different excision processes must occur to yield correct repair of deletion intermediates and expansion intermediates, respectively. Correct repair of deletion intermediates involves excision across slip outs, whereas correct repair of expansion intermediates involves excision of the slip out. The pattern of radio-incorporation repair products of expansion intermediates was strikingly different from that of deletion intermediates: in addition to correctly repaired homoduplex 30-repeat-containing products, a series of faster electrophoretic ladder DNA bands extended from the starting SI-DNA to the correctly repaired duplex with 30 repeats in both strands.

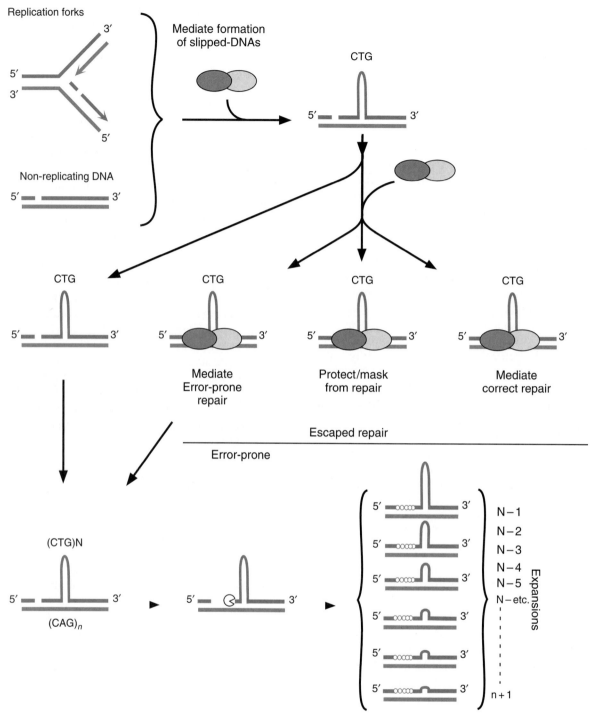

FIGURE 41-6 Proposed mechanism for the role of repair proteins in repeat expansions. Repair proteins may be involved in the formation of expansion intermediates. Once formed, the expansion intermediate might undergo error-prone repair to slipped products that have retained some excess repeats from the original slip out. Alternatively, repair proteins might be involved in mediating error-prone repair, or repair proteins may bind to the expansion intermediate and protect or mask it from repair, thereby permitting it to escape repair. Finally, repair proteins may be required to mediate the correct repair of the expansion intermediate. For a preformed expansion intermediate, the processes of error-prone repair, escaped repair, and correct repair are independent of the human mismatch repair proteins hMSH2, hMSH3, and hMLH1 and the nucleotide excision repair proteins XPF and XPG [71]. Thus, if the human MMR proteins are required for expansions, as are the murine proteins MSH2 and MSH3, the data suggest that their role may be limited to only mediating the formation of the expansion intermediates.

Error-prone repair products were detected only by radioincorporation, indicating that on a molar basis the production of any one species was modest, but the level of incorporation per molecule was considerable. Thus, whereas both deletion and expansion intermediates could be correctly repaired and escape repair, only expansion intermediates yielded error-prone repair. Like correct and escaped repair, error-prone repair showed no involvement of MMR and NER pathways because the cells lacking the MMR and NER proteins also generated similar error-prone repair products. Thus, if human MMR proteins are involved in (CTG)·(CAG) expansions, their role is distinct from the processing of slipped intermediates, but may be involved in their formation (Fig. 41-6). Although the DNA structure-specific endonucleases XPF and XPG are not involved, some other enodnuclease(s) or exonuclease(s) is likely to be responsible for the production of error-prone repair of slipped (CTG)·(CAG) repeats. Determination of what these are will require further experimentation.

G. Comparison of Slipped (CTG)·(CAG) Processing with Base–Base Mismatches and Random-Sequence Heteroduplexes

Processing of slipped CTG and CAG slip outs shows similarities and dissimilarities with the processing of base–base mismatches and random-sequence heteroduplexes, as summarized in Table 41-1. Similarities include the ability of nicks to direct repair to the nicked strand and to use the continuous (nonnicked) strand as a template for repair.

The increased repair efficiency of repeat slip outs with 5′ nicks over 3′ nicks may be a trait unique to slipped (CTG)·(CAG) repeat structures and not shared by base–base mismatches or random-sequence heteroduplexes [79]. Previous studies of random-sequence heteroduplexes claimed a strong preference for 5′ nicks [63, 81]; however, a more recent report found that this was artifactually due to the use of gpII endonuclease for substrate production [79]. Thus, the repair efficiency of either base–base mismatches or random-sequence heteroduplexes seems to depend only upon the nature of the unpaired sequence and not upon polarity of the nick [79]. In contrast, repair of slipped (CTG)·(CAG) DNAs did show increased repair efficiency for 5′ nicks over 3′ nicks (substrate preparations did not use gpII endonuclease). The poor repair of slip outs with 3′ nicks relative to 5′ nicks suggests that 3′→5′ excision is less processive than 5′→3′ excision, whereas the escaped repair of CTG slip outs opposite a 3′-nicked CAG strand may be due to inefficient excision.

The preferential repair of CAG slip outs over CTG slip outs may be due to differential recognition of the distinct structures assumed by each. The striking biophysical differences between slipped DNAs with slip outs of CAG or CTG repeats, including junctions and slip-out conformations, may subject them to very different processing efficiencies and fidelities. The poor repair of CTG slip outs, which assume intrastrand hairpin conformations, may be related to the inability of perfectly paired hairpins to be repaired in yeast [114]. Different base–base mismatches, which vary in repairability [79] assume different biophysical structures, and such structural differences correlate with differential abilities to be recognized and processed by mismatch repair [79, 105, 106, 118–120]. For randomsequence heteroduplex repair, it is unknown whether varying the single-strand conformation (i.e., the same length but different sequence) will affect repair outcome.

Slipped (CTG)·(CAG) repeats do not undergo loop-directed (loop excision) repair as do random-sequence heteroduplexes [55, 63, 81]. The inability of slipped DNAs to undergo loop-directed repair may seem surprising, particularly for CAG slip outs, which have been shown to assume predominantly single-stranded random coil conformations. However, its absence is not surprising because loop-directed repair would be expected to yield a repeat deletion bias, contrary to the expansion bias observed in disease families. The inability of slipped (CTG)·(CAG) repeats to undergo loop-directed repair is consistent with the observations in yeast by Lahue and colleagues, who found that hairpins with and without CTG repeats were subject to reduced levels of loop-directed repair relative to random-sequence heteroduplexes [55].

Slipped (CTG)·(CAG) repeat expansion substrates do undergo error-prone repair (incomplete excision of the slipped-out repeats). Only expansion intermediates, slip outs in the same strand as the nicks, are subject to this process. Slip outs with nicks located either 5′ or 3′ of the slip outs were equally capable of producing error-prone repair products, but the nick must be in the same strand as the slip out. Seemingly, excision of the slipped-out repeats is not the same as excision across the slip out. Whether this phenomenon is unique to slipped (CTG)·(CAG) repeats is not presently known. However, a distinct form of error-prone repair has been reported for random-sequence heteroduplexes transformed into primate cells [74]. There the sequence alterations, mostly from multilooped heteroduplexes, included point mutation, sequence insertions, duplications, and deletions, all of which occurred at or near the base of the looped-out sequence. Excision of random-sequence heteroduplexes and slipped (CTG)·(CAG) repeats may both be error-prone during similar steps but yield distinct products.

Slipped-strand repair is sensitive to aphidicolin, whereas dideoxythymidine triphosphate had no effect, suggesting an involvement of polymerases (pol) α, δ, and/or ϵ, but not polβ or many of the translesion synthesis polymerases such as polη, polι, polκ, polλ or polθ. The lack of a role of the translesion polymerases is consistent with an observation in yeast, which failed to detect altered repeat instability in the absence of lesion-bypass polymerases [121]. The sensitivity and resistance to aphidicolin and ddNTPs are shared by repair of base–base mismatches and random-sequence heteroduplexes. Whether slipped-DNA repair is mediated by polymerase δ, required for mismatch repair [122], is not known.

All slipped-DNA repair outcomes were independent of MMR and NER proteins, which contrasts with the absolute dependence of base–base mismatch repair on the MMR pathway [105, 106]. Similarly, repair of random-sequence heteroduplexes containing 2–12 nucleotides involves both MMR-dependent and MMR-independent components. The MMR-dependent repair is gradually reduced as loop size increased (from ~33% for 2 nt to ~3% for 12 nt) [81]. However, longer random-sequence heteroduplexes with loops from 17 up to 216 nucleotides are efficiently processed independent of MMR and NER proteins [63, 81]. It is not known whether a length limitation of slipped (CTG)·(CAG) repeats exists, for which MMR is and is not required.

Repair of slipped (CTG)·(CAG) repeats could be mediated by extracts of human neuroblastoma SH-SY5Y cells as well as by their neuronlike differentiated (retinoic acid treated) forms [71]. These same cells could also mediate base–base mismatch repair [123]. It is not known how random-sequence heteroduplexes may be processed by these cell extracts.

H. Role of Repair Proteins in (CTG)·(CAG) Instability?

Many repair proteins have been suggested to contribute to (CTG)·(CAG) expansions. These include polymerase β, and several MMR and NER proteins, and in several mice the spontaneous repeat expansions (not deletions) require these proteins. As outlined in Fig. 41-6, these proteins may act at various steps of (CTG)·(CAG) instability. The fact that repair proteins are required for the mutation process argues against the possibility that they are actually acting as repair proteins. The manner through which any repair protein may contribute to (CTG)·(CAG) instability is completely unknown. In contrast to the correction of base–base mismatches, where there is a plethora of biochemical *in vitro* repair data, a mechanistic role for any repair protein in (CTG)·(CAG) mutations can only be hypothesized in the absence of direct processing results. The fact that repair proteins are required for the mutation (not repair) process suggests that the proteins may not be functioning within the repair pathway. Data indicate that MSH2-MSH3 may bind CAG hairpins but does not hydrolyze ATP, the supposed signal for repair [57]. It is possible that binding is sufficient to induce instability. Data also reveal that these proteins are not required for either the correct, escaped, or error-prone processing of slipped (CTG)·(CAG) repeats [71]. Thus, if the role of MMR proteins is similar between mice and humans, the role of these repair proteins must precede that of slipped-DNA formation. Mediation of the formation of expansion (not deletion) intermediates seems to be an attractive role (Fig. 41-6).

It has been suggested that in bacteria the NER system may induce or protect from large (CTG)·(CAG) deletions [56, 124]. Nonrepeat DNA hairpins [59] and triplex DNAs [125] act as substrates for human NER proteins, which may suggest the potential to process slipped DNAs. However, this study did not reveal a requirement for human XPF or XPG in any of the repair outcomes of slipped (CTG)·(CAG) repeats. This suggests that, if human NER proteins are participating, their role must precede that of slip-out processing, possibly facilitating the formation of slipped-DNA intermediates during strand separation step damaged base repair [126] (Fig. 41-6). Polymerase β may contribute in a similar manner. Although polβ does not contribute to the processing of slipped DNAs, it may still be involved in the process of the formation of slipped (CTG)·(CAG) repeats via its strand-displacement activity (Fig. 41-6), as previously postulated [127].

Lack of involvement of MMR proteins in any of the repair outcomes [the correct, escaped, or error-prone repair of slipped (CTG)·(CAG) repeats] is intriguing because several MMR proteins are required for (CTG)·(CAG) expansions in transgenic mice. It is important to note that the escaped repair of certain CTG or CAG slip outs does not require either hMSH2 or hMSH3, thereby ruling out the hypothesis that binding by these proteins protects them from repair [57] (Fig. 41-6). These repair results, in association with the *in vivo* mouse studies [95–98, 101, 102], suggest that the role of MSH2, MSH3, and PMS2 in expansions may mediate the formation of expansion intermediates. This may be possible by displacing the newly replicated or nicked strands (nonmitotic cells), which may occur during error scanning that involves acute kinking of the homoduplex DNA [128] by MutS and yMSH2/yMSH6

(Fig. 41-6). An alternative explanation is that MMR proteins may actively form slipped intermediates at unwound repeats—a process that may be facilitated through their direct interaction with DNA helicases during the unwinding or annealing of DNA strands. MMR proteins are known to interact directly with several helicases, including WRN, BLM, and RecQ1, and these interactions may affect strand separation or reannealing [129–132]. Results suggest that error-prone repair of nicked slip outs with incomplete excision of the excess repeats will yield expansions. That MSH2/3 may actively participate in the formation of the (CTG)·(CAG) expansion intermediates (Fig. 41-6), but not deletion intermediates, could explain the expansion bias and stabilization–deletions observed in MMR-proficient and -deficient (CTG)·(CAG) mice, respectively [95–98, 101, 102]. A similar situation may arise in patients suffering from (CTG)·(CAG) expansion diseases, as they are repair-proficient as well.

On the other hand, even if mice show similar features with respect to instability diseases, they may handle the repair process in an entirely different manner than human proteins, a possibility suggested by the vast dissimilarity in the effects (or lack thereof) of MMR deficiencies on (CTG)·(CAG) instability between species (bacteria, yeast, fruit flies, and mice) [51, 133]. Thus, it is concluded that if, human MMR proteins are participating in the process of (CTG)·(CAG) instability, as in mice, their role must precede that of slip-out processing and may be at the initial step of slip-out processing.

I. Slipped-DNA Processing and Repeat Instability in Human Disease

How does slipped-DNA processing relate to repeat instability in human diseases? Using slipped DNAs containing the structural predicted features of true slipped DNAs, three distinct repair outcomes were observed. Correct repair, which occurred at surprisingly high efficiencies, escaped repair, which was expected, also occurred but only for certain substrates, and a novel form of error-prone repair, which was unexpected, led to products that retained excess repeats. These repair outcomes can, in part, explain how repeats may be stably maintained and display contraction biases or expansion biases. Distinct mutation levels, deletion biases, and expansion biases exist between disease loci or patient tissues.

These distinct repair outcomes may be related to the various mutation levels, deletion biases, and expansion biases observed in different diseases (reviewed in refs. [1a, 4] and [12]) or patient tissues [1a, 6, 8, 10, 11] showing different mutation types (deletion and expansion biases). The three distinct repair outcomes observed may be related to the mutation patterns occurring in patient tissues (Fig. 41-5). For example, in a given Huntington's disease patient, depending upon the tissue, expansion biases (striatum and sperm), deletion biases (cerebellum), and stable repeat maintenance (blood) occur [3, 8, 110]. The (CTG)·(CAG) repeats at different disease loci show different mutation types [2]. Expanded repeats can be stable (as in spinal and bulbar muscular atrophy) [7] or show deletion biases (as in the sperm of SCA8 patients [9] or the cerebellum of Huntington's disease [3, 8, 110], DM1 [10], spinocerebellar ataxia types 1–3, and dentatorubral pallidoluysian atrophy patients) (reviewed in ref. [12]) or expansion biases (as in the muscle, blood, and skin of myotonic dystrophy patients and the germline of many diseases) [1, 6, 10, 11]. Thus, the observation of three distinct repair results explains the origin of different mutation patterns observed in human diseases at the various genetic loci and tissues.

Correct repair of slip outs would maintain the integrity of the repeat tract. The efficiency of slipped-DNA repair was comparable to that of G–T mismatch repair, indicating the importance of this repair process. Slipped CAG or CTG repeats, particularly those with 5′ nicks, can be repaired with high efficiency and fidelity, an observation that contrasts with presumed escaped repair of these DNAs [57]. The G–T mismatch is the most efficiently repaired base–base mismatch by the same extracts used to process slipped (CTG)·(CAG) repeats. Mutagenic events may constantly be occurring but are never detected as they are continuously repaired with a high degree of fidelity. Thus, the apparent "stability" of certain expanded repeat loci, like the SBMA locus [7], may reflect the correct repair of mutagenic slipped intermediates, rather than the absence of their formation. Similarly, the stability and instability of the same expanded repeat in different tissues of the same patient or transgenic mouse may reflect differences in tissue-specific repairability rather than differences in the formation of slipped intermediates. In contrast to stable maintenance, both expansions and deletions may arise by escaped repair, and this may occur more frequently with CTG slip outs, particularly those with 3′ nicks.

The expansion bias observed in affected families and affected tissues may be explained by various aspects of slipped-DNA repair. Error-prone repair may be a source for the expansion bias observed in patients, as only expansion intermediates not deletion intermediates yielded error-prone products (Fig. 41-4). An expansion bias is further supported by the generally lower repair efficiency of expansion intermediates over deletion intermediates (Fig. 41-4). Both of these tendencies favor error-prone repair as a process to mediate expansions rather than deletions, which may argue for it as a candidate process to cause the expansion bias mutations occurring in humans. Error-prone repair did not require either MMR or NER proteins (Fig. 41-4), thereby revealing that this process, like correct repair and escaped repair, is independent of these proteins.

In many of the neurodegenerative diseases, the CAG repeats expand (striatum) or contract (cerebellum) or can be stable depending upon the brain region. MSH2-dependent CAG expansions can contribute to brain pathology evident as a delay in mutant Huntingtin protein accumulation [101]. How these mutations arise in nonmitotic neurons is unknown. It has been shown that processes of correct repair, escaped repair, and error-prone repair can contribute to these processes [71]. Human proteins from neural-like stem cells and neurally differentiated cells, with the coincident alterations in repair [116], can mediate these repair outcomes. Thus, this aberrant repair can contribute to the instability actively occurring in patient brains.

V. SUMMARY AND FUTURE STUDIES

Disease-associated repeat instability can occur in various patient tissues, including germ cells, proliferating somatic tissues, and nonproliferating tissues like the brain. Various DNA metabolic processes could potentially give rise to slipped DNAs at the repeats—the supposed mutagenic intermediate of repeat instability. Transgenic mice with (CTG)·(CAG) repeats have revealed a requirement for several mismatch repair genes for the spontaneous expansion of the repeats. Results with an *in vitro* repair assay using human cell extracts and structurally defined slipped DNAs have revealed several distinct repair outcomes: correct repair, escaped repair, and error-prone repair, some of which can give rise to the expansion bias occurring in patients. These processes can be mediated by neuronlike cells, supporting their contribution to instability in patient brains. Importantly, the mismatch and nucleotide excision repair proteins such as hMSH2, hMSH3, hMLH1, XPF, or XPG were not involved in processing of the slip outs. This suggests that, if these proteins are involved in expansion, their role precedes the step of slip-out processing. The formation of slipped DNAs seems attractive but will require further investigation.

Future studies include identifying proteins and pathways that contribute to the correct maintenance (repair) of slipped DNAs, as well as those factors (possibly proteins) that may contribute to escaped repair and error-prone repair. Careful consideration of both patient data and experimental models, *in vivo* and biochemical, will lead to a better understanding of this serious mutation. After over a decade of research, researchers are moving closer toward an understanding of the complexities and dynamics of this mutation.

Acknowledgments

We thank John D. Cleary for comments on the manuscript and for his excellent support in its preparation. We also thank the members of the Pearson lab for intellectual support. We are indebted to Arthur Kornberg for Commandment VIII ("Respect the personality of DNA") of his Ten Commandments of DNA (2003). *In* "Biology of DNA." Cold Spring Harbor Laboratory Press, Cold Spring Harbor, NY. This work was supported by the Canadian Institutes of Health Research (CIHR) and The Muscular Dystrophy Association (USA). SEM and MLR were supported by a Research Training Studentship (The Hospital for Sick Children) and Ontario Graduate Scholarships. GBP was supported by a Premier's Research Excellence Award to CEP. CEP is a CIHR Scholar and a Canadian Genetic Disease Scholar.

References

1a. Pearson, C. E., Nichol Edamura, K. and Cleary, J. D. (2005). Repeat insatbility: Mechanisms of dynamic mutations. *Nature Reviews Genetics*. **6**, 729–742.
1. Pearson, C. E. (2003). Slipping while sleeping? Trinucleotide repeat expansions in germ cells. *Trends Mol. Med.* **9**, 490–495.
2. Cleary, J. D., and Pearson, C. E. (2003). The contribution of cis-elements to disease-associated repeat instability: Clinical and experimental evidence. *Cytogenet. Genome Res.* **100**, 25–55.
3. Yoon, S. R., Dubeau, L., de Young, M., Wexler, N. S., and Arnheim, N. (2003). Huntington disease expansion mutations in humans can occur before meiosis is completed. *Proc. Natl. Acad. Sci. USA* **100**, 8834–8838.
4. Cleary J. D., Nichol Edamura, K., and Pearson, C. E. (2005). The complex nature of trinucleotide repeat instability. *Chem Tracts: Biochem. Mol. Biol.* **17**, 663–676.
5. Sinden, R. R. (2001). Neurodegenerative diseases. Origins of instability. *Nature* **411**, 757–758.
6. Yang, Z., Lau, R., Marcadier, J. L., Chitayat, D., and Pearson, C. E. (2003). Replication inhibitors modulate instability of an expanded trinucleotide repeat at the myotonic dystrophy type 1 disease locus in human cells. *Am. J. Hum. Genet.* **73**, 1092–1105.
7. Biancalana, V., Serville, F., Pommier, J., Julien, J., Hanauer, A., and Mandel, J. L. (1992). Moderate instability of the trinucleotide repeat in spino bulbar muscular atrophy. *Hum. Mol. Genet.* **1**, 255–258.
8. Telenius, H., Kremer, B., Goldberg, Y. P., Theilmann, J., Andrew, S. E., Zeisler, J., Adam, S., Greenberg, C., Ives, E. J., Clarke, L. A., et al. (1994). Somatic and gonadal mosaicism of the Huntington disease gene CAG repeat in brain and sperm. *Nat. Genet.* **6**, 409–414.
9. Moseley, M. L., Schut, L. J., Bird, T. D., Koob, M. D., Day, J. W., and Ranum, L. P. (2000). SCA8 CTG repeat: *en masse* contractions in sperm and intergenerational sequence changes may play a role in reduced penetrance. *Hum. Mol. Genet.* **9**, 2125–2130.
10. Thornton, C. A., Johnson, K., and Moxley, R. T., III (1994). Myotonic dystrophy patients have larger CTG expansions in skeletal muscle than in leukocytes. *Ann. Neurol.* **35**, 104–107.

11. Martorell, L., Johnson, K., Boucher, C. A., and Baiget, M. (1997). Somatic instability of the myotonic dystrophy (CTG)n repeat during human fetal development. *Hum. Mol. Genet.* **6**, 877–880.

11a. Cleary, J. D., La Spada, A. R., and Pearson, C. E. (2006). The Contribution of DNA replication to disease-associated repeat instability. In: *DNA Replication and Human Disease* (DePamphilis, M. L., Ed.) Cold Spring Harbour Laboratory Press, *in press*.

12. Cleary, J. D., Nichol, K., Wang, Y. H., and Pearson, C. E. (2002). Evidence of *cis*-acting factors in replication-mediated trinucleotide repeat instability in primate cells. *Nat. Genet.* **31**, 37–46.

13. Panigrahi, G. B., Cleary, J. D., and Pearson, C. E. (2002). *In vitro* (CTG)·(CAG) expansions and deletions by human cell extracts. *J. Biol. Chem.* **277**, 13926–13934.

14. Marcadier, J. L., and Pearson, C. E. (2003). Fidelity of primate cell repair of a double-strand break within a (CTG)·(CAG) tract. Effect of slipped DNA structures. *J. Biol. Chem.* **278**, 33848–33856.

15. Nichol Edamura, K., Leonard, M. R., and Pearson, C. E. (2005). Role of replication and CpG methylation in fragile X syndrome CGG deletions in primate cells. *Am. J. Hum. Genet.* **76**, 302–311.

16. Cleary J. D., and Pearson, C. E. (2005). Replication fork dynamics and dynamic mutations: The forks-shift model of repeat instability. *Trends Genet.* **21**, 272–280.

17. Samadashwily, G. M., Raca, G., and Mirkin, S. M. (1997). Trinucleotide repeats affect DNA replication *in vivo*. *Nat. Genet.* **17**, 298–304.

18. Nenguke, T., Aladjem, M. I., Gusella, J. F., Wexler, N. S., and Arnheim, N. (2003). Candidate DNA replication initiation regions at human trinucleotide repeat disease loci. *Hum. Mol. Genet.* **12**, 1021–1028.

19. Pelletier, R., Krasilnikova, M. M., Samadashwily, G. M., Lahue, R., and Mirkin, S. M. (2003). Replication and expansion of trinucleotide repeats in yeast. *Mol. Cell Biol.* **23**, 1349–1357.

20. Dere, R., Napierala, M., Ranum, L. P., and Wells, R. D. (2004). Hairpin structure-forming propensity of the (CCTG5CAGG) tetranucleotide repeats contributes to the genetic instability associated with myotonic dystrophy type 2. *J. Biol. Chem.* **279**, 41715–41726.

21. Krasilnikova, M. M., and Mirkin, S. M. (2004). Replication stalling at Friedreich's ataxia (GAA)n repeats *in vivo*. *Mol. Cell Biol.* **24**, 2286–2295.

22. Lahiri, M., Gustafson, T. L., Majors, E. R., and Freudenreich, C. H. (2004). Expanded CAG repeats activate the DNA damage checkpoint pathway. *Mol. Cell* **15**, 287–293.

23. Meservy, J. L., Sargent, R. G., Iyer, R. R., Chan, F., McKenzie, G. J., Wells, R. D., and Wilson, J. H. (2003). Long CTG tracts from the myotonic dystrophy gene induce deletions and rearrangements during recombination at the APRT locus in CHO cells. *Mol. Cell Biol.* **23**, 3152–3162.

24. Chetsanga, C. J., Boyd, V., Peterson, L., and Rushlow, K. (1975). Single-stranded regions in DNA of old mice. *Nature* **253**, 130–131.

25. Nakanishi, K., Shima, A., Fukuda, M., and Fujita, S. (1979). Age associated increase of single-stranded regions in the DNA of mouse brain and liver cells. *Mech. Ageing Dev.* **10**, 273–281.

26. Price, G. B., Modak, S. P., and Makinodan, T. (1971). Age-associated changes in the DNA of mouse tissue. *Science* **171**, 917–920.

27. Lilley, D. M., and White, M. F. (2001). The junction-resolving enzymes. *Nat. Rev. Mol. Cell Biol.* **2**, 433–443.

28. Fry, M., and Loeb, L. A. (1994). The fragile X syndrome d(CGG)n nucleotide repeats form a stable tetrahelical structure. *Proc. Natl. Acad. Sci. USA* **91**, 4950–4954.

29. LeProust, E. M., Pearson, C. E., Sinden, R. R., and Gao, X. (2000). Unexpected formation of parallel duplex in GAA and TTC trinucleotide repeats of Friedreich's ataxia. *J. Mol. Biol.* **302**, 1063–1080.

30. Vetcher, A. A., Napierala, M., and Wells, R. D. (2002). Sticky DNA: Effect of the polypurine·polypyrimidine sequence. *J. Biol. Chem.* **277**, 39228–39234.

31. Pearson, C. E., and Sinden, R. R. (1996). Alternative structures in duplex DNA formed within the trinucleotide repeats of the myotonic dystrophy and fragile X loci. *Biochemistry* **35**, 5041–5053.

32. Sinden, R. R. (1999). Biological implications of the DNA structures associated with disease-causing triplet repeats. *Am. J. Hum. Genet.* **64**, 346–353.

33. Pearson, C. E., Ewel, A., Acharya, S., Fishel, R. A., and Sinden, R. R. (1997). Human MSH2 binds to trinucleotide repeat DNA structures associated with neurodegenerative diseases. *Hum. Mol. Genet.* **6**, 1117–1123.

34. Pearson, C. E., Wang, Y. H., Griffith, J. D., and Sinden, R. R. (1998). Structural analysis of slipped-strand DNA (S-DNA) formed in (CTG)n·(CAG)n repeats from the myotonic dystrophy locus. *Nucleic Acids Res.* **26**, 816–823.

35. Pearson, C. E., Eichler, E. E., Lorenzetti, D., Kramer, S. F., Zoghbi, H. Y., Nelson, D. L., and Sinden, R. R. (1998). Interruptions in the triplet repeats of SCA1 and FRAXA reduce the propensity and complexity of slipped-strand DNA (S-DNA) formation. *Biochemistry* **37**, 2701–2708.

36. Pearson, C. E., Tam, M., Wang, Y. H., Montgomery, S. E., Dar, A., Cleary, J. D., and Nichol, K. (2002). Slipped-strand DNAs formed by long (CAG)·(CTG) repeats: Slipped-out repeats and slip-out junctions. *Nucleic Acids Res.* **30**, 4534–4547.

37. Tam, M., Montgomery, S. E., Kekis, M., David Stollar, B., Price, G. B., and Pearson, C. E. (2003). Slipped (CTG)·(CAG) repeats of the myotonic dystrophy locus: Surface probing with anti-DNA antibodies. *J. Mol. Biol.* **332**, 585–600.

38. Gacy, A. M., Goellner, G., Juranic, N., Macura, S., and McMurray, C. T. (1995). Trinucleotide repeats that expand in human disease form hairpin structures *in vitro*. *Cell* **81**, 533–540.

39. Mariappan, S. V., Garcoa, A. E., and Gupta, G. (1996). Structure and dynamics of the DNA hairpins formed by tandemly repeated CTG triplets associated with myotonic dystrophy. *Nucleic Acids Res.* **24**, 775–783.

40. Petruska, J., Arnheim, N., and Goodman, M. F. (1996). Stability of intrastrand hairpin structures formed by the CAG/CTG class of DNA triplet repeats associated with neurological diseases. *Nucleic Acids Res.* **24**, 1992–1998.

41. McMurray, C. T. (1999). DNA secondary structure: A common and causative factor for expansion in human disease. *Proc. Natl. Acad. Sci. USA* **96**, 1823–1825.

42. Hartenstine, M. J., Goodman, M. F., and Petruska, J. (2000). Base stacking and even/odd behavior of hairpin loops in DNA triplet repeat slippage and expansion with DNA polymerase. *J. Biol. Chem.* **275**, 18382–18390.

43. Zheng, M., Huang, X., Smith, G. K., Yang, X., and Gao, X. (1996). Genetically unstable CXG repeats are structurally dynamic and have a high propensity for folding. An NMR and UV spectroscopic study. *J. Mol. Biol.* **264**, 323–336.

44. Chi, L. M., and Lam, S. L. (2005). Structural roles of CTG repeats in slippage expansion during DNA replication. *Nucleic Acids Res.* **33**, 1604–1617.

45. Mitchell, J. E., Newbury, S. F., and McClellan, J. A. (1995). Compact structures of d(CNG)n oligonucleotides in solution and their possible relevance to fragile X and related human genetic diseases. *Nucleic Acids Res.* **23**, 1876–1881.

46. Volker, J., Makube, N., Plum, G. E., Klump, H. H., and Breslauer, K. J. (2002). Conformational energetics of stable

and metastable states formed by DNA triplet repeat oligonucleotides: Implications for triplet expansion diseases. *Proc. Natl. Acad. Sci. USA* **99**, 14700–14705.

47. Baldwin, R. L. (1968). Kinetics of helix formation and slippage of the dAT copolymer. *In* "Symposium on Molecular Associations in Biology" (B. Pullman, Ed.), pp. 145–162. Academic Press, New York.

48. Olivera, B. M., and Lehman, I. R. (1968). Enzymic joining of polynucleotides. 3. The polydeoxyadenylate–polydeoxythymidylate homopolymer pair. *J. Mol. Biol.* **36**, 261–274.

49. Woodson, S. A., and Crothers, D. M. (1988). Structural model for an oligonucleotide containing a bulged guanosine by NMR and energy minimization. *Biochemistry* **27**, 3130–3141.

50. Harvey, S. C. (1997). Slipped structures in DNA triplet repeat sequences: Entropic contributions to genetic instabilities. *Biochemistry* **36**, 3047–3049.

51. Lahue, R. S., and Slater, D. L. (2003). DNA repair and trinucleotide repeat instability. *Front. Biosci.* **8**, S653–665.

52. Moore, H., Greenwell, P. W., Liu, C. P., Arnheim, N., and Petes, T. D. (1999). Triplet repeats form secondary structures that escape DNA repair in yeast. *Proc. Natl. Acad. Sci. USA* **96**, 1504–1509.

53. Spiro, C., Pelletier, R., Rolfsmeier, M. L., Dixon, M. J., Lahue, R. S., Gupta, G., Park, M. S., Chen, X., Mariappan, S. V., and McMurray, C. T. (1999). Inhibition of FEN-1 processing by DNA secondary structure at trinucleotide repeats. *Mol. Cell* **4**, 1079–1085.

54. Henricksen, L. A., Tom, S., Liu, Y., and Bambara, R. A. (2000). Inhibition of flap endonuclease 1 by flap secondary structure and relevance to repeat sequence expansion. *J. Biol. Chem.* **275**, 16420–16427.

55. Corrette-Bennett, S. E., Mohlman, N. L., Rosado, Z., Miret, J. J., Hess, P. M., Parker, B. O., and Lahue, R. S. (2001). Efficient repair of large DNA loops in *Saccharomyces cerevisiae*. *Nucleic Acids Res.* **29**, 4134–4143.

56. Oussatcheva, E. A., Hashem, V. I., Zou, Y., Sinden, R. R., and Potaman, V. N. (2001). Involvement of the nucleotide excision repair protein UvrA in instability of CAG·CTG repeat sequences in *Escherichia coli*. *J. Biol. Chem.* **276**, 30878–30884.

57. Owen, B. A., Yang, Z., Lai, M., Gajek, M., Badger, J. D., Hayes, J. J., Edelmann, W., Kucherlapati, R., Wilson, T. M., and McMurray, C. T. (2005). (CAG)(n)-hairpin DNA binds to Msh2-Msh3 and changes properties of mismatch recognition. *Nat. Struct. Mol. Biol.* **12**, 663–670.

58. Pearson, C. E., Tam, M., Wang, Y. H., Montgomery, S. E., Dar, A. C., Cleary, J. D., and Nichol, K. (2002). Slipped-strand DNAs formed by long (CAG)·(CTG) repeats: Slipped-out repeats and slip-out junctions. *Nucleic Acids Res.* **30**, 4534–4547.

59. de Laat, W. L., Appeldoorn, E., Jaspers, N. G., and Hoeijmakers, J. H. (1998). DNA structural elements required for ERCC1-XPF endonuclease activity. *J. Biol. Chem.* **273**, 7835–7842.

60. Sancar, A. (1996). DNA excision repair. *Annu. Rev. Biochem.* **65**, 43–81.

61. Sancar, G. B., Siede, W., and van Zeeland, A. A. (1996). Repair and processing of DNA damage: A summary of recent progress. *Mutat. Res.* **362**, 127–146.

62. Wood, R. D. (1997). Nucleotide excision repair in mammalian cells. *J. Biol. Chem.* **272**, 23465–23468.

63. Littman, S. J., Fang, W. H., and Modrich, P. (1999). Repair of large insertion/deletion heterologies in human nuclear extracts is directed by a 5′ single-strand break and is independent of the mismatch repair system. *J. Biol. Chem.* **274**, 7474–7481.

64. Kirkpatrick, D. T., and Petes, T. D. (1997). Repair of DNA loops involves DNA-mismatch and nucleotide-excision repair proteins. *Nature* **387**, 929–931.

65. Vasquez, K. M., Christensen, J., Li, L., Finch, R. A., and Glazer, P. M. (2002). Human XPA and RPA DNA repair proteins participate in specific recognition of triplex-induced helical distortions. *Proc. Natl. Acad. Sci. USA* **99**, 5848–5853.

66. Parniewski, P., Bacolla, A., Jaworski, A., and Wells, R. D. (1999). Nucleotide excision repair affects the stability of long transcribed (CTG·CAG) tracts in an orientation-dependent manner in *Escherichia coli*. *Nucleic Acids Res.* **27**, 616–623.

67. Bowater, R. P., Jaworski, A., Larson, J. E., Parniewski, P., and Wells, R. D. (1997). Transcription increases the deletion frequency of long CTG·CAG triplet repeats from plasmids in *Escherichia coli*. *Nucleic Acids Res.* **25**, 2861–2868.

68. Schumacher, S., Pinet, I., and Bichara, M. (2001). Modulation of transcription reveals a new mechanism of triplet repeat instability in *Escherichia coli*. *J. Mol. Biol.* **307**, 39–49.

69. Freudenreich, C. H., Kantrow, S. M., and Zakian, V. A. (1998). Expansion and length-dependent fragility of CTG repeats in yeast. *Science*. **279**, 853–856.

70. Grewal, R. P. (1999). Neurodegeneration in Xeroderma pigmentosum: a trinucleotide repeat mutation analysis. *J. Neurol. Sci.* **163**, 183–186.

71. Panigrahi, G. B., Lau, R., Montgomery, S. E., Leonard, M. R., and Pearson, C. E. (2005). Slipped (CTG)·(CAG) repeats can be correctly repaired, escape repair or undergo error-prone repair. *Nat. Struct. Mol. Biol.*, **12**, 654–662.

72. Ayares, D., Ganea, D., Chekuri, L., Campbell, C. R., and Kucherlapati, R. (1987). Repair of single-stranded DNA nicks, gaps, and loops in mammalian cells. *Mol. Cell Biol.* **7**, 1656–1662.

73. Weiss, U., and Wilson, J. H. (1989). Effects of nicks on repair of single-stranded loops in heteroduplex DNA in mammalian cells. *Somat. Cell Mol. Genet.* **15**, 13–18.

74. Weiss, U., and Wilson, J. H. (1988). Heteroduplex-induced mutagenesis in mammalian cells. *Nucleic Acids Res.* **16**, 2313–2322.

75. Campbell, C. R., Ayares, D., Watkins, K., Wolski, R., and Kucherlapati, R. (1989). Single-stranded DNA gaps, tails and loops are repaired in *Escherichia coli*. *Mutat. Res.* **211**, 181–188.

76. Genschel, J., Littman, S. J., Drummond, J. T., and Modrich, P. (1998). Isolation of MutSbeta from human cells and comparison of the mismatch repair specificities of MutSbeta and MutSalpha. *J. Biol. Chem.* **273**, 19895–19901.

77. Corrette-Bennett, S. E., Parker, B. O., Mohlman, N. L., and Lahue, R. S. (1999). Correction of large mispaired DNA loops by extracts of *Saccharomyces cerevisiae*. *J. Biol. Chem.* **274**, 17605–17611.

78. McCulloch, S. D., Gu, L., and Li, G. M. (2003). Nick-dependent and -independent processing of large DNA loops in human cells. *J. Biol. Chem.* **278**, 50803–50809.

79. Huang, Y. M., Chen, S. U., Goodman, S. D., Wu, S. H., Kao, J. T., Lee, C. N., Cheng, W. C., Tsai, K. S., and Fang, W. H. (2004). Interaction of nick-directed DNA mismatch repair and loop repair in human cells. *J. Biol. Chem.* **279**, 30228–30235.

80. Fang, W. H., Wang, B. J., Wang, C. H., Lee, S. J., Chang, Y. T., Chuang, Y. K., and Lee, C. N. (2003). DNA loop repair by *Escherichia coli* cell extracts. *J. Biol. Chem.* **270**, 22446–22452.

81. McCulloch, S. D., Gu, L., and Li, G. M. (2003). Bi-directional processing of DNA loops by mismatch repair-dependent and -independent pathways in human cells. *J. Biol. Chem.* **278**, 3891–3896.

82. Umar, A., Boyer, J. C., and Kunkel, T. A. (1994). DNA loop repair by human cell extracts. *Science* **266**, 814–816.

83. Jaworski, A., Rosche, W. A., Gellibolian, R., Kang, S., Shimizu, M., Bowater, R. P., Sinden, R. R., and Wells, R. D. (1995). Mismatch repair in *Escherichia coli* enhances instability of (CTG)n triplet

repeats from human hereditary diseases. *Proc. Natl. Acad. Sci. USA* **92**, 11019–11023.
84. Schumacher, S., Fuchs, R. P., and Bichara, M. (1998). Expansion of CTG repeats from human disease genes is dependent upon replication mechanisms in *Escherichia coli*: The effect of long patch mismatch repair revisited. *J. Mol. Biol.* **279**, 1101–1110.
85. Miret, J. J., Pessoa-Brandao, L., and Lahue, R. S. (1997). Instability of CAG and CTG trinucleotide repeats in *Saccharomyces cerevisiae*. *Mol. Cell Biol.* **17**, 3382–3387.
86. Miret, J. J., Pessoa-Brandao, L., and Lahue, R. S. (1998). Orientation-dependent and sequence-specific expansions of CTG/CAG trinucleotide repeats in *Saccharomyces cerevisiae*. *Proc. Natl. Acad. Sci. USA* **95**, 12438–12443.
87. Schweitzer, J. K., and Livingston, D. M. (1997). Destabilization of CAG trinucleotide repeat tracts by mismatch repair mutations in yeast. *Hum. Mol. Genet.* **6**, 349–355.
88. Kolodner, R. (1996). Biochemistry and genetics of eukaryotic mismatch repair. *Genes Dev.* **10**, 1433–1442.
89. Acharya, S., Wilson, T., Gradia, S., Kane, M. F., Guerrette, S., Marsischky, G. T., Kolodner, R., and Fishel, R. (1996). hMSH2 forms specific mispair-binding complexes with hMSH3 and hMSH6. *Proc. Natl. Acad. Sci. USA* **93**, 13629–13634.
90. Wilson, T. E., Guerrette, S., and Fishel, R. (1999). Disassociation of mismatch recognition and ATPase activity by hMSH2-hMSH3. *J. Biol. Chem.* **274**, 21659–21664.
91. Kondo, E., Horii, A., and Fukushige, S. (2001). The interacting domains of three MutL heterodimers in man: hMLH1 interacts with 36 homologous amino acid residues within hMLH3, hPMS1 and hPMS2. *Nucleic Acids Res.* **29**, 1695–1702.
92. Raschle, M., Marra, G., Nystrom-Lahti, M., Schar, P., and Jiricny, J. (1999). Identification of hMutLbeta, a heterodimer of hMLH1 and hPMS1. *J. Biol. Chem.* **274**, 32368–32375.
93. Plotz, G., Raedle, J., Brieger, A., Trojan, J., and Zeuzem, S. (2002). hMutSalpha forms an ATP-dependent complex with hMutLalpha and hMutLbeta on DNA. *Nucleic Acids Res.* **30**, 711–718.
94. Buermeyer, A. B., Deschenes, S. M., Baker, S. M., and Liskay, R. M. (1999). Mammalian DNA mismatch repair. *Annu. Rev. Genet.* **33**, 533–564.
95. Manley, K., Shirley, T. L., Flaherty, L., and Messer, A. (1999). Msh2 deficiency prevents *in vivo* somatic instability of the CAG repeat in Huntington disease transgenic mice. *Nat. Genet.* **23**, 471–473.
96. van Den Broek, W. J., Nelen, M. R., Wansink, D. G., Coerwinkel, M. M., te Riele, H., Groenen, P. J., and Wieringa, B. (2002). Somatic expansion behaviour of the (CTG)(n) repeat in myotonic dystrophy knock-in mice is differentially affected by Msh3 and Msh6 mismatch-repair proteins. *Hum. Mol. Genet.* **11**, 191–198.
97. Savouret, C., Brisson, E., Essers, J., Kanaar, R., Pastink, A., te Riele, H., Junien, C., and Gourdon, G. (2003). CTG repeat instability and size variation timing in DNA repair-deficient mice. *EMBO J.* **22**, 2264–2273.
98. Savouret, C., Garcia-Cordier, C., Megret, J., te Riele, H., Junien, C., and Gourdon, G. (2004). MSH2-dependent germinal CTG repeat expansions are produced continuously in spermatogonia from DM1 transgenic mice. *Mol. Cell Biol.* **24**, 629–637.
99. Kovtun, I. V., and McMurray, C. T. (2001). Trinucleotide expansion in haploid germ cells by gap repair. *Nat. Genet.* **27**, 407–411.
100. Kovtun, I. V., Thornhill, A. R., and McMurray, C. T. (2004). Somatic deletion events occur during early embryonic development and modify the extent of CAG expansion in subsequent generations. *Hum. Mol. Genet.* **13**, 3057–3068.
101. Wheeler, V. C., Lebel, L. A., Vrbanac, V., Teed, A., te Riele, H., and MacDonald, M. E. (2003). Mismatch repair gene Msh2 modifies the timing of early disease in Hdh(Q111) striatum. *Hum. Mol. Genet.* **12**, 273–281.
102. Gomes-Pereira, M., Fortune, M. T., Ingram, L., McAbney, J. P., and Monckton, D. G. (2004). Pms2 is a genetic enhancer of trinucleotide CAG·CTG repeat somatic mosaicism: Implications for the mechanism of triplet repeat expansion. *Hum. Mol. Genet.* **13**, 1815–1825.
103. Yao, X., Buermeyer, A. B., Narayanan, L., Tran, D., Baker, S. M., Prolla, T. A., Glazer, P. M., Liskay, R. M., and Arnheim, N. (1999). Different mutator phenotypes in Mlh1- versus Pms2-deficient mice. *Proc. Natl. Acad. Sci. USA* **96**, 6850–6855.
104. Fishel, R., Acharya, S., Berardini, M., Bocker, T., Charbonneau, N., Cranston, A., Gradia, S., Guerrette, S., Heinen, C. D., Mazurek, A., Snowden, T., Schmutte, C., Shim, K. S., Tombline, G., and Wilson, T. (2000). Signaling mismatch repair: The mechanics of an adenosine-nucleotide molecular switch. *Cold Spring Harb. Symp. Quant. Biol.* **65**, 217–224.
105. Holmes, J., Jr., Clark, S., and Modrich, P. (1990). Strand-specific mismatch correction in nuclear extracts of human and *Drosophila melanogaster* cell lines. *Proc. Natl. Acad. Sci. USA* **87**, 5837–5841.
106. Thomas, D. C., Roberts, J. D., and Kunkel, T. A. (1991). Heteroduplex repair in extracts of human HeLa cells. *J. Biol. Chem.* **266**, 3744–3751.
107. Fang, W. H., and Modrich, P. (1993). Human strand-specific mismatch repair occurs by a bidirectional mechanism similar to that of the bacterial reaction. *J. Biol. Chem.* **268**, 11838–11844.
108. Lin, D. P., Wang, Y., Scherer, S. J., Clark, A. B., Yang, K., Avdievich, E., Jin, B., Werling, U., Parris, T., Kurihara, N., Umar, A., Kucherlapati, R., Lipkin, M., Kunkel, T. A., and Edelmann, W. (2004). An Msh2 point mutation uncouples DNA mismatch repair and apoptosis. *Cancer Res.* **64**, 517–522.
109. Pearson, C. E., and Sinden, R.R. (1997). Trinucleotide repeat DNA structures: Dynamic mutations from dynamic DNA. *Curr. Opin. Struct. Biol.* **38**, 321–330.
110. Kennedy, L., Evans, E., Chen, C. M., Craven, L., Detloff, P. J., Ennis, M., and Shelbourne, P. F. (2003). Dramatic tissue-specific mutation length increases are an early molecular event in Huntington disease pathogenesis. *Hum. Mol. Genet.* **12**, 3359–3367.
111. Lindahl, T., and Wood, R. D. (1999). Quality control by DNA repair. *Science* **286**, 1897–1905.
112. Lopez, B., and Coppey, J. (1987). Promotion of double-strand break repair by human nuclear extracts preferentially involves recombination with intact homologous DNA. *Nucleic Acids Res.* **15**, 6813–6826.
113. North, P., Ganesh, A., and Thacker, J. (1990). The rejoining of double-strand breaks in DNA by human cell extracts. *Nucleic Acids Res.* **18**, 6205–6210.
114. Nag, D. K., White, M. A., and Petes, T. D. (1989). Palindromic sequences in heteroduplex DNA inhibit mismatch repair in yeast. *Nature* **340**, 318–320.
115. Manley, K., Shirley, T. L., Flaherty, L., and Messer, A. (1999). Msh2 deficiency prevents *in vivo* somatic instability of the CAG repeat in Huntington disease transgenic mice. *Nat. Genet.* **23**, 471–473.
116. Nouspikel, T., and Hanawalt, P. C. (2002). DNA repair in terminally differentiated cells. *DNA Repair (Amsterdam)* **1**, 59–75.
117. David, P., Tocco, G., Krauss, S. W., and Goodman, M. F. (1997). DNA replication and postreplication mismatch repair in cell-free extracts from cultured human neuroblastoma and fibroblast cells. *J. Neurosci.* **17**, 8711–8720.
118. Werntges, H., Steger, G., Riesner, D., and Fritz, H. J. (1986). Mismatches in DNA double strands: Thermodynamic parameters and their correlation to repair efficiencies. *Nucleic Acids Res.* **14**, 3773–3790.

119. Fazakerley, G. V., Quignard, E., Woisard, A., Guschlbauer, W., van der Marel, G. A., van Boom, J. H., Jones, M., and Radman, M. (1986). Structures of mismatched base pairs in DNA and their recognition by the *Escherichia coli* mismatch repair system. *EMBO J.* **5**, 3697–3703.
120. Hunter, W. N., Brown, T., and Kennard, O. (1986). Structural features and hydration of d(C-G-C-G-A-A-T-T-A-G-C-G); a double helix containing two G·A mispairs. *J. Biomol. Struct. Dyn.* **4**, 173–191.
121. Dixon, M. J., and Lahue, R. S. (2002). Examining the potential role of DNA polymerases eta and zeta in triplet repeat instability in yeast. *DNA Repair (Amsterdam)* **1**, 763–770.
122. Longley, M. J., Pierce, A. J., and Modrich, P. (1997). DNA polymerase delta is required for human mismatch repair *in vitro*. *J. Biol. Chem.* **272**, 10917–10921.
123. David, G., Abbas, N., Stevanin, G., Durr, A., Yvert, G., Cancel, G., Weber, C., Imbert, G., Saudou, F., Antoniou, E., Drabkin, H., Gemmill, R., Giunti, P., Benomar, A., Wood, N., Ruberg, M., Agid, Y., Mandel, J. L., and Brice, A. (1997). Cloning of the SCA7 gene reveals a highly unstable CAG repeat expansion. *Nat. Genet.* **17**, 65–70.
124. Parniewski, P., Jaworski, A., Wells, R. D., and Bowater, R. P. (2000). Length of CTG·CAG repeats determines the influence of mismatch repair on genetic instability. *J. Mol. Biol.* **299**, 865–874.
125. Faruqi, A. F., Datta, H. J., Carroll, D., Seidman, M. M., and Glazer, P. M. (2000). Triple-helix formation induces recombination in mammalian cells via a nucleotide excision repair-dependent pathway. *Mol. Cell Biol.* **20**, 990–1000.
126. Patrick, S. M., and Turchi, J. J. (2002). Xeroderma pigmentosum complementation group A protein (XPA) modulates RPA–DNA interactions via enhanced complex stability and inhibition of strand separation activity. *J. Biol. Chem.* **277**, 16096–16101.
127. Hartenstine, M. J., Goodman, M. F., and Petruska, J. (2002). Weak strand displacement activity enables human DNA polymerase beta to expand CAG/CTG triplet repeats at strand breaks. *J. Biol. Chem.* **277**, 41379–41389.
128. Wang, H., Schofield, M. J., Du, C., Fridman, Y., Lee, S. D., Larson, E. D., Drummond, J. T., Alani, E., Hsieh, P., and Erie, D. A. (2003). DNA bending and unbending by MutS govern mismatch recognition and specificity. *Proc. Natl. Acad. Sci. USA* **100**, 14822–14827.
129. Yang, Q., Zhang, R., Wang, X. W., Linke, S. P., Sengupta, S., Hickson, I. D., Pedrazzi, G., Perrera, C., Stagljar, I., Littman, S. J., Modrich, P., and Harris, C. C. (2004). The mismatch DNA repair heterodimer, hMSH2/6, regulates BLM helicase. *Oncogene* **23**, 3749–3756.
130. Cheok, C. F., Wu, L., Garcia, P. L., Janscak, P., and Hickson, I. D. (2005). The Bloom's syndrome helicase promotes the annealing of complementary single-stranded DNA. *Nucleic Acids Res.* **33**, 3932–3941.
131. Doherty, K. M., Sharma, S., Uzdilla, L. A., Wilson, T. M., Cui, S., Vindigni, A., and Brosh, R. M., Jr. (2005). RECQ1 helicase interacts with human mismatch repair factors that regulate genetic recombination. *J. Biol. Chem.* **280**, 28085–28094.
132. Sharma, S., Sommers, J. A., Choudhary, S., Faulkner, J. K., Cui, S., Andreoli, L., Muzzolini, L., Vindigni, A., and Brosh, R. M., Jr. (2005). Biochemical analysis of the DNA unwinding and strand annealing activities catalyzed by human RECQ1. *J. Biol. Chem.* **280**, 28072–28084.
133. Jackson, S. M., Whitworth, A. J., Greene, J. C., Libby, R. T., Baccam, S. L., Pallanck, L. J., and La Spada, A. R. (2005). A SCA7 CAG/CTG repeat expansion is stable in *Drosophila melanogaster* despite modulation of genomic context and gene dosage. *Gene* **347**, 35–41.

CHAPTER 42

DNA Repair Models for Understanding Triplet Repeat Instability

YUAN LIU, RAJENDRA PRASAD, AND SAMUEL H. WILSON

Laboratory of Structural Biology, National Institute of Environmental Health Sciences, National Institutes of Health, Research Triangle Park, North Carolina 27709

I. Introduction and Background
II. Results and Discussion
III. Concluding Remarks
References

Trinucleotide repeat (TNR) expansion is a causative factor in many hereditary neurodegenerative diseases, and it has been proposed that formation of unusual DNA structures, such as the hairpin, within TNR sequences may lead to TNR expansion. DNA base damage-induced single-strand DNA breaks that occur in TNR sequences could allow DNA slippage and the formation of hairpin structures. To explore TNR expansion as a function of base excision repair (BER), we describe an *in vitro* BER model system involving several of the major human BER enzymes and a CAG repeat-containing DNA with an oxidative base lesion, 8-hydroxy-7, 8 dihydro-guanine (8-oxoG). Our results with this system are consistent with ssDNA break-initiated triplet repeat slippage and hairpin formation during BER. We propose that the triplet repeat track can form various sizes of hairpins and gaps during the process of BER. This model system enables further understanding roles of individual DNA repair proteins in maintaining triplet repeat stability.

I. INTRODUCTION AND BACKGROUND

Triplet repeat instability, in particular trinucleotide repeat (TNR) expansion, has been identified as a causative factor in over 20 hereditary human neurodegenerative diseases [1]. A molecular basis underlying the expansion is the formation of the triplet-repeat-invoked non-B-DNA structures, such as hairpins (CNG/CNG), tetraplexes (CGG/CCG), and sticky DNA (GAA/TTC) [1] during DNA replication and repair, as well as recombination [1, 2]. These noncanonical DNA structures can substantially defeat the cellular DNA repair enzymes, thereby protecting themselves from being repaired [3]. Therefore, the persistence of these structures allows the triplet repeats to be dynamic within the human genome and in some cases causes triplet repeat expansion at a coding or a noncoding region, leading to neuropathological conditions.

Multiple DNA metabolic pathways including DNA replication [4, 5], recombination [6–8], mismatch repair [9–12], and nucleotide excision repair [13, 14] have been shown to play active roles in triplet repeat instability. However, the importance of the base excision repair (BER) pathway and of single-strand DNA (ssDNA) break repair also must be emphasized, because the ssDNA break intermediate in BER is important in promoting genomic instability. A single-strand DNA breakage that specifically occurs within a TNR region, if not repaired efficiently, could allow DNA slippage and misalignment of the single-strand triplet repeat DNA. Consequently, this would promote the formation of various sizes of hairpin structures in either the damaged or the undamaged DNA strand. In the case of hairpins in the damaged strand, the gaps associated with such hairpin structures could be filled by a gap-filling DNA polymerase like DNA polymerase β (Pol β), and then the hairpins could be realigned to generate ligatable nicks that would be sealed by a DNA ligase. In this manner, a ssDNA break-induced triplet repeat expansion would be generated. Thus, the repair pathways for removing ssDNA breaks, such as BER, should be as critical in maintaining triplet repeat stability *in vivo* as the other DNA metabolic pathways.

A DNA single-strand break may be generated through both exogenous and endogenous stress. DNA base oxidation and single-strand-break damage can result directly from ionizing radiation, ultraviolet (UV) light, and environmental genotoxicants [15], as well as from endogenous reactive oxidative species (ROS) [16, 17]. Spontaneous cytosine deamination to uracil, endogenous base alkylation, and spontaneous base loss also can lead to strand-break damage [18–21]. The single-strand breaks also can result indirectly from removal of a damaged base or nucleotide by base excision repair enzymes, such as the bifunctional DNA glycosylase/apurinic/apyrimidinic (AP) lyases [22], AP endonucleases (APE) [21, 23], and the AP lyases, some of which are associated with DNA polymerases [24, 25]. Due to the broad range and abundance of damage sources, the ssDNA break may occur with a high frequency, and in the case of clustered base lesions, multiple ssDNA breaks and double-strand breaks can be generated. It has been estimated that the single-base lesion resulting from spontaneous hydrolytic depurination alone can occur at a frequency of 2×10^3–10^4 per human cell per day [16]. Thus, the ssDNA break associated with these processes may account for a very large proportion of the total cellular DNA damage.

The repair of the ssDNA break is mainly accomplished by BER proteins, along with a break sensor protein, poly(ADP-ribose) polymerase-1 (PARP-1) [26, 27], a scaffolding protein, X-ray repair cross-complementing 1 (XRCC1) [28–30], and polynucleotide kinase [29]. Pol β has been observed to accumulate in the nucleus at the site of a ssDNA break [31], indicating its importance in ssDNA repair. In addition, the Pol-β-mediated ssDNA repair is known to be XRCC1-dependent [31], suggesting that BER proteins coordinate with XRCC1 and the other repair and cell cycle signaling proteins to achieve efficient repair and to maintain genomic stability.

A number of *in vivo* studies have suggested that ssDNA break repair is involved in maintaining repeat sequence stability in genomic DNA. Oxidative DNA damage that can cause direct or indirect ssDNA breaks has already been implicated in promoting triplet repeat instability. An increase in the oxidative DNA damage products including 8-hydroxy-7,8 dihydroguanine has been identified in transgenic mouse models of Huntington's disease (HD) [32–34]. Furthermore, it has been proposed that an expansion-biased instability of CAG repeats in the striatal neurons of the HD model mouse may be due to multiple rounds of DNA damage and repair triggered by the age-related oxidative DNA damage and the reduced antioxidant defense in the striatum [35, 36]. It is possible that the frequency of ssDNA breaks from oxidative stress will significantly increase with age, thereby allowing multiple rounds of formation and integration of CAG hairpins into the genome.

In *Escherichia coli*, an increased deletion rate of CTG·CAG repeats was observed when the cells were subjected to oxidative damage from hydrogen peroxide [37]. Bacterial cells with a deficiency in superoxide dismutase (SOD) displayed a higher deletion rate than wild-type cells [37]. Because SOD deficiency and hydrogen peroxide treatment can increase oxidative DNA damage in *E. coli* [38–39], the notion emerges that ssDNA breaks in triplet repeats following the oxidative damage allow the formation of hairpins, specifically leading to triplet repeat deletion or expansion. Cellular antioxidant mechanisms, including SOD, can neutralize the oxidative damage potential and, therefore, reduce the amount of ssDNA breaks and prevent the deletion–expansion.

A functional deficiency or imbalance in repair proteins also can cause the accumulation of ssDNA that in turn triggers genomic instability. Under the challenge from an alkylating agent, PARP-1$^{-/-}$ mice displayed genomic instability involving the deletion of a fragment larger than 1 kb, as well as a small insertion at the junction of the large deletion [40]. Interestingly, the small insertion was later identified to be a duplication of a palindromic sequence. Because PARP-1 is a nick sensor protein and is critical for efficient BER [41] and ssDNA break repair [42], the compromised repair resulting from PARP-1 deficiency led to an accumulation of ssDNAs. The unrepaired ssDNAs were proposed to initiate double-strand DNA breaks (DSB) and activated nonhomologous end joining (NHEJ) repair, where a hairpin was generated at a single-strand palindromic sequence and this was incorporated into the genome leading to sequence duplication. This proposed mechanism for involving NHEJ may also be applicable to the situation in which a palindromic ssDNA can directly form a hairpin through DNA slippage and misalignment, if the ssDNA break cannot be repaired efficiently due to a functional defect in BER.

The importance of BER proteins in maintaining triplet repeat stability also has been implicated by *in vitro* evidence. Introduction of tetrahydrofuran (THF), an abasic site analogue, at the 3′ terminus of a triplet repeat tract used as a primer hindered primer extension by Pol β and significantly increased the expansion of the repeats [43]. Furthermore, insertion of a THF residue into the template strand of the GAA/TTC repeat track hinders Pol β synthesis, and this can also increase the amount of expanded products [44]. The structural characterization of the expanded triplet repeats by electron microscopy and endonuclease probing clearly demonstrated the formation of hairpin structures during Pol-β-mediated DNA synthesis [45]. These results suggest that a DNA base lesion at the repeated sequence, if not removed by BER, could promote DNA slippage and the formation of hairpin structures, subsequently causing sequence expansion. Therefore, the direct involvement of BER is indicated in triplet repeat instability initiated by a DNA base lesion. Thus far, the proteins identified that may affect the length of triplet repeats during BER and ssDNA break repair include Pol β, flap endonuclease 1 (FEN1), and DNA ligase I. However, their exact roles in triplet repeat instability initiated by base damage or a ssDNA break need to be further elucidated.

Pol β is critical in filling small gaps and removing damaged sugar phosphate during base excision repair (BER) [24, 25] and ssDNA break repair [31]. The efficient Pol β 1-nt gap-filling reaction [46] should guarantee a ligatable nick and the effective fulfillment of BER and ssDNA break repair, when DNA damage occurs at a TNR sequence. However, under certain circumstances, the nonprocessive DNA synthesis mediated by Pol β [47] during DNA repair may provide more opportunity for DNA slippage than the processive synthesis mediated by replicative DNA polymerases [47, 48]. This property allows Pol β to be used to produce triplet repeat expansion products *in vitro* [49]. Another mechanism underlying Pol-β-mediated TNR expansion is associated with its poor strand-displacement synthesis [50]. The ability of Pol β to displace a downstream strand of DNA is one-hundredfold lower than that of the Klenow fragment [50]. Thus, Pol β readily stalls and falls off the template after it fills in a gap and hits the 5′ end of a downstream strand of DNA during BER and ssDNA break repair. This would facilitate CAG- or CTG-repeat-mediated DNA slippage at the end of the extended primer strand that could further lead to the formation of a primer hairpin and the generation of a gap. Subsequently, Pol β gap-filling DNA synthesis extends the slipped primer and expands the TNR primer even more [50].

FEN1 is an essential enzyme for Okazaki fragment processing during DNA lagging strand synthesis [51, 52]. It is also critical for removing a DNA flap with a modified sugar phosphate residue during DNA long-patch base excision repair (LP-BER) [53–55]. In addition, FEN1 is well-known in maintaining genomic stability and preventing repeat sequence expansion. The specific function of FEN1 has been demonstrated *in vivo* in bacteria [56, 57], budding yeast [13, 58–64], and mouse [65, 66]. However, the connection between Huntington's disease and a FEN1 functional defect in the human population has not been established [67].

It has been proposed that, in most cases, FEN1 employs its 5′ endonuclease activity rather than its 5′–3′ exonuclease to efficiently remove a short triplet repeat flap before it can grow long enough to form a hairpin during DNA replication [64]. The long triplet repeat flaps that form stable hairpin and bubble structures can significantly inhibit FEN1 endonucleolytic cleavage [61, 64, 68]. Addition of FEN1 to a reaction in which a DNA polymerase was preincubated with GAA/TTC repeat tracks dramatically compromised the ability of FEN1 to suppress the TNR expansion [69]. Thus, preformed, stable hairpin and bubble structures cannot be removed effectively by FEN1 endonuclease activity and will remain in the configuration that favors sequence expansion. Interestingly, one study [70] indicated that FEN1 may manage to realign the TNR flap even in the situation where a long 5′ TNR flap is generated. It has been proposed that FEN1 can take advantage of reannealing of the long flap to the template strand, creating a "double-flap" structure with both 3′ and 5′ flaps [70]. Subsequently, FEN1 is loaded onto the 5′ flap and allows flap equilibration, while it is tracking along the full length of the flap, until it ultimately captures a double-flap intermediate with a 1-nt 3′ flap and a long 5′ triplet repeat flap; it then removes the 5′ TNR-containing flap [70]. Thus, in this scenario, FEN1 employs a unique tracking mechanism to remove a long TNR flap [70].

The specific mechanism of FEN1 involvement in sustaining the stability of triplet repeats during BER is not fully understood, as yet. Unlike replicative polymerases, the weak strand-displacement DNA synthesis of Pol β may not be strong enough [50, 71] to generate an authentic triplet repeat flap for FEN1 to cleave. It appears that a different mechanism might be adopted by FEN1 in preventing damage-induced triplet repeat instability during BER or ssDNA repair. Studies on FEN1 inhibition of Pol-β-mediated TNR instability [69] and characterization of Pol β–FEN1 coordination may provide some clues for defining the mechanism. The sequential coordination of Pol β and FEN1 activities during long-patch BER (LP-BER) has been identified, wherein Pol β fills in a 1-nt gap intermediate in BER leaving a nicked sugar phosphate flap. Removal of the flap by FEN1 starts the LP-BER process and creates another 1-nt gap for Pol β to fill in creating a nick. FEN1 then removes a nucleotide from this nick, again leaving a 1-nt gap [71], the preferred substrate for Pol β. This "hit and run" mechanism allows the BER intermediates to be passed from one enzyme to the other [71], as proposed several years ago, in the form of the "passing the baton" mechanism [72], wherein the baton is a BER intermediate and the specificity of the enzymes involved is the key mediator underlying the process of coordination in the BER mechanism. Thus, FEN1 may manage to maintain the stability of a DNA-damage-induced TNR expansion by keeping the gaps and nicks bound by the repair proteins and preventing the initiation of DNA slippage. This can be accomplished by sequential coordination of FEN1 and Pol β. Thus, the protein-to-protein coordination, rather than the removal of a TNR flap, may be the mechanism for FEN1 to sustain triplet repeat stability in LP-BER. FEN1 has been shown to physically interact with Werner syndrome protein (WRN) [73, 74] and Bloom protein [75, 76]. These Rec Q family proteins have helicase activity that can displace double-strand DNA, including hairpins, bubbles, and CGG/CCG tetraplex [76, 77]. It has been found that Bloom protein helicase activity can resolve the hairpins and bubbles into flaps, so that they ultimately can be removed by FEN1 [76]. In contrast, WRN cannot resolve a hairpin structure into a flap for FEN1 cleavage [74]. The importance of WNR in the LP-BER process has also been suggested [78]. WRN can stimulate FEN1 cleavage [73], suggesting that it may facilitate FEN1 activity in maintaining the stability of triplet repeat during BER through a protein–protein interaction as described earlier.

Another key protein that determines the triplet repeat stability during BER and ssDNA break repair is DNA ligase. Thus far, DNA ligase I (Lig I) [79–83] and ligase III (Lig III) [83, 84] have been identified as the major players in these pathways. DNA ligases may perform two opposite roles in determining the fate of the triplet repeat stability question, depending upon different situations of the substrate DNA. Under normal circumstances, DNA ligases seal nicks that are generated during the last step of BER and ssDNA break repair. The efficient nick sealing by ligase is critical for reducing the probability for promoting the production of ssDNA and DNA slippage, particularly at repeated sequences. Therefore, efficient ligase activity favors the maintenance of repeated sequence stability. On the other hand, ligase can also seal any ligatable nick created by DNA slippage and formation of hairpins and bubbles when they occur in the repeated sequence. However, an additional realignment step may be required to generate the ligatable nick associated with the hairpins as DNA ligase I cannot directly seal the nick adjacent to a hairpin, as demonstrated in Fig. 42-1 [64]. The ligation efficiency of ligase in these cases may vary depending on the stability of hairpins and bubbles [64], and in some cases ligase may promote TNR expansion. DNA Lig I has been shown to compete with FEN1 to lead to triplet repeat expansion [85]. In fact, Lig I cannot effectively compete with FEN1 on an extremely unstable hairpin or a bubble [64]; Lig I predominates at a stable hairpin or bubble and readily leads to sequence expansion [64]. For the hairpins and bubbles with intermediate stability, the competition between Lig I and FEN1 determines the path these structures will enter. Thus, the balance between these two enzymes may be the key: more Lig I will favor the ligation of the hairpins, whereas more FEN1 will lead to the removal of the hairpins [64]. The basis underlying the competition is equilibration between the hairpin

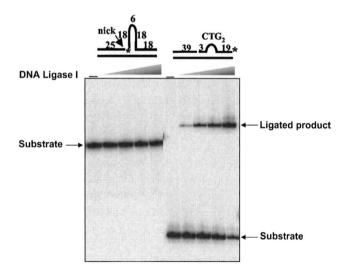

FIGURE 42-1 The nick adjacent to a DNA hairpin cannot be sealed by DNA ligase I. Reprinted with permission from *The Journal of Biological Chemistry* **278**, 13728–13739, copyright 2003, The American Society for Biochemistry and Molecular Biology, Inc. Substrates used in the assay for human DNA ligase I activity are illustrated at the top of the gels. The substrate shown in the left-hand portion of the figure was not ligated, whereas the substrate shown in the right-hand portion was ligated.

conformation and the flap conformation. More efficient ligation will shift the equilibration toward the formation of hairpin and expansion of the repeated sequence, whereas more efficient FEN1 cleavage will drive the equilibration toward the opposite direction, preventing the expansion [64]. DNA ligase I can also directly interact with Pol β [79, 86]. The biological significance of this interaction in BER has not yet been identified. It is conceivable that the interaction may allow the efficient loading of Lig I onto a nick generated by Pol β. Although this may not be so important for BER per se, it may be critical for preventing DNA slippage and subsequent repeated sequence expansion. XRCC1 also has been identified to physically interact with both Lig III [87, 88] and Pol β [89, 90]. Interestingly, XRCC1–Pol β interaction stimulates Lig III ligation efficiency [91], suggesting that formation of a complex having all three proteins is necessary for achieving an effective nick sealing. The importance of the protein interaction among XRCC1, Pol β, and Lig III or Lig I in maintaining triplet repeat during BER remains to be demonstrated.

The roles of the other essential BER enzymes in maintaining repeat sequence stability may be important to characterize as well. These include bifunctional DNA glycosylases and AP endonuclease. Bifunctional glycosylases can directly create a ssDNA break by its AP lyase activity, whereas AP endonuclease nicks DNA at the 5′ side of an abasic site. Thus, it is obvious that their function is critical for the stability of repeated sequence during BER. Their roles in triplet repeat expansion will be interesting topics to be explored further.

Despite the progress noted earlier concerning BER-related triplet repeat expansion, previous studies on the mechanisms underlying triplet repeat expansion mainly focused on DNA-replication-related expansion and used proliferating cells as model systems. Yet, the replication-mediated mechanisms cannot fully explain why triplet repeat expansion still occurs in some somatic cells, such as nonproliferating neurons, and increases along with aging [35]. Evidently an alternative mechanism underlies the tissue- or cell-type-specific sequence expansions. Striatal neurons are vulnerable to oxidative DNA damage [36], and ssDNA breaks occur at a high frequency [16]. The effects of these insults accumulate as a function of increasing age. It is apparent that DNA damage could be an initiator of sequence expansion.

It may be speculated that DNA damage, accompanied by direct or indirect single-strand break, may emerge as a theme that will bridge BER, ssDNA break repair, sequence expansion, and human neurodegenerative diseases. It is important to understand how ssDNA breaks can lead to triplet repeat instability, and how BER and ssDNA break repair may defend against or even promote the instability. Hence, research in this area may further help to develop the strategies for prevention and treatment of TNR-related human neurodegenerative diseases. By taking advantage of the experience of our laboratory in BER biochemistry, the authors established an *in vitro* system to characterize the roles of the major BER proteins in triplet repeat stability. In this system, purified BER proteins and designed substrates can be introduced to study the roles of individual BER proteins, as well as the multiple protein–protein coordination questions in maintaining triplet repeat stability. Here, experiments are described that were designed to establish a system to identify the roles of DNA base damage and BER specifically in CAG repeat expansion, because this is associated with Huntington's disease. Data suggest an interesting model for triplet repeat expansion as a function of Pol β activity during BER.

II. RESULTS AND DISCUSSION

To mimic the situation in which DNA base damage occurs at a CAG repeat, an oligonucleotide substrate was designed that has 20 CAG repeats flanked by a random DNA sequence at both 5′ and 3′ sides of the repeat track. An 8-hydroxy-7, 8 dihydro-guanine residue (8-oxoG) was introduced at the 5′ CAG triplet (Fig. 42-2). The control oligonucleotide used in these experiments is composed of a random sequence. In both cases, the 8-oxoG is in the same strand that will be the primer strand during Pol β gap-filling DNA synthesis associated with BER (Fig. 42-2). This oxidative base lesion is the 23rd nucleotide from the 5′ end of the oligonucleotide (Fig. 42-2). For both the repeat and random sequence substrates, the damaged strand is annealed to its corresponding template strand. The substrates were radiolabeled at the 5′ end of the damaged strand (Figs. 42-2 and 42-3). This substrate DNA was designed with the idea that it would allow the specific study of Pol β gap filling, reflecting the DNA slippage events occurring downstream of the damage site after the ssDNA break is generated from glycosylase and AP endonuclease reactions. In addition, to further understand the effect of downstream DNA on Pol-β-mediated DNA synthesis, a set of open template substrates was constructed by annealing a 22-nt primer to the same template strands used earlier, repeat sequence template, and the random sequence template, respectively (Fig. 42-4). These substrates were radiolabeled at the 5′ end of the 22-nt primer (Fig. 42-4). Purified human BER enzymes, including human 8-oxo-guanine DNA glycosylase (OGG1), human APE, and human Pol β, were employed in the system. The enzymatic reaction mixtures were incubated at 37°C for 15 min. The products were then separated from the substrates by making use of 15–18% urea polyacrylamide denaturing gel electrophoresis, and radiolabeled DNA molecules were detected by a PhosphorImager.

Incubation of OGG1 with the CAG repeat substrate produced three species of products, two of which are presumably the OGG1AP lyase product, with a "blocked" 3' phosphate or phosphoglycosylate terminus. These products are shown as the upper two bands of lane 1 in Fig. 42-2. The appearance of the lower band corresponding to the 22-nt molecule with 3' oxygen indicates either 5' cleavage of the abasic site after monofunctional removal of 8-oxoG or processing of the 3'-blocked molecules; this reaction product was probably due to a contaminating bacterial analogue of APE such as exonuclease III and endonuclease IV [92] in our purified OGG1 sample.

Addition of both purified APE and OGG1 to the reaction mixture yielded mainly a 22-nt product (Fig. 42-2, lane 2), indicating that APE was effective in cleaving at the 5' side of the AP site after 8-oxoG removal and/or processing the blocked 3' terminus, generating an unblocked 3' oxygen. With the random sequence substrate, OGG1 generated similar species of products as with the CAG repeat substrate (Fig. 42-3, lane 1). Incubation of APE and OGG1 with the random sequence substrate mainly produced the 22-nt product (Fig. 42-3, lane 2). These results indicate that the purified OGG1 and APE samples were effective in removing 8-oxoG and efficiently generating the BER intermediates with a ssDNA break.

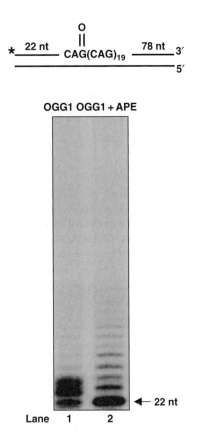

FIGURE 42-2 Removal of 8-oxoG from the CAG repeat track. The CAG repeat substrate containing 8-oxoG is illustrated schematically above the gel. OGG1 (8.3 nM) was incubated with 25 nM substrate in the absence (lane 1) or presence (lane 2) of 50 nM APE at 37°C for 15 min. The substrate was ^{32}P-radiolabeled at the 5' end of the damaged strand. The radiolabeled nucleotide is represented as an asterisk. The size of the DNA fragments and the APE cleavage product is indicated as nt. The upper portion of the gel is also shown. The substrate DNA migrated in the upper portion of the gel as a single band (not shown).

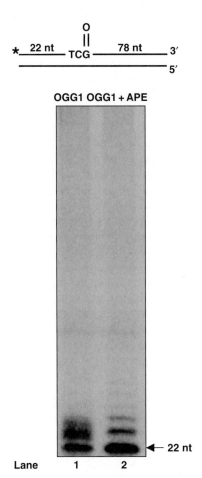

FIGURE 42-3 Removal of 8-oxoG from random sequence DNA. The DNA substrate is illustrated schematically above the gel. OGG1 (8.3 nM) was incubated with 25 nM substrate in the absence (lane 1) or presence (lane 2) of 50 nM APE at 37°C for 15 min. The substrate was ^{32}P-radiolabeled at the 5' end of the damaged strand. The radiolabeled nucleotide is represented as an asterisk. The size of the DNA fragments and the APE cleavage product is indicated as nt. The upper portion of the gel is also shown. The substrate DNA migrated in the upper portion of the gel as a single band (not shown).

To examine the DNA slippage mediated by (CAG)$_n$ repeats after a ssDNA break was generated by OGG1 and APE reactions, the authors took advantage of the fact that Pol β has strong gap-filling activity [93], but weak strand-displacement synthesis activity [50, 71]. Typically, at nanomolar and lower enzyme concentrations, Pol-β-mediated DNA synthesis on the BER intermediates results in insertion of only one nucleotide into a single-nucleotide gap. Alternatively, another enzyme such as FEN1 can continuously create the 1-nt gap for Pol β to fill, eventually allowing the enzyme to repair a longer track [71]. The other situations in which a low nanomolar level of Pol β is known to insert multiple nucleotides would be (1) the short gap of 6 nucleotides or less that are preferred substrates for filling by a processive mechanism; (2) the open template DNA without a downstream strand, where Pol β will insert nucleotides by a distributive mechanism; and (3) a large gap resulting from DNA slippage and misalignment, which Pol β will fill distributively. It was reasoned that a gap could form after the single-strand-break event due to DNA slippage and realignment in the triplet repeat track. Thus, if CAG-repeat-mediated DNA slippage occurs during BER of the 8-oxoG lesion, Pol β DNA synthesis resulting in multiple nucleotide insertions should be detected. On the other hand, if no slippage occurs, Pol β should perform 1-nt insertion only.

It was found that Pol β produced many more multinucleotide insertion products with the CAG repeat substrate than with the random sequence substrate (Fig. 42-4, compare lanes 1 and 2). Almost all of the 22-nt-long substrate

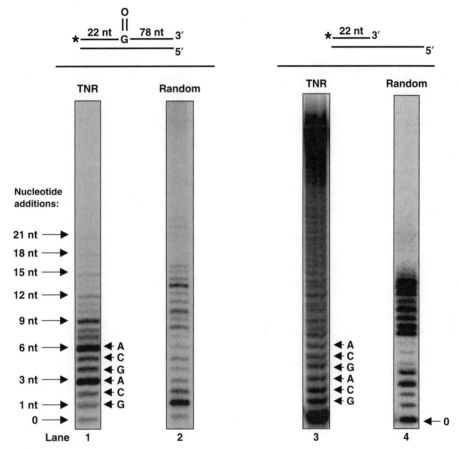

FIGURE 42-4 Pol-β-mediated DNA synthesis with the "closed" and "open" template DNA substrates. The closed substrates with 8-oxoG at either the CAG repeat track (TNR) (lane 1) or the random sequence (random, lane 2) are represented schematically above the gels. OGG1 (8.3 nM) and 50 nM APE along with 1 nM Pol β were incubated with 25 nM substrate at 37 °C for 15 min. For the open template substrates, each contains a 22-nt upstream primer that is annealed to the same template as their closed template counterpart (lane 3 for TNR, lane 4 for random). The reactions were performed by incubating 1 nM Pol β with 25 nM substrate at 37 °C for 15 min. The number of nucleotide additions by Pol β and the sizes of the DNA fragments are indicated by the arrows as nt. For CAG repeat substrates, the nucleotides inserted by Pol β are indicated by the arrows on the right side of lanes 1 and 3, respectively. Substrates were ^{32}P-labeled at the 5′ end of either the damaged DNA strands or the 22-nt upstream primers. The radiolabeled nucleotide is represented as an asterisk.

molecules (designated as zero-nucleotide additions in Fig. 42-4) were consumed during the incubation. The majority of Pol β DNA synthetic products with the repeat sequence substrate represented 3-, 6-, and 9-nucleotide insertions. Some products representing more than 9-nt insertions were also detected, and the amount of these was incrementally lower with an increase in the number of insertions. A distinct periodicity in product length corresponding to 3-nt insertions was evident. Inspection of the sequence indicated that A was the pausing site of Pol β DNA synthesis. Thus, Pol β insertion was slower with each attempt to insert an incoming G. In contrast, the major Pol β synthesis product with the random sequence was the 1-nt insertion product (Fig. 42-4, lane 2). Apparently, the presence of the downstream DNA strand blocks further nucleotide insertion by Pol β after it has filled in the 1-nt gap.

These results clearly suggested that the CAG repeat strand can undergo DNA slippage to create gaps and hairpins during BER. Thus, the 3-nt periodicity of Pol β gap filling on CAG substrate suggests several features of the proposed CAG repeat DNA slippage events. For example, the synthesis pattern is consistent with the formation of various sizes of gaps and hairpins, i.e., a 3-nt gap generated along with a 1-CAG-repeat hairpin and a 6-nt gap created by the formation of a 2-CAG-repeat hairpin. It is known that Pol β can fill gaps of these sizes processively on random DNA [94]. Next, the periodicity of Pol β pausing at A may also suggest the base pairing between the slipped 5′ end nucleotide C of a hairpin with G, the complementary nucleotide in the other strand that resulted from the realignment of the hairpin. From the enzyme's point of view, formation of a 1- or 2-triplet hairpin allows Pol β processive synthesis to fill the resulting short gap and also blocks further Pol β DNA synthesis after the gap has been filled.

To further understand whether the Pol β pausing pattern noted in Fig. 42-3 resulted from the inhibitory effect of a hairpin structure, Pol β DNA synthesis was measured with the "open template" substrates. Surprisingly, with the CAG repeat open template substrate, Pol β was able to perform DNA synthesis all the way to the end of the template to generate a significant amount of template length products (Fig. 42-4, lane 3). No strong 3-nt periodicity of pausing was observed with this substrate, suggesting that insertion bias or features of the CTG repeats in the template strand were not strong enough to pause Pol β. Pol β DNA synthesis with the random sequence was blocked at certain points up to about 15-nt additions, presumably due to the secondary structure in the template strand of this substrate (Fig. 42-4, lane 4).

Next, a substrate was employed that had a fixed 25-CAG-repeat hairpin with a 6-nt 5′-annealed region and a 6-nt gap, as illustrated schematically in Fig. 42-5. This was used to examine the prediction of processive synthesis on a hairpin-containing substrate with a 6-nt gap along with an annealed 5′ margin. It was found that Pol β processively filled in the 6-nt gap, but it failed to insert nucleotides farther downstream (Fig. 42-5).

These results allow for discussion of a model for triplet repeat expansion when DNA base damage or ssDNA break occurs in the triplet repeat track. The model is illustrated schematically in Fig. 42-6. DNA glycosylase (i.e., OGG1 in this case) initiates BER by removing a damaged base (8-oxoG), generating an abasic site (AP). Subsequently, APE cleaves at the AP site, creating a ssDNA break. The ssDNA breakage promotes DNA slippage within the triplet repeat track (CAG repeat) due to the self-base-pairing propensity of this repeat sequence. Consequently, the single-strand slipped DNA forms various sizes of hairpins and gaps. Pol β then fills these gaps, leaving a ligatable (right-hand side of the diagram) nick or in other cases a nick at the bottom of the hairpin (left-hand side of the diagram). Because DNA ligase cannot directly seal the nick at the bottom of a hairpin (Fig. 42-1) [64] and FEN1 cannot remove the hairpin [64, 68], the final integration of this newly synthesized DNA with the hairpin into the genome depends on the creation of a ligatable nick. This can be achieved by dynamic

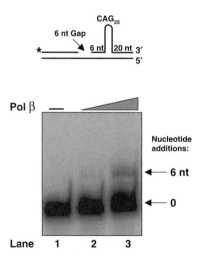

FIGURE 42-5 Pol-β-mediated DNA synthesis on the DNA substrate with a fixed CAG hairpin. The substrate has a fixed 25-CAG-repeat hairpin with a 6-nt 5′-annealed margin and a 6-nt gap, as illustrated schematically above the gel. Increasing concentrations of Pol β (1 and 2.5 nM) (lanes 2 and 3) were incubated with 25 nM substrate at 37°C for 15 min. The enzyme was omitted from the reaction mixture run in lane 1. The number of nucleotide additions by Pol β and the size of the DNA fragments and the gap are indicated as nt. The radioactively labeled nucleotide is illustrated as an asterisk.

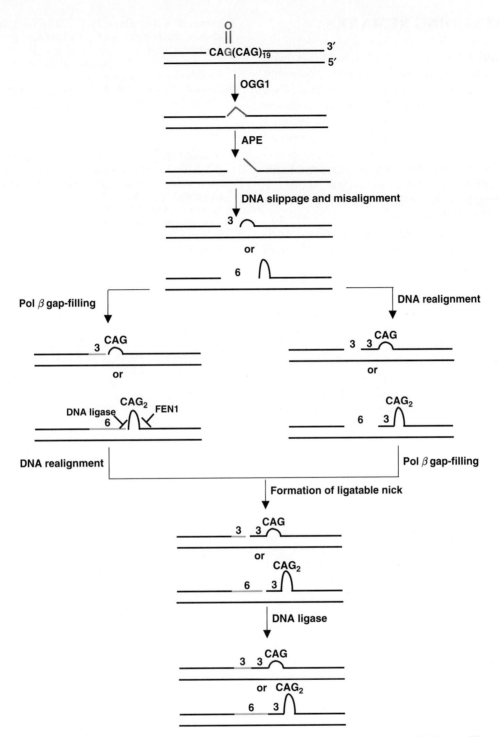

FIGURE 42-6 Model for CAG expansion by three and six nucleotides as a function of Pol β gap filling. Red represents DNA damage, whereas green represents newly synthesized DNA products resulting from Pol β gap filling. The number represents the length of a gap, a newly synthesized DNA fragment, or a 5′-annealed margin. See CD-ROM for color image.

reannealing of a portion of the 5′ end of the hairpin to the template, as illustrated in Fig. 42-6; a ligatable nick should have an annealed region at least 3 nt long (one triplet repeat unit) at the downstream side of the nick [64]. It should be noted that the question of whether Pol β can conduct processive synthesis on a gap with a hairpin at the 5′ margin has not yet been examined.

III. CONCLUDING REMARKS

In conclusion, an *in vitro* model system has been described to understand triplet repeat expansion reflecting events in base excision repair and ssDNA break repair. By employing this system, evidence obtained was consistent with ssDNA-break-initiated triplet repeat DNA slippage during BER. It is proposed that the slipped triplet repeat track can form various sizes of hairpins and gaps. Pol β may be responsible for filling these gaps and promoting repeat sequence expansion during BER or ssDNA repair. However, Pol β gap-filling DNA synthesis is weak on gaps of 9 nt and longer, suggesting that this enzyme may facilitate small expansions. Nevertheless, multiple rounds of small expansions could create larger expansions. Because the extent to which the individual BER proteins, such as glycosylases, APE, Pol β, FEN1, and DNA ligases, influence triplet repeat stability during DNA base damage repair is unknown, the model system described here enables further understanding of these individual enzymes. In the future, the combination of experiments with both *in vivo* and *in vitro* systems hopefully will lead to an understanding of the roles and mechanistic aspects of individual BER factors in maintaining triplet repeat stability.

Acknowledgments

This research was fostered through collaboration with Cynthia T. McMurray. We thank her for discussions and for assisting with conceptualization of the experiments described. We thank Esther W. Hou for samples of purified recombinant human BER enzymes. This research was supported by the Intramural Research Program of the NIH, National Institute of Environmental Health Sciences.

References

1. Wells, R. D., Dere, R., Hebert, M. L., Napierala, M., and Son, L. S. (2005). Advances in mechanisms of genetic instability related to hereditary neurological diseases. *Nucleic Acids Res.* **33**, 3785–3798.
2. Bacolla, A., and Wells, R. D. (2004). Non-B DNA conformations, genomic rearrangements, and human disease. *J. Biol. Chem.* **279**, 47411–47414.
3. Moore, H., Greenwell, P. W., Liu, C. P., Arnheim, N., and Petes, T. D. (1999). Triplet repeats form secondary structures that escape DNA repair in yeast. *Proc. Natl. Acad. Sci. USA* **96**, 1504–1509.
4. Wells, R. D. (1996). Molecular basis of genetic instability of triplet repeats. *J. Biol. Chem.* **271**, 2875–2878.
5. Gordenin, D. A., Kunkel, T. A., and Resnick, M. A. (1997). Repeat expansion—all in a flap? *Nat. Genet.* **16**, 116–118.
6. Jakupciak, J., and Wells, R. D. (2000). Gene conversion (recombination) mediates expansions of CTG·CAG repeats. *J. Biol. Chem.* **275**, 40003–40013.
7. Lahue, R. S., and Slater, D. L. (2003). DNA repair and trinucleotide repeat instability. *Front Biosci.* **8**, S653–S665.
8. Lenzmeier, B. A., and Freudenreich, C. H. (2003). Trinucleotide repeat instability: A hairpin curve at the crossroads of replication, recombination, and repair. *Cytogenet. Genome Res.* **100**, 7–24.
9. Jaworski, A., Rosche, W. A., Gellibolian, R., Kang, S., Shimizu, M., Bowater, R. P., Sinden, R. R., and Wells, R. D. (1995). Mismatch repair in *Escherichia coli* enhances instability of $(CTG)_n$ triplet repeats from human hereditary diseases. *Proc. Natl. Acad. Sci. USA* **92**, 11019–11023.
10. Pearson, C. E., Ewel, A., Acharya, S., Fishel, R. A., and Sinden, R. R. (1997). Human MSH2 binds to trinucleotide repeat DNA structures associated with neurodegenerative diseases. *Hum. Mol. Genet.* **6**, 1117–1123.
11. Manley, K., Shirley, T. L., Flaherty, L., and Messer, A. (1999). Msh2 deficiency prevents in vivo somatic instability of the CAG repeat in Huntington disease transgenic mice. *Nat. Genet.* **23**, 471–473.
12. Kovtun, I. V., and McMurray, C. T. (2001). Trinucleotide expansion in haploid germ cells by gap repair. *Nat. Genet.* **27**, 407–411.
13. Freudenreich, C. H., Kantrow, S. M., and Zakian, V. A. (1998). Expansion and length-dependent fragility of CTG repeats in yeast. *Science* **279**, 853–856.
14. Parniewski, P., Bacolla, A., Jaworski, A., and Wells, R. D. (1999). Nucleotide excision repair affects the stability of long transcribed (CTG·CAG) tracts in an orientation-dependent manner in *Escherichia coli*. *Nucleic Acids Res.* **27**, 616–623.
15. Caldecott, K. W. (2003). Protein-protein interactions during mammalian DNA single-strand break repair. *Biochem. Soc. Trans.* **31**, 247–251.
16. Lindahl, T. (1993). Instability and decay of the primary structure of DNA. *Nature* **362**, 709–715.
17. Ward, J. F. (1998). DNA repair in higher eukaryotes. *In* "DNA Damage and Repair" (J. A. Nivkoloff, and M. F. Hoekstra, Eds.), Vol. 2, pp. 65–84. Human Press, Totowa, NJ.
18. Lindahl, T., and Karlstrom, O. (1973). Heat-induced depyrimidination of deoxyribonucleic acid in neutral solution. *Biochemistry* **12**, 5151–5154.
19. Lindahl, T. (1974). An N-glycosylase from *Escherichia coli* that releases free uracil from DNA containing deaminated cytosine residues. *Proc. Natl. Acad. Sci. USA* **71**, 3649–3653.
20. Lindahl, T., and Ljungquist, S. (1975). Apurinic and apyrimidinic sites in DNA. *Basic Life Sci.* **5A**, 31–38.
21. Lindahl, T. (1979). DNA glycosylases, endonucleases for apurinic/apyrimidinic sites and base excision repair. *Prog. Nucleic Acid Res. Mol. Biol.* **22**, 135–192.
22. Friedberg, E. C., Walker, G. C., and Siede, W. (1995). "DNA Repair and Mutagenesis," pp. 154–169. ASM Press.
23. Lindahl, T. (1982). DNA repair enzymes. *Annu. Rev. Biochem.* **51**, 61–87.
24. Matsumoto, Y., and Kim, K. (1995). Excision of deoxyribose phosphate residues by DNA polymerase β during DNA repair. *Science* **269**, 699–702.
25. Prasad, R., Beard, W. A., Strauss, P. R., and Wilson, S. H. (1998). Human DNA polymerase beta deoxyribose phosphate lyase. Substrate specificity and catalytic mechanism. *J. Biol. Chem.* **273**, 15263–15270.
26. de Murcia, G., and Menissier de Murcia, J. (1994). Poly(ADP-ribose) polymerase: A molecular nick-sensor. *Trends Biochem. Sci.* **19**, 250.
27. Lindahl, T., Satoh, M. S., Poirier, G. G., and Klungland, A. (1995). Post-translational modification of poly(ADP-ribose) polymerase induced by DNA strand breaks. *Trends Biochem. Sci.* **20**, 405–411.
28. Moore, D. J., Taylor, R. M., Clements, P., and Caldecott, K. W. (2000). Mutation of a BRCT domain selectively disrupts DNA

single-strand break repair in noncycling Chinese hamster ovary cells. *Proc. Natl. Acad. Sci. USA* **97**, 13649–13654.
29. Whitehouse, C. J., Taylor, R. M., Thistlethwaite, A., Zhang, H., Karimi-Busheri, F., Lasko, D. D., Weinfeld, M., and Caldecott, K. W. (2001). XRCC1 stimulates human polynucleotide kinase activity at damaged DNA termini and accelerates DNA single-strand break repair. *Cell* **104**, 107–117.
30. Taylor, R. M., Thistlethwaite, A., and Caldecott, K. W. (2002). Central role for the XRCC1 BRCT I domain in mammalian DNA single-strand break repair. *Mol. Cell. Biol.* **22**, 2556–2563.
31. Li, L., Nakajima, S., Oohata, Y., Takao, M., Okano, S., Masutani, M., Wilson, S. H., and Yasui, A. (2004). In situ analysis of repair processes for oxidative DNA damage in mammalian cells. *Proc. Natl. Acad. Sci. USA* **101**, 13738–13743.
32. Perez-Severiano, F., Rios, C., and Segovia, J. (2000). Striatal oxidative damage parallels the expression of a neurological phenotype in mice transgenic for the mutation of Huntington's disease. *Brain Res.* **862**, 234–237.
33. Butterfield, D. A., Howard, B. J., and LaFontaine, M. A. (2001). Brain oxidative stress in animal models of accelerated aging and the age-related neurodegenerative disorders, Alzheimer's disease and Huntington's disease. *Curr. Med. Chem.* **8**, 815–828.
34. Bogdanov, M. B., Andreassen, O. A., Dedeoglu, A., Ferrante, R. J., and Beal, M. F. (2001). Increased oxidative damage to DNA in a transgenic mouse model of Huntington's disease. *J. Neurochem.* **79**, 1246–1249.
35. Kennedy, L., and Shelbourne, P. F. (2000). Dramatic mutation instability in HD mouse striatum: Does polyglutamine load contribute to cell-specific vulnerability in Huntington's disease? *Hum. Mol. Genet.* **9**, 2539–2544.
36. Cardozo-Pelaez, F., Song, S., Parthasarathy, A., Hazzi, C., Naidu, K., and Sanchez-Ramos, J. (1999). Oxidative DNA damage in the aging mouse brain. *Mov. Disorders* **14**, 972–980.
37. Hashem, V. I., and Sinden, R. R. (2002). Chemotherapeutically induced deletion of expanded triplet repeats. *Mutat. Res.* **508**, 107–119.
38. Nunoshiba, T., Obata, F., Boss, A. C., Oikawa, S., Mori, T., Kawanishi, S., and Yamamoto, K. (1999). Role of iron and superoxide for generation of hydroxyl radical, oxidative DNA lesions, and mutagenesis in *Escherichia coli*. *J. Biol. Chem.* **274**, 34832–34837.
39. Alhama, J., Ruiz-Laguna, J., Rodriguez-Ariza, A., Toribio, F., Lopez-Barea, J., and Pueyo, C. (1998). Formation of 8-oxoguanine in cellular DNA of *Escherichia coli* strains defective in different antioxidant defences. *Mutagenesis* **13**, 589–594.
40. Shibata, A., Kamada, N., Masumura, K., Nohmi, T., Kobayashi, S., Teraoka, H., Nakagama, H., Sugimura, T., Suzuki, H., and Masutani, M. (2005). Parp-1 deficiency causes an increase of deletion mutations and insertions/rearrangements in vivo after treatment with an alkylating agent. *Oncogene* **24**, 1328–1337.
41. Caldecott, K. W., Aoufouchi, S., Johnson, P., and Shall, S. (1996). XRCC1 polypeptide interacts with DNA polymerase beta and possibly poly (ADP-ribose) polymerase, and DNA ligase III is a novel molecular 'nick-sensor' in vitro. *Nucleic Acids Res.* **24**, 4387–4394.
42. Okano, S., Lan, L., Caldecott, K. W., Mori, T., and Yasui, A. (2003). Spatial and temporal cellular responses to single-strand breaks in human cells. *Mol. Cell. Biol.* **23**, 3974–3981 [Erratum (2003). *Mol. Cell. Biol.* **23**, 5472].
43. Lyons-Darden, T., and Topal, M. D. (1999). Abasic sites induce triplet-repeat expansion during DNA replication in vitro. *J. Biol. Chem.* **274**, 25975–25978.
44. Heidenfelder, B. L., and Topal, M. D. (2003). Effects of sequence on repeat expansion during DNA replication. *Nucleic Acids Res.* **31**, 7159–7164.
45. Heidenfelder, B. L., Makhov, A. M., and Topal, M. D. (2003). Hairpin formation in Friedreich's ataxia triplet repeat expansion. *J. Biol. Chem.* **278**, 2425–2431.
46. Srivastava, D. K., Vande Berg, B. J., Prasad, R., Molina, J. T., Beard, W. A., Tomkinson, A. E., and Wilson, S. H. (1998). Mammalian abasic site base excision repair. Identification of the reaction sequence and rate-determining steps. *J. Biol. Chem.* **273**, 21203–21209.
47. Wilson, S. H., Singhal, R. K., and Zmudzka, B. Z. (1998). Studies of DNA polymerases in replication-based repeat expansion. *In* "Genetic Instabilities and Hereditary Neurological Diseases" (R. D. Wells, and S. T., Warren, Eds.), pp. 693–698. Academic Press.
48. Kunkel, T. A., Alexander, P. S., Liu, J. C., and Fox, J. M. (1986). Mutagenesis in vitro by DNA polymerases α, β and γ. *Prog. Clin. Biol. Res.* **209A**, 441–447.
49. Ji, J., Clegg, N. J., Peterson, K. R., Jackson, A. L., Laird, C. D., and Loeb, L. A. (1996). In vitro expansion of GGC:GCC repeats: Identification of the preferred strand of expansion. *Nucleic Acids Res.* **24**, 2835–2840.
50. Hartenstine, M. J., Goodman, M. F., and Petruska, J. (2002). Weak strand displacement activity enables human DNA polymerase β to expand CAG/CTG triplet repeats at strand breaks. *J. Biol. Chem.* **277**, 41379–41389.
51. Bambara, R. A., Murante, R. S., and Henricksen, L. A. (1997). Enzymes and reactions at the eukaryotic DNA replication fork. *J. Biol. Chem.* **272**, 4647–4650.
52. Liu, Y., Kao, H.-I., and Bambara, R. A. (2004). Flap endonuclease 1: A central component of DNA metabolism. *Annu. Rev. Biochem.* **73**, 589–615.
53. DeMott, M. S., Shen, B., Park, M. S., Bambara, R. A., and Zigman, S. (1996). Human RAD2 homolog 1 5'- to 3'-exo/endonuclease can efficiently excise a displaced DNA fragment containing a 5'-terminal abasic lesion by endonuclease activity. *J. Biol. Chem.* **271**, 30068–30076.
54. Klungland, A., and Lindahl, T. (1997). Second pathway for completion of human DNA base excision-repair: Recombination with purified proteins and requirement for DNase IV (FEN1). *EMBO J.* **16**, 3341–3348.
55. Prasad, R., Lavrik, O. I., Kim, S. J., Kedar, P., Yang, X. P., Vande Berg, B. J., and Wilson, S. H. (2001). DNA polymerase β-mediated long patch base excision repair. Poly(ADP-ribose) polymerase-1 stimulates strand displacement DNA synthesis. *J. Biol. Chem.* **276**, 32411–32414.
56. Morel, P., Reverdy, C., Michel, B., Ehrlich, S. D., and Cassuto, E. (1998). The role of SOS and flap processing in microsatellite instability in *Escherichia coli*. *Proc. Natl. Acad. Sci. USA* **95**, 10003–10008.
57. Nagata, Y., Mashimo, K., Kawata, M., and Yamamoto K. (2002). The roles of Klenow processing and flap processing activities of DNA polymerase I in chromosome instability in *Escherichia coli* K12 strains. *Genetics* **160**, 13–23.
58. Johnson, R. E., Kovvali, G. K., Prakash, L., and Prakash, S. (1995). Requirement of the yeast RTH1 5' to 3' exonuclease for the stability of simple repetitive DNA. *Science* **269**, 238–240.
59. Tishkoff, D. X., Filosi, N., Gaida, G. M., and Kolodner, R. D. (1997). A novel mutation avoidance mechanism dependent on *S. cerevisiae* RAD27 is distinct from DNA mismatch repair. *Cell* **88**, 253–263.
60. Kokoska, R. J., Stefanovic, L., Tran, H. T., Resnick, M. A., Gordenin, D. A., and Petes, T. D. (1998). Destabilization of yeast micro- and minisatellite DNA sequences by mutations affecting a nuclease involved in Okazaki fragment processing (rad27) and DNA polymerase delta (pol3-t). *Mol. Cell. Biol.* **18**, 2779–2788.

61. Spiro, C., Pelletier, R., Rolfsmeier, M. L., Dixon, M. J., Lahue, R. S., Gupta, G., Park, M. S., Chen, X., Mariappan, S. V., and McMurray, C. T. (1999). Inhibition of FEN-1 processing by DNA secondary structure at trinucleotide repeats. *Mol. Cell* **4**, 1079–1085.
62. Greene, A., Snipe, J. R, Gordenin, D. A, and Resnick, M. A. (1999). Functional analysis of human FEN1 in *Saccharomyces cerevisiae* and its role in genome stability. *Hum. Mol. Genet.* **8**, 2263–2273.
63. Xie, Y., Liu, Y., Argueso, J. L., Henricksen, L. A., Kao, H. I., Bambara, R. A., and Alani, E. (2001). Identification of rad27 mutations that confer differential defects in mutation avoidance, repeat tract instability, and flap cleavage. *Mol. Cell. Biol.* **21**, 4889–4899.
64. Liu, Y., and Bambara, R. (2003). Analysis of human flap endonuclease 1 mutants reveals a mechanism to prevent triplet repeat expansion. *J. Biol. Chem.* **278**, 13728–13739.
65. Kucherlapati, M., Yang, K., Kuraguchi, M., Zhao, J., Lia, M., Heyer, J., Kane, M. F., Fan, K., Russell, R., Brown, A. M., Kneitz, B., Edelmann, W., Kolodner, R. D., Lipkin, M., and Kucherlapati, R. (2002). Haploinsufficiency of flap endonuclease (Fen1) leads to rapid tumor progression. *Proc. Natl. Acad. Sci. USA* **99**, 9924–9929.
66. Spiro, C., and McMurray, C. T. (2003). Nuclease-deficient FEN-1 blocks Rad51/BRCA1-mediated repair and causes trinucleotide repeat instability. *Mol. Cell. Biol.* **23**, 6063–6074.
67. Otto, C. J., Almqvist, E., Hayden, M. R., and Andrew, S. E. (2001). The "flap" endonuclease gene FEN1 is excluded as a candidate gene implicated in the CAG repeat expansion underlying Huntington disease. *Clin. Genet.* **59**, 122–127.
68. Henricksen, L. A., Tom, S., Liu, Y., and Bambara, R. A. (2000). Inhibition of flap endonuclease 1 by flap secondary structure and relevance to repeat sequence expansion. *J. Biol. Chem.* **275**, 16420–16427.
69. Ruggiero, B. L., and Topal, M. D. (2004). Triplet repeat expansion generated by DNA slippage is suppressed by human flap endonuclease 1. *J. Biol. Chem.* **279**, 23088–23097.
70. Liu, Y., Zhang, H., Veeraraghavan, J., Bambara, R. A., and Freudenreich, C. H. (2004). *Saccharomyces cerevisiae* flap endonuclease 1 uses flap equilibration to maintain triplet repeat stability. *Mol. Cell. Biol.* **24**, 4049–4064.
71. Liu, Y., Beard, W. A., Shock, D. D., Prasad, R., Hou, E. W., and Wilson, S. H. (2005). DNA polymerase beta and flap endonuclease 1 enzymatic specificities sustain DNA synthesis for long patch base excision repair. *J. Biol. Chem.* **280**, 3665–3674.
72. Wilson, S. H., and Kunkel, T. A. (2000). Passing the baton in base excision repair. *Nat. Struct. Biol.* **7**, 176–178.
73. Brosh, R. J., von Kobbe, C., Sommers, J., Karmakar, P., Opresko, P., Piotrowski, J., Dianova, I., Dianov, G., and Bohr, V. (2001). Werner syndrome protein interacts with human flap endonuclease 1 and stimulates its cleavage activity. *EMBO J.* **20**, 5791–5801.
74. Brosh, R. J., Driscoll, H., Dianov, G., and Sommers, J. (2002). Biochemical characterization of the WRN-FEN-1 functional interaction. *Biochemistry* **41**, 12204–12216.
75. Imamura, O., and Campbell, J. (2003). The human Bloom syndrome gene suppresses the DNA replication and repair defects of yeast DNA2 mutants. *Proc. Natl. Acad. Sci. USA* **100**, 8193–8198.
76. Wang, W., and Bambara, R. A. (2005). Human Bloom protein stimulates flap endonuclease 1 activity by resolving DNA secondary structure. *J. Biol. Chem.* **280**, 5391–5399.
77. Fry, M., and Loeb, L. A. (1999). Human Werner syndrome DNA helicase unwinds tetrahelical structures of the fragile X syndrome repeat sequence d(CGG)n. *J. Biol. Chem.* **274**, 12797–12802.
78. Harrigan, J. A., Opreako, P. L., von Kobbe, C., Kedar, P., Prasad, R., Wilson, S. H., and Bohr, V. A. (2003). The Werner syndrome protein stimulates DNA polymerase β strand-displacement synthesis via its helicase activity. *J. Biol. Chem.* **278**, 22686–22695.
79. Prasad, R., Singhal, R. K., Srivastava, D. K., Molina, J. T., Tomkinson, A. E., and Wilson, S. H. (1996). Specific interaction of DNA polymerase β and DNA ligase I in a multiprotein base excision repair complex from bovine testis. *J. Biol. Chem.* **271**, 16000–16007.
80. Matsumoto, Y., Kim, K., Hurwitz, J., Gary, R., Levin, D. S., Tomkinson, A. E., and Park, M. S. (1999). Reconstitution of proliferating cell nuclear antigen-dependent repair of apurinic/apyrimidinic sites with purified human proteins. *J. Biol. Chem.* **274**, 33703–33708.
81. Winters, T. A., Russell, P. S., Kohli, M., Dar, M. E., Neumann, R. D., and Jorgensen, T. J. (1999). Determination of human DNA polymerase utilization for the repair of a model ionizing radiation-induced DNA strand break lesion in a defined vector substrate. *Nucleic Acids Res.* **27**, 2423–2433.
82. Pascucci, B., Stucki, M., Jonsson, Z. O., Dogliotti, E., and Hubscher, U. (1999). Long patch base excision repair with purified human proteins. DNA ligase I as patch size mediator for DNA polymerases δ and ε. *J. Biol. Chem.* **274**, 33696–33702.
83. Tomkinson, A. E., Chen, L., Dong, Z., Leppard, J. B., Levin, D. S., Mackey, Z. B., and Motycka, T. A. (2001). Completion of base excision repair by mammalian DNA ligases. *Prog. Nucleic Acid Res. Mol. Biol.* **68**, 151–164.
84. Cappelli, E., Taylor, R., Cevasco, M., Abbondandolo, A., Caldecott, K., and Frosina, G. (1997). Involvement of XRCC1 and DNA ligase III gene products in DNA base excision repair. *J. Biol. Chem.* **272**, 23970–23975.
85. Henricksen, L., Veeraraghavan, J., Chafin, D. R., and Bambara, R. A. (2002). DNA ligase I competes with FEN1 to expand repetitive DNA sequences in vitro. *J. Biol. Chem.* **277**, 22361–22369.
86. Dimitriadis, E. K., Prasad, R., Vaske, M. K., Chen, L., Tomkinson, A. E., Lewis, M. S., and Wilson, S. H. (1998). Thermodynamics of human DNA ligase I trimerization and association with DNA polymerase beta. *J. Biol. Chem.* **273**, 20540–20550.
87. Caldecott, K. W., McKeown, C. K., Tucker, J. D., Ljungquist, S., and Thompson, L. H. (1994). An interaction between the mammalian DNA repair protein XRCC1 and DNA ligase III. *Mol. Cell. Biol.* **14**, 68–76.
88. Nash, R. A., Caldecott, K. W., Barnes, D. E., and Lindahl, T. (1997). XRCC1 protein interacts with one of two distinct forms of DNA ligase III. *Biochemistry* **36**, 5207–5211.
89. Marintchev, A., Robertson, A., Dimitriadis, E. K., Prasad, R., Wilson, S. H., and Mullen, G. P. (2000). Domain specific interaction in the XRCC1-DNA polymerase beta complex. *Nucleic Acids Res.* **28**, 2049–2059.
90. Gryk, M. R., Marintchev, A., Maciejewski, M. W., Robertson, A., Wilson, S. H., and Mullen, G. P. (2002). Mapping of the interaction interface of DNA polymerase β with XRCC1. *Structure* **10**, 1709–1720.
91. Dianova, I., Sleeth, K. M., Allinson, S. L., Parsons, J. L., Breslin, C., Caldecott, K. W., and Dianov, G. L. (2004). XRCC1-DNA polymerase beta interaction is required for efficient base excision repair. *Nucleic Acids Res.* **32**, 2550–2555.
92. Friedberg, E. C., Walker, G. C., and Siede, W. (1995). "DNA Repair and Mutagenesis," pp. 169–171. ASM Press.
93. Wilson, S. H. (1998). Mammalian base excision repair and DNA polymerase β. *Mutat. Res.* **407**, 203–215.
94. Singhal, R. K., and Wilson, S. H. (1993). Short gap-filling synthesis by DNA polymerase β is processive. *J. Biol. Chem.* **268**, 15906–15911.

Models of Repair Underlying Trinucleotide DNA Expansion

I. V. KOVTUN AND C. T. MCMURRAY

Department of Molecular Pharmacology and Experimental Therapeutics, Department of Biochemistry and Molecular Biology, Molecular Neuroscience Program, Mayo Clinic, Rochester, Minnesota 55905

I. Introduction
II. General Considerations: Mechanisms of Expansion Based on Human Disease Data
III. Repair Rather Than Mitotic Replication
IV. Break Repair Mechanisms
  A. Rescue of a Stalled Replication Fork by Recombination or Polymerase Reversal
  B. Excision Repair
V. Emerging and Unresolved Issues: The Role of Mismatch Repair Proteins
References

Genetic instability observed in human disease is termed an "expansion mutation" when the repeat length in a mutant allele exceeds that in a normal allele. Although the mutation is well-characterized, most of the proposed mechanisms for expansion fall short of explaining the features of human disease. Therefore, the mechanism by which expansion occurs is not yet clear. Emerging evidence over the past 6 years has suggested that the mutation arises from repair at a DNA break rather than during mitosis. How, why, and when breaks are made in DNA are subjects of debate. Nonetheless, DNA repair has taken center stage as the source of the mutational mechanism. This chapter will focus on repair-dependent models for the expansion at a DNA break.

Initially, all expansion diseases were reported to involve increases in CAG, CGG, GAA, or their complements [1–3]. However, the list of expanding repeats has grown over the years to include other triplets [4], tetramers [5], pentamers [6], and dodecamers [7]. Additionally, there is a growing list of polyalanine diseases [8–10], which arise from frameshift at CAG repeats within the coding sequence of genes. Thus, the mechanism of DNA expansion is undoubtedly complex. Despite its complexity, the common ability for DNA to "grow" at repetitive tracts suggests that the mechanism of expansion underlying each repeat is likely to share at least some steps.

I. INTRODUCTION

It has been over 12 years since DNA expansion was recognized as a mutation. Although data from a number of experimental systems have shed light on features of the mechanism, the basic properties of the expansion observed in human disease have not been adequately explained.

II. GENERAL CONSIDERATIONS: MECHANISMS OF EXPANSION BASED ON HUMAN DISEASE DATA

In considering any model for triplet expansion, it is crucial that the mechanism be consistent with the features of expansion observed in human disease.

Therefore, a proposed mechanism must explain (1) why expansion, rather than contraction, is favored as the mutation is transmitted from generation to generation and (2) why these expansions increase in frequency and in size during transmission. It is generally accepted that loops comprising secondary structures are intermediates in the expansion mechanism. Thus, the features of expansion in human disease most likely reflect a mechanism of loop processing and loop incorporation into DNA. The biophysical properties of repeat-containing sequences and their associated loop-out structures have been well-described [11–13], and model organisms have provided critical *in vivo* data to support their involvement. However, the mechanisms for expansion remain undefined and fall short of fully explaining the features of disease.

The central issues are when and how looped structures are formed and how they are processed to generate expansions. Looped DNA structures may form during mitosis by polymerase slippage. However, emerging data favor the notion that loops occur during error-prone repair of a DNA break. The most studied cases of DNA expansion are at the CAG, CTG, and CGG repeats (referred to as CNG repeats). Therefore, models and proposed mechanisms of expansion will be discussed in the context of the CNG repeats with the expectation that they might provide insight into DNA expansion mechanisms at more complex sites.

III. REPAIR RATHER THAN MITOTIC REPLICATION

Expansion is mediated by misaligned pairing of repeats and the inappropriate formation of DNA secondary structure after the duplex unpairs [14, 15]. The first and most straightforward mechanism proposed for DNA opening and expansion was polymerase slippage during mitosis (Fig. 43-1). In this model, the polymerase dissociates from the repeat segment on the template strand during DNA synthesis and "slips" back to pair at a previously replicated triplet unit. The replicated repeat DNA forms an extrahelical loop that can be incorporated into the DNA. Loops on the leading strand template give rise to contractions, whereas loops on the nascent strand give rise to expansions. Although unpairing reactions in duplex DNA are energetically unfavorable, slippage during DNA synthesis is possible because the unpairing reaction occurs largely when polymerase falls off the DNA. Because hydrogen-bonded structures appear to be important intermediates in expan-

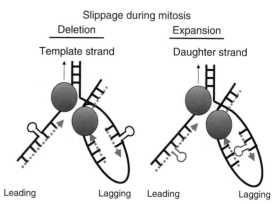

FIGURE 43-1 Model for expansion by polymerase slippage during replication. Expansion is mediated by misaligned pairing of repeats and the inappropriate formation of DNA secondary structure after the duplex unpairs during cell proliferation. The polymerase (red balls) dissociates from the repeat segment on the template strand during DNA synthesis and slips back to bind at a previously replicated triplet unit. The replicated repeat DNA forms an extrahelical loop that can be incorporated into the DNA. (Left) A loop can form on the template (solid lines) on either the leading or the lagging strand, as indicated. Loops on the template strand template give rise to deletions. (Right) A loop can form on the nascent daughter strand (dashed lines) on either the parental leading or the parental lagging strand, as indicated. Loops on the template strand template give rise to deletions. Loops on the nascent strand give rise to expansions.

sion, rapid and stable intrastrand hydrogen bonding is likely to allow extrahelical structures to prevail over duplex reannealing.

Slippage is known to occur at repetitive elements [16–21] and is, therefore, an attractive model for expansion. However, slippage, at least during mitosis, does not satisfactorily account for the features of expansion in human disease. Expansion in coding sequences of genes may be as large as 20 repeats, and in noncoding sequences there can be thousands within a single transmission [1]. However, slippage events as observed in bacteria and yeast are typically small [22, 23]. It is unlikely, then, that slips of 20–1000 occur in a single step. Much more likely is a model in which expansion arises from multiple slippage events when polymerase traverses a repeat tract. The number of probable slips is predicted to increase as the repeat tract grows, accounting for the increasing size of the expansion with the length of the repeat. However, a mitotic replication model predicts that slippage should occur in both the daughter and the template strands, and a relatively equal number of expansion and contraction events would be expected. This outcome, beyond the permutation range, is not borne out in human disease. In Huntington's disease (HD), 52.1% of parent–child transmissions of an intermediate allele result in expansion, whereas only 18% result in contraction (the remaining are stable transmissions) [24]. Once the permutation creeps into the full-mutation range, the expansion rate is reported to increase by nearly 175-fold over the contraction rate in expansion diseases [24–29]. When expansion is modeled in rapidly replicating systems such as yeast and bacteria, DNA instability indeed is observed at CAG/CTG repeats, but contractions, rather than expansions, prevail [17, 18, 21, 30]. *In vivo*, expanded products are present in the replicating system, but their detection requires a selection method. For example, Lahue and coworkers have quantified the expansion rate of CAG, CTG [19], GAA [30, 31], and CGG [32] repeats in yeast by using a novel 5-fluoroorotic acid selection system. Expansion of triplet repeats located within the 5′ untranslated region of the URA3 gene shuts off its expression, providing resistance to toxic effects of the drug. Surviving colonies displayed an increased rate of expansion only at structure-forming CNG repeats as compared to control CTA/TAG repeats [19]. In the absence of the selection system, contractions dominated.

In mammalian cell lines, the impact of replication on the fate of triplet repeats is far less clear. Human fibroblasts from patients with different triplet expansion diseases have shown little if any alteration at the corresponding disease locus upon proliferation in culture. These included fibroblasts from spinobulbar muscular atrophy [33], HD [34], and fragile X [35] patients. Similar results have been obtained in embryonic fibroblasts from R6/1 transgenic mice harboring an expanded version of the *hHD* gene (135 repeats) [34]. However, the length of the alleles and their tendency to expand are clearly related. Longer CAG repeat tracts (145–160 repeats) found in the immortalized fibroblasts from R6/2 *HD* transgenic mice show a slow tendency to expand when cultured over long periods (15 months) [36]. Similarly, synchronous expansion is observed at long CTG tracts [400 repeats] in proliferating cells isolated from myotonic dystrophy (DM) patients and from DM transgenic mice [35, 37]. Despite the tendency for long alleles to expand in culture, no correlation is found between expansion rate and the rate of cell division [37]. For example, in DM mouse embryonic fibroblasts (MEFS), lung cells isolated from DM animals stably maintained their CTG repeats in culture, whereas repeats in kidney cells were remarkably unstable. Thus, the degree of instability depended on tissue type rather than the rate of cell division. These data suggest that locus-specific differences and/or tissue-specific factors may be important components of the expansion process rather than replication per se.

It should also be noted that interpretation of *in vitro* data on triplet expansion in cultured mammalian cells is complicated by the fact that long triplet alleles can break during cell proliferation. Therefore, the effects of replication and repair on the stability of CAG and CTG tracts in cultured cells are difficult to separate. Expansion may arise in dividing cells, but through a break-dependent process.

To better distinguish between replication and repair, germ cells have been shown to be a useful system. Developing germ cells can be separated into distinct cell types, which divide (spermatogonia), undergo meiotic recombination (spermatocytes), or differentiate and allow repair without replication (haploid spermatids) (Fig. 43-2A). Several groups have measured the degree of expansion at each stage of sperm development to determine which cell type is associated with the mutation. In R6/1 mice, expansion of a CAG tract of the *hHD* transgene has been observed primarily in the haploid cells. Because haploid cells contain no homologous chromosome, a break repair mechanism for expansion is implicated [38]. In testicular cells from HD patients, expansion of CAG repeats, in addition to the postmeiotic cell population (spermatids), has been found in meiotic cells (spermatocytes), suggesting that expansions continue to occur in at least some cells during meiosis and/or after meiosis is complete [39]. In general, these data favor the idea that CAG expansion at the HD locus is likely to be a break-dependent process.

DM animals containing much longer repeat tracts, on the other hand, display a different profile. In DM mice, expansion could be observed in dividing spermatogonia and in terminally differentiated spermatids [40]. Repeat lengths in the germ cells of these DM mice continued to increase with age, indicating that expansions were continuously produced during the proliferative phase throughout life. Data from humans have revealed that the repeat tracts in oocytes of female DM patients were longer than those in blood, providing strong evidence that expansions were present in the oocyte precursor cells before oocyte development. Expansions were not observed at any developmental stage of the oocyte (Fig. 43-2B), but they were observed at postfertilization steps in dividing cells of early embryo [41]. These data from both DM patients and mice suggest that expansion can occur during proliferative stages of the development in somatic and germ cells [40, 41]. The difference in mutation behavior in DM and HD might be due to (1) the fact that repeat stretches are longer in the DM allele (favoring breakage), (2) locus-specific differences and/or differential processing of untranslated versus translated regions of DNA, or (3) entirely different mechanisms that are utilized. Neither somatic nor germ

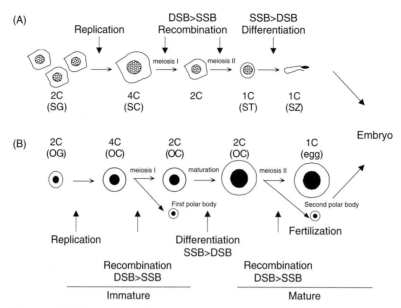

FIGURE 43-2 Stages of germ cell development in males and females. Stages of repair and replication occur during different steps of germ cell development. (A) Development of the male germ cells. Distinct cell types divide (spermatogonia, SG), undergo meiotic recombination (spermatocytes, SC), or differentiate and allow repair without replication (haploid spermatids, ST, or spermatozoa, SZ). Depending on the developmental stage, processes of double-strand-break repair, single-strand-break repair, or simple gap repair can be envisioned. The ploidy of cells in each developmental stage is shown (1C, 2C, and 4C represent the complement of DNA). Germ cell precursors, spermatogonia (SG), are 2C and undergo mitotic replication to produce 4C primary spermatocytes (SC). SC then undergo a rapid first division to produce secondary spermatocytes (2C), which, in turn, generate round spermatids (ST) in the second meiotic division, which undergo differentiation into mature SZ. (B) Schematic representation of oogenesis. Diploid (2C) oogonia (OG), upon replication, give rise to tetraploid primary oocytes (OC, 4C) which undergo first meiotic reduction to produce diploid secondary oocytes (OC, 2C). Diploid oocytes that are arrested in the metaphase of meiosis II go through the process of maturation, increasing in size. The second meiosis, generating a mature egg, occurs after fertilization. Fusion of the egg and sperm begins embryogenesis (embryo).

cells have clarified these issues. The importance of DNA breaks in causing expansion, however, has been directly examined by exposing proliferating cells to DNA-damaging genotoxic agents. The authors and others have reported that disease-length triplet repeats in cultured cells are more unstable after exposure [34, 42, 43]. Importantly, proliferating HD cells stably maintain their repeats in culture, but expand them after treatment with peroxide, indirectly implying the involvement of repair machinery in expansion [34].

A break-repair-dependent mechanism for expansion is supported by several lines of evidence. In yeast, when breaks are generated by HO endonuclease cleavage, repair can occur by gene conversion from a plasmid containing homologous sequences [44, 45]. Expansions in this system have been observed exclusively at the break sites, whereas contractions are seen only on the replicating plasmids [45]. It has been demonstrated that the occurrence of expansions and contractions in replicating mammalian systems depends on the proximity of the repeat tract to the origin of replication [46]. This dependence could be due to the presence of both proximal breaks and replication, where unligated ends are created during DNA synthesis at bidirectional origins of replication proximal to the repeats [46]. Several pieces of evidence demonstrate that contractions occur in cells that proliferate *in vivo*. For example, as they divide throughout a lifetime, testicular cells of both R6/1 HD transgenic mice [34] and HD patients show contractions rather than expansions. Expansion, on the other hand, is observed in the somatic tissues of transgenic animals as they age. The inherited repeat tract is maintained stably from birth; expansion begins at midlife and continues for the lifetime of the animal. Thus, expansion is associated with adult cells, which are terminally differentiated, and not with early stages of development when cells are proliferating. As a whole, a simple polymerase slippage during mitosis does not predict that expansions should be increasingly favored over contractions. Mitotic replication is likely to play a role in expansion, but in the context of DNA break repair.

IV. BREAK REPAIR MECHANISMS

A. Rescue of a Stalled Replication Fork by Recombination or Polymerase Reversal

At long arrays, replicative polymerase may fail to fully traverse the repeats, resulting in stalling of the replication fork and/or incomplete replication and chromosome breakage. Polymerase pausing and/or blocks have been demonstrated at CGG [32, 47, 48], CTG [32, 47, 48], and GAA [31], both *in vivo* and *in vitro*. Arrested replication forks may then be targeted by nucleases, thereby providing a substrate for double-strand-break (DSB) repair (Fig. 43-3A). For example, in bacteria, breakage was proposed to occur at replication forks blocked by a replication terminator sequence, a specific site that arrests bacterial chromosome replication [49], or at repeating sequences [50]. Alternatively, repair of triplet repeats may be induced by replication pausing at impassable sequences (bulky structures), or because the progressing fork encounters a nick or a lesion (Fig. 43-3B). Elements interfering with the progression of replication forks have been reported to induce rearrangements and/or render homologous recombination essential for viability in all organisms, from bacteria to humans [51]. If polymerase pausing at triplet repeats caused DSB, then repair would likely occur by homologous recombination.

An overwhelming amount of data has been accumulated in support of a recombination-based mechanism for expansion. It has not yet been demonstrated whether such a mechanism depends on preformed secondary structure at the repeats. However, it is clear from studies in simple model organisms that DSB can be a source of amplification of repeats once a break has occurred. Genes involved in recombination and DNA synthesis have been shown to influence the stability of trinucleotide repeats in both *Escherichia coli* and yeast. By using a genetic assay for repeat contraction in *E. coli*, the effect of mutations in the recA, recB, and lexA genes on the rate of instability of CTG/CAG

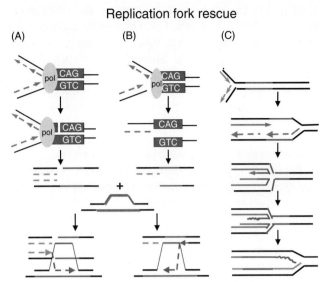

FIGURE 43-3 Rescue of a stalled replication fork by recombination or polymerase reversal. (A) Recombination repair of broken replication forks at stalled polymerases. Blockage at the replication fork leads to a break in the lagging strand template, thereby resulting in DSB. Arrested replication forks are targeted by nucleases, subsequently providing a substrate for repair by recombination. The processed end invades a homologous chromosome and uses its CAG/CTG stretch (in red) as a hybridization template to initiate repair. Depending on the extent of branch migration and the position for the hybridization, different size expansions can occur. DSB on the lagging strand template is shown, but a similar model can apply to breakage and repair of the leading strand. Increases in CAG/CTG length can also occur by polymerase slippage during the extension reaction. (B) Replication fork collapse. The progressing fork encounters a single-strand nick in the leading strand template. The replication fork collapses, and the broken free end invades a homologous chromosome and initiates recombination repair as in panel A. (C) Repeat instability by attempts to restart replication at a lesion or impassable sequence. In this scenario, the replication fork encounters the lesion or impassable sequence and backs up to recopy DNA by using the newly synthesized strands as replication templates. The resulting chicken foot structure allows repeat expansion if the recopied sequence is a CAG/CTG repeat (shown in red). The chicken foot structure is resolved and the block is bypassed by the recopied sequences. Replication restarts. In all cases, solid and dashed lines represent the template and the newly synthesized DNA, respectively; the arrowheads indicate the 3′ end of the growing strands. In all models described here, increases in CAG/CTG length can also occur by polymerase slippage during the extension reaction.

repeats of varying lengths was examined. The results show a decrease in the rate of contractions in the cells with mutated recA and recB as compared to recombination-proficient cells. These data indicate that recombination proficiency correlates with high rates of triplet repeat instability [52]. Similarly, recombination induced by DSB in yeast frequently results in contraction and expansion [44, 45, 53–55]. CTG tract instability in this system has been shown to depend on both Rad52p and

Rad1p [44]. In another set of experiments, overexpression of the Mre11–Rad50p–Xrs2p complex, crucial for strand processing at DSB, resulted in an increase of expansion products [45]. Moore and coworkers [56] have shown that, during meiosis, CNG hairpins can escape repair and are carried stably, predominantly causing expansion during a subsequent round of replication. Together, studies in yeast [44, 55, 57], bacteria [58–61], and mammalian cells [62] strongly indicate that recombination destabilizes triplet repeat tracts.

Interestingly, induction of the SOS system, a response to polymerase stalling, did not appear to play a significant role in repeat instability, nor did the presence of triplet repeats in cells turn on the SOS response [60]. Thus, it is also possible that repeat instability during the proliferative stage is a result of attempts to restart replication when the polymerase is blocked at triplet repeats. In this scenario, the replication fork forms a "chicken foot structure" in which the newly synthesized strand is available to serve as a replication template (Fig. 43-3C). The resolution of the "chicken foot" then can lead to fork cleavage, generating broken ends, the repair of which by homologous recombination results in instability at triplet repeats. In bacteria, this formation was proposed to occur in replication mutants, in which fork arrest is caused by a defect in a replication protein and by UV irradiation [49, 63, 64]. The chicken foot structure may also form during replication in eukaryotic organisms [65]. One study has shown that expansions and contractions of triplet repeats in proliferating yeast can occur if checkpoint machinery is defective. However, loss of MEC1, RAD24, RAD17, and RAD53 most often resulted in increased frequency of contractions [66]. The effect was attributed to the inability of these mutants to carry out faithful repair after the replication fork has stalled or a lesion leading to DSB has occurred at triplet repeats.

Because expansion has been shown to occur readily in yeast, which typically repair DNA by homologous recombination, a recombination-based mechanism has been proposed as likely in mammals. However, several pieces of evidence argue against this hypothesis. First, triplet repeats can undergo expansion by unequal sister chromatid exchange in yeast [67]. However, in human disease, one allele is typically stable [24, 68]. In Friedreich's ataxia, the only recessive trinucleotide repeat disorder [1], instability does not depend on the number of long alleles, and no change occurs in repeat number in the normal allele after unstable transmission [69]. These data indicate that the process of instability is likely to be intraallelic. Consistent with this finding, it has been observed that expansion in germ cells in HD transgenic R6/1 mice is largely restricted to haploid, terminally differentiated spermatozoa in which there is no homologous chromosome or sister chromatid [38]. Second, because some exchange of genetic material between chromosomes must be present as a consequence of recombination, one would expect that, at least some of the time, exchange would be observed in the sequence that flanks the repeat. However, no evidence exists for such an exchange in HD patients [70]. Third, in all tested recombination models, as in replication-based models (discussed previously), contractions are favored over expansions. This is in contrast to the expansion that is known to dominate in disease-affected families. Finally, the most compelling evidence arguing against recombination as a major mechanism in mammalian cells comes from experiments in transgenic animals. Gourdon and colleagues have demonstrated that, in mice harboring a long CTG tract in the human *DM* transgene, loss of proteins that are essential for homologous recombination (Rad52 and Rad54) and end joining (DNA-PK) has no effect on age-dependent expansion in these animals [71]. Thus, expansion by a double-strand-break repair mechanism does not appear to be well-supported in mammals. However, a single-strand-break repair mechanism is possible. Single breaks *in vivo* are generated as intermediates upon DNA damage repair by a base excision or nucleotide excision mechanism.

B. Excision Repair

In mammalian cells, hairpin structures may be considered as bulky adducts, which are typically repaired by the nucleotide excision repair (NER) pathway (Fig. 43-4). NER can correct a wide array of DNA damage, including intrastrand cross-links, bulky chemical lesions, cyclodeoxyadenosine, cyclopurines, and, infrequently, 8-oxoguanine thymine glycol [72, 73]. Removal of these lesions occurs by excision of approximately 28 nucleotides around the lesion on one strand, which is followed by gap repair synthesis using the other strand as a template.

Evidence of a role for NER in the expansion mechanism has been reported in bacteria [74, 75], yeast, and mammalian cells [76]. Many researchers have previously shown that stable hydrogen bonds within the hairpin can prevent FEN-1-mediated flap processing [30, 76] and leave a bulky lesion in DNA. Therefore, at the disease locus, FEN-1 is essentially "nuclease-defective" and leaves a precursor for expansion (Fig. 43-5). In mammalian cells, a nuclease dead state has been modeled by evaluating the properties of repeat instability in response to a mutant FEN-1, which cannot cleave

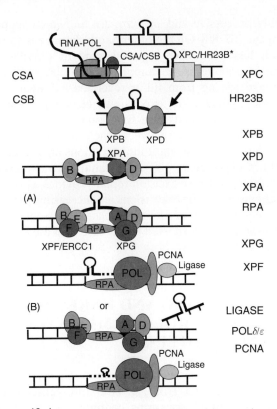

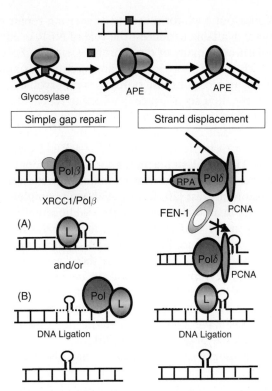

FIGURE 43-4 Expansion by a nucleotide excision repair mechanism. The nucleotide excision repair pathway (NER) can correct a wide array of DNA damage, including intrastrand cross-links, bulky chemical lesions, and secondary structures. One possibility is that the CAG hairpin itself can be the block (shown). Alternatively, lesions present in the DNA can be detected and hairpins can form during removal. Removal of these lesions occurs by excision of approximately 28 nucleotides around the lesion on one strand, which is followed by gap repair synthesis using the other strand as a template. DNA damage in the form of bulky adducts is recognized by two distinct pathways, depending on whether the damage resides in an active gene. If the gene is actively transcribed, a stalled RNA polymerase recruits two proteins: Cockayne proteins A (CSA) and B (CSB). This pathway is called transcription-coupled repair (TCR). CSA and CSB interact with two helicases within the TFIIH complex, XPD and XPB, which unwind the DNA. The open bubble structure is stabilized by XPA and a single-stranded binding protein replication factor A (RPA). The complex sequentially recruits two nucleases. XPG binds first and cleaves the 3′ side of the lesion, whereas XPF in a complex with ERCC1 binds next and cleaves the 5′ side of the lesion. The gap (around 28 nucleotides) is filled in by DNA polymerase δ or ε together with the processivity factor, proliferating cell nuclear antigen (PCNA), and ligated by DNA ligase 1. If the lesion is in nontranscribed DNA, then a separate recognition pathway is used involving XPC in a complex with HR23B. This pathway is called global genome repair (GGR). After recognition by the XPC complex, the other steps of GGR merge with those of TCR.

FIGURE 43-5 Expansion by a base excision repair model. In BER, damaged bases are removed from DNA and replaced with the correct base to restore Watson–Crick pairing. In the first step of BER, the damaged base is recognized by a DNA glycosylase. It binds specifically to a target base and hydrolyzes the N-glycosylic bond, releasing the inappropriate base while keeping the DNA backbone intact. The abasic site (i.e., the site of base loss) that is formed by DNA glycosylase activity is subsequently recognized by APE1 (the major AP endonuclease), which incises the phosphodiester backbone immediately 5′ to the lesion, leaving behind a strand break with a normal 3′-hydroxyl group and an abnormal 5′ abasic terminus. "Short-patch" BER proceeds with DNA Polβ removing the 5′ abasic residue via its 5′-deoxyribose-phosphodiesterase activity and filling in the single nucleotide gap. To complete the process, the nick is sealed by DNA ligase I or a complex of XRCC1 and ligase III. Additionally, an alternative BER pathway exists that involves the replacement of more than a single nucleotide and requires the FEN-1 protein to excise the flaplike structure that is produced by DNA polymerase strand displacement. The action of FEN-1 is prevented if hairpins form. This "long-patch" process has been divided into two subpathways: a PCNA-stimulated, Polβ-directed pathway and a PCNA-dependent, Polδ/ε-directed pathway. Removal of a damaged base produces a nick in the DNA backbone and can lead to hairpin formation and expansion by a strand displacement mechanism.

DNA. When flap cleavage by FEN-1 is inefficient, an alternative pathway involving ERCC1, which can also function in recombination, was recruited [76]. If expansion arises in the process of NER, then there are two likely mechanisms. An NER complex may be recruited, but it fails to excise the hairpin. In this case, the hairpin is incorporated into DNA after repair synthesis (Fig. 43-4). Alternatively, an NER complex may succeed in excising the hairpin; then extra repeats are incorporated after slippage and during repair synthesis (Fig. 43-4). Polymerase slippage can occur in either of these mechanisms during repair synthesis and may be solely responsible for expansion.

Knock-out mice for a variety of excision repair proteins are available to address the role of NER in expansion. However, many of these proteins, such as ERCC1 and XPF [77, 78], for example, are essential for animal viability, and loss of these proteins is embryonic lethal. Therefore, their significance in affecting repeat stability is difficult to discern *in vivo*. Nevertheless, heterozygous animal models can be used to test the role of dosage effect. For example, deletion of FEN-1/Rad27, which is involved in the fill-in reaction after excision repair (Fig. 43-5), has been shown to have a major impact on triplet expansion in yeast [20, 30, 44]. Although complete loss of FEN-1 in animals results in embryonic lethality [79], changes in the CAG repeat tract at the *hHD* locus in mice with a reduced level of FEN-1 (heterozygous) have been informative. Although there was no visible effect on age-dependent expansion in the somatic tissues of FEN-1(+/−) mice compared to their wild-type littermates, a decrease in deletions and an increase in expansions have been observed in the offspring of males heterozygous for FEN-1 [76].

The presence of oxidized bases may also be associated with the repeat expansion process. Oxidative lesions and genetic instability are associated with a host of neurodegenerative disorders, including HD [80], Alzheimer's disease [81], amyotrophic lateral sclerosis [82], Parkinson's disease [83], and ataxia telangiectasia [84]. Consistent with such a model, alterations in the somatic repeat length have been observed in mouse tissues as animals age, notably, in the brain. Somatic mosaicism is also observed in human disease [85, 86]. The observation that repeat expansion in terminally differentiated tissues increases with age suggests that a mechanism may involve repair of endogenous oxidative damage, which commonly arises during mitochondrial respiration and increases with age. Although NER can remove some oxidative lesions, the vast majority of thymine glycols and 8-oxoguanine are repaired by another type of excision process that is largely initiated by the removal of a single modified base. This pathway is called base excision repair (BER) (Fig. 43-5).

In BER, damaged bases are removed from DNA and replaced with the correct base in order to restore Watson–Crick pairing. As with NER, expansion in the process of BER occurs by an intrastrand gap repair process. The gap can occur due to the formation of a stable structure at a break site. Alternatively, a looped structure can form during strand displacement in a long-patch repair pathway (Fig. 43-5). Expansion by BER, therefore, may arise by polymerase slippage during repair synthesis or as a result of both slippage and structure formation during strand displacement. This mechanism for expansion is consistent with known properties of polymerases. Replicative polymerases are highly processive and, *in vitro*, are known to suppress slippage [87, 88]. On the other hand, slippage is favored when less processive, repair-dependent polymerases, such as polymerase-β, complete repair [87, 88]. Thus, slippage arising from fill-in reactions by repair polymerases would favor expansion over contractions, as observed in human disease. Excision repair mechanisms for expansion are attractive. In the process of a single-strand-break repair, gap-filling reactions are directional and are expected to give rise primarily to expansions. This feature of base excision repair provides a potential mechanism for why expansions are increasingly favored over contractions in human disease. The frequency of expansion is also expected to increase with the length of the repeat stretch.

V. EMERGING AND UNRESOLVED ISSUES: THE ROLE OF MISMATCH REPAIR PROTEINS

Much of what is known concerning repeat instability has been derived from *E. coli* and yeast models. Although simple organisms continue to provide key information regarding expansion, attention has focused on mouse *in vivo* models in more recent years. A mammalian *in vivo* system provides clear benefits. A mouse model is the only system that can recapitulate the molecular features of mutation associated with development and aging. Although transgenic models have been particularly important in providing powerful insights into key proteins that influence expansion, the role of such proteins is not always clear. For example, one of the first players shown to be essential for expansion *in vivo* is the mismatch repair (MMR) protein MSH2 [89]. Yet the role of MSH2 in expansion mechanisms is unresolved despite the fact that a causative role for MSH2 in expansion has been verified in several mouse models for HD [90–92] and DM [93, 94]. In all cases, the loss of MSH2 prevented expansion in both somatic tissues with age [71, 89, 95] and germ cells during development [38, 71].

On the surface, the involvement of MSH2 was not surprising given the fact that loss of MSH2 function was known to cause microsatellite instability in familial cases of hereditary polyposis colon cancer (HNPCC) [96–98]. However, a causative role for MSH2 in triplet expansion was not anticipated. In both HD and DM mouse models, expansion appears to depend primarily on a mismatch repair complex involving MSH2/MSH3 rather than its well-studied relative MSH2/MSH6 [99, 100]. A role for an MSH2/MSH3 complex in expansion is confirmed for both CAG and CTG repeats, indicating that the requirement for this complex is general and not influenced by C(A/T)G orientation, differences in base composition, or the context (coding versus noncoding) of the C(A/T)G repeat tract [1, 3, 5]. The generality of the MSH2/MSH3 requirement suggests that at least some steps of expansion among the different triplet repeats are similar and argues for a unified mechanism.

Evidence of a role for MMR machinery in trinucleotide repeat (TNR) instability appears to be consistent with the notion that the behavior of triplet repeats in MMR-deficient background differs from that of microsatellites observed in cancer. On the other hand, these data, when analyzed in detail, may suggest that the presence of functional MMR proteins simply eliminates bias toward deletions. For example, most of the changes at microsatellite loci in tumor cells that lack MSH2 are reported to be deletions [101–103]. Similarly, triplet repeats in bacteria and yeast show an overall higher frequency of instability in MSH2(−/−) background compared to wild-type cells, the majority of which are deletions [59, 104–106]. *In vivo*, it is widely observed that the absence of MSH2 in mouse models for DM and HD increases the incidence of deletions at expanded disease loci [34, 71, 95]. Thus, all studies agree that the loss of MSH2 greatly enhances the rate of deletion and inhibits expansion at a CNG repeat.

A model can be envisioned by considering the role of MSH2 in replication and repair separately. In the first scenario, slipped structures, such as loops and hairpins, can form during replication. If left unrepaired (no MMR), the repeats are bypassed by polymerase, which, in the next round of replication, leads to deletion (Fig. 43-6B). Alternatively, when hairpins form at breaks, the MSH2/MSH3 complex may recognize the mispaired bases of the hairpin, bind the hairpin, and prevent reannealing of the flap to its complementary strand. Expansion would be produced after polymerase fills in the gap and traps extra DNA after ligation (Fig. 43-6A). The latter model rests on the assumption that a hairpin is formed at the site of breakage. Biochemical data confirm that the MSH2/MSH3 complex binds CAG/CTG preformed hairpins. Upon binding, its enzymatic activity is altered in such a way that it prevents normal repair. Specifically, binding of purified human MSH2/MSH3 complex to the CAG hairpin has been demonstrated to inhibit the ATPase activity and nucleotide affinity for both ADP and ATP [100]. An MSH2–DNA complex is known to recruit the accessory MLH/PMS complex if the repair pathway proceeds through canonical MMR and to facilitate downstream steps of excision and resynthesis. Transgenic animals harboring the DM gene and lacking PMS2 have been shown to display a decrease in the rate of somatic expansion and an increase in large deletions [107]. The accumulated data on the role of MMR in expansion mutation are consistent with the model that

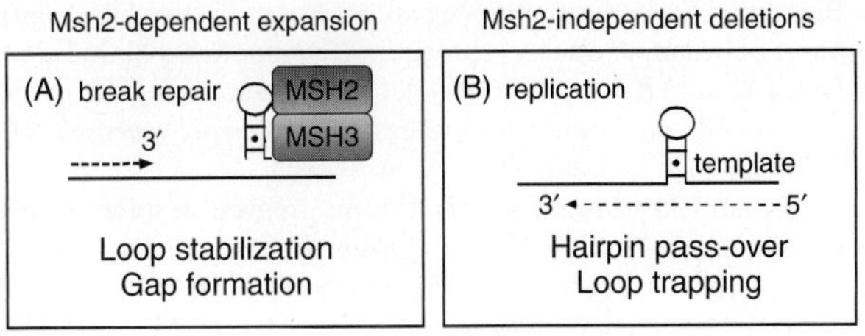

FIGURE 43-6 Possible roles for MSH2/MSH3 in expansion and deletion. MSH2-dependent expansions and MSH2-independent deletions in repeat tracts contribute to alterations at microsatellite loci. (A) Model for the dependence of expansion on MSH2/MSH3. Gap filling at a break, which can occur near or within the CAG tract or in the last step in a recombination-based mechanism. (B) Loops can form during DNA synthesis. In the absence of the MSH2/MSH3 complex, polymerase bypasses the loop to give rise to deletions.

MSH2/MSH3 complex stabilizes a hairpin, leading to altered states and/or aberrant repair. This model does not exclude the possibility that MSH2/MSH3 also functions in recombination and by cooperating with the repair machinery in BER and NER pathways. Future studies to identify factors that operate together with MSH2/MSH3 to cause expansion will be important in unraveling the expansion mechanism.

Mouse models have provided much insight into the expansion mechanism, yet many issues remained unresolved and will need to be answered in the coming years. For example, how are alternative structures such as hairpins allowed to form *in vivo*? Which proteins, in addition to MMR, are involved in hairpin recognition? For all of these questions, it is anticipated that mouse models will continue to be useful. Although a thorough understanding of the mechanism has not been achieved, the search is well worth the effort. A mechanistic solution to the problem of instability is likely to expose the very nature of heritable traits and evolution.

Acknowledgments

We thank S. M. Liang for help in figure preparation and C. Spiro, B. Owen, E. Trushina, and N. Kinzel for careful reading of the manuscript. This work was supported by the Mayo Foundation and National Institutes of Health grant numbers NS40738 and GM066359 (to CTM).

References

1. Cummings, C. J., and Zoghbi, H. Y. (2000). Trinucleotide repeats: Mechanisms and pathophysiology. *Ann. Rev. Genom. Hum. Genet.* **1**, 281–328.
2. Alper, G., and Narayanan, V. (2003). Friedreich's ataxia. *Ped. Neurol.* **28**, 335–341.
3. Manto, M. U. (2005). The wide spectrum of spinocerebellar ataxias (SCAs). *Cerebellum* **4**, 2–6.

4. Brais, B. (2003). Oculopharyngeal muscular dystrophy: A late-onset polyalanine disease. *Cytogenet. Genome Res.* **100**, 252–260.
5. Day, J. W., and Ranum, L. P. (2005). Genetics and molecular pathogenesis of the myotonic dystrophies. *Curr. Neurol. Neurosci. Rep.* **5**, 55–59.
6. Lin, X., and Ashizawa, T. (2005). Recent progress in spinocerebellar ataxia type-10 (SCA10). *Cerebellum* **4**, 37–42.
7. Chan, E. M., Andrade, D. M., Franceschetti, S., and Minassian, B. (2005). Progressive myoclonus epilepsies: EPM1, EPM2A, EPM2B. *Adv. Neurol.* **95**, 47–57.
8. Amiel, J., Trochet, D., Clement-Ziza, M., Munnich, A., Lyonnet, S. (2004). Polyalanine expansions in human. *Hum. Mol. Genet.* **2**, R235–R243.
9. Brown, L. Y., and Brown, S. A. (2004). Alanine tracts: The expanding story of human illness and trinucleotide repeats. *Trends Genet.* **20**, 51–58.
10. Albrecht, A., and Mundlos, S. (2005). The other trinucleotide repeat: Polyalanine expansion disorders. *Curr. Opin. Genet. Dev.* **15**, 285–293.
11. Gacy, A. M., Goellner, G., Juranic, N., Macura, S., and McMurray, C. T. (1995). DNA hairpin structure determines the threshold for expansion in genetic disease. *Cell* **81**, 533–540.
12. Gacy, A. M., and McMurray, C. T. (1998). Influence of hairpins on template reannealing at trinucleotide repeat duplexes: A model for slipped DNA. *Biochemistry* **37**, 9426–9434.
13. Paiva, A. M., and Sheardy, R. D. (2004). Influence of sequence context and length on the structure and stability of triplet repeat DNA oligomers. *Biochemistry* **43**, 14218–14227.
14. Parniewski, P., and Staczek, P. (2002). Molecular mechanisms of TRS instability. *Adv. Exp. Med. Biol.* **516**, 1–25.
15. Lenzmeier, B. A., and Freudenreich, C. H. (2003). Trinucleotide repeat instability: A hairpin curve at the crossroads of replication, recombination, and repair. *Cytogenet. Genome Res.* **100**, 7–24.
16. Kang, S., Jaworski, A., Ohshima, K., and Wells, R. D. (1995). Expansion and deletion of CTG repeats from human disease genes are determined by the direction of replication in *E. coli*. *Nat. Genet.* **10**, 213–218.
17. Freudenreich, C. H., Stavenhagen, J. B., and Zakian, V. A. (1997). Stability of a CTG/CAG trinucleotide repeat in yeast is dependent on its orientation in the genome. *Mol. Cell Biol.* **17**, 2090–2098.
18. Sia, E. A., Kokoska, R. J., Dominska, M., Greenwell, P., and Petes, T. D. (1997). Microsatellite instability in yeast: Dependence on repeat unit size and DNA mismatch repair genes. *Mol. Cell Biol.* **17**, 2851–2858.
19. Miret, J. J., Pessoabrandao, L., and Lahue, R. S. (1998). Orientation-dependent and sequence specific expansions of CTG/CAG trinucleotide repeats in *Saccharomyces cerevisiae*. *Proc. Natl. Acad. Sci. USA* **95**, 12438–12443.
20. Schweitzer, J. K., and Livingston, D. M. (1998). Expansions of CAG repeat tracts are frequent in a yeast mutant defective in Okazaki fragment maturation. *Hum. Mol. Genet.* **7**, 69–74.
21. Iyer, R. R., and Wells, R. D. (1999). Expansion and deletion of triplet repeat sequences in *Escherichia coli* occur on the leading strand of DNA replication. *J. Biol. Chem.* **274**, 3865–3877.
22. Levinson, G., and Gutman, G. A. (1987). Slipped-strand mispairing: a major mechanism for DNA sequence evolution. *Mol. Biol. Evol.* **4**, 203–221.
23. Modrich, P. (1994). Mismatch repair, genetic stability, and cancer. *Science* **266**, 1959–1960.
24. Kremer, B., Almqvist, E., Theilmann, J., Spence, N., Telenius, H., Goldberg, Y. P., and Hayden, M. R. (1995). Sex-dependent mechanisms for expansions and contractions of the CAG repeat on affected Huntington disease chromosomes. *Am. J. Hum. Genet.* **57**, 343–350.
25. Duyao, M., Ambrose, C., Myers, R., Novelletto, A., Persichetti, F., Frontali, M., Folstein, S., Ross, C., Franz, M., Abbott, M., et al. (1993). Trinucleotide repeat length instability and age of onset in Huntington's disease. *Nat. Genet.* **4**, 387–392.
26. Snow, K., Doud, L. K., Hagerman, R., Pergolizzi, R. G., Erster, S. H., and Thibodeau, S. N. (1993). Analysis of CGG sequence at the FMR-1 locus in fragile X families and in the general population. *Am. J. Hum. Genet.* **53**, 1217–1228.
27. McDonald, M. E., Barnes, G., Srinidhi, J., Duyao, M. P., Ambrose, C. M., Myers, R. H., Gray, J., Conneally, P. M., Young, A., Penney, J., et al. (1993). Gametic but not somatic instability of CAG repeat length in Huntington's disease. *J. Med. Genet.* **30**, 982–986.
28. Zhang, L., Fischbeck, K. H., and Arnheim, N. (1994). CAG repeat length variation in sperm from a patient with Kennedy's disease. *Hum. Mol. Genet.* **4**, 303–305.
29. Kunst, C. B., and Warren, S. T. (1994). Cryptic and polar variation of the fragile X repeat could result in predisposing normal alleles. *Cell* **77**, 853–861.
30. Spiro, C., Pelletier, R., Rolfsmeier, M. L., Dixon, M. J., Lahue, R. S., Gupta, G., Park, M. S., Chen, X., Mariappan, S. V. S., and McMurray, C. T. (1999). Inhibition of FEN-1 processing by DNA secondary structure at trinucleotide repeats. *Mol. Cell* **4**, 1079–1085.
31. Krasilnikova, M. M., and Mirkin, S. M. (2004). Replication stalling at Friedreich's ataxia (GAA)n repeats in vivo. *Mol. Cell Biol.* **24**, 2286–2295.
32. Pelletier, R., Krasilnikova, M. M., Samadashwily, G. M., Lahue, R., and Mirkin, S. M. (2003). Replication and expansion of trinucleotide repeats in yeast. *Mol. Cell Biol.* **23**, 1349–1357.
33. Spiegel, R., La Spada, A. R., Kress, W., Fischbeck, K. H., and Schmid, W. (1996). Somatic stability of the expanded CAG trinucleotide repeat in X-linked spinal and bulbar muscular atrophy. *Hum. Mutat.* **8**, 32–37.
34. Kovtun, I. V., Thornhill, A. R., and McMurray, C. T. (2004). Somatic deletion events occur during early embryonic development and modify the extent of CAG expansion in subsequent generations. *Hum. Mol. Genet.* **13**, 3057–3068.
35. Wohrle, D., Kennerknecht, I., Wolf, M., Enders, H., Schwemmle, S., and Steinbach, P. (1995). Heterogeneity of DM kinase repeat expansion in different fetal tissues and further expansion during cell proliferation in vitro: Evidence for a casual involvement of methyl-directed DNA mismatch repair in triplet repeat stability. *Hum. Mol. Genet.* **4**, 1147–1153.
36. Manley, K., Pugh, J., and Messer, A. (1999). Instability of the CAG repeat in immortalized fibroblast cell cultures from Huntington's disease transgenic mice. *Brain Res.* **835**, 74–79.
37. Gomes-Pereira, M., Fortune, M. T., and Monckton, D. G. (2001). Mouse tissue culture models of unstable triplet repeats: In vitro selection for larger alleles, mutational expansion bias and tissue specificity, but no association with cell division rates. *Hum. Mol. Genet.* **10**, 845–854.
38. Kovtun, I. V., and McMurray, C. T. (2001). Trinucleotide expansion in haploid germ cells by gap repair. *Nat. Genet.* **27**, 407–411.
39. Yoon, S. R., Dubeau, L., de Young, M., Wexler, N. S., and Arnheim, N. (2003). Huntington disease expansion mutations in humans can occur before meiosis is completed. *Proc. Natl. Acad. Sci. USA* **100**, 8834–8838.
40. Savouret, C., Garcia-Cordier, C., Megret, J., te Riele, H., Junien, C., and Gourdon, G. (2004). MSH2-dependent germinal CTG repeat expansions are produced continuously in spermatogonia from DM1 transgenic mice. *Mol. Cell Biol.* **24**, 629–637.
41. De Temmerman, N., Sermon, K., Seneca, S., De Rycke, M., Hilven, P., Lissens, W., Van Steirteghem, A., and Liebaers, I. (2004). Intergenerational instability of the expanded CTG repeat in the DMPK gene: Studies in human gametes and preimplantation embryos. *Am. J. Hum. Genet.* **75**, 325–329.

42. Yang, Z., Lau, R., Marcadier, J. L., Chitayat, D., and Pearson, C. E. (2003). Replication inhibitors modulate instability of an expanded trinucleotide repeat at the myotonic dystrophy type 1 disease locus in human cells. *Am. J. Hum. Genet.* **73**, 1092–1010.
43. Gomes-Pereira, M., and Monckton, D. G. (2004). Chemically induced increases and decreases in the rate of expansion of a CAG*CTG triplet repeat. *Nucleic Acids Res.* **32**, 2865–2872.
44. Freudenreich, C. H., Kantrow, S. M., and Zakian, V. A. (1998). Expansion and length-dependent fragility of CTG repeats in yeast. *Science* **270**, 853–856.
45. Richard, G.-F., Goellner, G. M., McMurray, C. T., and Haber, J. E. (2000). Recombination-induced CAG trinucleotide repeat expansions in yeast involve the MRE11/RAD50/XRS2 complex. *EMBO J.* **19**, 2381–2390.
46. Cleary, J. D., Nichol, K., Wang, Y. H., and Pearson, C. E. (2002). Evidence of cis-acting factors in replication-mediated trinucleotide repeat instability in primate cells. *Nat. Genet.* **31**, 37–46.
47. Kang, S., Ohshima, K., Shimizu, M., Amirhaeri, S., and Wells, R. D. (1995). Pausing of DNA synthesis in vitro at specific loci in CTG and CGG triplet repeats from human hereditary disease genes. *J. Biol. Chem.* **270**, 27014–27021.
48. Samadashwily, G. M., Raca, G., and Mirkin, S. M. (1997). Trinucleotide repeats affect DNA replication *in vivo. Nat. Genet.* **17**, 298–304.
49. Bidnenko, V., Ehrlich, S. D., and Michel, B. (2002). Replication fork collapse at replication terminator sequences. *EMBO J.* **21**, 3898–3907.
50. Bruand, C., Bidnenko, V., and Ehrlich, S. D. (2001). Replication mutations differentially enhance RecA-dependent and RecA-independent recombination between tandem repeats in *Bacillus subtilis. Mol. Microbiol.* **39**, 1248–1258.
51. Haber, J. E. (2000). Recombination: A frank view of exchanges and vice versa. *Curr. Opin. Cell Biol.* **12**, 286–292.
52. Jakupciak, J. P., and Wells, R. D. (1999). Genetic instabilities in CTG–CAG repeats occur by recombination. *J. Biol. Chem.* **274**, 23468–23479.
53. Paques, F., Leung, W. Y., and Haber, J. E. (1998). Expansions and contractions in a tandem repeat induced by double-strand break repair. *Mol. Cell Biol.* **18**, 2045–2054.
54. Richard, G. F., Dujon, B., and Haber, J. E. (1999). Double-strand break repair can lead to high frequencies of deletions within short CAG/CTG trinucleotide repeats. *Mol. Gen. Genet.* **261**, 871–882.
55. Jankowski, C., Nasar, F., and Nag, D. K. (2000). Meiotic instability of CAG repeat tracts occurs by double-strand break repair in yeast. *Proc. Natl. Acad. Sci. USA* **97**, 2134–21399.
56. Moore, H., Greenwell, P. W., Liu, C. P., Arnheim, N., and Petes, T. D. (1999). Triplet repeats form secondary structures that escape DNA repair in yeast. *Proc. Natl. Acad. Sci. USA* **96**, 1504–1509.
57. Jankowski, C., and Nag D. K. (2002). Most meiotic CAG repeat tract-length alterations in yeast are SPO11 dependent. *Mol. Genet. Genomics* **267**, 64–70.
58. Pluciennik, A., Iyer, R. R., Napierala, M., Larson, J. E., Filutowicz, M., and Wells, R. D. (2002). Long CTG–CAG repeats from myotonic dystrophy are preferred sites for intermolecular recombination. *J. Biol. Chem.* **277**, 34074–34086.
59. Hashem, V. I., Rosche, W. A., and Sinden, R. R. (2002). Genetic assays for measuring rates of (CAG/CTG) repeat instability in *Escherichia coli. Mutat. Res.* **502**, 25–37.
60. Hashem, V. I., Rosche, W. A., and Sinden, R. R. (2004). Genetic recombination destabilizes (CTG)n·(CAG)n repeats in *E. coli. Mutat. Res.* **554**, 95–109.
61. Hebert, M. L., Spitz, L. A., and Wells, R. D. (2004). DNA double-strand breaks induce deletion of CTG·CAG repeats in an orientation-dependent manner in *Escherichia coli. J. Mol. Biol.* **336**, 655–672.
62. Meservy, J. L., Sargent, R. G., Iyer, R. R., Chan, F., McKenzie, G. J., Wells, R. D., and Wilson, J. H. (2003). Long CTG tracts from the myotonic dystrophy gene induce deletions and rearrangements during recombination at the APRT locus in CHO cells. *Mol. Cell Biol.* **23**, 3152–3162.
63. Michel, B., Flores, M. J., Viguera, E., Grompone, G., Seigneur, M., and Bidnenko, V. (2001). Rescue of arrested replication forks by homologous recombination. *Proc. Natl. Acad. Sci. USA* **98**, 8181–8188.
64. Chow, K. H., and Courcelle, J. (2004). RecO acts with RecF and RecR to protect and maintain replication forks blocked by UV-induced DNA damage in *Escherichia coli. J. Biol. Chem.* **279**, 3492–3296.
65. Sharma, S., Otterlei, M., Sommers, J. A., Driscoll, H. C., Dianov, G. L., Kao, H. I., Bambara, R. A., and Brosh, R. M., Jr. (2004). WRN helicase and FEN-1 form a complex upon replication arrest and together process branch migrating DNA structures associated with the replication fork. *Mol. Biol. Cell* **15**, 734–750.
66. Lahiri, M., Gustafson, T. L., Majors, E. R., and Freudenreich, C. H. (2004). Expanded CAG repeats activate the DNA damage checkpoint pathway. *Mol. Cell* **15**, 287–293.
67. Nag, D. K., Suri, M., and Stenson, E. K. (2004). Both CAG repeats and inverted DNA repeats stimulate spontaneous unequal sister-chromatid exchange in *Saccharomyces cerevisiae. Nucleic Acids Res.* **32**, 5677–5684.
68. Leeflang, E. P., Zhang, L., Tavare, S., Hubert, R., Srinidhi, J., MacDonald, M. E., Myers, R. H., de Young, M., Wexler, N. S., Gusella, J. F., et al. (1995). Single sperm analysis of the trinucleotide repeats in the Huntington's disease gene: Quantification of the mutation frequency spectrum. *Hum. Mol. Genet.* **4**, 1519–1526.
69. Gacy, A. M., Goellner, G. M., Chen, X., Gupta, G., Spiro, C., Dyer, R., Yao, J., Mikesell, M., Johnson, A. J., Malancon, S. B., Richter, A., and McMurray, C. T. (1998). GAA instability in Friedreich's ataxia shares a common, DNA-directed and intra-allelic mechanism with other trinucleotide repeats. *Mol. Cell* **1**, 583–593.
70. Sutherland, G. R., and Richards, R. I. (1995). Simple tandem DNA repeats and human genetic disease. *Proc. Natl. Acad. Sci. USA* **92**, 3636–3641.
71. Savouret, C., Brisson, E., Essers, J., Kanaar, R., Pastink, A., te Riele, H., Junien C., and Gourdon, G. (2003). CTG repeat instability and size variation timing in DNA repair-deficient mice. *EMBO J.* **22**, 2264–2273.
72. Friedberg, E. C. (2003). DNA damage and repair. *Nature* **421**, 436–440.
73. Hanawalt, P. C., Ford, J. M., and Lloyd, D. R. (2003). Functional characterization of global genomic DNA repair and its implications for cancer. *Mutat. Res.* **544**, 107–114.
74. Parniewski, P., Bacolla, A., Jaworski, A., and Wells, R. D. (1999). Nucleotide excision repair affects the stability of long transcribed (CTG*CAG) tracts in an orientation-dependent manner in *Escherichia coli. Nucleic Acids Res.* **27**, 616–623.
75. Oussatcheva, E. A., Hashem, V. I., Zou, Y., Sinden, R. R., and Potaman, V. N. (2001). Involvement of the nucleotide excision repair protein UvrA in instability of CAG–CTG repeat sequences in *Escherichia coli. J. Biol. Chem.* **276**, 30878–30884.
76. Spiro, C., and McMurray, C. T. (2003). Nuclease-deficient FEN-1 blocks Rad51/BRCA1-mediated repair and causes trinucleotide repeat instability. *Mol. Cell Biol.* **23**, 6063–6074.
77. McWhir, J., Selfridge, J., Harrison, D. J., Squires, S., and Melton, D. W. (1993). Mice with DNA repair gene (ERCC-1) deficiency have elevated levels of p53, liver nuclear abnormalities and die before weaning. *Nat. Genet.* **5**, 217–224.
78. Weeda, G., Donker, I., de Wit, J., Morreau, H., Janssens, R., Vissers, C. J., Nigg, A., van Steeg, H., Bootsma, D., and Hoeijmakers, J. H. (1997). Disruption of mouse ERCC1 results in a novel repair

syndrome with growth failure, nuclear abnormalities and senescence. *Curr. Biol.* **7**, 427–439.
79. Larsen, E., Gran, C., Saether, B. E., Seeberg, E., and Klungland, A. (2003). Proliferation failure and gamma radiation sensitivity of Fen1 null mutant mice at the blastocyst stage. *Mol. Cell Biol.* **23**, 5346–5353.
80. Browne, S. E., Ferrante, R. J., and Beal, M. F. (1999). Oxidative stress in Huntington's disease. *Brain Pathol.* **9**, 147–163.
81. Butterfield, D. A., Drake, J., Pocernich, C., and Castegna, A. (2001). Evidence of oxidative damage in Alzheimer's disease brain: Central role for amyloid beta-peptide. *Trends Mol. Med.* **7**, 548–554.
82. Carri, M. T., Ferri, A., Cozzolino, M., Calabrese, L., and Rotilio, G. (2003). Neurodegeneration in amyotrophic lateral sclerosis: The role of oxidative stress and altered homeostasis of metals, *Brain Res. Bull.* **61**, 365–374.
83. Beal, M. F. (2003). Mitochondria, oxidative damage, and inflammation in Parkinson's disease. *Ann. NY Acad. Sci.* **991**, 120–131.
84. Watters, D. J. (2003). Oxidative stress in ataxia telangiectasia. *Redox Rep.* **8**, 23–29.
85. Kennedy, L., Evans, E., Chen, C. M., Craven, L., Detloff, P. J., Ennis, M., and Shelbourne, P. F. (2003). Dramatic tissue-specific mutation length increases are an early molecular event in Huntington disease pathogenesis. *Hum. Mol. Genet.* **12**, 3359–3367.
86. Telenius, H., Kremer, B., Goldberg, Y. P., Theilmann, J., Andrew, S. E., Zeisler, J., Adam, S., Greenberg, C., Ives, E. J., Clarke, L. A., et al. (1994). Somatic and gonadal mosaicism of the Huntington disease gene CAG repeat in brain and sperm. *Nat. Genet.* **6**, 409–414.
87. Hartenstine, M. J., Goodman, M. F., and Petruska, J. (2000). Base stacking and even/odd behavior of hairpin loops in DNA triplet repeat slippage and expansion with DNA polymerase. *J. Biol. Chem.* **275**, 18382–18390.
88. Hartenstine, M. J., Goodman, M. F., and Petruska, J. (2002). Weak strand displacement activity enables human DNA polymerase beta to expand CAG/CTG triplet repeats at strand breaks. *J. Biol. Chem.* **277**, 41379–41389.
89. Manley, K., Shirley, T. L., Flaherty, L., and Messer, A. (1999). MSH2 deficiency prevents in vivo somatic instability of the CAG repeat in Huntington disease transgenic mice. *Nat. Genet.* **23**, 471–473.
90. Mangiarini, L., Sathasivam, K., Seller, M., Cozens, B., Harper, A., Hetherington, C., Lawton, Trottier, Y., Lehrach, H., Davies, S. W., et al. (1996). Exon 1 of the HD gene with an expanded CAG repeat is sufficient to cause a progressive neurological phenotype in transgenic mice. *Cell* **87**, 493–506.
91. Mangiarini, L., Sathasivam, K., Mahal, A., Mott, R., Seller, M., and Bates, G. P. (1997). Instability of highly expanded CAG repeats in mice transgenic for the Huntington's disease mutation. *Nat. Genet.* **15**, 197–200.
92. Wheeler V. C., Auerbach, W., White, J. K., Srinidhi, J., Auerbach, A., Ryan, A., Duyao, M. P., Vrbanac, V., Weaver, M., Gusella, J. F., Joyner, A. L., and MacDonald, M. E. (1999). Length-dependent gametic CAG repeat instability in the Huntington's disease knock-in mouse. *Hum. Mol. Genet.* **8**, 115–122.
93. Gourdon, G., Radvanyi, F., Lia, A. S., Duros, C., Blanche, M., Abitbol, M., Junien, C., and Hofmann-Radvanyi, H. (1997). Moderate intergenerational and somatic instability of a 55-CTG repeat in transgenic mice. *Nat. Genet.* **15**, 190–192.
94. Monckton, D. G., Coolbaugh, M. I., Ashizawa, K. T., Siciliano, M. J., and Caskey, C. T. (1997). Hypermutable myotonic dystrophy CTG repeats in transgenic mice. *Nat. Genet.* **15**, 193–196.
95. Wheeler, V. C, Lebel, L. A., Vrbanac, V., Teed, A., te Riele, H., and MacDonald M. E. (2003). Mismatch repair gene Msh2 modifies the timing of early disease in Hdh(Q111) striatum. *Hum. Mol. Genet.* **12**, 273–281.
96. Verma, L., Kane, M. F., Brassett, C., Schmeits, J., Evans, D. G., Kolodner, R. D., and Maher, E. R. (1999). Mononucleotide microsatellite instability and germline MSH6 mutation analysis in early onset colorectal cancer. *J. Med. Genet.* **36**, 678–682.
97. Wagner, A., Barrows, A., Wijnen, J. T., van der Klift, H., Franken, P. F., Verkuijlen, P., Nakagawa, H., Geugien, M., Jaghmohan-Changur, S., Breukel, C., et al. (2003). Molecular analysis of hereditary nonpolyposis colorectal cancer in the United States: High mutation detection rate among clinically selected families and characterization of an American founder genomic deletion of the MSH2 gene. *Am. J. Hum. Genet.* **72**, 1088–1100.
98. Wahlberg, S. S., Schmeits, J., Thomas, G., Loda, M., Garber, J., Syngal, S., Kolodner, R. D., and Fox, E. (2002). Evaluation of microsatellite instability and immunohistochemistry for the prediction of germ-line MSH2 and MLH1 mutations in hereditary nonpolyposis colon cancer families. *Cancer Res.* **62**, 3485–3492.
99. van den Broek, W. J., Nelen, M. R., Wansink, D. G., Coerwinkel, M. M., te Riele, H., Groenen, P. J., and Wieringa, B. (2002). Somatic expansion behavior of the (CTG)n repeat in myotonic dystrophy knock-in mice is differentially affected by Msh3 and Msh6 mismatch-repair proteins. *Hum. Mol. Genet.* **11**, 191–198.
100. Owen, B. A., Yang, Z., Lai M., Gajek, M., Badger, J. D., II, Hayes, J. J., Edelmann, W., Kucherlapati, R., Wilson, T. M., and McMurray, C. T. (2005). (CAG)(n)-hairpin DNA binds to Msh2-Msh3 and changes properties of mismatch recognition. *Nat. Struct. Mol. Biol.* **12**, 663–670.
101. Goellner, G. M., Tester, D., Thibodeau, S., Almqvist, E., Goldberg, Y. P., Hayden, M. R., and McMurray, C. T. (1997). Different mechanisms underlie DNA instability in Huntington disease and colorectal cancer. *Am. J. Hum. Genet.* **60**, 879–890.
102. Tsao, J. L., Yatabe, Y., Salovaara, R., Jarvinen, H. J., Mecklin, J. P., Aaltonen, L. A., Tavare, S., and Shibata, D. (2000). Genetic reconstruction of individual colorectal tumor histories. *Proc. Natl. Acad. Sci. USA* **97**, 1236–1241.
103. Blake, C., Tsao, J.-L., and Shibata, D. (2001). Stepwise deletions of polyA sequences in mismatch repair-deficient colorectal cancers. *Am. J. Pathol.* **158**, 1867–1870.
104. Schweitzer, J. K., and Livingston, D. M. (1997). Destabilization of CAG trinucleotide repeat tracts by mismatch repair mutations in yeast. *Hum. Mol. Genet.* **6**, 349–355.
105. Parniewski, P., Jaworski, A., Wells R. D., and Bowater, R. P. (2000). Length of CTG/CAG repeats determines the influence of mismatch repair on genetic instability. *J. Mol. Biol.* **299**, 865–874.
106. Schmidt, K. H., Abbott, C. M., and Leach, D. R. F. (2000). Two opposing effects on mismatch repair on CTG repeat instability in *Escherichia coli*. *Mol. Microbiol.* **35**, 463–471.
107. Gomes-Pereira, M., Fortune, M. T., Ingram, L., McAbney, J. P., and Monckton, D. G. (2004). Pms2 is a genetic enhancer of trinucleotide CAG·CTG repeat somatic mosaicism: Implications for the mechanism of triplet repeat expansion. *Hum. Mol. Genet.* **13**, 1815–1825.

CHAPTER 44

Transcription and Triplet Repeat Instability

YUNFU LIN, VINCENT DION, AND JOHN H. WILSON

Verna and Marrs McLean Department of Biochemistry and Molecular Biology,
Baylor College of Medicine, Houston, Texas 77030

I. Introduction
II. Tissue-Specific Variation in Repeat Tract Lengths in Humans and Mice
   A. Genes Associated with Triplet Repeat Expansions Are Widely Expressed
   B. Tissue-Specific Variation in Repeat Tract Length Is Common
   C. Age-Dependent Repeat Instability Is Common, Even in Terminally Differentiated Neurons
III. Transcription and Repeat Instability in Bacteria, Yeast, and Human Cells
   A. Transcription Destabilizes Triplet Repeat Sequences in Bacteria
   B. Transcription Destabilizes Simple Sequence Repeats in Yeast
   C. Transcription Destabilizes Triplet Repeats in Human Cells
IV. Concluding Remarks
References

At several loci in the human genome, expansions of CAG·CTG trinucleotide repeats cause neurological diseases. The complex tissue-specific patterns of germline and somatic instability present a challenge to understanding the underlying mechanisms. Studies in bacteria and yeast have provided critical insights into the potential mechanisms of instability, but they cannot identify with certainty the pathways responsible for instability in humans. DNA replication-based models, for example, cannot account for the dramatic, age-dependent instability in neurons of the striatum, which do not divide. Thus, additional pathways must contribute to repeat instability in specific tissues. Here, we summarize the data that link transcription to triplet repeat instability in bacteria, yeast, mammalian cells, and mice. Transcription exposes single-strands of DNA, potentially allowing the repeats to form hairpins and slipped-strand structures. Interaction of these aberrant DNA structures with repair processes such as nucleotide excision repair and mismatch repair may bring about changes in the length of the repeat tract in nondividing cells.

I. INTRODUCTION

Short repetitive DNA sequences—termed microsatellite and minisatellite repeats—are unstable in all genomes, but at several loci in the human genome repeat instability is associated with disease [1–4]. Expansions of CAG·CTG trinucleotide (triplet) repeats are the cause of several human neurological diseases, including myotonic dystrophy, Huntington's disease, and a number of spinocerebellar ataxias [5, 6]. These diseases are characterized by the expansion of a triplet repeat beyond a threshold of about 25–35 repeats to a length that has pathologic consequences [1, 6].

Inheritance of repeat-sequence diseases typically shows a progressive worsening of the disease phenotype in subsequent generations as the repeat tract continues to expand, indicating a critical period of instability in the germline. It is the propensity of repeated sequences to expand in the germline that is the defining characteristic of this group of diseases. However, the somatic tissues of affected individuals also display a characteristic pattern of repeat instability; for example, CAG repeats in Huntington's disease typically are highly unstable in striatum, moderately unstable in liver and kidney, and stable in heart and muscle [7]. The complexity of the tissue-specific patterns of repeat instability—from the germline to various somatic tissues—presents a challenge to understanding the underlying mechanisms. Why does repeat instability vary from tissue to tissue? Do these patterns of instability arise by modulation of one fundamental mechanism, or do distinct mechanisms operate in different tissues? Are the same types of repeat at different locations in the genome destabilized by the same mechanism or by different ones?

The basis for the repeat instability in humans has been investigated by using model systems, including *Escherichia coli*, yeast, mammalian cells, and mice. In bacteria and yeast, virtually every process that exposes single strands of DNA has been shown to destabilize triplet repeats, including DNA replication, homologous recombination, DNA repair, and transcription, with replication and recombination showing the most dramatic effects [1, 4]. It is thought that exposure of single-stranded DNA allows CAG·CTG repeats to form hairpins and slipped-strand DNA duplexes, as they do *in vitro* [1, 3, 4, 8, 9]. These secondary structures interfere with normal DNA metabolic processes or trigger aberrant ones, ultimately leading to changes in repeat tract length. Thus, triplet repeat instability likely arises via a pathway in which a normal DNA metabolic process exposes single strands, allowing them to form a secondary structure, which in turn calls into play a normal or aberrant DNA repair process that brings about the change in repeat tract length.

Studies in bacteria and yeast have provided critical insights into the potential pathways that lead to repeat instability, but they do not identify those pathways that are responsible for instability in humans. For example, the processes identified as being most important in bacteria and yeast—DNA replication and homologous recombination—do not account for certain key observations in mammalian cells and mice, which arguably provide the most relevant models for the mitotic instability of repeats that is observed in the human germline and somatic tissues. The most difficult observation to accommodate in replication-based models is the ongoing instability that occurs over time in slowly dividing cells (for example, in the liver) and nondividing cells like the neurons in the striatum [10–13]. In addition, the degree of instability does not correlate with tissue-specific cell proliferation rates [14–16]. Moreover, instability has been reported to occur during meiotic arrest in the female germline [17] and in nonproliferating sperm precursors [18, 19]. Studies in mice also suggest a minor role for homologous recombination at best, as repeat stability was unaffected in mice deficient for the recombination proteins RAD52 and RAD54 [20]. Also, the ongoing instability in terminally differentiated neurons [10–13, 21] must occur in the absence of a sister chromatid, the vastly preferred partners for homologous recombination in mammalian cells [22]. Thus, replication and recombination are unlikely sources for the instability seen in somatic cells, especially in G1/G0-arrested neurons.

These considerations suggest that other pathways—in addition to or instead of replication and recombination—are likely to contribute to triplet repeat instability in specific tissues. This chapter, will examine the possibility that transcription through a triplet repeat triggers instability. The effects of transcription have been studied in bacteria [23–27], but the pathway that leads from transcription to repeat length change is not yet defined. Most results in bacteria have been interpreted in terms of interplay between transcription and replication [23–26], which seems unlikely to apply to the instability observed in nondividing neurons, for example. One transcription-induced, replication-independent pathway has been proposed [26], and a modified version of that pathway is shown in Fig. 44-1. Transcription is likely to trigger repeat instability by exposing single-stranded DNA as the RNA polymerase moves through the repeats, allowing the repeats to form hairpins and slipped-strand structures [1, 4, 8, 9]. These aberrant structures could engage DNA repair processes during transcription or in a process physically unlinked to transcription, as a way of dealing with structures left in the wake of a passing polymerase. It is the repair processes themselves that would bring about changes in the length of the repeat tract.

Two DNA repair processes—nucleotide excision repair (NER) and mismatch repair (MMR), or components thereof—are logical candidates to participate in a transcription-induced pathway of repeat instability. NER has a well-defined connection to transcription through a subpathway known as transcription-coupled repair (TC-NER) [28, 29]. In addition, the stability of triplet repeats is significantly altered in bacterial strains with mutations in NER [25, 30], as well as in human cells by knocking down NER components using siRNA (Lin et al.). MMR has been linked in some reports to TC-NER in bacteria [31] and in human cells [32–34]. Moreover, Msh2 has been shown to interact with components of NER in yeast [35] and to bind to CAG·CTG slipped-strand structures [36], and MMR has been linked to triplet repeat instability in bacteria [27, 37–39], yeast [40, 41], human cells (Lin et al.) and mice [10, 18, 20, 42–48].

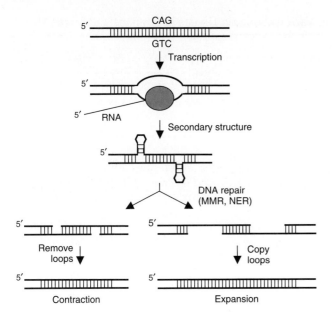

FIGURE 44-1 Pathway for transcription-induced triplet repeat instability. The passage of RNA polymerase separates the strands of the duplex allowing secondary structure to form in the nontranscribed strand. If the structure is present when the two strands reanneal, a slipped-strand structure such as the one shown here can form. Although equivalent structures are shown in each strand, CTG and CAG do not form hairpins with equal facility and thus the two strands may not have the same structure. DNA repair processes such as MMR and NER are proposed to recognize the hairpins and initiate repair. Removal of the loops would lead to contraction. Cleavage of the strands opposite the loops, coupled with DNA repair synthesis using the loops as templates, would cause expansions.

This chapter examines the basis for considering that a transcription-mediated pathway might contribute to the CAG·CTG triplet repeat instability observed in humans. The chapter reviews existing data on the transcription-induced instability and presents new data showing that elements of the pathway outlined in Fig. 44-1 operate in human cells.

II. TISSUE-SPECIFIC VARIATION IN REPEAT TRACT LENGTHS IN HUMANS AND MICE

Mice provide the most relevant model for the instability of triplet repeats observed in the human germline and somatic tissues. Transgenic and knock-in mouse models of triplet repeat diseases [49, 50] mimic the molecular pathology of the human diseases reasonably well, although not exactly [49–51]. Stability of triplet repeats in mice differs from that in humans in two important ways: (1) longer repeats are required to observe significant instability in mice, and (2) large jumps in tract length are much rarer [10, 49, 52, 53]. Nevertheless, studies in mice, cells derived from mice, modified mammalian cells, and cells from human patients have provided important insights into the mechanisms of triplet repeat instability [1, 4, 49].

A. Genes Associated with Triplet Repeat Expansions Are Widely Expressed

A common feature of the pathology of diseases associated with triplet repeat expansions is the degeneration of neurons in disease-specific regions of the central nervous system (CNS). Although many of the disease symptoms involve the CNS, the disease-associated genes are, in general, expressed in many tissues in humans and animals [54–60]. The gene causing Huntington's disease (HD), for example, is widely expressed, with its highest expression in brain and testis and lowest in liver [54, 55, 60]. Even its expression in the brain varies considerably in different regions, with the highest levels in the cerebral cortex and cerebellum and the lowest in the thalamus and caudate nucleus [54, 55]. Interestingly, HD expression levels do not correlate with the source of neuropathology, which arises from characteristic atrophy of the caudate nucleus. Similarly, other CAG·CTG triplet-repeat-associated disease genes, including those for spinocerebellar ataxia 1 (SCA1), dentatorubral pallidoluysian atrophy (DRPLA), spinal and bulbar muscular atrophy (SBMA), and myotonic dystrophy (DM), are also widely expressed, but in patterns that do not always match the neuropathology [56–59, 61–63].

These observations establish the main prerequisite for a transcription-induced pathway for triplet repeat instability, namely, that the affected gene be transcribed in tissues that display repeat instability. Because many of these disease genes are widely expressed, transcription has the potential to contribute to repeat instability in many tissues.

B. Tissue-Specific Variation in Repeat Tract Length Is Common

The various CAG·CTG triplet repeat diseases display repeat instability in the germline—the basis for intergenerational expansion—but they also display a characteristic pattern of instability in somatic tissues in the adult. Somatic instability is commonly observed in tissues that are affected by the disease, leading to the suggestion that somatic instability may exacerbate the disease process [64]. Table 44-1 summarizes the repeat instability that is observed in selected tissues from human patients and mouse models of HD, DM1, SCA1, SCA3, SCA7, and DRPLA.

These data highlight three aspects of tissue-specific triplet repeat instability. First, in somatic tissues, as in the

TABLE 44-1 Variation in Repeat Tract Length in Various Somatic Tissues in Humans and Mice[a]

Disease	Repeat	LI	KI	MU	H	LK	BR	ST	CE	TH	CC	HC	SC	Reference numbers
DM1/M	55	+	+	−	−	−	+							129
DM1/M	55	++	++	−	−		++		−	+				16
DM1/M	84	++	+++	+	−	−	++							46
DM1/M	162	++	+++	+	−		++	+++	+		++	+++	++	15, 47
DM1/M	>300	+++	++	+	−	−	++		−					20, 53
HD/P	40	+	+	−	−		++							7
HD/P	41			−	−	−	−							130
HD/M	48	−	−		−			−	−		−			128
HD/M	72	+	++		−			+++	+		++	++	++	13, 64
HD/M	90	+	−		−			++	−		+			128
HD/M	109	++	+		−			+++			++			128
HD/M	112	++			−			+++	−	++	++	++		48
HD/M	115	+++	+		−		+++	++	−	+	+			11, 18, 68
HD/M	~150	+++						+++	+		++			64
SCA1/M	154	+	++	+	−			+++	+		++	++	++	21
SCA3/P								++	−	++	−			131
SCA3/P	78	++	++	−			+							132
SCA7/M	92	+++	++	−	−			++						74
DRPLA/P	~60	++	+	+	−	−	++			++	+		++	133
DRPLA/P	62	++					++							132, 134
DRPLA/M	78	++	++		−				−		++			135

[a] Abbreviations: LI, liver; KI, kidney; MU, skeletal muscle; H, heart; LK leukocytes (blood); BR, brain; ST, striatum; CE, cerebellum; TH, thalamus; CC, cerebral cortex; HC, hippocampus; SC, spinal cord. Repeat instability as indicated by +'s and −'s, with more +'s indicating more instability.

germline, longer repeats give higher levels of instability. Second, tissues such as heart and leukocytes typically show very little instability, whereas others—liver and brain, especially the striatum—show high instability. The lack of correlation between cell proliferation rates and repeat instability has led to the suggestion that somatic instability can occur by mechanisms that are not replication-based [10–16, 46, 48]. Parallel observations of ongoing instability in unfertilized oocytes (which are arrested after meiotic DNA replication) and in nonproliferating sperm precursors have led to proposals that instability in these instances may be primarily due to DNA repair [17–19]. The common theme in these alternative proposals is that DNA damage in or near the repeats, including nicks, breaks, and oxidative damage, triggers a repair process that changes repeat tract length [13, 17, 18]. Third, the patterns of instability across tissues seem remarkably similar in the various diseases. This impression, if correct, implies that key elements of repeat instability may be tissue-specific, rather than disease-gene-specific.

A potential link between transcription, or open chromatin structure, and repeat instability was initially suggested by analyses of mouse lines that were transgenic for a portion of the huntingtin gene carrying 55 CAG repeats [11]. In three lines in which the transgene was expressed, there was clear germline and somatic repeat instability; however, in a fourth line with a silent transgene, the repeat was stable [11]. To address the possible relationship between repeat stability and transcription, Lia et al. [16] analyzed a transgenic DM1 mouse model that carried a human *DMPK* gene with a $(CTG)_{55}$ tract. They showed that *DMPK* mRNA was lowest in liver and kidney, about fourfold higher in gastrocnemius muscle, and about twentyfold higher in heart muscle. Repeat instability, however, was higher in liver and kidney than it was in heart or gastrocnemius muscle. Thus, *DMPK* mRNA levels and instability of the *DMPK* repeat are not correlated in these tissues [16].

The lack of a direct correlation between transcript levels and repeat instability is often cited as arguing against a role for transcription in triplet repeat instability [16, 47]. However, two considerations recommend caution in discounting transcription. First, the rates of transcription were not measured in this study, only the levels of stable mRNA. It is conceivable that the rates of *DMPK* transcription actually do correlate with repeat instability in these tissues, but that that correlation is obscured by different, tissue-specific rates of *DMPK* mRNA degradation. Second, the underlying assumption that higher transcription rates should be correlated with higher levels of repeat instability is not an essential feature of a transcription-induced pathway for repeat instability, as depicted in Fig. 44-1. For example, it could be that even low levels of transcription saturate the pathway. Under such conditions, tissue-specific differences in repeat stability would correlate with the efficiency of downstream DNA repair processes, rather than with differences in transcription rates.

C. Age-Dependent Repeat Instability Is Common, Even in Terminally Differentiated Neurons

One of the most striking observations associated with triplet repeat instability in mice and humans is the ongoing instability that occurs with age in germline and somatic tissues [10–13, 15–17, 21, 45–47, 53, 64–68]. Figure 44-2, which is taken from Kennedy et al. [64], shows a particularly clear example of age-dependent, expansion-biased repeat instability, as it occurs in several regions of the brain in an HD transgenic mouse model. Two aspects of these data are especially noteworthy. First, at 3 months of age there is minimal variability in repeat tract length. Thus, the many cell divisions that were required to establish the brain must have occurred without significantly altering the CAG repeat tract length. Second, at 24 months more than 80% of the cells in the striatum have an altered repeat tract length, with a strong bias toward expansion [13, 64]. This observation indicates that most of the cells in the striatum, as opposed to a small, replication-competent subset, display repeat instability. This is an important point because the striatum, like other tissues, contains a mixture of cell types. More than 65% of the cells in the mouse striatum are terminally differentiated neurons [69], and most of the remainder are glial cells that do not proliferate in the HD mouse striatum [70]. Furthermore, Kennedy et al. [13] showed that the small region of replication-competent cells in the striatum, the subventricular zone, does not contribute preferentially to repeat instability by demonstrating that tract length variation was indistinguishable in striata dissected to include or exclude the subventricular zone. Finally, repeat instability in neurons has been directly demonstrated by laser-captured microdissection of neurons from human patients with DRPLA [71]. Collectively, these observations indicate that there is at least one robust pathway for repeat instability that operates in terminally differentiated neurons.

Ongoing repeat instability in cells such as neurons, which are arrested in the G1/G0 phase of the cell cycle, eliminates DNA replication from consideration because the cells

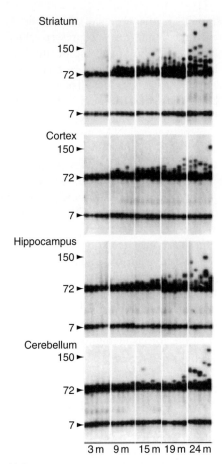

FIGURE 44-2 Age-dependent triplet repeat instability in tissues of the brain from HD mice. HD mice heterozygous for a knock-in allele carrying a $(CAG)_{72}$ repeat tract were analyzed for instability as a function of age. Illustrated here is small-pool PCR from samples that have about 10 cells worth of DNA. Numbers on the left indicate sizes of amplification products in terms of numbers of CAG repeats. (Figure is from ref. [64]: Kennedy et al. (2003). *Human Molecular Genetics* **12**, 3359–3367, with permission.)

do not cycle through S phase. Replication-independent pathways for repeat instability propose that DNA repair processes, acting on aberrant repeat-dependent secondary structures, alter the length of repeat tracts [17, 18, 26, 47]. In one class of models, DNA repair is initiated at sites of DNA damage (nicks [18] or oxidative damage [17], for example), which exposes single strands and allows secondary structures to form in repeat tracts. In the other class, secondary structures form as a consequence of duplex unwinding (during transcription [26] or by random breathing [47], for example), and DNA repair processes are engaged to deal with aberrant DNA structures. In both classes of model, tract lengths are changed when the repair processes encounter the aberrant DNA structure, as illustrated for the transcription-induced pathway in Fig. 44-1.

One set of observations that is difficult to reconcile with any of these pathways for replication-independent repeat instability is the absence of somatic instability in transgenic mice that carry expressed cDNAs with expanded repeats [72–76]. (The mouse models cited in Table 44-1 all carry transgenic genomic fragments or knock-ins at the endogenous loci.) The observations with cDNAs include transgenic mouse models for HD, SCA1, SCA7, and SBMA with a range of repeat lengths, and thus the difference between cDNAs and genomic fragments may be broadly valid. In one comprehensive study, Libby et al. [74] compared the stability of a $(CAG)_{92}$ tract in a full-length *SCA7* cDNA or in a 13.5-kb *SCA7* genomic fragment. In two transgenic lines, the cDNA construct displayed minimal instability (9–17%) in the germline and a lack of instability in somatic tissues. In contrast, four transgenic lines carrying the genomic fragment, with the same length repeat, were much more unstable in the germline (66–84%) and displayed a robust instability in several somatic tissues (see Table 44-1). Repeats in the genomic fragment could be partially stabilized by deleting most of the 3′ flanking sequences, which provides evidence of a role for cis elements in generating repeat instability [74].

Libby et al. directly address a potential role for transcription in their experiments [74]. Although the cDNA transgenes were expressed at roughly endogenous levels [77], reverse-transcriptase PCR (RT-PCR) analysis of the genomic transgenes failed to detect transcripts containing exons 3 and 4, the only exons in the fragment [74]. These observations are difficult to square with a transcription-induced pathway for repeat instability. Although transcription through the genomic fragment might produce an unstable transcript, and thus escape detection, the more difficult observation is that transcription through the cDNA does not destabilize the repeat.

One possible explanation is that transcription through the repeat is affected by the surrounding *cis* elements. For example, it may be that in the cDNA transgenes, in the absence of associated *cis* elements, RNA polymerase progresses through the repeat more rapidly, allowing less time for secondary structures to form [78]. A second possibility is that the key transcription event for repeat instability may generate antisense transcripts. It has become clear that antisense transcription is much more common than previously recognized and may play critical roles in gene regulation [79–82]. A recent report has identified antisense transcription through the repeat at the DM1 locus [137]. Antisense transcription could explain the need for *cis* elements and might also account for the different stabilities of the same length repeat tract in cDNA and genomic transgenes. For the disease genes whose cDNA transgenes are stable, *HD*, *SCA1*, *SCA7*, and *SBMA*, the nontranscribed strand (the strand not bound by RNA polymerase) contains the CAG repeat during normal transcription, but it would contain the CTG repeat if transcribed in the antisense direction. Because CTG repeats form a more stable secondary structure than CAG repeats [3, 8, 83], antisense transcription might be associated with a higher level of instability. Alternatively, collisions between oppositely directed polymerases may trigger repeat instability. Whatever the source of the stability differences between cDNAs and genomic fragments, it strongly supports a role for *cis* elements [1] and presents a challenge for all models of replication-independent triplet repeat instability.

III. TRANSCRIPTION AND REPEAT INSTABILITY IN BACTERIA, YEAST, AND HUMAN CELLS

A. Transcription Destabilizes Triplet Repeat Sequences in Bacteria

The effects of transcription through triplet repeat sequences have been reported in bacteria [23–27]. In all cases, triplet repeat sequences were cloned into a pUC plasmid in the 5′ end of a *lacZ* transcription unit, so that transcription could be controlled by addition of the inducer, IPTG, which increased transcription by up to tenfold [23]. Bacterial cells carrying the plasmids were grown for 15–180 cell divisions in the presence or absence of isopropyl-beta-D-thiogalacto-pyranoside (IPTG), and the resulting spectra of repeat tract lengths were analyzed by gel electrophoresis.

This elegantly simple experimental paradigm has revealed several effects of transcription on triplet repeat stability. First, the effects of transcription depend on the length and composition of the repeat sequences. Transcription did not detectably alter the stability of CAG·CTG repeats that were 50 repeat units or fewer in length [25, 27], but it did destabilize repeat tracts of 64 and 175 repeats [23–25]. GAC·GTC repeats displayed a similar sensitivity to transcription, with tract lengths of 27, 30, and 53 being unaffected, whereas tracts of 49, 69, and 74 were destabilized [26]. Second, orientation of the repeat relative to the direction of transcription can alter the stability of the repeat [23, 25, 26]. For example, $(GAC)_{49}$ in the transcribed strand was much more unstable than $(GTC)_{53}$ in the transcribed strand [26]. This orientation dependence disappeared at longer lengths: $(GAC)_{74}$ or $(GTC)_{69}$ in the transcribed strand showed similar levels of transcription-induced instability [26]. A similar lack of orientation dependence was demonstrated for $(CAG·CTG)_{64}$ repeats [24]. Curiously, a $(CAG·CTG)_{175}$ tract, which carries two G-to-A interruptions, displayed a dramatic orientation dependence, once again with the more susceptible orientation carrying the

CAG sequence in the transcribed strand [23, 25]. Overall, the sensitivity of transcription effects to the length, composition, and orientation of repeat tracts is consistent with secondary structure formation by the repeat.

The magnitude of the destabilization brought about by transcription through the repeats is modest. For (CAG·CTG)$_{64}$ tracts, which were analyzed after 15 generations, transcription caused a subtle shift in the spectrum of repeat lengths: the proportion of tracts 63–65 units long (about 90% of the total in the absence of transcription) was reduced by 5–10%, the proportion of larger expansions was reduced, and the proportion of larger deletions was increased by two- to fourfold [24]. When (CAG·CTG)$_{175}$ stability, in the sensitive orientation, was followed for 180 generations in log-phase growth, the rate of loss of the initial tract length was shown to be twofold greater when the repeat was transcribed [23]. A much more dramatic twenty-fivefold difference was apparent when cultures were cycled through stationary phase (about every 23 generations) [23]. This stationary-phase effect was shown to depend on both transcription and orientation of the repeat; it was interpreted to mean that stationary-phase growth or survival was advantageous for cells carrying plasmids with shorter repeats [23]. An alternative explanation might be that transcription through the repeat in stationary-phase cells dramatically destabilizes the repeat by a replication-independent pathway.

For the most part, the effects of transcription on repeat stability in bacteria have been interpreted in terms of interactions between transcription and replication [23–25]. In these systems, replication and transcription occur contemporaneously, and several studies have reported that DNA polymerase complexes can slow down or stall when they encounter an RNA polymerase [84–87]. These problems may be exacerbated by long triplet repeat sequences, which cause DNA polymerase pausing [88, 89]. To account for the orientation dependence of the transcriptional instability of (CAG·CTG)$_{175}$ and the preponderance of deletions, Parniewski et al. speculated that the CTG hairpins formed on the nontranscribed strand during transcription were subsequently bypassed by the DNA polymerase complex, thereby causing deletions [25]. On the basis of the involvement of NER, as described next, Parniewski et al. extended this model to one analogous to that depicted in Fig. 44-1 [25].

Of the DNA repair processes indicated in Fig. 44-1, only NER has been examined in bacteria in the context of transcription-induced triplet repeat instability [25]. In the absence of transcription, (CAG·CTG)$_{175}$ repeats were unaffected by mutations in the NER repair genes, *uvrA* or *uvrB*; however, transcription-induced instability was enhanced by mutations in *uvrA*, but reduced by mutations in *uvrB* [25]. In another study that used a similar plasmid system to selectively assay deletions of repeats in a transcribed gene, mutations in *uvrA* and *uvrB* were both shown to reduce the rates of deletion formation in (CAG·CTG)$_{79}$ repeats [30]. Although the conclusions of these two studies differ somewhat, which may derive from differences between the systems, they both indicate that NER can influence triplet repeat stability.

The role, if any, of MMR in transcription-induced triplet repeat instability has not been studied in bacteria; however, several groups have reported the effects of mutations in MMR genes on other aspects of repeat stability [27, 37–39, 90]. In *mutS* strains of bacteria, which are defective in MMR, the frequency of repeat tracts that have increased or decreased by a single repeat unit is dramatically increased [27, 37]. These frequent +1 and −1 changes are consistent with a role for MMR in protecting against slippage during DNA replication [91, 92] and are consistent with slippage models for repeat instability [93–96]. In addition to these frequent single-unit changes, there are significant effects on the frequencies of the rarer large deletions; however, there is disagreement on the direction of the effect. Two groups show that *mutS* strains have higher frequencies of large deletions [37, 90], whereas two other groups show that they have lower frequencies [27, 38, 39]. These studies agree that wild-type MMR functions to prevent single-unit changes, but they disagree on whether MMR normally promotes or prevents the formation of larger deletions.

B. Transcription Destabilizes Simple Sequence Repeats in Yeast

Although the effects of transcription on triplet repeat stability have not been examined in yeast, transcription has been shown to destabilize tracts of GT dinucleotides [97]. Changes in the lengths of GT tracts were monitored as a function of transcription by using a selectable gene driven by the galactose-inducible *GAL1–10* promoter. The fusion gene encodes Ura3 at its C terminus and confers a selectable Ura$^+$ phenotype. Frameshift mutations in a chromosomal copy of the fusion gene were engineered by placing 31 or 35 GT repeats into the N-terminal portion. Reversion of these frameshift mutations was assayed by selection for the Ura$^+$ phenotype, which occurred most commonly by deletion or addition of one GT repeat [97]. In this system, induction of transcription increased the rates of frameshift reversion by three- to ninefold [97].

The relationship between transcription-induced repeat instability and NER has not been examined, but the role of MMR has been investigated in this frameshift-reversion assay [97]. In yeast, as opposed to bacteria, the effects of MMR on repeat sequences are well-defined and noncontroversial. In simple sequence repeats, including triplet repeats, MMR-defective cells display a dramatically higher frequency of additions or deletions of 1–2 repeat units, consistent with the normal role of MMR in correcting replication-slippage events [41, 97–101]. The frequency of larger deletions, which are common in tracts of triplet repeats, is unaffected by mutations in MMR genes [41, 102, 103]. In the frameshift-reversion assay, defects in MMR genes (*MSH2* or *PMS1*) increase the rate of reversion by one-hundredfold, a rate that was stimulated by an additional two- to threefold upon the induction of transcription [97]. It was suggested that transcription has two effects: (1) to hinder the progress of DNA polymerase and thereby increase its tendency to form slipped primers, and (2) to interfere with the MMR machinery, decreasing the efficiency with which it recognizes and repairs slipped primers [97].

C. Transcription Destabilizes Triplet Repeats in Human Cells

Numerous publications have examined various aspects of triplet repeat biology in mammalian cells [14, 104–116], but only recently has the potential role of transcription in destabilizing triplet repeat sequences been studied [136]. To determine whether transcription was capable of destabilizing CAG triplet repeats in human cells, a direct-selection assay for repeat contraction [113] was used that employs an *HPRT* minigene driven by a Tet-ON inducible promoter [117]. The *HPRT* minigene was modified to carry a long tract of CAG repeats in its intron, oriented with the CAG repeat on the transcribed strand [113]. The direct-selection assay works because long CAG repeats are included in the mRNA as exons, thereby rendering the encoded protein nonfunctional [113]. Because long CAG tracts interfere with normal splicing, whereas repeat tracts of less than 39 repeats do not, triplet repeat contraction events can be readily monitored by selecting for HPRT+ colonies.

The modified *HPRT* minigene carrying a (CAG)$_{95}$ repeat tract was introduced into human HT1080 cells expressing the reverse tetracycline transcription activator (rtTA), and cell lines with single-copy integrants were established [136]. The effect of transcription through the repeat tracts in these cell lines was tested by growing the cells in the presence or absence of doxycycline, which binds to rtTA and activates it (Fig. 44-3). Real-time PCR measurements showed that doxycycline increased transcription through the *HPRT* minigene by twentyfivefold over a low background. Rates of contraction were measured by fluctuation analysis in the presence and absence of doxycycline. Transcription increased the rate of

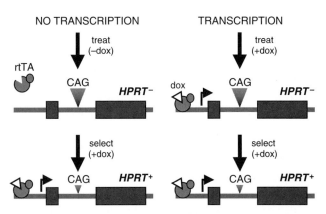

FIGURE 44-3 Selection assay for transcription-induced CAG repeat contraction. In this assay, transcription through the triplet repeat is controlled from a Tet-ON inducible promoter. When doxycycline is absent (left), rtTA does not bind its enhancer and transcription is suppressed. When doxycycline is present (right), rtTA is active, binds to its enhancer, and stimulates transcription. Cells were treated for various periods of time in the presence or absence of doxycycline and then replated in the presence of doxycycline to reveal the status of the CAG repeat. Tracts that have contracted to fewer than 39 CAG repeats will generate colonies under HPRT+ selection.

contraction by an average of fifteenfold to about 6×10^{-6} [136]. Analysis of individual HPRT+ colonies from these experiments showed that in all cases the repeat tracts had undergone contraction to a length of less than 39 repeats.

To address the potential role of replication, transcription-induced contraction frequencies were compared in proliferating and confluent cells, which differ by more than a factor of 10 in their rates of cell division [136]. In the presence of doxycycline, the two populations of cells accumulated HPRT+ repeat contractions at the same rate: for both proliferating and confluent cells the frequency of contractions was about 7×10^{-6} after 3 days of doxycycline treatment and about 15×10^{-6} after 7 days of treatment. The lack of correlation of contraction frequencies with cell division rates suggests that transcription-induced repeat instability is independent of DNA replication [136].

To investigate which DNA repair processes might participate in the pathway of transcription-induced repeat instability, expression of individual NER and MMR components was knocked down by using siRNA [136]. Multiple siRNAs were designed for each component to guard against nonspecific, off-target effects and then selected for testing on the basis of their ability to knock down expression of the target gene by at least 70%, as judged by Western blotting. Relative to a control siRNA that had no effect on doxycycline-induced repeat contraction, three siRNAs against XPA, three siRNAs against MSH2, and two siRNAs against MSH3 significantly reduced the frequency of contractions ($p < 0.005$). In contrast, siRNAs against MSH6 and XPC had no effect.

These results suggest that elements of both NER and MMR are involved in the pathway for transcription-induced triplet repeat instability. Specifically, they suggest that the MSH2/MSH3 complex (MutSβ)—but not the MSH2/MSH6 complex (MutSα)—is required to generate contractions and that TC-NER, which requires XPA but not XPC, may be involved in the pathway for contractions. Finally, these components of NER and MMR may collaborate in the same pathway because mixtures of siRNAs against XPA and MSH2 gave reductions that were indistinguishable from those obtained with either component alone [136].

Interactions between components of MMR and NER have been documented in several aspects of DNA repair. In yeast, Msh2 and Rad10 (the yeast homologue of mammalian NER protein, ERCC1) physically interact and *MSH2* mutations increase the ultraviolet (UV) sensitivity of NER-deficient yeast strains [35]. A functional link *between MMR and TC-NER* has been reported in bacteria and human cells [31–33], but this connection is controversial and is not supported by other studies [118–122]. MutSβ (MSH2/MSH3) and the ERCC1/XPF nuclease component of NER work together in the removal of psoralen cross-links from mammalian cell extracts [123]. During double-strand-break repair in yeast, the removal of 3′ nonhomologous tails requires both Msh2/Msh3 and Rad1/Rad10 (the counterpart of mammalian ERCC1/XPF) [124–126]. Perhaps most relevant, Msh2 and Rad1 have been shown to be required, as part of the same pathway, for the repair of 26-bp loops formed as heteroduplex intermediates in recombination in yeast [127]. Thus, there is ample precedent for the involvement of elements of MMR and NER in the pathway for transcription-induced triplet repeat instability proposed in Fig. 44-1.

The role of MMR, but not that of NER, has been investigated in mice [18, 20, 45–48]. When two models of HD—an exon 1 transgene with $(CAG)_{117}$ [12] and a knock-in with $(CAG)_{109}$ [128]—were bred onto an $Msh2^{-/-}$ background, the instability normally evident in sperm and various somatic tissues was absent, leading the authors to conclude that the normal function of MMR was required to generate germline and somatic instability [18, 48]. Studies in two models of DM1—a transgene fragment with $(CTG)_{152}$ [15] and a knock-in with $(CTG)_{84}$ [46]—extended those conclusions by analyzing repeat instability on $Msh3^{-/-}$, $Msh6^{-/-}$, and $Pms2^{-/-}$ backgrounds [46, 47]. On an $Msh3^{-/-}$ background, instability was eliminated, suggesting that the normal function of the MSH2/MSH3 complex (MutSβ) destabilizes triplet repeats [46]. Surprisingly, triplet repeat instability, especially expansion, was markedly enhanced on an $Msh6^{-/-}$ background [46]. Competition between MSH3 and MSH6 for binding to MSH2 could mean that in the absence of MSH6 additional MSH2/MSH3 complex is formed, which may account for the increased instability [46]. On a $Pms2^{-/-}$ background, repeat stability was also significantly reduced, suggesting that MMR-mediated processing events downstream of MSH2/MSH3 binding may normally be required to destabilize triplet repeats [47]. Note that these results support a dramatically different role for MMR in repeat instability in mice—to promote the whole range of instability, large jumps and small—whereas in bacteria and yeast MMR mainly prevents changes of one or two repeats. These results also match results in human cells, where MMR normally functions to promote repeat instability associated with transcription [136].

This tidy picture of MMR and repeat instability is challenged somewhat by results with a third DM1 mouse model, which carries a 45-kb human transgene with $(CTG)_{>300}$ [20, 45]. On an $Msh2^{-/-}$ background, the distribution of repeat tract lengths in germline and somatic tissues is somewhat narrowed, suggesting decreased instability, but the major effect is an overall shift toward shorter lengths, which is interpreted to indicate that the absence of MSH2 promotes contractions; that is, that a normal function of MSH2 is to prevent contractions [20, 45]. Whether the different effects of MMR deficiency in this model relate to the size of the CTG repeat, which was 2–3 times longer than in the mouse models discussed earlier, or to other factors is unclear. Taken together, however, all of the results in mouse models indicate that MMR plays a critical role in triplet repeat instability in the germline and in somatic tissues. Given the studies identifying a role for NER in transcription-induced repeat instability in cultured cells [136], it would be of great interest to examine the effects of NER deficiency in mouse models.

IV. CONCLUDING REMARKS

Expansion of triplet repeat sequences in the human genome occupies a position of growing importance in human disease. Originally termed dynamic mutations to denote their extreme instability [94], these novel genomic alterations have motivated intense scrutiny of two key questions: How do expanded repeats cause disease? Why are triplet repeat sequences unstable? Clear answers to either of these questions might provide promising routes to treatment of human patients. Both avenues of investigation have revealed a depth of complexity and subtlety that was completely unexpected. Arguably, at this point researchers have learned more about cell biology than about disease.

In the realm of triplet repeat instability, it is reasonably certain that their ability to form aberrant secondary structures, including hairpins and slipped-strand DNA, lies at the heart of their dynamic nature [1, 3, 4]. It is also clear that triplet repeats are rendered unstable through their interactions with the normal cellular machinery engaged in DNA transactions: no additional mutations are required. Repeat-induced secondary structures apparently confuse the normal processes or frequently send them down aberrant pathways. What is not known at this point, for humans anyway, are the identities of the processes that expose single strands to allow aberrant structure formation or those that mishandle the problem of structure removal. Nor is it known whether the actual events that lead to germline and somatic instability in humans are the outcome of one major pathway or a common result of multiple distinct or overlapping pathways.

These questions have been explored most intensively in bacteria and yeast, where the enormous power of genetics can be brought to bear on the problem. In those model organisms, replication has been identified as perhaps the most important modulator of repeat stability and, thus, the most likely candidate for the critical process in humans [1, 4, 96]. In mammalian cells and mice, however, proliferation rates correlate poorly with repeat instability, as is nowhere more evident than in the neurons of the striatum, which do not divide but display extreme repeat instability [10–16, 46, 48]. These observations have prompted the development of several replication-independent pathways for repeat stability [17–19], one of which—the transcription-induced pathway for repeat instability—has been the focus of this chapter.

The relevance of transcription to the triplet repeat instability observed in human germline and somatic tissues is not yet clear. The disease genes are widely transcribed, which is the main prerequisite for such a pathway. It is also attractive because it is an ongoing process that could account naturally for the age-dependent accumulation of repeat length changes. In addition, in human cells transcription through a CAG·CTG tract has been shown to destabilize the repeat in a way that does not depend on DNA replication [136]. Moreover, destabilization appears to require normal function of the MSH2/MSH3 complex [136], a complex that has been implicated in the repeat instability observed in several mouse models [18, 20, 45–48]. An additional requirement for the involvement of NER in transcription-induced instability has also been demonstrated in human cells [136], but it has not been tested in any animal model. The relevance of transcription-induced pathways to the instability observed in humans can only be tested adequately by additional experiments in mammalian cells and in mice.

Acknowledgments

Work on triplet repeat instability is supported by a grant from the NIH to JHW. (GM38219). VD was supported by an NSERC postgraduate scholarship.

References

1. Cleary, J. D., and Pearson, C. E. (2003). The contribution of cis-elements to disease-associated repeat instability: Clinical and experimental evidence. *Cytogenet. Genome Res.* **100**, 25–55.
2. Ellegren, H. (2004). Microsatellites: Simple sequences with complex evolution. *Nat. Rev. Genet.* **5**, 435–445.
3. Sinden, R. R., Potaman, V. N., Oussatcheva, E. A., Pearson, C. E., Lyubchenko, Y. L., and Shlyakhtenko, L. S. (2002). Triplet repeat DNA structures and human genetic disease: Dynamic mutations from dynamic DNA. *J. Biosci.* **27**, 53–65.
4. Lenzmeier, B. A., and Freudenreich, C. H. (2003). Trinucleotide repeat instability: A hairpin curve at the crossroads of replication, recombination, and repair. *Cytogenet. Genome Res.* **100**, 7–24.
5. Cummings, C. J., and Zoghbi, H. Y. (2000). Trinucleotide repeats: Mechanisms and pathophysiology. *Annu. Rev. Genomics Hum. Genet.* **1**, 281–328.
6. Usdin, K., and Grabczyk, E. (2000). DNA repeat expansions and human disease. *Cell. Mol. Life Sci.* **57**, 914–931.
7. Telenius, H., Kremer, B., Goldberg, Y. P., Theilmann, J., Andrew, S. E., Zeisler, J., Adam, S., Greenberg, C., Ives, E. J., Clarke, L. A., et al. (1994). Somatic and gonadal mosaicism of the Huntington disease gene CAG repeat in brain and sperm. *Nat. Genet.* **6**, 409–414.
8. Gacy, A. M., Goellner, G., Juranic, N., Macura, S., and McMurray, C. T. (1995). Trinucleotide repeats that expand in human disease form hairpin structures in vitro. *Cell* **81**, 533–540.
9. Pearson, C. E., and Sinden, R. R. (1996). Alternative structures in duplex DNA formed within the trinucleotide repeats of the myotonic dystrophy and fragile X loci. *Biochemistry* **35**, 5041–5053.
10. Wheeler, V. C., Auerbach, W., White, J. K., Srinidhi, J., Auerbach, A., Ryan, A., Duyao, M. P., Vrbanac, V., Weaver, M., Gusella, J. F., et al. (1999). Length-dependent gametic CAG repeat instability in the Huntington's disease knock-in mouse. *Hum. Mol. Genet.* **8**, 115–122.
11. Mangiarini, L., Sathasivam, K., Mahal, A., Mott, R., Seller, M., and Bates, G. P. (1997). Instability of highly expanded CAG repeats in mice transgenic for the Huntington's disease mutation. *Nat. Genet.* **15**, 197–200.
12. Mangiarini, L., Sathasivam, K., Seller, M., Cozens, B., Harper, A., Hetherington, C., Lawton, M., Trottier, Y., Lehrach, H., Davies, S. W., et al. (1996). Exon 1 of the HD gene with an expanded CAG repeat is sufficient to cause a progressive neurological phenotype in transgenic mice. *Cell* **87**, 493–506.
13. Kennedy, L., and Shelbourne, P. F. (2000). Dramatic mutation instability in HD mouse striatum: Does polyglutamine load contribute to cell-specific vulnerability in Huntington's disease? *Hum. Mol. Genet.* **9**, 2539–2544.
14. Gomes-Pereira, M., Fortune, M. T., and Monckton, D. G. (2001). Mouse tissue culture models of unstable triplet repeats: In vitro selection for larger alleles, mutational expansion bias and tissue specificity, but no association with cell division rates. *Hum. Mol. Genet.* **10**, 845–854.
15. Fortune, M. T., Vassilopoulos, C., Coolbaugh, M. I., Siciliano, M. J., and Monckton, D. G. (2000). Dramatic, expansion-biased, age-dependent, tissue-specific somatic mosaicism in a transgenic mouse model of triplet repeat instability. *Hum. Mol. Genet.* **9**, 439–445.
16. Lia, A. S., Seznec, H., Hofmann-Radvanyi, H., Radvanyi, F., Duros, C., Saquet, C., Blanche, M., Junien, C., and Gourdon, G. (1998). Somatic instability of the CTG repeat in mice transgenic for the myotonic dystrophy region is age dependent but not correlated to the relative intertissue transcription levels and proliferative capacities. *Hum. Mol. Genet.* **7**, 1285–1291.
17. Kaytor, M. D., Burright, E. N., Duvick, L. A., Zoghbi, H. Y., and Orr, H. T. (1997). Increased trinucleotide repeat instability with advanced maternal age. *Hum. Mol. Genet.* **6**, 2135–2139.

18. Kovtun, I. V., and McMurray, C. T. (2001). Trinucleotide expansion in haploid germ cells by gap repair. *Nat. Genet.* **27**, 407–411.
19. McMurray, C. T., and Kortun, I. V. (2003). Repair in haploid male germ cells occurs late in differentiation as chromatin is condensing. *Chromosoma* **111**, 505–508.
20. Savouret, C., Brisson, E., Essers, J., Kanaar, R., Pastink, A., te Riele, H., Junien, C., and Gourdon, G. (2003). CTG repeat instability and size variation timing in DNA repair-deficient mice. *EMBO J.* **22**, 2264–2273.
21. Watase, K., Venken, K. J., Sun, Y., Orr, H. T., and Zoghbi, H. Y. (2003). Regional differences of somatic CAG repeat instability do not account for selective neuronal vulnerability in a knock-in mouse model of SCA1. *Hum. Mol. Genet.* **12**, 2789–2795.
22. Godwin, A. R., Bollag, R. J., Christie, D. M., and Liskay, R. M. (1994). Spontaneous and restriction enzyme-induced chromosomal recombination in mammalian cells. *Proc. Natl. Acad. Sci. USA* **91**, 12554–12558.
23. Bowater, R. P., Jaworski, A., Larson, J. E., Parniewski, P., and Wells, R. D. (1997). Transcription increases the deletion frequency of long CTG·CAG triplet repeats from plasmids in *Escherichia coli*. *Nucleic Acids Res.* **1997**, 2861–2868.
24. Schumacher, S., Pinet, I., and Bichara, M. (2001). Modulation of transcription reveals a new mechanism of triplet repeat instability in *Escherichia coli*. *J. Mol. Biol.* **307**, 39–49.
25. Parniewski, P., Bacolla, A., Jaworski, A., and Wells, R. D. (1999). Nucleotide excision repair affects the stability of long transcribed (CTG*CAG) tracts in an orientation-dependent manner in *Escherichia coli*. *Nucleic Acids Res.* **27**, 616–623.
26. Mochmann, L. H., and Wells, R. D. (2004). Transcription influences the types of deletion and expansion products in an orientation-dependent manner from GAC*GTC repeats. *Nucleic Acids Res.* **32**, 4469–4479.
27. Schmidt, K. H., Abbott, C. M., and Leach, D. R. (2000). Two opposing effects of mismatch repair on CTG repeat instability in *Escherichia coli*. *Mol. Microbiol.* **35**, 463–471.
28. Lehmann, A. R. (2003). DNA repair-deficient diseases, xeroderma pigmentosum, Cockayne syndrome and trichothiodystrophy. *Biochimie* **85**, 1101–1111.
29. Citterio, E., Vermeulen, W., and Hoeijmakers, J. H. (2000). Transcriptional healing. *Cell* **101**, 447–450.
30. Oussatcheva, E. A., Hashem, V. I., Zou, Y., Sinden, R. R., and Potaman, V. N. (2001). Involvement of the nucleotide excision repair protein UvrA in instability of CAG*CTG repeat sequences in *Escherichia coli*. *J. Biol. Chem.* **276**, 30878–30884.
31. Mellon, I., Rajpal, D. K., Koi, M., Boland, C. R., and Champe, G. N. (1996). Transcription-coupled repair deficiency and mutations in human mismatch repair genes. *Science* **272**, 557–560.
32. Mellon, I., and Champe, G. N. (1996). Products of DNA mismatch repair genes mutS and mutL are required for transcription-coupled nucleotide-excision repair of the lactose operon in *Escherichia coli*. *Proc. Natl. Acad. Sci. USA* **93**, 1292–1297.
33. Lee, D. F., Drouin, R., Pitsikas, P., and Rainbow, A. J. (2004). Detection of an involvement of the human mismatch repair genes hMLH1 and hMSH2 in nucleotide excision repair is dependent on UVC fluence to cells. *Cancer Res.* **64**, 3865–3870.
34. Nara, K., Nagashima, F., and Yasui, A. (2001). Highly elevated ultraviolet-induced mutation frequency in isolated Chinese hamster cell lines defective in nucleotide excision repair and mismatch repair proteins. *Cancer Res.* **61**, 50–52.
35. Bertrand, P., Tishkoff, D. X., Filosi, N., Dasgupta, R., and Kolodner, R. D. (1998). Physical interaction between components of DNA mismatch repair and nucleotide excision repair. *Proc. Natl. Acad. Sci. USA* **95**, 14278–14283.
36. Pearson, C. E., Ewel, A., Acharya, S., Fishel, R. A., and Sinden, R. R. (1997). Human MSH2 binds to trinucleotide repeat DNA structures associated with neurodegenerative diseases. *Hum. Mol. Genet.* **6**, 1117–1123.
37. Schumacher, S., Fuchs, R. P., and Bichara, M. (1998). Expansion of CTG repeats from human disease genes is dependent upon replication mechanisms in *Escherichia coli*: The effect of long patch mismatch repair revisited. *J. Mol. Biol.* **279**, 1101–1110.
38. Parniewski, P., Jaworski, A., Wells, R. D., and Bowater, R. P. (2000). Length of CTG·CAG repeats determines the influence of mismatch repair on genetic instability. *J. Mol. Biol.* **299**, 865–874.
39. Jaworski, A., Rosche, W. A., Gellibolian, R., Kang, S., Shimizu, M., Bowater, R. P., Sinden, R. R., and Wells, R. D. (1995). Mismatch repair in *Escherichia coli* enhances instability of (CTG)n triplet repeats from human hereditary diseases. *Proc. Natl. Acad. Sci. USA* **92**, 11019–11023.
40. Rolfsmeier, M. L., Dixon, M. J., and Lahue, R. S. (2000). Mismatch repair blocks expansions of interrupted trinucleotide repeats in yeast. *Mol. Cell* **6**, 1501–1507.
41. Schweitzer, J. K., and Livingston, D. M. (1997). Destabilization of CAG trinucleotide repeat tracts by mismatch repair mutations in yeast. *Hum. Mol. Genet.* **6**, 349–355.
42. Kovtun, I. V., Spiro, C., and McMurray, C. T. (2004). Triplet repeats and DNA repair: Germ cell and somatic cell instability in transgenic mice. *Methods Mol. Biol.* **277**, 309–319.
43. Kovtun, I. V., Thornhill, A. R., and McMurray, C. T. (2004). Somatic deletion events occur during early embryonic development and modify the extent of CAG expansion in subsequent generations. *Hum. Mol. Genet.* **13**, 3057–3068.
44. Kovtun, I. V., Welch, G., Guthrie, H. D., Hafner, K. L., and McMurray, C. T. (2004). CAG repeat lengths in X- and Y-bearing sperm indicate that gender bias during transmission of Huntington's disease gene is determined in the embryo. *J. Biol. Chem.* **279**, 9389–9391.
45. Savouret, C., Garcia-Cordier, C., Megret, J., te Riele, H., Junien, C., and Gourdon, G. (2004). MSH2-dependent germinal CTG repeat expansions are produced continuously in spermatogonia from DM1 transgenic mice. *Mol. Cell Biol.* **24**, 629–637.
46. van den Broek, W. J., Nelen, M. R., Wansink, D. G., Coerwinkel, M. M., te Riele, H., Groenen, P., and Wieringa, B. (2002). Somatic expansion behaviour of the (CTG)n repeat in myotonic dystrophy knock-in mice is differentially affected by Msh3 and Msh6 mismatch-repair proteins. *Hum. Mol. Genet.* **11**, 191–198.
47. Gomes-Pereira, M., Fortune, M. T., Ingram, L., McAbney, J. P., and Monckton, D. G. (2004). Pms2 is a genetic enhancer of trinucleotide CAG·CTG repeat somatic mosaicism: Implications for the mechanism of triplet repeat expansion. *Hum. Mol. Genet.* **13**, 1815–1825.
48. Manley, K., Shirley, T. L., Flaherty, L., and Messer, A. (1999). Msh2 deficiency prevents in vivo somatic instability of the CAG repeat in Huntington disease transgenic mice. *Nat. Genet.* **23**, 471–473.
49. Bates, G. P., and Hay, D. G. (2004). Mouse models of triplet repeat diseases. *Methods Mol. Biol.* **277**, 3–15.
50. Everett, C. M., and Wood, N. W. (2004). Trinucleotide repeats and neurodegenerative disease. *Brain* **127**, 2385–2405.
51. Orr, H. T. (2002). Microarrays and polyglutamine disorders: Reports from the Hereditary Disease Array Group. *Hum. Mol. Genet.* **11**, 1909–1910.
52. Zhang, Y., Monckton, D. G., Siciliano, M. J., Connor, T. H., and Meistrich, M. L. (2002). Age and insertion site dependence of repeat number instability of a human DM1 transgene in individual mouse sperm. *Hum. Mol. Genet.* **11**, 791–798.
53. Seznec, H., Lia-Baldini, A. S., Duros, C., Fouquet, C., Lacroix, C., Hofmann-Radvanyi, H., Junien, C., and Gourdon, G. (2000). Transgenic mice carrying large human genomic sequences with expanded CTG repeat mimic closely the DM CTG repeat intergenerational and somatic instability. *Hum. Mol. Genet.* **9**, 1185–1194.

54. Strong, T. V., Tagle, D. A., Valdes, J. M., Elmer, L. W., Boehm, K., Swaroop, M., Kaatz, K. W., Collins, F. S., and Albin, R. L. (1993). Widespread expression of the human and rat Huntington's disease gene in brain and nonneural tissues. *Nat. Genet.* **5**, 259–265.
55. Li, S. H., Schilling, G., Young, W. S., III, Li, X. J., Margolis, R. L., Stine, O. C., Wagster, M. V., Abbott, M. H., Franz, M. L., Ranen, N. G., et al. (1993). Huntington's disease gene (IT15) is widely expressed in human and rat tissues. *Neuron* **11**, 985–993.
56. Nagafuchi, S., Yanagisawa, H., Ohsaki, E., Shirayama, T., Tadokoro, K., Inoue, T., and Yamada, M. (1994). Structure and expression of the gene responsible for the triplet repeat disorder, dentatorubral and pallidoluysian atrophy (DRPLA). *Nat. Genet.* **8**, 177–182.
57. Doyu, M., Sobue, G., Kimata, K., Yamamoto, K., and Mitsuma, T. (1994). Androgen receptor mRNA with increased size of tandem CAG repeat is widely expressed in the neural and nonneural tissues of X-linked recessive bulbospinal neuronopathy. *J. Neurol. Sci.* **127**, 43–47.
58. Yazawa, I., Nukina, N., Hashida, H., Goto, J., Yamada, M., and Kanazawa, I. (1995). Abnormal gene product identified in hereditary dentatorubral-pallidoluysian atrophy (DRPLA) brain. *Nat. Genet.* **10**, 99–103.
59. Banfi, S., Servadio, A., Chung, M. Y., Kwiatkowski, T. J., Jr., McCall, A. E., Duvick, L. A., Shen, Y., Roth, E. J., Orr, H. T., and Zoghbi, H. Y. (1994). Identification and characterization of the gene causing type 1 spinocerebellar ataxia. *Nat. Genet.* **7**, 513–520.
60. Dixon, K. T., Cearley, J. A., Hunter, J. M., and Detloff, P. J. (2004). Mouse Huntington's disease homolog mRNA levels: Variation and allele effects. *Gene Expr.* **11**, 221–231.
61. Gourfinkel-An, I., Cancel, G., Trottier, Y., Devys, D., Tora, L., Lutz, Y., Imbert, G., Saudou, F., Stevanin, G., Agid, Y., et al. (1997). Differential distribution of the normal and mutated forms of huntingtin in the human brain. *Ann. Neurol.* **42**, 712–719.
62. Trottier, Y., Cancel, G., An-Gourfinkel, I., Lutz, Y., Weber, C., Brice, A., Hirsch, E., and Mandel, J. L. (1998). Heterogeneous intracellular localization and expression of ataxin-3. *Neurobiol. Dis.* **5**, 335–347.
63. Servadio, A., Koshy, B., Armstrong, D., Antalffy, B., Orr, H. T., and Zoghbi, H. Y. (1995). Expression analysis of the ataxin-1 protein in tissues from normal and spinocerebellar ataxia type 1 individuals. *Nat. Genet.* **10**, 94–98.
64. Kennedy, L., Evans, E., Chen, C. M., Craven, L., Detloff, P. J., Ennis, M., and Shelbourne, P. F. (2003). Dramatic tissue-specific mutation length increases are an early molecular event in Huntington disease pathogenesis. *Hum. Mol. Genet.* **12**, 3359–3367.
65. Martorell, L., Monckton, D. G., Gamez, J., Johnson, K. J., Gich, I., de Munain, A. L., and Baiget, M. (1998). Progression of somatic CTG repeat length heterogeneity in the blood cells of myotonic dystrophy patients. *Hum. Mol. Genet.* **7**, 307–312.
66. Lorenzetti, D., Watase, K., Xu, B., Matzuk, M. M., Orr, H. T., and Zoghbi, H. Y. (2000). Repeat instability and motor incoordination in mice with a targeted expanded CAG repeat in the Sca1 locus. *Hum. Mol. Genet.* **9**, 779–785.
67. Wong, L. J., Ashizawa, T., Monckton, D. G., Caskey, C. T., and Richards, C. S. (1995). Somatic heterogeneity of the CTG repeat in myotonic dystrophy is age and size dependent. *Am. J. Hum. Genet.* **56**, 114–122.
68. Bates, G. P., Mangiarini, L., Mahal, A., and Davies, S. W. (1997). Transgenic models of Huntington's disease. *Hum. Mol. Genet.* **6**, 1633–1637.
69. Sturrock, R. R. (1986). A quantitative histological study of the indusium griseum and neostriatum in elderly mice. *J. Anat.* **149**, 195–203.
70. Shelbourne, P. F., Killeen, N., Hevner, R. F., Johnston, H. M., Tecott, L., Lewandoski, M., Ennis, M., Ramirez, L., Li, Z., Iannicola, C., et al. (1999). A Huntington's disease CAG expansion at the murine Hdh locus is unstable and associated with behavioural abnormalities in mice. *Hum. Mol. Genet.* **8**, 763–774.
71. Watanabe, H., Tanaka, F., Doyu, M., Riku, S., Yoshida, M., Hashizume, Y., and Sobue, G. (2000). Differential somatic CAG repeat instability in variable brain cell lineage in dentatorubral pallidoluysian atrophy (DRPLA): A laser-captured microdissection (LCM)-based analysis. *Hum. Genet.* **107**, 452–457.
72. Burright, E. N., Clark, H. B., Servadio, A., Matilla, T., Feddersen, R. M., Yunis, W. S., Duvick, L. A., Zoghbi, H. Y., and Orr, H. T. (1995). SCA1 transgenic mice: A model for neurodegeneration caused by an expanded CAG trinucleotide repeat. *Cell* **82**, 937–948.
73. Goldberg, Y. P., Kalchman, M. A., Metzler, M., Nasir, J., Zeisler, J., Graham, R., Koide, H. B., O'Kusky, J., Sharp, A. H., Ross, C. A., et al. (1996). Absence of disease phenotype and intergenerational stability of the CAG repeat in transgenic mice expressing the human Huntington disease transcript. *Hum. Mol. Genet.* **5**, 177–185.
74. Libby, R. T., Monckton, D. G., Fu, Y. H., Martinez, R. A., McAbney, J. P., Lau, R., Einum, D. D., Nichol, K., Ware, C. B., Ptacek, L. J., et al. (2003). Genomic context drives SCA7 CAG repeat instability, while expressed SCA7 cDNAs are intergenerationally and somatically stable in transgenic mice. *Hum. Mol. Genet.* **12**, 41–50.
75. Bingham, P. M., Scott, M. O., Wang, S., McPhaul, M. J., Wilson, E. M., Garbern, J. Y., Merry, D. E., and Fischbeck, K. H. (1995). Stability of an expanded trinucleotide repeat in the androgen receptor gene in transgenic mice. *Nat. Genet.* **9**, 191–196.
76. Zu, T., Duvick, L. A., Kaytor, M. D., Berlinger, M. S., Zoghbi, H. Y., Clark, H. B., and Orr, H. T. (2004). Recovery from polyglutamine-induced neurodegeneration in conditional SCA1 transgenic mice. *J. Neurosci.* **24**, 8853–8861.
77. La Spada, A. R., Fu, Y. H., Sopher, B. L., Libby, R. T., Wang, X., Li, L. Y., Einum, D. D., Huang, J., Possin, D. E., Smith, A. C., et al. (2001). Polyglutamine-expanded ataxin-7 antagonizes CRX function and induces cone-rod dystrophy in a mouse model of SCA7. *Neuron* **31**, 913–927.
78. Aguilera, A. (2002). The connection between transcription and genomic instability. *EMBO J.* **21**, 195–201.
79. Mattick, J. S. (2004). RNA regulation: A new genetics? *Nat. Rev. Genet.* **5**, 316–323.
80. Lavorgna, G., Dahary, D., Lehner, B., Sorek, R., Sanderson, C. M., and Casari, G. (2004). In search of antisense. *Trends Biochem. Sci.* **29**, 88–94.
81. Cawley, S., Bekiranov, S., Ng, H. H., Kapranov, P., Sekinger, E. A., Kampa, D., Piccolboni, A., Sementchenko, V., Cheng, J., Williams, A. J., et al. (2004). Unbiased mapping of transcription factor binding sites along human chromosomes 21 and 22 points to widespread regulation of noncoding RNAs. *Cell* **116**, 499–509.
82. Kampa, D., Cheng, J., Kapranov, P., Yamanaka, M., Brubaker, S., Cawley, S., Drenkow, J., Piccolboni, A., Bekiranov, S., Helt, G., et al. (2004). Novel RNAs identified from an in-depth analysis of the transcriptome of human chromosomes 21 and 22. *Genome Res.* **14**, 331–342.
83. Petruska, J., Arnheim, N., and Goodman, M. F. (1996). Stability of intrastrand hairpin structures formed by the CAG/CTG class of DNA triplet repeats associated with neurological diseases. *Nucleic Acids Res.* **24**, 1992–1998.
84. Liu, B., and Alberts, B. M. (1995). Head-on collision between a DNA replication apparatus and RNA polymerase transcription complex. *Science* **267**, 1131–1137.
85. French, S. (1992). Consequences of replication fork movement through transcription units in vivo. *Science* **258**, 1362–1365.
86. Deshpande, A. M., and Newlon, C. S. (1996). DNA replication fork pause sites dependent on transcription. *Science* **272**, 1030–1033.

87. Vilette, D., Ehrlich, S. D., and Michel, B. (1995). Transcription-induced deletions in *Escherichia coli* plasmids. *Mol. Microbiol.* **17**, 493–504.
88. Kang, S., Ohshima, K., Shimizu, M., Amirhaeri, S., and Wells, R. D. (1995). Pausing of DNA synthesis in vitro at specific loci in CTG and CGG triplet repeats from human hereditary disease genes. *J. Biol. Chem.* **270**, 27014–27021.
89. Ohshima, K., and Wells, R. D. (1997). Hairpin formation during DNA synthesis primer realignment in vitro in triplet repeat sequences from human hereditary disease genes. *J. Biol. Chem.* **272**, 16798–16806.
90. Hashem, V. I., Rosche, W. A., and Sinden, R. R. (2002). Genetic assays for measuring rates of (CAG)·(CTG) repeat instability in *Escherichia coli. Mutat. Res.*, **2002**, 25–37.
91. Fresco, J., and Alberts, B. M. (1960). The accomodation of noncomplementary bases in helical polyribonucleotides and deoxyribonucleic acid. *Proc. Natl. Acad. Sci. USA* **46**, 311–321.
92. Streisinger, G., Okada, Y., Emrich, J., Newton, J., Tsugita, A., Terzaghi, E., and Inouye, M. (1966). Frameshift mutations and the genetic code. This paper is dedicated to Professor Theodosius Dobzhansky on the occasion of his 66th birthday. *Cold Spring Harb. Symp. Quant. Biol.* **31**, 77–84.
93. Sinden, R. R., and Wells, R. D. (1992). DNA structure, mutations, and human genetic disease. *Curr. Opin. Biotechnol.* **3**, 612–622.
94. Richards, R. I., and Sutherland, G. R. (1992). Dynamic mutations: A new class of mutations causing human disease. *Cell* **70**, 709–712.
95. McMurray, C. T. (1995). Mechanisms of DNA expansion. *Chromosoma* **104**, 2–13.
96. Cleary, J. D., and Pearson, C. E. (2005). Replication fork dynamics and dynamic mutations: The fork-shift model of repeat instability. *Trends Genet.* **21**, 272–280.
97. Wierdl, M., Greene, C. N., Datta, A., Jinks-Robertson, S., and Petes, T. D. (1996). Destabilization of simple repetitive DNA sequences by transcription in yeast. *Genetics* **143**, 713–721.
98. Sia, E. A., Kokoska, R. J., Dominska, M., Greenwell, P., and Petes, T. D. (1997). Microsatellite instability in yeast: Dependence on repeat unit size and DNA mismatch repair genes. *Mol. Cell Biol.* **17**, 2851–2858.
99. Strand, M., Earley, M. C., Crouse, G. F., and Petes, T. D. (1995). Mutations in the MSH3 gene preferentially lead to deletions within tracts of simple repetitive DNA in *Saccharomyces cerevisiae. Proc. Natl. Acad. Sci. USA* **92**, 10418–10421.
100. Strand, M., Prolla, T. A., Liskay, R. M., and Petes, T. D. (1993). Destabilization of tracts of simple repetitive DNA in yeast by mutations affecting DNA mismatch repair. *Nature* **365**, 274–276.
101. Greene, C. N., and Jinks-Robertson, S. (1997). Frameshift intermediates in homopolymer runs are removed efficiently by yeast mismatch repair proteins. *Mol. Cell Biol.* **17**, 2844–2850.
102. Miret, J. J., Pessoa-Brandao, L., and Lahue, R. S. (1997). Instability of CAG and CTG trinucleotide repeats in *Saccharomyces cerevisiae. Mol. Cell Biol.* **17**, 3382–3387.
103. White, P. J., Borts, R. H., and Hirst, M. C. (1999). Stability of the human fragile X (CGG)(n) triplet repeat array in *Saccharomyces cerevisiae* deficient in aspects of DNA metabolism. *Mol. Cell Biol.* **19**, 5675–5684.
104. Khajavi, M., Tari, A. M., Patel, N. B., Tsuji, K., Siwak, D. R., Meistrich, M. L., Terry, N. H., and Ashizawa, T. (2001). "Mitotic drive" of expanded CTG repeats in myotonic dystrophy type 1 (DM1). *Hum. Mol. Genet.* **10**, 855–863.
105. Spiro, C., Pelletier, R., Rolfsmeier, M. L., Dixon, M. J., Lahue, R. S., Gupta, G., Park, M. S., Chen, X., Mariappan, S. V., and McMurray, C. T. (1999). Inhibition of FEN-1 processing by DNA secondary structure at trinucleotide repeats. *Mol. Cell* **4**, 1079–1085.
106. Cleary, J. D., Nichol, K., Wang, Y. H., and Pearson, C. E. (2002). Evidence of cis-acting factors in replication-mediated repeat instability in primate cells. *Nat. Genet.* **31**, 37–46.
107. Pineiro, E., Fernandez-Lopez, L., Gamez, J., Marcos, R., Surralles, J., and Velazquez, A. (2003). Mutagenic stress modulates the dynamics of CTG repeat instability associated with myotonic dystrophy type 1. *Nucleic Acids Res.* **31**, 6733–6740.
108. Yang, Z., Lau, R., Marcadier, J. L., Chitayat, D., and Pearson, C. E. (2003). Replication inhibitors modulate instability of an expanded trinucleotide repeat at the myotonic dystrophy type 1 disease locus in human cells. *Am. J. Hum. Genet.* **73**, 1092–1105.
109. Gomes-Pereira, M., and Monckton, D. G. (2004). Chemically induced increases and decreases in the rate of expansion of a CAG*CTG triplet repeat. *Nucleic Acids Res.* **32**, 2865–2872.
110. Dere, R., Napierala, M., Ranum, L. P., and Wells, R. D. (2004). Hairpin structure-forming propensity of the (CCTG·CAGG) tetranucleotide repeats contributes to the genetic instability associated with myotonic dystrophy type 2. *J. Biol. Chem.* **279**, 41715–41726.
111. Hashem, V. I., Pytlos, M. J., Klysik, E. A., Tsuji, K., Khajavi, M., Ashizawa, T., and Sinden, R. R. (2004). Chemotherapeutic deletion of CTG repeats in lymphoblast cells from DM1 patients. *Nucleic Acids Res.* **32**, 6334–6346.
112. Meservy, J. L., Sargent, R. G., Iyer, R. R., Chan, F., McKenzie, G. J., Wells, R. D., and Wilson, J. H. (2003). Long CTG tracts from the myotonic dystrophy gene induce deletions and rearrangements during recombination at the APRT locus in CHO cells. *Mol. Cell Biol.* **23**, 3152–3162.
113. Gorbunova, V., Seluanov, A., Dion, V., Sandor, Z., Meservy, J. L., and Wilson, J. H. (2003). Selectable system for monitoring the instability of ctg/cag triplet repeats in mammalian cells. *Mol. Cell Biol.* **23**, 4485–4493.
114. Gorbunova, V., Seluanov, A., Mittelman, D., and Wilson, J. H. (2004). Genome-wide demethylation destabilizes CTG·CAG trinucleotide repeats in mammalian cells. *Hum. Mol. Genet.* **13**, 2979–2989.
115. Ashizawa, T., Monckton, D. G., Vaishnav, S., Patel, B. J., Voskova, A., and Caskey, C. T. (1996). Instability of the expanded (CTG)n repeats in the myotonin protein kinase gene in cultured lymphoblastoid cell lines from patients with myotonic dystrophy. *Genomics* **36**, 47–53.
116. Manley, K., Pugh, J., and Messer, A. (1999). Instability of the CAG repeat in immortalized fibroblast cell cultures from Huntington's disease transgenic mice. *Brain Res.* **835**, 74–79.
117. Kringstein, A. M., Rossi, F. M., Hofmann, A., and Blau, H. M. (1998). Graded transcriptional response to different concentrations of a single transactivator. *Proc. Natl. Acad. Sci. USA* **95**, 13670–13675.
118. Adimoolam, S., Lin, C. X., and Ford, J. M. (2001). The p53-regulated cyclin-dependent kinase inhibitor, p21 (cip1, waf1, sdi1), is not required for global genomic and transcription-coupled nucleotide excision repair of UV-induced DNA photoproducts. *J. Biol. Chem.* **276**, 25813–25822.
119. Sweder, K. S., Verhage, R. A., Crowley, D. J., Crouse, G. F., Brouwer, J., and Hanawalt, P. C. (1996). Mismatch repair mutants in yeast are not defective in transcription-coupled DNA repair of UV-induced DNA damage. *Genetics* **143**, 1127–1135.
120. Kobayashi, K., O'Driscoll, M., Macpherson, P., Mullenders, L., Vreeswijk, M., and Karran, P. (2004). XPC lymphoblastoid cells defective in the hMutSalpha DNA mismatch repair complex exhibit normal sensitivity to UVC radiation and normal transcription-coupled excision repair of DNA cyclobutane pyrimidine dimers. *DNA Repair (Amsterdam)* **3** 649–657.

121. Sonneveld, E., Vrieling, H., Mullenders, L. H., and van Hoffen, A. (2001). Mouse mismatch repair gene Msh2 is not essential for transcription-coupled repair of UV-induced cyclobutane pyrimidine dimers. *Oncogene* **20**, 538–541.

122. Rochette, P. J., Bastien, N., McKay, B. C., Therrien, J. P., Drobetsky, E. A., and Drouin, R. (2002). Human cells bearing homozygous mutations in the DNA mismatch repair genes hMLH1 or hMSH2 are fully proficient in transcription-coupled nucleotide excision repair. *Oncogene* **21**, 5743–5752.

123. Zhang, N., Lu, X., Zhang, X., Peterson, C. A., and Legerski, R. J. (2002). hMutSbeta is required for the recognition and uncoupling of psoralen interstrand cross-links in vitro. *Mol. Cell Biol.* **22**, 2388–2397.

124. Paques, F., and Haber, J. E. (1997). Two pathways for removal of nonhomologous DNA ends during double-strand break repair in *Saccharomyces cerevisiae*. *Mol. Cell Biol.* **17**, 6765–6771.

125. Sugawara, N., Paques, F., Colaiacovo, M., and Haber, J. E. (1997). Role of *Saccharomyces cerevisiae* Msh2 and Msh3 repair proteins in double-strand break-induced recombination. *Proc. Natl. Acad. Sci. USA* **94**, 9214–9219.

126. Saparbaev, M., Prakash, L., and Prakash, S. (1996). Requirement of mismatch repair genes MSH2 and MSH3 in the RAD1-RAD10 pathway of mitotic recombination in *Saccharomyces cerevisiae*. *Genetics* **142**, 727–736.

127. Kirkpatrick, D. T., and Petes, T. D. (1997). Repair of DNA loops involves DNA-mismatch and nucleotide-excision repair proteins. *Nature* **387**, 929–931.

128. Wheeler, V. C., Auerbach, W., White, J. K., Srinidhi, J., Auerbach, A., Ryan, A., Duyao, M. P., Vrbanac, V., Weaver, M., Gusella, J. F., et al. (1999). Length-dependent gametic CAG repeat instability in the Huntington's disease knock-in mouse. *Hum. Mol. Genet.* **8**, 115–122.

129. Gourdon, G., Radvanyi, F., Lia, A. S., Duros, C., Blanche, M., Abitbol, M., Junien, C., and Hofmann-Radvanyi, H. (1997). Moderate intergenerational and somatic instability of a 55-CTG repeat in transgenic mice. *Nat. Genet.* **15**, 190–192.

130. Zuhlke, C., Riess, O., Bockel, B., Lange, H., and Thies, U. (1993). Mitotic stability and meiotic variability of the (CAG)n repeat in the Huntington disease gene. *Hum. Mol. Genet.* **2**, 2063–2067.

131. Lopes-Cendes, I., Maciel, P., Kish, S., Gaspar, C., Robitaille, Y., Clark, H. B., Koeppen, A. H., Nance, M., Schut, L., Silveira, I., et al. (1996). Somatic mosaicism in the central nervous system in spinocerebellar ataxia type 1 and Machado-Joseph disease. *Ann. Neurol.* **40**, 199–206.

132. Tanaka, F., Sobue, G., Doyu, M., Ito, Y., Yamamoto, M., Shimada, N., Yamamoto, K., Riku, S., Hshizume, Y., and Mitsuma, T. (1996). Differential pattern in tissue-specific somatic mosaicism of expanded CAG trinucleotide repeats in dentatorubral-pallidoluysian atrophy, Machado-Joseph disease, and X-linked recessive spinal and bulbar muscular atrophy. *J. Neurol. Sci.* **135**, 43–50.

133. Ueno, S., Kondoh, K., Kotani, Y., Komure, O., Kuno, S., Kawai, J., Hazama, F., and Sano, A. (1995). Somatic mosaicism of CAG repeat in dentatorubral-pallidoluysian atrophy (DRPLA). *Hum. Mol. Genet.* **4**, 663–666.

134. Takano, H., Onodera, O., Takahashi, H., Igarashi, S., Yamada, M., Oyake, M., Ikeuchi, T., Koide, R., Tanaka, H., Iwabuchi, K., et al. (1996). Somatic mosaicism of expanded CAG repeats in brains of patients with dentatorubral-pallidoluysian atrophy: cellular population-dependent dynamics of mitotic instability. *Am. J. Hum. Genet.* **58**, 1212–1222.

135. Sato, T., Oyake, M., Nakamura, K., Nakao, K., Fukusima, Y., Onodera, O., Igarashi, S., Takano, H., Kikugawa, K., Ishida, Y., et al. (1999). Transgenic mice harboring a full-length human mutant DRPLA gene exhibit age-dependent intergenerational and somatic instabilities of CAG repeats comparable with those in DRPLA patients. *Hum. Mol. Genet.* **8**, 99–106.

136. Lin, Y., Dion, V., and Wilson, J. H. (2006) Transcription promotes contraction of CAG repeat tracts in human cells. Nature Struct. Mol. Biol. (in press).

137. Cho, D. H., Thienes, C. P., Mahoney, S. E., Analau, E., Filippova, G. N. and Tapscott, S. J. (2005) Antisense transcription and heterochromatin at the DM1 CTG repeats are constrained by CTCF. *Mol. Cell* **20**, 483–489.

CHAPTER 45

Structural Characteristics of Trinucleotide Repeats in Transcripts

Wlodzimierz J. Krzyzosiak,[1] Krzysztof Sobczak,[1] and Marek Napierala[2]

[1]Institute of Bioorganic Chemistry, Polish Academy of Sciences, Noskowskiego 12/14, 61-704 Poznan, Poland, and [2]Institute of Biosciences and Technology, Center for Genome Research, Texas A&M University System Health Science Center, Texas Medical Center, Houston, Texas 77030-3303

I. Introduction
II. Simple Sequence Repeats in Genes and Proteins
III. Triplet Repeats in Transcripts
IV. Triplet Repeats in the Human Transcriptome
V. RNA Structures of Triplet Repeats
VI. RNA Structures of Triplet Repeats and Their Flanking Sequences
VII. Structural Role of the Repeat Interruptions
VIII. Cell Defense Systems Against dsRNA
IX. Concluding Remarks
References

This chapter reviews structural foundations fot the mechanism of RNA-mediated pathogenesis in Triplet Repeat Expansion Diseases. The occurrence of triplet repeats in human genes and transcripts is analyzed from evolutionary and functional perspectives. RNA structures formed by different types of triplet repeats and the structural roles of various naturally occuring repeat interruptions are discussed in relevance to their physiological function and role in RNA toxicity. The possible involvement of various cell defense systems against long dsRNA in the mechanism of pathogenesis triggered by expanded repeats in transcripts is also discussed.

I. INTRODUCTION

This chapter discusses structural features of triplet repeats in transcripts relevant to the more prominent role of RNA in the pathogenesis of triplet repeat expansion diseases (TREDs). The approach to this problem does not focus on any single disease, gene, transcript, or repeat type. Instead, all kinds of repeated sequences are analyzed collectively in a comparative manner. Similarities and differences are noted between structures formed by various triplet repeats in RNA, as well as between properties of these structures, as are the features that correlate well with pathogenesis. The rationale is that, in order to better understand the RNA-mediated processes that lead to cell dysfunction and disease, it is necessary to learn about the roles played by normal triplet repeats in transcripts. To gather this knowledge, both the occurrence and polymorphism of triplet repeats in transcripts are analyzed, as well as the expression and abundance of such transcripts in human transcriptomes. The repeats are also approached from the genomic and evolutionary perspectives to better characterize the sequences that may possibly participate in the RNA repeat-mediated pathomechanisms in cells.

II. SIMPLE SEQUENCE REPEATS IN GENES AND PROTEINS

Hundreds of thousands of simple sequence repeat (SSR) loci are present in the human genome. Their evolution is driven by two opposing mutational forces: length mutations, which tend to increase the repeat number, and point mutations, which slow down this process by breaking the regularity of the repeat tracts [1]. The mutations that are advantageous undergo positive selection. Thus, the evolutionary life cycle of the SSR tract very likely begins with point mutations that give rise to short repeat arrays. With time, these repeats become increasingly longer, most likely due to DNA replication slippage. Then, they again acquire point mutations and expand more slowly. Further point mutations stabilize these sequences as the end products of the SSR evolution [2]. The mutation rate at SSRs is several orders of magnitude higher than at other sequences and shows strong variation both between and within loci [3]. In agreement with this scenario, the plethora of SSRs present in the human genome are at various stages of this evolutionary process. Both short and long pure repeat tracts are present, as well as imperfect repeats containing different numbers and types of interruptions. Long pure repeats are often polymorphic in length [4], whereas interrupted repeats either show lower polymorphism or are monomorphic in the population.

In the human genome, the mononucleotides and dinucleotides are the most copious classes of SSRs. They occur predominantly in noncoding regions. In contrast, the trinucleotide repeats are more frequent in coding sequences of the human genes than in introns and intergenic regions [5]. They do not cause frameshifts when their repeat number is changed; thus, they are better tolerated than other SSRs in translated sequences [6]. The problem of whether the SSRs are only nonfunctional forerunners of the evolutionary processes or whether some of them had enough time to acquire specific functions in cells is a matter of continuous debate [7, 8]. The positive selection in exons implies some function for triplet repeats, but they are better known at present from the dysfunctions they cause. When these repeats are long enough, they tend to form unusual DNA structures that affect chromatin organization and DNA function [9]. Among the cellular processes impaired by the expanded triplet repeats are DNA recombination, cell cycle, replication, DNA mismatch repair, and transcription [7].

Length variation of triplet repeats within the open reading frame (ORF) is thought to be the fast evolutionary track to generate novel protein sequences. Some functional classes of proteins, such as transcription factors and protein kinases, took advantage of this rapid means of change more often than others [2]. Altogether about 20% of different human proteins may contain the single amino acid runs of Glu, Leu, Ala, and Pro predominantly [10], which are believed to be involved in the protein–protein interactions [11]. The variable lengths of amino acid repeats are thought to affect both protein structure and its interaction networks [12, 13]. The repeats are harmless to the cell within their normal length range. However, in the expanded form they become harmful, result in diseases, and undergo negative, purifying selection. In addition to the so-called "polyglutamine" diseases [14] are also the lesser known group of "polyalanine" disorders [15]. These two groups share at least some features of disease pathology, i.e., protein misfolding and aggregation [16].

III. TRIPLET REPEATS IN TRANSCRIPTS

Another interesting problem regarding triplet repeats is the role they possibly play in transcripts. Several relevant questions have already been asked, including the following. Do the repeats have any functions in the primary transcripts undergoing processing in the cell nucleus? Do they have any functions in mature RNAs present in the cytoplasm? If so, how do they realize these functions? Are there any specific repeat binding proteins involved in these activities? Was there enough time in the evolutionary scale for such proteins to coevolve with the repeats to perform these functions in tandem? Some partial answers to these questions already exist. With regard to cytosolic functions, the CUG [17] and CGG repeats [18] were shown to impede translation at its initiation step. Supporting the role of triplet repeats in the regulation of translation are their more frequent occurrence in the 5′ UTR than in the 3′ UTR and the strong predominance of GC-rich repeats (CGG and CCG) in the leader sequence [19]. In the cell nucleus, CUG [20], CCG [21], and CGG [22] repeats were shown to modulate the efficiency and accuracy of pre-mRNA splicing, including splice site selection. Proteins interacting with either single-stranded CUG repeats, which belong to the CELF family [23], or double-stranded CUG repeats, which belong to muscleblind family [24], were identified and well-characterized due to their participation in the RNA-mediated pathomechanism of myotonic dystrophy [25, 26]. As these proteins lack sequence specificity with respect to CUG repeats [27], it is unlikely that they have evolved to bind only these sequences. The same may apply to other repeated motifs and the putative proteins they bind. Thus, the likely scenario is that the repeats in transcripts, especially the longer repeats that appeared more recently in evolution, show some affinity for proteins that have been designed to perform other functions in cells. It was demonstrated earlier that the longer the CUG repeat, the more stable the structure it forms [28] and the more efficient the binding of proteins of the muscleblind family, which show affinity to the expanded repeats [24]. These proteins may be sequestered from their normal binding sites in other transcripts. At the same time, by an as yet unknown mechanism, the level of the CUG-BP protein becomes elevated [29, 30]. The end result of these pathogenic processes in DM1 cells is aberrant splicing of numerous primary transcripts [31]. Mutant transcripts containing the expanded CUG repeats were also reported to bind some transcription factors and impair the process of transcription in the DM1 cells [32]. The mechanism of pathogenesis in which RNA toxicity plays the primary role was also shown to trigger fragile-X-associated tremor ataxia syndrome (FXTAS) [33, 34].

IV. TRIPLET REPEATS IN THE HUMAN TRANSCRIPTOME

In the context of the previously outlined mechanism of RNA-mediated pathogenesis of some TREDs, and the possible contribution of this type of mechanism in other human diseases as well [19], the following questions have been asked: how many different transcripts contain triplet repeats in their sequences, and at what frequency do certain types of repeats occur? These questions seemed important because among the genes containing the longest repeat tracts could be candidates for involvement in diseases, for which the causative genes have not yet been identified. Among their transcripts should also be those whose function is compromised in the presence of mutant transcripts containing expanded repeats. It turned out that more than 600 human mRNAs harbor more than 700 triplet repeat tracts composed of 6 or more repeated motifs [19]. It was also revealed that CAG, CGG, CCG, CUG, AGG, and ACC are the most frequently occurring triplet repeat motifs and that ACG, AUC, CUU, AGU, CGU, and ACU repeats are very poorly represented. A rather small proportion (13%) of tracts were composed of more than 10 repeated motifs, and only 2% contained more than 20 repeated triplets. However, these numbers were derived from the bioinformatics survey of human gene sequences deposited in GenBank, in which information regarding repeat length polymorphism in the human population was scarce. Therefore, more comprehensive information was gathered experimentally. The genotyping project included 100 human genes selected to contain the longest runs of the most frequently occurring CAG and CUG repeats. The results demonstrated, however, that in genes other than those already known to be associated with TREDs, the very long and highly polymorphic repeat tracts were rare, which was in agreement with the results of the bioinformatic survey. It could be concluded from that analysis that, unless some interesting genes containing the CAG and CUG repeats were missing from the genotyping experiments, the RNAs bearing long repeat tracts were indeed infrequent among the corresponding transcripts.

It should also be stressed that not only the number of different transcripts harboring triplet repeats but also their expression levels in different cell types and tissues influence the cellular balance of the repeat containing transcripts and repeat binding proteins, which might be important for the mechanism of RNA-mediated pathogenesis. Bearing that in mind, the available gene expression data have been analyzed, by using tools of bioinformatics, to characterize the abundance of a large group of the repeat containing transcripts in different human tissues: brain, prostate, liver, kidney, pancreas, and lung. The CNG repeat containing transcripts were shown to belong to moderate and low abundance classes. Their expression levels were 1–2 orders of magnitude lower than those of the highly abundant reference transcripts. Among the 17 TRED-related transcripts analyzed, as many as 11 show moderate expression in at least one of the analyzed tissues. Interestingly, the ZNF9 transcript implicated in DM2 is the most abundant TRED transcript in most of the tissues analyzed.

V. RNA STRUCTURES OF TRIPLET REPEATS

To provide a basis for functional analysis of triplet repeats in RNA, it was necessary to determine their structural features. These structures were analyzed systematically, first out of the sequence context of their host transcripts. The transcripts investigated were composed of all 20 different triplet motifs reiterated 17 times. Transcripts composed of 6 repeated motifs, CGU, CGA, CAG, CUG, CCG, and CGG, were shown to form stable hairpins. The structures of 4 transcripts of the CNG type are shown in Fig. 45-1. The hairpin stem shows the same type of structure in all of these transcripts, as revealed by similar patterns of cleavage generated by ribonucleases T1, T2, and V1, nuclease S1, and lead ions [35]. The stem is composed of the periodically occurring base pairs and single base mismatches. Depending on the nature of bases involved in these mismatches, the stem structures show different thermodynamic stabilities. The most stable are stems containing the G–G mismatches, and the least stable are those with A–A. All repeats of the CNG type, with the exception of CGG repeats, show a tendency to assume several variant alignments, in which different central repeats form terminal loops and different 3′-terminal repeats form single-stranded protruding ends [35]. The hairpin variants having a 4-nt terminal loop and the longest possible stem strongly predominate (70–80%), and the contribution from slipped hairpins decreases with the degree of slippage. This kind of effect was observed for the first time in the DMPK transcript containing CUG repeats [28]. In case of the CGG repeat transcript, only two different hairpin alignments could be distinguished and these hairpins were present in equal proportions. This property of the CGG repeats could be explained by the strong helix-stabilizing effect of the G–G mismatches [35].

It should also be emphasized that details of the CNG repeat stem and loop structures could be revealed after these transcripts were forced to assume a single alignment, as the structure probing results of the uniformly aligned hairpins were much easier to interpret. The single structure was generated with help of the suitable GC clamp. Five G–C and C–G base pairs were sufficient to achieve that effect [35]. The artificial GC clamp was a good model for natural clamps existing in some of the human transcripts harboring triplet repeats.

VI. RNA STRUCTURES OF TRIPLET REPEATS AND THEIR FLANKING SEQUENCES

The CNG repeat regions in transcripts may form hairpin structures of different molecular architecture. This can be illustrated by different structural contributions from sequences flanking the repeat in transcripts of genes involved in TREDs. The secondary structures of 10 such transcripts were predicted by using the Mfold computer program [19]. It turned out that in five transcripts, FMR2, AR, SCA6, SCA7, and SCA12, the nearest flanking sequences contribute to a hairpin formed by the repeats by extending and strengthening its structure. On the other hand, this kind of effect was not observed in DMPK, DRPLA, SCA3, SCA8, and IT15 transcripts. According to the results of experimental analysis, which included most of these transcripts, the predictions turned out to be basically correct [28, 36–40]. One may ask the question, what is the functional significance of different types of interactions between the repeats and their flanks? It seems that tightly base-paired flanking sequences may strongly increase the ability of the expanded repeat hairpin to interact with the putative double-stranded repeat binding proteins and promote perturbations in cell functions. For example, hairpins of the same length formed by pure repeats in different host transcripts may trigger different effects, depending on the degree of additional stabilization of their structures by flanking sequences. By using the same reasoning, hairpins composed of a lower number of repeats but well-stabilized by the flanking sequences may exert effects similar to those composed of a larger number of repeats but lacking any additional stabilization.

In the SCA3 transcript, the 3′-terminal CAG repeats were shown to form several alternative base pairs with the so-called "pseudo-repeats" present in the 5′ flanking sequence [38]. This gives the impression of unusual repeat slippage in the 5′ direction. In addition, the frequent SNP located at the very 3′ end of the repeated sequence was demonstrated to influence the structure of that region at both of its sides. Interestingly, the SNP variant C usually accompanies longer CAG repeat tracts [41, 42] and is present in almost all *SCA3* mutant alleles described thus

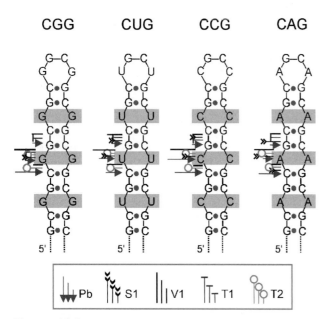

FIGURE 45-1 Schematic representation of the stem and loop structures formed by different types of CNG repeats in transcripts. The structures are ordered from left to right according to decreasing thermodynamic stability. The representative cleavage sites in the stem structures are indicated for the ribonucleases (T1, T2, and V1), nuclease S1, and lead ions (Pb). The size of the symbols corresponds to cleavage intensity.

far. The C variant is common in the human population, rare in chimpanzees, and absent in other primates [43]. It is therefore likely that the selection of the C variant in hominoid evolution facilitated increased variability at the SCA3 locus at the cost of an increased risk of pathogenic repeat expansion in humans [38].

VII. STRUCTURAL ROLE OF THE REPEAT INTERRUPTIONS

In four TRED-related genes (*FMR1, SCA1, SCA2,* and *SCA17*), most of the normal alleles contain specific interruptions located within the repeat tracts (Fig. 45-2). These are AGG interruptions, which break the regularity of the CGG repeat tracts in the *FMR1* gene, CAT triplets, which disrupt the homogeneity of CAG repeats in *SCA1*, and CAA interruptions, which break the continuity of CAG repeats in both *SCA2* and *SCA17* genes. These base substitutions may function in DNA as protective elements, preventing pathogenic repeat expansions. They decrease the opportunity of DNA strands to slip, inhibit interstrand slippage, and suppress intrastrand interaction [44]. But what could be the function of the repeat interspersions in transcripts? To answer this question, their structural effects were analyzed.

The structure probing experiments revealed that even a single AGG interruption within the CGG repeat of the FMR1 transcript prevents single hairpin structure formation by the repeats [40] (Fig. 45-3). The presence of the C→A substitution causes hairpins having A–G oppositions within their quasi-stable stem to be strongly disfavored. Instead, branched hairpins are formed that have the substituted base either in the side loop or in an enlarged terminal loop, depending on the location of the interruption. Even more strongly, the split hairpin structures are formed by FMR1 repeats harboring two and three base substitutions [40].

The *FMR1* premutations, i.e., the CGG repeat expansions in the 55–200 range, are associated with several syndromes, and for one of them, FXTAS, RNA-mediated pathogenesis is documented [34]. In some of the *FMR1* premutation alleles, the interruptions are retained and, as demonstrated for normal variants, they prevent the formation of single, long hairpins by repeats in transcripts. Thus, it is likely that the interruptions may protect some permutation carriers from being prone to FXTAS and related syndromes by shortening the effective length of the hairpin composed of pure CGG repeats. In this way, the status of the AGG interruptions, which determines

FMR1
30 repeats: 10-**1**-9-**1**-9

*Gggcgtgcggcagcg*cggcggcggcggcggcggcggcggcggcgg**AGG**cggcggcggcggcggcggcggcggcgg**AGG**cggcggcggcggcggcggcggcggcgg*ctgggcctcgagcgc*

SCA1
29 repeats: 12-**1**-1-**1**-14

*cgggacacaaggctgag*cagcagcagcagcagcagcagcagcagcagcagcag**CAT**cag**CAT**cagcagcagcagcagcagcagcagcagcagcagcagcagcagcag*caccctcagcaggg*

SCA2
22 repeats: 8-**1**-4-**1**-8

*gggcccctcaccatgtcgctgaagccc*cagcagcagcagcagcagcagcag**CAA**cagcagcagcag**CAA**cagcagcagcagcagcagcag*ccgccgcccgcggctgccaatgtccgc*

TBP
36 repeats: 3-**3**-9-**1**-1-**1**-16-**1**-1

*caaagg*cagcagcag**CAACAACAA**cagcagcagcagcagcagcagcag**CAA**cag**CAA**cagcagcagcagcagcagcagcagcagcagcagcagcagcagcag**CAA**cag*gcagtg*

FIGURE 45-2 Interruption patterns of the repeat region found in the predominant allelic variants of four TRED-related genes: *FMR1, SCA1, SCA2,* and *TBP*. The repeat regions are bold, interruptions are underlined, and repeat flanking regions are italicized.

the RNA structure of the repeat tract, may influence the correlation between repeat length and clinical outcome [40].

In addition, the CAU interruptions present in CAG repeats of the SCA1 transcripts either enlarge the hairpin terminal loop, nucleate internal loops or bulges, or force the repeats to form several smaller hairpins [37] (Fig. 45-3). Thus, the role of these G→U substitutions is to shorten and destabilize the CAG repeat hairpins, presumably to reduce their ability to interact with the dsCAG repeat binding proteins that were described earlier [45]. Whether their role is also to prevent CAG hairpin involvement in the RNA-mediated pathogenesis mechanism analogous to that documented for DM1, DM2, and FXTAS remains to be revealed. The fact that RNA structures formed by the repeat regions in some rare carriers of expanded interrupted repeats correlate better with pathogenesis than with total repeat length supports the role of RNA

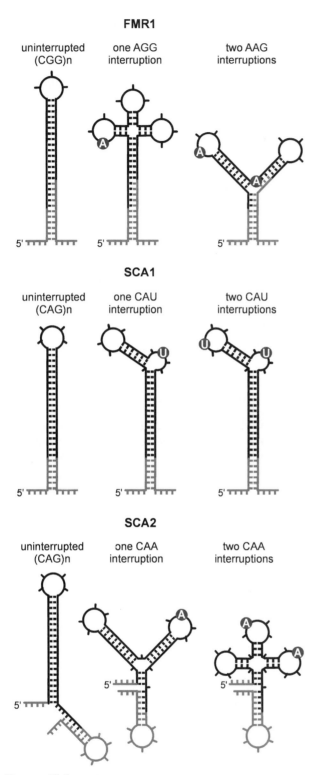

FIGURE 45-3 Structural organization of triplet repeat regions in the FMR1, SCA1, and SCA2 transcripts. The simplified secondary structure models presented are based on the results of experimental structure probing. Transcripts containing pure CAG (or CGG) repeats, RNAs having an asymmetrically located single interruption, and repeats harboring two interruptions are shown. Black and gray lines schematically depict the repeated sequences and specific flanking sequences, respectively. Positions of nucleotide substitutions are indicated by gray circles.

in SCA1 pathogenesis [37]. The pathogenic threshold of repeat length is 39 repeats for SCA1, which means that the RNA hairpin stem composed of 38 repeats corresponds to that threshold. In agreement with the RNA structure hypothesis for SCA1 pathogenesis, carriers of interrupted 39 and 43 repeats, in which branched hairpin structures are formed that do not reach the pathogenic length threshold, did not develop disease [37, 46].

In SCA2 the pathogenic threshold is 36 repeats, which correspond to the RNA hairpin stem composed of 32 pure CAG repeats. This is because four nucleotides are engaged in terminal loop formation and three repeats are involved in base pairing with specific flanking sequences [39] (Fig. 45-3). Here again, in some rare carriers of the interrupted "mutant" repeats who did not develop SCA2 symptoms, the total number of repeats is equal to or higher than the pathogenic threshold. However, the criterion of pathogenic RNA structure is not fulfilled [39]. Moreover, the RNA structure better correlates with the pathogenesis of SCA2 than the length of pure polyQ tract. The CAA interruptions also code for glutamine, and a stretch of 39 Q's present in the ataxin-2 protein is insufficient to develop typical SCA2 symptoms [47]. Taking this a step further, the examples of SCA17 and the genes unrelated to TREDs should also be recalled. In the mutated *TBP* gene implicated in SCA17, the CAA interruptions are present at locations that allow long hairpins composed of pure CAG repeats to be formed in transcripts [39]. On the other hand, in the normal population the *TBP* repeat region contains 29–42 interrupted repeats encoding 29Q–42Q [48]. Thus, it appears that even 42 consecutive glutamines are insufficient to cause SCA17 in the absence of the putative "toxic" RNA structures formed in the corresponding transcripts. Another instructive example is the *FOXP2* gene, implicated in human speech and language [49, 50]. The repeat region of this gene ended its evolutionary life cycle a long time ago [2] and is monomorphic in the human population. This region encodes 40 Q's and uses 26 CAG and 14 CAA triplets for this purpose. Unlike the case of TBP transcript, the more numerous and more dispersed CAA interruptions in *FOXP2* split the CAG repeats in such a way that long toxic hairpins cannot be formed [39]. Accordingly, none of the billions of carriers was reported to develop neurological disease due to this gene.

The correlations presented here between the RNA structures of the repeat regions and the appearance of the corresponding diseases speak in favor of a more general role for RNA in the pathogenesis of TREDs. One may argue, however, that these are only correlations and that direct evidence is needed to prove the hypothesis that RNA may also be involved in the pathogenesis of polyglutamine diseases. The appearance of cellular inclusions that either contain the mutated RNA [51] or do not is insufficient as these inclusions may be irrelevant to other pathways of pathogenesis. Some of these putative pathways are discussed briefly in the next section.

VIII. CELL DEFENSE SYSTEMS AGAINST dsRNA

It is now well-established that long hairpin structures are formed by triplet repeats in transcripts from most of the TRED-causative genes. The stem structures of these expanded hairpins behave in the structure probing experiments as if they were nearly perfect duplexes [28, 37–40]. These structures are formed within the cell nucleus, where they are present in the primary transcripts undergoing processing. The mature mRNAs are then exported to cytoplasm and translated. As endogenous long dsRNAs are rather rare in the cytoplasm of eukaryotic cells, the question arises as to whether the long repeat hairpins are recognized as targets of the cellular defense systems developed to combat the double-stranded viral RNA.

In vertebrate cells, long RNA duplexes induce several protective responses such as stimulation of dsRNA-activated protein kinase (PKR) and oligoadenylate synthetase (2-5A), induction of type 1 interferon (IFN), and activation of Dicer-related RNA interference pathways (RNAi) [52]. The PKR and 2-5A systems function to shut down protein translation in the infected cells. IFN induction also results in translation inhibition, which is achieved by the up-regulation of IFN-stimulated genes that include PKR and 2-5A. The antiviral defense network is thought to be activated by the dsRNA composed of at least 30 bp [52]. Only in rough accordance with this threshold is the fact that numerous microRNA precursors normally processed by Dicer in the cytoplasm fall into the 60- to 70-nt length range [53]. So what happens to the triplet repeat hairpins of TRED-related transcripts that are known to be present and translated in the cytoplasm? They form duplexes even 2–5 times longer than the 30-bp threshold, which calls the antiviral response systems on alert. But the repeat duplexes are somewhat different. They contain periodically occurring mismatches of different types [35]. Do these unusual types of duplex structures matter for PKR activation? According to the results of earlier studies, the enzyme seems to be fairly tolerant of various irregularities in RNA duplex structures [54], and hairpins composed of 15 CUG repeats were sufficient to activate PKR *in vitro* [55]. The activation was more efficient when longer repeats were used. The CAG repeats from the mutant IT15 transcript involved in Huntington's disease were also demonstrated to activate PKR [56]. More recently, the CGG repeats in the FMR1 premutation transcript were shown to be unable to activate PKR, but they were reportedly cleaved by Dicer [57].

Ribonuclease Dicer functions in cell cytoplasm to excise microRNAs from their precursors. The precursors are the stem and loop structures that have two protruding nucleotides at the 3' end [53]. These overhanging single-stranded nucleotides anchor the PAZ domain of Dicer [58], which cleaves RNA about 20 nt apart by using two RNaseIII domains [59]. According to results, the Dicer cleavages that occur not only in CGG but also in other CNG repeat hairpins residing in the interior of their host mRNAs must be executed in a fashion that takes advantage of different anchoring sites. Whether Dicer in some way controls the level of long triplet repeat hairpins in cells remains to be established.

Long RNA hairpins are more frequent in the cell nucleus. In addition to the giant CUG and CCUG hairpins formed in the cells of DM1 and DM2 patients and other expanded CNG repeat hairpins present in the cells of other TRED patients, structured Alu repeats and transcripts containing pri-microRNA hairpins are also present. The latter are substrates for Drosha ribonuclease, which generates the pre-microRNAs [60]. The question is whether the repeat hairpins are also Drosha substrates. If so, do the different CNG repeats differ in their substrate activity? These questions, which deal with problems that are at the crossroads of the triplet-repeat-induced pathogenesis and microRNA biogenesis pathways, also need to be answered.

IX. CONCLUDING REMARKS

After years of research focused on the structural characteristics of triplet repeats in transcripts of TRED-associated genes, most of the basic facts have been revealed. It is now known that normal CNG repeats stay single-stranded when they are short and those that are longer form hairpin structures. The rigidity and stability of these hairpins increase with repeat length. Transcripts from four genes having interrupted repeats are the exceptions as they form branched hairpin structures. The interruption systems, different in each gene, exert the same effect: they prevent single, long hairpin formation. The very long hairpins that are present in mutant transcripts may be toxic to cells, and such an RNA-mediated pathomechanism has been demonstrated for three diseases at this point. The kinds of perturbations to cell functions caused by the repeat hairpins are only beginning to be revealed, and splicing aberrations caused by the expanded CUG repeats are the best characterized so far. Transcriptomewide gene expression profiling using microarrays and proteomic approaches should provide new clues regarding the transcripts and proteins on which further research should focus.

Acknowledgments

This work was supported by the State Committee for Scientific Research (Grant Nos. 2P05A-08826 and PBZ-KBN-040/P04/2001) and the Foundation for Polish Science (Grant No. 8/2000).

References

1. Santibanez-Koref, M. F., Gangeswaran, R., and Hancock, J. M. (2001). A relationship between lengths of microsatellites and nearby substitution rates in mammalian genomes. *Mol. Biol. Evol.* **18**, 2119–2123.
2. Hancock, J. M., and Simon, M. (2005). Simple sequence repeats in proteins and their significance for network evolution. *Gene* **345**, 113–118.
3. Ellegren, H. (2004). Microsatellites: Simple sequences with complex evolution. *Nat. Rev. Genet.* **5**, 435–445.
4. Wren, J. D., Forgacs, E., Fondon, J. W., III, Pertsemlidis, A., Cheng, S. Y., Gallardo, T., Williams, R. S., Shohet, R. V., Minna, J. D., and Garner, H. R. (2000). Repeat polymorphisms within gene regions: Phenotypic and evolutionary implications. *Am. J. Hum. Genet.* **67**, 345–356.
5. Subramanian, S., Mishra, R. K., and Singh, L. (2003). Genome-wide analysis of microsatellite repeats in humans: Their abundance and density in specific genomic regions. *Genome Biol.* **4**, R13.
6. Jasinska, A., and Krzyzosiak, W. J. (2004). Repetitive sequences that shape the human transcriptome. *FEBS Lett.* **567**, 136–141.
7. Li, Y. C., Korol, A. B., Fahima, T., Beiles, A., and Nevo, E. (2002). Microsatellites: Genomic distribution, putative functions and mutational mechanisms: A review. *Mol. Ecol.* **11**, 2453–2465.
8. Li, Y. C., Korol, A. B., Fahima, T., and Nevo, E. (2004). Microsatellites within genes: Structure, function, and evolution. *Mol. Biol. Evol.* **21**, 991–1007.
9. Bacolla, A., and Wells, R. D. (2004). Non-B DNA conformations, genomic rearrangements, and human disease. *J. Biol. Chem.* **279**, 47411–47414.
10. Karlin, S., Chen, C., Gentles, A. J., and Cleary, M. (2002). Associations between human disease genes and overlapping gene groups and multiple amino acid runs. *Proc. Natl. Acad. Sci. USA* **99**, 17008–17013.
11. Hancock, J. M. (1993). Evolution of sequence repetition and gene duplications in the TATA-binding protein TBP (TFIID). *Nucleic Acids Res.* **21**, 2823–2830.
12. Mitchell, P. J., and Tjian, R. (1989). Transcriptional regulation in mammalian cells by sequence-specific DNA binding proteins. *Science* **245**, 371–378.
13. Perutz, M. F., Johnson, T., Suzuki, M., and Finch, J. T. (1994). Glutamine repeats as polar zippers: Their possible role in inherited neurodegenerative diseases. *Proc. Natl. Acad. Sci. USA* **91**, 5355–5358.
14. Zoghbi, H. Y., and Orr, H. T. (1999). Polyglutamine diseases: Protein cleavage and aggregation. *Curr. Opin. Neurobiol.* **9**, 566–570.
15. Brown, L. Y., and Brown, S. A. (2004). Alanine tracts: The expanding story of human illness and trinucleotide repeats. *Trends Genet.* **20**, 51–58.
16. Albrecht, A., and Mundlos, S. (2005). The other trinucleotide repeat: Polyalanine expansion disorders. *Curr. Opin. Genet. Dev.* **15**, 285–293.
17. Raca, G., Siyanova, E. Y., McMurray, C. T., and Mirkin, S. M. (2000). Expansion of the (CTG)(n) repeat in the 5′-UTR of a reporter gene impedes translation. *Nucleic Acids Res.* **28**, 3943–3949.
18. Kenneson, A., Zhang, F., Hagedorn, C. H., and Warren, S. T. (2001). Reduced FMRP and increased FMR1 transcription is proportionally associated with CGG repeat number in intermediate-length and premutation carriers. *Hum. Mol. Genet.* **10**, 1449–1454.
19. Jasinska, A., Michlewski, G., de Mezer, M., Sobczak, K., Kozlowski, P., Napierala, M., and Krzyzosiak, W. J. (2003). Structures of trinucleotide repeats in human transcripts and their functional implications. *Nucleic Acids Res.* **31**, 5463–5468.
20. Ho, T. H., Charlet, B. N., Poulos, M. G., Singh, G., Swanson, M. S., and Cooper, T. A. (2004). Muscleblind proteins regulate alternative splicing. *EMBO J.* **23**, 3103–3112.
21. Coleman, T. P., and Roesser, J. R. (1998). RNA secondary structure: An important cis-element in rat calcitonin/CGRP pre-messenger RNA splicing. *Biochemistry* **37**, 15941–15950.
22. Sirand-Pugnet, P., Durosay, P., Brody, E., and Marie, J. (1995). An intronic (A/U)GGG repeat enhances the splicing of an alternative intron of the chicken beta-tropomyosin pre-mRNA. *Nucleic Acids Res.* **23**, 3501–3507.
23. Ladd, A. N., Charlet, N., and Cooper, T. A. (2001). The CELF family of RNA binding proteins is implicated in cell-specific and developmentally regulated alternative splicing. *Mol. Cell Biol.* **21**, 1285–1296.
24. Miller, J. W., Urbinati, C. R., Teng-Umnuay, P., Stenberg, M. G., Byrne, B. J., Thornton, C. A., and Swanson, M. S. (2000). Recruitment of human muscleblind proteins to (CUG)(n) expansions associated with myotonic dystrophy. *EMBO J.* **19**, 4439–4448.
25. Tapscott, S. J., and Thornton, C. A. (2001). Biomedicine. Reconstructing myotonic dystrophy. *Science* **293**, 816–817.
26. Ranum, L. P., and Day, J. W. (2004). Myotonic dystrophy: RNA pathogenesis comes into focus. *Am. J. Hum. Genet.* **74**, 793–804.
27. Kino, Y., Mori, D., Oma, Y., Takeshita, Y., Sasagawa, N., and Ishiura, S. (2004). Muscleblind protein, MBNL1/EXP, binds specifically to CHHG repeats. *Hum. Mol. Genet.* **13**, 495–507.
28. Napierala, M., and Krzyzosiak, W. J. (1997). CUG repeats present in myotonin kinase RNA form metastable "slippery" hairpins. *J. Biol. Chem.* **272**, 31079–31085.
29. Timchenko, L. T., Miller, J. W., Timchenko, N. A., DeVore, D. R., Datar, K. V., Lin, L., Roberts, R., Caskey, C. T., and Swanson, M. S. (1996). Identification of a (CUG)n triplet repeat RNA-binding protein and its expression in myotonic dystrophy. *Nucleic Acids Res.* **24**, 4407–4414.
30. Faustino, N. A., and Cooper, T. A. (2003). Pre-mRNA splicing and human disease. *Genes Dev.* **17**, 419–437.
31. Day, J. W., and Ranum, L. P. (2005). RNA pathogenesis of the myotonic dystrophies. *Neuromuscul. Disord.* **15**, 5–16.
32. Ebralidze, A., Wang, Y., Petkova, V., Ebralidse, K., and Junghans, R. P. (2004). RNA leaching of transcription factors disrupts transcription in myotonic dystrophy. *Science* **303**, 383–387.
33. Hagerman, R. J., Leehey, M., Heinrichs, W., Tassone, F., Wilson, R., Hills, J., Grigsby, J., Gage, B., and Hagerman, P. J. (2001). Intention tremor, parkinsonism, and generalized brain atrophy in male carriers of fragile X. *Neurology* **57**, 127–130.

34. Hagerman, P. J., and Hagerman, R. J. (2004). The fragile-X premutation: A maturing perspective. *Am. J. Hum. Genet.* **74**, 805–816.
35. Sobczak, K., de Mezer, M., Michlewski, G., Krol, J., and Krzyzosiak, W. J. (2003). RNA structure of trinucleotide repeats associated with human neurological diseases. *Nucleic Acids Res.* **31**, 5469–5482.
36. Krzyzosiak, W. J., Napierala, M., and Drozdz, M. (1999). RNA structure modules with trinucleotide repeat motifs. *In* "RNA Biochemistry and Biotechnology" (J. Barciszewki and B. F. C. Clark, Eds.), pp. 303–314. Kluwer, Dordrecht, The Netherlands.
37. Sobczak, K., and Krzyzosiak, W. J. (2004). Imperfect CAG repeats form diverse structures in SCA1 transcripts. *J. Biol. Chem.* **279**, 41563–41572.
38. Michlewski, G., and Krzyzosiak, W. J. (2004). Molecular architecture of CAG repeats in human disease related transcripts. *J. Mol. Biol.* **340**, 665–679.
39. Sobczak, K., and Krzyzosiak, W. J. (2005). CAG repeats containing CAA interruptions form branched hairpin structures in spinocerebellar ataxia type 2 transcripts. *J. Biol. Chem.* **280**, 3898–3910.
40. Napierala, M., Michalowski, D., de Mezer, M., and Krzyzosiak, W. J. (2005). Facile FMR1 mRNA structure regulation by interruptions in CGG repeats. *Nucleic Acids Res.* **33**, 451–463.
41. Matsumura, R., Takayanagi, T., Murata, K., Futamura, N., Hirano, M., and Ueno, S. (1996). Relationship of (CAG)nC configuration to repeat instability of the Machado-Joseph disease gene. *Hum. Genet.* **98**, 643–645.
42. Igarashi, S., Takiyama, Y., Cancel, G., Rogaeva, E. A., Sasaki, H., Wakisaka, A., Zhou, Y. X., Takano, H., Endo, K., Sanpei, K., Oyake, M., Tanaka, H., Stevanin, G., Abbas, N., Durr, A., Rogaev, E. I., Sherrington, R., Tsuda, T., Ikeda, M., Cassa, E., Nishizawa, M., Benomar, A., Julien, J., Weissenbach, J., Wang, G.-X., Agid, Y., St. George-Hyslop, P. H., Brice, A., and Tsuji, S. (1996). Intergenerational instability of the CAG repeat of the gene for Machado-Joseph disease (MJD1) is affected by the genotype of the normal chromosome: implications for the molecular mechanisms of the instability of the CAG repeat. *Hum. Mol. Genet.* **5**, 923–932.
43. Limprasert, P., Nouri, N., Heyman, R. A., Nopparatana, C., Kamonsilp, M., Deininger, P. L., and Keats, B. J. (1996). Analysis of CAG repeat of the Machado-Joseph gene in human, chimpanzee and monkey populations: A variant nucleotide is associated with the number of CAG repeats. *Hum. Mol. Genet.* **5**, 207–213.
44. Pearson, C. E., Eichler, E. E., Lorenzetti, D., Kramer, S. F., Zoghbi, H. Y., Nelson, D. L., and Sinden, R. R. (1998). Interruptions in the triplet repeats of SCA1 and FRAXA reduce the propensity and complexity of slipped strand DNA (S-DNA) formation. *Biochemistry* **37**, 2701–2708.
45. McLaughlin, B. A., Spencer, C., and Eberwine, J. (1996). CAG trinucleotide RNA repeats interact with RNA-binding proteins. *Am. J. Hum. Genet.* **59**, 561–569.
46. Sobczak, K., and Krzyzosiak, W. J. (2004). Patterns of CAG repeat interruptions in SCA1 and SCA2 genes in relation to repeat instability. *Hum. Mutat.* **24**, 236–247.
47. Furtado, S., Farrer, M., Tsuboi, Y., Klimek, M. L., de la Fuente-Fernandez, R., Hussey, J., Lockhart, P., Calne, D. B., Suchowersky, O., Stoessl, A. J., and Wszolek, Z. K. (2002). SCA-2 presenting as parkinsonism in an Alberta family: Clinical, genetic, and PET findings. *Neurology* **59**, 1625–1627.
48. Nakamura, K., Jeong, S. Y., Uchihara, T., Anno, M., Nagashima, K., Nagashima, T., Ikeda, S., Tsuji, S., and Kanazawa, I. (2001). SCA17, a novel autosomal dominant cerebellar ataxia caused by an expanded polyglutamine in TATA-binding protein. *Hum. Mol. Genet.* **1**, 1441–1448.
49. Marcus, G. F., and Fisher, S. E. (2003). FOXP2 in focus: What can genes tell us about speech and language? *Trends Cogn. Sci.* **7**, 257–262.
50. Vargha-Khadem, F., Gadian, D. G., Copp, A., and Mishkin, M. (2005). FOXP2 and the neuroanatomy of speech and language. *Nat. Rev. Neurosci.* **6**, 131–138.
51. Tassone, F., Iwahashi, C., and Hagerman, P. J. (2004). FMR1 RNA within the intranuclear inclusions of fragile X-associated tremor/ataxia syndrome (FXTAS). *RNA Biol.* **1**, 103–105.
52. Karpala, A. J., Doran, T. J., and Bean, A. G. (2005). Immune responses to dsRNA: Implications for gene silencing technologies. *Immunol. Cell Biol.* **83**, 211–216.
53. Lee, Y., Ahn, C., Han, J., Choi, H., Kim, J., Yim, J., Lee, J., Provost, P., Radmark, O., Kim, S., and Kim, V. N. (2003). The nuclear RNase III Drosha initiates microRNA processing. *Nature* **425**, 415–419.
54. Peel, A. L. (2004). PKR activation in neurodegenerative disease. *J. Neuropathol. Exp. Neurol.* **63**, 97–105.
55. Tian, B., White, R. J., Xia, T., Welle, S., Turner, D. H., Mathews, M. B., and Thornton, C. A. (2000). Expanded CUG repeat RNAs form hairpins that activate the double-stranded RNA-dependent protein kinase PKR. *RNA* **6**, 79–87.
56. Peel, A. L., Rao, R. V., Cottrell, B. A., Hayden, M. R., Ellerby, L. M., and Bredesen, D. E. (2001). Double-stranded RNA-dependent protein kinase, PKR, binds preferentially to Huntington's disease (HD) transcripts and is activated in HD tissue. *Hum. Mol. Genet.* **10**, 1531–1538.
57. Handa, V., Saha, T., and Usdin, K. (2003). The fragile X syndrome repeats form RNA hairpins that do not activate the interferon-inducible protein kinase, PKR, but are cut by Dicer. *Nucleic Acids Res.* **31**, 6243–6248.
58. Lingel, A., Simon, B., Izaurralde, E., and Sattler, M. (2003). Structure and nucleic-acid binding of the Drosophila Argonaute 2 PAZ domain. *Nature* **426**, 465–469.
59. Zhang, H., Kolb, F. A., Jaskiewicz, L., Westhof, E., and Filipowicz, W. (2004). Single processing center models for human Dicer and bacterial RNase III. *Cell* **118**, 57–68.
60. Zeng, Y., Yi, R., and Cullen, B. R. (2005). Recognition and cleavage of primary microRNA precursors by the nuclear processing enzyme Drosha. *EMBO J.* **24**, 138–148.

Part XIII

Mutations in Flanking Sequences

Gross Rearrangements Caused by Long Triplet and Other Repeat Sequences

ALBINO BACOLLA, MARZENA WOJCIECHOWSKA, BEATA KOSMIDER, JACQUELYNN E. LARSON, AND ROBERT D. WELLS

Institute of Biosciences and Technology, Center for Genome Research, The Texas A&M University System Health Science Center, Texas Medical Center, Houston, Texas 77030

I. Introduction
II. Non-B-DNA Conformations Adopted by Triplet and Other Repeat Sequences
   A. Slipped and Hairpin Structures
   B. Cruciform DNA
   C. Triplex DNA
   D. Tetraplex DNA and *i*-Motifs
   E. Left-Handed Z-DNA
III. Non-B-DNA and Gross Rearrangements in Model Systems
   A. The 2.5-kb Poly(R·Y) Sequence of the Human *PKD1* Gene
   B. Myotonic Dystrophy Type 1 Triplet Repeat Sequence
   C. Other Non-B-DNA-Forming Sequences
IV. DNA Structure and Human Disease
   A. Statistical Analyses of Breakpoint Junction Sequences
   B. Triplex DNA in Follicular Lymphomas
   C. Segmental Duplications and Genomic Disorders
   D. DNA Repair and Chromosomal Rearrangements
V. Summary and Concluding Remarks
References

This chapter reviews current understanding of the roles of non-B-DNA conformations adopted by certain types of DNA sequences in genetic instabilities, especially gross rearrangements. Conclusions are as follows: (1) slipped (hairpin) structures, cruciforms, triplexes, tetraplexes and *i*-motifs, and left-handed Z-DNA are formed in chromosomes and elicit profound genetic consequences via recombination repair; (2) repeating sequences, probably in their non-B conformations, cause gross genomic rearrangements (deletions, insertions, inversions, translocations, and duplications); and (3) these rearrangements are the genetic basis for scores of human diseases, including polycystic kidney disease, adrenoleukodystrophy, follicular lymphomas, and spermatogenic failure.

I. INTRODUCTION

This chapter focuses on the relationships between the presence of repeating DNA motifs, including triplet repeat sequences (TRSs) in genomes, and the occurrence of gross rearrangements. The central thesis presented herein, which stems primarily from work conducted with TRSs and other repeat sequences in model systems [1–5], states that repeating DNA motifs adopt non-B conformations, which may serve as breakpoints for rearrangements, such as deletions, inversions, insertions, duplications, and translocations. The data support the presence of novel mechanisms of instabilities that differ radically from the well-known behavior of TRSs, i.e., the capacity of these repeats to expand and contract without the involvement of flanking sequences (reviewed in ref. [6]), and may impinge heavily on genome instability and human disease.

The chapter is organized into three sections. First, the basic non-B-DNA conformations and their underlying properties are reviewed. Second, the results that led to the formulation of the central thesis are described. Third, a number of human disease cases are presented where the pathogenic mechanisms are likely to be associated with non-B-DNA conformations. It should be emphasized that the evidence supporting the formation of non-B conformations *in vivo* is mostly based on the behavior of the DNA sequences *in vitro*, their relationships with DNA topology, and in certain cases antibody binding [7–12]. Direct demonstration that mutational reactions (i.e., gross deletions) are initiated specifically by non-B-DNA conformations will require a characterization of the structures *in situ*, which so far has been difficult to achieve.

II. NON-B-DNA CONFORMATIONS ADOPTED BY TRIPLET AND OTHER REPEAT SEQUENCES

The most fascinating and unique feature of TRSs and other repeat sequences in DNA is perhaps their ability to adopt alternative conformations that differ dramatically from the commonly known, right-handed, antiparallel, double helix, generally referred to as B-DNA (reviewed in ref. [6]). These non-B conformations, which include hairpins, cruciforms, triplexes, tetraplexes, and Z-DNA, are extremely polymorphic with respect to the number of strands involved (2–4), their orientations (parallel and antiparallel), the handedness of the helices (right-handed and left-handed), the orientation of the bases (*syn* and *anti*), and the types of hydrogen bonds (Watson–Crick, reverse Watson–Crick, Hoogsteen, reverse Hoogsteen, and others) that stabilize the structures [9, 10, 13, 14].

The TRS field was the first to consolidate the concept that non-B-DNA conformations were, at least in part, responsible for the instabilities (mostly expansions) associated with human neurological diseases (reviewed in ref. [15]). In addition, the realization that chromosomal abnormalities such as deletions, translocations, inversions, and duplications are also associated with repetitive sequences contributed to establishing a firm relationship between the polymorphic structural behavior of the DNA and human disease (reviewed in ref. [6]). Therefore, the pioneering work that laid the foundations for the characterization of DNA conformations and their properties [8, 16, 17], enabled the field of DNA structure to spread in the past few years from the bench of the biochemist to the attention of the human genetic and medical communities.

What are the forces that drive the transition from B-DNA to the less stable non-B conformations? First, there is a growing appreciation that the topological state in which the DNA normally exists in the cell is a key determinant [11, 18]. Negative supercoiling is the topological condition in which chromosomal DNA is organized *in vivo* and is similar to a twisted telephone cord obtained by rotating the receiver. In physical terms, the number of times the two strands of DNA revolve about one another *in vivo* is smaller than it would be *in vitro* under unconstrained conditions [9, 19]. It is this topological strain that represents the basic source of energy for the transition of segments of repeating tracts to non-B-DNA conformations [20]. Second, cellular processes that entail the separation of the complementary strands, such as transcription and replication, generate a competition between the reconstitution of B-DNA through strand reannealing and the formation of alternative structures at repeat sequences, which may be occasionally present on the displaced strands [21].

In addition, more recent studies suggest that other mechanisms, which are still unknown, may exist that promote the formation of non-B structures because, for example, large inverted repeats in the human genome appear to be organized into stable cruciforms as part of their normal chromosomal state [22, 23]. Herein will be reviewed the basic features of the most relevant non-B-DNA conformations and the findings that support their involvement in rearrangements and human disease.

A. Slipped and Hairpin Structures

Direct repeats (DRs) aligning out of register give rise to slipped structures with looped-out bases [Fig. 46-1, red loop]. When DRs involve several repeating motifs, like the TRSs, the GGGA·TCCC repeat, or telomeric sequences, the looped-out bases may form duplexes stabilized by non-Watson–Crick pairs [Fig. 46-1, blue loop], such as T·T in $(CTG)_n$, G·G and A·A in $(GGGA)_n$, or shared paired motifs [G·A in $(GGA)_n$ and the human TTAGGG telomeric sequence and GC·AA in other telomeric sequences], a specific arrangement characterized by interstrand, rather than intrastrand, stacking interactions [6, 24, 25]. In addition, nuclear magnetic resonance (NMR) solution structures of single-stranded oligomers composed of triplet repeat motifs, such as GNC (N = A or T) and CNG (N = A, C, T or G), showed the existence of unique arrangements, like the

Name	Conformation	Sequence Requirements	Model Sequence
Slipped and Hairpin Structures	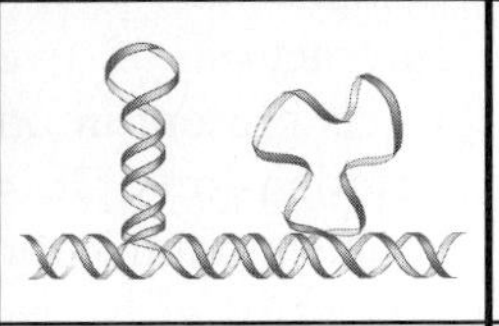	Direct Repeats	TCGGTTCGGT AGCCAAGCCA
Cruciform	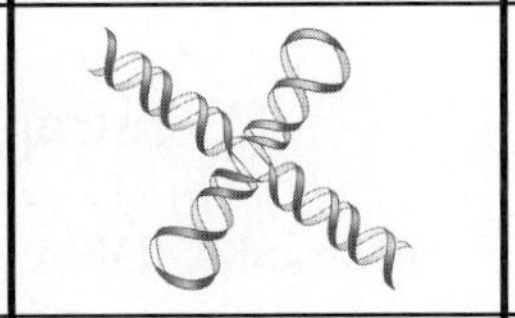	Inverted Repeats	TCGGTACCGA AGCCATGGCT
Triplex	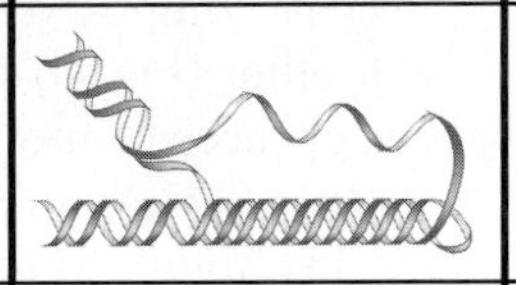	(R•Y)$_n$ Mirror Repeats	AAGAGG GGAGAA TTCTCC CCTCTT
Tetraplex	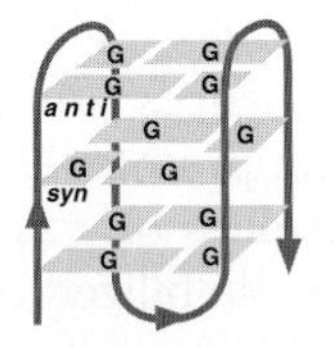	Four Oligo (G)	AG$_3$(T$_2$AG$_3$)$_3$ single strand
Left-handed Z - DNA	 B-Z Junctions	(YR•YR)$_n$	CGCGTGCGTGTG GCGCACGCACAC

FIGURE 46-1 Non-B-DNA conformations formed by triplet and other repeat sequences. See CD-ROM for color image. Adapted with permission from Bacolla and Wells (2004). *J. Biol. Chem.* **279**, 47411–47414 [6].

e motif formed by the CCG oligonucleotide [26]. In this case, two CCG-containing strands pair in a staggered antiparallel alignment stabilized by C·G dinucleotide units separated by nonbase-paired C residues located diagonally across the minor groove. The correlation between the CCG *e*-motif architecture and cytosine hypermethylation in the expanded CCG repeats of fragile X alleles [27] is intriguing, because the cytosine residues need to be brought into an extrahelical configuration during the methylation reaction [28]. Therefore, these NMR data, in conjunction with the finding that an increase in negative superhelical tension accelerates the methylation rate on (CGG·CCG)$_n$ templates [29], support the hypothesis that the formation of hairpin structures participates in the etiology of fragile X syndrome [30]. Detailed NMR studies on single-stranded CTG oligonucleotides indicated that hairpin stability is also contributed by the odd versus even number of repeats. Specifically, even numbers of CTG repeats enabled the formation of a TGCT loop, in which the two thymines formed hydrogen bonds and stacked with the loop-closing base pair. These interactions were absent from the odd-numbered CTG hairpins, which contained a 3-nt CTG loop instead. In addition, the TGCT loops were closed by a C·G pair (underlined in 5′-CTGTCG-3′) rather than a G·C pair (underlined in 5′-GCTGC-3′) in the case of the CTG loops. This 5′-to-3′ Y·R (Y, pyrimidine; R, purine) arrangement is particularly favorable for loop closing, and thereby contributed to the increased stability of hairpins with the TGCT loop and even numbers of CTGs, relative to their odd-numbered counterparts [31].

A computer search of all TRSs composed of at least 9 units in the human genome indicated the presence of well over 8000 elements, and a similar search for all microsatellites, from the 2-mer to the 16-mer perfect repeats, showed a total of more than 150,000 [32]. When one considers that most pericentromeric regions, which are composed almost exclusively of repeating sequences, have yet to be sequenced and that sequence gaps of millions of base pairs exist at these locations, the total number of direct repeats emphasizes the enormous potential for slipped and hairpin structure formation in the human genome.

B. Cruciform DNA

Cruciform DNA (Fig. 46-1) occurs at inverted repeats (IRs), which are defined as identical sequences on the complementary strands. Each strand folds at the IR center of symmetry and reconstitutes an intramolecular B helix capped by a single-stranded loop, which may extend from a few base pairs to several kilobase pairs [6]. A total of ~22,600 inverted repeats (with at least 75% sequence identity) were identified in the human genome, with arm sizes ranging from 25 bp to >100 kb and spacer lengths up to 100 kb [33]. Interestingly, most of the largest (>8 kb) IRs occurred in the sex chromosomes, where they shared >99% sequence identity. Also, most large IR sequences included testis-specific genes, which suggests that cruciform formation is a physiologic structural state of human chromosomes, most likely relevant to male germline gene expression and/or sequence integrity, which is believed to be maintained locally by gene conversion events [23].

C. Triplex DNA

Triplex DNA (triple helix, H-DNA) (Fig. 46-1) occurs at oligo(R·Y) tracts [6, 9, 10, 34, 35] and is favored by motifs containing a mirror repeat symmetry. The purine strand of the Watson–Crick duplex engages the third strand through Hoogsteen hydrogen bonds in the major groove while maintaining the original duplex structure in a B-like conformation. The third strand may be either pyrimidine- or purine-rich and parallel or antiparallel, respectively, to the complementary R strand. Moreover, this third strand may originate from within a single R·Y tract (intramolecular triplex) or from a separate tract (intermolecular triplex) [6]. With respect to (GAA·TTC)$_n$ expansions in Friedreich's ataxia, the sequence contains both the direct repeat symmetry required to form slipped conformations and the R·Y mirror symmetry characteristic of triplex DNA. In addition, biophysical, biochemical, and genetic studies revealed that this sequence is able to adopt additional structural polymorphisms. Indeed, the NMR solution structures of single-stranded TTC oligonucleotide mixed with equimolar single-stranded GAA oligonucleotides indicated the formation of parallel GAA·TTC duplexes stabilized by reverse Watson–Crick hydrogen bonds [36]. At higher molar ratios, the Y + R·Y triplex was detected (+ indicates the Hoogsteen interactions), but not the R+R·Y isomer, a result that was supported by chemical modifications on plasmid DNA *in vitro* [37]. Conversely, biochemical, electron microscopy, and genetic studies indicated the formation *in vivo* of R+R·Y triplex structures (sticky DNA) stabilized by magnesium ions that resulted from the intramolecular interaction of two distantly located (GAA·TTC)$_n$ tracts present in one DNA molecule in the direct orientation (refs. [38–40] and L. Son and R. D. Wells, personal communication).

D. Tetraplex DNA and *i*-Motifs

Tetraplex DNA (four-stranded DNA, G quadruplex) is composed of guanine tetrads stacked upon each other (Fig. 46-1) and therefore is favored by four runs of three or more guanines [6]. Originally reported for the G-rich single-stranded telomeric sequences [13], the ability to form tetrads was observed for other residues as well. In particular, studies on C-rich sequences led to the discovery of base pairing between hemiprotonated cytosines of one duplex and cytosine residues of a second duplex to form a tetraplex structure stabilized by intercalated C–C$^+$ pairs [41], known as the *i*-motif. Either strand of the (CGG·CCG)$_n$ duplex was shown to fold into tetraplex structures. Specifically, circular dichroism (CD) and ultraviolet (UV) spectroscopy indicated that CCG oligonucleotides formed a bimolecular tetraplex composed of two pairs of parallel hairpins stabilized by hemiprotonated intermolecular C–C$^+$ pairs, at both acidic and neutral pH [42]. At lower pH (<5), full cytosine protonation enabled the formation of unimolecular tetraplexes composed exclusively of *i*-motifs [42]. The ability of the CGG oligonucleotides to form tetraplexes was confirmed by NMR, which showed that two hairpins dimerized through their major groove edges using bifurcated hydrogen bonds to form internal G(*anti*)·C(*anti*)·G(*anti*)·C(*anti*) tetrads sandwiched between terminal G(*syn*)·G(*anti*)·G(*syn*)·G(*anti*) G tetrads [43]. Interestingly, circular dichroism and UV absorption spectroscopy analyses showed that the assembly of CGG oligonucleotides into the tetraplex conformation was slow when compared with the folding into homoduplexes and the tetraplex-forming CCG repeat [44], suggesting that these alternative conformations may be in competition *in vivo*.

Of particular interest were the observations that the Bloom's and Werner's syndrome helicases displayed strong enzymatic activities in the disassembly of non-B-DNA structures, particularly triplexes and tetraplexes [45–47]. These results suggest that the increased frequency of chromosomal abnormalities seen in these syndromes, which are defective for either helicase, is contributed to by the persistence of altered DNA conformations [48].

E. Left-Handed Z-DNA

This structure (Fig. 46-1) is adopted by alternating (YR·YR)$_n$ sequences, such as (CG·CG)$_n$ and (CA·TG)$_n$ [6, 8, 49–51]. The transition from the right-handed B to the left-handed Z form is accomplished by a 180° flip (upside down) of the base pair through a rotation of every other purine from the *anti* to the *syn* conformation and a corresponding change in the sugar puckering mode from the C2'-*endo* to the C3'-*endo* conformation [11, 51]. Although Z-DNA is a higher energy conformation than B-DNA, the B- to Z-DNA transition is promoted *in vivo* by negative supercoiling [20]. Two reports implicated left-handed Z-DNA formation under certain conditions for (GAC·GTC)$_n$ repeats [52] and for (CGG·CCG)$_n$ repeats [53] from spectroscopic and immunological studies on synthetic oligonucleotides. The (GAC·GTC)$_n$ repeats [54] are associated with skeletal dysplasias [55], and the expanded (CGG·CCG)$_n$ tracts are found in the 5' UTR of the fragile X gene (reviewed in ref. [15]). Hopefully, future studies will provide further insights into these surprising results.

In summary, these discoveries show the remarkable ability of simple repeating sequences to fold into a rich, and often unpredictable, variety of conformations, which opens additional avenues for the roles of these polymorphisms *in vivo*. The biological functions of these conformations are described next.

III. NON-B-DNA AND GROSS REARRANGEMENTS IN MODEL SYSTEMS

A. The 2.5-kb Poly(R·Y) Sequence of the Human *PKD1* Gene

Studies with a segment of the *PKD1* gene are described because they are representative of R·Y-type repeat sequences and are prefatory to additional work on other repeat sequences that are involved in hereditary neurological diseases. Mutations in the polycystic kidney disease 1 gene account for ~85% of cases with polycystic kidney, a condition that affects 1 out of every 500–1000 people [56]. Although most mutations are single point substitutions, recombination between paralogous sequences on chromosome 16 has been observed in some cases [57], and an unusually long (2.5 kb) poly(R·Y) tract located in intron 21 of the *PKD1* gene was postulated to play a role [57]. To address this possibility, the *PKD1* poly(R·Y) tract was first analyzed for the presence of repeat motifs with the potential of adopting non-B conformations. A total of 23 mirror repeats [58] were clustered in three distinct regions that could form intramolecular triplexes [18, 58]. Indeed, the ability of some of these repeats to undergo supercoil-dependent structural transitions was confirmed *in vitro* [59, 60]. In addition, a search for direct repeats indicated the presence of more than 1000 [18]. The TC·GA and CT·AG dinucleotides were the most prevalent; however, they were excluded from the 5' end of the tract, where the mirror repeats predominated. Additionally, tri-, tetra-, penta-, and longer nucleotide direct repeats were found, also clustered in specific regions of the 2.5-kb segment. Finally, closely spaced G runs were present throughout its length. It was concluded that the *PKD1* tract could adopt a large variety of different conformations, including slipped structures, triplexes, and tetraplexes [18].

Next, the 2.5-kb segment was cloned in plasmids and transformed *Escherichia coli* cells [18]. Cultures started from these cells exhibited a drop of up to 99% in the number of viable colony-forming units (CFUs) within the first few hours. This phenomenon was dependent on drugs that bound to the DNA and on the equilibrium distribution of negative superhelical tension of the cells. Specifically, actinomycin D, which intercalated in DNA and reduced the superhelical tension, alleviated cell death, whereas netropsin, which also bound to the DNA but increased the superhelical tension, further increased the fraction of inviable CFUs [18]. Also, the use of novobiocin, an inhibitor of DNA gyrase (an enzyme that introduces negative supercoil tension in *E. coli*), completely abrogated cell death. Similar experiments with genetic mutants of *E. coli* strains deficient in topoisomerase enzymes (topoisomerase I, which removes DNA supercoils, and DNA gyrase) showed that the extent of cell death directly correlated with the average number of negative supercoils of the plasmids. Because negative supercoiling is a primary driving force in promoting structural transitions between the B helix and alternative conformations [9, 20], it was concluded that the formation of non-B-DNA conformations by the poly(R·Y) tract, rather than its primary DNA sequence, was eliciting the biological responses.

Additional experiments indicated that the loss of viability required proficient DNA repair functions, because cells lacking either the UvrA or UvrB proteins of the nucleotide excision repair pathway did not exhibit the *PKD1*-dependent drop in viability even in the presence of netropsin. These results suggested that the poly(R·Y) non-B conformations were recognized as lesions, and hence induced DNA repair functions, that led to the loss of cell viability.

The poly(R·Y) tract was also cloned downstream of a reporter gene (*GFP*, which codes for the green fluorescent protein in the jellyfish *Aequorea victoria*) to determine whether the *PKD1* sequence was able to induce mutations in *cis* (Fig. 46-2) [1]. It was found that the poly(R·Y) tract increased the frequency of mutation events, including intermolecular recombinations between plasmids and intramolecular deletions, which disrupted or ablated the *GFP* gene. Analysis of the deletion mutants indicated that several breakpoint junctions were located within the poly(R·Y) tract, which confirmed that non-B-DNA-forming motifs were involved. Complexity analyses [1] showed that direct repeats or mirror repeats were usually present just adjacent to or encompassing the breakpoints, both in the poly(R·Y) tract and in the vector. Furthermore, in all cases the breakpoints occurred at homologous nucleotides, generally between 2 and 8 bp in length. Figure 46-3A shows an example where the *PKD1* tract induced a 1.6-kb deletion that removed part of the *GFP* gene. The breakpoint in the poly(R·Y) tract occurred at a CC·GG dinucleotide that was part of a mirror repeat, as well as two direct repeats. Additional direct repeats were also present upstream of the breakpoint. Hence, the DNA could fold into multiple non-B conformations at this site, including a triplex, slipped structures, or a tetraplex, which is shown. The CC·GG in the *GFP* gene was also adjacent to two pairs of direct repeats and therefore slipped conformations, one of which is shown.

Because similar features were found in all mutant clones, it was concluded that non-B-DNA conformations played a direct role in these rearrangements by specifically defining the locations of breakpoints. Because non-B-DNA conformations are expected to increase the rate of DNA breakage [20], this stimulation of gross rearrangements may have been a manifestation of an increased rate in double-strand break (DSB) formation [1].

B. Myotonic Dystrophy Type 1 Triplet Repeat Sequence

Myotonic dystrophy type 1 (DM1), an autosomal dominant neuromuscular disease, is caused by massive expansions of a CTG·CAG triplet repeat in the 3′ UTR region of the DM protein kinase gene (*DMPK*) [15]. Instability of CTG·CAG repeats is thought to arise from their inherent flexibility [61], which may facilitate the formation of slipped and hairpin structures [62, 63].

How these types of DNA conformational behaviors interact with the replication, repair, and recombination apparatuses and lead to instability has been the subject of a review [21]. These studies were extended by analyzing the propensity of noncanonical DNA structures formed by the long $(CTG \cdot CAG)_n$ repeats to induce gross rearrangements (2), which is a previously unrecognized type of genetic instability for this sequence. The capacity of DM1 TRSs of various lengths ($n = 17, 98$, and 175) to induce mutations in DNA flanking regions was evaluated by cloning the repeats in plasmids and transforming *E. coli* cells. The TRSs were cloned in the region of the plasmid that was adjacent to and downstream of the *GFP* gene (Fig. 46-2), so that repair of the putative non-B-DNA structures could disrupt the integrity of the reporter gene and abolish the fluorescent phenotype. "Wild-type" *E. coli* strains with different genetic backgrounds were transformed with plasmids containing TRSs of different lengths and grown for several generations. This strategy allowed for mutant cells to arise in the culture and to replicate along with their wild-type progenitor cells. To score for mutants, small fractions of these cultures were diluted and plated on solid medium under conditions where isolated colonies arose from single founder cells. Green–white screening indicated that the fraction of white colony-forming units (CFUs), e.g., colonies with the nonfunctional reporter gene, increased in the presence of $(CTG \cdot CAG)_n$ tracts. Specifically, the longer the tracts, the higher the frequency of the mutant white phenotype. For example, when the KMBL1001 strain was transformed with the plasmid containing the $(CTG \cdot CAG)_{98}$ insert, the fraction of white CFUs was 31%. In contrast, when the plasmid carried the $(CTG \cdot CAG)_{17}$ insert, this fraction was only 0.04%

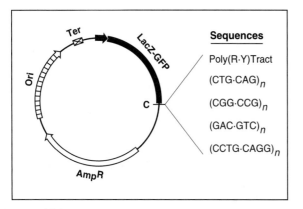

FIGURE 46-2 Repeat sequences cloned in plasmid DNA: striped arrow, *ColE1* origin of replication; crossed box, transcription terminator; solid arrow, lacZ promoter–operator; solid segment, *lacZ-GFP* fusion gene; open arrow, ampicillin resistance gene; C, cloning site.

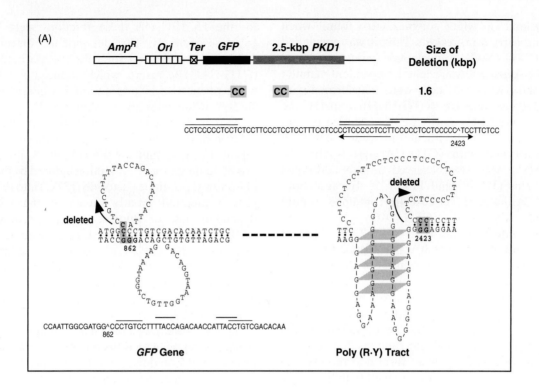

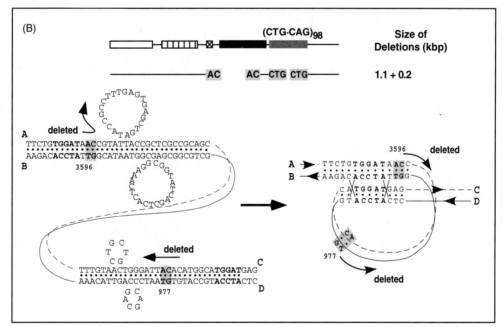

FIGURE 46-3 Rearrangements induced by non-B-DNA structure-forming sequences. The top parts of panels A and B are schematic diagrams of the plasmid maps with the same symbols from Fig. 46-2. Lower down is a scheme of the deletions showing the homologous nucleotides (gray boxes) at breakpoints, followed by the size of the deletions. Next are examples of predicted non-B-DNA conformations formed near the breakpoints, with arrows pointing toward the deleted DNA. Panel A: caret in linear sequence, breakpoint position; lines above linear sequence, direct repeats; arrows below linear sequence, mirror repeats. Adapted with permission from Bacolla et al. (2004). *Proc. Natl. Acad. Sci. USA* **101**, 14162–14167 [1] and Wojciechowska et al. (2005). *J. Biol. Chem.* **280**, 941–952 [2].

(4/10,000), and no white colonies were found when the plasmid carried no repeats. Thus, it was concluded that the (CTG·CAG)$_n$ trinucleotide repeat mediated plasmid DNA instability in a length-dependent manner.

When the KMBL1001 cells were transformed with a plasmid containing the (CTG·CAG)$_{175}$ insert, the fraction of white CFUs was only 13%. A similar two- to threefold decrease in the fraction of white colonies in cells transformed with (CTG·CAG)$_{175}$ relative to (CTG·CAG)$_{98}$ was also observed in two additional *E. coli* strains (JJC510 and JTT1) [2]. Because both (CTG·CAG)$_{98}$ and (CTG·CAG)$_{17}$ contained a pure repeat tract, whereas the (CTG·CAG)$_{175}$ insert contained two interruptions (two CTA·TAG units at repeats 28 and 69 [21, 64]), it was concluded that the interruptions caused a destabilization of slipped structures, due to mismatches in the duplex DNA [65], which reduced the mutagenic potential of the tract. Indeed, a correlation between the length and purity of a TRS and its capacity to adopt non-B-DNA conformations is well-documented [64, 66, 67], and interruptions that decrease the propensity of the repeat to form unorthodox secondary structures also increase its stability [68].

The (CTG·CAG)$_n$ inserts caused a disruption of the *GFP* gene in a manner consistent with their orientation relative to the direction of an incoming replication fork. Replication started from the *ColE1* origin (Fig. 46-2) and proceeded in a unidirectional fashion, such that one strand always provided the template for leading DNA synthesis, whereas the complementary strand served as a template for lagging strand synthesis. All (CTG·CAG)$_n$ inserts were cloned in both orientations so as to determine the mutagenic potential of each individual repeat when either the CTG (orientation I) or CAG (orientation II) tract was present on the lagging strand template. This difference was shown to be relevant in repeat instability, because deletions–expansions of (CTG·CAG) repeat units occurred predominantly in orientation II [64]. Because NMR solution structural data indicated that hairpins formed by the CTG repeats were more stable than those formed by the CAG repeats (due to a stronger interaction between T·T pairs than between A·A pairs [69]), the orientation dependence was explained by the more stable hairpins on the lagging strand template. In turn, stronger hairpins were proposed to cause replication fork collapse and, hence, mutagenesis, which was more likely than for the thermodynamically weaker CAG hairpins.

Consistent with these previous observations, the fractions of white colonies were greater when the (CTG·CAG)$_n$ repeats were cloned in orientation II than when they were cloned in orientation I [2]. For example, the fraction of white KMBL1001 cells transformed with the (CTG·CAG)$_{98}$ insert was 46% in orientation II and only 23% in orientation I, whereas for the JJC510 cells these fractions were 36% and 18%, respectively. The dependence on orientation was most pronounced for cells transformed with the (CTG·CAG)$_{175}$ insert, which induced practically no white phenotype in orientation I, whereas it induced 20–50% white colonies in orientation II.

Also investigated was whether transcription from the *GFP* promoter, which ran through the downstream triplet repeats to the terminator site (Fig. 46-2), stimulated the loss of fluorescence. Transcription induced by the addition of isopropyl β-D-thiogalactoside (IPTG) into the bacterial growth medium greatly increased the CTG·CAG-dependent loss of fluorescence. For example, when KMBL1001 cells were transformed with the plasmid carrying the (CTG·CAG)$_{98}$ insert, the fraction of white CFUs was 0.06% (6/10,000) in the absence of transcription and 59% after the induction of transcription. This dramatic effect revealed the important role of transcription as a biological process for stimulating mutagenesis, which is consistent with the known relationships between the translocation of an RNA polymerase complex and the formation of twin domains of supercoiling (i.e., a domain of positive supercoiling ahead of the complex and a domain of negative supercoiling behind) [70, 71].

Extensive sequence analyses of the plasmids recovered from the white colonies showed that the loss of fluorescence was caused by gross deletions (either single deletions or two separate deletions) that disrupted the integrity of the *GFP* gene (Fig. 46-3B). In general, the single deletions occurred between the terminator region and either the TRS or downstream locations, whereas the double deletions occurred within the TRS and in the upstream *GFP* reporter gene. The occurrence of multiple deletions was intriguing, because these were not observed in the absence of the TRS. Also, all mutant plasmids in which the (CTG·CAG)$_n$ tract was cloned in orientation II were found to contain the tract in the opposite orientation as a result of an inversion, which, in addition to the deletion, flipped the repeat and a few flanking nucleotides. Given the more stable condition of orientation I, it was tempting to propose that the inversion generated more viable plasmids, which conferred a growth advantage relative to the noninverted plasmids. In this regard, the culture experiments may have simulated "evolutionary" changes that in other biological systems would only be observed over extended periods of time. In summary, these results indicated that the TRSs stimulated the occurrence of gross rearrangements, involving both the TRSs and other sequences of the plasmid.

Analyses of the deletion–inversion breakpoints shed considerable insight into the mechanisms of TRS-mediated rearrangements. First, microhomologies of 1–8 bp were present at breakpoints (Fig. 46-3B). Second, in all cases the breakpoint positions occurred at or near repeat sequences that were able to adopt non-B-DNA conformations (both in the TRS and in the vector). Figure 46-3B depicts the representative DNA structures for one such mutation. Two deletions took place, one within the TRS tract of ~0.2 kb and the other in the vector, which removed ~1.1 kb. This latter deletion occurred between positions 3596 in the terminator region and 977 in the *GFP* gene at two homologous AC·GT dinucleotides located 5 bp away from two direct repeats and bracketed by two 5-bp TGGAT·ATCCA motifs. This arrangement suggested that the deletion was mediated by the two slipped structures of 21 and 6 bp, respectively, and by DNA looping stabilized by the interaction between the two TGGAT·ATCCA motifs, which exposed the homologous AC·GT sites and brought them in close proximity for the cut-and-paste deletion reaction.

Similar features were observed in most other mutant clones, which indicated that the rearrangements were triggered by the formation of non-B-DNA conformations. How might the (CTG·CAG)$_n$ sequence induce multiple rearrangements? Figure 46-4 (single deletion, left pathway) shows a DSB (A) introduced in the TRS as a consequence of slippage. An additional break (D) occurs at some distance at a second non-B-DNA structure that contains a sequence homologous to A. The two homologous sequences serve as a substrate for repair, which leads to the deletion of the intervening sequence (D–A repair). Alternatively (single deletion, right pathway), the

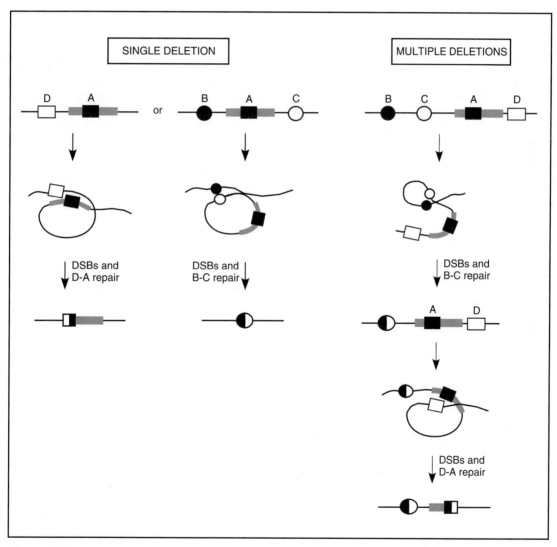

FIGURE 46-4 Model for non-B-DNA structure-induced deletions involving microhomology at breakpoints: gray rectangle, non-B-DNA-forming repeat sequence [poly(R·Y) tract or (CTG·CAG)$_n$]; solid box A, non-B-DNA conformation formed by the repeat sequence; open box D, non-B-DNA conformation containing few base pairs homologous to A; solid circle B, non-B-DNA conformation lacking base pairs homologous to A; open circle C, non-B-DNA conformation containing few base pairs homologous to B. Reprinted with permission from Wojciechowska et al. (2005). *J. Biol. Chem.* **280**, 941–952 [2].

DSB (A) occurring within the repeat sequence is followed by a DSB (B) at a second non-B-DNA structure that, because of lack of homology with A, is not an efficient template for repair. Thus, a third DSB at a novel non-B-DNA structure (C) contains nucleotides homologous to site B, and therefore DSB repair may occur between sites B and C, hence removing the original lesion (B–C repair). In the multiple deletions pathway, the third DSB (C) is located upstream from the A site rather than downstream. As a result, B–C repair will not remove the lesion at the A site. A fourth cleavage at a non-B-DNA structure at site D, which now contains nucleotides homologous to the A site, will enable A–D repair, thereby producing two deletion events.

The ability of long CTG·CAG repeats to cause long-range instability was also observed in CHO and COS1 cells [3, 4], and ongoing experiments in the authors' laboratory indicate that the results described earlier for *E. coli* are also found in mammalian COS7 cells (M. Wojciechowska and R. D. Wells, unpublished data). Therefore, these data support the conclusion that the non-B conformations adopted by triplet repeats and other sequences may have deleterious consequences because nucleotides prone to breakage are generated, whose repair results in rearrangements.

C. Other Non-B-DNA-Forming Sequences

Oligo(R·Y) tracts have been proposed to be involved in the regulation of expression of several disease-associated genes [72, 73], and an oligo(R·Y) tract with mirror repeat symmetry is present in the promoter region of the *c-myc* protooncogene. Eighty percent of Burkitt's lymphoma cases carry a t(8;14) translocation, resulting in the juxtaposition of the *c-myc* gene on chromosome 8 with *IgH* enhancer elements on chromosome 14, which lead to c-Myc mRNA overexpression and cancer (reviewed in ref. [74]). In the remaining 20% of cases, translocations occurring between chromosomes 2 and 8 or between chromosomes 8 and 22 place the *c-myc* gene adjacent to either κ or λ light chain loci and enhancer elements, respectively, again conferring overexpression. Several breakpoints were identified close to the H-DNA-forming sequence of *c-myc* (reviewed in ref. [5]), which prompted speculation as to whether alternative DNA conformations at this site may be involved in the mutagenic process.

To address this possibility, the *c-myc* H-DNA-forming sequence was cloned in a shuttle vector upstream of the suppressor tRNA gene *supF* (as the target for mutagenesis) and inserted in COS7 cells. After a period of growth, plasmids were harvested and transformed in *E. coli* MBM7070, a specific strain that carries a *lacZ* amber mutation. In the presence of isopropylthio-β-D-galactoside (IPTG), 5-bromo-4-chloro-3-indolyl-13-D-galactoside (X-Gal), and ampicillin, *E. coli* MBM7070 forms blue colonies if the *supF* gene contains no mutation. White colonies are formed, however, in the presence of a mutation within the *supF* region [5]. The *c-myc* oligo(R·Y) tract increased the mutation frequency by approximately twenty-fold in a manner consistent with the introduction of DSBs, which were located near the H-DNA structure adopted by the promoter sequence as determined by enzymatic structural probes. Sequence analyses showed that mutations included deletions, insertions, inversions, and duplications. In addition, microhomologies were present at the breakpoints in most cases, which indicated that the repair process proceeded by a nonhomologous end-joining pathway. Finally, the mutation frequencies depended on the orientation of the *c-myc* promoter tract as well as the integrity of the cellular repair functions, similar to *E. coli*. Hence, these results show that the *c-myc* promoter sequence forms a triplex DNA, which in primate COS7 cells is able to induce gross rearrangements, hence supporting the involvement of such a DNA structure in B-cell translocations.

Furthermore, the investigations were extended [1, 2, 6] to consider the capacity of other repeat sequences [(CGG·CCG)$_n$, (GTC·GAC)$_n$, and (CCTG·CAGG)$_n$ (Fig. 46-2)] to induce gross rearrangements, both in *E. coli* and in mammalian COS7 cells. The data indicate that, although all repeats studied to date are proficient in mutagenesis, profound differences exist (B. Kosmider, M. Wojciechowska, and R. D. Wells, unpublished data) in the frequencies and types of rearrangements observed. Taken together, these studies have revealed the surprising capacities of various kinds of non-B-DNA conformations to induce the cellular repair processes in both the prokaryotic and eukaryotic kingdoms, which triggered genomic instabilities [1].

As stated previously, this propensity of a TRS to cause gross rearrangements, either adjacent to the tract or at some distance away, is a novel and previously unrecognized mutagenic process.

IV. DNA STRUCTURE AND HUMAN DISEASE

A. Statistical Analyses of Breakpoint Junction Sequences

Metaanalysis consists of a statistical evaluation of pooled data from diverse sources as a way to overcome the problem of reduced statistical power in studies with small sample sizes. The Gross Rearrangement

Breakpoint Database (GRaBD at www.uwcm.ac.uk/uwcm/mg/grabd.html) and the Human Gene Mutation Database (HGMD at www.hgmd.org) represent large repositories of information for the analyses of DNA sequences involved in chromosomal rearrangements, microinsertions and microdeletions, and single base pair substitutions [75, 76]. A total of 397 germline and somatic DNA breakpoint junction sequences from 219 chromosomal rearrangements underlying human inherited diseases and cancers were deposited in GRaBD in 2004, and 39,415 mutation data were entered into HGMD in 2003 with a rate of new entries exceeding 5000 per year. Therefore, these databases represent invaluable resources for the application of metaanalytical studies.

Analysis of the DNA sequences close to gross rearrangement breakpoints in the GRaBD revealed that R·Y sequences and other repetitive elements were overrepresented [75, 77], suggesting that DNA structural features were involved in the rearrangements, as in the model systems described previously. Furthermore, statistical analyses aimed at evaluating the distance of the R·Y sequences or the YR·YR tracts from the exact breakpoint positions showed that the non-B-DNA-forming motifs were closer to the junctions than expected by chance, supporting their primary role in the mutation process [1]. The ability of repetitive DNA motifs to fold into non-B-DNA conformations was confirmed by a detailed analysis of the sequences present at the junctions prior to the mutation events in 11 cases. These consisted of reported deletions in patients with autosomal dominant polycystic kidney disease, early onset parkinsonism, Menkes' syndrome, α^+-thalassemia, adrenoleukodystrophy, and hydrocephalus [1]. In all cases, the most significant underlying feature was the location of the breakpoint junctions at, or very near, nucleotides that were part of alternative DNA structures, such as cruciforms, slipped direct repeats, and tetraplexes, as predicted from the DNA repeat composition. Therefore, these features were indistinguishable from those observed in *E. coli* cells (Fig. 46-3), strongly implying a direct mutational role of non-B-DNA conformations in humans. Finally, a metaanalysis conducted on all gross insertions, ranging from 21 bp to ~10 kb, in the HGMD [78] (158 total) revealed that misalignment of direct repeat sequences during DNA replication accounted for ~70% of the insertions, attesting to the commonality of slipped conformations in mutagenesis.

B. Triplex DNA in Follicular Lymphomas

The most extensively studied non-B-DNA-forming sequence shown to be directly implicated in human diseases to date is perhaps the *Bcl-2* major breakpoint region, involved in the t(14;18) translocation [12, 79, 80]. During this rearrangement, the *Bcl-2* gene on chromosome 18 becomes juxtaposed to the intron enhancer of the immunoglobulin heavy chain locus on chromosome 14, which is generated in the course of normal V(D)J recombination in pre-B cells. The resulting overexpression of the antiapoptotic bcl-2 protein is believed to be responsible for the high incidence of follicular lymphomas associated with this rearrangement, which account for nearly half of all non-Hodgkin's lymphomas [12]. Although the *Bcl-2* locus encompasses more than 200 kb of genomic DNA, most breakpoints are clustered within a 150-bp region in the 3' untranslated region of the gene, named Bcl-2 Mbr. Within these 150 bp, DNA breakages peak at three distinct sites, composed predominantly of imperfect R·Y motifs 15–20 bp in length, called peaks I, II, and III (Table 46-1).

In episomal DNA containing the isolated Bcl-2 Mbr and a pair of V(D)J heptamer–nonamer recombination signal sequences, recombination between the Bcl-2 Mbr and the D or J broken DNA ends was shown to depend on the RAG complex, supporting the ectopic nature of the reaction and the intrinsic susceptibility of the region to DNA breakage. A combination of structural studies *in vitro*, including bisulfite modification, circular dichroism, monoclonal antibody binding, gel shift assays, and electron microscopy, indicated the formation of two triplex structures in competition with one another at peak I. Although the nature of the structures at peaks II and III is still under investigation, the present data strongly support the involvement of triplex structures in the Bcl-2 t(14;18) translocation.

C. Segmental Duplications and Genomic Disorders

During the past few years, sequence analyses have revealed the prominent role of large duplicated segments of the human genome in chromosomal rearrangements (including deletions, inversions, duplications, and translocations [81]) associated with several genomic disorders and cancer (reviewed in ref. [6] and Table 46-1). These duplications (also called segmental duplications or low-copy repeats, LCRs) are specific to primates, and although their origin and evolutionary mechanisms are unclear, they appear to have contributed to genome reorganization and the acquisition of novel gene functions [82]. Two of the main mechanisms proposed for LCR-mediated mutagenesis are (1) homologous recombination between allelic sites (AHR) during meiosis and (2) nonallelic homologous recombination (NAHR) between paralogous LCRs, either on the same chromosome or on a separate chromosome [83]. An additional

TABLE 46-1 Selected Genetic Alterations, Diseases, and DNA Repeat Motifs[a]

Genetic alteration	Syndrome or metabolic event	DNA motifs
t(14;18)(q32.3;q21.3)	Follicular (B cells) lymphomas	(R·Y)-rich
t(11;22)(q23;q11.2)	Supernumerary der(22) syndrome	IR
t(17;22)(q11.2;q11.2)	Neurofibromatosis type 1	IR
t(1;22)(p21.2;q11.2)	Ependymoma	IR
t(4;22)(q35.1;q11.2)	Velocardiofacial syndrome	IR
del(22)(q11.2q11.2)	Di George, velocardiofacial, conotruncal anomaly face syndromes	IR, DR
dup(22)(q11.2q11.2)	Cat-eye syndrome	IR, DR
del(Yq)	Spermatogenic failure	IR, DR
i(17q)	Hematologic malignancies (chronic myeloid leukemia)	IR
del(17)(p11.2p11.2)	Smith–Magenis syndrome	IR, DR
dup(17)(p11.2p11.2)	Smith–Magenis syndrome	IR, DR
dup(17)(p12p12)	Charcot–Marie–Tooth type 1	IR, DR
del(17)(p12p12)	Hereditary neuropathy with liability to pressure palsies	IR, DR
del(15)(q11q13)	Prader–Willi and Angelman's syndromes	IR, DR
inv(15)(q11q13)	Predisposition to Angelman's syndrome	IR, DR
del(7)(q11.23q11.23)	Williams–Beuren syndrome	IR, DR
inv(7)(q11.23q11.23)	Predisposition to Williams–Beuren syndrome	IR
t(X;22)(q27;q11)	Myeloschizis and lumbosacral spina bifida	IR
del(5)(q35q35)	Sotos' syndrome	DR
inv(5)(q35q35)	Predisposition to Sotos' syndrome	IR, DR
del(17)(q11.2q11.2)	Neurofibromatosis type 1	DR
Expanded (CTG·CAG)$_n$ (coding)	Expanded polyglutamine diseases [Haw River, Huntington's, Hutington's disease-like 2, spinobulbar muscular atrophy (Kennedy), spinocerebellar ataxias (SCA) 1, SCA2, SCA3, SCA6, SCA7, SCA17]	DR
Expanded (CTG·CAG)$_n$ (noncoding)	Myotonic dystrophy type 1, SCA8, SCA12	DR
Expanded (CGG·CCG)$_n$ (coding)	Expanded polyalanine diseases [infantile spasm, cleidocranial dysplasia, blepharophimosis–ptosis–epicanthus inversus type B, hand–foot–genital, synpolydactyly, oculopharyngeal muscular dystrophy, holoprosencephaly, oculopharyngeal muscular dystrophy]	DR
Expanded (CGG·CCG)$_n$ (noncoding)	Fragile XA, fragile XE, fragile XF, Jacobsen (FRA11B)	DR
Expanded (GAA·TTC)$_n$	Friedreich's ataxia	DR with (R·Y)
Expanded (GAC·GTC)$_n$	Pseudoachondroplasia, multiple epiphyseal dysplasia	DR
Expanded (CCTG·CAGG)$_n$	Myotonic dystrophy type 2	DR
Expanded (ATTCT·AGAAT)$_n$	SCA10	DR
Expanded (CCCCGCCCCGCG)$_n$	Progressive myoclonus epilepsy type 1	DR
Expanded 24-mer	Creutzfeldt–Jakob syndrome	DR
Contracted 3.3-kb D4Z4 repeat	Facioscapulohumeral muscular dystrophy	DR

[a] Abbreviations: DR, direct repeat; IR, inverted repeat; DR with (R·Y), direct repeat composed of (R·Y)$_n$ with mirror repeat symmetry. Adapted with permission from Bacolla and Wells (2004). *J. Biol. Chem.* **279**, 47411–47414 [6].

relevant feature is the presence, within the LCRs, of hot spots for recombination where breakpoint junctions cluster preferentially [6, 83]. The existence of such hot spots is intriguing, given that LCRs comprise tens to hundreds of kilobase pairs of near-perfect sequence identity. Support is growing that other features in addition to sequence identity, such as the ability of internal motifs to fold into secondary DNA conformations, may play a role (reviewed in ref. [6] and Table 46-1).

The involvement of DNA secondary structures formed by LCRs and other extremely large inverted repeats of the Y chromosome in recurrent translocations and deletions of chromosome 22q11, isochromosome 17q, spermatogenic failure, Smith–Magenis syndrome, dup(17)(p11.2p11.2) syndrome, Charcot–Marie–Tooth type 1 disease, hereditary neuropathy with liability to pressure palsies, and other conditions has been reviewed [6]. More recently, Sotos' syndrome (SoS) has been shown to be caused by recombination between LCRs [84, 85]. This condition is an overgrowth disorder characterized by mental retardation, specific craniofacial features, and advanced bone age. Haploinsufficiency of the *NSD1* gene due to either intragenic mutations or a ~2.0-Mb common deletion is known to be the major cause of SoS. Sequence analyses identified the breakpoints for the ~2.0-Mb common deletion within two LCR blocks, a proximal LCR (PLCR, Sos-PREP) comprising ~390 kb of DNA and a distal LCR (DLCR, SOS-DREP) of ~429 kb. These LCRs are themselves composed of internal repetitive blocks, such that blocks A–H of the proximal LCR are repeated in the inverted orientation in the distal LCR but in a different order. A second block B is additionally present downstream of the distal LCR in the same orientation as block B of the proximal LCR, and most deletions are consistent with recombination events between these two directly oriented B blocks. Also, breakpoints took place mostly within a ~3- to 3.5-kb hot-spot region of the B blocks, which share >98% identity over more than 50 kb [84, 85]. Therefore, these mutagenic features conform to those observed previously with other LCR-mediated rearrangements.

However, most intriguing in these studies were the differences reported between Caucasian and Japanese patients. In the Caucasian population, intragenic *NSD1* mutations account for ~76% of SoS cases, whereas this fraction is only ~10% in the Japanese population [86]. Also, the common 2-Mb deletion between LCRs was typically observed in the Japanese cases [84, 85], whereas much more heterogeneous deletions were reported in the Caucasian patients [86]. Finally, all fathers of the Japanese children carrying the SoS 2-Mb common deletion in the paternally derived chromosome were heterozygous for an inversion between the proximal and distal LCRs. In at least two other conditions, Williams–Beuren syndrome [87, 88] and Angelman's syndrome [89], which are also caused by LCR-mediated deletions, heterozygous inversions in the parental chromosomes have been suggested to predispose to instability. This outcome would stem from the lack of homology at the site of inversion during meiotic interchromosomal pairing from the polymorphic carriers, which would favor unequal crossing over events and subsequent deletions [87]. Hence, these important studies suggest a novel mechanism for mutagenesis, whereby orientation polymorphism for large chromosomal segments in the population may be a predisposing factor for genetic instability.

Two studies have furthered understanding of the processes adopted by cells to repair DNA breaks occurring at unstable LCR regions. First, analyses of nonrecurrent deletions of 17p11.2, involving the Smith–Magenis syndrome region, indicated that these deletions originated at Alu-derived sequences through nonhomologous recombination reactions at locations sharing limited or no homology [90]. Analyses of the DNA sequences at breakpoints indicated the presence of short direct or inverted repeats and, hence, the formation of potential non-B-DNA conformations. These data are in conformity with the metaanalysis results and the selected cases analyzed previously, which strongly support the model of DSB formation at sites of altered DNA conformation as a risk factor for mutagenesis [1]. Second, a study on a patient afflicted with neurofibromatosis type 1 indicated that the condition was caused by a 2-Mb deletion at 17q11.2 [91], in the region normally involved in the recurrent cases. However, whereas one breakpoint junction was located within the recurrent region of the *JJAZ1* gene (adjacent to the distal NF1-LCR) [92], the proximal breakpoint occurred just downstream of an AT-rich segment that shared no homology with the distal site sandwiched between three directly oriented AluSx repeats. Hence, the mechanism proposed involved the occurrence of duplex DNA destabilization, promoted by slippage between the AluSx tracts and the low-melting AT-rich region, followed by repair through nonhomologous recombination.

D. DNA Repair and Chromosomal Rearrangements

Herein, it was reported that rearrangements may result from the repair of DSB through at least three pathways: (1) aberrant V(D)J recombination, (2) homologous recombination, and (3) nonhomologous recombination. How do these pathways compare in their ability to repair DNA? One study has addressed the question in the context of chromosomal translocations mediated by Alu repeats (reviewed in ref. [93]). A set of ES cells were constructed that carried two cassettes, one on chromosome 17 and the other on chromosome 14 [93]. The cassette on chromosome 17 (p5 allele) consisted of a 5′ neomycin fragment with a splice donor site followed by intronic sequences, including an Alu element from intron 1 of the *MLL* gene, an I-SceI endonuclease recognition site for DSB formation, and a 3′ puromycin fragment. The cassette on chromosome 14 (pF allele) consisted of a 5′ puromycin fragment, an I-SceI site, the same *MLL* Alu element, and a 3′ neomycin fragment with a splice acceptor site. The two Alu elements shared 290 bp of identity, whereas the puromycin fragments overlapped over a 265-bp segment. Depending on whether single-strand annealing (SSA, a subpathway of homologous recombination) or nonhomologous end joining (NHEJ) took place to repair the I-SceI-induced DSB, the chromosome 14 derivative would confer puromycin resistance (SSA) or puromycin sensitivity (NHEJ). Phenotype selection and sequence analyses showed that ~80% of the repair events were carried out by the SSA pathway. Hence, in the presence of sufficient homology, homologous recombination was the predominant pathway for DSB repair. However, a second type of experiment was conducted, whereby the two Alu elements were substituted so as to allow for only short (up to 25 bp) homologous stretches. Interestingly, Alu substitution did not cause a decrease in the frequency of translocations, although NHEJ became the predominant pathway. These results clearly demonstrate that the absence of homology is not rate-limiting in the repair of DSB. Most likely, it is the frequency of DSB formation or the simultaneous occurrence of two DSBs that governs the rate of repair. Hence, if DNA sequence motifs favor the transition to alternative conformations, thus lowering the activation energy for DNA hydrolysis, this will increase the frequency of DSB and, through repair, chromosome instability.

V. SUMMARY AND CONCLUDING REMARKS

This chapter reviews the current understanding of the roles of non-B-DNA conformations adopted by certain types of DNA sequences in genetic instabilities, especially gross rearrangements. Conclusions are as follows: (1) slipped (hairpin) structures, cruciforms, triplexes, tetraplexes and i-motifs, and left-handed Z-DNA are formed in chromosomes and elicit profound genetic consequences via recombination repair; (2) repeating sequences, probably in their non-B conformations, cause gross genomic rearrangements (deletions, insertions, inversions, translocations, and duplications); and (3) these rearrangements are the genetic basis for scores of human diseases, including polycystic kidney disease, adrenoleukodystrophy, follicular lymphomas, and spermatogenic failure.

Future work will focus on further characterization of all structures and metabolic processes involved, because this field of investigation is in its infancy. For the past 40 years, biochemists have characterized the conformations and properties of non-B-DNAs (reviewed in refs. [6, 8–10, 13–15, 17, 20]), and human geneticists have made excellent progress in mining the basis of genetic diseases. However, with the establishment of the human partial genome sequence, these disciplines can now be merged to gain deeper insights into the molecular basis of disease. Hopefully, this will be beneficial for developing effective therapeutic strategies in the future.

Acknowledgments

We thank Marek Napierala for helpful discussions. This work was supported by grants from the National Institutes of Health (NS37554 and ES11347), the Robert A. Welch Foundation, Friedreich's Ataxia Research Alliance, and the Seek-a-Miracle Foundation (Muscular Dystrophy Foundation).

References

1. Bacolla, A., Jaworski, A., Larson, J. E., Jakupciak, J. P., Chuzhanova, N., Abeysinghe, S. S., O'Connell, C. D., Cooper, D. N., and Wells, R. D. (2004). Breakpoints of gross deletions coincide with non-B DNA conformations. *Proc. Natl. Acad. Sci. USA* **101**, 14162–14167.
2. Wojciechowska, M., Bacolla, A., Larson, J. E., and Wells, R. D. (2005). The myotonic dystrophy type 1 triplet repeat sequence induces gross deletions and inversions. *J. Biol. Chem.* **280**, 941–952.
3. Meservy, J. L., Sargent, R. G., Iyer, R. R., Chan, F., McKenzie, G. J., Wells, R. D., and Wilson, J. H. (2003). Long CTG tracts from the myotonic dystrophy gene induce deletions and rearrangements during recombination at the APRT locus in CHO cells. *Mol. Cell Biol.* **23**, 3152–3162.
4. Marcadier, J. L., and Pearson, C. E. (2003). Fidelity of primate cell repair of a double-strand break within a (CTG)·(CAG) tract. Effect of slipped DNA structures. *J. Biol. Chem.* **278**, 33848–33856.
5. Wang, G., and Vasquez, K. M. (2004). Naturally occurring H-DNA-forming sequences are mutagenic in mammalian cells. *Proc. Natl. Acad. Sci. USA* **101**, 13448–13453.
6. Bacolla, A., and Wells, R. D. (2004). Non-B DNA conformations, genomic rearrangements, and human disease. *J. Biol. Chem.* **279**, 47411–47414.
7. Mirkin, S. M., and Frank-Kamenetskii, M. D. (1994). H-DNA and related structures. *Annu. Rev. Biophys. Biomol. Struct.* **23**, 541–576.
8. Wells, R. D., and Harvey, S. C. (1988). "Unusual DNA Structures." Springer-Verlag, New York.
9. Sinden, R. R. (1994). "DNA Structure and Function." Academic Press, San Diego, CA.
10. Soyfer, V. N., and Potaman, V. N. (1996). "Triple-Helical Nucleic Acids." Springer-Verlag, New York.
11. Rich, A., and Zhang, S. (2003). Timeline: Z-DNA: The long road to biological function. *Nat. Rev. Genet.* **4**, 566–572.
12. Raghavan, S. C., Chastain, P., Lee, J. S., Hegde, B. G., Houston, S., Langen, R., Hsieh, C. L., Haworth, I. S., and Lieber, M. R. (2005). Evidence for a triplex DNA conformation at the Bcl-2 major breakpoint region of the t(14;18) translocation. *J. Biol. Chem.* **280**, 22749–22760.
13. Neidle, S., and Parkinson, G. N. (2003). The structure of telomeric DNA. *Curr. Opin. Struct. Biol.* **13**, 275–283.

14. Majumdar, A., and Patel, D. J. (2002). Identifying hydrogen bond alignments in multistranded DNA architectures by NMR. *Acc. Chem. Res.* **35**, 1–11.
15. Wells, R. D., and Warren, S. T. (1998). "Genetic Instabilities and Hereditary Neurological Diseases." Academic Press, San Diego, CA.
16. Lilley, D. M., and Dahlberg, J. E. (1992). DNA structures. Part B. Chemical and electrophoretic analysis of DNA. *Methods Enzymol.* **212**, 3–458.
17. Lilley, D. M., and Dahlberg, J. E. (1992). DNA structures. Part A. Synthesis and physical analysis of DNA. *Methods Enzymol.* **211**, 3–567.
18. Bacolla, A., Jaworski, A., Connors, T. D., and Wells, R. D. (2001). *PKD1* unusual DNA conformations are recognized by nucleotide excision repair. *J. Biol. Chem.* **276**, 18597–18604.
19. Cozzarelli, N. R., Boles, T. C., and White, J. H. (1990). Primer on the topology and geometry of DNA supercoiling. In "DNA Topology and Its Biological Effects" (N. R. Cozzarelli and J. C. Wang, Eds.), pp. 139–184. Cold Spring Harbor Laboratory Press, Cold Spring Harbor, NY.
20. Wells, R. D. (1988). Unusual DNA structures. *J. Biol. Chem.* **263**, 1095–1098.
21. Wells, R. D., Dere, R., Hebert, M. L., Napierala, M., and Son, L. S. (2005). Advances in mechanisms of genetic instability related to hereditary neurological diseases. *Nucleic Acids Res.* **33**, 3785–3798.
22. Warburton, P. E., Giordano, J., Cheung, F., Gelfand, Y., and Benson, G. (2004). Inverted repeat structure of the human genome: The X-chromosome contains a preponderance of large, highly homologous inverted repeats that contain testes genes. *Genome Res.* **14**, 1861–1869.
23. Rozen, S., Skaletsky, H., Marszalek, J. D., Minx, P. J., Cordum, H. S., Waterston, R. H., Wilson, R. K., and Page, D. C. (2003). Abundant gene conversion between arms of palindromes in human and ape Y chromosomes. *Nature* **423**, 873–876.
24. Chou, S. H., Chin, K. H., and Wang, A. H. (2003). Unusual DNA duplex and hairpin motifs. *Nucleic Acids Res.* **31**, 2461–2474.
25. Huertas, D., and Azorin, F. (1996). Structural polymorphism of homopurine DNA sequences. d(GGA)n and d(GGGA)n repeats form intramolecular hairpins stabilized by different base-pairing interactions. *Biochemistry* **35**, 13125–13135.
26. Zheng, M., Huang, X., Smith, G. K., Yang, X., and Gao, X. (1996). Genetically unstable CXG repeats are structurally dynamic and have a high propensity of folding: An NMR and UV spectroscopic study. *J. Mol. Biol.* **264**, 323–336.
27. Oberle, I., Rousseau, F., Heitz, D., Kretz, C., Devys, D., Hanauer, A., Boue, J., Bertheas, M. F., and Mandel, J. L. (1991). Instability of a 550-base pair DNA segment and abnormal methylation in fragile X syndrome. *Science* **252**, 1097–1102.
28. Reinisch, K. M., Chen, L., Verdine, G. L., and Lipscomb, W. N. (1995). The crystal structure of HaeIII methyltransferase convalently complexed to DNA: An extrahelical cytosine and rearranged base pairing. *Cell* **82**, 143–153.
29. Bacolla, A., Pradhan, S., Larson, J. E., Roberts, R. J., and Wells, R. D. (2001). Recombinant human DNA (cytosine-5) methyltransferase. III. Allosteric control, reaction order, and influence of plasmid topology and triplet repeat length on methylation of the fragile X CGG·CCG sequence. *J. Biol. Chem.* **276**, 18605–18613.
30. Smith, S. S., Laayoun, A., Lingeman, R. G., Baker, D. J., and Riley, J. (1994). Hypermethylation of telomere-like foldbacks at codon 12 of the human c-Ha-ras gene and the trinucleotide repeat of the FMR-1 gene of fragile X. *J. Mol. Biol.* **243**, 143–151.
31. Chi, L. M., and Lam, S. L. (2005). Structural roles of CTG repeats in slippage expansion during DNA replication. *Nucleic Acids Res.* **33**, 1604–1617.
32. Collins, J. R., Stephens, R. M., Gold, B., Long, B., Dean, M., and Burt, S. K. (2003). An exhaustive DNA micro-satellite map of the human genome using high performance computing. *Genomics* **82**, 10–19.
34. Wells, R. D., Collier, D. A., Hanvey, J. C., Shimizu, M., and Wohlrab, F. (1988). The chemistry and biology of unusual DNA structures adopted by oligopurine·oligopyrimidine sequences. *FASEB J.* **2**, 2939–2949.
35. Frank-Kamenetskii, M. D., and Mirkin, S. M. (1995). Triplex DNA structures. *Annu. Rev. Biochem.* **64**, 65–95.
36. LeProust, E. M., Pearson, C. E., Sinden, R. R., and Gao, X. (2000). Unexpected formation of parallel duplex in GAA and TTC trinucleotide repeats of Friedreich's ataxia. *J. Mol. Biol.* **302**, 1063–1080.
37. Potaman, V. N., Oussatcheva, E. A., Lyubchenko, Y. L., Shlyakhtenko, L. S., Bidichandani, S. I., Ashizawa, T., and Sinden, R. R. (2004). Length-dependent structure formation in Friedreich ataxia (GAA)n*(TTC)n repeats at neutral pH. *Nucleic Acids Res.* **32**, 1224–1231.
38. Vetcher, A. A., Napierala, M., Iyer, R. R., Chastain, P. D., Griffith, J. D., and Wells, R. D. (2002). Sticky DNA, a long GAA·GAA·TTC triplex that is formed intramolecularly, in the sequence of intron 1 of the frataxin gene. *J. Biol. Chem.* **277**, 39217–39227.
39. Vetcher, A. A., Napierala, M., and Wells, R. D. (2002). Sticky DNA: Effect of the polypurine·polypyrimidine sequence. *J. Biol. Chem.* **277**, 39228–39234.
40. Napierala, M., Dere, R., Vetcher, A., and Wells, R. D. (2004). Structure-dependent recombination hot spot activity of GAA·TTC sequences from intron 1 of the Friedreich's ataxia gene. *J. Biol. Chem.* **279**, 6444–6454.
41. Mills, M., Lacroix, L., Arimondo, P. B., Leroy, J. L., Francois, J. C., Klump, H., and Mergny, J. L. (2002). Unusual DNA conformations: Implications for telomeres. *Curr. Med. Chem. Anti-Canc. Agents* **2**, 627–644.
42. Fojtik, P., and Vorlickova, M. (2001). The fragile X chromosome (GCC) repeat folds into a DNA tetraplex at neutral pH. *Nucleic Acids Res.* **29**, 4684–4690.
43. Kettani, A., Kumar, R. A., and Patel, D. J. (1995). Solution structure of a DNA quadruplex containing the fragile X syndrome triplet repeat. *J. Mol. Biol.* **254**, 638–656.
44. Fojtik, P., Kejnovska, I., and Vorlickova, M. (2004). The guanine-rich fragile X chromosome repeats are reluctant to form tetraplexes. *Nucleic Acids Res.* **32**, 298–306.
45. Sun, H., Karow, J. K., Hickson, I. D., and Maizels, N. (1998). The Bloom's syndrome helicase unwinds G4 DNA. *J. Biol. Chem.* **273**, 27587–27592.
46. Fry, M., and Loeb, L. A. (1999). Human Werner syndrome DNA helicase unwinds tetrahelical structures of the fragile X syndrome repeat sequence d(CGG)n. *J. Biol. Chem.* **274**, 12797–12802.
47. Brosh, R. M., Jr., Majumdar, A., Desai, S., Hickson, I. D., Bohr, V. A., and Seidman, M. M. (2001). Unwinding of a DNA triple helix by the Werner and Bloom syndrome helicases. *J. Biol. Chem.* **276**, 3024–3030.
48. Franchitto, A., and Pichierri, P. (2002). Protecting genomic integrity during DNA replication: Correlation between Werner's and Bloom's syndrome gene products and the MRE11 complex. *Hum. Mol. Genet.* **11**, 2447–2453.
49. Naylor, L. H., and Clark, E. M. (1990). d(TG)n·d(CA)n sequences upstream of the rat prolactin gene form Z-DNA and inhibit gene transcription. *Nucleic Acids Res.* **18**, 1595–1601.
50. Kladde, M. P., Kohwi, Y., Kohwi-Shigematsu, T., and Gorski, J. (1994). The non-B-DNA structure of d(CA/TG)n differs from that of Z-DNA. *Proc. Natl. Acad. Sci. USA* **91**, 1898–1902.
51. Rich, A., Nordheim, A., and Wang, A. H. (1984). The chemistry and biology of left-handed Z-DNA. *Annu. Review Biochem.* **53**, 791–846.

52. Vorlickova, M., Kejnovska, I., Tumova, M., and Kypr, J. (2001). Conformational properties of DNA fragments containing GAC trinucleotide repeats associated with skeletal displasias. *Eur. Biophys. J.* **30**, 179–185.
53. Latha, K. S., Anitha, S., Rao, K. S., and Viswamitra, M. A. (2002). Molecular understanding of aluminum-induced topological changes in $(CCG)_{12}$ triplet repeats: Relevance to neurological disorders. *Biochim. Biophys. Acta* **1588**, 56–64.
54. Mochmann, L. H., and Wells, R. D. (2004). Transcription influences the types of deletion and expansion products in an orientation-dependent manner from GAC*GTC repeats. *Nucleic Acids Res.* **32**, 4469–4479.
55. Delot, E., King, L. M., Briggs, M. D., Wilcox, W. R., and Cohn, D. H. (1999). Trinucleotide expansion mutations in the cartilage oligomeric matrix protein (COMP) gene. *Hum. Mol. Genet.* **8**, 123–128.
56. Rossetti, S., Chauveau, D., Walker, D., Saggar-Malik, A., Winearls, C. G., Torres, V. E., and Harris, P. C. (2002). A complete mutation screen of the ADPKD genes by DHPLC. *Kidney Int.* **61**, 1588–1599.
57. Watnick, T. J., Gandolph, M. A., Weber, H., Neumann, H. P. H., and Germino, G. G. (1998). Gene conversion is a likely cause of mutation in *PKD1*. *Hum. Mol. Genet.* **7**, 1239–1243.
58. Van Raay, T. J., Burn, T. C., Connors, T. D., Petri, L. R., Germino, G. G., Klinger, K. W., and Landes, G. M. (1996). A 2.5 kb polypyrimidine tract in the *PKD1* gene contains at least 23 H-DNA-forming sequences. *Microb. Comp. Genomics* **1**, 317–327.
59. Blaszak, R. T., Potaman, V., Sinden, R. R., and Bissler, J. J. (1999). DNA structural transitions within the *PKD1* gene. *Nucleic Acids Res.* **27**, 2610–2617.
60. Patel, H. P., Lu, L., Blaszak, R. T., and Bissler, J. J. (2004). *PKD1* intron 21: Triplex DNA formation and effect on replication. *Nucleic Acids Res.* **32**, 1460–1468.
61. Bacolla, A., Gellibolian, R., Shimizu, M., Amirhaeri, S., Kang, S., Ohshima, K., Larson, J. E., Harvey, S. C., Stollar, B. D., and Wells, R. D. (1997). Flexible DNA: Genetically unstable CTG·CAG and CGG·CCG from human hereditary neuromuscular disease genes. *J. Biol. Chem.* **272**, 16783–16792.
62. Bowater, R. P., and Wells, R. D. (2000). The intrinsically unstable life of DNA triplet repeats associated with human hereditary disorders. *Prog. Nucleic Acid Res. Mol. Biol.* **66**, 159–202.
63. Gellibolian, R., and Bacolla, A. (1998). Biophysical and structural studies on triplet repeat sequences: Duplex triplet repeat structures. In "Genetic Instabilities and Hereditary Neurological Diseases" (R. D. Wells and S. T. Warren, Eds.), pp. 561–583. Academic Press, San Diego, CA.
64. Kang, S., Jaworski, A., Ohshima, K., and Wells, R. D. (1995). Expansion and deletion of CTG repeats from human disease genes are determined by the direction of replication in *E. coli*. *Nat. Genet.* **10**, 213–218.
65. Wells, R. D., Parniewski, P., Pluciennik, A., Bacolla, A., Gellibolian, R., and Jaworski, A. (1998). Small slipped register genetic instabilities in *Escherichia coli* in triplet repeat sequences associated with hereditary neurological diseases. *J. Biol. Chem.* **273**, 19532–19541.
66. Kang, S., Ohshima, K., Jaworski, A., and Wells, R. D. (1996). CTG triplet repeats from the myotonic dystrophy gene are expanded in *Escherichia coli* distal to the replication origin as a single large event. *J. Mol. Biol.* **258**, 543–547.
67. Bowater, R. P., Jaworski, A., Larson, J. E., Parniewski, P., and Wells, R. D. (1997). Transcription increases the deletion frequency of long CTG·CAG triplet repeats from plasmids in *Escherichia coli*. *Nucleic Acids Res.* **25**, 2861–2868.
68. Kunst, C. B., Leeflang, E. P., Iber, J. C., Arnheim, N., and Warren, S. T. (1997). The effect of FMR1 CGG repeat interruptions on mutation frequency as measured by sperm typing. *J. Med. Genet.* **34**, 627–631.
69. Gao, X., Huang, X., Smith, G. K., and Zheng, M. (1998). Structure and dynamics of single-stranded nucleic acids containing trinucleotide repeats. In "Genetic Instabilities and Hereditary Neurological Diseases" (R. D. Wells and S. T. Warren, Eds.), pp. 623–646. Academic Press, San Diego, CA.
70. Liu, L. F., and Wang, J. C. (1987). Supercoiling of the DNA template during transcription. *Proc. Natl. Acad. Sci. USA* **84**, 7024–7027.
71. Rahmouni, A. R., and Wells, R. D. (1992). Direct evidence for the effect of transcription on local DNA supercoiling in vivo. *J. Mol. Biol.* **223**, 131–144.
72. Pestov, D. G., Dayn, A., Siyanova, E., George, D. L., and Mirkin, S. M. (1991). H-DNA and Z-DNA in the mouse c-Ki-ras promoter. *Nucleic Acids Res.* **19**, 6527–6532.
73. Kinniburgh, A. J. (1989). A cis-acting transcription element of the c-myc gene can assume an H-DNA conformation. *Nucleic Acids Res.* **17**, 7771–7778.
74. Blum, K. A., Lozanski, G., and Byrd, J. C. (2004). Adult Burkitt leukemia and lymphoma. *Blood* **104**, 3009–3020.
75. Abeysinghe, S. S., Chuzhanova, N., Krawczak, M., Ball, E. V., and Cooper, D. N. (2003). Translocation and gross deletion breakpoints in human inherited disease and cancer I: Nucleotide composition and recombination-associated motifs. *Hum. Mutat.* **22**, 229–244.
76. Stenson, P. D., Ball, E. V., Mort, M., Phillips, A. D., Shiel, J. A., Thomas, N. S., Abeysinghe, S., Krawczak, M., and Cooper, D. N. (2003). Human Gene Mutation Database (HGMD): 2003 update. *Hum. Mutat.* **21**, 577–581.
77. Chuzhanova, N., Abeysinghe, S. S., Krawczak, M., and Cooper, D. N. (2003). Translocation and gross deletion breakpoints in human inherited disease and cancer II: Potential involvement of repetitive sequence elements in secondary structure formation between DNA ends. *Hum. Mutat.* **22**, 245–251.
78. Chen, J. M., Chuzhanova, N., Stenson, P. D., Ferec, C., and Cooper, D. N. (2005). Meta-analysis of gross insertions causing human genetic disease: Novel mutational mechanisms and the role of replication slippage. *Hum. Mutat.* **25**, 207–221.
79. Raghavan, S. C., Houston, S., Hegde, B. G., Langen, R., Haworth, I. S., and Lieber, M. R. (2004). Stability and strand asymmetry in the non-B DNA structure at the Bcl-2 major breakpoint region. *J. Biol. Chem.* **279**, 46213–46225.
80. Raghavan, S. C., Swanson, P. C., Wu, X., Hsieh, C. L., and Lieber, M. R. (2004). A non-B-DNA structure at the *Bcl-*2 major breakpoint region is cleaved by the RAG complex. *Nature* **428**, 88–93.
81. Stankiewicz, P., and Lupski, J. R. (2002). Genome architecture, rearrangements and genomic disorders. *Trends Genet.* **18**, 74–82.
82. Samonte, R. V., and Eichler, E. E. (2002). Segmental duplications and the evolution of the primate genome. *Nat. Rev. Genet.* **3**, 65–72.
83. Lupski, J. R. (2004). Hotspots of homologous recombination in the human genome: Not all homologous sequences are equal. *Genome Biol.* **5**, 242.
84. Kurotaki, N., Stankiewicz, P., Wakui, K., Niikawa, N., and Lupski, J. R. (2005). Sotos syndrome common deletion is mediated by directly oriented subunits within inverted Sos-REP low-copy repeats. *Hum. Mol. Genet.* **14**, 535–542.
85. Visser, R., Shimokawa, O., Harada, N., Kinoshita, A., Ohta, T., Niikawa, N., and Matsumoto, N. (2005). Identification of a 3.0-kb major recombination hotspot in patients with sotos syndrome who carry a common 1.9-Mb microdeletion. *Am. J. Hum. Genet.* **76**, 52–67.

86. Tatton-Brown, K., Douglas, J., Coleman, K., Baujat, G., Chandler, K., Clarke, A., Collins, A., Davies, S., Faravelli, F., Firth, H., et al. (2005). Multiple mechanisms are implicated in the generation of 5q35 microdeletions in Sotos syndrome. *J. Med. Genet.* **42**, 307–313.
87. Bayes, M., Magano, L. F., Rivera, N., Flores, R., and Perez Jurado, L. A. (2003). Mutational mechanisms of Williams-Beuren syndrome deletions. *Am. J. Hum. Genet.* **73**, 131–151.
88. Osborne, L. R., Li, M., Pober, B., Chitayat, D., Bodurtha, J., Mandel, A., Costa, T., Grebe, T., Cox, S., Tsui, L. C., et al. (2001). A 1.5 million-base pair inversion polymorphism in families with Williams-Beuren syndrome. *Nat. Genet.* **29**, 321–325.
89. Gimelli, G., Pujana, M. A., Patricelli, M. G., Russo, S., Giardino, D., Larizza, L., Cheung, J., Armengol, L., Schinzel, A., Estivill, X., et al. (2003). Genomic inversions of human chromosome 15q11-q13 in mothers of Angelman syndrome patients with class II (BP2/3) deletions. *Hum. Mol. Genet.* **12**, 849–858.
90. Shaw, C. J., and Lupski, J. R. (2005). Non-recurrent 17p11.2 deletions are generated by homologous and non-homologous mechanisms. *Hum. Genet.* **116**, 1–7.
91. Lopez-Correa, C., Dorschner, M., Brems, H., Lazaro, C., Clementi, M., Upadhyaya, M., Dooijes, D., Moog, U., Kehrer-Sawatzki, H., Rutkowski, J. L., et al. (2001). Recombination hotspot in NF1 microdeletion patients. *Hum. Mol. Genet.* **10**, 1387–1392.
92. Kehrer-Sawatzki, H., Kluwe, L., Funsterer, C., and Mautner, V. F. (2005). Extensively high load of internal tumors determined by whole body MRI scanning in a patient with neurofibromatosis type 1 and a non-LCR-mediated 2-Mb deletion in 17q11.2. *Hum. Genet.* **116**, 466–475.
93. Elliott, B., Richardson, C., and Jasin, M. (2005). Chromosomal translocation mechanisms at intronic alu elements in mammalian cells. *Mol. Cell* **17**, 885–894.

PART XIV

Cancer and Genetic Instability

CHAPTER 47

Microsatellite Instability in Cancer

MICHAEL J. SICILIANO
Department of Molecular Genetics, The University of Texas M. D. Anderson Cancer Center,
Houston Texas 77030

I. Introduction
II. Microsatellite Instability (MSI) in Hereditary Nonpolyposis Color Cancer (HNPCC)
III. Downstream Impact of MSI on Cancer
IV. MSI-High HNPCC
V. HNPCC Tumors That Are Not MSI-High
VI. Alternate Bases for HNPCC in Non-MSI-High Tumors
    A. Impact of Different DNA Repair Genes on MSI
    B. Polymorphisms of Unknown Significance and HNPCC
VII. Increased Sensitivity of MSI Detection
    A. Sensitive Technology Detects MSI in PBLs of MMR Mutation Carriers
    B. Analysis of CRC Material by SP-PCR
    C. Mutation in Constitutive Tissue of Carriers of DNA Repair Mutations
    D. MSI Increases with Age in Normal Individuals
VIII. Conclusion
References

Microsatellite instability (MSI) in cancer is a genome wide event. As such, it is very different from the locus-specific phenomenon seen in inherited neurological diseases. Null mismatch repair (MMR) genes lead to such high levels of MSI that it can be seen by simple polymerase chain reaction (PCR) in many hereditary nonpolyposis colon cancer (HNPCC) tumors, where such germline MMR mutations have been shown to be the basis of this inherited cancer. After MSI analysis, immunohistochemical (IHC) staining has been used to determine the particular MMR protein missing so that the affected gene could be sequenced and the offending mutation identified. This has enabled the targeting of other family members carrying the mutation, so that appropriate preventive and early diagnostic procedures can be applied. Because the frequency of mutant MSI fragments must be >25% to be detectable by PCR, a more sensitive and quantitative method, small-pool PCR (SP-PCR), has been developed. Although SP-PCR has yet to be applied to the 30–70% of HNPCC families in which MSI has not been observed in tumors by traditional means, results have now been published on MSI frequency in the peripheral blood lymphocytes (PBLs) in normal individuals versus those carrying germline MMR mutations. Low, but measurable levels of MSI (1–3%) have been shown to increase in the PBLs of normal individuals with age, and levels significantly above age-matched controls have now been verified in the PBLs of MMR gene mutation carriers. Implications of these findings and this new approach are discussed.

I. INTRODUCTION

As a result of microsatellite instability (MSI), the number of repeats at microsatellite loci in the DNA become modified and the modifications are inherited. Therefore, the manifestations of these instabilities are mutations. Mutation provides the variation that drives evolution, and that is considered a good thing. However, mutation, being a random change in an orderly DNA array, is generally considered troublemaking within an organism. Throughout most of the chapters of this volume, the mayhem induced by MSI upon neurological function has been well-described. However, when one thinks of mutation and health, it is typical to think of cancer because mutagenic agents are generally carcinogenic. Curiously, the link between MSI and inherited cancer surfaced after unstable repeats were seen as the genetic bases of neurological disorders such as fragile X syndrome [1] and myotonic dystrophy [2]. However, the spectra of MSI in cancer versus those in neurological diseases are quite different. In cancer, there appears to be broad spectrum instability in which all of the many hundreds of thousands of microsatellite loci are subject to instability. In the different neurological syndromes, instability at specific microsatellite loci accounts for the pathologies. This has enabled MSI in cancer to be attributed to inherited defects in mismatch repair (MMR) genes, whereas the bases for the instability in neurological disorders remain problematic. Contrary to what is seen in cancer, deficiency of Msh2 (the product of a major MMR gene) has been shown to prevent MSI of the CAG repeat in Huntington's disease transgenic mice [3].

In this chapter, the understanding of MSI in the cancer in which it was first observed, hereditary nonpolyposis colon cancer (HNPCC), will be reviewed. New insights into the phenomenon as a result of techniques capable of determining the frequency of microsatellite mutants at individual loci will be considered, and how such analysis may broaden our view of the role of MSI will be evaluated.

II. MICROSATELLITE INSTABILITY (MSI) IN HEREDITARY NONPOLYPOSIS COLON CANCER (HNPCC)

The first observation of MSI in cancer was purely serendipitous. While the genomes of tumor DNA from HNPCC patients by polymerase chain reaction (PCR) were being scanned across polymorphic microsatellite loci for loss of heterozygosity (LOH), a remarkable observation was made [4–6]: the presence of new alleles (different numbers of repeats) in addition to the progenitor alleles with which the patients were born (Fig. 47-1). It soon was determined that the enabling events giving rise to such phenotypes were either mutation [7] or epigenetic silencing [8] of MMR genes. Such events were hypothesized to have severe clinical consequences in that the inability to repair replication errors could result in accelerated tumor initiation and progression of the mutator phenotype [9]. This was consistent with the observation that HNPCC patients present with disease symptoms 20 years earlier than the general population and also develop other epithelial tumors [10].

HNPCC is one of the two major colorectal cancers (CRCs) showing a dominant form of inheritance, the other being familial adenomatous polyposis (FAP) (see review in ref. [11]). The etiology of FAP has been traced to the inheritance of the tumor suppressor gene *APC* on chromosome 5 and leads to the formation of thousands of adenomatous polyps in the bowel, which advance to cancer as mutations in additional loci accumulate [12]. MSI is not believed to be a factor in FAP and it will not be discussed further here.

HNPCC, also called Lynch's syndrome, has a high penetrance (80–90%), has been characterized by early age of onset (40s), the lack of florid polyposis seen in FAP, and multiple tumors (35% of patients), and has a better prognosis than sporadic CRCs [13–15]. As one might expect in view of the MMR gene defects responsible for the disease, HNPCC families are at increased risk for malignant tumors at a number of extracolonic sites: endometrium, ovary, stomach, small bowel, hepatobiliary tract, pancreas, ureter, renal pelvis, and breast [16, 17].

HNPCC families are distinguished from sporadic CRCs by a number of features that have come to be known as the Amsterdam criteria: (1) at least three

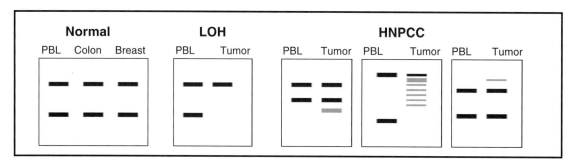

FIGURE 47-1 Cartoon depicting PCR fragments in loss of heterozygosity (LOH) and MSI. Several normal tissues indicate the pattern from a heterozygote. LOH is seen in tumors adjacent to the PBL sample having both alleles. In the three HNPCC examples, the PBL lanes indicate the heterozygous pattern for the individual. However, in the tumor lanes one sees the presence of "new" fragments—not in the normal genotype—and therefore MSI.

family members affected, two being first-degree relatives; (2) at least two generations affected; and (3) one of those affected being diagnosed before age 45 or an affected family member with more than one colon tumor [18, 19]. Criteria have been widened to include extracolonic tumors, as mentioned earlier, instead of just colon tumors [20]. HNPCC individuals carry germline MMR mutations (inherited and present in every cell), and either loss or loss of function of the wild-type alleles results in MSI and cancer in the tissue where that occurs (another affirmation of the Knudson hypothesis [21]). That HNPCC is an important system in which to study this phenomenon is underscored by the fact that at 5–8% of all colon cancers it is one of the most common forms of hereditary malignancy [11]. A significant proportion (10%) of sporadic CRC individuals may have MMR defects and MSI confined to the tumor tissue as a result of methylation of MMR genes, usually *MLH1* [22]. An important distinction between HNPCC and sporadic CRC for this discussion is that HNPCC individuals have germline cancer predisposition mutations in every cell and sporadic cases do not, even though both may have MSI in their tumors.

III. DOWNSTREAM IMPACT OF MSI ON CANCER

One may ask the same question here that is an issue in neurological disorders. How might MSI contribute to carcinogenesis because such repeats are commonly located in noncoding regions of the genome? An exhaustive *in silico* study of microsatellite loci [23] revealed over 128,000 dinucleotide, 8000 trinucleotide, 23,000 tetranucleotide, and 4000 pentanucleotide repeats consisting of at least 9 repeat units scattered across the chromosomes. Therefore, elevation of mutant frequencies at such loci because of some inherited condition would be expected to have an enormous impact on the genome. Collins et al. [23] then scanned for the presence of trinucleotide repeats in coding regions. A total of 62 genes—all of the expected neuromuscular genes (such as Huntington's chorea and the various spinocerebellar ataxias) as well as many that could affect carcinogenesis, such as homeobox genes, oncogenes, and cell cycle control genes—were identified.

However, instability in neuromuscular diseases has taught us that affected microsatellites in introns such as Friedreich's ataxia [24], in 5′ regions such as fragile X syndrome [1], and in 3′ regions such as myotonic dystrophy [2] can also cause disease. And that is just the trinucleotide repeats! There are hundreds of thousands more mono- and dinucleotide repeats. Therefore, it is reasonable to assume that even modest increases in MSI perhaps initiated by the most sublime alterations in MMR genes would have a profound effect on carcinogenesis over time.

Specific genes have been identified containing microsatellites, which when unstable have altered function contributing the carcinogenic process in HNPCC. The first of these caused a frameshift mutation in the *BAX* gene [25]. Summaries and references to various downstream genes that have been shown to be affected by such genome instability and that have been shown to have an effect on cancer initiation and/or progression are available [26, 27]. In addition, *in silico* genome scans to search databases for genes having microsatellites that, if mutated, would give rise to frameshift mutations rendering them ineffective and therefore likely candidates as downstream genes in a MSI scenario have identified new candidates [28, 29]. Finally, by turning the big guns of genomics on the problem, 81 genes were seen to be selectively down-regulated in MSI-high cancers using cDNA microarray of primary colon cancers [30]. Although this latter study was focused more on identifying genes that might be more affected by methylation than frameshift mutation, the list of expression changes downstream of genetic and epigenetic events associated with MSI is growing, and therefore the impact of even minor changes in MSI may have a major effect on disease. The downstream genes associated with cancer based on MSI will not be discussed further here.

IV. MSI-HIGH HNPCC

In what has become the classic quantification method to evaluate MSI levels in HNPCC [18], at least five of several recommended microsatellite loci are studied, and if new fragments are seen in at least two (or 40%) of those loci the sample is to be considered MSI-high, whereas failing to achieve that, tumors are lumped together into an MSI-low (where mutant fragments are observed at only one locus of the five) or MSS (stable, no mutant fragments seen at any of the loci screened) class. Though not statistically rigorous, this categorization has proved useful as the MSI-high phenotype has come to be recognized as a distinct class resulting from serious mutations or expression changes in at least one of the major mismatch repair genes.

The MSI-high class has been thoroughly and comprehensively studied by the de la Chapelle laboratory [22]. Tumors from 1066 CRC patients were screened for MSI without regard to family history. A total of 135 tumors were MSI-high and 73 were MSI-low. All showing MSI were then studied by immunohistochemical staining and sequencing for the major MMR genes—*MSH2*, *MLH1*, *MSH6*, and *PMS2*. Only 23 (all MSI-high) had what could be interpreted as deleterious mutations in one of those genes. Only these were considered Lynch's syndrome. None of the MSI-low tumors showed any such mutations. Most of all of the remaining MSI-high tumors had methylation of *MLH1* and had no mutations in the MMR genes that were considered deleterious. It should be noted that all tumor DNAs sequenced showed a great many mutations that were considered polymorphisms of unknown significance (PUS). These were tabulated and have been made available online. To put the value of this type of study in perspective, earlier studies [31, 32] revealed that, when family members were enrolled in prevention programs, significant reductions in CRC mortality (65%) and morbidity (62%) occurred. The identification of deleterious mutations makes it possible to specifically identify at-risk individuals for such prevention. Therefore, it may be concluded from this section that the identification of MSI-high individuals by the methods in place has great value as an initial step (subsequent steps involve sequencing) in identifying the genetic basis of the abnormality and developing tools and reagents to study others at risk.

What about MSI-high tumors caused by methylation of *MLH1*? These are not considered Lynch's syndrome [22] because they were not proved to be the result of germline mutations that lead to a family history of the disease. They seem to occur in older persons generally considered to be sporadic cases. However, the age line may be becoming obscure. In the Hamper et al. [22] study, 10 of the 23 probands with Lynch's syndrome were older than 50 years at the time of diagnosis. As shall be seen next, older CRC cases with clear germline mutations accounting for their syndrome are beginning to be identified. Furthermore, evidence is beginning to accumulate, through family studies, that there may be germline components to so-called "epigenetic" methylation events [31, 32, 33]. In fact, such a mutation was present in spermatozoa of one of the subjects, indicating not only the potential of passing such on to offspring but also that it can be a germline event [32].

V. HNPCC TUMORS THAT ARE NOT MSI-HIGH

Null mutations in MMR genes leading to the MSI-high phenotype have been shown to be present in only 30–70% of cancers that fit into the most encompassing set of criteria for HNPCC [34–36]. Therefore, a great percentage of HNPCC cases are not approachable by classical methods. Like their MSI-high counterparts, these patients, in addition to an autosomal dominant mode of inheritance, have a phenotype similar to those cancers likely arising as a result of an inherited mutator [10]. This large class of patients may have germline mutations predisposing them to MSI at a different, yet clinically significant level. Identification of those mutations would then make it possible to extend to such family members the identification of individuals at risk and subsequent prevention and early diagnosis procedures as are now available to the MSI-high families.

However, these cases do not come to light because where no MSI is detected, sequencing for mutations in MMR genes is not done. The frequency of mutant fragments at any microsatellite locus detectable by simple PCR must be >25% in order to be seen and not overwhelmed by progenitor fragments. By any somatic cell genetic standard, such a mutant frequency is indeed *enormous*. What if there were 5% MSI frequency, or 15%, at a locus? These would go undetected by standard PCR. Yet, from what has been seen previously on the number of target microsatellite loci in the genome and the growing number of potential downstream genes that could be affected by such instability, such relatively modest (compared to MSI-high) elevations in mutant frequencies would indeed be expected to have a clinical impact. Therefore, more sublime mutations of MMR genes that cause MSI, but at levels too low to be seen by classical PCR, might be the basis of the large number of families meeting Amsterdam-like criteria for HNPCC who do not have detectable MSI and who produce immunologically detectable, though modified, MMR protein as determined by IHC.

VI. ALTERNATE BASES FOR HNPCC IN NON-MSI-HIGH TUMORS

There are two ways to think about how mutations may have an intermediate or attenuated impact on MSI and, therefore, HNPCC: variable impact of the different genes of the MMR pathway and various attenuating mutations in high-impact MMR genes. Evidence exists for both types of events possibly playing roles in HNPCC.

A. Impact of Different DNA Repair Genes on MSI

The MMR pathway is complex and has many genes—perhaps over 50 as identified so far in *Saccharomyces cerevisiae* [37]. Some of the major ones seen to be involved in HNPCC have been studied for differential impact. By employing a procedure of diluting DNA to less than single molecule equivalents and conducting PCR on each of many pools of these in order to identify and quantify low-frequency mutants [38], Yao et al. [39] were able to show in transgenic mice that the frequency of mononucleotide repeat mutants was three times higher in $Mlh1^{-/-}$ mice than in $Pms2^{-/-}$ animals. Both of these genes are considered major MMR genes. It was the opinion of Yao et al. that such differences would have dramatic effects on tumor development. This was supported by an earlier observation from that laboratory of higher tumor burden in the $Mlh1$ null mice than in the $Pms2$ nulls [40]. It has been observed that $MSH6$ mutations are associated with late onset disease and low MSI [41, 42], and $MLH3$ mutations have been associated with variable degrees of MSI in tumor material [43]. In addition to different genes in the MMR pathway impacting MSI in different ways, Guo and Loeb [44] review the role of genes in other DNA repair pathways that have now been shown to affect MSI and cancer.

B. Polymorphisms of Unknown Significance and HNPCC

Missense mutations that often appear as simple polymorphisms have been shown to have functional consequences and be contributory to human pathology. A dramatic presentation of that has been made for the various mutations of the $PRODH$ gene on 22q11 in the region deleted in the velocardiofacial syndrome–DiGeorge syndrome, where missense mutations were associated with hyperprolinemia and schizophrenia [45]. Many missense mutations have been observed in major MMR genes in CRC patients [22]. Such mutations are generally considered polymorphisms of unknown significance (PUS). However, they may be attenuating mutations in major MMR genes, reducing the activity of the coded protein rather than completely destroying it, such that MSI would be produced at levels too low to be seen by simple PCR.

Lipkin et al. [46] have a very significant report on this topic. Hypothesizing that attenuated alleles might underlie susceptibility to colon cancer, they studied Israeli probands who did not have MSI as measured by traditional PCR for gene variants in MMR genes. Ages of onset in these patients were also older than those seen typically in HNPCC patients. An $MLH1$ variant resulting in a D132H amino acid substitution was identified as conferring clinically significant susceptibility to CRC. Functional analysis of the resultant protein indicated reduced, but not eliminated, ATPase activity. One might predict that such tumors will show statistically significant MSI by using a more sensitive method of detection.

In another case, attenuated alleles having reduced function came to light through consanguinity, whereby these alleles reach a homozygous state and produce an overt clinical syndrome [47]. Here three children were produced homozygous for a missense mutant $MLH1$ gene (Arg687Trp), the product of which was immunologically detected. The children had full-blown CRC with MSI-high. As yet, the parents are without symptoms even though they are carriers of the germline mutation. However, a more sensitive approach to MSI might reveal microsatellite mutations in them as well.

VII. INCREASED SENSITIVITY OF MSI DETECTION

All of the scenarios of lower level MSI playing a role in inherited cancer have one thing in common: lower level MSI. Consequently, the development of a technology capable of quantifying low, yet above background levels of MSI cuts right to the chase. Such technology hopefully would not only be able to quantify more accurately the level of MSI in tumors but also identify MSI in cells that can be monitored before individuals develop symptoms of neoplastic disease. This section covers those issues.

A. Sensitive Technology Detects MSI in PBLs of MMR Mutation Carriers

By cloning out and expanding individual lymphocytes, as well as by conducting PCR on highly diluted PBL DNA from HNPCC patients with MSI-high in their tumors, Parsons et al. [48] were able to detect mutant alleles at microsatellite loci in the normal tissues. Results came from family members containing certain truncating mutations in MMR genes $PMS2$ and $MLH1$. Later [49] it was indicated that for one of those patients—one having a nonsense mutation at codon 134 of the $PMS2$ gene—there was a reduced effect of the wild-type gene in controlling MSI. Because the product of $MLH1$ binds to the product of $PMS2$ as part of the normal function of this complex, truncating mutations in alleles of either of these were seen as participating in such an effect. This result suggested that for certain mutations a dominant negative (gain of function) or haploinsufficiency effect might result in MSI in constitutive or nontumor tissue. Therefore, identification of MSI in such cells could be an indication that the individual might carry such a mutant gene and therefore be susceptible to cancer. This very exciting finding had very little follow-up over the 10 years since the initial observation.

The preceding work suggested the possibility of addressing cancer family members whose tumor suppressor defect had not been identified with such a technology so that carriers of such a defect could be presymptomatically distinguished from those not carrying the mutation. The DNA dilution method introduced by Mockton and Jeffries [38], used in mice as described by Yao et al. [39], that yielded those initial results by Parsons et al. [48] in HNPCC patients was seen as a path to the sensitivity and quantification problems raised in evaluating MSI in the many HNPCC issues left unresolved. Some modifications were made [50]: use of fluorescently labeled probes and multiplexed loci for fragment analysis on a DNA sequencer [51] and employment of various quality-control [52] measures to guard against contamination and PCR artifact. In order to make the procedure more efficient, multiple single molecule PCRs, as practiced in the Arnheim laboratory [39], were not used; instead, each DNA pool was diluted to single genome equivalents (e.g., mean of 2 molecules/well). In doing that, it was necessary to describe a model in which the number of alleles in replicate pools were in a Poisson distribution, and in which particular allele frequencies constituted a fixed proportion of the total. Maximum likelihood estimates of the mean number of alleles in each pool and the frequencies of each allele were derived. The mutant frequencies were compared between groups for significance by using the arcsin transformed mutant frequencies and the bootstrap standard error. The procedure is called small-pool PCR (SP-PCR). By conducting it on multiple (approximately 100) small pools, if the frequency of mutant fragments is over 1% there is a high probability of trapping such fragments in some of the small pools. Such fragments within small pools are then no longer "overwhelmed" by the presence of the more frequent progenitors and can be identified and counted after amplification (see Fig. 47-2 for an example of what the chromatograms look like after amplifying the *DM1* microsatellite from the PBLs and tumor tissue of an HNPCC patient; from ref. [50]).

B. Analysis of CRC Material by SP-PCR

Six microsatellite loci were studied [50] for MSI in the constitutive tissue (PBLs and normal adjacent colon) and tumor of a 40-year-old HNPCC MSI-high patient who carried a germline null mutation for *MSH2* and was immunologically negative for the protein. These data were compared with similar tissues from a sporadic colon cancer patient having an MSI-high phenotype in his tumor due to methylated alleles of *MLH1*. The data are displayed in Table 47-1.

Results verified that loci seen to be unstable in MSI-high tumors by traditional PCR indeed had in excess of 25% mutant fragments; see the frequencies at each locus in the colon CA samples from both patients in Table 47-1. Supporting the initial observation of Parsons et al. [48], low, but significantly higher levels of MSI (0.06, the mean of all six loci) were detectable in the PBLs of the patient carrying the germline MMR mutation compared to the frequency in the PBLs of his or her age-matched control (0.01) ($p < 0.01$). The mean level of MSI in the PBLs of the sporadic patient was also elevated, but no higher than that observed in PBLs of his age-matched control. The results supported the notion that the presence of the germline MMR mutation contributed to the increased MSI in the PBLs of the mutation carriers. If verified on a wider scale, this procedure should help to identify individuals at risk where specific mutations in MMR genes have not been identified.

C. Mutation in Constitutive Tissue of Carriers of DNA Repair Mutations

HPRT mutation frequencies are increased in the PBLs of individuals homozygous for different rare genetic instability syndromes: xeroderma pigmentosum, ataxia telengiectasia, Bloom's, Werner's, and Fanconi anemia [53]. MSI has also been shown to exist in normal tissues of children with neurofibromatosis type 1 symptoms who are homozygous for mutations in the MMR gene *MLH1* [47, 54, 55]. MSI after traditional PCR was not reported in children with a similar syndrome but homozygous for a different MMR gene, *MSH2* [56–58]. This might reflect the relative impacts of *MLH1* versus *MSH2* on MSI, and the application of SP-PCR to the PBL DNA of *MSH2* homozygotes might quantify MSI.

However, the carriers of germline MMR mutations are more interesting (heterozygotes, not homozygotes). As indicated earlier, 6% MSI was identified in a multiple locus screen of PBL of a patient carrying a germline *MSH2* mutation [50]. Also using SP-PCR, 2.5–2.7% MSI at *BAT26* in the PBLs of the mother of one of the *MLH1* homozygous children discussed earlier was observed [47]. By using a completely different technique, MSI in the PBLs of MMR mutation carriers was observed by cloning out the *BAT26* alleles [59]. A total

CHAPTER 47 Microsatellite Instability in Cancer

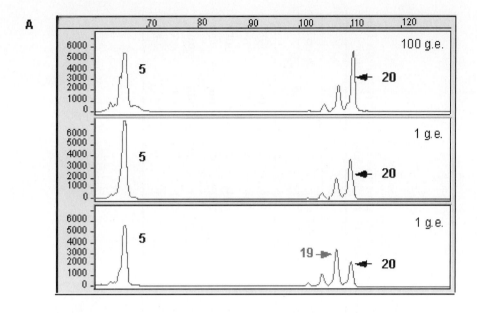

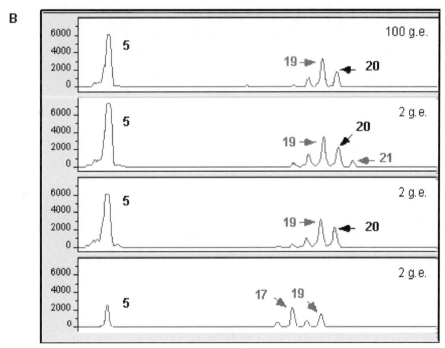

FIGURE 47-2 Chromatograms showing fluorescently labeled PCR products of the microsatellite repeat, *DM1*. Sizes (number of repeats) of progenitor alleles are labeled in black (5 and 20 repeats), whereas mutant alleles are labeled in red. (A) Tissue is the dissected normal colon from the MSI-high HNPCC patient. The top panel had over 100 ge of DNA amplified and indicates that the sample is from a heterozygous individual at this locus, 5 and 20 repeats. These peaks are clear and have the ever smaller "stutter" bands common to this analysis. The bottom two panels are two of the many small pools (<2 ge). Most pools had either one, the other, or both (as in the middle panel), or no progenitor fragments. In the bottom panel, in addition to the two progenitor fragments, a mutant fragment (19 repeats) at the size of the stutter band from the progenitor 20-repeat fragment is visible. (B) Tissue is colon tumor from the same patient. The top panel is a traditional PCR showing the progenitor fragments (5 and 20 repeats). In this case, the 19-repeat mutant is present in such high frequency as to be visible by traditional PCR. The bottom three panels are selected small pools (2 ge) where mutant fragments (17 and 21 repeats) are visible in addition to the common 19-repeat mutant and the progenitor fragments. The bottom panel shows that the 20-repeat progenitor fragment need not be present for the mutant 19-repeat fragment to be seen. See CD-ROM for color image. Reproduced with permission from ref. [49].

TABLE 47-1 Number of Alleles Screened and Mutant Frequencies at Six Microsatellite Loci Studied in Constitutive and Tumor Tissues from Two MSI-High Colon Cancer Patients[a]—One an HNPCC Carrying a Germline *MSH2* Mutation and the Other a Sporadic with *MLH1* m

Colon cancer		\multicolumn{12}{c}{Loci[b]}											
Patient	Tissue	*DM1*		*D2S123*		*D5S346*		*D17S250*		*BAT25*		*BAT26*		6 loci mean
		n	f	n	f	n	f	n	f	n	f	n	f	f
MSI-high, HNPCC	Control PBLs	462	<0.01	267	<0.01	381	<0.01	219	0.04	273	0.01	137	<0.01	0.01
	Patient PBLs	148	0.03*	241	0.08*	179	0.06*	164	0.11*	100	0.04	104	0.02	0.06*
	Patient Colon	201	0.03*	133	0.09*	194	0.01	135	0.09	138	0.04	278	<0.01	0.03*
	Colon CA	972	0.26*	169	0.31*	176	0.53*	136	0.59*	183	0.57*	278	<0.01	0.20*
MSI-high, sporadic	Control PBLs	86	0.09	128	0.06	179	0.04	150	0.04	153	0.03	126	<0.01	0.04
	Patient PBLs	165	0.08	130	0.04	123	0.00	144	0.04	137	<0.01	118	<0.01	0.03
	Colon CA	150	0.35*	114	0.25*	147	0.53*	130	0.46*	120	0.65*	141	<0.01	0.32*

[a] One was an HNPCC patient carrying a germline *MSH2* mutation and the other was a sporadic case with *MLH1* mutation.
[b] An asterisk denotes significantly different from control PBLs.

of 5.6% of *BAT26* clones had nonprogenitor fragments. No variant fragments were detected in clones from family members that did not carry an MMR mutation. Therefore, combination of those results with the original Parsons et al. [48] data shows that there may be a range of MSI in the PBL DNA of MMR gene mutations. This may be a way to identify individuals carrying MMR mutations that are predisposed to cancer before any such specific mutations are identified. Any proper evaluation of such data, however, will necessarily involve the use and understanding of the range in age-matched controls.

D. MSI Increases with Age in Normal Individuals

A general expectation exists that mutant cells will accumulate with age. One might expect that, as individuals age, various aspects of the DNA repair undergo mutation, leading to an increase in mutations in subsequent years (see review in ref. [60]). An increase of mutant genes in somatic cells with age has been demonstrated in mice and human cell culture systems [61, 62]. An increase in the frequency ($5-10 \times 10^{-6}$) of *HPRT* mutations with age in human lymphocyte DNA has also been observed [63]. From a different perspective, increased hypermethylation of *MLH1* with age has been reported [64].

The issue has some importance when considering the hypothesis that individuals carrying germline MMR mutations will have a higher frequency of MSI in their constitutive tissue than those without. In the single comparison made earlier [50], the 40-year-old *MSH2* mutation carrier had a significantly higher MSI level than his simultaneously run, age-matched control. This supports the hypothesis. However, the older CRC sporadic case with MSI-high due to *MLH1* methylation in his tumor had a somewhat elevated MSI in his PBLs (Table 47-1). Perhaps this is due to the fact that he is older. Therefore, to test that concept, SP-PCR using 6 microsatellite loci was directed (65) to PBL DNA from 17 "normal" blood bank donors varying in age from 20 to 67 years. MSI phenotypes were plotted against age in a regression analysis. A positive slope (Fig. 47-3) indicated a correlation between age and MSI phenotype ($p = 0.0006$). The mean weighted average mutant frequencies across all loci for all individuals in the age groups (0.009 for 20–30 years, 0.019 for 35–50 years, 0.034 for 60–70 years) were also significantly different from each other ($p < 0.01$). This was completely consistent with the results of Kendall et al. [66], who observed an MSI mutant frequency in young healthy children of 0.66–0.78%. The major significance of these results is not so much that MSI increases with age as that the identification of MSI levels in cells is capable of being monitored. Further data on MMR mutation carriers having higher MSI in their PBLs than age-matched controls may give a direct approach for identifying individuals at risk for cancer.

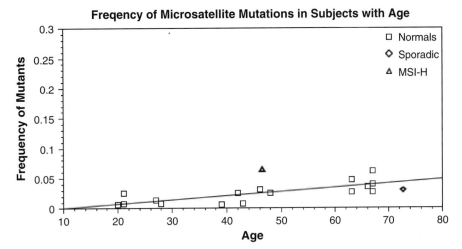

FIGURE 47-3 Mean frequency of mutant small-pool PCR fragments over 6 loci in the screen observed from the PBL DNA from 17 normal individual (squares) in three different age groups from 20 to 68 years. The linear regression line had a statistically significant positive slope ($p < 0.0006$), indicating that MSI increased with age. Triangle indicates the mean mutant frequency (0.06, $p < 0.01$) from age-matched controls in the DNA from the PBLs of an MSI-high HNPCC patient carrying an *MSH2* germline mutation. Mutant frequency in the PBLs of the older sporadic patient who had high-level MSI in tumor tissue is indicated by the triangle—a level clearly no higher than that of the age-matched cohort.

VIII. CONCLUSION

At this point in the evolution of the impact of genetics on human cancer, it can be said that the most important contributions are in the identification of individuals at risk so that they can be subject to prevention and early detection. The work of those who have studied HNPCC with MSI-high and the identification of germline mutations associated with the disease have made it possible to do just that in families where such genes have been shown to be segregating. Everyone in the field should be proud of that achievement. However, many less obvious genes and mutations are out there that render individuals susceptible to cancer. Researchers will continue to develop new technology to make the less obvious cases identifiable so that they may also benefit from the models developed from the MSI-high cases. Hopefully, the greater ability to detect and quantify MSI will also enable the resolution of problems in areas where MSI has been suggested but has not been so robust, such as in hematological [67, 68], prostate [69], and central nervous system [70] cancers.

Acknowledgments

Supported in part by NIH Grant CA-112588 and a gift from Mr. Kenneth D. Muller.

References

1. Richards, R. I., and Sutherland, G. R. (1992). Dynamic mutations: A new class of mutations causing human disease. *Cell* **70**, 709–712.
2. Fu, Y. H., Pizzuti, A., Fenwick, R. G., Jr., King, J., Rajnarayan, S., Dunne, P. W., Dubel, J., Nasser, G. A., Ashizawa, T., de Jong, P., Epstein, H., Perryman, B., and Caskey, T. C. (1992). An unstable triplet repeat in a gene related to myotonic muscular dystrophy. *Science* **255**, 1256–1258.
3. Manley, K., Shirley, T. L., Flaherty, L., and Messer, A. (1999). Msh2 deficiency prevents *in vivo* somatic instability of the CAG repeat in Huntington disease transgenic mice. *Nat. Genet.* **23**, 471–473.
4. Ionov, Y., Peinado, M. A., Malkhosyan, S., Shibata, D., and Perucho, M. (1993). Ubiquitous somatic mutations in simple repeated sequences reveal a new mechanism for colonic carcinogenesis. *Nature* **363**, 558–561.
5. Aaltonen, L. A., Peltomaki, P., Leach, F. S., Sistonen, P., Pylkkanen, L., Mecklin, J. P., Jarvinen, H., Powell, S. M., Jen, J., Hamilton, S. R., Peterson, G. M., Kinzler, K. W., Vogelstein, B., and de la Chapelle, A. (1993). Clues to the pathogenesis of familial colorectal cancer. *Science* **260**, 812–816.
6. Thibodeau, S. N., Bren, G., and Schaid, D. (1993). Microsatellite instability in cancer of the proximal colon. *Science* **260**, 816–819.
7. Fishel, R., Lescoe, M. K., Rao, M. R., Copeland, N. G., Jenkins, N. A., Garber, J., Kane, M., and Kolodner, R. (1993). The human mutator gene homolog MSH2 and its association with hereditary nonpolyposis colon cancer. *Cell* **75**, 1027–1038.
8. Kane, M. F., Loda, M., Gaida, G. M., Lipman, J., Mishra, R., Goldman, H., Jessup, J. M., and Kolodner, R. (1997). Methylation of the hMLH1 promoter correlates with lack of expression of hMLH1 in sporadic colon tumors and mismatch repair-defective human tumor cell lines. *Cancer Res.* **57**, 808–811.
9. Loeb, L. A. (2001). A mutator phenotype in cancer. *Cancer Res.* **61**, 3230–3239.
10. Lynch, H. T. (1993). Genetics, natural history, tumor spectrum, and pathology of hereditary nonpolyposis colorectal cancer: An updated review. *Gastroenterology* **104**, 1535–1549.
11. Lynch, H. T., and de la Chapelle, A. (1999). Genetic susceptibility to non-polyposis colorectal cancer. *J. Med. Genet.* **36**, 801–818.
12. Kinzler, D. W., and Vogelstein, B. (1996). Lessons from hereditary colorectal cancer. *Cell* **87**, 159–170.
13. Sankila, R., Aaltonen, L. A., Jarvinen, H. J., and Mecklin, J. P. (1996). Better survival rates in patients with MLH1-associated hereditary colorectal cancer. *Gastroenterology* **110**, 682–687.
14. Watanabe, T., Wu, T. T., Catalano, P. J., Ueki, T., Satriano, R., Haller, D. G., Benson, A. B., III, and Hamilton, S. R. (2001). Molecular predictors of survival after adjuvant chemotherapy for colon cancer. *New Engl. J. Med.* **344**, 1196–1206.
15. Watson, P., Lin, K. M., Rodriguez-Bigas, M. A., Smyrk, T., Lemon, S., Shashidharan, M., Franklin, B., Karr, B., Thorson, A., and Lynch, H. T. (1998). Colorectal carcinoma survival among hereditary nonpolyposis colorectal carcinoma family members. *Cancer* **83**, 259–266.
16. Risinger, J. I., Barrett, J. C., Watson, P., Lynch, H. T., and Boyd, J. (1996). Molecular genetic evidence of the occurrence of breast cancer as an integral tumor in patients with the hereditary nonpolyposis colorectal carcinoma syndrome. *Cancer* **77**, 1836–1843.
17. Watson, P., and Lynch, H. T. (1993). Extracolonic cancer in hereditary nonpolyposis colorectal cancer. *Cancer* **71**, 677–685.
18. Boland, C. R., Thibodeau, S. N., Hamilton, S. R., Sidransky, D., Eshleman, J. R., Burt, R. W., Meltzer, S. J., Rodriguez-Bigas, M. A., Fodde, R., Ranzani, G. N., and Srivastava, S. (1998). A National Cancer Institute Workshop on Microsatellite Instability for cancer detection and familial predisposition: Development of international criteria for the determination of microsatellite instability in colorectal cancer. *Cancer Res.* **58**, 5248–5257.
19. Vasen, H. F., Mecklin, J. P., Khan, P. M., and Lynch, H. T. (1991). The International Collaborative Group on Hereditary Non-Polyposis Colorectal Cancer (ICG-HNPCC). *Dis. Colon Rectum* **34**, 424–425.
20. Rodriguez-Bigas, M. A., Boland, C. R., Hamilton, S. R., Henson, D. E., Jass, J. R., Khan, P. M., Lynch, H., Perucho, M., Smyrk, T., Sobin, L., and Srivastava, S. (1997). A National Cancer Institute Workshop on Hereditary Nonpolyposis Colorectal Cancer Syndrome: Meeting highlights and Bethesda guidelines. *J. Natl. Cancer Inst.* **89**, 1758–1762.
21. Strong, L. C., and Amos, C. (1996). *In* "Cancer Epidemiology and Prevention," 2nd ed. (J. F. Fraumeni and D. Schottenfeld, Eds.), pp. 559–583. Oxford University Press, New York.
22. Hampel, H., Frankel, W. L., Martin, E., Arnold, M., Khanduja, K., Kuebler, P., Nakagawa, H., Sotamaa, K., Prior, T. W., Westman, J., Panescu, J., Fix, D., Lockman, J., Comeras, I., and de la Chapelle, A. (2005). Screen for the Lynch syndrome (hereditary nonpolyposis colon cancer). *New Engl. J. Med.* **352**, 1851–1860.
23. Collins, J. R., Stephens, R. M., Gold, B., Long, B., Dean, M., and Burt, S. K. (2003). An exhaustive DNA micro-satellite map of the human genome using high performance computing. *Genomics* **82**, 10–19.
24. Campuzano, V., Montermini, L., Molto, M. D., Pianese, L., Cossee, M., Cavalcanti, F., Monros, E., Rodius, F., Duclos, F., Monticelli, A., Zara, F., Canizares, J., Koutnikova, H., Bidichandani, S. I., Gellera, C., Brice, A., Trouillas, P., De Michele, G., Filla, A., De Frutos, R., Palau, F., Patel, P. I., Di Donato, S., Mandel, J.-L., Cocozza, S., Koenig, M., and Pandolfo, M. (1996). Friedreich's ataxia: Autosomal

recessive disease caused by an intronic GAA triplet repeat expansion. *Science* **271**, 1423–1427.
25. Rampino, N., Yamamoto, H., Ionov, Y., Li, Y., Sawai, H., Reed, J. C., and Perucho, M. (1997). Somatic frameshift mutations in the BAX gene in colon cancers of the microsatellite mutator phenotype. *Science* **275**, 967–969.
26. Peltomaki, P. (2001). Deficient DNA mismatch repair: A common etiologic factor for colon cancer. *Hum. Mol. Genet.* **10**, 735–740.
27. Woerner, S. M., Benner, A., Sutter, C., Schiller, M., Yuan, Y. P., Keller, G., Bork, P., Doeberitz, M. K., and Gebert, J. F. (2003). Pathogenesis of DNA repair-deficient cancers: A statistical meta-analysis of putative real common target genes. *Oncogene* **22**, 2226–2235.
28. Kim, N.-G., Rhee, H., Li, S. L., Kim, H., Lee, J.-S., Kim, J.-H., Kim, N. K., and Kim, H. (2002). Identification of *MARCKS*, *FLJ11383* and *TAF1B* as putative novel target genes in colorectal carcinomas with microsatellite instability. *Oncogene* **21**, 5081–5087.
29. Park, J., Betel, D., Gryfe, R., Michalickova, K., Di Nicola, N., Gallinger, S., Hogue, C. W. V., and Redston, M. (2002). Mutation profiling of mismatch repair-deficient colorectal cancers using an *in silico* genome scan to identify coding microsatellites. *Cancer Res.* **62**, 1284–1288.
30. Mori, Y., Yin, J., Sato, F. L., Sterian, A., Simms, L. A., Selaru, F. M., Schulmann, K., Xu, Y., Olaru, A., Wang, S., Deacu, E., Abraham, J. M., Young, J., Leggett, B. A., and Meltzer, S. J. (2004). Identification of genes uniquely involved in frequent microsatellite instability colon carcinogenesis by expression profiling combined with epigenetic scanning. *Cancer Res.* **64**, 2434–2438.
31. Jarvinen, H. J., Aarnio, M., Mustonen, H., Aktan-Collan, K., Aaltonen, L. A., Peltomaki, P., De La Chapelle, A., and Mecklin, J. P. (2000). Controlled 15-year trial on screening for colorectal cancer in families with hereditary nonpolyposis colorectal cancer. *Gastroenterology* **118**, 829–834.
32. Suter, C. M., Martin, D. I. K., and Ward, R. L. (2004). Germline epimutation of *MLH1* in individuals with multiple cancers. *Nat. Genet.* **36**, 497–501.
33. Frazier, M. L., Xi, L., Zong, J., Viscofsky, N., Rashid, A., Wu, E. F., Lynch, P. M., Amos, C. I., and Issa, J. P. (2003). Association of the CpG island methylator phenotype with family history of cancer in patients with colorectal cancer. *Cancer Res.* **63**, 4805–4808.
34. Benatti, P., Roncucci, L., Ganazzi, D., Percesepe, A., Di Gregorio, C., Pedroni, M., Borghi, F., Sala, E., Scarselli, A., Menigatti, M., Rossi, G., Genuardi, M., Viel, A., and Ponz De Leon, M. (2001). Clinical and biologic heterogeneity of hereditary nonpolyposis colorectal cancer. *Int. J. Cancer (Pred. Oncol.)* **95**, 323–328.
35. Park, J. G., Vasen, H. F., Park, Y. J., Park, K. J., Peltomaki, P., de Leon, M. P., Rodriguez-Bigas, M. A., Lubinski, J., Beck, N. E., Bisgaard, M. L., Miyaki, M., Wignen, J. T., Baba, S., Lindbolm, A., Madlenski, L., and Lynch, H. T. (2002). Suspected HNPCC and Amsterdam criteria II: Evaluation of mutation detection rate, an international collaborative study. *Int. J. Colorectal Dis.* **17**, 109–114.
36. Wiesner, G. L., Daley, D., Lewis, S., Ticknor, C., Platzer, P., Lutterbaugh, J., MacMillen, M., Baliner, B., Willis, J., Elston, R. C., and Markowitz, S. D. (2003). A subset of familial colorectal neoplasia kindreds linked to chromosome 9q22.2–31.2. *Proc. Natl. Acad. Sci. USA* **100**, 12961–12965.
37. Kolodner, R. D., Putnam, C. D., and Myung, K. (2002). Maintenance of genome stability in *Saccharomyces cerevisiae*. *Science* **297**, 552–557.
38. Monckton, D. G., and Jeffreys, A. J. (1991). Minisatellite "isoallele" discrimination in pseudohomozygotes by single molecule PCR and variant repeat mapping. *Genomics* **11**, 465–467.
39. Yao, X., Buermeyer, A. B., Narayanan, L., Tran, D., Baker, S. M., Prolla, T. A., Glazer, P. M., Liskay, R. M., and Arnheim, N. (1999). Different mutator phenotypes in *Mlh1* versus *Pms2* deficient mice. *Proc. Natl. Acad. Sci. USA* **96**, 6850–6855.
40. Prolla, T. A., Baker, S. M., Harris, A. C., Tsao, J. L., Yao, X., Bronner, C. E., Zheng, B., Gordon, M., Reneker, J., Arnheim, N., Shibata, D., Bradley, A., and Liskay, R. M. (1998). Tumour susceptibility and spontaneous mutation in mice deficient in Mlh1, Pms1 and Pms2 DNA mismatch repair. *Nat. Genet.* **18**, 276–279.
41. Hendriks, Y. M., Wagner, A., Morreau, H., Menko, F., Stormorken, A., Quehenberger, F., Sandkuijl, L., Moller, P., Genuardi, M., Van Houwelingen, H., Tops, C., Van Puijenbroek, M., Verkuijlen, P., Kenter, G., Van Mil, A., Meijers-Heijboer, H. Tan, G. B., Breuning, M. H., Fodde, R., Wijnen, J. T., Brocker-Vriends, A. H., and Vasen, H. (2004). Cancer risk in hereditary nonpolyposis colorectal cancer due to MSH6 mutations: Impact on counseling and surveillance. *Gastroenterology* **127**, 17–25.
42. Kolodner, R. D., Tytell, J. D., Schmeits, J. L., Kane, M. F., Gupta, R. D., Weger, J., Wahlberg, S., Fox, E. A., Peel, D., Ziogas, A., *et al.* (1999). Germ-line MSH6 mutations in colorectal cancer families. *Cancer Res.* **59**, 5068–5074.
43. Wu, Y., Berends, M. J. W., Mensink, R. G. J., Kempinga, C., Sijmons, R. H., van der Zee, A. G. J., Hollema, H., Kleibeuker, J. H., Buys, C. H. C. M., and Hofstra, R. M. W. (2000). Germline *hMLH3* mutations in patients with suspected HNPCC. *Am. J. Hum. Genet.* **67(Suppl.)**, 17.
44. Guo, H. H., and Loeb, S. A. (2003). Tumbling down a different pathway to genetic instability. *J. Clin. Invest.* **112**, 1793–1795.
45. Bender, H.-U., Almashanu, S., Steel, G., Hu, C.-A., Lin, W.-W., Willis, A., Pulver, A., and Valle, D. (2005). Functional consequences of *PRODH* missense mutations. *Am. J. Hum. Genet.* **76**, 409–420.
46. Lipkin, S. M., Rozek, L. S., Rennert, G., Yang, W., Chen, P. C., Hacia, J., Hunt, N., Shin, B., Fodor, S., Kokoris, M., Greenson, J. K., Fearon, E., Lynch, H., Collins, R., and Gruber, S. B. (2004). The MLH1 D132H variant is associated with susceptibility to sporadic colorectal cancer. *Nat. Genet.* **36**, 694–699.
47. Gallinger, S., Aronson, M., Shayan, K., Ratcliffe, E. M., Gerstle, J. T., Parkin, P. C., Rothenmund, H., Croitoru, M., Baumann, E., Durie, P. R., Weksberg, R., Pollett, R., Riddell, R. H., Ngan, B. Y., Cutz, E., Lagarde, A. E., and Chan, H. S. L. (2004). Gastrointestinal cancers and neurofibromatosis type 1 features in children with a germline homozygous MLH1 mutation. *Gastroenterology* **126**, 576–585.
48. Parsons, R., Li, G. M., Longley, M., Modrich, P., Liu, B., Berk, T., Hamilton, S. R., Kinzler, K. W., and Vogelstein, B. (1995). Mismatch repair deficiency in phenotypically normal human cells. *Science* **268**, 738–740.
49. Nicolaides, N. C., Littman, S. J., Modrich, P., Kinzler, K. W., and Vogelstein, B. (1998). A naturally occurring hPMS2 mutation can confer a dominant negative mutator phenotype. *Mol. Cell Biol.* **18**, 1635–1641.
50. Coolbaugh-Murphy, M., Maleki, A., Ramagli, L., Frazier, M., Lichtiger, B., Monckton, D. G., Siciliano, M. J., and Brown, B. W. (2004). Estimating mutant microsatellite allele frequencies in somatic cells by small-pool PCR. *Genomics* **84**, 419–430.
51. Canzian, F., Salovaara, R., Hemminki, A., Kristo, P., Chadwick, R. B., Aaltonen, L. A., and de la Chapelle, A. (1996). Semiautomated assessment of loss of heterozygosity and replication error in tumors. *Cancer Res.* **56**, 3331–3337.
52. Zhang, L., Leeflang, E. P., Yu, J., and Arnheim, N. (1994). Studying human mutations by sperm typing: Instability of CAG trinucleotide repeats in the human androgen receptor gene. *Nat. Genet.* **7**, 531–535.
53. Albertini, R. J., and Hayes, R. B. (1997). Somatic cell mutations in cancer epidemiology. *IARC Sci. Publ.*, 159–184.
54. Vilkki, S., Tsao, J. L., Loukola, A., Poyhonen, M., Vierimaa, O., Herva, R., Aaltonen, L. A., and Shibata, D. (2001). Extensive

somatic microsatellite mutations in normal human tissue. *Cancer Res.* **61**, 4541–4544.
55. Wang, Q., Lasset, C., Desseigne, F., Frappaz, D., Bergeron, C., Navarro, C., Ruano, E., and Puisieux, A. (1999). Neurofibromatosis and early onset of cancers in hMLH1-deficient children. *Cancer Res.* **59**, 294–297.
56. Bougeard, G., Charbonnier, F., Moerman, A., Martin, C., Ruchoux, M. M., Drouot, N., and Frebourg, T. (2003). Early onset brain tumor and lymphoma in MSH2-deficient children. *Am. J. Hum. Genet.* **72**, 213–216.
57. Ricciardone, M. D., Ozcelik, T., Cevher, B., Ozdag, H., Tuncer, M., Gurgey, A., Uzunalimoglu, O., Cetinkaya, H., Tanyeli, A., Erken, E., and Ozturk, M. (1999). Human MLH1 deficiency predisposes to hematological malignancy and neurofibromatosis type 1. *Cancer Res.* **59**, 290–293.
58. Whiteside, D., McLeod, R., Graham, G., Steckley, J. L., Booth, K., Somerville, M. J., and Andrew, S. E. (2002). A homozygous germ-line mutation in the human MSH2 gene predisposes to hematological malignancy and multiple cafe-au-lait spots. *Cancer Res.* **62**, 359–362.
59. Alazzouzi, H., Domingo, E., Gonzalez, S., Blanco, I., Armengol, M., Espin, E., Plaja, A., Schwartz, S., Capella, G., and Schwartz, S., Jr. (2005). Low levels of microsatellite instability characterize MLH1 and MSH2 HNPCC carriers before tumor diagnosis. *Hum. Mol. Genet.* **14**, 235–239.
60. Vijg, J. (2000). Somatic mutations and aging: A re-evaluation. *Mutat. Res.* **447**, 117–135.
61. Ono, T., Ikehata, H., Pithani, V. P., Uehara, Y., Chen, Y., Kinouchi, Y., Shimosegawa, T., and Hosoi, Y. (2004). Spontaneous mutations in digestive tract of old mice show tissue-specific patterns of genomic instability. *Cancer Res.* **64**, 6919–6923.
62. Geigl, J. B., Langer, S., Barwisch, S., Pfleghaar, K., Lederer, G., and Speicher, M. R. (2004). Analysis of gene expression patterns and chromosomal changes associated with aging. *Cancer Res.* **64**, 8550–8557.
63. Finette, B. A., Sullivan, L. M., O'Neill, J. P., Nicklas, J. A., Vacek, P. M., and Albertini, R. J. (1994). Determination of hprt mutant frequencies in T-lymphocytes from a healthy pediatric population: Statistical comparison between newborn, children and adult mutant frequencies, cloning efficiency and age. *Mutat. Res.* **308**, 223–231.
64. Nakagawa, H., Nuovo, G. J., Zervos, E. E., Martin, E. W., Jr., Salovaara, R., Aaltonen, L. A., and de la Chapelle, A. (2001). Age-related hypermethylation of the 5′ region of MLH1 in normal colonic mucosa is associated with microsatellite-unstable colorectal cancer development. *Cancer Res.* **61**, 6991–6995.
65. Coolbaugh-Murphy, M. I., Xu, J., Ramagli, S. S., Brown, B. W., and Siciliano, M. J. (2005). Microsatellite instability (MSI) increases with age in normal somatic cells. *Mech. Aging Dev.* **126**, 1051–1059.
66. Kendall, H. E., Vacek, P. M., and Finette, B. A. (2004). Analysis of microsatellite instability in children treated for acute lymphocytic leukemia with elevated *HPRT* mutant frequencies. *Mutagenesis* **19**, 409–412.
67. Teruya-Feldstein, J., Greene, J., Cohen, L., Popplewell, L., Ellis, N. A., and Offit, K. (2002). Analysis of mismatch repair defects in the familial occurrence of lymphoma and colorectal cancer. *Leuk. Lymphoma* **43**, 1619–1626.
68. Niv, E., Bomstein, Y., Yuklea, M., and Lishner, M. (2005). Microsatellite instability in patients with chronic B-cell lymphocytic leukaemia. *Br. J. Cancer* **92**, 1517–1523.
69. Chen, Y., Wang, J., Fraig, M. M., Metcalf, J., Turner, W. R., Bissada, N. K., Watson, D. K., and Schweinfest, C. W. (2001). Defects of DNA mismatch repair in human prostate cancer. *Cancer Res.* **61**, 4112–4121.
70. Alonso, M., Hamelin, R., Kim, M., Porwancher, K., Sung, T., Parhar, P., Miller, D. C., and Newcomb, E. W. (2001). Microsatellite instability occurs in distinct subtypes of pediatric but not adult central nervous system tumors. *Cancer Res.* **61**, 2124–2128.

Index

14-3-3 protein family, and SCA1, 344
4qter nuclear organization, and FSHD, 157

A

Acadian FRDA, 278
Acetazolamide, and SCA6, 382
Action tremors, 463, 464–465
ADCAs. *see* Autosomal dominant cerebellar ataxia
Adeno-associated virus serotype 2 (AAV2), 411
Adenovirus 12, and fragile site expression, 200
ADHD (attention deficit hyperactivity disorder), and fragile X syndrome, 166–167
Age of onset
    and allele length measurements, 545–546
    for DM1, 57
    for DM2, 59, 134–135
    for FSHD, 153
    for HDL2, 266*f*
    for Machado-Joseph disease, 364, 366
    for polyQ and polyAla diseases, 499
    and repeat expansion size, 517
    for SCA10, 340, 438*f*
    for SCA12, 463
    for SCA2, 352–354
    for SCA6, 380
    for SCA7, 402
Aggregation. *see* Polyglutamine aggregation
AKT signaling, and SCA1, 344–345
Alternative splicing
    disruption in myotonic dystrophy, 44–46
    misregulation and CELF proteins, 46–47
    misregulation and MBNL proteins, 46–47
    misregulation in myotonic dystrophy, 44*f*
    pathways, 44*f*
    and RNA-mediated disease process, 43–44
Alzheimer's disease
    and amyloid precursor protein (APP), 46
    and neurofibrillary tangles (NFTs), 58
    treatment with idebenone, 298
Amino acid repeats, 706
Amorphous aggregates, 524
Amplification of DNA by replication, 449
Amplifications at fragile sites and cancer, 200
Amyloid fibrils, 524
Amyloid-like aggregates, 524
Amyloid precursor protein (APP), 46, 143
    abnormal splicing in DM1, 62

Amylotrophic lateral sclerosis (ALS), SBMA misdiagnosed as, 212
Androgen insensitivity with SBMA, 212
Androgen receptor
    activation, 213–214
    function in nervous system, 214–215
    gene and protein structure, 214*f*
    ligands, 214
    repeat length effects on, 521–522
    and SBMA, 592
    structure, 213
Angelman's syndrome, 729
ANT1 gene, and FSHD, 156
Anti-androgens, and SBMA, 217
Anticipation, 24
    and age-of sampling effects, 546
    and CTG repeat instability, 57
    description, 56
    and Friedreich's ataxia, 311
    and Machado-Joseph disease, 366
    and microsatellite expansion diseases, 11–12
    molecular explanations for, 25
    and prevalence of DM1, 27–28
    and SCA12, 463
    and SCA2, 352–353, 356*f*
    and SCA7, 400, 402
AP endonuclease, 671
Aphidicolin, 660
    and fragile site expression, 199
Aphidicolin-sensitive common fragile sites, 197
Apoptosis
    in OPDM and polyQ diseases, 504
    role in glutamine repeat disorders, 592
Arrestins, 408
Ashizawa paradox, 546
Ataxia
    definition, 400
    and FXTAS, 168
    with Machado-Joseph disease, 364
    and SCA2, 354
Ataxin-1
    modifiers of phosphorylation, 345–347
    normal function, 340–341
    phorphorylation sites, 342–344
Ataxin-10 protein, 444
Ataxin-3, 366–368

Ataxin-7 (ATXN7)
　aggregation in nuclear inclusions, 391–392
　and cell type specificity with polyglutamine diseases, 410–411
　effect on TFTC/STAGA function, 390
　gene family, 389f
　incorporation into TFTC-type complexes, 390
　normal function, 389, 408–410
　paralogs, 389–390
　proteolytic processing of, 392
　and spinocerebellar ataxia type 7, 387
　stabilization, 392
　subcellular localization, 388–389
ATP2A1 genes, splicing abnormalities, 143
ATR deficiency, and fragile site expression, 198
(ATTCT)$_n$ repeats, 28, 434
　expansion identification, 434–436
　instability in lymphoblastoid cell lines (LBCLs), 440–441
　intergenerational changes in size, 437–438
　repeat-size mosaicism, 439–440
　somatic and germline instability, 439
　structure, 436
(ATTCT)$_n$·(AGAAT)$_n$ repeats
　and DNA unwinding, 450–454
　and DNA unwinding elements, 448–449
　instability in E. coli, 454–455
　models of complex expansion mutations, 455–456
Autism and fragile X syndrome, 166–167
Autosomal dominant cerebellar ataxias (ADCAs)
　identification, 22
　spinocerebellar ataxia type 7, 388–389
Axonal transport and Huntington's disease, 229

B

B-family units of PP2A
　expression, 468
　structure, 469
Base-base mismatches
　in human cell extracts, 652f
　processing vs. slipped (CTG)·(CAG) and random-sequence heteroduplexes, 659–660
　repair and human disease, 661–662
　repair and MMR and NER proteins, 660
　repair efficiencies, 655
　in slipped DNA, 648f
Base excision repair (BER), 649, 685f, 686
　and repeat instability, 671–675, 685f
　and triplet repeat instability, 668–671
Beta-blockers, 463
Beta-lactams, and Huntington's disease, 257
beta-sheet structures
　and polyglutamine aggregates, 525
　in polyQ and polyAla diseases, 498
Bifunctional DNA glycosylases, 671
Blank allele, 119, 121f
Blepharophimosis-ptosis-epicanthus inversus syndrome, 496
Bloom's syndrome, 720

Brain
　androgen receptor expression, 214
　DMPK levels, 105
　and fragile X syndrome premutation carriers, 167
　of Huntington's disease patients, 226
　loss of volume with FXTAS, 170
　neuroanatomy and physiology with fragile X syndrome, 179–180
　neuropathology with HDL2, 264
　neuropathology with SCA2, 355–356
　neuropathology with SCA6, 380–381
　postmortem analysis with Huntington's disease, 256
　synaptic plasticity and DMPK, 106–107
Brain imaging
　and Friedreich's ataxia, 279, 282–283
　MRI results for Machado-Joseph disease, 365
　MRI results for SCA6, 382
　MRI results for SCA7, 400
　MRI results for SCA8, 419
　and SCA12, 464
　and SCA17, 479
BRCA1, and fragile site expression, 198–199
Breakpoint junctions, 726–727
Breast cancer, 201
Bromodeoxyuridine-sensitive fragile sites, 197
Burkitt's lymphoma, 726

C

C2-8, 232
Caffeine, and fragile site expression, 198
CAG·CTG repeats, 475
CAG repeats
　aggregate varieties, 524
　base excision repair and repeat instability, 671–675
　expansions and Pol-beta gap filling, 675f
　and Huntington's disease, 223, 251
　in Machado-Joseph disease, 365–366
　mismatch repair and microsatellite instability, 4604
　in polyglutamine diseases, 488
　and SBMA, 4, 213
　and SCA6 pathogenesis, 379–381
　and SCA7, 399
　in spinocerebellar ataxia type 7, 388
CAG slip-outs, 647
　repair efficiency, 646, 655
CAG·CTG repeats
　with HDL2, 261, 263
　instability and gender of transmitting parent, 571–572
　loci, 538
　number of disorders, 538
　tissue-specific somatic instability, 572–574
Calcium channel
　and polyglutamine diseases, 492
　and SCA6 pathogenesis, 380
Cancer. see also Hereditary nonpolyposis colon cancer (HNPCC)
　Burkitt's lymphoma and oligo(R-Y) tract, 726
　downstream impact of microsatellite instability, 4605

drug resistance in tumor cells, 449
  follicular lymphomas, 727
  and fragile site expression, 200–203
  WWOX as tumor suppressor, 202
Canine progressive myoclonus epilepsy type 2, 620
Cannabinoids, 231
Cardiac troponin T RNA splicing, 62, 124, 143
Cardiopulmonary problems
  and cTNT splice variants, 45
  with DM1, 23, 58
  with DM2, 60, 123, 133
  and DMPK function in heart, 106
  and DMPK loss in mice, 68–70
  and Friedreich's ataxia, 280, 281–282, 299, 322
  with SCA7, 400
  and SIX5 deficiency, 71
Carnitine, and Friedreich's ataxia, 301
Caspase inhibitors, 231
Cataracts
  with DM1, 58
  with DM2, 122–123, 133
  and SIX5 deficiency, 29, 72, 123, 142
CBP (CREB-binding protein), and mutant ataxin-7, 392
CCTG repeat expansion
  with DM2, 59, 116
  size and disease severity, 122
  somatic instability, 117
(CCTG)·(CAGG) repeats, 604–607
(CCTG)$_n$ repeat expansion, 21
  mechanism of instability, 26–27
CCUG repeats
  and DM2, 42
  toxicity vs. CUG repeats, 63
Celastrol, 232
  and Huntington's disease, 227
CELF proteins
  and misregulated alternative splicing, 46–47
  and RNA splicing regulation, 47–48
Cell specificity, 505–506
Central nervous system. see CNS dysfunction
  and Friedreich's ataxia, 279–280, 282–283
Cerebellum
  and mutant ataxin-7, 392
  repeat stability, 547
CGG repeat expansion
  fly (drosophila) studies, 189
  and fragile X syndrome, 4, 165
Chaperones
  and drosophila polyglutamine models, 592–593
  and mutant ataxin-7 expression, 392–393
  in OPDM and polyQ diseases, 502–503
  and polyglutamine aggregate formation, 530
  and polyglutamine diseases, 491
  as therapeutic target in Huntington's disease, 226–227
Charcot-Marie-Tooth type 1 disease, 729
Chicken foot structure, 684
Chloride channel splicing alterations, 125, 143

Chromosomal fragile sites
  and cancer, 200–203
  and CGG·CCG repeats, 625–626
  common, 197–203
  components in expression, 199f
  folate-sensitive rare fragile sites, 196–197
  genes at common fragile site loci, 201–203
  historical aspects, 196
  rare, 196–197
  rare and common, 195–196
Chromosome 12q24.1, 352
Chromosome 15q21-24, 22
Chromosome 15q21-q24, 131
Chromosome 16p, 131
Chromosome 19q13.3, 56
  structure of *Dm-1* locus, 100
Chromosome 22q13-qter, 434
Chromosome 3p12-21.1, 402
Chromosome 3q21, 59
Chromosome 4q, and FSHD, 155
Chromosome 4q35, and FSHD, 153
Chromosome 9q13, and Friedreich's ataxia, 301–302
Circadian rhythm, and fragile X syndrome, 185
cis Effects of CAG expansion in SCA7, 403
cis Effects of CTG expansion, 56, 63–64, 547–548, 567
Cis modifiers of repeat instability, 567–571
Cleidocranial dysplasia, 496
Clonazepam, and Machado-Joseph disease, 372
CNS dysfunction
  and androgen receptor expression, 214–215
  and ataxin-7, 405
  with DM1, 58
  with DM2, 60
  with DM2 and DM1, 123
  with fragile X syndrome, 187–188
  and glutamate, 251–252
Coding repeat diseases, 488
Coenzyme Q, and Huntington's disease, 258
Coenzyme Q10, and Friedreich's ataxia, 300–301
Cognitive dysfunction
  and amyloid precursor protein (APP), 46
  dementia with SCA2, 355
  with DM1, 23
  and fragile X syndrome, 166–167
  and FXTAS, 168
  and NMDARs, 46
  with SCA12, 463
  and tau splicing abnormalities, 45
Colon cancer. see also Hereditary nonpolyposis colon cancer (HNPCC)
  types, 4604
Common fragile sites, 195–196
  cis-acting elements, 198
  mechanism of cytogenetic formation, 197–203
  trans-acting factors, 198–199
Cone opsins, and SCA7 pathogenesis, 393
Congenital central hypoventilation syndrome, 496

Congenital myotonic dystrophy, 24
   absence in mouse studies, 49
   and anticipation, 27–28
   molecular explanations for, 25
   severity and expansion length, 126
Congo red, and Huntington's disease, 240
Creatine, and Friedreich's ataxia, 301
Creatine kinase, and SBMA, 212
CREB binding protein, 491
Creutzfeldt-Jakob disease, 475
CRIK proteins, 84, 85f
Cruciform structures, 720
   adopted by triplet sequences, 719f
Crx, and SCA7 pathogenesis, 394
CTG expansion
   cis effects at DM1 locus, 63–64
   effects and RNA mechanism of DM1 pathology, 74f
   and heterochromatin spread, 74
   and SCA8, 418–419
CTG repeats
   and DM1, 56
   location, 56
   mechanism of instability, 57
CTG slip-outs, 647
   repair efficiency, 646, 655
CTG*CAG expansions
   caused by gene conversion, 599–600
   instability and mismatch repair (MMR), 11
   instability and repair proteins, 660–661
   intermolecular recombination, 602, 604
   interruptions and repeat stability, 724
   intramolecular recombination, 602
   orientation and repeat stability, 724
   and plasmid DNA instability, 724
   processing by human cell extracts, 654–662
   processing vs. base-base mismatches and random-sequence heteroduplexes, 659–660
   recombination capacity, 600–602
   role of MMR proteins, 653–654
   and role of MSH2 and MSH3, 652
$(CTG)_n$ repeat expansion, 21
   effect on adjacent genes, 29
   mechanism of instability, 26
   position within DMPK gene, 38
CUG-BP
   in normal and DM skeletal muscle, 144
   overexpression in DM1, 143
   overexpression in DM1 and DM2, 124–125
CUG BP protein, 62
   overexpression in DM1, 62
CUG repeat expansion, 62–63
   and DM phenotype, 143
   and DM1 molecular pathophysiology, 80–81
   effects on intracellular transcript localization, 40
   mechanistic basis of toxicity, 61–62
   overexpression and myoblast fusion, 40
   proteins interacting with, 62–63
   and RNA-mediated disease process, 39
   toxicity vs. CCUG repeats, 61, 63
   transgenic mouse studies of overexpression, 40–42, 49

D

D4Z4 repeat
   and FSHD, 153, 154–155
   hypomethylation with FSHD, 156–159
Databases for rearrangement breakpoints and mutations, 727
Deletion intermediates, 646–653, 654
   formation, 647f
Deletion-inversion breakpoints, 724–725
Deletions at fragile sites, and cancer, 200
Dentatorubral pallidoluysian atrophy (DRPLA). *see* DRPLA
Diabetes mellitus, 280, 282
Diagnostic methods for DM2, 137–138, 140f, 141f
Dideoxythymidine, 660
Dinucleotide repeats, 13
Distamycin A-sensitive fragile sites, 197
Dm-1 locus, 100
DM1
   aberrantly spliced genes, 143
   age of onset and disease course, 57
   allele length measurements, 545–546
   and alternative RNA splicing, 108–109
   anticipation, 24
   chromatograph of PCR products of microsatellite repeat, 743f
   clinical presentation, 22–23, 57–58
   comparison of symptoms to DM2, 134t
   congenital form, 24
   evidence against conventional mechanisms for genetic dominance, 38–39
   gene splicing abnormalities, 143
   genetics, 56–57, 80–81
   genetics and etiology, 61
   and haploinsufficiency of DMPK, 109
   identification, 21
   local chromatin perturbation and DMPK, 108
   locus specific effects, 63–64
   molecular parallels with SCA8, 428
   multisystem phenotype, 22–23
   mutation identification and mapping, 24–25
   as noncoding repeat disease, 488
   origin of the mutation, 27
   pathogenic mechanisms, 28–29, 80f
   prevalence, 27
   recombination capacity of repeat expansion, 601–604
   repeat location, 619
   RNA gain-of-function model, 418
   role of genetic recombination, 599
   somatic mosaicism, 541, 564–565
   somatic mosaicism and disease pathology, 546–551
   summary, 5

Index

timing of somatic repeat instability, 577
triplet repeat sequence, 722–726
DM2
  age of onset, 134–135
  age of onset and disease course, 59
  clinical presentation, 23, 59–60
  comparison of symptoms to DM1, 134t
  diagnostic methods, 117, 119–120, 137–138, 140f, 141f
  disease severity and size of mutant expansion, 135
  evolutionary conservation of the DM2 repeat, 137
  expansion region, 117f
  expansion repeat, 4
  gene splicing abnormalities, 143
  genetics, 59, 80–81, 116, 133, 136–142
  genetics and etiology, 61
  genotype-phenotype correlation, 120–121
  haplotype analysis, 116–117, 138f
  haplotypes, 102f
  homozygosity for DM2 mutation, 135
  identification, 132–133
  instability of CCGT expansion, 120
  intergenerational changes, 120–121
  and IR (insulin receptor) splicing abnormalities, 45
  linkage disequilibrium around the mutation, 137
  molecular pathogenesis, 29, 142–144
  mouse models, 144
  muscle biopsy data, 135–136
  mutation identification, 24
  as noncoding repeat disease, 488
  origin of the mutation, 137–138
  population studies, 137–138
  premutation allele, 119f
  prevalence, 27
  recombination properties of tetranucleotide repeats, 604–606
  repeat characteristics, 136–137
  repeat location, 619
  repeat tracts in different species, 119f
  RNA gain-of-function model, 418
  somatic mosaicism, 117
  summary, 5
  symptoms, 133
  and untranslated CCUG repeats, 42
  variability of phenotypic expression, 135
DMPK
  activators, 90f
  and the actomyosin cytoskeleton, 91
  AGC serine/threonone protein kinase group, 84, 85f
  alternative splicing, 79, 81–84, 101
  amino-terminal region, 100–101
  c-terminal tails, 88–89
  carboxy-terminal region, 101, 106
  catalytic domain, 101
  coiled-coil region, 87–88, 101
  in cytosol, 106
  and depolarization-mediated calcium efflux from SR, 66
  Dm-1 pathogenesis, 108–109
  in endoplasmic reticulum, 106
  and etiology of DM1, 64–70
  family members, 90f, 103–104
  function in heart, 106
  functional biochemical properties, 101–103
  haploinsufficiency, 109, 123
  homodimerization through coiled-coil region, 102
  human amino acid sequence, 100f
  inactivation in mice, 65
  interaction with other regulatory proteins, 102
  and ion homeostasis, 89–91
  isoforms, 80–81, 101
  loss of function and pathogenesis, 28–29
  in mitochondria, 106
  overexpression in Tg26-hDMPK model, 92
  phosphorylation of myosin phosphatase target subunit 1, 103
  phosphorylation of phospholamban, 103
  phosphorylation of serum response factor, 103
  role of N-terminal leucine-rich domain, 84–85
  role of serine/threonine protein kinase domain, 86–87
  site of genetic mutation for DM1, 21
  structural domains, 100–101
  subcellular localization, 88f, 105–106
  substrate specificity, 103
  substrates, 90f
  substrates and function, 89–92
  and synaptic plasticity in brain, 107
  Tg26-hDMPK, 92
  tissue expression, 104–105
  tissue expression and *in situ* localization, 81–84
  transcription and repeat instability, 694
  transgenic mouse studies, 92
  VSGGG motif, 87, 101
DMPK haploinsufficiency, 38–39
  and DM1 pathogenesis, 108f, 142
DMPK loss
  and cardiac conduction disorders, 68–70
  and DM1 pathology, 70f
  and hippocampal function, 70
  and skeletal muscle effects of DM1, 66–68
DMPK mRNA gain of function, 28
DMPK mutant RNA, 42–43
DMPK RNA expression vs. ZNF9 RNA, 61
DMWD gene, 142
DMWD transcript levels, 64
DNA duplexes and gene expression, 626
DNA expansion. *see* Repeat expansions
DNA instability and cancer, 200–203
DNA lesion recognizing proteins, 651f
DNA ligase and ssDNA break repair, 670
DNA repair. *see* Repair
DNA replication slippage, 549–550
DNA structures, 620–622
  breakpoint junctions, 726–727
  and human disease, 726–730
  segmental duplications, 727–729
DNA unwinding, 448–449
  and SCA10 repeats, 450–454

Dopaminergic agents, and Machado-Joseph disease, 372
Double-strand break repair, 7–10, 684
    and non-B-DNA structures, 725–726
    pathways, 729–730
Downbeat positioning nystagmus, and SCA6, 381
Doxycycline
    and Huntington's disease, 240
    and OPDM, 505
Drosophila models
    homologues for human genes, 588
    of Huntington's disease, 232, 590
    Huntington's disease (HD) and axonal transport, 229
    Huntington's disease (HD) and rapamycin, 228
    of Machado-Joseph disease, 369, 589–590
    neurodegeneration and polQ protein expression, 234f
    neurodegeneration studies, 588
    nontoxicity of $(CUG)_n$ repeat RNAs, 144
    overexpression of Hsp70, 226
    of polyglutamine diseases, 588–592
    polyglutamine modifiers, 592–593
    pure polyglutamine models, 590–591
    of SBMA, 591–592
    of SCA1, 341, 345, 591
    of SCA8, 428–429
    of SMBA, 215
    SUMOylation, 492
    WWOX gene studies, 202
DRPLA
    and CAG expansions, 657
    Drosophila homologues of genes, 588
    identification, 22
    mutation identification, 24
    pathogenic mechanisms, 28
    and repeat instability, 646, 661–662
    size of expanded repeats, 26
    somatic mosaicism, 542
    summary, 5
dsRNA, cellular defense against, 711
DUE (DNA unwinding elements), 448–449
DUX4 protein, and FSHD, 155–156
Dystroophia myotonia, 22

E

E. coli studies
    cell death and negative supercoils, 721
    CTG expansion, 57
    gene conversion and repeat expansions, 599–600
    recombination and repeat instability, 8–10, 683
    recombinational potential of different TRSs, 603
    ssDNA break repair and repeat instability, 668
e-motif structures, 620, 621f, 719f
Endocrine pathology, 58
    with DM2, 60
    with DM2 and DM1, 123
    with SBMA, 212
EPI-A0001, 301
EPM1, 618
    repeat location, 619

Epstein-Barr virus, 543
Error-prone repair, 655
    mechanisms, 657–659
    of random-sequence heteroduplexes, 659
    of slipped (CTG)·(CAG) repeat expansions, 659
Escaped repair, 649–650, 655
    mechanisms, 657, 658f
Eukaryotic DNA replication, 449
Exons, 43
Expansion intermediates, 646–653, 654
    formation, 647f
Eye pathology
    with DM1, 58
    with DM2, 60
    with Machado-Joseph disease, 364
    retinal degeneration with SCA7, 407–408, 410f
    with SCA7, 400–401
    slowed saccades with SCA2, 351

F

FA. see Friedreich's ataxia
Facioscapulohumeral Muscular Dystrophy. see FSHD
Familial adenomatous polyposis (FAP), 783
Fast skeletal muscle troponin T splicing, 143
FEN-1, 623–624, 669, 685
FHIT gene, 201
FMR1
    knockout mice studies, 177–185
    paralogs, 184–185
    repeat location, 619
    role of genetic recombination, 599
    schematic representation, 178f
    structural role of interruptions, 709, 710f
FMRP
    and fragile X syndrome, 176
    and hippocampal-related memory loss in Fmr1 knockout mice, 180–181
FNCA, and fragile site expression, 199
Folate-senitive rare fragile sites, 196–197
Follicular lymphomas, 727
FOR gene, 201–202
FRA-associated genes, 201f, 202
FRA4F fragile site, 202
FRA6E fragile site, 202
Fragile sites. see Chromosomal fragile sites
Fragile X associated premature ovarian failure, 619
Fragile X-associated tremor/ataxia syndrome (FXTAS)
    clinical features, 168
    diagnostic criteria, 168t
    epidemiology, 168–170
    molecular pathogenesis, 170–171
    neuropathology, 170
    overview, 167–168
    screening studies, 169t
Fragile X syndrome
    animals used for studying, 177
    anticipation, 24
    characteristics, 175–177

clinical involvement in premutation carriers, 167
clinical presentation, 22, 166–167
and hairpin structures, 719
identification, 22
loss of function and pathogenesis, 28
macroorchidism, 178–179
medications for, 167
molecular pathophysiology, 142
murine model, 177
mutation identification, 24
as noncoding repeat disease, 488
premutation carriers, 176
summary, 5
Fragile X syndrome fly studies, 185–190
behavior phenotypes, 187
biochemistry, 189
FXTAS model, 189
modifying phenotypes with genes and drugs, 188–189
neuronal phenotypes, 187–188
Fragile X syndrome mouse studies
behavior, 181–183
brain neuroanatomy and physiology, 179–180
CGG repeat instability, 183–184
environment effects on FMRP levels, 183
Fmr1 knockout mice, 177–178
LTP/LTD and hippocampal-related memory deficits, 180–181
macroorchidism, 178–179
structural abnormalities, 180
Fragile XE mental retardation
as noncoding repeat disease, 488
summary, 5
Frataxin, 297–298
expression and expanded GAA repeat, 288–289
point mutations, 285
FRDA gene
expression, 284–285
mapping and cloning, 284
mutations in Friedreich's ataxia, 286–287
structure, 284
FRDA1. *see also* Friedreich's ataxia
pathogenic mechanisms, 28
FRG1 gene, and FSHD, 156
FRG2 gene, and FSHD, 156
Friedreich's ataxia (FA). *see also* GAA repeats
age of onset, 290*f*
and *Alu* elements, 306
borderline alleles, 315
clinical presentation, 22, 280–281, 306, 321
conditional mouse models, 322–323
diagnostic criteria, 278
epidemiology, 278–279
expression of FRDA gene, 284–285
founder effect, 307
FRDA upregulation, 301–302
GAA-TR allele, 306–309
GAA-TR allele length and disease severity, 306
(GAGGAA) hexanucleotide interruption, 315
gene identification, 297–298

geographic distribution and population genetics, 309–310
identification, 22, 277
intergenerational instability of GAA-TR alleles, 311–312
laboratory investigations, 282
mapping FRDA gene, 284
molecular pathophysiology, 142
mouse models, 322–324
multistep hypothesis for allele class generation, 309
mutation identification, 24
neuroimaging, 282–283
neurophysiological investigations, 283
as noncoding repeat disease, 488
onset, 280
pathology, 279–280
phenotype-genotype correlations, 290–291
prevalence, 278
prognosis, 283–284
recombination of GAA·TTC repeats, 606–607
reproduction of molecular defect in mice, 324
somatic instability of GAA-TR alleles, 312–316
structure of FRDA gene, 284
summary, 5
therapeutic approaches, 291
treatments, 298–301
and triplex DNA, 720
uniqueness of GAA-TR sequence on FRDA locus, 316
FSH levels
and DM2, 123
and SIX5 deficiency, 72–73
FSHD
candidate region genes, 155–156
chromatin remodeling, 156–159
clinical characteristics, 152
D4Z4 rearrangement, 155
description, 151
disease models, 158*f*
genetic diagnosis, 154–155
genetic/linkage heterogeneity, 154
identification, 151–152
linkage analysis, 152–154
muscle biopsy characteristics, 152
myoblast studies, 159
primate specificity, 159
repeat expansion, 4
FXTAS (Fragile X-associated tremor/ataxia syndrome)
fly (drosophila) studies, 189
mouse model, 184
pathogenic mechanisms, 28
repeat location, 619
RNA gain-of-function model, 418

G

G-tetraplex structures, 621*f*
GAA repeats
detection and diagnostic value, 286–287
expansion, 285–286

GAA repeats (*Continued*)
  and Friedreich's ataxia, 278–279, 302
  instability in myotonic dystrophies, 287
  origin, 287–288
  pathogenic mechanisms, 288–290
GAA-TR allele
  intergenerational instability, 311–312
  length and Friedreich's ataxia severity, 306
  origin and evolution at FRDA locus, 306–309
  somatic instability, 312–316
  uniqueness to FRDA locus, 316
GAA·TTC repeats
  and Friedreich's ataxia, 327
  structure-dependent recombination activity, 606–607
  tissue-specific somatic instability, 572–574
Gain-of-function model, 12–13, 28
  for polyglutamine diseases, 489, 518
  for polyQ and polyAla diseases, 500
  for SCA7, 409–410
Gap repair, 669
Gastric carcinoma, 201
Gastrointestinal tract involvement, 58
Gender
  and FSHD, 155
  and symptoms of fragile X syndrome, 166
Gender of parent
  and congenital myotonic dystrophies, 24
  and intergenerational repeat instability, 564, 571–572
  molecular explanations for effects, 25
  and repeat instability, 4
  and timing of repeat instability, 576
Genes, at common fragile site loci, 201–203
Genetic alterations and diseases, 728t
Genetic anticipation. *see* Anticipation
Genetic recombination and repeat instability, 598–599
Germ cells
  development, 682f
  and repeat expansions, 681–682
Glucose tolerance, and SBMA, 212
Glutamate
  and excitotoxicity, 252–254
  and neurotransmission, 251–252
  receptor families and subtypes, 252f
Glutamine, chemical and physical properties, 518–519
Glutathione peroxidase mimetics, and Friedreich's ataxia, 301
GRID2 gene, 202
Gross Rearrangement Breakpoint Database, 727

H

H-DNA, 637. *see* Triplex structures
Hairpin structure, 622
  adopted by triplet sequences, 718–719
  and $(CCTG)_n$ instability, 27
  DNA-mediated effects, 626
  repair efficiency, 655
  in RNA, 622
  RNA-mediated effects, 626–629
  in slipped DNA, 647–648, 648f
  and somatic mosaicism, 551
  of trinucleotide repeats, 707–708
Hand-foot-genital syndrome, 493
Haploinsufficiency
  definition, 38
  of DMPK, 38–39, 80–81, 108f, 123, 142
  of SIX5, 80–81, 123
  of ZNF9, 145
Haw River syndrome, 5
HD. *see* Huntington's disease
HDL2 (Huntington's disease-like 2)
  CAG·CTG repeat expansions, 263
  clinical presentation, 263
  discovery, 261–262
  disease pathology, 620
  distinguished from polyglutamine diseases, 266
  epidemiology, 264–265
  and intranuclear protein aggregates, 269–270
  locus and junctophilin-3, 266–269
  and neuroacanthocytes, 263–264
  neuropathology, 264
  phenotype-genotype relationship, 265
  prevalence, 265
  and toxic transcripts, 269
Heart. *see* Cardiopulmonary problems
Heat shock inducers
  and Huntington's disease, 227
  and SBMA, 592
Heat shock proteins, and SCA17, 480
Hereditary nonpolyposis colon cancer (HNPCC), 737
  microsatellite-high, 739–740
  and microsatellite instability, 738–739
  microsatellite-low, 740
  polymorphisms of unknown significance, 741
Heterochromatin, 74
Heteroduplexes, 647
  of $(CTG)_x \cdot (CAG)_y$, 647
  processing vs. slipped (CTG)·(CAG) and base mismatches, 659–660
  random-sequence in human cells extracts, 652f
  repair, 646–653, 649–650
  repair of DNA loops, 650
  structure, 648f
Hippocampus
  and DMPK loss in mice, 70
  expression of DMPK in rats, 105
  with fragile X syndrome, 179
  long-term potentiation (LTP) of Fmr1 knockout mice, 180–181
  and mutant ataxin-7, 392
Histone deacetylase inhibitors, 553
  and Huntington's disease, 230
  and polyglutamine diseases, 411
  and SBMA, 217
HNPCC (hereditary nonpolyposis colorectal carcinoma), 650–651
Holliday junctions, and slipped DNA formation, 647
Holoprosencephaly, 496

Homoduplex slipped DNAs, 647
HPRT minigene, 698
HSA transgenic mice, 40–42
   and SERCA gene expression, 46
Hsp40 chaperones
   and Huntington's disease, 226
   and polyglutamine diseases, 592
Hsp70 chaperones
   and Huntington's disease, 226
   and polyglutamine diseases, 592
Human Gene Mutation Database, 727
Huntingtin
   and axonal transport defect in HD, 229
   Drosophila homologue, 588
   N-terminal fragments, 225–226
   repeat coil structure, 520
   and SCA7 pathogenesis, 394
   and transcription regulation, 411
Huntington's disease (HD), 391
   alleviating transcriptional repression, 240–241
   and CAG expansions, 657
   calcium signaling disruption and neurotoxicity, 253, 257*f*
   clinical presentation, 22
   drosophila models, 590
   and excitotoxicity, 254–255
   genetic models, 234–235
   glutamate and excitotoxicity, 252–254
   glutamate and neurotransmission, 251–252
   glutamate receptor signaling, 257*f*
   identification, 22
   and inclusion bodies, 224
   interpretation of study results, 240–241
   length of polyglutamine tract and age of onset, 223
   mitochondria and excitotoxicity, 253–254, 256
   modulating polyglutamic aggregation, 240
   mutation identification, 24
   neurotoxic lesion models, 233–234
   NMDA receptors and calcium-induced cytotoxicity, 252–253
   preclinical testing in mouse models, 235, 240–242
   and proteolytic cleavage and turnover, 410–411
   published therapeutic approaches in mammalian models, 236–239
   and repeat instability, 646, 661–662
   role of environment, 235, 240
   role of InsP3R1 receptors, 255–256
   role of mGluR5 receptors, 255–256
   size of expanded repeats, 26
   somatic mosaicism, 541–542
   somatic mosaicism and disease pathology, 546–547
   summary, 5
   therapeutic approaches, 256–258
   tissue specificity of somatic mosaicism, 543–544
   treatment with idebebone, 298
   YAC72 transgenic mouse models, 254
Huntington's disease (HD) experimental therapeutics
   invertebrate models, 232–233
   mammalian cell culture, 231–232
   rodent models, 233–235
   in vitro filter retardation assay, 231
   yeast-based models, 231–232
Huntington's disease (HD) pathogenesis
   autophagy-lysosome pathway, 228
   axonal transport impairment, 229
   cellular protein quality control mechanisms, 226
   mechanisms, 28
   molecular chaperones as a therapeutic target, 226–227
   toxicity of N-terminal fragments, 225–226
   transcriptional dysregulation, 229–230
   ubiquitin-proteasome system (UPS), 228
Huntington's disease-like 2, 5
Huntington's disease-like 4. *see* SCA17
Hyperhidrosis, 123
Hyperlipidemia, and SBMA, 212
Hypogammaglobulinemia, 123
Hypogonadism with DM2, 133

I

i Motifs, 720–721
i-tetraplex structures, 621*f*
ICF syndrome, 154
   and hypomethylation of D4Z4, 157
Idebenone, 298–300, 323
Inclusion bodies
   and Huntington's disease, 225, 226
   and SCA17, 475, 480
   and SCA7, 401–402
Inclusion, definition, 524
Inheritance of repeat-sequence diseases, 691
Insertions at fragile sites and cancer, 200
INSR (insulin receptor) splicing alterations, 125
Insulin receptor RNA splicing, 62, 143
Insulin resistance
   with DM2, 123, 133
   and misregulated splicing of insulin receptor (IR), 45
Intergenerational contraction, 26
Intranuclear inclusions
   and OPDM, 497
   in polyQ and polyALA diseases, 498
Introns, 43
Inverted repeats (IRs), 720

J

Jacobsen syndrome, 197
JPH3 repeat length and HDL2, 265
Juglone, 232
Junctophilins, 266–269

K

Kainate receptors, 251–252
Kennedy's disease. *see* SBMA (spinal and bulbar muscular atrophy)

Knockin mouse studies
  of FXTAS, 170
  of Huntington's disease, 234–235
  of SCA7, 408
  of somatic instability of GAA-TR alleles, 313
Knockout mice studies
  absence of congenital DM, 49
  cardiac conduction defects and DMPK activity, 38–39
  (CTG)$_n$ repeat expansion and adjacent genes, 29
  DMPK loss, 92
  Fmr1 (fragile X syndrome), 176f
  for Friedreich's ataxia, 323–324
  haploinsufficiency of DMPK, 80–81, 142
  haploinsufficiency of SIX5, 80–81
  and MBNL protein sequestration, 49
  of somatic mosaicism, 543–544
Knudson hypothesis, 4605

L

Lamotrigine, and Huntington's disease, 257
Late-onset Friedrich's ataxia (LOFA), 278
Learning disabilities, 166–167
Left-handed Z-DNA, 719f, 721
Lens, DMPK levels, 104
Leydig cell proliferation, and SIX5 deficiency, 72–73
Linkage disequlibrium of DM2 expansion mutation, 137
Lipinski's rule of five, 231
Loop-directed repair, 650
Loop repair, 650
  and repeat expansions, 680
  role of MSH2 and MLH1, 651
Loops in slipped DNA, 648f
Loss of function model, 13
  for OPDM and polyQ diseases, 506
  for polyglutamine diseases, 518
  for polyQ and polyAla diseases, 500
  for SCA10, 444
Low-copy repeats, 727–729
Lynch's syndrome, 738

M

Machado-Joseph disease
  identification, 22
  pathogenic mechanisms, 28
  role of genetic recombination, 599
  summary, 6
Machado-Joseph disease (MJD)
  and ataxin-3, 366–368
  clinical features, 364–365
  drosophila models, 589–590
  identification, 22
  insights from animal models, 370–372
  molecular genetic features, 365–366
  neuropathology, 365
  pathogenic mechanisms, 28
  prevalence, 364
  protein misfolding and pathogenesis, 368–370
  role of genetic recombination, 599
  and segregation distortion, 27
  size of expanded repeats, 26
  somatic mosaicism, 542
  subtypes, 366
  therapeutic approaches, 372
Macular degeneration and SCA7, 388
Mammalian MMR proteins, 651f
MBLL proteins, 62
MBNL genes, 48, 49f
MBNL proteins
  inactivation in DM1, 62
  interaction with CUG repeat sequences, 62
  and misregulated alternative splicing, 46–47
  in normal and DM skeletal muscle, 144
  role in DM expression, 48, 124–125
MBXL proteins, 62
McLeod syndrome, 263
Meiotic drive, and anticipation, 27
Memantine, and Huntington's disease, 257–258
Metal ions, and triplex stabilization, 328–329
Microaggregate, 524
Microsatellite instability
  and age, 745
  contributing factors, 610f
  detection methods, 741–742
  downstream impact on cancer, 739
  and hereditary nonpolyposis colon cancer (HNPCC), 738–740
  impact of DNA repair, 740–741
  role of non-B DNA structures, 600–601
  role of non-B RNA structures, 600–601
Microsatellites
  definition, 597
  genotype-phenotype correlation, 11–12
  influence on genetic recombination, 601t
Microtubule-associated protein tau, 143
Minocycline, and Huntington's disease, 240
Mismatch repair (MMR), 650–651
  and (CTG)·(CAG) instability, 650–653
  and expansion mechanisms, 624, 686–687
  and microsatellite instability, 740–741
  and microsatellite instability in cancer, 738
  mutations in constitutive tissue of carriers, 742
  and recombination intermediates destabilization, 604
  and repeat instability, 10–11, 574–576, 692, 699
  and repeat instability in bacteria, 697
  and repeat instability in yeast, 697–698
  and somatic mosaicism, 548–549, 551–552
  as therapeutic target, 553
  and triplet repeat instability, 668
Mithramycin, 240
Mitochondria
  dysfunction and polyglutamine diseases, 492
  and excitotoxicity with Huntington's disease, 253–254, 256
  and frataxin deficiency, 277
  and Machado-Joseph disease, 371

MitoQ, 301
Mitotic drive, 26
MJD. *see* Machado-Joseph disease
MMR. *see* Mismatch repair (MMR)
MMR proteins
   and base-base mismatch repair, 660
   and non-MMR DNA metabolic processes, 651
   and slipped DNA repair, 656–657, 660
   and trinucleotide repair, 653–654
Modafinil, and Machado-Joseph disease, 372
Morris water maze test, 181–182
Mouse studies
   acoustic startle reflex test, 182
   of ataxin-7 and SCA7 pathogenesis, 411
   audiogenic seizures, 182–183
   behavior and fragile X syndrome, 181–183
   brain anatomy/physiology with fragile X syndrome, 179–180
   DMPK deficiencies, 28–29
   environment effects on FMRP levels, 183
   eyeblink conditioning test, 182
   fear test, 182
   of fragile X syndrome, 177–185
   frataxin deficiency and Friedreich's ataxia, 298
   for Friedreich's ataxia, 322–324
   FXTAS model, 184
   hippocampal-related memory loss in Fmr1 knockout mice, 180–181
   intergenerational instability, 568t
   junctophilin deficiency, 268
   Machado-Joseph disease, 368–370
   macroorchidism with fragile X syndrome, 178–179
   models of Huntington's disease, 233–235
   Morris water maze test, 181–182
   motor activity and anxiety test, 182
   overcoming *frda* null mice lethality, 312
   overexpression of Hsp70, 227
   recreation of somatic mosaicism, 578–579
   of repeat instability, 565–568
   of repeat instability and tissue specificity, 693–696
   rotarod apparatus for muscle testing, 235f
   SAHA treatment for Huntington's disease, 240
   of SCA1, 341–342
   of SCA2, 358
   of SCA8, 428–429
   somatic mosaicism dynamics, 545
   transcriptional dysregulation and Huntington's disease, 229–230
MRCK proteins, 84, 85f
MRI of brain. *see* Brain imaging
mRNA, degradation by NMD pathway, 43
MSH2
   binding to trinucleotide repeats, 653–654
   and repair efficiencies, 646
MSH2-MSH3, 651, 687
MSH2-MSH6, 651
MSH3 and repair efficiency, 646
MSH6 deficiency and CAG stabilization, 652
MTMR1 (myotubularin-related 1) gene, 46

Muscleblind proteins
   affinity for expanded repeats, 707
   interaction with CUG repeat sequences, 62
Muscleblind sequestration model of DM, 48–49, 62, 124–125
Mycophenolic acid, 231
Myoclonic epilepsy (EPM1), 4
Myogenesis, and overexpression of CUG repeats, 40
Myotonia
   causal role of CUG repeats, 61
   in DM1, 23, 57
   in DM2, 23, 59–60, 122, 134
   and DMPK loss in mice, 65
   in HSA transgenic mice, 41
   and misregulated splicing of muscle-specific chloride channel (ClC-1), 45
Myotonic atrophica, 22
Myotonic dystrophies
   additional loci, 139
   comparison of symptoms for DM1 and DM2, 60, 126, 134t
   congenital, 24
   diagnostics and therapeutics, 29–30
   genetic mutations, 21–22
   pathogenic models prior to DM2, 123
   phenotypic differences between DM1 and DM2, 29
   prevalence, 132
Myotonic dystrophy protein kinase. *see* DMPK
Myotubularin NMDA NR1 receptor splicing, 125, 143
Myotubularin receptor splicing, 62

N

Negative supercoiling, 718
Nematode worm studies of Huntington's disease, 232, 233f
NER. *see* Nucleotide excision repair
NER proteins
   and base-base mismatch repair, 660
   and slipped DNA repair, 656–657, 660
Nervous system. *see* Central nervous system
Neuroacanthocytes, 263–264
Neurodegeneration studies, using drosophila, 588
Neurofibromatosis type 1, 729
Neuroimaging. *see* Brain imaging
Nick-directed repair, 650
   and PARP-1, 668–669
   sealing by ligase, 670
   and slippage of repeats, 646
Nifedipine, 66
Nipecotic acid, 231
NMDA NR1 receptor splicing, 62, 143
NMDA receptors
   and calcium-induced excitotoxicity, 252–253
   and excitotoxicity with Huntington's disease, 254–255
   and neurotransmission, 251–252
NMDARs, 46
Nomenclature guidelines, 22
   for aggregates, 524
   for CELF proteins, 47f

Non-B-DNA sequences, 328
    and 2.5-kb poly (R-Y) sequence of PKD1 gene, 721–722
    adopted by triplet sequences, 718–721, 719f
    and double-strand break repair, 725–726
    rearrangements induced by, 723f
Non-Watson-Crick pairs, 718–719
Noncoding trinucleotide repeat diseases, 488
Novobiocin and Friedreich's ataxia, 302
Nuclear inclusions
    of mutant ataxin-7, 391–392
    and polyglutamine diseases, 411
    and SCA7, 401–402
Nuclear retention of DMPK/ZNF9 RNA, 42–43
Nuclei per muscle fiber, 41
Nucleotide excision repair (NER), 650
    and expansion mechanisms, 624
    pathways, 685f
    and PKD1-dependent drop in viability, 721–722
    and repeat instability, 11, 650, 684–685, 692
    and repeat instability in bacteria, 697

O

Ocular problems
    with DM1, 58
    with DM2, 60
    downbeat positioning nystagmus with SCA6, 381
    extraocular muscle remodeling in OPDM, 505
    eye movement abnormalities and SCA12, 463
    and Friedreich's ataxia, 281
    with FSHD, 152
    with Machado-Joseph disease, 364
    retinal degeneration with SCA7, 407–408, 410f
    with SCA7, 400–401
    slowed saccades with SCA2, 351
Oculopharyngeal muscular dystrophy. see OPDM
Oligoadenylate synthetase, 711
Oligomerization
    and non-B-DNA structures, 726
    in OPDM and polyQ diseases, 501–502
Oligomers, 524
Oligospermia, 123
OMIM number, 22
OPDM
    apoptosis, 504
    background information, 500–501
    cell specificity, 505–506
    and HDL2 pathogenesis, 269
    molecular mechanisms, 501
    neurological involvement, 504–505
    oligomerization, 501–502
    pathogenic mechanisms, 28
    polyA binding protein nuclear 1, 501
    protein aggregates, 501
    role of nucleus, 504
    role of RNA, 503–504
    role of soluble and aggregate proteins, 505
    summary, 5
    transcriptional dysregulation, 504

    treatment, 494–495, 505
    ubiquitin-proteasome system, 502–503
Oscillopsia, and SCA6, 381
Oxidative stress, with Friedreich's ataxia, 298

P

Parabenzoquinones, and Friedreich's ataxia, 298–301
Parallel structures, 647
Parental gender
    and congenital myotonic dystrophies, 24
    molecular explanations for effects, 25
    and transmission of Friedreich's ataxia, 311–312
    and transmission of SCA7, 403
Parkin gene, 202
Pcp2 promoters, and SCA7 pathogenesis, 390
PCR assay for DM2 diagnosis, 119, 121f
Pentanucleotide repeats, 4
Peripheral nervous system, and Friedreich's ataxia, 280
pH
    and H-DNA structure, 637
    and sticky DNA formation, 329
    and tetraplex DNA formation, 720
Phosphorylation of ataxin-1, 342–344
PKD1 gene, 721–722
PMS2, in MMR pathway, 651
Pol beta
    ang gap repair, 669
    and DNA synthesis, 673–674
Polar zipper hypothesis, 404, 523
Poly-Q disorders
    role of transcriptional dysregulation, 393
    spinocerebellar ataxia type 7, 390–394
Poly (R-Y) tract, 721
Polyalanine diseases
    blepharophimosis-ptosis-epicanthus inversus syndrome, 496
    cleidocranial dysplasia, 496
    comparison to polyglutamine diseases, 498–500
    congenital central hypoventilation syndrome, 496
    description, 493
    gene structure and function, 499
    hand-foot-genital syndrome, 493
    holoprosencephaly, 496
    list of, 488
    summary of, 490
    synpolydactyly syndrome (SPD), 493
    x-linked infantile spasm syndrome (West syndrome), 497
    x-linked mental retardation with growth hormone deficit, 497
Polyglutamic diseases. see also specific diseases
    and adeno-associated virus setorype 2, 411
    apoptosis, 504
    and calcium signaling, 492
    caused by microsatellite instability, 598
    common features, 488–489
    comparison to polyalanine diseases, 498–500
    drosophila models, 588–592

gain of function model, 12–13
gene structure and function, 499
genotype-phenotype correlation, 11–12
and histone deacetylase inhibitors, 411
list of, 340, 488
loss of function model, 13
and mitochondrial dysfunction, 492
neurological involvement, 504–505
nuclear inclusions, 411
oligomerization, 501–502
overlapping features, 518
pathogenic mechanisms, 12
and phosphorylation, 492
proposed mechanisms, 489, 491–493
protein folding mechanism, 518
and protein misfolding, 491
recent treatment advances, 493
repeat length, 488–489
and RNA interference, 411
role of nucleus, 504
role of RNA, 503–504
role of soluble and aggregate proteins, 505
similarities to OPDM, 501–505
summary, 5–6, 489t
and SUMOylation, 492
toxic fragment hypothesis, 410–411
and transcription regulation, 491–492
transcriptional dysregulation, 504
and transport defects, 492
treatments, 494–495, 505
type I and type II, 599
and ubiquitin-proteasome pathway, 491, 502–503
and ubiquitization, 492
Polyglutamine
aggregate varieties, 524
as aggregation domain, 522–523
molecular rheostat role, 522
repeat length and protein folding and stability, 522
as solubilizing domain, 522
Polyglutamine aggregates
amyloid-like, 524–529
cellular factors influencing, 530
effects of flanking sequences, 528–529
elongation, 527–529
elongation and nucleation acceleration, 527–528
formation by Josephin domain, 529–530
influence of molecular chaperones, 530
and normal repeat length, 519
nucleation of assembly of amyloid-like structures, 525–527
protein recruitment, 527
specificity of amyloid-like elongation, 529
stability, 529
structure, 525
Polyglutamine expansion
aggregation, 523–531
conformation, 520
evolutionary trends in length, 520–521
in Huntington's disease, 251

length thresholds, 518
normal function, 520–523
normal repeat length, 519–520
repeat length and protein function, 521–522
Polymerase beta, and (CTG)·(CAG) instability, 660
Polymerase-triplex interplay, 638
Polypurine-polypyrimidine (R-Y) sequences, and triplex formation, 328–329
PP2A (protein phosphatase 2A)
function, 467
subcellular localization, 469
PPP2R2B gene
function of protein phosphatase 2A, 467–468
genomic structure, 466f
and SCA12, 465
pre-mRNAs, alternative splicing, 43–44
Primidone, 463
Promiscuous replication hypothesis, 449–450
Proteasomes, and mutant ataxin-7 expression, 392–393
Protein aggregates. *see also* nuclear inclusions
and HDL2, 269–270
in OPDM and polyQ diseases, 501
Protein folding, and mutant ataxin-7 expression, 393
Protein kinase (PKR), 711
Protein misfolding
and polyglutamine diseases, 491
and RNA hairpin structures, 628
Protein sequestration model, 48–49
Protofibril, 524
Proximal myotonic dystrophy (PMD), 22
Proximal myotonic myopathy (PROMM), 22, 132
mutation identification, 25
Purkinje cells
and SCA17, 480
and SCA6 pathogenesis, 380
and SCA7 pathogenesis, 401–402, 406

Q

Quadruplexes, 647

R

(R-Y) sequences, 328
Rad52 proteins, 575
Radicicol, 227
Random coil structure of polyglutamine, 519–520
Random-sequence heteroduplexes, 659–660
Rapamycin, 228
RAPID cloning, 418
Rare fragile sites, 195, 196–197
non-folate sensitive, 197
Rasagiline, 258
Reciprocal exchange, 8
Recombination
effect of sticky DNA, 333
and repeat instability, 683, 684, 692
and triplet repeat instability, 668
Renal carcinoma, 201

Repair, 655
   and aphidicolin, 660
   arrested replication forks, 683
   BER repair and repeat instability, 671–675
   and chromosomal rearrangements, 729–730
   and dideoxythymidine, 660
   double-strand break repair, 7–10
   efficiencies, 655
   of heteroduplexes, 649–650
   impact on microsatellite instability, 740–741
   mechanisms, 657–659, 658f
   mechanisms and repeat instability, 683–686
   mismatch repair, 650–653
   mismatch repair and somatic mosaicism, 548–549
   nucleotide excision repair (NER), 650, 684–685
   outcomes, 655
   proposed mechanisms, 656f
   random-sequence heteroduplex repair, 650
   and repeat expansions, 680–683
   and repeat instability in human disease, 661–662
   role of MMR proteins, 653–654
   and somatic mosaicism, 549
   stalled replication forks, 683f
   and transcription-induced repeat instability, 698–699

Repeat expansion
   associated diseases, 679–680
   and base excision repair (BER), 667
   diseases with repeat outside coding sequence, 618
   and DNA base damage of ssDNA break, 674
   future directions for research, 13
   and gene expression, 693
   and human disease, 4
   and interruptions, 622
   length variability, 4
   location, 4
   mechanisms based on human disease data, 680
   possible mechanisms, 623–624
   and replication events, 681
   replication model, 641–642
   and replication slowing, 639–640
   spectrum in human diseases, 4

Repeat instability. *see also* Microsatellite instability
   age dependence, 695–696
   and base excision repair (BER), 668–675
   and CG-rich environment, 570–571
   cis-acting factors affecting, 570
   cis modifiers, 567–571
   and DNA mismatch repair, 574–575
   and DNA repair, 649–650
   and double-strand break repair in mammals, 10
   and double-strand breaks, 607–608
   and double-strand repair break, 7–10
   factors affecting, 567
   and flanking sequences, 570–571
   and gender of parent, 571–572
   and gender of transmitting parent, 4
   and gene conversion, 598
   gene conversion as mechanism, 598–599
   genetic modifiers, 574–575
   and genetic recombination, 7–10
   and human disease, 5–6
   in humans, 692
   inital mouse studies, 564–565
   intergenerational, 564
   intertissue variability, 572–574
   and mismatch repair (MMR), 10–11, 686–687
   during mitotic and meiotic recombination, 9–10
   molecular mechanisms, 7–11
   mouse model summary, 568t
   and non-B-DNA structures, 667, 726
   and nucleotide excision repair, 11, 668
   and orientation polymorphism, 729
   overview, 4, 7
   in patient-derived tissues, 7
   and population genetics, 11
   and recombination, 668, 705
   and recombination in mammals, 10
   and repair mechanisms, 683–686
   and replication events, 623, 668
   and replication-repair, 10
   and replication slippage, 598
   and single-strand DNA break repair, 668
   and single-strand gaps, 608–610
   and single-strand nicks, 608–610
   and site of transgene integration, 569–570
   and strain effects, 549
   recreation in transgenic mice, 577–578
   timing, 575–576
   timing of somatic mosaicism, 577
   trans-acting factors, 571–574
   and transcription in bacteria, 696–697
   and transcription in human cells, 698–699
   and transcription in yeast, 697–698

Repeat motifs and associated diseases, 728t

Replication
   chicken foot structure, 684
   effect of sticky DNA, 332–333
   model for repeat expansions, 641–642
   promiscuous replication hypothesis, 449–450
   and repeat instability, 692
   slowing by repeat expansions, 638–640
   and triplet repeat instability, 668

Replication fork collapse, 683f

Replication slippage model
   for CCTG·CAGG repeats, 605
   and somatic mosaicism, 549–550

Reproductive disturbances
   androgen insensitivity with SBMA, 212
   premature ovarian failure, 197

Respiratory problems
   with DM1, 23
   with FSHD, 152

Retina, and mutant ataxin-7, 392
Rhodopsin, and SCA7 pathogenesis, 390, 391, 393, 394
Ribosome stalling, 628
Riboswitches, 39
Riluzole, and Huntington's disease, 240

RNA gain-of-function model, 13, 28, 123–124, 124f, 142, 143
  for DM1, 418
  for DM2, 418
  for FXTAS, 170–171, 418
  for SCA10, 418, 444
  for SCA12, 418
  for SCA8, 418
RNA interference, 411
RNA-mediated disease process, 64f, 123
  abnormal splicing in DM1, 62
  alternative splicing disruption, 44–46
  alternative splicing of pre-mRNAs, 43–44
  and CTG expansion, 74f
  evidence for, 39–40
  evidence for mechanism in DM1, 38–42
  as incomplete explanation of DM1 and DM2 etiology, 61
  mechanisms for misregulated alternative splicing, 47–48
  muscleblind sequestration model of DM, 48–49
  nuclear retention of mutant DMPK/ZNF9 RNA, 42–43
  and skeletal muscle effects of DM1, 61
  summary, 80–81
  toxic gain-of-function effect, 79
RNA structures, 622
ROCK proteins, 84, 85f
Ryanodine, 66
Ryanodine receptor 1 splicing, 143
RyR1 (ryanodine receptor) gene, 45–46

S

S776 and ataxin-1 phosphorylation, 342–344
SAGA-like complexes and SCA7, 389
SAHA treatment, 553
  of Huntington's disease, 240
SbcC protein, 57
SBMA
  aggregation of mutant protein, 215–216
  androgen receptor structure and activation, 213–214
  clinical features, 212
  clinical presentation, 22
  differential diagnosis, 212–213
  drosophila models, 591–592
  gender specificity, 592
  genetics, 213
  identification, 22
  intracellular inclusions, 215
  laboratory studies, 212
  ligand-dependent effects, 216–217
  management, 213
  model of cellular pathogenesis, 216f
  mutation identification, 24
  pathogenic mechanisms, 28
  pathological mechanisms, 215–217
  somatic mosaicism, 542
  summary, 5
  therapeutic approaches, 217
  toxic gain of androgen receptor function, 215

SCA1
  and 14-3-3 protein family, 344
  and AKT signaling, 344–345
  and ataxin-1 phosphorylation modifiers, 345–347
  and CAG expansions, 657
  characteristics, 340
  mutation identification, 24
  and normal ataxin-1 function, 340–341
  and nuclear inclusions, 341
  pathogenic mechanisms, 28, 341–342
  phenotype comparison with SCA2, 354f
  and repeat instability, 661–662
  size of expanded repeats, 26
  slipped DNA structure, 649–650
  somatic mosaicism, 542
  structural role of interruptions, 709, 710f
  summary, 5
SCA10. see also $(ATTCT)_n \cdot (AGAAT)_n$ repeats
  age of onset and repeat size, 438f
  age of onset and size of of repeat expansion, 436
  $(ATTCT)_n$ repeat expansion instability, 438
  clinical presentation, 22
  drosophila models, 591
  epilepsy type age of onset and repeat size, 443f
  expansion mutations, 437f
  expansion repeat, 4
  gain of function model, 444–445
  genotype-phenotype correlation, 443
  intergenerational changes in repeat size, 437–438
  loss of function and pathogenesis, 444
  molecular diagnosis, 443
  pathogenic mechanisms, 28, 443–445
  pentanucleotide repeat, 434
  phenotype and ethnicity, 442–443
  repeat expansion identification, 434–436
  repeat instability in E. coli, 454–455
  repeat location, 619
  repeats form unwound DNA, 450–454
  RNA gain-of-function model, 418
  structure of $(ATTCT)_n$ repeats, 436
  summary, 6
SCA12
  brain imaging, 464
  clinical presentation, 463–464
  diagnosis, 464–465
  disease pathogenesis, 469–471
  epidemiology and genetics, 462–463
  molecular basis, 465–467
  as noncoding repeat disease, 488
  not a polyglutamine disease, 467
  overexpression of Bbeta1, 461
  pathogenic mechanisms, 28
  PPP2R2B gene products, 467–468
  repeat length polymorphism, 462f
  RNA gain-of-function model, 418
  summary, 6
  treatment with beta-blockers, 463
  treatment with primidone, 463

SCA17, 6
   epidemiology and frequency, 476
   expansion origin and instability, 476
   identification, 476
   intergenerational instability, 475
   neurological signs and frequency, 478t
   neuropathology, 480
   pathogenic mechanisms, 28
   pathophysiology, 480
   phenotype, 478–480
   phenotype-genotype correlations, 479
   repeat expansion behavior, 476
   repeat size, 476
SCA2
   anticipation, 352–353, 356f
   and CAG expansions, 657
   CAG repeat range, 352
   dementia, 355
   eye movements and retinal changes, 354–355
   frequency and phenotype, 354–355
   gene function, 356–358
   gene identification, 351–352
   genetic modifiers of age of onset, 353–354
   infantile onset, 24
   mouse models, 358
   movement disorders, 355
   neuropathology, 355–356
   pathogenic mechanisms, 28
   phenotype comparison with SCA1, 354f
   prevalence, 351
   slipped DNA structure, 649–650
   somatic mosaicism, 542
   structural role of interruptions, 709, 710f
   summary, 5
SCA3. *see* Machado-Joseph disease (MJD)
SCA6
   clinical presentation, 381–382
   diagnosis, 382
   etiology and neuropathology, 379–381
   summary, 6
   treatment and prognosis, 382–383
SCA7
   about, 387–388, 399–401
   age of onset and disease course, 402
   anticipation, 568
   ataxin-7 function and disease pathogenesis, 408–410
   ataxin-7 proteolytic cleavage and turnover, 410–411
   causes, 388
   cis modifiers of repeat instability, 570
   clinical presentation, 22
   disease progression, 391
   electrophysiological abnormalities, 394f
   gene identification, 402
   genotype-phenotype correlation, 402f
   histological abnormalities, 394f
   infantile onset, 24
   mouse models, 388t
   neurodegeneration models, 404–407
   neuropathology, 401–402
   pathogenic mechanisms, 28
   polar zipper hypothesis, 404
   repeat instability, 403–404. *see* DRPLA
   retinal degeneration, 407–408, 410f
   retinal pathology in mice, 391
   somatic mosaicism, 542
   summary, 6
   transcriptional abnormalities, 394f
   transcriptional dysregulation, 393–394
SCA8
   clinical presentation, 419
   disease penetrance in MN-A family, 421
   expansions on control chromosomes, 423
   fly (drosophila) studies, 428–429
   gene organization, 419f
   haplotype analysis, 423–425
   identification of repeat expansion, 418–419
   inheritance patterns, 421–423
   (MN-A) family analysis, 419
   (MN-A) family haplotype analysis, 420f, 426f
   molecular parallels with DM1, 428
   mouse models, 428–429
   MRI of brain, 419, 421f
   neuropathology, 419, 421
   as noncoding repeat disease, 488
   repeat contractions in sperm, 427–428
   repeat instability, 425–428
   repeat location, 619
   similarity to DM1, 145
   summary, 6
Scapuloperoneal syndrome, 152
Segmental duplications, 727–729
Selenium, and Friedreich's ataxia, 301
SERCA genes
   and splicing abnormalities, 45–46
   splicing abnormalities, 143
Serine/threonine protein kinases, 85f
Sherman Paradox, 24
Single-strand DNA break repair, 668
SIX5 gene
   deficiency, 29
   deletion in mice, 70–73
   and DM1, 56, 142
   haploinsufficiency, 123
   inactivation in DM1, 64
Skeletal defects with DM1, 58
Skeletal muscle
   abnormal gating of sodium channels with DMPK loss, 67
   biopsy data for DM2, 135–136
   biopsy features with DM1 and DM2, 122, 143–144
   depolarization-mediated calcium efflux from SR with DMPK loss, 66
   DMPK levels, 104–105
   excitation-contraction coupling, 66
   and Friedreich's ataxia, 280
   molecular mechanisms for weakness in DM1, 66–68
   and SIX5 gene deletion, 71
   weakness with DM2, 122
   weakness with FSHD, 152

Slipped (CTG)·(CAG) repeats, 652f
Slipped DNAs
  adopted by triplet sequences, 718–719
  and CAG repeat expansion, 674
  expansion and deletion intermediate formation, 647f
  as mutagenic intermediates, 646–653
  processing by human cell extracts, 654–662
  repair and MMR/NER proteins, 656–657
  and repeat expansions, 680–681
  and repeat instability, 10
  and repeat instability in human disease, 661–662
  structure, 622, 646–649, 648f
Small-pool polymerase chain reaction, 4608
SMC1, and fragile site expression, 199
Smith-Magenis syndrome, 729
Sodium butyrate, 553
Somatic instability
  definition, 312
  of GAA-TR alleles at FRDA locus, 312–316
Somatic mosaicism, 564–565
  in CAG·CTG repeat disorders, 542
  cis-acting modifiers, 547–548
  in DM1, 541
  dynamics in human tissue culture, 543–544
  dynamics in mouse tissue culture, 545
  expansion length and disease pathology, 546–547
  genotyping and measuring, 538–541
  and hairpin structures, 551
  in humans, 542–543
  in Huntington's disease, 541–542
  and mismatch repair (MMR), 551–552
  mouse models, 573t
  recreation in transgenic mice, 578–579
  and replication slippage, 549–550
  as therapeutic target, 552–554
  and timing of repeat instability, 577
  tissue specificity, 543–544, 572–574
  trans-acting modifiers, 548–549
  in transgenic mouse models, 543–544
Sotos' syndrome, 729
Southern blot analysis, 120, 121f, 139
  detection of mutant alleles, 539
  and FSHD diagnosis, 154–155
  of GAA-TR alleles at FRDA locus, 313
Spermatogenic cell death, and SIX5 deficiency, 72–73
Spherical oligomers, 524
Spheroids, 524
Spinocerebellar ataxia
  repeat instability, 646
  subgroups, 462
Splicing of pre-mRNAs, 43–44
STAGA complex, and SCA7 pathogenesis, 408–409
Steinert's disease, 22
Sticky DNA
  detection methods, 329
  effect on recombination, 333
  effect on replication, 332–333
  effects on transcription, 330–332
  electron microscopy, 332f

formation requirements, 329
and Friedreich's ataxia, 289–290, 302, 328
model, 328f
structures, 647
and triplet repeats, 720
SUMOylation, 492
Surgical treatment, of OPDM, 505
Synpolydactyly syndrome (SPD), 493

T

Taltirelin hydrate, and SCA6, 382
TATA box binding protein, 491
Tau, 143
  abnormal RNA splicing, 62, 125
  implication in other neurological disorders, 45
TBP, structural role of interruptions, 709
Testicular atrophy, and SIX5 deficiency, 72–73
Tetradotoxin, 66
Tetranucleotide repeats, 4
Tetraplexes, 620, 621f
  adopted by triplet sequences, 719f, 720–721
Thymidylate levels, and rare fragile sites, 197
Tissue specificity
  and repeat instability, 681, 692
  and repeat tract lengths in humans and mice, 693–696
  and somatic mosaicism, 547, 572–574
Titin gene splicing abnormalities, 143
TNNT3 gene splicing abnormalities, 46
Toxic fragment hypothesis, 410–411
Toxicity criteria, 300
Trans-acting modifiers of repeat instability, 571–574
Trans-acting modifiers of somatic mosaicism, 548–549
trans-Dominant RNA-gain-of-function theory, 25
Transcription
  and age-dependent repeat instability, 695–696
  effect of sticky DNA, 330–332
  in OPDM and polyQ diseases, 504
  and polyglutamine diseases, 491–492
  regulation by TAF paralogs, 389
  and repeat instability, 692, 693f, 694
  and repeat instability in bacteria, 696–697
  and repeat instability in human cells, 698–699
  and repeat instability in yeast, 697–698
  triplex-mediated effects, 629
Transcription-coupled repair, 692
Transcriptional healing, 200
Transducins, and SCA7 pathogenesis, 393
Transgenic mouse studies
  abnormal chloride channel splicing, 62
  absence of congenital DM, 49
  CAG expansions and repair, 657
  $(CTG)_n$ repeat expansion size, 26
  CUG repeats and skeletal muscle disease, 61
  for DM2, 144
  DMPK inactivation, 65–66
  DMPK overexpression, 80–81
  expressing CUG-BP1, 48
  Huntington's disease (HD) and rapamycin, 228

Transgenic mouse studies (*Continued*)
    Machado-Joseph disease, 369–370
    macroorchidism with fragile X syndrome, 179
    MBNL protein inactivation, 62–63
    MSH2 and MSH3 and (CTG)·(CAG) repeat stability, 652
    overexpression of CUG expansion RNA, 40–42, 49
    overexpression of human DMPK, 39
    PMS2 deficiency, 653
    recreation of somatic mosaicism, 578–579
    repeat instability and DSB repair, 10
    repeat instability and transcription, 694
    retinal degeneration with SCA7, 407–408
    S776 and ataxin-1 phosphorylation with SCA1, 343*f*
    of SCA7, 406
    SIX5 gene deletion, 70–73
    SIX5 inactivation, 64
    of somatic instability of GAA-TR alleles, 313
    of somatic mosaicism, 543–544
    YAC72 model of Huntington's disease, 254
Trehalose, and Huntington's disease, 240
Tremor, and FXTAS, 168
Trinucleotide repeat diseases. *see also* specific disease
    noncoding and coding diseases, 488
    somatic mosaicism, 564–565
Trinucleotide repeats. *see also* Repeat instability
    dynamics, 564–565
    flanking sequences and RNA structure, 708–709
    function, 706
    in human transcriptions, 707
    length and tissue specificity, 693–696
    length variability, 706
    RNA structures, 707–708
    simple sequence repeat (SSR) loci, 706
    somatic instability, 693–694
    structural role of interruptions, 709–711
    in transcripts, 706–707
Triplet expansion. *see* Repeat expansion
Triplet repeats
    cloned in plasmid DNA, 722*f*
    cruciform DNA structures, 720
    and deletion-inversion breakpoints, 724–725
    for DM1, 722–726
    i motifs, 720–721
    and left-handed Z-DNA structures, 721
    non-B-DNA conformations, 718–721
    tetraplex DNA structures, 720–721
    triplex DNA structures, 720
Triplex structures, 622, 647
    adopted by triplet sequences, 719*f*, 720
    detection methods, 329
    and expanded GAA repeats, 289
    in follicular lymphomas, 727
    and Friedreich's ataxia, 328
    graphic representation, 621*f*
    inter- and intramolecular, 720
    models, 328*f*
    and polymerization arrest, 638
    types of structures, 328–329
Tubulin-related gene (TUBB4q), and FSHD, 156
Type 1 interferon, 711

U

Ubiquitin-proteasome system (UPS), 228
    and OPDM, 501
    in OPDM and polyQ diseases, 502–503
    and polyglutamine diseases, 491
Unpaired DNA, nucleotide excision repair (NER), 650
Unwound DNA. *see* DNA unwinding

V

Vertigo, and SCA6, 381
Vision
    and Friedreich's ataxia, 279
    and spinocerebellar ataxia type 7, 388
Vitamin E, and Friedreich's ataxia, 300–301

W

Web sites
    Gross Rearrangement Breakpoint Database, 727
    Human Gene Mutation Database, 727
    human genes with fly homologues, 588
    NCI toxicity criteria, 300
Werner's syndrome, 720
Williams-Beuren syndrome, 729
WWOX gene, 201–202

X

X-linked infantile spasm syndrome, 497
X-linked mental retardation with growth hormone deficit, 497
Xeroderma pigmentosum (XP), 650

Y

Y-27632, 232
Yeast studies
    break-repair mechanism for repeat expansion, 682
    of chromosomal fragile sites, 626
    of DNA repair of heteroduplexes, 649–650
    of expansion mechanisms, 624
    recombination and repeat instability, 683, 684
    repeat instability during mitotic and meiotic recombination, 9–10
    transcription and repeat instability, 697–698

Z

Z-disk-associated genes, 143
Zinc finger 9 gene. *see* ZNF9
ZNF9 gene
    genomic organization, 116
    loss of function and pathogenesis, 29
    site of genetic mutation for DM2, 21
ZNF9 haploinsufficiency, 145
ZNF9 mutant RNA, 42–43
ZNF9 RNA, expression vs. DMPK RNA, 61